Electrical Circuit Theory and Technology

A fully comprehensive text for courses in electrical principles, circuit theory and electrical technology, providing 800 worked examples and over 1,350 further problems for students to work through at their own pace. This book is ideal for students studying engineering for the first time as part of BTEC National and other pre-degree vocational courses, as well as Higher Nationals, Foundation Degrees and first-year undergraduate modules.

John Bird, BSc (Hons), CEng, CSci, CMath, FITE, FIMA, FCollT, is the former Head of Applied Electronics in the Faculty of Technology at Highbury College, Portsmouth, UK. More recently he has combined freelance lecturing and examining, and is the author of over 130 textbooks on engineering and mathematical subjects with worldwide sales of over one million copies. He is currently lecturing at the Defence School of Marine and Air Engineering in the Defence College of Technical Training at HMS Sultan, Gosport, Hampshire, UK.

D0303145

WITHDRAWN FROM STOCKPORT COLLEGE LEARNING CENTRE

127853

In Memory of Elizabeth

Electrical Circuit Theory and Technology

Sixth edition

John Bird

LONDON AND NEW YORK

STOCKPORT COLLEGE
LIBRARY+
I-1839367 2/5/18
13956
127853 621.31
4wk BIR

Sixth edition published 2017
by Routledge
2 Park Square, Milton Park, Abingdon, Oxon OX14 4RN

and by Routledge
711 Third Avenue, New York, NY 10017

Routledge is an imprint of the Taylor & Francis Group, an informa business

© 2017 John Bird

The right of John Bird to be identified as author of this work has been asserted by him in accordance with sections 77 and 78 of the Copyright, Designs and Patents Act 1988.

All rights reserved. No part of this book may be reprinted or reproduced or utilized in any form or by any electronic, mechanical, or other means, now known or hereafter invented, including photocopying and recording, or in any information storage or retrieval system, without permission in writing from the publishers.

Trademark notice: Product or corporate names may be trademarks or registered trademarks, and are used only for identification and explanation without intent to infringe.

First edition published by Newnes 1997
Fifth edition published by Routledge 2014

British Library Cataloguing in Publication Data
A catalogue record for this book is available from the British Library

Library of Congress Cataloging in Publication Data
Names: Bird, J. O., author.
Title: Electrical circuit theory and technology / John Bird.
Description: 6th ed. | New York : Routledge, [2017] | Includes index.
Identifiers: LCCN 2016038154| ISBN 9781138673496 | ISBN 9781315561929
Subjects: LCSH: Electric circuits. | Electrical engineering.
Classification: LCC TK454 .B48 2017 | DDC 621.319/2–dc23
LC record available at https://lccn.loc.gov/2016038154

ISBN: 978-1-138-67349-6 (pbk)
ISBN: 978-1-315-56192-9 (ebk)

Typeset in Times by
Servis Filmsetting Ltd, Stockport, Cheshire
Printed and bound in Great Britain by Bell & Bain Ltd, Glasgow

Visit the companion website: www.routledge.com/cw/bird

Contents

Preface xii

Part 1 Revision of some basic mathematics 1

1 Some mathematics revision 3
- 1.1 Use of calculator and evaluating formulae 4
- 1.2 Fractions 7
- 1.3 Percentages 8
- 1.4 Ratio and proportion 10
- 1.5 Laws of indices 13
- 1.6 Brackets 16
- 1.7 Solving simple equations 16
- 1.8 Transposing formulae 19
- 1.9 Solving simultaneous equations 21

2 Further mathematics revision 23
- 2.1 Radians and degrees 24
- 2.2 Measurement of angles 25
- 2.3 Trigonometry revision 26
- 2.4 Logarithms and exponentials 28
- 2.5 Straight line graphs 33
- 2.6 Gradients, intercepts and equation of a graph 35
- 2.7 Practical straight line graphs 37
- 2.8 Calculating areas of common shapes 38

Main formulae for Part 1 Revision of some basic mathematics 44

Part 2 Basic electrical engineering principles 47

3 Units associated with basic electrical quantities 49
- 3.1 SI units 49
- 3.2 Charge 50
- 3.3 Force 50
- 3.4 Work 51
- 3.5 Power 52
- 3.6 Electrical potential and e.m.f. 53
- 3.7 Resistance and conductance 53
- 3.8 Electrical power and energy 54
- 3.9 Summary of terms, units and their symbols 55

4 An introduction to electric circuits 56
- 4.1 Standard symbols for electrical components 57
- 4.2 Electric current and quantity of electricity 57
- 4.3 Potential difference and resistance 58
- 4.4 Basic electrical measuring instruments 58
- 4.5 Linear and non-linear devices 59
- 4.6 Ohm's law 59
- 4.7 Multiples and sub-multiples 59
- 4.8 Conductors and insulators 61
- 4.9 Electrical power and energy 61
- 4.10 Main effects of electric current 64
- 4.11 Fuses 64
- 4.12 Insulation and the dangers of constant high current flow 64

5 Resistance variation 65
- 5.1 Resistor construction 66
- 5.2 Resistance and resistivity 66
- 5.3 Temperature coefficient of resistance 68
- 5.4 Resistor colour coding and ohmic values 70

6 Batteries and alternative sources of energy 73
- 6.1 Introduction to batteries 74
- 6.2 Some chemical effects of electricity 74
- 6.3 The simple cell 75
- 6.4 Corrosion 76
- 6.5 E.m.f. and internal resistance of a cell 76
- 6.6 Primary cells 78
- 6.7 Secondary cells 79
- 6.8 Lithium-ion batteries 81
- 6.9 Cell capacity 84
- 6.10 Safe disposal of batteries 84
- 6.11 Fuel cells 84
- 6.12 Alternative and renewable energy sources 85
- 6.13 Solar energy 86

Revision Test 1 89

7 Series and parallel networks 90
7.1 Series circuits 91
7.2 Potential divider 92
7.3 Parallel networks 94
7.4 Current division 96
7.5 Loading effect 99
7.6 Potentiometers and rheostats 100
7.7 Relative and absolute voltages 103
7.8 Earth potential and short circuits 104
7.9 Wiring lamps in series and in parallel 104

8 Capacitors and capacitance 106
8.1 Introduction to capacitors 107
8.2 Electrostatic field 107
8.3 Electric field strength 108
8.4 Capacitance 108
8.5 Capacitors 109
8.6 Electric flux density 110
8.7 Permittivity 110
8.8 The parallel plate capacitor 111
8.9 Capacitors connected in parallel and series 112
8.10 Dielectric strength 116
8.11 Energy stored 117
8.12 Practical types of capacitor 117
8.13 Supercapacitors 119
8.14 Discharging capacitors 121

9 Magnetic circuits 122
9.1 Introduction to magnetism and magnetic circuits 123
9.2 Magnetic fields 124
9.3 Magnetic flux and flux density 125
9.4 Magnetomotive force and magnetic field strength 125
9.5 Permeability and *B–H* curves 126
9.6 Reluctance 127
9.7 Composite series magnetic circuits 129
9.8 Comparison between electrical and magnetic quantities 132
9.9 Hysteresis and hysteresis loss 132

Revision Test 2 134

10 Electromagnetism 135
10.1 Magnetic field due to an electric current 136
10.2 Electromagnets 137
10.3 Force on a current-carrying conductor 139
10.4 Principle of operation of a simple d.c. motor 142
10.5 Principle of operation of a moving-coil instrument 143
10.6 Force on a charge 143

11 Electromagnetic induction 145
11.1 Introduction to electromagnetic induction 146
11.2 Laws of electromagnetic induction 147
11.3 Rotation of a loop in a magnetic field 150
11.4 Inductance 151
11.5 Inductors 152
11.6 Energy stored 153
11.7 Inductance of a coil 153
11.8 Mutual inductance 155

12 Electrical measuring instruments and measurements 158
12.1 Introduction 159
12.2 Analogue instruments 159
12.3 Shunts and multipliers 159
12.4 Electronic instruments 161
12.5 The ohmmeter 161
12.6 Multimeters 162
12.7 Wattmeters 162
12.8 Instrument 'loading' effect 162
12.9 The oscilloscope 164
12.10 Virtual test and measuring instruments 169
12.11 Virtual digital storage oscilloscopes 170
12.12 Waveform harmonics 173
12.13 Logarithmic ratios 174
12.14 Null method of measurement 176
12.15 Wheatstone bridge 177
12.16 D.c. potentiometer 177
12.17 A.c. bridges 178
12.18 Measurement errors 179

13 Semiconductor diodes 182
13.1 Types of material 183
13.2 Semiconductor materials 183
13.3 Conduction in semiconductor materials 185
13.4 The p–n junction 185
13.5 Forward and reverse bias 186
13.6 Semiconductor diodes 189
13.7 Characteristics and maximum ratings 190
13.8 Rectification 190
13.9 Zener diodes 190
13.10 Silicon controlled rectifiers 192
13.11 Light emitting diodes 193
13.12 Varactor diodes 193
13.13 Schottky diodes 193

14 Transistors 195
14.1 Transistor classification 196
14.2 Bipolar junction transistors (BJTs) 196
14.3 Transistor action 197
14.4 Leakage current 198
14.5 Bias and current flow 199
14.6 Transistor operating configurations 199

14.7 Bipolar transistor characteristics 200
14.8 Transistor parameters 201
14.9 Current gain 202
14.10 Typical BJT characteristics and maximum ratings 203
14.11 Field effect transistors 204
14.12 Field effect transistor characteristics 205
14.13 Typical FET characteristics and maximum ratings 206
14.14 Transistor amplifiers 206
14.15 Load lines 208

Revision Test 3 **213**

Main formulae for Part 2 Basic electrical and electronic principles **215**

Part 3 Electrical principles and technology **217**

15 D.c. circuit theory **219**
15.1 Introduction 219
15.2 Kirchhoff's laws 220
15.3 The superposition theorem 224
15.4 General d.c. circuit theory 226
15.5 Thévenin's theorem 228
15.6 Constant-current source 233
15.7 Norton's theorem 233
15.8 Thévenin and Norton equivalent networks 236
15.9 Maximum power transfer theorem 239

16 Alternating voltages and currents **242**
16.1 Introduction 243
16.2 The a.c. generator 243
16.3 Waveforms 244
16.4 A.c. values 245
16.5 Electrical safety – insulation and fuses 248
16.6 The equation of a sinusoidal waveform 248
16.7 Combination of waveforms 251
16.8 Rectification 254
16.9 Smoothing of the rectified output waveform 255

Revision Test 4 **257**

17 Single-phase series a.c. circuits **258**
17.1 Purely resistive a.c. circuit 259
17.2 Purely inductive a.c. circuit 259
17.3 Purely capacitive a.c. circuit 260
17.4 *R–L* series a.c. circuit 261
17.5 *R–C* series a.c. circuit 264
17.6 *R–L–C* series a.c. circuit 266
17.7 Series resonance 269
17.8 Q-factor 270
17.9 Bandwidth and selectivity 272
17.10 Power in a.c. circuits 272
17.11 Power triangle and power factor 274

18 Single-phase parallel a.c. circuits **277**
18.1 Introduction 278
18.2 *R–L* parallel a.c. circuit 278
18.3 *R–C* parallel a.c. circuit 279
18.4 *L–C* parallel a.c. circuit 280
18.5 *LR–C* parallel a.c. circuit 282
18.6 Parallel resonance and Q-factor 285
18.7 Power factor improvement 289

19 D.c. transients **294**
19.1 Introduction 295
19.2 Charging a capacitor 295
19.3 Time constant for a *C–R* circuit 296
19.4 Transient curves for a *C–R* circuit 296
19.5 Discharging a capacitor 300
19.6 Camera flash 302
19.7 Current growth in an *L–R* circuit 302
19.8 Time constant for an *L–R* circuit 303
19.9 Transient curves for an *L–R* circuit 303
19.10 Current decay in an *L–R* circuit 305
19.11 Switching inductive circuits 307
19.12 The effect of time constant on a rectangular waveform 307

20 Operational amplifiers **309**
20.1 Introduction to operational amplifiers 310
20.2 Some op amp parameters 311
20.3 Op amp inverting amplifier 312
20.4 Op amp non-inverting amplifier 314
20.5 Op amp voltage-follower 315
20.6 Op amp summing amplifier 315
20.7 Op amp voltage comparator 316
20.8 Op amp integrator 317
20.9 Op amp differential amplifier 318
20.10 Digital to analogue (D/A) conversion 320
20.11 Analogue to digital (A/D) conversion 320

Revision Test 5 **322**

21 Ways of generating electricity – the present and the future **323**
21.1 Introduction 324
21.2 Generating electrical power using coal 324
21.3 Generating electrical power using oil 326
21.4 Generating electrical power using natural gas 327
21.5 Generating electrical power using nuclear energy 328

21.6 Generating electrical power using hydro power 329
21.7 Generating electrical power using pumped storage 330
21.8 Generating electrical power using wind 331
21.9 Generating electrical power using tidal power 331
21.10 Generating electrical power using biomass 333
21.11 Generating electrical power using solar energy 333
21.12 Harnessing the power of wind, tide and sun on an 'energy island' – a future possibility? 334

22 Three-phase systems 336
22.1 Introduction 337
22.2 Three-phase supply 337
22.3 Star connection 337
22.4 Delta connection 340
22.5 Power in three-phase systems 342
22.6 Measurement of power in three-phase systems 343
22.7 Comparison of star and delta connections 348
22.8 Advantages of three-phase systems 348

23 Transformers 349
23.1 Introduction 350
23.2 Transformer principle of operation 350
23.3 Transformer no-load phasor diagram 352
23.4 E.m.f. equation of a transformer 354
23.5 Transformer on-load phasor diagram 356
23.6 Transformer construction 357
23.7 Equivalent circuit of a transformer 358
23.8 Regulation of a transformer 359
23.9 Transformer losses and efficiency 360
23.10 Resistance matching 363
23.11 Auto transformers 365
23.12 Isolating transformers 367
23.13 Three-phase transformers 367
23.14 Current transformers 368
23.15 Voltage transformers 369

Revision Test 6 370

24 D.c. machines 371
24.1 Introduction 372
24.2 The action of a commutator 372
24.3 D.c. machine construction 373
24.4 Shunt, series and compound windings 373
24.5 E.m.f. generated in an armature winding 374
24.6 D.c. generators 375
24.7 Types of d.c. generator and their characteristics 376
24.8 D.c. machine losses 380
24.9 Efficiency of a d.c. generator 380
24.10 D.c. motors 381
24.11 Torque of a d.c. machine 382
24.12 Types of d.c. motor and their characteristics 383
24.13 The efficiency of a d.c. motor 387
24.14 D.c. motor starter 389
24.15 Speed control of d.c. motors 390
24.16 Motor cooling 392

25 Three-phase induction motors 393
25.1 Introduction 394
25.2 Production of a rotating magnetic field 394
25.3 Synchronous speed 396
25.4 Construction of a three-phase induction motor 397
25.5 Principle of operation of a three-phase induction motor 397
25.6 Slip 398
25.7 Rotor e.m.f. and frequency 399
25.8 Rotor impedance and current 400
25.9 Rotor copper loss 400
25.10 Induction motor losses and efficiency 401
25.11 Torque equation for an induction motor 402
25.12 Induction motor torque–speed characteristics 404
25.13 Starting methods for induction motors 405
25.14 Advantages of squirrel-cage induction motors 406
25.15 Advantages of wound rotor induction motor 407
25.16 Double cage induction motor 407
25.17 Uses of three-phase induction motors 407

Revision Test 7 408

Main formulae for Part 3 Electrical principles and technology 409

Part 4 Advanced circuit theory and technology 411

26 Revision of complex numbers 413
26.1 Introduction 413
26.2 Operations involving Cartesian complex numbers 415
26.3 Complex equations 417
26.4 The polar form of a complex number 418

26.5 Multiplication and division using complex numbers in polar form 419
26.6 De Moivre's theorem – powers and roots of complex numbers 420

27 Application of complex numbers to series a.c. circuits 423
27.1 Introduction 423
27.2 Series a.c. circuits 424
27.3 Further worked problems on series a.c. circuits 430

28 Application of complex numbers to parallel a.c. networks 435
28.1 Introduction 435
28.2 Admittance, conductance and susceptance 436
28.3 Parallel a.c. networks 439
28.4 Further worked problems on parallel a.c. networks 443

29 Power in a.c. circuits 446
29.1 Introduction 446
29.2 Determination of power in a.c. circuits 447
29.3 Power triangle and power factor 449
29.4 Use of complex numbers for determination of power 450
29.5 Power factor improvement 454

Revision Test 8 459

30 A.c. bridges 460
30.1 Introduction 461
30.2 Balance conditions for an a.c. bridge 461
30.3 Types of a.c. bridge circuit 462
30.4 Worked problems on a.c. bridges 467

31 Series resonance and Q-factor 471
31.1 Introduction 472
31.2 Series resonance 472
31.3 Q-factor 474
31.4 Voltage magnification 476
31.5 Q-factors in series 478
31.6 Bandwidth 479
31.7 Small deviations from the resonant frequency 483

32 Parallel resonance and Q-factor 486
32.1 Introduction 486
32.2 The *LR–C* parallel network 487
32.3 Dynamic resistance 488
32.4 The *LR–CR* parallel network 488
32.5 Q-factor in a parallel network 489
32.6 Further worked problems on parallel resonance and Q-factor 493

Revision Test 9 496

33 Introduction to network analysis 497
33.1 Introduction 497
33.2 Solution of simultaneous equations using determinants 498
33.3 Network analysis using Kirchhoff's laws 499

34 Mesh-current and nodal analysis 507
34.1 Mesh-current analysis 507
34.2 Nodal analysis 511

35 The superposition theorem 518
35.1 Introduction 518
35.2 Using the superposition theorem 518
35.3 Further worked problems on the superposition theorem 523

36 Thévenin's and Norton's theorems 528
36.1 Introduction 528
36.2 Thévenin's theorem 529
36.3 Further worked problems on Thévenin's theorem 535
36.4 Norton's theorem 539
36.5 Thévenin and Norton equivalent networks 546

Revision Test 10 551

37 Delta–star and star–delta transformations 552
37.1 Introduction 552
37.2 Delta and star connections 552
37.3 Delta–star transformation 553
37.4 Star–delta transformation 561

38 Maximum power transfer theorems and impedance matching 565
38.1 Maximum power transfer theorems 566
38.2 Impedance matching 571

Revision Test 11 574

39 Complex waveforms 575
39.1 Introduction 576
39.2 The general equation for a complex waveform 576
39.3 Harmonic synthesis 577
39.4 Fourier series of periodic and non-periodic functions 585
39.5 Even and odd functions and Fourier series over any range 590
39.6 R.m.s. value, mean value and the form factor of a complex wave 594
39.7 Power associated with complex waves 597
39.8 Harmonics in single-phase circuits 599
39.9 Further worked problems on harmonics in single-phase circuits 602
39.10 Resonance due to harmonics 606
39.11 Sources of harmonics 608

40 A numerical method of harmonic analysis **612**
- 40.1 Introduction 612
- 40.2 Harmonic analysis on data given in tabular or graphical form 612
- 40.3 Complex waveform considerations 616

41 Magnetic materials **619**
- 41.1 Revision of terms and units used with magnetic circuits 620
- 41.2 Magnetic properties of materials 621
- 41.3 Hysteresis and hysteresis loss 622
- 41.4 Eddy current loss 626
- 41.5 Separation of hysteresis and eddy current losses 629
- 41.6 Non-permanent magnetic materials 631
- 41.7 Permanent magnetic materials 633

Revision Test 12 **634**

42 Dielectrics and dielectric loss **635**
- 42.1 Electric fields, capacitance and permittivity 635
- 42.2 Polarization 636
- 42.3 Dielectric strength 636
- 42.4 Thermal effects 637
- 42.5 Mechanical properties 638
- 42.6 Types of practical capacitor 638
- 42.7 Liquid dielectrics and gas insulation 638
- 42.8 Dielectric loss and loss angle 638

43 Field theory **642**
- 43.1 Field plotting by curvilinear squares 643
- 43.2 Capacitance between concentric cylinders 646
- 43.3 Capacitance of an isolated twin line 651
- 43.4 Energy stored in an electric field 654
- 43.5 Induced e.m.f. and inductance 656
- 43.6 Inductance of a concentric cylinder (or coaxial cable) 656
- 43.7 Inductance of an isolated twin line 659
- 43.8 Energy stored in an electromagnetic field 662

44 Attenuators **665**
- 44.1 Introduction 666
- 44.2 Characteristic impedance 666
- 44.3 Logarithmic ratios 668
- 44.4 Symmetrical T- and π-attenuators 670
- 44.5 Insertion loss 675
- 44.6 Asymmetrical T- and π-sections 678
- 44.7 The L-section attenuator 681
- 44.8 Two-port networks in cascade 683
- 44.9 *ABCD* parameters 686
- 44.10 *ABCD* parameters for networks 689
- 44.11 Characteristic impedance in terms of *ABCD* parameters 695

Revision Test 13 **697**

45 Filter networks **698**
- 45.1 Introduction 698
- 45.2 Basic types of filter sections 699
- 45.3 The characteristic impedance and the attenuation of filter sections 701
- 45.4 Ladder networks 702
- 45.5 Low-pass filter sections 703
- 45.6 High-pass filter sections 709
- 45.7 Propagation coefficient and time delay in filter sections 714
- 45.8 '*m*-derived' filter sections 720
- 45.9 Practical composite filters 725

46 Magnetically coupled circuits **728**
- 46.1 Introduction 728
- 46.2 Self-inductance 728
- 46.3 Mutual inductance 729
- 46.4 Coupling coefficient 730
- 46.5 Coils connected in series 731
- 46.6 Coupled circuits 734
- 46.7 Dot rule for coupled circuits 739

47 Transmission lines **746**
- 47.1 Introduction 746
- 47.2 Transmission line primary constants 747
- 47.3 Phase delay, wavelength and velocity of propagation 748
- 47.4 Current and voltage relationships 749
- 47.5 Characteristic impedance and propagation coefficient in terms of the primary constants 751
- 47.6 Distortion on transmission lines 755
- 47.7 Wave reflection and the reflection coefficient 757
- 47.8 Standing-waves and the standing-wave ratio 760

48 Transients and Laplace transforms **765**
- 48.1 Introduction 766
- 48.2 Response of *R–C* series circuit to a step input 766
- 48.3 Response of *R–L* series circuit to a step input 768
- 48.4 *L–R–C* series circuit response 771
- 48.5 Introduction to Laplace transforms 774
- 48.6 Inverse Laplace transforms and the solution of differential equations 779
- 48.7 Laplace transform analysis directly from the circuit diagram 784

48.8 L–R–C series circuit using Laplace transforms 794
48.9 Initial conditions 797

Revision Test 14 **801**

Main formulae for Part 4 Advanced circuit theory and technology **802**

Part 5 General reference **807**

Standard electrical quantities – their symbols and units **809**

Greek alphabet **812**

Common prefixes **813**

Resistor colour coding and ohmic values **814**

Answers to Practice Exercises **815**

Index **837**

On the Website

Some practical laboratory experiments

1 Ohm's law 2
2 Series–parallel d.c. circuit 3
3 Superposition theorem 4
4 Thévenin's theorem 6
5 Use of a CRO to measure voltage, frequency and phase 8
6 Use of a CRO with a bridge rectifier circuit 9
7 Measurement of the inductance of a coil 10
8 Series a.c. circuit and resonance 11
9 Parallel a.c. circuit and resonance 13
10 Charging and discharging a capacitor 15

To download and edit go to:
www.routledge.com/cw/bird

Preface

Electrical Circuit Theory and Technology 6th Edition provides coverage for a wide range of courses that contain electrical principles, circuit theory and technology in their syllabuses, from **introductory to degree level** – and including Edexcel BTEC Levels 2 to 5 National Certificate/Diploma, Higher National Certificate/Diploma and Foundation degree in Engineering

In this new sixth edition, **new material added** includes some mathematics revision needed for electrical and electronic principles, ways of generating electricity – the present and the future (including more on renewable energy), more on lithium-ion batteries, along with other minor modifications.

The text is set out in **five parts** as follows:

PART 1, comprising chapters 1 to 12, involves **Revision of some Basic Mathematics** needed for Electrical and Electronic Principles.

PART 2, involving chapters 3 to 14, contains **Basic Electrical Engineering Principles** which any student wishing to progress in electrical engineering would need to know. An introduction to units, electrical circuits, resistance variation, batteries and alternative sources of energy, series and parallel circuits, capacitors and capacitance, magnetic circuits, electromagnetism, electromagnetic induction, electrical measuring instruments and measurements, semiconductor diodes and transistors are all included in this section.

PART 3, involving chapters 15 to 25, contains **Electrical Principles and Technology** suitable for National Certificate, National Diploma and City and Guilds courses in electrical and electronic engineering. D.c. circuit theory, alternating voltages and currents, single-phase series and parallel circuits, d.c. transients, operational amplifiers, ways of generating electricity, three-phase systems, transformers, d.c. machines and three-phase induction motors are all included in this section.

PART 4, involving chapters 26 to 48, contains **Advanced Circuit Theory and Technology** suitable for Degree, Foundation degree, Higher National Certificate/Diploma and City and Guilds courses in electrical and electronic/telecommunications engineering. The three earlier sections of the book will provide a valuable reference/revision for students at this level.

Complex numbers and their application to series and parallel networks, power in a.c. circuits, a.c. bridges, series and parallel resonance and Q-factor, network analysis involving Kirchhoff's laws, mesh and nodal analysis, the superposition theorem, Thévenin's and Norton's theorems, delta-star and star-delta transforms, maximum power transfer theorems and impedance matching, complex waveforms, Fourier series, harmonic analysis, magnetic materials, dielectrics and dielectric loss, field theory, attenuators, filter networks, magnetically coupled circuits, transmission line theory and transients and Laplace transforms are all included in this section.

PART 5 provides a short **General Reference** for standard electrical quantities – their symbols and units, the Greek alphabet, common prefixes and resistor colour coding and ohmic values.

At the beginning of each of the 48 chapters a brief explanation as to why it is important to understand the material contained within that chapter is included, together with a list of **learning objectives**.

At the end of each of the first four parts of the text is a handy reference of the **main formulae** used.

There are a number of Internet downloads freely available to both students and lecturers/instructors; these are listed on page xiii.

It is not possible to acquire a thorough understanding of electrical principles, circuit theory and technology without working through a large number of numerical problems. It is for this reason that *Electrical Circuit Theory and Technology 6th Edition* contains nearly **800 detailed worked problems**, together with some **1350 further problems (with answers at the back of the book)**, arranged within **202 Practice Exercises** that appear every few pages throughout the text. Some **1153 line diagrams** further enhance the understanding of the theory.

Fourteen Revision Tests have been included, interspersed within the text every few chapters. For example, Revision Test 1 tests understanding of chapters 3 to 6, Revision Test 2 tests understanding of chapters 7 to 9, Revision Test 3 tests understanding of chapters 10 to 14, and so on. These Revision Tests do not have answers given since it is envisaged that lecturers/instructors could set the Revision Tests for students to attempt as part of their course structure. Lecturers/instructors may obtain a complimentary set of solutions of the Revision Tests in an **Instructor's Manual** available from the publishers via the internet – see below.

Learning by example is at the heart of *Electrical Circuit Theory and Technology 6th Edition*.

JOHN BIRD
Royal Naval Defence College of Marine and Air Engineering, HMS Sultan,
formerly University of Portsmouth
and Highbury College, Portsmouth

John Bird is the former Head of Applied Electronics in the Faculty of Technology at Highbury College, Portsmouth, UK. More recently, he has combined freelance lecturing at the University of Portsmouth with Examiner responsibilities for Advanced Mathematics with City and Guilds, and examining for the International Baccalaureate. He is the author of some 130 textbooks on engineering and mathematical subjects with worldwide sales of over one million copies. He is currently lecturing at the Defence School of Marine and Air Engineering in the Defence College of Technical Training at HMS Sultan, Gosport, Hampshire, UK.

Free Web downloads

The following support material is available from www.routledge.com/cw/bird

For Students:

1. **Full solutions to all 1350 further questions in the Practice Exercises**
2. **A set of formulae for each of the first four sections of the text**
3. **Multiple choice questions**
4. **Information on 38 Engineers/Scientists mentioned in the text**

For Lecturers/Instructors:

1–4. **As per students 1–4 above**

5. **Full solutions and marking scheme for each of the 14 Revision Tests; also, each test may be downloaded.**
6. **Lesson Plans and revision material**. Typical 30-week lesson plans for 'Electrical and Electronic Principles', Unit 6, and 'Further Electrical Principles', Unit 64, are included, together with two practice examination question papers (with solutions) for each of the modules.
7. **Ten practical Laboratory Experiments** are available. It may be that tutors will want to edit these experiments to suit their own equipment/component availability.
8. **All 1153 illustrations used in the text may be downloaded for use in PowerPoint Presentations.**

Part 1

Revision of some basic mathematics

Chapter 1

Some mathematics revision

***Why it is important to understand:* Some mathematics revision**

Mathematics is a vital tool for professional and chartered engineers. It is used in electrical and electronic engineering, in mechanical and manufacturing engineering, in civil and structural engineering, in naval architecture and marine engineering and in aeronautical and rocket engineering. In these various branches of engineering, it is very often much cheaper and safer to design your artefact with the aid of mathematics – rather than through guesswork. 'Guesswork' may be reasonably satisfactory if you are designing an exactly similar artefact as one that has already proven satisfactory; however, the classification societies will usually require you to provide the calculations proving that the artefact is safe and sound. Moreover, these calculations may not be readily available to you and you may have to provide fresh calculations, to prove that your artefact is 'roadworthy'. For example, if you design a tall building or a long bridge by 'guesswork', and the building or bridge do not prove to be structurally reliable, it could cost you a fortune to rectify the deficiencies. This cost may dwarf the initial estimate you made to construct these structures, and cause you to go bankrupt. Thus, without mathematics, the prospective professional or chartered engineer is very severely disadvantaged.

Knowledge of mathematics provides the basis for all engineering.

At the end of this chapter you should be able to:

- use a calculator and evaluate formulae
- manipulate fractions
- understand and perform calculations with percentages
- appreciate ratios and direct and inverse proportion
- understand and use the laws of indices
- expand equations containing brackets
- solve simple equations
- transpose formulae
- solve simultaneous equations in two unknowns

Electrical Circuit Theory and Technology. 978-1-138-67349-6, © 2017 John Bird. Published by Taylor & Francis. All rights reserved.

1.1 Use of calculator and evaluating formulae

In engineering, calculations often need to be performed. For simple numbers it is useful to be able to use mental arithmetic. However, when numbers are larger an electronic calculator needs to be used.

In engineering calculations it is essential to have a **scientific notation calculator** which will have all the necessary functions needed, and more. This chapter assumes you have a **CASIO fx-991ES PLUS calculator**, or similar. If you can accurately use a calculator, your confidence with engineering calculations will improve.

Check that you can use a calculator in the following Practice Exercise.

Practice Exercise 1 Use of calculator (Answers on page 815)

1. Evaluate $378.37 - 298.651 + 45.64 - 94.562$
2. Evaluate $\dfrac{17.35 \times 34.27}{41.53 \div 3.76}$ correct to 3 decimal places
3. Evaluate $\dfrac{(4.527 + 3.63)}{(452.51 \div 34.75)} + 0.468$ correct to 5 significant figures
4. Evaluate $52.34 - \dfrac{(912.5 \div 41.46)}{(24.6 - 13.652)}$ correct to 3 decimal places
5. Evaluate $\dfrac{52.14 \times 0.347 \times 11.23}{19.73 \div 3.54}$ correct to 4 significant figures
6. Evaluate 6.85^2 correct to 3 decimal places
7. Evaluate $(0.036)^2$ in engineering form
8. Evaluate 1.3^3
9. Evaluate $(0.38)^3$ correct to 4 decimal places
10. Evaluate $(0.018)^3$ in engineering form
11. Evaluate $\dfrac{1}{0.00725}$ correct to 1 decimal place
12. Evaluate $\dfrac{1}{0.065} - \dfrac{1}{2.341}$ correct to 4 significant figures
13. Evaluate 2.1^4
14. Evaluate $(0.22)^5$ correct to 5 significant figures in engineering form
15. Evaluate $(1.012)^7$ correct to 4 decimal places
16. Evaluate $1.1^3 + 2.9^4 - 4.4^2$ correct to 4 significant figures
17. Evaluate $\sqrt{34528}$ correct to 2 decimal places
18. Evaluate $\sqrt[3]{17}$ correct to 3 decimal places
19. Evaluate $\sqrt[6]{2451} - \sqrt[4]{46}$ correct to 3 decimal places

Express the answers to questions 20 to 23 in engineering form.

20. Evaluate $5 \times 10^{-3} \times 7 \times 10^8$
21. Evaluate $\dfrac{6 \times 10^3 \times 14 \times 10^{-4}}{2 \times 10^6}$
22. Evaluate $\dfrac{56.43 \times 10^{-3} \times 3 \times 10^4}{8.349 \times 10^3}$ correct to 3 decimal places
23. Evaluate $\dfrac{99 \times 10^5 \times 6.7 \times 10^{-3}}{36.2 \times 10^{-4}}$ correct to 4 significant figures
24. Evaluate $\dfrac{4}{5} - \dfrac{1}{3}$ as a decimal, correct to 4 decimal places
25. Evaluate $\dfrac{2}{3} - \dfrac{1}{6} + \dfrac{3}{7}$ as a fraction
26. Evaluate $2\dfrac{5}{6} + 1\dfrac{5}{8}$ as a decimal, correct to 4 significant figures
27. Evaluate $5\dfrac{6}{7} - 3\dfrac{1}{8}$ as a decimal, correct to 4 significant figures
28. Evaluate $\dfrac{3}{4} \times \dfrac{4}{5} - \dfrac{2}{3} \div \dfrac{4}{9}$ as a fraction
29. Evaluate $8\dfrac{8}{9} \div 2\dfrac{2}{3}$ as a mixed number
30. Evaluate $3\dfrac{1}{5} \times 1\dfrac{1}{3} - 1\dfrac{7}{10}$ as a decimal, correct to 3 decimal places

31. Evaluate $\dfrac{\left(4\frac{1}{5}-1\frac{2}{3}\right)}{\left(3\frac{1}{4}\times 2\frac{3}{5}\right)}-\dfrac{2}{9}$ as a decimal, correct to 3 significant figures

In questions 32 to 38, evaluate correct to 4 decimal places.

32. Evaluate $\sin 67°$
33. Evaluate $\tan 71°$
34. Evaluate $\cos 63.74°$
35. Evaluate $\tan 39.55° - \sin 52.53°$
36. Evaluate $\sin(0.437\,\text{rad})$
37. Evaluate $\tan(5.673\,\text{rad})$
38. Evaluate $\dfrac{(\sin 42.6°)(\tan 83.2°)}{\cos 13.8°}$

In questions 39 to 45, evaluate correct to 4 significant figures.

39. 1.59π
40. $2.7(\pi - 1)$
41. $\pi^2\left(\sqrt{13}-1\right)$
42. $8.5e^{-2.5}$
43. $3e^{(2\pi-1)}$
44. $\sqrt{\left[\dfrac{5.52\pi}{2e^{-2}\times\sqrt{26.73}}\right]}$
45. $\sqrt{\left[\dfrac{e^{\left(2-\sqrt{3}\right)}}{\pi\times\sqrt{8.57}}\right]}$

Evaluation of formulae

The statement $\mathbf{y = mx + c}$ is called a **formula** for y in terms of m, x and c.
y, m, x and c are called **symbols**.
When given values of m, x and c we can evaluate y.
There are a large number of formulae used in engineering and in this section we will insert numbers in place of symbols to evaluate engineering quantities.
Here are some practical examples. Check with your calculator that you agree with the working and answers.

Problem 1. In an electrical circuit the voltage V is given by Ohm's law, i.e. V = IR. Find, correct to 4 significant figures, the voltage when I = 5.36 A and R = 14.76 Ω

$$V = IR = I \times R = 5.36 \times 14.76$$

Hence, **voltage V = 79.11 V, correct to 4 significant figures**

Problem 2. Velocity v is given by v = u + at. If u = 9.54 m/s, a = 3.67 m/s^2 and t = 7.82 s, find v, correct to 3 significant figures.

$$\begin{aligned} v &= u + at = 9.54 + 3.67 \times 7.82 \\ &= 9.54 + 28.6994 = 38.2394 \end{aligned}$$

Hence, **velocity v = 38.2 m/s, correct to 3 significant figures**

Problem 3. The area, A, of a circle is given by $A = \pi r^2$. Determine the area correct to 2 decimal places, given radius r = 5.23 m.

$$A = \pi r^2 = \pi(5.23)^2 = \pi(27.3529)$$

Hence, **area, A = 85.93 m^2, correct to 2 decimal places**

Problem 4. Density $= \dfrac{\text{mass}}{\text{volume}}$. Find the density when the mass is 6.45 kg and the volume is $300 \times 10^{-6}\,\text{m}^3$.

$$\textbf{Density} = \frac{\text{mass}}{\text{volume}} = \frac{6.45\,\text{kg}}{300\times10^{-6}\,\text{m}^3} = \mathbf{21500\,kg/m^3}$$

Problem 5. The power, P watts, dissipated in an electrical circuit is given by the formula $P = \dfrac{V^2}{R}$. Evaluate the power, correct to 4 significant figures, given that V = 230 V and R = 35.63 Ω

$$P = \frac{V^2}{R} = \frac{(230)^2}{35.63} = \frac{52900}{35.63} = 1484.70390\ldots$$

Press ENG and $1.48470390.. \times 10^3$ appears on the screen
Hence, **power, P = 1485 W or 1.485 kW correct to 4 significant figures.**

Problem 6. Resistance, R Ω, varies with temperature according to the formula $R = R_0(1+\alpha t)$. Evaluate R, correct to 3 significant figures, given $R_0 = 14.59$, $\alpha = 0.0043$ and $t = 80$

$$R = R_0(1+\alpha t) = 14.59[1+(0.0043)(80)]$$
$$= 14.59(1+0.344) = 14.59(1.344)$$

Hence, **resistance, R = 19.6 Ω, correct to 3 significant figures**

Problem 7. The current, I amperes, in an a.c. circuit is given by: $I = \dfrac{V}{\sqrt{(R^2+X^2)}}$ Evaluate the current, correct to 2 decimal places, when V = 250 V, R = 25.0 Ω and X = 18.0 Ω

$$I = \frac{V}{\sqrt{(R^2+X^2)}} = \frac{250}{\sqrt{(25.0^2+18.0^2)}} = 8.11534341\ldots$$

Hence, **current, I = 8.12 A, correct to 2 decimal places**

Now try the following Practice Exercise

Practice Exercise 2 Evaluation of formulae (Answers on page 815)

1. The area A of a rectangle is given by the formula $A = l \times b$. Evaluate the area, correct to 2 decimal places, when l = 12.4 cm and b = 5.37 cm
2. The circumference C of a circle is given by the formula $C = 2\pi r$. Determine the circumference, correct to 2 decimal places, given r = 8.40 mm
3. A formula used in connection with gases is $R = \dfrac{PV}{T}$. Evaluate R when P = 1500, V = 5 and T = 200
4. The velocity of a body is given by $v = u + at$. The initial velocity u is measured when time t is 15 seconds and found to be 12 m/s. If the acceleration a is 9.81 m/s^2 calculate the final velocity v
5. Calculate the current I in an electrical circuit, correct to 3 significant figures, when I = V/R amperes when the voltage V is measured and found to be 7.2 V and the resistance R is 17.7 Ω
6. Find the distance s, given that $s = \dfrac{1}{2}gt^2$. Time t = 0.032 seconds and acceleration due to gravity g = 9.81 m/s^2. Give the answer in millimetres correct to 3 significant figures.
7. The energy stored in a capacitor is given by $E = \dfrac{1}{2}CV^2$ joules. Determine the energy when capacitance $C = 5 \times 10^{-6}$ farads and voltage V = 240 V
8. Find the area A of a triangle, correct to 1 decimal place, given $A = \dfrac{1}{2}bh$, when the base length b is 23.42 m and the height h is 53.7 m
9. Resistance R_2 is given by $R_2 = R_1(1+\alpha t)$. Find R_2, correct to 4 significant figures, when $R_1 = 220$, $\alpha = 0.00027$ and t = 75.6
10. Density $= \dfrac{\text{mass}}{\text{volume}}$. Find the density, correct to 4 significant figures, when the mass is 2.462 kg and the volume is 173 cm^3. Give the answer in units of kg/m^3. Note that 1 cm^3 = 10^{-6}m^3
11. Evaluate resistance R_T, correct to 4 significant figures, given $\dfrac{1}{R_T} = \dfrac{1}{R_1} + \dfrac{1}{R_2} + \dfrac{1}{R_3}$ when $R_1 = 5.5\,\Omega$, $R_2 = 7.42\,\Omega$ and $R_3 = 12.6\,\Omega$
12. The potential difference, V volts, available at battery terminals is given by $V = E - Ir$. Evaluate V when E = 5.62, I = 0.70 and R = 4.30
13. The current I amperes flowing in a number of cells is given by $I = \dfrac{nE}{R+nr}$. Evaluate the current, correct to 3 significant figures, when n = 36. E = 2.20, R = 2.80 and r = 0.50
14. Energy, E joules, is given by the formula $E = \dfrac{1}{2}LI^2$. Evaluate the energy when L = 5.5 H and I = 1.2 A
15. The current I amperes in an a.c. circuit is given by $I = \dfrac{V}{\sqrt{(R^2+X^2)}}$. Evaluate the

current, correct to 4 significant figures, when $V = 250\,V$, $R = 11.0\,\Omega$ and $X = 16.2\,\Omega$

1.2 Fractions

An example of a fraction is $\frac{2}{3}$ where the top line, i.e. the 2, is referred to as the **numerator** and the bottom line, i.e. the 3, is referred to as the **denominator**.

A **proper fraction** is one where the numerator is smaller than the denominator, examples being $\frac{2}{3}, \frac{1}{2}, \frac{3}{8}, \frac{5}{16}$, and so on.

An **improper fraction** is one where the denominator is smaller than the numerator, examples being $\frac{3}{2}, \frac{2}{1}, \frac{8}{3}, \frac{16}{5}$, and so on.

Addition of fractions is demonstrated in the following worked problems.

Problem 8. Evaluate A, given $A = \frac{1}{2} + \frac{1}{3}$

The lowest common denominator of the two denominators 2 and 3 is 6, i.e. 6 is the lowest number that both 2 and 3 will divide into.

Then $\frac{1}{2} = \frac{3}{6}$ and $\frac{1}{3} = \frac{2}{6}$ i.e. both $\frac{1}{2}$ and $\frac{1}{3}$ have the common denominator, namely 6.

The two fractions can therefore be added as:

$$\mathbf{A} = \frac{\mathbf{1}}{\mathbf{2}} + \frac{\mathbf{1}}{\mathbf{3}} = \frac{3}{6} + \frac{2}{6} = \frac{3+2}{6} = \frac{\mathbf{5}}{\mathbf{6}}$$

Problem 9. Evaluate A, given $A = \frac{2}{3} + \frac{3}{4}$

A common denominator can be obtained by multiplying the two denominators together, i.e. the common denominator is $3 \times 4 = 12$

The two fractions can now be made equivalent, i.e. $\frac{2}{3} = \frac{8}{12}$ and $\frac{3}{4} = \frac{9}{12}$

so that they can be easily added together, as follows:

$$A = \frac{2}{3} + \frac{3}{4} = \frac{8}{12} + \frac{9}{12} = \frac{8+9}{12} = \frac{17}{12}$$

i.e. $\mathbf{A} = \frac{\mathbf{2}}{\mathbf{3}} + \frac{\mathbf{3}}{\mathbf{4}} = \mathbf{1}\frac{\mathbf{5}}{\mathbf{12}}$

Problem 10. Evaluate A, given $A = \frac{1}{6} + \frac{2}{7} + \frac{3}{2}$

A suitable common denominator can be obtained by multiplying $6 \times 7 = 42$, and all three denominators divide exactly into 42.

Thus, $\frac{1}{6} = \frac{7}{42}, \frac{2}{7} = \frac{12}{42}$ and $\frac{3}{2} = \frac{63}{42}$

Hence, $A = \frac{1}{6} + \frac{2}{7} + \frac{3}{2} = \frac{7}{42} + \frac{12}{42} + \frac{63}{42}$

$= \frac{7+12+63}{42} = \frac{82}{42} = \frac{41}{21}$

i.e. $\mathbf{A} = \frac{\mathbf{1}}{\mathbf{6}} + \frac{\mathbf{2}}{\mathbf{7}} + \frac{\mathbf{3}}{\mathbf{2}} = \mathbf{1}\frac{\mathbf{20}}{\mathbf{21}}$

Problem 11. Determine A as a single fraction, given $A = \frac{1}{x} + \frac{2}{y}$

A common denominator can be obtained by multiplying the two denominators together, i.e. xy

Thus, $\frac{1}{x} = \frac{y}{xy}$ and $\frac{2}{y} = \frac{2x}{xy}$

Hence, $A = \frac{1}{x} + \frac{2}{y} = \frac{y}{xy} + \frac{2x}{xy}$ i.e. $\mathbf{A} = \frac{\mathbf{y + 2x}}{\mathbf{xy}}$

Note that addition, subtraction, multiplication and division of fractions may be determined using a **calculator** (for example, the CASIO fx-991ES PLUS).

Locate the $\frac{\square}{\square}$ and $\square\frac{\square}{\square}$ functions on your calculator (the latter function is a shift function found above the $\frac{\square}{\square}$ function) and then check the following worked problems.

Problem 12. Evaluate $\frac{1}{4} + \frac{2}{3}$ using a calculator

(i) Press $\frac{\square}{\square}$ function

(ii) Type in 1

(iii) Press ↓ on the cursor key and type in 4

(iv) $\frac{1}{4}$ appears on the screen

(v) Press → on the cursor key and type in +

(vi) Press $\frac{\square}{\square}$ function

(vii) Type in 2

(viii) Press ↓ on the cursor key and type in 3

(ix) Press → on the cursor key

(x) Press = and the answer $\frac{11}{12}$ appears

(xi) Press S ⇔ D function and the fraction changes to a decimal 0.9166666….

Thus, $\mathbf{\frac{1}{4}+\frac{2}{3}=\frac{11}{12}=0.9167}$ as a decimal, correct to 4 decimal places.

It is also possible to deal with **mixed numbers** on the calculator.

Press Shift then the $\frac{\square}{\square}$ function and $\square\frac{\square}{\square}$ appears.

Problem 13. Evaluate $5\frac{1}{5}-3\frac{3}{4}$ using a calculator

(i) Press Shift then the $\frac{\square}{\square}$ function and $\square\frac{\square}{\square}$ appears on the screen

(ii) Type in 5 then → on the cursor key

(iii) Type in 1 and ↓ on the cursor key

(iv) Type in 5 and $5\frac{1}{5}$ appears on the screen

(v) Press → on the cursor key

(vi) Type in – and then press Shift then the $\frac{\square}{\square}$ function and $5\frac{1}{5}-\square\frac{\square}{\square}$ appears on the screen

(vii) Type in 3 then → on the cursor key

(viii) Type in 3 and ↓ on the cursor key

(ix) Type in 4 and $5\frac{1}{5}-3\frac{3}{4}$ appears on the screen

(x) Press = and the answer $\frac{29}{20}$ appears

(xi) Press shift and then $S \Leftrightarrow D$ function and $1\frac{9}{20}$ appears

(xii) Press S ⇔ D function and the fraction changes to a decimal 1.45

Thus, $\mathbf{5\frac{1}{5}-3\frac{3}{4}=\frac{29}{20}=1\frac{9}{20}=1.45}$ as a decimal

---

Now try the following Practice Exercise

Practice Exercise 3 Fractions (Answers on page 815)

In problems 1 to 3, evaluate the given fractions

1. $\frac{1}{3}+\frac{1}{4}$
2. $\frac{1}{5}+\frac{1}{4}$
3. $\frac{1}{6}+\frac{1}{2}-\frac{1}{5}$

In problems 4 and 5, use a calculator to evaluate the given expressions

4. $\frac{1}{3}-\frac{3}{4}\times\frac{8}{21}$
5. $\frac{3}{4}\times\frac{4}{5}-\frac{2}{3}\div\frac{4}{9}$
6. Evaluate $\frac{3}{8}+\frac{5}{6}-\frac{1}{2}$ as a decimal, correct to 4 decimal places.
7. Evaluate $8\frac{8}{9}\div 2\frac{2}{3}$ as a mixed number.
8. Evaluate $3\frac{1}{5}\times 1\frac{1}{3}-1\frac{7}{10}$ as a decimal, correct to 3 decimal places.
9. Determine $\frac{2}{x}+\frac{3}{y}$ as a single fraction.

1.3 Percentages

Percentages are used to give a common standard. The use of percentages is very common in many aspects of commercial life, as well as in engineering. Interest rates, sale reductions, pay rises, exams and VAT are all examples where percentages are used.

Percentages are fractions having 100 as their denominator.

For example, the fraction $\frac{40}{100}$ is written as 40% and is read as 'forty per cent'.

The easiest way to understand percentages is to go through some worked examples.

Problem 14. Express 0.275 as a percentage

$$0.275 = 0.275 \times 100\% = \mathbf{27.5\%}$$

Problem 15. Express 17.5% as a decimal number

$$17.5\% = \frac{17.5}{100} = \mathbf{0.175}$$

Problem 16. Express $\frac{5}{8}$ as a percentage

$$\frac{5}{8} = \frac{5}{8} \times 100\% = \frac{500}{8}\% = \mathbf{62.5\%}$$

Problem 17. In two successive tests a student gains marks of 57/79 and 49/67. Is the second mark better or worse than the first?

$$57/79 = \frac{57}{79} = \frac{57}{79} \times 100\% = \frac{5700}{79}\%$$

$$= \mathbf{72.15}\% \text{ correct to 2 decimal places.}$$

$$49/67 = \frac{49}{67} = \frac{49}{67} \times 100\% = \frac{4900}{67}\%$$

$$= \mathbf{73.13}\% \text{ correct to 2 decimal places}$$

Hence, **the second test mark is marginally better than the first test.**

This question demonstrates how much easier it is to compare two fractions when they are expressed as percentages.

Problem 18. Express 75% as a fraction

$$75\% = \frac{75}{100} = \mathbf{\frac{3}{4}}$$

The fraction $\frac{75}{100}$ is reduced to its simplest form by cancelling, i.e. dividing numerator and denominator by 25.

Problem 19. Express 37.5% as a fraction

$$37.5\% = \frac{37.5}{100}$$

$$= \frac{375}{1000} \text{ by multiplying numerator and denominator by 10}$$

$$= \frac{15}{40} \text{ by dividing numerator and denominator by 25}$$

$$= \mathbf{\frac{3}{8}} \text{ by dividing numerator and denominator by 5}$$

Problem 20. Find 27% of £65

$$27\% \text{ of } £65 = \frac{27}{100} \times 65 = \mathbf{£17.55} \text{ by calculator}$$

Problem 21. A 160 GB iPod is advertised as costing £190 excluding VAT. If VAT is added at 20%, what will be the total cost of the iPod?

$$\text{VAT} = 20\% \text{ of } £190 = \frac{20}{100} \times 190 = £38$$

$$\text{Total cost of iPod} = £190 + £38 = \mathbf{£228}$$

A quicker method to determine the total cost is: $1.20 \times £190 = \mathbf{£228}$

Problem 22. Express 23 cm as a percentage of 72 cm, correct to the nearest 1%

23 cm as a percentage of 72 cm

$$= \frac{23}{72} \times 100\% = 31.94444\ldots\ldots\%$$

$$= \mathbf{32}\% \text{ correct to the nearest } 1\%$$

Problem 23. A box of screws increases in price from £45 to £52. Calculate the percentage change in cost, correct to 3 significant figures.

$$\% \text{ change} = \frac{\text{new value} - \text{original value}}{\text{original value}} \times 100\%$$

$$= \frac{52 - 45}{45} \times 100\% = \frac{7}{45} \times 100 = \mathbf{15.6\%}$$

$$= \textbf{percentage change in cost}$$

Problem 24. A drilling speed should be set to 400 rev/min. The nearest speed available on the machine is 412 rev/min. Calculate the percentage over-speed.

% over-speed

$$= \frac{\text{available speed} - \text{correct speed}}{\text{correct speed}} \times 100\%$$

$$= \frac{412 - 400}{400} \times 100\% = \frac{12}{400} \times 100\% = \mathbf{3\%}$$

Now try the following Practice Exercise

Practice Exercise 4 Percentages (Answers on page 815)

In problems 1 and 2, express the given numbers as percentages.

1. 0.057
2. 0.374
3. Express 20% as a decimal number
4. Express $\frac{11}{16}$ as a percentage
5. Express $\frac{5}{13}$ as a percentage, correct to 3 decimal places
6. Place the following in order of size, the smallest first, expressing each as percentages, correct to 1 decimal place:
(a) $\frac{12}{21}$ (b) $\frac{9}{17}$ (c) $\frac{5}{9}$ (d) $\frac{6}{11}$
7. Express 65% as a fraction in its simplest form
8. Calculate 43.6% of 50 kg
9. Determine 36% of 27 mv
10. Calculate correct to 4 significant figures:
(a) 18% of 2758 tonnes
(b) 47% of 18.42 grams
(c) 147% of 14.1 seconds
11. Express:
(a) 140 kg as a percentage of 1 t
(b) 47 s as a percentage of 5 min
(c) 13.4 cm as a percentage of 2.5 m
12. A computer is advertised on the internet at £520, exclusive of VAT. If VAT is payable at 20%, what is the total cost of the computer?
13. Express 325 mm as a percentage of 867 mm, correct to 2 decimal places.
14. When signing a new contract, a Premiership footballer's pay increases from £15,500 to £21,500 per week. Calculate the percentage pay increase, correct to 3 significant figures.
15. A metal rod 1.80 m long is heated and its length expands by 48.6 mm. Calculate the percentage increase in length.

1.4 Ratio and proportion

Ratios

Ratio is a way of comparing amounts of something; it shows how much bigger one thing is than the other. Ratios are generally shown as numbers separated by a colon (:) so the ratio of 2 and 7 is written as 2:7 and we read it as a ratio of 'two to seven'.

Here are some worked examples to help us understand more about ratios.

Problem 25. In a class, the ratio of female to male students is 6:27. Reduce the ratio to its simplest form.

Both 6 and 27 can be divided by 3

Thus, 6:27 is the same as **2:9**

6:27 and 2:9 are called **equivalent ratios**.

It is normal to express ratios in their lowest, or simplest, form. In this example, the simplest form is **2:9** which means for every 2 females in the class there are 9 male students.

Problem 26. A gear wheel having 128 teeth is in mesh with a 48 tooth gear. What is the gear ratio?

Gear ratio = 128:48

A ratio can be simplified by finding common factors.

128 and 48 can both be divided by 2, i.e. 128:48 is the same as 64:24

64 and 24 can both be divided by 8, i.e. 64:24 is the same as 8:3

There is no number that divides completely into both 8 and 3 so 8:3 is the simplest ratio, i.e. **the gear ratio is 8:3**

128:48 is equivalent to 64:24 which is equivalent to 8:3

8:3 is the simplest form.

Problem 27. A wooden pole is 2.08 m long. Divide it in the ratio of 7 to 19.

Since the ratio is 7:19, the total number of parts is $7 + 19 = 26$ parts

26 parts corresponds to 2.08 m = 208 cm, hence, 1 part corresponds to $\frac{208}{26} = 8$

Thus, 7 parts corresponds to $7 \times 8 =$ **56 cm**,

and 19 parts corresponds to $19 \times 8 =$ **152 cm**

Hence, **2.08 m divides in the ratio of 7:19 as 56 cm to 152 cm**

(Check: 56 + 152 must add up to 208, otherwise an error would have been made.)

Problem 28. Express 45 p as a ratio of £7.65 in its simplest form.

Changing both quantities to the same units, i.e. to pence, gives a ratio of 45:765

Dividing both quantities by 5 gives: $45{:}765 \equiv 9{:}153$

Dividing both quantities by 3 gives: $9{:}153 \equiv 3{:}51$

Dividing both quantities by 3 again gives: $3{:}51 \equiv 1{:}17$

Thus, **45p as a ratio of £7.65 is 1:17**

45:765, 9:153, 3:51 and 1:17 are **equivalent ratios** and **1:17 is the simplest ratio**

Problem 29. A glass contains 30 ml of whisky which is 40% alcohol. If 45 ml of water is added and the mixture stirred, what is now the alcohol content?

The 30 ml of whisky contains 40% alcohol

$$= \frac{40}{100} \times 30 = 12\,\text{ml}$$

After 45 ml of water is added we have $30 + 45 = 75$ ml of fluid of which alcohol is 12 ml

Fraction of alcohol present $= \frac{12}{75}$

Percentage of alcohol present $= \frac{12}{75} \times 100\% =$ **16%**

Now try the following Practice Exercise

Practice Exercise 5 Ratios (Answers on page 815)

1. In a box of 333 paper clips, 9 are defective. Express the non-defective paper clips as a ratio of the defective paper clips, in its simplest form.
2. A gear wheel having 84 teeth is in mesh with a 24 tooth gear. Determine the gear ratio in its simplest form.
3. A metal pipe 3.36 m long is to be cut into two in the ratio 6 to 15. Calculate the length of each piece.
4. In a will, £6440 is to be divided between three beneficiaries in the ratio 4:2:1. Calculate the amount each receives.
5. A local map has a scale of 1:22,500. The distance between two motorways is 2.7 km. How far are they apart on the map?
6. Express 130 g as a ratio of 1.95 kg
7. In a laboratory, acid and water are mixed in the ratio 2:5. How much acid is needed to make 266 ml of the mixture?
8. A glass contains 30 ml of gin which is 40% alcohol. If 18 ml of water is added and the mixture stirred, determine the new percentage alcoholic content.
9. A wooden beam 4 m long weighs 84 kg. Determine the mass of a similar beam that is 60 cm long.
10. An alloy is made up of metals P and Q in the ratio 3.25:1 by mass. How much of P has to be added to 4.4 kg of Q to make the alloy.

Direct proportion

Two quantities are in **direct proportion** when they increase or decrease in the **same ratio.**

Here are some worked examples to help us understand more about direct proportion.

Problem 30. 3 energy saving light bulbs cost £7.80. Determine the cost of 7 such light bulbs.

If 3 light bulbs cost £7.80

then 1 light bulb cost $\frac{7.80}{3} = £2.60$

Hence, **7 light bulbs cost** $7 \times £2.60 = $ **£18.20**

Problem 31. If 56 litres of petrol costs £59.92, calculate the cost of 32 litres.

If 56 litres of petrol costs £59.92

then 1 litre of petrol costs $\frac{59.92}{56} = £1.07$

Hence, **32 litres cost** $32 \times 1.07 = $ **£34.24**

Problem 32. Hooke's law states that stress, σ, is directly proportional to strain, ε, within the elastic limit of a material. When, for mild steel, the stress is 63 MPa, the strain is 0.0003. Determine

(a) the value of strain when the stress is 42 MPa

(b) the value of stress when the strain is 0.00072

(a) Stress is directly proportional to strain.
When the stress is 63 MPa, the strain is 0.0003,

hence a stress of 1 MPa corresponds to a strain of $\frac{0.0003}{63}$

and **the value of strain when the stress is 42 MPa** $= \frac{0.0003}{63} \times 42 = \mathbf{0.0002}$

(b) If when the strain is 0.0003, the stress is 63 MPa,

then a strain of 0.0001 corresponds to $\frac{63}{3}$ MPa

and **the value of stress when the strain is 0.00072** $= \frac{63}{3} \times 7.2 = \mathbf{151.2\,MPa}$

Problem 33. Ohm's law state that the current flowing in a fixed resistance is directly proportional to the applied voltage. When 90 mV is applied across a resistor the current flowing is 3 A. Determine

(a) the current when the voltage is 60 mV

(b) the voltage when the current is 4.2 A

(a) Current is directly proportional to the voltage.
When voltage is 90 mV, the current is 3 A, hence a voltage of 1 mV corresponds to a current of $\frac{3}{90}$ A

and **when the voltage is 60 mV,**

the current $= 60 \times \frac{3}{90} = \mathbf{2\,A}$

(b) Voltage is directly proportional to the current.
When current is 3 A, the voltage is 90 mV, hence a current of 1 A corresponds to a voltage of $\frac{90}{3}$ mV = 30 mV

and **when the current is 4.2 A,**

the voltage $= 30 \times 4.2 = \mathbf{126\,mV}$

Now try the following Practice Exercise

Practice Exercise 6 Direct proportion (Answers on page 815)

1. 3 engine parts cost £208.50. Calculate the cost of 8 such parts.
2. If 9 litres of gloss white paint costs £24.75, calculate the cost of 24 litres of the same paint.
3. The total mass of 120 household bricks is 57.6 kg. Determine the mass of 550 such bricks.
4. Hooke's law states that stress is directly proportional to strain within the elastic limit of a material. When, for copper, the stress is 60 MPa, the strain is 0.000625. Determine (a) the strain when the stress is 24 MPa, and (b) the stress when the strain is 0.0005
5. Charles's law states that volume is directly proportional to thermodynamic temperature for a given mass of gas at constant pressure. A gas occupies a volume of 4.8 litres at 330 K. Determine (a) the temperature when the volume is 6.4 litres, and (b) the volume when the temperature is 396 K.
6. Ohm's law states that current is proportional to p.d. in an electrical circuit. When a p.d. of 60 mV is applied across a circuit a current of 24 μA flows. Determine:
   (a) the current flowing when the p.d. is 5 V, and
   (b) the p.d. when the current is 10 mA
7. If 2.2 lb = 1 kg, and 1 lb = 16 oz, determine the number of pounds and ounces in 38 kg (correct to the nearest ounce).
8. If 1 litre = 1.76 pints, and 8 pints = 1 gallon, determine (a) the number of litres in 35 gallons, and (b) the number of gallons in 75 litres.

Inverse proportion

Two variables, x and y, are in inverse proportion to one another if y is proportional to $\frac{1}{x}$, i.e. $y \, \alpha \frac{1}{x}$ or $y = \frac{k}{x}$ or $k = xy$ where k is a constant, called the **coefficient of proportionality**.

Inverse proportion means that as the value of one variable increases, the value of another decreases, and that their product is always the same.

Here are some worked examples on inverse proportion.

Problem 34. It is estimated that a team of four designers would take a year to develop an engineering process. How long would three designers take?

If 4 designers take 1 year, then 1 designer would take 4 years to develop the process.

Hence, 3 designers would take $\frac{4}{3}$ years, i.e. **1 year 4 months**

Problem 35. A team of five people can deliver leaflets to every house in a particular area in four hours. How long will it take a team of three people?

If 5 people take 4 hours to deliver the leaflets, then 1 person would take $5 \times 4 = 20$ hours

Hence, 3 people would take $\frac{20}{3}$ hours, i.e. $6\frac{2}{3}$ hours, i.e. **6 hours 40 minutes**

Problem 36. The electrical resistance R of a piece of wire is inversely proportional to the cross-sectional area A. When $A = 5\,\text{mm}^2$, $R = 7.2$ ohms. Determine

(a) the coefficient of proportionality and

(b) the cross-sectional area when the resistance is 4 ohms.

(a) $R \, \alpha \frac{1}{A}$ i.e. $R = \frac{k}{A}$ or $k = RA$. Hence, when $R = 7.2$ and $A = 5$, the

coefficient of proportionality, $k = (7.2)(5) = \mathbf{36}$

(b) Since $k = RA$ then $A = \frac{k}{R}$

When $R = 4$, the **cross sectional area,**

$$A = \frac{36}{4} = \mathbf{9\,mm^2}$$

Problem 37. Boyle's law states that at constant temperature, the volume V of a fixed mass of gas is inversely proportional to its absolute pressure p. If a gas occupies a volume of $0.08\,\text{m}^3$ at a pressure of 1.5×10^6 pascals, determine (a) the coefficient of proportionality and (b) the volume if the pressure is changed to 4×10^6 pascals.

(a) $V \alpha \frac{1}{p}$ i.e. $V = \frac{k}{p}$ or $k = pV$

Hence, **the coefficient of proportionality,** $k = (1.5 \times 10^6)(0.08) = \mathbf{0.12 \times 10^6}$

(b) **Volume,** $V = \frac{k}{p} = \frac{0.12 \times 10^6}{4 \times 10^6} = \mathbf{0.03 m^3}$

Now try the following Practice Exercise

Practice Exercise 7 Further inverse proportion (Answers on page 816)

1. A 10 kg bag of potatoes lasts for a week with a family of 7 people. Assuming all eat the same amount, how long will the potatoes last if there were only two in the family?
2. If 8 men take 5 days to build a wall, how long would it take 2 men?
3. If y is inversely proportional to x and $y = 15.3$ when $x = 0.6$, determine (a) the coefficient of proportionality, (b) the value of y when x is 1.5, and (c) the value of x when y is 27.2
4. A car travelling at 50 km/h makes a journey in 70 minutes. How long will the journey take at 70 km/h?
5. Boyle's law states that for a gas at constant temperature, the volume of a fixed mass of gas is inversely proportional to its absolute pressure. If a gas occupies a volume of $1.5\,\text{m}^3$ at a pressure of 200×10^3 Pascal's, determine (a) the constant of proportionality, (b) the volume when the pressure is 800×10^3 Pascals and (c) the pressure when the volume is $1.25\,\text{m}^3$

1.5 Laws of indices

The manipulation of indices, powers and roots is a crucial underlying skill needed in algebra.

Law 1: When multiplying two or more numbers having the same base, the indices are added.

For example, $2^2 \times 2^3 = 2^{2+3} = 2^5$

and $5^4 \times 5^2 \times 5^3 = 5^{4+2+3} = 5^9$

More generally, $\mathbf{a^m \times a^n = a^{m+n}}$

For example, $a^3 \times a^4 = a^{3+4} = a^7$

Law 2: When dividing two numbers having the same base, the index in the denominator is subtracted from the index in the numerator.

For example, $\frac{2^5}{2^3} = 2^{5-3} = 2^2$ and $\frac{7^8}{7^5} = 7^{8-5} = 7^3$

More generally, $\mathbf{\frac{a^m}{a^n} = a^{m-n}}$

For example, $\frac{c^5}{c^2} = c^{5-2} = c^3$

Law 3: When a number which is raised to a power is raised to a further power, the indices are multiplied.

For example, $(2^2)^3 = 2^{2\times3} = 2^6$ and $(3^4)^2 = 3^{4\times2} = 3^8$

More generally, $\mathbf{(a^m)^n = a^{mn}}$

For example, $(d^2)^5 = d^{2\times5} = d^{10}$

Law 4: When a number has an index of 0, its value is 1.

For example, $3^0 = 1$ and $17^0 = 1$

More generally, $\mathbf{a^0 = 1}$

Law 5: A number raised to a negative power is the reciprocal of that number raised to a positive power.

For example, $3^{-4} = \frac{1}{3^4}$ and $\frac{1}{2^{-3}} = 2^3$

More generally, $\mathbf{a^{-n} = \frac{1}{a^n}}$ For example, $a^{-2} = \frac{1}{a^2}$

Law 6: When a number is raised to a fractional power the denominator of the fraction is the root of the number and the numerator is the power.

For example, $8^{\frac{2}{3}} = \sqrt[3]{8^2} = (2)^2 = 4$

and $25^{\frac{1}{2}} = \sqrt[2]{25^1} = \sqrt{25^1} = \pm5$ (Note that $\sqrt{\ } \equiv \sqrt[2]{\ }$)

More generally, $\mathbf{a^{\frac{m}{n}} = \sqrt[n]{a^m}}$ For example, $x^{\frac{4}{3}} = \sqrt[3]{x^4}$

Problem 38. Evaluate in index form $5^3 \times 5 \times 5^2$

$5^3 \times 5 \times 5^2 = 5^3 \times 5^1 \times 5^2$ (Note that 5 means 5^1)

$= 5^{3+1+2} = \mathbf{5^6}$ from law 1

Problem 39. Evaluate $\frac{3^5}{3^4}$

From law 2: $\frac{3^5}{3^4} = 3^{5-4} = 3^1 = \mathbf{3}$

Problem 40. Evaluate $\frac{2^4}{2^4}$

$\frac{2^4}{2^4} = 2^{4-4}$ from law 2

$= 2^0 = \mathbf{1}$ from law 4

Any number raised to the power of zero equals 1

Problem 41. Evaluate $\frac{3 \times 3^2}{3^4}$

$$\frac{3 \times 3^2}{3^4} = \frac{3^1 \times 3^2}{3^4} = \frac{3^{1+2}}{3^4} = \frac{3^3}{3^4}$$

$= 3^{3-4} = 3^{-1}$ from laws 1 and 2

$= \mathbf{\frac{1}{3}}$ from law 5

Problem 42. Evaluate $\frac{10^3 \times 10^2}{10^8}$

$\frac{10^3 \times 10^2}{10^8} = \frac{10^{3+2}}{10^8} = \frac{10^5}{10^8}$ from law 1

$= 10^{5-8} = 10^{-3}$ from law 2

$= \frac{1}{10^{+3}} = \frac{1}{1000}$ from law 5

Hence, $\frac{10^3 \times 10^2}{10^8} = \mathbf{10^{-3}} = \mathbf{\frac{1}{1000}} = \mathbf{0.001}$

Problem 43. Simplify: (a) $(2^3)^4$ (b) $(3^2)^5$ expressing the answers in index form.

From law 3:

(a) $(2^3)^4 = 2^{3\times4} = \mathbf{2^{12}}$

(b) $(3^2)^5 = 3^{2\times5} = \mathbf{3^{10}}$

Problem 44. Evaluate: $\frac{(10^2)^3}{10^4 \times 10^2}$

From laws 1, 2, and 3: $\frac{(10^2)^3}{10^4 \times 10^2} = \frac{10^{(2\times3)}}{10^{(4+2)}}$

$= \frac{10^6}{10^6} = 10^{6-6} = 10^0 = \mathbf{1}$ from law 4

Problem 45. Evaluate (a) $4^{1/2}$ (b) $16^{3/4}$ (c) $27^{2/3}$ (d) $9^{-\frac{1}{2}}$

(a) $4^{1/2} = \sqrt[2]{4^1} = \sqrt{4} = \pm\mathbf{2}$

(b) $16^{3/4} = \sqrt[4]{16^3} = (2)^3 = \mathbf{8}$
(Note that it does not matter whether the 4th root of 16 is found first or whether 16 cubed is found first – the same answer will result)

(c) $27^{2/3} = \sqrt[3]{27^2} = (3)^2 = \mathbf{9}$

(d) $9^{-\frac{1}{2}} = \dfrac{1}{9^{\frac{1}{2}}} = \dfrac{1}{\sqrt{9}} = \dfrac{1}{\pm 3} = \pm\mathbf{\dfrac{1}{3}}$

Problem 46. Simplify $a^2b^3c \times ab^2c^5$

$$a^2b^3c \times ab^2c^5 = a^2 \times b^3 \times c \times a \times b^2 \times c^5$$
$$= a^2 \times b^3 \times c^1 \times a^1 \times b^2 \times c^5$$

Grouping together like terms gives:

$$a^2 \times a^1 \times b^3 \times b^2 \times c^1 \times c^5$$

Using law 1 of indices gives:

$$a^{2+1} \times b^{3+2} \times c^{1+5} = a^3 \times b^5 \times c^6$$

i.e. $\mathbf{a^2b^3c \times ab^2c^5 = a^3b^5c^6}$

Problem 47. Simplify $\dfrac{x^5y^2z}{x^2yz^3}$

$$\frac{x^5y^2z}{x^2yz^3} = \frac{x^5 \times y^2 \times z}{x^2 \times y \times z^3} = \frac{x^5}{x^2} \times \frac{y^2}{y^1} \times \frac{z}{z^3}$$
$$= x^{5-2} \times y^{2-1} \times z^{1-3} \text{ by law 2}$$
$$= x^3 \times y^1 \times z^{-2} = \mathbf{x^3yz^{-2}} \text{ or } \mathbf{\frac{x^3y}{z^2}} \text{ by law 5}$$

Now try the following Practice Exercise

Practice Exercise 8 Laws of indices (Answers on page 816)

In questions 1 to 18, evaluate without the aid of a calculator.

1. Evaluate $2^2 \times 2 \times 2^4$
2. Evaluate $3^5 \times 3^3 \times 3$ in index form
3. Evaluate $\dfrac{2^7}{2^3}$
4. Evaluate $\dfrac{3^3}{3^5}$
5. Evaluate 7^0
6. Evaluate $\dfrac{2^3 \times 2 \times 2^6}{2^7}$
7. Evaluate $\dfrac{10 \times 10^6}{10^5}$
8. Evaluate $10^4 \div 10$
9. Evaluate $\dfrac{10^3 \times 10^4}{10^9}$
10. Evaluate $5^6 \times 5^2 \div 5^7$
11. Evaluate $(7^2)^3$ in index form
12. Evaluate $(3^3)^2$
13. Evaluate $\dfrac{3^7 \times 3^4}{3^5}$ in index form
14. Evaluate $\dfrac{(9 \times 3^2)^3}{(3 \times 27)^2}$
15. Evaluate $\dfrac{(16 \times 4)^2}{(2 \times 8)^3}$
16. Evaluate $\dfrac{5^{-2}}{5^{-4}}$
17. Evaluate $\dfrac{3^2 \times 3^{-4}}{3^3}$
18. Evaluate $\dfrac{7^2 \times 7^{-3}}{7 \times 7^{-4}}$

In problems 19 to 36, simplify the following, giving each answer as a power:

19. $z^2 \times z^6$
20. $a \times a^2 \times a^5$
21. $n^8 \times n^{-5}$
22. $b^4 \times b^7$
23. $b^2 \div b^5$
24. $c^5 \times c^3 \div c^4$
25. $\dfrac{m^5 \times m^6}{m^4 \times m^3}$

26. $\frac{(x^2)(x)}{x^6}$

27. $(x^3)^4$

28. $(y^2)^{-3}$

29. $(t \times t^3)^2$

30. $(c^{-7})^{-2}$

31. $\left(\frac{a^2}{a^5}\right)^3$

32. $\left(\frac{1}{b^3}\right)^4$

33. $\left(\frac{b^2}{b^7}\right)^{-2}$

34. $\frac{1}{(s^3)^3}$

35. $p^3qr^2 \times p^2q^5r \times pqr^2$

36. $\frac{x^3y^2z}{x^5yz^3}$

1.6 Brackets

The use of brackets, which are used in many engineering equations, is explained through the following worked problems.

Problem 48. Expand the bracket to determine A, given $A = a(b + c + d)$

Multiplying each term in the bracket by 'a' gives:

$$\mathbf{A} = a(b + c + d) = \mathbf{ab + ac + ad}$$

Problem 49. Expand the brackets to determine A, given $A = a[b(c + d) - e(f - g)]$

When there is more than one set of brackets the innermost brackets are multiplied out first. Hence,
$A = a[b(c + d) - e(f - g)] = a[bc + bd - ef + eg]$
Note that $-e \times -g = +eg$

Now multiplying each term in the square brackets by 'a' gives:

$$\mathbf{A = abc + abd - aef + aeg}$$

Problem 50. Expand the brackets to determine A, given $A = a[b(c + d - e) - f(g - h\{j - k\})]$

The inner brackets are determined first, hence

$$\begin{aligned} A &= a[b(c + d - e) - f(g - h\{j - k\})] \\ &= a[b(c + d - e) - f(g - hj + hk)] \\ &= a[bc + bd - be - fg + fhj - fhk] \end{aligned}$$

i.e. $\mathbf{A = abc + abd - abe - afg + afhj - afhk}$

Problem 51. Evaluate A, given $A = 2[3(6 - 1) - 4(7\{2 + 5\} - 6)]$

$$\begin{aligned} \mathbf{A} &= 2[3(6 - 1) - 4(7\{2 + 5\} - 6)] \\ &= 2[3(6 - 1) - 4(7 \times 7 - 6)] \\ &= 2[3 \times 5 - 4 \times 43] \\ &= 2[15 - 172] = 2[-157] = \mathbf{-314} \end{aligned}$$

Now try the following Practice Exercise

Practice Exercise 9 Brackets (Answers on page 816)

In problems 1 to 2, evaluate A

1. $A = 3(2 + 1 + 4)$
2. $A = 4[5(2 + 1) - 3(6 - 7)]$

Expand the brackets in problems 3 to 7.

3. $2(x - 2y + 3)$
4. $(3x - 4y) + 3(y - z) - (z - 4x)$
5. $2x + [y - (2x + y)]$
6. $24a - [2\{3(5a - b) - 2(a + 2b)\} + 3b]$
7. $ab[c + d - e(f - g + h\{i + j\})]$

1.7 Solving simple equations

To '**solve an equation**' means '**to find the value of the unknown**'.

Here are some examples to demonstrate how simple equations are solved.

Problem 52. Solve the equation: $4x = 20$

Dividing each side of the equation by 4 gives: $\frac{4x}{4} = \frac{20}{4}$

i.e. $\mathbf{x = 5}$ by cancelling

which is the solution to the equation $4x = 20$

The same operation **must** be applied to both sides of an equation so that the equality is maintained.

We can do anything we like to an equation, **as long as we do the same to both sides.**

Problem 53. Solve the equation: $\frac{2x}{5} = 6$

Multiplying both sides by 5 gives: $5\left(\frac{2x}{5}\right) = 5(6)$

Cancelling and removing brackets gives: $2x = 30$

Dividing both sides of the equation by 2 gives: $\frac{2x}{2} = \frac{30}{2}$

Cancelling gives: $\mathbf{x = 15}$

which is the solution of the equation $\frac{2x}{5} = 6$

Problem 54. Solve the equation: $a - 5 = 8$

Adding 5 to both sides of the equation gives:
$a - 5 + 5 = 8 + 5$

i.e. $a = 8 + 5$

i.e. $\mathbf{a = 13}$

which is the solution of the equation $a - 5 = 8$

Note that adding 5 to both sides of the above equation results in the '-5' moving from the LHS to the RHS, but the sign is changed to '+'

Problem 55. Solve the equation: $x + 3 = 7$

Subtracting 3 from both sides gives: $x + 3 - 3 = 7 - 3$

i.e. $x = 7 - 3$

i.e. $\mathbf{x = 4}$

which is the solution of the equation $x + 3 = 7$

Note that subtracting 3 from both sides of the above equation results in the '+3' moving from the LHS to the RHS, but the sign is changed to '−'

So we can move straight from $x + 3 = 7$ to: $x = 7 - 3$

Thus a term can be moved from one side of an equation to the other **as long as a change in sign is made.**

Problem 56. Solve the equation: $6x + 1 = 2x + 9$

In such equations the terms containing x are grouped on one side of the equation and the remaining terms grouped on the other side of the equation. As in Problems 54 and 55, changing from one side of an equation to the other must be accompanied by a change of sign.

Since $6x + 1 = 2x + 9$

then $6x - 2x = 9 - 1$

i.e. $4x = 8$

Dividing both sides by 4 gives: $\frac{4x}{4} = \frac{8}{4}$

Cancelling gives: $\mathbf{x = 2}$

which is the solution of the equation $6x + 1 = 2x + 9$

In the above examples, the solutions can be checked. Thus, in problem 56, where $6x + 1 = 2x + 9$, if $x = 2$ then:

$$\text{LHS of equation} = 6(2) + 1 = 13$$

$$\text{RHS of equation} = 2(2) + 9 = 13$$

Since the left hand side equals the right hand side then $x = 2$ must be the correct solution of the equation.

When solving simple equations, always check your answers by substituting your solution back into the original equation.

Problem 57. Solve the equation: $3(x - 2) = 9$

Removing the bracket gives: $3x - 6 = 9$

Rearranging gives: $3x = 9 + 6$

i.e. $3x = 15$

Dividing both sides by 3 gives: $\mathbf{x = 5}$

which is the solution of the equation $3(x - 2) = 9$

The equation may be checked by substituting $x = 5$ back into the original equation.

Problem 58. Solve the equation:
$4(2r - 3) - 2(r - 4) = 3(r - 3) - 1$

Removing brackets gives:
$8r - 12 - 2r + 8 = 3r - 9 - 1$

Rearranging gives: $8r - 2r - 3r = -9 - 1 + 12 - 8$

i.e. $3r = -6$

Dividing both sides by 3 gives: $\mathbf{r} = \frac{-6}{3} = \mathbf{-2}$

which is the solution of the equation
$4(2r - 3) - 2(r - 4) = 3(r - 3) - 1$

Problem 59. Solve the equation: $\frac{4}{x} = \frac{2}{5}$

The lowest common multiple (LCM) of the denominators, i.e. the lowest algebraic expression that both x and 5 will divide into, is 5x

Multiplying both sides by 5x gives: $5x\left(\frac{4}{x}\right) = 5x\left(\frac{2}{5}\right)$

Cancelling gives: $5(4) = x(2)$

i.e. $\quad 20 = 2x \qquad (1)$

Dividing both sides by 2 gives: $\frac{20}{2} = \frac{2x}{2}$

Cancelling gives: $\mathbf{10 = x}$ or $\mathbf{x = 10}$

which is the solution of the equation $\frac{4}{x} = \frac{2}{5}$

When there is just one fraction on each side of the equation as in this example, there is a quick way to arrive at equation (1) without needing to find the LCM of the denominators.

We can move from $\frac{4}{x} = \frac{2}{5}$ to: $4 \times 5 = 2 \times x$

by what is called '**cross-multiplication**'.

In general, if $\frac{a}{b} = \frac{c}{d}$ then: $ad = bc$

We can use cross-multiplication when there is one fraction only on each side of the equation.

Problem 60. Solve the equation:

$$\frac{2y}{5} + \frac{3}{4} + 5 = \frac{1}{20} - \frac{3y}{2}$$

The lowest common multiple (LCM) of the denominators is 20, i.e. the lowest number that 4, 5, 20 and 2 will divide into.

Multiplying each term by 20 gives:

$$20\left(\frac{2y}{5}\right) + 20\left(\frac{3}{4}\right) + 20(5) = 20\left(\frac{1}{20}\right) - 20\left(\frac{3y}{2}\right)$$

Cancelling gives: $4(2y) + 5(3) + 100 = 1 - 10(3y)$

i.e. $\quad 8y + 15 + 100 = 1 - 30y$

Rearranging gives: $8y + 30y = 1 - 15 - 100$

i.e. $\quad 38y = -114$

Dividing both sides by 38 gives: $\frac{38y}{38} = \frac{-114}{38}$

Cancelling gives: $\mathbf{y = -3}$

which is the solution of the equation $\frac{2y}{5} + \frac{3}{4} + 5 = \frac{1}{20} - \frac{3y}{2}$

Problem 61. Solve the equation: $2\sqrt{d} = 8$

Whenever square roots are involved in an equation, the square root term needs to be isolated on its own before squaring both sides

Dividing both sides by 2 gives: $\sqrt{d} = \frac{8}{2}$

Cancelling gives: $\quad \sqrt{d} = 4$

Squaring both sides gives: $\left(\sqrt{d}\right)^2 = (4)^2$

i.e. $\quad \mathbf{d = 16}$

which is the solution of the equation $2\sqrt{d} = 8$

Problem 62. Solve the equation: $x^2 = 25$

Whenever a square term is involved, the square root of both sides of the equation must be taken.

Taking the square root of both sides gives: $\sqrt{x^2} = \sqrt{25}$

i.e. $\quad \mathbf{x = \pm 5}$

which is the solution of the equation $x^2 = 25$

Now try the following Practice Exercise

Practice Exercise 10 Solving simple equations (Answers on page 816)

Solve the following equations:

1. $2x + 5 = 7$
2. $8 - 3t = 2$
3. $\frac{2}{3}c - 1 = 3$
4. $2x - 1 = 5x + 11$
5. $2a + 6 - 5a = 0$
6. $3x - 2 - 5x = 2x - 4$
7. $20d - 3 + 3d = 11d + 5 - 8$
8. $2(x - 1) = 4$
9. $16 = 4(t + 2)$
10. $5(f - 2) - 3(2f + 5) + 15 = 0$
11. $2x = 4(x - 3)$
12. $6(2 - 3y) - 42 = -2(y - 1)$
13. $2(3g - 5) - 5 = 0$
14. $4(3x + 1) = 7(x + 4) - 2(x + 5)$

15. $11+3(r-7)=16-(r+2)$

16. $8+4(x-1)-5(x-3)=2(5-2x)$

17. $\frac{1}{5}d+3=4$

18. $2+\frac{3}{4}y=1+\frac{2}{3}y+\frac{5}{6}$

19. $\frac{1}{4}(2x-1)+3=\frac{1}{2}$

20. $\frac{1}{5}(2f-3)+\frac{1}{6}(f-4)+\frac{2}{15}=0$

21. $\frac{1}{3}(3m-6)-\frac{1}{4}(5m+4)+\frac{1}{5}(2m-9)=-3$

22. $\frac{x}{3}-\frac{x}{5}=2$

23. $\frac{2}{a}=\frac{3}{8}$

24. $\frac{1}{3n}+\frac{1}{4n}=\frac{7}{24}$

25. $\frac{x+3}{4}=\frac{x-3}{5}+2$

26. $\frac{3t}{20}=\frac{6-t}{12}+\frac{2t}{15}-\frac{3}{2}$

27. $\frac{y}{5}+\frac{7}{20}=\frac{5-y}{4}$

28. $\frac{v-2}{2v-3}=\frac{1}{3}$

29. $\frac{2}{a-3}=\frac{3}{2a+1}$

30. $3\sqrt{t}=9$

31. $2\sqrt{y}=5$

32. $10=5\sqrt{\left(\frac{x}{2}-1\right)}$

33. $16=\frac{t^2}{9}$

34. $\sqrt{\left(\frac{y+2}{y-2}\right)}=\frac{1}{2}$

35. $\frac{6}{a}=\frac{2a}{3}$

1.8 Transposing formulae

There are no new rules for transposing formulae. The same rules as were used for simple equations are used, i.e. **the balance of an equation must be maintained**.

Here are some worked examples to help understanding of transposing formulae.

Problem 63. Transpose $p=q+r+s$ to make r the subject

The object is to obtain r on its own on the left-hand side (LHS) of the equation. Changing the equation around so that r is on the LHS gives:

$$q+r+s=p \qquad (1)$$

From the previous chapter on simple equations, a term can be moved from one side of an equation to the other side as long as the sign is changed.

Rearranging gives: $\mathbf{r=p-q-s}$

Mathematically, we have subtracted $q+s$ from both sides of equation (1)

Problem 64. Transpose $v=f\lambda$ to make λ the subject

$v=f\lambda$ relates velocity v, frequency f and wavelength λ

Rearranging gives: $f\lambda=v$

Dividing both sides by f gives: $\frac{f\lambda}{f}=\frac{v}{f}$

Cancelling gives: $\boldsymbol{\lambda=\frac{v}{f}}$

Problem 65. When a body falls freely through a height h, the velocity v is given by $v^2=2gh$. Express this formula with h as the subject.

Rearranging gives: $2gh=v^2$

Dividing both sides by 2g gives: $\frac{2gh}{2g}=\frac{v^2}{2g}$

Cancelling gives: $\mathbf{h=\frac{v^2}{2g}}$

Problem 66. If $I=\frac{V}{R}$, rearrange to make V the subject

$I = \frac{V}{R}$ is Ohm's law, where I is the current, V is the voltage and R is the resistance.

Rearranging gives: $\frac{V}{R} = I$

Multiplying both sides by R gives: $R\left(\frac{V}{R}\right) = R(I)$

Cancelling gives: $\mathbf{V = IR}$

Problem 67. Rearrange the formula $R = \frac{\rho L}{A}$ to make (i) A the subject, and (ii) L the subject

$R = \frac{\rho L}{A}$ relates resistance R of a conductor, resistivity ρ, conductor length L and conductor cross-sectional area A.

(i) Rearranging gives: $\frac{\rho L}{A} = R$

Multiplying both sides by A gives:

$$A\left(\frac{\rho L}{A}\right) = A(R)$$

Cancelling gives: $\rho L = AR$

Rearranging gives: $AR = \rho l$

Dividing both sides by R gives: $\frac{AR}{R} = \frac{\rho L}{R}$

Cancelling gives: $\mathbf{A = \frac{\rho L}{R}}$

(ii) Multiplying both sides of $\frac{\rho L}{A} = R$ by A gives:

$\rho L = AR$

Dividing both sides by ρ gives: $\frac{\rho L}{\rho} = \frac{AR}{\rho}$

Cancelling gives: $\mathbf{L = \frac{AR}{\rho}}$

Problem 68. Transpose $y = mx + c$ to make m the subject

$y = mx + c$ is the equation of a straight line graph, where y is the vertical axis variable, x is the horizontal axis variable, m is the gradient of the graph and c is the y-axis intercept.

Subtracting c from both sides gives: $y - c = mx$

or $mx = y - c$

Dividing both sides by x gives: $\mathbf{m = \frac{y - c}{x}}$

Problem 69. The final length, L_2 of a piece of wire heated through θ°C is given by the formula $L_2 = L_1(1 + \alpha\theta)$ where L_1 is the original length. Make the coefficient of expansion, α, the subject.

Rearranging gives: $L_1(1 + \alpha\theta) = L_2$

Removing the bracket gives: $L_1 + L_1\alpha\theta = L_2$

Rearranging gives: $L_1\alpha\theta = L_2 - L_1$

Dividing both sides by $L_1\theta$ gives: $\frac{L_1\alpha\theta}{L_1\theta} = \frac{L_2 - L_1}{L_1\theta}$

Cancelling gives: $\boldsymbol{\alpha = \frac{L_2 - L_1}{L_1\theta}}$

An alternative method of transposing $L_2 = L_1(1 + \alpha\theta)$ for α is shown below.

Dividing both sides by L_1 gives: $\frac{L_2}{L_1} = 1 + \alpha\theta$

Subtracting 1 from both sides gives: $\frac{L_2}{L_1} - 1 = \alpha\theta$

or $\alpha\theta = \frac{L_2}{L_1} - 1$

Dividing both sides by θ gives: $\boldsymbol{\alpha = \frac{\frac{L_2}{L_1} - 1}{\theta}}$

The two answers $\alpha = \frac{L_2 - L_1}{L_1\theta}$ and $\alpha = \frac{\frac{L_2}{L_1} - 1}{\theta}$ look quite different. They are, however, equivalent. The first answer looks tidier but is no more correct than the second answer.

Problem 70. A formula for the distance s moved by a body is given by: $s = \frac{1}{2}(v + u)t$. Rearrange the formula to make u the subject.

Rearranging gives: $\frac{1}{2}(v + u)t = s$

Multiplying both sides by 2 gives: $(v + u)t = 2s$

Dividing both sides by t gives: $\frac{(v + u)t}{t} = \frac{2s}{t}$

Cancelling gives: $v + u = \frac{2s}{t}$

Rearranging gives: $\mathbf{u = \frac{2s}{t} - v}$ or $\mathbf{u = \frac{2s - vt}{t}}$

Problem 71. In a right angled triangle having sides x, y and hypotenuse z, Pythagoras' theorem states $z^2 = x^2 + y^2$. Transpose the formula to find x.

Rearranging gives: $x^2 + y^2 = z^2$

and $x^2 = z^2 - y^2$

Taking the square root of both sides gives: $\mathbf{x = \sqrt{z^2 - y^2}}$

Problem 72. The impedance Z of an a.c. circuit is given by: $Z=\sqrt{R^2+X^2}$ where R is the resistance. Make the reactance, X , the subject.

Rearranging gives: $\sqrt{R^2+X^2}=Z$

Squaring both sides gives: $R^2+X^2=Z^2$

Rearranging gives: $X^2=Z^2-R^2$

Taking the square root of both sides gives:

$$\mathbf{X=\sqrt{Z^2-R^2}}$$

Now try the following Practice Exercise

Practice Exercise 11 Transposing formulae (Answers on page 816)

Make the symbol indicated the subject of each of the formulae shown, and express each in its simplest form.

1. $a+b=c-d-e$ (d)
2. $y=7x$ (x)
3. $pv=c$ (v)
4. $v=u+at$ (a)
5. $V=IR$ (R)
6. $x+3y=t$ (y)
7. $c=2\pi r$ (r)
8. $y=mx+c$ (x)
9. $I=PRT$ (T)
10. $X_L=2\pi fL$ (L)
11. $I=\dfrac{E}{R}$ (R)
12. $y=\dfrac{x}{a}+3$ (x)
13. $F=\dfrac{9}{5}C+32$ (C)
14. $X_C=\dfrac{1}{2\pi fC}$ (f)
15. $S=\dfrac{a}{1-r}$ (r)
16. $y=\dfrac{\lambda(x-d)}{d}$ (x)
17. $A=\dfrac{3(F-f)}{L}$ (f)
18. $y=\dfrac{AB^2}{5CD}$ (D)
19. $R=R_0(1+\alpha t)$ (t)
20. $I=\dfrac{E-e}{R+r}$ (R)
21. $y=4ab^2c^2$ (b)
22. $t=2\pi\sqrt{\dfrac{L}{g}}$ (L)
23. $v^2=u^2+2as$ (u)
24. $N=\sqrt{\left(\dfrac{a+x}{y}\right)}$ (a)
25. Transpose $Z=\sqrt{R^2+(2\pi fL)^2}$ for L, and evaluate L when $Z=27.82$, $R=11.76$ and $f=50$.

1.9 Solving simultaneous equations

The solution of simultaneous equations is demonstrated in the following worked problems.

Problem 73. If 6 apples and 2 pears cost £1.80 and 8 apples and 6 pears cost £2.90, calculate how much an apple and a pear each cost.

Let an apple = A and a pear = P, then:

$$6A+2P=180 \quad (1)$$

$$8A+6P=290 \quad (2)$$

From equation (1), $6A=180-2P$

and $A=\dfrac{180-2P}{6}=30-0.3333P \quad (3)$

From equation (2), $8A=290-6P$

and $A=\dfrac{290-6P}{8}=36.25-0.75P \quad (4)$

Equating (3) and (4) gives:

$$30-0.3333P=36.25-0.75P$$

i.e. $0.75P-0.3333P=36.25-30$

and $0.4167P=6.25$

and $P=\dfrac{6.25}{0.4167}=15$

Substituting in (3) gives:

$$A = 30 - 0.3333(15) = 30 - 5 = 25$$

Hence, **an apple costs 25p and a pear costs 15p**

The above method of solving simultaneous equations is called the **substitution method**.

Problem 74. If 6 bananas and 5 peaches cost £3.45 and 4 bananas and 8 peaches cost £4.40, calculate how much a banana and a peach each cost.

Let a banana = B and a peach = P, then:

$$6B + 5P = 345 \qquad (1)$$
$$4B + 8P = 440 \qquad (2)$$

Multiplying equation (1) by 2 gives:

$$12B + 10P = 690 \qquad (3)$$

Multiplying equation (2) by 3 gives:

$$12B + 24P = 1320 \qquad (4)$$

Equation (4) – equation (3) gives: $14P = 630$

from which, $P = \frac{630}{14} = 45$

Substituting in (1) gives: $6B + 5(45) = 345$

i.e. $6B = 345 - 5(45)$

i.e. $6B = 120$

and $B = \frac{120}{6} = 20$

Hence, **a banana costs 20p and a peach costs 45p**

The above method of solving simultaneous equations is called the **elimination method**.

Problem 75. If 20 bolts and 2 spanners cost £10, and 6 spanners and 12 bolts cost £18, how much does a spanner and a bolt cost?

Let s = a spanner and b = a bolt.

Therefore, $2s + 20b = 10 \qquad (1)$

and $6s + 12b = 18 \qquad (2)$

Multiplying equation (1) by 3 gives:

$$6s + 60b = 30 \qquad (3)$$

Equation (3) – equations (2) gives: $48b = 12$

from which, $b = \frac{12}{48} = 0.25$

Substituting in (1) gives:

$$2s + 20(0.25) = 10$$

i.e. $2s = 10 - 20(0.25)$

i.e. $2s = 5$

and $s = \frac{5}{2} = 2.5$

Therefore, **a spanner costs £2.50 and a bolt costs £0.25 or 25p**

Now try the following Practice Exercises

Practice Exercise 12 Simultaneous equations (Answers on page 816)

1. If 5 apples and 3 bananas cost £1.45 and 4 apples and 6 bananas cost £2.42, determine how much an apple and a banana each cost.
2. If 7 apples and 4 oranges cost £2.64 and 3 apples and 3 oranges cost £1.35, determine how much an apple and a banana each cost.
3. Three new cars and four new vans supplied to a dealer together cost £93000, and five new cars and two new vans of the same models cost £99000. Find the respective costs of a car and a van.
4. In a system of forces, the relationship between two forces F_1 and F_2 is given by:
$$5F_1 + 3F_2 = -6$$
$$3F_1 + 5F_2 = -18$$
Solve for F_1 and F_2
5. Solve the simultaneous equations:
$$a + b = 7$$
$$a - b = 3$$
6. Solve the simultaneous equations:
$$8a - 3b = 51$$
$$3a + 4b = 14$$

For fully worked solutions to each of the problems in Practice Exercises 1 and 12 in this chapter, go to the website:
www.routledge.com/cw/bird

Chapter 2

Further mathematics revision

***Why it is important to understand:* Further mathematics revision**

There are an enormous number of uses of trigonometry; fields that use trigonometry include astronomy (especially for locating apparent positions of celestial objects, in which spherical trigonometry is essential) and hence navigation (on the oceans, in aircraft, and in space), electrical engineering, music theory, electronics, medical imaging (CAT scans and ultrasound), number theory (and hence cryptology), oceanography, land surveying and geodesy (a branch of earth sciences), architecture, mechanical engineering, civil engineering, computer graphics and game development. It is clear that a good knowledge of trigonometry is essential in many fields of engineering.

All types of engineers use natural and common logarithms. In electrical engineering, a dB (decibel) scale is very useful for expressing attenuations in radio propagation and circuit gains, and logarithms are used for implementing arithmetic operations in digital circuits. Exponential functions are used in engineering, physics, biology and economics. There are many quantities that grow exponentially; some examples are population, compound interest and charge in a capacitor. There is also exponential decay; some examples include radioactive decay, atmospheric pressure, Newton's law of cooling and linear expansion. Understanding and using logarithms and exponential functions are important in many branches of engineering.

Graphs have a wide range of applications in engineering and in physical sciences because of their inherent simplicity. A graph can be used to represent almost any physical situation involving discrete objects and the relationship among them. If two quantities are directly proportional and one is plotted against the other, a straight line is produced. Examples of this include an applied force on the end of a spring plotted against spring extension, the speed of a flywheel plotted against time, and strain in a wire plotted against stress (Hooke's law). In engineering, the straight line graph is the most basic graph to draw and evaluate.

When designing a new building, or seeking planning permission, it is often necessary to specify the total floor area of the building. In construction, calculating the area of a gable end of a building is important when determining the number of bricks and mortar to order. When using a bolt, the most important thing is that it is long enough for your particular application and it may also be necessary to calculate the shear area of the bolt connection. Arches are everywhere, from sculptures and monuments to pieces of architecture and strings on musical instruments; finding the height of an arch or its cross-sectional area is often required. Determining the cross-sectional areas of beam structures is vitally important in design engineering. There are thus a large number of situations in engineering where determining area is important. The floodlit area at a football ground, the area an automatic garden sprayer sprays and the angle of lap of a belt drive all rely on calculations involving the arc of a circle. The ability to handle calculations involving circles and their properties is clearly essential in several branches of engineering design.

Electrical Circuit Theory and Technology. 978-1-138-67349-6, © 2017 John Bird. Published by Taylor & Francis. All rights reserved.

Surveyors, farmers and landscapers often need to determine the area of irregularly shaped pieces of land to work with the land properly. There are many applications in all aspects of engineering, where finding the areas of irregular shapes and the lengths of irregular shaped curves are important applications. Typical earthworks include roads, railway beds, causeways, dams and canals. The mid-ordinate rule is a staple of scientific data analysis and engineering.

At the end of this chapter, you should be able to:

- change radians to degrees and vice versa
- calculate sine, cosine and tangent for large and small angles
- calculate unknown sides of a right-angled triangle
- use Pythagoras' theorem
- use the sine and cosine rules for acute-angled triangles
- define a logarithm
- state and use the laws of logarithms to simplify logarithmic expressions
- solve equations involving logarithms
- solve indicial equations
- solve equations using Napierian logarithms
- appreciate the many examples of laws of growth and decay in engineering and science
- perform calculations involving the laws of growth and decay
- understand rectangular axes, scales and co-ordinates
- plot given co-ordinates and draw the best straight line graph
- determine the gradient and vertical-axis intercept of a straight line graph
- state the equation of a straight line graph
- plot straight line graphs involving practical engineering examples
- calculate the areas of common shapes
- use the mid-ordinate rule to determine irregular areas

2.1 Radians and degrees

There are 2π radians or $360°$ in a complete circle, thus:

$$\boldsymbol{\pi} \textbf{ radians} = \mathbf{180°} \qquad \text{from which,}$$

$$\mathbf{1\ rad} = \frac{\mathbf{180°}}{\pi} \quad \text{or} \quad \mathbf{1°} = \frac{\pi}{\mathbf{180}}\text{rad}$$

where $\pi = 3.14159265358979323846\ldots$ to 20 decimal places!

Problem 1. Convert the following angles to degrees correct to 3 decimal places:
(a) 0.1 rad (b) 0.7 rad (c) 1.3 rad

(a) $0.1\,\text{rad} = 0.1\,\text{rad} \times \dfrac{180°}{\pi\ \text{rad}} = \mathbf{5.730°}$

(b) $0.7\,\text{rad} = 0.7\,\text{rad} \times \dfrac{180°}{\pi\ \text{rad}} = \mathbf{40.107°}$

(c) $1.3\,\text{rad} = 1.3\,\text{rad} \times \dfrac{180°}{\pi\ \text{rad}} = \mathbf{74.485°}$

Problem 2. Convert the following angles to radians correct to 4 decimal places:
(a) 5° (b) 40° (c) 85°

(a) $5° = 5° \times \dfrac{\pi\ \text{rad}}{180°} = \dfrac{\pi}{36}\text{rad} = \mathbf{0.0873\ rad}$

(b) $40° = 40° \times \dfrac{\pi\ \text{rad}}{180°} = \dfrac{4\pi}{18}\text{rad} = \mathbf{0.6981\ rad}$

(c) $85° = 85° \times \dfrac{\pi\ \text{rad}}{180°} = \dfrac{85\pi}{180}\text{rad} = \mathbf{1.4835\ rad}$

Now try the following Practice Exercise

Practice Exercise 13 Radians and degrees (Answers on page 816)

1. Convert the following angles to degrees correct to 3 decimal places (where necessary):
   (a) 0.6 rad (b) 0.8 rad (c) 2 rad
   (d) 3.14159 rad

2. Convert the following angles to radians correct to 4 decimal places:
   (a) 45° (b) 90° (c) 120° (d) 180°

2.2 Measurement of angles

Angles are measured starting from the horizontal 'x' axis, in an **anticlockwise direction**, as shown by θ_1 to θ_4 in Figure 2.1. An angle can also be measured in a **clockwise direction**, as shown by θ_5 in Figure 2.1, but in this case the angle has a negative sign before it. If, for example, $\theta_4 = 320°$ then $\theta_5 = -40°$

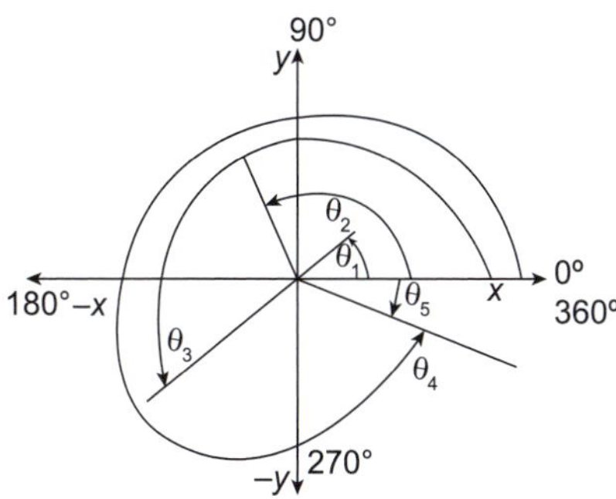

Figure 2.1

Problem 3. Use a calculator to determine the cosine, sine and tangent of the following angles, each measured anticlockwise from the horizontal 'x' axis, each correct to 4 decimal places:
(a) 30° (b) 120° (c) 250° (d) 320°
(e) 390° (f) 480°

(a) $\cos 30° = \mathbf{0.8660}$ $\sin 30° = \mathbf{0.5000}$
$\tan 30° = \mathbf{0.5774}$

(b) $\cos 120° = \mathbf{-0.5000}$ $\sin 120° = \mathbf{0.8660}$
$\tan 120° = \mathbf{-1.7321}$

(c) $\cos 250° = \mathbf{-0.3420}$ $\sin 250° = \mathbf{-0.9397}$
$\tan 250° = \mathbf{2.7475}$

(d) $\cos 320° = \mathbf{0.7660}$ $\sin 320° = \mathbf{-0.6428}$
$\tan 320° = \mathbf{-0.8391}$

(e) $\cos 390° = \mathbf{0.8660}$ $\sin 390° = \mathbf{0.5000}$
$\tan 390° = \mathbf{0.5774}$

(f) $\cos 480° = \mathbf{-0.5000}$ $\sin 480° = \mathbf{0.8660}$
$\tan 480° = \mathbf{-1.7321}$

These angles are now drawn in Figure 2.2. Note that the cosine and sine of angles always lie between −1 and +1, but that tangent values can be > 1 and < 1.

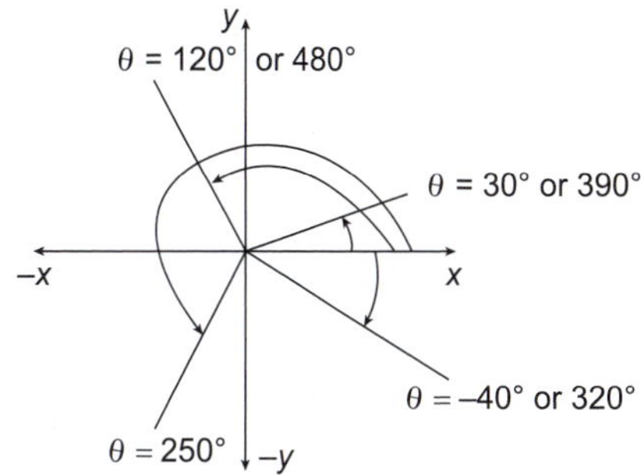

Figure 2.2

Note from Figure 2.2 that $\theta = 30°$ is the same as $\theta = 390°$ and so are their cosines, sines and tangents. Similarly, note that $\theta = 120°$ is the same as $\theta = 480°$ and so are their cosines, sines and tangents. Also, note that $\theta = -40°$ is the same as $\theta = +320°$ and so are their cosines, sines and tangents.

It is noted from above that

- in the **first quadrant**, i.e. where θ varies from 0° to 90°, all (A) values of cosine, sine and tangent are positive
- in the **second quadrant**, i.e. where θ varies from 90° to 180°, only values of sine (S) are positive
- in the **third quadrant**, i.e. where θ varies from 180° to 270°, only values of tangent (T) are positive
- in the **fourth quadrant**, i.e. where θ varies from 270° to 360°, only values of cosine (C) are positive.

These positive signs, A, S, T and C are shown in Figure 2.3.

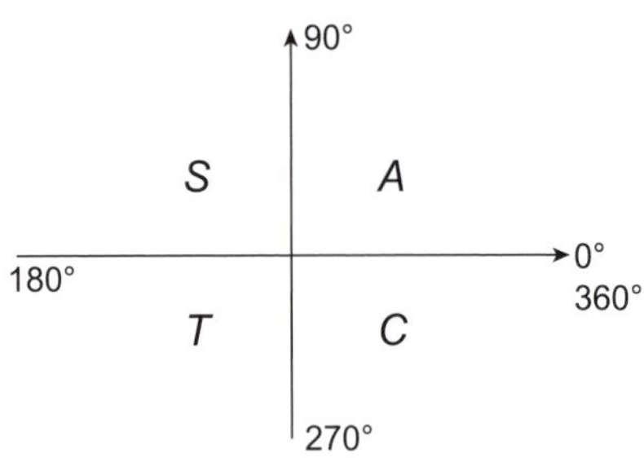

Figure 2.3

Now try the following Practice Exercise

Practice Exercise 14 Measurement of angles (Answers on page 816)

1. Find the cosine, sine and tangent of the following angles, where appropriate each correct to 4 decimal places:
(a) 60° (b) 90° (c) 150° (d) 180°
(e) 210° (f) 270° (g) 330° (h) −30°
(i) 420° (j) 450° (k) 510°

2.3 Trigonometry revision

(a) Sine, cosine and tangent

From Figure 2.4, $\sin\theta = \dfrac{BC}{AC}$ $\quad\cos\theta = \dfrac{AB}{AC}$

$\tan\theta = \dfrac{BC}{AB}$

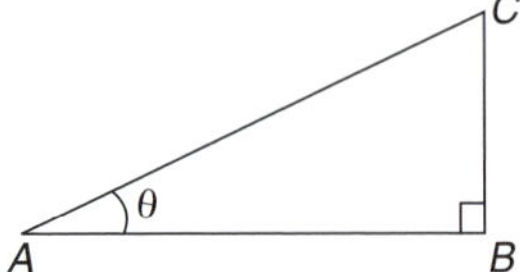

Figure 2.4

Problem 4. In Fig. 2.4, if AB = 2 and AC = 3, determine the angle θ.

It is convenient to use the expression for cos θ, since 'AB' and 'AC' are given.

Hence, $\cos\theta = \dfrac{AB}{AC} = \dfrac{2}{3} = 0.66667$

from which, $\theta = \cos^{-1}(0.66667) = \mathbf{48.19°}$

Problem 5. In Figure 2.4, if BC = 1.5 and AC = 2.2, determine the angle θ.

It is convenient to use the expression for sin θ, since 'BC' and 'AC' are given.

Hence, $\sin\theta = \dfrac{BC}{AC} = \dfrac{1.5}{2.2} = 0.68182$

from which, $\boldsymbol{\theta} = \sin^{-1}(0.68182) = \mathbf{42.99°}$

Problem 6. In Figure 2.4, if BC = 8 and AB = 1.3, determine the angle θ.

It is convenient to use the expression for tan θ, since 'BC' and 'AB' are given.

Hence, $\tan\theta = \dfrac{BC}{AB} = \dfrac{8}{1.3} = 6.1538$

from which, $\boldsymbol{\theta} = \tan^{-1}(6.1538) = \mathbf{80.77°}$

(b) Pythagoras' Theorem

Pythagoras' theorem* states that:
$(\text{hypotenuse})^2 = (\text{adjacent side})^2 + (\text{opposite side})^2$

*Who was **Pythagoras**? Pythagoras of Samos (c. 570 BC – c. 495 BC) was an Ionian Greek philosopher and mathematician, best known for the Pythagorean theorem. To find out more go to www.routledge.com/cw/bird

i.e. in the triangle of Figure 2.5, $\mathbf{AC^2 = AB^2 + BC^2}$

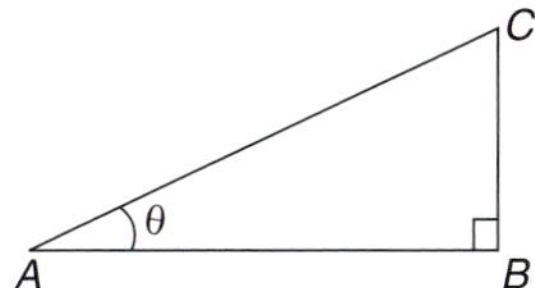

Figure 2.5

Problem 7. In Figure 2.5, if AB = 5.1 m and BC = 6.7 m, determine the length of the hypotenuse, AC.

From Pythagoras, $AC^2 = AB^2 + BC^2$

$$= 5.1^2 + 6.7^2 = 26.01 + 44.89$$
$$= 70.90$$

from which, $\mathbf{AC} = \sqrt{70.90} = \mathbf{8.42\,m}$

Now try the following Practice Exercise

Practice Exercise 15 Sines, cosines and tangents and Pythagoras' theorem (Answers on page 817)

In problems 1 to 5, refer to Figure 2.5.

1. If AB = 2.1 m and BC = 1.5 m, determine angle θ
2. If AB = 2.3 m and AC = 5.0 m, determine angle θ
3. If BC = 3.1 m and AC = 6.4 m, determine angle θ
4. If AB = 5.7 cm and BC = 4.2 cm, determine the length AC
5. If AB = 4.1 m and AC = 6.2 m, determine length BC

(c) The sine and cosine rules

For the triangle ABC shown in Figure 2.6,

the sine rule states: $\dfrac{\mathbf{a}}{\mathbf{\sin A}} = \dfrac{\mathbf{b}}{\mathbf{\sin B}} = \dfrac{\mathbf{c}}{\mathbf{\sin C}}$

and the cosine rule states: $\mathbf{a^2 = b^2 + c^2 - 2bc\cos A}$

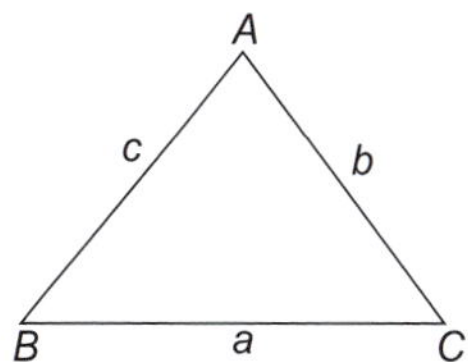

Figure 2.6

Problem 8. In Figure 2.6, if a = 3 m, A = 20° and B = 120°, determine lengths b, c and angle C.

Using the sine rule, $\dfrac{a}{\sin A} = \dfrac{b}{\sin B}$

i.e. $\dfrac{3}{\sin 20^\circ} = \dfrac{b}{\sin 120^\circ}$

from which, $\mathbf{b} = \dfrac{3\sin 120^\circ}{\sin 20^\circ} = \dfrac{3 \times 0.8660}{0.3420} = \mathbf{7.596\,m}$

Angle, $\mathbf{C} = 180^\circ - 20^\circ - 120^\circ = \mathbf{40^\circ}$

Using the sine rule again gives: $\dfrac{c}{\sin C} = \dfrac{a}{\sin A}$

i.e. $\mathbf{c} = \dfrac{a\sin C}{\sin A} = \dfrac{3 \times \sin 40^\circ}{\sin 20^\circ} = \mathbf{5.638\,m}$

Problem 9. In Figure 2.6, if b = 8.2 cm. c = 5.1 cm and A = 70°, determine the length a and angles B and C.

From the cosine rule,

$$a^2 = b^2 + c^2 - 2bc\cos A$$
$$= 8.2^2 + 5.1^2 - 2 \times 8.2 \times 5.1 \times \cos 70^\circ$$
$$= 67.24 + 26.01 - 2(8.2)(5.1)\cos 70^\circ$$
$$= 64.643$$

Hence, **length, a** $= \sqrt{64.643} = \mathbf{8.04\,cm}$

Using the sine rule: $\dfrac{a}{\sin A} = \dfrac{b}{\sin B}$

i.e. $\dfrac{8.04}{\sin 70^\circ} = \dfrac{8.2}{\sin B}$

from which, $8.04\sin B = 8.2\sin 70^\circ$

and $\sin B = \dfrac{8.2\sin 70^\circ}{8.04} = 0.95839$

and $\mathbf{B} = \sin^{-1}(0.95839) = \mathbf{73.41^\circ}$

Since A + B + C = 180°, then
$\mathbf{C} = 180^\circ - A - B = 180^\circ - 70^\circ - 73.41^\circ = \mathbf{36.59^\circ}$

Now try the following Practice Exercise

Practice Exercise 16 Sine and cosine rules (Answers on page 817)

In problems 1 to 4, refer to Figure 2.6.

1. If b = 6 m, c = 4 m and B = 100°, determine angles C and A and length a.
2. If a = 15 m, c = 23 m and B = 67°, determine length b and angles A and C.
3. If a = 4 m, b = 8 m and c = 6 m, determine angle A.
4. If a = 10.0 cm, b = 8.0 cm and c = 7.0 cm, determine angles A, B and C.
5. In Figure 2.7, PR represents the inclined jib of a crane and is 10.0 m long. PQ is 4.0 m long. Determine the inclination of the jib to the vertical (i.e. angle P) and the length of tie QR.

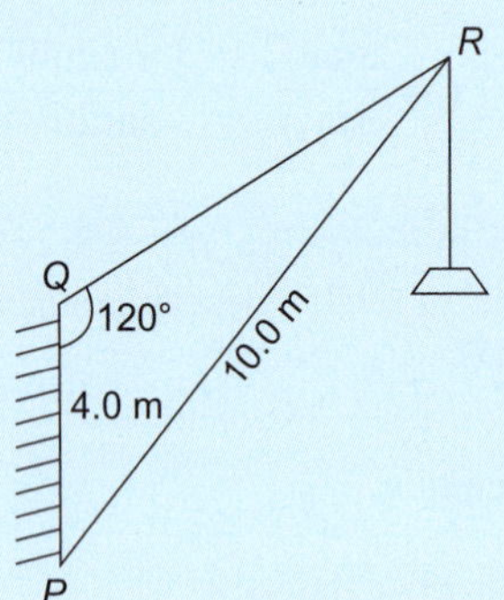

Figure 2.7

2.4 Logarithms and exponentials

In general, if a number y can be written in the form a^x, then the index x is called the 'logarithm of y to the base of a',

i.e. **if $y = a^x$ then $x = \log_a y$**

For example, the two statements: $\mathbf{16 = 2^4}$ and $\mathbf{\log_2 16 = 4}$ are equivalent.

Logarithms having a base of 10 are called **common logarithms** and $\log_{10}$ is usually abbreviated to lg.

Logarithms having a base of e (where 'e' is a mathematical constant approximately equal to 2.7183) are called **hyperbolic**, **Napierian** or **natural logarithms**, and $\log_e$ is usually abbreviated to ln.

(a) Laws of logarithms

(i) $\mathbf{\log(A \times B) = \log A + \log B}$

(ii) $\mathbf{\log\left(\frac{A}{B}\right) = \log A - \log B}$

(iii) $\mathbf{\log A^n = n\log A}$

Here are some worked problems to help understanding of the laws of logarithms.

Problem 10. Write log 4 + log 7 as the logarithm of a single number

$$\log 4 + \log 7 = \log(7 \times 4) \text{ by the first law of logarithms}$$
$$= \mathbf{\log 28}$$

Problem 11. Write log 16 − log 2 as the logarithm of a single number

$$\log 16 - \log 2 = \log\left(\frac{16}{2}\right) \text{ by the second law of logarithms}$$
$$= \mathbf{\log 8}$$

Problem 12. Write 2 log 3 as the logarithm of a single number

$$2\log 3 = \log 3^2 \text{ by the third law of logarithms}$$
$$= \mathbf{\log 9}$$

Problem 13. Write $\frac{1}{2}\log 25$ as the logarithm of a single number

$$\frac{1}{2}\log 25 = \log 25^{\frac{1}{2}} \text{ by the third law of logarithms}$$
$$= \log\sqrt{25} = \mathbf{\log 5}$$

Problem 14. Write $\frac{1}{2}\log 16 + \frac{1}{3}\log 27 - 2\log 5$ as the logarithm of a single number

$$\frac{1}{2}\log 16 + \frac{1}{3}\log 27 - 2\log 5$$

$$= \log 16^{\frac{1}{2}} + \log 27^{\frac{1}{3}} - \log 5^2$$

by the third law of logarithms

$$= \log\sqrt{16} + \log\sqrt[3]{27} - \log 25$$

by the laws of indices

$$= \log 4 + \log 3 - \log 25$$

$$= \log\left(\frac{4\times 3}{25}\right)$$

by the first and second laws of logarithms

$$= \log\left(\frac{12}{25}\right) = \mathbf{\log 0.48}$$

Problem 15. Solve the equation: $\log(x-1) + \log(x+8) = 2\log(x+2)$

$$\text{LHS} = \log(x-1) + \log(x+8)$$

$$= \log(x-1)(x+8)$$

from the first law of logarithms

$$= \log(x^2 + 7x - 8)$$

$$\text{RHS} = 2\log(x+2) = \log(x+2)^2$$

from the third law of logarithms

$$= \log(x^2 + 4x + 4)$$

Hence, $\log(x^2 + 7x - 8) = \log(x^2 + 4x + 4)$

from which, $x^2 + 7x - 8 = x^2 + 4x + 4$

i.e. $7x - 8 = 4x + 4$

i.e. $3x = 12$

and $\mathbf{x = 4}$

Problem 16. Solve the equation: $\log\left(x^2 - 3\right) - \log x = \log 2$

$$\log\left(x^2 - 3\right) - \log x = \log\left(\frac{x^2 - 3}{x}\right)$$

from the second law of logarithms

Hence, $\log\left(\dfrac{x^2 - 3}{x}\right) = \log 2$

from which, $\dfrac{x^2 - 3}{x} = 2$

Rearranging gives: $x^2 - 3 = 2x$

and $x^2 - 2x - 3 = 0$

Factorising gives: $(x-3)(x+1) = 0$

from which, $x = 3$ or $x = -1$ (or use the quadratic formula or a calculator)

$x = -1$ is not a valid solution since the logarithm of a negative number has no real root.

Hence, **the solution of the equation is**: $\mathbf{x = 3}$

Now try the following Practice Exercise

Practice Exercise 17 Laws of logarithms (Answers on page 817)

In Problems 1 to 10, write as the logarithm of a single number:

1. $\log 2 + \log 3$
2. $\log 3 + \log 5$
3. $\log 3 + \log 4 - \log 6$
4. $\log 7 + \log 21 - \log 49$
5. $2\log 2 + \log 3$
6. $2\log 2 + 3\log 5$
7. $2\log 5 - \frac{1}{2}\log 81 + \log 36$
8. $\frac{1}{3}\log 8 - \frac{1}{2}\log 81 + \log 27$
9. $\frac{1}{2}\log 4 - 2\log 3 + \log 45$
10. $\frac{1}{4}\log 16 + 2\log 3 - \log 18$

Solve the equations given in Problems 11 to 14:

11. $\log x^4 - \log x^3 = \log 5x - \log 2x$
12. $\log 2t^3 - \log t = \log 16 + \log t$
13. $2\log b^2 - 3\log b = \log 8b - \log 4b$
14. $\log(x+1) + \log(x-1) = \log 3$

(b) Indicial equations

To solve, say, $3^x = 27$, logarithms to a base of 10 are taken of both sides,

i.e. $\log_{10} 3^x = \log_{10} 27$

and $x\log_{10} 3 = \log_{10} 27$

by the third law of logarithms

Rearranging gives: $\mathbf{x} = \frac{\log_{10} 27}{\log_{10} 3} = \frac{1.43136\ldots}{0.47712\ldots} = \mathbf{3}$

which may be readily checked.

(Note, $\frac{\log 27}{\log 3}$ is **not** equal to $\log \frac{27}{3}$)

Problem 17. Solve the equation: $2^x = 5$, correct to 4 significant figures.

Taking logarithms to base 10 of both sides of $2^x = 5$ gives:

$$\log_{10} 2^x = \log_{10} 5$$

i.e. $$x \log_{10} 2 = \log_{10} 5$$

by the third law of logarithms

Rearranging gives:

$\mathbf{x} = \frac{\log_{10} 5}{\log_{10} 2} = \frac{0.6989700\ldots}{0.3010299\ldots} = \mathbf{2.322}$, correct to 4 significant figures.

Problem 18. Solve the equation: $x^{2.7} = 34.68$, correct to 4 significant figures.

Taking logarithms to base 10 of both sides gives:

$$\log_{10} x^{2.7} = \log_{10} 34.68$$

$$2.7 \log_{10} x = \log_{10} 34.68$$

Hence, $$\log_{10} x = \frac{\log_{10} 34.68}{2.7} = 0.57040$$

Thus, $\mathbf{x} = \text{antilog } 0.57040 = 10^{0.57040} = \mathbf{3.719}$, correct to 4 significant figures.

Now try the following Practice Exercise

Practice Exercise 18 Indicial equations (Answers on page 817)

In problems 1 to 6, solve the indicial equations for x, each correct to 4 significant figures:

1. $3^x = 6.4$
2. $2^x = 9$
3. $x^{1.5} = 14.91$
4. $25.28 = 4.2^x$
5. $x^{-0.25} = 0.792$
6. $0.027^x = 3.26$
7. The decibel gain n of an amplifier is given by: $n = 10 \log_{10}\left(\frac{P_2}{P_1}\right)$ where P_1 is the power input and P_2 is the power output. Find the power gain $\frac{P_2}{P_1}$ when n = 25 decibels.

(c) Solving equations involving exponential functions

It may be shown that: $\mathbf{\log_e e^x = x}$

For example, $\log_e e^2 = 2$ and $\log_e e^{5t} = 5t$

This is useful when solving equations involving exponential functions.

For example, to solve $e^{3x} = 7$, take Napierian logarithms of both sides,

which gives: $$\ln e^{3x} = \ln 7$$

i.e. $$3x = \ln 7$$

from which $\mathbf{x} = \frac{1}{3} \ln 7 = \mathbf{0.6486}$, correct to 4 decimal places.

Problem 19. Solve the equation: $9 = 4e^{-3x}$ to find x, correct to 4 significant figures.

Rearranging $9 = 4e^{-3x}$ gives: $\frac{9}{4} = e^{-3x}$

Taking Napierian logarithms of both sides gives:

$$\ln\left(\frac{9}{4}\right) = \ln(e^{-3x})$$

Since $\log_e e^\alpha = \alpha$, then $\ln\left(\frac{9}{4}\right) = -3x$

Hence, $\mathbf{x} = \frac{\ln\left(\frac{9}{4}\right)}{-3} = \mathbf{-0.2703}$, correct to 4 significant figures.

Problem 20. Given $32 = 70(1 - e^{-\frac{t}{2}})$ determine the value of t, correct to 3 significant figures.

Rearranging $32 = 70(1 - e^{-\frac{t}{2}})$ gives: $\frac{32}{70} = 1 - e^{-\frac{t}{2}}$

and $$e^{-\frac{t}{2}} = 1 - \frac{32}{70} = \frac{38}{70}$$

Taking Napierian logarithms of both sides gives:

$$\ln e^{-\frac{t}{2}} = \ln\left(\frac{38}{70}\right)$$

i.e. $$-\frac{t}{2} = \ln\left(\frac{38}{70}\right)$$

from which, $\mathbf{t} = -2 \ln\left(\frac{38}{70}\right) = \mathbf{1.22}$, correct to 3 significant figures.

Problem 21. Solve the equation:

$$2.68 = \ln\left(\frac{4.87}{x}\right) \text{ to find x}$$

From the definition of a logarithm, since $2.68 = \ln\left(\frac{4.87}{x}\right)$ then $e^{2.68} = \frac{4.87}{x}$

Rearranging gives: $x = \frac{4.87}{e^{2.68}} = 4.87e^{-2.68}$

i.e. $\mathbf{x = 0.3339}$, correct to 4 significant figures.

Now try the following Practice Exercise

Practice Exercise 19 Evaluating Napierian logarithms (Answers on page 817)

In Problems 1 to 8 solve the given equations, each correct to 4 significant figures.

1. $1.5 = 4e^{2t}$
2. $7.83 = 2.91e^{-1.7x}$
3. $16 = 24(1 - e^{-\frac{t}{2}})$
4. $5.17 = \ln\left(\frac{x}{4.64}\right)$
5. $3.72\ln\left(\frac{1.59}{x}\right) = 2.43$
6. $5 = 8\left(1 - e^{\frac{-x}{2}}\right)$
7. $\ln(x + 3) - \ln x = \ln(x - 1)$
8. $\ln(x - 1)^2 - \ln 3 = \ln(x - 1)$
9. If $\frac{P}{Q} = 10\log_{10}\left(\frac{R_1}{R_2}\right)$ find the value of R_1 when $P = 160$, $Q = 8$ and $R_2 = 5$
10. If $U_2 = U_1 e^{\left(\frac{W}{PV}\right)}$ make W the subject of the formula.

(d) Laws of growth and decay

Laws of exponential growth and decay are of the form $y = Ae^{-kx}$ and $y = A(1 - e^{-kx})$, where A and k are constants. When plotted, the form of these equations is as shown in Figure 2.8.

The laws occur frequently in engineering and science and examples of quantities related by a natural law include:

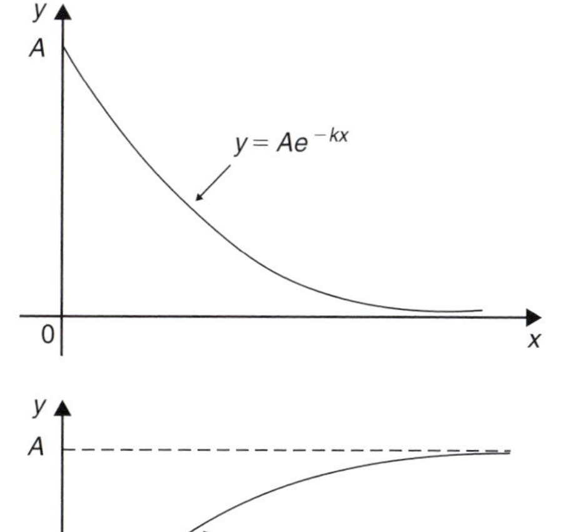

Figure 2.8

(i)	Linear expansion	$l = l_0 e^{\alpha\theta}$
(ii)	Change in electrical resistance with temperature	$R_\theta = R_0 e^{\alpha\theta}$
(iv)	Newton's law of cooling	$\theta = \theta_0 e^{-kt}$
(vi)	Discharge of a capacitor	$q = Qe^{-\frac{t}{CR}}$
(vii)	Atmospheric pressure	$p = p_0 e^{-h/c}$
(viii)	Radioactive decay	$N = N_0 e^{-\lambda t}$
(ix)	Decay of current in an inductive circuit	$i = Ie^{-\frac{Rt}{L}}$
(x)	Growth of current in a capacitive circuit	$i = I(1 - e^{-\frac{t}{CR}})$

Here are some worked problems to demonstrate the laws of growth and decay.

Problem 22. The resistance R of an electrical conductor at temperature θ°C is given by $R = R_0 e^{\alpha\theta}$, where α is a constant and $R_0 = 5\,k\Omega$. Determine the value of α correct to 4 significant figures, when $R = 6\,k\Omega$ and $\theta = 1500$°C. Also, find the temperature, correct to the nearest degree, when the resistance R is 5.4 kΩ.

Transposing $R = R_0 e^{\alpha\theta}$ gives: $\frac{R}{R_0} = e^{\alpha\theta}$

Taking Napierian logarithms of both sides gives:

$$\ln\frac{R}{R_0} = \ln e^{\alpha\theta} = \alpha\theta$$

Hence,

$$\alpha = \frac{1}{\theta} \ln \frac{R}{R_0} = \frac{1}{1500} \ln\left(\frac{6 \times 10^3}{5 \times 10^3}\right)$$

$$= \frac{1}{1500}(0.1823215\ldots) = 1.215477\ldots \times 10^{-4}$$

Hence, $\boldsymbol{\alpha = 1.215 \times 10^{-4}}$ correct to 4 significant figures.

From above, $\ln \frac{R}{R_0} = \alpha\theta$ hence $\theta = \frac{1}{\alpha} \ln \frac{R}{R_0}$

When $R = 5.4 \times 10^3$, $\alpha = 1.215477\ldots \times 10^{-4}$ and $R_0 = 5 \times 10^3$

$$\theta = \frac{1}{1.215477\ldots \times 10^{-4}} \ln\left(\frac{5.4 \times 10^3}{5 \times 10^3}\right)$$

$$= \frac{10^4}{1.215477\ldots}(7.696104\ldots \times 10^{-2})$$

$= \mathbf{633°C}$ correct to the nearest degree.

Problem 23. The current i amperes flowing in a capacitor at time t seconds is given by: $i = 8.0(1 - e^{-\frac{t}{CR}})$, where the circuit resistance R is 25 kΩ and capacitance C is 16 μF. Determine (a) the current i after 0.5 seconds and (b) the time, to the nearest millisecond, for the current to reach 6.0 A. Sketch the graph of current against time.

(a) Current
$$i = 8.0(1 - e^{-\frac{t}{CR}})$$
$$= 8.0[1 - e^{-0.5/(16\times10^{-6})(25\times10^3)}]$$
$$= 8.0(1 - e^{-1.25})$$
$$= 8.0(1 - 0.2865047\ldots)$$
$$= 8.0(0.7134952\ldots)$$
$= \mathbf{5.71\ amperes}$

(b) Transposing $i = 8.0(1 - e^{-\frac{t}{CR}})$

gives: $\frac{i}{8.0} = 1 - e^{-\frac{t}{CR}}$

from which, $e^{-\frac{t}{CR}} = 1 - \frac{i}{8.0} = \frac{8.0 - i}{8.0}$

Taking Napierian logarithms of both sides gives:

$$-\frac{t}{CR} = \ln\left(\frac{8.0 - i}{8.0}\right)$$

Hence,
$$t = -CR \ln\left(\frac{8.0 - i}{8.0}\right)$$

When $i = 6.0$ A,

$$t = -(16 \times 10^{-6})(25 \times 10^3) \ln\left(\frac{8.0 - 6.0}{8.0}\right)$$

i.e.

$$\mathbf{t} = -(0.40) \ln\left(\frac{2.0}{8.0}\right) = -0.4 \ln 0.25 = 0.5545\,s$$

$= \mathbf{555\ ms}$ correct to the nearest ms.

A graph of current against time is shown in Figure 2.9.

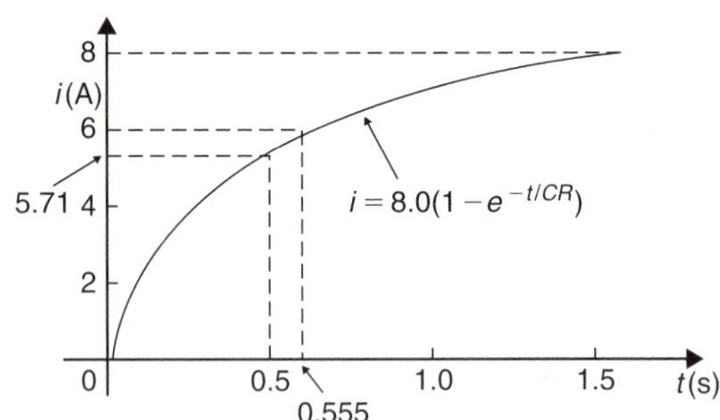

Figure 2.9

Now try the following Practice Exercise

Practice Exercise 20 Laws of growth and decay (Answers on page 817)

1. The temperature, T°C, of a cooling object varies with time, t minutes, according to the equation: $T = 150\,e^{-0.04t}$. Determine the temperature when (a) t = 0, (b) t = 10 minutes.
2. The voltage drop, v volts, across an inductor L henrys at time t seconds is given by: $v = 200e^{-\frac{Rt}{L}}$, where R = 150Ω and $L = 12.5 \times 10^{-3}$ H. Determine (a) the voltage when $t = 160 \times 10^{-6}$ s, and (b) the time for the voltage to reach 85 V.
3. The length l metres of a metal bar at temperature t°C is given by $l = l_0 e^{\alpha t}$, where l_0 and α are constants. Determine (a) the value of l when $l_0 = 1.894$, $\alpha = 2.038 \times 10^{-4}$ and t = 250°C, and (b) the value of l_0 when l = 2.416, t = 310°C and $\alpha = 1.682 \times 10^{-4}$.
4. The instantaneous current i at time t is given by: $i = 10e^{-t/CR}$ when a capacitor is being charged. The capacitance C is 7×10^{-6} farads

and the resistance R is 0.3×10^6 ohms. Determine:

(a) the instantaneous current when t is 2.5 seconds, and

(b) the time for the instantaneous current to fall to 5 amperes.

Sketch a curve of current against time from t = 0 to t = 6 seconds

5. The current i flowing in a capacitor at time t is given by: $i = 12.5(1 - e^{-t/CR})$ where resistance R is 30 kΩ and the capacitance C is 20 μF. Determine (a) the current flowing after 0.5 seconds, and (b) the time for the current to reach 10 amperes.

2.5 Straight line graphs

A graph is a visual representation of information, showing how one quantity varies with another related quantity.

The most common method of showing a relationship between two sets of data is to use a pair of reference axes – these are two lines drawn at right angles to each other, (often called **Cartesian** or **rectangular axes**), as shown in Figure 2.10.

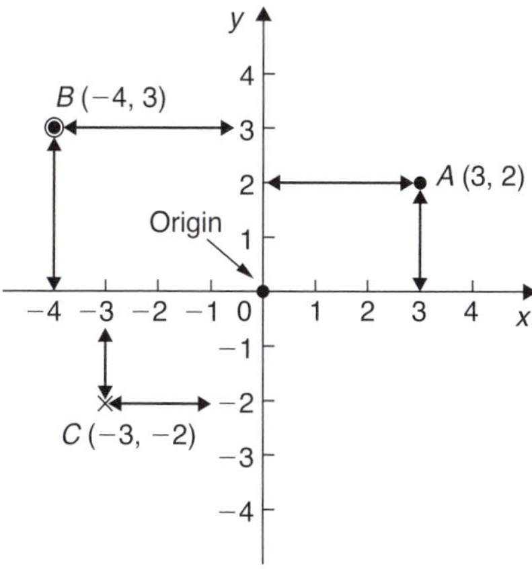

Figure 2.10

The horizontal axis is labelled the x-axis, and the vertical axis is labelled the y-axis.

The point where x is 0 and y is 0 is called the **origin**.

x values have **scales** that are positive to the right of the origin and negative to the left.

y values have scales that are positive up from the origin and negative down from the origin.

Co-ordinates are written with brackets and a comma in between two numbers. For example, point A is shown with co-ordinates (3, 2) and is located by starting at the origin and moving 3 units in the positive x direction (i.e. to the right) and then 2 units in the positive y direction (i.e. up).

When co-ordinates are stated, the first number is always the x value, and the second number is always the y value. Also in Figure 2.10, point B has co-ordinates (−4, 3) and point C has co-ordinates (−3, −2)

The following table gives the force F Newtons which, when applied to a lifting machine, overcomes a corresponding load of L Newtons.

F (Newtons)	19	35	50	93	125	147
L (Newtons)	40	120	230	410	540	680

1. Plot L horizontally and F vertically.
2. Scales are normally chosen such that the graph occupies as much space as possible on the graph paper. So in this case, the following scales are chosen:

   Horizontal axis (i.e. L): 1 cm = 50 N

   Vertical axis (i.e. F): 1 cm = 10 N

3. Draw the axes and label them L (Newtons) for the horizontal axis and F (newtons) for the vertical axis.
4. Label the origin as 0.
5. Write on the horizontal scaling 100, 200, 300, and so on, every 2 cm.
6. Write on the vertical scaling 10, 20, 30, and so on, every 1 cm.
7. Plot on the graph the co-ordinates (40, 19), (120, 35), (230, 50), (410, 93), (540, 125) and (680, 147) marking each with a cross or a dot.
8. Using a ruler, draw the best straight line through the points. You will notice that not all of the points lie exactly on a straight line. This is quite normal with experimental values. In a practical situation it would be surprising if all of the points lay exactly on a straight line.
9. Extend the straight line at each end.
10. From the graph, determine the force applied when the load is 325 N. It should be close to 75 N. This

process of finding an equivalent value within the given data is called **interpolation**.
Similarly, determine the load that a force of 45 N will overcome. It should be close to 170 N.

11. From the graph, determine the force needed to overcome a 750 N load. It should be close to 161 N. This process of finding an equivalent value outside the given data is called **extrapolation**. To extrapolate we need to have extended the straight line drawn. Similarly, determine the force applied when the load is zero. It should be close to 11 N. Where the straight line crosses the vertical axis is called the **vertical-axis intercept**. So in this case, the vertical-axis intercept = 11 N at co-ordinates (0, 11)

The graph you have drawn should look something like Figure 2.11.

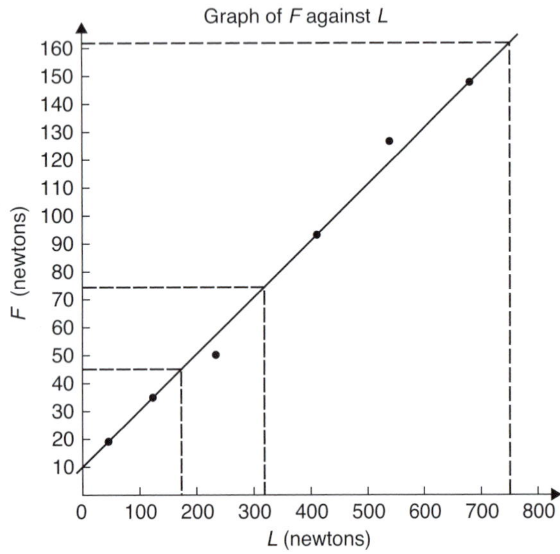

Figure 2.11

In another example, let the relationship between two variables x and y be $y = 3x + 2$

When $x = 0$, $y = 3 \times 0 + 2 = 0 + 2 = 2$

When $x = 1$, $y = 3 \times 1 + 2 = 3 + 2 = 5$

When $x = 2$, $y = 3 \times 2 + 2 = 6 + 2 = 8$, and so on.

The co-ordinates (0, 2), (1, 5) and (2, 8) have been produced and are plotted, with others, as shown in Figure 2.12.
When the points are joined together **a straight line graph results**, i.e. $y = 3x + 2$ is a straight line graph.

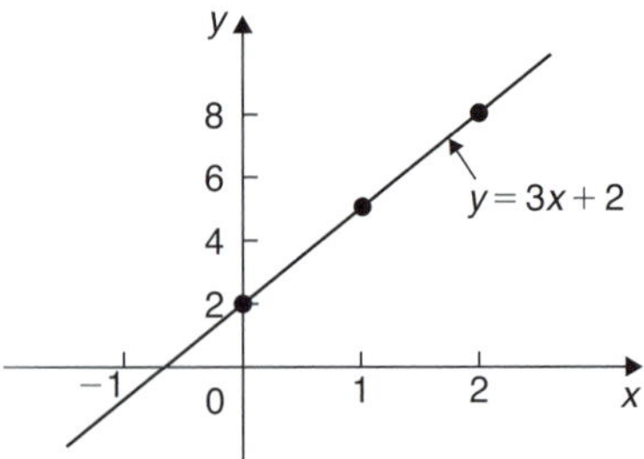

Figure 2.12

Now try the following Practice Exercise

Practice Exercise 21 Straight line graphs (Answers on page 817)

1. Corresponding values obtained experimentally for two quantities are:

x	−5	−3	−1	0	2	4
y	−13	−9	−5	−3	1	5

Plot a graph of y (vertically) against x (horizontally) to scales of 2 cm = 1 for the horizontal x- axis and 1 cm = 1 for the vertical y-axis. (This graph will need the whole of the graph paper with the origin somewhere in the centre of the paper).
From the graph find:

(a) the value of y when $x = 1$
(b) the value of y when $x = -2.5$
(c) the value of x when $y = -6$
(d) the value of x when $y = 5$

2. Corresponding values obtained experimentally for two quantities are:

x	−2.0	−0.5	0	1.0	2.5	3.0	5.0
y	−13.0	−5.5	−3.0	2.0	9.5	12.0	22.0

Use a horizontal scale for x of 1 cm = $\frac{1}{2}$ unit and a vertical scale for y of 1 cm = 2 units and draw a graph of x against y. Label the graph and each of its axes. By interpolation, find from the graph the value of y when x is 3.5

3. Draw a graph of $y - 3x + 5 = 0$ over a range of $x = -3$ to $x = 4$. Hence determine (a) the value of y when $x = 1.3$ and (b) the value of x when $y = -9.2$

4. The speed n rev/min of a motor changes when the voltage V across the armature is varied.

The results are shown in the following table:

n (rev/min)	560	720	900	1010	1240	1410
V (volts)	80	100	120	140	160	180

It is suspected that one of the readings taken of the speed is inaccurate. Plot a graph of speed (horizontally) against voltage (vertically) and find this value. Find also (a) the speed at a voltage of 132 V, and (b) the voltage at a speed of 1300 rev/min.

2.6 Gradients, intercepts and equation of a graph

Gradient

The **gradient or slope** of a straight line is the ratio of the change in the value of y to the change in the value of x between any two points on the line. If, as x increases, (→), y also increases, (↑), then the gradient is positive.
In Figure 2.13(a), a straight line graph $y = 2x + 1$ is shown. To find the gradient of this straight line, choose two points on the straight line graph, such as A and C.

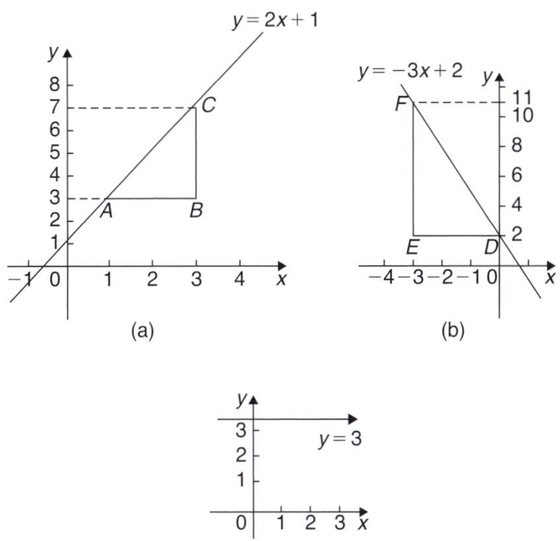

Figure 2.13

Then construct a right angled triangle, such as ABC, where BC is vertical and AB is horizontal.

Then,

$$\textbf{gradient of AC} = \frac{\text{change in y}}{\text{change in x}} = \frac{CB}{BA}$$

$$= \frac{7-3}{3-1} = \frac{4}{2} = \mathbf{2}$$

In Figure 2.13(b), a straight line graph $y = -3x + 2$ is shown. To find the gradient of this straight line, choose two points on the straight line graph, such as D and F. Then construct a right angled triangle, such as DEF, where EF is vertical and ED is horizontal.
Then,

$$\textbf{gradient of DF} = \frac{\text{change in y}}{\text{change in x}} = \frac{FE}{ED}$$

$$= \frac{11-2}{-3-0} = \frac{9}{-3} = \mathbf{-3}$$

Figure 2.13(c) shows a straight line graph $y = 3$. **Since the straight line is horizontal the gradient is zero**.

y-axis intercept

The value of y when $x = 0$ is called the **y-axis intercept**. In Figure 2.5(a) the y-axis intercept is 1 and in Figure 2.13(b) the y-axis intercept is 2

Equation of a straight line graph

The general equation of a straight line graph is:

$$\mathbf{y = mx + c}$$

where m is the gradient or slope, and c is the y-axis intercept
Thus, as we have found in Figure 2.13(a), $y = 2x + 1$ represents a straight line of gradient 2 and y-axis intercept 1. So, given an equation $y = 2x + 1$, we are able to state, on sight, that the gradient = 2 and the y-axis intercept = 1, without the need for any analysis.
Similarly, in Figure 2.13(b), $y = -3x + 2$ represents a straight line of gradient −3 and y-axis intercept 2.
In Figure 2.13(c), $y = 3$ may be re-written as $y = 0x + 3$ and therefore represents a straight line of gradient 0 and y-axis intercept 3.

Here are some worked problems to help understanding of gradients, intercepts and equation of a graph.

Problem 24. Determine for the straight line shown in Figure 2.14: (a) the gradient and (b) the equation of the graph

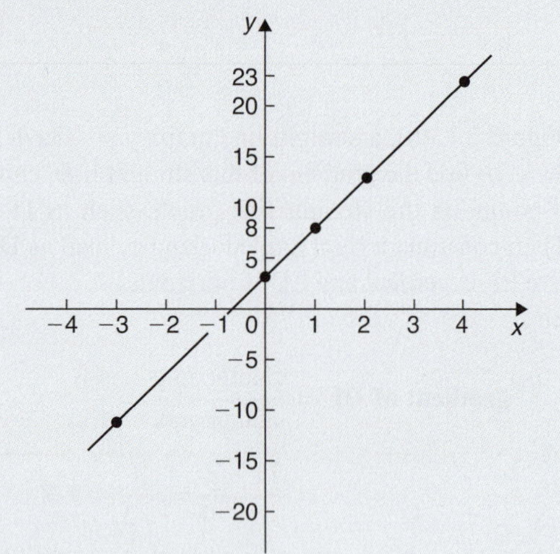

Figure 2.14

(a) A right angled triangle ABC is constructed on the graph as shown in Figure 2.15.

$$\textbf{Gradient} = \frac{AC}{CB} = \frac{23-8}{4-1} = \frac{15}{3} = \mathbf{5}$$

(b) The y-axis intercept at $x = 0$ is seen to be at $y = 3$
$y = mx + c$ is a straight line graph where m = gradient and c = y-axis intercept.
From above, $m = 5$ and $c = 3$.
Hence, equation of graph is: $\mathbf{y = 5x + 3}$

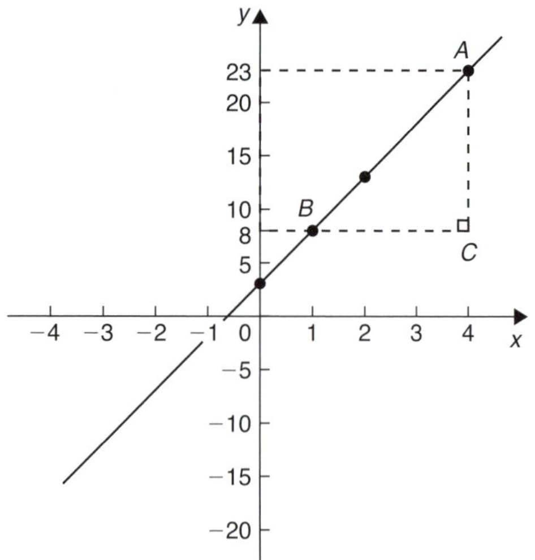

Figure 2.15

Problem 25. Determine the equation of the straight line shown in Figure 2.16.

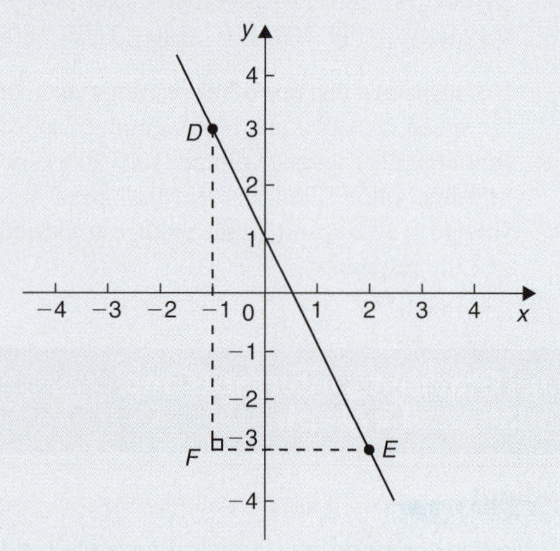

Figure 2.16

The triangle DEF is shown constructed in Figure 2.16.

$$\text{Gradient of DE} = \frac{DF}{FE} = \frac{3-(-3)}{-1-2} = \frac{6}{-3} = -2$$

and the y-axis intercept = 1
Hence, **the equation of the straight line is**:
$y = mx + c$ i.e. $\mathbf{y = -2x + 1}$

Now try the following Practice Exercise

Practice Exercise 22 Gradients, intercepts and equation of a graph (Answers on page 817)

1. The equation of a line is $4y = 2x + 5$. A table of corresponding values is produced and is shown below. Complete the table and plot a graph of y against x. Find the gradient of the graph.

x	−4	−3	−2	−1	0	1	2	3	4
y		−0.25			1.25				3.25

2. Determine the gradient and intercept on the y-axis for each of the following equations:
   (a) $y = 4x - 2$ (b) $y = -x$
   (c) $y = -3x - 4$ (d) $y = 4$

3. Draw on the same axes the graphs of $y = 3x - 5$ and $3y + 2x = 7$. Find the co-ordinates of the point of intersection.

4. A piece of elastic is tied to a support so that it hangs vertically, and a pan, on which weights can be placed, is attached to the free end. The length of the elastic is measured as various weights are added to the pan and the results obtained are as follows:

Load, W (N)	5	10	15	20	25
Length, l (cm)	60	72	84	96	108

Plot a graph of load (horizontally) against length (vertically) and determine: (a) the length when the load is 17 N, (b) the value of load when the length is 74 cm, (c) its gradient, and (d) the equation of the graph.

2.7 Practical straight line graphs

When a set of co-ordinate values are given or are obtained experimentally and it is believed that they follow a law of the form y = mx + c, then if a straight line can be drawn reasonably close to most of the co-ordinate values when plotted, this verifies that a law of the form y = mx + c exists. From the graph, constants m (i.e. gradient) and c (i.e. y-axis intercept) can be determined.
Here is a worked practical problems.

Problem 26. The following values of resistance R ohms and corresponding voltage V volts are obtained from a test on a filament lamp.

R ohms	30	48.5	73	107	128
V volts	16	29	52	76	94

Choose suitable scales and plot a graph with R representing the vertical axis and V the horizontal axis. Determine (a) the gradient of the graph, (b) the R axis intercept value, (c) the equation of the graph, (d) the value of resistance when the voltage is 60 V, and (e) the value of the voltage when the resistance is 40 ohms. (f) If the graph were to continue in the same manner, what value of resistance would be obtained at 110 V?

The co-ordinates (16, 30), (29, 48.5), and so on, are shown plotted in Figure 2.17 where the best straight line is drawn through the points.

(a) The slope or gradient of the straight line AC is given by:

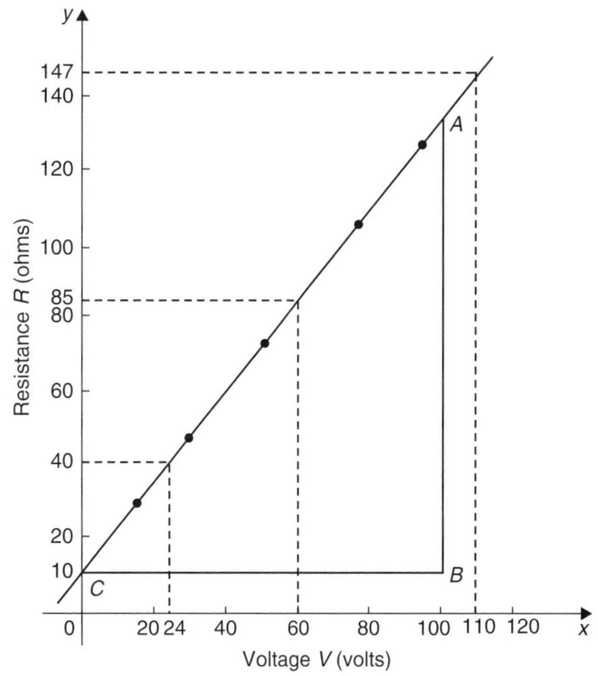

Figure 2.17

$$\frac{AB}{BC} = \frac{135 - 10}{100 - 0} = \frac{125}{100} = \mathbf{1.25}$$

(Note that the vertical line AB and the horizontal line BC may be constructed anywhere along the length of the straight line. However, calculations are made easier if the horizontal line BC is carefully chosen, in this case equal to 100)

(b) The R-axis intercept is at **R = 10 ohms** (by extrapolation)

(c) The equation of a straight line is y = mx + c, when y is plotted on the vertical axis and x on the horizontal axis. m represents the gradient and c the y-axis intercept. In this case, R corresponds to y, V corresponds to x, m = 1.25 and c = 10. Hence the equation of the graph is:

$$\mathbf{R = (1.25V + 10)\Omega}$$

From Figure 2.17,

(d) when the voltage is 60 V, the resistance is **85Ω**

(e) when the resistance is 40 ohms, the voltage is **24 V**, and

(f) by extrapolation, when the voltage is 110 V, the resistance is **147Ω**

Now try the following Practice Exercise

Practice Exercise 23 Practical problems involving straight line graphs (Answers on page 817)

1. The resistance R ohms of a copper winding is measured at various temperatures t°C and the results are as follows:

R ohms	112	120	126	131	134
t°C	20	36	48	58	64

Plot a graph of R (vertically) against t (horizontally) and find from it (a) the temperature when the resistance is 122 Ω and (b) the resistance when the temperature is 52°C

2. The following table gives the force F Newtons which, when applied to a lifting machine, overcomes a corresponding load of L Newtons.

Force F Newtons	25	47	64	120	149	187
Load L Newtons	50	140	210	430	550	700

Choose suitable scales and plot a graph of F (vertically) against L (horizontally). Draw the best straight line through the points. Determine from the graph (a) the gradient, (b) the F-axis intercept, (c) the equation of the graph, (d) the force applied when the load is 310 N, and (e) the load that a force of 160 N will overcome. (f) If the graph were to continue in the same manner, what value of force will be needed to overcome a 800 N load?

3. The speed of a motor varies with armature voltage as shown by the following experimental results:

n (rev/min)	285	517	615	750	917	1050
V volts	60	95	110	130	155	175

Plot a graph of speed (horizontally) against voltage (vertically) and draw the best straight line through the points. Find from the graph (a) the speed at a voltage of 145 V, and (b) the voltage at a speed of 400 rev/min.

4. An experiment with a set of pulley blocks gave the following results:

Effort, E (newtons)	9.0	11.0	13.6	17.4	20.8	23.6
Load, L (newtons)	15	25	38	57	74	88

Plot a graph of effort (vertically) against load (horizontally) and determine (a) the gradient, (b) the vertical axis intercept, (c) the law of the graph, (d) the effort when the load is 30 N and (e) the load when the effort is 19 N.

2.8 Calculating areas of common shapes

The formulae for the areas of common shapes are shown in Table 2.1.
Here are some worked problems to demonstrate how the formulae are used to determine the area of common shapes.

Problem 27. Calculate the area of the parallelogram shown in Figure 2.18.

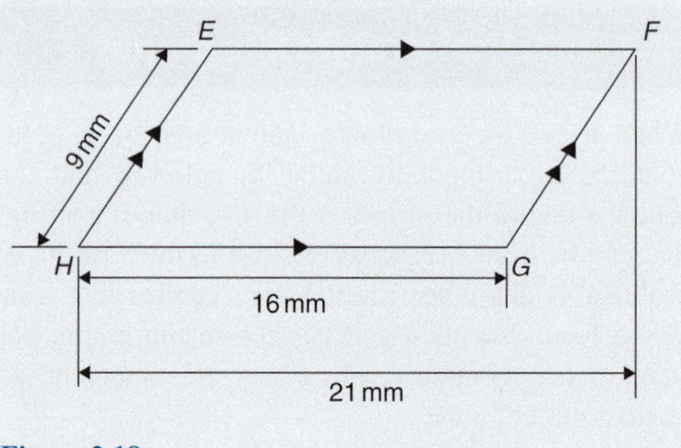

Figure 2.18

Area of a parallelogram = base × perpendicular height. The perpendicular height h is not shown on Figure 2.18 but may be found using Pythagoras's theorem (see Chapter 1).
From Figure 2.19, $9^2 = 5^2 + h^2$
from which, $h^2 = 9^2 - 5^2 = 81 - 25 = 56$
Hence, perpendicular height, $h = \sqrt{56} = 7.48$ mm

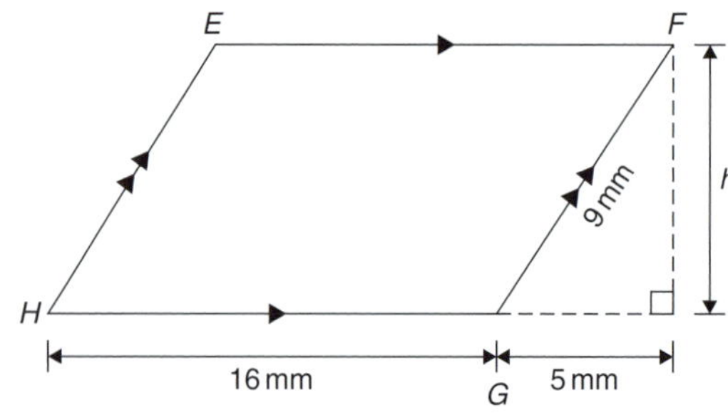

Figure 2.19

Hence, **area of parallelogram EFGH**
$= 16\,\text{mm} \times 7.48\,\text{mm} = \mathbf{120\,mm^2}$

Table 2.1 Formulae for the areas of common shapes

Area of plane figures	
Square	Area $= x^2$
Rectangle	Area $= l \times b$
Parallelogram	Area $= b \times h$
Triangle	Area $= \frac{1}{2} \times b \times h$
Trapezium	Area $= \frac{1}{2}(a+b)h$
Circle	Area $= \pi r^2$ or $\frac{\pi d^2}{4}$ Circumference $= 2\pi r$ Radian measure: 2π radians $= 360$ degrees
Sector of circle	Area $= \frac{\theta°}{360}(\pi r^2)$

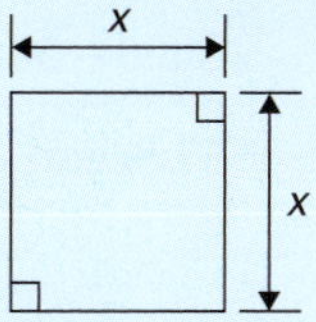

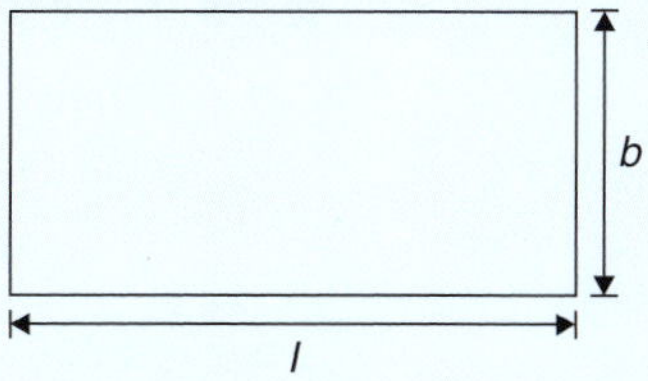

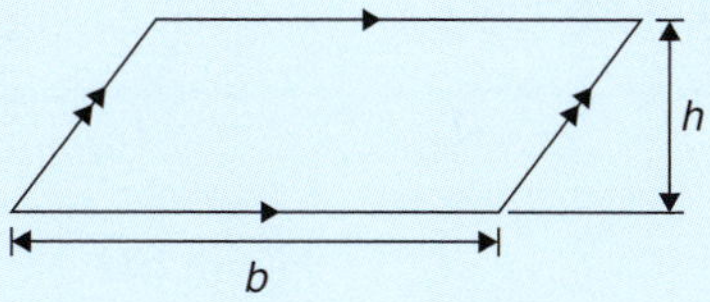

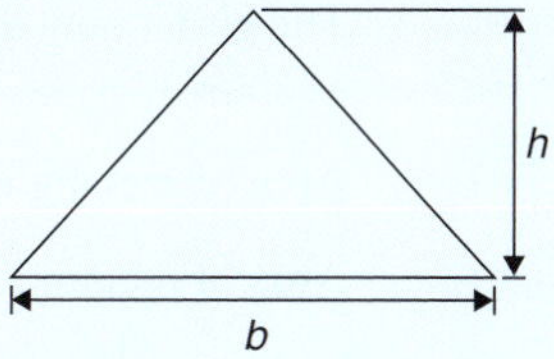

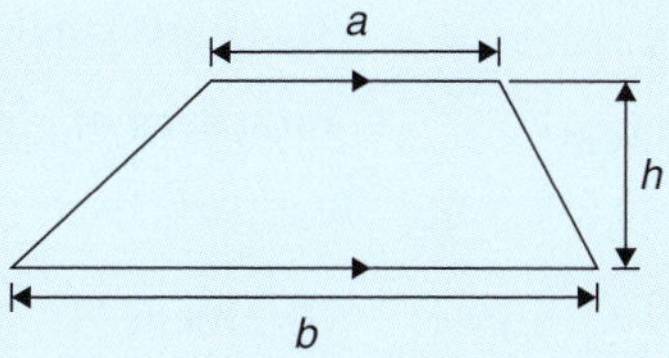

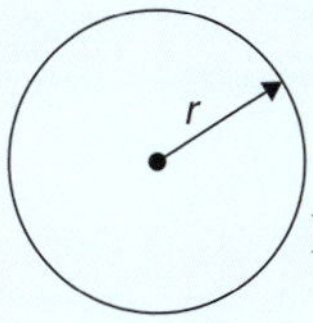

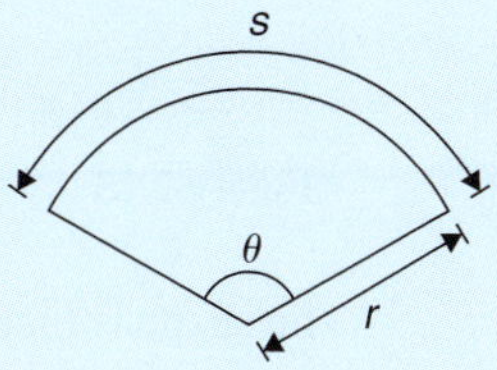

Problem 28. Calculate the area of the triangle shown in Figure 2.20.

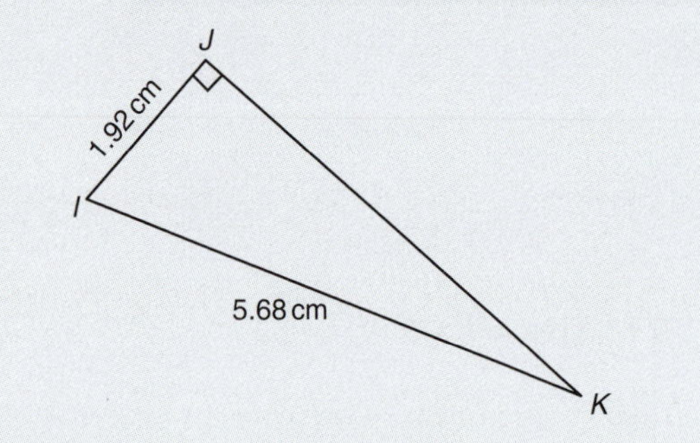

Figure 2.20

Area of triangle IJK $= \frac{1}{2} \times \text{base} \times \text{perpendicular height}$

$$= \frac{1}{2} \times IJ \times JK$$

To find JK, Pythagoras's theorem is used, i.e.

$$5.68^2 = 1.92^2 + JK^2$$

$$\text{from which, } JK = \sqrt{5.68^2 - 1.92^2} = 5.346\,\text{cm}$$

Hence, **area of triangle IJK**

$$= \frac{1}{2} \times 1.92 \times 5.346 = \mathbf{5.132\,cm^2}$$

Problem 29. The outside measurements of a picture frame are 100 cm by 50 cm. If the frame is 4 cm wide, find the area of the wood used to make the frame.

A sketch of the frame is shown shaded in Figure 2.21.

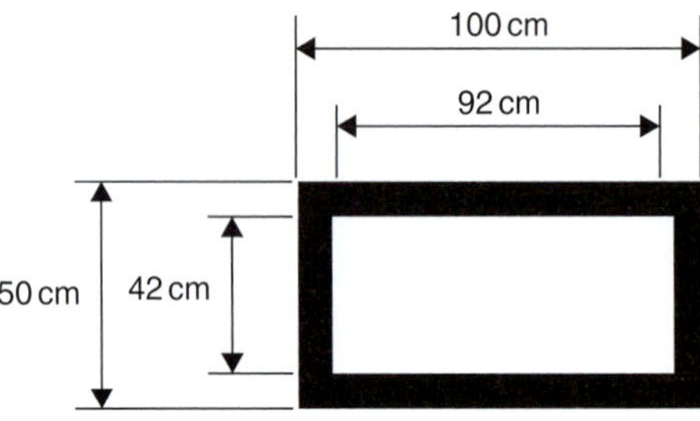

Figure 2.21

Area of wood

$= \text{area of large rectangle} - \text{area of small rectangle}$

$= (100 \times 50) - (92 \times 42)$

$= 5000 - 3864$

$= \mathbf{1136\,cm^2}$

Problem 30. Find the cross-sectional area of the girder shown in Figure 2.22.

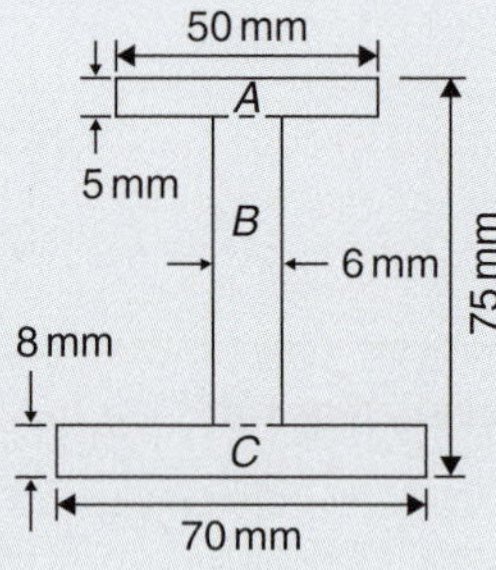

Figure 2.22

The girder may be divided into three separate rectangles as shown.

Area of rectangle A $= 50 \times 5 = 250\,\text{mm}^2$

Area of rectangle B $= (75 - 8 - 5) \times 6$
$= 62 \times 6 = 372\,\text{mm}^2$

Area of rectangle C $= 70 \times 8 = 560\,\text{mm}^2$

Total area of girder $= 250 + 372 + 560$
$= \mathbf{1182\,mm^2}$ or $\mathbf{11.82\,cm^2}$

Problem 31. Figure 2.23 shows the gable end of a building. Determine the area of brickwork in the gable end.

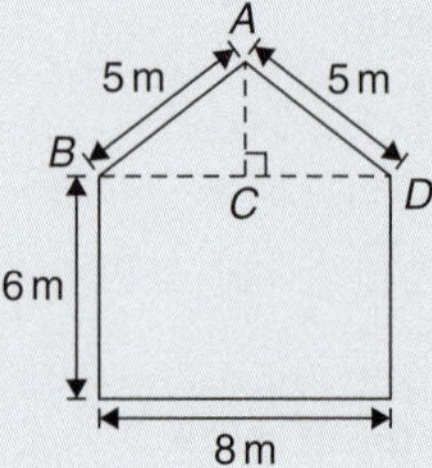

Figure 2.23

The shape is that of a rectangle and a triangle.

Area of rectangle $= 6 \times 8 = 48\,\text{m}^2$

Area of triangle $= \frac{1}{2} \times \text{base} \times \text{height}$

CD = 4 m, AD = 5 m, hence AC = 3 m (since it is a 3, 4, 5 triangle – or by Pythagoras)

Hence, area of triangle ABD $= \frac{1}{2} \times 8 \times 3 = 12\,\text{m}^2$

Total area of brickwork = 48 + 12 = **60 m²**

Problem 32. Find the area of the circle having a diameter of 15 mm.

Area of circle $= \frac{\pi d^2}{4} = \frac{\pi(15)^2}{4} = \frac{225\pi}{4}$
$= \mathbf{176.7\,mm^2}$

Problem 33. Find the area of the circle having a circumference of 70 mm.

Circumference, $c = 2\pi r$

hence radius, $r = \frac{c}{2\pi} = \frac{70}{2\pi} = \frac{35}{\pi}$ mm

Area of circle $= \pi r^2 = \pi\left(\frac{35}{\pi}\right)^2 = \frac{35^2}{\pi}$
$= \mathbf{389.9\,mm^2}$ or $\mathbf{3.899\,cm^2}$

Problem 34. Calculate the area of the sector a of circle having diameter 80 mm with angle subtended at centre 107°42'.

If diameter = 80 mm, then radius, r = 40 mm, and area

of sector $= \frac{107°42'}{360}(\pi 40^2) = \frac{107\frac{42}{60}}{360}(\pi 40^2)$
$= \frac{107.7}{360}(\pi 40^2) = \mathbf{1504\,mm^2}$ or $\mathbf{15.04\,cm^2}$

Problem 35. A hollow shaft has an outside diameter of 5.45 cm and an inside diameter of 2.25 cm. Calculate the cross-sectional area of the shaft.

The cross-sectional area of the shaft is shown by the shaded part in Figure 2.24 (often called an **annulus**).

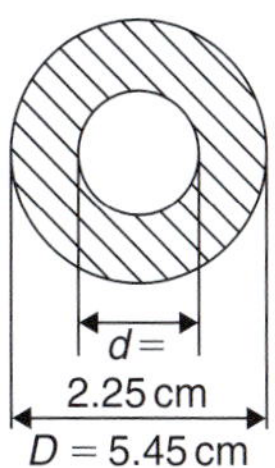

Figure 2.24

Area of shaded part

= area of large circle - area of small circle

$= \frac{\pi D^2}{4} - \frac{\pi d^2}{4} = \frac{\pi}{4}(D^2 - d^2) = \frac{\pi}{4}(5.45^2 - 2.25^2)$
$= \mathbf{19.35\,cm^2}$

Problem 36. A football stadium floodlight can spread its illumination over an angle of 45° to a distance of 55 m. Determine the maximum area that is floodlit.

Floodlit area = area of sector $= \frac{\theta°}{360}\left(\pi r^2\right)$
$= \frac{45}{360}\left(\pi \times 55^2\right)$ from Table 2.1
$= \mathbf{1188\,m^2}$

Problem 37. An automatic garden spray produces a spray to a distance of 1.8 m and revolves through an angle α which may be varied. If the desired spray catchment area is to be 2.5 m², to what should angle α be set, correct to the nearest degree.

Area of sector, $2.5 = \frac{\alpha}{360}\left(\pi r^2\right)$

from which,

$\alpha = \frac{2.5 \times 360}{\pi \times 1.8^2}$
$= 88.42°$

Hence, **angle** $\boldsymbol{\alpha}$ **= 88°**, correct to the nearest degree.

Now try the following Practice Exercise

Practice Exercise 24 Areas of common shapes (Answers on page 817)

1. A rectangular field has an area of 1.2 hectares and a length of 150 m. If 1 hectare = 10000 m² find (a) its width, and (b) the length of a diagonal.
2. Find the area of a triangle whose base is 8.5 cm and perpendicular height 6.4 cm.
3. A square has an area of 162 cm². Determine the length of a diagonal.
4. A rectangular picture has an area of 0.96 m². If one of the sides has a length of 800 mm, calculate, in millimetres, the length of the other side.

5. Determine the area of each of the angle iron sections shown in Figure 2.25.

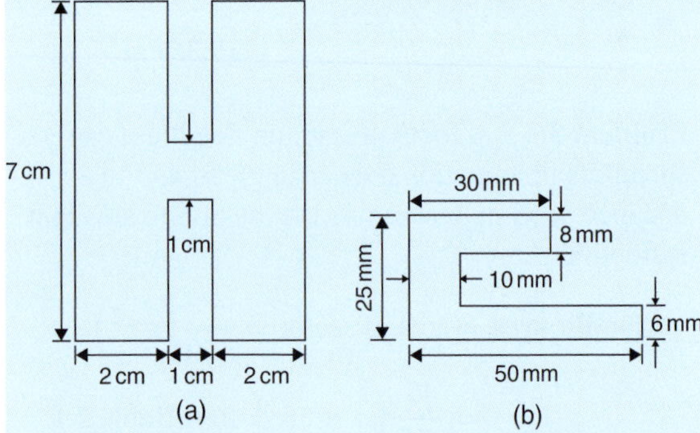

Figure 2.25

6. Figure 2.26 shows a 4 m wide path within the outside wall of a 41 m by 37 m garden. Calculate the area of the path.

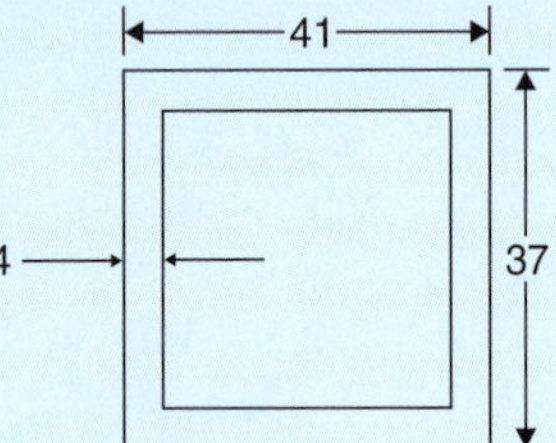

Figure 2.26

7. Calculate the area of the steel plate shown in Figure 2.27

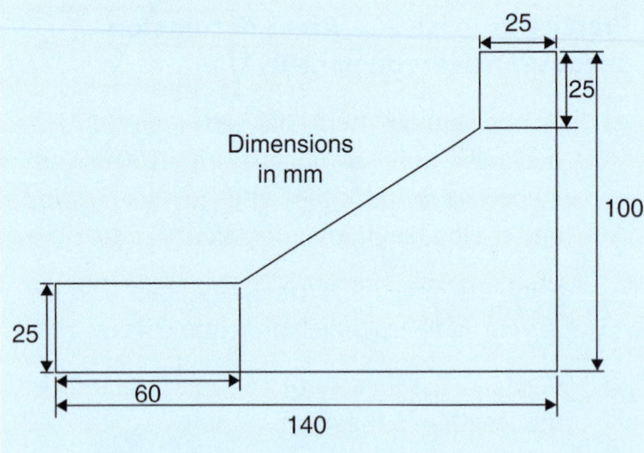

Figure 2.27

8. Determine the area of an equilateral triangle of side 10.0 cm.

9. If paving slabs are produced in 250 mm by 250 mm squares, determine the number of slabs required to cover an area of 2 m^2.

10. A rectangular garden measures 40 m by 15 m. A 1 m flower border is made round the two shorter sides and one long side. A circular swimming pool of diameter 8 m is constructed in the middle of the garden. Find, correct to the nearest square metre, the area remaining.

11. Determine the area of circles having (a) a radius of 4 cm (b) a diameter of 30 mm (c) a circumference of 200 mm.

12. Calculate the areas of the following sectors of circles:
    (a) radius 9 cm, angle subtended at centre 75°
    (b) diameter 35 mm, angle subtended at centre 48°37′

13. Determine the shaded area of the template shown in Figure 2.28.

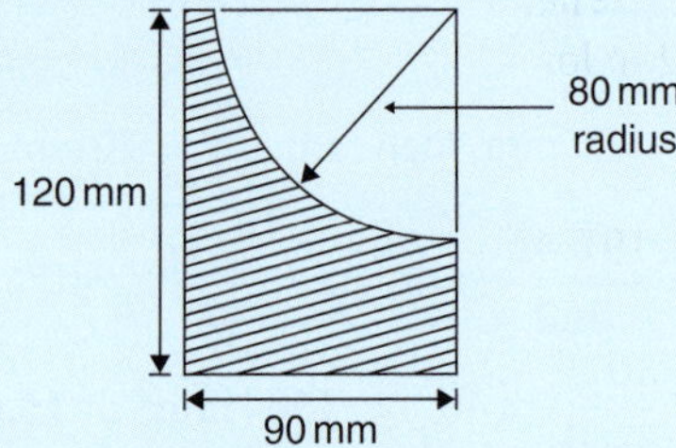

Figure 2.28

14. An archway consists of a rectangular opening topped by a semi-circular arch as shown in Figure 2.29. Determine the area of the opening if the width is 1 m and the greatest height is 2 m.

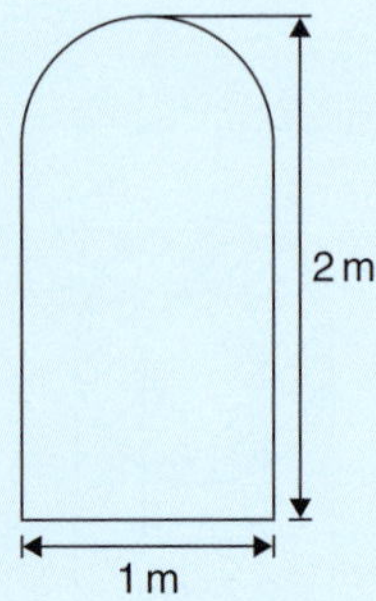

Figure 2.29

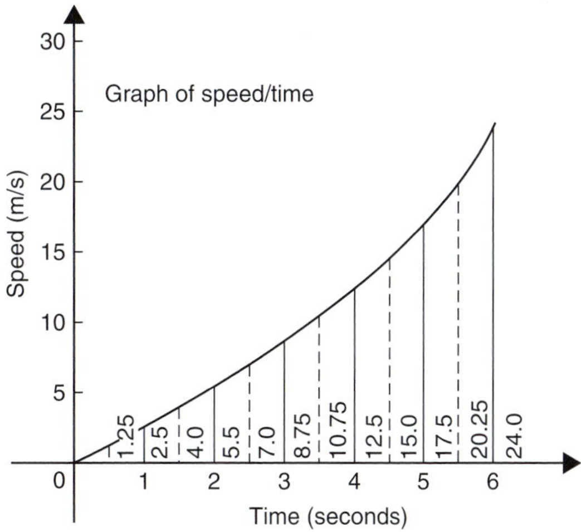

Figure 2.30

15. The floodlights at a sports ground spread its illumination over an angle of 40° to a distance of 48 m. Determine (a) the angle in radians, and (b) the maximum area that is floodlit.

16. Find the area swept out in 50 minutes by the minute hand of a large floral clock, if the hand is 2 m long.

Areas of irregular figures

Areas of irregular plane surfaces may be approximately determined by using the mid-ordinate rule. The **mid-ordinate rule** states:

Area = (width of interval)(sum of mid-ordinates)

Problem 38. A car starts from rest and its speed is measured every second for 6 s:

Time t(s)	0	1	2	3	4	5	6
Speed v (m/s)	0	2.5	5.5	8.75	12.5	17.5	24.0

Determine the distance travelled in 6 seconds (i.e. the area under the v/t graph), by the mid-ordinate rule.

A graph of speed/time is shown in Figure 2.30.

The time base is divided into 6 strips each of width 1 second.

Mid-ordinates are erected as shown in Figure 2.30 by the broken lines.

The length of each mid-ordinate is measured. Thus

area = (width of interval)(sum of mid-ordinates)

$= (1)\,[1.25 + 4.0 + 7.0 + 10.75 + 15.0 + 20.25]$

$=$ **58.25 m**

Now try the following Practice Exercise

Practice Exercise 25 Areas of irregular figures (Answers on page 818)

1. Plot a graph of $y = 3x - x^2$ by completing a table of values of y from $x = 0$ to $x = 3$. Determine the area enclosed by the curve, the x-axis and ordinate $x = 0$ and $x = 3$ by the mid-ordinate rule.

2. Plot the graph of $y = 2x^2 + 3$ between $x = 0$ and $x = 4$. Estimate the area enclosed by the curve, the ordinates $x = 0$ and $x = 4$, and the x-axis by the mid-ordinate rule.

For fully worked solutions to each of the problems in Practice Exercises 13 to 25 in this chapter, go to the website:
www.routledge.com/cw/bird

Main formulae for revision of some basic mathematics

Laws of indices

$$a^m \times a^n = a^{m+n} \quad \frac{a^m}{a^n} = a^{m-n} \quad (a^m)^n = a^{mn}$$

$$a^{\frac{m}{n}} = \sqrt[n]{a^m} \quad a^{-n} = \frac{1}{a^n} \quad a^0 = 1$$

Radian measure

2π radians $=$ 360 degrees

Theorem of Pythagoras

$b^2 = a^2 + c^2$

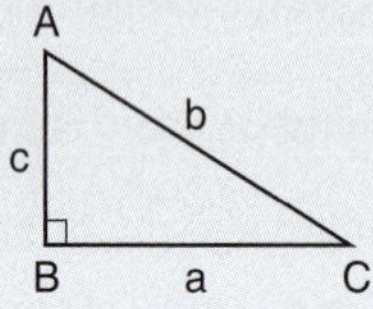

$$\sin A = \frac{a}{b} \quad \cos A = \frac{c}{b} \quad \tan A = \frac{a}{c}$$

Sine rule

$$\frac{a}{\sin A} = \frac{b}{\sin B} = \frac{c}{\sin C}$$

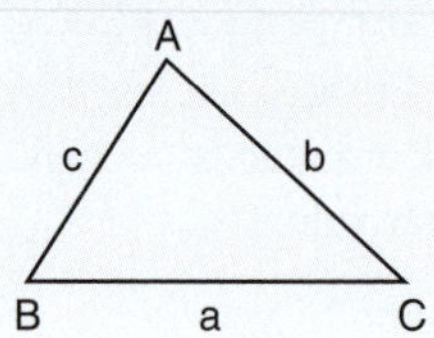

Cosine rule

$a^2 = b^2 + c^2 - 2bc\cos A$

Definition of a logarithm

If $y = a^x$ then $x = \log_a y$

Laws of logarithms

$$\log(A \times B) = \log A + \log B$$

$$\log\left(\frac{A}{B}\right) = \log A - \log B \qquad \log A^n = n \times \log A$$

Equation of a straight line

$y = mx + c$ where m is the gradient and c is the y-axis intercept

Part 1

Area of plane figures

Square Area $= x^2$

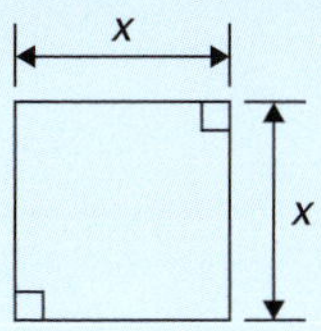

Rectangle Area $= l \times b$

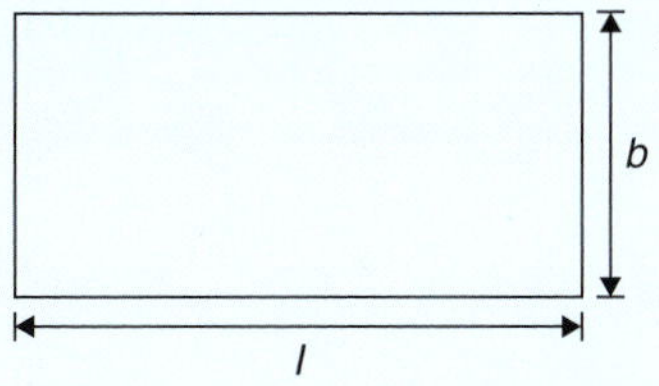

Parallelogram Area $= b \times h$

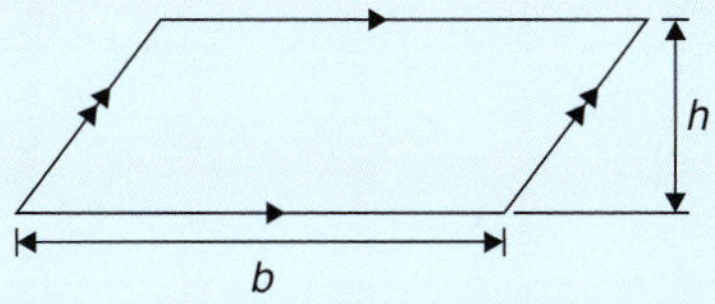

Triangle Area $= \frac{1}{2} \times b \times h$

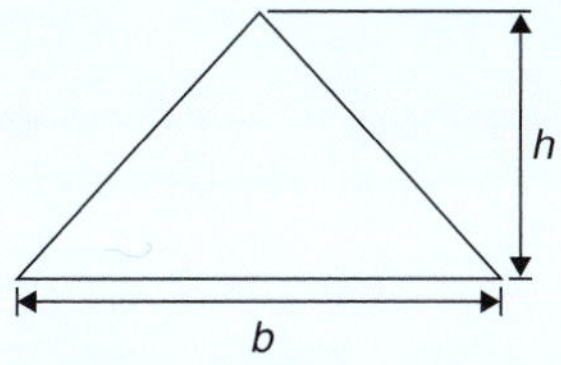

Trapezium Area $= \frac{1}{2}(a+b)h$

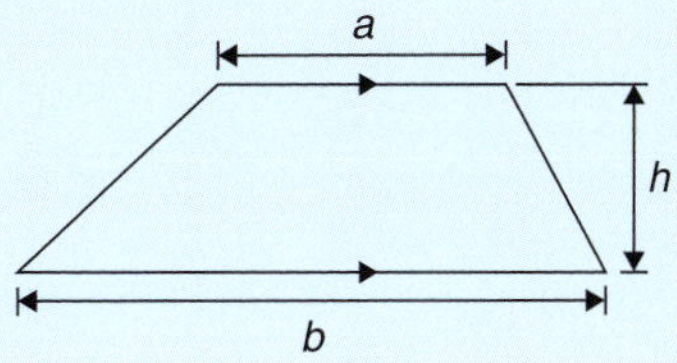

Circle Area $= \pi r^2$ or $\frac{\pi d^2}{4}$

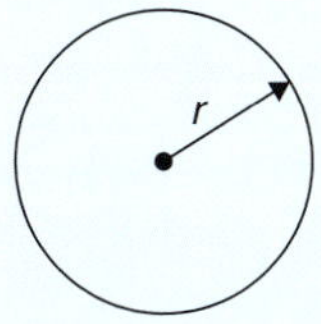

Circumference $= 2\pi r$

Radian measure: 2π radians = 360 degrees

Sector of circle Area $= \frac{\theta°}{360}(\pi r^2)$

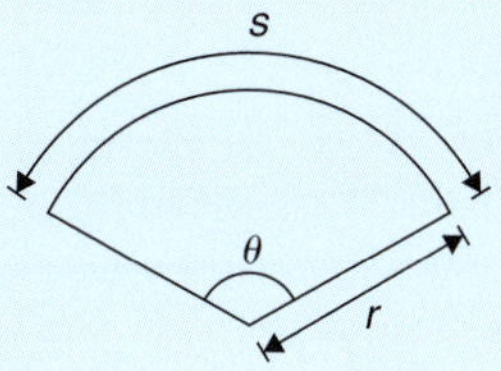

These formulae are available for downloading at the website:
www.routledge.com/cw/bird

Part 2

Basic electrical engineering principles

Chapter 3

Units associated with basic electrical quantities

***Why it is important to understand:* Units associated with basic electrical quantities**

The relationship between quantities can be written using words or symbols (letters), but symbols are normally used because they are much shorter; for example, *V* is used for voltage, *I* for current and *R* for resistance. Some of the units have a convenient size for electronics, but most are either too large or too small to be used directly so they are used with prefixes. The prefixes make the unit larger or smaller by the value shown; for example, 25 mA is read as 25 milliamperes and means 25×10^{-3}A = 25 x 0.001A = 0.025A. Knowledge of this chapter is essential for future studies and provides the basis of electrical units and prefixes; some simple calculations help understanding.

At the end of this chapter you should be able to:

- state the basic SI units
- recognize derived SI units
- understand prefixes denoting multiplication and division
- state the units of charge, force, work and power and perform simple calculations involving these units
- state the units of electrical potential, e.m.f., resistance, conductance, power and energy and perform simple calculations involving these units

3.1 SI units

The system of units used in engineering and science is the Système Internationale d'Unités (international system of units), usually abbreviated to SI units, and is based on the metric system. This was introduced in 1960 and is now adopted by the majority of countries as the official system of measurement.

The basic units in the SI system are listed with their symbols in Table 3.1.

Derived SI units use combinations of basic units and there are many of them. Two examples are:

- Velocity – metres per second (m/s)
- Acceleration – metres per second squared (m/s^2)

SI units may be made larger or smaller by using prefixes which denote multiplication or division by a particular amount. The six most common multiples, with their meaning, are listed in Table 3.2. For a more complete list of prefixes, see page 813.

Electrical Circuit Theory and Technology. 978-1-138-67349-6, © 2017 John Bird. Published by Taylor & Francis. All rights reserved.

Table 3.1 Basic SI units

Quantity	Unit
Length	metre, m
Mass	kilogram, kg
Time	second, s
Electric current	ampere, A
Thermodynamic temperature	kelvin, K
Luminous intensity	candela, cd
Amount of substance	mole, mol

3.2 Charge

The **unit of charge** is the **coulomb*** (C), where one coulomb is one ampere second (1 coulomb $= 6.24 \times 10^{18}$ electrons). The coulomb is defined as the quantity of electricity which flows past a given point in an electric circuit when a current of one **ampere*** is maintained for one second. Thus,

charge, in coulombs $\quad Q = It$

where I is the current in amperes and t is the time in seconds.

* **Who was** Coulomb? **Charles-Augustin de Coulomb** (14 June 1736–23 August 1806) was best known for developing Coulomb's law, the definition of the electrostatic force of attraction and repulsion. To find out more go to **www.routledge.com/cw/bird**

Problem 1. If a current of 5 A flows for 2 minutes, find the quantity of electricity transferred.

Quantity of electricity $Q = It$ coulombs

$$I = 5\,\text{A},\ t = 2 \times 60 = 120\,\text{s}$$

Hence $\quad Q = 5 \times 120 = \mathbf{600\ C}$

3.3 Force

The **unit of force** is the **newton*** (N) where one newton is one kilogram metre per second squared. The newton is defined as the force which, when applied to a mass of one kilogram, gives it an acceleration of one metre per second squared. Thus,

force, in newtons $\quad F = ma$

where m is the mass in kilograms and a is the acceleration in metres per second squared. **Gravitational force**, or **weight**, is mg, where $g = 9.81\,\text{m/s}^2$.

* **Who was** Ampére? **André-Marie Ampére** (1775–1836) is generally regarded as one of the founders of classical electromagnetism. To find out more go to **www.routledge.com/cw/bird**

Table 3.2

Prefix	Name	Meaning	
M	mega	multiply by 1 000 000	(i.e. $\times 10^6$)
k	kilo	multiply by 1000	(i.e. $\times 10^3$)
m	milli	divide by 1000	(i.e. $\times 10^{-3}$)
μ	micro	divide by 1 000 000	(i.e. $\times 10^{-6}$)
n	nano	divide by 1 000 000 000	(i.e. $\times 10^{-9}$)
p	pico	divide by 1 000 000 000 000	(i.e. $\times 10^{-12}$)

Problem 2. A mass of 5000 g is accelerated at $2\,\text{m/s}^2$ by a force. Determine the force needed.

Force = mass × acceleration

$$= 5\,\text{kg} \times 2\,\text{m/s}^2 = 10\frac{\text{kg m}}{\text{s}^2} = \mathbf{10\,N}$$

Problem 3. Find the force acting vertically downwards on a mass of 200 g attached to a wire.

Mass = 200 g = 0.2 kg and acceleration due to gravity, $g = 9.81\,\text{m/s}^2$

Force acting downwards = weight = mass × acceleration

$$= 0.2\,\text{kg} \times 9.81\,\text{m/s}^2$$

$$= \mathbf{1.962\,N}$$

3.4 Work

The **unit of work or energy** is the **joule*** (J), where one joule is one newton metre. The joule is defined as the

* **Who was** Newton? **Sir Isaac Newton** (25 December 1642–20 March 1727) was the English polymath who laid the foundations for much of classical mechanics used today. To find out more go to **www.routledge.com/cw/bird**

* **Who was** Joule? **James Prescott Joule** (24 December 1818–11 October 1889) was an English physicist and brewer. He studied the nature of heat, and discovered its relationship to mechanical work. To find out more go to **www.routledge.com/cw/bird**

work done or energy transferred when a force of one newton is exerted through a distance of one metre in the direction of the force. Thus

work done on a body, in joules $\quad W = Fs$

where F is the force in newtons and s is the distance in metres moved by the body in the direction of the force. Energy is the capacity for doing work.

3.5 Power

The **unit of power** is the **watt*** (W) where one watt is one joule per second. Power is defined as the rate of doing work or transferring energy. Thus,

power in watts, $\quad P = \dfrac{W}{t}$

where W is the work done or energy transferred in joules and t is the time in seconds. Thus

energy in joules, $\quad W = Pt$

* **Who was** Watt? **James Watt** (19 January 1736–25 August 1819) was a Scottish inventor and mechanical engineer whose radically improved both the power and efficiency of steam engines. To find out more go to **www.routledge.com/cw/bird**

Problem 4. A portable machine requires a force of 200 N to move it. How much work is done if the machine is moved 20 m and what average power is utilized if the movement takes 25 s?

$$\text{Work done} = \text{force} \times \text{distance} = 200\,\text{N} \times 20\,\text{m} = \mathbf{4000\,Nm\ or\ 4\,kJ}$$

$$\text{Power} = \frac{\text{work done}}{\text{time taken}} = \frac{4000\,\text{J}}{25\,\text{s}} = \mathbf{160\,J/s = 160\,W}$$

Problem 5. A mass of 1000 kg is raised through a height of 10 m in 20 s. What is (a) the work done and (b) the power developed?

(a) Work done = force × distance

force = mass × acceleration

$$\text{Hence, work done} = (1000\,\text{kg} \times 9.81\,\text{m/s}^2) \times (10\,\text{m}) = 98\,100\,\text{Nm} = \mathbf{98.1\,kNm\ or\ 98.1\,kJ}$$

(b) $$\text{Power} = \frac{\text{work done}}{\text{time taken}} = \frac{98\,100\,\text{J}}{20\,\text{s}} = 4905\,\text{J/s} = \mathbf{4905\,W\ or\ 4.905\,kW}$$

Now try the following Practice Exercise

Practice Exercise 26 Charge, force, work and power (Answers on page 818)

(Take $g = 9.81\,\text{m/s}^2$ where appropriate)

1. What force is required to give a mass of 20 kg an acceleration of $30\,\text{m/s}^2$?
2. Find the accelerating force when a car having a mass of 1.7 Mg increases its speed with a constant acceleration of $3\,\text{m/s}^2$.
3. A force of 40 N accelerates a mass at $5\,\text{m/s}^2$. Determine the mass.
4. Determine the force acting downwards on a mass of 1500 g suspended on a string.
5. A force of 4 N moves an object 200 cm in the direction of the force. What amount of work is done?

6. A force of 2.5 kN is required to lift a load. How much work is done if the load is lifted through 500 cm?

7. An electromagnet exerts a force of 12 N and moves a soft iron armature through a distance of 1.5 cm in 40 ms. Find the power consumed.

8. A mass of 500 kg is raised to a height of 6 m in 30 s. Find (a) the work done and (b) the power developed.

9. What quantity of electricity is carried by 6.24×10^{21} electrons?

10. In what time would a current of 1 A transfer a charge of 30 C?

11. A current of 3 A flows for 5 minutes. What charge is transferred?

12. How long must a current of 0.1 A flow so as to transfer a charge of 30 C?

13. Rewrite the following as indicated:
    (a) 1000 pF = ... nF
    (b) 0.02 μF = ... pF
    (c) 5000 kHz = ... MHz
    (d) 47 kΩ = ... MΩ
    (e) 0.32 mA = ... μA

3.6 Electrical potential and e.m.f.

The **unit of electric potential** is the volt (V), where one volt is one joule per coulomb. One volt is defined as the difference in potential between two points in a conductor which, when carrying a current of one ampere, dissipates a power of one watt, i.e.

$$\text{volts} = \frac{\text{watts}}{\text{amperes}} = \frac{\text{joules/second}}{\text{amperes}} = \frac{\text{joules}}{\text{ampere seconds}} = \frac{\text{joules}}{\text{coulombs}}$$

(The **volt** is named after the Italian physicist **Alessandro Volta**.*)

A change in electric potential between two points in an electric circuit is called a **potential difference**. The **electromotive force (e.m.f.)** provided by a source of energy such as a battery or a generator is measured in volts.

3.7 Resistance and conductance

The **unit of electric resistance** is the **ohm*** **(Ω)**, where one ohm is one volt per ampere. It is defined as the resistance between two points in a conductor when a constant electric potential of one volt applied at the two points produces a current flow of one ampere in the conductor. Thus,

$$\text{resistance in ohms,} \quad R = \frac{V}{I}$$

where V is the potential difference across the two points in volts and I is the current flowing between the two points in amperes.

The reciprocal of resistance is called **conductance** and is measured in siemens (S), named after the German inventor and industrialist **Ernst Siemens.*** Thus,

$$\text{conductance in siemens,} \quad G = \frac{1}{R}$$

where R is the resistance in ohms.

Problem 6. Find the conductance of a conductor of resistance (a) 10 Ω, (b) 5 kΩ and (c) 100 mΩ.

*** Who was Volta? Alessandro Giuseppe Antonio Anastasio Volta** (18 February 1745–5 March 1827) was the Italian physicist who invented the battery. To find out more go to **www.routledge.com/cw/bird**

Part 2

* **Who was** Ohm? **Georg Simon Ohm** (16 March 1789–6 July 1854) was a Bavarian physicist and mathematician who wrote a complete theory of electricity, in which he stated his law for electromotive force.To find out more go to **www.routledge.com/cw/bird**

* **Who was** Siemens? **Ernst Werner Siemens** (13 December 1816–6 December 1892) was a German inventor and industrialist, known world-wide for his advances in various technologies. To find out more go to **www.routledge.com/cw/bird**

(a) Conductance $G = \frac{1}{R} = \frac{1}{10}$ siemen $= \mathbf{0.1\,S}$

(b) $G = \frac{1}{R} = \frac{1}{5 \times 10^3}\,\text{S} = 0.2 \times 10^{-3}\,\text{S} = \mathbf{0.2\,mS}$

(c) $G = \frac{1}{R} = \frac{1}{100 \times 10^{-3}}\,\text{S} = \frac{10^3}{100}\,\text{S} = \mathbf{10\,S}$

3.8 Electrical power and energy

When a direct current of I amperes is flowing in an electric circuit and the voltage across the circuit is V volts, then

power in watts, $\boldsymbol{P = VI}$

Electrical energy $=$ Power $\times$ time

$= \boldsymbol{VIt}$ **joules**

Although the unit of energy is the joule, when dealing with large amounts of energy the unit used is the **kilowatt hour (kWh)** where

$$1\,\text{kWh} = 1000 \text{ watt hour}$$
$$= 1000 \times 3600 \text{ watt seconds or joules}$$
$$= 3\,600\,000\,\text{J}$$

Problem 7. A source e.m.f. of 5 V supplies a current of 3 A for 10 minutes. How much energy is provided in this time?

Energy = power × time and power = voltage × current
Hence

Energy $= VIt = 5 \times 3 \times (10 \times 60) = 9000$ Ws or J

$= \mathbf{9\,kJ}$

Problem 8. An electric heater consumes 1.8 MJ when connected to a 250 V supply for 30 minutes. Find the power rating of the heater and the current taken from the supply.

Energy = power × time, hence

$$\text{power} = \frac{\text{energy}}{\text{time}} = \frac{1.8 \times 10^6\,\text{J}}{30 \times 60\,\text{s}} = 1000\,\text{J/s} = 1000\,\text{W}$$

i.e. **Power rating of heater = 1 kW**

$$\text{Power } P = VI, \text{ thus } I = \frac{P}{V} = \frac{1000}{250} = 4\,\text{A}$$

Hence the current taken from the supply is 4 A

Now try the following Practice Exercise

Practice Exercise 27 E.m.f., resistance, conductance, power and energy (Answers on page 818)

1. Find the conductance of a resistor of resistance (a) 10 Ω, (b) 2 kΩ, (c) 2 mΩ.
2. A conductor has a conductance of 50 μS. What is its resistance?
3. An e.m.f. of 250 V is connected across a resistance and the current flowing through the resistance is 4 A. What is the power developed?
4. 450 J of energy are converted into heat in 1 minute. What power is dissipated?
5. A current of 10 A flows through a conductor and 10 W is dissipated. What p.d. exists across the ends of the conductor?
6. A battery of e.m.f. 12 V supplies a current of 5 A for 2 minutes. How much energy is supplied in this time?
7. A d.c. electric motor consumes 36 MJ when connected to a 250 V supply for 1 hour. Find the power rating of the motor and the current taken from the supply.

3.9 Summary of terms, units and their symbols

Quantity	Quantity symbol	Unit	Unit symbol
Length	*l*	metre	m
Mass	*m*	kilogram	kg
Time	*t*	second	s
Velocity	*v*	metres per second	m/s or m s^{-1}
Acceleration	*a*	metres per second squared	m/s^2 or m s^{-2}
Force	*F*	newton	N
Electrical charge or quantity	*Q*	coulomb	C
Electric current	*I*	ampere	A
Resistance	*R*	ohm	Ω
Conductance	*G*	siemen	S
Electromotive force	*E*	volt	V
Potential difference	*V*	volt	V
Work	*W*	joule	J
Energy	*E* (or W)	joule	J
Power	*P*	watt	W

As progress is made through *Electrical Circuit Theory and Technology* many more terms will be met. A full list of electrical quantities, together with their symbols and units are given in Part 5, page 809.

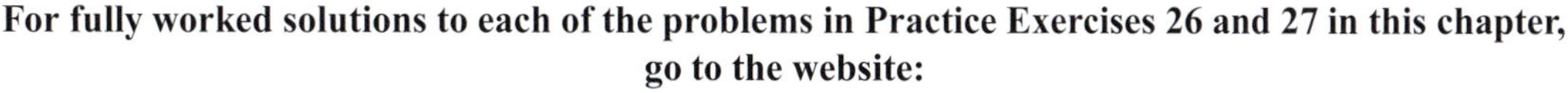

For fully worked solutions to each of the problems in Practice Exercises 26 and 27 in this chapter, go to the website:
www.routledge.com/cw/bird

Chapter 4

An introduction to electric circuits

***Why it is important to understand:* An introduction to electric circuits**

Electric circuits are a part of the basic fabric of modern technology. A circuit consists of electrical elements connected together, and we can use symbols to draw circuits. Engineers use electrical circuits to solve problems that are important in modern society, such as in the generation, transmission and consumption of electrical power and energy. The outstanding characteristics of electricity compared with other power sources are its mobility and flexibility. The elements in an electric circuit include sources of energy, resistors, capacitors, inductors, and so on. Analysis of electric circuits means determining the unknown quantities such as voltage, current and power associated with one or more elements in the circuit. Basic electric circuit analysis and laws are explained in this chapter and knowledge of these are essential in the solution of engineering problems.

At the end of this chapter you should be able to:

- recognize common electrical circuit diagram symbols
- understand that electric current is the rate of movement of charge and is measured in amperes
- appreciate that the unit of charge is the coulomb
- calculate charge or quantity of electricity Q from $Q=It$
- understand that a potential difference (p.d.) between two points in a circuit is required for current to flow
- appreciate that the unit of p.d. is the volt
- understand that resistance opposes current flow and is measured in ohms
- appreciate what an ammeter, a voltmeter, an ohmmeter, a multimeter, an oscilloscope, a wattmeter, a bridge megger, a tachometer and stroboscope measure
- distinguish between linear and non-linear devices
- state Ohm's law as $V=IR$ or $I=\frac{V}{R}$ or $R=\frac{V}{I}$
- use Ohm's law in calculations, including multiples and sub-multiples of units
- describe a conductor and an insulator, giving examples of each
- appreciate that electrical power P is given by $P=VI=I^2R=\frac{V^2}{R}$ watts

Electrical Circuit Theory and Technology. 978-1-138-67349-6, © 2017 John Bird. Published by Taylor & Francis. All rights reserved.

- calculate electrical power
- define electrical energy and state its unit
- calculate electrical energy
- state the three main effects of an electric current, giving practical examples of each
- explain the importance of fuses in electrical circuits
- appreciate the dangers of constant high current flow with insulation materials

4.1 Standard symbols for electrical components

Symbols are used for components in electrical circuit diagrams and some of the more common ones are shown in Figure 4.1.

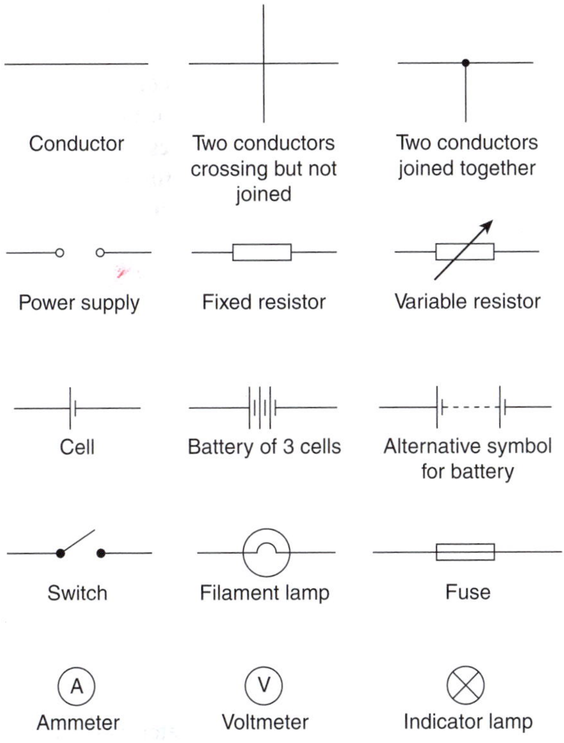

Figure 4.1

4.2 Electric current and quantity of electricity

All **atoms** consist of **protons, neutrons** and **electrons**. The protons, which have positive electrical charges, and the neutrons, which have no electrical charge, are contained within the **nucleus**. Removed from the nucleus are minute negatively charged particles called electrons. Atoms of different materials differ from one another by having different numbers of protons, neutrons and electrons. An equal number of protons and electrons exist within an atom and it is said to be electrically balanced, as the positive and negative charges cancel each other out. When there are more than two electrons in an atom the electrons are arranged into **shells** at various distances from the nucleus.

All atoms are bound together by powerful forces of attraction existing between the nucleus and its electrons. Electrons in the outer shell of an atom, however, are attracted to their nucleus less powerfully than are electrons whose shells are nearer the nucleus.

It is possible for an atom to lose an electron; the atom, which is now called an **ion**, is not now electrically balanced, but is positively charged and is thus able to attract an electron to itself from another atom. Electrons that move from one atom to another are called free electrons and such random motion can continue indefinitely. However, if an electric pressure or **voltage** is applied across any material there is a tendency for electrons to move in a particular direction. This movement of free electrons, known as **drift**, constitutes an electric current flow. **Thus current is the rate of movement of charge.**

Conductors are materials that contain electrons that are loosely connected to the nucleus and can easily move through the material from one atom to another.

Insulators are materials whose electrons are held firmly to their nucleus.

The unit used to measure the **quantity of electrical charge Q** is called the **coulomb*** (where 1 coulomb $= 6.24 \times 10^{18}$ electrons).

If the drift of electrons in a conductor takes place at the rate of one coulomb per second the resulting current is said to be a current of one **ampere**.*

*Who was **Coulomb**? For image and resume of Coulomb, see page 50. To find out more go to **www.routledge.com/cw/bird**

*Who was **Ampere**? For image and resume of Ampere, see page 50. To find out more go to **www.routledge.com/cw/bird**

Thus, 1 ampere = 1 coulomb per second or 1 A = 1 C/s. Hence, 1 coulomb = 1 ampere second or 1 C = 1 As. Generally, if I is the current in amperes and t the time in seconds during which the current flows, then $I \times t$ represents the quantity of electrical charge in coulombs, i.e.
quantity of electrical charge transferred,

$$Q = I \times t \text{ coulombs}$$

Problem 1. What current must flow if 0.24 coulombs is to be transferred in 15 ms?

Since the quantity of electricity, $Q = It$, then

$$I = \frac{Q}{t} = \frac{0.24}{15 \times 10^{-3}} = \frac{0.24 \times 10^3}{15} = \frac{240}{15} = \mathbf{16\ A}$$

Problem 2. If a current of 10 A flows for four minutes, find the quantity of electricity transferred.

Quantity of electricity, $Q = It$ coulombs

$I = 10$ A; $t = 4 \times 60 = 240$ s

Hence $Q = 10 \times 240 = \mathbf{2400\ C}$

Now try the following Practice Exercise

Practice Exercise 28 Electric current and charge (Answers on page 818)

1. In what time would a current of 10 A transfer a charge of 50 C?
2. A current of 6 A flows for 10 minutes. What charge is transferred?
3. How long must a current of 100 mA flow so as to transfer a charge of 80 C?

4.3 Potential difference and resistance

For a continuous current to flow between two points in a circuit a **potential difference (p.d.)** or **voltage,** V, is required between them; a complete conducting path is necessary to and from the source of electrical energy. The unit of p.d. is the **volt, V** (named in honour of the Italian physicist **Alessandro VoltaVolta***).

*Who was Volta? For image and resume of Volta, see page 53. To find out more go to **www.routledge.com/cw/bird**

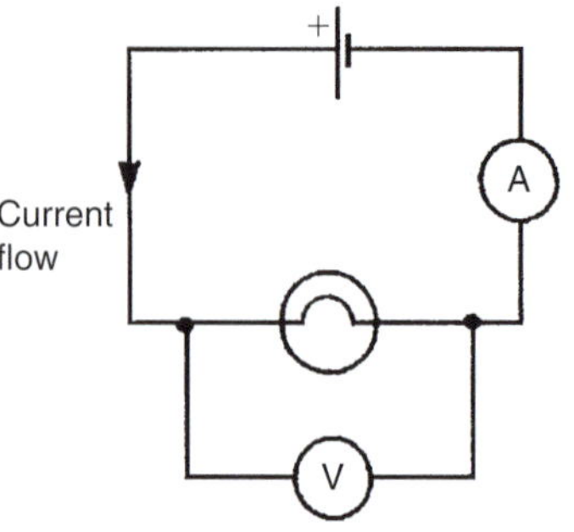

Figure 4.2

Figure 4.2 shows a cell connected across a filament lamp. Current flow, by convention, is considered as flowing from the positive terminal of the cell, around the circuit to the negative terminal.
The flow of electric current is subject to friction. This friction, or opposition, is called **resistance, *R***, and is the property of a conductor that limits current. The unit of resistance is the **ohm**; 1 ohm is defined as the resistance which will have a current of 1 ampere flowing through it when 1 volt is connected across it, i.e.

$$\textbf{resistance } R = \frac{\textbf{potential difference}}{\textbf{current}}$$

4.4 Basic electrical measuring instruments

An **ammeter** is an instrument used to measure current and must be connected **in series** with the circuit. Figure 4.2 shows an ammeter connected in series with the lamp to measure the current flowing through it. Since all the current in the circuit passes through the ammeter it must have a very **low resistance**.
A **voltmeter** is an instrument used to measure p.d. and must be connected **in parallel** with the part of the circuit whose p.d. is required. In Figure 4.2, a voltmeter is connected in parallel with the lamp to measure the p.d. across it. To avoid a significant current flowing through it a voltmeter must have a very **high resistance**.
An **ohmmeter** is an instrument for measuring resistance.
A **multimeter**, or universal instrument, may be used to measure voltage, current and resistance. An 'Avometer' and '**fluke**' are typical examples.
The **oscilloscope** may be used to observe waveforms and to measure voltages and currents. The display of an oscilloscope involves a spot of light moving across a screen. The amount by which the spot is deflected from its initial position depends on the p.d. applied to the terminals of the oscilloscope and the range selected.

The displacement is calibrated in 'volts per cm'. For example, if the spot is deflected 3 cm and the volts/cm switch is on 10 V/cm then the magnitude of the p.d. is 3 cm × 10 V/cm, i.e. 30 V.

A **wattmeter** is an instrument for the measurement of power in an electrical circuit.

A **BM80** or a **420 MIT megger** or a **bridge megger** may be used to measure both continuity and insulation resistance. **Continuity testing** is the measurement of the resistance of a cable to discover if the cable is continuous, i.e. that it has no breaks or high-resistance joints. **Insulation resistance testing** is the measurement of resistance of the insulation between cables, individual cables to earth or metal plugs and sockets, and so on. An insulation resistance in excess of 1 MΩ is normally acceptable.

A **tachometer** is an instrument that indicates the speed, usually in revolutions per minute, at which an engine shaft is rotating.

A **stroboscope** is a device for viewing a rotating object at regularly recurring intervals, by means of either (a) a rotating or vibrating shutter, or (b) a suitably designed lamp which flashes periodically. If the period between successive views is exactly the same as the time of one revolution of the revolving object, and the duration of the view very short, the object will appear to be stationary. See Chapter 12 for more detail about electrical measuring instruments and measurements.

4.5 Linear and non-linear devices

Figure 4.3 shows a circuit in which current I can be varied by the variable resistor R_2. For various settings of R_2, the current flowing in resistor R_1, displayed on the ammeter, and the p.d. across R_1, displayed on the voltmeter, are noted and a graph is plotted of p.d. against current. The result is shown in Figure 4.4(a), where the straight line graph passing through the origin indicates that current is directly proportional to the p.d. Since the gradient i.e. (p.d./current) is constant, resistance R_1 is constant. A resistor is thus an example of a **linear device**.

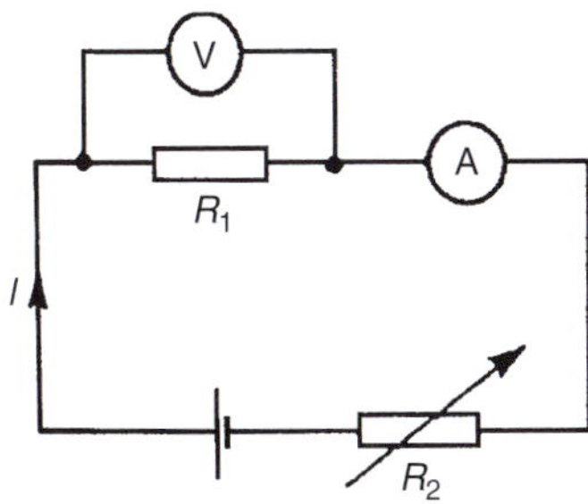

Figure 4.3

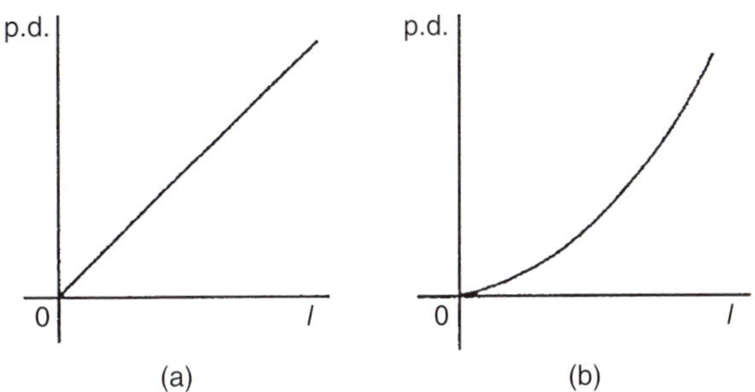

Figure 4.4

If the resistor R_1 in Figure 4.3 is replaced by a component such as a lamp, then the graph shown in Figure 4.4(b) results when values of p.d. are noted for various current readings. Since the gradient is changing, the lamp is an example of a **non-linear device**.

4.6 Ohm's law

Ohm's law* states that the current I flowing in a circuit is directly proportional to the applied voltage V and inversely proportional to the resistance R, provided the temperature remains constant. Thus,

$$I = \frac{V}{R} \quad \text{or} \quad V = IR \quad \text{or} \quad R = \frac{V}{I}$$

For a practical laboratory experiment on Ohm's law, see the website.

Problem 3. The current flowing through a resistor is 0.8 A when a p.d. of 20 V is applied. Determine the value of the resistance.

From Ohm's law,

resistance $R = \frac{V}{I} = \frac{20}{0.8} = \frac{200}{8} = \mathbf{25\,\Omega}$

4.7 Multiples and sub-multiples

Currents, voltages and resistances can often be very large or very small. Thus multiples and sub-multiples of units are often used, as stated in Chapter 3. The most common ones, with an example of each, are listed in Table 4.1.

*Who was **Ohm**? For image and resume of Ohm, see page 54. To find out more go to **www.routledge.com/cw/bird**

Part 2

Table 4.1

Prefix	Name	Meaning	Example
M	mega	multiply by 1 000 000 (i.e. $\times 10^6$)	$2\,M\Omega = 2\,000\,000$ ohms
k	kilo	multiply by 1000 (i.e. $\times 10^3$)	$10\,kV = 10\,000$ volts
m	milli	divide by 1000 (i.e. $\times 10^{-3}$)	$25\,mA = \frac{25}{1000}\,A = 0.025$ amperes
μ	micro	divide by 1 000 000 (i.e. $\times 10^{-6}$)	$50\,\mu V = \frac{50}{1\,000\,000}\,V = 0.000\,05$ volts

A more extensive list of common prefixes are given on page 813.

Problem 4. Determine the p.d. which must be applied to a 2 kΩ resistor in order that a current of 10 mA may flow.

Resistance $R = 2\,k\Omega = 2 \times 10^3 = 2000\,\Omega$

Current $I = 10\,mA$

$$= 10 \times 10^{-3}\,A \text{ or } \frac{10}{10^3} \text{ or } \frac{10}{1000}\,A$$

$$= 0.01\,A$$

From Ohm's law, potential difference,

$$V = IR = (0.01)(2000) = \mathbf{20\,V}$$

Problem 5. A coil has a current of 50 mA flowing through it when the applied voltage is 12 V. What is the resistance of the coil?

$$\text{Resistance, } R = \frac{V}{I} = \frac{12}{50 \times 10^{-3}} = \frac{12 \times 10^3}{50} = \frac{12\,000}{50} = \mathbf{240\,\Omega}$$

Problem 6. A 100 V battery is connected across a resistor and causes a current of 5 mA to flow. Determine the resistance of the resistor. If the voltage is now reduced to 25 V, what will be the new value of the current flowing?

$$\text{Resistance } R = \frac{V}{I} = \frac{100}{5 \times 10^{-3}} = \frac{100 \times 10^3}{5} = 20 \times 10^3 = \mathbf{20\,k\Omega}$$

Current when voltage is reduced to 25 V,

$$I = \frac{V}{R} = \frac{25}{20 \times 10^3} = \frac{25}{20} \times 10^{-3} = \mathbf{1.25\,mA}$$

Problem 7. What is the resistance of a coil which draws a current of (a) 50 mA and (b) 200 μA from a 120 V supply?

(a) $$\text{Resistance } R = \frac{V}{I} = \frac{120}{50 \times 10^{-3}} = \frac{120}{0.05} = \frac{12\,000}{5} = \mathbf{2400\,\Omega} \text{ or } \mathbf{2.4\,k\Omega}$$

(b) $$\text{Resistance } R = \frac{120}{200 \times 10^{-6}} = \frac{120}{0.0002} = \frac{1\,200\,000}{2} = \mathbf{600\,000\,\Omega} \text{ or } \mathbf{600\,k\Omega} \text{ or } \mathbf{0.6\,M\Omega}$$

Problem 8. The current/voltage relationship for two resistors A and B is as shown in Figure 4.5. Determine the value of the resistance of each resistor.

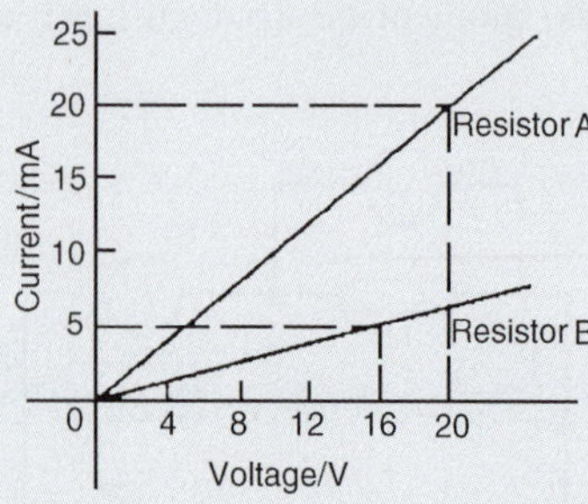

Figure 4.5

For resistor A,

$$R = \frac{V}{I} = \frac{20\,\text{A}}{20\,\text{mA}} = \frac{20}{0.02} = \frac{2000}{2} = \mathbf{1000\,\Omega} \text{ or } \mathbf{1\,k\Omega}$$

For resistor B,

$$R = \frac{V}{I} = \frac{16\,\text{V}}{5\,\text{mA}} = \frac{16}{0.005} = \frac{16000}{5} = \mathbf{3200\,\Omega} \text{ or } \mathbf{3.2\,k\Omega}$$

Now try the following Practice Exercise

Practice Exercise 29 Ohms law (Answers on page 818)

1. The current flowing through a heating element is 5 A when a p.d. of 35 V is applied across it. Find the resistance of the element.
2. A 60 W electric light bulb is connected to a 240 V supply. Determine (a) the current flowing in the bulb and (b) the resistance of the bulb.
3. Graphs of current against voltage for two resistors, P and Q, are shown in Figure 4.6. Determine the value of each resistor.

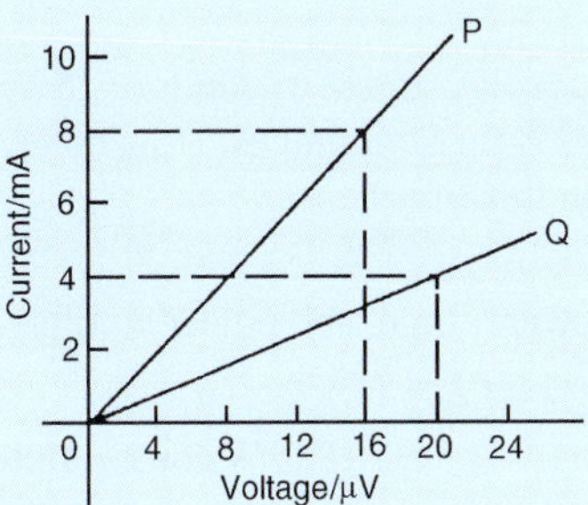

Figure 4.6

4. Determine the p.d. which must be applied to a 5 kΩ resistor such that a current of 6 mA may flow.
5. A 20 V source of e.m.f. is connected across a circuit having a resistance of 400 Ω. Calculate the current flowing.

4.8 Conductors and insulators

A **conductor** is a material having a low resistance which allows electric current to flow in it. All metals are conductors and some examples include copper, aluminium, brass, platinum, silver, gold and carbon.

An **insulator** is a material having a high resistance which does not allow electric current to flow in it. Some examples of insulators include plastic, rubber, glass, porcelain, air, paper, cork, mica, ceramics and certain oils.

4.9 Electrical power and energy

Electrical power

Power *P* in an electrical circuit is given by the product of potential difference V and current I, as stated in Chapter 1. The unit of power is the **watt***, **W**. Hence

$$\mathbf{P = V \times I \text{ watts}} \qquad (4.1)$$

From Ohm's law, $V = IR$

Substituting for V in equation (4.1) gives:

$$P = (IR) \times I$$

i.e. $$\mathbf{P = I^2 R \text{ watts}}$$

Also, from Ohm's law, $I = \frac{V}{R}$

Substituting for I in equation (4.1) gives:

$$P = V \times \frac{V}{R}$$

i.e. $$\mathbf{P = \frac{V^2}{R} \text{ watts}}$$

There are thus three possible formulae which may be used for calculating power.

Problem 9. A 100 W electric light bulb is connected to a 250 V supply. Determine (a) the current flowing in the bulb, and (b) the resistance of the bulb.

Power $P = V \times I$, from which current $I = \frac{P}{V}$

(a) Current $I = \frac{100}{250} = \frac{10}{25} = \frac{2}{5} = \mathbf{0.4\,A}$

(b) Resistance $R = \frac{V}{I} = \frac{250}{0.4} = \frac{2500}{4} = \mathbf{625\,\Omega}$

*Who was **Watt**? For image and resume of Watt, see page 52. To find out more go to **www.routledge.com/cw/bird**

Problem 10. Calculate the power dissipated when a current of 4 mA flows through a resistance of 5 kΩ.

Power $P = I^2 R = (4\times10^{-3})^2(5\times10^3)$

$= 16\times10^{-6}\times5\times10^3 = 80\times10^{-3}$

$= \mathbf{0.08\,W}$ or $\mathbf{80\,mW}$

Alternatively, since $I = 4\times10^{-3}$ and $R = 5\times10^3$, then from Ohm's law,
voltage $V = IR = 4\times10^{-3}\times5\times10^{-3} = 20\,V$
Hence, power $P = V\times I = 20\times4\times10^{-3} = \mathbf{80\,mW}$

Problem 11. An electric kettle has a resistance of 30 Ω. What current will flow when it is connected to a 240 V supply? Find also the power rating of the kettle.

Current, $I = \frac{V}{R} = \frac{240}{30} = \mathbf{8\,A}$

Power, $P = VI = 240\times8 = 1920\,W$

$= \mathbf{1.92\,kW}$

= power rating of kettle

Problem 12. A current of 5 A flows in the winding of an electric motor, the resistance of the winding being 100 Ω. Determine (a) the p.d. across the winding, and (b) the power dissipated by the coil.

(a) Potential difference across winding,
$V = IR = 5\times100 = \mathbf{500\,V}$

(b) Power dissipated by coil, $P = I^2R = 5^2\times100$

$= \mathbf{2500\,W}$ or $\mathbf{2.5\,kW}$

(Alternatively, $P = V\times I = 500\times5 = \mathbf{2500\,W}$ or $\mathbf{2.5\,kW}$)

Problem 13. The hot resistance of a 240 V filament lamp is 960 Ω. Find the current taken by the lamp and its power rating.

From Ohm's law,

current $I = \frac{V}{R} = \frac{240}{960} = \frac{24}{96} = \frac{\mathbf{1}}{\mathbf{4}}$ **A** or **0.25 A**

Power rating $P = VI = (240)\left(\frac{1}{4}\right) = \mathbf{60\,W}$

Electrical energy

Electrical energy = power × time

If the power is measured in watts and the time in seconds then the unit of energy is watt-seconds or **joules***. If the power is measured in kilowatts and the time in hours then the unit of energy is **kilowatt-hours**, often called the **'unit of electricity'**. The 'electricity meter' in the home records the number of kilowatt-hours used and is thus an energy meter.

Problem 14. A 12 V battery is connected across a load having a resistance of 40 Ω. Determine the current flowing in the load, the power consumed and the energy dissipated in 2 minutes.

Current $I = \frac{V}{R} = \frac{12}{40} = \mathbf{0.3\,A}$

Power consumed, $P = VI = (12)(0.3) = \mathbf{3.6\,W}$

Energy dissipated
= power × time
= (3.6 W)(2 × 60 s) = **432 J** (since 1 J = 1 Ws)

Problem 15. A source of e.m.f. of 15 V supplies a current of 2 A for six minutes. How much energy is provided in this time?

Energy = power × time, and power = voltage × current

Hence energy $= VIt = 15\times2\times(6\times60)$

$= 10\,800$ Ws or J = **10.8 kJ**

Problem 16. Electrical equipment in an office takes a current of 13 A from a 240 V supply. Estimate the cost per week of electricity if the equipment is used for 30 hours each week and 1 kWh of energy costs 13.56 p.

Power = VI watts = 240 × 13 = 3120 W = 3.12 kW

Energy used per week
= power × time
= (3.12 kW) × (30 h) = 93.6 kWh

Cost at 13.56 p per kWh = 93.6 × 13.56 = 1269.216 p

Hence **weekly cost of electricity = £12.69**

*Who was **Joules**? For image and resume of Joules, see page 51. To find out more go to **www.routledge.com/cw/bird**

Problem 17. An electric heater consumes 3.6 MJ when connected to a 250 V supply for 40 minutes. Find the power rating of the heater and the current taken from the supply.

$$\text{Power} = \frac{\text{energy}}{\text{time}} = \frac{3.6\times10^6}{40\times60}\frac{\text{J}}{\text{s}} \text{ (or W)} = 1500\text{ W}$$

i.e. Power rating of heater = **1.5 kW**

$$\text{Power } P = VI, \text{ thus } I = \frac{P}{V} = \frac{1500}{250} = 6\text{ A}$$

Hence the current taken from the supply is **6 A**

Problem 18. Determine the power dissipated by the element of an electric fire of resistance 20 Ω when a current of 10 A flows through it. If the fire is on for 6 hours determine the energy used and the cost if 1 unit of electricity costs 13 p.

Power $P = I^2R = 10^2 \times 20 = 100 \times 20 =$ **2000 W** or **2 kW** (Alternatively, from Ohm's law, $V = IR = 10\times 20 = 200$ V, hence power $P = V \times I = 200 \times 10 = 2000$ W = 2 kW)

Energy used in 6 hours
= power × time
= 2 kW × 6 h = **12 kWh**

1 unit of electricity = 1 kWh

Hence the number of units used is 12

Cost of energy = 12 × 13 p = **£1.56**

Problem 19. A business uses two 3 kW fires for an average of 20 hours each per week, and six 150 W lights for 30 hours each per week. If the cost of electricity is 14.25 p per unit, determine the weekly cost of electricity to the business.

Energy = power × time

Energy used by one 3 kW fire in 20 hours
= 3 kW × 20 h = 60 kWh

Hence weekly energy used by two 3 kW fires
= 2 × 60 = 120 kWh

Energy used by one 150 W light for 30 hours
= 150 W × 30 h
= 4500 Wh = 4.5 kWh

Hence weekly energy used by six 150 W lamps
= 6 × 4.5 = 27 kWh

Total energy used per week = 120 + 27 = 147 kWh

1 unit of electricity = 1 kWh of energy

Thus weekly cost of energy at
14.25 p per kWh = 14.25 × 147 = 2094.75 p
= **£20.95**

Now try the following Practice Exercise

Practice Exercise 30 Power and energy **(Answers on page 818)**

1. The hot resistance of a 250 V filament lamp is 625 Ω. Determine the current taken by the lamp and its power rating.
2. Determine the resistance of a coil connected to a 150 V supply when a current of (a) 75 mA, (b) 300 μA flows through it.
3. Determine the resistance of an electric fire which takes a current of 12 A from a 240 V supply. Find also the power rating of the fire and the energy used in 20 h.
4. Determine the power dissipated when a current of 10 mA flows through an appliance having a resistance of 8 kΩ.
5. 85.5 J of energy are converted into heat in nine seconds. What power is dissipated?
6. A current of 4 A flows through a conductor and 10 W is dissipated. What p.d. exists across the ends of the conductor?
7. Find the power dissipated when:
   (a) a current of 5 mA flows through a resistance of 20 kΩ
   (b) a voltage of 400 V is applied across a 120 kΩ resistor
   (c) a voltage applied to a resistor is 10 kV and the current flow is 4 mA.
8. A battery of e.m.f. 15 V supplies a current of 2 A for 5 min. How much energy is supplied in this time?
9. In a household during a particular week three 2 kW fires are used on average 25 h each and eight 100 W light bulbs are used on average 35 h each. Determine the cost of electricity for the week if 1 unit of electricity costs 15 p.
10. Calculate the power dissipated by the element of an electric fire of resistance 30 Ω when a current of 10 A flows in it. If the fire is on for 30 hours in a week determine the energy used. Determine also the weekly cost of energy if electricity costs 13.5 p per unit.

4.10 Main effects of electric current

The three main effects of an electric current are:

(a) magnetic effect

(b) chemical effect

(c) heating effect.

Some practical applications of the effects of an electric current include:

Magnetic effect: bells, relays, motors, generators, transformers, telephones, car-ignition and lifting magnets (see Chapter 10)

Chemical effect: primary and secondary cells and electroplating (see Chapter 6)

Heating effect: cookers, water heaters, electric fires, irons, furnaces, kettles and soldering irons

4.11 Fuses

If there is a fault in a piece of equipment then excessive current may flow. This will cause overheating and possibly a fire; **fuses** protect against this happening. Current from the supply to the equipment flows through the fuse. The fuse is a piece of wire which can carry a stated current; if the current rises above this value it will melt. If the fuse melts (blows) then there is an open circuit and no current can then flow – thus protecting the equipment by isolating it from the power supply.

The fuse must be able to carry slightly more than the normal operating current of the equipment to allow for tolerances and small current surges. With some equipment there is a very large surge of current for a short time at switch on. If a fuse is fitted to withstand this large current there would be no protection against faults which cause the current to rise slightly above the normal value. Therefore special anti-surge fuses are fitted. These can stand 10 times the rated current for 10 milliseconds. If the surge lasts longer than this the fuse will blow.

A circuit diagram symbol for a fuse is shown in Figure 4.1 on page 57.

Problem 20. If 5 A, 10 A and 13 A fuses are available, state which is most appropriate for the following appliances, which are both connected to a 240 V supply
(a) electric toaster having a power rating of 1 kW
(b) electric fire having a power rating of 3 kW.

Power $P = VI$, from which current $I = \frac{P}{V}$

(a) For the toaster,

$$\text{current } I = \frac{P}{V} = \frac{1000}{240} = \frac{100}{24} = 4.17\,\text{A}$$

Hence a **5 A fuse** is most appropriate

(b) For the fire,

$$\text{current } I = \frac{P}{V} = \frac{3000}{240} = \frac{300}{24} = 12.5\,\text{A}$$

Hence a **13 A fuse** is most appropriate

Now try the following Practice Exercise

Practice Exercise 31 Fuses (Answers on page 818)

1. A television set having a power rating of 120 W and electric lawn-mower of power rating 1 kW are both connected to a 240 V supply. If 3 A, 5 A and 10 A fuses are available state which is the most appropriate for each appliance.

4.12 Insulation and the dangers of constant high current flow

The use of insulation materials on electrical equipment, whilst being necessary, also has the effect of preventing heat loss, i.e. the heat is not able to dissipate, thus creating possible danger of fire. In addition, the insulating material has a maximum temperature rating – this is heat it can withstand without being damaged. The current rating for all equipment and electrical components is therefore limited to keep the heat generated within safe limits. In addition, the maximum voltage present needs to be considered when choosing insulation.

For fully worked solutions to each of the problems in Practice Exercises 28 to 31 in this chapter, go to the website:
www.routledge.com/cw/bird

Chapter 5

Resistance variation

***Why it is important to understand:* Resistance variation**

An electron travelling through the wires and loads of an electric circuit encounters resistance. Resistance is the hindrance to the flow of charge. The flow of charge through wires is often compared to the flow of water through pipes. The resistance to the flow of charge in an electric circuit is analogous to the frictional effects between water and the pipe surfaces as well as the resistance offered by obstacles that are present in its path. It is this resistance that hinders the water flow and reduces both its flow rate and its drift speed. Like the resistance to water flow, the total amount of resistance to charge flow within a wire of an electric circuit is affected by some clearly identifiable variables. Factors which affect resistance are length, cross-sectional area and type of material. The value of a resistor also changes with changing temperature, but this is not as we might expect, mainly due to a change in the dimensions of the component as it expands or contracts. It is due mainly to a change in the resistivity of the material caused by the changing activity of the atoms that make up the resistor. Resistance variation due to length, cross-sectional area, type of material and temperature variation are explained in this chapter, with calculations to aid understanding. In addition, the resistor colour coding/ohmic values are explained.

At the end of this chapter you should be able to:

- recognize three common methods of resistor construction
- appreciate that electrical resistance depends on four factors
- appreciate that resistance $R=\dfrac{\rho l}{a}$, where ρ is the resistivity
- recognize typical values of resistivity and its unit
- perform calculations using $R=\dfrac{\rho l}{a}$
- define the temperature coefficient of resistance, α
- recognize typical values for α
- perform calculations using $R_\theta=R_0(1+\alpha\theta)$
- determine the resistance and tolerance of a fixed resistor from its colour code
- determine the resistance and tolerance of a fixed resistor from its letter and digit code

Electrical Circuit Theory and Technology. 978-1-138-67349-6, © 2017 John Bird. Published by Taylor & Francis. All rights reserved.

5.1 Resistor construction

There is a wide range of resistor types. Four of the most common methods of construction are:

(i) Surface Mount Technology (SMT)

Many modern circuits use SMT resistors. Their manufacture involves depositing a film of resistive material such as tin oxide on a tiny ceramic chip. The edges of the resistor are then accurately ground or cut with a laser to give a precise resistance across the ends of the device. Tolerances may be as low as ±0.02% and SMT resistors normally have very low power dissipation. Their main advantage is that very high component density can be achieved.

(ii) Wire wound resistors

A length of wire such, as nichrome or manganin, whose resistive value per unit length is known, is cut to the desired value and wound around a ceramic former prior to being lacquered for protection. This type of resistor has a large physical size, which is a disadvantage; however, they can be made with a high degree of accuracy, and can have a **high power rating**.

Wire wound resistors are used in **power circuits** and **motor starters**.

(iii) Metal film resistors

Metal film resistors are made from small rods of ceramic coated with metal, such as a nickel alloy. The value of resistance is controlled firstly by the thickness of the coating layer (the thicker the layer, the lower the value of resistance), and secondly by cutting a fine spiral groove along the rod using a laser or diamond cutter to cut the metal coating into a long spiral strip, which forms the resistor.

Metal film resistors are low tolerance, precise resistors (±1% or less) and are used in **electronic circuits**.

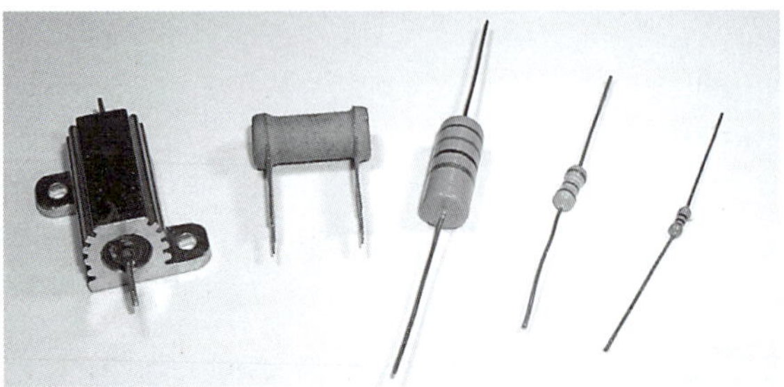

Figure 5.1

(iv) Carbon film resistors

Carbon film resistors have a similar construction to metal film resistors but generally with wider tolerance, typically ±5%. They are inexpensive, in common use, and are used in **electronic circuits**.
Some typical resistors are shown in Figure 5.1.

5.2 Resistance and resistivity

The resistance of an electrical conductor depends on four factors, these being: (a) the length of the conductor, (b) the cross-sectional area of the conductor, (c) the type of material and (d) the temperature of the material.

Resistance, R, is directly proportional to length, l, of a conductor, i.e. $R \propto l$. Thus, for example, if the length of a piece of wire is doubled, then the resistance is doubled.

Resistance, R, is inversely proportional to cross-sectional area, a, of a conductor, i.e. $R \propto 1/a$. Thus, for example, if the cross-sectional area of a piece of wire is doubled then the resistance is halved.

Since $R \propto l$ and $R \propto 1/a$ then $R \propto l/a$. By inserting a constant of proportionality into this relationship the type of material used may be taken into account. The constant of proportionality is known as the **resistivity** of the material and is given the symbol ρ (Greek rho). Thus,

resistance $\quad \mathbf{R = \dfrac{\rho l}{a}}$ **ohms**

ρ is measured in ohm metres (Ωm)
The value of the resistivity is that resistance of a unit cube of the material measured between opposite faces of the cube.
Resistivity varies with temperature and some typical values of resistivities measured at about room temperature are given below:

Copper	$1.7 \times 10^{-8}\,\Omega$m	(or $0.017\,\mu\Omega$m)
Aluminium	$2.6 \times 10^{-8}\,\Omega$m	(or $0.026\,\mu\Omega$m)
Carbon (graphite)	$10 \times 10^{-8}\,\Omega$m	(or $0.10\,\mu\Omega$m)
Glass	$1 \times 10^{10}\,\Omega$m	
Mica	$1 \times 10^{13}\,\Omega$m	

Note that good conductors of electricity have a low value of resistivity and good insulators have a high value of resistivity.

Problem 1. The resistance of a 5 m length of wire is 600 Ω. Determine (a) the resistance of an 8 m length of the same wire, and (b) the length of the same wire when the resistance is 420 Ω.

(a) Resistance, R, is directly proportional to length, l, i.e. $R \propto l$.

Hence, $600\,\Omega \propto 5\,\text{m}$ or $600=(k)(5)$, where k is the coefficient of proportionality. Hence,

$$k=\frac{600}{5}=120$$

When the length l is 8 m, then resistance

$$R=kl=(120)(8)=\mathbf{960\,\Omega}$$

(b) When the resistance is $420\,\Omega$, $420=kl$, from which,

$$\text{length } l=\frac{420}{k}=\frac{420}{120}=\mathbf{3.5\,m}$$

Problem 2. A piece of wire of cross-sectional area $2\,\text{mm}^2$ has a resistance of $300\,\Omega$. Find (a) the resistance of a wire of the same length and material if the cross-sectional area is $5\,\text{mm}^2$, (b) the cross-sectional area of a wire of the same length and material of resistance $750\,\Omega$.

Resistance, R, is inversely proportional to cross-sectional area, a, i.e. $R \propto \frac{1}{a}$.

Hence $300\,\Omega \propto \frac{1}{2\,\text{mm}^2}$ or $300=(k)\left(\frac{1}{2}\right)$

from which the coefficient of proportionality,

$$k=300\times 2=600$$

(a) When the cross-sectional area $a=5\,\text{mm}^2$

then $R=(k)\left(\frac{1}{5}\right)=(600)\left(\frac{1}{5}\right)=\mathbf{120\,\Omega}$

(Note that resistance has decreased as the cross-sectional area is increased.)

(b) When the resistance is $750\,\Omega$ then $750=(k)(1/a)$, from which cross-sectional area,

$$a=\frac{k}{750}=\frac{600}{750}=\mathbf{0.8\,mm^2}$$

Problem 3. A wire of length 8 m and cross-sectional area $3\,\text{mm}^2$ has a resistance of $0.16\,\Omega$. If the wire is drawn out until its cross-sectional area is $1\,\text{mm}^2$, determine the resistance of the wire.

Resistance, R, is directly proportional to length, l, and inversely proportional to the cross-sectional area, a, i.e. $R \propto \frac{l}{a}$ or $R=k\left(\frac{l}{a}\right)$, where k is the coefficient of proportionality.

Since $R=0.16$, $l=8$ and $a=3$, then $0.16=(k)\left(\frac{8}{3}\right)$ from which

$$k=0.16\times\frac{3}{8}=0.06$$

If the cross-sectional area is reduced to $\frac{1}{3}$ of its original area then the length must be tripled to 3×8, i.e. 24 m

New resistance, $R=k\left(\frac{l}{a}\right)=0.06\left(\frac{24}{1}\right)=\mathbf{1.44\,\Omega}$

Problem 4. Calculate the resistance of a 2 km length of aluminium overhead power cable if the cross-sectional area of the cable is $100\,\text{mm}^2$. Take the resistivity of aluminium to be $0.03\times 10^{-6}\,\Omega\text{m}$.

Length $l=2\text{ km}=2000\,\text{m}$;
area, $a=100\,\text{mm}^2=100\times 10^{-6}\,\text{m}^2$;
resistivity $\rho=0.03\times 10^{-6}\,\Omega\text{m}$

$$\text{Resistance,}\quad R=\frac{\rho l}{a}=\frac{(0.03\times 10^{-6}\,\Omega\text{m})(2000\,\text{m})}{(100\times 10^{-6}\,\text{m}^2)}$$

$$=\frac{0.03\times 2000}{100}\,\Omega$$

$$=\mathbf{0.6\,\Omega}$$

Problem 5. Calculate the cross-sectional area, in mm^2, of a piece of copper wire, 40 m in length and having a resistance of $0.25\,\Omega$. Take the resistivity of copper as $0.02\times 10^{-6}\,\Omega\text{m}$.

Resistance, $R=\frac{\rho l}{a}$ hence cross-sectional area $a=\frac{\rho l}{R}$

$$=\frac{(0.02\times 10^{-6}\,\Omega\text{m})(40\,\text{m})}{0.25\,\Omega}$$

$$=3.2\times 10^{-6}\,\text{m}^2$$

$$=(3.2\times 10^{-6})\times 10^{6}\,\text{mm}^2=\mathbf{3.2\,mm^2}$$

Problem 6. The resistance of 1.5 km of wire of cross-sectional area $0.17\,\text{mm}^2$ is $150\,\Omega$. Determine the resistivity of the wire.

Resistance, $R=\frac{\rho l}{a}$

hence, resistivity, $\rho=\frac{Ra}{l}=\frac{(150\,\Omega)(0.17\times 10^{-6}\,\text{m}^2)}{(1500\,\text{m})}$

$$=\mathbf{0.017\times 10^{-6}\,\Omega m} \text{ or } \mathbf{0.017\,\mu\Omega m}$$

Problem 7. Determine the resistance of 1200 m of copper cable having a diameter of 12 mm if the resistivity of copper is $1.7 \times 10^{-8}\,\Omega$m.

$$\text{Cross-sectional area of cable, } a = \pi r^2 = \pi\left(\frac{12}{2}\right)^2$$
$$= 36\pi \text{ mm}^2$$
$$= 36\pi \times 10^{-6}\,\text{m}^2$$

$$\text{Resistance, } R = \frac{\rho l}{a} = \frac{(1.7 \times 10^{-8}\,\Omega\text{m})(1200\,\text{m})}{(36\pi \times 10^{-6}\,\text{m}^2)}$$
$$= \frac{1.7 \times 1200 \times 10^6}{10^8 \times 36\pi}\,\Omega = \frac{1.7 \times 12}{36\pi}\,\Omega$$
$$= \mathbf{0.180\,\Omega}$$

Now try the following Practice Exercise

Practice Exercise 32 Resistance and resistivity (Answers on page 818)

1. The resistance of a 2 m length of cable is 2.5 Ω. Determine (a) the resistance of a 7 m length of the same cable and (b) the length of the same wire when the resistance is 6.25 Ω.
2. Some wire of cross-sectional area 1 mm^2 has a resistance of 20 Ω. Determine (a) the resistance of a wire of the same length and material if the cross-sectional area is 4 mm^2, and (b) the cross-sectional area of a wire of the same length and material if the resistance is 32 Ω.
3. Some wire of length 5 m and cross-sectional area 2 mm^2 has a resistance of 0.08 Ω. If the wire is drawn out until its cross-sectional area is 1 mm^2, determine the resistance of the wire.
4. Find the resistance of 800 m of copper cable of cross-sectional area 20 mm^2. Take the resistivity of copper as 0.02 μΩm.
5. Calculate the cross-sectional area, in mm^2, of a piece of aluminium wire 100 m long and having a resistance of 2 Ω. Take the resistivity of aluminium as $0.03 \times 10^{-6}\,\Omega$m.
6. (a) What does the resistivity of a material mean?
   (b) The resistance of 500 m of wire of cross-sectional area 2.6 mm^2 is 5 Ω. Determine the resistivity of the wire in μΩm.
7. Find the resistance of 1 km of copper cable having a diameter of 10 mm if the resistivity of copper is $0.017 \times 10^{-6}\,\Omega$m.

5.3 Temperature coefficient of resistance

In general, as the temperature of a material increases, most conductors increase in resistance, insulators decrease in resistance, whilst the resistances of some special alloys remain almost constant.

The **temperature coefficient of resistance** of a material is the increase in the resistance of a 1 Ω resistor of that material when it is subjected to a rise of temperature of 1°C. The symbol used for the temperature coefficient of resistance is α (Greek alpha). Thus, if some copper wire of resistance 1 Ω is heated through 1°C and its resistance is then measured as 1.0043 Ω then $\alpha = 0.0043\,\Omega/\Omega$°C for copper. The units are usually expressed only as 'per °C', i.e. $\alpha = 0.0043$/°C for copper. If the 1 Ω resistor of copper is heated through 100°C then the resistance at 100°C would be $1 + 100 \times 0.0043 = 1.43\,\Omega$.
Some typical values of temperature coefficient of resistance measured at 0°C are given below:

Copper	0.0043/°C	Aluminium	0.0038/°C
Nickel	0.0062/°C	Carbon	−0.000 48/°C
Constantan	0	Eureka	0.000 01/°C

(Note that the negative sign for carbon indicates that its resistance falls with increase of temperature.)
If the resistance of a material at 0°C is known, the resistance at any other temperature can be determined from:

$$\mathbf{R_\theta = R_0(1 + \alpha_0\theta)}$$

where R_0 = resistance at 0°C

R_θ = resistance at temperature θ°C

α_0 = temperature coefficient of resistance at 0°C.

Problem 8. A coil of copper wire has a resistance of 100 Ω when its temperature is 0°C. Determine its resistance at 70°C if the temperature coefficient of resistance of copper at 0°C is 0.0043/°C.

Resistance $R_\theta = R_0(1+\alpha_0\theta)$

Hence resistance at 70°C, $R_{70} = 100[1+(0.0043)(70)]$

$= 100[1+0.301]$

$= 100(1.301)$

$= \mathbf{130.1\,\Omega}$

Problem 9. An aluminium cable has a resistance of 27 Ω at a temperature of 35°C. Determine its resistance at 0°C. Take the temperature coefficient of resistance at 0°C to be 0.0038/°C.

Resistance at θ°C, $R_\theta = R_0(1+\alpha_0\theta)$

Hence resistance at 0°C, $R_0 = \dfrac{R_\theta}{(1+\alpha_0\theta)}$

$= \dfrac{27}{[1+(0.0038)(35)]}$

$= \dfrac{27}{1+0.133} = \dfrac{27}{1.133}$

$= \mathbf{23.83\,\Omega}$

Problem 10. A carbon resistor has a resistance of 1 kΩ at 0°C. Determine its resistance at 80°C. Assume that the temperature coefficient of resistance for carbon at 0°C is −0.0005/°C.

Resistance at temperature θ°C, $R_\theta = R_0(1+\alpha_0\theta)$

i.e. $R_\theta = 1000[1+(-0.0005)(80)]$

$= 1000[1-0.040] = 1000(0.96)$

$= \mathbf{960\,\Omega}$

If the resistance of a material at room temperature (approximately 20°C), R_{20}, and the temperature coefficient of resistance at 20°C, α_{20}, are known, then the resistance R_θ at temperature θ°C is given by:

$$R_\theta = R_{20}[1+\alpha_{20}(\theta-20)]$$

Problem 11. A coil of copper wire has a resistance of 10 Ω at 20°C. If the temperature coefficient of resistance of copper at 20°C is 0.004/°C, determine the resistance of the coil when the temperature rises to 100°C.

Resistance at θ°C, $R = R_{20}[1+\alpha_{20}(\theta-20)]$

Hence resistance at 100°C,

$R_{100} = 10[1+(0.004)(100-20)]$

$= 10[1+(0.004)(80)]$

$= 10[1+0.32]$

$= 10(1.32)$

$= \mathbf{13.2\,\Omega}$

Problem 12. The resistance of a coil of aluminium wire at 18°C is 200 Ω. The temperature of the wire is increased and the resistance rises to 240 Ω. If the temperature coefficient of resistance of aluminium is 0.0039/°C at 18°C, determine the temperature to which the coil has risen.

Let the temperature rise to θ°

Resistance at θ°C, $R_\theta = R_{18}[1+\alpha_{18}(\theta-18)]$

i.e. $240 = 200[1+(0.0039)(\theta-18)]$

$240 = 200+(200)(0.0039)(\theta-18)$

$240-200 = 0.78(\theta-18)$

$40 = 0.78(\theta-18)$

$\dfrac{40}{0.78} = \theta-18$

$51.28 = \theta-18$, from which,

$\theta = 51.28+18 = \mathbf{69.28°C}$

Hence the temperature of the coil increases to 69.28°C

If the resistance at 0°C is not known, but is known at some other temperature θ_1, then the resistance at any temperature can be found as follows:

$R_1 = R_0(1+\alpha_0\theta_1)$ and $R_2 = R_0(1+\alpha_0\theta_2)$

Dividing one equation by the other gives:

$$\frac{R_1}{R_2} = \frac{1+\alpha_0\theta_1}{1+\alpha_0\theta_2}$$

where R_2 = resistance at temperature θ_2

Problem 13. Some copper wire has a resistance of 200 Ω at 20°C. A current is passed through the wire and the temperature rises to 90°C. Determine the resistance of the wire at 90°C, correct to the nearest ohm, assuming that the temperature coefficient of resistance is 0.004/°C at 0°C.

$R_{20} = 200\,\Omega$, $\alpha_0 = 0.004/°C$

$$\frac{R_{20}}{R_{90}} = \frac{[1+\alpha_0(20)]}{[1+\alpha_0(90)]}$$

$$\text{Hence} \quad R_{90} = \frac{R_{20}[1+90\alpha_0]}{[1+20\alpha_0]}$$

$$= \frac{200[1+90(0.004)]}{[1+20(0.004)]}$$

$$= \frac{200[1+0.36]}{[1+0.08]}$$

$$= \frac{200(1.36)}{(1.08)} = \mathbf{251.85\,\Omega}$$

i.e. the resistance of the wire at 90°C is 252 Ω

Now try the following Practice Exercise

Practice Exercise 33 Temperature coefficient of resistance (Answers on page 818)

1. A coil of aluminium wire has a resistance of 50 Ω when its temperature is 0°C. Determine its resistance at 100°C if the temperature coefficient of resistance of aluminium at 0°C is 0.0038/°C.
2. A copper cable has a resistance of 30 Ω at a temperature of 50°C. Determine its resistance at 0°C. Take the temperature coefficient of resistance of copper at 0°C as 0.0043/°C.
3. The temperature coefficient of resistance for carbon at 0°C is −0.00048/°C. What is the significance of the minus sign? A carbon resistor has a resistance of 500 Ω at 0°C. Determine its resistance at 50°C.
4. A coil of copper wire has a resistance of 20 Ω at 18°C. If the temperature coefficient of resistance of copper at 18°C is 0.004/°C, determine the resistance of the coil when the temperature rises to 98°C.
5. The resistance of a coil of nickel wire at 20°C is 100 Ω. The temperature of the wire is increased and the resistance rises to 130 Ω. If the temperature coefficient of resistance of nickel is 0.006/°C at 20°C, determine the temperature to which the coil has risen.
6. Some aluminium wire has a resistance of 50 Ω at 20°C. The wire is heated to a temperature of 100°C. Determine the resistance of the wire at 100°C, assuming that the temperature coefficient of resistance at 0°C is 0.004/°C.
7. A copper cable is 1.2 km long and has a cross-sectional area of 5 mm^2. Find its resistance at 80°C if at 20°C the resistivity of copper is 0.02×10^{-6} Ωm and its temperature coefficient of resistance is 0.004/°C.

5.4 Resistor colour coding and ohmic values

(a) Colour code for fixed resistors

The colour code for fixed resistors is given in Table 5.1.

Table 5.1

Colour	Significant figures	Multiplier	Tolerance
Silver	–	10^{-2}	±10%
Gold	–	10^{-1}	±5%
Black	0	1	–
Brown	1	10	±1%
Red	2	10^2	±2%
Orange	3	10^3	–
Yellow	4	10^4	–
Green	5	10^5	±0.5%
Blue	6	10^6	±0.25%
Violet	7	10^7	±0.1%
Grey	8	10^8	–
White	9	10^9	–
None	–	–	±20%

(i) For a **four-band fixed resistor** (i.e. resistance values with two significant figures): yellow-violet-orange-red indicates 47 kΩ with a tolerance of ±2%
(Note that the first band is the one nearest the end of the resistor.)

(ii) For a **five-band fixed resistor** (i.e. resistance values with three significant figures): red-yellow-white-orange-brown indicates 249 kΩ with a tolerance of ±1%
(Note that the fifth band is 1.5 to 2 times wider than the other bands.)

Problem 14. Determine the value and tolerance of a resistor having a colour coding of: orange-orange-silver-brown.

The first two bands, i.e. orange-orange, give 33 from Table 5.1.
The third band, silver, indicates a multiplier of 10^2 from Table 5.1, which means that the value of the resistor is $33 \times 10^{-2} = 0.33\,\Omega$

The fourth band, i.e. brown, indicates a tolerance of ±1% from Table 5.1. Hence a colour coding of orange-orange-silver-brown represents a resistor of value **0.33 Ω with a tolerance of ±1%**

Problem 15. Determine the value and tolerance of a resistor having a colour coding of brown-black-brown.

The first two bands, i.e. brown-black, give 10 from Table 5.1.
The third band, brown, indicates a multiplier of 10 from Table 5.1, which means that the value of the resistor is $10 \times 10 = 100\,\Omega$

There is no fourth band colour in this case; hence, from Table 5.1, the tolerance is ±20%. Hence a colour coding of brown-black-brown represents a resistor of value **100 Ω with a tolerance of ±20%**

Problem 16. Between what two values should a resistor with colour coding brown-black-brown-silver lie?

From Table 5.1, brown-black-brown-silver indicates 10×10, i.e. 100 Ω, with a tolerance of ±10%
This means that the value could lie between

$$(100 - 10\% \text{ of } 100)\,\Omega$$

and $$(100 + 10\% \text{ of } 100)\,\Omega$$

i.e. brown-black-brown-silver indicates any value **between 90 Ω and 110 Ω**

Problem 17. Determine the colour coding for a 47 kΩ resistor having a tolerance of ±5%

From Table 5.1, $47\,\text{k}\Omega = 47 \times 10^3$ has a colour coding of yellow-violet-orange. With a tolerance of ±5%, the fourth band will be gold.
Hence 47 kΩ ± 5% has a colour coding of: **yellow-violet-orange-gold**.

Problem 18. Determine the value and tolerance of a resistor having a colour coding of orange-green-red-yellow-brown.

Orange-green-red-yellow-brown is a five-band fixed resistor and from Table 5.1 indicates: $352 \times 10^4\,\Omega$ with a tolerance of ±1%

$$352 \times 10^4\,\Omega = 3.52 \times 10^6\,\Omega, \text{ i.e. } 3.52\,\text{M}\Omega$$

Hence orange-green-red-yellow-brown indicates **3.52 MΩ ± 1%**

(b) Letter and digit code for resistors

Another way of indicating the value of resistors is the letter and digit code shown in Table 5.2.

Table 5.2

Resistance value	Marked as
0.47 Ω	R47
1 Ω	1R0
4.7 Ω	4R7
47 Ω	47R
100 Ω	100R
1 kΩ	1K0
10 kΩ	10 K
10 MΩ	10 M

Tolerance is indicated as follows: $F = \pm 1\%$, $G = \pm 2\%$, $J = \pm 5\%$, $K = \pm 10\%$ and $M = \pm 20\%$

Part 2

Thus, for example,

$$R33M = 0.33\,\Omega \pm 20\%$$
$$4R7K = 4.7\,\Omega \pm 10\%$$
$$390RJ = 390\,\Omega \pm 5\%$$

Problem 19. Determine the value of a resistor marked as 6K8F

From Table 5.2, 6K8F is equivalent to: **6.8 kΩ ± 1%**

Problem 20. Determine the value of a resistor marked as 4M7M

From Table 5.2, 4M7M is equivalent to: **4.7 MΩ ± 20%**

Problem 21. Determine the letter and digit code for a resistor having a value of 68 kΩ ± 10%

From Table 5.2, 68 kΩ ± 10% has a letter and digit code of: **68KK**

Now try the following Practice Exercise

Practice Exercise 34 Resistor colour coding and ohmic values (Answers on page 818)

1. Determine the value and tolerance of a resistor having a colour coding of: blue-grey-orange-red.
2. Determine the value and tolerance of a resistor having a colour coding of: yellow-violet-gold.
3. Determine the value and tolerance of a resistor having a colour coding of: blue-white-black-black-gold.
4. Determine the colour coding for a 51 kΩ four-band resistor having a tolerance of ±2%
5. Determine the colour coding for a 1 MΩ four-band resistor having a tolerance of ±10%
6. Determine the range of values expected for a resistor with colour coding: red-black-green-silver.
7. Determine the range of values expected for a resistor with colour coding: yellow-black-orange-brown.
8. Determine the value of a resistor marked as (a) R22G (b) 4K7F
9. Determine the letter and digit code for a resistor having a value of 100 kΩ ± 5%
10. Determine the letter and digit code for a resistor having a value of 6.8 MΩ ± 20%

For fully worked solutions to each of the problems in Practice Exercises 32 to 34 in this chapter, go to the website:
www.routledge.com/cw/bird

Chapter 6

Batteries and alternative sources of energy

Why it is important to understand: **Batteries and alternative sources of energy**

Batteries store electricity in a chemical form, inside a closed-energy system. They can be re-charged and re-used as a power source in small appliances, machinery and remote locations. Batteries can store d.c. electrical energy produced by renewable sources such as solar, wind and hydro power in chemical form. Because renewable energy-charging sources are often intermittent in their nature, batteries provide energy storage in order to provide a relatively constant supply of power to electrical loads regardless of whether the sun is shining or the wind is blowing. In an off-grid photovoltaic (PV) system, for example, battery storage provides a way to power common household appliances regardless of the time of day or the current weather conditions. In a grid-tie with battery backup PV system, batteries provide uninterrupted power in case of utility power failure. Energy causes movement; every time something moves, energy is being used. Energy moves cars, makes machines run, heats ovens, and lights our homes. One form of energy can be changed into another form. When petrol is burned in a vehicle engine, the energy stored in petrol is changed into heat energy. When we stand in the sun, light energy is changed into heat. When a torch or flashlight is turned on, chemical energy stored in the battery is changed into light and heat. To find energy, look for motion, heat, light, sound, chemical reactions or electricity. The sun is the source of all energy. The sun's energy is stored in coal, petroleum, natural gas, food, water and wind. While there are two types of energy, renewable and non-renewable, most of the energy we use comes from burning non-renewable fuels – coal, petroleum or oil or natural gas. These supply the majority of our energy needs because we have designed ways to transform their energy on a large scale to meet consumer needs. Regardless of the energy source, the energy contained in them is changed into a more useful form of electricity. This chapter explains the increasingly important area of battery use and briefly looks at some alternative sources of energy.

At the end of this chapter you should be able to:

- list practical applications of batteries
- understand electrolysis and its applications, including electroplating
- appreciate the purpose and construction of a simple cell
- explain polarization and local action
- explain corrosion and its effects
- define the terms e.m.f., E, and internal resistance, r, of a cell

Electrical Circuit Theory and Technology. 978-1-138-67349-6, © 2017 John Bird. Published by Taylor & Francis. All rights reserved.

- perform calculations using $V = E - Ir$
- determine the total e.m.f. and total internal resistance for cells connected in series and in parallel
- distinguish between primary and secondary cells
- explain the construction and practical applications of the Leclanché, mercury, lead–acid and alkaline cells
- list the advantages and disadvantages of alkaline cells over lead–acid cells
- understand the importance of lithium-ion batteries – applications, advantages and disadvantages
- understand the term 'cell capacity' and state its unit
- understand the importance of safe battery disposal
- appreciate advantages of fuel cells and their likely future applications
- understand the implications of alternative energy sources and state six examples
- appreciate the importance and uses of solar energy, its advantages, disadvantages and practical applications

6.1 Introduction to batteries

A battery is a device that **converts chemical energy to electricity**. If an appliance is placed between its terminals the current generated will power the device. Batteries are an indispensable item for many electronic devices and are essential for devices that require power when no mains power is available. For example, without the battery there would be no mobile phones or laptop computers.

The battery is now over 200 years old and batteries are found almost everywhere in consumer and industrial products. Some **practical examples** where batteries are used include:

> in laptops, in cameras, in mobile phones, in cars, in watches and clocks, for security equipment, in electronic meters, for smoke alarms, for meters used to read gas, water and electricity consumption at home, to power a camera for an endoscope looking internally at the body, and for transponders used for toll collection on highways throughout the world.

Batteries tend to be split into two categories – **primary**, which are not designed to be electrically re-charged, i.e. are disposable (see Section 6.6), and **secondary batteries**, which are designed to be re-charged, such as those used in mobile phones (see Section 6.7).

In more recent years it has been necessary to design batteries with reduced size, but with increased lifespan and capacity.

If an application requires small size and high power then the 1.5 V battery is used. If longer lifetime is required then the 3 to 3.6 V battery is used. In the 1970s the 1.5 V **manganese battery** was gradually replaced by the **alkaline battery**. **Silver oxide batteries** were gradually introduced in the 1960s and are still the preferred technology for watch batteries today.

Lithium-ion batteries were introduced in the 1970s because of the need for longer lifetime applications. Indeed, some such batteries have been known to last well over ten years before replacement, a characteristic that means that these batteries are still very much in demand today for digital cameras, and sometimes for watches and computer clocks. Lithium batteries are capable of delivering high currents but tend to be expensive. For more on lithium-ion batteries see Section 6.8, page 81. More types of batteries and their uses are listed in Table 6.2 on page 82.

6.2 Some chemical effects of electricity

A material must contain **charged particles** to be able to conduct electric current. In **solids**, the current is carried by **electrons**. Copper, lead, aluminium, iron and carbon are some examples of solid conductors. In **liquids and gases**, the current is carried by the part of a molecule which has acquired an electric charge, called **ions**. These can possess a positive or negative charge, and examples include hydrogen ion H^+, copper ion Cu^{++} and hydroxyl ion OH^-. Distilled water contains no ions and is a poor conductor of electricity, whereas salt water contains ions and is a fairly good conductor of electricity.

Electrolysis is the decomposition of a liquid compound by the passage of electric current through it. Practical applications of electrolysis include the electroplating of metals (see below), the refining of copper and the extraction of aluminium from its ore.

An **electrolyte** is a compound which will undergo electrolysis. Examples include salt water, copper sulphate and sulphuric acid.

The **electrodes** are the two conductors carrying current to the electrolyte. The positive-connected electrode is called the **anode** and the negative-connected electrode the **cathode**.

When two copper wires connected to a battery are placed in a beaker containing a salt water solution, current will flow through the solution. Air bubbles appear around the wires as the water is changed into hydrogen and oxygen by electrolysis.

Electroplating uses the principle of electrolysis to apply a thin coat of one metal to another metal. Some practical applications include the tin-plating of steel, silver-plating of nickel alloys and chromium-plating of steel. If two copper electrodes connected to a battery are placed in a beaker containing copper sulphate as the electrolyte it is found that the cathode (i.e. the electrode connected to the negative terminal of the battery) gains copper whilst the anode loses copper.

6.3 The simple cell

The purpose of an **electric cell** is to convert chemical energy into electrical energy.

A **simple cell** comprises two dissimilar conductors (electrodes) in an electrolyte. Such a cell is shown in Figure 6.1, comprising copper and zinc electrodes. An electric current is found to flow between the electrodes. Other possible electrode pairs exist, including zinc–lead and zinc–iron. The electrode potential (i.e. the p.d. measured between the electrodes) varies for each pair of metals. By knowing the e.m.f. of each metal with respect to some standard electrode, the e.m.f. of any pair of metals may be determined. The standard used is the hydrogen electrode. The **electrochemical series** is a way of listing elements in order of electrical potential, and Table 6.1 shows a number of elements in such a series.

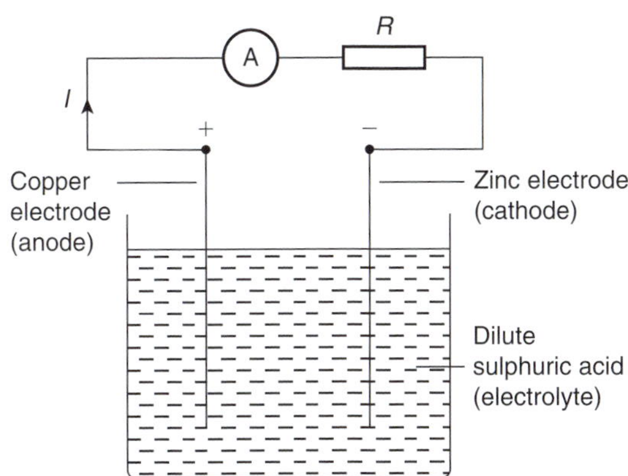

Figure 6.1

Table 6.1 Part of the electrochemical series

Potassium
Sodium
Aluminium
Zinc
Iron
Lead
Hydrogen
Copper
Silver
Carbon

In a simple cell two faults exist – those due to **polarization** and **local action**.

Polarization

If the simple cell shown in Figure 6.1 is left connected for some time, the current I decreases fairly rapidly. This is because of the formation of a film of hydrogen bubbles on the copper anode. This effect is known as the polarization of the cell. The hydrogen prevents full contact between the copper electrode and the electrolyte and this increases the internal resistance of the cell. The effect can be overcome by using a chemical depolarizing agent or depolarizer, such as potassium dichromate, which removes the hydrogen bubbles as they form. This allows the cell to deliver a steady current.

Local action

When commercial zinc is placed in dilute sulphuric acid, hydrogen gas is liberated from it and the zinc dissolves. The reason for this is that impurities, such as traces of iron, are present in the zinc which set up small primary cells with the zinc. These small cells are short-circuited by the electrolyte, with the result that localized currents flow, causing corrosion. This action is known as local action of the cell. This may be prevented by rubbing a small amount of mercury on the zinc surface, which forms a protective layer on the surface of the electrode.

When two metals are used in a simple cell the electrochemical series may be used to predict the behaviour of the cell:

Part 2

(i) The metal that is higher in the series acts as the negative electrode, and vice-versa. For example, the zinc electrode in the cell shown in Figure 6.1 is negative and the copper electrode is positive.

(ii) The greater the separation in the series between the two metals the greater is the e.m.f. produced by the cell.

The electrochemical series is representative of the order of reactivity of the metals and their compounds:

(i) The higher metals in the series react more readily with oxygen and vice-versa.

(ii) When two metal electrodes are used in a simple cell the one that is higher in the series tends to dissolve in the electrolyte.

6.4 Corrosion

Corrosion is the gradual destruction of a metal in a damp atmosphere by means of simple cell action. In addition to the presence of moisture and air required for rusting, an electrolyte, an anode and a cathode are required for corrosion. Thus, if metals widely spaced in the electrochemical series are used in contact with each other in the presence of an electrolyte, corrosion will occur. For example, if a brass valve is fitted to a heating system made of steel, corrosion will occur.

The **effects of corrosion** include the weakening of structures, the reduction of the life of components and materials, the wastage of materials and the expense of replacement.

Corrosion may be **prevented** by coating with paint, grease, plastic coatings and enamels, or by plating with tin or chromium. Also, iron may be galvanized, i.e. plated with zinc, the layer of zinc helping to prevent the iron from corroding.

6.5 E.m.f. and internal resistance of a cell

The **electromotive force (e.m.f.), *E***, of a cell is the p.d. between its terminals when it is not connected to a load (i.e. the cell is on 'no load').

The e.m.f. of a cell is measured by using a **high resistance voltmeter** connected in parallel with the cell. The voltmeter must have a high resistance otherwise it will pass current and the cell will not be on 'no-load'. For example, if the resistance of a cell is 1 Ω and that of a voltmeter 1 MΩ, then the equivalent resistance of the circuit is 1 MΩ + 1 Ω, i.e. approximately 1 MΩ, hence no current flows and the cell is not loaded.

The voltage available at the terminals of a cell falls when a load is connected. This is caused by the **internal resistance** of the cell which is the opposition of the material of the cell to the flow of current. The internal resistance acts in series with other resistances in the circuit. Figure 6.2 shows a cell of e.m.f. E volts and internal resistance, r, and XY represents the terminals of the cell.

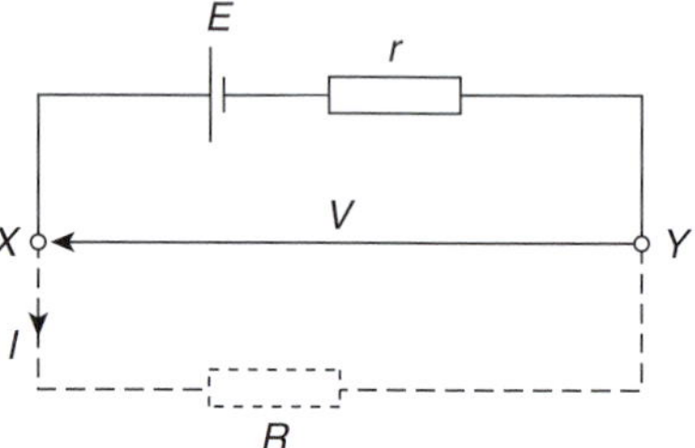

Figure 6.2

When a load (shown as resistance R) is not connected, no current flows and the terminal p.d., $V = E$. When R is connected a current I flows which causes a voltage drop in the cell, given by Ir. The p.d. available at the cell terminals is less than the e.m.f. of the cell and is given by:

$$V = E - Ir$$

Thus if a battery of e.m.f. 12 volts and internal resistance 0.01 Ω delivers a current of 100 A, the terminal p.d.,

$$V = 12 - (100)(0.01)$$
$$= 12 - 1 = 11\,\text{V}$$

When different values of potential difference V across a cell or power supply are measured for different values of current I, a graph may be plotted as shown in Figure 6.3. Since the e.m.f. E of the cell or power supply is the p.d. across its terminals on no load (i.e. when $I = 0$), then E is as shown by the broken line.

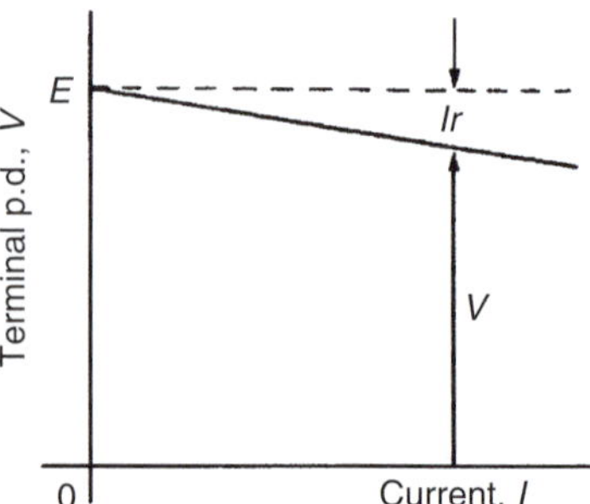

Figure 6.3

Since $V = E - Ir$ then the internal resistance may be calculated from

$$r = \frac{E - V}{I}$$

When a current is flowing in the direction shown in Figure 6.2 the cell is said to be **discharging** ($E > V$).
When a current flows in the opposite direction to that shown in Figure 6.2 the cell is said to be **charging** ($V > E$).
A **battery** is a combination of more than one cell. The cells in a battery may be connected in series or in parallel.

(i) **For cells connected in series:**

Total e.m.f. = sum of cells' e.m.f.s

Total internal resistance = sum of cells' internal resistances

(ii) **For cells connected in parallel:**

If each cell has the same e.m.f. and internal resistance:

Total e.m.f. = e.m.f. of one cell

Total internal resistance of n cells

$= \frac{1}{n} \times$ internal resistance of one cell

Problem 1. Eight cells, each with an internal resistance of 0.2 Ω and an e.m.f. of 2.2 V are connected (a) in series, (b) in parallel. Determine the e.m.f. and the internal resistance of the batteries so formed.

(a) When connected in series, total e.m.f.

= sum of cells' e.m.f.

$= 2.2 \times 8 = \mathbf{17.6\,V}$

Total internal resistance

= sum of cells' internal resistance

$= 0.2 \times 8 = \mathbf{1.6\,\Omega}$

(b) When connected in parallel, total e.m.f.

= e.m.f. of one cell

$= \mathbf{2.2\,V}$

Total internal resistance of 8 cells

$= \frac{1}{8} \times$ internal resistance of one cell

$= \frac{1}{8} \times 0.2 = \mathbf{0.025\,\Omega}$

Problem 2. A cell has an internal resistance of 0.02 Ω and an e.m.f. of 2.0 V. Calculate its terminal p.d. if it delivers (a) 5 A, (b) 50 A.

(a) Terminal p.d. $V = E - Ir$ where E = e.m.f. of cell, I = current flowing and r = internal resistance of cell

$E = 2.0\,\text{V}$, $I = 5\,\text{A}$ and $r = 0.02\,\Omega$

Hence **terminal p.d.**

$\mathbf{V} = 2.0 - (5)(0.02) = 2.0 - 0.1 = \mathbf{1.9\,V}$

(b) When the current is 50 A, terminal p.d.,

$V = E - Ir = 2.0 - 50(0.02)$

i.e. $\mathbf{V} = 2.0 - 1.0 = \mathbf{1.0\,V}$

Thus the terminal p.d. decreases as the current drawn increases.

Problem 3. The p.d. at the terminals of a battery is 25 V when no load is connected and 24 V when a load taking 10 A is connected. Determine the internal resistance of the battery.

When no load is connected the e.m.f. of the battery, E, is equal to the terminal p.d., V, i.e. $E = 25\,\text{V}$

When current $I = 10\,\text{A}$ and terminal p.d.

$V = 24\,\text{V}$, then $V = E - Ir$

i.e. $24 = 25 - (10)r$

Hence, rearranging gives

$10r = 25 - 24 = 1$

and the internal resistance,

$r = \frac{1}{10} = \mathbf{0.1\,\Omega}$

Problem 4. Ten 1.5 V cells, each having an internal resistance of 0.2 Ω, are connected in series to a load of 58 Ω. Determine (a) the current flowing in the circuit and (b) the p.d. at the battery terminals.

(a) For ten cells, battery e.m.f., $E = 10 \times 1.5 = 15\,\text{V}$, and the total internal resistance, $r = 10 \times 0.2 = 2\,\Omega$.

Part 2

Part 2

When connected to a 58 Ω load the circuit is as shown in Figure 6.4

$$\text{Current } I = \frac{\text{e.m.f.}}{\text{total resistance}} = \frac{15}{58+2} = \frac{15}{60} = \mathbf{0.25\,A}$$

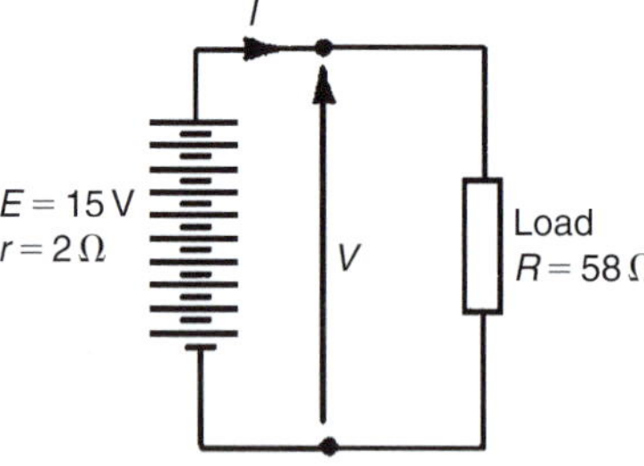

Figure 6.4

(b) P.d. at battery terminals, $V = E - Ir$

i.e. $V = 15 - (0.25)(2) = \mathbf{14.5\,V}$

Now try the following Practice Exercise

Practice Exercise 35 E.m.f. and internal resistance of a cell (Answers on page 818)

1. Twelve cells, each with an internal resistance of 0.24 Ω and an e.m.f. of 1.5 V are connected (a) in series, (b) in parallel. Determine the e.m.f. and internal resistance of the batteries so formed.
2. A cell has an internal resistance of 0.03 Ω and an e.m.f. of 2.2 V. Calculate its terminal p.d. if it delivers

   (a) 1 A (b) 20 A (c) 50 A
3. The p.d. at the terminals of a battery is 16 V when no load is connected and 14 V when a load taking 8 A is connected. Determine the internal resistance of the battery.
4. A battery of e.m.f. 20 V and internal resistance 0.2 Ω supplies a load taking 10 A. Determine the p.d. at the battery terminals and the resistance of the load.
5. Ten 2.2 V cells, each having an internal resistance of 0.1 Ω, are connected in series to a load of 21 Ω. Determine (a) the current flowing in the circuit, and (b) the p.d. at the battery terminals
6. For the circuits shown in Figure 6.5 the resistors represent the internal resistance of the batteries. Find, in each case:
   (i) the total e.m.f. across PQ
   (ii) the total equivalent internal resistances of the batteries.

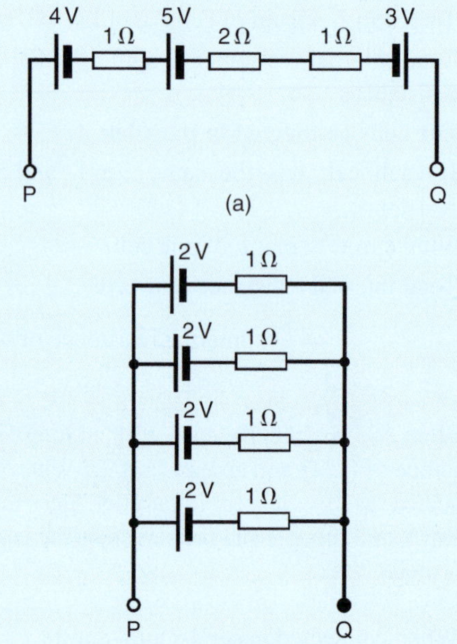

Figure 6.5

7. The voltage at the terminals of a battery is 52 V when no load is connected and 48.8 V when a load taking 80 A is connected. Find the internal resistance of the battery. What would be the terminal voltage when a load taking 20 A is connected?

6.6 Primary cells

Primary cells cannot be recharged, that is, the conversion of chemical energy to electrical energy is irreversible and the cell cannot be used once the chemicals are exhausted. Examples of primary cells include the Leclanché cell and the mercury cell.

Leclanché* cell

A typical dry Leclanché cell is shown in Figure 6.6. Such a cell has an e.m.f. of about 1.5 V when new, but this falls rapidly if in continuous use due to polarization. The hydrogen film on the carbon electrode forms faster than can be dissipated by the depolarizer. The Leclanché cell is suitable only for intermittent use; applications include torches, transistor radios, bells, indicator circuits, gas lighters, controlling switch-gear, and so on. The cell is the most commonly used of primary cells, is cheap, requires little maintenance and has a shelf life of about two years.

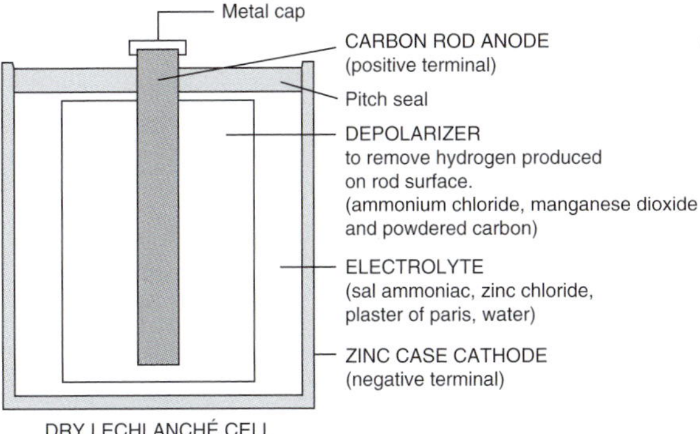

Figure 6.6

* **Who was** Leclanché? **Georges Leclanché** (1839–September 14, 1882) was the French electrical engineer who invented the Leclanché cell, the forerunner of the modern battery. To find out more go to **www.routledge.com/cw/bird**

Mercury cell

A typical mercury cell is shown in Figure 6.7. Such a cell has an e.m.f. of about 1.3 V which remains constant for a relatively long time. Its main advantages over the Leclanché cell is its smaller size and its long shelf life. Typical practical applications include hearing aids, medical electronics, cameras and guided missiles.

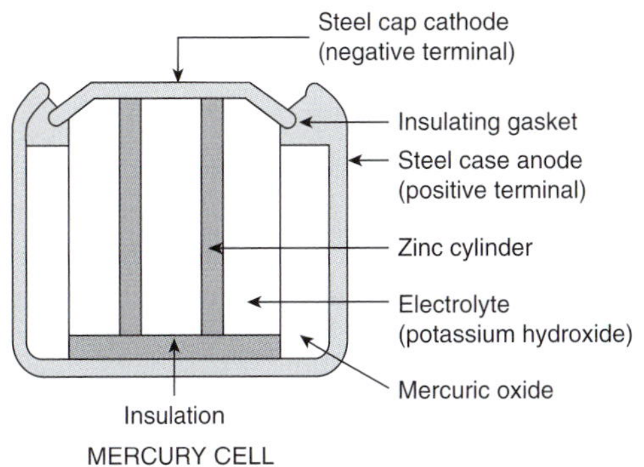

Figure 6.7

6.7 Secondary cells

Secondary cells can be recharged after use, that is, the conversion of chemical energy to electrical energy is reversible and the cell may be used many times. Examples of secondary cells include the lead–acid cell and the nickel cadmium and nickel–metal cells. Practical applications of such cells include car batteries, telephone circuits and for traction purposes – such as milk delivery vans and fork-lift trucks.

Lead–acid cell

A typical lead–acid cell is constructed of:

(i) A container made of glass, ebonite or plastic.

(ii) **Lead plates**

(a) the negative plate (cathode) consists of spongy lead

(b) the positive plate (anode) is formed by pressing lead peroxide into the lead grid.

The plates are interleaved as shown in the plan view of Figure 6.8 to increase their effective cross-sectional area and to minimize internal resistance.

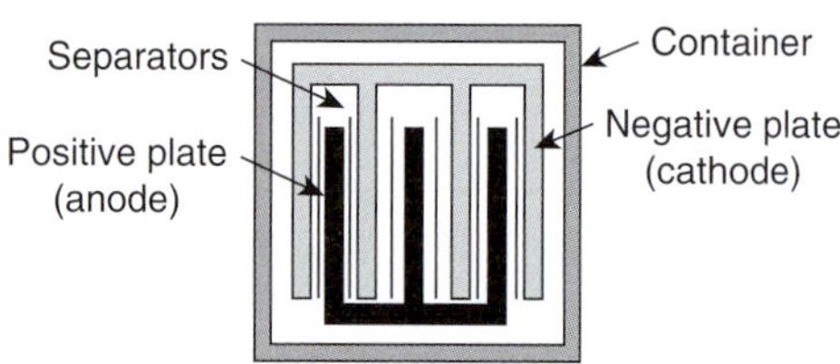

Figure 6.8

(iii) **Separators** made of glass, celluloid or wood.

(iv) An **electrolyte** which is a mixture of sulphuric acid and distilled water.

The relative density (or specific gravity) of a lead–acid cell, which may be measured using a hydrometer, varies between about 1.26 when the cell is fully charged to about 1.19 when discharged. The terminal p.d. of a lead–acid cell is about 2 V.
When a cell supplies current to a load it is said to be **discharging**. During discharge:

(i) the lead peroxide (positive plate) and the spongy lead (negative plate) are converted into lead sulphate, and

(ii) the oxygen in the lead peroxide combines with hydrogen in the electrolyte to form water. The electrolyte is therefore weakened and the relative density falls.

The terminal p.d. of a lead–acid cell when fully discharged is about 1.8 V. A cell is **charged** by connecting a d.c. supply to its terminals, the positive terminal of the cell being connected to the positive terminal of the supply. The charging current flows in the reverse direction to the discharge current and the chemical action is reversed. During charging:

(i) the lead sulphate on the positive and negative plates is converted back to lead peroxide and lead, respectively, and

(ii) the water content of the electrolyte decreases as the oxygen released from the electrolyte combines with the lead of the positive plate. The relative density of the electrolyte thus increases.

The colour of the positive plate when fully charged is dark brown and when discharged is light brown. The colour of the negative plate when fully charged is grey and when discharged is light grey.

To help maintain lead-acid cells, always store them in a charged condition, never let the open cell voltage drop much below 2.10 V, apply a topping charge every six months or when recommended, avoid repeated deep discharges, charge more often or use a larger battery, prevent sulphation and grid corrosion by choosing the correct charge and float voltages, and avoid operating them at elevated ambient temperatures.

Nickel cadmium and nickel–metal cells

In both types the positive plate is made of nickel hydroxide enclosed in finely perforated steel tubes, the resistance being reduced by the addition of pure nickel or graphite. The tubes are assembled into nickel–steel plates.

In the nickel–metal cell (sometimes called the **Edison*** **cell** or **nife cell**), the negative plate is made of iron oxide, with the resistance being reduced by a little mercuric oxide, the whole being enclosed in perforated steel tubes and assembled in steel plates. In the nickel–cadmium cell the negative plate is made of cadmium. The electrolyte in each type of cell is a solution of potassium hydroxide which does not undergo any

* **Who was** Edison? **Thomas Alva Edison** (11 February 1847–18 October 1931) was an American inventor and businessman. Edison is the fourth most prolific inventor in history, holding well over 1,000 US patents in his name, as well as many patents elsewhere. To find out more go to **www.routledge.com/cw/bird**

chemical change and thus the quantity can be reduced to a minimum. The plates are separated by insulating rods and assembled in steel containers which are then enclosed in a non-metallic crate to insulate the cells from one another. The average discharge p.d. of an alkaline cell is about 1.2 V.

Advantages of a nickel cadmium cell or a nickel–metal cell over a lead–acid cell include:

(i) more robust construction

(ii) capable of withstanding heavy charging and discharging currents without damage

(iii) has a longer life

(iv) for a given capacity is lighter in weight

(v) can be left indefinitely in any state of charge or discharge without damage

(vi) is not self-discharging.

Disadvantages of a nickel cadmium and nickel–metal cell over a lead–acid cell include:

(i) is relatively more expensive

(ii) requires more cells for a given e.m.f.

(iii) has a higher internal resistance

(iv) must be kept sealed

(v) has a lower efficiency.

Nickel cells may be used in extremes of temperature, in conditions where vibration is experienced or where duties require long idle periods or heavy discharge currents. Practical examples include traction and marine work, lighting in railway carriages, military portable radios and for starting diesel and petrol engines.

See also Table 6.2, page 82.

6.8 Lithium-ion batteries

Lithium-ion batteries are incredibly popular and may be found in laptops, hand-held pcs, mobile phones and iPods.

A **lithium-ion battery** (sometimes **Li-ion battery** or **LIB**) is a member of a family of rechargeable battery types in which lithium ions move from the negative electrode to the positive electrode during discharge and back when charging. Li-ion batteries use an intercalated lithium compound as one electrode material, compared to the metallic lithium used in a non-rechargeable lithium battery. The electrolyte, which allows for ionic movement, and the two electrodes are the constituent components of a lithium-ion battery cell.

Lithium-ion batteries are common in **consumer electronics**. They are one of the most popular types of rechargeable batteries for portable electronics, with a high energy density, small memory effect and only a slow loss of charge when not in use. Beyond consumer electronics, LIBs are also growing in popularity for **military**, **battery electric vehicle** and **aerospace applications**. Also, lithium-ion batteries are becoming a common replacement for the lead-acid batteries that have been used historically for **golf carts** and **utility vehicles**. Instead of heavy lead plates and acid electrolyte, the trend is to use lightweight lithium-ion battery packs that can provide the same voltage as lead-acid batteries, so no modification to the vehicle's drive system is required.

Chemistry, performance, cost and safety characteristics vary across LIB types. Handheld electronics mostly use LIBs based on lithium cobalt oxide (LiCoO2), which offers high energy density, but presents safety risks, especially when damaged. Lithium iron phosphate (LiFePO4), lithium manganese oxide (LMnO or LMO) and lithium nickel manganese cobalt oxide (LiNiMnCoO2 or NMC) offer lower energy density, but longer lives and inherent safety. Such batteries are widely used for **electric tools**, **medical equipment** and other roles. NMC in particular is a leading contender for **automotive applications**. Lithium nickel cobalt aluminium oxide (LiNiCoAlO2 or NCA) and lithium titanate (Li4Ti5O12 or LTO) are specialty designs aimed at particular niche roles. The new lithium sulphur batteries promise the highest performance to weight ratio.

Lithium-ion batteries can be dangerous under some conditions and can pose a safety hazard since they contain, unlike other rechargeable batteries, a flammable electrolyte and are also kept pressurized. Because of this the testing standards for these batteries are more stringent than those for acid-electrolyte batteries, requiring both a broader range of test conditions and additional battery-specific tests. This is in response to reported accidents and failures, and there have been battery-related recalls by some companies.

For many years, nickel-cadmium had been the only suitable battery for **portable equipment** from **wireless communications** to **mobile computing**. Nickel-metal-hydride and lithium-ion emerged in the early 1990s and today, lithium-ion is the fastest growing and most promising battery chemistry. Pioneer work with the lithium battery began in 1912 under G.N. Lewis but it was not until the early 1970s when the first non-rechargeable lithium batteries became commercially

Table 6.2

Type of battery	Common uses	Hazardous component	Disposal recycling options
Wet cell (i.e. a primary cell that has a liquid electrolyte)			
Lead–acid batteries	Electrical energy supply for vehicles including cars, trucks, boats, tractors and motorcycles. Small sealed lead–acid batteries are used for emergency lighting and uninterruptible power supplies	Sulphuric acid and lead	Recycle – most petrol stations and garages accept old car batteries, and council waste facilities have collection points for lead–acid batteries
Dry cell: Non-chargeable – single use (for example, AA, AAA, C, D, lantern and miniature watch sizes)			
Zinc carbon	Torches, clocks, shavers, radios, toys and smoke alarms	Zinc	Not classed as hazardous waste – can be disposed with household waste
Zinc chloride	Torches, clocks, shavers, radios, toys and smoke alarms	Zinc	Not classed as hazardous waste – can be disposed with household waste
Alkaline manganese	Personal stereos and radio/cassette players	Manganese	Not classed as hazardous waste – can be disposed with household waste
Primary button cells (i.e. a small flat battery shaped like a 'button' used in small electronic devices)			
Mercuric oxide	Hearing aids, pacemakers and cameras	Mercury	Recycle at council waste facility, if available
Zinc air	Hearing aids, pagers and cameras	Zinc	Recycle at council waste facility, if available
Silver oxide	Calculators, watches and cameras	Silver	Recycle at council waste facility, if available
Lithium	Computers, watches and cameras	Lithium (explosive and flammable)	Recycle at council waste facility, if available
Dry cell rechargeable – secondary batteries			
Nickel cadmium (NiCd)	Mobile phones, cordless power tools, laptop computers, shavers, motorized toys, personal stereos	Cadmium	Recycle at council waste facility, if available
Nickel–metal hydride (NiMH)	Alternative to NiCd batteries, but longer life	Nickel	Recycle at council waste facility, if available
Lithium-ion (Li-ion)	Alternative to NiCd and NiMH batteries, but greater energy storage capacity	Lithium	Recycle at council waste facility, if available

available. Lithium is the lightest of all metals, has the greatest electrochemical potential and provides the largest energy density for weight.

Attempts to develop rechargeable lithium batteries initially failed due to safety problems. Because of the inherent instability of lithium metal, especially during charging, research shifted to a non-metallic lithium battery using lithium ions. Although slightly lower in energy density than lithium metal, lithium-ion is safe, provided certain precautions are met when charging and discharging. In 1991, the Sony Corporation commercialized the first lithium-ion battery and other manufacturers followed suit.

The energy density of lithium-ion is typically twice that of the standard nickel-cadmium. There is potential for higher energy densities. The load characteristics are reasonably good and behave similarly to nickel-cadmium in terms of discharge. The high cell voltage of 3.6 volts allows battery pack designs with only one cell. Most of today's mobile phones run on a single cell. A nickel-based pack would require three 1.2-volt cells connected in series.

Lithium-ion is a low maintenance battery, an advantage that most other chemistries cannot claim. There is no memory and no scheduled cycling is required to prolong the battery's life. In addition, the self-discharge is less than half compared to nickel-cadmium, making lithium-ion well suited for modern fuel gauge applications. Lithium-ion cells cause little harm when disposed.

Despite its overall advantages, lithium-ion has its drawbacks. It is fragile and requires a protection circuit to maintain safe operation. Built into each pack, the protection circuit limits the peak voltage of each cell during charge and prevents the cell voltage from dropping too low on discharge. In addition, the cell temperature is monitored to prevent temperature extremes. The maximum charge and discharge current on most packs are limited to between 1° C and 2° C. With these precautions in place, the possibility of metallic lithium plating occurring due to overcharge is virtually eliminated.

Ageing is a concern with most lithium-ion batteries and many manufacturers remain silent about this issue. Some capacity deterioration is noticeable after one year, whether the battery is in use or not. The battery frequently fails after two or three years. It should be noted that other chemistries also have age-related degenerative effects. This is especially true for nickel-metal-hydride if exposed to high ambient temperatures. At the same time, lithium-ion packs are known to have served for five years in some applications.

Manufacturers are constantly improving lithium-ion. New and enhanced chemical combinations are introduced every six months or so. With such rapid progress, it is difficult to assess how well the revised battery will age.

Storage in a cool place slows the ageing process of lithium-ion (and other chemistries). Manufacturers recommend storage temperatures of 15 °C (59°F). In addition, the battery should be partially charged during storage; the manufacturer recommends a 40% charge.

The most economical lithium-ion battery in terms of cost-to-energy ratio is the cylindrical 18650 (size is 18mm x 65.2mm). This cell is used for mobile computing and other applications that do not demand ultra-thin geometry. If a slim pack is required, the prismatic lithium-ion cell is the best choice. These cells come at a higher cost in terms of stored energy.

Summary of advantages of lithium-ion batteries

(i) High energy density – potential for yet higher capacities.

(ii) Does not need prolonged priming when new; one regular charge is all that's needed.

(iii) Relatively low self-discharge – is less than half that of nickel-based batteries.

(iv) Low maintenance – no periodic discharge is needed; there is no memory.

(v) Speciality cells can provide very high current to applications such as power tools.

Summary of limitations of lithium-ion batteries

(i) Requires protection circuit to maintain voltage and current within safe limits.

(ii) Subject to ageing, even if not in use – storage in a cool place at 40% charge reduces the aging effect.

(iii) Expensive to manufacture – about 40% higher in cost than nickel-cadmium.

(iv) Not fully mature – metals and chemicals are changing on a continuing basis.

To extend the battery life of lithium-ion batteries:

(i) **Keep batteries at room temperature**, meaning between 20°C and 25°C; heat is by far the greatest factor in reducing lithium-ion battery life.

(ii) **Obtain a high-capacity lithium-ion battery, rather than carrying a spare**; batteries deteriorate over time, whether they're being used or not so a spare battery won't last much longer than the one in use.

(iii) **Allow partial discharges and avoid full ones**; unlike NiCad batteries, lithium-ion batteries do not have a charge memory, which means that deep-discharge cycles are not required – in fact, it's better for the battery to use partial-discharge cycles.

(iv) **Avoid completely discharging lithium-ion batteries**; if a lithium-ion battery is discharged below 2.5 volts per cell, a safety circuit built into the battery opens and the battery appears to be dead and the original charger will be of no use – only battery analyzers with the boost function have a chance of recharging the battery.

(v) **For extended storage, discharge a lithium-ion battery to about 40% and store it in a cool place.**

Lithium-ion batteries are a huge improvement over previous types of batteries and getting 500 charge/discharge cycles from a lithium-ion battery is becoming commonplace.

6.9 Cell capacity

The **capacity** of a cell is measured in ampere-hours (Ah). A fully charged 50 Ah battery rated for 10 h discharge can be discharged at a steady current of 5 A for 10 h, but if the load current is increased to 10 A then the battery is discharged in 3–4 h, since the higher the discharge current, the lower is the effective capacity of the battery. Typical discharge characteristics for a lead–acid cell are shown in Figure 6.9.

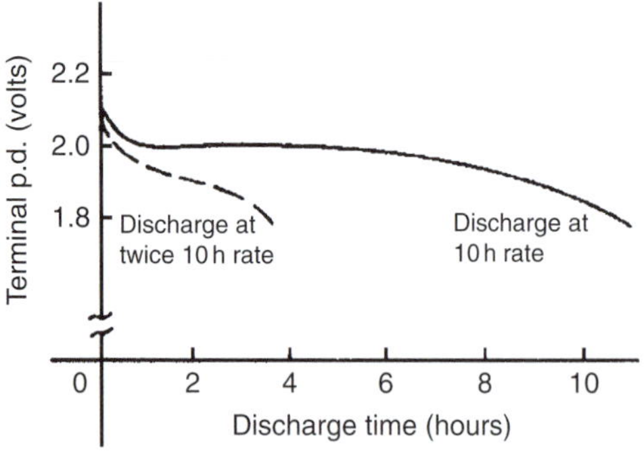

Figure 6.9

6.10 Safe disposal of batteries

Battery disposal has become a topical subject in the UK because of greater awareness of the dangers and implications of depositing up to 300 million batteries per annum – a waste stream of over 20 000 tonnes – into landfill sites.

Certain batteries contain substances which can be a hazard to humans, wildlife and the environment, as well as posing a fire risk. Other batteries can be recycled for their metal content.

Waste batteries are a concentrated source of toxic heavy metals such as mercury, lead and cadmium. If batteries containing heavy metals are disposed of incorrectly, the metals can leach out and pollute the soil and groundwater, endangering humans and wildlife. Long-term exposure to cadmium, a known human carcinogen (i.e. a substance producing cancerous growth), can cause liver and lung disease. Mercury can cause damage to the human brain, spinal system, kidneys and liver. Sulphuric acid in lead–acid batteries can cause severe skin burns or irritation upon contact. It is increasingly important to correctly dispose of all types of batteries.

Table 6.2 lists types of batteries, their common uses, their hazardous components and disposal recycling options.

Battery disposal has become more regulated since the Landfill Regulations 2002 and Hazardous Waste Regulations 2005. From the Waste Electrical and Electronic Equipment (WEEE) Regulations 2006, commencing July 2007 all producers (manufacturers and importers) of electrical and electronic equipment will be responsible for the cost of collection, treatment and recycling of obligated WEEE generated in the UK.

6.11 Fuel cells

A **fuel cell** is an electrochemical energy conversion device, similar to a battery, but differing from the latter in that it is designed for continuous replenishment of the reactants consumed, i.e. it produces electricity from an external source of fuel and oxygen, as opposed to the limited energy storage capacity of a battery. Also, the electrodes within a battery react and change as a battery is charged or discharged, whereas a fuel cell's electrodes are catalytic (i.e. not permanently changed) and relatively stable.

Typical reactants used in a fuel cell are hydrogen on the anode side and oxygen on the cathode side (i.e. a **hydrogen cell**). Usually, reactants flow in and reaction products flow out. Virtually continuous long-term operation is feasible as long as these flows are maintained.

Fuel cells are very attractive in modern applications for their high efficiency and ideally emission-free use, in contrast to currently more modern fuels such as methane or natural gas that generate carbon dioxide. The only by-product of a fuel cell operating on pure hydrogen is water vapour.

Currently, fuel cells are a very expensive alternative to internal combustion engines. However, continued research and development is likely to make fuel cell vehicles available at market prices within a few years.

Fuel cells are very useful as power sources in remote locations, such as spacecraft, remote weather stations, and in certain military applications. A fuel cell running on hydrogen can be compact, lightweight and has no moving parts.

6.12 Alternative and renewable energy sources

Alternative energy refers to energy sources which could replace coal, traditional gas and oil, all of which increase the atmospheric carbon when burned as fuel.

Renewable energy implies that it is derived from a source which is automatically replenished or one that is effectively infinite so that it is not depleted as it is used. Coal, gas and oil are not renewable because, although the fields may last for generations, their time span is finite and will eventually run out.

There are many means of harnessing energy which have less damaging impacts on our environment, including the following:

1. **Solar energy** is one of the most resourceful sources of energy for the future. The reason for this is that the total energy received each year from the sun is around 35 000 times the total energy used by man. However, about one-third of this energy is either absorbed by the outer atmosphere or reflected back into space. Solar energy could be used to run cars, power plants and space ships. **Solar panels** on roofs capture heat in water storage systems. **Photovoltaic cells**, when suitably positioned, convert sunlight to electricity. For more on solar energy, see Section 6.13 following.
2. **Wind power** is another alternative energy source that can be used without producing by-products that are harmful to nature. The fins of a windmill rotate in a vertical plane which is kept vertical to the wind by means of a tail fin and as wind flow crosses the blades of the windmill it is forced to rotate and can be used to generate electricity (see Chapter 11). Like solar power, harnessing the wind is highly dependent upon weather and location. The average wind velocity of Earth is around 9 m/s, and the power that could be produced when a windmill is facing a wind of 10 m.p.h. (i.e. around 4.5 m/s) is around 50 watts.
3. **Hydroelectricity** is achieved by the damming of rivers and utilizing the potential energy in the water. As the water stored behind a dam is released at high pressure, its kinetic energy is transferred onto turbine blades and used to generate electricity. The system has enormous initial costs but has relatively low maintenance costs and provides power quite cheaply.
4. **Tidal power** utilizes the natural motion of the tides to fill reservoirs which are then slowly discharged through electricity-producing turbines.
5. **Geothermal energy** is obtained from the internal heat of the planet and can be used to generate steam to run a steam turbine which, in turn, generates electricity. The radius of the Earth is about 4000 miles, with an internal core temperature of around 4000°C at the centre. Drilling 3 miles from the surface of the Earth, a temperature of 100 °C is encountered; this is sufficient to boil water to run a steam-powered electric power plant. Although drilling 3 miles down is possible, it is not easy. Fortunately, however, volcanic features called **geothermal hotspots** are found all around the world. These are areas which transmit excess internal heat from the interior of the Earth to the outer crust, which can be used to generate electricity.
   Advantages of geothermal power include being environmentally friendly, global warming effects are mitigated, has no fuel costs, gives predictable 24/7 power, has a high load factor and no pollution.
   Disadvantages of geothermal power include having a long gestation time leading to possible cost overruns, slow technology improvement, problems with financing, innumerable regulations, and having limited locations.
6. **Biomass** is fuel that is developed from organic materials, a renewable and sustainable source of energy used to create electricity or other forms of power. Some examples of materials that make up biomass fuels are scrap lumber, forest debris, certain crops, manure and some types of waste residues. With a constant supply of waste – from construction and demolition activities, to wood not used in papermaking, to municipal solid waste – green energy production can continue indefinitely. There are several methods to convert biomass

into electricity and these are briefly discussed in Chapter 21.

6.13 Solar energy

Solar energy – power from the sun – is a vast and inexhaustible resource. Once a system is in place to convert it into useful energy, the fuel is free and will never be subject to the variations of energy markets. Furthermore, it represents a clean alternative to the fossil fuels that currently pollute air and water, threaten public health and contribute to global warming. Given the abundance and the appeal of solar energy, this resource will play a prominent role in our energy future.

In the broadest sense, solar energy supports all life on Earth and is the basis for almost every form of energy used. The sun makes plants grow, which can be burned as 'biomass' fuel or, if left to rot in swamps and compressed underground for millions of years, in the form of coal and oil. Heat from the sun causes temperature differences between areas, producing wind that can power turbines. Water evaporates because of the sun, falls on high elevations and rushes down to the sea, spinning hydroelectric turbines as it passes. However, solar energy usually refers to ways the sun's energy can be used to directly generate heat, lighting and electricity.

The amount of energy from the sun that falls on the Earth's surface is enormous. All the energy stored in Earth's reserves of coal, oil and natural gas is matched by the energy from just 20 days of sunshine.

Originally developed for energy requirements for an orbiting earth satellite, solar power has expanded in recent years for use in domestic and industrial needs. Solar power is produced by collecting sunlight and converting it into electricity. This is achieved by using solar panels, which are large flat panels made up of many individual solar cells.

Some advantages of solar power

(a) The major advantage of solar power is that no pollution is created in the process of generating electricity. Environmentally it is the most clean and green energy. It is renewable (unlike gas, oil and coal) and sustainable, helping to protect the environment.

(b) Solar energy does not require any fuel.

(c) Solar energy does not pollute air by releasing carbon dioxide, nitrogen oxide, sulphur dioxide or mercury into the atmosphere like many traditional forms of electrical generation do.

(d) Solar energy does not contribute to global warming, acid rain or smog. It actively contributes to the decrease of harmful green house gas emissions.

(e) There is no on-going cost for the power solar energy generates, as solar radiation is free everywhere; once installed, there are no recurring costs.

(f) Solar energy can be flexibly applied to a variety of stationary or portable applications. Unlike most forms of electrical generation, the panels can be made small enough to fit pocket-size electronic devices, such as a calculator, or sufficiently large to charge an automobile battery or supply electricity to entire buildings.

(g) Solar energy offers much more self-reliance than depending upon a power utility for all electricity.

(h) Solar energy is quite economical in the long run. After the initial investment has been recovered, the energy from the sun is practically free. Solar energy systems are virtually maintenance free and will last for decades.

(i) Solar energy is unaffected by the supply and demand of fuel and is therefore not subjected to the ever-increasing price of fossil fuel.

(j) By not using any fuel, solar energy does not contribute to the cost and problems of the recovery and transportation of fuel or the storage of radioactive waste.

(k) Solar energy is generated where it is needed, hence large scale transmission cost is minimized.

(l) Solar energy can be utilized to offset utility-supplied energy consumption. It does not only reduce electricity bills, but will also continue to supply homes/businesses with electricity in the event of a power outage.

(m) A solar energy system can operate entirely independently, not requiring a connection to a power or gas grid at all. Systems can therefore be installed in remote locations, making it more practical and cost-effective than the supply of utility electricity to a new site.

(n) Solar energy projects operate silently, have no moving parts, do not release offensive smells and do not require the addition of any additional fuel.

(o) Solar energy projects support local job and wealth creation, fuelling local economies.

Some disadvantages of solar power

(a) The initial cost is the main disadvantage of installing a solar energy system, largely because of the high cost of the semi-conducting materials used in building solar panels.

(b) The cost of solar energy is also high compared to non-renewable utility-supplied electricity. As energy shortages are becoming more common, solar energy is becoming more price-competitive.

(c) Solar panels require quite a large area for installation to achieve a good level of efficiency.

(d) The efficiency of the system also relies on the location of the sun, although this problem can be overcome with the installation of certain components.

(e) The production of solar energy is influenced by the presence of clouds or pollution in the air. Similarly, no solar energy will be produced during the night although a battery backup system and/or net metering will solve this problem.

Some applications of solar energy

(i) **Concentrating Solar Power (CSP):** Concentrating solar power (CSP) plants are utility-scale generators that produce electricity using mirrors or lenses to efficiently concentrate the sun's energy. The four principal CSP technologies are parabolic troughs, dish-Stirling engine systems, central receivers and concentrating photovoltaic systems (CPV).

(ii) **Solar Thermal Electric Power Plants:** Solar thermal energy involves harnessing solar power for practical applications from solar heating to electrical power generation. Solar thermal collectors, such as solar hot water panels, are commonly used to generate **solar hot water for domestic and light industrial applications**. This energy system is also used in architecture and building design to **control heating and ventilation** in both active solar and passive solar designs.

(iii) **Photovoltaics:** Photovoltaic or PV technology employs solar cells or solar photovoltaic arrays to convert energy from the sun into electricity. Solar cells produce direct current electricity from the sun's rays, which can be used to power equipment or to recharge batteries. Many **pocket calculators** incorporate a single solar cell, but for larger applications, cells are generally grouped together to form PV modules that are in turn arranged in solar arrays. Solar arrays can be used to power **orbiting satellites** and other **spacecraft**, and in remote areas as a source of power for **roadside emergency telephones**, **remote sensing**, **school crossing warning signs** and **cathodic protection of pipelines**.

(iv) **Solar Heating Systems:** Solar hot water systems use sunlight to heat water. The systems are composed of solar thermal collectors and a storage tank, and they may be active, passive or batch systems.

(v) **Passive Solar Energy:** Building designers are concerned to maintain the building environment at a comfortable temperature through the sun's daily and annual cycles. This can be achieved by: (a) **direct gain** in which the positioning of windows, skylights, and shutters to control the amount of direct solar radiation reaching the interior and warming the air and surfaces within a building, (b) **indirect gain** in which solar radiation is captured by a part of the building envelope and then transmitted indirectly to the building through conduction and convection, and (c) **isolated gain** which involves passively capturing solar heat and then moving it passively into or out of the building via a liquid or air directly or using a thermal store. Sunspaces, greenhouses and solar closets are alternative ways of capturing isolated heat gain from which warmed air can be taken.

(vi) **Solar Lighting:** Also known as day-lighting, solar lighting is the use of natural light to provide illumination to offset energy use in electric lighting systems and reduce the cooling load on high voltage a.c. systems. Day-lighting features include building orientation, window orientation, exterior shading, saw-tooth roofs, clerestory windows, light shelves, skylights and light tubes. Architectural trends increasingly recognise day-lighting as a cornerstone of sustainable design.

(vii) **Solar Cars:** A solar car is an electric vehicle powered by energy obtained from solar panels on the surface of the car which convert the sun's energy directly into electrical energy. Solar cars are not currently a practical form of transportation. Although they can operate for limited distances without sun, the solar cells are generally very fragile. Development teams have focused their efforts on optimising the efficiency of the vehicle, but many have only enough room for one or two people.

(viii) **Solar Power Satellite:** A solar power satellite (SPS) is a proposed satellite built in high Earth orbit that uses microwave power transmission to beam solar power to a very large antenna on Earth where it can be used in place of conventional power sources. The advantage of placing the solar collectors in space is the unobstructed view of the sun, unaffected by the day/night cycle, weather or seasons. However, the costs of construction are very high, and SPSs will not be able to compete with conventional sources unless low launch costs can be achieved or unless a space-based manufacturing industry develops and they can be built in orbit from off-earth materials.

(ix) **Solar Updraft Tower:** A solar updraft tower is a proposed type of renewable-energy power plant. Air is heated in a very large circular greenhouse-like structure, and the resulting convection causes the air to rise and escape through a tall tower. The moving air drives turbines, which produce electricity. There are no solar updraft towers in operation at present. A research prototype operated in Spain in the 1980s, and EnviroMission is proposing to construct a full-scale power station using this technology in Australia.

(x) **Renewable Solar Power Systems with Regenerative Fuel Cell Systems:** NASA in the USA has long recognised the unique advantages of regenerative fuel cell (RFC) systems to provide energy storage for solar power systems in space. RFC systems are uniquely qualified to provide the necessary energy storage for solar surface power systems on the moon or Mars during long periods of darkness, i.e. during the 14-day lunar night or the 12-hour Martian night. The nature of the RFC and its inherent design flexibility enables it to effectively meet the requirements of space missions. In the course of implementing the NASA RFC Programme, researchers recognised that there are numerous applications in government, industry, transportation and the military for RFC systems as well.

For fully worked solutions to each of the problems in Practice Exercise 35 in this chapter, go to the website:
www.routledge.com/cw/bird

Revision Test 1

This revision test covers the material contained in Chapters 3 to 6. *The marks for each question are shown in brackets at the end of each question.*

1. An electromagnet exerts a force of 15 N and moves a soft iron armature through a distance of 12 mm in 50 ms. Determine the power consumed. (5)

2. A d.c. motor consumes 47.25 MJ when connected to a 250 V supply for 1 hour 45 minutes. Determine the power rating of the motor and the current taken from the supply. (5)

3. A 100 W electric light bulb is connected to a 200 V supply. Calculate (a) the current flowing in the bulb, and (b) the resistance of the bulb. (4)

4. Determine the charge transferred when a current of 5 mA flows for 10 minutes. (2)

5. A current of 12 A flows in the element of an electric fire of resistance 25 Ω. Determine the power dissipated by the element. If the fire is on for five hours every day, calculate for a one-week period (a) the energy used, and (b) cost of using the fire if electricity cost 13.5 p per unit. (6)

6. Calculate the resistance of 1200 m of copper cable of cross-sectional area 15 mm^2. Take the resistivity of copper as 0.02 $\mu\Omega$m. (5)

7. At a temperature of 40°C, an aluminium cable has a resistance of 25 Ω. If the temperature coefficientof resistance at 0°C is 0.0038/°C, calculate its resistance at 0°C. (5)

8. (a) Determine the values of the resistors with the following colour coding:
   (i) red-red-orange-silver
   (ii) orange-orange-black-blue-green.
   (b) What is the value of a resistor marked as 47KK? (6)

9. Four cells, each with an internal resistance of 0.40 Ω and an e.m.f. of 2.5 V, are connected in series to a load of 38.40 Ω. (a) Determine the current flowing in the circuit and the p.d. at the battery terminals. (b) If the cells are connected in parallel instead of in series, determine the current flowing and the p.d. at the battery terminals. (10)

10. (a) State six typical applications of primary cells.
    (b) State six typical applications of secondary cells.
    (c) State the advantages of a fuel cell over a conventional battery and state three practical applications. (12)

11. State for lithium-ion batteries
    (a) three typical practical applications
    (b) four advantages compared with other batteries
    (c) three limitations. (10)

12. Name six alternative, renewable energy sources, and give a brief description of each. (18)

13. For solar power, state
    (a) seven advantages
    (b) five disadvantages. (12)

For lecturers/instructors/teachers, fully worked solutions to each of the problems in Revision Test 1, together with a full marking scheme, are available at the website:
www.routledge.com/cw/bird

Chapter 7

Series and parallel networks

***Why it is important to understand:* Series and parallel networks**

There are two ways in which components may be connected together in an electric circuit. One way is 'in series' where components are connected 'end-to-end'; another way is 'in parallel' where components are connected 'across each other'. When a circuit is more complicated than two or three elements, it is very likely to be a network of individual series and parallel circuits. A firm understanding of the basic principles associated with series and parallel circuits is a sufficient background to begin an investigation of any single-source d.c. network having a combination of series and parallel elements or branches. Confidence in the analysis of series-parallel networks comes only through exposure, practice and experience. At first glance, these circuits may seem very complicated, but with a methodical analysis approach the functionality of the circuit can become obvious. This chapter explains with examples, series, parallel and series/parallel networks. The relationships between voltages, currents and resistances for these networks are considered through calculations.

At the end of this chapter you should be able to:

- calculate unknown voltages, currents and resistances in a series circuit
- understand voltage division in a series circuit
- calculate unknown voltages, currents and resistances in a parallel network
- calculate unknown voltages, currents and resistances in series-parallel networks
- understand current division in a two-branch parallel network
- appreciate the loading effect of a voltmeter
- understand the difference between potentiometers and rheostats
- perform calculations to determine load currents and voltages in potentiometers and rheostats
- understand and perform calculations on relative and absolute voltages
- state three causes of short circuits in electrical circuits
- describe the advantages and disadvantages of series and parallel connection of lamps

Electrical Circuit Theory and Technology. 978-1-138-67349-6, © 2017 John Bird. Published by Taylor & Francis. All rights reserved.

7.1 Series circuits

Figure 7.1 shows three resistors, R_1, R_2 and R_3, connected end to end, i.e. in series, with a battery source of V volts. Since the circuit is closed a current I will flow and the p.d. across each resistor may be determined from the voltmeter readings V_1, V_2 and V_3

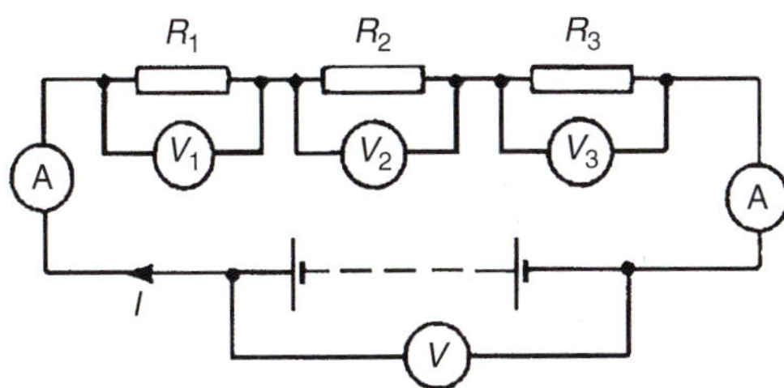

Figure 7.1

In a series circuit

(a) The current I is the same in all parts of the circuit and hence the same reading is found on each of the two ammeters shown.

(b) The sum of the voltages V_1, V_2 and V_3 is equal to the total applied voltage, V, i.e.

$$V = V_1 + V_2 + V_3$$

From Ohm's law:

$V_1 = IR_1$, $V_2 = IR_2$, $V_3 = IR_3$ and $V = IR$

where R is the total circuit resistance.

Since $V = V_1 + V_2 + V_3$

then $IR = IR_1 + IR_2 + IR_3$

Dividing throughout by I gives

$$R = R_1 + R_2 + R_3$$

Thus for a series circuit, the total resistance is obtained by adding together the values of the separate resistances.

Problem 1. For the circuit shown in Figure 7.2, determine (a) the battery voltage V, (b) the total resistance of the circuit, and (c) the values of resistance of resistors R_1, R_2 and R_3, given that the p.d.s across R_1, R_2 and R_3 are 5 V, 2 V and 6 V, respectively.

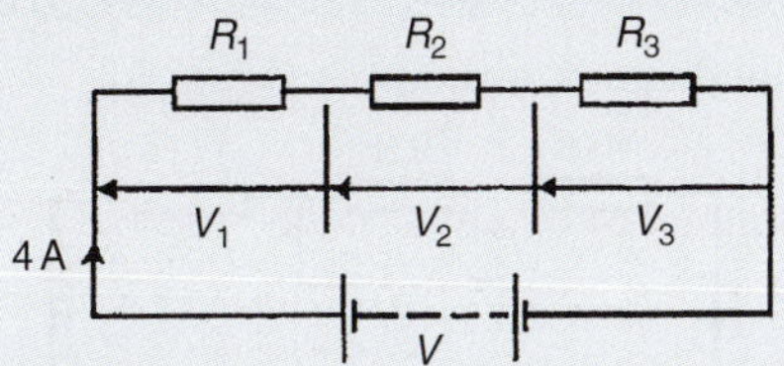

Figure 7.2

(a) Battery voltage $V = V_1 + V_2 + V_3$

$= 5 + 2 + 6 = \mathbf{13\,V}$

(b) Total circuit resistance $R = \frac{V}{I} = \frac{13}{4} = \mathbf{3.25\,\Omega}$

(c) Resistance $R_1 = \frac{V_1}{I} = \frac{5}{4} = \mathbf{1.25\,\Omega}$

Resistance $R_2 = \frac{V_2}{I} = \frac{2}{4} = \mathbf{0.5\,\Omega}$

Resistance $R_3 = \frac{V_3}{I} = \frac{6}{4} = \mathbf{1.5\,\Omega}$

(Check: $R_1 + R_2 + R_3 = 1.25 + 0.5 + 1.5$
$= 3.25\,\Omega = R$)

Problem 2. For the circuit shown in Figure 7.3, determine the p.d. across resistor R_3. If the total resistance of the circuit is 100 Ω, determine the current flowing through resistor R_1. Find also the value of resistor R_2.

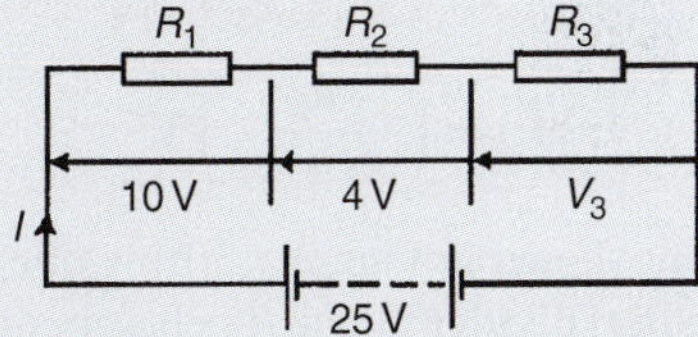

Figure 7.3

P.d. across R_3, $V_3 = 25 - 10 - 4 = \mathbf{11\,V}$

Current $I = \frac{V}{R} = \frac{25}{100} = \mathbf{0.25\,A}$, which is the current flowing in each resistor

Resistance $R_2 = \frac{V_2}{I} = \frac{4}{0.25} = \mathbf{16\,\Omega}$

Problem 3. A 12 V battery is connected in a circuit having three series-connected resistors having resistances of 4 Ω, 9 Ω and 11 Ω. Determine the current flowing through, and the p.d. across the

Part 2

9 Ω resistor. Find also the power dissipated in the 11 Ω resistor.

The circuit diagram is shown in Figure 7.4.

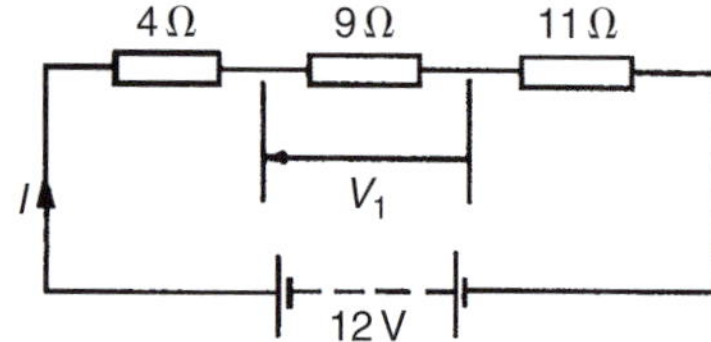

Figure 7.4

Total resistance $R = 4 + 9 + 11 = 24\,\Omega$

Current $I = \frac{V}{R} = \frac{12}{24} = \mathbf{0.5\,A}$, which is the current in the 9 Ω resistor.

P.d. across the 9 Ω resistor, $V_1 = I \times 9 = 0.5 \times 9$

$$= \mathbf{4.5\,V}$$

Power dissipated in the 11 Ω resistor,

$$P = I^2 R = 0.5^2(11)$$
$$= 0.25(11)$$
$$= \mathbf{2.75\,W}$$

7.2 Potential divider

The voltage distribution for the circuit shown in Figure 7.5(a) is given by:

$$V_1 = \left(\frac{R_1}{R_1 + R_2}\right) V$$

$$V_2 = \left(\frac{R_2}{R_1 + R_1}\right) V$$

The circuit shown in Figure 7.5(b) is often referred to as a **potential divider** circuit. Such a circuit can consist of a number of similar elements in series connected across a voltage source, voltages being taken from connections between the elements. Frequently the divider consists of two resistors, as shown in Figure 7.5(b), where

$$V_{OUT} = \left(\frac{R_2}{R_1 + R_2}\right) V_{IN}$$

A potential divider is the simplest way of producing a source of lower e.m.f. from a source of higher e.m.f., and is the basic operating mechanism of the **potentiometer**, a measuring device for accurately measuring potential differences (see page 177).

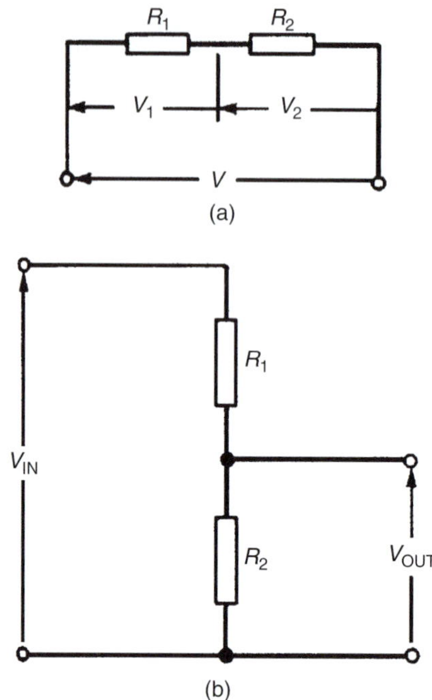

Figure 7.5

Problem 4. Determine the value of voltage V shown in Figure 7.6.

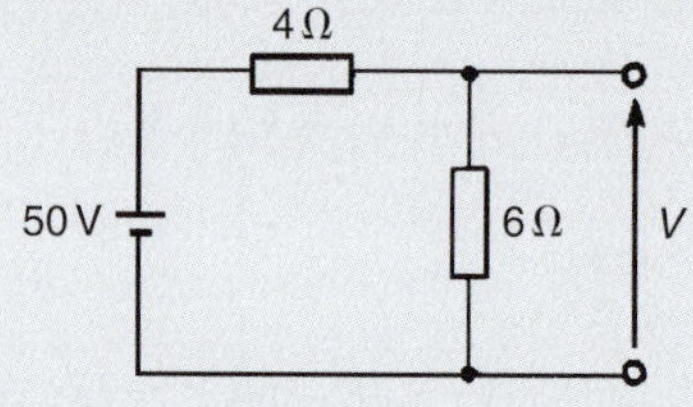

Figure 7.6

Figure 7.6 may be redrawn as shown in Figure 7.7, and voltage

$$V = \left(\frac{6}{6+4}\right)(50) = \mathbf{30\,V}$$

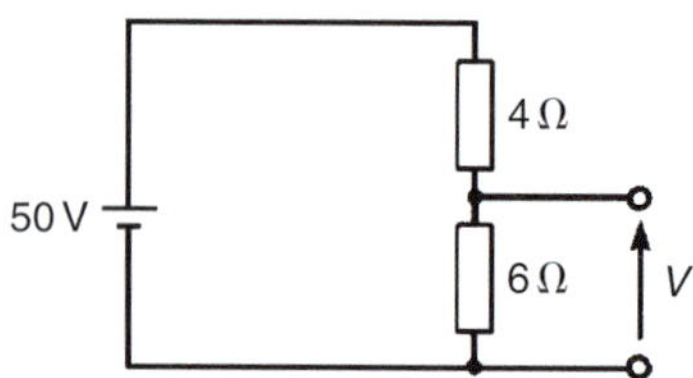

Figure 7.7

Problem 5. Two resistors are connected in series across a 24 V supply and a current of 3 A flows in the circuit. If one of the resistors has a resistance of 2 Ω, determine (a) the value of the other resistor, and (b) the p.d. across the 2 Ω resistor. If the circuit is connected for 50 hours, how much energy is used?

The circuit diagram is shown in Figure 7.8.

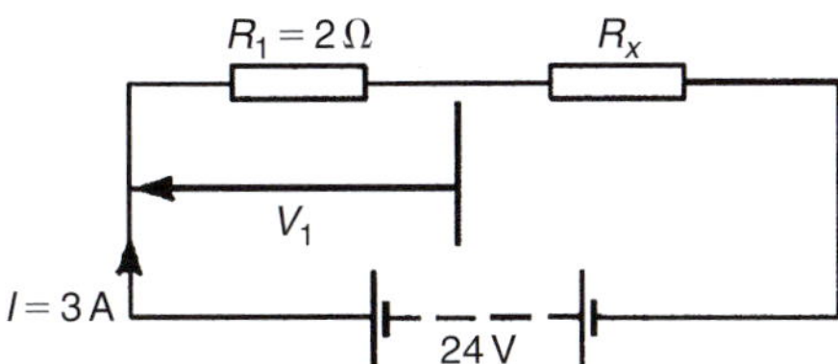

Figure 7.8

(a) Total circuit resistance $R = \frac{V}{I} = \frac{24}{3} = 8\,\Omega$

Value of unknown resistance, $R_x = 8 - 2 = \mathbf{6\,\Omega}$

(b) P.d. across 2 Ω resistor, $V_1 = IR_1 = 3 \times 2 = \mathbf{6\,V}$

Alternatively, from above,

$$V_1 = \left(\frac{R_1}{R_1 + R_x}\right) V = \left(\frac{2}{2+6}\right)(24) = 6\,\text{V}$$

Energy used = power × time

$$= V \times I \times t$$

$$= (24 \times 3\,\text{W})\ (50\,\text{h})$$

$$= 3600\,\text{Wh} = \mathbf{3.6\,kWh}$$

Now try the following Practice exercise

Practice Exercise 36 Series circuits (Answers on page 819)

1. The p.d.s measured across three resistors connected in series are 5 V, 7 V and 10 V, and the supply current is 2 A. Determine (a) the supply voltage, (b) the total circuit resistance and (c) the values of the three resistors.

2. For the circuit shown in Figure 7.9, determine the value of V_1. If the total circuit resistance is 36 Ω, determine the supply current and the value of resistors R_1, R_2 and R_3

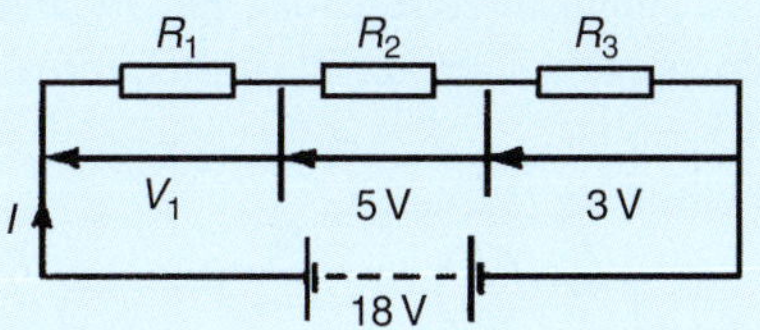

Figure 7.9

3. When the switch in the circuit in Figure 7.10 is closed the reading on voltmeter 1 is 30 V and that on voltmeter 2 is 10 V. Determine the reading on the ammeter and the value of resistor R_x

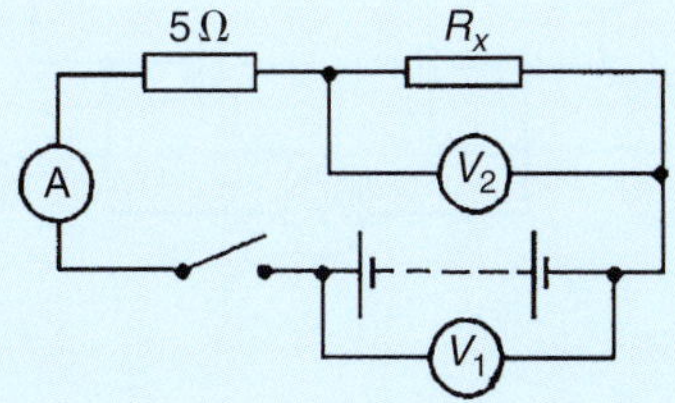

Figure 7.10

4. Calculate the value of voltage V in Figure 7.11.

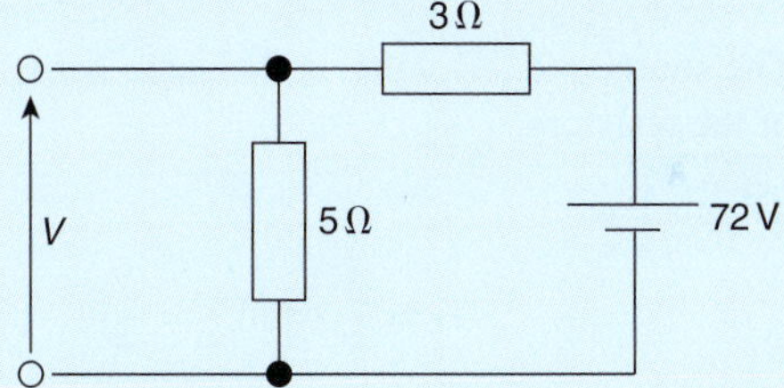

Figure 7.11

5. Two resistors are connected in series across an 18 V supply and a current of 5 A flows. If one of the resistors has a value of 2.4 Ω, determine (a) the value of the other resistor and (b) the p.d. across the 2.4 Ω resistor.

6. An arc lamp takes 9.6 A at 55 V. It is operated from a 120 V supply. Find the value of the stabilizing resistor to be connected in series.

7. An oven takes 15 A at 240 V. It is required to reduce the current to 12 A. Find (a) the resistor which must be connected in series, and (b) the voltage across the resistor.

7.3 Parallel networks

Figure 7.12 shows three resistors, R_1, R_2 and R_3, connected across each other, i.e. in parallel, across a battery source of V volts.

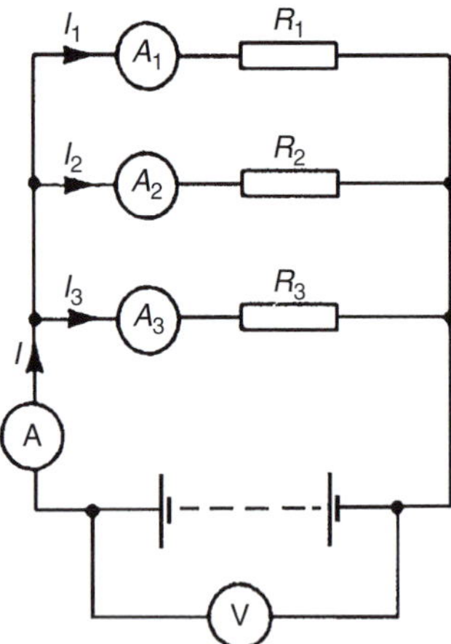

Figure 7.12

In a parallel circuit:

(a) The sum of the currents I_1, I_2 and I_3 is equal to the total circuit current, I, i.e. $\boldsymbol{I = I_1 + I_2 + I_3}$

(b) The source p.d., V volts, is the same across each of the resistors.

From Ohm's law:

$$I_1 = \frac{V}{R_1}, \quad I_2 = \frac{V}{R_2}, \quad I_3 = \frac{V}{R_3} \text{ and } I = \frac{V}{R}$$

where R is the total circuit resistance.

Since $\boldsymbol{I = I_1 + I_2 + I_3}$

then $\dfrac{V}{R} = \dfrac{V}{R_1} + \dfrac{V}{R_2} + \dfrac{V}{R_3}$

Dividing throughout by V gives:

$$\boldsymbol{\frac{1}{R} = \frac{1}{R_1} + \frac{1}{R_2} + \frac{1}{R_3}}$$

This equation must be used when finding the total resistance R of a parallel circuit. For the special case of **two resistors in parallel**

$$\frac{1}{R} = \frac{1}{R_1} + \frac{1}{R_2} = \frac{R_2 + R_1}{R_1 R_2}$$

Hence $\quad \boldsymbol{R = \frac{R_1 R_2}{R_1 + R_2}} \quad \left(\text{i.e. } \frac{\text{product}}{\text{sum}}\right)$

Problem 6. For the circuit shown in Figure 7.13, determine (a) the reading on the ammeter, and (b) the value of resistor R_2

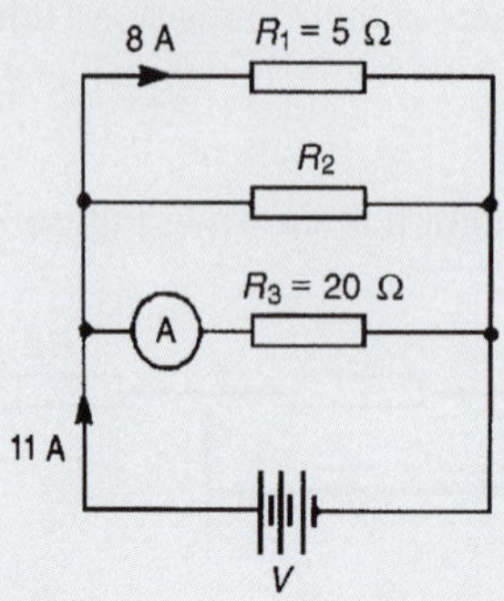

Figure 7.13

P.d. across R_1 is the same as the supply voltage V.

Hence supply voltage, $V = 8 \times 5 = 40\,\text{V}$

(a) Reading on ammeter, $I = \dfrac{V}{R_3} = \dfrac{40}{20} = \mathbf{2\,A}$

(b) Current flowing through $R_2 = 11 - 8 - 2 = 1\,\text{A}$

Hence, $R_2 = \dfrac{V}{I_2} = \dfrac{40}{1} = \mathbf{40\,\Omega}$

Problem 7. Two resistors, of resistance 3 Ω and 6 Ω, are connected in parallel across a battery having a voltage of 12 V. Determine (a) the total circuit resistance and (b) the current flowing in the 3 Ω resistor.

The circuit diagram is shown in Figure 7.14.

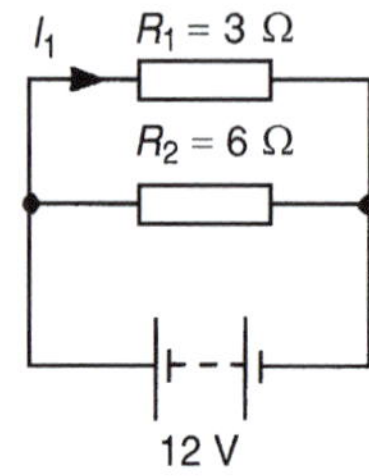

Figure 7.14

(a) The total circuit resistance R is given by

$$\frac{1}{R} = \frac{1}{R_1} + \frac{1}{R_2} = \frac{1}{3} + \frac{1}{6}$$

$$\frac{1}{R}=\frac{2+1}{6}=\frac{3}{6}$$

Hence, $R=\frac{6}{3}=\mathbf{2\,\Omega}$

$$\left(\text{Alternatively, } R=\frac{R_1R_2}{R_1+R_2}=\frac{3\times 6}{3+6}=\frac{18}{9}=\mathbf{2\,\Omega}\right)$$

(b) Current in the $3\,\Omega$ resistance, $I_1=\frac{V}{R_1}=\frac{12}{3}=\mathbf{4\,A}$

Problem 8. For the circuit shown in Figure 7.15, find (a) the value of the supply voltage V and (b) the value of current I.

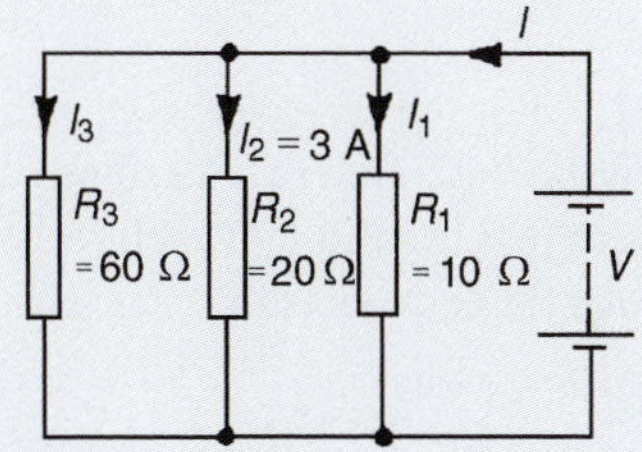

Figure 7.15

(a) P.d. across $20\,\Omega$ resistor $=I_2R_2=3\times 20=60\,\text{V}$, hence supply voltage $\mathbf{V=60\,V}$ since the circuit is connected in parallel.

(b) Current $I_1=\frac{V}{R_1}=\frac{60}{10}=6\,\text{A};\ I_2=3\,\text{A}$

$$I_3=\frac{V}{R_3}=\frac{60}{60}=1\,\text{A}$$

Current $I=I_1+I_2+I_3$ and hence $I=6+3+1$

$$=\mathbf{10\,A}$$

Alternatively,

$$\frac{1}{R}=\frac{1}{60}+\frac{1}{20}+\frac{1}{10}=\frac{1+3+6}{60}=\frac{10}{60}$$

Hence total resistance $R=\frac{60}{10}=6\,\Omega$

Current $I=\frac{V}{R}=\frac{60}{6}=\mathbf{10\,A}$

Problem 9. Given four $1\,\Omega$ resistors, state how they must be connected to give an overall resistance of (a) $\frac{1}{4}\,\Omega$ (b) $1\,\Omega$ (c) $1\frac{1}{3}\,\Omega$ (d) $2\frac{1}{2}\,\Omega$, all four resistors being connected in each case.

(a) **All four in parallel** (see Figure 7.16),

since $\frac{1}{R}=\frac{1}{1}+\frac{1}{1}+\frac{1}{1}+\frac{1}{1}=\frac{4}{1}$, i.e. $R=\frac{1}{4}\,\Omega$

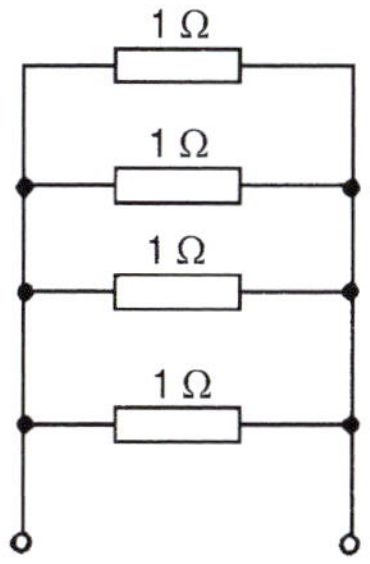

Figure 7.16

(b) **Two in series, in parallel with another two in series** (see Figure 7.17), since $1\,\Omega$ and $1\,\Omega$ in series gives $2\,\Omega$, and $2\,\Omega$ in parallel with $2\,\Omega$ gives:

$$\frac{2\times 2}{2+2}=\frac{4}{4}=1\,\Omega$$

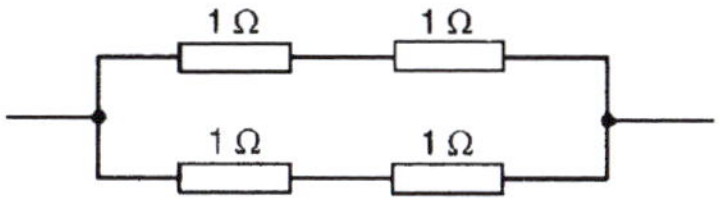

Figure 7.17

(c) **Three in parallel, in series with one** (see Figure 7.18), since for the three in parallel,

$\frac{1}{R}=\frac{1}{1}+\frac{1}{1}+\frac{1}{1}=\frac{3}{1}$, i.e. $R=\frac{1}{3}\,\Omega$ and $\frac{1}{3}\,\Omega$ in series with $1\,\Omega$ gives $1\frac{1}{3}\,\Omega$

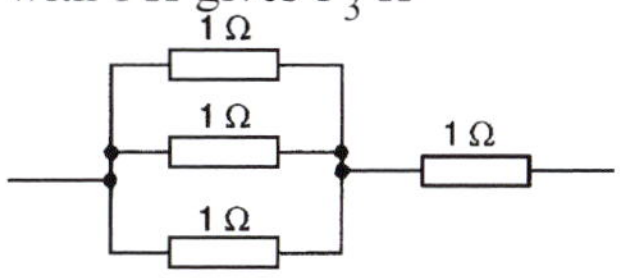

Figure 7.18

(d) **Two in parallel, in series with two in series** (see Figure 7.19), since for the two in parallel $R=\frac{1\times 1}{1+1}=\frac{1}{2}\,\Omega$, and $\frac{1}{2}\,\Omega$, $1\,\Omega$ and $1\,\Omega$ in series gives $2\frac{1}{2}\,\Omega$

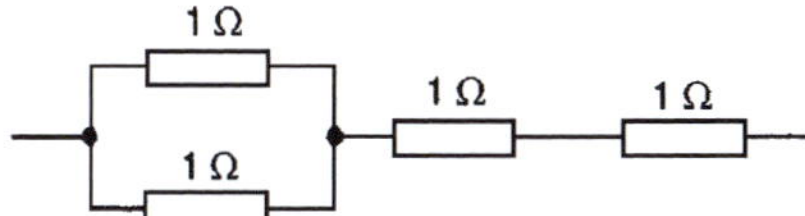

Figure 7.19

Part 2

Problem 10. Find the equivalent resistance for the circuit shown in Figure 7.20.

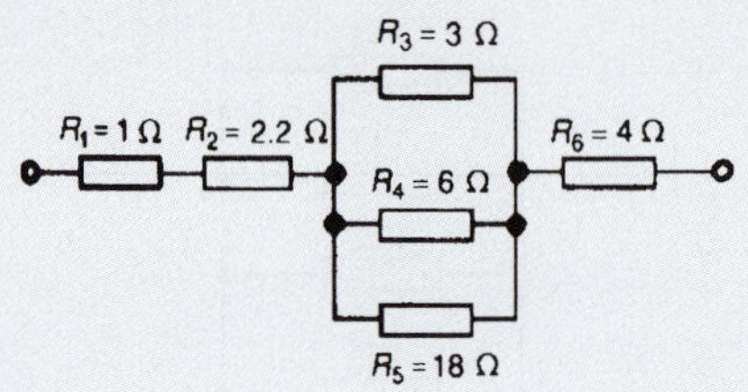

Figure 7.20

R_3, R_4 and R_5 are connected in parallel and their equivalent resistance R is given by:

$$\frac{1}{R} = \frac{1}{3} + \frac{1}{6} + \frac{1}{18} = \frac{6+3+1}{18} = \frac{10}{18}$$

$$\text{Hence } R = \frac{18}{10} = 1.8\,\Omega$$

The circuit is now equivalent to four resistors in series and the equivalent circuit resistance

$$= 1 + 2.2 + 1.8 + 4 = \mathbf{9\,\Omega}$$

7.4 Current division

For the circuit shown in Figure 7.21, the total circuit resistance, R_T, is given by:

$$R_T = \frac{R_1 R_2}{R_1 + R_2}$$

$$\text{and } V = IR_T = I\left(\frac{R_1 R_2}{R_1 + R_2}\right)$$

$$\text{Current } I_1 = \frac{V}{R_1} = \frac{I}{R_1}\left(\frac{R_1 R_2}{R_1 + R_2}\right) = \left(\frac{\boldsymbol{R_2}}{\boldsymbol{R_1 + R_2}}\right)(\boldsymbol{I})$$

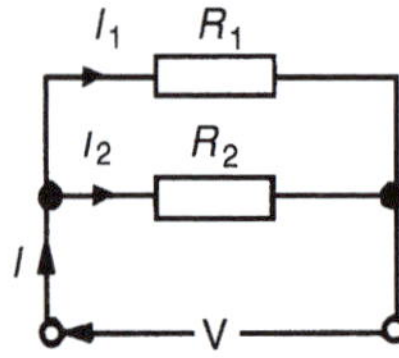

Figure 7.21

Similarly,

$$\text{current } I_2 = \frac{V}{R_2} = \frac{I}{R_2}\left(\frac{R_1 R_2}{R_1 + R_2}\right) = \left(\frac{\boldsymbol{R_1}}{\boldsymbol{R_1 + R_2}}\right)(\boldsymbol{I})$$

Summarizing, with reference to Figure 7.21

$$\boldsymbol{I_1} = \left(\frac{\boldsymbol{R_2}}{\boldsymbol{R_1 + R_2}}\right)(\boldsymbol{I}) \text{ and } \boldsymbol{I_2} = \left(\frac{\boldsymbol{R_1}}{\boldsymbol{R_1 + R_2}}\right)(\boldsymbol{I})$$

It is important to note that current division can only be applied to **two** parallel resistors. If there are more than two parallel resistors, then current division cannot be determined using the above formulae.

Problem 11. For the series-parallel arrangement shown in Figure 7.22, find (a) the supply current, (b) the current flowing through each resistor and (c) the p.d. across each resistor.

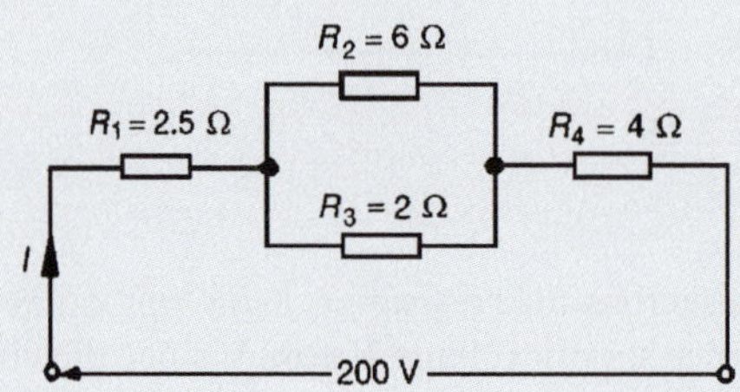

Figure 7.22

(a) The equivalent resistance R_x of R_2 and R_3 in parallel is:

$$R_x = \frac{6 \times 2}{6+2} = \frac{12}{8} = 1.5\,\Omega$$

The equivalent resistance R_T of R_1, R_x and R_4 in series is:

$$R_T = 2.5 + 1.5 + 4 = 8\,\Omega$$

$$\text{Supply current } I = \frac{V}{R_T} = \frac{200}{8} = \mathbf{25\,A}$$

(b) The current flowing through R_1 and R_4 is 25 A
The current flowing through R_2

$$= \left(\frac{R_3}{R_2 + R_3}\right) I = \left(\frac{2}{6+2}\right) 25$$

$$= \mathbf{6.25\,A}$$

The current flowing through R_3

$$= \left(\frac{R_2}{R_2 + R_3}\right) I = \left(\frac{6}{6+2}\right) 25$$

$$= \mathbf{18.75\,A}$$

(Note that the currents flowing through R_2 and R_3 must add up to the total current flowing into the parallel arrangement, i.e. 25 A)

(c) The equivalent circuit of Figure 7.22 is shown in Figure 7.23.

p.d. across R_1, i.e. $V_1 = IR_1 = (25)(2.5) = \mathbf{62.5\,V}$

p.d. across R_x, i.e. $V_x = IR_x = (25)(1.5) = \mathbf{37.5\,V}$

p.d. across R_4, i.e. $V_4 = IR_4 = (25)(4) = \mathbf{100\,V}$

Hence the p.d. across R_2 = p.d. across $R_3 = \mathbf{37.5\,V}$

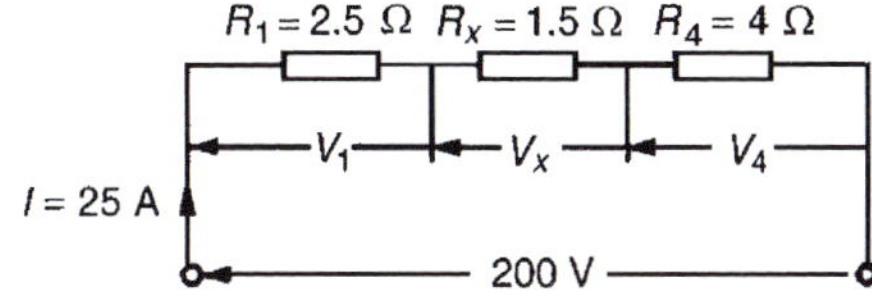

Figure 7.23

Problem 12. For the circuit shown in Figure 7.24 calculate (a) the value of resistor R_x such that the total power dissipated in the circuit is 2.5 kW, and (b) the current flowing in each of the four resistors.

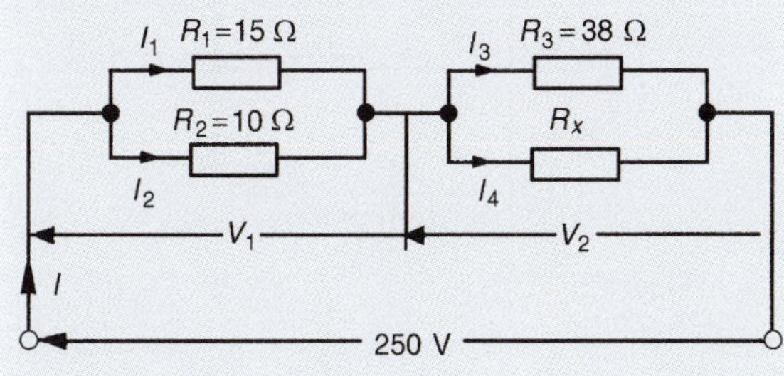

Figure 7.24

(a) Power dissipated $P = VI$ watts, hence $2500 = (250)(I)$

i.e. $I = \dfrac{2500}{250} = 10\,\text{A}$

From Ohm's law, $R_T = \dfrac{V}{I} = \dfrac{250}{10} = 25\,\Omega$, where R_T is the equivalent circuit resistance.

The equivalent resistance of R_1 and R_2 in parallel is

$$\frac{15 \times 10}{15 + 10} = \frac{150}{25} = 6\,\Omega$$

The equivalent resistance of resistors R_3 and R_x in parallel is equal to $25\,\Omega - 6\,\Omega$, i.e. $19\,\Omega$.
There are three methods whereby R_x can be determined.

Method 1

The voltage $V_1 = IR$, where R is $6\,\Omega$, from above, i.e. $V_1 = (10)(6) = 60\,\text{V}$

Hence $V_2 = 250\,\text{V} - 60\,\text{V} = 190\,\text{V}$ = p.d. across R_3
= p.d. across R_x

$I_3 = \dfrac{V_2}{R_3} = \dfrac{190}{38} = 5\,\text{A}$. Thus $I_4 = 5\,\text{A}$ also,

since $I = 10\,\text{A}$

Thus $\boldsymbol{R_x} = \dfrac{V_2}{I_4} = \dfrac{190}{5} = \mathbf{38\,\Omega}$

Method 2

Since the equivalent resistance of R_3 and R_x in parallel is $19\,\Omega$,

$$19 = \frac{38R_x}{38 + R_x} \quad \left(\text{i.e. } \frac{\text{product}}{\text{sum}}\right)$$

Hence $19(38 + R_x) = 38R_x$

$$722 + 19R_x = 38R_x$$

$$722 = 38R_x - 19R_x = 19R_x$$

Thus $\boldsymbol{R_x} = \dfrac{722}{19} = \mathbf{38\,\Omega}$

Method 3

When two resistors having the same value are connected in parallel the equivalent resistance is always half the value of one of the resistors. Thus, in this case, since $R_T = 19\,\Omega$ and $R_3 = 38\,\Omega$, then $\mathbf{R_x = 38\,\Omega}$ could have been deduced on sight.

(b) Current $I_1 = \left(\dfrac{R_2}{R_1 + R_2}\right) I = \left(\dfrac{10}{15 + 10}\right)(10)$

$$= \left(\frac{2}{5}\right)(10) = \mathbf{4\,A}$$

$$\text{Current } I_2 = \left(\frac{R_1}{R_1+R_2}\right) I = \left(\frac{15}{15+10}\right)(10)$$

$$= \left(\frac{3}{5}\right)(10) = \mathbf{6\,A}$$

From part (a), method 1, $\boldsymbol{I_3 = I_4 = 5\,A}$

Problem 13. For the arrangement shown in Figure 7.25, find the current I_x.

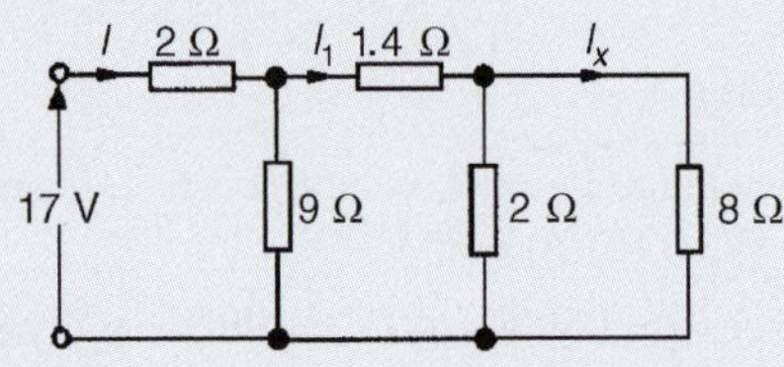

Figure 7.25

Commencing at the right-hand side of the arrangement shown in Figure 7.25, the circuit is gradually reduced in stages as shown in Figure 7.26(a)–(d).

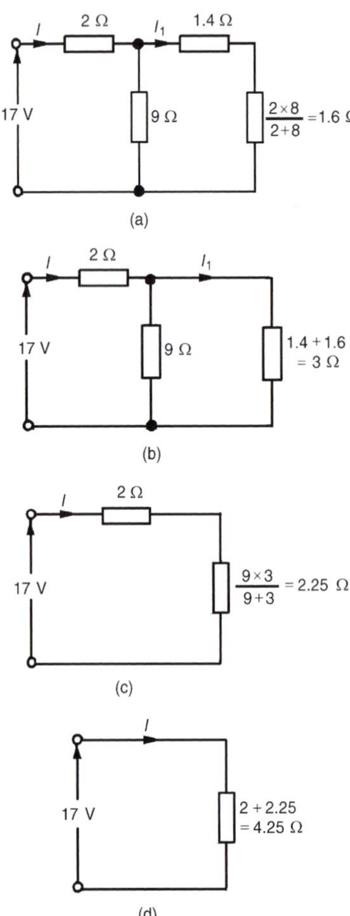

Figure 7.26

From Figure 7.26(d), $I = \dfrac{17}{4.25} = 4\,\text{A}$

From Figure 7.26(b), $I_1 = \left(\dfrac{9}{9+3}\right)(I) = \left(\dfrac{9}{12}\right)(4)$

$= 3\,\text{A}$

From Figure 7.25, $I_x = \left(\dfrac{2}{2+8}\right)(I_1) = \left(\dfrac{2}{10}\right)(3)$

$= \mathbf{0.6\,A}$

For a practical laboratory experiment on series-parallel d.c. circuits, see the website.

Now try the following Practice Exercise

Practice Exercise 37 Parallel networks (Answers on page 819)

1. Resistances of 4 Ω and 12 Ω are connected in parallel across a 9 V battery. Determine (a) the equivalent circuit resistance, (b) the supply current and (c) the current in each resistor.

2. For the circuit shown in Figure 7.27 determine (a) the reading on the ammeter, and (b) the value of resistor R.

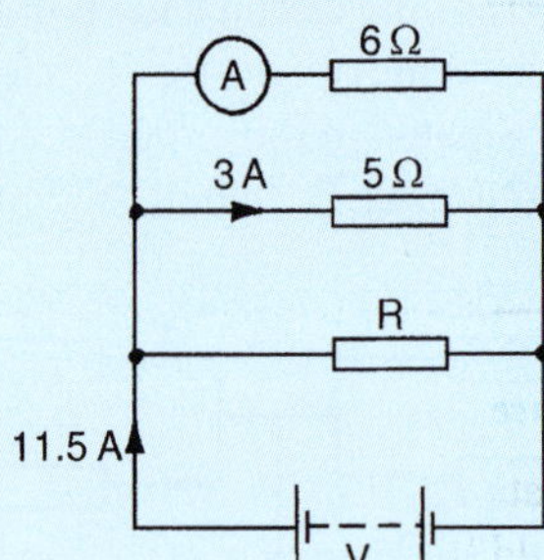

Figure 7.27

3. Find the equivalent resistance when the following resistances are connected (a) in series, (b) in parallel.
(i) 3 Ω and 2 Ω (ii) 20 kΩ and 40 kΩ
(iii) 4 Ω, 8 Ω and 16 Ω
(iv) 800 Ω, 4 kΩ and 1500 Ω

4. Find the total resistance between terminals A and B of the circuit shown in Figure 7.28(a).

5. Find the equivalent resistance between terminals C and D of the circuit shown in Figure 7.28(b).

Part 2

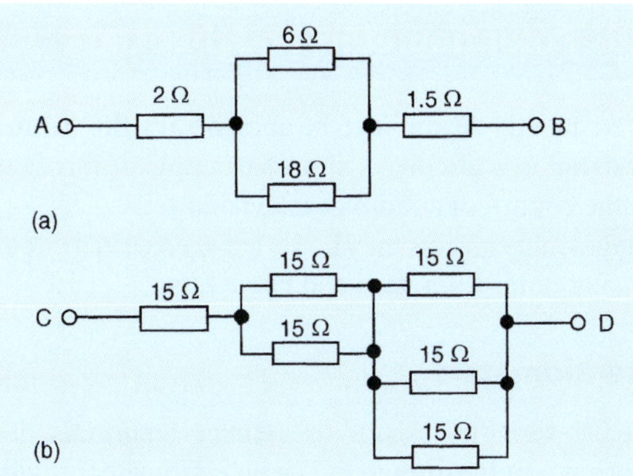

Figure 7.28

6. Resistors of 20 Ω, 20 Ω and 30 Ω are connected in parallel. What resistance must be added in series with the combination to obtain a total resistance of 10 Ω. If the complete circuit expends a power of 0.36 kW, find the total current flowing.

7. (a) Calculate the current flowing in the 30 Ω resistor shown in Figure 7.29.

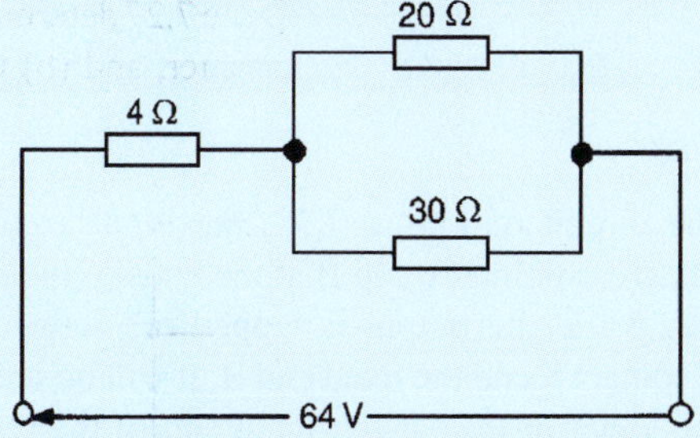

Figure 7.29

(b) What additional value of resistance would have to be placed in parallel with the 20 Ω and 30 Ω resistors to change the supply current to 8 A, the supply voltage remaining constant.

8. Determine the currents and voltages indicated in the circuit shown in Figure 7.30.

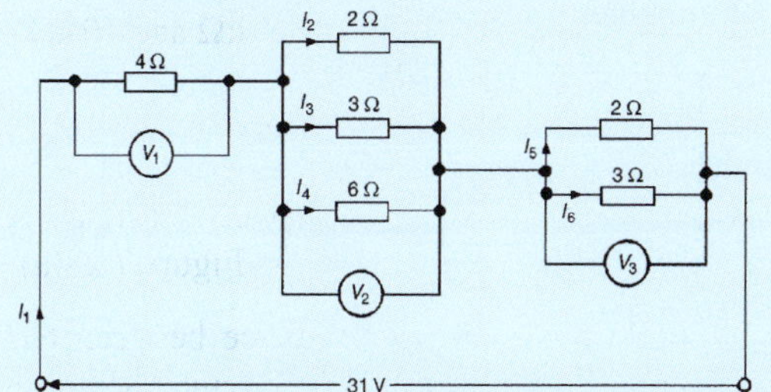

Figure 7.30

9. Find the current I in Figure 7.31.

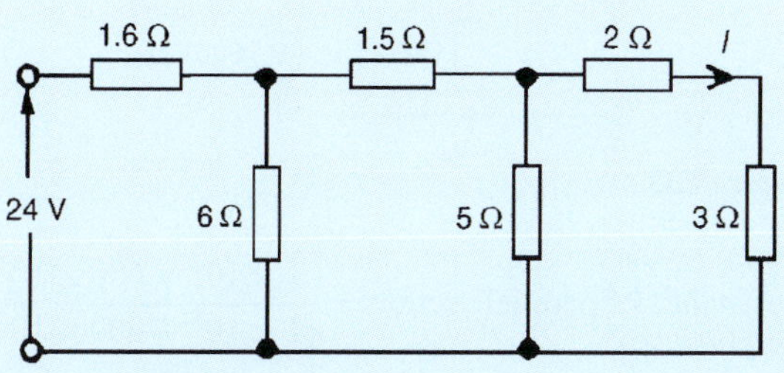

Figure 7.31

10. A resistor of 2.4 Ω is connected in series with another of 3.2 Ω. What resistance must be placed across the one of 2.4 Ω so that the total resistance of the circuit shall be 5 Ω.

11. A resistor of 8 Ω is connected in parallel with one of 12 Ω and the combination is connected in series with one of 4 Ω. A p.d. of 10 V is applied to the circuit. The 8 Ω resistor is now placed across the 4 Ω resistor. Find the p.d. required to send the same current through the 8 Ω resistor.

7.5 Loading effect

Loading effect is the terminology used when a measuring instrument such as an oscilloscope or voltmeter is connected across a component and the current drawn by the instrument upsets the circuit under test. The best way of demonstrating loading effect is by a numerical example.

In the simple circuit of Figure 7.32, the voltage across each of the resistors can be calculated using voltage division, or by inspection. In this case, the voltage shown as V should be 20 V.

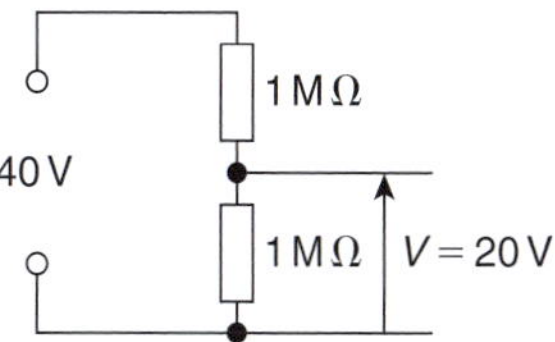

Figure 7.32

Using a voltmeter having a resistance of, say, 600 kΩ places 600 kΩ in parallel with the 1 MΩ resistor, as shown in Figure 7.33.

Part 2

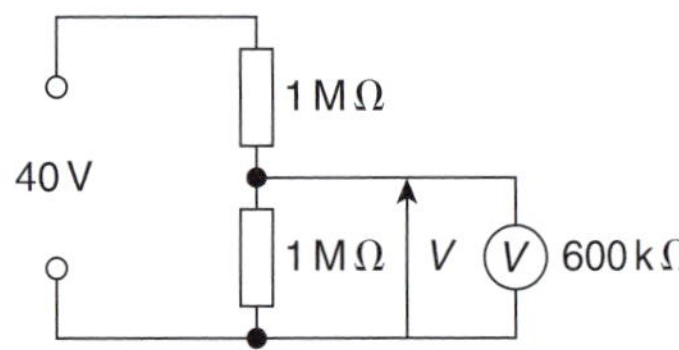

Figure 7.33

$$\text{Resistance of parallel section} = \frac{1\times10^6\times600\times10^3}{(1\times10^6+600\times10^3)}$$

$$= 375\ \text{k}\Omega\ \text{(using product/sum)}$$

$$\text{The voltage } V \text{ now equals } \frac{375\times10^3}{(1\times10^6+375\times10^3)}\times40$$

$$= \mathbf{10.91\ V}\ \text{(by voltage division)}$$

The voltmeter has loaded the circuit by drawing current for its operation, and by so doing, reduces the voltage across the 1 MΩ resistor from the correct value of 20 V to 10.91 V.

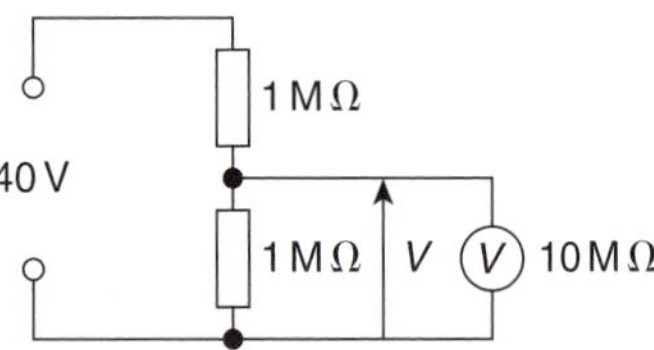

Figure 7.34

Using a Fluke (or multimeter) which has a set internal resistance of, say, 10 MΩ, as shown in Figure 7.34, produces a much better result and the loading effect is minimal, as shown below.

$$\text{Resistance of parallel section} = \frac{1\times10^6\times10\times10^6}{(1\times10^6+10\times10^6)}$$

$$= 0.91\text{M}\Omega$$

$$\text{The voltage } V \text{ now equals } \frac{0.91\times10^6}{(0.91\times10^6+1\times10^6)}\times40$$

$$= \mathbf{19.06\ V}$$

When taking measurements, it is vital that the loading effect is understood and kept in mind at all times. An incorrect voltage reading may be due to this loading effect rather than the equipment under investigation being defective. Ideally, **the resistance of a voltmeter should be infinite**.

7.6 Potentiometers and rheostats

It is frequently desirable to be able to **vary the value of a resistor** in a circuit. A simple example of this is the volume control of a radio or television set.
Voltages and currents may be varied in electrical circuits by using **potentiometers** and **rheostats**.

Potentiometers

When a variable resistor uses **three terminals**, it is known as a **potentiometer**. The potentiometer provides an adjustable voltage divider circuit, which is useful as a means of obtaining **various voltages** from a fixed potential difference. Consider the potentiometer circuit shown in Figure 7.35 incorporating a lamp and supply voltage V.

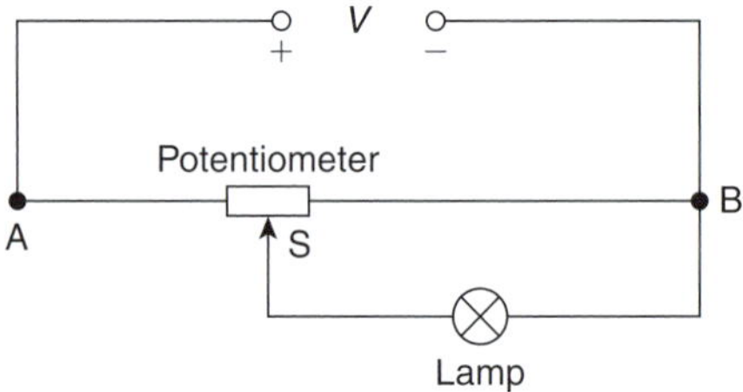

Figure 7.35

In the circuit of Figure 7.35, the input voltage is applied across points A and B at the ends of the potentiometer, while the output is tapped off between the sliding contact S and the fixed end B. It will be seen that with the slider at the far left-hand end of the resistor, the full voltage will appear across the lamp, and as the slider is moved towards point B the lamp brightness will reduce. When S is at the far right of the potentiometer, the lamp is short-circuited, no current will flow through it, and the lamp will be fully off.

Problem 14. Calculate the volt drop across the 60 Ω load in the circuit shown in Figure 7.36 when the slider S is at the halfway point of the 200 Ω potentiometer.

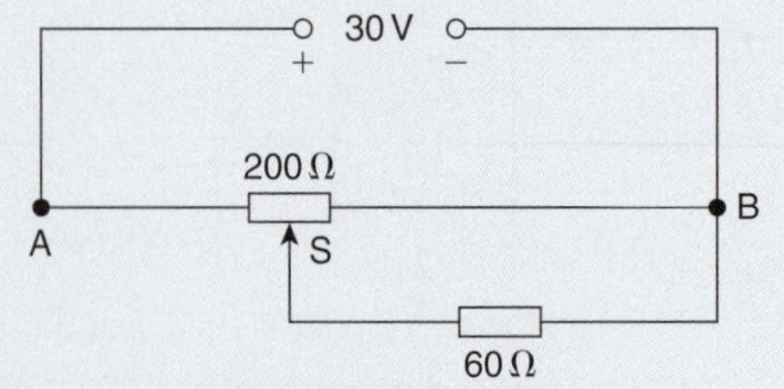

Figure 7.36

With the slider halfway, the equivalent circuit is shown in Figure 7.37.

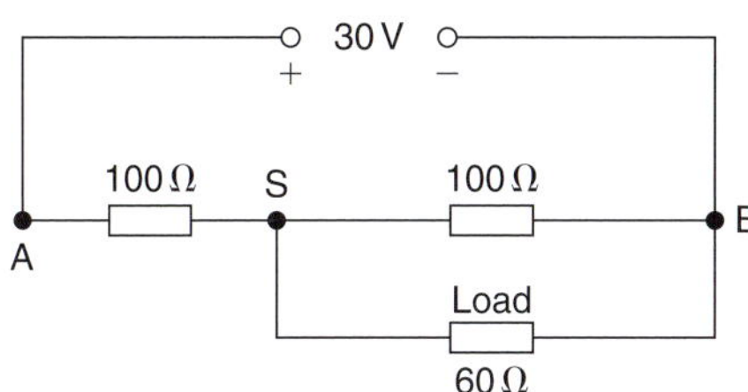

Figure 7.37

For the parallel resistors, total resistance,

$$R_P = \frac{100 \times 60}{100 + 60} = \frac{100 \times 60}{160} = 37.5\ \Omega$$

(or use $\frac{1}{R_P} = \frac{1}{100} + \frac{1}{60}$ to determine R_p)

The equivalent circuit is now as shown in Figure 7.38.

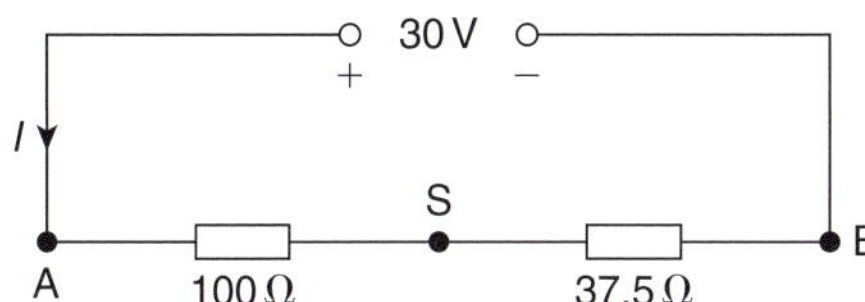

Figure 7.38

The volt drop across the 37.5 Ω resistor in Figure 7.38 is the same as the volt drop across both of the parallel resistors in Figure 7.37.

There are two methods for determining the volt drop, V_{SB}:

Method 1

Total circuit resistance, $R_T = 100 + 37.5 = 137.5\,\Omega$

Hence, supply current, $I = \frac{30}{137.5} = 0.2182\,\text{A}$

Thus, volt drop, $V_{SB} = I \times 37.5 = 0.2182 \times 37.5 = \mathbf{8.18\ V}$

Method 2

By the principle of voltage division,

$$V_{SB} = \left(\frac{37.5}{100 + 37.5}\right)(30) = 8.18\,\text{V}$$

Hence, **the volt drop across the 60 Ω load of Figure 7.36 is 8.18 V**

Rheostats

A variable resistor where only **two terminals** are used, one fixed and one sliding, is known as a **rheostat**. The rheostat circuit, shown in Figure 7.39, similar in construction to the potentiometer, is used to **control current flow**. The rheostat also acts as a dropping resistor, reducing the voltage across the load, but is more effective at controlling current.

For this reason **the resistance of the rheostat should be greater than that of the load**, otherwise it will have little or no effect. Typical uses are in a train set or Scalextric. Another practical example is in varying the brilliance of the panel lighting controls in a car.

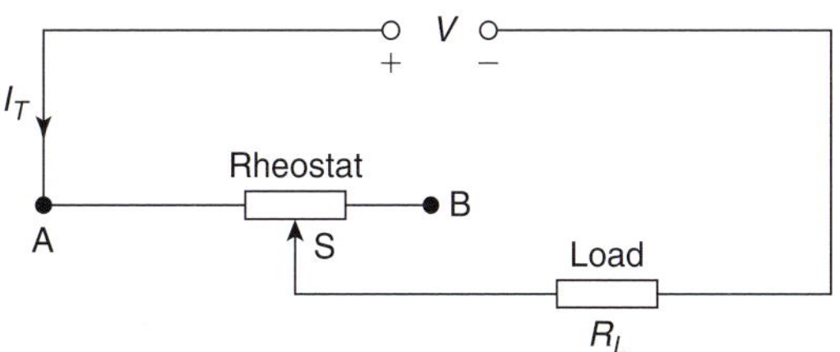

Figure 7.39

The rheostat resistance is connected in series with the load circuit, R_L, with the slider arm tapping off an amount of resistance (i.e. that between A and S) to provide the current flow required. With the slider at the far left-hand end, the load receives maximum current; with the slider at the far right-hand end, minimum current flows. The current flowing can be calculated by finding the total resistance of the circuit (i.e. $R_T = R_{AS} + R_L$), then by applying Ohm's law, $\boldsymbol{I_T = \frac{V}{R_{AS} + R_L}}$

Calculations involved with the rheostat circuit are simpler than those for the potentiometer circuit.

Problem 15. In the circuit of Figure 7.40, calculate the current flowing in the 100 Ω load, when the sliding point S is 2/3 of the way from A to B.

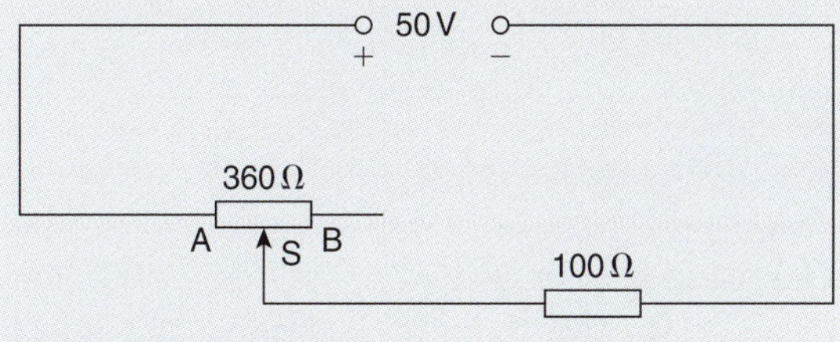

Figure 7.40

Resistance, $R_{AS} = \frac{2}{3} \times 360 = 240\,\Omega$

Total circuit resistance, $R_T = R_{AS} + R_L$
$= 240 + 100 = 340\,\Omega$

Current flowing in load,

$$I = \frac{V}{R_T} = \frac{50}{340} = \mathbf{0.147\,A} \text{ or } \mathbf{147\,mA}$$

Summary

A **potentiometer** (a) has three terminals, and (b) is used for voltage control.
A **rheostat** (a) has two terminals, (b) is used for current control.
A rheostat is not suitable if the load resistance is higher than the rheostat resistance; rheostat resistance must be higher than the load resistance to be able to influence current flow.

Now try the following Practice Exercise

Practice Exercise 38 Potentiometers and rheostats (Answers on page 819)

1. For the circuit shown in Figure 7.41, AS is 3/5 of AB. Determine the voltage across the 120 Ω load. Is this a potentiometer or a rheostat circuit?

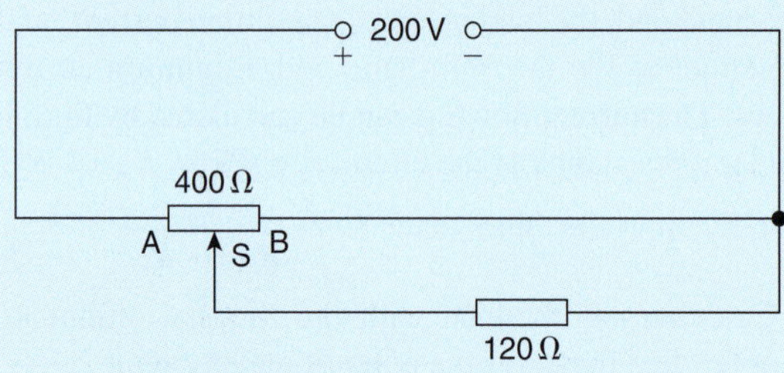

Figure 7.41

2. For the circuit shown in Figure 7.42, calculate the current flowing in the 25 Ω load and the voltage drop across the load when (a) AS is half of AB, (b) point S coincides with point B. Is this a potentiometer or a rheostat?

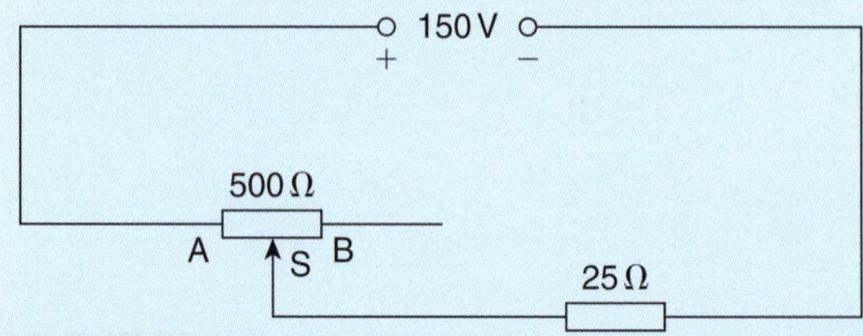

Figure 7.42

3. For the circuit shown in Figure 7.43, calculate the voltage across the 600 Ω load when point S splits AB in the ratio 1:3

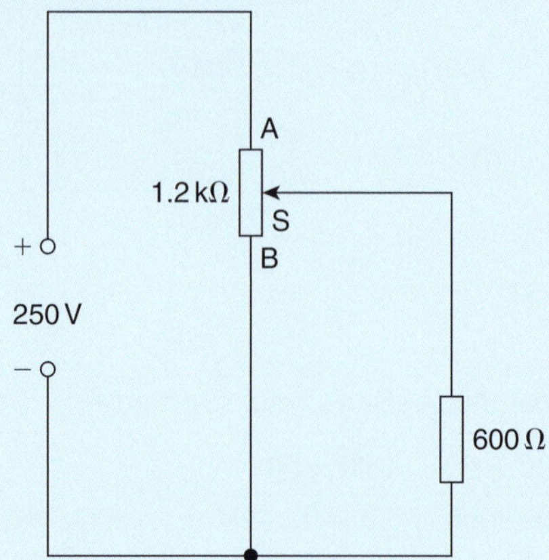

Figure 7.43

4. For the circuit shown in Figure 7.44, the slider S is set at halfway. Calculate the voltage drop across the 120 Ω load.

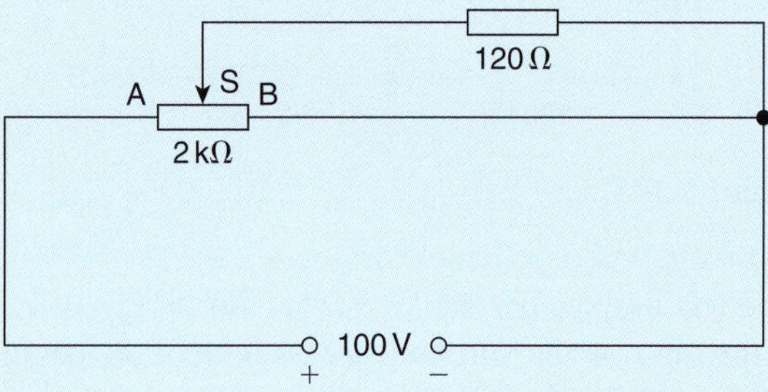

Figure 7.44

5. For the potentiometer circuit shown in Figure 7.45, AS is 60% of AB. Calculate the voltage across the 70 Ω load.

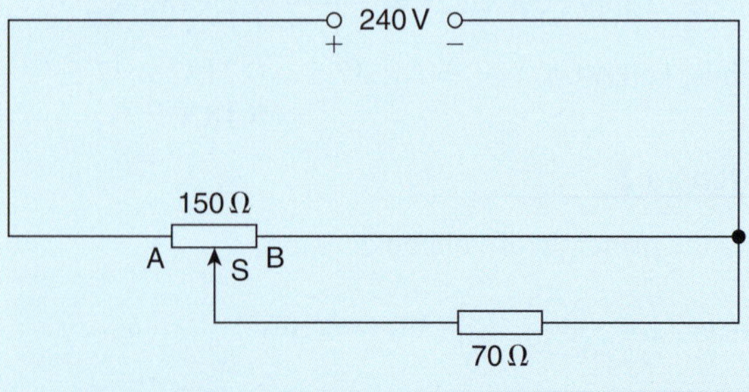

Figure 7.45

Part 2

7.7 Relative and absolute voltages

In an electrical circuit, the voltage at any point can be quoted as being 'with reference to' (w.r.t.) any other point in the circuit. Consider the circuit shown in Figure 7.46. The total resistance,

$$R_T = 30 + 50 + 5 + 15 = 100\,\Omega$$

$$\text{and current, } I = \frac{200}{100} = 2\,\text{A}$$

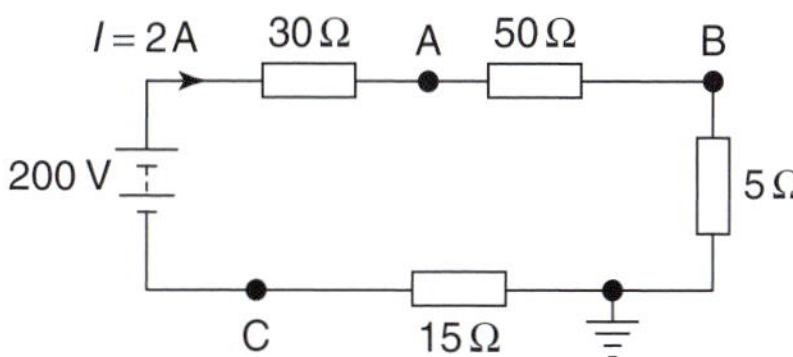

Figure 7.46

If a voltage at point A is quoted with reference to point B then the voltage is written as V_{AB}. This is known as a **'relative voltage'**. In the circuit shown in Figure 7.46, the voltage at A w.r.t. B is $I \times 50$, i.e. $2 \times 50 = 100\,\text{V}$ and is written as $V_{AB} = 100\,\text{V}$

It must also be indicated whether the voltage at A w.r.t. B is closer to the positive terminal or the negative terminal of the supply source. Point A is nearer to the positive terminal than B so is written as $V_{AB} = 100\,\text{V}$ or $V_{AB} = +100\,\text{V}$ or $V_{AB} = 100\,\text{V}$ +ve.

If no positive or negative is included, then the voltage is always taken to be positive.

If the voltage at B w.r.t. A is required, then V_{BA} is negative and written as $V_{BA} = -100\,\text{V}$ or $V_{BA} = 100\,\text{V}$ −ve.

If the reference point is changed to the **earth point** then any voltage taken w.r.t. the earth is known as an **'absolute potential'**. If the absolute voltage of A in Figure 7.46 is required, then this will be the sum of the voltages across the 50 Ω and 5 Ω resistors, i.e. $100 + 10 = 110\,\text{V}$ and is written as $V_A = 110\,\text{V}$ or $V_A = +110\,\text{V}$ or $V_A = 110\,\text{V}$ +ve, positive since moving from the earth point to point A is moving towards the positive terminal of the source. If the voltage is negative w.r.t. earth then this must be indicated; for example, $V_C = 30\,\text{V}$ negative w.r.t. earth, and is written as $V_C = -30\,\text{V}$ or $V_C = 30\,\text{V}$ −ve.

Problem 16. For the circuit shown in Figure 7.47, calculate (a) the voltage drop across the 4 kΩ resistor, (b) the current through the 5 kΩ resistor, (c) the power developed in the 1.5 kΩ resistor, (d) the voltage at point X w.r.t. earth, and (e) the absolute voltage at point X.

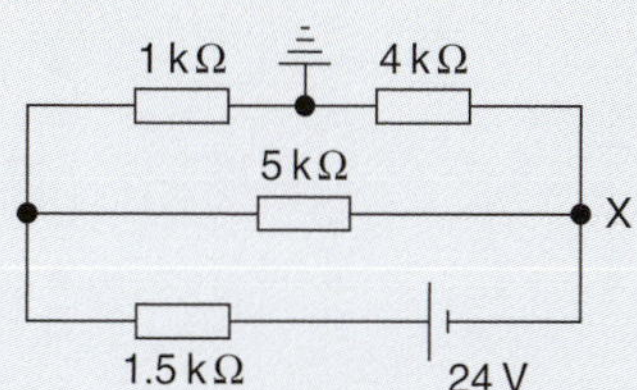

Figure 7.47

(a) Total circuit resistance, $R_T = [(1 + 4)\text{k}\Omega$ in parallel with $5\,\text{k}\Omega]$ in series with $1.5\,\text{k}\Omega$ Total circuit current, $I_T = \frac{V}{R_T} = \frac{24}{4 \times 10^3} = 6\,\text{mA}$

By current division, current in top branch

$$= \left(\frac{5}{5 + 1 + 4}\right) \times 6 = 3\,\text{mA}$$

Hence, **volt drop across 4 kΩ resistor**
$$= 3 \times 10^{-3} \times 4 \times 10^3 = \mathbf{12\,V}$$

(b) **Current through the 5 kΩ resistor**

$$= \left(\frac{1 + 4}{5 + 1 + 4}\right) \times 6 = \mathbf{3\,mA}$$

(c) **Power in the 1.5 kΩ resistor**
$= I_T^2 R = (6 \times 10^{-3})^2 (1.5 \times 10^3) = \mathbf{54\,mW}$

(d) The voltage at the earth point is 0 volts. The volt drop across the 4 kΩ is 12 V, from part (a). Since moving from the earth point to point X is moving towards the negative terminal of the voltage source, the voltage at point X w.r.t. earth is **−12 V**.

(e) The 'absolute voltage at point X' means the 'voltage at point X w.r.t. earth', hence **the absolute voltage at point X is −12 V**. Questions (d) and (e) mean the same thing.

Now try the following Practice Exercise

Practice Exercise 39 Relative and absolute voltages (Answers on page 819)

1. For the circuit of Figure 7.48, calculate (a) the absolute voltage at points A, B and C,

Part 2

(b) the voltage at A relative to B and C, and (c) the voltage at D relative to B and A.

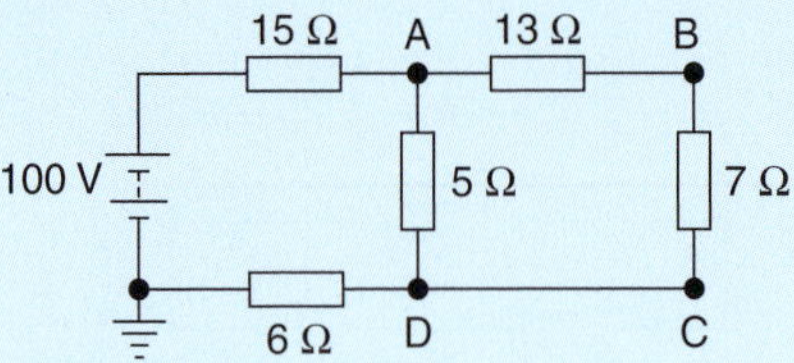

Figure 7.48

2. For the circuit shown in Figure 7.49, calculate (a) the voltage drop across the 7 Ω resistor, (b) the current through the 30 Ω resistor, (c) the power developed in the 8 Ω resistor, (d) the voltage at point X w.r.t. earth, and (e) the absolute voltage at point X.

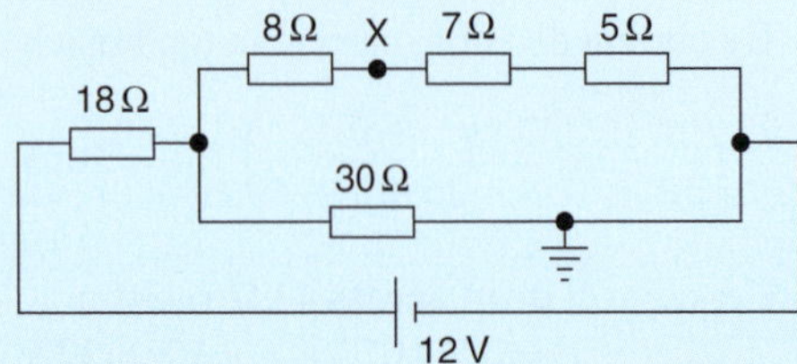

Figure 7.49

3. In the bridge circuit of Figure 7.50 calculate (a) the absolute voltages at points A and B, and (b) the voltage at A relative to B.

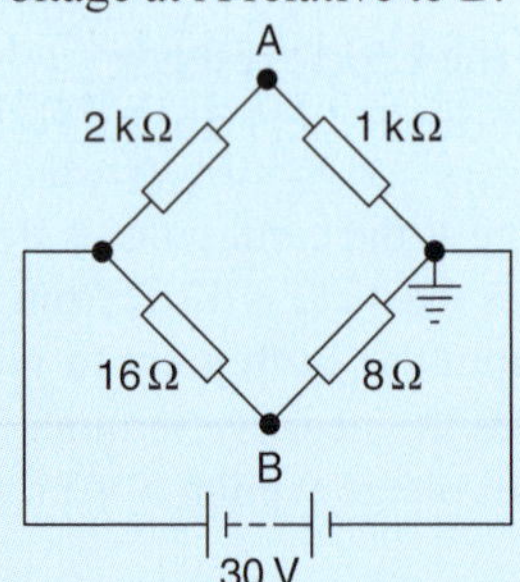

Figure 7.50

7.8 Earth potential and short circuits

The earth, and hence the sea, is at a potential of zero volts. Items connected to the earth (or sea), i.e. circuit wiring and electrical components, are said to be earthed or at earth potential. This means that there is no difference of potential between the item and earth. A ship's hull, being immersed in the sea, is at earth potential and therefore at zero volts. Earth faults, or short circuits, are caused by low resistance between the current-carrying conductor and earth. This occurs when the insulation resistance of the circuit wiring decreases, and is normally caused by:

1. dampness
2. insulation becoming hard or brittle with age or heat
3. accidental damage.

7.9 Wiring lamps in series and in parallel

Series connection

Figure 7.51 shows three lamps, each rated at 240 V, connected in series across a 240 V supply.

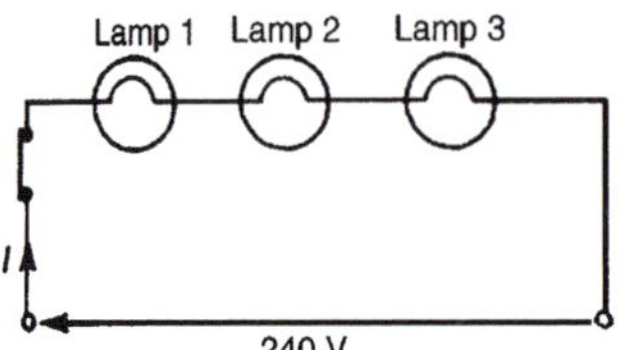

Figure 7.51

(i) Each lamp has only $\frac{240}{3}$ V, i.e. 80 V across it and thus each lamp glows dimly.

(ii) If another lamp of similar rating is added in series with the other three lamps then each lamp now has $\frac{240}{4}$ V, i.e. 60 V across it and each now glows even more dimly.

(iii) If a lamp is removed from the circuit or if a lamp develops a fault (i.e. an open circuit) or if the switch is opened, then the circuit is broken, no current flows, and the remaining lamps will not light up.

(iv) Less cable is required for a series connection than for a parallel one.

The series connection of lamps is usually limited to decorative lighting such as for Christmas tree lights.

Parallel connection

Figure 7.52 shows three similar lamps, each rated at 240 V, connected in parallel across a 240 V supply.

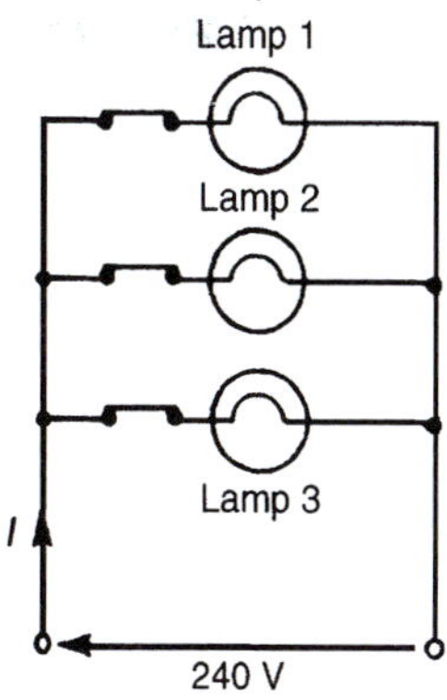

Figure 7.52

(i) Each lamp has 240 V across it and thus each will glow brilliantly at their rated voltage.

(ii) If any lamp is removed from the circuit or develops a fault (open circuit) or a switch is opened, the remaining lamps are unaffected.

(iii) The addition of further similar lamps in parallel does not affect the brightness of the other lamps.

(iv) More cable is required for a parallel connection than for a series one.

The parallel connection of lamps is the most widely used in electrical installations.

Problem 17. If three identical lamps are connected in parallel and the combined resistance is 150 Ω, find the resistance of one lamp.

Let the resistance of one lamp be R, then,

$$\frac{1}{150}=\frac{1}{R}+\frac{1}{R}+\frac{1}{R}=\frac{3}{R}, \text{ from which, } R=3\times 150$$
$$=\mathbf{450\,\Omega}$$

Problem 18. Three identical lamps A, B and C, are connected in series across a 150 V supply. State (a) the voltage across each lamp, and (b) the effect of lamp C failing.

(a) Since each lamp is identical and they are connected in series there is $\frac{150}{3}$ V, i.e. **50 V across each**.

(b) If lamp C fails, i.e. open circuits, no current will flow and **lamps A and B will not operate**.

Now try the following Practice Exercise

Practice Exercise 40 Wiring lamps in series and parallel (Answers on page 819)

1. If four identical lamps are connected in parallel and the combined resistance is 100 Ω, find the resistance of one lamp.

2. Three identical filament lamps are connected (a) in series, (b) in parallel across a 210 V supply. State for each connection the p.d. across each lamp.

Part 2

For fully worked solutions to each of the problems in Practice Exercises 36 to 40 in this chapter, go to the website:
www.routledge.com/cw/bird

Chapter 8

Capacitors and capacitance

***Why it is important to understand:* Capacitance and capacitors**

The capacitor is a widely used electrical component and it has several features that make it useful and important. A capacitor can store energy, so capacitors are often found in power supplies. Capacitors are used for timing – for example with a 555 timer IC controlling the charging and discharging; for smoothing – for example in a power supply; for coupling – for example between stages of an audio system and a loudspeaker; for filtering – for example in the tone control of an audio system; for tuning – for example in a radio system; and for storing energy – for example in a camera flash circuit. Capacitors find uses in virtually every form of electronics circuit from analogue circuits, including amplifiers and power supplies, through to oscillators, integrators and many more. Capacitors are also used in logic circuits, primarily for providing decoupling to prevent spikes and ripple on the supply lines which could cause spurious triggering of the circuits. Capacitors are often used in car stereo systems and hooked up to the sub-woofer speaker because the loud bass 'boom' sounds require a lot of power: the capacitors store a steady amount of charge and then quickly release it when the sub needs it for its boom. In some applications a capacitor can be used to reduce the spark created by the opening of the points on a switch or relay. A capacitor has a voltage that is proportional to the charge that is stored in the capacitor, so a capacitor can be used to perform interesting computations in op-amp circuits. Circuits with capacitors exhibit frequency-dependent behaviour so that circuits that amplify certain frequencies selectively can be built. Capacitors are very important components in electrical and electronic circuits, and this chapter introduces the terminology and calculations to aid understanding.

At the end of this chapter you should be able to:

- appreciate some applications of capacitors
- describe an electrostatic field
- define electric field strength E and state its unit
- define capacitance and state its unit
- describe a capacitor and draw the circuit diagram symbol
- perform simple calculations involving $C = \frac{Q}{V}$ and $Q = It$
- define electric flux density D and state its unit
- define permittivity, distinguishing between ε_0, ε_r and ε
- perform simple calculations involving $D = \frac{Q}{A}$, $E = \frac{V}{D}$ and $\frac{D}{E} = \varepsilon_0\varepsilon_r$
- understand that for a parallel plate capacitor, $C = \frac{\varepsilon_0\varepsilon_r A(n-1)}{d}$

Electrical Circuit Theory and Technology. 978-1-138-67349-6, © 2017 John Bird. Published by Taylor & Francis. All rights reserved.

- perform calculations involving capacitors connected in parallel and in series
- define dielectric strength and state its unit
- state that the energy stored in a capacitor is given by $W = \frac{1}{2} CV^2$ joules
- describe practical types of capacitor
- understand the precautions needed when discharging capacitors

8.1 Introduction to capacitors

A capacitor is an electrical device that is used to store electrical energy. Next to the resistor, the capacitor is the most commonly encountered component in electrical circuits. Capacitors are used extensively in electrical and electronic circuits. For example, capacitors are used to smooth rectified a.c. outputs, they are used in telecommunication equipment – such as radio receivers – for tuning to the required frequency, they are used in time delay circuits, in electrical filters, in oscillator circuits and in magnetic resonance imaging (MRI) in medical body scanners, to name but a few practical applications.
Some typical small capacitors are shown in Figure 8.1.

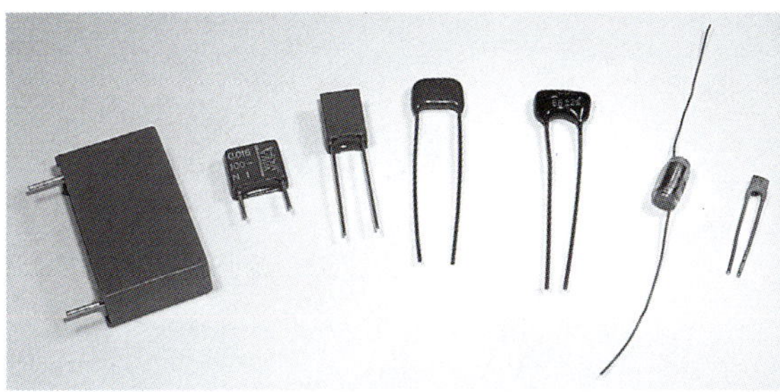

Figure 8.1

8.2 Electrostatic field

Figure 8.2 represents two parallel metal plates, A and B, charged to different potentials. If an electron that has a negative charge is placed between the plates, a force will act on the electron, tending to push it away from the negative plate B towards the positive plate, A. Similarly, a positive charge would be acted on by a force tending to move it towards the negative plate. Any region such as that shown between the plates in Figure 8.2, in which an electric charge experiences a force, is called an **electrostatic field**. The direction of the field is defined as that of the force acting on a positive charge placed in the field. In Figure 8.2, the direction of the force is from the positive plate to the negative plate.

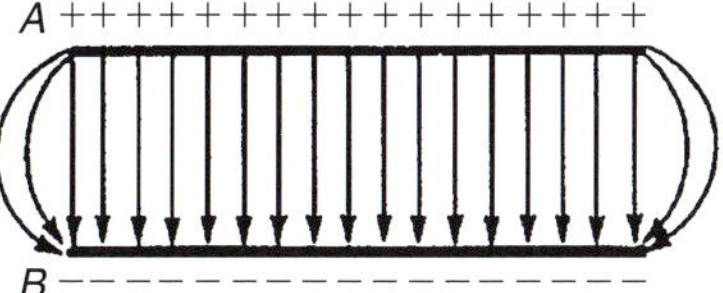

Figure 8.2 Electrostatic field

Such a field may be represented in magnitude and direction by **lines of electric force** drawn between the charged surfaces. The closeness of the lines is an indication of the field strength. Whenever a p.d. is established between two points, an electric field will always exist. Figure 8.3(a) shows a typical field pattern for an isolated point charge, and Figure 8.3(b) shows the field pattern for adjacent charges of opposite polarity. Electric lines of force (often called electric flux lines) are continuous and start and finish on point charges. Also, the lines cannot cross each other. When a charged body is placed close to an uncharged body, an induced charge of opposite sign appears on the surface of the uncharged body. This is because lines of force from the charged body terminate on its surface.

The concept of field lines or lines of force is used to illustrate the properties of an electric field. However, it should be remembered that they are only aids to the imagination.
The **force of attraction or repulsion** between two electrically charged bodies is proportional to the magnitude of their charges and inversely proportional to the square of the distance separating them,

i.e. force $\propto \frac{q_1 q_2}{d^2}$ or $\quad$ **force** $= k\frac{q_1 q_2}{d^2}$

where constant $k \approx 9 \times 10^9$ in air
This is known as **Coulomb's law***.

***Who was** Coulomb? For image and resume of Coulomb, see page 50. To find out more go to **www.routledge.com/cw/bird**

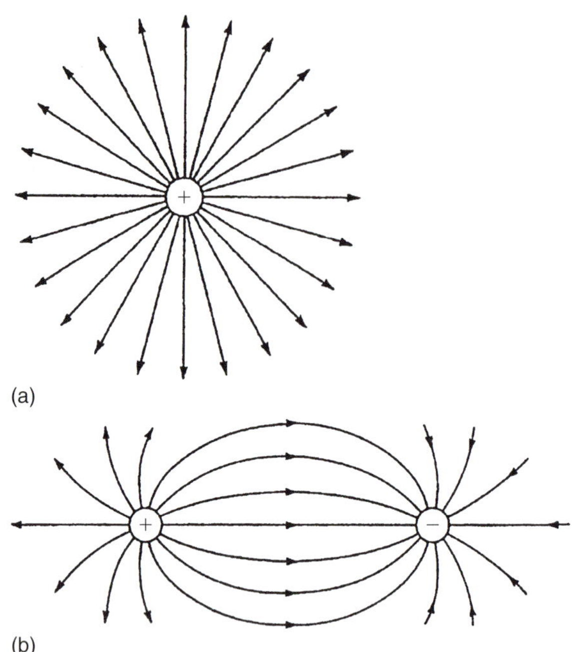

Figure 8.3 (a) Isolated point charge; (b) adjacent charges of opposite polarity

Hence the force between two charged spheres in air with their centres 16 mm apart and each carrying a charge of $+1.6\,\mu C$ is given by:

$$\text{force} = k\frac{q_1 q_2}{d^2} \approx (9\times 10^9)\frac{(1.6\times 10^{-6})^2}{(16\times 10^{-3})^2}$$

$$= \mathbf{90\ newtons}$$

8.3 Electric field strength

Figure 8.4 shows two parallel conducting plates separated from each other by air. They are connected to opposite terminals of a battery of voltage V volts.

There is therefore an electric field in the space between the plates. If the plates are close together, the electric lines of force will be straight and parallel and equally spaced, except near the edge where fringing will occur (see Figure 8.2). Over the area in which there is negligible fringing,

$$\textbf{Electric field strength, } E = \frac{V}{d} \textbf{ volts/metre}$$

where d is the distance between the plates. Electric field strength is also called **potential gradient**.

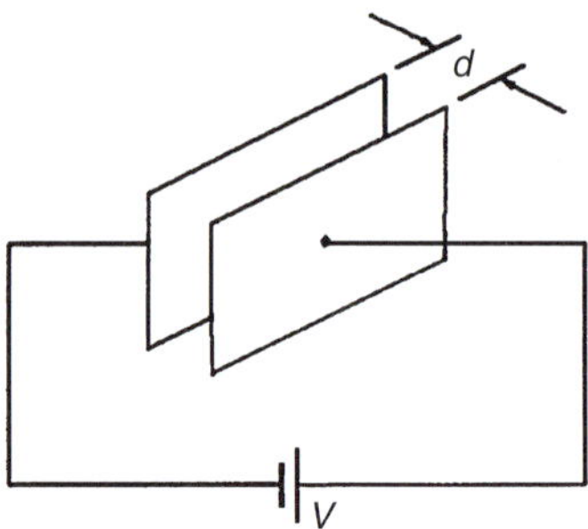

Figure 8.4

8.4 Capacitance

Static electric fields arise from electric charges, electric field lines beginning and ending on electric charges. Thus the presence of the field indicates the presence of equal positive and negative electric charges on the two plates of Figure 8.4. Let the charge be $+Q$ coulombs on one plate and $-Q$ coulombs on the other. The property of this pair of plates which determines how much charge corresponds to a given p.d. between the plates is called their **capacitance**:

$$\textbf{capacitance } C = \frac{Q}{V}$$

The **unit of capacitance** is the **farad**, **F** (or more usually $\mu F = 10^{-6}\,F$ or $pF = 10^{-12}\,F$), which is defined as the capacitance when a p.d. of one volt appears across the plates when charged with one coulomb. The unit farad is named after **Michael Faraday**.*

***Who was** Faraday? **Michael Faraday** (22 September 1791–25 August 1867) was an English scientist whose main discoveries include electromagnetic induction, diamagnetism and electrolysis. To find out more go to **www.routledge.com/cw/bird**

8.5 Capacitors

Every system of electrical conductors possesses capacitance. For example, there is capacitance between the conductors of overhead transmission lines and also between the wires of a telephone cable. In these examples the capacitance is undesirable but has to be accepted, minimized or compensated for. There are other situations where capacitance is a desirable property.

Devices specially constructed to possess capacitance are called **capacitors** (or condensers, as they used to be called). In its simplest form a capacitor consists of two plates which are separated by an insulating material known as a **dielectric**. A capacitor has the ability to store a quantity of static electricity.

The symbols for a fixed capacitor and a variable capacitor used in electrical circuit diagrams are shown in Figure 8.5.

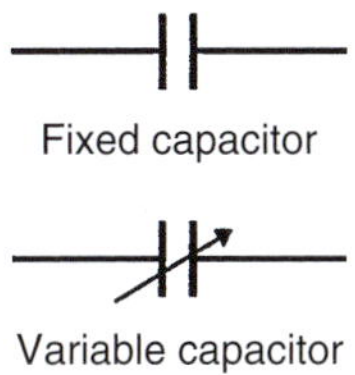

Figure 8.5

The **charge Q** stored in a capacitor is given by:

$$Q = I \times t \text{ coulombs}$$

where I is the current in amperes and t the time in seconds.

Problem 1. (a) Determine the p.d. across a 4 μF capacitor when charged with 5 mC.
(b) Find the charge on a 50 pF capacitor when the voltage applied to it is 2 kV.

(a) $C = 4\,\mu F = 4 \times 10^{-6} F$; $Q = 5\,mC = 5 \times 10^{-3}\,C$

Since $C = \frac{Q}{V}$ then $V = \frac{Q}{C} = \frac{5 \times 10^{-3}}{4 \times 10^{-6}} = \frac{5 \times 10^{6}}{4 \times 10^{3}}$

$= \frac{5000}{4}$

Hence p.d. = 1250 V or 1.25 kV

(b) $C = 50\,pF = 50 \times 10^{-12}\,F$; $V = 2\,kV = 2000\,V$

$Q = CV = 50 \times 10^{-12} \times 2000 = \frac{5 \times 2}{10^8}$

$= 0.1 \times 10^{-6}$

Hence charge = 0.1 μC

Problem 2. A direct current of 4 A flows into a previously uncharged 20 μF capacitor for 3 ms. Determine the p.d. between the plates.

$I = 4\,A$; $C = 20\,\mu F = 20 \times 10^{-6}\,F$

$t = 3\,ms = 3 \times 10^{-3}\,s$

$Q = It = 4 \times 3 \times 10^{-3}\,C$

$V = \frac{Q}{C} = \frac{4 \times 3 \times 10^{-3}}{20 \times 10^{-6}} = \frac{12 \times 10^{6}}{20 \times 10^{3}} = 0.6 \times 10^{3}$

$= 600\,V$

Hence, the p.d. between the plates is 600 V

Problem 3. A 5 μF capacitor is charged so that the p.d. between its plates is 800 V. Calculate how long the capacitor can provide an average discharge current of 2 mA.

$C = 5\,\mu F = 5 \times 10^{-6}\,F$; $V = 800V$;

$I = 2\,mA = 2 \times 10^{-3}\,A$

$Q = CV = 5 \times 10^{-6} \times 800 = 4 \times 10^{-3}\,C$

Also, $Q = It$. Thus, $t = \frac{Q}{I} = \frac{4 \times 10^{-3}}{2 \times 10^{-3}} = 2\,s$

Hence the capacitor can provide an average discharge current of 2 mA for 2 s

Now try the following Practice Exercise

Practice Exercise 41 Charge and capacitance (Answers on page 819)

1. Find the charge on a 10 μF capacitor when the applied voltage is 250 V.
2. Determine the voltage across a 1000 pF capacitor to charge it with 2 μC.
3. The charge on the plates of a capacitor is 6 mC when the potential between them is 2.4 kV. Determine the capacitance of the capacitor.
4. For how long must a charging current of 2 A be fed to a 5 μF capacitor to raise the p.d. between its plates by 500 V.
5. A steady current of 10 A flows into a previously uncharged capacitor for 1.5 ms when the p.d. between the plates is 2 kV. Find the capacitance of the capacitor.

8.6 Electric flux density

Unit flux is defined as emanating from a positive charge of 1 coulomb. Thus electric flux Ψ is measured in coulombs, and for a charge of Q coulombs, the flux $\Psi = Q$ coulombs.
Electric flux density D is the amount of flux passing through a defined area A that is perpendicular to the direction of the flux:

$$\textbf{electric flux density, } D = \frac{Q}{A} \textbf{ coulombs/metre}^2$$

Electric flux density is also called **charge density,** σ

8.7 Permittivity

At any point in an electric field, the electric field strength E maintains the electric flux and produces a particular value of electric flux density D at that point. For a field established in **vacuum** (or for practical purposes in air), the ratio D/E is a constant ε_0, i.e.

$$\frac{D}{E} = \varepsilon_0$$

where ε_0 is called the **permittivity of free space** or the free space constant. The value of ε_0 is 8.85×10^{-12} F/m.

When an insulating medium, such as mica, paper, plastic or ceramic, is introduced into the region of an electric field the ratio of D/E is modified:

$$\frac{D}{E} = \varepsilon_0 \varepsilon_r$$

where ε_r, the **relative permittivity** of the insulating material, indicates its insulating power compared with that of vacuum:

$$\text{relative permittivity } \varepsilon_r = \frac{\text{flux density in material}}{\text{flux density in vacuum}}$$

ε_r has no unit. Typical values of ε_r include:
air, 1.00; polythene, 2.3; mica, 3–7; glass, 5–10; water, 80; ceramics, 6–1000
The product $\varepsilon_0 \varepsilon_r$ is called the **absolute permittivity**, ε, i.e.

$$\varepsilon = \varepsilon_0 \varepsilon_r$$

The insulating medium separating charged surfaces is called a **dielectric**. Compared with conductors, dielectric materials have very high resistivities. They are therefore used to separate conductors at different potentials, such as capacitor plates or electric power lines.

Problem 4. Two parallel rectangular plates measuring 20 cm by 40 cm carry an electric charge of 0.2 μC. Calculate the electric flux density. If the plates are spaced 5 mm apart and the voltage between them is 0.25 kV, determine the electric field strength.

Charge $Q = 0.2\,\mu\text{C} = 0.2 \times 10^{-6}\,C$;

Area $A = 20\,\text{cm} \times 40\,\text{cm} = 800\,\text{cm}^2 = 800 \times 10^{-4}\,\text{m}^2$

$$\textbf{Electric flux density } D = \frac{Q}{A} = \frac{0.2 \times 10^{-6}}{800 \times 10^{-4}} = \frac{0.2 \times 10^4}{800 \times 10^6}$$

$$= \frac{2000}{800} \times 10^{-6} = \mathbf{2.5\,\mu C/m^2}$$

Voltage $V = 0.25\,\text{kV} = 250\,\text{V}$;

plate spacing, $d = 5\,\text{mm} = 5 \times 10^{-3}\,\text{m}$

$$\textbf{Electric field strength } E = \frac{V}{d} = \frac{250}{5 \times 10^{-3}} = \mathbf{50\,kV/m}$$

Problem 5. The flux density between two plates separated by mica of relative permittivity 5 is $2\,\mu\text{C/m}^2$. Find the voltage gradient between the plates.

Flux density $D = 2\,\mu\text{C/m}^2 = 2 \times 10^{-6}\,\text{C/m}^2$;
$\varepsilon_0 = 8.85 \times 10^{-12}\,\text{F/m}$; $\varepsilon_r = 5$.

$$\frac{D}{E} = \varepsilon_0 \varepsilon_r$$

hence **voltage gradient** $E = \dfrac{D}{\varepsilon_0 \varepsilon_r}$

$$= \frac{2 \times 10^{-6}}{8.85 \times 10^{-12} \times 5}\,\text{V/m}$$

$$= \mathbf{45.2\,kV/m}$$

Problem 6. Two parallel plates having a p.d. of 200 V between them are spaced 0.8 mm apart. What is the electric field strength? Find also the flux density when the dielectric between the plates is (a) air, and (b) polythene of relative permittivity 2.3

$$\textbf{Electric field strength } E = \frac{V}{D} = \frac{200}{0.8 \times 10^{-3}}$$

$$= \mathbf{250\,kV/m}$$

(a) For air: $\varepsilon_r = 1$

$\frac{D}{E} = \varepsilon_0 \varepsilon_r$. Hence

electric flux density $D = E\varepsilon_0\varepsilon_r$

$= (250 \times 10^3 \times 8.85 \times 10^{-12} \times 1)$ C/m^2

$= \mathbf{2.213\,\mu C/m^2}$

(b) For polythene, $\varepsilon_r = 2.3$

Electric flux density $D = E\varepsilon_0\varepsilon_r$

$= (250 \times 10^3 \times 8.85 \times 10^{-12} \times 2.3)$ C/m^2

$= \mathbf{5.089\,\mu C/m^2}$

Now try the following Practice Exercise

Practice Exercise 42 Electric field strength, electric flux density and permittivity (Answers on page 819)

(Where appropriate take ε_0 as 8.85×10^{-12} F/m.)

1. A capacitor uses a dielectric 0.04 mm thick and operates at 30 V. What is the electric field strength across the dielectric at this voltage?
2. A two-plate capacitor has a charge of 25 C. If the effective area of each plate is 5 cm^2 find the electric flux density of the electric field.
3. A charge of 1.5 μC is carried on two parallel rectangular plates each measuring 60 mm by 80 mm. Calculate the electric flux density. If the plates are spaced 10 mm apart and the voltage between them is 0.5 kV determine the electric field strength.
4. The electric flux density between two plates separated by polystyrene of relative permittivity 2.5 is 5 μC/m^2. Find the voltage gradient between the plates.
5. Two parallel plates having a p.d. of 250 V between them are spaced 1 mm apart. Determine the electric field strength. Find also the electric flux density when the dielectric between the plates is (a) air and (b) mica of relative permittivity 5

8.8 The parallel plate capacitor

For a parallel plate capacitor, as shown in Figure 8.6(a), experiments show that capacitance C is proportional to

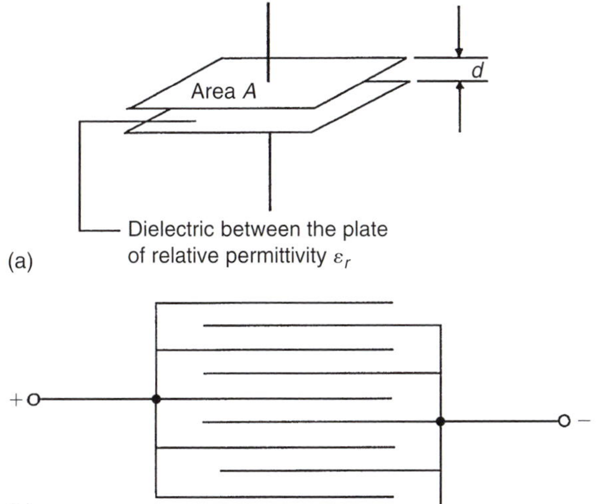

Figure 8.6

the area A of a plate, inversely proportional to the plate spacing d (i.e. the dielectric thickness) and depends on the nature of the dielectric:

$$\textbf{Capacitance, } C = \frac{\varepsilon_0\varepsilon_r A}{d} \textbf{ farads}$$

where $\varepsilon_0 = 8.85 \times 10^{-12}$ F/m (constant)

ε_r = relative permittivity

A = area of one of the plates in m^2 and

d = thickness of dielectric in m.

Another method used to increase the capacitance is to interleave several plates as shown in Figure 8.6(b). Ten plates are shown, forming nine capacitors with a capacitance nine times that of one pair of plates.
If such an arrangement has n plates then capacitance $C \propto (n-1)$.

$$\text{Thus capacitance } \quad C = \frac{\varepsilon_0\varepsilon_r A(n-1)}{d} \textbf{ farads}$$

Problem 7. (a) A ceramic capacitor has an effective plate area of 4 cm^2 separated by 0.1 mm of ceramic of relative permittivity 100. Calculate the capacitance of the capacitor in picofarads. (b) If the capacitor in part (a) is given a charge of 1.2 μC, what will be the p.d. between the plates?

(a) Area $A = 4\,\text{cm}^2 = 4 \times 10^{-4}\,\text{m}^2$;
$d = 0.1\,\text{mm} = 0.1 \times 10^{-3}$ m;
$\varepsilon_0 = 8.85 \times 10^{-12}$ F/m; $\varepsilon_r = 100$

Capacitance $C = \frac{\varepsilon_0 \varepsilon_r A}{d}$ farads

$$= \frac{8.85 \times 10^{-12} \times 100 \times 4 \times 10^{-4}}{0.1 \times 10^{-3}} \text{ F}$$

$$= \frac{8.85 \times 4}{10^{10}} \text{ F} = \frac{8.85 \times 4 \times 10^{12}}{10^{10}} \text{ pF}$$

$$= \mathbf{3540\,pF}$$

(b) $Q = CV$ thus $V = \frac{Q}{C} = \frac{1.2 \times 10^{-6}}{3540 \times 10^{-12}} V = \mathbf{339\,V}$

Problem 8. A waxed paper capacitor has two parallel plates, each of effective area 800 cm^2. If the capacitance of the capacitor is 4425 pF, determine the effective thickness of the paper if its relative permittivity is 2.5

$A = 800\,\text{cm}^2 = 800 \times 10^{-4}\,\text{m}^2 = 0.08\,\text{m}^2;$

$C = 4425\,\text{pF} = 4425 \times 10^{-12}\,\text{F};$

$\varepsilon_0 = 8.85 \times 10^{-12}\,\text{F/m}; \varepsilon_r = 2.5$

Since $C = \frac{\varepsilon_0 \varepsilon_r A}{d}$ then $d = \frac{\varepsilon_0 \varepsilon_r A}{C}$

Hence, $d = \frac{8.85 \times 10^{-12} \times 2.5 \times 0.08}{4425 \times 10^{-12}} = 0.0004\,\text{m}$

Hence the thickness of the paper is 0.4 mm

Problem 9. A parallel plate capacitor has 19 interleaved plates each 75 mm by 75 mm separated by mica sheets 0.2 mm thick. Assuming the relative permittivity of the mica is 5, calculate the capacitance of the capacitor.

$n = 19; n - 1 = 18;$

$A = 75 \times 75 = 5625\,\text{mm}^2 = 5625 \times 10^{-6}\,\text{m}^2;$

$\varepsilon_r = 5; \varepsilon_0 = 8.85 \times 10^{-12}\,\text{F/m};$

$d = 0.2\,\text{mm} = 0.2 \times 10^{-3}\,\text{m}$

Capacitance $C = \frac{\varepsilon_0 \varepsilon_r A(n-1)}{d}$

$$= \frac{8.85 \times 10^{-12} \times 5 \times 5625 \times 10^{-6} \times 18}{0.2 \times 10^{-3}} \text{ F}$$

$= \mathbf{0.0224\,\mu F}$ or $\mathbf{22.4\,nF}$

Now try the following Practice Exercise

Practice Exercise 43 Parallel plate capacitors (Answers on page 819)

(Where appropriate take ε_0 as 8.85×10^{-12} F/m)

1. A capacitor consists of two parallel plates each of area 0.01 m^2, spaced 0.1 mm in air. Calculate the capacitance in picofarads.
2. A waxed paper capacitor has two parallel plates, each of effective area 0.2 m^2. If the capacitance is 4000 pF, determine the effective thickness of the paper if its relative permittivity is 2
3. Calculate the capacitance of a parallel plate capacitor having five plates, each 30 mm by 20 mm and separated by a dielectric 0.75 mm thick having a relative permittivity of 2.3
4. How many plates has a parallel plate capacitor having a capacitance of 5 nF, if each plate is 40 mm by 40 mm and each dielectric is 0.102 mm thick with a relative permittivity of 6
5. A parallel plate capacitor is made from 25 plates, each 70 mm by 120 mm interleaved with mica of relative permittivity 5. If the capacitance of the capacitor is 3000 pF, determine the thickness of the mica sheet.
6. The capacitance of a parallel plate capacitor is 1000 pF. It has 19 plates, each 50 mm by 30 mm separated by a dielectric of thickness 0.40 mm. Determine the relative permittivity of the dielectric.
7. A capacitor is to be constructed so that its capacitance is 4250 pF and to operate at a p.d. of 100 V across its terminals. The dielectric is to be polythene ($\varepsilon_r = 2.3$) which, after allowing a safety factor, has a dielectric strength of 20 MV/m. Find (a) the thickness of the polythene needed, and (b) the area of a plate.

8.9 Capacitors connected in parallel and series

(a) Capacitors connected in parallel

Figure 8.7 shows three capacitors, C_1, C_2 and C_3, connected in parallel with a supply voltage V applied across the arrangement.

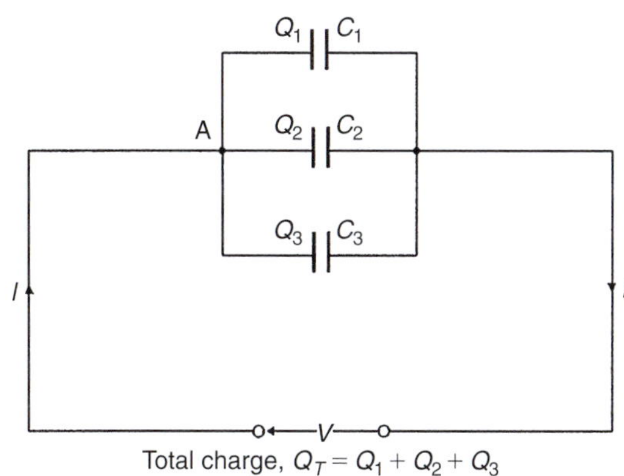

Total charge, $Q_T = Q_1 + Q_2 + Q_3$

Figure 8.7

When the charging current I reaches point A it divides, some flowing into C_1, some flowing into C_2 and some into C_3. Hence the total charge $Q_T(=I \times t)$ is divided between the three capacitors. The capacitors each store a charge and these are shown as Q_1, Q_2 and Q_3, respectively. Hence

$$Q_T = Q_1 + Q_2 + Q_3$$

But $Q_T = CV$, $Q_1 = C_1 V$, $Q_2 = C_2 V$ and $Q_3 = C_3 V$. Therefore $CV = C_1 V + C_2 V + C_3 V$, where C is the total equivalent circuit capacitance,

i.e. $\boldsymbol{C = C_1 + C_2 + C_3}$

It follows that for n parallel-connected capacitors,

$$\boldsymbol{C = C_1 + C_2 + C_3 + \cdots + C_n}$$

i.e. the equivalent capacitance of a group of parallel-connected capacitors is the sum of the capacitances of the individual capacitors. (Note that this formula is similar to that used for **resistors** connected in **series**.)

(b) Capacitors connected in series

Figure 8.8 shows three capacitors, C_1, C_2 and C_3, connected in series across a supply voltage V. Let the p.d. across the individual capacitors be V_1, V_2 and V_3, respectively, as shown.

Let the charge on plate 'a' of capacitor C_1 be $+Q$ coulombs. This induces an equal but opposite charge of $-Q$ coulombs on plate 'b'. The conductor between plates 'b' and 'c' is electrically isolated from the rest of the circuit so that an equal but opposite charge of $+Q$ coulombs must appear on plate 'c', which, in turn, induces an equal and opposite charge of $-Q$ coulombs on plate 'd', and so on.

Hence when capacitors are connected in series the charge on each is the same.

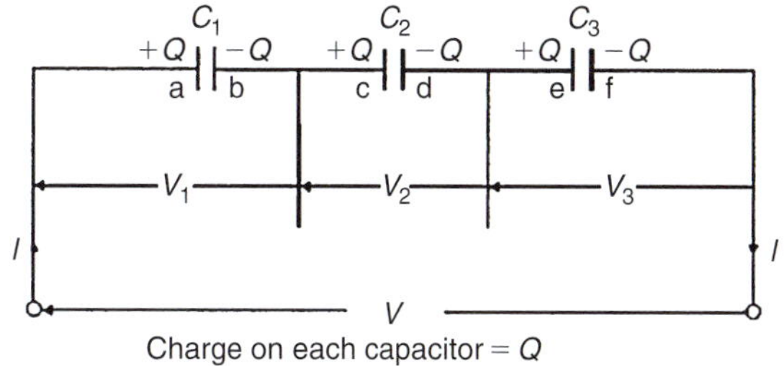

Charge on each capacitor $= Q$

Figure 8.8

In a series circuit: $V = V_1 + V_2 + V_3$

Since $V = \frac{Q}{C}$ then $\frac{Q}{C} = \frac{Q}{C_1} + \frac{Q}{C_2} + \frac{Q}{C_3}$

where C is the total equivalent circuit capacitance,

i.e. $\boldsymbol{\frac{1}{C} = \frac{1}{C_1} + \frac{1}{C_2} + \frac{1}{C_3}}$

It follows that for n series-connected capacitors:

$$\boldsymbol{\frac{1}{C} = \frac{1}{C_1} + \frac{1}{C_2} + \frac{1}{C_3} + \cdots + \frac{1}{C_n}}$$

i.e. for series-connected capacitors, the reciprocal of the equivalent capacitance is equal to the sum of the reciprocals of the individual capacitances. (Note that this formula is similar to that used for **resistors** connected in **parallel**.)

For the special case of **two capacitors in series**:

$$\frac{1}{C} = \frac{1}{C_1} + \frac{1}{C_2} = \frac{C_2 + C_1}{C_1 C_2}$$

Hence $\boldsymbol{C = \frac{C_1 C_2}{C_1 + C_2}}$ $\left(\text{i.e. } \frac{\text{product}}{\text{sum}}\right)$

Problem 10. Calculate the equivalent capacitance of two capacitors of 6 μF and 4 μF connected (a) in parallel and (b) in series.

(a) In parallel, equivalent capacitance

$C = C_1 + C_2 = 6\,\mu\text{F} + 4\,\mu\text{F} = \mathbf{10\,\mu F}$

(b) In series, equivalent capacitance C is given by:

$$C = \frac{C_1 C_2}{C_1 + C_2}$$

This formula is used for the special case of **two** capacitors in series.

Thus $C = \frac{6 \times 4}{6 + 4} = \frac{24}{10} = \mathbf{2.4\,\mu F}$

Problem 11. What capacitance must be connected in series with a 30 μF capacitor for the equivalent capacitance to be 12 μF?

Let $C = 12\,\mu F$ (the equivalent capacitance), $C_1 = 30\,\mu F$ and C_2 be the unknown capacitance.

For two capacitors in series $\frac{1}{C} = \frac{1}{C_1} + \frac{1}{C_2}$

Hence $\frac{1}{C_2} = \frac{1}{C} - \frac{1}{C_1} = \frac{C_1 - C}{CC_1}$

and $C_2 = \frac{CC_1}{C_1 - C} = \frac{12 \times 30}{30 - 12}$

$= \frac{360}{18} = \mathbf{20\,\mu F}$

Problem 12. Capacitances of 1 μF, 3 μF, 5 μF and 6 μF are connected in parallel to a direct voltage supply of 100 V. Determine (a) the equivalent circuit capacitance, (b) the total charge and (c) the charge on each capacitor.

(a) The equivalent capacitance C for four capacitors in parallel is given by:

$$C = C_1 + C_2 + C_3 + C_4$$

i.e. $C = 1+3+5+6 = \mathbf{15\,\mu F}$

(b) Total charge $Q_T = CV$ where C is the equivalent circuit capacitance

i.e. $Q_T = 15 \times 10^{-6} \times 100 = 1.5 \times 10^{-3} C = \mathbf{1.5\,mC}$

(c) The charge on the 1 μF capacitor

$Q_1 = C_1 V = 1 \times 10^{-6} \times 100$

$= \mathbf{0.1\,mC}$

The charge on the 3 μF capacitor

$Q_2 = C_2 V = 3 \times 10^{-6} \times 100$

$= \mathbf{0.3\,mC}$

The charge on the 5 μF capacitor

$Q_3 = C_3 V = 5 \times 10^{-6} \times 100$

$= \mathbf{0.5\,mC}$

The charge on the 6 μF capacitor

$Q_4 = C_4 V = 6 \times 10^{-6} \times 100$

$= \mathbf{0.6\,mC}$

[Check: in a parallel circuit

$Q_T = Q_1 + Q_2 + Q_3 + Q_4$

$Q_1 + Q_2 + Q_3 + Q_4 = 0.1 + 0.3 + 0.5 + 0.6$

$= 1.5\,mC = Q_T$]

Problem 13. Capacitances of 3 μF, 6 μF and 12 μF are connected in series across a 350 V supply. Calculate (a) the equivalent circuit capacitance, (b) the charge on each capacitor and (c) the p.d. across each capacitor.

The circuit diagram is shown in Figure 8.9.

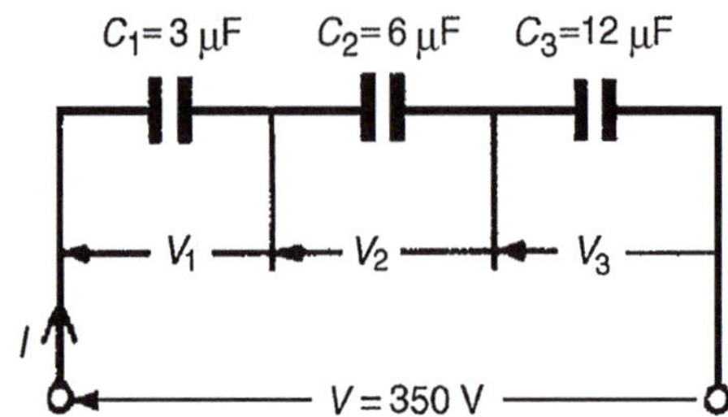

Figure 8.9

(a) The equivalent circuit capacitance C for three capacitors in series is given by:

$$\frac{1}{C} = \frac{1}{C_1} + \frac{1}{C_2} + \frac{1}{C_3}$$

i.e. $\frac{1}{C} = \frac{1}{3} + \frac{1}{6} + \frac{1}{12} = \frac{4+2+1}{12} = \frac{7}{12}$

Hence the equivalent circuit capacitance $\mathbf{C = \frac{12}{7} = 1\frac{5}{7}\,\mu F}$

(b) Total charge $Q_T = CV$,

hence

$Q_T = \frac{12}{7} \times 10^{-6} \times 350 = 600\,\mu C$ or $0.6\,mC$

Since the capacitors are connected in series 0.6 mC is the charge on each of them.

(c) The voltage across the 3 μF capacitor,

$V_1 = \frac{Q}{C_1} = \frac{0.6 \times 10^{-3}}{3 \times 10^{-6}}$

$= \mathbf{200\,V}$

The voltage across the 6 μF capacitor,

$V_2 = \frac{Q}{C_2} = \frac{0.6 \times 10^{-3}}{6 \times 10^{-6}}$

$= \mathbf{100\,V}$

The voltage across the 12 μF capacitor,

$$V_3 = \frac{Q}{C_3} = \frac{0.6 \times 10^{-3}}{12\times 10^{-6}}$$
$$= \mathbf{50\,V}$$

[Check: in a series circuit

$$V = V_1 + V_2 + V_3$$

$$V_1 + V_2 + V_3 = 200 + 100 + 50 = 350\,\text{V}$$

$$= \text{supply voltage}]$$

In practice, capacitors are rarely connected in series unless they are of the same capacitance. The reason for this can be seen from the above problem where the lowest valued capacitor (i.e. 3 μF) has the highest p.d. across it (i.e. 200 V), which means that if all the capacitors have an identical construction they must all be rated at the highest voltage.

Problem 14. For the arrangement shown in Figure 8.10 find (a) the equivalent capacitance of the circuit, (b) the voltage across QR, and (c) the charge on each capacitor.

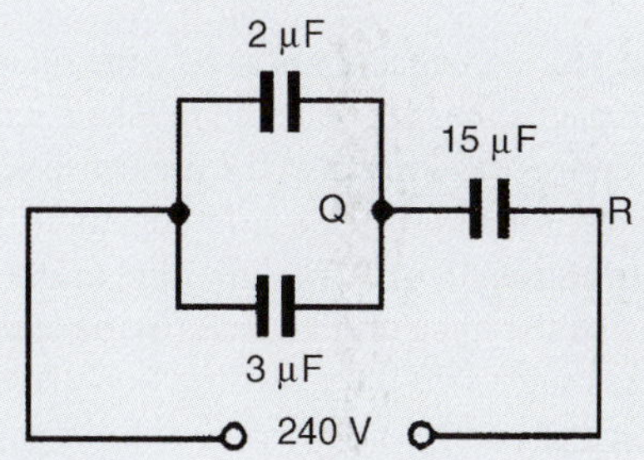

Figure 8.10

(a) 2 μF in parallel with 3 μF gives an equivalent capacitance of 2 μF + 3 μF = 5 μF. The circuit is now as shown in Figure 8.11.

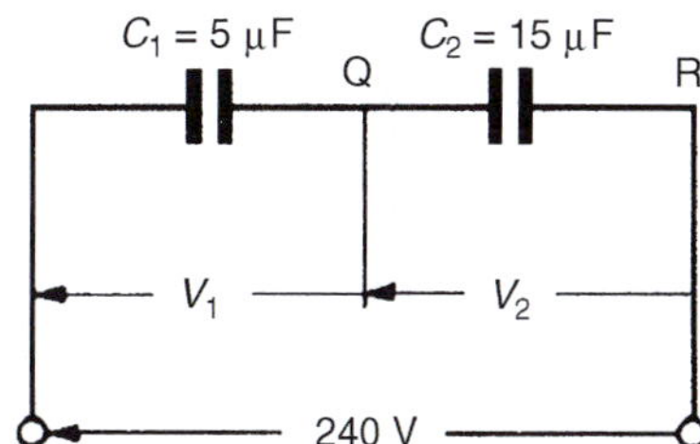

Figure 8.11

The **equivalent capacitance** of 5 μF in series with 15 μF is given by

$$\frac{5 \times 15}{5 + 15}\,\mu\text{F} = \frac{75}{20}\,\mu\text{F} = \mathbf{3.75\,\mu F}$$

(b) The charge on each of the capacitors shown in Figure 8.11 will be the same since they are connected in series. Let this charge be Q coulombs.

Then $\quad Q = C_1 V_1 = C_2 V_2$

i.e. $\quad 5V_1 = 15V_2$

$$V_1 = 3V_2 \qquad (1)$$

Also $\quad V_1 + V_2 = 240\,\text{V}$

Hence $\quad 3V_2 + V_2 = 240\,\text{V}$ from equation (1)

Thus $\quad V_2 = 60\,\text{V}$ and $V_1 = 180\,\text{V}$

Hence the voltage across QR is 60 V

(c) The charge on the 15 μF capacitor is

$C_2V_2 = 15 \times 10^{-6} \times 60 = \mathbf{0.9\,mC}$

The charge on the 2 μF capacitor is

$2 \times 10^{-6} \times 180 = \mathbf{0.36\,mC}$

The charge on the 3 μF capacitor is

$3 \times 10^{-6} \times 180 = \mathbf{0.54\,mC}$

Now try the following Practice Exercise

Practice Exercise 44 Capacitors in parallel and series (Answers on page 819)

1. Capacitors of 2 μF and 6 μF are connected (a) in parallel and (b) in series. Determine the equivalent capacitance in each case.
2. Find the capacitance to be connected in series with a 10 μF capacitor for the equivalent capacitance to be 6 μF
3. Two 6 μF capacitors are connected in series with one having a capacitance of 12 μF. Find the total equivalent circuit capacitance. What capacitance must be added in series to obtain a capacitance of 1.2 μF?

4. Determine the equivalent capacitance when the following capacitors are connected (a) in parallel and (b) in series:

   (i) 2 μF, 4 μF and 8 μF
   (ii) 0.02 μF, 0.05 μF and 0.10 μF
   (iii) 50 pF and 450 pF
   (iv) 0.01 μF and 200 pF

5. For the arrangement shown in Figure 8.12 find (a) the equivalent circuit capacitance and (b) the voltage across a 4.5 μF capacitor.

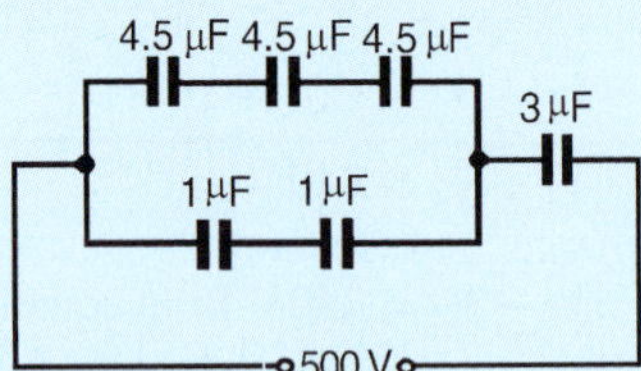

Figure 8.12

6. Three 12 μF capacitors are connected in series across a 750 V supply. Calculate (a) the equivalent capacitance, (b) the charge on each capacitor and (c) the p.d. across each capacitor.

7. If two capacitors having capacitances of 3 μF and 5 μF respectively are connected in series across a 240 V supply, determine (a) the p.d. across each capacitor and (b) the charge on each capacitor.

8. In Figure 8.13 capacitors P, Q and R are identical and the total equivalent capacitance of the circuit is 3 μF. Determine the values of P, Q and R.

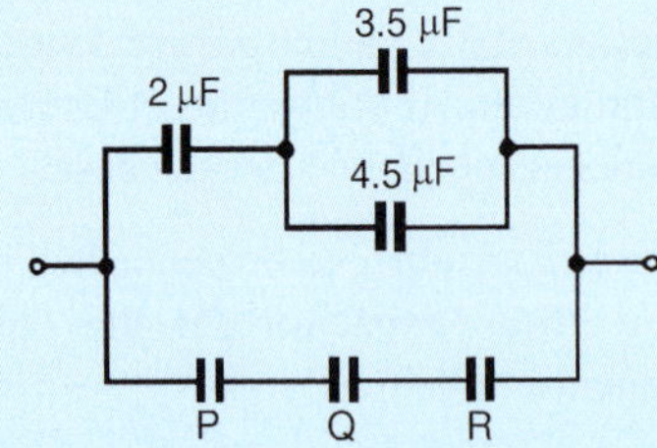

Figure 8.13

9. For the circuit shown in Figure 8.14, determine (a) the total circuit capacitance, (b) the total energy in the circuit and (c) the charges in the capacitors shown as C_1 and C_2

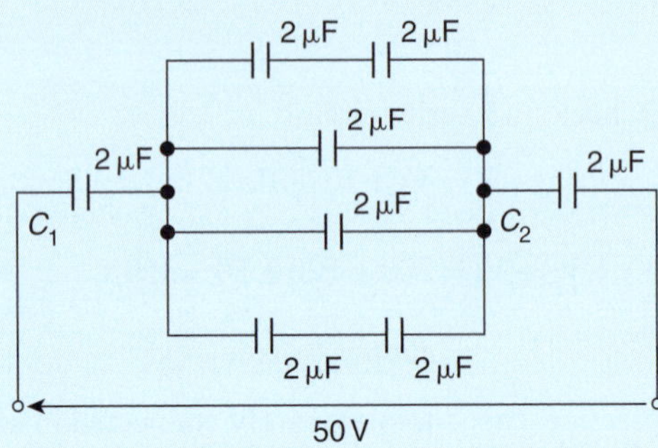

Figure 8.14

8.10 Dielectric strength

The maximum amount of field strength that a dielectric can withstand is called the dielectric strength of the material.

Dielectric strength, $E_m = \dfrac{V_m}{d}$

Problem 15. A capacitor is to be constructed so that its capacitance is 0.2 μF and to take a p.d. of 1.25 kV across its terminals. The dielectric is to be mica which, after allowing a safety factor of 2, has a dielectric strength of 50 MV/m. Find (a) the thickness of the mica needed, and (b) the area of a plate assuming a two-plate construction. (Assume ε_r for mica to be 6)

(a) Dielectric strength, $E = \dfrac{V}{d}$, i.e. $d = \dfrac{V}{E}$

$$= \frac{1.25 \times 10^3}{50 \times 10^6}\ \text{m}$$

$$= \mathbf{0.025\ mm}$$

(b) Capacitance, $C = \dfrac{\varepsilon_0 \varepsilon_r A}{d}$

hence area $A = \dfrac{Cd}{\varepsilon_0 \varepsilon_r}$

$$= \frac{0.2 \times 10^{-6} \times 0.025 \times 10^{-3}}{8.85 \times 10^{-12} \times 6}\ \text{m}^2$$

$$= 0.09416\ \text{m}^2 = \mathbf{941.6\ cm^2}$$

8.11 Energy stored

The energy, W, stored by a capacitor is given by

$$W = \frac{1}{2}CV^2 \text{ joules}$$

Problem 16. (a) Determine the energy stored in a 3 μF capacitor when charged to 400 V. (b) Find also the average power developed if this energy is dissipated in a time of 10 μs.

(a) **Energy stored** $W = \frac{1}{2}CV^2$ joules

$$= \frac{1}{2} \times 3 \times 10^{-6} \times 400^2$$
$$= \frac{3}{2} \times 16 \times 10^{-2}$$
$$= \mathbf{0.24\ J}$$

(b) **Power** $= \dfrac{\text{energy}}{\text{time}} = \dfrac{0.24}{10 \times 10^{-6}}\ \text{W} = \mathbf{24\ kW}$

Problem 17. A 12 μF capacitor is required to store 4 J of energy. Find the p.d. to which the capacitor must be charged.

Energy stored $W = \dfrac{1}{2}CV^2$ hence $V^2 = \dfrac{2W}{C}$

and $V = \sqrt{\left(\dfrac{2W}{C}\right)} = \sqrt{\left(\dfrac{2 \times 4}{12 \times 10^{-6}}\right)} = \sqrt{\left(\dfrac{2 \times 10^6}{3}\right)}$

$$= \mathbf{816.5\ V}$$

Problem 18. A capacitor is charged with 10 mC. If the energy stored is 1.2 J, find (a) the voltage and (b) the capacitance.

Energy stored $W = \dfrac{1}{2}CV^2$ and $C = \dfrac{Q}{V}$

Hence $\quad W = \dfrac{1}{2}\left(\dfrac{Q}{V}\right)V^2 = \dfrac{1}{2}QV$

from which $\quad V = \dfrac{2W}{Q}$

$Q = 10\,\text{mC} = 10 \times 10^{-3}C$ and $W = 1.2\,\text{J}$

(a) **Voltage** $V = \dfrac{2W}{Q} = \dfrac{2 \times 1.2}{10 \times 10^{-3}} = \mathbf{0.24\ kV}$ or $\mathbf{240\ V}$

(b) **Capacitance** $C = \dfrac{Q}{V} = \dfrac{10 \times 10^{-3}}{240}\ \text{F}$

$$= \frac{10 \times 10^6}{240 \times 10^3}\ \mu\text{F} = \mathbf{41.67\ \mu F}$$

Now try the following Practice Exercise

Practice Exercise 45 Energy stored in capacitors (Answers on page 819)

(Where appropriate take ε_0 as 8.85×10^{-12} F/m)

1. When a capacitor is connected across a 200 V supply the charge is 4 μC. Find (a) the capacitance and (b) the energy stored.
2. Find the energy stored in a 10 μF capacitor when charged to 2 kV.
3. A 3300 pF capacitor is required to store 0.5 mJ of energy. Find the p.d. to which the capacitor must be charged.
4. A capacitor, consisting of two metal plates each of area 50 cm^2 and spaced 0.2 mm apart in air, is connected across a 120 V supply. Calculate (a) the energy stored, (b) the electric flux density and (c) the potential gradient.
5. A bakelite capacitor is to be constructed to have a capacitance of 0.04 μF and to have a steady working potential of 1 kV maximum. Allowing a safe value of field stress of 25 MV/m, find (a) the thickness of bakelite required, (b) the area of plate required if the relative permittivity of bakelite is 5, (c) the maximum energy stored by the capacitor and (d) the average power developed if this energy is dissipated in a time of 20 μs.

8.12 Practical types of capacitor

Practical types of capacitor are characterized by the material used for their dielectric. The main types include: variable air, mica, paper, ceramic, plastic, titanium oxide and electrolytic.

1. **Variable air capacitors.** These usually consist of two sets of metal plates (such as aluminium), one fixed, the other variable. The set of moving plates rotate on a spindle as shown by the end view of Figure 8.15.

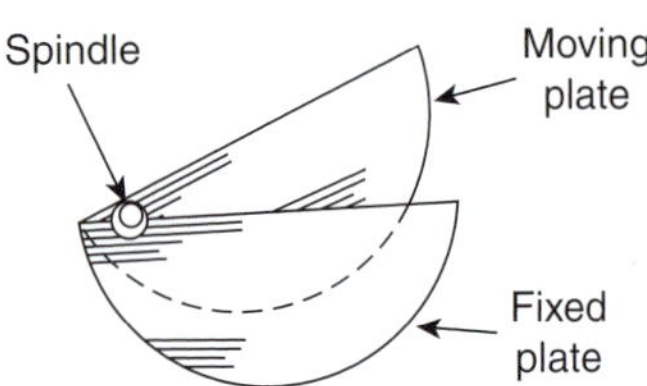

Figure 8.15

As the moving plates are rotated through half a revolution, the meshing, and therefore the capacitance, varies from a minimum to a maximum value. Variable air capacitors are used in radio and electronic circuits where very low losses are required, or where a variable capacitance is needed. The maximum value of such capacitors is between 500 pF and 1000 pF.

A typical practical variable air capacitor is shown in Figure 8.16.

Figure 8.16

2. **Mica capacitors.** A typical older type construction is shown in Figure 8.17.

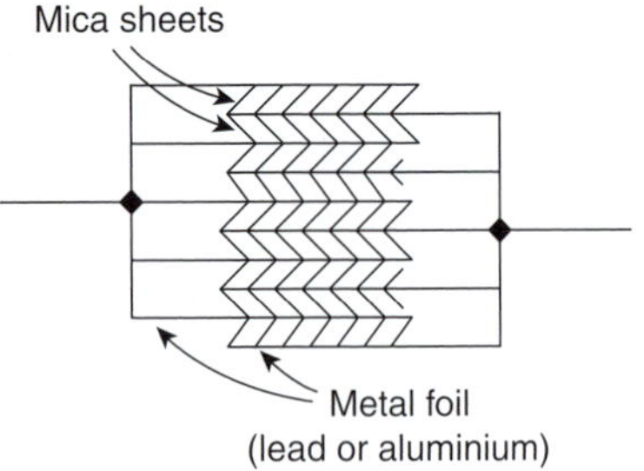

Figure 8.17

Usually the whole capacitor is impregnated with wax and placed in a bakelite case. Mica is easily obtained in thin sheets and is a good insulator. However, mica is expensive and is not used in capacitors above about 0.2 μF. A modified form of mica capacitor is the silvered mica type. The mica is coated on both sides with a thin layer of silver which forms the plates. Capacitance is stable and less likely to change with age. Such capacitors have a constant capacitance with change of temperature, a high working voltage rating and a long service life and are used in high-frequency circuits with fixed values of capacitance up to about 1000 pF.

3. **Paper capacitors.** A typical paper capacitor is shown in Figure 8.18 where the length of the roll corresponds to the capacitance required. The whole is usually impregnated with oil or wax to exclude moisture, and then placed in a plastic or aluminium container for protection. Paper capacitors are made in various working voltages up to about 150 kV and are used where loss is not very important. The maximum value of this type of capacitor is between 500 pF and 10 μF. Disadvantages of paper capacitors include variation in capacitance with temperature change and a shorter service life than most other types of capacitor.

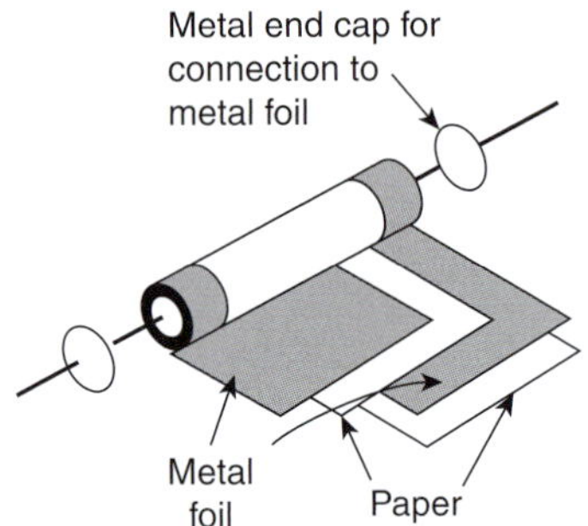

Figure 8.18

4. **Ceramic capacitors.** These are made in various forms, each type of construction depending on the value of capacitance required. For high values, a tube of ceramic material is used as shown in the cross-section of Figure 8.19. For smaller values the cup construction is used as shown in Figure 8.20, and for still smaller values the disc construction shown in Figure 8.21 is used. Certain ceramic materials have a very high permittivity and this enables capacitors of high capacitance to be made which are of small physical size with a high working voltage rating. Ceramic capacitors are available in the

range 1 pF to 0.1 μF and may be used in high-frequency electronic circuits subject to a wide range of temperatures.

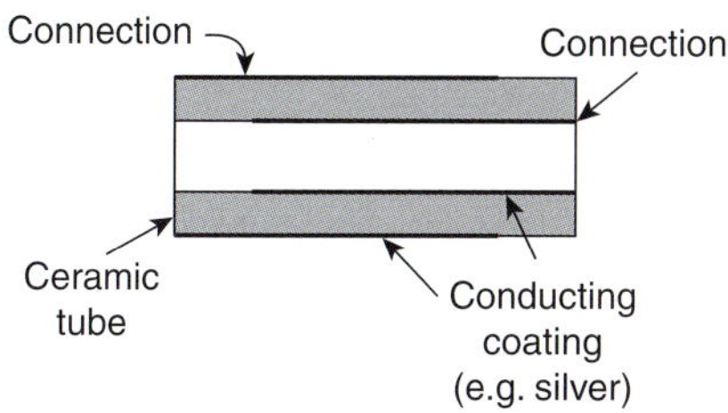

Figure 8.19

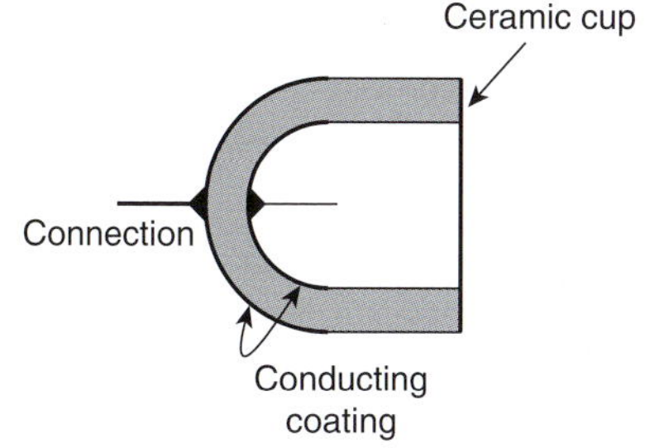

Figure 8.20

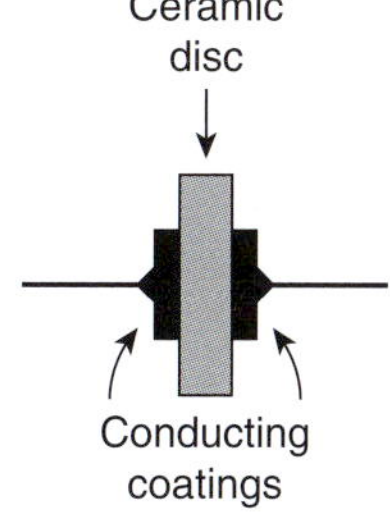

Figure 8.21

5. **Plastic capacitors.** Some plastic materials such as polystyrene and Teflon can be used as dielectrics. Construction is similar to the paper capacitor but using a plastic film instead of paper. Plastic capacitors operate well under conditions of high temperature, provide a precise value of capacitance, a very long service life and high reliability.
6. **Titanium oxide capacitors** have a very high capacitance with a small physical size when used at a low temperature.
7. **Electrolytic capacitors.** Construction is similar to the paper capacitor, with aluminium foil used for the plates and with a thick absorbent material, such as paper, impregnated with an electrolyte (ammonium borate), separating the plates. The finished capacitor is usually assembled in an aluminium container and hermetically sealed. Its operation depends on the formation of a thin aluminium oxide layer on the positive plate by electrolytic action when a suitable direct potential is maintained between the plates. This oxide layer is very thin and forms the dielectric. (The absorbent paper between the plates is a conductor and does not act as a dielectric.) Such capacitors **must always be used on d.c.** and must be connected with the correct polarity; if this is not done the capacitor will be destroyed since the oxide layer will be destroyed. Electrolytic capacitors are manufactured with working voltage from 6 V to 600 V, although accuracy is generally not very high. These capacitors possess a much larger capacitance than other types of capacitors of similar dimensions due to the oxide film being only a few microns thick. The fact that they can be used only on d.c. supplies limits their usefulness.
   Some typical electrolytic capacitors are shown in Figure 8.22.

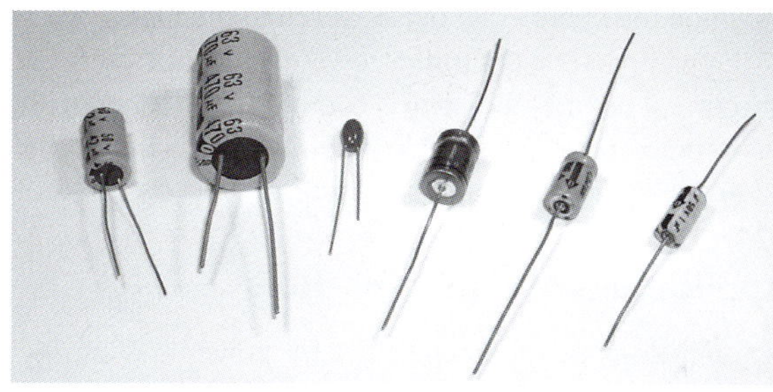

Figure 8.22

8.13 Supercapacitors

Electrical double-layer capacitors (EDLC) are, together with pseudocapacitors, part of a new type of electrochemical capacitor called **supercapacitors**, also known as **ultracapacitors**. Supercapacitors do not have a conventional solid dielectric. The capacitance value of an electrochemical capacitor is determined by two storage principles:

(a) **Double-layer capacitance** is the electrostatic storage of the electrical energy achieved by separation of charge in a Helmholtz double layer at the interface between the surface of a conductor electrode and an electrolytic solution electrolyte. The separation of charge distance in a double-layer is of the order of a few Angstroms (0.3–0.8 nm) and is static in origin.

(b) **Pseudocapacitance** is the electrochemical storage of the electrical energy, achieved by redox reactions on the surface of the electrode or by specifically adsorped ions that results in a reversible faradaic charge-transfer on the electrode.

Double-layer capacitance and pseudocapacitance both contribute to the total capacitance value of a supercapacitor.

Supercapacitors have the **highest available capacitance** values per unit volume and the greatest energy density of all capacitors. Supercapacitors support up to 12 000 F/1.2 V, with specific capacitance values up to 10 000 times that of electrolytic capacitors. Supercapacitors bridge the gap between capacitors and batteries. In terms of specific energy as well as in terms of specific power this gap covers several orders of magnitude. However, batteries still have about 10 times the capacity of supercapacitors. While existing supercapacitors have energy densities that are approximately 10% of a conventional battery, their power density is generally 10 to 100 times as great. Power density combines energy density with the speed at which the energy can be delivered to the load.

Supercapacitors are **polarized** and must operate with the correct polarity. Polarity is controlled by design with asymmetric electrodes, or, for symmetric electrodes, by a voltage applied during manufacture.

Applications of supercapacitors

Applications of supercapacitors for power and energy requirements include long, small currents for static memory (SRAM) in electronic equipment, power electronics that require very short, high currents as in the KERS system in Formula 1 cars, and recovery of braking energy in vehicles.

Advantages of supercapacitors include:

(i) Long life, with little degradation over hundreds of thousands of charge cycles. Due to the capacitor's high number of charge–discharge cycles (millions or more compared to 200 to 1000 for most commercially available rechargeable batteries) it will last for the entire lifetime of most devices, which makes the device environmentally friendly. Rechargeable batteries wear out typically over a few years and their highly reactive chemical electrolytes present a disposal and safety hazard. Battery lifetime can be optimized by charging only under favourable conditions, at an ideal rate and, for some chemistries, as infrequently as possible. EDLCs can help in conjunction with batteries by acting as a charge conditioner, storing energy from other sources for load balancing purposes and then using any excess energy to charge the batteries at a suitable time.

(ii) Low cost per cycle.

(iii) Good reversibility.

(iv) Very high rates of charge and discharge.

(v) Extremely low internal resistance (ESR) and consequent high cycle efficiency (95% or more) and extremely low heating levels.

(vi) High output power.

(vii) High specific power; the specific power of electric double-layer capacitors can exceed 6 kW/kg at 95% efficiency.

(viii) Improved safety, no corrosive electrolyte and low toxicity of materials.

(ix) Simple charge methods – no full-charge detection is needed; no danger of overcharging.

(x) When used in conjunction with rechargeable batteries, in some applications the EDLC can supply energy for a short time, reducing battery cycling duty and extending life.

Disadvantages of supercapacitors include:

(i) The amount of energy stored per unit weight is generally lower than that of an electrochemical battery (3 to 5 Wh/kg for a standard ultracapacitor, although 85 Wh/kg has been achieved in the lab compared to 30 to 40 Wh/kg for a lead–acid battery, 100 to 250 Wh/kg for a lithium-ion battery and about 0.1% of the volumetric energy density of gasoline.

(ii) Has the highest dielectric absorption of any type of capacitor.

(iii) High self-discharge – the rate is considerably higher than that of an electrochemical battery.

(iv) Low maximum voltage – series connections are needed to obtain higher voltages and voltage balancing may be required.

(v) Unlike practical batteries, the voltage across any capacitor, including EDLCs, drops significantly as it discharges. Effective storage and recovery of energy requires complex electronic control and switching equipment, with consequent energy loss.

(vi) Very low internal resistance allows extremely rapid discharge when shorted, resulting in a spark hazard similar to any other capacitor of similar voltage and capacitance (generally much higher than electrochemical cells).

Summary

Ultracapacitors are some of the best devices available for delivering a quick surge of power. Because an ultracapacitor stores energy in an electric field, rather than in a chemical reaction, it can survive hundreds of thousands more charge and discharge cycles than a battery can.

8.14 Discharging capacitors

When a capacitor has been disconnected from the supply it may still be charged and it may retain this charge for some considerable time. Thus precautions must be taken to ensure that the capacitor is automatically discharged after the supply is switched off. This is done by connecting a high-value resistor across the capacitor terminals.

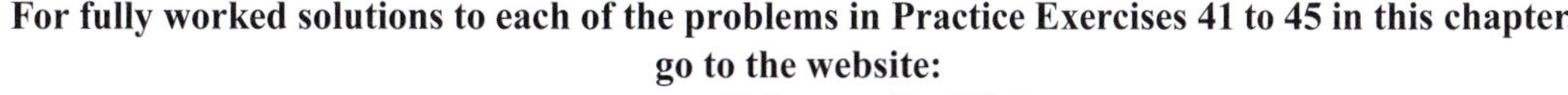

For fully worked solutions to each of the problems in Practice Exercises 41 to 45 in this chapter, go to the website:
www.routledge.com/cw/bird

Chapter 9

Magnetic circuits

Why it is important to understand: **Magnetic circuits**

Many common devices rely on magnetism. Familiar examples include computer disk drives, tape recorders, VCRs, transformers, motors, generators and so on. Practically all transformers and electric machinery uses magnetic material for shaping and directing the magnetic fields which act as a medium for transferring and connecting energy. It is therefore important to be able to analyse and describe magnetic field quantities for understanding these devices. Magnetic materials are significant in determining the properties of a piece of electromagnetic equipment or an electric machine, and affect its size and efficiency. To understand their operation, knowledge of magnetism and magnetic circuit principles is required. In this chapter, we look at fundamentals of magnetism, relationships between electrical and magnetic quantities, magnetic circuit concepts and methods of analysis.

At the end of this chapter you should be able to:

- appreciate some applications of magnets
- describe the magnetic field around a permanent magnet
- state the laws of magnetic attraction and repulsion for two magnets in close proximity
- define magnetic flux, Φ, and magnetic flux density, B, and state their units
- perform simple calculations involving $B = \dfrac{\Phi}{A}$
- define magnetomotive force, F_m, and magnetic field strength, H, and state their units
- perform simple calculations involving $F_m = NI$ and $H = \dfrac{NI}{l}$
- define permeability, distinguishing between μ_0, μ_r and μ
- understand the B–H curves for different magnetic materials
- appreciate typical values of μ_r
- perform calculations involving $B = \mu_0 \mu_r H$
- define reluctance, S, and state its units
- perform calculations involving $S = \dfrac{\text{mmf}}{\Phi} = \dfrac{l}{\mu_0 \mu_r A}$
- perform calculations on composite series magnetic circuits
- compare electrical and magnetic quantities
- appreciate how a hysteresis loop is obtained and that hysteresis loss is proportional to its area

Electrical Circuit Theory and Technology. 978-1-138-67349-6, © 2017 John Bird. Published by Taylor & Francis. All rights reserved.

9.1 Introduction to magnetism and magnetic circuits

The study of magnetism began in the thirteenth century with many eminent scientists and physicists such as William Gilbert,* Hans Christian Oersted*, Michael Faraday*, James Maxwell*, André Ampère* and Wilhelm Weber* all having some input on the subject since. The association between electricity and magnetism is a fairly recent finding in comparison with the very first understanding of basic magnetism.

Today, magnets have **many varied practical applications**. For example, they are used in motors and generators, telephones, relays, loudspeakers, computer hard drives and floppy disks, anti-lock brakes, cameras, fishing reels, electronic ignition systems, keyboards, TV and radio components and in transmission equipment.

Hans Christian Örsted.

* **Who was** Oersted? **Hans Christian Oersted** (14 August 1777–9 March 1851) was the Danish physicist and chemist who discovered that electric currents create magnetic fields. To find out more go to **www.routledge.com/cw/bird**

* Who was Faraday? For image and resume of Faraday, see page 108. To find out more go to **www.routledge.com/cw/bird**

***Who was** Gilbert? **William Gilbert** (24 May 1544–30 November 1603) was an English physician, physicist and natural philosopher who is credited as one of the originators of the term *electricity*. To find out more go to **www.routledge.com/cw/bird**

James Clerk Maxwell.

* **Who was** Maxwell? **James Clerk Maxwell** (13 June 1831–5 November 1879) was a Scottish theoretical physicist who formulated classical electromagnetic theory.

* **Who was** Ampère? For image and resume of Ampère, see page 50. To find out more go to **www.routledge.com/cw/bird**

The full theory of magnetism is one of the most complex of subjects; this chapter provides an introduction to the topic.

9.2 Magnetic fields

A **permanent magnet** is a piece of ferromagnetic material (such as iron, nickel or cobalt) which has properties of attracting other pieces of these materials. A permanent magnet will position itself in a north and south direction when freely suspended. The north-seeking end of the magnet is called the **north pole**, ***N***, and the south-seeking end the **south pole**, ***S***.

The area around a magnet is called the **magnetic field** and it is in this area that the effects of the **magnetic force** produced by the magnet can be detected. A magnetic field cannot be seen, felt, smelled or heard and therefore is difficult to represent. Michael Faraday suggested that the magnetic field could be represented pictorially, by imagining the field to consist of **lines of magnetic flux**, which enables investigation of the distribution and density of the field to be carried out.

The distribution of a magnetic field can be investigated by using some iron filings. A bar magnet is placed on a flat surface covered by, say, cardboard, upon which is sprinkled some iron filings. If the cardboard is gently tapped the filings will assume a pattern similar to that shown in Figure 9.1. If a number of magnets of different strength are used, it is found that the stronger the field the closer are the lines of magnetic flux and vice-versa. Thus a magnetic field has the property of exerting a force, demonstrated in this case by causing the iron filings to move into the pattern shown. The strength of the magnetic field decreases as we move away from the magnet. It should be realized, of course, that the magnetic field is three dimensional in its effect, and not acting in one plane as appears to be the case in this experiment.

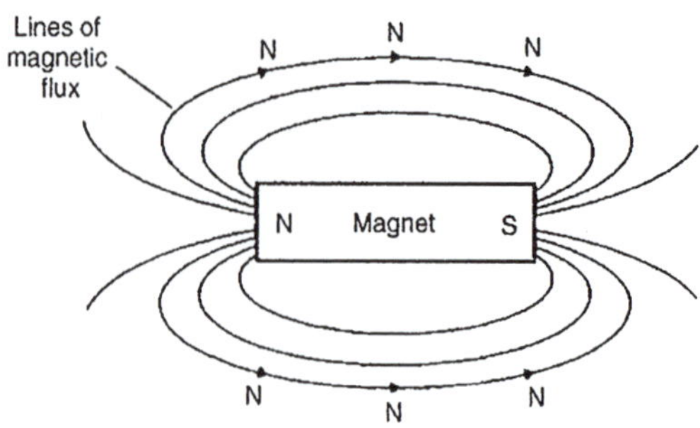

Figure 9.1

If a compass is placed in the magnetic field in various positions, the direction of the lines of flux may be determined by noting the direction of the compass pointer. The direction of a magnetic field at any point is taken as that in which the north-seeking pole of a compass needle points when suspended in the field. The direction of a line of flux is from the north pole to the south pole on the outside of the magnet and is then assumed to continue through the magnet back to the point at which it emerged at the north pole. Thus such lines of flux always form complete closed loops or paths, they never intersect and always have a definite direction. The laws of magnetic attraction and repulsion can be demonstrated by using two bar magnets. In Figure 9.2(a), **with unlike poles adjacent, attraction takes place**. Lines of flux are

* **Who was** Weber? **Wilhelm Eduard Weber** (24 October 1804–23 June 1891) was a German physicist who, together with Gauss, invented the first electromagnetic telegraph. To find out more go to **www.routledge.com/cw/bird**

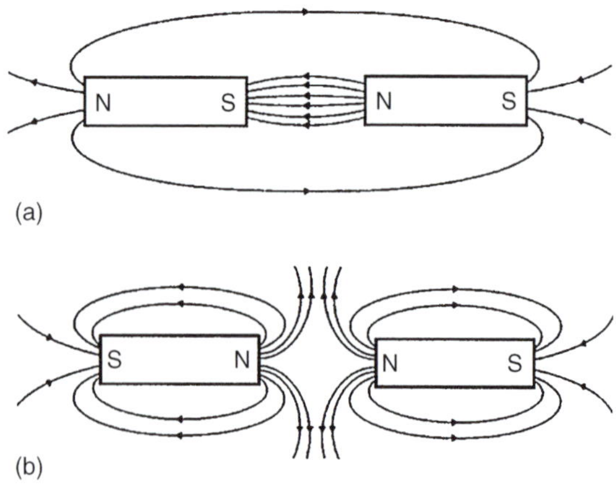

Figure 9.2

imagined to contract and the magnets try to pull together. The magnetic field is strongest in between the two magnets, shown by the lines of flux being close together. In Figure 9.2(b), **with similar poles adjacent (i.e. two north poles), repulsion occurs**, i.e. the two north poles try to push each other apart, since magnetic flux lines running side by side in the same direction repel.

9.3 Magnetic flux and flux density

Magnetic flux is the amount of magnetic field (or the number of lines of force) produced by a magnetic source. The symbol for magnetic flux is Φ (Greek letter 'phi'). The unit of magnetic flux is the **weber,*** **Wb**.

Magnetic flux density is the amount of flux passing through a defined area that is perpendicular to the direction of the flux:

$$\textbf{Magnetic flux density} = \frac{\textbf{magnetic flux}}{\textbf{area}}$$

The symbol for magnetic flux density is B. The unit of magnetic flux density is the **tesla,*** **T**, where $1\,\text{T} = 1\,\text{Wb/m}^2$. Hence

$$B = \frac{\Phi}{A}\ \textbf{tesla} \quad \text{where } A(\text{m}^2) \text{ is the area}$$

Problem 1. A magnetic pole face has a rectangular section having dimensions 200 mm by 100 mm. If the total flux emerging from the pole is 150 μWb, calculate the flux density.

Flux $\Phi = 150\,\mu\text{Wb} = 150 \times 10^{-6}\,\text{Wb}$

Cross-sectional area $A = 200 \times 100 = 20\,000\,\text{mm}^2$

$= 20\,000 \times 10^{-6}\,\text{m}^2$

$$\text{Flux density } B = \frac{\Phi}{A} = \frac{150 \times 10^{-6}}{20\,000 \times 10^{-6}}$$

$$= \mathbf{0.0075\ T\ or\ 7.5\ mT}$$

Problem 2. The maximum working flux density of a lifting electromagnet is 1.8 T and the effective area of a pole face is circular in cross-section. If the total magnetic flux produced is 353 mWb, determine the radius of the pole face.

Flux density $B = 1.8\,\text{T}$;
flux $\Phi = 353\,\text{mWb} = 353 \times 10^{-3}\,\text{Wb}$

Since $B = \frac{\Phi}{A}$, cross-sectional area $A = \frac{\Phi}{B}$

$$= \frac{353 \times 10^{-3}}{1.8}\,\text{m}^2$$

$$= 0.1961\,\text{m}^2$$

The pole face is circular, hence area $= \pi r^2$, where r is the radius.
Hence $\pi r^2 = 0.1961$

from which $r^2 = \frac{0.1961}{\pi}$ and radius $r = \sqrt{\left(\frac{0.1961}{\pi}\right)}$

$$= 0.250\,\text{m}$$

i.e. **the radius of the pole face is 250 mm**

9.4 Magnetomotive force and magnetic field strength

Magnetomotive force (mmf) is the cause of the existence of a magnetic flux in a magnetic circuit,

$$\textbf{mmf, } F_m = NI \textbf{ amperes}$$

* **Who was** Tesla? **Nikola Tesla** (10 July 1856–7 January 1943), after emigrating to the United States in 1884 to work for Thomas Edison, set up a number of laboratories and companies to develop a range of electrical devices. To find out more go to **www.routledge.com/cw/bird**

* **Who was** Weber? For image and resume of Weber, see page 124. To find out more go to **www.routledge.com/cw/bird**

Part 2

where N is the number of conductors (or turns) and I is the current in amperes. The unit of mmf is sometimes expressed as 'ampere-turns'. However, since 'turns' have no dimensions, the SI unit of mmf is the ampere. **Magnetic field strength** (or **magnetizing force**),

$$H = NI/l \text{ ampere per metre}$$

where l is the mean length of the flux path in metres. Thus **mmf = NI = Hl amperes**.

Problem 3. A magnetizing force of 8000 A/m is applied to a circular magnetic circuit of mean diameter 30 cm by passing a current through a coil wound on the circuit. If the coil is uniformly wound around the circuit and has 750 turns, find the current in the coil.

$H = 8000$ A/m; $l = \pi d = \pi \times 30 \times 10^{-2}$ m;

$N = 750$ turns

Since $H = \frac{NI}{l}$ then, $I = \frac{Hl}{N} = \frac{8000 \times \pi \times 30 \times 10^{-2}}{750}$

Thus, **current I = 10.05 A**

Now try the following Practice Exercise

Practice Exercise 46 Flux, flux density, mmf and magnetic field strength (Answers on page 820)

1. What is the flux density in a magnetic field of cross-sectional area 20 cm^2 having a flux of 3 mWb?
2. Determine the total flux emerging from a magnetic pole face having dimensions 5 cm by 6 cm, if the flux density is 0.9 T
3. The maximum working flux density of a lifting electromagnet is 1.9 T and the effective area of a pole face is circular in cross-section. If the total magnetic flux produced is 611 mWb, determine the radius of the pole face.
4. A current of 5 A is passed through a 1000-turn coil wound on a circular magnetic circuit of radius 120 mm. Calculate (a) the magnetomotive force, and (b) the magnetic field strength.

9.5 Permeability and *B–H* curves

For air, or any non-magnetic medium, the ratio of magnetic flux density to magnetizing force is a constant, i.e. B/H = a constant. This constant is μ_0, the **permeability of free space** (or the magnetic space constant) and is equal to $4\pi \times 10^{-7}$ H/m, i.e. **for air, or any non-magnetic medium**, the ratio $B/H = \mu_0$. (Although all non-magnetic materials, including air, exhibit slight magnetic properties, these can effectively be neglected.)

For all media other than free space, $B/H = \mu_0 \mu_r$

where u_r is the relative permeability, and is defined as

$$\mu_r = \frac{\text{flux density in material}}{\text{flux density in a vacuum}}$$

μ_r varies with the type of magnetic material and, since it is a ratio of flux densities, it has no unit. From its definition, μ_r for a vacuum is 1.

$\mu_0 \mu_r = \mu$, called the **absolute permeability**.

By plotting measured values of flux density B against magnetic field strength H, a **magnetization curve** (or ***B–H* curve**) is produced. For non-magnetic materials this is a straight line. Typical curves for four magnetic materials are shown in Figure 9.3.

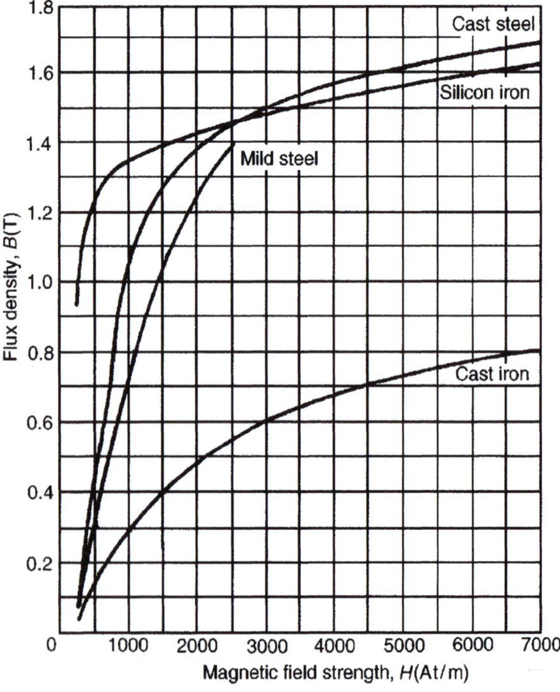

Figure 9.3 *B–H* curves for four materials

The **relative permeability** of a ferromagnetic material is proportional to the slope of the B–H curve and thus varies with the magnetic field strength. The approximate range of values of relative permeability μ_r for some common magnetic materials are:

Cast iron	$\mu_r = 100$–250	Mild steel	$\mu_r = 200$–800
Silicon iron	$\mu_r = 1000$–5000	Cast steel	$\mu_r = 300$–900
Mumetal	$\mu_r = 200$–5000	Stalloy	$\mu_r = 500$–6000

Problem 4. A flux density of 1.2 T is produced in a piece of cast steel by a magnetizing force of 1250 A/m. Find the relative permeability of the steel under these conditions.

For a magnetic material:

$$B = \mu_0 \mu_r H$$

i.e. $$u_r = \frac{B}{\mu_0 H} = \frac{1.2}{(4\pi \times 10^{-7})(1250)} = \mathbf{764}$$

Problem 5. Determine the magnetic field strength and the mmf required to produce a flux density of 0.25 T in an air gap of length 12 mm.

For air: $B = \mu_0 H$ (since $\mu_r = 1$)

Magnetic field strength $$H = \frac{B}{\mu_0} = \frac{0.25}{4\pi \times 10^{-7}} = \mathbf{198\,940\ A/m}$$

$$\mathbf{mmf} = Hl = 198\,940 \times 12 \times 10^{-3} = \mathbf{2387\ A}$$

Problem 6. A coil of 300 turns is wound uniformly on a ring of non-magnetic material. The ring has a mean circumference of 40 cm and a uniform cross-sectional area of 4 cm^2. If the current in the coil is 5 A, calculate (a) the magnetic field strength, (b) the flux density and (c) the total magnetic flux in the ring.

(a) Magnetic field strength $$H = \frac{NI}{l} = \frac{300 \times 5}{40 \times 10^{-2}} = \mathbf{3750\ A/m}$$

(b) For a non-magnetic material $\mu_r = 1$, thus flux density $B = \mu_0 H$

i.e. $\mathbf{B} = 4\pi \times 10^{-7} \times 3750 = \mathbf{4.712\ mT}$

(c) Flux $$\Phi = BA = (4.712 \times 10^{-3})(4 \times 10^{-4}) = \mathbf{1.885\ \mu Wb}$$

Problem 7. An iron ring of mean diameter 10 cm is uniformly wound with 2000 turns of wire. When a current of 0.25 A is passed through the coil a flux density of 0.4 T is set up in the iron. Find (a) the magnetizing force and (b) the relative permeability of the iron under these conditions.

$l = \pi d = \pi \times 10\,\text{cm} = \pi \times 10 \times 10^{-2}\,\text{m}$;
$N = 2000$ turns; $I = 0.25$ A; $B = 0.4$ T

(a) $$\mathbf{H} = \frac{NI}{l} = \frac{2000 \times 0.25}{\pi \times 10 \times 10^{-2}} = \frac{5000}{\pi} = \mathbf{1592\,A/m}$$

(b) $$B = \mu_0 \mu_r H, \text{ hence } \boldsymbol{\mu_r} = \frac{B}{\mu_0 H} = \frac{0.4}{(4\pi \times 10^{-7})(1592)} = \mathbf{200}$$

Problem 8. A uniform ring of cast iron has a cross-sectional area of 10 cm^2 and a mean circumference of 20 cm. Determine the mmf necessary to produce a flux of 0.3 mWb in the ring. The magnetization curve for cast iron is shown in Figure 9.3.

$A = 10\ \text{cm}^2 = 10 \times 10^{-4}\,\text{m}^2$; $l = 20$ cm $= 0.2$ m;
$\Phi = 0.3 \times 10^{-3}$ Wb

$$\text{Flux density } B = \frac{\Phi}{A} = \frac{0.3 \times 10^{-3}}{10 \times 10^{-4}} = 0.3\,\text{T}$$

From the magnetization curve for cast iron in Figure 9.3, when $B = 0.3$ T, $H = 1000$ A/m, hence
mmf$= Hl = 1000 \times 0.2 = \mathbf{200\ A}$
A tabular method could have been used in this problem. Such a solution is shown at the bottom of page 128.

9.6 Reluctance

Reluctance $\boldsymbol{S}$ (or R_M) is the 'magnetic resistance' of a magnetic circuit to the presence of magnetic flux.

Reluctance,

$$\boldsymbol{S} = \frac{F_M}{\Phi} = \frac{NI}{\Phi} = \frac{Hl}{BA} = \frac{l}{(B/H)A} = \frac{\boldsymbol{l}}{\boldsymbol{\mu_0 \mu_r A}}$$

The unit of reluctance is 1/H (or H^{-1}) or A/Wb.
Ferromagnetic materials have a low reluctance and can be used as **magnetic screens** to prevent magnetic fields affecting materials within the screen.

Problem 9. Determine the reluctance of a piece of mumetal of length 150 mm and cross-sectional area 1800 mm^2 when the relative permeability is 4000. Find also the absolute permeability of the mumetal.

$$\textbf{Reluctance } S = \frac{l}{\mu_0 \mu_r A}$$

$$= \frac{150 \times 10^{-3}}{(4\pi \times 10^{-7})(4000)(1800 \times 10^{-6})}$$

$$= \mathbf{16\,580/H} \text{ or } \mathbf{16\,580\ A/Wb} \text{ or } \mathbf{16.58\ kA/Wb}$$

$$\textbf{Absolute permeability, } \boldsymbol{\mu} = \mu_0 \mu_r$$

$$= (4\pi \times 10^{-7})(4000)$$

$$= \mathbf{5.027 \times 10^{-3}\ H/m}$$

Problem 10. A mild steel ring has a radius of 50 mm and a cross-sectional area of 400 mm^2. A current of 0.5 A flows in a coil wound uniformly around the ring and the flux produced is 0.1 mWb. If the relative permeability at this value of current is 200, find (a) the reluctance of the mild steel and (b) the number of turns on the coil.

$l = 2\pi r = 2 \times \pi \times 50 \times 10^{-3}$ m; $A = 400 \times 10^{-6}$ m^2;

$I = 0.5$ A; $\Phi = 0.1 \times 10^{-3}$ Wb; $\mu_r = 200$

(a) $$\textbf{Reluctance } S = \frac{l}{\mu_0 \mu_r A}$$

$$= \frac{2 \times \pi \times 50 \times 10^{-3}}{(4\pi \times 10^{-7})(200)(400 \times 10^{-6})}$$

$$= \mathbf{3.125 \times 10^6/H}$$

(b) $S = \dfrac{\text{mmf}}{\Phi}$ i.e. mmf $= S\Phi$

so that $NI = S\Phi$ and

$$\text{hence } N = \frac{S\Phi}{I} = \frac{3.125 \times 10^6 \times 0.1 \times 10^{-3}}{0.5}$$

$$= \mathbf{625\ turns}$$

Now try the following Practice Exercise

Practice Exercise 47 Magnetic circuits (Answers on page 820)

(Where appropriate assume: $\mu_0 = 4\pi \times 10^{-7}$ H/m.)

1. Find the magnetic field strength and the magnetomotive force needed to produce a flux density of 0.33 T in an air gap of length 15 mm.
2. An air gap between two pole pieces is 20 mm in length and the area of the flux path across the gap is 5 cm^2. If the flux required in the air gap is 0.75 mWb, find the mmf necessary.
3. Find the magnetic field strength applied to a magnetic circuit of mean length 50 cm when a coil of 400 turns is applied to it, carrying a current of 1.2 A.
4. A solenoid 20 cm long is wound with 500 turns of wire. Find the current required to establish a magnetizing force of 2500 A/m inside the solenoid.
5. A magnetic field strength of 5000 A/m is applied to a circular magnetic circuit of mean diameter 250 mm. If the coil has 500 turns find the current in the coil.
6. Find the relative permeability of a piece of silicon iron if a flux density of 1.3 T is produced by a magnetic field strength of 700 A/m
7. Part of a magnetic circuit is made from steel of length 120 mm, cross-sectional area 15 cm^2 and relative permeability 800. Calculate (a) the reluctance and (b) the absolute permeability of the steel.
8. A steel ring of mean diameter 120 mm is uniformly wound with 1500 turns of wire. When a current of 0.30 A is passed through the coil a flux density of 1.5 T is set up in the steel. Find the relative permeability of the steel under these conditions.

Part of circuit	Material	Φ (Wb)	A (m^2)	$B = \dfrac{\Phi}{A}$ (T)	H from graph	l (m)	mmf= Hl (A)
Ring	Cast iron	0.3×10^{-3}	10×10^{-4}	0.3	1000	0.2	200

9. A mild steel closed magnetic circuit has a mean length of 75 mm and a cross-sectional area of 320.2 mm^2. A current of 0.40 A flows in a coil wound uniformly around the circuit and the flux produced is 200 μWb. If the relative permeability of the steel at this value of current is 400, find (a) the reluctance of the material and (b) the number of turns of the coil.

10. A uniform ring of cast steel has a cross-sectional area of 5 cm^2 and a mean circumference of 15 cm. Find the current required in a coil of 1200 turns wound on the ring to produce a flux of 0.8 mWb. (Use the magnetization curve for cast steel shown on page 126.)

11. (a) A uniform mild steel ring has a diameter of 50 mm and a cross-sectional area of 1 cm^2. Determine the mmf necessary to produce a flux of 50 μWb in the ring. (Use the B–H curve for mild steel shown on page 126.)
(b) If a coil of 440 turns is wound uniformly around the ring in part (a), what current would be required to produce the flux?

9.7 Composite series magnetic circuits

For a series magnetic circuit having n parts, the **total reluctance** $\boldsymbol{S}$ is given by:

$$S = S_1 + S_2 + \cdots + S_n$$

(This is similar to resistors connected in series in an electrical circuit.)

Problem 11. A closed magnetic circuit of cast steel contains a 6 cm long path of cross-sectional area 1 cm^2 and a 2 cm path of cross-sectional area 0.5 cm^2. A coil of 200 turns is wound around the 6 cm length of the circuit and a current of 0.4 A flows. Determine the flux density in the 2 cm path, if the relative permeability of the cast steel is 750.

For the 6 cm long path:

$$\text{Reluctance } S_1 = \frac{l_1}{\mu_0\mu_r A_1} = \frac{6\times10^{-2}}{(4\pi\times10^{-7})(750)(1\times10^{-4})} = 6.366\times10^5/\text{H}$$

For the 2 cm long path:

$$\text{Reluctance } S_2 = \frac{l_2}{\mu_0\mu_r A_2} = \frac{2\times10^{-2}}{(4\pi\times10^{-7})(750)(0.5\times10^{-4})} = 4.244\times10^5/\text{H}$$

$$\text{Total circuit reluctance } S = S_1 + S_2 = (6.366+4.244)\times10^5 = 10.61\times10^5/\text{H}$$

$$S = \frac{\text{mmf}}{\Phi}, \text{ i.e. } \Phi = \frac{\text{mmf}}{S} = \frac{NI}{S} = \frac{200\times0.4}{10.61\times10^5} = 7.54\times10^{-5}\text{ Wb}$$

$$\text{Flux density in the 2 cm path, } B = \frac{\Phi}{A} = \frac{7.54\times10^{-5}}{0.5\times10^{-4}} = \mathbf{1.51\ T}$$

Problem 12. A silicon iron ring of cross-sectional area 5 cm^2 has a radial air gap of 2 mm cut into it. If the mean length of the silicon iron path is 40 cm, calculate the magnetomotive force to produce a flux of 0.7 mWb. The magnetization curve for silicon is shown on page 126.

There are two parts to the circuit – the silicon iron and the air gap. The total mmf will be the sum of the mmfs of each part.

For the silicon iron: $B = \dfrac{\Phi}{A} = \dfrac{0.7\times10^{-3}}{5\times10^{-4}} = 1.4\text{T}$

From the B–H curve for silicon iron on page 126, when B = 1.4 T, H = 1650 At/m.

Hence the mmf for the iron path $= Hl = 1650\times0.4 = 660$ A

For the air gap:

The flux density will be the same in the air gap as in the iron, i.e. 1.4 T. (This assumes no leakage or fringing occurring.)

$$\text{For air, } H = \frac{B}{\mu_0} = \frac{1.4}{4\pi\times10^{-7}} = 1\,114\,000\text{ A/m}$$

Hence the mmf for the air gap $= Hl$

$$= 1\,114\,000 \times 2 \times 10^{-3}$$
$$= 2228\,\text{A}$$

Total mmf to produce a flux of 0.7 mWb

$$= 660 + 2228$$
$$= \mathbf{2888\ A}$$

A tabular method could have been used as shown at the bottom of the page.

Part of circuit	Material	Φ (Wb)	A (m^2)	B (T)	H (A/m)	l (m)	mmf = Hl (A)
Ring	Silicon iron	0.7×10^{-3}	5×10^{-4}	1.4	1650 (from graph)	0.4	660
Air gap	Air	0.7×10^{-3}	5×10^{-4}	1.4	$\frac{1.4}{4\pi \times 10^{-7}}$ $= 1\,114\,000$	2×10^{-3}	2228
							Total: **2888 A**

Problem 13. Figure 9.4 shows a ring formed with two different materials – cast steel and mild steel. The dimensions are:

	Mean length	Cross-sectional area
Mild steel	400 mm	500 mm^2
Cast steel	300 mm	312.5 mm^2

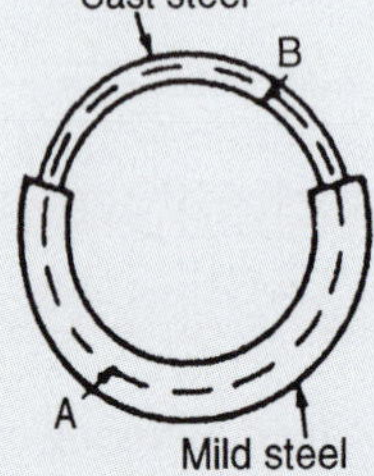

Figure 9.4

Find the total mmf required to cause a flux of 500 μWb in the magnetic circuit. Determine also the total circuit reluctance.

A tabular solution is shown at the top of page 131.

$$\textbf{Total circuit reluctance } S = \frac{\text{mmf}}{\Phi} = \frac{2000}{500 \times 10^{-6}}$$
$$= \mathbf{4 \times 10^6/H}$$

Problem 14. A section through a magnetic circuit of uniform cross-sectional area 2 cm^2 is shown in Figure 9.5. The cast steel core has a mean length of 25 cm. The air gap is 1 mm wide and the coil has 5000 turns. The B–H curve for cast steel is shown on page 126. Determine the current in the coil to produce a flux density of 0.80 T in the air gap, assuming that all the flux passes through both parts of the magnetic circuit.

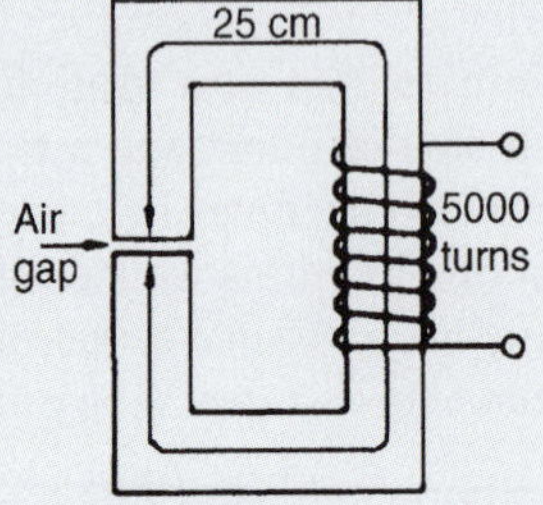

Figure 9.5

For the cast steel core, when $B = 0.80$ T, $H = 750$ A/m (from page 126)

Reluctance of core $S_1 = \dfrac{l_1}{\mu_0 \mu_r A_1}$ and since $B = \mu_0 \mu_r H$,

then $\mu_r = \dfrac{B}{\mu_0 H}$. Thus $S_1 = \dfrac{l_1}{\mu_0 \left(\dfrac{B}{\mu_0 H}\right) A} = \dfrac{l_1 H}{BA}$

$$= \frac{(25 \times 10^{-2})(750)}{(0.8)(2 \times 10^{-4})}$$
$$= 1\,172\,000/\text{H}$$

For the air gap:

Reluctance, $S_2 = \dfrac{l_2}{\mu_0 \mu_r A_2} = \dfrac{l_2}{\mu_0 A_2}$

(since $\mu_r = 1$ for air)

$$= \frac{1 \times 10^{-3}}{(4\pi \times 10^{-7})(2 \times 10^{-4})}$$
$$= 3\,979\,000/\text{H}$$

Part of circuit	Material	Φ (Wb)	A (m^2)	B(T) (= Φ/A)	H (A/m) (from graphs p 69)	l (m)	mmf= Hl (A)
A	Mild steel	500×10^{-6}	500×10^{-6}	1.0	1400	400×10^{-3}	560
B	Cast steel	500×10^{-6}	312.5×10^{-6}	1.6	4800	300×10^{-3}	1440
							Total: **2000 A**

$$\text{Total circuit reluctance } S = S_1 + S_2 = 1\,172\,000 + 3\,979\,000 = 5\,151\,000/\text{H}$$

$$\text{Flux } \Phi = BA = 0.80 \times 2 \times 10^{-4} = 1.6 \times 10^{-4}\,\text{Wb}$$

$$S = \frac{\text{mmf}}{\Phi}, \text{ thus mmf} = S\Phi$$

Hence $NI = S\Phi$

and current $I = \dfrac{S\Phi}{N}$

$$= \frac{(5\,151\,000)(1.6\times10^{-4})}{5000} = \mathbf{0.165\,A}$$

Now try the following Practice Exercise

Practice Exercise 48 Composite series magnetic circuits (Answers on page 820)

(Where appropriate assume $\mu_0 = 4\pi \times 10^{-7}$ H/m)

1. A magnetic circuit of cross-sectional area $0.4\,\text{cm}^2$ consists of one part 3 cm long, of material having relative permeability 1200, and a second part 2 cm long of material having relative permeability 750. With a 100-turn coil carrying 2 A, find the value of flux existing in the circuit.
2. (a) A cast steel ring has a cross-sectional area of $600\,\text{mm}^2$ and a radius of 25 mm. Determine the mmf necessary to establish a flux of 0.8 mWb in the ring. Use the B–H curve for cast steel shown on page 126.
   (b) If a radial air gap 1.5 mm wide is cut in the ring of part (a), find the mmf now necessary to maintain the same flux in the ring.
3. For the magnetic circuit shown in Figure 9.6 find the current I in the coil needed to produce a flux of 0.45 mWb in the air gap. The silicon iron magnetic circuit has a uniform cross-sectional area of $3\,\text{cm}^2$ and its magnetization curve is as shown on page 126.

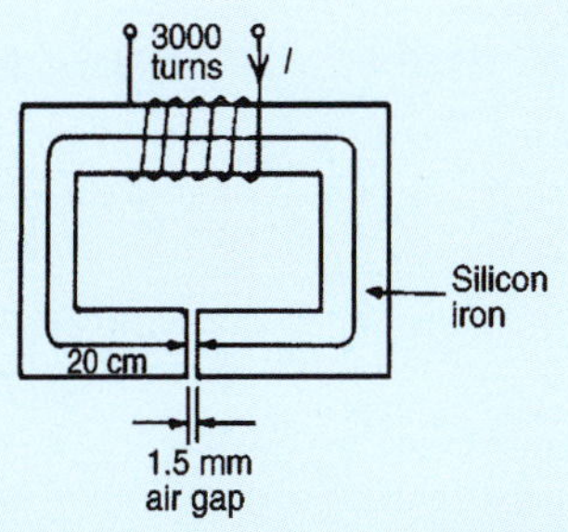

Figure 9.6

4. A ring forming a magnetic circuit is made from two materials; one part is mild steel of mean length 25 cm and cross-sectional area $4\,\text{cm}^2$, and the remainder is cast iron of mean length 20 cm and cross-sectional area $7.5\,\text{cm}^2$. Use a tabular approach to determine the total mmf required to cause a flux of 0.30 mWb in the magnetic circuit. Find also the total reluctance of the circuit. Use the magnetization curves shown on page 126.
5. Figure 9.7 shows the magnetic circuit of a relay. When each of the air gaps are 1.5 mm wide find the mmf required to produce a flux density of 0.75 T in the air gaps. Use the B–H curves shown on page 126.

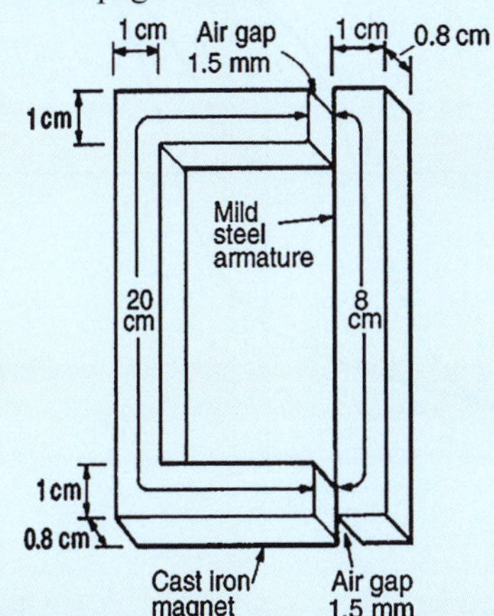

Figure 9.7

9.8 Comparison between electrical and magnetic quantities

Electrical circuit		Magnetic circuit	
e.m.f. E	(V)	mmf F_m	(A)
current I	(A)	flux Φ	(Wb)
resistance R	(Ω)	reluctance S (H^{-1})	
$I=\frac{E}{R}$		$\Phi=\frac{\text{mmf}}{S}$	
$R=\frac{\rho l}{A}$		$S=\frac{l}{\mu_0\mu_r A}$	

9.9 Hysteresis and hysteresis loss

Hysteresis loop

Let a ferromagnetic material which is completely demagnetized, i.e. one in which $B=H=0$ be subjected to increasing values of magnetic field strength H and the corresponding flux density B measured. The resulting relationship between B and H is shown by the curve Oab in Figure 9.8. At a particular value of H, shown as Oy, it becomes difficult to increase the flux density any further. The material is said to be saturated. Thus ***by*** is the **saturation flux density**.

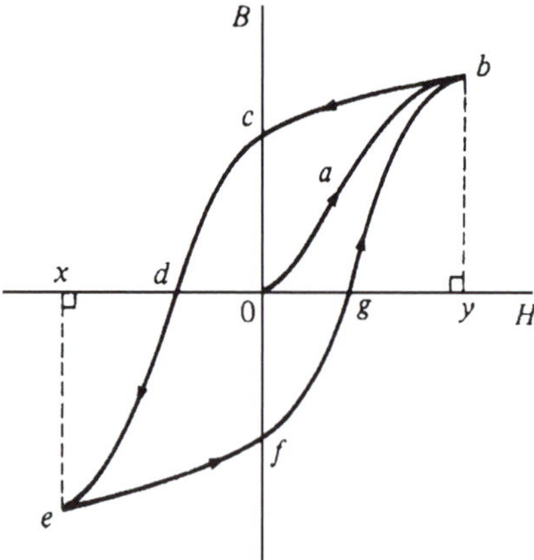

Figure 9.8

If the value of H is now reduced it is found that the flux density follows curve ***bc***. When H is reduced to zero, flux remains in the iron. This **remanent flux density** or **remanence** is shown as ***Oc*** in Figure 9.8. When H is increased in the opposite direction, the flux density decreases until, at a value shown as ***Od***, the flux density has been reduced to zero. The magnetic field strength ***Od*** required to remove the residual magnetism, i.e. reduce B to zero, is called the **coercive force**.

Further increase of H in the reverse direction causes the flux density to increase in the reverse direction until saturation is reached, as shown by curve ***de***. If H is varied backwards from ***Ox*** to ***Oy***, the flux density follows the curve ***efgb***, similar to curve ***bcde***.

It is seen from Figure 9.8 that the flux density changes lag behind the changes in the magnetic field strength. This effect is called **hysteresis**. The closed figure ***bcdefgb*** is called the **hysteresis loop** (or the *B*/*H* loop).

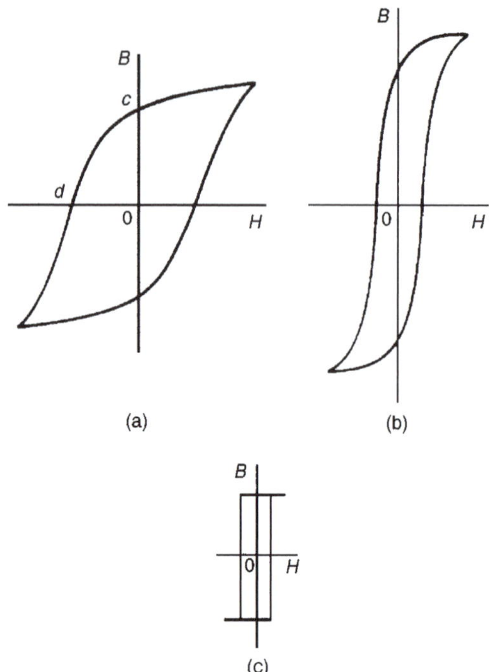

Figure 9.9

Hysteresis loss

A disturbance in the alignment of the domains (i.e. groups of atoms) of a ferromagnetic material causes energy to be expended in taking it through a cycle of magnetization. This energy appears as heat in the specimen and is called the **hysteresis loss**.

The energy loss associated with hysteresis is proportional to the area of the hysteresis loop.

The area of a hysteresis loop varies with the type of material. The area, and thus the energy loss, is much greater for hard materials than for soft materials.

Figure 9.9 shows typical hysteresis loops for:

(a) **hard material**, which has a high remanence *Oc* and a large coercivity ***Od***

(b) **soft steel**, which has a large remanence and small coercivity

(c) **ferrite**, this being a ceramic-like magnetic substance made from oxides of iron, nickel, cobalt, magnesium, aluminium and mangenese; the hysteresis of ferrite is very small.

For a.c.-excited devices the hysteresis loop is repeated every cycle of alternating current. Thus a hysteresis loop with a large area (as with hard steel) is often unsuitable since the energy loss would be considerable. Silicon steel has a narrow hysteresis loop, and thus small hysteresis loss, and is suitable for transformer cores and rotating machine armatures.

For fully worked solutions to each of the problems in Practice Exercises 46 to 48 in this chapter, go to the website:
www.routledge.com/cw/bird

Revision Test 2

This revision test covers the material contained in Chapters 7 to 9. *The marks for each question are shown in brackets at the end of each question.*

1. Resistances of 5 Ω, 7 Ω and 8 Ω are connected in series. If a 10 V supply voltage is connected across the arrangement, determine the current flowing through and the p.d. across the 7 Ω resistor. Calculate also the power dissipated in the 8 Ω resistor. (6)
2. For the series-parallel network shown in Figure RT2.1, find (a) the supply current, (b) the current flowing through each resistor, (c) the p.d. across each resistor, (d) the total power dissipated in the circuit, (e) the cost of energy if the circuit is connected for 80 hours. Assume electrical energy costs 14p per unit. (15)
3. The charge on the plates of a capacitor is 8 mC when the potential between them is 4 kV. Determine the capacitance of the capacitor. (2)
4. Two parallel rectangular plates measuring 80 mm by 120 mm are separated by 4 mm of mica and carry an electric charge of 0.48 C. The voltage between the plates is 500 V. Calculate (a) the electric flux density, (b) the electric field strength and (c) the capacitance of the capacitor, in picofarads, if the relative permittivity of mica is 5 (7)
5. A 4 F capacitor is connected in parallel with a 6 F capacitor. This arrangement is then connected in series with a 10 F capacitor. A supply p.d. of 250 V is connected across the circuit. Find (a) the equivalent capacitance of the circuit, (b) the voltage across the 10 F capacitor and (c) the charge on each capacitor. (7)
6. A coil of 600 turns is wound uniformly on a ring of non-magnetic material. The ring has a uniform cross-sectional area of 200 mm^2 and a mean circumference of 500 mm. If the current in the coil is 4 A, determine (a) the magnetic field strength, (b) the flux density and (c) the total magnetic flux in the ring. (5)
7. A mild steel ring of cross-sectional area 4 cm^2 has a radial air-gap of 3 mm cut into it. If the mean length of the mild steel path is 300 mm, calculate the magnetomotive force to produce a flux of 0.48 mWb. (Use the B–H curve on page 126.) (8)
8. In the circuit shown in Figure RT2.2, the slider S is at the halfway point.

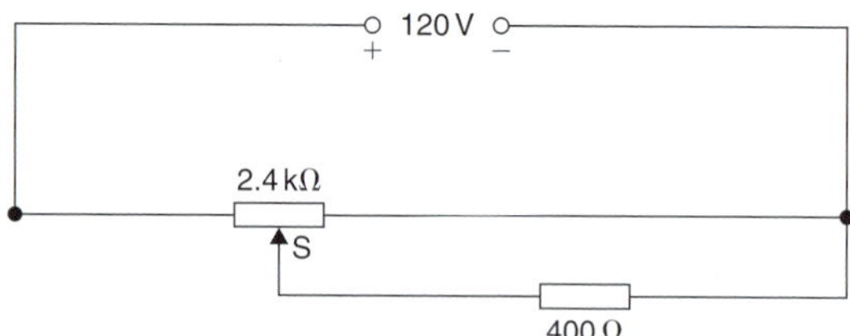

Figure RT2.2

   (a) Calculate the p.d. across and the current flowing in the 400 Ω load resistor.
   (b) Is the circuit a potentiometer or a rheostat? (5)
9. For the circuit shown in Figure RT2.3, calculate the current flowing in the 50 Ω load and the voltage drop across the load when
   (a) *XS* is 3/5 of *XY*
   (b) point S coincides with point *Y* (5)

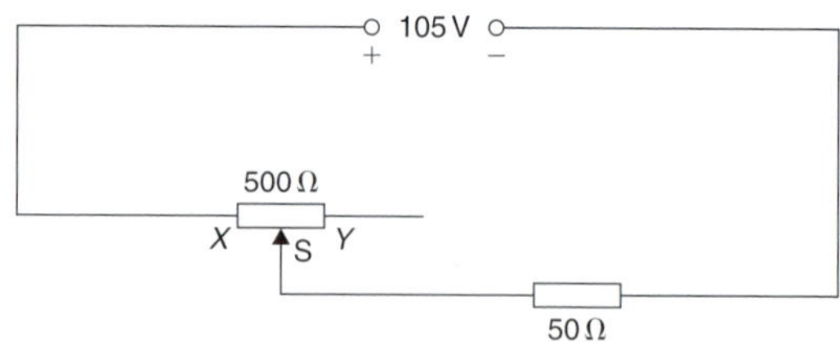

Figure RT2.3

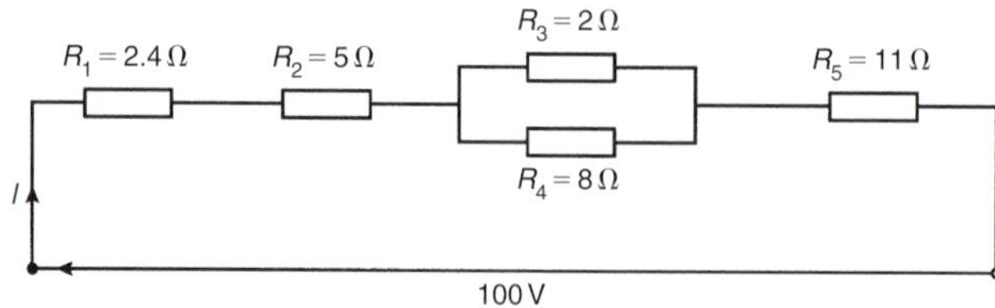

Figure RT2.1

For lecturers/instructors/teachers, fully worked solutions to each of the problems in Revision Test 2, together with a full marking scheme, are available at the website:
www.routledge.com/cw/bird

Chapter 10

Electromagnetism

Why it is important to understand: **Electromagnetism**

While the basic facts about magnetism have been known since ancient times, it was not until the early 1800s that the connection between electricity and magnetism was made and the foundations of modern electromagnetic theory established. In 1819, Hans Christian Oersted, a Danish scientist, demonstrated that electricity and magnetism were related when he showed that a compass needle was deflected by a current-carrying conductor. The following year, Andre Ampère showed that current-carrying conductors attract or repel each other just like magnets. However, it was Michael Faraday who developed our present concept of the magnetic field as a collection of flux lines in space that conceptually represent both the intensity and the direction of the field. It was this concept that led to an understanding of magnetism and the development of important practical devices such as the transformer and the electric generator. In electrical machines, ferromagnetic materials may form the magnetic circuits (as in transformers), or by ferromagnetic materials in conjunction with air (as in rotating machines). In most electrical machines, the magnetic field is produced by passing an electric current through coils wound on ferromagnetic material. In this chapter important concepts of electromagnetism are explained and simple calculations performed.

At the end of this chapter you should be able to:

- understand that magnetic fields are produced by electric currents
- apply the screw rule to determine direction of a magnetic field
- recognize that the magnetic field around a solenoid is similar to a magnet
- apply the screw rule or grip rule to a solenoid to determine magnetic field direction
- recognize and describe practical applications of an electromagnet, i.e. electric bell, relay, lifting magnet, telephone receiver
- appreciate factors upon which the force F on a current-carrying conductor depends
- perform calculations using $F = BIl$ and $F = BIl \sin\theta$
- recognize that a loudspeaker is a practical application of force F
- use Fleming's left-hand rule to pre-determine direction of force in a current-carrying conductor
- describe the principle of operation of a simple d.c. motor
- describe the principle of operation and construction of a moving coil instrument
- appreciate the force F on a charge in a magnetic field is given by $F = QvB$
- perform calculations using $F = QvB$

Electrical Circuit Theory and Technology. 978-1-138-67349-6, © 2017 John Bird. Published by Taylor & Francis. All rights reserved.

10.1 Magnetic field due to an electric current

Magnetic fields can be set up not only by permanent magnets, as shown in Chapter 9, but also by electric currents.

Let a piece of wire be arranged to pass vertically through a horizontal sheet of cardboard, on which is placed some iron filings, as shown in Figure 10.1(a).

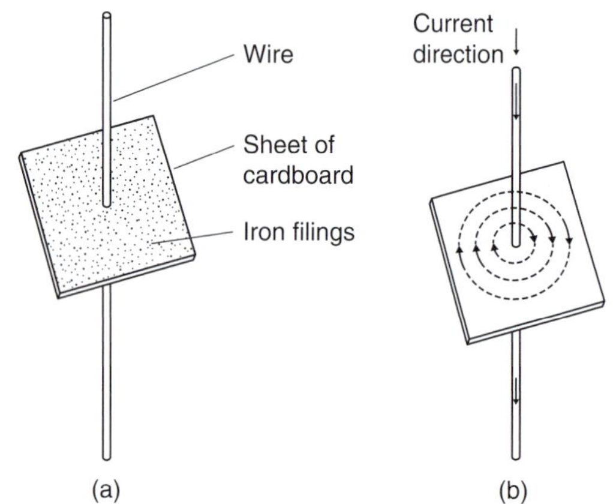

Figure 10.1

If a current is now passed through the wire, then the iron filings will form a definite circular field pattern with the wire at the centre when the cardboard is gently tapped. By placing a compass in different positions the lines of flux are seen to have a definite direction, as shown in Figure 10.1(b). If the current direction is reversed, the direction of the lines of flux is also reversed. The effect on both the iron filings and the compass needle disappears when the current is switched off. The magnetic field is thus produced by the electric current. The magnetic flux produced has the same properties as the flux produced by a permanent magnet. If the current is increased the strength of the field increases and, as for the permanent magnet, the field strength decreases as we move away from the current-carrying conductor.

In Figure 10.1 the effect of only a small part of the magnetic field is shown.

If the whole length of the conductor is similarly investigated it is found that the magnetic field around a straight conductor is in the form of concentric cylinders as shown in Figure 10.2, the field direction depending on the direction of the current flow.

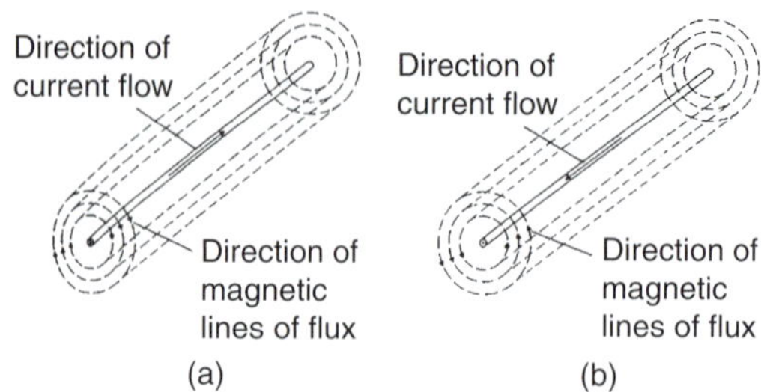

Figure 10.2

When dealing with magnetic fields formed by electric current it is usual to portray the effect as shown in Figure 10.3. The convention adopted is:

(i) Current flowing away from the viewer, i.e. into the paper, is indicated by ⊕. This may be thought of as the feathered end of the shaft of an arrow. See Figure 10.3(a).

(ii) Current flowing towards the viewer, i.e. out of the paper, is indicated by ⊙. This may be thought of as the tip of an arrow. See Figure 10.3(b).

The direction of the magnetic lines of flux is best remembered by the **screw rule**. This states that:

'If a normal right-hand thread screw is screwed along the conductor in the direction of the current, the direction of rotation of the screw is in the direction of the magnetic field.'

For example, with current flowing away from the viewer (Figure 10.3(a)) a right-hand thread screw driven into the paper has to be rotated clockwise. Hence the direction of the magnetic field is clockwise.

A magnetic field set up by a long coil, or **solenoid**, is shown in Figure 10.4(a) and is seen to be similar to that of a bar magnet. If the solenoid is wound on an iron bar, as shown in Figure 10.4(b), an even stronger magnetic field is produced, the iron becoming magnetized and behaving like a permanent magnet.

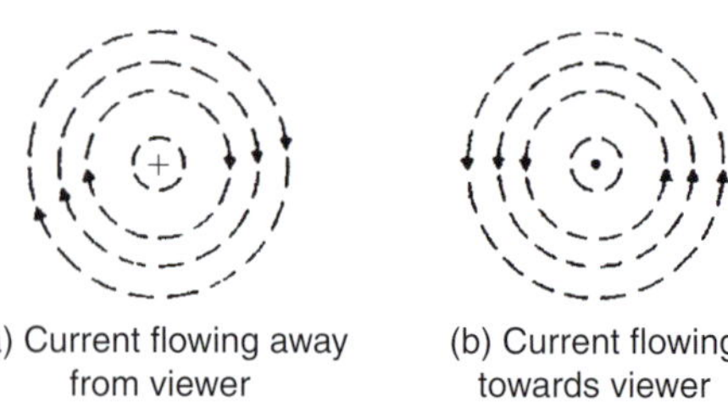

Figure 10.3

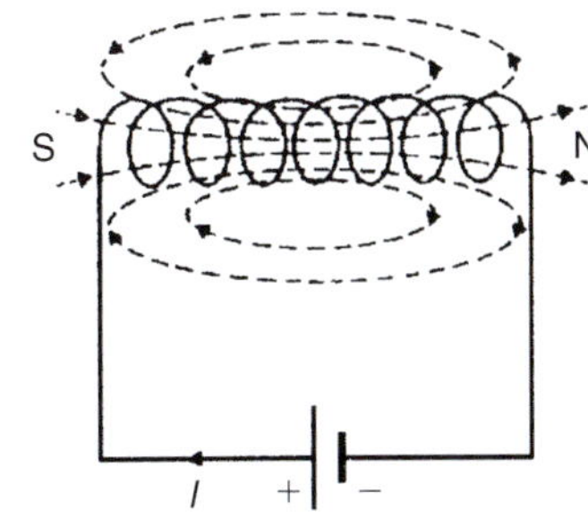

(a) Magnetic field of a solenoid

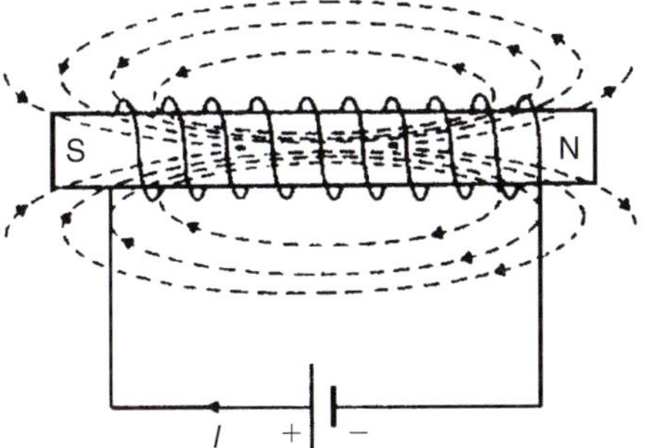

(b) Magnetic field of an iron-cored solenoid

Figure 10.4

The direction of the magnetic field produced by the current I in the solenoid may be found by either of two methods, i.e. the screw rule or the grip rule.

(a) **The screw rule** states that if a normal right-hand thread screw is placed along the axis of the solenoid and is screwed in the direction of the current it moves in the direction of the magnetic field **inside** the solenoid. The direction of the magnetic field **inside** the solenoid is from south to north. Thus in Figures 10.4(a) and (b) the north pole is to the right.

(b) **The grip rule** states that if the coil is gripped with the **right** hand, with the fingers pointing in the direction of the current, then the thumb, outstretched parallel to the axis of the solenoid, points in the direction of the magnetic field **inside** the solenoid.

Problem 1. Figure 10.5 shows a coil of wire wound on an iron core connected to a battery. Sketch the magnetic field pattern associated with the current-carrying coil and determine the polarity of the field.

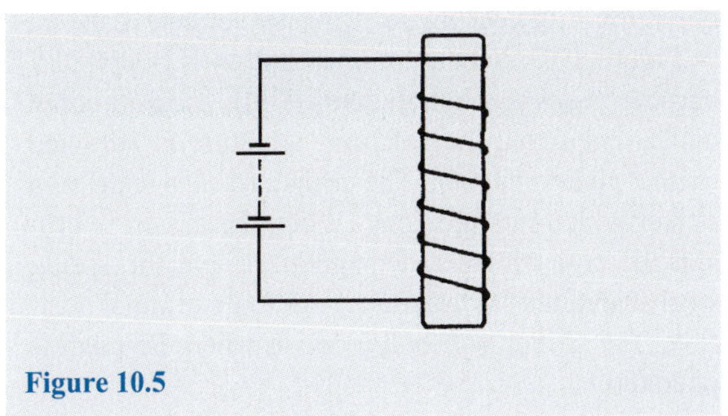

Figure 10.5

The magnetic field associated with the solenoid in Figure 10.5 is similar to the field associated with a bar magnet and is as shown in Figure 10.6. The polarity of the field is determined either by the screw rule or by the grip rule. Thus the north pole is at the bottom and the south pole at the top.

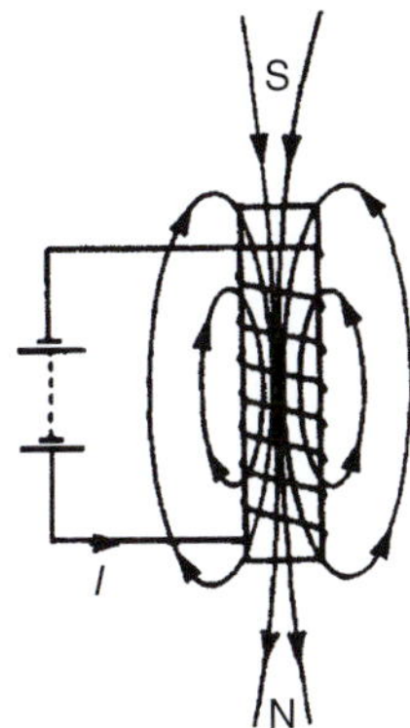

Figure 10.6

10.2 Electromagnets

The solenoid is very important in electromagnetic theory since the magnetic field inside the solenoid is practically uniform for a particular current, and is also versatile, inasmuch that a variation of the current can alter the strength of the magnetic field. An electromagnet, based on the solenoid, provides the basis of many items of electrical equipment, examples of which include electric bells, relays, lifting magnets and telephone receivers.

(i) Electric bell

There are various types of electric bell, including the single-stroke bell, the trembler bell, the buzzer and a continuously ringing bell, but all depend on the attraction exerted by an electromagnet on a soft iron armature. A typical single-stroke bell circuit is shown in Figure 10.7. When the push button is operated a current passes through the coil. Since the iron-cored coil is energized the soft iron armature is attracted to the electromagnet. The armature also carries a striker which hits the gong. When the circuit is broken the coil becomes demagnetized and the spring steel strip pulls the armature back to its original position. The striker will only operate when the push is operated.

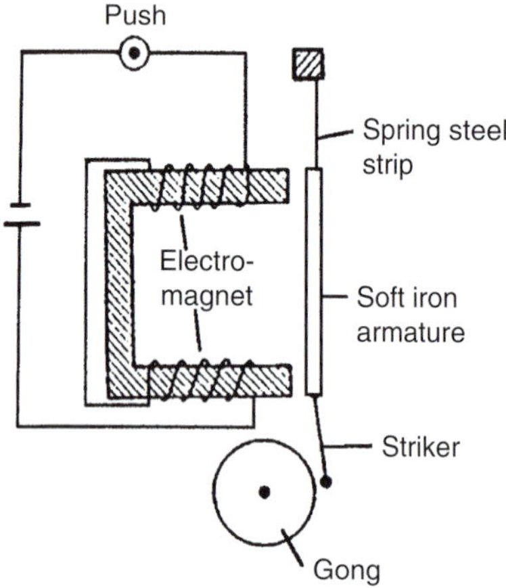

Figure 10.7

(ii) Relay

A relay is similar to an electric bell except that contacts are opened or closed by operation instead of a gong being struck. A typical simple relay is shown in Figure 10.8, which consists of a coil wound on a soft iron core. When the coil is energized the hinged soft iron armature is attracted to the electromagnet and pushes against two fixed contacts so that they are connected together, thus closing some other electrical circuit.

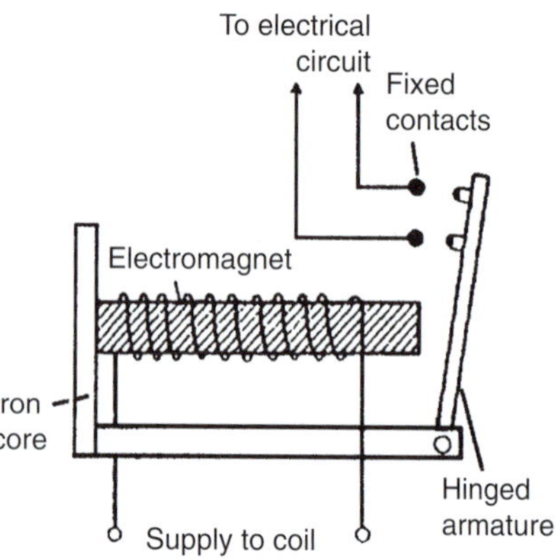

Figure 10.8

(iii) Lifting magnet

Lifting magnets, incorporating large electromagnets, are used in iron and steel works for lifting scrap metal. A typical robust lifting magnet, capable of exerting large attractive forces, is shown in the elevation and plan view of Figure 10.9, where a coil, C, is wound round a central core, P, of the iron casting. Over the face of the electromagnet is placed a protective non-magnetic sheet of material, R. The load, Q, which must be of magnetic material, is lifted when the coils are energized, the magnetic flux paths, M, being shown by the broken lines.

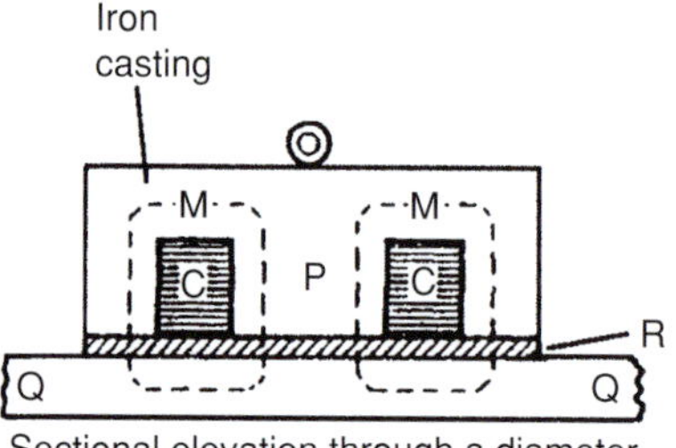

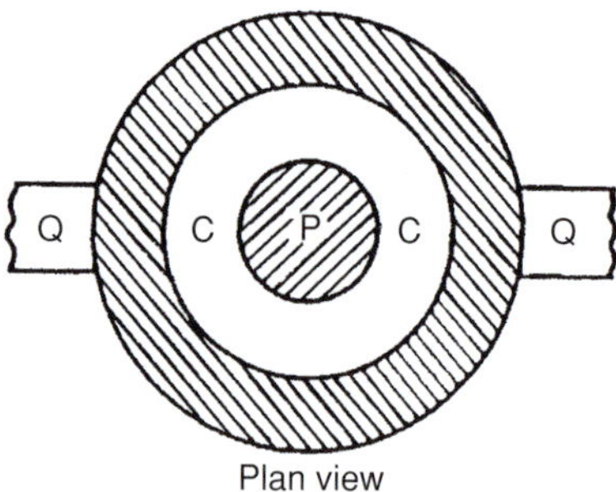

Figure 10.9

(iv) Telephone receiver

Whereas a transmitter or microphone changes sound waves into corresponding electrical signals, a telephone receiver converts the electrical waves back into sound waves. A typical telephone receiver is shown in Figure 10.10 and consists of a permanent magnet with coils wound on its poles. A thin, flexible diaphragm of magnetic material is held in position near to the magnetic poles but not touching them. Variation in current from the transmitter varies the magnetic field and the diaphragm consequently vibrates. The vibration produces sound variations corresponding to those transmitted.

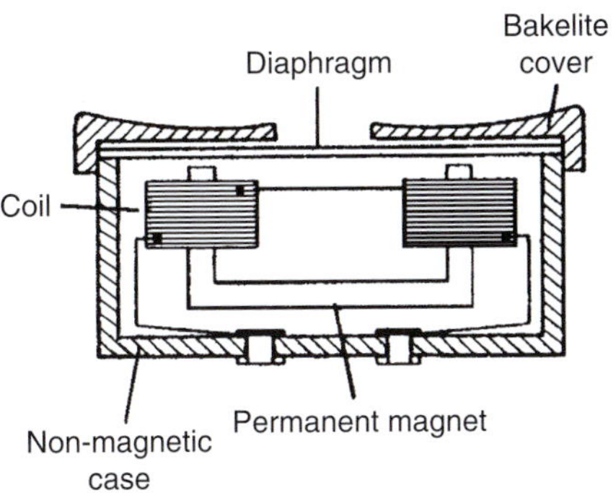

Figure 10.10

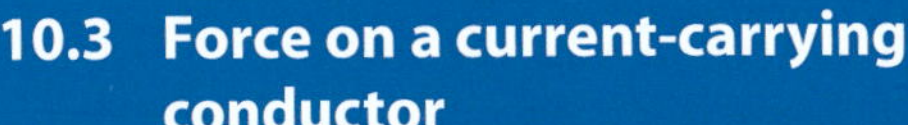

10.3 Force on a current-carrying conductor

If a current-carrying conductor is placed in a magnetic field produced by permanent magnets, then the fields due to the current-carrying conductor and the permanent magnets interact and cause a force to be exerted on the conductor. The force on the current-carrying conductor in a magnetic field depends upon:

(a) the flux density of the field, B teslas

(b) the strength of the current, I amperes

(c) the length of the conductor perpendicular to the magnetic field, l metres, and

(d) the directions of the field and the current.

When the magnetic field, the current and the conductor are mutually at right-angles, then:

Force $F = BIl$ newtons

When the conductor and the field are at an angle $\theta°$ to each other then:

Force $F = BIl \sin\theta$ newtons

Since when the magnetic field, current and conductor are mutually at right angles, $F = BIl$, the magnetic flux density B may be defined by $B = F/Il$, i.e. the flux density is 1 T if the force exerted on 1 m of a conductor when the conductor carries a current of 1 A is 1 N.

Loudspeaker

A simple application of the above force is the moving-coil loudspeaker. The loudspeaker is used to convert electrical signals into sound waves.

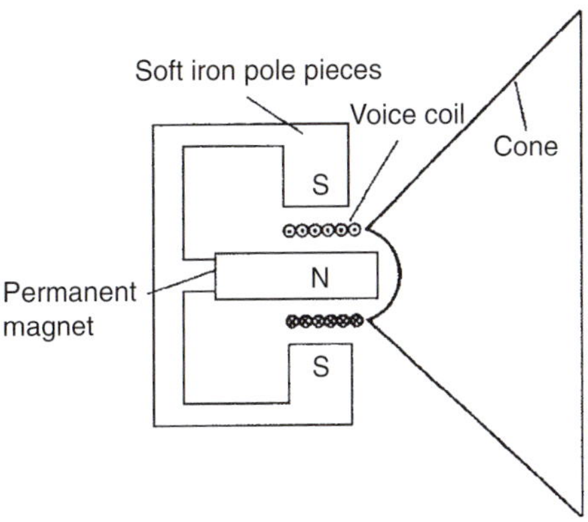

Figure 10.11

Figure 10.11 shows a typical loudspeaker having a magnetic circuit comprising a permanent magnet and soft iron pole pieces so that a strong magnetic field is available in the short cylindrical air gap. A moving coil, called the voice or speech coil, is suspended from the end of a paper or plastic cone so that it lies in the gap. When an electric current flows through the coil it produces a force which tends to move the cone backwards and forwards according to the direction of the current. The cone acts as a piston, transferring this force to the air, and producing the required sound waves.

Problem 2. A conductor carries a current of 20 A and is at right-angles to a magnetic field having a flux density of 0.9 T. If the length of the conductor in the field is 30 cm, calculate the force acting on the conductor. Determine also the value of the force if the conductor is inclined at an angle of 30° to the direction of the field.

$B = 0.9\,\text{T}$; $I = 20\,\text{A}$; $l = 30\,\text{cm} = 0.30\,\text{m}$

Force $F = BIl = (0.9)(20)(0.30)$ newtons when the conductor is at right-angles to the field, as shown in Figure 10.12(a), i.e. $\boldsymbol{F = 5.4\,\text{N}}$

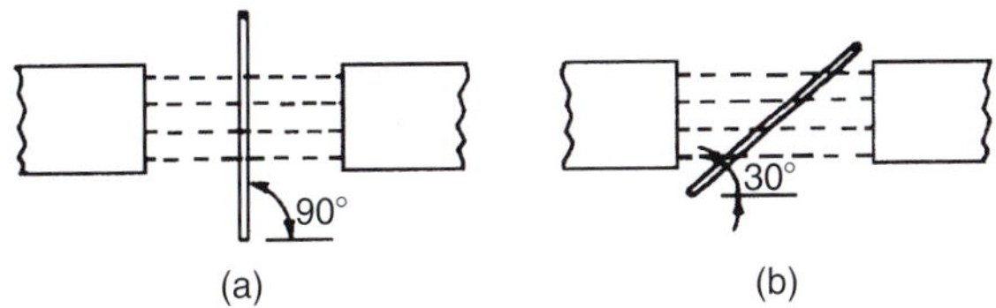

Figure 10.12

Part 2

When the conductor is inclined at 30° to the field, as shown in Figure 10.12(b), then force $F = BIl \sin\theta$

$$= (0.9)(20)(0.30) \sin 30°$$

i.e. $\mathbf{F = 2.7\,N}$

If the current-carrying conductor shown in Figure 10.3(a) is placed in the magnetic field shown in Figure 10.13(a), then the two fields interact and cause a force to be exerted on the conductor as shown in Figure 10.13(b). The field is strengthened above the conductor and weakened below, thus tending to move the conductor downwards. This is the basic principle of operation of the electric motor (see Section 10.4) and the moving-coil instrument (see Section 10.5).

The direction of the force exerted on a conductor can be pre-determined by using **Fleming's* left-hand rule** (often called the motor rule), which states:

Let the thumb, first finger and second finger of the left hand be extended such that they are all at right-angles to each other (as shown in Figure 10.14). If the first finger points in the direction of the magnetic field, the second finger points in the direction of the current, then the thumb will point in the direction of the motion of the conductor.

* **Who was** Fleming? **Sir John Ambrose Fleming** (29 November 1849–18 April 1945) was the English electrical engineer and physicist best known for inventing the vacuum tube. To find out more go to **www.routledge.com/cw/bird**

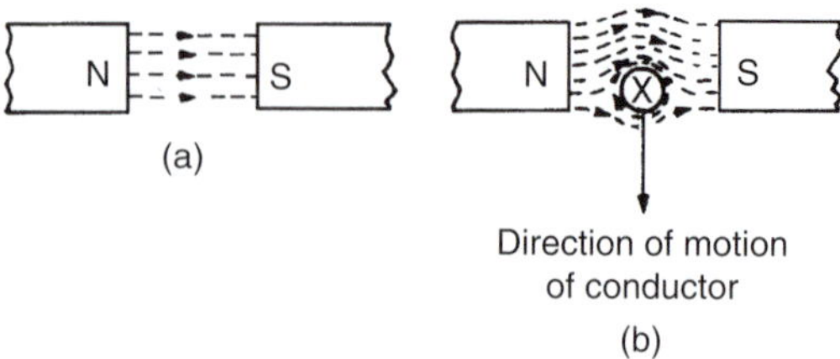

Figure 10.13

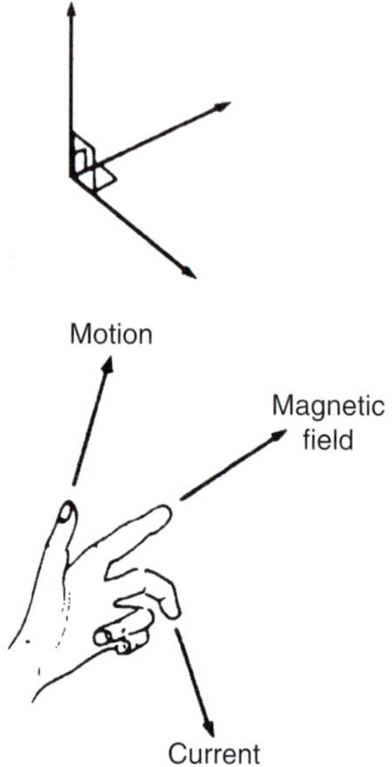

Figure 10.14

Summarizing:

First finger – Field

SeCond finger – Current

ThuMb – Motion

Problem 3. Determine the current required in a 400 mm length of conductor of an electric motor, when the conductor is situated at right-angles to a magnetic field of flux density 1.2 T, if a force of 1.92 N is to be exerted on the conductor. If the conductor is vertical, the current flowing downwards and the direction of the magnetic field is from left to right, what is the direction of the force?

Force $= 1.92\,N$; $l = 400\,mm = 0.40\,m$; $B = 1.2\,T$

Since $F = BIl$, then $I = \dfrac{F}{Bl}$

hence current $I = \dfrac{1.92}{(1.2)(0.4)} = \mathbf{4\,A}$

If the current flows downwards, the direction of its magnetic field due to the current alone will be clockwise when viewed from above. The lines of flux will reinforce (i.e. strengthen) the main magnetic field at the back of the conductor and will be in opposition in the front (i.e. weaken the field).
Hence the force on the conductor will be from back to front (i.e. towards the viewer). This direction may also have been deduced using Fleming's left-hand rule.

Problem 4. A conductor 350 mm long carries a current of 10 A and is at right-angles to a magnetic field lying between two circular pole faces each of radius 60 mm. If the total flux between the pole faces is 0.5 mWb, calculate the magnitude of the force exerted on the conductor.

$l = 350\,\text{mm} = 0.35\,\text{m}$; $I = 10\,\text{A}$;

Area of pole face $A = \pi r^2 = \pi(0.06)^2\,\text{m}^2$;

$\Phi = 0.5\,\text{mWb} = 0.5 \times 10^{-3}\,\text{Wb}$

Force $F = BIl$, and $B = \dfrac{\Phi}{A}$

$$\text{hence force } F = \left(\frac{\Phi}{A}\right) Il$$

$$= \frac{(0.5 \times 10^{-3})}{\pi(0.06)^2}(10)(0.35)\,\text{newtons}$$

i.e. force = 0.155 N

Problem 5. With reference to Figure 10.15, determine (a) the direction of the force on the conductor in Figure 10.15(a), (b) the direction of the force on the conductor in Figure 10.15(b), (c) the direction of the current in Figure 10.15(c), (d) the polarity of the magnetic system in Figure 10.15(d).

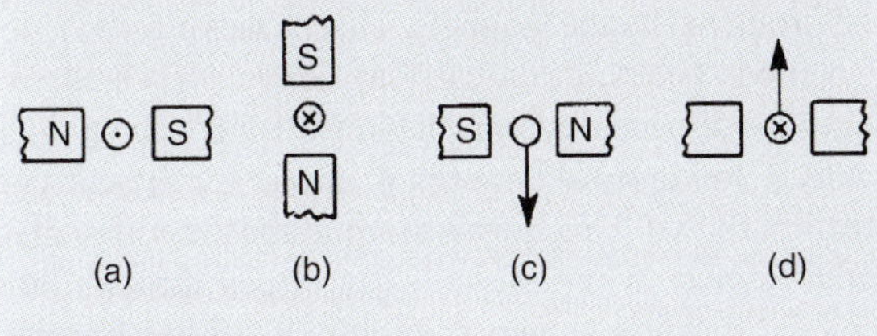

Figure 10.15

(a) The direction of the main magnetic field is from north to south, i.e. left to right. The current is flowing towards the viewer, and using the screw rule, the direction of the field is anticlockwise. Hence either by Fleming's left-hand rule, or by sketching the interacting magnetic field as shown in Figure 10.16(a), the direction of the force on the conductor is seen to be upward.

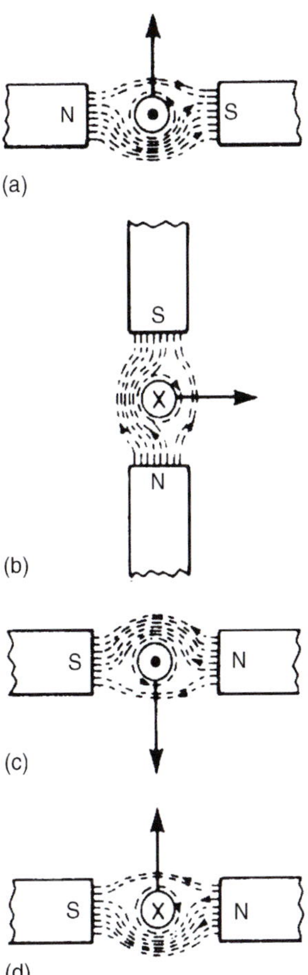

Figure 10.16

(b) Using a similar method to part (a) it is seen that the force on the conductor is to the right – see Figure 10.16(b).

(c) Using Fleming's left-hand rule, or by sketching as in Figure 10.16(c), it is seen that the current is towards the viewer, i.e. out of the paper.

(d) Similar to part (c), the polarity of the magnetic system is as shown in Figure 10.16(d).

Problem 6. A coil is wound on a rectangular former of width 24 mm and length 30 mm. The former is pivoted about an axis passing through the

Part 2

middle of the two shorter sides and is placed in a uniform magnetic field of flux density 0.8 T, the axis being perpendicular to the field. If the coil carries a current of 50 mA, determine the force on each coil side (a) for a single-turn coil, (b) for a coil wound with 300 turns.

(a) Flux density $B = 0.8\,\text{T}$; length of conductor lying at right-angles to field $l = 30\,\text{mm} = 30 \times 10^{-3}\,\text{m}$; current $I = 50\,\text{mA} = 50 \times 10^{-3}\,\text{A}$
For a single-turn coil, force on each coil side
$F = BIl = 0.8 \times 50 \times 10^{-3} \times 30 \times 10^{-3}$
$= \mathbf{1.2 \times 10^{-3}\,N}$ or $\mathbf{0.0012\,N}$

(b) When there are 300 turns on the coil there are effectively 300 parallel conductors each carrying a current of 50 mA. Thus the total force produced by the current is 300 times that for a single-turn coil. Hence force on coil side
$F = 300\,BIl = 300 \times 0.0012 = \mathbf{0.36\,N}$

Now try the following Practice Exercise

Practice Exercise 49 The force on a current-carrying conductor (Answers on page 820)

1. A conductor carries a current of 70 A at right-angles to a magnetic field having a flux density of 1.5 T. If the length of the conductor in the field is 200 mm, calculate the force acting on the conductor. What is the force when the conductor and field are at an angle of 45°?

2. Calculate the current required in a 240 mm length of conductor of a d.c. motor when the conductor is situated at right-angles to the magnetic field of flux density 1.25 T, if a force of 1.20 N is to be exerted on the conductor.

3. A conductor 30 cm long is situated at right-angles to a magnetic field. Calculate the flux density of the magnetic field if a current of 15 A in the conductor produces a force on it of 3.6 N

4. A conductor 300 mm long carries a current of 13 A and is at right-angles to a magnetic field between two circular pole faces, each of diameter 80 mm. If the total flux between the pole faces is 0.75 mWb, calculate the force exerted on the conductor.

5. (a) A 400 mm length of conductor carrying a current of 25 A is situated at right-angles to a magnetic field between two poles of an electric motor. The poles have a circular cross-section. If the force exerted on the conductor is 80 N and the total flux between the pole faces is 1.27 mWb, determine the diameter of a pole face.

   (b) If the conductor in part (a) is vertical, the current flowing downwards and the direction of the magnetic field is from left to right, what is the direction of the 80 N force?

6. A coil is wound uniformly on a former having a width of 18 mm and a length of 25 mm. The former is pivoted about an axis passing through the middle of the two shorter sides and is placed in a uniform magnetic field of flux density 0.75 T, the axis being perpendicular to the field. If the coil carries a current of 120 mA, determine the force exerted on each coil side, (a) for a single-turn coil, (b) for a coil wound with 400 turns.

10.4 Principle of operation of a simple d.c. motor

A rectangular coil which is free to rotate about a fixed axis is shown placed inside a magnetic field produced by permanent magnets in Figure 10.17. A direct current is fed into the coil via carbon brushes bearing on a commutator, which consists of a metal ring split into two halves separated by insulation.

When current flows in the coil a magnetic field is set up around the coil which interacts with the magnetic field produced by the magnets. This causes a force F to be exerted on the current-carrying conductor which, by Fleming's left-hand rule, is downwards between points A and B and upward between C and D for the current direction shown. This causes a torque and the coil rotates anticlockwise. When the coil has turned through 90° from the position shown in Figure 10.17 the brushes connected to the positive and negative terminals of the supply make contact with different halves of the commutator ring, thus reversing the direction of the current flow in the conductor. If the current is not reversed and the coil rotates past this position the forces acting on it

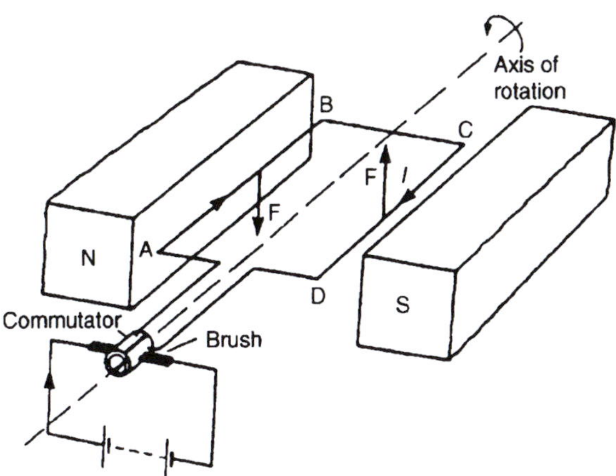

Figure 10.17

change direction and it rotates in the opposite direction, thus never making more than half a revolution. The current direction is reversed every time the coil swings through the vertical position and thus the coil rotates anticlockwise for as long as the current flows. This is the principle of operation of a d.c. motor which is thus a device that takes in electrical energy and converts it into mechanical energy.

10.5 Principle of operation of a moving-coil instrument

A moving-coil instrument operates on the motor principle. When a conductor carrying current is placed in a magnetic field, a force F is exerted on the conductor, given by $F = BIl$. If the flux density B is made constant (by using permanent magnets) and the conductor is a fixed length (say, a coil) then the force will depend only on the current flowing in the conductor.

In a moving-coil instrument a coil is placed centrally in the gap between shaped pole pieces as shown by the front elevation in Figure 10.18(a). (The air gap is kept as small as possible, although for clarity it is shown exaggerated in Figure 10.18.) The coil is supported by steel pivots, resting in jewel bearings, on a cylindrical iron core. Current is led into and out of the coil by two phosphor bronze spiral hairsprings which are wound in opposite directions to minimize the effect of temperature change and to limit the coil swing (i.e. to **control** the movement) and return the movement to zero position when no current flows. Current flowing in the coil produces forces as shown in Figure 10.18(b), the directions being obtained by Fleming's left-hand rule. The two forces, F_A and F_B, produce a torque which will move the coil in a clockwise direction, i.e. move the

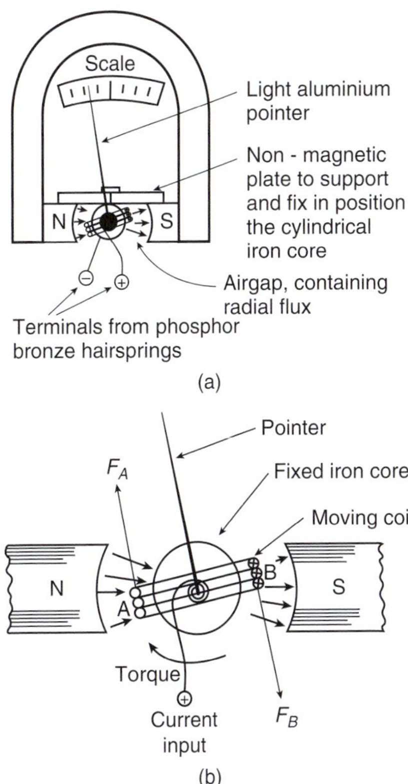

Figure 10.18

pointer from left to right. Since force is proportional to current the scale is linear.

When the aluminium frame, on which the coil is wound, is rotated between the poles of the magnet, small currents (called eddy currents) are induced into the frame, and this provides automatically the necessary **damping** of the system due to the reluctance of the former to move within the magnetic field. The moving-coil instrument will measure only direct current or voltage and the terminals are marked positive and negative to ensure that the current passes through the coil in the correct direction to deflect the pointer 'up the scale'.

The range of this sensitive instrument is extended by using shunts and multipliers (see Chapter 12).

10.6 Force on a charge

When a charge of Q coulombs is moving at a velocity of v m/s in a magnetic field of flux density B teslas, the charge moving perpendicular to the field, then the magnitude of the force F exerted on the charge, is given by:

$$F = QvB \text{ newtons}$$

Problem 7. An electron in a television tube has a charge of 1.6×10^{-19} coulombs and travels at 3×10^{7} m/s perpendicular to a field of flux density 18.5 μT. Determine the force exerted on the electron in the field.

From above, force $F = QvB$ newtons, where

Q = charge in coulombs $= 1.6\times10^{-19}$ C;

v = velocity of charge $= 3\times10^{7}$ m/s;

and B = flux density $= 18.5\times10^{-6}$ T

Hence force on electron

$$F = 1.6\times10^{-19}\times3\times10^{7}\times18.5\times10^{-6}$$
$$= 1.6\times3\times18.5\times10^{-18}$$
$$= 88.8\times10^{-18}$$
$$= \mathbf{8.88\times10^{-17}\ N}$$

Now try the following Practice Exercise

Practice Exercise 50 The force on a charge (Answers on page 820)

1. Calculate the force exerted on a charge of 2×10^{-18}C travelling at 2×10^{6} m/s perpendicular to a field of density 2×10^{-7}T.
2. Determine the speed of a 10^{-19} C charge travelling perpendicular to a field of flux density 10^{-7} T, if the force on the charge is 10^{-20} N.

Part 2

For fully worked solutions to each of the problems in Practice Exercises 49 and 50 in this chapter, go to the website:
www.routledge.com/cw/bird

Chapter 11

Electromagnetic induction

Why it is important to understand: **Electromagnetic induction**

Electromagnetic induction is the production of a potential difference (voltage) across a conductor when it is exposed to a varying magnetic field. Michael Faraday is generally credited with the discovery of induction in the 1830s. Faraday's law of induction is a basic law of electromagnetism that predicts how a magnetic field will interact with an electric circuit to produce an electromotive force (e.m.f.). It is the fundamental operating principle of transformers, inductors and many types of electrical motors, generators and solenoids. A.c. generators use Faraday's law to produce rotation and thus convert electrical and magnetic energy into rotational kinetic energy. This idea can be used to run all kinds of motors. Probably one of the greatest inventions of all time is the transformer. Alternating current from the primary coil moves quickly back and forth across the secondary coil. The moving magnetic field caused by the changing field (flux) induces a current in the secondary coil. This chapter explains electromagnetic induction, Faraday's laws, Lenz's law and Fleming's rule and develops various calculations to help understanding of the concepts.

At the end of this chapter you should be able to:

- understand how an e.m.f. may be induced in a conductor
- state Faraday's laws of electromagnetic induction
- state Lenz's law
- use Fleming's right-hand rule for relative directions
- appreciate that the induced e.m.f., $E = Blv$ or $E = Blv \sin\theta$
- calculate induced e.m.f. given B, l, v and θ and determine relative directions
- understand and perform calculations on rotation of a loop in a magnetic field
- define inductance L and state its unit
- define mutual inductance

Electrical Circuit Theory and Technology. 978-1-138-67349-6, © 2017 John Bird. Published by Taylor & Francis. All rights reserved.

- appreciate that e.m.f. $E = -N\dfrac{d\Phi}{dt} = -L\dfrac{dI}{dt}$
- calculate induced e.m.f. given N, t, L, change of flux or change of current
- appreciate factors which affect the inductance of an inductor
- draw the circuit diagram symbols for inductors
- calculate the energy stored in an inductor using $W = \frac{1}{2}LI^2$ joules
- calculate inductance L of a coil, given $L = \dfrac{N\Phi}{I}$ and $L = \dfrac{N^2}{S}$
- calculate mutual inductance using $E_2 = -M\dfrac{dI_1}{dt}$ and $M = \dfrac{N_1 N_2}{S}$

11.1 Introduction to electromagnetic induction

When a conductor is moved across a magnetic field so as to cut through the lines of force (or flux), an electromotive force (e.m.f.) is produced in the conductor. If the conductor forms part of a closed circuit then the e.m.f. produced causes an electric current to flow round the circuit. Hence an e.m.f. (and thus current) is 'induced' in the conductor as a result of its movement across the magnetic field. This effect is known as **'electromagnetic induction'**.

Figure 11.1(a) shows a coil of wire connected to a centre-zero galvanometer, which is a sensitive ammeter with the zero-current position in the centre of the scale.

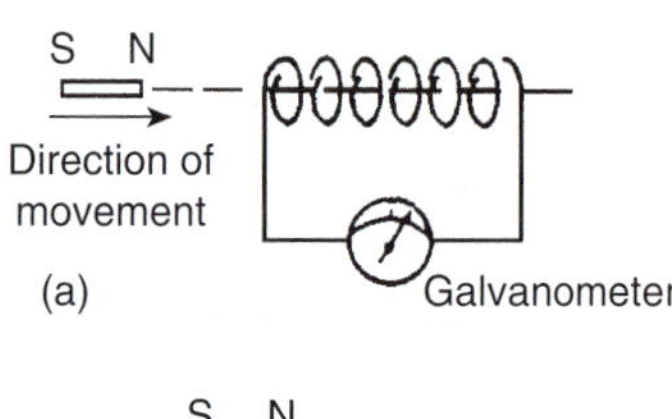

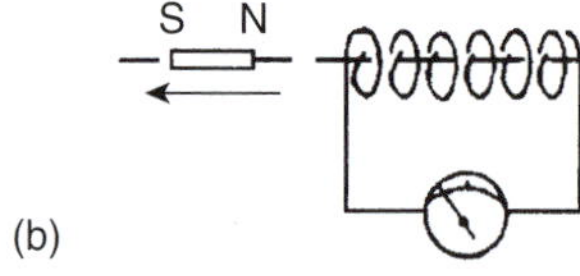

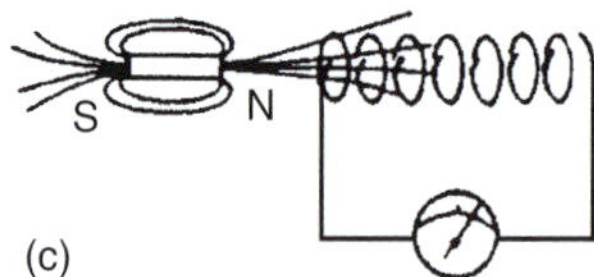

Figure 11.1

(a) When the magnet is moved at constant speed towards the coil (Figure 11.1(a)), a deflection is noted on the galvanometer showing that a current has been produced in the coil.

(b) When the magnet is moved at the same speed as in (a) but away from the coil the same deflection is noted but is in the opposite direction (see Figure 11.1(b)).

(c) When the magnet is held stationary, even within the coil, no deflection is recorded.

(d) When the coil is moved at the same speed as in (a) and the magnet held stationary the same galvanometer deflection is noted.

(e) When the relative speed is, say, doubled, the galvanometer deflection is doubled.

(f) When a stronger magnet is used, a greater galvanometer deflection is noted.

(g) When the number of turns of wire of the coil is increased, a greater galvanometer deflection is noted.

Figure 11.1(c) shows the magnetic field associated with the magnet. As the magnet is moved towards the coil, the magnetic flux of the magnet moves across, or cuts, the coil. **It is the relative movement of the magnetic flux and the coil that causes an e.m.f., and thus current, to be induced in the coil.** This effect is known as electromagnetic induction. The laws of electromagnetic induction stated in Section 11.2 evolved from experiments such as those described above.

11.2 Laws of electromagnetic induction

Faraday's* laws of electromagnetic induction state:

(i) *An induced e.m.f. is set up whenever the magnetic field linking that circuit changes.*

(ii) *The magnitude of the induced e.m.f. in any circuit is proportional to the rate of change of the magnetic flux linking the circuit.*

Lenz's* law states:
The direction of an induced e.m.f. is always such that it tends to set up a current opposing the motion or the change of flux responsible for inducing that e.m.f.

An alternative method to Lenz's law of determining relative directions is given by **Fleming's* Right-hand rule** (often called the gene*R*ator rule) which states:

* Who was **Lenz**? **Heinrich Friedrich Emil Lenz** (12 February 1804–10 February 1865) was the Russian physicist remembered for formulating Lenz's law in electrodynamics. To find out more go to **www.routledge.com/cw/bird**

* Who was **Faraday**? For image and resume of Faraday, see page 108. To find out more go to **www.routledge.com/cw/bird**

*Who was **Fleming**? For image and resume of Fleming, see page 140. To find out more go to **www.routledge.com/cw/bird**

Let the thumb, first finger and second finger of the right hand be extended such that they are all at right-angles to each other (as shown in Figure 11.2).

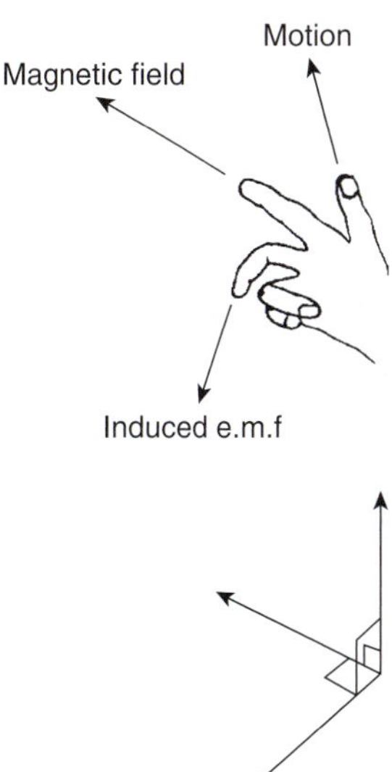

Figure 11.2

If the first finger points in the direction of the magnetic field, the thumb points in the direction of motion of the conductor relative to the magnetic field, then the second finger will point in the direction of the induced e.m.f.

Summarizing:

First finger – Field

ThuMb – Motion

SEcond finger – E.m.f.

In a generator, conductors forming an electric circuit are made to move through a magnetic field. By Faraday's law an e.m.f. is induced in the conductors and thus a source of e.m.f. is created. A generator converts mechanical energy into electrical energy. (The action of a simple a.c. generator is described in Chapter 16.) The **induced e.m.f.** E set up between the ends of the conductor shown in Figure 11.3 is given by:

$$E = Blv \text{ volts}$$

where B, the flux density, is measured in teslas, l, the length of conductor in the magnetic field, is measured in metres, and v, the conductor velocity, is measured in metres per second.

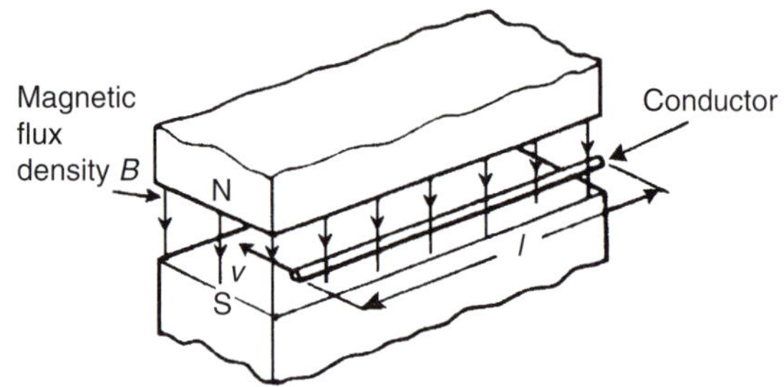

Figure 11.3

If the conductor moves at an angle $\theta°$ to the magnetic field (instead of at 90°, as assumed above) then

$$E = Blv \sin\theta \text{ volts}$$

Problem 1. A conductor 300 mm long moves at a uniform speed of 4 m/s at right-angles to a uniform magnetic field of flux density 1.25 T. Determine the current flowing in the conductor when (a) its ends are open-circuited, (b) its ends are connected to a load of 20 Ω resistance.

When a conductor moves in a magnetic field it will have an e.m.f. induced in it, but this e.m.f. can only produce a current if there is a closed circuit.

Induced e.m.f. $E = Blv = (1.25)\left(\dfrac{300}{1000}\right)(4) = 1.5\,\text{V}$

(a) If the ends of the conductor are open-circuited **no current will flow** even though 1.5 V has been induced.

(b) From Ohm's law, $I = \dfrac{E}{R} = \dfrac{1.5}{20} = \mathbf{0.075\,A}$ or **75 mA**

Problem 2. At what velocity must a conductor 75 mm long cut a magnetic field of flux density 0.6 T if an e.m.f. of 9 V is to be induced in it? Assume the conductor, the field and the direction of motion are mutually perpendicular.

Induced e.m.f. $E = Blv$, hence velocity $v = \dfrac{E}{Bl}$

Hence $v = \dfrac{9}{(0.6)(75 \times 10^{-3})} = \dfrac{9 \times 10^3}{0.6 \times 75} = \mathbf{200\,m/s}$

Problem 3. A conductor moves with a velocity of 15 m/s at an angle of (a) 90°, (b) 60° and (c) 30° to a magnetic field produced between two square-faced poles of side length 2 cm. If the flux leaving a pole face is 5 µWb, find the magnitude of the induced e.m.f. in each case.

$v = 15\,\text{m/s}$; length of conductor in magnetic field, $l = 2\,\text{cm} = 0.02\,\text{m}$; $A = 2 \times 2\,\text{cm}^2 = 4 \times 10^{-4}\,\text{m}^2$, $\Phi = 5 \times 10^{-6}\,\text{Wb}$

(a) $E_{90} = Blv \sin 90° = \left(\dfrac{\Phi}{A}\right) lv \sin 90°$

$$= \frac{(5 \times 10^{-6})}{(4 \times 10^{-4})}(0.02)(15)(1)$$

$$= \mathbf{3.75\,mV}$$

(b) $E_{60} = Blv \sin 60° = E_{90} \sin 60° = 3.75 \sin 60°$

$$= \mathbf{3.25\,mV}$$

(c) $E_{30} = Blv \sin 30° = E_{90} \sin 30° = 3.75 \sin 30°$

$$= \mathbf{1.875\,mV}$$

Problem 4. The wing span of a metal aeroplane is 36 m. If the aeroplane is flying at 400 km/h, determine the e.m.f. induced between its wing tips. Assume the vertical component of the earth's magnetic field is 40 µT.

Induced e.m.f. across wing tips, $E = Blv$

$B = 40\,\mu\text{T} = 40 \times 10^{-6}\,\text{T}$; $l = 36\,\text{m}$

$$v = 400\frac{\text{km}}{\text{h}} \times 1000\frac{\text{m}}{\text{km}} \times \frac{1\,\text{h}}{60 \times 60\,\text{s}} = \frac{(400)(1000)}{3600}$$

$$= \frac{4000}{36}\,\text{m/s}$$

Hence $E = Blv = (40 \times 10^{-6})(36)\left(\dfrac{4000}{36}\right)$

$$= \mathbf{0.16\,V}$$

Problem 5. The diagram shown in Figure 11.4 represents the generation of e.m.f.s. Determine (i) the direction in which the conductor has to be moved in Figure 11.4(a), (ii) the direction of the induced e.m.f. in Figure 11.4(b), (iii) the polarity of the magnetic system in Figure 11.4(c).

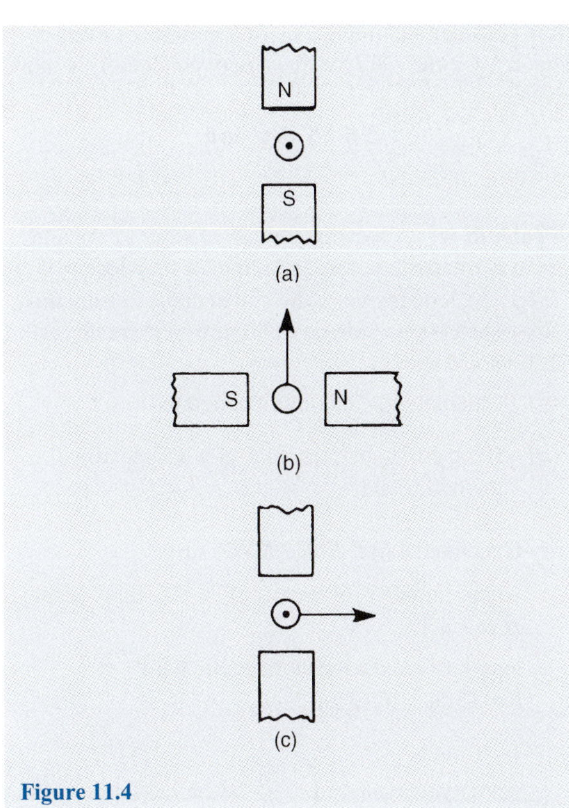

Figure 11.4

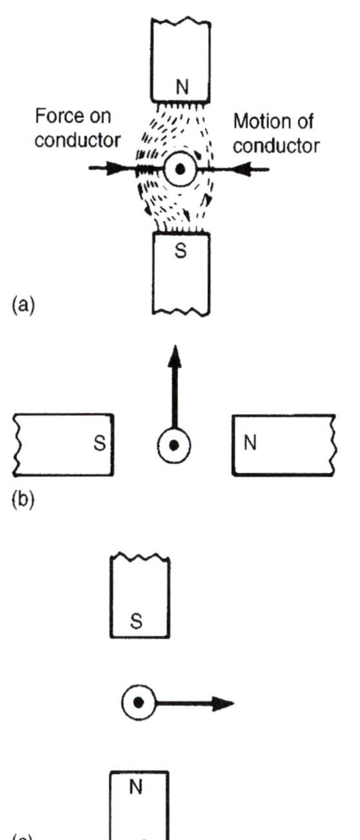

Figure 11.5

The direction of the e.m.f., and thus the current due to the e.m.f., may be obtained by either Lenz's law or Fleming's Right-hand rule (i.e. GeneRator rule).

(i) Using Lenz's law: the field due to the magnet and the field due to the current-carrying conductor are shown in Figure 11.5(a) and are seen to reinforce to the left of the conductor. Hence the force on the conductor is to the right. However, Lenz's law states that the direction of the induced e.m.f. is always such as to oppose the effect producing it. **Thus the conductor will have to be moved to the left.**

(ii) Using Fleming's right-hand rule:

First finger – Field, i.e. $N \rightarrow S$, or right to left;

ThuMb – Motion, i.e. upwards;

SEcond finger – E.m.f., i.e. **towards the viewer or out of the paper**, as shown in Figure 11.5(b)

(iii) The polarity of the magnetic system of Figure 11.4(c) is shown in Figure 11.5(c) and is obtained using Fleming's right-hand rule.

Now try the following Practice Exercise

Practice Exercise 51 Induced e.m.f. (Answers on page 820)

1. A conductor of length 15 cm is moved at 750 mm/s at right-angles to a uniform flux density of 1.2 T. Determine the e.m.f. induced in the conductor.
2. Find the speed that a conductor of length 120 mm must be moved at right-angles to a magnetic field of flux density 0.6 T to induce in it an e.m.f. of 1.8 V
3. A 25 cm long conductor moves at a uniform speed of 8 m/s through a uniform magnetic field of flux density 1.2 T. Determine the current flowing in the conductor when (a) its ends are open-circuited, (b) its ends are connected to a load of 15 ohms resistance.

4. A car is travelling at 80 km/h. Assuming the back axle of the car is 1.76 m in length and the vertical component of the earth's magnetic field is 40 μT, find the e.m.f. generated in the axle due to motion.

5. A conductor moves with a velocity of 20 m/s at an angle of (a) 90° (b) 45° (c) 30°, to a magnetic field produced between two square-faced poles of side length 2.5 cm. If the flux on the pole face is 60 mWb, find the magnitude of the induced e.m.f. in each case.

6. A conductor 400 mm long is moved at 70° to a 0.85 T magnetic field. If it has a velocity of 115 km/h, calculate (a) the induced voltage, and (b) force acting on the conductor if connected to an 8 Ω resistor.

11.3 Rotation of a loop in a magnetic field

Figure 11.6 shows a view of a looped conductor whose sides are moving across a magnetic field.

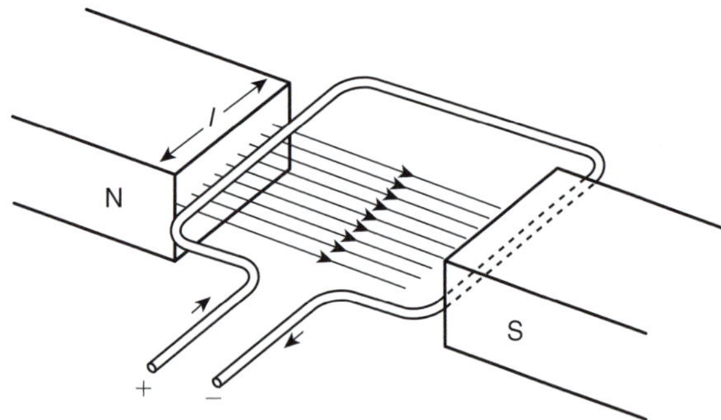

Figure 11.6

The left-hand side is moving in an upward direction (check using Fleming's right-hand rule), with length l cutting the lines of flux which are travelling from left to right. By definition, the induced e.m.f. will be equal to $Blv \sin\theta$ and flowing into the page.

The right-hand side is moving in a downward direction (again, check using Fleming's right-hand rule), with length l cutting the same lines of flux as above. The induced e.m.f. will also be equal to $Blv \sin\theta$ but flowing out of the page.

Therefore the total e.m.f. for the loop conductor $= 2\,Blv \sin\theta$

Now consider a coil made up of a number of turns N. The total e.m.f. E for the loop conductor is now given by:

$$\boldsymbol{E = 2N\,Blv\,\sin\theta}$$

Problem 6. A rectangular coil of sides 12 cm and 8 cm is rotated in a magnetic field of flux density 1.4 T, the longer side of the coil actually cutting this flux. The coil is made up of 80 turns and rotates at 1200 rev/min.

(a) Calculate the maximum generated e.m.f.

(b) If the coil generates 90 V, at what speed will the coil rotate?

(a) Generated e.m.f. $E = 2\,N\,Blv \sin\theta$

where number of turns, $N = 80$, flux density, $B = 1.4$ T,

length of conductor in magnetic field, $l = 12\text{ cm} = 0.12$ m,

$$\text{velocity, } v = \omega r = \left(\frac{1200}{60} \times 2\pi \text{ rad/s}\right)\left(\frac{0.08}{2}\text{ m}\right)$$

$$= 1.6\pi \text{ m/s},$$

and for maximum e.m.f. induced, $\theta = 90°$, from which, $\sin\theta = 1$

Hence, **maximum e.m.f. induced,**

$$\boldsymbol{E} = 2\,N\,Blv \sin\theta$$

$$= 2 \times 80 \times 1.4 \times 0.12 \times 1.6\pi \times 1 = \mathbf{135.1\ volts}$$

(b) Since $E = 2\,N\,Blv \sin\theta$

then $90 = 2 \times 80 \times 1.4 \times 0.12 \times v \times 1$

from which, $v = \dfrac{90}{2 \times 80 \times 1.4 \times 0.12}$

$$= 3.348 \text{ m/s}$$

$v = \omega r$ hence, angular velocity,

$$\omega = \frac{v}{r} = \frac{3.348}{\frac{0.08}{2}} = 83.7 \text{ rad/s}$$

$$\textbf{Speed of coil in rev/min} = \frac{83.7 \times 60}{2\pi}$$
$$= \textbf{799 rev/min}$$

An **alternative method** of determining (b) is by **direct proportion**.

Since $E = 2N\,Blv\sin\theta$, then with N, B, l and θ being constant, $\boldsymbol{E \propto v}$

If from (a), 135.1 V is produced by a speed of 1200 rev/min,

then 1 V would be produced by a speed of $\dfrac{1200}{135.1}$
$= 8.88$ rev/min

Hence, 90 V would be produced by a speed of $90 \times 8.88 =$ **799 rev/min**

Now try the following Practice Exercise

Practice Exercise 52 Induced e.m.f. in a coil (Answers on page 820)

1. A rectangular coil of sides 8 cm by 6 cm is rotating in a magnetic field such that the longer sides cut the magnetic field. Calculate the maximum generated e.m.f. if there are 60 turns on the coil, the flux density is 1.6 T and the coil rotates at 1500 rev/min.
2. A generating coil on a former 100 mm long has 120 turns and rotates in a 1.4 T magnetic field. Calculate the maximum e.m.f. generated if the coil, having a diameter of 60 mm, rotates at 450 rev/min.
3. If the coils in Problems 1 and 2 generate 60 V, calculate (a) the new speed for each coil, and (b) the flux density required if the speed is unchanged.

11.4 Inductance

Inductance is the name given to the property of a circuit whereby there is an e.m.f. induced into the circuit by the change of flux linkages produced by a current change.
When the e.m.f. is induced in the same circuit as that in which the current is changing, the property is called **self inductance**, ***L***.

When the e.m.f. is induced in a circuit by a change of flux due to current changing in an adjacent circuit, the property is called **mutual inductance**, ***M***.
The unit of inductance is the **henry,*** **H**.

A circuit has an inductance of one henry when an e.m.f. of one volt is induced in it by a current changing at the rate of one ampere per second.

Induced e.m.f. in a coil of N turns,

$$\boldsymbol{E = -N\frac{d\Phi}{dt}\ \text{volts}}$$

where $d\Phi$ is the change in flux in Webers, and dt is the time taken for the flux to change in seconds (i.e. $d\Phi/dt$ is the rate of change of flux).

Induced e.m.f. in a coil of inductance L henrys,

$$\boldsymbol{E = -L\frac{dI}{dt}\ \text{volts}}$$

where dI is the change in current in amperes and dt is the time taken for the current to change in seconds (i.e. dI/dt is the rate of change of current). The minus sign in each of the above two equations remind us of its direction (given by Lenz's law).

*Who was **Henry**? **Joseph Henry** (17 December 1797–13 May 1878) was an American scientist who discovered the electromagnetic phenomenon of self-inductance. To find out more go to **www.routledge.com/cw/bird**

Part 2

Problem 7. Determine the e.m.f. induced in a coil of 200 turns when there is a change of flux of 25 mWb linking with it in 50 ms.

$$\text{Induced e.m.f. } E = -N\frac{d\Phi}{dt} = -(200)\left(\frac{25\times10^{-3}}{50\times10^{-3}}\right)$$

$$= \mathbf{-100\ volts}$$

Problem 8. A flux of 400 μWb passing through a 150-turn coil is reversed in 40 ms. Find the average e.m.f. induced.

Since the flux reverses, the flux changes from +400 μWb to −400 μWb, a total change of flux of 800 μWb

$$\text{Induced e.m.f. } E = -N\frac{d\Phi}{dt} = -(150)\left(\frac{800\times10^{-6}}{40\times10^{-3}}\right)$$

$$= -\left(\frac{150\times800\times10^{3}}{40\times10^{6}}\right)$$

Hence **the average e.m.f. induced $E = -3$ volts**

Problem 9. Calculate the e.m.f. induced in a coil of inductance 12 H by a current changing at the rate of 4 A/s.

$$\text{Induced e.m.f. } E = -L\frac{dI}{dt} = -(12)(4) = \mathbf{-48\ volts}$$

Problem 10. An e.m.f. of 1.5 kV is induced in a coil when a current of 4 A collapses uniformly to zero in 8 ms. Determine the inductance of the coil.

Change in current, $dI = (4-0) = 4$ A;
$dt = 8\text{ ms} = 8\times10^{-3}$ s;

$$\frac{dI}{dt} = \frac{4}{8\times10^{-3}} = \frac{4000}{8} = 500\text{ A/s};$$

$$E = 1.5\text{ kV} = 1500\text{ V}$$

$$\text{Since } |E| = L\left(\frac{dI}{dt}\right)$$

$$\text{inductance, } L = \frac{|E|}{(dI/dt)} = \frac{1500}{500} = \mathbf{3\ H}$$

(Note that $|E|$ means the 'magnitude of E', which disregards the minus sign.)

Now try the following Practice Exercise

Practice Exercise 53 Inductance (Answers on page 820)

1. Find the e.m.f. induced in a coil of 200 turns when there is a change of flux of 30 mWb linking with it in 40 ms.
2. An e.m.f. of 25 V is induced in a coil of 300 turns when the flux linking with it changes by 12 mWb. Find the time, in milliseconds, in which the flux makes the change.
3. An ignition coil having 10 000 turns has an e.m.f. of 8 kV induced in it. What rate of change of flux is required for this to happen?
4. A flux of 0.35 mWb passing through a 125-turn coil is reversed in 25 ms. Find the magnitude of the average e.m.f. induced.

11.5 Inductors

A component called an inductor is used when the property of inductance is required in a circuit. The basic form of an inductor is simply a coil of wire.
Factors which affect the inductance of an inductor include:

(i) the number of turns of wire – the more turns the higher the inductance

(ii) the cross-sectional area of the coil of wire – the greater the cross-sectional area the higher the inductance

(iii) the presence of a magnetic core – when the coil is wound on an iron core the same current sets up a more concentrated magnetic field and the inductance is increased

(iv) the way the turns are arranged – a short, thick coil of wire has a higher inductance than a long, thin one.

Two examples of practical inductors are shown in Figure 11.7, and the standard electrical circuit diagram symbols for air-cored and iron-cored inductors are shown in Figure 11.8.

An iron-cored inductor is often called a **choke** since, when used in a.c. circuits, it has a choking effect, limiting the current flowing through it. Inductance is

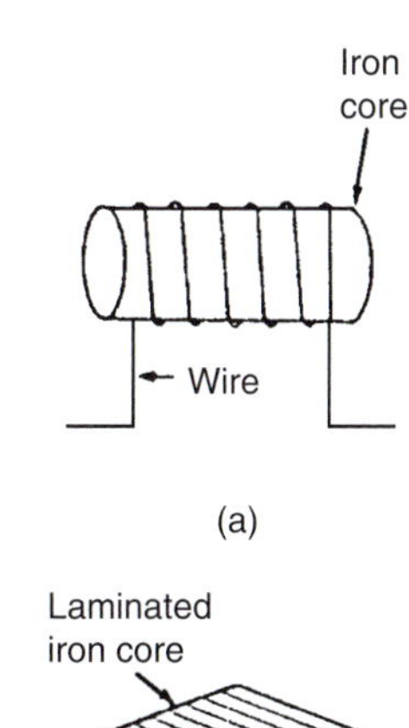

(a)

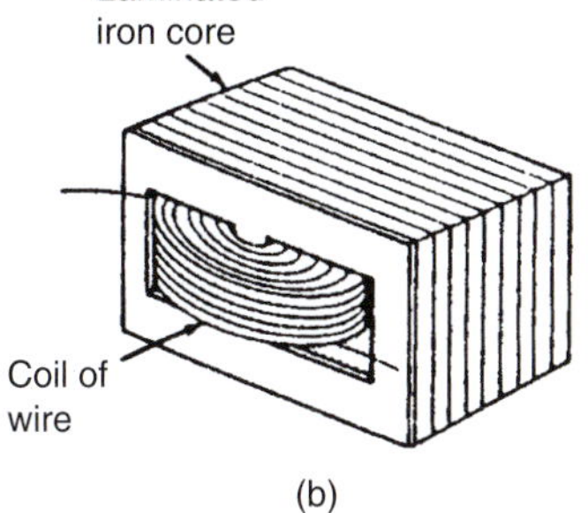

(b)

Figure 11.7

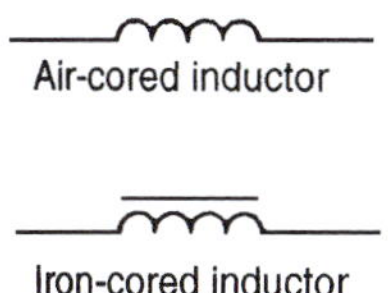

Figure 11.8

often undesirable in a circuit. To reduce inductance to a minimum the wire may be bent back on itself, as shown in Figure 11.9, so that the magnetizing effect of one conductor is neutralized by that of the adjacent conductor. The wire may be coiled around an insulator, as shown, without increasing the inductance. Standard resistors may be non-inductively wound in this manner.

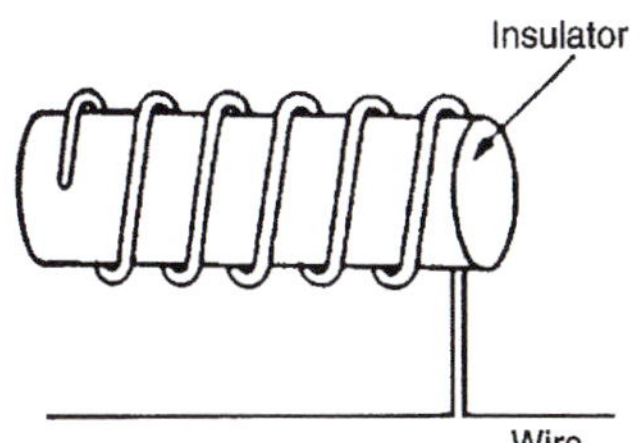

Figure 11.9

Some typical small inductors are shown in Figure 11.9

11.6 Energy stored

An inductor possesses an ability to store energy. The energy stored, W, in the magnetic field of an inductor is given by:

$$W = \frac{1}{2}LI^2 \text{ joules}$$

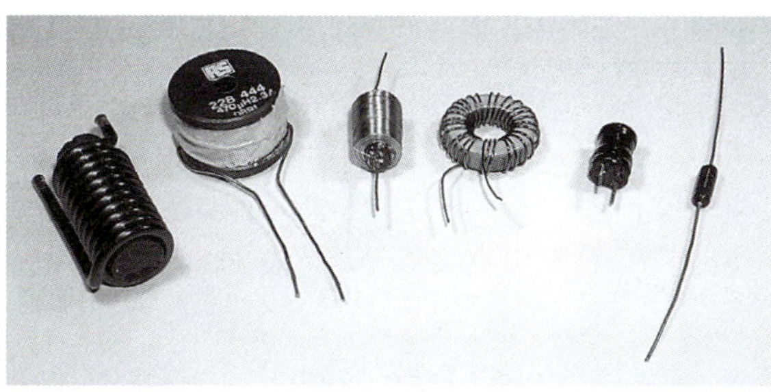

Figure 11.10

Problem 11. An 8 H inductor has a current of 3 A flowing through it. How much energy is stored in the magnetic field of the inductor?

Energy stored, $W = \frac{1}{2}LI^2 = \frac{1}{2}(8)(3)^2 = \mathbf{36\ joules}$

Now try the following Practice Exercise

Practice Exercise 54 Energy stored (Answers on page 820)

1. Calculate the value of the energy stored when a current of 30 mA is flowing in a coil of inductance 400 mH.
2. The energy stored in the magnetic field of an inductor is 80 J when the current flowing in the inductor is 2 A. Calculate the inductance of the coil.

11.7 Inductance of a coil

If a current changing from 0 to I amperes produces a flux change from 0 to Φ Webers, then $dI = I$ and $d\Phi = \Phi$. Then, from Section 11.4, induced e.m.f. $E = N\Phi/t = LI/t$, from which

inductance of coil, $L = \frac{N\Phi}{I}$ **henrys**

Since $E = -L\dfrac{dI}{dt} = -N\dfrac{d\Phi}{dt}$ then $L = N\dfrac{d\Phi}{dt}\left(\dfrac{dt}{dI}\right)$

i.e. $$L = N\frac{d\Phi}{dI}$$

From Chapter 9, mmf $= \Phi S$ from which, $\Phi = \dfrac{\text{mmf}}{S}$

Substituting into $L = N\dfrac{d\Phi}{dI}$

gives $$L = N\frac{d}{dI}\left(\frac{\text{mmf}}{S}\right)$$

i.e. $$L = \frac{N}{S}\frac{d(NI)}{dI} \quad \text{since mmf} = NI$$

i.e. $$L = \frac{N^2}{S}\frac{dI}{dI} \quad \text{and since } \frac{dI}{dI} = 1,$$

$$\boldsymbol{L = \frac{N^2}{S}} \textbf{ henrys}$$

Problem 12. Calculate the coil inductance when a current of 4 A in a coil of 800 turns produces a flux of 5 mWb linking with the coil.

For a coil, inductance $L = \dfrac{N\Phi}{I}$

$$= \frac{(800)(5\times 10^{-3})}{4} = \mathbf{1\,H}$$

Problem 13. A flux of 25 mWb links with a 1500 turn coil when a current of 3 A passes through the coil. Calculate (a) the inductance of the coil, (b) the energy stored in the magnetic field and (c) the average e.m.f. induced if the current falls to zero in 150 ms.

(a) **Inductance,** $\boldsymbol{L} = \dfrac{N\Phi}{I} = \dfrac{(1500)(25\times 10^{-3})}{3}$

$= \mathbf{12.5\,H}$

(b) **Energy stored,** $\boldsymbol{W} = \frac{1}{2}LI^2 = \frac{1}{2}(12.5)(3)^2$

$= \mathbf{56.25\,J}$

(c) **Induced e.m.f.,** $\boldsymbol{E} = -L\dfrac{dI}{dt}$

$= -(12.5)\left(\dfrac{3-0}{150\times 10^{-3}}\right)$

$= \mathbf{-250\,V}$

(Alternatively, $E = -N\left(\dfrac{d\Phi}{dt}\right)$

$= -(1500)\left(\dfrac{25\times 10^{-3}}{150\times 10^{-3}}\right)$

$= \mathbf{-250\,V}$

since if the current falls to zero so does the flux.)

Problem 14. A 750-turn coil of inductance 3 H carries a current of 2 A. Calculate the flux linking the coil and the e.m.f. induced in the coil when the current collapses to zero in 20 ms.

Coil inductance, $L = \dfrac{N\Phi}{I}$ from which,

flux $\Phi = \dfrac{LI}{N} = \dfrac{(3)(2)}{750} = 8\times 10^{-3} = \mathbf{8\,mWb}$

Induced e.m.f. $E = -L\left(\dfrac{dI}{dt}\right) = -3\left(\dfrac{2-0}{20\times 10^{-3}}\right)$

$= \mathbf{-300\,V}$

(Alternatively, $E = -N\dfrac{d\Phi}{dt} = -(750)\left(\dfrac{8\times 10^{-3}}{20\times 10^{-3}}\right)$

$= \mathbf{-300\,V}$)

Problem 15. A silicon iron ring is wound with 800 turns, the ring having a mean diameter of 120 mm and a cross-sectional area of 400 mm^2. If when carrying a current of 0.5 A the relative permeability is found to be 3000, calculate (a) the self inductance of the coil, (b) the induced e.m.f. if the current is reduced to zero in 80 ms.

The ring is shown sketched in Figure 11.11.

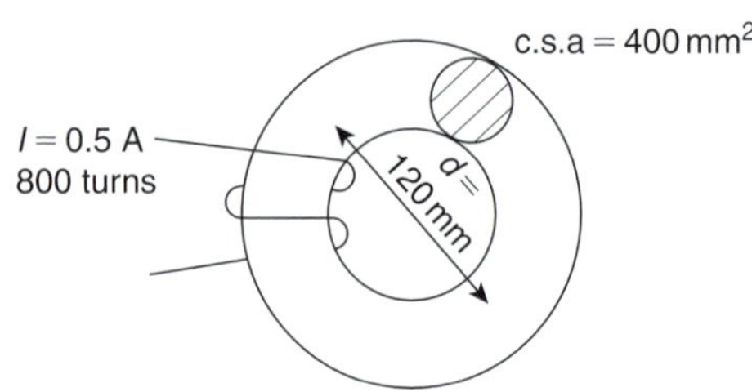

Figure 11.11

(a) Inductance, $L=\frac{N^2}{S}$ and from Chapter 9,

reluctance, $S=\frac{l}{\mu_0\mu_r A}$

i.e. $S=\frac{\pi\times 120\times 10^{-3}}{4\pi\times 10^{-7}\times 3000\times 400\times 10^{-6}}$

$=250\times 10^3$ A/Wb

Hence, **self inductance,** $\mathbf{L}=\frac{N^2}{S}=\frac{800^2}{250\times 10^3}$

$= \mathbf{2.56\ H}$

(b) **Induced e.m.f.,** $\mathbf{E}=-L\frac{dI}{dt}$

$=-(2.56)\frac{(0.5-0)}{80\times 10^{-3}}$

$= \mathbf{-16\ V}$

Now try the following Practice Exercise

Practice Exercise 55 Inductance of a coil (Answers on page 820)

1. A flux of 30 mWb links with a 1200-turn coil when a current of 5 A is passing through the coil. Calculate (a) the inductance of the coil, (b) the energy stored in the magnetic field and (c) the average e.m.f. induced if the current is reduced to zero in 0.20 s
2. An e.m.f. of 2 kV is induced in a coil when a current of 5 A collapses uniformly to zero in 10 ms. Determine the inductance of the coil.
3. An average e.m.f. of 60 V is induced in a coil of inductance 160 mH when a current of 7.5 A is reversed. Calculate the time taken for the current to reverse.
4. A coil of 2500 turns has a flux of 10 mWb linking with it when carrying a current of 2 A. Calculate the coil inductance and the e.m.f. induced in the coil when the current collapses to zero in 20 ms.
5. A coil is wound with 600 turns and has a self inductance of 2.5 H. What current must flow to set up a flux of 20 mWb?
6. When a current of 2 A flows in a coil, the flux linking with the coil is 80 μWb. If the coil inductance is 0.5 H, calculate the number of turns of the coil.
7. A steady current of 5 A when flowing in a coil of 1000 turns produces a magnetic flux of 500 μWb. Calculate the inductance of the coil. The current of 5 A is then reversed in 12.5 ms. Calculate the e.m.f. induced in the coil.
8. An iron ring has a cross-sectional area of 500 mm^2 and a mean length of 300 mm. It is wound with 100 turns and its relative permeability is 1600. Calculate (a) the current required to set up a flux of 500 μWb in the coil, (b) the inductance of the system and (c) the induced e.m.f. if the field collapses in 1 ms.

11.8 Mutual inductance

Mutually induced e.m.f. in the second coil,

$$E_2=-M\frac{dI_1}{dt}\text{ volts}$$

where M is the **mutual inductance** between two coils, in henrys, and dI_1/dt is the rate of change of current in the first coil.

The phenomenon of mutual inductance is used in **transformers** (see Chapter 23, page 349). Mutual inductance is developed further in Chapter 46 on magnetically coupled circuits (see page 728).

Another expression for *M*

Let an iron ring have two coils, A and B, wound on it. If the fluxes Φ_1 and Φ_2 are produced from currents I_1 and I_2 in coils A and B respectively, then the reluctance could be expressed as:

$$S=\frac{I_1N_1}{\Phi_1}=\frac{I_2N_2}{\Phi_2}$$

If the flux in coils A and B are the same and produced from the current I_1 in coil A only, assuming 100% coupling, then the mutual inductance can be expressed as:

$$M=\frac{N_2\Phi_1}{I_1}$$

Multiplying by $\left(\frac{N_1}{N_1}\right)$ gives: $M = \frac{N_2\Phi_1 N_1}{I_1 N_1}$

However, $S = \frac{I_1 N_1}{\Phi_1}$

Thus, mutual inductance, $\boldsymbol{M = \frac{N_1 N_2}{S}}$

Problem 16. Calculate the mutual inductance between two coils when a current changing at 200 A/s in one coil induces an e.m.f. of 1.5 V in the other.

Induced e.m.f. $|E_2| = M\frac{dI_1}{dt}$, i.e. $1.5 = M(200)$

Thus **mutual inductance,** $\boldsymbol{M} = \frac{1.5}{200}$

$= \mathbf{0.0075\ H}$ or $\mathbf{7.5\ mH}$

Problem 17. The mutual inductance between two coils is 18 mH. Calculate the steady rate of change of current in one coil to induce an e.m.f. of 0.72 V in the other.

Induced e.m.f., $|E_2| = M\frac{dI_1}{dt}$

Hence rate of change of current, $\frac{dI_1}{dt} = \frac{|E_2|}{M}$

$= \frac{0.72}{0.018} = \mathbf{40\ A/s}$

Problem 18. Two coils have a mutual inductance of 0.2 H. If the current in one coil is changed from 10 A to 4 A in 10 ms, calculate (a) the average induced e.m.f. in the second coil, (b) the change of flux linked with the second coil if it is wound with 500 turns.

(a) Induced e.m.f. $E_2 = -M\frac{dI_1}{dt}$

$= -(0.2)\left(\frac{10-4}{10\times10^{-3}}\right)$

$= \mathbf{-120\ V}$

(b) Induced e.m.f. $|E_2| = N\frac{d\Phi}{dt}$, hence $d\Phi = \frac{|E_2|dt}{N}$

Thus the change of flux, $d\Phi = \frac{120(10\times10^{-3})}{500}$

$= \mathbf{2.4\ mWb}$

Problem 19. In the device shown in Figure 11.12, when the current in the primary coil of 1000 turns increases linearly from 1 A to 6 A in 200 ms, an e.m.f. of 15 V is induced into the secondary coil of 480 turns, which is left open circuited. Determine (a) the mutual inductance of the two coils, (b) the reluctance of the former and (c) the self inductance of the primary coil.

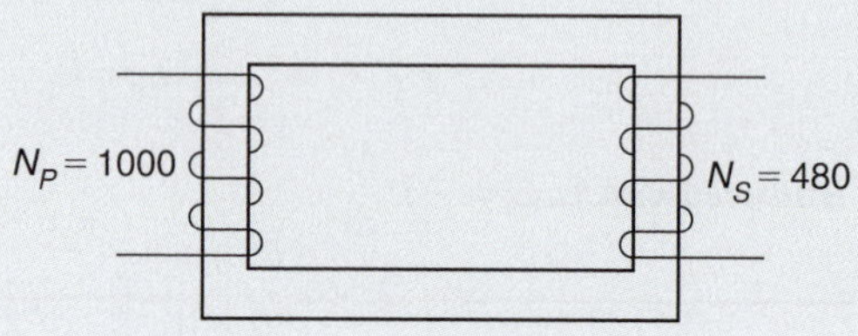

Figure 11.12

(a) $E_S = M\frac{dI_P}{dt}$ from which,

mutual inductance, $\mathbf{M} = \frac{E_S}{\frac{dI_P}{dt}} = \frac{15}{\left(\frac{6-1}{200\times10^{-3}}\right)}$

$= \frac{15}{25} = \mathbf{0.60\ H}$

(b) $M = \frac{N_P N_S}{S}$ from which,

reluctance, $S = \frac{N_P N_S}{M} = \frac{(1000)(480)}{0.60}$

$= \mathbf{800\ 000\ A/Wb}$ or $\mathbf{800\ kA/Wb}$

(c) Primary self inductance, $L_P = \frac{N_P^2}{S} = \frac{(1000)^2}{800\,000}$

$= \mathbf{1.25\ H}$

Now try the following Practice Exercise

Practice Exercise 56 Mutual inductance (Answers on page 820)

1. The mutual inductance between two coils is 150 mH. Find the magnitude of the e.m.f. induced in one coil when the current in the other is increasing at a rate of 30 A/s.
2. Determine the mutual inductance between two coils when a current changing at 50 A/s in

one coil induces an e.m.f. of 80 mV in the other.

3. Two coils have a mutual inductance of 0.75 H. Calculate the magnitude of the e.m.f. induced in one coil when a current of 2.5 A in the other coil is reversed in 15 ms.

4. The mutual inductance between two coils is 240 mH. If the current in one coil changes from 15 A to 6 A in 12 ms, calculate (a) the average e.m.f. induced in the other coil, (b) the change of flux linked with the other coil if it is wound with 400 turns.

5. When the current in the primary coil of 400 turns of a magnetic circuit increases linearly from 10 mA to 35 mA in 100 ms, an e.m.f. of 75 mV is induced into the secondary coil of 240 turns, which is left open circuited. Determine (a) the mutual inductance of the two coils, (b) the reluctance of the former and (c) the self inductance of the secondary coil.

For fully worked solutions to each of the problems in Practice Exercises 51 to 56 in this chapter, go to the website:
www.routledge.com/cw/bird

COMPANION @ WEBSITE

Chapter 12

Electrical measuring instruments and measurements

Why it is important to understand: **Electrical measuring instruments and measurements**

Future electrical engineers need to be able to appreciate basic measurement techniques, instruments and methods used in everyday practice. This chapter covers both analogue and digital instruments, measurement errors, bridges, oscilloscopes, data acquisition, instrument controls and measurement systems. Accurate measurements are central to virtually every scientific and engineering discipline. Electrical measurements often come down to either measuring current or measuring voltage. Even if you are measuring frequency, you will be measuring the frequency of a current signal or a voltage signal and you will need to know how to measure either voltage or current. Many times you will use a digital multimeter – a DMM – to measure either voltage or current; actually, a DMM will also usually measure frequency (of a voltage signal) and resistance. The quality of a measuring instrument is assessed from its accuracy, precision, reliability, durability and so on, all of which are related to its cost.

At the end of this chapter you should be able to:

- recognize the importance of testing and measurements in electric circuits
- appreciate the essential devices comprising an analogue instrument
- calculate values of shunts for ammeters and multipliers for voltmeters
- understand the advantages of electronic instruments
- understand the operation of an ohmmeter/megger
- appreciate the operation of multimeters/Avometers/Flukes
- understand the operation of a wattmeter
- appreciate instrument 'loading' effect
- understand the operation of an oscilloscope for d.c. and a.c. measurements

Electrical Circuit Theory and Technology. 978-1-138-67349-6, © 2017 John Bird. Published by Taylor & Francis. All rights reserved.

- calculate periodic time, frequency, peak to peak values from waveforms on an oscilloscope
- appreciate virtual test and measuring instruments
- recognize harmonics present in complex waveforms
- determine ratios of powers, currents and voltages in decibels
- understand null methods of measurement for a Wheatstone bridge and d.c. potentiometer
- understand the operation of a.c. bridges
- appreciate the most likely source of errors in measurements
- appreciate calibration accuracy of instruments

12.1 Introduction

Tests and measurements are important in designing, evaluating, maintaining and servicing electrical circuits and equipment. In order to detect electrical quantities such as current, voltage, resistance or power, it is necessary to transform an electrical quantity or condition into a visible indication. This is done with the aid of instruments (or meters) that indicate the magnitude of quantities either by the position of a pointer moving over a graduated scale (called an analogue instrument) or in the form of a decimal number (called a digital instrument).

The digital instrument has, in the main, become the instrument of choice in recent years; in particular, computer-based instruments are rapidly replacing items of conventional test equipment, with the virtual storage test instrument, the **digital storage oscilloscope**, being the most common. This is explained later in this chapter, but before that some analogue instruments, which are still used in some installations, are explored.

12.2 Analogue instruments

All analogue electrical indicating instruments require three essential devices:

(a) A **deflecting or operating device**. A mechanical force is produced by the current or voltage which causes the pointer to deflect from its zero position.

(b) **A controlling device**. The controlling force acts in opposition to the deflecting force and ensures that the deflection shown on the meter is always the same for a given measured quantity. It also prevents the pointer always going to the maximum deflection. There are two main types of controlling device – spring control and gravity control.

(c) **A damping device**. The damping force ensures that the pointer comes to rest in its final position quickly and without undue oscillation. There are three main types of damping used – eddy-current damping, air-friction damping and fluid-friction damping.

There are basically two types of scale – linear and non-linear.

A **linear scale** is shown in Figure 12.1(a), where the divisions or graduations are evenly spaced. The voltmeter shown has a range 0–100 V, i.e. a full-scale deflection (f.s.d.) of 100 V. A **non-linear scale** is shown in Figure 12.1(b). The scale is cramped at the beginning and the graduations are uneven throughout the range. The ammeter shown has a f.s.d. of 10 A.

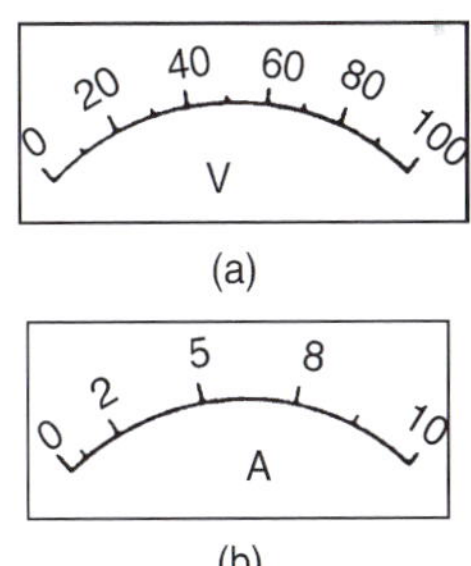

Figure 12.1

12.3 Shunts and multipliers

An **ammeter**, which measures current, has a low resistance (ideally zero) and must be connected in series with the circuit.

A **voltmeter**, which measures p.d., has a high resistance (ideally infinite) and must be connected in parallel with the part of the circuit whose p.d. is required.

There is no difference between the basic instrument used to measure current and voltage since both use a milliammeter as their basic part. This is a sensitive instrument which gives f.s.d. for currents of only a few milliamperes. When an ammeter is required to measure currents of larger magnitude, a proportion of the current is diverted through a low-value resistance connected in parallel with the meter. Such a diverting resistor is called a **shunt**.

From Figure 12.2(a), $V_{PQ} = V_{RS}$. Hence $I_a r_a = I_S R_S$

Thus the value of the shunt, $\boldsymbol{R_s = \frac{I_a r_a}{I_s}}$ **ohms**

The milliammeter is converted into a voltmeter by connecting a high-value resistance (called a **multiplier**) in series with it, as shown in Figure 12.2(b). From Figure 12.2(b), $V = V_a + V_M = Ir_a + IR_M$

Thus the value of the multiplier, $\boldsymbol{R_M = \frac{V - Ir_a}{I}}$ **ohms**

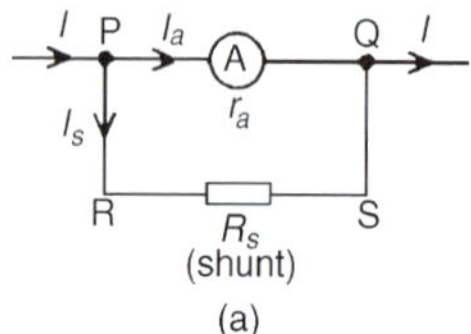

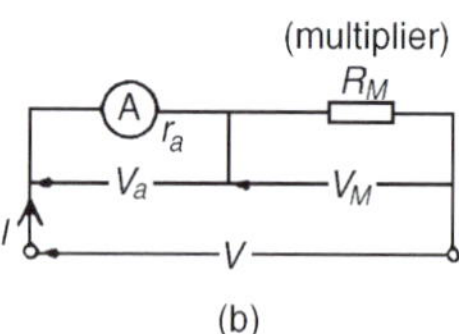

Figure 12.2

Problem 1. A moving-coil instrument gives an f.s.d. when the current is 40 mA and its resistance is 25 Ω. Calculate the value of the shunt to be connected in parallel with the meter to enable it to be used as an ammeter for measuring currents up to 50 A.

The circuit diagram is shown in Figure 12.3,

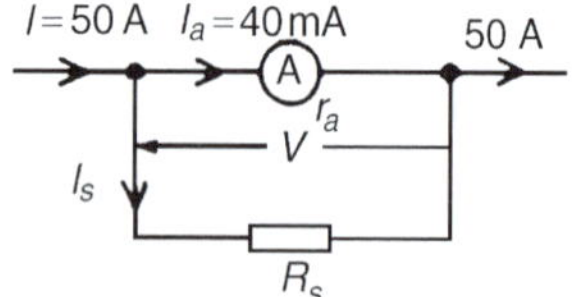

Figure 12.3

where r_a = resistance of instrument = 25 Ω,

R_s = resistance of shunt,

I_a = maximum permissible current flowing in instrument = 40 mA = 0.04 A,

I_s = current flowing in shunt,

I = total circuit current required to give f.s.d. = 50 A

Since $I = I_a + I_s$ then $I_s = I - I_a = 50 - 0.04 = 49.96$ A

$V = I_a r_a = I_s R_s$

Hence $R_s = \frac{I_a r_a}{I_s} = \frac{(0.04)(25)}{49.96} = 0.02002\,\Omega$

$= \mathbf{20.02\,m\Omega}$

Thus for the moving-coil instrument to be used as an ammeter with a range 0–50 A, a resistance of value 20.02 mΩ needs to be connected in parallel with the instrument.

Problem 2. A moving-coil instrument having a resistance of 10 Ω gives an f.s.d. when the current is 8 mA. Calculate the value of the multiplier to be connected in series with the instrument so that it can be used as a voltmeter for measuring p.d.s up to 100 V.

The circuit diagram is shown in Figure 12.4,

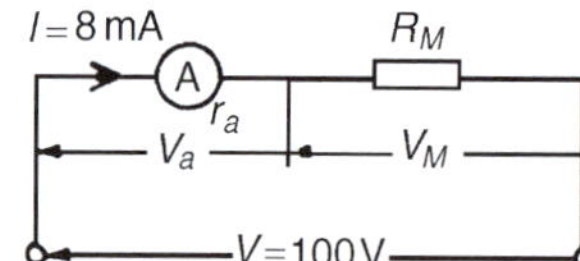

Figure 12.4

where r_a = resistance of instrument = 10 Ω,

R_M = resistance of multiplier,

I = total permissible instrument current = 8 mA = 0.008 A,

V = total p.d. required to give f.s.d. = 100 V

$V = V_a + V_M = Ir_a + IR_M$

i.e. $100 = (0.008)(10) + (0.008)\,R_M$

or $100 - 0.08 = 0.008\,R_M$

thus $R_M = \dfrac{99.92}{0.008} = 12\,490\,\Omega = \mathbf{12.49\,k\Omega}$

Hence for the moving-coil instrument to be used as a voltmeter with a range 0–100 V, a resistance of value 12.49 kΩ needs to be connected in series with the instrument.

Now try the following Practice Exercise

Practice Exercise 57 Shunts and multipliers (Answers on page 820)

1. A moving-coil instrument gives f.s.d. for a current of 10 mA. Neglecting the resistance of the instrument, calculate the approximate value of series resistance needed to enable the instrument to measure up to (a) 20 V, (b) 100 V, (c) 250 V.
2. A meter of resistance 50 Ω has an f.s.d. of 4 mA. Determine the value of shunt resistance required in order that f.s.d. should be (a) 15 mA, (b) 20 A, (c) 100 A.
3. A moving-coil instrument having a resistance of 20 Ω gives an f.s.d. when the current is 5 mA. Calculate the value of the multiplier to be connected in series with the instrument so that it can be used as a voltmeter for measuring p.d.s up to 200 V.
4. A moving-coil instrument has a f.s.d. of 20 mA and a resistance of 25 Ω. Calculate the values of resistance required to enable the instrument to be used (a) as a 0–10 A ammeter, and (b) as a 0–100 V voltmeter. State the mode of resistance connection in each case.

12.4 Electronic instruments

Electronic measuring instruments have advantages over instruments such as the moving-iron or moving-coil meters, in that they have a much higher input resistance (some as high as 1000 MΩ) and can handle a much wider range of frequency (from d.c. up to MHz).

The digital voltmeter (DVM) is one which provides a digital display of the voltage being measured. Advantages of a DVM over analogue instruments include higher accuracy and resolution, no observational or parallax errors (see Section 12.18) and a very high input resistance, constant on all ranges.

A digital multimeter is a DVM with additional circuitry which makes it capable of measuring a.c. voltage, d.c. and a.c. current and resistance.

Instruments for a.c. measurements are generally calibrated with a sinusoidal alternating waveform to indicate r.m.s. values when a sinusoidal signal is applied to the instrument. Some instruments, such as the moving-iron and electro-dynamic instruments, give a true r.m.s. indication. With other instruments the indication is either scaled up from the mean value (such as with the rectifier moving-coil instrument) or scaled down from the peak value.

Sometimes quantities to be measured have complex waveforms (see Section 12.12), and whenever a quantity is non-sinusoidal, errors in instrument readings can occur if the instrument has been calibrated for sine waves only.

Such waveform errors can be largely eliminated by using electronic instruments.

12.5 The ohmmeter

An **ohmmeter** is an instrument for measuring electrical resistance.

A simple ohmmeter circuit is shown in Figure 12.5(a). Unlike the ammeter or voltmeter, the ohmmeter circuit does not receive the energy necessary for its operation from the circuit under test. In the ohmmeter this energy is supplied by a self-contained source of voltage, such as a battery. Initially, terminals XX are short-circuited and R adjusted to give f.s.d. on the milliammeter. If current I is at a maximum value and voltage E is constant, then resistance $R = E/I$ is at a minimum value. Thus f.s.d. on the milliammeter is made zero on the resistance scale. When terminals XX are open circuited no current flows and $R(=E/O)$ is infinity, ∞.

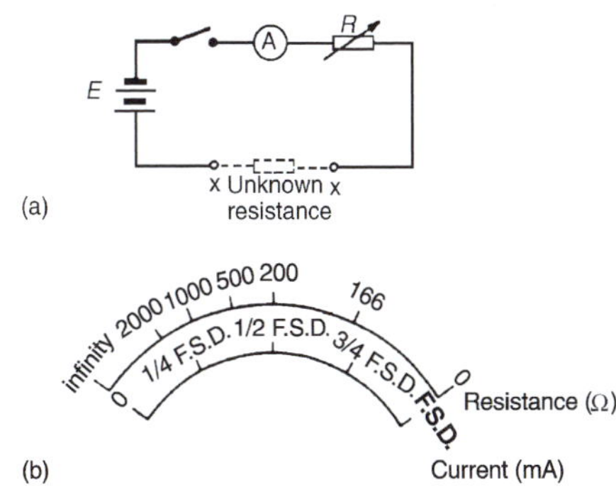

Figure 12.5

The milliammeter can thus be calibrated directly in ohms. A cramped (non-linear) scale results and is 'back to front', as shown in Figure 12.5(b). When calibrated, an unknown resistance is placed between terminals XX and its value determined from the position of the pointer on the scale. An ohmmeter designed for measuring low values of resistance is called a **continuity tester**. An ohmmeter designed for measuring high values of resistance (i.e. megohms) is called an **insulation resistance tester** (e.g. **'Megger'**).

12.6 Multimeters

Instruments are manufactured that combine a moving-coil meter with a number of shunts and series multipliers, to provide a range of readings on a single scale graduated to read current and voltage. If a battery is incorporated then resistance can also be measured. Such instruments are called **multimeters** or **universal instruments** or **multirange instruments**. An 'Avometer' is a typical example. A particular range may be selected either by the use of separate terminals or by a selector switch. Only one measurement can be performed at a time. Often such instruments can be used in a.c. as well as d.c. circuits when a rectifier is incorporated in the instrument.

Digital multimeters (DMM) are now almost universally used, the **Fluke Digital Multimeter** being an industry leader for performance, accuracy, resolution, ruggedness, reliability and safety. These instruments measure d.c. currents and voltages, resistance and continuity, a.c. (r.m.s.) currents and voltages, temperature and much more.

12.7 Wattmeters

A **wattmeter** is an instrument for measuring electrical power in a circuit. Figure 12.6 shows typical connections of a wattmeter used for measuring power supplied to a load. The instrument has two coils:

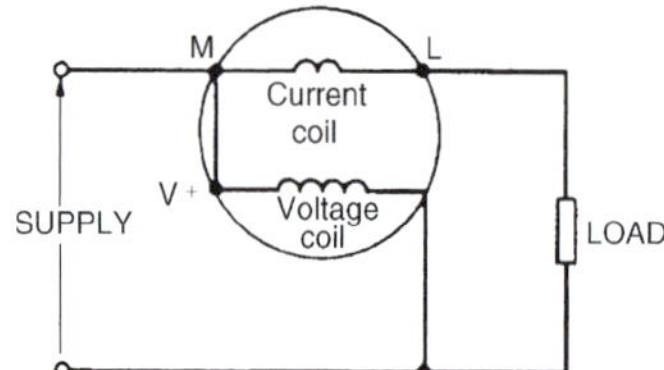

Figure 12.6

(i) a current coil, which is connected in series with the load, like an ammeter, and

(ii) a voltage coil, which is connected in parallel with the load, like a voltmeter.

12.8 Instrument 'loading' effect

Some measuring instruments depend for their operation on power taken from the circuit in which measurements are being made. Depending on the 'loading' effect of the instrument (i.e. the current taken to enable it to operate), the prevailing circuit conditions may change.

The resistance of voltmeters may be calculated since each have a stated sensitivity (or 'figure of merit'), often stated in 'kΩ per volt' of f.s.d. A voltmeter should have as high a resistance as possible – ideally infinite.

In a.c. circuits the impedance of the instrument varies with frequency and thus the loading effect of the instrument can change.

Problem 3. Calculate the power dissipated by the voltmeter and by resistor R in Figure 12.7 when (a) $R = 250\,\Omega$, (b) $R = 2\,M\Omega$. Assume that the voltmeter sensitivity (sometimes called figure of merit) is $10\,k\Omega/V$.

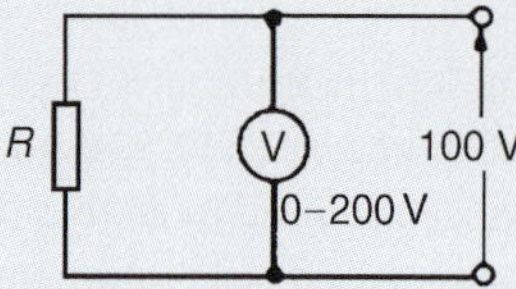

Figure 12.7

(a) Resistance of voltmeter, $R_v = \text{sensitivity} \times \text{f.s.d.}$

Hence, $R_v = (10\,k\Omega/V) \times (200\,V) = 2000\,k\Omega$

$= 2\,M\Omega$

Current flowing in voltmeter, $I_v = \dfrac{V}{R_v} = \dfrac{100}{2 \times 10^6}$

$= 50 \times 10^{-6}\,A$

Power dissipated by voltmeter $= VI_v$

$= (100)(50 \times 10^{-6})$

$= \mathbf{5\,mW}$

When $R = 250\,\Omega$, current in resistor, $I_R = \frac{V}{R}$

$$= \frac{100}{250} = \mathbf{0.4\,A}$$

Power dissipated in load resistor $R = VI_R$

$$= (100)(0.4) = \mathbf{40\,W}$$

Thus the power dissipated in the voltmeter is insignificant in comparison with the power dissipated in the load.

(b) When $R = 2\,M\Omega$, current in resistor,

$$I_R = \frac{V}{R} = \frac{100}{2 \times 10^6} = 50 \times 10^{-6}\,A$$

Power dissipated in load resistor

$$R = VI_R = 100 \times 50 \times 10^{-6} = \mathbf{5\,mW}$$

In this case the higher load resistance reduced the power dissipated such that the voltmeter is using as much power as the load.

Problem 4. An ammeter has an f.s.d. of 100 mA and a resistance of 50 Ω. The ammeter is used to measure the current in a load of resistance 500 Ω when the supply voltage is 10 V. Calculate (a) the ammeter reading expected (neglecting its resistance), (b) the actual current in the circuit, (c) the power dissipated in the ammeter and (d) the power dissipated in the load.

From Figure 12.8,

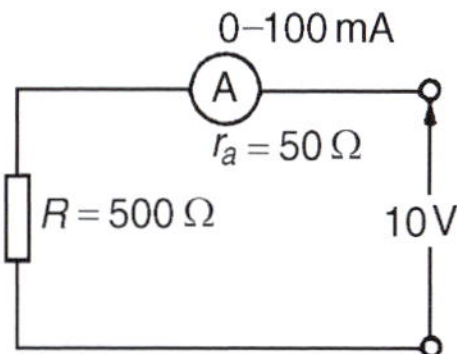

Figure 12.8

(a) expected ammeter reading $= \frac{V}{R} = \frac{10}{500} = \mathbf{20\,mA}$

(b) Actual ammeter reading $= \frac{V}{R + r_a} = \frac{10}{500 + 50}$

$$= \mathbf{18.18\,mA}$$

Thus the ammeter itself has caused the circuit conditions to change from 20 mA to 18.18 mA

(c) Power dissipated in the ammeter

$$= I^2 r_a = (18.18 \times 10^{-3})^2(50) = \mathbf{16.53\,mW}$$

(d) Power dissipated in the load resistor

$$= I^2 R = (18.18 \times 10^{-3})^2(500) = \mathbf{165.3\,mW}$$

Problem 5. A voltmeter having an f.s.d. of 100 V and a sensitivity of 1.6 kΩ/V is used to measure voltage V_1 in the circuit of Figure 12.9. Determine (a) the value of voltage V_1 with the voltmeter not connected, and (b) the voltage indicated by the voltmeter when connected between A and B.

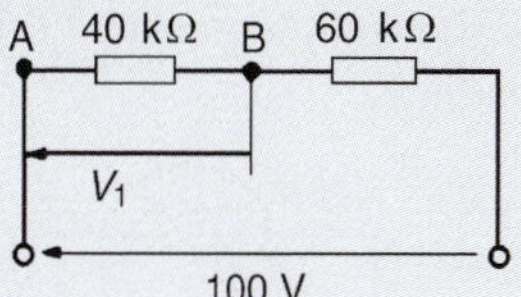

Figure 12.9

(a) By voltage division, $V_1 = \left(\frac{40}{40 + 60}\right) 100 = \mathbf{40\,V}$

(b) The resistance of a voltmeter having a 100 V f.s.d. and sensitivity 1.6 kΩ/V is 100 V × 1.6 kΩ/V = 160 kΩ.
When the voltmeter is connected across the 40 kΩ resistor the circuit is as shown in Figure 12.10(a) and the equivalent resistance of the parallel network is given by

$$\left(\frac{40 \times 160}{40 + 160}\right) k\Omega \text{ i.e. } \left(\frac{40 \times 160}{200}\right) \Omega = 32\,k\Omega$$

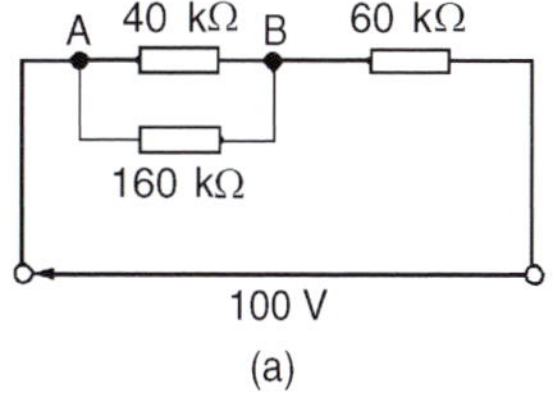

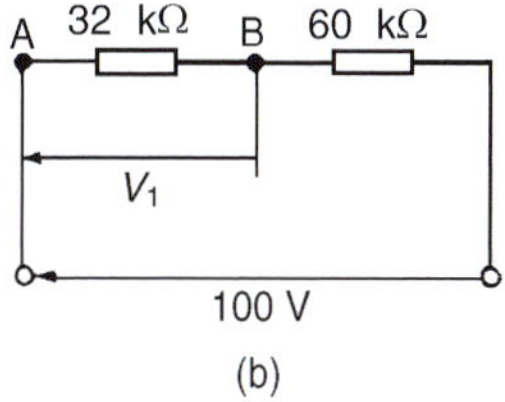

Figure 12.10

The circuit is now effectively as shown in Figure 12.10(b).
Thus the voltage indicated on the voltmeter is

$$\left(\frac{32}{32+60}\right)100\,\text{V} = \mathbf{34.78\,V}$$

A considerable error is thus caused by the loading effect of the voltmeter on the circuit. The error is reduced by using a voltmeter with a higher sensitivity.

Problem 6. (a) A current of 20 A flows through a load having a resistance of 2 Ω. Determine the power dissipated in the load. (b) A wattmeter, whose current coil has a resistance of 0.01 Ω, is connected as shown in Figure 12.11. Determine the wattmeter reading.

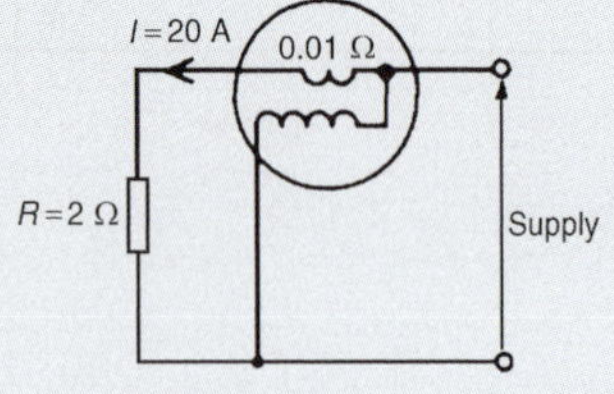

Figure 12.11

(a) Power dissipated in the load, $P = I^2R = (20)^2(2)$ $= \mathbf{800\,W}$

(b) With the wattmeter connected in the circuit the total resistance R_T is $2 + 0.01 = 2.01\,\Omega$

The wattmeter reading is thus $I^2R_T = (20)^2(2.01)$ $= \mathbf{804\,W}$

Now try the following Practice Exercise

Practice Exercise 58 Instrument 'loading' effects (Answers on page 820)

1. A 0–1 A ammeter having a resistance of 50 Ω is used to measure the current flowing in a 1 kΩ resistor when the supply voltage is 250 V. Calculate: (a) the approximate value of current (neglecting the ammeter resistance), (b) the actual current in the circuit, (c) the power dissipated in the ammeter, (d) the power dissipated in the 1 kΩ resistor.

2. (a) A current of 15 A flows through a load having a resistance of 4 Ω. Determine the power dissipated in the load. (b) A wattmeter, whose current coil has a resistance of 0.02 Ω, is connected (as shown in Figure 12.13) to measure the power in the load. Determine the wattmeter reading assuming the current in the load is still 15 A.

3. A voltage of 240 V is applied to a circuit consisting of an 800 Ω resistor in series with a 1.6 kΩ resistor. What is the voltage across the 1.6 kΩ resistor? The p.d. across the 1.6 kΩ resistor is measured by a voltmeter of f.s.d. 250 V and sensitivity 100 Ω/V. Determine the voltage indicated.

4. A 240 V supply is connected across a load of resistance R. Also connected across R is a voltmeter having an f.s.d. of 300 V and a figure of merit (i.e. sensitivity) of 8 kΩ/V. Calculate the power dissipated by the voltmeter and by the load resistance if (a) $R = 100\,\Omega$ (b) $R = 1\,\text{M}\Omega$. Comment on the results obtained.

12.9 The oscilloscope

The oscilloscope is basically a graph-displaying device – it draws a graph of an electrical signal. In most applications the graph shows how signals change over time. From the graph it is possible to:

- determine the time and voltage values of a signal
- calculate the frequency of an oscillating signal
- see the 'moving parts' of a circuit represented by the signal
- tell if a malfunctioning component is distorting the signal
- find out how much of a signal is d.c. or a.c.
- tell how much of the signal is noise and whether the noise is changing with time

Oscilloscopes are used by everyone from television repair technicians to physicists. They are indispensable for anyone designing or repairing electronic equipment. The usefulness of an oscilloscope is not limited to the world of electronics. With the proper transducer (i.e. a device that creates an electrical signal in response to physical stimuli, such as sound, mechanical stress, pressure, light or heat), an oscilloscope can measure any kind of phenomena. An automobile engineer uses an oscilloscope to measure engine vibrations; a medical researcher uses an oscilloscope to measure brain waves, and so on.

Oscilloscopes are available in both analogue and digital types. An **analogue oscilloscope** works by directly applying a voltage being measured to an electron beam moving across the oscilloscope screen. The voltage deflects the beam up or down proportionally, tracing the waveform on the screen. This gives an immediate picture of the waveform.

In contrast, a **digital oscilloscope** samples the waveform and uses an analogue-to-digital converter (see Section 20.11, page 320) to convert the voltage being measured into digital information. It then uses this digital information to reconstruct the waveform on the screen.

For many applications either an analogue or digital oscilloscope is appropriate. However, each type does possess some unique characteristics making it more or less suitable for specific tasks.

Analogue oscilloscopes are often preferred when it is important to display rapidly varying signals in 'real time' (i.e. as they occur).

Digital oscilloscopes allow the capture and viewing of events that happen only once. They can process the digital waveform data or send the data to a computer for processing. Also, they can store the digital waveform data for later viewing and printing. Digital storage oscilloscopes are explained in Section 12.11.

Analogue oscilloscopes

When an oscilloscope probe is connected to a circuit, the voltage signal travels through the probe to the vertical system of the oscilloscope. Figure 12.12 shows a simple block diagram that shows how an analogue oscilloscope displays a measured signal.

Depending on how the vertical scale (volts/division control) is set, an attenuator reduces the signal voltage or an amplifier increases the signal voltage. Next, the signal travels directly to the vertical deflection plates of the cathode ray tube (CRT). Voltage applied to these deflection plates causes a glowing dot to move. (An electron beam hitting phosphor inside the CRT creates the glowing dot.) A positive voltage causes the dot to move up while a negative voltage causes the dot to move down. The signal also travels to the trigger system to start or trigger a 'horizontal sweep'. Horizontal sweep is a term referring to the action of the horizontal system causing the glowing dot to move across the screen. Triggering the horizontal system causes the horizontal time base to move the glowing dot across the screen from left to right within a specific time interval. Many sweeps in rapid sequence cause the movement of the glowing dot to blend into a solid line. At higher speeds, the dot may sweep across the screen up to 500 000 times each second.

Together, the horizontal sweeping action (i.e. the X direction) and the vertical deflection action (i.e. the Y

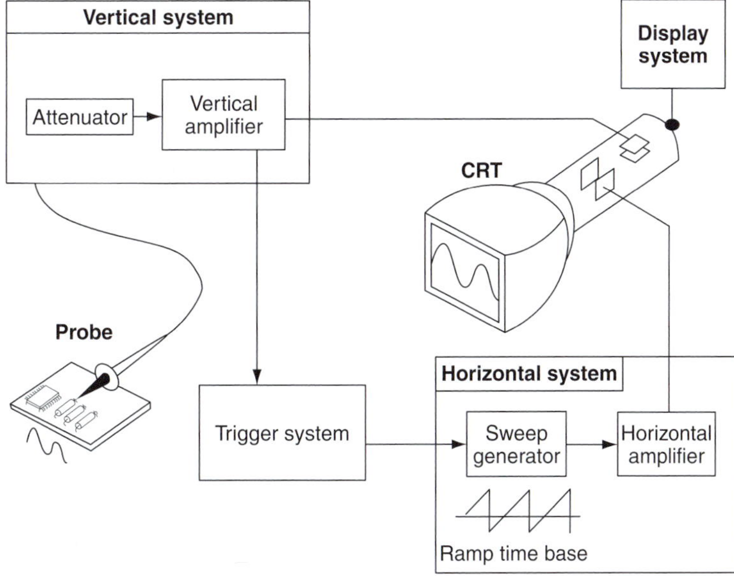

Figure 12.12

Part 2

direction), trace a graph of the signal on the screen. The trigger is necessary to stabilize a repeating signal. It ensures that the sweep begins at the same point of a repeating signal, resulting in a clear picture.

In conclusion, to use an analogue oscilloscope, three basic settings to accommodate an incoming signal need to be adjusted:

- the attenuation or amplification of the signal – use the volts/division control to adjust the amplitude of the signal before it is applied to the vertical deflection plates
- the time base – use the time/division control to set the amount of time per division represented horizontally across the screen
- the triggering of the oscilloscope – use the trigger level to stabilize a repeating signal, as well as triggering on a single event.

Also, adjusting the focus and intensity controls enable a sharp, visible display to be created.

(i) With **direct voltage measurements**, only the Y amplifier 'volts/cm' switch on the oscilloscope is used. With no voltage applied to the Y plates the position of the spot trace on the screen is noted. When a direct voltage is applied to the Y plates the new position of the spot trace is an indication of the magnitude of the voltage. For example, in Figure 12.13(a), with no voltage applied to the Y plates, the spot trace is in the centre of the screen (initial position) and then the spot trace moves 2.5 cm to the final position shown, on application of a d.c. voltage. With the 'volts/cm' switch on 10 volts/cm the magnitude of the direct voltage is 2.5 cm × 10 volts/cm, i.e. 25 volts.

(ii) With **alternating voltage measurements**, let a sinusoidal waveform be displayed on an oscilloscope screen as shown in Figure 12.13(b). If the time/cm switch is on, say, 5 ms/cm then the **periodic time *T*** of the sinewave is 5 ms/cm × 4 cm, i.e. **20 ms** or **0.02 s**

Since frequency $f=\frac{1}{T}$, **frequency** $=\frac{1}{0.02}=$ **50 Hz**

If the 'volts/cm' switch is on, say, 20 volts/cm then the **amplitude** or **peak value** of the sinewave shown is 20 volts/cm × 2 cm, i.e. 40 V.

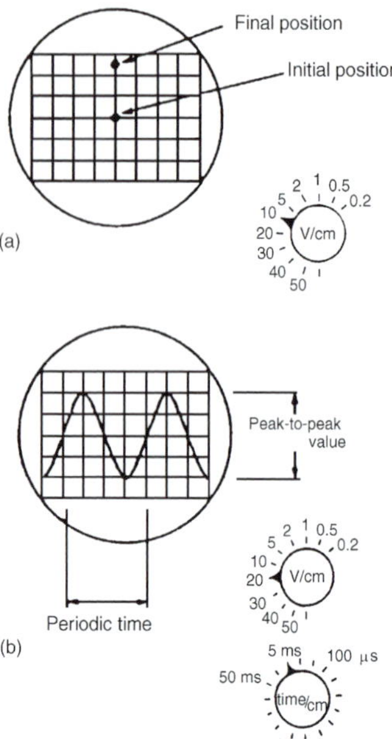

Figure 12.13

$$\text{Since r.m.s. voltage} = \frac{\text{peak voltage}}{\sqrt{2}},$$

(see Chapter 16),

$$\textbf{r.m.s. voltage} = \frac{40}{\sqrt{2}} = \textbf{28.28 volts}$$

Double beam oscilloscopes are useful whenever two signals are to be compared simultaneously.

The oscilloscope demands reasonable skill in adjustment and use. However, its greatest advantage is in observing the shape of a waveform – a feature not possessed by other measuring instruments.

Digital oscilloscopes

Some of the systems that make up digital oscilloscopes are the same as those in analogue oscilloscopes; however, digital oscilloscopes contain additional data processing systems – as shown in the block diagram of Figure 12.14. With the added systems, the digital oscilloscope collects data for the entire waveform and then displays it.

When a digital oscilloscope probe is attached to a circuit, the vertical system adjusts the amplitude of the signal, just as in the analogue oscilloscope. Next, the analogue-to-digital converter (ADC) in the acquisition system samples the signal at discrete points in time and

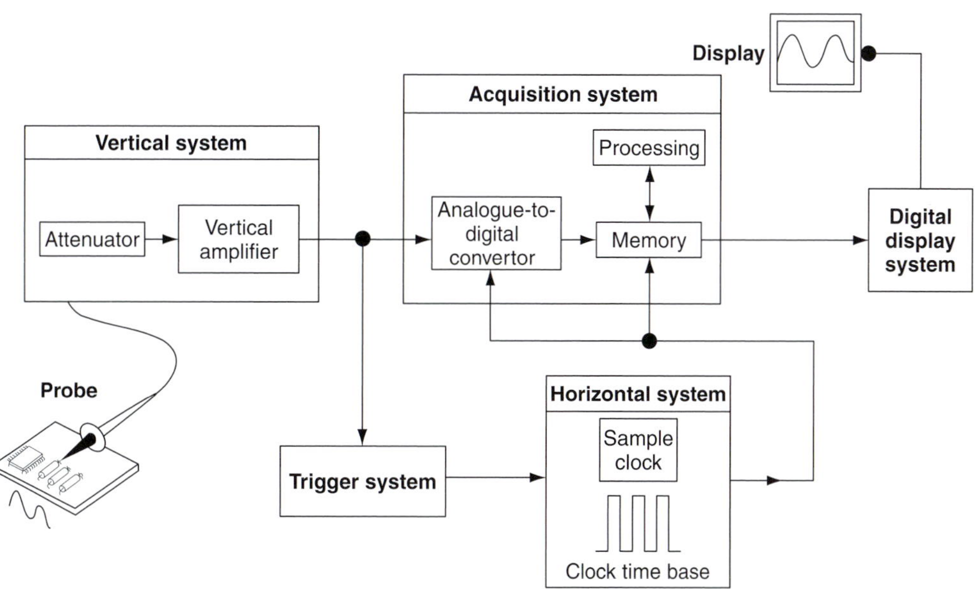

Figure 12.14

converts the signals' voltage at these points to digital values called *sample points*. The horizontal systems' sample clock determines how often the ADC takes a sample. The rate at which the clock 'ticks' is called the sample rate and is measured in samples per second.

The sample points from the ADC are stored in memory as *waveform points*. More than one sample point may make up one waveform point.

Together, the waveform points make up one waveform *record*. The number of waveform points used to make a waveform record is called a *record length*. The trigger system determines the start and stop points of the record. The display receives these record points after being stored in memory.

Depending on the capabilities of an oscilloscope, additional processing of the sample points may take place, enhancing the display. Pre-trigger may be available, allowing events to be seen before the trigger point.

Fundamentally, with a digital oscilloscope as with an analogue oscilloscope, there is a need to adjust vertical, horizontal and trigger settings to take a measurement. A typical double-beam digital fluke oscilloscope is shown in Figure 12.15.

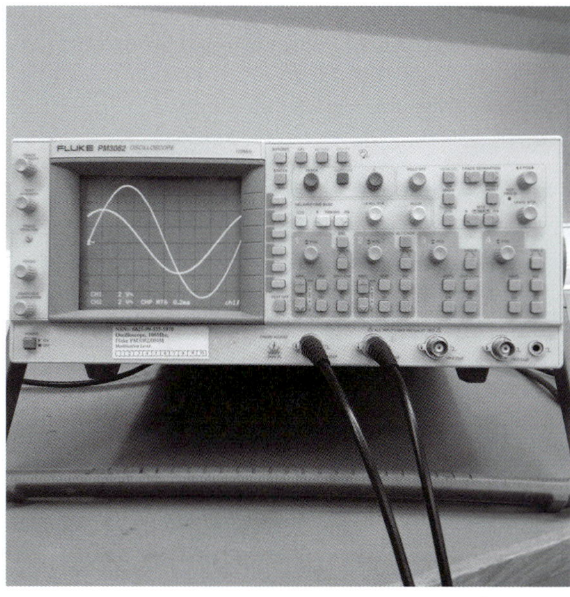

Figure 12.15

Problem 7. For the oscilloscope square voltage waveform shown in Figure 12.16, determine (a) the periodic time, (b) the frequency and (c) the peak-to-peak voltage. The 'time/cm' (or timebase control) switch is on 100 μs/cm and the 'volts/cm' (or signal amplitude control) switch is on 20 V/cm.

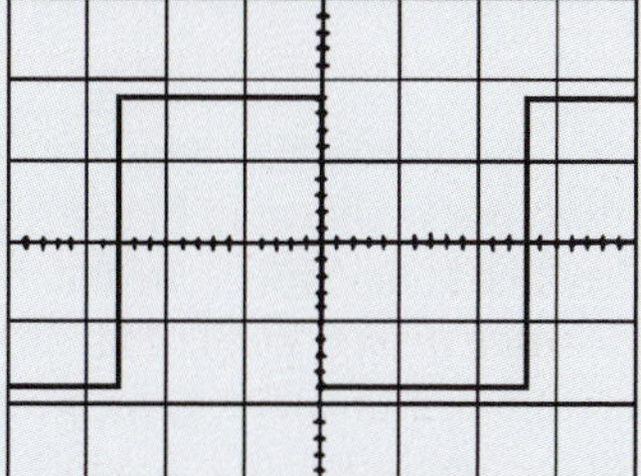

Figure 12.16

(In Figures 12.16 to 12.19 assume that the squares shown are 1 cm by 1 cm.)

(a) The width of one complete cycle is 5.2 cm

Hence the periodic time,
$T = 5.2\,\text{cm} \times 100 \times 10^{-6}\,\text{s/cm} = \mathbf{0.52\,ms}$

(b) Frequency, $f = \frac{1}{T} = \frac{1}{0.52 \times 10^{-3}} = \mathbf{1.92\,kHz}$

(c) The peak-to-peak height of the display is 3.6 cm, hence the peak-to-peak voltage $= 3.6\,\text{cm} \times 20\,\text{V/cm} = \mathbf{72\,V}$

Problem 8. For the oscilloscope display of a pulse waveform shown in Figure 12.17 the 'time/cm' switch is on 50 ms/cm and the 'volts/cm' switch is on 0.2 V/cm. Determine (a) the periodic time, (b) the frequency, (c) the magnitude of the pulse voltage.

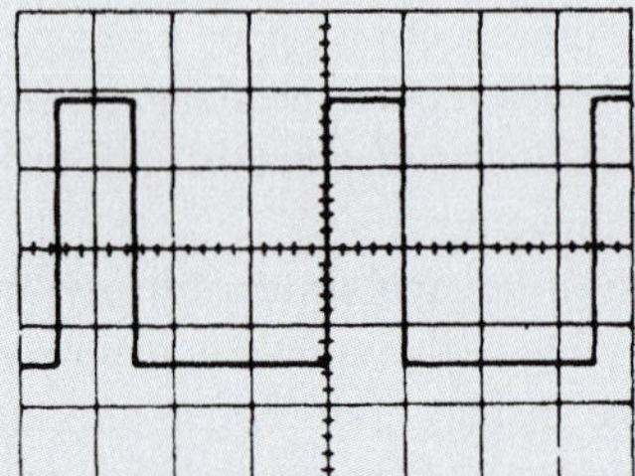

Figure 12.17

(a) The width of one complete cycle is 3.5 cm

Hence the periodic time, $T = 3.5\,\text{cm} \times 50\,\text{ms/cm} = \mathbf{175\,ms}$

(b) Frequency, $f = \frac{1}{T} = \frac{1}{0.175} = \mathbf{5.71\,Hz}$

(c) The height of a pulse is 3.4 cm, hence the magnitude of the pulse voltage $= 3.4\,\text{cm} \times 0.2\,\text{V/cm} = \mathbf{0.68\,V}$

Problem 9. A sinusoidal voltage trace displayed by an oscilloscope is shown in Figure 12.18. If the 'time/cm' switch is on 500 μs/cm and the 'volts/cm' switch is on 5 V/cm, find, for the waveform, (a) the frequency, (b) the peak-to-peak voltage, (c) the amplitude, (d) the r.m.s. value.

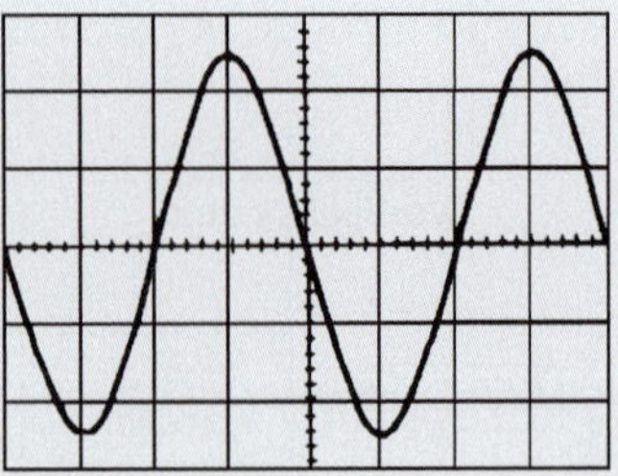

Figure 12.18

(a) The width of one complete cycle is 4 cm. Hence the periodic time, T is $4\,\text{cm} \times 500\,\mu\text{s/cm}$, i.e. 2 ms

Frequency, $f = \frac{1}{T} = \frac{1}{2 \times 10^{-3}} = \mathbf{500\,Hz}$

(b) The peak-to-peak height of the waveform is 5 cm. Hence the peak-to-peak voltage $= 5\,\text{cm} \times 5\,\text{V/cm} = \mathbf{25\,V}$

(c) Amplitude $\frac{1}{2} \times 25\,V = \mathbf{12.5\,V}$

(d) The peak value of voltage is the amplitude, i.e. 12.5 V.

r.m.s. voltage $= \frac{\text{peak voltage}}{\sqrt{2}} = \frac{12.5}{\sqrt{2}} = \mathbf{8.84\,V}$

Problem 10. For the double-beam oscilloscope displays shown in Figure 12.19, determine (a) their frequency, (b) their r.m.s. values, (c) their phase difference. The 'time/cm' switch is on 100 μs/cm and the 'volts/cm' switch on 2 V/cm.

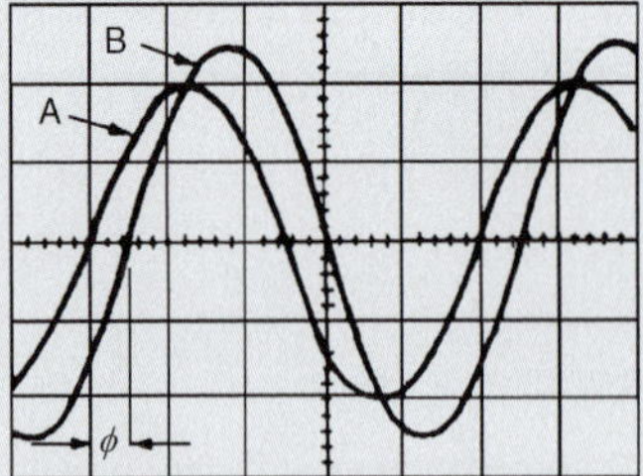

Figure 12.19

(a) The width of each complete cycle is 5 cm for both waveforms. Hence the periodic time, T, of each waveform is $5\,\text{cm} \times 100\,\mu\text{s/cm}$, i.e. 0.5 ms.

Frequency of each waveform, $f = \frac{1}{T} = \frac{1}{0.5 \times 10^{-3}}$

$= \mathbf{2\,kHz}$

(b) The peak value of waveform A is $2\,\text{cm} \times 2\,\text{V/cm} = \mathbf{4\,V}$,

hence the r.m.s. value of waveform A

$$= \frac{4}{\sqrt{2}} = \mathbf{2.83\,V}$$

The peak value of waveform B is $2.5\,\text{cm} \times 2\,\text{V/cm} = 5\,\text{V}$,

hence the r.m.s. value of waveform B

$$= \frac{5}{\sqrt{2}} = \mathbf{3.54\,V}$$

(c) Since 5 cm represents 1 cycle, then 5 cm represents $360°$,

i.e. 1 cm represents $\frac{360}{5} = 72°$

The phase angle $\phi = 0.5\,\text{cm} = 0.5\,\text{cm} \times 72°/\text{cm} = 36°$

Hence waveform A leads waveform B by 36°

Now try the following Practice Exercise

Practice Exercise 59 The oscilloscope (Answers on page 820)

1. For the square voltage waveform displayed on an oscilloscope shown in Figure 12.20, find (a) its frequency, (b) its peak-to-peak voltage.

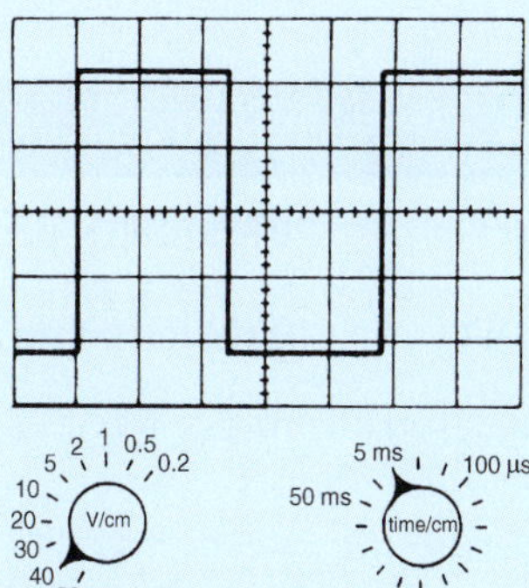

Figure 12.20

2. For the pulse waveform shown in Figure 12.21, find (a) its frequency, (b) the magnitude of the pulse voltage.

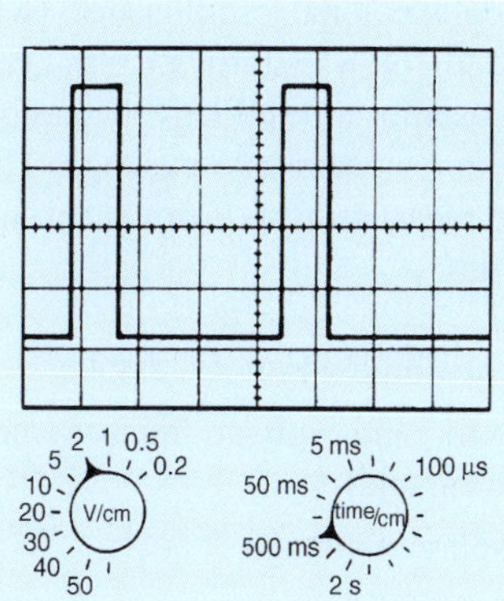

Figure 12.21

3. For the sinusoidal waveform shown in Figure 12.22, determine (a) its frequency, (b) the peak-to-peak voltage, (c) the r.m.s. voltage.

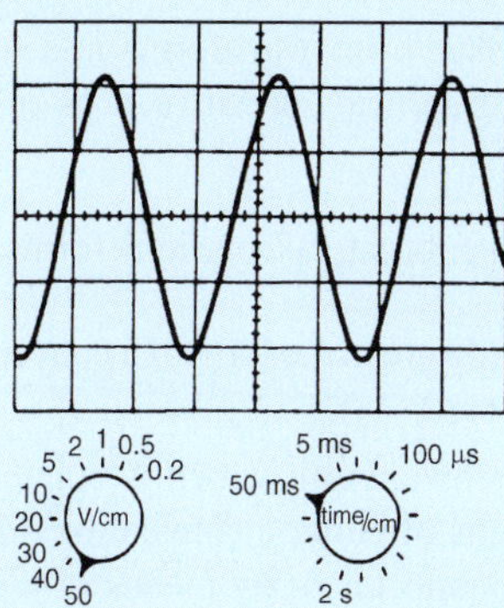

Figure 12.22

12.10 Virtual test and measuring instruments

Computer-based instruments are rapidly replacing items of conventional test equipment in many of today's test and measurement applications. Probably the most commonly available virtual test instrument is the digital storage oscilloscope (DSO). Because of the processing power available from the PC coupled with the mass storage capability, a computer-based virtual DSO is able to provide a variety of additional functions, such as spectrum analysis and digital display of both frequency and voltage. In addition, the ability to save waveforms and captured measurement data for future analysis or for comparison purposes can be extremely valuable, particularly where evidence of conformance with standards or specifications is required.

Unlike a conventional oscilloscope (which is primarily intended for waveform display), a computer-based virtual oscilloscope effectively combines several test instruments in one single package. The functions and available measurements from such an instrument usually includes:

- real-time or stored waveform display
- precise time and voltage measurement (using adjustable cursors)
- digital display of voltage
- digital display of frequency and/or periodic time
- accurate measurement of phase angle
- frequency spectrum display and analysis
- data logging (stored waveform data can be exported in formats that are compatible with conventional spreadsheet packages, e.g. as .xls files)
- ability to save/print waveforms and other information in graphical format (e.g. as .jpg or .bmp files).

Virtual instruments can take various forms, including:

- internal hardware in the form of a conventional PCI expansion card
- external hardware unit which is connected to the PC by means of either a conventional 25-pin parallel port connector or by means of a serial USB connector.

The software (and any necessary drivers) is invariably supplied on CD-ROM or can be downloaded from the manufacturer's website. Some manufacturers also supply software drivers together with sufficient accompanying documentation in order to allow users to control virtual test instruments from their own software developed using popular programming languages such as VisualBASIC or C++.

12.11 Virtual digital storage oscilloscopes

Several types of virtual DSO are currently available. These can be conveniently arranged into three different categories according to their application:

- low-cost DSO
- high-speed DSO
- high-resolution DSO.

Unfortunately, there is often some confusion between the last two categories. A high-speed DSO is designed for examining waveforms that are rapidly changing. Such an instrument does not necessarily provide high-resolution measurement. Similarly, a high-resolution DSO is useful for displaying waveforms with a high degree of precision, but it may not be suitable for examining fast waveforms. The difference between these two types of DSO should become a little clearer later on.

Low-cost DSO are primarily designed for low-frequency signals (typically signals up to around 20 kHz) and are usually able to sample their signals at rates of between 10 K and 100 K samples per second. Resolution is usually limited to either 8-bits or 12-bits (corresponding to 256 and 4096 discrete voltage levels, respectively).

High-speed DSOs are rapidly replacing CRT-based oscilloscopes. They are invariably dual-channel instruments and provide all the features associated with a conventional 'scope', including trigger selection, time-base and voltage ranges, and an ability to operate in X–Y mode.

Additional features available with a computer-based instrument include the ability to capture transient signals (as with a conventional digital storage 'scope') and save waveforms for future analysis. The ability to analyse a signal in terms of its frequency spectrum is yet another feature that is only possible with a DSO (see later).

Upper frequency limit

The upper signal frequency limit of a DSO is determined primarily by the rate at which it can sample an incoming signal. Typical sampling rates for different types of virtual instrument are:

Type of DSO	Typical sampling rate
Low-cost DSO	20 K to 100 K per second
High-speed DSO	100 M to 1000 M per second
High-resolution DSO	20 M to 100 M per second

In order to display waveforms with reasonable accuracy it is normally suggested that the sampling rate should be *at least* twice and *preferably more* than five times the highest signal frequency. Thus, in order to display a 10 MHz signal with any degree of accuracy a sampling rate of 50 M samples per second will be required.

The 'five times rule' merits a little explanation. When sampling signals in a digital-to-analogue converter we usually apply the **Nyquist*** criterion that the sampling frequency must be at least twice the highest analogue signal frequency. Unfortunately, this no longer applies in the case of a DSO where we need to sample at an even faster rate if we are to accurately display the signal. In practise we would need a minimum of about five points within a single cycle of a sampled waveform in order to reproduce it with approximate fidelity. Hence the sampling rate should be at least five times that of the highest signal frequency in order to display a waveform reasonably faithfully.

A special case exists with dual channel DSOs. Here the sampling rate may be shared between the two channels. Thus an effective sampling rate of 20 M samples per second might equate to 10 M samples per second for *each* of the two channels. In such a case the upper frequency limit would not be 4 MHz but only a mere 2 MHz.

The approximate bandwidth required to display different types of signals with reasonable precision is given in the table below:

Signal	Bandwidth required (approx.)
Low-frequency and power	d.c. to 10 kHz
Audio frequency (general)	d.c. to 20 kHz
Audio frequency (high-quality)	d.c. to 50 kHz
Square and pulse waveforms (up to 5 kHz)	d.c. to 100 kHz
Fast pulses with small rise-times	d.c. to 1 MHz
Video	d.c. to 10 MHz
Radio (LF, MF and HF)	d.c. to 50 MHz

The general rule is that, for sinusoidal signals, the bandwidth should ideally be at least double that of the highest signal frequency whilst for square wave and pulse signals, the bandwidth should be at least ten times that of the highest signal frequency.

It is worth noting that most manufacturers define the bandwidth of an instrument as the frequency at which a sine wave input signal will fall to 0.707 of its true amplitude (i.e. the −3 dB point). To put this into context, at the cut-off frequency the displayed trace will be in error by a whopping 29%!

Resolution

The relationship between resolution and signal accuracy (not bandwidth) is simply that the more bits used in the conversion process the more discrete voltage levels can be resolved by the DSO. The relationship is as follows:

$$x = 2^n$$

where x is the number of discrete voltage levels and n is the number of bits. Thus, each time we use an additional bit in the conversion process we double the resolution of the DSO, as shown in the table below:

Number of bits, n	Number of discrete voltage levels, x
8-bit	256
10-bit	1024
12-bit	4096
16-bit	65 536

Buffer memory capacity

A DSO stores its captured waveform samples in a buffer memory. Hence, for a given sampling rate, the size of this memory buffer will determine for how long the DSO can capture a signal before its buffer memory becomes full.

The relationship between sampling rate and buffer memory capacity is important. A DSO with a high sampling rate but small memory will only be able to use its full sampling rate on the top few time base ranges.

To put this into context, it's worth considering a simple example. Assume that we need to display 10 000 cycles of a 10 MHz square wave. This signal will occur in a time frame of 1 ms. If applying the 'five times rule' we would need a bandwidth of at least 50 MHz to display this signal accurately.

To reconstruct the square wave we would need a minimum of about five samples per cycle so a minimum

*__Who was__ Nyquist? **Harry Theodor Nyquist** (7 February 1889–4 April 1976) received the IRE Medal of Honour in 1960 for contributions to thermal noise, data transmission and negative feedback. To find out more go to **www.routledge.com/cw/bird**

sampling rate would be 5×10 MHz = 50 M samples per second. To capture data at the rate of 50 M samples per second for a time interval of 1 ms requires a memory that can store 50 000 samples. If each sample uses 16-bits we would require 100 kbyte of extremely fast memory.

Accuracy

The measurement resolution or measurement accuracy of a DSO (in terms of the smallest voltage change that can be measured) depends on the actual range that is selected. So, for example, on the 1 V range an 8-bit DSO is able to detect a voltage change of one two hundred and fifty sixth of a volt or (1/256) V or about 4 mV. For most measurement applications this will prove to be perfectly adequate as it amounts to an accuracy of about 0.4% of full-scale.

Figure 12.23 depicts a PicoScope software display showing multiple windows providing conventional oscilloscope waveform display, spectrum analyser display, frequency display and voltmeter display.

Adjustable cursors make it possible to carry out extremely accurate measurements. In Figure 12.24 the peak value of the (nominal 10 V peak) waveform is measured at precisely 9625 mV (9.625 V). The time to reach the peak value (from 0 V) is measured as 246.7 μs (0.2467 ms).

The addition of a second time cursor makes it possible to measure the time accurately between two events. In Figure 12.25, event 'o' occurs 131 ns before the trigger point whilst event 'x' occurs 397 ns after the trigger point. The elapsed time between these two events is 528 ns. The two cursors can be adjusted by means of the mouse (or other pointing device) or, more accurately, using the PC's cursor keys.

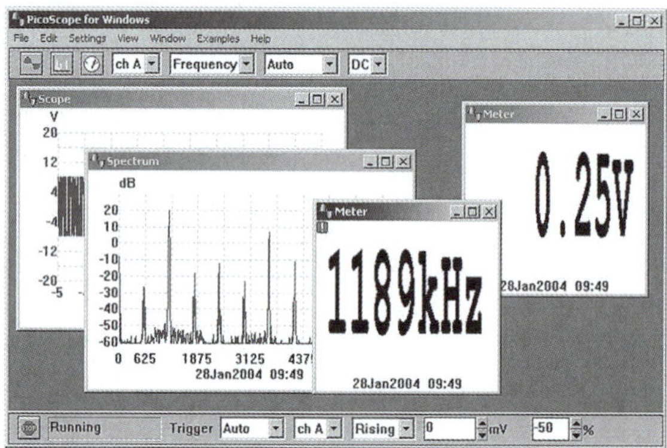

Figure 12.23

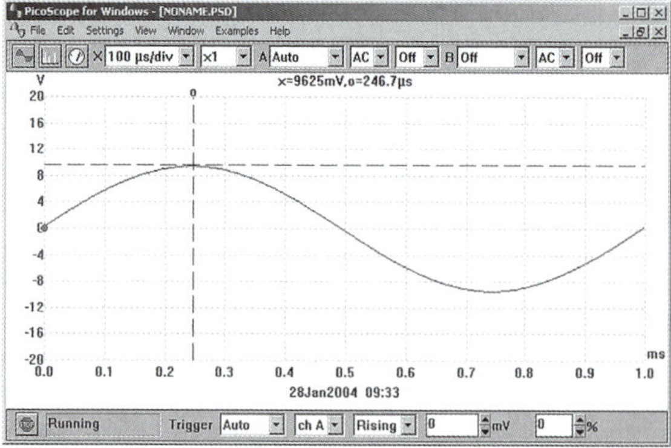

Figure 12.24

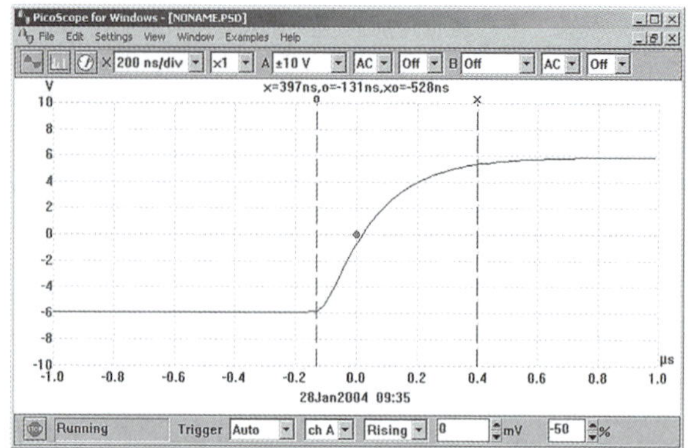

Figure 12.25

Autoranging

Autoranging is another very useful feature that is often provided with a virtual DSO. If you regularly use a conventional 'scope' for a variety of measurements you will know only too well how many times you need to make adjustments to the vertical sensitivity of the instrument.

High-resolution DSO

High-resolution DSOs are used for precision applications where it is necessary to faithfully reproduce a waveform and also to be able to perform an accurate analysis of noise floor and harmonic content. Typical applications include small-signal work and high-quality audio.

Unlike the low-cost DSO, which typically has 8-bit resolution and poor d.c. accuracy, these units are usually accurate to better than 1% and have either 12-bit or 16-bit resolution. This makes them ideal for audio, noise and vibration measurements.

The increased resolution also allows the instrument to be used as a spectrum analyser with very wide dynamic range (up to 100 dB). This feature is ideal for performing noise and distortion measurements on low-level analogue circuits.

Bandwidth alone is not enough to ensure that a DSO can accurately capture a high-frequency signal. The goal of manufacturers is to achieve a flat frequency response. This response is sometimes referred to as a Maximally Flat Envelope Delay (MFED). A frequency response of this type delivers excellent pulse fidelity with minimum overshoot, undershoot and ringing.

It is important to remember that if the input signal is not a pure sine wave it will contain a number of higher-frequency harmonics. For example, a square wave will contain odd harmonics that have levels that become progressively reduced as their frequency increases. Thus, to display a 1 MHz square wave accurately you need to take into account the fact that there will be signal components present at 3 MHz, 5 MHz, 7 MHz, 9 MHz, 11 MHz and so on.

Spectrum analysis

The technique of Fast Fourier Transformation (FFT) calculated using software algorithms using data captured by a virtual DSO has made it possible to produce frequency spectrum displays. Such displays can be used to investigate the harmonic content of waveforms as well as the relationship between several signals within a composite waveform.

Figure 12.26 shows the frequency spectrum of the 1 kHz sine wave signal from a low-distortion signal generator. Here the virtual DSO has been set to capture samples at a rate of 4096 per second within a frequency range of d.c. to 12.2 kHz. The display clearly shows the second harmonic (at a level of −50 dB or −70 dB relative to the fundamental), plus further harmonics at 3 kHz, 5 kHz and 7 kHz (all of which are greater than 75 dB down on the fundamental).

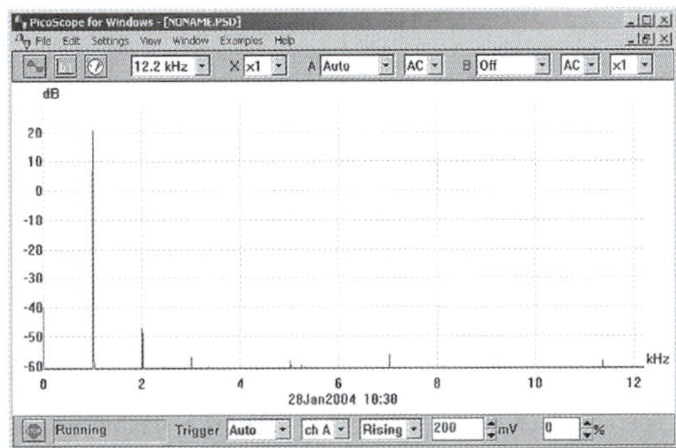

Figure 12.26

Problem 11. Figure 12.27 shows the frequency spectrum of a signal at 1184 kHz displayed by a high-speed virtual DSO. Determine (a) the harmonic relationship between the signals marked 'o' and 'x', (b) the difference in amplitude (expressed in dB) between the signals marked 'o' and 'x' and (c) the amplitude of the second harmonic relative to the fundamental signal 'o'

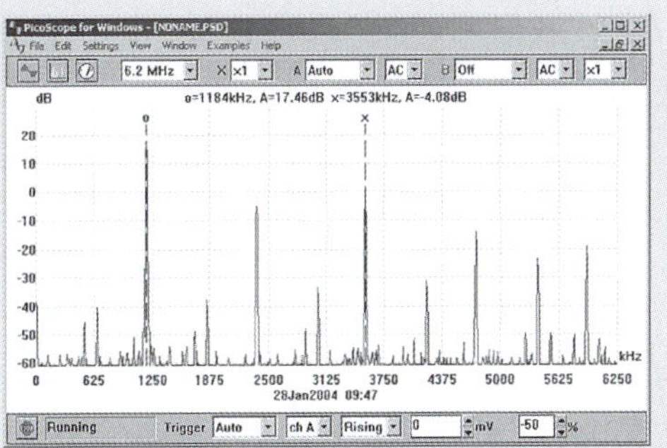

Figure 12.27

(a) The signal x is at a frequency of 3553 kHz. This is three times the frequency of the signal at 'o' which is at 1184 kHz. Thus, **x is the third harmonic of the signal 'o'**

(b) The signal at 'o' has an amplitude of +17.46 dB whilst the signal at 'x' has an amplitude of −4.08 dB. Thus, **the difference in level** $= (+17.46) - (-4.08) =$ **21.54 dB**

(c) **The amplitude of the second harmonic** (shown at approximately 2270 kHz) = **−5 dB**

12.12 Waveform harmonics

(i) Let an instantaneous voltage v be represented by $v = V_m \sin 2\pi ft$ volts. This is a waveform which varies sinusoidally with time t, has a frequency f, and a maximum value V_m. Alternating voltages are usually assumed to have wave shapes which are sinusoidal, where only one frequency is present. If the waveform is not sinusoidal it is called a **complex wave**, and, whatever its shape, it may be split up mathematically into components called the **fundamental** and a number of **harmonics**. This process is called harmonic analysis. The fundamental (or first harmonic) is sinusoidal and has the supply frequency, f; the other harmonics are also sine waves having frequencies which are integer multiples of f. Thus, if the supply frequency is 50 Hz, then the third harmonic frequency is 150 Hz, the fifth 250 Hz, and so on.

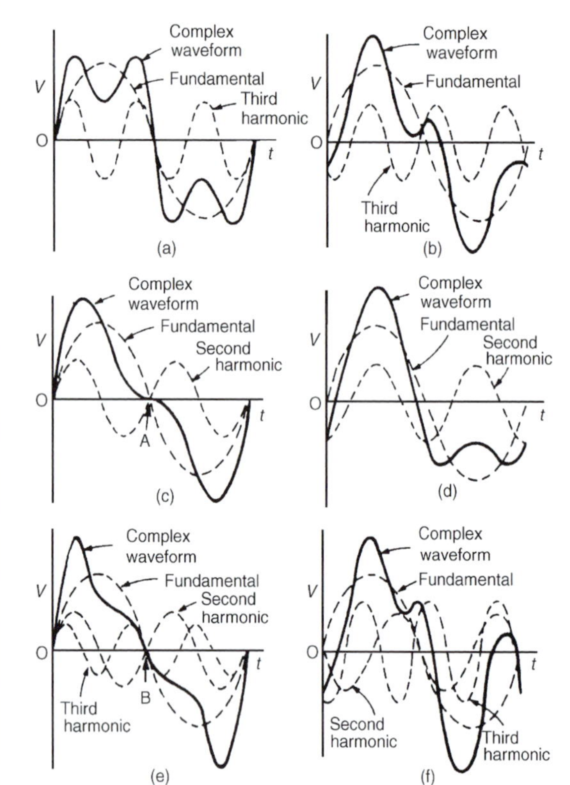

Figure 12.28

(ii) A complex waveform comprising the sum of the fundamental and a third harmonic of about half the amplitude of the fundamental is shown in Figure 12.28(a), both waveforms being initially in phase with each other. If further odd harmonic waveforms of the appropriate amplitudes are added, a good approximation to a square wave results. In Figure 12.28(b), the third harmonic is shown having an initial phase displacement from the fundamental. The positive and negative half cycles of each of the complex waveforms shown in Figures 12.28(a) and (b) are identical in shape, and this is a feature of waveforms containing the fundamental and only odd harmonics.

(iii) A complex waveform comprising the sum of the fundamental and a second harmonic of about half the amplitude of the fundamental is shown in Figure 12.28(c), each waveform being initially in phase with each other. If further even harmonics of appropriate amplitudes are added a good approximation to a triangular wave results. In Figure 12.28(c) the negative cycle appears as a mirror image of the positive cycle about point A. In Figure 12.28(d) the second harmonic is shown with an initial phase displacement from the fundamental and the positive and negative half cycles are dissimilar.

(iv) A complex waveform comprising the sum of the fundamental, a second harmonic and a third harmonic is shown in Figure 12.28(e), each waveform being initially 'in-phase'. The negative half cycle appears as a mirror image of the positive cycle about point B. In Figure 12.28(f), a complex waveform comprising the sum of the fundamental, a second harmonic and a third harmonic are shown with initial phase displacement. The positive and negative half cycles are seen to be dissimilar.

The features mentioned relative to Figures 12.28(a) to (f) make it possible to recognize the harmonics present in a complex waveform displayed on an oscilloscope.

More on complex waveforms may be found in Chapter 39, page 575.

12.13 Logarithmic ratios

In electronic systems, the ratio of two similar quantities measured at different points in the system are often expressed in logarithmic units. By definition, if the ratio of two powers P_1 and P_2 is to be expressed in **decibel (dB) units**, then the number of decibels, X, is given by:

$$\mathbf{X = 10\ lg\left(\frac{P_2}{P_1}\right)\ dB} \qquad (1)$$

A decibel is one-tenth of a bel, the bel being a unit named in honour of **Alexander Graham Bell**.*

Thus, when the power ratio,

$\frac{P_2}{P_1} = 1$ then the decibel power ratio

$= 10 \lg 1 = 0$

when the power ratio,

$\frac{P_2}{P_1} = 100$ then the decibel power ratio

$= 10 \lg 100 = +20$

(i.e. a power gain),

*__Who was__ Bell? **Alexander Graham Bell** (3 March 1847–2 August 1922) is credited with inventing the first practical telephone. To find out more go to **www.routledge.com/cw/bird**

and when the power ratio,

$\frac{P_2}{P_1} = \frac{1}{100}$ then the decibel power ratio

$= 10 \lg \frac{1}{100} = -20$

(i.e. a power loss or attenuation).

Logarithmic units may also be used for voltage and current ratios.

Power, P, is given by $P = I^2R$ or $P = V^2/R$

Substituting in equation (1) gives:

$$X = 10 \lg \left(\frac{I_2^2 R_2}{I_1^2 R_1}\right) \text{dB or } X = 10 \lg \left(\frac{V_2^2/R_2}{V_1^2/R_1}\right) \text{dB}$$

If $R_1 = R_2$ then $X = 10 \lg \left(\frac{I_2^2}{I_1^2}\right)$ dB or $X = 10 \lg \left(\frac{V_2^2}{V_1^2}\right)$ dB

i.e. $\mathbf{X = 20 \lg \left(\frac{I_2}{I_1}\right) dB}$ or $\mathbf{X = 20 \lg \left(\frac{V_2}{V_1}\right) dB}$

(from the laws of logarithms – see Chapter 2).

From equation (1), X decibels is a logarithmic ratio of two similar quantities and is not an absolute unit of measurement. It is therefore necessary to state a **reference level** to measure a number of decibels above or below that reference. The most widely used reference level for power is 1 mW, and when power levels are expressed in decibels, above or below the 1 mW reference level, the unit given to the new power level is dBm.

A voltmeter can be re-scaled to indicate the power level directly in decibels. The scale is generally calibrated by taking a reference level of 0 dB when a power of 1 mW is dissipated in a 600 Ω resistor (this being the natural impedance of a simple transmission line). The reference voltage V is then obtained from

$P = \frac{V^2}{R}$, i.e. $1 \times 10^{-3} = \frac{V^2}{600}$ from which,

$V = 0.775$ volts

In general, the number of dBm, $X = 20 \lg \left(\frac{V}{0.775}\right)$

Thus $V = 0.20$ V corresponds to $20 \lg \left(\frac{0.20}{0.775}\right)$

$= -11.77$ dBm and

$V = 0.90$ V corresponds to $20 \lg \left(\frac{0.90}{0.775}\right)$

$= +1.3$ dBm, and so on.

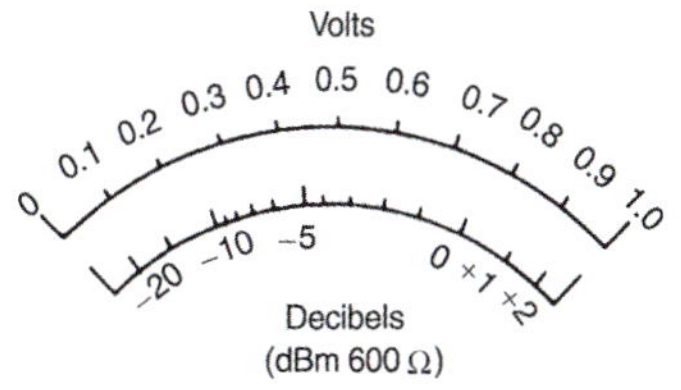

Figure 12.29

A typical **decibelmeter**, or **dB meter**, scale is shown in Figure 12.29. Errors are introduced with dB meters when the circuit impedance is not 600 Ω.

Problem 12. The ratio of two powers is (a) 3, (b) 20, (c) 400, (d) $\frac{1}{20}$. Determine the decibel power ratio in each case.

From above, the power ratio in decibels, X, is given by:

$X = 10 \lg \left(\frac{P_2}{P_1}\right)$

(a) When $\frac{P_2}{P_1} = 3$, $X = 10 \lg(3) = 10(0.477)$
$= \mathbf{4.77\,dB}$

(b) When $\frac{P_2}{P_1} = 20$, $X = 10 \lg(20) = 10(1.30)$
$= \mathbf{13.0\,dB}$

(c) When $\frac{P_2}{P_1} = 400$, $X = 10 \lg(400) = 10(2.60)$
$= \mathbf{26.0\,dB}$

(d) When $\frac{P_2}{P_1} = \frac{1}{20} = 0.05$, $X = 10 \lg(0.05)$
$= 10(-1.30) = \mathbf{-13.0\,dB}$

(a), (b) and (c) represent power gains and (d) represents a power loss or attenuation.

Problem 13. The current input to a system is 5 mA and the current output is 20 mA. Find the decibel current ratio, assuming the input and load resistances of the system are equal.

From above, the decibel current ratio is

$$20 \lg \left(\frac{I_2}{I_1}\right) = 20 \lg \left(\frac{20}{5}\right)$$
$$= 20 \lg 4$$
$$= 20(0.60)$$
$$= \mathbf{12\,dB\ gain}$$

Problem 14. 6% of the power supplied to a cable appears at the output terminals. Determine the power loss in decibels.

If $P_1 =$ input power and $P_2 =$ output power then $\frac{P_2}{P_1} = \frac{6}{100} = 0.06$

$$\text{Decibel power ratio} = 10\lg\left(\frac{P_2}{P_1}\right) = 10\lg(0.06)$$
$$= 10(-1.222) = -12.22\,\text{dB}$$

Hence the decibel power loss, or attenuation, is 12.22 dB

Problem 15. An amplifier has a gain of 14 dB. Its input power is 8 mW. Find its output power.

$$\text{Decibel power ratio} = 10\lg\left(\frac{P_2}{P_1}\right)$$

where $P_1 =$ input power $= 8$ mW,

and $P_2 =$ output power

$$\text{Hence } 14 = 10\lg\left(\frac{P_2}{P_1}\right)$$
$$1.4 = \lg\left(\frac{P_2}{P_1}\right)$$

and $10^{1.4} = \frac{P_2}{P_1}$ from the definition of a logarithm

i.e. $25.12 = \frac{P_2}{P_1}$

Output power, $P_2 = 25.12$, $P_1 = (25.12)(8) =$ **201 mW** or **0.201 W**

Problem 16. The output voltage from an amplifier is 4 V. If the voltage gain is 27 dB, calculate the value of the input voltage assuming that the amplifier input resistance and load resistance are equal.

$$\text{Voltage gain in decibels} = 27 = 20\lg\left(\frac{V_2}{V_1}\right)$$
$$= 20\lg\left(\frac{4}{V_1}\right)$$

$$\text{Hence } \frac{27}{20} = \lg\left(\frac{4}{V_1}\right)$$
$$1.35 = \lg\left(\frac{4}{V_1}\right)$$

$10^{1.35} = \frac{4}{V_1}$, from which

$$V_1 = \frac{4}{10^{1.35}} = \frac{4}{22.39} = 0.179\,\text{V}$$

Hence the input voltage V_1 is 0.179 V

Now try the following Practice Exercise

Practice Exercise 60 Logarithmic ratios (Answers on page 820)

1. The ratio of two powers is (a) 3, (b) 10, (c) 20, (d) 10 000. Determine the decibel power ratio for each.
2. The ratio of two powers is (a) $\frac{1}{10}$, (b) $\frac{1}{3}$, (c) $\frac{1}{40}$, (d) $\frac{1}{100}$ Determine the decibel power ratio for each.
3. The input and output currents of a system are 2 mA and 10 mA, respectively. Determine the decibel current ratio of output to input current assuming input and output resistances of the system are equal.
4. 5% of the power supplied to a cable appears at the output terminals. Determine the power loss in decibels.
5. An amplifier has a gain of 24 dB. Its input power is 10 mW. Find its output power.
6. The output voltage from an amplifier is 7 mV. If the voltage gain is 25 dB, calculate the value of the input voltage assuming that the amplifier input resistance and load resistance are equal.
7. The scale of a voltmeter has a decibel scale added to it, which is calibrated by taking a reference level of 0 dB when a power of 1 mW is dissipated in a 600 Ω resistor. Determine the voltage at (a) 0 dB, (b) 1.5 dB, and (c) −15 dB (d) What decibel reading corresponds to 0.5 V?

12.14 Null method of measurement

A **null method of measurement** is a simple, accurate and widely used method which depends on an instrument reading being adjusted to read zero current only. The method assumes:

(i) if there is any deflection at all, then some current is flowing;

(ii) if there is no deflection, then no current flows (i.e. a null condition).

Hence it is unnecessary for a meter sensing current flow to be calibrated when used in this way. A sensitive milliammeter or microammeter with centre-zero position setting is called a **galvanometer**. Examples where the method is used are in the Wheatstone bridge (see Section 12.15), in the d.c. potentiometer (see Section 12.16) and with a.c. bridges (see Section 12.17).

12.15 Wheatstone bridge

Figure 12.30 shows a **Wheatstone*** bridge circuit which compares an unknown resistance R_x with others of known values, i.e. R_1 and R_2, which have fixed values, and R_3, which is variable. R_3 is varied until zero deflection is obtained on the galvanometer G. No current then flows through the meter, $V_A = V_B$, and the bridge is said to be 'balanced'.

At balance, $R_1R_x = R_2R_3$, i.e. $\mathbf{R_x = \frac{R_2R_3}{R_1}}$ **ohms**

***Who was** Wheatstone? **Sir Charles Wheatstone** (6 February 1802–19 October 1875), was an English scientist and inventor of the concertina, the stereoscope, and the Playfair cipher. To find out more go to **www.routledge.com/cw/bird**

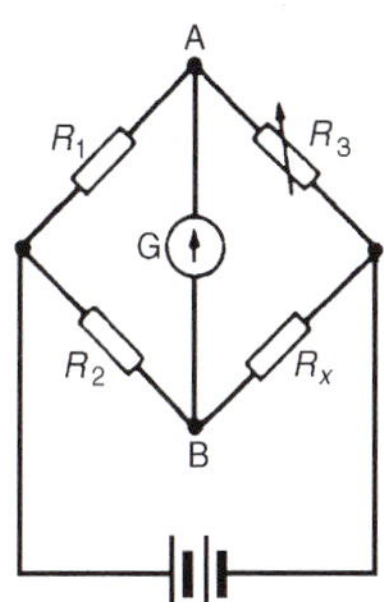

Figure 12.30

Problem 17. In a Wheatstone bridge *ABCD*, a galvanometer is connected between A and C, and a battery between B and D. A resistor of unknown value is connected between A and B. When the bridge is balanced, the resistance between B and C is 100 Ω, that between C and D is 10 Ω and that between D and A is 400 Ω. Calculate the value of the unknown resistance.

The Wheatstone bridge is shown in Figure 12.31 where R_x is the unknown resistance. At balance, equating the products of opposite ratio arms gives:

$$(R_x)(10) = (100)(400)$$

and $R_x = \frac{(100)(400)}{10} = 4000\,\Omega$

Hence the unknown resistance, $R_x = 4\,k\Omega$

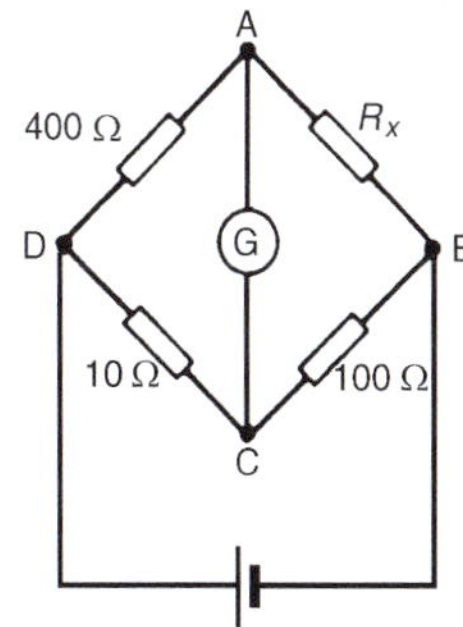

Figure 12.31

12.16 D.c. potentiometer

The **d.c. potentiometer** is a null-balance instrument used for determining values of e.m.f.s and p.d.s. by comparison with a known e.m.f. or p.d. In Figure 12.32(a), using a standard cell of known e.m.f. E_1, the slider S is

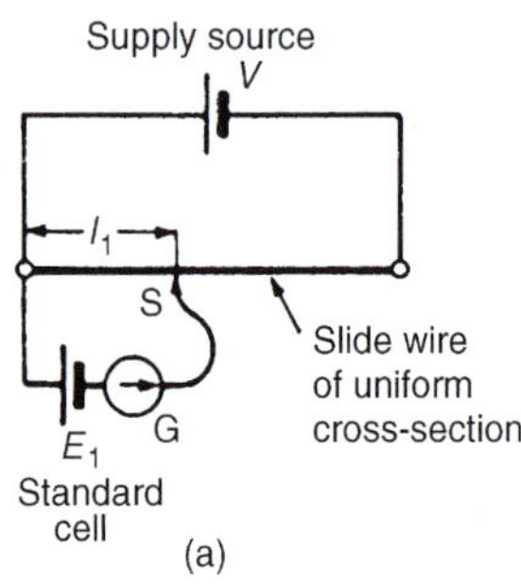

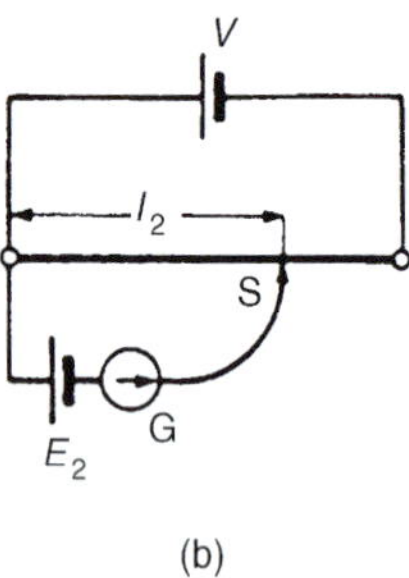

Figure 12.32

moved along the slide wire until balance is obtained (i.e. the galvanometer deflection is zero), shown as length l_1

The standard cell is now replaced by a cell of unknown e.m.f., E_2 (see Figure 12.32(b)), and again balance is obtained (shown as l_2).

Since $E_1 \alpha l_1$ and $E_2 \alpha l_2$ then $\frac{E_1}{E_2} = \frac{l_1}{l_2}$

and $$\boldsymbol{E_2 = E_1 \left(\frac{l_2}{l_1}\right) \textbf{ volts}}$$

A potentiometer may be arranged as a resistive two-element potential divider in which the division ratio is adjustable to give a simple variable d.c. supply. Such devices may be constructed in the form of a resistive element carrying a sliding contact which is adjusted by a rotary or linear movement of the control knob.

Problem 18. In a d.c. potentiometer, balance is obtained at a length of 400 mm when using a standard cell of 1.0186 volts. Determine the e.m.f. of a dry cell if balance is obtained with a length of 650 mm.

$E_1 = 1.0186\,\text{V}$, $l_1 = 400\,\text{mm}$, $l_2 = 650\,\text{mm}$

With reference to Figure 12.32, $\frac{E_1}{E_2} = \frac{l_1}{l_2}$

from which, $E_2 = E_1\left(\frac{l_2}{l_1}\right) = (1.0186)\left(\frac{650}{400}\right)$
$= \mathbf{1.655\ volts}$

Now try the following Practice Exercise

Practice Exercise 61 Wheatstone bridge and d.c. potentiometer (Answers on page 821)

1. In a Wheatstone bridge *PQRS*, a galvanometer is connected between Q and S and a voltage source between P and R. An unknown resistor R_x is connected between P and Q. When the bridge is balanced, the resistance between Q and R is 200 Ω, that between R and S is 10 Ω and that between S and P is 150 Ω. Calculate the value of R_x

2. Balance is obtained in a d.c. potentiometer at a length of 31.2 cm when using a standard cell of 1.0186 volts. Calculate the e.m.f. of a dry cell if balance is obtained with a length of 46.7 cm.

12.17 A.c. bridges

A Wheatstone bridge type circuit, shown in Figure 12.33, may be used in a.c. circuits to determine unknown values of inductance and capacitance, as well as resistance.

When the potential differences across Z_3 and Z_x (or across Z_1 and Z_2) are equal in magnitude and phase, then the current flowing through the galvanometer, G, is zero.

At balance, $Z_1 Z_x = Z_2 Z_3$, from which,

$$\boldsymbol{Z_x = \frac{Z_2 Z_3}{Z_1}\,\Omega}$$

There are many forms of a.c. bridge, and these include: the Maxwell, Hay, Owen and Heaviside bridges for measuring inductance, and the De Sauty, Schering and Wien

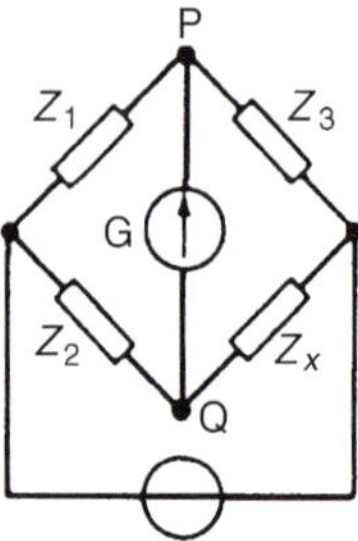

Figure 12.33

bridges for measuring capacitance. A **commercial or universal bridge** is one which can be used to measure resistance, inductance or capacitance.

A.c. bridges require a knowledge of complex numbers, as explained in Chapter 26, and such bridges are discussed in detail in Chapter 30.

12.18 Measurement errors

Errors are always introduced when using instruments to measure electrical quantities. The errors most likely to occur in measurements are those due to:

(i) the limitations of the instrument

(ii) the operator

(iii) the instrument disturbing the circuit

(i) Errors in the limitations of the instrument

The **calibration accuracy** of an instrument depends on the precision with which it is constructed. Every instrument has a margin of error which is expressed as a percentage of the instrument's full-scale deflection.

For example, industrial-grade instruments have an accuracy of ±2% of f.s.d. Thus if a voltmeter has an f.s.d. of 100 V and it indicates 40 V, say, then the actual voltage may be anywhere between 40±(2% of 100), or 40±2, i.e. between 38 V and 42 V.

When an instrument is calibrated, it is compared against a standard instrument and a graph is drawn of 'error' against 'meter deflection'.

A typical graph is shown in Figure 12.34 where it is seen that the accuracy varies over the scale length. Thus a meter with a ±2% f.s.d. accuracy would tend to have an accuracy which is much better than ±2% f.s.d. over much of the range. Calibration is usually carried out by ISO accredited organisations following appropriate quality standards. (ISO is the International Organisation for Standardisation.)

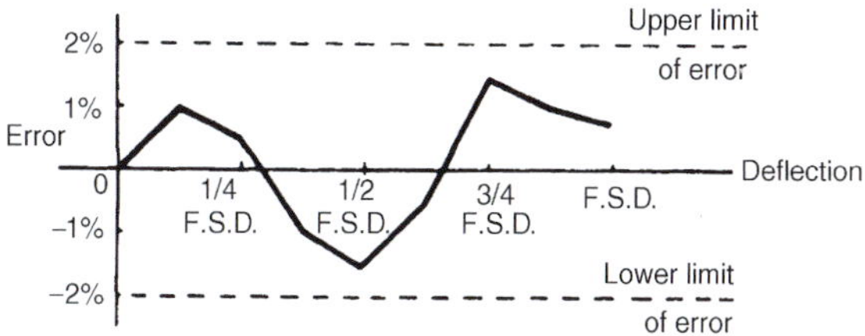

Figure 12.34

(ii) Errors by the operator

It is easy for an operator to misread an instrument. With linear scales the values of the sub-divisions are reasonably easy to determine; non-linear scale graduations are more difficult to estimate. Also, scales differ from instrument to instrument and some meters have more than one scale (as with multimeters) and mistakes in reading indications are easily made. When reading a meter scale it should be viewed from an angle perpendicular to the surface of the scale at the location of the pointer; a meter scale should not be viewed 'at an angle'. Errors by the operator are largely eliminated using digital instruments.

(iii) Errors due to the instrument disturbing the circuit

Any instrument connected into a circuit will affect that circuit to some extent. Meters require some power to operate, but provided this power is small compared with the power in the measured circuit, then little error will result. Incorrect positioning of instruments in a circuit can be a source of errors. For example, let a resistance be measured by the voltmeter-ammeter method as shown in Figure 12.35. Assuming 'perfect' instruments, the resistance should be given by the voltmeter reading divided by the ammeter reading (i.e. $R = V/I$).

However, in Figure 12.35(a), $V/I = R + r_a$ and in Figure 12.35(b) the current through the ammeter is that through the resistor plus that

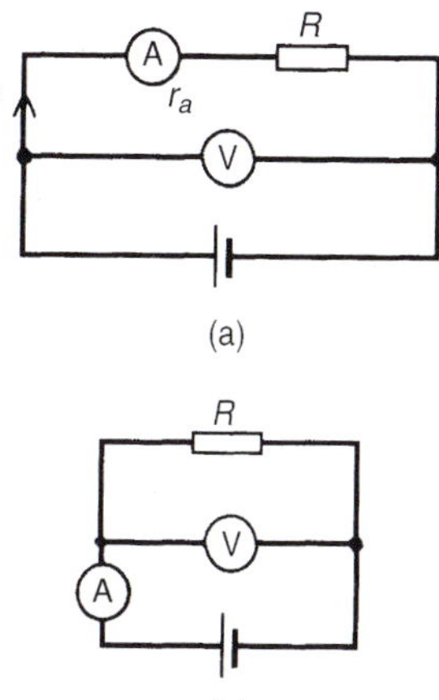

Figure 12.35

through the voltmeter. Hence the voltmeter reading divided by the ammeter reading will not give the true value of the resistance R for either method of connection.

Problem 19. The current flowing through a resistor of $5\,k\Omega \pm 0.4\%$ is measured as 2.5 mA with an accuracy of measurement of $\pm 0.5\%$. Determine the nominal value of the voltage across the resistor and its accuracy.

Voltage, $V = IR = (2.5 \times 10^{-3})(5 \times 10^{3}) = 12.5\,V$.

The maximum possible error is $0.4\% + 0.5\% = 0.9\%$

Hence the voltage, $V = 12.5\,V \pm 0.9\%$ of 12.5 V $= 0.9/100 \times 12.5 = 0.1125\,V = 0.11\,V$ correct to 2 significant figures. Hence the voltage V may also be expressed as **12.5 ± 0.11 volts** (i.e. a voltage lying between 12.39 V and 12.61 V).

Problem 20. The current I flowing in a resistor R is measured by a 0–10 A ammeter which gives an indication of 6.25 A. The voltage V across the resistor is measured by a 0–50 V voltmeter, which gives an indication of 36.5 V. Determine the resistance of the resistor, and its accuracy of measurement if both instruments have a limit of error of 2% of f.s.d. Neglect any loading effects of the instruments.

$$\text{Resistance, } R = \frac{V}{I} = \frac{36.5}{6.25} = 5.84\,\Omega$$

Voltage error is $\pm 2\%$ of $50\,V = \pm 1.0\,V$ and expressed as a percentage of the voltmeter reading gives $\frac{\pm 1}{36.5} \times 100\% = \pm 2.74\%$

Current error is $\pm 2\%$ of $10\,A = \pm 0.2\,A$ and expressed as a percentage of the ammeter reading gives $\frac{\pm 0.2}{6.25} \times 100\% = \pm 3.2\%$

Maximum relative error = sum of errors $= 2.74\% + 3.2\% = \pm 5.94\%$, and 5.94% of $5.84\,\Omega = 0.347\,\Omega$

Hence the resistance of the resistor may be expressed as: **5.84 Ω ± 5.94%**, or **5.84 ± 0.35 Ω** (rounding off).

Problem 21. The arms of a Wheatstone bridge $ABCD$ have the following resistances: AB: $R_1 = 1000\,\Omega \pm 1.0\%$; BC: $R_2 = 100\,\Omega \pm 0.5\%$; CD: unknown resistance R_x; DA: $R_3 = 432.5\,\Omega \pm 0.2\%$. Determine the value of the unknown resistance and its accuracy of measurement.

The Wheatstone bridge network is shown in Figure 12.36 and at balance:

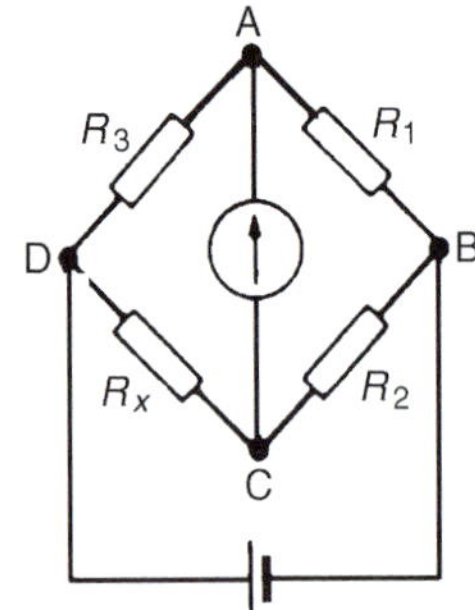

Figure 12.36

$$R_1 R_x = R_2 R_3, \text{ i.e. } R_x = \frac{R_2 R_3}{R_1} = \frac{(100)(432.5)}{1000} = 43.25\,\Omega$$

The maximum relative error of R_x is given by the sum of the three individual errors, i.e.

$1.0\% + 0.5\% + 0.2\% = 1.7\%$

Hence $R_x = 43.25\,\Omega \pm 1.7\%$

1.7% of $43.25\,\Omega = 0.74\,\Omega$ (rounding off).

Thus R_x may also be expressed as $\boldsymbol{R_x = 43.25 \pm 0.74\,\Omega}$

Now try the following Practice Exercise

Practice Exercise 62 Measurement errors (Answers on page 821)

1. The p.d. across a resistor is measured as 37.5 V with an accuracy of $\pm 0.5\%$. The value of the resistor is $6\,k\Omega \pm 0.8\%$. Determine the current flowing in the resistor and its accuracy of measurement.

2. The voltage across a resistor is measured by a 75 V f.s.d. voltmeter which gives an indication of 52 V. The current flowing in the resistor is measured by a 20 A f.s.d. ammeter which gives an indication of 12.5 A. Determine the resistance of the resistor and its accuracy if both instruments have an accuracy of ±2% of f.s.d.

3. A Wheatstone bridge *PQRS* has the following arm resistances:
PQ, 1 kΩ ± 2%; *QR*, 100 Ω ± 0.5%; *RS*, unknown resistance; *SP*, 273.6 Ω ± 0.1%. Determine the value of the unknown resistance, and its accuracy of measurement.

For fully worked solutions to each of the problems in Practice Exercises 57 to 62 in this chapter, go to the website:
www.routledge.com/cw/bird

Chapter 13

Semiconductor diodes

Why it is important to understand: **Semiconductor diodes**

Semiconductors have had a monumental impact on our society. Semiconductors are found at the heart of microprocessor chips as well as transistors. Anything that's computerized or uses radio waves depends on semiconductors. Today, most semiconductor chips and transistors are created with silicon; semiconductors are the foundation of modern electronics. Semiconductor-based electronic components include transistors, solar cells, many kinds of diodes including the light-emitting diode (LED), the silicon controlled rectifier, photo-diodes, and digital and analogue integrated circuits. A diode is the simplest possible semiconductor device. The ability of the diode to conduct current easily in one direction, but not in the reverse direction, is very useful. For example, in a car, diodes allow current from the alternator to charge the battery when the engine is running. However, when the engine stops, the diode prevents the battery from discharging through the alternator (preventing damage). Diodes are widely used in power supplies and battery chargers to convert the mains a.c. voltage to a d.c. level (rectifiers). They are also used to protect elements and systems from excessive voltages or currents, polarity reversals, arcing and shorting. Diodes are one of the most fundamental devices that are used in electronics. This chapter explains the operation of the p–n junction, and the characteristics and applications of various types of diode.

At the end of this chapter you should be able to:

- classify materials as conductors, semiconductors or insulators
- appreciate the importance of silicon and germanium
- understand n-type and p-type materials
- understand the p–n junction
- appreciate forward and reverse bias of p–n junctions
- recognize the symbols used to represent diodes in circuit diagrams
- understand the importance of diode characteristics and maximum ratings
- know the characteristics and applications of various types of diode – signal diodes, rectifiers, zener diodes, silicon controlled rectifiers, light emitting diodes, varactor diodes and Schottky diodes

Electrical Circuit Theory and Technology. 978-1-138-67349-6, © 2017 John Bird. Published by Taylor & Francis. All rights reserved.

13.1 Types of material

Materials may be classified as conductors, semiconductors or insulators. The classification depends on the value of resistivity of the material. Good conductors are usually metals and have resistivities in the order of 10^{-7} to $10^{-8}\,\Omega$m, semiconductors have resistivities in the order of 10^{-3} to $3\times10^{3}\,\Omega$m, and the resistivities of insulators are in the order of 10^{4} to $10^{14}\,\Omega$m. Some typical approximate values at normal room temperatures are:

Conductors	
Aluminium	$2.7\times10^{-8}\,\Omega$m
Brass (70 Cu/30 Zn)	$8\times10^{-8}\,\Omega$m
Copper (pure annealed)	$1.7\times10^{-8}\,\Omega$m
Steel (mild)	$15\times10^{-8}\,\Omega$m

Semiconductors (at 27°C)	
Silicon	$2.3\times10^{3}\,\Omega$m
Germanium	$0.45\,\Omega$m

Insulators	
Glass	$\geq10^{10}\,\Omega$m
Mica	$\geq10^{11}\,\Omega$m
PVC	$\geq10^{13}\,\Omega$m
Rubber (pure)	10^{12} to $10^{14}\,\Omega$m

In general, over a limited range of temperatures, the resistance of a conductor increases with temperature increase, the resistance of insulators remains approximately constant with variation of temperature and the resistance of semiconductor materials decreases as the temperature increases. For a specimen of each of these materials, having the same resistance (and thus completely different dimensions) at, say, 15°C, the variation for a small increase in temperature to t°C is as shown in Figure 13.1.

As the temperature of semiconductor materials is raised above room temperature, the resistivity is reduced and ultimately a point is reached where they effectively become conductors. For this reason, silicon should not operate at a working temperature in excess of 150°C to 200°C, depending on its purity, and germanium should not operate at a working temperature in excess of 75°C to 90°C, depending on its purity. As the temperature of a semiconductor is reduced below normal room temperature, the resistivity increases until, at very low temperatures, the semiconductor becomes an insulator.

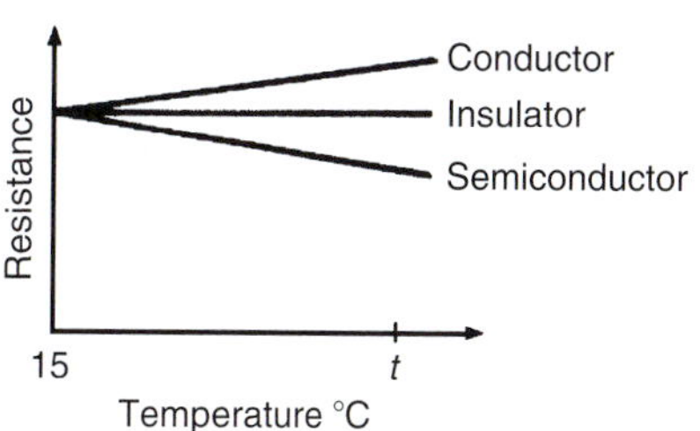

Figure 13.1

13.2 Semiconductor materials

From Chapter 4, it was stated that an atom contains both negative charge carriers **(electrons)** and positive charge carriers **(protons)**. Electrons each carry a single unit of negative electric charge while protons each exhibit a single unit of positive charge. Since atoms normally contain an equal number of electrons and protons, the net charge present will be zero. For example, if an atom has 11 electrons, it will also contain 11 protons. The end result is that the negative charge of the electrons will be exactly balanced by the positive charge of the protons.

Electrons are in constant motion as they orbit around the nucleus of the atom. Electron orbits are organized into **shells**. The maximum number of electrons present in the first shell is two, in the second shell eight, and in the third, fourth and fifth shells it is 18, 32 and 50, respectively. In electronics, only the electron shell furthermost from the nucleus of an atom is important. It is important to note that the movement of electrons between atoms only involves those present in the outer **valence shell**.

If the valence shell contains the maximum number of electrons possible, the electrons are rigidly bonded together and the material has the properties of an insulator (see Figure 13.2). If, however, the valence shell does not have its full complement of electrons, the electrons can be easily detached from their orbital bonds, and the material has the properties associated with an electrical conductor.

In its pure state, silicon is an insulator because the covalent bonding rigidly holds all of the electrons, leaving no free (easily loosened) electrons to conduct

Part 2

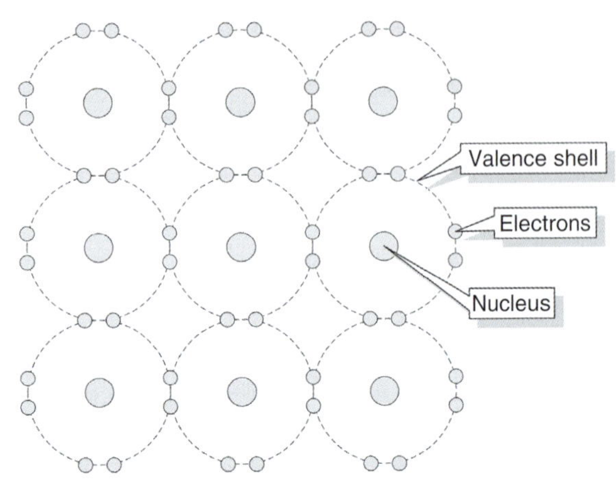

Figure 13.2

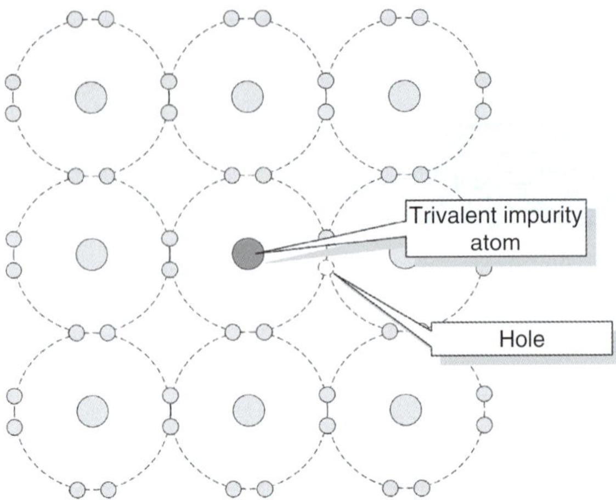

Figure 13.4

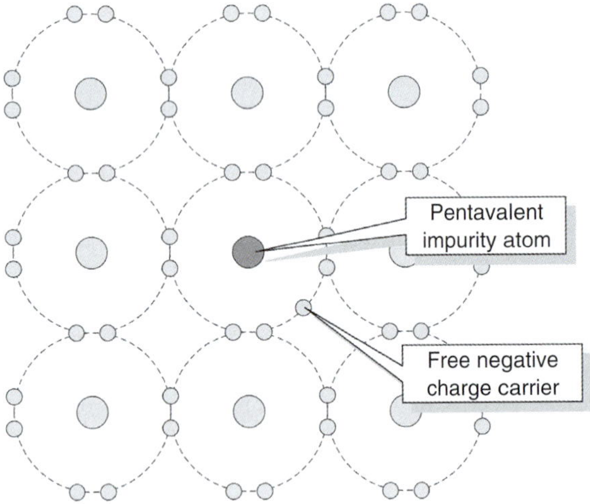

Figure 13.3

current. If, however, an atom of a different element (i.e. an **impurity**) is introduced that has five electrons in its valence shell, a surplus electron will be present (see Figure 13.3). These free electrons become available for use as charge carriers and they can be made to move through the lattice by applying an external potential difference to the material.

Similarly, if the impurity element introduced into the pure silicon lattice has three electrons in its valence shell, the absence of the fourth electron needed for proper covalent bonding will produce a number of spaces into which electrons can fit (see Figure 13.4). These spaces are referred to as **holes**. Once again, current will flow when an external potential difference is applied to the material.

Regardless of whether the impurity element produces surplus electrons or holes, the material will no longer behave as an insulator, neither will it have the properties that we normally associate with a metallic conductor. Instead, we call the material a **semiconductor** – the term simply serves to indicate that the material is no longer a good insulator nor is it a good conductor but is somewhere in between. Examples of semiconductor materials include **silicon (Si), germanium (Ge), gallium arsenide (GaAs)** and **indium arsenide (InAs)**.

Antimony, arsenic and **phosphorus** are **n-type impurities** and form an n-type material when any of these impurities are added to pure semiconductor material such as silicon or germanium. The amount of impurity added usually varies from 1 part impurity in 10^5 parts semiconductor material to 1 part impurity to 10^8 parts semiconductor material, depending on the resistivity required. **Indium, aluminium** and **boron** are all **p-type impurities** and form a p-type material when any of these impurities are added to a pure semiconductor.

The process of introducing an atom of another (impurity) element into the lattice of an otherwise pure material is called **doping**. When the pure material is doped with an impurity with five electrons in its valence shell (i.e. a **pentavalent impurity**) it will become an **n-type** (i.e. negative type) semiconductor material. If, however, the pure material is doped with an impurity having three electrons in its valence shell (i.e. a **trivalent impurity**) it will become a **p-type** (i.e. positive type) semiconductor material. Note that n-type semiconductor material contains an excess of negative charge carriers, and p-type material contains an excess of positive charge carriers.

In semiconductor materials, there are very few charge carriers per unit volume free to conduct. This is because the 'four electron structure' in the outer shell of the atoms (called **valency electrons**), form strong **covalent bonds** with neighbouring atoms, resulting in a tetrahedral (i.e. four-sided) structure with the electrons held fairly rigidly in place.

13.3 Conduction in semiconductor materials

Arsenic, antimony and phosphorus have five valency electrons and when a semiconductor is doped with one of these substances, some impurity atoms are incorporated in the tetrahedral structure. The 'fifth' valency electron is not rigidly bonded and is free to conduct, the impurity atom donating a charge carrier.

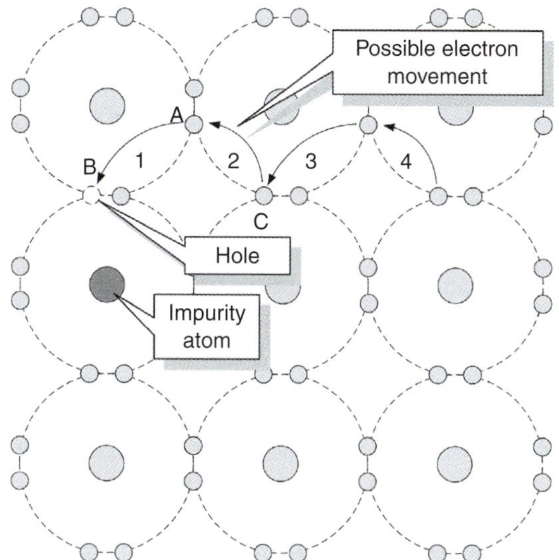

Figure 13.5

Indium, aluminium and boron have three valency electrons, and when a semiconductor is doped with one of these substances, some of the semiconductor atoms are replaced by impurity atoms. One of the four bonds associated with the semiconductor material is deficient by one electron and this deficiency is called a **hole**. Holes give rise to conduction when a potential difference exists across the semiconductor material due to movement of electrons from one hole to another, as shown in Figure 13.5. In this diagram, an electron moves from A to B, giving the appearance that the hole moves from B to A. Then electron C moves to A, giving the appearance that the hole moves to C, and so on.

13.4 The p–n junction

A p–n junction is a piece of semiconductor material in which part of the material is p-type and part is n-type.

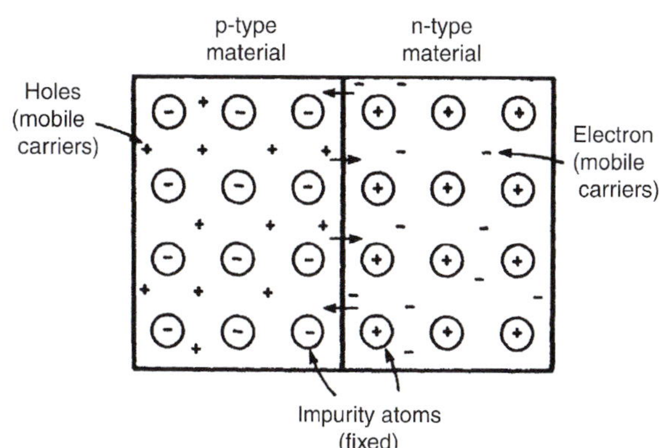

Figure 13.6

In order to examine the charge situation, assume that separate blocks of p-type and n-type materials are pushed together. Also assume that a hole is a positive charge carrier and that an electron is a negative charge carrier.

At the junction, the donated electrons in the n-type material, called **majority carriers**, diffuse into the p-type material (diffusion is from an area of high density to an area of lower density) and the acceptor holes in the p-type material diffuse into the n-type material, as shown by the arrows in Figure 13.6. Because the n-type material has lost electrons, it acquires a positive potential with respect to the p-type material and thus tends to prevent further movement of electrons. The p-type material has gained electrons and becomes negatively charged with respect to the n-type material and hence tends to retain holes. Thus after a short while, the movement of electrons and holes stops due to the potential difference across the junction, called the **contact potential**. The area in the region of the junction becomes depleted of holes and electrons due to electron-hole recombination, and is called a **depletion layer**, as shown in Figure 13.7.

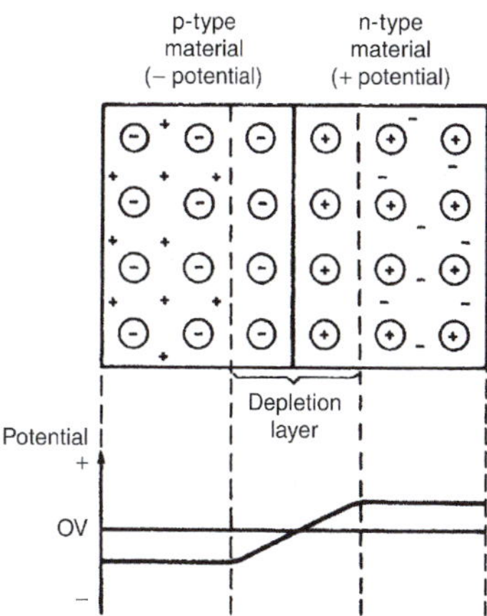

Figure 13.7

Part 2

Problem 1. Explain briefly the terms given below when they are associated with a p–n junction:
(a) conduction in intrinsic semiconductors,
(b) majority and minority carriers, and
(c) diffusion.

(a) Silicon or germanium with no doping atoms added are called **intrinsic semiconductors**. At room temperature, some of the electrons acquire sufficient energy for them to break the covalent bond between atoms and become free mobile electrons. This is called **thermal generation of electron-hole pairs**. Electrons generated thermally create a gap in the crystal structure called a hole, the atom associated with the hole being positively charged, since it has lost an electron. This positive charge may attract another electron released from another atom, creating a hole elsewhere. When a potential is applied across the semiconductor material, holes drift towards the negative terminal (unlike charges attract), and electrons towards the positive terminal, and hence a small current flows.

(b) When additional mobile electrons are introduced by doping a semiconductor material with pentavalent atoms (atoms having five valency electrons), these mobile electrons are called **majority carriers**. The relatively few holes in the n-type material produced by intrinsic action are called **minority carriers**.

For p-type materials, the additional holes are introduced by doping with trivalent atoms (atoms having three valency electrons). The holes are apparently positive mobile charges and are majority carriers in the p-type material. The relatively few mobile electrons in the p-type material produced by intrinsic action are called minority carriers.

(c) Mobile holes and electrons wander freely within the crystal lattice of a semiconductor material. There are more free electrons in n-type material than holes and more holes in p-type material than electrons. Thus, in their random wanderings, on average, holes pass into the n-type material and electrons into the p-type material. This process is called **diffusion**.

Problem 2. Explain briefly why a junction between p-type and n-type materials creates a contact potential.

Intrinsic semiconductors have resistive properties, in that when an applied voltage across the material is reversed in polarity, a current of the same magnitude flows in the opposite direction. When a p–n junction is formed, the resistive property is replaced by a rectifying property, that is, current passes more easily in one direction than the other.

An n-type material can be considered to be a stationary crystal matrix of fixed positive charges together with a number of mobile negative charge carriers (electrons). The total number of positive and negative charges are equal. A p-type material can be considered to be a number of stationary negative charges together with mobile positive charge carriers (holes).

Again, the total number of positive and negative charges are equal and the material is neither positively nor negatively charged. When the materials are brought together, some of the mobile electrons in the n-type material diffuse into the p-type material. Also, some of the mobile holes in the p-type material diffuse into the n-type material.

Many of the majority carriers in the region of the junction combine with the opposite carriers to complete covalent bonds and create a region on either side of the junction with very few carriers. This region, called the **depletion layer**, acts as an insulator and is in the order of 0.5 μm thick. Since the n-type material has lost electrons, it becomes positively charged. Also, the p-type material has lost holes and becomes negatively charged, creating a potential across the junction, called the **barrier** or **contact potential**.

13.5 Forward and reverse bias

When an external voltage is applied to a p–n junction making the p-type material positive with respect to the n-type material, as shown in Figure 13.8, the p–n junction is **forward biased**. The applied voltage opposes the contact potential, and, in effect, closes the depletion layer. Holes and electrons can now cross the junction and a current flows. An increase in the applied voltage above that required to narrow the depletion layer (about 0.2 V for germanium and 0.6 V for silicon), results in a rapid rise in the current flow.

When an external voltage is applied to a p–n junction making the p-type material negative with respect to the n-type material, as shown in Figure 13.9, the p–n junction is **reverse biased**. The applied voltage is now in the same sense as the contact potential and opposes the

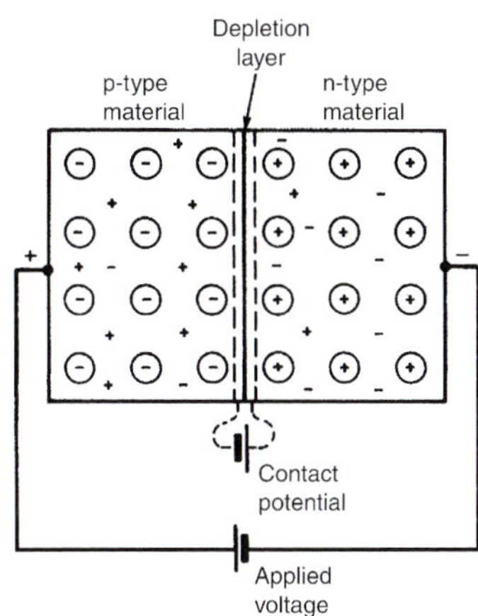

Figure 13.8

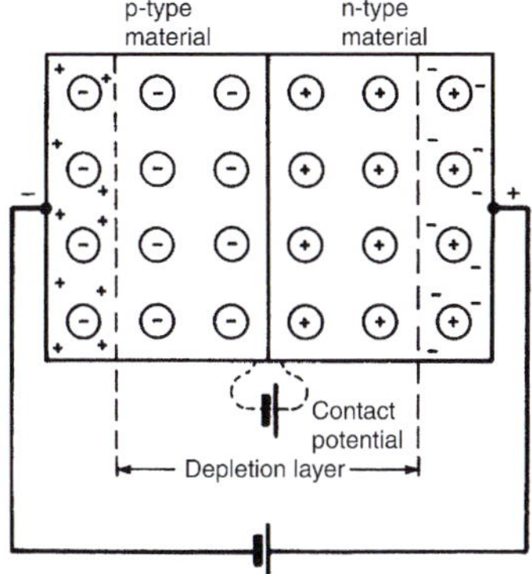

Figure 13.9

movement of holes and electrons due to opening up the depletion layer. Thus, in theory, no current flows. However, at normal room temperature certain electrons in the covalent bond lattice acquire sufficient energy from the heat available to leave the lattice, generating mobile electrons and holes. This process is called **electron-hole generation by thermal excitation**.

The electrons in the p-type material and holes in the n-type material caused by thermal excitation are called minority carriers and these will be attracted by the applied voltage. Thus, in practice, a small current of a few microamperes for germanium and less than one microampere for silicon, at normal room temperature, flows under reverse bias conditions.

Graphs depicting the current–voltage relationship for forward and reverse biased p–n junctions, for both germanium and silicon, are shown in Figure 13.10.

Problem 3. Sketch the forward and reverse characteristics of a silicon p–n junction diode and describe the shapes of the characteristics drawn.

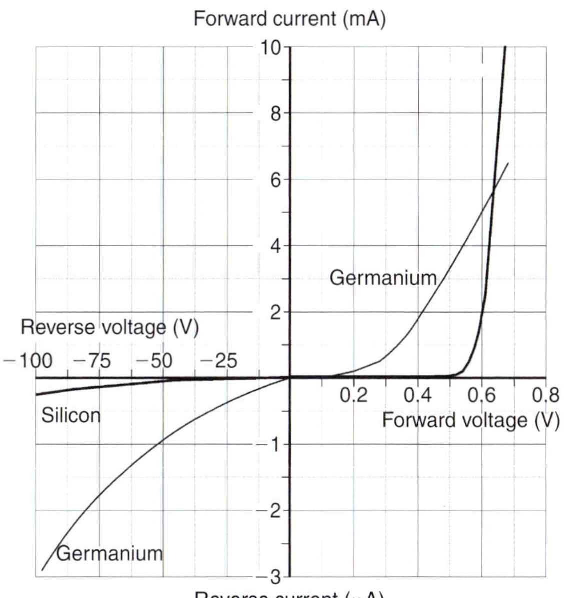

Figure 13.10

A typical characteristic for a silicon p–n junction is shown in Figure 13.10. When the positive terminal of the battery is connected to the p-type material and the negative terminal to the n-type material, the diode is forward biased. Due to like charges repelling, the holes in the p-type material drift towards the junction. Similarly, the electrons in the n-type material are repelled by the negative bias voltage and also drift towards the junction. The width of the depletion layer and size of the contact potential are reduced. For applied voltages from 0 to about 0.6 V, very little current flows. At about 0.6 V, majority carriers begin to cross the junction in large numbers and current starts to flow. As the applied voltage is raised above 0.6 V, the current increases exponentially (see Figure 13.10).

When the negative terminal of the battery is connected to the p-type material and the positive terminal to the n-type material the diode is reverse biased. The holes in the p-type material are attracted towards the negative terminal and the electrons in the n-type material are attracted towards the positive terminal (unlike charges attract). This drift increases the magnitude of both the contact potential and the thickness of the depletion layer, so that only very few majority carriers have sufficient energy to surmount the junction.

The thermally excited minority carriers, however, can cross the junction since it is, in effect, forward biased for these carriers. The movement of minority carriers results in a small constant current flowing. As the magnitude of the reverse voltage is increased a point will be reached where a large current suddenly starts to flow. The voltage at which this occurs is called the **breakdown voltage**. This current is due to two effects:

(i) the **zener effect**, resulting from the applied voltage being sufficient to break some of the covalent bonds, and

(ii) the **avalanche effect**, resulting from the charge carriers moving at sufficient speed to break covalent bonds by collision.

Problem 4. The forward characteristic of a diode is shown in Figure 13.11. Use the characteristic to determine (a) the current flowing in the diode when a forward voltage of 0.4 V is applied, (b) the voltage dropped across the diode when a forward current of 9 mA is flowing in it, (c) the resistance of the diode when the forward voltage is 0.6 V, and (d) whether the diode is a Ge or Si type.

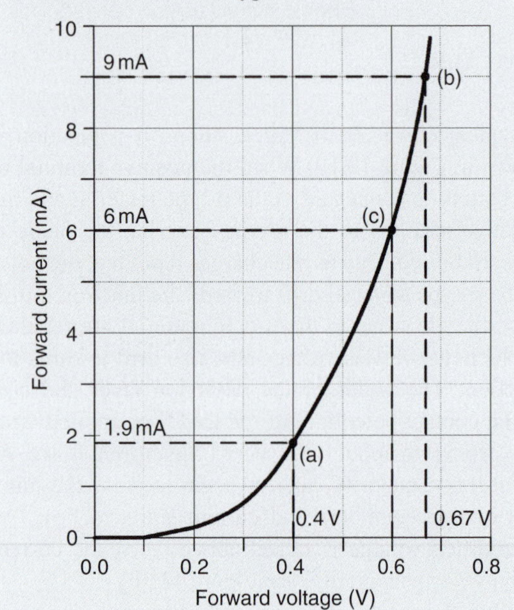

Figure 13.11

(a) From Figure 13.11, when $V = 0.4$ V, **current flowing, $I = 1.9$ mA**

(b) When $I = 9$ mA, **the voltage dropped across the diode, $V = 0.67$ V**

(c) From the graph, when $V = 0.6$ V, $I = 6$ mA.

Thus, **resistance of the diode,** $R = \frac{V}{I} = \frac{0.6}{6 \times 10^{-3}}$
$= 0.1 \times 10^3 = \mathbf{100\,\Omega}$

(d) The onset of conduction occurs at approximately 0.2 V. This suggests that the diode is a **Ge type**.

Problem 5. Corresponding readings of current, I, and voltage, V, for a semiconductor device are given in the table:

V_f (V)	0	0.1	0.2	0.3	0.4	0.5	0.6	0.7	0.8
I_f (mA)	0	0	0	0	0	1	9	24	50

Plot the I/V characteristic for the device and identify the type of device.

The I/V characteristic is shown in Figure 13.12. Since the device begins to conduct when a potential of approximately 0.6 V is applied to it we can infer that **the semiconductor material is silicon** rather than germanium.

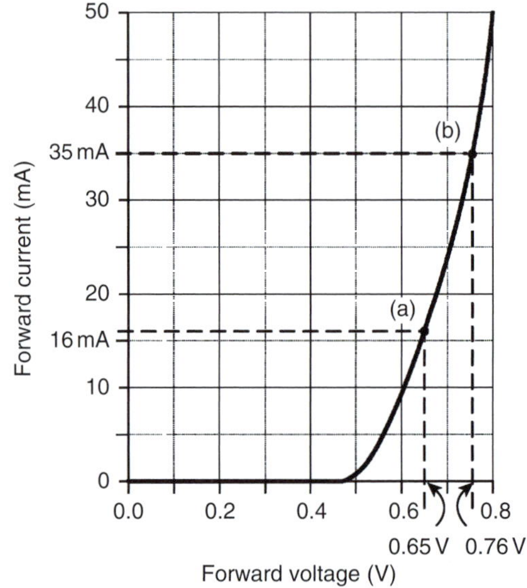

Figure 13.12

Problem 6. For the characteristic of Figure 13.12, determine for the device (a) the forward current when the forward voltage is 0.65 V, and (b) the forward voltage when the forward current is 35 mA.

(a) From Figure 13.12, when the forward voltage is 0.65 V, **the forward current = 16 mA**

(b) When the forward current is 35 mA, **the forward voltage = 0.76 V**

Now try the following Practice Exercise

Practice Exercise 63 Semiconductor materials and p–n junctions (Answers on page 821)

1. Explain what you understand by the term intrinsic semiconductor and how an intrinsic semiconductor is turned into either a p-type or an n-type material.
2. Explain what is meant by minority and majority carriers in an n-type material and state whether the numbers of each of these carriers are affected by temperature.
3. A piece of pure silicon is doped with (a) pentavalent impurity and (b) trivalent impurity. Explain the effect these impurities have on the form of conduction in silicon.
4. With the aid of simple sketches, explain how pure germanium can be treated in such a way that conduction is predominantly due to (a) electrons and (b) holes.
5. Explain the terms given below when used in semiconductor terminology: (a) covalent bond, (b) trivalent impurity, (c) pentavalent impurity, (d) electron-hole pair generation.
6. Explain briefly why although both p-type and n-type materials have resistive properties when separate, they have rectifying properties when a junction between them exists.
7. The application of an external voltage to a junction diode can influence the drift of holes and electrons. With the aid of diagrams explain this statement and also how the direction and magnitude of the applied voltage affects the depletion layer.
8. State briefly what you understand by the terms: (a) reverse bias, (b) forward bias, (c) contact potential, (d) diffusion, (e) minority carrier conduction.
9. Explain briefly the action of a p–n junction diode: (a) on open-circuit, (b) when provided with a forward bias and (c) when provided with a reverse bias. Sketch the characteristic curves for both forward and reverse bias conditions.
10. Draw a diagram illustrating the charge situation for an unbiased p–n junction. Explain the change in the charge situation when compared with that in isolated p-type and n-type materials. Mark on the diagram the depletion layer and the majority carriers in each region.
11. The graph shown in Figure 13.13 was obtained during an experiment on a diode. (a) What type of diode is this? Give reasons. (b) Determine the forward current for a forward voltage of 0.5 V. (c) Determine the forward voltage for a forward current of 30 mA. (d) Determine the resistance of the diode when the forward voltage is 0.4 V.

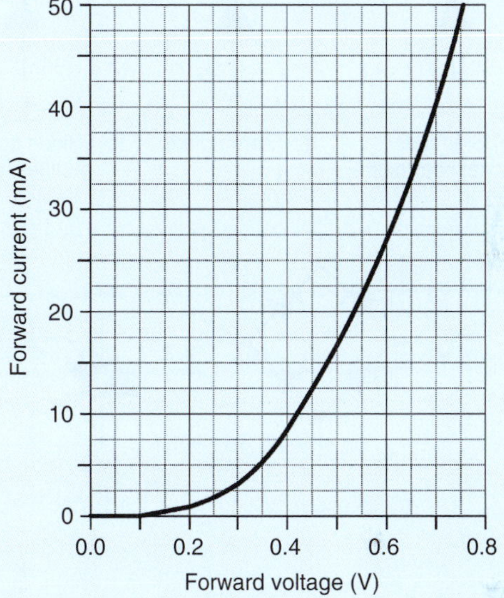

Figure 13.13

13.6 Semiconductor diodes

When a junction is formed between p-type and n-type semiconductor materials, the resulting device is called a **semiconductor diode**. This component offers an extremely low resistance to current flow in one direction and an extremely high resistance to current flow in the other. This property allows diodes to be used in applications that require a circuit to behave differently according to the direction of current flowing in it. Note that an ideal diode would pass an infinite current in one direction and no current at all in the other direction.

A semiconductor diode is an encapsulated p–n junction fitted with connecting leads or tags for connection to external circuitry. Where an appreciable current is present (as is the case with many rectifier circuits) the diode may be mounted in a metal package designed to conduct heat away from the junction. The connection to the p-type material is referred to as the **anode** while that to the n-type material is called the **cathode**.

Various different types of diode are available for different applications. These include **rectifier diodes** for use in power supplies, **zener diodes** for use as voltage reference sources, **light emitting diodes** and **varactor diodes**. Figure 13.14 shows the symbols used to represent diodes in electronic circuit diagrams, where 'a' is the anode and 'k' the cathode.

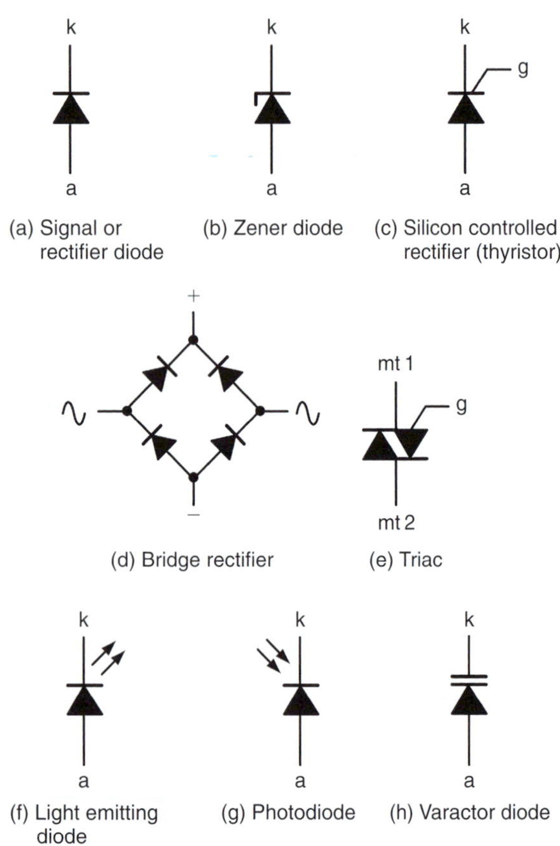

Figure 13.14

Some typical practical diodes are shown in Figure 13.15

Figure 13.15

13.7 Characteristics and maximum ratings

Signal diodes require consistent forward characteristics with low forward voltage drop. Rectifier diodes need to be able to cope with high values of reverse voltage and large values of forward current, and consistency of characteristics is of secondary importance in such applications. Table 13.1 summarizes the characteristics of some common semiconductor diodes. It is worth noting that diodes are limited by the amount of forward current and reverse voltage they can withstand. This limit is based on the physical size and construction of the diode.

A typical general-purpose diode may be specified as having a forward threshold voltage of 0.6 V and a reverse breakdown voltage of 200 V. If the latter is exceeded, the diode may suffer irreversible damage. Typical values of **maximum repetitive reverse voltage** (V_{RRM}) or **peak inverse voltage** (PIV) range from about 50 V to over 500 V. The reverse voltage may be increased until the maximum reverse voltage for which the diode is rated is reached. If this voltage is exceeded the junction may break down and the diode may suffer permanent damage.

13.8 Rectification

The process of obtaining unidirectional currents and voltages from alternating currents and voltages is called **rectification**. Semiconductor diodes are commonly used to convert alternating current (a.c.) to direct current (d.c.), in which case they are referred to as **rectifiers**. The simplest form of rectifier circuit makes use of a single diode and, since it operates on only either positive or negative halfcycles of the supply, it is known as a **half-wave rectifier**. Four diodes are connected as a **bridge rectifier** – see Figure 13.14(d) – and are often used as a **full-wave rectifier**. Note that in both cases, automatic switching of the current is carried out by the diode(s). For methods of half-wave and full-wave rectification, see Section 16.8, page 254.

13.9 Zener diodes

Zener diodes are heavily doped silicon diodes that, unlike normal diodes, exhibit an abrupt reverse breakdown at relatively low voltages (typically less than

Table 13.1 Characteristics of some typical signal and rectifier diodes

Device code	Material	Max repetitive reverse voltage (V_{RRM})	Max forward current ($I_{F(max)}$)	Max reverse current ($I_{R(max)}$)	Application
1N4148	Silicon	100 V	75 mA	25 nA	General purpose
1N914	Silicon	100 V	75 mA	25 nA	General purpose
AA113	Germanium	60 V	10 mA	200 μA	RF detector
OA47	Germanium	25 V	110 mA	100 μA	Signal detector
OA91	Germanium	115 V	50 mA	275 μA	General purpose
1N4001	Silicon	50 V	1 A	10 μA	Low voltage rectifier
1N5404	Silicon	400 V	3 A	10 μA	High voltage rectifier
BY127	Silicon	1250 V	1 A	10 μA	High voltage rectifier

6 V). A similar effect, called **avalanche breakdown**, occurs in less heavily doped diodes. These avalanche diodes also exhibit a rapid breakdown with negligible current flowing below the avalanche voltage and a relatively large current flowing once the avalanche voltage has been reached. For avalanche diodes, this breakdown voltage usually occurs at voltages above 6 V. In practice, however, both types of diode are referred to as **Zener*** **diodes**. The symbol for a Zener diode is shown in Figure 13.14(b) whilst a typical Zener diode characteristic is shown in Figure 13.16.

Whereas reverse breakdown is a highly undesirable effect in circuits that use conventional diodes, it can be extremely useful in the case of Zener diodes where the breakdown voltage is precisely known. When a diode is undergoing reverse breakdown and provided its maximum ratings are not exceeded, the voltage appearing across it will remain substantially constant (equal to the nominal Zener voltage) regardless of the current flowing. This property makes the Zener diode ideal for use as a **voltage regulator**.

*Who was Zener? **Clarence Melvin Zener** (1 December 1905–15 July 1993) was the first person to describe the breakdown of electrical insulators. To find out more go to **www.routledge.com/cw/bird**

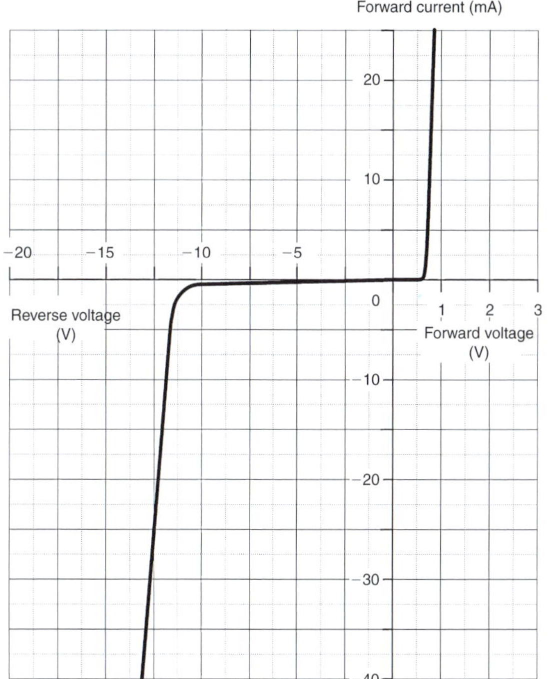

Figure 13.16

Zener diodes are available in various families (according to their general characteristics, encapsulations and power ratings) with reverse breakdown (Zener) voltages in the range 2.4 V to 91 V.

Problem 7. The characteristic of a Zener diode is shown in Figure 13.17. Use the characteristic to determine (a) the current flowing in the diode when a reverse voltage of 30 V is applied, (b) the voltage dropped across the diode when a reverse current of 5 mA is flowing in it, (c) the voltage rating for the Zener diode, and (d) the power dissipated in the Zener diode when a reverse voltage of 30 V appears across it.

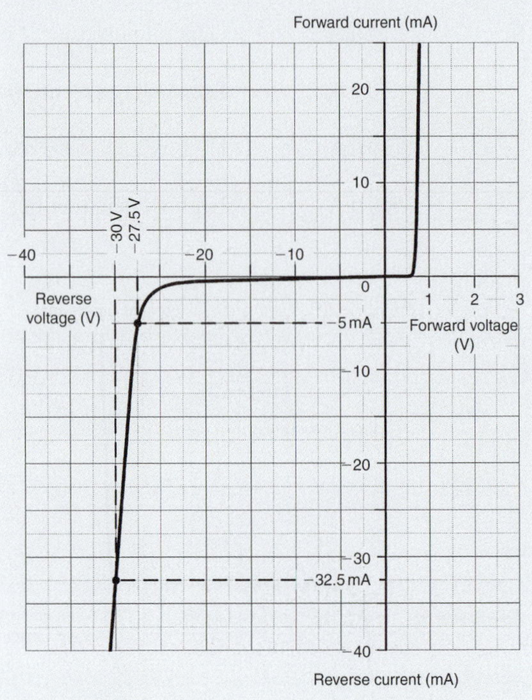

Figure 13.17

(a) When $V = -30$ V, **the current flowing in the diode,** $\boldsymbol{I = -32.5}$ **mA**

(b) When $I = -5$ mA, **the voltage dropped across the diode,** $\boldsymbol{V = -27.5}$ **V**

(c) The characteristic shows the onset of Zener action at 27 V; this would suggest a **Zener voltage rating of 27 V**

(d) Power, $P = V \times I$, from which, **power dissipated when the reverse voltage is 30 V,**

$$P = 30 \times (32.5 \times 10^{-3}) = 0.975\,\text{W} = \mathbf{975\,mW}$$

13.10 Silicon controlled rectifiers

Silicon controlled rectifiers (or **thyristors**) are three-terminal devices which can be used for switching and a.c. power control. Silicon controlled rectifiers can switch very rapidly from conducting to a non-conducting state. In the off state, the silicon controlled rectifier exhibits negligible leakage current, while in the on state the device exhibits very low resistance. This results in very little power loss within the silicon controlled rectifier even when appreciable power levels are being controlled.

Once switched into the conducting state, the silicon controlled rectifier will remain conducting (i.e. it is latched in the on state) until the forward current is removed from the device. In d.c. applications this necessitates the interruption (or disconnection) of the supply before the device can be reset into its non-conducting state. Where the device is used with an alternating supply, the device will automatically become reset whenever the main supply reverses. The device can then be triggered on the next half cycle having correct polarity to permit conduction.

Like their conventional silicon diode counterparts, silicon controlled rectifiers have anode and cathode connections; control is applied by means of a gate terminal, g. The symbol for a silicon controlled rectifier is shown in Figure 13.14(c).

In normal use, a silicon controlled rectifier (SCR) is triggered into the conducting (on) state by means of the application of a current pulse to the gate terminal – see Figure 13.18. The effective triggering of a silicon controlled rectifier requires a gate trigger pulse having a fast rise time derived from a low-resistance source. Triggering can become erratic when insufficient gate current is available or when the gate current changes slowly.

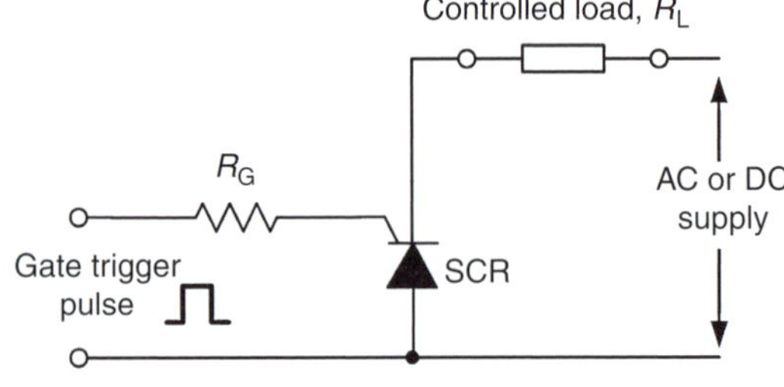

Figure 13.18

A typical silicon controlled rectifier for mains switching applications will require a gate trigger pulse of about 30 mA at 2.5 V to control a current of up to 5 A.

13.11 Light emitting diodes

Light emitting diodes (LED) can be used as general-purpose indicators and, compared with conventional filament lamps, operate from significantly smaller voltages and currents. LEDs are also very much more reliable than filament lamps. Most LEDs will provide a reasonable level of light output when a forward current of between 5 mA and 20 mA is applied.

LEDs are available in various formats, with the round types being most popular. Round LEDs are commonly available in the 3 mm and 5 mm (0.2 inch) diameter plastic packages and also in a 5 mm × 2 mm rectangular format. The viewing angle for round LEDs tends to be in the region of 20° to 40°, whereas for rectangular types this is increased to around 100°. The peak wavelength of emission depends on the type of semiconductor employed, but usually lies in the range 630 to 690 nm. The symbol for an LED is shown in Figure 13.14(f).

13.12 Varactor diodes

It was shown earlier that when a diode is operated in the reverse biased condition, the width of the depletion region increases as the applied voltage increases. Varying the width of the depletion region is equivalent to varying the plate separation of a very small capacitor such that the relationship between junction capacitance and applied reverse voltage will look something like that shown in Figure 13.19. The typical variation of capacitance provided by a varactor is from about 50 pF to 10 pF as the reverse voltage is increased from 2 V to 20 V. The symbol for a varactor diode is shown in Figure 13.14(h).

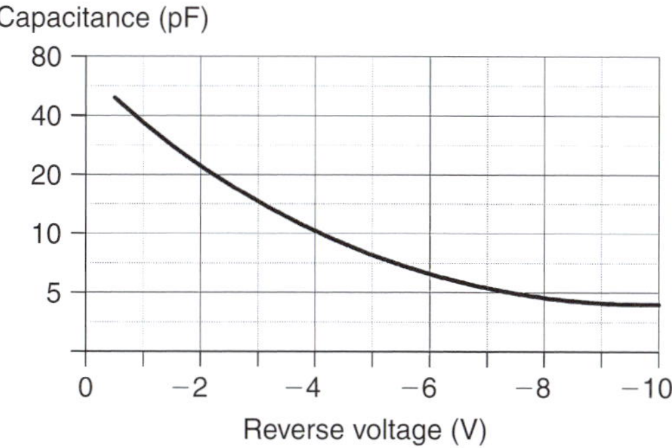

Figure 13.19

13.13 Schottky diodes

The conventional p–n junction diode explained in Section 13.4 operates well as a rectifier and switching device at relatively low frequencies (i.e. 50 Hz to 400 Hz) but its performance as a rectifier becomes seriously impaired at high frequencies due to the presence of stored charge carriers in the junction. These have the effect of momentarily allowing current to flow in the reverse direction when reverse voltage is applied. This problem becomes increasingly more problematic as the frequency of the a.c. supply is increased and the periodic time of the applied voltage becomes smaller.

To avoid these problems a diode that uses a metal–semiconductor contact rather than a p–n junction (see Figure 13.20) is employed. When compared with conventional silicon junction diodes, these **Schottky*** **diodes** have a lower forward voltage (typically 0.35 V) and a slightly reduced maximum reverse voltage rating (typically 50 V to 200 V). Their main advantage, however, is that they operate with high efficiency in **switched-mode power supplies** (SMPS) at frequencies of up to 1 MHz. Schottky diodes are also extensively used in the construction of **integrated circuits** designed for high-speed digital logic applications.

*Who was Schottky? **Walter Hermann Schottky** (23 July 1886–4 March 1976) played a major role in developing the theory of electron and ion emission phenomena, invented the screen-grid vacuum tube and the pentode and co-invented the Ribbon microphone and loudspeaker with Dr. Gerwin Erlach To find out more go to **www.routledge.com/cw/bird**

Part 2

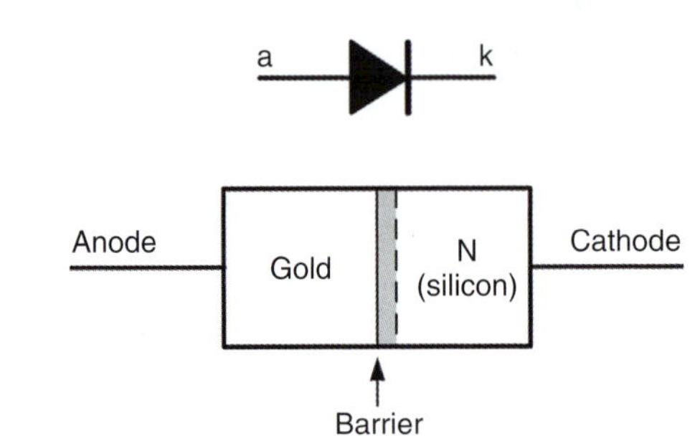

Figure 13.20

Now try the following Practice Exercise

Practice Exercise 64 Semiconductor diodes (Answers on page 821)

1. Identify the types of diodes shown in Figure 13.21.

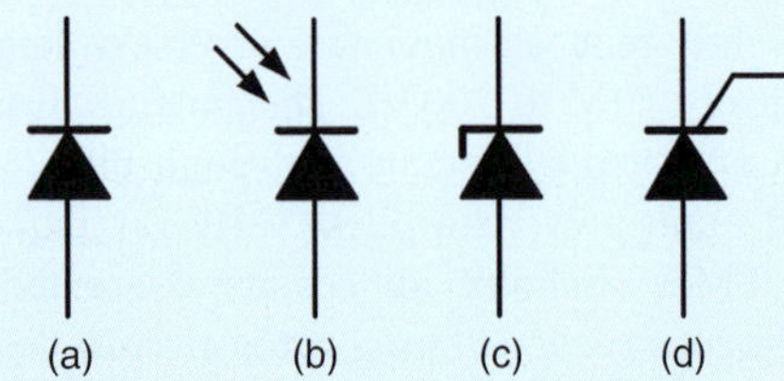

Figure 13.21

2. Sketch a circuit to show how a thyristor can be used as a controlled rectifier.
3. Sketch a graph showing how the capacitance of a varactor diode varies with applied reverse voltage.
4. State TWO advantages of light emitting diodes when compared with conventional filament indicating lamps.
5. State TWO applications for Schottky diodes.
6. The graph shown in Figure 13.22 was obtained during an experiment on a zener diode.

   (a) Estimate the zener voltage for the diode. (b) Determine the reverse voltage for a reverse current of −20 mA. (c) Determine the reverse current for a reverse voltage of −5.5 V. (d) Determine the power dissipated by the diode when the reverse voltage is −6 V.

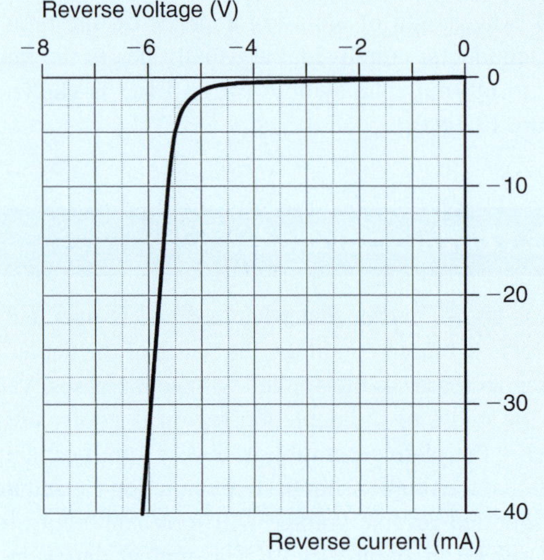

Figure 13.22

For fully worked solutions to each of the problems in Practice Exercises 63 and 64 in this chapter, go to the website:
www.routledge.com/cw/bird

Chapter 14

Transistors

Why it is important to understand: **Transistors**

The invention of the bipolar transistor in 1948 started a revolution in electronics. Technical feats previously requiring relatively large, mechanically fragile, power-hungry vacuum tubes were suddenly achievable with tiny, mechanically rugged, power-thrifty specks of crystalline silicon. This revolution made possible the design and manufacture of lightweight, inexpensive electronic devices that we now take for granted. Understanding how transistors function is of paramount importance to anyone interested in understanding modern electronics. A transistor is a three-terminal semiconductor device that can perform two functions that are fundamental to the design of electronic circuits – amplification and switching. Put simply, amplification consists of magnifying a signal by transferring energy to it from an external source, whereas a transistor switch is a device for controlling a relatively large current between or voltage across two terminals by means of a small control current or voltage applied at a third terminal. Transistors can be mass produced at very low costs, and transistors are the reason that computers keep getting smaller yet more powerful every day. There are more than 60 million transistors built every year for every man, woman and child on Earth. Transistors are the key to our modern world. This chapter explains the structure and operation of the transistor, incorporating some simple calculations.

At the end of this chapter you should be able to:

- understand the structure of bipolar junction transistors (BJT) and junction gate field effect transistors (JFET)
- understand the action of BJT and JFET devices
- appreciate different classes and applications for BJT and JFET devices
- draw the circuit symbols for BJT and JFET devices
- appreciate common base, common emitter and common collector connections
- appreciate common gate, common source and common drain connections
- interpret characteristics for BJT and JFET devices
- appreciate how transistors are used as Class-A amplifiers
- use a load line to determine the performance of a transistor amplifier
- estimate quiescent operating conditions and gain from transistor characteristics and other data

Electrical Circuit Theory and Technology. 978-1-138-67349-6, © 2017 John Bird. Published by Taylor & Francis. All rights reserved.

Table 14.1 Transistor classification

Low-frequency	Transistors designed specifically for audio low-frequency applications (below 100 kHz)
High-frequency	Transistors designed specifically for high radio-frequency applications (100 kHz and above)
Switching	Transistors designed for switching applications
Low-noise	Transistors that have low-noise characteristics and which are intended primarily for the amplification of low-amplitude signals
High-voltage	Transistors designed specifically to handle high voltages
Driver	Transistors that operate at medium power and voltage levels and which are often used to precede a final (power) stage which operates at an appreciable power level
Small-signal	Transistors designed for amplifying small voltages in amplifiers and radio receivers
Power	Transistor designed to handle high currents and voltages

14.1 Transistor classification

Transistors fall into **two main classes** – **bipolar** and **field effect**. They are also classified according to semiconductor material employed – silicon or germanium, and to their field of application (for example, general purpose, switching, high frequency, and so on). Transistors are also classified according to the application that they are designed for, as shown in Table 14.1. Note that these classifications can be combined so that it is possible, for example, to classify a transistor as a 'low-frequency power transistor' or as a 'low-noise high-frequency transistor'.

14.2 Bipolar junction transistors (BJTs)

Bipolar transistors generally comprise n–p–n or p–n–p junctions of either silicon (Si) or germanium (Ge) material. The junctions are, in fact, produced in a single slice of silicon by diffusing impurities through a photographically reduced mask. Silicon transistors are superior when compared with germanium transistors in the vast majority of applications (particularly at high temperatures) and thus germanium devices are very rarely encountered in modern electronic equipment.

The construction of typical n–p–n and p–n–p transistors is shown in Figures 14.1 and 14.2. In order to conduct the heat away from the junction (important in medium- and high-power applications) the collector is connected to the metal case of the transistor.

The **symbols** and simplified junction models for n–p–n and p–n–p transistors are shown in Figure 14.3. It is important to note that the base region (p-type material in the case of an n–p–n transistor or n-type material in the case of a p–n–p transistor) is extremely narrow.

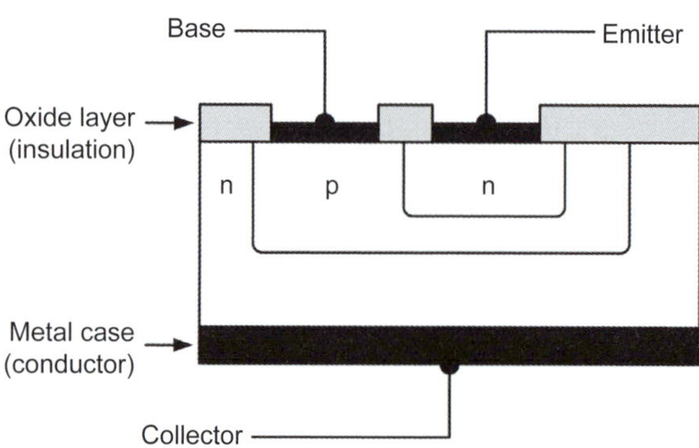

Figure 14.1

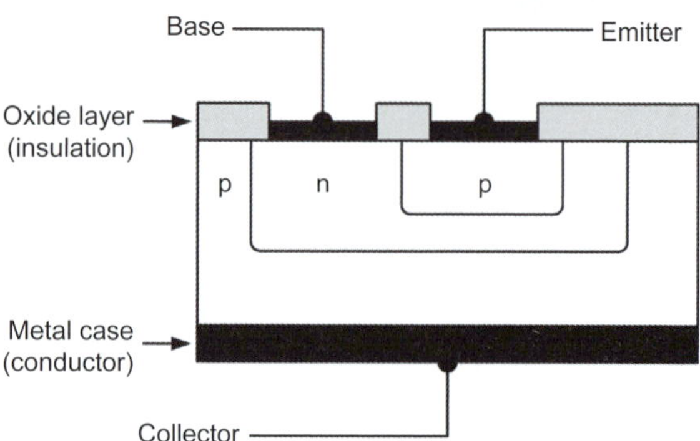

Figure 14.2

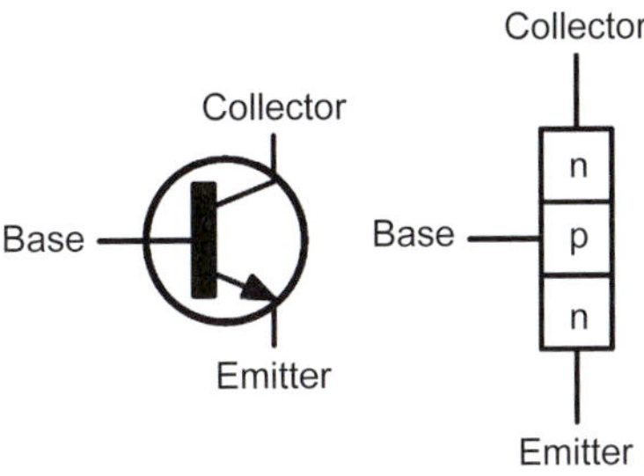

(a) n–p–n bipolar junction transistor (BJT)

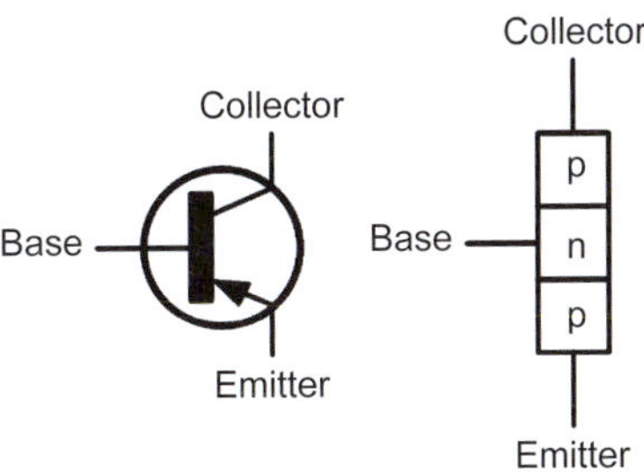

(b) p–n–p bipolar junction transistor (BJT)

Figure 14.3

Some typical practical transistors are shown in Figure 14.4.

Figure 14.4

14.3 Transistor action

In the **n–p–n transistor**, connected as shown in Figure 14.5(a), transistor action is accounted for as follows:

(a) the majority carriers in the n-type emitter material are electrons

(b) the base–emitter junction is forward biased to these majority carriers and electrons cross the junction and appear in the base region

(c) the base region is very thin and only lightly doped with holes, so some recombination with holes occurs but many electrons are left in the base region

(d) the base–collector junction is reverse biased to holes in the base region and electrons in the collector region, but is forward biased to electrons in the base region; these electrons are attracted by the positive potential at the collector terminal

(e) a large proportion of the electrons in the base region cross the base collector junction into the collector region, creating a collector current.

The **transistor action** for an n–p–n device is shown diagrammatically in Figure 14.6(a). Conventional current flow is taken to be in the direction of the motion of holes, that is, in the opposite direction to electron flow. Around 99.5% of the electrons leaving the emitter will cross the base collector junction and only 0.5% of the electrons will recombine with holes in the narrow base region.

In the **p–n–p transistor**, connected as shown in Figure 14.5(b), transistor action is accounted for as follows:

(a) the majority carriers in the emitter p-type material are holes

(b) the base–emitter junction is forward biased to the majority carriers and the holes cross the junction and appear in the base region

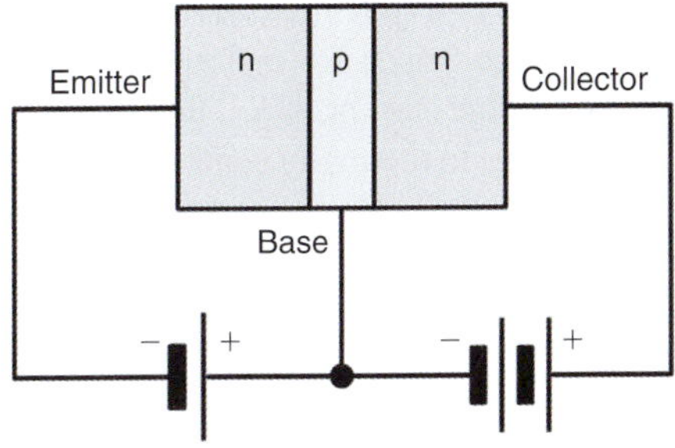

(a) n–p–n bipolar junction transistor

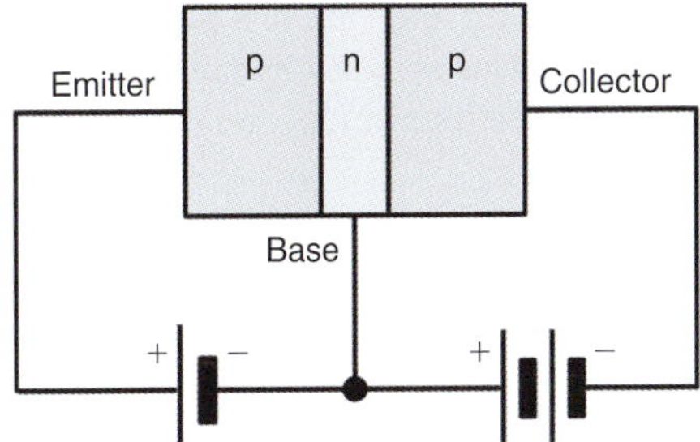

(b) p–n–p bipolar junction transistor

Figure 14.5

(c) the base region is very thin and is only lightly doped with electrons so although some electron–hole pairs are formed, many holes are left in the base region

(d) the base–collector junction is reverse biased to electrons in the base region and holes in the collector region, but forward biased to holes in the base region; these holes are attracted by the negative potential at the collector terminal

(e) a large proportion of the holes in the base region cross the base–collector junction into the collector region, creating a collector current; conventional current flow is in the direction of hole movement.

The **transistor action** for a p–n–p device is shown diagrammatically in Figure 14.6(b). Around 99.5% of the holes leaving the emitter will cross the base–collector junction and only 0.5% of the holes will recombine with electrons in the narrow base region.

14.4 Leakage current

For an **n–p–n transistor**, the base–collector junction is reversed biased for majority carriers, but a small leakage current, I_{CBO}, flows from the collector to the base due to thermally generated minority carriers (holes in the collector and electrons in the base) being present. The base–collector junction is forward biased to these minority carriers.

Similarly, for a **p–n–p transistor**, the base–collector junction is reverse biased for majority carriers. However, a small leakage current, I_{CBO}, flows from the base to the collector due to thermally generated minority carriers (electrons in the collector and holes in the base), being present. Once again, the base–collector junction is forward biased to these minority carriers.

With modern transistors, leakage current is usually very small (typically less than 100 nA) and in most applications it can be ignored.

Problem 1. With reference to a p–n–p transistor, explain briefly what is meant by the term 'transistor action' and why a bipolar junction transistor is so named.

For the transistor as depicted in Figure 14.5(b), the emitter is relatively heavily doped with acceptor atoms (holes). When the emitter terminal is made sufficiently positive with respect to the base, the base–emitter junction is forward biased to the majority carriers. The majority carriers are holes in the emitter and these drift from the emitter to the base.

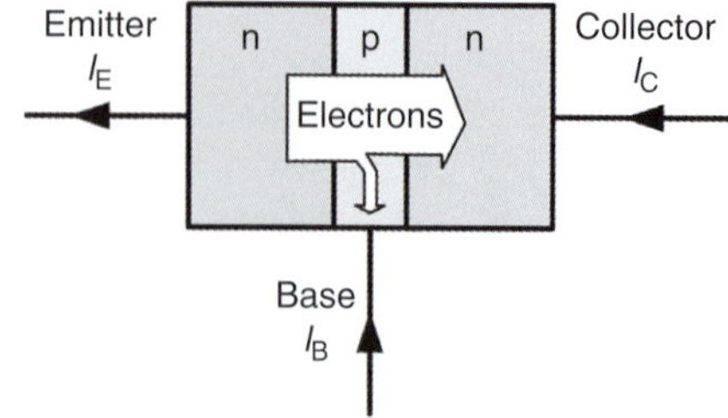

(a) n–p–n bipolar junction transistor

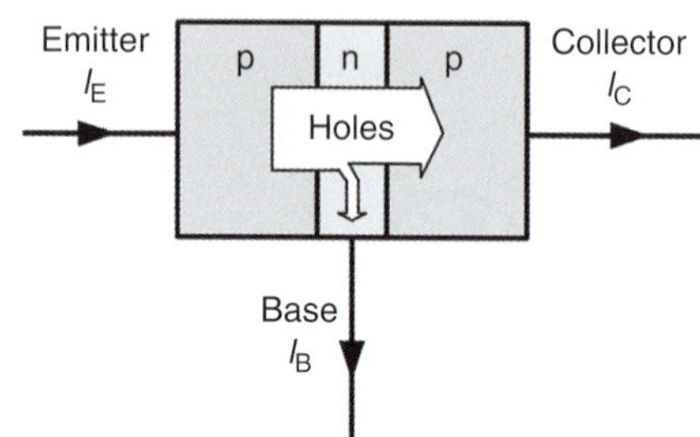

(b) p–n–p bipolar junction transistor

Figure 14.6

The base region is relatively lightly doped with donor atoms (electrons) and although some electron–hole recombinations take place, perhaps 0.5%, most of the holes entering the base do not combine with electrons.

The base–collector junction is reverse biased to electrons in the base region, but forward biased to holes in

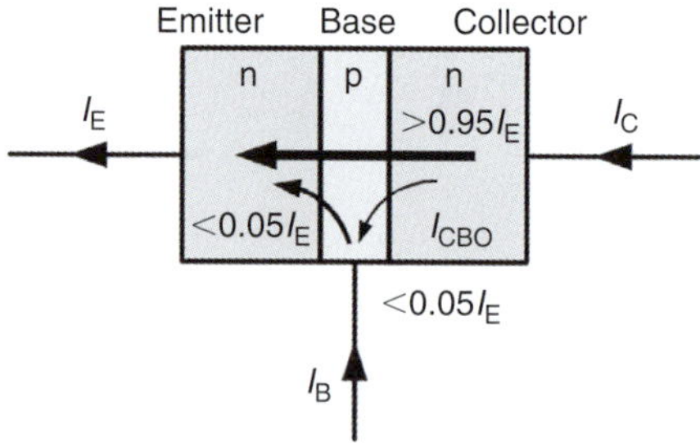

(a) n–p–n bipolar junction transistor

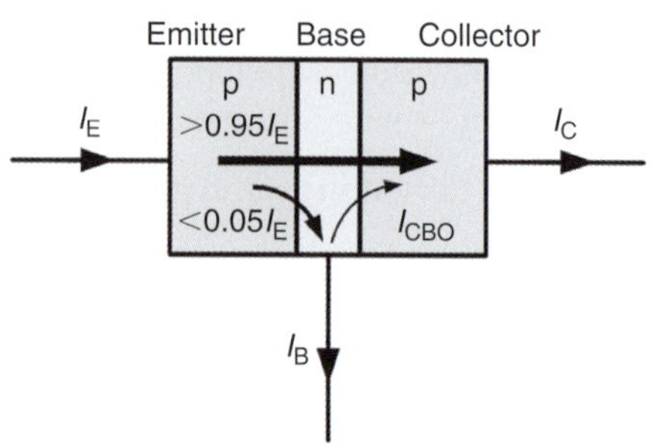

(b) p–n–p bipolar junction transistor

Figure 14.7

the base region. Since the base is very thin and now is packed with holes, these holes pass the base–emitter junction towards the negative potential of the collector terminal. The control of current from emitter to collector is largely independent of the collector–base voltage and almost wholly governed by the emitter–base voltage.

The essence of transistor action is this current control by means of the base–emitter voltage. In a p–n–p transistor, holes in the emitter and collector regions are majority carriers, but are minority carriers when in the base region. Also thermally generated electrons in the emitter and collector regions are minority carriers as are holes in the base region. However, both majority and minority carriers contribute towards the total current flow (see Figure 14.7). It is because a transistor makes use of both types of charge carriers (holes and electrons) that they are called **bipolar**. The transistor also comprises two p–n junctions and for this reason it is a **junction transistor**; hence the name – **bipolar junction transistor**.

14.5 Bias and current flow

In normal operation (i.e. for operation as a linear amplifier) the base–emitter junction of a transistor is forward biased and the collector–base junction is reverse biased. The base region is, however, made very narrow so that carriers are swept across it from emitter to collector so that only a relatively small current flows in the base. To put this into context, the current flowing in the emitter circuit is typically 100 times greater than that flowing in the base. The direction of conventional current flow is from emitter to collector in the case of a p–n–p transistor, and collector to emitter in the case of an n–p–n device, as shown in Figure 14.8.

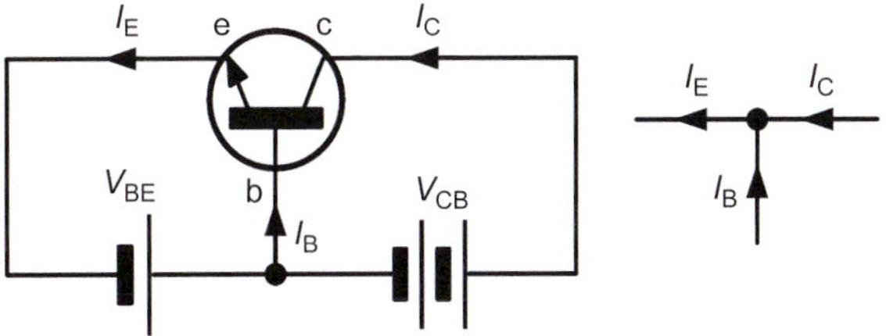

(a) n–p–n bipolar junction transistor (BJT)

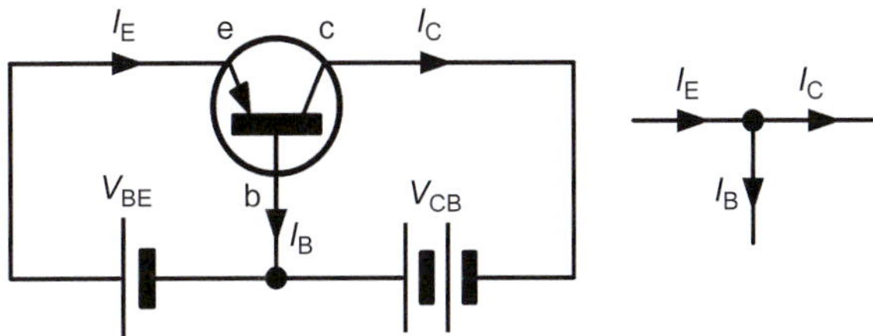

(b) p–n–p bipolar junction transistor (BJT)

Figure 14.8

The equation that relates current flow in the collector, base, and emitter circuits (see Figure 14.8) is:

$$I_E = I_B + I_C$$

where I_E is the emitter current, I_B is the base current and I_C is the collector current (all expressed in the same units).

Problem 2. A transistor operates with a collector current of 100 mA and an emitter current of 102 mA. Determine the value of base current.

Emitter current, $I_E = I_B + I_C$
from which, base current, $I_B = I_E - I_C$
Hence, **base current,** $\boldsymbol{I_B} = 102 - 100 = \mathbf{2\ mA}$

14.6 Transistor operating configurations

Three basic circuit configurations are used for transistor amplifiers. These three circuit configurations depend upon which one of the three transistor connections is made common to both the input and the output. In the case of bipolar junction transistors, the configurations are known as **common emitter**, **common collector** (or **emitter follower**) and **common base**, as shown in Figure 14.9.

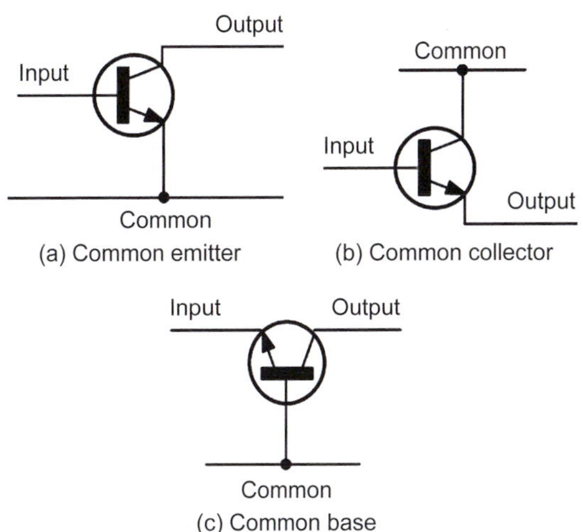

Figure 14.9

Part 2

14.7 Bipolar transistor characteristics

The characteristics of a bipolar junction transistor are usually presented in the form of a set of graphs relating voltage and current present at the transistor terminals. Figure 14.10 shows a typical **input characteristic** (I_B plotted against V_{BE}) for an n–p–n bipolar junction transistor operating in common-emitter mode. In this mode, the input current is applied to the base and the output current appears in the collector (the emitter is effectively **common** to both the input and output circuits, as shown in Figure 14.9(a)).

The input characteristic shows that very little base current flows until the base–emitter voltage V_{BE} exceeds 0.6 V. Thereafter, the base current increases rapidly – this characteristic bears a close resemblance to the forward part of the characteristic for a silicon diode.

Figure 14.11 shows a typical set of **output (collector) characteristics** (I_C plotted against V_{CE}) for an n–p–n bipolar transistor. Each curve corresponds to a different value of base current. Note the 'knee' in the characteristic below $V_{CE} = 2$ V. Also note that the curves are quite flat. For this reason (i.e. since the collector current does not change very much as the collector–emitter voltage changes) we often refer to this as a **constant current characteristic**.

Figure 14.12 shows a typical **transfer characteristic** for an n–p–n bipolar junction transistor. Here I_C is plotted against I_B for a small-signal general-purpose transistor. The slope of this curve (i.e. the ratio of I_C to I_B) is the common-emitter current gain of the transistor which is explored further in Section 14.9.

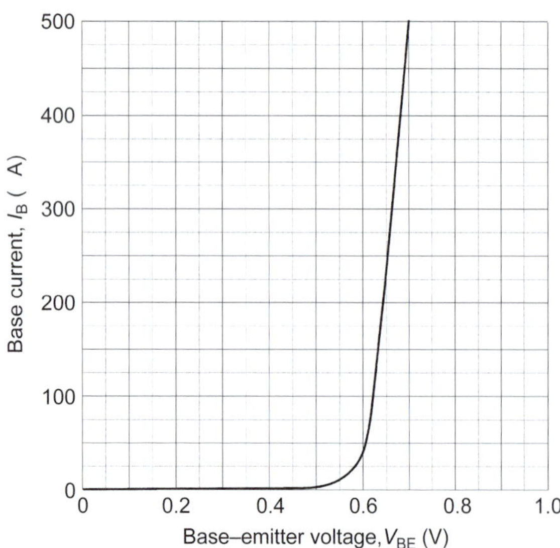

Figure 14.10

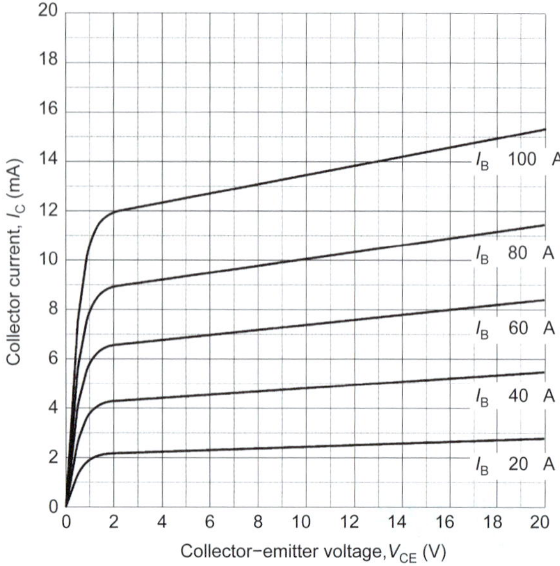

Figure 14.11

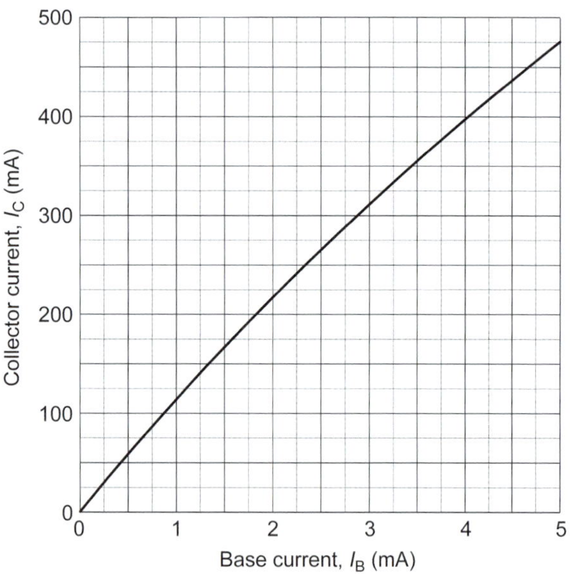

Figure 14.12

A circuit that can be used for obtaining the common-emitter characteristics of an n–p–n BJT is shown in Figure 14.13. For the input characteristic, VR1 is set at a particular value and the corresponding values of V_{BE} and I_B are noted. This is repeated for various settings of VR1 and plotting the values gives the typical input characteristic of Figure 14.10.

For the output characteristics, VR1 is varied so that I_B is, say, 20 μA. Then VR2 is set at various values and corresponding values of V_{CE} and I_C are noted. The graph

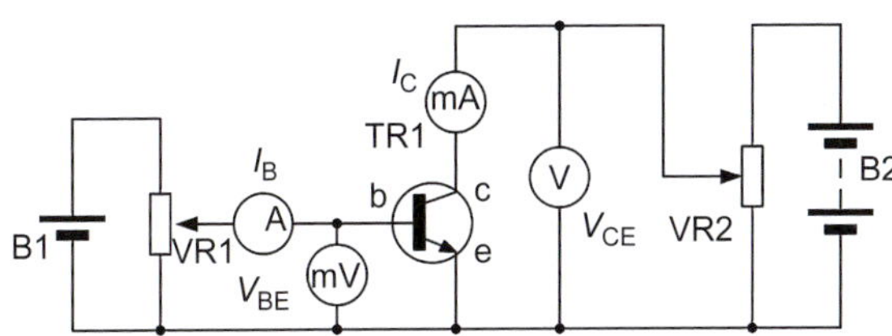

Figure 14.13

of V_{CE}/I_C is then plotted for $I_B = 20\,\mu A$. This is repeated for, say, $I_B = 40\,\mu A$, $I_B = 60\,\mu A$ and so on. Plotting the values gives the typical output characteristics of Figure 14.11.

14.8 Transistor parameters

The transistor characteristics met in the previous section provide us with some usefule information that can help us to model the behaviour of a transistor. In particular, the three characteristic graphs can be used to determine the following parameters for operation in common-emitter mode:

Input resistance (from the input characteristic, Figure 14.10)

Static (or d.c.) input resistance $= \dfrac{V_{BE}}{I_B}$ (from corresponding points on the graph).

Dynamic (or a.c.) input resistance $= \dfrac{\Delta V_{BE}}{\Delta I_B}$ (from the slope of the graph).

(Note that ΔV_{BE} means 'change of V_{BE}' and ΔI_B means 'change of I_B'.)

Output resistance (from the output characteristic, Figure 14.11)

Static (or d.c.) output resistance $= \dfrac{V_{CE}}{I_C}$ (from corresponding points on the graph).

Dynamic (or a.c.) output resistance $= \dfrac{\Delta V_{CE}}{\Delta I_C}$ (from the slope of the graph).

(Note that ΔV_{CE} means 'change of V_{CE}' and ΔI_C means 'change of I_C'.)

Current gain (from the transfer characteristic, Figure 14.12)

Static (or d.c.) current gain $= \dfrac{I_C}{I_B}$ (from corresponding points on the graph).

Dynamic (or a.c.) current gain $= \dfrac{\Delta I_C}{\Delta I_B}$ (from the slope of the graph).

(Note that ΔI_C means 'change of I_C' and ΔI_B means 'change of I_B'.)

The method for determining these parameters from the relevant characteristic is illustrated in the following worked problems.

Problem 3. Figure 14.14 shows the input characteristic for an n–p–n silicon transistor. When the base–emitter voltage is 0.65 V, determine (a) the value of base current, (b) the static value of input resistance and (c) the dynamic value of input resistance.

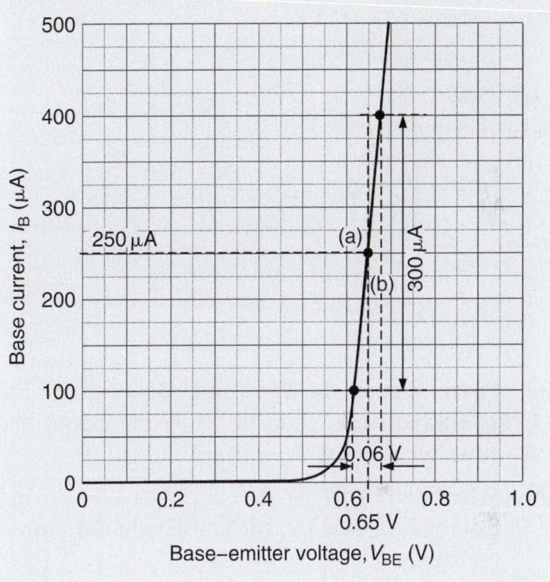

Figure 14.14

(a) From Figure 14.14, when $V_{BE} = 0.65$ V, **base current, $I_B = 250\,\mu A$** (shown as (a) on the graph).

(b) When $V_{BE} = 0.65$ V, $I_B = 250\,\mu A$, hence,

the static value of input resistance

$$= \frac{V_{BE}}{I_B} = \frac{0.65}{250 \times 10^{-6}} = \mathbf{2.6\,k\Omega}$$

(c) From Figure 14.14, V_{BE} changes by 0.06 V when I_B changes by 300 μ A (as shown by (b) on the graph). Hence,

dynamic value of input resistance

$$= \frac{\Delta V_{BE}}{\Delta I_B} = \frac{0.06}{300 \times 10^{-6}} = \mathbf{200\,\Omega}$$

Problem 4. Figure 14.15 shows the output characteristic for an n–p–n silicon transistor. When the collector–emitter voltage is 10 V and the base current is 80 μA, determine (a) the value of collector current, (b) the static value of output resistance and (c) the dynamic value of output resistance.

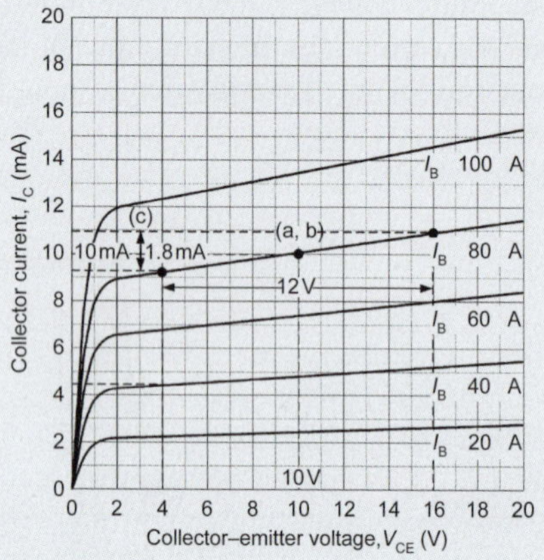

Figure 14.15

(a) From Figure 14.15, when $V_{CE}=10\,\text{V}$ and $I_B=80\,\mu\text{A}$, (i.e. point (a, b) on the graph), the **collector current, $I_C=10\,\text{mA}$**

(b) When $V_{CE}=10\,\text{V}$ and $I_B=80\,\mu\text{A}$ then $I_C=10\,\text{mA}$ from part (a). Hence, **the static value of output resistance**

$$= \frac{V_{CE}}{I_C} = \frac{10}{10 \times 10^{-3}} = \mathbf{1\,k\Omega}$$

(c) When the change in V_{CE} is 12 V, the change in I_C is 1.8 mA (shown as point (c) on the graph). Hence,

the dynamic value of output resistance

$$= \frac{\Delta V_{CE}}{\Delta I_C} = \frac{12}{1.8 \times 10^{-3}} = \mathbf{6.67\,k\Omega}$$

Problem 5. Figure 14.16 shows the transfer characteristic for an n–p–n silicon transistor. When the base current is 2.5 mA, determine (a) the value of collector current, (b) the static value of current gain and (c) the dynamic value of current gain.

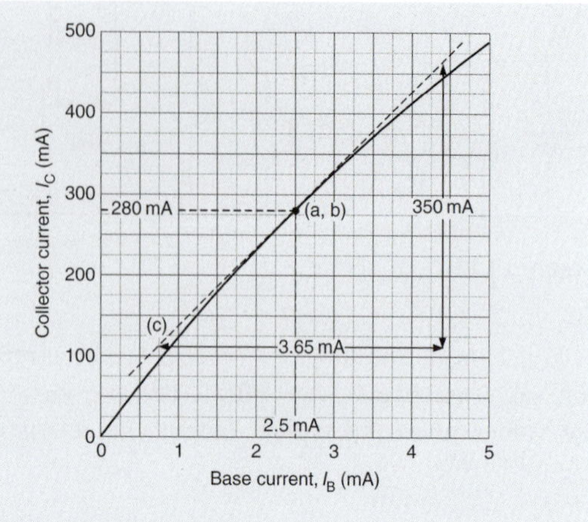

Figure 14.16

(a) From Figure 14.16, when $I_B=2.5\,\text{mA}$, **collector current, $I_C=280\,\text{mA}$** (see point (a, b) on the graph).

(b) From part (a), when $I_B=2.5\,\text{mA}$, $I_C=280\,\text{mA}$ hence,

the static value of current gain

$$= \frac{I_C}{I_B} = \frac{280 \times 10^{-3}}{2.5 \times 10^{-3}} = \mathbf{112}$$

(c) In Figure 14.16, the tangent through the point (a, b) is shown by the broken straight line (c). Hence,

the dynamic value of current gain

$$= \frac{\Delta I_C}{\Delta I_B} = \frac{(460 - 110) \times 10^{-3}}{(4.4 - 0.75) \times 10^{-3}} = \frac{350}{3.65} = \mathbf{96}$$

14.9 Current gain

As stated earlier, the common-emitter current gain is given by the ratio of collector current, I_C, to base current, I_B. We use the symbol h_{FE} to represent the static value of common-emitter current gain, thus:

$$h_{FE} = \frac{I_C}{I_B}$$

Similarly, we use h_{fe} to represent the dynamic value of common-emitter current gain, thus:

$$h_{fe} = \frac{\Delta I_C}{\Delta I_B}$$

As we showed earlier, values of h_{FE} and h_{fe} can be obtained from the transfer characteristic (I_C plotted against I_B). Note that h_{FE} is found from corresponding static values while h_{fe} is found by measuring the slope of the graph. Also note that, if the transfer characteristic is linear, there is little (if any) difference between h_{FE} and h_{fe}.

It is worth noting that current gain (h_{fe}) varies with collector current. For most small-signal transistors, h_{fe} is a maximum at a collector current in the range 1 mA and 10 mA. Current gain also falls to very low values for power transistors when operating at very high values of collector current. Furthermore, most transistor parameters (particularly common-emitter current gain, h_{fe}) are liable to wide variation from one device to the next. It is, therefore, important to design circuits on the basis of the minimum value for h_{fe} in order to ensure successful operation with a variety of different devices.

Problem 6. A bipolar transistor has a common-emitter current gain of 125. If the transistor operates with a collector current of 50 mA, determine the value of base current.

Common-emitter current gain, $h_{FE} = \frac{I_C}{I_B}$

from which, **base current**, $I_B = \frac{I_C}{h_{FE}} = \frac{50 \times 10^{-3}}{125}$

$= \mathbf{400\ \mu A}$

14.10 Typical BJT characteristics and maximum ratings

Table 14.2 on page 204 summarizes the characteristics of some typical bipolar junction transistors for different applications, where I_C max is the maximum collector current, V_{CE} max is the maximum collector–emitter voltage, P_{TOT} max is the maximum device power dissipation, and h_{fe} is the typical value of common-emitter current gain.

Note that the BC548 is a general-purpose n-p-n bipolar junction transistor commonly used in European electronic equipment, low in cost and widely available. The BC548 is the modern plastic-packaged BC108 and can be used in any circuit designed for the BC108 or BC148.

Problem 7. Which of the bipolar transistors listed in Table 14.2 would be most suitable for each of the following applications: (a) the input stage of a radio receiver, (b) the output stage of an audio amplifier and (c) generating a 5 V square wave pulse.

(a) **BF180**, since this transistor is designed for use in radio frequency (RF) applications.

(b) **2N3055**, since this is the only device in the list that can operate at a sufficiently high power level.

(c) **2N3904**, since switching transistors are designed for use in pulse and square wave applications.

Now try the following Practice Exercise

Practice Exercise 65 Bipolar junction transistors (Answers on page 821)

1. Explain, with the aid of sketches, the operation of an n–p–n transistor and also explain why the collector current is very nearly equal to the emitter current.
2. Describe the basic principle of operation of a bipolar junction transistor, including why majority carriers crossing into the base from the emitter pass to the collector and why the collector current is almost unaffected by the collector potential.
3. Explain what is meant by 'leakage current' in a bipolar junction transistor and why this can usually be ignored.
4. For a transistor connected in common-emitter configuration, sketch the typical output characteristics relating collector current and the collector-emitter voltage, for various values of base current. Explain the shape of the characteristics.
5. Sketch the typical input characteristic relating base current and the base–emitter voltage for a transistor connected in common-emitter configuration and explain its shape.
6. With the aid of a circuit diagram, explain how the input and output characteristic of a common-emitter n–p–n transistor may be produced.
7. Define the term 'current gain' for a bipolar junction transistor operating in common-emitter mode.
8. A bipolar junction transistor operates with a collector current of 1.2 A and a base current of 50 mA. What will the value of emitter current be?
9. What is the value of common-emitter current gain for the transistor in Problem 8?

Table 14.2 Transistor characteristics and maximum ratings

Device	Type	I_C max	V_{CE} max	P_{TOT} max	h_{FE} typical	Application
BC548	n–p–n	100 mA	20 V	300 mW	125	General-purpose small-signal amplifier
BCY70	n–p–n	200 mA	−40 V	360 mW	150	General-purpose small-signal amplifier
2N3904	n–p–n	200 mA	40 V	310 mW	150	Switching
BF180	n–p–n	20 mA	20 V	150 mW	100	RF amplifier
2N3053	n–p–n	700 mA	40 V	800 mW	150	Low-frequency amplifier/driver
2N3055	n–p–n	15 A	60 V	115 W	50	Low-frequency power

10. Corresponding readings of base current, I_B, and base–emitter voltage, V_{BE}, for a bipolar junction transistor are given in the table below:

V_{BE} (V)	0	0.1	0.2	0.3	0.4	0.5	0.6	0.7	0.8
I_B (μA)	0	0	0	0	1	3	19	57	130

Plot the I_B/V_{BE} characteristic for the device and use it to determine (a) the value of I_B when $V_{BE} = 0.65$ V, (b) the static value of input resistance when $V_{BE} = 0.65$ V and (c) the dynamic value of input resistance when $V_{BE} = 0.65$ V

11. Corresponding readings of base current, I_B, and collector current, I_C, for a bipolar junction transistor are given in the table below:

I_B (μA)	0	10	20	30	40	50	60	70	80
I_C (mA)	0	1.1	2.1	3.1	4.0	4.9	5.8	6.7	7.6

Plot the I_C/I_B characteristic for the device and use it to determine the static value of common-emitter current gain when $I_B = 45$ μA.

14.11 Field effect transistors

Field effect transistors are available in two basic forms; junction gate and insulated gate. The gate source junction of a **junction gate field effect transistor (JFET)** is effectively a reverse-biased p–n junction. The gate connection of an **insulated gate field effect transistor (IGFET)**, on the other hand, is insulated from the channel and charge is capacitively coupled to the channel. To keep things simple, we will consider only JFET devices. Figure 14.17 shows the basic construction of an n-channel JFET.

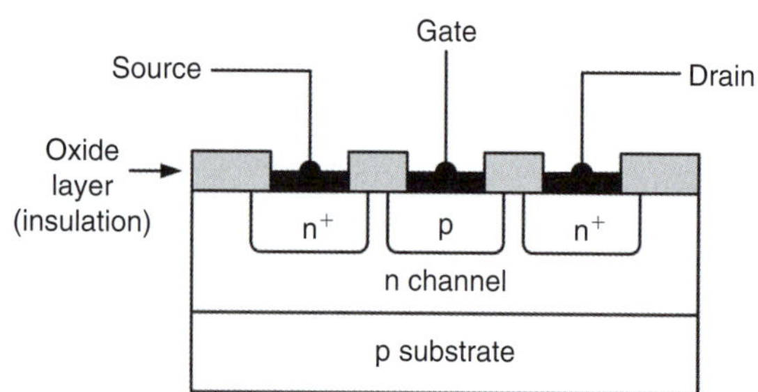

Figure 14.17

JFET transistors comprise a channel of p-type or n-type material surrounded by material of the opposite polarity. The ends of the channel (in which conduction takes place) form electrodes known as the source and drain. The effective width of the channel (in which conduction takes place) is controlled by a charge placed on the third (gate) electrode. The effective resistance between the source and drain is thus determined by the voltage present at the gate. (The + signs in Figure 14.17 are used to indicate a region of heavy doping thus n^+ simply indicates a heavily doped n-type region.)

JFETs offer a very much higher input resistance when compared with bipolar transistors. For example, the input resistance of a bipolar transistor operating in common-emitter mode is usually around 2.5 kΩ.

A JFET transistor operating in equivalent common-source mode would typically exhibit an input resistance of 100 MΩ! This feature makes JFET devices ideal for use in applications where a very high input resistance is desirable.

As with bipolar transistors, the characteristics of a FET are often presented in the form of a set of graphs relating voltage and current present at the transistor's terminals.

14.12 Field effect transistor characteristics

A typical **mutual characteristic** (I_D plotted against V_{GS}) for a small-signal general-purpose n-channel field effect transistor operating in common-source mode is shown in Figure 14.18. This characteristic shows that the drain current is progressively reduced as the gate–source voltage is made more negative. At a certain value of V_{GS} the drain current falls to zero and the device is said to be cut-off.

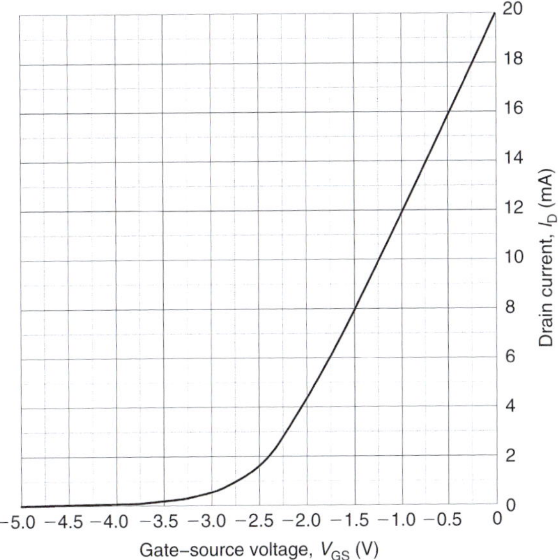

Figure 14.18

Figure 14.19 shows a typical family of **output characteristics** (I_D plotted against V_{DS}) for a small-signal general-purpose n-channel FET operating in common-source mode. This characteristic comprises a family of curves, each relating to a different value of gate–source voltage V_{GS}. You might also like to compare this characteristic with the output characteristic for a transistor operating in common-emitter mode that you met earlier in Figure 14.11.

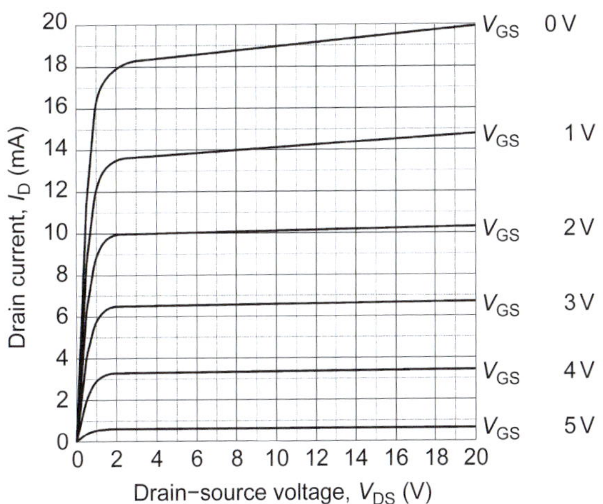

Figure 14.19

As in the case of the bipolar junction transistor, the output characteristic curves for an n-channel FET have a 'knee' that occurs at low values of V_{DS}. Also, note how the curves become flattened above this value with the drain current I_D not changing very significantly for a comparatively large change in drain–source voltage V_{DS}. These characteristics are, in fact, even flatter than those for a bipolar transistor. Because of their flatness, they are often said to represent a constant current characteristic.

The gain offered by a field effect transistor is normally expressed in terms of its **forward transconductance** (g_{fs} or Y_{fs}) in common-source mode. In this mode, the input voltage is applied to the gate and the output current appears in the drain (the source is effectively common to both the input and output circuits).

In common-source mode, **the static (or d.c.) forward transfer conductance** is given by:

$$g_{fs} = \frac{I_D}{V_{GS}} \text{ (from corresponding points on the graph)}$$

whilst **the dynamic (or a.c.) forward transfer conductance** is given by:

$$g_{fs} = \frac{\Delta I_D}{\Delta V_{GS}} \text{ (from the slope of the graph)}$$

(Note that ΔI_D means 'change of I_D' and ΔV_{GS} means 'change of V_{GS}'.)

The method for determining these parameters from the relevant characteristic is illustrated in worked Problem 8 below.

Forward transfer conductance (g_{fs}) varies with drain current. For most small-signal devices, g_{fs}, is quoted for values of drain current between 1 mA and 10 mA. Most FET parameters (particularly forward transfer conductance) are liable to wide variation from one device to the next. It is, therefore, important to design circuits on the basis of the minimum value for g_{fs}, in order to ensure successful operation with a variety of different devices. The experimental circuit for obtaining the common-source characteristics of an n-channel JFET transistor is shown in Figure 14.20.

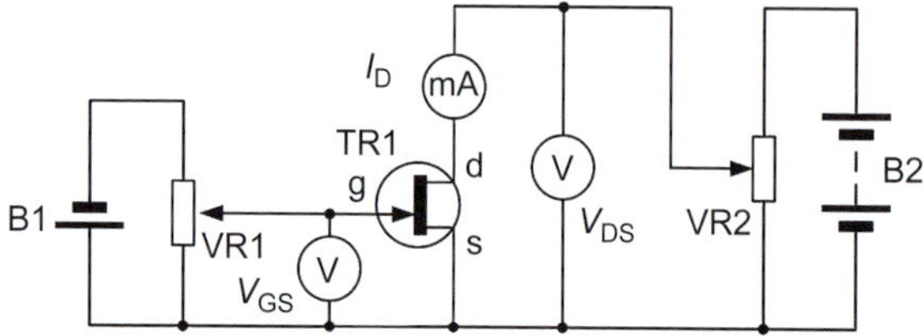

Figure 14.20

Problem 8. Figure 14.21 shows the mutual characteristic for a junction gate field effect transistor. When the gate–source voltage is −2.5 V, determine (a) the value of drain current, (b) the dynamic value of forward transconductance.

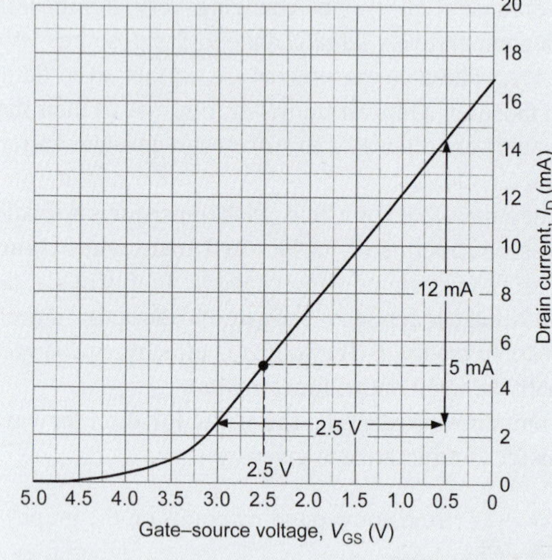

Figure 14.21

(a) From Figure 14.21, when $V_{GS} = -2.5$ V, the **drain current, $I_D = 5$ mA**

(b) From Figure 14.21,

$$g_{fs} = \frac{\Delta I_D}{\Delta V_{GS}} = \frac{(14.5 - 2.5) \times 10^{-3}}{2.5}$$

i.e. **the dynamic value of forward transconductance** $= \frac{12 \times 10^{-3}}{2.5} =$ **4.8 mS** (note the unit – **siemens, S**).

Problem 9. A field effect transistor operates with a drain current of 100 mA and a gate–source bias of −1 V. The device has a g_{fs} value of 0.25. If the bias voltage decreases to −1.1 V, determine (a) the change in drain current and (b) the new value of drain current.

(a) The change in gate–source voltage (V_{GS}) is −0.1 V and the resulting change in drain current can be determined from: $g_{fs} = \frac{\Delta I_D}{\Delta V_{GS}}$

Hence, **the change in drain current,**

$\mathbf{\Delta I_D} = g_{fs} \times \Delta V_{GS} = 0.25 \times -0.1 = -0.025$ A $=$ **−25 mA**

(b) The **new value of drain current** $= (100 - 25) =$ **75 mA**

14.13 Typical FET characteristics and maximum ratings

Table 14.3 summarizes the characteristics of some typical field effect transistors for different applications, where I_D max is the maximum drain current, V_{DS} max is the maximum drain–source voltage, P_D max is the maximum drain power dissipation and g_{fs} typ is the typical value of forward transconductance for the transistor. The list includes both depletion and enhancement types as well as junction and insulated gate types.

Problem 10. Which of the field effect transistors listed in Table 14.3 would be most suitable for each of the following applications: (a) the input stage of a radio receiver, (b) the output stage of a transmitter and (c) switching a load connected to a high-voltage supply.

(a) **BF244A**, since this transistor is designed for use in radio frequency (RF) applications.

(b) **MRF171A**, since this device is designed for RF power applications.

(c) **IRF830**, since this device is intended for switching applications and can operate at up to 500 V

14.14 Transistor amplifiers

Three basic circuit arrangements are used for transistor amplifiers and these are based on the three circuit configurations that we met earlier (i.e. they depend upon which

Table 14.3 FET characteristics and maximum ratings

Device	Type	I_D max	V_{DS} max	P_D max	g_{fs} typ	Application
2N2819	n-chan.	10 mA	25 V	200 mW	4.5 mS	General purpose
2N5457	n-chan.	10 mA	25 V	310 mW	1.2 mS	General purpose
2N7000	n-chan.	200 mA	60 V	400 mW	0.32 S	Low-power switching
BF244A	n-chan.	100 mA	30 V	360 mW	3.3 mS	RF amplifier
BSS84	p-chan.	−130 mA	−50 V	360 mW	0.27 S	Low-power switching
IRF830	n-chan.	4.5 A	500 V	75 W	3.0 S	Power switching
MRF171A	n-chan.	4.5 A	65 V	115 W	1.8 S	RF power amplifier

+V
R_L
TR
Output
Input
Common, 0 V
(a) Common emitter
+V
TR
Input
R_L
Output
Common, 0 V
(b) Common collector
+V
R_L
TR
Input
Output
Common, 0 V
(c) Common base

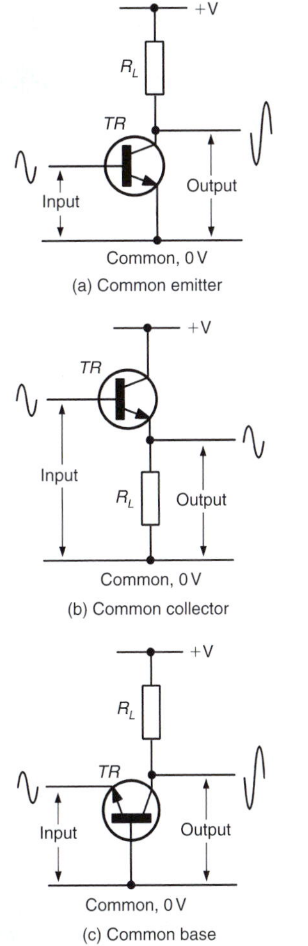

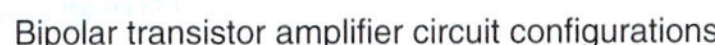

Bipolar transistor amplifier circuit configurations

Figure 14.22

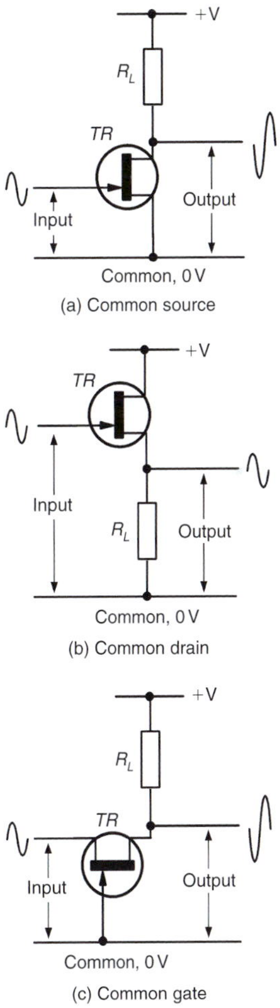

+V
R_L
TR
Output
Input
Common, 0 V
(a) Common source
+V
TR
Input
R_L
Output
Common, 0 V
(b) Common drain
+V
R_L
TR
Input
Output
Common, 0 V
(c) Common gate

Field effect transistor amplifier circuit configurations

Figure 14.23

one of the three transistor connections is made common to both the input and the output). In the case of **bipolar transistors**, the configurations are known as **common emitter, common collector** (or emitter follower) and **common base**.

Where **field effect transistors** are used, the corresponding configurations are **common source, common drain** (or source follower) and **common gate**.

These basic circuit configurations, depicted in Figures 14.22 and 14.23, exhibit quite different performance characteristics, as shown in Tables 14.4 and 14.5, respectively.

A requirement of most amplifiers is that the output signal should be a faithful copy of the input signal or be somewhat larger in amplitude. Other types of amplifier are 'non-linear', in which case their input and output waveforms will not necessarily be similar. In practice, the degree of linearity provided by an amplifier can be affected by a number of factors, including the amount of bias applied and the amplitude of the input signal. It is also worth noting that a linear amplifier will become non-linear when the applied input signal exceeds a threshold value. Beyond this value the amplifier is said to be overdriven and the output will become increasingly distorted if the input signal is further increased.

The optimum value of bias for **linear (Class A) amplifiers** is that value which ensures that the active devices are operated at the mid-point of their characteristics. In practice, this means that a static value of collector current will flow even when there is no signal present. Furthermore, the collector current will flow throughout the complete cycle of an input signal (i.e. conduction will take place over an angle of 360°). At no stage should the transistor be **saturated** ($V_{CE} \approx 0$ V or $V_{DS} \approx 0$ V) nor should it be **cut-off** ($V_{CE} \approx V_{CC}$ or $V_{DS} \approx V_{DD}$).

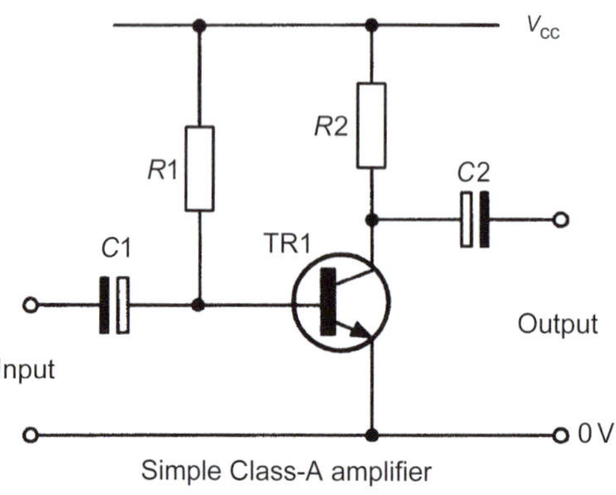

Figure 14.24

In order to ensure that a static value of collector current flows in a transistor, a small current must be applied to the base of the transistor. This current can be derived from the same voltage rail that supplies the collector circuit (via the **collector load**). Figure 14.24 shows a simple Class-A common-emitter circuit in which the **base bias resistor**, R1, and **collector load resistor**, R2, are connected to a common positive supply rail.

The a.c. signal is applied to the base terminal of the transistor via a coupling capacitor, C1. This capacitor removes the d.c. component of any signal applied to the input terminals and ensures that the base bias current delivered by R1 is unaffected by any device connected to the input. C2 couples the signal out of the stage and also prevents d.c. current flow appearing at the output terminals.

14.15 Load lines

The a.c. performance of a transistor amplifier stage can be predicted using a **load line** superimposed on the relevant set of output characteristics. For a bipolar transistor operating in common-emitter mode the required characteristics are I_C plotted against V_{CE}. One end of the load line corresponds to the supply voltage (V_{CC}) while the other end corresponds to the value of collector or drain current that would flow with the device totally saturated ($V_{CE} = 0$ V). In this condition:

$$I_C = \frac{V_{CC}}{R_L}$$

where R_L is the value of collector or drain load resistance.

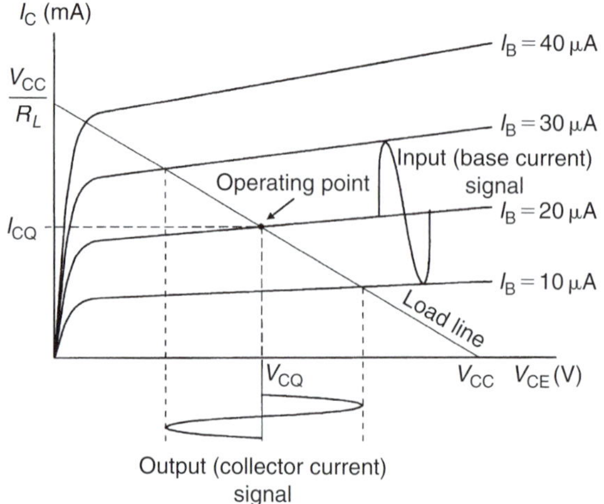

Figure 14.25

Table 14.4 Characteristics of BJT amplifiers

Bipolar transistor amplifiers (see Figure 14.22)			
Parameter	**Common emitter**	**Common collector**	**Common base**
Voltage gain	Medium/high (40)	Unity (1)	High (200)
Current gain	High (200)	High (200)	Unity (1)
Power gain	Very high (8000)	High (200)	High (200)
Input resistance	Medium (2.5 kΩ)	High (100 kΩ)	Low (200 Ω)
Output resistance	Medium/high (20 kΩ)	Low (100 Ω)	High (100 kΩ)
Phase shift	180°	0°	0°
Typical applications	General purpose, AF and RF amplifiers	Impedance matching, input and output stages	RF and VHF amplifiers

Table 14.5 Characteristics of FET amplifiers

Field effect transistor amplifiers (see Figure 14.23)			
Parameter	**Common source**	**Common drain**	**Common gate**
Voltage gain	Medium/high (40)	Unity (1)	High (250)
Current gain	Very high (200 000)	Very high (200 000)	Unity (1)
Power gain	Very high (8 000 000)	Very high (200 000)	High (250)
Input resistance	Very high (1 MΩ)	Very high (1 MΩ)	Low (500 Ω)
Output resistance	Medium/high (50 kΩ)	Low (200 Ω)	High (150 kΩ)
Phase shift	180°	0°	0°
Typical applications	General purpose, AF and RF amplifiers	Impedance matching stages	RF and VHF amplifiers

Figure 14.25 shows a load line superimposed on a set of output characteristics for a bipolar transistor operating in common-emitter mode. The quiescent point (or operating point) is the point on the load line that corresponds to the conditions that exist when no signal is applied to the stage. In Figure 14.25, the base bias current is set at 20 μA so that the **quiescent point** effectively sits roughly halfway along the load line. This position ensures that the collector voltage can swing both positively (above) and negatively (below) its quiescent value (V_{CQ}).

The effect of superimposing an alternating base current (of 20 μA peak–peak) to the d.c. bias current (of 20 μA) can be clearly seen. The corresponding collector current signal can be determined by simply moving up and down the load line.

Problem 11. The characteristic curves shown in Figure 14.26 relate to a transistor operating in common-emitter mode. If the transistor is operated with $I_B = 30$ μA, a load resistor of 1.2 kΩ and an 18 V supply, determine (a) the quiescent values of collector voltage and current (V_{CQ} and I_{CQ}), and (b) the peak–peak output voltage that would be produced by an input signal of 40 μA peak–peak.

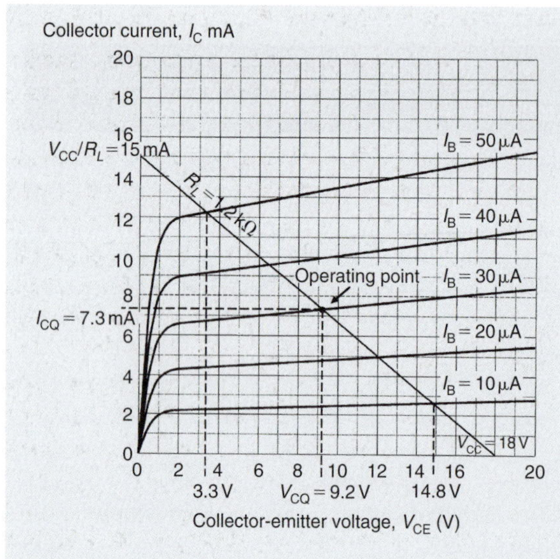

Figure 14.26

(a) First we need to construct the load line on Figure 14.26. The two ends of the load line will correspond to V_{CC}, the 18 V supply, on the collector–emitter voltage axis and 18 V/1.2 kΩ or 15 mA on the collector current axis.

Next we locate the **operating point** (or **quiescent point**) from the point of intersection of the $I_B = 30\,\mu A$ characteristic and the load line. Having located the operating point we can read off the **quiescent values**, i.e. the no-signal values, of collector–emitter voltage (V_{CQ}) and collector current (I_{CQ}). Hence, $\boldsymbol{V_{CQ} = 9.2\,V}$ and $\boldsymbol{I_{CQ} = 7.3\,mA}$.

(b) Next we can determine the maximum and minimum values of collector–emitter voltage by locating the appropriate intercept points on Figure 14.26. Note that the maximum and minimum values of base current will be $(30\,\mu A + 20\,\mu A) = 50\,\mu A$ on positive peaks of the signal and $(30\,\mu A - 20\,\mu A) = 10\,\mu A$ on negative peaks of the signal. The maximum and minimum values of V_{CE} are, respectively, 14.8 V and 3.3 V. Hence,

the output voltage swing

$= (14.8\,V - 3.3\,V) = \mathbf{11.5\,V\ peak–peak}$

Problem 12. An n–p–n transistor has the following characteristics, which may be assumed to be linear between the values of collector voltage stated.

Base current (μA)	Collector current (mA) for collector voltages of: 1 V	5 V
30	1.4	1.6
50	3.0	3.5
70	4.6	5.2

The transistor is used as a common-emitter amplifier with load resistor $R_L = 1.2\,k\Omega$ and a collector supply of 7 V. The signal input resistance is 1 kΩ. If an input current of 20 μA peak varies sinusoidally about a mean bias of 50 μA, estimate (a) the quiescent values of collector voltage and current, (b) the output voltage swing, (c) the voltage gain (d) the dynamic current gain and (e) the power gain.

The characteristics are drawn as shown in Figure 14.27.

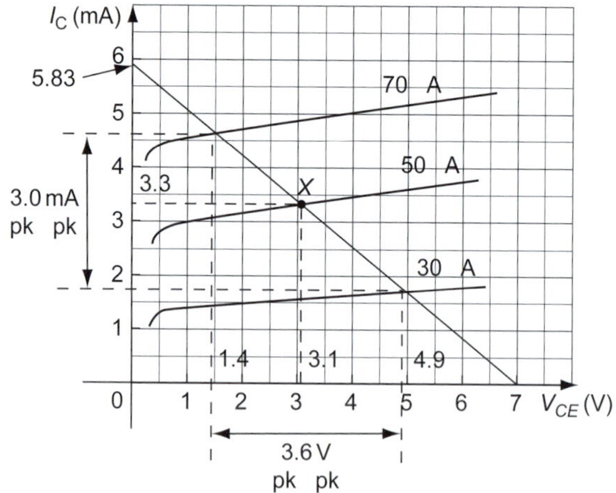

Figure 14.27

The two ends of the load line will correspond to V_{CC}, the 7 V supply, on the collector–emitter voltage axis and 7 V/1.2 kΩ = 5.83 mA on the collector current axis.

(a) The operating point (or quiescent point), X, is located from the point of intersection of the $I_B = 50\,\mu A$ characteristic and the load line. Having located the operating point we can read off the **quiescent values**, i.e. the no-signal values, of collector–emitter voltage (V_{CQ}) and collector current (I_{CQ}). Hence, $\boldsymbol{V_{CQ} = 3.1\,V}$ and $\boldsymbol{I_{CQ} = 3.3\,mA}$.

(b) The maximum and minimum values of collector–emitter voltage may be determined by locating the appropriate intercept points on Figure 14.27. Note that the maximum and minimum values of base current will be $(50\,\mu A + 20\,\mu A) = 70\,\mu A$ on positive peaks of the signal and $(50\,\mu A - 20\,\mu A) = 30\,\mu A$ on negative peaks of the signal. The maximum and minimum values of V_{CE} are, respectively, 4.9 V and 1.4 V. Hence,

the output voltage swing

$$= (4.9\,V - 1.4\,V) = \mathbf{3.5\,V\ peak\text{–}peak}$$

(c) $$\text{Voltage gain} = \frac{\text{change in collector voltage}}{\text{change in base voltage}}$$

The change in collector voltage $= 3.5\,V$ from part (b).

The input voltage swing is given by: $i_b R_i$,

where i_b is the base current swing $= (70 - 30) = 40\,\mu A$ and R_i is the input resistance $= 1\,k\Omega$.

$$\text{Hence, input voltage swing} = 40 \times 10^{-6} \times 1 \times 10^3 = 40\,mV = \text{change in base voltage.}$$

$$\text{Thus, } \mathbf{voltage\ gain} = \frac{\text{change in collector voltage}}{\text{change in base voltage}} = \frac{\Delta V_C}{\Delta V_B} = \frac{3.5}{40 \times 10^{-3}} = \mathbf{87.5}$$

(d) $$\text{Dynamic current gain, } h_{fe} = \frac{\Delta I_C}{\Delta I_B}$$

From Figure 14.27, the output current swing, i.e. the change in collector current, $\Delta I_C = 3.0\,mA$ peak to peak. The input base current swing, the change in base current, $\Delta I_B = 40\,\mu A$.

Hence, **the dynamic current gain**,

$$h_{fe} = \frac{\Delta I_C}{\Delta I_B} = \frac{3.0 \times 10^{-3}}{40 \times 10^{-6}} = \mathbf{75}$$

(e) For a resistive load, the power gain is given by:

$$\mathbf{power\ gain} = \text{voltage gain} \times \text{current gain} = 87.5 \times 75 = \mathbf{6562.5}$$

Now try the following Practice Exercise

Practice Exercise 66 Transistors (Answers on page 821)

1. State whether the following statements are true or false:
   (a) The purpose of a transistor amplifier is to increase the frequency of the input signal.
   (b) The gain of an amplifier is the ratio of the output signal amplitude to the input signal amplitude.
   (c) The output characteristics of a transistor relate the collector current to the base current.
   (d) If the load resistor value is increased the load line gradient is reduced.
   (e) In a common-emitter amplifier, the output voltage is shifted through 180° with reference to the input voltage.
   (f) In a common-emitter amplifier, the input and output currents are in phase.
   (g) The dynamic current gain of a transistor is always greater than the static current gain.
2. In relation to a simple transistor amplifier stage, explain what is meant by the terms: (a) Class-A, (b) saturation, (c) cut-off, (d) quiescent point.
3. Sketch the circuit of a simple Class-A BJT amplifier and explain the function of the components.
4. Explain, with the aid of a labelled sketch, how a load line can be used to determine the operating point of a simple Class-A transistor amplifier.

5. Sketch circuits showing how a JFET can be connected as an amplifier in: (a) common source configuration, (b) common drain configuration, (c) common gate configuration. State typical values of voltage gain and input resistance for each circuit.

6. The output characteristics for a BJT are shown in Figure 14.28. If this device is used in a common-emitter amplifier circuit operating from a 12 V supply with a base bias of 60 μA and a load resistor of 1 kΩ, determine (a) the quiescent values of collector–emitter voltage and collector current, and (b) the peak–peak collector voltage when an 80 μA peak–peak signal current is applied.

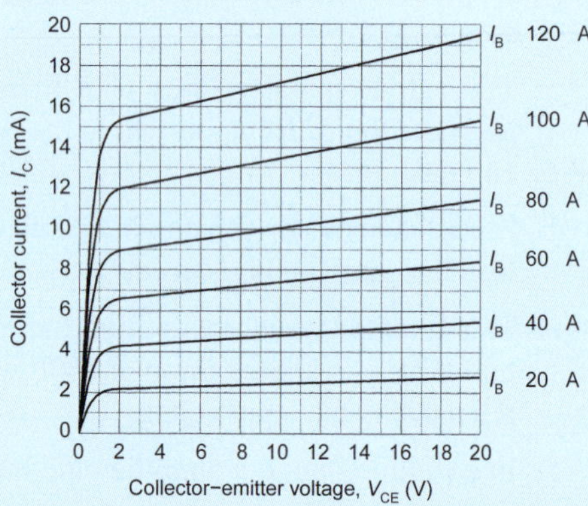

Figure 14.28

7. The output characteristics of a JFET are shown in Figure 14.29. If this device is used in an amplifier circuit operating from an 18 V supply with a gate–source bias voltage of −3 V and a load resistance of 900 Ω, determine (a) the quiescent values of drain–source voltage and drain current, (b) the peak–peak output voltage when an input voltage of 2 V peak–peak is applied and (c) the voltage gain of the stage.

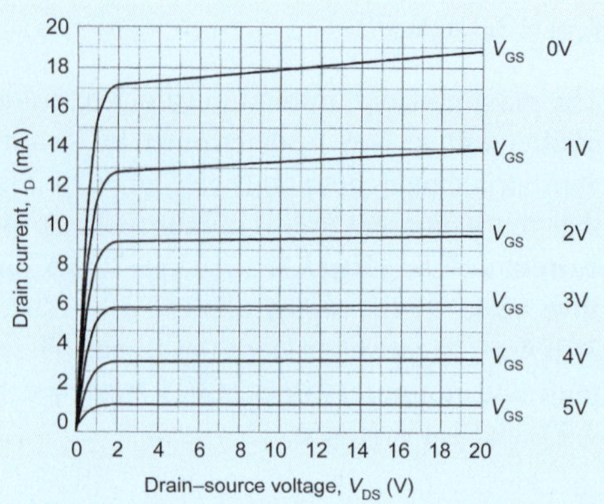

Figure 14.29

8. An amplifier has a current gain of 40 and a voltage gain of 30. Determine the power gain.

9. The output characteristics of a transistor in common-emitter mode configuration can be regarded as straight lines connecting the following points.

	$I_B = 20\,\mu A$		$50\,\mu A$		$80\,\mu A$	
V_{CE} (V)	1.0	8.0	1.0	8.0	1.0	8.0
I_C (mA)	1.2	1.4	3.4	4.2	6.1	8.1

Plot the characteristics and superimpose the load line for a 1 kΩ load, given that the supply voltage is 9 V and the d.c. base bias is 50 μA. The signal input resistance is 800 Ω. When a peak input current of 30 μA varies sinusoidally about a mean bias of 50 μA, determine (a) the quiescent values of collector voltage and current, (b) the output voltage swing, (c) the voltage gain, (d) the dynamic current gain and (e) the power gain.

For fully worked solutions to each of the problems in Practice Exercises 65 and 66 in this chapter, go to the website:
www.routledge.com/cw/bird

Revision Test 3

This revision test covers the material contained in Chapters 10 to 14. *The marks for each question are shown in brackets at the end of each question.*

1. A conductor, 25 cm long, is situated at right-angles to a magnetic field. Determine the strength of the magnetic field if a current of 12 A in the conductor produces a force on it of 4.5 N. (3)
2. An electron in a television tube has a charge of 1.5×10^{-19} C and travels at 3×10^{7} m/s perpendicular to a field of flux density 20 μT. Calculate the force exerted on the electron in the field. (3)
3. A lorry is travelling at 100 km/h. Assuming the vertical component of the earth's magnetic field is 40 μT and the back axle of the lorry is 1.98 m, find the e.m.f. generated in the axle due to motion. (4)
4. An e.m.f. of 2.5 kV is induced in a coil when a current of 2 A collapses to zero in 5 ms. Calculate the inductance of the coil. (4)
5. Two coils, P and Q, have a mutual inductance of 100 mH. If a current of 3 A in coil P is reversed in 20 ms, determine (a) the average e.m.f. induced in coil Q, and (b) the flux change linked with coil Q if it wound with 200 turns. (5)
6. A moving coil instrument gives an f.s.d. when the current is 50 mA and has a resistance of 40 Ω. Determine the value of resistance required to enable the instrument to be used (a) as a 0–5 A ammeter, and (b) as a 0–200 V voltmeter. State the mode of connection in each case. (8)
7. An amplifier has a gain of 20 dB. Its input power is 5 mW. Calculate its output power. (3)
8. A sinusoidal voltage trace displayed on an oscilloscope is shown in Figure RT3.1; the 'time/cm' switch is on 50 ms and the 'volts/cm' switch is on 2 V/cm. Determine for the waveform (a) the frequency, (b) the peak-to-peak voltage, (c) the amplitude, (d) the r.m.s. value. (7)
9. With reference to a p–n junction, briefly explain the terms: (a) majority carriers, (b) contact potential, (c) depletion layer, (d) forward bias, (e) reverse bias. (10)
10. Briefly describe each of the following, drawing their circuit diagram symbol and stating typical applications: (a) zenor diode, (b) silicon controlled rectifier, (c) light emitting diode, (d) varactor diode, (e) Schottky diode. (20)

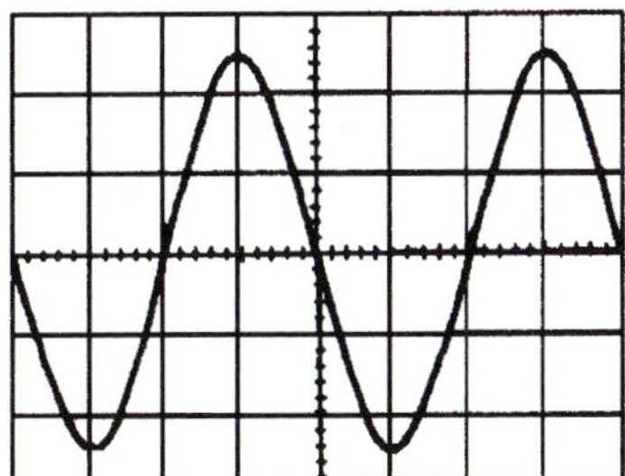

Figure RT3.1

11. The following values were obtained during an experiment on a varactor diode.

Voltage, V	5	10	15	20	25
Capacitance, pF	42	28	18	12	8

Plot a graph showing the variation of capacitance with voltage for the varactor. Label your axes clearly and use your graph to determine (a) the capacitance when the reverse voltage is −17.5 V, (b) the reverse voltage for a capacitance of 35 pF and (c) the change in capacitance when the voltage changes from −2.5 V to −22.5 V. (8)

12. Briefly describe, with diagrams, the action of an n–p–n transistor. (7)
13. The output characteristics of a common-emitter transistor amplifier are given below. Assume that the characteristics are linear between the values of collector voltage stated.

	$I_B = 10\,\mu A$		40 μA		70 μA	
V_{CE} (V)	1.0	7.0	1.0	7.0	1.0	7.0
I_C (mA)	0.6	0.7	2.5	2.9	4.6	5.35

Plot the characteristics and superimpose the load line for a 1.5 kΩ load and collector supply voltage of 8 V. The signal input resistance is 1.2 kΩ. When a peak input current of 30 μA varies sinusoidally about a mean bias of 40 μA, determine (a) the quiescent values of collector voltage and current, (b) the output voltage swing, (c) the voltage gain, (d) the dynamic current gain and (e) the power gain. (18)

For lecturers/instructors/teachers, fully worked solutions to each of the problems in Revision Test 3, together with a full marking scheme, are available at the website:
www.routledge.com/cw/bird

Main formulae for Part 2 Basic electrical and electronic principles

General

Charge $Q = It$ Force $F = ma$

Work $W = Fs$ Power $P = \frac{W}{t}$

Energy $W = Pt$

Ohm's law $V = IR$ or $I = \frac{V}{R}$ or $R = \frac{V}{I}$

Conductance $G = \frac{1}{R}$

Power $P = VI = I^2R = \frac{V^2}{R}$ Resistance $R = \frac{\rho l}{a}$

Resistance at θ°C, $R_\theta = R_0(1 + \alpha_0\theta)$

Terminal p.d. of source, $V = E - Ir$

Series circuit $R = R_1 + R_2 + R_3 + \cdots$

Parallel network $\frac{1}{R} = \frac{1}{R_1} + \frac{1}{R_2} + \frac{1}{R_3} + \cdots$

Capacitors and capacitance

$E = \frac{V}{d}$ $C = \frac{Q}{V}$ $Q = It$ $D = \frac{Q}{A}$

$\frac{D}{E} = \varepsilon_0\varepsilon_r$

$C = \frac{\varepsilon_0\varepsilon_r A(n-1)}{d}$

Capacitors in parallel $C = C_1 + C_2 + C_3 + \cdots$

Capacitors in series $\frac{1}{C} = \frac{1}{C_1} + \frac{1}{C_2} + \frac{1}{C_3} + \cdots$

$W = \frac{1}{2}CV^2$

Magnetic circuits

$B = \frac{\Phi}{A}$ $F_m = NI$ $H = \frac{NI}{l}$ $\frac{B}{H} = \mu_0\mu_r$

$S = \frac{\text{mmf}}{\Phi} = \frac{l}{\mu_0\mu_r A}$

Electromagnetism

$F = BIl\sin\theta$ $F = QvB$

Electromagnetic induction

$E = Blv\sin\theta$ $E = -N\frac{d\Phi}{dt} = -L\frac{dI}{dt}$

$W = \frac{1}{2}LI^2$

$L = \frac{N\Phi}{I} = \frac{N^2}{S}$ $E_2 = -M\frac{dI_1}{dt}$ $M = \frac{N_1N_2}{S}$

Measurements

Shunt $R_s = \frac{I_a r_a}{I_s}$ Multiplier $R_M = \frac{V - Ir_a}{I}$

Power in decibels $= 10\log\frac{P_2}{P_1} = 20\log\frac{I_2}{I_1} = 20\log\frac{V_2}{V_1}$

Wheatstone bridge $R_x = \frac{R_2R_3}{R_1}$

Potentiometer $E_2 = E_1\left(\frac{l_2}{l_1}\right)$

These formulae are available for download at the website:
www.routledge.com/cw/bird

Part 3

Electrical principles and technology

Chapter 15

D.c. circuit theory

***Why it is important to understand:* D.c circuit theory**

In earlier chapters it was seen that a single equivalent resistance can be found when two or more resistors are connected together in either series, parallel or combinations of both, and that these circuits obey Ohm's law. However, sometimes in more complex circuits we cannot simply use Ohm's law alone to find the voltages or currents circulating within the circuit. For these types of calculations we need certain rules which allow us to obtain the circuit equations and for this we can use Kirchhoff's laws. In addition, there are a number of circuit theorems – superposition theorem, Thévenin's theorem, Norton's theorem – which allow us to analyse more complex circuits. In addition, the maximum power transfer theorem enables us to determine maximum power in a d.c. circuit. In this chapter Kirchhoff's laws and the circuit theorems are explained in detail using many numerical worked examples. An electrical/electronic engineer often needs to be able to analyse an electrical network to determine currents flowing in each branch and the voltage across each branch.

At the end of this chapter you should be able to:

- state and use Kirchhoff's laws to determine unknown currents and voltages in d.c. circuits
- understand the superposition theorem and apply it to find currents in d.c. circuits
- understand general d.c. circuit theory
- understand Thévenin's theorem and apply a procedure to determine unknown currents in d.c. circuits
- recognize the circuit diagram symbols for ideal voltage and current sources
- understand Norton's theorem and apply a procedure to determine unknown currents in d.c. circuits
- appreciate and use the equivalence of the Thévenin and Norton equivalent networks
- state the maximum power transfer theorem and use it to determine maximum power in a d.c. circuit

15.1 Introduction

The laws which determine the currents and voltage drops in d.c. networks are: (a) Ohm's law (see Chapter 4), (b) the laws for resistors in series and in parallel (see Chapter 7) and (c) Kirchhoff's laws (see Section 15.2 following). In addition, there are a number of circuit theorems which have been developed for solving problems in electrical networks. These include:

(i) the superposition theorem (see Section 15.3),
(ii) Thévenin's theorem (see Section 15.5),
(iii) Norton's theorem (see Section 15.7) and

Electrical Circuit Theory and Technology. 978-1-138-67349-6, © 2017 John Bird. Published by Taylor & Francis. All rights reserved.

(iv) the maximum power transfer theorem (see Section 15.9).

15.2 Kirchhoff's* laws

Kirchhoff's* **laws** state:

(a) **Current Law.** *At any junction in an electric circuit the total current flowing towards that junction is equal to the total current flowing away from the junction,* i.e. $\Sigma I = 0$

Thus, referring to Figure 15.1:
$I_1 + I_2 = I_3 + I_4 + I_5$ or

$I_1 + I_2 - I_3 - I_4 - I_5 = 0$

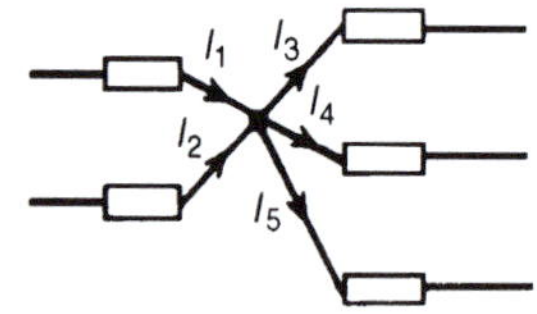

Figure 15.1

*Who was **Kirchhoff**? Gustav Robert Kirchhoff (12 March 1824–17 October 1887) was a German physicist. Concepts in circuit theory and thermal emission are named 'Kirchhoff's laws' after him, as well as a law of thermochemistry. To find out more go to **www.routledge.com/cw/bird**

(b) **Voltage Law.** *In any closed loop in a network, the algebraic sum of the voltage drops (i.e. products of current and resistance) taken around the loop is equal to the resultant e.m.f. acting in that loop.*

Thus, referring to Figure 15.2:

$E_1 - E_2 = IR_1 + IR_2 + IR_3$

(Note that if current flows away from the positive terminal of a source, that source is considered by convention to be positive. Thus moving anticlockwise around the loop of Figure 15.2, E_1 is positive and E_2 is negative.)

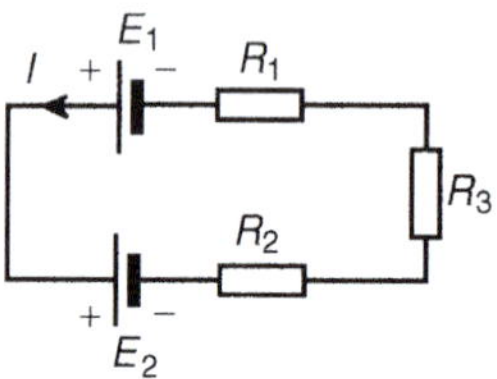

Figure 15.2

Problem 1. (a) Find the unknown currents marked in Figure 15.3(a). (b) Determine the value of e.m.f. E in Figure 15.3(b).

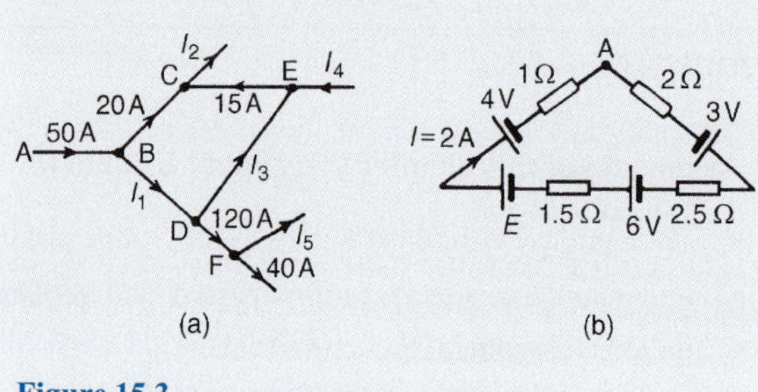

Figure 15.3

(a) Applying Kirchhoff's current law:

For junction B: $50 = 20 + I_1$. Hence $\boldsymbol{I_1 = 30\,A}$

For junction C: $20 + 15 = I_2$. Hence $\boldsymbol{I_2 = 35\,A}$

For junction D: $I_1 = I_3 + 120$

i.e. $30 = I_3 + 120$. Hence $\boldsymbol{I_3 = -90\,A}$

(i.e. in the opposite direction to that shown in Figure 15.3(a).)

For junction E: $I_4 + I_3 = 15$

i.e. $I_4 = 15 - (-90)$.

Hence $\boldsymbol{I_4 = 105\,A}$

For junction F: $120 = I_5 + 40$. Hence $\boldsymbol{I_5 = 80\,A}$

(b) Applying Kirchhoff's voltage law and moving clockwise around the loop of Figure 15.3(b) starting at point A:

$$3+6+E-4=(I)(2)+(I)(2.5)+(I)(1.5)+(I)(1)$$
$$=I(2+2.5+1.5+1)$$

i.e. $5+E=2(7)$, since $I=2$ A

Hence $E=14-5=\mathbf{9\,V}$

Problem 2. Use Kirchhoff's laws to determine the currents flowing in each branch of the network shown in Figure 15.4.

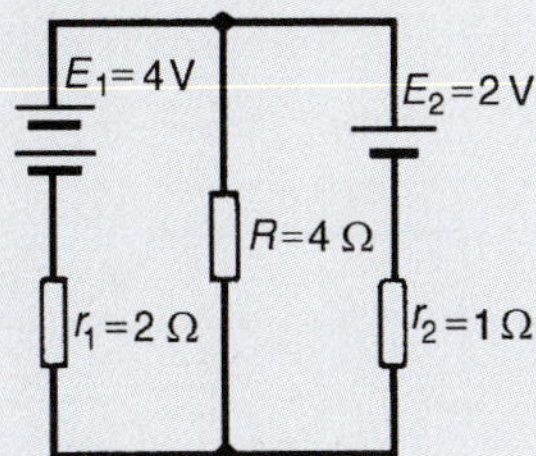

Figure 15.4

Procedure

1. Use Kirchhoff's current law and label current directions on the original circuit diagram. The directions chosen are arbitrary, but it is usual, as a starting point, to assume that current flows from the positive terminals of the batteries. This is shown in Figure 15.5, where the three branch currents are expressed in terms of I_1 and I_2 only, since the current through R is I_1+I_2

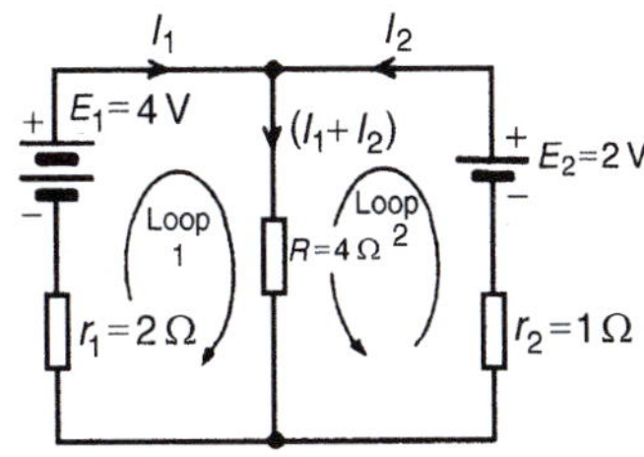

Figure 15.5

2. Divide the circuit into two loops and apply Kirchhoff's voltage law to each. From loop 1 of Figure 15.5, and moving in a clockwise direction as indicated (the direction chosen does not matter), gives

$$E_1=I_1r_1+(I_1+I_2)R,\ \text{i.e. } 4=2I_1+4(I_1+I_2)$$
$$\text{i.e. } 6I_1+4I_2=4 \qquad (1)$$

From loop 2 of Figure 15.5, and moving in an anticlockwise direction as indicated (once again, the choice of direction does not matter; it does not have to be in the same direction as that chosen for the first loop), gives:

$$E_2=I_2r_2+(I_1+I_2)R,\ \text{i.e. } 2=I_2+4(I_1+I_2)$$
$$\text{i.e. } 4I_1+5I_2=2 \qquad (2)$$

3. Solve equations (1) and (2) for I_1 and I_2

$2\times(1)$ gives: $12I_1+8I_2=8$ (3)

$3\times(2)$ gives: $12I_1+15I_2=6$ (4)

$(3)-(4)$ gives: $-7I_2=2$ hence $I_2=-\dfrac{2}{7}$
$=\mathbf{-0.286\,A}$

(i.e. I_2 is flowing in the opposite direction to that shown in Figure 15.5.)

From (1) $6I_1+4(-0.286)=4$

$6I_1=4+1.144$

Hence $I_1=\dfrac{5.144}{6}=\mathbf{0.857\,A}$

Current flowing through resistance R is

$$I_1+I_2=0.857+(-0.286)=\mathbf{0.571\,A}$$

Note that a third loop is possible, as shown in Figure 15.6, giving a third equation which can be used as a check:

$$E_1-E_2=I_1r_1-I_2r_2$$
$$4-2=2I_1-I_2$$
$$2=2I_1-I_2$$

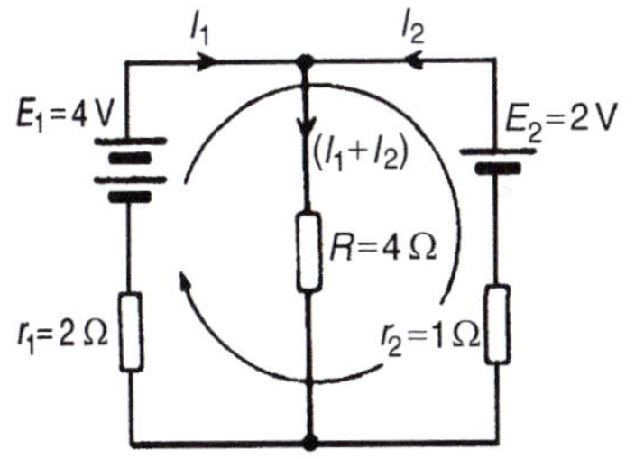

Figure 15.6

[Check: $2I_1-I_2=2(0.857)-(-0.286)=2$]

Part 3

Problem 3. Determine, using Kirchhoff's laws, each branch current for the network shown in Figure 15.7.

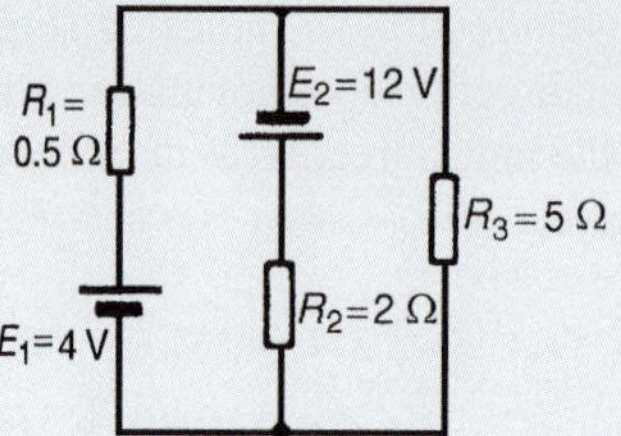

Figure 15.7

1. Currents and their directions are shown labelled in Figure 15.8 following Kirchhoff's current law. It is usual, although not essential, to follow conventional current flow with current flowing from the positive terminal of the source.

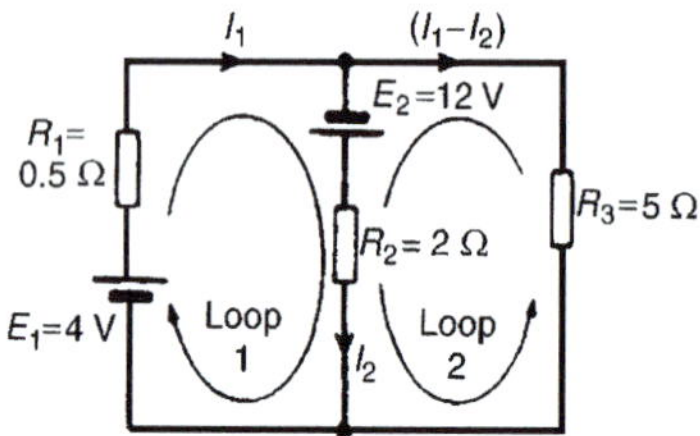

Figure 15.8

2. The network is divided into two loops as shown in Figure 15.8. Applying Kirchhoff's voltage law gives:

For loop 1:

$$E_1 + E_2 = I_1 R_1 + I_2 R_2$$

i.e. $$16 = 0.5I_1 + 2I_2 \qquad (1)$$

For loop 2:

$$E_2 = I_2 R_2 - (I_1 - I_2)R_3$$

Note that since loop 2 is in the opposite direction to current $(I_1 - I_2)$, the volt drop across R_3 (i.e. $(I_1 - I_2)(R_3)$) is by convention negative.

Thus $12 = 2I_2 - 5(I_1 - I_2)$

i.e. $$12 = -5I_1 + 7I_2 \qquad (2)$$

3. Solving equations (1) and (2) to find I_1 and I_2:

$$10 \times (1) \text{ gives } 160 = 5I_1 + 20I_2 \qquad (3)$$

$$(2) + (3) \text{ gives } 172 = 27I_2 \text{ hence } I_2 = \frac{172}{27}$$

$$= \mathbf{6.37\ A}$$

From (1): $16 = 0.5I_1 + 2(6.37)$

$$I_1 = \frac{16 - 2(6.37)}{0.5} = \mathbf{6.52\ A}$$

Current flowing in R_3 $= I_1 - I_2 = 6.52 - 6.37$

$$= \mathbf{0.15\ A}$$

Problem 4. For the bridge network shown in Figure 15.9 determine the currents in each of the resistors.

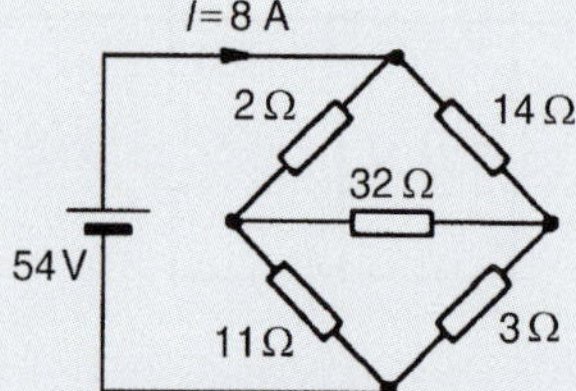

Figure 15.9

Let the current in the 2 Ω resistor be I_1, then by Kirchhoff's current law, the current in the 14 Ω resistor is $(I - I_1)$. Let the current in the 32 Ω resistor be I_2 as shown in Figure 15.10. Then the current in the 11 Ω resistor is $(I_1 - I_2)$ and that in the 3 Ω resistor is $(I - I_1 + I_2)$. Applying Kirchhoff's voltage law to loop 1 and moving in a clockwise direction as shown in Figure 15.10 gives:

$$54 = 2I_1 + 11(I_1 - I_2)$$

i.e. $$13I_1 - 11I_2 = 54 \qquad (1)$$

Applying Kirchhoff's voltage law to loop 2 and moving in an anticlockwise direction as shown in Figure 15.10 gives:

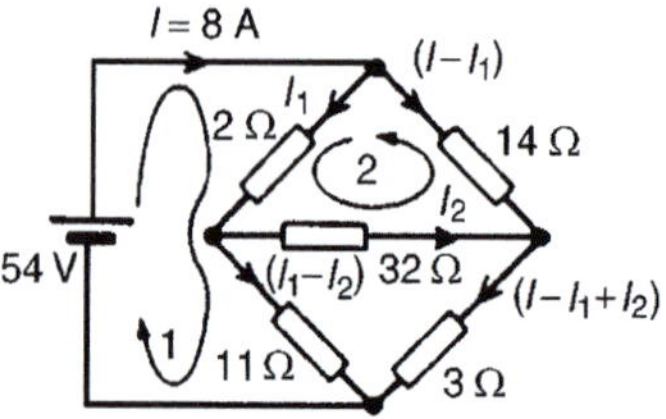

Figure 15.10

$$0 = 2I_1 + 32I_2 - 14(I - I_1)$$

However $\quad I = 8\text{ A}$

Hence $\quad 0 = 2I_1 + 32I_2 - 14(8 - I_1)$

i.e. $\quad 16I_1 + 32I_2 = 112 \qquad (2)$

Equations (1) and (2) are simultaneous equations with two unknowns, I_1 and I_2.

$16 \times (1)$ gives : $\quad 208I_1 - 176I_2 = 864 \qquad (3)$

$13 \times (2)$ gives : $\quad 208I_1 + 416I_2 = 1456 \qquad (4)$

$(4) - (3)$ gives : $\quad 592I_2 = 592$

$$I_2 = 1\text{ A}$$

Substituting for I_2 in (1) gives:

$$13I_2 - 11 = 54$$

$$I_1 = \frac{65}{13} = 5\text{ A}$$

Hence,
the current flowing in the 2 Ω resistor $= I_1 = \mathbf{5\text{ A}}$

the current flowing in the 14 Ω resistor $= I - I_1 = 8 - 5 = \mathbf{3\text{ A}}$

the current flowing in the 32 Ω resistor $= I_2 = \mathbf{1\text{ A}}$

the current flowing in the 11 Ω resistor $= I_1 - I_2 = 5 - 1 = \mathbf{4\text{ A}}$

the current flowing in the 3 Ω resistor $= I - I_1 + I_2 = 8 - 5 + 1 = \mathbf{4\text{ A}}$

Now try the following Practice Exercise

Practice Exercise 67 Kirchhoff's laws (Answers on page 821)

1. Find currents I_3, I_4 and I_6 in Figure 15.11

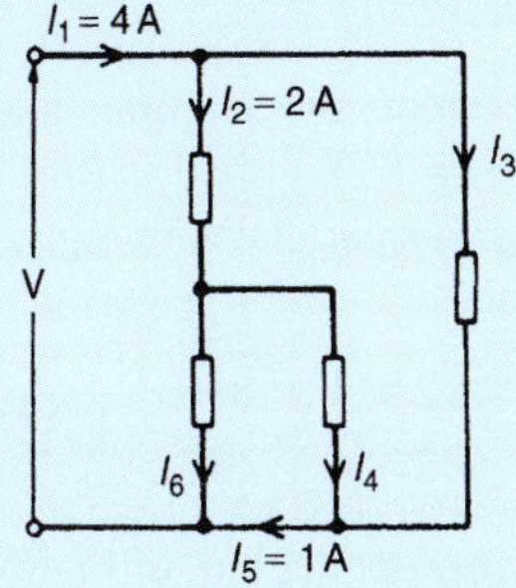

Figure 15.11

2. For the networks shown in Figure 15.12, find the values of the currents marked.

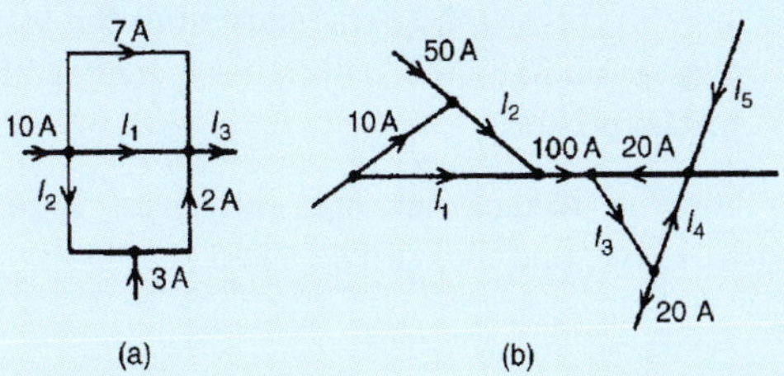

Figure 15.12

3. Calculate the currents I_1 and I_2 in Figure 15.13.

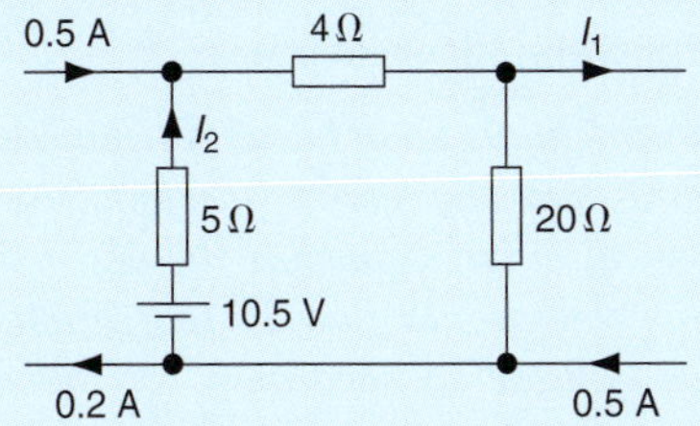

Figure 15.13

4. Use Kirchhoff's laws to find the current flowing in the 6 Ω resistor of Figure 15.14 and the power dissipated in the 4 Ω resistor.

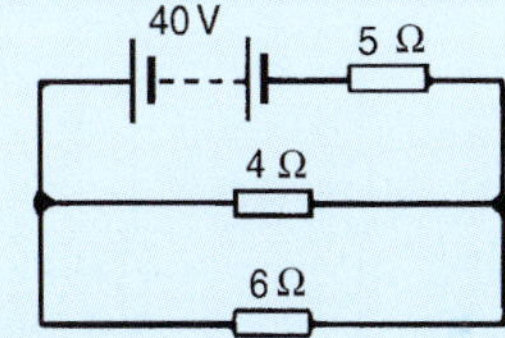

Figure 15.14

5. Find the current flowing in the 3 Ω resistor for the network shown in Figure 15.15(a). Find also the p.d. across the 10 Ω and 2 Ω resistors.

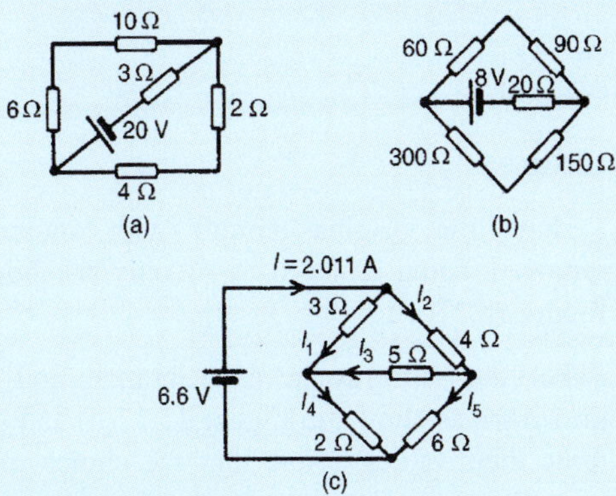

Figure 15.15

6. For the network shown in Figure 15.15(b) find: (a) the current in the battery, (b) the current in the 300 Ω resistor, (c) the current in the 90 Ω resistor and (d) the power dissipated in the 150 Ω resistor.
7. For the bridge network shown in Figure 15.15 (c), find the currents I_1 to I_5

15.3 The superposition theorem

The **superposition theorem** states:

'In any network made up of linear resistances and containing more than one source of e.m.f., the resultant current flowing in any branch is the algebraic sum of the currents that would flow in that branch if each source was considered separately, all other sources being replaced at that time by their respective internal resistances.'

Problem 5. Figure 15.16 shows a circuit containing two sources of e.m.f., each with their

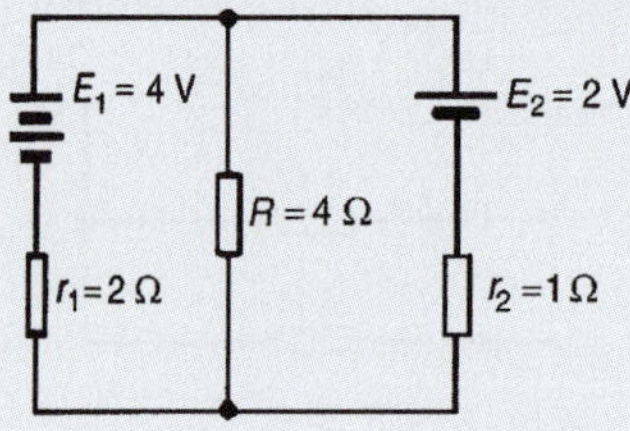

Figure 15.16

internal resistance. Determine the current in each branch of the network by using the superposition theorem.

Procedure:

1. Redraw the original circuit with source E_2 removed, being replaced by r_2 only, as shown in Figure 15.17(a).
2. Label the currents in each branch and their directions as shown in Figure 15.17(a) and determine their values. (Note that the choice of current directions depends on the battery polarity, which, by convention is taken as flowing from the positive battery terminal as shown.)

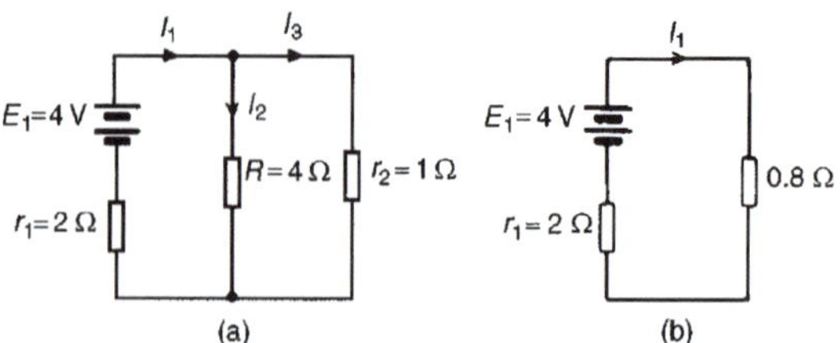

Figure 15.17

R in parallel with r_2 gives an equivalent resistance of:

$$\frac{4\times 1}{4+1}=0.8\,\Omega$$

From the equivalent circuit of Figure 15.17(b)

$$I_1=\frac{E_1}{r_1+0.8}=\frac{4}{2+0.8}=1.429\,\text{A}$$

From Figure 15.17(a)

$$I_2=\left(\frac{1}{4+1}\right)I_1=\frac{1}{5}(1.429)=0.286\,\text{A}$$

and

$$I_3=\left(\frac{4}{4+1}\right)I_1=\frac{4}{5}(1.429)$$

$$=1.143\,\text{A by current division}$$

3. Redraw the original circuit with source E_1 removed, being replaced by r_1 only, as shown in Figure 15.18(a).

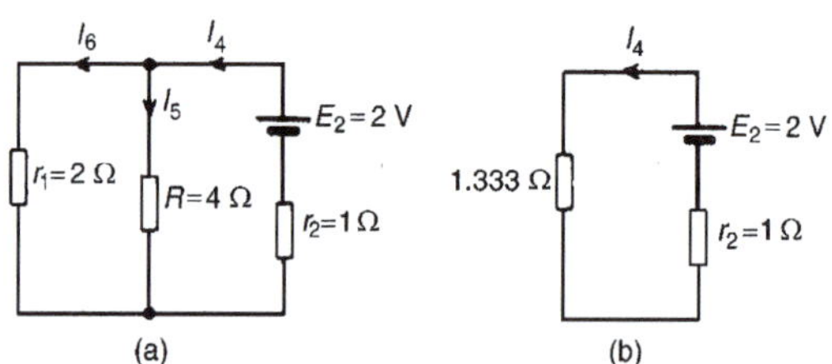

Figure 15.18

4. Label the currents in each branch and their directions as shown in Figure 15.18(a) and determine their values. r_1 in parallel with R gives an equivalent resistance of:

$$\frac{2\times 4}{2+4}=\frac{8}{6}=1.333\,\Omega$$

From the equivalent circuit of Figure 15.18(b)

$$I_4=\frac{E_2}{1.333+r_2}=\frac{2}{1.333+1}=0.857\,\text{A}$$

From Figure 15.18(a)

$$I_5 = \left(\frac{2}{2+4}\right) I_4 = \frac{2}{6}(0.857) = 0.286\,\text{A}$$

$$I_6 = \left(\frac{4}{2+4}\right) I_4 = \frac{4}{6}(0.857) = 0.571\,\text{A}$$

5. Superimpose Figure 15.18(a) on to Figure 15.17(a), as shown in Figure 15.19.

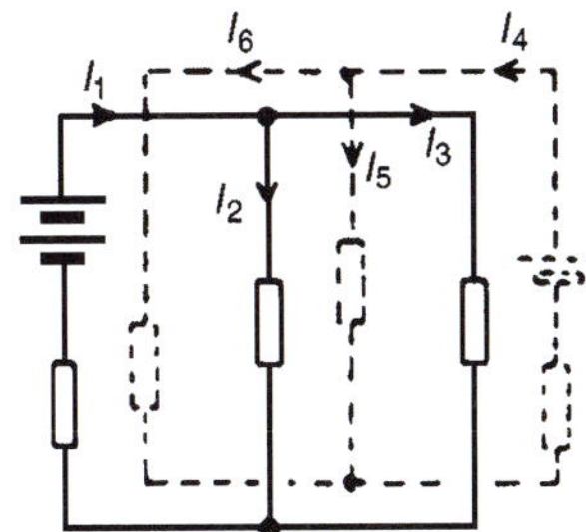

Figure 15.19

6. Determine the algebraic sum of the currents flowing in each branch.
Resultant current flowing through source 1, i.e.

$$I_1 - I_6 = 1.429 - 0.571$$
$$= \mathbf{0.858\ A\ (discharging)}$$

Resultant current flowing through source 2, i.e.

$$I_4 - I_3 = 0.857 - 1.143$$
$$= \mathbf{-0.286\ A\ (charging)}$$

Resultant current flowing through resistor R, i.e.

$$I_2 + I_5 = 0.286 + 0.286$$
$$= \mathbf{0.572\ A}$$

The resultant currents with their directions are shown in Figure 15.20.

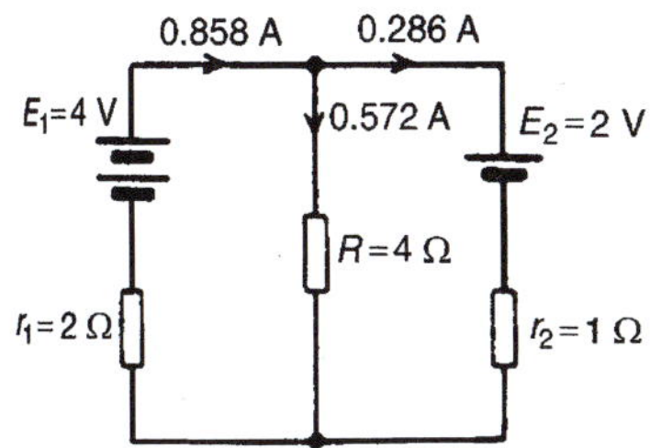

Figure 15.20

Problem 6. For the circuit shown in Figure 15.21, find, using the superposition theorem, (a) the current flowing in and the p.d. across the 18 Ω resistor, (b) the current in the 8 V battery and (c) the current in the 3 V battery.

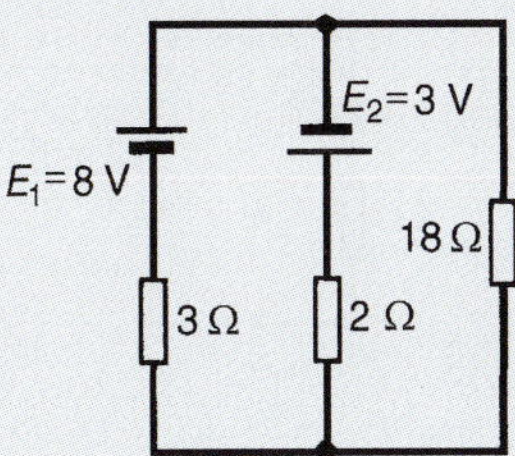

Figure 15.21

1. Removing source E_2 gives the circuit of Figure 15.22(a).

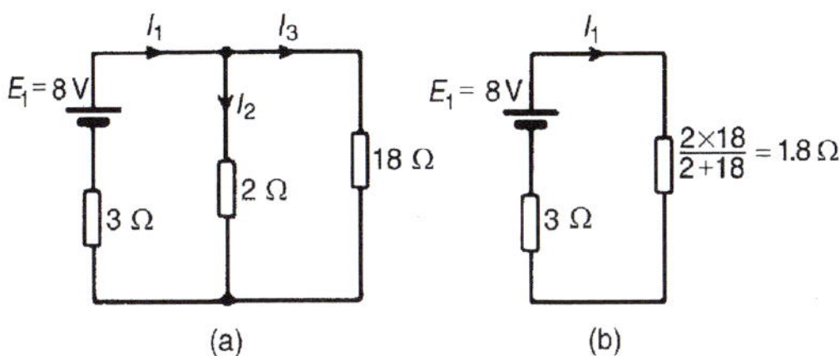

Figure 15.22

2. The current directions are labelled as shown in Figure 15.22(a), I_1 flowing from the positive terminal of E_1

From Figure 15.22(b),

$$I_1 = \frac{E_1}{3+1.8} = \frac{8}{4.8} = 1.667\,\text{A}$$

From Figure 15.22(a), $I_2 = \left(\frac{18}{2+18}\right) I_1$

$$= \frac{18}{20}(1.667) = 1.500\,\text{A}$$

and $\quad I_3 = \left(\frac{2}{2+18}\right) I_1$

$$= \frac{2}{20}(1.667) = 0.167\,\text{A}$$

3. Removing source E_1 gives the circuit of Figure 15.23(a) (which is the same as Figure 15.23(b)).

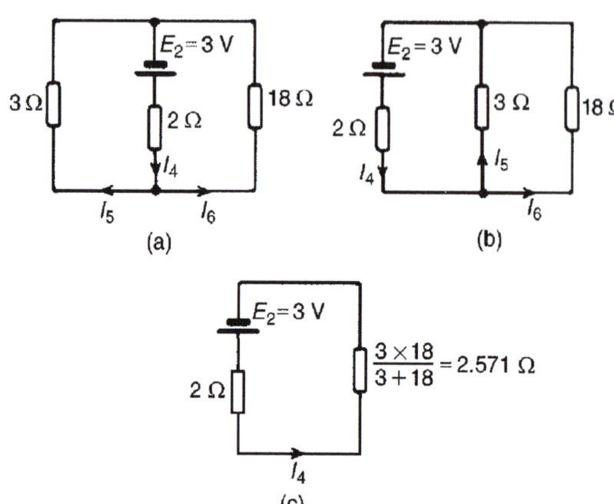

Figure 15.23

4. The current directions are labelled as shown in Figures 15.23(a) and 15.23(b), I_4 flowing from the positive terminal of E_2

From Figure 15.23(c), $I_4 = \frac{E_2}{2+2.571} = \frac{3}{4.571}$

$$= 0.656\,\text{A}$$

From Figure 15.23(b), $I_5 = \left(\frac{18}{3+18}\right) I_4$

$$= \frac{18}{21}(0.656) = 0.562\,\text{A}$$

$$I_6 = \left(\frac{3}{3+18}\right) I_4 = \frac{3}{21}(0.656)$$

$$= 0.094\,\text{A}$$

5. Superimposing Figure 15.23(a) on to Figure 15.22(a) gives the circuit in Figure 15.24.

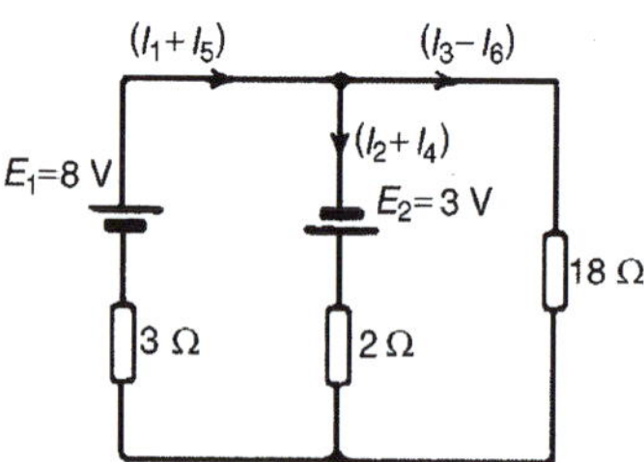

Figure 15.24

6. (a) Resultant current in the 18 Ω resistor

$= I_3 - I_6$

$= 0.167 - 0.094$

$= \mathbf{0.073\,A}$

P.d. across the 18 Ω resistor

$= 0.073 \times 18 = \mathbf{1.314\,V}$

(b) Resultant current in the 8 V battery

$= I_1 + I_5 = 1.667 + 0.562$

$= \mathbf{2.229\,A}$ **(discharging)**

(c) Resultant current in the 3 V battery

$= I_2 + I_4 = 1.500 + 0.656$

$= \mathbf{2.156\,A}$ **(discharging)**

For a practical laboratory experiment on the superposition theorem, see the website.

Now try the following Practice Exercise

Practice Exercise 68 Superposition theorem (Answers on page 821)

1. Use the superposition theorem to find currents I_1, I_2 and I_3 of Figure 15.25(a).

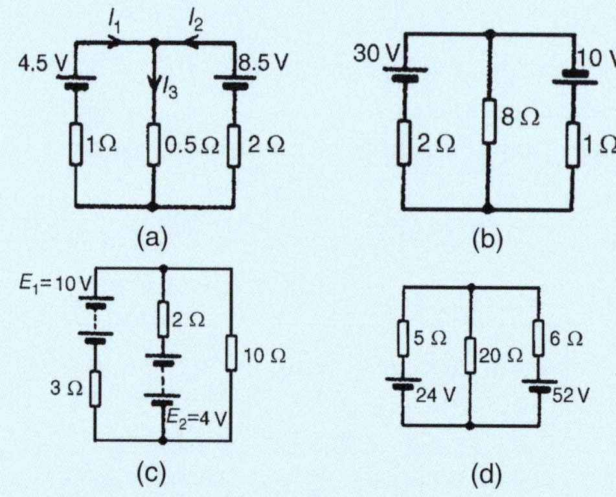

Figure 15.25

2. Use the superposition theorem to find the current in the 8 Ω resistor of Figure 15.25(b).
3. Use the superposition theorem to find the current in each branch of the network shown in Figure 15.25(c).
4. Use the superposition theorem to determine the current in each branch of the arrangement shown in Figure 15.25(d).

15.4 General d.c. circuit theory

The following points involving d.c. circuit analysis need to be appreciated before proceeding with problems using Thévenin's and Norton's theorems:

(i) The open-circuit voltage, E, across terminals AB in Figure 15.26 is equal to 10 V, since no current flows through the 2 Ω resistor and hence no voltage drop occurs.

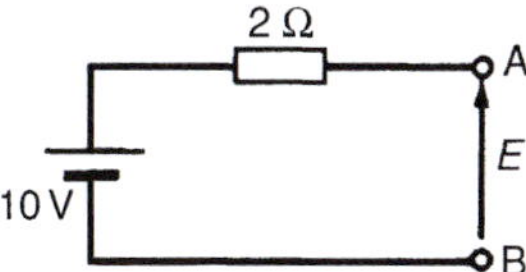

Figure 15.26

(ii) The open-circuit voltage, E, across terminals AB in Figure 15.27(a) is the same as the voltage across the 6 Ω resistor. The circuit may be redrawn as shown in Figure 15.27(b).

$$E = \left(\frac{6}{6+4}\right)(50)$$

by voltage division in a series circuit, i.e. $\boldsymbol{E = 30\,\mathrm{V}}$

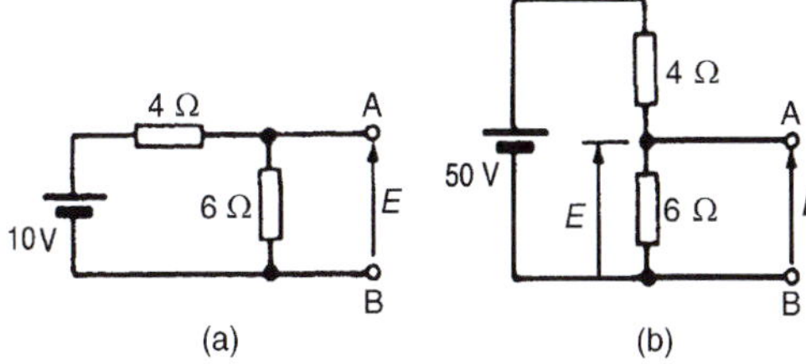

Figure 15.27

(iii) For the circuit shown in Figure 15.28(a) representing a practical source supplying energy, $V = E - Ir$, where E is the battery e.m.f., V is the battery terminal voltage and r is the internal resistance of the battery (as shown in Section 6.5). For the circuit shown in Figure 15.28(b), $V = E - (-I)r$, i.e. $V = E + Ir$

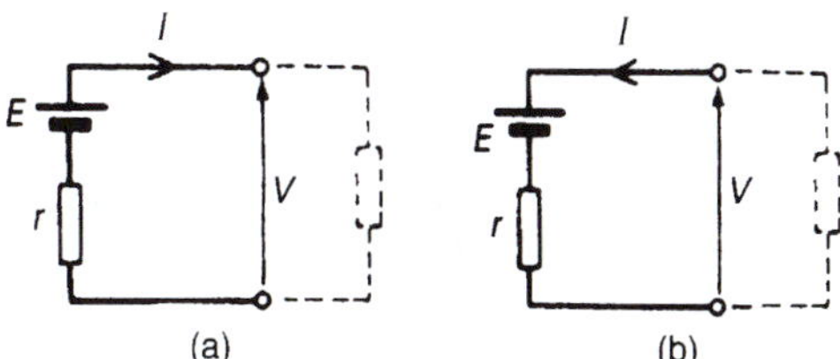

Figure 15.28

(iv) The resistance 'looking-in' at terminals AB in Figure 15.29(a) is obtained by reducing the circuit in stages as shown in Figures 15.29(b) to (d). Hence the equivalent resistance across AB is 7 Ω

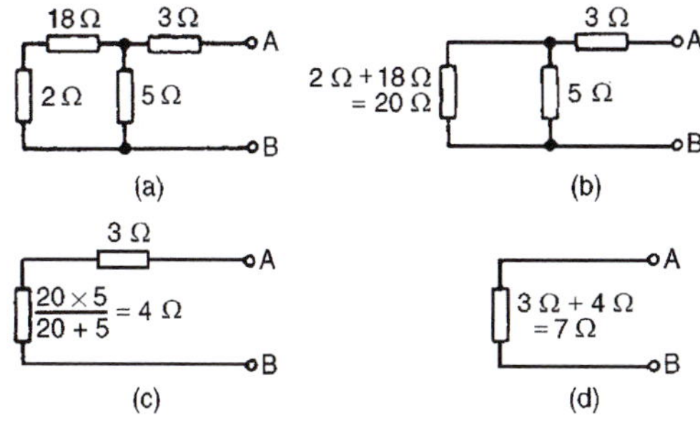

Figure 15.29

(v) For the circuit shown in Figure 15.30(a), the 3 Ω resistor carries no current and the p.d. across the 20 Ω resistor is 10 V. Redrawing the circuit gives Figure 15.30(b), from which

$$E = \left(\frac{4}{4+6}\right) \times 10 = \mathbf{4\,V}$$

(vi) If the 10 V battery in Figure 15.30(a) is removed and replaced by a short-circuit, as shown in Figure 15.30(c), then the 20 Ω resistor may be removed. The reason for this is that a short-circuit has zero resistance, and 20 Ω in parallel with zero ohms gives an equivalent resistance of: $(20 \times 0/20 + 0)$, i.e. 0 Ω. The circuit is then as shown in Figure 15.30(d), which is redrawn in Figure 15.30(e).

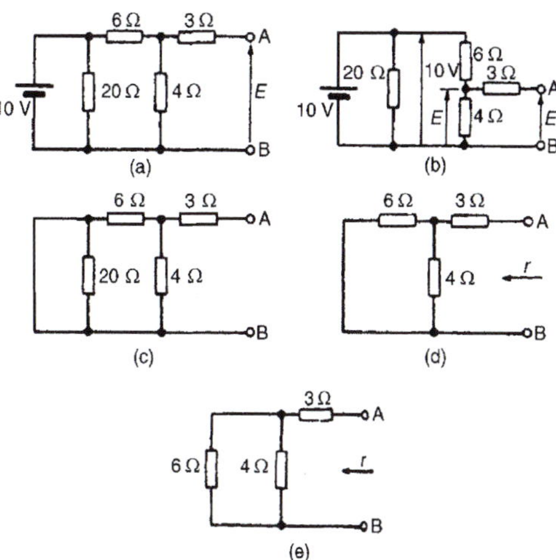

Figure 15.30

From Figure 15.30(e), the equivalent resistance across AB,

$$r = \frac{6 \times 4}{6+4} + 3 = 2.4 + 3 = \mathbf{5.4\,\Omega}$$

(vii) To find the voltage across AB in Figure 15.31: since the 20 V supply is across the 5 Ω and 15 Ω resistors in series then, by voltage division, the voltage drop across AC,

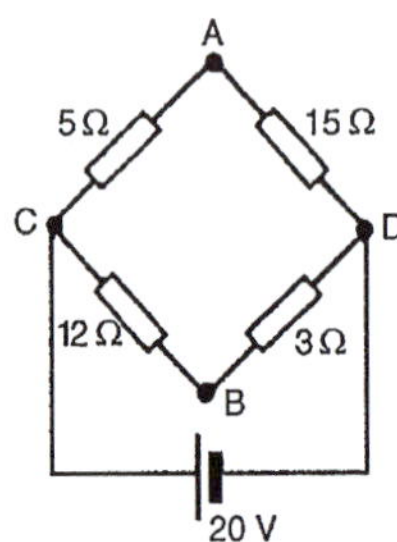

Figure 15.31

$$V_{AC} = \left(\frac{5}{5+15}\right)(20) = 5\,V$$

Similarly, $V_{CB} = \left(\frac{12}{12+3}\right)(20) = 16\,V$

V_C is at a potential of +20 V.

$V_A = V_C - V_{AC} = +20 - 5 = 15\,V$ and

$V_B = V_C - V_{BC} = +20 - 16 = 4\,V$

Hence the voltage between AB is $V_A - V_B = 15 - 4 = 11\,V$ and current would flow from A to B since A has a higher potential than B.

(viii) In Figure 15.32(a), to find the equivalent resistance across AB the circuit may be redrawn as in Figures 15.32(b) and (c). From Figure 15.32(c), the equivalent resistance across AB

$$= \frac{5 \times 15}{5+15} + \frac{12 \times 3}{12+3} = 3.75 + 2.4 = \mathbf{6.15\,\Omega}$$

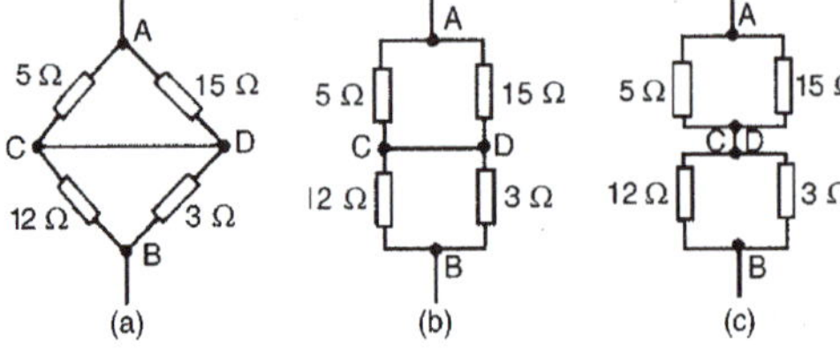

Figure 15.32

(ix) In the worked problems in Sections 15.5 and 15.7 following, it may be considered that Thévenin's and Norton's theorems have no obvious advantages compared with, say, Kirchhoff's laws. However, these theorems can be used to analyse part of a circuit, and in much more complicated networks the principle of replacing the supply by a constant voltage source in series with a resistance (or impedance) is very useful.

15.5 Thévenin's theorem

Thévenin's* **theorem** states:

'The current in any branch of a network is that which would result if an e.m.f. equal to the p.d. across a break made in the branch, were introduced into the branch, all other e.m.f.s being removed and represented by the internal resistances of the sources.'

The procedure adopted when using Thévenin's theorem is summarized below. To determine the current in any branch of an active network (i.e. one containing a source of e.m.f.):

(i) remove the resistance R from that branch,

(ii) determine the open-circuit voltage, E, across the break,

(iii) remove each source of e.m.f. and replace them by their internal resistances and then determine the resistance, r, 'looking-in' at the break,

(iv) determine the value of the current from the equivalent circuit shown in Figure 15.33, i.e.

$$\boldsymbol{I = \frac{E}{R+r}}$$

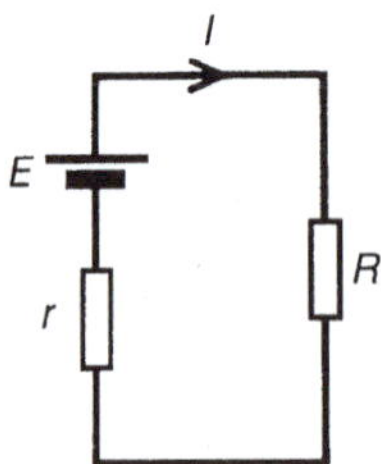

Figure 15.33

*Who was **Thévenin**? Léon Charles Thévenin (30 March 1857–21 September 1926) extended Ohm's law to the analysis of complex electrical circuits. To find out more go to **www.routledge.com/cw/bird**

Part 3

Problem 7. Use Thévenin's theorem to find the current flowing in the 10 Ω resistor for the circuit shown in Figure 15.34(a).

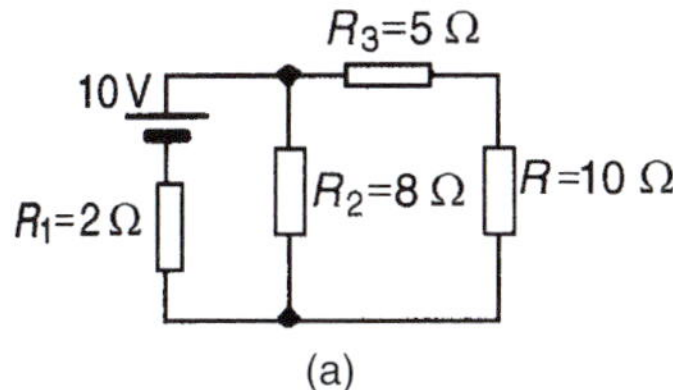

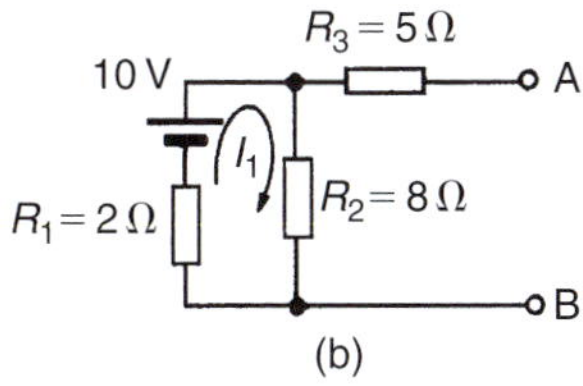

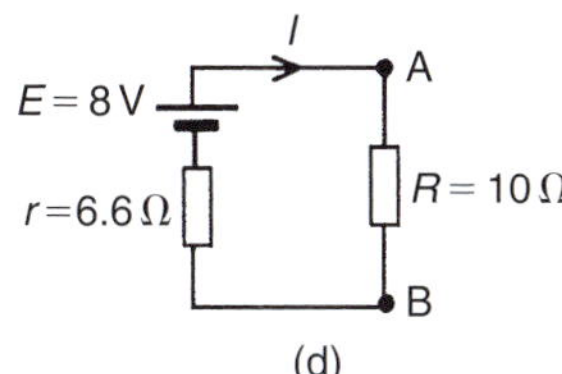

Figure 15.34

Following the above procedure:

(i) The 10 Ω resistance is removed from the circuit as shown in Figure 15.34(b)

(ii) There is no current flowing in the 5 Ω resistor and current I_1 is given by:

$$I_1 = \frac{10}{R_1 + R_2} = \frac{10}{2+8} = 1\,\text{A}$$

P.d. across $R_2 = I_1 R_2 = 1 \times 8 = 8\,\text{V}$
Hence p.d. across AB, i.e. the open-circuit voltage across the break, $E = 8\,\text{V}$

(iii) Removing the source of e.m.f. gives the circuit of Figure 15.34(c).

$$\text{Resistance, } r = R_3 + \frac{R_1 R_2}{R_1 + R_2} = 5 + \frac{2 \times 8}{2+8}$$

$$= 5 + 1.6 = 6.6\,\Omega$$

(iv) The equivalent Thévenin's circuit is shown in Figure 15.34(d).

$$\text{Current } I = \frac{E}{R+r} = \frac{8}{10+6.6} = \frac{8}{16.6} = 0.482\,\text{A}$$

Hence the current flowing in the 10 Ω resistor of Figure 15.34(a) is **0.482 A**

Problem 8. For the network shown in Figure 15.35(a) determine the current in the 0.8 Ω resistor using Thévenin's theorem.

Following the procedure:

(i) The 0.8 Ω resistor is removed from the circuit as shown in Figure 15.35(b).

(ii) $$\text{Current } I_1 = \frac{12}{1+5+4} = \frac{12}{10} = 1.2\,\text{A}$$

P.d. across 4 Ω resistor $= 4I_1 = (4)\,(1.2) = 4.8\,\text{V}$
Hence p.d. across AB, i.e. the open-circuit voltage across AB, $E = 4.8\,\text{V}$

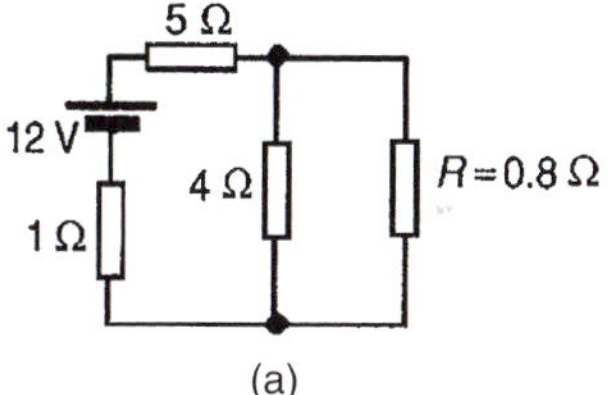

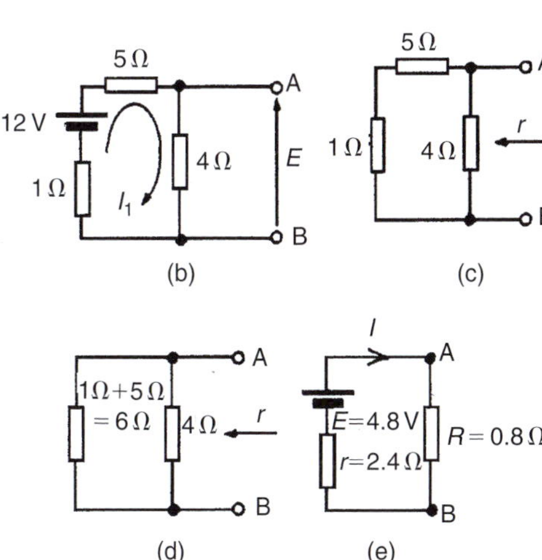

Figure 15.35

(iii) Removing the source of e.m.f. gives the circuit shown in Figure 15.35(c). The equivalent circuit of Figure 15.35(c) is shown in Figure 15.35(d), from which,

$$\text{resistance } r = \frac{4\times 6}{4+6} = \frac{24}{10} = 2.4\,\Omega$$

(iv) The equivalent Thévenin's circuit is shown in Figure 15.35(e), from which,

$$\text{current } I = \frac{E}{r+R} = \frac{4.8}{2.4+0.8} = \frac{4.8}{3.2}$$

$$I = \mathbf{1.5\,A = current\ in\ the\ 0.8\,\Omega\ resistor.}$$

Problem 9. Use Thévenin's theorem to determine the current I flowing in the $4\,\Omega$ resistor shown in Figure 15.36(a). Find also the power dissipated in the $4\,\Omega$ resistor.

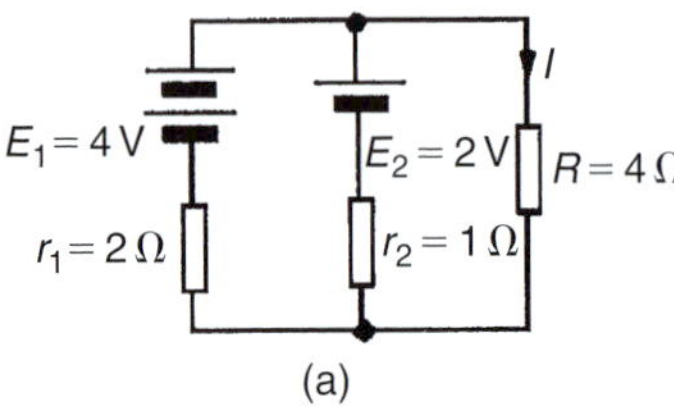

(a)

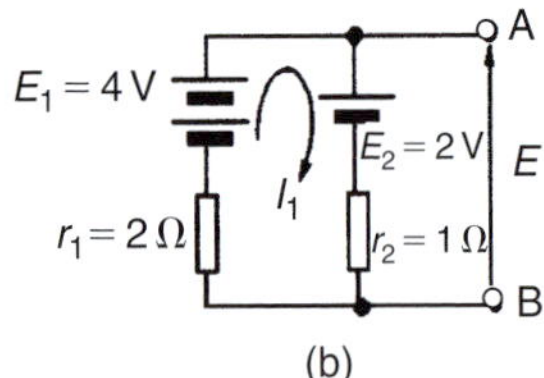

(b)

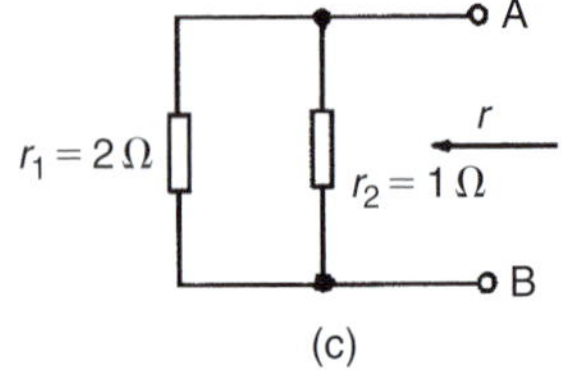

(c)

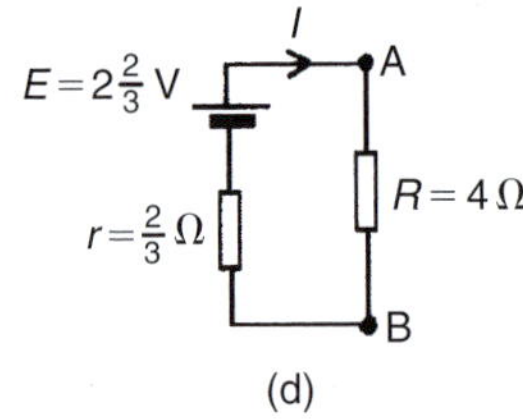

(d)

Figure 15.36

Following the procedure:

(i) The $4\,\Omega$ resistor is removed from the circuit as shown in Figure 15.36(b).

(ii) $$\text{Current } I_1 = \frac{E_1 - E_2}{r_1 + r_2} = \frac{4-2}{2+1} = \frac{2}{3}\,\text{A}$$

P.d. across AB, $E = E_1 - I_1 r_1 = 4 - \left(\frac{2}{3}\right)(2) = 2\frac{2}{3}\,\text{V}$

(see Section 15.4(iii)).

(Alternatively, p.d. across AB,

$$E = E_2 - I_1 r_2$$
$$= 2 - \left(\tfrac{2}{3}\right)(1) = 2\tfrac{2}{3}\,\text{V})$$

(iii) Removing the sources of e.m.f. gives the circuit shown in Figure 15.36(c), from which resistance

$$r = \frac{2\times 1}{2+1} = \frac{2}{3}\,\Omega$$

(iv) The equivalent Thévenin's circuit is shown in Figure 15.36(d), from which,

$$\text{current, } I = \frac{E}{r+R} = \frac{2\frac{2}{3}}{\frac{2}{3}+4} = \frac{8/3}{14/3}$$
$$= \frac{8}{14} = \mathbf{0.571\,A}$$
$$= \textbf{current in the } \mathbf{4\,\Omega} \textbf{ resistor}$$

Power dissipated in $4\,\Omega$ resistor,

$$P = I^2R = (0.571)^2(4) = \mathbf{1.304\,W}$$

Problem 10. Use Thévenin's theorem to determine the current flowing in the $3\,\Omega$ resistance of the network shown in Figure 15.37(a). The voltage source has negligible internal resistance.

(Note the symbol for an ideal voltage source in Figure 15.37(a) – from BS EN 60617-2: 1996, which superseded BS 3939-2: 1985 – which may be used as an alternative to the battery symbol.)

Following the procedure.

(i) The $3\,\Omega$ resistance is removed from the circuit as shown in Figure 15.37(b).

Part 3

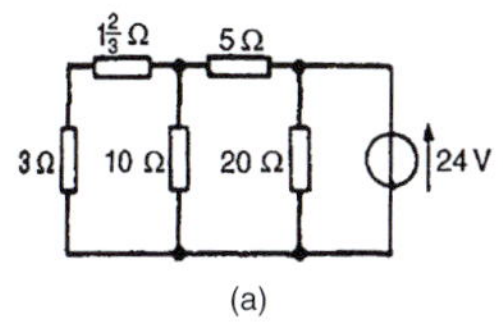

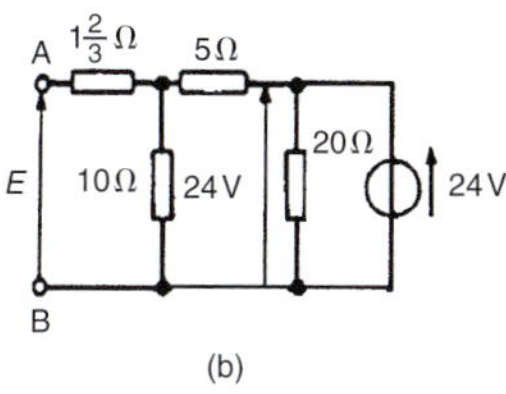

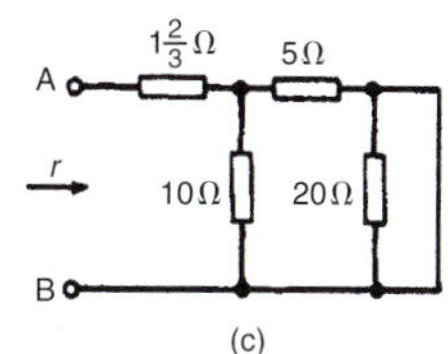

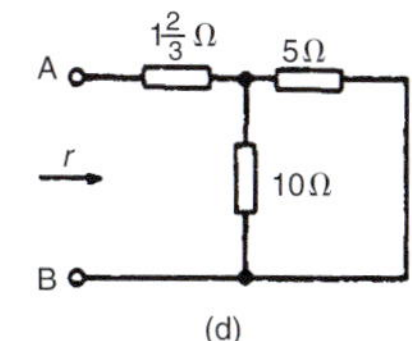

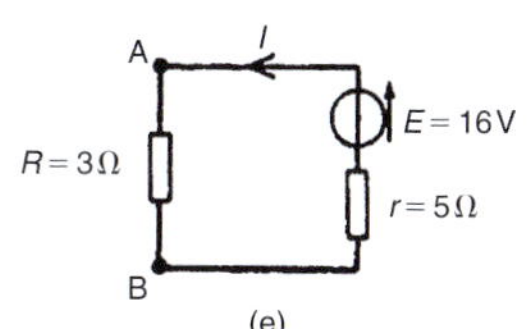

Figure 15.37

(ii) The $1\frac{2}{3}\,\Omega$ resistance now carries no current.

$$\text{P.d. across } 10\,\Omega \text{ resistor} = \left(\frac{10}{10+5}\right)(24)$$

$$= \mathbf{16\,V} \text{ (see Section 15.4(v)).}$$

Hence p.d. across AB, $E = 16\,\text{V}$

(iii) Removing the source of e.m.f. and replacing it by its internal resistance means that the $20\,\Omega$ resistance is short-circuited, as shown in Figure 15.37(c) since its internal resistance is zero. The $20\,\Omega$ resistance may thus be removed as shown in Figure 15.37(d) (see Section 15.4 (vi)).

From Figure 15.37(d), resistance,

$$r = 1\frac{2}{3} + \frac{10\times 5}{10+5}$$
$$= 1\frac{2}{3} + \frac{50}{15} = 5\,\Omega$$

(iv) The equivalent Thévenin's circuit is shown in Figure 15.37(e), from which

$$\text{current, } I = \frac{E}{r+R} = \frac{16}{3+5} = \frac{16}{8} = \mathbf{2\,A}$$
$$= \textbf{current in the } \mathbf{3\,\Omega} \textbf{ resistance}$$

Problem 11. A Wheatstone bridge network is shown in Figure 15.38(a). Calculate the current flowing in the $32\,\Omega$ resistor, and its direction, using Thévenin's theorem. Assume the source of e.m.f. to have negligible resistance.

Following the procedure:

(i) The $32\,\Omega$ resistor is removed from the circuit as shown in Figure 15.38(b)

(ii) The p.d. between A and C,

$$V_{AC} = \left(\frac{R_1}{R_1+R_4}\right)(E) = \left(\frac{2}{2+11}\right)(54) = 8.31\,\text{V}$$

The p.d. between B and C,

$$V_{BC} = \left(\frac{R_2}{R_2+R_3}\right)(E) = \left(\frac{14}{14+3}\right)(54) = 44.47\,\text{V}$$

Hence the p.d. between A and B $= 44.47 - 8.31 = \mathbf{36.16\,V}$

Point C is at a potential of +54 V. Between C and A is a voltage drop of 8.31 V. Hence the voltage at point A is $54 - 8.31 = 45.69$ V. Between C and B is a voltage drop of 44.47 V. Hence the voltage at point B is $54 - 44.47 = 9.53$ V. Since the voltage at A is greater than at B, current must flow in the direction A to B (see Section 15.4 (vii)).

(iii) Replacing the source of e.m.f. with a short-circuit (i.e. zero internal resistance) gives the circuit shown in Figure 15.38(c). The circuit is redrawn and simplified as shown in Figure 15.38(d) and (e), from which the resistance between terminals A and B,

$$r = \frac{2\times 11}{2+11} + \frac{14\times 3}{14+3} = \frac{22}{13} + \frac{42}{17}$$
$$= 1.692 + 2.471 = \mathbf{4.163\,\Omega}$$

(iv) The equivalent Thévenin's circuit is shown in Figure 15.38(f), from which,

$$\text{current } I = \frac{E}{r+R_5} = \frac{36.16}{4.163+32} = 1\,\text{A}$$

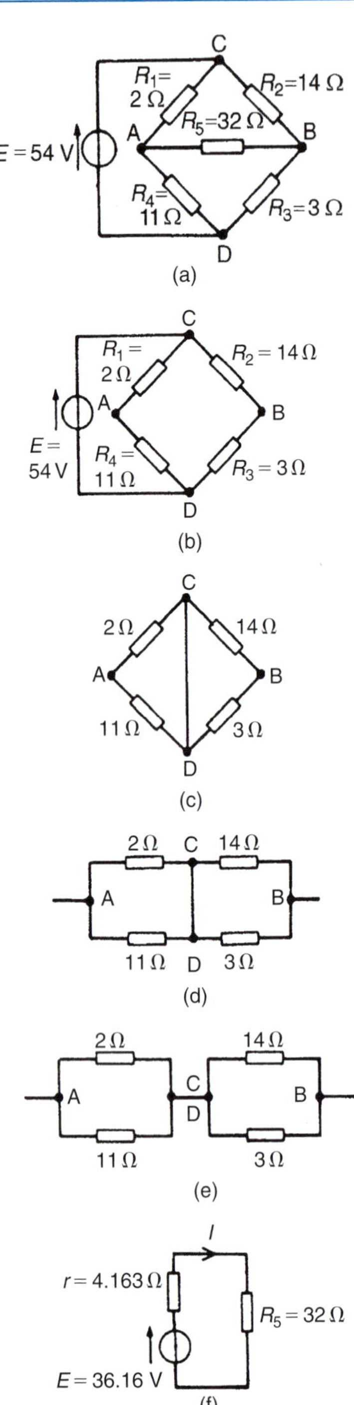

Figure 15.38

Hence the current in the 32 Ω resistor of Figure 15.38(a) is 1 A, flowing from A to B.

For a practical laboratory experiment on Thévenin's theorem, see the website.

Now try the following Practice Exercise

Practice Exercise 69 Thévenin's theorem (Answers on page 821)

1. Use Thévenin's theorem to find the current flowing in the 14 Ω resistor of the network shown in Figure 15.39. Find also the power dissipated in the 14 Ω resistor.

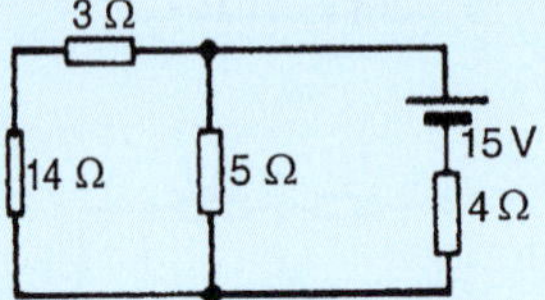

Figure 15.39

2. Use Thévenin's theorem to find the current flowing in the 6 Ω resistor shown in Figure 15.40 and the power dissipated in the 4 Ω resistor.

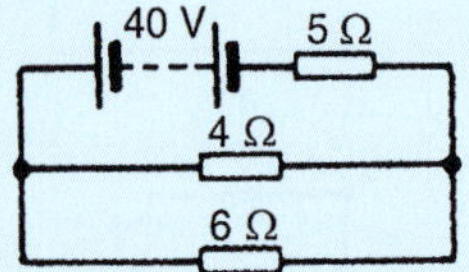

Figure 15.40

3. Repeat Problems 1 to 4 of Exercise 68 on page 226 using Thévenin's theorem.

4. In the network shown in Figure 15.41, the battery has negligible internal resistance. Find, using Thévenin's theorem, the current flowing in the 4 Ω resistor.

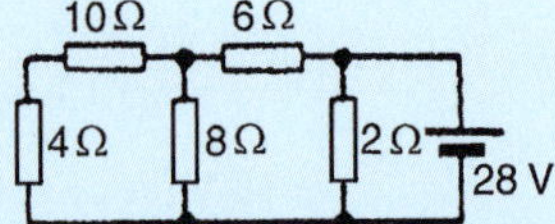

Figure 15.41

5. For the bridge network shown in Figure 15.42, find the current in the 5 Ω resistor, and its direction, by using Thévenin's theorem.

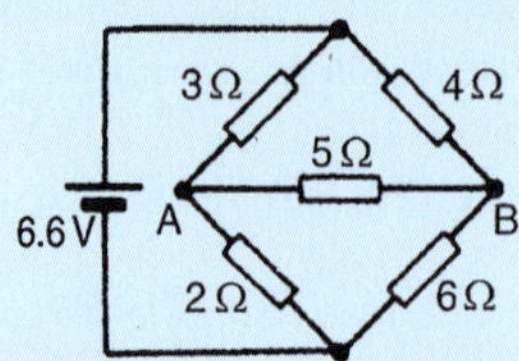

Figure 15.42

15.6 Constant-current source

A source of electrical energy can be represented by a source of e.m.f. in series with a resistance. In Section 15.5, the Thévenin constant-voltage source consisted of a constant e.m.f. E in series with an internal resistance r. However, this is not the only form of representation. A source of electrical energy can also be represented by a constant-current source in parallel with a resistance. It may be shown that the two forms are equivalent. An **ideal constant-voltage generator** is one with zero internal resistance so that it supplies the same voltage to all loads. An **ideal constant-current generator** is one with infinite internal resistance so that it supplies the same current to all loads.
Note the symbol for an ideal current source (from BS EN 60617-2: 1996, which superseded BS 3939-2: 1985), shown in Figure 15.43.

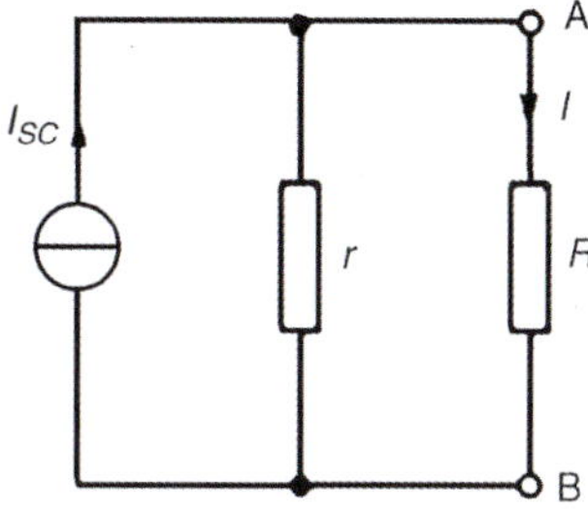

Figure 15.43

15.7 Norton's theorem

Norton's* **theorem** states:

'The current that flows in any branch of a network is the same as that which would flow in the branch if it were connected across a source of electrical energy, the short-circuit current of which is equal to the current that would flow in a short-circuit across the branch, and the internal resistance of which is equal to the resistance which appears across the open-circuited branch terminals.'

The procedure adopted when using Norton's theorem is summarized below.

*Who was **Norton**? Edward Lawry Norton (28 July 1898–28 January 1983) is best remembered for development of the dual of Thevenin's equivalent circuit, now referred to as Norton's equivalent circuit. To find out more go to **www.routledge.com/cw/bird**

To determine the current flowing in a resistance R of a branch AB of an active network:

(i) short-circuit branch AB

(ii) determine the short-circuit current I_{SC} flowing in the branch

(iii) remove all sources of e.m.f. and replace them by their internal resistance (or, if a current source exists, replace with an open-circuit), then determine the resistance r, 'looking-in' at a break made between A and B

(iv) determine the current I flowing in resistance R from the Norton equivalent network shown in Figure 15.43, i.e.

$$\boldsymbol{I = \left(\frac{r}{r+R}\right) I_{SC}}$$

Problem 12. Use Norton's theorem to determine the current flowing in the 10 Ω resistance for the circuit shown in Figure 15.44(a).

Following the above procedure:

(i) The branch containing the 10 Ω resistance is short-circuited as shown in Figure 15.44(b).

(ii) Figure 15.44(c) is equivalent to Figure 15.44(b). Hence

$$I_{SC} = \frac{10}{2} = 5\,\text{A}$$

(iii) If the 10 V source of e.m.f. is removed from Figure 15.44(b) the resistance 'looking-in' at a break made between A and B is given by:

$$r = \frac{2 \times 8}{2+8} = 1.6\,\Omega$$

(iv) From the Norton equivalent network shown in Figure 15.44(d) the current in the 10 Ω resistance, by current division, is given by:

$$I = \left(\frac{1.6}{1.6+5+10}\right)(5) = \mathbf{0.482\,A}$$

as obtained previously in Problem 7 using Thévenin's theorem.

Problem 13. Use Norton's theorem to determine the current I flowing in the 4 Ω resistance shown in Figure 15.45(a).

Part 3

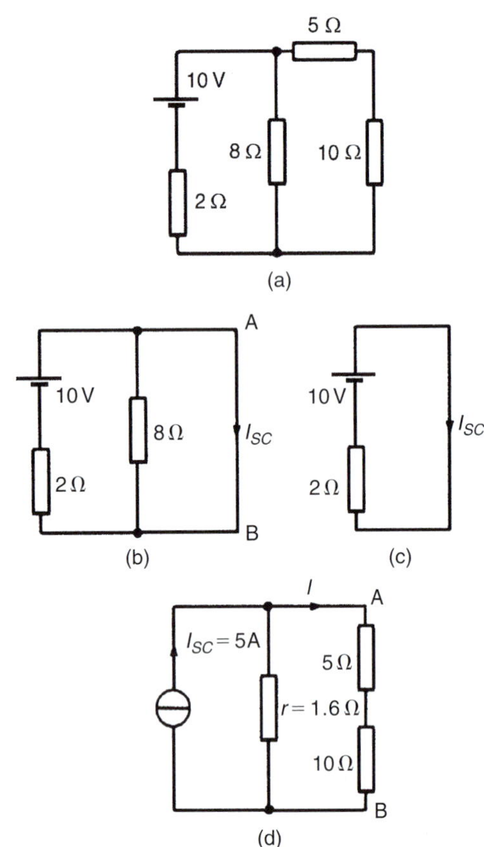

Figure 15.44

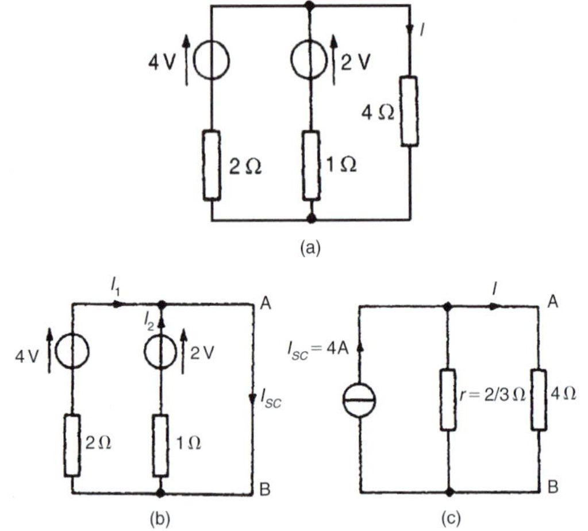

Figure 15.45

Following the procedure:

(i) The 4 Ω branch is short-circuited as shown in Figure 15.45(b).

(ii) From Figure 15.45(b), $I_{SC} = I_1 + I_2 = \frac{4}{2} + \frac{2}{1} = 4\,\text{A}$

(iii) If the sources of e.m.f. are removed the resistance 'looking-in' at a break made between A and B is given by:

$$r = \frac{2 \times 1}{2 + 1} = \frac{2}{3}\,\Omega$$

(iv) From the Norton equivalent network shown in Figure 15.45(c) the current in the 4 Ω resistance is given by:

$$I = \left[\frac{2/3}{(2/3) + 4}\right](4) = \mathbf{0.571\,A},$$

as obtained previously in Problems 2, 5 and 9 using Kirchhoff's laws and the theorems of superposition and Thévenin.

Problem 14. Use Norton's theorem to determine the current flowing in the 3 Ω resistance of the network shown in Figure 15.46(a). The voltage source has negligible internal resistance.

Following the procedure:

(i) The branch containing the 3 Ω resistance is short-circuited as shown in Figure 15.46(b).

(ii) From the equivalent circuit shown in Figure 15.46(c),

$$I_{SC} = \frac{24}{5} = 4.8\,\text{A}$$

(iii) If the 24 V source of e.m.f. is removed the resistance 'looking-in' at a break made between A and B is obtained from Figure 15.46(d) and its equivalent circuit shown in Figure 15.46(e) and is given by:

$$r = \frac{10 \times 5}{10 + 5} = \frac{50}{15} = 3\frac{1}{3}\,\Omega$$

(iv) From the Norton equivalent network shown in Figure 15.46(f) the current in the 3 Ω resistance is given by:

$$I = \left[\frac{3\frac{1}{3}}{3\frac{1}{3} + 1\frac{2}{3} + 3}\right](4.8) = \mathbf{2\,A},$$

as obtained previously in Problem 10 using Thévenin's theorem.

Part 3

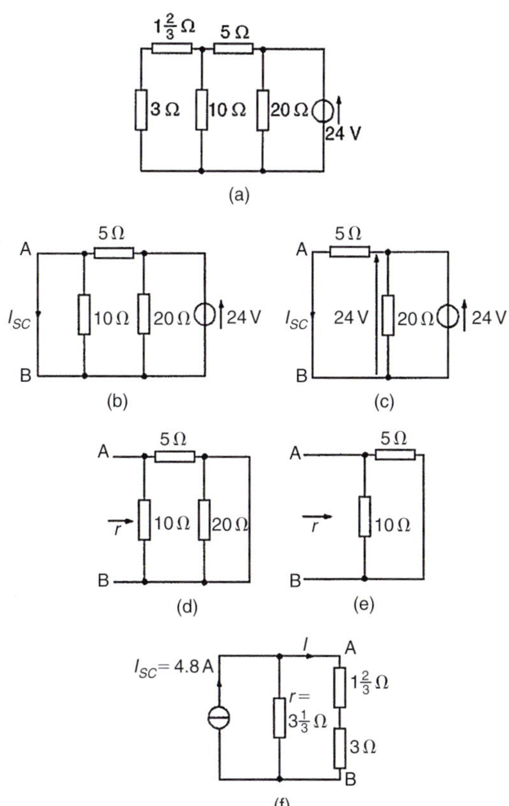

Figure 15.46

Problem 15. Determine the current flowing in the 2 Ω resistance in the network shown in Figure 15.47(a).

Following the procedure:

(i) The 2 Ω resistance branch is short-circuited as shown in Figure 15.47(b).

(ii) Figure 15.47(c) is equivalent to Figure 15.47(b).

Hence $I_{SC} = \left(\frac{6}{6+4}\right)(15) = \mathbf{9\,A}$ by current division.

(iii) If the 15 A current source is replaced by an open-circuit then from Figure 15.47(d) the resistance 'looking-in' at a break made between A and B is given by (6+4) Ω in parallel with (8+7) Ω, i.e.

$$r = \frac{(10)(15)}{10+15} = \frac{150}{25} = 6\,\Omega$$

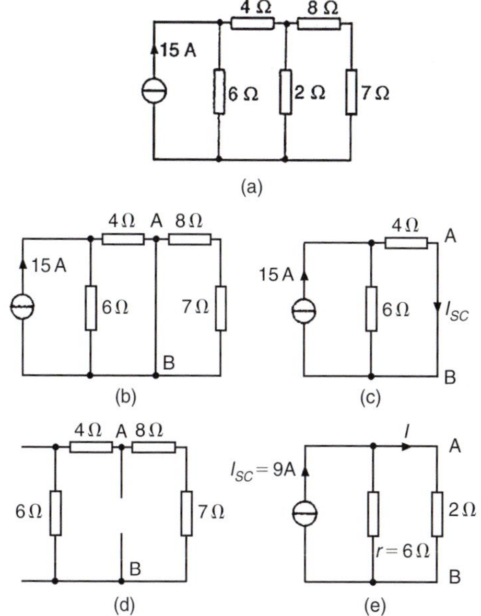

Figure 15.47

(iv) From the Norton equivalent network shown in Figure 15.47(e) the current in the 2 Ω resistance is given by:

$$I = \left(\frac{6}{6+2}\right)(9) = \mathbf{6.75\,A}$$

Now try the following Practice Exercise

Practice Exercise 70 Norton's theorem (Answers on page 821)

1. Repeat Problems 1 to 4 of Exercise 68 on page 226 using Norton's theorem.
2. Repeat Problems 1, 2, 4 and 5 of Exercise 69 on page 232 using Norton's theorem.
3. Determine the current flowing in the 6 Ω resistance of the network shown in Figure 15.48 by using Norton's theorem.

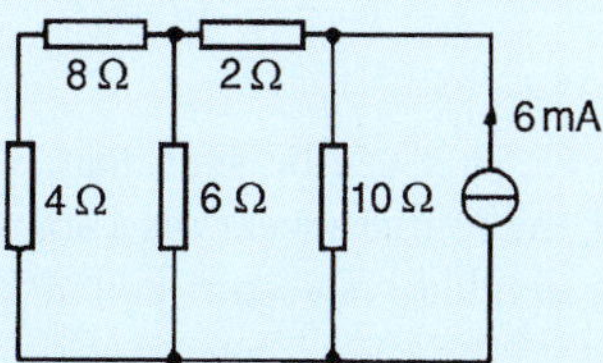

Figure 15.48

15.8 Thévenin and Norton equivalent networks

The Thévenin and Norton networks shown in Figure 15.49 are equivalent to each other. The resistance 'looking-in' at terminals AB is the same in each of the networks, i.e. r

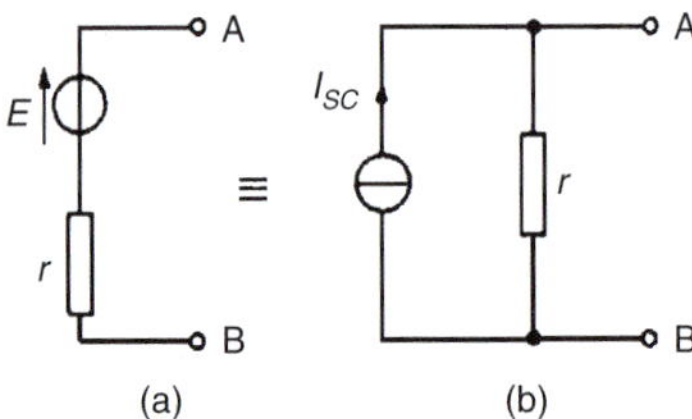

Figure 15.49

If terminals AB in Figure 15.49(a) are short-circuited, the short-circuit current is given by E/r. If terminals AB in Figure 15.49(b) are short-circuited, the short-circuit current is I_{SC}. For the circuit shown in Figure 15.49(a) to be equivalent to the circuit in Figure 15.49(b) the same short-circuit current must flow. Thus $I_{SC}=E/r$

Figure 15.50 shows a source of e.m.f. E in series with a resistance r feeding a load resistance R

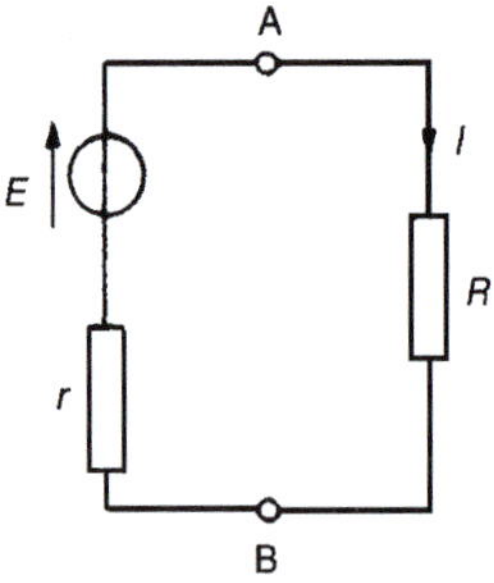

Figure 15.50

From Figure 15.50, $I=\dfrac{E}{r+R}=\dfrac{E/r}{(r+R)/r}=\left(\dfrac{r}{r+R}\right)\dfrac{E}{r}$

i.e. $$I=\left(\frac{r}{r+R}\right)I_{SC}$$

From Figure 15.51 it can be seen that, when viewed from the load, the source appears as a source of current I_{SC} which is divided between r and R connected in parallel.

Thus the two representations shown in Figure 15.49 are equivalent.

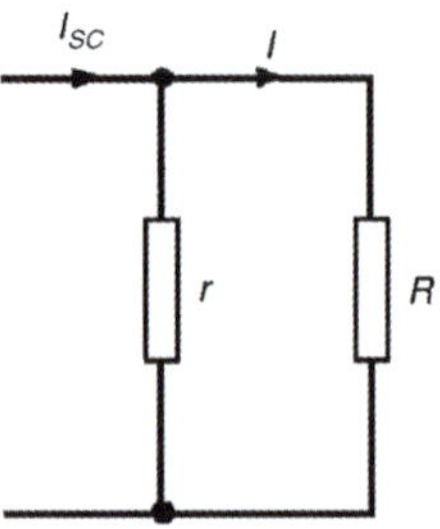

Figure 15.51

Problem 16. Convert the circuit shown in Figure 15.52 to an equivalent Norton network.

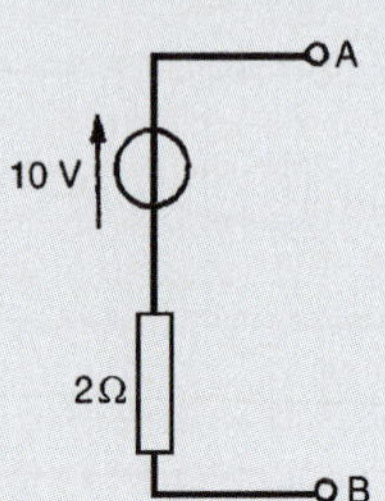

Figure 15.52

If terminals AB in Figure 15.52 are short-circuited, the short-circuit current $I_{SC}=\frac{10}{2}=5\,\text{A}$.

The resistance 'looking-in' at terminals AB is $2\,\Omega$. Hence the equivalent Norton network is as shown in Figure 15.53.

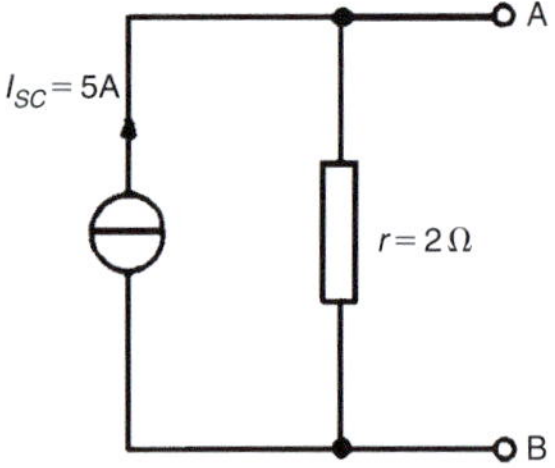

Figure 15.53

Problem 17. Convert the network shown in Figure 15.54 to an equivalent Thévenin circuit.

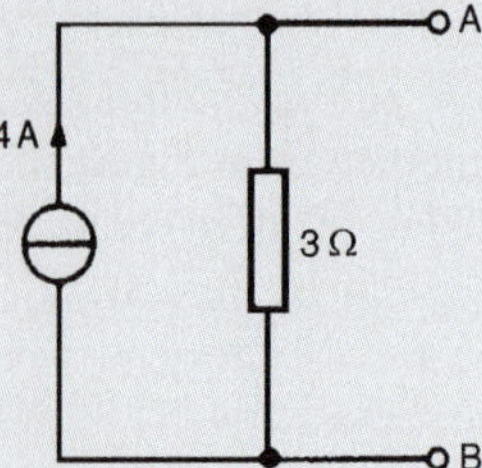

Figure 15.54

The open-circuit voltage E across terminals AB in Figure 15.54 is given by:

$E = (I_{SC})(r) = (4)(3) = 12\,V$.

The resistance 'looking-in' at terminals AB is $3\,\Omega$. Hence the equivalent Thévenin circuit is as shown in Figure 15.55.

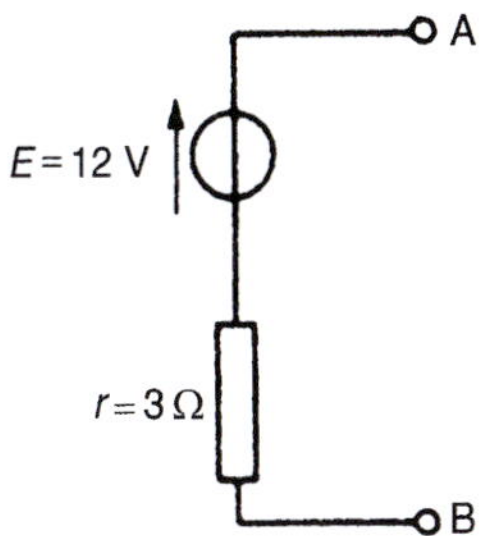

Figure 15.55

Problem 18. (a) Convert the circuit to the left of terminals AB in Figure 15.56(a) to an equivalent Thévenin circuit by initially converting to a Norton equivalent circuit. (b) Determine the current flowing in the 1.8 Ω resistor.

(a) For the branch containing the 12 V source, converting to a Norton equivalent circuit gives $I_{SC} = 12/3 = 4\,A$ and $r_1 = 3\,\Omega$. For the branch containing the 24 V source, converting to a Norton equivalent circuit gives $I_{SC2} = 24/2 = 12\,A$ and $r_2 = 2\,\Omega$.

Thus Figure 15.56(b) shows a network equivalent to Figure 15.56(a).

From Figure 15.56(b) the total short-circuit current is $4 + 12 = 16\,A$ and the total resistance is given by:

$$\frac{3 \times 2}{3 + 2} = 1.2\,\Omega$$

Thus Figure 15.56(b) simplifies to Figure 15.56(c).

The open-circuit voltage across AB of Figure 15.56(c), $E = (16)(1.2) = 19.2\,V$, and the resistance 'looking-in' at AB is $1.2\,\Omega$. Hence the Thévenin equivalent circuit is as shown in Figure 15.56(d).

(b) When the 1.8 Ω resistance is connected between terminals A and B of Figure 15.56(d) the current I flowing is given by:

$$I = \frac{19.2}{1.2 + 1.8} = \mathbf{6.4\,A}$$

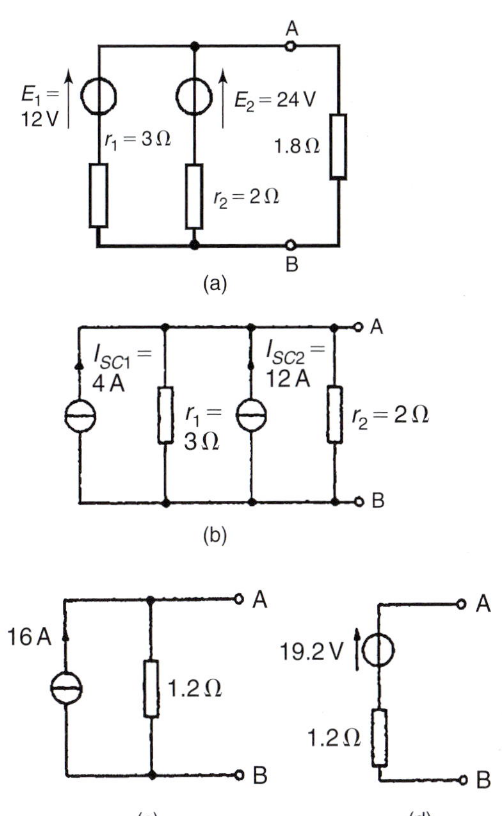

Figure 15.56

Problem 19. Determine by successive conversions between Thévenin and Norton equivalent networks a Thévenin equivalent circuit for terminals AB of Figure 15.57(a). Hence determine the current flowing in the 200 Ω resistance.

For the branch containing the 10 V source, converting to a Norton equivalent network gives

$$I_{SC} = \frac{10}{2000} = 5\,mA \text{ and } r_1 = 2\,k\Omega$$

For the branch containing the 6 V source, converting to a Norton equivalent network gives

$$I_{SC} = \frac{6}{3000} = 2\,mA \text{ and } r_2 = 3\,k\Omega$$

Thus the network of Figure 15.57(a) converts to Figure 15.57(b).

Combining the 5 mA and 2 mA current sources gives the equivalent network of Figure 15.57(c), where the short-circuit current for the original two branches considered is 7 mA and the resistance is

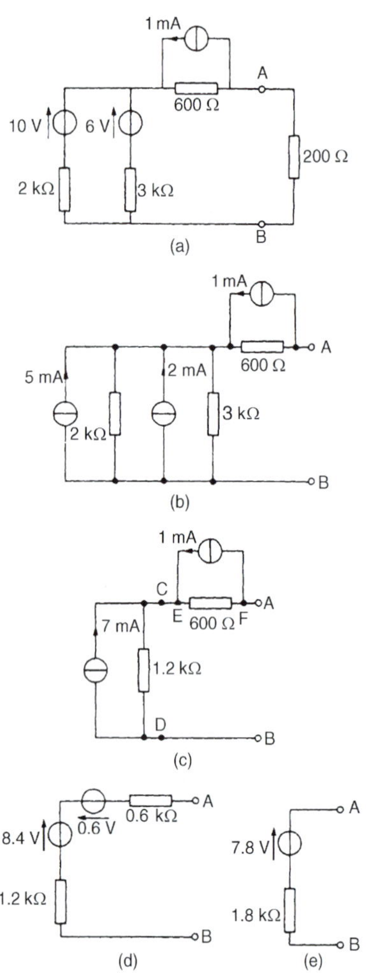

Figure 15.57

$$\frac{2\times 3}{2+3}=1.2\,\text{k}\Omega$$

Both of the Norton equivalent networks shown in Figure 15.57(c) may be converted to Thévenin equivalent circuits. The open-circuit voltage across CD is: $(7\times 10^{-3})(1.2\times 10^{3})=8.4\,\text{V}$ and the resistance 'looking-in' at CD is $1.2\,\text{k}\Omega$.
The open-circuit voltage across EF is $(1\times 10^{-3})(600)=0.6\,\text{V}$ and the resistance 'looking-in' at EF is $0.6\,\text{k}\Omega$. Thus Figure 15.57(c) converts to Figure 15.57(d). Combining the two Thévenin circuits gives

$$E=8.4-0.6=\mathbf{7.8\,V}\text{ and the resistance}$$
$$r=(1.2+0.6)\,\text{k}\Omega=\mathbf{1.8\,k\Omega}$$

Thus the Thévenin equivalent circuit for terminals AB of Figure 15.57(a) is as shown in Figure 15.57(e).

Hence the current I flowing in a $200\,\Omega$ resistance connected between A and B is given by:

$$I=\frac{7.8}{1800+200}=\frac{7.8}{2000}=\mathbf{3.9\,mA}$$

Now try the following Practice Exercise

Practice Exercise 71 Thévenin and Norton equivalent networks (Answers on page 821)

1. Convert the circuits shown in Figure 15.58 to Norton equivalent networks.

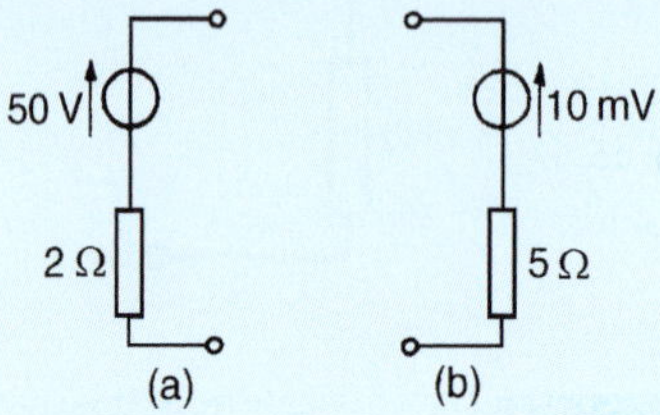

Figure 15.58

2. Convert the networks shown in Figure 15.59 to Thévenin equivalent circuits.

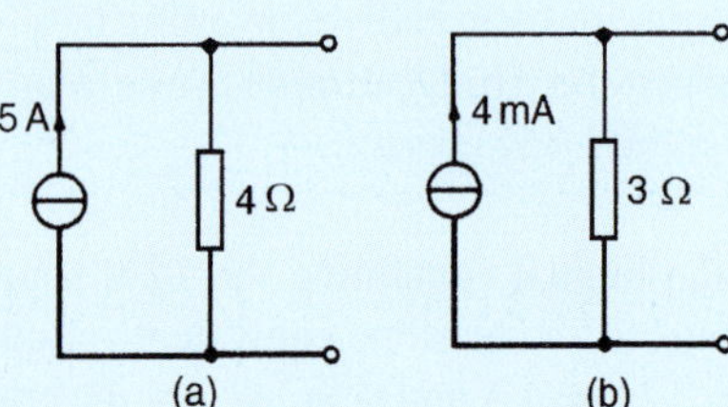

Figure 15.59

3. (a) Convert the network to the left of terminals AB in Figure 15.60 to an equivalent Thévenin circuit by initially converting to a Norton equivalent network.
   (b) Determine the current flowing in the $1.8\,\Omega$ resistance connected between A and B in Figure 15.60.

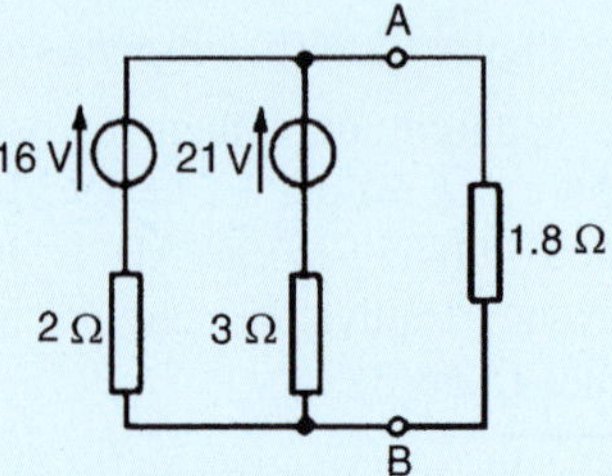

Figure 15.60

4. Determine, by successive conversions between Thévenin and Norton equivalent networks, a

Thévenin equivalent circuit for terminals AB in Figure 15.61. Hence determine the current flowing in a 6 Ω resistor connected between A and B.

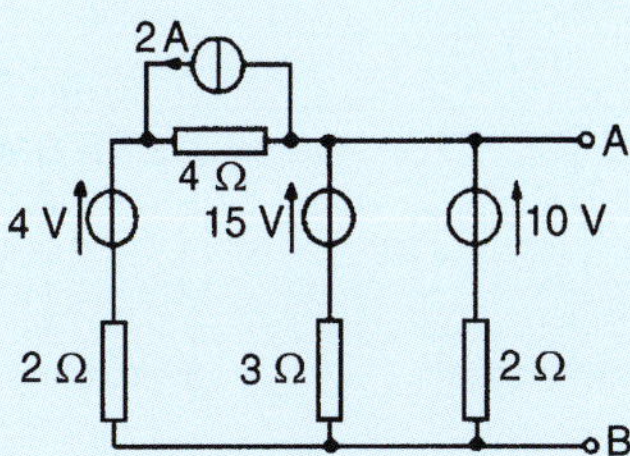

Figure 15.61

15.9 Maximum power transfer theorem

The **maximum power transfer theorem** states:

'The power transferred from a supply source to a load is at its maximum when the resistance of the load is equal to the internal resistance of the source.'

Hence, in Figure 15.62, when $R=r$ the power transferred from the source to the load is a maximum.

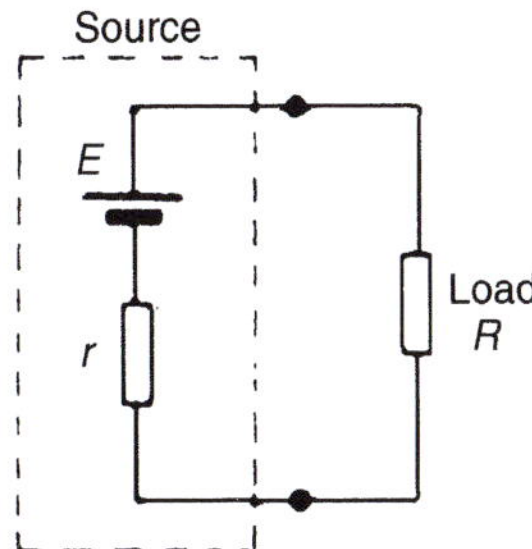

Figure 15.62

Typical practical applications of the maximum power transfer theorem are found in stereo amplifier design, seeking to maximize power delivered to speakers, and in electric vehicle design, seeking to maximize power delivered to drive a motor.

Problem 20. The circuit diagram of Figure 15.63 shows dry cells of source e.m.f. 6 V, and internal resistance 2.5 Ω. If the load resistance R_L is varied from 0 to 5 Ω in 0.5 Ω steps, calculate the power dissipated by the load in each case. Plot a graph of R_L (horizontally) against power (vertically) and determine the maximum power dissipated.

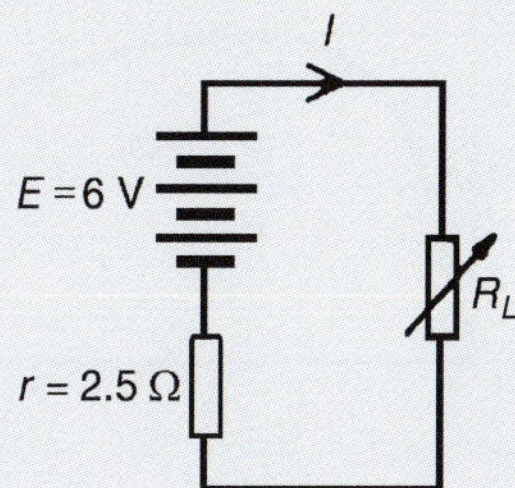

Figure 15.63

When $R_L=0$, current $I=\frac{E}{r+R_L}=\frac{6}{2.5}=2.4\,\text{A}$ and power dissipated in R_L, $P=I^2R_L$

i.e. $$P=(2.4)^2\,(0)=0\,\text{W}$$

When $R_L=0.5\,\Omega$, current $I=\frac{E}{r+R_L}=\frac{6}{2.5+0.5}=2\,\text{A}$

and $P=I^2R_L=(2)^2\,(0.5)=2\,\text{W}$

When $R_L=1.0\,\Omega$, current $I=\frac{6}{2.5+1.0}=1.714\,\text{A}$

and $P=(1.714)^2\,(1.0)=2.94\,\text{W}$

With similar calculations the following table is produced:

$R_L(\Omega)$	$I=\frac{E}{r+R_L}$	$P=I^2R_L(W)$
0	2.4	0
0.5	2.0	2.00
1.0	1.714	2.94
1.5	1.5	3.38
2.0	1.333	3.56
2.5	1.2	3.60
3.0	1.091	3.57
3.5	1.0	3.50
4.0	0.923	3.41
4.5	0.857	3.31
5.0	0.8	3.20

A graph of R_L against P is shown in Figure 15.64. **The maximum value of power is 3.60 W**, which occurs when R_L is 2.5 Ω, i.e. **maximum power occurs when $R_L = r$**, which is what the maximum power transfer theorem states.

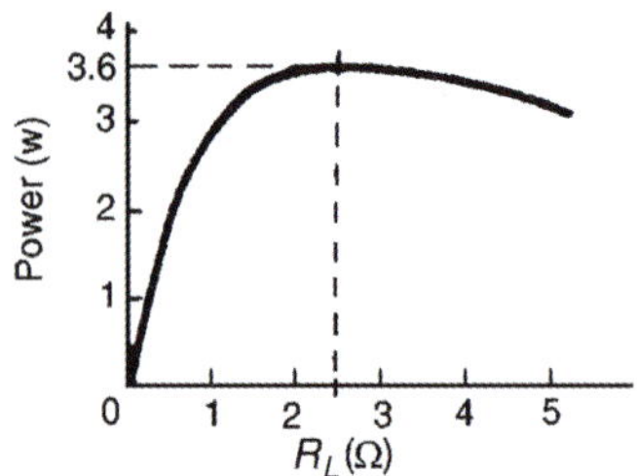

Figure 15.64

Problem 21. A d.c. source has an open-circuit voltage of 30 V and an internal resistance of 1.5 Ω. State the value of load resistance that gives maximum power dissipation and determine the value of this power.

The circuit diagram is shown in Figure 15.65. From the maximum power transfer theorem, for maximum power dissipation,

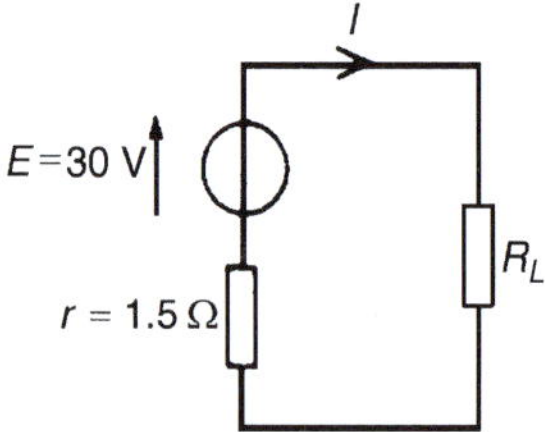

Figure 15.65

$$\boldsymbol{R_L = r = 1.5\,\Omega}$$

From Figure 15.65, current

$$I = \frac{E}{r + R_L} = \frac{30}{1.5 + 1.5} = 10\,\text{A}$$

Power $P = I^2 R_L = (10)^2(1.5) = \mathbf{150\,W = maximum}$ **power dissipated**.

Problem 22. Find the value of the load resistor R_L shown in Figure 15.66(a) that gives maximum power dissipation and determine the value of this power.

Using the procedure for Thévenin's theorem:

(i) Resistance R_L is removed from the circuit as shown in Figure 15.66(b).

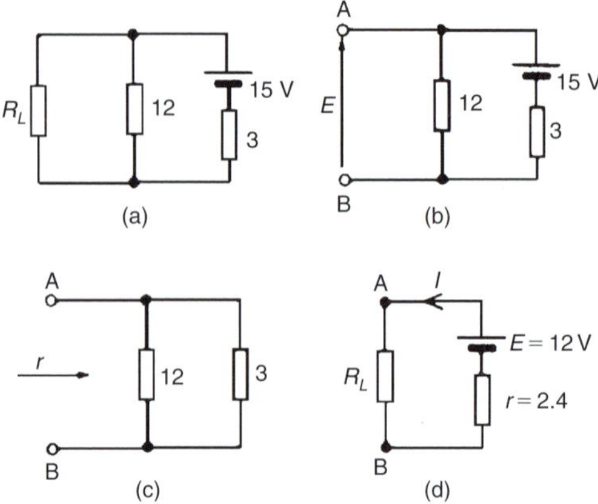

Figure 15.66

(ii) The p.d. across AB is the same as the p.d. across the 12 Ω resistor.

Hence $E = \left(\frac{12}{12+3}\right)(15) = 12\,\text{V}$

(iii) Removing the source of e.m.f. gives the circuit of Figure 15.66(c),

from which resistance, $r = \frac{12 \times 3}{12+3} = \frac{36}{15} = 2.4\,\Omega$

(iv) The equivalent Thévenin's circuit supplying terminals AB is shown in Figure 15.66(d), from which, current, $I = E/(r + R_L)$

For maximum power, $\boldsymbol{R_L = r = 2.4\,\Omega}$.

Thus current, $I = \frac{12}{2.4 + 2.4} = 2.5\,\text{A}$.

Power, P, dissipated in load R_L,
$P = I^2 R_L = (2.5)^2\,(2.4) = \mathbf{15\,W}$

Now try the following Practice Exercise

Practice Exercise 72 Maximum power transfer theorem (Answers on page 821)

1. A d.c. source has an open-circuit voltage of 20 V and an internal resistance of 2 Ω. Determine the value of the load resistance that gives maximum power dissipation. Find the value of this power.
2. A d.c. source having an open-circuit voltage of 42 V and an internal resistance of 3 Ω is

connected to a load of resistance R_L. Determine the maximum power dissipated by the load.

3. A voltage source comprising six 2 V cells, each having an internal resistance of 0.2 Ω, is connected to a load resistance R. Determine the maximum power transferred to the load.

4. The maximum power dissipated in a 4 Ω load is 100 W when connected to a d.c. voltage V and internal resistance r. Calculate (a) the current in the load, (b) internal resistance r and (c) voltage V.

5. Determine the value of the load resistance R_L shown in Figure 15.67 that gives maximum power dissipation and find the value of the power.

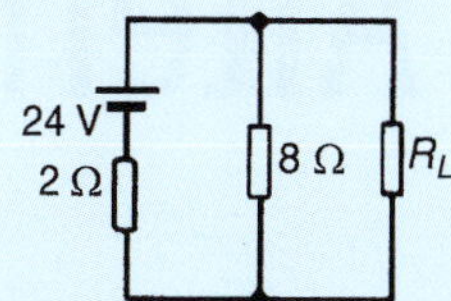

Figure 15.67

For fully worked solutions to each of the problems in Practice Exercises 67 to 72 in this chapter, go to the website:
www.routledge.com/cw/bird

Chapter 16

Alternating voltages and currents

***Why it is important to understand:* Alternating voltages and currents**

With alternating current (a.c.), the flow of electric charge periodically reverses direction, whereas with direct current (d.c.), the flow of electric charge is only in one direction. In a power station, electricity can be made most easily by using a gas or steam turbine or water impeller to drive a generator consisting of a spinning magnet inside a set of coils. The resultant voltage is always 'alternating' by virtue of the magnet's rotation. Alternating voltage can be carried around the country via cables far more effectively than direct current because a.c. can be passed through a transformer and a high voltage can be reduced to a low voltage, suitable for use in homes. The electricity arriving at your home is alternating voltage. Electric light bulbs and toasters can operate perfectly from 230 volts a.c. Other equipment such as televisions has an internal power supply which converts the 230 volts a.c. to a low d.c. voltage for the electronic circuits. How is this done? There are several ways, but the simplest is to use a transformer to reduce the voltage to, say, 12 volts a.c. This lower voltage can be fed through a 'rectifier' which combines the negative and positive alternating cycles so that only positive cycles emerge. A.c. is the form in which electric power is delivered to businesses and residences. The usual waveform of an a.c. power circuit is a sine wave. In certain applications, different waveforms are used, such as triangular or square waves. Audio and radio signals carried on electrical wires are also examples of alternating current. The frequency of the electrical system varies by country; most electric power is generated at either 50 or 60 hertz. Some countries have a mixture of 50 Hz and 60 Hz supplies, notably Japan. A low frequency eases the design of electric motors, particularly for hoisting, crushing and rolling applications, and commutator-type traction motors for applications such as railways. However, low frequency also causes noticeable flicker in arc lamps and incandescent light bulbs. The use of lower frequencies also provides the advantage of lower impedance losses, which are proportional to frequency. 16.7 Hz frequency is still used in some European rail systems, such as in Austria, Germany, Norway, Sweden and Switzerland. Off-shore, military, textile industry, marine, computer mainframe, aircraft, and spacecraft applications sometimes use 400 Hz, for benefits of reduced weight of apparatus or higher motor speeds. This chapter introduces alternating current and voltages, with its terminology and values and its sinusoidal expression. Also, the addition of two sine waves is explained, as are rectifiers and their smoothing.

Electrical Circuit Theory and Technology. 978-1-138-67349-6, © 2017 John Bird. Published by Taylor & Francis. All rights reserved.

At the end of this chapter you should be able to:

- appreciate why a.c. is used in preference to d.c.
- describe the principle of operation of an a.c. generator
- distinguish between unidirectional and alternating waveforms
- define cycle, period or periodic time T and frequency f of a waveform
- perform calculations involving $T = \dfrac{1}{f}$
- define instantaneous, peak, mean and r.m.s. values, and form and peak factors for a sine wave
- calculate mean and r.m.s. values and form and peak factors for given waveforms
- understand and perform calculations on the general sinusoidal equation $v = V_m \sin(\omega t \pm \phi)$
- understand lagging and leading angles
- combine two sinusoidal waveforms (a) by plotting graphically, (b) by drawing phasors to scale and (c) by calculation
- understand rectification, and describe methods of obtaining half-wave and full-wave rectification
- appreciate methods of smoothing a rectified output waveform

16.1 Introduction

Electricity is produced by generators at power stations and then distributed by a vast network of transmission lines (called the National Grid system) to industry and for domestic use. It is easier and cheaper to generate alternating current (a.c.) than direct current (d.c.) and a.c. is more conveniently distributed than d.c. since its voltage can be readily altered using transformers. Whenever d.c. is needed in preference to a.c., devices called rectifiers are used for conversion (see Section 16.7).

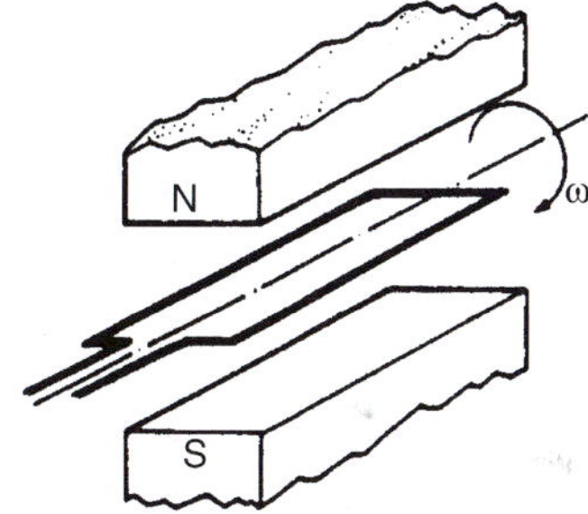

Figure 16.1

16.2 The a.c. generator

Let a single turn coil be free to rotate at constant angular velocity symmetrically between the poles of a magnet system, as shown in Figure 16.1.

An e.m.f. is generated in the coil (from Faraday's laws) which varies in magnitude and reverses its direction at regular intervals. The reason for this is shown in Figure 16.2. In positions (a), (e) and (i) the conductors of the loop are effectively moving along the magnetic field, no flux is cut and hence no e.m.f. is induced. In position (c) maximum flux is cut and hence maximum e.m.f. is induced. In position (g), maximum flux is cut and hence maximum e.m.f. is again induced. However, using Fleming's right-hand rule, the induced e.m.f. is

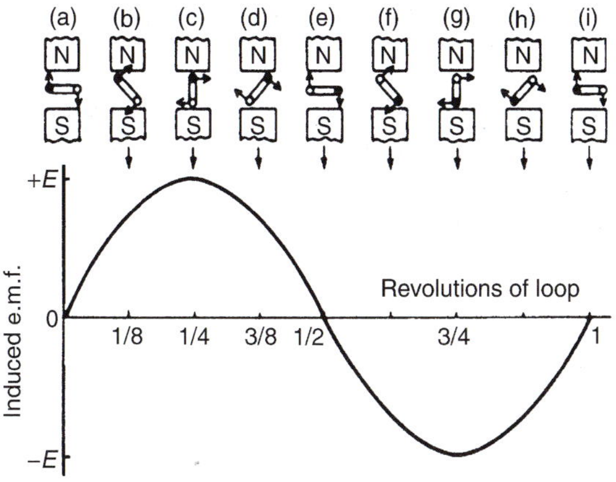

Figure 16.2

in the opposite direction to that in position (c) and is thus shown as $-E$. In positions (b), (d), (f) and (h) some flux is cut and hence some e.m.f. is induced. If all such positions of the coil are considered, in one revolution of the coil one cycle of alternating e.m.f. is produced as shown. This is the principle of operation of the a.c. generator (i.e. the alternator).

16.3 Waveforms

If values of quantities which vary with time t are plotted to a base of time, the resulting graph is called a **waveform**. Some typical waveforms are shown in Figure 16.3. Waveforms (a) and (b) are **unidirectional waveforms**, for, although they vary considerably with time, they flow in one direction only (i.e. they do not cross the time axis and become negative). Waveforms (c) to (g) are called **alternating waveforms** since their quantities are continually changing in direction (i.e. alternately positive and negative).

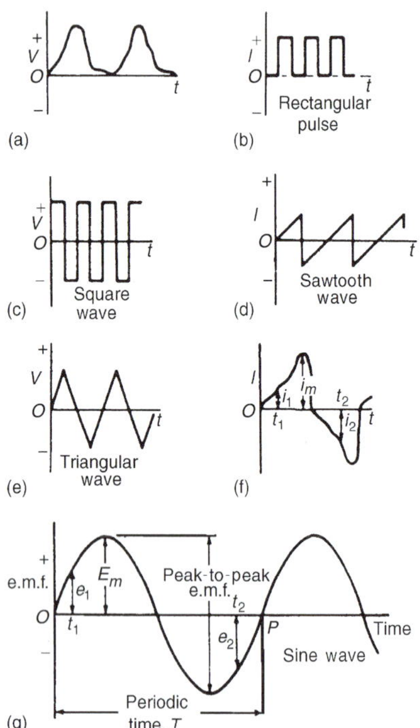

Figure 16.3

A waveform of the type shown in Figure 16.3(g) is called a **sine wave**. It is the shape of the waveform of e.m.f. produced by an alternator and thus the mains electricity supply is of 'sinusoidal' form.

One complete series of values is called a **cycle** (i.e. from O to P in Figure 16.3(g)).

The time taken for an alternating quantity to complete one cycle is called the **period** or the **periodic time,** T, of the waveform.

The number of cycles completed in one second is called the **frequency,** f, of the supply and is measured in **hertz,*** **Hz**. The standard frequency of the electricity supply in Great Britain is 50 Hz.

$$T=\frac{1}{f} \text{ or } f=\frac{1}{T}$$

Problem 1. Determine the periodic time for frequencies of (a) 50 Hz and (b) 20 kHz.

(a) Periodic time $T=\frac{1}{f}=\frac{1}{50}=\mathbf{0.02\,s}$ or $\mathbf{20\,ms}$

(b) Periodic time $T=\frac{1}{f}=\frac{1}{20\,000}=\mathbf{0.00005\,s}$ or $\mathbf{50\,\mu s}$

Problem 2. Determine the frequencies for periodic times of (a) 4 ms, (b) 4 μs.

(a) Frequency $f=\frac{1}{T}=\frac{1}{4\times10^{-3}}=\frac{1000}{4}=\mathbf{250\,Hz}$

*Who was **Hertz**? Heinrich Rudolf Hertz (22 February 1857–1 January 1894) was the first person to conclusively prove the existence of electromagnetic waves. To find out more go to **www.routledge.com/cw/bird**

Part 3

(b) Frequency $f=\frac{1}{T}=\frac{1}{4\times10^{-6}}=\frac{1\,000\,000}{4}$

$=\mathbf{250\,000\,Hz}$ or **250 kHz** or **0.25 MHz**

Problem 3. An alternating current completes 5 cycles in 8 ms. What is its frequency?

Time for 1 cycle $=\frac{8}{5}$ ms $=1.6$ ms $=$ periodic time T

Frequency $f=\frac{1}{T}=\frac{1}{1.6\times10^{-3}}=\frac{1000}{1.6}=\frac{10\,000}{16}$

$=\mathbf{625\,Hz}$

Now try the following Practice Exercise

Practice Exercise 73 Frequency and periodic time (Answers on page 822)

1. Determine the periodic time for the following frequencies:
(a) 2.5 Hz (b) 100 Hz (c) 40 kHz
2. Calculate the frequency for the following periodic times:
(a) 5 ms (b) 50 μs (c) 0.2 s
3. An alternating current completes 4 cycles in 5 ms. What is its frequency?

16.4 A.c. values

Instantaneous values are the values of the alternating quantities at any instant of time. They are represented by small letters, i, v, e, etc. (see Figures 16.3(f) and (g)).

The largest value reached in a half cycle is called the **peak value** or the **maximum value** or the **amplitude** of the waveform. Such values are represented by V_m, I_m, etc. (see Figures 16.3(f) and (g)). A **peak-to-peak** value of e.m.f. is shown in Figure 16.3(g) and is the difference between the maximum and minimum values in a cycle.

The **average** or **mean value** of a symmetrical alternating quantity (such as a sine wave) is the average value measured over a half cycle (since over a complete cycle the average value is zero).

$$\textbf{Average or mean value} = \frac{\textbf{area under the curve}}{\textbf{length of base}}$$

The area under the curve is found by approximate methods such as the trapezoidal rule, the mid-ordinate rule or Simpson's rule. Average values are represented by V_{AV}, I_{AV}, etc.

For a sine wave,

average value = 0.637 × maximum value

(i.e. 2/π × maximum value)

The **effective value** of an alternating current is that current which will produce the same heating effect as an equivalent direct current. The effective value is called the **root mean square (r.m.s.) value** and whenever an alternating quantity is given, it is assumed to be the r.m.s. value. For example, the domestic mains supply in Great Britain is 240 V and is assumed to mean '240 V r.m.s'. The symbols used for r.m.s. values are I, V, E, etc. For a non-sinusoidal waveform as shown in Figure 16.4 the r.m.s. value is given by:

$$I=\sqrt{\left(\frac{i_1^2+i_2^2+\cdots+i_n^2}{n}\right)}$$

where n is the number of intervals used.

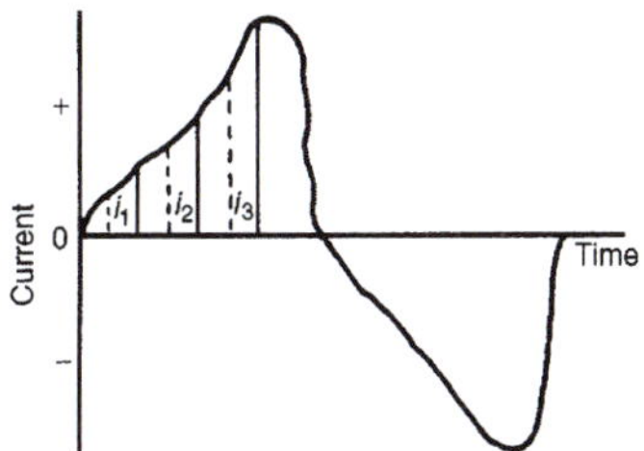

Figure 16.4

For a sine wave,

r.m.s. value = 0.707 × maximum value

(i.e. $1/\sqrt{2}$ × maximum value)

$$\textbf{Form factor} = \frac{\textbf{r.m.s. value}}{\textbf{average value}}$$ For a sine wave, form factor $=1.11$

$$\textbf{Peak factor} = \frac{\textbf{maximum value}}{\textbf{r.m.s. value}}$$ For a sine wave, peak factor $=1.41$

The values of form and peak factors give an indication of the shape of waveforms.

Problem 4. For the periodic waveforms shown in Figure 16.5 determine for each: (i) frequency (ii) average value over half a cycle (iii) r.m.s. value (iv) form factor and (v) peak factor.

Part 3

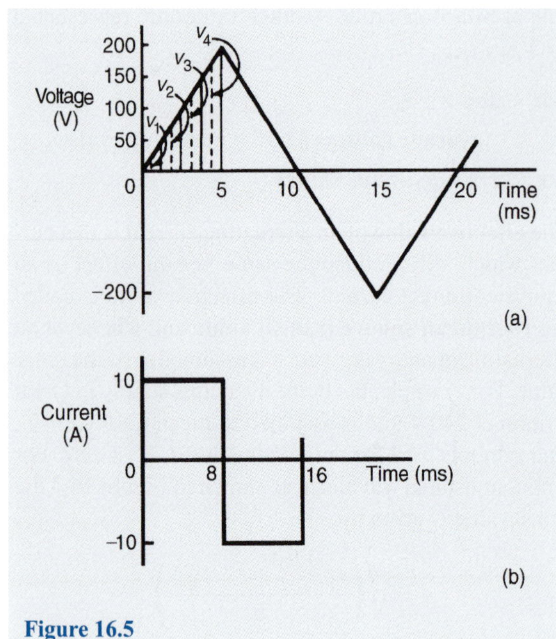

Figure 16.5

(a) **Triangular waveform (Figure 16.5(a))**

(i) Time for 1 complete cycle = 20 ms = periodic time, T

Hence frequency $f = \frac{1}{T} = \frac{1}{20 \times 10^{-3}} = \frac{1000}{20}$

$= \mathbf{50\,Hz}$

(ii) Area under the triangular waveform for a half cycle

$= \frac{1}{2} \times \text{base} \times \text{height} = \frac{1}{2} \times (10 \times 10^{-3}) \times 200$

$= 1$ volt second

Average value of waveform

$= \frac{\text{area under curve}}{\text{length of base}} = \frac{1\text{ volt second}}{10 \times 10^{-3}\text{ second}}$

$= \frac{1000}{10} = \mathbf{100\,V}$

(iii) In Figure 16.5(a), the first 1/4 cycle is divided into 4 intervals.

Thus r.m.s. value

$$= \sqrt{\left(\frac{v_1^2 + v_2^2 + v_3^2 + v_4^2}{4}\right)}$$

$$= \sqrt{\left(\frac{25^2 + 75^2 + 125^2 + 175^2}{4}\right)}$$

$= \mathbf{114.6\,V}$

(Note that the greater the number of intervals chosen, the greater the accuracy of the result. For example, if twice the number of ordinates as that chosen above are used, the r.m.s. value is found to be 115.6 V)

(iv) Form factor $= \frac{\text{r.m.s. value}}{\text{average value}} = \frac{114.6}{100} = \mathbf{1.15}$

(v) Peak factor $= \frac{\text{maximum value}}{\text{r.m.s. value}} = \frac{200}{114.6}$

$= \mathbf{1.75}$

(b) **Rectangular waveform (Figure 16.5(b))**

(i) Time for 1 complete cycle = 16 ms = periodic time, T

Hence frequency, $f = \frac{1}{T} = \frac{1}{16 \times 10^{-3}}$

$= \frac{1000}{16} = \mathbf{62.5\,Hz}$

(ii) Average value over half a cycle

$= \frac{\text{area under curve}}{\text{length of base}}$

$= \frac{10 \times (8 \times 10^{-3})}{8 \times 10^{-3}}$

$= \mathbf{10\,A}$

(iii) The r.m.s. value $= \sqrt{\left(\frac{i_1^2 + i_2^2 + \cdots + i_n^2}{n}\right)}$

$= \mathbf{10\,A}$

however many intervals are chosen, since the waveform is rectangular.

(iv) Form factor $= \frac{\text{r.m.s. value}}{\text{average value}} = \frac{10}{10} = \mathbf{1}$

(v) Peak factor $= \frac{\text{maximum value}}{\text{r.m.s. value}} = \frac{10}{10} = \mathbf{1}$

Problem 5. The following table gives the corresponding values of current and time for a half cycle of alternating current.

time t (ms)	0	0.5	1.0	1.5	2.0	2.5
current i (A)	0	7	14	23	40	56

time t (ms)	3.0	3.5	4.0	4.5	5.0
current i (A)	68	76	60	5	0

Assuming the negative half cycle is identical in shape to the positive half cycle, plot the waveform and find (a) the frequency of the supply, (b) the instantaneous values of current after 1.25 ms and 3.8 ms, (c) the peak or maximum value, (d) the mean or average value and (e) the r.m.s. value of the waveform.

The half cycle of alternating current is shown plotted in Figure 16.6

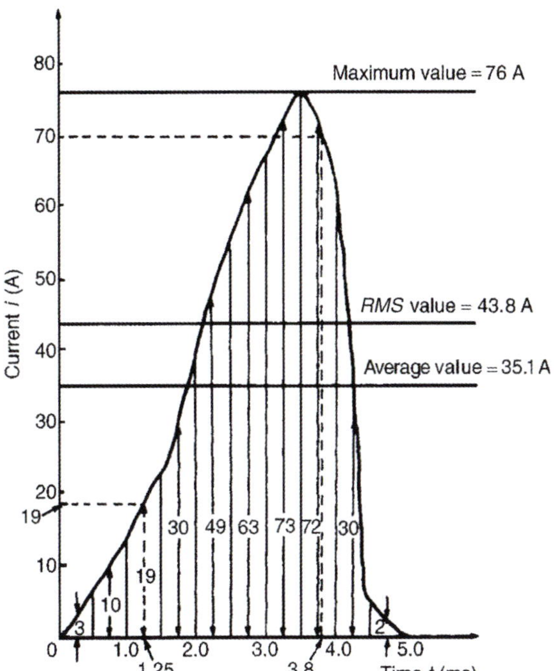

Figure 16.6

(a) Time for a half cycle = 5 ms. Hence the time for 1 cycle, i.e. the periodic time, $T = 10$ ms or 0.01 s

$$\text{Frequency, } f = \frac{1}{T} = \frac{1}{0.01} = \mathbf{100\ Hz}$$

(b) Instantaneous value of current after 1.25 ms is **19 A**, from Figure 16.6

Instantaneous value of current after 3.8 ms is **70 A**, from Figure 16.6

(c) Peak or maximum value = **76 A**

(d) Mean or average value $= \dfrac{\text{area under curve}}{\text{length of base}}$

Using the mid-ordinate rule with 10 intervals, each of width 0.5 ms gives:

area under curve

$$= (0.5 \times 10^{-3})[3 + 10 + 19 + 30 + 49 + 63 + 73 + 72 + 30 + 2] \text{ (see Figure 16.6)}$$

$$= (0.5 \times 10^{-3})(351)$$

$$\text{Hence mean or average value} = \frac{(0.5 \times 10^{-3})(351)}{5 \times 10^{-3}} = \mathbf{35.1\ A}$$

(e) r.m.s. value

$$= \sqrt{\left(\frac{3^2 + 10^2 + 19^2 + 30^2 + 49^2 + 63^2 + 73^2 + 72^2 + 30^2 + 2^2}{10}\right)}$$

$$= \sqrt{\left(\frac{19\,157}{10}\right)} = \mathbf{43.8\ A}$$

Problem 6. Calculate the r.m.s. value of a sinusoidal current of maximum value 20 A.

For a sine wave, r.m.s. value = 0.707 × maximum value

$$= 0.707 \times 20 = \mathbf{14.14\ A}$$

Problem 7. Determine the peak and mean values for a 240 V mains supply.

For a sine wave, r.m.s. value of voltage $V = 0.707 \times V_m$. A 240 V mains supply means that 240 V is the r.m.s. value, hence

$$V_m = \frac{V}{0.707} = \frac{240}{0.707} = \mathbf{339.5\ V = peak\ value}$$

Mean value $V_{AV} = 0.637\,V_m = 0.637 \times 339.5 = \mathbf{216.3\ V}$

Problem 8. A supply voltage has a mean value of 150 V. Determine its maximum value and its r.m.s. value.

For a sine wave, mean value = 0.637 × maximum value.

$$\text{Hence maximum value} = \frac{\text{mean value}}{0.637} = \frac{150}{0.637} = \mathbf{235.5\ V}$$

$$\text{r.m.s. value} = 0.707 \times \text{maximum value} = 0.707 \times 235.5 = \mathbf{166.5\ V}$$

Now try the following Practice Exercise

Practice Exercise 74 A.c. values of waveforms (Answers on page 822)

1. An alternating current varies with time over half a cycle as follows:

current (A)	0	0.7	2.0	4.2	8.4	8.2
time (ms)	0	1	2	3	4	5

current (A)	2.5	1.0	0.4	0.2	0
time (ms)	6	7	8	9	10

   The negative half cycle is similar. Plot the curve and determine: (a) the frequency, (b) the instantaneous values at 3.4 ms and 5.8 ms, (c) its mean value and (d) its r.m.s. value

2. For the waveforms shown in Figure 16.7 determine for each (i) the frequency, (ii) the average value over half a cycle, (iii) the r.m.s. value, (iv) the form factor, (v) the peak factor.

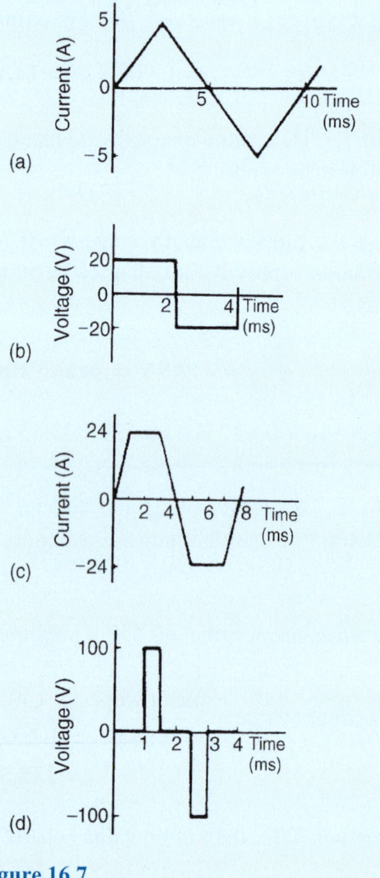

Figure 16.7

3. An alternating voltage is triangular in shape, rising at a constant rate to a maximum of 300 V in 8 ms and then falling to zero at a constant rate in 4 ms. The negative half cycle is identical in shape to the positive half cycle. Calculate (a) the mean voltage over half a cycle, and (b) the r.m.s. voltage.

4. Calculate the r.m.s. value of a sinusoidal curve of maximum value 300 V

5. Find the peak and mean values for a 200 V mains supply.

6. A sinusoidal voltage has a maximum value of 120 V. Calculate its r.m.s. and average values.

7. A sinusoidal current has a mean value of 15.0 A. Determine its maximum and r.m.s. values.

16.5 Electrical safety – insulation and fuses

Insulation is used to prevent 'leakage', and when determining what type of insulation should be used, the maximum voltage present must be taken into account. For this reason, **peak values are always considered when choosing insulation materials**.

Fuses are the weak link in a circuit and are used to break the circuit if excessive current is drawn. Excessive current could lead to a fire. Fuses rely on the heating effect of the current, and for this reason **r.m.s values must always be used when calculating the appropriate fuse size**.

16.6 The equation of a sinusoidal waveform

In Figure 16.8, OA represents a vector that is free to rotate anticlockwise about 0 at an angular velocity of ω rad/s. A rotating vector is known as a **phasor**.

After time t seconds the vector OA has turned through an angle ωt. If the line BC is constructed perpendicular to OA as shown, then

$$\sin\omega t = \frac{BC}{OB} \quad \text{i.e. } BC = OB\sin\omega t$$

Part 3

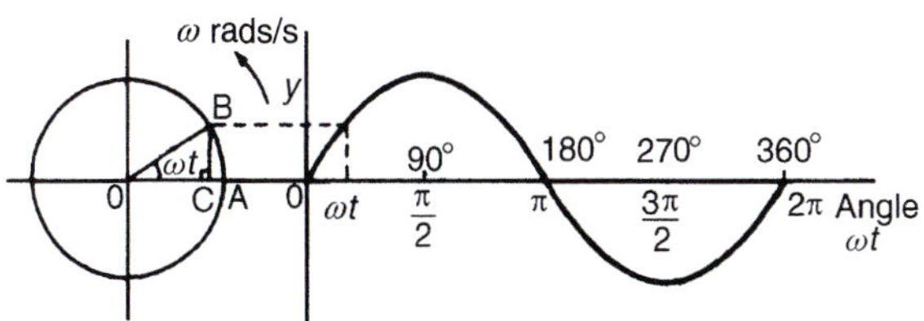

Figure 16.8

If all such vertical components are projected on to a graph of y against angle ωt (in radians), a sine curve results of maximum value OA. Any quantity which varies sinusoidally can thus be represented as a phasor.

A sine curve may not always start at $0°$. To show this a periodic function is represented by $y=\sin(\omega t \pm \phi)$, where ϕ is the phase (or angle) difference compared with $y=\sin\omega t$. In Figure 16.9(a), $y_2=\sin(\omega t+\phi)$ starts ϕ radians earlier than $y_1=\sin\omega t$ and is thus said to **lead** y_1 by ϕ radians. Phasors y_1 and y_2 are shown in Figure 16.9(b) at the time when $t=0$

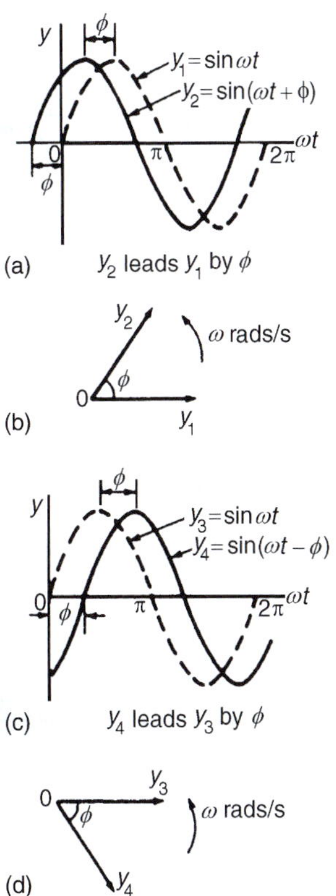

Figure 16.9

In Figure 16.9(c), $y_4=\sin(\omega t-\phi)$ starts ϕ radians later than $y_3=\sin\omega t$ and is thus said to **lag** y_3 by ϕ radians. Phasors y_3 and y_4 are shown in Figure 16.9(d) at the time when $t=0$.

Given the general sinusoidal voltage, $v=V_m\sin(\omega t \pm \phi)$, then

(i) Amplitude or maximum value $=V_m$

(ii) Peak-to-peak value $=2\,V_m$

(iii) Angular velocity $=\omega$ rad/s

(iv) Periodic time, $T=2\pi/\omega$ seconds

(v) Frequency, $f=\omega/2\pi$ Hz (since $\omega=2\pi f$)

(vi) $\phi=$ angle of lag or lead (compared with $v=V_m\sin\omega t$)

Problem 9. An alternating voltage is given by $v=282.8\sin 314t$ volts. Find (a) the r.m.s. voltage, (b) the frequency and (c) the instantaneous value of voltage when $t=4$ ms.

(a) The general expression for an alternating voltage is $v=V_m\sin(\omega t \pm \phi)$

Comparing $v=282.8\sin 314t$ with this general expression gives the peak voltage as 282.8 V

Hence the r.m.s. voltage $=0.707\times$ maximum value

$=0.707\times 282.8=$ **200 V**

(b) Angular velocity, $\omega=314$ rad/s, i.e. $2\pi f=314$

Hence frequency, $f=\dfrac{314}{2\pi}=$ **50 Hz**

(c) When $t=4$ ms, $v=282.8\sin(314\times 4\times 10^{-3})$

$=282.8\sin(1.256)=$ **268.9 V**

(Note that 1.256 radians $=\left[1.256\times\dfrac{180}{\pi}\right]^\circ$

$=71.96°$

Hence $v=282.8\sin 71.96°=$ **268.9 V**)

Problem 10. An alternating voltage is given by

$v=75\sin(200\pi t-0.25)$ volts.

Find (a) the amplitude, (b) the peak-to-peak value, (c) the r.m.s. value, (d) the periodic time, (e) the frequency and (f) the phase angle (in degrees and minutes) relative to $75\sin 200\pi t$.

Comparing $v=75\sin(200\pi t-0.25)$ with the general expression $v=V_m\sin(\omega t \pm \phi)$ gives:

(a) Amplitude, or peak value $=$ **75 V**

(b) Peak-to-peak value $= 2 \times 75 = \mathbf{150\,V}$

(c) The r.m.s. value $= 0.707 \times$ maximum value $= 0.707 \times 75 = \mathbf{53\,V}$

(d) Angular velocity, $\omega = 200\pi$ rad/s

Hence periodic time, $T = \dfrac{2\pi}{\omega} = \dfrac{2\pi}{200\pi} = \dfrac{1}{100} = \mathbf{0.01\,s\ or\ 10\,ms}$

(e) Frequency, $f = \dfrac{1}{T} = \dfrac{1}{0.01} = \mathbf{100\,Hz}$

(f) Phase angle, $\phi = 0.25$ radians lagging $75\sin 200\pi t$

$0.25 \text{ rads} = \left(0.25 \times \dfrac{180}{\pi}\right)^\circ = 14.32^\circ$

Hence phase angle $= \mathbf{14.32^\circ}$ **lagging**

Problem 11. An alternating voltage, v, has a periodic time of 0.01 s and a peak value of 40 V. When time t is zero, $v = -20$ V. Express the instantaneous voltage in the form $v = V_m \sin(\omega t \pm \phi)$

Amplitude, $V_m = 40$ V

Periodic time, $T = \dfrac{2\pi}{\omega}$ hence angular velocity,

$$\omega = \frac{2\pi}{T} = \frac{2\pi}{0.01} = 200\pi \text{ rad/s}$$

$v = V_m \sin(\omega t + \phi)$ thus becomes

$$v = 40\sin(200\pi t + \phi)\text{ V}$$

When time $t = 0$, $v = -20$ V

i.e. $-20 = 40\sin\phi$

so that $\sin\phi = \dfrac{-20}{40} = -0.5$

Hence $\phi = \sin^{-1}(-0.5) = -30^\circ = \left(-30 \times \dfrac{\pi}{180}\right)$ rads $= -\dfrac{\pi}{6}$ rads

Thus $\mathbf{v = 40\sin\left(200\pi t - \dfrac{\pi}{6}\right)\ V}$

Problem 12. The current in an a.c. circuit at any time t seconds is given by: $i = 120\sin(100\pi t + 0.36)$ amperes. Find:

(a) the peak value, the periodic time, the frequency and phase angle relative to $120\sin 100\pi t$

(b) the value of the current when $t = 0$

(c) the value of the current when $t = 8$ ms

(d) the time when the current first reaches 60 A

(e) the time when the current is first a maximum.

(a) Peak value $= \mathbf{120\,A}$

Periodic time, $T = \dfrac{2\pi}{\omega} = \dfrac{2\pi}{100\pi}$ (since $\omega = 100\pi$) $= \dfrac{1}{50} = \mathbf{0.02\,s}$ or $\mathbf{20\,ms}$

Frequency, $f = \dfrac{1}{T} = \dfrac{1}{0.02} = \mathbf{50\,Hz}$

Phase angle $= 0.36 \text{ rads} = \left(0.36 \times \dfrac{180}{\pi}\right)^\circ = \mathbf{20.63^\circ}$ **leading**

(b) When $t = 0$, $i = 120\sin(0 + 0.36) = 120\sin 20.63^\circ = \mathbf{49.3\,A}$

(c) When $t = 8$ ms, $i = 120\sin\left[100\pi\left(\dfrac{8}{10^3}\right) + 0.36\right]$

$= 120\sin 2.8733 = \mathbf{31.8\,A}$

(d) When $i = 60$ A, $60 = 120\sin(100\pi t + 0.36)$

thus $\dfrac{60}{120} = \sin(100\pi t + 0.36)$

so that $(100\pi t + 0.36) = \sin^{-1} 0.5 = 30^\circ = \dfrac{\pi}{6}$ rads $= 0.5236$ rads

Hence time, $t = \dfrac{0.5236 - 0.36}{100\pi} = \mathbf{0.521\,ms}$

(e) When the current is a maximum, $i = 120$ A

Thus $120 = 120\sin(100\pi t + 0.36)$

$1 = \sin(100\pi t + 0.36)$

$(100\pi t + 0.36) = \sin^{-1} 1 = 90^\circ = \dfrac{\pi}{2}$ rads $= 1.5708$ rads

Hence time, $t = \dfrac{1.5708 - 0.36}{100\pi} = \mathbf{3.85\,ms}$

For a practical laboratory experiment on the use of the CRO to measure voltage, frequency and phase, see the website.

Now try the following Practice Exercise

Practice Exercise 75 The equation of a sinusoidal waveform (Answers on page 822)

1. An alternating voltage is represented by $v = 20\sin 157.1t$ volts. Find (a) the maximum value, (b) the frequency, (c) the periodic time. (d) What is the angular velocity of the phasor representing this waveform?
2. Find the peak value, the r.m.s. value, the frequency, the periodic time and the phase angle (in degrees and minutes) of the following alternating quantities:
   (a) $v = 90\sin 400\pi t$ volts
   (b) $i = 50\sin(100\pi t + 0.30)$ amperes
   (c) $e = 200\sin(628.4t - 0.41)$ volts
3. A sinusoidal current has a peak value of 30 A and a frequency of 60 Hz. At time $t = 0$, the current is zero. Express the instantaneous current i in the form $i = I_m \sin\omega t$
4. An alternating voltage v has a periodic time of 20 ms and a maximum value of 200 V. When time $t = 0$, $v = -75$ volts. Deduce a sinusoidal expression for v and sketch one cycle of the voltage showing important points.
5. The instantaneous value of voltage in an a.c. circuit at any time t seconds is given by:

   $$v = 100\sin(50\pi t - 0.523)\text{ V}$$

   Find:

   (a) the peak-to-peak voltage, the frequency, the periodic time and the phase angle
   (b) the voltage when $t = 0$
   (c) the voltage when $t = 8$ ms
   (d) the times in the first cycle when the voltage is 60 V
   (e) the times in the first cycle when the voltage is -40 V
   (f) the first time when the voltage is a maximum.

   Sketch the curve for one cycle showing relevant points.

16.7 Combination of waveforms

The resultant of the addition (or subtraction) of two sinusoidal quantities may be determined either:

(a) by plotting the periodic functions graphically (see worked Problems 13 and 16), or

(b) by resolution of phasors by drawing or calculation (see worked Problems 14 and 15).

Problem 13. The instantaneous values of two alternating currents are given by $i_1 = 20\sin\omega t$ amperes and $i_2 = 10\sin(\omega t + \pi/3)$ amperes. By plotting i_1 and i_2 on the same axes, using the same scale, over one cycle, and adding ordinates at intervals, obtain a sinusoidal expression for $i_1 + i_2$

$i_1 = 20\sin\omega t$ and $i_2 = 10\sin\left(\omega t + \dfrac{\pi}{3}\right)$ are shown plotted in Figure 16.10.

Ordinates of i_1 and i_2 are added at, say, 15° intervals (a pair of dividers are useful for this).

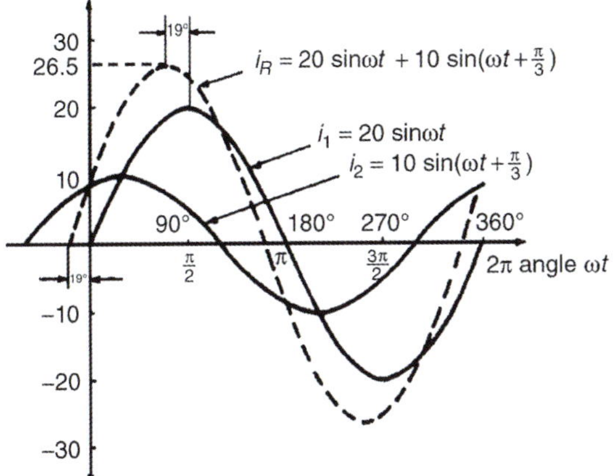

Figure 16.10

For example,

at 30°, $i_1 + i_2 = 10 + 10 = 20$ A

at 60°, $i_1 + i_2 = 8.7 + 17.3 = 26$ A

at 150°, $i_1 + i_2 = 10 + (-5) = 5$ A, and so on.

The resultant waveform for $i_1 + i_2$ is shown by the broken line in Figure 16.10. It has the same period, and hence frequency, as i_1 and i_2. The amplitude or peak value is 26.5 A.

Part 3

The resultant waveform leads the curve $i_1 = 20 \sin \omega t$ by 19°,

i.e. $\left(19 \times \frac{\pi}{180}\right)$ rads $= 0.332$ rads

Hence the sinusoidal expression for the resultant $i_1 + i_2$ is given by:

$$\mathbf{i_R = i_1 + i_2 = 26.5 \sin(\omega t + 0.332)\,A}$$

Problem 14. Two alternating voltages are represented by $v_1 = 50 \sin \omega t$ volts and $v_2 = 100 \sin(\omega t - \pi/6)$ V. Draw the phasor diagram and find, by calculation, a sinusoidal expression to represent $v_1 + v_2$

Phasors are usually drawn at the instant when time $t = 0$. Thus v_1 is drawn horizontally 50 units long and v_2 is drawn 100 units long, lagging v_1 by $\pi/6$ rads, i.e. 30°. This is shown in Figure 16.11(a), where 0 is the point of rotation of the phasors.

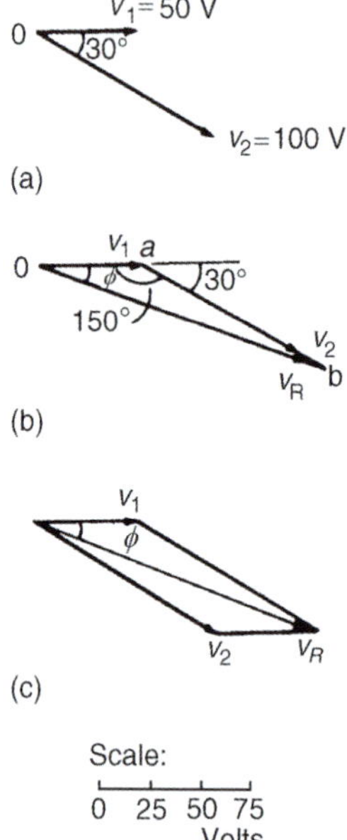

Figure 16.11

Procedure to draw phasor diagram to represent $v_1 + v_2$:

(i) Draw v_1 horizontal 50 units long, i.e. Oa of Figure 16.11(b)

(ii) Join v_2 to the end of v_1 at the appropriate angle, i.e. ab of Figure 16.11(b)

(iii) The resultant $v_R = v_1 + v_2$ is given by the length Ob and its phase angle ϕ may be measured with respect to v_1

Alternatively, when two phasors are being added the resultant is always the diagonal of the parallelogram, as shown in Figure 16.11(c).

From the drawing, by measurement, $v_R = 145$ V and angle $\phi = 20°$ lagging v_1

A more accurate solution is obtained by calculation, using the cosine and sine rules. Using the cosine rule on triangle Oab of Figure 16.11(b) gives:

$$v_R^2 = v_1^2 + v_2^2 - 2v_1v_2 \cos 150°$$
$$= 50^2 + 100^2 - 2(50)(100) \cos 150°$$
$$= 2500 + 10\,000 - (-8660)$$
$$v_R = \sqrt{(21\,160)} = 145.5\,\text{V}$$

Using the sine rule, $\frac{100}{\sin \phi} = \frac{145.5}{\sin 150°}$

from which $\sin \phi = \frac{100 \sin 150°}{145.5} = 0.3436$

and $\phi = \sin^{-1} 0.3436 = 0.35$ radians, and lags v_1

Hence $v_R = v_1 + v_2 =$ $\mathbf{145.5 \sin(\omega t - 0.35)\,V}$

Problem 15. Find a sinusoidal expression for $(i_1 + i_2)$ of worked Problem 13, (a) by drawing phasors, (b) by calculation.

(a) The relative positions of i_1 and i_2 at time $t = 0$ are shown as phasors in Figure 16.12(a). The phasor diagram in Figure 16.12(b) shows the resultant i_R, and i_R is measured as 26 A and angle ϕ as 19° or 0.33 rads leading i_1

Hence, by drawing, $i_R = 26 \sin(\omega t + 0.33)$ A

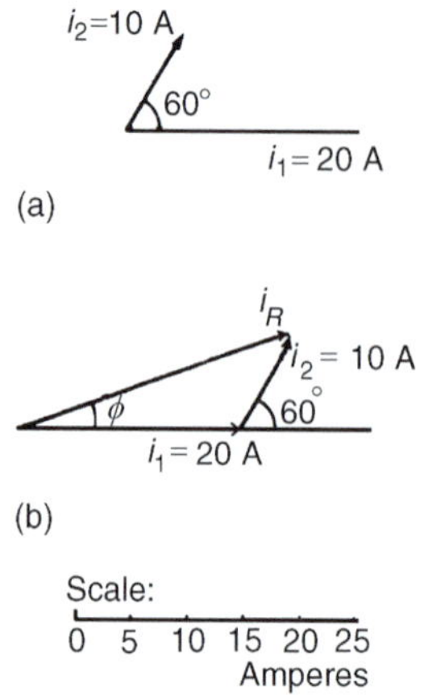

Figure 16.12

(b) From Figure 16.12(b), by the cosine rule:

$i_R^2 = 20^2 + 10^2 - 2(20)(10)(\cos 120°)$

from which $i_R = \mathbf{26.46\,A}$

By the sine rule: $\dfrac{10}{\sin\phi} = \dfrac{26.46}{\sin 120°}$

from which $\phi = 19.10°$ (i.e. 0.333 rads)

Hence, by calculation

$$i_R = 26.46\sin(\omega t + 0.333)\,\mathrm{A}$$

An alternative method of calculation is to use **complex numbers** (see Chapter 26).

Then $i_1 + i_2 = 20\sin\omega t + 10\sin\left(\omega t + \dfrac{\pi}{3}\right)$

$\equiv 20\angle 0 + 10\angle\dfrac{\pi}{3}\text{rad}$

or $20\angle 0° + 10\angle 60°$

$= (20 + j0) + (5 + j8.66)$

$= (25 + j8.66) = 26.46\angle 19.106°$

or $26.46\angle 0.333\,\text{rad}$

$\equiv \mathbf{26.46\sin(\omega t + 0.333)\,A}$

Problem 16. Two alternating voltages are given by $v_1 = 120\sin\omega t$ volts and $v_2 = 200\sin(\omega t - \pi/4)$ volts. Obtain sinusoidal expressions for $v_1 - v_2$ (a) by plotting waveforms, and (b) by resolution of phasors.

(a) $v_1 = 120\sin\omega t$ and $v_2 = 200\sin(\omega t - \pi/4)$ are shown plotted in Figure 16.13. Care must be taken when subtracting values of ordinates, especially when at least one of the ordinates is negative. For example

at 30°, $v_1 - v_2 = 60 - (-52) = 112\,\mathrm{V}$

at 60°, $v_1 - v_2 = 104 - 52 = 52\,\mathrm{V}$

at 150°, $v_1 - v_2 = 60 - 193 = -133\,\mathrm{V}$ and so on.

The resultant waveform, $v_R = v_1 - v_2$, is shown by the broken line in Figure 16.13. The maximum value of v_R is 143 V and the waveform is seen to lead v_1 by 99° (i.e. 1.73 radians).

Hence, by drawing, $\boldsymbol{v_R = v_1 - v_2}$

$= \mathbf{143\sin(\omega t + 1.73)\,volts}$

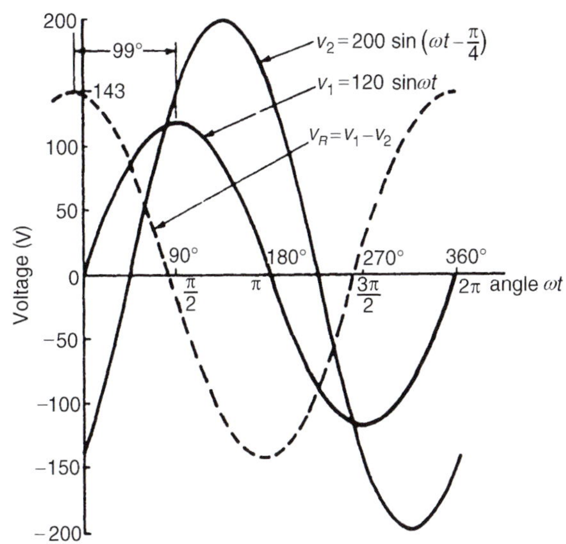

Figure 16.13

(b) The relative positions of v_1 and v_2 are shown at time $t = 0$ as phasors in Figure 16.14(a). Since the resultant of $v_1 - v_2$ is required, $-v_2$ is drawn in the opposite direction to $+v_2$ and is shown by the broken line in Figure 16.14(a). The phasor diagram with the resultant is shown in Figure 16.14(b), where $-v_2$ is added phasorially to v_1

By resolution:

Sum of horizontal components of v_1 and v_2

$= 120\cos 0° + 200\cos 135° = -21.42$

Sum of vertical components of v_1 and v_2

$= 120\sin 0° + 200\sin 135° = 141.4$

From Figure 16.14(c), resultant

$v_R = \sqrt{[(-21.42)^2 + (141.4)^2]} = 143.0,$

and $\tan\phi' = \dfrac{141.4}{21.42} = \tan 6.6013$, from which

$\phi' = \tan^{-1} 6.6013 = 81.39°$ and

$\phi = 98.61°$ or 1.721 radians

Hence, by resolution of phasors,

$$\boldsymbol{v_R = v_1 - v_2 = 143.0\sin(\omega t + 1.721)\,\textbf{volts}}$$

(By complex number: $v_R = v_1 - v_2$

$= 120\angle 0 - 200\angle -\dfrac{\pi}{4}$

$= (120 + j0)$

$-(141.42 - j141.42)$

$= -21.42 + j141.42$

$= 143.0\angle 98.61°$

or $143.9\angle 1.721\,\text{rad}$

Hence, $\boldsymbol{v_R = v_1 - v_2 =}$ **143.0 sin(**$\boldsymbol{\omega t}$ **+ 1.721) volts)**

Part 3

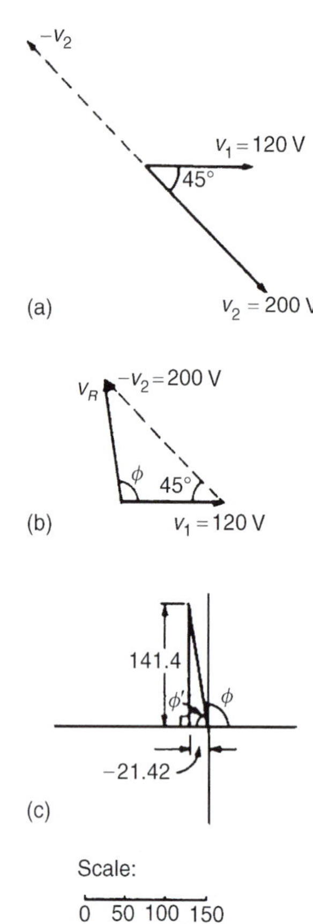

Figure 16.14

Now try the following Practice Exercise

Practice Exercise 76 The combination of periodic waveforms (Answers on page 822)

1. The instantaneous values of two alternating voltages are given by $v_1 = 5\sin\omega t$ and $v_2 = 8\sin(\omega t - \pi/6)$. By plotting v_1 and v_2 on the same axes, using the same scale, over one cycle, obtain expressions for (a) $v_1 + v_2$ and (b) $v_1 - v_2$
2. Repeat Problem 1 by calculation.
3. Construct a phasor diagram to represent $i_1 + i_2$ where $i_1 = 12\sin\omega t$ and $i_2 = 15\sin(\omega t + \pi/3)$. By measurement, or by calculation, find a sinusoidal expression to represent $i_1 + i_2$
4. Determine, either by plotting graphs and adding ordinates at intervals, or by calculation, the following periodic functions in the form $v = V_m \sin(\omega t \pm \phi)$:

   (a) $10\sin\omega t + 4\sin(\omega t + \pi/4)$

   (b) $80\sin(\omega t + \pi/3) + 50\sin(\omega t - \pi/6)$

   (c) $100\sin\omega t - 70\sin(\omega t - \pi/3)$
5. The voltage drops across two components when connected in series across an a.c. supply are $v_1 = 150\sin 314.2t$ and $v_2 = 90\sin(314.2t - \pi/5)$ volts, respectively. Determine (a) the voltage of the supply, in trigonometric form, (b) the r.m.s. value of the supply voltage and (c) the frequency of the supply.
6. If the supply to a circuit is $25\sin 628.3t$ volts and the voltage drop across one of the components is $18\sin(628.3t - 0.52)$ volts, calculate (a) the voltage drop across the remainder of the circuit, (b) the supply frequency and (c) the periodic time of the supply.
7. The voltages across three components in a series circuit when connected across an a.c. supply are:

   $v_1 = 30\sin\left(300\pi t - \dfrac{\pi}{6}\right)$ volts,

   $v_2 = 40\sin\left(300\pi t + \dfrac{\pi}{4}\right)$ volts and

   $v_3 = 50\sin\left(300\pi t + \dfrac{\pi}{3}\right)$ volts.

   Calculate (a) the supply voltage, in sinusoidal form, (b) the frequency of the supply, (c) the periodic time and (d) the r.m.s. value of the supply.

16.8 Rectification

The process of obtaining unidirectional currents and voltages from alternating currents and voltages is called rectification. Automatic switching in circuits is achieved using diodes (see Chapter 13).

Half-wave rectification

Using a single diode, D, as shown in Figure 16.15, **half-wave rectification** is obtained. When P is sufficiently positive with respect to Q, diode D is switched on and current i flows. When P is negative with respect to Q, diode D is switched off. Transformer T isolates the equipment from direct connection with the mains supply and enables the mains voltage to be changed.

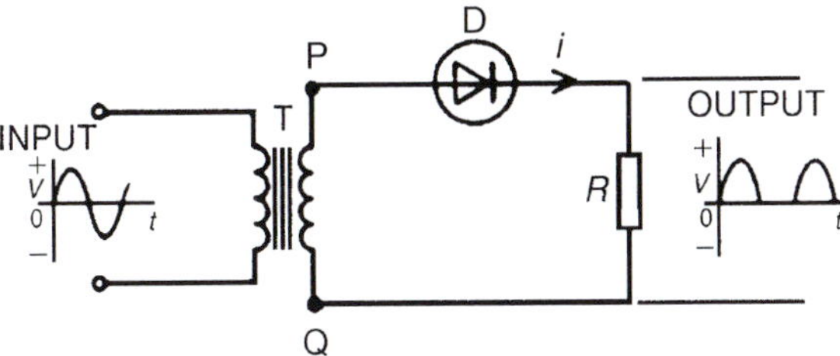

Figure 16.15

Thus, an alternating, sinusoidal waveform applied to the transformer primary is rectified into a unidirectional waveform. Unfortunately, the output waveform shown in Figure 16.15 is not constant (i.e. steady), and as such, would be unsuitable as a d.c. power supply for electronic equipment. It would, however, be satisfactory as a battery charger. In Section 16.9, methods of smoothing the output waveform are discussed.

Full-wave rectification using a centre-tapped transformer

Two diodes may be used as shown in Figure 16.16 to obtain **full-wave rectification** where a centre-tapped transformer T is used. When P is sufficiently positive with respect to Q, diode D_1 conducts and current flows (shown by the broken line in Figure 16.16). When S is positive with respect to Q, diode D_2 conducts and current flows (shown by the continuous line in Figure 16.16).

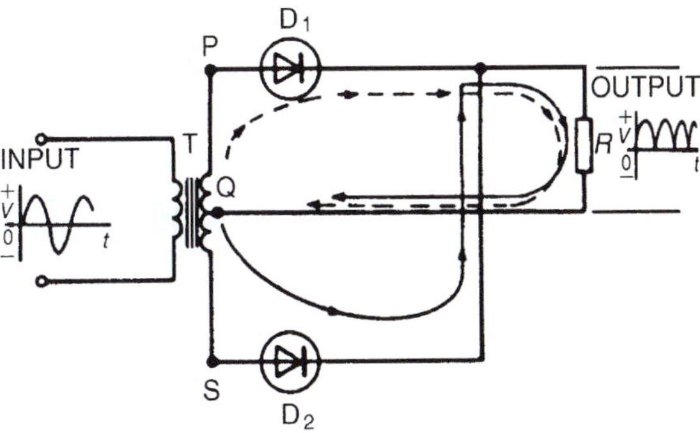

Figure 16.16

The current flowing in the load R is in the same direction for both half-cycles of the input. The output waveform is thus as shown in Figure 16.16. The output is unidirectional, but is not constant; however, it is better than the output waveform produced with a half-wave rectifier. Section 16.8 explains how the waveform may be improved so as to be of more use.

A **disadvantage** of this type of rectifier is that centre-tapped transformers are expensive.

Full-wave bridge rectification

Four diodes may be used in a **bridge rectifier** circuit, as shown in Figure 16.17, to obtain **full-wave rectification**. (Note, the term 'bridge' means a network of four elements connected to form a square, the input being applied to two opposite corners and the output being taken from the remaining two corners.) As for the rectifier shown in Figure 16.16, the current flowing in load R is in the same direction for both half cycles of the input giving the output waveform shown.

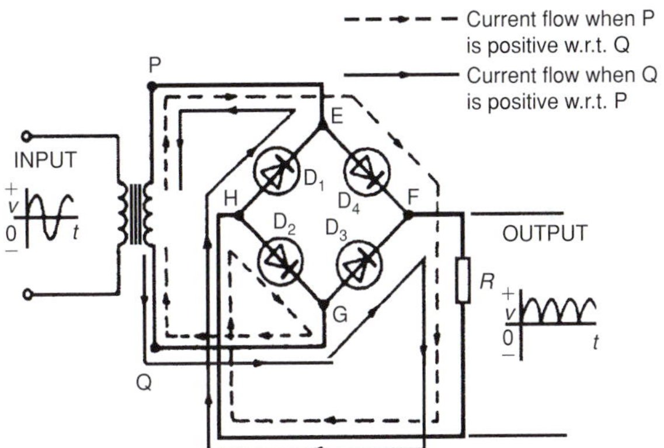

Figure 16.17

Following the broken line in Figure 16.17:
When P is positive with respect to Q, current flows from the transformer to point E, through diode D_4 to point F, then through load R to point H, through D_2 to point G, and back to the transformer.

Following the full line in Figure 16.17:
When Q is positive with respect to P, current flows from the transformer to point G, through diode D_3 to point F, then through load R to point H, through D_1 to point E, and back to the transformer. The output waveform is not steady and needs improving; a method of smoothing is explained in the next section.

16.9 Smoothing of the rectified output waveform

The pulsating outputs obtained from the half- and full-wave rectifier circuits are not suitable for the operation of equipment that requires a steady d.c. output, such as would be obtained from batteries. For example, for applications such as audio equipment, a supply with a large variation is unacceptable since it produces 'hum' in the output. **Smoothing** is the process of removing the worst of the output waveform variations.

To smooth out the pulsations a large capacitor, C, is connected across the output of the rectifier, as shown in Figure 16.18; the effect of this is to maintain the output voltage at a level which is very near to the peak of the output waveform. The improved waveforms for half-wave and full-wave rectifiers are shown in more detail in Figure 16.19.

During each pulse of output voltage, the capacitor C charges to the same potential as the peak of the waveform, as shown as point X in Figure 16.19. As the waveform dies away, the capacitor discharges across the load, as shown by XY. The output voltage is then restored to the peak value the next time the rectifier conducts, as shown by YZ. This process continues as shown in Figure 16.19.

Capacitor C is called a **reservoir capacitor** since it stores and releases charge between the peaks of the rectified waveform.

The variation in potential between points X and Y is called **ripple**, as shown in Figure 16.19; the object is to reduce ripple to a minimum. Ripple may be reduced even further by the addition of inductance and another capacitor in a **'filter'** circuit arrangement, as shown in Figure 16.20.

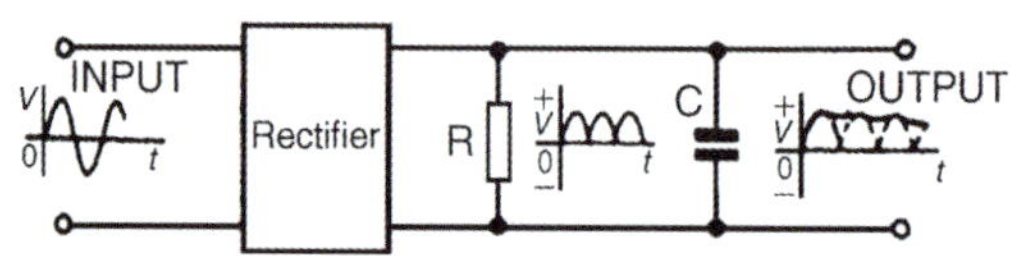

Figure 16.18

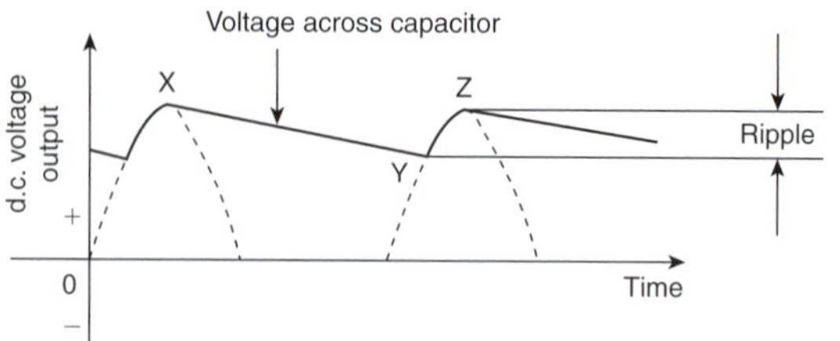

(a) Half-wave rectifier

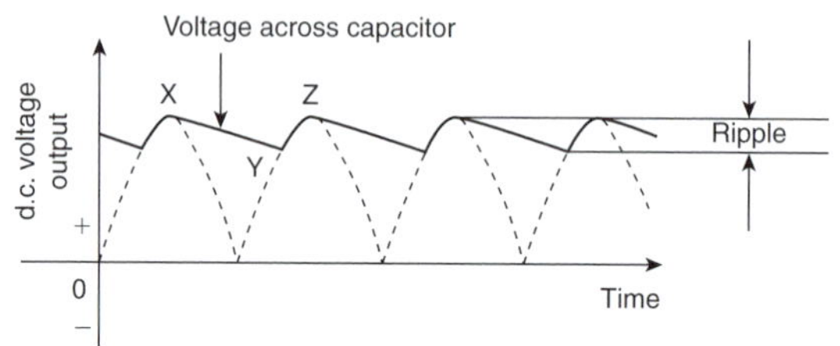

(b) Full-wave rectifier

Figure 16.19

The output voltage from the rectifier is applied to capacitor C_1 and the voltage across points AA is shown in Figure 16.20, similar to the waveforms of Figure 16.19. The load current flows through the inductance L; when current is changing, e.m.f.s are induced, as explained in Chapter 11. By Lenz's law, the induced voltages will oppose those causing the current changes.

As the ripple voltage increases and the load current increases, the induced e.m.f. in the inductor will oppose the increase. As the ripple voltage falls and the load current falls, the induced e.m.f. will try to maintain the current flow.

The voltage across points BB in Figure 16.20 and the current in the inductance are almost ripple-free. A further capacitor, C_2, completes the process.

For a practical laboratory experiment on the use of the CRO with a bridge rectifier circuit, see the website.

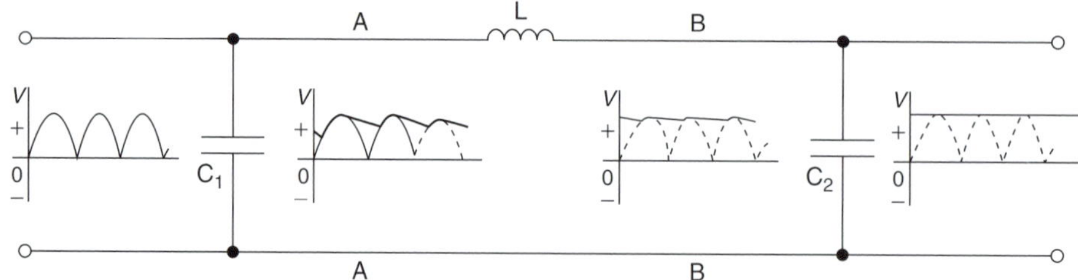

Figure 16.20

For fully worked solutions to each of the problems in Practice Exercises 73 to 76 in this chapter, go to the website:

www.routledge.com/cw/bird

Part 3

Revision Test 4

This revision test covers the material contained in Chapters 15 and 16. *The marks for each question are shown in brackets at the end of each question.*

1. Find the current flowing in the 5 Ω resistor of the circuit shown in Figure RT4.1 using (a) Kirchhoff's laws, (b) the superposition theorem, (c) Thévenin's theorem, (d) Norton's theorem. Demonstrate that the same answer results from each method. Find also the current flowing in each of the other two branches of the circuit. (27)

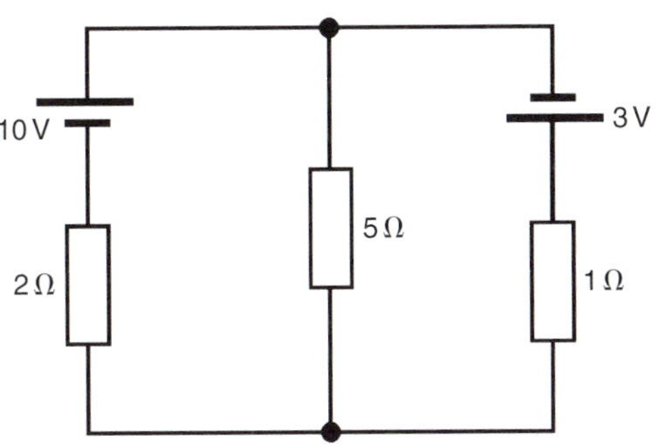

Figure RT4.1

2. A d.c. voltage source has an internal resistance of 2 Ω and an open-circuit voltage of 24 V. State the value of load resistance that gives maximum power dissipation and determine the value of this power. (5)

3. A sinusoidal voltage has a mean value of 3.0 A. Determine its maximum and r.m.s. values. (4)

4. The instantaneous value of current in an a.c. circuit at any time t seconds is given by:

$$i = 50\sin(100\pi t - 0.45)\,\text{mA}$$

Determine
(a) the peak-to-peak current, the frequency, the periodic time and the phase angle (in degrees and minutes)
(b) the current when $t = 0$
(c) the current when $t = 8$ ms
(d) the first time when the current is a maximum.

Sketch the current for one cycle showing relevant points. (14)

Part 3

For lecturers/instructors/teachers, fully worked solutions to each of the problems in Revision Test 4, together with a full marking scheme, are available at the website:
www.routledge.com/cw/bird

Chapter 17

Single-phase series a.c. circuits

***Why it is important to understand:* Single-phase series a.c. circuits**

The analysis of basic a.c. electric circuits containing impedances and ideal a.c. supplies are presented in this chapter. Series circuits containing pure resistance, *R*, pure inductance, *L*, and pure capacitance, *C*, are initially explained. Then series *R–L*, *R–C* and *R–L–C* series circuits are explored using phasors which greatly simplifies the analysis. When capacitors or inductors are involved in an a.c. circuit, the current and voltage do not peak at the same time. The fraction of a period difference between the peaks expressed in degrees is said to be the phase difference. The phase difference is less than or equal to 90°. Calculations of current, voltage, reactance, impedance and phase are explained via many worked examples. The important phenomena of resonance are explored in an *R–L–C* series circuit – there are many applications for this circuit – together with Q-factor, bandwidth and selectivity. Resonance is used in many different types of oscillator circuits; another important application is for tuning, such as in radio receivers or television sets, where it is used to select a narrow range of frequencies from the ambient radio waves. In this role the circuit is often referred to as a tuned circuit. Finally, power in a.c. circuits is explained, together with the terms true power, apparent power, reactive power and power factor. Single-phase series a.c. circuit theory is of great importance in electrical/electronic engineering.

At the end of this chapter you should be able to:

- draw phasor diagrams and current and voltage waveforms for (a) purely resistive, (b) purely inductive and (c) purely capacitive a.c. circuits
- perform calculations involving $X_L = 2\pi fL$ and $X_C = \dfrac{1}{2\pi f C}$
- draw circuit diagrams, phasor diagrams and voltage and impedance triangles for *R–L*, *R–C* and *R–L–C* series a.c. circuits and perform calculations using Pythagoras' theorem, trigonometric ratios and $Z = \dfrac{V}{I}$

Electrical Circuit Theory and Technology. 978-1-138-67349-6, © 2017 John Bird. Published by Taylor & Francis. All rights reserved.

- understand resonance
- derive the formula for resonant frequency and use it in calculations
- understand Q-factor and perform calculations using $\dfrac{V_L(\text{or } V_C)}{V}$ or $\dfrac{\omega_r L}{R}$ or $\dfrac{1}{\omega_r CR}$ or $\dfrac{1}{R}\sqrt{\left(\dfrac{L}{C}\right)}$
- understand bandwidth and half-power points
- perform calculations involving $(f_2 - f_1) = \dfrac{f_r}{Q}$
- understand selectivity and typical values of Q-factor
- appreciate that power P in an a.c. circuit is given by $P = VI\cos\phi$ or $I_R^2 R$ and perform calculations using these formulae
- understand true, apparent and reactive power and power factor and perform calculations involving these quantities

17.1 Purely resistive a.c. circuit

In a purely resistive a.c. circuit, the current I_R and applied voltage V_R are in phase. See Figure 17.1.

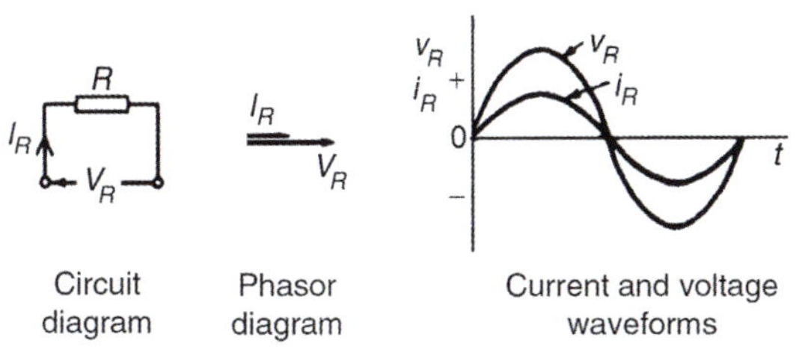

Figure 17.1

17.2 Purely inductive a.c. circuit

In a purely inductive a.c. circuit, the current I_L **lags** the applied voltage V_L by 90° (i.e. $\pi/2$ rads). See Figure 17.2.

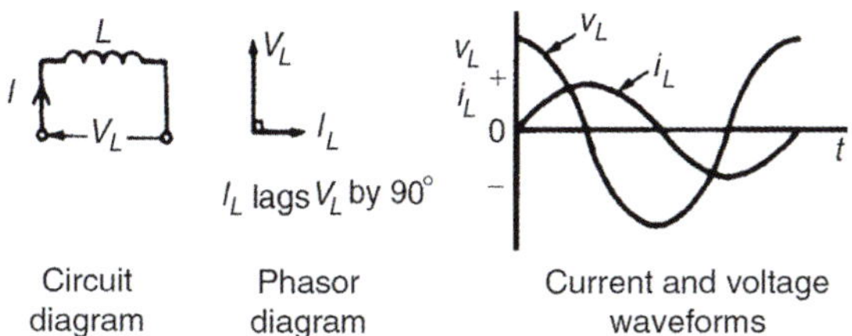

Figure 17.2

In a purely inductive circuit the opposition to the flow of alternating current is called the **inductive reactance,** X_L

$$\boldsymbol{X_L = \frac{V_L}{I_L} = 2\pi f L\ \Omega}$$

where f is the supply frequency, in hertz, and L is the inductance, in henrys.
X_L is proportional to f, as shown in Figure 17.3.

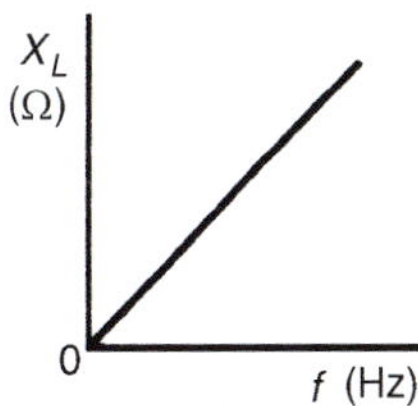

Figure 17.3

Problem 1. (a) Calculate the reactance of a coil of inductance 0.32 H when it is connected to a 50 Hz supply. (b) A coil has a reactance of 124 Ω in a circuit with a supply of frequency 5 kHz. Determine the inductance of the coil.

(a) Inductive reactance, $X_L = 2\pi f L = 2\pi(50)(0.32)$

$$= \mathbf{100.5\ \Omega}$$

(b) Since $X_L = 2\pi f L$, inductance

$$L = \frac{X_L}{2\pi f} = \frac{124}{2\pi(5000)}\text{H}$$
$$= \mathbf{3.95\ mH}$$

Problem 2. A coil has an inductance of 40 mH and negligible resistance. Calculate its inductive reactance and the resulting current if connected to (a) a 240 V, 50 Hz supply, and (b) a 100 V, 1 kHz supply.

(a) Inductive reactance, $X_L = 2\pi fL$

$$= 2\pi(50)(40\times10^{-3})$$

$$= \mathbf{12.57\,\Omega}$$

$$\text{Current, } I = \frac{V}{X_L} = \frac{240}{12.57} = \mathbf{19.09\,A}$$

(b) Inductive reactance, $X_L = 2\pi(1000)(40\times10^{-3})$

$$= \mathbf{251.3\,\Omega}$$

$$\text{Current, } I = \frac{V}{X_L} = \frac{100}{251.3} = \mathbf{0.398\,A}$$

17.3 Purely capacitive a.c. circuit

In a purely capacitive a.c. circuit, the current I_C **leads** the applied voltage V_C by 90° (i.e. $\pi/2$ rads). See Figure 17.4.

In a purely capacitive circuit the opposition to the flow of alternating current is called the **capacitive reactance, X_C**

$$X_C = \frac{V_C}{I_C} = \frac{1}{2\pi fC}\,\Omega$$

where C is the capacitance in farads.

X_C varies with frequency f, as shown in Figure 17.5.

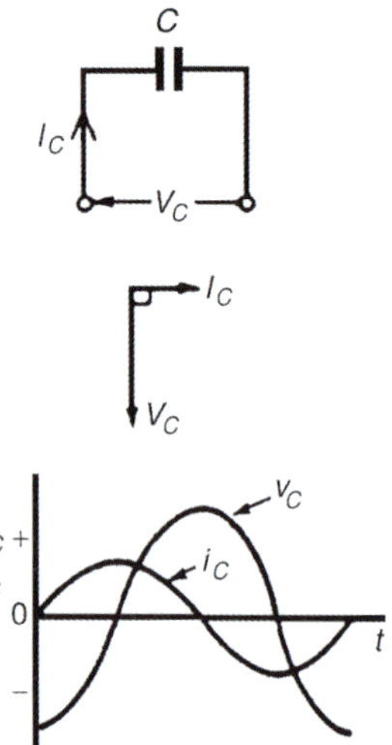

Figure 17.4

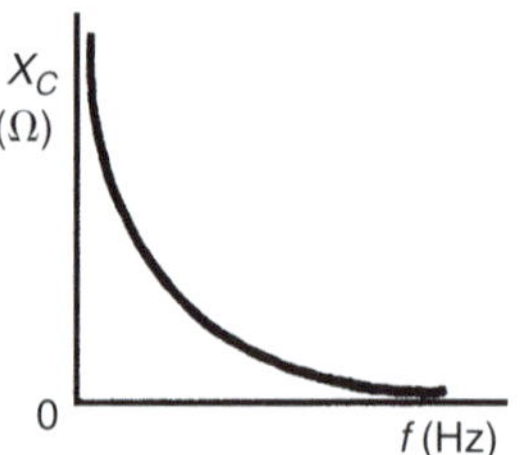

Figure 17.5

Problem 3. Determine the capacitive reactance of a capacitor of 10 μF when connected to a circuit of frequency (a) 50 Hz, (b) 20 kHz.

(a) Capacitive reactance $X_C = \dfrac{1}{2\pi fC}$

$$= \frac{1}{2\pi(50)(10\times10^{-6})}$$

$$= \frac{10^6}{2\pi(50)(10)}$$

$$= \mathbf{318.3\,\Omega}$$

(b) $X_C = \dfrac{1}{2\pi fC} = \dfrac{1}{2\pi(20\times10^3)(10\times10^{-6})}$

$$= \frac{10^6}{2\pi(20\times10^3)(10)}$$

$$= \mathbf{0.796\,\Omega}$$

Hence as the frequency is increased from 50 Hz to 20 kHz, X_C decreases from 318.3 Ω to 0.796 Ω (see Figure 17.5).

Problem 4. A capacitor has a reactance of 40 Ω when operated on a 50 Hz supply. Determine the value of its capacitance.

Since $X_C = \dfrac{1}{2\pi fC}$, capacitance $C = \dfrac{1}{2\pi f X_C}$

$$= \frac{1}{2\pi(50)(40)}\,\text{F}$$

$$= \frac{10^6}{2\pi(50)(40)}\,\mu\text{F}$$

$$= \mathbf{79.58\,\mu F}$$

Problem 5. Calculate the current taken by a 23 μF capacitor when connected to a 240 V, 50 Hz supply.

$$\text{Current } I = \frac{V}{X_C} = \frac{V}{\left(\dfrac{1}{2\pi fC}\right)}$$

$$= 2\pi fCV = 2\pi(50)(23\times10^{-6})(240)$$

$$= \mathbf{1.73\,A}$$

CIVIL

The relationship between voltage and current for the inductive and capacitive circuits can be summarized using the word 'CIVIL', which represents the following: **in a capacitor (C) the current (I) is ahead of the voltage (V), and the voltage (V) is ahead of the current (I) for the inductor (L).**

Now try the following Practice Exercise

Practice Exercise 77 Purely inductive and capacitive a.c. circuits (Answers on page 822)

1. Calculate the reactance of a coil of inductance 0.2 H when it is connected to (a) a 50 Hz, (b) a 600 Hz and (c) a 40 kHz supply.
2. A coil has a reactance of 120 Ω in a circuit with a supply frequency of 4 kHz. Calculate the inductance of the coil.
3. A supply of 240 V, 50 Hz is connected across a pure inductance and the resulting current is 1.2 A. Calculate the inductance of the coil.
4. An e.m.f. of 200 V at a frequency of 2 kHz is applied to a coil of pure inductance 50 mH. Determine (a) the reactance of the coil, and (b) the current flowing in the coil.
5. Calculate the capacitive reactance of a capacitor of 20 μF when connected to an a.c. circuit of frequency (a) 20 Hz, (b) 500 Hz, (c) 4 kHz
6. A capacitor has a reactance of 80 Ω when connected to a 50 Hz supply. Calculate the value of its capacitance.
7. A capacitor has a capacitive reactance of 400 Ω when connected to a 100 V, 25 Hz supply. Determine its capacitance and the current taken from the supply.
8. Two similar capacitors are connected in parallel to a 200 V, 1 kHz supply. Find the value of each capacitor if the circuit current is 0.628 A.

17.4 *R–L* series a.c. circuit

In an a.c. circuit containing inductance L and resistance R, the applied voltage V is the phasor sum of V_R and V_L (see Figure 17.6), and thus the current I lags the applied voltage V by an angle lying between 0° and 90° (depending on the values of V_R and V_L), shown as angle ϕ. In any a.c. series circuit the current is common to each component and is thus taken as the reference phasor.

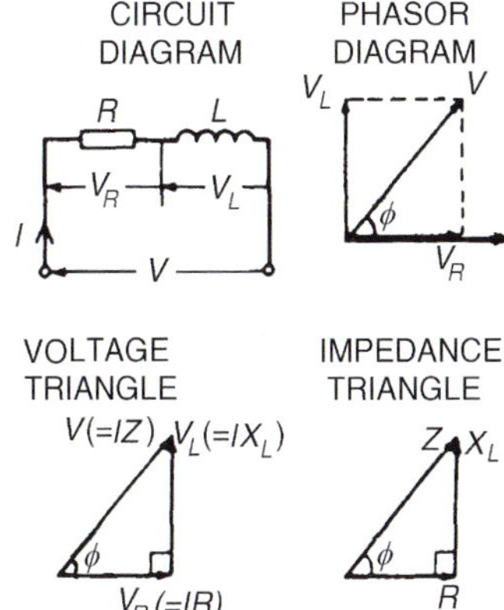

Figure 17.6

From the phasor diagram of Figure 17.6, the **'voltage triangle'** is derived.

For the *R–L* circuit:

$$V = \sqrt{(V_R^2 + V_L^2)} \quad \text{(by Pythagoras' theorem)}$$

$$\text{and } \tan\phi = \frac{V_L}{V_R} \quad \text{(by trigonometric ratios)}$$

In an a.c. circuit, the ratio $\dfrac{\text{applied voltage } V}{\text{current } I}$ is called the **impedance *Z***, i.e.

$$\boldsymbol{Z = \frac{V}{I}\,\Omega}$$

If each side of the voltage triangle in Figure 17.6 is divided by current I then the **'impedance triangle'** is derived.

For the *R–L* circuit: $Z = \sqrt{(R^2 + X_L^2)}$

$$\tan\phi = \frac{X_L}{R}, \sin\phi = \frac{X_L}{Z} \text{ and } \cos\phi = \frac{R}{Z}$$

Problem 6. In a series *R–L* circuit the p.d. across the resistance R is 12 V and the p.d. across the

Part 3

inductance L is 5 V. Find the supply voltage and the phase angle between current and voltage.

From the voltage triangle of Figure 17.6,
supply voltage $V=\sqrt{(12^2+5^2)}$ i.e. $\boldsymbol{V=13\,V}$
(Note that in a.c. circuits, the supply voltage is **not** the arithmetic sum of the p.d.s across components. It is, in fact, the **phasor sum**.)

$$\tan\phi=\frac{V_L}{V_R}=\frac{5}{12}, \text{ from which } \phi=\tan^{-1}\left(\frac{5}{12}\right)$$
$$= \mathbf{22.62^\circ\ lagging}$$

('Lagging' infers that the current is 'behind' the voltage, since phasors revolve anticlockwise.)

Problem 7. A coil has a resistance of 4 Ω and an inductance of 9.55 mH. Calculate (a) the reactance, (b) the impedance, and (c) the current taken from a 240 V, 50 Hz supply. Determine also the phase angle between the supply voltage and current.

$R=4\,\Omega$; $L=9.55\,\text{mH}=9.55\times10^{-3}\,\text{H}$;
$f=50\,\text{Hz}$; $V=240\,\text{V}$

(a) Inductive reactance, $X_L=2\pi fL$
$$=2\pi(50)(9.55\times10^{-3})$$
$$=\mathbf{3\,\Omega}$$

(b) Impedance, $Z=\sqrt{(R^2+X_L^2)}=\sqrt{(4^3+3^2)}=\mathbf{5\,\Omega}$

(c) Current, $I=\dfrac{V}{Z}=\dfrac{240}{5}=\mathbf{48\,A}$

The circuit and phasor diagrams and the voltage and impedance triangles are as shown in Figure 17.6.

Since $\tan\phi=\dfrac{X_L}{R}$, $\phi=\tan^{-1}\dfrac{X_L}{R}=\tan^{-1}\dfrac{3}{4}$
$$=\mathbf{36.87^\circ\ lagging}$$

Problem 8. A coil takes a current of 2 A from a 12 V d.c. supply. When connected to a 240 V, 50 Hz supply the current is 20 A. Calculate the resistance, impedance, inductive reactance and inductance of the coil.

$$\text{Resistance } R=\frac{\text{d.c. voltage}}{\text{d.c. current}}=\frac{12}{2}=6\,\Omega$$

$$\text{Impedance } Z=\frac{\text{a.c. voltage}}{\text{a.c. current}}=\frac{240}{20}=12\,\Omega$$

Since $Z=\sqrt{(R^2+X_L^2)}$, inductive reactance,
$$X_L=\sqrt{(Z^2-R^2)}$$
$$=\sqrt{(12^2-6^2)}$$
$$=10.39\,\Omega$$

Since $X_L=2\pi fL$, inductance $L=\dfrac{X_L}{2\pi f}=\dfrac{10.39}{2\pi(50)}$
$$=\mathbf{33.1\,mH}$$

This problem indicates a simple method for finding the inductance of a coil, i.e. firstly to measure the current when the coil is connected to a d.c. supply of known voltage, and then to repeat the process with an a.c. supply.

For a practical laboratory experiment on the measurement of inductance of a coil, see the website.

Problem 9. A coil of inductance 318.3 mH and negligible resistance is connected in series with a 200 Ω resistor to a 240 V, 50 Hz supply. Calculate (a) the inductive reactance of the coil, (b) the impedance of the circuit, (c) the current in the circuit, (d) the p.d. across each component and (e) the circuit phase angle.

$L=318.3\,\text{mH}=0.3183\,\text{H}$; $R=200\,\Omega$;
$V=240\,\text{V}$; $f=50\,\text{Hz}$

The circuit diagram is as shown in Figure 17.6.

(a) Inductive reactance $X_L=2\pi fL=2\pi(50)(0.3183)$
$$=\mathbf{100\,\Omega}$$

(b) Impedance $Z=\sqrt{(R^2+X_L^2)}=\sqrt{[(200)^2+(100)^2]}$
$$=\mathbf{223.6\,\Omega}$$

(c) Current $I=\dfrac{V}{Z}=\dfrac{240}{223.6}=\mathbf{1.073\,A}$

(d) The p.d. across the coil, $V_L=IX_L=1.073\times100$
$$=\mathbf{107.3\,V}$$

The p.d. across the resistor, $V_R=IR=1.073\times200$
$$=\mathbf{214.6\,V}$$

[Check: $\sqrt{(V_R^2+V_L^2)}=\sqrt{[(214.6)^2+(107.3)^2]}=240\,\text{V}$, the supply voltage]

(e) From the impedance triangle, angle $\phi=\tan^{-1}\dfrac{X_L}{R}$
$$=\tan^{-1}\left(\frac{100}{200}\right)$$

Hence the phase angle $\boldsymbol{\phi=26.57^\circ}$ lagging

Problem 10. A coil consists of a resistance of 100 Ω and an inductance of 200 mH. If an alternating voltage, v, given by $v = 200 \sin 500t$ volts is applied across the coil, calculate (a) the circuit impedance, (b) the current flowing, (c) the p.d. across the resistance, (d) the p.d. across the inductance and (e) the phase angle between voltage and current.

Since $v = 200 \sin 500t$ volts then $V_m = 200$ V and $\omega = 2\pi f = 500$ rad/s

Hence r.m.s. voltage $V = 0.707 \times 200 = 141.4$ V

Inductive reactance, $X_L = 2\pi fL = \omega L$

$$= 500 \times 200 \times 10^{-3} = 100\,\Omega$$

(a) Impedance $Z = \sqrt{(R^2 + X_L^2)}$

$$= \sqrt{(100^2 + 100^2)} = \mathbf{141.4\,\Omega}$$

(b) Current $I = \dfrac{V}{Z} = \dfrac{141.4}{141.4} = \mathbf{1\,A}$

(c) p.d. across the resistance $V_R = IR = 1 \times 100 = \mathbf{100\,V}$

(d) p.d. across the inductance $V_L = IX_L = 1 \times 100 = \mathbf{100\,V}$

(e) Phase angle between voltage and current is given by: $\tan\phi = \left(\dfrac{X_L}{R}\right)$

from which, $\phi = \tan^{-1}(100/100)$, hence $\boldsymbol{\phi = 45°}$ or $\boldsymbol{\dfrac{\pi}{4}}$ **rads**

Problem 11. A pure inductance of 1.273 mH is connected in series with a pure resistance of 30 Ω. If the frequency of the sinusoidal supply is 5 kHz and the p.d. across the 30 Ω resistor is 6 V, determine the value of the supply voltage and the voltage across the 1.273 mH inductance. Draw the phasor diagram.

The circuit is shown in Figure 17.7(a).

Supply voltage, $V = IZ$

Current $I = \dfrac{V_R}{R} = \dfrac{6}{30} = 0.20$ A

Inductive reactance $X_L = 2\pi fL$

$$= 2\pi(5 \times 10^3)(1.273 \times 10^{-3})$$

$$= 40\,\Omega$$

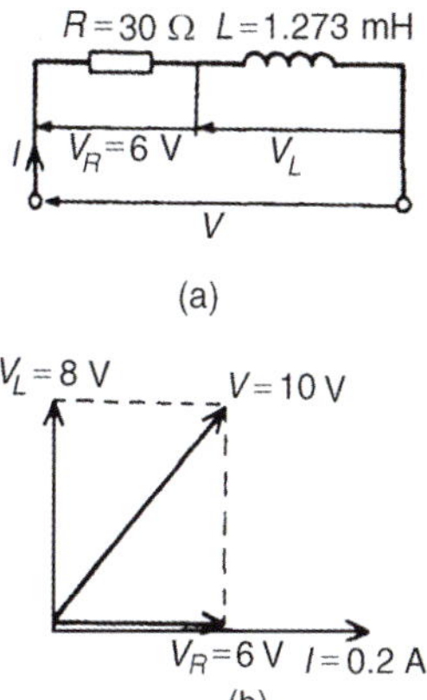

Figure 17.7

Impedance, $Z = \sqrt{(R^2 + X_L^2)} = \sqrt{(30^2 + 40^2)} = 50\,\Omega$

Supply voltage $V = IZ = (0.20)(50) = \mathbf{10\,V}$

Voltage across the 1.273 mH inductance, $V_L = IX_L = (0.2)(40) = \mathbf{8\,V}$

The phasor diagram is shown in Figure 17.7(b).

(Note that in a.c. circuits, the supply voltage is **not** the arithmetic sum of the p.d.s across components but the **phasor sum**.)

Problem 12. A coil of inductance 159.2 mH and resistance 20 Ω is connected in series with a 60 Ω resistor to a 240 V, 50 Hz supply. Determine (a) the impedance of the circuit, (b) the current in the circuit, (c) the circuit phase angle, (d) the p.d. across the 60 Ω resistor and (e) the p.d. across the coil. (f) Draw the circuit phasor diagram showing all voltages.

The circuit diagram is shown in Figure 17.8(a). When impedances are connected in series the individual resistances may be added to give the total circuit resistance. The equivalent circuit is thus shown in Figure 17.8(b).

Inductive reactance $X_L = 2\pi fL$

$$= 2\pi(50)(159.2 \times 10^{-3})$$

$$= 50\,\Omega$$

(a) Circuit impedance, $Z = \sqrt{(R_T^2 + X_L^2)}$

$$= \sqrt{(80^2 + 50^2)}$$

$$= \mathbf{94.34\,\Omega}$$

(b) Circuit current, $I = \dfrac{V}{Z} = \dfrac{240}{94.34} = \mathbf{2.544\,A}$

Part 3

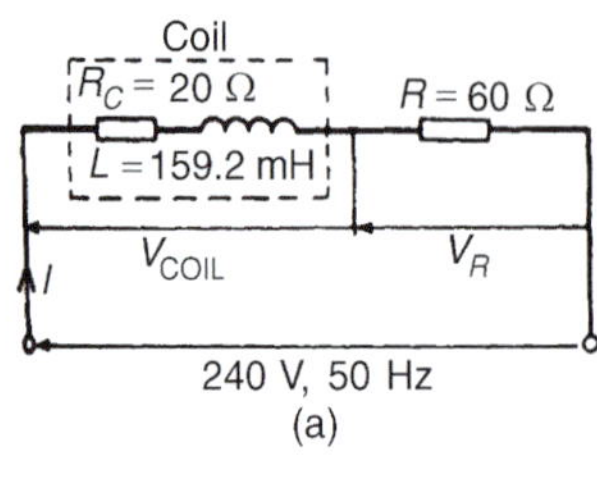

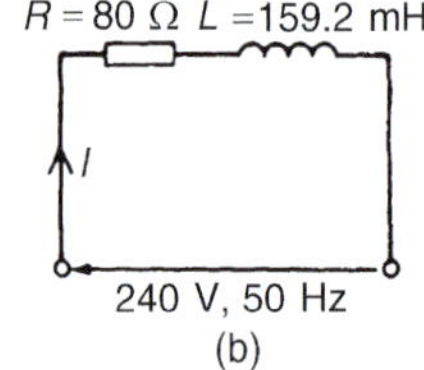

Figure 17.8

(c) Circuit phase angle $\phi = \tan^{-1}\left(\frac{X_L}{R}\right)$

$= \tan^{-1}(50/80)$

$= \mathbf{32° \text{ lagging}}$

From Figure 17.8(a):

(d) $V_R = IR = (2.544)(60) = \mathbf{152.6\,V}$

(e) $V_{COIL} = IZ_{COIL}$, where $Z_{COIL} = \sqrt{(R_C^2 + X_L^2)}$

$= \sqrt{(20^2 + 50^2)}$

$= 53.85\,\Omega$

Hence $V_{COIL} = (2.544)(53.85) = \mathbf{137.0\,V}$

(f) For the phasor diagram, shown in Figure 17.9,

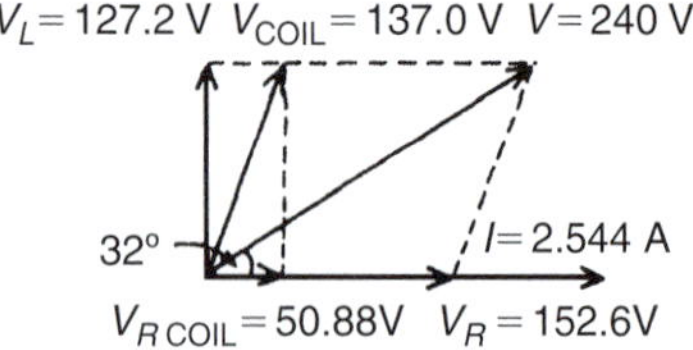

Figure 17.9

$V_L = IX_L = (2.544)(50) = 127.2\,V$

$V_{R\,COIL} = IR_C = (2.544)(20) = 50.88\,V$

The 240 V supply voltage is the phasor sum of V_{COIL} and V_R

Now try the following Practice Exercise

Practice Exercise 78 *R–L* series a.c. circuits (Answers on page 822)

1. Determine the impedance of a coil which has a resistance of 12 Ω and a reactance of 16 Ω.
2. A coil of inductance 80 mH and resistance 60 Ω is connected to a 200 V, 100 Hz supply. Calculate the circuit impedance and the current taken from the supply. Find also the phase angle between the current and the supply voltage.
3. An alternating voltage given by $v = 100\sin 240t$ volts is applied across a coil of resistance 32 Ω and inductance 100 mH. Determine (a) the circuit impedance, (b) the current flowing, (c) the p.d. across the resistance and (d) the p.d. across the inductance.
4. A coil takes a current of 5 A from a 20 V d.c. supply. When connected to a 200 V, 50 Hz a.c. supply the current is 25 A. Calculate (a) the resistance, (b) impedance and (c) inductance of the coil.
5. A coil of inductance 636.6 mH and negligible resistance is connected in series with a 100 Ω resistor to a 250 V, 50 Hz supply. Calculate (a) the inductive reactance of the coil, (b) the impedance of the circuit, (c) the current in the circuit, (d) the p.d. across each component and (e) the circuit phase angle.

17.5 *R–C* series a.c. circuit

In an a.c. series circuit containing capacitance C and resistance R, the applied voltage V is the phasor sum of V_R and V_C (see Figure 17.10) and thus the current I leads the applied voltage V by an angle lying between 0° and 90° (depending on the values of V_R and V_C), shown as angle α.

From the phasor diagram of Figure 17.10, the **'voltage triangle'** is derived. For the R–C circuit:

$$V = \sqrt{(V_R^2 + V_C^2)} \quad \text{(by Pythagoras' theorem)}$$

and $\tan\alpha = \frac{V_C}{V_R}$ (by trigonometric ratios)

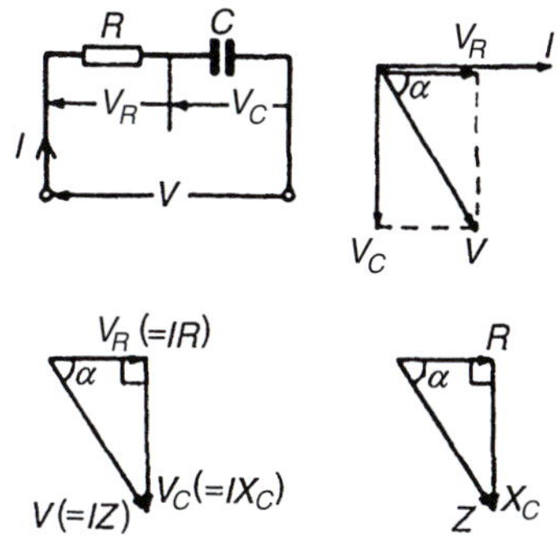

Figure 17.10

As stated in Section 17.4, in an a.c. circuit, the ratio (applied voltage V)/(current I) is called the **impedance Z**, i.e. $Z = \frac{V}{I}\,\Omega$

If each side of the voltage triangle in Figure 17.10 is divided by current I then the **'impedance triangle'** is derived.

For the R–C circuit: $Z = \sqrt{(R^2 + X_C^2)}$

$$\tan\alpha = \frac{X_C}{R},\quad \sin\alpha = \frac{X_C}{Z}\quad\text{and}\quad \cos\alpha = \frac{R}{Z}$$

Problem 13. A resistor of 25 Ω is connected in series with a capacitor of 45 μF. Calculate (a) the impedance and (b) the current taken from a 240 V, 50 Hz supply. Find also the phase angle between the supply voltage and the current.

$R = 25\,\Omega$; $C = 45\,\mu F = 45\times10^{-6}\,F$; $V = 240\,V$; $f = 50\,Hz$

The circuit diagram is as shown in Figure 17.10

$$\text{Capacitive reactance, } X_C = \frac{1}{2\pi fC} = \frac{1}{2\pi(50)(45\times10^{-6})} = 70.74\,\Omega$$

(a) Impedance $Z = \sqrt{(R^2 + X_C^2)} = \sqrt{[(25)^2 + (70.74)^2]} = \mathbf{75.03\,\Omega}$

(b) Current $I = \frac{V}{Z} = \frac{240}{75.03} = \mathbf{3.20\,A}$

Phase angle between the supply voltage and current, $\alpha = \tan^{-1}\left(\frac{X_C}{R}\right)$

hence $\alpha = \tan^{-1}\left(\frac{70.74}{25}\right) = \mathbf{70.54^\circ}$ **leading**

('Leading' infers that the current is 'ahead' of the voltage, since phasors revolve anticlockwise.)

Problem 14. A capacitor C is connected in series with a 40 Ω resistor across a supply of frequency 60 Hz. A current of 3 A flows and the circuit impedance is 50 Ω. Calculate: (a) the value of capacitance, C, (b) the supply voltage, (c) the phase angle between the supply voltage and current, (d) the p.d. across the resistor and (e) the p.d. across the capacitor. Draw the phasor diagram.

(a) Impedance $Z = \sqrt{(R^2 + X_C^2)}$

Hence $X_C = \sqrt{(Z^2 - R^2)} = \sqrt{(50^2 - 40^2)} = 30\,\Omega$

$$X_C = \frac{1}{2\pi fC}$$

$$\text{hence } C = \frac{1}{2\pi f X_C} = \frac{1}{2\pi(60)30}\,F = \mathbf{88.42\,\mu F}$$

(b) Since $Z = \frac{V}{I}$ then $V = IZ = (3)(50) = \mathbf{150\,V}$

(c) Phase angle, $\alpha = \tan^{-1}\frac{X_C}{R} = \tan^{-1}\left(\frac{30}{40}\right) = \mathbf{36.87^\circ}$ **leading**

(d) P.d. across resistor, $V_R = IR = (3)(40) = \mathbf{120\,V}$

(e) P.d. across capacitor, $V_C = IX_C = (3)(30) = \mathbf{90\,V}$

The phasor diagram is shown in Figure 17.11, where the supply voltage V is the phasor sum of V_R and V_C

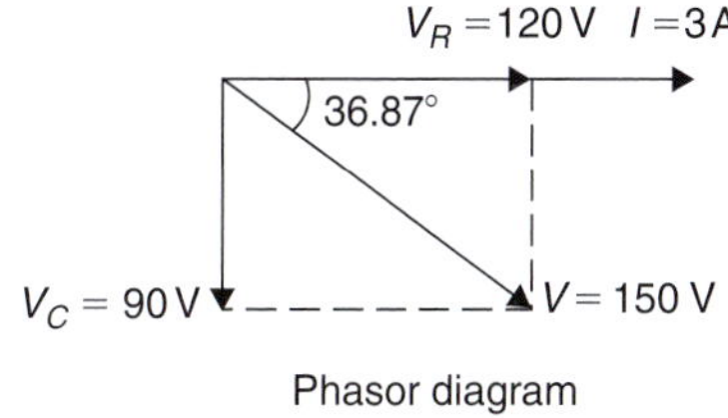

Figure 17.11

Now try the following Practice Exercise

Practice Exercise 79 *R–C* series a.c. circuits (Answers on page 822)

1. A voltage of 35 V is applied across a C–R series circuit. If the voltage across the resistor is 21 V, find the voltage across the capacitor.

2. A resistance of 50 Ω is connected in series with a capacitance of 20 μF. If a supply of

Part 3

200 V, 100 Hz is connected across the arrangement, find (a) the circuit impedance, (b) the current flowing and (c) the phase angle between voltage and current.

3. An alternating voltage $v = 250 \sin 800t$ volts is applied across a series circuit containing a 30 Ω resistor and 50 μF capacitor. Calculate (a) the circuit impedance, (b) the current flowing, (c) the p.d. across the resistor, (d) the p.d. across the capacitor and (e) the phase angle between voltage and current.

4. A 400 Ω resistor is connected in series with a 2358 pF capacitor across a 12 V a.c. supply. Determine the supply frequency if the current flowing in the circuit is 24 mA.

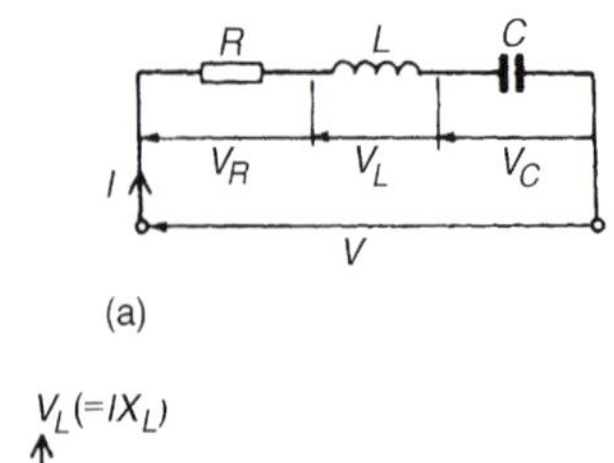

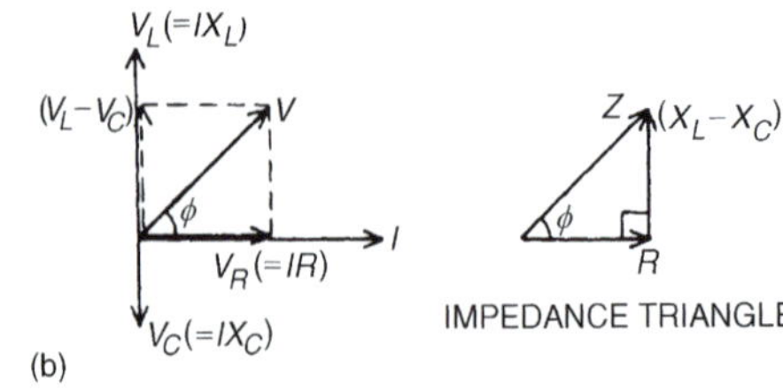

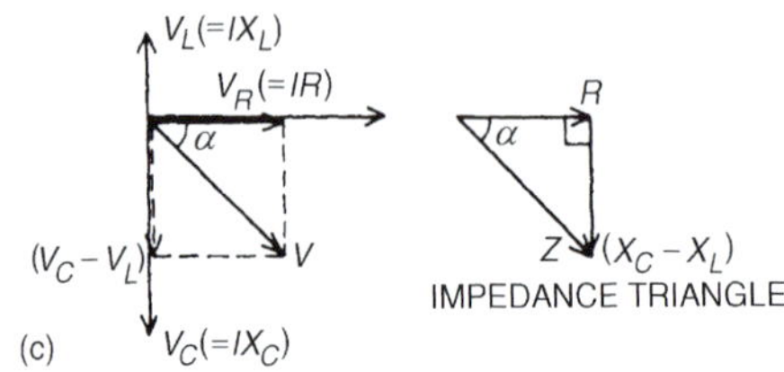

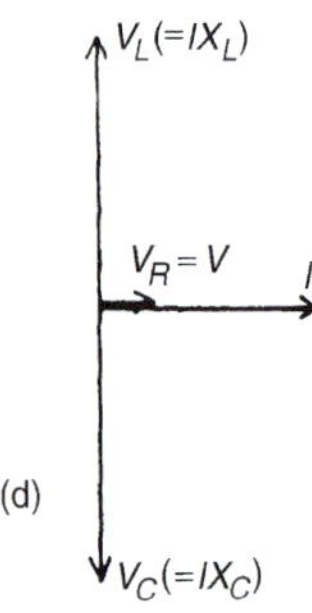

Figure 17.12

17.6 *R–L–C* series a.c. circuit

In an a.c. series circuit containing resistance R, inductance L and capacitance C, the applied voltage V is the phasor sum of V_R, V_L and V_C (see Figure 17.12). V_L and V_C are anti-phase, i.e. displaced by 180°, and there are three phasor diagrams possible – each depending on the relative values of V_L and V_C

When $X_L > X_C$ (Figure 17.12(b)):

$$Z = \sqrt{[R^2 + (X_L - X_C)^2]}$$

$$\text{and } \tan\phi = \frac{(X_L - X_C)}{R}$$

When $X_C > X_L$ (Figure 17.12(c)):

$$Z = \sqrt{[R^2 + (X_C - X_L)^2]}$$

$$\text{and} \quad \tan\alpha = \frac{(X_C - X_L)}{R}$$

When $X_L = X_C$ (Figure 17.12(d)), the applied voltage V and the current I are in phase. This effect is called **series resonance** (see Section 17.7).

Problem 15. A coil of resistance 5 Ω and inductance 120 mH in series with a 100 μF capacitor is connected to a 300 V, 50 Hz supply. Calculate (a) the current flowing, (b) the phase difference between the supply voltage and current, (c) the voltage across the coil and (d) the voltage across the capacitor.

The circuit diagram is shown in Figure 17.13

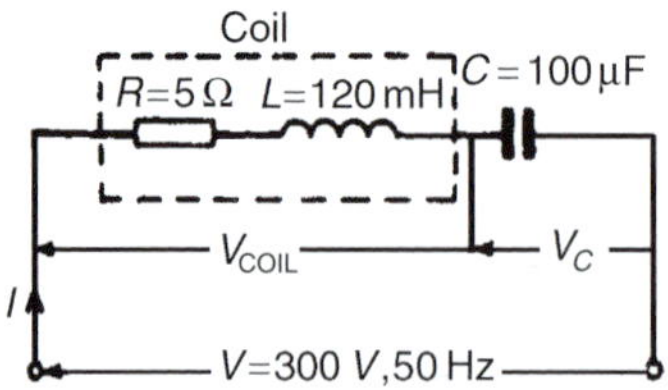

Figure 17.13

$$X_L = 2\pi fL = 2\pi(50)(120 \times 10^{-3}) = \mathbf{37.70\,\Omega}$$

$$X_C = \frac{1}{2\pi fC} = \frac{1}{2\pi(50)(100 \times 10^{-6})} = \mathbf{31.83\,\Omega}$$

Since X_L is greater than X_C the circuit is inductive.

$X_L - X_C = 37.70 - 31.83 = 5.87\,\Omega$

Impedance $Z = \sqrt{[R^2 + (X_L - X_C)^2]}$

$= \sqrt{[(5)^2 + (5.87)^2]}$

$= 7.71\,\Omega$

(a) Current $I = \frac{V}{Z} = \frac{300}{7.71} = \mathbf{38.91\,A}$

(b) Phase angle $\phi = \tan^{-1}\left(\frac{X_L - X_C}{R}\right) = \tan^{-1}\frac{5.87}{5}$

$= \mathbf{49.58°}$

(c) Impedance of coil, Z_{COIL}

$= \sqrt{(R^2 + X_L^2)} = \sqrt{[(5)^2 + (37.70)^2]} = 38.03\,\Omega$

Voltage across coil $V_{COIL} = IZ_{COIL}$

$= (38.91)(38.03)$

$= \mathbf{1480\,V}$

Phase angle of coil $= \tan^{-1}\frac{X_L}{R} = \tan^{-1}\left(\frac{37.70}{5}\right)$

$= 82.45°$ lagging

(d) Voltage across capacitor

$V_C = IX_C = (38.91)(31.83)$

$= \mathbf{1239\,V}$

The phasor diagram is shown in Figure 17.14. The supply voltage V is the phasor sum of V_{COIL} and V_C

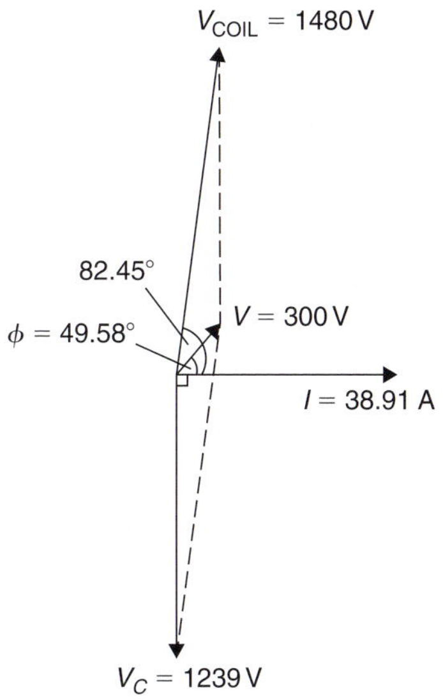

Figure 17.14

Series connected impedances

For series connected impedances the total circuit impedance can be represented as a single L–C–R circuit by combining all values of resistance together, all values of inductance together and all values of capacitance together (remembering that for series connected capacitors $\frac{1}{C} = \frac{1}{C_1} + \frac{1}{C_2} + \cdots$).

For example, the circuit of Figure 17.15(a) showing three impedances has an equivalent circuit of Figure 17.15(b).

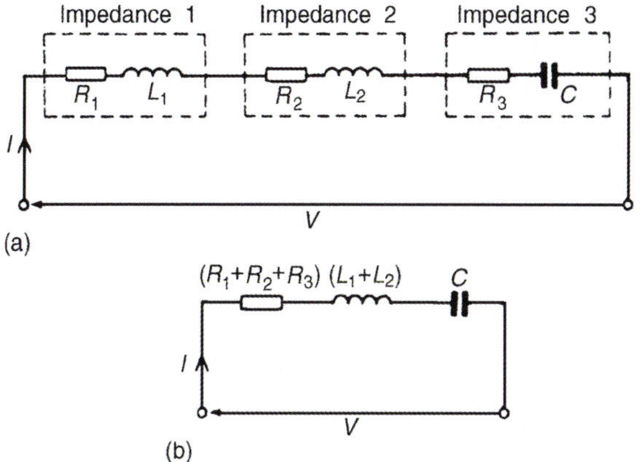

Figure 17.15

Problem 16. The following three impedances are connected in series across a 40 V, 20 kHz supply: (i) a resistance of 8 Ω, (ii) a coil of inductance 130 μH and 5 Ω resistance and (iii) a 10 Ω resistor in series with a 0.25 μF capacitor. Calculate (a) the circuit current, (b) the circuit phase angle and (c) the voltage drop across each impedance.

The circuit diagram is shown in Figure 17.16(a). Since the total circuit resistance is $8 + 5 + 10$, i.e. 23 Ω, an

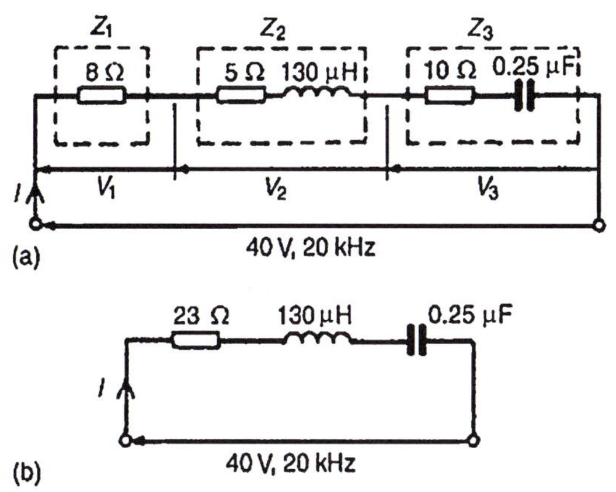

Figure 17.16

equivalent circuit diagram may be drawn, as shown in Figure 17.16(b)

Inductive reactance, $X_L = 2\pi fL$

$$= 2\pi(20\times10^3)(130\times10^{-6})$$
$$= 16.34\,\Omega$$

Capacitive reactance,

$$X_C = \frac{1}{2\pi fC} = \frac{1}{2\pi(20\times10^3)(0.25\times10^{-6})} = 31.83\,\Omega$$

Since $X_C > X_L$, the circuit is capacitive (see phasor diagram in Figure 17.12(c)).

$X_C - X_L = 31.83 - 16.34 = 15.49\,\Omega$.

(a) Circuit impedance, $Z = \sqrt{[R^2+(X_C-X_L)^2]}$

$$= \sqrt{[23^2+15.49^2]}$$
$$= 27.73\,\Omega$$

Circuit current, $I = \frac{V}{Z} = \frac{40}{27.73} = \mathbf{1.442\,A}$

(b) From Figure 17.12(c), circuit phase angle

$$\phi = \tan^{-1}\left(\frac{X_C-X_L}{R}\right)$$

i.e. $\phi = \tan^{-1}\left(\frac{15.49}{23}\right) = \mathbf{33.96^\circ\ leading}$

(c) From Figure 17.16(a), $V_1 = IR_1 = (1.442)(8) = \mathbf{11.54\,V}$

$V_2 = IZ_2 = I\sqrt{(5^2+16.34^2)} = (1.442)(17.09) = \mathbf{24.64\,V}$

$V_3 = IZ_3 = I\sqrt{(10^2+31.83^2)} = (1.442)(33.36) = \mathbf{48.11\,V}$

The 40 V supply voltage is the phasor sum of V_1, V_2 and V_3

Problem 17. Determine the p.d.s V_1 and V_2 for the circuit shown in Figure 17.17 if the frequency of the supply is 5 kHz. Draw the phasor diagram and hence determine the supply voltage V and the circuit phase angle.

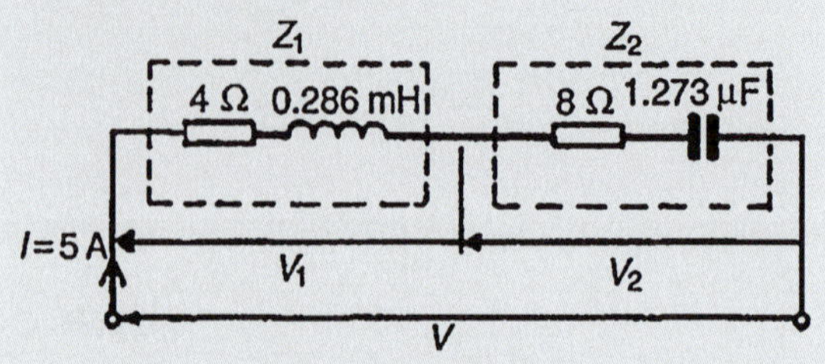

Figure 17.17

For impedance Z_1:

$R_1 = 4\Omega$ and $X_L = 2\pi fL$

$$= 2\pi(5\times10^3)(0.286\times10^{-3})$$
$$= 8.985\,\Omega$$

$V_1 = IZ_1 = I\sqrt{(R^2+X_L^2)} = 5\sqrt{(4^2+8.985^2)}$

$$= 49.18\,V$$

Phase angle $\phi_1 = \tan^{-1}\left(\frac{X_L}{R}\right)$

$$= \tan^{-1}\left(\frac{8.985}{4}\right) = 66.0^\circ \text{ lagging}$$

For impedance Z_2:

$R_2 = 8\,\Omega$ and $X_C = \frac{1}{2\pi fC}$

$$= \frac{1}{2\pi(5\times10^3)(1.273\times10^{-6})}$$
$$= 25.0\,\Omega$$

$V_2 = IZ_2 = I\sqrt{(R^2+X_C^2)} = 5\sqrt{(8^2+25.0^2)}$

$$= 131.2\,V$$

Phase angle $\phi_2 = \tan^{-1}\left(\frac{X_C}{R}\right)$

$$= \tan^{-1}\left(\frac{25.0}{8}\right) = 72.26^\circ \text{ leading}$$

The phasor diagram is shown in Figure 17.18.
The phasor sum of V_1 and V_2 gives the supply voltage V of 100 V at a phase angle of **53.13° leading**. These values may be determined by drawing or by calculation – either by resolving into horizontal and vertical components or by the cosine and sine rules.

Part 3

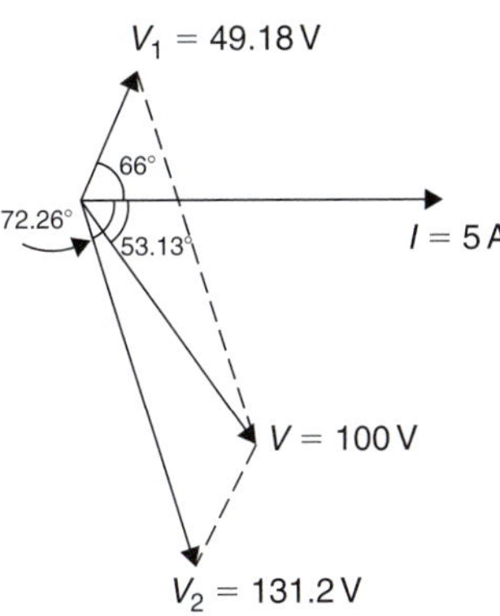

Figure 17.18

Now try the following Practice Exercise

Practice Exercise 80 *R–L–C* series a.c. circuits (Answers on page 822)

1. A 40 μF capacitor in series with a coil of resistance 8 Ω and inductance 80 mH is connected to a 200 V, 100 Hz supply. Calculate (a) the circuit impedance, (b) the current flowing, (c) the phase angle between voltage and current, (d) the voltage across the coil and (e) the voltage across the capacitor.

2. Find the values of resistance *R* and inductance *L* in the circuit of Figure 17.19.

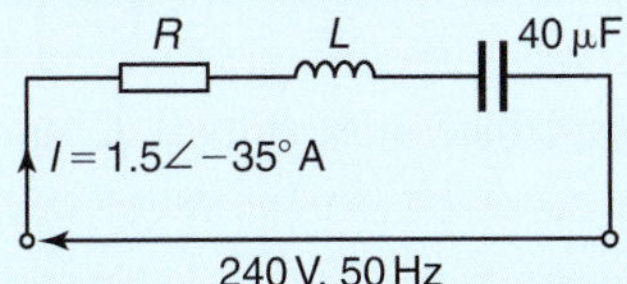

Figure 17.19

3. Three impedances are connected in series across a 100 V, 2 kHz supply. The impedances comprise:
   (i) an inductance of 0.45 mH and 2 Ω resistance,
   (ii) an inductance of 570 μH and 5 Ω resistance and
   (iii) a capacitor of capacitance 10 μF and resistance 3 Ω.

   Assuming no mutual inductive effects between the two inductances, calculate (a) the circuit impedance, (b) the circuit current, (c) the circuit phase angle and (d) the voltage across each impedance.

4. For the circuit shown in Figure 17.20 determine the voltages V_1 and V_2 if the supply frequency is 1 kHz. Draw the phasor diagram and hence determine the supply voltage V and the circuit phase angle.

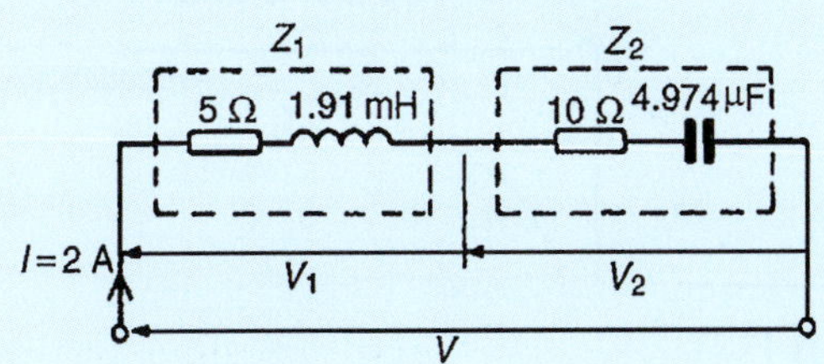

Figure 17.20

17.7 Series resonance

As stated in Section 17.6, for an *R–L–C* series circuit, when $X_L = X_C$ (Figure 17.12(d)), the applied voltage V and the current I are in phase. This effect is called **series resonance**. At resonance:

(i) $V_L = V_C$

(ii) $Z = R$ (i.e. the minimum circuit impedance possible in an *L–C–R* circuit).

(iii) $I = \dfrac{V}{R}$ (i.e. the maximum current possible in an *L–C–R* circuit).

(iv) Since $X_L = X_C$, then $2\pi f_r L = \dfrac{1}{2\pi f_r C}$

from which, $f_r^2 = \dfrac{1}{(2\pi)^2 LC}$

and, $$\boldsymbol{f_r = \frac{1}{2\pi\sqrt{(LC)}}\ \textbf{Hz}}$$

where f_r is the resonant frequency.

(v) The series resonant circuit is often described as an **acceptor circuit** since it has its minimum impedance, and thus maximum current, at the resonant frequency.

(vi) Typical graphs of current I and impedance Z against frequency are shown in Figure 17.21.

Problem 18. A coil having a resistance of 10 Ω and an inductance of 125 mH is connected in series with a 60 μF capacitor across a 120 V supply. At what frequency does resonance occur? Find the current flowing at the resonant frequency.

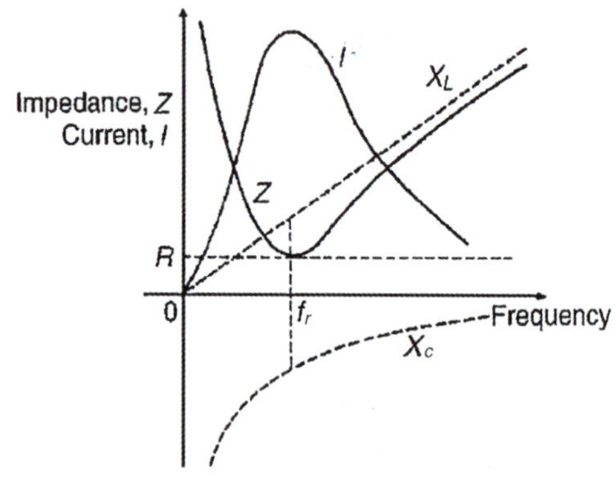

Figure 17.21

$$\text{Resonant frequency, } f_r = \frac{1}{2\pi\sqrt{(LC)}}\text{ Hz}$$

$$= \frac{1}{2\pi\sqrt{\left[\left(\frac{125}{10^3}\right)\left(\frac{60}{10^6}\right)\right]}}\text{ Hz}$$

$$= \frac{1}{2\pi\sqrt{\left(\frac{125\times 6}{10^8}\right)}}$$

$$= \frac{1}{2\pi\frac{\sqrt{[(125)(6)]}}{10^4}}$$

$$= \frac{10^4}{2\pi\sqrt{[(125)(6)]}} = \mathbf{58.12\ Hz}$$

At resonance, $X_L = X_C$ and impedance $Z = R$

Hence current, $I = \frac{V}{R} = \frac{120}{10} = \mathbf{12\ A}$

Problem 19. The current at resonance in a series L–C–R circuit is $100\,\mu$A. If the applied voltage is 2 mV at a frequency of 200 kHz, and the circuit inductance is $50\,\mu$H, find (a) the circuit resistance, and (b) the circuit capacitance.

(a) $I = 100\,\mu A = 100\times 10^{-6}$ A;
$V = 2\,mV = 2\times 10^{-3}$ V

At resonance, impedance Z = resistance R

$$\text{Hence } R = \frac{V}{I} = \frac{20\times 10^{-3}}{100\times 10^{-6}} = \frac{2\times 10^6}{100\times 10^3} = \mathbf{20\ \Omega}$$

(b) At resonance $X_L = X_C$

$$\text{i.e. } 2\pi fL = \frac{1}{2\pi fC}$$

Hence capacitance

$$C = \frac{1}{(2\pi f)^2 L}$$

$$= \frac{1}{(2\pi\times 200\times 10^3)^2(50\times 10^{-6})}\text{ F}$$

$$= \frac{(10^6)(10^6)}{(4\pi)^2(10^{10})(50)}\,\mu\text{F}$$

$$= \mathbf{0.0127\,\mu F} \text{ or } \mathbf{12.7\,nF}$$

17.8 Q-factor

At resonance, if R is small compared with X_L and X_C, it is possible for V_L and V_C to have voltages many times greater than the supply voltage (see Figure 17.12(d)).

$$\textbf{Voltage magnification at resonance} = \frac{\textbf{voltage across } \boldsymbol{L} \textbf{ (or } \boldsymbol{C}\textbf{)}}{\textbf{supply voltage } \boldsymbol{V}}$$

This ratio is a measure of the quality of a circuit (as a resonator or tuning device) and is called the **Q-factor**.

$$\text{Hence Q-factor} = \frac{V_L}{V} = \frac{IX_L}{IR} = \frac{X_L}{R} = \frac{\mathbf{2\pi f_r L}}{\mathbf{R}}$$

$$\text{Alternatively, Q-factor} = \frac{V_C}{V} = \frac{IX_C}{IR} = \frac{X_C}{R} = \frac{\mathbf{1}}{\mathbf{2\pi f_r CR}}$$

$$\text{At resonance } f_r = \frac{1}{2\pi\sqrt{(LC)}} \text{ i.e. } 2\pi f_r = \frac{1}{\sqrt{(LC)}}$$

$$\text{Hence } Q\text{-factor} = \frac{2\pi f_r L}{R} = \frac{1}{\sqrt{(LC)}}\left(\frac{L}{R}\right) = \frac{\mathbf{1}}{\mathbf{R}}\sqrt{\left(\frac{L}{C}\right)}$$

(Q-factor is explained more fully in Chapter 31, page 471)

Problem 20. A coil of inductance 80 mH and negligible resistance is connected in series with a capacitance of $0.25\,\mu$F and a resistor of resistance $12.5\,\Omega$ across a 100 V, variable frequency supply. Determine (a) the resonant frequency and (b) the current at resonance. How many times greater than the supply voltage is the voltage across the reactances at resonance?

(a) Resonant frequency f_r

$$= \frac{1}{2\pi\sqrt{\left[\left(\frac{80}{10^3}\right)\left(\frac{0.25}{10^6}\right)\right]}} = \frac{1}{2\pi\sqrt{\left[\frac{(8)(0.25)}{10^8}\right]}}$$

$$= \frac{10^4}{2\pi\sqrt{2}}$$

$$= \mathbf{1125.4\ Hz = 1.1254\ kHz}$$

(b) Current at resonance $I = \frac{V}{R} = \frac{100}{12.5} = \mathbf{8\,A}$

Voltage across inductance, at resonance,

$$V_L = IX_L = (I)(2\pi fL) = (8)(2\pi)(1125.4)(80 \times 10^{-3}) = 4525.5\,\text{V}$$

(Also, voltage across capacitor,

$$V_C = IX_C = \frac{I}{2\pi fC} = \frac{8}{2\pi(1125.4)(0.25 \times 10^{-6})} = 4525.5\,\text{V})$$

$$\text{Voltage magnification at resonance} = \frac{V_L}{V} \text{ or } \frac{V_c}{V} = \frac{4525.5}{100} = \mathbf{45.255\,V}$$

i.e. at resonance, the voltage across the reactances are 45.255 times greater than the supply voltage. Hence Q-factor of circuit is 45.255

Problem 21. A series circuit comprises a coil of resistance 2 Ω and inductance 60 mH, and a 30 μF capacitor. Determine the Q-factor of the circuit at resonance.

$$\text{At resonance, Q-factor} = \frac{1}{R}\sqrt{\left(\frac{L}{C}\right)} = \frac{1}{2}\sqrt{\left(\frac{60 \times 10^{-3}}{30 \times 10^{-6}}\right)} = \frac{1}{2}\sqrt{\left(\frac{60 \times 10^{6}}{30 \times 10^{3}}\right)} = \frac{1}{2}\sqrt{(2000)} = \mathbf{22.36}$$

Problem 22. A coil of negligible resistance and inductance 100 mH is connected in series with a capacitance of 2 μF and a resistance of 10 Ω across a 50 V, variable frequency supply. Determine (a) the resonant frequency, (b) the current at resonance, (c) the voltages across the coil and the capacitor at resonance and (d) the Q-factor of the circuit.

(a) Resonant frequency,

$$f_r = \frac{1}{2\pi\sqrt{(LC)}} = \frac{1}{2\pi\sqrt{\left[\left(\frac{100}{10^3}\right)\left(\frac{2}{10^6}\right)\right]}} = \frac{1}{2\pi\sqrt{\left(\frac{20}{10^8}\right)}} = \frac{1}{\left(\frac{2\pi\sqrt{20}}{10^4}\right)} = \frac{10^4}{2\pi\sqrt{20}} = \mathbf{355.9\,Hz}$$

(b) Current at resonance $I = \frac{V}{R} = \frac{50}{10} = \mathbf{5\,A}$

(c) Voltage across coil at resonance,

$$V_L = IX_L = I(2\pi f_r L) = (5)(2\pi \times 355.9 \times 100 \times 10^{-3}) = \mathbf{1118\,V}$$

Voltage across capacitance at resonance,

$$V_C = IX_C = \frac{I}{2\pi f_r C} = \frac{5}{2\pi(355.9)(2 \times 10^{-6})} = \mathbf{1118\,V}$$

(d) Q-factor (i.e. voltage magnification at resonance)

$$= \frac{V_L}{V} \text{ or } \frac{V_C}{V} = \frac{1118}{50} = \mathbf{22.36}$$

Q-factor may also have been determined by $\frac{2\pi f_r L}{R}$ or $\frac{1}{2\pi f_r CR}$ or $\frac{1}{R}\sqrt{\left(\frac{L}{C}\right)}$

Now try the following Practice Exercise

Practice Exercise 81 Series resonance and Q-factor (Answers on page 822)

1. Find the resonant frequency of a series a.c. circuit consisting of a coil of resistance 10 Ω and

Part 3

inductance 50 mH and capacitance 0.05 μF. Find also the current flowing at resonance if the supply voltage is 100 V.

2. The current at resonance in a series L–C–R circuit is 0.2 mA. If the applied voltage is 250 mV at a frequency of 100 kHz and the circuit capacitance is 0.04 μF, find the circuit resistance and inductance.

3. A coil of resistance 25 Ω and inductance 100 mH is connected in series with a capacitance of 0.12 μF across a 200 V, variable frequency supply. Calculate (a) the resonant frequency, (b) the current at resonance and (c) the factor by which the voltage across the reactance is greater than the supply voltage.

4. Calculate the inductance which must be connected in series with a 1000 pF capacitor to give a resonant frequency of 400 kHz.

5. A series circuit comprises a coil of resistance 20 Ω and inductance 2 mH and a 500 pF capacitor. Determine the Q-factor of the circuit at resonance. If the supply voltage is 1.5 V, what is the voltage across the capacitor?

17.9 Bandwidth and selectivity

Figure 17.22 shows how current I varies with frequency in an R–L–C series circuit. At the resonant frequency f_r, current is a maximum value, shown as I_r. Also shown are the points A and B where the current is 0.707 of the maximum value at frequencies f_1 and f_2. The power delivered to the circuit is I^2R. At $I = 0.707I_r$ the power is $(0.707I_r)^2R = 0.5I_r^2R$, i.e. half the power that occurs at frequency f_r. The points corresponding to f_1 and f_2 are called the **half-power points**. The distance between these points, i.e. $(f_2 - f_1)$, is called the **bandwidth**.

It may be shown that

$$Q = \frac{f_r}{f_2 - f_1} \quad \text{or} \quad (f_2 - f_1) = \frac{f_r}{Q}$$

(This formula is proved in Chapter 31, page 480–481)

Problem 23. A filter in the form of a series L–R–C circuit is designed to operate at a resonant frequency of 5 kHz. Included within the filter is a 20 mH inductance and 10 Ω resistance. Determine the bandwidth of the filter.

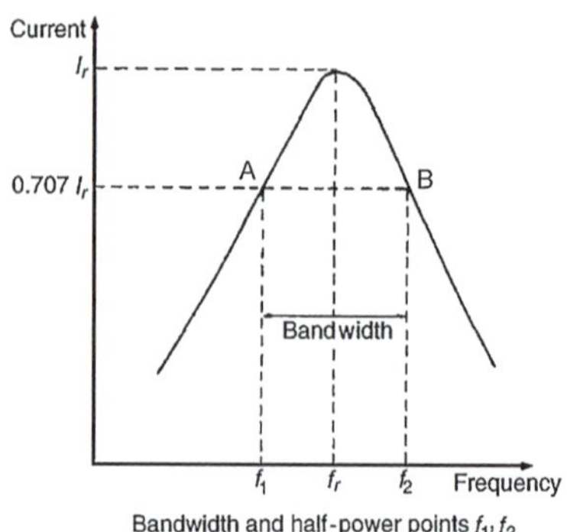

Figure 17.22

Q-factor at resonance is given by

$$Q_r = \frac{\omega_r L}{R} = \frac{(2\pi 5000)(20 \times 10^{-3})}{10} = 62.83$$

Since $Q_r = f_r/(f_2 - f_1)$

bandwidth, $(f_2 - f_1) = \dfrac{f_r}{Q_r} = \dfrac{5000}{62.83} = \mathbf{79.6\,Hz}$

Selectivity is the ability of a circuit to respond more readily to signals of a particular frequency to which it is tuned than to signals of other frequencies. The response becomes progressively weaker as the frequency departs from the resonant frequency. The higher the Q-factor, the narrower the bandwidth and the more selective is the circuit. Circuits having high Q-factors (say, in the order of 100 to 300) are therefore useful in communications engineering. A high Q-factor in a series power circuit has disadvantages in that it can lead to dangerously high voltages across the insulation and may result in electrical breakdown.

(For more on bandwidth and selectivity see Chapter 31, pages 479–483)

For a practical laboratory experiment on series a.c. circuits and resonance, see the website.

17.10 Power in a.c. circuits

In Figures 17.23(a)–(c), the value of power at any instant is given by the product of the voltage and current at that instant, i.e. the instantaneous power, $p = vi$, as shown by the broken lines.

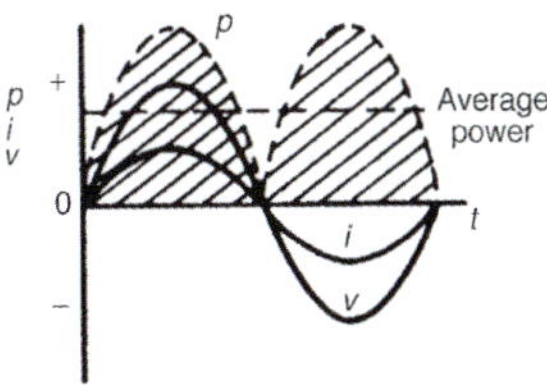

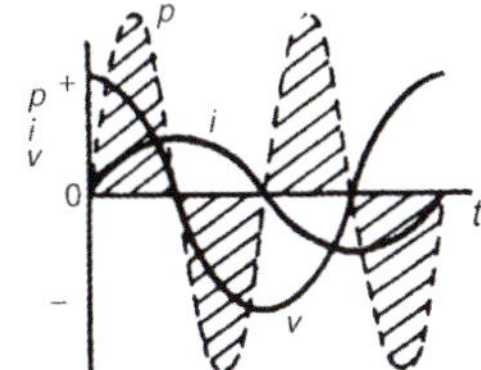

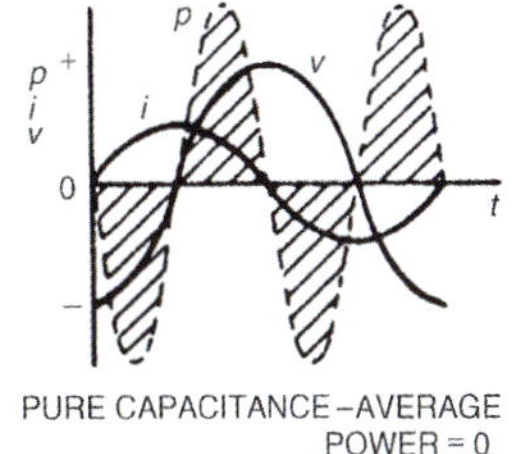

Figure 17.23

(a) For a purely resistive a.c. circuit, the average power dissipated, P, is given by:

$$P = VI = I^2R = \frac{V^2}{R} \text{ watts}$$

(V and I being r.m.s. values).

See Figure 17.23(a). The unit 'watt' is named after **James Watt***

(b) For a purely inductive a.c. circuit, the average power is zero. See Figure 17.23(b).

(c) For a purely capacitive a.c. circuit, the average power is zero. See Figure 17.23(c).

Figure 17.24 shows current and voltage waveforms for an R–L circuit where the current lags the voltage by angle ϕ. The waveform for power (where $p = vi$) is shown by the broken line, and its shape, and hence average power depends on the value of angle ϕ.

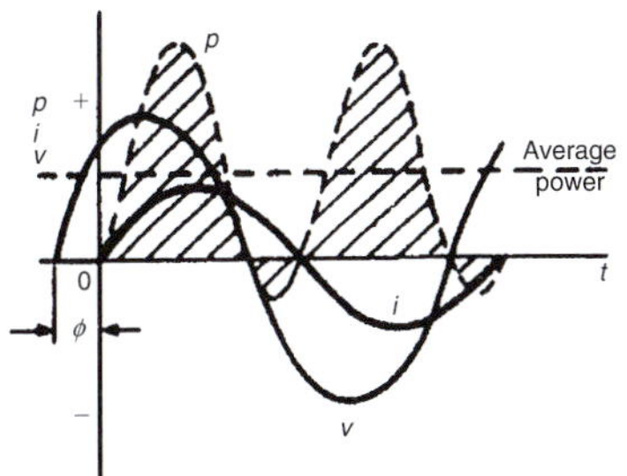

Figure 17.24

For an R–L, R–C or R–L–C series a.c. circuit, the average power P is given by:

$$P = VI\cos\phi \text{ watts}$$

or

$$P = I^2R \text{ watts} \quad (V \text{ and } I \text{ being r.m.s. values})$$

The formulae for power are proved in Chapter 29, page 446.

Problem 24. An instantaneous current, $i = 250\sin\omega t$ mA flows through a pure resistance of 5 kΩ. Find the power dissipated in the resistor.

Power dissipated, $P = I^2R$ where I is the r.m.s. value of current.

If $i = 250\sin\omega t$ mA, then $I_m = 0.250$ A and r.m.s. current, $I = (0.707 \times 0.250)$ A.

Hence power $P = (0.707 \times 0.250)^2(5000)$
$= \mathbf{156.2\ watts}$

Problem 25. A series circuit of resistance 60 Ω and inductance 75 mH is connected to a 110 V, 60 Hz supply. Calculate the power dissipated.

Inductive reactance,

$$X_L = 2\pi fL = 2\pi(60)(75 \times 10^{-3}) = 28.27\,\Omega$$

Impedance,

$$Z = \sqrt{(R^2 + X_L^2)} = \sqrt{[(60)^2 + (28.27)^2]} = 66.33\,\Omega$$

$$\text{Current, } I = \frac{V}{Z} = \frac{100}{66.33} = 1.658\text{ A}$$

To calculate power dissipation in an a.c. circuit two formulae may be used:

(i) $P = I^2R = (1.658)^2(60) = \mathbf{165\ W}$

* Who was **Watt**? For image and resume of Watt, see page 52. To find out more go to **www.routledge.com/cw/bird**

Part 3

or (ii) $P = VI\cos\phi$ where $\cos\phi = \frac{R}{Z}$

$$= \frac{60}{66.33} = 0.9046$$

Hence $P = (110)(1.658)(0.9046) = \mathbf{165\,W}$

17.11 Power triangle and power factor

Figure 17.25(a) shows a phasor diagram in which the current I lags the applied voltage V by angle ϕ. The horizontal component of V is $V\cos\phi$ and the vertical component of V is $V\sin\phi$. If each of the voltage phasors is multiplied by I, Figure 17.25(b) is obtained and is known as the **'power triangle'**.

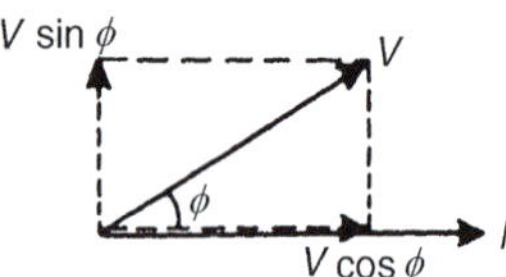

(a) PHASOR DIAGRAM

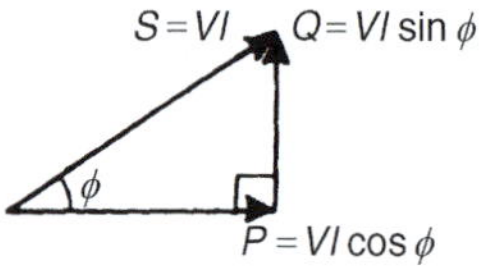

(b) POWER TRIANGLE

Figure 17.25

Apparent power, $S = VI$ voltamperes (VA)
True or active power, $P = VI\cos\phi$ watts (W)
Reactive power, $Q = VI\sin\phi$ reactive voltamperes (var)

$$\textbf{Power factor} = \frac{\textbf{true power } P}{\textbf{apparent power } S}$$

For sinusoidal voltages and currents,

power factor $= \frac{P}{S} = \frac{VI\cos\phi}{VI}$, i.e.

$$\textbf{p.f.} = \cos\phi = \frac{R}{Z} \text{ (from Figure 17.6)}$$

The relationships stated above are also true when current I leads voltage V. More on the power triangle and power factor is contained in Chapter 29, page 449.

Problem 26. A pure inductance is connected to a 150 V, 50 Hz supply, and the apparent power of the circuit is 300 VA. Find the value of the inductance.

Apparent power $S = VI$

Hence current $I = \frac{S}{V} = \frac{300}{150} = 2\,\text{A}$

Inductive reactance $X_L = \frac{V}{I} = \frac{150}{2} = 75\,\Omega$

Since $X_L = 2\pi fL$, inductance $L = \frac{X_L}{2\pi f} = \frac{75}{2\pi(50)}$

$$= \mathbf{0.239\,H}$$

Problem 27. A transformer has a rated output of 200 kVA at a power factor of 0.8. Determine the rated power output and the corresponding reactive power.

$VI = 200\,\text{kVA} = 200\times10^3$; p.f. $= 0.8 = \cos\phi$

Power output, $P = VI\cos\phi = (200\times10^3)(0.8)$

$$= \mathbf{160\,kW}$$

Reactive power, $Q = VI\sin\phi$

If $\cos\phi = 0.8$, then $\phi = \cos^{-1}0.8 = 36.87^\circ$

Hence $\sin\phi = \sin 36.87^\circ = 0.6$

Hence **reactive power**, $Q = (200\times10^3)(0.6)$

$$= \mathbf{120\,kvar}$$

Problem 28. The power taken by an inductive circuit when connected to a 120 V, 50 Hz supply is 400 W and the current is 8 A. Calculate (a) the resistance, (b) the impedance, (c) the reactance, (d) the power factor and (e) the phase angle between voltage and current.

(a) Power $P = I^2R$. Hence $R = \frac{P}{I^2} = \frac{400}{(8)^2} = \mathbf{6.25\,\Omega}$

(b) Impedance $Z = \frac{V}{I} = \frac{120}{8} = \mathbf{15\,\Omega}$

(c) Since $Z = \sqrt{(R^2 + X_L^2)}$,

then $X_L = \sqrt{(Z^2 - R^2)}$

$= \sqrt{[(15)^2 - (6.25)^2]}$

$= \mathbf{13.64\,\Omega}$

(d) Power factor $= \dfrac{\text{true power}}{\text{apparent power}}$

$$= \frac{VI\cos\phi}{VI}$$

$$= \frac{400}{(120)(8)} = \mathbf{0.4167}$$

(e) p.f. $= \cos\phi = 0.4167$. Hence phase angle,

$$\phi = \cos^{-1}0.4167$$
$$= \mathbf{65.37^\circ\ lagging}$$

Problem 29. A circuit consisting of a resistor in series with a capacitor takes 100 watts at a power factor of 0.5 from a 100 V, 60 Hz supply. Find (a) the current flowing, (b) the phase angle, (c) the resistance, (d) the impedance and (e) the capacitance.

(a) Power factor $= \dfrac{\text{true power}}{\text{apparent power}}$

i.e. $0.5 = \dfrac{100}{(100)(I)}$. Hence $I = \dfrac{100}{(0.5)(100)} = \mathbf{2\,A}$

(b) Power factor $= 0.5 = \cos\phi$. Hence phase angle,

$$\phi = \cos^{-1}0.5 = \mathbf{60^\circ\ leading}$$

(c) Power $P = I^2R$. Hence resistance $R = \dfrac{P}{I^2} = \dfrac{100}{(2)^2}$

$$= \mathbf{25\,\Omega}$$

(d) Impedance $Z = \dfrac{V}{I} = \dfrac{100}{2} = \mathbf{50\,\Omega}$

(e) Capacitive reactance, $X_C = \sqrt{(Z^2 - R^2)}$

$$= \sqrt{(50^2 - 25^2)}$$
$$= 43.30\,\Omega$$

$X_C = \dfrac{1}{2\pi fC}$ hence capacitance,

$$C = \frac{1}{2\pi f X_c} = \frac{1}{2\pi(60)(43.30)}\,\text{F}$$
$$= \mathbf{61.26\,\mu F}$$

Now try the following Practice Exercise

Practice Exercise 82 Power in a.c. circuits (Answers on page 822)

1. A voltage $v = 200\sin\omega t$ volts is applied across a pure resistance of 1.5 kΩ. Find the power dissipated in the resistor.
2. A 50 μF capacitor is connected to a 100 V, 200 Hz supply. Determine the true power and the apparent power.
3. A motor takes a current of 10 A when supplied from a 250 V a.c. supply. Assuming a power factor of 0.75 lagging, find the power consumed. Find also the cost of running the motor for 1 week continuously if 1 kWh of electricity costs 12.20 p.
4. A motor takes a current of 12 A when supplied from a 240 V a.c. supply. Assuming a power factor of 0.70 lagging, find the power consumed.
5. A substation is supplying 200 kVA and 150 kvar. Calculate the corresponding power and power factor.
6. A load takes 50 kW at a power factor of 0.8 lagging. Calculate the apparent power and the reactive power.
7. A coil of resistance 400 Ω and inductance 0.20 H is connected to a 75 V, 400 Hz supply. Calculate the power dissipated in the coil.
8. An 80 Ω resistor and a 6 μF capacitor are connected in series across a 150 V, 200 Hz supply. Calculate (a) the circuit impedance, (b) the current flowing and (c) the power dissipated in the circuit.
9. The power taken by a series circuit containing resistance and inductance is 240 W when connected to a 200 V, 50 Hz supply. If the current flowing is 2 A, find the values of the resistance and inductance.

10. A circuit consisting of a resistor in series with an inductance takes 210 W at a power factor of 0.6 from a 50 V, 100 Hz supply. Find (a) the current flowing, (b) the circuit phase angle, (c) the resistance, (d) the impedance and (e) the inductance.

11. A 200 V, 60 Hz supply is applied to a capacitive circuit. The current flowing is 2 A and the power dissipated is 150 W. Calculate the values of the resistance and capacitance.

For fully worked solutions to each of the problems in Practice Exercises 77 to 82 in this chapter, go to the website:
www.routledge.com/cw/bird

Chapter 18

Single-phase parallel a.c. circuits

***Why it is important to understand:* Single-phase parallel a.c. circuits**

The analysis of basic a.c. parallel electric circuits containing impedances and ideal a.c. supplies are presented in this chapter. Parallel networks containing *R–L*, *R–C*, *L–C* and *R–L–C* parallel circuits are explored using phasors, which greatly simplifies the analysis. Calculations of current, voltage, reactance, impedance and phase are explained via many worked examples. The important phenomena of resonance are explored in an *RL–C* parallel circuit, together with Q-factor, bandwidth and selectivity. Resonance is used in many different types of oscillator and filter circuits. A method of power factor improvement is explained. If a network is 100% efficient, its power factor is 1 or unity. This is the ideal for power transmission, but is practically impossible to attain. Variation in power factor is caused by different types of electrical devices connected to the grid that consume or generate reactive power. Unless this variation is corrected, higher currents are drawn from the grid, leading to grid instability, higher costs and reduced transmission capacity. A poor power factor results in additional costs for the electricity supplier. These costs are passed on to the customer as a 'reactive power charge' or 'exceeded capacity charge'. All UK electricity suppliers impose a reactive penalty charge when the average power factor falls below around 0.95. The causes of poor power factor include inductive loads on equipment such as a.c. motors, arc welders, furnaces, fluorescent lighting and air conditioning. The more inductive loads there are on the network, the greater the possibility there is of a poor power factor. Single-phase parallel a.c. circuit theory is of great importance in electrical/electronic engineering.

At the end of this chapter you should be able to:

- calculate unknown currents, impedances and circuit phase angle from phasor diagrams for (a) *R–L*, (b) *R–C*, (c) *L–C*, (d) *LR–C* parallel a.c. circuits
- state the condition for parallel resonance in an *LR–C* circuit
- derive the resonant frequency equation for an *LR–C* parallel a.c. circuit
- determine the current and dynamic resistance at resonance in an *LR–C* parallel circuit
- understand and calculate Q-factor in an *LR–C* parallel circuit
- understand how power factor may be improved

Electrical Circuit Theory and Technology. 978-1-138-67349-6, © 2017 John Bird. Published by Taylor & Francis. All rights reserved.

18.1 Introduction

In parallel circuits, such as those shown in Figures 18.1 and 18.2, the voltage is common to each branch of the network and is thus taken as the reference phasor when drawing phasor diagrams.

CIRCUIT DIAGRAM

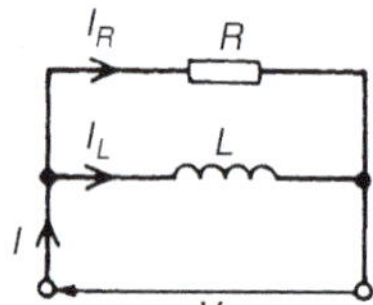

PHASOR DIAGRAM

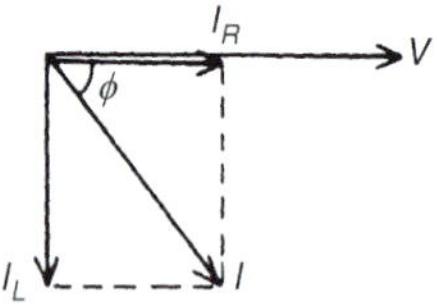

Figure 18.1

For any parallel a.c. circuit:

True or active power, $P = VI\cos\phi$ watts (W)

or $P = I_R^2 R$ watts

Apparent power, $S = VI$ voltamperes (VA)

Reactive power, $Q = VI\sin\phi$ reactive voltamperes (var)

$$\text{Power factor} = \frac{\text{true power}}{\text{apparent power}} = \frac{P}{S} = \cos\phi$$

(These formulae are the same as for series a.c. circuits as used in Chapter 17.)

18.2 *R–L* parallel a.c. circuit

In the two-branch parallel circuit containing resistance R and inductance L shown in Figure 18.1, the current flowing in the resistance, I_R, is in-phase with the supply voltage V and the current flowing in the inductance, I_L, lags the supply voltage by 90°. The supply current I is the phasor sum of I_R and I_L and thus the current I lags the applied voltage V by an angle lying between 0° and 90° (depending on the values of I_R and I_L), shown as angle ϕ in the phasor diagram.

From the phasor diagram:

$$I = \sqrt{(I_R^2 + I_L^2)} \text{ (by Pythagoras' theorem)}$$

$$\text{where } I_R = \frac{V}{R} \text{ and } I_L = \frac{V}{X_L}$$

$$\tan\phi = \frac{I_L}{I_R},\ \sin\phi = \frac{I_L}{I} \text{ and}$$

$$\cos\phi = \frac{I_R}{I} \text{ (by trigonometric ratios)}$$

$$\text{Circuit impedance, } Z = \frac{V}{I}$$

Problem 1. A 20 Ω resistor is connected in parallel with an inductance of 2.387 mH across a 60 V, 1 kHz supply. Calculate (a) the current in each branch, (b) the supply current, (c) the circuit phase angle, (d) the circuit impedance, and (e) the power consumed.

The circuit and phasor diagrams are as shown in Figure 18.1.

(a) Current flowing in the resistor $I_R = \frac{V}{R} = \frac{60}{20} = \mathbf{3\,A}$

Current flowing in the inductance

$$I_L = \frac{V}{X_L} = \frac{V}{2\pi f L} = \frac{60}{2\pi(1000)(2.387\times10^{-3})} = \mathbf{4\,A}$$

(b) From the phasor diagram, supply current,

$$I = \sqrt{(I_R^2 + I_L^2)} = \sqrt{(3^2+4^2)} = \mathbf{5\,A}$$

(c) Circuit phase angle, $\phi = \tan^{-1}\frac{I_L}{I_R} = \tan^{-1}\left(\frac{4}{3}\right)$

$= \mathbf{53.13°\ lagging}$

(d) Circuit impedance, $Z = \frac{V}{I} = \frac{60}{5} = \mathbf{12\,\Omega}$

(e) Power consumed $P = VI\cos\phi$

$= (60)(5)(\cos 53.13°)$

$= \mathbf{180\,W}$

(Alternatively, power consumed $P = I_R^2 R = (3)^2(20) = \mathbf{180\,W}$)

Part 3

Now try the following Practice Exercise

Practice Exercise 83 *R–L* parallel a.c. circuits (Answers on page 822)

1. A 30 Ω resistor is connected in parallel with a pure inductance of 3 mH across a 110 V, 2 kHz supply. Calculate (a) the current in each branch, (b) the circuit current, (c) the circuit phase angle, (d) the circuit impedance, (e) the power consumed, and (f) the circuit power factor.
2. A 40 Ω resistance is connected in parallel with a coil of inductance L and negligible resistance across a 200 V, 50 Hz supply and the supply current is found to be 8 A. Sketch a phasor diagram and determine the inductance of the coil.

18.3 *R–C* parallel a.c. circuit

In the two-branch parallel circuit containing resistance R and capacitance C shown in Figure 18.2, I_R is in-phase with the supply voltage V and the current flowing in the capacitor, I_C, leads V by 90°. The supply current I is the phasor sum of I_R and I_C and thus the current I leads the applied voltage V by an angle lying between 0° and 90° (depending on the values of I_R and I_C), shown as angle α in the phasor diagram.

From the phasor diagram:

$$I = \sqrt{(I_R^2 + I_C^2)} \text{ (by Pythagoras' theorem)}$$

where $I_R = \dfrac{V}{R}$ and $I_C = \dfrac{V}{X_C}$

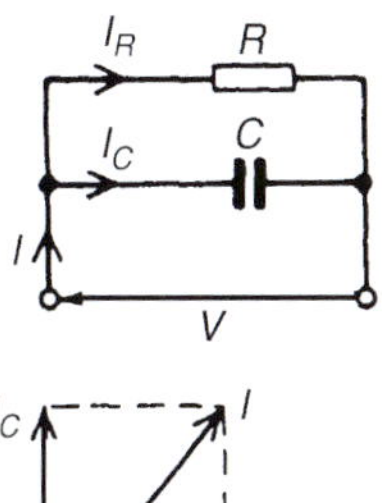

Figure 18.2

$$\tan\alpha = \frac{I_C}{I_R},\ \sin\alpha = \frac{I_C}{I} \text{ and } \cos\alpha = \frac{I_R}{I}$$

(by trigonometric ratios)

Circuit impedance $Z = \dfrac{V}{I}$

Problem 2. A 30 μF capacitor is connected in parallel with an 80 Ω resistor across a 240 V, 50 Hz supply. Calculate (a) the current in each branch, (b) the supply current, (c) the circuit phase angle, (d) the circuit impedance, (e) the power dissipated and (f) the apparent power.

The circuit and phasor diagrams are as shown in Figure 18.2.

(a) Current in resistor, $I_R = \dfrac{V}{R} = \dfrac{240}{80} = \mathbf{3\ A}$

Current in capacitor, $I_C = \dfrac{V}{X_C} = \dfrac{V}{\left(\dfrac{1}{2\pi f C}\right)}$

$$= 2\pi f C V$$

$$= 2\pi(50)(30\times10^{-6})(240)$$

$$= \mathbf{2.262\ A}$$

(b) Supply current, $I = \sqrt{(I_R^2 + I_C^2)} = \sqrt{(3^2 + 2.262^2)}$

$$= \mathbf{3.757\ A}$$

(c) Circuit phase angle, $\alpha = \tan^{-1}\dfrac{I_C}{I_R} = \tan^{-1}\left(\dfrac{2.262}{3}\right)$

$$= \mathbf{37.02°\ leading}$$

(d) Circuit impedance, $Z = \dfrac{V}{I} = \dfrac{240}{3.757} = \mathbf{63.88\ \Omega}$

(e) True or active power dissipated,

$$P = VI\cos\alpha$$

$$= 240(3.757)\cos 37.02°$$

$$= \mathbf{720\ W}$$

(Alternatively, true power $P = I_R^2 R = (3)^2(80)$
$= 720$ W)

(f) Apparent power, $S = VI = (240)(3.757)$

$$= \mathbf{901.7\ VA}$$

Problem 3. A capacitor C is connected in parallel with a resistor R across a 120 V, 200 Hz supply. The supply current is 2 A at a power factor of 0.6 leading. Determine the values of C and R.

The circuit diagram is shown in Figure 18.3(a).

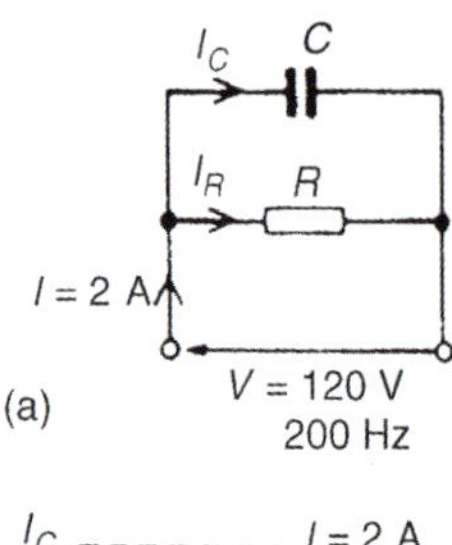

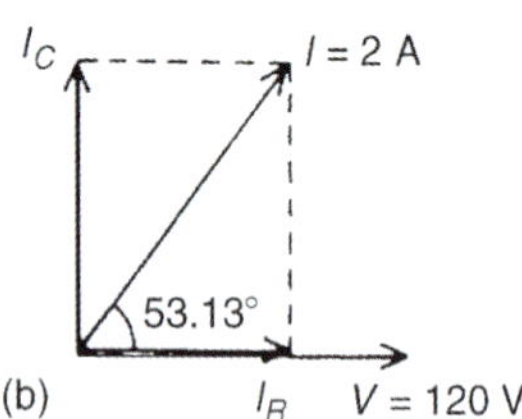

Figure 18.3

Power factor $= \cos\phi = 0.6$ leading, hence $\phi = \cos^{-1} 0.6 = 53.13°$ leading.

From the phasor diagram shown in Figure 18.3(b),

$$I_R = I\cos 53.13° = (2)(0.6)$$
$$= \mathbf{1.2\,A}$$

and $I_C = I\sin 53.13° = (2)(0.8)$

$$= \mathbf{1.6\,A}$$

(Alternatively, I_R and I_C can be measured from the scaled phasor diagram.)

From the circuit diagram,

$$I_R = \frac{V}{R} \text{ from which } R = \frac{V}{I_R} = \frac{120}{1.2} = \mathbf{100\,\Omega}$$

and $I_C = \dfrac{V}{X_C} = 2\pi fCV$, from which,

$$C = \frac{I_C}{2\pi fV}$$
$$= \frac{1.6}{2\pi(200)(120)}$$
$$= \mathbf{10.61\,\mu F}$$

Now try the following Practice Exercise

Practice Exercise 84 *R–C* parallel a.c. circuits (Answers on page 823)

1. A 1500 nF capacitor is connected in parallel with a 16 Ω resistor across a 10 V, 10 kHz supply. Calculate (a) the current in each branch, (b) the supply current, (c) the circuit phase angle, (d) the circuit impedance, (e) the power consumed, (f) the apparent power and (g) the circuit power factor. Sketch the phasor diagram.
2. A capacitor C is connected in parallel with a resistance R across a 60 V, 100 Hz supply. The supply current is 0.6 A at a power factor of 0.8 leading. Calculate the values of R and C.

18.4 *L–C* parallel a.c. circuit

In the two-branch parallel circuit containing inductance L and capacitance C shown in Figure 18.4, I_L lags V by 90° and I_C leads V by 90°.

Theoretically there are three phasor diagrams possible – each depending on the relative values of I_L and I_C:

(i) $I_L > I_C$ (giving a supply current, $I = I_L - I_C$ lagging V by 90°)

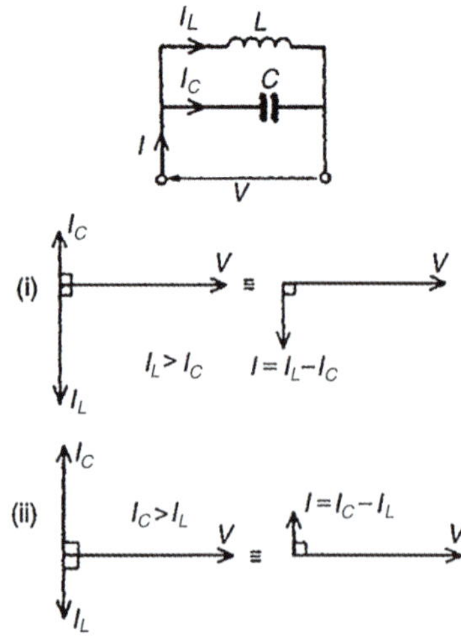

Figure 18.4

(ii) $I_C > I_L$ (giving a supply current, $I = I_C - I_L$ leading V by 90°)

(iii) $I_L = I_C$ (giving a supply current, $I = 0$)

The latter condition is not possible in practice due to circuit resistance inevitably being present (as in the circuit described in Section 18.5).

For the L–C parallel circuit, $I_L = \dfrac{V}{X_L}$, $I_C = \dfrac{V}{X_C}$

I = phasor difference between I_L and I_C, and $Z = \dfrac{V}{I}$

Problem 4. A pure inductance of 120 mH is connected in parallel with a 25 μF capacitor and the network is connected to a 100 V, 50 Hz supply. Determine (a) the branch currents, (b) the supply current and its phase angle, (c) the circuit impedance and (d) the power consumed.

The circuit and phasor diagrams are as shown in Figure 18.4.

(a) Inductive reactance, $X_L = 2\pi fL$

$$= 2\pi(50)(120 \times 10^{-3})$$

$$= 37.70\,\Omega$$

Capacitive reactance, $X_C = \dfrac{1}{2\pi fC}$

$$= \frac{1}{2\pi(50)(25 \times 10^{-6})}$$

$$= 127.3\,\Omega$$

Current flowing in inductance, $I_L = \dfrac{V}{X_L} = \dfrac{100}{37.70}$

$$= \mathbf{2.653\,A}$$

Current flowing in capacitor, $I_C = \dfrac{V}{X_C} = \dfrac{100}{127.3}$

$$= \mathbf{0.786\,A}$$

(b) I_L and I_C are anti-phase. Hence supply current,

$I = I_L - I_C = 2.653 - 0.786 =$ **1.867 A and the current lags the supply voltage V by 90°** (see Figure 18.4(i)).

(c) Circuit impedance, $Z = \dfrac{V}{I} = \dfrac{100}{1.867} = \mathbf{53.56\,\Omega}$

(d) Power consumed, $P = VI\cos\phi$

$$= (100)(1.867)(\cos 90°)$$

$$= \mathbf{0\,W}$$

Problem 5. Repeat worked Problem 4 for the condition when the frequency is changed to 150 Hz.

(a) Inductive reactance, $X_L = 2\pi(150)(120 \times 10^{-3})$

$$= 113.1\,\Omega$$

Capacitive reactance, $X_C = \dfrac{1}{2\pi(150)(25 \times 10^{-6})}$

$$= 42.44\,\Omega$$

Current flowing in inductance, $I_L = \dfrac{V}{X_L} = \dfrac{100}{113.1}$

$$= \mathbf{0.844\,A}$$

Current flowing in capacitor, $I_C = \dfrac{V}{X_C} = \dfrac{100}{42.44}$

$$= \mathbf{2.356\,A}$$

(b) Supply current, $I = I_C - I_L = 2.356 - 0.884$

= **1.472 A leading *V* by 90°** (see Figure 18.4(ii))

(c) Circuit impedance, $Z = \dfrac{V}{I} = \dfrac{100}{1.472} = \mathbf{67.93\,\Omega}$

(d) Power consumed, $P = VI\cos\phi = \mathbf{0\,W}$ (since $\phi = 90°$)

From Problems 4 and 5:

(i) When $X_L < X_C$ then $I_L > I_C$ and I lags V by 90°

(ii) When $X_L > X_C$ then $I_L < I_C$ and I leads V by 90°

(iii) In a parallel circuit containing no resistance the power consumed is zero

---

Now try the following Practice Exercise

Practice Exercise 85 *L–C* parallel a.c. circuits (Answers on page 823)

1. An inductance of 80 mH is connected in parallel with a capacitance of 10 μF across a 60 V,

100 Hz supply. Determine (a) the branch currents, (b) the supply current, (c) the circuit phase angle, (d) the circuit impedance and (e) the power consumed.

2. Repeat Problem 1 for a supply frequency of 200 Hz.

18.5 *LR–C* parallel a.c. circuit

In the two-branch circuit containing capacitance C in parallel with inductance L and resistance R in series (such as a coil) shown in Figure 18.5(a), the phasor diagram for the LR branch alone is shown in Figure 18.5(b) and the phasor diagram for the C branch is shown alone in Figure 18.5(c). Rotating each and superimposing on one another gives the complete phasor diagram shown in Figure 18.5(d).

The current I_{LR} of Figure 18.5(d) may be resolved into horizontal and vertical components. The horizontal component, shown as op, is $I_{LR}\cos\phi_1$ and the vertical component, shown as pq, is $I_{LR}\sin\phi_1$. There are three possible conditions for this circuit:

(i) $I_C > I_{LR}\sin\phi_1$ (giving a supply current I leading V by angle ϕ – as shown in Figure 18.5(e))

(ii) $I_{LR}\sin\phi_1 > I_C$ (giving I lagging V by angle ϕ – as shown in Figure 18.5(f))

(iii) $I_C = I_{LR}\sin\phi_1$ (this is called parallel resonance, see Section 18.6).

There are two methods of finding the phasor sum of currents I_{LR} and I_C in Figures 18.5(e) and (f). These are: (i) by a scaled phasor diagram, or (ii) by resolving each current into their **'in-phase'** (i.e. horizontal) and **'quadrature'** (i.e. vertical) **components**, as demonstrated in Problems 6 and 7.

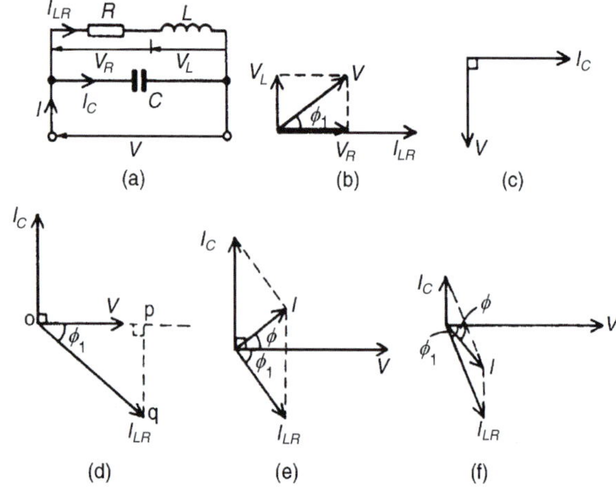

Figure 18.5

With reference to the phasor diagrams of Figure 18.5:

Impedance of LR branch, $Z_{LR} = \sqrt{(R^2 + X_L^2)}$

Current, $I_{LR} = \dfrac{V}{Z_{LR}}$ and $I_C = \dfrac{V}{X_C}$

Supply current I = phasor sum of I_{LR} and I_C (by drawing)

$$= \sqrt{\{(I_{LR}\cos\phi_1)^2 + (I_{LR}\sin\phi_1 \sim I_C)^2\}}$$ (by calculation)

where $\sim$ means 'the difference between'.

Circuit impedance $Z = \dfrac{V}{I}$

$$\tan\phi_1 = \frac{V_L}{V_R} = \frac{X_L}{R},\ \sin\phi_1 = \frac{X_L}{Z_{LR}} \text{ and } \cos\phi_1 = \frac{R}{Z_{LR}}$$

$$\tan\phi = \frac{I_{LR}\sin\phi_1 \sim I_C}{I_{LR}\cos\phi_1} \text{ and } \cos\phi = \frac{I_{LR}\cos\phi_1}{I}$$

Problem 6. A coil of inductance 159.2 mH and resistance 40 Ω is connected in parallel with a 30 μF capacitor across a 240 V, 50 Hz supply. Calculate (a) the current in the coil and its phase angle, (b) the current in the capacitor and its phase angle, (c) the supply current and its phase angle, (d) the circuit impedance, (e) the power consumed, (f) the apparent power and (g) the reactive power. Draw the phasor diagram.

The circuit diagram is shown in Figure 18.6(a).

(a) For the coil, inductive reactance

$$X_L = 2\pi f L = 2\pi(50)(159.2\times10^{-3}) = 50\,\Omega$$

Impedance $Z_1 = \sqrt{(R^2 + X_L^2)} = \sqrt{(40^2 + 50^2)} = 64.03\,\Omega$

Current in coil, $I_{LR} = \dfrac{V}{Z_1} = \dfrac{240}{64.03} = \mathbf{3.748\ A}$

Branch phase angle $\phi_1 = \tan^{-1}\dfrac{X_L}{R} = \tan^{-1}\left(\dfrac{50}{40}\right) = \tan^{-1}1.25 = \mathbf{51.34°\ lagging}$

(see phasor diagram in Figure 18.6(b))

(b) Capacitive reactance, $X_C = \dfrac{1}{2\pi f C}$

$$= \frac{1}{2\pi(50)(30\times 10^{-6})}$$

$$= 106.1\,\Omega$$

Current in capacitor, $I_C = \dfrac{V}{X_C} = \dfrac{240}{106.1}$

$= \mathbf{2.262\,A}$ **leading the supply voltage by 90°**

(see phasor diagram of Figure 18.6(b)).

(c) The supply current I is the phasor sum of I_{LR} and I_C. This may be obtained by drawing the phasor diagram to scale and measuring the current I and its phase angle relative to V. (Current I will always be the diagonal of the parallelogram formed as in Figure 18.6(a))

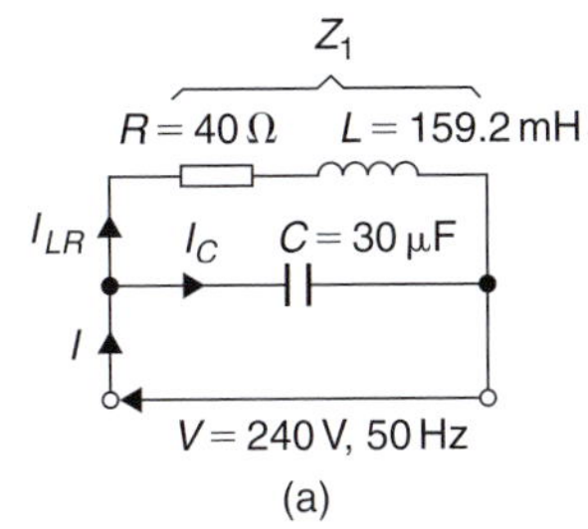

(a)

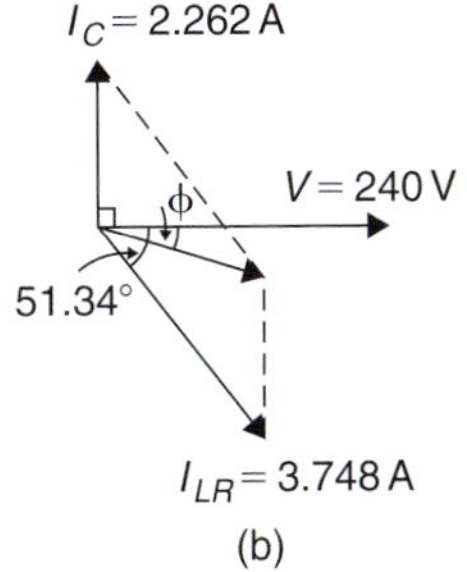

(b)

Figure 18.6

Alternatively the current I_{LR} and I_C may be resolved into their horizontal (or 'in-phase') and vertical (or 'quadrature') components. The horizontal component of I_{LR} is

$$I_{LR}\cos(51.34°) = 3.748\cos 51.34° = 2.342\,\text{A}$$

The horizontal component of I_C is $I_C \cos 90° = 0$

Thus the total horizontal component, $I_H = \mathbf{2.342\,A}$

The vertical component of

$$I_{LR} = -I_{LR}\sin(51.34°)$$
$$= -3.748\sin 51.34°$$
$$= -2.926\,\text{A}$$

The vertical component of $I_C = I_C \sin 90°$

$$= 2.262\sin 90°$$
$$= 2.262\,\text{A}$$

Thus the total vertical component,

$$I_V = -2.926 + 2.262$$
$$= \mathbf{-0.664\,A}$$

I_H and I_V are shown in Figure 18.7, from which,

$$I = \sqrt{[2.342]^2 + (-0.664)^2]} = 2.434\,\text{A}$$

$I_H = 2.342$ A
φ
$I_V = -0.664$ A
I

Figure 18.7

$$\text{Angle } \phi = \tan^{-1}\left(\frac{0.664}{2.342}\right) = 15.83° \text{ lagging}$$

Hence the supply current $I = 2.434$ A lagging V by 15.83°

(d) Circuit impedance, $Z = \dfrac{V}{I} = \dfrac{240}{2.434} = \mathbf{98.60\,\Omega}$

(e) Power consumed, $P = VI\cos\phi$

$$= (240)(2.434)\cos 15.83°$$
$$= \mathbf{562\,W}$$

(Alternatively, $P = I_R^2 R = I_{LR}^2 R$ (in this case)

$$= (3.748)^2 (40) = \mathbf{562\,W})$$

(f) Apparent power, $S = VI = (240)(2.434)$

$$= \mathbf{584.2\,VA}$$

(g) Reactive power, $Q = VI\sin\phi$

$$= (240)(2.434)(\sin 15.83°)$$
$$= \mathbf{159.4\,var}$$

Problem 7. A coil of inductance 0.12 H and resistance 3 kΩ is connected in parallel with a 0.02 μF capacitor and is supplied at 40 V at a

Part 3

frequency of 5 kHz. Determine (a) the current in the coil, and (b) the current in the capacitor. (c) Draw to scale the phasor diagram and measure the supply current and its phase angle; check the answer by calculation. Determine (d) the circuit impedance and (e) the power consumed.

The circuit diagram is shown in Figure 18.8(a).

$R = 3\,\text{k}\Omega$ $L = 0.12\,\text{H}$

I_{LR} $C = 0.02\,\mu\text{F}$

I_C

I

$V = 40\,\text{V}$, 5 kHz

(a)

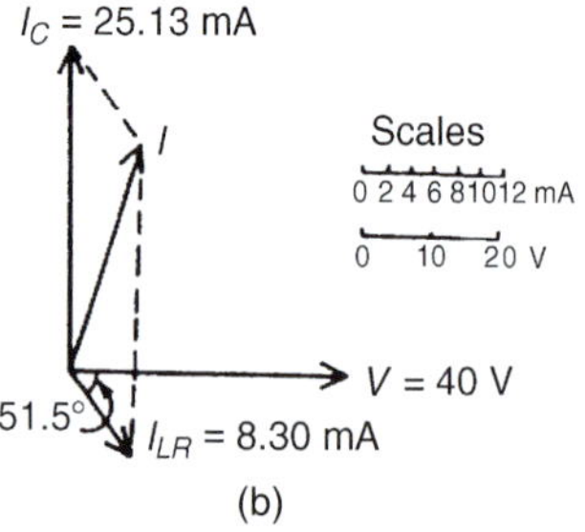

(b)

Figure 18.8

(a) Inductive reactance, $X_L = 2\pi fL = 2\pi(5000)(0.12)$
$= 3770\,\Omega$

Impedance of coil, $Z_1 = \sqrt{(R^2 + X_L^2)}$
$= \sqrt{[(3000)^2 + (3770)^2]}$
$= 4818\,\Omega$

Current in coil, $I_{LR} = \dfrac{V}{Z_1} = \dfrac{40}{4818} = \mathbf{8.30\,mA}$

Branch phase angle $\phi = \tan^{-1}\dfrac{X_L}{R} = \tan^{-1}\dfrac{3770}{3000}$
$= \mathbf{51.5°}$ **lagging**

(b) Capacitive reactance, $X_C = \dfrac{1}{2\pi fC}$

$= \dfrac{1}{2\pi(5000)(0.02 \times 10^{-6})}$

$= 1592\,\Omega$

Capacitor current, $I_C = \dfrac{V}{X_C} = \dfrac{40}{1592}$
$= \mathbf{25.13\,mA}$ **leading** $\boldsymbol{V}$ **by 90°**

(c) Currents I_{LR} and I_C are shown in the phasor diagram of Figure 18.8(b). The parallelogram is completed as shown and the supply current is given by the diagonal of the parallelogram. The current I is measured as **19.3 mA** leading voltage V by **74.5°**

By calculation,

$I = \sqrt{[(I_{LR}\cos 51.5°)^2 + (I_C - I_{LR}\sin 51.5°)^2]}$
$= \mathbf{19.34\,mA}$

and $\phi = \tan^{-1}\left(\dfrac{I_C - I_{LR}\sin 51.5°}{I_{LR}\cos 51.5°}\right) = \mathbf{74.50°}$

(d) Circuit impedance, $Z = \dfrac{V}{I} = \dfrac{40}{19.34 \times 10^{-3}}$
$= \mathbf{2.068\,k\Omega}$

(e) Power consumed, $P = VI\cos\phi$
$= (40)(19.34 \times 10^{-3})(\cos 74.50°)$
$= \mathbf{206.7\,mW}$

(Alternatively, $P = I_R^2 R = I_{LR}^2 R$
$= (8.30 \times 10^{-3})^2\,(3000)$
$= \mathbf{206.7\,mW}$)

Now try the following Practice Exercise

Practice Exercise 86 *L–R–C* parallel a.c. circuits (Answers on page 823)

1. A coil of resistance 60 Ω and inductance 318.4 mH is connected in parallel with a 15 μF capacitor across a 200 V, 50 Hz supply. Calculate (a) the current in the coil, (b) the current in the capacitor, (c) the supply current and its phase angle, (d) the circuit impedance, (e) the power consumed, (f) the apparent power and (g) the reactive power. Sketch the phasor diagram.
2. A 25 nF capacitor is connected in parallel with a coil of resistance 2 kΩ and inductance 0.20 H across a 100 V, 4 kHz supply. Determine (a) the current in the coil, (b) the current in the capacitor, (c) the supply current and its phase angle (by drawing a phasor diagram to scale, and also by calculation), (d) the circuit impedance and (e) the power consumed.

18.6 Parallel resonance and Q-factor

Parallel resonance

Resonance occurs in the two-branch network containing capacitance C in parallel with inductance L and resistance R in series (see Figure 18.5(a)) when the quadrature (i.e. vertical) component of current I_{LR} is equal to I_C. At this condition the supply current I is in-phase with the supply voltage V.

Resonant frequency

When the quadrature component of I_{LR} is equal to I_C then: $I_C = I_{LR}\sin\phi_1$ (see Figure 18.9)

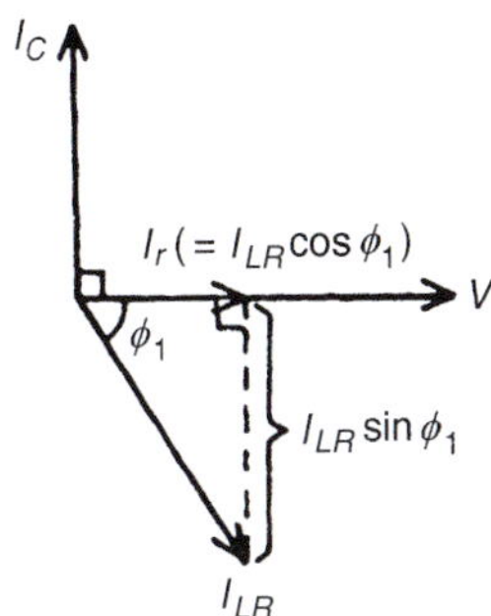

Figure 18.9

Hence $\dfrac{V}{X_c} = \left(\dfrac{V}{Z_{LR}}\right)\left(\dfrac{X_L}{Z_{LR}}\right)$ (from Section 18.5)

from which, $Z_{LR}^2 = X_C X_L = (2\pi f_r L)\left(\dfrac{1}{2\pi f_r C}\right) = \dfrac{L}{C}$ (1)

Hence $[\sqrt{(R^2+X_L^2)}]^2 = \dfrac{L}{C}$ and $R^2 + X_L^2 = \dfrac{L}{C}$

Thus $(2\pi f_r L)^2 = \dfrac{L}{C} - R^2$ and

$$2\pi f_r L = \sqrt{\left(\frac{L}{C} - R^2\right)}$$

and

$$f_r = \frac{1}{2\pi L}\sqrt{\left(\frac{L}{C} - R^2\right)}$$

$$= \frac{1}{2\pi}\sqrt{\left(\frac{L}{L^2 C} - \frac{R^2}{L^2}\right)}$$

i.e. parallel resonant frequency,

$$\boldsymbol{f_r = \frac{1}{2\pi}\sqrt{\left(\frac{1}{LC} - \frac{R^2}{L^2}\right)}\ \text{Hz}}$$

(When R is negligible, then $f_r = \dfrac{1}{2\pi\sqrt{(LC)}}$, which is the same as for series resonance.)

Current at resonance

Current at resonance, $I_r = I_{LR}\cos\phi_1$ (from Figure 18.9)

$$= \left(\frac{V}{Z_{LR}}\right)\left(\frac{R}{Z_{LR}}\right)$$

(from Section 18.5)

$$= \frac{VR}{Z_{LR}^2}$$

However, from equation (18.1), $Z_{LR}^2 = \dfrac{L}{C}$

hence $\boldsymbol{I_r = \dfrac{VR}{\frac{L}{C}} = \dfrac{VRC}{L}}$ (18.2)

The current is at a **minimum** at resonance.

Dynamic resistance

Since the current at resonance is in-phase with the voltage the impedance of the circuit acts as a resistance. This resistance is known as the **dynamic resistance, R_D** (or sometimes, the dynamic impedance).

From equation (18.2), impedance at resonance

$$= \frac{V}{I_r} = \frac{V}{\left(\frac{VRC}{L}\right)} = \frac{L}{RC}$$

i.e. dynamic resistance, $\boldsymbol{R_D = \dfrac{L}{RC}}$ **ohms**

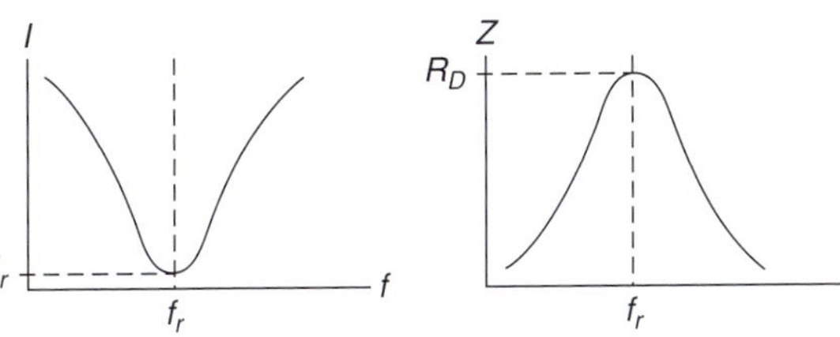

Figure 18.10

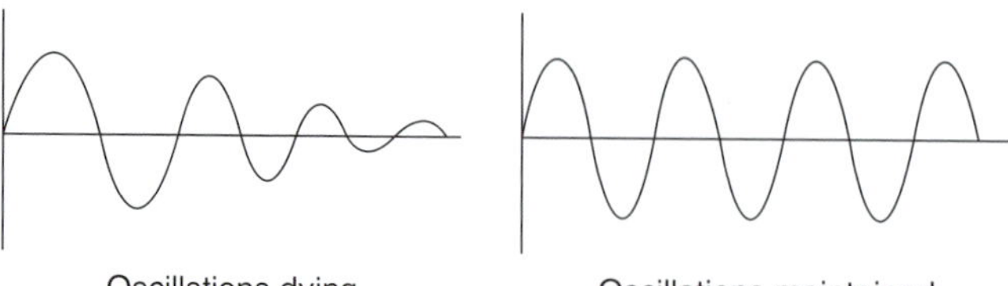

Figure 18.11

Graphs of current and impedance against frequency near to resonance for a parallel circuit are shown in Figure 18.10, and are seen to be the reverse of those in a series circuit (from page 270).

Rejector circuit

The parallel resonant circuit is often described as a **rejector** circuit since it presents its maximum impedance at the resonant frequency and the resultant current is a minimum.

Mechanical analogy

Electrical resonance for the parallel circuit can be likened to a mass hanging on a spring which, if pulled down and released, will oscillate up and down but due to friction the oscillations will slowly die. To maintain the oscillation the mass would require a small force applied each time it reaches its point of maximum travel and this is exactly what happens with the electrical circuit. A small current is required to overcome the losses and maintain the oscillations of current. Figure 18.11 shows the two cases.

Applications of resonance

One use for resonance is to establish a condition of **stable frequency** in circuits designed to produce a.c. signals. Usually, a parallel circuit is used for this purpose, with the capacitor and inductor directly connected together, exchanging energy between each other. Just as a pendulum can be used to stabilize the frequency of a clock mechanism's oscillations, so can a parallel circuit be used to stabilize the electrical frequency of an a.c. oscillator circuit.

Another use for resonance is in applications where the effects of greatly increased or decreased impedance at a particular frequency is desired. A resonant circuit can be used to 'block' (i.e. present high impedance towards) a frequency or range of frequencies, thus acting as a sort of frequency **'filter'** to strain certain frequencies out of a mix of others. In fact, these particular circuits are called filters, and their design is considered in Chapter 45. In essence, this is how analogue radio receiver tuner circuits work to filter, or select, one station frequency out of the mix of different radio station frequency signals intercepted by the antenna.

Q-factor

Currents higher than the supply current can circulate within the parallel branches of a parallel resonant circuit, the current leaving the capacitor and establishing the magnetic field of the inductor, this then collapsing and recharging the capacitor, and so on. The **Q-factor** of a parallel resonant circuit is the ratio of the current circulating in the parallel branches of the circuit to the supply current, i.e. the current magnification.

Q-factor at resonance = current magnification

$$= \frac{\text{circulating current}}{\text{supply current}}$$

$$= \frac{I_C}{I_r} = \frac{I_{LR}\sin\phi_1}{I_r}$$

$$= \frac{I_{LR}\sin\phi_1}{I_{LR}\cos\phi_1} = \frac{\sin\phi_1}{\cos\phi_1}$$

$$= \tan\phi_1 = \frac{X_L}{R}$$

i.e. **Q-factor at resonance** $= \boldsymbol{\frac{2\pi f_r L}{R}}$

(which is the same as for a series circuit)

Note that in a **parallel** circuit the Q-factor is a measure of **current magnification**, whereas in a **series** circuit it is a measure of **voltage magnification**.

At mains frequencies the Q-factor of a parallel circuit is usually low, typically less than 10, but in radio-frequency circuits the Q-factor can be very high.

Problem 8. A pure inductance of 150 mH is connected in parallel with a 40 μF capacitor across a 50 V, variable frequency supply. Determine (a) the resonant frequency of the circuit and (b) the current circulating in the capacitor and inductance at resonance.

The circuit diagram is shown in Figure 18.12.

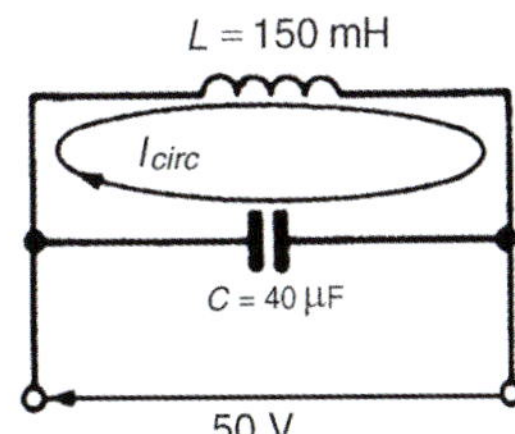

Figure 18.12

(a) Parallel resonant frequency,

$$f_r = \frac{1}{2\pi}\sqrt{\left(\frac{1}{LC} - \frac{R^2}{L^2}\right)}$$

However, resistance $R = 0$. Hence,

$$f_r = \frac{1}{2\pi}\sqrt{\left(\frac{1}{LC}\right)}$$

$$= \frac{1}{2\pi}\sqrt{\left[\frac{1}{(150\times10^{-3})(40\times10^{-6})}\right]}$$

$$= \frac{1}{2\pi}\sqrt{\left(\frac{10^7}{(15)(4)}\right)}$$

$$= \frac{10^3}{2\pi}\sqrt{\left(\frac{1}{6}\right)} = \mathbf{64.97\,Hz}$$

(b) Current circulating in L and C at resonance,

$$I_{CIRC} = \frac{V}{X_C} = \frac{V}{\left(\dfrac{1}{2\pi f_r C}\right)} = 2\pi f_r CV$$

Hence $I_{CIRC} = 2\pi(64.97)(40\times10^{-6})(50)$
$= \mathbf{0.816\,A}$

(Alternatively, $I_{CIRC} = \dfrac{V}{X_L} = \dfrac{V}{2\pi f_r L}$

$= \dfrac{50}{2\pi(64.97)(0.15)} = \mathbf{0.817\,A}$)

Problem 9. A coil of inductance 0.20 H and resistance 60 Ω is connected in parallel with a 20 μF capacitor across a 20 V, variable frequency supply. Calculate (a) the resonant frequency, (b) the dynamic resistance, (c) the current at resonance and (d) the circuit Q-factor at resonance.

(a) Parallel resonant frequency,

$$f_r = \frac{1}{2\pi}\sqrt{\left(\frac{1}{LC} - \frac{R^2}{L^2}\right)}$$

$$= \frac{1}{2\pi}\sqrt{\left(\frac{1}{(0.20)(20\times10^{-6})} - \frac{(60)^2}{(0.2)^2}\right)}$$

$$= \frac{1}{2\pi}\sqrt{(250\,000 - 90\,000)}$$

$$= \frac{1}{2\pi}\sqrt{(160\,000)} = \frac{1}{2\pi}(400)$$

$$= \mathbf{63.66\,Hz}$$

(b) Dynamic resistance, $R_D = \dfrac{L}{RC} = \dfrac{0.20}{(60)(20\times10^{-6})}$

$= \mathbf{166.7\,\Omega}$

(c) Current at resonance, $I_r = \dfrac{V}{R_D} = \dfrac{20}{166.7} = \mathbf{0.12\,A}$

(d) Circuit Q-factor at resonance $= \dfrac{2\pi f_r L}{R}$

$= \dfrac{2\pi(63.66)(0.2)}{60}$

$= \mathbf{1.33}$

Alternatively, Q-factor at resonance = current magnification (for a parallel circuit) $= \dfrac{I_c}{I_r}$

$$I_c = \frac{V}{X_c} = \frac{V}{\left(\dfrac{1}{2\pi f_r C}\right)} = 2\pi f_r CV$$

$= 2\pi(63.66)(20\times10^{-6})(20)$

$= 0.16\,A$

Hence Q-factor $= \dfrac{I_c}{I_r} = \dfrac{0.16}{0.12} = \mathbf{1.33}$, as obtained above

Problem 10. A coil of inductance 100 mH and resistance 800 Ω is connected in parallel with a variable capacitor across a 12 V, 5 kHz supply. Determine for the condition when the supply current is a minimum: (a) the capacitance of the capacitor, (b) the dynamic resistance, (c) the supply current and (d) the Q-factor.

(a) The supply current is a minimum when the parallel circuit is at resonance.

Resonant frequency, $f_r = \dfrac{1}{2\pi}\sqrt{\left(\dfrac{1}{LC} - \dfrac{R^2}{L^2}\right)}$

Transposing for C gives: $(2\pi f_r)^2 = \dfrac{1}{LC} - \dfrac{R^2}{L^2}$

$$(2\pi f_r)^2 + \frac{R^2}{L^2} = \frac{1}{LC}$$

$$C = \frac{1}{L\left\{(2\pi f_r)^2 + \dfrac{R^2}{L^2}\right\}}$$

When $L = 100\,\text{mH}$, $R = 800\,\Omega$ and $f_r = 5000\,\text{Hz}$,

$$C = \frac{1}{100\times 10^{-3}\left\{(2\pi 5000)^2 + \dfrac{800^2}{(100\times 10^{-3})^2}\right\}}$$

$$= \frac{1}{0.1[\pi^2 10^8 + (0.64)10^8]}\,\text{F}$$

$$= \frac{10^6}{0.1(10.51\times 10^8)}\,\mu\text{F}$$

$= \mathbf{0.009515\,\mu F}$ or $\mathbf{9.515\,nF}$

(b) Dynamic resistance, $R_D = \dfrac{L}{CR}$

$$= \frac{100\times 10^{-3}}{(9.515\times 10^{-9})(800)}$$

$= \mathbf{13.14\,k\Omega}$

(c) Supply current at resonance, $I_r = \dfrac{V}{R_D}$

$$= \frac{12}{13.14\times 10^3}$$

$= \mathbf{0.913\,mA}$

(d) Q-factor at resonance $= \dfrac{2\pi f_r L}{R}$

$$= \frac{2\pi(5000)(100\times 10^{-3})}{800}$$

$= \mathbf{3.93}$

Alternatively, Q-factor at resonance

$$= \frac{I_c}{I_r} = \frac{V/X_c}{I_r} = \frac{2\pi f_r CV}{I_r}$$

$$= \frac{2\pi(5000)(9.515\times 10^{-9})(12)}{0.913\times 10^{-3}}$$

$= \mathbf{3.93}$

For a practical laboratory experiment on parallel a.c. circuits and resonance, see the website.

Now try the following Practice Exercise

Practice Exercise 87 Parallel resonance and Q-factor (Answers on page 823)

1. A $0.15\,\mu$F capacitor and a pure inductance of 0.01 H are connected in parallel across a 10 V, variable frequency supply. Determine (a) the resonant frequency of the circuit, and (b) the current circulating in the capacitor and inductance.

2. A $30\,\mu$F capacitor is connected in parallel with a coil of inductance 50 mH and unknown resistance R across a 120 V, 50 Hz supply. If the circuit has an overall power factor of 1, find (a) the value of R, (b) the current in the coil and (c) the supply current.

3. A coil of resistance $25\,\Omega$ and inductance 150 mH is connected in parallel with a $10\,\mu$F capacitor across a 60 V, variable frequency supply. Calculate (a) the resonant frequency, (b) the dynamic resistance, (c) the current at resonance and (d) the Q-factor at resonance.

4. A coil of resistance $1.5\,\text{k}\Omega$ and 0.25 H inductance is connected in parallel with a variable capacitance across a 10 V, 8 kHz supply. Calculate (a) the capacitance of the capacitor when the supply current is a minimum, (b) the dynamic resistance and (c) the supply current.

5. A parallel circuit as shown in Figure 18.13 is tuned to resonance by varying capacitance C. Resistance, $R = 30\,\Omega$, inductance, $L = 400\,\mu$H and the supply voltage, $V = 200$ V, 5 MHz.

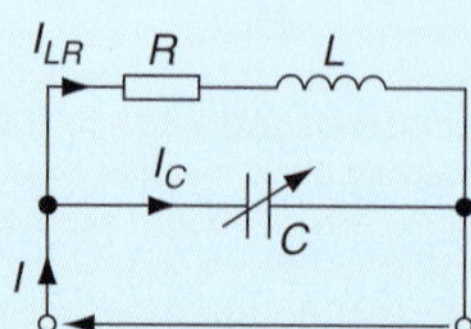

Figure 18.13

Calculate (a) the value of C to give resonance at 5 MHz, (b) the dynamic resistance, (c) the Q-factor, (d) the bandwidth, (e) the current in each branch, (f) the supply current and (g) the power dissipated at resonance.

18.7 Power factor improvement

From page 278, in any a.c. circuit, **power factor = cos ϕ**, where ϕ is the phase angle between supply current and supply voltage.

Industrial loads such as a.c. motors are essentially inductive (i.e. R–L) and may have a low power factor.

For example, let a motor take a current of 50 A at a power factor of 0.6 lagging from a 240 V, 50 Hz supply, as shown in the circuit diagram of Figure 18.14(a).

If power factor = 0.6 lagging, then:

$$\cos\phi = 0.6 \text{ lagging}$$

Hence, phase angle, $\phi = \cos^{-1} 0.6 = 53.13^\circ$ lagging

Lagging means that I lags V (remember CIVIL), and the phase diagram is as shown in Figure 18.14(b).

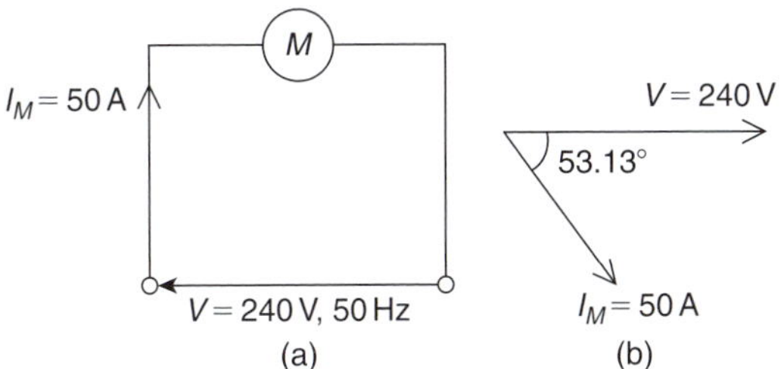

Figure 18.14

How can this power factor of 0.6 be 'improved' or 'corrected' to, say, unity?

Unity power factor means: $\cos\phi = 1$ from which, $\phi = 0$

So how can the circuit of Figure 18.14(a) be modified so that the circuit phase angle is changed from 53.13° to 0°?

The answer is to connect a capacitor in parallel with the motor as shown in Figure 18.15(a). When a capacitor is connected in parallel with the inductive load, it takes a current shown as I_C. In the phasor diagram of Figure 18.15(b), current I_C is shown leading the voltage V by 90° (again, remember CIVIL).

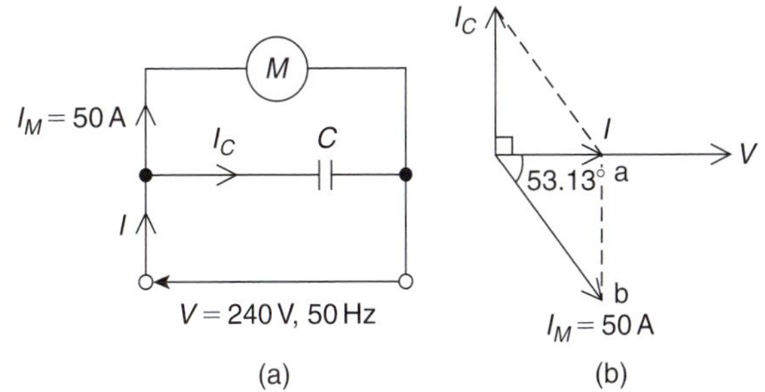

Figure 18.15

The supply current in Figure 18.15(a) is shown as I and is now the phasor sum of I_M and I_C

In the phasor diagram of Figure 18.15(b), current I is shown as the phasor sum of I_M and I_C and is in phase with V, i.e. the circuit phase angle is 0°, which means that the power factor is $\cos 0^\circ = 1$

Thus, by connecting a capacitor in parallel with the motor, the power factor has been improved from 0.6 lagging to unity.

From right-angle triangles,

$$\cos 53.13^\circ = \frac{\text{adjacent}}{\text{hypotenuse}} = \frac{I}{50}$$

from which, **supply current,**

$$I = 50\cos 53.13^\circ = \mathbf{30\ A}$$

Before the capacitor was connected, the supply current was 50 A. Now it is 30 A.

Herein lies **the advantage of power factor improvement – the supply current has been reduced**.

When power factor is improved, the **supply current is reduced, the supply system has lower losses** (i.e. lower I^2R losses) and therefore **cheaper running costs**.

Problem 11. In the circuit of Figure 18.16, what value of capacitor is needed to improve the power factor from 0.6 lagging to unity?

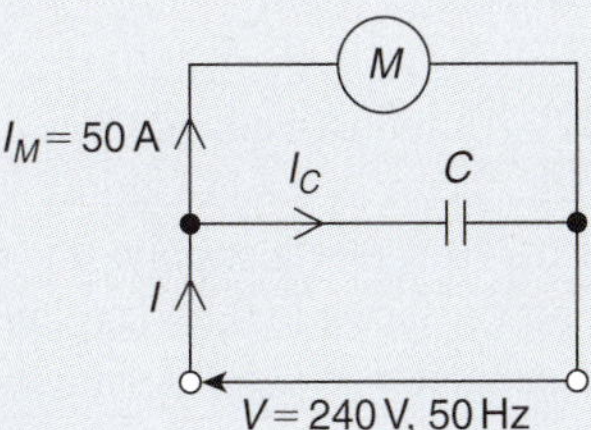

Figure 18.16

This is the same circuit as used above where the supply current was reduced from 50 A to 30 A by power factor improvement. In the phasor diagram of Figure 18.17, current I_C needs to equal ab if I is to be in phase with V.

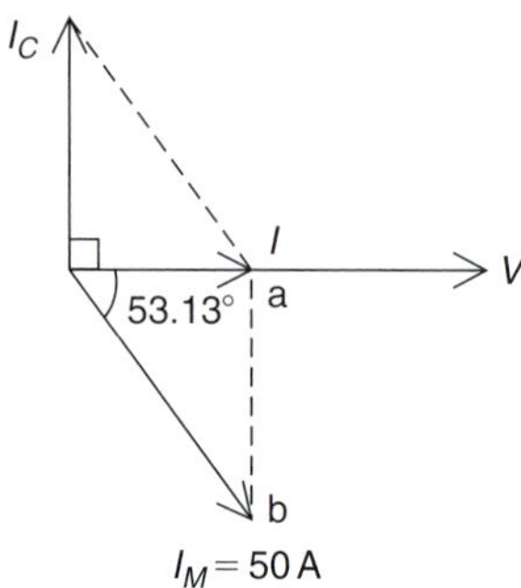

Figure 18.17

From right-angle triangles,

$$\sin 53.13° = \frac{\text{opposite}}{\text{hypotenuse}} = \frac{ab}{50}$$

from which, $ab = 50 \sin 53.13° = 40\,\text{A}$

Hence, **a capacitor has to be of such a value as to take 40 A** for the power factor to be improved from 0.6 to 1

From a.c. theory, in the circuit of Figure 18.16,

$$I_C = \frac{V}{X_C} = \frac{V}{\left(\frac{1}{2\pi fC}\right)} = 2\pi fCV$$

from which, **capacitance,**

$$\boldsymbol{C} = \frac{I_C}{2\pi fV} = \frac{40}{2\pi(50)(240)} = \mathbf{530.5\,\mu F}$$

In **practical situations** a power factor of 1 is not normally required but a power factor in the region of **0.8** or better is usually aimed for. (Actually, a power factor of 1 means resonance!)

Problem 12. An inductive load takes a current of 60 A at a power factor of 0.643 lagging when connected to a 240 V, 60 Hz supply. It is required to improve the power factor to 0.80 lagging by connecting a capacitor in parallel with the load. Calculate (a) the new supply current, (b) the capacitor current and (c) the value of the power factor correction capacitor.

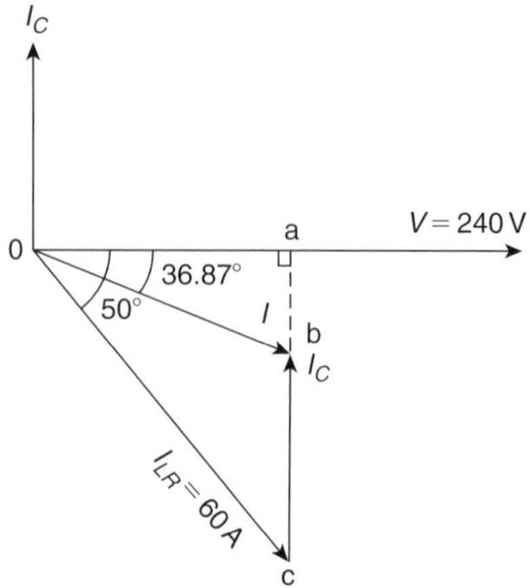

Figure 18.18

(a) A power factor of 0.643 means $\cos\phi_1 = 0.643$

from which, $\phi_1 = \cos^{-1} 0.643 = \mathbf{50°}$

A power factor of 0.80 means $\cos\phi_2 = 0.80$

from which, $\phi_2 = \cos^{-1} 0.80 = \mathbf{36.87°}$

The phasor diagram is shown in Figure 18.18, where the new supply current I is shown by length $0b$

From triangle $0ac$, $\cos 50° = \frac{0a}{60}$

from which, $0a = 60 \cos 50° = 38.57\,\text{A}$

From triangle $0ab$, $\cos 36.87° = \frac{0a}{0b} = \frac{38.57}{I}$

from which, **new supply current**,

$$\boldsymbol{I} = \frac{38.57}{\cos 36.87°} = \mathbf{48.21\,A}$$

(b) The new supply current I is the phasor sum of I_C and I_{LR}

Thus, if $I = I_C + I_{LR}$ then $I_C = I - I_{LR}$

i.e. **capacitor current**,

$$\begin{aligned}\boldsymbol{I_C} &= 48.21\angle -36.87° - 60\angle -50° \\ &= (38.57 - j28.93) - (38.57 - j45.96) \\ &= (0 + j17.03)\,\text{A or } \mathbf{17.03\angle 90°\,A}\end{aligned}$$

(c) Current, $I_C = \frac{V}{X_C} = \frac{V}{\left(\frac{1}{2\pi fC}\right)} = 2\pi fCV$

from which, **capacitance**,

$$\boldsymbol{C} = \frac{I_C}{2\pi fV} = \frac{17.03}{2\pi(60)(240)} = \mathbf{188.2\,\mu F}$$

Problem 13. A motor has an output of 4.8 kW, an efficiency of 80% and a power factor of 0.625 lagging when operated from a 240 V, 50 Hz supply. It is required to improve the power factor to 0.95 lagging by connecting a capacitor in parallel with the motor. Determine (a) the current taken by the motor, (b) the supply current after power factor correction, (c) the current taken by the capacitor, (d) the capacitance of the capacitor and (e) the kvar rating of the capacitor.

(a) $\text{Efficiency} = \dfrac{\text{power output}}{\text{power input}}$

hence $\dfrac{80}{100} = \dfrac{4800}{\text{power input}}$

$\text{Power input} = \dfrac{4800}{0.8} = 6000\,\text{W}$

Hence, $6000 = VI_M \cos\phi = (240)(I_M)(0.625)$,

since $\cos\phi = \text{p.f.} = 0.625$

Thus current taken by the motor,

$I_M = \dfrac{6000}{(240)(0.625)} = \mathbf{40\,A}$

The circuit diagram is shown in Figure 18.19(a).

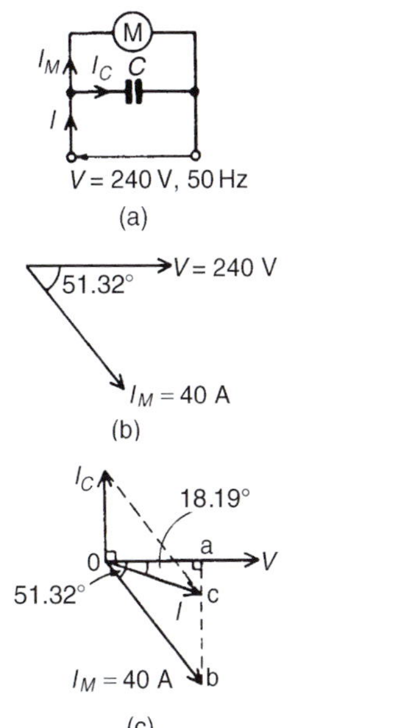

Figure 18.19

The phase angle between I_M and V is given by:

$\phi = \cos^{-1} 0.625 = 51.32°$, hence the phasor diagram is as shown in Figure 18.19(b).

(b) When a capacitor C is connected in parallel with the motor a current I_C flows which leads V by 90°. The phasor sum of I_M and I_C gives the supply current I, and has to be such as to change the circuit power factor to 0.95 lagging, i.e. a phase angle of $\cos^{-1} 0.95$ or 18.19° lagging, as shown in Figure 18.19(c).

The horizontal component of I_M (shown as $0a$)

$$= I_M \cos 51.32° = 40 \cos 51.32° = 25\,\text{A}$$

The horizontal component of I (also given by $0a$)

$$= I \cos 18.19° = 0.95\,I$$

Equating the horizontal components gives:

$$25 = 0.95 I$$

Hence the supply current after p.f. correction,

$$I = \frac{25}{0.95} = \mathbf{26.32\,A}$$

(c) The vertical component of I_M (shown as ab)

$$= I_M \sin 51.32° = 40 \sin 51.32° = 31.22\,\text{A}$$

The vertical component of I (shown as ac)

$$= I \sin 18.19° = 26.32 \sin 18.19° = 8.22\,\text{A}$$

The magnitude of the capacitor current I_C (shown as bc) is given by $ab - ac$, i.e. $31.22 - 8.22 = \mathbf{23\,A}$

(d) Current $I_C = \dfrac{V}{X_c} = \dfrac{V}{\left(\dfrac{1}{2\pi f C}\right)} = 2\pi f C V$,

from which $C = \dfrac{I_C}{2\pi f V} = \dfrac{23}{2\pi(50)(240)}$

$$= \mathbf{305\,\mu F}$$

(e) kvar rating of the capacitor $= \dfrac{VI_c}{1000} = \dfrac{(240)(23)}{1000}$

$$= \mathbf{5.52\,kvar}$$

In this problem the supply current has been reduced from 40 A to 26.32 A without altering the current or power taken by the motor. This means that the I^2R losses are reduced, with an obvious saving in cost.

Part 3

Problem 14. A 250 V, 50 Hz single-phase supply feeds the following loads: (i) incandescent lamps taking a current of 10 A at unity power factor, (ii) fluorescent lamps taking 8 A at a power factor of 0.7 lagging, (iii) a 3 kVA motor operating at full load and at a power factor of 0.8 lagging and (iv) a static capacitor. Determine, for the lamps and motor, (a) the total current, (b) the overall power factor and (c) the total power. (d) Find the value of the static capacitor to improve the overall power factor to 0.975 lagging.

A phasor diagram is constructed as shown in Figure 18.20(a), where 8 A is lagging voltage V by $\cos^{-1}$ 0.7, i.e. 45.57°, and the motor current is 3000/250, i.e. 12 A lagging V by $\cos^{-1}$ 0.8, i.e. 36.87°.

(a) The horizontal component of the currents

$$= 10\cos 0^\circ + 12\cos 36.87^\circ + 8\cos 45.57^\circ$$
$$= 10 + 9.6 + 5.6 = 25.2\,\text{A}$$

The vertical component of the currents

$$= 10\sin 0^\circ - 12\sin 36.87^\circ - 8\sin 45.57^\circ$$
$$= 0 - 7.2 - 5.713 = -12.91\,\text{A}$$

From Figure 18.20(b), total current,

$$I_L = \sqrt{[(25.2)^2 + (12.91)^2]}$$
$$= \mathbf{28.31\,A}$$

at a phase angle of $\phi = \tan^{-1}\left(\frac{12.91}{25.2}\right)$, i.e. **27.13° lagging**

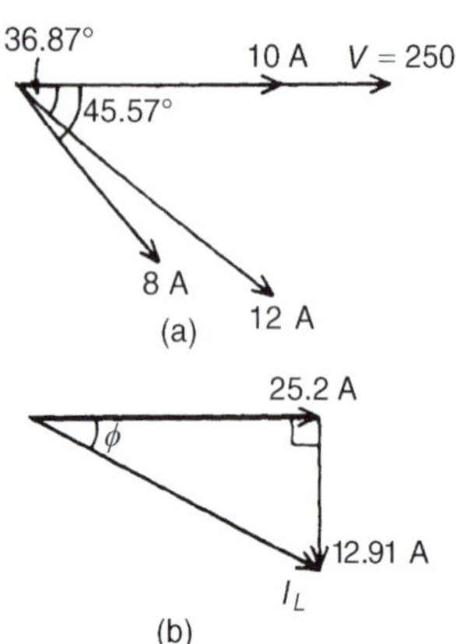

Figure 18.20

(b) Power factor $= \cos\phi = \cos 27.13^\circ$
$= \mathbf{0.890\ lagging}$

(c) Total power, $P = VI_L\cos\phi$

$$= (250)(28.31)(0.890)$$
$$= \mathbf{6.3\,kW}$$

(d) To improve the power factor, a capacitor is connected in parallel with the loads. The capacitor takes a current I_C such that the supply current falls from 28.31 A to I, lagging V by $\cos^{-1}$ 0.975, i.e. 12.84°. The phasor diagram is shown in Figure 18.21.

$$oa = 28.31\cos 27.13^\circ = I\cos 12.84^\circ$$

$$\text{Hence } I = \frac{28.31\cos 27.13^\circ}{\cos 12.84^\circ} = 25.84\,\text{A}$$

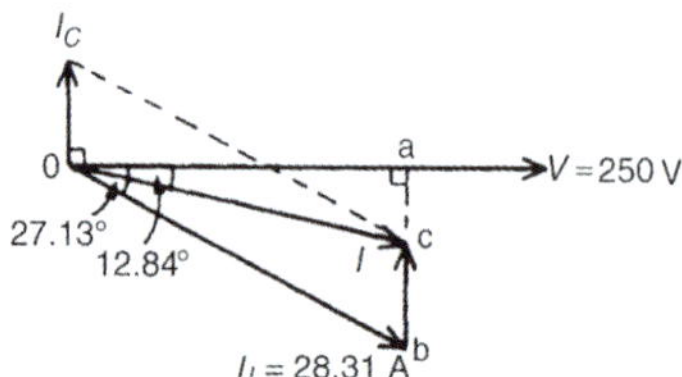

Figure 18.21

Current $I_C = bc = (ab - ac)$

$$= 28.31\sin 27.13^\circ - 25.84\sin 12.84^\circ$$
$$= 12.91 - 5.742$$
$$= 7.168\,\text{A}$$

$$I_c = \frac{V}{X_c} = \frac{V}{\left(\frac{1}{2\pi fC}\right)} = 2\pi fCV$$

$$\text{Hence capacitance } C = \frac{I_c}{2\pi fV}$$
$$= \frac{7.168}{2\pi(50)(250)}$$
$$= \mathbf{91.27\,\mu F}$$

Thus to improve the power factor from 0.890 to 0.975 lagging a 91.27 μF capacitor is connected in parallel with the loads.

Now try the following Practice Exercise

Practice Exercise 88 Power factor improvement (Answers on page 823)

1. A 415 V alternator is supplying a load of 55 kW at a power factor of 0.65 lagging. Calculate (a) the kVA loading and (b) the current taken from the alternator. (c) If the power factor is now raised to unity, find the new kVA loading.
2. A single-phase motor takes 30 A at a power factor of 0.65 lagging from a 240 V, 50 Hz supply. Determine (a) the current taken by the capacitor connected in parallel to correct the power factor to unity, and (b) the value of the supply current after power factor correction.
3. A 20 Ω non-reactive resistor is connected in series with a coil of inductance 80 mH and negligible resistance. The combined circuit is connected to a 200 V, 50 Hz supply. Calculate (a) the reactance of the coil, (b) the impedance of the circuit, (c) the current in the circuit, (d) the power factor of the circuit, (e) the power absorbed by the circuit, (f) the value of a power factor correction capacitor to produce a power factor of unity and (g) the value of a power factor correction capacitor to produce a power factor of 0.9.
4. A motor has an output of 6 kW, an efficiency of 75% and a power factor of 0.64 lagging when operated from a 250 V, 60 Hz supply. It is required to raise the power factor to 0.925 lagging by connecting a capacitor in parallel with the motor. Determine (a) the current taken by the motor, (b) the supply current after power factor correction, (c) the current taken by the capacitor, (d) the capacitance of the capacitor and (e) the kvar rating of the capacitor.
5. A 200 V, 50 Hz single-phase supply feeds the following loads: (i) fluorescent lamps taking a current of 8 A at a power factor of 0.9 leading, (ii) incandescent lamps taking a current of 6 A at unity power factor, (iii) a motor taking a current of 12 A at a power factor of 0.65 lagging. Determine the total current taken from the supply and the overall power factor. Find also the value of a static capacitor connected in parallel with the loads to improve the overall power factor to 0.98 lagging.

For fully worked solutions to each of the problems in Practice Exercises 83 to 88 in this chapter, go to the website:
www.routledge.com/cw/bird

Chapter 19

D.c. transients

***Why it is important to understand:* D.c. transients**

The study of transient and steady state responses of a circuit is very important as they form the building blocks of most electrical circuits. This chapter explores the response of capacitors and inductors to sudden changes in d.c. voltage, called a transient voltage, when connected in series with a resistor. Unlike resistors, which respond instantaneously to applied voltage, capacitors and inductors react over time as they absorb and release energy. Each time a switch is made or an input is connected, circuit conditions change, but only for a short time; during these brief transient events, components and circuits may behave differently to the way they behave under normal 'static' conditions. The voltage across a capacitor cannot change instantaneously as some time is required for the electric charge to build up on, or leave the capacitor plates. Series R–C networks have many practical uses. They are often used in timing circuits to control events that must happen repeatedly at a fixed time interval. One example is a circuit that causes an LED to blink on and off once every second. There are several ways to design a circuit to do this, but one of the most common ways uses a series R–C circuit. By adjusting the value of the resistor or the capacitor, the designer can cause the LED to blink faster or slower. An understanding of d.c. transients is thus important and these are explained in this chapter.

At the end of this chapter you should be able to:

- understand the term 'transient'
- describe the transient response of capacitor and resistor voltages, and current in a series C–R d.c. circuit
- define the term 'time constant'
- calculate time constant in a C–R circuit
- draw transient growth and decay curves for a C–R circuit
- use equations $v_C = V(1 - \mathrm{e}^{-t/\tau})$, $v_R = V\mathrm{e}^{-t/\tau}$ and $i = I\mathrm{e}^{-t/\tau}$ for a C–R circuit
- describe the transient response when discharging a capacitor
- describe the transient response of inductor and resistor voltages, and current in a series L–R d.c. circuit
- calculate time constant in an L–R circuit
- draw transient growth and decay curves for an L–R circuit
- use equations $v_L = V\mathrm{e}^{-t/\tau}$, $v_R = V(1 - \mathrm{e}^{-t/\tau})$ and $i = I(1 - \mathrm{e}^{-t/\tau})$
- describe the transient response for current decay in an L–R circuit
- understand the switching of inductive circuits
- describe the effects of time constant on a rectangular waveform via integrator and differentiator circuits

Electrical Circuit Theory and Technology. 978-1-138-67349-6, © 2017 John Bird. Published by Taylor & Francis. All rights reserved.

19.1 Introduction

When a d.c. voltage is applied to a capacitor C and resistor R connected in series, there is a short period of time immediately after the voltage is connected, during which the current flowing in the circuit and voltages across C and R are changing.

Similarly, when a d.c. voltage is connected to a circuit having inductance L connected in series with resistance R, there is a short period of time immediately after the voltage is connected, during which the current flowing in the circuit and the voltages across L and R are changing.

These changing values are called **transients**.

19.2 Charging a capacitor

(a) The circuit diagram for a series-connected C–R circuit is shown in Figure 19.1. When switch S is closed then by Kirchhoff's voltage law:

$$V = v_C + v_R \qquad (1)$$

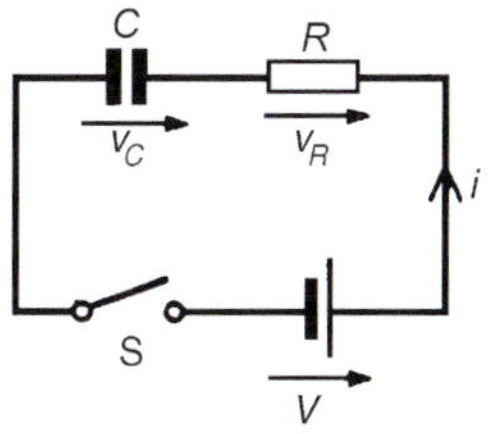

Figure 19.1

(b) The battery voltage V is constant. The capacitor voltage v_C is given by q/C, where q is the charge on the capacitor. The voltage drop across R is given by iR, where i is the current flowing in the circuit. Hence at all times:

$$V = \frac{q}{C} + iR \qquad (2)$$

At the instant of closing S (initial circuit condition), assuming there is no initial charge on the capacitor, q_0 is zero, hence v_{Co} is zero. Thus from equation (1), $V = 0 + v_{Ro}$, i.e. $v_{Ro} = V$. This shows that the resistance to current is solely due to R, and the initial current flowing, $i_o = I = V/R$

(c) A short time later at time t_1 seconds after closing S, the capacitor is partly charged to, say, q_1 coulombs because current has been flowing. The voltage v_{C1} is now q_1/C volts. If the current flowing is i_1 amperes, then the voltage drop across R has fallen to i_1R volts. Thus, equation (2) is now $V = (q_1/C) + i_1R$

(d) A short time later still, say at time t_2 seconds after closing the switch, the charge has increased to q_2 coulombs and v_C has increased to q_2/C volts. Since $V = v_C + v_R$ and V is a constant, then v_R decreases to i_2R. Thus v_C is increasing and i and v_R are decreasing as time increases.

(e) Ultimately, a few seconds after closing S (i.e. at the final or **steady state** condition), the capacitor is fully charged to, say, Q coulombs, current no longer flows, i.e. $i = 0$, and hence $v_R = iR = 0$. It follows from equation (1) that $v_C = V$

(f) Curves showing the changes in v_C, v_R and i with time are shown in Figure 19.2.

The curve showing the variation of v_C with time is called an **exponential growth curve** and the graph is called the 'capacitor voltage/time' characteristic. The curves showing the variation of v_R and i with time are called **exponential**

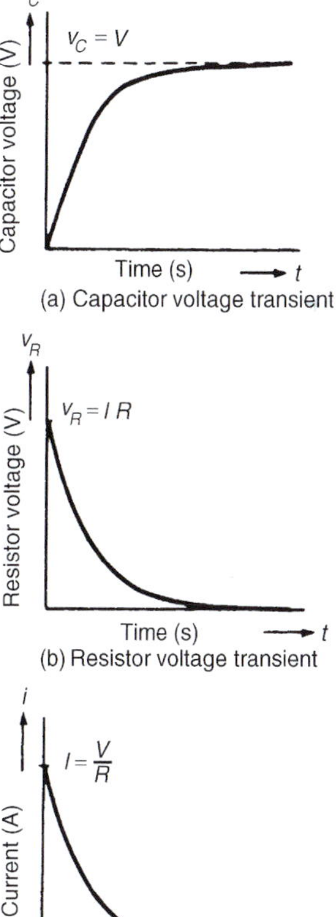

Figure 19.2

decay curves, and the graphs are called 'resistor voltage/time' and 'current/time' characteristics, respectively. (The name 'exponential' shows that the shape can be expressed mathematically by an exponential mathematical equation, as shown in Section 19.4.)

19.3 Time constant for a *C–R* circuit

(a) If a constant d.c. voltage is applied to a series-connected C–R circuit, a transient curve of capacitor voltage v_C is as shown in Figure 19.2(a).

(b) With reference to Figure 19.3, let the constant voltage supply be replaced by a variable voltage supply at time t_1 seconds. Let the voltage be varied so that the **current** flowing in the circuit is **constant**.

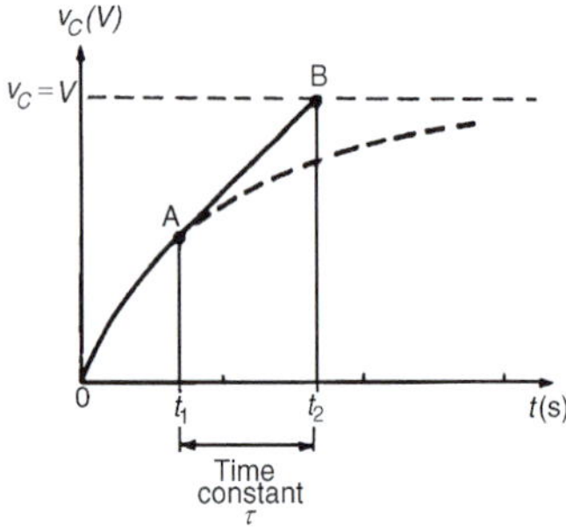

Figure 19.3

(c) Since the current flowing is a constant, the curve will follow a tangent, AB, drawn to the curve at point A.

(d) Let the capacitor voltage v_C reach its final value of V at time t_2 seconds.

(e) The time corresponding to $(t_2 - t_1)$ seconds is called the **time constant** of the circuit, denoted by the Greek letter 'tau', τ. The value of the time constant is CR seconds, i.e. for a series connected C–R circuit,

time constant $\tau =$ CR seconds

Since the variable voltage mentioned in para. (b) above can be applied to any instant during the transient change, it may be applied at $t=0$, i.e. at the instant of connecting the circuit to the supply. If this is done, then the time constant of the circuit may be defined as:

'the time taken for a transient to reach its final state if the initial rate of change is maintained'.

19.4 Transient curves for a *C–R* circuit

There are two main methods of drawing transient curves graphically, these being:

(a) the **tangent method** – this method is shown in worked Problem 1 below and

(b) the **initial slope and three point method**, which is shown in worked Problem 2, and is based on the following properties of a transient exponential curve:

  (i) for a growth curve, the value of a transient at a time equal to one time constant is 0.632 of its steady state value (usually taken as 63% of the steady state value), at a time equal to two and a half time constants is 0.918 of its steady state value (usually taken as 92% of its steady state value) and at a time equal to five time constants is equal to its steady state value,

  (ii) for a decay curve, the value of a transient at a time equal to one time constant is 0.368 of its initial value (usually taken as 37% of its initial value), at a time equal to two and a half time constants is 0.082 of its initial value (usually taken as 8% of its initial value) and at a time equal to five time constants is equal to zero.

The transient curves shown in Figure 19.2 have **mathematical equations**, obtained by solving the differential equations representing the circuit. The equations of the curves are:

growth of capacitor voltage, $v_C = V(1 - e^{-t/CR})$
$= V(1 - e^{-t/\tau})$

decay of resistor voltage, $v_R = Ve^{-t/CR}$
$= Ve^{-t/\tau}$ **and**

decay of current flowing, $i = Ie^{-t/CR} = Ie^{-t/\tau}$

These equations are derived analytically in Chapter 48.

Problem 1. A 15 µF uncharged capacitor is connected in series with a 47 kΩ resistor across a 120 V d.c. supply. Use the tangential graphical method to draw the capacitor voltage/time characteristic of the circuit. From the characteristic, determine the capacitor voltage at a time equal to one time constant after being connected to the supply, and also two seconds after being connected to the supply. Also, find the time for the capacitor voltage to reach one half of its steady state value.

To construct an exponential curve, the time constant of the circuit and steady state value need to be determined.

$$\text{Time constant} = CR = 15\,\mu\text{F} \times 47\,\text{k}\Omega$$
$$= 15 \times 10^{-6} \times 47 \times 10^{3}$$
$$= 0.705\,\text{s}$$

Steady state value of $v_C = V$,

i.e. $v_C = 120\,\text{V}$

With reference to Figure 19.4, the scale of the horizontal axis is drawn so that it spans at least five time constants, i.e. 5 × 0.705 or about 3.5 seconds. The scale of the vertical axis spans the change in the capacitor voltage, that is, from 0 to 120 V. A broken line AB is drawn corresponding to the final value of v_C

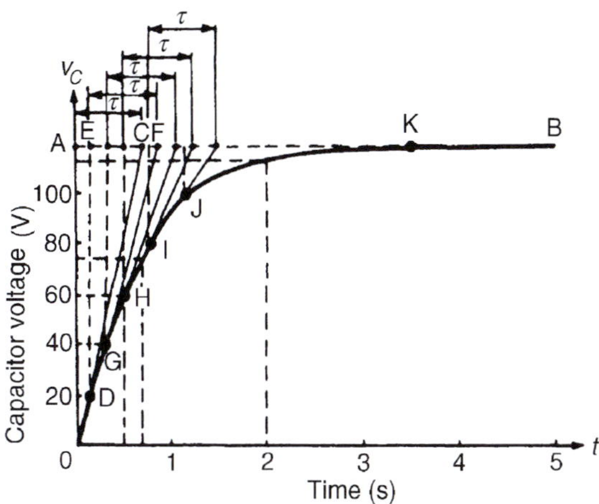

Figure 19.4

Point C is measured along AB so that AC is equal to 1τ, i.e. AC = 0.705 s. Straight line $0C$ is drawn. Assuming that about five intermediate points are needed to draw the curve accurately, a point D is selected on $0C$ corresponding to a v_C value of about 20 V. DE is drawn vertically. EF is made to correspond to 1τ, i.e. $EF = 0.705$ s. A straight line is drawn joining DF. This procedure of

(a) drawing a vertical line through point selected,

(b) at the steady state value, drawing a horizontal line corresponding to 1τ, and

(c) joining the first and last points,

is repeated for v_C values of 40, 60, 80 and 100 V, giving points G, H, I and J.

The capacitor voltage effectively reaches its steady state value of 120 V after a time equal to five time constants, shown as point K. Drawing a smooth curve through points O, D, G, H, I, J and K gives the exponential growth curve of capacitor voltage.

From the graph, the value of capacitor voltage at a time equal to the time constant is about **75 V**. It is a characteristic of all exponential growth curves that after a time equal to one time constant, the value of the transient is 0.632 of its steady state value. In this problem, 0.632 × 120 = 75.84 V. Also from the graph, when t is two seconds, v_C is about **115 V**. [This value may be checked using the equation $v_C(1 - e^{-t/\tau})$, where $V = 120$ V, $\tau = 0.705$ s and $t = 2$ s. This calculation gives $v_C = 112.97$ V.]

The time for v_C to rise to one half of its final value, i.e. 60 V, can be determined from the graph and is about **0.5 s**. [This value may be checked using $v_C = V(1 - \text{e}^{-t/\tau})$ where $V = 120$ V, $v_C = 60$ V and $\tau = 0.705$ s, giving $t = 0.489$ s]

Problem 2. A 4 μF capacitor is charged to 24 V and then discharged through a 220 kΩ resistor. Use the 'initial slope and three point' method to draw: (a) the capacitor voltage/time characteristic, (b) the resistor voltage/time characteristic and (c) the current/time characteristic, for the transients which occur. From the characteristics determine the value of capacitor voltage, resistor voltage and current one and a half seconds after discharge has started.

To draw the transient curves, the time constant of the circuit and steady state values are needed.

$$\text{Time constant, } \tau = CR = 4 \times 10^{-6} \times 220 \times 10^{3}$$
$$= 0.88\,\text{s}$$

Initially, capacitor voltage $v_C = v_R = 24$ V,

$$i = \frac{V}{R} = \frac{24}{220 \times 10^{3}}$$
$$= 0.109\,\text{mA}$$

Finally, $v_C = v_R = i = 0$

(a) The exponential decay of capacitor voltage is from 24 V to 0 V in a time equal to five time constants, i.e. 5 × 0.88 = 4.4 s. With reference to Figure 19.5, to construct the decay curve:

  (i) the horizontal scale is made so that it spans at least five time constants, i.e. 4.4 s,

  (ii) the vertical scale is made to span the change in capacitor voltage, i.e. 0 to 24 V,

  (iii) point A corresponds to the initial capacitor voltage, i.e. 24 V,

(iv) OB is made equal to one time constant and line AB is drawn. This gives the initial slope of the transient,

(v) the value of the transient after a time equal to one time constant is 0.368 of the initial value, i.e. $0.368 \times 24 = 8.83$ V; a vertical line is drawn through B and distance BC is made equal to 8.83 V,

(vi) the value of the transient after a time equal to two and a half time constants is 0.082 of the initial value, i.e. $0.082 \times 24 = 1.97$ V, shown as point D in Figure 19.5,

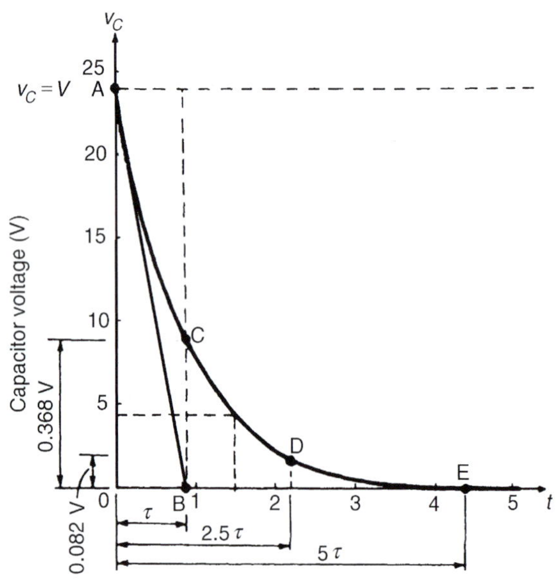

Figure 19.5

(vii) the transient effectively dies away to zero after a time equal to five time constants, i.e. 4.4 s, giving point E.

The smooth curve drawn through points A, C, D and E represents the decay transient. At $1\frac{1}{2}$ s after decay has started, $v_C \approx$ **4.4 V**. [This may be checked using $v_C = V e^{-t/\tau}$, where $V = 24$, $t = 1\frac{1}{2}$ and $\tau = 0.88$, giving $v_C = 4.36$ V]

(b) The voltage drop across the resistor is equal to the capacitor voltage when a capacitor is discharging through a resistor, thus the resistor voltage/time characteristic is identical to that shown in Figure 19.5.

Since $v_R = v_C$, then at $1\frac{1}{2}$ seconds after decay has started, $v_R \approx$ **4.4 V** (see (vii) above).

(c) The current/time characteristic is constructed in the same way as the capacitor voltage/time characteristic, shown in part (a) of this problem, and is as shown in Figure 19.6. The values are:

point A: initial value of current $= 0.109$ mA

point C: at 1τ, $i = 0.368 \times 0.109 = 0.040$ mA

point D: at 2.5τ, $i = 0.082 \times 0.109 = 0.009$ mA

point E: at 5τ, $i = 0$

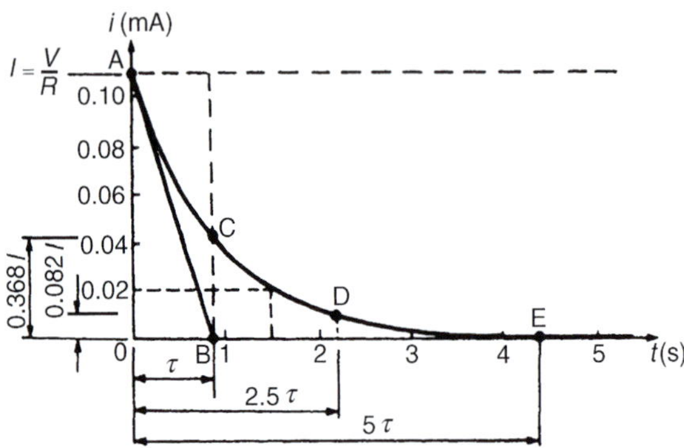

Figure 19.6

Hence the current transient is as shown. At a time of $1\frac{1}{2}$ seconds, the value of current, from the characteristic is **0.02 mA**. [This may be checked using $i = I e^{-t/\tau}$ where $I = 0.109$, $t = 1\frac{1}{2}$ and $\tau = 0.88$, giving $i = 0.0198$ mA or 19.8 μA]

Problem 3. A 20 μF capacitor is connected in series with a 50 kΩ resistor and the circuit is connected to a 20 V d.c. supply. Determine

(a) the initial value of the current flowing,

(b) the time constant of the circuit,

(c) the value of the current one second after connection,

(d) the value of the capacitor voltage two seconds after connection and

(e) the time after connection when the resistor voltage is 15 V

Parts (c), (d) and (e) may be determined graphically, as shown in worked Problems 1 and 2 or by calculation as shown below.

$V = 20\,\text{V}, C = 20\,\mu\text{F} = 20 \times 10^{-6}$ F,

$R = 50\,\text{k}\Omega = 50 \times 10^{3}$ V

(a) The initial value of the current flowing is

$$I = \frac{V}{R} = \frac{20}{50 \times 10^3} = \mathbf{0.4\,mA}$$

(b) From Section 19.3 the time constant,

$$\tau = CR = (20 \times 10^{-6}) \times (50 \times 10^3) = \mathbf{1\,s}$$

(c) Current, $i = I\mathrm{e}^{-t/\tau}$

Working in mA units, $i = 0.4\mathrm{e}^{-1/1} = 0.4 \times 0.368$
$= \mathbf{0.147\,mA}$

(d) Capacitor voltage, $v_C = V(1 - \mathrm{e}^{-t/\tau})$

$$= 20(1 - \mathrm{e}^{-2/1})$$
$$= 20(1 - 0.135)$$
$$= 20 \times 0.865$$
$$= \mathbf{17.3\,V}$$

(e) Resistor voltage, $v_R = V\mathrm{e}^{-t/\tau}$

Thus, $15 = 20\mathrm{e}^{-t/1}$, $\frac{15}{20} = \mathrm{e}^{-t}$, i.e. $\mathrm{e}^t = \frac{20}{15} = \frac{4}{3}$

Taking natural logarithms of each side of the equation gives

$$t = \ln\frac{4}{3} = \ln 1.3333$$

i.e. time, t = 0.288 s

Problem 4. A circuit consists of a resistor connected in series with a 0.5 μF capacitor and has a time constant of 12 ms. Determine (a) the value of the resistor, and (b) the capacitor voltage 7 ms after connecting the circuit to a 10 V supply.

(a) The time constant $\tau = CR$, hence $R = \frac{\tau}{C}$

i.e. $R = \frac{12 \times 10^{-3}}{0.5 \times 10^{-6}} = 24 \times 10^3 = \mathbf{24\,k\Omega}$

(b) The equation for the growth of capacitor voltage is:

$$v_C = V(1 - \mathrm{e}^{-t/\tau})$$

Since $\tau = 12\,\text{ms} = 12 \times 10^{-3}$ s, $V = 10$ V and

$$t = 7\,\text{ms} = 7 \times 10^{-3}\,\text{s},$$

then $v_C = 10\left[1 - \mathrm{e}^{-\frac{7\times10^{-3}}{12\times10^{-3}}}\right] = 10(1 - \mathrm{e}^{-0.583})$

$$= 10(1 - 0.558) = \mathbf{4.42\,V}$$

Alternatively, the value of v_C when t is 7 ms may be determined using the growth characteristic as shown in worked Problem 1.

Problem 5. A circuit consists of a 10 μF capacitor connected in series with a 25 kΩ resistor with a switchable 100 V d.c. supply. When the supply is connected, calculate (a) the time constant, (b) the maximum current, (c) the voltage across the capacitor after 0.5 s, (d) the current flowing after one time constant, (e) the voltage across the resistor after 0.1 s, (f) the time for the capacitor voltage to reach 45 V and (g) the initial rate of voltage rise.

(a) **Time constant,** $\tau = C \times R = 10 \times 10^{-6} \times 25 \times 10^3$
$= \mathbf{0.25\,s}$

(b) Current is a maximum when the circuit is first connected and is only limited by the value of resistance in the circuit, i.e.

$$I_m = \frac{V}{R} = \frac{100}{25 \times 10^3} = \mathbf{4\,mA}$$

(c) Capacitor voltage, $v_C = V_m(1 - \mathrm{e}^{-\frac{t}{\tau}})$

When time, $t = 0.5$ s, then

$v_C = 100(1 - \mathrm{e}^{-\frac{0.5}{0.25}}) = 100(0.8647) = \mathbf{86.47\,V}$

(d) Current, $i = I_m\mathrm{e}^{-\frac{t}{\tau}}$

and when $t = \tau$, **current,** $i = 4\mathrm{e}^{-\frac{\tau}{\tau}} = 4\mathrm{e}^{-1}$
$= \mathbf{1.472\,mA}$

Alternatively, after one time constant the capacitor voltage will have risen to 63.2% of the supply voltage and the current will have fallen to 63.2% of its final value, i.e. 36.8% of I_m

Hence, $i = 36.8\%$ of $4 = 0.368 \times 4 = \mathbf{1.472\,mA}$

(e) The voltage across the resistor, $v_R = V\mathrm{e}^{-\frac{t}{\tau}}$

When $t = 0.1$ s, **resistor voltage,**

$$v_R = 100\mathrm{e}^{-\frac{0.1}{0.25}} = \mathbf{67.03\,V}$$

(f) Capacitor voltage, $v_C = V_m(1 - \mathrm{e}^{-\frac{t}{\tau}})$

When the capacitor voltage reaches 45 V, then:

$$45 = 100\left(1 - \mathrm{e}^{-\frac{t}{0.25}}\right)$$

from which,

$\frac{45}{100} = 1 - \mathrm{e}^{-\frac{t}{0.25}}$ and $\mathrm{e}^{-\frac{t}{0.25}} = 1 - \frac{45}{100} = 0.55$

Hence,

$-\frac{t}{0.25} = \ln 0.55$ and **time,** $t = -0.25 \ln 0.55$
$= \mathbf{0.149\,s}$

(g) **Initial rate of voltage rise** $= \dfrac{V}{\tau} = \dfrac{100}{0.25} =$ **400 V/s**
(i.e. gradient of the tangent at $t = 0$)

19.5 Discharging a capacitor

When a capacitor is charged (i.e. with the switch in position A in Figure 19.7), and the switch is then moved to position B, the electrons stored in the capacitor keep the current flowing for a short time. Initially, at the instant of moving from A to B, the current flow is such that the capacitor voltage v_C is balanced by an equal and opposite voltage $v_R = iR$. Since initially $v_C = v_R = V$, then $i = I = V/R$. During the transient decay, by applying Kirchhoff's voltage law to Figure 19.7, $v_C = v_R$. Finally the transients decay exponentially to zero, i.e. $v_C = v_R = 0$. The transient curves representing the voltages and current are as shown in Figure 19.8.

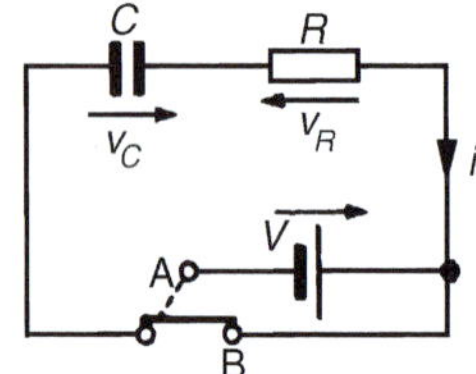

Figure 19.7

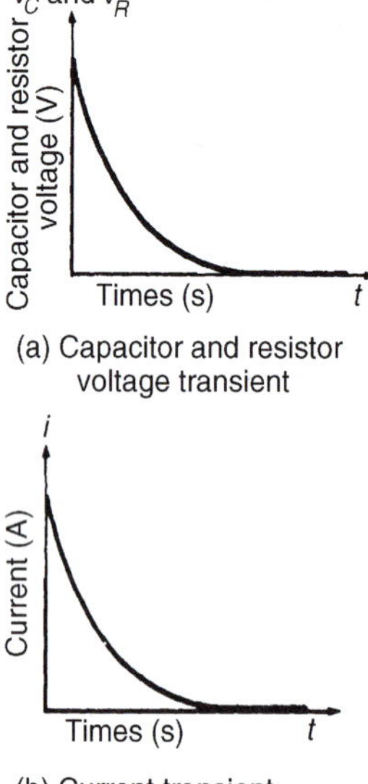

Figure 19.8

The equations representing the transient curves during the discharge period of a series-connected C–R circuit are:

decay of voltage, $v_C = v_R = V e^{(-t/CR)} = V e^{(-t/\tau)}$

decay of current, $i = I e^{(-t/CR)} = I e^{(-t/\tau)}$

When a capacitor has been disconnected from the supply it may still be charged and it may retain this charge for some considerable time. Thus precautions must be taken to ensure that the capacitor is automatically discharged after the supply is switched off. This is done by connecting a high-value resistor across the capacitor terminals.

Problem 6. A capacitor is charged to 100 V and then discharged through a 50 kΩ resistor. If the time constant of the circuit is 0.8 s, determine: (a) the value of the capacitor, (b) the time for the capacitor voltage to fall to 20 V, (c) the current flowing when the capacitor has been discharging for 0.5 s, and (d) the voltage drop across the resistor when the capacitor has been discharging for one second.

Parts (b), (c) and (d) of this problem may be solved graphically, as shown in worked Problems 1 and 2 or by calculation as shown below.

$V = 100\,\text{V},\ \tau = 0.08\,\text{s},\ R = 50\,\text{k}\Omega = 50 \times 10^3\,\Omega$

(a) Since time constant, $\tau = CR$, $C = \tau/R$

$$\text{i.e. } C = \frac{0.8}{50 \times 10^3} = \mathbf{16\,\mu F}$$

(b) $v_C = V e^{-t/\tau}$

$20 = 100 e^{-t/0.8}$, i.e. $\frac{1}{5} = e^{-t/0.8}$

Thus $e^{t/0.8} = 5$ and taking natural logarithms of each side, gives

$$\frac{t}{0.8} = \ln 5, \text{ i.e. } t = 0.8 \ln 5$$

Hence $t = 1.29$ s

(c) $i = I e^{-t/\tau}$

The initial current flowing, $I = \dfrac{V}{R}$

$$= \frac{100}{50 \times 10^3} = 2\,\text{mA}$$

Working in mA units, $i = I e^{-t/\tau} = 2e^{(-0.5/0.8)}$
$$= 2e^{-0.625}$$
$$= 2 \times 0.535 = \mathbf{1.07\,mA}$$

(d) $v_R = v_C = V e^{-t/\tau}$
$$= 100e^{-1/0.8} = 100e^{-1.25}$$
$$= 100 \times 0.287 = \mathbf{28.7\,V}$$

Problem 7. A 0.1 μF capacitor is charged to 200 V before being connected across a 4 kΩ resistor. Determine (a) the initial discharge current, (b) the time constant of the circuit and (c) the minimum time required for the voltage across the capacitor to fall to less than 2 V.

(a) Initial discharge current,
$$i = \frac{V}{R} = \frac{200}{4 \times 10^3} = \mathbf{0.05\,A} \text{ or } \mathbf{50\,mA}$$

(b) Time constant $\tau = CR = 0.1 \times 10^{-6} \times 4 \times 10^3$
$$= \mathbf{0.0004\,s} \text{ or } \mathbf{0.4\,ms}$$

(c) The minimum time for the capacitor voltage to fall to less than 2 V, i.e. less than $\frac{2}{200}$ or 1% of the initial value is given by 5τ
$$5\tau = 5 \times 0.4 = \mathbf{2\,ms}$$

In a d.c. circuit, a capacitor blocks the current except during the times that there are changes in the supply voltage.
For a practical laboratory experiment on the charging and discharging of a capacitor, see the website.

Now try the following Practice Exercise

Practice Exercise 89 Transients in series connected C–R circuits (Answers on page 823)

1. An uncharged capacitor of 0.2 μF is connected to a 100 V d.c. supply through a resistor of 100 kΩ. Determine the capacitor voltage 10 ms after the voltage has been applied.

2. A circuit consists of an uncharged capacitor connected in series with a 50 kΩ resistor and has a time constant of 15 ms. Determine (a) the capacitance of the capacitor and (b) the voltage drop across the resistor 5 ms after connecting the circuit to a 20 V d.c. supply.

3. A 10 μF capacitor is charged to 120 V and then discharged through a 1.5 MΩ resistor. Determine either graphically or by calculation the capacitor voltage 2 s after discharging has commenced. Also find how long it takes for the voltage to fall to 25 V.

4. A capacitor is connected in series with a voltmeter of resistance 750 kΩ and a battery. When the voltmeter reading is steady the battery is replaced with a shorting link. If it takes 17 s for the voltmeter reading to fall to two-thirds of its original value, determine the capacitance of the capacitor.

5. When a 3 μF charged capacitor is connected to a resistor, the voltage falls by 70% in 3.9 s. Determine the value of the resistor.

6. A 50 μF uncharged capacitor is connected in series with a 1 kΩ resistor and the circuit is switched to a 100 V d.c. supply. Determine:
   (a) the initial current flowing in the circuit,
   (b) the time constant,
   (c) the value of current when t is 50 ms and
   (d) the voltage across the resistor 60 ms after closing the switch.

7. An uncharged 5 μF capacitor is connected in series with a 30 kΩ resistor across a 110 V d.c. supply. Determine the time constant of the circuit and the initial charging current. Determine the current flowing 120 ms after connecting to the supply.

8. An uncharged 80 μF capacitor is connected in series with a 1 kΩ resistor and is switched across a 110 V supply. Determine the time constant of the circuit and the initial value of current flowing. Determine the value of current flowing after (a) 40 ms and (b) 80 ms.

9. A 60 μF capacitor is connected in series with a 10 kΩ resistor and connected to a 120 V d.c. supply. Calculate (a) the time constant, (b) the initial rate of voltage rise, (c) the initial charging current and (d) the time for the capacitor voltage to reach 50 V.

10. A 200 V d.c. supply is connected to a 2.5 MΩ resistor and a 2 μF capacitor in series. Calculate (a) the current flowing 4 s after connecting, (b) the voltage across the resistor after 4 s and (c) the energy stored in the capacitor after 4 s.

11. (a) In the circuit shown in Figure 19.9, with the switch in position 1, the capacitor is uncharged. If the switch is moved to position 2 at time $t = 0$ s, calculate (i) the initial current through the 0.5 MΩ, (ii) the voltage across the capacitor when $t = 1.5$ s and (iii) the time taken for the voltage across the capacitor to reach 12 V.

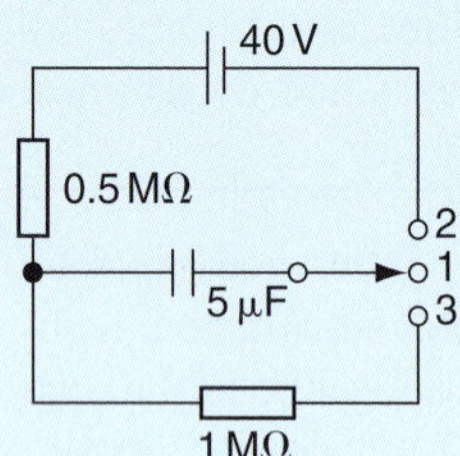

Figure 19.9

(b) If at the time $t = 1.5$ s, the switch is moved to position 3, calculate (i) the initial current through the 1 MΩ resistor, (ii) the energy stored in the capacitor 3.5 s later (i.e. when $t = 5$ s).

(c) Sketch a graph of the voltage across the capacitor against time from $t = 0$ to $t = 5$ s, showing the main points.

19.6 Camera flash

The internal workings of a camera flash are an example of the application of C–R circuits. When a camera is first switched on, a battery slowly charges a capacitor to its full potential via a C–R circuit. When the capacitor is fully charged, an indicator (red light) typically lets the photographer know that the flash is ready for use. Pressing the shutter button quickly discharges the capacitor through the flash (i.e. a resistor). The current from the capacitor is responsible for the bright light that is emitted. The flash rapidly draws current in order to emit the bright light. The capacitor must then be discharged before the flash can be used again.

19.7 Current growth in an L–R circuit

(a) The circuit diagram for a series-connected L–R circuit is shown in Figure 19.10. When switch S is closed, then by Kirchhoff's voltage law:

$$V = v_L + v_R \qquad (3)$$

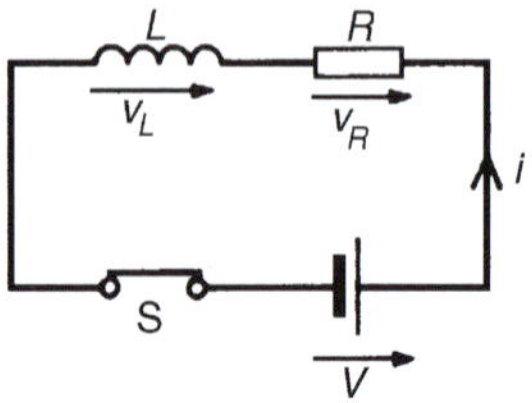

Figure 19.10

(b) The battery voltage V is constant. The voltage across the inductance is the induced voltage, i.e.

$$v_L = L \times \frac{\text{change of current}}{\text{change of time}} = L\frac{\mathrm{d}i}{\mathrm{d}t}$$

The voltage drop across R, v_R is given by iR. Hence, at all times:

$$V = L(\mathrm{d}i/\mathrm{d}t) + iR \qquad (4)$$

(c) At the instant of closing the switch, the rate of change of current is such that it induces an e.m.f. in the inductance which is equal and opposite to V, hence $V = v_L + 0$, i.e. $v_L = V$. From equation (3), because $v_L = V$, then $v_R = 0$ and $i = 0$.

(d) A short time later at time t_1 seconds after closing S, current i_1 is flowing since there is a rate of change of current initially, resulting in a voltage drop of $i_1 R$ across the resistor. Since V (constant) $= v_L + v_R$ the induced e.m.f. is reduced, and equation (4) becomes:

$$V = L\frac{\mathrm{d}i_1}{\mathrm{d}t_1} + i_1 R$$

(e) A short time later still, say at time t_2 seconds after closing the switch, the current flowing is i_2, and the voltage drop across the resistor increases to $i_2 R$. Since v_R increases, v_L decreases.

(f) Ultimately, a few seconds after closing S, the current flow is entirely limited by R, the rate of change of current is zero and hence v_L is zero. Thus

$V = iR$. Under these conditions, steady state current flows, usually signified by I. Thus, $I = V/R$, $v_R = IR$ and $v_L = 0$ at steady state conditions.

(g) Curves showing the changes in v_L, v_R and i with time are shown in Figure 19.11 and indicate that v_L is a maximum value initially (i.e equal to V), decaying exponentially to zero, whereas v_R and i grow exponentially from zero to their steady state values of V and $I = V/R$, respectively.

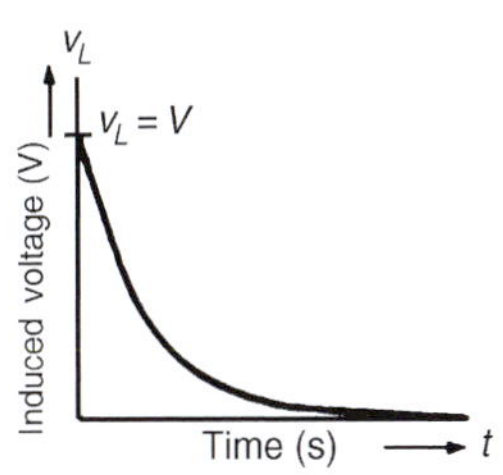

(a) Induced voltage transient

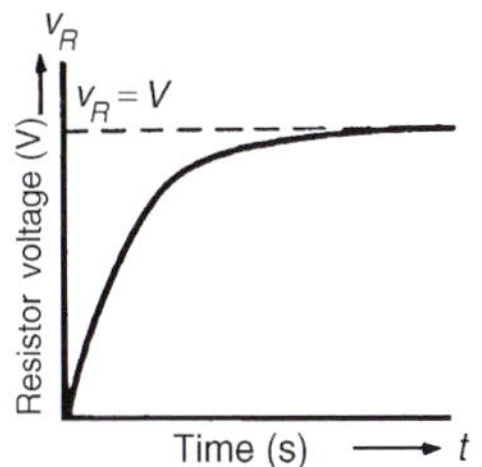

(b) Resistor voltage transient

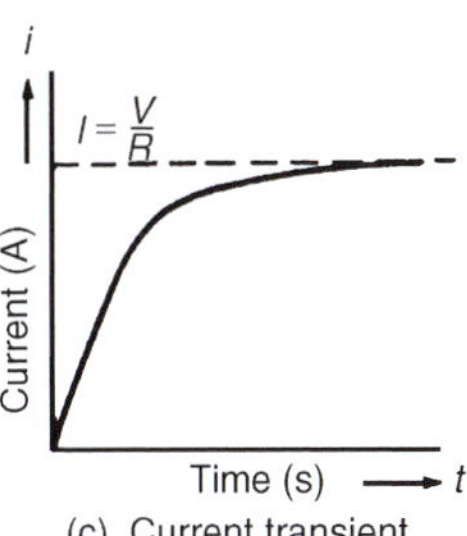

(c) Current transient

Figure 19.11

19.8 Time constant for an L–R circuit

With reference to Section 19.3, the time constant of a series-connected L–R circuit is defined in the same way as the time constant for a series-connected C–R circuit, i.e. it is the time taken to reach its final value if the initial rate of change is maintained. Its value is given by:

time constant, $\tau = L/R$ seconds

19.9 Transient curves for an L–R circuit

Transient curves representing the induced voltage/time, resistor voltage/time and current/time characteristics may be drawn graphically, as outlined in Section 19.4. A method of construction is shown in worked Problem 8. Each of the transient curves shown in Figure 19.11 have mathematical equations, and these are:

decay of induced voltage, $v_L = Ve^{(-Rt/L)} = Ve^{(-t/\tau)}$

growth of resistor voltage, $v_R = V(1 - e^{-Rt/L})$
$= V(1 - e^{-t/\tau})$

growth of current flow, $i = I(1 - e^{-Rt/L})$
$= I(1 - e^{-t/\tau})$

These equations are derived analytically in Chapter 48. The application of these equations is shown in worked Problem 10.

Problem 8. A relay has an inductance of 100 mH and a resistance of 20 Ω. It is connected to a 60 V d.c. supply. Use the 'initial slope and three point' method to draw the current/time characteristic and hence determine the value of current flowing at a time equal to two time constants and the time for the current to grow to 1.5 A.

Before the current/time characteristic can be drawn, the time constant and steady state value of the current have to be calculated.

$$\text{Time constant, } \tau = \frac{L}{R} = \frac{100 \times 10^{-3}}{20} = 5\,\text{ms}$$

$$\text{Final value of current, } I = \frac{V}{R} = \frac{60}{20} = 3\,\text{A}$$

The method used to construct the characteristic is the same as that used in worked Problem 2.

(a) The scales should span at least five time constants (horizontally), i.e. 25 ms, and 3 A (vertically).

(b) With reference to Figure 19.12, the initial slope is obtained by making AB equal to 1 time constant, (5 ms), and joining $0B$.

(c) At a time of 1 time constant, CD is $0.632 \times I = 0.632 \times 3 = 1.896$ A
At a time of 2.5 time constants, EF is $0.918 \times I = 0.918 \times 3 = 2.754$ A
At a time of 5 time constants, $GH = 3$ A

Part 3

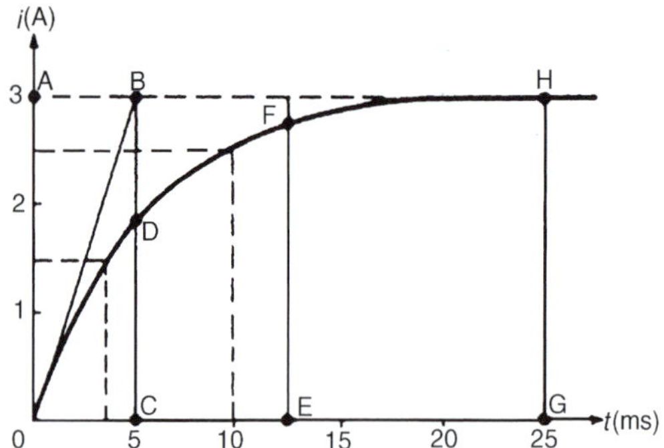

Figure 19.12

(d) A smooth curve is drawn through points 0, D, F and H and this curve is the current/time characteristic.

From the characteristic, when $t=2\tau$, $i \approx$ **2.6 A**.
[This may be checked by calculation using $i=I(1-e^{-t/\tau})$, where $I=3$ and $t=2\tau$, giving $i=2.59$ A]
Also, when the current is 1.5 A, the corresponding time is about **3.6 ms**.
[This may be checked by calculation, using $i=I(1-e^{-t/\tau})$ where $i=1.5$, $I=3$ and $\tau=5$ms, giving $t=3.466$ ms]

Problem 9. A coil of inductance 0.04 H and resistance 10 Ω is connected to a 120 V d.c. supply. Determine (a) the final value of current, (b) the time constant of the circuit, (c) the value of current after a time equal to the time constant from the instant the supply voltage is connected, and (d) the expected time for the current to rise to within 1% of its final value.

(a) Final steady current, $I=\dfrac{V}{R}=\dfrac{120}{10}=$ **12 A**

(b) Time constant of the circuit, $\tau=\dfrac{L}{R}=\dfrac{0.04}{10}$

$=$ **0.004 s** or **4 ms**

(c) In the time τ s the current rises to 63.2% of its final value of 12 A, i.e. in 4 ms the current rises to $0.632\times 12=$ **7.58 A**

(d) The expected time for the current to rise to within 1% of its final value is given by 5τ s, i.e. $5\times 4=$ **20 ms**

Problem 10. The winding of an electromagnet has an inductance of 3 H and a resistance of 15 Ω. When it is connected to a 120 V d.c. supply, calculate:

(a) the steady state value of current flowing in the winding,

(b) the time constant of the circuit,

(c) the value of the induced e.m.f. after 0.1 s,

(d) the time for the current to rise to 85% of its final value and

(e) the value of the current after 0.3 s

(a) The steady state value of current is $I=V/R$, i.e. $I=120/15=$ **8 A**

(b) The time constant of the circuit, $\tau=L/R=3/15=$ **0.2 s**

Parts (c), (d) and (e) of this problem may be determined by drawing the transients graphically, as shown in worked Problem 7 or by calculation as shown below.

(c) The induced e.m.f., v_L is given by $v_L=Ve^{-t/\tau}$ The d.c. voltage V is 120 V, t is 0.1 s and τ is 0.2 s, hence

$$v_L=120e^{-0.1/0.2}=120e^{-0.5}=120\times 0.6065$$

i.e. $v_L=$ **72.78 V**

(d) When the current is 85% of its final value, $i=0.85I$.

Also, $i=I(1-e^{-t/\tau})$, thus $0.85I=I(1-e^{-t/\tau})$

$$0.85=1-e^{-t/\tau} \text{ and since } \tau=0.2,$$

$$0.85=1-e^{-t/0.2}$$

$$e^{-t/0.2}=1-0.85=0.15$$

$$e^{t/0.2}=\frac{1}{0.15}=6.\dot{6}$$

Taking natural logarithms of each side of this equation gives:

$\ln e^{t/0.2}=\ln 6.\dot{6}$, and by the laws of logarithms

$\dfrac{t}{0.2}\ln e=\ln 6.\dot{6}$. But $\ln e=1$, hence

$t=0.2\ln 6.\dot{6}$ i.e. $t=$ **0.379 s**

(e) The current at any instant is given by $i=I(1-e^{-t/\tau})$

Part 3

When $I = 8, t = 0.3$ and $\tau = 0.2$, then

$$i = 8(1 - e^{-0.3/0.2}) = 8(1 - e^{-1.5})$$
$$= 8(1 - 0.2231) = 8 \times 0.7769$$

i.e. $i =$ **6.215 A**

19.10 Current decay in an *L–R* circuit

When a series-connected L–R circuit is connected to a d.c. supply as shown with S in position A of Figure 19.13, a current $I = V/R$ flows after a short time, creating a magnetic field ($\Phi \propto I$) associated with the inductor. When S is moved to position B, the current value decreases, causing a decrease in the strength of the magnetic field. Flux linkages occur, generating a voltage v_L, equal to $L(di/dt)$. By Lenz's law, this voltage keeps current i flowing in the circuit, its value being limited by R. Since $V = v_L + v_R, 0 = v_L + v_R$ and $v_L = -v_R$, i.e. v_L and v_R are equal in magnitude but opposite in direction. The current decays exponentially to zero and since v_R is proportional to the current flowing, v_R decays exponentially to zero. Since $v_L = v_R$, v_L also decays exponentially to zero. The curves representing these transients are similar to those shown in Figure 19.8.

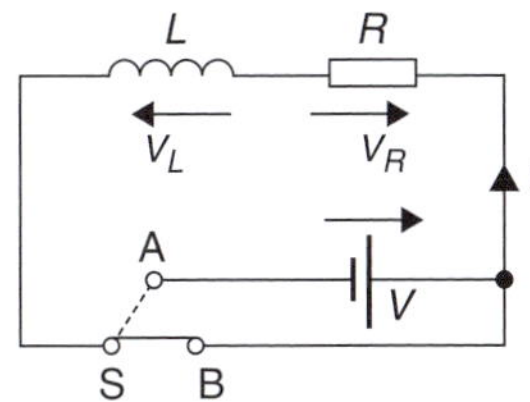

Figure 19.13

The equations representing the decay transient curves are:

decay of voltages, $v_L = v_R = Ve^{(-Rt/L)} = Ve^{(-t/\tau)}$

decay of current, $i = Ie^{(-Rt/L)} = Ie^{(-t/\tau)}$

Problem 11. The field winding of a 110 V d.c. motor has a resistance of 15 Ω and a time constant of 2 s. Determine the inductance and use the tangential method to draw the current/time characteristic when the supply is removed and replaced by a shorting link. From the characteristic, determine (a) the current flowing in the winding 3 s after being shorted-out and (b) the time for the current to decay to 5 A.

Since the time constant, $\tau = \frac{L}{R}$, $L = R\tau$

i.e. inductance $L = 15 \times 2 =$ **30 H**

The current/time characteristic is constructed in a similar way to that used in worked Problem 1.

(i) The scales should span at least five time constants horizontally, i.e. 10 s, and $I = V/R = 110/15 = 7.\dot{3}$ A vertically.

(ii) With reference to Figure 19.14, the initial slope is obtained by making *0B* equal to 1 time constant, (2 s), and joining *AB*.

(iii) At, say, $i = 6$ A, let C be the point on *AB* corresponding to a current of 6 A. Make *DE* equal to 1 time constant (2 s) and join *CE*.

(iv) Repeat the procedure given in (iii) for current values of, say, 4 A, 2 A and 1 A, giving points F, G and H.

(v) Point J is at five time constants, when the value of current is zero.

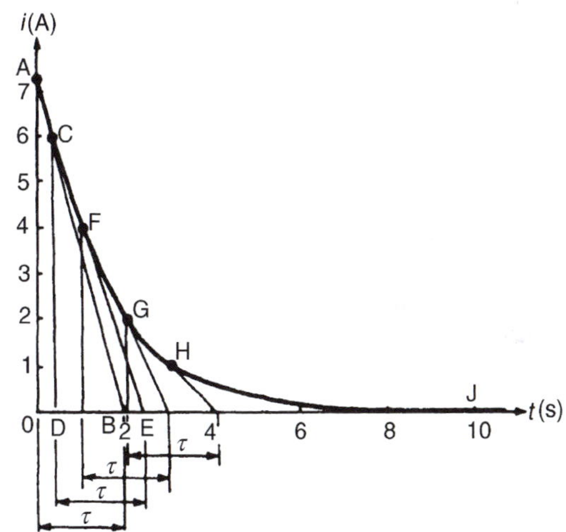

Figure 19.14

(vi) Join points A, C, F, G, H and J with a smooth curve. This curve is the current/time characteristic.

(a) From the current/time characteristic, when $t = 3$ s, $i =$ **1.5 A**. [This may be checked by calculation using $i = Ie^{-t/\tau}$, where $I = 7.\dot{3}$, $t = 3$ and $\tau = 2$, giving $i = 1.64$ A] The discrepancy between the two results is due to relatively few values, such as C, F, G and H, being taken.

(b) From the characteristic, when $i = 5$ A, $t =$ **0.70 s**. [This may be checked by

Part 3

calculation using $i = Ie^{-t/\tau}$, where $i = 5$, $I = 7.\dot{3}$, $\tau = 2$, giving $t = 0.766$ s.] Again, the discrepancy between the graphical and calculated values is due to relatively few values, such as C, F, G and H, being taken.

Problem 12. A coil having an inductance of 6 H and a resistance of R Ω is connected in series with a resistor of 10 Ω to a 120 V d.c. supply. The time constant of the circuit is 300 ms. When steady state conditions have been reached, the supply is replaced instantaneously by a short-circuit. Determine: (a) the resistance of the coil, (b) the current flowing in the circuit one second after the shorting link has been placed in the circuit and (c) the time taken for the current to fall to 10% of its initial value.

(a) The time constant, $\tau = \dfrac{\text{circuit inductance}}{\text{total circuit resistance}} = \dfrac{L}{R+10}$

Thus $R = \dfrac{L}{\tau} - 10 = \dfrac{6}{0.3} = 10 = \mathbf{10\,\Omega}$

Parts (b) and (c) may be determined graphically as shown in worked Problems 8 and 11 or by calculation as shown below.

(b) The steady state current, $I = \dfrac{V}{R} = \dfrac{120}{10+10} = 6\text{ A}$

The transient current after 1 second,

$i = Ie^{-t/\tau} = 6e^{-1/0.3}$

Thus $i = 6e^{-3.\dot{3}} = 6 \times 0.03567 = \mathbf{0.214\,A}$

(c) 10% of the initial value of the current is $(10/100) \times 6$, i.e. 0.6 A

Using the equation $i = Ie^{-t/\tau}$ gives

$$0.6 = 6e^{-t/0.3}$$

i.e.
$$\frac{0.6}{6} = e^{-t/0.3}$$

or
$$e^{t/0.3} = \frac{6}{0.6} = 10$$

Taking natural logarithms of each side of this equation gives:

$$\frac{t}{0.3} = \ln 10$$

$$t = 0.3 \ln 10 = \mathbf{0.691\,s}$$

Problem 13. An inductor has a negligible resistance and an inductance of 200 mH and is connected in series with a 1 kΩ resistor to a 24 V d.c. supply. Determine the time constant of the circuit and the steady state value of the current flowing in the circuit. Find (a) the current flowing in the circuit at a time equal to one time constant, (b) the voltage drop across the inductor at a time equal to two time constants and (c) the voltage drop across the resistor after a time equal to three time constants.

The time constant, $\tau = \dfrac{L}{R} = \dfrac{0.2}{1000} = \mathbf{0.2\,ms}$

The steady state current $I = \dfrac{V}{R} = \dfrac{24}{1000} = \mathbf{24\,mA}$

(a) The transient current, $i = I(1 - e^{-t/\tau})$ and $t = 1\tau$

Working in mA units gives, $i = 24(1 - e^{-(1\tau/\tau)})$

$$= 24(1 - e^{-1})$$

$$= 24(1 - 0.368)$$

$$= \mathbf{15.17\,mA}$$

(b) The voltage drop across the inductor, $v_L = Ve^{-t/\tau}$

When $t = 2\tau$, $v_L = 24e^{-2\tau/\tau} = 24e^{-2}$

$$= \mathbf{3.248\,V}$$

(c) The voltage drop across the resistor,

$$v_R = V(1 - e^{-t/\tau})$$

When $t = 3\tau$, $v_R = 24(1 - e^{-3\tau/\tau}) = 24(1 - e^{-3})$

$$= \mathbf{22.81\,V}$$

Now try the following Practice Exercise

Practice Exercise 90 Transients in series *L–R* circuits (Answers on page 823)

1. A coil has an inductance of 1.2 H and a resistance of 40 Ω and is connected to a 200 V d.c. supply. Either by drawing the current/

Part 3

time characteristic or by calculation, determine the value of the current flowing 60 ms after connecting the coil to the supply.

2. A 25 V d.c. supply is connected to a coil of inductance 1 H and resistance 5 Ω. Either by using a graphical method to draw the exponential growth curve of current or by calculation, determine the value of the current flowing 100 ms after being connected to the supply.

3. An inductor has a resistance of 20 Ω and an inductance of 4 H. It is connected to a 50 V d.c. supply. Calculate (a) the value of current flowing after 0.1 s and (b) the time for the current to grow to 1.5 A.

4. The field winding of a 200 V d.c. machine has a resistance of 20 Ω and an inductance of 500 mH. Calculate:
   (a) the time constant of the field winding,
   (b) the value of current flow one time constant after being connected to the supply and
   (c) the current flowing 50 ms after the supply has been switched on.

5. A circuit comprises an inductor of 9 H of negligible resistance connected in series with a 60 Ω resistor and a 240 V d.c. source. Calculate (a) the time constant, (b) the current after 1 time constant, (c) the time to develop maximum current, (d) the time for the current to reach 2.5 A and (e) the initial rate of change of current.

6. In the inductive circuit shown in Figure 19.15, the switch is moved from position A to position B until maximum current is flowing. Calculate (a) the time taken for the voltage across the resistance to reach 8 volts, (b) the time taken for maximum current to flow in the circuit, (c) the energy stored in the inductor when maximum current is flowing and (d) the time for current to drop to 750 mA after switching to position C.

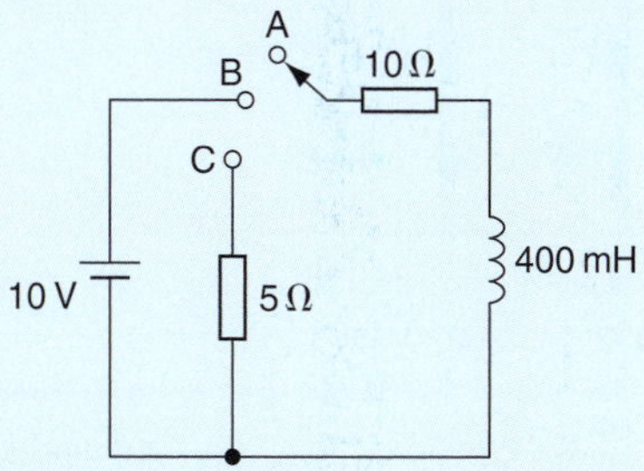

Figure 19.15

19.11 Switching inductive circuits

Energy stored in the magnetic field of an inductor exists because a current provides the magnetic field. When the d.c. supply is switched off the current falls rapidly, the magnetic field collapses causing a large induced e.m.f. which will either cause an arc across the switch contacts or will break down the insulation between adjacent turns of the coil. The high induced e.m.f. acts in a direction which tends to keep the current flowing, i.e. in the same direction as the applied voltage. The energy from the magnetic field will thus be aided by the supply voltage in maintaining an arc, which could cause severe damage to the switch. To reduce the induced e.m.f. when the supply switch is opened, a discharge resistor R_D is connected in parallel with the inductor as shown in Figure 19.16. The magnetic field energy is dissipated as heat in R_D and R and arcing at the switch contacts is avoided.

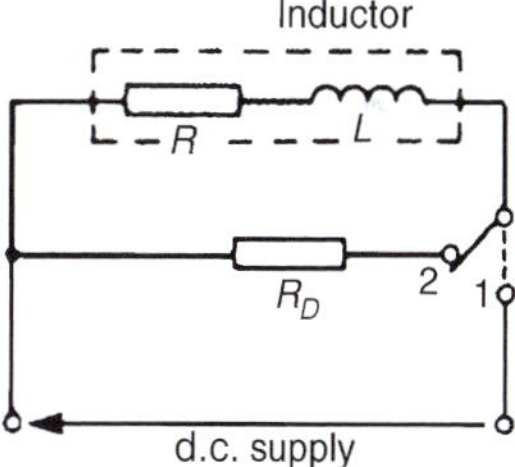

Figure 19.16

19.12 The effect of time constant on a rectangular waveform

Integrator circuit

By varying the value of either C or R in a series-connected C–R circuit, the time constant ($\tau = CR$), of a circuit can be varied. If a rectangular waveform varying from $+E$ to $-E$ is applied to a C–R circuit as shown in Figure 19.17, output waveforms of the capacitor voltage have various shapes, depending on the value

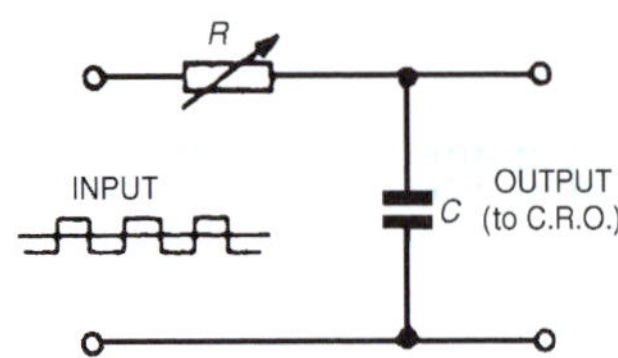

Figure 19.17

of R. When R is small, $t = CR$ is small and an output waveform such as that shown in Figure 19.18(a) is obtained. As the value of R is increased, the waveform changes to that shown in Figure 19.18(b). When R is large, the waveform is as shown in Figure 19.18(c), the circuit then being described as an **integrator circuit**.

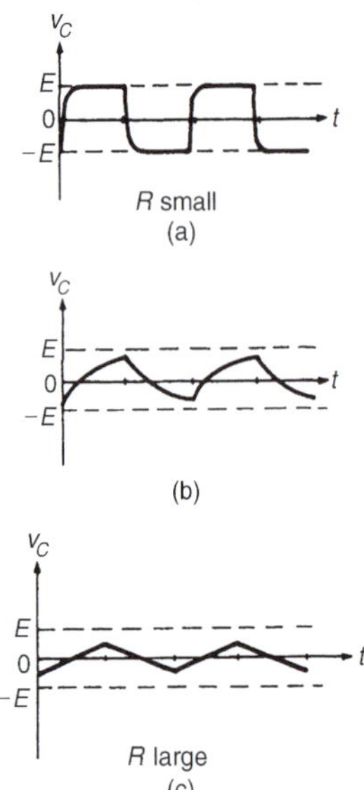

Figure 19.18

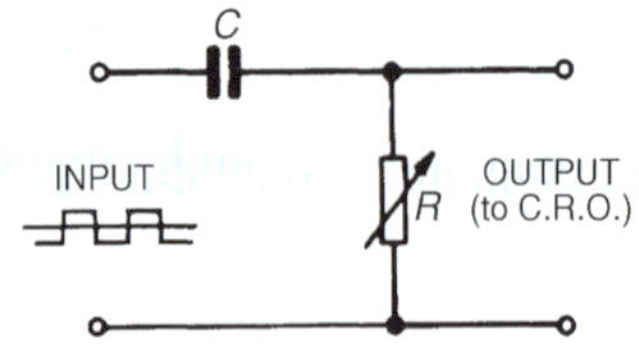

Figure 19.19

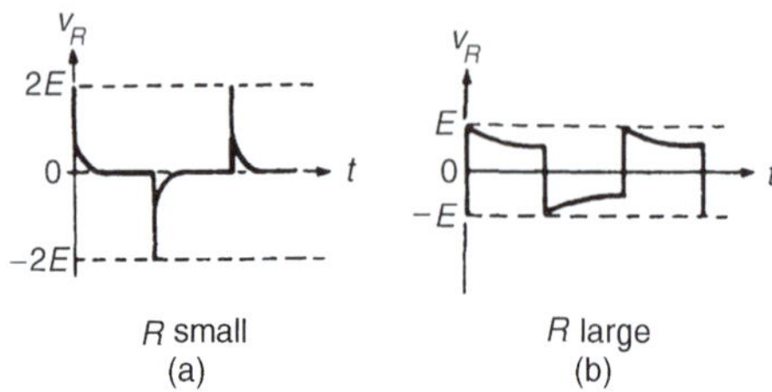

Figure 19.20

Differentiator circuit

If a rectangular waveform varying from $+E$ to $-E$ is applied to a series-connected C–R circuit and the waveform of the voltage drop across the resistor is observed, as shown in Figure 19.19, the output waveform alters as R is varied due to the time constant, ($\tau = CR$) altering. When R is small, the waveform is as shown in Figure 19.20(a), the voltage being generated across R by the capacitor discharging fairly quickly. Since the change in capacitor voltage is from $+E$ to $-E$, the change in discharge current is $2E/R$, resulting in a change in voltage across the resistor of $2E$. This circuit is called a **differentiator circuit**. When R is large, the waveform is as shown in Figure 19.20(b).

For fully worked solutions to each of the problems in Practice Exercises 89 and 90 in this chapter, go to the website:
www.routledge.com/cw/bird

Chapter 20

Operational amplifiers

***Why it is important to understand:* Operational amplifiers**

The term operational amplifier, abbreviated to op amp, was coined in the 1940s to refer to a special kind of amplifier that, by proper selection of external components, could be configured to perform a variety of mathematical operations. Early op amps were made from vacuum tubes consuming lots of space and energy. Later op amps were made smaller by implementing them with discrete transistors. Today, op amps are monolithic integrated circuits, highly efficient and cost effective. Operational amplifiers can be used to perform mathematical operations on voltage signals such as inversion, addition, subtraction, integration, differentiation and multiplication by a constant. The op amp is one of the most useful and important components of analogue electronics and they are widely used. This chapter explains the main properties of op amps and explains the principle of operation of the inverter, non-inverter, voltage follower, summing, voltage comparator, integrator and differentiator op amps. In addition digital to analogue and analogue to digital conversions are explained.

At the end of this chapter you should be able to:

- recognize the main properties of an operational amplifier
- understand op amp parameters input bias current and offset current and voltage
- define and calculate common-mode rejection ratio
- appreciate slew rate
- explain the principle of operation, draw the circuit diagram symbol and calculate gain for the following operational amplifiers:
  - inverter
  - non-inverter
  - voltage follower (or buffer)
  - summing
  - voltage comparator
  - integrator
  - differentiator
- understand digital to analogue conversion
- understand analogue to digital conversion

Electrical Circuit Theory and Technology. 978-1-138-67349-6, © 2017 John Bird. Published by Taylor & Francis. All rights reserved.

20.1 Introduction to operational amplifiers

Operational amplifiers (usually called **'op amps'**) were originally made from discrete components, being designed to solve mathematical equations electronically, by performing operations such as addition and division in analogue computers. Now produced in integrated-circuit (IC) form, op amps have many uses, with one of the most important being as a high-gain d.c. and a.c. voltage amplifier.

The **main properties** of an op amp include:

(i) a very high open-loop voltage gain A_o of around 10^5 for d.c. and low frequency a.c., which decreases with frequency increase

(ii) a very high input impedance, typically $10^6\,\Omega$ to $10^{12}\,\Omega$, such that current drawn from the device, or the circuit supplying it, is very small and the input voltage is passed on to the op amp with little loss

(iii) a very low output impedance, around $100\,\Omega$, such that its output voltage is transferred efficiently to any load greater than a few kilohms

The **circuit diagram symbol** for an op amp is shown in Figure 20.1. It has one output, $\boldsymbol{V_o}$, and two inputs; the **inverting input, $\boldsymbol{V_1}$**, is marked – and the **non-inverting input, $\boldsymbol{V_2}$**, is marked +

The operation of an op amp is most convenient from a dual balanced d.c. power supply $\pm V_S$ (i.e. $+V_S$, $0, -V_S$); the centre point of the supply, i.e. 0 V, is common to the input and output circuits and is taken as their voltage reference level. The power supply connections are not usually shown in a circuit diagram.

An op amp is basically a **differential** voltage amplifier, i.e. it amplifies the difference between input voltages V_1 and V_2. Three situations are possible:

(i) if $V_2 > V_1$, V_o is positive

(ii) if $V_2 < V_1$, V_o is negative

(iii) if $V_2 = V_1$, V_o is zero.

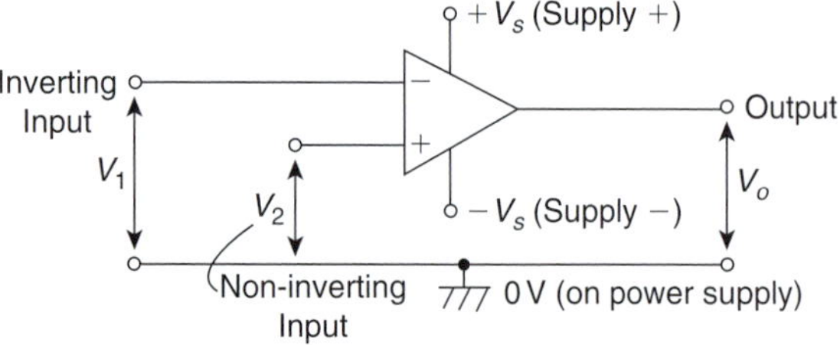

Figure 20.1

In general, $$\boldsymbol{V_o = A_o(V_2 - V_1)}$$

or $$\boldsymbol{A = \frac{V_o}{V_2 - V_1}} \qquad (1)$$

where A_o is the open-loop voltage gain

Problem 1. A differential amplifier has an open-loop voltage gain of 120. The input signals are 2.45 V and 2.35 V. Calculate the output voltage of the amplifier.

From equation (1), **output voltage,**

$$\boldsymbol{V_o} = A_o(V_2 - V_1) = 120(2.45 - 2.35)$$
$$= (120)(0.1) = \mathbf{12\ V}$$

Transfer characteristic

A typical **voltage characteristic** showing how the output V_o varies with the input $(V_2 - V_1)$ is shown in Figure 20.2.

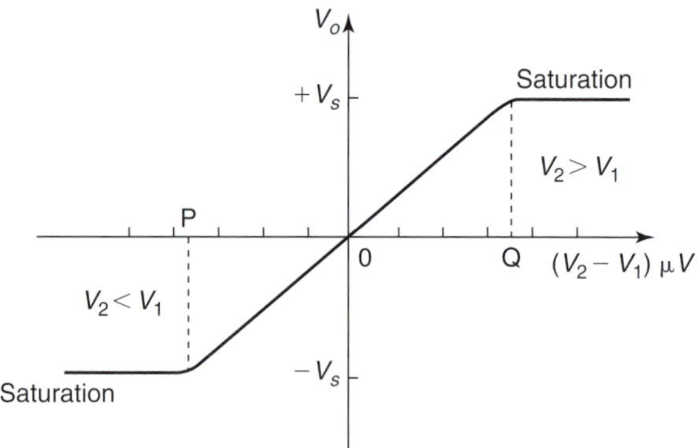

Figure 20.2

It is seen from Figure 20.2 that only within the very small input range $P0Q$ is the output directly proportional to the input; it is in this range that the op amp behaves linearly and there is minimum distortion of the amplifier output. Inputs outside the linear range cause saturation and the output is then close to the maximum value, i.e. $+V_S$ or $-V_S$. The limited linear behaviour is due to the very high open-loop gain A_o, and the higher it is the greater is the limitation.

Part 3

Negative feedback

Operational amplifiers nearly always use **negative feedback**, obtained by feeding back some, or all, of the output to the inverting (−) input (as shown in Figure 20.5 later). The feedback produces an output voltage that opposes the one from which it is taken. This reduces the new output of the amplifier and the resulting closed-loop gain A is then less than the open-loop gain A_o. However, as a result, a wider range of voltages can be applied to the input for amplification. As long as $A_o >> A$, negative feedback gives:
(i) a constant and predictable voltage gain A, (ii) reduced distortion of the output, and (iii) better frequency response.

The advantages of using negative feedback outweigh the accompanying loss of gain which is easily increased by using two or more op amp stages.

Bandwidth

The open-loop voltage gain of an op amp is not constant at all frequencies; because of capacitive effects it falls at high frequencies. Figure 20.3 shows the gain/bandwidth characteristic of a 741 op amp. At frequencies below 10 Hz the gain is constant, but at higher frequencies the gain falls at a constant rate of 6 dB/octave (equivalent to a rate of 20 dB per decade) to 0 dB

The gain-bandwidth product for any amplifier is the linear voltage gain multiplied by the bandwidth at that gain. The value of frequency at which the open-loop gain has fallen to unity is called the transition frequency f_T

$$f_T = \text{closed-loop voltage gain} \times \text{bandwidth} \qquad (2)$$

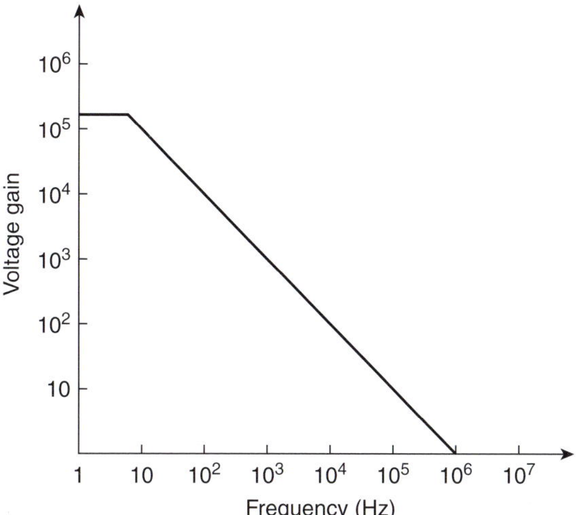

Figure 20.3

In Figure 20.3, $f_T = 10^6$ Hz or 1 MHz; a gain of 20 dB (i.e. 20 $\log_{10}$ 10) gives a 100 kHz bandwidth, whilst a gain of 80 dB (i.e. 20 $\log_{10} 10^4$) restricts the bandwidth to 100 Hz.

20.2 Some op amp parameters

Input bias current

The input bias current, I_B, is the average of the currents into the two input terminals with the output at zero volts, which is typically around 80 nA (i.e. 80×10^{-9} A) for a 741 op amp. The input bias current causes a volt drop across the equivalent source impedance seen by the op amp input.

Input offset current

The input offset current, I_{os}, of an op amp is the difference between the two input currents with the output at zero volts. In a 741 op amp, I_{os} is typically 20 nA.

Input offset voltage

In the ideal op amp, with both inputs at zero there should be zero output. Due to imbalances within the amplifier this is not always the case and a small output voltage results. The effect can be nullified by applying a small offset voltage, V_{os}, to the amplifier. In a 741 op amp, V_{os} is typically 1 mV.

Common-mode rejection ratio

The output voltage of an op amp is proportional to the difference between the voltages applied to its two input terminals. Ideally, when the two voltages are equal, the output voltages should be zero. A signal applied to both input terminals is called a common-mode signal and it is usually an unwanted noise voltage. The ability of an op amp to suppress common-mode signals is expressed in terms of its common-mode rejection ratio (CMRR), which is defined by:

$$\text{CMRR} = 20\log_{10}\left(\frac{\text{differential voltage gain}}{\text{common mode gain}}\right)\text{dB} \qquad (3)$$

In a 741 op amp, the CMRR is typically 90 dB.
The common-mode gain, A_{com}, is defined as:

$$A_{com} = \frac{V_o}{V_{com}} \qquad (4)$$

where V_{com} is the common input signal

Part 3

Problem 2. Determine the common-mode gain of an op amp that has a differential voltage gain of 150×10^3 and a CMRR of 90 dB.

From equation (3),

$$\text{CMRR} = 20\log_{10}\left(\frac{\text{differential voltage gain}}{\text{common-mode gain}}\right)\text{dB}$$

Hence $$90 = 20\log_{10}\left(\frac{150 \times 10^3}{\text{common-mode gain}}\right)$$

from which $$4.5 = \log_{10}\left(\frac{150 \times 10^3}{\text{common-mode gain}}\right)$$

and $$10^{4.5} = \left(\frac{150 \times 10^3}{\text{common-mode gain}}\right)$$

Hence, **common-mode gain** $= \dfrac{150 \times 10^3}{10^{4.5}} = \mathbf{4.74}$

Problem 3. A differential amplifier has an open-loop voltage gain of 120 and a common input signal of 3.0 V to both terminals. An output signal of 24 mV results. Calculate the common-mode gain and the CMRR.

From equation (4), the common-mode gain,

$$A_{com} = \frac{V_o}{V_{com}} = \frac{24 \times 10^{-3}}{3.0} = 8 \times 10^{-3} = \mathbf{0.008}$$

From equation (3), the

$$\textbf{CMRR} = 20\log_{10}\left(\frac{\text{differential voltage gain}}{\text{common-mode gain}}\right)\text{dB}$$

$$= 20\log_{10}\left(\frac{120}{0.008}\right) = 20\log_{10} 15\,000 = \mathbf{83.52\ dB}$$

Slew rate

The slew rate of an op amp is the maximum rate of change of output voltage following a step input voltage. Figure 20.4 shows the effects of slewing; it causes the output voltage to change at a slower rate than the input, such that the output waveform is a distortion of the input waveform. 0.5 V/μs is a typical value for the slew rate.

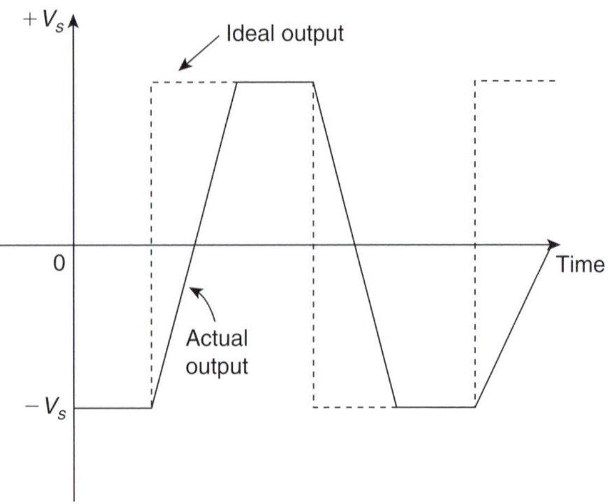

Figure 20.4

20.3 Op amp inverting amplifier

The basic circuit for an inverting amplifier is shown in Figure 20.5, where the input voltage V_i (a.c. or d.c.) to be amplified is applied via resistor R_i to the inverting (−) terminal; the output voltage V_o is therefore in anti-phase with the input. The non-inverting (+) terminal is held at 0 V. Negative feedback is provided by the feedback resistor, R_f, feeding back a certain fraction of the output voltage to the inverting terminal.

Amplifier gain

In an **ideal op amp** two assumptions are made, these being that:

(i) each input draws zero current from the signal source, i.e. their input impedances are infinite, and

(ii) the inputs are both at the same potential if the op amp is not saturated, i.e. $V_A = V_B$ in Figure 20.5.

In Figure 20.5, $V_B = 0$, hence $V_A = 0$ and point X is called a **virtual earth**.

Thus, $I_1 = \dfrac{V_i - 0}{R_i}$ and $I_2 = \dfrac{0 - V_o}{R_f}$

However, $I_1 = I_2$ from assumption (i) above.

Hence $$\frac{V_i}{R_i} = -\frac{V_o}{R_f}$$

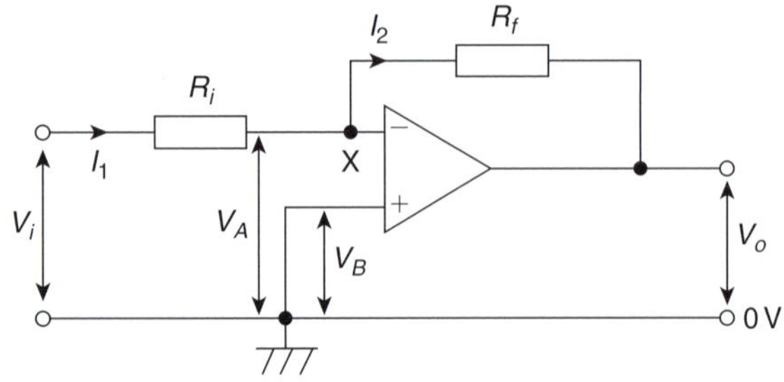

Figure 20.5

Part 3

the negative sign showing that V_o is negative when V_i is positive, and vice versa.
The **closed-loop gain A** is given by:

$$A = \frac{V_o}{V_i} = \frac{-R_f}{R_i} \qquad (5)$$

This shows that the gain of the amplifier depends only on the two resistors, which can be made with precise values, and not on the characteristics of the op amp, which may vary from sample to sample.
For example, if $R_i = 10\,\text{k}\Omega$ and $R_f = 100\,\text{k}\Omega$, then the closed-loop gain,

$$\mathbf{A} = \frac{-R_f}{R_i} = \frac{-100 \times 10^3}{10 \times 10^3} = \mathbf{-10}$$

Thus an input of 100 mV will cause an output change of 1 V.

Input impedance

Since point X is a virtual earth (i.e. at 0 V), R_i may be considered to be connected between the inverting (−) input terminal and 0 V. The input impedance of the circuit is therefore R_i in parallel with the much greater input impedance of the op amp, i.e. effectively R_i. The circuit input impedance can thus be controlled by simply changing the value of R_i

Problem 4. In the inverting amplifier of Figure 20.5, $R_i = 1\,\text{k}\Omega$ and $R_f = 2\,\text{k}\Omega$. Determine the output voltage when the input voltage is: (a) +0.4 V, (b) −1.2 V.

From equation (5), $V_o = \left(\frac{-R_f}{R_i}\right) V_i$

(a) When $V_i = +0.4\,\text{V}$, $\boldsymbol{V_o} = \left(\frac{-2000}{1000}\right)(+0.4) = \mathbf{-0.8\,V}$

(b) When $V_i = -1.2\,\text{V}$, $\boldsymbol{V_o} = \left(\frac{-2000}{1000}\right)(-1.2) = \mathbf{+2.4\,V}$

Problem 5. The op amp shown in Figure 20.6 has an input bias current of 100 nA at 20°C. Calculate (a) the voltage gain, and (b) the output offset voltage due to the input bias current. (c) How can the effect of input bias current be minimized?

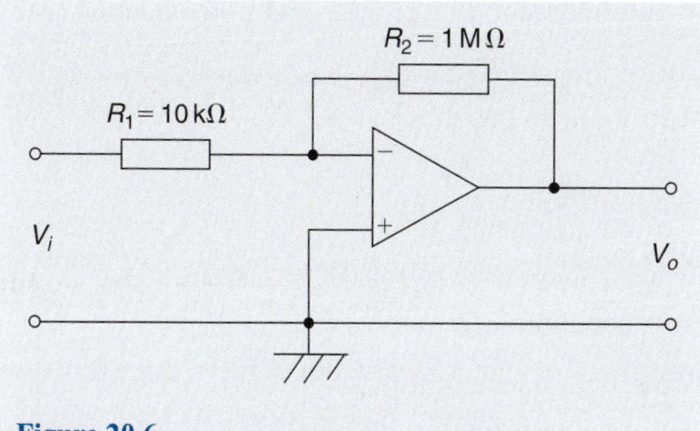

Figure 20.6

Comparing Figure 20.6 with Figure 20.5 gives $R_i = 10\,\text{k}\Omega$ and $R_f = 1\,\text{M}\Omega$

(a) From equation (5), **voltage gain,**

$$A = \frac{-R_f}{R_i} = \frac{-1 \times 10^6}{10 \times 10^3} = \mathbf{-100}$$

(b) The input bias current, I_B, causes a volt drop across the equivalent source impedance seen by the op amp input, in this case, R_i and R_f in parallel. Hence, the offset voltage, V_{os}, at the input due to the 100 nA input bias current, I_B, is given by:

$$V_{os} = I_B \left(\frac{R_i R_f}{R_i + R_f}\right)$$
$$= (100 \times 10^{-9})\left(\frac{10 \times 10^3 \times 1 \times 10^6}{(10 \times 10^3) + (1 \times 10^6)}\right)$$
$$= (10^{-7})(9.9 \times 10^3) = 9.9 \times 10^{-4} = \mathbf{0.99\,mV}$$

(c) The effect of input bias current can be minimized by ensuring that both inputs 'see' the same driving resistance. This means that **a resistance of value 9.9 kΩ** (from part (b)) **should be placed between the non-inverting (+) terminal and earth** in Figure 20.6.

Problem 6. Design an inverting amplifier to have a voltage gain of 40 dB, a closed-loop bandwidth of 5 kHz and an input resistance of 10 kΩ.

The voltage gain of an op amp, in decibels, is given by:

gain in decibels $= 20\log_{10}$ (voltage gain)
(from Chapter 12)

Hence $\quad 40 = 20\log_{10} A$

from which, $\quad 2 = \log_{10} A$

and $\quad A = 10^2 = \mathbf{100}$

With reference to Figure 20.5, and from equation (5),

$$A = \left|\frac{R_f}{R_i}\right|$$

i.e. $$100 = \frac{R_f}{10 \times 10^3}$$

Hence $\boldsymbol{R_f} = 100 \times 10 \times 10^3 = \mathbf{1\,M\Omega}$

From equation (2),

frequency $=$ gain $\times$ bandwidth $= 100 \times 5 \times 10^3$

$= \mathbf{0.5\,MHz}$ or **500 kHz**

Now try the following Practice Exercise

Practice Exercise 91 Introduction to operational amplifiers (Answers on page 823)

1. A differential amplifier has an open-loop voltage gain of 150 when the input signals are 3.55 V and 3.40 V. Determine the output voltage of the amplifier.
2. Calculate the differential voltage gain of an op amp that has a common-mode gain of 6.0 and a CMRR of 80 dB.
3. A differential amplifier has an open-loop voltage gain of 150 and a common input signal of 4.0 V to both terminals. An output signal of 15 mV results. Determine the common-mode gain and the CMRR.
4. In the inverting amplifier of Figure 20.5 (on page 256), $R_i = 1.5\,k\Omega$ and $R_f = 2.5\,k\Omega$. Determine the output voltage when the input voltage is: (a) $+0.6$ V, (b) -0.9 V
5. The op amp shown in Figure 20.7 has an input bias current of 90 nA at 20°C. Calculate (a) the voltage gain and (b) the output offset voltage due to the input bias current.

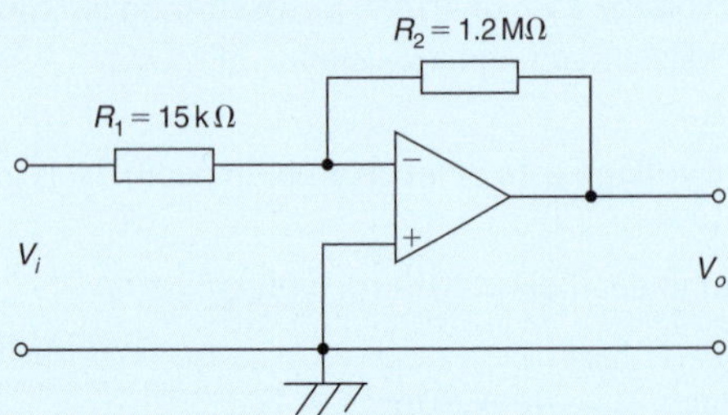

Figure 20.7

6. Determine (a) the value of the feedback resistor and (b) the frequency for an inverting amplifier to have a voltage gain of 45 dB, a closed-loop bandwidth of 10 kHz and an input resistance of 20 kΩ.

20.4 Op amp non-inverting amplifier

The basic circuit for a non-inverting amplifier is shown in Figure 20.8, where the input voltage V_i (a.c. or d.c.) is applied to the non-inverting (+) terminal of the op amp. This produces an output V_o that is in phase with the input. Negative feedback is obtained by feeding back to the inverting (−) terminal, the fraction of V_o developed across R_i in the voltage divider formed by R_f and R_i across V_o

Amplifier gain

In Figure 20.8, let the feedback factor,

$$\beta = \frac{R_i}{R_i + R_f}$$

It may be shown that for an amplifier with open-loop gain A_o, the closed-loop voltage gain A is given by:

$$A = \frac{A_o}{1 + \beta A_o}$$

For a typical op amp, $A_o = 10^5$, thus βA_o is large compared with 1, and the above expression approximates to:

$$A = \frac{A_o}{\beta A_o} = \frac{1}{\beta} \qquad (6)$$

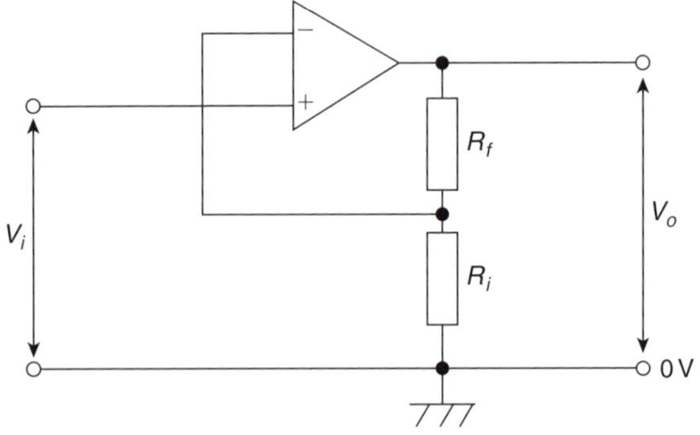

Figure 20.8

Hence $$\boldsymbol{A = \frac{V_o}{V_i} = \frac{R_i + R_f}{R_i} = 1 + \frac{R_f}{R_i}} \quad (7)$$

For example, if $R_i = 10\,\text{k}\Omega$ and $R_f = 100\,\text{k}\Omega$,

then $A = 1 + \dfrac{100 \times 10^3}{10 \times 10^3} = 1 + 10 = \mathbf{11}$

Again, the gain depends only on the values of R_i and R_f and is independent of the open-loop gain A_o

Input impedance

Since there is no virtual earth at the non-inverting (+) terminal, the input impedance is much higher (typically 50 MΩ) than that of the inverting amplifier. Also, it is unaffected if the gain is altered by changing R_f and/or R_i. This non-inverting amplifier circuit gives good matching when the input is supplied by a high impedance source.

Problem 7. For the op amp shown in Figure 20.9, $R_1 = 4.7\,\text{k}\Omega$ and $R_2 = 10\,\text{k}\Omega$. If the input voltage is -0.4 V, determine (a) the voltage gain (b) the output voltage.

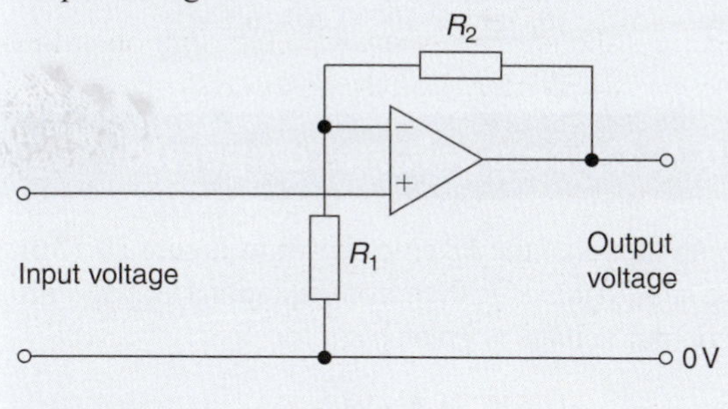

Figure 20.9

The op amp shown in Figure 20.9 is a non-inverting amplifier, similar to Figure 20.8.

(a) From equation (7), **voltage gain,**

$$A = 1 + \frac{R_f}{R_i} = 1 + \frac{R_2}{R_1} = 1 + \frac{10 \times 10^3}{4.7 \times 10^3}$$
$$= 1 + 2.13 = \mathbf{3.13}$$

(b) Also from equation (7), **output voltage,**

$$V_o = \left(1 + \frac{R_2}{R_1}\right) V_i$$
$$= (3.13)(-0.4) = \mathbf{-1.25\,V}$$

20.5 Op amp voltage-follower

The **voltage-follower** is a special case of the non-inverting amplifier in which 100% negative feedback is obtained by connecting the output directly to the inverting (−) terminal, as shown in Figure 20.10. Thus R_f in Figure 20.8 is zero and R_i is infinite.

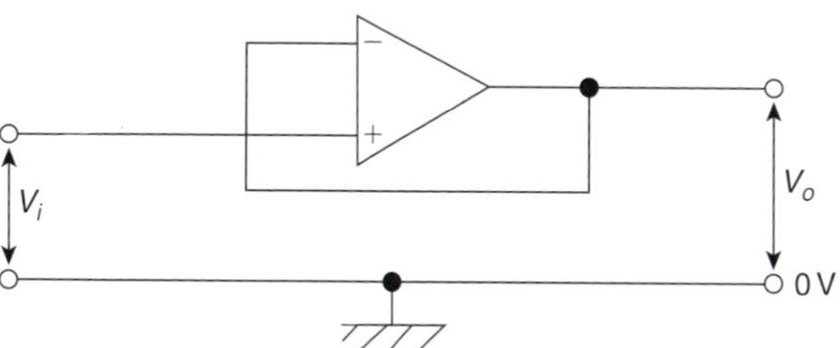

Figure 20.10

From equation (6), $A = 1/\beta$ (when A_o is very large). Since all of the output is fed back, $\beta = 1$ and $A \approx 1$. Thus the voltage gain is nearly 1 and $V_o = V_i$ to within a few millivolts.

The circuit of Figure 20.10 is called a voltage-follower since, as with its transistor emitter-follower equivalent, V_o follows V_i. It has an extremely high input impedance and a low output impedance. Its main use is as a **buffer amplifier**, giving current amplification, to match a high impedance source to a low impedance load. For example, it is used as the input stage of an analogue voltmeter where the highest possible input impedance is required so as not to disturb the circuit under test; the output voltage is measured by a relatively low impedance moving-coil meter.

20.6 Op amp summing amplifier

Because of the existence of the virtual earth point, an op amp can be used to add a number of voltages (d.c. or a.c.) when connected as a multi-input inverting amplifier. This, in turn, is a consequence of the high value of the open-loop voltage gain A_o. Such circuits may be used as 'mixers' in audio systems to combine the outputs of microphones, electric guitars, pick-ups, etc. They are also used to perform the mathematical process of addition in analogue computing.

The circuit of an op amp summing amplifier having three input voltages V_1, V_2 and V_3 applied via input resistors R_1, R_2 and R_3 is shown in Figure 20.11. If it is assumed that the inverting (−) terminal of the op amp

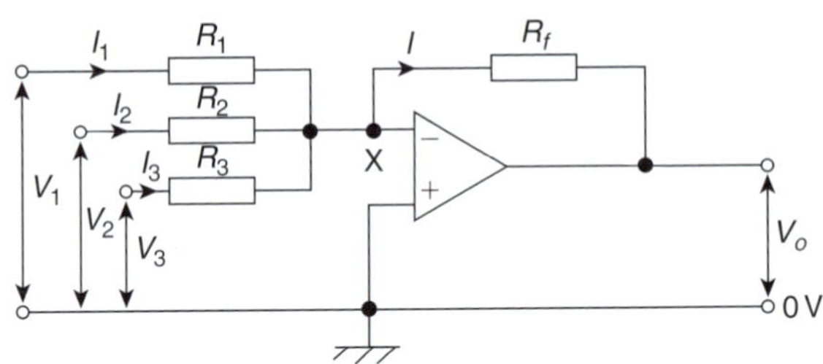

Figure 20.11

draws no input current, all of it passing through R_f, then:

$$I = I_1 + I_2 + I_3$$

Since X is a virtual earth (i.e. at 0 V), it follows that:

$$\frac{-V_o}{R_f} = \frac{V_1}{R_1} + \frac{V_2}{R_2} + \frac{V_3}{R_3}$$

Hence

$$\mathbf{V_o = -\left(\frac{R_f}{R_1}V_1 + \frac{R_f}{R_2}V_2 + \frac{R_f}{R_3}V_3\right) = -R_f\left(\frac{V_1}{R_1} + \frac{V_2}{R_2} + \frac{V_3}{R_3}\right)} \quad (8)$$

The three input voltages are thus added and amplified if R_f is greater than each of the input resistors; 'weighted' summation is said to have occurred.

Alternatively, the input voltages are added and attenuated if R_f is less than each input resistor.

For example, if $\frac{R_f}{R_1} = 4$, $\frac{R_f}{R_2} = 3$ and $\frac{R_f}{R_3} = 1$ and $V_1 = V_2 = V_3 = +1$ V, then

$$V_o = \left(\frac{R_f}{R_1}V_1 + \frac{R_f}{R_2}V_2 + \frac{R_f}{R_3}V_3\right)$$
$$= -(4 + 3 + 1) = \mathbf{-8V}$$

If $R_1 = R_2 = R_3 = R_i$, the input voltages are amplified or attenuated equally, and

$$V_o = -\frac{R_f}{R_i}(V_1 + V_2 + V_3)$$

If, also, $R_i = R_f$ then $V_o = -(V_1 + V_2 + V_3)$

The virtual earth is also called the **summing point** of the amplifier. It isolates the inputs from one another so that each behaves as if none of the others existed and none feeds any of the other inputs even though all the resistors are connected at the inverting (−) input.

Problem 8. For the summing op amp shown in Figure 20.12, determine the output voltage, V_o.

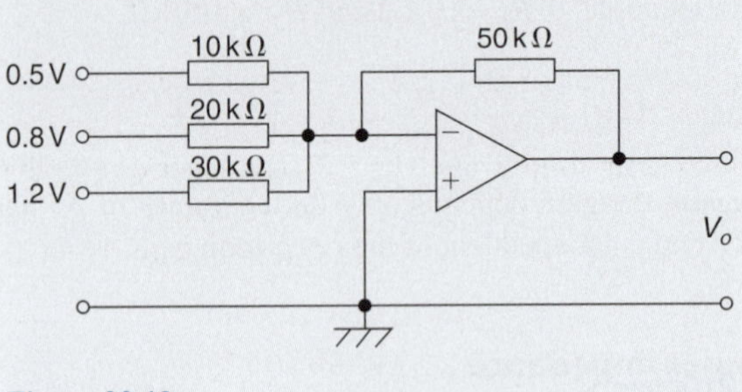

Figure 20.12

From equation (8),

$$V_o = -R_f\left(\frac{V_1}{R_1} + \frac{V_2}{R_2} + \frac{V_3}{R_3}\right)$$
$$= -(50 \times 10^3)\left(\frac{0.5}{10 \times 10^3} + \frac{0.8}{20 \times 10^3} + \frac{1.2}{30 \times 10^3}\right)$$
$$= -(50 \times 10^3)(5 \times 10^{-5} + 4 \times 10^{-5} + 4 \times 10^{-5})$$
$$= -(50 \times 10^3)(13 \times 10^{-5}) = \mathbf{-6.5V}$$

20.7 Op amp voltage comparator

If both inputs of the op amp shown in Figure 20.13 are used simultaneously, then from equation (1), page 310, the output voltage is given by:

$$V_o = A_o(V_2 - V_1)$$

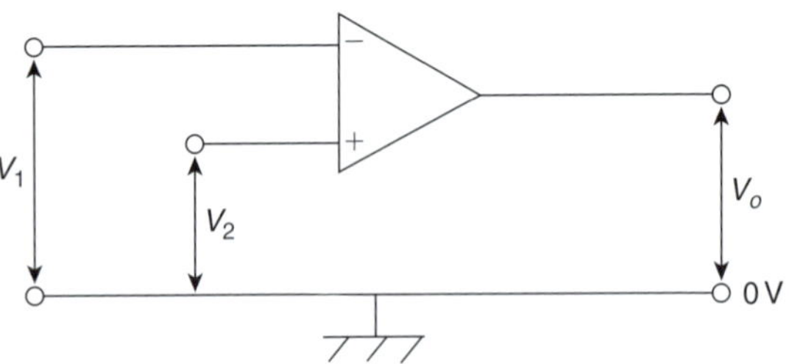

Figure 20.13

When $V_2 > V_1$ then V_o is positive, its maximum value being the positive supply voltage $+V_s$, which it has when $(V_2 - V_1) \geq V_s/A_o$. The op amp is then saturated. For example, if $V_s = +9$V and $A_o = 10^5$, then saturation

occurs when $(V_2 - V_1) \geq 9/10^5$ i.e. when V_2 exceeds V_1 by $90\,\mu$V and $V_o \approx 9$ V.
When $V_1 > V_2$, then V_o is negative and saturation occurs if V_1 exceeds V_2 by V_s/A_o i.e. around $90\,\mu$V in the above example; in this case, $V_o \approx -V_s = -9$ V
A small change in $(V_2 - V_1)$ therefore causes V_o to switch between near $+V_s$ and near to $-V_s$ and enables the op amp to indicate when V_2 is greater or less than V_1, i.e. to act as a **differential amplifier** and compare two voltages. It does this in an electronic digital voltmeter.

Problem 9. Devise a light-operated alarm circuit using an op amp, an LDR, an LED and a ±15 V supply.

A typical light-operated alarm circuit is shown in Figure 20.14.

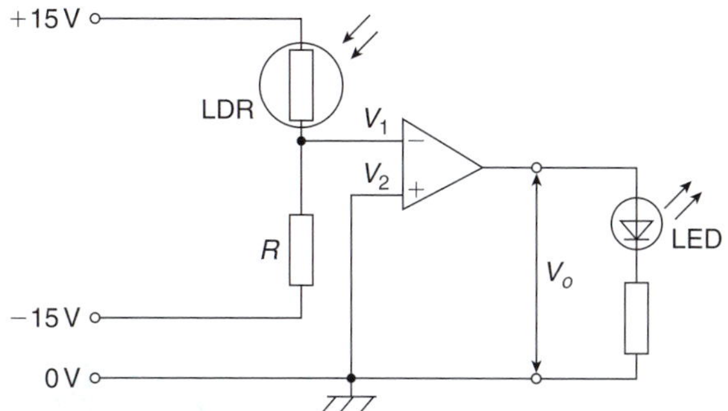

Figure 20.14

Resistor R and the light-dependent resistor (LDR) form a voltage divider across the +15/0/−15 V supply. The op amp compares the voltage V_1 at the voltage divider junction, i.e. at the inverting (−) input, with that at the non-inverting (+) input, i.e. with V_2, which is 0 V. In the dark the resistance of the LDR is much greater than that of R, so more of the 30 V across the voltage divider is dropped across the LDR, causing V_1 to fall below 0 V. Now $V_2 > V_1$ and the output voltage V_o switches from near −15 V to near +15 V and the light emitting diode (LED) lights.

20.8 Op amp integrator

The circuit for the op amp integrator shown in Figure 20.15 is the same as for the op amp inverting amplifier shown in Figure 20.5, but feedback occurs via a capacitor, C, rather than via a resistor.

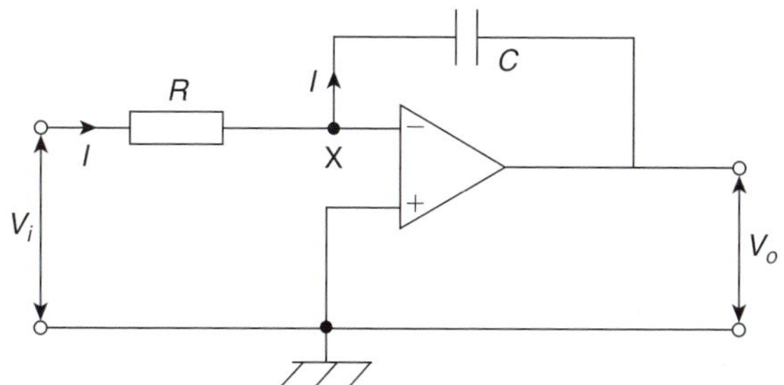

Figure 20.15

The output voltage is given by:

$$V_o = -\frac{1}{CR}\int V_i\,dt \quad (9)$$

Since the inverting (−) input is used in Figure 20.15, V_o is negative if V_i is positive, and vice-versa, hence the negative sign in equation (9).

Since X is a virtual earth in Figure 20.15, i.e. at 0 V, the voltage across R is V_i and that across C is V_o. Assuming again that none of the input current I enters the op amp inverting (−) input, then all of current I flows through C and charges it up. If V_i is constant, I will be a constant value given by $I = V_i/R$. Capacitor C therefore charges at a constant rate and the potential of the output side of C ($= V_o$, since its input side is zero) charges so that the feedback path absorbs I. If Q is the charge on C at time t and the p.d. across it (i.e. the output voltage) changes from 0 to V_o in that time then:

$Q = -V_oC = It$ (from Chapter 8)

i.e. $-V_oC = \frac{V_i}{R}t$

i.e. $V_o = -\frac{1}{CR}V_it$

This result is the same as would be obtained from $V_o = -\frac{1}{CR}\int V_i\,dt$ if V_i is a constant value.
For example, if the input voltage $V_i = -2$ V and, say, $CR = 1$ s, then

$V_o = -(-2)t = 2t$

A graph of V_o/t will be a ramp function as shown in Figure 20.16 ($V_o = 2t$ is of the straight line form $y = mx + c$; in this case $y = V_o$ and $x = t$, gradient, $m = 2$ and vertical axis intercept $c = 0$). V_o rises steadily by +2 V/s in Figure 20.16, and if the power supply is, say, ±9 V, then V_o reaches +9 V after 4.5 s when the op amp saturates.

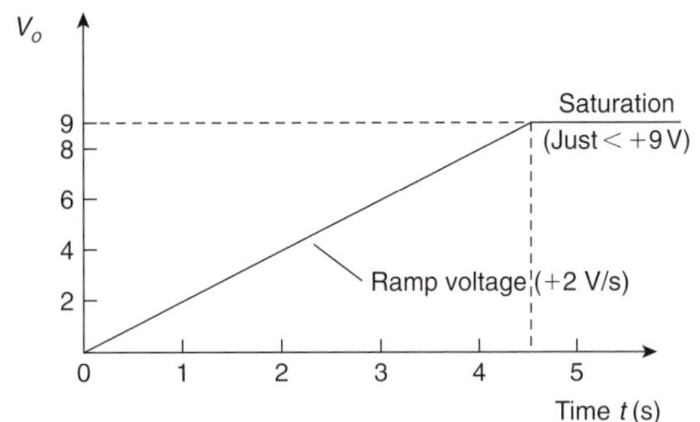

Figure 20.16

Problem 10. A steady voltage of −0.75 V is applied to an op amp integrator having component values of $R = 200\,\text{k}\Omega$ and $C = 2.5\,\mu\text{F}$. Assuming that the initial capacitor charge is zero, determine the value of the output voltage 100 ms after application of the input.

From equation (9), output voltage,

$$V_o = -\frac{1}{CR}\int V_i\,\text{d}t$$

$$= -\frac{1}{(2.5\times10^{-6})(200\times10^{3})}\int(-0.75)\,\text{d}t$$

$$= -\frac{1}{0.5}\int(-0.75)\,\text{d}t = -2[-0.75t] = +1.5t$$

When time $t = 100$ ms, **output voltage,**

$$\boldsymbol{V_o} = (1.5)(100\times10^{-3}) = \mathbf{0.15\,V}$$

20.9 Op amp differential amplifier

The circuit for an op amp differential amplifier is shown in Figure 20.17 where voltages V_1 and V_2 are applied to its two input terminals and the difference between these voltages is amplified.

(i) Let V_1 volts be applied to terminal 1 and 0 V be applied to terminal 2. The difference in the potentials at the inverting (−) and non-inverting (+) op amp inputs is practically zero and hence the inverting terminal must be at zero potential. Then $I_1 = V_1/R_1$. Since the op amp input resistance is high, this current flows through the feedback resistor R_f. The volt drop across R_f, which is the output voltage $V_o = (V_1/R_1)R_f$; hence, the closed-loop voltage gain A is given by:

$$\boldsymbol{A = \frac{V_o}{V_1} = -\frac{R_f}{R_1}} \qquad (10)$$

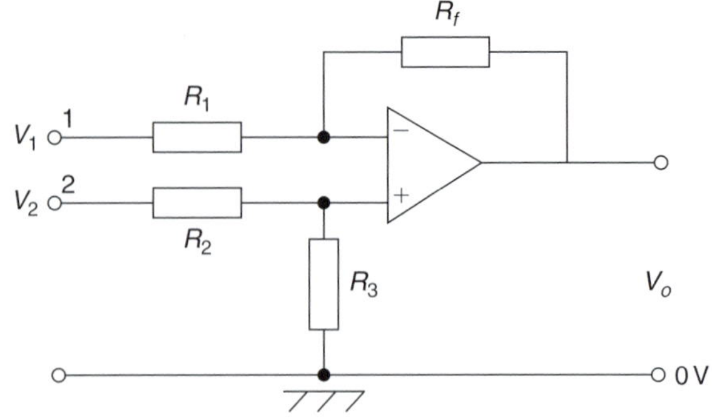

Figure 20.17

(ii) By similar reasoning, if V_2 is applied to terminal 2 and 0 V to terminal 1, then the voltage appearing at the non-inverting terminal will be $(R_3/(R_2+R_3))V_2$ volts. This voltage will also appear at the inverting (−) terminal and thus the voltage across R_1 is equal to $-(R_3/(R_2+R_3))V_2$ volts.
Now the output voltage,

$$V_o = \left(\frac{R_3}{R_2+R_3}\right)V_2 + \left[-\left(\frac{R_3}{R_2+R_3}\right)V_2\right]\left(-\frac{R_f}{R_1}\right)$$

and the voltage gain,

$$A = \frac{V_o}{V_2} = \left(\frac{R_3}{R_2+R_3}\right) + \left[-\left(\frac{R_3}{R_2+R_3}\right)\right]\left(-\frac{R_f}{R_1}\right)$$

i.e. $$\boldsymbol{A = \frac{V_o}{V_2} = \left(\frac{R_3}{R_2+R_3}\right)\left(1+\frac{R_f}{R_1}\right)} \qquad (11)$$

(iii) Finally, if the voltages applied to terminals 1 and 2 are V_1 and V_2, respectively, then the difference between the two voltages will be amplified.

If $V_1 > V_2$, then:

$$\boldsymbol{V_o = (V_1 - V_2)\left(-\frac{R_f}{R_1}\right)} \qquad (12)$$

If $V_2 > V_1$, then:

$$\boldsymbol{V_o = (V_2 - V_1)\left(\frac{R_3}{R_2+R_3}\right)\left(1+\frac{R_f}{R_1}\right)} \qquad (13)$$

Problem 11. In the differential amplifier shown in Figure 20.17, $R_1 = 10\,\text{k}\Omega$, $R_2 = 10\,\text{k}\Omega$, $R_3 = 100\,\text{k}\Omega$ and $R_f = 100\,\text{k}\Omega$. Determine the output voltage V_o if:

(a) $V_1 = 5$ mV and $V_2 = 0$

(b) $V_1 = 0$ and $V_2 = 5$ mV

(c) $V_1 = 50$ mV and $V_2 = 25$ mV

(d) $V_1 = 25$ mV and $V_2 = 50$ mV

(a) From equation (10),

$$V_o = -\frac{R_f}{R_1}V_1 = -\left(\frac{100 \times 10^3}{10 \times 10^3}\right)(5)\,\text{mV}$$

$$= \mathbf{-50\,mV}$$

(b) From equation (11),

$$V_o = \left(\frac{R_3}{R_2 + R_3}\right)\left(1 + \frac{R_f}{R_1}\right)V_2$$

$$= \left(\frac{100}{110}\right)\left(1 + \frac{100}{10}\right)(5)\text{ mV} = \mathbf{+50\,mV}$$

(c) $V_1 > V_2$ hence from equation (12),

$$V_o = (V_1 - V_2)\left(-\frac{R_f}{R_1}\right)$$

$$= (50 - 25)\left(-\frac{100}{10}\right)\text{mV} = \mathbf{-250\,mV}$$

(d) $V_2 > V_1$ hence from equation (13),

$$V_o = (V_2 - V_1)\left(\frac{R_3}{R_2 + R_3}\right)\left(1 + \frac{R_f}{R_1}\right)$$

$$= (50 - 25)\left(\frac{100}{100 + 10}\right)\left(1 + \frac{100}{10}\right)\text{mV}$$

$$= (25)\left(\frac{100}{110}\right)(11) = \mathbf{+250\,mV}$$

Now try the following Practice Exercise

Practice Exercise 92 Operational amplifiers (Answers on page 823)

1. If the input voltage for the op amp shown in Figure 20.18 is −0.5 V, determine (a) the voltage gain, (b) the output voltage.

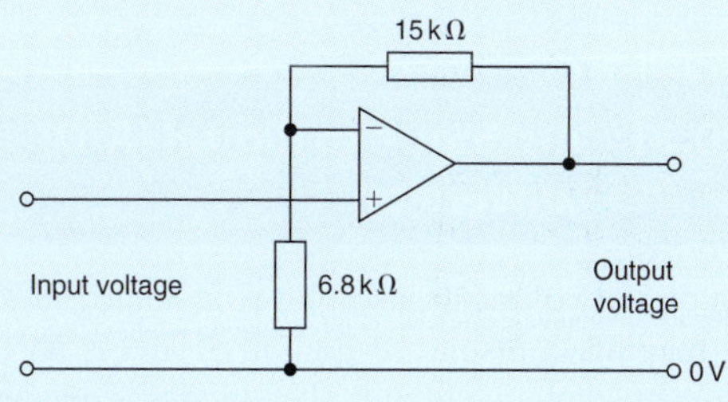

Figure 20.18

2. In the circuit of Figure 20.19, determine the value of the output voltage, V_o, when (a) $V_1 = +1$ V and $V_2 = +3$ V_o (b) $V_1 = +1$ V and $V_2 = -3$ V.

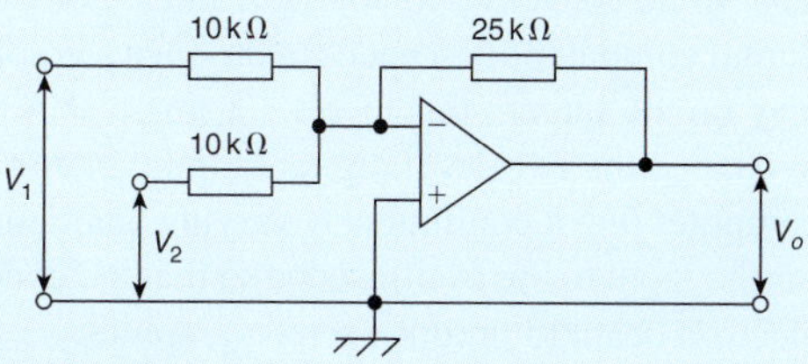

Figure 20.19

3. For the summing op amp shown in Figure 20.20, determine the output voltage, V_o

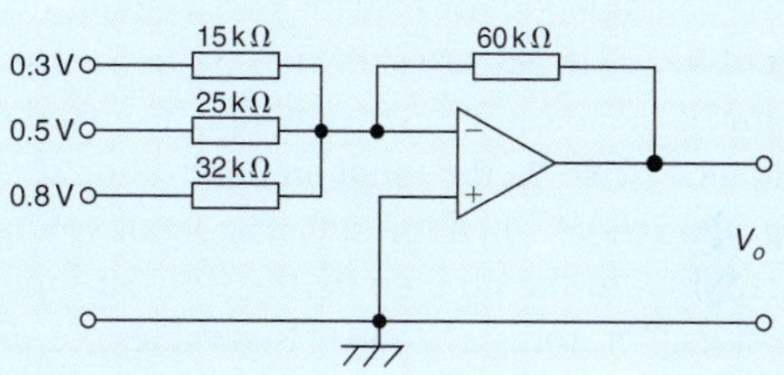

Figure 20.20

4. A steady voltage of −1.25 V is applied to an op amp integrator having component values of $R = 125$ kΩ and $C = 4.0$ μF. Calculate the value of the output voltage 120 ms after applying the input, assuming that the initial capacitor charge is zero.

5. In the differential amplifier shown in Figure 20.21, determine the output voltage, V_o, if: (a) $V_1 = 4$ mV and $V_2 = 0$, (b) $V_1 = 0$ and $V_2 = 6$ mV, (c) $V_1 = 40$ mV and $V_2 = 30$ mV, (d) $V_1 = 25$ mV and $V_2 = 40$ mV.

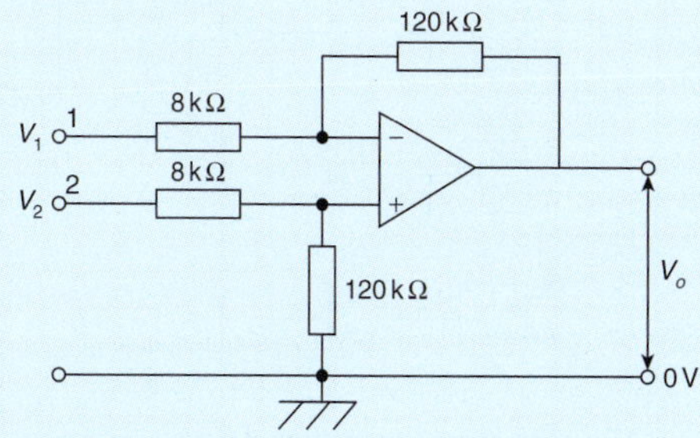

Figure 20.21

20.10 Digital to analogue (D/A) conversion

There are a number of situations when digital signals have to be converted to analogue ones. For example, a digital computer often needs to produce a graphical display on the screen; this involves using a D/A converter to change the two-level digital output voltage from the computer into a continuously varying analogue voltage for the input to the oscilloscope so that it can deflect the electron beam to produce screen graphics.

A binary weighted resistor D/A converter is shown in Figure 20.22 for a four-bit input. The values of the resistors, $R, 2R, 4R, 8R$ increase according to the binary scale – hence the name of the converter. The circuit uses an op amp as a **summing amplifier** (see Section 20.6) with a feedback resistor R_f. Digitally controlled electronic switches are shown as S_1 to S_4. Each switch connects the resistor in series with it to a fixed reference voltage V_{ref} when the input bit controlling it is a 1 and to ground (0 V) when it is a 0. The input voltages V_1 to V_4 applied to the op amp by the four-bit input via the resistors therefore have one of two values, i.e. either V_{ref} or 0 V.

From equation (8), page 316, the analogue output voltage V_o is given by:

$$V_o = -\left(\frac{R_f}{R}V_1 + \frac{R_f}{2R}V_2 + \frac{R_f}{4R}V_3 + \frac{R_f}{8R}V_4\right)$$

Let $R_f = R = 1\,\mathrm{k\Omega}$, then:

$$V_o = \left(V_1 + \frac{1}{2}V_2 + \frac{1}{4}V_3 + \frac{1}{8}V_4\right)$$

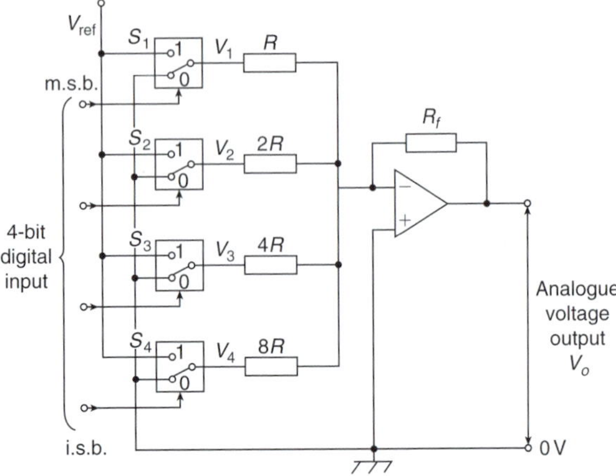

Figure 20.22

With a four-bit input of 0001 (i.e. decimal 1), S_4 connects $8R$ to V_{ref}, i.e. $V_4 = V_{ref}$, and S_1, S_2 and S_3 connect R, $2R$ and $4R$ to 0V, making $V_1 = V_2 = V_3 = 0$. Let $V_{ref} = -8$V, then output voltage,

$$V_o = -\left(0+0+0+\frac{1}{8}(-8)\right) = \mathbf{+1\,V}$$

With a four-bit input of 0101 (i.e. decimal 5), S_2 and S_4 connects $2R$ and $4R$ to V_{ref}, i.e. $V_2 = V_4 = V_{ref}$, and S_1 and S_3 connect R and $4R$ to 0 V, making $V_1 = V_3 = 0$. Again, if $V_{ref} = -8$ V, then output voltage,

$$V_o = -\left(0+\frac{1}{2}(-8)+0+\frac{1}{8}(-8)\right) = \mathbf{+5\,V}$$

If the input is 0111 (i.e. decimal 7), the output voltage will be 7 V, and so on. From these examples, it is seen that the analogue output voltage, V_o, is directly proportional to the digital input.

V_o has a 'stepped' waveform, the waveform shape depending on the binary input. A typical waveform is shown in Figure 20.23.

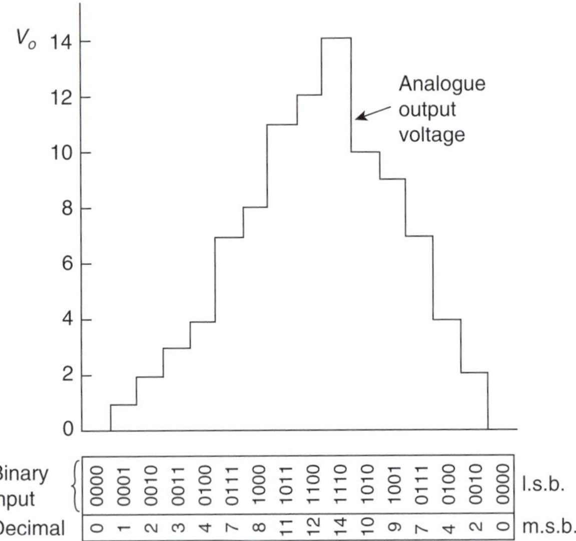

Figure 20.23

20.11 Analogue to digital (A/D) conversion

In a digital voltmeter, its input is in analogue form and the reading is displayed digitally. This is an example where an analogue to digital converter is needed.

A block diagram for a four-bit counter type A/D conversion circuit is shown in Figure 20.24. An op amp

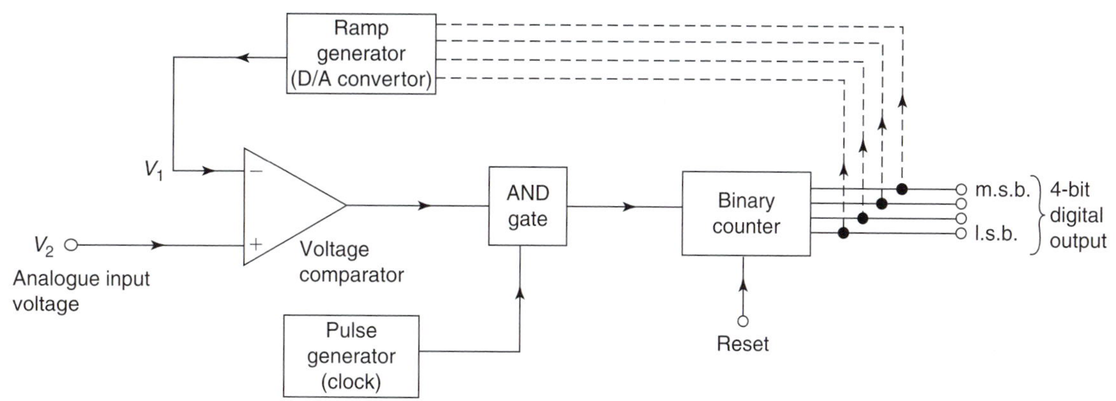

Figure 20.24

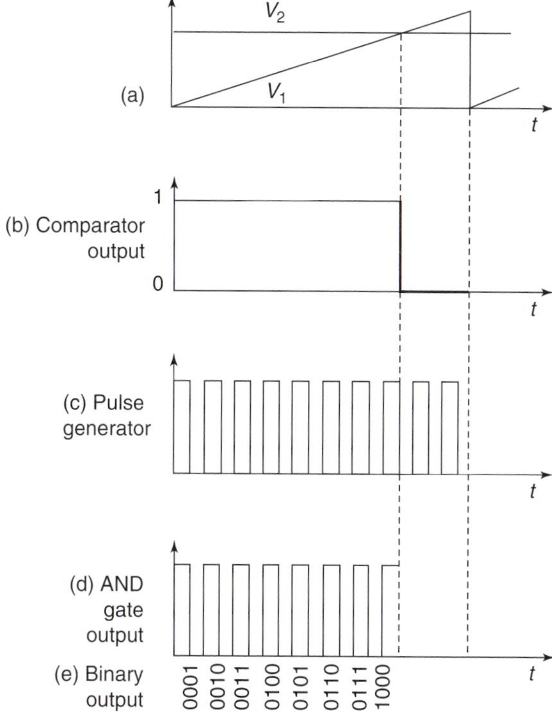

Figure 20.25

is again used, in this case as a **voltage comparator** (see Section 20.7). The analogue input voltage V_2, shown in Figure 20.25(a) as a steady d.c. voltage, is applied to the non-inverting (+) input, whilst a sawtooth voltage V_1 supplies the inverting (−) input.

The output from the comparator is applied to one input of an AND gate and is a 1 (i.e. 'high') until V_1 equals or exceeds V_2, when it then goes to 0 (i.e. 'low') as shown in Figure 20.25(b). The other input of the AND gate is fed by a steady train of pulses from a pulse generator, as shown in Figure 20.25(c). When both inputs to the AND gate are 'high', the gate 'opens' and gives a 'high' output, i.e. a pulse, as shown in Figure 20.25(d). The time taken by V_1 to reach V_2 is proportional to the analogue voltage if the ramp is linear. The output pulses from the AND gate are recorded by a binary counter and, as shown in Figure 20.25(e), are the digital equivalent of the analogue input voltage V_2. In practice, the ramp generator is a D/A converter which takes its digital input from the binary counter, shown by the broken lines in Figure 20.24. As the counter advances through its normal binary sequence, a staircase waveform with equal steps (i.e. a ramp) is built up at the output of the D/A converter (as shown by the first few steps in Figure 20.23).

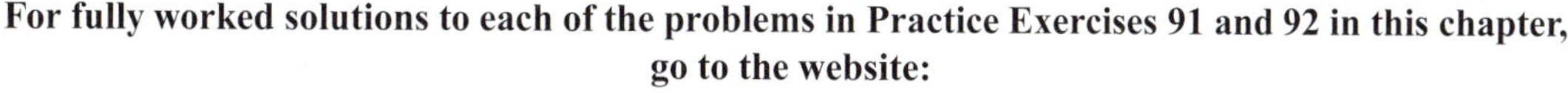
For fully worked solutions to each of the problems in Practice Exercises 91 and 92 in this chapter, go to the website:
www.routledge.com/cw/bird

Revision Test 5

This revision test covers the material contained in Chapters 17 to 20. *The marks for each question are shown in brackets at the end of each question.*

1. The power taken by a series inductive circuit when connected to a 100 V, 100 Hz supply is 250 W and the current is 5 A. Calculate (a) the resistance, (b) the impedance, (c) the reactance, (d) the power factor and (e) the phase angle between voltage and current. (9)

2. A coil of resistance 20 Ω and inductance 200 mH is connected in parallel with a 4 μF capacitor across a 50 V, variable frequency supply. Calculate (a) the resonant frequency, (b) the dynamic resistance, (c) the current at resonance and (d) the Q-factor at resonance. (10)

3. A series circuit comprises a coil of resistance 30 Ω and inductance 50 mH, and a 2500 pF capacitor. Determine the Q-factor of the circuit at resonance. (4)

4. The winding of an electromagnet has an inductance of 110 mH and a resistance of 5.5 Ω. When it is connected to a 110 V d.c. supply, calculate (a) the steady state value of current flowing in the winding, (b) the time constant of the circuit, (c) the value of the induced e.m.f. after 0.01 s, (d) the time for the current to rise to 75% of it's final value and (e) the value of the current after 0.02 s (11)

5. A single-phase motor takes 30 A at a power factor of 0.65 lagging from a 300 V, 50 Hz supply. Calculate (a) the current taken by a capacitor connected in parallel with the motor to correct the power factor to unity, and (b) the value of the supply current after power factor correction. (7)

6. For the summing operational amplifier shown in Figure RT5.1, determine the value of the output voltage, V_o (3)

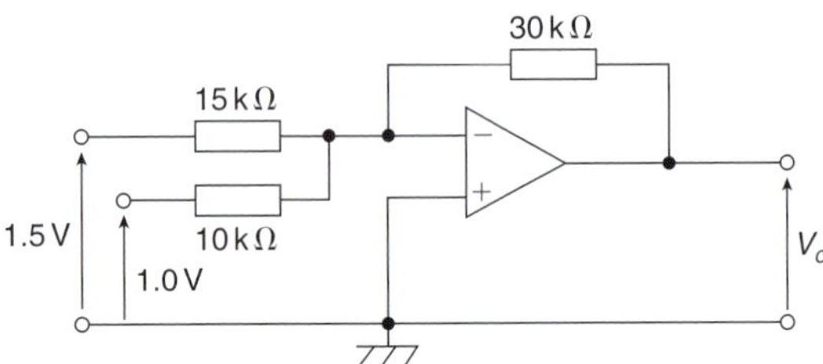

Figure RT5.1

7. In the differential amplifier shown in Figure RT5.2, determine the output voltage, V_o, when: (a) $V_1 = 4$ mV and $V_2 = 0$, (b) $V_1 = 0$ and $V_2 = 5$ mV, (c) $V_1 = 20$ mV and $V_2 = 10$ mV (6)

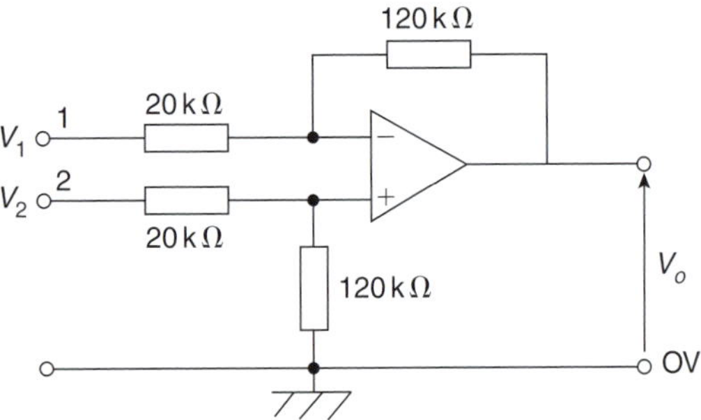

Figure RT5.2

For lecturers/instructors/teachers, fully worked solutions to each of the problems in Revision Test 5, together with a full marking scheme, are available at the website:
www.routledge.com/cw/bird

Chapter 21

Ways of generating electricity – the present and the future

Why it is important to understand: **Ways of generating electricity - the present and the future**

In 1831, Michael Faraday devised a machine that generated electricity from rotary motion – but it took almost 50 years for the technology to reach a commercially viable stage. In 1878, in the USA, Thomas Edison developed and sold a commercially viable replacement for gas lighting and heating using locally generated and distributed direct current electricity. The world's first public electricity supply was provided in late 1881, when the streets of the Surrey town of Godalming in the UK were lit with electric light. This system was powered from a water wheel on the River Wey, which drove a Siemens alternator that supplied a number of arc lamps within the town.

In the 1920s the use of electricity for home lighting increased and by the mid-1930s electrical appliances were standard in the homes of the better-off; after the Second World War they became common in all households. Now, some 80 years later, we have become almost totally dependent on electricity! Without electricity we have no lighting, no communication via mobile phones/smartphones, internet or computer, television, iPods or radios, no air-conditioning, no fans, no electric heating, no refrigerator or freezer, no coffee maker, no kitchen appliances, no dishwasher, no electric stoves, ovens or microwaves, no washing machines or tumble dryers, problems with drinking water if it comes from a system dependent on electrical pumps, ... Our whole lifestyles have become totally dependent on a reliable electricity supply.

In the future, civilization will be forced to research and develop alternative energy sources. Our current rate of fossil fuel usage will lead to an energy crisis this century. In order to survive the energy crisis many companies in the energy industry are inventing new ways to extract energy from renewable sources and while the rate of development is slow, mainstream awareness and government pressures are growing. Traditional methods of generating electricity are unsustainable, and new energy sources must be found that do not produce as much carbon. The recognized need for alternative power sources is not new. Massive solar arrays are seen unveiled in vast deserts, enormous on-and-offshore wind-farms, wave-beams converting the power of our oceans, and a host of biomass solutions arrive and disappear. One of the most important of current engineering problems centres upon the realization that fossil fuels are unsustainable, and therefore the challenge of how to generate enough electricity using clean, renewable sources.

Electrical Circuit Theory and Technology. 978-1-138-67349-6, © 2017 John Bird. Published by Taylor & Francis. All rights reserved.

At the end of this chapter, you should be able to:

- briefly describe the process of generating electricity using:

(a) coal	(b) oil
(c) natural gas	(d) nuclear energy
(e) hydro power	(f) pumped storage
(g) wind	(h) tidal power
(i) biomass	(j) solar energy

- appreciate the advantages and disadvantages of each of the methods of generating electricity
- understand a future possibility of harnessing the power of wind, tide and sun on an 'energy island'

21.1 Introduction

In this chapter, methods of generating electricity using **coal, oil, natural gas, nuclear energy, hydro power, pumped storage, wind, tidal power, biomass** and **solar energy** are explained.

Coal, oil and gas are called '**fossil fuels**' because they have been formed from the organic remains of prehistoric plants and animals. Historically, the transition from one energy system to another, as from wood to coal or coal to oil, has proven an enormously complicated process, requiring decades to complete. In similar fashion, it will be many years before renewable forms of energy – wind, solar, tidal, geothermal and others still in development – replace fossil fuels as the world's leading energy providers. This chapter explains some ways of generating electricity for the present and for the future.

21.2 Generating electrical power using coal

A coal power station turns the chemical energy in coal into electrical energy that can be used in homes and businesses.

In Figure 21.1, the coal (1) is ground to a fine powder and blown into the boiler (2), where it is burned, converting its chemical energy into heat energy. Grinding the coal into powder increases its surface area, which helps it to burn faster and hotter, producing as much heat and as little waste as possible.

As well as heat, burning coal produces ash and exhaust gases. The ash falls to the bottom of the boiler and is removed by the ash systems (3). It is usually then sold to the building industry and used as an ingredient in various building materials, like concrete.

The gases enter the exhaust stack (4), which contains equipment that filters out any dust and ash, before venting into the atmosphere. The exhaust stacks of coal power stations are built tall so that the exhaust plume (5) can disperse before it touches the ground. This ensures that it does not affect the quality of the air around the station.

Burning the coal heats water in pipes coiled around the boiler, turning it into steam. The hot steam expands in the pipes, so when it emerges it is under high pressure. The pressure drives the steam over the blades of the steam turbine (6), causing it to spin, converting the heat energy released in the boiler into mechanical energy.

A shaft connects the steam turbine to the turbine generator (7), so when the turbine spins, so does the generator. The generator uses an electromagnetic field to convert this mechanical energy into electrical energy (as described in Chapter 11).

After passing through the turbine, the steam comes into contact with pipes full of cold water (8); in coastal stations this water is pumped straight from the sea. The cold pipes cool the steam so that it condenses back into water. It is then piped back to the boiler, where it can be heated up again, and turned into steam to keep the turbine turning.

Finally, a transformer converts the electrical energy from the generator to a high voltage. The national grid uses high voltages to transmit electricity efficiently through the power lines (9) to the homes and businesses that need it (10). Here, other transformers reduce the voltage back down to a usable level.

Coal has been a reliable source of energy for many decades; however coal is considered to produce the highest amount of carbon emissions of any form of electricity generation. A new technology called Carbon Capture and Storage (CCS) is being developed to remove up to 90% carbon dioxide from power station emissions and store it underground. CCS will be an important way to cut emissions and meet international low carbon targets.

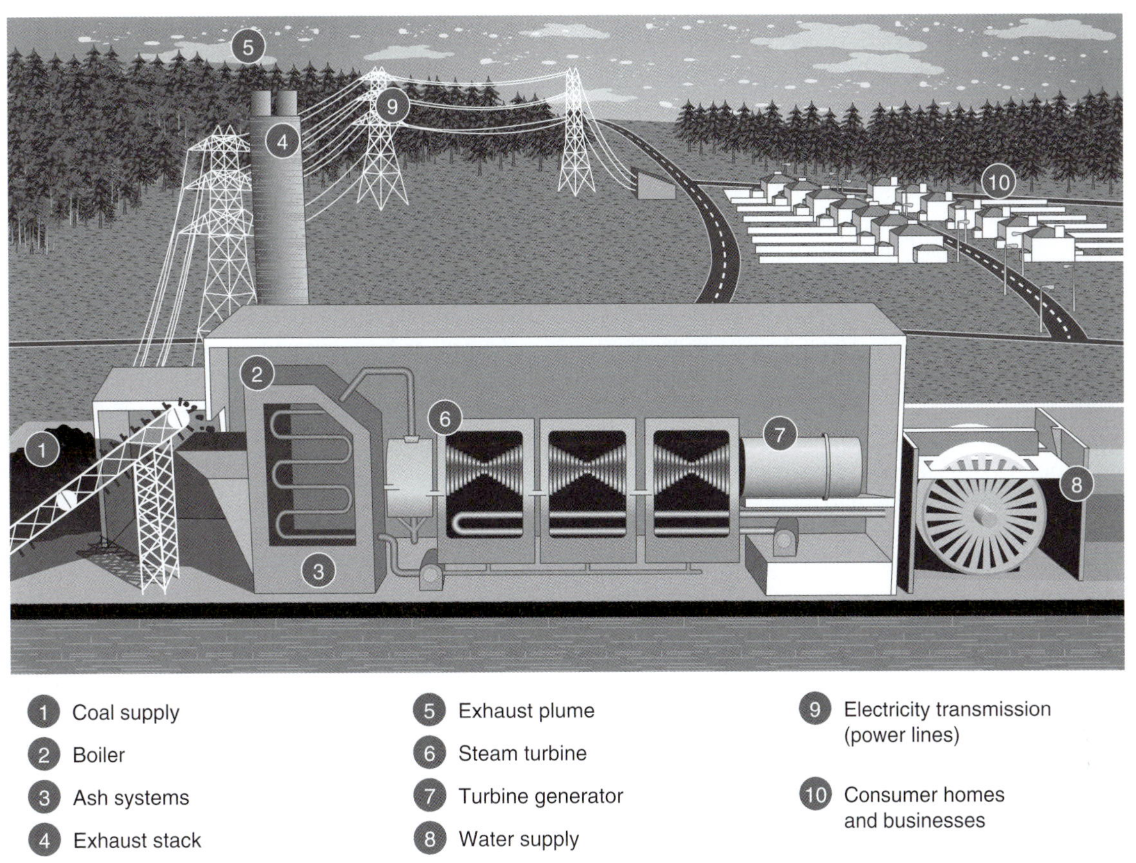

Figure 21.1

Advantages of coal

Coal is easily combustible, and burns at low temperatures, making coal-fired boilers cheaper and simpler than many others. It is widely and easily distributed all over the world, is comparatively inexpensive to buy on the open market due to large reserves and easy accessibility, has good availability for much of the world (i.e. coal is found in many more places than other fossil fuels) and is mainly simple to mine, making it by far the least expensive fossil fuel to actually obtain. It is economically possible to build a wide variety of sizes of generation plants, and a fossil-fuelled power station can be built almost anywhere, so long as you can get large quantities of fuel to it – most coal fired power stations have dedicated rail links to supply the coal.

Disadvantages of coal

Coal is non-renewable and fast depleting, has the lowest energy density of any fossil fuel, i.e. it produces the least energy per ton of fuel, and has the lowest energy density per unit volume, meaning that the amount of energy generated per cubic metre is lower than any other fossil fuel. Coal has high transportation costs due to the bulk of coal, coal dust is an extreme explosion hazard, so transportation and storage must take special precautions to mitigate this danger. Storage costs are high, especially if required to have enough stock for a few years to assure power production availability. Burning fossil fuels releases carbon dioxide, a powerful greenhouse gas, that had been stored in the earth for millions of years – contributing to global warming – and coal leaves behind harmful by-products upon combustion (both airborne and in solid-waste form), thereby causing a lot of pollution (air pollution due to burning coal is much worse than any other form of power generation, and very expensive 'scrubbers' must be installed to remove a significant amount of it; even then, a non-trivial amount escapes into the air). Mining of coal leads to irreversible damage to the adjoining environment, coal will eventually run out, cannot be recycled, and prices for all fossil fuels are rising, especially if the real cost of their carbon is included.

21.3 Generating electrical power using oil

An oil power station turns the chemical energy in oil into electrical energy that can be used in homes and businesses.

In Figure 21.2, the oil (1) is piped into the boiler (2), where it is burned, converting its chemical energy into heat energy. This heats water in pipes coiled around the boiler, turning it into steam. The hot steam expands in the narrow pipes, so when it emerges it is under high pressure.

The pressure drives the steam over the blades of the steam turbine (3), causing it to spin, converting the heat energy released in the boiler into mechanical energy. A shaft connects the steam turbine to the turbine generator (4), so when the turbine spins, so does the generator. The generator uses an electromagnetic field to convert this mechanical energy into electrical energy.

After passing through the turbine, the steam comes into contact withpipes full of cold water (5). The cold pipes cool the steam so that it condenses back into water. It is then piped back to the boiler, where it can be heated up again, and turned into steam again to keep the turbine turning.

Finally, a transformer converts the electrical energy from the generator to a high voltage. The national grid uses high voltages to transmit electricity efficiently through the power lines (6) to the homes and businesses that need it (7). Here, other transformers reduce the voltage back down to a usable level.

As well as heat, burning oil produces exhaust gases. These are piped from the boiler to the exhaust stack (8), which contains equipment that filters out any particles, before venting into the atmosphere. The stack is built tall so that the exhaust gas plume (9) can disperse before it touches the ground. This ensures that it does not affect the quality of the air around the station.

Advantages of oil

Oil has a high energy density (i.e. a small amount of oil can produce a large amount of energy), has easy

Figure 21.2

availability and infrastructure for transport, is easy to use, is crucial for a wide variety of industries, is relatively easy to produce and refine, is a constant power source and is highly reliable.

Disadvantages of oil

Oil produces greenhouse gas emissions (GHG), can cause pollution of water and earth, emits harmful substances like sulphur dioxide, carbon monoxide, and acid rain, and can lead to production of very harmful and toxic materials during refining (plastic being one of the most harmful of substances).

21.4 Generating electrical power using natural gas

A gas power station turns the chemical energy in natural gas into electrical energy that can be used in homes and businesses.

In Figure 21.3, natural gas (1) is pumped into the gas turbine (2), where it is mixed with air (3) and burned, converting its chemical energy into heat energy. As well as heat, burning natural gas produces a mixture of gases called the combustion gas. The heat makes the combustion gas expand. In the enclosed gas turbine, this causes a build-up of pressure.

The pressure drives the combustion gas over the blades of the gas turbine, causing it to spin, converting some of the heat energy into mechanical energy. A shaft connects the gas turbine to the gas turbine generator (4), so when the turbine spins, the generator does too. The generator uses an electromagnetic field to convert this mechanical energy into electrical energy.

After passing through the gas turbine, the still-hot combustion gas is piped to the heat recovery steam generator (5). Here it is used to heat pipes full of water, turning the water to steam, before escaping through the exhaust stack (6). Natural gas burns very cleanly, but the stack is still built tall so that the exhaust gas plume (7)

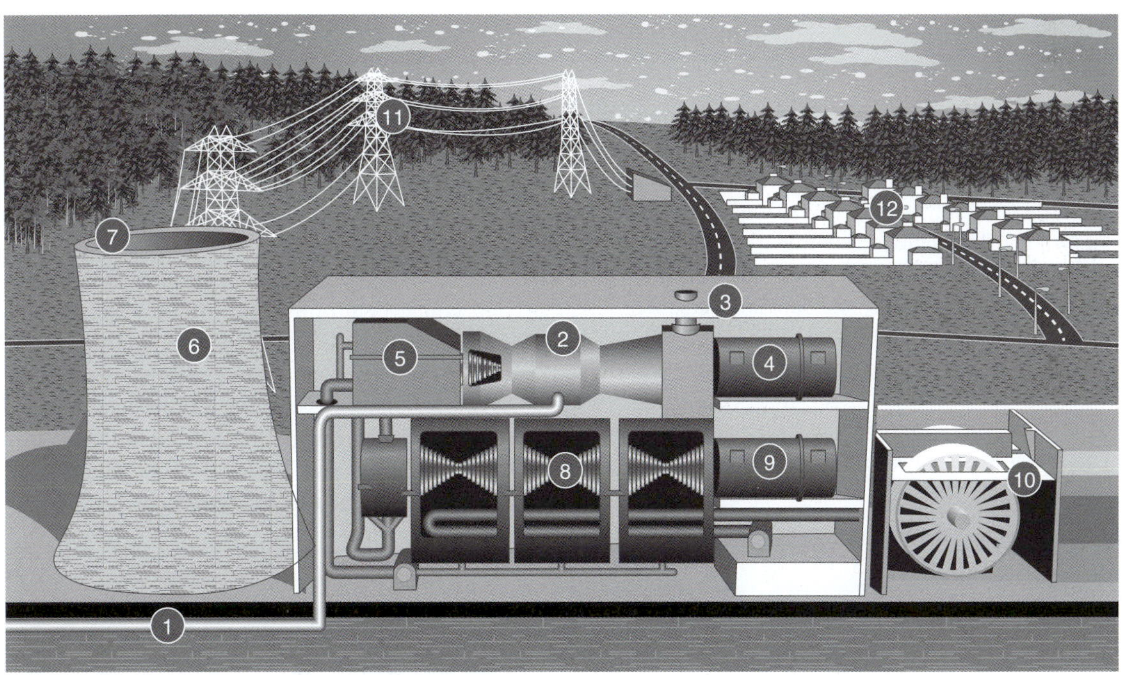

1 Gas line in
2 Gas turbine
3 Air in
4 Gas turbing generator
5 Heat recovery steam generator
6 Stack for exhaust gases
7 Exhaust gas plume
8 Steam turbine
9 Steam turbine generator
10 Cooling water supply
11 Electricity transmission (power lines)
12 Consumer homes and businesses

Figure 21.3

Part 3

can disperse before it touches the ground. This ensures that it does not affect the quality of the air around the station.

The hot steam expands in the pipes, so when it emerges it is under high pressure. These high-pressure steam jets spin the steam turbine (8), just like the combustion gas spins the gas turbine. The steam turbine is connected by a shaft to the steam turbine generator (9), which converts the turbine's mechanical energy into electrical energy.

After passing through the turbine, the steam comes into contact with pipes full of cold water (10). In coastal stations this water is pumped straight from the sea. The cold pipes cool the steam so that it condenses back into water. It is then piped back to the heat recovery steam generator to be re-used.

Finally, a transformer converts the electrical energy from the generator to a high voltage. The national grid uses high voltages to transmit electricity efficiently through the power lines (11) to the homes and businesses that need it (12). Here, other transformers reduce the voltage back down to a usable level.

Advantages of natural gas

Gas is used to produce electricity, is less harmful than coal or oil, is easy to store and transport, has much residential use, is used for vehicle fuel, burns cleaner without leaving any smell, ash or smoke, is an instant energy, ideal in kitchens, has much industrial use for producing hydrogen, ammonia for fertilizers and some paints and plastics, and is abundant and versatile.

Disadvantages of natural gas

Gas is toxic and flammable, can damage the environment, is non-renewable, and is expensive to install.

21.5 Generating electrical power using nuclear energy

A nuclear power station turns the nuclear energy in uranium atoms into electrical energy that can be used in homes and businesses.

In Figure 21.4, the reactor vessel (1) is a tough steel capsule that houses the fuel rods – sealed metal cylinders containing pellets of uranium oxide. When a neutron – a neutrally charged subatomic particle – hits a uranium atom, the atom sometimes splits, releasing two or three more neutrons. This process converts the nuclear energy that binds the atom together into heat energy.

The fuel assemblies are arranged in such a way that when atoms in the fuel split, the neutrons they release are likely to hit other atoms and make them split as well. This chain reaction produces large quantities of heat.

Water flows through the reactor vessel, where the chain reaction heats it to around 300 °C. The water needs to stay in liquid form for the power station to work, so the pressuriser (2) subjects it to around 155 times atmospheric pressure, which stops it boiling.

The reactor coolant pump (3) circulates the hot pressurised water from the reactor vessel to the steam generator (4). Here, the water flows through thousands of looped pipes before circulating back to the reactor vessel. A second stream of water flows through the steam generator, around the outside of the pipes. This water is under much less pressure, so the heat from the pipes boils it into steam.

The steam then passes through a series of turbines (5), causing them to spin, converting the heat energy produced in the reactor into mechanical energy. A shaft connects the turbines to a generator, so when the turbines spin, so does the generator. The generator uses an electromagnetic field to convert this mechanical energy into electrical energy.

A transformer converts the electrical energy from the generator to a high voltage. The national grid uses high voltages to transmit electricity efficiently through the power lines (6) to the homes and businesses that need it (7). Here, other transformers reduce the voltage back down to a usable level.

After passing through the turbines, the steam comes into contact with pipes full of cold water pumped in from the sea (8). The cold pipes cool the steam so that it condenses back into water. It is then piped back to the steam generator, where it can be heated up again, and turned into steam again to keep the turbines turning.

Advantages of nuclear energy

Nuclear energy is reliable, has low fuel costs, low electricity costs, no greenhouse gas emissions/air pollution, has a high load factor and huge potential.

Disadvantages of nuclear energy

Disadvantages of nuclear energy includes the fear of nuclear and radiation accidents, the problems of nuclear waste disposal, the low level of radioactivity from normal operations, the fear of nuclear proliferation, high capital investment, cost overruns and long gestation time, the many regulations for nuclear energy power plants, and fuel danger – uranium is limited to only a few countries and suppliers.

1 Reactor vessel
2 Pressuriser
3 Reactor coolant pump
4 Steam generator
5 Turbine generator and turbines
6 Electricity transmission (power lines)
7 Consumer homes and businesses
8 Cooling via sea water

Figure 21.4

21.6 Generating electrical power using hydro power

A hydroelectric power station converts the kinetic – or moving – energy in flowing or falling water into electrical energy that can be used in homes and businesses. Hydroelectric power can be generated on a small scale with a 'run-of-river' installation, which uses naturally flowing river water to turn one or more turbines, or on a large scale with a hydroelectric dam.

A hydroelectric dam straddles a river, blocking the water's progress downstream. With reference to Figure 21.5, water collects on the upstream side of the dam, forming an artificial lake known as a reservoir (1). Damming the river converts the water's kinetic energy into potential energy: the reservoir becomes a sort of battery, storing energy that can be released a little at a time. As well as being a source of energy, some reservoirs are used as boating lakes or drinking water supplies.

The reservoir's potential energy is converted back into kinetic energy by opening underwater gates, or intakes (2), in the dam. When an intake opens, the immense weight of the reservoir forces water through a channel called the penstock (3) towards a turbine. The water rushes past the turbine, hitting its blades and causing it to spin, converting some of the water's kinetic energy into mechanical energy. The water then finally flows out of the dam and continues its journey downstream.

A shaft connects the turbine to a generator (4), so when the turbine spins, so does the generator. The generator uses an electromagnetic field to convert this mechanical energy into electrical energy.

As long as there is plenty of water in the reservoir, a hydroelectric dam can respond quickly to changes in demand for electricity. Opening and closing the intakes directly controls the amount of water flowing through the penstock, which determines the amount of electricity the dam is generating.

The turbine and generator are located in the dam's power house (5), which also houses a transformer. The transformer converts the electrical energy from the generator to a high voltage. The national grid uses high voltages to transmit electricity efficiently through the power lines (6) to the homes and businesses that need it (7). Here, other transformers reduce the voltage back down to a usable level.

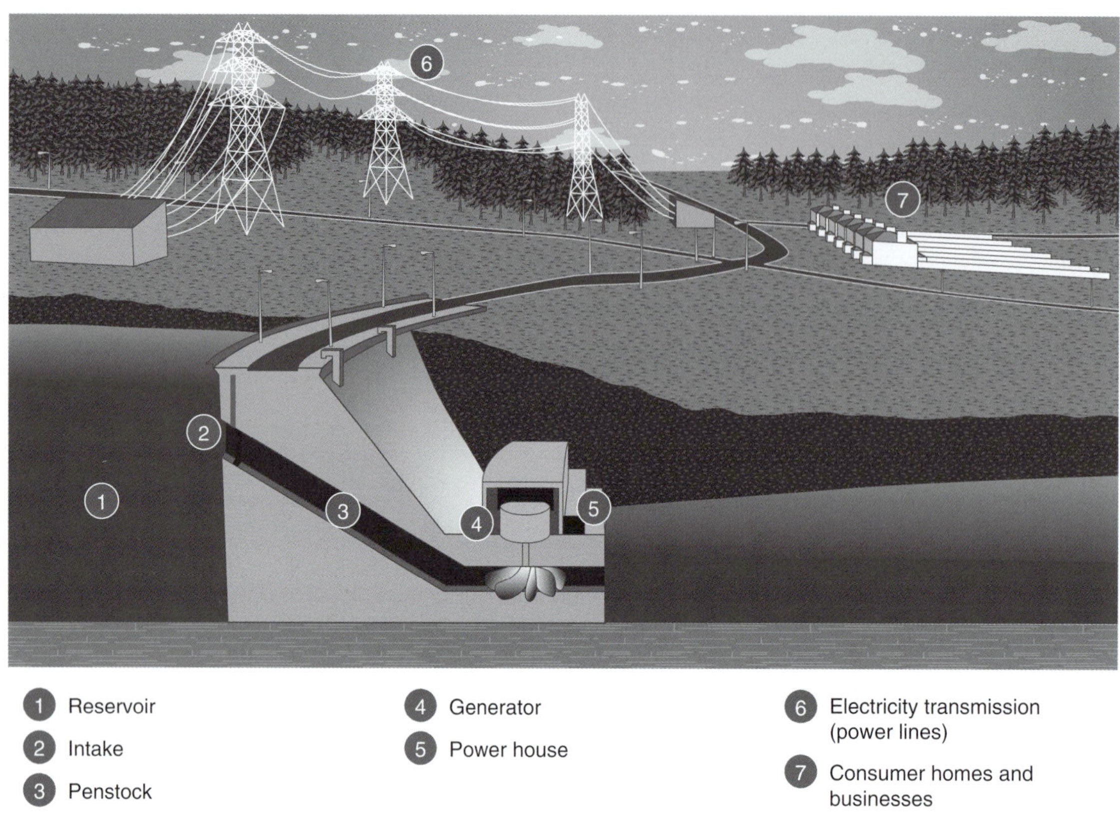

Figure 21.5

Advantages of hydro power

With hydro power there are no fuel costs, low operating costs and little maintenance, low electricity costs, no greenhouse gas emissions/air pollution, energy storage possibilities, small size hydro plants possible, reliability, a high load factor and long life.

Disadvantages of hydro power

Disadvantages of hydro power includes environmental, dislocation and tribal rights difficulties, wildlife and fish being affected, the possibility of earthquake vulnerability, siltation, dam failure due to poor construction or terrorism, the fact that plants cannot be built anywhere, and long gestation times.

21.7 Generating electrical power using pumped storage

Pumped storage reservoirs provide a place to store energy until it's needed. There are fluctuations in demand for electricity throughout the day. For example when a popular TV programme finishes, many people put the kettle on, causing a peak in demand for electrical power.

When electricity is suddenly demanded, a way is needed of producing power which can go from producing no power to full power immediately, and keep generating power for half an hour or so until other power stations can catch up with the demand for energy. This is why pumped storage reservoirs are so useful.

A pumped storage plant has two separate reservoirs, an upper and a lower one. When electricity is in low demand, for example at night, water is pumped into the upper reservoir.

When there is a sudden demand for power, giant taps known as the head gates are opened. This allows water from the upper reservoir to flow through pipes, powering a turbine, into the lower reservoir.

The movement of the turbine turns a generator which creates electricity. The electricity is created in the generator by using powerful magnets and coils of wire. When the coils are spun quickly inside the magnets, they produce electricity.

Water exiting from the pipe flows into the lower reservoir rather than re-entering a river and flowing downstream. At night, the water in the lower reservoir can be pumped back up into the upper reservoir to be used again.

In terms of how pumped storage and dammed water generate electricity, the methods are the same. The difference is that in pumped storage, the water is continually reused, whereas in hydroelectric dams, the water which generates electricity continues flowing downriver after use.

Pumped storage in the UK: Most pumped storage plants are located in Scotland, except the largest of all, Dinorwig, which is in North Wales. Dinorwig, built in 1984, produces 1728 MW – which is enough electricity to power nearly 7 million desktop computers. Dinorwig has the fastest 'response time' of any pumped storage plant in the world – it can provide 1320 MW in 12 seconds.

Advantages of pumped storage

Pumped storage provides a way to generate electricity instantly and quickly, no pollution or waste is created, and there is little effect on the landscape, as typically pumped storage plants are made from existing lakes in mountains.

Disadvantages of pumped storage

Pumped storage facilities are expensive to build, and once the pumped storage plant is used, it cannot be used again until the water is pumped back to the upper reservoir.

21.8 Generating electrical power using wind

Wind turbines use the wind's kinetic energy to generate electrical energy that can be used in homes and businesses. Individual wind turbines can be used to generate electricity on a small scale – to power a single home, for example. A large number of wind turbines grouped together, sometimes known as a wind farm or wind park, can generate electricity on a much larger scale.

A wind turbine works like a high-tech version of an old-fashioned windmill. The wind blows on the angled blades of the rotor, causing it to spin, converting some of the wind's kinetic energy into mechanical energy. Sensors in the turbine detect how strongly the wind is blowing and from which direction. The rotor automatically turns to face the wind, and automatically brakes in dangerously high winds to protect the turbine from damage.

With reference to Figure 21.6, a shaft and gearbox connect the rotor to a generator (1), so when the rotor spins, so does the generator. The generator uses an electromagnetic field to convert this mechanical energy into electrical energy.

The electrical energy from the generator is transmitted along cables to a substation (2). Here, the electrical energy generated by all the turbines in the wind farm is combined and converted to a high voltage. The national grid uses high voltages to transmit electricity efficiently through the power lines (3) to the homes and businesses that need it (4). Here, other transformers reduce the voltage back down to a usable level.

Advantages of wind energy

Wind energy has no pollution and global warming effects, low costs, a large industrial base, no fuel costs, and offshore advantages.

Disadvantages of wind energy

Wind energy has low persistent noise, can cause a loss of scenery, requires land usage, and is intermittent in nature.

21.9 Generating electrical power using tidal power

The tide moves a huge amount of water twice each day, and harnessing it could provide a great deal of energy – for example, around 20% of Britain's needs. Although the energy supply is reliable and plentiful, converting it into useful electrical power is not easy.

There are eight main sites around Britain where tidal power stations could usefully be built, including the Severn, Dee, Solway and Humber estuaries. Only around 20 sites in the world have been identified as possible tidal power stations.

Tidal energy is produced through the use of tidal energy generators, as shown in Figure 21.7. These large underwater turbines are placed in areas with high tidal movements, and are designed to capture the kinetic motion of the ebbing and surging of ocean tides in order to produce electricity. Tidal power has great potential for future power and electricity generation because of the massive size of the oceans.

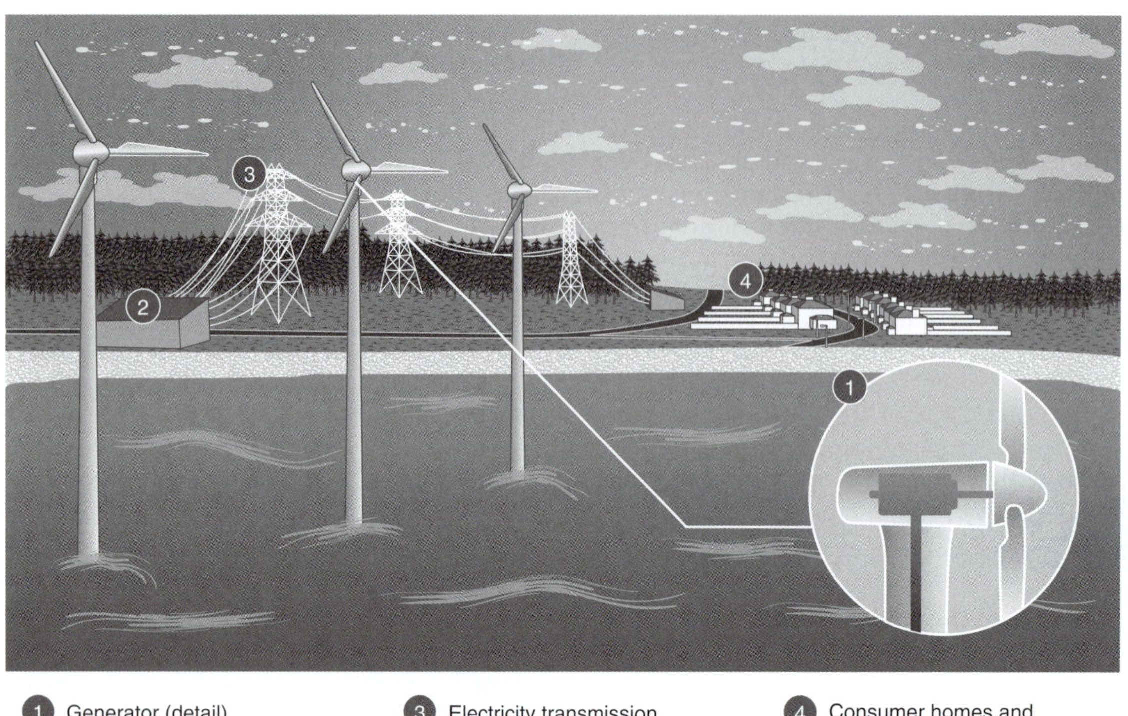

Figure 21.6

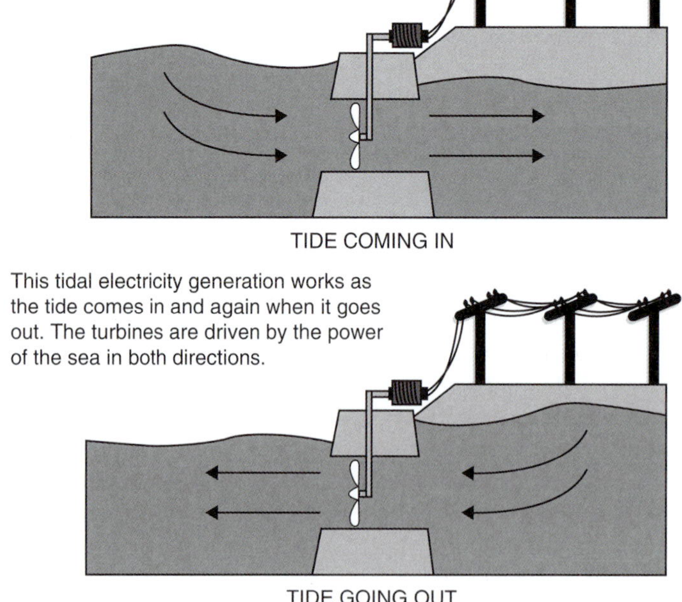

Figure 21.7

Advantages of tidal power

Tidal power is renewable, non-polluting and carbon negative, predictable, needs no fuel, has low costs, long life, high energy density and high load factor.

Disadvantages of tidal power

Tidal power has high initial capital investment, limited locations, detrimental effects on marine life, immature technology, long gestation time, some difficulties in transmission of tidal electricity, and weather effects that can damage tidal power equipment.

21.10 Generating electrical power using biomass

Biomass is fuel that is developed from organic materials, a renewable and sustainable source of energy used to create electricity or other forms of power, as stated in para 6, page 85.

There are several methods to convert biomass into electricity:

One way is to simply burn biomass directly, heat water to steam, and sending it through a steam turbine, which then generates electricity. The second way requires gasification of biomass. A biomass gasifier takes dry biomass, such as agriculture waste, and with the absence of oxygen and high temperatures produces synthesis gas (CO + H2), also known as pyrolysis of biomass. The gasification process turns wet biomass, such as food waste and manure, into methane (CH4) in a digestion tank. Both methane and synthesis gas (syngas) can be used in a gas engine or a gas turbine for electricity production. A third way to produce electricity from gasified biomass is by using fuel cells. If biogas/bio-syngas is available with high enough purity fuel cells can be used to produce bio-electricity. However, the fuel cells break down quickly if the gas in any way contains impurities. This technology is not yet commercial.

Biofuels, like ethanol, biodiesel and bio-oil can be also be used for power production in most types of power generators built for gasoline or diesel.

The generation of electricity from biomass results not only in electricity, but also a lot of heat. A traditional gas engine will have an efficiency of 30–35%. Gas turbines and steam turbines will end up at around 50%.

It is common to also use the heat from these processes, thus increasing the overall energy efficiency. There are many plants that combine heat and power production from biomass – the biogas plant cost increases, but the long term savings that come with better energy efficiency is in most cases worth it.

Advantages of biomass energy

Biomass energy is carbon neutral, uses waste efficiently, is a continuous source of power, can use a large variety of feedstock, has a low capital investment, can be built in remote areas and on a small scale, reduces methane, is easily available and is a low cost resource.

Disadvantages of biomass energy

With biomass energy, pollution can occur where poor technology is used, continuous feedstock is needed for efficiency, good management of biomass plants are required, has limited potential compared to other forms of energy like solar, hydro etc, and biomass plants are unpopular if constructed near homes.

21.11 Generating electrical power using solar energy

Solar panels turn energy from the sun's rays directly into useful energy that can be used in homes and businesses. There are two main types: solar thermal and photovoltaic, or PV. Solar thermal panels use the sun's energy to heat water that can be used in washing and heating. PV panels use the photovoltaic effect to turn the sun's energy directly into electricity, which can supplement or replace a building's usual supply.

A PV panel is made up of a semiconducting material, usually silicon-based, sandwiched between two electrical contacts. Referring to Figure 21.8, to generate as much electricity as possible, PV panels need to spend as much time as possible in direct sunlight (1a). A sloping, south-facing roof is the ideal place to mount a solar panel.

A sheet of glass (1b) protects the semiconductor sandwich from hail, grit blown by the wind, and wildlife. The semiconductor is also coated in an antireflective substance (1c), which makes sure that it absorbs the sunlight it needs instead of scattering it uselessly away.

When sunlight strikes the panel and is absorbed, it knocks loose electrons from some of the atoms that make up the semiconductor (1d). The semiconductor is positively charged on one side and negatively charged on the other side, which encourages all these loose electrons to travel in the same direction, creating an electric current. The contacts (1e and 1f) capture this current (1g) in an electrical circuit.

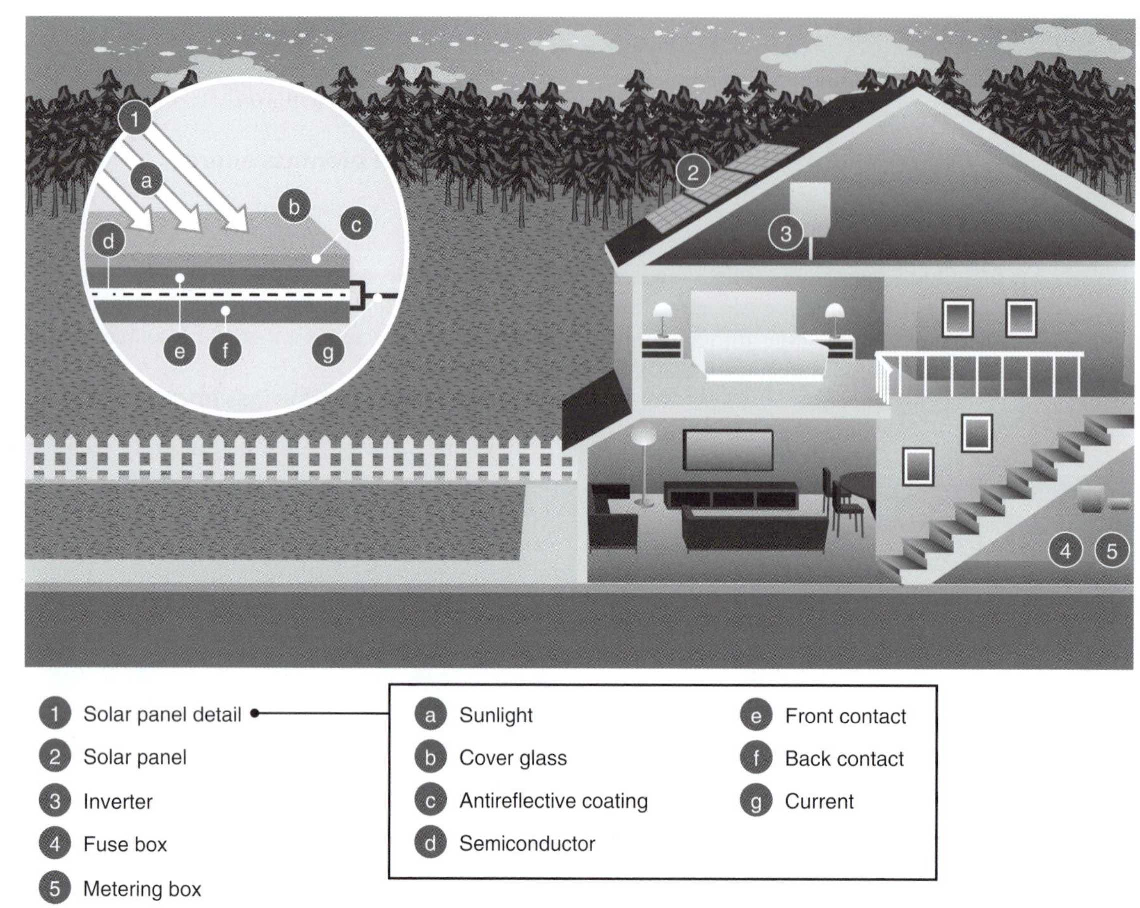

Figure 21.8

The electricity PV panels (2) generate direct current (d.c.). Before it can be used in homes and businesses, it has to be changed into alternating current (a.c.) electricity using an inverter (3). The inverted current then travels from the inverter to the building's fuse box (4) and from there to the appliances that need it.

PV systems installed in homes and businesses can include a dedicated metering box (5) that measures how much electricity the panels are generating. As an incentive to generate renewable energy, energy suppliers pay the system's owner a fixed rate for every unit of electricity it generates – plus a bonus for units the owner doesn't use, because these can help supply the national grid. Installing a PV system is not cheap, but this deal can help the owner to earn back the cost more quickly – and potentially even make a profit one day.

Advantages of solar power

Solar power is environmentally friendly, has declining costs, no fuel, low maintenance, no pollution, has almost unlimited potential, has the advantage that installations can be any size, and installation can be quick.

Disadvantages of solar power

Solar power has higher initial costs than fossil energy forms, is intermittent in nature, and has high capital investment.

21.12 Harnessing the power of wind, tide and sun on an 'energy island' – a future possibility?

Man-made 'energy islands' anchored, say, in the North Sea and the English Channel could help the world meet increasing power demands – and tackle the problem of expanding population by providing homes for people. The artificial structures – the brainchild of mechanical engineer Professor Carl Ross, of the University of Portsmouth – would produce energy by harnessing the power of the wind, tide and sun, and could be towed far out to sea to avoid complaints of noise and unsightliness.

A particular problem, especially in smaller overcrowded countries, such as those found in Europe and Asia, is the NIMBY syndrome – that is, the 'Not In My Back Yard' reaction of people. Complaints that using these renewable methods of producing energy takes up valuable land space, is unsightly, and causes noise are common. Putting these three renewable energy-producing forms on a floating island, all these negative points can be avoided. Two-thirds of the Earth's surface is covered by water, so dry land is not being 'lost' – in fact, the oceans are being colonised. Professor Ross feels that once this technology is matured, humans can even start living on the floating island to help counter over-population.

In 2011 the United Nations announced that the global population had reached seven billion, and the momentum of a growing population does not show any signs of slowing down.

The floating islands would be attached to the sea bed by tubular pillars with vacuum chambers in their bases, similar to offshore drilling rigs. The islands would support wind turbines and solar panels on their upper surface, while underneath, tidal turbines would harness the power of the oceans.

Professor Ross proposes that the floating energy islands be deployed in locations including the deep sea region of the North Sea, the west coast of Scotland and the opening of the English Channel. Each one would supply power for some 119,500 homes.

The islands come with a hefty price tag – an estimated £1.7 billion – but Professor Ross believes the initial outlay would be recovered through household energy bills after 11 years. However, this estimate is based mostly on wind farms situated on land, and not at sea. For wind farms at sea, the wind is about twice that on dry land. Moreover the kinetic energy produced on the wind farm is proportional to the mass of the wind, times its velocity squared; but the mass of the wind hitting the wind turbine is proportional to its velocity; thus the kinetic energy produced on the wind turbine at sea is proportional to the velocity cubed of the wind hitting it, and NOT on its velocity squared! This means that a wind turbine at sea is likely to produce about 8 times the kinetic energy than that produced on dry land, so that instead of the wind farm paying for itself in about 11 years it will pay for itself in less than 2 years! This revelation is likely to favourably change the economic argument in favour of wind turbines at sea!

[For more, see *The Journal of Ocean Technology* Vol 10, No 4, 2015, pages 45 to 52, 'Floating Energy Islands – producing multiple forms of renewable energy' by Carl T F Ross and Tien Y Sien]

Chapter 22

Three-phase systems

Why it is important to understand: **Three-phase systems**

A three-phase circuit is an electrical distribution method that uses three alternating currents to supply power. This type of power distribution is the most widely used in the world for transferring power from generating systems to electrical supply grids. A three-phase circuit is also commonly used on large motors, pumps and other pieces of mechanical equipment. Most households receive electricity in the form of single-phase circuits, though some may have special three-phase circuit breakers installed for appliances such as washing machines or stoves. A three-phase system is usually more economical than an equivalent single-phase or two-phase system at the same voltage because it uses less conductor material to transmit electrical power. The three currents, together, deliver a balanced load, something not possible with single-phase alternating current. With alternating current, the current direction alternates, flowing back and forth in the circuit; this means that the voltage alternates as well, constantly changing from maximum to minimum. Three-phase power combines the three wires to offset the maximum and minimum oscillations, so that a device receiving this type of power does not experience such a wide variation in voltage. This makes three-phase power a very efficient form of electrical power distribution. Consequently, a three-phase electric motor uses less electricity and normally lasts longer than a single-phase motor of the same voltage and rating. This chapter describes a three-phase system with star and delta connections and explains how power is calculated.

At the end of this chapter you should be able to:

- describe a single-phase supply
- describe a three-phase supply
- understand a star connection and recognize that $I_L = I_p$ and $V_L = \sqrt{3}V_p$
- draw a complete phasor diagram for a balanced, star-connected load
- understand a delta connection and recognize that $V_L = V_p$ and $I_L = \sqrt{3}I_p$
- draw a phasor diagram for a balanced, delta-connected load
- calculate power in three-phase systems using $P = \sqrt{3}V_L I_L \cos\phi$
- appreciate how power is measured in a three-phase system, by the one, two and three-wattmeter methods
- compare star and delta connections
- appreciate the advantages of three-phase systems

Electrical Circuit Theory and Technology. 978-1-138-67349-6, © 2017 John Bird. Published by Taylor & Francis. All rights reserved.

22.1 Introduction

Generation, transmission and distribution of electricity via the National Grid system is accomplished by three-phase alternating currents.

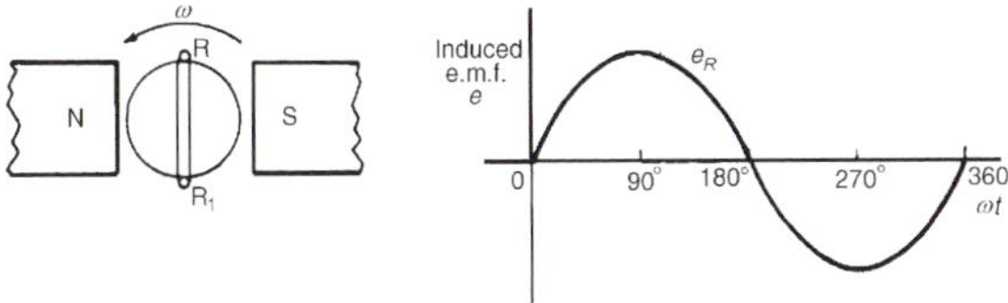

Figure 22.1

The voltage induced by a single coil when rotated in a uniform magnetic field is shown in Figure 22.1 and is known as a **single-phase voltage**. Most consumers are fed by means of a single-phase a.c. supply. Two wires are used, one called the live conductor (usually coloured red) and the other is called the neutral conductor (usually coloured black). The neutral is usually connected via protective gear to earth, the earth wire being coloured green. The standard voltage for a single-phase a.c. supply is 240 V. The majority of single-phase supplies are obtained by connection to a three-phase supply (see Figure 22.5, page 338).

22.2 Three-phase supply

A **three-phase supply** is generated when three coils are placed 120° apart and the whole rotated in a uniform magnetic field as shown in Figure 22.2(a). The result is three independent supplies of equal voltages which are each displaced by 120° from each other, as shown in Figure 22.2(b).

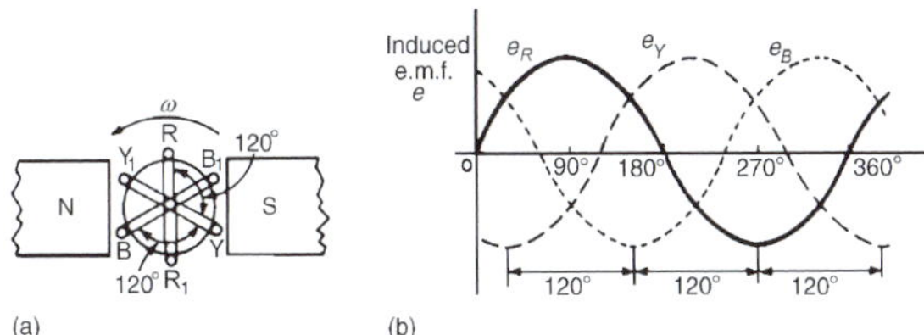

Figure 22.2

(i) The convention adopted to identify each of the phase voltages is: R – red, Y – yellow, and B – blue, as shown in Figure 22.2.

(ii) The **phase-sequence** is given by the sequence in which the conductors pass the point initially taken by the red conductor. The national standard phase sequence is R, Y, B.

A three-phase a.c. supply is carried by three conductors, called **'lines'** which are coloured red, yellow and blue. The currents in these conductors are known as line currents (I_L) and the p.d.s between them are known as line voltages (V_L). A fourth conductor, called the **neutral** (coloured black, and connected through protective devices to earth) is often used with a three-phase supply.

If the three-phase windings shown in Figure 22.2 are kept independent then six wires are needed to connect a supply source (such as a generator) to a load (such as a motor). To reduce the number of wires it is usual to interconnect the three phases. There are two ways in which this can be done, these being: (a) a **star connection**, and (b) a **delta**, or **mesh, connection**. Sources of three-phase supplies, i.e. alternators, are usually connected in star, whereas three-phase transformer windings, motors and other loads may be connected either in star or delta.

22.3 Star connection

(i) A **star-connected load** is shown in Figure 22.3, where the three line conductors are each connected to a load and the outlets from the loads are joined together at N to form what is termed the **neutral point** or the **star point**.

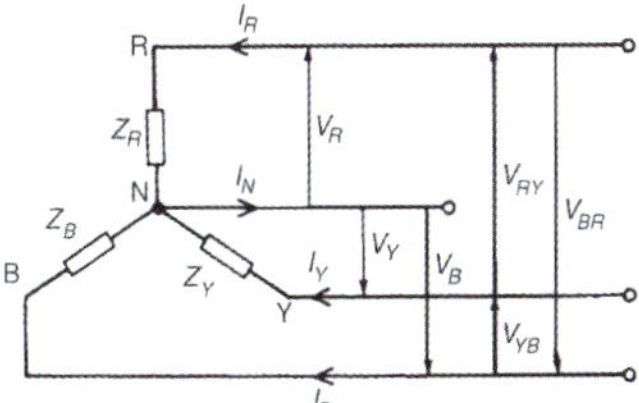

Figure 22.3

(ii) The voltages, V_R, V_Y and V_B are called **phase voltages** or line to neutral voltages. Phase voltages are generally denoted by V_p

(iii) The voltages, V_{RY}, V_{YB} and V_{BR} are called **line voltages**.

(iv) From Figure 22.3 it can be seen that the phase currents (generally denoted by I_p) are equal to their respective line currents I_R, I_Y and I_B, i.e. for a star connection:

$$I_L = I_p$$

(v) For a balanced system:
$I_R = I_Y = I_B$, $V_R = V_Y = V_B$
$V_{RY} = V_{YB} = V_{BR}$, $Z_R = Z_Y = Z_B$
and the current in the neutral conductor, $I_N = 0$. When a star-connected system is balanced, then the neutral conductor is unnecessary and is often omitted.

(vi) The line voltage, V_{RY}, shown in Figure 22.4(a) is given by $\boldsymbol{V_{RY} = V_R - V_Y}$ (V_Y is negative since it is in the opposite direction to V_{RY}). In the phasor diagram of Figure 22.4(b), phasor V_Y is reversed (shown by the broken line) and then added phasorially to V_R (i.e. $\boldsymbol{V_{RY} = V_R + (-V_Y)}$). By trigonometry, or by measurement, $V_{RY} = \sqrt{3}V_R$, i.e. for a balanced star connection:

$$V_L = \sqrt{3}\, V_p$$

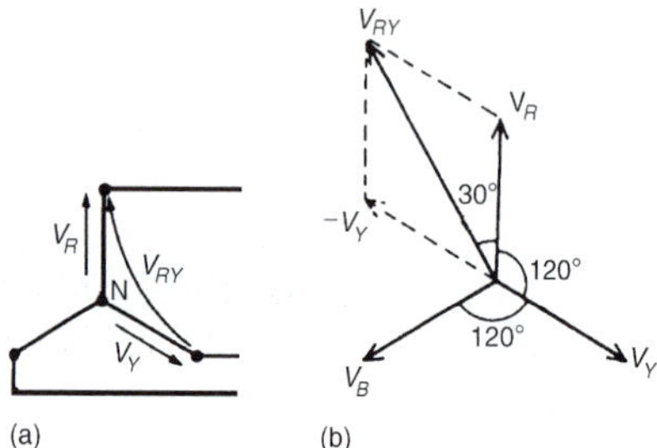

Figure 22.4

(See worked Problem 3 following for a complete phasor diagram of a star-connected system.)

(vii) The star connection of the three phases of a supply, together with a neutral conductor, allows the use of two voltages – the phase voltage and the line voltage. A four-wire system is also used when the load is not balanced. The standard electricity supply to consumers in Great Britain is 415/240 V, 50 Hz, three-phase, four-wire alternating current A diagram of connections is shown in Figure 22.5.

For most of the twentieth century, the **supply voltage in the UK in domestic premises has been 240 V a.c.** (r.m.s.) at 50 Hz. In 1988, a European-wide agreement was reached to change the various national voltages, which ranged at the time from 220 V to 240 V, to a common European standard of **230 V**.

As a result, the standard nominal supply voltage in domestic single-phase 50 Hz installations in

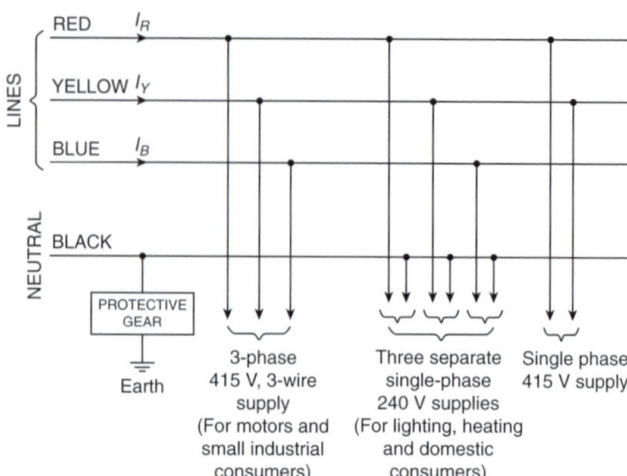

Figure 22.5

the UK has been 230 V since 1995. However, as an interim measure, electricity suppliers can work with an asymmetric voltage tolerance of 230 V +10%/−6% (i.e. 216.2 V to 253 V). The old standard was 240 V ±6% (i.e. 225.6 V to 254.4 V), which is mostly contained within the new range, and so in practice suppliers have had no reason to actually change voltages.

Similarly, the **three-phase voltage** in the UK had been for many years **415 V** ±6% (i.e. 390 V to 440 V). European harmonization required this to be changed to **400 V** +10%/−6% (i.e. 376 V to 440 V). Again, since the present supply voltage of 415 V lies within this range, supply companies are unlikely to reduce their voltages in the near future.

Many of the calculations following are based on the 240 V/415 V supply voltages which have applied for many years and are likely to continue to do so.

Problem 1. Three loads, each of resistance 30 Ω, are connected in star to a 415 V, three-phase supply. Determine (a) the system phase voltage, (b) the phase current and (c) the line current.

A '415 V, three-phase supply' means that 415 V is the line voltage, V_L

(a) For a star connection, $V_L = \sqrt{3}V_p$

Hence **phase voltage,** $\boldsymbol{V_p} = \dfrac{V_L}{\sqrt{3}} = \dfrac{415}{\sqrt{3}}$

$= \mathbf{239.6\,V}$ or $\mathbf{240\,V}$ correct to 3 significant figures

(b) **Phase current,** $\boldsymbol{I_p = \frac{V_p}{R_p} = \frac{240}{30} = 8\,A}$

(c) For a star connection, $I_p = I_L$

Hence the **line current,** $\boldsymbol{I_L = 8\,A}$

Problem 2. A star-connected load consists of three identical coils each of resistance 30 Ω and inductance 127.3 mH. If the line current is 5.08 A, calculate the line voltage if the supply frequency is 50 Hz.

Inductive reactance $X_L = 2\pi fL$
$= 2\pi(50)(127.3 \times 10^{-3})$
$= 40\,\Omega$

Impedance of each phase $Z_p = \sqrt{(R^2 + X_L^2)}$
$= \sqrt{(30^2 + 40^2)} = 50\,\Omega$

For a star connection $I_L = I_p = \frac{V_p}{Z_p}$

Hence phase voltage $V_p = I_p Z_p = (5.08)(50)$
$= 254\,V$

Line voltage $\boldsymbol{V_L} = \sqrt{3}V_p = \sqrt{3}(254)$
$= \mathbf{440\,V}$

Problem 3. A balanced, three-wire, star-connected, three-phase load has a phase voltage of 240 V, a line current of 5 A and a lagging power factor of 0.966. Draw the complete phasor diagram.

The phasor diagram is shown in Figure 22.6.

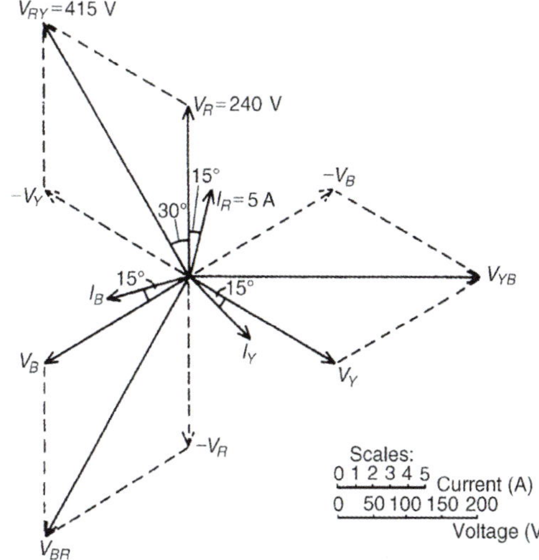

Figure 22.6

Procedure to construct the phasor diagram:

(i) Draw $V_R = V_Y = V_B = 240\,V$ and spaced 120° apart. (Note that V_R is shown vertically upwards – this, however, is immaterial for it may be drawn in any direction.)

(ii) Power factor $= \cos\phi = 0.966$ lagging. Hence the load phase angle is given by $\cos^{-1} 0.966$, i.e. 15° lagging. Hence $I_R = I_Y = I_B = 5\,A$, lagging V_R, V_Y and V_B, respectively, by 15°

(iii) $V_{RY} = V_R - V_Y$ (phasorially). Hence V_Y is reversed and added phasorially to V_R. By measurement, $V_{RY} = 415\,V$ (i.e. $\sqrt{3}(240)$) and leads V_R by 30°. Similarly, $V_{YB} = V_Y - V_B$ and $V_{BR} = V_B - V_R$

Problem 4. A 415 V, three-phase, four wire, star-connected system supplies three resistive loads as shown in Figure 22.7. Determine (a) the current in each line and (b) the current in the neutral conductor.

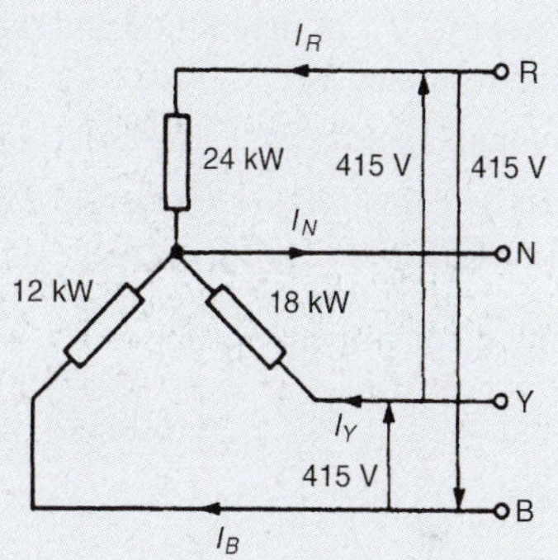

Figure 22.7

(a) For a star-connected system $V_L = \sqrt{3}V_p$

Hence $V_p = \frac{V_L}{\sqrt{3}} = \frac{415}{\sqrt{3}} = 240\,V$

Since current $I = \frac{\text{power } P}{\text{voltage } V}$ for a resistive load

then $I_R = \frac{P_R}{V_R} = \frac{24\,000}{240} = \mathbf{100\,A}$

$I_Y = \frac{P_Y}{V_Y} = \frac{18\,000}{240} = \mathbf{75\,A}$

and $I_B = \frac{P_B}{V_B} = \frac{12\,000}{240} = \mathbf{50\,A}$

(b) The three line currents are shown in the phasor diagram of Figure 22.8. Since each load is resistive the currents are in phase with the phase voltages and are hence mutually displaced by 120°. The current in the neutral conductor is given by:

$I_N = I_R + I_Y + I_B$ phasorially.

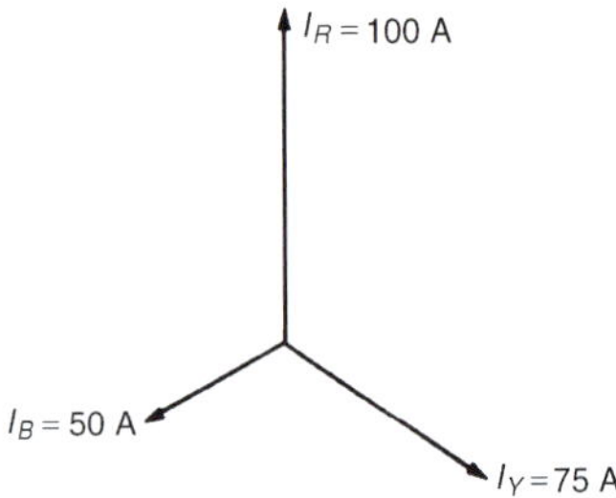

Figure 22.8

Figure 22.9 shows the three line currents added phasorially. *Oa* represents I_R in magnitude and direction. From the nose of *Oa*, *ab* is drawn representing I_Y in magnitude and direction. From the nose of *ab*, *bc* is drawn representing I_B in magnitude and direction. *Oc* represents the resultant, I_N

By measurement, $\boldsymbol{I_N = 43}$ **A**

Alternatively, by calculation, considering I_R at 90°, I_B at 210° and I_Y at 330°:

$$\text{Total horizontal component} = 100\cos 90^\circ + 75\cos 330^\circ + 50\cos 210^\circ = 21.65$$

$$\text{Total vertical component} = 100\sin 90^\circ + 75\sin 330^\circ + 50\sin 210^\circ = 37.50$$

$$\text{Hence magnitude of } I_N = \sqrt{(21.65^2 + 37.50^2)} = \mathbf{43.3\ A}$$

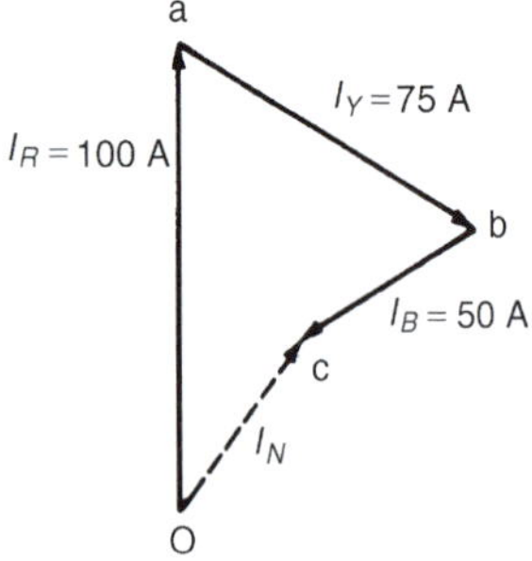

Figure 22.9

22.4 Delta connection

(i) A **delta (or mesh) connected load** is shown in Figure 22.10 where the end of one load is connected to the start of the next load.

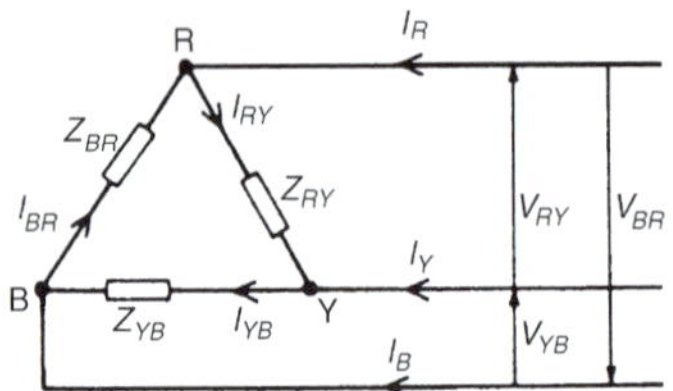

Figure 22.10

(ii) From Figure 22.10, it can be seen that the line voltages V_{RY}, V_{YB} and V_{BR} are the respective phase voltages, i.e. for a delta connection:

$$V_L = V_p$$

(iii) Using Kirchhoff's current law in Figure 22.10, $I_R = I_{RY} - I_{BR} = I_{RY} + (-I_{BR})$. From the phasor diagram shown in Figure 22.11, by trigonometry or by measurement, $I_R = \sqrt{3} I_{RY}$, i.e. for a delta connection:

$$I_L = \sqrt{3} I_p$$

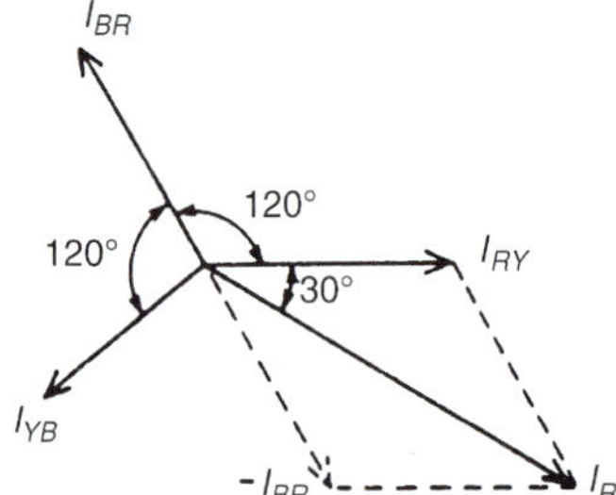

Figure 22.11

Problem 5. Three identical coils each of resistance 30 Ω and inductance 127.3 mH are connected in delta to a 440 V, 50 Hz, three-phase supply. Determine (a) the phase current, and (b) the line current.

Phase impedance, $Z_p = 50\,\Omega$ (from worked Problem 2) and for a delta connection, $V_p = V_L$

(a) Phase current, $$I_p = \frac{V_p}{Z_p} = \frac{V_L}{Z_p} = \frac{440}{50} = \mathbf{8.8\ A}$$

(b) For a delta connection, $I_L = \sqrt{3}I_p = \sqrt{3}(8.8)$
$= \mathbf{15.24\,A}$

Thus when the load is connected in delta, three times the line current is taken from the supply than is taken if connected in star.

Problem 6. Three identical capacitors are connected in delta to a 415 V, 50 Hz, three-phase supply. If the line current is 15 A, determine the capacitance of each of the capacitors.

For a delta connection $I_L = \sqrt{3}I_p$

Hence phase current $I_p = \frac{I_L}{\sqrt{3}} = \frac{15}{\sqrt{3}} = 8.66\,\text{A}$

Capacitive reactance per phase, $X_C = \frac{V_p}{I_p} = \frac{V_L}{I_p}$ (since for a delta connection $V_L = V_p$)

Hence $X_C = \frac{415}{8.66} = 47.92\,\Omega$

$X_C = \frac{1}{2\pi fC}$, from which capacitance,

$$C = \frac{1}{2\pi f X_C} = \frac{1}{2\pi(50)(47.92)}\,\text{F} = \mathbf{66.43\,\mu F}$$

Problem 7. Three coils each having resistance 3 Ω and inductive reactance 4 Ω are connected (i) in star and (ii) in delta to a 415 V, three-phase supply. Calculate for each connection (a) the line and phase voltages and (b) the phase and line currents.

(i) **For a star connection:** $I_L = I_p$ and $V_L = \sqrt{3}V_p$

(a) A 415 V, three-phase supply means that the

line voltage, $V_L = \mathbf{415\,V}$

Phase voltage, $V_p = \frac{V_L}{\sqrt{3}} = \frac{415}{\sqrt{3}} = \mathbf{240\,V}$

(b) Impedance per phase, $Z_p = \sqrt{(R^2 + X_L^2)}$
$= \sqrt{(3^2 + 4^2)} = 5\,\Omega$

Phase current, $I_p = \frac{V_p}{Z_p} = \frac{240}{5} = \mathbf{48\,A}$

Line current, $I_L = I_p = \mathbf{48\,A}$

(ii) **For a delta connection:** $V_L = V_p$ and $I_L = \sqrt{3}I_p$

(a) Line voltage, $V_L = 415\,\text{V}$

Phase voltage, $V_p = V_L = \mathbf{415\,V}$

(b) Phase current, $I_p = \frac{V_p}{Z_p} = \frac{415}{5} = \mathbf{83\,A}$

Line current, $I_L = \sqrt{3}I_p = \sqrt{3}(83) = \mathbf{144\,A}$

Now try the following Practice Exercise

Practice Exercise 93 Star and delta connections (Answers on page 823)

1. Three loads, each of resistance 50 Ω are connected in star to a 400 V, three-phase supply. Determine (a) the phase voltage, (b) the phase current and (c) the line current.
2. If the loads in question 1 are connected in delta to the same supply, determine (a) the phase voltage, (b) the phase current and (c) the line current.
3. A star-connected load consists of three identical coils, each of inductance 159.2 mH and resistance 50 Ω. If the supply frequency is 50 Hz and the line current is 3 A, determine (a) the phase voltage and (b) the line voltage.
4. Three identical capacitors are connected (a) in star, (b) in delta to a 400 V, 50 Hz, three-phase supply. If the line current is 12 A determine in each case the capacitance of each of the capacitors.
5. Three coils each having resistance 6 Ω and inductance *LH* are connected (a) in star and (b) in delta to a 415 V, 50 Hz, three-phase supply. If the line current is 30 A, find for each connection the value of *L*.
6. A 400 V, three-phase, four wire, star-connected system supplies three resistive loads of 15 kW, 20 kW and 25 kW in the red, yellow and blue phases, respectively. Determine the current flowing in each of the four conductors.
7. A three-phase, star-connected alternator delivers a line current of 65 A to a balanced delta-connected load at a line voltage of 380 V. Calculate (a) the phase voltage of the alternator, (b) the alternator phase current and (c) the load phase current.
8. Three 24 μF capacitors are connected in star across a 400 V, 50 Hz, three-phase supply. What value of capacitance must be connected in delta in order to take the same line current?

22.5 Power in three-phase systems

The power dissipated in a three-phase load is given by the sum of the power dissipated in each phase. If a load is balanced then the total power P is given by: $P=3\times$ power consumed by one phase.

The power consumed in one phase $=I_p^2 R_p$ or $V_p I_p \cos\phi$ (where ϕ is the phase angle between V_p and I_p)

For a star connection, $V_p = \dfrac{V_L}{\sqrt{3}}$ and $I_p = I_L$ hence

$$P = 3\left(\frac{V_L}{\sqrt{3}}\right) I_L \cos\phi$$
$$= \sqrt{3} V_L I_L \cos\phi$$

For a delta connection, $V_p = V_L$ and $I_p = \dfrac{I_L}{\sqrt{3}}$ hence

$$P = 3V_L\left(\frac{I_L}{\sqrt{3}}\right)\cos\phi$$
$$= \sqrt{3} V_L I_L \cos\phi$$

Hence for either a star or a delta balanced connection the total power P is given by:

$$\boldsymbol{P = \sqrt{3}\, V_L I_L \cos\phi \text{ watts}} \quad \text{or} \quad \boldsymbol{P = 3I_p^2 R_p \text{ watts}}$$

Total volt-amperes, $\boldsymbol{S = \sqrt{3}\, V_L I_L}$ **volt-amperes**

Problem 8. Three 12 Ω resistors are connected in star to a 415 V, three-phase supply. Determine the total power dissipated by the resistors.

Power dissipated, $P=\sqrt{3} V_L I_L \cos\phi$ or $P=3I_p^2 R_p$

Line voltage, $V_L=415$ V and

phase voltage $V_p=\dfrac{415}{\sqrt{3}}=240$ V

(since the resistors are star-connected)

Phase current, $I_p=\dfrac{V_p}{Z_p}=\dfrac{V_p}{R_p}=\dfrac{240}{12}=20$ A

For a star connection $I_L=I_p=20$ A

For a purely resistive load, the power factor $=\cos\phi=1$

Hence power $P=\sqrt{3} V_L I_L \cos\phi=\sqrt{3}(415)(20)(1)$

$=\mathbf{14.4\,kW}$

or power $P=3I_p^2 R_p=3(20)^2(12)=\mathbf{14.4\,kW}$

Problem 9. The input power to a three-phase a.c. motor is measured as 5 kW. If the voltage and current to the motor are 400 V and 8.6 A, respectively, determine the power factor of the system.

Power, $P=5000$ W; line voltage $V_L=400$ V; line current, $I_L=8.6$ A

Power, $P=\sqrt{3} V_L I_L \cos\phi$

Hence power factor $=\cos\phi=\dfrac{P}{\sqrt{3} V_L I_L}$

$=\dfrac{5000}{\sqrt{3}(400)(8.6)}$

$=\mathbf{0.839}$

Problem 10. Three identical coils, each of resistance 10 Ω and inductance 42 mH are connected (a) in star and (b) in delta to a 415 V, 50 Hz, three-phase supply. Determine the total power dissipated in each case.

(a) **Star connection**

Inductive reactance $X_L=2\pi f L$

$=2\pi(50)(42\times10^{-3})$

$=13.19\,\Omega$

Phase impedance $Z_p=\sqrt{(R^2+X_L^2)}$

$=\sqrt{(10^2+13.19^2)}$

$=16.55\,\Omega$

Line voltage $V_L=415$ V and

phase voltage, $V_p=\dfrac{V_L}{\sqrt{3}}=\dfrac{415}{\sqrt{3}}=240$ V

Phase current, $I_p=\dfrac{V_p}{Z_p}=\dfrac{240}{16.55}=14.50$ A

Line current, $I_L=I_p=14.50$ A

Power factor $=\cos\phi=\dfrac{R_p}{Z_p}=\dfrac{10}{16.55}$

$=0.6042$ lagging

Power dissipated, $P=\sqrt{3} V_L I_L \cos\phi$

$=\sqrt{3}(415)(14.50)(0.6042)$

$=\mathbf{6.3\,kW}$

(Alternatively, $P=3I_p^2 R_p=3(14.50)^2(10)$

$=\mathbf{6.3\,kW}$)

(b) **Delta connection**

$$V_L = V_p = 415\,\text{V},\ Z_p = 16.55\,\Omega,$$
$$\cos\phi = 0.6042 \text{ lagging (from above).}$$

$$\text{Phase current, } I_p = \frac{V_p}{Z_p} = \frac{415}{16.55} = 25.08\,\text{A}$$

$$\text{Line current, } I_L = \sqrt{3} I_p = \sqrt{3}(25.08)$$
$$= 43.44\,\text{A}$$

Power dissipated, $P = \sqrt{3} V_L I_L \cos\phi$

$$= \sqrt{3}(415)(43.44)(0.6042)$$
$$= \mathbf{18.87\,kW}$$

(Alternatively, $P = 3I_p^2 R_p = 3(25.08)^2(10)$
$= \mathbf{18.87\,kW}$)

Hence loads connected in delta dissipate three times the power than when connected in star, and also take a line current three times greater.

Problem 11. A 415 V three-phase a.c. motor has a power output of 12.75 kW and operates at a power factor of 0.77 lagging and with an efficiency of 85%. If the motor is delta-connected, determine (a) the power input, (b) the line current and (c) the phase current.

(a) $\text{Efficiency} = \dfrac{\text{power output}}{\text{power input}}$

hence $\dfrac{85}{100} = \dfrac{12\,750}{\text{power input}}$

from which, **power input** $= \dfrac{12\,750 \times 100}{85}$

$= \mathbf{15\,000\,W}$ or $\mathbf{15\,kW}$

(b) Power, $P = \sqrt{3} V_L I_L \cos\phi$, hence

line current, $I_L = \dfrac{P}{\sqrt{3} V_L \cos\phi} = \dfrac{15\,000}{\sqrt{3}(415)(0.77)}$

$= \mathbf{27.10\,A}$

(c) For a delta connection, $I_L = \sqrt{3} I_p$

hence **phase current**, $I_p = \dfrac{I_L}{\sqrt{3}} = \dfrac{27.10}{\sqrt{3}}$

$= \mathbf{15.65\,A}$

22.6 Measurement of power in three-phase systems

Power in three-phase loads may be measured by the following methods:

(i) **One-wattmeter method for a balanced load**

Wattmeter connections for both star and delta are shown in Figure 22.12.

Total power = 3 x wattmeter reading

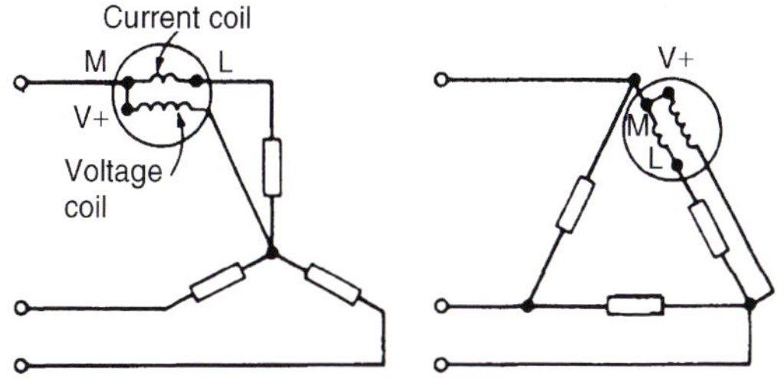

Figure 22.12

(ii) **Two-wattmeter method for balanced or unbalanced loads**

A connection diagram for this method is shown in Figure 22.13 for a star-connected load. Similar connections are made for a delta-connected load.

Total power = sum of wattmeter readings
$$= P_1 + P_2$$

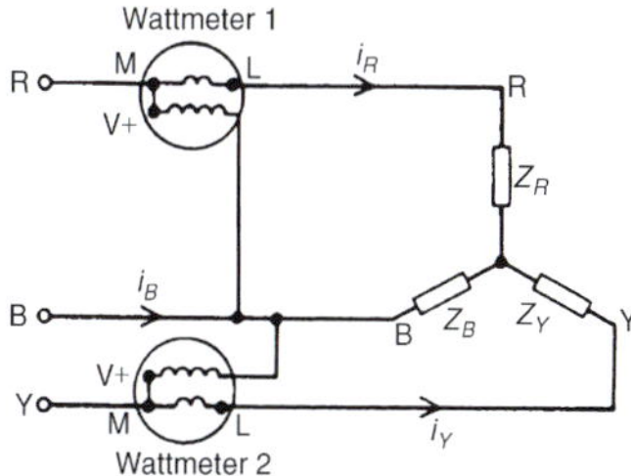

Figure 22.13

The power factor may be determined from:

$$\tan\phi = \sqrt{3}\left(\frac{P_1 - P_2}{P_1 + P_2}\right) \quad \text{(see Problems 12 and 15 to 18)}$$

It is possible, depending on the load power factor, for one wattmeter to have to be 'reversed' to obtain a reading. In this case it is taken as a negative reading (see worked Problem 17).

(iii) **Three-wattmeter method for a three-phase, four-wire system for balanced and unbalanced loads** (see Figure 22.14).

Total power = $P_1 + P_2 + P_3$

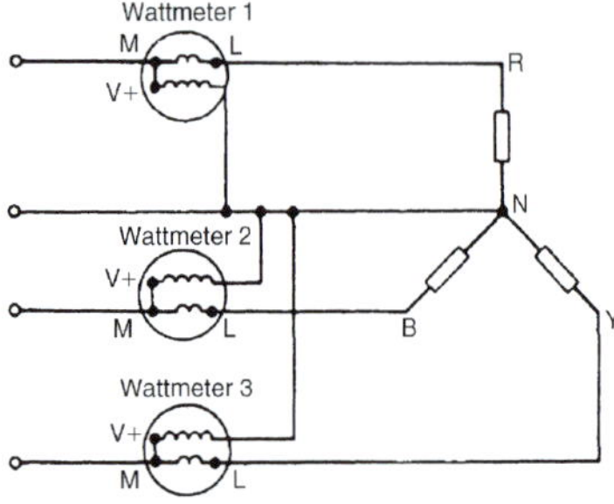

Figure 22.14

Problem 12. (a) Show that the total power in a three-phase, three-wire system using the two-wattmeter method of measurement is given by the sum of the wattmeter readings. Draw a connection diagram. (b) Draw a phasor diagram for the two-wattmeter method for a balanced load. (c) Use the phasor diagram of part (b) to derive a formula from which the power factor of a three-phase system may be determined using only the wattmeter readings.

(a) A connection diagram for the two-wattmeter method of a power measurement is shown in Figure 22.15 for a star-connected load.

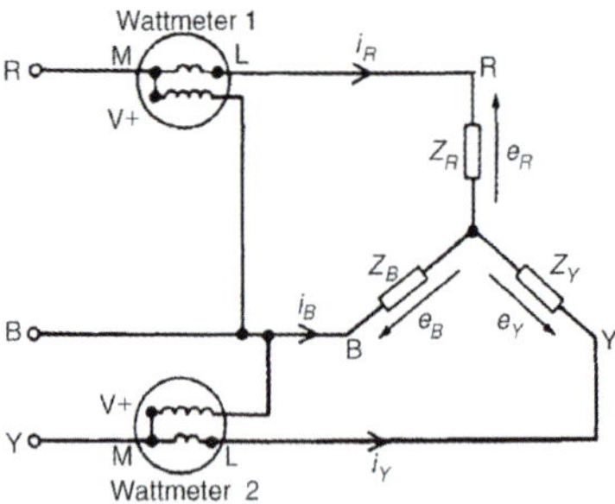

Figure 22.15

Total instantaneous power, $p = e_R i_R + e_Y i_Y + e_B i_B$ and in any three-phase system $i_R + i_Y + i_B = 0$. Hence $i_B = -i_R - i_Y$

Thus, $p = e_R i_R + e_Y i_Y + e_B(-i_R - i_Y)$

$$= (e_R - e_B)i_R + (e_Y - e_B)i_Y$$

However, $(e_R - e_B)$ is the p.d. across wattmeter 1 in Figure 22.15 and $(e_Y - e_B)$ is the p.d. across wattmeter 2.

Hence total instantaneous power,

$$p = (\text{wattmeter 1 reading}) + (\text{wattmeter 2 reading})$$
$$= p_1 + p_2$$

The moving systems of the wattmeters are unable to follow the variations which take place at normal frequencies and they indicate the mean power taken over a cycle. Hence the total power, $\boldsymbol{P = P_1 + P_2}$ for balanced or unbalanced loads.

(b) The phasor diagram for the two-wattmeter method for a balanced load having a lagging current is shown in Figure 22.16, where $V_{RB} = V_R - V_B$ and $V_{YB} = V_Y - V_B$ (phasorially).

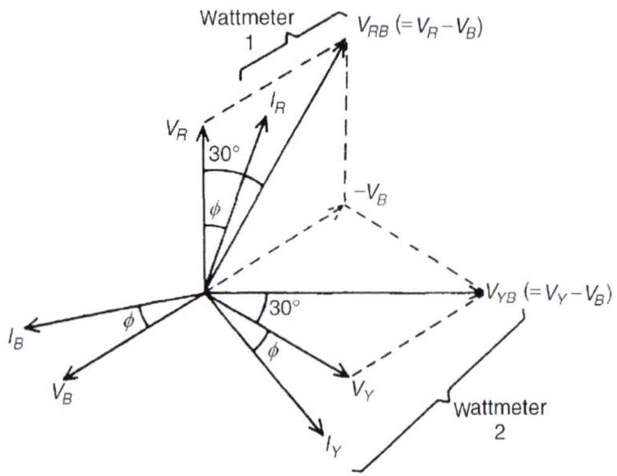

Figure 22.16

(c) Wattmeter 1 reads $V_{RB} I_R \cos(30° - \phi) = P_1$

Wattmeter 2 reads $V_{YB} I_Y \cos(30° + \phi) = P_2$

$$\frac{P_1}{P_2} = \frac{V_{RB} I_R \cos(30° - \phi)}{V_{YB} I_Y \cos(30° + \phi)} = \frac{\cos(30° - \phi)}{\cos(30° + \phi)}$$

since $I_R = I_Y$ and $V_{RB} = V_{YB}$ for a balanced load.

Hence $\dfrac{P_1}{P_2} = \dfrac{\cos 30° \cos\phi + \sin 30° \sin\phi}{\cos 30° \cos\phi - \sin 30° \sin\phi}$

(from compound angle formulae, see *Higher Engineering Mathematics*, J.O. Bird, 2017, 8th edition, Taylor & Francis.)

Dividing throughout by cos 30° cos ϕ gives:

$$\frac{P_1}{P_2} = \frac{1 + \tan 30° \tan\phi}{1 - \tan 30° \tan\phi}$$

$$= \frac{1 + \frac{1}{\sqrt{3}}\tan\phi}{1 - \frac{1}{\sqrt{3}}\tan\phi} \quad \left(\text{since } \frac{\sin\phi}{\cos\phi} = \tan\phi\right)$$

Part 3

Cross-multiplying gives:

$$P_1 - \frac{P_1}{\sqrt{3}} \tan\phi = P_2 + \frac{P_2}{\sqrt{3}} \tan\phi$$

Hence $$P_1 - P_2 = (P_1 + P_2)\frac{\tan\phi}{\sqrt{3}}$$

from which $$\boldsymbol{\tan\phi = \sqrt{3}\left(\frac{P_1 - P_2}{P_1 + P_2}\right)}$$

ϕ, $\cos\phi$ and thus power factor can be determined from this formula.

Problem 13. A 400 V, three-phase star-connected alternator supplies a delta-connected load, each phase of which has a resistance of 30 Ω and inductive reactance 40 Ω. Calculate (a) the current supplied by the alternator and (b) the output power and the kVA of the alternator, neglecting losses in the line between the alternator and load.

A circuit diagram of the alternator and load is shown in Figure 22.17.

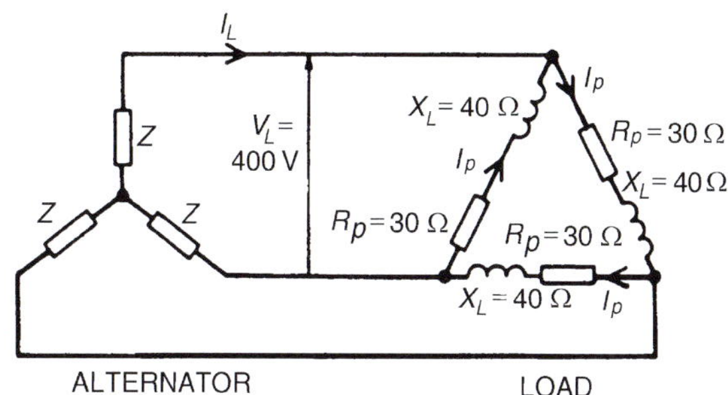

Figure 22.17

(a) Considering the load: phase current, $I_p = \dfrac{V_p}{Z_p}$

$V_p = V_L$ for a delta connection. Hence $V_p = 400$ V

Phase impedance, $Z_p = \sqrt{(R_p^2 + X_L^2)}$

$$= \sqrt{(30^2 + 40^2)} = 50\,\Omega$$

Hence $I_p = \dfrac{V_p}{Z_p} = \dfrac{400}{50} = 8$ A

For a delta connection, line current, $I_L = \sqrt{3}I_p = \sqrt{3}(8) = 13.86$ A

Hence 13.86 A is the current supplied by the alternator.

(b) Alternator output power is equal to the power dissipated by the load.

i.e. $P = \sqrt{3}V_L I_L \cos\phi$, where $\cos\phi = \dfrac{R_p}{Z_p}$

$$= \frac{30}{50} = 0.6$$

Hence $P = \sqrt{3}(400)(13.86)(0.6) = \mathbf{5.76\,kW}$

Alternator output kVA, $S = \sqrt{3}V_L I_L$

$$= \sqrt{3}(400)(13.86)$$
$$= \mathbf{9.60\,kVA}$$

Problem 14. Each phase of a delta-connected load comprises a resistance of 30 Ω and an 80 μF capacitor in series. The load is connected to a 400 V, 50 Hz, three-phase supply. Calculate (a) the phase current, (b) the line current, (c) the total power dissipated and (d) the kVA rating of the load. Draw the complete phasor diagram for the load.

(a) Capacitive reactance, $X_C = \dfrac{1}{2\pi f C}$

$$= \frac{1}{2\pi(50)(80 \times 10^{-6})}$$
$$= 39.79\,\Omega$$

Phase impedance, $Z_p = \sqrt{(R_p^2 + X_C^2)}$

$$= \sqrt{(30^2 + 39.79^2)}$$
$$= 49.83\,\Omega$$

Power factor $= \cos\phi = \dfrac{R_p}{Z_p} = \dfrac{30}{49.83} = 0.602$

Hence $\phi = \cos^{-1} 0.602 = 52.99°$ leading.

Phase current, $I_p = \dfrac{V_p}{Z_p}$ and $V_p = V_L$ for a delta connection

Hence $I_p = \dfrac{400}{49.83} = \mathbf{8.027\,A}$

(b) Line current $I_L = \sqrt{3}I_p$ for a delta connection

Hence $I_L = \sqrt{3}(8.207) = \mathbf{13.90\,A}$

(c) Total power dissipated, $P = \sqrt{3}V_L I_L \cos\phi$

$$= \sqrt{3}(400)(13.90)(0.602)$$
$$= \mathbf{5.797\,kW}$$

(d) Total kVA, $S = \sqrt{3}V_L I_L = \sqrt{3}(400)(13.90)$

$$= \mathbf{9.630\,kVA}$$

The phasor diagram for the load is shown in Figure 22.18.

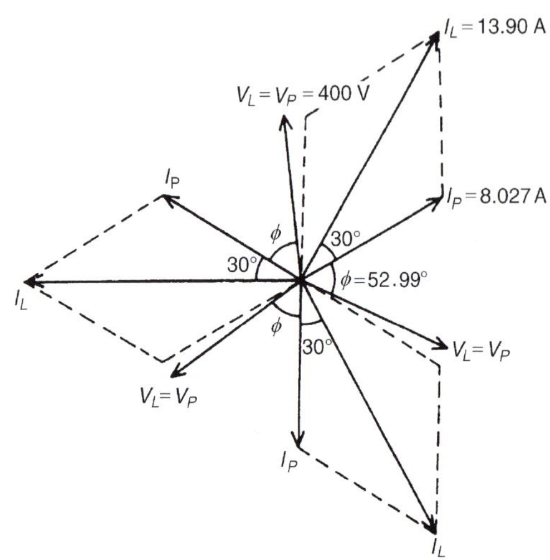

Figure 22.18

Problem 15. Two wattmeters are connected to measure the input power to a balanced three-phase load by the two-wattmeter method. If the instrument readings are 8 kW and 4 kW, determine (a) the total power input and (b) the load power factor.

(a) Total input power, $P = P_1 + P_2 = 8 + 4 = \mathbf{12\,kW}$

(b) $\tan\phi = \sqrt{3}\left(\dfrac{P_1 - P_2}{P_1 + P_2}\right) = \sqrt{3}\left(\dfrac{8-4}{8+4}\right)$

$$= \sqrt{3}\left(\frac{4}{12}\right) = \sqrt{3}\left(\frac{1}{3}\right) = \frac{1}{\sqrt{3}}$$

Hence $\phi = \tan^{-1}\dfrac{1}{\sqrt{3}} = 30°$

Power factor $= \cos\phi = \cos 30° = \mathbf{0.866}$

Problem 16. Two wattmeters connected to a three-phase motor indicate the total power input to be 12 kW. The power factor is 0.6. Determine the readings of each wattmeter.

If the two wattmeters indicate P_1 and P_2, respectively, then $P_1 + P_2 = 12\,\text{kW}$ (1)

$\tan\phi = \sqrt{3}\left(\dfrac{P_1 - P_2}{P_1 + P_2}\right)$ and power factor $= 0.6 = \cos\phi$

Angle $\phi = \cos^{-1} 0.6 = 53.13°$ and $\tan 53.13° = 1.3333$

Hence $1.3333 = \dfrac{\sqrt{3}(P_1 - P_2)}{12}$, from which,

$$P_1 - P_2 = \frac{12(1.3333)}{\sqrt{3}}$$

i.e. $P_1 - P_2 = 9.237\,\text{kW}$ (2)

Adding equations (1) and (2) gives: $2P_1 = 21.237$

i.e. $P_1 = \dfrac{21.237}{2} = 10.62\,\text{kW}$

Hence **wattmeter 1 reads 10.62 kW**
From equation (1), **wattmeter 2 reads (12 − 10.62) = 1.38 kW**

Problem 17. Two wattmeters indicate 10 kW and 3 kW, respectively, when connected to measure the input power to a three-phase balanced load, the reverse switch being operated on the meter indicating the 3 kW reading. Determine (a) the input power and (b) the load power factor.

Since the reversing switch on the wattmeter had to be operated, the 3 kW reading is taken as −3 kW.

(a) Total input power, $P = P_1 + P_2 = 10 + (-3) = \mathbf{7\,kW}$

(b) $\tan\phi = \sqrt{3}\left(\dfrac{P_1 - P_2}{P_1 + P_2}\right) = \sqrt{3}\left(\dfrac{10-(-3)}{10+(-3)}\right)$

$$= \sqrt{3}\left(\frac{13}{7}\right) = 3.2167$$

Angle $\phi = \tan^{-1} 3.2167 = 72.73°$

Power factor $= \cos\phi = \cos 72.73° = \mathbf{0.297}$

Problem 18. Three similar coils, each having a resistance of 8 Ω and an inductive reactance of 8 Ω, are connected (a) in star and (b) in delta, across a 415 V, three-phase supply. Calculate for each connection the readings on each of two wattmeters connected to measure the power by the two-wattmeter method.

(a) **Star connection:** $V_L = \sqrt{3}V_p$ and $I_L = I_p$

Phase voltage, $V_p = \dfrac{V_L}{\sqrt{3}} = \dfrac{415}{\sqrt{3}}$ and

phase impedance, $Z_p = \sqrt{(R_p^2 + X_L^2)}$

$$= \sqrt{(8^2 + 8^2)} = 11.31\,\Omega$$

Hence phase current, $I_p = \dfrac{V_p}{Z_p} = \dfrac{415/\sqrt{3}}{11.31} = 21.18\,\text{A}$

Total power, $P = 3I_p^2 R_p = 3(21.18)^2(8) = 10\,766\,\text{W}$

If wattmeter readings are P_1 and P_2, then

$$P_1 + P_2 = 10\,766 \qquad (1)$$

Since $R_p = 8\,\Omega$ and $X_L = 8\,\Omega$, then phase angle $\phi = 45°$ (from impedance triangle)

$$\tan\phi = \sqrt{3}\left(\frac{P_1 - P_2}{P_1 + P_2}\right) \text{ hence}$$

$$\tan 45° = \frac{\sqrt{3}(P_1 - P_2)}{10\,766}$$

$$\text{from which } P_1 - P_2 = \frac{10\,766(1)}{\sqrt{3}} = 6216\,\text{W} \qquad (2)$$

Adding equations (1) and (2) gives:

$$2P_1 = 10\,766 + 6216$$
$$= 16\,982\,\text{W}$$

Hence $P_1 = 8491\,\text{W}$

From equation (1), $P_2 = 10\,766 - 8491 = 2275\,\text{W}$

When the coils are star-connected the wattmeter readings are thus 8.491 kW and 2.275 kW.

(b) **Delta connection:** $V_L = V_p$ and $I_L = \sqrt{3} I_p$

$$\text{Phase current, } I_p = \frac{V_p}{Z_p} = \frac{415}{11.31} = 36.69\,\text{A}$$

Total power, $P = 3I_p^2 R_p = 3(36.69)^2(8) = 32\,310\,\text{W}$

$$\text{Hence} \quad P_1 + P_2 = 32\,310\,\text{W} \qquad (3)$$

$$\tan\phi = \sqrt{3}\left(\frac{P_1 - P_2}{P_1 + P_2}\right) \text{ thus } 1 = \frac{\sqrt{3}(P_1 - P_2)}{32\,310}$$

$$\text{from which, } P_1 - P_2 = \frac{32\,310}{\sqrt{3}} = 18\,650\,\text{W} \qquad (4)$$

Adding equations (3) and (4) gives:

$2P_1 = 50\,960$, from which $P_1 = 25\,480\,\text{W}$

From equation (3), $P_2 = 32\,310 - 25\,480 = 6830\,\text{W}$

When the coils are delta-connected the wattmeter readings are thus 25.48 kW and 6.83 kW.

Now try the following Practice Exercise

Practice Exercise 94 Power in three-phase systems (Answers on page 823)

1. Determine the total power dissipated by three 20 Ω resistors when connected (a) in star and (b) in delta to a 440 V, three-phase supply.
2. Determine the power dissipated in the circuit of Problem 3 of Exercise 93, page 341.
3. A balanced delta-connected load has a line voltage of 400 V, a line current of 8 A and a lagging power factor of 0.94. Draw a complete phasor diagram of the load. What is the total power dissipated by the load?
4. Three inductive loads, each of resistance 4 Ω and reactance 9 Ω, are connected in delta. When connected to a three-phase supply the loads consume 1.2 kW. Calculate (a) the power factor of the load, (b) the phase current, (c) the line current and (d) the supply voltage.
5. The input voltage, current and power to a motor is measured as 415 V, 16.4 A and 6 kW, respectively. Determine the power factor of the system.
6. A 440 V, three-phase a.c. motor has a power output of 11.25 kW and operates at a power factor of 0.8 lagging and with an efficiency of 84%. If the motor is delta connected, determine (a) the power input, (b) the line current and (c) the phase current.
7. Two wattmeters are connected to measure the input power to a balanced three-phase load. If the wattmeter readings are 9.3 kW and 5.4 kW, determine (a) the total output power, and (b) the load power factor.
8. 8 kW is found by the two-wattmeter method to be the power input to a three-phase motor. Determine the reading of each wattmeter if the power factor of the system is 0.85
9. Three similar coils, each having a resistance of 4.0 Ω and an inductive reactance of 3.46 Ω, are connected (a) in star and (b) in delta across a 400 V, three-phase supply. Calculate for each connection the readings on each

of two wattmeters connected to measure the power by the two-wattmeter method.

10. A three-phase, star-connected alternator supplies a delta-connected load, each phase of which has a resistance of 15 Ω and inductive reactance of 20 Ω. If the line voltage is 400 V, calculate (a) the current supplied by the alternator and (b) the output power and kVA rating of the alternator, neglecting any losses in the line between the alternator and the load.

11. Each phase of a delta-connected load comprises a resistance of 40 Ω and a 40 μF capacitor in series. Determine, when connected to a 415 V, 50 Hz, three-phase supply (a) the phase current, (b) the line current, (c) the total power dissipated and (d) the kVA rating of the load.

22.7 Comparison of star and delta connections

(i) Loads connected in delta dissipate three times more power than when connected in star to the same supply.

(ii) For the same power, the phase currents must be the same for both delta and star connections (since power $=3I_p^2R_p$), hence the line current in the delta-connected system is greater than the line current in the corresponding star-connected system. To achieve the same phase current in a star-connected system as in a delta-connected system, the line voltage in the star system is $\sqrt{3}$ times the line voltage in the delta system.

Thus for a given power transfer, a delta system is associated with larger line currents (and thus larger conductor cross-sectional area) and a star system is associated with a larger line voltage (and thus greater insulation).

22.8 Advantages of three-phase systems

Advantages of three-phase systems over single-phase supplies include:

(i) For a given amount of power transmitted through a system, the three-phase system requires conductors with a smaller cross-sectional area. This means a saving of copper (or aluminium) and thus the original installation costs are less.

(ii) Two voltages are available (see Section 22.3(vii)).

(iii) Three-phase motors are very robust, relatively cheap, generally smaller, have self-starting properties, provide a steadier output and require little maintenance compared with single-phase motors.

For fully worked solutions to each of the problems in Practice Exercises 93 and 94 in this chapter, go to the website:
www.routledge.com/cw/bird

Chapter 23

Transformers

Why it is important to understand: **Transformers**

The transformer is one of the simplest of electrical devices. Its basic design, materials, and principles have changed little over the last 100 years, yet transformer designs and materials continue to be improved. Transformers are essential in high-voltage power transmission, providing an economical means of transmitting power over large distances. A major application of transformers is to increase voltage before transmitting electrical energy over long distances through cables. Cables have resistance and so dissipate electrical energy. By transforming electrical power to a high-voltage, and therefore low-current form, for transmission and back again afterwards, transformers enable economical transmission of power over long distances. Consequently, transformers have shaped the electricity supply industry, permitting generation to be located remotely from points of demand. All but a tiny fraction of the world's electrical power has passed through a series of transformers by the time it reaches the consumer. Transformers are also used extensively in electronic products to step down the supply voltage to a level suitable for the low-voltage circuits they contain. The transformer also electrically isolates the end user from contact with the supply voltage. Signal and audio transformers are used to couple stages of amplifiers and to match devices such as microphones and record players to the input of amplifiers. Audio transformers allowed telephone circuits to carry on a two-way conversation over a single pair of wires. This chapter explains the principle of operation of a transformer, its construction and associated calculations, including losses and efficiency. Resistance matching, the auto transformer, the three-phase transformer and current and voltage transformers are also discussed.

At the end of this chapter you should be able to:

- understand the principle of operation of a transformer
- understand the term 'rating' of a transformer
- use $\frac{V_1}{V_2} = \frac{N_1}{N_2} = \frac{I_2}{I_1}$ in calculations on transformers
- construct a transformer no-load phasor diagram and calculate magnetizing and core loss components of the no-load current
- state the e.m.f. equation for a transformer $E = 4.44 f \Phi_m N$ and use it in calculations
- construct a transformer on-load phasor diagram for an inductive circuit assuming the volt drop in the windings is negligible
- describe transformer construction
- derive the equivalent resistance, reactance and impedance referred to the primary of a transformer
- understand voltage regulation

Electrical Circuit Theory and Technology. 978-1-138-67349-6, © 2017 John Bird. Published by Taylor & Francis. All rights reserved.

- describe losses in transformers and calculate efficiency
- appreciate the concept of resistance matching and how it may be achieved
- perform calculations using $R_1 = \left(\frac{N_1}{N_2}\right)^2 R_L$
- describe an auto transformer, its advantages/disadvantages and uses
- describe an isolating transformer, stating uses
- describe a three-phase transformer
- describe current and voltage transformers

23.1 Introduction

A transformer is a device which uses the phenomenon of mutual induction (see Chapter 11) to change the values of alternating voltages and currents. In fact, one of the main advantages of a.c. transmission and distribution is the ease with which an alternating voltage can be increased or decreased by transformers.

Losses in transformers are generally low and thus efficiency is high. Being static, they have a long life and are very stable.

Transformers range in size from the miniature units used in electronic applications to the large power transformers used in power stations. The principle of operation is the same for each.

A transformer is represented in Figure 23.1(a) as consisting of two electrical circuits linked by a common ferromagnetic core. One coil is termed the **primary winding**, which is connected to the supply of electricity, and the other the **secondary winding**, which may be connected to a load. A circuit diagram symbol for a transformer is shown in Figure 23.1(b).

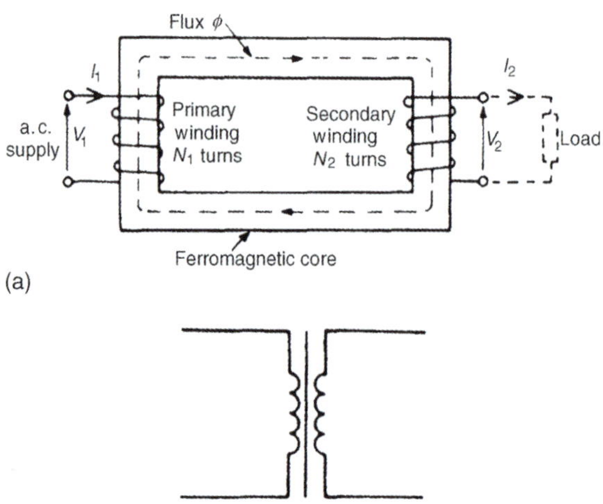

Figure 23.1

Some typical practical transformers are shown in Figure 23.2.

Figure 23.2

23.2 Transformer principle of operation

When the secondary is an open-circuit and an alternating voltage V_1 is applied to the primary winding, a small current – called the no-load current I_0 – flows, which sets up a magnetic flux in the core. This alternating flux links with both primary and secondary coils and induces in them e.m.f.s of E_1 and E_2, respectively, by mutual induction.

The induced e.m.f. E in a coil of N turns is given by

$$E = -N\frac{d\Phi}{dt} \text{ volts,}$$

where $d\Phi/dt$ is the rate of change of flux. In an ideal transformer, the rate of change of flux is the same for both primary and secondary and thus $E_1/N_1 = E_2/N_2$, i.e. **the induced e.m.f. per turn is constant**.

Assuming no losses, $E_1 = V_1$ and $E_2 = V_2$. Hence

$$\frac{V_1}{N_1} = \frac{V_2}{N_2} \quad \text{or} \quad \frac{V_1}{V_2} = \frac{N_1}{N_2} \qquad (1)$$

V_1/V_2 is called the **voltage ratio** and N_1/N_2 the **turns ratio**, or the **'transformation ratio'** of the transformer. If N_2 is less than N_1 then V_2 is less than V_1 and the device is termed a **step-down transformer**. If N_2 is greater then N_1 then V_2 is greater than V_1 and the device is termed a **step-up transformer**.

When a load is connected across the secondary winding, a current I_2 flows. In an ideal transformer losses are neglected and a transformer is considered to be 100% efficient.

Hence input power = output power, or $V_1 I_1 = V_2 I_2$, i.e. in an ideal transformer, the **primary and secondary volt-amperes are equal**.

Thus $$\frac{V_1}{V_2} = \frac{I_2}{I_1} \qquad (2)$$

Combining equations (1) and (2) gives:

$$\frac{V_1}{V_2} = \frac{N_1}{N_2} = \frac{I_2}{I_1} \qquad (3)$$

The **rating** of a transformer is stated in terms of the volt-amperes that it can transform without overheating. With reference to Figure 23.1(a), the transformer rating is either $V_1 I_1$ or $V_2 I_2$, where I_2 is the full load secondary current.

Problem 1. A transformer has 500 primary turns and 3000 secondary turns. If the primary voltage is 240 V, determine the secondary voltage, assuming an ideal transformer.

For an ideal transformer, voltage ratio = turns ratio, i.e.

$$\frac{V_1}{V_2} = \frac{N_1}{N_2}, \text{ hence } \frac{240}{V_2} = \frac{500}{3000}$$

Thus secondary voltage $V_2 = \dfrac{(3000)(240)}{(500)}$

$= \mathbf{1440\ V}$ or $\mathbf{1.44\ kV}$

Problem 2. An ideal transformer with a turns ratio of 2:7 is fed from a 240 V supply. Determine its output voltage.

A turns ratio of 2:7 means that the transformer has 2 turns on the primary for every 7 turns on the secondary (i.e. a step-up transformer). Thus,

$$\frac{N_1}{N_2} = \frac{2}{7}$$

For an ideal transformer, $\dfrac{N_1}{N_2} = \dfrac{V_1}{V_2}$; hence $\dfrac{2}{7} = \dfrac{240}{V_2}$

Thus the secondary voltage $V_2 = \dfrac{(240)(7)}{(2)} = \mathbf{840\ V}$

Problem 3. An ideal transformer has a turns ratio of 8:1 and the primary current is 3 A when it is supplied at 240 V. Calculate the secondary voltage and current.

A turns ratio of 8:1 means $\dfrac{N_1}{N_2} = \dfrac{8}{1}$, i.e. a step-down transformer.

$\dfrac{N_1}{N_2} = \dfrac{V_1}{V_2}$ or secondary voltage $V_2 = V_1\left(\dfrac{N_2}{N_1}\right)$

$= 240\left(\dfrac{1}{8}\right)$

$= \mathbf{30\ volts}$

Also, $\dfrac{N_1}{N_2} = \dfrac{I_2}{I_1}$; hence secondary current $I_2 = I_1\left(\dfrac{N_1}{N_2}\right)$

$= 3\left(\dfrac{8}{1}\right)$

$= \mathbf{24\ A}$

Problem 4. An ideal transformer, connected to a 240 V mains, supplies a 12 V, 150 W lamp. Calculate the transformer turns ratio and the current taken from the supply.

$V_1 = 240$ V, $V_2 = 12$ V, $I_2 = \dfrac{P}{V_2} = \dfrac{150}{12} = 12.5$ A

Turns ratio $= \dfrac{N_1}{N_2} = \dfrac{V_1}{V_2} = \dfrac{240}{12} = \mathbf{20}$

$\dfrac{V_1}{V_2} = \dfrac{I_2}{I_1}$, from which, $I_1 = I_2\left(\dfrac{V_2}{V_1}\right) = 12.5\left(\dfrac{12}{240}\right)$

Hence current taken from the supply,

$$I_1 = \frac{12.5}{20} = \mathbf{0.625\ A}$$

Problem 5. A 5 kVA single-phase transformer has a turns ratio of 10:1 and is fed from a 2.5 kV supply. Neglecting losses, determine (a) the full load secondary current, (b) the minimum load resistance which can be connected across the secondary winding to give full load kVA and (c) the primary current at full load kVA.

(a) $\frac{N_1}{N_2} = \frac{10}{1}$ and $V_1 = 2.5\,\text{kV} = 2500\,\text{V}$

Since $\frac{N_1}{N_2} = \frac{V_1}{V_2}$, secondary voltage

$$V_2 = V_1\left(\frac{N_2}{N_1}\right) = 2500\left(\frac{1}{10}\right) = 250\,\text{V}$$

The transformer rating in volt-amperes $= V_2 I_2$ (at full load), i.e. $5000 = 250 I_2$

Hence full load secondary current $I_2 = \frac{5000}{250}$

$$= \mathbf{20A}$$

(b) Minimum value of load resistance, $R_L = \frac{V_2}{I_2}$

$$= \frac{250}{20}$$

$$= \mathbf{12.5\,\Omega}$$

(c) $\frac{N_1}{N_2} = \frac{I_2}{I_1}$, from which primary current,

$$I_1 = I_2\left(\frac{N_2}{N_1}\right)$$

$$= 20\left(\frac{1}{10}\right)$$

$$= \mathbf{2\,A}$$

Now try the following Practice Exercise

Practice Exercise 95 Transformer principle of operation (Answers on page 824)

1. A transformer has 600 primary turns connected to a 1.5 kV supply. Determine the number of secondary turns for a 240 V output voltage, assuming no losses.
2. An ideal transformer with a turns ratio of 2:9 is fed from a 220 V supply. Determine its output voltage.
3. A transformer has 800 primary turns and 2000 secondary turns. If the primary voltage is 160 V, determine the secondary voltage, assuming an ideal transformer.
4. An ideal transformer has a turns ratio of 12:1 and is supplied at 192 V. Calculate the secondary voltage.
5. An ideal transformer has a turns ratio of 15:1 and is supplied at 180 V when the primary current is 4 A. Calculate the secondary voltage and current.
6. A step-down transformer having a turns ratio of 20:1 has a primary voltage of 4 kV and a load of 10 kW. Neglecting losses, calculate the value of the secondary current.
7. A transformer has a primary to secondary turns ratio of 1:15. Calculate the primary voltage necessary to supply a 240 V load. If the load current is 3 A, determine the primary current. Neglect any losses.
8. A 10 kVA, single-phase transformer has a turns ratio of 12:1 and is supplied from a 2.4 kV supply. Neglecting losses, determine (a) the full load secondary current, (b) the minimum value of load resistance which can be connected across the secondary winding without the kVA rating being exceeded and (c) the primary current.
9. A 20 Ω resistance is connected across the secondary winding of a single-phase power transformer whose secondary voltage is 150 V. Calculate the primary voltage and the turns ratio if the supply current is 5 A, neglecting losses.

23.3 Transformer no-load phasor diagram

(i) The core flux is common to both primary and secondary windings in a transformer and is thus taken as the reference phasor in a phasor diagram. On no-load the primary winding takes a small no-load current I_0 and since, with losses neglected, the primary winding is a pure inductor, this current lags the applied voltage V_1 by 90°. In the phasor diagram, assuming no losses, shown in Figure 23.3(a), current I_0 produces the flux and is drawn in phase with the flux. The primary induced e.m.f. E_1 is in phase opposition to V_1 (by Lenz's law) and is shown 180° out of phase with V_1 and equal in magnitude. The secondary induced e.m.f. is shown for a 2:1 turns ratio transformer.

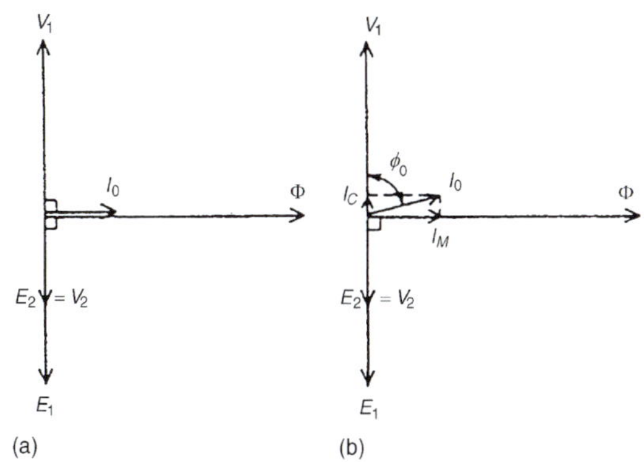

Figure 23.3

(ii) A no-load phasor diagram for a practical transformer is shown in Figure 23.3(b). If current flows then losses will occur. When losses are considered then the no-load current I_0 is the phasor sum of two components – (a) $\boldsymbol{I_M}$**, the magnetizing component**, in phase with the flux, and (b) $\boldsymbol{I_C}$**, the core loss component** (supplying the hysteresis and eddy current losses). From Figure 23.3(b):

No-load current, $\boldsymbol{I_0 = \sqrt{(I_M^2 + I_C^2)}}$, where

$$\boldsymbol{I_M = I_0 \sin \phi_0}$$

and $$\boldsymbol{I_C = I_0 \cos \phi_0}$$

Power factor on no-load $= \cos \phi_0 = \dfrac{I_C}{I_0}$

The total core losses (i.e. iron losses) $= V_1 I_0 \cos \phi_0$

Problem 6. A 2400 V/400 V single-phase transformer takes a no-load current of 0.5 A and the core loss is 400 W. Determine the values of the magnetizing and core loss components of the no-load current. Draw to scale the no-load phasor diagram for the transformer.

$V_1 = 2400\,\text{V},\ V_2 = 400\,\text{V},\ I_0 = 0.5\,\text{A}$

Core loss (i.e. iron loss) $= 400 = V_1 I_0 \cos\phi_0$

i.e. $400 = (2400)(0.5)\cos\phi_0$

$$\text{Hence } \cos\phi_0 = \frac{400}{(2400)(0.5)} = 0.3333$$

$$\phi_0 = \cos^{-1} 0.3333 = 70.53^\circ$$

The no-load phasor diagram is shown in Figure 23.4.

Magnetizing component, $\boldsymbol{I_M} = I_0 \sin\phi_0$
$= 0.5 \sin 70.53^\circ$
$= \mathbf{0.471\ A}$

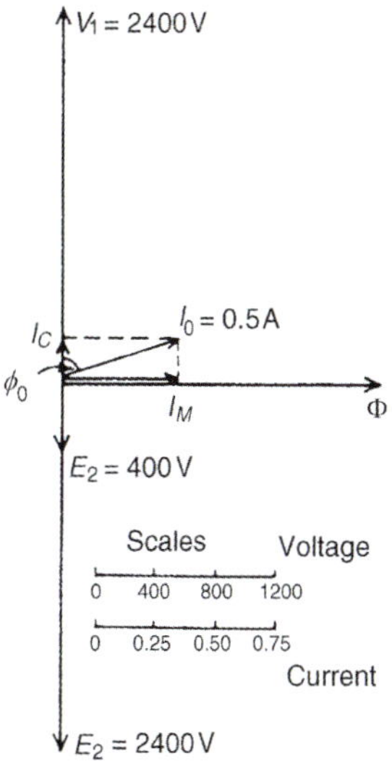

Figure 23.4

Core loss component, $\boldsymbol{I_C} = I_0 \cos\phi_0 = 0.5 \cos 70.53^\circ$
$= \mathbf{0.167\ A}$

Problem 7. A transformer takes a current of 0.8 A when its primary is connected to a 240 volt, 50 Hz supply, the secondary being on open circuit. If the power absorbed is 72 watts, determine (a) the iron loss current, (b) the power factor on no-load and (c) the magnetizing current.

$I_0 = 0.8\,\text{A},\ V_1 = 240\,\text{V}$

(a) Power absorbed = total core loss = 72
$= V_1 I_0 \cos\phi_0$

Hence $72 = 240\, I_0 \cos\phi_0$

and iron loss current, $I_C = I_0 \cos\phi_0 = \dfrac{72}{240}$
$= \mathbf{0.30\ A}$

(b) Power factor at no load, $\cos\phi_0 = \dfrac{I_C}{I_0} = \dfrac{0.30}{0.80}$
$= \mathbf{0.375}$

(c) From the right-angled triangle in Figure 23.3(b) and using Pythagoras' theorem, $I_0^2 = I_C^2 + I_M^2$, from which magnetizing current,

$$I_M = \sqrt{(I_0^2 - I_C^2)} = \sqrt{(0.80^2 - 0.30^2)}$$
$$= \mathbf{0.74\ A}$$

Now try the following Practice Exercise

Practice Exercise 96 The no-load phasor diagram (Answers on page 824)

1. (a) Draw the phasor diagram for an ideal transformer on no-load.
   (b) A 500 V/100 V, single-phase transformer takes a full load primary current of 4 A. Neglecting losses, determine (i) the full load secondary current, and (ii) the rating of the transformer.
2. A 3300 V/440 V, single-phase transformer takes a no-load current of 0.8 A and the iron loss is 500 W. Draw the no-load phasor diagram and determine the values of the magnetizing and core loss components of the no-load current.
3. A transformer takes a current of 1 A when its primary is connected to a 300 V, 50 Hz supply, the secondary being on open-circuit. If the power absorbed is 120 watts, calculate (a) the iron loss current, (b) the power factor on no-load and (c) the magnetizing current.

23.4 E.m.f. equation of a transformer

The magnetic flux Φ set up in the core of a transformer when an alternating voltage is applied to its primary winding is also alternating and is sinusoidal.

Let Φ_m be the maximum value of the flux and f be the frequency of the supply. The time for 1 cycle of the alternating flux is the periodic time T, where $T=1/f$ seconds.

The flux rises sinusoidally from zero to its maximum value in $\frac{1}{4}$ cycle, and the time for $\frac{1}{4}$ cycle is $1/4f$ seconds.

Hence the average rate of change of flux $=\dfrac{\Phi_m}{(1/4f)}=4f\Phi_m$ Wb/s, and since 1 Wb/s = 1 volt, the average e.m.f. induced in each turn $=4f\Phi_m$ volts.

As the flux Φ varies sinusoidally, then a sinusoidal e.m.f. will be induced in each turn of both primary and secondary windings.

For a sine wave, form factor $=\dfrac{\text{r.m.s. value}}{\text{average value}}$ = 1.11 (see Chapter 16)

Hence r.m.s. value = form factor × average value = 1.11 × average value

Thus r.m.s. e.m.f. induced in each turn

$$= 1.11 \times 4f\Phi_m \text{ volts}$$
$$= 4.44f\Phi_m \text{ volts}$$

Therefore, r.m.s. value of e.m.f. induced in primary,

$$\mathbf{E_1 = 4.44\, f\Phi_m N_1 \text{ volts}} \qquad (4)$$

and r.m.s. value of e.m.f. induced in secondary,

$$\mathbf{E_2 = 4.44\, f\Phi_m N_2 \text{ volts}} \qquad (5)$$

Dividing equation (4) by equation (5) gives:

$\dfrac{E_1}{E_2}=\dfrac{N_1}{N_2}$, as previously obtained in Section 23.2.

Problem 8. A 100 kVA, 4000 V/200 V, 50 Hz single-phase transformer has 100 secondary turns. Determine (a) the primary and secondary current, (b) the number of primary turns and (c) the maximum value of the flux.

$V_1=4000$ V, $V_2=200$ V, $f=50$ Hz, $N_2=100$ turns

(a) Transformer rating $=V_1I_1=V_2I_2=100\,000$ VA

Hence primary current, $I_1=\dfrac{100\,000}{V_1}=\dfrac{100\,000}{4000}$ $=\mathbf{25\ A}$

and secondary current, $I_2=\dfrac{100\,000}{V_2}=\dfrac{100\,000}{200}$ $=\mathbf{500\ A}$

(b) From equation (3), $\dfrac{V_1}{V_2}=\dfrac{N_1}{N_2}$

from which, primary turns, $N_1=\left(\dfrac{V_1}{V_2}\right)(N_2)=\left(\dfrac{4000}{200}\right)(100)$

i.e. $\mathbf{N_1=2000}$ **turns**

Part 3

(c) From equation (5), $E_2 = 4.44 f \Phi_m N_2$

from which, maximum flux Φ_m

$$= \frac{E_2}{4.44 f N_2} = \frac{200}{4.44(50)(100)}$$

(assuming $E_2 = V_2$)

$$= \mathbf{9.01 \times 10^{-3}\ Wb} \text{ or } \mathbf{9.01\ mWb}$$

[Alternatively, equation (4) could have been used, where $E_1 = 4.44 f \Phi_m N_1$,

from which $\Phi_m = \frac{E_1}{4.44 f N_1} = \frac{4000}{4.44(50)(2000)}$ (assuming $E_1 = V_1$)

$= 9.01$ mWb, as above]

Problem 9. A single-phase, 50 Hz transformer has 25 primary turns and 300 secondary turns. The cross-sectional area of the core is 300 cm^2. When the primary winding is connected to a 250 V supply, determine (a) the maximum value of the flux density in the core, and (b) the voltage induced in the secondary winding.

(a) From equation (4),

e.m.f. $E_1 = 4.44 f \Phi_m N_1$ volts i.e.

$$250 = 4.44(50)\Phi_m(25)$$

from which, maximum flux density,

$$\Phi_m = \frac{250}{(4.44)(50)(25)} \text{ Wb}$$
$$= 0.04505 \text{ Wb}$$

However, $\Phi_m = B_m \times A$, where B_m = maximum flux density in the core and A = cross-sectional area of the core (see Chapter 9).

Hence $B_m \times 300 \times 10^{-4} = 0.04505$

from which, **maximum flux density**,

$$\boldsymbol{B_m} = \frac{0.04505}{300 \times 10^{-4}} = \mathbf{1.50\ T}$$

(b) $\frac{V_1}{V_2} = \frac{N_1}{N_2}$, from which, $V_2 = V_1\left(\frac{N_2}{N_1}\right)$

i.e. voltage induced in the secondary winding,

$$V_2 = (250)\left(\frac{300}{25}\right) = \mathbf{3000\ V} \text{ or } \mathbf{3\ kV}$$

Problem 10. A single-phase 500 V/100 V, 50 Hz transformer has a maximum core flux density of 1.5 T and an effective core cross-sectional area of 50 cm^2. Determine the number of primary and secondary turns.

The e.m.f. equation for a transformer is $E = 4.44 f \Phi_m N$

and maximum flux, $\Phi_m = B \times A = (1.5)(50 \times 10^{-4})$

$$= 75 \times 10^{-4} \text{ Wb}$$

Since $E_1 = 4.44 f \Phi_m N_1$

then primary turns, $N_1 = \frac{E_1}{4.44 f \Phi_m}$

$$= \frac{500}{4.44(50)(75 \times 10^{-4})}$$
$$= \mathbf{300\ turns}$$

Since $E_2 = 4.44 f \Phi_m N_2$

then secondary turns, $N_2 = \frac{E_2}{4.44 f \Phi_m}$

$$= \frac{100}{4.44(50)(75 \times 10^{-4})}$$
$$= \mathbf{60\ turns}$$

Problem 11. A 4500 V/225 V, 50 Hz single-phase transformer is to have an approximate e.m.f. per turn of 15 V and operate with a maximum flux density of 1.4 T. Calculate (a) the number of primary and secondary turns and (b) the cross-sectional area of the core.

(a) E.m.f. per turn $= \frac{E_1}{N_1} = \frac{E_2}{N_2} = 15$

Hence primary turns, $N_1 = \frac{E_1}{15} = \frac{4500}{15} = \mathbf{300}$

and secondary turns, $N_2 = \frac{E_2}{15} = \frac{225}{15} = \mathbf{15}$

(b) E.m.f. $E_1 = 4.44 f \Phi_m N_1$

from which, $\Phi_m = \frac{E_1}{4.44 f N_1} = \frac{4500}{4.44(50)(300)}$

$$= 0.0676 \text{ Wb}$$

Now flux $\Phi_m = B_m \times A$, where A is the cross-sectional area of the core, hence

area $A = \frac{\Phi_m}{B_m} = \frac{0.0676}{1.4} = \mathbf{0.0483\ m^2}$ or $\mathbf{483\ cm^2}$

Now try the following Practice Exercise

Practice Exercise 97 The transformer e.m.f. equation (Answers on page 824)

1. A 60 kVA, 1600 V/100 V, 50 Hz, single-phase transformer has 50 secondary windings. Calculate (a) the primary and secondary current, (b) the number of primary turns and (c) the maximum value of the flux.
2. A single-phase, 50 Hz transformer has 40 primary turns and 520 secondary turns. The cross-sectional area of the core is 270 cm^2. When the primary winding is connected to a 300 volt supply, determine (a) the maximum value of flux density in the core, and (b) the voltage induced in the secondary winding.
3. A single-phase 800 V/100 V, 50 Hz transformer has a maximum core flux density of 1.294 T and an effective cross-sectional area of 60 cm^2. Calculate the number of turns on the primary and secondary windings.
4. A 3.3 kV/110 V, 50 Hz, single-phase transformer is to have an approximate e.m.f. per turn of 22 V and operate with a maximum flux of 1.25 T. Calculate (a) the number of primary and secondary turns, and (b) the cross-sectional area of the core.

23.5 Transformer on-load phasor diagram

If the voltage drop in the windings of a transformer are assumed negligible, then the terminal voltage V_2 is the same as the induced e.m.f. E_2 in the secondary. Similarly, $V_1 = E_1$. Assuming an equal number of turns on primary and secondary windings, then $E_1 = E_2$, and let the load have a lagging phase angle ϕ_2

In the phasor diagram of Figure 23.5, current I_2 lags V_2 by angle ϕ_2. When a load is connected across the secondary winding a current I_2 flows in the secondary winding. The resulting secondary e.m.f. acts so as to tend to reduce the core flux. However this does not happen since reduction of the core flux reduces E_1, hence a reflected increase in primary current I_1' occurs which provides a restoring mmf. Hence at all loads, primary and secondary mmfs are equal, but in opposition, and the core flux remains constant. I_1' is sometimes called the 'balancing' current and is equal, but in the opposite direction, to current I_2, as shown in Figure 23.5. I_0, shown at a phase angle ϕ_0 to V_1, is the no-load current of the transformer (see Section 23.3).

The phasor sum of I_1' and I_0 gives the supply current I_1, and the phase angle between V_1, and I_1 is shown as ϕ_1

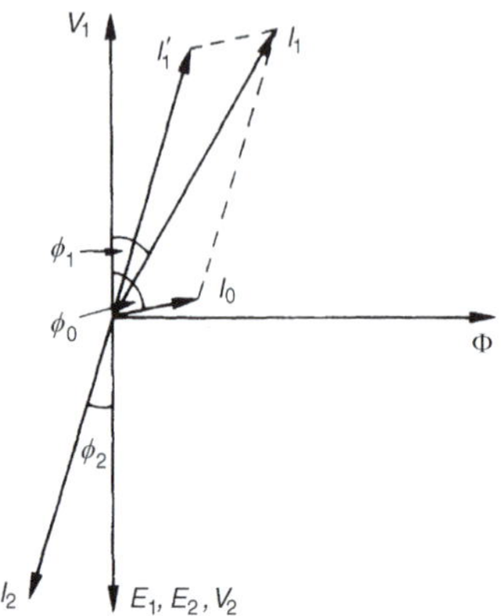

Figure 23.5

Problem 12. A single-phase transformer has 2000 turns on the primary and 800 turns on the secondary. Its no-load current is 5 A at a power factor of 0.20 lagging. Assuming the volt drop in the windings is negligible, determine the primary current and power factor when the secondary current is 100 A at a power factor of 0.85 lagging.

Let I_1' be the component of the primary current which provides the restoring mmf. Then

$$I_1' N_1 = I_2 N_2$$

i.e. $I_1'(2000) = (100)(800)$

from which $I_1' = \dfrac{(100)(800)}{2000} = 40\,\text{A}$

If the power factor of the secondary is 0.85

then $\cos\phi_2 = 0.85$, from which, $\phi_2 = \cos^{-1} 0.85 = 31.8°$

If the power factor on no-load is 0.20

then $\cos\phi_0 = 0.2$ and $\phi_0 = \cos^{-1} 0.2 = 78.5°$

In the phasor diagram shown in Figure 23.6, $I_2 = 100$ A is shown at an angle of $\phi_2 = 31.8°$ to V_2 and $I_1' = 40$ A is shown in anti-phase to I_2

The no-load current $I_0 = 5$ A is shown at an angle of $\phi_0 = 78.5°$ to V_1.

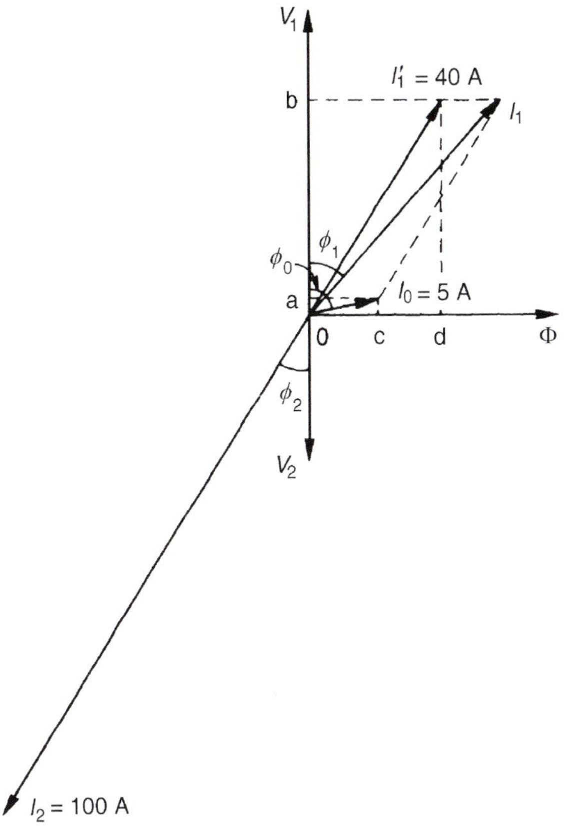

Figure 23.6

Current I_1 is the phasor sum of I_1' and I_0 and by drawing to scale, $I_1 = 44$ A and angle $\phi_1 = 37°$.

By calculation, $I_1 \cos\phi_1 = 0a + 0b$

$$= I_0 \cos\phi_0 + I_1' \cos\phi_2$$
$$= (5)(0.2) + (40)(0.85)$$
$$= 35.0\ \text{A}$$

and $I_1 \sin\phi_1 = 0c + 0d$

$$= I_0 \sin\phi_0 + I_1' \sin\phi_2$$
$$= (5)\sin 78.5° + (40)\sin 31.8°$$
$$= 25.98\ \text{A}$$

Hence the magnitude of $I_1 = \sqrt{(35.0^2 + 25.98^2)}$

$$= \mathbf{43.59\ A}$$

and $\tan\phi_1 = \left(\frac{25.98}{35.0}\right)$, from which,

$$\phi_1 = \tan^{-1}\left(\frac{25.98}{35.0}\right)$$
$$= \mathbf{36.59°}$$

Hence the power factor of the primary $= \cos\phi_1$

$$= \cos 36.59°$$
$$= \mathbf{0.80}$$

Now try the following Practice Exercise

Practice Exercise 98 The transformer on-load (Answers on page 824)

1. A single-phase transformer has 2400 turns on the primary and 600 turns on the secondary. Its no-load current is 4 A at a power factor of 0.25 lagging. Assuming the volt drop in the windings is negligible, calculate the primary current and power factor when the secondary current is 80 A at a power factor of 0.8 lagging.

23.6 Transformer construction

(i) There are broadly two types of single-phase double-wound transformer constructions – the **core type** and the **shell type**, as shown in Figure 23.7. The low- and high-voltage windings are wound as shown to reduce leakage flux.

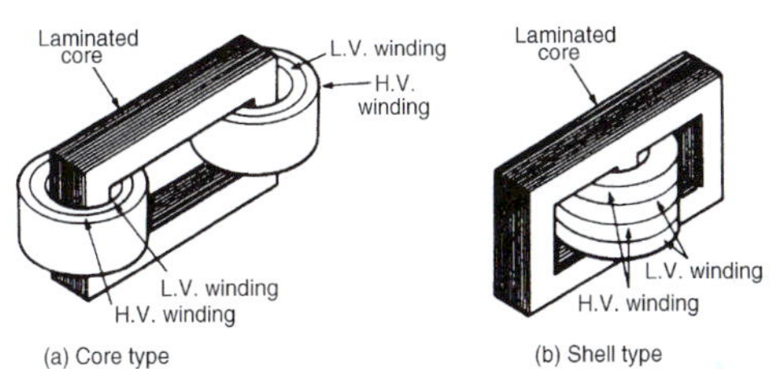

Figure 23.7

(ii) For **power transformers**, rated possibly at several MVA and operating at a frequency of 50 Hz in Great Britain, the core material used is usually laminated silicon steel or stalloy, the laminations reducing eddy currents and the silicon steel keeping hysteresis loss to a minimum.

Large power transformers are used in the main distribution system and in industrial supply circuits. Small power transformers have many applications, examples including welding and rectifier supplies, domestic bell circuits, imported washing machines, and so on.

(iii) For **audio frequency (a.f.) transformers**, rated from a few mVA to no more than 20 VA, and operating at frequencies up to about 15 kHz, the small

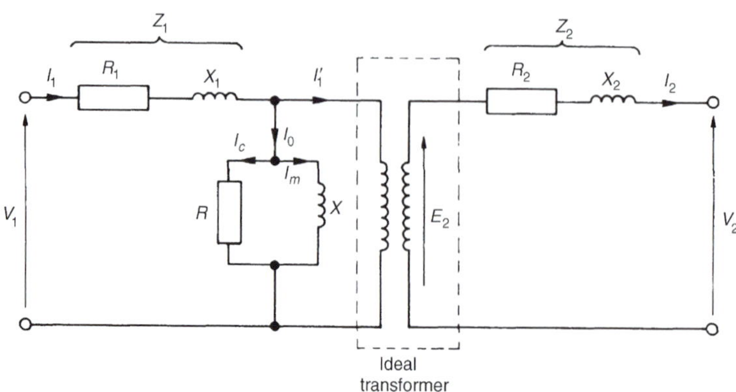

Figure 23.8

core is also made of laminated silicon steel. A typical application of a.f. transformers is in an audio amplifier system.

(iv) **Radio frequency (r.f.) transformers**, operating in the MHz frequency region, have either an air core, a ferrite core or a dust core. Ferrite is a ceramic material having magnetic properties similar to silicon steel, but having a high resistivity. Dust cores consist of fine particles of carbonyl iron or permalloy (i.e. nickel and iron), each particle of which is insulated from its neighbour. Applications of r.f. transformers are found in radio and television receivers.

(v) Transformer **windings** are usually of enamel-insulated copper or aluminium.

(vi) **Cooling** is achieved by air in small transformers and oil in large transformers.

23.7 Equivalent circuit of a transformer

Figure 23.8 shows an equivalent circuit of a transformer. R_1 and R_2 represent the resistances of the primary and secondary windings and X_1 and X_2 represent the reactances of the primary and secondary windings, due to leakage flux.

The core losses due to hysteresis and eddy currents are allowed for by resistance R which takes a current I_C, the core loss component of the primary current. Reactance X takes the magnetizing component I_M.

In a simplified equivalent circuit shown in Figure 23.9, R and X are omitted since the no-load current I_0 is normally only about 3–5% of the full load primary current.

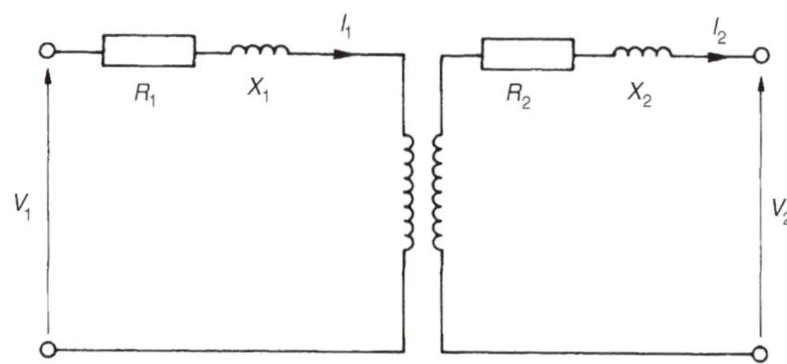

Figure 23.9

It is often convenient to assume that all of the resistance and reactance is on one side of the transformer.

Resistance R_2 in Figure 23.9 can be replaced by inserting an additional resistance R_2' in the primary circuit such that the power absorbed in R_2' when carrying the primary current is equal to that in R_2 due to the secondary current, i.e. $I_1^2 R_2' = I_2^2 R_2$ from which,

$$R_2' = R_2\left(\frac{I_2}{I_1}\right)^2 = R_2\left(\frac{V_1}{V_2}\right)^2$$

Then the total equivalent resistance in the primary circuit R_e is equal to the primary and secondary resistances of the actual transformer. Hence

$$R_e = R_1 + R_2', \text{ i.e. } \boldsymbol{R_e = R_1 + R_2\left(\frac{V_1}{V_2}\right)^2} \qquad (6)$$

By similar reasoning, the equivalent reactance in the primary circuit is given by

$$X_e = X_1 + X_2', \text{ i.e. } \boldsymbol{X_e = X_1 + X_2\left(\frac{V_1}{V_2}\right)^2} \qquad (7)$$

The equivalent impedance Z_e of the primary and secondary windings referred to the primary is given by

$$\boldsymbol{Z_e = \sqrt{(R_e^2 + X_e^2)}} \qquad (8)$$

If ϕ_e is the phase angle between I_1 and the volt drop $I_1 Z_e$ then

$$\cos\phi_e = \frac{R_e}{Z_e} \qquad (9)$$

The simplified equivalent circuit of a transformer is shown in Figure 23.10.

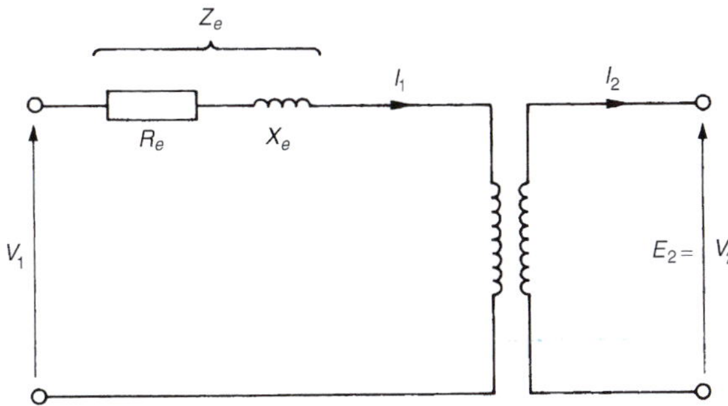

Figure 23.10

Problem 13. A transformer has 600 primary turns and 150 secondary turns. The primary and secondary resistances are 0.25 Ω and 0.01 Ω, respectively, and the corresponding leakage reactances are 1.0 Ω and 0.04 Ω, respectively. Determine (a) the equivalent resistance referred to the primary winding, (b) the equivalent reactance referred to the primary winding, (c) the equivalent impedance referred to the primary winding and (d) the phase angle of the impedance.

(a) From equation (6), equivalent resistance

$$R_e = R_1 + R_2\left(\frac{V_1}{V_2}\right)^2$$

i.e. $\boldsymbol{R_e} = 0.25 + 0.01\left(\frac{600}{150}\right)^2$ since $\frac{V_1}{V_2} = \frac{N_1}{N_2}$

$= \mathbf{0.41\,\Omega}$

(b) From equation (7), equivalent reactance,

$$X_e = X_1 + X_2\left(\frac{V_1}{V_2}\right)^2$$

i.e. $X_e = 1.0 + 0.04\left(\frac{600}{150}\right)^2 = \mathbf{1.64\,\Omega}$

(c) From equation (8), equivalent impedance,

$$Z_e = \sqrt{(R_e^2 + X_e^2)} = \sqrt{(0.41^2 + 1.64^2)} = \mathbf{1.69\,\Omega}$$

(d) From equation (9), $\cos\phi_e = \frac{R_e}{Z_e} = \frac{0.41}{1.69}$

Hence $\phi_e = \cos^{-1}\left(\frac{0.41}{1.69}\right) = \mathbf{75.96^\circ}$

Now try the following Practice Exercise

Practice Exercise 99 The equivalent circuit of a transformer (Answers on page 824)

1. A transformer has 1200 primary turns and 200 secondary turns. The primary and secondary resistances are 0.2 Ω and 0.02 Ω, respectively, and the corresponding leakage reactances are 1.2 Ω and 0.05 Ω, respectively. Calculate (a) the equivalent resistance, reactance and impedance referred to the primary winding, and (b) the phase angle of the impedance.

23.8 Regulation of a transformer

When the secondary of a transformer is loaded, the secondary terminal voltage, V_2, falls. As the power factor decreases, this voltage drop increases. This is called the **regulation of the transformer** and it is usually expressed as a percentage of the secondary no-load voltage, E_2. For full-load conditions:

$$\textbf{Regulation} = \left(\frac{E_2 - V_2}{E_2}\right) \times 100\% \qquad (10)$$

The fall in voltage, $(E_2 - V_2)$, is caused by the resistance and reactance of the windings.
Typical values of voltage regulation are about 3% in small transformers and about 1% in large transformers.

Problem 14. A 5 kVA, 200 V/400 V, single-phase transformer has a secondary terminal voltage of 387.6 volts when loaded. Determine the regulation of the transformer.

From equation (10):

$$\text{regulation} = \frac{\text{(No-load secondary voltage} - \text{terminal voltage on load)}}{\text{no-load secondary voltage}} \times 100\%$$

$$= \left[\frac{400 - 387.6}{400}\right] \times 100\%$$

$$= \left(\frac{12.4}{400}\right) \times 100\% = \mathbf{3.1\%}$$

Problem 15. The open-circuit voltage of a transformer is 240 V. A tap-changing device is set to operate when the percentage regulation drops below 2.5%. Determine the load voltage at which the mechanism operates.

$$\text{Regulation} = \frac{\text{(no load voltage} - \text{terminal load voltage)}}{\text{no load voltage}} \times 100\%$$

Hence $\quad 2.5 = \left[\frac{240 - V_2}{240}\right] 100\%$

Therefore $\frac{(2.5)(240)}{100} = 240 - V_2$

i.e. $\quad 6 = 240 - V_2$

from which, **load voltage, $V_2 = 240 - 6 = 234$ volts**

Now try the following Practice Exercise

Practice Exercise 100 Transformer regulation (Answers on page 824)

1. A 6 kVA, 100 V/500 V, single-phase transformer has a secondary terminal voltage of 487.5 volts when loaded. Determine the regulation of the transformer.

2. A transformer has an open-circuit voltage of 110 volts. A tap-changing device operates when the regulation falls below 3%. Calculate the load voltage at which the tap-changer operates.

23.9 Transformer losses and efficiency

There are broadly two sources of **losses in transformers** on load, these being copper losses and iron losses.

(a) **Copper losses** are variable and result in a heating of the conductors, due to the fact that they possess resistance. If R_1 and R_2 are the primary and secondary winding resistances then the total copper loss is $I_1^2 R_1 + I_2^2 R_2$

(b) **Iron losses** are constant for a given value of frequency and flux density and are of two types – hysteresis loss and eddy current loss.

   (i) **Hysteresis loss** is the heating of the core as a result of the internal molecular structure reversals which occur as the magnetic flux alternates. The loss is proportional to the area of the hysteresis loop and thus low loss nickel iron alloys are used for the core since their hysteresis loops have small areas. (See Chapters 9 and 41)

   (ii) **Eddy current loss** is the heating of the core due to e.m.f.s being induced not only in the transformer windings but also in the core. These induced e.m.f.s set up circulating currents, called eddy currents. Owing to the low resistance of the core, eddy currents can be quite considerable and can cause a large power loss and excessive heating of the core. Eddy current losses can be reduced by increasing the resistivity of the core material or, more usually, by laminating the core (i.e. splitting it into layers or leaves) when very thin layers of insulating material can be inserted between each pair of laminations. This increases the resistance of the eddy current path, and reduces the value of the eddy current.

$$\textbf{Transformer efficiency}, \eta = \frac{\text{output power}}{\text{input power}} = \frac{\text{input power} - \text{losses}}{\text{input power}}$$

$$\boldsymbol{\eta = 1 - \frac{\textbf{losses}}{\textbf{input power}}} \qquad (11)$$

and is usually expressed as a percentage. It is not uncommon for power transformers to have efficiencies of between 95% and 98%.

Output power $= V_2 I_2 \cos\phi_2$

total losses = copper loss + iron losses

and input power = output power + losses

Problem 16. A 200 kVA rated transformer has a full-load copper loss of 1.5 kW and an iron loss of 1 kW. Determine the transformer efficiency at full load and 0.85 power factor.

$$\text{Efficiency } \eta = \frac{\text{output power}}{\text{input power}} = \frac{\text{input power} - \text{losses}}{\text{input power}}$$

$$= 1 - \frac{\text{losses}}{\text{input power}}$$

Full-load output power $= VI\cos\phi = (200)(0.85)$

$= 170\,\text{kW}$

Total losses $= 1.5 + 1.0 = 2.5\,\text{kW}$

Input power = output power + losses $= 170 + 2.5$

$= 172.5\,\text{kW}$

$$\text{Hence efficiency} = \left(1 - \frac{2.5}{172.5}\right) = 1 - 0.01449$$

$= 0.9855$ or **98.55%**

Problem 17. Determine the efficiency of the transformer in worked Problem 16 at half full load and 0.85 power factor.

Half full-load power output $= \frac{1}{2}(200)(0.85) = 85\,\text{kW}$

Copper loss (or I^2R loss) is proportional to current squared.

Hence the copper loss at half full load is

$$\left(\tfrac{1}{2}\right)^2 (1500) = 375\,\text{W}$$

Iron loss = 1000 W (constant)

Total losses $= 375 + 1000 = 1375$ W or 1.375 kW

Input power at half full load = output power at half full load + losses

$= 85 + 1.375 = 86.375\,\text{kW}$

$$\text{Hence efficiency} = \left(1 - \frac{\text{losses}}{\text{input power}}\right) = \left(1 - \frac{1.375}{86.375}\right)$$

$= 1 - 0.01592 = 0.9841$ or **98.41%**

Problem 18. A 400 kVA transformer has a primary winding resistance of 0.5 Ω and a secondary winding resistance of 0.001 Ω. The iron loss is 2.5 kW and the primary and secondary voltages are 5 kV and 320 V, respectively. If the power factor of the load is 0.85, determine the efficiency of the transformer (a) on full load, and (b) on half load.

(a) Rating $= 400\,\text{kVA} = V_1 I_1 = V_2 I_2$

$$\text{Hence primary current, } I_1 = \frac{400 \times 10^3}{V_1} = \frac{400 \times 10^3}{5000} = 80\,\text{A}$$

$$\text{and secondary current, } I_2 = \frac{400 \times 10^3}{V_2} = \frac{400 \times 10^3}{320} = 1250\,\text{A}$$

Total copper loss $= I_1^2 R_1 + I_2^2 R_2$,

(where $R_1 = 0.5\,\Omega$ and $R_2 = 0.001\,\Omega$)

$= (80)^2(0.5) + (1250)^2(0.001)$

$= 3200 + 1562.5 = 4762.5$ watts

On full load, total loss = copper loss + iron loss

$= 4762.5 + 2500$

$= 7262.5\,\text{W} = 7.2625\,\text{kW}$

Total output power on full load $= V_2 I_2 \cos\phi_2$

$= (400 \times 10^3)(0.85)$

$= 340\,\text{kW}$

Input power = output power + losses

$= 340\,\text{kW} + 7.2625\,\text{kW}$

$= 347.2625\,\text{kW}$

$$\text{Efficiency, } \eta = \left[1 - \frac{\text{losses}}{\text{input power}}\right] \times 100\%$$

$$= \left[1 - \frac{7.2625}{347.2625}\right] \times 100\%$$

$=$ **97.91%**

(b) Since the copper loss varies as the square of the current, then total copper loss on half load $= 4762.5 \times \left(\frac{1}{2}\right)^2 = 1190.625\,\text{W}$

Hence total loss on half load $= 1190.625 + 2500$

$= 3690.625\,\text{W}$ or

$3.691\,\text{kW}$

Output power on half full load $= \frac{1}{2}(340) = 170\,\text{kW}$

Input power on half full load

$= \text{output power} + \text{losses}$

$= 170\,\text{kW} + 3.691\,\text{kW} = 173.691\,\text{kW}$

Hence efficiency at half full load,

$$\eta = \left[1 - \frac{\text{losses}}{\text{input power}}\right] \times 100\%$$

$$= \left[1 - \frac{3.691}{173.691}\right] \times 100\% = \mathbf{97.87\%}$$

Maximum efficiency

It may be shown that the efficiency of a transformer is a maximum when the variable copper loss (i.e. $I_1^2 R_1 + I_2^2 R_2$) is equal to the constant iron losses.

Problem 19. A 500 kVA transformer has a full load copper loss of 4 kW and an iron loss of 2.5 kW. Determine (a) the output kVA at which the efficiency of the transformer is a maximum, and (b) the maximum efficiency, assuming the power factor of the load is 0.75

(a) Let x be the fraction of full load kVA at which the efficiency is a maximum.

The corresponding total copper loss $= (4\,\text{kW})(x^2)$

At maximum efficiency, copper loss = iron loss. Hence

$$4x^2 = 2.5$$

from which $x^2 = \frac{2.5}{4}$ and $x = \sqrt{\left(\frac{2.5}{4}\right)} = 0.791$

Hence **the output kVA at maximum efficiency**

$= 0.791 \times 500 = \mathbf{395.5\,kVA}$

(b) Total loss at maximum efficiency $= 2 \times 2.5 = 5\,\text{kW}$

Output power $= 395.5\,\text{kVA} \times \text{p.f.} = 395.5 \times 0.75$

$= 296.625\,\text{kW}$

Input power = output power + losses

$= 296.625 + 5 = 301.625\,\text{kW}$

Maximum efficiency,

$$\eta = \left[1 - \frac{\text{losses}}{\text{input power}}\right] \times 100\%$$

$$= \left[1 - \frac{5}{301.625}\right] \times 100\%$$

$$= \mathbf{98.34\%}$$

Now try the following Practice Exercise

Practice Exercise 101 Transformer losses and efficiency (Answers on page 824)

1. A single-phase transformer has a voltage ratio of 6:1 and the h.v. winding is supplied at 540 V. The secondary winding provides a full load current of 30 A at a power factor of 0.8 lagging. Neglecting losses, find (a) the rating of the transformer, (b) the power supplied to the load, (c) the primary current.
2. A single-phase transformer is rated at 40 kVA. The transformer has full load copper losses of 800 W and iron losses of 500 W. Determine the transformer efficiency at full load and 0.8 power factor.
3. Determine the efficiency of the transformer in Problem 2 at half full load and 0.8 power factor.
4. A 100 kVA, 2000 V/400 V, 50 Hz, single-phase transformer has an iron loss of 600 W and a full load copper loss of 1600 W. Calculate its efficiency for a load of 60 kW at 0.8 power factor.
5. (a) What are eddy currents? State how their effect is reduced in transformers.
   (b) Determine the efficiency of a 15 kVA transformer for the following conditions:
      (i) full load, unity power factor
      (ii) 0.8 full load, unity power factor
      (iii) half full load, 0.8 power factor.

Assume that iron losses are 200 W and the full load copper loss is 300 W.

6. A 250 kVA transformer has a full load copper loss of 3 kW and an iron loss of 2 kW. Calculate (a) the output kVA at which the efficiency of the transformer is a maximum, and (b) the maximum efficiency, assuming the power factor of the load is 0.80

23.10 Resistance matching

Varying a load resistance to be equal, or almost equal, to the source internal resistance is called **matching**. Examples where resistance matching is important include coupling an aerial to a transmitter or receiver, or in coupling a loudspeaker to an amplifier, where coupling transformers may be used to give maximum power transfer.

With d.c. generators or secondary cells, the internal resistance is usually very small. In such cases, if an attempt is made to make the load resistance as small as the source internal resistance, overloading of the source results.

A method of achieving maximum power transfer between a source and a load (see Section 15.9, page 239), is to adjust the value of the load resistance to 'match' the source internal resistance. A transformer may be used as a **resistance matching device** by connecting it between the load and the source.

The reason why a transformer can be used for this is shown below. With reference to Figure 23.11:

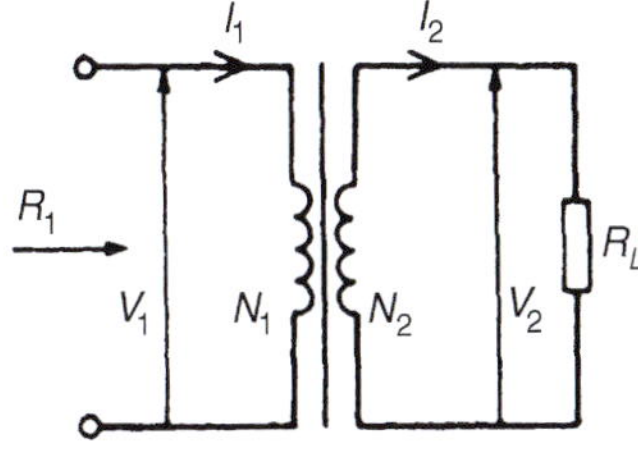

Figure 23.11

$$R_L = \frac{V_2}{I^2} \quad \text{and} \quad R_1 = \frac{V_1}{I_1}$$

For an ideal transformer, $V_1 = \left(\frac{N_1}{N_2}\right) V_2$ and

$$I_1 = \left(\frac{N_2}{N_1}\right) I_2$$

Thus the equivalent input resistance R_1 of the transformer is given by:

$$R_1 = \frac{V_1}{I_1} = \frac{\left(\frac{N_1}{N_2}\right) V_2}{\left(\frac{N_2}{N_1}\right) I_2}$$

$$= \left(\frac{N_1}{N_2}\right)^2 \left(\frac{V_2}{I_2}\right) = \left(\frac{N_1}{N_2}\right)^2 R_L$$

i.e. $$\boldsymbol{R_1 = \left(\frac{N_1}{N_2}\right)^2 R_L}$$

Hence by varying the value of the turns ratio, the equivalent input resistance of a transformer can be 'matched' to the internal resistance of a load to achieve maximum power transfer.

Problem 20. A transformer having a turns ratio of 4:1 supplies a load of resistance 100 Ω. Determine the equivalent input resistance of the transformer.

From above, the equivalent input resistance,

$$R_1 = \left(\frac{N_1}{N_2}\right)^2 R_L = \left(\frac{4}{1}\right)^2 (100) = \mathbf{1600\,\Omega}$$

Problem 21. The output stage of an amplifier has an output resistance of 112 Ω. Calculate the optimum turns ratio of a transformer which would match a load resistance of 7 Ω to the output resistance of the amplifier.

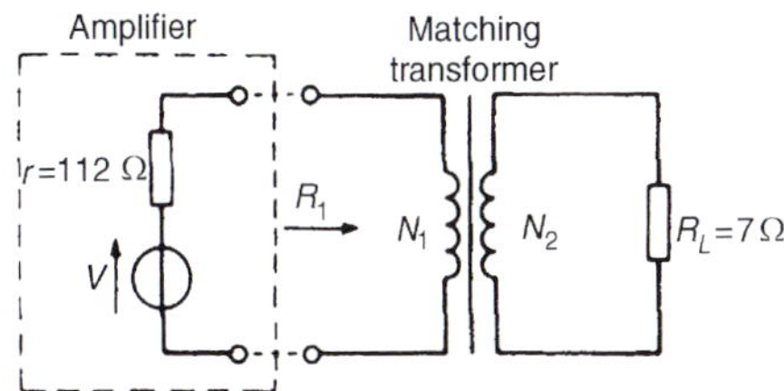

Figure 23.12

The circuit is shown in Figure 23.12. The equivalent input resistance, R_1 of the transformer needs to be 112 Ω for maximum power transfer.

Part 3

$$R_1 = \left(\frac{N_1}{N_2}\right)^2 R_L$$

Hence $\left(\frac{N_1}{N_2}\right)^2 = \frac{R_1}{R_L} = \frac{112}{7} = 16$

i.e. $\frac{N_1}{N_2} = \sqrt{(16)} = 4$

Hence the optimum turns ratio is 4:1

Problem 22. Determine the optimum value of load resistance for maximum power transfer if the load is connected to an amplifier of output resistance 150 Ω through a transformer with a turns ratio of 5:1

The equivalent input resistance R_1 of the transformer needs to be 150 Ω for maximum power transfer.

$$R_1 = \left(\frac{N_1}{N_2}\right)^2 R_L, \text{ from which, } R_L = R_1\left(\frac{N_2}{N_1}\right)^2$$

$$= 150\left(\frac{1}{5}\right)^2 = \mathbf{6\,\Omega}$$

Problem 23. A single-phase, 220 V/1760 V ideal transformer is supplied from a 220 V source through a cable of resistance 2 Ω. If the load across the secondary winding is 1.28 kΩ, determine (a) the primary current flowing and (b) the power dissipated in the load resistor.

The circuit diagram is shown in Figure 23.13.

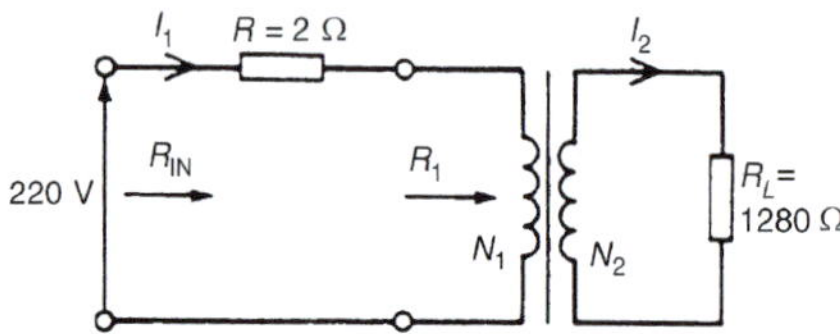

Figure 23.13

(a) Turns ratio $\frac{N_1}{N_2} = \frac{V_1}{V_2} = \frac{220}{1760} = \frac{1}{8}$

Equivalent input resistance of the transformer,

$$R_1 = \left(\frac{N_1}{N_2}\right)^2 R_L$$

$$= \left(\frac{1}{8}\right)^2 (1.28 \times 10^3) = 20\,\Omega$$

Total input resistance, $R_{IN} = R + R_1$

$$= 2 + 20 = 22\,\Omega$$

Primary current, $I_1 = \frac{V_1}{R_{IN}} = \frac{220}{22} = \mathbf{10\,A}$

(b) For an ideal transformer $\frac{V_1}{V_2} = \frac{I_2}{I_1}$, from which

$$I_2 = I_1\left(\frac{V_1}{V_2}\right)$$

$$= 10\left(\frac{220}{1760}\right) = 1.25\,\text{A}$$

Power dissipated in load resistor R_L,

$$P = I_2^2 R_L = (1.25)^2 (1.28 \times 10^3)$$

$$= \textbf{2000 watts or 2 kW}$$

Problem 24. An a.c. source of 24 V and internal resistance 15 kΩ is matched to a load by a 25:1 ideal transformer. Determine (a) the value of the load resistance and (b) the power dissipated in the load.

The circuit diagram is shown in Figure 23.14.

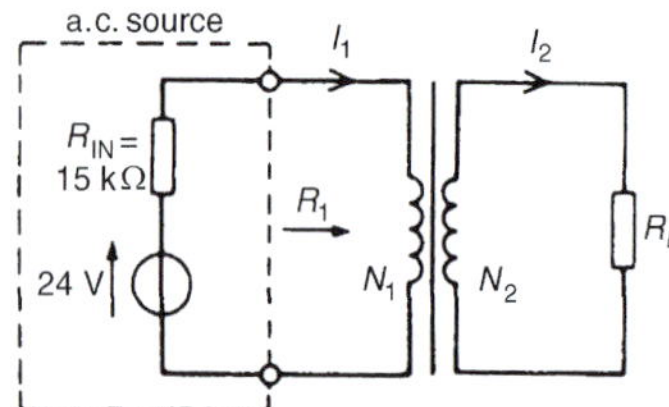

Figure 23.14

(a) For maximum power transfer R_1 needs to be equal to 15 kΩ

$$R_1 = \left(\frac{N_1}{N_2}\right)^2 R_L, \text{ from which load resistance,}$$

$$R_L = R_1\left(\frac{N_2}{N_1}\right)^2$$

$$= (15000)\left(\frac{1}{25}\right)^2 = \mathbf{24\,\Omega}$$

(b) The total input resistance when the source is connected to the matching transformer is $R_{IN} + R_1$ i.e. 15 kΩ + 15 kΩ = 30 kΩ

$$\text{Primary current, } I_1 = \frac{V}{30\,000} = \frac{24}{30\,000} = 0.8\,\text{mA}$$

$$\frac{N_1}{N_2} = \frac{I_2}{I_1}, \text{ from which, } I_2 = I_1\left(\frac{N_1}{N_2}\right)$$

$$= (0.8 \times 10^{-3})\left(\frac{25}{1}\right)$$

$$= 20 \times 10^{-3}\,\text{A}$$

Power dissipated in the load R_L,

$$P = I_2^2 R_L = (20 \times 10^{-3})^2(24) = 9600 \times 10^{-6}\,\text{W}$$

$$= \mathbf{9.6\,mW}$$

Now try the following Practice Exercise

Practice Exercise 102 Resistance matching (Answers on page 824)

1. A transformer having a turns ratio of 8:1 supplies a load of resistance 50 Ω. Determine the equivalent input resistance of the transformer.
2. What ratio of transformer turns is required to make a load of resistance 30 Ω appear to have a resistance of 270 Ω?
3. A single-phase, 240 V/2880 V ideal transformer is supplied from a 240 V source through a cable of resistance 3 Ω. If the load across the secondary winding is 720 Ω, determine (a) the primary current flowing and (b) the power dissipated in the load resistance.
4. A load of resistance 768 Ω is to be matched to an amplifier which has an effective output resistance of 12 Ω. Determine the turns ratio of the coupling transformer.
5. An a.c. source of 20 V and internal resistance 20 kΩ is matched to a load by a 16:1 single-phase transformer. Determine (a) the value of the load resistance and (b) the power dissipated in the load.

23.11 Auto transformers

An auto transformer is a transformer which has part of its winding common to the primary and secondary circuits. Figure 23.15(a) shows the circuit for a double-wound transformer and Figure 23.15(b) that for an auto transformer. The latter shows that the secondary is actually part of the primary, the current in the secondary being $(I_2 - I_1)$. Since the current is less in this section, the cross-sectional area of the winding can be reduced, which reduces the amount of material necessary.

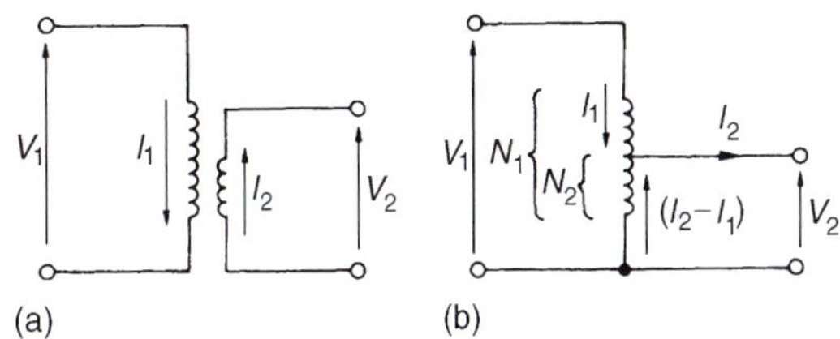

Figure 23.15

Figure 23.16 shows the circuit diagram symbol for an auto transformer.

Figure 23.16

Problem 25. A single-phase auto transformer has a voltage ratio 320 V:250 V and supplies a load of 20 kVA at 250 V. Assuming an ideal transformer, determine the current in each section of the winding.

$$\text{Rating} = 20\,\text{kVA} = V_1 I_1 = V_2 I_2$$

$$\text{Hence primary current, } I_1 = \frac{20 \times 10^3}{V_1} = \frac{20 \times 10^3}{320}$$

$$= \mathbf{62.5\,A}$$

$$\text{and secondary current, } I_2 = \frac{20 \times 10^3}{V_2} = \frac{20 \times 10^3}{250}$$

$$= \mathbf{80\,A}$$

Hence current in common part of the winding

$$= 80 - 62.5 = \mathbf{17.5\,A}$$

The current flowing in each section of the transformer is shown in Figure 23.17.

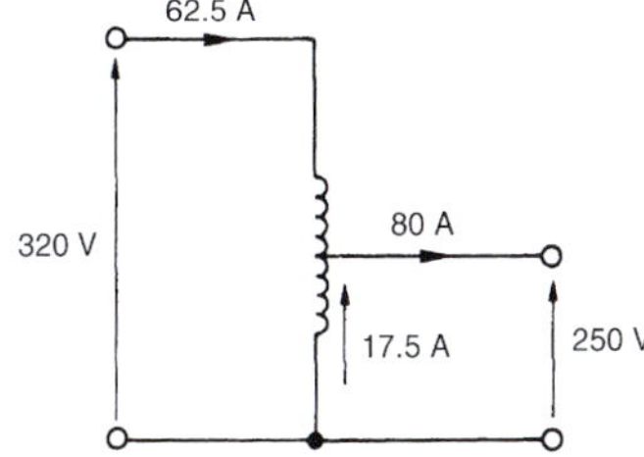

Figure 23.17

Saving of copper in an auto transformer

For the same output and voltage ratio, the auto transformer requires less copper than an ordinary double-wound transformer. This is explained below.

The volume, and hence weight, of copper required in a winding is proportional to the number of turns and to the cross-sectional area of the wire. In turn, this is proportional to the current to be carried, i.e. volume of copper is proportional to NI

Volume of copper in an auto transformer

$$\propto (N_1 - N_2)I_1 + N_2(I_2 - I_1) \qquad \text{see Figure 23.15(b)}$$

$$\propto N_1I_1 - N_2I_1 + N_2I_2 - N_2I_1$$

$$\propto N_1I_1 + N_2I_2 - 2N_2I_1$$

$$\propto 2N_1I_1 - 2N_2I_1$$

(since $N_2I_2 = N_1I_1$)

Volume of copper in a double-wound transformer $\propto N_1I_1 + N_2I_2$
$\propto 2N_1I_1$

(again, since $N_2I_2 = N_1I_1$)

$$\text{Hence } \frac{\text{volume of copper in an auto transformer}}{\text{volume of copper in a double-wound transformer}} = \frac{2N_1I_1 - 2N_2I_1}{2N_1I_1} = \frac{2N_1I_1}{2N_1I_1} - \frac{2N_2I_1}{2N_1I_1} = 1 - \frac{N_2}{N_1}$$

If $\frac{N_2}{N_1} = x$ then

(volume of copper in auto transformer)

$=(1-x)$ (volume of copper in a double-wound transformer) (12)

If, say, $x = \frac{4}{5}$ then

(volume of copper in auto transformer)

$= \left(1 - \frac{4}{5}\right)$ (volume of copper in a double-wound transformer)

$= \frac{1}{5}$(volume in double-wound transfomer)

i.e. a saving of 80%

Similarly, if $x = \frac{1}{4}$, the saving is 25%, and so on.

The closer N_2 is to N_1, the greater the saving in copper.

Problem 26. Determine the saving in the volume of copper used in an auto transformer compared with a double-wound transformer for (a) a 200 V:150 V transformer, and (b) a 500 V:100 V transformer.

(a) For a 200 V:150 V transformer, $x = \frac{V_2}{V_1} = \frac{150}{200} = 0.75$

Hence from equation (12), (volume of copper in auto transformer)

$=(1-0.75)$ (volume of copper in double-wound transformer)

$=(0.25)$ (volume of copper in double-wound transformer)

$=25\%$ of copper in a double-wound transformer

Hence the saving is 75%

(b) For a 500 V:100 V transformer, $x = \frac{V_2}{V_1} = \frac{100}{500} = 0.2$

Hence (volume of copper in auto transformer)

$=(1-0.2)$ (volume of copper in double-wound transformer)

$=(0.8)$ (volume in double-wound transformer)

$=80\%$ of copper in a double-wound transformer

Hence the saving is 20%

Now try the following Practice Exercise

Practice Exercise 103 Auto transformers (Answers on page 824)

1. A single-phase auto transformer has a voltage ratio of 480 V:300 V and supplies a load of 30 kVA at 300 V. Assuming an ideal transformer, calculate the current in each section of the winding.

Part 3

2. Calculate the saving in the volume of copper used in an auto transformer compared with a double-wound transformer for (a) a 300 V:240 V transformer, and (b) a 400 V:100 V transformer.

Advantages of auto transformers

The advantages of auto transformers over double-wound transformers include:

1. a saving in cost since less copper is needed (see above)
2. less volume, hence less weight
3. a higher efficiency, resulting from lower I^2R losses
4. a continuously variable output voltage is achievable if a sliding contact is used
5. a smaller percentage voltage regulation.

Disadvantages of auto transformers

The primary and secondary windings are not electrically separate, hence if an open-circuit occurs in the secondary winding the full primary voltage appears across the secondary.

Uses of auto transformers

Auto transformers are used for reducing the voltage when starting induction motors (see Chapter 25) and for interconnecting systems that are operating at approximately the same voltage.

23.12 Isolating transformers

Transformers not only enable current or voltage to be transformed to some different magnitude, but provide a means of isolating electrically one part of a circuit from another when there is no electrical connection between primary and secondary windings. An **isolating transformer** is a 1:1 ratio transformer with several important applications, including bathroom shaver-sockets, portable electric tools, model railways and so on.

23.13 Three-phase transformers

Three-phase, double-wound transformers are mainly used in power transmission and are usually of the core type. They basically consist of three pairs of single-phase windings mounted on one core, as shown in Figure 23.18, which gives a considerable saving in the amount of iron used. The primary and secondary windings in Figure 23.18 are wound on top of each other in the form of concentric cylinders, similar to that shown in Figure 23.7(a). The windings may be with the primary delta-connected and the secondary star-connected, or star–delta, star–star or delta–delta, depending on its use.

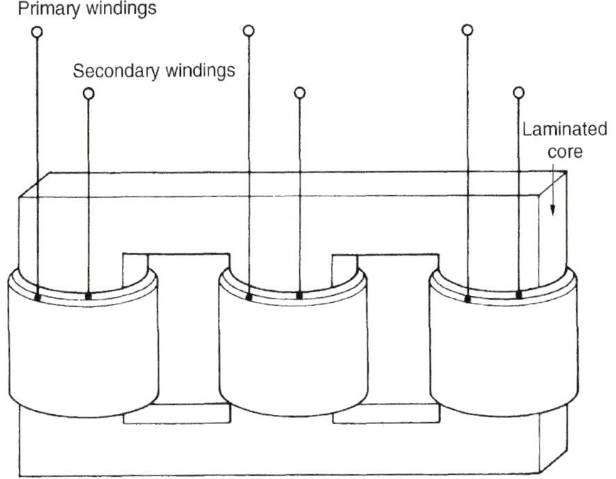

Figure 23.18

A delta connection is shown in Figure 23.19(a) and a star connection in Figure 23.19(b).

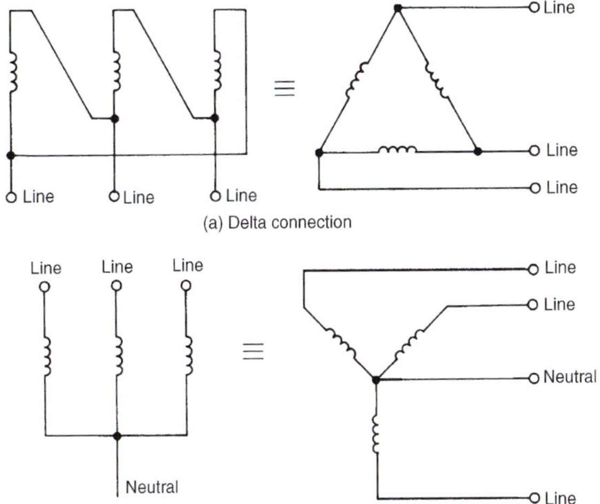

Figure 23.19

Problem 27. A three-phase transformer has 500 primary turns and 50 secondary turns. If the supply voltage is 2.4 kV, find the secondary line voltage on no-load when the windings are connected (a) star–delta, (b) delta–star.

(a) For a star connection, $V_L = \sqrt{3} V_p$ (see Chapter 22).

Primary phase voltage, $V_{p1} = \frac{V_{L1}}{\sqrt{3}} = \frac{2400}{\sqrt{3}}$
$= 1385.64$ volts

For a delta connection, $V_L = V_p$

$\frac{N_1}{N_2} = \frac{V_1}{V_2}$, from which,

secondary phase voltage, $V_{p2} = V_{p1}\left(\frac{N_2}{N_1}\right)$
$= (1385.64)\left(\frac{50}{500}\right)$
$=$ **138.6 volts**

(b) For a delta connection, $V_L = V_p$
hence primary phase voltage $V_{p1} = 2.4$ kV
$= 2400$ volts

Secondary phase voltage, $V_{p2} = V_{p1}\left(\frac{N_2}{N_1}\right)$
$= (2400)\left(\frac{50}{500}\right)$
$= 240$ V

For a star connection, $V_L = \sqrt{3} V_p$

hence the secondary line voltage $= \sqrt{3}(240)$
$=$ **416 volts**

Now try the following Practice Exercise

Practice Exercise 104 Three-phase transformer (Answers on page 824)

1. A three-phase transformer has 600 primary turns and 150 secondary turns. If the supply voltage is 1.5 kV, determine the secondary line voltage on no-load when the windings are connected (a) delta–star, (b) star–delta.

23.14 Current transformers

For measuring currents in excess of about 100 A, a current transformer is normally used. With a d.c. moving-coil ammeter the current required to give full-scale deflection is very small – typically a few milliamperes. When larger currents are to be measured a shunt resistor is added to the circuit (see Chapter 12). However, even with shunt resistors added it is not possible to measure very large currents. When a.c. is being measured a shunt cannot be used since the proportion of the current which flows in the meter will depend on its impedance, which varies with frequency.

In a double-wound transformer: $\frac{I_1}{I_2} = \frac{N_2}{N_1}$

from which, **secondary current** $\boldsymbol{I_2 = I_1\left(\frac{N_1}{N_2}\right)}$

In current transformers the primary usually consists of one or two turns whilst the secondary can have several hundred turns. A typical arrangement is shown in Figure 23.20.

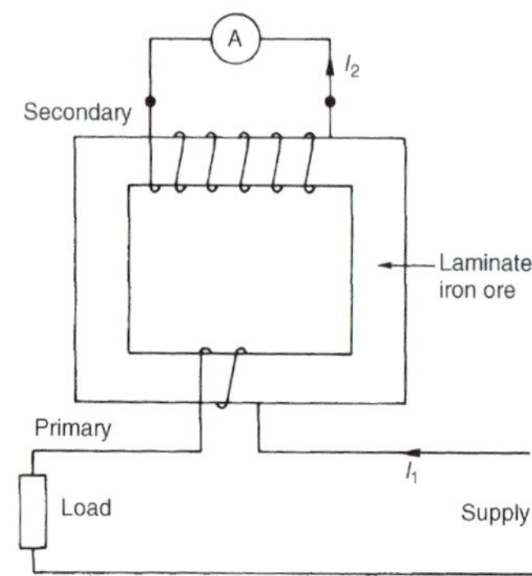

Figure 23.20

If, for example, the primary has 2 turns and the secondary 200 turns, then if the primary current is 500 A,

secondary current, $I_2 = I_1\left(\frac{N_1}{N_2}\right) = (500)\left(\frac{2}{200}\right) = 5$ A

Current transformers isolate the ammeter from the main circuit and allow the use of a standard range of ammeters giving full-scale deflections of 1 A, 2 A or 5 A.

For very large currents the transformer core can be mounted around the conductor or bus-bar. Thus the primary then has just one turn. It is very important to short-circuit the secondary winding before removing the ammeter. This is because if current is flowing in the primary, dangerously high voltages could be induced in the secondary should it be open-circuited.

Current transformer circuit diagram symbols are shown in Figure 23.21.

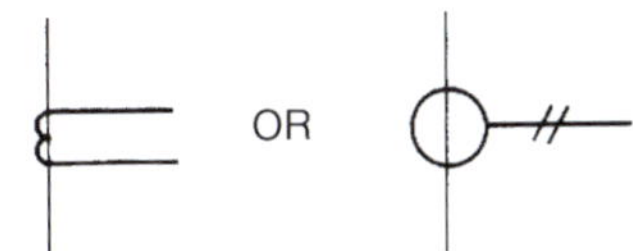

Figure 23.21

Problem 28. A current transformer has a single turn on the primary winding and a secondary winding of 60 turns. The secondary winding is connected to an ammeter with a resistance of 0.15 Ω. The resistance of the secondary winding is 0.25 Ω. If the current in the primary winding is 300 A, determine (a) the reading on the ammeter, (b) the potential difference across the ammeter and (c) the total load (in VA) on the secondary.

(a) Reading on the ammeter,

$$I_2 = I_1\left(\frac{N_1}{N_2}\right)$$

$$= 300\left(\frac{1}{60}\right) = \mathbf{5\,A}$$

(b) P.d. across the ammeter

$= I_2 R_A$, where R_A is the ammeter resistance

$= (5)(0.15) = \mathbf{0.75\ volts}$

(c) Total resistance of secondary circuit $= 0.15 + 0.25$

$= 0.40\,\Omega$

Induced e.m.f. in secondary $= (5)(0.40) = 2.0\,V$

Total load on secondary $= (2.0)(5) = \mathbf{10\,VA}$

Now try the following Practice Exercise

Practice Exercise 105 Current transformer (Answers on page 824)

1. A current transformer has two turns on the primary winding and a secondary winding of 260 turns. The secondary winding is connected to an ammeter with a resistance of 0.2 Ω. The resistance of the secondary winding is 0.3 Ω. If the current in the primary winding is 650 A, determine (a) the reading on the ammeter, (b) the potential difference across the ammeter and (c) the total load in VA on the secondary.

23.15 Voltage transformers

For measuring voltages in excess of about 500 V it is often safer to use a voltage transformer. These are normal double-wound transformers with a large number of turns on the primary, which is connected to a high voltage supply, and a small number of turns on the secondary. A typical arrangement is shown in Figure 23.22.

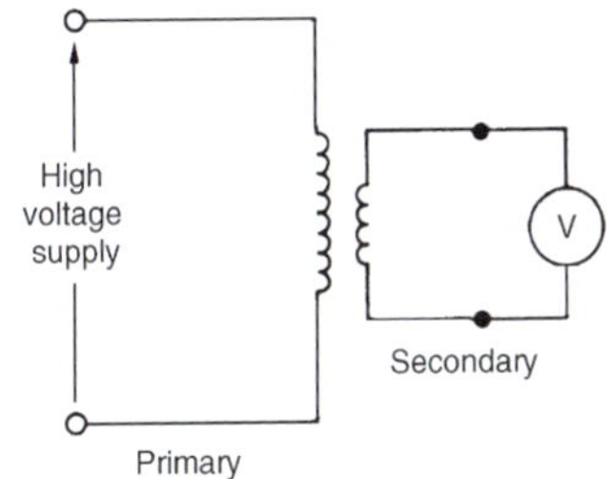

Figure 23.22

$$\text{Since } \frac{V_1}{V_2} = \frac{N_1}{N_2}$$

the **secondary voltage,** $\mathbf{V_2 = V_1\left(\frac{N_2}{N_1}\right)}$

Thus, if the arrangement in Figure 23.22 has 4000 primary turns and 20 secondary turns then for a voltage of 22 kV on the primary, the voltage on the secondary,

$$V_2 = V_1\left(\frac{N_2}{N_1}\right) = 22\,000\left(\frac{20}{4000}\right) = \mathbf{110\ volts}$$

For fully worked solutions to each of the problems in Practice Exercises 95 to 105 in this chapter, go to the website:
www.routledge.com/cw/bird

Part 3

Revision Test 6

This revision test covers the material contained in Chapters 21 to 23. *The marks for each question are shown in brackets at the end of each question.*

1. (a) Briefly explain how fossil fuels are used to generate electricity.
   (b) State four advantages of coal, oil and natural gas.
   (c) State four disadvantages of coal, oil and natural gas. (17)

2. List five advantages and five disadvantages of using nuclear power to generate electricity. (10)

3. (a) State five renewable methods used to generate electricity. (b) State for each renewable method listed in part (a), two advantages and two disadvantages. (25)

4. Three identical coils, each of resistance 40 Ω and inductive reactance 30 Ω, are connected (i) in star, and (ii) in delta to a 400 V, three-phase supply. Calculate for each connection (a) the line and phase voltages, (b) the phase and line currents and (c) the total power dissipated. (12)

5. Two wattmeters are connected to measure the input power to a balanced three-phase load by the two-wattmeter method. If the instrument readings are 10 kW and 6 kW, determine (a) the total power input and (b) the load power factor. (5)

6. An ideal transformer connected to a 250 V mains, supplies a 25 V, 200 W lamp. Calculate the transformer turns ratio and the current taken from the supply. (4)

7. A 200 kVA, 8000 V/320 V, 50 Hz single-phase transformer has 120 secondary turns. Determine (a) the primary and secondary currents, (b) the number of primary turns and (c) the maximum value of flux. (8)

8. Determine the percentage regulation of an 8 kVA, 100 V/200 V, single-phase transformer when its secondary terminal voltage is 194 V when loaded. (3)

9. A 500 kVA rated transformer has a full load copper loss of 4 kW and an iron loss of 3 kW. Determine the transformer efficiency (a) at full load and 0.80 power factor and (b) at half full load and 0.80 power factor. (10)

10. Determine the optimum value of load resistance for maximum power transfer if the load is connected to an amplifier of output resistance 288 Ω through a transformer with a turns ratio 6:1 (3)

11. A single-phase auto transformer has a voltage ratio of 250 V:200 V and supplies a load of 15 kVA at 200 V. Assuming an ideal transformer, determine the current in each section of the winding. (3)

For lecturers/instructors/teachers, fully worked solutions to each of the problems in Revision Test 6, together with a full marking scheme, are available at the website:
www.routledge.com/cw/bird

Chapter 24

D.c. machines

***Why it is important to understand:* D.c. machines**

A machine which converts d.c. electrical power into mechanical power is known as a d.c. motor. A machine which converts mechanical power into electrical power is called a d.c. generator. From a construction point of view there is no difference between a d.c. motor and generator. D.c. motors have been available for nearly 100 years; in fact the first electric motors were designed and built for operation from direct current power. D.c. motors have a wide speed range, good speed regulation, compact size and light weight (relative to mechanical variable speed), ease of control, low maintenance and low cost. The armature and field in a d.c. motor can be connected three different ways to provide varying amounts of torque or different types of speed control. The armature and field windings are designed slightly differently for different types of d.c. motors. The three basic types of d.c. motors are the series motor, the shunt motor and the compound motor. The series motor is designed to move large loads with high starting torque in applications such as a crane motor or lift hoist. The shunt motor is designed slightly differently, since it is made for applications such as pumping fluids, where constant-speed characteristics are important. The compound motor is designed with some of the series motor's characteristics and some of the shunt motor's characteristics. This allows the compound motor to be used in applications where high starting torque and controlled operating speed are both required. In this chapter, types of d.c. motor and generator are described, together with associated calculations. The motor starter and methods of speed control are also considered.

At the end of this chapter you should be able to:

- distinguish between the function of a motor and a generator
- describe the action of a commutator
- describe the construction of a d.c. machine
- distinguish between wave and lap windings
- understand shunt, series and compound windings of d.c. machines
- understand armature reaction
- calculate generated e.m.f. in an armature winding using $E = \dfrac{2p\Phi nZ}{c}$
- describe types of d.c. generator and their characteristics
- calculate generated e.m.f. for a generator using $E = V + I_a R_a$
- state typical applications of d.c. generators
- list d.c. machine losses and calculate efficiency
- calculate back e.m.f. for a d.c. motor using $E = V - I_a R_a$

Electrical Circuit Theory and Technology. 978-1-138-67349-6, © 2017 John Bird. Published by Taylor & Francis. All rights reserved.

- calculate the torque of a d.c. motor using $T = \frac{EI_a}{2\pi n}$ and $T = \frac{p\Phi ZI_a}{\pi c}$
- describe types of d.c. motor and their characteristics
- state typical applications of d.c. motors
- describe a d.c. motor starter
- describe methods of speed control of d.c. motors
- list types of enclosure for d.c. motors

24.1 Introduction

When the input to an electrical machine is electrical energy (seen as applying a voltage to the electrical terminals of the machine), and the output is mechanical energy (seen as a rotating shaft), the machine is called an electric **motor**. Thus an electric motor converts electrical energy into mechanical energy.
The principle of operation of a motor is explained in Section 10.4, page 142.
When the input to an electrical machine is mechanical energy (seen as, say, a diesel motor, coupled to the machine by a shaft), and the output is electrical energy (seen as a voltage appearing at the electrical terminals of the machine), the machine is called a **generator**. Thus, a generator converts mechanical energy to electrical energy.
The principle of operation of a generator is explained in Section 11.2, page 147.

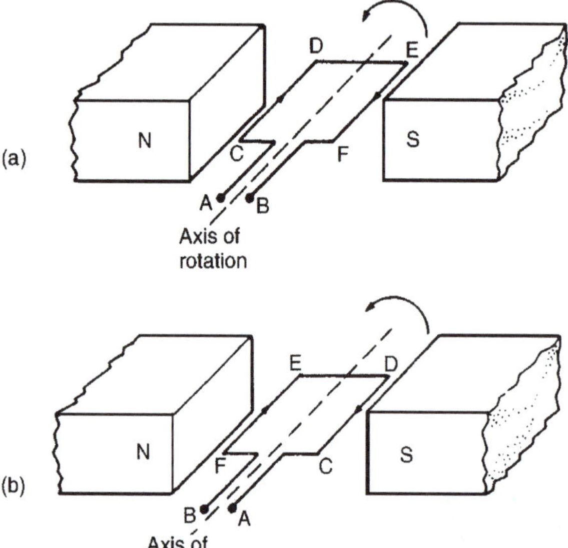

Figure 24.1

24.2 The action of a commutator

In an electric motor, conductors rotate in a uniform magnetic field. A single-loop conductor mounted between permanent magnets is shown in Figure 24.1. A voltage is applied at points A and B in Figure 24.1(a).

A force, F, acts on the loop due to the interaction of the magnetic field of the permanent magnets and the magnetic field created by the current flowing in the loop. This force is proportional to the flux density, B, the current flowing, I, and the effective length of the conductor, l, i.e. $F = BIl$. The force is made up of two parts, one acting vertically downwards due to the current flowing from C to D and the other acting vertically upwards due to the current flowing from E to F (from Fleming's left-hand rule). If the loop is free to rotate, then when it has rotated through 180°, the conductors are as shown in Figure 24.1(b). For rotation to continue in the same direction, it is necessary for the current flow to be as shown in Figure 24.1(b), i.e. from D to C and from F to E. This apparent reversal in the direction of current flow is achieved by a process called **commutation**. With reference to Figure 24.2(a), when a direct voltage is applied at A and B, then as the single-loop conductor rotates, current flow will always be away from the commutator for the part of the conductor adjacent to the N-pole and towards the commutator for the part of the conductor adjacent to the S-pole. Thus the forces act to give continuous rotation in an anticlockwise direction. The arrangement shown in Figure 24.2(a) is called a 'two-segment' commutator and the voltage is applied to the rotating segments by stationary **brushes** (usually carbon blocks), which slide on the commutator material (usually copper), when rotation takes place.

In practice, there are many conductors on the rotating part of a d.c. machine and these are attached to many

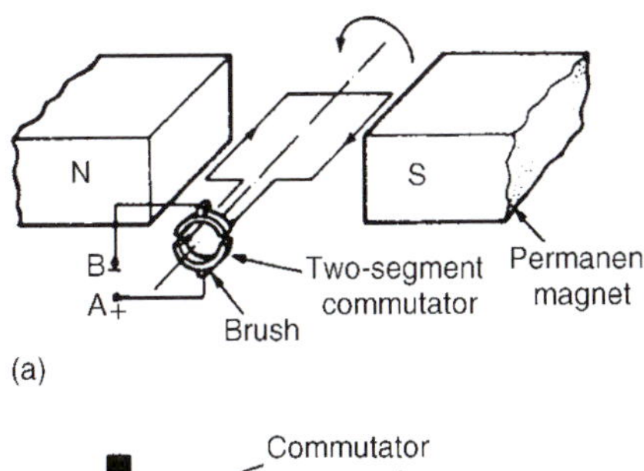

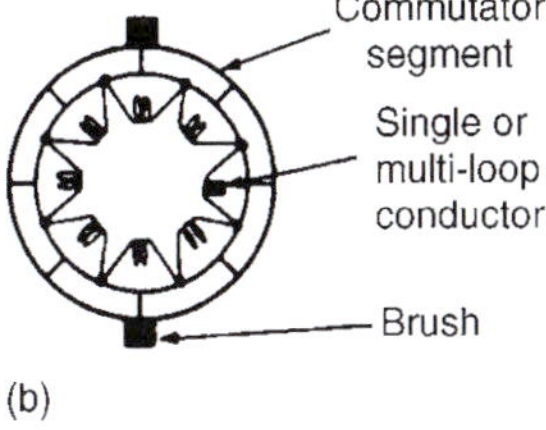

Figure 24.2

commutator segments. A schematic diagram of a multi-segment commutator is shown in Figure 24.2(b).

Poor commutation results in sparking at the trailing edge of the brushes. This can be improved by using **interpoles** (situated between each pair of main poles), high resistance brushes, or using brushes spanning several commutator segments.

24.3 D.c. machine construction

The basic parts of any d.c. machine are shown in Figure 24.3, and comprise:

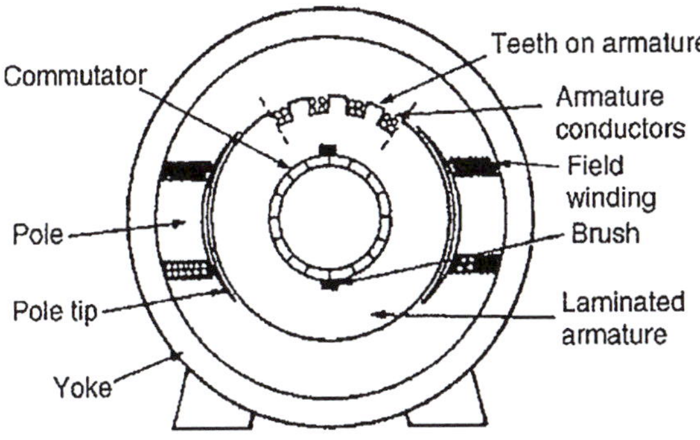

Figure 24.3

(a) A stationary part called the **stator** having,

 (i) a steel ring called the **yoke**, to which are attached

 (ii) the magnetic **poles**, around which are the

 (iii) **field windings**, i.e. many turns of a conductor wound round the pole core; current passing through this conductor creates an electromagnet (rather than the permanent magnets shown in Figures 24.1 and 24.2).

(b) A rotating part called the **armature** mounted in bearings housed in the stator and having,

 (iv) a laminated cylinder of iron or steel called the **core**, on which teeth are cut to house the

 (v) **armature winding**, i.e. a single or multi-loop conductor system and

 (vi) the **commutator** (see Section 24.2).

Armature windings can be divided into two groups, depending on how the wires are joined to the commutator. These are called **wave windings** and **lap windings**.

(a) In **wave windings** there are two paths in parallel, irrespective of the number of poles, each path supplying half the total current output. Wave-wound generators produce high-voltage, low-current outputs.

(b) In **lap windings** there are as many paths in parallel as the machine has poles. The total current output divides equally between them. Lap-wound generators produce high-current, low-voltage output.

24.4 Shunt, series and compound windings

When the field winding of a d.c. machine is connected in parallel with the armature, as shown in Figure 24.4(a), the machine is said to be **shunt** wound. If the field winding is connected in series with the armature, as shown in Figure 24.4(b), then the machine is said to be **series** wound. A **compound** wound machine has a combination of series and shunt windings.

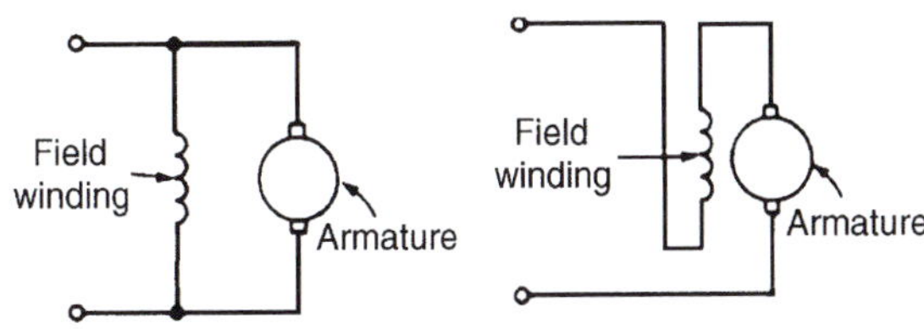

Figure 24.4

Depending on whether the electrical machine is series wound, shunt wound or compound wound, it behaves differently when a load is applied. The behaviour of a d.c. machine under various conditions is shown by means of graphs, called characteristic curves or just **characteristics**. The characteristics shown in the following sections are theoretical, since they neglect the effects of armature reaction.

Armature reaction is the effect that the magnetic field produced by the armature current has on the magnetic field produced by the field system. In a generator, armature reaction results in a reduced output voltage, and in a motor, armature reaction results in increased speed.

A way of overcoming the effect of armature reaction is to fit compensating windings, located in slots in the pole face.

24.5 E.m.f. generated in an armature winding

Let Z = number of armature conductors,

Φ = useful flux per pole, in webers

p = number of **pairs** of poles

and n = armature speed in rev/s

The e.m.f. generated by the armature is equal to the e.m.f. generated by one of the parallel paths. Each conductor passes $2p$ poles per revolution and thus cuts $2p\Phi$ webers of magnetic flux per revolution. Hence flux cut by one conductor per second $= 2p\Phi n$ Wb and so the average e.m.f. E generated per conductor is given by:

$E = 2p\Phi n$ volts (since 1 volt = 1 weber per second)

Let c = number of parallel paths through the winding between positive and negative brushes

$c = 2$ **for a wave winding**

$c = 2p$ **for a lap winding**

The number of conductors in series in each path $= \dfrac{Z}{c}$

The total e.m.f. between brushes

= (average e.m.f./conductor)(number of conductors in series per path)

$= 2p\Phi n \dfrac{Z}{c}$

i.e. **generated e.m.f.,** $\boldsymbol{E = \dfrac{2p\Phi n Z}{c}}$ **volts** (1)

Since Z, p and c are constant for a given machine, then $E \propto \Phi n$. However $2\pi n$ is the angular velocity ω in radians per second, hence the generated e.m.f. is proportional to Φ and ω, i.e.

generated e.m.f., $\boldsymbol{E \propto \Phi\,\omega}$ (2)

Problem 1. An 8-pole, wave-connected armature has 600 conductors and is driven at 625 rev/min. If the flux per pole is 20 mWb, determine the generated e.m.f.

$Z = 600$, $c = 2$ (for a wave winding), $p = 4$ pairs

$n = \dfrac{625}{60}$ rev/s, $\Phi = 20 \times 10^{-3}$ Wb

$$\text{Generated e.m.f., } E = \frac{2p\Phi nZ}{c} = \frac{2(4)(20\times10^{-3})\left(\frac{625}{60}\right)(600)}{2}$$

$= \mathbf{500\ volts}$

Problem 2. A 4-pole generator has a lap-wound armature with 50 slots with 16 conductors per slot. The useful flux per pole is 30 mWb. Determine the speed at which the machine must be driven to generate an e.m.f. of 240 V.

$E = 240$ V, $c = 2p$ (for a lap winding),
$Z = 50 \times 16 = 800$, $\Phi = 30 \times 10^{-3}$ Wb.

$$\text{Generated e.m.f. } E = \frac{2p\Phi nZ}{c} = \frac{2p\Phi nZ}{2p} = \Phi nZ$$

$$\text{Rearranging gives, speed, } n = \frac{E}{\Phi Z} = \frac{240}{(30\times10^{-3})(800)}$$

$= \mathbf{10\ rev/s}$ or $\mathbf{600\ rev/min}$

Problem 3. An 8-pole, lap-wound armature has 1200 conductors and a flux per pole of 0.03 Wb. Determine the e.m.f. generated when running at 500 rev/min.

Generated e.m.f., $E = \dfrac{2p\Phi nZ}{c} = \dfrac{2p\Phi nZ}{2p}$ for a lap-wound machine, i.e.

$E = \Phi nZ = (0.03)\left(\dfrac{500}{60}\right)(1200) = \mathbf{300\ volts}$

Problem 4. Determine the generated e.m.f. in Problem 3 if the armature is wave-wound.

Part 3

Generated e.m.f., $E = \dfrac{2p\Phi nZ}{c} = \dfrac{2p\Phi nZ}{2}$

(since $c = 2$ for wave-wound)

$= p\Phi nZ = (4)(\Phi nZ)$

$= (4)(300)$ from Problem 3,

$=$ **1200 volts**

Problem 5. A d.c. shunt-wound generator running at constant speed generates a voltage of 150 V at a certain value of field current. Determine the change in the generated voltage when the field current is reduced by 20%, assuming the flux is proportional to the field current.

The generated e.m.f. E of a generator is proportional to $\Phi\omega$, i.e. is proportional to Φn, where Φ is the flux and n is the speed of rotation.

It follows that $E = k\Phi n$, where k is a constant.

At speed n_1 and flux Φ_1, $E_1 = k\Phi_1 n_1$

At speed n_2 and flux Φ_2, $E_2 = k\Phi_2 n_2$

Thus, by division:

$$\frac{E_1}{E_2} = \frac{k\Phi_1 n_1}{k\Phi_2 n_2} = \frac{\Phi_1 n_1}{\Phi_2 n_2}$$

The initial conditions are $E_1 = 150\,\text{V}$, $\Phi = \Phi_1$ and $n = n_1$. When the flux is reduced by 20%, the new value of flux is 80/100 or 0.8 of the initial value, i.e. $\Phi_2 = 0.8\Phi_1$. Since the generator is running at constant speed, $n_2 = n_1$

Thus $\dfrac{E_1}{E_2} = \dfrac{\Phi_1 n_1}{\Phi_2 n_2} = \dfrac{\Phi_1 n_1}{0.8\Phi_1 n_1} = \dfrac{1}{0.8}$

that is, $E_2 = 150 \times 0.8 = 120\,\text{V}$

Thus, a reduction of 20% in the value of the flux **reduces the generated voltage to 120 V** at constant speed.

Problem 6. A d.c. generator running at 30 rev/s generates an e.m.f. of 200 V. Determine the percentage increase in the flux per pole required to generate 250 V at 20 rev/s.

From equation (2), generated e.m.f., $E \propto \Phi\omega$ and since $\omega = 2\pi n$, $E \propto \Phi n$

Let $E_1 = 200\,\text{V}$, $n_1 = 30$ rev/s and flux per pole at this speed be Φ_1

Let $E_2 = 250\,\text{V}$, $n_1 = 20$ rev/s and flux per pole at this speed be Φ_2

Since $E \propto \Phi n$, then $\dfrac{E_1}{E_2} = \dfrac{\Phi_1 n_1}{\Phi_2 n_2}$

Hence $\dfrac{200}{250} = \dfrac{\Phi_1(30)}{\Phi_2(20)}$

from which $\Phi_2 = \dfrac{\Phi_1(30)(250)}{(20)(200)} = 1.875\,\Phi_1$

Hence the increase in flux per pole needs to be **87.5%**

Now try the following Practice Exercise

Practice Exercise 106 Generated e.m.f. (Answers on page 824)

1. A 4-pole, wave-connected armature of a d.c. machine has 750 conductors and is driven at 720 rev/min. If the useful flux per pole is 15 mWb, determine the generated e.m.f.
2. A 6-pole generator has a lap-wound armature with 40 slots with 20 conductors per slot. The flux per pole is 25 mWb. Calculate the speed at which the machine must be driven to generate an e.m.f. of 300 V.
3. A 4-pole armature of a d.c. machine has 1000 conductors and a flux per pole of 20 mWb. Determine the e.m.f. generated when running at 600 rev/min when the armature is (a) wave-wound, (b) lap-wound.
4. A d.c. generator running at 25 rev/s generates an e.m.f. of 150 V. Determine the percentage increase in the flux per pole required to generate 180 V at 20 rev/s.

24.6 D.c. generators

D.c. generators are classified according to the method of their field excitation. These groupings are:

(i) **Separately excited generators**, where the field winding is connected to a source of supply other than the armature of its own machine.

(ii) **Self-excited generators**, where the field winding receives its supply from the armature of its own machine, and which are sub-divided into (a) shunt, (b) series and (c) compound wound generators.

24.7 Types of d.c. generator and their characteristics

(a) Separately excited generator

A typical separately excited generator circuit is shown in Figure 24.5.

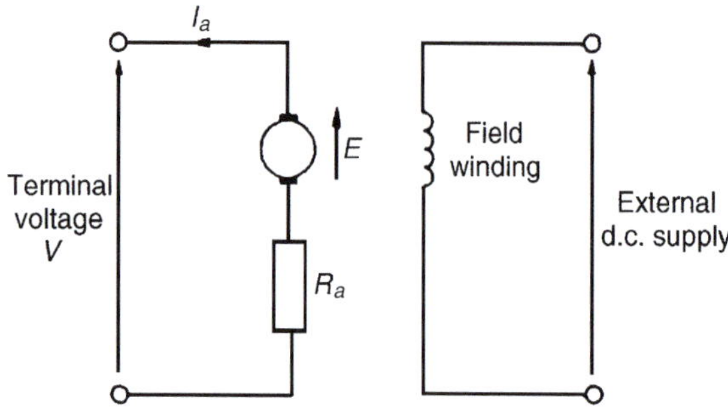

Figure 24.5

When a load is connected across the armature terminals, a load current I_a will flow. The terminal voltage V will fall from its open-circuit e.m.f. E due to a volt drop caused by current flowing through the armature resistance, shown as R_a, i.e.

terminal voltage, $V = E - I_a R_a$

or **generated e.m.f., $E = V + I_a R_a$** (3)

Problem 7. Determine the terminal voltage of a generator which develops an e.m.f. of 200 V and has an armature current of 30 A on load. Assume the armature resistance is 0.30 Ω.

With reference to Figure 24.5, terminal voltage,

$$V = E - I_a R_a = 200 - (30)(0.30) = 200 - 9$$
$$= \mathbf{191\ volts}$$

Problem 8. A generator is connected to a 60 Ω load and a current of 8 A flows. If the armature resistance is 1 Ω, determine (a) the terminal voltage and (b) the generated e.m.f.

(a) Terminal voltage, $V = I_a R_L = (8)(60) = \mathbf{480\ volts}$

(b) Generated e.m.f.,

$$E = V + I_a R_a \quad \text{from equation (3)}$$
$$= 480 + (8)(1) = 480 + 8 = \mathbf{488\ volts}$$

Problem 9. A separately excited generator develops a no-load e.m.f. of 150 V at an armature speed of 20 rev/s and a flux per pole of 0.10 Wb. Determine the generated e.m.f. when (a) the speed increases to 25 rev/s and the pole flux remains unchanged, (b) the speed remains at 20 rev/s and the pole flux is decreased to 0.08 Wb and (c) the speed increases to 24 rev/s and the pole flux is decreased to 0.07 Wb.

(a) From Section 24.5, generated e.m.f. $E \propto \Phi n$

$$\text{from which} \quad \frac{E_1}{E_2} = \frac{\Phi_1 n_1}{\Phi_2 n_2}$$

$$\text{Hence} \quad \frac{150}{E_2} = \frac{(0.10)(20)}{(0.10)(25)}$$

$$\text{from which} \quad E_2 = \frac{(150)(0.10)(25)}{(0.10)(20)}$$
$$= \mathbf{187.5\ volts}$$

(b) $$\frac{150}{E_3} = \frac{(0.10)(20)}{(0.08)(20)}$$

$$\text{from which e.m.f.,} \quad E_3 = \frac{(150)(0.08)(20)}{(0.10)(20)}$$
$$= \mathbf{120\ volts}$$

(c) $$\frac{150}{E_4} = \frac{(0.10)(20)}{(0.07)(24)}$$

$$\text{from which e.m.f.,} \quad E_4 = \frac{(150)(0.07)(24)}{(0.10)(20)}$$
$$= \mathbf{126\ volts}$$

Characteristics

The two principal generator characteristics are the generated voltage/field current characteristics, called the **open-circuit characteristic** and the terminal voltage/load current characteristic, called the **load characteristic**. A typical separately excited generator **open-circuit characteristic** is shown in Figure 24.6(a) and a typical **load characteristic** is shown in Figure 24.6(b).

A separately excited generator is used only in special cases, such as when a wide variation in terminal p.d. is required, or when exact control of the field current is necessary. Its disadvantage lies in requiring a separate source of direct current.

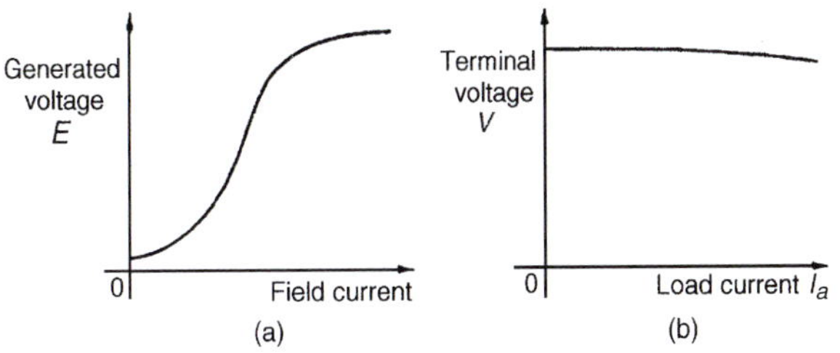

Figure 24.6

(b) Shunt-wound generator

In a shunt-wound generator the field winding is connected in parallel with the armature, as shown in Figure 24.7. The field winding has a relatively high resistance and therefore the current carried is only a fraction of the armature current.

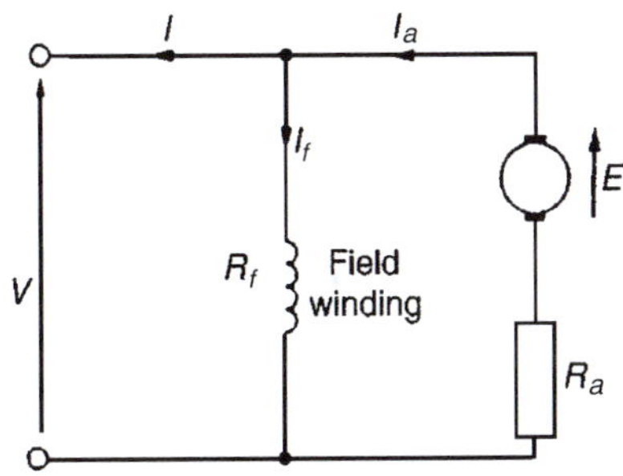

Figure 24.7

For the circuit shown in Figure 24.7,

terminal voltage $V = E - I_a R_a$

or generated e.m.f., $E = V + I_a R_a$

$I_a = I_f + I$, from Kirchhoff's current law,

where I_a = armature current

I_f = field current $\left(= \dfrac{V}{R_f}\right)$

and I = load current

Problem 10. A shunt generator supplies a 20 kW load at 200 V through cables of resistance, $R = 100\,\text{m}\Omega$. If the field winding resistance, $R_f = 50\,\Omega$ and the armature resistance, $R_a = 40\,\text{m}\Omega$, determine (a) the terminal voltage and (b) the e.m.f. generated in the armature.

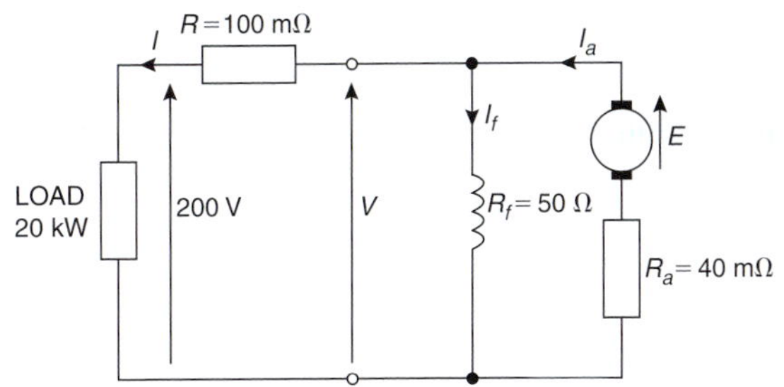

Figure 24.8

(a) The circuit is as shown in Figure 24.8.

$$\text{Load current, } I = \frac{20\,000 \text{ watts}}{200 \text{ volts}} = 100\,\text{A}$$

Volt drop in the cables to the load

$$= IR = (100)(100 \times 10^{-3})$$
$$= 10\,\text{V}$$

Hence terminal voltage, $V = 200 + 10 = $ **210 volts**

(b) Armature current $I_a = I_f + I$

$$\text{Field current, } I_f = \frac{V}{R_f} = \frac{210}{50} = 4.2\,\text{A}$$

Hence $I_a = I_f + I = 4.2 + 100 = 104.2\,\text{A}$

Generated e.m.f., $E = V + I_a R_a$

$$= 210 + (104.2)(40 \times 10^{-3})$$
$$= 210 + 4.168$$
$$= \mathbf{214.17 \text{ volts}}$$

Characteristics

The generated e.m.f., E, is proportional to $\Phi\omega$ (see Section 24.5), hence at constant speed, since $\omega = 2\pi n$, $E \propto \Phi$. Also the flux Φ is proportional to field current I_f until magnetic saturation of the iron circuit of the generator occurs. Hence the open circuit characteristic is as shown in Figure 24.9(a).

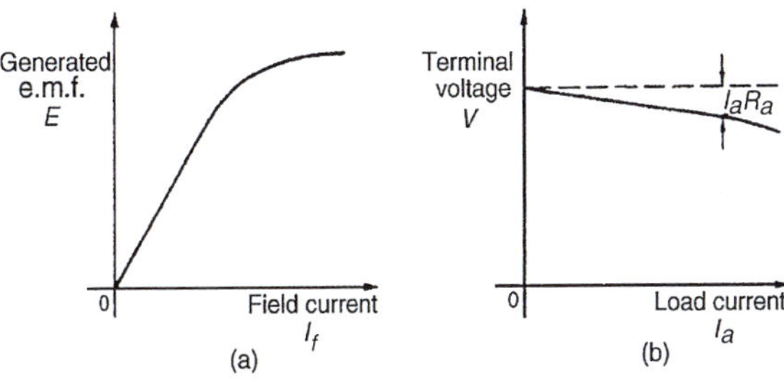

Figure 24.9

Part 3

As the load current on a generator having constant field current and running at constant speed increases, the value of armature current increases, hence the armature volt drop, $I_a R_a$ increases. The generated voltage E is larger than the terminal voltage V and the voltage equation for the armature circuit is $V = E - I_a R_a$. Since E is constant, V decreases with increasing load. The load characteristic is as shown in Figure 24.9(b). In practice, the fall in voltage is about 10% between no-load and full-load for many d.c. shunt-wound generators.

The shunt-wound generator is the type most used in practice, but the load current must be limited to a value that is well below the maximum value. This then avoids excessive variation of the terminal voltage. Typical applications are with battery charging and motor car generators.

(c) Series-wound generator

In the series-wound generator the field winding is connected in series with the armature as shown in Figure 24.10.

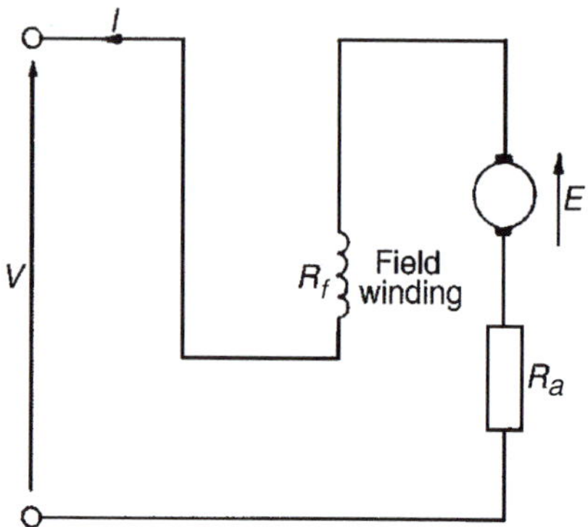

Figure 24.10

Characteristics

The load characteristic is the terminal voltage/current characteristic. The generated e.m.f., E, is proportional to $\Phi\omega$ and at constant speed $\omega(=2\pi n)$ is a constant. Thus E is proportional to Φ. For values of current below magnetic saturation of the yoke, poles, air gaps and armature core, the flux Φ is proportional to the current, hence $E \propto I$. For values of current above those required for magnetic saturation, the generated e.m.f. is approximately constant. The values of field resistance and armature resistance in a series-wound machine are small, hence the terminal voltage V is very nearly equal to E. A typical load characteristic for a series generator is shown in Figure 24.11.

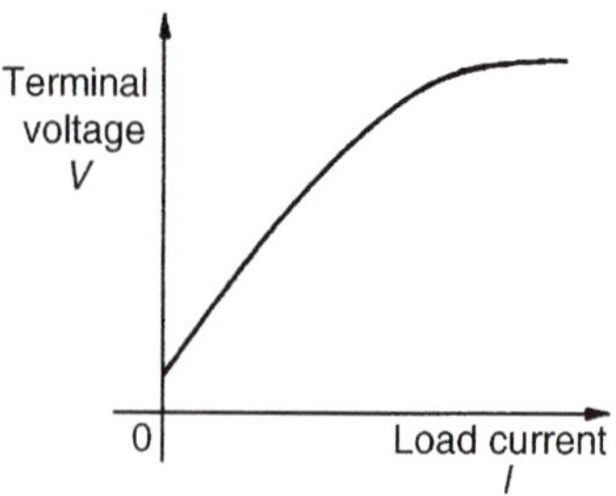

Figure 24.11

In a series-wound generator, the field winding is in series with the armature and it is not possible to have a value of field current when the terminals are open circuited, thus it is not possible to obtain an open-circuit characteristic.

Series-wound generators are rarely used in practice, but can be used as a 'booster' on d.c. transmission lines.

(d) Compound-wound generator

In the compound-wound generator two methods of connection are used, both having a mixture of shunt and series windings, designed to combine the advantages of each. Figure 24.12(a) shows what is termed a **long-shunt** compound generator, and Figure 24.12(b) shows a **short-shunt** compound generator. The latter is the most generally used form of d.c. generator.

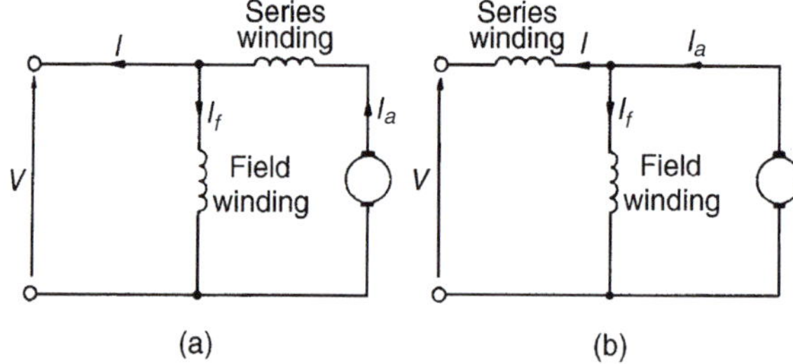

Figure 24.12

Problem 11. A short-shunt compound generator supplies 80 A at 200 V. If the field resistance, $R_f = 40\,\Omega$, the series resistance, $R_{Se} = 0.02\,\Omega$ and the armature resistance, $R_a = 0.04\,\Omega$, determine the e.m.f. generated.

The circuit is shown in Figure 24.13.

Volt drop in series winding $= IR_{Se} = (80)(0.02) = 1.6\,\text{V}$

P.d. across the field winding = p.d. across armature

$$= V_1 = 200 + 1.6 = 201.6\,\text{V}$$

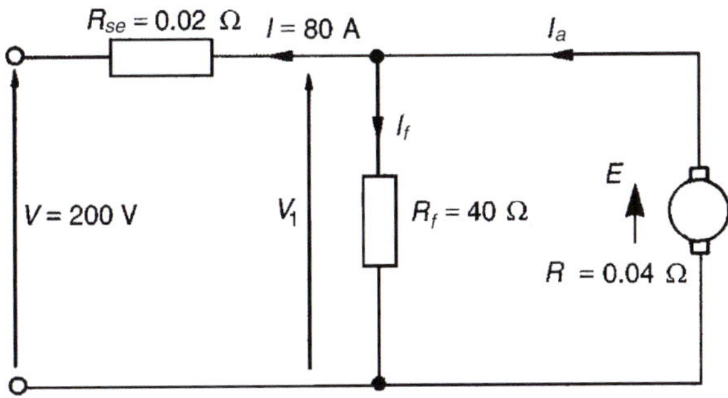

Figure 24.13

$$\text{Field current, } I_f = \frac{V_1}{R_f} = \frac{201.6}{40} = 5.04\,\text{A}$$

$$\text{Armature current, } I_a = I + I_f = 80 + 5.04 = 85.04\,\text{A}$$

$$\begin{aligned}\text{Generated e.m.f., } E &= V_1 + I_a R_a \\ &= 201.6 + (85.04)(0.04) \\ &= 201.6 + 3.4016 \\ &= \mathbf{205\,volts}\end{aligned}$$

Characteristics

In cumulative-compound machines the magnetic flux produced by the series and shunt fields are additive. Included in this group are **over-compounded, level-compounded** and **under-compounded machines** – the degree of compounding obtained depending on the number of turns of wire on the series winding.

A large number of series winding turns results in an over-compounded characteristic, as shown in Figure 24.14, in which the full-load terminal voltage exceeds the no-load voltage. A level-compound machine gives a full-load terminal voltage which is equal to the no-load voltage, as shown in Figure 24.14.

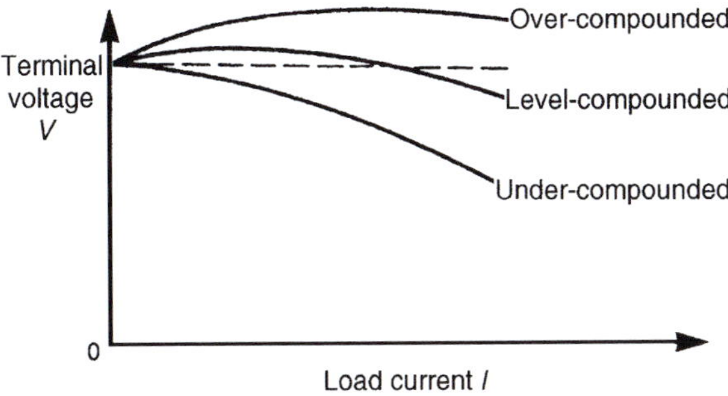

Figure 24.14

An under-compounded machine gives a full-load terminal voltage which is less than the no-load voltage, as shown in Figure 24.14. However, even this latter characteristic is a little better than that for a shunt generator alone.

Compound-wound generators are used in electric arc welding, with lighting sets and with marine equipment.

Now try the following Practice Exercise

Practice Exercise 107 The d.c. generator (Answers on page 824)

1. Determine the terminal voltage of a generator which develops an e.m.f. of 240 V and has an armature current of 50 A on load. Assume the armature resistance is 40 mΩ.
2. A generator is connected to a 50 Ω load and a current of 10 A flows. If the armature resistance is 0.5 Ω, determine (a) the terminal voltage and (b) the generated e.m.f.
3. A separately excited generator develops a no-load e.m.f. of 180 V at an armature speed of 15 rev/s and a flux per pole of 0.20 Wb. Calculate the generated e.m.f. when
   (a) the speed increases to 20 rev/s and the flux per pole remains unchanged,
   (b) the speed remains at 15 rev/s and the pole flux is decreased to 0.125 Wb,
   (c) the speed increases to 25 rev/s and the pole flux is decreased to 0.18 Wb.
4. A shunt generator supplies a 50 kW load at 400 V through cables of resistance 0.2 Ω. If the field winding resistance is 50 Ω and the armature resistance is 0.05 Ω, determine (a) the terminal voltage, (b) the e.m.f. generated in the armature.
5. A short-shunt compound generator supplies 50 A at 300 V. If the field resistance is 30 Ω, the series resistance 0.03 Ω and the armature resistance 0.05 Ω, determine the e.m.f. generated.
6. A d.c. generator has a generated e.m.f. of 210 V when running at 700 rev/min and the flux per pole is 120 mWb. Determine the generated e.m.f. (a) at 1050 rev/min, assuming the flux remains constant, (b) if the flux is reduced by one-sixth at constant speed and (c) at a speed of 1155 rev/min and a flux of 132 mWb.

7. A 250 V d.c. shunt-wound generator has an armature resistance of 0.1 Ω. Determine the generated e.m.f. when the generator is supplying 50 kW, neglecting the field current of the generator.

24.8 D.c. machine losses

As stated in Section 24.1, a generator is a machine for converting mechanical energy into electrical energy and a motor is a machine for converting electrical energy into mechanical energy. When such conversions take place, certain losses occur which are dissipated in the form of heat.

The principal **losses of machines** are:

(i) **Copper loss**, due to I^2R heat losses in the armature and field windings.

(ii) **Iron (or core) loss**, due to hysteresis and eddy-current losses in the armature. This loss can be reduced by constructing the armature of silicon steel laminations having a high resistivity and low hysteresis loss. At constant speed, the iron loss is assumed constant.

(iii) **Friction and windage losses**, due to bearing and brush contact friction and losses due to air resistance against moving parts (called windage). At constant speed, these losses are assumed to be constant.

(iv) **Brush contact loss** between the brushes and commutator. This loss is approximately proportional to the load current.

The total losses of a machine can be quite significant and operating efficiencies of between 80% and 90% are common.

24.9 Efficiency of a d.c. generator

The efficiency of an electrical machine is the ratio of the output power to the input power and is usually expressed as a percentage. The Greek letter 'η' (eta) is used to signify efficiency and since the units are power/power, then efficiency has no units. Thus

$$\textbf{efficiency, } \eta = \left(\frac{\textbf{output power}}{\textbf{input power}}\right) \times 100\%$$

If the total resistance of the armature circuit (including brush contact resistance) is R_a, then **the total loss in the armature circuit is $I_a^2 R_a$**

If the terminal voltage is V and the current in the shunt circuit is I_f, then **the loss in the shunt circuit is $I_f V$**

If the sum of the iron, friction and windage losses is C then **the total losses is given by**:

$$I_a^2 R_a + I_f V + C$$

$$(I_a^2 R_a + I_f V \text{ is, in fact, the 'copper loss'})$$

If the output current is I, then **the output power is VI**.

Total input power $= VI + I_a^2 R_a + I_f V + C$. Hence

$$\textbf{efficiency, } \eta = \frac{\textbf{output}}{\textbf{input}}$$

$$= \left(\frac{VI}{VI + I_a^2 R_a + I_f V + C}\right) \times 100\% \quad (4)$$

The **efficiency of a generator is a maximum** when the load is such that:

$$I_a^2 R_a = VI_f + C$$

i.e. when the variable loss = the constant loss

Problem 12. A 10 kW shunt generator having an armature circuit resistance of 0.75 Ω and a field resistance of 125 Ω generates a terminal voltage of 250 V at full load. Determine the efficiency of the generator at full load, assuming the iron, friction and windage losses amount to 600 W.

The circuit is shown in Figure 24.15.

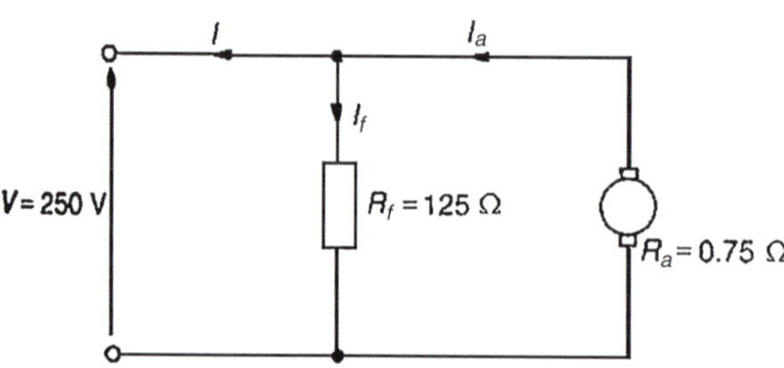

Figure 24.15

Part 3

Output power $= 10\,000\,W = VI$

from which, load current $I = \frac{10\,000}{V} = \frac{10\,000}{250} = 40\,\text{A}$

$$\text{Field current, } I_f = \frac{V}{R_f} = \frac{250}{125} = 2\,\text{A}$$

$$\text{Armature current, } I_a = I_f + I = 2 + 40 = 42\,\text{A}$$

Efficiency,

$$\eta = \left(\frac{VI}{VI + I_a^2 R_a + I_f V + C}\right) \times 100\%$$

$$= \left(\frac{10\,000}{10\,000 + (42)^2(0.75) + (2)(250) + 600}\right) \times 100\%$$

$$= \frac{10\,000}{12\,423} \times 100\% = \mathbf{80.50\%}$$

Now try the following Practice Exercise

Practice Exercise 108 Efficiency of a d.c. generator (Answers on page 824)

1. A 15 kW shunt generator having an armature circuit resistance of 0.4 Ω and a field resistance of 100 Ω generates a terminal voltage of 240 V at full load. Determine the efficiency of the generator at full load, assuming the iron, friction and windage losses amount to 1 kW.

24.10 D.c. motors

The construction of a d.c. motor is the same as a d.c. generator. The only difference is that in a generator the generated e.m.f. is greater than the terminal voltage, whereas in a motor the generated e.m.f. is less than the terminal voltage.

D.c. motors are often used in power stations to drive emergency stand-by pump systems which come into operation to protect essential equipment and plant should the normal a.c. supplies or pumps fail.

Back e.m.f.

When a d.c. motor rotates, an e.m.f. is induced in the armature conductors. By Lenz's law this induced e.m.f. E opposes the supply voltage V and is called a **back e.m.f.**, and the supply voltage V is given by:

$$\boldsymbol{V = E + I_a R_a} \quad \text{or} \quad \boldsymbol{E = V - I_a R_a} \qquad (5)$$

Problem 13. A d.c. motor operates from a 240 V supply. The armature resistance is 0.2 Ω. Determine the back e.m.f. when the armature current is 50 A.

For a motor, $V = E + I_a R_a$

hence back e.m.f., $E = V - I_a R_a$

$$= 240 - (50)(0.2) = 240 - 10$$

$$= \mathbf{230\ volts}$$

Problem 14. The armature of a d.c. machine has a resistance of 0.25 Ω and is connected to a 300 V supply. Calculate the e.m.f. generated when it is running: (a) as a generator giving 100 A, and (b) as a motor taking 80 A.

(a) As a generator, generated e.m.f.,

$E = V + I_a R_a$, from equation (3),

$= 300 + (100)(0.25)$

$= 300 + 25 = \mathbf{325\ volts}$

(b) As a motor, generated e.m.f. (or back e.m.f.),

$E = V - I_a R_a$, from equation (5),

$= 300 - (80)(0.25) = \mathbf{280\ volts}$

Now try the following Practice Exercise

Practice Exercise 109 Back e.m.f. (Answers on page 824)

1. A d.c. motor operates from a 350 V supply. If the armature resistance is 0.4 Ω, determine the back e.m.f. when the armature current is 60 A.
2. The armature of a d.c. machine has a resistance of 0.5 Ω and is connected to a 200 V supply. Calculate the e.m.f. generated when it is running (a) as a motor taking 50 A and (b) as a generator giving 70 A.
3. Determine the generated e.m.f. of a d.c. machine if the armature resistance is 0.1 Ω and it (a) is running as a motor connected to a 230 V supply, the armature current being 60 A, and (b) is running as a generator with a terminal voltage of 230 V, the armature current being 80 A.

24.11 Torque of a d.c. machine

From equation (5), for a d.c. motor, the supply voltage V is given by

$$V = E + I_a R_a$$

Multiplying each term by current I_a gives:

$$VI_a = EI_a + I_a^2 R_a$$

The term $\boldsymbol{VI_a}$ is the **total electrical power supplied to the armature**, the term $\boldsymbol{I_a^2 R_a}$ is the **loss due to armature resistance** and the term $\boldsymbol{EI_a}$ is the **mechanical power developed by the armature**.

If T is the torque, in newton metres, then the mechanical power developed is given by $T\omega$ watts (see *Science for Engineering*, 5th edition, Taylor & Francis).

Hence $T\omega = 2\pi nT = EI_a$, from which

$$\textbf{torque } T = \frac{EI_a}{2\pi n} \textbf{ newton metres} \qquad (6)$$

From Section 24.5, equation (1), the e.m.f. E generated is given by

$$E = \frac{2p\Phi nZ}{c}$$

Hence $2\pi nT = EI_a = \left(\frac{2p\Phi nZ}{c}\right) I_a$

and torque $T = \frac{\left(\frac{2p\Phi nZ}{c}\right) I_a}{2\pi n}$

i.e. $$T = \frac{p\Phi ZI_a}{\pi c} \textbf{ newton metres} \qquad (7)$$

For a given machine, Z, c and p are fixed values

Hence torque, $$T \propto \Phi I_a \qquad (8)$$

Problem 15. An 8-pole d.c. motor has a wave-wound armature with 900 conductors. The useful flux per pole is 25 mWb. Determine the torque exerted when a current of 30 A flows in each armature conductor.

$p = 4$, $c = 2$ for a wave winding, $\Phi = 25 \times 10^{-3}$ Wb, $Z = 900$, $I_a = 30$ A

From equation (7),

$$\text{torque } T = \frac{p\Phi ZI_a}{\pi c} = \frac{(4)(25 \times 10^{-3})(900)(30)}{\pi(2)} = \mathbf{429.7\,Nm}$$

Problem 16. Determine the torque developed by a 350 V d.c. motor having an armature resistance of 0.5 Ω and running at 15 rev/s. The armature current is 60 A.

$V = 350$ V, $R_a = 0.5\,\Omega$, $n = 15$ rev/s, $I_a = 60$ A

Back e.m.f. $E = V - I_a R_a = 350 - (60)(0.5) = 320$ V

From equation (6), torque $T = \frac{EI_a}{2\pi n} = \frac{(320)(60)}{2\pi(15)} = \mathbf{203.7\,Nm}$

Problem 17. A 6-pole lap-wound motor is connected to a 250 V d.c. supply. The armature has 500 conductors and a resistance of 1 Ω. The flux per pole is 20 mWb. Calculate (a) the speed and (b) the torque developed when the armature current is 40 A.

$V = 250$ V, $Z = 500$, $R_a = 1\,\Omega$, $\Phi = 20 \times 10^{-3}$ Wb, $I_a = 40$ A, $c = 2p$ for a lap winding

(a) Back e.m.f. $E = V - I_a R_a = 250 - (40)(1) = 210$ V

E.m.f. $E = \frac{2p\Phi nZ}{c}$

i.e. $210 = \frac{2p(20 \times 10^{-3})n(500)}{2p}$

Hence speed $n = \frac{210}{(20 \times 10^{-3})(500)} = \mathbf{21\,rev/s}$

or $(21 \times 60) = \mathbf{1260\,rev/min}$

(b) Torque $T = \frac{EI_a}{2\pi n} = \frac{(210)(40)}{2\pi(21)} = \mathbf{63.66\,Nm}$

Problem 18. The shaft torque of a diesel motor driving a 100 V d.c. shunt-wound generator is 25 Nm. The armature current of the generator is 16 A at this value of torque. If the shunt field regulator is adjusted so that the flux is reduced by 15%, the torque increases to 35 Nm. Determine the armature current at this new value of torque.

From equation (8), the shaft torque T of a generator is proportional to ΦI_a, where Φ is the flux and I_a is the armature current. Thus, $T = k\Phi I_a$, where k is a constant.

The torque at flux Φ_1 and armature current I_{a1} is $T_1 = k\Phi_1 I_{a1}$.

Similarly, $T_2 = k\Phi_2 I_{a2}$

By division $\dfrac{T_1}{T_2} = \dfrac{k\Phi_1 I_{a1}}{k\Phi_2 I_{a2}} = \dfrac{\Phi_1 I_{a1}}{\Phi_2 I_{a2}}$

Hence $\dfrac{25}{35} = \dfrac{\Phi_1 \times 16}{0.85\Phi_1 \times I_{a2}}$

i.e. $I_{a2} = \dfrac{16 \times 35}{0.85 \times 25} = 26.35\text{A}$

That is, **the armature current at the new value of torque is 26.35 A**

Problem 19. A 100 V d.c. generator supplies a current of 15 A when running at 1500 rev/min. If the torque on the shaft driving the generator is 12 N m, determine (a) the efficiency of the generator and (b) the power loss in the generator.

(a) From Section 24.9, the efficiency of a

$$\text{generator} = \frac{\text{output power}}{\text{input power}} \times 100\%$$

The output power is the electrical output, i.e. VI watts. The input power to a generator is the mechanical power in the shaft driving the generator, i.e. $T\omega$ or $T(2\pi n)$ watts, where T is the torque in Nm and n is speed of rotation in rev/s. Hence, for a generator

$$\text{efficiency,} \quad \eta = \frac{VI}{T(2\pi n)} \times 100\%$$

$$\text{i.e.} \quad \eta = \frac{(100)(15)(100)}{(12)(2\pi)\left(\frac{1500}{60}\right)}$$

i.e. **efficiency = 79.6%**

(b) The input power = output power + losses

Hence, $T(2\pi n) = VI + \text{losses}$

$$\text{i.e.} \quad \text{losses} = T(2\pi n) - VI$$

$$= \left[(12)(2\pi)\left(\frac{1500}{60}\right)\right] - [(100)(15)]$$

i.e. **power loss** = 1885 − 1500 = **385 W**

Now try the following Practice Exercise

Practice Exercise 110 Losses, efficiency and torque (Answers on page 824)

1. The shaft torque required to drive a d.c. generator is 18.7 Nm when it is running at 1250 rev/min. If its efficiency is 87% under these conditions and the armature current is 17.3 A, determine the voltage at the terminals of the generator.
2. A 220 V d.c. generator supplies a load of 37.5 A and runs at 1550 rev/min. Determine the shaft torque of the diesel motor driving the generator, if the generator efficiency is 78%.
3. A 4-pole d.c. motor has a wave-wound armature with 800 conductors. The useful flux per pole is 20 mWb. Calculate the torque exerted when a current of 40 A flows in each armature conductor.
4. Calculate the torque developed by a 240 V d.c. motor whose armature current is 50 A, armature resistance is 0.6 Ω and is running at 10 rev/s.
5. An 8-pole lap-wound d.c. motor has a 200 V supply. The armature has 800 conductors and a resistance of 0.8 Ω. If the useful flux per pole is 40 mWb and the armature current is 30 A, calculate (a) the speed and (b) the torque developed.
6. A 150 V d.c. generator supplies a current of 25 A when running at 1200 rev/min. If the torque on the shaft driving the generator is 35.8 Nm, determine (a) the efficiency of the generator, and (b) the power loss in the generator.

24.12 Types of d.c. motor and their characteristics

(a) Shunt-wound motor

In the shunt-wound motor the field winding is in parallel with the armature across the supply, as shown in Figure 24.16.

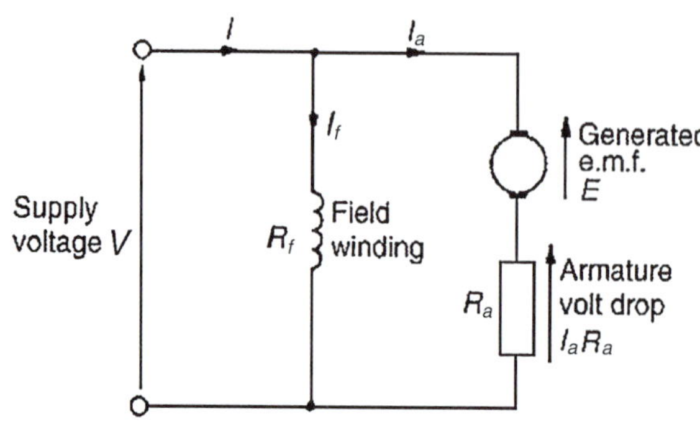

Figure 24.16

For the circuit shown in Figure 24.16,

Supply voltage, $V = E + I_a R_a$

or generated e.m.f., $E = V - I_a R_a$

Supply current, $I = I_a + I_f$, from Kirchhoff's current law.

Problem 20. A 240 V shunt motor takes a total current of 30 A. If the field winding resistance $R_f = 150\,\Omega$ and the armature resistance $R_a = 0.4\,\Omega$, determine (a) the current in the armature, and (b) the back e.m.f.

(a) Field current, $I_f = \dfrac{V}{R_f} = \dfrac{240}{150} = 1.6\,\text{A}$

Supply current, $I = I_a + I_f$

Hence armature current, $I_a = I - I_f = 30 - 1.6$
$= \mathbf{28.4\,A}$

(b) Back e.m.f., $E = V - I_a R_a$
$= 240 - (28.4)(0.4)$
$= \mathbf{228.64\ volts}$

Characteristics

The two principal characteristics are the torque/armature current and speed/armature current relationships. From these, the torque/speed relationship can be derived.

(i) The theoretical torque/armature current characteristic can be derived from the expression $T \propto \Phi I_a$ (see Section 24.11). For a shunt-wound motor, the field winding is connected in parallel with the armature circuit and thus the applied voltage gives a constant field current, i.e. a shunt-wound motor is a constant flux machine. Since Φ is constant, it follows that $T \propto I_a$, and the characteristic is as shown in Figure 24.17.

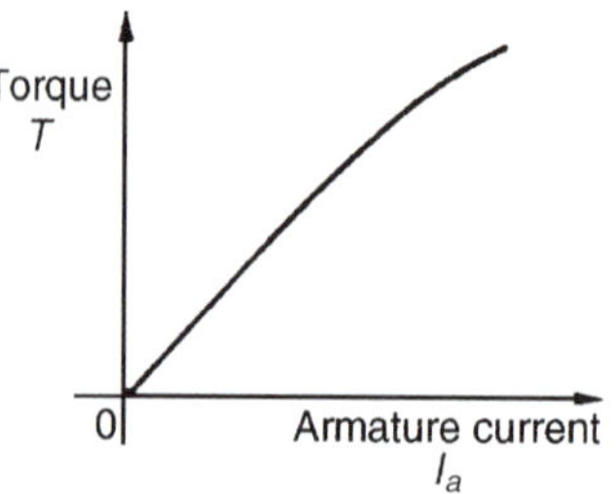

Figure 24.17

(ii) The armature circuit of a d.c. motor has resistance due to the armature winding and brushes, R_a ohms, and when armature current I_a is flowing through it, there is a voltage drop of $I_a R_a$ volts. In Figure 24.16 the armature resistance is shown as a separate resistor in the armature circuit to help understanding. Also, even though the machine is a motor, because conductors are rotating in a magnetic field, a voltage, $E \propto \Phi\omega$, is generated by the armature conductors. From equation (5), $V = E + I_a R_a$ or $E = V - I_a R_a$

However, from Section 24.5, $E \propto \Phi n$, hence $n \propto E/\Phi$, i.e.

$$\textbf{speed of rotation, } \boldsymbol{n \propto \frac{E}{\Phi} \propto \frac{V - I_a R_a}{\Phi}} \qquad (9)$$

For a shunt motor, V, Φ and R_a are constants, hence as armature current I_a increases, $I_a R_a$ increases and $V - I_a R_a$ decreases, and the speed is proportional to a quantity which is decreasing and is as shown in Figure 24.18. As the load on the shaft of the motor increases, I_a increases and the speed drops slightly. In practice, the speed falls by about 10% between no-load and full-load on many d.c. shunt-wound motors. Due to this relatively small drop in

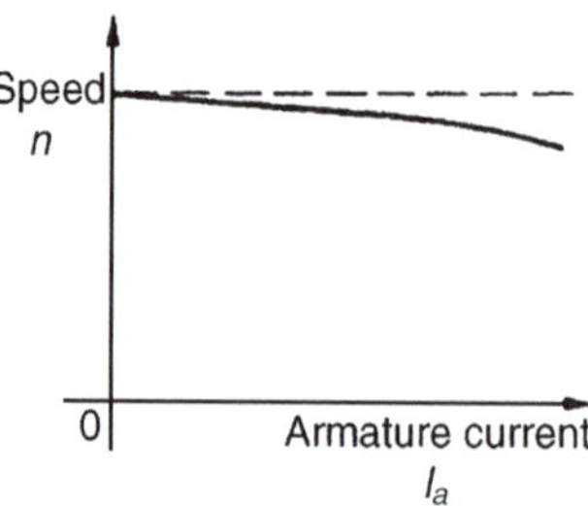

Figure 24.18

Part 3

speed, the d.c. shunt-wound motor is taken as basically being a constant-speed machine and may be used for driving lathes, lines of shafts, fans, conveyor belts, pumps, compressors, drilling machines and so on.

(iii) Since torque is proportional to armature current (see (i) above), the theoretical speed/torque characteristic is as shown in Figure 24.19.

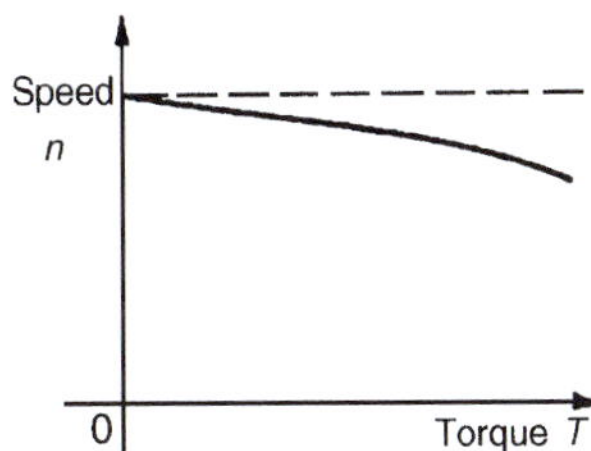

Figure 24.19

Problem 21. A 200 V d.c. shunt-wound motor has an armature resistance of 0.4 Ω and at a certain load has an armature current of 30 A and runs at 1350 rev/min. If the load on the shaft of the motor is increased so that the armature current increases to 45 A, determine the speed of the motor, assuming the flux remains constant.

The relationship $E \propto \Phi n$ applies to both generators and motors. For a motor,

$$E = V - I_a R_a \text{ (see equation 5)}$$

Hence $E_1 = 200 - 30 \times 0.4 = 188\,\text{V}$

and $E_2 = 200 - 45 \times 0.4 = 182\,\text{V}$

The relationship, $\dfrac{E_1}{E_2} = \dfrac{\Phi_1 n_1}{\Phi_2 n_2}$

applies to both generators and motors. Since the flux is constant, $\Phi_1 = \Phi_2$.

$$\text{Hence } \frac{188}{182} = \frac{\Phi_1 \times \left(\frac{1350}{60}\right)}{\Phi_1 \times n_2}, \text{ i.e. } n_2 = \frac{22.5 \times 182}{188} = 21.78\,\text{rev/s}$$

Thus the speed of the motor when the armature current is 45 A is 21.78 × 60 rev/min, i.e. **1307 rev/min**.

Problem 22. A 220 V d.c. shunt-wound motor runs at 800 rev/min and the armature current is 30 A. The armature circuit resistance is 0.4 Ω. Determine (a) the maximum value of armature current if the flux is suddenly reduced by 10% and (b) the steady state value of the armature current at the new value of flux, assuming the shaft torque of the motor remains constant.

(a) For a d.c. shunt-wound motor, $E = V - I_a R_a$. Hence initial generated e.m.f., $E_1 = 220 - 30 \times 0.4 = 208\,\text{V}$. The generated e.m.f. is also such that $E \propto \Phi n$, so at the instant the flux is reduced, the speed has not had time to change, and $E = 208 \times 90/100 = 187.2\,\text{V}$.
Hence, the voltage drop due to the armature resistance is 220 − 187.2, i.e. 32.8 V. The **instantaneous value of the current** is 32.8/0.4, i.e. **82 A**. This increase in current is about three times the initial value and causes an increase in torque, $(T \propto \Phi I_a)$. The motor accelerates because of the larger torque value until steady state conditions are reached.

(b) $T \propto \Phi I_a$ and since the torque is constant,

$\Phi_1 I_{a1} = \Phi_2 I_{a2}$. The flux Φ is reduced by 10%, hence

$$\Phi_2 = 0.9\Phi_1$$

Thus, $\Phi_1 \times 30 = 0.9\Phi_1 \times I_{a2}$

i.e. the steady state value of armature current,

$$I_{a2} = \frac{30}{0.9} = \mathbf{33\frac{1}{3}\,A}$$

(b) Series-wound motor

In the series-wound motor the field winding is in series with the armature across the supply as shown in Figure 24.20.

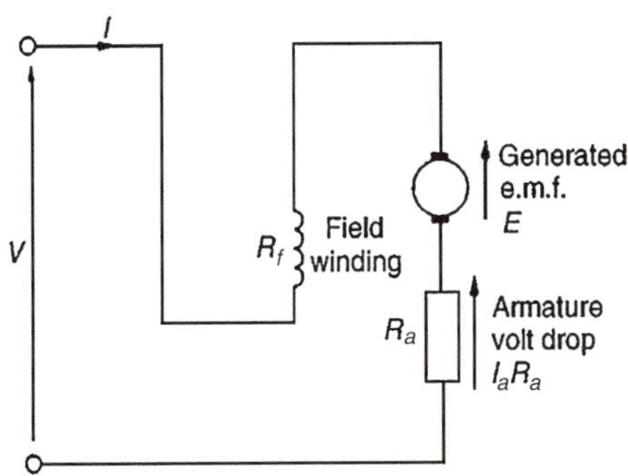

Figure 24.20

For the series motor shown in Figure 24.20,

Supply voltage, $V = E + I(R_a + R_f)$

or generated e.m.f., $E = V - I(R_a + R_f)$

Characteristics

In a series motor, the armature current flows in the field winding and is equal to the supply current, I.

(i) **The torque/current characteristic**
It is shown in Section 24.11 that torque $T \propto \Phi I_a$. Since the armature and field currents are the same current, I, in a series machine, then $T \propto \Phi I$ over a limited range, before magnetic saturation of the magnetic circuit of the motor is reached (i.e. the linear portion of the B–H curve for the yoke, poles, air gap, brushes and armature in series). Thus $\Phi \propto I$ and $T \propto I^2$. After magnetic saturation, Φ almost becomes a constant and $T \propto I$. Thus the theoretical torque/current characteristic is as shown in Figure 24.21.

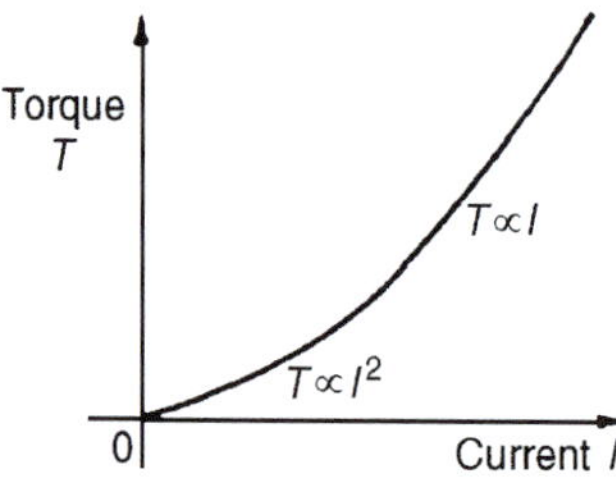

Figure 24.21

(ii) **The speed/current characteristic**
It is shown in equation (9) that $n \propto (V - I_a R_a)/\Phi$. In a series motor, $I_a = I$ and below the magnetic saturation level, $\Phi \propto I$. Thus $n \propto (V - IR)/I$ where R is the combined resistance of the series field and armature circuit. Since IR is small compared with V, then an approximate relationship for the speed is $n \propto V/I \propto 1/I$ since V is constant. Hence the theoretical speed/current characteristic is as shown in Figure 24.22. The high speed at small values of current indicate that this type of motor must not be run on very light loads and, invariably, such motors are permanently coupled to their loads.

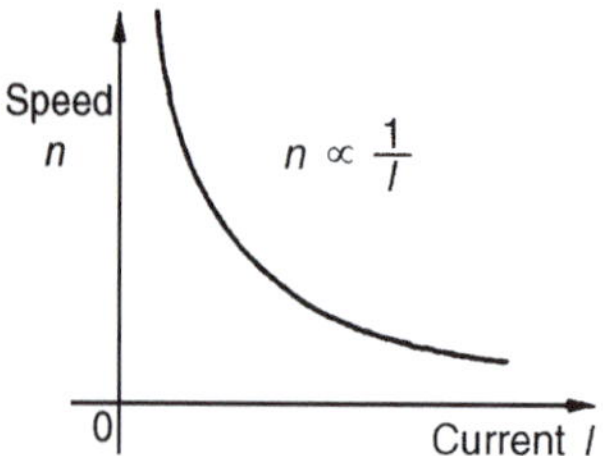

Figure 24.22

(iii) The theoretical speed/torque characteristic may be derived from (i) and (ii) above by obtaining the torque and speed for various values of current and plotting the co-ordinates on the speed/torque characteristics. A typical speed/torque characteristic is shown in Figure 24.23.

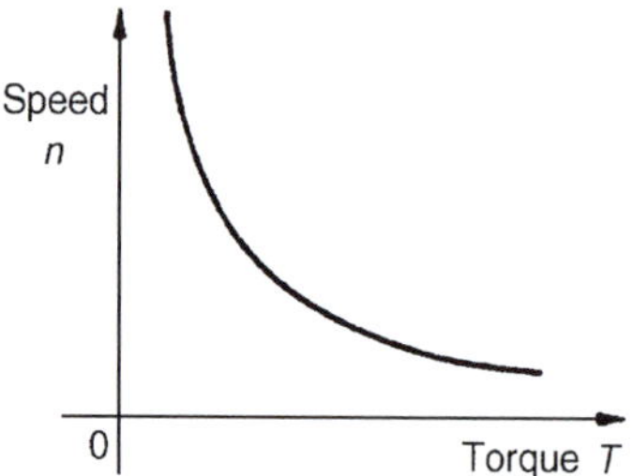

Figure 24.23

A d.c. series motor takes a large current on starting and the characteristic shown in Figure 24.21 shows that the series-wound motor has a large torque when the current is large. Hence these motors are used for traction (such as trains, milk delivery vehicles, etc.), driving fans and for cranes and hoists, where a large initial torque is required.

Problem 23. A series motor has an armature resistance of 0.2 Ω and a series field resistance of 0.3 Ω. It is connected to a 240 V supply and at a particular load runs at 24 rev/s when drawing 15 A from the supply.

(a) Determine the generated e.m.f. at this load.

(b) Calculate the speed of the motor when the load is changed such that the current is increased to 30 A. Assume that this causes a doubling of the flux.

(a) With reference to Figure 24.20, generated e.m.f., E, at initial load, is given by

$$E_1 = V - I_a(R_a + R_f)$$
$$= 240 - (15)(0.2 + 0.3) = 240 - 7.5$$
$$= \mathbf{232.5\ volts}$$

(b) When the current is increased to 30 A, the generated e.m.f. is given by:

Part 3

$$E_2 = V - I_a(R_a + R_f)$$
$$= 240 - (30)(0.2 + 0.3) = 240 - 15$$
$$= 225 \text{ volts}$$

Now e.m.f. $E \propto \Phi n$

thus $\dfrac{E_1}{E_2} = \dfrac{\Phi_1 n_1}{\Phi_2 n_2}$

i.e. $\dfrac{232.5}{225} = \dfrac{\Phi_1(24)}{(2\Phi_1)(n_2)}$ since $\Phi_2 = 2\Phi_1$

Hence **speed of motor,** $n_2 = \dfrac{(24)(225)}{(232.5)(2)}$

$= \mathbf{11.6\ rev/s}$

As the current has been increased from 15 A to 30 A, the speed has decreased from 24 rev/s to 11.6 rev/s. Its speed/current characteristic is similar to Figure 24.22.

(c) Compound-wound motor

There are two types of compound-wound motor:

(i) **Cumulative compound**, in which the series winding is so connected that the field due to it assists that due to the shunt winding.

(ii) **Differential compound**, in which the series winding is so connected that the field due to it opposes that due to the shunt winding.

Figure 24.24(a) shows a **long-shunt** compound motor and Figure 24.24(b) a **short-shunt** compound motor.

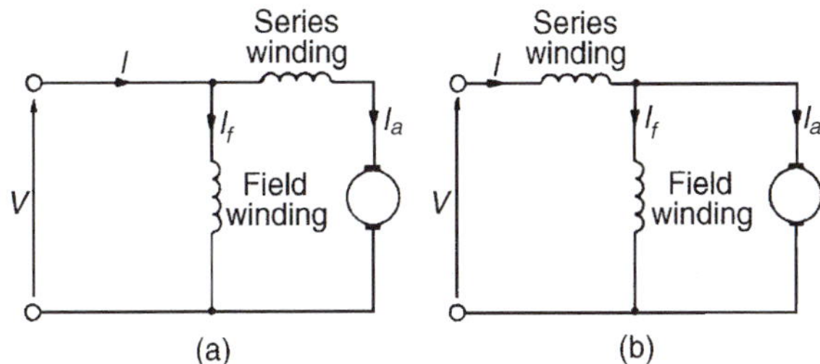

Figure 24.24

Characteristics

A compound-wound motor has both a series and a shunt field winding (i.e. one winding in series and one in parallel with the armature), and is usually wound to have a characteristic similar in shape to a series-wound motor (see Figures 24.21–24.23). A limited amount of shunt winding is present to restrict the no-load speed to a safe value. However, by varying the number of turns on the series and shunt windings and the directions of the magnetic fields produced by these windings (assisting or opposing), families of characteristics may be obtained to suit almost all applications. Generally, compound-wound motors are used for heavy duties, particularly in applications where sudden heavy load may occur, such as for driving plunger pumps, presses, geared lifts, conveyors, hoists and so on.

Typical compound motor torque and speed characteristics are shown in Figure 24.25.

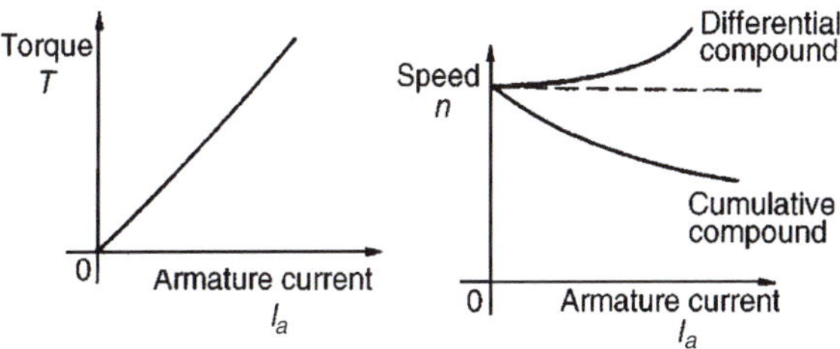

Figure 24.25

24.13 The efficiency of a d.c. motor

It was stated in Section 24.9, that the efficiency of a d.c. machine is given by:

$$\text{efficiency, } \eta = \frac{\text{output power}}{\text{input power}} \times 100\%$$

Also, the total losses $= I_a^2 R_a + I_f V + C$ (for a shunt motor) where C is the sum of the iron, friction and windage losses.

For a motor, the input power $= VI$

and the output power $= VI -$ losses

$$= VI - I_a^2 R_a - I_f V - C$$

Hence

$$\textbf{efficiency } \eta = \left(\frac{VI - I_a^2 R_a - I_f V - C}{VI}\right) \times 100\% \qquad (10)$$

The **efficiency of a motor is a maximum** when the load is such that:

$$I_a^2 R_a = I_f V + C$$

Problem 24. A 320 V shunt motor takes a total current of 80 A and runs at 1000 rev/min. If the iron, friction and windage losses amount to 1.5 kW, the shunt field resistance is 40 Ω and the armature resistance is 0.2 Ω, determine the overall efficiency of the motor.

The circuit is shown in Figure 24.26.

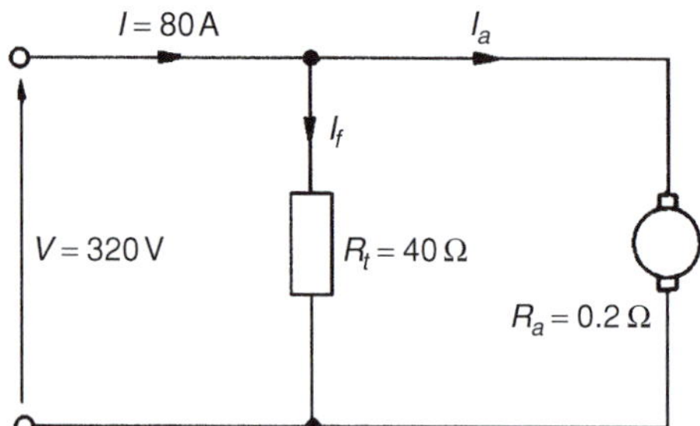

Figure 24.26

Field current, $I_f = \frac{V}{R_f} = \frac{320}{40} = 8\,\text{A}$

Armature current $I_a = I - I_f = 80 - 8 = 72\,\text{A}$

C = iron, friction and windage losses = 1500 W

Efficiency,

$$\eta = \left(\frac{VI - I_a^2 R_a - I_f V - C}{VI}\right) \times 100\%$$

$$= \left(\frac{(320)(80) - (72)^2(0.2) - (8)(320) - 1500}{(320)(80)}\right) \times 100\%$$

$$= \left(\frac{25\,600 - 1036.8 - 2560 - 1500}{25\,600}\right) \times 100\%$$

$$= \left(\frac{20\,503.2}{25\,600}\right) \times 100\% = \mathbf{80.1\%}$$

Problem 25. A 250 V series motor draws a current of 40 A. The armature resistance is 0.15 Ω and the field resistance is 0.05 Ω. Determine the maximum efficiency of the motor.

The circuit is as shown in Figure 24.27.
From equation 10, efficiency,

$$\eta = \left(\frac{VI - I_a^2 R_a - I_f V - C}{VI}\right) \times 100\%$$

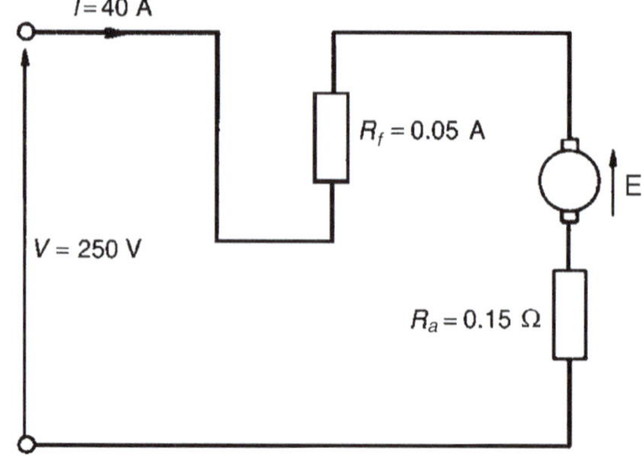

Figure 24.27

However, for a series motor, $I_f = 0$ and the $I_a^2 R_a$ loss needs to be $I^2(R_a + R_f)$

Hence efficiency, $\eta = \left(\frac{VI - I^2(R_a + R_f) - C}{VI}\right) \times 100\%$

For maximum efficiency $I^2(R_a + R_f) = C$

Hence efficiency,

$$\eta = \left(\frac{VI - 2I^2(R_a + R_f)}{VI}\right) \times 100\%$$

$$= \left(\frac{(250)(40) - 2(40)^2(0.15 + 0.05)}{(250)(40)}\right) \times 100\%$$

$$= \left(\frac{10\,000 - 640}{10\,000}\right) \times 100\%$$

$$= \left(\frac{9360}{10\,000}\right) \times 100\% = \mathbf{93.6\%}$$

Problem 26. A 200 V d.c. motor develops a shaft torque of 15 Nm at 1200 rev/min. If the efficiency is 80%, determine the current supplied to the motor.

The efficiency of a motor $= \frac{\text{output power}}{\text{input power}} \times 100\%$

The output power of a motor is the power available to do work at its shaft and is given by $T\omega$ or $T(2\pi n)$ watts, where T is the torque in Nm and n is the speed of rotation in rev/s. The input power is the electrical power in watts supplied to the motor, i.e. VI watts.
Thus for a motor, efficiency,

$$\eta = \frac{T(2\pi n)}{VI} \times 100\%$$

i.e.

$$80 = \left[\frac{(15)(2\pi)(1200/60)}{(200)(I)}\right](100)$$

Thus the current supplied, $I = \dfrac{(15)(2\pi)(20)(100)}{(200)(80)}$

$= \mathbf{11.8\,A}$

Problem 27. A d.c. series motor drives a load at 30 rev/s and takes a current of 10 A when the supply voltage is 400 V. If the total resistance of the motor is 2 Ω and the iron, friction and windage losses amount to 300 W, determine the efficiency of the motor.

$$\text{Efficiency, } \eta = \left(\frac{VI - I^2R - C}{VI}\right) \times 100\%$$

$$= \left(\frac{(400)(10) - (10)^2(2) - 300}{(400)(10)}\right) \times 100\%$$

$$= \left(\frac{4000 - 200 - 300}{4000}\right) \times 100\%$$

$$= \left(\frac{3500}{4000}\right) \times 100\% = \mathbf{87.5\%}$$

Now try the following Practice Exercise

Practice Exercise 111 D.c. motors (Answers on page 825)

1. A 240 V shunt motor takes a total current of 80 A. If the field winding resistance is 120 Ω and the armature resistance is 0.4 Ω, determine (a) the current in the armature and (b) the back e.m.f.
2. A d.c. motor has a speed of 900 rev/min when connected to a 460 V supply. Find the approximate value of the speed of the motor when connected to a 200 V supply, assuming the flux decreases by 30% and neglecting the armature volt drop.
3. A series motor having a series field resistance of 0.25 Ω and an armature resistance of 0.15 Ω is connected to a 220 V supply and at a particular load runs at 20 rev/s when drawing 20 A from the supply. Calculate the e.m.f. generated at this load. Determine also the speed of the motor when the load is changed such that the current increases to 25 A. Assume the flux increases by 25%.
4. A 500 V shunt motor takes a total current of 100 A and runs at 1200 rev/min. If the shunt field resistance is 50 Ω, the armature resistance is 0.25 Ω and the iron, friction and windage losses amount to 2 kW, determine the overall efficiency of the motor.
5. A 250 V, series-wound motor is running at 500 rev/min and its shaft torque is 130 Nm. If its efficiency at this load is 88%, find the current taken from the supply.
6. In a test on a d.c. motor, the following data was obtained. Supply voltage: 500 V. Current taken from the supply: 42.4 A. Speed: 850 rev/min. Shaft torque: 187 Nm. Determine the efficiency of the motor correct to the nearest 0.5%
7. A 300 V series motor draws a current of 50 A. The field resistance is 40 mΩ and the armature resistance is 0.2 Ω. Determine the maximum efficiency of the motor.
8. A series motor drives a load at 1500 rev/min and takes a current of 20 A when the supply voltage is 250 V. If the total resistance of the motor is 1.5 Ω and the iron, friction and windage losses amount to 400 W, determine the efficiency of the motor.
9. A series-wound motor is connected to a d.c. supply and develops full load torque when the current is 30 A and speed is 1000 rev/min. If the flux per pole is proportional to the current flowing, find the current and speed at half full load torque, when connected to the same supply.

24.14 D.c. motor starter

If a d.c. motor whose armature is stationary is switched directly to its supply voltage, it is likely that the fuses protecting the motor will burn out. This is because the armature resistance is small, frequently being less than one ohm. Thus, additional resistance must be added to the armature circuit at the instant of closing the switch to start the motor.

As the speed of the motor increases, the armature conductors are cutting flux and a generated voltage, acting in opposition to the applied voltage, is produced, which limits the flow of armature current. Thus the value of the additional armature resistance can then be reduced.

When at normal running speed, the generated e.m.f. is such that no additional resistance is required in the armature circuit. To achieve this varying resistance in the armature circuit on starting, a d.c. motor starter is used, as shown in Figure 24.28.

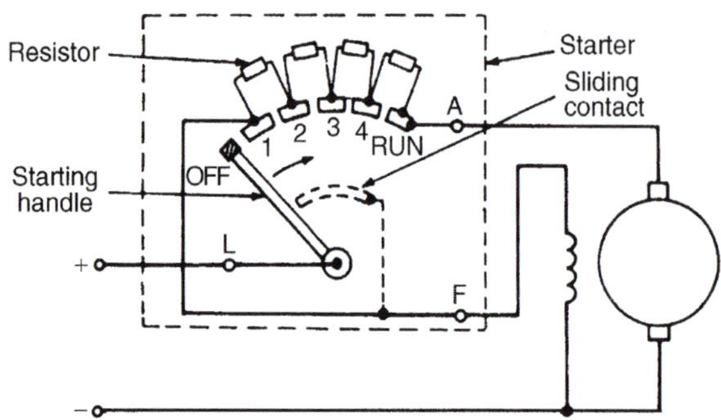

Figure 24.28

The starting handle is moved **slowly** in a clockwise direction to start the motor. For a shunt-wound motor, the field winding is connected to stud 1 or to L via a sliding contact on the starting handle, to give maximum field current, hence maximum flux, hence maximum torque on starting, since $T \propto \Phi I_a$
A similar arrangement without the field connection is used for series motors.

24.15 Speed control of d.c. motors

Shunt-wound motors

The speed of a shunt-wound d.c. motor, n, is proportional to $(V - I_a R_a)/\Phi$ (see equation (9)). The speed is varied either by varying the value of flux, Φ, or by varying the value of R_a. The former is achieved by using a variable resistor in series with the field winding, as shown in Figure 24.29(a) and such a resistor is called the **shunt field regulator**. As the value of resistance of the shunt field regulator is increased, the value of the field current, I_f, is decreased.

This results in a decrease in the value of flux, Φ, and hence an increase in the speed, since $n \propto 1/\Phi$. Thus only speeds **above** that given without a shunt field regulator can be obtained by this method. Speeds **below** those given by $(V - I_a R_a)/\Phi$ are obtained by increasing the resistance in the armature circuit, as shown in Figure 24.29(b), where

$$n \propto \frac{V - I_a(R_a + R)}{\Phi}$$

Since resistor R is in series with the armature, it carries the full armature current and results in a large

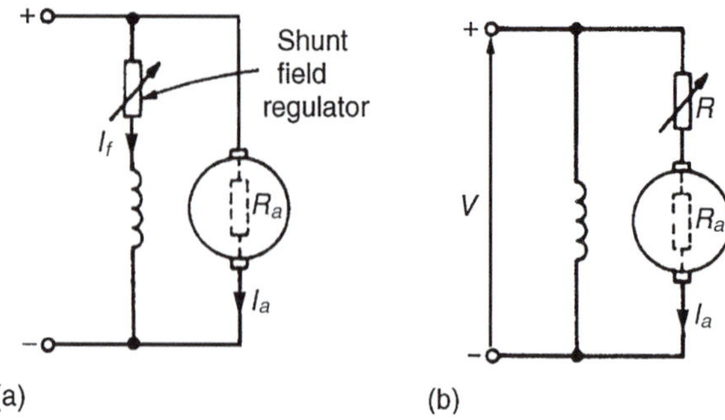

Figure 24.29

power loss in large motors where a considerable speed reduction is required for long periods.
These methods of speed control are demonstrated in the following worked problem.

Problem 28. A 500 V shunt motor runs at its normal speed of 10 rev/s when the armature current is 120 A. The armature resistance is 0.2 Ω.

(a) Determine the speed when the current is 60 A and a resistance of 0.5 Ω is connected in series with the armature, the shunt field remaining constant.

(b) Determine the speed when the current is 60 A and the shunt field is reduced to 80% of its normal value by increasing resistance in the field circuit.

(a) With reference to Figure 24.29(b), back e.m.f. at 120 A,

$$E_1 = V - I_a R_a = 500 - (120)(0.2) = 500 - 24 = 476 \text{ volts}$$

When $I_a = 60$ A, $E_2 = 500 - (60)(0.2 + 0.5)$
$= 500 - (60)(0.7)$
$= 500 - 42 = 458$ volts

Now $\frac{E_1}{E_2} = \frac{\Phi_1 n_1}{\Phi_2 n_2}$

i.e. $\frac{476}{458} = \frac{\Phi_1(10)}{\Phi_1(n_2)}$ since $\Phi_2 = \Phi_1$

from which, speed $n_2 = \frac{(10)(458)}{(476)} = $ **9.62 rev/s**

(b) Back e.m.f. when $I_a = 60$ A,

$$E_2 = 500 - (60)(0.2) = 500 - 12 = 488 \text{ volts}$$

Now $\frac{E_1}{E_2} = \frac{\Phi_1 n_1}{\Phi_2 n_2}$

Part 3

i.e. $\frac{476}{488}=\frac{(\Phi_1)(10)}{(0.8\Phi_1)(n_3)}$, since $\Phi_2=0.8\Phi_1$

from which, speed $n_3=\frac{(10)(488)}{(0.8)(476)}=\mathbf{12.82\ rev/s}$

Series-wound motors

The speed control of series-wound motors is achieved using either (a) field resistance or (b) armature resistance techniques.

(a) The speed of a d.c. series-wound motor is given by:

$$n=k\left(\frac{V-IR}{\Phi}\right)$$

where k is a constant, V is the terminal voltage, R is the combined resistance of the armature and series field and Φ is the flux.

Thus, a reduction in flux results in an increase in speed. This is achieved by putting a variable resistance in parallel with the field winding and reducing the field current, and hence flux, for a given value of supply current. A circuit diagram of this arrangement is shown in Figure 24.30(a). A variable resistor connected in parallel with the series-wound field to control speed is called a **diverter**. Speeds above those given with no diverter are obtained by this method. Worked Problem 29 below demonstrates this method.

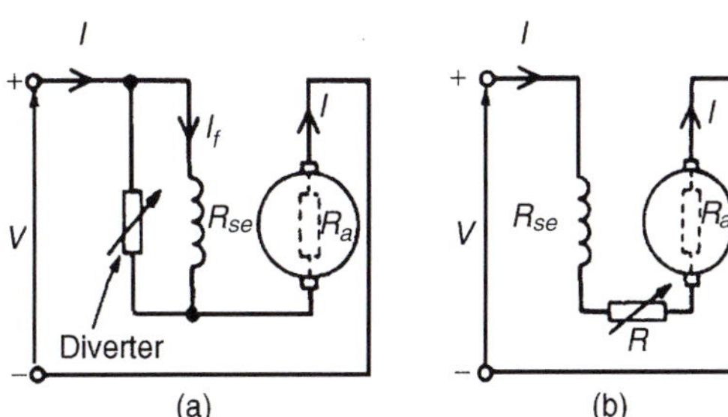

Figure 24.30

(b) Speeds below normal are obtained by connecting a variable resistor in series with the field winding and armature circuit, as shown in Figure 24.30(b). This effectively increases the value of R in the equation

$$n=k\left(\frac{V-IR}{\Phi}\right)$$

and thus reduces the speed. Since the additional resistor carries the full supply current, a large power loss is associated with large motors in which a considerable speed reduction is required for long periods. This method is demonstrated in worked Problem 30.

Problem 29. On full-load a 300 V series motor takes 90 A and runs at 15 rev/s. The armature resistance is 0.1 Ω and the series winding resistance is 50 mΩ. Determine the speed when developing full load torque but with a 0.2 Ω diverter in parallel with the field winding. (Assume that the flux is proportional to the field current.)

At 300 V, e.m.f., $E_1=V-IR$

$$\begin{aligned}&=V-I(R_a+R_{se})\\&=300-(90)(0.1+0.05)\\&=300-(90)(0.15)\\&=300-13.5=286.5\text{ volts}\end{aligned}$$

With the 0.2 Ω diverter in parallel with R_{se} (see Figure 24.30(a)), the equivalent resistance,

$$R=\frac{(0.2)(0.05)}{(0.2)+(0.05)}=\frac{(0.2)(0.05)}{(0.25)}=0.04\,\Omega$$

By current division, current I_f (in Figure 24.30(a))

$$=\left(\frac{0.2}{0.2+0.05}\right)I=0.8I$$

Torque, $T\propto I_a\Phi$ and for full load torque, $I_{a1}\Phi_1=I_{a2}\Phi_2$

Since flux is proportional to field current $\Phi_1\propto I_{a1}$ and $\Phi_2\propto 0.8I_{a2}$ then $(90)(90)=(I_{a2})(0.8I_{a2})$

from which, $I_{a2}^2=\frac{(90)^2}{0.8}$ and $I_{a2}=\frac{90}{\sqrt{(0.8)}}=100.62\text{ A}$

Hence e.m.f. $E_2=V-I_{a2}(R_a+R)$

$$\begin{aligned}&=300-(100.62)(0.1+0.04)\\&=300-(100.62)(0.14)\\&=300-14.087=285.9\text{ volts}\end{aligned}$$

Now e.m.f., $E\propto\Phi n$, from which

$$\frac{E_1}{E_2}=\frac{\Phi_1 n_1}{\Phi_2 n_2}=\frac{I_{a1}n_1}{0.8I_{a2}n_2}$$

Hence $\frac{286.5}{285.9}=\frac{(90)(15)}{(0.8)(100.62)n_2}$

and new speed, $n_2=\frac{(285.9)(90)(15)}{(286.5)(0.8)(100.62)}=\mathbf{16.74\ rev/s}$

Thus the speed of the motor has increased from 15 rev/s (i.e. 900 rev/min) to 16.74 rev/s (i.e. 1004 rev/min) by inserting a 0.2 Ω diverter resistance in parallel with the series winding.

Part 3

Problem 30. A series motor runs at 800 rev/min when the voltage is 400 V and the current is 25 A. The armature resistance is 0.4 Ω and the series field resistance is 0.2 Ω. Determine the resistance to be connected in series to reduce the speed to 600 rev/min with the same current.

With reference to Figure 24.30(b), at 800 rev/min,

$$\text{e.m.f., } E_1 = V - I(R_a + R_{se}) = 400 - (25)(0.4 + 0.2)$$

$$= 400 - (25)(0.6)$$

$$= 400 - 15 = 385 \text{ volts}$$

At 600 rev/min, since the current is unchanged, the flux is unchanged.

Thus $E \propto \Phi n$, or $E \propto n$, and $\dfrac{E_1}{E_2} = \dfrac{n_1}{n_2}$

Hence $\dfrac{385}{E_2} = \dfrac{800}{600}$

from which, $E_2 = \dfrac{(385)(600)}{(800)} = 288.75$ volts

and $E_2 = V - I(R_a + R_{se} + R)$

Hence $288.75 = 400 - 25(0.4 + 0.2 + R)$

Rearranging gives: $0.6 + R = \dfrac{400 - 288.75}{25} = 4.45$

from which, extra series resistance, $R = 4.45 - 0.6$

i.e. $\boldsymbol{R = 3.85\,\Omega}$

Thus the addition of a series resistance of 3.85 Ω has reduced the speed from 800 rev/min to 600 rev/min.

Now try the following Practice Exercise

Practice Exercise 112 Speed control of d.c. motors (Answers on page 825)

1. A 350 V shunt motor runs at its normal speed of 12 rev/s when the armature current is 90 A. The resistance of the armature is 0.3 Ω. (a) Find the speed when the current is 45 A and a resistance of 0.4 Ω is connected in series with the armature, the shunt field remaining constant. (b) Find the speed when the current is 45 A and the shunt field is reduced to 75% of its normal value by increasing resistance in the field circuit.
2. A series motor runs at 900 rev/min when the voltage is 420 V and the current is 40 A. The armature resistance is 0.3 Ω and the series field resistance is 0.2 Ω. Calculate the resistance to be connected in series to reduce the speed to 720 rev/min with the same current.
3. A 320 V series motor takes 80 A and runs at 1080 rev/min at full load. The armature resistance is 0.2 Ω and the series winding resistance is 0.05 Ω. Assuming the flux is proportional to the field current, calculate the speed when developing full-load torque, but with a 0.15 Ω diverter in parallel with the field winding.

24.16 Motor cooling

Motors are often classified according to the type of enclosure used, the type depending on the conditions under which the motor is used and the degree of ventilation required.

The most common type of protection is the **screen-protected type**, where ventilation is achieved by fitting a fan internally, with the openings at the end of the motor fitted with wire mesh.

A **drip-proof type** is similar to the screen-protected type but has a cover over the screen to prevent drips of water entering the machine.

A **flame-proof type** is usually cooled by the conduction of heat through the motor casing.

With a **pipe-ventilated type**, air is piped into the motor from a dust-free area, and an internally fitted fan ensures the circulation of this cool air.

Part 3

For fully worked solutions to each of the problems in Practice Exercises 106 to 112 in this chapter, go to the website:
www.routledge.com/cw/bird

Chapter 25

Three-phase induction motors

Why it is important to understand: **Three-phase induction motors**

The induction motor is a three-phase a.c. motor and is the most widely used machine in industrial applications. Its characteristic features are simple and rugged construction, low cost and minimum maintenance, high reliability and sufficiently high efficiency, and needs no extra starting motor and need not be synchronized. An induction motor, also called an asynchronous motor, is an a.c. motor in which all electromagnetic energy is transferred by inductive coupling from a primary winding to a secondary winding, the two windings being separated by an air gap. In three-phase induction motors that are inherently self-starting, energy transfer is usually from the stator to either a wound rotor or a short-circuited squirrel cage rotor. Three-phase cage rotor induction motors are widely used in industrial drives because they are rugged, reliable and economical. Single-phase induction motors are also used extensively for smaller loads. Although most a.c. motors have long been used in fixed-speed load drive service, they are increasingly being used in variable-frequency drive (VFD) service, variable-torque centrifugal fan, pump and compressor loads being by far the most important energy-saving applications for VFD service. Squirrel cage induction motors are most commonly used in both fixed-speed and VFD applications. This chapter explains the principle of operation of the different types of three-phase induction motors. Calculations of slip and losses and efficiency are also included. Methods of starting induction motors are also discussed.

At the end of this chapter you should be able to:

- appreciate the merits of three-phase induction motors
- understand how a rotating magnetic field is produced
- state the synchronous speed, $n_s = (f/p)$ and use in calculations
- describe the principle of operation of a three-phase induction motor
- distinguish between squirrel-cage and wound rotor types of motor
- understand how a torque is produced causing rotor movement
- understand and calculate slip
- derive expressions for rotor e.m.f., frequency, resistance, reactance, impedance, current and copper loss, and use them in calculations

Electrical Circuit Theory and Technology. 978-1-138-67349-6, © 2017 John Bird. Published by Taylor & Francis. All rights reserved.

- state the losses in an induction motor and calculate efficiency
- derive the torque equation for an induction motor, state the condition for maximum torque, and use in calculations
- describe torque-speed and torque-slip characteristics for an induction motor
- state and describe methods of starting induction motors
- state advantages of cage rotor and wound rotor types of induction motor
- describe the double cage induction motor
- state typical applications of three-phase induction motors

25.1 Introduction

In d.c. motors, introduced in Chapter 24, conductors on a rotating armature pass through a stationary magnetic field. In a **three-phase induction motor**, the magnetic field rotates and this has the advantage that no external electrical connections to the rotor need be made. Its name is derived from the fact that the current in the rotor is **induced** by the magnetic field instead of being supplied through electrical connections to the supply. The result is a motor which: (i) is cheap and robust, (ii) is explosion proof, due to the absence of a commutator or slip-rings and brushes with their associated sparking, (iii) requires little or no skilled maintenance and (iv) has self-starting properties when switched to a supply with no additional expenditure on auxiliary equipment. The principal disadvantage of a three-phase induction motor is that its speed cannot be readily adjusted.

25.2 Production of a rotating magnetic field

When a three-phase supply is connected to symmetrical three-phase windings, the currents flowing in the windings produce a magnetic field. This magnetic field is constant in magnitude and rotates at constant speed as shown below, and is called the **synchronous speed**.

With reference to Figure 25.1, the windings are represented by three single-loop conductors, one for each phase, marked $R_S R_F$, $Y_S Y_F$ and $B_S B_F$, the S and F signifying start and finish. In practice, each phase winding comprises many turns and is distributed around the stator; the single-loop approach is for clarity only.

When the stator windings are connected to a three-phase supply, the current flowing in each winding varies with time and is as shown in Figure 25.1(a). If the value

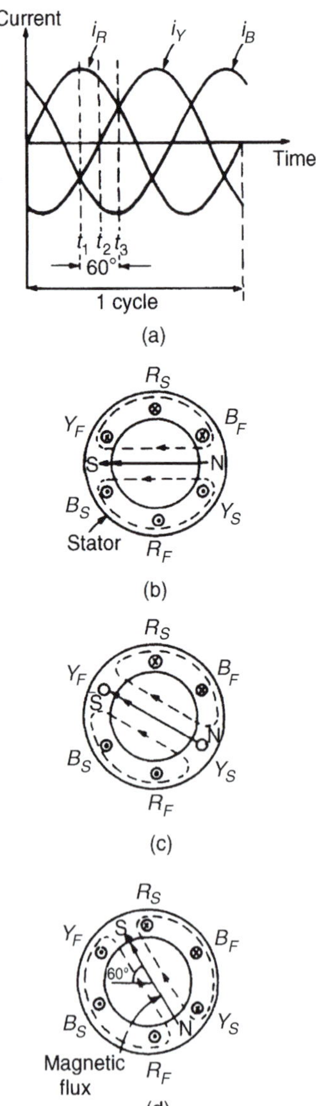

Figure 25.1

of current in a winding is positive, the assumption is made that it flows from start to finish of the winding, i.e. if it is the red phase, current flows from R_S to R_F, i.e. away from the viewer in R_S and towards the viewer in R_F. When the value of current is negative, the assumption is made that it flows from finish to start, i.e. towards the viewer in an 'S' winding and away from the viewer in an 'F' winding. At time, say t_1, shown in Figure 25.1(a), the current flowing in the red phase is a maximum positive value. At the same time, t_1, the currents flowing in the yellow and blue phases are both 0.5 times the maximum value and are negative.

The current distribution in the stator windings is therefore as shown in Figure 25.1(b), in which current flows away from the viewer (shown as ⊗) in R_S since it is positive, but towards the viewer (shown as ⊙) in Y_S and B_S, since these are negative. The resulting magnetic field is as shown, due to the 'solenoid' action and application of the corkscrew rule.

A short time later at time t_2, the current flowing in the red phase has fallen to about 0.87 times its maximum value and is positive, the current in the yellow phase is zero and the current in the blue phase is about 0.87 times its maximum value and is negative. Hence the currents and resultant magnetic field are as shown in Figure 25.1(c). At time t_3, the currents in the red and yellow phases are 0.5 of their maximum values and the current in the blue phase is a maximum negative value. The currents and resultant magnetic field are as shown in Figure 25.1(d).

Similar diagrams to Figure 25.1(b), (c) and (d) can be produced for all time values and these would show that the magnetic field travels through one revolution for each cycle of the supply voltage applied to the stator windings. By considering the flux values rather than the current values, it is shown below that the rotating magnetic field has a constant value of flux. The three coils shown in Figure 25.2(a) are connected in star to a three-phase supply. Let the positive directions of the fluxes produced by currents flowing in the coils, be ϕ_A, ϕ_B and ϕ_C, respectively. The directions of ϕ_A, ϕ_B and ϕ_C do not alter, but their magnitudes are proportional to the currents flowing in the coils at any particular time. At time t_1, shown in Figure 25.2(b), the currents flowing in the coils are:

i_B, a maximum positive value, i.e. the flux is towards point P;

i_A and i_C, half the maximum value and negative, i.e. the flux is away from point P.

These currents give rise to the magnetic fluxes ϕ_A, ϕ_B and ϕ_C, whose magnitudes and directions are as shown in Figure 25.2(c). The resultant flux is the phasor sum of ϕ_A, ϕ_B and ϕ_C, shown as Φ in Figure 25.2(c). At time t_2, the currents flowing are:

i_B, 0.866 × maximum positive value, i_C, zero, and i_A, 0.866 × maximum negative value.

The magnetic fluxes and the resultant magnetic flux are as shown in Figure 25.2(d).

At time t_3, i_B is 0.5 × maximum value and is positive
i_A is a maximum negative value, and
i_C is 0.5 × maximum value and is positive.

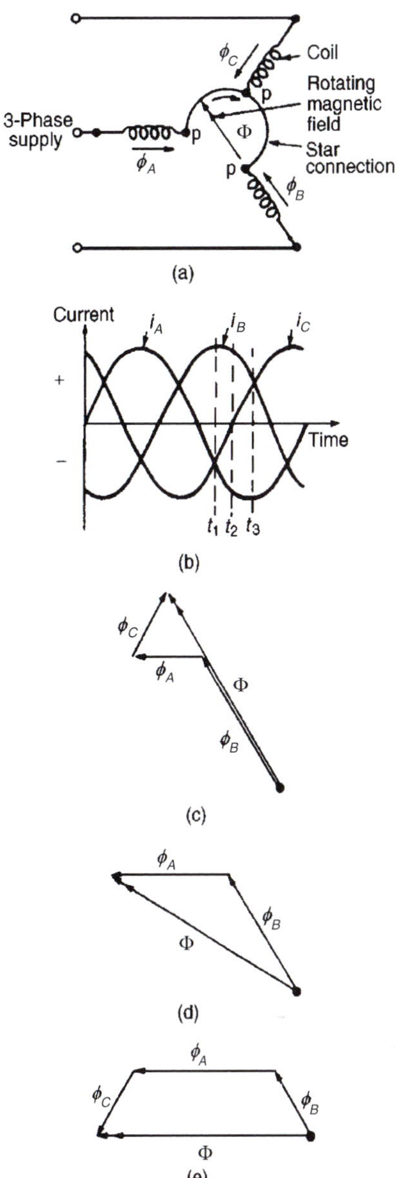

Figure 25.2

The magnetic fluxes and the resultant magnetic flux are as shown in Figure 25.2(e).

Inspection of Figures 25.2(c), (d) and (e) shows that the magnitude of the resultant magnetic flux, Φ, in each case is constant and is $1\frac{1}{2} \times$ the maximum value of ϕ_A, ϕ_B or ϕ_C, but that its direction is changing. The process of determining the resultant flux may be repeated for all values of time and shows that the magnitude of the resultant flux is constant for all values of time and also that it rotates at constant speed, making one revolution for each cycle of the supply voltage.

25.3 Synchronous speed

The rotating magnetic field produced by three-phase windings could have been produced by rotating a permanent magnet's north and south pole at synchronous speed, (shown as N and S at the ends of the flux phasors in Figures 25.1(b), (c) and (d)). For this reason, it is called a two-pole system and an induction motor using three-phase windings only is called a two-pole induction motor.

If six windings displaced from one another by 60° are used, as shown in Figure 25.3(a), by drawing the current and resultant magnetic field diagrams at various time values, it may be shown that one cycle of the supply current to the stator windings causes the magnetic field to move through half a revolution. The current distribution in the stator windings are shown in Figure 25.3(a), for the time t shown in Figure 25.3(b).

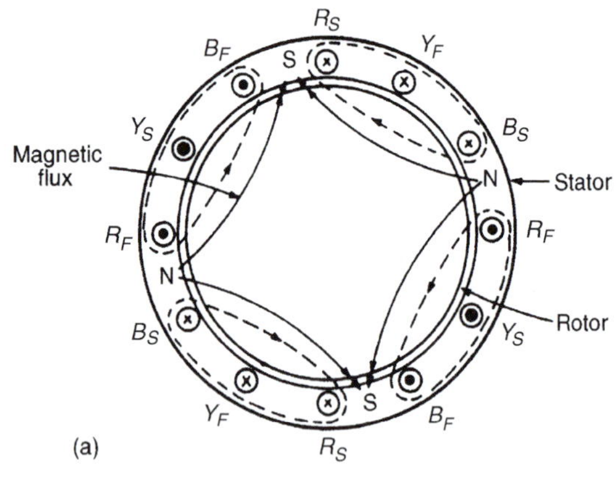

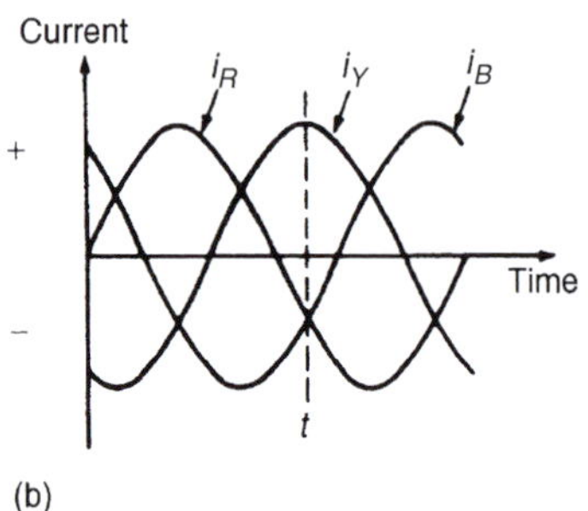

Figure 25.3

It can be seen that for six windings on the stator, the magnetic flux produced is the same as that produced by rotating two permanent magnet north poles and two permanent magnet south poles at synchronous speed. This is called a four-pole system and an induction motor using six-phase windings is called a four-pole induction motor. By increasing the number of phase windings the number of poles can be increased to any even number.

In general, if f is the frequency of the currents in the stator windings and the stator is wound to be equivalent to p **pairs** of poles, the speed of revolution of the rotating magnetic field, i.e. the synchronous speed, n_s is given by:

$$\boldsymbol{n_s = \frac{f}{p}\ \text{rev/s}}$$

Problem 1. A three-phase two-pole induction motor is connected to a 50 Hz supply. Determine the synchronous speed of the motor in rev/min.

From above, $n_s = f/p$ rev/s, where n_s is the synchronous speed, f is the frequency in hertz of the supply to the stator and p is the number of **pairs** of poles. Since the motor is connected to a 50 hertz supply, $f = 50$. The motor has a two-pole system, hence p, the number of pairs of poles is one.

Thus, synchronous speed, $n_s = \dfrac{50}{1} = 50$ rev/s

$= 50 \times 60$ rev/min

$=$ **3000 rev/min**

Problem 2. A stator winding supplied from a three-phase 60 Hz system is required to produce a magnetic flux rotating at 900 rev/min. Determine the number of poles.

Synchronous speed, $n_s = 900\,\text{rev/min} = \dfrac{900}{60}$ rev/s

$= 15$ rev/s

Since $n_s = \dfrac{f}{p}$ then $p = \dfrac{f}{n_s} = \dfrac{60}{15} = 4$

Hence the number of pole pairs is 4 and thus **the number of poles is 8**

Problem 3. A three-phase two-pole motor is to have a synchronous speed of 6000 rev/min. Calculate the frequency of the supply voltage.

Since $n_s = \frac{f}{p}$ then **frequency,** $f = (n_s)(p)$

$$= \left(\frac{6000}{60}\right)\left(\frac{2}{2}\right)$$

$$= \mathbf{100\,Hz}$$

Now try the following Practice Exercise

Practice Exercise 113 Synchronous speed (Answers on page 825)

1. The synchronous speed of a three-phase, four-pole induction motor is 60 rev/s. Determine the frequency of the supply to the stator windings.
2. The synchronous speed of a three-phase induction motor is 25 rev/s and the frequency of the supply to the stator is 50 Hz. Calculate the equivalent number of pairs of poles of the motor.
3. A six-pole, three-phase induction motor is connected to a 300 Hz supply. Determine the speed of rotation of the magnetic field produced by the stator.

25.4 Construction of a three-phase induction motor

The stator of a three-phase induction motor is the stationary part corresponding to the yoke of a d.c. machine. It is wound to give a two-pole, four-pole, six-pole, ... rotating magnetic field, depending on the rotor speed required. The rotor, corresponding to the armature of a d.c. machine, is built up of laminated iron, to reduce eddy currents.

In the type most widely used, known as a **squirrel-cage rotor**, copper or aluminium bars are placed in slots cut in the laminated iron, the ends of the bars being welded or brazed into a heavy conducting ring (see Figure 25.4(a)). A cross-sectional view of a three-phase induction motor is shown in Figure 25.4(b).

The conductors are placed in slots in the laminated iron rotor core. If the slots are skewed, better starting

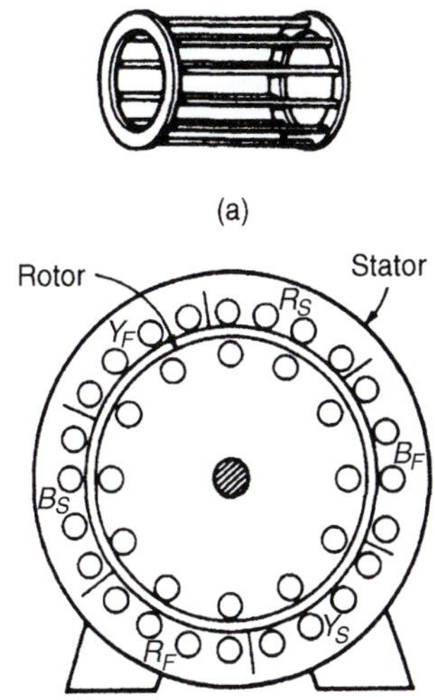

Figure 25.4

and quieter running is achieved. This type of rotor has no external connections, which means that slip-rings and brushes are not needed. The squirrel-cage motor is cheap, reliable and efficient.

Another type of rotor is the **wound rotor**. With this type there are phase windings in slots, similar to those in the stator. The windings may be connected in star or delta and the connections made to three slip-rings. The slip-rings are used to add external resistance to the rotor circuit, particularly for starting (see Section 25.13), but for normal running the slip-rings are short circuited.

The principle of operation is the same for both the squirrel-cage and the wound rotor machines.

25.5 Principle of operation of a three-phase induction motor

When a three-phase supply is connected to the stator windings, a rotating magnetic field is produced. As the magnetic flux cuts a bar on the rotor, an e.m.f. is induced in it and since it is joined, via the end conducting rings, to another bar one pole pitch away, a current flows in the bars. The magnetic field associated with this current flowing in the bars interacts with the rotating magnetic field and a force is produced, tending to turn the rotor in the same direction as the rotating magnetic field, (see Figure 25.5). Similar forces are applied to all the conductors on the rotor, so that a torque is produced, causing the rotor to rotate.

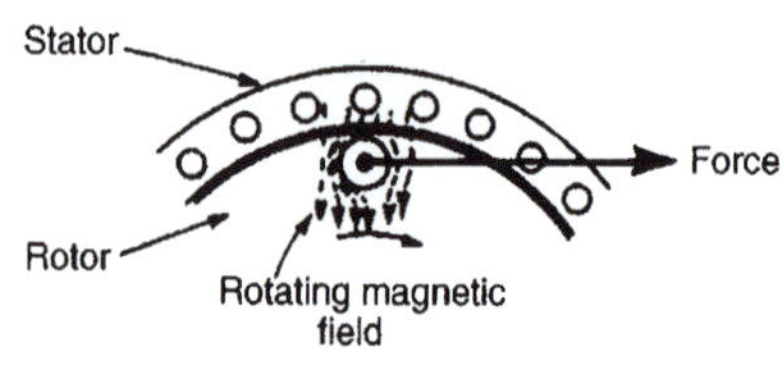

Figure 25.5

25.6 Slip

The force exerted by the rotor bars causes the rotor to turn in the direction of the rotating magnetic field. As the rotor speed increases, the rate at which the rotating magnetic field cuts the rotor bars is less and the frequency of the induced e.m.f.s in the rotor bars is less. If the rotor runs at the same speed as the rotating magnetic field, no e.m.f.s are induced in the rotor, hence there is no force on them and no torque on the rotor. Thus the rotor slows down. For this reason the rotor can never run at synchronous speed.

When there is no load on the rotor, the resistive forces due to windage and bearing friction are small and the rotor runs very nearly at synchronous speed. As the rotor is loaded, the speed falls and this causes an increase in the frequency of the induced e.m.f.s in the rotor bars and hence the rotor current, force and torque increase. The difference between the rotor speed, n_r, and the synchronous speed, n_s, is called the **slip speed**, i.e.

$$\textbf{slip speed} = \boldsymbol{n_s - n_r} \textbf{ rev/s}$$

The ratio $(n_s - n_r)/n_s$ is called the **fractional slip** or just the **slip**, s, and is usually expressed as a percentage. Thus

$$\textbf{slip, } s = \left(\frac{n_s - n_r}{n_s}\right) \times 100\%$$

Typical values of slip between no load and full load are about 4 to 5% for small motors and 1.5 to 2% for large motors.

Problem 4. The stator of a three-phase, four-pole induction motor is connected to a 50 Hz supply. The rotor runs at 1455 rev/min at full load. Determine (a) the synchronous speed and (b) the slip at full load.

(a) The number of pairs of poles, $p = 4/2 = 2$

The supply frequency $f = 50\,\text{Hz}$

The **synchronous speed,** $n_s = \frac{f}{p} = \frac{50}{2} = \textbf{25 rev/s}$

(b) The rotor speed, $n_r = \frac{1455}{60} = 24.25\,\text{rev/s}$

The slip, $s = \left(\frac{n_s - n_r}{n_s}\right) \times 100\%$

$$= \left(\frac{25 - 24.25}{25}\right) \times 100\% = \textbf{3\%}$$

Problem 5. A three-phase, 60 Hz induction motor has two poles. If the slip is 2% at a certain load, determine (a) the synchronous speed, (b) the speed of the rotor and (c) the frequency of the induced e.m.f.s in the rotor.

(a) $f = 60\,\text{Hz}$, $p = \frac{2}{2} = 1$

Hence **synchronous speed,** $n_s = \frac{f}{p} = \frac{60}{1}$

$$= \textbf{60 rev/s}$$

or $60 \times 60 = \textbf{3600 rev/min}$

(b) Since slip, $s = \left(\frac{n_s - n_r}{n_s}\right) \times 100\%$

$$2 = \left(\frac{60 - n_r}{60}\right) \times 100$$

Hence $\frac{2 \times 60}{100} = 60 - n_r$

i.e. $n_r = 60 - \frac{20 \times 60}{100} = 58.8\,\text{rev/s}$

i.e. the rotor runs at $58.8 \times 60 = \textbf{3528 rev/min}$

(c) Since the synchronous speed is 60 rev/s and that of the rotor is 58.8 rev/s, the rotating magnetic field cuts the rotor bars at (60 − 58.8), i.e. 1.2 rev/s.

Thus the frequency of the e.m.f.s induced in the rotor bars, $\boldsymbol{f = n_s p = (1.2)\left(\frac{2}{2}\right) = 1.2\,\textbf{Hz}}$

Problem 6. A three-phase induction motor is supplied from a 50 Hz supply and runs at 1200 rev/min when the slip is 4%. Determine the synchronous speed.

Slip, $s = \left(\frac{n_s - n_r}{n_s}\right) \times 100\%$

Rotor speed, $n_r = \frac{1200}{60} = 20\,\text{rev/s}$, and $s = 4$

Hence $\quad 4 = \left(\frac{n_s - 20}{n_s}\right) \times 100\%$

or $\quad 0.04 = \frac{n_s - 20}{n_s}$

from which, $n_s(0.04) = n_s - 20$

and $\quad 20 = n_s - 0.04 n_s = n_s(1 - 0.04)$

Hence **synchronous speed,** $\boldsymbol{n_s} = \left(\frac{20}{1 - 0.04}\right)$

$= 20.8\dot{3}\,\text{rev/s}$

$= (20.8\dot{3} \times 60)\,\text{rev/min}$

$= \mathbf{1250\,rev/min}$

Now try the following Practice Exercise

Practice Exercise 114 Slip (Answers on page 825)

1. A six-pole, three-phase induction motor runs at 970 rev/min at a certain load. If the stator is connected to a 50 Hz supply, find the percentage slip at this load.
2. A three-phase, 50 Hz induction motor has eight poles. If the full load slip is 2.5%, determine (a) the synchronous speed, (b) the rotor speed, and (c) the frequency of the rotor e.m.f.s.
3. A three-phase induction motor is supplied from a 60 Hz supply and runs at 1710 rev/min when the slip is 5%. Determine the synchronous speed.
4. A four-pole, three-phase, 50 Hz induction motor runs at 1440 rev/min at full load. Calculate (a) the synchronous speed, (b) the slip and (c) the frequency of the rotor-induced e.m.f.s.

25.7 Rotor e.m.f. and frequency

Rotor e.m.f.

When an induction motor is stationary, the stator and rotor windings form the equivalent of a transformer, as shown in Figure 25.6.

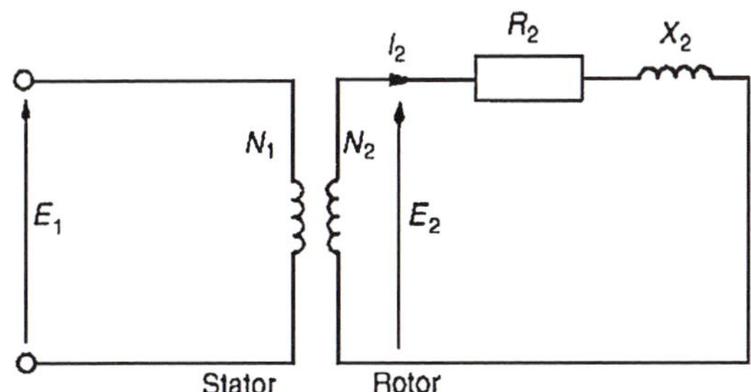

Figure 25.6

The rotor e.m.f. at standstill is given by

$$E_2 = \left(\frac{N_2}{N_1}\right) E_1 \qquad (1)$$

where E_1 is the supply voltage per phase to the stator.

When an induction motor is running, the induced e.m.f. in the rotor is less since the relative movement between conductors and the rotating field is less. The induced e.m.f. is proportional to this movement, hence it must be proportional to the slip, s.

Hence when running, rotor e.m.f. per phase

$$= E_r = sE_2 = s\left(\frac{N_2}{N_1}\right) E_1 \qquad (2)$$

Rotor frequency

The rotor e.m.f. is induced by an alternating flux and the rate at which the flux passes the conductors is the slip speed. Thus the frequency of the rotor e.m.f. is given by:

$$f_r = (n_s - n_r)p = \frac{(n_s - n_r)}{n_s}(n_s p)$$

However $\left(\frac{n_s - n_r}{n_s}\right)$ is the slip s and $n_s p$ is the supply frequency f, hence

$$f_r = sf \qquad (3)$$

Problem 7. The frequency of the supply to the stator of an eight-pole induction motor is 50 Hz and the rotor frequency is 3 Hz. Determine (a) the slip and (b) the rotor speed.

(a) From equation (3), $f_r = sf$

Hence $\quad 3 = (s)(50)$

from which, **slip,** $s = \frac{3}{50} = \mathbf{0.06}$ or $\mathbf{6\%}$

(b) Synchronous speed, $n_s = \dfrac{f}{p} = \dfrac{50}{4} = 12.5\,\text{rev/s}$

or $(12.5 \times 60) = 750\,\text{rev/min}$

Slip, $s = \left(\dfrac{n_s - n_r}{n_s}\right)$, hence $0.06 = \left(\dfrac{12.5 - n_r}{12.5}\right)$

$(0.06)(12.5) = 12.5 - n_r$

and **rotor speed,** $\boldsymbol{n_r} = 12.5 - (0.06)(12.5)$

$= \mathbf{11.75\,rev/s}$ or $\mathbf{705\,rev/min}$

Now try the following Practice Exercise

Practice Exercise 115 Rotor frequency (Answers on page 825)

1. A 12-pole, three-phase, 50 Hz induction motor runs at 475 rev/min. Determine (a) the slip speed, (b) the percentage slip and (c) the frequency of rotor currents.
2. The frequency of the supply to the stator of a six-pole induction motor is 50 Hz and the rotor frequency is 2 Hz. Determine (a) the slip and (b) the rotor speed in rev/min.

25.8 Rotor impedance and current

Rotor resistance

The rotor resistance R_2 is unaffected by frequency or slip, and hence remains constant.

Rotor reactance

Rotor reactance varies with the frequency of the rotor current.

At standstill, reactance per phase, $X_2 = 2\pi fL$

When running, reactance per phase,

$X_r = 2\pi f_r L$

$= 2\pi(sf)L$ from equation (3)

$= s(2\pi fL)$

i.e. $$X_r = sX_2 \qquad (4)$$

Figure 25.7 represents the rotor circuit when running.

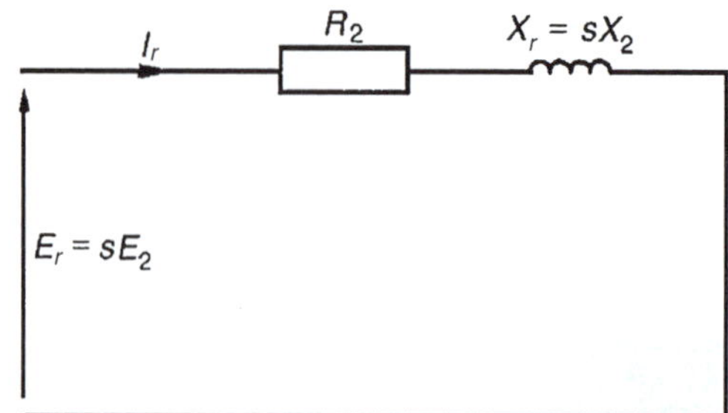

Figure 25.7

Rotor impedance

Rotor impedance per phase, $\boldsymbol{Z_r = \sqrt{[R_2^2 + (sX_2)^2]}}$ (5)

At standstill, slip $s = 1$, then $\boldsymbol{Z_2 = \sqrt{[R_2^2 + X_2^2]}}$ (6)

Rotor current

From Figures 25.6 and 25.7,

at standstill, starting current, $$I_2 = \frac{E_2}{Z_2} = \frac{\left(\frac{N_2}{N_1}\right)E_1}{\sqrt{[R_2^2 + X_2^2]}} \qquad (7)$$

and when running, current, $$I_r = \frac{E_r}{Z_r} = \frac{s\left(\frac{N_2}{N_1}\right)E_1}{\sqrt{[R_2^2 + (sX_2)^2]}} \qquad (8)$$

25.9 Rotor copper loss

Power $P = 2\pi nT$, where T is the torque in newton metres, hence torque $T = (P/2\pi n)$

If P_2 is the power input to the rotor from the rotating field, and P_m is the mechanical power output (including friction losses)

then $$T = \frac{P_2}{2\pi n_s} = \frac{P_m}{2\pi n_r}$$

from which, $\dfrac{P_2}{n_s} = \dfrac{P_m}{n_r}$ or $\dfrac{P_m}{P_2} = \dfrac{n_r}{n_s}$

Hence $1 - \dfrac{P_m}{P_2} = 1 - \dfrac{n_r}{n_s}$

$$\frac{P_2 - P_m}{P_2} = \frac{n_s - n_r}{n_s} = s$$

$P_2 - P_m$ is the electrical or copper loss in the rotor, i.e. $P_2 - P_m = I_r^2 R_2$

Hence **slip,** $\boldsymbol{s = \dfrac{\text{rotor copper loss}}{\text{rotor input}} = \dfrac{I_r^2 R_2}{P_2}}$ (9)

or power input to the rotor, $\boldsymbol{P_2 = \dfrac{I_r^2 R_2}{s}}$ (10)

25.10 Induction motor losses and efficiency

Figure 25.8 summarizes losses in induction motors.

Motor efficiency, $\boldsymbol{\eta = \dfrac{\text{output power}}{\text{input power}} = \dfrac{P_m}{P_1} \times 100\%}$

Problem 8. The power supplied to a three-phase induction motor is 32 kW and the stator losses are 1200 W. If the slip is 5%, determine (a) the rotor copper loss, (b) the total mechanical power developed by the rotor, (c) the output power of the motor if friction and windage losses are 750 W, and (d) the efficiency of the motor, neglecting rotor iron loss.

(a) Input power to rotor = stator input power − stator losses

$$= 32\,\text{kW} - 1.2\,\text{kW}$$
$$= 30.8\,\text{kW}$$

From equation (9), slip $= \dfrac{\text{rotor copper loss}}{\text{rotor input}}$

i.e. $\dfrac{5}{100} = \dfrac{\text{rotor copper loss}}{30.8}$

from which, **rotor copper loss** $= (0.05)(30.8)$
$= \mathbf{1.54\,kW}$

(b) Total mechanical power developed by the rotor

$$= \text{rotor input power} - \text{rotor losses}$$
$$= 30.8 - 1.54 = \mathbf{29.26\,kW}$$

(c) Output power of motor

$$= \text{power developed by the rotor} - \text{friction and windage losses}$$
$$= 29.26 - 0.75 = \mathbf{28.51\,kW}$$

(d) Efficiency of induction motor,

$$\eta = \left(\frac{\text{output power}}{\text{input power}}\right) \times 100\%$$
$$= \left(\frac{28.51}{32}\right) \times 100\% = \mathbf{89.10\%}$$

Problem 9. The speed of the induction motor of Problem 8 is reduced to 35% of its synchronous speed by using external rotor resistance. If the torque and stator losses are unchanged, determine (a) the rotor copper loss, and (b) the efficiency of the motor.

(a) Slip, $s = \left(\dfrac{n_s - n_r}{n_s}\right) \times 100\%$

$$= \left(\frac{n_s - 0.35 n_s}{n_s}\right) \times 100\%$$
$$= (0.65)(100) = 65\%$$

Input power to rotor = 30.8 kW (from Problem 8)

Since $s = \dfrac{\text{rotor copper loss}}{\text{rotor input}}$

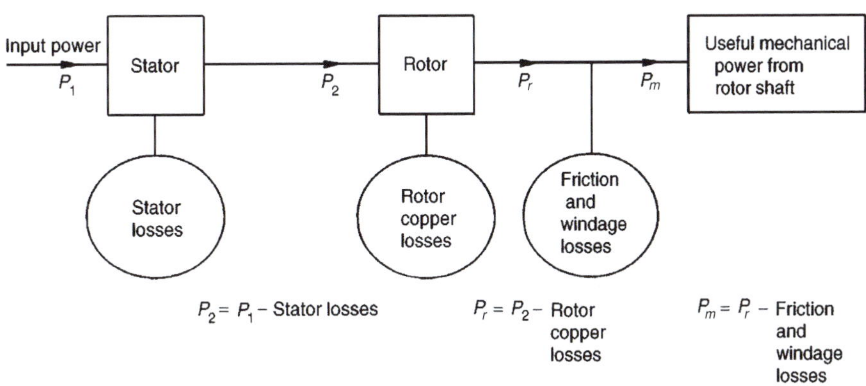

Figure 25.8

then **rotor copper loss**

$$= (s)(\text{rotor input})$$

$$= \left(\frac{65}{100}\right)(30.8) = \mathbf{20.02\ kW}$$

(b) Power developed by rotor

$$= \text{input power to rotor} - \text{rotor copper loss}$$

$$= 30.8 - 20.02 = 10.78\,\text{kW}$$

Output power of motor

$$= \text{power developed by rotor} - \text{friction and windage losses}$$

$$= 10.78 - 0.75 = 10.03\,\text{kW}$$

$$\text{Efficiency, } \eta = \frac{\text{output power}}{\text{input power}} \times 100\%$$

$$= \left(\frac{10.03}{32}\right) \times 100\%$$

$$= \mathbf{31.34\%}$$

Now try the following Practice Exercise

Practice Exercise 116 Losses and efficiency (Answers on page 825)

1. The power supplied to a three-phase induction motor is 50 kW and the stator losses are 2 kW. If the slip is 4%, determine (a) the rotor copper loss, (b) the total mechanical power developed by the rotor, (c) the output power of the motor if friction and windage losses are 1 kW and (d) the efficiency of the motor, neglecting rotor iron losses.
2. By using external rotor resistance, the speed of the induction motor in Problem 1 is reduced to 40% of its synchronous speed. If the torque and stator losses are unchanged, calculate (a) the rotor copper loss and (b) the efficiency of the motor.

25.11 Torque equation for an induction motor

$$\text{Torque } T = \frac{P_2}{2\pi n_s} = \left(\frac{1}{2\pi n_s}\right)\left(\frac{I_r^2 R_2}{s}\right) \text{ (from equation (10))}$$

From equation (8), $I_r = \dfrac{s(N_2/N_1)E_1}{\sqrt{[R_2^2 + (sX_2)^2]}}$

Hence torque per phase,

$$T = \left(\frac{1}{2\pi n_s}\right)\left[\frac{s^2(N_2/N_1)^2 E_1^2}{R_2^2 + (sX_2)^2}\right]\left(\frac{R_2}{s}\right)$$

$$\text{i.e. } T = \left(\frac{1}{2\pi n_s}\right)\left[\frac{s(N_2/N_1)^2 E_1^2 R_2}{R_2^2 + (sX_2)^2}\right]$$

If there are m phases then

$$\text{torque, } T = \left(\frac{m}{2\pi n_s}\right)\left[\frac{s(N_2/N_1)^2 E_1^2 R_2}{R_2^2 + (sX_2)^2}\right]$$

$$\text{i.e. } \boldsymbol{T = \left[\frac{m(N_2/N_1)^2}{2\pi n_s}\right]\left[\frac{sE_1^2 R_2}{R_2^2 + (sX_2)^2}\right]} \quad (11)$$

$$= k\left(\frac{sE_1^2 R_2}{R_2^2 + (sX_2)^2}\right) \quad \text{where } k \text{ is a constant for a particular machine,}$$

$$\text{i.e. } \quad \textbf{torque } \boldsymbol{T \propto \frac{sE_1^2 R_2}{R_2^2 + (sX_2)^2}} \quad (12)$$

Under normal conditions, the supply voltage is usually constant, hence equation (12) becomes:

$$T \propto \frac{sR_2}{R_2^2 + (sX_2)^2} \propto \frac{R_2}{\dfrac{R_2^2}{s} + sX_2^2}$$

The torque will be a maximum when the denominator is a minimum and this occurs when $R_2^2/s = sX_2^2$

i.e. when $s = \dfrac{R_2}{X_2}$ or $R_2 = sX_2 = X_r$

(from equation (4))

Thus **maximum torque** occurs when rotor resistance and rotor reactance are equal, i.e. $\boldsymbol{R_2 = X_r}$.

Problems 10 to 13 following illustrate some of the characteristics of three-phase induction motors.

Problem 10. A 415 V, three-phase, 50 Hz, four-pole, star-connected induction motor runs at 24 rev/s on full load. The rotor resistance and reactance per phase are 0.35 Ω and 3.5 Ω, respectively, and the effective rotor–stator turns ratio is 0.85:1. Calculate (a) the synchronous speed, (b) the slip, (c) the full load torque, (d) the power output if mechanical losses amount to 770 W, (e) the maximum torque, (f) the speed at which maximum torque occurs and (g) the starting torque.

(a) Synchronous speed, $n_s = \frac{f}{p} = \frac{50}{2}$

$= \mathbf{25\,rev/s}$ or (25×60)

$= \mathbf{1500\,rev/min}$

(b) Slip, $s = \left(\frac{n_s - n_r}{n_s}\right) = \frac{25-24}{25} = \mathbf{0.04}$ or $\mathbf{4\%}$

(c) Phase voltage, $E_1 = \frac{415}{\sqrt{3}} = 239.6$ volts

Full load torque,

$$T = \left[\frac{m(N_2/N_1)^2}{2\pi n_s}\right]\left[\frac{sE_1^2 R_2}{R_2^2 + (sX_2)^2}\right]$$

(from equation (11))

$$= \left[\frac{3(0.85)^2}{2\pi(25)}\right]\left[\frac{0.04\,(239.6)^2 0.35}{(0.35)^2 + (0.04 \times 3.5)^2}\right]$$

$$= (0.01380)\left(\frac{803.71}{0.1421}\right) = \mathbf{78.05\,N\,m}$$

(d) Output power, including friction losses,

$P_m = 2\pi n_r T$

$= 2\pi(24)(78.05)$

$= 11\,770$ watts

Hence **power output** $= P_m$ − mechanical losses

$= 11\,770 - 770 = 11\,000$ W

$= \mathbf{11\,kW}$

(e) Maximum torque occurs when $R_2 = X_r = 0.35\,\Omega$

Slip, $s = \frac{R_2}{X_2} = \frac{0.35}{3.5} = 0.1$

Hence **maximum torque,**

$$\boldsymbol{T_m} = (0.01380)\left[\frac{sE_1^2 R_2}{R_2^2 + (sX_2)^2}\right] \quad \text{(from part (c))}$$

$$= (0.01380)\left[\frac{0.1(239.6)^2 0.35}{0.35^2 + 0.35^2}\right]$$

$$= (0.01380)\left[\frac{2009.29}{0.245}\right]$$

$= \mathbf{113.18\,N\,m}$

(f) For maximum torque, slip $s = 0.1$

Slip, $s = \left(\frac{n_s - n_r}{n_s}\right)$ i.e. $0.1 = \left(\frac{25 - n_r}{25}\right)$

Hence $(0.1)(25) = 25 - n_r$ and $n_r = 25 - (0.1)(25)$

Thus speed at which maximum torque occurs,

$n_r = 25 - 2.5$

$= \mathbf{22.5\,rev/s}$ or $\mathbf{1350\,rev/min}$

(g) At the start, i.e. at standstill, slip $s = 1$

$$\text{Hence starting torque} = \left[\frac{m(N_2/N_1)^2}{2\pi n_s}\right]\left[\frac{E_1^2 R_2}{R_2^2 + X_2^2}\right]$$

from equation (11) with $s = 1$

$$= (0.01380)\left[\frac{(239.6)^2 0.35}{0.35^2 + 3.5^2}\right]$$

$$= (0.01380)\left(\frac{20092.86}{12.3725}\right)$$

i.e. **starting torque = 22.41 N m**

(Note that the full load torque (from part (c)) is 78.05 N m but the starting torque is only 22.41 N m)

Problem 11. Determine for the induction motor in Problem 10 at full load, (a) the rotor current, (b) the rotor copper loss and (c) the starting current.

(a) From equation (8), **rotor current,**

$$I_r = \frac{s\left(\frac{N_2}{N_1}\right)E_1}{\sqrt{[R_2^2 + (sX_2)^2]}}$$

$$= \frac{(0.04)(0.85)(239.6)}{\sqrt{[0.35^2 + (0.04 \times 3.5)^2]}}$$

$$= \frac{8.1464}{0.37696} = \mathbf{21.61\,A}$$

(b) Rotor copper loss per phase $= I_r^2 R_2$

$= (21.61)^2(0.35)$

$= 163.45$ W

Total copper loss (for 3 phases) $= 3 \times 163.45$

$= \mathbf{490.35\,W}$

(c) From equation (7), starting current,

$$I_2 = \frac{\left(\frac{N_2}{N_1}\right)E_1}{\sqrt{[R_2^2 + X_2^2]}} = \frac{(0.85)(239.6)}{\sqrt{[0.35^2 + 3.5^2]}} = \mathbf{57.90\,A}$$

(Note that the starting current of 57.90 A is considerably higher than the full load current of 21.61 A)

Problem 12. For the induction motor in Problems 10 and 11, if the stator losses are 650 W, determine (a) the power input at full load, (b) the efficiency of the motor at full load and (c) the current taken from the supply at full load, if the motor runs at a power factor of 0.87 lagging.

(a) Output power $P_m = 11.770\,\text{kW}$ from part (d), Problem 10

Rotor copper loss $= 490.35\,\text{W} = 0.49035\,\text{kW}$ from part (b), Problem 11

Stator input power,

$$P_1 = P_m + \text{rotor copper loss} + \text{rotor stator loss}$$
$$= 11.770 + 0.49035 + 0.650$$
$$= \mathbf{12.910\,kW}$$

(b) Net power output $= 11\,\text{kW}$ from part (d), Problem 10

$$\text{Hence efficiency, } \eta = \frac{\text{output}}{\text{input}} \times 100\%$$
$$= \left(\frac{11}{12.910}\right) \times 100\%$$
$$= \mathbf{85.21\%}$$

(c) Power input, $P_1 = \sqrt{3} V_L I_L \cos\phi$ (see Chapter 22) and $\cos\phi = \text{p.f.} = 0.87$

$$\text{hence, } \textbf{supply current, } I_L = \frac{P_1}{\sqrt{3} V_L \cos\phi}$$
$$= \frac{12.910 \times 1000}{\sqrt{3}(415)\,0.87}$$
$$= \mathbf{20.64\,A}$$

Problem 13. For the induction motor of Problems 10 to 12, determine the resistance of the rotor winding required for maximum starting torque.

From equation (4), rotor reactance $X_r = sX_2$
At the moment of starting, slip, $s = 1$
Maximum torque occurs when rotor reactance equals rotor resistance; hence for **maximum torque,** $R_2 = X_r = sX_2 = X_2 = \mathbf{3.5\,\Omega}$
Thus if the induction motor was a wound rotor type with slip-rings then an external star-connected resistance of $(3.5 - 0.35)\,\Omega = 3.15\,\Omega$ per phase could be added to the rotor resistance to give maximum torque at starting (see Section 25.13).

Now try the following Practice Exercise

Practice Exercise 117 The torque equation (Answers on page 825)

1. A 400 V, three-phase, 50 Hz, two-pole, star-connected induction motor runs at 48.5 rev/s on full load. The rotor resistance and reactance per phase are 0.4 Ω and 4.0 Ω, respectively, and the effective rotor–stator turns ratio is 0.8:1. Calculate (a) the synchronous speed, (b) the slip, (c) the full load torque, (d) the power output if mechanical losses amount to 500 W, (e) the maximum torque, (f) the speed at which maximum torque occurs, and (g) the starting torque.
2. For the induction motor in Problem 1, calculate at full load (a) the rotor current, (b) the rotor copper loss and (c) the starting current.
3. If the stator losses for the induction motor in Problem 1 are 525 W, calculate at full load (a) the power input, (b) the efficiency of the motor and (c) the current taken from the supply if the motor runs at a power factor of 0.84
4. For the induction motor in Problem 1, determine the resistance of the rotor winding required for maximum starting torque.

25.12 Induction motor torque–speed characteristics

From Problem 10, parts (c) and (g), it is seen that the normal starting torque may be less than the full load torque. Also, from Problem 10, parts (e) and (f), it is seen that the speed at which maximum torque occurs is determined by the value of the rotor resistance. At synchronous speed, slip $s = 0$ and torque is zero. From these observations, the torque-speed and torque-slip characteristics of an induction motor are as shown in Figure 25.9.

The rotor resistance of an induction motor is usually small compared with its reactance (for example, $R_2 = 0.35\,\Omega$ and $X_2 = 3.5\,\Omega$ in the above problems), so that maximum torque occurs at a high speed, typically about 80% of synchronous speed.

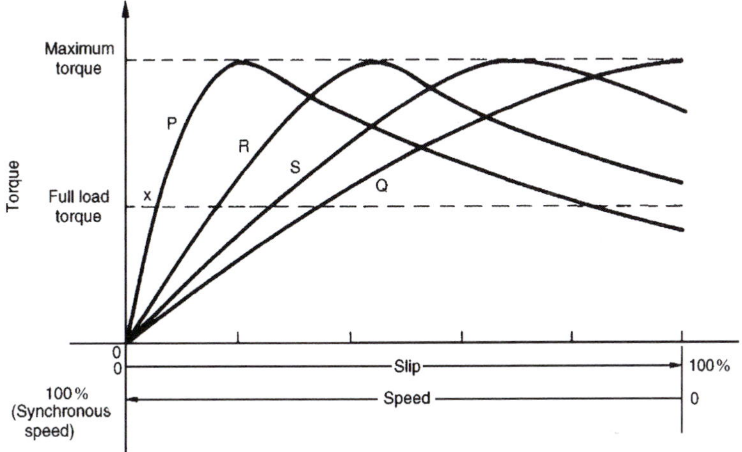

Figure 25.9

Curve P in Figure 25.9 is a typical characteristic for an induction motor. The curve P cuts the full load torque line at point X, showing that at full load the slip is about 4–5%. The normal operating conditions are between 0 and X, thus it can be seen that for normal operation the speed variation with load is quite small – the induction motor is an almost constant-speed machine. Redrawing the speed–torque characteristic between 0 and X gives the characteristic shown in Figure 25.10, which is similar to a d.c. shunt motor as shown in Chapter 24.

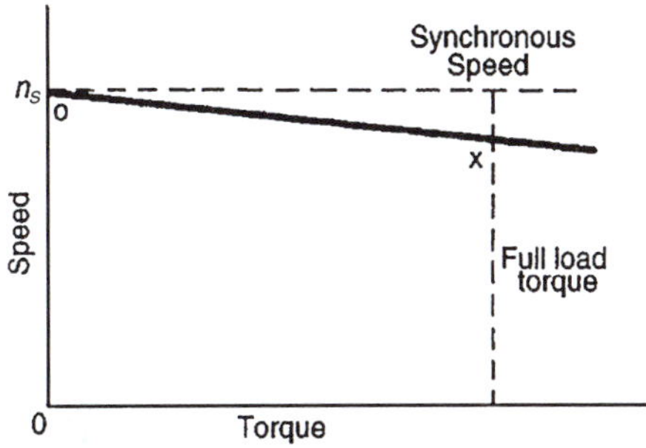

Figure 25.10

If maximum torque is required at starting then a high resistance rotor is necessary, which gives characteristic Q in Figure 25.9. However, as can be seen, the motor has a full load slip of over 30%, which results in a drop in efficiency. Also, such a motor has a large speed variation with variations of load. Curves R and S of Figure 25.9 are characteristics for values of rotor resistances between those of P and Q. Better starting torque than for curve P is obtained, but with lower efficiency and with speed variations under operating conditions.

A **squirrel-cage induction motor** would normally follow characteristic P. This type of machine is highly efficient and about constant-speed under normal running conditions. However, it has a poor starting torque and must be started off-load or very lightly loaded (see Section 25.13 below). Also, on starting, the current can be four or five times the normal full load current, due to the motor acting like a transformer with secondary short-circuited. In Problem 11, for example, the current at starting was nearly three times the full load current.

A **wound-rotor induction motor** would follow characteristic P when the slip-rings are short-circuited, which is the normal running condition. However, the slip-rings allow for the addition of resistance to the rotor circuit externally and, as a result, for starting, the motor can have a characteristic similar to curve Q in Figure 25.9 and the high starting current experienced by the cage induction motor can be overcome.

In general, for three-phase induction motors, the power factor is usually between about 0.8 and 0.9 lagging, and the full load efficiency is usually about 80–90%

From equation (12) it is seen that torque is proportional to the square of the supply voltage. Any voltage variations therefore would seriously affect the induction motor performance.

25.13 Starting methods for induction motors

Squirrel-cage rotor

(i) **Direct-on-line starting**

With this method, starting current is high and may cause interference with supplies to other consumers.

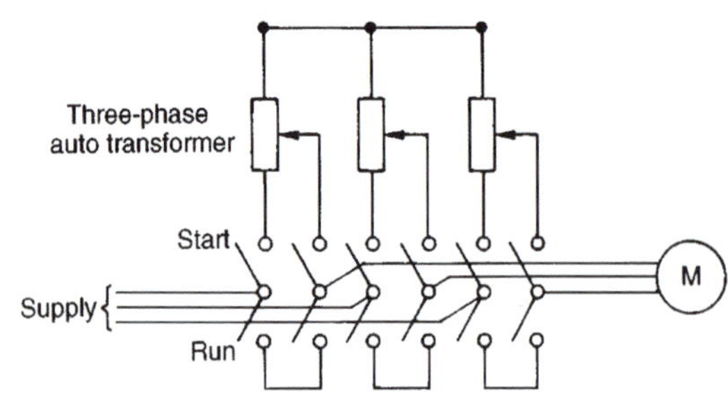

Figure 25.11

(ii) **Auto transformer starting**

With this method, an auto transformer is used to reduce the stator voltage, E_1, and thus the starting current (see equation (7)). However, the starting torque is seriously reduced (see equation (12)), so the voltage is reduced only sufficiently to give the required reduction of the starting current. A typical arrangement is shown in Figure 25.11. A double-throw switch connects the auto transformer in circuit for starting, and when the motor is up to speed the switch is moved to the run position, which connects the supply directly to the motor.

(iii) **Star–delta starting**

With this method for starting, the connections to the stator phase winding are star-connected, so that the voltage across each phase winding is $1/\sqrt{3}$ (i.e. 0.577) of the line voltage. For running, the windings are switched to delta connection. A typical arrangement is shown in Figure 25.12. This method of starting is less expensive than by auto transformer.

Wound rotor

When starting on load is necessary, a wound rotor induction motor must be used. This is because maximum torque at starting can be obtained by adding external resistance to the rotor circuit via slip-rings (see Problem 13). A face-plate type starter is used, and as the resistance is gradually reduced, the machine characteristics at each stage will be similar to Q, S, R and P of Figure 25.13. At each resistance step, the motor operation will transfer from one characteristic to the next so that the overall starting characteristic will be as shown by the bold line in Figure 25.13. For very large induction motors, very gradual and smooth starting is achieved by a liquid-type resistance.

25.14 Advantages of squirrel-cage induction motors

The advantages of squirrel-cage motors compared with the wound rotor type are that they:

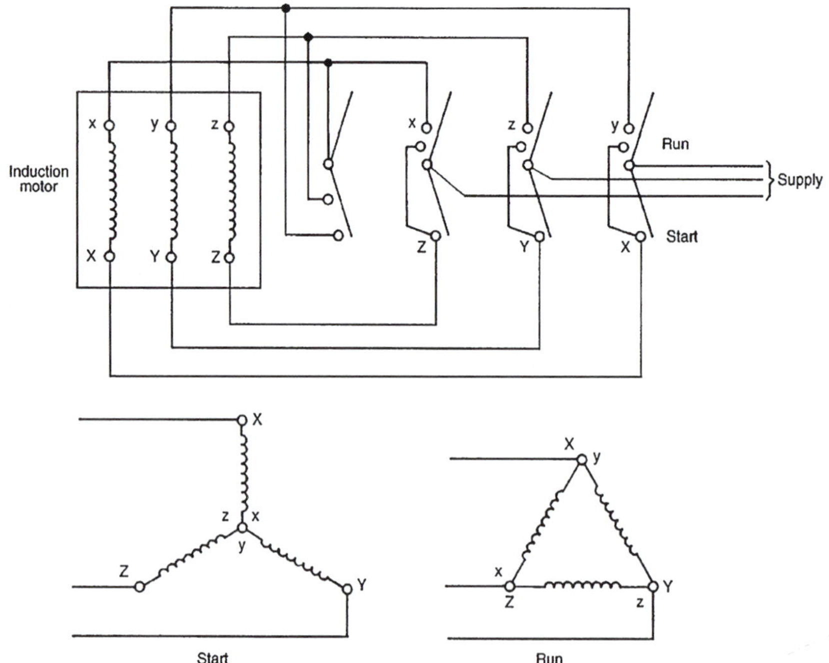

Figure 25.12

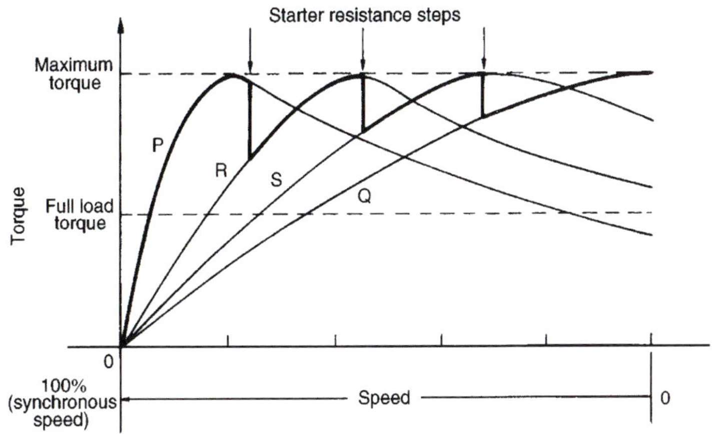

Figure 25.13

(i) are cheaper and more robust

(ii) have slightly higher efficiency and power factor

(iii) are explosion-proof, since the risk of sparking is eliminated by the absence of slip-rings and brushes.

25.15 Advantages of wound rotor induction motor

The advantages of the wound rotor motor compared with the cage type are that they:

(i) have a much higher starting torque

(ii) have a much lower starting current

(iii) have a means of varying speed by use of external rotor resistance.

25.16 Double cage induction motor

The advantages of squirrel-cage and wound rotor induction motors are combined in the double cage induction motor. This type of induction motor is specially constructed with the rotor having two cages, one inside the other. The outer cage has high-resistance conductors so that maximum torque is achieved at or near starting. The inner cage has normal low-resistance copper conductors but high reactance since it is embedded deep in the iron core. The torque–speed characteristic of the inner cage is that of a normal induction motor, as shown in Figure 25.14. At starting, the outer cage produces the torque, but when running the inner cage produces the torque. The combined characteristic of inner and outer cages is shown in Figure 25.14. The double cage induction motor is highly efficient when running.

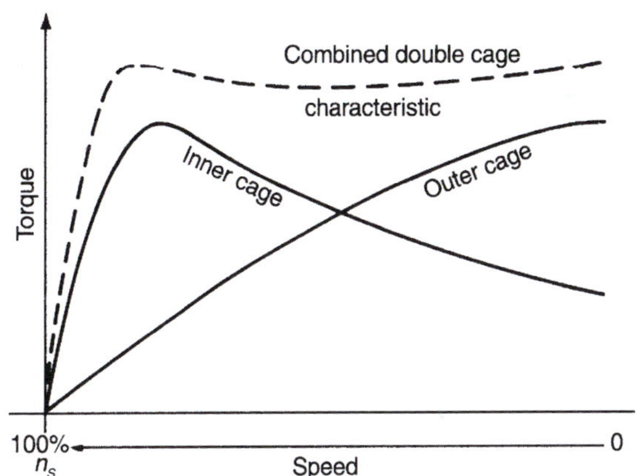

Figure 25.14

25.17 Uses of three-phase induction motors

Three-phase induction motors are widely used in industry and constitute almost all industrial drives where a nearly constant speed is required, from small workshops to the largest industrial enterprises.

Typical applications are with machine tools, pumps and mill motors. The squirrel cage rotor type is the most widely used of all a.c. motors.

For fully worked solutions to each of the problems in Practice Exercises 113 to 117 in this chapter, go to the website:
www.routledge.com/cw/bird

Revision Test 7

This revision test covers the material contained in Chapters 24 and 25. *The marks for each question are shown in brackets at the end of each question.*

1. A six-pole armature has 1000 conductors and a flux per pole of 40 mWb. Determine the e.m.f. generated when running at 600 rev/min when (a) lap wound, (b) wave wound. (6)

2. The armature of a d.c. machine has a resistance of 0.3 Ω and is connected to a 200 V supply. Calculate the e.m.f. generated when it is running (a) as a generator giving 80 A, (b) as a motor taking 80 A. (4)

3. A 15 kW shunt generator having an armature circuit resistance of 1 Ω and a field resistance of 160 Ω generates a terminal voltage of 240 V at full load. Determine the efficiency of the generator at full load assuming the iron, friction and windage losses amount to 544 W. (6)

4. A four-pole d.c. motor has a wave-wound armature with 1000 conductors. The useful flux per pole is 40 mWb. Calculate the torque exerted when a current of 25 A flows in each armature conductor. (4)

5. A 400 V shunt motor runs at its normal speed of 20 rev/s when the armature current is 100 A. The armature resistance is 0.25 Ω. Calculate the speed, in rev/min, when the current is 50 A and a resistance of 0.40 Ω is connected in series with the armature, the shunt field remaining constant. (7)

6. The stator of a three-phase, six-pole induction motor is connected to a 60 Hz supply. The rotor runs at 1155 rev/min at full load. Determine (a) the synchronous speed and (b) the slip at full load. (6)

7. The power supplied to a three-phase induction motor is 40 kW and the stator losses are 2 kW. If the slip is 4%, determine (a) the rotor copper loss, (b) the total mechanical power developed by the rotor, (c) the output power of the motor if frictional and windage losses are 1.48 kW and (d) the efficiency of the motor, neglecting rotor iron loss. (9)

8. A 400 V, three-phase, 100 Hz, eight-pole induction motor runs at 24.25 rev/s on full load. The rotor resistance and reactance per phase are 0.2 Ω and 2 Ω, respectively, and the effective rotor–stator turns ratio is 0.80:1. Calculate (a) the synchronous speed, (b) the percentage slip and (c) the full load torque. (8)

For lecturers/instructors/teachers, fully worked solutions to each of the problems in Revision Test 7, together with a full marking scheme, are available at the website:
www.routledge.com/cw/bird

Main formulae for Part 3 Electrical principles and technology

A.c. theory

$T=\frac{1}{f}$ or $f=\frac{1}{T}$ $\qquad I=\sqrt{\left(\frac{i_1^2+i_2^2+i_3^2+\cdots+i_n}{n}\right)}$

For a sine wave: $I_{AV}=\frac{2}{\pi}I_m$ or $0.637I_m$

$I=\frac{1}{\sqrt{2}}I_m$ or $0.707I_m$

Form factor $=\frac{\text{r.m.s}}{\text{average}}$ Peak factor $=\frac{\text{maximum}}{\text{r.m.s}}$

General sinusoidal voltage: $v=V_m\sin(\omega t\pm\phi)$

Single-phase circuits

$X_L=2\pi fL \qquad X_C=\frac{1}{2\pi fC}$

$Z=\frac{V}{I}=\sqrt{(R^2+X^2)}$

Series resonance: $f_r=\frac{1}{2\pi\sqrt{LC}}$

$Q=\frac{V_L}{V}$ or

$\frac{V_C}{V}=\frac{2\pi f_r L}{R}=\frac{1}{2\pi f_r CR}=\frac{1}{R}\sqrt{\frac{L}{C}}$

$Q=\frac{f_r}{f_2-f_1}$ or

$(f_2-f_1)=\frac{f_r}{Q}$

Parallel resonance (LR–C circuit):

$f_r=\frac{1}{2\pi}\sqrt{\frac{1}{LC}-\frac{R^2}{L^2}}$

$I_r=\frac{VRC}{L} \qquad R_D=\frac{L}{CR}$

$Q=\frac{2\pi f_r L}{R}=\frac{I_C}{I_r}$

$P=VI\cos\phi$ or

$I^2R \qquad S=VI$

$Q=VI\sin\phi \qquad$ power factor $=\cos\phi=\frac{R}{Z}$

D.c. transients

***C–R* circuit** $\tau=CR$

Charging: $v_C=V(1-\mathrm{e}^{-(t/CR)})$

$v_r=V\mathrm{e}^{-(t/CR)}$

$i=I\mathrm{e}^{-(t/CR)}$

Discharging: $v_C=v_R=V\mathrm{e}^{-(t/CR)}$

$i=I\mathrm{e}^{-(t/CR)}$

***L–R* circuit** $\tau=\frac{L}{R}$

Current growth: $v_L=V\mathrm{e}^{-(Rt/L)}$

$v_R=V(1-\mathrm{e}^{-(Rt/L)})$

$i=I(1-\mathrm{e}^{-(Rt/L)})$

Current decay: $v_L=v_R=V\mathrm{e}^{-(Rt/L)}$

$i=I\mathrm{e}^{-(Rt/L)}$

Operational amplifiers

$\text{CMRR}=20\log_{10}\left(\frac{\text{differential voltage gain}}{\text{common mode gain}}\right)\text{dB}$

Inverter: $A=\frac{V_o}{V_i}=\frac{-R_f}{R_i}$

Non-inverter: $A=\frac{V_o}{V_i}=1+\frac{R_f}{R_i}$

Summing: $V_o=-R_f\left(\frac{V_1}{R_1}+\frac{V_2}{R_2}+\frac{V_3}{R_3}\right)$

Integrator: $V_o=-\frac{1}{CR}\int V_i\,\mathrm{d}t$

Differential If $V_1>V_2: V_o=(V_1-V_2)\left(-\frac{R_f}{R_1}\right)$

If $V_2>V_1$:

$V_o=(V_2-V_1)\left(\frac{R_3}{R_2+R_3}\right)\left(1+\frac{R_f}{R_1}\right)$

Three-phase systems

Star: $I_L=I_p \quad V_L=\sqrt{3}V_p$

Delta: $V_L=V_p \quad I_L=\sqrt{3}I_p$

$P=\sqrt{3}V_L I_L\cos\phi$ or $P=3I_p^2R_p$

Two-wattmeter method $P=P_1+P_2$

$\tan\phi=\sqrt{3}\frac{(P_1-P_2)}{(P_1+P_2)}$

Transformers

$$\frac{V_1}{V_2}=\frac{N_1}{N_2}=\frac{I_2}{I_1} \qquad I_0=\sqrt{(I_M^2+I_C^2)}$$

$$I_M=I_0\sin\phi_0 \qquad I_C=I_0\cos\phi_0$$

$$E=4.44f\Phi_m N \quad \text{Regulation}=\left(\frac{E_2-E_1}{E_2}\right)\times 100\%$$

Equivalent circuit: $R_e=R_1+R_2\left(\frac{V_1}{V_2}\right)^2$

$$X_e=X_1+X_2\left(\frac{V_1}{V_2}\right)^2$$

$$Z_e=\sqrt{(R_e^2+X_e^2)}$$

Efficiency, $\eta=1-\frac{\text{losses}}{\text{input power}}$

Output power $=V_2I_2\cos\phi_2$

Total loss = copper loss + iron loss

Input power = output power + losses

Resistance matching: $R_1=\left(\frac{N_1}{N_2}\right)^2 R_L$

D.c. machines

General e.m.f. $E=\frac{2p\Phi nZ}{c}\propto\Phi\omega$

($c=2$ for wave winding, $c=2p$ for lap winding)

Generator: $E=V+I_aR_a$

Efficiency, $\eta=\left(\frac{VI}{VI+I_a^2R_a+I_fV+C}\right)\times 100\%$

Motor: $E=V-I_aR_a$

Efficiency, $\eta=\left(\frac{VI-I_a^2R_a-I_fV-C}{VI}\right)\times 100\%$

$$\text{Torque}=\frac{EI_a}{2\pi n}=\frac{p\Phi ZI_a}{\pi c}\propto I_a\Phi$$

Three-phase induction motors

$$n_s=\frac{f}{p} \qquad s=\left(\frac{n_s-n_r}{n_s}\right)\times 100$$

$$f_r=sf \qquad X_r=sX_2$$

$$I_r=\frac{E_r}{Z_r}=\frac{s\left(\frac{N_2}{N_1}\right)E_1}{\sqrt{[R_2^2+(sX_2)^2]}} \qquad s=\frac{I_r^2R_2}{P_2}$$

Efficiency, $\eta=\frac{P_m}{P_l}$

$$=\frac{\text{input} - \text{stator loss} - \text{rotor copper loss} - \text{friction and windage loss}}{\text{input power}}$$

Torque, $T=\left(\frac{m(N_2/N_1)^2}{2\pi n_s}\right)\left(\frac{sE_1^2R_2}{R_2^2+(sX_2)^2}\right)$

$$\propto\frac{sE_1^2R_2}{R_2^2+(sX_2)^2}$$

These formulae are available for download at the website:
www.routledge.com/cw/bird

Part 4

Advanced circuit theory and technology

Chapter 26

Revision of complex numbers

***Why it is important to understand:* Complex numbers**

Complex numbers are used in many scientific fields, including engineering, electromagnetism, quantum physics and applied mathematics, such as chaos theory. Any physical motion which is periodic, such as an oscillating beam, string, wire, pendulum, electronic signal or electromagnetic wave can be represented by a complex number function. This can make calculations with the various components simpler than with real numbers and sines and cosines. In control theory, systems are often transformed from the time domain to the frequency domain using the Laplace transform. In fluid dynamics, complex functions are used to describe potential flow in two dimensions. In electrical engineering, the Fourier transform is used to analyse varying voltages and currents. Complex numbers are used in signal analysis and other fields for a convenient description for periodically varying signals. This use is also extended into digital signal processing and digital image processing, which utilize digital versions of Fourier analysis (and wavelet analysis) to transmit, compress, restore and otherwise process digital audio signals, still images and video signals. Knowledge of complex numbers is clearly absolutely essential for further studies in so many engineering disciplines and is used extensively in many of the ensuing chapters.

At the end of this chapter you should be able to:

- define a complex number
- understand the Argand diagram
- perform calculations on addition, subtraction, multiplication and division in Cartesian and polar forms
- use De Moivre's theorem for powers and roots of complex numbers

26.1 Introduction

A **complex number** is of the form $(a+jb)$ where a is a **real number** and jb is an **imaginary number**. Hence $(1+j2)$ and $(5-j7)$ are examples of complex numbers.

By definition, $\boldsymbol{j=\sqrt{-1}}$ and $\boldsymbol{j^2=-1}$

Complex numbers are widely used in the analysis of series, parallel and series–parallel electrical networks supplied by alternating voltages (see Chapters 27 to 29), in deriving balance equations with a.c. bridges (see Chapter 30), in analysing a.c. circuits using Kirchhoff's laws (Chapter 33), mesh and nodal analysis (Chapter 34), the superposition theorem (Chapter 35), with Thévenin's and Norton's theorems (Chapter 36) and with delta–star and star–delta transforms (Chapter 37) and in many other aspects of higher electrical engineering. The advantage of the use of complex numbers is that the manipulative processes become simply algebraic processes.

Electrical Circuit Theory and Technology. 978-1-138-67349-6, © 2017 John Bird. Published by Taylor & Francis. All rights reserved.

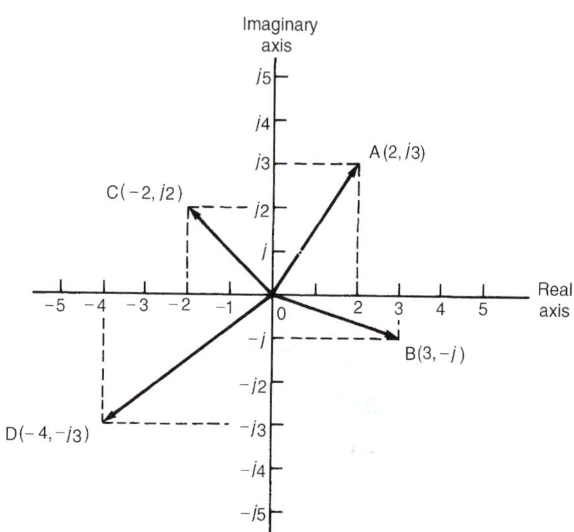

Figure 26.1 The Argand diagram

A complex number can be represented pictorially on an **Argand*** **diagram**. In Figure 26.1, the line $0A$ represents the complex number $(2+j3)$, $0B$ represents $(3-j)$, $0C$ represents $(-2+j2)$ and $0D$ represents $(-4-j3)$

A complex number of the form $a+jb$ is called a **Cartesian or rectangular complex number** (Cartesian being named after **Descartes***). The significance of the j operator is shown in Figure 26.2. In Figure 26.2(a) the number 4 (i.e. $4+j0$) is shown drawn as a phasor horizontally to the right of the origin on the real axis. (Such a phasor could represent, for example, an alternating current, $i=4\sin\omega t$ amperes, when time t is zero.)

The number $j4$ (i.e. $0+j4$) is shown in Figure 26.2(b) drawn vertically upwards from the origin on the imaginary axis. Hence multiplying the number 4 by the operator j results in an anticlockwise phase-shift of 90° without altering its magnitude.

Multiplying $j4$ by j gives $j^2 4$, i.e. −4, and is shown in Figure 26.2(c) as a phasor four units long on the horizontal real axis to the left of the origin – an anticlockwise phase-shift of 90° compared with the position shown in Figure 26.2(b). Thus multiplying by j^2 reverses the original direction of a phasor.

Multiplying $j^2 4$ by j gives $j^3 4$, i.e. $-j4$, and is shown in Figure 26.2(d) as a phasor four units long on the vertical, imaginary axis downward from the origin – an anticlockwise phase-shift of 90° compared with the position shown in Figure 26.2(c).

Multiplying $j^3 4$ by j gives $j^4 4$, i.e. 4, which is the original position of the phasor shown in Figure 26.2(a). **Summarizing**, application of the operator j to any number rotates it 90° anticlockwise on the Argand

* Who was **Argand**? **Jean-Robert Argand** (July 18 1768–August 13 1822) was a highly influential mathematician who published the first complete proof of the fundamental theorem of algebra To find out more go to **www.routledge.com/cw/bird**

* Who was **Descartes**? **René Descartes** (31 March 1596–11 February 1650) was a French philosopher, mathematician and writer. He wrote many influential texts including *Meditations on First Philosophy*. Descartes is best known for the philosophical statement 'Cogito ergo sum' (I think, therefore I am), found in part IV of *Discourse on the Method*. To find out more go to **www.routledge.com/cw/bird**

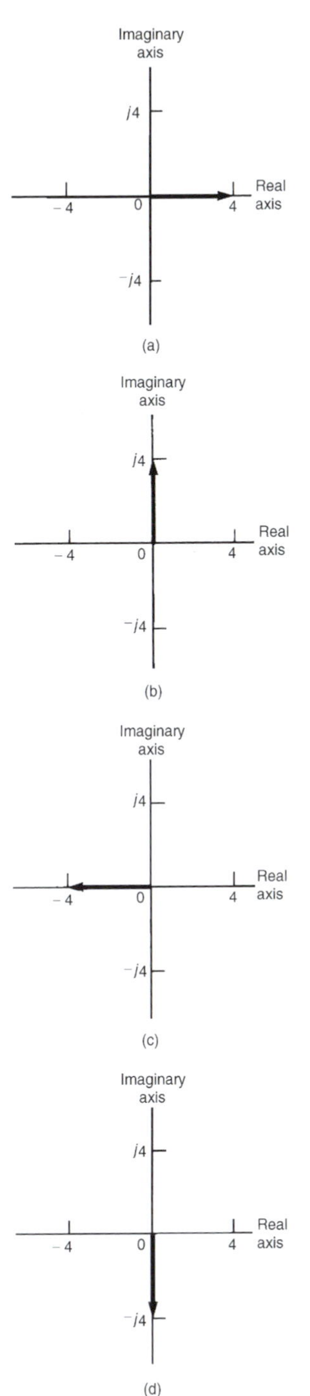

Figure 26.2

diagram, multiplying a number by j^2 rotates it 180° anticlockwise, multiplying a number by j^3 rotates it 270° anticlockwise and multiplication by j^4 rotates it 360° anticlockwise, i.e. back to its original position. In each case the phasor is unchanged in its magnitude.

By similar reasoning, if a phasor is operated on by $-j$ then a phase shift of $-90°$ (i.e. clockwise direction) occurs, again without change of magnitude.
In electrical circuits, 90° phase shifts occur between voltage and current with pure capacitors and inductors; this is the key to why j notation is used so much in the analysis of electrical networks. This is explained in Chapter 27.

26.2 Operations involving Cartesian complex numbers

(a) Addition and subtraction

$$(a+jb)+(c+jd)=(a+c)+j(b+d)$$

and $$(a+jb)-(c+jd)=(a-c)+j(b-d)$$

Thus, $(3+j2)+(2-j4)=3+j2+2-j4=\mathbf{5-j2}$

and $(3+j2)-(2-j4)=3+j2-2+j4=\mathbf{1+j6}$

(b) Multiplication

$$\begin{aligned}(a+jb)(c+jd)&=ac+a(jd)+(jb)c+(jb)(jd)\\&=ac+jad+jbc+j^2bd\end{aligned}$$

But $j^2=-1$, thus

$$(a+jb)(c+jd)=(ac-bd)+j(ad+bc)$$

For example,

$$\begin{aligned}(3+j2)(2-j4)&=6-j12+j4-j^28\\&=(6-(-1)8)+j(-12+4)\\&=14+j(-8)=\mathbf{14-j8}\end{aligned}$$

(c) Complex conjugate

The **complex conjugate** of $(a+jb)$ is $(a-jb)$. For example, the conjugate of $(3-j2)$ is $(3+j2)$

The product of a complex number and its complex conjugate is always a real number, and this is an important property used when dividing complex numbers. Thus

$$\begin{aligned}(a+jb)(a-jb)&=a^2-jab+jab-j^2b^2\\&=a^2-(-b^2)\\&=a^2+b^2 \text{ (i.e. a real number)}\end{aligned}$$

For example, $(1+j2)(1-j2) = 1^2 + 2^2 = \mathbf{5}$

and $(3-j4)(3+j4) = 3^2 + 4^2 = \mathbf{25}$

(d) Division

The expression of one complex number divided by another, in the form $a+jb$, is accomplished by multiplying the numerator and denominator by the complex conjugate of the denominator. This has the effect of making the denominator a real number. Hence, for example,

$$\frac{2+j4}{3-j4} = \frac{2+j4}{3-j4} \times \frac{3+j4}{3+j4} = \frac{6+j8+j12+j^2 16}{3^2+4^2}$$

$$= \frac{6+j8+j12-16}{25}$$

$$= \frac{-10+j20}{25}$$

$$= \frac{\mathbf{-10}}{\mathbf{25}} + j\frac{\mathbf{20}}{\mathbf{25}} \text{ or } \mathbf{-0.4+j0.8}$$

The elimination of the imaginary part of the denominator by multiplying both the numerator and denominator by the conjugate of the denominator is often termed **'rationalizing'**.

Problem 1. In an electrical circuit the total impedance Z_T is given by

$$Z_T = \frac{Z_1 Z_2}{Z_1 + Z_2} + Z_3$$

Determine Z_T in $(a+jb)$ form, correct to two decimal places, when $Z_1 = 5 - j3$, $Z_2 = 4 + j7$ and $Z_3 = 3.9 - j6.7$

$$Z_1 Z_2 = (5-j3)(4+j7) = 20 + j35 - j12 - j^2 21$$

$$= 20 + j35 - j12 + 21 = 41 + j23$$

$$Z_1 + Z_2 = (5-j3) + (4+j7) = 9 + j4$$

Hence $$\frac{Z_1 Z_2}{Z_1+Z_2} = \frac{41+j23}{9+j4} = \frac{(41+j23)(9-j4)}{(9+j4)(9-j4)}$$

$$= \frac{369 - j164 + j207 - j^2 92}{9^2+4^2}$$

$$= \frac{369 - j164 + j207 + 92}{97}$$

$$= \frac{461 + j43}{97} = 4.753 + j0.443$$

Thus $$\frac{Z_1 Z_2}{Z_1+Z_2} + Z_3 = (4.753 + j0.443) + (3.9 - j6.7)$$

$= \mathbf{8.65 - j6.26}$, correct to two decimal places.

Problem 2. Given $Z_1 = 3 + j4$ and $Z_2 = 2 - j5$ determine in Cartesian form correct to three decimal places:

(a) $\frac{1}{Z_1}$, (b) $\frac{1}{Z_2}$, (c) $\frac{1}{Z_1} + \frac{1}{Z_2}$, (d) $\frac{1}{(1/Z_1)+(1/Z_2)}$

(a) $$\frac{1}{Z_1} = \frac{1}{3+j4} = \frac{3-j4}{(3+j4)(3-j4)} = \frac{3-j4}{3^2+4^2}$$

$$= \frac{3-j4}{25} = \frac{3}{25} - j\frac{4}{25} = \mathbf{0.120 - j0.160}$$

(b) $$\frac{1}{Z_2} = \frac{1}{2-j5} = \frac{2+j5}{(2-j5)(2+j5)} = \frac{2+j5}{2^2+5^2}$$

$$= \frac{2+j5}{29} = \frac{2}{29} + j\frac{5}{29} = \mathbf{0.069 + j0.172}$$

(c) $$\frac{1}{Z_1} + \frac{1}{Z_2} = (0.120 - j0.160) + (0.069 + j0.172)$$

$$= \mathbf{0.189 + j0.012}$$

(d) $$\frac{1}{(1/Z_1)+(1/Z_2)} = \frac{1}{0.189 + j0.012}$$

$$= \frac{0.189 - j0.012}{(0.189 + j0.012)(0.189 - j0.012)}$$

$$= \frac{0.189 - j0.012}{0.189^2 + 0.012^2}$$

$$= \frac{0.189 - j0.012}{0.03587}$$

$$= \frac{0.189}{0.03587} - \frac{j0.012}{0.03587}$$

$$= \mathbf{5.269 - j0.335}$$

Part 4

Now try the following Practice Exercise

Practice Exercise 118 Operations involving Cartesian complex numbers (Answers on page 825)

In Problems 1 to 5, evaluate in $a+jb$ form assuming that $Z_1=2+j3$, $Z_2=3-j4$, $Z_3=-1+j2$ and $Z_4=-2-j5$

1. (a) Z_1-Z_2, (b) $Z_2+Z_3-Z_4$
2. (a) Z_1Z_2, (b) Z_3Z_4
3. (a) $Z_1Z_3Z_4$, (b) $Z_2Z_3+Z_4$
4. (a) $\frac{Z_1}{Z_2}$, (b) $\frac{Z_1+Z_2}{Z_3+Z_4}$
5. (a) $\frac{Z_1Z_2}{Z_1+Z_2}$, (b) $Z_1+\frac{Z_2}{Z_3}+Z_4$
6. Evaluate $\left[\frac{(1+j)^2-(1-j)^2}{j}\right]$
7. If $Z_1=4-j3$ and $Z_2=2+j$, evaluate x and y given

   $$x+jy=\frac{1}{Z_1-Z_2}+\frac{1}{Z_1Z_2}$$

8. Evaluate (a) $(1+j)^4$, (b) $\frac{2-j}{2+j}$, (c) $\frac{1}{2+j3}$
9. If $Z=\frac{1+j3}{1-j2}$ evaluate Z^2 in $a+jb$ form.
10. In an electrical circuit the equivalent impedance Z is given by

    $$Z=Z_1+\frac{Z_2Z_3}{Z_2+Z_3}$$

    Determine Z in rectangular form, correct to two decimal places, when $Z_1=5.91+j3.15$, $Z_2=5+j12$ and $Z_3=8-j15$
11. Given $Z_1=5-j9$ and $Z_2=7+j2$, determine in $(a+jb)$ form, correct to four decimal places

    (a) $\frac{1}{Z_1}$, (b) $\frac{1}{Z_2}$, (c) $\frac{1}{Z_1}+\frac{1}{Z_2}$, (d) $\frac{1}{(1/Z_1)+(1/Z_2)}$

26.3 Complex equations

If two complex numbers are equal, then their real parts are equal and their imaginary parts are equal. Hence, if $a+jb=c+jd$ then $a=c$ and $b=d$. This is a useful property, since equations having two unknown quantities can be solved from one equation. Complex equations are used when deriving balance equations with a.c. bridges (see Chapter 30).

Problem 3. Solve the following complex equations:

(a) $3(a+jb)=9-j2$

(b) $(2+j)(-2+j)=x+jy$

(c) $(a-j2b)+(b-j3a)=5+j2$

(a) $3(a+jb)=9-j2$. Thus $3a+j3b=9-j2$

Equating real parts gives: $3a=9$, i.e. $\boldsymbol{a=3}$

Equating imaginary parts gives:

$3b=-2$, i.e. $\boldsymbol{b=-2/3}$

(b) $(2+j)(-2+j)=x+jy$

Thus $-4+j2-j2+j^2=x+jy$

$$-5+j0=x+jy$$

Equating real and imaginary parts gives: $\boldsymbol{x=-5}$, $\boldsymbol{y=0}$

(c) $(a-j2b)+(b-j3a)=5+j2$

Thus $(a+b)+j(-2b-3a)=5+j2$

Hence $\quad a+b=5 \quad (1)$

and $\quad -2b-3a=2 \quad (2)$

We have two simultaneous equations to solve. Multiplying equation (1) by (2) gives:

$$2a+2b=10 \quad (3)$$

Adding equations (2) and (3) gives $-a=12$, i.e. $\boldsymbol{a=-12}$

From equation (1), $\boldsymbol{b=17}$

Problem 4. An equation derived from an a.c. bridge network is given by

$$R_1 R_3 = (R_2 + j\omega L_2)\left[\frac{1}{(1/R_4) + j\omega C}\right]$$

R_1, R_3, R_4 and C_4 are known values. Determine expressions for R_2 and L_2 in terms of the known components.

Multiplying both sides of the equation by $(1/R_4 + j\omega C_4)$ gives

$$(R_1 R_3)(1/R_4 + j\omega C_4) = R_2 + j\omega L_2$$

i.e. $$R_1 R_3/R_4 + jR_1 R_3 \omega C_4 = R_2 + j\omega L_2$$

Equating the real parts gives: $\boldsymbol{R_2 = R_1 R_3/R_4}$

Equating the imaginary parts gives:

$\omega L_2 = R_1 R_3 \omega C_4$, from which, $\boldsymbol{L_2 = R_1 R_3 C_4}$

Now try the following Practice Exercise

Practice Exercise 119 Complex equations (Answers on page 825)

In Problems 1 to 4 solve the given complex equations.

1. $4(a + jb) = 7 - j3$
2. $(3 + j4)(2 - j3) = x + jy$
3. $(a - j3b) + (b - j2a) = 4 + j6$
4. $5 + j2 = \sqrt{(e + jf)}$
5. An equation derived from an a.c. bridge circuit is given by

$$(R_3)\left[\frac{-j}{\omega C_1}\right] = \left[R_x - \frac{j}{\omega C_x}\right]\left[\frac{R_4(-j/(\omega C_4))}{R_4 - (j/(\omega C_4))}\right]$$

Components R_3, R_4, C_1 and C_4 have known values. Determine expressions for R_x and C_x in terms of the known components.

26.4 The polar form of a complex number

In Figure 26.3(a), $Z = x + jy = r\cos\theta + jr\sin\theta$ from trigonometry,

$$= r(\cos\theta + j\sin\theta)$$

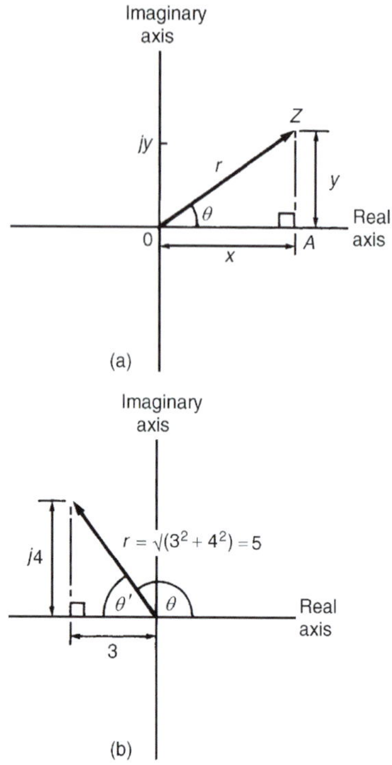

Figure 26.3

This latter form is usually abbreviated to $\boldsymbol{Z = r\angle\theta}$, and is called the **polar form** of a complex number.

r is called the **modulus** (or magnitude of Z) and is written as mod Z or $|Z|$. r is determined from Pythagoras' theorem on triangle OAZ, i.e.

$$|Z| = r = \sqrt{(x^2 + y^2)}$$

The modulus is represented on the Argand diagram by the distance OZ. θ is called the **argument** (or amplitude) of Z and is written as arg Z. θ is also deduced from triangle OAZ: $\textbf{arg}\ \boldsymbol{Z = \theta = \tan^{-1} y/x}$.

For example, the Cartesian complex number $(3 + j4)$ is equal to $r\angle\theta$ in polar form, where $\boldsymbol{r} = \sqrt{(3^2 + 4^2)} = \mathbf{5}$ and $\boldsymbol{\theta} = \tan^{-1}\frac{4}{3} = \mathbf{53.13°}$

Hence $\mathbf{(3 + j4) = 5\angle 53.13°}$

Similarly, $(-3 + j4)$ is shown in Figure 26.3(b), where $r = \sqrt{(3^2 + 4^2)} = 5$, $\theta' = \tan^{-1}\frac{4}{3} = 53.13°$

and $\theta = 180° - 53.13° = 126.87°$

Hence $\mathbf{(-3 + j4) = 5\angle 126.87°}$

Part 4

26.5 Multiplication and division using complex numbers in polar form

(a) Multiplication

$$(r_1\angle\theta_1)(r_2\angle\theta_2) = r_1 r_2\angle(\theta_1+\theta_2)$$

Thus $3\angle 25^\circ \times 2\angle 32^\circ = 6\angle 57^\circ$,

$4\angle 11^\circ \times 5\angle -18^\circ = 20\angle -7^\circ$,

$2\angle(\pi/3) \times 7\angle(\pi/6) = 14\angle(\pi/2)$, and so on.

(b) Division

$$\frac{r_1\angle\theta_1}{r_2\angle\theta_2} = \frac{r_1}{r_2}\angle(\theta_1+\theta_2)$$

Thus $\dfrac{8\angle 58^\circ}{2\angle 11^\circ} = 4\angle 47^\circ$,

$\dfrac{9\angle 136^\circ}{3\angle -60^\circ} = 3\angle(136^\circ - -60^\circ)$

$= 3\angle 196^\circ$ or $3\angle -164^\circ$,

and $\dfrac{10\angle(\pi/2)}{5\angle(-\pi/4)} = 2\angle(3\pi/4)$, and so on.

Conversion from Cartesian or rectangular form to polar form, and vice-versa, may be achieved by using the $R\rightarrow P$ and $P\rightarrow R$ conversion facility which is available on most calculators with scientific notation. This allows, of course, a saving of time.

Problem 5. Convert $5\angle -132^\circ$ into $a+jb$ form correct to four significant figures.

Figure 26.4 indicates that the polar complex number $5\angle -132^\circ$ lies in the third quadrant of the Argand diagram.

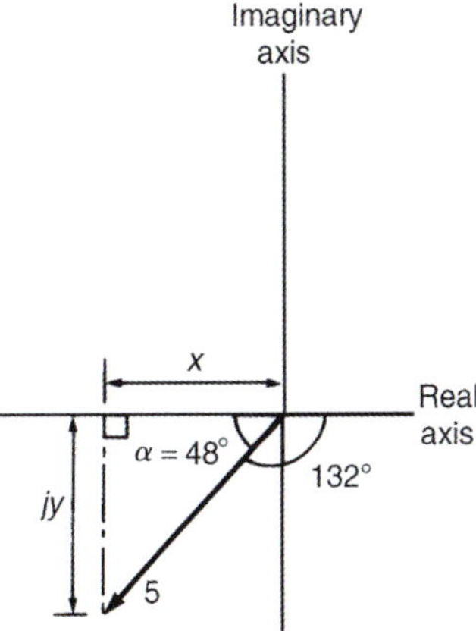

Figure 26.4

Using trigonometrical ratios,

$x = 5\cos 48^\circ = 3.346$ and $y = 5\sin 48^\circ = 3.716$

Hence $\mathbf{5\angle -132^\circ = -3.346 - j3.716}$

Alternatively,

$$\begin{aligned} 5\angle -132^\circ &= 5(\cos -132^\circ + j\sin -132^\circ) \\ &= 5\cos(-132^\circ) + j5\sin(-132^\circ) \\ &= \mathbf{-3.346 - j3.716}, \text{ as above} \end{aligned}$$

With this latter method the real and imaginary parts are obtained directly, using a calculator.

Problem 6. Two impedances in an electrical network are given by $Z_1 = 4.7\angle 35^\circ$ and $Z_2 = 7.3\angle -48^\circ$. Determine in polar form the total impedance Z_T given that $Z_T = Z_1 Z_2/(Z_1+Z_2)$

$$\begin{aligned} Z_1 &= 4.7\angle 35^\circ = 4.7\cos 35^\circ + j4.7\sin 35^\circ \\ &= 3.85 + j2.70 \end{aligned}$$

$$\begin{aligned} Z_2 &= 7.3\angle -48^\circ = 7.3\cos(-48^\circ) + j7.3\sin(-48^\circ) \\ &= 4.88 - j5.42 \end{aligned}$$

$$\begin{aligned} Z_1 + Z_2 &= (3.85 + j2.70) + (4.88 - j5.42) \\ &= 8.73 - j2.72 \\ &= \sqrt{(8.73^2 + 2.72^2)}\angle\tan^{-1}\left(\frac{-2.72}{8.73}\right) \\ &= 9.14\angle -17.31^\circ \end{aligned}$$

Hence

$$\begin{aligned} Z_T = Z_1 Z_2/(Z_1+Z_2) &= \frac{4.7\angle 35^\circ \times 7.3\angle -48^\circ}{9.14\angle -17.31^\circ} \\ &= \frac{4.7\times 7.3}{9.14}\angle[35^\circ - 48^\circ - (-17.31^\circ)] \\ &= \mathbf{3.75\angle 4.31^\circ} \text{ or } \mathbf{3.75\angle 4^\circ 19'} \end{aligned}$$

Now try the following Practice Exercise

Practice Exercise 120 The polar form of complex numbers (Answers on page 825)

In Problems 1 and 2 determine the modulus and the argument of each of the complex numbers given.

1. (a) $3+j4$, (b) $2-j5$
2. (a) $-4+j$, (b) $-5-j3$

Part 4

In Problems 3 and 4 express the given Cartesian complex numbers in polar form, leaving answers in surd form.

3. (a) $6+j5$, (b) $3-j2$, (c) -3

4. (a) $-5+j$, (b) $-4-j3$, (c) $-j2$

In Problems 5 to 7 convert the given polar complex numbers into $(a+jb)$ form, giving answers correct to four significant figures.

5. (a) $6\angle 30°$, (b) $4\angle 60°$, (c) $3\angle 45°$

6. (a) $2\angle \pi/2$, (b) $3\angle \pi$, (c) $5\angle(5\pi/6)$

7. (a) $8\angle 150°$, (b) $4.2\angle -120°$, (c) $3.6\angle -25°$

In Problems 8 to 10, evaluate in polar form.

8. (a) $2\angle 40° \times 5\angle 20°$, (b) $2.6\angle 72° \times 4.3\angle 45°$

9. (a) $5.8\angle 35° \div 2\angle -10°$

   (b) $4\angle 30° \times 3\angle 70° \div 2\angle -15°$

10. (a) $\dfrac{4.1\angle 20° \times 3.2\angle -62°}{1.2\angle 150°}$

    (b) $6\angle 25° + 3\angle -36° - 4\angle 72°$

11. Solve the complex equations, giving answers correct to four significant figures.

    (a) $\dfrac{12\angle(\pi/2) \times 3\angle(3\pi/4)}{2\angle -(\pi/3)} = x + jy$

    (b) $15\angle \pi/3 + 12\angle \pi/2 - 6\angle -\pi/3 = r\angle\theta$

12. The total impedance Z_T of an electrical circuit is given by

$$Z_T = \frac{Z_1 \times Z_2}{Z_1 + Z_2} + Z_3$$

Determine Z_T in polar form correct to three significant figures when $Z_1 = 3.2\angle -56°$, $Z_2 = 7.4\angle 25°$ and $Z_3 = 6.3\angle 62°$

13. A star-connected impedance Z_1 is given by

$$Z_1 = \frac{Z_A Z_B}{Z_A + Z_B + Z_C}$$

Evaluate Z_1, in both Cartesian and polar form, given $Z_A = (20 + j0)\Omega$, $Z_B = (0 - j20)\Omega$ and $Z_C = (10 + j10)\Omega$

14. The current I flowing in an impedance is given by

$$I = \frac{(8\angle 60°)(10\angle 0°)}{(8\angle 60° + 5\angle 30°)}\,\text{A}$$

Determine the value of current in polar form, correct to two decimal places.

15. A delta-connected impedance Z_A is given by

$$Z_A = \frac{Z_1 Z_2 + Z_2 Z_3 + Z_3 Z_1}{Z_2}$$

Determine Z_A, in both Cartesian and polar form, given $Z_1 = (10 + j0)\Omega$, $Z_2 = (0 - j10)\Omega$ and $Z_3 = (10 + j10)\Omega$

26.6 De Moivre's theorem* – powers and roots of complex numbers

De Moivre's theorem,* states:

$$[r\angle\theta]^n = r^n\angle n\theta$$

*Who was **De Moivre**? **Abraham de Moivre** (26 May 1667–27 November 1754) was a French mathematician famous for de Moivre's formula, which links complex numbers and trigonometry, and for his work on the normal distribution and probability theory. To find out more go to **www.routledge.com/cw/bird**

This result is true for all positive, negative or fractional values of n. De Moivre's theorem is thus useful in determining powers and roots of complex numbers. For example,

$$[2\angle 15°]^6 = 2^6\angle(6\times 15°) = \mathbf{64\angle 90° = 0 + j64}$$

A square root of a complex number is determined as follows:

$$\sqrt{[r\angle\theta]} = [r\angle\theta]^{1/2} = r^{1/2}\angle\tfrac{1}{2}\theta$$

However, it is important to realize that a real number has two square roots, equal in size but opposite in sign. On an Argand diagram the roots are 180° apart (see worked Problem 8 following).

Problem 7. Determine $(-2+j3)^5$ in polar and in Cartesian form.

$Z = -2 + j3$ is situated in the second quadrant of the Argand diagram.

Thus $r = \sqrt{[(2)^2 + (3)^2]} = \sqrt{13}$ and $\alpha = \tan^{-1} 3/2 = 56.31°$

Hence the argument $\theta = 180° - 56.31° = 123.69°$

Thus $-2 + j3$ in polar form is $\sqrt{13}\angle 123.69°$

$$(-2+j3)^5 = [\sqrt{13}\angle 123.69°]^5$$
$$= (\sqrt{13})^5\angle(5\times 123.69°)$$

from De Moivre's theorem

$$= 13^{5/2}\angle 618.45°$$
$$= 13^{5/2}\angle 258.45°$$
$$(\text{since } 618.45° \equiv 618.45° - 360°)$$
$$= 13^{5/2}\angle -101.55° = \mathbf{609.3\angle -101.55°}$$

In Cartesian form,

$$609.3\angle -101.55° = 609.3\cos(-101.55°) + j609.3\sin(-101.55°)$$
$$= \mathbf{-122 - j597}$$

Problem 8. Determine the two square roots of the complex number $(12+j5)$ in Cartesian and polar form, correct to three significant figures. Show the roots on an Argand diagram.

In polar form $12 + j5 = \sqrt{(12^2 + 5^2)}\angle\tan^{-1}(5/12)$, since $12 + j5$ is in the first quadrant of the Argand diagram, i.e. $12 + j5 = 13\angle 22.62°$

Since we are finding the square roots of $13\angle 22.62°$ there will be two solutions. To obtain the second solution it is helpful to express $13\angle 22.62°$ also as $13\angle(360° + 22.62°)$, i.e. $13\angle 382.62°$ (we have merely rotated one revolution to obtain this result). The reason for doing this is that when we divide the angles by 2 we still obtain angles less than 360°, as shown below.

Hence $\sqrt{(12+j5)} = \sqrt{[13\angle 22.62°]}$ or $\sqrt{[13\angle 382.62°]}$

$$= [13\angle 22.62°]^{1/2} \text{ or } [13\angle 382.62°]^{1/2}$$
$$= 13^{1/2}\angle\left(\tfrac{1}{2}\times 22.62°\right) \text{ or } 13^{1/2}\angle\left(\tfrac{1}{2}\times 382.62°\right)$$

from De Moivre's theorem,

$$= \sqrt{13}\angle 11.31° \text{ or } \sqrt{13}\angle 191.31°$$

i.e. $= \mathbf{3.61\angle 11.31°}$ or $\mathbf{3.61\angle -168.69°}$

These two solutions of $\sqrt{(12+j5)}$ are shown in the Argand diagram of Figure 26.5. $3.61\angle 11.31°$ is in the first quadrant of the Argand diagram.

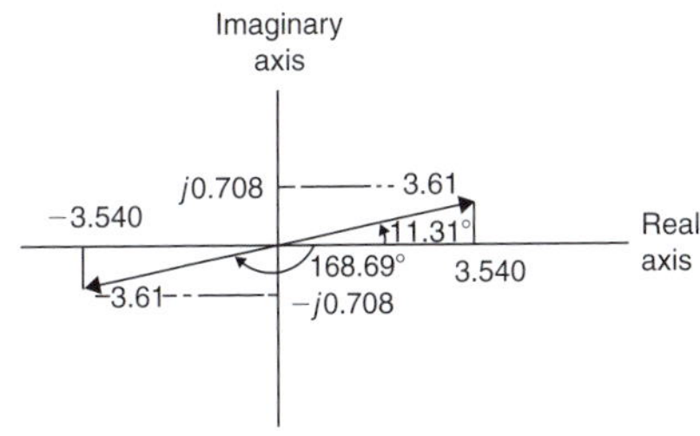

Figure 26.5

Thus $3.61\angle 11.31° = 3.61(\cos 11.31° + j\sin 11.31°) = 3.540 + j0.708$

$3.61\angle -168.69°$ is in the third quadrant of the Argand diagram.

Thus $3.61\angle -168.69° = 3.61[\cos(-168.69°) + j\sin(-168.69°)] = -3.540 - j0.708$

Thus in Cartesian form the two roots are

$$\mathbf{\pm(3.540 + j0.708)}$$

From the Argand diagram the roots are seen to be 180° apart, i.e. they lie on a straight line. This is always true when finding square roots of complex numbers.

Part 4

Now try the following Practice Exercise

Practice Exercise 121 Powers and roots of complex numbers (Answers on page 825)

In Problems 1 to 4, evaluate in Cartesian and in polar form.

1. (a) $(2+j3)^2$, (b) $(4-j5)^2$
2. (a) $(-3+j2)^5$, (b) $(-2-j)^3$
3. (a) $(4\angle 32°)^4$, (b) $(2\angle 125°)^5$
4. (a) $(3\angle -\pi/3)^3$, (b) $1.5\angle -160°)^4$

In Problems 5 to 7, determine the two square roots of the given complex numbers in Cartesian form and show the results on an Argand diagram.

5. (a) $2+j$, (b) $3-j2$
6. (a) $-3+j4$, (b) $-1-j3$
7. (a) $5\angle 36°$, (b) $14\angle 3\pi/2$
8. Convert $2-j$ into polar form and hence evaluate $(2-j)^7$ in polar form.

With a calculator, such as the CASIO fx-991ES PLUS, it is possible, using the Complex mode, to achieve all of the calculations in Practice Exercises 118 to 121 very much more quickly than the methods shown in this chapter. Since complex numbers are used so extensively with a.c. circuit calculations it is important to be able to use the quickest and most accurate method of solution – and this is by using a calculator.

For fully worked solutions to each of the problems in Practice Exercises 118 to 121 in this chapter, go to the website:
www.routledge.com/cw/bird

Chapter 27

Application of complex numbers to series a.c. circuits

Why it is important to understand: **Application of complex numbers to series a.c. circuits**

Complex numbers are useful for a.c. circuit analysis because they provide a convenient method of symbolically denoting phase shift between a.c. quantities like voltage and current. When analysing alternating current circuits, it is found that quantities of voltage, current and even resistance (called impedance in a.c. circuits) are not the familiar one-dimensional quantities used in d.c. circuits. Rather, these quantities, because they're dynamic, i.e. they are alternating in direction and amplitude, possess other dimensions that must be taken into account. Frequency and phase shift are two of these dimensions. Even with relatively simple a.c. circuits, dealing with a single frequency, there is still phase shift to contend with in addition to the amplitude. A complex number is a single mathematical quantity able to express these two dimensions of amplitude and phase shift at once. Oscillating currents and voltages are complex values that have a real part we can measure and an imaginary part which we cannot. At first it seems pointless to create something we can't see or measure, but it is actually useful in a number of ways. Firstly, it helps us understand the behaviour of circuits which contain reactance, produced by capacitors or inductors, when we apply a.c. signals and, secondly, it gives us a new way to think about oscillations. This is useful when we want to apply concepts like the conservation of energy to understanding the behaviour of systems which range from simple mechanical pendulums to a quartz-crystal oscillator. Knowledge of complex numbers makes the analysis of a.c. series circuits straightforward.

At the end of this chapter you should be able to:

- appreciate the use of complex numbers in a.c. circuits
- perform calculations on series a.c. circuits using complex numbers

27.1 Introduction

Simple a.c. circuits may be analysed by using phasor diagrams. However, when circuits become more complicated analysis is considerably simplified by using complex numbers. It is essential that the basic operations used with complex numbers, as outlined in Chapter 26, are thoroughly understood before proceeding with a.c. circuit analysis. The theory introduced in Chapter 17 is relevant; in this chapter similar circuits will be analysed using j notation and Argand diagrams.

Electrical Circuit Theory and Technology. 978-1-138-67349-6, © 2017 John Bird. Published by Taylor & Francis. All rights reserved.

27.2 Series a.c. circuits

(a) Pure resistance

In an a.c. circuit containing resistance R only (see Figure 27.1(a)), the current I_R is **in phase** with the applied voltage V_R as shown in the phasor diagram of Figure 27.1(b). The phasor diagram may be superimposed on the Argand diagram as shown in Figure 27.1(c). The impedance $\boldsymbol{Z}$ of the circuit is given by

$$\boldsymbol{Z = \frac{V_R\angle 0^\circ}{I_R\angle 0^\circ} = R}$$

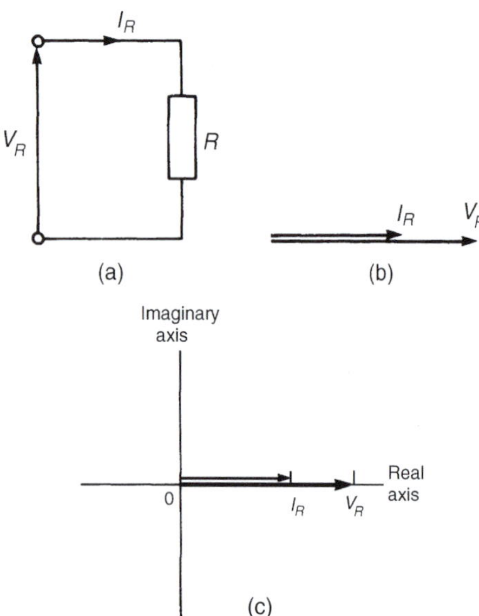

Figure 27.1 (a) Circuit diagram (b) phasor diagram (c) Argand diagram

(b) Pure inductance

In an a.c. circuit containing pure inductance L only (see Figure 27.2(a)), the current I_L **lags** the applied voltage V_L by 90°, as shown in the phasor diagram of Figure 27.2(b). The phasor diagram may be superimposed on the Argand diagram as shown in Figure 27.2(c). The impedance Z of the circuit is given by

$$Z = \frac{V_L\angle 90^\circ}{I_L\angle 0^\circ} = \frac{V_L}{I_L}\angle 90^\circ = \boldsymbol{X_L\angle 90^\circ} \text{ or } \boldsymbol{jX_L}$$

where X_L is the **inductive reactance** given by

$$\boldsymbol{X_L = \omega L = 2\pi f L \text{ ohms}}$$

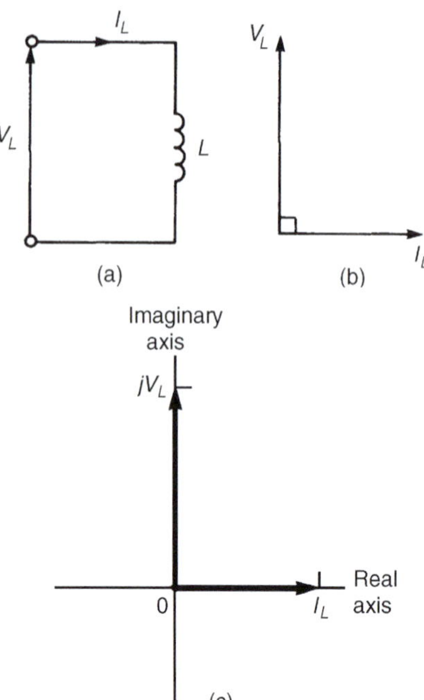

Figure 27.2 (a) Circuit diagram (b) phasor diagram (c) Argand diagram

where f is the frequency in hertz and L is the inductance in henrys.

(c) Pure capacitance

In an a.c. circuit containing pure capacitance only (see Figure 27.3(a)), the current I_C **leads** the applied voltage V_C by 90° as shown in the phasor diagram of Figure 27.3(b). The phasor diagram may be superimposed on the Argand diagram as shown in Figure 27.3(c). The impedance Z of the circuit is given by

$$Z = \frac{V_C\angle -90^\circ}{I_C\angle 0^\circ} = \frac{V_C}{I_C}\angle -90^\circ = \boldsymbol{X_C\angle -90^\circ} \text{ or } \boldsymbol{-jX_C}$$

where X_C is the **capacitive reactance** given by

$$\boldsymbol{X_C = \frac{1}{\omega C} = \frac{1}{2\pi f C} \text{ ohms}}$$

where C is the capacitance in farads.

$$\left[\text{Note: } -jX_C = \frac{-j}{\omega C} = \frac{-j(j)}{\omega C(j)} = \frac{-j^2}{j\omega C} = \frac{-(-1)}{j\omega C} = \frac{1}{j\omega C}\right]$$

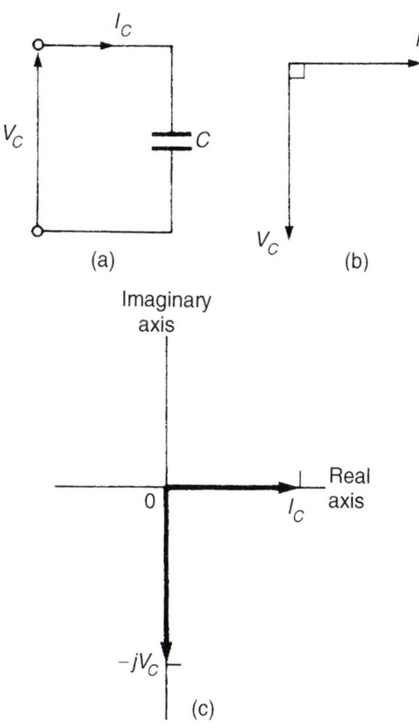

Figure 27.3 (a) Circuit diagram (b) phasor diagram (c) Argand diagram

(d) *R–L* series circuit

In an a.c. circuit containing resistance R and inductance L in series (see Figure 27.4(a)), the applied voltage V is the phasor sum of V_R and V_L as shown in the phasor diagram of Figure 27.4(b). The current I lags the applied voltage V by an angle lying between $0°$ and $90°$ – the actual value depending on the values of V_R and V_L, which depend on the values of R and L. The circuit phase angle, i.e. the angle between the current and the applied voltage, is shown as angle ϕ in the phasor diagram. In any series circuit the current is common to all components and is thus taken as the reference phasor in Figure 27.4(b). The phasor diagram may be superimposed on the Argand diagram as shown in Figure 27.4(c), where it may be seen that in complex form the supply voltage V is given by:

$$V = V_R + jV_L$$

Figure 27.5(a) shows the voltage triangle that is derived from the phasor diagram of Figure 27.4(b) (i.e. triangle *Oab*). If each side of the voltage triangle is divided by current I then the impedance triangle of Figure 27.5(b) is derived. The impedance triangle may be superimposed on the Argand diagram, as shown in

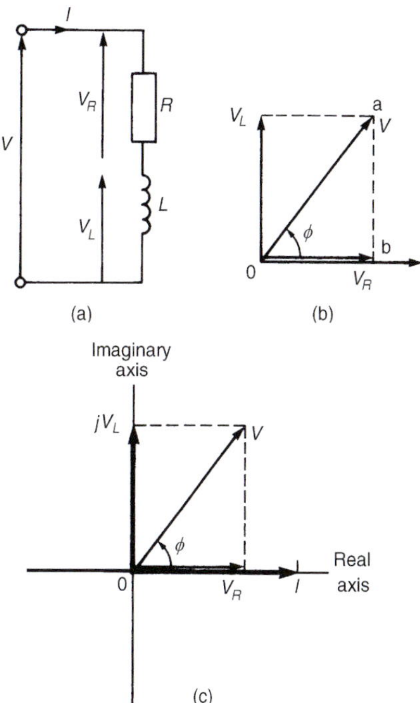

Figure 27.4 (a) Circuit diagram (b) phasor diagram (c) Argand diagram

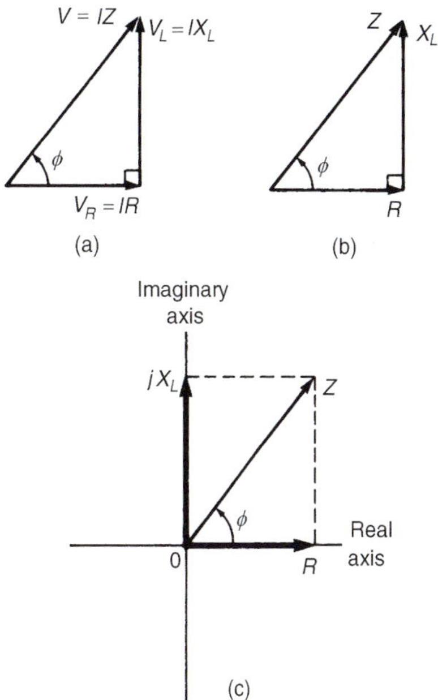

Figure 27.5 (a) Voltage triangle (b) impedance triangle (c) Argand diagram

Figure 27.5(c), where it may be seen that in complex form the impedance Z is given by:

$$Z = R + jX_L$$

Thus, for example, an impedance expressed as $(3+j4)\,\Omega$ means that the resistance is $3\,\Omega$ and the inductive reactance is $4\,\Omega$.
In polar form, $Z=|Z|\angle\phi$ where, from the impedance triangle, the modulus of impedance $|Z|=\sqrt{(R^2+X_L^2)}$ and the circuit phase angle $\phi=\tan^{-1}(X_L/R)$ lagging.

(e) *R–C* series circuit

In an a.c. circuit containing resistance R and capacitance C in series (see Figure 27.6(a)), the applied voltage V is the phasor sum of V_R and V_C, as shown in the phasor diagram of Figure 27.6(b). The current I leads the applied voltage V by an angle lying between 0° and 90° – the actual value depending on the values of V_R and V_C, which depend on the values of R and C. The circuit phase angle is shown as angle ϕ in the phasor diagram. The phasor diagram may be superimposed on the Argand diagram as shown in Figure 27.6(c), where it may be seen that in complex form the supply voltage V is given by:

$$\boldsymbol{V=V_R-jV_C}$$

Figure 27.7(a) shows the voltage triangle that is derived from the phasor diagram of Figure 27.6(b). If each side of the voltage triangle is divided by current I, the impedance triangle is derived as shown in Figure 27.7(b). The impedance triangle may be superimposed on the Argand diagram as shown in Figure 27.7(c), where it may be seen that in complex form the impedance Z is given by

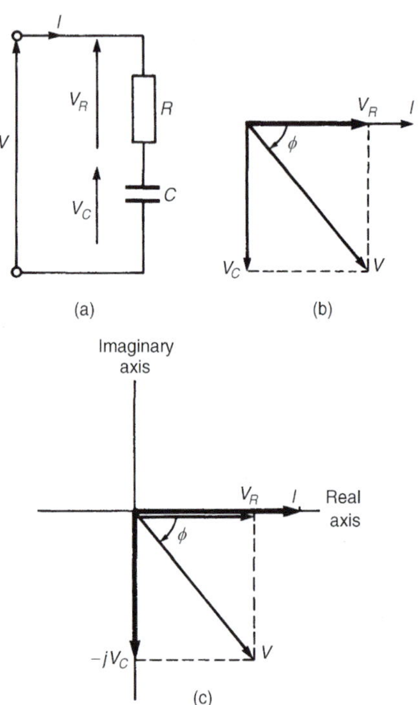

Figure 27.6 (a) Circuit diagram (b) phasor diagram (c) Argand diagram

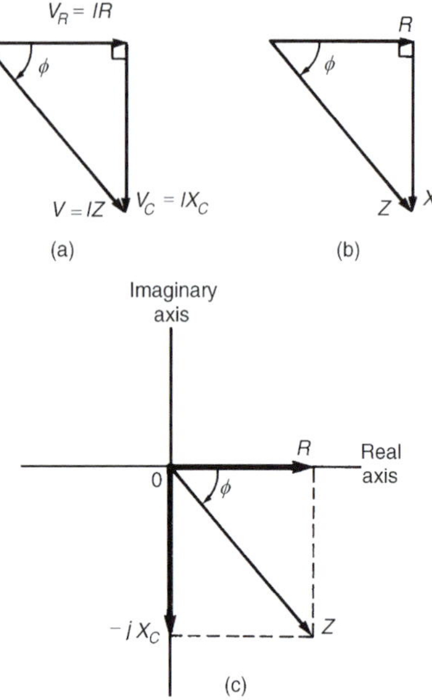

Figure 27.7 (a) Voltage triangle (b) impedance triangle (c) argand diagram

$$\boldsymbol{Z=R-jX_C}$$

Thus, for example, an impedance expressed as $(9-j14)\,\Omega$ means that the resistance is $9\,\Omega$ and the capacitive reactance $\boldsymbol{X_C}$ is $14\,\Omega$
In polar form, $Z=|Z|\angle\phi$ where, from the impedance triangle, $|Z|=\sqrt{(R^2+X_C^2)}$ and $\phi=\tan^{-1}(X_C/R)$ leading

(f) *R–L–C* series circuit

In an a.c. circuit containing resistance R, inductance L and capacitance C in series (see Figure 27.8(a)), the applied voltage V is the phasor sum of V_R, V_L and V_C, as shown in the phasor diagram of Figure 27.8(b) (where the condition $V_L > V_C$ is shown). The phasor diagram may be superimposed on the Argand diagram as shown in Figure 27.8(c), where it may be seen that in complex form the supply voltage V is given by:

$$\boldsymbol{V=V_R+j(V_L-V_C)}$$

From the voltage triangle the impedance triangle is derived and superimposing this on the Argand diagram gives, in complex form,

impedance $\boldsymbol{Z=R+j(X_L-X_C)}$ **or** $\boldsymbol{Z=|Z|\angle\phi}$

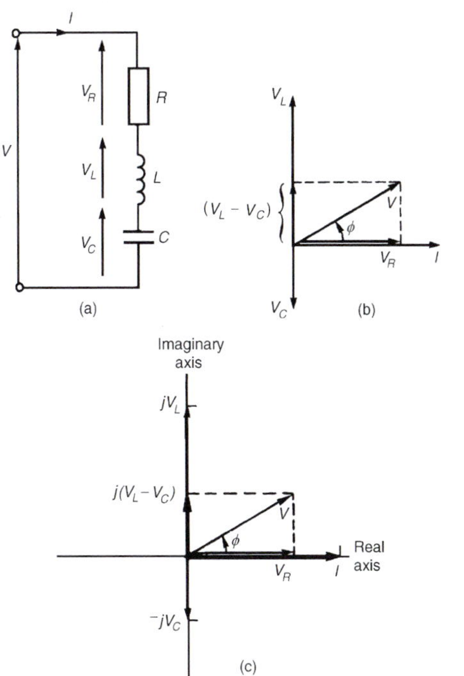

Figure 27.8 (a) Circuit diagram (b) phasor diagram (c) Argand diagram

where,

$|Z| = \sqrt{[R^2 + (X_L - X_C)^2]}$ and

$\phi = \tan^{-1}(X_L - X_C)/R$

When $V_L = V_C, X_L = X_C$ and the applied voltage V and the current I are in phase. This effect is called **series resonance** and is discussed separately in Chapter 31.

(g) General series circuit

In an a.c. circuit containing several impedances connected in series, say, $Z_1, Z_2, Z_3, \ldots, Z_n$, then the total equivalent impedance Z_T is given by

$$Z_T = Z_1 + Z_2 + Z_3 + \cdots + Z_n$$

Problem 1. Determine the values of the resistance and the series-connected inductance or capacitance for each of the following impedances: (a) $(12 + j5)\,\Omega$, (b) $-j40\,\Omega$, (c) $30\angle 60^\circ\,\Omega$, (d) $2.20 \times 10^6 \angle -30^\circ\,\Omega$. Assume for each a frequency of 50 Hz.

(a) From Section 27.2(d), for an R–L series circuit, impedance $Z = R + jX_L$

Thus $Z = (12 + j5)\,\Omega$ represents a resistance of $12\,\Omega$ and an inductive reactance of $5\,\Omega$ in series.

Since inductive reactance $X_L = 2\pi f L$

$$\text{inductance } L = \frac{X_L}{2\pi f} = \frac{5}{2\pi(50)} = 0.0159\,\text{H}$$

i.e. the inductance is 15.9 mH

Thus an impedance $(12 + j5)\,\Omega$ represents a resistance of $12\,\Omega$ in series with an inductance of 15.9 mH

(b) From Section 27.2(c), for a purely capacitive circuit, impedance $Z = -jX_C$

Thus $Z = -j40\,\Omega$ represents zero resistance and a capacitive reactance of $40\,\Omega$.

Since capacitive reactance $X_C = 1/(2\pi f C)$

$$\text{capacitance } C = \frac{1}{2\pi f X_C} = \frac{1}{2\pi(50)(40)}\,\text{F}$$
$$= \frac{10^6}{2\pi(50)(40)}\,\mu\text{F} = 79.6\,\mu\text{F}$$

Thus an impedance $-j40\,\Omega$ represents a pure capacitor of capacitance $79.6\,\mu$F

(c) $30\angle 60^\circ = 30(\cos 60^\circ + j\sin 60^\circ) = 15 + j25.98$

Thus $Z = 30\angle 60^\circ\,\Omega = (15 + j25.98)\,\Omega$ represents a resistance of $15\,\Omega$ and an inductive reactance of $25.98\,\Omega$ in series (from Section 27.2(d)).

Since $X_L = 2\pi f L$

$$\text{inductance } L = \frac{X_L}{2\pi f} = \frac{25.98}{2\pi(50)}$$
$$= 0.0827\text{ H or } 82.7\,\text{mH}$$

Thus an impedance $30\angle 60^\circ\,\Omega$ represents a resistance of $15\,\Omega$ in series with an inductance of 82.7 mH

(d) $2.20 \times 10^6 \angle -30^\circ$

$= 2.20 \times 10^6[\cos(-30^\circ) + j\sin(-30^\circ)]$

$= 1.905 \times 10^6 - j1.10 \times 10^6$

Thus $Z = 2.20 \times 10^6 \angle -30^\circ\,\Omega$

$= (1.905 \times 10^6 - j1.10 \times 10^6)\,\Omega$

Part 4

represents a resistance of $1.905 \times 10^6\,\Omega$ (i.e. $1.905\,\text{M}\Omega$) and a capacitive reactance of $1.10 \times 10^6\,\Omega$ in series (from Section 27.2(e)).

Since capacitive reactance $X_C = 1/(2\pi f C)$,

$$\text{capacitance } C = \frac{1}{2\pi f X_C}$$

$$= \frac{1}{2\pi(50)(1.10 \times 10^6)}\,\text{F}$$

$$= 2.894 \times 10^{-9}\,\text{F or } 2.894\,\text{nF}$$

Thus an impedance $2.2 \times 10^6\angle -30°\,\Omega$ represents a resistance of $1.905\,\text{M}\Omega$ in series with a 2.894 nF capacitor.

Problem 2. Determine, in polar and rectangular forms, the current flowing in an inductor of negligible resistance and inductance 159.2 mH when it is connected to a 250 V, 50 Hz supply.

Inductive reactance

$$X_L = 2\pi f L = 2\pi(50)(159.2 \times 10^{-3}) = 50\,\Omega$$

Thus circuit impedance $Z = (0 + j50)\,\Omega = 50\angle 90°\,\Omega$

Supply voltage, $V = 250\angle 0°$ V (or $(250 + j0)$V)

(Note that since the voltage is given as 250 V, this is assumed to mean $250\angle 0°$ V or $(250 + j0)$V)

$$\text{Hence current } I = \frac{V}{Z} = \frac{250\angle 0°}{50\angle 90°} = \frac{250}{50}\angle(0° - 90°)$$

$$= \mathbf{5\angle -90°\,A}$$

$$\text{Alternatively, } I = \frac{V}{Z} = \frac{(250 + j0)}{(0 + j50)} = \frac{250(-j50)}{j50(-j50)}$$

$$= \frac{-j(50)(250)}{50^2} = \mathbf{-j5\,A}$$

which is the same as $5\angle -90°$ A

Problem 3. A $3\,\mu$F capacitor is connected to a supply of frequency 1 kHz and a current of $2.83\angle 90°$ A flows. Determine the value of the supply p.d.

$$\text{Capacitive reactance } X_C = \frac{1}{2\pi f C}$$

$$= \frac{1}{2\pi(1000)(3 \times 10^{-6})}$$

$$= 53.05\,\Omega$$

Hence circuit impedance

$$Z = (0 - j53.05)\,\Omega = 53.05\angle -90°\,\Omega$$

$$\text{Current } I = 2.83\angle 90°\,\text{A (or } (0 + j2.83)\text{A)}$$

Supply p.d., $V = IZ = (2.83\angle 90°)(53.05\angle -90°)$

i.e. **p.d. = $150\angle 0°$ V**

Alternatively, $V = IZ = (0 + j2.83)(0 - j53.05)$

$$= -j^2(2.83)(53.05) = \mathbf{150\,V}$$

Problem 4. The impedance of an electrical circuit is $(30 - j50)$ ohms. Determine (a) the resistance, (b) the capacitance, (c) the modulus of the impedance and (d) the current flowing and its phase angle, when the circuit is connected to a 240 V, 50 Hz supply.

(a) Since impedance $Z = (30 - j50)\,\Omega$, **the resistance is 30 ohms** and the capacitive reactance is $50\,\Omega$

(b) Since $X_C = 1/(2\pi f C)$, **capacitance**,

$$C = \frac{1}{2\pi f X_c} = \frac{1}{2\pi(50)(50)} = \mathbf{63.66\,\mu F}$$

(c) The modulus of impedance,

$$|Z| = \sqrt{(R^2 + X_C^2)} = \sqrt{(30^2 + 50^2)}$$

$$= \mathbf{58.31\,\Omega}$$

(d) Impedance $Z = (30 - j50)\,\Omega$

$$= 58.31\angle \tan^{-1}\frac{X_C}{R}$$

$$= 58.31\angle -59.04°\,\Omega$$

$$\text{Hence current } I = \frac{V}{Z} = \frac{240\angle 0°}{58.31\angle -59.04°}$$

$$= \mathbf{4.12\angle 59.04°\,A}$$

Part 4

Problem 5. A 200 V, 50 Hz supply is connected across a coil of negligible resistance and inductance 0.15 H connected in series with a 32 Ω resistor. Determine (a) the impedance of the circuit, (b) the current and circuit phase angle, (c) the p.d. across the 32 Ω resistor and (d) the p.d. across the coil.

(a) Inductive reactance $X_L = 2\pi fL = 2\pi(50)(0.15) = 47.1\,\Omega$

Impedance $Z = R + jX_L$
$= \mathbf{(32 + j47.1)\,\Omega}$ or $\mathbf{57.0\angle 55.81°\,\Omega}$

The circuit diagram is shown in Figure 27.9

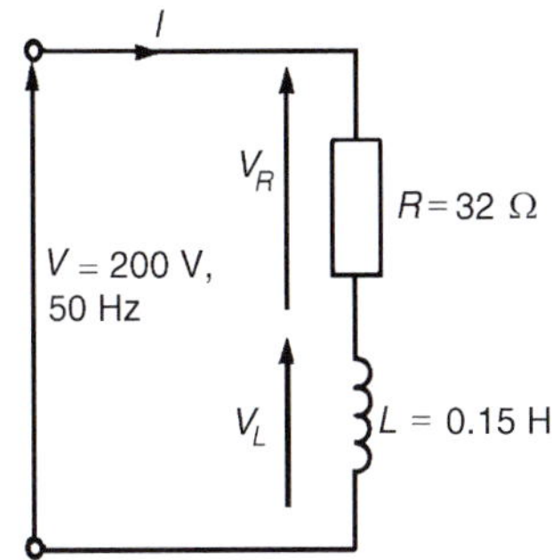

Figure 27.9

(b) Current $I = \frac{V}{Z} = \frac{200\angle 0°}{57.0\angle 55.81°}$
$= \mathbf{3.51\angle -55.81°\,A}$

i.e. **the current is 3.51 A lagging the voltage by 55.81°**

(c) P.d. across the 32 Ω resistor,
$V_R = IR = (3.51\angle -55.81°)(32\angle 0°)$
i.e. $\mathbf{V_R = 112.3\angle -55.81°\,V}$

(d) P.d. across the coil,
$V_L = IX_L = (3.51\angle -55.81°)(47.1\angle 90°)$
i.e. $\mathbf{V_L = 165.3\angle 34.19°\,V}$

The phasor sum of V_R and V_L is the supply voltage V as shown in the phasor diagram of Figure 27.10.

$V_R = 112.3\angle -55.81° = (63.11 - j92.89)\,V$

$V_L = 165.3\angle 34.19°\,V = (136.73 + j92.89)\,V$

Hence

$V = V_R + V_L = (63.11 - j92.89) + (136.73 + j92.89)$
$= (200 + j0)V$ or $200\angle 0°$ V, correct to three significant figures.

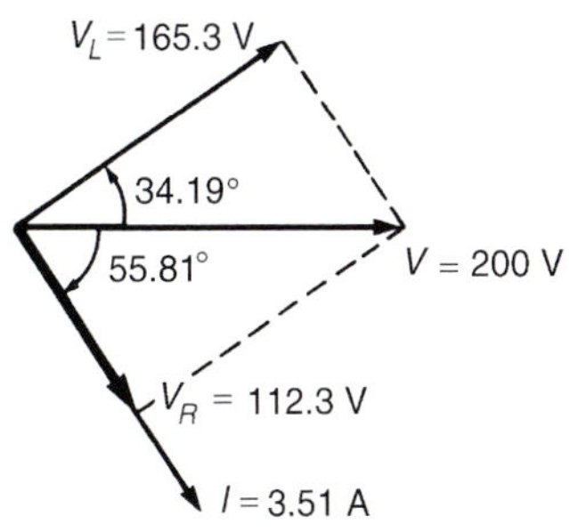

Figure 27.10

Problem 6. Determine the value of impedance if a current of $(7 + j16)$ A flows in a circuit when the supply voltage is $(120 + j200)$V. If the frequency of the supply is 5 MHz, determine the value of the components forming the series circuit.

Impedance $Z = \frac{V}{I} = \frac{(120 + j200)}{(7 + j16)}$
$= \frac{233.24\angle 59.04°}{17.464\angle 66.37°}$
$= 13.36\angle -7.33\,\Omega$ or $(13.25 - j1.705)\,\Omega$

The series circuit thus consists of a **13.25 Ω resistor** and a capacitor of capacitive reactance **1.705 Ω**

Since $X_C = \frac{1}{2\pi fC}$

capacitance $C = \frac{1}{2\pi f X_C}$
$= \frac{1}{2\pi(5 \times 10^6)(1.705)}$
$= 1.867 \times 10^{-8}\,\text{F} = \mathbf{18.67\,nF}$

Now try the following Practice Exercise

Practice Exercise 122 Series a.c. circuits (Answers on page 826)

1. Determine the resistance R and series inductance L (or capacitance C) for each of the following impedances, assuming the frequency to be 50 Hz. (a) $(4 + j7)\,\Omega$, (b) $(3 - j20)\,\Omega$, (c) $j10\,\Omega$, (d) $-j3\,\text{k}\Omega$, (e) $15\angle(\pi/3)\,\Omega$, (f) $6\angle -45°\,\text{M}\Omega$

2. A 0.4 μF capacitor is connected to a 250 V, 2 kHz supply. Determine the current flowing.

3. Two voltages in a circuit are represented by $(15+j10)$V and $(12-j4)$V. Determine the magnitude of the resultant voltage when these voltages are added.

4. A current of $2.5\angle-90°$ A flows in a coil of inductance 314.2 mH and negligible resistance when connected across a 50 Hz supply. Determine the value of the supply p.d.

5. A voltage $(75+j90)$ V is applied across an impedance and a current of $(5+j12)$ A flows. Determine (a) the value of the circuit impedance, and (b) the values of the components comprising the circuit if the frequency is 1 kHz.

6. A 30 μF capacitor is connected in series with a resistance R at a frequency of 200 Hz. The resulting current leads the voltage by 30°. Determine the magnitude of R.

7. A coil has a resistance of 40 Ω and an inductive reactance of 75 Ω. The current in the coil is $1.70\angle0°$ A. Determine the value of (a) the supply voltage, (b) the p.d. across the 40 Ω resistance, (c) the p.d. across the inductive part of the coil and (d) the circuit phase angle. Draw the phasor diagram.

8. An alternating voltage of 100 V, 50 Hz is applied across an impedance of $(20-j30)\Omega$. Calculate (a) the resistance, (b) the capacitance, (c) the current and (d) the phase angle between current and voltage.

9. A capacitor C is connected in series with a coil of resistance R and inductance 30 mH. The current flowing in the circuit is $2.5\angle-40°$ A when the supply p.d. is 200 V at 400 Hz. Determine the value of (a) resistance R, (b) capacitance C, (c) the p.d. across C and (d) the p.d., across the coil. Draw the phasor diagram.

10. If the p.d. across a coil is $(30+j20)$V at 60 Hz and the coil consists of a 50 mH inductance and 10 Ω resistance, determine the value of current flowing (in polar and Cartesian forms).

27.3 Further worked problems on series a.c. circuits

Problem 7. For the circuit shown in Figure 27.11, determine the value of impedance Z_2

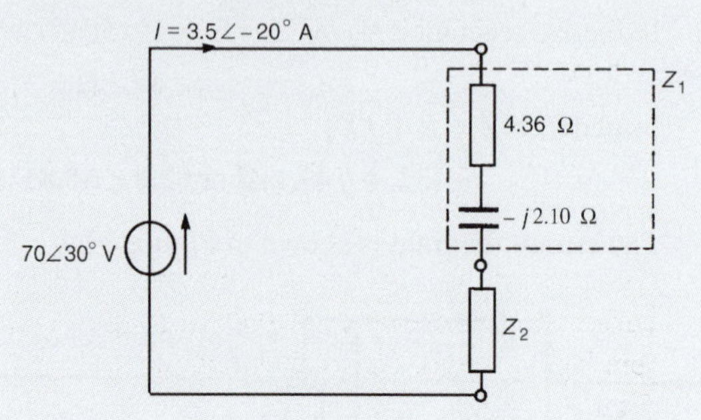

Figure 27.11

Total circuit impedance

$$Z=\frac{V}{I}=\frac{70\angle30°}{3.5\angle-20°}$$
$$=20\angle50°\ \Omega \text{ or } (12.86+j15.32)\Omega$$

Total impedance $Z=Z_1+Z_2$ (see Section 27.2(g)).

Hence $(12.86+j15.32)=(4.36-j2.10)+Z_2$

from which, impedance

$$Z_2=(12.86+j15.32)-(4.36-j2.10)$$
$$=\mathbf{(8.50+j\,17.42)\,\Omega} \text{ or } \mathbf{19.38\angle63.99°\,\Omega}$$

Problem 8. A circuit comprises a resistance of 90 Ω in series with an inductor of inductive reactance 150 Ω. If the supply current is $1.35\angle0°$ A, determine (a) the supply voltage, (b) the voltage across the 90 Ω resistance, (c) the voltage across the inductance, and (d) the circuit phase angle. Draw the phasor diagram.

The circuit diagram is shown in Figure 27.12

(a) Circuit impedance $Z=R+jX_L=(90+j150)\Omega$ or $174.93\angle59.04°\Omega$

Supply voltage

$$V=IZ=(1.35\angle0°)(174.93\angle59.04°)$$
$$=\mathbf{236.2\angle59.04°\ V} \text{ or } \mathbf{(121.5+j202.5)\ V}$$

(b) Voltage across 90 Ω resistor, $\boldsymbol{V_R}=\mathbf{121.5\ V}$ (since $V=V_R+jV_L$)

Part 4

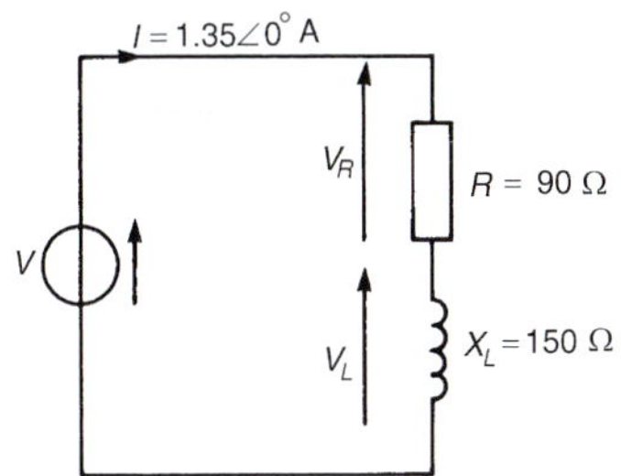

Figure 27.12

(c) Voltage across inductance, $V_L = 202.5$ **V** leading V_R by 90°

(d) Circuit phase angle is the angle between the supply current and voltage, i.e. **59.04° lagging** (i.e. current lags voltage). The phasor diagram is shown in Figure 27.13.

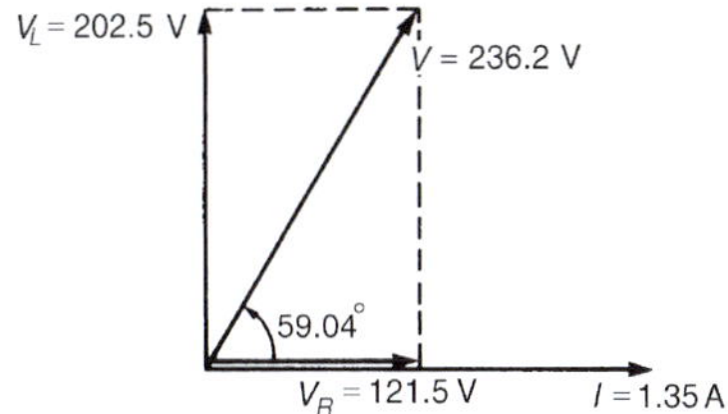

Figure 27.13

Problem 9. A coil of resistance 25 Ω and inductance 20 mH has an alternating voltage given by $v = 282.8 \sin(628.4t + (\pi/3))$ volts applied across it. Determine (a) the r.m.s. value of voltage (in polar form), (b) the circuit impedance, (c) the r.m.s. current flowing and (d) the circuit phase angle.

(a) Voltage $v = 282.8 \sin(628.4t + (\pi/3))$ volts means $V_m = 282.8$ V, hence r.m.s. voltage

$$V = 0.707 \times 282.8 \left[\text{or } \frac{1}{\sqrt{2}} \times 282.8\right]$$

i.e. $V = 200$ V

In complex form the r.m.s. voltage may be expressed as **200∠π/3 V** or **200∠60° V**

(b) $\omega = 2\pi f = 628.4$ rad/s, hence frequency

$f = 628.4/(2\pi) = 100$ Hz

Inductive reactance

$X_L = 2\pi f L = 2\pi(100)(20 \times 10^{-3}) = 12.57\,\Omega$

Hence circuit impedance

$Z = R + jX_L = (\mathbf{25 + j12.57})\,\boldsymbol{\Omega}$ or

$\mathbf{27.98\angle 26.69°\,\Omega}$

(c) R.m.s current, $I = \dfrac{V}{Z} = \dfrac{200\angle 60°}{27.98\angle 26.69°}$

$= \mathbf{7.148\angle 33.31°\ A}$

(d) Circuit phase angle is the angle between current I and voltage V, i.e. $60° - 33.31° =$ **26.69° lagging**.

Problem 10. A 240 V, 50 Hz voltage is applied across a series circuit comprising a coil of resistance 12 Ω and inductance 0.10 H, and 120 μF capacitor. Determine the current flowing in the circuit.

The circuit diagram is shown in Figure 27.14.

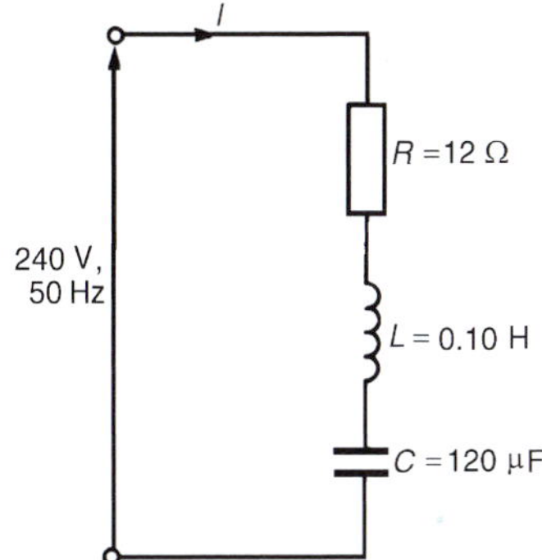

Figure 27.14

Inductive reactance, $X_L = 2\pi f L = 2\pi(50)(0.10)$

$= 31.4\,\Omega$

Capacitive reactance,

$$X_C = \frac{1}{2\pi f C} = \frac{1}{2\pi(50)(120 \times 10^{-6})} = 26.5\,\Omega$$

Impedance $Z = R + j(X_L - X_C)$ (see Section 27.2(f))

i.e. $Z = 12 + j(31.4 - 26.5)$

$= (12 + j4.9)\,\Omega$ or $13.0\angle 22.2°\,\Omega$

Current flowing, $I = \dfrac{V}{Z} = \dfrac{240\angle 0°}{13.0\angle 22.2°}$

$= \mathbf{18.5\angle -22.2°\ A}$

i.e. the current flowing is 18.5 A, lagging the voltage by 22.2°

The phasor diagram is shown on the Argand diagram in Figure 27.15

Part 4

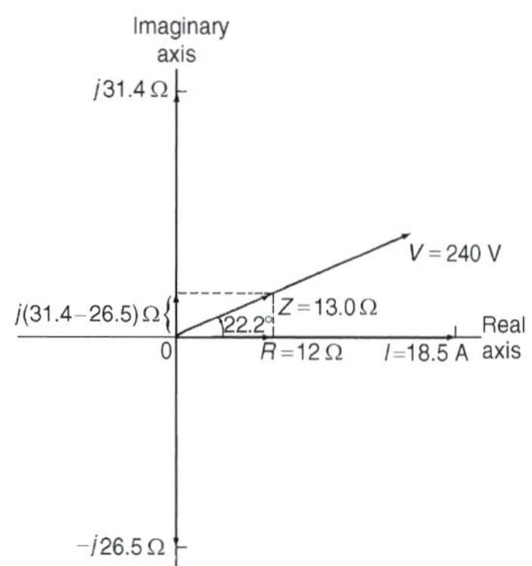

Figure 27.15

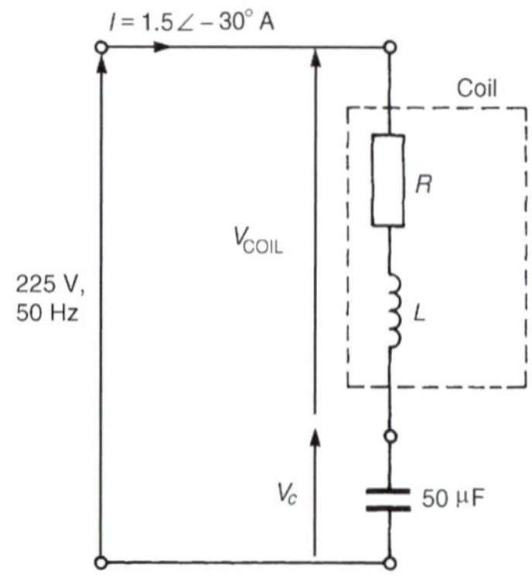

Figure 27.16

Problem 11. A coil of resistance R ohms and inductance L henrys is connected in series with a 50 μF capacitor. If the supply voltage is 225 V at 50 Hz and the current flowing in the circuit is $1.5\angle -30°$ A, determine the values of R and L. Determine also the voltage across the coil and the voltage across the capacitor.

Circuit impedance,

$$Z = \frac{V}{Z} = \frac{225\angle 0°}{1.5\angle -30°}$$

$$= 150\angle 30°\,\Omega \text{ or } (129.9 + j75.0)\,\Omega$$

Capacitive reactance,

$$X_C = \frac{1}{2\pi f C} = \frac{1}{2\pi(50)(50\times 10^{-6})} = 63.66\,\Omega$$

Circuit impedance $Z = R + j(X_L - X_C)$

i.e. $129.9 + j75.0 = R + j(X_L - 63.66)$

Equating the real parts gives: **resistance $R = 129.9\Omega$.**

Equating the imaginary parts gives: $75.0 = X_L - 63.66$,

from which, $X_L = 75.0 + 63.66 = 138.66\,\Omega$

Since $X_L = 2\pi f L$, **inductance** $L = \frac{X_L}{2\pi f} = \frac{138.66}{2\pi(50)}$

$$= \mathbf{0.441\ H}$$

The circuit diagram is shown in Figure 27.16.

Voltage across coil, $V_{COIL} = IZ_{COIL}$

$Z_{COIL} = R + jX_L$

$= (129.9 + j138.66)\Omega$ or $190\angle 46.87°\,\Omega$

Hence $V_{COIL} = (1.5\angle -30°)(190\angle 46.87°)$

$= \mathbf{285\angle 16.87°\ V}$ or $\mathbf{(272.74 + j82.71)\ V}$

Voltage across capacitor,

$V_C = IX_C = (1.5\angle -30°)(63.66\angle -90°)$

$= \mathbf{95.49\angle -120°\ V}$ or $\mathbf{(-47.75 - j82.70)\ V}$

[Check: Supply voltage,

$V = V_{COIL} + V_C$

$= (272.74 + j82.71) + (-47.75 - j82.70)$

$= (225 + j0)$ V or $225\angle 0°$ V]

Problem 12. For the circuit shown in Figure 27.17, determine the values of voltages V_1 and V_2 if the supply frequency is 4 kHz. Determine also the value of the supply voltage V and the circuit phase angle. Draw the phasor diagram.

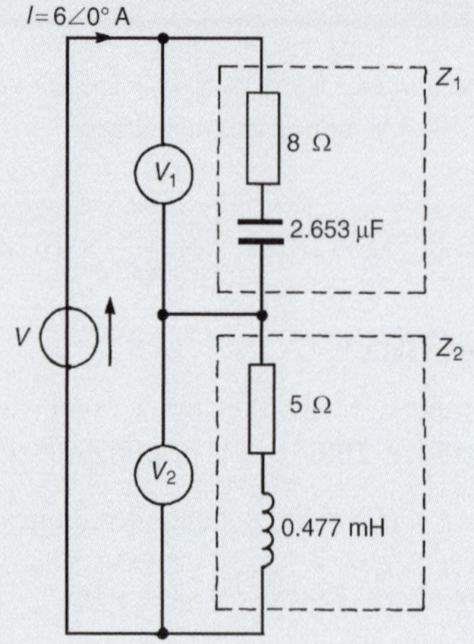

Figure 27.17

Part 4

For impedance Z_1,

$$X_C = \frac{1}{2\pi fC} = \frac{1}{2\pi(4000)(2.653\times 10^{-6})} = 15\,\Omega$$

Hence $Z_1 = (8 - j15)\Omega$ or $17\angle{-61.93}°\,\Omega$

and **voltage** $V_1 = IZ_1$

$$= (6\angle 0°)(17\angle{-61.93}°)$$

$$= \mathbf{102\angle{-61.93}°\ V} \text{ or } \mathbf{(48 - j90)V}$$

For impedance Z_2,

$$X_L = 2\pi fL = 2\pi(4000)(0.477\times 10^{-3}) = 12\,\Omega$$

Hence $Z_2 = (5 + j12)\Omega$ or $13\angle 67.38°\,\Omega$

and **voltage** $V_2 = IZ_2 = (6\angle 0°)(13\angle 67.38°)$

$$= \mathbf{78\angle 67.38°\ V} \text{ or } \mathbf{(30 + j72)V}$$

Supply voltage, $V = V_1 + V_2 = (48 - j90) + (30 + j72)$

$$= \mathbf{(78 - j18)V} \text{ or } \mathbf{80\angle{-13}°\ V}$$

Circuit phase angle, $\boldsymbol{\phi = 13°}$ **leading**. The phasor diagram is shown in Figure 27.18.

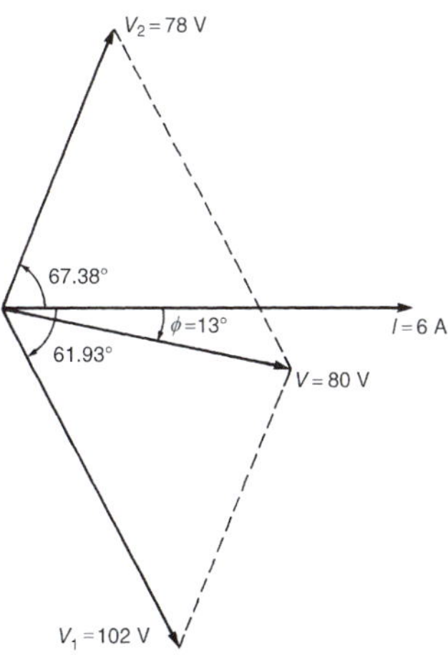

Figure 27.18

Now try the following Practice Exercise

Practice Exercise 123 Series a.c. circuits (Answers on page 826)

1. Determine, in polar form, the complex impedances for the circuits shown in Figure 27.19 if the frequency in each case is 50 Hz.

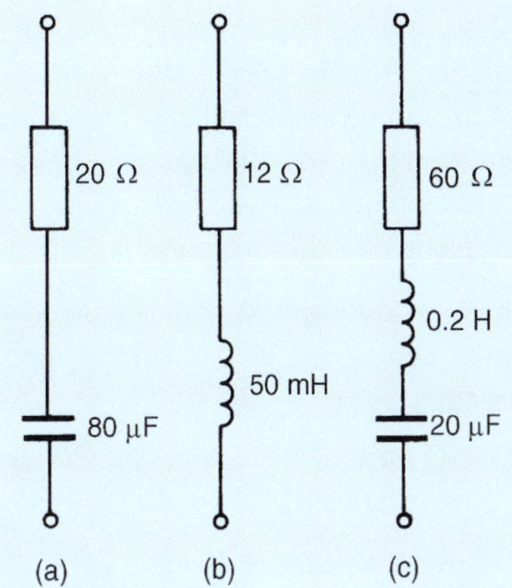

Figure 27.19

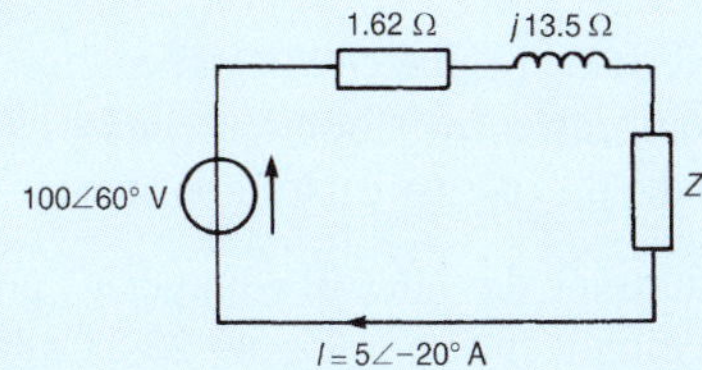

Figure 27.20

2. For the circuit shown in Figure 27.20, determine the impedance Z in polar and rectangular forms.
3. A series circuit consists of a 10 Ω resistor, a coil of inductance 0.09 H and negligible resistance, and a 150 µF capacitor, and is connected to a 100 V, 50 Hz supply. Calculate the current flowing and its phase relative to the supply voltage.
4. A 150 mV, 5 kHz source supplies an a.c. circuit consisting of a coil of resistance 25 Ω and inductance 5 mH connected in series with a capacitance of 177 nF. Determine the current flowing and its phase angle relative to the source voltage.
5. Two impedances, $Z_1 = 5\angle 30°\,\Omega$ and $Z_2 = 10\angle 45°\,\Omega$, draw a current of 3.36 A when connected in series to a certain a.c. supply. Determine (a) the supply voltage, (b) the phase angle between the voltage and current, (c) the p.d. across Z_1 and (d) the p.d. across Z_2
6. A 4500 pF capacitor is connected in series with a 50 Ω resistor across an alternating

Part 4

voltage $v = 212.1\sin(\pi 10^6 t + \pi/4)$ volts. Calculate (a) the r.m.s. value of the voltage, (b) the circuit impedance, (c) the r.m.s. current flowing, (d) the circuit phase angle, (e) the voltage across the resistor and (f) the voltage across the capacitor.

7. Three impedances are connected in series across a 120 V, 10 kHz supply. The impedances are:
   (i) Z_1, a coil of inductance 200 μH and resistance 8 Ω
   (ii) Z_2, a resistance of 12 Ω
   (iii) Z_3, a 0.50 μF capacitor in series with a 15 Ω resistor.

   Determine (a) the circuit impedance, (b) the circuit current, (c) the circuit phase angle and (d) the p.d. across each impedance.

8. Determine the value of voltages V_1 and V_2 in the circuit shown in Figure 27.21, if the frequency of the supply is 2.5 kHz. Find also the value of the supply voltage V and the circuit phase angle. Draw the phasor diagram.

9. A circuit comprises a coil of inductance 40 mH and resistance 20 Ω in series with a variable capacitor. The supply voltage is 120 V at 50 Hz. Determine the value of capacitance needed to cause a current of 2.0 A to flow in the circuit.

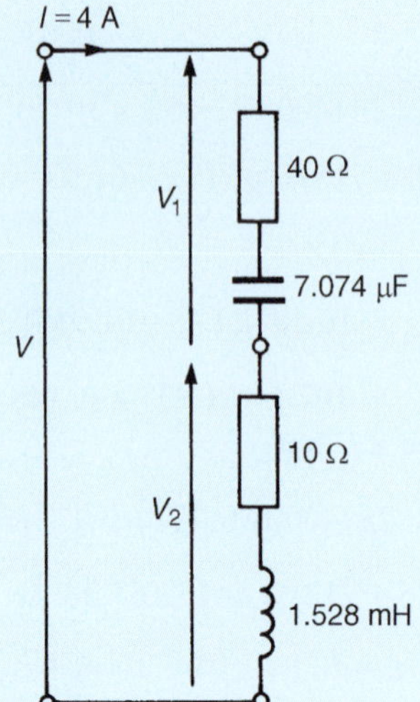

Figure 27.21

10. For the circuit shown in Figure 27.22, determine (i) the circuit current I flowing, and (ii) the p.d. across each impedance.

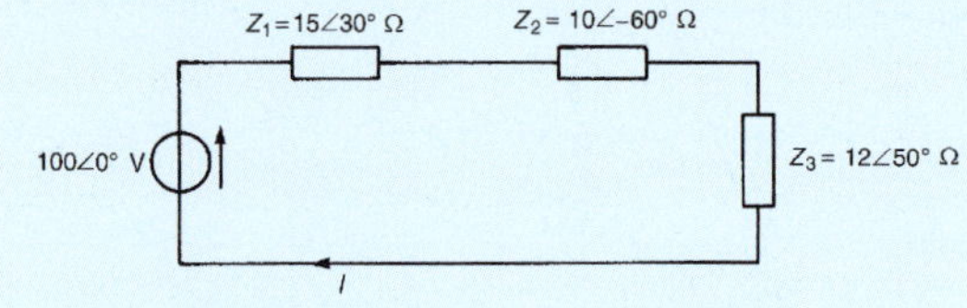

Figure 27.22

Part 4

For fully worked solutions to each of the problems in Practice Exercises 122 and 123 in this chapter, go to the website:
www.routledge.com/cw/bird

Chapter 28

Application of complex numbers to parallel a.c. networks

Why it is important to understand: **Application of complex numbers to parallel a.c. circuits**

As with series circuits, parallel networks may be analysed by using phasor diagrams. However, with parallel networks containing more than two branches, this can become very complicated. It is with parallel a.c. network analysis in particular that the full benefit of using complex numbers may be appreciated. Parallel a.c. circuits are like parallel d.c. circuits, except that phasors need to be used throughout the analysis. Also, the same rules that are needed in analysing parallel d.c. circuits, such as Kirchhoff's current law and the current division rule, are also needed for parallel a.c. networks. Before analysing such networks, admittance, conductance and susceptance are defined. Knowledge of complex numbers makes the analysis of a.c. parallel networks straightforward.

At the end of this chapter you should be able to:

- determine admittance, conductance and susceptance in a.c. circuits
- perform calculations on parallel a.c. circuits using complex numbers

28.1 Introduction

As with series circuits, parallel networks may be analysed by using phasor diagrams. However, with parallel networks containing more than two branches this can become very complicated. It is with parallel a.c. network analysis in particular that the full benefit of using complex numbers may be appreciated. The theory for parallel a.c. networks introduced in Chapter 18 is relevant; more advanced networks will be analysed in this chapter using j notation. Before analysing such networks admittance, conductance and susceptance are defined.

Electrical Circuit Theory and Technology. 978-1-138-67349-6, © 2017 John Bird. Published by Taylor & Francis. All rights reserved.

28.2 Admittance, conductance and susceptance

Admittance is defined as the current I flowing in an a.c. circuit divided by the supply voltage V (i.e. it is the reciprocal of impedance Z). The symbol for admittance is Y. Thus

$$Y=\frac{I}{V}=\frac{1}{Z}$$

The unit of admittance is the **siemen**,* **S**.

An impedance may be resolved into a real part R and an imaginary part X, giving $Z=R\pm jX$. Similarly, an admittance may be resolved into two parts – the real part being called the **conductance** G, and the imaginary part being called the **susceptance** B – and expressed in complex form. Thus admittance

$$Y=G\pm jB$$

When an a.c. circuit contains:

(a) **pure resistance**, then

$$Z=R \text{ and } Y=\frac{1}{Z}=\frac{1}{R}=G$$

(b) **pure inductance**, then

$$Z=jX_L \text{ and } Y=\frac{1}{Z}=\frac{1}{jX_L}=\frac{-j}{(jX_L)(-j)}$$

$$=\frac{-j}{X_L}=-jB_L$$

thus a negative sign is associated with inductive susceptance, B_L

(c) **pure capacitance**, then

$$Z=-jX_C \text{ and } Y=\frac{1}{Z}=\frac{1}{-jX_C}=\frac{j}{(-jX_C)(j)}$$

$$=\frac{j}{X_C}=+jB_C$$

thus a positive sign is associated with capacitive susceptance, B_C

(d) **resistance and inductance in series**, then

$$Z=R+jX_L \text{ and } Y=\frac{1}{Z}=\frac{1}{R+jX_L}$$

$$=\frac{(R-jX_L)}{R^2+X_L^2}$$

i.e. $Y=\dfrac{R}{R^2+X_L^2}-j\dfrac{X_L}{R^2+X_L^2}$ or $\boldsymbol{Y=\dfrac{R}{|Z|^2}-j\dfrac{X_L}{|Z|^2}}$

Thus conductance, $G=R/|Z|^2$ and inductive susceptance, $B_L=-X_L/|Z|^2$

(Note that in an inductive circuit, the imaginary term of the impedance, X_L, is positive, whereas the imaginary term of the admittance, B_L, is negative.)

(e) **resistance and capacitance in series**, then

$$Z=R-jX_C \text{ and } Y=\frac{1}{Z}=\frac{1}{R-jX_C}=\frac{R+jX_C}{R^2+X_C^2}$$

i.e. $Y=\dfrac{R}{R^2+X_C^2}+j\dfrac{X_C}{R^2+X_C^2}$ or

$$\boldsymbol{Y=\frac{R}{|Z|^2}+j\frac{X_C}{|Z|^2}}$$

Thus conductance, $G=R/|Z|^2$ and capacitive susceptance, $B_C=X_C/|Z|^2$

(Note that in a capacitive circuit, the imaginary term of the impedance, X_C, is negative, whereas the imaginary term of the admittance, B_C, is positive.)

(f) **resistance and inductance in parallel**, then

$$\frac{1}{Z}=\frac{1}{R}+\frac{1}{jX_L}=\frac{jX_L+R}{(R)(jX_L)}$$

from which, $\boldsymbol{Z=\dfrac{(R)(jX_L)}{R+jX_L}}$ $\left(\text{i.e. } \dfrac{\text{product}}{\text{sum}}\right)$

and $Y=\dfrac{1}{Z}=\dfrac{R+jX_L}{jRX_L}=\dfrac{R}{jRX_L}+\dfrac{jX_L}{jRX_L}$

i.e. $Y=\dfrac{1}{jX_L}+\dfrac{1}{R}=\dfrac{(-j)}{(jX_L)(-j)}+\dfrac{1}{R}$

or $\boldsymbol{Y=\dfrac{1}{R}-\dfrac{j}{X_L}}$

Thus conductance, $\boldsymbol{G=1/R}$ and inductive susceptance, $\boldsymbol{B_L=-1/X_L}$

(g) **resistance and capacitance in parallel**, then

$$Z=\frac{(R)(-jX_C)}{R-jX_C}\ \left(\text{i.e. } \frac{\text{product}}{\text{sum}}\right)$$

and $Y=\dfrac{1}{Z}=\dfrac{R-jX_C}{-jRX_C}=\dfrac{R}{-jRX_C}-\dfrac{jX_C}{-jRX_C}$

*Who was **Siemens**? For image and resume of Siemens, see page 54. To find out more go to **www.routledge.com/cw/bird**

i.e. $$Y=\frac{1}{-jX_C}+\frac{1}{R}=\frac{(j)}{(-jX_C)(j)}+\frac{1}{R}$$

or $$\boldsymbol{Y=\frac{1}{R}+\frac{j}{X_C}} \qquad (1)$$

Thus conductance, $\boldsymbol{G=1/R}$ and capacitive susceptance, $\boldsymbol{B_C=1/X_C}$

The conclusions that may be drawn from Sections (d) to (g) above are:

(i) that a **series** circuit is more easily represented by an **impedance**,

(ii) that a **parallel** circuit is often more easily represented by an **admittance** especially when more than two parallel impedances are involved.

Problem 1. Determine the admittance, conductance and susceptance of the following impedances: (a) $-j5\,\Omega$, (b) $(25+j40)\,\Omega$, (c) $(3-j2)\,\Omega$, (d) $50\angle 40°\,\Omega$

(a) If impedance $Z=-j5\,\Omega$, then

$$\text{admittance } Y=\frac{1}{Z}=\frac{1}{-j5}=\frac{j}{(-j5)(j)}=\frac{j}{5}$$

$$=\boldsymbol{j\,0.2\text{ S}} \text{ or } \boldsymbol{0.2\angle 90°\text{ S}}$$

Since there is no real part, **conductance,** $\boldsymbol{G=0}$, and **capacitive susceptance,** $\boldsymbol{B_C=0.2\text{ S}}$.

(b) If impedance $Z=(25+j40)\Omega$ then

$$\text{admittance } Y=\frac{1}{Z}=\frac{1}{(25+j40)}=\frac{25-j40}{25^2+40^2}$$

$$=\frac{25}{2225}-\frac{j40}{2225}$$

$$=\boldsymbol{(0.0112-j0.0180)\text{ S}}$$

Thus **conductance,** $\boldsymbol{G=0.0112\text{ S}}$ and **inductive susceptance,** $\boldsymbol{B_L=0.0180\text{ S}}$.

(c) If impedance $Z=(3-j2)\Omega$, then

$$\text{admittance } Y=\frac{1}{Z}=\frac{1}{(3-j2)}=\frac{3+j2}{3^2+2^2}$$

$$=\left(\frac{3}{13}+j\frac{2}{13}\right)\text{S or}$$

$$\boldsymbol{(0.231+j0.154)\text{ S}}$$

Thus **conductance,** $\boldsymbol{G=0.231\text{ S}}$ and **capacitive susceptance,** $\boldsymbol{B_C=0.154\text{ S}}$.

(d) If impedance $Z=50\angle 40°\,\Omega$, then

$$\text{admittance } Y=\frac{1}{Z}=\frac{1}{50\angle 40°}=\frac{1\angle 0°}{50\angle 40°}$$

$$=\frac{1}{50}\angle -40°=\boldsymbol{0.02\angle -40°\text{ S}} \text{ or}$$

$$\boldsymbol{(0.0153-j0.0129)\text{ S}}$$

Thus **conductance,** $\boldsymbol{G=0.0153\text{ S}}$ and **inductive susceptance,** $\boldsymbol{B_L=0.0129\text{ S}}$.

Problem 2. Determine expressions for the impedance of the following admittances: (a) $0.004\angle 30°\text{ S}$, (b) $(0.001-j0.002)\text{ S}$, (c) $(0.05+j\,0.08)\text{ S}$.

(a) Since admittance $Y=1/Z$, impedance $Z=1/Y$

$$\text{Hence impedance } Z=\frac{1}{0.004\angle 30°}=\frac{1\angle 0°}{0.004\angle 30°}$$

$$=\boldsymbol{250\angle -30°\,\Omega} \text{ or}$$

$$\boldsymbol{(216.5-j125)\,\Omega}$$

(b) $$\text{Impedance } Z=\frac{1}{(0.001-j0.002)}$$

$$=\frac{0.001+j0.002}{(0.001)^2+(0.002)^2}$$

$$=\frac{0.001+j0.002}{0.000\,005}$$

$$=\boldsymbol{(200+j400)\,\Omega} \text{ or } \boldsymbol{447.2\angle 63.43°\,\Omega}$$

(c) Admittance $Y=(0.05+j0.08)\text{ S}$

$$=0.094\angle 57.99°\text{ S}$$

$$\text{Hence impedance } Z=\frac{1}{0.0094\angle 57.99°}$$

$$=\boldsymbol{10.64\angle -57.99°\,\Omega} \text{ or}$$

$$\boldsymbol{(5.64-j9.02)\,\Omega}$$

Problem 3. The admittance of a circuit is $(0.040+j0.025)\text{ S}$. Determine the values of the resistance and the capacitive reactance of the circuit if they are connected (a) in parallel, (b) in series. Draw the phasor diagram for each of the circuits.

(a) Parallel connection

Admittance $Y=(0.040+j0.025)$ S, therefore conductance, $G=0.040$ S and capacitive susceptance, $B_C=0.025$ S. From equation (1) when a circuit consists of resistance R and capacitive reactance in parallel, then $Y=(1/R)+(j/X_C)$.

Hence resistance $R=\dfrac{1}{G}=\dfrac{1}{0.040}=\mathbf{25\,\Omega}$

and capacitive reactance $X_C=\dfrac{1}{B_C}=\dfrac{1}{0.025}=\mathbf{40\,\Omega}$

The circuit and phasor diagrams are shown in Figure 28.1.

(b) Series connection

Admittance $Y=(0.040+j0.025)$ S, therefore

$$\text{impedance } Z=\frac{1}{Y}=\frac{1}{0.040+j0.025}$$

$$=\frac{0.040-j0.025}{(0.040)^2+(0.025)^2}$$

$$=(17.98-j11.24)\,\Omega$$

Thus the **resistance, $R=17.98\,\Omega$** and **capacitive reactance, $X_C=11.24\,\Omega$**

The circuit and phasor diagrams are shown in Figure 28.2.

The circuits shown in Figures 28.1(a) and 28.2(a) are equivalent in that they take the same supply current

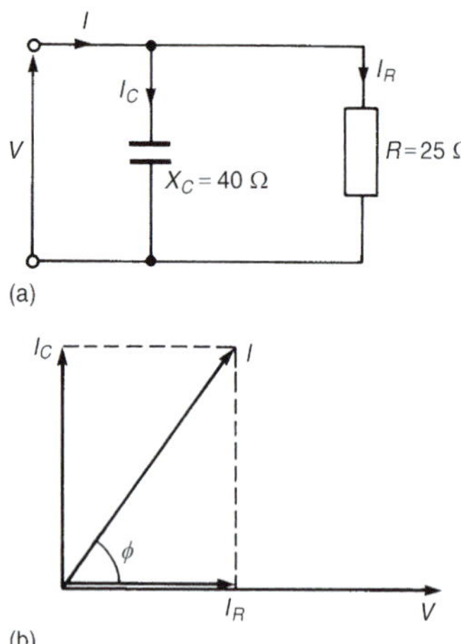

Figure 28.1 (a) Circuit diagram (b) phasor diagram

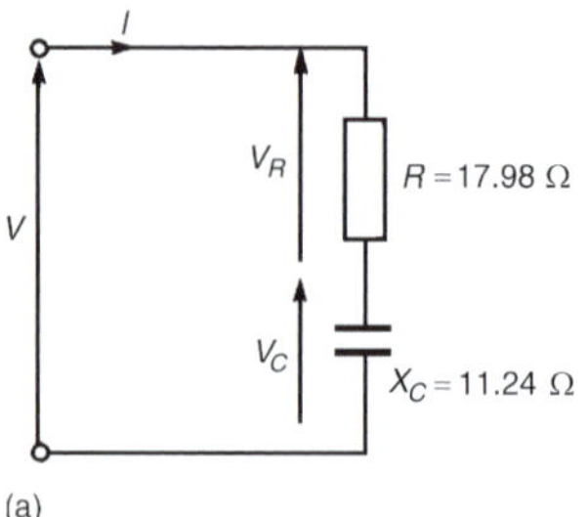

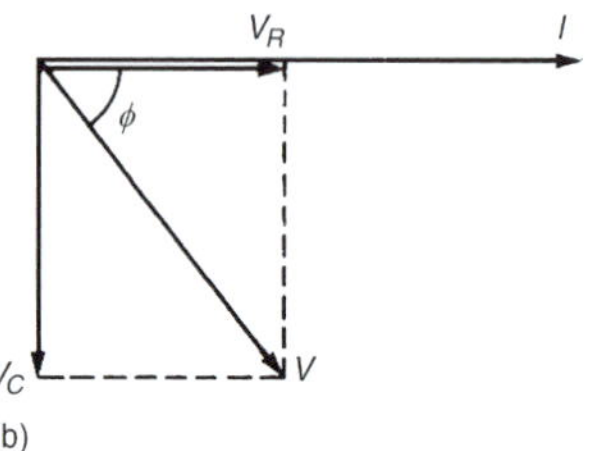

Figure 28.2 (a) Circuit diagram (b) phasor diagram

I for a given supply voltage V; the phase angle ϕ between the current and voltage is the same in each of the phasor diagrams shown in Figures 28.1(b) and 28.2(b).

Now try the following Practice Exercise

Practice Exercise 124 Admittance, conductance and susceptance (Answers on page 826)

1. Determine the admittance (in polar form), conductance and susceptance of the following impedances: (a) $j10\,\Omega$, (b) $-j40\,\Omega$, (c) $32\angle-30°\,\Omega$, (d) $(5+j9)\Omega$, (e) $(16-j10)\Omega$
2. Derive expressions, in polar form, for the impedances of the following admittances: (a) $0.05\angle40°$ S, (b) $0.0016\angle-25°$ S, (c) $(0.1+j0.4)$ S, (d) $(0.025-j0.040)$ S
3. The admittance of a series circuit is $(0.010-j0.004)$ S. Determine the values of the circuit components if the frequency is 50 Hz.
4. The admittance of a network is $(0.05-j0.08)$ S. Determine the values of resistance and reactance in the circuit if they are connected (a) in series, (b) in parallel.

5. The admittance of a two-branch parallel network is $(0.02 + j0.05)$ S. Determine the circuit components if the frequency is 1 kHz.

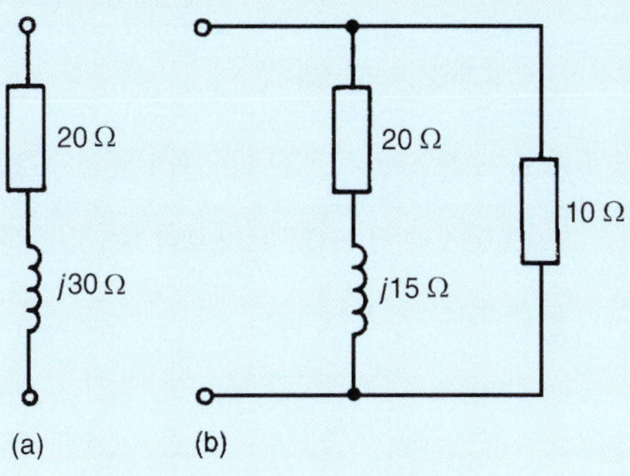

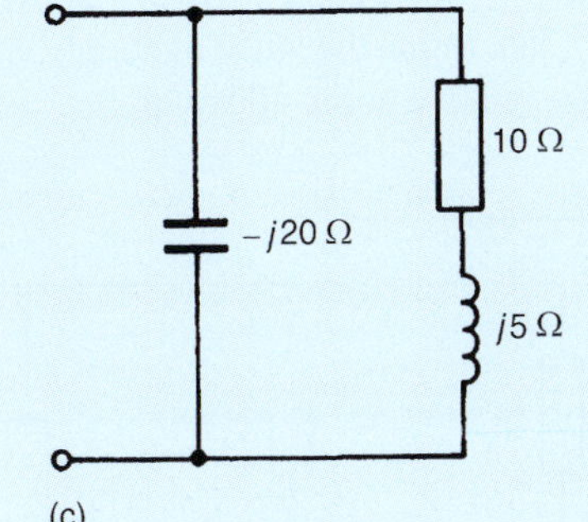

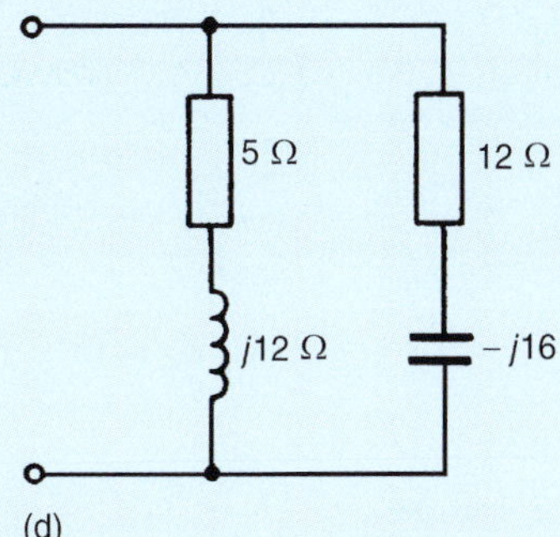

Figure 28.3

6. Determine the total admittance, in rectangular and polar forms, of each of the networks shown in Figure 28.3.

28.3 Parallel a.c. networks

Figure 28.4 shows a circuit diagram containing three impedances, Z_1, Z_2 and Z_3, connected in parallel. The potential difference across each impedance is the same, i.e. the supply voltage V. Current $I_1 = V/Z_1$, $I_2 = V/Z_2$ and $I_3 = V/Z_3$. If Z_T is the total equivalent impedance of the circuit then $I = V/Z_T$. The supply current, $I = I_1 + I_2 + I_3$ (phasorially).

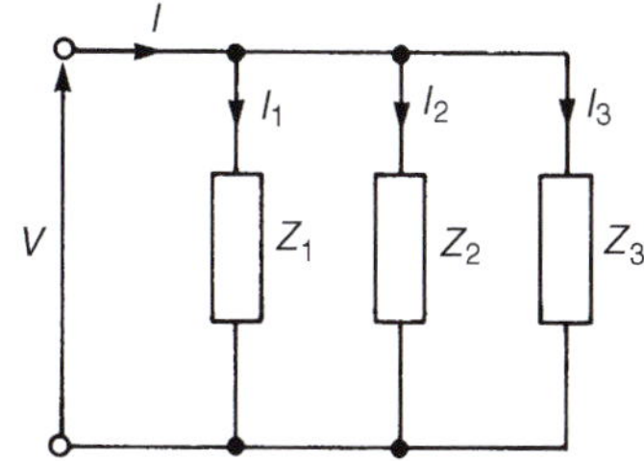

Figure 28.4

Thus $\frac{V}{Z_T} = \frac{V}{Z_1} + \frac{V}{Z_2} + \frac{V}{Z_3}$ and

$$\mathbf{\frac{1}{Z_T} = \frac{1}{Z_1} + \frac{1}{Z_2} + \frac{1}{Z_3}}$$

or total admittance, $\mathbf{Y_T = Y_1 + Y_2 + Y_3}$

In general, for n impedances connected in parallel,

$$\mathbf{Y_T = Y_1 + Y_2 + Y_3 + \cdots + Y_n} \quad \text{(phasorially)}$$

It is in parallel circuit analysis that the use of admittance has its greatest advantage.

Current division in a.c. circuits

For the special case of two impedances, Z_1 and Z_2, connected in parallel (see Figure 28.5),

$$\frac{1}{Z_T} = \frac{1}{Z_1} + \frac{1}{Z_2} = \frac{Z_2 + Z_1}{Z_1 Z_2}$$

The total impedance, $\mathbf{Z_T = Z_1 Z_2/(Z_1 + Z_2)}$ (i.e. product/sum).

From Figure 28.5,

supply voltage, $V = IZ_T = I\left(\frac{Z_1 Z_2}{Z_1 + Z_2}\right)$

Also, $V = I_1 Z_1$ (and $V = I_2 Z_2$)

Thus, $I_1 Z_1 = I\left(\frac{Z_1 Z_2}{Z_1 + Z_2}\right)$

i.e. $\textbf{current } \mathbf{I_1 = I\left(\frac{Z_2}{Z_1 + Z_2}\right)}$

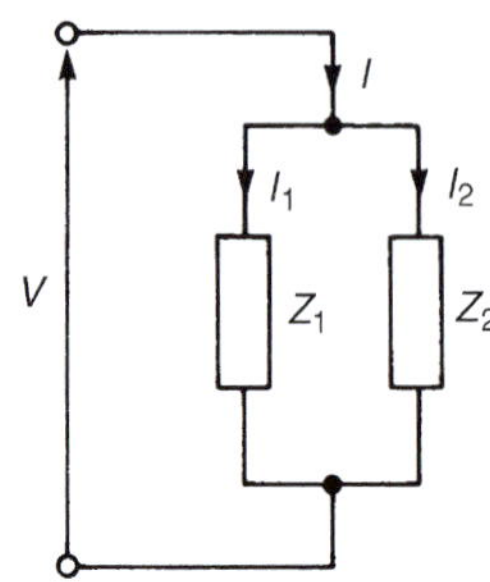

Figure 28.5

Similarly, **current** $\boldsymbol{I_2 = I\left(\frac{Z_1}{Z_1 + Z_2}\right)}$

Note that all of the above circuit symbols infer complex quantities either in Cartesian or polar form.
The following problems show how complex numbers are used to analyse parallel a.c. networks.

Problem 4. Determine the values of currents I, I_1 and I_2 shown in the network of Figure 28.6.

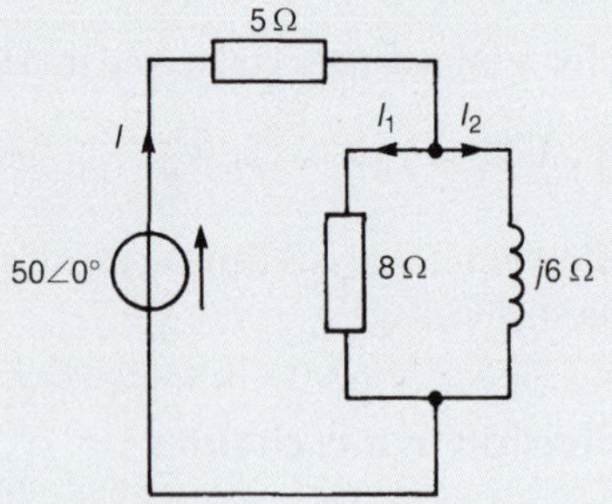

Figure 28.6

Total circuit impedance,

$$Z_T = 5 + \frac{(8)(j6)}{8 + j6} = 5 + \frac{(j48)(8 - j6)}{8^2 + 6^2}$$

$$= 5 + \frac{j384 + 288}{100}$$

$$= (7.88 + j3.84)\ \Omega \text{ or } 8.77\angle 25.98^\circ\ \Omega$$

$$\text{Current } I = \frac{V}{Z_T} = \frac{50\angle 0^\circ}{8.77\angle 25.98^\circ} = \mathbf{5.70\angle -25.98^\circ\ A}$$

$$\text{Current } I_1 = I\left(\frac{j6}{8 + j6}\right)$$

$$= (5.70\angle -25.98^\circ)\left(\frac{6\angle 90^\circ}{10\angle 36.87^\circ}\right)$$

$$= \mathbf{3.42\angle 27.15^\circ\ A}$$

$$\text{Current } I_2 = I\left(\frac{8}{8 + j6}\right)$$

$$= (5.70\angle -25.98^\circ)\left(\frac{8\angle 0^\circ}{10\angle 36.87^\circ}\right)$$

$$= \mathbf{4.56\angle -62.85^\circ\ A}$$

[Note: $I = I_1 + I_2 = 3.42\angle 27.15^\circ + 4.56\angle -62.85^\circ$

$$= (3.043 + j1.561) + (2.081 - j4.058)$$

$$= (5.124 - j2.497)\text{A}$$

$$= 5.70\angle -25.98^\circ\text{A}]$$

Problem 5. For the parallel network shown in Figure 28.7, determine the value of supply current I and its phase relative to the 40 V supply.

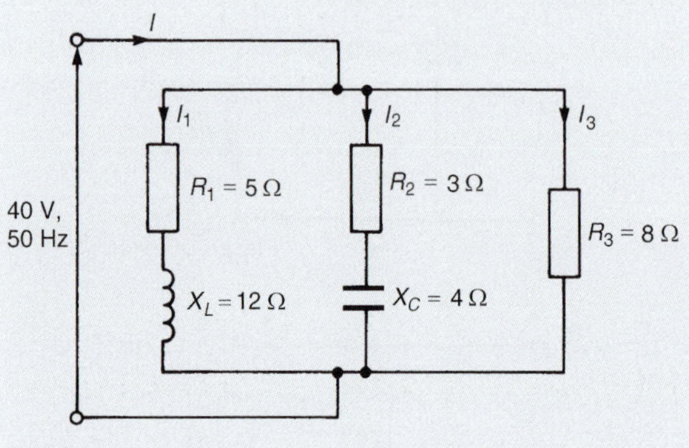

Figure 28.7

Impedance $Z_1 = (5 + j12)\ \Omega$, $Z_2 = (3 - j4)\ \Omega$ and $Z_3 = 8\ \Omega$

Supply current $I = \frac{V}{Z_T} = VY_T$ where $Z_T =$ total circuit impedance, and $Y_T =$ total circuit admittance.

$$Y_T = Y_1 + Y_2 + Y_3$$

$$= \frac{1}{Z_1} + \frac{1}{Z_2} + \frac{1}{Z_3} = \frac{1}{(5 + j12)} + \frac{1}{(3 - j4)} + \frac{1}{8}$$

$$= \frac{5 - j12}{5^2 + 12^2} + \frac{3 + j4}{3^2 + 4^2} + \frac{1}{8}$$

$$= (0.0296 - j0.0710) + (0.1200 + j0.1600) + (0.1250)$$

i.e. $Y_T = (0.2746 + j0.0890)\text{ S or } 0.2887\angle 17.96^\circ\text{S}$

Current $I = VY_T = (40\angle 0^\circ)(0.2887\angle 17.96^\circ)$

$= 11.55\angle 17.96^\circ$ A

Hence the current I is 11.55 A and is leading the 40 V supply by 17.96°

Alternatively, current $I = I_1 + I_2 + I_3$

Current $I_1 = \dfrac{40\angle 0^\circ}{5 + j12} = \dfrac{40\angle 0^\circ}{13\angle 67.38^\circ}$

$= 3.077\angle -67.38^\circ$ A or $(1.183 - j2.840)$ A

Current $I_2 = \dfrac{40\angle 0^\circ}{3 - j4} = \dfrac{40\angle 0^\circ}{5\angle -53.13^\circ} = 8\angle 53.13^\circ$ A or $(4.80 + j6.40)$ A

Current $I_3 = \dfrac{40\angle 0^\circ}{8\angle 0^\circ} = 5\angle 0^\circ$ A or $(5 + j0)$ A

Thus current $I = I_1 + I_2 + I_3$

$= (1.183 - j2.840) + (4.80 + j6.40) + (5 + j0)$

$= 10.983 + j3.560 = \mathbf{11.55\angle 17.96^\circ\ A}$, as previously obtained.

Problem 6. An a.c. network consists of a coil, of inductance 79.58 mH and resistance 18 Ω, in parallel with a capacitor of capacitance 64.96 μF. If the supply voltage is 250∠0° V at 50 Hz, determine (a) the total equivalent circuit impedance, (b) the supply current, (c) the circuit phase angle, (d) the current in the coil and (e) the current in the capacitor.

The circuit diagram is shown in Figure 28.8.

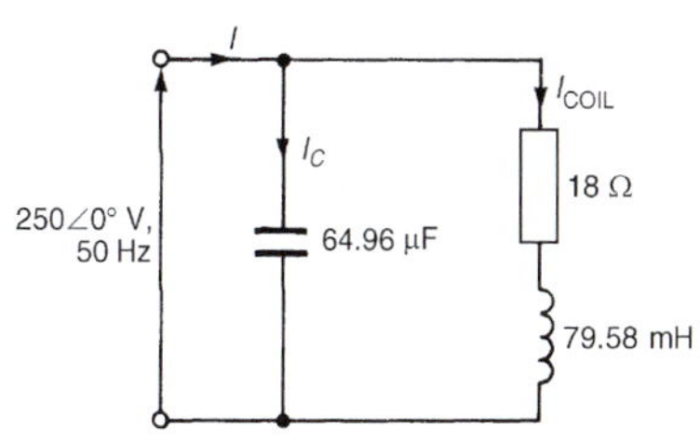

Figure 28.8

Inductive reactance, $X_L = 2\pi fL$

$= 2\pi(50)(79.58 \times 10^{-3}) = 25\Omega$

Hence the impedance of the coil,

$Z_{COIL} = (R + jX_L) = (18 + j25)\ \Omega$ or $30.81\angle 54.25^\circ\ \Omega$

Capacitive reactance, $X_C = \dfrac{1}{2\pi fC}$

$= \dfrac{1}{2\pi(50)(64.96 \times 10^{-6})}$

$= 49\,\Omega$

In complex form, the impedance presented by the capacitor, Z_C is $-jX_C$, i.e. $-j49\,\Omega$ or $49\angle -90^\circ\,\Omega$.

(a) Total equivalent circuit impedance,

$$\boldsymbol{Z_T} = \frac{Z_{COIL}X_C}{Z_{COIL} + Z_C}\left(\text{i.e. } \frac{\text{product}}{\text{sum}}\right)$$

$$= \frac{(30.81\angle 54.25^\circ)(49\angle -90^\circ)}{(18 + j25) + (-j49)}$$

$$= \frac{(30.81\angle 54.25^\circ)(49\angle -90^\circ)}{18 - j24}$$

$$= \frac{(30.81\angle 54.25^\circ)(49\angle -90^\circ)}{30\angle -53.13^\circ}$$

$$= 50.32\angle(54.25^\circ - 90^\circ - (-53.13^\circ))$$

$$= \mathbf{50.32\angle 17.38^\circ} \text{ or } \mathbf{(48.02 + j15.03)\Omega}$$

(b) Supply current $\boldsymbol{I} = \dfrac{V}{Z_T} = \dfrac{250\angle 0^\circ}{50.32\angle 17.38^\circ}$

$= \mathbf{4.97\angle -17.38^\circ\ A}$

(c) Circuit phase angle $=$ **17.38° lagging**, i.e. the current I lags the voltage V by 17.38°

(d) Current in the coil, $\boldsymbol{I_{COIL}} = \dfrac{V}{Z_{COIL}}$

$= \dfrac{250\angle 0^\circ}{30.81\angle 54.25^\circ}$

$= \mathbf{8.11\angle -54.25^\circ\ A}$

(e) Current in the capacitor, $\boldsymbol{I_C} = \dfrac{V}{Z_C} = \dfrac{250\angle 0^\circ}{49\angle -90^\circ}$

$= \mathbf{5.10\angle 90^\circ\ A}$

Now try the following Practice Exercise

Practice Exercise 125 Parallel a.c. circuits (Answers on page 826)

1. Determine the equivalent circuit impedances of the parallel networks shown in Figure 28.9.

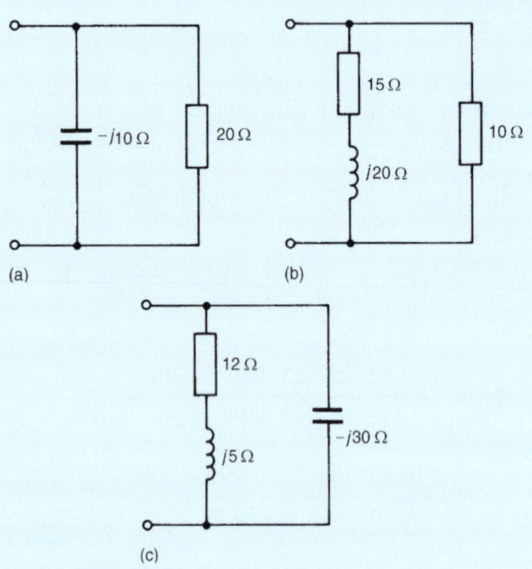

Figure 28.9

2. Determine the value and phase of currents I_1 and I_2 in the network shown in Figure 28.10.

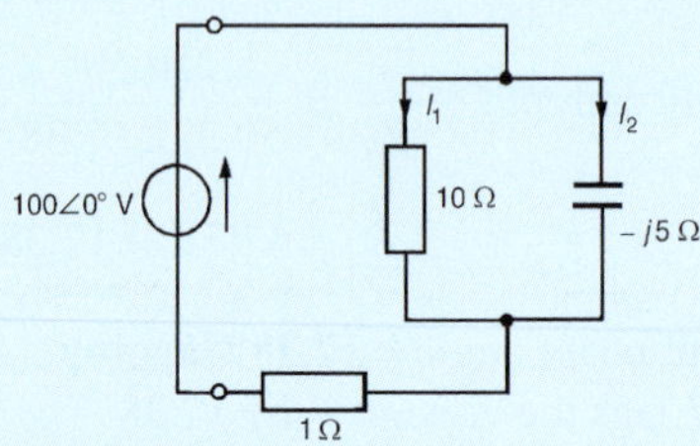

Figure 28.10

3. For the series–parallel network shown in Figure 28.11, determine (a) the total network impedance across AB, and (b) the supply current flowing if a supply of alternating voltage $30\angle 20°$ V is connected across AB.

4. For the parallel network shown in Figure 28.12, determine (a) the equivalent circuit impedance, (b) the supply current I, (c) the circuit phase angle and (d) currents I_1 and I_2

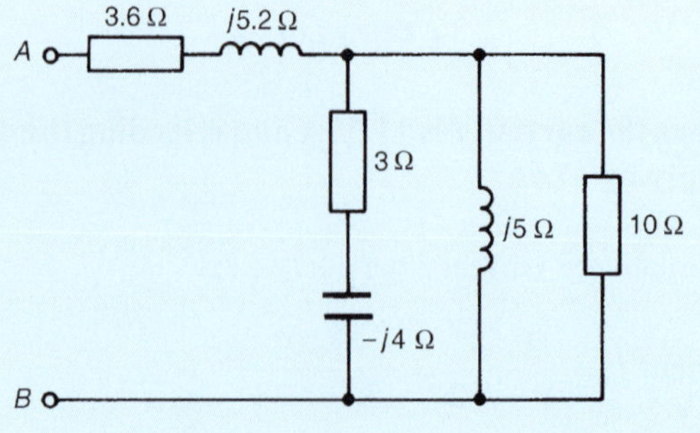

Figure 28.11

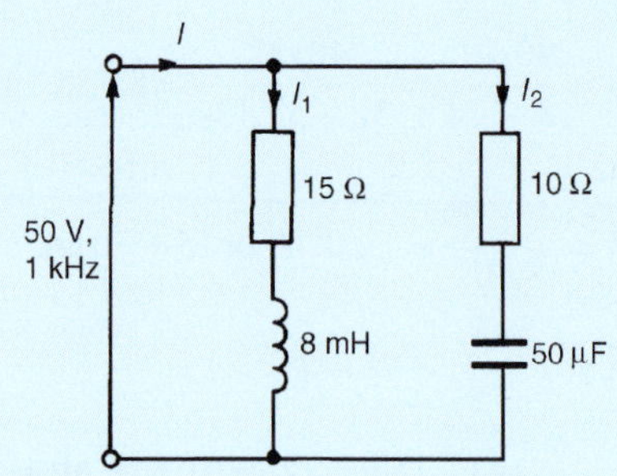

Figure 28.12

5. For the network shown in Figure 28.13, determine (a) current I_1, (b) current I_2, (c) current I, (d) the equivalent input impedance and (e) the supply phase angle.

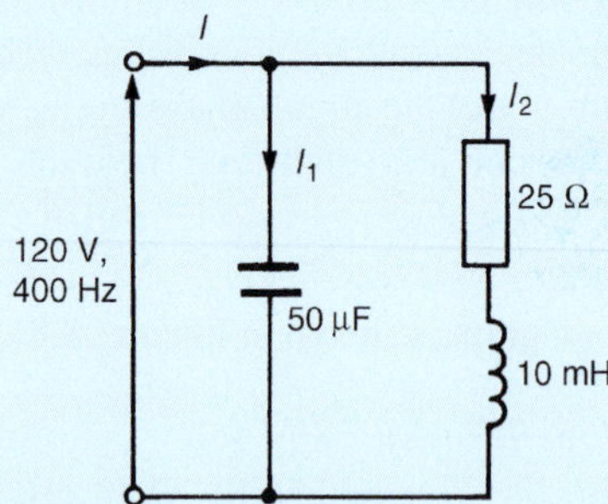

Figure 28.13

6. Determine, for the network shown in Figure 28.14, (a) the total network admittance, (b) the total network impedance, (c) the supply current I, (d) the network phase angle and (e) currents I_1, I_2, I_3 and I_4

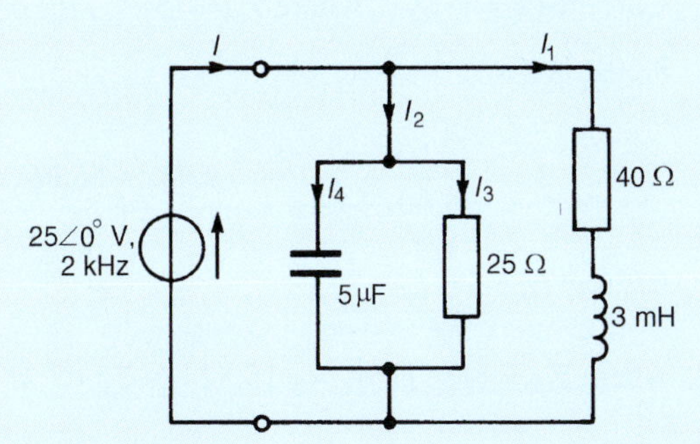

Figure 28.14

7. Four impedances of $(10-j20)\,\Omega$, $(30+j0)\Omega$, $(2-j15)\,\Omega$ and $(25+j12)\Omega$ are connected in parallel across a 250 V a.c. supply. Find the supply current and its phase angle.

8. In the network shown in Figure 28.15, the voltmeter indicates 24 V. Determine the reading on the ammeter.

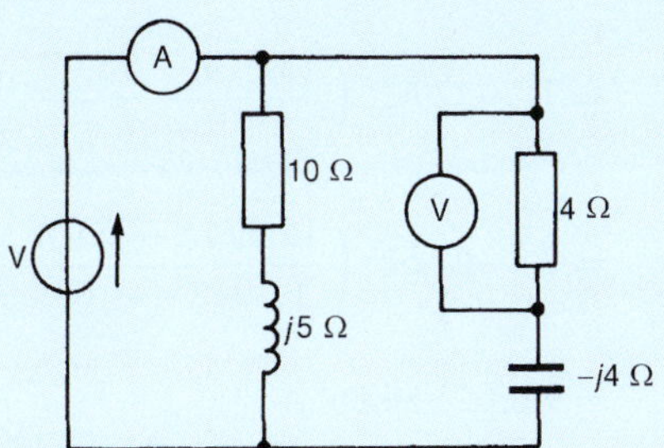

Figure 28.15

28.4 Further worked problems on parallel a.c. networks

Problem 7. (a) For the network diagram of Figure 28.16, determine the value of impedance Z_1 (b) If the supply frequency is 5 kHz, determine the value of the components comprising impedance Z_1

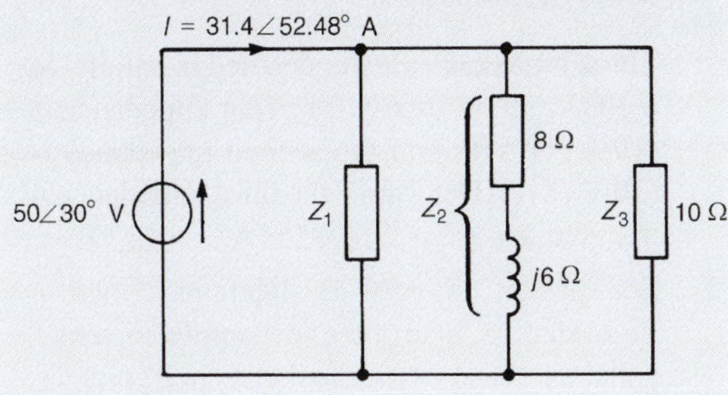

Figure 28.16

(a) Total circuit admittance,

$$Y_T = \frac{I}{V} = \frac{31.4\angle 52.48^\circ}{50\angle 30^\circ}$$

$$= \mathbf{0.628\angle 25.48^\circ\,S} \text{ or } \mathbf{(0.58+j0.24)\,S}$$

$$Y_T = Y_1 + Y_2 + Y_3$$

$$\text{Thus } (0.58+j0.24) = Y_1 + \frac{1}{(8+j6)} + \frac{1}{10}$$

$$= Y_1 + \frac{8-j6}{8^2+6^2} + 0.1$$

$$\text{i.e. } 0.58+j0.24 = Y_1 + 0.08 - j0.06 + 0.1$$

$$\text{Hence } Y_1 = (0.58-0.08-0.1) + j(0.24+j0.06)$$

$$= (0.4+j0.3)\text{S or } 0.5\angle 36.87^\circ\text{S}$$

$$\text{Thus impedance, } Z_1 = \frac{1}{Y_1} = \frac{1}{0.5\angle 36.87^\circ}$$

$$= \mathbf{2\angle -36.87^\circ\,\Omega} \text{ or } \mathbf{(1.6-j1.2)\Omega}$$

(b) Since $Z_1 = (1.6 - j1.2)\,\Omega$, **resistance = 1.6 Ω** and capacitive reactance, $X_C = 1.2\,\Omega$.

Since $X_C = \dfrac{1}{2\pi f C}$

$$\text{capacitance } C = \frac{1}{2\pi f X_C}$$

$$= \frac{1}{2\pi(5000)(1.2)}\text{F}$$

i.e. **capacitance = 26.53 μF**

Problem 8. For the series–parallel arrangement shown in Figure 28.17, determine (a) the equivalent series circuit impedance, (b) the supply current I, (c) the circuit phase angle, (d) the values of voltages V_1 and V_2 and (e) the values of currents I_A and I_B

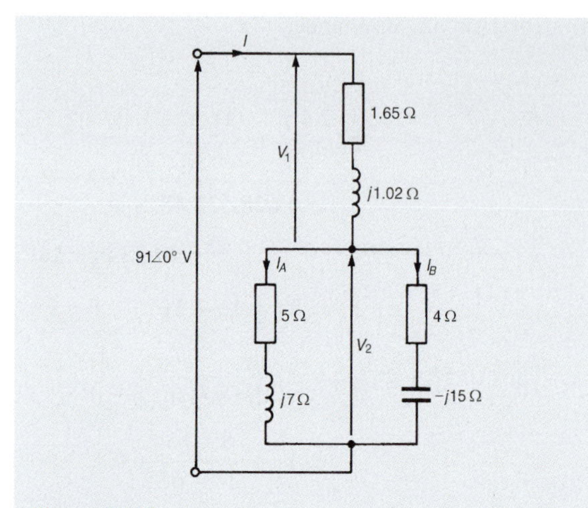

Figure 28.17

(a) The impedance, Z, of the two branches connected in parallel is given by:

$$Z = \frac{(5+j7)(4-j15)}{(5+j7)+(4-j15)}$$

$$= \frac{20 - j75 + j28 - j^2 105}{9 - j8}$$

$$= \frac{125 - j47}{9 - j8} = \frac{133.54\angle -20.61^\circ}{12.04\angle -41.63^\circ}$$

$$= 11.09\angle -21.02^\circ\,\Omega \text{ or } (10.35 + j3.98)\Omega$$

Equivalent series circuit impedance,

$$\boldsymbol{Z_T} = (1.65 + j1.02) + (10.35 + j3.98)$$
$$= \mathbf{(12 + j5)\,\Omega} \text{ or } \mathbf{13\angle 22.62^\circ\,\Omega}$$

(b) Supply current, $I = \frac{V}{Z} = \frac{91\angle 0^\circ}{13\angle 22.62^\circ}$

$$= \mathbf{7\angle -22.62^\circ\ A}$$

(c) Circuit phase angle = **22.62° lagging**

(d) Voltage $V_1 = IZ_1$, where $Z_1 = (1.65 + j1.02)\,\Omega$ or $1.94\angle 31.72^\circ\,\Omega$

Hence $\boldsymbol{V_1} = (7\angle -22.62^\circ)(1.94\angle 31.72^\circ)$

$$= \mathbf{13.58\angle 9.10^\circ\ V}$$

Voltage $V_2 = IZ$, where Z is the equivalent impedance of the two branches connected in parallel.

Hence $\boldsymbol{V_2} = (7\angle -22.62^\circ)\,(11.09\angle 21.02^\circ)$

$$= \mathbf{77.63\angle -1.60^\circ\ V}$$

(e) Current $I_A = V_2/Z_A$, where $Z_A = (5 + j7)\,\Omega$ or $8.60\angle 54.46^\circ\Omega$

Thus $\quad \boldsymbol{I_A} = \frac{77.63\angle -1.60^\circ}{8.60\angle 54.46^\circ} = \mathbf{9.03\angle -56.06^\circ\,A}$

Current $I_B = V_2/Z_B$

where $Z_B = (4 - j15)\Omega$ or $15.524\angle -75.07^\circ\Omega$

Thus $\boldsymbol{I_B} = \frac{77.63\angle -1.60^\circ}{15.524\angle -75.07^\circ} = \mathbf{5.00\angle 73.47^\circ\ A}$

[Alternatively, by current division,

$$I_A = I\left(\frac{Z_B}{Z_A + Z_B}\right)$$

$$= 7\angle -22.62^\circ\left(\frac{15.524\angle -75.07^\circ}{(5 + j7) + (4 - j15)}\right)$$

$$= 7\angle -22.62^\circ\left(\frac{15.524\angle -75.07^\circ}{9 - j8}\right)$$

$$= 7\angle -22.62^\circ\left(\frac{15.524\angle -75.07^\circ}{12.04\angle -41.63^\circ}\right)$$

$$= \mathbf{9.03\angle -56.06^\circ\,A}$$

$$I_B = I\left(\frac{Z_A}{Z_A + Z_B}\right)$$

$$= 7\angle -22.62^\circ\left(\frac{8.60\angle 54.46^\circ}{12.04\angle -41.63^\circ}\right)$$

$$= \mathbf{5.00\angle 73.47^\circ A}]$$

Now try the following Practice Exercise

Practice Exercise 126 Parallel a.c. circuits (Answers on page 826)

1. Three impedances are connected in parallel to a 100 V, 50 Hz supply. The first impedance is $(10 + j12.5)\,\Omega$ and the second impedance is $(20 + j8)\,\Omega$. Determine the third impedance if the total current is $20\angle -25^\circ$ A.
2. For each of the network diagrams shown in Figure 28.18, determine the supply current I and their phase relative to the applied voltages.

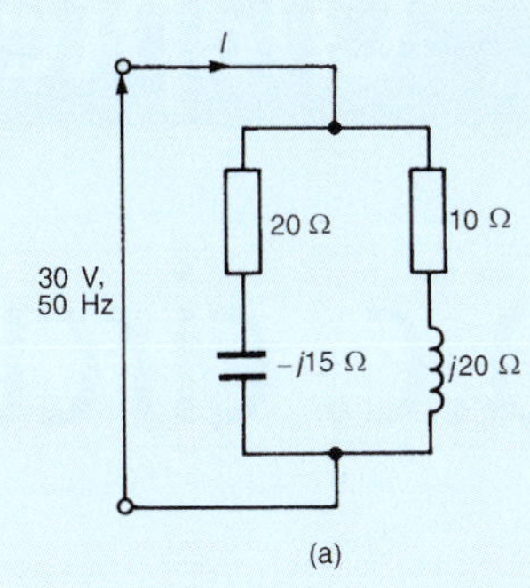

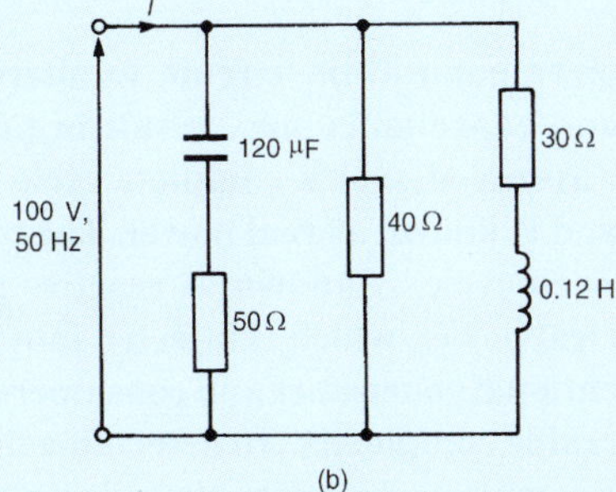

Figure 28.18

3. Determine the value of current flowing in the $(12+j9)\,\Omega$ impedance in the network shown in Figure 28.19.

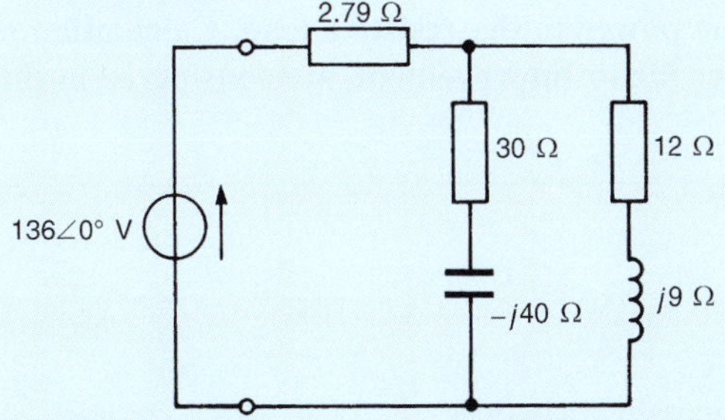

Figure 28.19

4. In the series–parallel network shown in Figure 28.20 the p.d. between points A and B is $50\angle{-68.13°}$ V. Determine (a) the supply current I, (b) the equivalent input impedance, (c) the supply voltage V, (d) the supply phase angle, (e) the p.d. across points B and C and (f) the value of currents I_1 and I_2

5. For the network shown in Figure 28.21, determine (a) the value of impedance Z_2, (b) the current flowing in Z_2 and (c) the components comprising Z_2 if the supply frequency is 2 kHz.

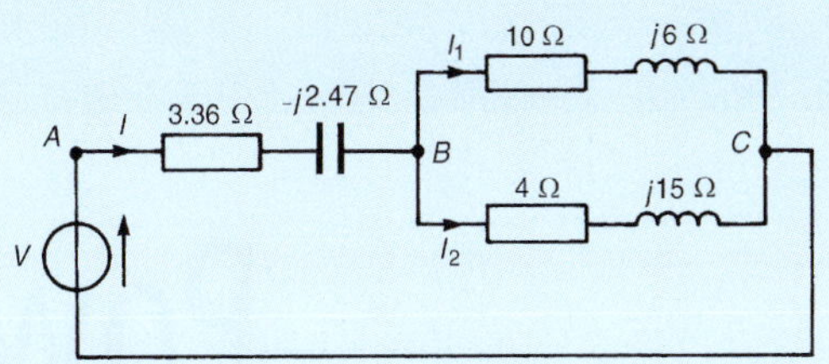

Figure 28.20

6. Coils of impedance $(5+j8)\,\Omega$ and $(12+j16)\,\Omega$ are connected in parallel. In series with this combination is an impedance of $(15-j40)\,\Omega$. If the alternating supply p.d. is $150\angle 0°$ V, determine (a) the equivalent network impedance, (b) the supply current, (c) the supply phase angle, (d) the current in the $(5+j8)\,\Omega$ impedance, and (e) the current in the $(12+j16)\,\Omega$ impedance.

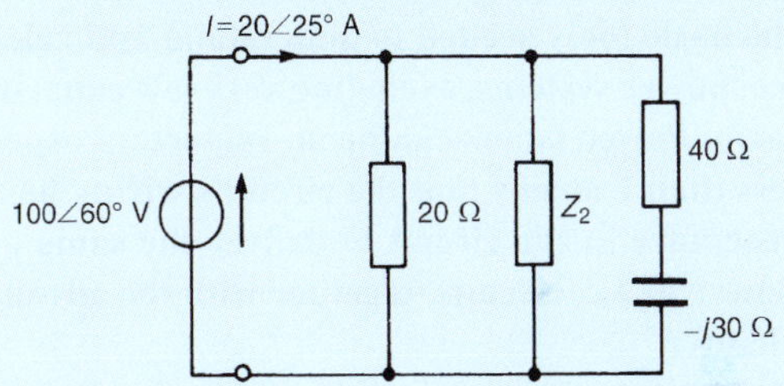

Figure 28.21

7. For the circuit shown in Figure 28.22, determine (a) the input impedance, (b) the source voltage V, (c) the p.d. between points A and B and (d) the current in the 10 Ω resistor.

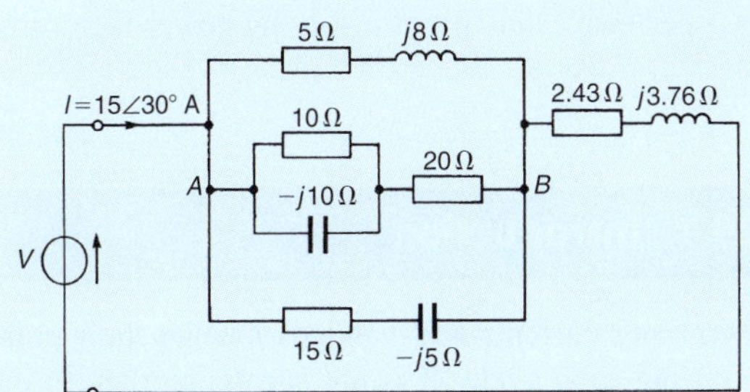

Figure 28.22

Part 4

For fully worked solutions to each of the problems in Practice Exercises 124 to 126 in this chapter, go to the website:
www.routledge.com/cw/bird

Chapter 29

Power in a.c. circuits

***Why it is important to understand:* Power in a.c. circuits**

Power in an electric circuit is the rate of flow of energy past a given point of the circuit. In alternating current circuits, energy storage elements such as inductance and capacitance may result in periodic reversals of the direction of energy flow. The portion of power averaged over a complete cycle of the a.c. waveform results in net transfer of energy in one direction, and is known as real power. The portion of power due to stored energy, which returns to the source in each cycle, is known as reactive power. Renewable energy sources are used mainly to generate a.c. electrical power, which is injected into power networks consisting of a large number of transmission lines, conventional generators and consumers. Such power networks, especially in developed countries, are of considerable complexity. To determine the way these injected powers flow from generators to consumers requires complex calculations based on network analysis. This chapter introduces the basic concepts of active, reactive and apparent power. These are the basic tools needed to understand and calculate flows of energy from generators to consumers. In all a.c. power systems, excluding very low capacity ones, power is generated and transported in three-phase form. Power factor can be an important aspect to consider in an a.c. circuit, because any power factor less than 1 means that the circuit's wiring has to carry more current than would be necessary with zero reactance in the circuit to deliver the same amount of true power to the resistive load. Calculation of power in a.c. circuits, together with the advantages of power factor improvement, are considered in this chapter.

At the end of this chapter you should be able to:

- determine active, apparent and reactive power in a.c. series/parallel networks
- appreciate the need for power factor improvement
- perform calculations involving power factor improvement

29.1 Introduction

Alternating currents and voltages change their polarity during each cycle. It is not surprising therefore to find that power also pulsates with time. The product of voltage v and current i at any instant of time is calledinstantaneous power p, and is given by:

$$p = vi$$

The unit of power is the **watt***

* Who was **Watt**? For image and resume of Watt, see page 52. To find out more go to **www.routledge.com/cw/bird**

Electrical Circuit Theory and Technology. 978-1-138-67349-6, © 2017 John Bird. Published by Taylor & Francis. All rights reserved.

29.2 Determination of power in a.c. circuits

(a) Purely resistive a.c. circuits

Let a voltage $v = V_m \sin\omega t$ be applied to a circuit comprising resistance only. The resulting current is $i = I_m \sin\omega t$, and the corresponding instantaneous power, p, is given by:

$$p = vi = (V_m \sin\omega t)(I_m \sin\omega t)$$

i.e. $p = V_m I_m \sin^2\omega t$

From trigonometrical double angle formulae, $\cos 2A = 1 - 2\sin^2 A$, from which,

$$\sin^2 A = \tfrac{1}{2}(1 - \cos 2A)$$

Thus $\sin^2\omega t = \tfrac{1}{2}(1 - \cos 2\omega t)$

Then power $p = V_m I_m \left[\tfrac{1}{2}(1 - \cos 2\omega t)\right]$

i.e. $\boldsymbol{p = \tfrac{1}{2} V_m I_m (1 - \cos 2\omega t)}$

The waveforms of v, i and p are shown in Figure 29.1. The waveform of power repeats itself after π/ω seconds and hence the power has a frequency twice that of voltage and current. The power is always positive, having a maximum value of $V_m I_m$. The average or mean value of the power is $\frac{1}{2} V_m I_m$

The r.m.s. value of voltage $V = 0.707 V_m$, i.e. $V = V_m/\sqrt{2}$, from which, $V_m = \sqrt{2}\,V$. Similarly, the r.m.s. value of current, $I = I_m/\sqrt{2}$, from which, $I_m = \sqrt{2}\,I$. Hence the average power, P, developed in a purely resistive a.c. circuit is given by

$P = \frac{1}{2} V_m I_m = \frac{1}{2}(\sqrt{2}\,V)(\sqrt{2}\,I) = VI$ watts

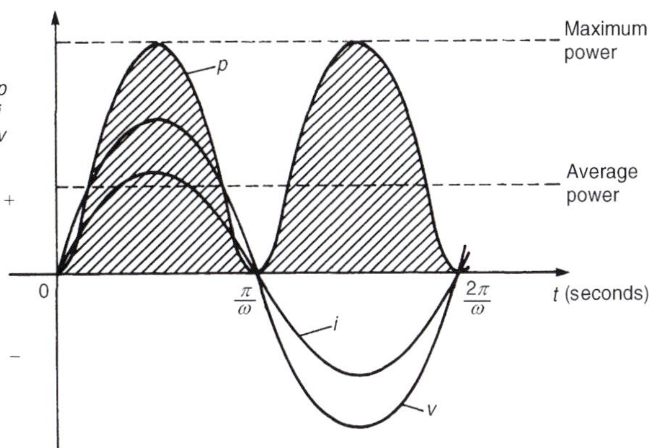

Figure 29.1 The waveforms of v, i and p

Also, power $P = I^2 R$ or V^2/R as for a d.c. circuit, since $V = IR$.

Summarizing, the average power P in a purely resistive a.c. circuit is given by

$$\boldsymbol{P = VI = I^2 R = \frac{V^2}{R} \text{ watts}}$$

where V and I are r.m.s. values.

(b) Purely inductive a.c. circuits

Let a voltage $v = V_m \sin\omega t$ be applied to a circuit containing pure inductance (theoretical case). The resulting current is $i = I_m \sin(\omega t - (\pi/2))$ since current lags voltage by 90° in a purely inductive circuit, and the corresponding instantaneous power, p, is given by:

$$p = vi = (V_m \sin\omega t) I_m \sin(\omega t - (\pi/2))$$

i.e. $p = V_m I_m \sin\omega t \sin(\omega t - (\pi/2))$

However, $\sin(\omega t - (\pi/2)) = -\cos\omega t$

Thus $p = -V_m I_m \sin\omega t \cos\omega t$

Rearranging gives: $p = -\dfrac{1}{2} V_m I_m \quad (2\sin\omega t \cos\omega t)$.
However, from the double angle formulae, $2\sin\omega t \cos\omega t = \sin 2\omega t$

Thus **power,** $\boldsymbol{p = -\dfrac{1}{2} V_m I_m \sin 2\omega t}$

The waveforms of v, i and p are shown in Figure 29.2. The frequency of power is twice that of voltage and current. For the power curve shown in Figure 29.2, the area above the horizontal axis is equal to the area below, thus over a complete cycle the average power P is zero.

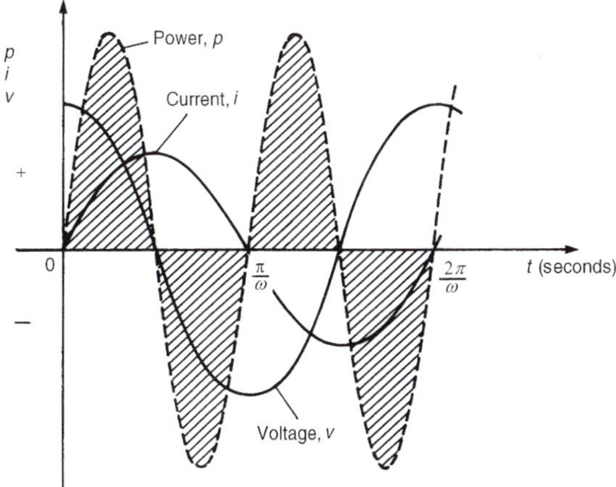

Figure 29.2 Power in a purely inductive a.c. circuit

Part 4

It is noted that when v and i are both positive, power p is positive and energy is delivered from the source to the inductance; when v and i have opposite signs, power p is negative and energy is returned from the inductance to the source.
In general, when the current through an inductance is increasing, energy is transferred from the circuit to the magnetic field, but this energy is returned when the current is decreasing.

Summarizing, the average power P in a purely inductive a.c. circuit is zero.

(c) Purely capacitive a.c. circuits

Let a voltage $v = V_m \sin\omega t$ be applied to a circuit containing pure capacitance. The resulting current is $i = I_m \sin(\omega t + (\pi/2))$, since current leads voltage by $90°$ in a purely capacitive circuit, and the corresponding instantaneous power, p, is given by:

$$p = vi = (V_m \sin\omega t) I_m \sin(\omega t + (\pi/2))$$

i.e. $p = V_m I_m \sin\omega t \sin(\omega t + (\pi/2))$

However, $\sin(\omega t + (\pi/2)) = \cos\omega t$

Thus $P = V_m I_m \sin\omega t \cos\omega t$

Rearranging gives $p = \frac{1}{2} V_m I_m (2 \sin\omega t \cos\omega t)$

Thus **power, $p = \frac{1}{2} V_m I_m \sin 2\omega t$**

The waveforms of v, i and p are shown in Figure 29.3. Over a complete cycle the average power P is zero. When the voltage across a capacitor is increasing, energy is transferred from the circuit to the electric field, but this energy is returned when the voltage is decreasing.

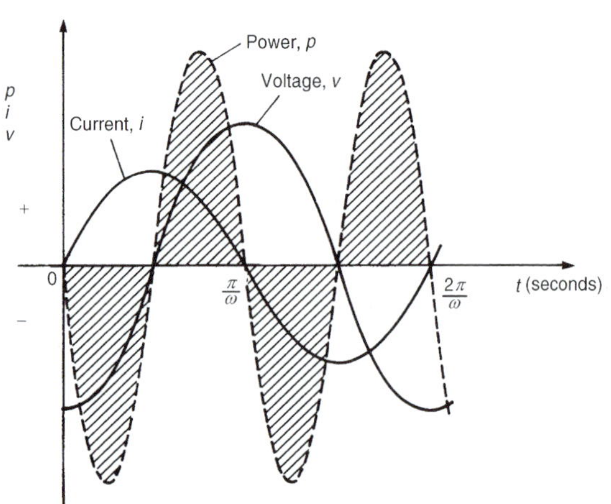

Figure 29.3 Power in a purely capacitive a.c. circuit

Summarizing, the average power P in a purely capacitive a.c. circuit is zero.

(d) R–L or R–C a.c. circuits

Let a voltage $v = V_m \sin\omega t$ be applied to a circuit containing resistance and inductance or resistance and capacitance. Let the resulting current be $i = I_m \sin(\omega t + \phi)$, where phase angle ϕ will be positive for an R–C circuit and negative for an R–L circuit. The corresponding instantaneous power, p, is given by:

$$p = vi = (V_m \sin\omega t)(I_m \sin(\omega t + \phi))$$

i.e. $p = V_m I_m \sin\omega t \sin(\omega t + \phi)$

Products of sine functions may be changed into differences of cosine functions by using:

$\sin A \sin B = -\frac{1}{2}[\cos(A + B) - \cos(A - B)]$

Substituting $\omega t = A$ and $(\omega t + \phi) = B$ gives:

$$\text{power, } p = V_m I_m \left\{-\tfrac{1}{2}[\cos(\omega t + \omega t + \phi) - \cos(\omega t - (\omega t + \phi))]\right\}$$

i.e. $p = \frac{1}{2} V_m I_m [\cos(-\phi) - \cos(2\omega t + \phi)]$

However, $\cos(-\phi) = \cos\phi$

Thus $\boldsymbol{p = \frac{1}{2} V_m I_m [\cos\phi - \cos(2\omega t + \phi)]}$

The instantaneous power p thus consists of

(i) a sinusoidal term, $-\frac{1}{2} V_m I_m \cos(2\omega t + \phi)$, which has a mean value over a cycle of zero, and

(ii) a constant term, $\frac{1}{2} V_m I_m \cos\phi$ (since ϕ is constant for a particular circuit).

Thus the average value of power, $P = \frac{1}{2} V_m I_m \cos\phi$

Since $V_m = \sqrt{2}\, V$ and $I_m = \sqrt{2}\, I$

average power, $P = \frac{1}{2}(\sqrt{2}\, V)(\sqrt{2}\, I) \cos\phi$

i.e. $\boldsymbol{P = VI \cos\phi}$ **watts**

The waveforms of v, i and p are shown in Figure 29.4 for an R–L circuit. The waveform of power is seen to pulsate at twice the supply frequency. The areas of the power curve (shown shaded) above the horizontal time axis represent power supplied to the load; the small areas below the axis represent power being returned to the supply from the inductance as the magnetic field collapses.

Part 4

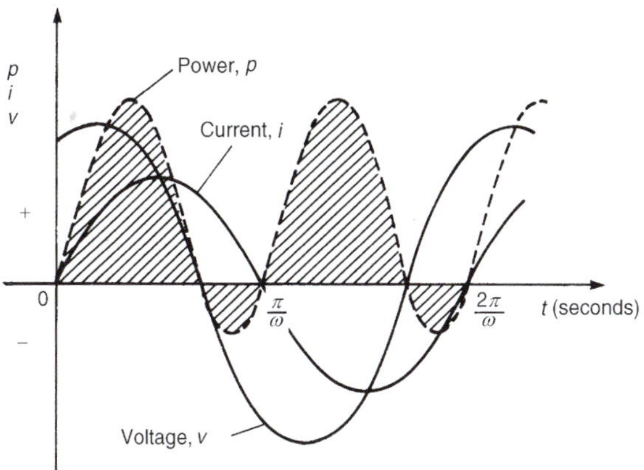

Figure 29.4 Power in a.c. circuit containing resistance and inductive reactance

A similar shape of power curve is obtained for an R–C circuit, the small areas below the horizontal axis representing power being returned to the supply from the charged capacitor. The difference between the areas above and below the horizontal axis represents the heat loss due to the circuit resistance. Since power is dissipated only in a pure resistance, the alternative equations for power, $P=I_R^2 R$, may be used, where I_R is the r.m.s. current flowing through the resistance.

Summarizing, the average power P in a circuit containing resistance and inductance and/or capacitance, whether in series or in parallel, is given by $P=VI\cos\phi$ or $P=I_R^2\,R$ (V, I and I_R being r.m.s. values).

29.3 Power triangle and power factor

A phasor diagram in which the current I lags the applied voltage V by angle ϕ (i.e. an inductive circuit) is shown in Figure 29.5(a). The horizontal component of V is $V\cos\phi$, and the vertical component of V is $V\sin\phi$. If each of the voltage phasors of triangle $0ab$ is multiplied by I, Figure 29.5(b) is produced and is known as the **'power triangle'**. Each side of the triangle represents a particular type of power:

True or active power $P=VI\cos\phi$ watts (W)

Apparent power $S=VI$ voltamperes (VA)

Reactive power $Q=VI\sin\phi$ vars (var)

The power triangle is **not** a phasor diagram since quantities P, Q and S are mean values and not r.m.s. values of sinusoidally varying quantities.

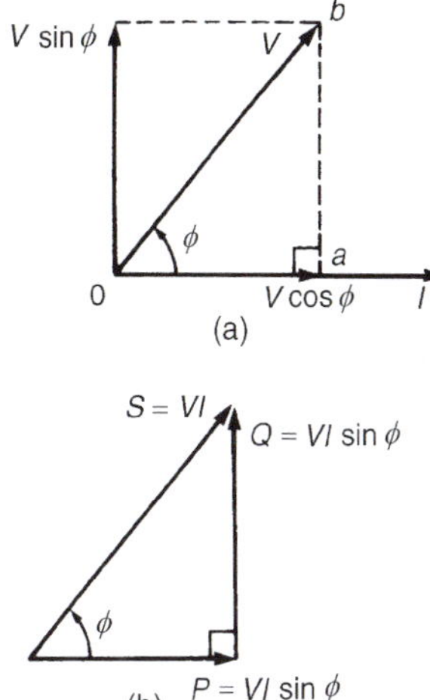

Figure 29.5 (a) Phasor diagram, (b) power triangle for inductive circuit

Superimposing the power triangle on an Argand diagram produces a relationship between P, S and Q in complex form, i.e.

$$S=P+jQ$$

Apparent power, S, is an important quantity since a.c. apparatus, such as generators, transformers and cables, is usually rated in voltamperes rather than in watts. The allowable output of such apparatus is usually limited not by mechanical stress but by temperature rise, and hence by the losses in the device. The losses are determined by the voltage and current and are almost independent of the power factor. Thus the amount of electrical equipment installed to supply a certain load is essentially determined by the voltamperes of the load rather than by the power alone.

The **rating** of a machine is defined as the maximum apparent power that it is designed to carry continuously without overheating.

The **reactive power, Q,** contributes nothing to the net energy transfer and yet it causes just as much loading of the equipment as if it did so. Reactive power is a term much used in power generation, distribution and utilization of electrical energy.

Inductive reactive power, by convention, is defined as positive reactive power; capacitive reactive power, by convention, is defined as negative reactive power. The above relationships derived from the phasor diagram of an inductive circuit may be shown to be true for a capacitive circuit, the power triangle being as shown in Figure 29.6.

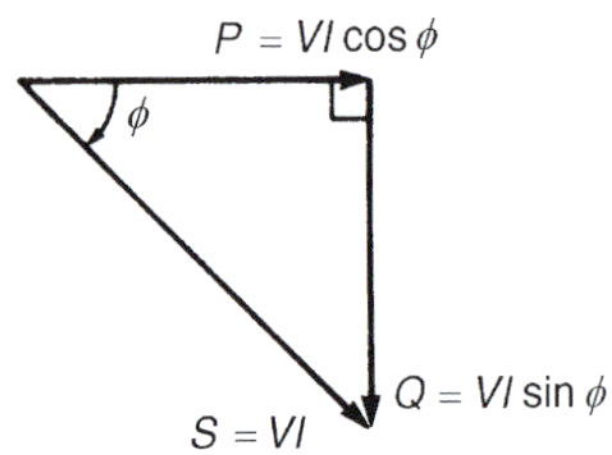

Figure 29.6 Power triangle for capacitive circuit

Power factor is defined as:

$$\textbf{power factor} = \frac{\textbf{active power } P}{\textbf{apparent power } S}$$

For sinusoidal voltages and currents,

$$\text{power factor} = \frac{P}{S} = \frac{VI\cos\phi}{VI}$$

$$= \cos\phi = \frac{R}{Z} \text{ (from the impedance triangle)}$$

A circuit in which current lags voltage (i.e. an inductive circuit) is said to have a lagging power factor, and indicates a lagging reactive power Q
A circuit in which current leads voltage (i.e. a capacitive circuit) is said to have a leading power factor, and indicates a leading reactive power Q

29.4 Use of complex numbers for determination of power

Let a circuit be supplied by an alternating voltage $V\angle\alpha$, where

$$V\angle\alpha = V(\cos\alpha + j\sin\alpha) = V\cos\alpha + jV\sin\alpha$$
$$= a + jb \qquad (1)$$

Let the current flowing in the circuit be $I\angle\beta$, where

$$I\angle\beta = I(\cos\beta + j\sin\beta) = I\cos\beta + jI\sin\beta$$
$$= c + jd \qquad (2)$$

From Sections 29.2 and 29.3, power $P = VI\cos\phi$, where ϕ is the angle between the voltage V and current I. If the voltage is $V\angle\alpha^\circ$ and the current is $I\angle\beta^\circ$, then the angle between voltage and current is $(\alpha - \beta)^\circ$

Thus power, $P = VI\cos(\alpha - \beta)$

From compound angle formulae,

$\cos(\alpha - \beta) = \cos\alpha\cos\beta + \sin\alpha\sin\beta$

Hence power, $P = VI[\cos\alpha\cos\beta + \sin\alpha\sin\beta]$

Rearranging gives

$P = (V\cos\alpha)(I\cos\beta) + (V\sin\alpha)(I\sin\beta)$, i.e.

$P = (a)(c) + (b)(d)$ from equations (1) and (2)

Summarizing, if $V = (a + jb)$ and $I = (c + jd)$, then

$$\textbf{power, } P = ac + bd \qquad (3)$$

Thus power may be calculated from the sum of the products of the real components and imaginary components of voltage and current.

Reactive power, $Q = VI\sin(\alpha - \beta)$

From compound angle formulae,

$\sin(\alpha - \beta) = \sin\alpha\cos\beta - \cos\alpha\sin\beta$

Thus $Q = VI[\sin\alpha\cos\beta - \cos\alpha\sin\beta]$

Rearranging gives

$Q = (V\sin\alpha)(I\cos\beta) - (V\cos\alpha)(I\sin\beta)$, i.e.

$Q = (b)(c) - (a)(d)$ from equations (1) and (2)

Summarizing, if $V = (a + jb)$ and $I = (c + jd)$, then

$$\textbf{reactive power, } Q = bc - ad \qquad (4)$$

Expressions (3) and (4) provide an alternative method of determining true power P and reactive power Q when the voltage and current are complex quantities. From Section 29.3, apparent power $S = P + jQ$. However, merely multiplying V by I in complex form will not give this result, i.e. (from above)

$$S = VI = (a + jb)(c + jd) = (ac - bd) + j(bc + ad)$$

Here the real part is not the expression for power as given in equation (3) and the imaginary part is not the expression of reactive power given in equation (4).

The correct expression may be derived by multiplying the voltage V by the conjugate of the current, i.e. $(c - jd)$, denoted by I^*. Thus

$$\textbf{apparent power } S = VI^* = (a + jb)(c - jd)$$
$$= (ac + bd) + j(bc - ad)$$

i.e. $\quad S = P + jQ$ from equations (3) and (4)

Thus the active and reactive powers may be determined if, and only if, the voltage V is multiplied by the conjugate of current I. As stated in Section 29.3, a positive value of Q indicates an inductive circuit, i.e. a circuit having a lagging power factor, whereas a negative value of Q indicates a capacitive circuit, i.e. a circuit having a leading power factor.

Problem 1. A coil of resistance 5 Ω and inductive reactance 12 Ω is connected across a supply voltage of $52\angle 30^\circ$ volts. Determine the active power in the circuit.

The circuit diagram is shown in Figure 29.7.

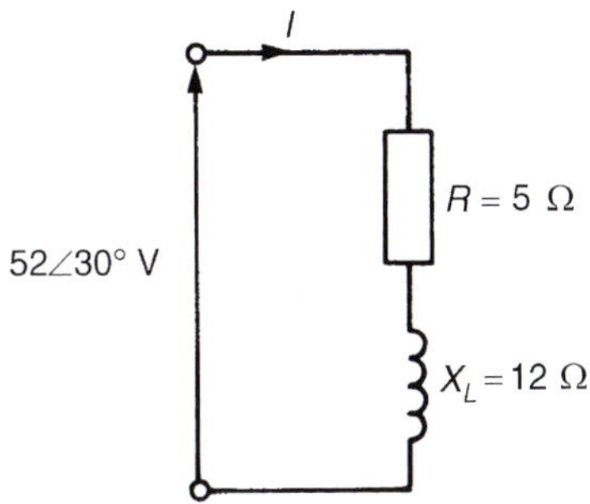

Figure 29.7

Impedance $Z = (5 + j12)\Omega$ or $13\angle 67.38^\circ\,\Omega$

Voltage $V = 52\angle 30^\circ$ V or $(45.03 + j26.0)$ V

$$\text{Current } I = \frac{V}{Z} = \frac{52\angle 30^\circ}{13\angle 67.38^\circ}$$

$$= 4\angle -37.38^\circ \text{ A or } (3.18 - j2.43)\text{ A}$$

There are three methods of calculating power.

Method 1. Active power, $P = VI\cos\phi$, where ϕ is the angle between voltage V and current I. Hence

$$P = (52)(4)\cos[30^\circ - (-37.38^\circ)]$$

$$= (52)(4)\cos 67.38^\circ = \mathbf{80\ W}$$

Method 2. Active power, $P = I_R^2 R = (4)^2(5) = \mathbf{80\ W}$

Method 3. Since $V = (45.03 + j26.0)$ V and $I = (3.18 - j2.43)$ A, then active power,

$$P = (45.03)(3.18) + (26.0)(-2.43) \quad \text{from equation (3)}$$

i.e. $P = 143.2 - 63.2 = \mathbf{80\ W}$

Problem 2. A current of $(15 + j8)$ A flows in a circuit whose supply voltage is $(120 + j200)$ V. Determine (a) the active power and (b) the reactive power.

(a) *Method 1.* Active power $P = (120)(15) + (200)(8)$, from equation (3), i.e.

$P = 1800 + 1600 = \mathbf{3400\ W}$ or $\mathbf{3.4\ kW}$

Method 2. Current $I = (15 + j8)\text{A} = 17\angle 28.07^\circ$ A and

Voltage $V = (120 + j200)\,\text{V} = 233.24\angle 59.04^\circ\,V$

Angle between voltage and current

$$= 59.04^\circ - 28.07^\circ$$

$$= 30.97^\circ$$

Hence power, $P = VI\cos\phi$

$$= (233.24)(17)\cos 30.97^\circ$$

$$= \mathbf{3.4\ kW}$$

(b) *Method 1.* Reactive power,

$Q = (200)(15) - (120)(8)$

from equation (4), i.e.

$Q = 3000 - 960 = \mathbf{2040\ var}$ or $\mathbf{2.04\ kvar}$

Method 2. Reactive power,

$Q = VI\sin\phi = (233.24)(17)\sin 30.97^\circ$

$= \mathbf{2.04\ kvar}$

Alternatively, parts (a) and (b) could have been obtained directly, using

Apparent power, $S = VI^*$

$$= (120 + j200)(15 - j8)$$

$$= (1800 + 1600) + j(3000 - 960)$$

$$= 3400 + j2040 = P + jQ$$

from which, **power $P = 3400$ W** and **reactive power, $Q = 2040$ var**

Problem 3. A series circuit possesses resistance R and capacitance C. The circuit dissipates a power of 1.732 kW and has a power factor of 0.866 leading. If the applied voltage is given by $v = 141.4\sin(10^4 t + (\pi/9))$ volts, determine (a) the current flowing and its phase, (b) the value of resistance R and (c) the value of capacitance C.

(a) Since $v = 141.4\sin(10^4t + (\pi/9))$ volts, then 141.4 V represents the maximum value, from which the r.m.s. voltage, $V = 141.4/\sqrt{2} = 100$ V, and the phase angle of the voltage $= +\pi/9$ rad or 20° leading. Hence as a phasor the voltage V is written as **100∠20° V**

Power factor = 0.866 = $\cos\phi$, from which $\phi = \cos^{-1} 0.866 = 30°$

Hence the angle between voltage and current is 30°

Power $P = VI\cos\phi$

Hence $1732 = (100)I\cos 30°$ from which,

$$\text{current, } |I| = \frac{1732}{(100)(0.866)} = \mathbf{20\,A}$$

Since the power factor is leading, the current phasor leads the voltage – in this case by 30°. Since the voltage has a phase angle of 20°,

current, $\boldsymbol{I} = 20\angle(20° + 30°)\,\text{A} = \mathbf{20\angle 50°\,A}$

(b) Impedance

$$Z = \frac{V}{I} = \frac{100\angle 20°}{20\angle 50°}$$

$$= 5\angle -30°\,\Omega \text{ or } (4.33 - j2.5)\,\Omega$$

Hence the **resistance, $\boldsymbol{R} = 4.33\,\Omega$** and the capacitive reactance, $X_C = 2.5\,\Omega$.

Alternatively, the resistance may be determined from active power, $P = I^2R$. Hence $1732 = (20)^2R$, from which,

$$\textbf{resistance } \boldsymbol{R} = \frac{1732}{(20)^2} = \mathbf{4.33\,\Omega}$$

(c) Since $v = 141.4\sin(10^4t + (\pi/9))$ volts, angular velocity $\omega = 10^4$ rad/s.
Capacitive reactance, $X_C = 2.5\,\Omega$, thus

$$2.5 = \frac{1}{2\pi fC} = \frac{1}{\omega C}$$

from which, **capacitance,** $\boldsymbol{C} = \dfrac{1}{2.5\omega}$

$$= \frac{1}{(2.5)(10^4)}\,\text{F}$$

$$= \mathbf{40\,\mu F}$$

Problem 4. For the circuit shown in Figure 29.8, determine the active power developed between points (a) A and B, (b) C and D, (c) E and F

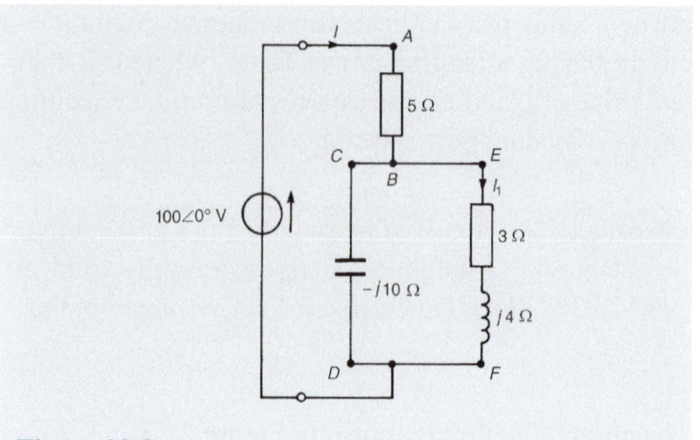

Figure 29.8

Circuit impedance,

$$Z = 5 + \frac{(3 + j4)(-j10)}{(3 + j4 - j10)} = 5 + \frac{(40 - j30)}{(3 - j6)}$$

$$= 5 + \frac{50\angle -36.87°}{6.71\angle -63.43°} = 5 + 7.45\angle 26.56°$$

$$= 5 + 6.66 + j3.33 = (11.66 + j3.33)\,\Omega \text{ or } 12.13\angle 15.94°\,\Omega$$

$$\text{Current } I = \frac{V}{Z} = \frac{100\angle 0°}{12.13\angle 15.94°} = 8.24\angle -15.94°\,\text{A}$$

(a) Active power developed between points A and B $= I^2R = (8.24)^2(5) = \mathbf{339.5\,W}$

(b) Active power developed between points C and D **is zero**, since no power is developed in a pure capacitor.

(c) Current, $I_1 = I\left(\dfrac{Z_{CD}}{Z_{CD} + Z_{EF}}\right)$

$$= 8.24\angle -15.94°\left(\frac{-j10}{3 - j6}\right)$$

$$= 8.24\angle -15.94°\left(\frac{10\angle -90°}{6.71\angle -63.43°}\right)$$

$$= 12.28\angle -42.51°\,\text{A}$$

Hence the active power developed between points E and F $= I_1^2R = (12.28)^2(3) = \mathbf{452.4\,W}$

[Check: Total active power developed = 339.5 + 452.4 = 791.9 W or 792 W, correct to three significant figures

Total active power, $P = I^2R_T = (8.24)^2(11.66) = 792$ W (since 11.66 Ω is the total circuit equivalent resistance)

or $P = VI\cos\phi = (100)(8.24)\cos 15.94° = 792$ W]

Problem 5. The circuit shown in Figure 29.9 dissipates an active power of 400 W and has a power factor of 0.766 lagging. Determine (a) the apparent power, (b) the reactive power, (c) the value and phase of current I and (d) the value of impedance Z.

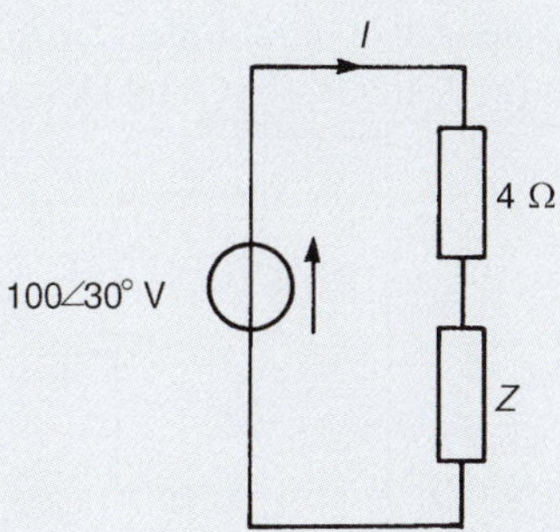

Figure 29.9

Since power factor = 0.766 lagging, the circuit phase angle $\phi = \cos^{-1}0.766$, i.e. $\phi = 40°$ lagging which means that the current I lags voltage V by 40°.

(a) Since power, $P = VI\cos\phi$, the magnitude of apparent power,

$$\boldsymbol{S} = VI = \frac{P}{\cos\phi} = \frac{400}{0.766} = \mathbf{522.2\ VA}$$

(b) Reactive power, $Q = VI\sin\phi = (522.2)(\sin 40°) =$ **335.7 var lagging**. (The reactive power is lagging since the circuit is inductive, which is indicated by the lagging power factor.) The power triangle is shown in Figure 29.10.

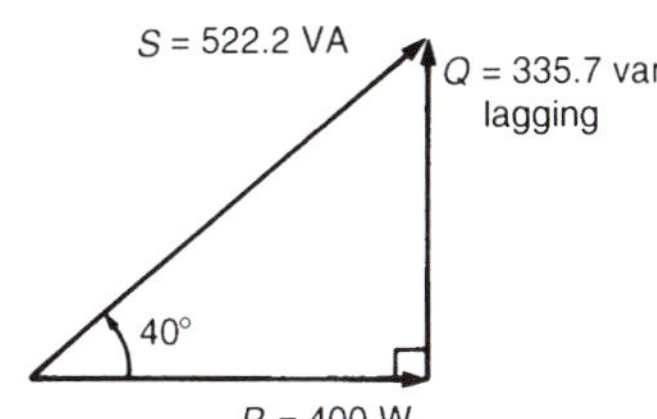

Figure 29.10

(c) Since $VI = 522.2$ VA,

$$\text{magnitude of current } |I| = \frac{522.2}{V} = \frac{522.2}{100} = \mathbf{5.222\ A}$$

Since the voltage is at a phase angle of 30° (see Figure 29.9) and current lags voltage by 40°, the phase angle of current is $30° - 40° = -10°$

Hence **current $\boldsymbol{I} = 5.222\angle{-10°}$ A**

(d) Total circuit impedance

$$Z_T = \frac{V}{I} = \frac{100\angle 30°}{5.222\angle{-10°}} = 19.15\angle 40°\ \Omega \text{ or } (14.67 + j12.31)\ \Omega$$

Hence **impedance**

$$\boldsymbol{Z} = Z_T - 4 = (14.67 + j12.31) - 4$$

$$= \mathbf{(10.67 + j12.31)\ \Omega} \text{ or } \mathbf{16.29\angle 49.08°\ \Omega}$$

Now try the following Practice Exercise

Practice Exercise 127 Power in a.c. circuits (Answers on page 827)

1. When the voltage applied to a circuit is given by $(2 + j5)$V, the current flowing is given by $(8 + j4)$A. Determine the power dissipated in the circuit.
2. A current of $(12 + j5)$A flows in a circuit when the supply voltage is $(150 + j220)$V. Determine (a) the active power, (b) the reactive power and (c) the apparent power. Draw the power triangle.
3. A capacitor of capacitive reactance 40 Ω and a resistance of 30 Ω are connected in series to a supply voltage of $200\angle 60°$ V. Determine the active power in the circuit.
4. The circuit shown in Figure 29.11 takes 81 VA at a power factor of 0.8 lagging. Determine the value of impedance Z.

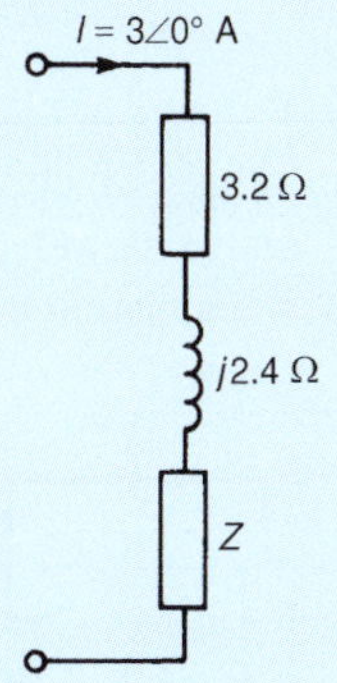

Figure 29.11

5. A series circuit possesses inductance L and resistance R. The circuit dissipates a power of 2.898 kW and has a power factor of 0.966 lagging. If the applied voltage is given by $v = 169.7\sin(100t - (\pi/4))$ volts, determine

Part 4

(a) the current flowing and its phase, (b) the value of resistance R and (c) the value of inductance L.

6. The p.d. across and the current in a certain circuit are represented by $(190+j40)$ V and $(9-j4)$ A, respectively. Determine the active power and the reactive power, stating whether the latter is leading or lagging.

7. Two impedances, $Z_1 = 6\angle 40°\,\Omega$ and $Z_2 = 10\angle 30°\,\Omega$, are connected in series and have a total reactive power of 1650 var lagging. Determine (a) the average power, (b) the apparent power and (c) the power factor.

8. A current $i = 7.5\sin(\omega t - (\pi/4))$ A flows in a circuit which has an applied voltage $v = 180\sin(\omega t + (\pi/12))$V. Determine (a) the circuit impedance, (b) the active power, (c) the reactive power and (d) the apparent power. Draw the power triangle.

9. The circuit shown in Figure 29.12 has a power of 480 W and a power factor of 0.8 leading. Determine (a) the apparent power, (b) the reactive power and (c) the value of impedance Z.

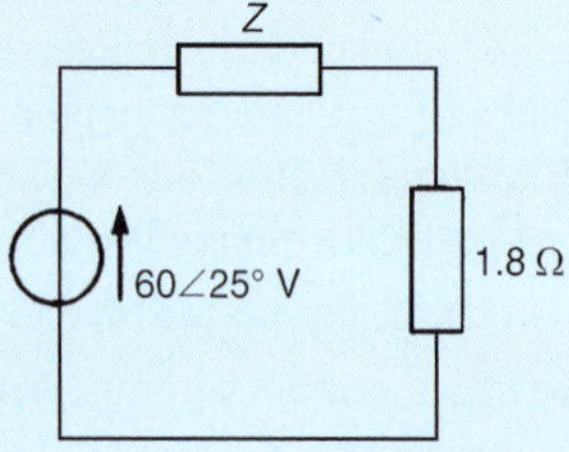

Figure 29.12

10. For the network shown in Figure 29.13, determine (a) the values of currents I_1 and I_2, (b) the total active power, (c) the reactive power and (d) the apparent power.

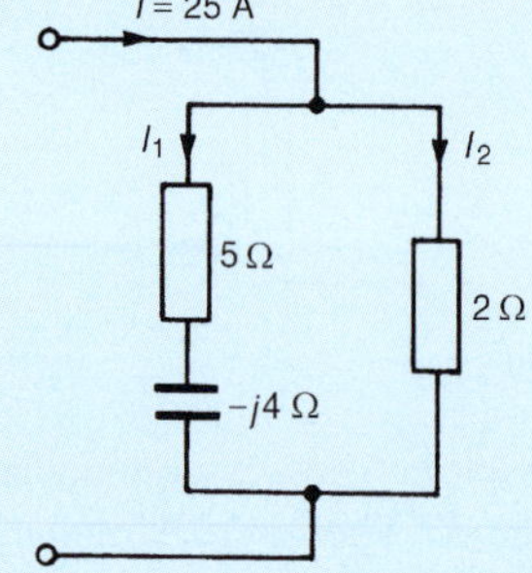

Figure 29.13

11. A circuit consists of an impedance $5\angle -45°\,\Omega$ in parallel with a resistance of 10 Ω. The supply current is 4 A. Determine for the circuit (a) the active power, (b) the reactive power and (c) the power factor.

12. For the network shown in Figure 29.14, determine the active power developed between points (a) A and B, (b) C and D, (c) E and F.

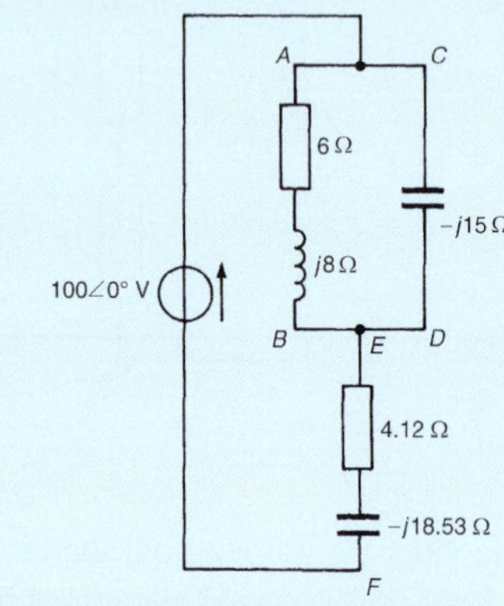

Figure 29.14

29.5 Power factor improvement

For a particular power supplied, a high power factor reduces the current flowing in a supply system, which consequently lowers losses (i.e. I^2R losses) and hence results in cheaper running costs (as stated in Section 18.7, page 289). Supply authorities use tariffs which encourage consumers to operate at a reasonably high power factor. One method of improving the power factor of an inductive load is to connect a bank of capacitors in parallel with the load. Capacitors are rated in reactive voltamperes and the effect of the capacitors is to reduce the reactive power of the system without changing the active power. Most residential and industrial loads on a power system are inductive, i.e. they operate at a lagging power factor.

A simplified circuit diagram is shown in Figure 29.15(a) where a capacitor C is connected across an inductive load. Before the capacitor is connected the circuit current is I_{LR} and is shown lagging voltage V by angle ϕ_1 in the phasor diagram of Figure 29.15(b). When the capacitor C is connected it takes a current I_C which is shown in the phasor diagram leading voltage V by 90°. The supply current I in Figure 29.15(a) is now the phasor sum of currents I_{LR} and I_C as shown in Figure 29.15(b). The circuit phase angle, i.e. the angle

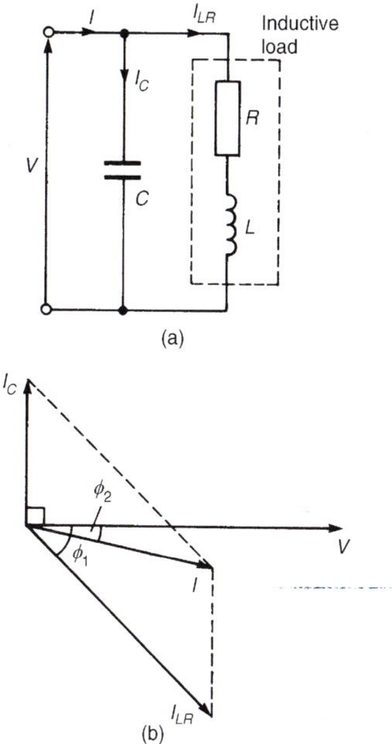

Figure 29.15 (a) Circuit diagram (b) phasor diagram

between V and I, has been reduced from ϕ_1 to ϕ_2 and the power factor has been improved from cos ϕ_1 to cos ϕ_2

Figure 29.16(a) shows the power triangle for an inductive circuit with a lagging power factor of cos ϕ_1. In Figure 29.16(b), the angle ϕ_1 has been reduced to ϕ_2, i.e. the power factor has been improved from cos ϕ_1 to cos ϕ_2 by introducing leading reactive voltamperes (shown as length *ab*) which is achieved by connecting capacitance in parallel with the inductive load. The power factor has been improved by reducing the reactive voltamperes; the active power P has remained unaffected.

Power factor correction results in the apparent power S decreasing (from *0a* to *0b* in Figure 29.16(b)) and thus the current decreasing, so that the power distribution system is used more efficiently.

Another method of power factor improvement, besides the use of static capacitors, is by using synchronous motors; such machines can be made to operate at leading power factors.

Problem 6. A 300 kVA transformer is at full load with an overall power factor of 0.70 lagging. The power factor is improved by adding capacitors in parallel with the transformer until the overall power factor becomes 0.90 lagging. Determine the rating (in kilovars) of the capacitors required.

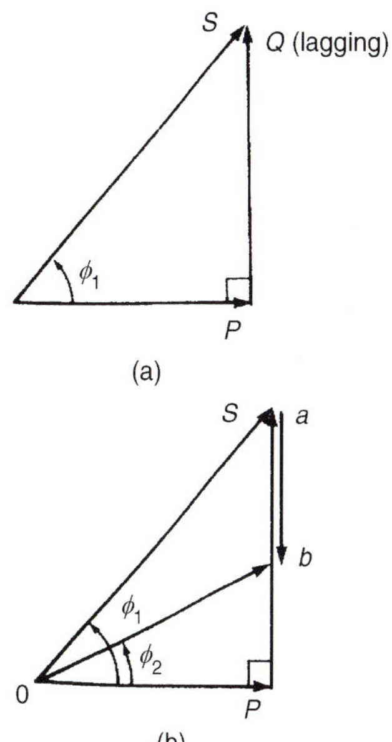

Figure 29.16 Effect of connecting capacitance in parallel with the inductive load

At full load, active power, $P = VI\cos\phi = (300)(0.70) = 210\,\text{kW}$

Circuit phase angle $\phi = \cos^{-1}0.70 = 45.57°$

Reactive power, $Q = VI\sin\phi = (300)(\sin 45.57°) = 214.2\,\text{kvar lagging}$

The power triangle is shown as triangle *0ab* in Figure 29.17. When the power factor is 0.90, the circuit phase angle $\phi = \cos^{-1}0.90 = 25.84°$. The capacitor rating needed to improve the power factor to 0.90 is given by length *bd* in Figure 29.17.

Tan $25.84° = ad/210$, from which, $ad = 210\tan 25.84° = 101.7\,\text{kvar}$. Hence the capacitor rating, i.e. $bd = ab - ad = 214.2 - 101.7 =$ **112.5 kvar leading**.

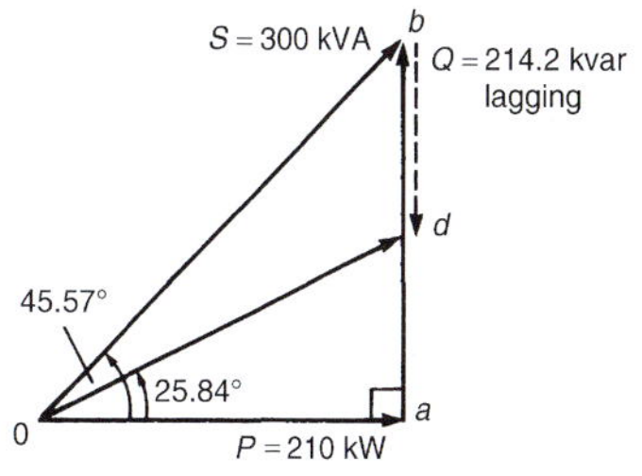

Figure 29.17

Problem 7. A circuit has an impedance $Z = (3 + j4)\,\Omega$ and a source p.d. of $50\angle 30°\,\text{V}$ at a frequency of 1.5 kHz. Determine (a) the supply current, (b) the active, apparent and reactive power,

Part 4

(c) the rating of a capacitor to be connected in parallel with impedance Z to improve the power factor of the circuit to 0.966 lagging and (d) the value of capacitance needed to improve the power factor to 0.966 lagging.

(a) Supply current, $I=\frac{V}{Z}=\frac{50\angle 30^\circ}{(3+j4)}=\frac{50\angle 30^\circ}{5\angle 53.13^\circ}$

$$=\mathbf{10\angle -23.13^\circ\ A}$$

(b) Apparent power, $S=VI^*=(50\angle 30^\circ)(10\angle 23.13^\circ)$

$$=500\angle 53.13^\circ\ \text{VA}$$
$$=(300+j400)\ \text{VA}=P+jQ$$

Hence **active power, $P=300$ W**

apparent power, $S=500$ VA and

reactive power, $Q=400$ var lagging

The power triangle is shown in Figure 29.18.

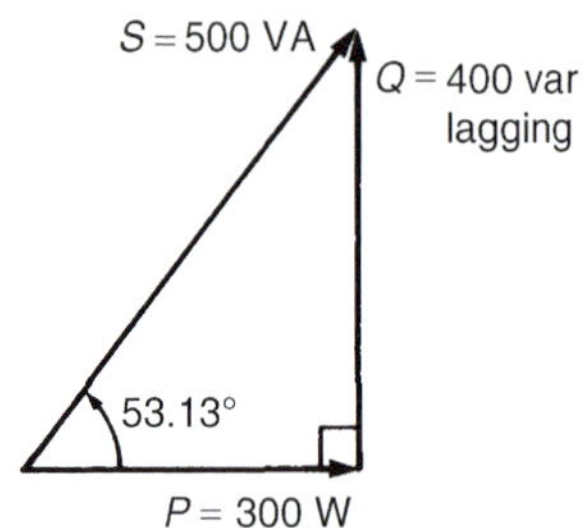

Figure 29.18

(c) A power factor of 0.966 means that $\cos\phi=0.966$.

Hence angle $\phi=\cos^{-1} 0.966=15^\circ$

To improve the power factor from cos 53.13°, i.e. 0.60, to 0.966, the power triangle will need to change from $0cb$ (see Figure 29.19) to $0ab$, the length ca representing the rating of a capacitor connected in parallel with the circuit. From Figure 29.19, $\tan 15^\circ=ab/300$, from which, $ab=300\tan 15^\circ=80.38$ var.

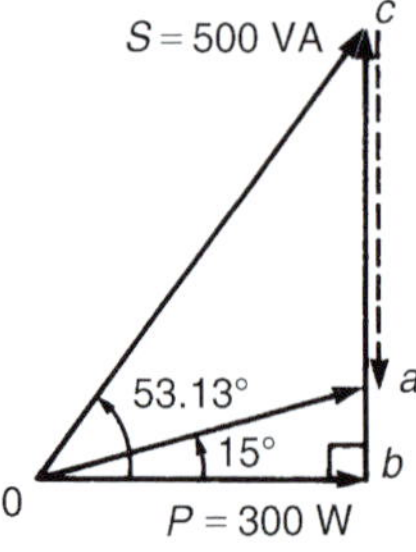

Figure 29.19

Hence the **rating of the capacitor**,

$$ca=cb-ab$$
$$=400-80.38$$
$$=\mathbf{319.6\ var\ leading}.$$

(d) Current in capacitor, $I_C=\frac{Q}{V}=\frac{319.6}{50}=6.39$ A

Capacitive reactance, $X_C=\frac{V}{I_C}=\frac{50}{6.39}=7.82\,\Omega$

Thus $7.82=1/(2\pi fC)$, from which,

required capacitance $C=\frac{1}{2\pi(1500)(7.82)}$ F

$$\equiv\mathbf{13.57\,\mu F}$$

Problem 8. A 30 Ω non-reactive resistor is connected in series with a coil of inductance 100 mH and negligible resistance. The combined circuit is connected to a 300 V, 50 Hz supply. Calculate (a) the reactance of the coil, (b) the impedance of the circuit, (c) the current in the circuit, (d) the power factor of the circuit, (e) the power absorbed by the circuit and (f) the value of the power factor correction capacitor to produce a power factor of 0.85

(a) **Inductive reactance**,

$$X_L=2\pi fL=2\pi(50)(100\times 10^{-3})=\mathbf{31.42\,\Omega}$$

(b) **Impedance**, $Z=R+jX_L=(30+j31.42)$

$$=\mathbf{43.44\angle 46.32^\circ\ \Omega}$$

(c) **Current**, $I=\frac{V}{Z}=\frac{300}{43.44\angle 46.32^\circ}$

$$=\mathbf{6.906\angle -46.32^\circ\ A}$$

(d) **Power factor** $=\cos\phi=\cos 46.32^\circ=\mathbf{0.691}$

(e) **Power**, $P=I^2R=(6.906)^2(30)=\mathbf{1431\ W}$

or $P=VI\cos\phi=(300)(6.906)\cos 46.32^\circ$

$$=\mathbf{1431\ W}$$

(f) To improve the power factor, **a capacitor C is connected in parallel with the $R-L$ circuit** as shown in Figure 29.20. In the phasor diagram of Figure 29.21, current I_{LR} is shown as 6.906 A at 46.32° lagging.
If power factor is to be improved to 0.85, then $\cos\phi=0.85$ and $\phi=\cos^{-1}0.85=31.79^\circ$

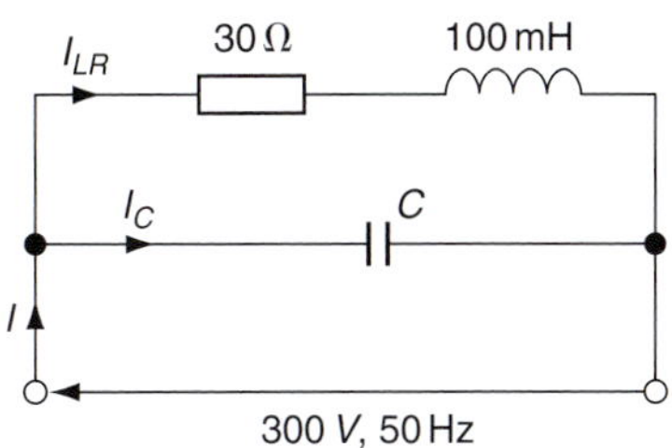

Figure 29.20

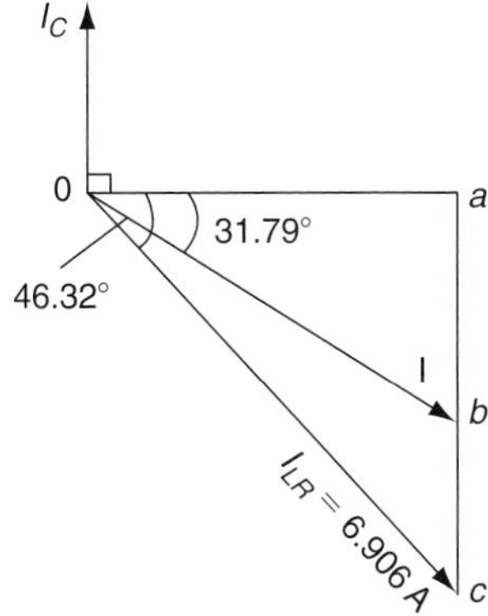

Figure 29.21

$\cos 46.32° = \frac{0a}{6.906}$ from which,

$0a = 6.906 \cos 46.32° = 4.769\,\text{A}$

To improve the power factor to 0.85, a capacitor is connected in parallel with the R–L circuit such that the capacitor takes a current of I_C which is given by the length bc in the phasor diagram. Length $bc = ac - ab$

$\tan 31.79° = \frac{ab}{4.769}$ from which,

$ab = 4.769 \tan 31.79° = 2.956\,\text{A}$

$\tan 46.32° = \frac{ac}{4.769}$ from which,

$ac = 4.769 \tan 46.32° = 4.994\,\text{A}$

Hence, length $bc = ac - ab = 4.994 - 2.956$
$= 2.038\,\text{A}$

Thus, capacitor current, $\boldsymbol{I_C = 2.038\,\text{A}}$

$$\text{Now } I_C = \frac{V}{X_C} = \frac{V}{\frac{1}{2\pi fC}} = 2\pi fCV$$

from which, **capacitance,**

$$C = \frac{I_C}{2\pi fV} = \frac{2.038}{2\pi(50)(300)} = \mathbf{21.62\,\mu F}$$

[In the phasor diagram, current I is the phasor sum of I_{LR} and I_C. Thus, an alternative method of determining I_C is as follows:

$\cos 31.79° = \frac{0a}{0b} = \frac{4.769}{0b}$ from which,

$$0b = \frac{4.769}{\cos 31.79°} = 5.611\,\text{A},$$

i.e. $I = 5.611\angle -31.79°\,\text{A}$

Now $I = I_{LR} + I_C$ i.e. $I_C = I - I_{LR}$ and in complex number form:

$\boldsymbol{I_C} = 5.611\angle -31.79° - 6.906\angle -46.32°$

$= (0 + j2.038)\,\text{A}$ or $\mathbf{2.038\angle 90°\,A}$,

the magnitude of which is the same as that obtained above]

Now try the following Practice Exercise

Practice Exercise 128 Power factor improvement (Answers on page 827)

1. A 600 kVA transformer is at full load with an overall power factor of 0.64 lagging. The power factor is improved by adding capacitors in parallel with the transformer until the overall power factor becomes 0.95 lagging. Determine the rating (in kvars) of the capacitors needed.
2. A source p.d. of $130\angle 40°\,\text{V}$ at 2 kHz is applied to a circuit having an impedance of $(5 + j12)\,\Omega$. Determine (a) the supply current, (b) the active, apparent and reactive powers, (c) the rating of the capacitor to be connected in parallel with the impedance to improve the power factor of the circuit to 0.940 lagging and (d) the value of the capacitance of the capacitor required.
3. The network shown in Figure 29.22 has a total active power of 2253 W. Determine (a) the total impedance, (b) the supply current, (c) the apparent power, (d) the reactive power, (e) the circuit power factor, (f) the capacitance of the capacitor to be connected in parallel with the network to improve the power factor to 0.90 lagging, if the supply frequency is 50 Hz.

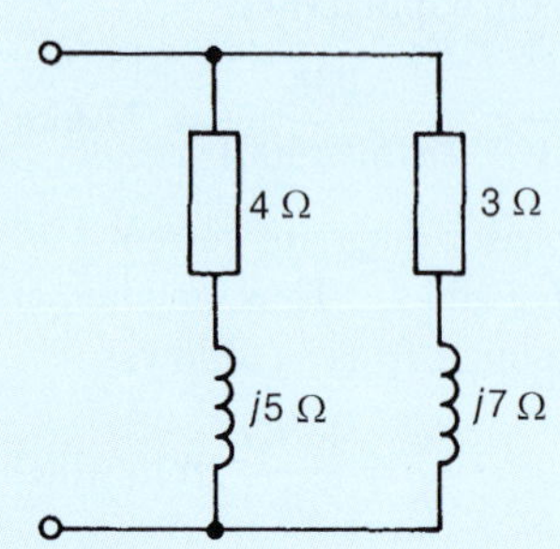

Figure 29.22

4. The power factor of a certain load is improved to 0.92 lagging with the addition of a 30 kvar bank of capacitors. If the resulting supply apparent power is 200 kVA, determine (a) the active power, (b) the reactive power before power factor correction and (c) the power factor before correction.

5. A 15 Ω non-reactive resistor is connected in series with a coil of inductance 75 mH and negligible resistance. The combined circuit is connected to a 200 V, 50 Hz supply. Calculate (a) the reactance of the coil, (b) the impedance of the circuit, (c) the current in the circuit, (d) the power factor of the circuit, (e) the power absorbed by the circuit and (f) the value of the power factor correction capacitor to produce a power factor of 0.92

For fully worked solutions to each of the problems in Practice Exercises 127 and 128 in this chapter, go to the website:
www.routledge.com/cw/bird

Revision Test 8

This revision test covers the material contained in Chapters 26 to 29. *The marks for each question are shown in brackets at the end of each question.*

1. The total impedance Z_T of an electrical circuit is given by:

$$Z_T = Z_1 + \frac{Z_2 \times Z_3}{Z_2 + Z_3}$$

Determine Z_T in polar form, correct to 3 significant figures, when

$Z_1 = 5.5\angle -21^\circ\ \Omega$, $Z_2 = 2.6\angle 30^\circ\ \Omega$ and

$Z_3 = 4.8\angle 71^\circ\ \Omega$ (8)

2. For the network shown in Figure RT8.1, determine
   (a) the equivalent impedance of the parallel branches
   (b) the total circuit equivalent impedance
   (c) current I
   (d) the circuit phase angle
   (e) currents I_1 and I_2
   (f) the p.d. across points A and B
   (g) the p.d. across points B and C
   (h) the active power developed in the inductive branch
   (i) the active power developed across the $-j\,10\,\Omega$ capacitor
   (j) the active power developed between points B and C
   (k) the total active power developed in the network
   (l) the total apparent power developed in the network
   (m) the total reactive power developed in the network. (27)

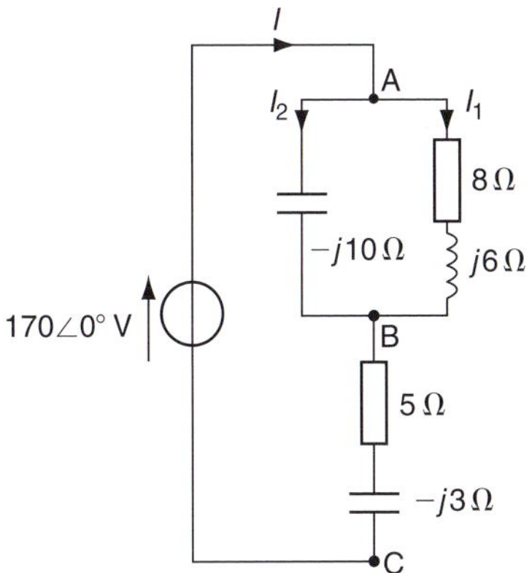

Figure RT8.1

3. An inductive load takes a current of 60 A at a power factor of 0.643 lagging when connected to a 240 V, 50 Hz supply. It is required to improve the power factor to 0.80 lagging by connecting a capacitor in parallel with the load. Calculate (a) the new supply current, (b) the capacitor current and (c) the value of the power factor correction capacitor. Draw the circuit phasor diagram. (15)

Part 4

For lecturers/instructors/teachers, fully worked solutions to each of the problems in Revision Test 8, together with a full marking scheme, are available at the website:
www.routledge.com/cw/bird

Chapter 30

A.c. bridges

Why it is important to understand: **A.c. bridges**

A.c. bridges are used for measuring the values of inductors and capacitors or for converting the signals measured from inductive or capacitive components into a suitable form such as a voltage. Inductors and capacitors can also be measured using an approximate method of voltage division. These methods are discussed in this chapter for a number of bridge circuits for which complex numbers are needed for the determination of unknown component values. Impedance bridges work the same way as a Wheatstone bridge, which measures resistance, only the balance equation is with complex quantities, as both magnitude and phase across the components of the two dividers must be equal in order for the null detector to indicate zero. The null detector must be a device capable of detecting very small a.c. voltages. An oscilloscope is often used for this, although very sensitive electromechanical meter movements and even headphones (small speakers) may be used if the source frequency is within audio range.

At the end of this chapter you should be able to:

- derive the balance equations of any a.c. bridge circuit
- state types of a.c. bridge circuit
- calculate unknown components when using an a.c. bridge circuit

Electrical Circuit Theory and Technology. 978-1-138-67349-6, © 2017 John Bird. Published by Taylor & Francis. All rights reserved.

30.1 Introduction

A.C. bridges are electrical networks, based upon an extension of the **Wheatstone*** **bridge principle,** used for the determination of an unknown impedance by comparison with known impedances and for the determination of frequency. In general, they contain four impedance arms, an a.c. power supply and a balance detector which is sensitive to alternating currents. It is more difficult to achieve balance in an a.c. bridge than in a d.c. bridge because both the magnitude and the phase angle of impedances are related to the balance condition. Balance equations are derived by using complex numbers. A.c. bridges provide precise methods of measurement of inductance and capacitance, as well as resistance.

30.2 Balance conditions for an a.c. bridge

The majority of well known a.c. bridges are classified as four-arm bridges and consist of an arrangement of four impedances (in complex form, $Z = R \pm jX$), as shown in Figure 30.1. As with the d.c. Wheatstone bridge circuit, an a.c. bridge is said to be 'balanced' when the current through the detector is zero (i.e. when no current flows between B and D of Figure 30.1). If the current through the detector is zero, then the current I_1 flowing in impedance Z_1 must also flow in impedance Z_2. Also, at balance, the current I_4 flowing in impedance Z_4, must also flow through Z_3

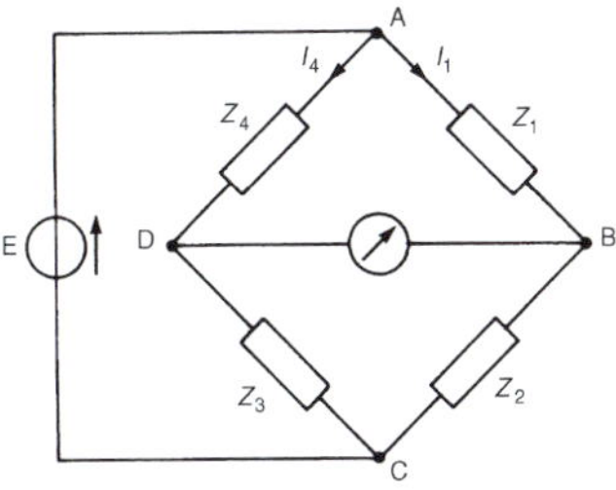

Figure 30.1 Four-arm bridge

*Who was **Wheatstone**? For image and resume of Wheatstone, see page 177. To find out more go to **www.routledge.com/cw/bird**

At balance:

(i) the volt drop between A and B is equal to the volt drop between A and D,

i.e. $V_{AB} = V_{AD}$

i.e. $I_1 Z_1 = I_4 Z_4$ (both in magnitude and in phase) (1)

(ii) the volt drop between B and C is equal to the volt drop between D and C,

i.e. $V_{BC} = V_{DC}$

i.e. $I_1 Z_2 = I_4 Z_3$ (both in magnitude and in phase) (2)

Dividing equation (1) by equation (2) gives

$$\frac{I_1 Z_1}{I_1 Z_2} = \frac{I_4 Z_4}{I_4 Z_3}$$

from which $\dfrac{Z_1}{Z_2} = \dfrac{Z_4}{Z_3}$

or $$\mathbf{Z_1 Z_3 = Z_2 Z_4} \qquad (3)$$

Equation (3) shows that at balance the products of the impedances of opposite arms of the bridge are equal. If in polar form, $Z_1 = |Z_1|\angle\alpha_1$, $Z_2 = |Z_2|\angle\alpha_2$, $Z_3 = |Z_3|\angle\alpha_3$, and $Z_4 = |Z_4|\angle\alpha_4$, then from equation (3), $(|Z_1|\angle\alpha_1)(|Z_3|\angle\alpha_3) = (|Z_2|\angle\alpha_2)(|Z_4|\angle\alpha_4)$, which shows that there are **two** conditions to be satisfied simultaneously for balance in an a.c. bridge, i.e.

$$\mathbf{|Z_1|\,|Z_3| = |Z_2||Z_4| \quad \text{and} \quad \alpha_1 + \alpha_3 = \alpha_2 + \alpha_4}$$

When deriving balance equations of a.c. bridges, where at least two of the impedances are in complex form, it is important to appreciate that for a complex equation $a+jb=c+jd$ the real parts are equal, i.e. $a=c$, and the imaginary parts are equal, i.e. $b=d$

Usually one arm of an a.c. bridge circuit contains the unknown impedance while the other arms contain known fixed or variable components. Normally only two components of the bridge are variable. When balancing a bridge circuit, the current in the detector is gradually reduced to zero by successive adjustments of the two variable components. At balance, the unknown impedance can be expressed in terms of the fixed and variable components.

Procedure for determining the balance equations of any a.c. bridge circuit

(i) Determine for the bridge circuit the impedance in each arm in complex form and write down the balance equation as in equation (3). Equations are usually easier to manipulate if L and C are initially expressed as X_L and X_C, rather than ωL or $1/(\omega C)$

(ii) Isolate the unknown terms on the left-hand side of the equation in the form $a+jb$

(iii) Manipulate the terms on the right-hand side of the equation into the form $c+jd$

(iv) Equate the real parts of the equation, i.e. $a=c$, and equate the imaginary parts of the equation, i.e. $b=d$

(v) Substitute ωL for X_L and $1/(\omega C)$ for X_c where appropriate and express the final equations in their simplest form.

Types of detector used with a.c. bridges vary with the type of bridge and with the frequency at which it is operated. Common detectors used include:

(i) an oscilloscope, which is suitable for use with a very wide range of frequencies;

(ii) earphones (or telephone headsets), which are suitable for frequencies up to about 10 kHz and are used often at about 1 kHz, in which region the human ear is very sensitive;

(iii) various electronic detectors, which use tuned circuits to detect current at the correct frequency; and

(iv) vibration galvanometers, which are usually used for mains-operated bridges. This type of detector consists basically of a narrow moving coil which is suspended on a fine phosphor bronze wire between the poles of a magnet. When a current of the correct frequency flows through the coil, it is set into vibration. This is because the mechanical resonant frequency of the suspension is purposely made equal to the electrical frequency of the coil current. A mirror attached to the coil reflects a spot of light on to a scale, and when the coil is vibrating the spot appears as an extended beam of light. When the band reduces to a spot the bridge is balanced. Vibration galvanometers are available in the frequency range 10 Hz to 300 Hz.

30.3 Types of a.c. bridge circuit

A large number of bridge circuits have been developed, each of which has some particular advantage under certain conditions. Some of the most important a.c. bridges include the Maxwell, Hay, Owen and Maxwell-Wien bridges for measuring inductance, the De Sauty and Schering bridges for measuring capacitance, and the Wien bridge for measuring frequency. Obviously a large number of combinations of components in bridges is possible.

In many bridges it is found that two of the balancing impedances will be of the same nature, and often consist of standard non-inductive resistors.

For a bridge to balance quickly the requirement is either:

(i) the adjacent arms are both pure components (i.e. either both resistors, or both pure capacitors, or one of each) – this type of bridge being called a **ratio-arm bridge** (see, for example, paras (a), (c), (e) and (g) below); or

(ii) a pair of opposite arms are pure components – this type of bridge being called a **product-arm bridge** (see, for example, paras (b), (d) and (f) below).

A ratio-arm bridge can only be used to measure reactive quantities of the same type. When using a product-arm bridge the reactive component of the balancing impedance must be of opposite sign to the unknown reactive component.

A commercial or universal bridge is available and can be used to measure resistance, inductance or capacitance.

(a) The simple Maxwell* bridge

This bridge is used to measure the resistance and inductance of a coil having a high Q-factor (where Q-factor$=\omega L/R$, see Chapters 17 and 31).

A coil having unknown resistance R_x and inductance L_x is shown in the circuit diagram of a simple Maxwell bridge in Figure 30.2. R_4 and L_4 represent a standard coil having known variable values. At balance, expressions for R_x and L_x may be derived in terms of known components R_2, R_3, R_4 and L_4

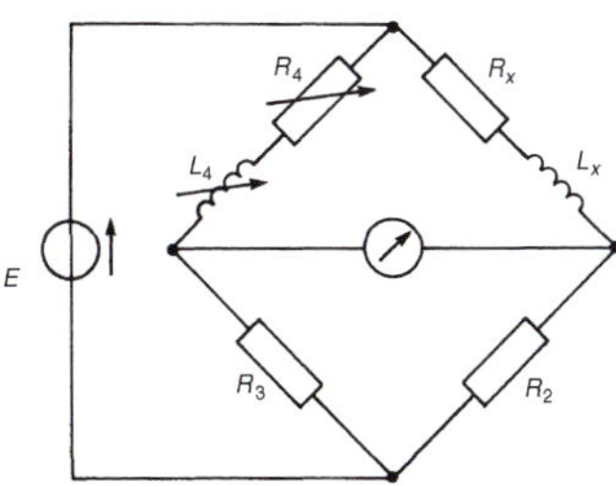

Figure 30.2 Simple Maxwell bridge

The procedure for determining the balance equations given in Section 30.2 may be followed.

(i) From Figure 30.2, $Z_x=R_x+jX_{L_x}$, $Z_2=R_2$, $Z_3=R_3$ and $Z_4=R_4+jX_{L_4}$

At balance,

$$(Z_x)(Z_3)=(Z_2)(Z_4),$$

from equation (3),

i.e. $(R_x+jX_{L_x})(R_3)=(R_2)(R_4+jX_{L_4})$

(ii) Isolating the unknown impedance on the left-hand side of the equation gives

$$(R_x+jX_{L_x})=\frac{R_2}{R_3}(R_4+jX_{L_4})$$

(iii) Manipulating the right-hand side of the equation into $(a+jb)$ form gives

$$(R_x+jX_{L_x})=\frac{R_2R_4}{R_3}+j\frac{R_2X_{L4}}{R_3}$$

(iv) Equating the real parts gives $R_x=\dfrac{R_2R_4}{R_3}$

Equating the imaginary parts gives $X_{L_x}=\dfrac{R_2X_{L4}}{R_3}$

(v) Since $X_L=\omega L$, then

$$\omega L_x=\frac{R_2(\omega L_4)}{R_3}\text{ from which }L_x=\frac{R_2L_4}{R_3}$$

Thus at balance the unknown components in the simple Maxwell bridge are given by

$$\boldsymbol{R_x=\frac{R_2R_4}{R_3}}\quad\text{and}\quad\boldsymbol{L_x=\frac{R_2L_4}{R_3}}$$

These are known as the **'balance equations'** for the bridge.

(b) The Hay bridge

This bridge is used to measure the resistance and inductance of a coil having a very high Q-factor. A coil having unknown resistance R_x and inductance L_x is shown in the circuit diagram of a Hay bridge in Figure 30.3.

Following the procedure of Section 30.2 gives:

(i) From Figure 30.3, $Z_x=R_x+jX_{L_x}$, $Z_2=R_2, Z_3=R_3-jX_{C_3}$, and $Z_4=R_4$

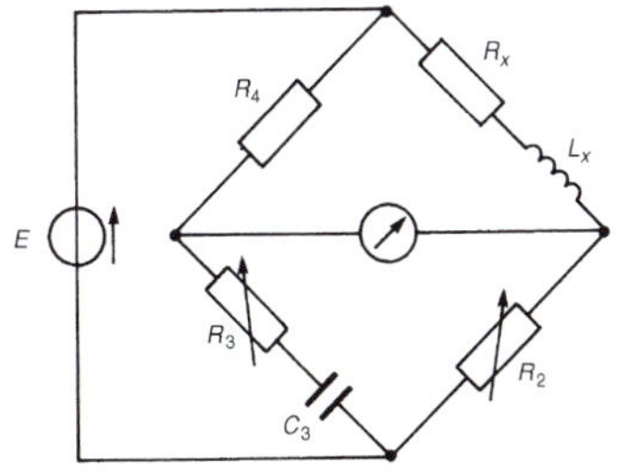

Figure 30.3 Hay bridge

*Who was **Maxwell**? For image and resume of Maxwell, see page 123. To find out more go to **www.routledge.com/cw/bird**

At balance $(Z_x)(Z_3)=(Z_2)(Z_4)$, from equation (3),

i.e. $(R_x+jX_{L_x})(R_3-jX_{C_3})=(R_2)(R_4)$

(ii) $(R_x+jX_{L_x})=\dfrac{R_2R_4}{R_3-jX_{C_3}}$

(iii) Rationalizing the right-hand side gives

$$(R_x+jX_{L_x})=\frac{R_2R_4(R_3+jX_{C_3})}{(R_3-jX_{C_3})(R_3+jX_{C_3})}$$

$$=\frac{R_2R_4(R_3+jX_{C_3})}{R_3^2+X_{C_3}^2}$$

i.e. $(R_x+jX_{L_x})=\dfrac{R_2R_3R_4}{R_3^2+X_{C_3}^2}+j\dfrac{R_2R_4X_{C_3}}{R_3^2+X_{C_3}^2}$

(iv) Equating the real parts gives $R_x=\dfrac{R_2R_3R_4}{R_3^2+X_{C_3}^2}$

Equating the imaginary parts gives

$$X_{L_x}=\frac{R_2R_4X_{C_3}}{R_3^2+X_{C_3}^2}$$

(v) Since $X_{C_3}=\dfrac{1}{\omega C_3}$

$$R_x=\frac{R_2R_3R_4}{R_3^2+(1/(\omega^2C_3^2))}$$

$$=\frac{R_2R_3R_4}{(\omega^2C_3^2R_3^2+1)/(\omega^2C_3^2)}$$

i.e. $R_x=\dfrac{\omega^2C_3^2R_2R_3R_4}{1+\omega^2C_3^2R_3^2}$

Since $X_{L_x}=\omega L_x$

$$\omega L_x=\frac{R_2R_4(1/(\omega C_3))}{(\omega^2C_3^2R_3^2+1)/(\omega^2C_3^2)}$$

$$=\frac{\omega^2C_3^2R_2R_4}{\omega C_3(1+\omega^2C_3^2R_3^2)}$$

i.e. $L_x=\dfrac{C_3R_2R_4}{(1+\omega^2C_3^2R_3^2)}$ by cancelling.

Thus at balance the unknown components in the Hay bridge are given by

$$\mathbf{R_x=\frac{\omega^2C_3^2R_2R_3R_4}{(1+\omega^2C_3^2R_3^2)}} \quad \text{and} \quad \mathbf{L_x=\frac{C_3R_2R_4}{(1+\omega^2C_3^2R_3^2)}}$$

Since $\omega(=2\pi f)$ appears in the balance equations, the bridge is **frequency-dependent**.

(c) The Owen bridge

This bridge is used to measure the resistance and inductance of coils possessing a large value of inductance. A coil having unknown resistance R_x and inductance L_x is shown in the circuit diagram of an Owen bridge in Figure 30.4, from which $Z_x=R_x+jX_{L_x}$, $Z_2=R_2-jX_{C_2}$, $Z_3=-jX_{C_3}$ and $Z_4=R_4$

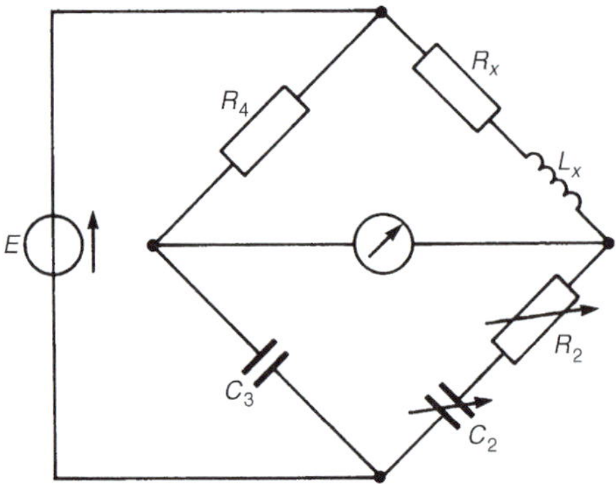

Figure 30.4 Owen bridge

At balance $(Z_x)(Z_3)=(Z_2)(Z_4)$, from equation (3), i.e.

$$(R_x+jX_{L_x})(-jX_{C_3})=(R_2-jX_{C_2})(R_4)$$

Rearranging gives $R_x+jX_{L_x}=\dfrac{(R_2-jX_{C_2})R_4}{-1jX_{C_3}}$

By rationalizing and equating real and imaginary parts it may be shown that at balance the unknown components in the Owen bridge are given by

$$\mathbf{R_x=\frac{R_4C_3}{C_2}} \quad \text{and} \quad \mathbf{L_x=R_2R_4C_3}$$

(d) The Maxwell-Wien bridge

This bridge is used to measure the resistance and inductance of a coil having a low Q-factor. A coil having unknown resistance R_x and inductance L_x is shown in the circuit diagram of a Maxwell-Wien bridge in Figure 30.5, from which $Z_x=R_x+jX_{L_x}$, $Z_2=R_2$ and $Z_4=R_4$

Part 4

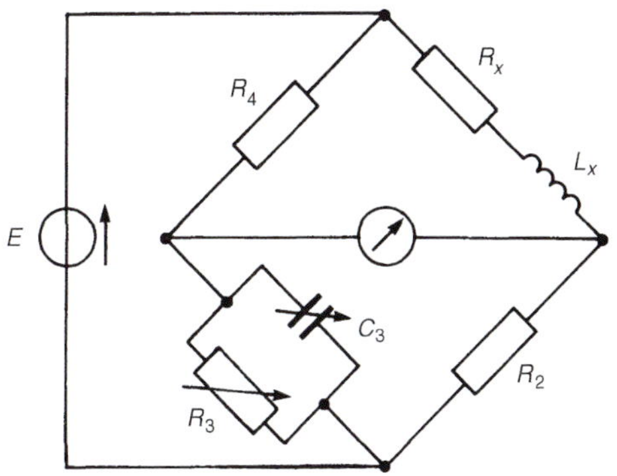

Figure 30.5 Maxwell-Wien bridge

Arm 3 consists of two parallel-connected components. The equivalent impedance Z_3, is given either

(i) by $\frac{\text{product}}{\text{sum}}$ i.e. $Z_3 = \frac{(R_3)(-jX_{C_3})}{(R_3 - jX_{C_3})}$ or

(ii) by using the reciprocal impedance expression,

$$\frac{1}{Z_3} = \frac{1}{R_3} + \frac{1}{-jX_{C_3}}$$

$$\text{from which } Z_3 = \frac{1}{(1/R_3) + (1/(-jX_{C_3}))}$$

$$= \frac{1}{(1/R_3) + (j/X_{C_3})}$$

$$\text{or} \quad Z_3 = \frac{1}{\frac{1}{R_3} + j\omega C_3} \text{ since } X_{C_3} = \frac{1}{\omega C_3}$$

Whenever an arm of an a.c. bridge consists of two branches in parallel, either method of obtaining the equivalent impedance may be used.

For the Maxwell-Wien bridge of Figure 30.5, at balance

$$(Z_x)(Z_3) = (Z_2)(Z_4), \text{ from equation (3)}$$

$$\text{i.e. } (R_x + jX_{L_x})\frac{(R_3)(-jX_{C_3})}{(R_3 - jX_{C_3})} = R_2 R_4$$

using method (i) for Z_3. Hence

$$(R_x + jX_{L_x}) = R_2 R_4 \frac{(R_3 - jX_{C_3})}{(R_3)(-jX_{C_3})}$$

By rationalizing and equating real and imaginary parts it may be shown that at balance the unknown components in the Maxwell-Wien bridge are given by

$$\boldsymbol{R_x = \frac{R_2 R_4}{R_3}} \quad \text{and} \quad \boldsymbol{L_x = C_3 R_2 R_4}$$

(e) The de Sauty bridge

This bridge provides a very simple method of measuring a capacitance by comparison with another known capacitance. In the de Sauty bridge shown in Figure 30.6, C_x is an unknown capacitance and C_4 is a standard capacitor.

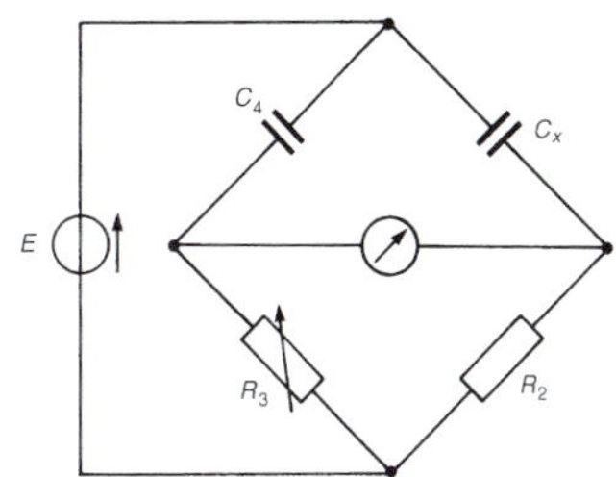

Figure 30.6 De Sauty bridge

$$\text{At balance} \quad (Z_x)(Z_3) = (Z_2)(Z_4)$$

$$\text{i.e.} \quad (-jX_{C_x})R_3 = (R_2)(-jX_{C_4})$$

$$\text{Hence} \quad (X_{C_x})(R_3) = (R_2)(X_{C_4})$$

$$\left(\frac{1}{\omega C_x}\right)(R_3) = (R_2)\left(\frac{1}{\omega C_4}\right)$$

$$\text{from which } \frac{R_3}{C_x} = \frac{R_2}{C_4} \quad \text{or} \quad \boldsymbol{C_x = \frac{R_3 C_4}{R_2}}$$

This simple bridge is usually inadequate in most practical cases. The power factor of the capacitor under test is significant because of internal dielectric losses – these losses being the dissipation within a dielectric material when an alternating voltage is applied to a capacitor.

(f) The Schering bridge

This bridge is used to measure the capacitance and equivalent series resistance of a capacitor. From the measured values the power factor of insulating materials and dielectric losses may be determined. In the circuit diagram of a Schering bridge shown in Figure 30.7, C_x is the unknown capacitance and R_x its equivalent series resistance.

From Figure 30.7, $Z_x = R_x - jX_{C_x}$, $Z_2 = -jX_{C_2}$

$$Z_3 = \frac{(R_3)(-jX_{C_3})}{(R_3 - jX_{C_3})} \text{ and } Z_4 = R_4$$

At balance, $(Z_x)(Z_3) = (Z_2)(Z_4)$ from equation (3),

$$\text{i.e.} \quad (R_x - jX_{C_x})\frac{(R_3)(-jX_{C_3})}{R_3 - jX_{C_3}} = (-jX_{C_2})(R_4)$$

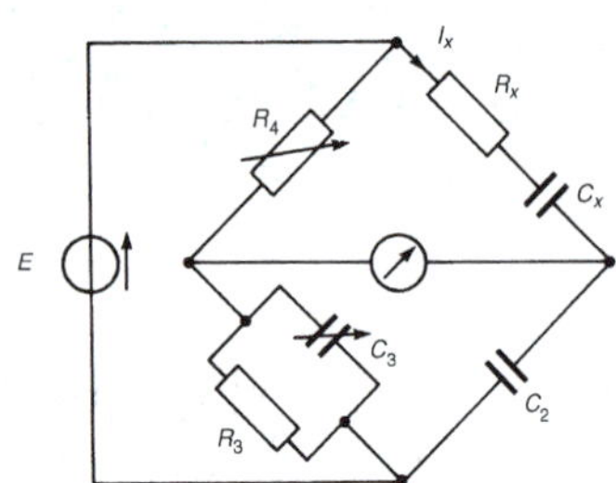

Figure 30.7 Schering bridge

from which $$(R_x - jX_{C_x}) = \frac{(-jX_{C_2}R_4)(R_3 - jX_{C_3})}{-jX_{C_3}R_3}$$

$$= \frac{X_{C_2}R_4}{X_{C_3}R_3}(R_3 - jX_{C_3})$$

Equating the real parts gives

$$R_x = \frac{X_{C_2}R_4}{X_{C_3}} = \frac{(1/\omega C_2)R_4}{(1/\omega C_3)} = \frac{C_3 R_4}{C_2}$$

Equating the imaginary parts gives

$$-X_{C_x} = \frac{-X_{C_2}R_2}{R_3}$$

i.e. $$\frac{1}{\omega C_x} = \frac{(1/\omega C_2)R_4}{R_3} = \frac{R_4}{\omega C_2 R_3}$$

from which $$C_x = \frac{C_2 R_3}{R_4}$$

Thus at balance the unknown components in the Schering bridge are given by

$$\boldsymbol{R_x = \frac{C_3 R_4}{C_2}} \quad \textbf{and} \quad \boldsymbol{C_x = \frac{C_2 R_3}{R_4}}$$

The loss in a dielectric may be represented by either (a) a resistance in parallel with a capacitor or (b) a lossless capacitor in series with a resistor.

If the dielectric is represented by an R–C circuit, as shown by R_x and C_x in Figure 30.7, the phasor diagram for the unknown arm is as shown in Figure 30.8.

Angle ϕ is given by

$$\phi = \tan^{-1}\frac{V_{C_x}}{V_{R_x}} = \tan^{-1}\frac{I_x X_{C_x}}{I_x R_x}$$

i.e. $$\phi = \tan^{-1}\left(\frac{1}{\omega C_x R_x}\right)$$

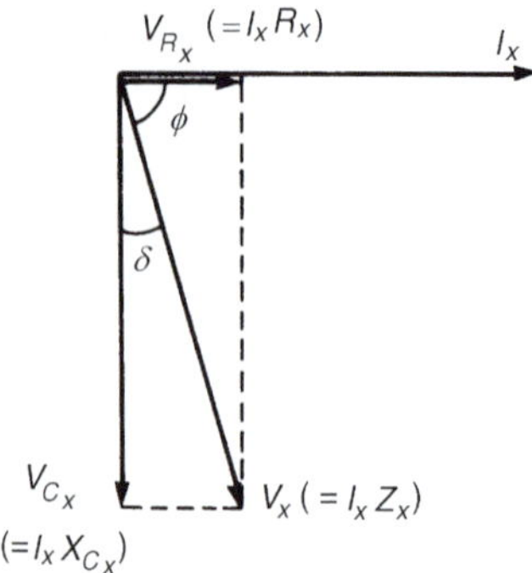

Figure 30.8 Phasor diagram for the unknown arm in the Schering bridge

The power factor of the unknown arm is given by $\cos\phi$. The angle δ $(= 90° - \phi)$ is called the **loss angle** and is given by

$$\delta = \tan^{-1}\frac{V_{R_x}}{V_{C_x}} = \boldsymbol{\tan^{-1}\omega C_x R_x} \text{ and}$$

$$\delta = \tan^{-1}\left[\omega\left(\frac{C_2 R_3}{R_4}\right)\left(\frac{C_3 R_4}{C_2}\right)\right] \text{ from above}$$

$$= \boldsymbol{\tan^{-1}(\omega R_3 C_3)}$$

(See also Chapter 42, page 635.)

(g) The Wien bridge

This bridge is used to measure frequency in terms of known components (or, alternatively, to measure capacitance if the frequency is known). It may also be used as a frequency-stabilizing network.

A typical circuit diagram of a Wien bridge is shown in Figure 30.9, from which

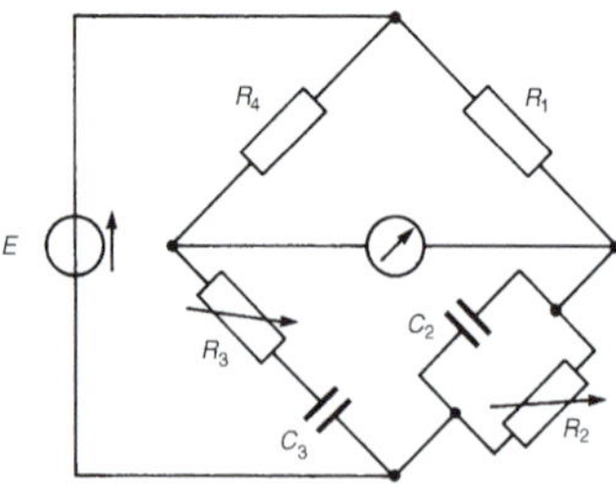

Figure 30.9 Wien bridge

$$Z_1 = R_1, Z_2 = \frac{1}{(1/R_2) + j\omega C_2}$$ (see (ii), para. (d), page 465),

$$Z_3 = R_3 - jX_{C_3} \quad \text{and} \quad Z_4 = R_4$$

At balance, $(Z_1)(Z_3)=(Z_2)(Z_4)$ from equation (3), i.e.

$$(R_1)(R_3-jX_{C_3})=\left(\frac{1}{(1/R_2)+j\omega C_2}\right)(R_4)$$

Rearranging gives

$$\left(R_3-\frac{j}{\omega C_3}\right)\left(\frac{1}{R_2}+j\omega C_2\right)=\frac{R_4}{R_1}$$

$$\frac{R_3}{R_2}+\frac{C_2}{C_3}-j\left(\frac{1}{\omega C_3 R_2}\right)+j\omega C_2 R_3=\frac{R_4}{R_1}$$

Equating real parts gives

$$\boldsymbol{\frac{R_3}{R_2}+\frac{C_2}{C_3}=\frac{R_4}{R_1}} \qquad (4)$$

Equating imaginary parts gives

$$-\frac{1}{\omega C_3 R_2}+\omega C_2 R_3=0$$

i.e. $\omega C_2 R_3=\dfrac{1}{\omega C_3 R_2}$

from which $\omega^2=\dfrac{1}{C_2 C_3 R_2 R_3}$

Since $\omega=2\pi f$, **frequency,** $\boldsymbol{f=\dfrac{1}{2\pi\sqrt{(C_2C_3R_2R_3)}}}$ (5)

Note that if $C_2=C_3=C$ and $R_2=R_3=R$,

frequency, $f=\dfrac{1}{2\pi\sqrt{(C^2R^2)}}=\dfrac{1}{2\pi CR}$

30.4 Worked problems on a.c. bridges

Problem 1. The a.c. bridge shown in Figure 30.10 is used to measure the capacitance C_x and resistance R_x. (a) Derive the balance equations of the bridge. (b) Given $R_3=R_4$, $C_2=0.2\,\mu F$, $R_2=2.5\,k\Omega$ and the frequency of the supply is 1 kHz, determine the values of R_x and C_x at balance.

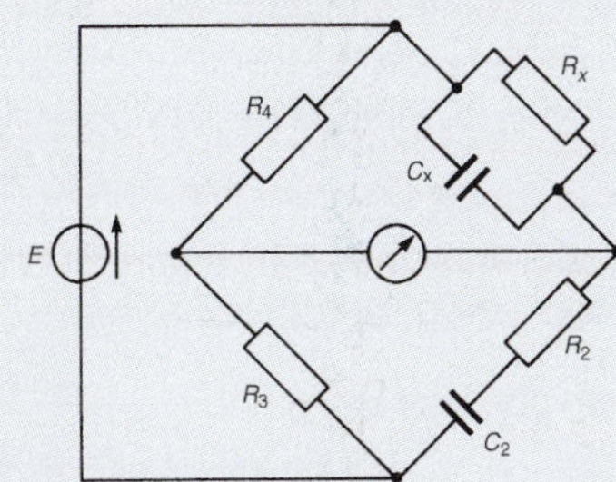

Figure 30.10

(a) Since C_x and R_x are the unknown values and are connected in parallel, it is easier to use the reciprocal impedance form for this branch $\left(\text{rather than } \dfrac{\text{product}}{\text{sum}}\right)$

i.e. $\dfrac{1}{Z_x}=\dfrac{1}{R_x}+\dfrac{1}{-jX_{C_x}}=\dfrac{1}{R_x}+\dfrac{j}{X_{C_x}}$

from which $Z_x=\dfrac{1}{(1/R_x)+j\omega C_x}$

From Figure 30.10, $Z_2=R_2-jX_{C_2}$, $Z_3=R_3$ and $Z_4=R_4$

At balance, $(Z_x)(Z_3)=(Z_2)(Z_4)$

$$\left(\frac{1}{(1/R_x)+j\omega C_x}\right)(R_3)=(R_2-j\omega X_{C_2})(R_4)$$

hence $\dfrac{R_3}{R_4(R_2-jX_{C_2})}=\dfrac{1}{R_x}+j\omega C_x$

Rationalizing gives $\dfrac{R_3(R_2+jX_{C_2})}{R_4(R_2^2+X_{C_2}^2)}=\dfrac{1}{R_x}+j\omega C_x$

$$\text{Hence } \frac{1}{R_x}+j\omega C_x=\frac{R_3R_2}{R_4(R_2^2+(1/\omega^2C_2^2))}+\frac{jR_3(1/\omega C_2)}{R_3(R_2^2+(1/\omega^2C_2^2))}$$

Equating the real parts gives

$$\frac{1}{R_x}=\frac{R_3R_2}{R_4(R_2^2+(1/\omega^2C_2^2))}$$

i.e. $R_x=\dfrac{R_4}{R_2R_3}\left(\dfrac{R_2^2\omega^2C_2^2+1}{\omega^2C_2^2}\right)$

and $\boldsymbol{R_x=\dfrac{R_4(1+\omega^2 C_2^2R_2^2)}{R_2R_3\omega^2C_2^2}}$

Equating the imaginary parts gives

$$\omega C_x=\frac{R_3(1/\omega C_2)}{R_4(R_2^2+(1/\omega^2C_2^2))}=\frac{R_3}{\omega C_2R_4((R_2^2\omega^2C_2^2+1)/\omega^2C_2^2)}$$

i.e. $\omega C_x=\dfrac{R_3\omega^2C_2^2}{\omega C_2R_4(1+\omega^2C_2^2R_2^2)}$

and $\boldsymbol{C_x=\dfrac{R_3C_2}{R_4(1+\omega^2 C_2^2R_2^2)}}$

Part 4

(b) Substituting the given values gives

$$R_x = \frac{(1+\omega^2 C_2^2 R_2^2)}{R_2\omega^2 C_2^2} \quad \text{since } R_3 = R_4$$

$$\text{i.e. } R_x = \frac{1+(2\pi 1000)^2(0.2\times10^{-6})^2(2.5\times10^3)^2}{(2.5\times10^3)(2\pi 1000)^2(0.2\times10^{-6})^2}$$

$$= \frac{1+9.8696}{3.9478\times10^{-3}} \equiv 2.75\,\text{k}\Omega$$

$$C_x = \frac{C_2}{(1+\omega^2 C_2^2 R_2^2)} \quad \text{since } R_3 = R_4$$

$$= \frac{(0.2\times10^{-6})}{1+9.8696}\text{F} = 0.01840\,\mu\text{F or } 18.40\,\text{nF}$$

Hence at balance $R_x = 2.75\,\text{k}\Omega$ and $C_x = 18.40\,\text{nF}$

Problem 2. For the Wien bridge shown in Figure 30.9, $R_2 = R_3 = 30\,\text{k}\Omega$, $R_4 = 1\,\text{k}\Omega$ and $C_2 = C_3 = 1\,\text{nF}$. Determine, when the bridge is balanced, (a) the value of resistance R_1, and (b) the frequency of the bridge.

(a) From equation (4)

$$\frac{R_3}{R_2} + \frac{C_2}{C_3} = \frac{R_4}{R_1}$$

i.e. $1+1 = 1000/R_1$, since $R_2 = R_3$ and $C_2 = C_3$, from which

$$\textbf{resistance } R_1 = \frac{1000}{2} = \mathbf{500\,\Omega}$$

(b) From equation (5),

$$\text{frequency, } f = \frac{1}{2\pi\sqrt{(C_2C_3R_2R_3)}}$$

$$= \frac{1}{2\pi\sqrt{[(10^{-9})^2(30\times10^3)^2]}}$$

$$= \frac{1}{2\pi(10^{-9})(30\times10^3)} \equiv \mathbf{5.305\,kHz}$$

Problem 3. A Schering bridge network is as shown in Figure 30.7, page 466. Given $C_2 = 0.2\,\mu\text{F}$, $R_4 = 200\,\Omega$, $R_3 = 600\,\Omega$, $C_3 = 4000\,\text{pF}$ and the supply frequency is 1.5 kHz, determine, when the bridge is balanced, (a) the value of resistance R_x, (b) the value of capacitance C_x, (c) the phase angle of the unknown arm, (d) the power factor of the unknown arm and (e) its loss angle.

From para. (f), the equations for R_x and C_x at balance are given by

$$R_x = \frac{R_4C_3}{C_2} \quad \text{and} \quad C_x = \frac{C_2R_3}{R_4}$$

(a) Resistance, $R_x = \dfrac{R_4C_3}{C_2} = \dfrac{(200)(4000\times10^{-12})}{0.2\times10^{-6}}$

$= \mathbf{4\,\Omega}$

(b) Capacitance, $C_x = \dfrac{C_2R_3}{R_4} = \dfrac{(0.2\times10^{-6})(600)}{(200)}\text{F}$

$= \mathbf{0.6\,\mu F}$

(c) The phasor diagram for R_x and C_x in series is shown in Figure 30.11.

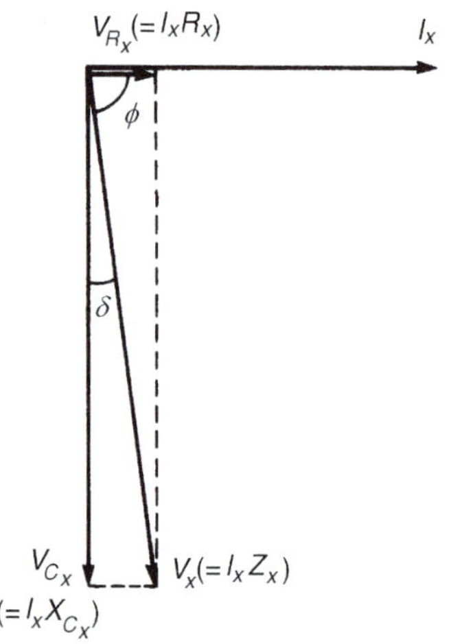

Figure 30.11

$$\text{Phase angle, } \phi = \tan^{-1}\frac{V_{C_x}}{V_{R_x}} = \tan^{-1}\frac{I_xX_{C_x}}{I_xR_x}$$

$$= \tan^{-1}\frac{1}{\omega C_xR_x}$$

$$\text{i.e.} \quad \phi = \tan^{-1}\left(\frac{1}{(2\pi 1500)(0.6\times10^{-6})(4)}\right)$$

$$= \tan^{-1}44.21 = \mathbf{88.7^\circ\ lead}$$

(d) Power factor of capacitor $= \cos\phi = \cos 88.7^\circ$ $= \mathbf{0.0227}$

(e) Loss angle, shown as δ in Figure 30.11, is given by $\delta = 90^\circ - 88.7^\circ = \mathbf{1.3^\circ}$

Alternatively, loss angle

$\delta = \tan^{-1}\omega C_xR_x$ (see para. (f), page 466)

Part 4

$$= \tan^{-1}\left(\frac{1}{44.21}\right) \text{ from (c) above,}$$

i.e. $\delta = 1.3°$

Now try the following Practice Exercise

Practice Exercise 129 A.c. bridges (Answers on page 827)

1. A Maxwell-Wien bridge circuit *ABCD* has the following arm impedances: *AB*, 250 Ω resistance; *BC*, 2 μF capacitor in parallel with a 10 kΩ resistor; *CD*, 400 Ω resistor; *DA*, unknown inductor having inductance *L* in series with resistance *R*. Determine the values of *L* and *R* if the bridge is balanced.

2. In a four-arm de Sauty a.c. bridge, arm 1 contains a 2 kΩ non-inductive resistor, arm 3 contains a loss-free 2.4 μF capacitor and arm 4 contains a 5 kΩ non-inductive resistor. When the bridge is balanced, determine the value of the capacitor contained in arm 2.

3. A four-arm bridge *ABCD* consists of: *AB* – fixed resistor R_1; *BC* – variable resistor R_2 in series with a variable capacitor C_2; *CD* – fixed resistor R_3; *DA* – coil of unknown resistance *R* and inductance *L*. Determine the values of *R* and *L* if, at balance, $R_1 = 1\,k\Omega$, $R_2 = 2.5\,k\Omega$, $C_2 = 4000\,pF$, $R_3 = 1\,k\Omega$ and the supply frequency is 1.6 kHz

4. The bridge shown in Figure 30.12 is used to measure capacitance C_x and resistance R_x. Derive the balance equations of the bridge and determine the values of C_x and R_x when $R_1 = R_4$, $C_2 = 0.1\,\mu F$, $R_2 = 2\,k\Omega$ and the supply frequency is 1 kHz

5. In a Schering bridge network *ABCD*, the arms are made up as follows: *AB* – a standard capacitor C_1; *BC* – a capacitor C_2 in parallel with a resistor R_2; *CD* – a resistor R_3; *DA* – the capacitor under test, represented by a capacitor C_x in series with a resistor R_x. The detector is connected between B and D and the a.c. supply is connected between A and C. Derive the equations for R_x and C_x when the bridge is balanced. Evaluate R_x and C_x if, at balance, $C_1 = 1\,nF$, $R_2 = 100\,\Omega$, $R_3 = 1\,k\Omega$ and $C_2 = 10\,nF$.

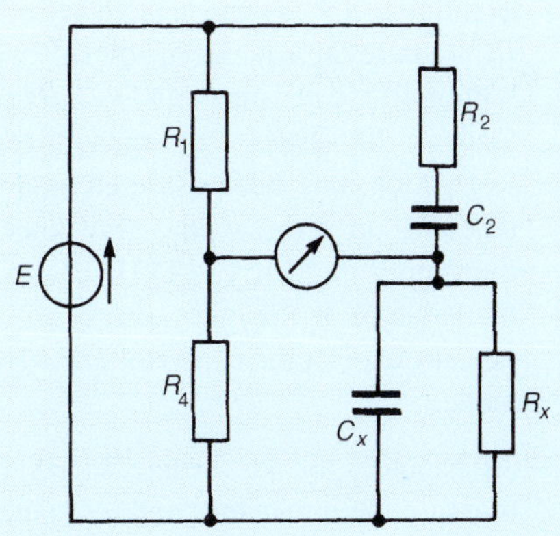

Figure 30.12

6. The a.c. bridge shown in Figure 30.13 is balanced when the values of the components are as shown. Determine at balance the values of R_x and L_x

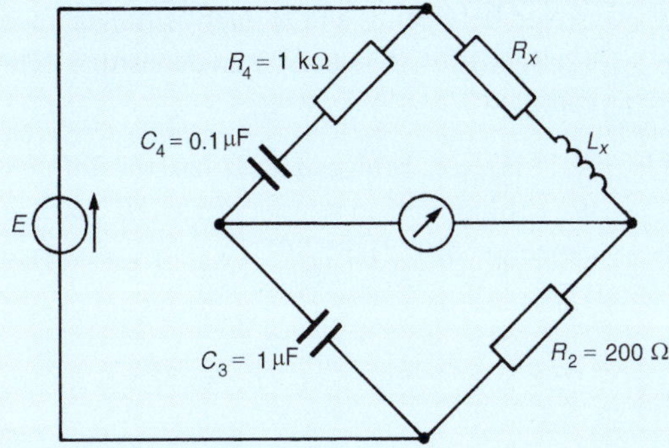

Figure 30.13

7. An a.c. bridge has, in arm *AB*, a pure capacitor of 0.4 μF; in arm *BC*, a pure resistor of 500 Ω; in arm *CD*, a coil of 50 Ω resistance and 0.1 H inductance; in arm *DA*, an unknown impedance comprising resistance R_x and capacitance C_x in series. If the frequency of the bridge at balance is 800 Hz, determine the values of R_x and C_x

8. When the Wien bridge shown in Figure 30.9 is balanced, the components have the following values: $R_2 = R_3 = 20\,k\Omega$, $R_4 = 500\,\Omega$, $C_2 = C_3 = 800\,pF$. Determine for the balance condition (a) the value of resistance R_1 and (b) the frequency of the bridge supply.

9. The conditions at balance of a Schering bridge *ABCD* used to measure the capacitance and loss angle of a paper capacitor are

Part 4

as follows: AB – a pure capacitance of 0.2 μF; BC – a pure capacitance of 3000 pF in parallel with a 400 Ω resistance; CD – a pure resistance of 200 Ω; DA – the capacitance under test which may be considered as a capacitance C_x in series with a resistance R_x. If the supply frequency is 1 kHz determine (a) the value of R_x, (b) the value of C_x, (c) the power factor of the capacitor and (d) its loss angle.

10. At balance, an a.c. bridge $PQRS$ used to measure the inductance and resistance of an inductor has the following values: PQ – a non-inductive 400 Ω resistor; QR – the inductor with unknown inductance L_x in series with resistance R_x; RS – a 3 μF capacitor in series with a non-inductive 250 Ω resistor; SP – a 15 μF capacitor. A detector is connected between Q and S and the a.c. supply is connected between P and R. Derive the balance equations for R_x and L_x and determine their values.

11. A 1 kHz a.c. bridge $ABCD$ has the following components in its four arms: AB – a pure capacitor of 0.2 μF; BC – a pure resistance of 500 Ω; CD – an unknown impedance; DA – a 400 Ω resistor in parallel with a 0.1 μF capacitor. If the bridge is balanced, determine the series components comprising the impedance in arm CD.

12. An a.c. bridge $ABCD$ has in arm AB a standard lossless capacitor of 200 pF; arm BC, an unknown impedance, represented by a lossless capacitor C_x in series with a resistor R_x; arm CD, a pure 5 kΩ resistor; arm DA, a 6 kΩ resistor in parallel with a variable capacitor set at 250 pF. The frequency of the bridge supply is 1500 Hz. Determine for the condition when the bridge is balanced (a) the values of R_x and C_x and (b) the loss angle.

13. An a.c. bridge $ABCD$ has the following components: AB – a 1 kΩ resistance in parallel with a 0.2 μF capacitor; BC – a 1.2 kΩ resistance; CD – a 750 Ω resistance; DA – a 0.8 μF capacitor in series with an unknown resistance. Determine (a) the frequency for which the bridge is in balance and (b) the value of the unknown resistance in arm DA to produce balance.

Part 4

For fully worked solutions to each of the problems in Practice Exercise 129 in this chapter, go to the website:
www.routledge.com/cw/bird

Chapter 31

Series resonance and Q-factor

***Why it is important to understand:* Series resonance and Q-factor**

Series resonance circuits are one of the most important circuits used in electrical and electronic engineering. They can be found in various forms, such as in a.c. mains filters, noise filters and also in radio and television tuning circuits, producing a very selective tuning circuit for the receiving of the different frequency channels. Resonance may be achieved either by varying the frequency or by varying one of the components, usually the capacitance, until the inductive and capacitive reactances are equal. Resonance makes the circuit respond in a markedly different manner at the resonance frequency than at other frequencies, a behaviour which is a basis of the selection process used in many communications systems, such as the tuning of a radio. Resonance, Q-factor, bandwidth and selectivity in a.c. series circuits are explained in this chapter

At the end of this chapter you should be able to:

- state the conditions for resonance in an a.c. series circuit
- calculate the resonant frequency in an a.c. series circuit,
$$f_r = \frac{1}{2\pi\sqrt{(LC)}}$$
- define Q-factor as $\frac{X}{R}$ and as $\frac{V_L}{V}$ or $\frac{V_C}{V}$
- determine the maximum value of V_C and V_{COIL} and the frequency at which this occurs
- determine the overall Q-factor for two components in series
- define bandwidth and selectivity
- calculate Q-factor and bandwidth in an a.c. series circuit
- determine the current and impedance when the frequency deviates from the resonant frequency

Electrical Circuit Theory and Technology. 978-1-138-67349-6, © 2017 John Bird. Published by Taylor & Francis. All rights reserved.

31.1 Introduction

When the voltage V applied to an electrical network containing resistance, inductance and capacitance is in phase with the resulting current I, the circuit is said to be **resonant**. The phenomenon of **resonance** is of great value in all branches of radio, television and communications engineering, since it enables small portions of the communications frequency spectrum to be selected for amplification independently of the remainder.

At resonance, the equivalent network impedance Z is purely resistive since the supply voltage and current are in phase. The power factor of a resonant network is unity (i.e. power factor $= \cos\phi = \cos 0 = 1$)

In electrical work there are two types of resonance – one associated with series circuits (which was introduced in Chapter 17), when the input impedance is a minimum (which is discussed further in this chapter), and the other associated with simple parallel networks, when the input impedance is a maximum (which was discussed in Chapter 18 and is further discussed in Chapter 32).

31.2 Series resonance

Figure 31.1 shows a circuit comprising a coil of inductance L and resistance R connected in series with a capacitor C. The R–L–C series circuit has a total impedance Z given by $Z = R + j(X_L - X_C)$ ohms, or $Z = R + j(\omega L - 1/\omega C)$ ohms where $\omega = 2\pi f$. The circuit is at resonance when $(X_L - X_C) = 0$, i.e. when $X_L = X_C$ or $\omega L = 1/(\omega C)$. The phasor diagram for this condition is shown in Figure 31.2, where $|V_L| = |V_C|$

Since at resonance

$$\omega_r L = \frac{1}{\omega_r C}, \quad \omega_r^2 = \frac{1}{LC} \quad \text{and} \quad \omega = \frac{1}{\sqrt{(LC)}}$$

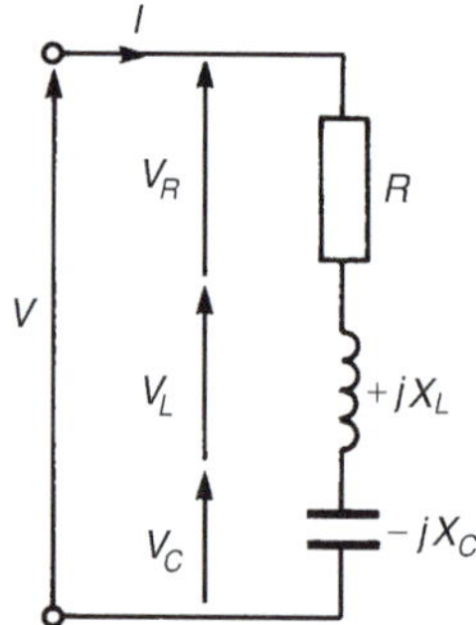

Figure 31.1 R–L–C series circuit

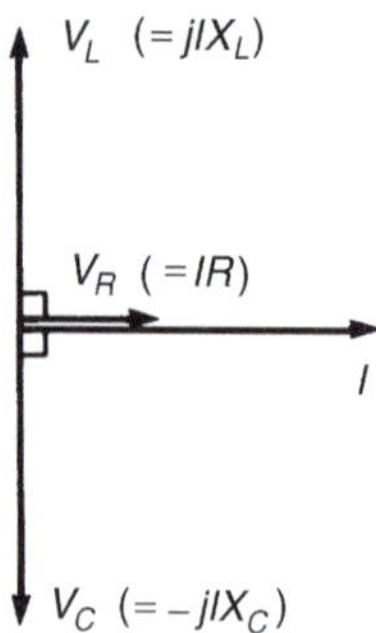

Figure 31.2 Phasor diagram $|V_L| = |V_C|$

Thus resonant frequency,

$$f_r = \frac{1}{2\pi\sqrt{(LC)}} \textbf{ hertz}, \text{ since } \omega_r = 2\pi f_r$$

Figure 31.3 shows how inductive reactance X_L and capacitive reactance X_C vary with the frequency. At the resonant frequency f_r, $|X_L| = |X_C|$. Since impedance $Z = R + j(X_L - X_C)$ and, at resonance, $(X_L - X_C) = 0$, then **impedance $Z = R$ at resonance**. This is the **minimum** value possible for the impedance as shown in the graph of the modulus of impedance, $|Z|$, against frequency in Figure 31.4.

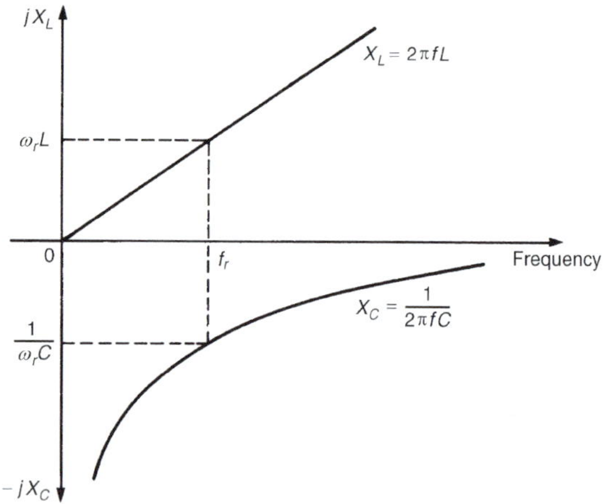

Figure 31.3 Variation of X_L and X_C with frequency

At frequencies less than f_r, $X_L < X_C$ and the circuit is capacitive; at frequencies greater than f_r, $X_L > X_C$ and the circuit is inductive.

Current $I = V/Z$. Since impedance Z is a minimum value at resonance, the **current I has a maximum value**. At resonance, current $I = V/R$. A graph of current against frequency is shown in Figure 31.4.

Part 4

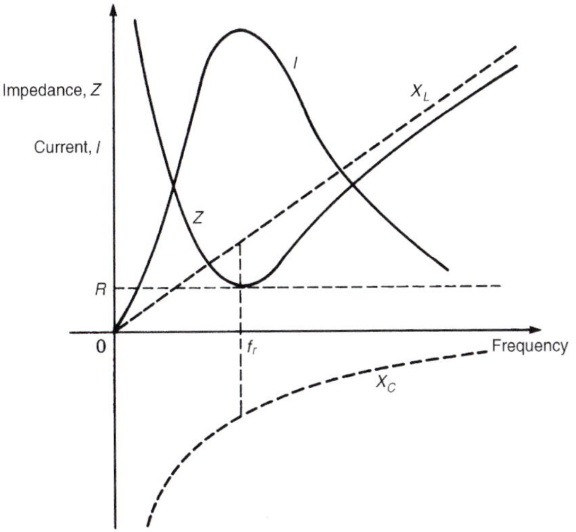

Figure 31.4 $|Z|$ and I plotted against frequency

Problem 1. A coil having a resistance of $10\,\Omega$ and an inductance of 75 mH is connected in series with a $40\,\mu F$ capacitor across a 200 V a.c. supply. Determine (a) at what frequency resonance occurs, and (b) the current flowing at resonance.

(a) Resonant frequency,

$$f_r = \frac{1}{2\pi\sqrt{(LC)}}$$

$$= \frac{1}{2\pi\sqrt{[(75\times 10^{-3})(40\times 10^{-6})]}}$$

i.e. $\boldsymbol{f_r = 91.9\,\text{Hz}}$

(b) Current at resonance, $\boldsymbol{I} = \dfrac{V}{R} = \dfrac{200}{10} = \mathbf{20\,A}$

Problem 2. An R–L–C series circuit is comprised of a coil of inductance 10 mH and resistance $8\,\Omega$ and a variable capacitor C. The supply frequency is 1 kHz. Determine the value of capacitor C for series resonance.

At resonance, $\omega_r L = 1/(\omega_r C)$, from which, capacitance, $C = 1/(\omega_r^2 L)$

Hence **capacitance**, $\boldsymbol{C} = \dfrac{1}{(2\pi 1000)^2(10\times 10^{-3})}$

$$= \mathbf{2.53\,\mu F}$$

Problem 3. A coil having inductance L is connected in series with a variable capacitor C. The circuit possesses stray capacitance C_S which is assumed to be constant and effectively in parallel with the variable capacitor C. When the capacitor is set to 1000 pF the resonant frequency of the circuit is 92.5 kHz, and when the capacitor is set to 500 pF the resonant frequency is 127.8 kHz. Determine the values of (a) the stray capacitance C_S, and (b) the coil inductance L.

For a series R–L–C circuit the resonant frequency f_r is given by:

$$f_r = \frac{1}{2\pi\sqrt{(LC)}}$$

The total capacitance of C in parallel with C_S is given by $(C + C_S)$

At 92.5 kHz, $C = 1000$ pF. Hence

$$92.5\times 10^3 = \frac{1}{2\pi\sqrt{[L(1000+C_S)10^{-12}]}} \qquad (1)$$

At 127.8 kHz, $C = 500$ pF. Hence

$$127.8\times 10^3 = \frac{1}{2\pi\sqrt{[L(500+C_S)10^{-12}]}} \qquad (2)$$

(a) Dividing equation (2) by equation (1) gives:

$$\frac{127.8\times 10^3}{92.5\times 10^3} = \frac{\dfrac{1}{2\pi\sqrt{[L(500+C_S)10^{-12}]}}}{\dfrac{1}{2\pi\sqrt{[L(1000+C_S)10^{-12}]}}}$$

i.e. $$\frac{127.8}{92.5} = \frac{\sqrt{[L(1000+C_S)10^{-12}]}}{\sqrt{[L(500+C_S)10^{-12}]}} = \sqrt{\left(\frac{1000+C_S}{500+C_S}\right)}$$

where C_S is in picofarads, from which,

$$\left(\frac{127.8}{92.5}\right)^2 = \frac{100+C_S}{500+C_S}$$

i.e. $$1.909 = \frac{1000+C_S}{500+C_S}$$

Hence $1.909\,(500+C_S) = 1000 + C_S$

$$954.5 + 1.909 C_S = 1000 + C_S$$

$$1.909\,C_S - C_S = 1000 - 954.5$$

$$0.909\,C_S = 45.5$$

Thus **stray capacitance** $\boldsymbol{C_S} = 45.5/0.909$

$$= \mathbf{50\,pF}$$

Part 4

(b) Substituting $C_S = 50\,\text{pF}$ in equation (1) gives:

$$92.5 \times 10^3 = \frac{1}{2\pi\sqrt{[L(1050 \times 10^{-12})]}}$$

Hence $(92.5 \times 10^3 \times 2\pi)^2 = \dfrac{1}{L(1050 \times 10^{-12})}$

from which, **inductance**

$$\boldsymbol{L} = \frac{1}{(1050 \times 10^{-12})(92.5 \times 10^3 \times 2\pi)^2}\,\text{H}$$

$$= \mathbf{2.82\ mH}$$

Now try the following Practice Exercise

Practice Exercise 130 Series resonance (Answers on page 827)

1. A coil having an inductance of 50 mH and resistance 8.0 Ω is connected in series with a 25 μF capacitor across a 100 V a.c. supply. Determine (a) the resonant frequency of the circuit, and (b) the current flowing at resonance.

2. The current at resonance in a series R–L–C circuit is 0.12 mA. The circuit has an inductance of 0.05 H and the supply voltage is 24 mV at a frequency of 40 kHz. Determine (a) the circuit resistance and (b) the circuit capacitance.

3. A coil of inductance 2.0 mH and resistance 4.0 Ω is connected in series with a 0.3 μF capacitor. The circuit is connected to a 5.0 V, variable frequency supply. Calculate (a) the frequency at which resonance occurs, (b) the voltage across the capacitance at resonance and (c) the voltage across the coil at resonance.

4. A series R–L–C circuit having an inductance of 0.40 H has an instantaneous voltage, $v = 60\sin(4000t - (\pi/6))$ volts and an instantaneous current, $i = 2.0\sin 4000t$ amperes. Determine (a) the values of the circuit resistance and capacitance and (b) the frequency at which the circuit will be resonant.

5. A variable capacitor C is connected in series with a coil having inductance L. The circuit possesses stray capacitance C_S which is assumed to be constant and effectively in parallel with the variable capacitor C. When the capacitor is set to 2.0 nF the resonant frequency of the circuit is 86.85 kHz, and when the capacitor is set to 1.0 nF the resonant frequency is 120 kHz. Determine the values of (a) the stray circuit capacitance C_S and (b) the coil inductance L.

31.3 Q-factor

Q-factor is a figure of merit for a resonant device such as an L–C–R circuit.

Such a circuit resonates by cyclic interchange of stored energy, accompanied by energy dissipation due to the resistance.

By definition, at resonance

$$Q = 2\pi\left(\frac{\text{maximum energy stored}}{\text{energy loss per cycle}}\right)$$

Since the energy loss per cycle is equal to (the average power dissipated) × (periodic time),

$$Q = 2\pi\left(\frac{\text{maximum energy stored}}{\text{average power dissipated} \times \text{periodic time}}\right)$$

$$= 2\pi\left(\frac{\text{maximum energy stored}}{\text{average power dissipated} \times (1/f_r)}\right)$$

since the periodic time $T = 1/f_r$

Thus $Q = 2\pi f_r\left(\dfrac{\text{maximum energy stored}}{\text{average power dissipated}}\right)$

i.e. $Q = \omega_r\left(\dfrac{\text{maximum energy stored}}{\text{average power dissipated}}\right)$

where ω_r is the angular frequency at resonance.

In an L–C–R circuit both of the reactive elements store energy during a quarter cycle of the alternating supply input and return it to the circuit source during the following quarter cycle. An inductor stores energy in its magnetic field, then transfers it to the electric field of the capacitor and then back to the magnetic field, and so on. Thus the inductive and capacitive elements transfer energy from one to the other successively with the source of supply ideally providing no additional energy at all. Practical reactors both store and dissipate energy.

Q-factor is an abbreviation for **quality factor** and refers to the 'goodness' of a reactive component.

For an **inductor**, $Q=\omega_r\left(\dfrac{\text{maximum energy stored}}{\text{average power dissipated}}\right)$

$$=\omega_r\left(\frac{\frac{1}{2}LI_m^2}{I^2R}\right)=\frac{\omega_r\left(\frac{1}{2}LI_m^2\right)}{(I_m/\sqrt{2})^2R}$$

$$=\frac{\omega_r L}{R} \qquad (1)$$

For a **capacitor**, $Q=\dfrac{\omega_r\left(\frac{1}{2}CV_m^2\right)}{(I_m/\sqrt{2})^2R}=\dfrac{\omega_r\frac{1}{2}C(I_mX_C)^2}{(I_m/\sqrt{2})^2R}$

$$=\frac{\omega_r\frac{1}{2}CI_m^2(1/\omega_r C)^2}{(I_m/\sqrt{2})^2R}$$

i.e. $$Q=\frac{1}{\omega_r CR} \qquad (2)$$

From expressions (1) and (2) it can be deduced that

$$Q=\frac{X_L}{R}=\frac{X_C}{R}=\frac{\text{reactance}}{\text{resistance}}$$

In fact, Q-factor can also be defined as

$$\text{Q-factor}=\frac{\text{reactance power}}{\text{resistance}}=\frac{Q}{P}$$

where Q is the reactive power which is also the peak rate of energy storage, and P is the average energy dissipation rate. Hence

$$\text{Q-factor}=\frac{Q}{P}=\frac{I^2X_L(\text{or } I^2X_C)}{I^2R}=\frac{X_L}{R}\left(\text{or }\frac{X_C}{R}\right)$$

i.e. $$\boldsymbol{Q=\frac{\textbf{reactance}}{\textbf{resistance}}}$$

In an R–L–C series circuit the amount of energy stored at resonance is constant.
When the capacitor voltage is a maximum, the inductor current is zero, and vice versa, i.e. $\frac{1}{2}LI_m{}^2=\frac{1}{2}CV_m{}^2$
Thus the Q-factor at resonance, Q_r, is given by

$$\boldsymbol{Q_r=\frac{\omega_r L}{R}=\frac{1}{\omega_r CR}} \qquad (3)$$

However, at resonance $\omega_r=1/\sqrt{(LC)}$

Hence $$Q_r=\frac{\omega_r L}{R}=\frac{1}{\sqrt{(LC)}}\left(\frac{L}{R}\right)$$

i.e. $$\boldsymbol{Q_r=\frac{1}{R}\sqrt{\left(\frac{L}{C}\right)}}$$

It should be noted that when Q-factor is referred to, it is nearly always assumed to mean 'the Q-factor at resonance'.
With reference to Figures 31.1 and 31.2, at resonance, $V_L=V_C$

$$V_L=IX_L=I\omega_r L=\frac{V}{R}\omega_r L=\left(\frac{\omega_r L}{R}\right)V=Q_rV$$

and $$V_C=IX_C=\frac{I}{\omega_r C}=\frac{V/R}{\omega_r C}=\left(\frac{1}{\omega_r CR}\right)V=Q_rV$$

Hence, at resonance, $V_L=V_C=Q_rV$

or $$\boldsymbol{Q_r=\frac{V_L\ (\text{or } V_C)}{V}}$$

The voltages V_L and V_C at resonance may be much greater than that of the supply voltage V. For this reason Q is often called the **circuit magnification factor**. It represents a measure of the number of times V_L or V_C is greater than the supply voltage.

The Q-factor at resonance can have a value of several hundreds. Resonance is usually of interest only in circuits of Q-factor greater than about 10; circuits having Q considerably below this value are effectively merely operating at unity power factor.

Problem 4. A series circuit comprises a 10 Ω resistance, a 5 μF capacitor and a variable inductance L. The supply voltage is $20\angle 0°$ volts at a frequency of 318.3 Hz. The inductance is adjusted until the p.d. across the 10 Ω resistance is a maximum. Determine for this condition (a) the value of inductance L, (b) the p.d. across each component and (c) the Q-factor.

(a) The maximum voltage across the resistance occurs at resonance when the current is a maximum. At resonance, $\omega_r L=1/(\omega_r C)$, from which

$$\textbf{inductance } L=\frac{1}{\omega_r^2 C}=\frac{1}{(2\pi 318.3)^2(5\times 10^{-6})}$$
$$=\mathbf{0.050\ H} \text{ or } \mathbf{50\ mH}$$

(b) Current at resonance $$I_r=\frac{V}{R}=\frac{20\angle 0°}{10\angle 0°}=2.0\angle 0°\ \text{A}$$

p.d. across resistance, $V_R=I_rR=(2.0\angle 0°)(10)=\mathbf{20\angle 0°\ V}$

p.d. across inductance, $V_L=IX_L$

$$X_L=2\pi(318.3)(0.050)=100\ \Omega$$

Hence $V_L=(2.0\angle 0°)(100\angle 90°)=\mathbf{200\angle 90°\ V}$

p.d. across capacitor, $\boldsymbol{V_C} = IX_C$

$$= (2.0\angle 0°)(100\angle -90°)$$

$$= \mathbf{200\angle -90° \ V}$$

(c) Q-factor at resonance, $\boldsymbol{Q_r} = \dfrac{V_L \text{ (or } V_C)}{V} = \dfrac{200}{20} = \mathbf{10}$

[Alternatively, $Q_r = \dfrac{\omega_r L}{R} = \dfrac{100}{10} = \mathbf{10}$

or $Q_r = \dfrac{1}{\omega_r CR}$

$$= \frac{1}{2\pi(318.3)(5\times 10^{-6})(10)}$$

$$= \mathbf{10}$$

or $Q_r = \dfrac{1}{R}\sqrt{\left(\dfrac{L}{C}\right)}$

$$= \frac{1}{10}\sqrt{\left(\frac{0.050}{5\times 10^{-6}}\right)} = \mathbf{10}]$$

31.4 Voltage magnification

For a circuit with a high value of Q (say, exceeding 100), the maximum volt drop across the coil, V_{COIL}, and the maximum volt drop across the capacitor, V_C, coincide with the maximum circuit current at the resonant frequency, f_r, as shown in Figure 31.5(a). However, if a circuit of low Q (say, less than 10) is used, it may be shown experimentally that the maximum value of V_C occurs at a frequency less than f_r while the maximum value of V_{COIL} occurs at a frequency higher than f_r, as shown in Figure 31.5(b). The maximum current, however, still occurs at the resonant frequency with low Q. This is analysed below.

Since $Q_r = \dfrac{V_C}{V}$ then $V_C = VQ_r$

However, $V_C = IX_C = I\left(\dfrac{-j}{\omega C}\right) = I\left(\dfrac{1}{j\omega C}\right)$ and since

$$I = \frac{V}{Z}$$

$$V_C = \frac{V}{Z}\left(\frac{1}{j\omega C}\right) = \frac{V}{(j\omega C)Z}$$

$$Z = R + j\left(\omega L - \frac{1}{\omega C}\right)$$

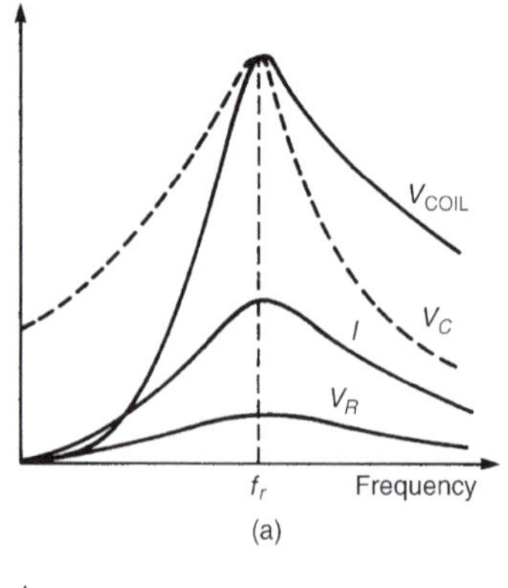

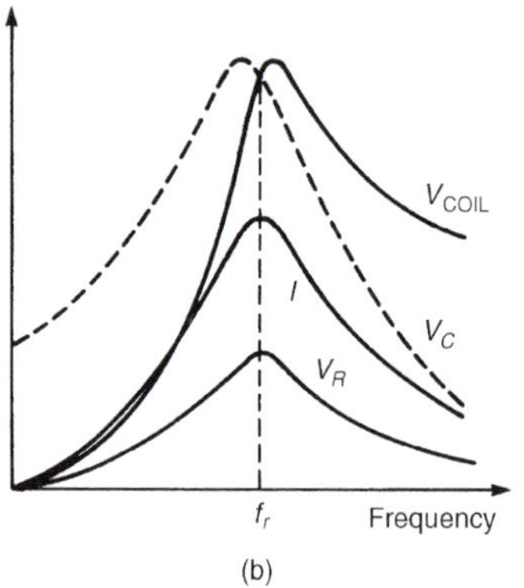

Figure 31.5 (a) High Q-factor (b) low Q-factor

thus $V_C = \dfrac{V}{(j\omega C)\left[R + j\left(\omega L - \dfrac{1}{\omega C}\right)\right]}$

$$= \frac{V}{j\omega CR + j^2\omega^2 CL - j^2\dfrac{\omega C}{\omega C}}$$

$$= \frac{V}{j\omega CR - \omega^2 LC + 1} = \frac{V}{(1-\omega^2 LC) + j\omega CR}$$

$$= \frac{V[(1-\omega^2 LC) - j\omega CR]}{[(1-\omega^2 LC,) + j\omega CR][(1-\omega^2 LC) - j\omega CR]}$$

$$= \frac{V[(1-\omega^2 LC) - j\omega CR\ ]}{[(1-\omega^2 LC)^2 + (\omega CR)^2]}$$

The magnitude of V_C,

$$|V_C| = \frac{V\sqrt{[(1-\omega^2 LC)^2 + (\omega CR)^2]}}{[(1-\omega^2 LC)^2 + (\omega CR)^2]}$$

from the Argand diagram

$$= \frac{V}{\sqrt{[(1-\omega^2 LC)^2 + (\omega CR)^2]}} \qquad (4)$$

To find the maximum value of V_C, equation (4) is differentiated with respect to ω, equated to zero and then

Part 4

solved – this being the normal procedure for maximum/minimum problems. Thus, using the quotient and function of a function rules:

$$\frac{dV_C}{d\omega} = \frac{\sqrt{[(1-\omega^2LC)^2+(\omega CR)^2]}[0] - [V]\frac{1}{2}[(1-\omega^2LC)^2+(\omega CR)^2]^{-1/2} \times[2(1-\omega^2LC)(-2\omega LC)+2\omega C^2R^2]}{\{\sqrt{[(1-\omega^2LC)^2+(\omega CR)^2]}\}^2}$$

$$= \frac{0-\frac{V}{2}[(1-\omega^2LC)^2+(\omega CR)^2]^{-1/2} \times[2(1-\omega^2LC)(-2\omega LC)+2\omega C^2R^2]}{(1-\omega^2LC)^2+(\omega CR)^2}$$

$$= \frac{-\frac{V}{2}[2(1-\omega^2LC)^2(-2\omega LC)+2\omega C^2R^2]}{[(1-\omega^2LC)^2+(\omega CR)^2]^{3/2}} = 0$$

for a maximum value.

Hence $-\dfrac{V}{2}[2(1-\omega^2LC)(-2\omega LC)+2\omega C^2R^2]=0$

and $-V[(1-\omega^2LC)(-2\omega LC)+\omega C^2R^2]=0$

and $(1-\omega^2LC)(-2\omega LC)+\omega C^2R^2=0$

from which, $\omega C^2R^2=(1-\omega^2LC)(2\omega LC)$

i.e. $C^2R^2 = 2LC(1-\omega^2LC)$

$$\frac{C^2R^2}{LC} = 2-2\omega^2LC \quad \text{and} \quad 2\omega^2LC = 2-\frac{CR^2}{L}$$

Hence $\omega^2 = \dfrac{2}{2LC} - \dfrac{\frac{CR^2}{L}}{2LC} = \dfrac{1}{LC} - \dfrac{1}{2}\left(\dfrac{R}{L}\right)^2$

The resonant frequency, $\omega_r = \dfrac{1}{\sqrt{(LC)}}$ from which, $\omega_r^2 = \dfrac{1}{LC}$

Thus $\omega^2 = \omega_r^2 - \dfrac{1}{2}\left(\dfrac{R}{L}\right)^2 \qquad (5)$

$Q = \dfrac{\omega_r L}{R}$ from which $\dfrac{R}{L} = \dfrac{\omega_r}{Q}$ and $\left(\dfrac{R}{L}\right)^2 = \dfrac{\omega_r^2}{Q^2}$

Hence, from equation (5) $\omega^2 = \omega_r^2 - \dfrac{1}{2}\dfrac{\omega_r^2}{Q^2}$

i.e. $$\omega^2 = \omega_r^2\left(1-\frac{1}{2Q^2}\right) \qquad (6)$$

or $$\omega = \omega_r\sqrt{\left(1-\frac{1}{2Q^2}\right)}$$

or $$\boldsymbol{f = f_r\sqrt{\left(1-\frac{1}{2Q^2}\right)}} \qquad (7)$$

Hence the maximum p.d. across the capacitor does not occur at the resonant frequency, but at a frequency slightly less than f_r, as shown in Figure 31.5(b). If Q is large, then $f \approx f_r$ as shown in Figure 31.5(a).

From equation (4),

$$|V_C| = \frac{V}{\sqrt{[(1-\omega^2LC)^2+(\omega CR)^2]}}$$

and substituting $\omega^2 = \omega_r^2\left(1-\dfrac{1}{2Q^2}\right)$ from equation (6) gives maximum value of V_c,

$$V_{C_m} = \frac{V}{\sqrt{\left[\left(1-\omega_r^2\left(1-\frac{1}{2Q^2}\right)LC\right)^2 + \omega_r^2\left(1-\frac{1}{2Q^2}\right)C^2R^2\right]}}$$

$\omega_r^2 = \dfrac{1}{LC}$ hence

$$V_{C_m} = \frac{V}{\sqrt{\left[\left(1-\frac{1}{LC}\left(1-\frac{1}{2Q^2}\right)LC\right)^2 + \frac{1}{LC}\left(1-\frac{1}{2Q^2}\right)C^2R^2\right]}}$$

$$= \frac{V}{\sqrt{\left[\left(1-\left(1-\frac{1}{2Q^2}\right)\right)^2 + \frac{CR^2}{L}\left(1-\frac{1}{2Q^2}\right)\right]}}$$

$$= \frac{V}{\sqrt{\left[\frac{1}{4Q^4}+\frac{CR^2}{L}-\frac{CR^2}{L}\left(\frac{1}{2Q^2}\right)\right]}} \qquad (8)$$

$Q = \dfrac{\omega_r L}{R} = \dfrac{1}{\omega_r CR}$ hence

$$Q^2 = \left(\frac{\omega_r L}{R}\right)\left(\frac{1}{\omega_r CR}\right) = \frac{L}{CR^2}$$

from which, $\dfrac{CR^2}{L} = \dfrac{1}{Q^2}$

Substituting in equation (8),

$$V_{C_m} = \frac{V}{\sqrt{\left(\frac{1}{4Q^4}+\frac{1}{Q^2}-\frac{1}{2Q^4}\right)}}$$

$$= \frac{V}{\sqrt{\left(\frac{1}{Q^2}\left[\frac{1}{4Q^2}+1-\frac{1}{2Q^2}\right]\right)}}$$

$$= \frac{V}{\frac{1}{Q}\sqrt{\left[1 - \frac{1}{4Q^2}\right]}}$$

i.e. $$V_{C_m} = \frac{QV}{\sqrt{\left[1 - \left(\frac{1}{2Q}\right)^2\right]}} \qquad (9)$$

From equation (9), when Q is large, $V_{C_m} \approx QV$

If a similar exercise is undertaken for the voltage across the inductor it is found that the maximum value is given by:

$$V_{L_m} = \frac{QV}{\sqrt{\left[1 - \left(\frac{1}{2Q}\right)^2\right]}}$$

i.e. the same equation as for V_{C_m}, and frequency,

$$f = \frac{f_r}{\sqrt{\left[\left(1 - \frac{1}{2Q^2}\right)\right]}}$$

showing that the maximum p.d. across the coil does not occur at the resonant frequency but at a value slightly greater than f_r, as shown in Figure 31.5(b).

Problem 5. A series L–R–C circuit has a sinusoidal input voltage of maximum value 12 V. If inductance, $L = 20$ mH, resistance, $R = 80\,\Omega$ and capacitance, $C = 400$ nF, determine (a) the resonant frequency, (b) the value of the p.d. across the capacitor at the resonant frequency, (c) the frequency at which the p.d. across the capacitor is a maximum and (d) the value of the maximum voltage across the capacitor.

(a) The resonant frequency,

$$f_r = \frac{1}{2\pi\sqrt{(LC)}}$$

$$= \frac{1}{2\pi\sqrt{[(20\times10^{-3})(400\times10^{-9})]}}$$

$$= \mathbf{1779.4\,Hz}$$

(b) $V_C = QV$ and

$$Q = \frac{\omega_r L}{R}\left(\text{or } \frac{1}{\omega_r CR} \text{ or } \frac{1}{R}\sqrt{\frac{L}{C}}\right)$$

Hence $$Q = \frac{(2\pi\,1779.4)(20\times10^{-3})}{80} = 2.80$$

Thus $V_C = QV = (2.80)(12) = \mathbf{33.60\,V}$

(c) From equation (7), the frequency f at which V_C is a maximum value,

$$f = f_r\sqrt{\left(1 - \frac{1}{2Q^2}\right)}$$

$$= (1779.4)\sqrt{\left(1 - \frac{1}{2(2.80)^2}\right)}$$

$$= \mathbf{1721.7\,Hz}$$

(d) From equation (9), the maximum value of the p.d. across the capacitor is given by:

$$V_{C_m} = \frac{QV}{\sqrt{\left[1 - \left(\frac{1}{2Q}\right)^2\right]}}$$

$$= \frac{(2.80)(12)}{\sqrt{\left[1 - \left(\frac{1}{2(2.80)}\right)^2\right]}}$$

$$= \mathbf{34.15\,V}$$

31.5 Q-factors in series

If the losses of a capacitor are not considered as negligible, the overall Q-factor of the circuit will depend on the Q-factor of the individual components. Let the Q-factor of the inductor be Q_L and that of the capacitor be Q_C

The overall Q-factor, $Q_T = \frac{1}{R_T}\sqrt{\frac{L}{C}}$ from Section 31.3,

where $R_T = R_L + R_C$

Since $Q_L = \frac{\omega_r L}{R_L}$ then $R_L = \frac{\omega_r L}{Q_L}$ and since

$Q_C = \frac{1}{\omega_r C R_C}$ then $R_C = \frac{1}{Q_C \omega_r C}$

Hence

$$Q_T = \frac{1}{R_L + R_C}\sqrt{\frac{L}{C}}$$

$$= \frac{1}{\left(\frac{\omega_r L}{Q_L} + \frac{1}{Q_C \omega_r C}\right)}\sqrt{\frac{L}{C}}$$

Part 4

$$= \frac{1}{\left[\frac{\left(\frac{1}{\sqrt{(LC)}}\right)L}{Q_L} + \frac{1}{Q_C\left(\frac{1}{\sqrt{(LC)}}\right)C}\right]}\sqrt{\frac{L}{C}}$$

$$\text{since } \omega_r = \frac{1}{\sqrt{(LC)}}$$

$$= \frac{1}{\frac{L}{Q_L L^{1/2}C^{1/2}} + \frac{L^{1/2}C^{1/2}}{Q_C C}}\sqrt{\frac{L}{C}}$$

$$= \frac{1}{\frac{1}{Q_L}\frac{L^{1/2}}{C^{1/2}} + \frac{1}{Q_C}\frac{L^{1/2}}{C^{1/2}}}\sqrt{\frac{L}{C}}$$

$$= \frac{1}{\frac{1}{Q_L}\sqrt{\frac{L}{C}} + \frac{1}{Q_C}\sqrt{\frac{L}{C}}}\sqrt{\frac{L}{C}}$$

$$= \frac{1}{\sqrt{\frac{L}{C}}\left(\frac{1}{Q_L} + \frac{1}{Q_C}\right)}\sqrt{\frac{L}{C}}$$

$$= \frac{1}{\frac{1}{Q_L} + \frac{1}{Q_C}} = \frac{1}{\frac{Q_C + Q_L}{Q_L Q_C}}$$

i.e. the overall Q-factor,

$$\mathbf{Q_T = \frac{Q_L Q_C}{Q_L + Q_C}}$$

Problem 6. An inductor of Q-factor 60 is connected in series with a capacitor having a Q-factor of 390. Determine the overall Q-factor of the circuit.

From above, overall Q-factor,

$$Q_T = \frac{Q_L Q_C}{Q_L + Q_C} = \frac{(60)(390)}{60 + 390} = \frac{23\,400}{450} = \mathbf{52}$$

Now try the following Practice Exercise

Practice Exercise 131 Q-factor (Answers on page 827)

1. A series R–L–C circuit comprises a 5 μF capacitor, a 4 Ω resistor and a variable inductance L. The supply voltage is $10\angle 0°$ V at a frequency of 159.1 Hz. The inductance is adjusted until the p.d. across the 4 Ω resistance is a maximum. Determine for this condition (a) the value of inductance, (b) the p.d. across each component and (c) the Q-factor of the circuit.
2. A series L–R–C circuit has a supply input of 5 volts. Given that inductance, $L = 5$ mH, resistance, $R = 75\,\Omega$ and capacitance, $C = 0.2\,\mu$F, determine (a) the resonant frequency, (b) the value of voltage across the capacitor at the resonant frequency, (c) the frequency at which the p.d. across the capacitance is a maximum and (d) the value of the maximum voltage across the capacitor.
3. A capacitor having a Q-factor of 250 is connected in series with a coil which has a Q-factor of 80. Calculate the overall Q-factor of the circuit.
4. An R–L–C series circuit has a maximum current of 2 mA flowing in it when the frequency of the 0.1 V supply is 4 kHz. The Q-factor of the circuit under these conditions is 90. Determine (a) the voltage across the capacitor and (b) the values of the circuit resistance, inductance and capacitance.
5. Calculate the inductance of a coil which must be connected in series with a 4000 pF capacitor to give a resonant frequency of 200 kHz. If the coil has a resistance of 12 Ω, determine the circuit Q-factor.

31.6 Bandwidth

Figure 31.6 shows how current I varies with frequency f in an R–L–C series circuit. At the resonant frequency, f_r, current is a maximum value, shown as I_r. Also shown are the points A and B where the current is 0.707 of the maximum value at frequencies f_1 and f_2. The power delivered to the circuit is I^2R. At $I = 0.707I_r$, the power is $(0.707I_r)^2R = 0.5\,I_r^2R$, i.e. half the power that occurs at frequency f_r. The points corresponding to f_1 and f_2 are called the **half-power points**. The distance between these points, i.e. $(f_2 - f_1)$, is called the **bandwidth**.

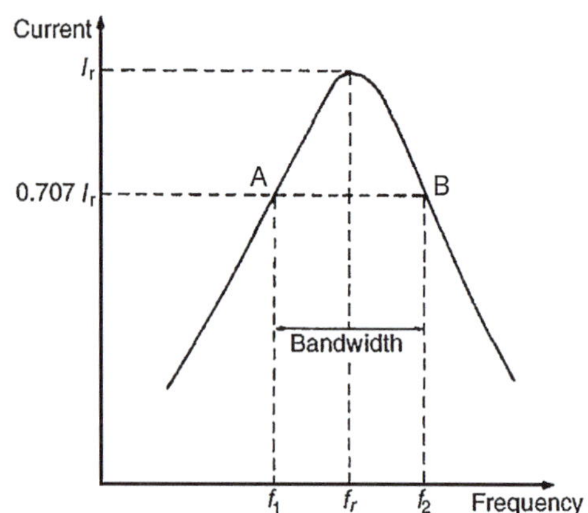

Figure 31.6 Bandwidth and half-power points f_1 and f_2

When the ratio of two powers P_1 and P_2 is expressed in decibel units, the number of decibels X is given by:

$$X = 10\lg\left(\frac{P_2}{P_1}\right) \text{ dB (see Section 12.13, page 174)}$$

Let the power at the half-power points be $(0.707I_r)^2 R = (I_r^2 R)/2$ and let the peak power be $I_r^2 R$, then the ratio of the power in decibels is given by:

$$10\lg\left[\frac{I_r^2 R/2}{I_r^2 R}\right] = 10\lg\frac{1}{2} = \mathbf{-3\,dB}$$

It is for this reason that the half-power points are often referred to as **'the −3 dB points'**.

At the half-power frequencies, $I = 0.707I_r$, thus impedance

$$Z = \frac{V}{I} = \frac{V}{0.707I_r} = 1.414\left(\frac{V}{I_r}\right) = \sqrt{2}Z_r = \sqrt{2}R$$

(since at resonance $Z_r = R$)

Since $Z = \sqrt{2}R$, an isosceles triangle is formed by the impedance triangles, as shown in Figure 31.7, where $ab = bc$. From the impedance triangles it can be seen that the equivalent circuit reactance is equal to the circuit resistance at the half-power points.

At f_1, the lower half-power frequency $|X_C| > |X_L|$ (see Figure 31.4).

Thus $\dfrac{1}{2\pi f_1 C} - 2\pi f_1 L = R$

from which, $1 - 4\pi^2 f_1^2 LC = 2\pi f_1 CR$

i.e. $(4\pi^2 LC)f_1^2 + (2\pi CR)f_1 - 1 = 0$

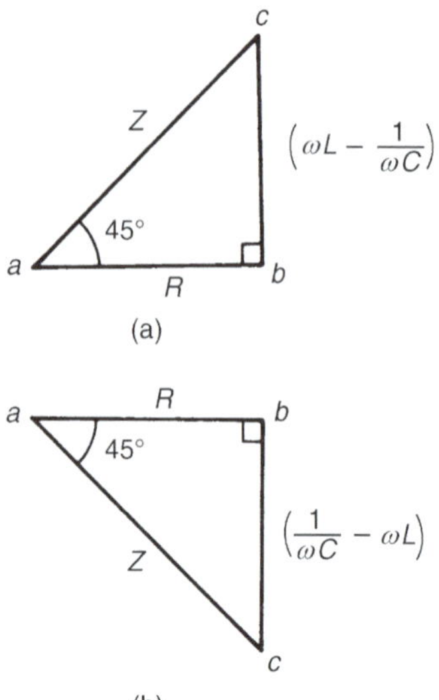

Figure 31.7 (a) Inductive impedance triangle (b) capacitive impedance triangle

This is a quadratic equation in f_1. Using the quadratic formula gives:

$$f_1 = \frac{-(2\pi CR) \pm \sqrt{[(2\pi CR)^2 - 4(4\pi^2 LC)(-1)]}}{2(4\pi^2 LC)}$$

$$= \frac{-(2\pi CR) \pm \sqrt{[4\pi^2 C^2 R^2 + 16\pi^2 LC]}}{8\pi^2 LC}$$

$$= \frac{-(2\pi CR) \pm \sqrt{[4\pi^2 C^2 (R^2 + (4L/C))]}}{8\pi^2 LC}$$

$$= \frac{-(2\pi CR) \pm 2\pi C\sqrt{[R^2 + (4L/C)]}}{8\pi^2 LC}$$

Hence $\boldsymbol{f_1 = \dfrac{-R \pm \sqrt{[R^2 + (4L/C)]}}{4\pi L}}$

$$\boldsymbol{= \frac{-R + \sqrt{[R^2 + (4L/C)]}}{4\pi L}}$$

(since $\sqrt{[R^2 + (4L/C)]} > R$ and f_1 cannot be negative).

At f_2, the upper half-power frequency $|X_L| > |X_C|$ (see Figure 31.4)

Thus $2\pi f_2 L - \dfrac{1}{2\pi f_2 C} = R$

from which, $4\pi^2 f_2^2 LC - 1 = R(2\pi f_2 C)$

i.e. $(4\pi^2 LC)f_2^2 - (2\pi CR)f_2 - 1 = 0$

This is a quadratic equation in f_2 and may be solved using the quadratic formula as for f_1, giving:

$$f_2 = \frac{R + \sqrt{[R^2 + (4L/C)]}}{4\pi L}$$

Bandwidth $= (f_2 - f_1)$

$$= \left\{\frac{R + \sqrt{[R^2 + (4L/C)]}}{4\pi L}\right\} - \left\{\frac{-R + \sqrt{[R^2 + (4L/C)]}}{4\pi L}\right\}$$

$$= \frac{2R}{4\pi L} = \frac{R}{2\pi L} = \frac{1}{2\pi L/R}$$

$$= \frac{f_r}{2\pi f_r L/R} = \frac{f_r}{Q_r}$$

from equation (3). Hence for a series R–L–C circuit

$$Q_r = \frac{f_r}{f_2 - f_1} \qquad (10)$$

Problem 7. A filter in the form of a series L–R–C circuit is designed to operate at a resonant frequency of 20 kHz. Included within the filter is a 20 mH inductance and 8 Ω resistance. Determine the bandwidth of the filter.

Q-factor at resonance is given by

$$Q_r = \frac{\omega_r L}{R} = \frac{(2\pi 20\,000)(10 \times 10^{-3})}{8} = 157.08$$

Since $Q_r = f_r/(f_2 - f_1)$

bandwidth, $(f_2 - f_1) = \frac{f_r}{Q_r} = \frac{20\,000}{157.08} = \mathbf{127.3\,Hz}$

An alternative equation involving f_r

At the lower half-power frequency f_1: $\frac{1}{\omega_1 C} - \omega_1 L = R$

At the higher half-power frequency f_2: $\omega_2 L - \frac{1}{\omega_2 C} = R$

Equating gives: $\frac{1}{\omega_1 C} - \omega_1 L = \omega_2 L - \frac{1}{\omega_2 C}$

Multiplying throughout by C gives:

$$\frac{1}{\omega_1} - \omega_1 LC = \omega_2 LC - \frac{1}{\omega_2}$$

However, for series resonance, $\omega_r^2 = 1/(LC)$

Hence $\frac{1}{\omega_1} - \frac{\omega_1}{\omega_r^2} = \frac{\omega_2}{\omega_r^2} - \frac{1}{\omega_2}$

i.e. $\frac{1}{\omega_1} + \frac{1}{\omega_2} = \frac{\omega_2}{\omega_r^2} + \frac{\omega_1}{\omega_r^2} = \frac{\omega_1 + \omega_2}{\omega_r^2}$

Therefore $\frac{\omega_2 + \omega_1}{\omega_1 \omega_2} = \frac{\omega_1 + \omega_2}{\omega_r^2}$

from which, $\omega_r^2 = \omega_1 \omega_2$ or $\omega_r = \sqrt{(\omega_1 \omega_2)}$

Hence $2\pi f_r = \sqrt{[(2\pi f_1)(2\pi f_2)]}$ and $f_r = \sqrt{(f_1 f_2)}$ (11)

Selectivity is the ability of a circuit to respond more readily to signals of a particular frequency to which it is tuned than to signals of other frequencies. The response becomes progressively weaker as the frequency departs from the resonant frequency. Discrimination against other signals becomes more pronounced as circuit losses are reduced, i.e. as the Q-factor is increased. Thus $Q_r = f_r/(f_2 - f_1)$ is a measure of the circuit selectivity in terms of the points on each side of resonance where the circuit current has fallen to 0.707 of its maximum value reached at resonance. The higher the Q-factor, the narrower the bandwidth and the more selective is the circuit. Circuits having high Q-factors (say, in the order 300) are therefore useful in communications engineering. A high Q-factor in a series power circuit has disadvantages in that it can lead to dangerously high voltages across the insulation and may result in electrical breakdown.

For example, suppose that the working voltage of a capacitor is stated as 1 kV and is used in a circuit having a supply voltage of 240 V. The maximum value of the supply will be $\sqrt{2}(240)$, i.e. 340 V. The working voltage of the capacitor would appear to be ample. However, if the Q-factor is, say, 10, the voltage across the capacitor will reach 2.4 kV. Since the capacitor is rated only at 1 kV, dielectric breakdown is more than likely to occur.

Low Q-factors, say, in the order of 5 to 25, may be found in power transformers using laminated iron cores.

A capacitor-start induction motor, as used in domestic appliances such as washing machines and vacuum-cleaners, having a Q-factor as low as 1.5 at starting would result in a voltage across the capacitor 1.5 times that of the supply voltage; hence the cable joining the capacitor to the motor would require extra insulation.

Problem 8. An R–L–C series circuit has a resonant frequency of 1.2 kHz and a Q-factor at resonance of 30. If the impedance of the circuit at resonance is 50 Ω determine the values of (a) the inductance and (b) the capacitance. Find also (c) the bandwidth, (d) the lower and upper half-power frequencies and (e) the value of the circuit impedance at the half-power frequencies.

(a) At resonance the circuit impedance, $Z = R$, i.e. $R = 50\,\Omega$.

Q-factor at resonance, $Q_r = \omega_r L/R$

Hence **inductance,** $\boldsymbol{L} = \dfrac{Q_r R}{\omega_r} = \dfrac{(30)(50)}{(2\pi 1200)}$

$= \mathbf{0.199\,H}$ or $\mathbf{199\,mH}$

(b) At resonance $\omega_r L = 1/(\omega_r C)$

Hence **capacitance,** $\boldsymbol{C} = \dfrac{1}{\omega_r^2 L}$

$= \dfrac{1}{(2\pi 1200)^2(0.199)}$

$= \mathbf{0.088\,\mu F}$ or $\mathbf{88\,nF}$

(c) Q-factor at resonance is also given by $Q_r = f_r/(f_2 - f_1)$, from which,

bandwidth, $\boldsymbol{(f_2 - f_1)} = \dfrac{f_r}{Q_r} = \dfrac{1200}{30} = \mathbf{40\,Hz}$

(d) From equation (11), resonant frequency, $f_r = \sqrt{(f_1 f_2)}$, i.e. $1200 = \sqrt{(f_1 f_2)}$
from which, $f_1 f_2 = (1200)^2 = 1.44 \times 10^6$ (12)

From part (c), $f_2 - f_1 = 40$ (13)
From equation (12), $f_1 = (1.44 \times 10^6)/f_2$
Substituting in equation 13 gives:

$$f_2 - \frac{1.44 \times 10^2}{f_2} = 40$$

Multiplying throughout by f_2 gives:

$f_2^2 - 1.44 \times 10^6 = 40 f_2$

i.e. $f_2^2 - 40 f_2 - 1.44 \times 10^6 = 0$

This is a quadratic equation in f_2. Using the quadratic formula gives:

$$f_2 = \frac{40 \pm \sqrt{[(40)^2 - 4(1.44 \times 10^6)]}}{2}$$

$$= \frac{40 \pm 2400}{2}$$

$$= \frac{40 + 2400}{2} \text{ (since } f_2 \text{ cannot be negative)}$$

Hence **the upper half-power frequency,**

$\boldsymbol{f_2} = \mathbf{1220\,Hz}$

From equation (12), **the lower half-power frequency,**

$f_1 = f_2 - 40 = 1220 - 40 = \mathbf{1180\,Hz}$

Note that the upper and lower half-power frequency values are symmetrically placed about the resonance frequency. This is usually the case when the Q-factor has a high value (say, >10).

(e) At the half-power frequencies, current $I = 0.707\,I_r$

Hence impedance,

$$Z = \frac{V}{I} = \frac{V}{0.707\,I_r} = 1.414\left(\frac{V}{I_r}\right) = \sqrt{2}Zr$$

$$= \sqrt{2}R$$

Thus **impedance at the half-power frequencies,**

$Z = \sqrt{2}R = \sqrt{2}(50) = \mathbf{70.71\,\Omega}$

Problem 9. A series $R-L-C$ circuit is connected to a 0.2 V supply and the current is at its maximum value of 4 mA when the supply frequency is adjusted to 3 kHz. The Q-factor of the circuit under these conditions is 100. Determine the value of (a) the circuit resistance, (b) the circuit inductance, (c) the circuit capacitance and (d) the voltage across the capacitor.

Since the current is at its maximum, the circuit is at resonance and the resonant frequency is 3 kHz.

(a) At resonance, impedance, $Z = R = \dfrac{V}{I}$

$$= \frac{0.2}{4 \times 10^{-3}} = 50\,\Omega$$

Hence **the circuit resistance is 50 Ω**

(b) Q-factor at resonance is given by $Q_r = \omega_r L/R$ from which,

inductance, $\boldsymbol{L} = \dfrac{Q_r R}{\omega_r} = \dfrac{(100)(50)}{2\pi 3000}$

$= \mathbf{0.265\,H}$ or $\mathbf{265\,mH}$

(c) Q-factor at resonance is also given by $Q_r = 1/(\omega_r CR)$, from which,

capacitance, $\boldsymbol{C} = \dfrac{1}{\omega_r R Q_r} = \dfrac{1}{(2\pi 3000)(50)(100)}$

$= \mathbf{0.0106\,\mu F}$ or $\mathbf{10.6\,nF}$

(d) Q-factor at resonance in a series circuit represents the voltage magnification, i.e. $Q_r = V_C/V$, from which, $V_C = Q_r V = (100)(0.2) = 20\,\text{V}$

Hence **the voltage across the capacitor is 20 V**

(Alternatively, $V_C = IX_C = \dfrac{I}{\omega_r C}$

$$= \frac{4\times 10^{-3}}{(2\pi 3000)(0.0106\times 10^{-6})}$$

$= \mathbf{20\,V}$)

Problem 10. A coil of inductance 351.8 mH and resistance 8.84 Ω is connected in series with a 20 μF capacitor. Determine (a) the resonant frequency, (b) the Q-factor at resonance, (c) the bandwidth and (d) the lower and upper −3 dB frequencies.

(a) Resonant frequency,

$$f_r = \frac{1}{2\pi\sqrt{(LC)}} = \frac{1}{2\pi\sqrt{[(0.3518)(20\times 10^{-6})]}}$$

$$= \mathbf{60.0\,Hz}$$

(b) Q-factor at resonance,

$$Q_r = \frac{1}{R}\sqrt{\frac{L}{C}} = \frac{1}{8.84}\sqrt{\left(\frac{0.3518}{20\times 10^{-6}}\right)} = \mathbf{15}$$

[Alternatively, $Q_r = \dfrac{\omega_r L}{R} = \dfrac{2\pi(60.0)(0.3518)}{8.84}$

$= \mathbf{15}$

or $Q_r = \dfrac{1}{\omega_r CR}$

$$= \frac{1}{(2\pi 60.0)(20\times 10^{-6})(8.84)}$$

$= \mathbf{15}$]

(c) Bandwidth, $(f_2 - f_1) = \dfrac{f_r}{Q_r} = \dfrac{60.0}{15} = \mathbf{4\,Hz}$

(d) With a Q-factor of 15 it may be assumed that the lower and upper −3 dB frequencies, f_1 and f_2, are symmetrically placed about the resonant frequency of 60.0 Hz. Hence **the lower −3 dB frequency, $f_1 = 58$ Hz, and the upper −3 dB frequency, $f_2 = 62$ Hz.**

[This may be checked by using $(f_2 - f_1) = 4$ and $f_r = \sqrt{(f_1 f_2)}$]

31.7 Small deviations from the resonant frequency

Let ω_1 be a frequency below the resonant frequency ω_r in an L–R–C series circuit, and ω_2 be a frequency above ω_r by the same amount as ω_1 is below, i.e. $\omega_r - \omega_1 = \omega_2 - \omega_r$

Let the fractional deviation from the resonant frequency be δ where

$$\delta = \frac{\omega_r - \omega_1}{\omega_r} = \frac{\omega_2 - \omega_r}{\omega_r}$$

Hence $\omega_r\delta = \omega_r - \omega_1$ and $\omega_r\delta = \omega_2 - \omega_r$

from which, $\omega_1 = \omega_r - \omega_r\delta$ and $\omega_2 = \omega_r + \omega_r\delta$

i.e. $\omega_1 = \omega_r(1-\delta)$ (14)

and $\omega_2 = \omega_r(1+\delta)$ (15)

At resonance, $I_r = \dfrac{V}{R}$ and at other frequencies, $I = \dfrac{V}{Z}$ where Z is the circuit impedance.

Hence $\dfrac{I}{I_r} = \dfrac{V/Z}{V/R} = \dfrac{R}{Z} = \dfrac{R}{R + j\left(\omega L - \dfrac{1}{\omega C}\right)}$

From equation (15), **at frequency ω_2**

$$\frac{I}{I_r} = \frac{R}{R + j\left[\omega_r(1+\delta)L - \dfrac{1}{\omega_r(1+\delta)C}\right]}$$

$$= \frac{R/R}{\dfrac{R}{R} + j\left[\dfrac{\omega_r L}{R}(1+\delta) - \dfrac{1}{\omega_r RC(1+\delta)}\right]}$$

At resonance, $\dfrac{1}{\omega_r C} = \omega_r L$ hence

$$\frac{I}{I_r} = \frac{1}{1 + j\left[\dfrac{\omega_r L}{R}(1+\delta) - \dfrac{\omega_r L}{R(1+\delta)}\right]}$$

$$= \frac{1}{1 + j\dfrac{\omega_r L}{R}\left[(1+\delta) - \dfrac{1}{(1+\delta)}\right]}$$

Since $\dfrac{\omega_r L}{R} = Q$ then

$$\frac{I}{I_r} = \frac{1}{1 + jQ\left[\dfrac{(1+\delta)^2 - 1}{(1+\delta)}\right]}$$

$$= \frac{1}{1 + jQ\left[\frac{1 + 2\delta + \delta^2 - 1}{(1+\delta)}\right]}$$

$$= \frac{1}{1 + jQ\left[\frac{2\delta + \delta^2}{1+\delta}\right]} = \frac{1}{1 + j\delta Q\left[\frac{2+\delta}{1+\delta}\right]}$$

If the deviation from the resonant frequency δ is very small such that $\delta \ll 1$

then $$\frac{I}{I_r} \approx \frac{1}{1 + j\delta Q\left[\frac{2}{1}\right]} = \mathbf{\frac{1}{1 + j2\delta Q}} \quad (16)$$

and $$\frac{I}{I_r} = \frac{V/Z}{V/Z_r} = \frac{Z_r}{Z} = \frac{1}{1 + j2\delta Q}$$

from which, $$\mathbf{\frac{Z}{Z_r} = 1 + j2\delta Q} \quad (17)$$

It may be shown that **at frequency ω_1**, $\frac{I}{I_r} = \mathbf{\frac{1}{1 - j2\delta Q}}$

and

$$\mathbf{\frac{Z}{Z_r} = 1 - j2\delta Q}$$

Problem 11. In an L–R–C series network, the inductance, $L = 8$ mH, the capacitance, $C = 0.3\,\mu$F and the resistance, $R = 15\,\Omega$. Determine the current flowing in the circuit when the input voltages $7.5\angle 0°$ V and the frequency is (a) the resonant frequency, (b) a frequency 3% above the resonant frequency. Find also (c) the impedance of the circuit when the frequency is 3% above the resonant frequency.

(a) At resonance, $Z_r = R = 15\,\Omega$

Current at resonance, $$\boldsymbol{I_r} = \frac{V}{Z_r} = \frac{7.5\angle 0°}{15\angle 0°} = \mathbf{0.5\angle 0°\ A}$$

(b) If the frequency is 3% above the resonant frequency, then $\delta = 0.03$

From equation (16), $$\frac{I}{I_r} = \frac{1}{1 + j2\delta Q}$$

$$Q = \frac{1}{R}\sqrt{\frac{L}{C}} = \frac{1}{15}\sqrt{\left(\frac{8 \times 10^{-3}}{0.3 \times 10^{-6}}\right)} = 10.89$$

Hence $$\frac{1}{0.5\angle 0°} = \frac{1}{1 + j2(0.03)(10.89)}$$

$$= \frac{1}{1 + j0.6534}$$

$$= \frac{1}{1.1945\angle 33.16°}$$

and $$\boldsymbol{I} = \frac{0.5\angle 0°}{1.1945\angle 33.16°} = \mathbf{0.4186\angle{-33.16°}\ A}$$

(c) From equation (17), $\frac{Z}{Z_r} = 1 + j2\delta Q$

hence $$\boldsymbol{Z} = Z_r(1 + j2\delta Q) = R(1 + j2\delta Q)$$
$$= 15(1 + j2(0.03)(10.89))$$
$$= 15(1 + j0.6534)$$
$$= 15(1.1945\angle 33.16°)$$
$$= \mathbf{17.92\angle 33.16°\ \Omega}$$

Alternatively, $$\boldsymbol{Z} = \frac{V}{I} = \frac{7.5\angle 0°}{0.4186\angle{-33.16°}} = \mathbf{17.92\angle 33.16°\ \Omega}$$

Now try the following Practice Exercise

Practice Exercise 132 Bandwidth (Answers on page 827)

1. A coil of resistance 10.05 Ω and inductance 400 mH is connected in series with a 0.396 μF capacitor. Determine (a) the resonant frequency, (b) the resonant Q-factor, (c) the bandwidth and (d) the lower and upper half-power frequencies.

2. An R–L–C series circuit has a resonant frequency of 2 kHz and a Q-factor at resonance of 40. If the impedance of the circuit at resonance is 30 Ω, determine the values of (a) the inductance and (b) the capacitance. Find also (c) the bandwidth, (d) the lower and upper −3 dB frequencies and (e) the impedance at the −3 dB frequencies.

3. A filter in the form of a series L–C–R circuit is designed to operate at a resonant frequency of 20 kHz and incorporates a 20 mH inductor and 30 Ω resistance. Determine the bandwidth of the filter.

4. A circuit consists of a coil of inductance 200 μH and resistance 8.0 Ω in series with a lossless 500 pF capacitor. Determine (a) the resonant Q-factor, and (b) the bandwidth of the circuit.

5. A coil of inductance 200 μH and resistance 50.27 Ω and a variable capacitor are connected in series to a 5 mV supply of frequency 2 MHz. Determine (a) the value of capacitance to tune the circuit to resonance, (b) the supply current at resonance, (c) the p.d. across the capacitor at resonance, (d) the bandwidth and (e) the half-power frequencies.

6. A supply voltage of 3 V is applied to a series *R–L–C* circuit whose resistance is 12 Ω, inductance is 7.5 mH and capacitance is 0.5 μF. Determine (a) the current flowing at resonance, (b) the current flowing at a frequency 2.5% below the resonant frequency and (c) the impedance of the circuit when the frequency is 1% lower than the resonant frequency.

For fully worked solutions to each of the problems in Practice Exercises 130 to 132 in this chapter, go to the website:
www.routledge.com/cw/bird

COMPANION @ WEBSITE

Chapter 32

Parallel resonance and Q-factor

Why it is important to understand: **Parallel resonance and Q-factor**

There are three basic conditions in a parallel circuit; dependent on frequency and component values, the circuit will be operating below, above or at resonance. This chapter investigates the conditions for parallel resonance, and explains dynamic impedance, Q-factor and bandwidth. One use for resonance is to establish a condition of stable frequency in circuits designed to produce a.c. signals. Usually, a parallel circuit is used for this purpose, with the capacitor and inductor directly connected together, exchanging energy between each other. Just as a pendulum can be used to stabilise the frequency of a clock mechanism's oscillations, so can a parallel circuit be used to stabilise the electrical frequency of an a.c. oscillator circuit. Another use for resonance is in applications where the effects of greatly increased or decreased impedance at a particular frequency are desired. A resonant circuit can be used to block a frequency or range of frequencies, thus acting as a sort of frequency filter to strain certain frequencies out of a mix of others; these particular circuits are called filters.

At the end of this chapter you should be able to:

- state the condition for resonance in an a.c. parallel network
- calculate the resonant frequency in a.c. parallel networks
- calculate dynamic resistance $R_D = \frac{L}{CR}$ in an a.c. parallel network
- calculate Q-factor and bandwidth in an a.c. parallel network
- determine the overall Q-factor for capacitors connected in parallel
- determine the impedance when the frequency deviates from the resonant frequency

32.1 Introduction

A parallel network containing resistance R, pure inductance L and pure capacitance C connected in parallel is shown in Figure 32.1. Since the inductance and capacitance are considered as pure components, this circuit is something of an 'ideal' circuit. However, it may be used to highlight some important points regarding resonance which are applicable to any parallel circuit.

From Figure 32.1,

the admittance of the resistive branch, $G = \frac{1}{R}$

the admittance of the inductive branch, $B_L = \frac{1}{jX_L} = \frac{-j}{\omega L}$

Electrical Circuit Theory and Technology. 978-1-138-67349-6, © 2017 John Bird. Published by Taylor & Francis. All rights reserved.

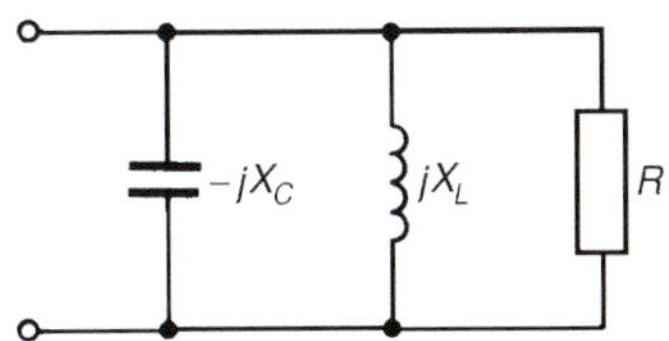

Figure 32.1 Parallel R–L–C circuit

the admittance of the capacitive branch,

$$B_C = \frac{1}{-jX_C} = \frac{j}{1/\omega C} = j\omega C$$

Total circuit admittance, $Y = G + j(B_C - B_L)$

i.e. $$Y = \frac{1}{R} + j\left(\omega C - \frac{1}{\omega L}\right)$$

The circuit is at resonance when the imaginary part is zero, i.e. when $\omega C - (1/\omega L) = 0$. Hence at resonance $\omega_r C = 1/(\omega_r L)$ and $\omega_r^2 = 1/(LC)$, from which $\omega_r = 1/\sqrt{(LC)}$ and the resonant frequency

$$\mathbf{f_r = \frac{1}{2\pi\sqrt{(LC)}}\ hertz}$$

the same expression as for a series R–L–C circuit.

Figure 32.2 shows typical graphs of B_C, B_L, G and Y against frequency f for the circuit shown in Figure 32.1. At resonance, $B_C = B_L$ and admittance $Y = G = 1/R$. This represents the condition of **minimum admittance** for the circuit and thus **maximum impedance**.

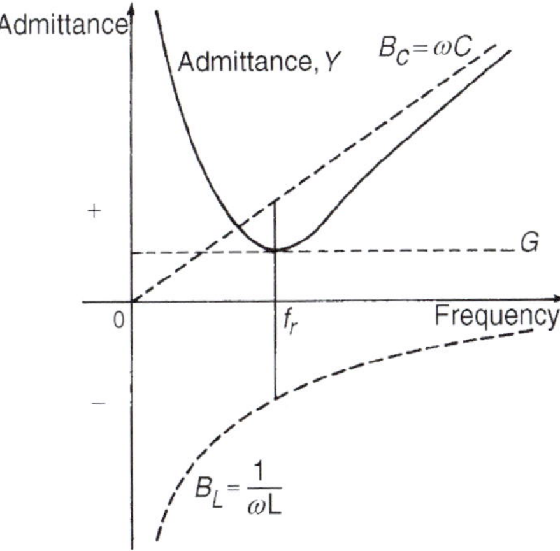

Figure 32.2 $|Y|$ plotted against frequency

Since current $I = V/Z = VY$, the **current** is at a **minimum** value at resonance in a parallel network.

From the ideal circuit of Figure 32.1 we have therefore established the following facts which apply to any parallel circuit. At resonance:

(i) admittance Y is a minimum
(ii) impedance Z is a maximum
(iii) current I is a minimum
(iv) an expression for the resonant frequency f_r may be obtained by making the 'imaginary' part of the complex expression for admittance equal to zero.

32.2 The *LR–C* parallel network

A more practical network, containing a coil of inductance L and resistance R in parallel with a pure capacitance C, is shown in Figure 32.3.

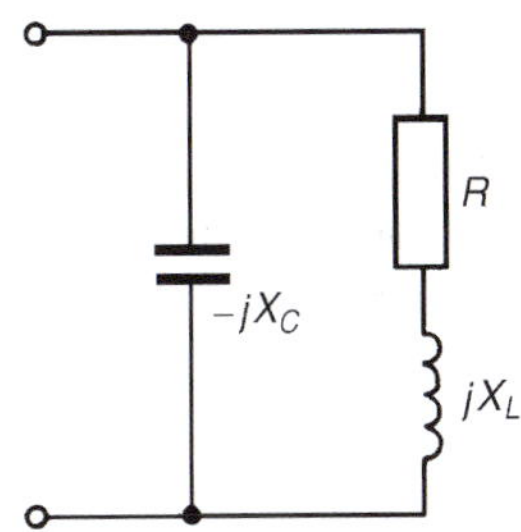

Figure 32.3

Admittance of coil, $Y_{COIL} = \dfrac{1}{R + jX_L} = \dfrac{R - jX_L}{R^2 + X_L^2}$

$$= \frac{R}{R^2 + \omega^2 L^2} - \frac{j\omega L}{R^2 + \omega^2 L^2}$$

Admittance of capacitor, $Y_C = \dfrac{1}{-jX_C} = \dfrac{j}{X_c} = j\omega C$

Total circuit admittance, $Y = Y_{COIL} + Y_C$

$$= \frac{R}{R^2 + \omega^2 L^2} - \frac{j\omega L}{R^2 + \omega^2 L^2} + j\omega C \quad (1)$$

At resonance, the total circuit admittance Y is real $(Y = R/(R^2 + \omega^2 L^2))$, i.e. the imaginary part is zero. Hence, at resonance:

$$\frac{-\omega_r L}{R^2 + \omega_r^2 L^2} + \omega_r C = 0$$

Therefore $\dfrac{\omega_r L}{R^2 + \omega_r^2 L^2} = \omega_r C$ and $\dfrac{L}{C} = R^2 + \omega_r^2 L^2$

Part 4

Thus $\omega_r^2 L^2 = \frac{L}{C} - R^2$

and $$\omega_r^2 = \frac{L}{CL^2} - \frac{R^2}{L^2} = \frac{1}{LC} - \frac{R^2}{L^2} \qquad (2)$$

Hence $\omega_r = \sqrt{\left(\frac{1}{LC} - \frac{R^2}{L^2}\right)}$

and **resonant frequency,** $\boldsymbol{f_r = \frac{1}{2\pi}\sqrt{\left(\frac{1}{LC} - \frac{R^2}{L^2}\right)}}$ (3)

Note that when $R^2/L^2 \ll 1/(LC)$ then $f_r = 1/2\pi\sqrt{(LC)}$, as for the series R–L–C circuit. Equation (3) is the same as obtained in Chapter 18, page 285; however, the above method may be applied to any parallel network, as demonstrated in Section 32.4 below.

32.3 Dynamic resistance

Since the current at resonance is in phase with the voltage, the impedance of the network acts as a resistance. This resistance is known as the **dynamic resistance,** $\boldsymbol{R_D}$. Impedance at resonance, $R_D = V/I_r$, where I_r is the current at resonance.

$$I_r = VY_r = V\left(\frac{R}{R^2 + \omega_r^2 L^2}\right)$$

from equation (1) with the j terms equal to zero.

Hence $$R_D = \frac{V}{I_r} = \frac{V}{VR/(R^2 + \omega_r^2 L^2)} = \frac{R^2 + \omega_r^2 L^2}{R}$$

$$= \frac{R^2 + L^2(1/LC) - (R^2/L^2)}{R}$$

from equation (2)

$$= \frac{R^2 + (L/C) - R^2}{R} = \frac{L/C}{R} = \frac{L}{CR}$$

Hence **dynamic resistance,** $\boldsymbol{R_D = \frac{L}{CR}}$ (4)

32.4 The *LR–CR* parallel network

A more general network comprising a coil of inductance L and resistance R_L in parallel with a capacitance C and resistance R_C in series is shown in Figure 32.4. Admittance of inductive branch,

$$Y_L = \frac{1}{R_L + jX_L} = \frac{R_L - jX_L}{R_L^2 + X_L^2} = \frac{R_L}{R_L^2 + X_L^2} - \frac{jX_L}{R_L^2 + X_L^2}$$

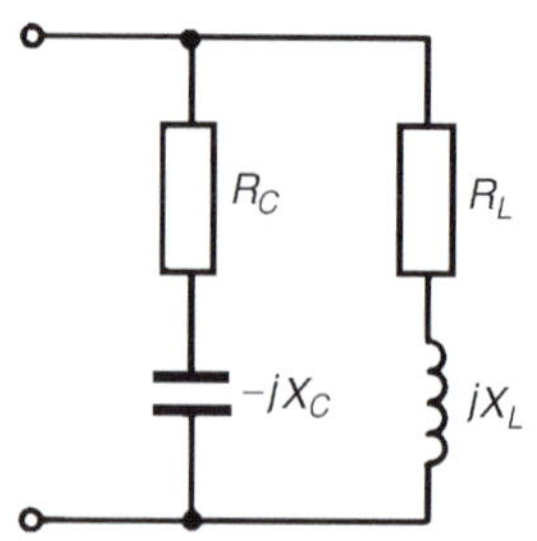

Figure 32.4

Admittance of capacitive branch,

$$Y_C = \frac{1}{R_C - jX_C} = \frac{R_C + jX_C}{R_C^2 + X_C^2} = \frac{R_C}{R_C^2 + X_C^2} + \frac{jX_C}{R_C^2 + X_C^2}$$

Total network admittance,

$$Y = Y_L + Y_C = \frac{R_L}{R_L^2 + X_L^2} - \frac{jX_L}{R_L^2 + X_L^2} + \frac{R_C}{R_C^2 + X_C^2} + \frac{jX_C}{R_C^2 + X_C^2}$$

At resonance the admittance is a minimum, i.e. when the imaginary part of Y is zero. Hence, at resonance,

$$\frac{-X_L}{R_L^2 + X_L^2} + \frac{X_C}{R_C^2 + X_C^2} = 0$$

i.e. $$\frac{\omega_r L}{R_L^2 + \omega^2 L^2} = \frac{1/(\omega_r C)}{R_C^2 + 1/\omega_r^2 C^2} \qquad (5)$$

Rearranging gives:

$$\omega_r L\left(R_C^2 + \frac{1}{\omega_r^2 C^2}\right) = \frac{1}{\omega_r C}(R_L^2 + \omega_r^2 L^2)$$

$$\omega_r L R_C^2 + \frac{L}{\omega_r C^2} = \frac{R_L^2}{\omega_r C} + \frac{\omega_r L^2}{C}$$

Multiplying throughout by $\omega_r C^2$ gives:

$$\omega_r^2 C^2 L R_C^2 + L = R_L^2 C + \omega_r^2 L^2 C$$
$$\omega_r^2(C^2 L R_C^2 - L^2 C) = R_L^2 C - L$$
$$\omega_r^2 CL(CR_C^2 - L) = R_L^2 C - L$$

Hence $$\omega_r^2 = \frac{(CR_L^2 - L)}{LC(CR_C^2 - L)}$$

i.e. $$\omega_r = \frac{1}{\sqrt{(LC)}}\sqrt{\left(\frac{R_L^2 - (L/C)}{R_C^2 - (L/C)}\right)}$$

Hence

$$\textbf{resonant frequency}, f_r = \frac{1}{2\pi\sqrt{(LC)}}\sqrt{\left(\frac{R_L^2-(L/C)}{R_C^2-(L/C)}\right)} \quad (6)$$

It is clear from equation (5) that parallel resonance may be achieved in such a circuit in several ways – by varying either the frequency f, the inductance L, the capacitance C, the resistance R_L or the resistance R_C

32.5 Q-factor in a parallel network

The Q-factor in the series R–L–C circuit is a measure of the voltage magnification. In a parallel circuit, currents higher than the supply current can circulate within the parallel branches of a parallel resonant network, the current leaving the capacitor and establishing the magnetic field of the inductance, this then collapsing and recharging the capacitor, and so on. The Q-factor of a parallel resonant circuit is the ratio of the current circulating in the parallel branches of the circuit to the supply current, i.e. in a parallel circuit, Q-factor is a measure of the **current magnification**.

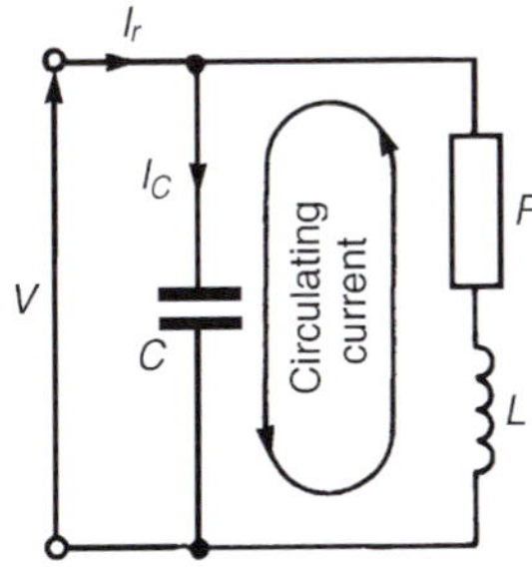

Figure 32.5

Circulating currents may be several hundreds of times greater than the supply current at resonance. For the parallel network of Figure 32.5, the Q-factor at resonance is given by:

$$Q_r = \frac{\textbf{circulating current}}{\textbf{current at resonance}} = \frac{\textbf{capacitor current}}{\textbf{current at resonance}} = \frac{I_C}{I_r}$$

Current in capacitor, $I_C = V/X_C = V\omega_r C$

Current at resonance, $I_r = \dfrac{V}{R_D} = \dfrac{V}{L/CR} = \dfrac{VCR}{L}$

Hence $Q_r = \dfrac{I_C}{I_r} = \dfrac{V\omega_r C}{VCR/L}$ i.e. $\boldsymbol{Q_r = \dfrac{\omega_r L}{R}}$

the same expression as for series resonance.

The difference between the resonant frequency of a series circuit and that of a parallel circuit can be quite small. The resonant frequency of a coil in parallel with a capacitor is shown in equation (3); however, around the closed loop comprising the coil and capacitor the energy would naturally resonate at a frequency given by that for a series R–L–C circuit, as shown in Chapter 31. This latter frequency is termed the **natural frequency, f_n**, and the frequency of resonance seen at the terminals of Figure 32.5 is often called the **forced resonant frequency, f_r**. (For a series circuit, the forced and natural frequencies coincide.)

From the coil–capacitor loop of Figure 32.5,

$$f_n = \frac{1}{2\pi\sqrt{(LC)}}$$

and the forced resonant frequency,

$$f_r = \frac{1}{2\pi}\sqrt{\left(\frac{1}{LC}-\frac{R^2}{L^2}\right)}$$

Thus $$\frac{f_r}{f_n} = \frac{\frac{1}{2\pi}\sqrt{\left(\frac{1}{LC}-\frac{R^2}{L^2}\right)}}{\frac{1}{2\pi\sqrt{(LC)}}} = \frac{\sqrt{\left(\frac{1}{LC}-\frac{R^2}{L^2}\right)}}{\frac{1}{\sqrt{(LC)}}}$$

$$= \sqrt{\left(\frac{1}{LC}-\frac{R^2}{L^2}\right)}\sqrt{(LC)} = \sqrt{\left(\frac{LC}{LC}-\frac{LCR^2}{L^2}\right)}$$

$$= \sqrt{\left(1-\frac{R^2C}{L}\right)}$$

From Chapter 31, $Q = \dfrac{1}{R}\sqrt{\left(\dfrac{L}{C}\right)}$ from which

$$Q^2 = \frac{1}{R^2}\left(\frac{L}{C}\right) \text{ or } \frac{R^2C}{L} = \frac{1}{Q^2}$$

Hence $$\frac{f_r}{f_n} = \sqrt{\left(1-\frac{R^2C}{L}\right)} = \sqrt{\left(1-\frac{1}{Q^2}\right)}$$

i.e. $$\boldsymbol{f_r = f_n\sqrt{\left(1-\frac{1}{Q^2}\right)}}$$

Thus it is seen that even with small values of Q the difference between f_r and f_n tends to be very small. A high value of Q makes the parallel resonant frequency tend to the same value as that of the series resonant frequency.

The expressions already obtained in Chapter 31 for bandwidth and resonant frequency also apply to parallel circuits,

i.e. $$Q_r = f_r/(f_2 - f_1) \qquad (7)$$

and $$f_r = \sqrt{(f_1 f_2)} \qquad (8)$$

The overall Q-factor Q_T of two parallel components having different Q-factors is given by:

$$Q_T = \frac{Q_L Q_C}{Q_L + Q_C} \qquad (9)$$

as for the series circuit.

By similar reasoning to that of the series R–L–C circuit, it may be shown that at the half-power frequencies the admittance is $\sqrt{2}$ times its minimum value at resonance and, since $Z = 1/Y$, the value of impedance at the half-power frequencies is $1/\sqrt{2}$ or 0.707 times its maximum value at resonance.

By similar analysis to that given in Chapter 31, it may be shown that for a parallel network:

$$\frac{Y}{Y_r} = \frac{R_D}{Z} = 1 + j2\delta Q \qquad (10)$$

where Y is the circuit admittance, Y_r is the admittance at resonance, Z is the network impedance and R_D is the dynamic resistance (i.e. the impedance at resonance) and δ is the fractional deviation from the resonant frequency.

Problem 1. A coil of inductance 5 mH and resistance 10 Ω is connected in parallel with a 250 nF capacitor across a 50 V variable-frequency supply. Determine (a) the resonant frequency, (b) the dynamic resistance, (c) the current at resonance and (d) the circuit Q-factor at resonance.

(a) Resonance frequency

$$f_r = \frac{1}{2\pi}\sqrt{\left(\frac{1}{LC} - \frac{R^2}{L^2}\right)} \quad \text{from equation (3),}$$

$$= \frac{1}{2\pi}\sqrt{\left(\frac{1}{5\times 10^{-3}\times 250\times 10^{-9}} - \frac{10^2}{(5\times 10^{-3})^2}\right)}$$

$$= \frac{1}{2\pi}\sqrt{(800\times 10^6 - 4\times 10^6)} = \frac{1}{2\pi}\sqrt{(796\times 10^6)}$$

$$= \mathbf{4490\ Hz}$$

(b) From equation (4), dynamic resistance,

$$R_D = \frac{L}{CR} = \frac{5\times 10^{-3}}{(250\times 10^{-9})(10)} = \mathbf{2000\ \Omega}$$

(c) Current at resonance, $I_r = \frac{V}{R_D} = \frac{50}{2000} = \mathbf{25\ mA}$

(d) Q-factor at resonance, $Q_r = \frac{\omega_r L}{R}$

$$= \frac{(2\pi\, 4490)(5\times 10^{-3})}{10}$$

$$= \mathbf{14.1}$$

Problem 2. In the parallel network of Figure 32.6, inductance, $L = 100$ mH and capacitance, $C = 40\ \mu$F. Determine the resonant frequency for the network if (a) $R_L = 0$ and (b) $R_L = 30\ \Omega$

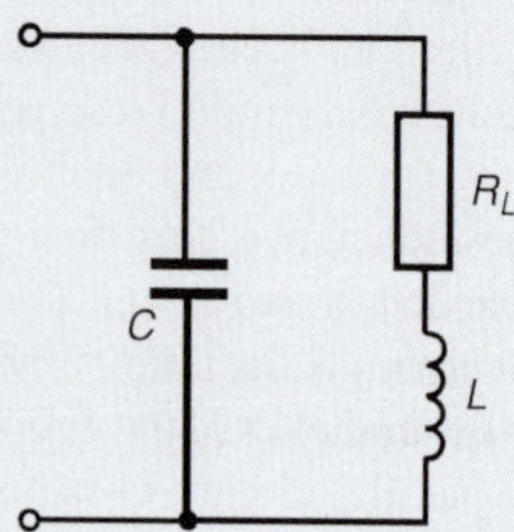

Figure 32.6

Total circuit admittance,

$$Y = \frac{1}{R_L + jX_L} + \frac{1}{-jX_C} = \frac{R_L - jX_L}{R_L^2 + X_L^2} + \frac{j}{X_C}$$

$$= \frac{R_L}{R_L^2 + X_L^2} - \frac{jX_L}{R_L^2 + X_L^2} + \frac{j}{X_C}$$

The network is at resonance when the admittance is at a minimum value, i.e. when the imaginary part is zero. Hence, at resonance,

$$\frac{-X_L}{R_L^2 + X_L^2} + \frac{1}{X_C} = 0 \quad \text{or} \quad \omega_r C = \frac{\omega_r L}{R_L^2 + \omega_r^2 L^2} \qquad (11)$$

(a) When $R_L = 0$, $\omega_r C = \frac{\omega_r L}{\omega_r^2 L^2}$

from which, $\omega_r^2 = \frac{1}{LC}$ and $\omega_r = \frac{1}{\sqrt{(LC)}}$

Hence resonant frequency,

$$f_r = \frac{1}{2\pi\sqrt{(LC)}} = \frac{1}{2\pi\sqrt{(100\times 10^{-3}\times 40\times 10^{-6})}}$$

$$= \mathbf{79.6\ Hz}$$

(b) When $R_L = 30\,\Omega$, $\omega_r C = \dfrac{\omega_r L}{30^2 + \omega_r^2 L^2}$ from equation (11) above

from which, $$30^2 + \omega_r^2 L^2 = \frac{L}{C}$$

i.e. $$\omega_r^2 (100 \times 10^{-3})^2 = \frac{100 \times 10^{-3}}{40 \times 10^{-6}} - 900$$

i.e. $$\omega_r^2 (0.01) = 2500 - 900 = 1600$$

Thus, $\omega_r^2 = 1600/0.01 = 160\,000$ and
$$\omega_r = \sqrt{160\,000} = 400\,\text{rad/s}$$

Hence resonant frequency, $\boldsymbol{f_r} = \dfrac{400}{2\pi} = \mathbf{63.7\,Hz}$

[Alternatively, from equation (3),

$$f_r = \frac{1}{2\pi}\sqrt{\left(\frac{1}{LC} - \frac{R^2}{L^2}\right)}$$
$$= \frac{1}{2\pi}\sqrt{\left(\frac{1}{(100 \times 10^{-3})(40 \times 10^{-6})} - \frac{30^2}{(100 \times 10^{-3})^2}\right)}$$
$$= \frac{1}{2\pi}\sqrt{(250\,000 - 90\,000)} = \frac{1}{2\pi}\sqrt{160\,000}$$
$$= \frac{1}{2\pi}(400) = \mathbf{63.7\,Hz}]$$

Hence, as the resistance of a coil increases, the resonant frequency decreases in the circuit of Figure 32.6.

Problem 3. A coil of inductance 120 mH and resistance 150 Ω is connected in parallel with a variable capacitor across a 20 V, 4 kHz supply. Determine for the condition when the supply current is a minimum, (a) the capacitance of the capacitor, (b) the dynamic resistance, (c) the supply current, (d) the Q-factor, (e) the bandwidth, (f) the upper and lower −3 dB frequencies and (g) the value of the circuit impedance at the −3 dB frequencies.

(a) The supply current is a minimum when the parallel network is at resonance.

Resonant frequency, $\boldsymbol{f_r} = \dfrac{1}{2\pi}\sqrt{\left(\dfrac{1}{LC} - \dfrac{R^2}{L^2}\right)}$ from equation (3),

from which, $(2\pi f_r)^2 = \dfrac{1}{LC} - \dfrac{R^2}{L^2}$

Hence $$\frac{1}{LC} = (2\pi f_r)^2 + \frac{R^2}{L^2}$$ and

capacitance

$$C = \frac{1}{L[(2\pi f_r)^2 + (R^2/L^2)]}$$
$$= \frac{1}{120 \times 10^{-3}[(2\pi 4000)^2 + (150^2/(120 \times 10^{-3})^2)]}$$
$$= \frac{1}{0.12(631.65 \times 10^6 + 1.5625 \times 10^6)}$$
$$= \mathbf{0.01316\,\mu F} \text{ or } \mathbf{13.16\,nF}$$

(b) Dynamic resistance,

$$\mathbf{R_D} = \frac{L}{CR} = \frac{120 \times 10^{-3}}{(13.16 \times 10^{-9})(150)} = \mathbf{60.79\,k\Omega}$$

(c) Supply current at resonance,

$$\boldsymbol{I_r} = \frac{V}{R_D} = \frac{20}{60.79 \times 10^{-3}} = \mathbf{0.329\,mA} \text{ or } \mathbf{329\,\mu A}$$

(d) Q-factor at resonance,

$$Q_r = \frac{\omega_r L}{R} = \frac{(2\pi 4000)(120 \times 10^{-3})}{150} = \mathbf{20.11}$$

[Note that the expressions $Q_r = \dfrac{1}{\omega_r CR}$ or

$$Q_r = \frac{1}{R}\sqrt{\left(\frac{L}{C}\right)}$$

used for the R–L–C series circuit may also be used in parallel circuits when the resistance of the coil is much smaller than the inductive reactance of the coil.

In this case $R = 150\,\Omega$
and $X_L = 2\pi(4000)(120 \times 10^{-3}) = 3016\,\Omega$.
Hence, alternatively,

$$\boldsymbol{Q_r} = \frac{1}{\omega_r CR} = \frac{1}{(2\pi 4000)(13.16 \times 10^{-9})(150)} = \mathbf{20.16}$$

or $$\boldsymbol{Q_r} = \frac{1}{R}\sqrt{\left(\frac{L}{C}\right)} = \frac{1}{150}\sqrt{\left(\frac{120 \times 10^{-3}}{13.16 \times 10^{-9}}\right)} = \mathbf{20.13}]$$

(e) If the lower and upper −3 dB frequencies are f_1 and f_2, respectively, then the bandwidth is $(f_2 - f_1)$. Q-factor at resonance is given by

$Q_r = f_r/(f_2 - f_1)$, from which, bandwidth,

$$(f_2 - f_1) = \frac{f_r}{Q_r} = \frac{4000}{20.11} = \mathbf{199\,Hz}$$

(f) Resonant frequency, $f_r = \sqrt{(f_1\, f_2)}$, from which

$$f_1\, f_2 = f_r^2 = (4000)^2 = 16 \times 10^6 \qquad (12)$$

Also, from part (e), $f_2 - f_1 = 199 \qquad (13)$

From equation (12), $f_1 = \dfrac{16 \times 10^6}{f_2}$

Substituting in equation (13) gives:

$$f_2 - \frac{16 \times 10^6}{f_2} = 199$$

i.e. $f_2^2 - 16 \times 10^6 = 199\, f_2$ from which,

$$f_2^2 - 199\, f_2 - 16 \times 10^6 = 0$$

Solving this quadratic equation gives:

$$f_2 = \frac{199 \pm \sqrt{[(199)^2 - 4(-16 \times 10^6)]}}{2}$$

$$= \frac{199 \pm 8002.5}{2}$$

i.e. **the upper 3 dB frequency, f_2 = 4100 Hz** (neglecting the negative answer).

From equation (12),

the lower −3 dB frequency, $f_1 = \dfrac{10 \times 10^6}{f_2}$

$$= \frac{16 \times 10^6}{4100}$$

$$= \mathbf{3900\,Hz}$$

(Note that f_1 and f_2 are equally displaced about the resonant frequency, f_r, as they always will be when Q is greater than about 10 – just as for a series circuit.)

(g) The value of the circuit impedance, Z, at the −3 dB frequencies is given by

$$Z = \frac{1}{\sqrt{2}} Z_r$$

where Z_r is the impedance at resonance.

The impedance at resonance $Z_r = R_D$, the dynamic resistance.

Hence **impedance at the −3 dB frequencies**

$$= \frac{1}{\sqrt{2}}(60.79 \times 10^3)$$

$$= \mathbf{42.99\,k\Omega}$$

Figure 32.7 shows impedance plotted against frequency for the circuit in the region of the resonant frequency.

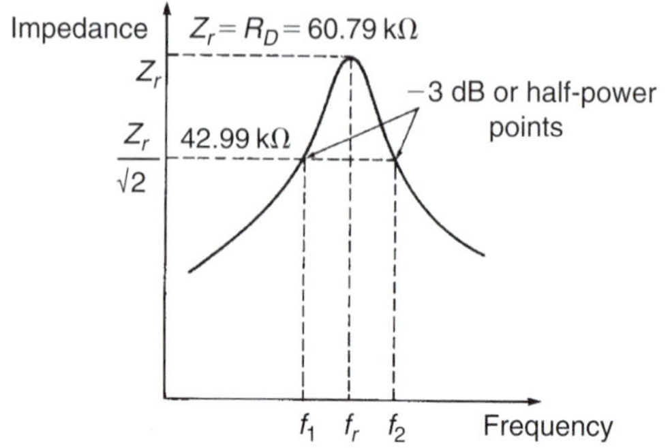

Figure 32.7

Now try the following Practice Exercise

Practice Exercise 133 Parallel resonance and Q-factor (Answers on page 827)

1. A coil of resistance 20 Ω and inductance 100 mH is connected in parallel with a 50 μF capacitor across a 30 V variable-frequency supply. Determine (a) the resonant frequency of the circuit, (b) the dynamic resistance, (c) the current at resonance, and (d) the circuit Q-factor at resonance.

2. A 25 V, 2.5 kHz supply is connected to a network comprising a variable capacitor in parallel with a coil of resistance 250 Ω and inductance 80 mH. Determine for the condition when the supply current is a minimum (a) the capacitance of the capacitor, (b) the dynamic resistance, (c) the supply current, (d) the Q-factor, (e) the bandwidth, (f) the upper and lower half-power frequencies and (g) the value of the circuit impedance at the −3 dB frequencies.

3. A 0.1 μF capacitor and a pure inductance of 0.02 H are connected in parallel across a 12 V variable-frequency supply. Determine (a) the resonant frequency of the circuit and (b) the current circulating in the capacitance and inductance at resonance.

4. A coil of resistance 300 Ω and inductance 100 mH and a 4000 pF capacitor are connected

(i) in series and (ii) in parallel. Find for each connection (a) the resonant frequency, (b) the Q-factor and (c) the impedance at resonance.

5. A network comprises a coil of resistance 100 Ω and inductance 0.8 H and a capacitor having capacitance 30 μF. Determine the resonant frequency of the network when the capacitor is connected (a) in series with, and (b) in parallel with the coil.

6. Determine the value of capacitor C shown in Figure 32.8 for which the resonant frequency of the network is 1 kHz.

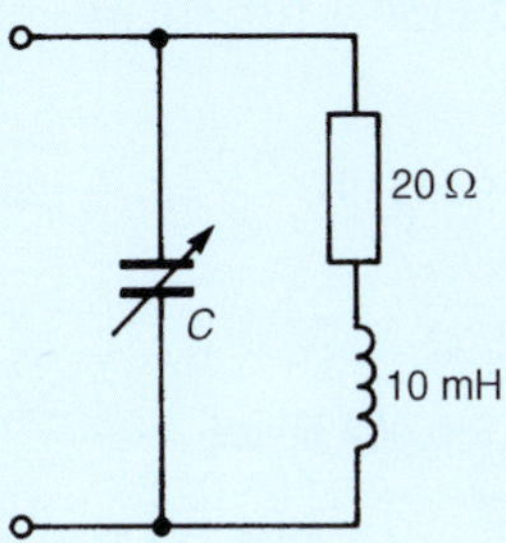

Figure 32.8

7. In the parallel network shown in Figure 32.9, inductance L is 40 mH and capacitance C is 5 μF. Determine the resonant frequency of the circuit if (a) $R_L = 0$ and (b) $R_L = 40$ Ω.

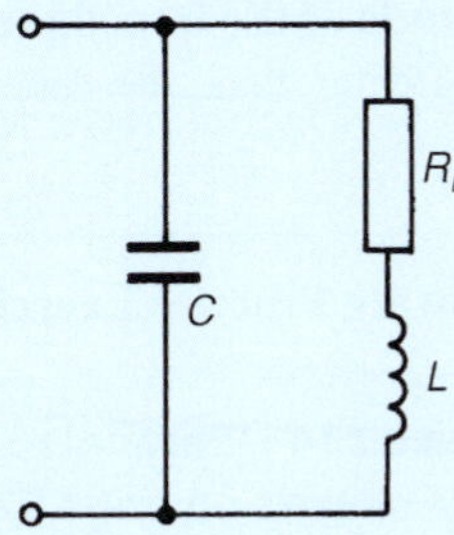

Figure 32.9

32.6 Further worked problems on parallel resonance and Q-factor

Problem 4. A two-branch parallel network is shown in Figure 32.10. Determine the resonant frequency of the network.

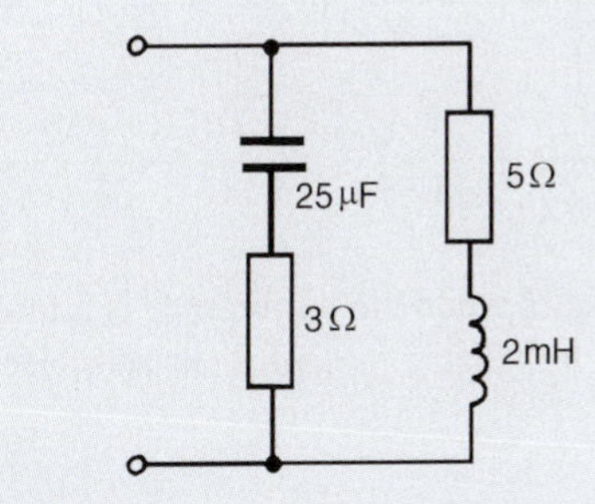

Figure 32.10

From equation (6),

$$\text{resonant frequency, } f_r = \frac{1}{2\pi\sqrt{(LC)}}\sqrt{\left(\frac{R_L^2 - (L/C)}{R_C^2 - (L/C)}\right)}$$

where $R_L = 5$ Ω, $R_C = 3$ Ω, $L = 2$ mH and $C = 25$ μF. Thus

$$f_r = \frac{1}{2\pi\sqrt{[(2\times10^{-3})(25\times10^{-6})]}}\sqrt{\left(\frac{5^2 - ((2\times10^{-3})/(25\times10^{-6}))}{3^2 - ((2\times10^{-3})/(25\times10^{-6}))}\right)}$$

$$= \frac{1}{2\pi\sqrt{(5\times10^{-8})}}\sqrt{\left(\frac{25-80}{9-80}\right)}$$

$$= \frac{10^4}{2\pi\sqrt{5}}\sqrt{\left(\frac{-55}{-71}\right)} = \mathbf{626.5\,Hz}$$

Problem 5. Determine for the parallel network shown in Figure 32.11 the values of inductance L for which the network is resonant at a frequency of 1 kHz.

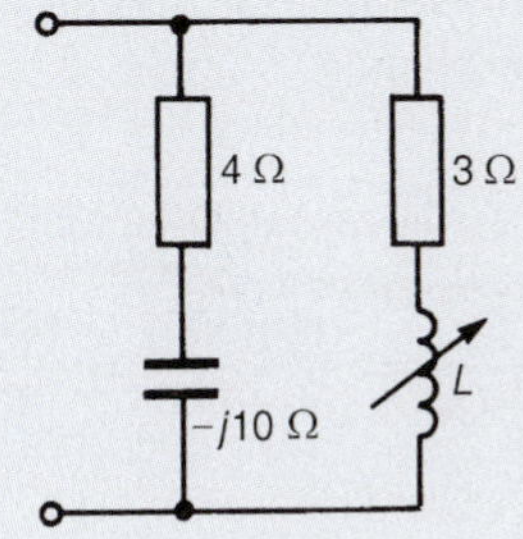

Figure 32.11

The total network admittance, Y, is given by

$$Y = \frac{1}{3 + jX_L} + \frac{1}{4 - j10} = \frac{3 - jX_L}{3^2 + X_L^2} + \frac{4 + j10}{4^2 + 10^2}$$

$$= \frac{3}{3^2+X_L^2} - \frac{jX_L}{3^2+X_L^2} + \frac{4}{116} + \frac{j10}{116}$$

$$= \left(\frac{3}{3^2+X_L^2} + \frac{4}{116}\right) + j\left(\frac{10}{116} - \frac{X_L}{3^2+X_L^2}\right)$$

Resonance occurs when the admittance is a minimum, i.e. when the imaginary part of Y is zero. Hence, at resonance,

$$\frac{10}{116} - \frac{X_L}{3^2+X_L^2} = 0 \quad \text{i.e.} \quad \frac{10}{116} = \frac{X_L}{3^2+X_L^2}$$

Therefore $\qquad 10(9+X_L^2)=116X_L$

i.e. $\qquad 10X_L^2 - 116X_L + 90 = 0$

from which, $X_L^2 - 11.6X_L + 9 = 0$
Solving the quadratic equation gives:

$$X_L = \frac{11.6 \pm \sqrt{[(-11.6)^2 - 4(9)]}}{2} = \frac{11.6 \pm 9.93}{2}$$

i.e. $X_L = 10.765\,\Omega$ or $0.835\,\Omega$
Hence $10.765 = 2\pi f_r L_1$, from which,

inductance $L_1 = \dfrac{10.765}{2\pi(1000)} = 1.71\,\text{mH}$

and $0.835 = 2\pi f_r L_2$ from which,

inductance, $L_2 = \dfrac{0.835}{2\pi(1000)} = 0.13\,\text{mH}$

Thus the conditions for the circuit of Figure 32.11 to be resonant are that inductance L is either 1.71 mH or 0.13 mH

Problem 6. A capacitor having a Q-factor of 300 is connected in parallel with a coil having a Q-factor of 60. Determine the overall Q-factor of the parallel combination.

From equation (9), the overall Q-factor is given by:

$$Q_T = \frac{Q_L Q_C}{Q_L + Q_C} = \frac{(60)(300)}{60+300} = \frac{18\,000}{360} = \mathbf{50}$$

Problem 7. In an LR–C network, the capacitance is 10.61 nF, the bandwidth is 500 Hz and the resonant frequency is 150 kHz. Determine for the circuit (a) the Q-factor, (b) the dynamic resistance and (c) the magnitude of the impedance when the supply frequency is 0.4% greater than the tuned frequency.

(a) From equation (7), $\boldsymbol{Q} = \dfrac{f_r}{f_2 - f_1} = \dfrac{150 \times 10^3}{500}$

$$= \mathbf{300}$$

(b) From equation (4), dynamic resistance, $R_D = \dfrac{L}{CR}$

Also, in an LR–C network, $Q = \dfrac{\omega_r L}{R}$, from which $R = \dfrac{\omega_r L}{Q}$

Hence, $\boldsymbol{R_D} = \dfrac{L}{CR} = \dfrac{L}{C\left(\dfrac{\omega_r L}{Q}\right)} = \dfrac{LQ}{C\omega_r L} = \dfrac{Q}{\omega_r C}$

$$= \frac{300}{(2\pi 150 \times 10^3)(10.61 \times 10^{-9})} = \mathbf{30\,k\Omega}$$

(c) From equation (10), $\dfrac{R_D}{Z} = 1 + j2\delta Q$, from which $Z = \dfrac{R_D}{1 + j2\delta Q}$

$\delta = 0.4\% = 0.004$ hence

$$Z = \frac{30 \times 10^3}{1 + j2(0.004)(300)}$$

$$= \frac{30 \times 10^3}{1 + j2.4} = \frac{30 \times 10^3}{2.6\angle 67.38^\circ}$$

$$= 11.54\angle -67.38^\circ\,\text{k}\Omega$$

Hence **the magnitude of the impedance** when the frequency is 0.4% greater than the tuned frequency is **11.54 kΩ**

Now try the following Practice Exercise

Practice Exercise 134 Parallel resonance and Q-factor (Answers on page 828)

1. A capacitor of reactance 5 Ω is connected in series with a 10 Ω resistor. The whole circuit is then connected in parallel with a coil of inductive reactance 20 Ω and a variable resistor. Determine the value of this resistance for which the parallel network is resonant.
2. Determine, for the parallel network shown in Figure 32.12, the values of inductance L for which the circuit is resonant at a frequency of 600 Hz.

Part 4

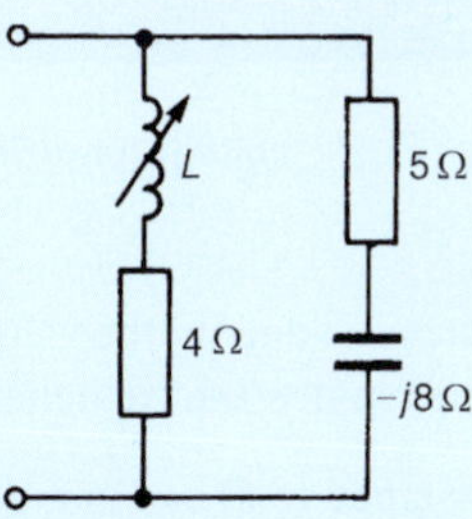

Figure 32.12

3. Find the resonant frequency of the two-branch parallel network shown in Figure 32.13.

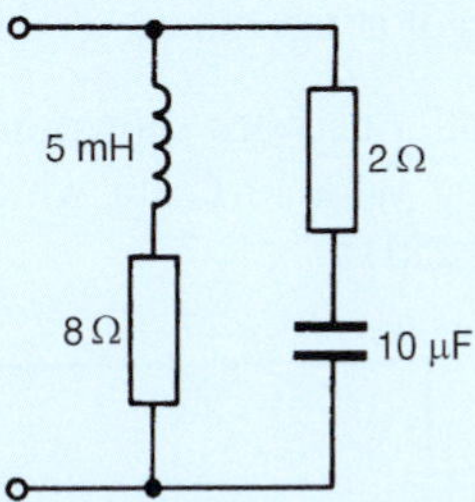

Figure 32.13

4. Determine the value of the variable resistance R in Figure 32.14 for which the parallel network is resonant.

5. For the parallel network shown in Figure 32.15, determine the resonant frequency. Find also the value of resistance to be connected in series with the 10 μF capacitor to change the resonant frequency to 1 kHz.

6. Determine the overall Q-factor of a parallel arrangement consisting of a capacitor having a Q-factor of 410 and an inductor having a Q-factor of 90.

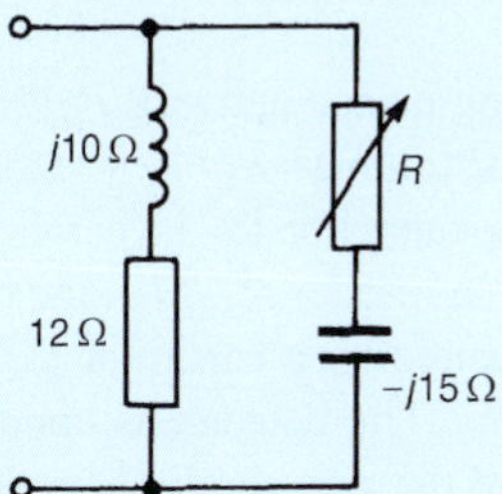

Figure 32.14

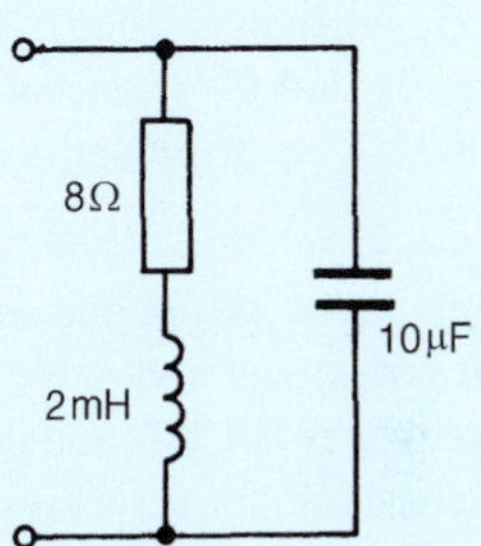

Figure 32.15

7. The value of capacitance in an *LR–C* parallel network is 49.74 nF. If the resonant frequency of the circuit is 200 kHz and the bandwidth is 800 Hz, determine for the network (a) the Q-factor, (b) the dynamic resistance and (c) the magnitude of the impedance when the supply frequency is 0.5% smaller than the tuned frequency.

Part 4

For fully worked solutions to each of the problems in Practice Exercises 133 and 134 in this chapter, go to the website:
www.routledge.com/cw/bird

Revision Test 9

This revision test covers the material contained in Chapters 30 to 32. *The marks for each part of the question are shown in brackets at the end of each question.*

1. In a Schering bridge network *PQRS*, the arms are made up as follows: *PQ* – a standard capacitor C_1, *QR* – a capacitor C_2 in parallel with a resistor R_2, *RS* – a resistor R_3, *SP* – the capacitor under test, represented by a capacitor C_x in series with a resistor R_x. The detector is connected between Q and S and the a.c. supply is connected between P and R.

   (a) Sketch the bridge and derive the equations for R_x and C_x when the bridge is balanced.
   (b) Evaluate R_x and C_x if, at balance $C_1 = 5\,\text{nF}$, $R_2 = 300\,\Omega$, $C_2 = 30\,\text{nF}$ and $R_3 = 1.5\,\text{k}\Omega$. (16)

2. A coil of inductance 25 mH and resistance 5 Ω is connected in series with a variable capacitor C. If the supply frequency is 1 kHz and the current flowing is 2 A, determine, for series resonance, (a) the value of capacitance C, (b) the supply p.d. and (c) the p.d. across the capacitor. (8)

3. An L–R–C series circuit has a peak current of 5 mA flowing in it when the frequency of the 200 mV supply is 5 kHz. The Q-factor of the circuit under these conditions is 75. Determine (a) the voltage across the capacitor and (b) the values of the circuit resistance, inductance and capacitance. (8)

4. A coil of resistance 15 Ω and inductance 150 mH is connected in parallel with a 4 μF capacitor across a 50 V variable-frequency supply. Determine (a) the resonant frequency of the circuit, (b) the dynamic resistance (c) the current at resonance and (d) the circuit Q-factor at resonance. (10)

5. For the parallel network shown in Figure RT9.1, determine the value of C for which the resonant frequency is 2 kHz. (8)

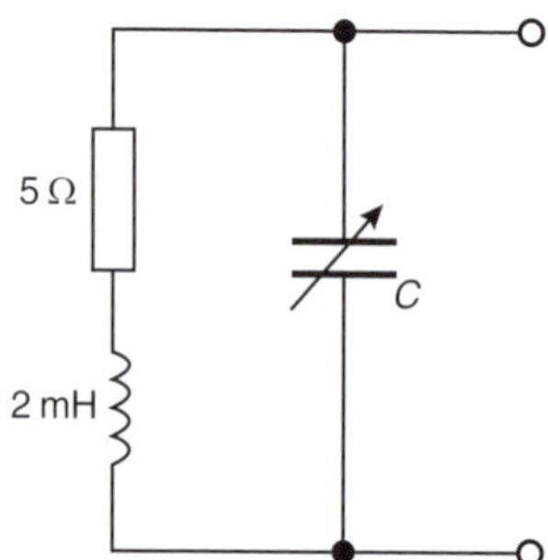

Figure RT9.1

For lecturers/instructors/teachers, fully worked solutions to each of the problems in Revision Test 9, together with a full marking scheme, are available at the website: www.routledge.com/cw/bird

Chapter 33

Introduction to network analysis

Why it is important to understand: **Introduction to network analysis**

Network analysis is any structured technique used to mathematically analyse a circuit or a network of interconnected components. Quite often a technician or engineer will encounter circuits containing multiple sources of power or component configurations which defy simplification by series/parallel analysis techniques using Ohm's law. In such cases, other means are required. This and the following chapters present some techniques useful in analysing such complex circuits. These are a collection of techniques for finding the voltages and currents in every component of the network. Network analysis is the process by which designers and manufacturers measure the electrical performance of the components and circuits used in more complex systems. When these systems are conveying signals with information content, there is a concern with getting the signal from one point to another with maximum efficiency and minimum distortion. This chapter introduces the techniques of network analysis, solves simultaneous equations in two and three unknowns, and uses Kirchhoff's laws to determine unknown values of branch currents.

At the end of this chapter you should be able to:

- appreciate available methods of analysing networks
- solve simultaneous equations in two and three unknowns using determinants
- analyse a.c. networks using Kirchhoff's laws

33.1 Introduction

Voltage sources in series–parallel networks cause currents to flow in each branch of the circuit and corresponding volt drops occur across the circuit components. A.c. circuit (or network) analysis involves the determination of the currents in the branches and/or the voltages across components.

The laws which determine the currents and voltage drops in a.c. networks are:

(a) **current, $I = V/Z$,** where Z is the complex impedance and V the voltage across the impedance;

(b) **the laws for impedances in series and parallel**, i.e. total impedance,

Electrical Circuit Theory and Technology. 978-1-138-67349-6, © 2017 John Bird. Published by Taylor & Francis. All rights reserved.

$Z_T = Z_1 + Z_2 + Z_3 + \cdots + Z_n$ for n impedances connected in series,

and $\frac{1}{Z_T} = \frac{1}{Z_1} + \frac{1}{Z_2} + \frac{1}{Z_3} + \cdots + \frac{1}{Z_n}$ for n impedances connected in parallel; and

(c) **Kirchhoff's laws**, which may be stated as:
  (i) *'At any point in an electrical circuit the phasor sum of the currents flowing towards that junction is equal to the phasor sum of the currents flowing away from the junction.'*
  (ii) *'In any closed loop in a network, the phasor sum of the voltage drops (i.e. the products of current and impedance) taken around the loop is equal to the phasor sum of the e.m.f.s acting in that loop.'*

In any circuit the currents and voltages at any point may be determined by applying Kirchhoff's laws (as demonstrated in this chapter), or by extensions of Kirchhoff's laws, called mesh-current analysis and nodal analysis (see Chapter 34).

However, for more complicated circuits, a number of circuit theorems have been developed as alternatives to the use of Kirchhoff's laws to solve problems involving both d.c. and a.c. electrical networks. These include:

(a) the superposition theorem (see Chapter 35)
(b) Thévenin's theorem (see Chapter 36)
(c) Norton's theorem (see Chapter 36)
(d) the maximum power transfer theorems (see Chapter 38).

In addition to these theorems, and often used as a preliminary to using circuit theorems, star–delta (or $T-\pi$) and delta–star (or $\pi - T$) transformations provide a method for simplifying certain circuits (see Chapter 37).

In a.c. circuit analysis involving Kirchhoff's laws or circuit theorems, the use of complex numbers is essential.

The above laws and theorems apply to linear circuits, i.e. circuits containing impedances whose values are independent of the direction and magnitude of the current flowing in them.

33.2 Solution of simultaneous equations using determinants

When Kirchhoff's laws are applied to electrical circuits, simultaneous equations result which require solution. If two loops are involved, two simultaneous equations containing two unknowns need to be solved; if three loops are involved, three simultaneous equations containing three unknowns need to be solved and so on. The elimination and substitution methods of solving simultaneous equations may be used to solve such equations. However, a more convenient method is to use **determinants**.

Two unknowns

When solving linear simultaneous equations in two unknowns using determinants:

(i) the equations are initially written in the form:

$$a_1x + b_1y + c_1 = 0$$

$$a_2x + b_2y + c_2 = 0$$

(ii) the solution is given by:

$$\frac{x}{D_x} = \frac{-y}{D_y} = \frac{1}{D}$$

where $D_x = \begin{vmatrix} b_1 & c_1 \\ b_2 & c_2 \end{vmatrix}$

i.e. the determinant of the coefficients left when the x-column is 'covered up',

$$D_y = \begin{vmatrix} a_1 & c_1 \\ a_2 & c_2 \end{vmatrix}$$

i.e. the determinant of the coefficients left when the y-column is 'covered up',

and $D = \begin{vmatrix} a_1 & b_1 \\ a_2 & b_2 \end{vmatrix}$

i.e. the determinant of the coefficients left when the constants-column is 'covered up'.

A '2×2' determinant $\begin{vmatrix} \boldsymbol{a} & \boldsymbol{c} \\ \boldsymbol{b} & \boldsymbol{d} \end{vmatrix}$ is evaluated as $ad - bc$

Three unknowns

When solving linear simultaneous equations in three unknowns using determinants:

(i) the equations are initially written in the form:

$$a_1x + b_1y + c_1z + d_1 = 0$$

$$a_2x + b_2y + c_2z + d_2 = 0$$

$$a_3x + b_3y + c_3z + d_3 = 0$$

(ii) the solution is given by:

$$\frac{x}{D_x}=\frac{-y}{D_y}=\frac{z}{D_z}=\frac{-1}{D}$$

where $D_x=\begin{vmatrix} b_1 & c_1 & d_1\\ b_2 & c_2 & d_2\\ b_3 & c_3 & d_3\end{vmatrix}$

i.e. the determinant of the coefficients left when the x-column is 'covered up',

$$D_y=\begin{vmatrix} a_1 & c_1 & d_1\\ a_2 & c_2 & d_2\\ a_3 & c_3 & d_3\end{vmatrix}$$

i.e. the determinant of the coefficients left when the y-column is 'covered up',

$$D_z=\begin{vmatrix} a_1 & b_1 & d_1\\ a_2 & b_2 & d_2\\ a_3 & b_3 & d_3\end{vmatrix}$$

i.e. the determinant of the coefficients left when the z-column is 'covered up',

and $D=\begin{vmatrix} a_1 & b_1 & c_1\\ a_2 & b_2 & c_2\\ a_3 & b_3 & c_3\end{vmatrix}$

i.e. the determinant of the coefficients left when the constants-column is 'covered up'.

To evaluate a 3 × 3 determinant:

(a) The **minor** of an element of a 3 by 3 matrix is the value of the 2 by 2 determinant obtained by covering up the row and column containing that element.

Thus for the matrix $\begin{pmatrix} 1 & 2 & 3\\ 4 & 5 & 6\\ 7 & 8 & 9\end{pmatrix}$ the minor of element 4 is the determinant $\begin{vmatrix} 2 & 3\\ 8 & 9\end{vmatrix}$

i.e. $(2\times 9)-(3\times 8)=18-24=-6$

Similarly, the minor of element 3 is $\begin{vmatrix} 4 & 5\\ 7 & 8\end{vmatrix}$

i.e. $(4\times 8)-(5\times 7)=32-35=-3$

(b) The sign of the minor depends on its position within the matrix, the sign pattern being $\begin{pmatrix} + & - & +\\ - & + & -\\ + & - & +\end{pmatrix}$. Thus the signed minor of element 4 in the above matrix is $-\begin{vmatrix} 2 & 3\\ 8 & 9\end{vmatrix}=-(-6)=6$

The signed-minor of an element is called the **cofactor** of the element.

Thus the cofactor of element 2 is

$$-\begin{vmatrix} 4 & 6\\ 7 & 9\end{vmatrix}=-(36-42)=6$$

(c) **The value of a 3 by 3 determinant is the sum of the products of the elements and their cofactors of any row or any column of the corresponding 3 by 3 matrix.**

Thus a 3 by 3 determinant $\begin{vmatrix} a & b & c\\ d & e & f\\ g & h & j\end{vmatrix}$ is evaluated as

$a\begin{vmatrix} e & f\\ h & j\end{vmatrix}-b\begin{vmatrix} d & f\\ g & j\end{vmatrix}+c\begin{vmatrix} d & e\\ g & h\end{vmatrix}$ using the top row,

or $-b\begin{vmatrix} d & f\\ g & j\end{vmatrix}+e\begin{vmatrix} a & c\\ g & j\end{vmatrix}-h\begin{vmatrix} a & c\\ d & f\end{vmatrix}$ using the second column.

There are thus six ways of evaluating a 3 by 3 determinant.
Determinants are used to solve simultaneous equations in some of the following problems and in Chapter 34.

33.3 Network analysis using Kirchhoff's* laws

Kirchhoff's* **laws** may be applied to both d.c. and a.c. circuits. The laws are introduced in Chapter 15 for d.c. circuits. To demonstrate the method of analysis, consider the d.c. network shown in Figure 33.1. If the current flowing in each branch is required, the following three-step procedure may be used:

(i) Label branch currents and their directions on the circuit diagram. The directions chosen are arbitrary but, as a starting point, a useful guide

*Who was **Kirchhoff**? For image and resume of Kirchhoff, see page 220. To find out more go to **www.routledge.com/cw/bird**

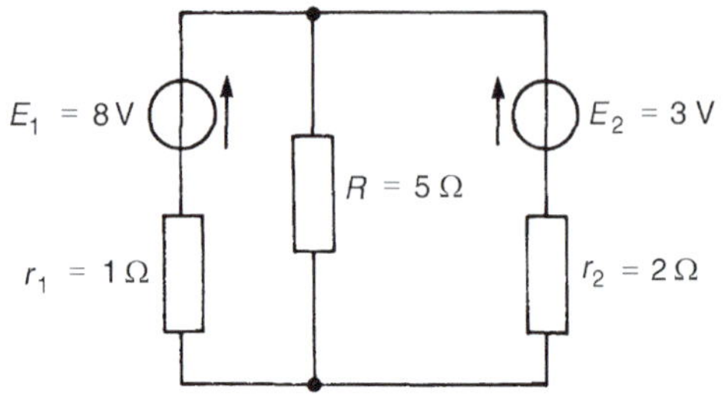

Figure 33.1

is to assume that current flows from the positive terminals of the voltage sources. This is shown in Figure 33.2 where the three branch currents are expressed in terms of I_1 and I_2 only, since the current through resistance R, by Kirchhoff's current law, is $(I_1 + I_2)$.

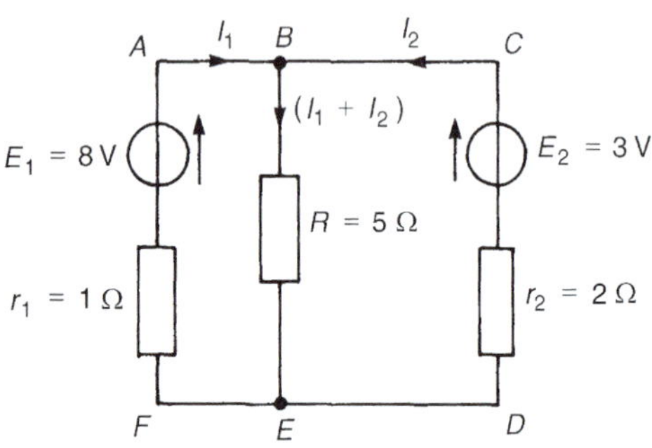

Figure 33.2

(ii) Divide the circuit into loops – two in this case (see Figure 33.2) and then apply Kirchhoff's voltage law to each loop in turn. From loop *ABEF*, and moving in a clockwise direction (the choice of loop direction is arbitrary), $E_1 = I_1 r + (I_1 + I_2)R$ (note that the two voltage drops are positive since the loop direction is the same as the current directions involved in the volt drops). Hence

$$8 = I_1 + 5(I_1 + I_2)$$

or $$6I_1 + 5I_2 = 8 \qquad (1)$$

From loop *BCDE* in Figure 33.2, and moving in an anticlockwise direction (note that the direction does not have to be the same as that used for the first loop), $E_2 = I_2 r_2 + (I_1 + I_2)R$,

i.e. $$3 = 2I_2 + 5(I_1 + I_2)$$

or $$5I_1 + 7I_2 = 3 \qquad (2)$$

(iii) Solve simultaneous equations (1) and (2) for I_1 and I_2.

Multiplying equation (1) by 7 gives:

$$42I_1 + 35I_2 = 56 \qquad (3)$$

Multiplying equation (2) by 5 gives:

$$25I_1 + 35I_2 = 15 \qquad (4)$$

Equation (3) − equation (4) gives:

$$17I_1 = 41$$

from which, current $\boldsymbol{I_1} = 41/17 = 2.412\,A =$ **2.41 A**, correct to two decimal places.

From equation (1): $6(2.412) + 5I_2 = 8$, from which,

$$\text{current } \boldsymbol{I_2} = \frac{8 - 6(2.412)}{5} = -1.294\,\text{A}$$

$= \mathbf{-1.29\,A}$, correct to two decimal places.

The minus sign indicates that current I_2 flows in the opposite direction to that shown in Figure 33.2.

The current flowing through resistance R is

$$(I_1 + I_2) = 2.412 + (-1.294) = 1.118\,\text{A}$$

$= \mathbf{1.12\,A}$, correct to two decimal places.

[A third loop may be selected in Figure 33.2 (just as a check), moving clockwise around the outside of the network.
Then $E_1 - E_2 = I_1 r_1 - I_2 r_2$ i.e. $8 - 3 = I_1 - 2I_2$. Thus $5 = 2.412 - 2(-1.294) = 5$]

An alternative method of solving equations (1) and (2) is shown below using determinants. Since

$$6I_1 + 5I_2 - 8 = 0 \qquad (1)$$

$$5I_1 + 7I_2 - 3 = 0 \qquad (2)$$

then $$\frac{I_1}{\begin{vmatrix} 5 & -8 \\ 7 & -3 \end{vmatrix}} = \frac{-I_2}{\begin{vmatrix} 6 & -8 \\ 5 & -3 \end{vmatrix}} = \frac{1}{\begin{vmatrix} 6 & 5 \\ 5 & 7 \end{vmatrix}}$$

i.e. $$\frac{I_1}{-15 + 56} = \frac{-I_2}{-18 + 40} = \frac{1}{42 - 25}$$

and $$\frac{I_1}{41} = \frac{-I_2}{22} = \frac{1}{17}$$

from which, $\boldsymbol{I_1} = 41/17 =$ **2.41 A** and $\boldsymbol{I_2} = -22/17 = \mathbf{-1.29\,A}$, as obtained previously.

The above procedure is shown for a simple d.c. circuit having two unknown values of current. The procedure, however, applies equally well to a.c. networks and/or to circuits where three unknown currents are involved. This is illustrated in the following problems.

Problem 1. Use Kirchhoff's laws to find the current flowing in each branch of the network shown in Figure 33.3.

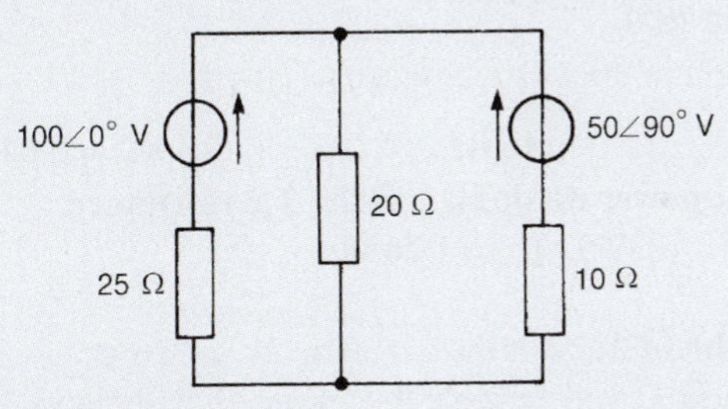

Figure 33.3

(i) The branch currents and their directions are labelled as shown in Figure 33.4

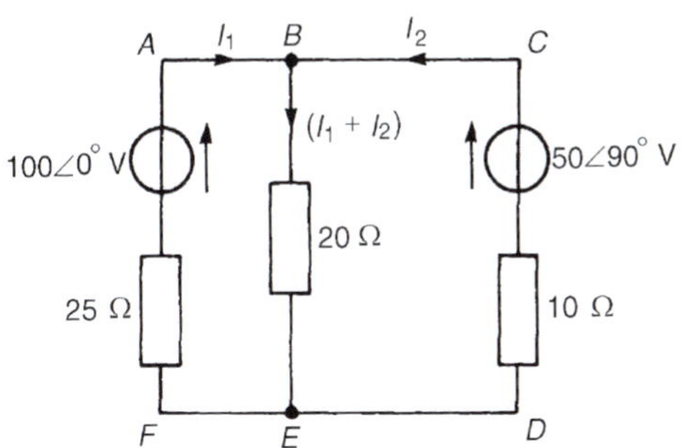

Figure 33.4

(ii) Two loops are chosen. From loop $ABEF$, and moving clockwise,

$$25I_1 + 20(I_1 + I_2) = 100\angle 0^\circ$$

i.e. $$45I_1 + 20I_2 = 100 \quad (1)$$

From loop $BCDE$, and moving anticlockwise,

$$10I_2 + 20(I_1 + I_2) = 50\angle 90^\circ$$

i.e. $$20I_1 + 30I_2 = j50 \quad (2)$$

3 × equation (1) gives: $135I_1 + 60I_2 = 300$ (3)

2 × equation (2) gives: $40I_1 + 60I_2 = j100$ (4)

Equation (3) − equation (4) gives:

$$95I_1 = 300 - j100$$

from which, current $I_1 = \dfrac{300 - j100}{95}$

$= \mathbf{3.329\angle -18.43^\circ}$ **A** or **(3.158 − j1.052) A**

Substituting in equation (1) gives:

$45(3.158 - j1.052) + 20I_2 = 100$, from which,

$$I_2 = \frac{100 - 45(3.158 - j1.052)}{20}$$

$= \mathbf{(-2.106 + j2.367)}$ **A** or $\mathbf{3.168\angle 131.66^\circ}$ **A**

Thus

$$I_1 + I_2 = (3.158 - j1.052) + (-2.106 + j2.367)$$

$= \mathbf{(1.052 + j1.315)}$ **A** or $\mathbf{1.684\angle 51.34^\circ}$ **A**

Problem 2. Determine the current flowing in the 2 Ω resistor of the circuit shown in Figure 33.5 using Kirchhoff's laws. Find also the power dissipated in the 3 Ω resistance.

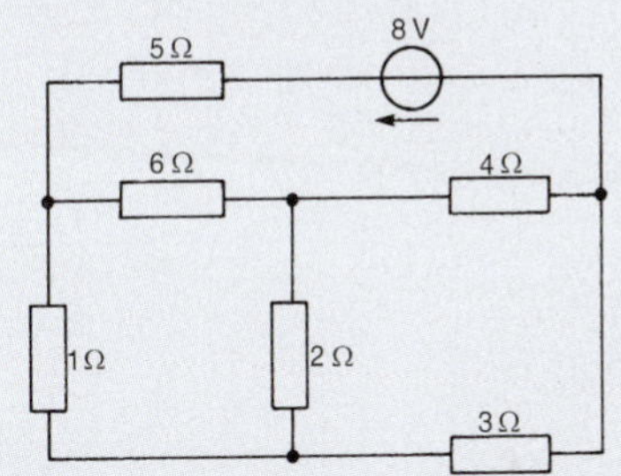

Figure 33.5

(i) Currents and their directions are assigned as shown in Figure 33.6.

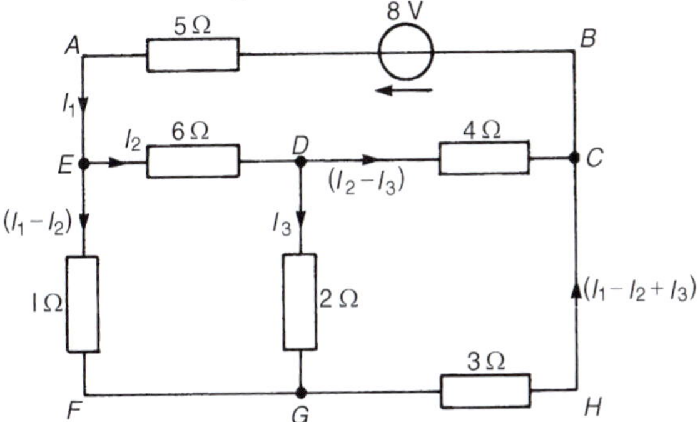

Figure 33.6

(ii) Three loops are chosen since three unknown currents are required. The choice of loop directions is arbitrary. From loop $ABCDE$, and moving anticlockwise,

$$5I_1+6I_2+4(I_2-I_3)=8$$

i.e. $$5I_1+10I_2-4I_3=8 \qquad (1)$$

From loop *EDGF*, and moving clockwise,

$$6I_2+2I_3-1(I_1-I_2)=0$$

i.e. $$-I_1+7I_2+2I_3=0 \qquad (2)$$

From loop *DCHG*, and moving anticlockwise,

$$2I_3+3(I_1-I_2+I_3)-4(I_2-I_3)=0$$

i.e. $$3I_1-7I_2+9I_3=0$$

(iii) Thus $5I_1+10I_2-4I_3-8=0$

$$-I_1+7I_2+2I_3+0=0$$

$$3I_1-7I_2+9I_3+0=0$$

Hence, using determinants,

$$\frac{I_1}{\begin{vmatrix}10 & -4 & -8\\ 7 & 2 & 0\\ -7 & 9 & 0\end{vmatrix}}=\frac{-I_2}{\begin{vmatrix}5 & -4 & -8\\ -1 & 2 & 0\\ 3 & 9 & 0\end{vmatrix}}=\frac{I_3}{\begin{vmatrix}5 & 10 & -8\\ -1 & 7 & 0\\ 3 & -7 & 0\end{vmatrix}}=\frac{-1}{\begin{vmatrix}5 & 10 & -4\\ -1 & 7 & 2\\ 3 & -7 & 9\end{vmatrix}}$$

Thus

$$\frac{I_1}{-8\begin{vmatrix}7 & 2\\ -7 & 9\end{vmatrix}}=\frac{-I_2}{-8\begin{vmatrix}-1 & 2\\ 3 & 9\end{vmatrix}}=\frac{I_3}{-8\begin{vmatrix}-1 & 7\\ 3 & -7\end{vmatrix}}$$

$$=\frac{-1}{5\begin{vmatrix}7 & 2\\ -7 & 9\end{vmatrix}-10\begin{vmatrix}-1 & 2\\ 3 & 9\end{vmatrix}-4\begin{vmatrix}-1 & 7\\ 3 & -7\end{vmatrix}}$$

$$\frac{I_1}{-8(63+14)}=\frac{-I_2}{-8(-9-6)}=\frac{I_3}{-8(7-21)}$$

$$=\frac{-1}{5(63+14)-10(-9-6)-4(7-21)}$$

$$\frac{I_1}{-616}=\frac{-I_2}{120}=\frac{I_3}{112}=\frac{-1}{591}$$

Hence $$I_1=\frac{616}{591}=1.042\,\text{A},$$

$$I_2=\frac{120}{591}=0.203\,\text{A and}$$

$$I_3=\frac{-112}{591}=-0.190\,\text{A}$$

Thus **the current flowing in the 2 Ω resistance is 0.190 A** in the opposite direction to that shown in Figure 33.6.

Current in the 3 Ω resistance $=I_1-I_2+I_3$

$$=1.042-0.203+(-0.190)=0.649\,\text{A}$$

Hence **power dissipated in the 3 Ω resistance,** $I^2(3)=(0.649)^2(3)=$ **1.26 W**

Problem 3. For the a.c. network shown in Figure 33.7, determine the current flowing in each branch using Kirchhoff's laws.

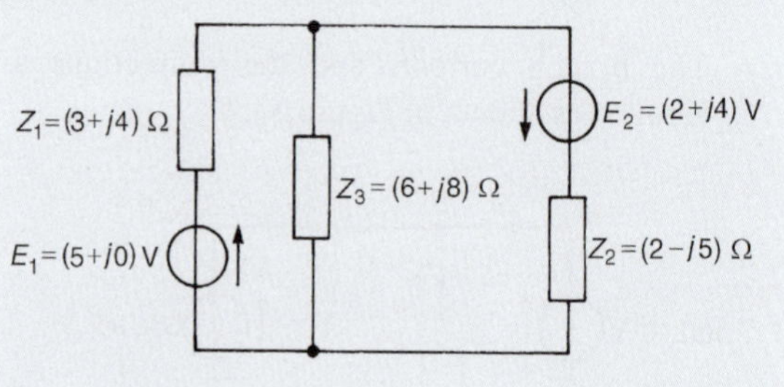

Figure 33.7

(i) Currents I_1 and I_2 with their directions are shown in Figure 33.8.

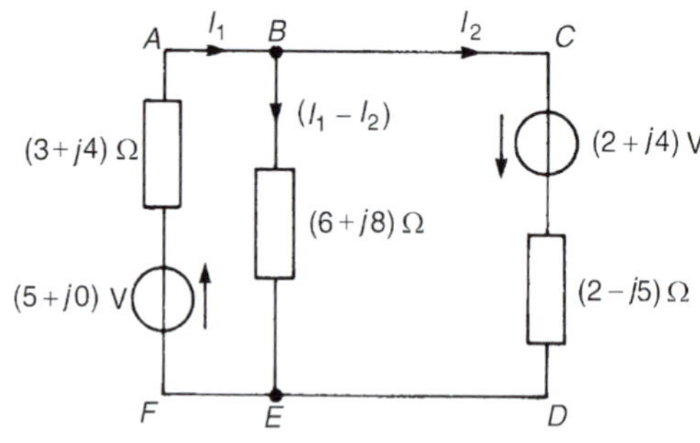

Figure 33.8

(ii) Two loops are chosen with their directions both clockwise.

From loop *ABEF*,

$$(5+j0)=I_1(3+j4)+(I_1-I_2)(6+j8)$$

i.e. $$5=(9+j12)I_1-(6+j8)I_2 \qquad (1)$$

From loop $BCDE$,

$$(2+j4)=I_2(2-j5)-(I_1-I_2)(6+j8)$$

i.e. $(2+j4)=-(6+j8)I_1+(8+j3)I_2$ (2)

(iii) Multiplying equation (1) by $(8+j3)$ gives:

$$5(8+j3)=(8+j3)(9+j12)I_1 -(8+j3)(6+j8)I_2 \quad (3)$$

Multiplying equation (2) by $(6+j8)$ gives:

$$(6+j8)(2+j4)=-(6+j8)(6+j8)I_1 +(6+j8)(8+j3)I_2 \quad (4)$$

Adding equations (3) and (4) gives:

$$5(8+j3)+(6+j8)(2+j4)=[(8+j3)(9+j12) - (6+j8)(6+j8)]I_1$$

i.e. $(20+j55)=(64+j27)I_1$

from which, $\boldsymbol{I_1}=\dfrac{20+j55}{64+j27}=\dfrac{58.52\angle 70.02^\circ}{69.46\angle 22.87^\circ}$

$=\mathbf{0.842\angle 47.15^\circ\ A}$

$=(0.573+j0.617)\,\text{A}$

$=\mathbf{(0.57+j0.62)\ A}$, correct to two decimal places.

From equation (1), $5=(9+j12)(0.573+j0.617) - (6+j8)I_2$

$5=(-2.247+j12.429) - (6+j8)I_2$

from which, $\boldsymbol{I_2}=\dfrac{-2.247+j12.429-5}{6+j8}$

$=\dfrac{14.39\angle 120.25^\circ}{10\angle 53.13^\circ}$

$=\mathbf{1.439\angle 67.12^\circ\ A}$

$=(0.559+j1.326)\,\text{A}$

$=\mathbf{(0.56+j1.33)\ A}$, correct to two decimal places.

The current in the $(6+j8)\,\Omega$ impedance,

$\boldsymbol{I_1-I_2}=(0.573+j0.617)-(0.559+j1.326)$

$=\mathbf{(0.014-j0.709)\ A}$ or $\mathbf{0.709\angle -88.87^\circ\ A}$

An alternative method of solving equations (1) and (2) is shown below, using determinants.

$$(9+j12)I_1-(6+j8)I_2-5=0 \quad (1)$$

$$-(6+j8)I_1+(8+j3)I_2-(2+j4)=0 \quad (2)$$

Thus

$$\frac{I_1}{\begin{vmatrix}-(6+j8) & -5\\ (8+j3) & -(2+j4)\end{vmatrix}}=\frac{-I_2}{\begin{vmatrix}(9+j12) & -5\\ -(6+j8) & -(2+j4)\end{vmatrix}}=\frac{1}{\begin{vmatrix}(9+j12) & -(6+j8)\\ -(6+j8) & (8+j3)\end{vmatrix}}$$

$$\frac{I_1}{(-20+j40)+(40+j15)}=\frac{-I_2}{(30-j60)-(30+j40)}=\frac{1}{(36+j123)-(-28+j96)}$$

$$\frac{I_1}{20+j55}=\frac{-I_2}{-j100}=\frac{1}{64+j27}$$

Hence $\quad \boldsymbol{I_1}=\dfrac{20+j55}{64+j27}=\dfrac{58.52\angle 70.02^\circ}{69.46\angle 22.87^\circ}$

$=\mathbf{0.842\angle 47.15^\circ\ A}$

and $\quad \boldsymbol{I_2}=\dfrac{100\angle 90^\circ}{69.46\angle 22.87^\circ}=\mathbf{1.440\angle 67.13^\circ\ A}$

The current flowing in the $(6+j8)\,\Omega$ impedance is given by:

$\boldsymbol{I_1-I_2}=0.842\angle 47.15^\circ-1.440\angle 67.13^\circ\,\text{A}$

$=\mathbf{(0.013-j0.709)\ A}$ or $\mathbf{0.709\angle -88.95^\circ\ A}$

Problem 4. For the network shown in Figure 33.9, use Kirchhoff's laws to determine the magnitude of the current in the $(4+j3)\,\Omega$ impedance.

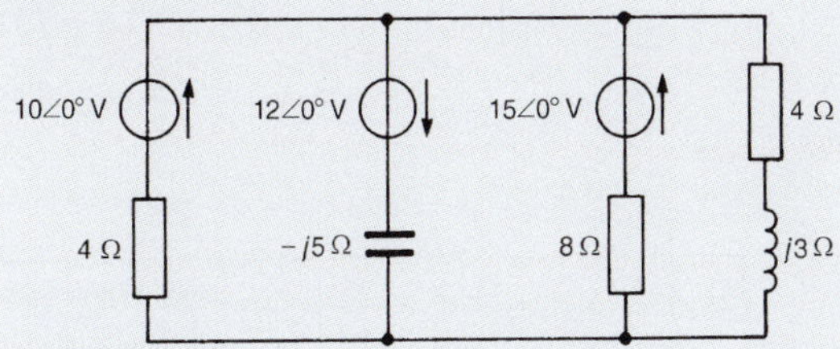

Figure 33.9

(i) Currents I_1, I_2 and I_3 with their directions are shown in Figure 33.10. The current in the

Part 4

$(4+j3)\,\Omega$ impedance is specified by one symbol only (i.e. I_3), which means that the three equations formed need to be solved for only one unknown current.

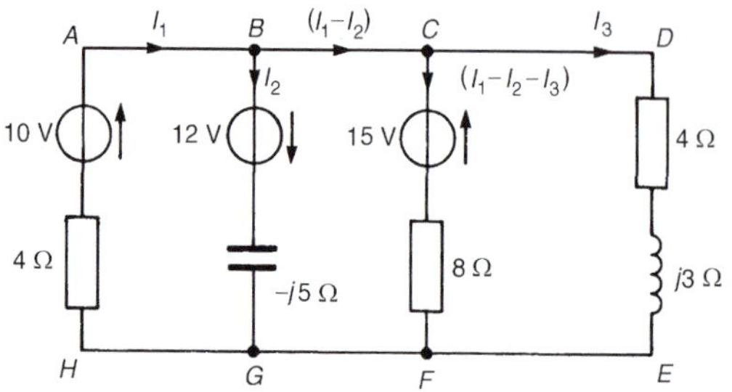

Figure 33.10

(ii) Three loops are chosen. From loop *ABGH*, and moving clockwise,

$$4I_1 - j5I_2 = 10 + 12 \quad (1)$$

From loop *BCFG*, and moving anticlockwise,

$$-j5I_2 - 8(I_1 - I_2 - I_3) = 15 + 12 \quad (2)$$

From loop *CDEF*, and moving clockwise,

$$-8(I_1 - I_2 - I_3) + (4 + j3)(I_3) = 15 \quad (3)$$

Hence

$$4I_1 - j5I_2 + 0I_3 - 22 = 0$$

$$-8I_1 + (8 - j5)I_2 + 8I_3 - 27 = 0$$

$$-8I_1 + 8I_2 + (12 + j3)I_3 - 15 = 0$$

Solving for I_3 using determinants gives:

$$\frac{I_3}{\begin{vmatrix} 4 & -j5 & -22 \\ -8 & (8-j5) & -27 \\ -8 & 8 & -15 \end{vmatrix}} = \frac{-1}{\begin{vmatrix} 4 & -j5 & 0 \\ -8 & (8-j5) & 8 \\ -8 & 8 & (12+j3) \end{vmatrix}}$$

Thus

$$\frac{I_3}{4\begin{vmatrix} (8-j5) & -27 \\ 8 & -15 \end{vmatrix} + j5\begin{vmatrix} -8 & -27 \\ -8 & -15 \end{vmatrix} - 22\begin{vmatrix} -8 & (8-j5) \\ -8 & 8 \end{vmatrix}}$$

$$= \frac{-1}{4\begin{vmatrix} (8-j5) & 8 \\ 8 & (12+j3) \end{vmatrix} + j5\begin{vmatrix} -8 & 8 \\ -8 & (12+j3) \end{vmatrix}}$$

Hence

$$\frac{I_3}{384 + j700} = \frac{-1}{308 - j304} \text{ from which,}$$

$$I_3 = \frac{-(384 + j700)}{(308 - j304)}$$

$$= \frac{798.41\angle -118.75}{432.76\angle -44.63^\circ}$$

$$= 1.85\angle -74.12^\circ\,\text{A}$$

Hence the magnitude of the current flowing in the $(4+j3)\Omega$ impedance is 1.85 A

Now try the following Practice Exercise

Practice Exercise 135 Network analysis using Kirchhoff's laws (Answers on page 828)

1. For the network shown in Figure 33.11, determine the current flowing in each branch.

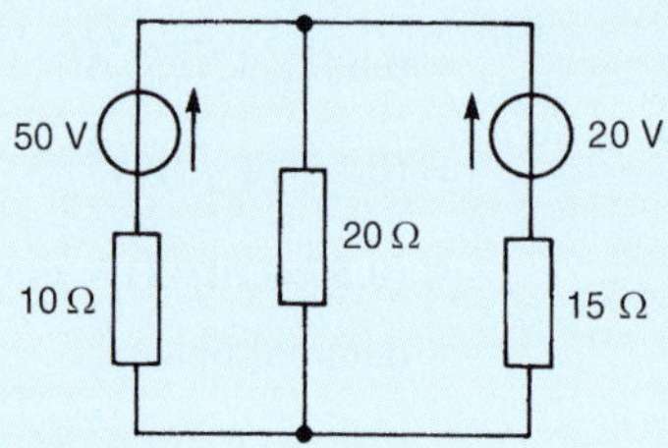

Figure 33.11

2. Determine the value of currents I_A, I_B and I_C for the network shown in Figure 33.12.

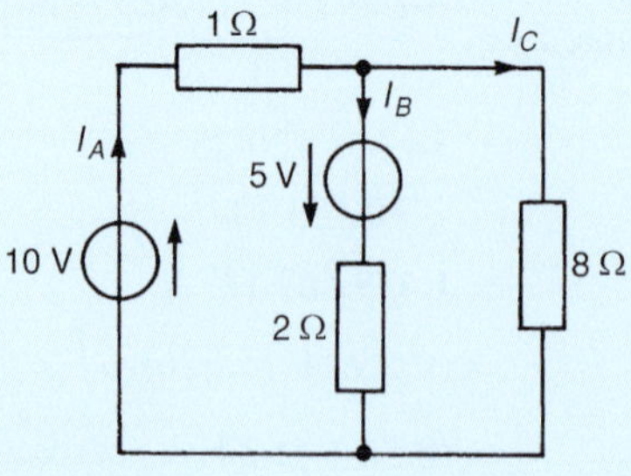

Figure 33.12

3. For the bridge shown in Figure 33.13, determine the current flowing in (a) the 5 Ω

resistance, (b) the 22 Ω resistance and (c) the 2 Ω resistance.

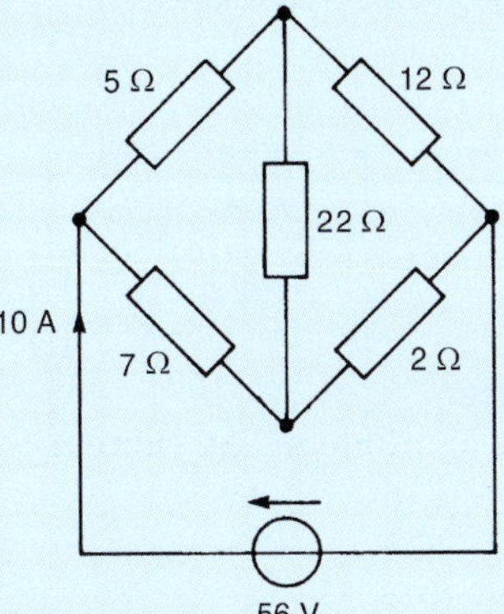

Figure 33.13

4. For the circuit shown in Figure 33.14, determine (a) the current flowing in the 10 V source, (b) the p.d. across the 6 Ω resistance and (c) the active power dissipated in the 4 Ω resistance.

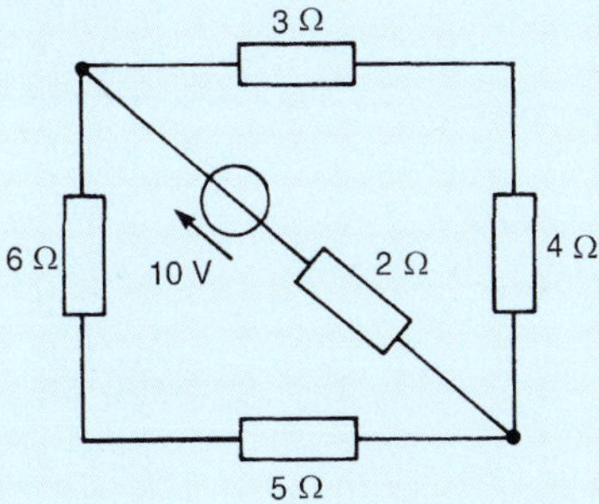

Figure 33.14

5. Use Kirchhoff's laws to determine the current flowing in each branch of the network shown in Figure 33.15.

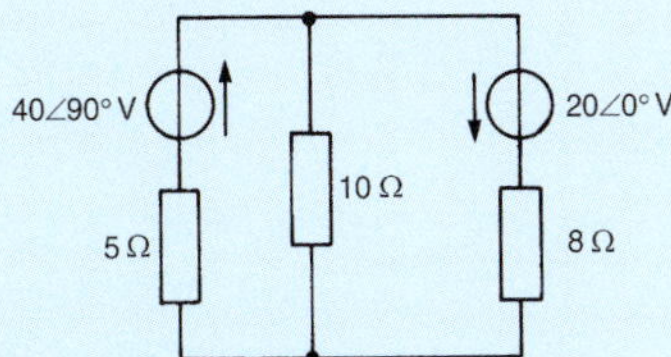

Figure 33.15

6. For the network shown in Figure 33.16, use Kirchhoff's laws to determine the current flowing in the capacitive branch.

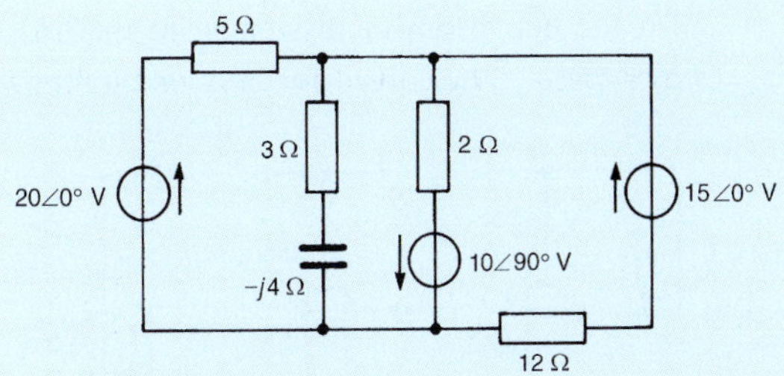

Figure 33.16

7. Use Kirchhoff's laws to determine, for the network shown in Figure 33.17, the current flowing in (a) the 20 Ω resistance and (b) the 4 Ω resistance. Determine also (c) the p.d. across the 8 Ω resistance and (d) the active power dissipated in the 10 Ω resistance.

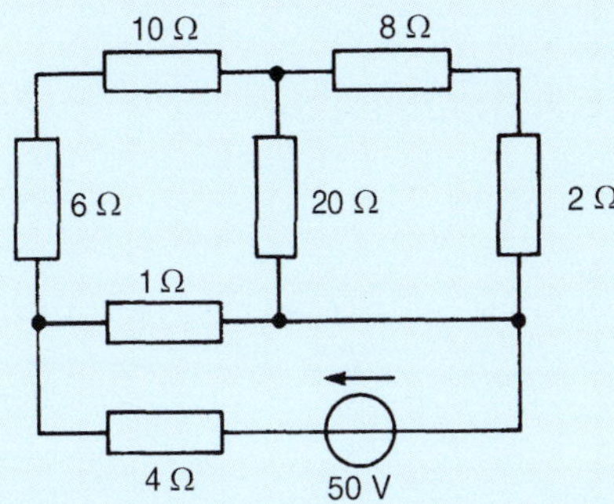

Figure 33.17

8. Determine the value of currents I_A, I_B and I_C shown in the network of Figure 33.18, using Kirchhoff's laws.

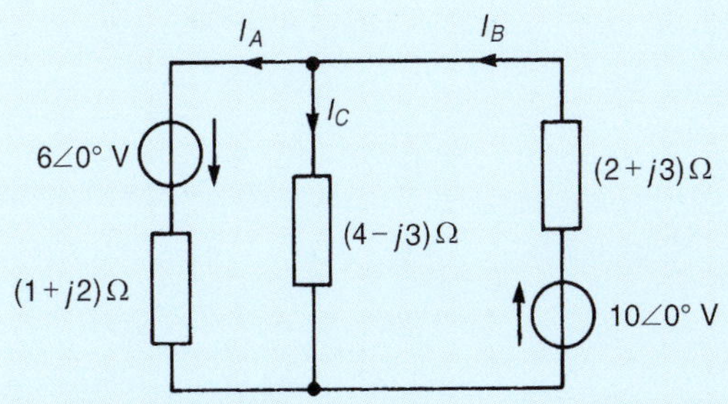

Figure 33.18

9. Use Kirchhoff's laws to determine the currents flowing in (a) the 3 Ω resistance, (b) the 6 Ω resistance and (c) the 4 V source of the network shown in Figure 33.19. Determine

also the active power dissipated in the 5 Ω resistance.

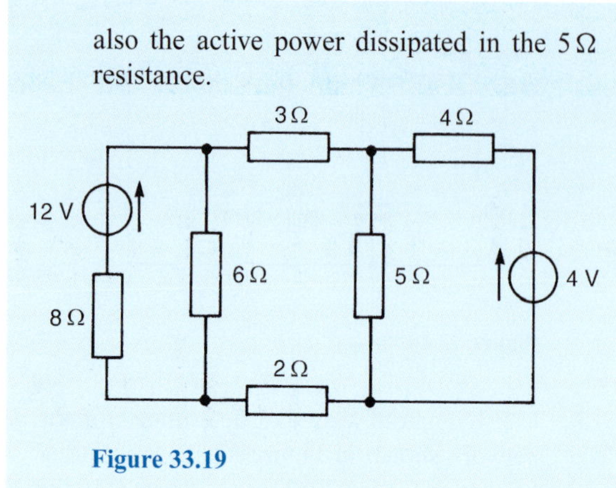

Figure 33.19

10. Determine the magnitude of the p.d. across the $(8 + j6)\,\Omega$ impedance shown in Figure 33.20 by using Kirchhoff's laws.

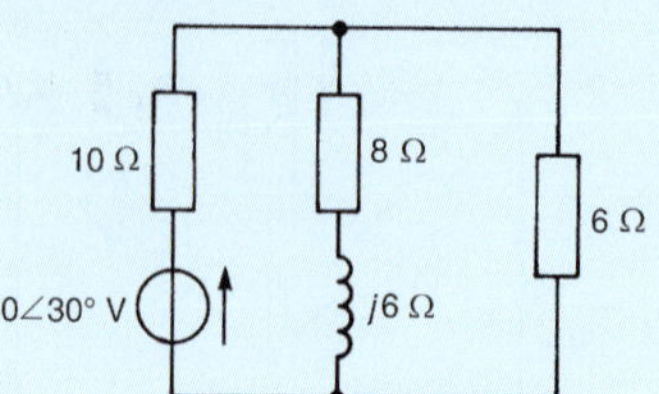

Figure 33.20

For fully worked solutions to each of the problems in Practice Exercise 135 in this chapter, go to the website:
www.routledge.com/cw/bird

Chapter 34

Mesh-current and nodal analysis

Why it is important to understand: **Mesh-current and nodal analysis**

When circuits become large and complicated, it is useful to have various methods for simplifying and analysing the circuit. There is no perfect formula for solving a circuit; depending on the type of circuit, there are different methods that can be employed to solve the circuit. Some methods might not work, and some methods may be very difficult in terms of long mathematical problems. Two of the most important methods for solving circuits are mesh current analysis and nodal analysis. Mesh analysis is the application of Kirchhoff's voltage law to solve for mesh currents, whereas nodal analysis is the application of Kirchhoff's current law to solve for the voltages at each node in a network. Mesh analysis is often easier as it requires fewer unknowns. This chapter demonstrates how to solve d.c. and a.c. networks using mesh and nodal analysis.

At the end of this chapter you should be able to:

- solve d.c. and a.c. networks using mesh-current analysis
- solve d.c. and a.c. networks using nodal analysis

34.1 Mesh-current analysis

Mesh-current analysis is merely an extension of the use of Kirchhoff's laws, explained in Chapter 33. Figure 34.1 shows a network whose circulating currents I_1, I_2 and I_3 have been assigned to closed loops in the circuit rather than to branches. Currents I_1, I_2 and I_3 are called **mesh-currents** or **loop-currents**.

In mesh-current analysis the loop-currents are all arranged to circulate in the same direction (in

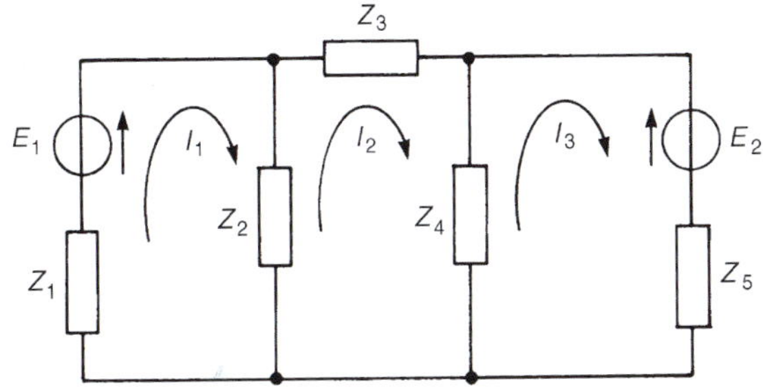

Figure 34.1

Electrical Circuit Theory and Technology. 978-1-138-67349-6, © 2017 John Bird. Published by Taylor & Francis. All rights reserved.

Figure 34.1, shown as clockwise direction). Kirchhoff's second law is applied to each of the loops in turn, which in the circuit of Figure 34.1 produces three equations in three unknowns which may be solved for I_1, I_2 and I_3. The three equations produced from Figure 34.1 are:

$$I_1(Z_1+Z_2)-I_2Z_2=E_1$$
$$I_2(Z_2+Z_3+Z_4)-I_1Z_2-I_3Z_4=0$$
$$I_3(Z_4+Z_5)-I_2Z_4=-E_2$$

The branch currents are determined by taking the phasor sum of the mesh currents common to that branch. For example, the current flowing in impedance Z_2 of Figure 34.1 is given by (I_1-I_2) phasorially. The method of mesh-current analysis, called **Maxwell's theorem**, is demonstrated in the following problems.

Problem 1. Use mesh-current analysis to determine the current flowing in (a) the 5 Ω resistance and (b) the 1 Ω resistance of the d.c. circuit shown in Figure 34.2.

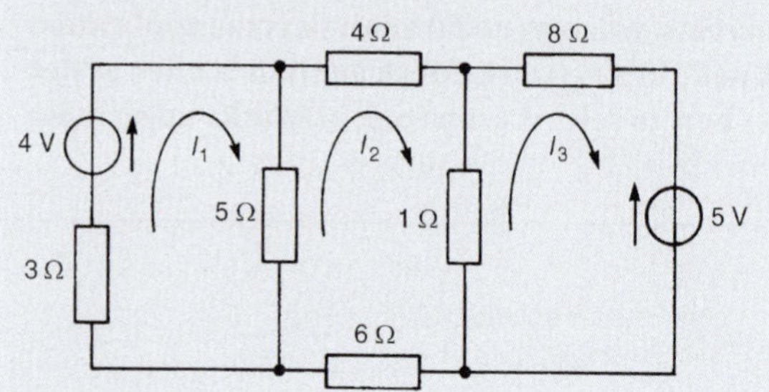

Figure 34.2

The mesh currents I_1, I_2 and I_3 are shown in Figure 34.2. Using Kirchhoff's voltage law:

For loop 1, $(3+5)I_1-5I_2=4$ (1)

For loop 2, $(4+1+6+5)I_2-(5)I_1-(1)I_3=0$ (2)

For loop 3, $(1+8)I_3-(1)I_2=-5$ (3)

Thus

$$8I_1-5I_2 \quad -4=0 \qquad (1')$$
$$-5I_1+16I_2-I_3 \quad =0 \qquad (2')$$
$$-I_2 \quad +9I_3+5=0 \qquad (3')$$

Using determinants,

$$\frac{I_1}{\begin{vmatrix}-5 & 0 & -4\\ 16 & -1 & 0\\ -1 & 9 & 5\end{vmatrix}}=\frac{-I_2}{\begin{vmatrix}8 & 0 & -4\\ -5 & -1 & 0\\ 0 & 9 & 5\end{vmatrix}}=\frac{I_3}{\begin{vmatrix}8 & -5 & -4\\ -5 & 16 & 0\\ 0 & -1 & 5\end{vmatrix}}=\frac{-1}{\begin{vmatrix}8 & -5 & 0\\ -5 & 16 & -1\\ 0 & -1 & 9\end{vmatrix}}$$

$$\frac{I_1}{-5\begin{vmatrix}-1 & 0\\ 9 & 5\end{vmatrix}-4\begin{vmatrix}16 & -1\\ -1 & 9\end{vmatrix}}=\frac{-I_2}{8\begin{vmatrix}-1 & 0\\ 9 & 5\end{vmatrix}-4\begin{vmatrix}-5 & -1\\ 0 & 9\end{vmatrix}}$$
$$=\frac{I_3}{-4\begin{vmatrix}-5 & 16\\ 0 & -1\end{vmatrix}+5\begin{vmatrix}8 & -5\\ -5 & 16\end{vmatrix}}$$
$$=\frac{-1}{8\begin{vmatrix}16 & -1\\ -1 & 9\end{vmatrix}+5\begin{vmatrix}-5 & -1\\ 0 & 9\end{vmatrix}}$$

$$\frac{I_1}{-5(-5)-4(143)}=\frac{-I_2}{8(-5)-4(-45)}$$
$$=\frac{I_3}{-4(5)+5(103)}$$
$$=\frac{-1}{8(143)+5(-45)}$$

$$\frac{I_1}{-547}=\frac{-I_2}{140}=\frac{I_3}{495}=\frac{-1}{919}$$

Hence $I_1=\dfrac{547}{919}=0.595\,\text{A}$,

$$I_2=\frac{140}{919}=0.152\,\text{A}$$

$$I_3=\frac{-495}{919}=-0.539\,\text{A}$$

(a) **Current in the 5 Ω resistance** $=I_1-I_2$
$=0.595-0.152$
$=\mathbf{0.44\,A}$

(b) **Current in the 1 Ω resistance** $=I_2-I_3$
$=0.152-(-0.539)$
$=\mathbf{0.69\,A}$

Part 4

Problem 2. For the a.c. network shown in Figure 34.3 determine, using mesh-current analysis, (a) the mesh currents I_1 and I_2, (b) the current flowing in the capacitor and (c) the active power delivered by the $100\angle 0°$ V voltage source.

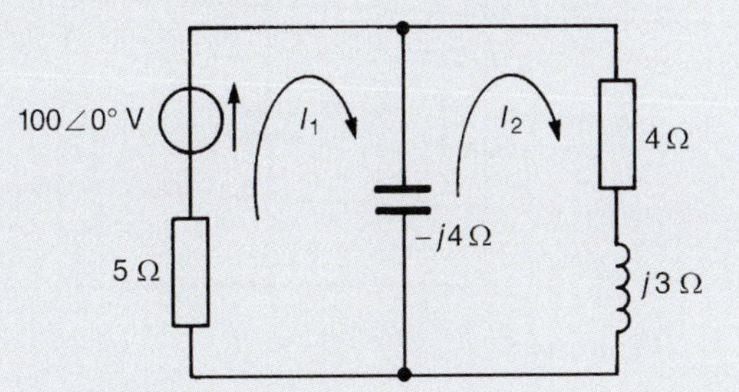

Figure 34.3

(a) For the first loop $(5-j4)I_1-(-j4I_2)=100\angle 0°$ (1)

For the second loop

$$(4+j3-j4)I_2-(-j4I_1)=0 \quad (2)$$

Rewriting equations (1) and (2) gives:

$$(5-j4)I_1+j4I_2-100=0 \quad (1')$$

$$j4I_1+(4-j)I_2+0=0 \quad (2')$$

Thus, using determinants,

$$\frac{I_1}{\begin{vmatrix} j4 & -100 \\ (4-j) & 0 \end{vmatrix}} = \frac{-I_2}{\begin{vmatrix} (5-j4) & -100 \\ j4 & 0 \end{vmatrix}} = \frac{1}{\begin{vmatrix} (5-j4) & j4 \\ j4 & (4-j) \end{vmatrix}}$$

$$\frac{I_1}{(400-j100)} = \frac{-I_2}{j400} = \frac{1}{(32-j21)}$$

Hence $\boldsymbol{I_1} = \dfrac{(400-j100)}{(32-j21)} = \dfrac{412.31\angle -14.04°}{38.28\angle -33.27°}$

$= 10.77\angle 19.23°\text{ A} = \mathbf{10.8\angle -19.2°\ A}$, correct to one decimal place

$\boldsymbol{I_2} = \dfrac{400\angle -90°}{38.28\angle -33.27°} = 10.45\angle -56.73°\text{ A}$

$= \mathbf{10.5\angle -56.7°\ A}$, correct to one decimal place

(b) Current flowing in capacitor

$= I_1 - I_2$

$= 10.77\angle 19.23° - 10.45\angle -56.73°$

$= 4.44 + j12.28 = 13.1\angle 70.12°\text{ A}$,

i.e. **the current in the capacitor is 13.1 A**

(c) Source power $P = VI\cos\phi$

$= (100)(10.77)\cos 19.23°$

$= \mathbf{1016.9\ W} = 1020\text{ W}$, correct to three significant figures.

[Check: power in 5 Ω resistor

$= I_1^2(5) = (10.77)^2(5) = 579.97\text{ W}$

and power in 4 Ω resistor

$= I_2^2(4) = (10.45)^2(4) = 436.81\text{ W}$

Thus total power dissipated

$= 579.97 + 436.81$

$= 1016.8\text{ W} = 1020\text{ W}$, correct to three significant figures.]

Problem 3. A balanced star-connected three-phase load is shown in Figure 34.4. Determine the value of the line currents I_R, I_Y and I_B using mesh-current analysis.

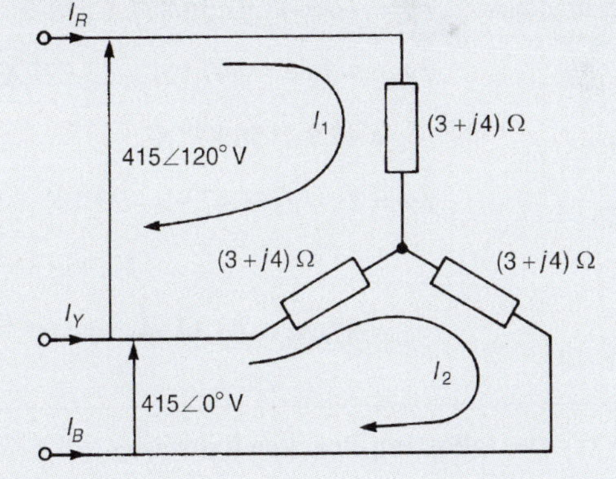

Figure 34.4

Two mesh currents I_1 and I_2 are chosen as shown in Figure 34.4.

From loop 1,

$$I_1(3+j4)+I_1(3+j4)-I_2(3+j4)=415\angle 120°$$

i.e. $(6+j8)I_1-(3+j4)I_2-415\angle 120°=0$ (1)

Part 4

From loop 2,

$$I_2(3+j4) - I_1(3+j4) + I_2(3+j4) = 415\angle 0^\circ$$

i.e. $-(3+j4)I_1 + (6+j8)I_2 - 415\angle 0^\circ = 0$ (2)

Solving equations (1) and (2) using determinants gives:

$$\frac{I_1}{\begin{vmatrix} -(3+j4) & -415\angle 120^\circ \\ (6+j8) & -415\angle 0^\circ \end{vmatrix}} = \frac{-I_2}{\begin{vmatrix} (6+j8) & -415\angle 120^\circ \\ -(3+j4) & -415\angle 0^\circ \end{vmatrix}}$$

$$= \frac{1}{\begin{vmatrix} (6+j8) & -(3+j4) \\ -(3+j4) & (6+j8) \end{vmatrix}}$$

$$\frac{I_1}{2075\angle 53.13^\circ + 4150\angle 173.13^\circ}$$

$$= \frac{-I_2}{-4150\angle 53.13^\circ - 2075\angle 173.13^\circ}$$

$$= \frac{1}{100\angle 106.26^\circ - 25\angle 106.26^\circ}$$

$$\frac{I_1}{3594\angle 143.13^\circ} = \frac{I_2}{3594\angle 83.13^\circ} = \frac{1}{75\angle 106.26^\circ}$$

Hence $I_1 = \dfrac{3594\angle 143.13^\circ}{75\angle 106.26^\circ} = 47.9\angle 36.87^\circ\,\text{A}$

and $I_2 = \dfrac{3594\angle 83.13^\circ}{75\angle 106.26^\circ} = 47.9\angle -23.13^\circ\,\text{A}$

Thus line current $\boldsymbol{I_R = I_1 = 47.9\angle 36.87^\circ\,\text{A}}$

$$I_B = -I_2 = -(47.9\angle -23.23^\circ\,\text{A})$$

$$= \mathbf{47.9\angle 156.87^\circ\,A}$$

and

$$I_Y = I_2 - I_1 = 47.9\angle{-23.13^\circ} - 47.96\angle 36.87^\circ$$

$$= \mathbf{47.9\angle -83.13^\circ\,A}$$

Now try the following Practice Exercise

Practice Exercise 136 Mesh-current analysis (Answers on page 828)

1. Repeat Problems 1 to 10 of Exercise 135, page 504 using mesh-current analysis.
2. For the network shown in Figure 34.5, use mesh-current analysis to determine the value of current I and the active power output of the voltage source.

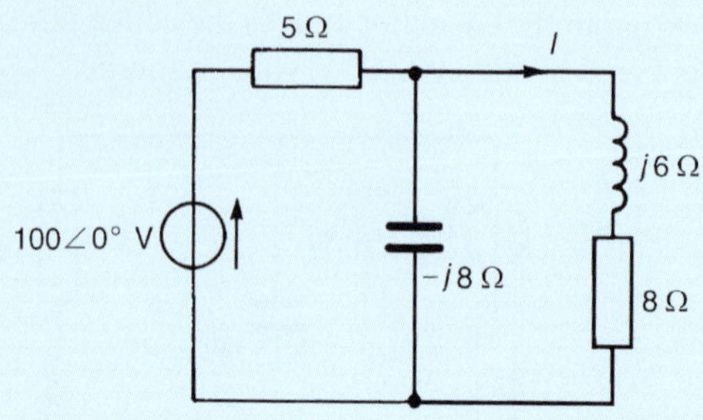

Figure 34.5

3. Use mesh-current analysis to determine currents I_1, I_2 and I_3 for the network shown in Figure 34.6.

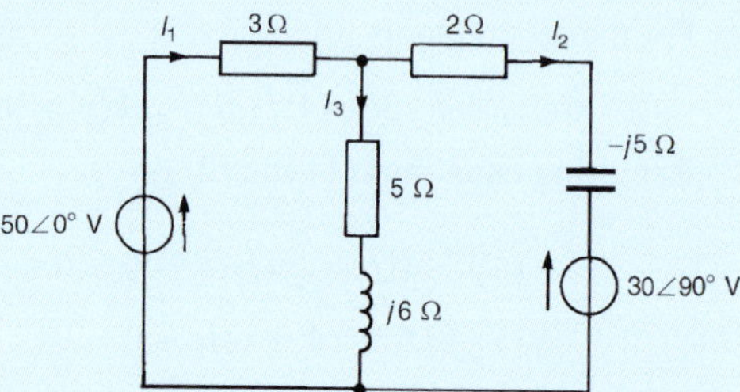

Figure 34.6

4. For the network shown in Figure 34.7, determine the current flowing in the $(4+j3)\,\Omega$ impedance.

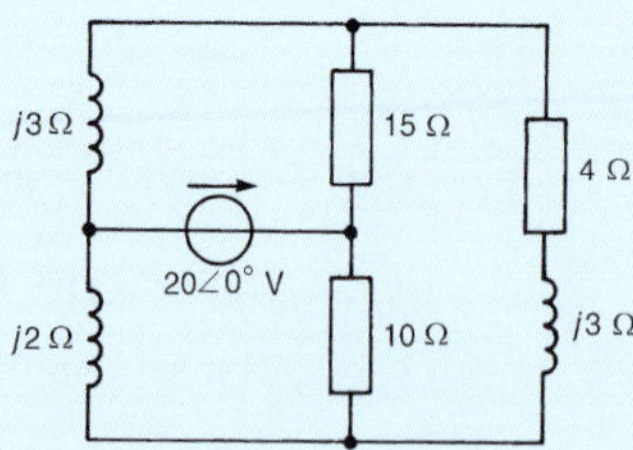

Figure 34.7

5. For the network shown in Figure 34.8, use mesh-current analysis to determine (a) the current in the capacitor, I_C, (b) the current in the inductance, I_L, (c) the p.d. across the $4\,\Omega$ resistance and (d) the total active circuit power.

Part 4

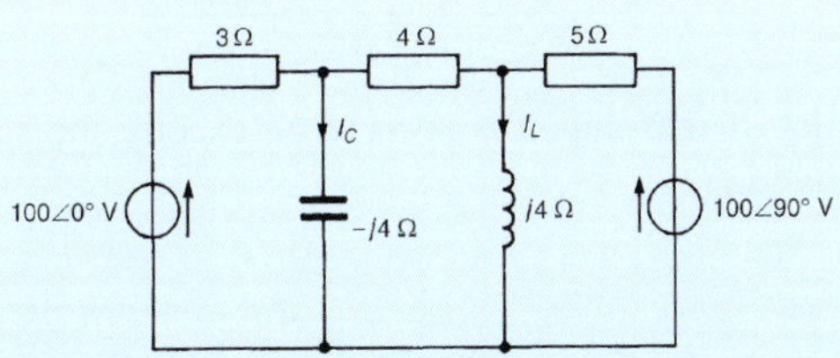

Figure 34.8

6. Determine the value of the currents I_R, I_Y and I_B in the network shown in Figure 34.9 by using mesh-current analysis.

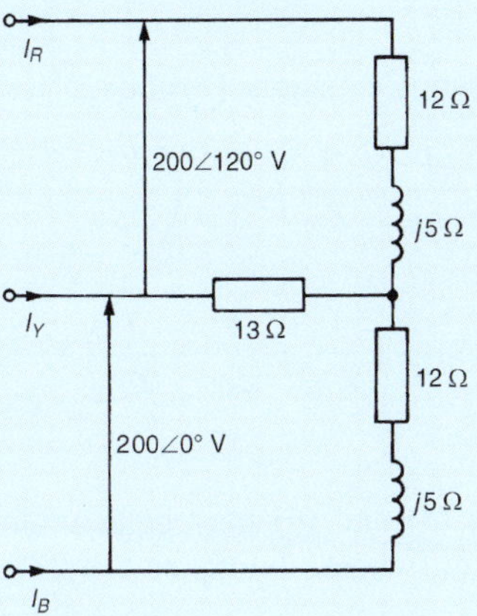

Figure 34.9

7. In the network of Figure 34.10, use mesh-current analysis to determine (a) the current in the capacitor, (b) the current in the 5 Ω resistance, (c) the active power output of the 15∠0° V source and (d) the magnitude of the p.d. across the $j2$ Ω inductance.

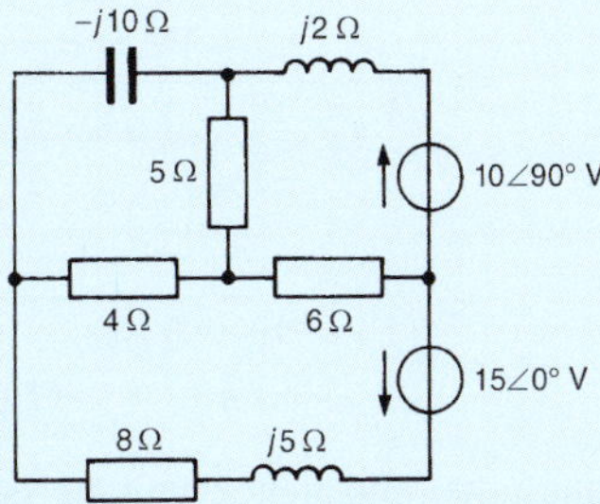

Figure 34.10

8. A balanced three-phase delta-connected load is shown in Figure 34.11. Use mesh-current analysis to determine the values of mesh currents I_1, I_2 and I_3 shown and hence find the line currents I_R, I_Y and I_B.

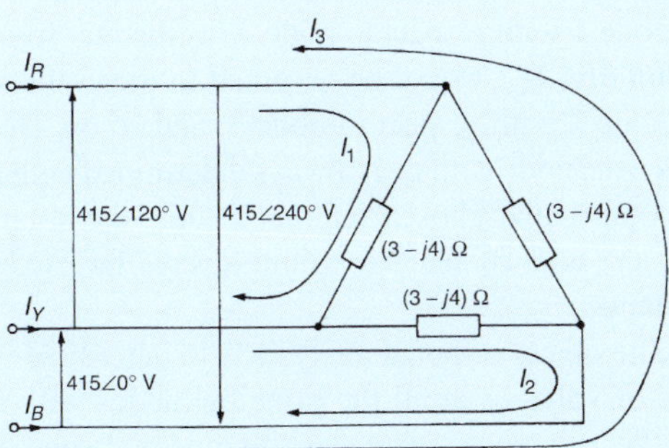

Figure 34.11

9. Use mesh-circuit analysis to determine the value of currents I_A to I_E in the circuit shown in Figure 34.12.

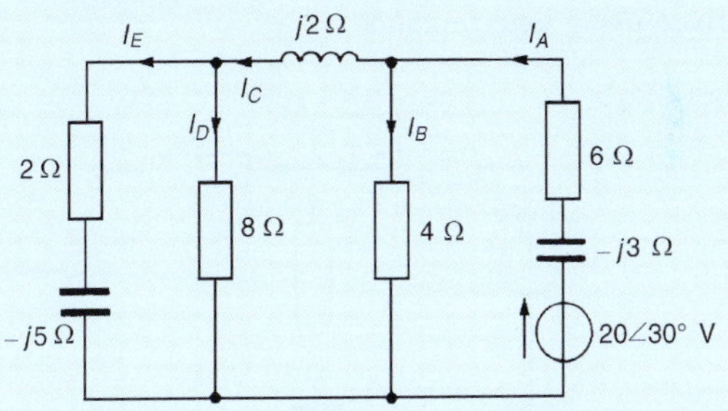

Figure 34.12

34.2 Nodal analysis

A **node** of a network is defined as a point where two or more branches are joined. If three or more branches join at a node, then that node is called a **principal node** or **junction**. In Figure 34.13, points 1, 2, 3, 4 and 5 are nodes, and points 1, 2 and 3 are principal nodes.

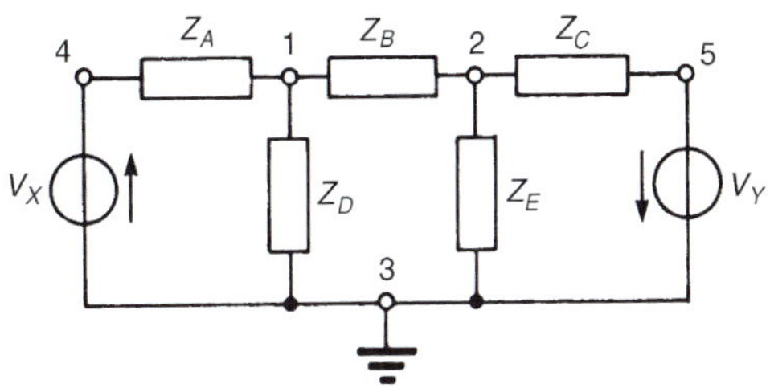

Figure 34.13

Part 4

A node voltage is the voltage of a particular node with respect to a node called the reference node. If in Figure 34.13, for example, node 3 is chosen as the reference node then V_{13} is assumed to mean the voltage at node 1 with respect to node 3 (as distinct from V_{31}). Similarly, V_{23} would be assumed to mean the voltage at node 2 with respect to node 3, and so on. However, since the node voltage is always determined with respect to a particular chosen reference node, the notation V_1 for V_{13} and V_2 for V_{23} would always be used in this instance.

The object of nodal analysis is to determine the values of voltages at all the principal nodes with respect to the reference node, e.g., to find voltages V_1 and V_2 in Figure 34.13. When such voltages are determined, the currents flowing in each branch can be found.

Kirchhoff's current law is applied to nodes 1 and 2 in turn in Figure 34.13 and two equations in unknowns V_1 and V_2 are obtained which may be simultaneously solved using determinants.

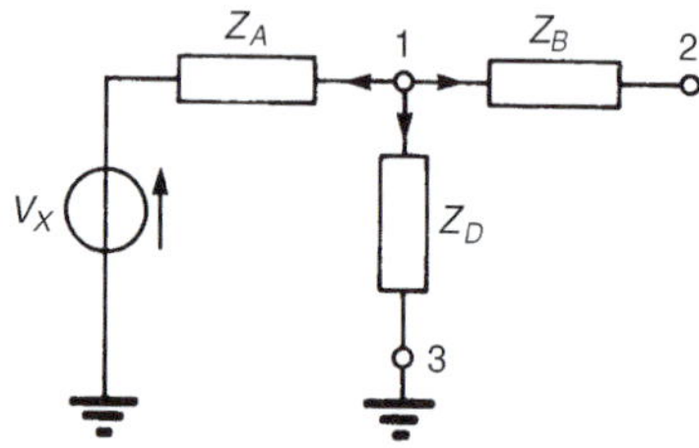

Figure 34.14

The branches leading to node 1 are shown separately in Figure 34.14. Let us assume that all branch currents are leaving the node as shown. Since the sum of currents at a junction is zero,

$$\frac{V_1 - V_x}{Z_A} + \frac{V_1}{Z_D} + \frac{V_1 - V_2}{Z_B} = 0 \qquad (1)$$

Similarly, for node 2, assuming all branch currents are leaving the node as shown in Figure 34.15,

$$\frac{V_2 - V_1}{Z_B} + \frac{V_2}{Z_E} + \frac{V_2 + V_Y}{Z_C} = 0 \qquad (2)$$

In equations (1) and (2), the currents are all assumed to be leaving the node. In fact, any selection in the direction of the branch currents may be made – the resulting equations will be identical. (For example, if for node 1 the current flowing in Z_B is considered as flowing towards node 1 instead of away, then the equation for node 1 becomes

$$\frac{V_1 - V_x}{Z_A} + \frac{V_1}{Z_D} = \frac{V_2 - V_1}{Z_B}$$

which if rearranged is seen to be exactly the same as equation (1).

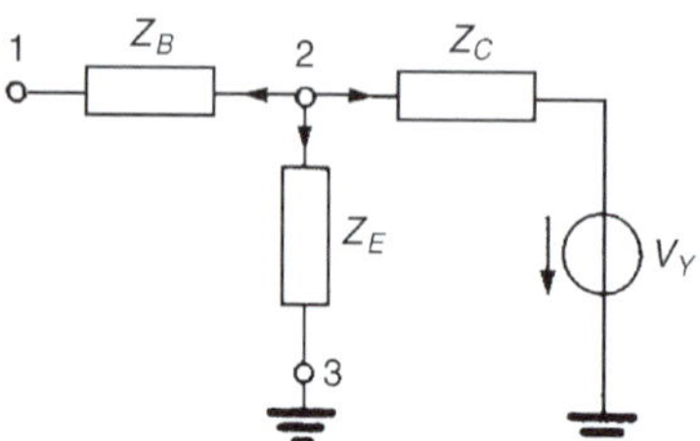

Figure 34.15

Rearranging equations (1) and (2) gives:

$$\left(\frac{1}{Z_A} + \frac{1}{Z_B} + \frac{1}{Z_D}\right) V_1 - \left(\frac{1}{Z_B}\right) V_2 - \left(\frac{1}{Z_A}\right) V_x = 0 \qquad (3)$$

$$-\left(\frac{1}{Z_B}\right) V_1 + \left(\frac{1}{Z_B} + \frac{1}{Z_C} + \frac{1}{Z_E}\right) V_2 + \left(\frac{1}{Z_C}\right) V_Y = 0 \qquad (4)$$

Equations (3) and (4) may be rewritten in terms of admittances (where admittance $Y = l/Z$):

$$(Y_A + Y_B + Y_D)V_1 - Y_B V_2 - Y_A V_x = 0 \qquad (5)$$

$$-Y_B V_1 + (Y_B + Y_C + Y_E)V_2 + Y_C V_Y = 0 \qquad (6)$$

Equations (5) and (6) may be solved for V_1 and V_2 by using determinants. Thus

$$\frac{V_1}{\begin{vmatrix} -Y_B & -Y_A \\ (Y_B + Y_C + Y_E) & Y_C \end{vmatrix}} = \frac{-V_2}{\begin{vmatrix} (Y_A + Y_B + Y_D) & -Y_A \\ -Y_B & Y_C \end{vmatrix}} = \frac{1}{\begin{vmatrix} (Y_A + Y_B + Y_D) & -Y_B \\ -Y_B & (Y_B + Y_C + Y_E) \end{vmatrix}}$$

Current equations, and hence voltage equations, may be written at each principal node of a network with the exception of a reference node. The number of equations necessary to produce a solution for a circuit is, in fact, always one less than the number of principal nodes.

Whether mesh-current analysis or nodal analysis is used to determine currents in circuits depends on the number of loops and nodes the circuit contains.

Basically, the method that requires the least number of equations is used. The method of nodal analysis is demonstrated in the following problems.

Problem 4. For the network shown in Figure 34.16, determine the voltage V_{AB}, by using nodal analysis.

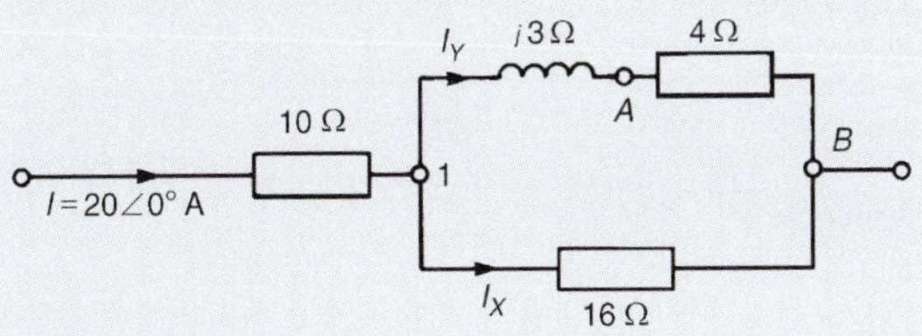

Figure 34.16

Figure 34.16 contains two principal nodes (at 1 and B) and thus only one nodal equation is required. B is taken as the reference node and the equation for node 1 is obtained as follows. Applying Kirchhoff's current law to node 1 gives:

$$I_X + I_Y = I$$

i.e.
$$\frac{V_1}{16} + \frac{V_1}{(4+j3)} = 20\angle 0°$$

Thus
$$V_1\left(\frac{1}{16} + \frac{1}{4+j3}\right) = 20$$

$$V_1\left(0.0625 + \frac{4-j3}{4^2+3^2}\right) = 20$$

$$V_1(0.0625 + 0.16 - j0.12) = 20$$

$$V_1(0.2225 - j0.12) = 20$$

from which
$$V_1 = \frac{20}{(0.2225 - j0.12)}$$

$$= \frac{20}{0.2528\angle -28.34°}$$

i.e. voltage
$$V_1 = 79.1\angle 28.34° \text{ V}$$

The current through the $(4+j3)\,\Omega$ branch, $I_Y = V_1/(4+j3)$

Hence the voltage drop between points A and B, i.e. across the $4\,\Omega$ resistance, is given by:

$$V_{AB} = (I_Y)(4) = \frac{V_1(4)}{(4+j3)} = \frac{79.1\angle 28.34°}{5\angle 36.87°}(4)$$

$$= \mathbf{63.3\angle -8.53°\ V}$$

Problem 5. Determine the value of voltage V_{XY} shown in the circuit of Figure 34.17.

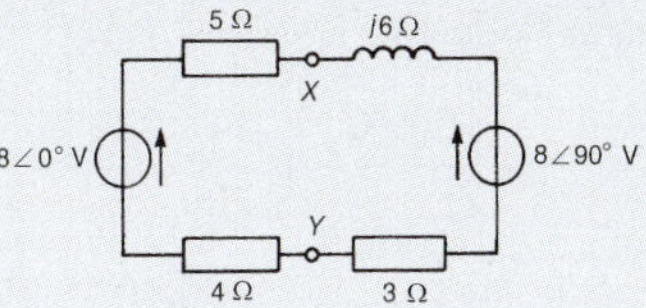

Figure 34.17

The circuit contains no principal nodes. However, if point Y is chosen as the reference node then an equation may be written for node X assuming that current leaves point X by both branches.

Thus
$$\frac{V_X - 8\angle 0°}{(5+4)} + \frac{V_x - 8\angle 90°}{(3+j6)} = 0$$

from which,
$$V_X\left(\frac{1}{9} + \frac{1}{3+j6}\right) = \frac{8}{9} + \frac{j8}{3+j6}$$

$$V_X\left(\frac{1}{9} + \frac{3-j6}{3^2+6^2}\right) = \frac{8}{9} + \frac{j8(3-j6)}{3^2+6^2}$$

$$V_X(0.1778 - j0.1333) = 0.8889 + \frac{48+j24}{45}$$

$$V_X(0.2222\angle -36.86°) = 1.9556 + j0.5333$$

$$= 2.027\angle 15.25°$$

Since point Y is the reference node,

$$\text{voltage } V_X = \boldsymbol{V_{XY}} = \frac{2.027\angle 15.25°}{0.2222\angle -36.86°}$$

$$= \mathbf{9.12\angle 52.11°\ V}$$

Problem 6. Use nodal analysis to determine the current flowing in each branch of the network shown in Figure 34.18.

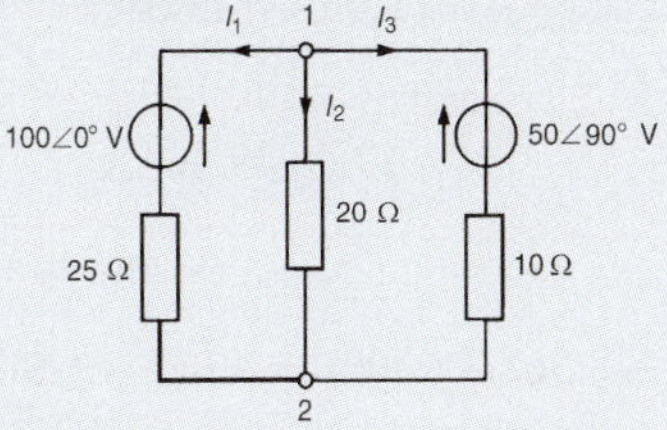

Figure 34.18

This is the same problem as Problem 1 of Chapter 33, page 501, which was solved using Kirchhoff's laws. A comparison of methods can be made.

Part 4

There are only two principal nodes in Figure 34.18 so only one nodal equation is required. Node 2 is taken as the reference node.

The equation at node 1 is $I_1+I_2+I_3=0$

i.e. $$\frac{V_1-100\angle 0^\circ}{25}+\frac{V_1}{20}+\frac{V_1-50\angle 90^\circ}{10}=0$$

i.e. $$\left(\frac{1}{25}+\frac{1}{20}+\frac{1}{10}\right)V_1-\frac{100\angle 0^\circ}{25}-\frac{50\angle 90^\circ}{10}=0$$

$$0.19V_1=4+j5$$

Thus the voltage at node 1, $V_1=\dfrac{4+j5}{0.19}$

$$=33.70\angle 51.34^\circ\,\text{V}$$

or $(21.05+j26.32)$ V

Hence the current in the 25 Ω resistance,

$$I_1=\frac{V_1-100\angle 0^\circ}{25}=\frac{21.05+j26.32-100}{25}$$

$$=\frac{-78.95+j26.32}{25}$$

$=\mathbf{3.33\angle 161.56^\circ\,A}$ flowing away from node 1

(or $3.33\angle(161.56^\circ-180^\circ)$ A $=\mathbf{3.33\angle -18.44^\circ\,A}$ flowing toward node 1)

The current in the 20 Ω resistance,

$$I_2=\frac{V_1}{20}=\frac{33.70\angle 51.34^\circ}{20}=\mathbf{1.69\angle 51.34^\circ\,A}$$

flowing from node 1 to node 2

The current in the 10 Ω resistor,

$$I_3=\frac{V_1-50\angle 90^\circ}{10}=\frac{21.05+j26.32-j50}{10}$$

$$=\frac{21.05-j23.68}{10}$$

$=\mathbf{3.17\angle -48.36^\circ\,A}$

away from node 1

(or $3.17\angle(-48.36^\circ-180^\circ)=3.17\angle -228.36^\circ$ A

$=\mathbf{3.17\angle 131.64^\circ\,A}$

toward node 1)

Problem 7. In the network of Figure 34.19 use nodal analysis to determine (a) the voltage at nodes 1 and 2, (b) the current in the $j4\,\Omega$ inductance, (c) the current in the 5 Ω resistance and (d) the magnitude of the active power dissipated in the 2.5 Ω resistance.

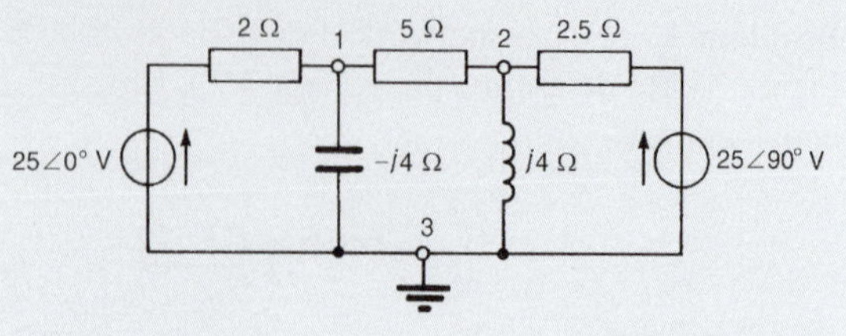

Figure 34.19

(a) At node 1, $$\frac{V_1-25\angle 0^\circ}{2}+\frac{V_1}{-j4}+\frac{V_1-V_2}{5}=0$$

Rearranging gives:

$$\left(\frac{1}{2}+\frac{1}{-j4}+\frac{1}{5}\right)V_1-\left(\frac{1}{5}\right)V_2-\frac{25\angle 0^\circ}{2}=0$$

i.e. $$(0.7+j0.25)V_1-0.2V_2-12.5=0 \quad (1)$$

At node 2, $$\frac{V_2-25\angle 90^\circ}{2.5}+\frac{V_2}{j4}+\frac{V_2-V_1}{5}=0$$

Rearranging gives:

$$-\left(\frac{1}{5}\right)V_1+\left(\frac{1}{2.5}+\frac{1}{j4}+\frac{1}{5}\right)V_2-\frac{25\angle 90^\circ}{2.5}=0$$

i.e. $$-0.2V_1+(0.6-j0.25)V_2-j10=0 \quad (2)$$

Thus two simultaneous equations have been formed with two unknowns, V_1 and V_2. Using determinants, if

$$(0.7+j0.25)V_1-0.2V_2-12.5=0 \quad (1)$$

and $$-0.2V_1+(0.6-j0.25)V_2-j10=0 \quad (2)$$

then $$\frac{V_1}{\begin{vmatrix}-0.2 & -12.5\\(0.6-j0.25) & -j10\end{vmatrix}}=\frac{-V_2}{\begin{vmatrix}(0.7+j0.25) & -12.5\\-0.2 & -j10\end{vmatrix}}=\frac{1}{\begin{vmatrix}(0.7+j0.25) & -0.2\\-0.2 & (0.6-j0.25)\end{vmatrix}}$$

i.e.

$$\frac{V_1}{(j2+7.5-j3.125)}=\frac{-V_2}{(-j7+2.5-2.5)}$$

$$=\frac{1}{(0.42-j0.175+j0.15+0.0625-0.04)}$$

and $$\frac{V_1}{7.584\angle -8.53^\circ}=\frac{-V_2}{-7\angle 90^\circ}$$

$$=\frac{1}{0.443\angle -3.23^\circ}$$

Thus **voltage,** $\boldsymbol{V_1}=\dfrac{7.584\angle -8.53^\circ}{0.443\angle -3.23^\circ}$

$$=17.12\angle -5.30^\circ\,\text{V}$$

$=\mathbf{17.1\angle -5.3^\circ\,V}$, correct to one decimal place.

and **voltage,** $\boldsymbol{V_2}=\dfrac{7\angle 90^\circ}{0.443\angle -3.23^\circ}$

$$=15.80\angle 93.23^\circ\,\text{V}$$

$=\mathbf{15.8\angle 93.2^\circ\,V}$, correct to one decimal place.

(b) The current in the $j4\,\Omega$ inductance is given by:

$$\frac{V_2}{j4}=\frac{15.80\angle 93.23^\circ}{4\angle 90^\circ}$$

$=\mathbf{3.95\angle 3.23^\circ\,A}$ flowing away from node 2

(c) The current in the $5\,\Omega$ resistance is given by:

$$I_5=\frac{V_1-V_2}{5}$$

$$=\frac{17.12\angle -5.30^\circ-15.80\angle 93.23^\circ}{5}$$

i.e. $$I_5=\frac{(17.05-j1.58)-(-0.89+j15.77)}{5}$$

$$=\frac{17.94-j17.35}{5}=\frac{24.96\angle -44.04^\circ}{5}$$

$=\mathbf{4.99\angle -44.04^\circ\,A}$ flowing from node 1 to node 2

(d) The active power dissipated in the $2.5\,\Omega$ resistor is given by

$$P_{2.5}=(I_{2.5})^2(2.5)=\left(\frac{V_2-25\angle 90^\circ}{2.5}\right)^2(2.5)$$

$$=\frac{(0.89+j15.77-j25)^2}{2.5}$$

$$=\frac{(9.273\angle -95.51^\circ)^2}{2.5}$$

$$=\frac{85.99\angle -191.02^\circ}{2.5}\text{ by de Moivre's theorem}$$

$$=34.4\angle 169^\circ\,\text{W}$$

Thus the magnitude of the active power dissipated in the 2.5 Ω resistance is 34.4 W

Problem 8. In the network shown in Figure 34.20 determine the voltage V_{XY} using nodal analysis.

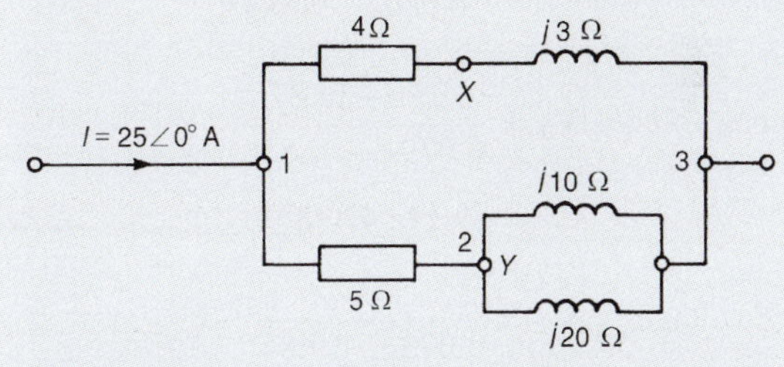

Figure 34.20

Node 3 is taken as the reference node.

At node 1, $$25\angle 0^\circ=\frac{V_1}{4+j3}+\frac{V_1-V_2}{5}$$

i.e. $$\left(\frac{4-j3}{25}+\frac{1}{5}\right)V_1-\frac{1}{5}V_2-25=0$$

or $$(0.379\angle -18.43^\circ)V_1-0.2V_2-25=0 \quad (1)$$

At node 2, $$\frac{V_2}{j10}+\frac{V_2}{j20}+\frac{V_2-V_1}{5}=0$$

i.e. $$-0.2V_1+\left(\frac{1}{j10}+\frac{1}{j20}+\frac{1}{5}\right)V_2=0$$

or $$-0.2V_1+(-j0.1-j0.05+0.2)V_2=0$$

i.e. $$-0.2V_1+(0.25\angle -36.87^\circ)V_2+0=0 \quad (2)$$

Simultaneous equations (1) and (2) may be solved for V_1 and V_2 by using determinants. Thus,

$$\frac{V_1}{\begin{vmatrix} -0.2 & -25 \\ 0.25\angle -36.87^\circ & 0 \end{vmatrix}} = \frac{-V_2}{\begin{vmatrix} 0.379\angle -18.43^\circ & -25 \\ -0.2 & 0 \end{vmatrix}}$$

$$= \frac{1}{\begin{vmatrix} 0.379\angle -18.43^\circ & -0.2 \\ -0.2 & 0.25\angle -36.87^\circ \end{vmatrix}}$$

i.e. $$\frac{V_1}{6.25\angle -36.87^\circ} = \frac{-V_2}{-5}$$

$$= \frac{1}{0.09475\angle -55.30^\circ - 0.04}$$

$$= \frac{1}{0.079\angle -79.85^\circ}$$

Hence voltage, $$\mathbf{V_1} = \frac{6.25\angle -36.87^\circ}{0.079\angle -79.85^\circ} = \mathbf{79.11\angle 42.98^\circ\ V}$$

and voltage, $$\mathbf{V_2} = \frac{5}{0.079\angle -79.85^\circ} = \mathbf{63.29\angle 79.85^\circ\ V}$$

The current flowing in the $(4 + j3)\ \Omega$ branch is $V_1/(4 + j3)$. Hence the voltage between point X and node 3 is:

$$\frac{V_1}{(4+j3)}(j3) = \frac{(79.11\angle 42.98^\circ)(3\angle 90^\circ)}{5\angle 36.87^\circ} = 47.47\angle 96.11^\circ\ V$$

Thus the voltage

$$V_{XY} = V_X - V_Y = V_X - V_2 = 47.47\angle 96.11^\circ - 63.29\angle 79.85^\circ$$

$$= -16.21 - j15.10 = \mathbf{22.15\angle -137^\circ\ V}$$

Problem 9. Use nodal analysis to determine the voltages at nodes 2 and 3 in Figure 34.21 and hence determine the current flowing in the 2 Ω resistor and the power dissipated in the 3 Ω resistor.

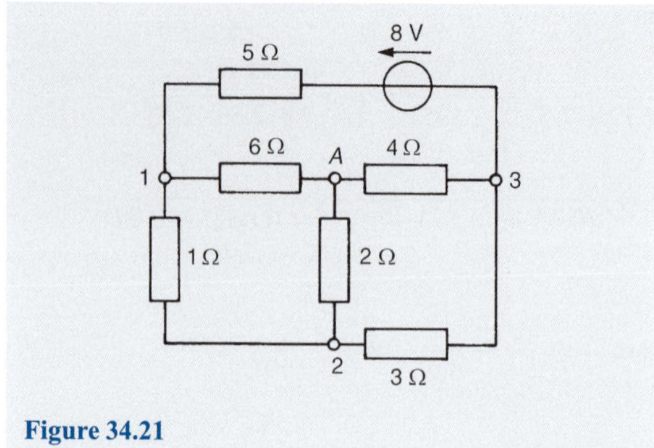

Figure 34.21

This is the same problem as Problem 2 of Chapter 33, page 501, which was solved using Kirchhoff's laws. In Figure 34.21, the reference node is shown at point A.

At node 1, $$\frac{V_1 - V_2}{1} + \frac{V_1}{6} + \frac{V_1 - 8 - V_3}{5} = 0$$

i.e. $$1.367V_1 - V_2 - 0.2V_3 - 1.6 = 0 \qquad (1)$$

At node 2, $$\frac{V_2}{2} + \frac{V_2 - V_1}{1} + \frac{V_2 - V_3}{3} = 0$$

i.e. $$-V_1 + 1.833V_2 - 0.333V_3 + 0 = 0 \qquad (2)$$

At node 3, $$\frac{V_3}{4} + \frac{V_3 - V_2}{3} + \frac{V_3 + 8 - V_1}{5} = 0$$

i.e. $$-0.2V_1 - 0.333V_2 + 0.783V_3 + 1.6 = 0 \qquad (3)$$

Equations (1) to (3) can be solved for V_1, V_2 and V_3 by using determinants. Hence

$$\frac{V_1}{\begin{vmatrix} -1 & -0.2 & -1.6 \\ 1.833 & -0.333 & 0 \\ -0.333 & 0.783 & 1.6 \end{vmatrix}} = \frac{-V_2}{\begin{vmatrix} 1.367 & -0.2 & -1.6 \\ -1 & -0.333 & 0 \\ -0.2 & 0.783 & 1.6 \end{vmatrix}}$$

$$= \frac{V_3}{\begin{vmatrix} 1.367 & -1 & -1.6 \\ -1 & 1.833 & 0 \\ -0.2 & -0.333 & 1.6 \end{vmatrix}} = \frac{-1}{\begin{vmatrix} 1.367 & -1 & -0.2 \\ -1 & 1.833 & -0.333 \\ -0.2 & -0.333 & 0.783 \end{vmatrix}}$$

Solving for V_2 gives:

$$\frac{-V_2}{-1.6(-0.8496) + 1.6(-0.6552)}$$

$$= \frac{-1}{1.367(1.3244) + 1(-0.8496) - 0.2(0.6996)}$$

hence $\dfrac{-V_2}{0.31104} = \dfrac{-1}{0.82093}$, from which

voltage, $\boldsymbol{V_2} = \dfrac{0.31104}{0.82093} = \mathbf{0.3789\,V}$

Thus the current in the 2 Ω resistor

$= \dfrac{V_2}{2} = \dfrac{0.3789}{2} = \mathbf{0.19\,A}$, flowing from node 2 to node A.

Solving for V_3 gives:

$$\frac{V_3}{-1.6(0.6996) + 1.6(1.5057)} = \frac{-1}{0.82093}$$

hence $\dfrac{V_3}{1.2898} = \dfrac{-1}{0.82093}$ from which,

voltage, $\boldsymbol{V_3} = \dfrac{-1.2898}{0.82093} = \mathbf{-1.571\,V}$

Power in the 3 Ω resistor

$$= (I_3)^2(3) = \left(\frac{V_2 - V_3}{3}\right)^2(3)$$

$$= \frac{(0.3789 - (-1.571))^2}{3} = \mathbf{1.27\,W}$$

Now try the following Practice Exercise

Practice Exercise 137 Nodal analysis (Answers on page 828)

1. Repeat Problems 1, 2, 5, 8 and 10 of Exercise 135, page 504 using nodal analysis.
2. Repeat Problems 2, 3, 5 and 9 of Exercise 136, page 510 using nodal analysis.
3. Determine for the network shown in Figure 34.22 the voltage at node 1 and the voltage V_{AB}

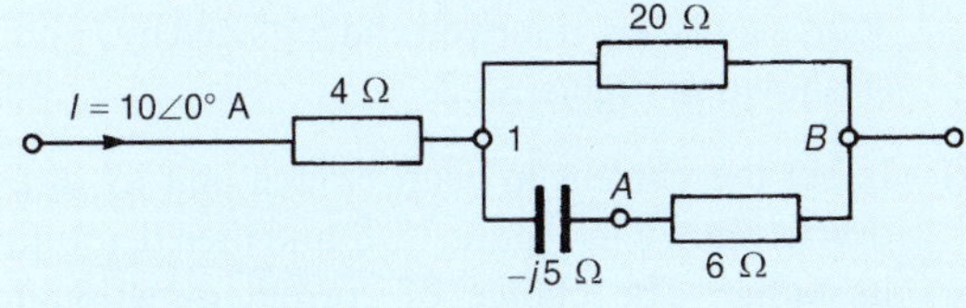

Figure 34.22

4. Determine the voltage V_{PQ} in the network shown in Figure 34.23.

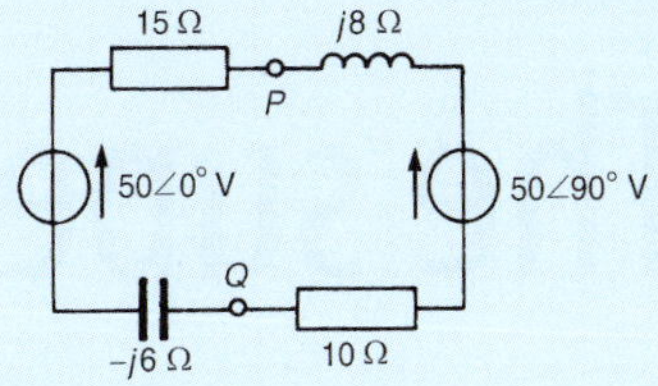

Figure 34.23

5. Use nodal analysis to determine the currents I_A, I_B and I_C shown in the network of Figure 34.24.

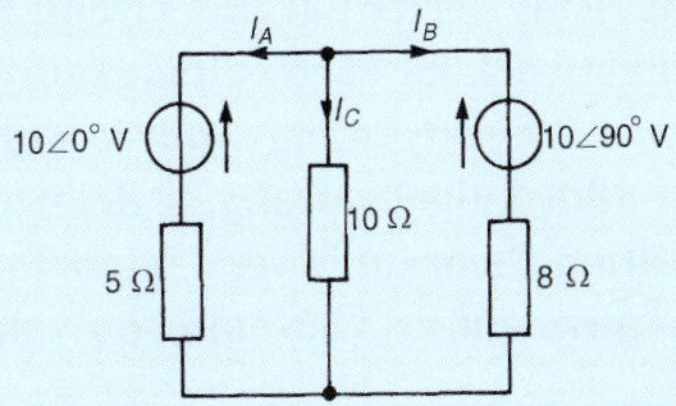

Figure 34.24

6. For the network shown in Figure 34.25 determine (a) the voltages at nodes 1 and 2, (b) the current in the 40 Ω resistance, (c) the current in the 20 Ω resistance and (d) the magnitude of the active power dissipated in the 10 Ω resistance.

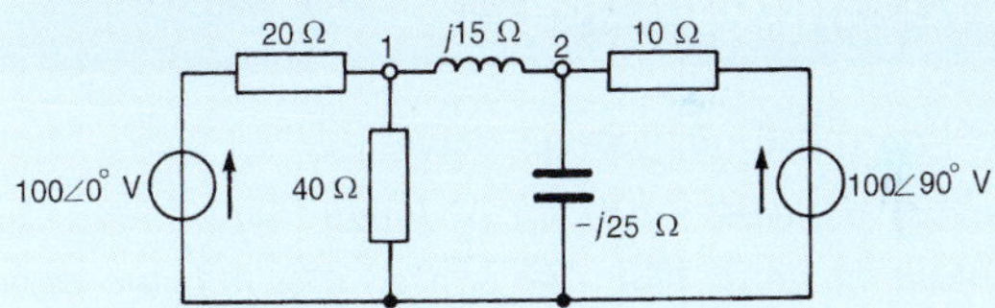

Figure 34.25

7. Determine the voltage V_{AB} in the network of Figure 34.26, using nodal analysis.

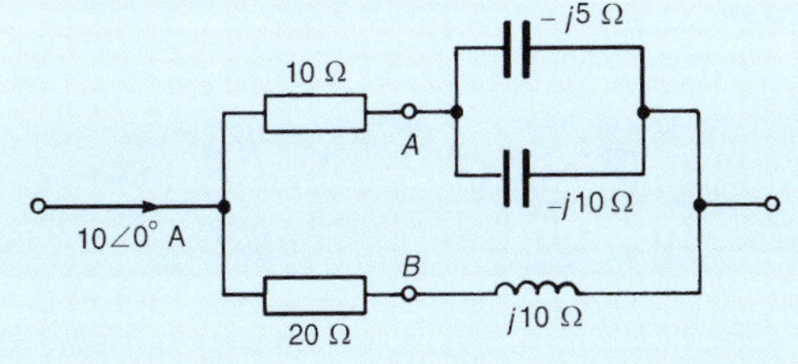

Figure 34.26

For fully worked solutions to each of the problems in Practice Exercises 136 and 137 in this chapter, go to the website:
www.routledge.com/cw/bird

Part 4

Chapter 35

The superposition theorem

***Why it is important to understand:* The superposition theorem**

The strategy used in the superposition theorem is to eliminate all but one voltage source within a network at a time, using series/parallel circuit analysis to determine voltage drops, and/or currents, within the modified network for each voltage source separately. Then, once voltage drops and/or currents have been determined for each voltage source working separately, the values are all 'superimposed' on top of each other, i.e. added algebraically, to determine the actual voltage drops/currents with all sources active. The superposition theorem is very important in circuit analysis; it is used in converting any circuit into its Norton equivalent or Thévenin equivalent (which will be explained in the next chapter).

At the end of this chapter you should be able to:

- solve d.c. and a.c. networks using the superposition theorem

35.1 Introduction

The superposition theorem states:

'In any network made up of linear impedances and containing more than one source of e.m.f. the resultant current flowing in any branch is the phasor sum of the currents that would flow in that branch if each source were considered separately, all other sources being replaced at that time by their respective internal impedances.'

35.2 Using the superposition theorem

The superposition theorem, which was introduced in Chapter 15 for d.c. circuits, may be applied to both d.c. and a.c. networks. A d.c. network is shown in Figure 35.1 and will serve to demonstrate the principle of application of the superposition theorem.

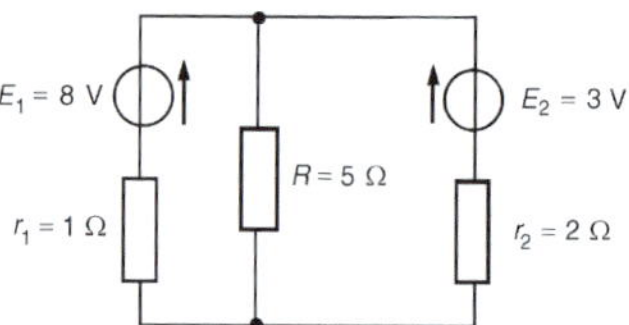

Figure 35.1

To find the current flowing in each branch of the circuit, the following six-step procedure can be adopted:

(i) Redraw the original network with one of the sources, say E_2, removed and replaced by r_2 only, as shown in Figure 35.2.

(ii) Label the current in each branch and its direction as shown in Figure 35.2, and then determine its value. The choice of current direction for I_1 depends on the source polarity which, by convention, is taken as flowing from the positive terminal as shown.

Electrical Circuit Theory and Technology. 978-1-138-67349-6, © 2017 John Bird. Published by Taylor & Francis. All rights reserved.

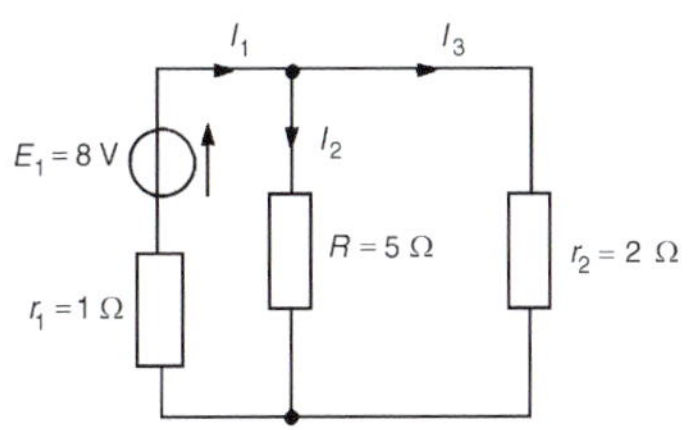

Figure 35.2

R in parallel with r_2 gives an equivalent resistance of

$(5 \times 2)/(5+2) = 10/7 = 1.429\,\Omega$

as shown in the equivalent network of Figure 35.3. From Figure 35.3,

$$\text{current } I_1 = \frac{E_1}{(r_1 + 1.429)} = \frac{8}{2.429} = 3.294\,\text{A}$$

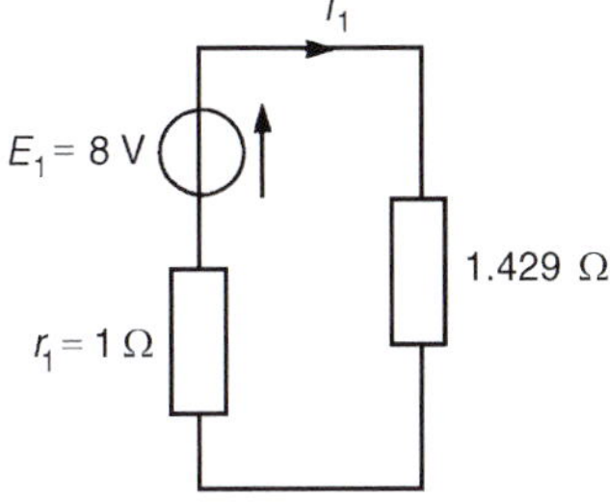

Figure 35.3

From Figure 35.2,

$$\text{current } I_2 = \left(\frac{r_2}{R+r_2}\right)(I_1) = \left(\frac{2}{5+2}\right)(3.294)$$

$$= 0.941\,\text{A}$$

$$\text{and current } I_3 = \left(\frac{5}{5+2}\right)(3.294) = 2.353\,\text{A}$$

(iii) Redraw the original network with source E_1 removed and replaced by r_1 only, as shown in Figure 35.4.

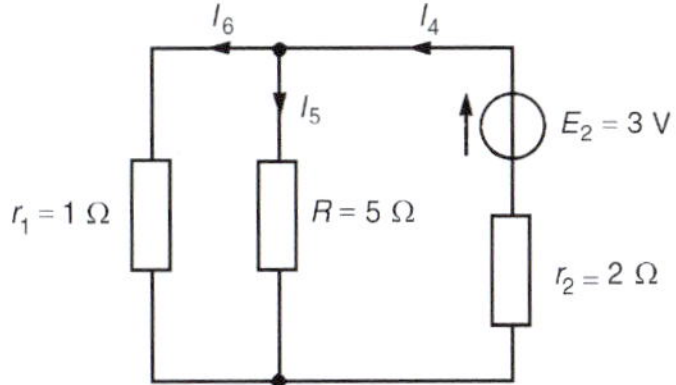

Figure 35.4

(iv) Label the currents in each branch and their directions as shown in Figure 35.4, and determine their values.

R and r_1 in parallel gives an equivalent resistance of

$(5 \times 1)/(5+1) = 5/6\,\Omega$ or $0.833\,\Omega$,

as shown in the equivalent network of Figure 35.5.

From Figure 35.5,

$$\text{current } I_4 = \frac{E_2}{r_2 + 0.833} = \frac{3}{2.833} = 1.059\,\text{A}$$

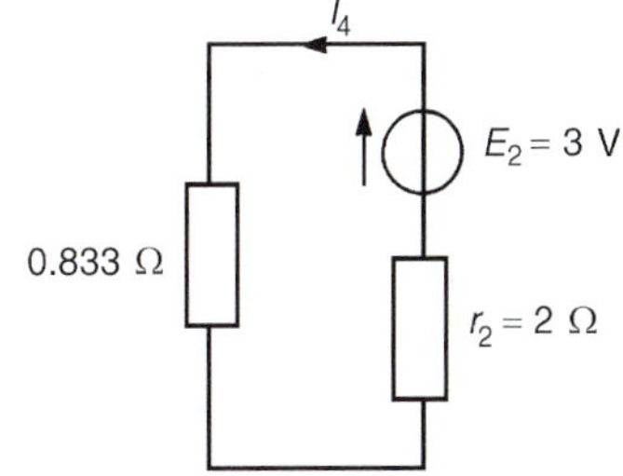

Figure 35.5

From Figure 35.4,

$$\text{current } I_5 = \left(\frac{1}{1+5}\right)(1.059) = 0.177\,\text{A}$$

$$\text{and current } I_6 = \left(\frac{5}{1+5}\right)(1.059) = 0.8825\,\text{A}$$

(v) Superimpose Figure 35.2 on Figure 35.4, as shown in Figure 35.6.

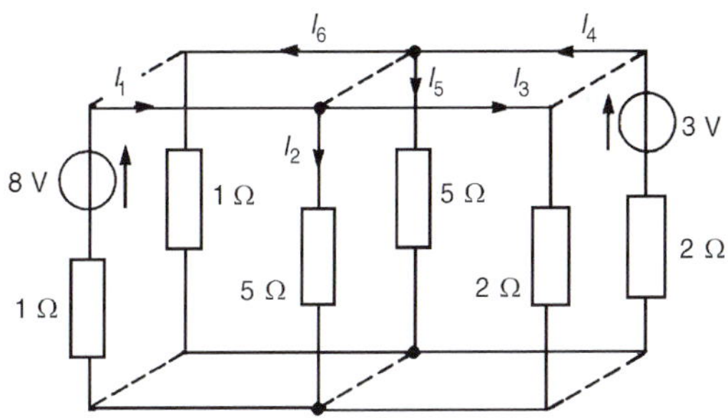

Figure 35.6

(vi) Determine the algebraic sum of the currents flowing in each branch. (Note that in an a.c. circuit it is the phasor sum of the currents that is required.)

From Figure 35.6, the resultant current flowing through the 8 V source is given by

$I_1 - I_6 = 3.294 - 0.8825 =$ **2.41 A** (discharging, i.e. flowing from the positive terminal of the source).

The resultant current flowing in the 3 V source is given by

$I_3 - I_4 = 2.353 - 1.059 = \mathbf{1.29\,A}$ (charging, i.e. flowing into the positive terminal of the source).

The resultant current flowing in the 5 Ω resistance is given by

$I_2 + I_5 = 0.941 + 0.177 = \mathbf{1.12\,A}$

The values of current are the same as those obtained on page 500 by using Kirchhoff's laws.

The following problems demonstrate further the use of the superposition theorem in analysing a.c. as well as d.c. networks. The theorem is straightforward to apply, but is lengthy. Thévenin's and Norton's theorems (described in Chapter 36) produce results more quickly.

Problem 1. A.c. sources of $100\angle 0°$V and internal resistance 25 Ω, and $50\angle 90°$V and internal resistance 10 Ω, are connected in parallel across a 20 Ω load. Determine using the superposition theorem, the current in the 20 Ω load and the current in each voltage source.

(This is the same problem as Problem 1 on page 501 and Problem 6 on page 513 and a comparison of methods may be made.)
The circuit diagram is shown in Figure 35.7. Following the above procedure:

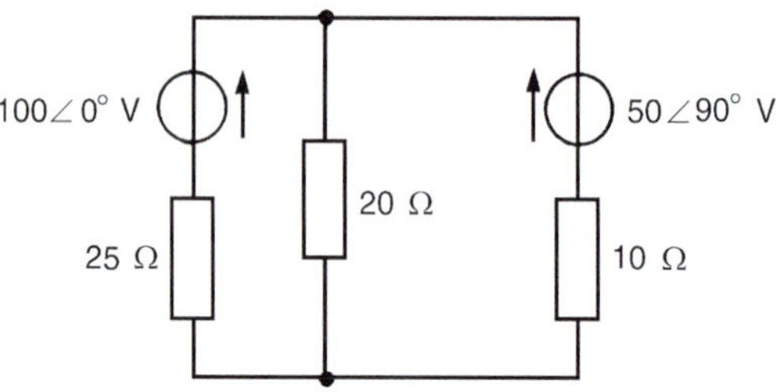

Figure 35.7

(i) The network is redrawn with the $50\angle 90°$ V source removed as shown in Figure 35.8

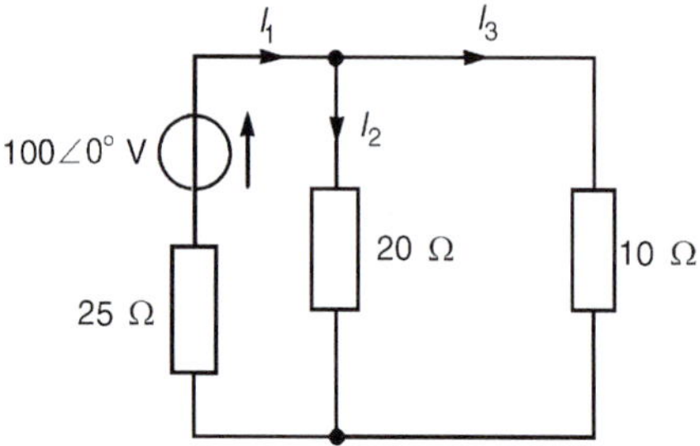

Figure 35.8

(ii) Currents I_1, I_2 and I_3 are labelled as shown in Figure 35.8.

$$I_1 = \frac{100\angle 0°}{25 + (10 \times 20)/(10+20)} = \frac{100\angle 0°}{25 + 6.667}$$
$$= 3.518\angle 0°\,\text{A}$$

$$I_2 = \left(\frac{10}{10+20}\right)(3.158\angle 0°) = 1.053\angle 0°\,\text{A}$$

$$I_3 = \left(\frac{20}{10+20}\right)(3.158\angle 0°) = 2.105\angle 0°\,\text{A}$$

(iii) The network is redrawn with the $100\angle 0°$ V source removed as shown in Figure 35.9

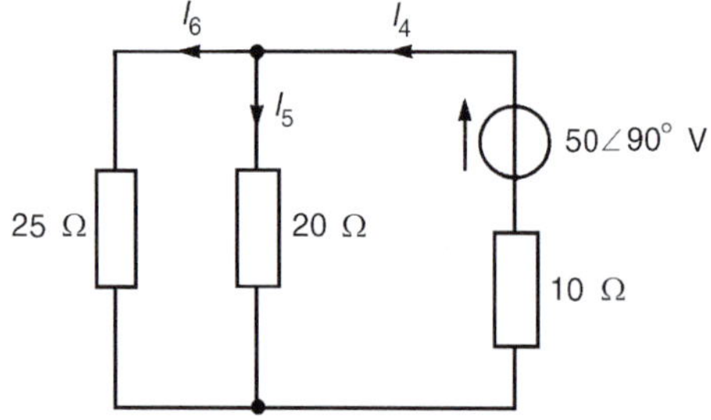

Figure 35.9

(iv) Currents I_4, I_5 and I_6 are labelled as shown in Figure 35.9.

$$I_4 = \frac{50\angle 90°}{10 + (25 \times 20)/(25+20)} = \frac{50\angle 90°}{10 + 11.111}$$
$$= 2.368\angle 90°\,\text{A or } j2.368\,\text{A}$$

$$I_5 = \left(\frac{25}{20+25}\right)(j2.368) = j1.316\,\text{A}$$

$$I_6 = \left(\frac{20}{20+25}\right)(j2.368) = j1.052\,\text{A}$$

(v) Figure 35.10 shows Figure 35.9 superimposed on Figure 35.8, giving the currents shown.

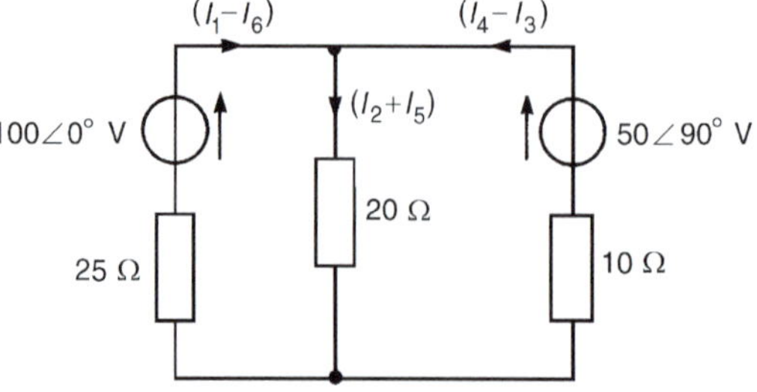

Figure 35.10

(vi) Current in the 20 Ω load,
$I_2 + I_5 = (1.053 + j1.316)$ A or **1.69∠51.33°A**

Current in the 100∠0° V source,
$I_1 - I_6 = (3.158 - j1.052)$ A or **3.33∠−18.42° A**

Current in the 50∠90° V source,
$I_4 - I_3 = (j2.368 - 2.105)$ or **3.17∠131.64° A**

Problem 2. Use the superposition theorem to determine the current in the 4 Ω resistor of the network shown in Figure 35.11.

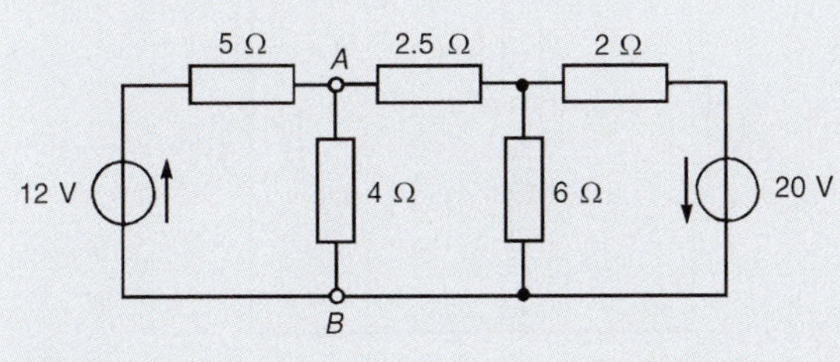

Figure 35.11

(i) Removing the 20 V source gives the network shown in Figure 35.12.

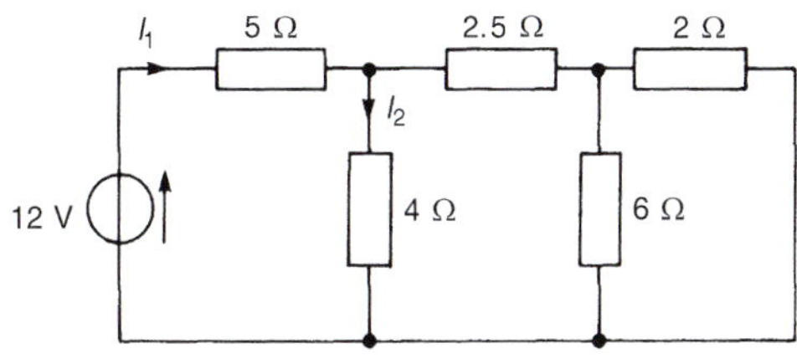

Figure 35.12

(ii) Currents I_1 and I_2 are shown labelled in Figure 35.12. It is unnecessary to determine the currents in all the branches since only the current in the 4 Ω resistance is required.

From Figure 35.12, 6 Ω in parallel with 2 Ω gives $(6 \times 2)/(6+2) = 1.5\,\Omega$, as shown in Figure 35.13.

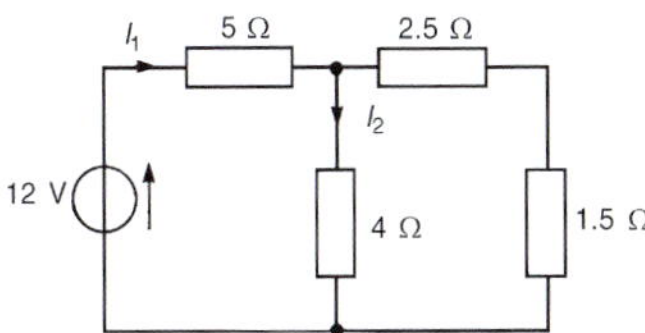

Figure 35.13

2.5 Ω in series with 1.5 Ω gives 4 Ω, 4 Ω in parallel with 4 Ω gives 2 Ω, and 2 Ω in series with 5 Ω gives 7 Ω.

Thus current $I_1 = \dfrac{12}{7} = 1.714$ A and

$$\text{current } I_2 = \left(\frac{4}{4+4}\right)(1.714) = 0.857\,\text{A}$$

(iii) Removing the 12 V source from the original network gives the network shown in Figure 35.14.

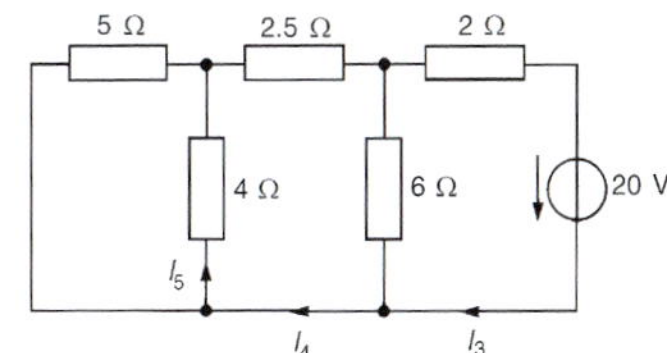

Figure 35.14

(iv) Currents I_3, I_4 and I_5 are shown labelled in Figure 35.14.

From Figure 35.14, 5 Ω in parallel with 4 Ω gives $(5 \times 4)/(5+4) = 20/9 = 2.222\,\Omega$, as shown in Figure 35.15, 2.222 Ω in series with 2.5 Ω gives 4.722 Ω, 4.722 Ω in parallel with 6 Ω gives $(4.722 \times 6)/(4.722+6) = 2.642\,\Omega$, 2.642 Ω in series with 2 Ω gives 4.642 Ω.

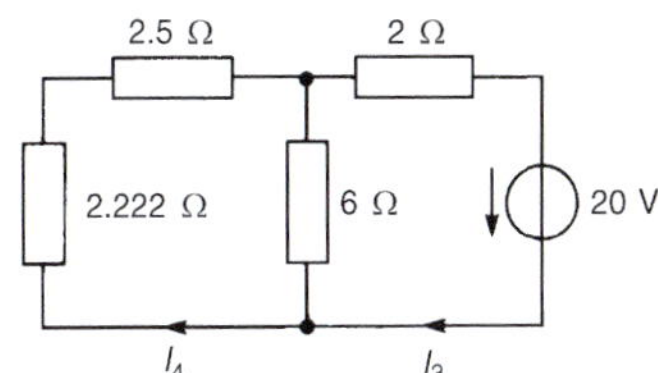

Figure 35.15

$$\text{Hence } I_3 = \frac{20}{4642} = 4.308\,\text{A}$$

$$I_4 = \left(\frac{6}{6+4.722}\right)(4.308) = 2.411\,\text{A},$$

from Figure 35.15

$$I_5 = \left(\frac{5}{4+5}\right)(2.411) = 1.339\,\text{A},$$

from Figure 35.14

(v) Superimposing Figure 35.14 on Figure 35.12 shows that the current flowing in the 4 Ω resistor is given by $I_5 - I_2$

(vi) $I_5 - I_2 = 1.339 - 0.857 =$ **0.48 A, flowing from B toward A** (see Figure 35.11).

Problem 3. Use the superposition theorem to obtain the current flowing in the $(4+j3)\,\Omega$ impedance of Figure 35.16.

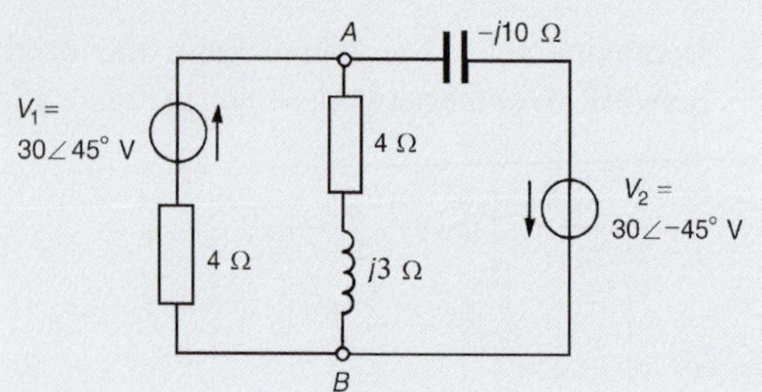

Figure 35.16

(i) The network is redrawn with V_2 removed, as shown in Figure 35.17.

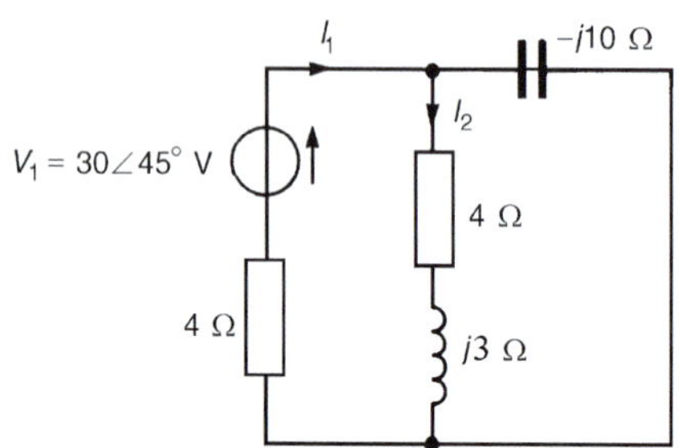

Figure 35.17

(ii) Current I_1 and I_2 are shown in Figure 35.17. From Figure 35.17, $(4+j3)\,\Omega$ in parallel with $-j10\,\Omega$ gives an equivalent impedance of

$$\frac{(4+j3)(-j10)}{(4+j3-j10)}=\frac{30-j40}{4-j7}$$

$$=\frac{50\angle-53.13^\circ}{8.062\angle-60.26^\circ}$$

$$=6.202\angle 7.13^\circ \text{ or}$$

$$(6.154+j0.770)\,\Omega$$

Total impedance of Figure 35.17 is

$$6.154+j0.770+4=(10.154+j0.770)\,\Omega \text{ or}$$

$$10.183\angle 4.34^\circ\Omega$$

$$\text{Hence current } I_1=\frac{30\angle 45^\circ}{10.183\angle 4.34^\circ}$$

$$=2.946\angle 40.66^\circ\text{ A}$$

$$\text{and current } I_2=\left(\frac{-j10}{4-j7}\right)(2.946\angle 40.66^\circ)$$

$$=\frac{(10\angle-90^\circ)(2.946\angle 40.66^\circ)}{8.062\angle-60.26^\circ}$$

$$=3.654\angle 10.92^\circ\text{ A or}$$

$$(3.588+j0.692)\text{ A}$$

(iii) The original network is redrawn with V_1 removed, as shown in Figure 35.18.

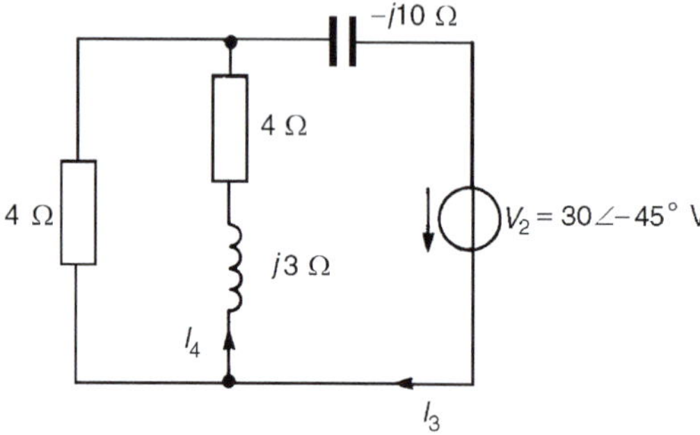

Figure 35.18

(iv) Currents I_3 and I_4 are shown in Figure 35.18. From Figure 35.18, $4\,\Omega$ in parallel with $(4+j3)\,\Omega$ gives an equivalent impedance of

$$\frac{4(4+j3)}{4+4+j3}=\frac{16+j12}{8+j3}=\frac{20\angle 36.87^\circ}{8.544\angle 20.56^\circ}$$

$$=2.341\angle 16.31^\circ\Omega \text{ or}$$

$$(2.247+j0.657)\,\Omega$$

Total impedance of Figure 35.18 is

$$2.247+j0.657-j10=(2.247-j9.343)\,\Omega \text{ or}$$

$$9.609\angle-76.48^\circ\Omega$$

$$\text{Hence current } I_3=\frac{30\angle-45^\circ}{9.609\angle-76.48^\circ}$$

$$=3.122\angle 31.48^\circ\text{ A}$$

$$\text{and current } I_4=\left(\frac{4}{8+j3}\right)(3.122\angle 31.48^\circ)$$

$$=\frac{(4\angle 0^\circ)(3.122\angle 31.48^\circ)}{8.544\angle 20.56^\circ}$$

$$=1.462\angle 10.92^\circ\text{ A or}$$

$$(1.436+j0.277)\text{ A}$$

(v) If the network of Figure 35.18 is superimposed on the network of Figure 35.17, it can be seen that the current in the $(4 + j3)\,\Omega$ impedance is given by $I_2 - I_4$

(vi) $I_2 - I_4 = (3.588 + j0.692) - (1.436 + j0.277)$

$= \mathbf{(2.152 + j0.415)A}$ or $\mathbf{2.192\angle 10.92°\,A}$,

flowing from **A to B** in Figure 35.16.

Now try the following Practice Exercise

Practice Exercise 138 The superposition theorem (Answers on page 828)

1. Repeat Problems 1, 5, 8 and 9 of Exercise 135, page 504 using the superposition theorem.
2. Repeat Problems 3 and 5 of Exercise 136, page 510 using the superposition theorem.
3. Repeat Problem 5 of Exercise 137, page 517 using the superposition theorem.
4. Two batteries each of e.m.f. 15 V are connected in parallel to supply a load of resistance 2.0 Ω. The internal resistances of the batteries are 0.5 Ω and 0.3 Ω. Determine, using the superposition theorem, the current in the load and the current supplied by each battery.
5. Use the superposition theorem to determine the magnitude of the current flowing in the capacitive branch of the network shown in Figure 35.19.

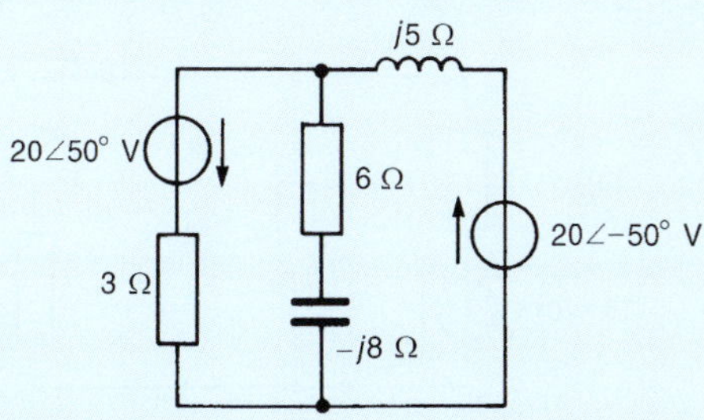

Figure 35.19

6. A.c. sources of 20∠90° V and internal resistance 10 Ω and 30∠0° V and internal resistance 12 Ω are connected in parallel across an 8 Ω load. Use the superposition theorem to determine (a) the current in the 8 Ω load, and (b) the current in each voltage source.
7. Use the superposition theorem to determine current I_x flowing in the 5 Ω resistance of the network shown in Figure 35.20.

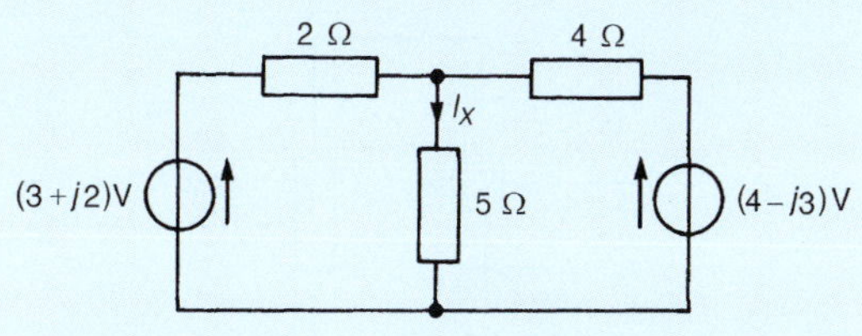

Figure 35.20

35.3 Further worked problems on the superposition theorem

Problem 4. For the a.c. network shown in Figure 35.21 determine, using the superposition theorem, (a) the current in each branch, (b) the magnitude of the voltage across the $(6 + j8)\Omega$ impedance and (c) the total active power delivered to the network.

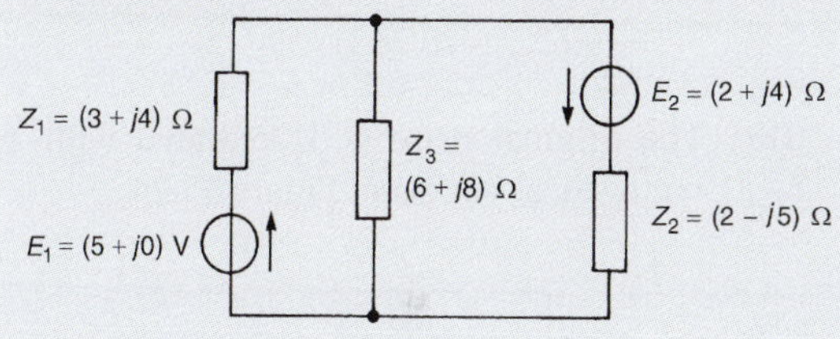

Figure 35.21

(a) (i) The original network is redrawn with E_2 removed, as shown in Figure 35.22.

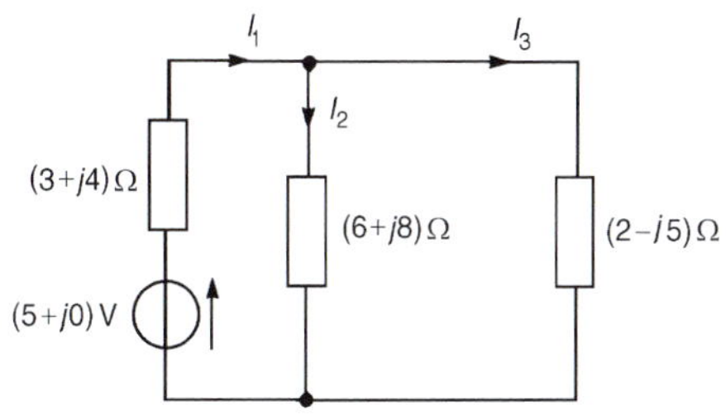

Figure 35.22

(ii) Currents I_1, I_2 and I_3 are labelled as shown in Figure 35.22. From Figure 35.22, $(6 + j8)\Omega$ in parallel with $(2 - j5)\Omega$ gives an equivalent impedance of

Part 4

$$\frac{(6+j8)(2-j5)}{(6+j8)+(2-j5)}=(5.123-j3.671)\,\Omega$$

From the equivalent network of Figure 35.23,

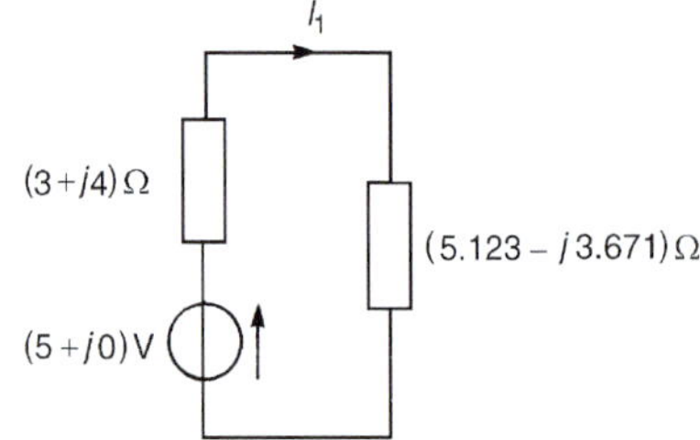

Figure 35.23

$$\text{current } I_1=\frac{5+j0}{(3+j4)+(5.123-j3.671)}$$
$$=(0.614-j0.025)\,\text{A}$$

current

$$I_2=\left[\frac{(2-j5)}{(6+j8)+(2-j5)}\right](0.614-j0.025)$$
$$=(-0.00731-j0.388)\,\text{A}$$

and current

$$I_3=\left[\frac{(6+j8)}{(6+j8)+(2-j5)}\right](0.614-j0.025)$$
$$=(0.622+j0.363)\,\text{A}$$

(iii) The original network is redrawn with E_1 removed, as shown in Figure 35.24.

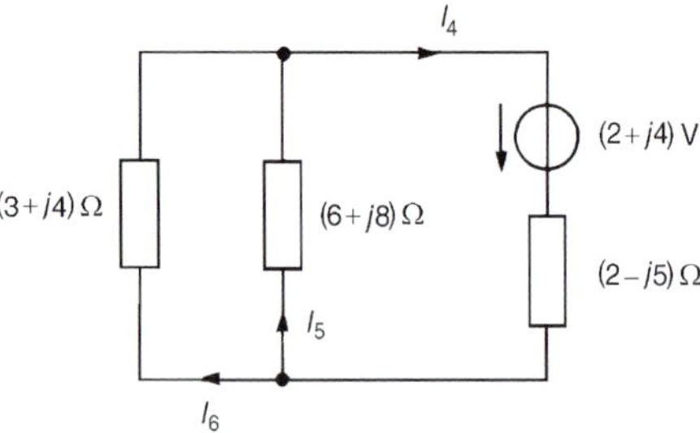

Figure 35.24

(iv) Currents I_4, I_5 and I_6 are shown labelled in Figure 35.24 with I_4 flowing away from the positive terminal of the $(2+j4)$V source.

From Figure 35.24, $(3+j4)\,\Omega$ in parallel with $(6+j8)\,\Omega$ gives an equivalent impedance of

$$\frac{(3+j4)(6+j8)}{(3+j4)+(6+j8)}=(2.00+j2.667)\,\Omega$$

From the equivalent network of Figure 35.25,

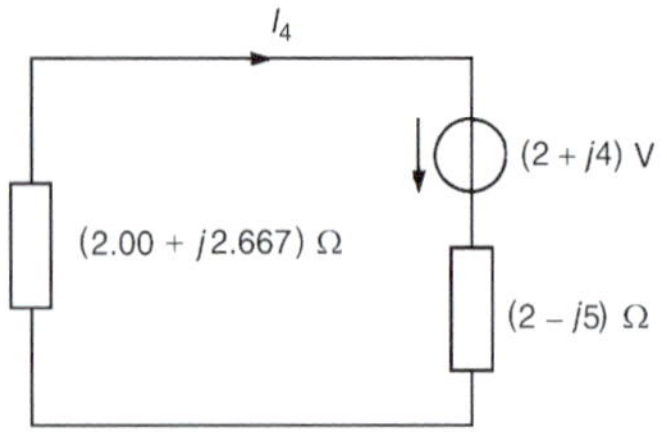

Figure 35.25

$$\text{current } I_4=\frac{(2+j4)}{(2.00+j2.667)+(2-j5)}$$
$$=(-0.062+j0.964)\,\text{A}$$

From Figure 35.24,
current $I_5=$

$$\left[\frac{(3+j4)}{(3+j4)+(6+j8)}\right](-0.062+j0.964)$$
$$=(-0.0207+j0.321)\,\text{A}$$

and current

$I_6=$

$$\left[\frac{6+j8}{(3+j4)+(6+j8)}\right](-0.062+j0.964)$$
$$=(-0.041+j0.643)\,\text{A}$$

(v) If Figure 35.24 is superimposed on Figure 35.22, the resultant currents are as shown in Figure 35.26.

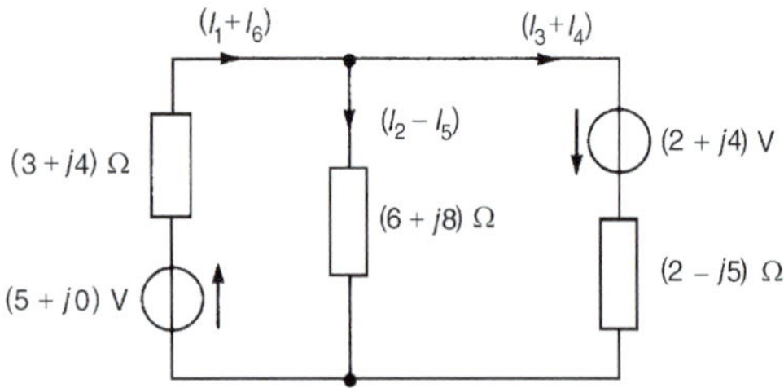

Figure 35.26

(vi) Resultant current flowing from $(5+j0)$V source is given by

$$I_1+I_6=(0.614-j0.025)+(-0.041+j0.643)$$
$$=\mathbf{(0.573+j0.618)\,A} \text{ or } \mathbf{0.843\angle 47.16^\circ\,A}$$

Part 4

Resultant current flowing from $(2+j4)$ V source is given by

$$I_3 + I_4 = (0.622 + j0.363) + (-0.062 + j0.964)$$
$$= \mathbf{(0.560 + j1.327)\,A} \text{ or } \mathbf{1.440\angle 67.12^\circ\,A}$$

Resultant current flowing through the $(6+j8)\Omega$ impedance is given by

$$I_2 - I_5 = (-0.00731 - j0.388) - (-0.0207 + j0.321)$$
$$= \mathbf{(0.0134 - j0.709)\,A} \text{ or } \mathbf{0.709\angle -88.92^\circ\,A}$$

(b) Voltage across $(6+j8)\,\Omega$ impedance is given by

$$(I_2 - I_5)(6+j8) = (0.709\angle -88.92^\circ)(10\angle 53.13^\circ) = 7.09\angle -35.79^\circ\,\text{V}$$

i.e. the magnitude of the voltage across the $(6+j8)\,\Omega$ impedance is **7.09 V**

(c) Total active power P delivered to the network is given by

$$P = E_1(I_1 + I_6)\cos\phi_1 + E_2(I_3 + I_4)\cos\phi_2$$

where ϕ_1 is the phase angle between E_1 and $(I_1 + I_6)$ and ϕ_2 is the phase angle between E_2 and $(I_3 + I_4)$, i.e.

$$P = (5)\,(0.843)\cos(47.16^\circ - 0^\circ) + (\sqrt{(2^2+4^2)})(1.440)\cos(67.12^\circ - \tan^{-1}\tfrac{4}{2})$$
$$= 2.866 + 6.427 = 9.293\,\text{W}$$
$$= \mathbf{9.3\,W}\text{, correct to one decimal place.}$$

(This value may be checked since total active power dissipated is given by:

$$P = (I_1 + I_6)^2(3) + (I_2 - I_5)^2(6) + (I_3 + I_4)^2(2)$$
$$= (0.843)^2(3) + (0.709)^2(6) + (1.440)^2(2)$$
$$= 2.132 + 3.016 + 4.147 = 9.295\,\text{W}$$
$$= \mathbf{9.3\,W}\text{, correct to one decimal place.)}$$

Problem 5. Use the superposition theorem to determine, for the network shown in Figure 35.27, (a) the magnitude of the current flowing in the capacitor, (b) the p.d. across the 5 Ω resistance, (c) the active power dissipated in the 20 Ω resistance and (d) the total active power taken from the supply.

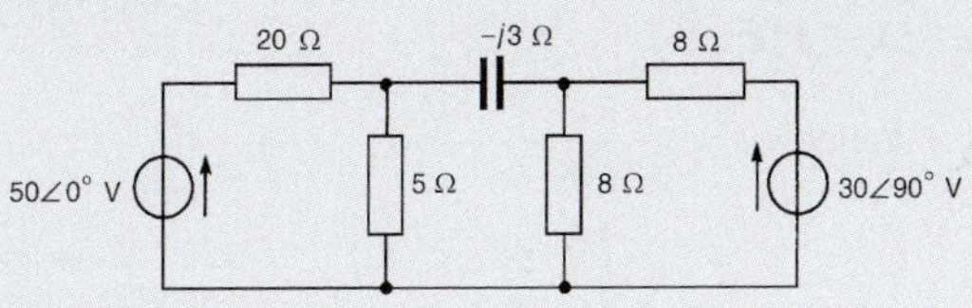

Figure 35.27

(i) The network is redrawn with the $30\angle 90^\circ$ V source removed, as shown in Figure 35.28.

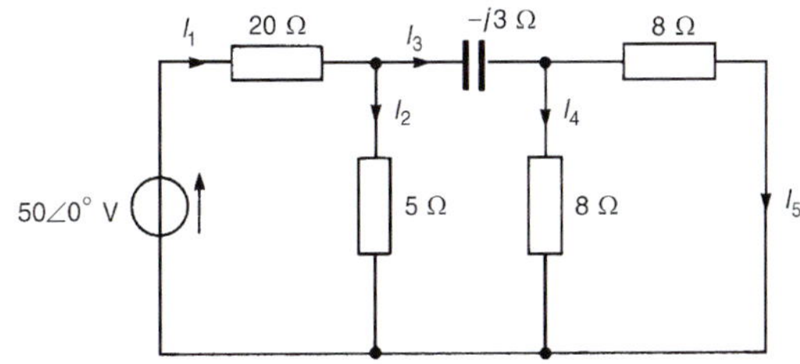

Figure 35.28

(ii) Currents I_1 to I_5 are shown labelled in Figure 35.28. From Figure 35.28, two 8 Ω resistors in parallel give an equivalent resistance of 4 Ω.

$$\text{Hence } I_1 = \frac{50\angle 0^\circ}{20 + (5(4-j3)/(5+4-j3))}$$
$$= 2.220\angle 2.12^\circ\,\text{A}$$
$$I_2 = \frac{(4-j3)}{(5+4-j3)} I_1$$
$$= 1.170\angle -16.32^\circ\,\text{A}$$
$$I_3 = \left(\frac{5}{5+4-j3}\right) I_1 = 1.170\angle 20.55^\circ\,\text{A}$$
$$I_4 = \left(\frac{8}{8+8}\right) I_3 = 0.585\angle 20.55^\circ\,\text{A} = I_5$$

(iii) The original network is redrawn with the $50\angle 0^\circ$ V source removed, as shown in Figure 35.29.

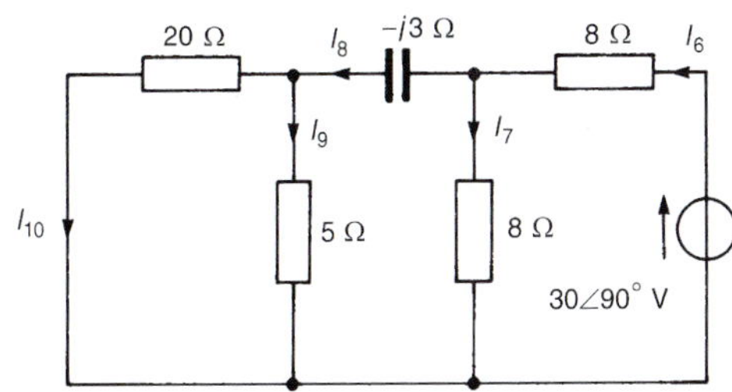

Figure 35.29

(iv) Currents I_6 to I_{10} are shown labelled in Figure 35.29. From Figure 35.29, 20 Ω in parallel with 5 Ω gives an equivalent resistance of (20 × 5)/(20 + 5) = 4 Ω.

$$\text{Hence } I_6 = \frac{30\angle 90^\circ}{8+(8(4-j3)/(8+4-j3))}$$

$$= 2.715\angle 96.52^\circ \text{ A}$$

$$I_7 = \frac{(4-j3)}{(8+4-j3)} I_6 = 1.097\angle 73.69^\circ \text{ A}$$

$$I_8 = \left(\frac{8}{8+4-j3}\right) I_6 = 1.756\angle 110.56^\circ \text{ A}$$

$$I_9 = \left(\frac{20}{20+5}\right) I_8 = 1.405\angle 110.56^\circ \text{ A}$$

$$\text{and} \quad I_{10} = \left(\frac{5}{20+5}\right) I_8 = 0.351\angle 110.56^\circ \text{ A}$$

(a) The current flowing in the capacitor is given by

$$(I_3 - I_8) = 1.170\angle 20.55^\circ - 1.756\angle 110.56^\circ$$
$$= (1.712 - j1.233) \text{ A or } 2.11\angle -35.76^\circ \text{ A}$$

i.e. **the magnitude of the current in the capacitor is 2.11 A**

(b) The p.d. across the 5 Ω resistance is given by $(I_2 + I_9)$ (5)

$$(I_2 + I_9) = 1.170\angle -16.32^\circ + 1.405\angle 110.56^\circ$$
$$= (0.629 + j0.987) \text{ A or } 1.17\angle 57.49^\circ \text{ A}$$

Hence **the magnitude of the p.d. across the 5 Ω resistance** is (1.17) (5) = **5.85 V**

(c) Active power dissipated in the 20 Ω resistance is given by $(I_1 - I_{10})^2(20)$

$$(I_1 - I_{10}) = 2.220\angle 2.12^\circ - 0.351\angle 110.56^\circ$$
$$= (2.342 - j0.247) \text{ A or } 2.355\angle -6.02^\circ \text{ A}$$

Hence **the active power dissipated in the 20 Ω resistance** is $(2.355)^2(20)$ = **111 W**

(d) Active power developed by the 50∠0° V source

$$P_1 = V(I_1 - I_{10})\cos\phi_1$$
$$= (50)(2.355)\cos(6.02^\circ - 0^\circ)$$
$$= 117.1 \text{ W}$$

Active power developed by 30∠90° V source,

$$P_2 = 30(I_6 - I_5)\cos\phi_2$$

$$(I_6 - I_5) = 2.7156\angle 96.52^\circ - 0.585\angle 20.55^\circ$$
$$= (-0.856 + j2.492) \text{ A or } 2.635\angle 108.96^\circ \text{ A}$$

$$\text{Hence } P_2 = (30)(2.635)\cos(108.96^\circ - 90^\circ)$$
$$= 74.8 \text{ W}$$

Total power developed, $P = P_1 + P_2$
$$= 117.1 + 74.8$$
$$= \mathbf{191.9\ W}$$

(This value may be checked by summing the I^2R powers dissipated in the four resistors.)

Now try the following Practice Exercise

Practice Exercise 139 Superposition theorem (Answers on page 828)

1. For the network shown in Figure 35.30, determine, using the superposition theorem, (a) the current flowing in the capacitor, (b) the current flowing in the 2 Ω resistance, (c) the p.d. across the 5 Ω resistance and (d) the total active circuit power.

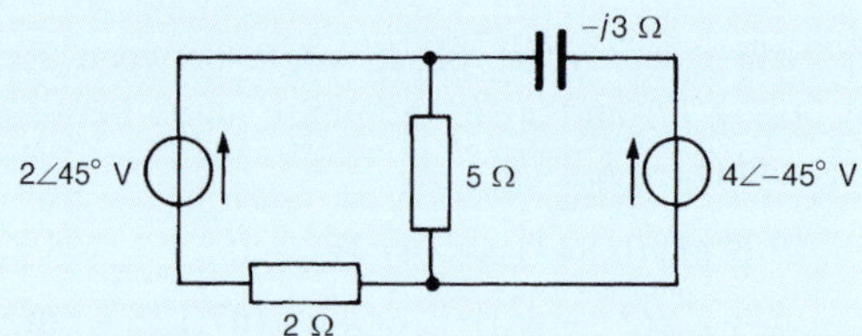

Figure 35.30

2. (a) Use the superposition theorem to determine the current in the 12 Ω resistance of the network shown in Figure 35.31. Determine also the p.d. across the 8 Ω resistance and the power dissipated in the 20 Ω resistance.

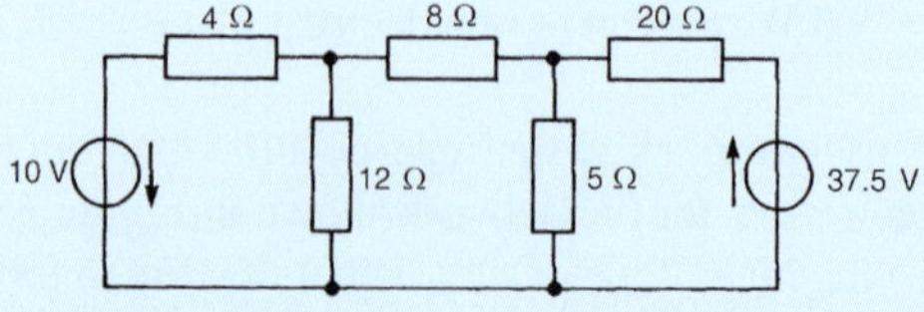

Figure 35.31

Part 4

(b) If the 37.5 V source in Figure 35.31 is reversed in direction, determine the current in the 12 Ω resistance.

3. For the network shown in Figure 35.32, use the superposition theorem to determine (a) the current in the capacitor, (b) the p.d. across the 10 Ω resistance, (c) the active power dissipated in the 20 Ω resistance and (d) the total active circuit power.

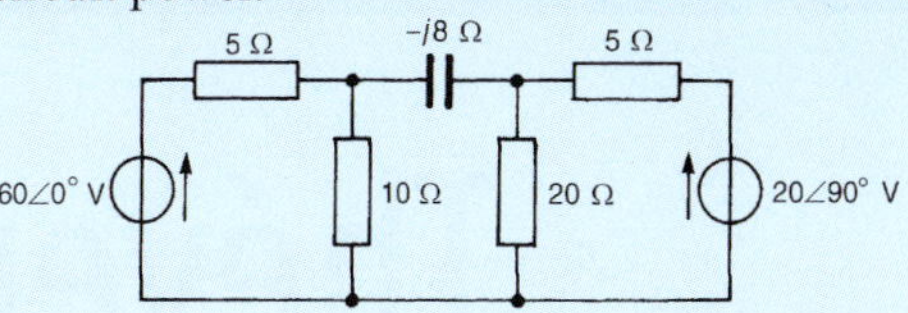

Figure 35.32

For fully worked solutions to each of the problems in Practice Exercises 138 and 139 in this chapter, go to the website:
www.routledge.com/cw/bird

Chapter 36

Thévenin's and Norton's theorems

Why it is important to understand: **Thévenin's and Norton's theorems**

Two powerful circuit analysis techniques are Thévenin's theorem and Norton's theorem. Both theorems convert a complex circuit to a simpler series or parallel equivalent circuit for easier analysis. Analysis involves removing part of the circuit across two terminals to aid calculation, later combining the circuit with the Thévenin or Norton equivalent circuit. Thévenin's theorem was independently derived in 1853 by the German scientist Hermann von Helmholtz and in 1883 by Léon Charles Thévenin (1857–1926), an electrical engineer with France's national Postes et Télégraphes telecommunications organization. Norton's theorem is an extension of Thévenin's theorem and was introduced in 1926 separately by two people: Hause-Siemens researcher Hans Ferdinand Mayer (1895–1980) and Bell Labs engineer Edward Lawry Norton (1898–1983). This chapter explains both theorems, analysing both d.c. and a.c. networks through calculations.

At the end of this chapter you should be able to:

- understand and use Thévenin's theorem to analyse a.c. and d.c. networks
- understand and use Norton's theorem to analyse a.c. and d.c. networks
- appreciate and use the equivalence of Thévenin and Norton networks

36.1 Introduction

Many of the networks analysed in Chapters 33, 34 and 35 using Kirchhoff's laws, mesh-current and nodal analysis and the superposition theorem can be analysed more quickly and easily by using Thévenin's or Norton's theorems. Each of these theorems involves replacing what may be a complicated network of sources and linear impedances with a simple equivalent circuit. A set procedure may be followed when using each theorem, the procedures themselves requiring a knowledge of basic circuit theory. (It may be worth checking some general d.c. circuit theory in Section 15.4. page 226, before proceeding.)

Electrical Circuit Theory and Technology. 978-1-138-67349-6, © 2017 John Bird. Published by Taylor & Francis. All rights reserved.

36.2 Thévenin's theorem

Thévenin's* theorem states:

'The current which flows in any branch of a network is the same as that which would flow in the branch if it were connected across a source of electrical energy, the e.m.f. of which is equal to the potential difference which would appear across the branch if it were open-circuited, and the internal impedance of which is equal to the impedance which appears across the open-circuited branch terminals when all sources are replaced by their internal impedances.'

The theorem applies to any linear active network ('linear' meaning that the measured values of circuit components are independent of the direction and magnitude of the current flowing in them, and 'active' meaning that it contains a source, or sources, of e.m.f.).

The above statement of Thévenin's theorem simply means that a complicated network with output terminals AB, as shown in Figure 36.1(a), can be replaced by a single voltage source E in series with an impedance z, as shown in Figure 36.1(b). E is the open-circuit voltage measured at terminals AB and z is the equivalent impedance of the network at the terminals AB when all internal sources of e.m.f. are made zero. The polarity of voltage E is chosen so that the current flowing through an impedance connected between A and B will have the same direction as would result if the impedance had been connected between A and B of the original network. Figure 36.1(b) is known as the **Thévenin equivalent circuit**, and was initially introduced in Section 15.5, page 228 for d.c. networks.

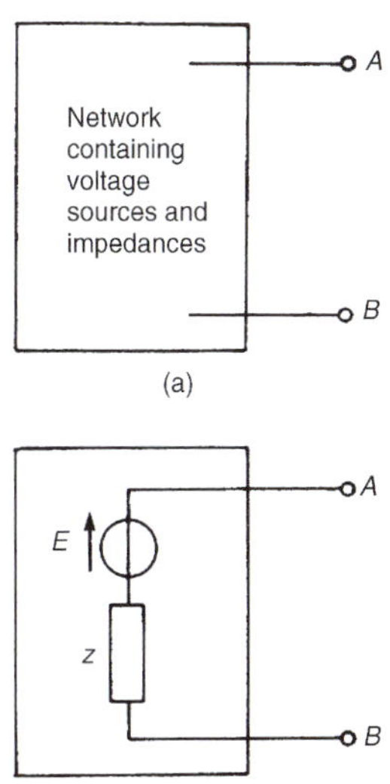

Figure 36.1 The Thévenin equivalent circuit

The following four-step **procedure** can be adopted when determining, by means of Thévenin's theorem, the current flowing in a branch containing impedance Z_L of an active network:

(i) remove the impedance Z_L from that branch;

(ii) determine the open-circuit voltage E across the break;

(iii) remove each source of e.m.f. and replace it by its internal impedance (if it has zero internal impedance then replace it by a short-circuit), and then determine the internal impedance, z, 'looking in' at the break;

(iv) determine the current from the Thévenin equivalent circuit shown in Figure 36.2, i.e.

$$\textbf{current } \boldsymbol{i_L} = \frac{\boldsymbol{E}}{\boldsymbol{Z_L + z}}$$

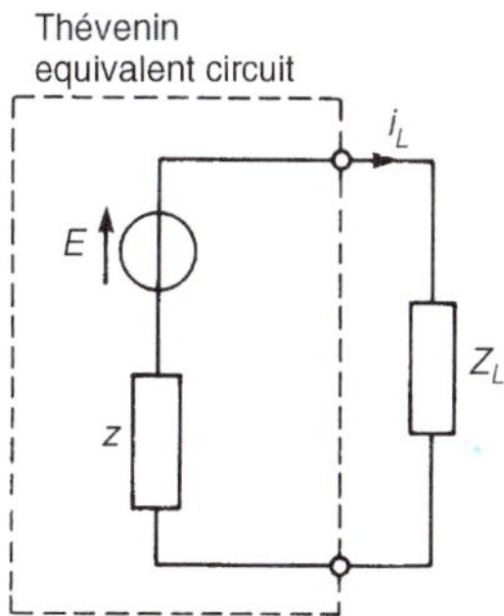

Figure 36.2

A simple d.c. network (Figure 36.3) serves to demonstrate how the above procedure is applied to determine the current flowing in the 5 Ω resistance by using Thévenin's theorem. This is the same network as used in Chapter 33 when it was solved using Kirchhoff's laws (see page 500), and by means of the superposition

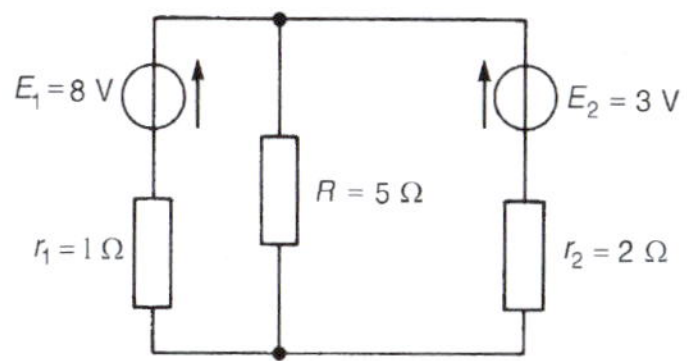

Figure 36.3

*Who was **Thévenin**? For resume of Thévenin, see page 228. To find out more go to **www.routledge.com/cw/bird**

Part 4

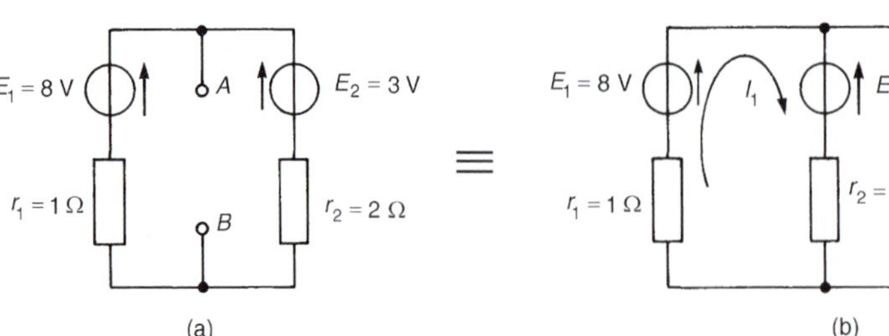

Figure 36.4

theorem in Chapter 35 (see page 518). A comparison of methods may be made.

Using the above procedure:

(i) the 5 Ω resistor is removed, as shown in Figure 36.4(a).

(ii) The open-circuit voltage E across the break is now required. The network of Figure 36.4(a) is redrawn for convenience as shown in Figure 36.4(b), where current,

$$I_1 = \frac{E_1 - E_2}{r_1 + r_2} = \frac{8-3}{1+2} = \frac{5}{3} \quad \text{or} \quad 1\frac{2}{3}\,\text{A}$$

Hence the open-circuit voltage E is given by

$$E = E_1 - I_1 r_1$$

i.e. $E = 8 - \left(1\tfrac{2}{3}\right)(1) = 6\tfrac{1}{3}\,\text{V}$

(Alternatively, $E = E_2 - (-I_1) r_2$
$= 3 + \left(1\tfrac{2}{3}\right)(2) = 6\tfrac{1}{3}\,\text{V})$

(iii) Removing each source of e.m.f. gives the network of Figure 36.5. The impedance, z, 'looking in' at the break AB is given by

$$z = (1 \times 2)/(1+2) = \tfrac{2}{3}\,\Omega$$

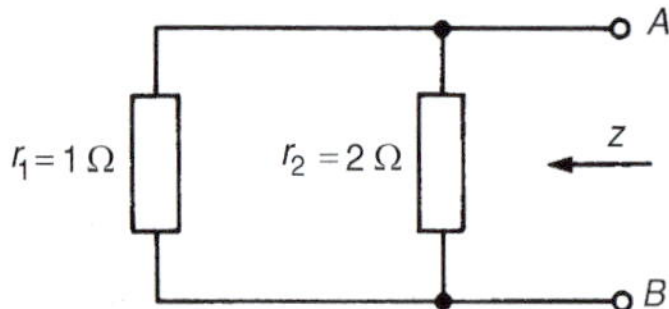

Figure 36.5

(iv) The Thévenin equivalent circuit is shown in Figure 36.6, where current i_L is given by

$$i_L = \frac{E}{Z_L + z} = \frac{6\frac{1}{3}}{5 + \frac{2}{3}} = 1.1177$$

$= \mathbf{1.12\,A}$, correct to two decimal places.

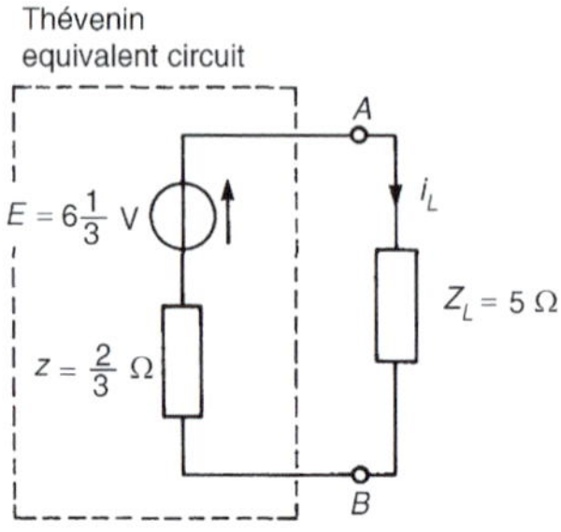

Figure 36.6

To determine the currents flowing in the other two branches of the circuit of Figure 36.3, basic circuit theory is used. Thus, from Figure 36.7, voltage $V = (1.1177)(5) = 5.5885\,\text{V}$

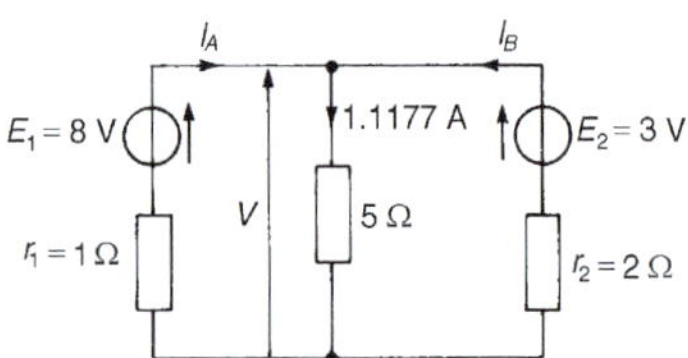

Figure 36.7

Then $V = E_1 - I_A r_1$, i.e. $5.5885 = 8 - I_A(1)$,

from which

current $I_A = 8 - 5.5885 = \mathbf{2.41\,A}$

Similarly, $V = E_2 - I_B r_2$, i.e. $5.5885 = 3 - I_B(2)$,

from which

$$\text{current } I_B = \frac{3 - 5.5885}{2} = \mathbf{-1.29\,A}$$

(i.e. flowing in the direction opposite to that shown in Figure 36.7).

The Thévenin theorem procedure used above may be applied to a.c. as well as d.c. networks, as shown below.

An a.c. network is shown in Figure 36.8, where it is required to find the current flowing in the $(6+j8)\,\Omega$ impedance by using Thévenin's theorem.

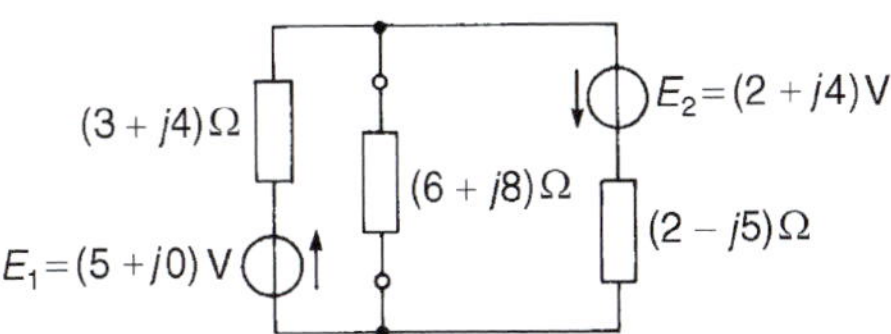

Figure 36.8

Using the above procedure

(i) The $(6+j8)\,\Omega$ impedance is removed, as shown in Figure 36.9(a).

(ii) The open-circuit voltage across the break is now required. The network is redrawn for convenience as shown in Figure 36.9(b), where current

$$I_1 = \frac{(5+j0)+(2+j4)}{(3+j4)+(2-j5)} = \frac{(7+j4)}{(5-j)} = 1.581\angle 41.05^\circ\,\text{A}$$

Hence open-circuit voltage across AB,

$$E = E_1 - I_1(3+j4), \text{ i.e.}$$

$$E = (5+j0) - (1.581\angle 41.05^\circ)(5\angle 53.13^\circ)$$

from which $E = 9.657\angle -54.73^\circ\,\text{V}$

(iii) From Figure 36.10, the impedance z 'looking in' at terminals AB is given by

$$z = \frac{(3+j4)(2-j5)}{(3+j4)+(2-j5)} = 5.281\angle -3.76^\circ\,\Omega \quad \text{or} \quad (5.270-j0.346)\,\Omega$$

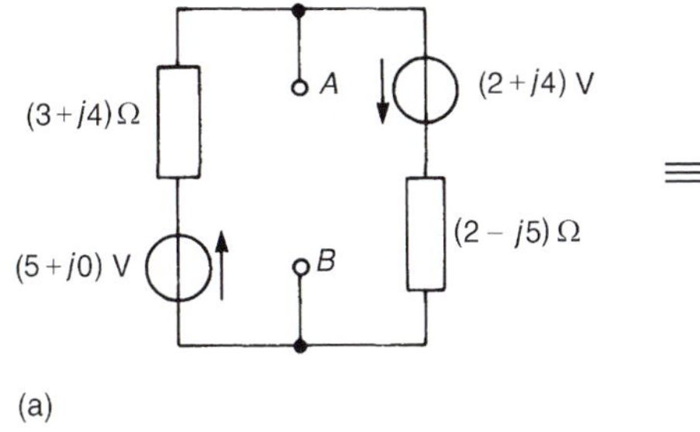

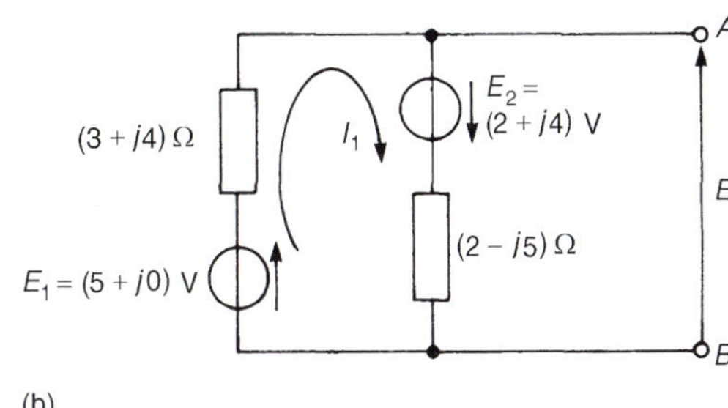

Figure 36.9

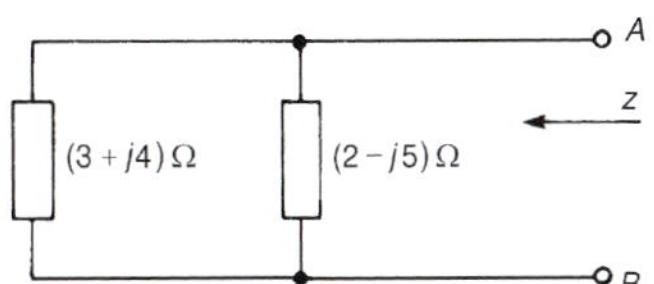

Figure 36.10

(iv) The Thévenin equivalent circuit is shown in Figure 36.11, from which current

$$i_L = \frac{E}{Z_L + z} = \frac{9.657\angle -54.73^\circ}{(6+j8)+(5.270-j0.346)}$$

Thus, current in $(6+j8)\,\Omega$ impedance,

$$i_L = \frac{9.657\angle -54.73^\circ}{13.623\angle 34.18^\circ} = \mathbf{0.71\angle -88.91^\circ\,A}$$

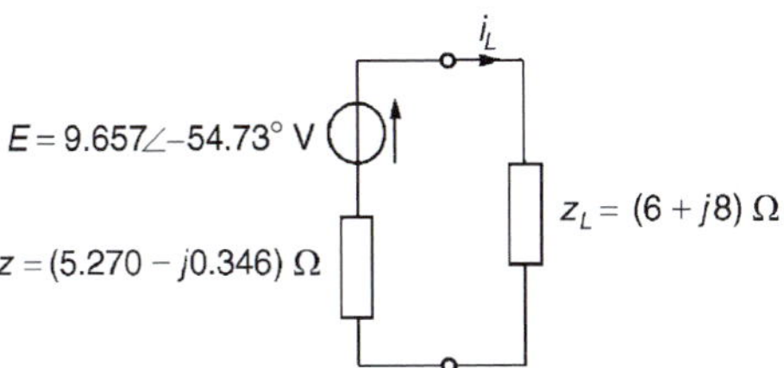

Figure 36.11

The network of Figure 36.8 is analysed using Kirchhoff's laws in Problem 3, page 502, and by the superposition theorem in Problem 4, page 523. The above analysis using Thévenin's theorem is seen to be much quicker.

Problem 1. For the circuit shown in Figure 36.12, use Thévenin's theorem to determine (a) the current flowing in the capacitor and (b) the p.d. across the $150\,\text{k}\Omega$ resistor.

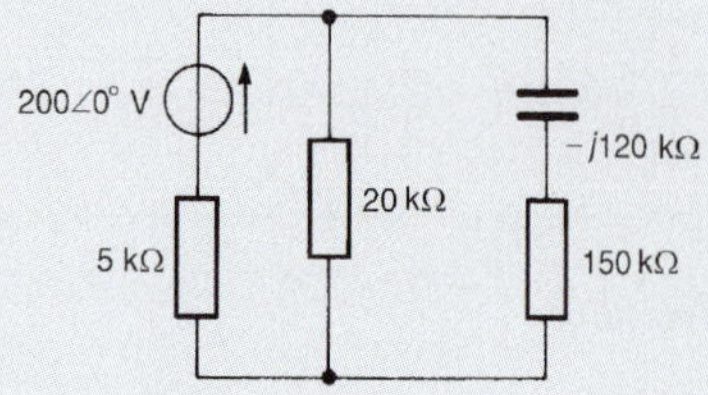

Figure 36.12

(a) (i) Initially the $(150 - j120)\,\text{k}\Omega$ impedance is removed from the circuit as shown in Figure 36.13.

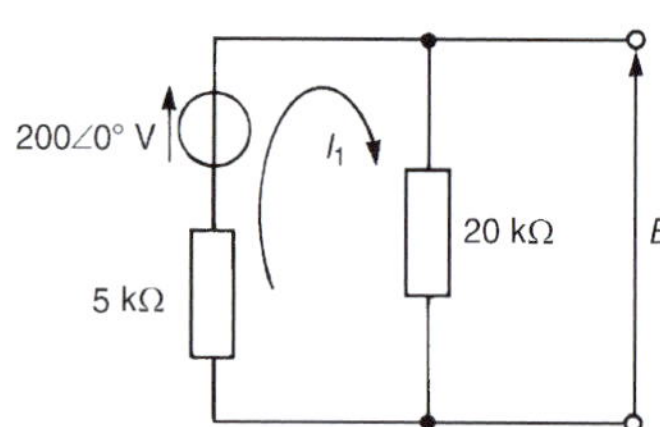

Figure 36.13

Note that, to find the current in the capacitor, only the capacitor need have been initially removed from the circuit. However, removing each of the components from the branch through which the current is required will often result in a simpler solution.

(ii) From Figure 36.13,

$$\text{current } I_1 = \frac{200\angle 0^\circ}{(5000 + 20\,000)} = 8\,\text{mA}$$

The open-circuit e.m.f. E is equal to the p.d. across the $20\,\text{k}\Omega$ resistor, i.e.

$$E = (8 \times 10^{-3})(20 \times 10^{3}) = \mathbf{160\,V}$$

(iii) Removing the $200\angle 0^\circ$ V source gives the network shown in Figure 36.14.

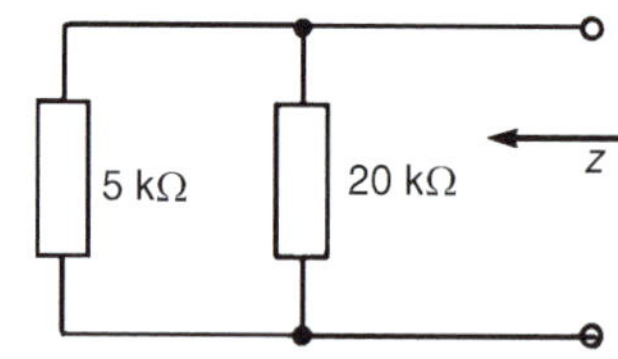

Figure 36.14

The impedance, z, 'looking in' at the open-circuited terminals is given by

$$z = \frac{5 \times 20}{5 + 20}\,\text{k}\Omega = \mathbf{4\,k\Omega}$$

(iv) The Thévenin equivalent circuit is shown in Figure 36.15, where current i_L is given by

$$i_L = \frac{E}{Z_L + z} = \frac{160}{(150 - j120) \times 10^3 + 4 \times 10^3}$$

$$= \frac{160}{195.23 \times 10^3\angle -37.93^\circ}$$

$$= 0.82\angle 37.93^\circ\,\text{mA}$$

Thus the current flowing in the capacitor is 0.82 mA

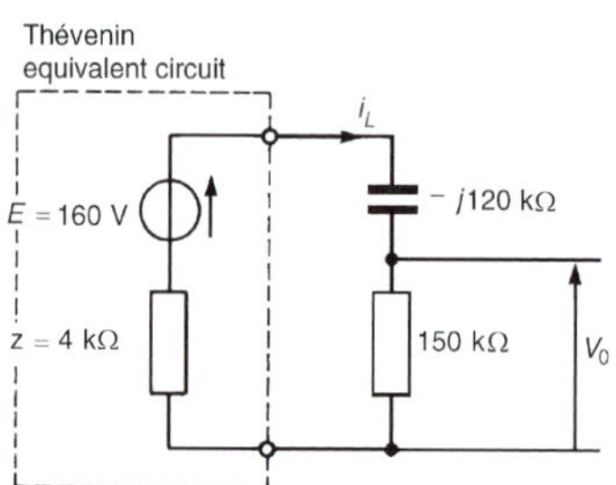

Figure 36.15

(b) P.d. across the $150\,\text{k}\Omega$ resistor,

$$V_0 = i_L R = (0.82 \times 10^{-3})(150 \times 10^{3}) = \mathbf{123\,V}$$

Problem 2. Determine, for the network shown in Figure 36.16, the value of current I. Each of the voltage sources has a frequency of 2 kHz.

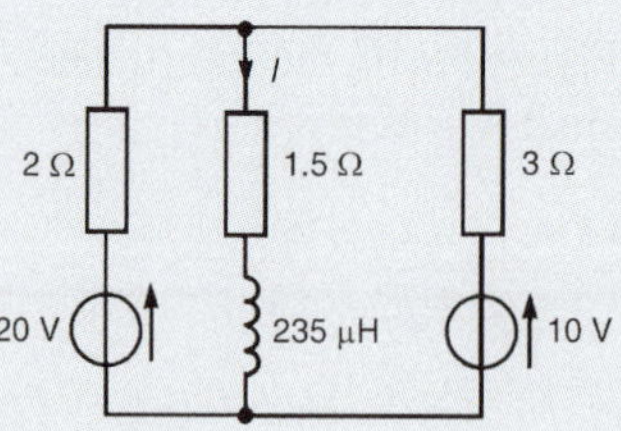

Figure 36.16

(i) The impedance through which current I is flowing is initially removed from the network, as shown in Figure 36.17.

(ii) From Figure 36.17,

$$\text{current, } I_1 = \frac{20 - 10}{2 + 3} = 2\,\text{A}$$

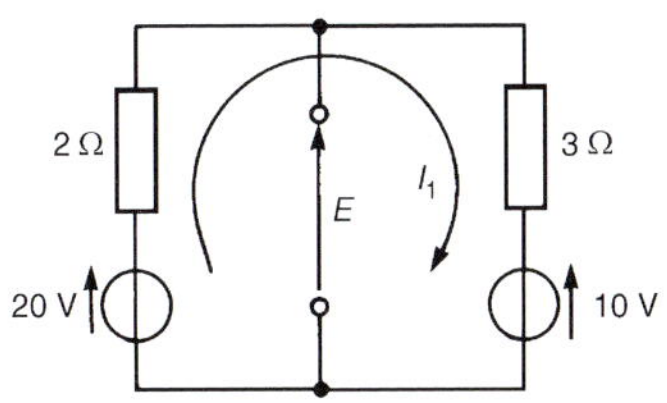

Figure 36.17

Hence the open circuit e.m.f.

$$\boldsymbol{E} = 20 - I_1(2) = 20 - 2(2) = \mathbf{16\,V}$$

(Alternatively, $E = 10 + I_1(3) = 10 + (2)(3) = 16\,\text{V}$)

(iii) When the sources of e.m.f. are removed from the circuit, the impedance, z, 'looking in' at the break is given by

$$z = \frac{2 \times 3}{2+3} = \mathbf{1.2\,\Omega}$$

(iv) The Thévenin equivalent circuit is shown in Figure 36.18, where inductive reactance,

$$X_L = 2\pi f L = 2\pi(2000)(235 \times 10^{-6}) = 2.95\,\Omega$$

Hence current

$$I = \frac{16}{(1.2 + 1.5 + j2.95)} = \frac{16}{4.0\angle 47.53^\circ}$$
$$= \mathbf{4.0\angle{-47.53^\circ}\,A} \text{ or } \mathbf{(2.70 - j2.95)\,A}$$

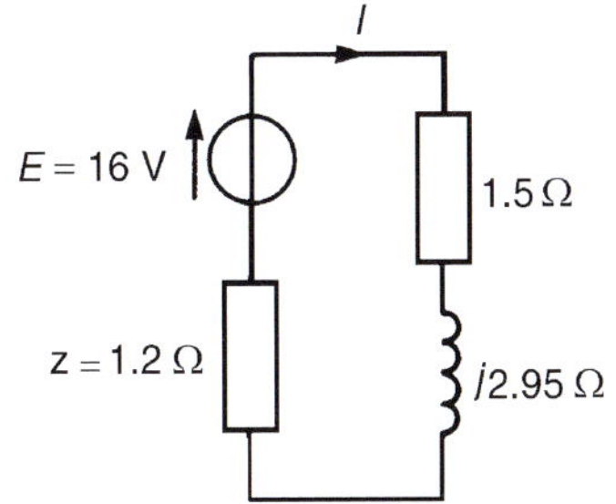

Figure 36.18

Problem 3. Use Thévenin's theorem to determine the power dissipated in the 48 Ω resistor of the network shown in Figure 36.19.

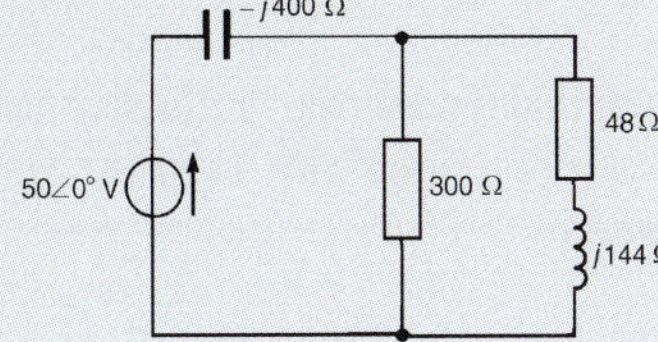

Figure 36.19

The power dissipated by a current I flowing through a resistor R is given by I^2R, hence initially the current flowing in the 48 Ω resistor is required.

(i) The $(48 + j144)\Omega$ impedance is initially removed from the network as shown in Figure 36.20.

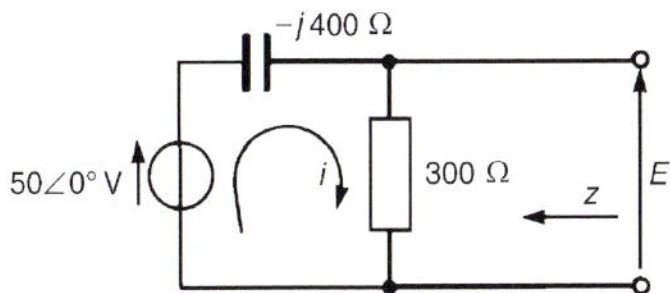

Figure 36.20

(ii) From Figure 36.20,

$$\text{current, } i = \frac{50\angle 0^\circ}{(300 - j400)} = 0.1\angle 53.13^\circ\,\text{A}$$

Hence the open-circuit voltage

$$\boldsymbol{E} = i(300) = (0.1\angle 53.13^\circ)(300) = \mathbf{30\angle 53.13^\circ\,V}$$

(iii) When the $50\angle 0^\circ$ V source shown in Figure 36.20 is removed, the impedance, z, is given by

$$z = \frac{(-j400)(300)}{(300 - j400)} = \frac{(400\angle{-90^\circ})(300)}{500\angle{-53.13^\circ}}$$
$$= 240\angle{-36.87^\circ}\,\Omega \text{ or } \mathbf{(192 - j144)\,\Omega}$$

(iv) The Thévenin equivalent circuit is shown in Figure 36.21 connected to the $(48 + j144)\,\Omega$ load.

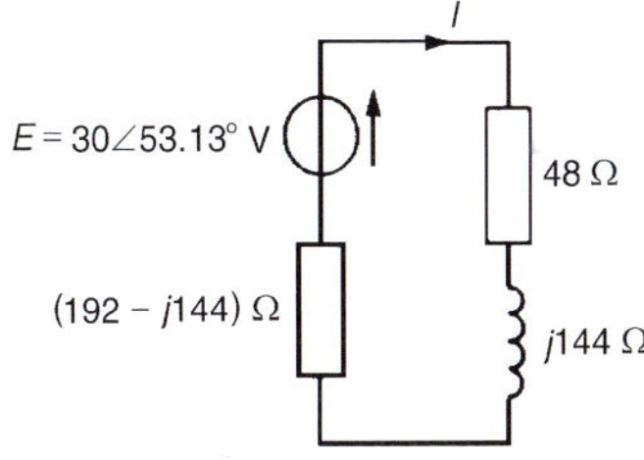

Figure 36.21

$$\text{Current } I = \frac{30\angle 53.13^\circ}{(192 - j144) + (48 + j144)}$$
$$= \frac{30\angle 53.13^\circ}{240\angle 0^\circ} = 0.125\angle 53.13^\circ\,\text{A}$$

Hence the power dissipated in the 48 Ω resistor $= I^2R = (0.125)^2(48) = \mathbf{0.75\,W}$

Problem 4. For the network shown in Figure 36.22, use Thévenin's theorem to determine the current flowing in the 80 Ω resistor.

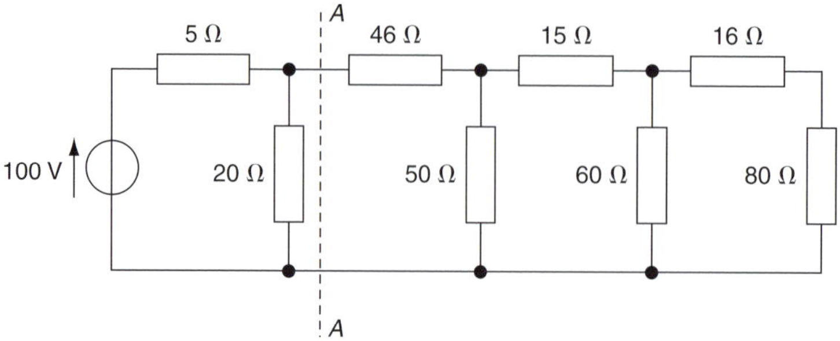

Figure 36.22

One method of analysing a multi-branch network as shown in Figure 36.22 is to use Thévenin's theorem on one part of the network at a time. For example, the part of the circuit to the left of AA may be reduced to a Thévenin equivalent circuit.
From Figure 36.23,

$$E_1 = \left(\frac{20}{20+5}\right)100 = 80\,\text{V, by voltage division}$$

$$\text{and } z_1 = \frac{20\times 5}{20+5} = 4\,\Omega$$

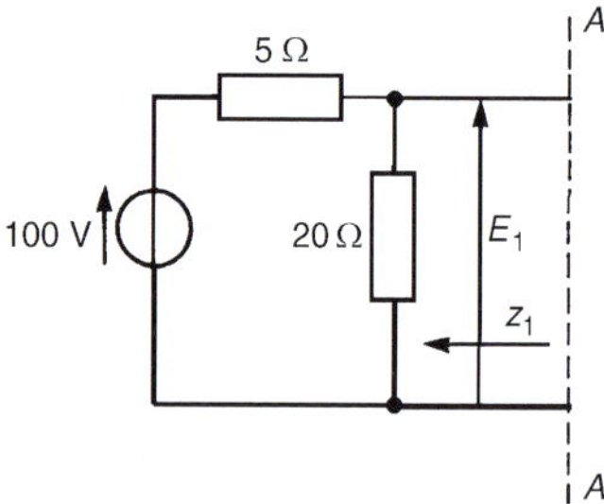

Figure 36.23

Thus the network of Figure 36.22 reduces to that of Figure 36.24. The part of the network shown in

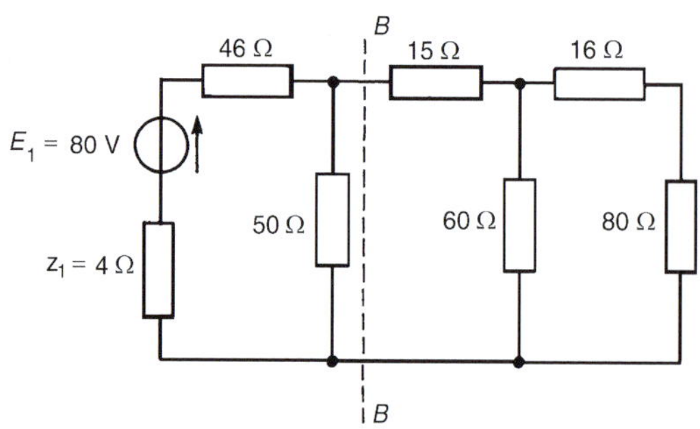

Figure 36.24

Figure 36.24 to the left of BB may be reduced to a Thévenin equivalent circuit, where

$$E_2 = \left(\frac{50}{50+46+4}\right)(80) = 40\,\text{V}$$

$$\text{and } z_2 = \frac{50\times 50}{50+50} = 25\,\Omega$$

Thus the original network reduces to that shown in Figure 36.25.

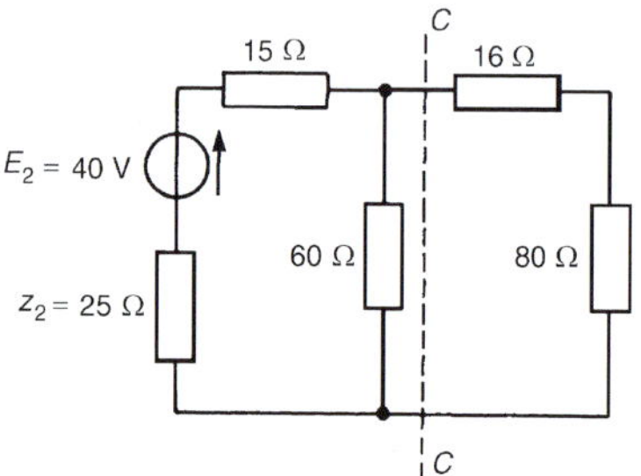

Figure 36.25

The part of the network shown in Figure 36.25 to the left of CC may be reduced to a Thévenin equivalent circuit, where

$$E_3 = \left(\frac{60}{60+25+15}\right)(40) = 24\,\text{V}$$

$$\text{and } z_3 = \frac{(60)(40)}{(60+40)} = 24\,\Omega$$

Thus the original network reduces to that of Figure 36.26, from which **the current in the 80 Ω resistor** is given by

$$I = \left(\frac{24}{80+16+24}\right) = \mathbf{0.20\,A}$$

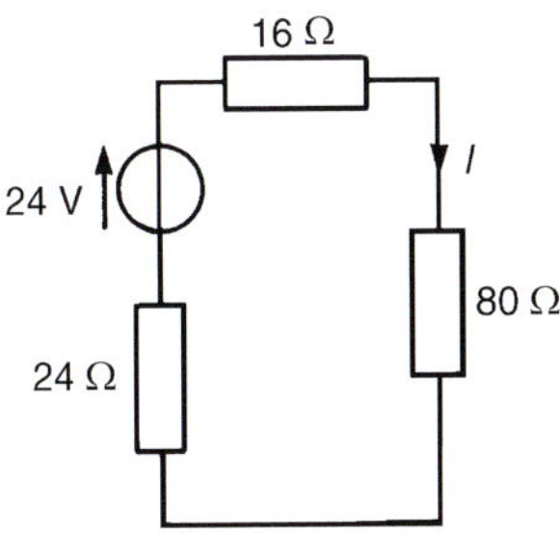

Figure 36.26

Now try the following Practice Exercise

Practice Exercise 140 Thévenin's theorem (Answers on page 829)

1. Use Thévenin's theorem to determine the current flowing in the 10 Ω resistor of the d.c. network shown in Figure 36.27.

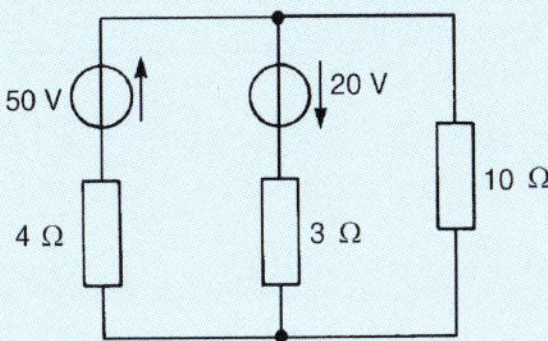

Figure 36.27

2. Determine, using Thévenin's theorem, the values of currents I_1, I_2 and I_3 of the network shown in Figure 36.28.

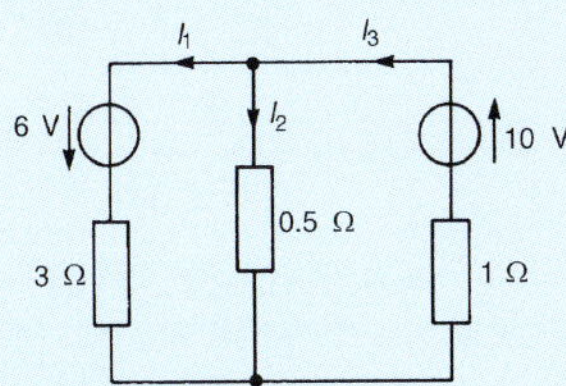

Figure 36.28

3. Determine the Thévenin equivalent circuit with respect to terminals AB of the network shown in Figure 36.29. Hence determine the magnitude of the current flowing in a $(4-j7)$ Ω impedance connected across terminals AB and the power delivered to this impedance.

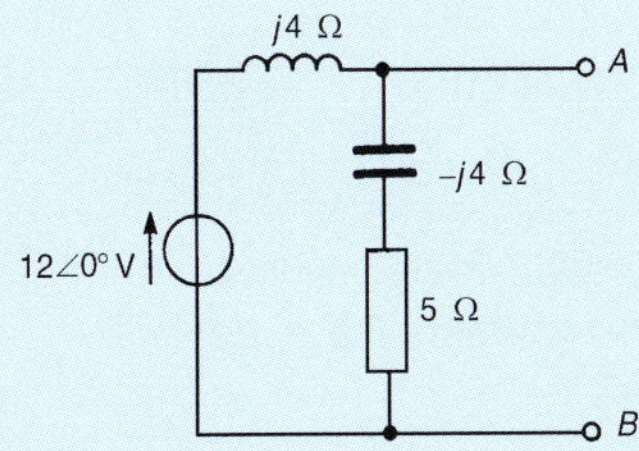

Figure 36.29

4. For the network shown in Figure 36.30, use Thévenin's theorem to determine the current flowing in the 3 Ω resistance.

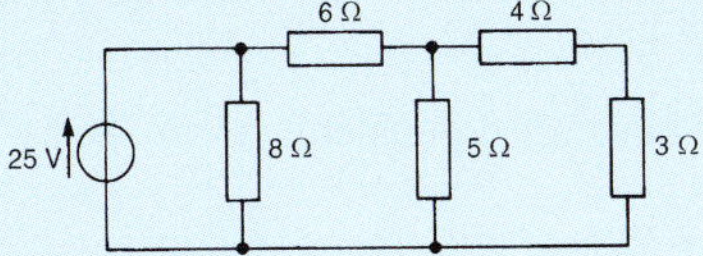

Figure 36.30

5. Derive for the network shown in Figure 36.31 the Thévenin equivalent circuit at terminals AB, and hence determine the current flowing in a 20 Ω resistance connected between A and B.

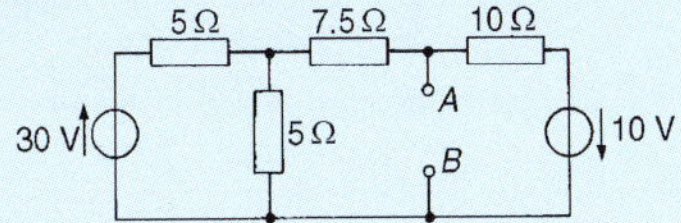

Figure 36.31

36.3 Further worked problems on Thévenin's theorem

Problem 5. Determine the Thévenin equivalent circuit with respect to terminals AB of the circuit shown in Figure 36.32. Hence determine (a) the magnitude of the current flowing in a $(3.75+j11)$ Ω impedance connected across terminals AB, and (b) the magnitude of the p.d. across the $(3.75+j11)$ Ω impedance.

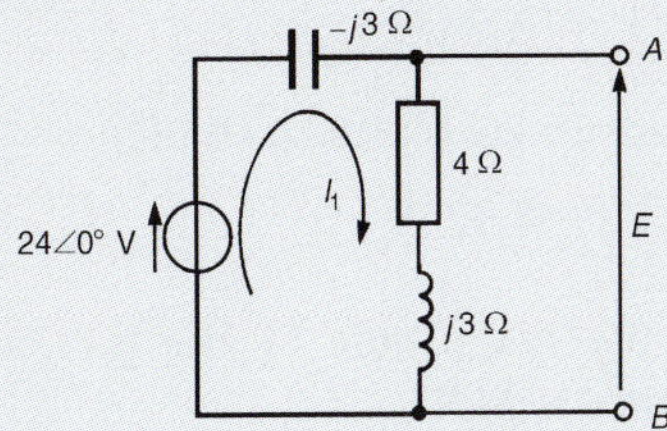

Figure 36.32

Current I_1 shown in Figure 36.32 is given by

$$I_1 = \frac{24\angle 0^\circ}{(4 + j3 - j3)} = \frac{24\angle 0^\circ}{4\angle 0^\circ} = 6\angle 0^\circ \text{ A}$$

The Thévenin equivalent voltage, i.e. the open-circuit voltage across terminals AB, is given by

$$\mathbf{E} = I_1(4 + j3) = (6\angle 0^\circ)(5\angle 36.87^\circ) = \mathbf{30\angle 36.87^\circ\ V}$$

When the $24\angle 0^\circ$ V source is removed, the impedance z 'looking in' at AB is given by

$$z = \frac{(4 + j3)(-j3)}{(4 + j3 - j3)} = \frac{9 - j12}{4} = \mathbf{(2.25 - j3.0)\ \Omega}$$

Thus the Thévenin equivalent circuit is as shown in Figure 36.33.

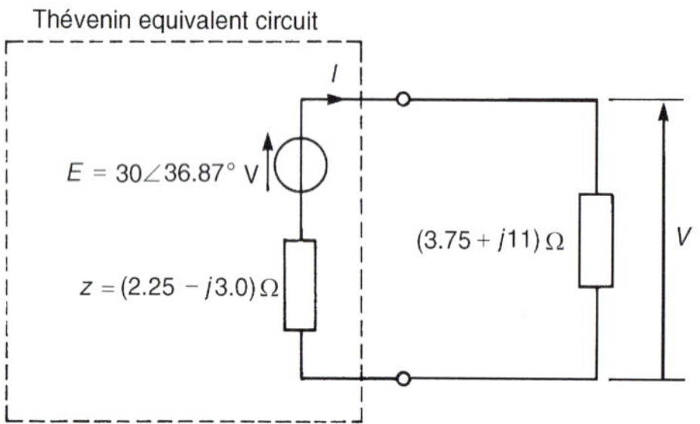

Figure 36.33

(a) When a $(3.75 + j11)\,\Omega$ impedance is connected across terminals AB, the current I flowing in the impedance is given by

$$I = \frac{30\angle 36.87^\circ}{(3.75 + j11) + (2.25 - j3.0)} = \frac{30\angle 36.87^\circ}{10\angle 53.13^\circ} = 3\angle -16.26^\circ \text{A}$$

Hence the current flowing in the $(3.75 + j11)\,\Omega$ impedance is 3 A

(b) P.d. across the $(3.75 + j11)\,\Omega$ impedance is given by

$$V = (3\angle -16.26^\circ)(3.75 + j11)$$
$$= (3\angle -16.26^\circ)(11.62\angle 71.18^\circ)$$
$$= 34.86\angle 54.92^\circ \text{ V}$$

Hence the magnitude of the p.d. across the impedance is 34.9 V

Problem 6. Use Thévenin's theorem to determine the current flowing in the capacitor of the network shown in Figure 36.34.

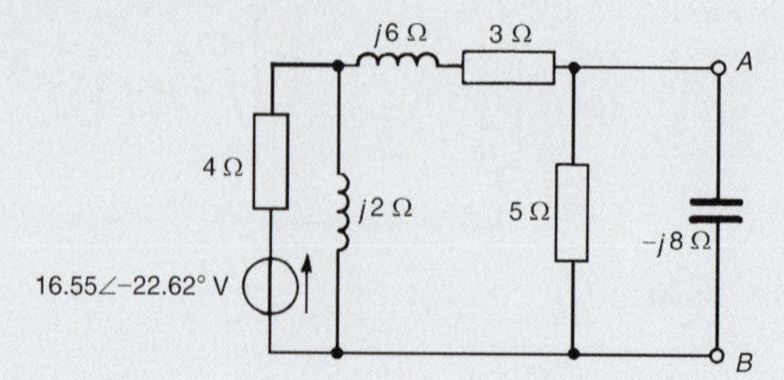

Figure 36.34

(i) The capacitor is removed from branch AB, as shown in Figure 36.35.

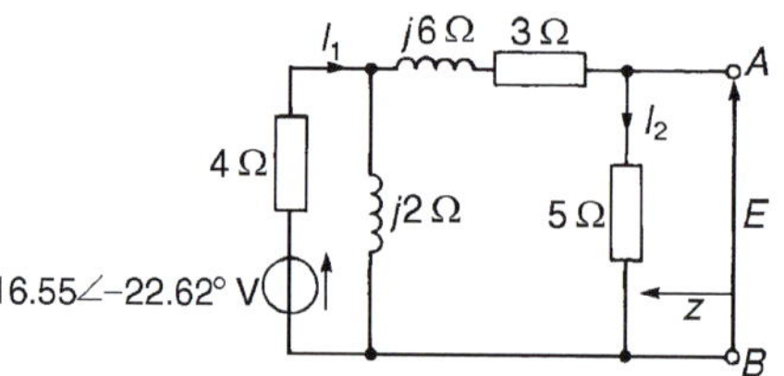

Figure 36.35

(ii) The open-circuit voltage, E, shown in Figure 36.35, is given by $(I_2)(5)$. I_2 may be determined by current division if I_1 is known. (Alternatively, E may be determined by the method used in Problem 4.)

Current $I_1 = V/Z$, where Z is the total circuit impedance and $V = 16.55\angle -22.62^\circ$ V.

$$\text{Impedance, } Z = 4 + \frac{(j2)(8 + j6)}{j2 + 8 + j6}$$
$$= 4 + \frac{-12 + j16}{8 + j8}$$
$$= 4.596\angle 22.38^\circ\ \Omega$$

$$\text{Hence } I_1 = \frac{16.55\angle -22.62^\circ}{4.596\angle 22.38^\circ}$$
$$= 3.60\angle -45^\circ \text{ A}$$

$$\text{and } \quad I_2 = \left(\frac{j2}{j2 + 3 + j6 + 5}\right) I_1$$
$$= \frac{(2\angle 90^\circ)(3.60\angle -45^\circ)}{11.314\angle 45^\circ}$$
$$= 0.636\angle 0^\circ \text{ A}$$

(An alternative method of finding I_2 is to use Kirchhoff's laws or mesh-current or nodal analysis on Figure 36.35.)

Hence $\boldsymbol{E} = (I_2)(5) = (0.636\angle 0°)(5) = \mathbf{3.18\angle 0°\ V}$

(iii) If the $16.55\angle -22.62°$ V source is removed from Figure 36.35, the impedance, z, 'looking in' at AB is given by

$$z = \frac{5[((4 \times j2)/(4 + j2)) + (3 + j6)]}{5 + [((4 \times j2)/(4 + j2)) + 3 + j6]} = \frac{5(3.8 + j7.6)}{8.8 + j7.6}$$

i.e. $z = 3.654\angle 22.61°\,\Omega$ or $\mathbf{(3.373 + j1.405)\,\Omega}$

(iv) The Thévenin equivalent circuit is shown in Figure 36.36, where the current flowing in the capacitor, I, is given by

$$I = \frac{3.18\angle 0°}{(3.373 + j1.405) - j8} = \frac{3.18\angle 0°}{7.408\angle -62.91°}$$

$= \mathbf{0.43\angle 62.91°\ A}$ **in the direction from A to B.**

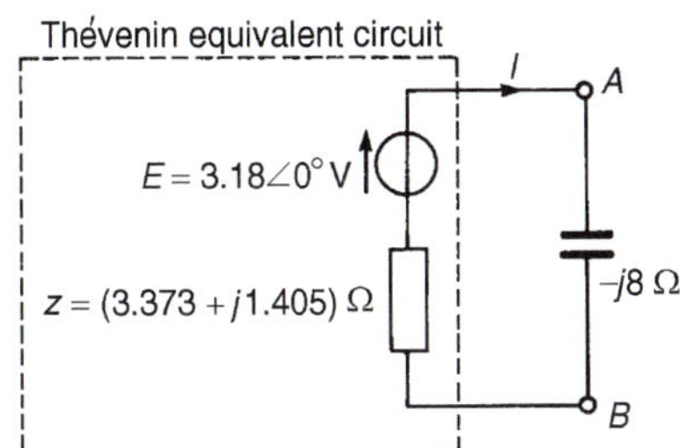

Figure 36.36

Problem 7. For the network shown in Figure 36.37, derive the Thévenin equivalent circuit with respect to terminals PQ, and hence determine the power dissipated by a 2 Ω resistor connected across PQ.

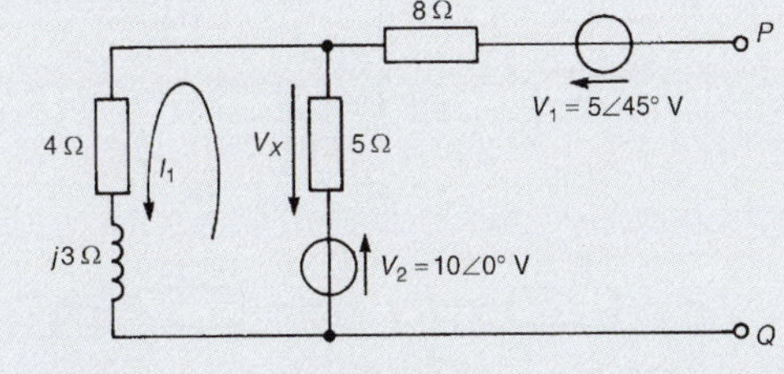

Figure 36.37

Current I_1 shown in Figure 36.37 is given by

$$I_1 = \frac{10\angle 0°}{(5 + 4 + j3)} = 1.054\angle -18.43°\ \text{A}$$

Hence the voltage drop across the 5 Ω resistor is given by $V_X = (I_1)(5) = 5.27\angle -18.43°$ V, and is in the direction shown in Figure 36.37, i.e. the direction opposite to that in which I_1 is flowing.

The open-circuit voltage E across PQ is the phasor sum of V_1, V_x and V_2, as shown in Figure 36.38.

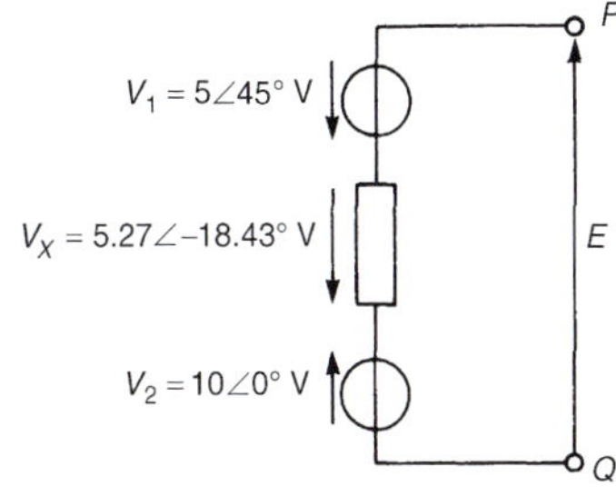

Figure 36.38

Thus $\boldsymbol{E} = 10\angle 0° - 5\angle 45° - 5.27\angle -18.43°$

$= (1.465 - j1.869)$ V or $\mathbf{2.375\angle -51.91°\ V}$

The impedance, z, 'looking in' at terminals PQ with the voltage sources removed is given by

$$z = 8 + \frac{5(4 + j3)}{(5 + 4 + j3)} = 8 + 2.635\angle 18.44°$$

$$= \mathbf{(10.50 + j0.833)\,\Omega}$$

The Thévenin equivalent circuit is shown in Figure 36.39 with the 2 Ω resistance connected across terminals PQ.

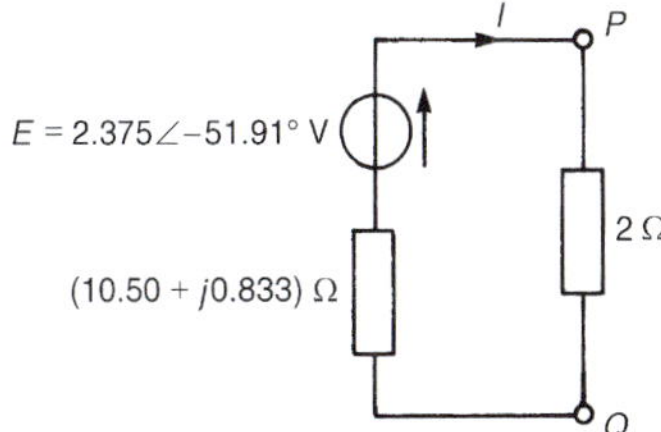

Figure 36.39

The current flowing in the 2 Ω resistance is given by

$$I = \frac{2.375\angle -51.91°}{(10.50 + j0.833) + 2} = 0.1896\angle -55.72°\ \text{A}$$

The power P dissipated in the 2 Ω resistor is given by

$P = I^2R = (0.1896)^2(2) = \mathbf{0.0719\ W \equiv 72\ mW,}$

correct to two significant figures.

Problem 8. For the a.c. bridge network shown in Figure 36.40, determine the current flowing in the capacitor, and its direction, by using Thévenin's theorem. Assume the $30\angle 0°$ V source to have negligible internal impedance.

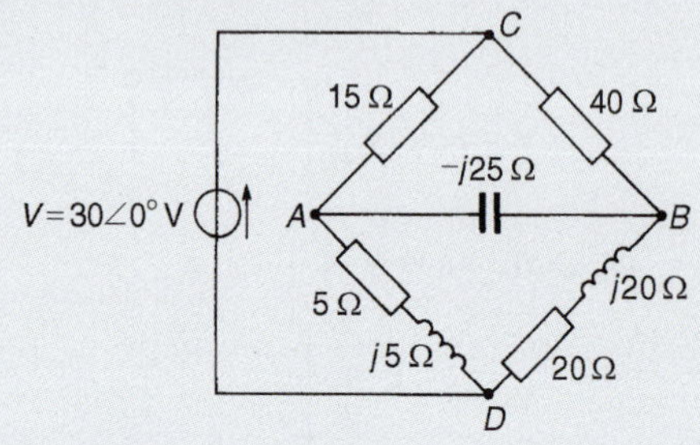

Figure 36.40

(i) The $-j25\,\Omega$ capacitor is initially removed from the network, as shown in Figure 36.41.

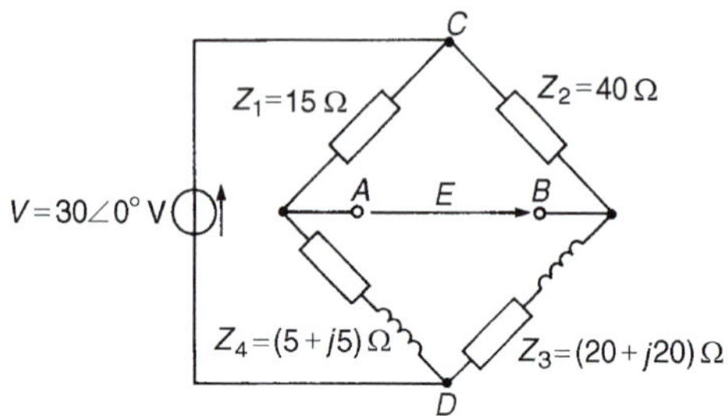

Figure 36.41

(ii) P.d. between A and C,

$$V_{AC} = \left(\frac{Z_1}{Z_1 + Z_4}\right) V = \left(\frac{15}{15 + 5 + j5}\right)(30\angle 0°)$$
$$= 21.83\angle -14.04°\ \text{V}$$

P.d. between B and C,

$$V_{BC} = \left(\frac{Z_2}{Z_2 + Z_3}\right) V = \left(\frac{40}{40 + 20 + j20}\right)(30\angle 0°)$$
$$= 18.97\angle -18.43°\ \text{V}$$

Assuming that point A is at a higher potential than point B, then the p.d. between A and B is

$$21.83\angle -14.04° - 18.97\angle -18.43°$$
$$= (3.181 + j0.701)\ \text{V or } 3.257\angle 12.43°\ \text{V},$$

i.e. the open-circuit voltage across AB is given by

$$E = 3.257\angle 12.43°\ \text{V}$$

Point C is at a potential of $30\angle 0°$ V. Between C and A is a volt drop of $21.83\angle -14.04°$ V. Hence the **voltage at point A** is

$$30\angle 0° - 21.83\angle -14.04° = \mathbf{10.29\angle 30.98°\ V}$$

Between points C and B is a voltage drop of $18.97\angle -18.43°$ V. Hence the **voltage at point B** is $30\angle 0° - 18.97\angle -18.43° = \mathbf{13.42\angle 26.55°\ V}$.

Since the magnitude of the voltage at B is higher than at A, current must flow in the direction B to A.

(iii) Replacing the $30\angle 0°$ V source with a short-circuit (i.e. zero internal impedance) gives the network shown in Figure 36.42(a). The network is shown redrawn in Figure 36.42(b) and simplified in Figure 36.42(c). Hence the impedance, z, 'looking in' at terminals AB is given by

$$z = \frac{(15)(5 + j5)}{(15 + 5 + j5)} + \frac{(40)(20 + j20)}{(40 + 20 + j20)}$$
$$= 5.145\angle 30.96° + 17.889\angle 26.57°$$

i.e. $z = (20.41 + j10.65)\,\Omega$

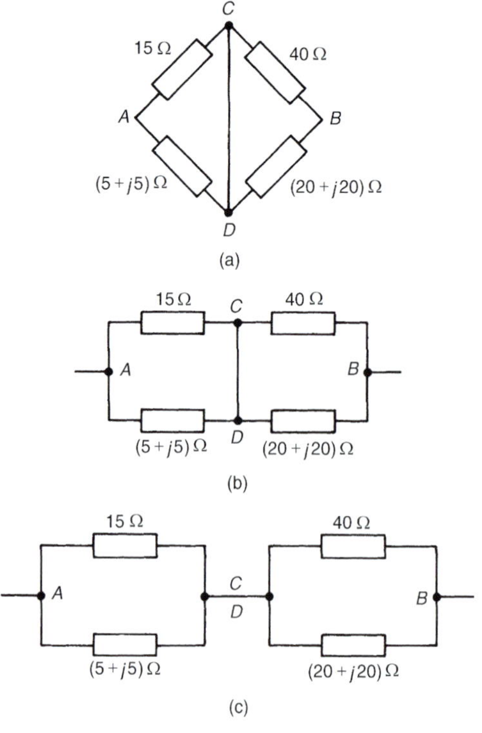

Figure 36.42

Part 4

(iv) The Thévenin equivalent circuit is shown in Figure 36.43, where current I is given by

$$I = \frac{3.257\angle 12.43^\circ}{(20.41 + j10.65) - j25} = \frac{3.257\angle 12.43^\circ}{24.95\angle -35.11^\circ} = 0.131\angle 47.54^\circ\,\text{A}$$

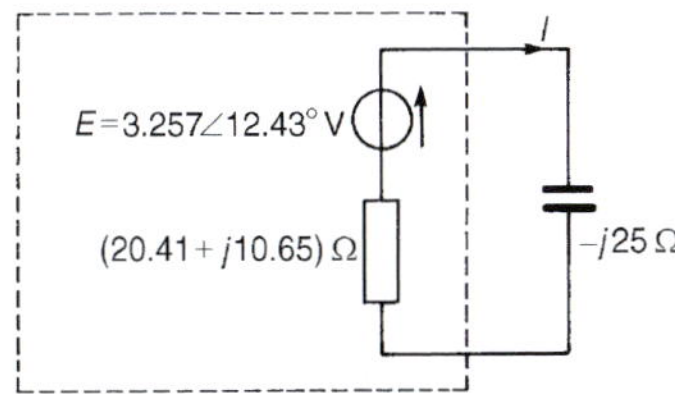

Figure 36.43

Thus a current of 131 mA flows in the capacitor in a direction from B to A.

Now try the following Practice Exercise

Practice Exercise 141 Thévenin's theorem (Answers on page 829)

1. Determine for the network shown in Figure 36.44 the Thévenin equivalent circuit with respect to terminals AB, and hence determine the current flowing in the $(5 + j6)\,\Omega$ impedance connected between A and B.

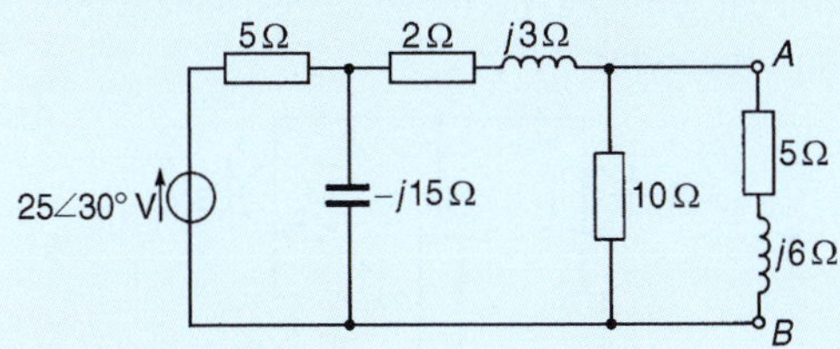

Figure 36.44

2. For the network shown in Figure 36.45, derive the Thévenin equivalent circuit with respect to terminals AB, and hence determine the magnitude of the current flowing in a $(2 + j13)\,\Omega$ impedance connected between A and B.

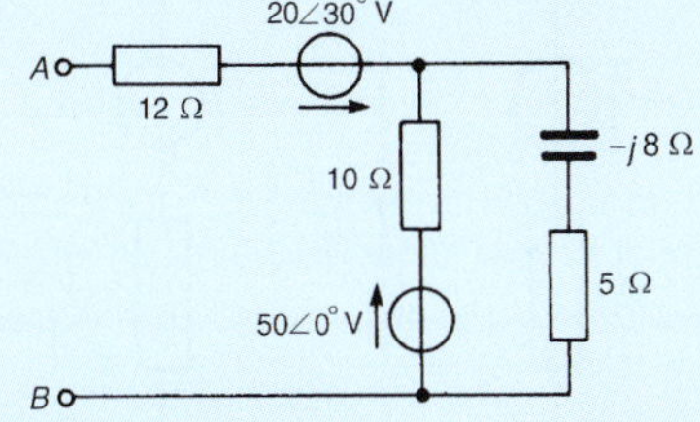

Figure 36.45

3. Use Thévenin's theorem to determine the power dissipated in the $4\,\Omega$ resistance of the network shown in Figure 36.46.

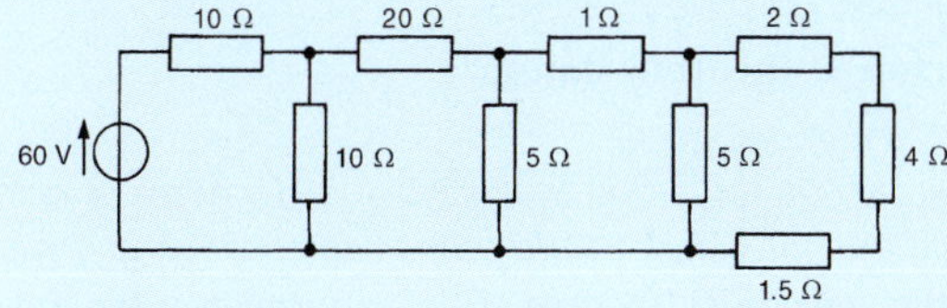

Figure 36.46

4. For the bridge network shown in Figure 36.47 use Thévenin's theorem to determine the current flowing in the $(4 + j3)\,\Omega$ impedance and its direction. Assume that the $20\angle 0^\circ\,\text{V}$ source has negligible internal impedance.

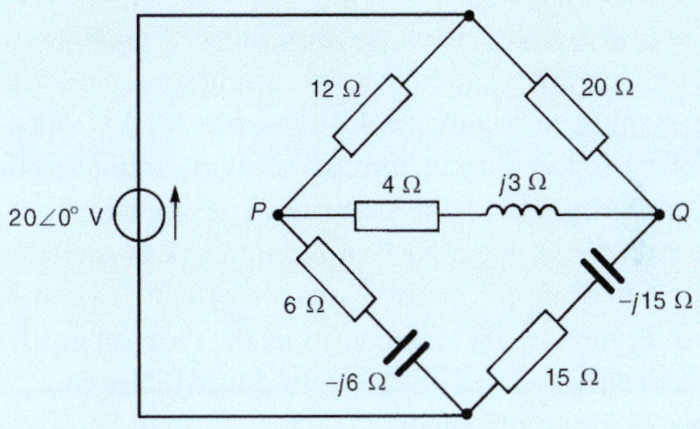

Figure 36.47

5. Repeat Problems 1 to 10 of Exercise 135, page 504 using Thévenin's theorem.
6. Repeat Problems 2 and 3 of Exercise 136, page 510 using Thévenin's theorem.
7. Repeat Problems 3 to 7 of Exercise 137, page 517 using Thévenin's theorem.
8. Repeat Problems 4 to 7 of Exercise 138, page 523 using Thévenin's theorem.
9. Repeat Problems 1 to 3 of Exercise 139, page 526 using Thévenin's theorem.

36.4 Norton's theorem

A source of electrical energy can be represented by a source of e.m.f. in series with an impedance. In Section 36.2, the Thévenin constant-voltage source consisted of a constant e.m.f., E, which may be alternating or direct, in series with an internal impedance, z.

However, this is not the only form of representation. A source of electrical energy can also be represented by a constant-current source, which may be alternating or direct, in parallel with an impedance. It is shown in Section 36.5 that the two forms are in fact equivalent.

Norton's* theorem states:

'The current that flows in any branch of a network is the same as that which would flow in the branch if it were connected across a source of electrical energy, the short-circuit current of which is equal to the current that would flow in a short-circuit across the branch, and the internal impedance of which is equal to the impedance which appears across the open-circuited branch terminals.'

The above statement simply means that any linear active network with output terminals AB, as shown in Figure 36.48(a), can be replaced by a current source in parallel with an impedance z as shown in Figure 36.48(b). The equivalent current source I_{SC} (note the symbol in Figure 36.48(b) as per BS EN 60617-2:1996) is the current through a short-circuit applied to the terminals of the network. The impedance z is the equivalent impedance of the network at the terminals AB when all internal sources of e.m.f. are made zero. Figure 36.48(b) is known as the **Norton equivalent circuit**, and was initially introduced in Section 15.7, page 233 for d.c. networks.

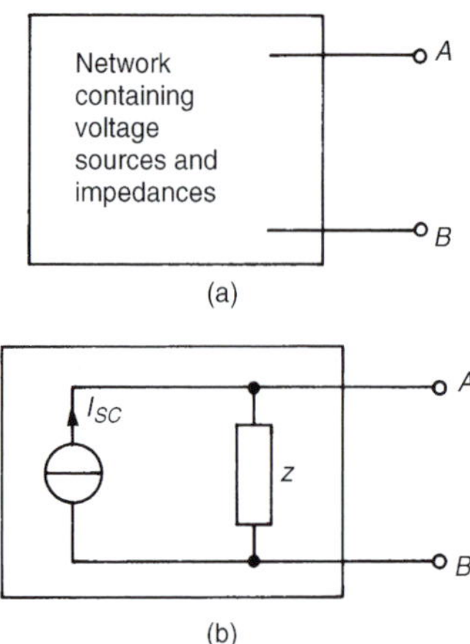

Figure 36.48 The Norton equivalent circuit

The following four-step procedure may be adopted when determining the current flowing in an impedance Z_L of a branch AB of an active network, using Norton's theorem:

(i) short-circuit branch AB

(ii) determine the short-circuit current I_{SC}

(iii) remove each source of e.m.f. and replace it by its internal impedance (or, if a current source exists, replace with an open circuit), then determine the impedance, z, 'looking in' at a break made between A and B

(iv) determine the value of the current i_L flowing in impedance Z_L from the Norton equivalent network shown in Figure 36.49, i.e.

$$i_L = \left(\frac{z}{Z_L + z}\right) I_{SC}$$

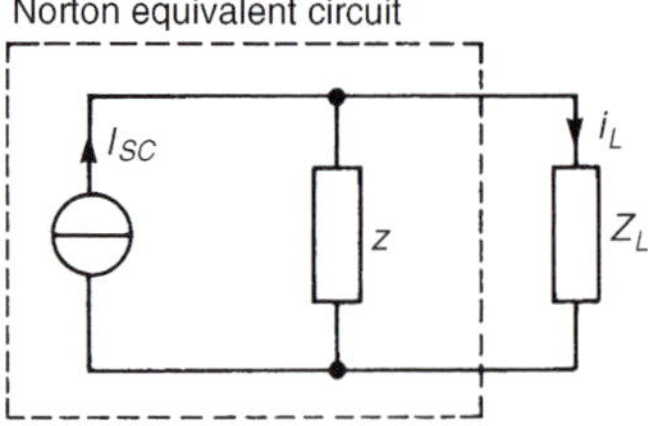

Figure 36.49

A simple d.c. network (Figure 36.50) serves to demonstrate how the above procedure is applied to determine the current flowing in the 5 Ω resistance by using Norton's theorem:

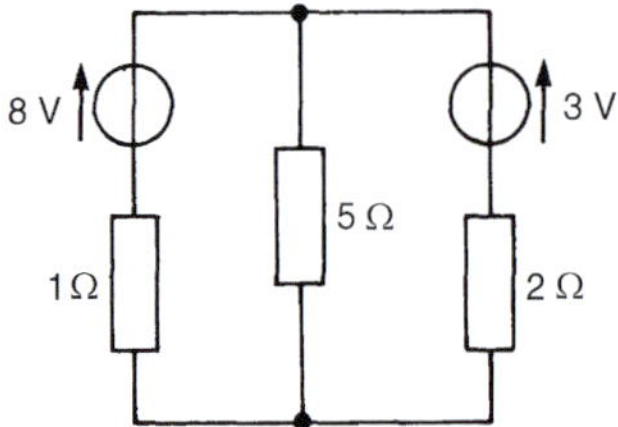

Figure 36.50

(i) The 5 Ω branch is short-circuited, as shown in Figure 36.51.

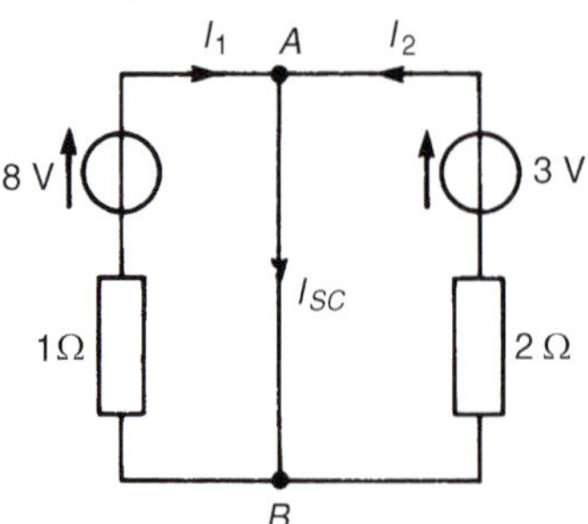

Figure 36.51

*Who was **Norton**? For resume of Norton, see page 233. To find out more go to **www.routledge.com/cw/bird**

(ii) From Figure 36.51,

$$I_{SC} = I_1 + I_2 = \frac{8}{1} + \frac{3}{2} = 9.5\,\text{A}$$

(iii) If each source of e.m.f. is removed the impedance 'looking in' at a break made between A and B is given by $z = (1 \times 2)/(1+2) = \frac{2}{3}\,\Omega$

(iv) From the Norton equivalent network shown in Figure 36.52, the current in the 5 Ω resistance is given by $I_L = \left(\frac{2}{3}/\left(5+\frac{2}{3}\right)\right) 9.5 = \mathbf{1.12\,A}$, as obtained previously using Kirchhoff's laws, the superposition theorem and by Thévenin's theorem.

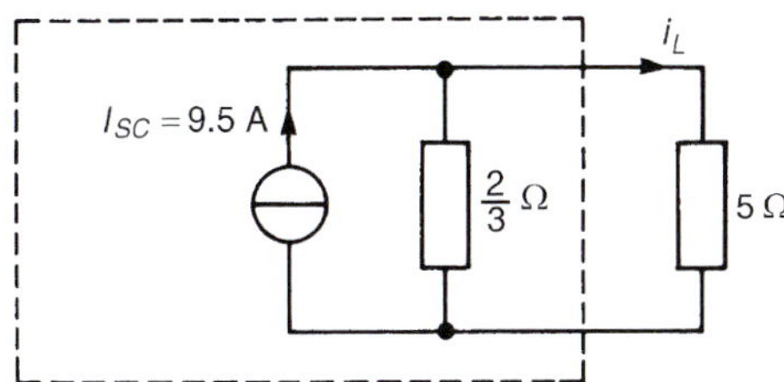

Figure 36.52

As with Thévenin's theorem, Norton's theorem may be used with a.c. as well as d.c. networks, as shown below.

An a.c. network is shown in Figure 36.53, where it is required to find the current flowing in the $(6+j8)\,\Omega$ impedance by using Norton's theorem.
Using the above procedure:

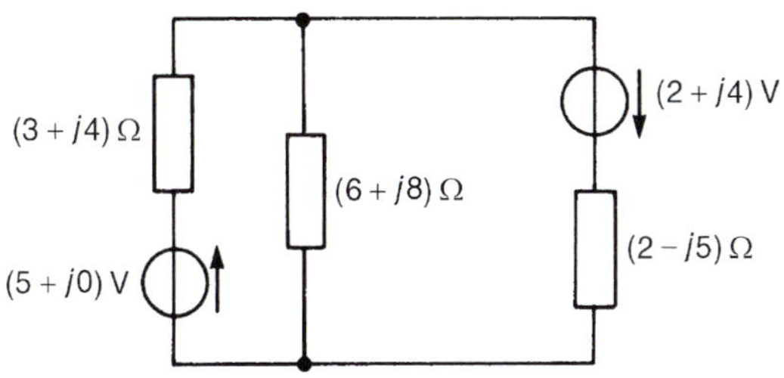

Figure 36.53

(i) Initially the $(6+j8)\,\Omega$ impedance is short-circuited, as shown in Figure 36.54.

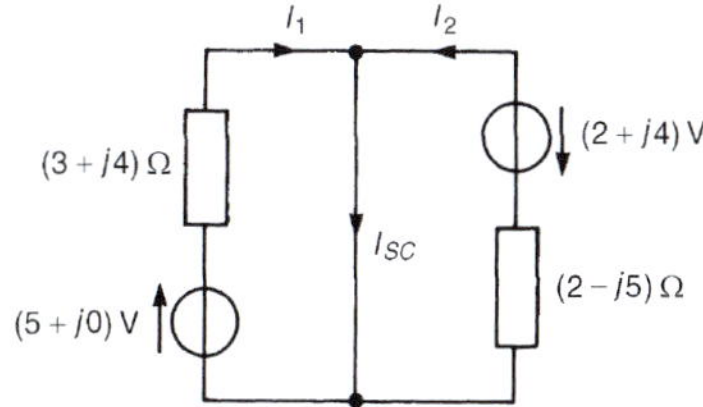

Figure 36.54

(ii) From Figure 36.54,

$$\boldsymbol{I_{SC}} = I_1 + I_2 = \frac{(5+j0)}{(3+j4)} + \frac{(-(2+j4))}{(2-j5)}$$

$$= 1\angle{-53.13^\circ} - \frac{4.472\angle 63.43^\circ}{5.385\angle{-68.20^\circ}}$$

$$= (1.152 - j1.421)\,\text{A or } \mathbf{1.829\angle{-50.97^\circ}\,A}$$

(iii) If each source of e.m.f. is removed, the impedance, z, 'looking in' at a break made between A and B is given by

$$z = \frac{(3+j4)(2-j5)}{(3+j4)+(2-j5)}$$

$$= \mathbf{5.28\angle{-3.76^\circ}\,\Omega} \quad \text{or} \quad \mathbf{(5.269 - j0.346)\,\Omega}$$

(iv) From the Norton equivalent network shown in Figure 36.55, the current is given by

$$i_L = \left(\frac{z}{Z_L + z}\right) I_{SC}$$

$$= \left(\frac{5.28\angle{-3.76^\circ}}{(6+j8)+(5.269-j0.346)}\right) 1.829\angle{-50.97^\circ}$$

i.e. **current in $(6+j8)\,\Omega$ impedance, $i_L = 0.71\angle{-88.91^\circ}$ A**

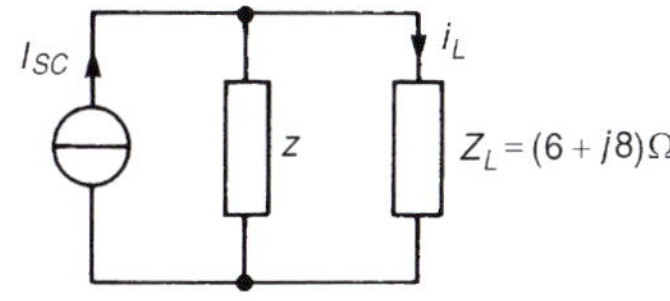

Figure 36.55

Problem 9. Use Norton's theorem to determine the value of current I in the circuit shown in Figure 36.56.

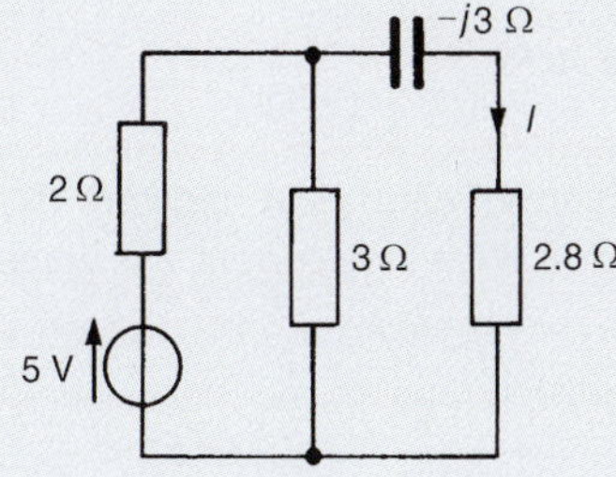

Figure 36.56

(i) The branch containing the 2.8 Ω resistor is short-circuited, as shown in Figure 36.57.

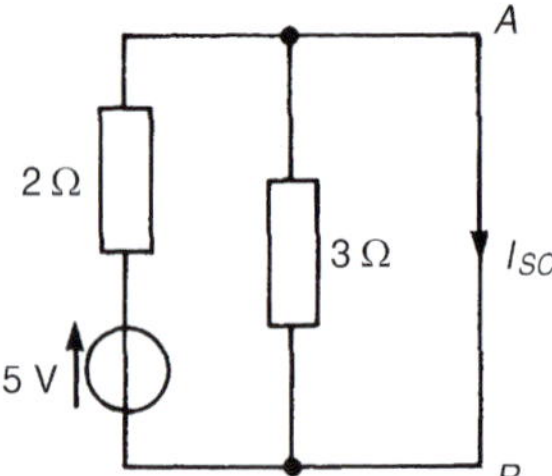

Figure 36.57

(ii) The 3 Ω resistor in parallel with a short-circuit is the same as 3 Ω in parallel with 0, giving an equivalent impedance of $(3\times 0)/(3+0)=0$. Hence the network reduces to that shown in Figure 36.58, where $\boldsymbol{I_{SC}=5/2=2.5\,A}$

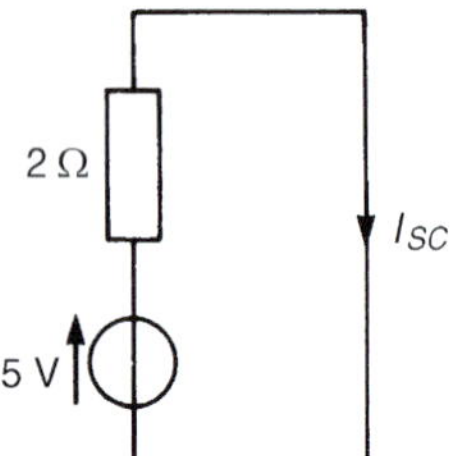

Figure 36.58

(iii) If the 5 V source is removed from the network the input impedance, z, 'looking-in' at a break made in AB of Figure 36.57 gives $z=(2\times 3)/(2+3)=\mathbf{1.2\,\Omega}$ (see Figure 36.59).

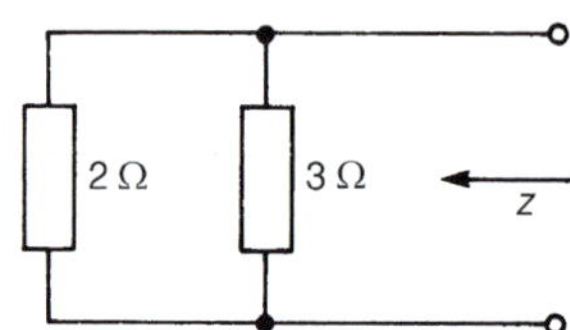

Figure 36.59

(iv) The Norton equivalent network is shown in Figure 36.60, where current I is given by

$$I=\left(\frac{1.2}{1.2+(2.8-j3)}\right)(2.5)=\frac{3}{4-j3}$$

$$=\mathbf{0.60\angle 36.87^\circ\,A}$$

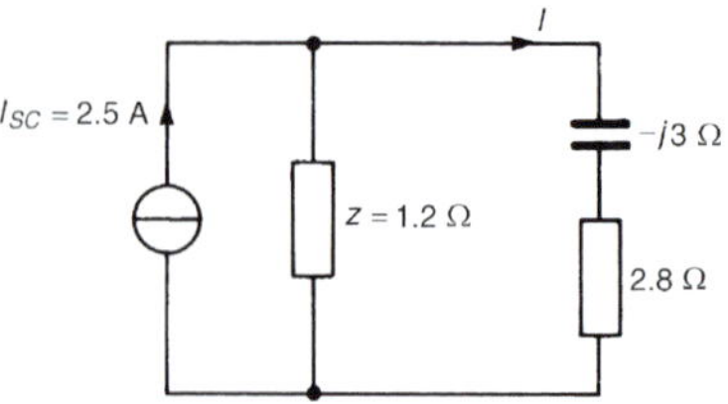

Figure 36.60

Problem 10. For the circuit shown in Figure 36.61, determine the current flowing in the inductive branch by using Norton's theorem.

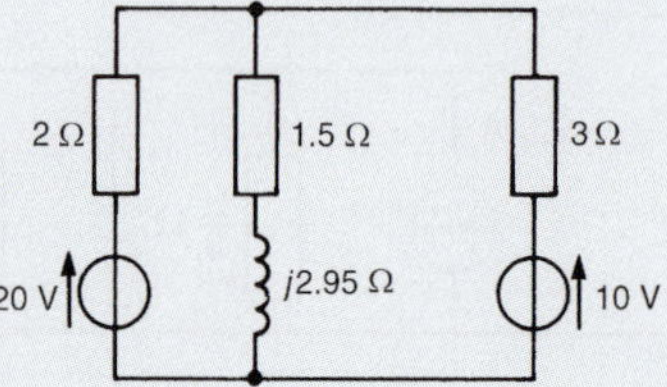

Figure 36.61

(i) The inductive branch is initially short-circuited, as shown in Figure 36.62.

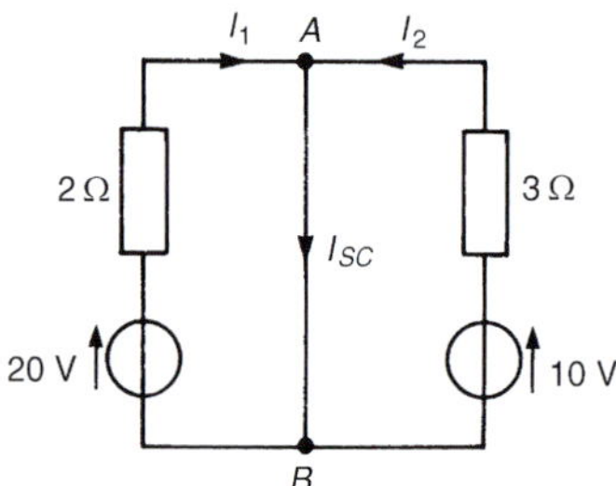

Figure 36.62

(ii) From Figure 36.62,

$$I_{SC}=I_1+I_2=\frac{20}{2}+\frac{10}{3}=\mathbf{13.\dot{3}\,A}$$

(iii) If the voltage sources are removed, the impedance, z, 'looking in' at a break made in AB is given by $z=(2\times 3)/(2+3)=\mathbf{1.2\,\Omega}$

(iv) The Norton equivalent network is shown in Figure 36.63, where current I is given by

$$I=\left(\frac{1.2}{1.2+1.5+j2.95}\right)(13.\dot{3})=\frac{16}{2.7+j2.95}$$

$$=\mathbf{4.0\angle -47.53^\circ A}\ \text{ or }\ \mathbf{(2.7-j2.95)\,A}$$

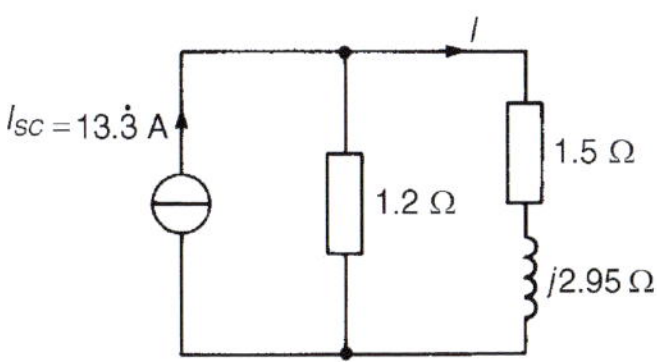

Figure 36.63

Problem 11. Use Norton's theorem to determine the magnitude of the p.d. across the 1 Ω resistance of the network shown in Figure 36.64.

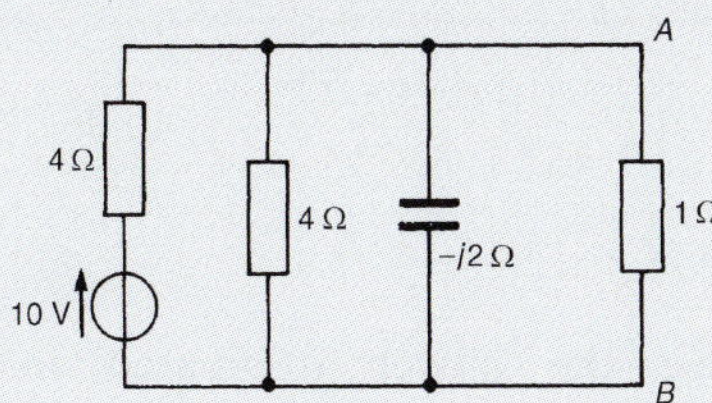

Figure 36.64

(i) The branch containing the 1 Ω resistance is initially short-circuited, as shown in Figure 36.65.

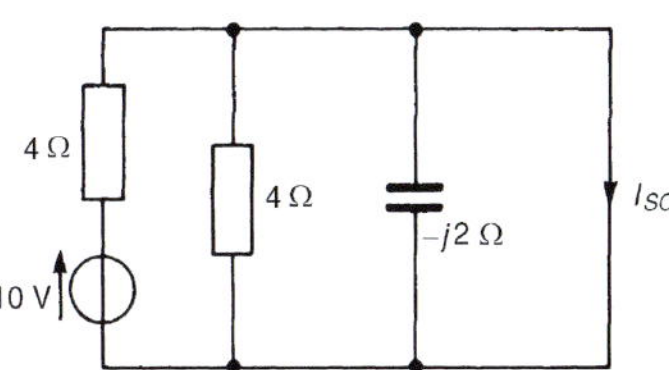

Figure 36.65

(ii) 4 Ω in parallel with $-j2\,\Omega$ in parallel with 0 Ω (i.e. the short-circuit) is equivalent to 0, giving the equivalent circuit of Figure 36.66. Hence $\boldsymbol{I_{SC}} = 10/4 = \mathbf{2.5\,A}$

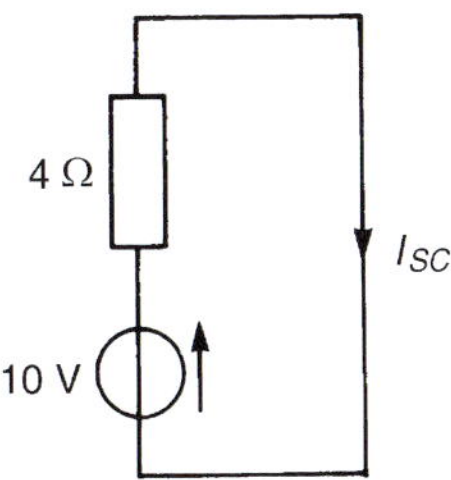

Figure 36.66

(iii) The 10 V source is removed from the network of Figure 36.64, as shown in Figure 36.67, and the impedance z, 'looking in' at a break made in AB is given by

$$\frac{1}{z} = \frac{1}{4} + \frac{1}{4} + \frac{1}{-j2} = \frac{-j - j + 2}{-j4} = \frac{2 - j2}{-j4}$$

from which

$$z = \frac{-j4}{2 - j2} = \frac{-j4(2 + j2)}{2^2 + 2^2} = \frac{8 - j8}{8} = \mathbf{(1 - j1)\,\Omega}$$

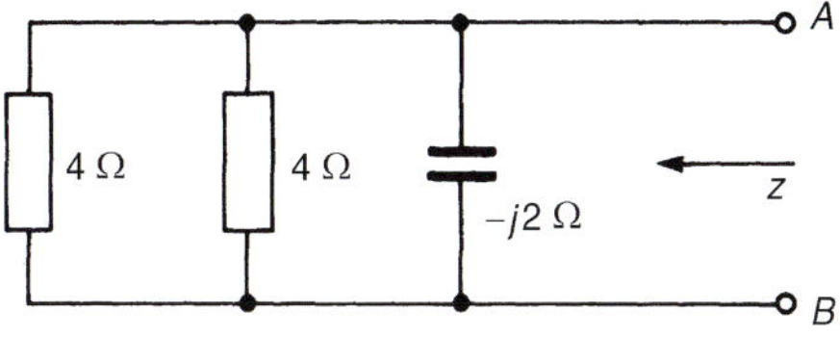

Figure 36.67

(iv) The Norton equivalent network is shown in Figure 36.68, from which current I is given by

$$I = \left(\frac{1 - j1}{(1 - j1) + 1}\right)(2.5) = 1.58\angle -18.43^\circ\,\text{A}$$

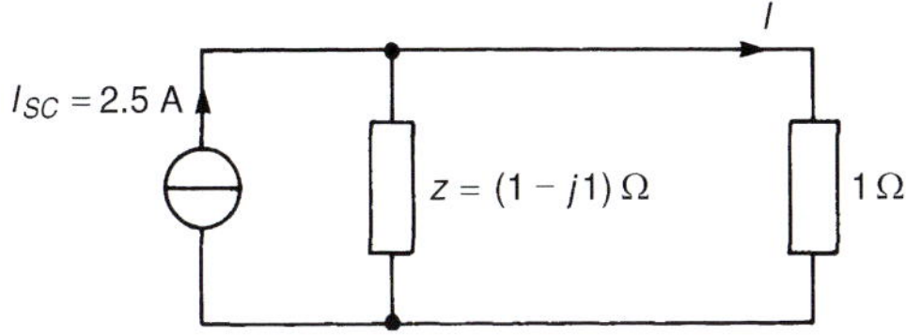

Figure 36.68

Hence the magnitude of the p.d. across the 1 Ω resistor is given by

$$IR = (1.58)(1) = \mathbf{1.58\,V}$$

Problem 12. For the network shown in Figure 36.69, obtain the Norton equivalent network at terminals AB. Hence determine the power dissipated in a 5 Ω resistor connected between A and B.

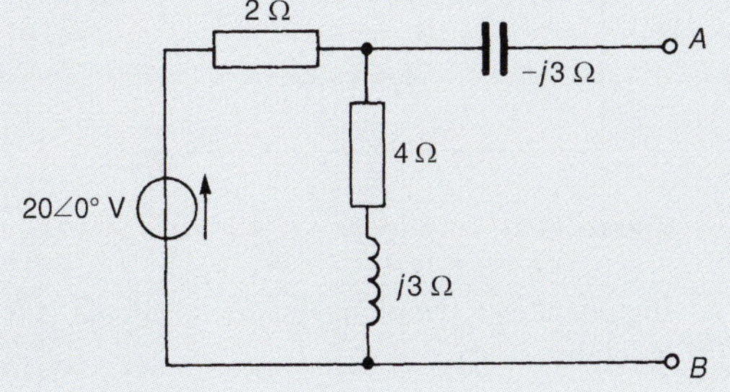

Figure 36.69

(i) Terminals AB are initially short-circuited, as shown in Figure 36.70.

(ii) The circuit impedance Z presented to the $20\angle 0°$ V source is given by

$$Z = 2 + \frac{(4+j3)(-j3)}{(4+j3)+(-j3)} = 2 + \frac{9-j12}{4}$$

$$= (4.25 - j3)\,\Omega \quad \text{or} \quad 5.202\angle -35.22°\,\Omega$$

Thus current I in Figure 36.70 is given by

$$I = \frac{20\angle 0°}{5.202\angle -35.22°} = 3.845\angle 35.22°\,\text{A}$$

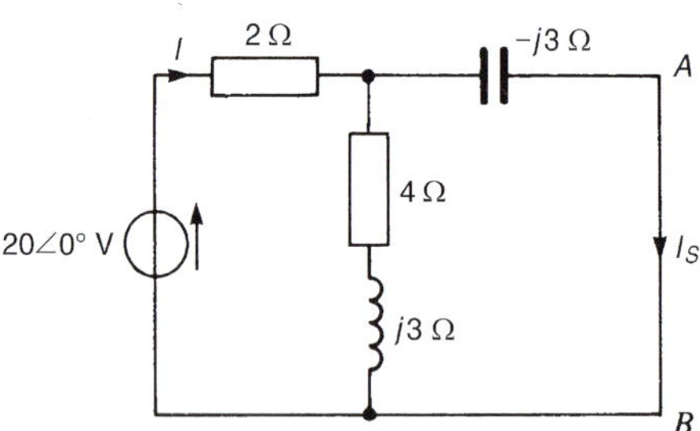

Figure 36.70

Hence

$$I_{SC} = \left(\frac{(4+j3)}{(4+j3)-j3}\right)(3.845\angle 35.22°)$$

$$= \mathbf{4.806\angle 72.09°\,A}$$

(iii) Removing the $20\angle 0°$ V source of Figure 36.69 gives the network of Figure 36.71.

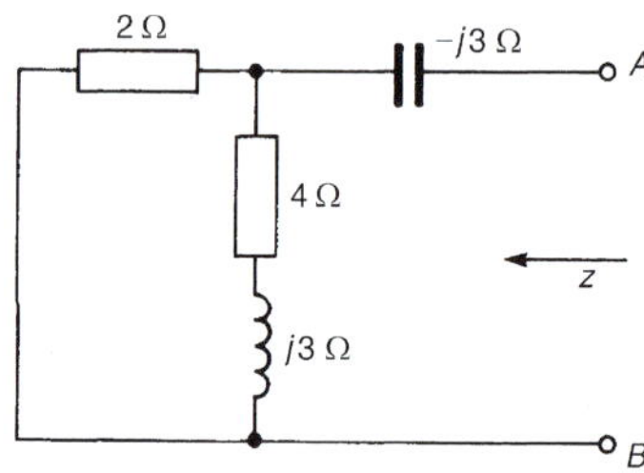

Figure 36.71

Impedance, z, 'looking in' at terminals AB is given by

$$z = -j3 + \frac{2(4+j3)}{2+4+j3} = -j3 + 1.491\angle 10.3°$$

$$= \mathbf{(1.467 - j2.733)\,\Omega} \quad \text{or} \quad \mathbf{3.102\angle -61.77°\,\Omega}$$

(iv) The Norton equivalent network is shown in Figure 36.72.

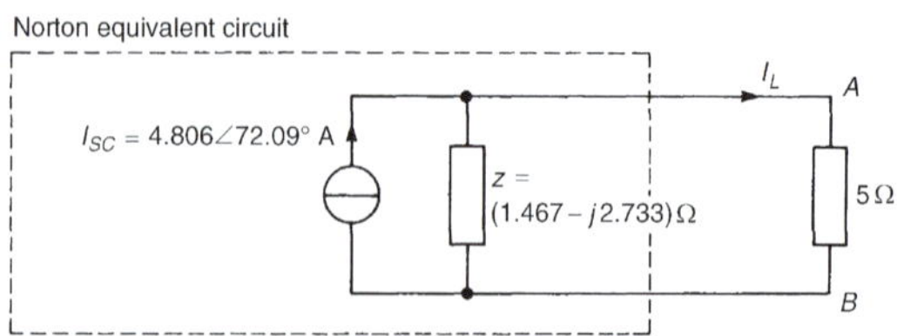

Figure 36.72

Current I_L

$$= \left(\frac{3.102\angle -61.77°}{1.467 - j2.733 + 5}\right)(4.806\angle 72.09°)$$

$$= 2.123\angle 33.23°\,\text{A}$$

Hence the power dissipated in the 5Ω resistor is

$$I_L^2 R = (2.123)^2(5) = \mathbf{22.5\,W}$$

Problem 13. Derive the Norton equivalent network with respect to terminals PQ for the network shown in Figure 36.73 and hence determine the magnitude of the current flowing in a 2 Ω resistor connected across PQ.

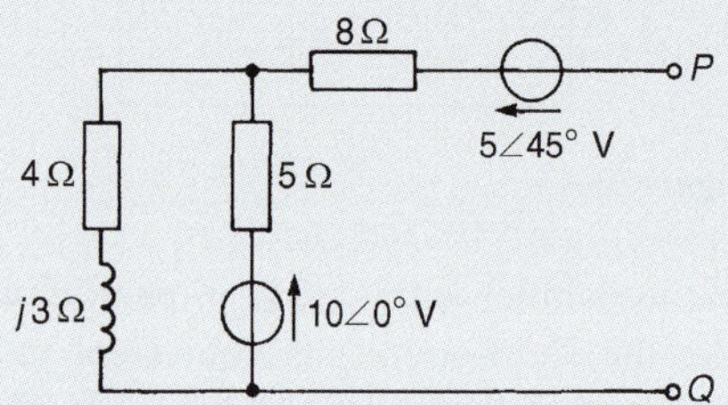

Figure 36.73

This is the same problem as Problem 7 on page 537 which was solved by Thévenin's theorem.
A comparison of methods may thus be made.

(i) Terminals PQ are initially short-circuited, as shown in Figure 36.74.

(ii) Currents I_1 and I_2 are shown labelled. Kirchhoff's laws are used. For loop $ABCD$, and moving anticlockwise,

$$10\angle 0° = 5I_1 + (4+j3)(I_1 + I_2)$$

i.e. $\quad (9+j3)I_1 + (4+j3)I_2 - 10 = 0 \quad (1)$

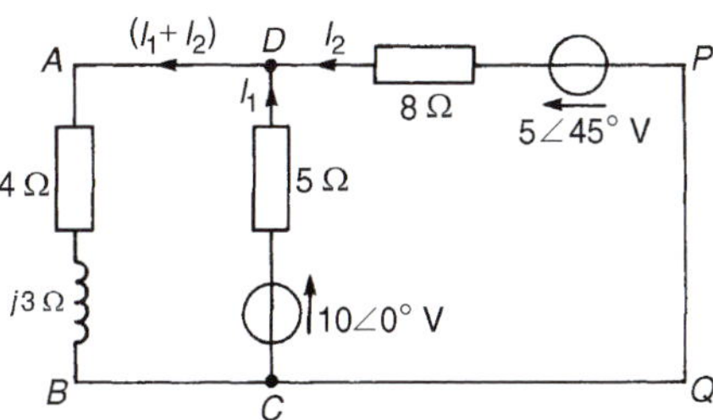

Figure 36.74

For loop *DPQC*, and moving clockwise,

$$10\angle 0^\circ - 5\angle 45^\circ = 5I_1 - 8I_2$$

i.e. $\quad 5I_1 - 8I_2 + (5\angle 45^\circ - 10) = 0 \qquad (2)$

Solving equations (1) and (2) by using determinants gives

$$\frac{I_1}{\begin{vmatrix}(4+j3) & -10 \\ -8 & (5\angle 45^\circ - 10)\end{vmatrix}} = \frac{-I_2}{\begin{vmatrix}(9+j3) & -10 \\ 5 & (5\angle 45^\circ - 10)\end{vmatrix}} = \frac{I}{\begin{vmatrix}(9+j3) & (4+j3) \\ 5 & -8\end{vmatrix}}$$

from which

$$I_2 = \frac{-\begin{vmatrix}(9+j3) & -10 \\ 5 & (5\angle 45^\circ - 10)\end{vmatrix}}{\begin{vmatrix}(9+j3) & (4+j3) \\ 5 & -8\end{vmatrix}}$$

$$= \frac{-[(9+j3)(5\angle 45^\circ - 10) + 50]}{[-72 - j24 - 20 - j15]}$$

$$= \frac{-[22.52\angle 146.50^\circ]}{[99.925\angle -157.03^\circ]}$$

$$= -0.225\angle 303.53^\circ \text{ or } -0.225\angle -56.47^\circ$$

Hence the short-circuit current $I_{SC} = 0.225\angle -56.47^\circ$ A flowing from P to Q.

(iii) The impedance, z, 'looking in' at a break made between P and Q is given by

$z = (10.50 + j0.833)\,\Omega$

(see Problem 7, page 537).

(iv) The Norton equivalent circuit is shown in Figure 36.75, where current I is given by

$$I = \left(\frac{10.50 + j0.833}{10.50 + j0.833 + 2}\right)(0.225\angle -56.47^\circ)$$

$$= 0.19\angle -55.74^\circ \text{ A}$$

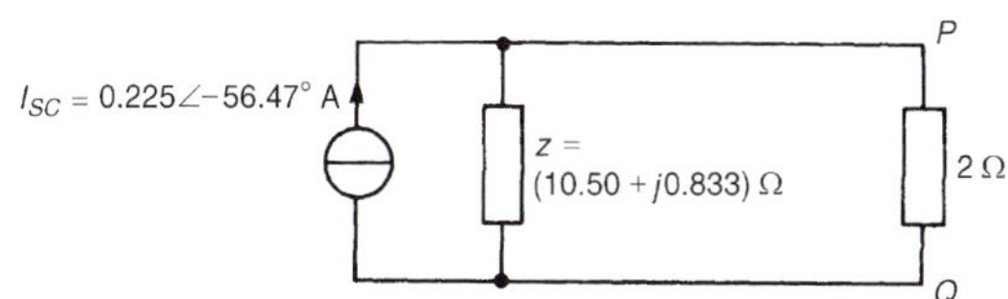

Figure 36.75

Hence the magnitude of the current flowing in the 2 Ω resistor is 0.19 A.

Now try the following Practice Exercise

Practice Exercise 142 Norton's theorem (Answers on page 829)

1. Repeat Problems 1 to 4 of Exercise 140, page 535 using Norton's theorem.
2. Repeat Problems 1 to 3 of Exercise 141, page 539 using Norton's theorem.
3. Determine the current flowing in the 10 Ω resistance of the network shown in Figure 36.76 by using Norton's theorem.

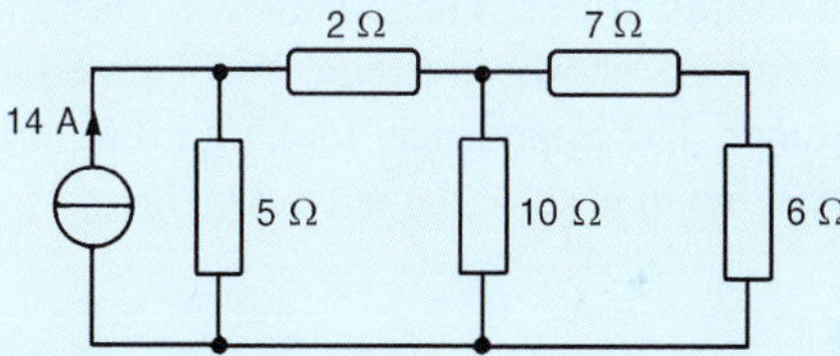

Figure 36.76

4. For the network shown in Figure 36.77, use Norton's theorem to determine the current flowing in the 10 Ω resistance.

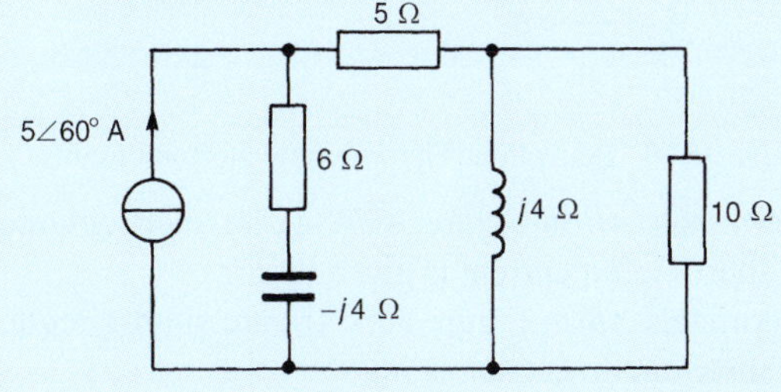

Figure 36.77

5. Determine for the network shown in Figure 36.78 the Norton equivalent network

at terminals AB. Hence determine the current flowing in a $(2+j4)\,\Omega$ impedance connected between A and B.

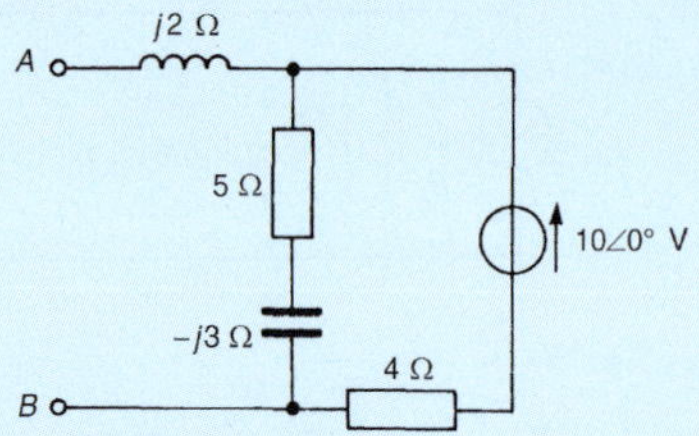

Figure 36.78

6. Repeat Problems 1 to 10 of Exercise 135, page 504 using Norton's theorem.
7. Repeat Problems 2 and 3 of Exercise 136, page 510 using Norton's theorem.
8. Repeat Problems 3 to 6 of Exercise 137, page 517 using Norton's theorem.

36.5 Thévenin and Norton equivalent networks

It is seen in Sections 36.2 and 36.4 that when Thévenin's and Norton's theorems are applied to the same circuit, identical results are obtained. Thus the Thévenin and Norton networks shown in Figure 36.79 are equivalent to each other. The impedance 'looking in' at terminals AB is the same in each of the networks, i.e. z.

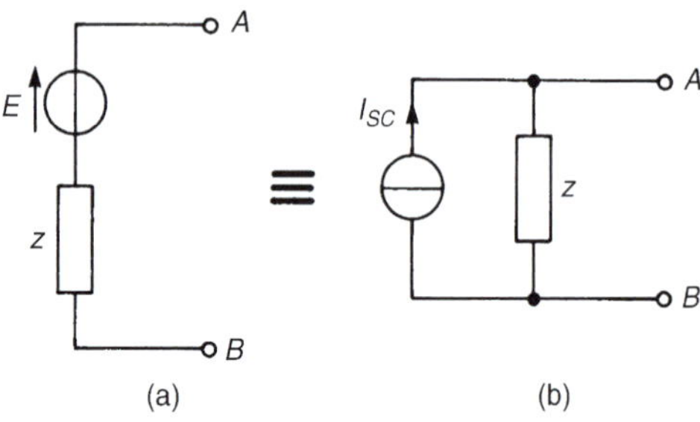

Figure 36.79 Equivalent Thévenin and Norton circuits

If terminals AB in Figure 36.79(a) are short-circuited, the short-circuit current is given by E/z

If terminals AB in Figure 36.79(b) are short-circuited, the short-circuit current is I_{SC}

Thus $\quad \boldsymbol{I_{SC} = E/z}$

Figure 36.80 shows a source of e.m.f. E in series with an impedance z feeding a load impedance Z_L. From Figure 36.80,

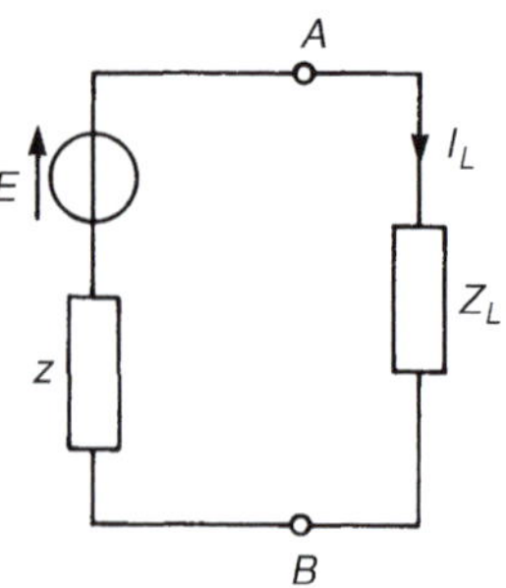

Figure 36.80

$$I_L = \frac{E}{z+Z_L} = \frac{E/z}{(z+Z_L)/z} = \left(\frac{z}{z+Z_L}\right)\frac{E}{z}$$

i.e. $\quad \boldsymbol{I_L = \left(\frac{z}{z+Z_L}\right) I_{SC}}$, from above.

From Figure 36.81 it can be seen that, when viewed from the load, the source appears as a source of current I_{SC} which is divided between z and Z_L connected in parallel.

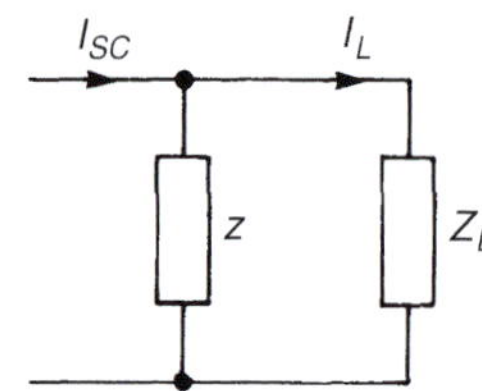

Figure 36.81

Thus it is shown that the two representations shown in Figure 36.79 are equivalent.

Problem 14. (a) Convert the circuit shown in Figure 36.82(a) to an equivalent Norton network. (b) Convert the network shown in Figure 36.82(b) to an equivalent Thévenin circuit.

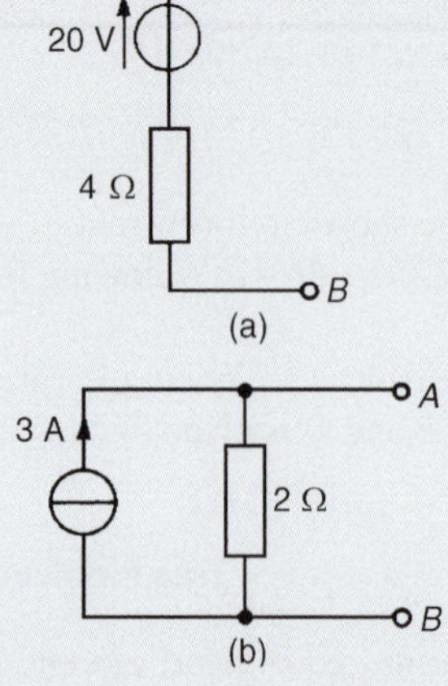

Figure 36.82

Part 4

(a) If the terminals AB of Figure 36.82(a) are short circuited, the short-circuit current, $I_{SC}=20/4=5$ A. The impedance 'looking in' at terminals AB is 4 Ω. Hence the equivalent Norton network is as shown in Figure 36.83(a).

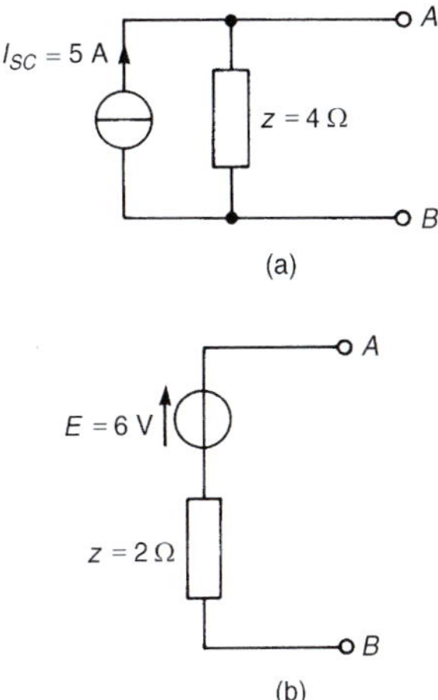

Figure 36.83

(b) The open-circuit voltage E across terminals AB in Figure 36.82(b) is given by $E=(I_{SC})(z)=(3)(2)=6$ V. The impedance 'looking in' at terminals AB is 2 Ω.

Hence the equivalent Thévenin circuit is as shown in Figure 36.83(b).

Problem 15. (a) Convert the circuit to the left of terminals AB in Figure 36.84 to an equivalent Thévenin circuit by initially converting to a Norton equivalent circuit. (b) Determine the magnitude of the current flowing in the $(1.8+j4)$ Ω impedance connected between terminals A and B of Figure 36.84.

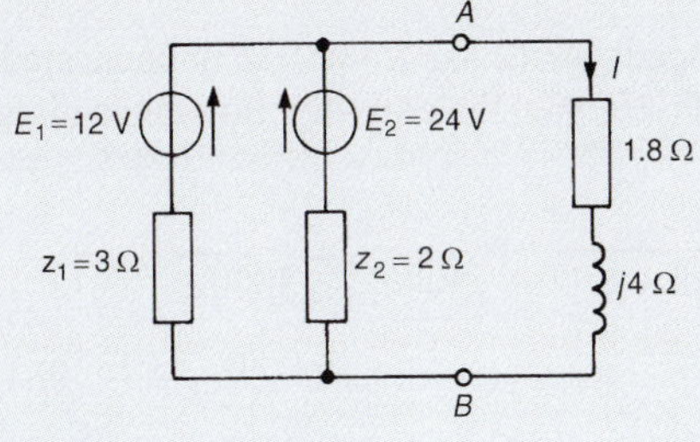

Figure 36.84

(a) For the branch containing the 12 V source, conversion to a Norton equivalent network gives $I_{SC_1}=12/3=4$ A and $z_1=3$ Ω. For the branch containing the 24 V source, conversion to a Norton equivalent circuit gives $I_{SC_2}=24/2=12$ A and $z_2=2$ Ω.

Thus Figure 36.85 shows a network equivalent to Figure 36.84. From Figure 36.85, the total short-circuit current is $4+12=16$ A, and the total impedance is given by $(3\times 2)/(3+2)=1.2$ Ω Thus Figure 36.85 simplifies to Figure 36.86.

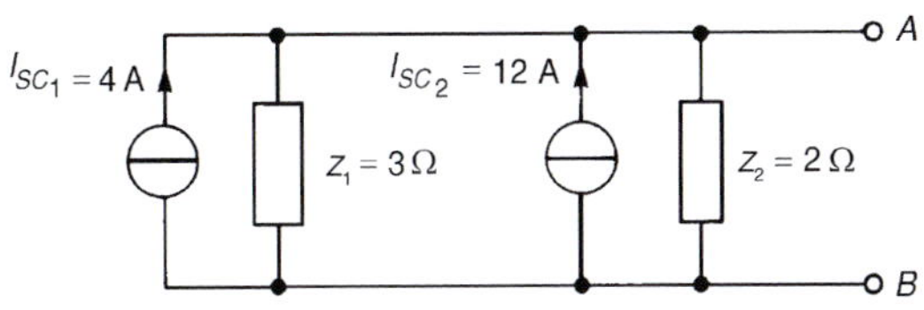

Figure 36.85

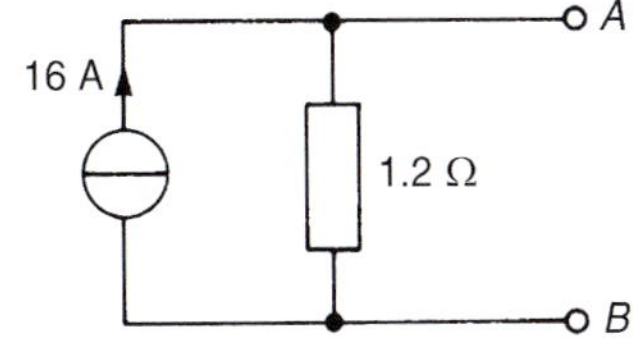

Figure 36.86

The open-circuit voltage across AB of Figure 36.86, $E=(16)(1.2)=19.2$ V, and the impedance 'looking in' at AB, $z=1.2$ Ω. Hence the Thévenin equivalent circuit is as shown in Figure 36.87.

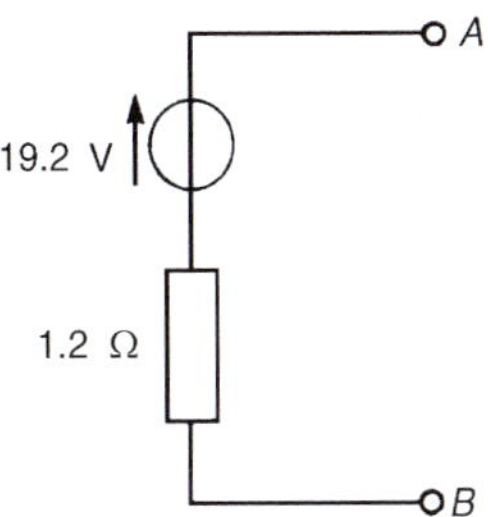

Figure 36.87

(b) When the $(1.8+j4)$ Ω impedance is connected to terminals AB of Figure 36.87, the current I flowing is given by

$$I=\frac{19.2}{(1.2+1.8+j4)}=3.84\angle -53.13^\circ \text{ A}$$

Hence the current flowing in the $(1.8+j4)$ Ω impedance is 3.84 A

Part 4

Problem 16. Determine, by successive conversions between Thévenin's and Norton's equivalent networks, a Thévenin equivalent circuit for terminals AB of Figure 36.88. Hence determine the magnitude of the current flowing in the capacitive branch connected to terminals AB.

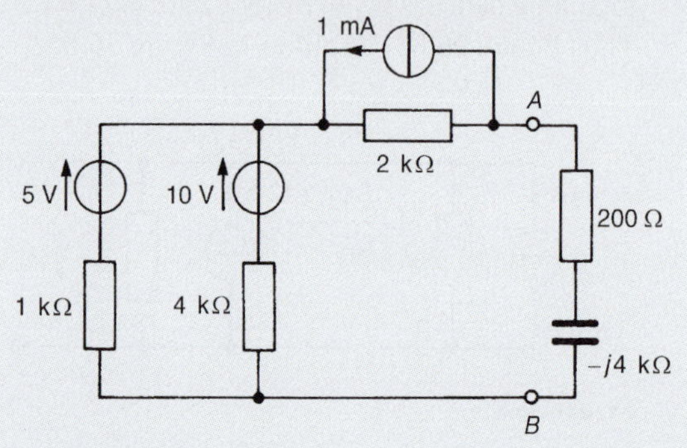

Figure 36.88

For the branch containing the 5 V source, converting to a Norton equivalent network gives $I_{SC}=5/1000=5$ mA and $z=1$ kΩ. For the branch containing the 10 V source, converting to a Norton equivalent network gives $I_{SC}=10/4000=2.5$ mA and $z=4$ kΩ. Thus the circuit of Figure 36.88 converts to that of Figure 36.89.

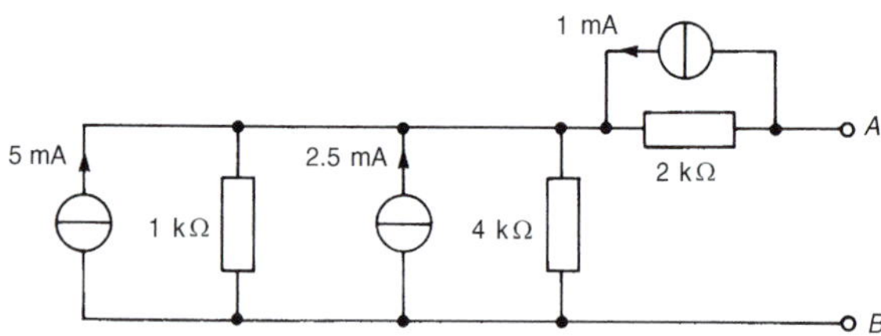

Figure 36.89

The two Norton equivalent networks shown in Figure 36.89 may be combined, since the total short-circuit current is $(5+2.5)=7.5$ mA and the total impedance z is given by $(1\times 4)/(1+4)=0.8$ kΩ. This results in the network of Figure 36.90.

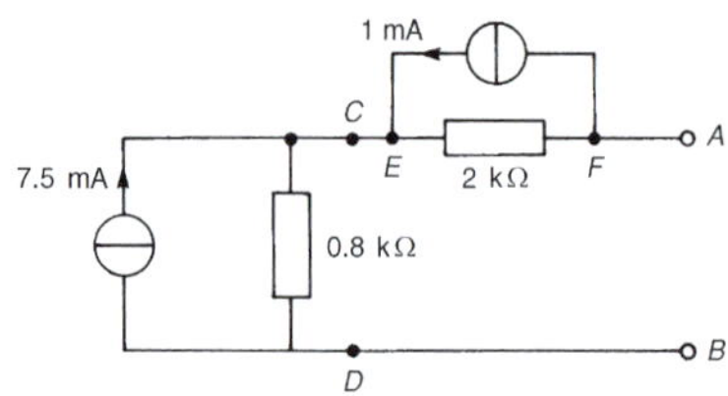

Figure 36.90

Both of the Norton equivalent networks shown in Figure 36.90 may be converted to Thévenin equivalent circuits. Open-circuit voltage across CD is

$$(7.5\times 10^{-3})(0.8\times 10^{3})=6\,\text{V}$$

and the impedance 'looking in' at CD is 0.8 kΩ. Open-circuit voltage across EF is $(1\times 10^{-3})(2\times 10^{2})=2$ V and the impedance 'looking in' at EF is 2 kΩ. Thus Figure 36.90 converts to Figure 36.91.

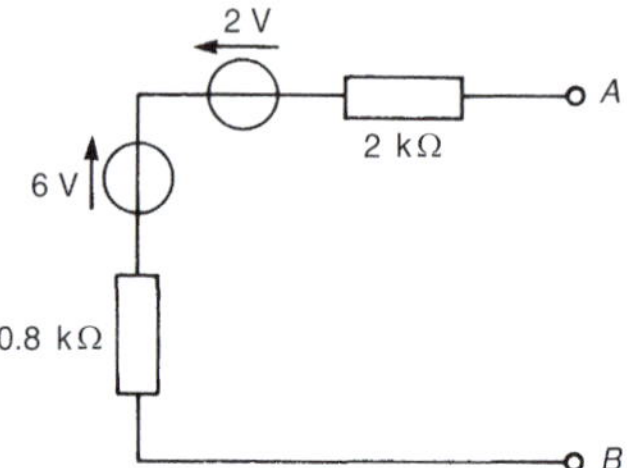

Figure 36.91

Combining the two Thévenin circuits gives e.m.f. $E=6-2=$ **4 V**, and impedance $z=(0.8+2)=$ **2.8 kΩ**. Thus the Thévenin equivalent circuit for terminals AB of Figure 36.88 is as shown in Figure 36.92.

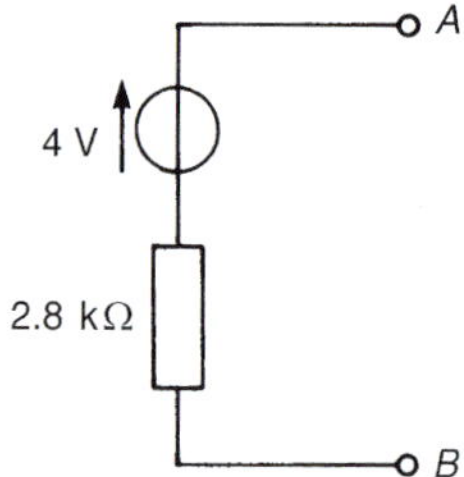

Figure 36.92

If an impedance $(200-j4000)\,\Omega$ is connected across terminals AB, then the current I flowing is given by

$$I=\frac{4}{2800+(200-j4000)}=\frac{4}{5000\angle -53.13^\circ}$$

$$=0.80\angle 53.13^\circ\,\text{mA}$$

i.e. **the current in the capacitive branch is 0.80 mA**

Problem 17. (a) Determine an equivalent Thévenin circuit for terminals AB of the network shown in Figure 36.93. (b) Calculate the power dissipated in a $(600 - j800)\,\Omega$ impedance connected between A and B of Figure 36.93.

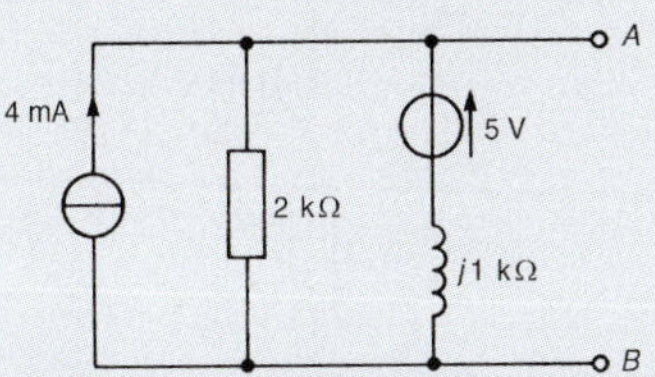

Figure 36.93

(a) Converting the Thévenin circuit to a Norton network gives

$$I_{SC} = \frac{5}{j1000} = -j5\,\text{mA} \quad \text{or} \quad 5\angle{-90^\circ}\,\text{mA and}$$

$$z = j1\,\text{k}\Omega$$

Thus Figure 36.93 converts to that shown in Figure 36.94. The two Norton equivalent networks may be combined, giving

$$I_{SC} = 4 + 5\angle{-90^\circ}$$
$$= (4 - j5)\,\text{mA or } 6.403\angle{-51.34^\circ}\,\text{mA}$$

and $$z = \frac{(2)(j1)}{(2 + j1)} = (0.4 + j0.8)\,\text{k}\Omega \quad \text{or}$$
$$0.894\angle 63.43^\circ\,\text{k}\Omega$$

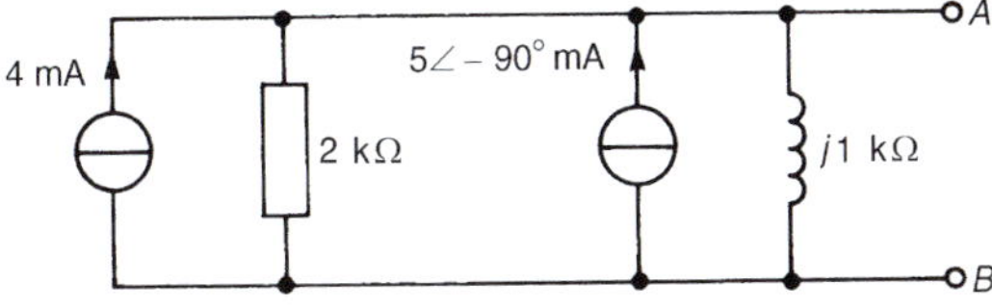

Figure 36.94

This results in the equivalent network shown in Figure 36.95. Converting to an equivalent Thévenin circuit gives open circuit e.m.f. across AB,

$$E = (6.403 \times 10^{-3}\angle{-51.34^\circ})(0.894 \times 10^{3}\angle 63.43^\circ)$$
$$= \mathbf{5.724\angle 12.09^\circ\ V}$$

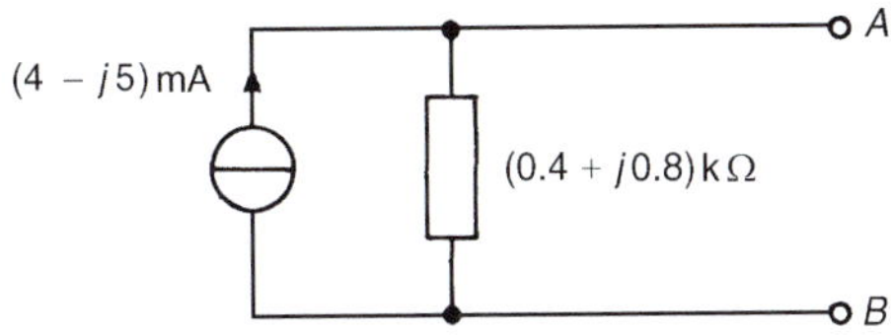

Figure 36.95

and

impedance $z = 0.894\angle 63.43^\circ\,\text{k}\Omega$ or

$$\mathbf{(400 + j800)\,\Omega}$$

Thus the Thévenin equivalent circuit is as shown in Figure 36.96.

(b) When a $(600 - j800)\,\Omega$ impedance is connected across AB, the current I flowing is given by

$$I = \frac{5.724\angle 12.09^\circ}{(400 + j800) + (600 - j800)}$$

$$= 5.724\angle 12.09^\circ\,\text{mA}$$

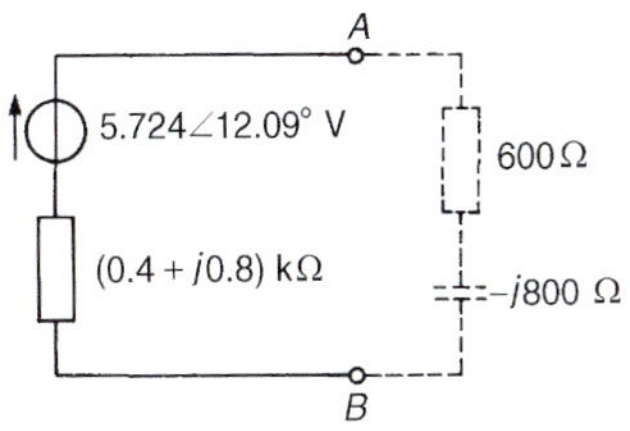

Figure 36.96

Hence the power P dissipated in the $(600 - j800)\,\Omega$ impedance is given by

$$P = I^2R = (5.724 \times 10^{-3})^2(600) = \mathbf{19.7\,mW}$$

Now try the following Practice Exercise

Practice Exercise 143 Thévenin and Norton equivalent networks (Answers on page 829)

1. Convert the circuits shown in Figure 36.97 to Norton equivalent networks.

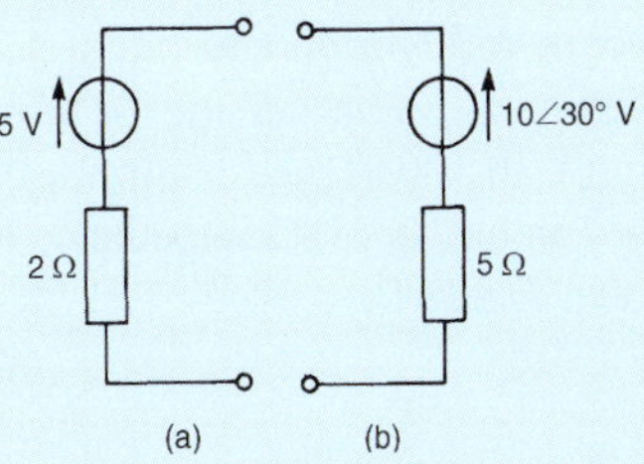

Figure 36.97

2. Convert the networks shown in Figure 36.98 to Thévenin equivalent circuits.

Part 4

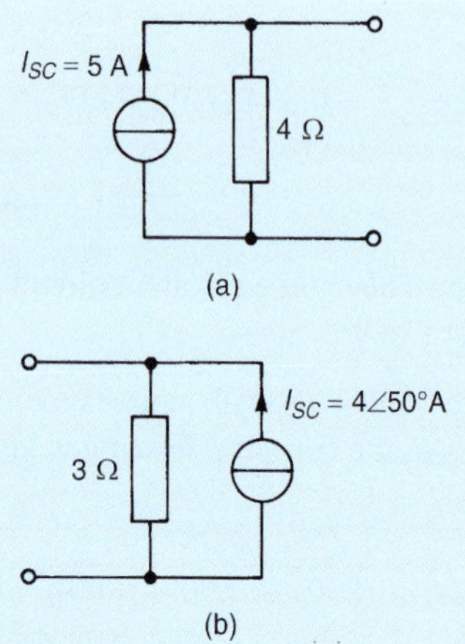

Figure 36.98

3. (a) Convert the network to the left of terminals AB in Figure 36.99 to an equivalent Thévenin circuit by initially converting to a Norton equivalent network.

   (b) Determine the current flowing in the $(2.8 - j3)\,\Omega$ impedance connected between A and B in Figure 36.99.

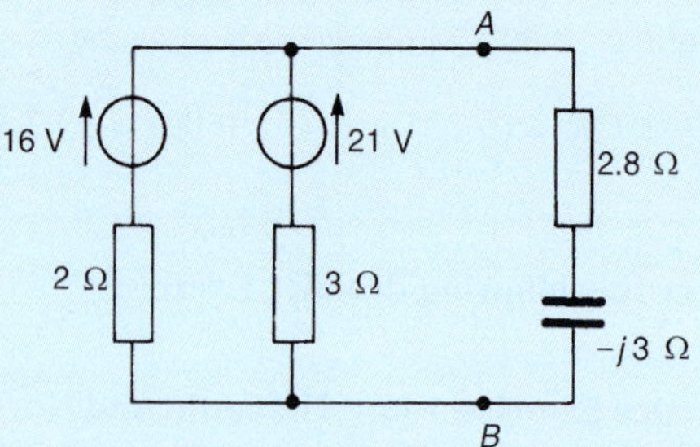

Figure 36.99

4. Determine, by successive conversions between Thévenin and Norton equivalent networks, a Thévenin equivalent circuit for terminals AB of Figure 36.100. Hence determine the current flowing in a $(2+j4)\,\Omega$ impedance connected between A and B.

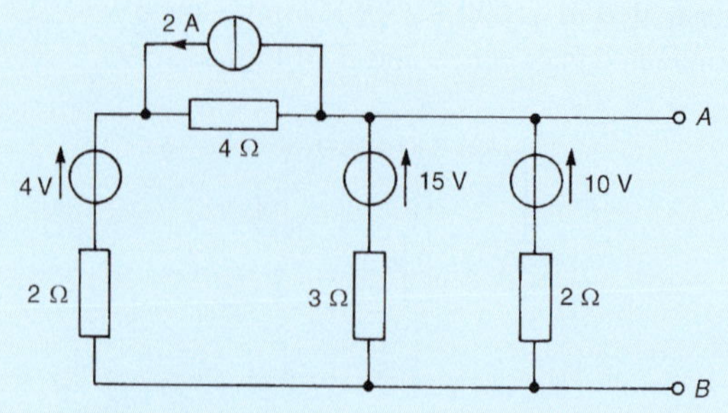

Figure 36.100

5. Derive an equivalent Thévenin circuit for terminals AB of the network shown in Figure 36.101. Hence determine the p.d. across AB when a $(3+j4)\,\mathrm{k}\Omega$ impedance is connected between these terminals.

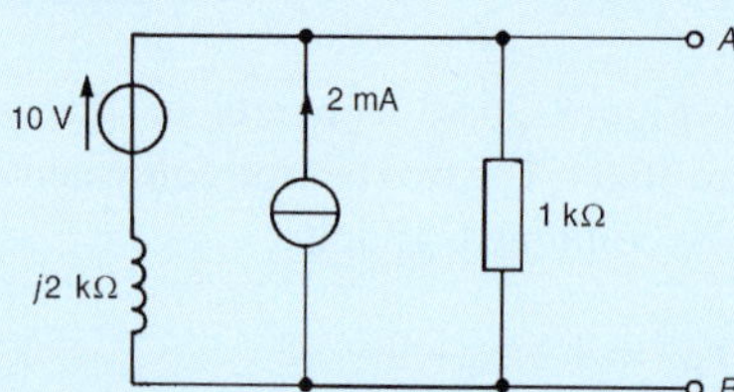

Figure 36.101

6. For the network shown in Figure 36.102, derive (a) the Thévenin equivalent circuit, and (b) the Norton equivalent network. (c) A 6 Ω resistance is connected between A and B. Determine the current flowing in the 6 Ω resistance by using both the Thévenin and Norton equivalent circuits.

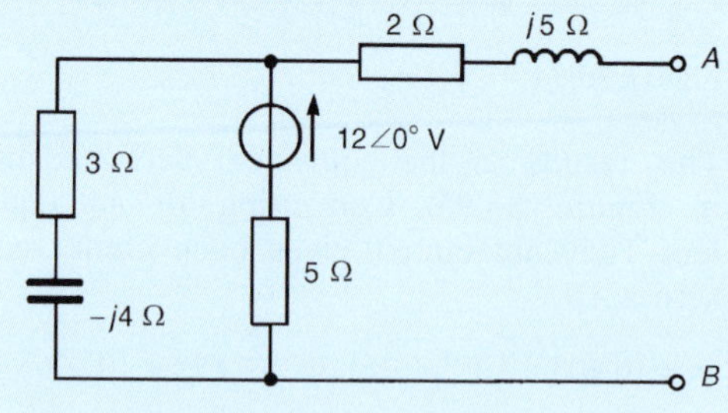

Figure 36.102

For fully worked solutions to each of the problems in Practice Exercises 140 to 143 in this chapter, go to the website:
www.routledge.com/cw/bird

Revision Test 10

This revision test covers the material contained in Chapters 33 to 36. *The marks for each question are shown in brackets at the end of each question.*

For the network shown in Figure RT10.1, determine the current flowing in each branch using:

(a) Kirchhoff's laws (10)

(b) Mesh-current analysis (12)

(c) Nodal analysis (12)

(d) The superposition theorem (22)

(e) Thévenin's theorem (14)

(f) Norton's theorem (10)

Demonstrate that each method gives the same value for each of the branch currents.

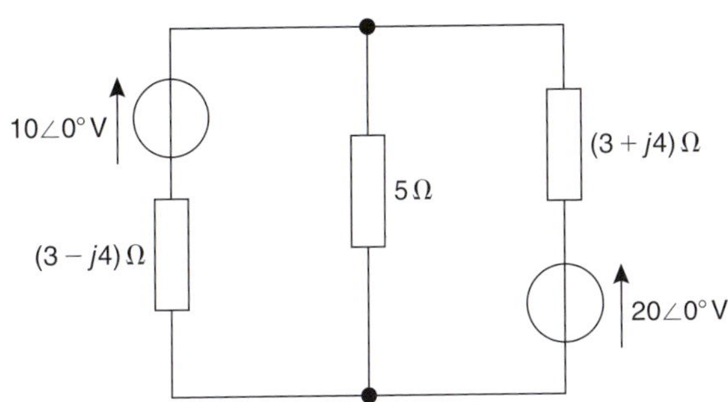

Figure RT10.1

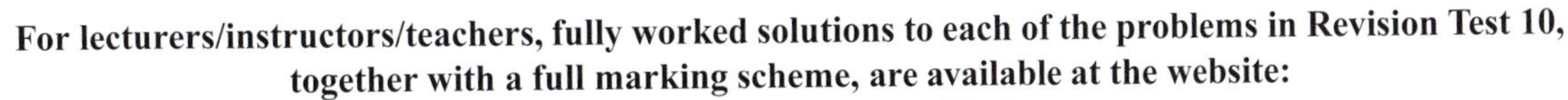
For lecturers/instructors/teachers, fully worked solutions to each of the problems in Revision Test 10, together with a full marking scheme, are available at the website:
www.routledge.com/cw/bird

Chapter 37

Delta–star and star–delta transformations

Why it is important to understand: **Delta–star and star–delta transforms**

There are many popular circuit configurations; two such commonly used configurations are star-connected networks and delta-connected networks. A star-connected network is also referred to as the T network. Similarly, the delta-connected network is also referred to as the π network. The star-connected and delta-connected configurations are used often and a complex circuit may contain such configurations within the circuit. In order to analyse such a complex circuit, it is at times necessary to transform a star-connected configuration within the circuit to a delta-connected configuration. In some cases, it may be the other way around. It is often easier to analyse the circuit after such a transformation. This chapter firstly explains how to recognize a delta and a star connection and then applies the delta–star and star–delta transformations to d.c. and a.c. networks through calculations.

At the end of this chapter you should be able to:

- recognize delta (or π) and star (or T) connections
- apply the delta–star and star–delta transformations in appropriate a.c. and d.c. networks

37.1 Introduction

By using Kirchhoff's laws, mesh-current analysis, nodal analysis or the superposition theorem, currents and voltages in many networks can be determined as shown in Chapters 33 to 35. Thévenin's and Norton's theorems, introduced in Chapter 36, provide an alternative method of solving networks and often with considerably reduced numerical calculations. Also, these latter theorems are especially useful when only the current in a particular branch of a complicated network is required. Delta–star and star–delta transformations may be applied in certain types of circuit to simplify them before application of circuit theorems.

37.2 Delta and star connections

The network shown in Figure 37.1(a) consisting of three impedances Z_A, Z_B and Z_C is said to be **π-connected**. This network can be redrawn as shown in Figure 37.1(b), where the arrangement is referred to as **delta-connected** or **mesh-connected**.

The network shown in Figure 37.2(a), consisting of three impedances, Z_1, Z_2 and Z_3, is said to be

Electrical Circuit Theory and Technology. 978-1-138-67349-6, © 2017 John Bird. Published by Taylor & Francis. All rights reserved.

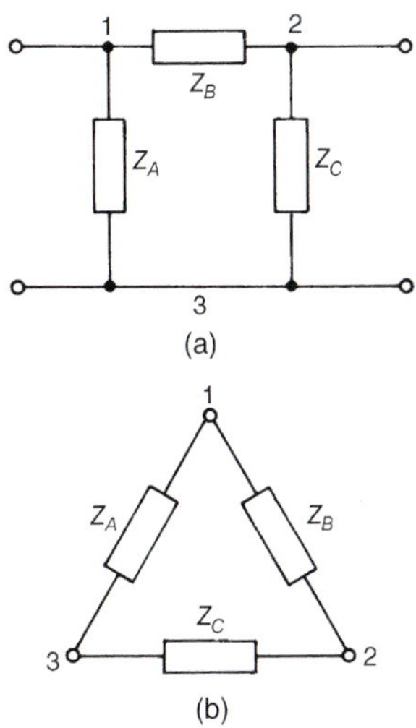

Figure 37.1 (a) π-connected network, (b) delta-connected network

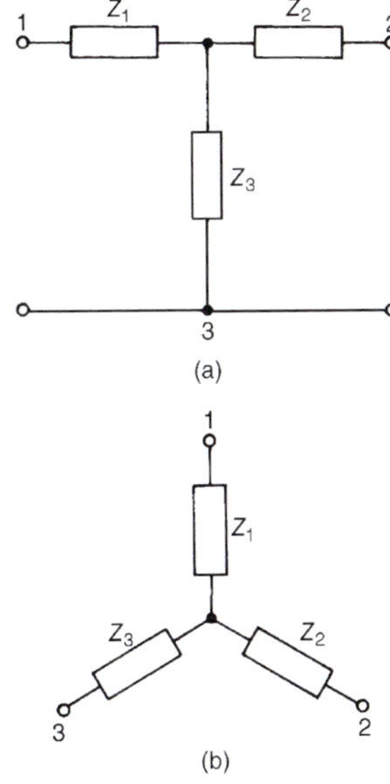

Figure 37.2 (a) T-connected network, (b) star-connected network

***T*-connected**. This network can be redrawn as shown in Figure 37.2(b), where the arrangement is referred to as **star-connected**.

37.3 Delta–star transformation

It is possible to replace the delta connection shown in Figure 37.3(a) by an equivalent star connection as shown in Figure 37.3(b), such that the impedance measured

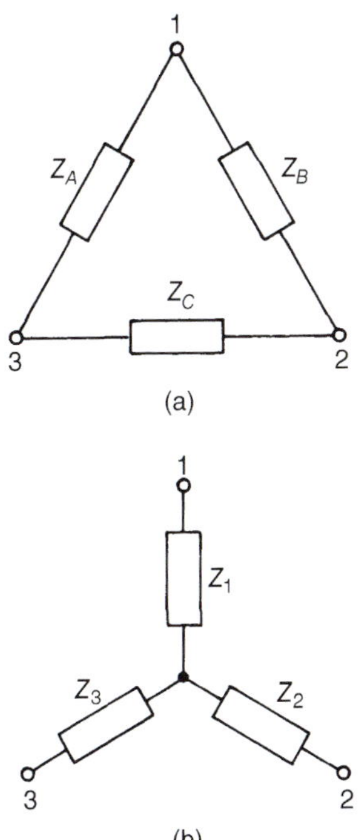

Figure 37.3

between any pair of terminals (1–2, 2–3 or 3–1) is the same in star as in delta. The equivalent star network will consume the same power and operate at the same power factor as the original delta network. A delta–star transformation may alternatively be termed 'π to T transformation'.

Considering terminals 1 and 2 of Figure 37.3(a), the equivalent impedance is given by the impedance Z_B in parallel with the series combination of Z_A and Z_C

i.e. $$\frac{Z_B(Z_A+Z_C)}{Z_B+Z_A+Z_C}$$

In Figure 37.3(b), the equivalent impedance between terminals 1 and 2 is Z_1 and Z_2 in series, i.e. Z_1+Z_2

Thus,

Delta Star

$$Z_{12}=\frac{Z_B(Z_A+Z_C)}{Z_B+Z_A+Z_C}=Z_1+Z_2 \quad (1)$$

By similar reasoning,

$$Z_{23}=\frac{Z_C(Z_A+Z_B)}{Z_C+Z_A+Z_B}=Z_2+Z_3 \quad (2)$$

and

$$Z_{31}=\frac{Z_A(Z_B+Z_C)}{Z_A+Z_B+Z_C}=Z_3+Z_1 \quad (3)$$

Hence we have three simultaneous equations to be solved for Z_1, Z_2 and Z_3

Equation (1) – equation (2) gives:

$$\frac{Z_A Z_B - Z_A Z_C}{Z_A + Z_B + Z_C} = Z_1 - Z_3 \qquad (4)$$

Equation (3) + equation (4) gives:

$$\frac{2Z_A Z_B}{Z_A + Z_B + Z_C} = 2Z_1$$

from which $$Z_1 = \frac{Z_A Z_B}{Z_A + Z_B + Z_C}$$

Similarly, equation (2) – equation (3) gives:

$$\frac{Z_B Z_C - Z_A Z_B}{Z_A + Z_B + Z_C} = Z_2 - Z_1 \qquad (5)$$

Equation (1) + equation (5) gives:

$$\frac{2Z_B Z_C}{Z_A + Z_B + Z_C} = 2Z_2$$

from which $$Z_2 = \frac{Z_B Z_C}{Z_A + Z_B + Z_C}$$

Finally, equation (3) – equation (1) gives:

$$\frac{Z_A Z_C - Z_B Z_C}{Z_A + Z_B + Z_C} = Z_3 - Z_2 \qquad (6)$$

Equation (2) + equation (6) gives:

$$\frac{2Z_A Z_C}{Z_A + Z_B + Z_C} = 2Z_3$$

from which $$Z_3 = \frac{Z_A Z_C}{Z_A + Z_B + Z_C}$$

Summarizing, the star section shown in Figure 37.3(b) is equivalent to the delta section shown in Figure 37.3(a) when

$$\mathbf{Z_1 = \frac{Z_A Z_B}{Z_A + Z_B + Z_C}} \qquad (7)$$

$$\mathbf{Z_2 = \frac{Z_B Z_C}{Z_A + Z_B + Z_C}} \qquad (8)$$

and $$\mathbf{Z_3 = \frac{Z_A Z_C}{Z_A + Z_B + Z_C}} \qquad (9)$$

It is noted that impedance Z_1 is given by the product of the two impedances in delta joined to terminal 1 (i.e. Z_A and Z_B), divided by the sum of the three impedances; impedance Z_2 is given by the product of the two impedances in delta joined to terminal 2 (i.e. Z_B and Z_C), divided by the sum of the three impedances; and impedance Z_3 is given by the product of the two impedances in delta joined to terminal 3 (i.e. Z_A and Z_C), divided by the sum of the three impedances.

Thus, for example, the star equivalent of the resistive delta network shown in Figure 37.4 is given by

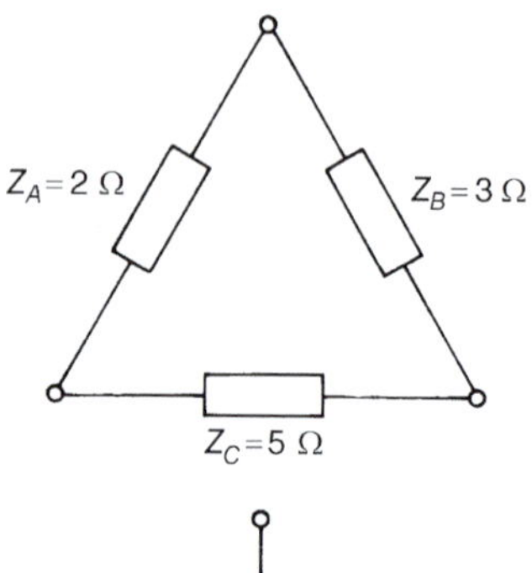

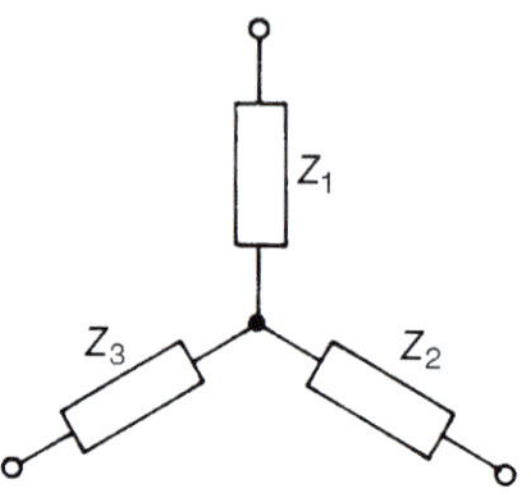

Figure 37.4

$$Z_1 = \frac{(2)(3)}{2+3+5} = \mathbf{0.6\,\Omega}$$

$$Z_2 = \frac{(3)(5)}{2+3+5} = \mathbf{1.5\,\Omega}$$

and $$Z_3 = \frac{(2)(5)}{2+3+5} = \mathbf{1.0\,\Omega}$$

Problem 1. Replace the delta-connected network shown in Figure 37.5 by an equivalent star connection.

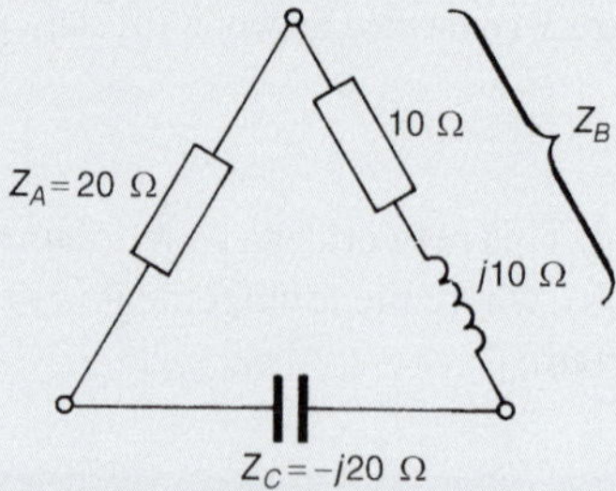

Figure 37.5

Let the equivalent star network be as shown in Figure 37.6. Then, from equation (7),

Part 4

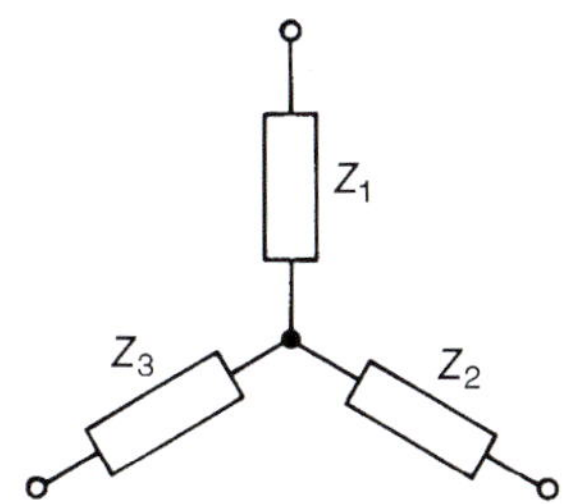

Figure 37.6

$$Z_1 = \frac{Z_A Z_B}{Z_A + Z_B + Z_C}$$

$$= \frac{(20)(10 + j10)}{20 + 10 + j10 - j20}$$

$$= \frac{(20)(10 + j10)}{(30 - j10)}$$

$$= \frac{(20)(1.414\angle 45°)}{31.62\angle -18.43°}$$

$$= \mathbf{8.944\angle 63.43°\ \Omega} \text{ or } \mathbf{(4 + j8)\ \Omega}$$

From equation (8),

$$Z_2 = \frac{Z_B Z_C}{Z_A + Z_B + Z_C}$$

$$= \frac{(10 + j10)(-j20)}{31.62\angle -18.43°}$$

$$= \frac{(1.414\angle 45°)(20\angle -90°)}{31.62\angle -18.43°}$$

$$= \mathbf{8.944\angle -26.57°\ \Omega} \text{ or } \mathbf{(8 - j4)\ \Omega}$$

From equation (9),

$$Z_3 = \frac{Z_A Z_C}{Z_A + Z_B + Z_C} = \frac{(20)(-j20)}{31.62\angle -18.43°}$$

$$= \frac{(400\angle -90°)}{31.62\angle -18.43°}$$

$$= \mathbf{12.650\angle -71.57°\ \Omega} \text{ or } \mathbf{(4 - j12)\ \Omega}$$

Problem 2. For the network shown in Figure 37.7, determine (a) the equivalent circuit impedance across terminals AB, (b) supply current I and (c) the power dissipated in the 10 Ω resistor.

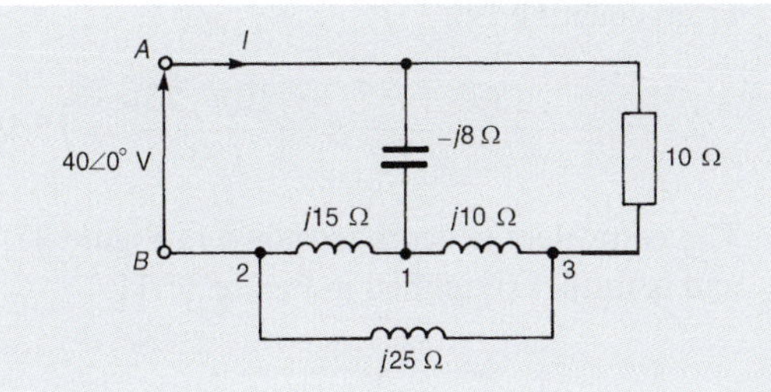

Figure 37.7

(a) The network of Figure 37.7 is redrawn, as in Figure 37.8, showing more clearly the part of the network 1, 2, 3 forming a delta connection. This may be transformed into a star connection as shown in Figure 37.9.

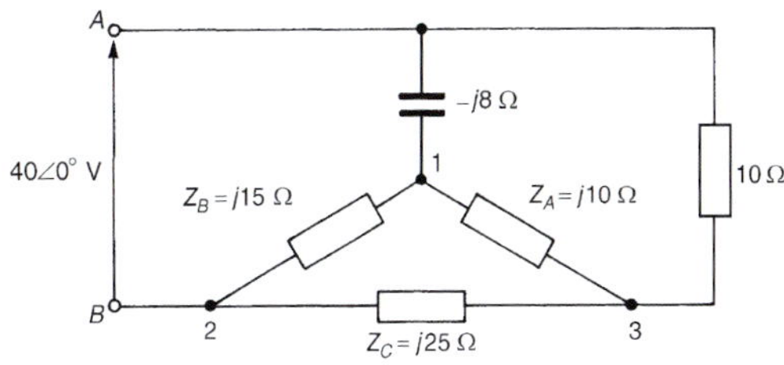

Figure 37.8

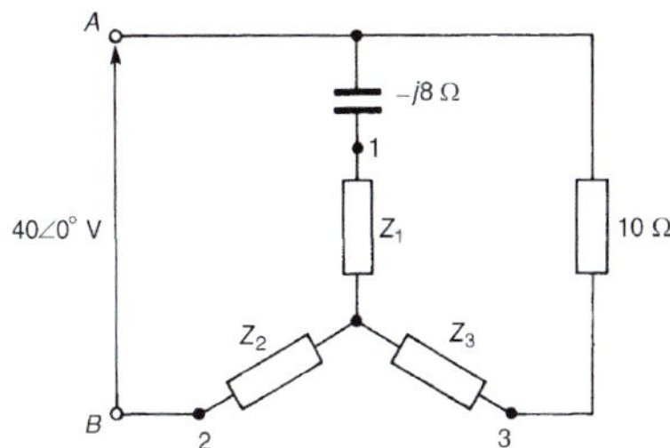

Figure 37.9

From equation (7),

$$Z_1 = \frac{Z_A Z_B}{Z_A + Z_B + Z_C} = \frac{(j10)(j15)}{j10 + j15 + j25}$$

$$= \frac{(j10)(j15)}{(j50)} = \mathbf{j3\ \Omega}$$

From equation (8),

$$Z_2 = \frac{Z_B Z_C}{Z_A + Z_B + Z_C} = \frac{(j15)(j25)}{j50} = \mathbf{j7.5\ \Omega}$$

Part 4

From equation (9),

$$Z_3 = \frac{Z_A Z_C}{Z_A + Z_B + Z_C} = \frac{(j10)(j25)}{j50} = \mathbf{j5\ \Omega}$$

The equivalent network is shown in Figure 37.10 and is further simplified in Figure 37.11.

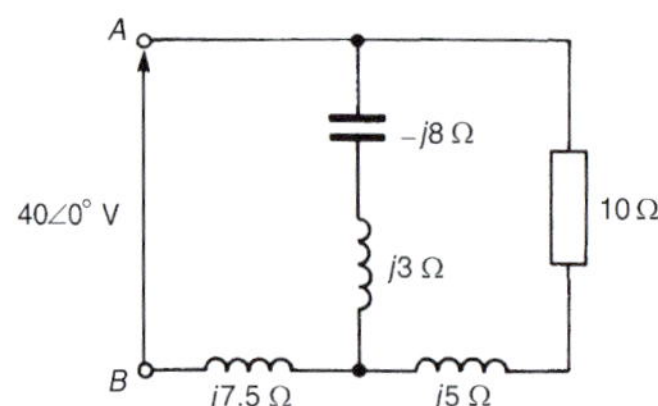

Figure 37.10

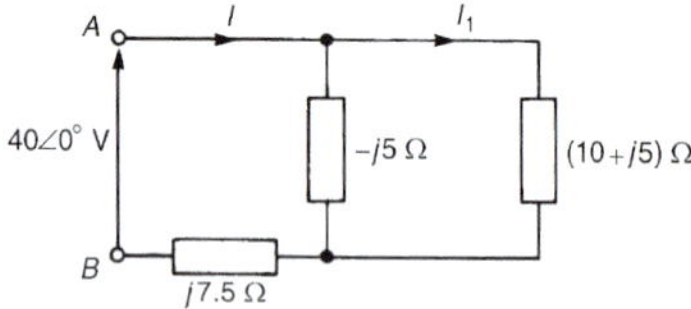

Figure 37.11

$(10+j5)\Omega$ in parallel with $-j5\,\Omega$ gives an equivalent impedance of

$$\frac{(10+j5)(-j5)}{(10+j5-j5)} = (2.5-j5)\,\Omega$$

Hence the total circuit equivalent impedance across terminals AB is given by

$$\boldsymbol{Z_{AB}} = (2.5-j5)+j7.5 = \mathbf{(2.5+j2.5)\ \Omega}$$

or $\mathbf{3.54\angle 45^\circ\ \Omega}$

(b) Supply current $\boldsymbol{I} = \frac{V}{Z_{AB}} = \frac{40\angle 0^\circ}{3.54\angle 45^\circ}$

$$= \mathbf{11.3\angle -45^\circ\,A}$$

(c) Power P dissipated in the $10\,\Omega$ resistance of Figure 37.7 is given by $(I_1)^2(10)$, where I_1 (see Figure 37.11) is given by:

$$I_1 = \left[\frac{-j5}{10+j5-j5}\right](11.3\angle -45^\circ)$$

$$= 5.65\angle -135^\circ\text{A}$$

Hence power $\boldsymbol{P} = (5.65)^2(10) = \mathbf{319\,W}$

Problem 3. Determine, for the bridge network shown in Figure 37.12, (a) the value of the single equivalent resistance that replaces the network between terminals A and B, (b) the current supplied by the 52 V source and (c) the current flowing in the 8 Ω resistance.

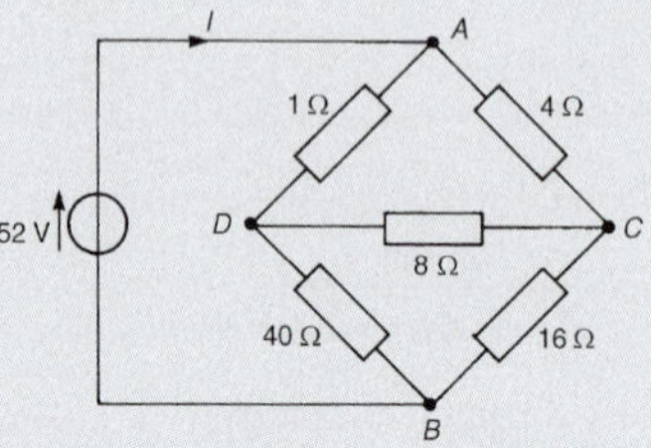

Figure 37.12

(a) In Figure 37.12, no resistances are directly in parallel or directly in series with each other. However, ACD and BCD are both delta connections and either may be converted into an equivalent star connection. The delta network BCD is redrawn in Figure 37.13(a) and is transformed into an equivalent star connection as shown in Figure 37.13(b), where

$$Z_1 = \frac{(8)(16)}{8+16+40} = 2\,\Omega \qquad \text{(from equation (7))}$$

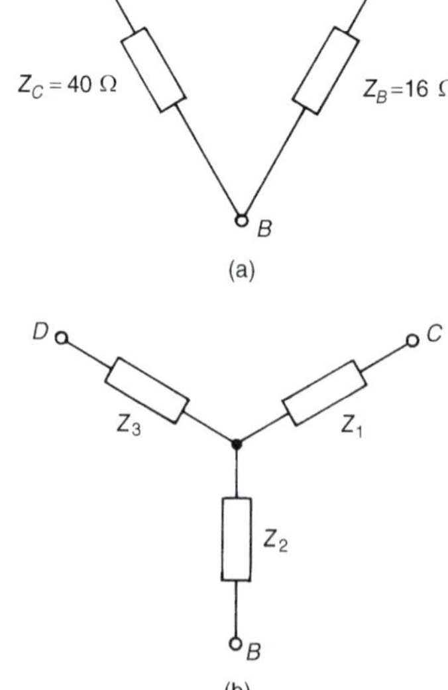

Figure 37.13

$$Z_2 = \frac{(16)(40)}{8+16+40} = 10\,\Omega \quad \text{(from equation (8))}$$

$$Z_3 = \frac{(8)(40)}{8+16+40} = 5\,\Omega \quad \text{(from equation (9))}$$

The network of Figure 37.12 may thus be redrawn as shown in Figure 37.14. The 4 Ω and 2 Ω resistances are in series with each other, as are the 1 Ω and 5 Ω resistors. Hence the equivalent network is as shown in Figure 37.15. The total equivalent resistance across terminals A and B is given by

$$\boldsymbol{R_{AB}} = \frac{(6)(6)}{(6)+(6)} + 10 = \mathbf{13\,\Omega}$$

(b) Current supplied by the 52 V source, i.e. current I in Figure 37.15, is given by

$$\boldsymbol{I} = \frac{V}{Z_{AB}} = \frac{52}{13} = \mathbf{4\,A}$$

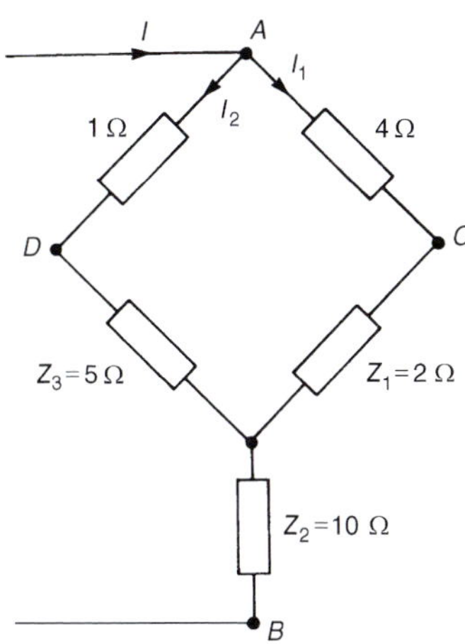

Figure 37.14

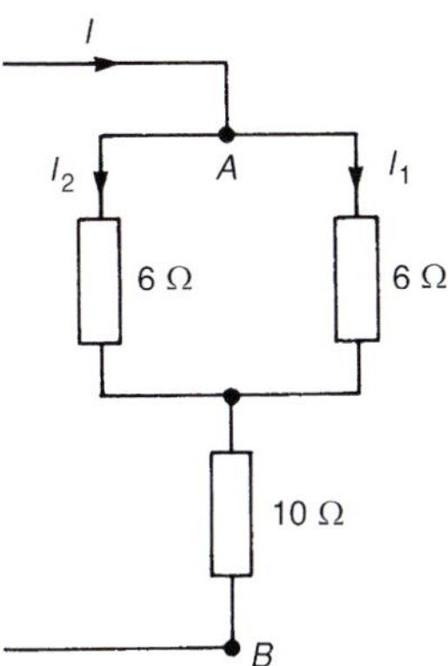

Figure 37.15

(c) From Figure 37.15,
current $I_1 = [6/(6+6)](I) = 2\,A$, and
current $I_2 = 2$ A also.
From Figure 37.14, p.d. across AC, $V_{AC} = (I_1)(4)$ $= 8$ V and p.d. across AD, $V_{AD} = (I_2)(1) = 2$ V. Hence p.d. between C and D (i.e. p.d. across the 8 Ω resistance of Figure 37.12) is given by $(8-2) = 6$ V

Thus **the current in the 8 Ω resistance** is given by $V_{CD}/8 = 6/8 = \mathbf{0.75\,A}$

Problem 4. Figure 37.16 shows an Anderson bridge used to measure, with high accuracy, inductance L_X and series resistance R_X.

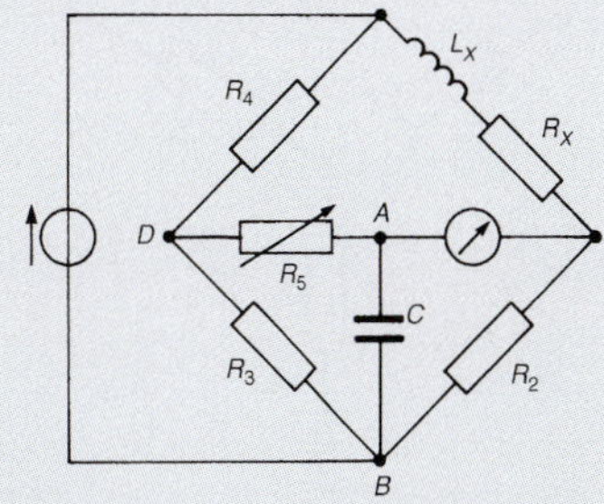

Figure 37.16

(a) Transform the delta ABD into its equivalent star connection and hence determine the balance equations for R_X and L_X

(b) If $R_2 = R_3 = 1\,k\Omega$, $R_4 = 500\,\Omega$, $R_5 = 200\,\Omega$ and $C = 2\,\mu F$, determine the values of R_X and L_X at balance.

(a) The delta ABD is redrawn separately in Figure 37.17, together with its equivalent star connection comprising impedances Z_1, Z_2 and Z_3

From equation (7),

$$Z_1 = \frac{(R_5)(-jX_C)}{R_5 - jX_C + R_3} = \frac{-jR_5X_C}{(R_3+R_5) - jX_C}$$

From equation (8),

$$Z_2 = \frac{(-jX_C)(R_3)}{R_5 - jX_C + R_3} = \frac{-jR_3X_C}{(R_3+R_5) - jX_C}$$

From equation (9),

$$Z_3 = \frac{R_5R_3}{(R_3+R_5) - jX_C}$$

Part 4

The network of Figure 37.16 is redrawn with the star replacing the delta as shown in Figure 37.18, and further simplified in Figure 37.19. (Note that impedance Z_1 does not affect the balance of the bridge since it is in series with the detector.)

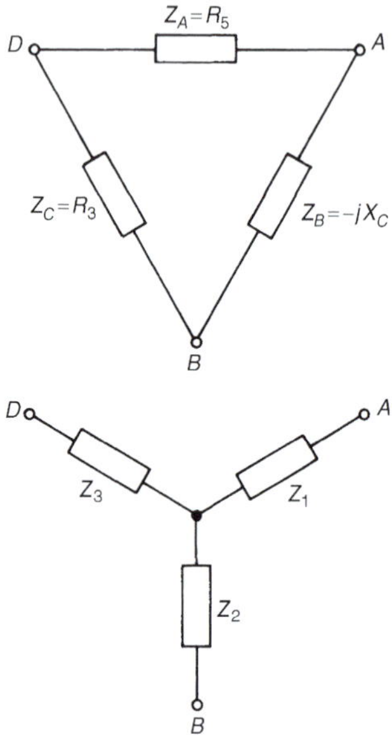

Figure 37.17

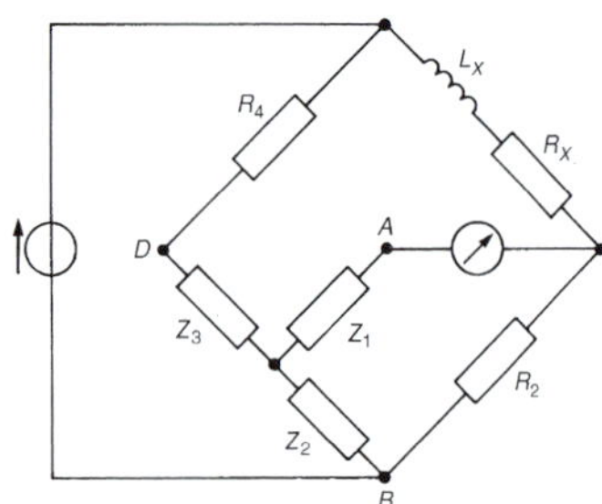

Figure 37.18

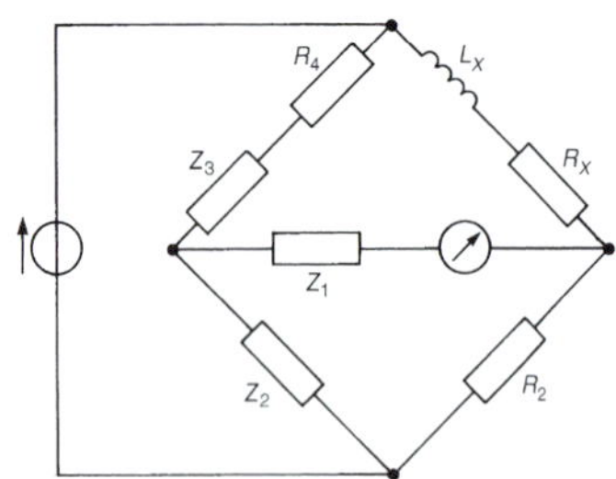

Figure 37.19

At balance,

$(R_X + jX_{L_X})(Z_2) = (R_2)(R_4 + Z_3)$

from Chapter 30,

from which,

$$(R_X + jX_{L_X}) = \frac{R_2}{Z_2}(R_4 + Z_3) = \frac{R_2 R_4}{Z_2} + \frac{R_2 Z_3}{Z_2}$$

$$= \frac{R_2 R_4}{-jR_3 X_C/((R_3 + R_5) - jX_C)}$$

$$+ \frac{R_2(R_5 R_3/((R_3 + R_5) - jX_C))}{-jR_3 X_C/((R_3 + R_5) - jX_C)}$$

$$= \frac{R_2 R_4((R_3 + R_5) - jX_C)}{-jR_3 X_C} + \frac{R_2 R_5 R_3}{-jR_3 X_C}$$

$$= \frac{jR_2 R_4((R_3 + R_5) - jX_C)}{R_3 X_C} + \frac{jR_2 R_5}{X_C}$$

i.e. $$(R_X + jX_{L_X}) = \frac{jR_2 R_4(R_3 + R_5)}{R_3 X_C}$$

$$+ \frac{R_2 R_4 X_C}{R_3 X_C} + \frac{jR_2 R_5}{X_C}$$

Equating the real parts gives:

$$\boldsymbol{R_X = \frac{R_2 R_4}{R_3}}$$

Equating the imaginary parts gives:

$$X_{L_X} = \frac{R_2 R_4(R_3 + R_5)}{R_3 X_C} + \frac{R_2 R_5}{X_C}$$

i.e. $$\omega L_X = \frac{R_2 R_4 R_3}{R_3(1/\omega C)} + \frac{R_2 R_4 R_5}{R_3(1/\omega C)}$$

$$+ \frac{R_2 R_5}{(1/\omega C)}$$

$$= \omega C R_2 R_4 + \frac{\omega C R_2 R_4 R_5}{R_3}$$

$$+ \omega C R_2 R_5$$

Hence $\boldsymbol{L_X = R_2 C\left(R_4 + \frac{R_4 R_5}{R_3} + R_5\right)}$

(b) When $R_2 = R_3 = 1\,\text{k}\Omega$, $R_4 = 500\,\Omega$, $R_5 = 200\,\Omega$ and $C = 2\,\mu\text{F}$, then, at balance

$$R_X = \frac{R_2 R_4}{R_3} = \frac{(1000)(500)}{(1000)} = \mathbf{500\,\Omega}$$

and

$$L_X = R_2 C\left(R_4 + \frac{R_4 R_5}{R_3} + R_5\right)$$

$$= (1000)(2 \times 10^{-6})\left[500 + \frac{(500)(200)}{(1000)} + 200\right]$$

$$= \mathbf{1.60\,H}$$

Part 4

Problem 5. For the network shown in Figure 37.20, determine (a) the current flowing in the $(0+j10)\,\Omega$ impedance and (b) the power dissipated in the $(20+j0)\,\Omega$ impedance.

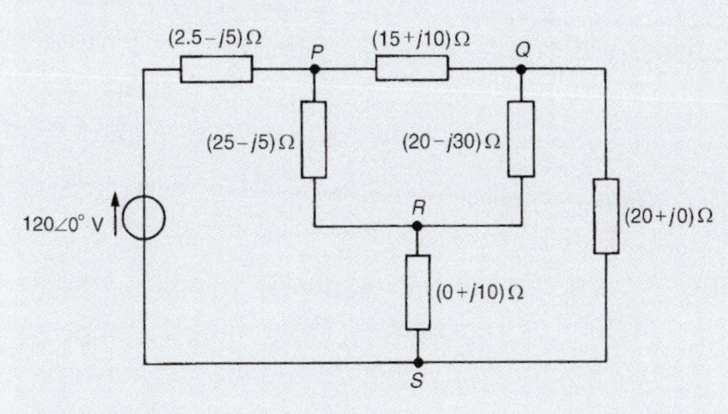

Figure 37.20

(a) The network may initially be simplified by transforming the delta *PQR* to its equivalent star connection as represented by impedances Z_1, Z_2 and Z_3 in Figure 37.21. From equation (7),

$$Z_1 = \frac{(15+j10)(25-j5)}{(15+j10)+(25-j5)+(20-j30)}$$

$$= \frac{(15+j10)(25-j5)}{(60-j25)}$$

$$= \frac{(18.03\angle 33.69^\circ)(25.50\angle -11.31^\circ)}{65\angle -22.62^\circ}$$

$$= 7.07\angle 45^\circ\,\Omega \quad \text{or} \quad (5+j5)\,\Omega$$

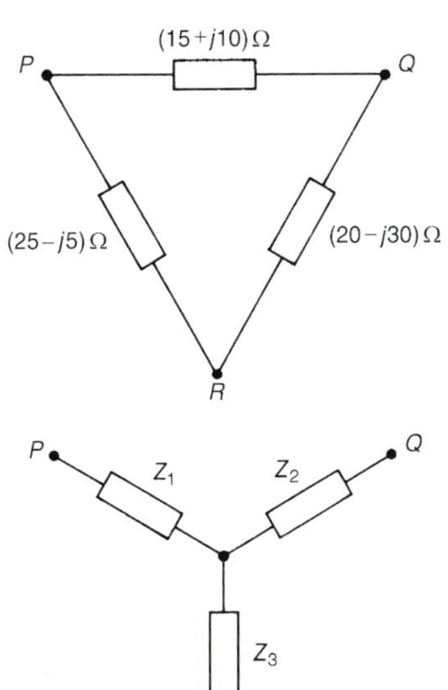

Figure 37.21

From equation (8),

$$Z_2 = \frac{(15+j10)(20-j30)}{(65\angle -22.62^\circ)}$$

$$= \frac{(18.03\angle 33.69^\circ)(36.06\angle -56.31^\circ)}{65\angle -22.62^\circ}$$

$$= 10.0\angle 0^\circ \quad \text{or} \quad (10+j0)\Omega$$

From equation (9),

$$Z_3 = \frac{(25-j5)(20-j30)}{(65\angle -22.62^\circ)}$$

$$= \frac{(25.50\angle -11.31^\circ)(36.06\angle -56.31^\circ)}{65\angle -22.62^\circ}$$

$$= 14.15\angle -45^\circ\Omega \quad \text{or} \quad (10-j10)\Omega$$

The network is shown redrawn in Figure 37.22 and further simplified in Figure 37.23, from which,

$$\text{current } I_1 = \frac{120\angle 0^\circ}{7.5+((10)(30)/(10+30))}$$

$$= \frac{120\angle 0^\circ}{15} = 8\,\text{A}$$

$$\text{current } I_2 = \left(\frac{10}{10+30}\right)(8) = 2\,A$$

$$\text{current } I_3 = \left(\frac{30}{10+30}\right)(8) = 6\,A$$

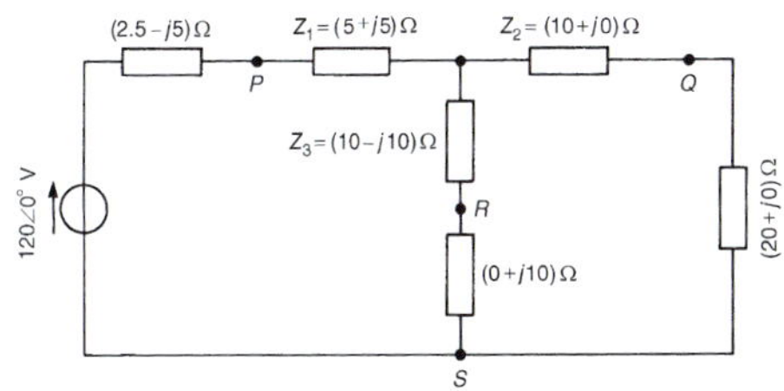

Figure 37.22

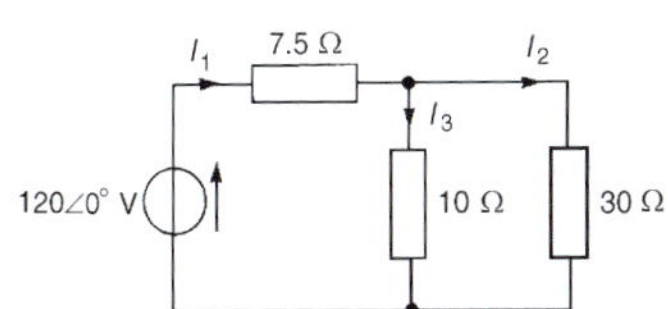

Figure 37.23

The current flowing in the $(0+j10)\Omega$ impedance of Figure 37.20 is the current I_3 shown in Figure 37.23, i.e. **6 A**.

(b) The power P dissipated in the $(20+j0)\Omega$ impedance of Figure 37.20 is given by

$$P=I_2^2(20)=(2)^2(20)=\mathbf{80\,W}$$

Now try the following Practice Exercise

Practice Exercise 144 Delta–star transformations (Answers on page 829)

1. Transform the delta-connected networks shown in Figure 37.24 to their equivalent star-connected networks.

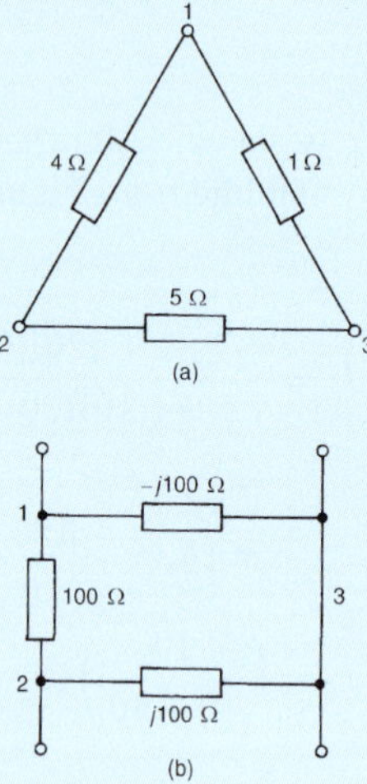

Figure 37.24

2. Transform the π network shown in Figure 37.25 to its equivalent star-connected network.

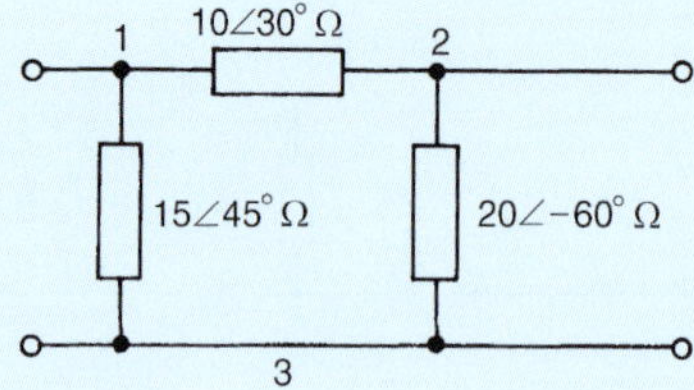

Figure 37.25

3. For the network shown in Figure 37.26 determine (a) current I and (b) the power dissipated in the $10\,\Omega$ resistance.

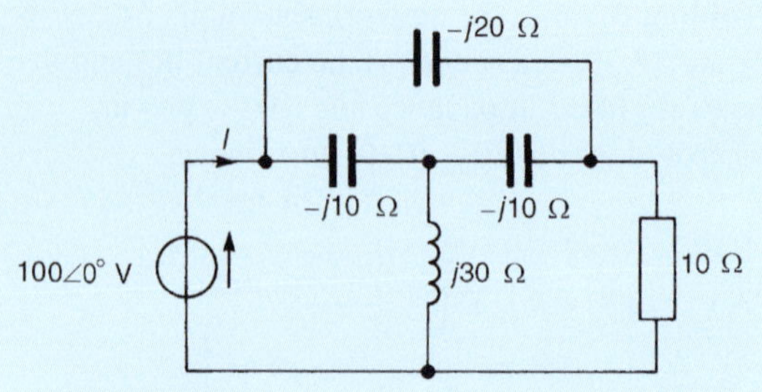

Figure 37.26

4. A delta-connected network contains three $24\angle 60^\circ\ \Omega$ impedances. Determine the impedances of the equivalent star-connected network.

5. For the a.c. bridge network shown in Figure 37.27, transform the delta-connected network ABC into an equivalent star, and hence determine the current flowing in the capacitor.

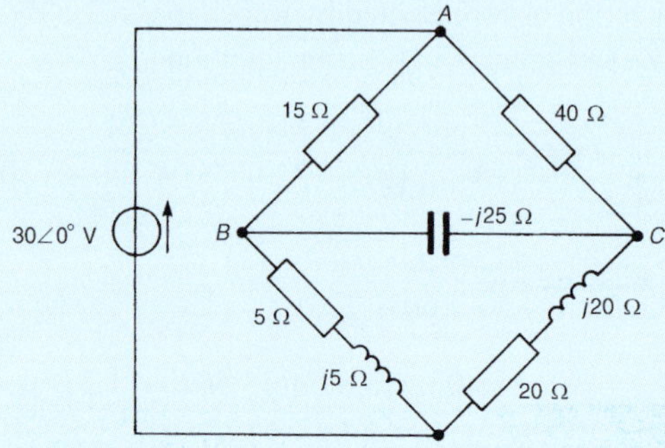

Figure 37.27

6. For the network shown in Figure 37.28 transform the delta-connected network ABC to an equivalent star-connected network, convert the 35 A, $2\,\Omega$ Norton circuit to an equivalent Thévenin circuit and hence determine the p.d. across the $12.5\,\Omega$ resistor.

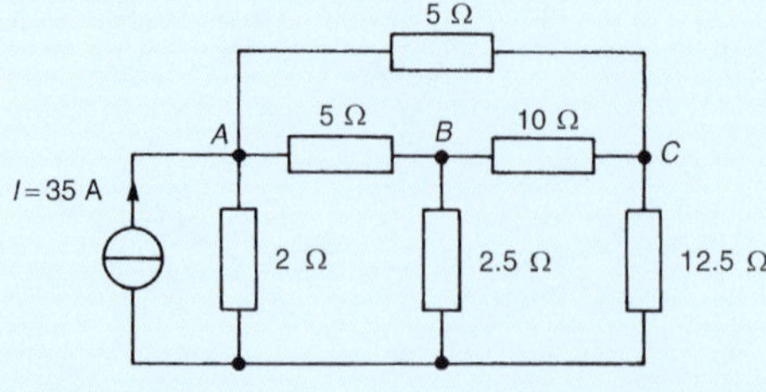

Figure 37.28

7. Transform the delta-connected network ABC shown in Figure 37.29 and hence determine the

magnitude of the current flowing in the 20 Ω resistance.

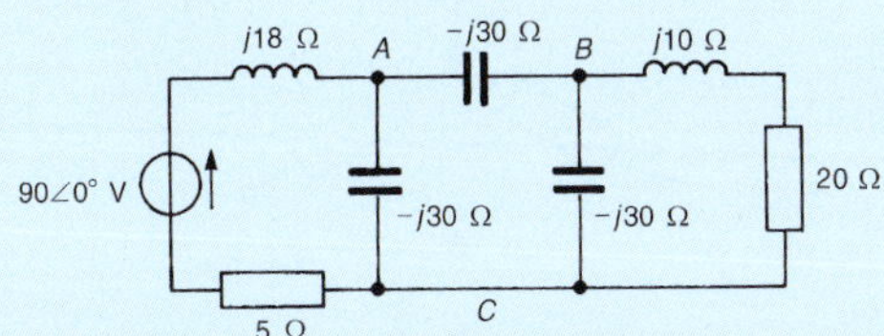

Figure 37.29

8. For the network shown in Figure 37.30 determine (a) the current supplied by the $80\angle 0°$ V source and (b) the power dissipated in the $(2.00 - j0.916)$ Ω impedance.

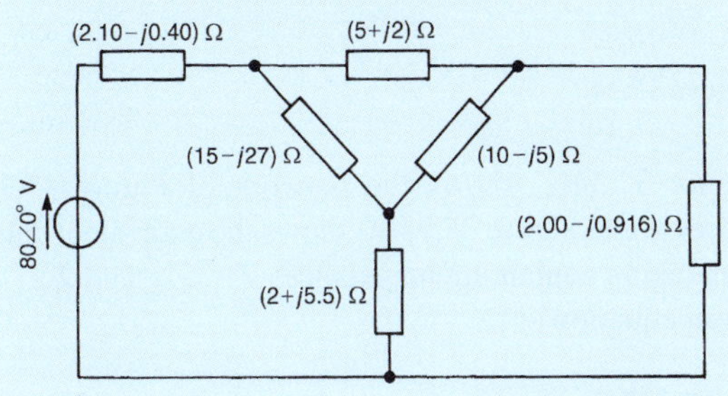

Figure 37.30

37.4 Star-delta transformation

It is possible to replace the star section shown in Figure 37.31(a) by an equivalent delta section, as shown in Figure 37.31(b). Such a transformation is also known as a 'T to π transformation'.

From equations (7), (8) and (9),

$$Z_1Z_2 + Z_2Z_3 + Z_3Z_1 = \frac{Z_AZ_B^2Z_C + Z_AZ_BZ_C^2 + Z_A^2Z_BZ_C}{(Z_A + Z_B + Z_C)^2}$$

$$= \frac{Z_AZ_BZ_C(Z_B + Z_C + Z_A)}{(Z_A + Z_B + Z_C)^2}$$

$$= \frac{Z_AZ_BZ_C}{(Z_A + Z_B + Z_C)} \quad (10)$$

i.e. $$Z_1Z_2 + Z_2Z_3 + Z_3Z_1 = Z_A\left(\frac{Z_BZ_C}{Z_A + Z_B + Z_C}\right)$$

$$= Z_A(Z_2)$$

from equation (8)

Hence $$Z_A = \frac{Z_1Z_2 + Z_2Z_3 + Z_3Z_1}{Z_2}$$

From equation (10),

$$Z_1Z_2 + Z_2Z_3 + Z_3Z_1 = Z_B\left(\frac{Z_AZ_C}{Z_A + Z_B + Z_C}\right)$$

$$= Z_B(Z_3) \quad \text{from equation (9)}$$

Hence $$Z_B = \frac{Z_1Z_2 + Z_2Z_3 + Z_3Z_1}{Z_3}$$

Also from equation (10),

$$Z_1Z_2 + Z_2Z_3 + Z_3Z_1 = Z_C\left(\frac{Z_AZ_B}{Z_A + Z_B + Z_C}\right)$$

$$= Z_C(Z_1) \quad \text{from equation (7)}$$

Hence $$Z_C = \frac{Z_1Z_2 + Z_2Z_3 + Z_3Z_1}{Z_1}$$

Summarizing, the delta section shown in Figure 37.31(b) is equivalent to the star section shown in Figure 37.31(a) when

$$\mathbf{Z_A = \frac{Z_1Z_2 + Z_2Z_3 + Z_3Z_1}{Z_2}} \quad (11)$$

$$\mathbf{Z_B = \frac{Z_1Z_2 + Z_2Z_3 + Z_3Z_1}{Z_3}} \quad (12)$$

and $$\mathbf{Z_C = \frac{Z_1Z_2 + Z_2Z_3 + Z_3Z_1}{Z_1}} \quad (13)$$

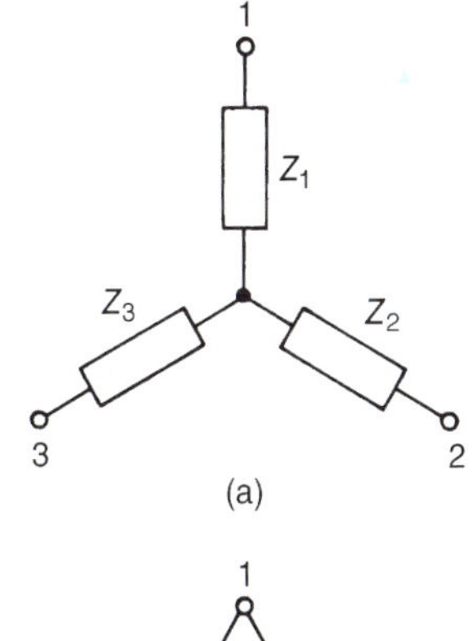

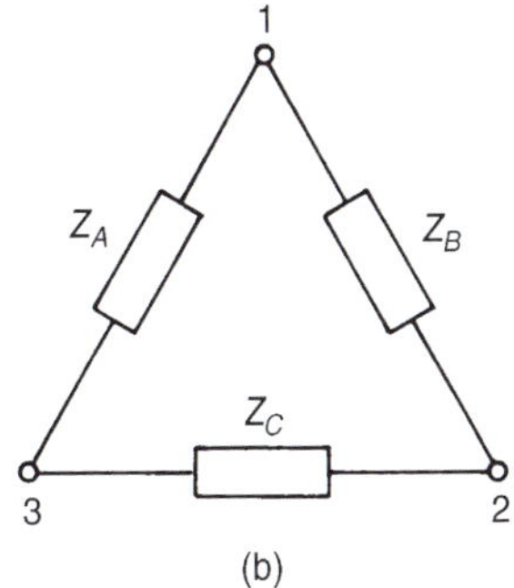

Figure 37.31

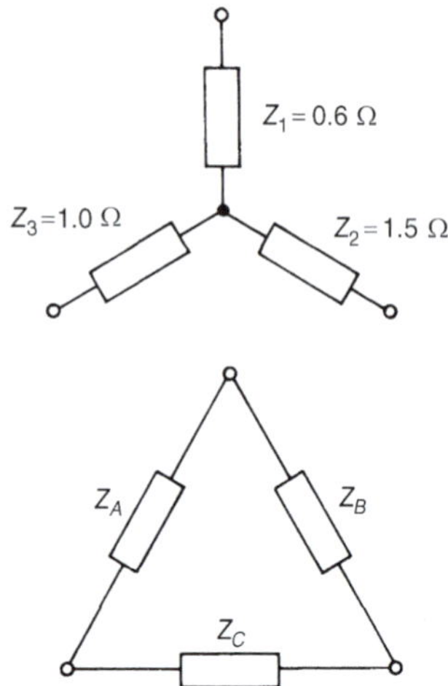

Figure 37.32

It is noted that the numerator in each expression is the sum of the products of the star impedances taken in pairs. The denominator of the expression for Z_A, which is connected between terminals 1 and 3 of Figure 37.31(b), is Z_2, which is connected to terminal 2 of Figure 37.31(a). Similarly, the denominator of the expression for Z_B, which is connected between terminals 1 and 2 of Figure 37.31(b), is Z_3, which is connected to terminal 3 of Figure 37.31(a). Also the denominator of the expression for Z_c, which is connected between terminals 2 and 3 of Figure 37.31(b), is Z_1, which is connected to terminal 1 of Figure 37.31(a).

Thus, for example, the delta equivalent of the resistive star circuit shown in Figure 37.32 is given by:

$$\mathbf{Z_A} = \frac{(0.6)(1.5)+(1.5)(1.0)+(1.0)(0.6)}{1.5}$$

$$= \frac{3.0}{1.5} = \mathbf{2\,\Omega},$$

$$\mathbf{Z_B} = \frac{3.0}{1.0} = \mathbf{3\,\Omega}, \quad \mathbf{Z_C} = \frac{3.0}{0.6} = \mathbf{5\,\Omega}$$

Problem 6. Determine the delta-connected equivalent network for the star-connected impedances shown in Figure 37.33

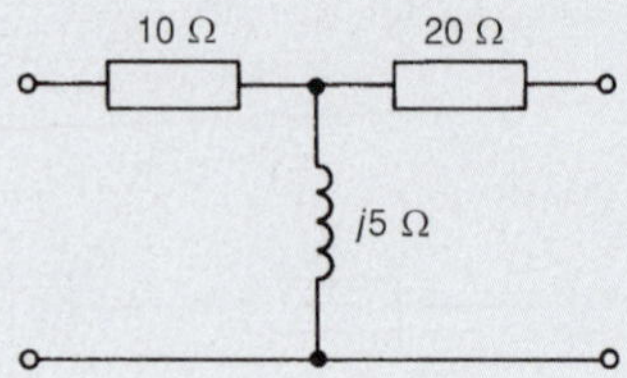

Figure 37.33

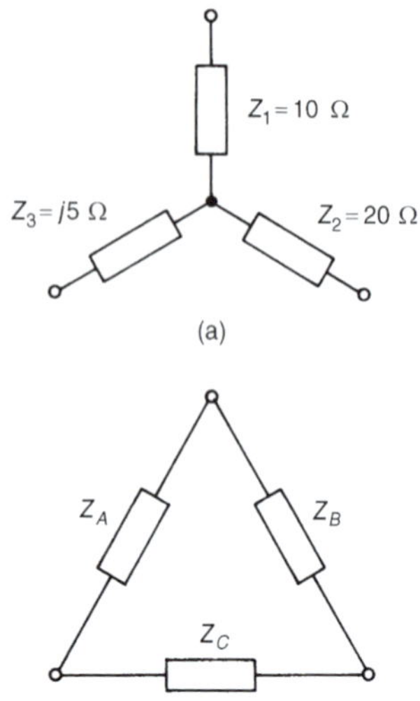

Figure 37.34

Figure 37.34(a) shows the network of Figure 37.33 redrawn and Figure 37.34(b) shows the equivalent delta connection containing impedances Z_A, Z_B and Z_c

From equation (11),

$$Z_A = \frac{Z_1Z_2 + Z_2Z_3 + Z_3Z_1}{Z_2}$$

$$= \frac{(10)(20)+(20)(j5)+(j5)(10)}{20}$$

$$= \frac{200+j150}{20} = \mathbf{(10+j7.5)\,\Omega}$$

From equation (12),

$$Z_B = \frac{(200+j150)}{Z_3} = \frac{(200+j150)}{j5}$$

$$= \frac{-j5(200+j150)}{25} = \mathbf{(30-j40)\,\Omega}$$

From equation (13),

$$Z_C = \frac{(200+j150)}{Z_1} = \frac{(200+j150)}{10} = \mathbf{(20+j15)\,\Omega}$$

Problem 7. Three impedances, $Z_1 = 100\angle 0°\,\Omega$, $Z_2 = 63.25\angle 18.43°\,\Omega$ and $Z_3 = 100\angle -90°\,\Omega$ are connected in star. Convert the star to an equivalent delta connection.

The star-connected network and the equivalent delta network comprising impedances Z_A, Z_B and Z_C are shown in Figure 37.35. From equation (11),

$$Z_A = \frac{Z_1Z_2 + Z_2Z_3 + Z_3Z_1}{Z_2}$$

$$= \frac{(100\angle 0^\circ)(63.25\angle 18.43^\circ) + (63.25\angle 18.43^\circ)(100\angle -90^\circ) + (100\angle -90^\circ)(100\angle 0^\circ)}{63.25\angle 18.43^\circ}$$

$$= \frac{6325\angle 18.43^\circ + 6325\angle -71.57^\circ + 10\,000\angle -90^\circ}{63.25\angle 18.43^\circ}$$

$$= \frac{6000 + j2000 + 2000 - j6000 - j10\,000}{63.25\angle 18.43^\circ}$$

$$= \frac{8000 - j14\,000}{63.25\angle 18.43^\circ} = \frac{16\,124.5\angle -60.26^\circ}{63.25\angle 18.43^\circ}$$

$$= \mathbf{254.93\angle -78.69^\circ\,\Omega} \text{ or } \mathbf{(50 - j250)\Omega}$$

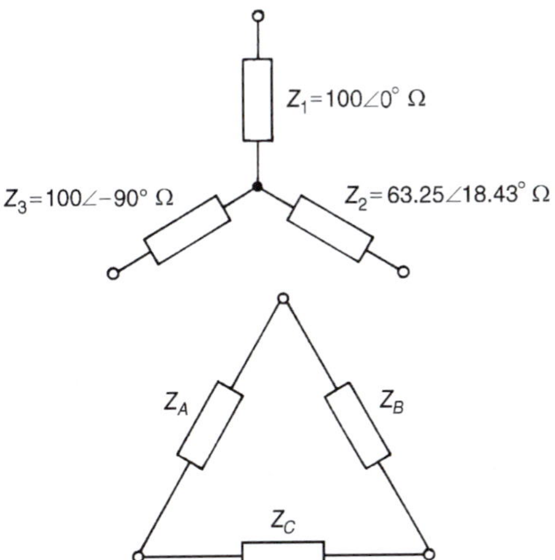

Figure 37.35

From equation (12),

$$Z_B = \frac{Z_1Z_2 + Z_2Z_3 + Z_3Z_1}{Z_3}$$

$$= \frac{16\,124.5\angle -60.26^\circ}{100\angle -90^\circ}$$

$$= \mathbf{161.25\angle 29.74^\circ\,\Omega} \text{ or } \mathbf{(140 + j80)\Omega}$$

From equation (13),

$$Z_C = \frac{Z_1Z_2 + Z_2Z_3 + Z_3Z_1}{Z_1} = \frac{16\,124.5\angle -60.26^\circ}{100\angle 0^\circ}$$

$$= \mathbf{161.25\angle -60.26^\circ\,\Omega} \text{ or } \mathbf{(80 - j140)\Omega}$$

Now try the following Practice Exercise

Practice Exercise 145 Star-delta transformations (Answers on page 829)

1. Determine the delta-connected equivalent networks for the star-connected impedances shown in Figure 37.36.
2. Change the *T*-connected network shown in Figure 37.37 to its equivalent delta-connected network.

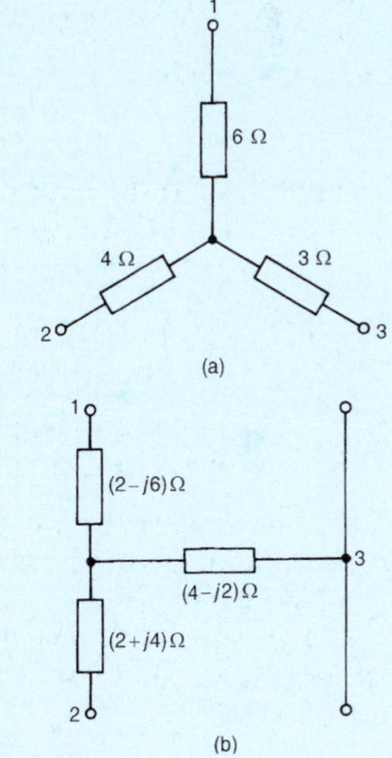

Figure 37.36

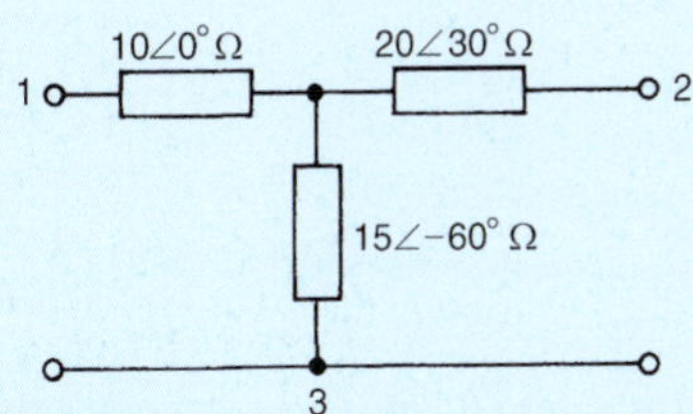

Figure 37.37

Part 4

3. Three impedances, each of $(2+j3)\,\Omega$, are connected in star. Determine the impedances of the equivalent delta-connected network.
4. (a) Derive the star-connected network of three impedances equivalent to the network shown in Figure 37.38.
   (b) Obtain the delta-connected equivalent network for Figure 37.38.

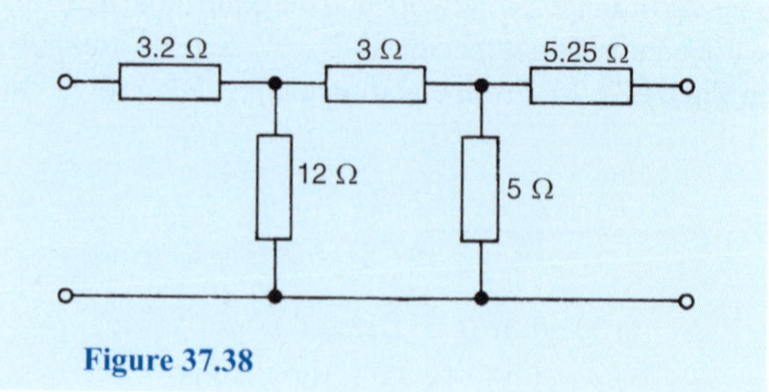

Figure 37.38

Part 4

For fully worked solutions to each of the problems in Practice Exercises 144 and 145 in this chapter, go to the website:
www.routledge.com/cw/bird

Chapter 38

Maximum power transfer theorems and impedance matching

Why it is important to understand: **Maximum power transfer theorems and impedance matching**

In electrical engineering, the maximum power transfer theorem states that to obtain maximum external power from a source with a finite internal resistance, the resistance of the load must equal the resistance of the source as viewed from its output terminals. Moritz von Jacobi published the maximum power (transfer) theorem around 1840; it is also referred to as Jacobi's law. A related concept is reflection-less impedance matching. In radio, transmission lines, and other electronics, there is often a requirement to match the source impedance (such as a transmitter) to the load impedance (such as an antenna) to avoid reflections in the transmission line. In electronics, impedance matching is the practice of designing the input impedance of an electrical load (or the output impedance of its corresponding signal source) to maximize the power transfer or minimize reflections from the load. Matching a load to a source for maximum power transfer is extremely important in microwaves, as well as all manner of lower frequency applications such as stereo sound systems, electrical generating plants, solar cells and hybrid electric cars. This chapter explains, with calculations, the maximum power transfer theorems and the importance of impedance matching.

At the end of this chapter you should be able to:

- appreciate typical applications of the maximum power transfer theorem
- appreciate the conditions for maximum power transfer in a.c. networks
- apply the maximum power transfer theorems to a.c. networks
- appreciate applications and advantages of impedance matching in a.c. networks
- perform calculations involving matching transformers for impedance matching in a.c. networks

Electrical Circuit Theory and Technology. 978-1-138-67349-6, © 2017 John Bird. Published by Taylor & Francis. All rights reserved.

38.1 Maximum power transfer theorems

As mentioned in Section 15.9 on page 239, the maximum power transfer theorem has applications in stereo amplifier design, where it is necessary to maximize power delivered to speakers, and in electric vehicle design, where it is necessary to maximize power delivered to drive a motor.

A network that contains linear impedances and one or more voltage or current sources can be reduced to a Thévenin equivalent circuit, as shown in Chapter 36. When a load is connected to the terminals of this equivalent circuit, power is transferred from the source to the load.

A Thévenin equivalent circuit is shown in Figure 38.1 with source internal impedance, $z=(r+jx)\,\Omega$ and complex load $Z=(R+jX)\Omega$

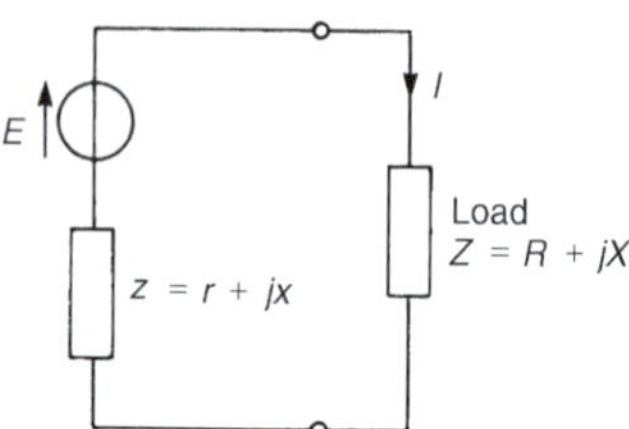

Figure 38.1

The maximum power transferred from the source to the load depends on the following four conditions.

Condition 1. Let the load consist of a pure variable resistance R (i.e. let $X=0$). Then current I in the load is given by:

$$I=\frac{E}{(r+R)+jx}$$

and the magnitude of current, $|I|=\dfrac{E}{\sqrt{[(r+R)^2+x^2]}}$

The active power P delivered to load R is given by

$$P=|I|^2R=\frac{E^2R}{(r+R)^2+x^2}$$

To determine the value of R for maximum power transferred to the load, P is differentiated with respect to R and then equated to zero (this being the normal procedure for finding maximum or minimum values using calculus). Using the quotient rule of differentiation,

$$\frac{dP}{dR}=E^2\left\{\frac{[(r+R)^2+x^2](1)-(R)(2)(r+R)}{[(r+R)^2+x^2]^2}\right\}$$

$= 0$ for a maximum (or minimum) value.

For $\dfrac{dP}{dR}$ to be zero, the numerator of the fraction must be zero.

Hence $\quad (r+R)^2+x^2-2R(r+R)=0$

i.e. $\quad r^2+2rR+R^2+x^2-2Rr-2R^2=0$

from which, $r^2+x^2=R^2 \qquad (1)$

or $\quad$ $\boldsymbol{R=\sqrt{(r^2+x^2)}=|z|}$

Thus, with a variable purely resistive load, the maximum power is delivered to the load if the load resistance R is made equal to the magnitude of the source impedance.

Condition 2. Let both the load and the source impedance be purely resistive (i.e. let $x=X=0$). From equation (1) it may be seen that the maximum power is transferred when $\boldsymbol{R=r}$ (this is, in fact, the d.c. condition explained in Chapter 15, page 239)

Condition 3. Let the load Z have both variable resistance R and variable reactance X. From Figure 38.1,

current $I=\dfrac{E}{(r+R)+j(x+X)}$ and

$$|I|=\frac{E}{\sqrt{[(r+R)^2+(x+x)^2]}}$$

The active power P delivered to the load is given by $P=|I|^2R$ (since power can only be dissipated in a resistance) i.e.

$$P=\frac{E^2R}{(r+R)^2+(x+x)^2}$$

If X is adjusted such that $X=-x$ then the value of power is a maximum.

If $X=-x$ then $P=\dfrac{E^2R}{(r+R)^2}$

$$\frac{dP}{dR}=E^2\left\{\frac{(r+R)^2(1)-(R)(2)(r+R)}{(r+R)^4}\right\}$$

$= 0$ for a maximum value

Hence $(r+R)^2 - 2R(r+R) = 0$

i.e. $r^2 + 2rR + R^2 - 2Rr - 2R^2 = 0$

from which, $r^2 - R^2 = 0$ and $R = r$

Thus with the load impedance Z consisting of variable resistance R and variable reactance X, maximum power is delivered to the load when $\mathbf{X = -x}$ **and** $\mathbf{R = r}$ i.e. when $R + jX = r - jx$. Hence maximum power is delivered to the load when the load impedance is the complex conjugate of the source impedance.

Condition 4. Let the load impedance Z have variable resistance R and fixed reactance X. From Figure 38.1, the magnitude of current,

$$|I| = \frac{E}{\sqrt{[(r+R)^2 + (x+X)^2]}}$$

and the power dissipated in the load,

$$P = \frac{E^2 R}{(r+R)^2 + (x+X)^2}$$

$$\frac{dP}{dR} = E^2 \left\{ \frac{[(r+R)^2 + (x+X)^2(1) - (R)(2)(r+R)]}{[(r+R)^2 + (x+X)^2]^2} \right\}$$

$= 0$ for a maximum value

Hence $(r+R)^2 + (x+X)^2 - 2R(r+R) = 0$

$r^2 + 2rR + R^2 + (x+X)^2 - 2Rr - 2R^2 = 0$

from which, $R^2 = r^2 + (x+X)^2$ and

$$\mathbf{R = \sqrt{[r^2 + (x+X)^2)]}}$$

Summary

With reference to Figure 38.1:

1. When the load is purely resistive (i.e. $X = 0$) and adjustable, maximum power transfer is achieved when

$$\mathbf{R = |z| = \sqrt{(r^2 + x^2)}}$$

2. When both the load and the source impedance are purely resistive (i.e. $X = x = 0$), maximum power transfer is achieved when $\mathbf{R = r}$

3. When the load resistance R and reactance X are both independently adjustable, maximum power transfer is achieved when

$$\mathbf{X = -x \text{ and } R = r}$$

4. When the load resistance R is adjustable with reactance X fixed, maximum power transfer is achieved when

$$\mathbf{R = \sqrt{[r^2 + (x+X)^2]}}$$

The maximum power transfer theorems are primarily important where a small source of power is involved – such as, for example, the output from a telephone system (see Section 38.2)

Problem 1. For the circuit shown in Figure 38.2 the load impedance Z is a pure resistance. Determine (a) the value of R for maximum power to be transferred from the source to the load and (b) the value of the maximum power delivered to R.

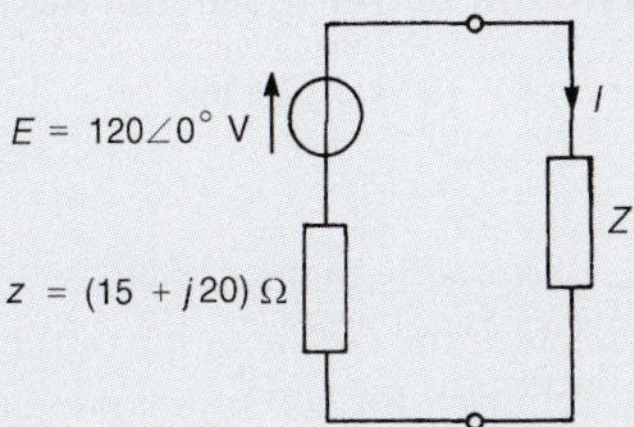

Figure 38.2

(a) From condition 1, maximum power transfer occurs when $R = |z|$, i.e. when

$$R = |15 + j20| = \sqrt{(15^2 + 20^2)} = \mathbf{25\,\Omega}$$

(b) Current I flowing in the load is given by $I = E/Z_T$, where the total circuit impedance

$Z_T = z + R = 15 + j20 + 25$

$= (40 + j20)\,\Omega$ or $44.72\angle 26.57°\,\Omega$

Hence $I = \dfrac{120\angle 0°}{44.72\angle 26.57°} = 2.683\angle -26.57°$ A

Thus **maximum power delivered,**

$\mathbf{P = I^2 R}$

$= (2.683)^2(25)$

$= \mathbf{180\,W}$

Problem 2. If the load impedance Z in Figure 38.2 of Problem 1 consists of variable resistance R and variable reactance X, determine (a) the value of Z that results in maximum power transfer and (b) the value of the maximum power.

Part 4

(a) From condition 3, maximum power transfer occurs when $X=-x$ and $R=r$

Thus if $z=r+jx=(15+j20)\,\Omega$ then

$\boldsymbol{Z=(15-j20)\,\Omega}$ or $\boldsymbol{25\angle-53.13^\circ\,\Omega}$

(b) Total circuit impedance at maximum power transfer condition, $Z_T=z+Z$, i.e.

$$Z_T=(15+j20)+(15-j20)=30\,\Omega$$

Hence current in load, $I=\dfrac{E}{Z_T}=\dfrac{120\angle0^\circ}{30}$
$=4\angle0^\circ$ A

and **maximum power** transfer in the load,

$P=I^2R=(4)^2\,(15)=\mathbf{240\,W}$

Problem 3. For the network shown in Figure 38.3, determine (a) the value of the load resistance R required for maximum power transfer, and (b) the value of the maximum power transferred.

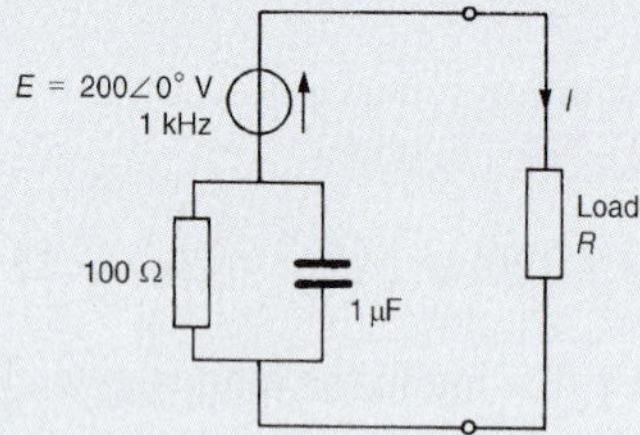

Figure 38.3

(a) This problem is an example of condition 1, where maximum power transfer is achieved when $R=|z|$. Source impedance z is composed of a $100\,\Omega$ resistance in parallel with a $1\,\mu$F capacitor.

Capacitive reactance, $X_C=\dfrac{1}{2\pi fC}$

$$=\frac{1}{2\pi(1000)(1\times10^{-6})}$$

$$=159.15\,\Omega$$

Hence source impedance,

$$z=\frac{(100)(-j159.15)}{(100-j159.15)}=\frac{159.15\angle-90^\circ}{187.96\angle-57.86^\circ}$$

$$=84.67\angle-32.14^\circ\,\Omega \quad\text{or}\quad (71.69-j45.04)\,\Omega$$

Thus the value of **load resistance** for maximum power transfer is **84.67 Ω** (i.e. $|z|$)

(b) With $z=(71.69-j45.04)\,\Omega$ and $R=84.67\,\Omega$ for maximum power transfer, the total circuit impedance,

$$Z_T=71.69+84.67-j45.04$$
$$=(156.36-j45.04)\,\Omega \text{ or } 162.72\angle-16.07^\circ\,\Omega$$

Current flowing in the load, $I=\dfrac{V}{Z_T}$

$$=\frac{200\angle0^\circ}{162.72\angle-16.07^\circ}$$

$$=1.23\angle16.07^\circ\text{ A}$$

Thus the **maximum power** transferred,

$P=I^2R=(1.23)^2(84.67)=\mathbf{128\,W}$

Problem 4. In the network shown in Figure 38.4 the load consists of a fixed capacitive reactance of 7 Ω and a variable resistance R. Determine (a) the value of R for which the power transferred to the load is a maximum, and (b) the value of the maximum power.

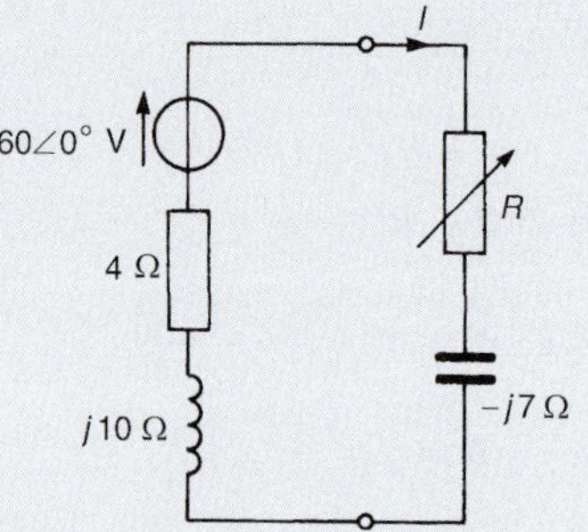

Figure 38.4

(a) From condition (4), maximum power transfer is achieved when

$$\boldsymbol{R}=\sqrt{[r^2+(x+X)^2]}=\sqrt{[4^2+(10-7)^2]}$$
$$=\sqrt{(4^2+3^2)}=\mathbf{5\,\Omega}$$

(b) Current $I=\dfrac{60\angle0^\circ}{(4+j10)+(5-j7)}=\dfrac{60\angle0^\circ}{(9+j3)}$

$$=\frac{60\angle0^\circ}{9.487\angle18.43^\circ}=6.324\angle-18.43^\circ\text{ A}$$

Thus the **maximum power** transferred,

$P=I^2R=(6.324)^2(5)=\mathbf{200\,W}$

Part 4

Problem 5. Determine the value of the load resistance R shown in Figure 38.5 that gives maximum power dissipation and calculate the value of this power.

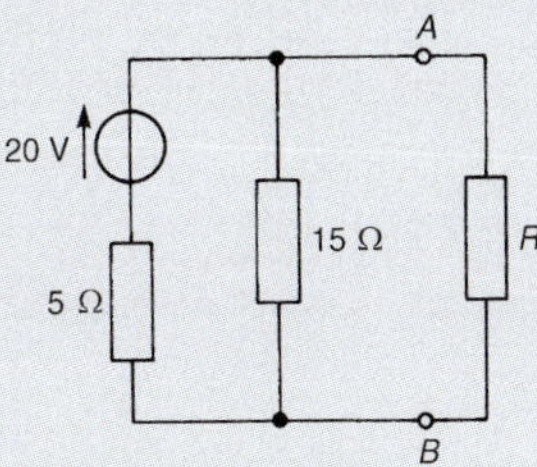

Figure 38.5

Using the procedure of Thévenin's theorem (see page 529):

(i) R is removed from the network as shown in Figure 38.6

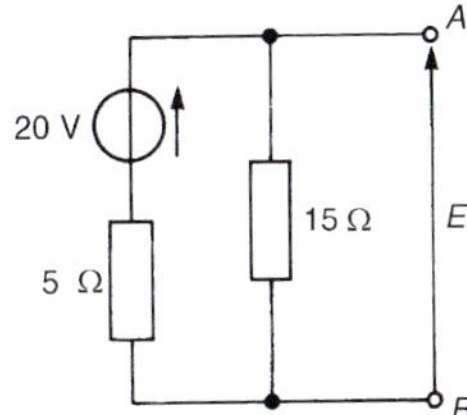

Figure 38.6

(ii) P.d. across AB, $E = (15/(15+5))(20) = 15\,\text{V}$

(iii) Impedance 'looking-in' at terminals AB with the 20 V source removed is given by $r = (5 \times 15)/(5+15) = 3.75\,\Omega$

(iv) The equivalent Thévenin circuit supplying terminals AB is shown in Figure 38.7. From condition (2), for maximum power transfer, $R = r$, i.e. $\boldsymbol{R = 3.75\,\Omega}$

$$\text{Current } I = \frac{E}{R+r} = \frac{15}{3.75+3.75} = 2\,\text{A}$$

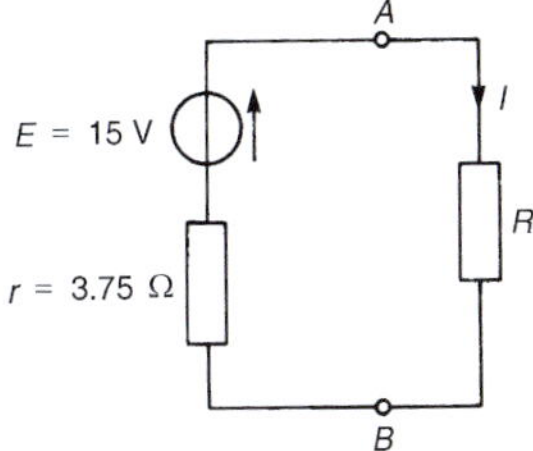

Figure 38.7

Thus the **maximum power** dissipated in the load,

$$P = I^2R = (2)^2(3.75) = \mathbf{15\,W}$$

Problem 6. Determine, for the network shown in Figure 38.8, (a) the values of R and X that will result in maximum power being transferred across terminals AB and (b) the value of the maximum power.

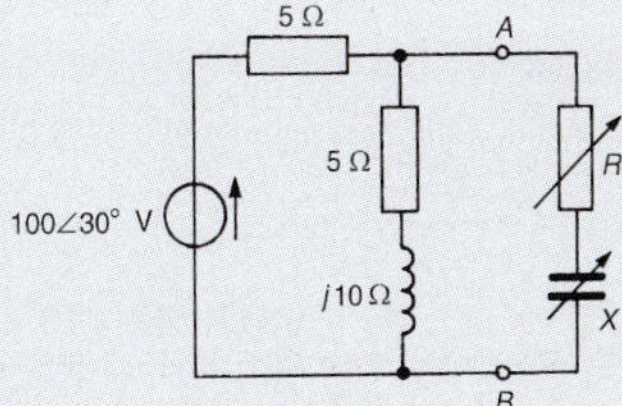

Figure 38.8

(a) Using the procedure for Thévenin's theorem:

(i) Resistance R and reactance X are removed from the network as shown in Figure 38.9

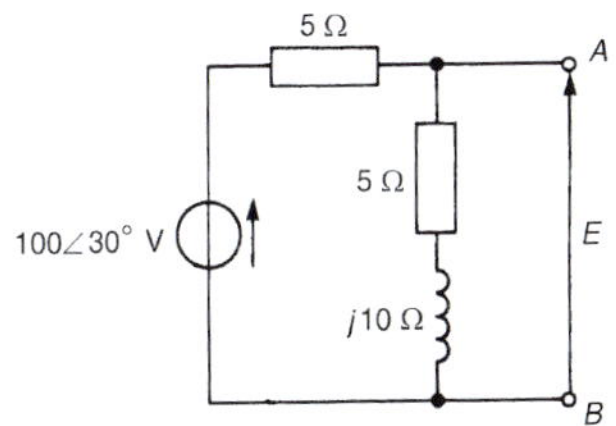

Figure 38.9

(ii) P.d. across AB,

$$E = \left(\frac{5+j10}{5+j10+5}\right)(100\angle 30^\circ)$$

$$= \frac{(11.18\angle 63.43^\circ)(100\angle 30^\circ)}{14.14\angle 45^\circ}$$

$$= 79.07\angle 48.43^\circ\,\text{V}$$

(iii) With the $100\angle 30^\circ$ V source removed the impedance, z, 'looking in' at terminals AB is given by:

$$z = \frac{(5)(5+j10)}{(5+5+j10)} = \frac{(5)(11.18\angle 63.43^\circ)}{(14.14\angle 45^\circ)}$$

$$= 3.953\angle 18.43^\circ\,\Omega \text{ or } (3.75 + j1.25)\,\Omega$$

(iv) The equivalent Thévenin circuit is shown in Figure 38.10. From condition 3, maximum power transfer is achieved when $X = -x$

Part 4

and $R = r$, i.e. in this case when
$\mathbf{X = -1.25\,\Omega}$ and $\mathbf{R = 3.75\,\Omega}$

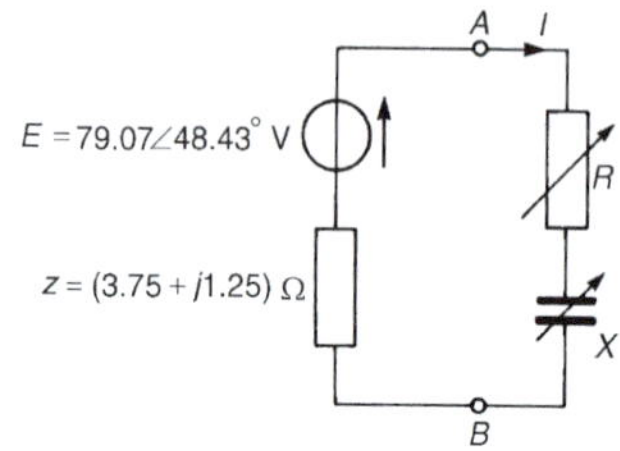

Figure 38.10

(b) Current $I = \dfrac{E}{z + Z}$

$$= \frac{79.07\angle 48.43^\circ}{(3.75 + j1.25) + (3.75 - j1.25)}$$

$$= \frac{79.07\angle 48.43^\circ}{7.5}$$

$$= 10.543\angle 48.43^\circ\,\text{A}$$

Thus the **maximum power** transferred,

$P = I^2R = (10.543)^2(3.75) = \mathbf{417\,W}$

Now try the following Practice Exercise

Practice Exercise 146 Maximum power transfer theorems (Answers on page 829)

1. For the circuit shown in Figure 38.11 determine the value of the source resistance r if the maximum power is to be dissipated in the 15 Ω load. Determine the value of this maximum power.

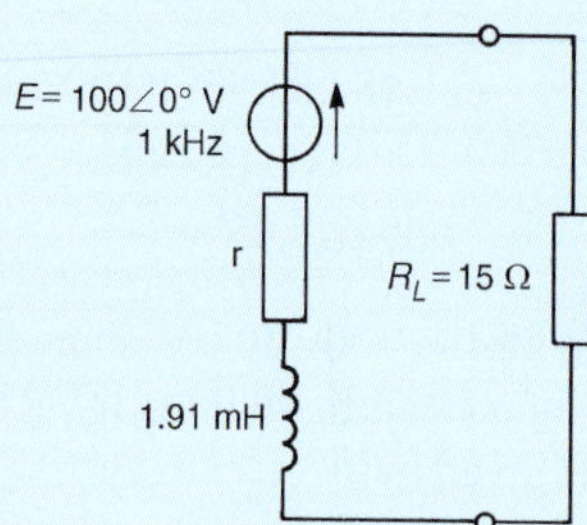

Figure 38.11

2. In the circuit shown in Figure 38.12 the load impedance Z_L is a pure resistance R. Determine (a) the value of R for maximum power to be transferred from the source to the load, and (b) the value of the maximum power delivered to R.

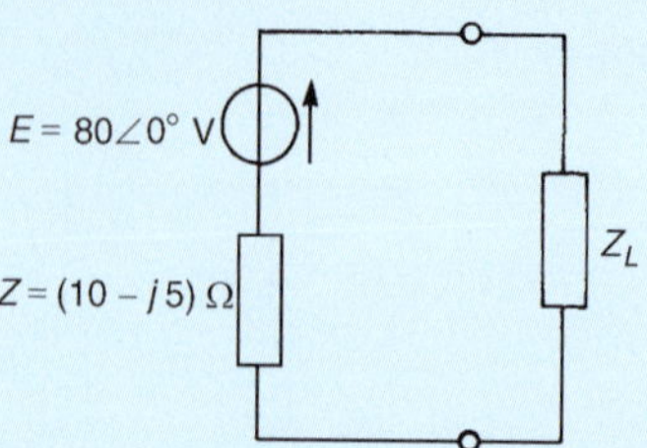

Figure 38.12

3. If the load impedance Z_L in Figure 38.12 of Problem 2 consists of a variable resistance R and variable reactance X, determine (a) the value of Z_L which results in maximum power transfer and (b) the value of the maximum power.

4. For the network shown in Figure 38.13 determine (a) the value of the load resistance R_L required for maximum power transfer and (b) the value of the maximum power.

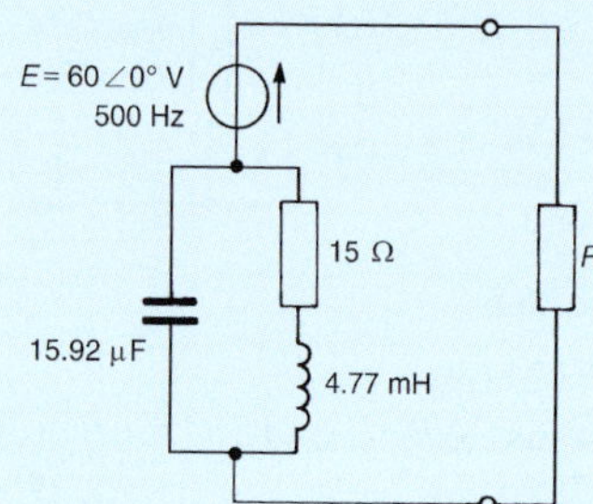

Figure 38.13

5. Find the value of the load resistance R_L shown in Figure 38.14 that gives maximum power dissipation, and calculate the value of this power.

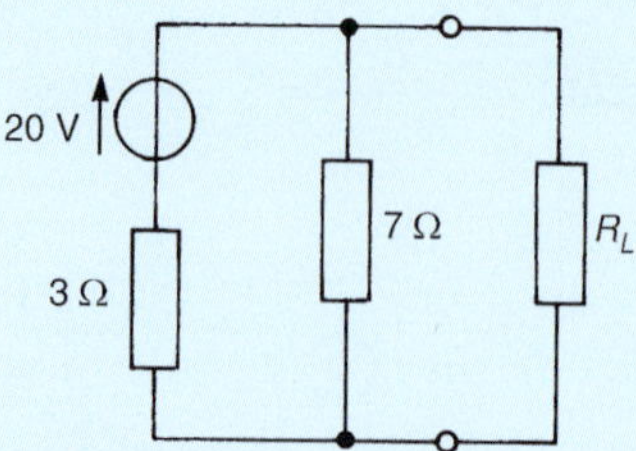

Figure 38.14

6. For the circuit shown in Figure 38.15 determine (a) the value of load resistance R_L which

results in maximum power transfer and (b) the value of the maximum power.

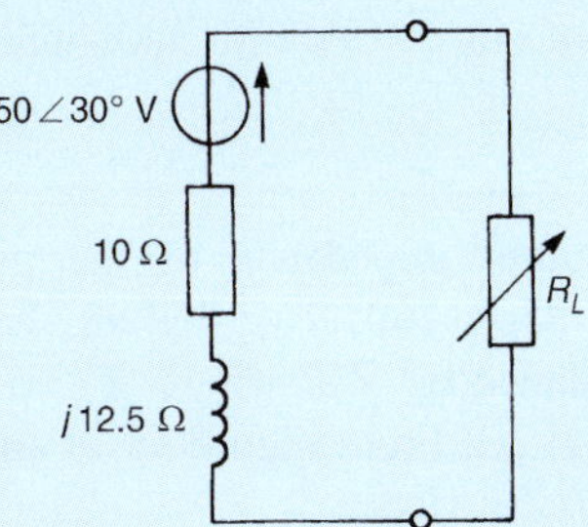

Figure 38.15

7. Determine, for the network shown in Figure 38.16, (a) the values of R and X which result in maximum power being transferred across terminals AB and (b) the value of the maximum power.

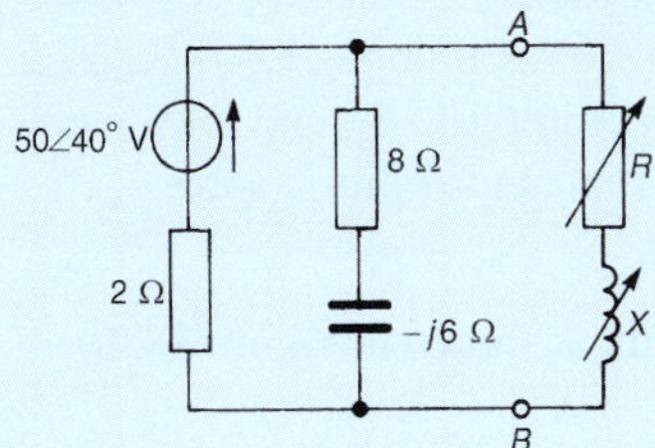

Figure 38.16

8. A source of $120\angle 0°$ V and impedance $(5+j3)\,\Omega$ supplies a load consisting of a variable resistor R in series with a fixed capacitive reactance of 8 Ω. Determine (a) the value of R to give maximum power transfer and (b) the value of the maximum power.

9. If the load Z_L between terminals A and B of Figure 38.17 is variable in both resistance and reactance, determine the value of Z_L such that it will receive maximum power. Calculate the value of the maximum power.

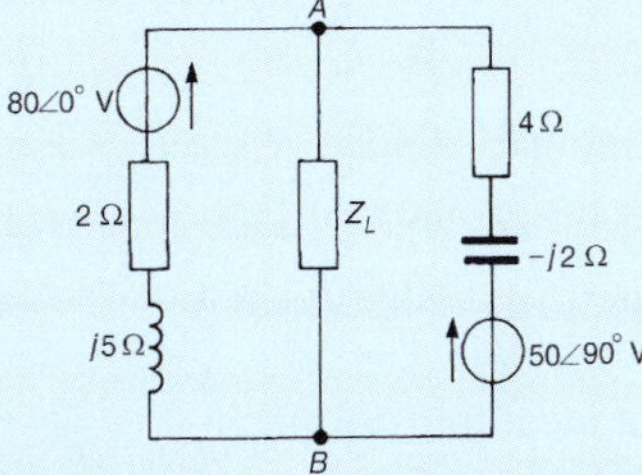

Figure 38.17

10. For the circuit of Figure 38.18, determine the value of load impedance Z_L for maximum load power if (a) Z_L comprises a variable resistance R and variable reactance X and (b) Z_L is a pure resistance R. Determine the values of load power in each case.

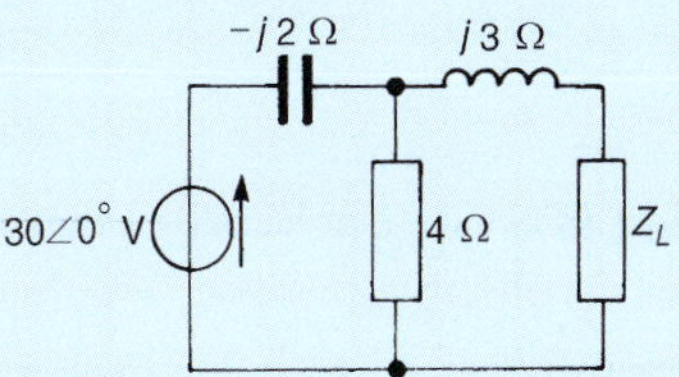

Figure 38.18

38.2 Impedance matching

It is seen from Section 38.1 that when it is necessary to obtain the maximum possible amount of power from a source, it is advantageous if the circuit components can be adjusted to give equality of impedances. This adjustment is called **'impedance matching'** and is an important consideration in electronic and communications devices which normally involve small amounts of power. Examples where matching is important include coupling an aerial to a transmitter or receiver, or coupling a loudspeaker to an amplifier. Also, matching a load to a source for maximum power transfer is extremely important in microwaves, as well as all manner of lower frequency applications such as electrical generating plants and solar cells.

The mains power supply is considered as infinitely large compared with the demand upon it, and under such conditions it is unnecessary to consider the conditions for maximum power transfer. With transmission lines (see Chapter 47), the lines are 'matched' ideally, i.e. terminated in their characteristic impedance.

With d.c. generators, motors or secondary cells, the internal impedance is usually very small and in such cases, if an attempt is made to make the load impedance as small as the source internal impedance, overloading of the source results.

A method of achieving maximum power transfer between a source and a load is to adjust the value of the load impedance to match the source impedance, which can be done using a **'matching-transformer'**.

A transformer is represented in Figure 38.19 supplying a load impedance Z_L

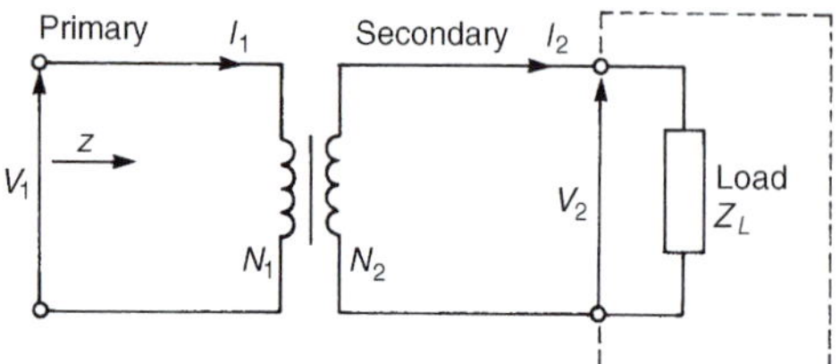

Figure 38.19 Matching impedance by means of a transformer

Small transformers used in low-power networks are usually regarded as ideal (i.e. losses are negligible), such that

$$\frac{V_1}{V_2}=\frac{N_1}{N_2}=\frac{I_2}{I_1}$$

From Figure 38.19, the primary input impedance $|z|$ is given by

$$|z|=\frac{V_1}{I_1}=\frac{(N_1/N_2)V_2}{(N_2/N_1)I_2}=\left(\frac{N_1}{N_2}\right)^2\frac{V_2}{I_2}$$

Since the load impedance $|Z_L|=V_2/I_2$

$$|z|=\left(\frac{N_1}{N_2}\right)^2|Z_L| \quad (2)$$

If the input impedance of Figure 38.19 is purely resistive (say, r) and the load impedance is purely resistive (say, R_L) then equation (2) becomes

$$r=\left(\frac{N_1}{N_2}\right)^2 R_L \quad (3)$$

(This is the case introduced in Section 23.10, page 363.) Thus by varying the value of the transformer turns ratio, the equivalent input impedance of the transformer can be 'matched' to the impedance of a source to achieve maximum power transfer.

Problem 7. Determine the optimum value of load resistance for maximum power transfer if the load is connected to an amplifier of output resistance 448 Ω through a transformer with a turns ratio of 8:1

The equivalent input resistance r of the transformer must be 448 Ω for maximum power transfer. From equation (3), $r=(N_1/N_2)^2R_L$, from which, load resistance $\boldsymbol{R_L}=r(N_2/N_1)^2=448(1/8)^2=\mathbf{7\,\Omega}$

Problem 8. A generator has an output impedance of $(450+j60)\,\Omega$. Determine the turns ratio of an ideal transformer necessary to match the generator to a load of $(40+j19)\,\Omega$ for maximum transfer of power.

Let the output impedance of the generator be z, where $z=(450+j60)\,\Omega$ or $453.98\angle 7.59°\,\Omega$, and the load impedance be Z_L, where $Z_L=(40+j19)\,\Omega$ or $44.28\angle 25.41°\,\Omega$. From Figure 38.19 and equation (2), $z=(N_1/N_2)^2Z_L$. Hence

$$\text{transformer turns ratio } \left(\frac{N_1}{N_2}\right)=\sqrt{\frac{z}{Z_L}}=\sqrt{\frac{453.98}{44.28}}=\sqrt{(10.25)}=\mathbf{3.20}$$

Problem 9. A single-phase, 240 V/1920 V ideal transformer is supplied from a 240 V source through a cable of resistance 5 Ω. If the load across the secondary winding is 1.60 kΩ, determine (a) the primary current flowing and (b) the power dissipated in the load resistance.

The network is shown in Figure 38.20.

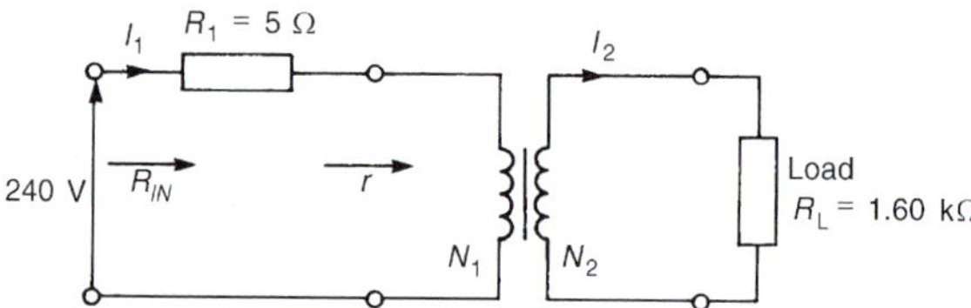

Figure 38.20

(a) Turns ratio, $\dfrac{N_1}{N_2}=\dfrac{V_1}{V_2}=\dfrac{240}{1920}=\dfrac{1}{8}$

Equivalent input resistance of the transformer,

$$r=\left(\frac{N_1}{N_2}\right)^2R_L=\left(\frac{1}{8}\right)^2(1600)=25\,\Omega$$

Total input resistance, $R_{IN}=R_1+r=5+25=30\,\Omega$. Hence the primary current,

$\boldsymbol{I_1}=V_1/R_{IN}=240/30=\mathbf{8\,A}$

(b) For an ideal transformer, $\dfrac{V_1}{V_2}=\dfrac{I_2}{I_1}$

from which, $I_2=I_1\left(\dfrac{V_1}{V_2}\right)=(8)\left(\dfrac{240}{1920}\right)=1\,\text{A}$

Part 4

Power dissipated in the load resistance,

$$P = I_2^2 R_L = (I)^2(1600) = \mathbf{1.6\,kW}$$

Problem 10. An a.c. source of $30\angle 0^\circ$ V and internal resistance 20 kΩ is matched to a load by a 20:1 ideal transformer. Determine for maximum power transfer (a) the value of the load resistance and (b) the power dissipated in the load.

The network diagram is shown in Figure 38.21.

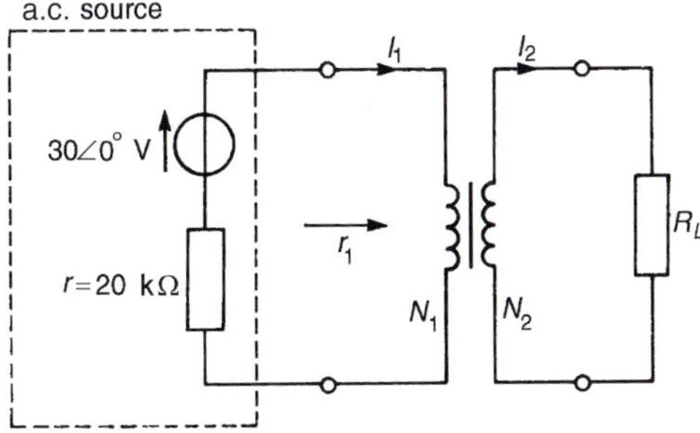

Figure 38.21

(a) For maximum power transfer, r_1 must be equal to 20 kΩ. From equation (3), $r_1 = (N_1/N_2)^2 R_L$ from which,

$$\text{load resistance } \boldsymbol{R_L} = r_1\left(\frac{N_2}{N_1}\right)^2 = (20\,000)\left(\frac{1}{20}\right)^2 = \mathbf{50\,\Omega}$$

(b) The total input resistance when the source is connected to the matching transformer is $(r + r_1)$, i.e. 20 kΩ + 20 kΩ = 40 kΩ. Primary current,

$$I_1 = V/40\,000 = 30/40\,000 = 0.75\,\text{mA}$$

$$\frac{N_1}{N_2} = \frac{I_2}{I_1} \text{ from which, } I_2 = I_1\left(\frac{N_1}{N_2}\right) = (0.75\times 10^{-3})\left(\frac{20}{1}\right) = 15\,\text{mA}$$

Power dissipated in load resistance R_L is given by

$$\boldsymbol{P} = I_2^2 R_L = (15\times 10^{-3})^2(50) = \mathbf{0.01125\,W} \text{ or } \mathbf{11.25\,mW}$$

Now try the following Practice Exercise

Practice Exercise 147 Impedance matching (Answers on page 829)

1. The output stage of an amplifier has an output resistance of 144 Ω. Determine the optimum turns ratio of a transformer that would match a load resistance of 9 Ω to the output resistance of the amplifier for maximum power transfer.
2. Find the optimum value of load resistance for maximum power transfer if a load is connected to an amplifier of output resistance 252 Ω through a transformer with a turns ratio of 6:1
3. A generator has an output impedance of $(300 + j45)\,\Omega$. Determine the turns ratio of an ideal transformer necessary to match the generator to a load of $(37 + j19)\,\Omega$ for maximum power transfer.
4. A single-phase, 240 V/2880 V ideal transformer is supplied from a 240 V source through a cable of resistance 3.5 Ω. If the load across the secondary winding is 1.8 kΩ, determine (a) the primary current flowing, and (b) the power dissipated in the load resistance.
5. An a.c. source of $20\angle 0^\circ$ V and internal resistance 10.24 kΩ is matched to a load for maximum power transfer by a 16:1 ideal transformer. Determine (a) the value of the load resistance and (b) the power dissipated in the load.

Part 4

For fully worked solutions to each of the problems in Practice Exercises 146 and 147 in this chapter, go to the website:
www.routledge.com/cw/bird

Revision Test 11

This revision test covers the material in Chapters 37 and 38. *The marks for each question are shown in brackets at the end of each question.*

1. Determine the delta-connected equivalent network for the star-connected impedances shown in Figure RT11.1 (9)

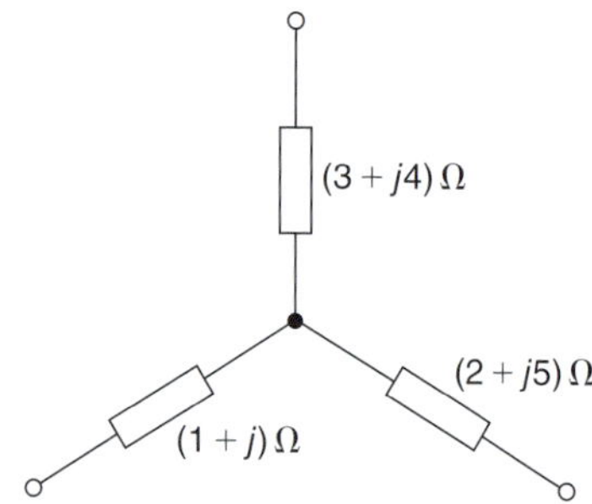

Figure RT11.1

2. Transform the delta-connection in Figure RT11.2 to its equivalent star connection. Hence determine the following for the network shown in Figure RT11.3:

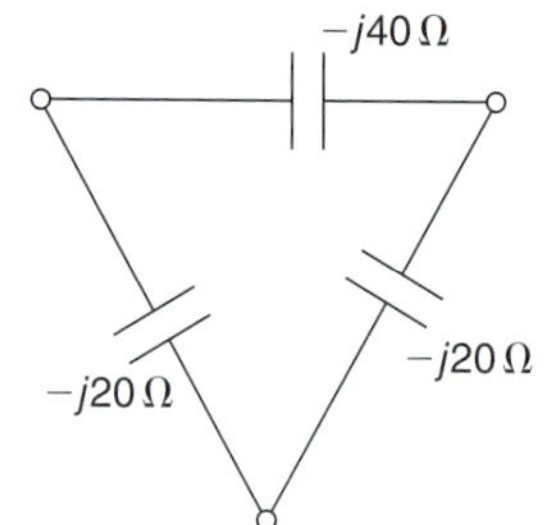

Figure RT11.2

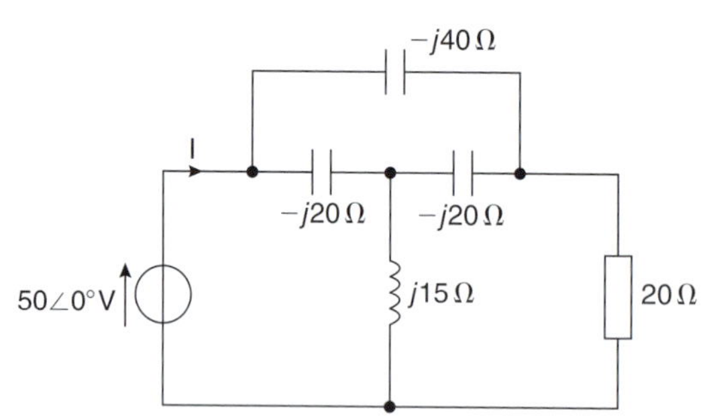

Figure RT11.3

(a) the total circuit impedance
(b) the current I
(c) the current in the 20 Ω resistor
(d) the power dissipated in the 20 Ω resistor. (17)

3. If the load impedance Z in Figure RT11.4 consists of variable resistance and variable reactance, find (a) the value of Z that results in maximum power transfer and (b) the value of the maximum power. (6)

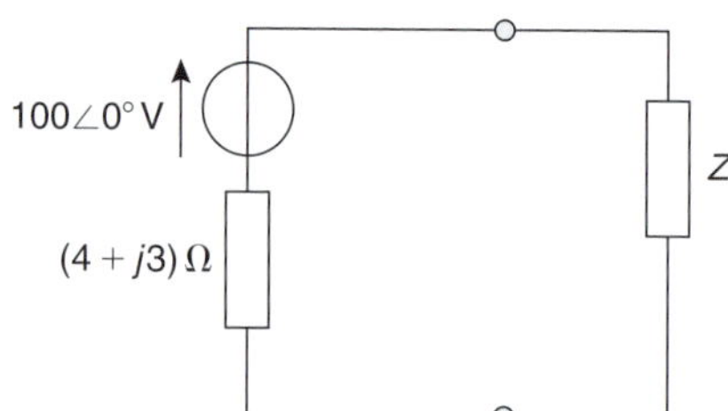

Figure RT11.4

4. Determine the value of the load resistance R in Figure RT11.5 that gives maximum power dissipation and calculate the value of power. (9)

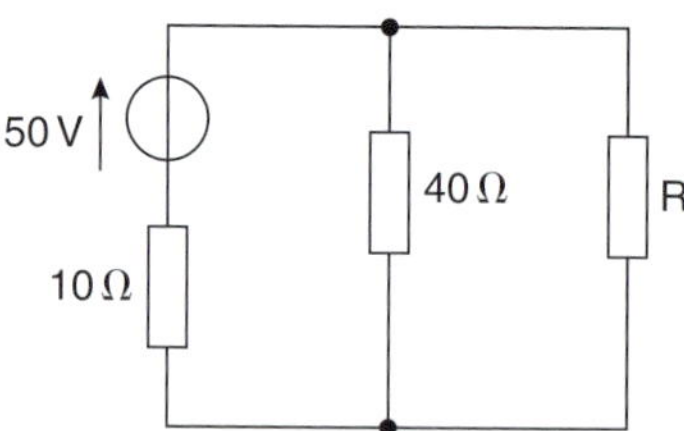

Figure RT11.5

5. An a.c. source of $10\angle 0^\circ$ V and internal resistance 5 kΩ is matched to a load for maximum power transfer by a 5:1 ideal transformer. Determine (a) the value of the load resistance and (b) the power dissipated in the load. (9)

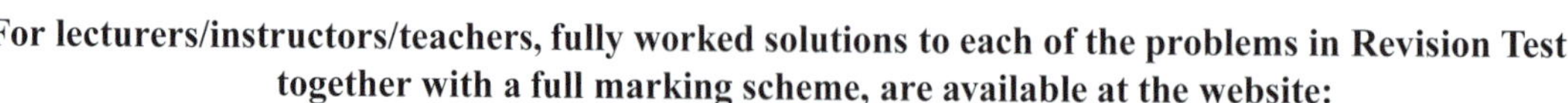

For lecturers/instructors/teachers, fully worked solutions to each of the problems in Revision Test 11, together with a full marking scheme, are available at the website:

www.routledge.com/cw/bird

Chapter 39

Complex waveforms

***Why it is important to understand:* Complex waveforms**

Alternating currents and voltages come in very many forms; for example, the electricity power supply, sound, light, video, radio, all produce signals that alternate – meaning they change their values over time, alternating above and below a particular value (often, but not always zero). Our bodies also produce alternating electrical signals, as do all sorts of natural and man-made objects and devices. These signals are what engineers and technicians are most often interested in when studying electronics, but signals come in many very different forms. To understand complex signals, there is often a need to simplify them; if the signal can be understood in its simplest form, then that understanding can be applied to the complex signal. A complex wave is a wave made up of a series of sine waves; it is therefore more complex than a single pure sine wave. This series of sine waves always contains a sine wave called the fundamental which has the same frequency (repetition rate) as the complex wave being created. This chapter explains harmonic synthesis, uses Fourier series for simple functions, calculates r.m.s., mean values and power with complex waves, and lists and explains some sources of harmonics.

At the end of this chapter you should be able to:

- define a complex wave
- recognize periodic functions
- recognize the general equation of a complex waveform
- use harmonic synthesis to build up a complex wave
- recognize characteristics of waveforms containing odd, even or odd and even harmonics, with or without phase change
- determine Fourier series for simple functions
- calculate r.m.s. and mean values, and form factor of a complex wave
- calculate power associated with complex waves
- perform calculations on single-phase circuits containing harmonics
- define and perform calculations on harmonic resonance
- list and explain some sources of harmonics

Electrical Circuit Theory and Technology. 978-1-138-67349-6, © 2017 John Bird. Published by Taylor & Francis. All rights reserved.

39.1 Introduction

In preceding chapters a.c. supplies have been assumed to be sinusoidal, this being a form of alternating quantity commonly encountered in electrical engineering. However, many supply waveforms are **not** sinusoidal. For example, sawtooth generators produce ramp waveforms, and rectangular waveforms may be produced by multivibrators. A waveform that is not sinusoidal is called a **complex wave**. Such a waveform may be shown to be composed of the sum of a series of sinusoidal waves having various interrelated periodic times.

A function $f(t)$ is said to be **periodic** if $f(t+T)=f(t)$ for all values of t, where T is the interval between two successive repetitions and is called the **period** of the function $f(t)$. A sine wave having a period of $2\pi/\omega$ is a familiar example of a periodic function.

A typical complex periodic-voltage waveform, shown in Figure 39.1, has period T seconds and frequency f hertz. A complex wave such as this can be resolved into the sum of a number of sinusoidal waveforms, and each of the sine waves can have a different frequency, amplitude and phase.

The initial, major sine wave component has a frequency f equal to the frequency of the complex wave and this frequency is called the **fundamental frequency**. The other sine wave components are known as **harmonics**, these having frequencies which are integer multiples of frequency f. Hence the second harmonic has a frequency of $2f$, the third harmonic has a frequency of $3f$, and so on. Thus if the fundamental (i.e. supply) frequency of a complex wave is 50 Hz, then the third harmonic frequency is 150 Hz, the fourth harmonic frequency is 200 Hz, and so on.

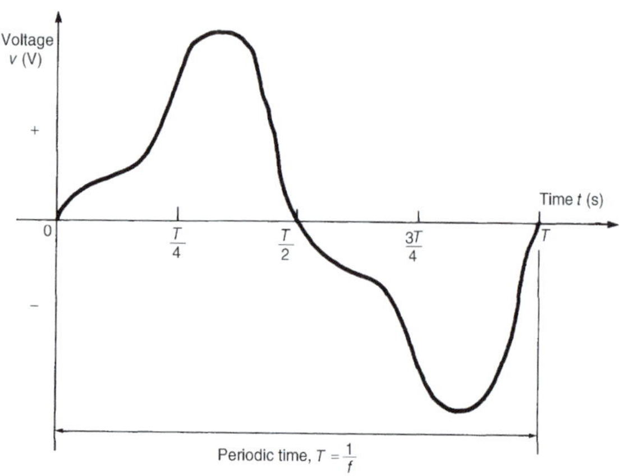

Figure 39.1 Typical complex periodic voltage waveform

39.2 The general equation for a complex waveform

The instantaneous value of a complex voltage wave v acting in a linear circuit may be represented by the general equation

$$\mathbf{v = V_m \sin(\omega t + \Psi_1) + V_{2m} \sin(2\omega t + \Psi_2) + \cdots + V_{nm} \sin(n\omega t + \Psi_n)\,volts} \quad (1)$$

Here $V_{1m}\sin(\omega t+\Psi_1)$ represents the fundamental component of which V_{1m} is the maximum or peak value, frequency, $f=\omega/2\pi$ and Ψ_1 is the phase angle with respect to time, $t=0$.

Similarly, $V_{2m}\sin(2\omega t+\Psi_2)$ represents the second harmonic component, and $V_{nm}\sin(n\omega t+\Psi_n)$ represents the nth harmonic component, of which V_{nm} is the peak value, frequency $= n\omega/2\pi(=nf)$ and Ψ_n is the phase angle.

In the same way, the instantaneous value of a complex current i may be represented by the general equation

$$\mathbf{i = I_{1m} \sin(\omega t + \theta_1) + I_{2m} \sin(2\omega t + \theta_2) + \cdots + I_{nm} \sin(n\omega t + \theta_n)\,amperes} \quad (2)$$

Where equations (1) and (2) refer to the voltage across and the current flowing through a given linear circuit, the phase angle between the fundamental voltage and current is $\phi_1=(\Psi_1-\theta_1)$, the phase angle between the second harmonic voltage and current is $\phi_2=(\Psi_2-\theta_2)$, and so on.

It often occurs that not all harmonic components are present in a complex waveform. Sometimes only the fundamental and odd harmonics are present, and in others only the fundamental and even harmonics are present.

The following worked problems help introduce complex waveform equations and revise a.c. values from Chapter 16.

Problem 1. A complex voltage wave is given by:

$v=200\sin 100\pi t+80\sin 300\pi t+40\sin 500\pi t$ volts

Determine (a) which harmonics are present, (b) the r.m.s. value of the fundamental, (c) the frequency of the fundamental, (d) the periodic time of the fundamental, (e) the frequencies of the harmonics.

(a) The first term, or fundamental, $200\sin 100\pi t$, has $\omega_1 = 100\pi$ rad/s and maximum value 200 V. The second term, $80\sin 300\pi t$, has an angular velocity of 300π rad/s, which is **three times** that for the fundamental.

Hence, **$80\sin 300\pi t$ is the third harmonic term**.

Similarly, **$40\sin 500\pi t$ is the fifth harmonic term**.

(b) R.m.s. value of the fundamental $= 0.707 \times 200$ $= \mathbf{141.4\,V}$

(c) Frequency of fundamental,
$$f_1 = \frac{\omega_1}{2\pi} = \frac{100\pi}{2\pi} = \mathbf{50\,Hz}$$

(d) Periodic time of fundamental,
$$T = \frac{1}{f_1} = \frac{1}{50} = \mathbf{0.02\,s} \text{ or } \mathbf{20\,ms}$$

(e) Frequency of third harmonic $= 3 \times 50 = \mathbf{150\,Hz}$

$\left(\text{or} = \frac{300\pi}{2\pi} = 150\,\text{Hz}\right)$

Frequency of fifth harmonic $= 5 \times 50 = \mathbf{250\,Hz}$

Problem 2. A complex current wave is represented by:
$$i = 60\sin 240\pi t + 24\sin\left(480\pi t - \frac{\pi}{4}\right) + 15\sin\left(720\pi t + \frac{\pi}{3}\right)\text{mA}$$
Determine (a) the frequency of the fundamental (b) the percentage second harmonic, (c) the percentage third harmonic, (d) the r.m.s. value of the second harmonic, (e) the phase angles of the harmonic components and (f) mean value of the third harmonic.

Since 480 is twice 240, and 720 is three times 240, the harmonics present in the given wave are the **second and third**.

(a) Frequency of fundamental, $f_1 = \frac{240\pi}{2\pi} = \mathbf{120\,Hz}$

(b) Percentage second harmonic means expressing the maximum value of the second harmonic as a percentage of the maximum value of the fundamental,

i.e. percentage second harmonic $= \frac{24}{60} \times 100\%$ $= \mathbf{40\%}$

(c) Percentage third harmonic $= \frac{15}{60} \times 100\% = \mathbf{25\%}$

(d) R.m.s. value of second harmonic $= 0.707 \times 24$ $= \mathbf{16.97\,mA}$

(e) The second harmonic has a phase angle of $\frac{\pi}{4}$ **rad lagging** (or 45° lagging).

The third harmonic has a phase angle of $\frac{\pi}{3}$ **rad leading** (or 60° leading).

(f) Mean or average value of third harmonic $= 0.637 \times 15 = \mathbf{9.56\,mA}$

Now try the following Practice Exercise

Practice Exercise 148 The equation of a complex waveform (Answers on page 830)

1. A complex voltage wave is given by:
$$v = 150\sin 200\pi t + 60\sin 400\pi t + 30\sin 800\pi t \text{ volts}$$
Determine (a) which harmonics are present, (b) the r.m.s. value of the fundamental, (c) the frequency of the fundamental, (d) the periodic time of the fundamental and (e) the frequencies of the harmonics.
2. A complex current wave is represented by:
$$i = 20\sin 160\pi t + 8\sin\left(480\pi t + \frac{\pi}{2}\right) + 2\sin\left(800\pi t - \frac{\pi}{5}\right)\text{A}$$
Determine (a) the frequency of the fundamental, (b) the percentage third harmonic, (c) the percentage fifth harmonic, (d) the r.m.s. value of the third harmonic, (e) the phase angles of the harmonic components and (f) the mean value of the fifth harmonic.
3. A complex waveform comprises a fundamental voltage with a peak value of 30 V and a frequency of 400 Hz together with a third harmonic having a peak value of 12 V leading by 60°. Write down an expression for the instantaneous value of the complex voltage.

39.3 Harmonic synthesis

Harmonic analysis is the process of resolving a complex periodic waveform into a series of sinusoidal components of ascending order of frequency. Many of the waveforms met in practice can be represented by mathematical expressions similar to those of equations (1) and (2), and the magnitude of their harmonic components together with their phase may be calculated using **Fourier series** (see Section 39.4 and Section 39.5). **Numerical methods** are used to analyse waveforms for which simple mathematical expressions cannot be obtained. A numerical method of harmonic analysis is explained in Chapter 40. In a laboratory, waveform analysis may be performed using a **waveform analyser** which produces a direct readout of the component waves present in a complex wave.

By adding the instantaneous values of the fundamental and progressive harmonics of a complex wave for given instants in time, the shape of a complex waveform can be gradually built up. This graphical procedure is known as **harmonic synthesis** (synthesis meaning 'the putting together of parts or elements so as to make up a complex whole').

A number of examples of harmonic synthesis will now be considered.

Example 1

Consider the complex voltage expression given by

$$v_a = 100\sin\omega t + 30\sin 3\omega t \text{ volts}$$

The waveform is made up of a fundamental wave of maximum value 100 V and frequency, $f = \omega/2\pi$ hertz and a third harmonic component of maximum value 30 V and frequency $= 3\omega/2\pi (= 3f)$, the fundamental and third harmonics being initially in phase with each other. Since the maximum value of the third harmonic is 30 V and that of the fundamental is 100 V, the resultant waveform v_a is said to contain 30/100, i.e. '30% third harmonic'. In Figure 39.2, the fundamental waveform is shown by the broken line plotted over one cycle, the periodic time being $2\pi/\omega$ seconds. On the same axis is plotted $30\sin 3\omega t$, shown by the dotted line, having a maximum value of 30 V and for which three cycles are completed in time T seconds. At zero time, $30\sin 3\omega t$ is in phase with $100\sin\omega t$.

The fundamental and third harmonic are combined by adding ordinates at intervals to produce the waveform for v_a as shown. For example, at time $T/12$ seconds, the fundamental has a value of 50 V and the third harmonic a value of 30 V. Adding gives a value of 80 V for waveform v_a, at time $T/12$ seconds. Similarly, at time $T/4$ seconds, the fundamental has a value of 100 V and the third harmonic a value of −30 V. After addition, the resultant waveform v_a is 70 V at time $T/4$. The procedure is continued between $t = 0$ and $t = T$ to produce the complex waveform for v_a. The negative half-cycle

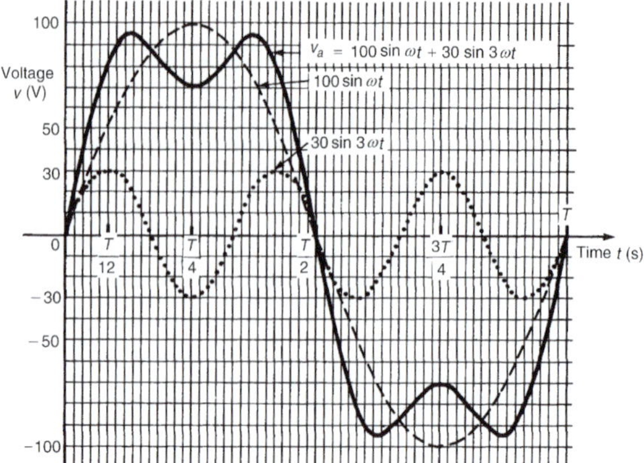

Figure 39.2

of waveform v_a is seen to be identical in shape to the positive half-cycle.

Example 2

Consider the addition of a fifth harmonic component to the complex waveform of Figure 39.2, giving a resultant waveform expression

$$v_b = 100\sin\omega t + 30\sin 3\omega t + 20\sin 5\omega t \text{ volts}$$

Figure 39.3 shows the effect of adding $(100\sin\omega t + 30\sin 3\omega t)$ obtained from Figure 39.2 to $20\sin 5\omega t$. The shapes of the negative and positive half-cycles are still identical. If further odd harmonics of the appropriate amplitude and phase were added to v_b, a good approximation to **a square wave** would result.

Example 3

Consider the complex voltage expression given by

$$v_c = 100\sin\omega t + 30\sin\left(3\omega t + \frac{\pi}{2}\right) \text{ volts}$$

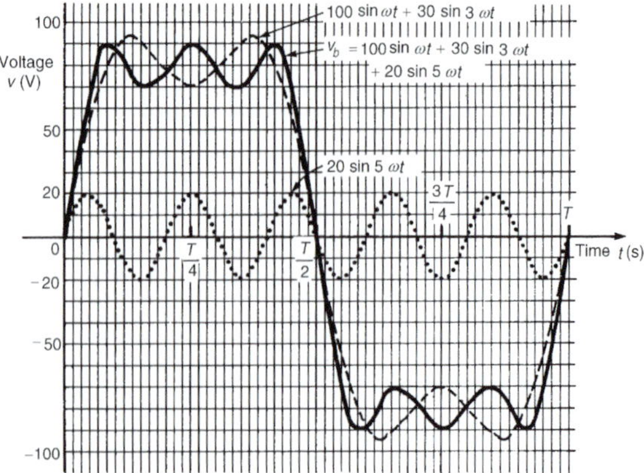

Figure 39.3

Part 4

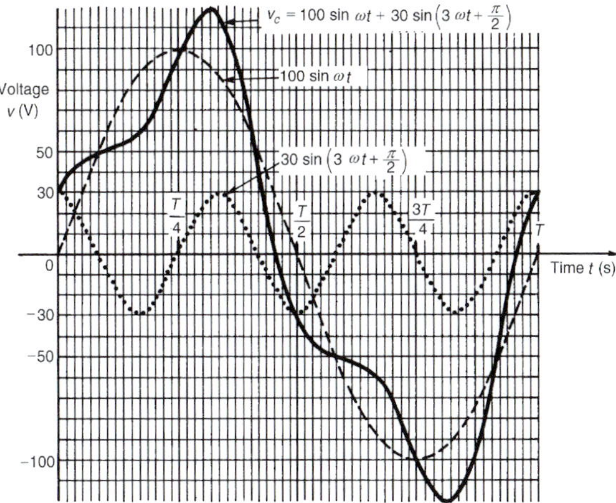

Figure 39.4

This expression is similar to voltage v_a in that the peak value of the fundamental and third harmonic are the same. However, the third harmonic has a phase displacement of $\pi/2$ radian leading (i.e. leading $30\sin 3\omega t$ by $\pi/2$ radian). Note that, since the periodic time of the fundamental is T seconds, the periodic time of the third harmonic is $T/3$ seconds, and a phase displacement of $\pi/2$ radian or $\frac{1}{4}$ cycle of the third harmonic represents a time interval of $(T/3)\div 4$, i.e. $T/12$ seconds.

Figure 39.4 shows graphs of $100\sin\omega t$ and $30\sin(3\omega t+\pi/2)$ over the time for one cycle of the fundamental. When ordinates of the two graphs are added at intervals, the resultant waveform v_c is as shown. The shape of the waveform v_c is quite different from that of waveform v_a shown in Figure 39.2, even though the percentage third harmonic is the same. If the negative half-cycle in Figure 39.4 is reversed it can be seen that the shape of the positive and negative half-cycles are identical.

Example 4

Consider the complex voltage expression given by

$$\boldsymbol{v_d = 100\sin\omega t + 30\sin\left(3\omega t - \frac{\pi}{2}\right)\ \text{volts}}$$

The fundamental, $100\sin\omega t$, and the third harmonic component, $30\sin(3\omega t - \pi/2)$, are plotted in Figure 39.5, the latter lagging $30\sin 3\omega t$ by $\pi/2$ radian or $T/12$ seconds. Adding ordinates at intervals gives the resultant waveform v_d as shown. The negative half-cycle of v_d is identical in shape to the positive half-cycle.

Example 5

Consider the complex voltage expression given by

$$\boldsymbol{v_e = 100\sin\omega t + 30\sin(3\omega t + \pi)\ \text{volts}}$$

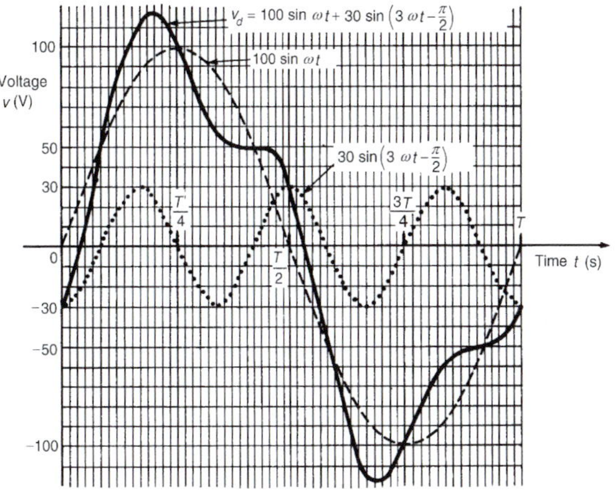

Figure 39.5

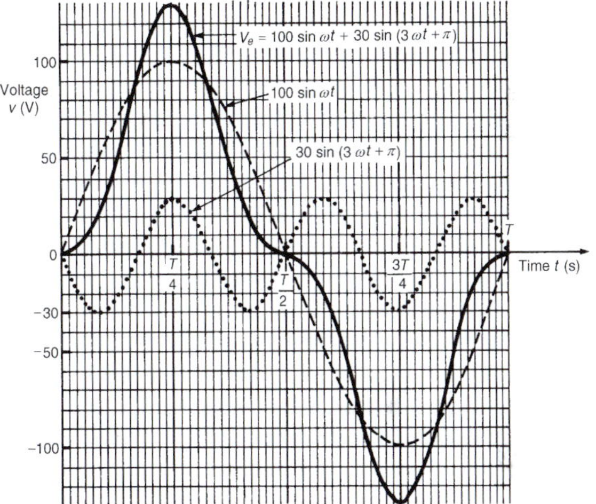

Figure 39.6

The fundamental, $100\sin\omega t$, and the third harmonic component, $30\sin(3\omega t+\pi)$, are plotted as shown in Figure 39.6, the latter leading $30\sin 3\omega t$ by π radian or $T/6$ seconds. Adding ordinates at intervals gives the resultant waveform v_e as shown. The negative half-cycle of v_e is identical in shape to the positive half-cycle.

Example 6

Consider the complex voltage expression given by

$$\boldsymbol{v_f = 100\sin\omega t - 30\sin\left(3\omega t + \frac{\pi}{2}\right)\ \text{volts}}$$

The phasor representing $30\sin(3\omega t+\pi/2)$ is shown in Figure 39.7(a) at time $t=0$. The phasor representing $-30\sin(3\omega t+\pi/2)$ is shown in Figure 39.7(b) where it is seen to be in the opposite direction to that shown in Figure 39.7(a).

$-30\sin(3\omega t+\pi/2)$ is the same as $30\sin(3\omega t-\pi/2)$

Thus

$$v_f = 100\sin\omega t - 30\sin\left(3\omega t+\frac{\pi}{2}\right)$$
$$= 100\sin\omega t + 30\sin\left(3\omega t-\frac{\pi}{2}\right)$$

The waveform representing this expression has already been plotted in Figure 39.5.

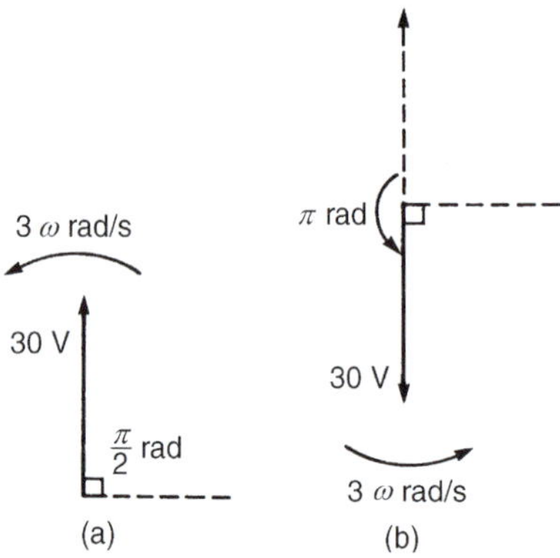

Figure 39.7

General conclusions on examples 1 to 6

Whenever odd harmonics are added to a fundamental waveform, whether initially in phase with each other or not, the positive and negative half-cycles of the resultant complex wave are identical in shape (i.e. in Figures 39.2 to 39.6, the values of voltage in the third quadrant – between $T/2$ seconds and $3T/4$ seconds – are identical to the voltage values in the first quadrant – between 0 and $T/4$ seconds, except that they are negative, and the values of voltage in the second and fourth quadrants are identical, except for the sign change. This is a feature of waveforms containing a fundamental and odd harmonics and is true whether harmonics are added or subtracted from the fundamental.

From Figures 39.2 to 39.6, it is seen that a waveform can change its shape considerably as a result of changes in both phase and magnitude of the harmonics.

Example 7

Consider the complex current expression given by

$$i_a = \mathbf{10\sin\omega t + 4\sin 2\omega t \text{ amperes}}$$

Current i_a consists of a fundamental component, $10\sin\omega t$, and a second harmonic component, $4\sin 2\omega t$, the components being initially in phase with each other. Current i_a contains 40% second harmonic. The fundamental and second harmonic are shown plotted separately in Figure 39.8. By adding ordinates at intervals, the complex waveform representing i_a is produced as shown. It is noted that if all the values in the negative half-cycle were reversed then this half-cycle would appear as a mirror image of the positive half-cycle about a vertical line drawn through time $t = T/2$

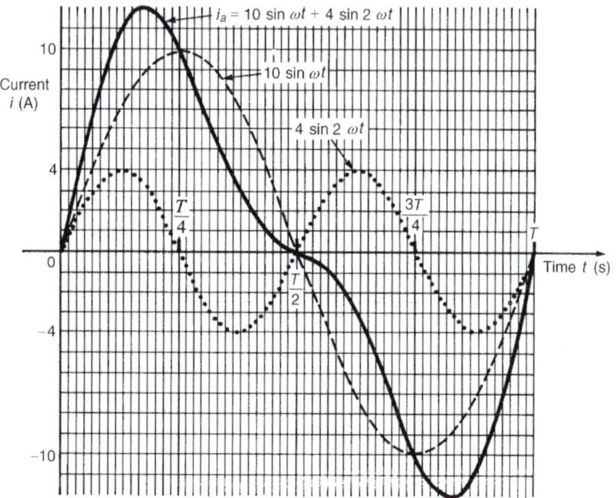

Figure 39.8

Example 8

Consider the complex current expression given by

$$i_b = \mathbf{10\sin\omega t + 4\sin 2\omega t + 3\sin 4\omega t \text{ amperes}}$$

The waveforms representing $(10\sin\omega t + 4\sin 2\omega t)$ and the fourth harmonic component, $3\sin 4\omega t$, are each shown separately in Figure 39.9, the former waveform having been produced in Figure 39.8. By adding ordinates at intervals, the complex waveform for i_b is produced as shown in Figure 39.9. If the half-cycle between times $T/2$ and T is reversed then it is seen to be a mirror image of the half-cycle lying between 0 and $T/2$ about a vertical line drawn through the time $t = T/2$

Example 9

Consider the complex current expressions given by

$$i_c = \mathbf{10\sin\omega t + 4\sin\left(2\omega t+\frac{\pi}{2}\right) \text{ amperes}}$$

The fundamental component, $10\sin\omega t$, and the second harmonic component, having an amplitude of 4 A

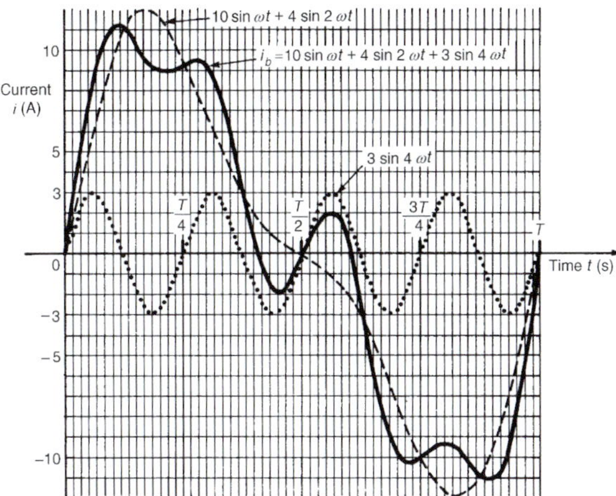

Figure 39.9

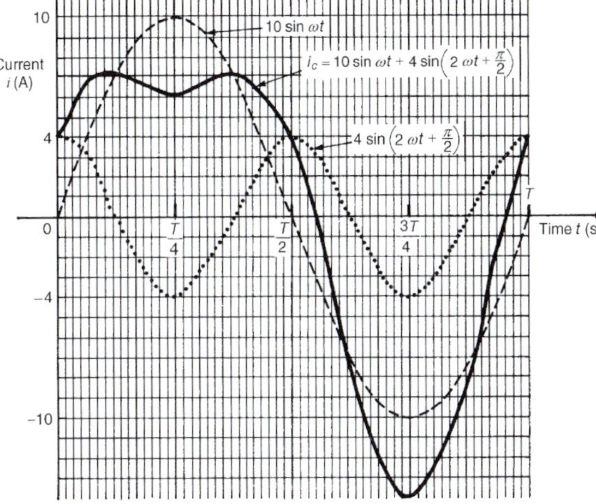

Figure 39.10

and a phase displacement of $\pi/2$ radian leading (i.e. leading $4\sin 2\omega t$ by $\pi/2$ radian or $T/8$ seconds), are shown plotted separately in Figure 39.10. By adding ordinates at intervals, the complex waveform for i_c is produced as shown. The positive and negative half-cycles of the resultant waveform i_c are seen to be quite dissimilar.

Example 10

Consider the complex current expression given by

$$i_d = \mathbf{10\sin\omega t + 4\sin(2\omega t + \pi)\ amperes}$$

The fundamental, $10\sin\omega t$, and the second harmonic component which leads $4\sin 2\omega t$ by π rad are shown separately in Figure 39.11. By adding ordinates at intervals, the resultant waveform i_d is produced as shown. If the negative half-cycle is reversed, it is seen to be a mirror image of the positive half-cycle about a line drawn vertically through time $t = T/2$

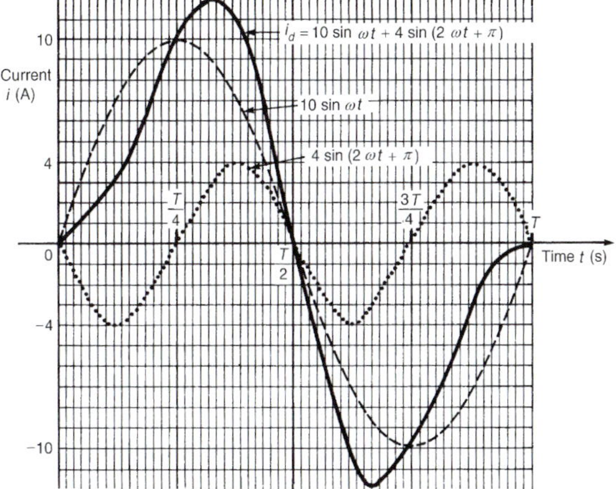

Figure 39.11

General conclusions on examples 7 to 10

Whenever even harmonics are added to a fundamental component:

(a) if the harmonics are initially in phase or if there is a phase shift of π rad, the negative half-cycle, when reversed, is a mirror image of the positive half-cycle about a vertical line drawn through time $t = T/2$;

(b) if the harmonics are initially out of phase with each other (i.e. other than π rad), the positive and negative half-cycles are dissimilar.

These are features of waveforms containing the fundamental and even harmonics.

Example 11

Consider the complex voltage expression given by

$$v_g = \mathbf{50\sin\omega t + 25\sin 2\omega t + 15\sin 3\omega t\ volts}$$

The fundamental and the second and third harmonics are each shown separately in Figure 39.12. By adding ordinates at intervals, the resultant waveform v_g is produced as shown. If the negative half-cycle is reversed, it appears as a mirror image of the positive half-cycle about a vertical line drawn through time $= T/2$

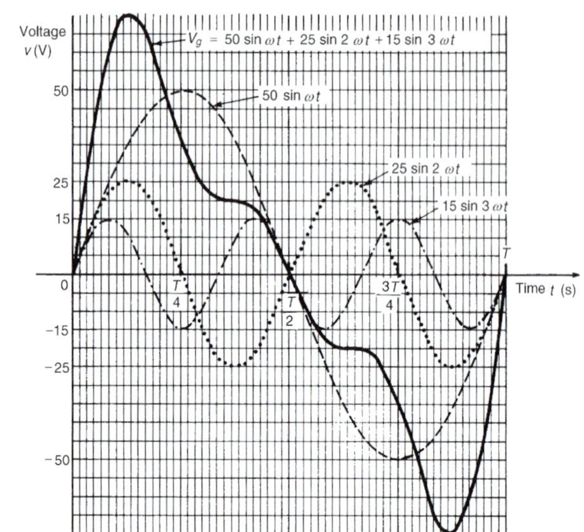

Figure 39.12

Example 12

Consider the complex voltage expression given by

$$v_h = 50 \sin \omega t + 25 \sin(2\omega t - \pi) + 15 \sin\left(3\omega t + \frac{\pi}{2}\right) \text{ volts}$$

The fundamental, the second harmonic lagging by π radian and the third harmonic leading by $\pi/2$ radian are initially plotted separately, as shown in Figure 39.13. Adding ordinates at intervals gives the resultant waveform v_h as shown. The positive and negative half-cycles are seen to be quite dissimilar.

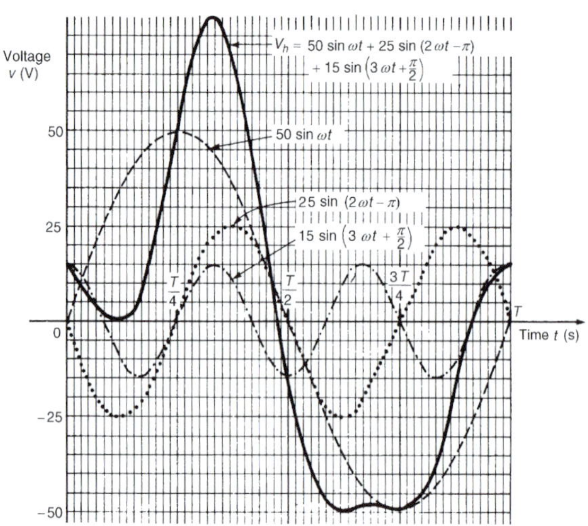

Figure 39.13

General conclusions on examples 11 and 12

Whenever a waveform contains both odd and even harmonics:

(a) if the harmonics are initially in phase with each other, the negative cycle, when reversed, is a mirror image of the positive half-cycle about a vertical line drawn through time $t = T/2$;

(b) if the harmonics are initially out of phase with each other, the positive and negative half-cycles are dissimilar.

Example 13

Consider the complex current expression given by

$$i = 32 + 50 \sin \omega t + 20 \sin\left(2\omega t - \frac{\pi}{2}\right) \text{ mA}$$

The current i comprises three components – a 32 mA d.c. component, a fundamental of amplitude 50 mA and a second harmonic of amplitude 20 mA, lagging by $\pi/2$ radian. The fundamental and second harmonic are shown separately in Figure 39.14. Adding ordinates at intervals gives the complex waveform $50 \sin \omega t + 20 \sin(2\omega t - \pi/2)$

This waveform is then added to the 32 mA d.c. component to produce the waveform i as shown. The effect of the d.c. component is seen to be to shift the whole wave 32 mA upward. The waveform approaches that expected from a **half-wave rectifier** (see Section 39.11).

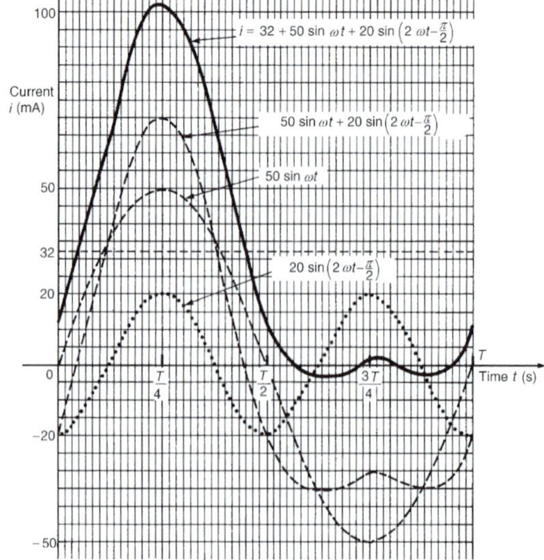

Figure 39.14

Problem 3. A complex waveform v comprises a fundamental voltage of 240 V r.m.s. and frequency 50 Hz, together with a 20% third harmonic which has a phase angle lagging by $3\pi/4$ rad at time = 0 (a) Write down an expression to represent voltage v. (b) Use harmonic synthesis to sketch the complex waveform representing voltage v over one cycle of the fundamental component.

(a) A fundamental voltage having an r.m.s. value of 240 V has a maximum value, or amplitude, of $(\sqrt{2})(240)$, i.e. 339.4 V

If the fundamental frequency is 50 Hz then angular velocity, $\omega = 2\pi f = 2\pi(50) = 100\pi$ rad/s. Hence the fundamental voltage is represented by $339.4 \sin 100\pi t$ volts. Since the fundamental frequency is 50 Hz, the time for one cycle of the fundamental is given by $T = 1/f = 1/50$ s or 20 ms.

The third harmonic has an amplitude equal to 20% of 339.4 V, i.e. 67.9 V. The frequency of the third harmonic component is $3 \times 50 = 150$ Hz, thus the angular velocity is 2π (150), i.e. 300π rad/s. Hence the third harmonic voltage is represented by $67.9 \sin(300\pi t - 3\pi/4)$ volts. Thus

$$\textbf{voltage, } v = \mathbf{339.4 \sin 100\pi t + 67.9 \sin\left(300\pi t - \frac{3\pi}{4}\right) \textbf{ volts}}$$

(b) One cycle of the fundamental, $339.4 \sin 100\pi t$, is shown sketched in Figure 39.15, together with three cycles of the third harmonic component, $67.9 \sin(300\pi t - 3\pi/4)$ initially lagging by $3\pi/4$ rad. By adding ordinates at intervals, the complex waveform representing voltage is produced as shown. If the negative half-cycle is reversed, it is seen to be identical to the positive half-cycle, which is a feature of waveforms containing the fundamental and odd harmonics.

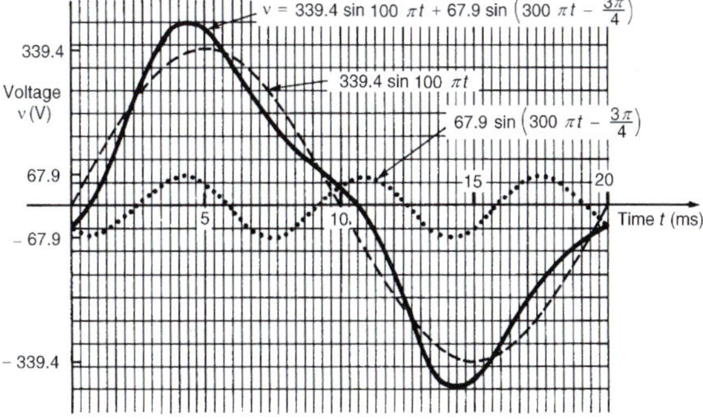

Figure 39.15

Problem 4. For each of the periodic complex waveforms shown in Figure 39.16, suggest whether odd or even harmonics (or both) are likely to be present.

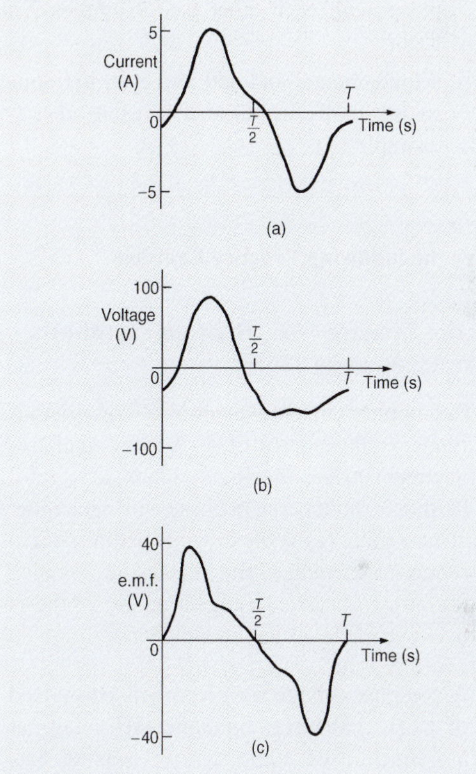

Figure 39.16

(a) If in Figure 39.16(a) the negative half-cycle is reversed, it is seen to be identical to the positive half-cycle. This feature indicates that the complex current waveform is composed of a fundamental and odd harmonics only (see examples 1 to 6).

(b) In Figure 39.16(b) the negative half-cycle is quite dissimilar to the positive half-cycle. This indicates that the complex voltage waveform comprises either

(i) a fundamental and even harmonics, initially out of phase with each other (see example 9), or

(ii) a fundamental and odd and even harmonics, one or more of the harmonics being initially out of phase (see example 12).

(c) If in Figure 39.16(c) the negative half-cycle is reversed, it is seen to be a mirror image of the positive half-cycle about a vertical line drawn through time $T/2$. This feature indicates that the complex e.m.f. waveform comprises either:

(i) a fundamental and even harmonics initially in phase with each other (see examples 7 and 8), or

(ii) a fundamental and odd and even harmonics, each initially in phase with each other (see example 11).

Now try the following Practice Exercise

Practice Exercise 149 Harmonic synthesis (Answers on page 830)

1. A complex current waveform i comprises a fundamental current of 50 A r.m.s. and frequency 100 Hz, together with a 24% third harmonic, both being in phase with each other at zero time. (a) Write down an expression to represent current, i. (b) Sketch the complex waveform of current using harmonic synthesis over one cycle of the fundamental.

2. A complex voltage waveform v is comprised of a 212.1 V r.m.s. fundamental voltage at a frequency of 50 Hz, a 30% second harmonic component lagging by $\pi/2$ rad, and a 10% fourth harmonic component leading by $\pi/3$ rad. (a) Write down an expression to represent voltage v. (b) Sketch the complex voltage waveform using harmonic synthesis over one cycle of the fundamental waveform.

3. A voltage waveform is represented by

$$v = 20 + 50\sin\omega t + 20\sin(2\omega t - \pi/2) \text{ volts}$$

Draw the complex waveform over one cycle of the fundamental by using harmonic synthesis.

4. Write down an expression representing a current having a fundamental component of amplitude 16 A and frequency 1 kHz, together with its third and fifth harmonics being respectively one-fifth and one-tenth the amplitude of the fundamental, all components being in phase at zero time. Sketch the complex current waveform for one cycle of the fundamental using harmonic synthesis.

5. For each of the waveforms shown in Figure 39.17, state which harmonics are likely to be present.

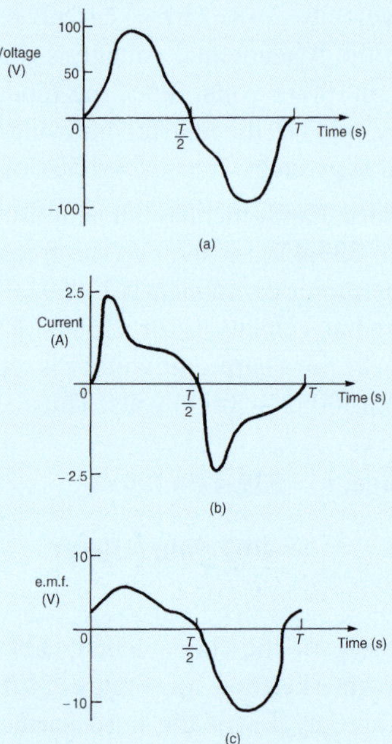

Figure 39.17

6. A voltage waveform is described by

$$v = 200\sin 377t + 80\sin(1131t + (\pi/4)) + 20\sin(1885t - (\pi/3)) \text{ volts}$$

Determine (a) the fundamental and harmonic frequencies of the waveform, (b) the percentage third harmonic and (c) the percentage fifth harmonic. Sketch the voltage waveform using harmonic synthesis over one cycle of the fundamental.

Part 4

39.4 Fourier series of periodic and non-periodic functions

Fourier* **series** provides a method of analysing periodic functions into their constituent components. Alternating currents and voltages, displacement, velocity and acceleration of slider-crank mechanisms and acoustic waves are typical practical examples in engineering and science where periodic functions are involved and often requiring analysis. (The main topics of Fourier series are covered here, but for more, see *Higher Engineering Mathematics 8th Edition*.)

Periodic functions

As stated earlier in this chapter, a function $f(x)$ is said to be **periodic** if $f(x+T)=f(x)$ for all values of x, where T is some positive number. T is the interval between two successive repetitions and is called the **period** of the functions $f(x)$.

For example, $y=\sin x$ is periodic in x with period 2π since $\sin x=\sin(x+2\pi)=\sin(x+4\pi)$, and so on. In general, if $y=\sin\omega t$ then the period of the waveform is $2\pi/\omega$. The function shown in Figure 39.18 is also periodic of period 2π and is defined by:

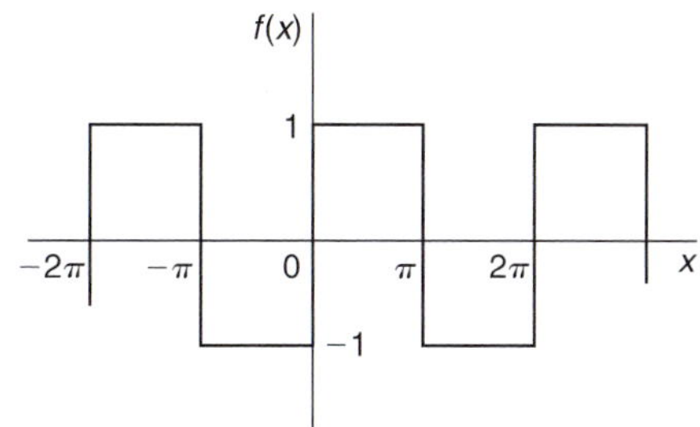

Figure 39.18

$$f(x)=\begin{cases}-1, & \text{when } -\pi<x<0\\ 1, & \text{when } \quad 0<x<\pi\end{cases}$$

If a graph of a function has no sudden jumps or breaks it is called a **continuous function**, examples being the graphs of sine and cosine functions. However, other graphs make finite jumps at a point or points in the interval. The square wave shown in Figure 39.18 has **finite discontinuities** at $x=\pi, 2\pi, 3\pi$, and so on. A great advantage of Fourier series over other series is that it can be applied to functions which are discontinuous as well as those which are continuous.

The basis of a Fourier series is that all functions of practical significance which are defined in the interval $-\pi\le x\le\pi$ can be expressed in terms of a **convergent trigonometric series** of the form:

$$f(x)=a_0+a_1\cos x+a_2\cos 2x+a_3\cos 3x+\cdots$$
$$+b_1\sin x+b_2\sin 2x+b_3\sin 3x+\cdots$$

when $a_0, a_1, a_2, \ldots b_1, b_2, \ldots$ are real constants,

i.e. $$\boldsymbol{f(x)=a_0+\sum_{n=1}^{\infty}(a_n\cos nx+b_n\sin nx)}, \qquad (3)$$

where for the range $-\pi$ to π:

$$\boldsymbol{a_0=\frac{1}{2\pi}\int_{-\pi}^{\pi}f(x)\,dx} \qquad (4)$$

$$\boldsymbol{a_n=\frac{1}{\pi}\int_{-\pi}^{\pi}f(x)\cos nx\,dx\ (n=1,2,3,\ldots)} \qquad (5)$$

and $$\boldsymbol{b_n=\frac{1}{\pi}\int_{-\pi}^{\pi}f(x)\sin nx\,dx\ (n=1,2,3,\ldots)} \qquad (6)$$

a_0, a_n and b_n are called the **Fourier coefficients** of the series and if these can be determined, the series of equation (3) is called the **Fourier series** corresponding to $f(x)$.

For the series of equation (3):

the term $(a_1\cos x+b_1\sin x)$ or $c_1\sin(x+\alpha_1)$ is called the **first harmonic** or the **fundamental**, the

***Who was** Fourier? **Jean Baptiste Joseph Fourier** (21 March 1768–16 May 1830) was a French mathematician and physicist best known for initiating the investigation of **Fourier series** and their applications to problems of heat transfer and vibrations. To find out more go to **www.routledge.com/cw/bird**

term $(a_2 \cos 2x + b_2 \sin 2x)$ or $c_2 \sin(2x + \alpha_2)$ is called the **second harmonic**, and so on.

For an exact representation of a complex wave, an infinite number of terms are, in general, required. In many practical cases, however, it is sufficient to take the first few terms only. Obtaining a Fourier series for a periodic function of period 2π is demonstrated in the following worked problems.

Problem 5. Obtain a Fourier series for the periodic function $f(x)$ defined as:

$$f(x) = \begin{cases} -k, & \text{when } -\pi < x < 0 \\ +k, & \text{when } \quad 0 < x < \pi \end{cases}$$

The function is periodic outside of this range with period 2π

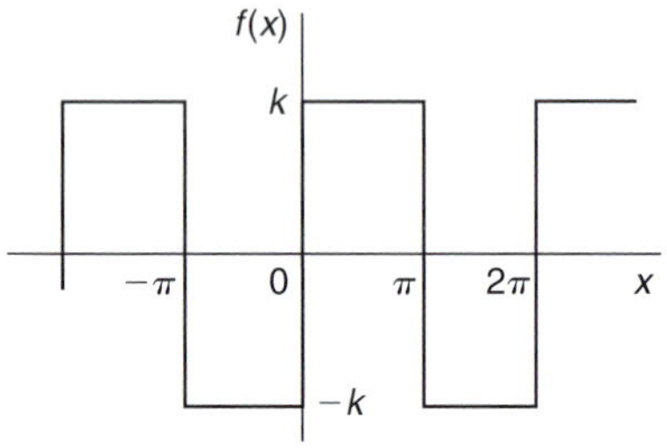

Figure 39.19

The square wave function defined is shown in Figure 39.19. Since $f(x)$ is given by two different expressions in the two halves of the range the integration is performed in two parts, one from $-\pi$ to 0 and the other from 0 to π

From equation (4):

$$a_0 = \frac{1}{2\pi}\int_{-\pi}^{\pi} f(x)\mathrm{d}x = \frac{1}{2\pi}\left[\int_{-\pi}^{0} -k\,\mathrm{d}x + \int_{0}^{\pi} k\,\mathrm{d}x\right]$$

$$= \frac{1}{2\pi}\{[-kx]_{-\pi}^{0} + [kx]_0^{\pi}\} = 0$$

[a_0 is in fact the **mean value** of the waveform over a complete period of 2π and this value could have been deduced on sight from Figure 39.19.]

From equation (5):

$$a_n = \frac{1}{\pi}\int_{-\pi}^{\pi} f(x)\cos nx\,\mathrm{d}x = \frac{1}{\pi}\left\{\int_{-\pi}^{0} -k\cos nx\,\mathrm{d}x + \int_{0}^{\pi} k\cos nx\,\mathrm{d}x\right\}$$

$$= \frac{1}{\pi}\left\{\left[\frac{-k\sin nx}{n}\right]_{-\pi}^{0} + \left[\frac{k\sin nx}{n}\right]_0^{\pi}\right\} = 0$$

Hence $a_1, a_2, a_3, \ldots$ are all zero (since $\sin 0 = \sin(-n\pi) = \sin n\pi = 0$), and therefore **no cosine terms** will appear in the Fourier series.

From equation (6):

$$b_n = \frac{1}{\pi}\int_{-\pi}^{\pi} f(x)\sin nx\,dx = \frac{1}{\pi}\left\{\int_{-\pi}^{0} -k\sin nx\,\mathrm{d}x + \int_{0}^{\pi} k\sin nx\,\mathrm{d}x\right\}$$

$$= \frac{1}{\pi}\left\{\left[\frac{k\cos nx}{n}\right]_{-\pi}^{0} + \left[\frac{-k\cos nx}{n}\right]_0^{\pi}\right\}$$

When n is odd:

$$b_n = \frac{k}{\pi n}\{[(1) - (-1)] + [-(-1) - (-1)]\}$$

$$= \frac{k}{\pi n}\{2 + 2\} = \frac{4k}{n\pi}$$

Hence, $b_1 = \dfrac{4k}{\pi}$, $b_3 = \dfrac{4k}{3\pi}$, $b_5 = \dfrac{4k}{5\pi}$, and so on

When n is even:

$$b_n = \frac{k}{\pi n}\{[1 - 1] + [-1 - (-1)]\} = 0$$

Hence, from equation (3), the Fourier series for the function shown in Figure 39.19 is given by:

$$f(x) = a_0 + \sum_{n=1}^{\infty}(a_n \cos nx + b_n \sin nx)$$

$$= 0 + \sum_{n=1}^{\infty}(0 + b_n \sin nx)$$

i.e. $f(x) = \dfrac{4k}{\pi}\sin x + \dfrac{4k}{3\pi}\sin 3x + \dfrac{4k}{5\pi}\sin 5x + \cdots\cdots$.

i.e. $\boldsymbol{f(x) = \dfrac{4k}{\pi}\left(\sin x + \dfrac{1}{3}\sin 3x + \dfrac{1}{5}\sin 5x + \cdots\cdots\right)}$

Problem 6. For the Fourier series of Problem 5, let $k = \pi$. Show by plotting the first three partial sums of this Fourier series that, as the series is added together term by term, the result approximates more and more closely to the function it represents.

Part 4

If $k=\pi$ in the above Fourier series, then

$$f(x)=4\left(\sin x+\frac{1}{3}\sin 3x+\frac{1}{5}\sin 5x+\cdots\cdots\right)$$

$4\sin x$ is termed the first partial sum of the Fourier series of $f(x)$,

$\left(4\sin x+\frac{4}{3}\sin 3x\right)$ is termed the second partial sum of the Fourier series, and

$\left(4\sin x+\frac{4}{3}\sin 3x+\frac{4}{5}\sin 5x\right)$ is termed the third partial sum, and so on.

Let $P_1=4\sin x$, $P_2=\left(4\sin x+\frac{4}{3}\sin 3x\right)$ and

$$P_3=\left(4\sin x+\frac{4}{3}\sin 3x+\frac{4}{5}\sin 5x\right).$$

Graphs of P_1, P_2 and P_3, obtained by drawing up tables of values, and adding waveforms, are shown in Figures 39.20(a) to (c) and they show that the series is convergent, i.e. continually approximating towards a definite limit as more and more partial sums are taken, and in the limit will have the sum $f(x)=\pi$

Even with just three partial sums, the waveform is starting to approach the **rectangular wave** the Fourier series is representing.

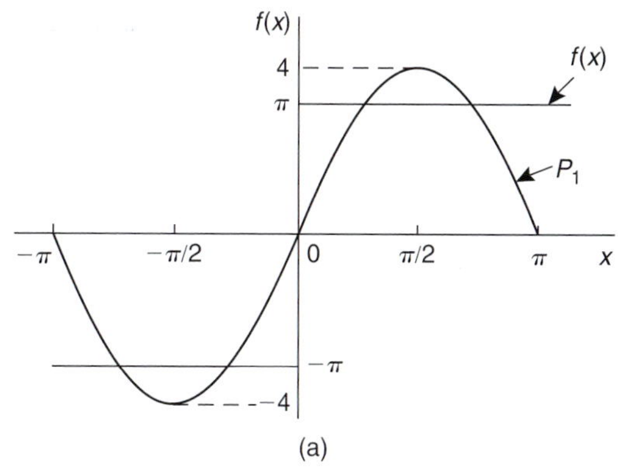

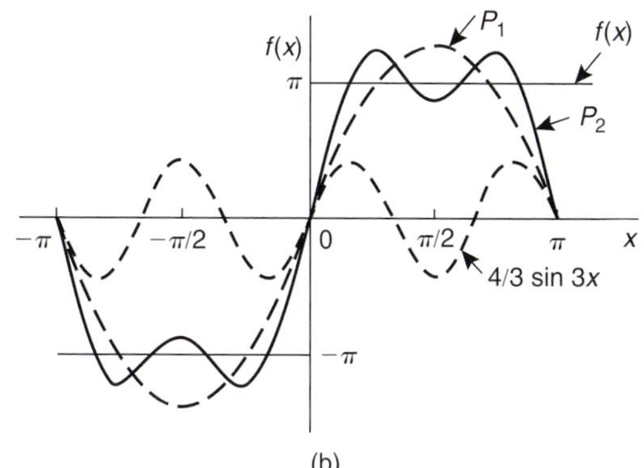

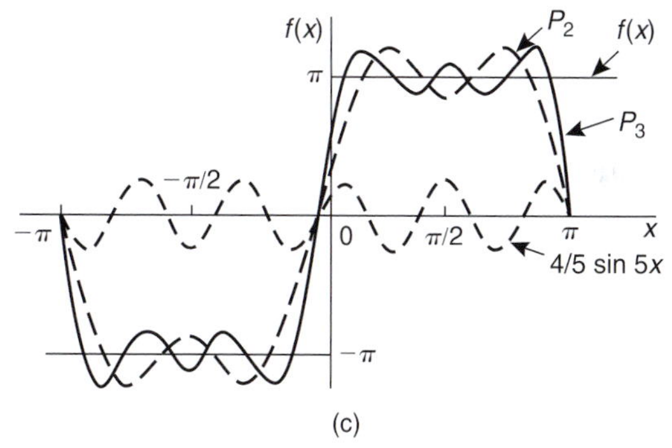

Figure 39.20

Expansion of non-periodic functions

If a function $f(x)$ is not periodic then it cannot be expanded in a Fourier series for **all** values of x. However, it is possible to determine a Fourier series to represent the function over any range of width 2π

Given a non-periodic function, a new function may be constructed by taking the values of $f(x)$ in the given range and then repeating them outside of the given range at intervals of 2π. Since this new function is, by construction, periodic with period 2π, it may then be expanded in a Fourier series for all values of x. For example, the function $f(x)=x$ is not a periodic function. However, if a Fourier series for $f(x)=x$ is required then the function is constructed outside of this range so that it is periodic with period 2π as shown by the broken lines in Figure 39.21.

For non-periodic functions, such as $f(x)=x$, the sum of the Fourier series is equal to $f(x)$ at all points in the given range but it is not equal to $f(x)$ at points outside of the range.

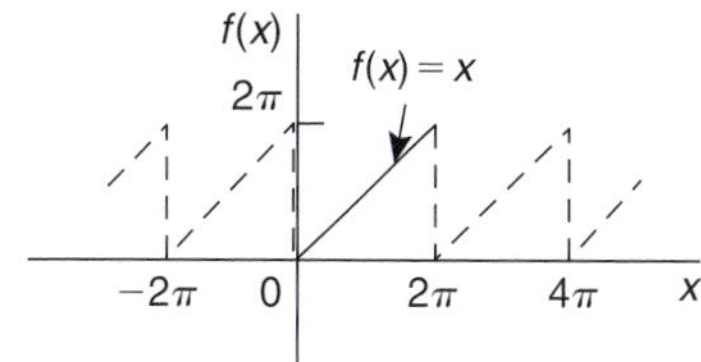

Figure 39.21

For determining a Fourier series of a non-periodic function over a range 2π, exactly the same formulae for the Fourier coefficients are used as previously, i.e. equations (4) to (6).

Part 4

Problem 7. Determine the Fourier series to represent the function $f(x) = 2x$ in the range $-\pi$ to $+\pi$

The function $f(x) = 2x$ is not periodic. The function is shown in the range $-\pi$ to π in Figure 39.22 and is then constructed outside of that range so that it is periodic of period 2π (see broken lines) with the resulting saw-tooth waveform.

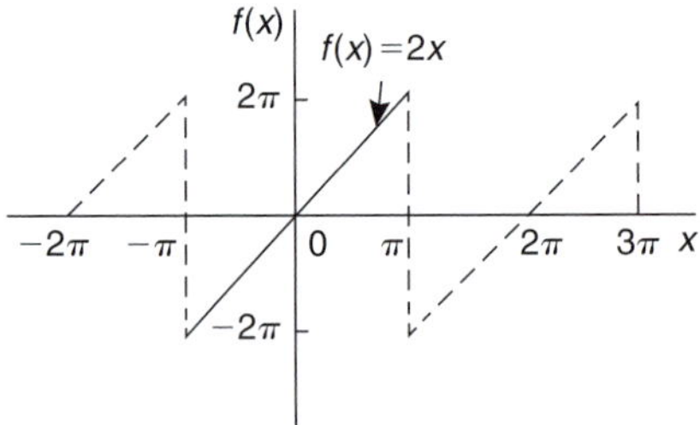

Figure 39.22

For a Fourier series:

$$f(x) = a_0 + \sum_{n=1}^{\infty}(a_n \cos nx + b_n \sin nx)$$

From equation (4),

$$a_0 = \frac{1}{2\pi}\int_{-\pi}^{\pi} f(x)\mathrm{d}x = \frac{1}{2\pi}\int_{-\pi}^{\pi} 2x\,\mathrm{d}x = \frac{2}{2\pi}\left[\frac{x^2}{2}\right]_{-\pi}^{\pi}$$
$$= 0 \text{ (i.e. the mean value)}$$

From equation (5),

$$a_n = \frac{1}{\pi}\int_{-\pi}^{\pi} f(x)\cos nx\,\mathrm{d}x = \frac{1}{\pi}\int_{-\pi}^{\pi} 2x\cos nx\,\mathrm{d}x$$
$$= \frac{2}{\pi}\left[\frac{x\sin nx}{n} - \int \frac{\sin nx}{n}\mathrm{d}x\right]_{-\pi}^{\pi}$$

by integration by parts

$$= \frac{2}{\pi}\left[\frac{x\sin nx}{n} + \frac{\cos nx}{n^2}\right]_{-\pi}^{\pi}$$
$$= \frac{2}{\pi}\left[\left(0 + \frac{\cos n\pi}{n^2}\right) - \left(0 + \frac{\cos n(-\pi)}{n^2}\right)\right] = 0$$

since $\cos n\pi = \cos(-n\pi)$

From equation (6),

$$b_n = \frac{1}{\pi}\int_{-\pi}^{\pi} f(x)\sin nx\,\mathrm{d}x = \frac{1}{\pi}\int_{-\pi}^{\pi} 2x\sin nx\,\mathrm{d}x$$
$$= \frac{2}{\pi}\left[\frac{-x\cos nx}{n} - \int\left(\frac{-\cos nx}{n}\right)\mathrm{d}x\right]_{-\pi}^{\pi} \quad \text{by parts}$$
$$= \frac{2}{\pi}\left[\frac{-x\cos nx}{n} + \frac{\sin nx}{n^2}\right]_{-\pi}^{\pi}$$
$$= \frac{2}{\pi}\left[\left(\frac{-\pi\cos n\pi}{n} + \frac{\sin n\pi}{n^2}\right) - \left(\frac{-(-\pi)\cos n(-\pi)}{n} + \frac{\sin n(-\pi)}{n^2}\right)\right]$$
$$= \frac{2}{\pi}\left[\frac{-\pi\cos n\pi}{n} - \frac{\pi\cos(-n\pi)}{n}\right]$$
$$= \frac{-4}{n}\cos n\pi \quad \text{since } \cos(-n\pi) = \cos n\pi$$

When n is odd, $b_n = \frac{4}{n}$. Thus, $b_1 = 4$, $b_3 = \frac{4}{3}$, $b_5 = \frac{4}{5}$, and so on.

When n is even, $b_n = \frac{-4}{n}$. Thus $b_2 = -\frac{4}{2}$, $b_4 = -\frac{4}{4}$, $b_6 = -\frac{4}{6}$, and so on.

Thus, $f(x) = 2x = 4\sin x - \frac{4}{2}\sin 2x + \frac{4}{3}\sin 3x - \frac{4}{4}\sin 4x + \frac{4}{5}\sin 5x - \frac{4}{6}\sin 6x + \cdots\cdots$

i.e. $$\mathbf{2x = 4\left(\sin x - \frac{1}{2}\sin 2x + \frac{1}{3}\sin 3x - \frac{1}{4}\sin 4x + \frac{1}{5}\sin 5x - \frac{1}{6}\sin 6x + \cdots\right)}$$

for values of $f(x)$ between $-\pi$ and π

For values of $f(x)$ outside the range $-\pi$ to $+\pi$ the sum of the series is not equal to $f(x)$.

Problem 8. Obtain a Fourier series for the function defined by:

$$f(x) = \begin{cases} x, & \text{when } 0 < x < \pi \\ 0, & \text{when } \pi < x < 2\pi \end{cases}$$

The defined function is shown in Figure 39.23 between 0 and 2π. The function is constructed outside of this range so that it is periodic of period 2π, as shown by the broken line in Figure 39.23.

For a Fourier series:

$$f(x) = a_0 + \sum_{n=1}^{\infty}(a_n \cos nx + b_n \sin nx)$$

It is more convenient in this case to take the limits from 0 to 2π instead of from $-\pi$ to $+\pi$. The value of the Fourier coefficients are unaltered by this change of limits. Hence,

Part 4

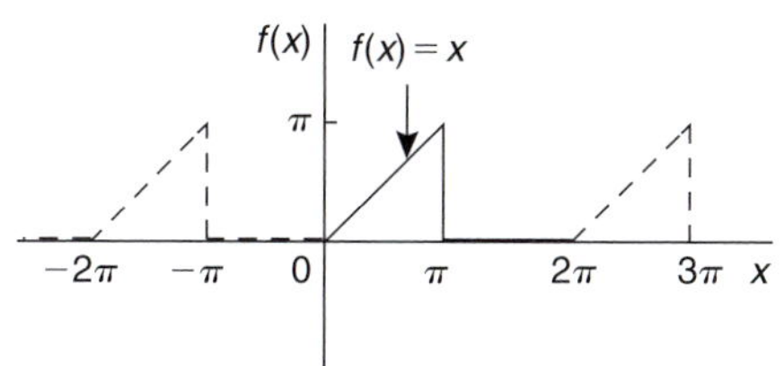

Figure 39.23

$$a_0 = \frac{1}{2\pi}\int_0^{2\pi} f(x)\mathrm{d}x = \frac{1}{2\pi}\left[\int_0^{\pi} x\,\mathrm{d}x + \int_{\pi}^{2\pi} 0\,\mathrm{d}x\right]$$

$$= \frac{1}{2\pi}\left[\frac{x^2}{2}\right]_0^{\pi} = \frac{1}{2\pi}\left(\frac{\pi^2}{2}\right) = \frac{\pi}{4}$$

$$a_n = \frac{1}{\pi}\int_0^{2\pi} f(x)\cos nx\,\mathrm{d}x$$

$$= \frac{1}{\pi}\left[\int_0^{\pi} x\cos nx\,\mathrm{d}x + \int_{\pi}^{2\pi} 0\,\mathrm{d}x\right]$$

$$= \frac{1}{\pi}\left[\frac{x\sin nx}{n} + \frac{\cos nx}{n^2}\right]_0^{\pi}$$

(from Problem 7, by parts)

$$= \frac{1}{\pi}\left\{\left[\frac{\pi\sin n\pi}{n} + \frac{\cos n\pi}{n^2}\right] - \left[0 + \frac{\cos 0}{n^2}\right]\right\}$$

$$= \frac{1}{\pi n^2}(\cos n\pi - 1)$$

When n is even, $a_n = 0$

When n is odd, $a_n = \frac{-2}{\pi n^2}$. Hence, $a_1 = \frac{-2}{\pi}$, $a_3 = \frac{-2}{3^2\pi}$, $a_5 = \frac{-2}{5^2\pi}$, and so on

$$b_n = \frac{1}{\pi}\int_0^{2\pi} f(x)\sin nx\,\mathrm{d}x$$

$$= \frac{1}{\pi}\left[\int_0^{\pi} x\sin nx\,\mathrm{d}x - \int_{\pi}^{2\pi} 0\,\mathrm{d}x\right]$$

$$= \frac{1}{\pi}\left[\frac{-x\cos nx}{n} + \frac{\sin nx}{n^2}\right]_0^{\pi}$$

(from Problem 7, by parts)

$$= \frac{1}{\pi}\left\{\left[\frac{-\pi\cos n\pi}{n} + \frac{\sin n\pi}{n^2}\right] - \left[0 + \frac{\sin 0}{n^2}\right]\right\}$$

$$= \frac{1}{\pi}\left[\frac{-\pi\cos n\pi}{n}\right] = \frac{-\cos n\pi}{n}$$

Hence $b_1 = -\cos\pi = 1$, $b_2 = -\frac{1}{2}$, $b_3 = \frac{1}{3}$, and so on.

Thus the Fourier series is:

$$f(x) = a_0 + \sum_{n=1}^{\infty}(a_n\cos nx + b_n\sin nx)$$

i.e. $f(x) = \frac{\pi}{4} - \frac{2}{\pi}\cos x - \frac{2}{3^2\pi}\cos 3x - \frac{2}{5^2\pi}\cos 5x - \cdots + \sin x - \frac{1}{2}\sin 2x + \frac{1}{3}\sin 3x - \cdots$

i.e. $\boldsymbol{f(x) = \frac{\pi}{4} - \frac{2}{\pi}\left(\cos x + \frac{\cos 3x}{3^2} + \frac{\cos 5x}{5^2} + \cdots\right) + \left(\sin x - \frac{1}{2}\sin 2x + \frac{1}{3}\sin 3x - \cdots\right)}$

Now try the following Practice Exercise

Practice Exercise 150 Fourier series of periodic and non-periodic functions (Answers on page 830)

1. Determine the Fourier series for the periodic function: $f(x) = \begin{cases} -2, \text{ when } & -\pi < x < 0 \\ +2, \text{ when } & 0 < x < \pi \end{cases}$ which is periodic outside this range of period 2π

2. Find the term representing the third harmonic for the periodic function of period 2π given by:
$$f(x) = \begin{cases} 0, & \text{when } -\pi < x < 0 \\ 1, & \text{when } 0 < x < \pi \end{cases}$$

3. Determine the Fourier series for the periodic function of period 2π defined by:
$$f(t) = \begin{cases} 0, & \text{when } -\pi < t < 0 \\ 1, & \text{when } 0 < t < \frac{\pi}{2} \\ -1, & \text{when } \frac{\pi}{2} < t < \pi \end{cases}$$
The function has a period of 2π

4. Show that the Fourier series for the function $f(x) = x$ over the range $x = 0$ to $x = 2\pi$ is given by:
$$f(x) = \pi - 2\left(\sin x + \frac{1}{2}\sin 2x + \frac{1}{3}\sin 3x + \frac{1}{4}\sin 4x + \cdots\right)$$

Part 4

5. Determine the Fourier series up to and including the third harmonic for the function defined by:

$$f(x) = \begin{cases} x, & \text{when } 0 \le x \le \pi \\ 2\pi - x, & \text{when } \pi \le x \le 2\pi \end{cases}$$

Sketch a graph of the function within and outside of the given range, assuming the period is 2π

6. Find the Fourier series for the function $f(x) = x + \pi$ within the range $-\pi < x < \pi$

39.5 Even and odd functions and Fourier series over any range

Even functions

A function $y = f(x)$ is said to be **even** if $f(-x) = f(x)$ for all values of x. Graphs of even functions are always **symmetrical about the y-axis** (i.e. is a mirror image). $y = \cos x$ is a typical example.

Fourier cosine series

The Fourier series of an **even periodic** function $f(x)$ having period 2π contains **cosine terms only** (i.e. contains no sine terms) and may contain a constant term.

Hence, $$\boldsymbol{f(x) = a_0 + \sum_{n=1}^{\infty} a_n \cos nx} \qquad (7)$$

where $$\boldsymbol{a_0 = \frac{1}{2\pi}\int_{-\pi}^{\pi} f(x)\mathrm{d}x = \frac{1}{\pi}\int_0^{\pi} f(x)\mathrm{d}x} \qquad (8)$$

(due to symmetry)

and $$\boldsymbol{a_n} = \frac{1}{\pi}\int_{-\pi}^{\pi} f(x)\cos nx\,\mathrm{d}x$$

$$= \boldsymbol{\frac{2}{\pi}\int_0^{\pi} f(x)\cos nx\,\mathrm{d}x} \qquad (9)$$

Problem 9. Determine the Fourier series for the periodic function of period 2π defined by:

$$f(x) = \begin{cases} -2, \text{ when } -\pi < x < -\dfrac{\pi}{2} \\ 2, \text{ when } -\dfrac{\pi}{2} < x < \dfrac{\pi}{2} \\ -2, \text{ when } \dfrac{\pi}{2} < x < \pi \end{cases}$$

The square wave shown in Figure 39.24 is an **even function** since it is symmetrical about the $f(x)$ axis.

Hence from equation (7), the Fourier series is given by:

$$f(x) = a_0 + \sum_{n=1}^{\infty} a_n \cos nx$$

(i.e. the series contains no sine terms)

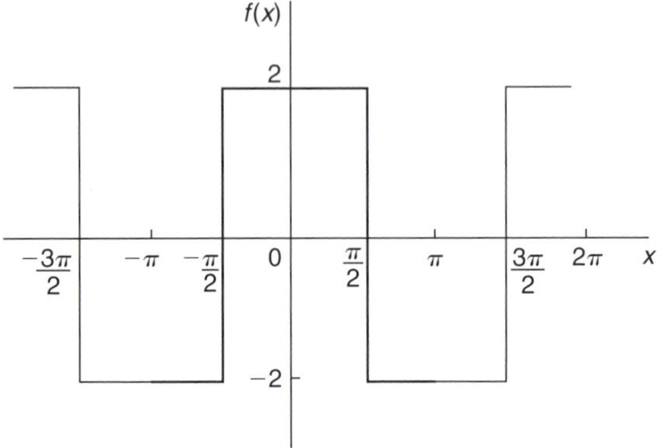

Figure 39.24

From equation (8),

$$a_0 = \frac{1}{\pi}\int_0^{\pi} f(x)\mathrm{d}x = \frac{1}{\pi}\left\{\int_0^{\pi/2} 2\mathrm{d}x + \int_{\pi/2}^{\pi} -2\mathrm{d}x\right\}$$

$$= \frac{1}{\pi}\{[2x]_0^{\pi/2} + [-2x]_{\pi 2}^{\pi}\}$$

$$= \frac{1}{\pi}\{(\pi) + [(-2\pi) - (-\pi)]\} = 0$$

(i.e. the mean value)

From equation (9),

$$a_n = \frac{2}{\pi}\int_0^{\pi} f(x)\cos nx\,\mathrm{d}x$$

$$= \frac{2}{\pi}\left\{\int_0^{\pi/2} 2\cos nx\,\mathrm{d}x + \int_{\pi/2}^{\pi} -2\cos nx\,\mathrm{d}x\right\}$$

$$= \frac{4}{\pi}\left\{\left[\frac{\sin nx}{n}\right]_0^{\pi/2} + \left[\frac{-\sin nx}{n}\right]_{\pi/2}^{\pi}\right\}$$

$$= \frac{4}{\pi}\left\{\left(\frac{\sin(\pi/2)n}{n} - 0\right) + \left(0 - \frac{-\sin(\pi/2)n}{n}\right)\right\}$$

$$= \frac{4}{\pi}\left(\frac{2\sin(\pi/2)n}{n}\right) = \frac{8}{\pi n}\left(\sin\frac{n\pi}{2}\right)$$

When n is even, $a_n = 0$

When n is odd, $a_n = \dfrac{8}{\pi n}$ for $n = 1, 5, 9, \ldots$

and $a_n = \dfrac{-8}{\pi n}$ for $n = 3, 7, 11, \ldots$

Hence, $a_1 = \frac{8}{\pi}, a_3 = \frac{-8}{3\pi}, a_5 = \frac{8}{5\pi}$, and so on.

Hence, the Fourier series for the waveform of Figure 39.24 is given by:

$$f(x) = \frac{8}{\pi}\left(\cos x - \frac{1}{3}\cos 3x + \frac{1}{5}\cos 5x - \frac{1}{7}\cos 7x + \cdots\right)$$

Odd functions

A function $y = f(x)$ is said to be **odd** if $f(-x) = -f(x)$ for all values of x. Graphs of odd functions are always **symmetrical about the origin**. $y = \sin x$ is a typical example.

Many functions are neither even nor odd.

Fourier sine series

The Fourier series of an **odd** periodic function $f(x)$ having period 2π contains **sine terms only** (i.e. contains no constant term and no cosine terms).

Hence, $$f(x) = \sum_{n=1}^{\infty} b_n \sin nx \qquad (10)$$

where $$b_n = \frac{1}{\pi}\int_{-\pi}^{\pi} f(x)\sin nx\,dx = \frac{2}{\pi}\int_0^{\pi} f(x)\sin nx\,dx \qquad (11)$$

Problem 10. Obtain the Fourier series for the square wave shown in Figure 39.25.

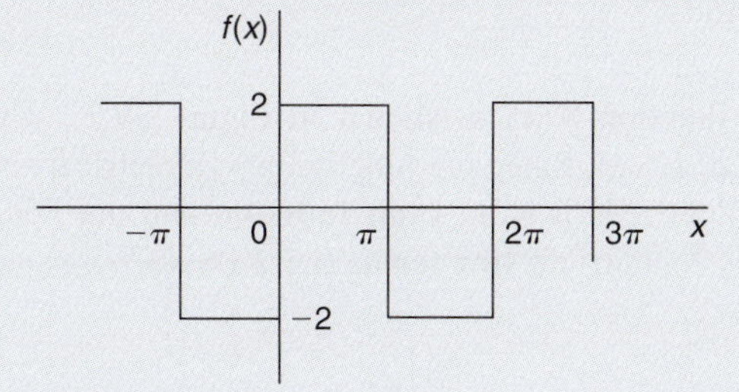

Figure 39.25

The square wave shown in Figure 39.25 is an **odd function** since it is symmetrical about the origin.

Hence, from equation (10), the Fourier series is given by:

$$f(x) = \sum_{n=1}^{\infty} b_n \sin nx$$

The function is defined by:

$$f(x) = \begin{cases} -2, & \text{when } -\pi < x < 0 \\ 2, & \text{when } 0 < x < \pi \end{cases}$$

From equation (11),

$$b_n = \frac{2}{\pi}\int_0^{\pi} f(x)\sin nx\,dx = \frac{2}{\pi}\int_0^{\pi} 2\sin nx\,dx$$

$$= \frac{4}{\pi}\left[\frac{-\cos nx}{n}\right]_0^{\pi}$$

$$= \frac{4}{\pi}\left[\left(\frac{-\cos n\pi}{n}\right) - \left(-\frac{1}{n}\right)\right] = \frac{4}{\pi n}(1 - \cos n\pi)$$

When n is even, $b_n = 0$

When n is odd, $b_n = \frac{4}{\pi n}[1-(-1)] = \frac{8}{\pi n}$

Hence, $b_1 = \frac{8}{\pi}, b_3 = \frac{8}{3\pi}, b_5 = \frac{8}{5\pi}$, and so on.

Hence the Fourier series is:

$$f(x) = \frac{8}{\pi}\left(\sin x + \frac{1}{3}\sin 3x + \frac{1}{5}\sin 5x + \frac{1}{7}\sin 7x + \cdots\right)$$

Expansion of a periodic function of period L

It may be shown that if $f(x)$ is a periodic function of period L, then the Fourier series is given by:

$$f(x) = a_0 + \sum_{n=1}^{\infty}\left[a_n\cos\left(\frac{2\pi nx}{L}\right) + b_n\sin\left(\frac{2\pi nx}{L}\right)\right] \qquad (12)$$

where, in the range $-\frac{L}{2}$ to $+\frac{L}{2}$:

$$a_0 = \frac{1}{L}\int_{-L/2}^{L/2} f(x)\,dx \qquad (13)$$

$$a_n = \frac{2}{L}\int_{-L/2}^{L/2} f(x)\cos\left(\frac{2\pi nx}{L}\right)dx \qquad (14)$$

and $$b_n = \frac{2}{L}\int_{-L/2}^{L/2} f(x)\sin\left(\frac{2\pi nx}{L}\right)dx \qquad (15)$$

The limits of integration may be replaced by any interval of length L, such as from 0 to L.

Problem 11. The voltage from a square wave generator is of the form:

$$v(t) = \begin{cases} 0, & -4 < t < 0 \\ 10, & 0 < t < 4 \end{cases} \quad \text{and has a period of 8 ms.}$$

Find the Fourier series for this periodic function.

The square wave is shown in Figure 39.26.

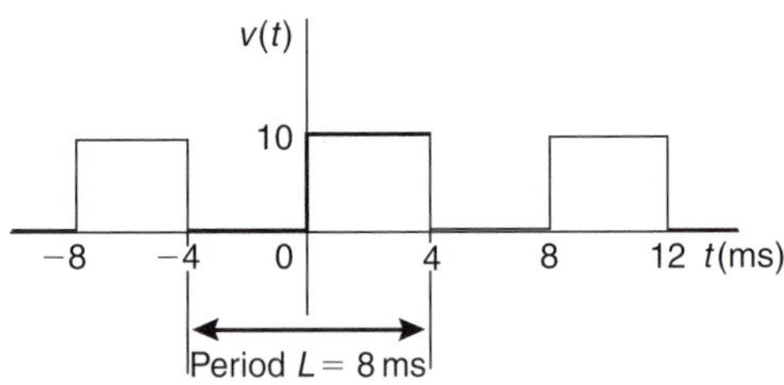

Figure 39.26

From equation (12), the Fourier series is of the form:

$$v(t) = a_0 + \sum_{n=1}^{\infty}\left[a_n \cos\left(\frac{2\pi nt}{L}\right) + b_n \sin\left(\frac{2\pi nt}{L}\right)\right]$$

From equation (13)

$$a_0 = \frac{1}{L}\int_{-L/2}^{L/2} v(t)\,dt = \frac{1}{8}\int_{-4}^{4} v(t)\,dt$$

$$= \frac{1}{8}\left\{\int_{-4}^{0} 0\,dt + \int_{0}^{4} 10\,dt\right\} = \frac{1}{8}[10t]_0^4 = 5$$

From equation (14),

$$a_n = \frac{2}{L}\int_{-L/2}^{L/2} v(t)\cos\left(\frac{2\pi nt}{L}\right)dt$$

$$= \frac{2}{8}\int_{-4}^{4} v(t)\cos\left(\frac{2\pi nt}{8}\right)dt$$

$$= \frac{1}{4}\left\{\int_{-4}^{0} 0\cos\left(\frac{\pi nt}{4}\right)dt + \int_{0}^{4} 10\cos\left(\frac{\pi nt}{4}\right)dt\right\}$$

$$= \frac{1}{4}\left[\frac{10\sin\left(\frac{\pi nt}{4}\right)}{\left(\frac{\pi n}{4}\right)}\right]_0^4 = \frac{10}{\pi n}[\sin \pi n - \sin 0] = 0$$

for $n = 1, 2, 3, ..$

From equation (15),

$$b_n = \frac{2}{L}\int_{-L/2}^{L/2} v(t)\sin\left(\frac{2\pi nt}{L}\right)dt$$

$$= \frac{2}{8}\int_{-4}^{4} v(t)\sin\left(\frac{2\pi nt}{8}\right)dt$$

$$= \frac{1}{4}\left\{\int_{-4}^{0} 0\sin\left(\frac{\pi nt}{4}\right)dt + \int_{0}^{4} 10\sin\left(\frac{\pi nt}{4}\right)dt\right\}$$

$$= \frac{1}{4}\left[\frac{-10\cos\left(\frac{\pi nt}{4}\right)}{\left(\frac{\pi n}{4}\right)}\right]_0^4 = \frac{-10}{\pi n}[\cos \pi n - \cos 0]$$

When n is even, $b_n = 0$

When n is odd, $b_1 = \frac{-10}{\pi}(-1-1) = \frac{20}{\pi}$,

$b_3 = \frac{-10}{3\pi}(-1-1) = \frac{20}{3\pi}$, $b_5 = \frac{20}{5\pi}$, and so on

Thus the Fourier series for the function $v(t)$ is given by:

$$\boldsymbol{v(t) = 5 + \frac{20}{\pi}\left[\sin\left(\frac{\pi t}{4}\right) + \frac{1}{3}\sin\left(\frac{3\pi t}{4}\right) + \frac{1}{5}\sin\left(\frac{5\pi t}{4}\right) + \cdots\right]}$$

Problem 12. Obtain the Fourier series for the function defined by:

$$f(x) = \begin{cases} 0, & \text{when } -2 < x < -1 \\ 5, & \text{when } -1 < x < 1 \\ 0, & \text{when } \quad 1 < x < 2 \end{cases}$$

The function is periodic outside of this range of period 4.

The function $f(x)$ is shown in Figure 39.27 where period, $L = 4$. Since the function is symmetrical about the $f(x)$ axis it is an **even function** and the Fourier series contains **no sine terms** (i.e. $b_n = 0$)

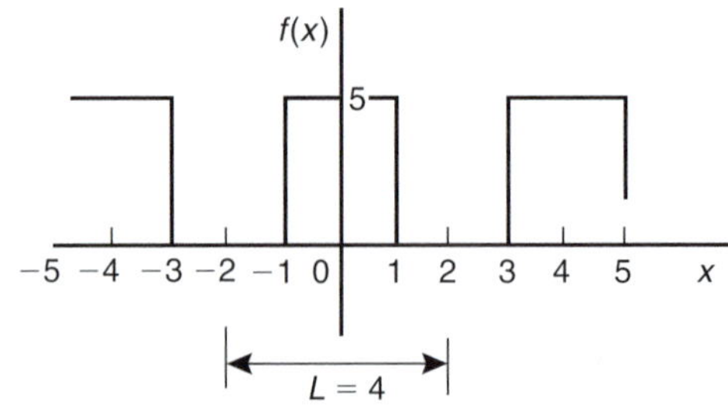

Figure 39.27

Thus, from equation (12),

$$f(x) = a_0 + \sum_{n=1}^{\infty} a_n \cos\left(\frac{2\pi n x}{L}\right)$$

From equation (13),

$$a_0 = \frac{1}{L}\int_{-L/2}^{L/2} f(x)\mathrm{d}x = \frac{1}{4}\int_{-2}^{2} f(x)\mathrm{d}x$$

$$= \frac{1}{4}\left\{\int_{-2}^{-1} 0\,\mathrm{d}x + \int_{-1}^{1} 5\mathrm{d}x + \int_{1}^{2} 0\,\mathrm{d}x\right\}$$

$$= \frac{1}{4}[5x]_{-1}^{1} = \frac{1}{4}[(5) - (-5)] = \frac{10}{4} = \frac{5}{2}$$

From equation (14),

$$a_n = \frac{2}{L}\int_{-L/2}^{L/2} f(x)\cos\left(\frac{2\pi n x}{L}\right)\mathrm{d}x$$

$$= \frac{2}{4}\int_{-2}^{2} f(x)\cos\left(\frac{2\pi n x}{4}\right)\mathrm{d}x$$

$$= \frac{1}{2}\left\{\int_{-2}^{-1} 0\cos\left(\frac{\pi n x}{2}\right)\mathrm{d}x + \int_{-1}^{1} 5\cos\left(\frac{\pi n x}{2}\right)\mathrm{d}x + \int_{1}^{2} 0\cos\left(\frac{\pi n x}{2}\right)\mathrm{d}x\right\}$$

$$= \frac{5}{2}\left[\frac{\sin\frac{\pi n x}{2}}{\frac{\pi n}{2}}\right]_{-1}^{1} = \frac{5}{\pi n}\left[\sin\left(\frac{\pi n}{2}\right) - \sin\left(\frac{-\pi n}{2}\right)\right]$$

When n is even, $a_n = 0$

When n is odd, $a_1 = \frac{5}{\pi}(1 - -1) = \frac{10}{\pi}$,

$a_3 = \frac{5}{3\pi}(-1 - 1) = \frac{-10}{3\pi}$, $a_5 = \frac{5}{5\pi}(1 - -1) = \frac{10}{5\pi}$,

and so on

Hence the Fourier series for the function $f(x)$ is given by:

$$\mathbf{f(x) = \frac{5}{2} + \frac{10}{\pi}\left[\cos\left(\frac{\pi x}{2}\right) - \frac{1}{3}\cos\left(\frac{3\pi x}{2}\right) + \frac{1}{5}\cos\left(\frac{5\pi x}{2}\right) - \frac{1}{7}\cos\left(\frac{7\pi x}{2}\right) + \cdots\right]}$$

Now try the following Practice Exercise

Practice Exercise 151 Even and odd functions and Fourier series over any range (Answers on page 830)

1. Determine the Fourier series for the function defined by:
$$f(x) = \begin{cases} -1, & -\pi < x < -\frac{\pi}{2} \\ 1, & -\frac{\pi}{2} < x < \frac{\pi}{2} \\ -1, & \frac{\pi}{2} < x < \pi \end{cases}$$
which is periodic outside of this range of period 2π.
2. Obtain the Fourier series of the function defined by:
$$f(t) = \begin{cases} t + \pi, & -\pi < t < 0 \\ t - \pi, & 0 < t < \pi \end{cases}$$
which is periodic of period 2π. Sketch the given function.
3. Determine the Fourier series defined by
$$f(x) = \begin{cases} 1 - x, & -\pi < x < 0 \\ 1 + x, & 0 < x < \pi \end{cases}$$
which is periodic of period 2π
4. The voltage from a square wave generator is of the form:
$$v(t) = \begin{cases} 0, & -10 < t < 0 \\ 5, & 0 < t < 10 \end{cases}$$
and is periodic of period 20.
Show that the Fourier series for the function is given by:
$$f(t) = \frac{5}{2} + \frac{10}{\pi}\left(\sin\left(\frac{\pi t}{10}\right) + \frac{1}{3}\sin\left(\frac{3\pi t}{10}\right) + \frac{1}{5}\sin\left(\frac{5\pi t}{10}\right) + \cdots\right)$$
5. Find the Fourier series for $f(x) = x$ in the range $x = 0$ to $x = 5$.
6. A periodic function of period 2π is defined by:
$$f(x) = \begin{cases} -3, & -2 < x < 0 \\ +3, & 0 < x < 2 \end{cases}$$
Sketch the function and obtain the Fourier series for the function.

39.6 R.m.s. value, mean value and the form factor of a complex wave

R.m.s. value

Let the instantaneous value of a complex current, i, be given by

$$i = I_{1m}\sin(\omega t+\theta_1) + I_{2m}\sin(2\omega t+\theta_2) + \cdots + I_{nm}\sin(n\omega t+\theta_n)\text{ amperes}$$

The effective or r.m.s. value of this current is given by

$$I = \sqrt{(\text{mean value of } i^2)}$$

$$i^2 = [I_{1m}\sin(\omega t+\theta_1) + I_{2m}\sin(2\omega t+\theta_2) + \cdots + I_{nm}\sin(n\omega t+\theta_n)]^2$$

i.e. $$i^2 = I_{1m}^2\sin^2(\omega t+\theta_1) + I_{2m}^2\sin^2(2\omega t+\theta_2) + \cdots + I_{nm}^2\sin^2(n\omega t+\theta_n) + 2I_{1m}I_{2m}\sin(\omega t+\theta_1)\sin(2\omega t+\theta_2) + \cdots \quad (16)$$

Without writing down all terms involved when squaring current i, it can be seen that two types of term result, these being:

(i) terms such as $I_{1m}^2\sin^2(\omega t+\theta_1)$, $I_{2m}^2\sin^2(2\omega t+\theta_2)$, and so on, and

(ii) terms such as $2I_{1m}I_{2m}\sin(\omega t+\theta_1)\sin(2\omega t+\theta_2)$, i.e. products of different harmonics.

The mean value of i^2 is the sum of the mean values of each term in equation (16).
Taking an example of the first type, say $I_{1m}^2\sin^2(\omega t+\theta_1)$, the mean value over one cycle of the fundamental is determined using integral calculus:

Mean value of $I_{1m}^2\sin^2(\omega t+\theta_1)$

$$= \frac{1}{2\pi}\int_0^{2\pi} I_{1m}^2\sin^2(\omega t+\theta_1)\,d(\omega t)$$

(since the mean value of $y = f(x)$ between $x = a$ and $x = b$ is given by $\frac{1}{b-a}\int_a^b y\,dx$)

$$= \frac{I_{1m}^2}{2\pi}\int_0^{2\pi}\left\{\frac{1-\cos 2(\omega t+\theta_1)}{2}\right\}d(\omega t),$$

(since $\cos 2x = 1 - 2\sin^2 x$, from which $\sin^2 x = (1-\cos 2x)/2$),

$$= \frac{I_{1m}^2}{4\pi}\left[\omega t - \frac{\sin 2(\omega t+\theta_1)}{2}\right]_0^{2\pi}$$

$$= \frac{I_{1m}^2}{4\pi}\left[\left(2\pi - \frac{\sin 2(2\pi+\theta_1)}{2}\right) - \left(0 - \frac{\sin 2(0+\theta_1)}{2}\right)\right]$$

$$= \frac{I_{1m}^2}{4\pi}\left[2\pi - \frac{\sin 2(2\pi+\theta_1)}{2} + \frac{\sin 2\theta_1}{2}\right] = \frac{I_{1m}^2}{4\pi}(2\pi)$$

$$= \mathbf{\frac{I_{1m}^2}{2}}$$

Hence it follows the mean value of $I_{nm}^2\sin^2(n\omega t+\theta_n)$ is given by $I_{nm}^2/2$

Taking an example of the second type, say,

$$2I_{1m}I_{2m}\sin(\omega t+\theta_1)\sin(2\omega t+\theta_2)$$

the mean value over one cycle of the fundamental is also determined using integration:

Mean value of $2I_{1m}I_{2m}\sin(\omega t+\theta_1)\sin(2\omega t+\theta_2)$

$$= \frac{1}{2\pi}\int_0^{2\pi} 2I_{1m}I_{2m}\sin(\omega t+\theta_1)\sin(2\omega t+\theta_2)d(\omega t)$$

$$= \frac{I_{1m}I_{2m}}{\pi}\int_0^{2\pi}\frac{1}{2}\{\cos(\omega t+\theta_2-\theta_1) - \cos(3\omega t+\theta_2+\theta_1)\}d(\omega t)$$

(since $\sin A \sin B = \frac{1}{2}[\cos(A-B) - \cos(A+B)]$, and taking $A = (2\omega t+\theta_2)$ and $B = (\omega t+\theta_1)$)

$$= \frac{I_{1m}I_{2m}}{2\pi}\left[\sin(\omega t+\theta_2-\theta_1) - \frac{\sin(3\omega t+\theta_2+\theta_1)}{3}\right]_0^{2\pi}$$

$$= \frac{I_{1m}I_{2m}}{2\pi}\left[\left(\sin(2\pi+\theta_2-\theta_1) - \frac{\sin(6\pi+\theta_2+\theta_1)}{3}\right) - \left(\sin(\theta_2-\theta_1) - \frac{\sin(\theta_2+\theta_1)}{3}\right)\right]$$

$$= \frac{I_{1m}I_{2m}}{2\pi}[0] = \mathbf{0} \quad (17)$$

Hence it follows that all such products of different harmonics will have a mean value of zero. Thus

$$\text{mean value of } i^2 = \frac{I_{1m}^2}{2} + \frac{I_{2m}^2}{2} + \cdots + \frac{I_{nm}^2}{2}$$

Hence the r.m.s. value of current,

$$I = \sqrt{\left(\frac{I_{1m}^2}{2} + \frac{I_{2m}^2}{2} + \cdots + \frac{I_{nm}^2}{2}\right)}$$

i.e. $$I=\sqrt{\left(\frac{I_{1m}^2+I_{2m}^2+\cdots+I_{nm}^2}{2}\right)} \qquad (18)$$

For a sine wave, r.m.s. value $= (1/\sqrt{2})$ maximum value, i.e. maximum value $= \sqrt{2}$ r.m.s. value. Hence, for example, $I_{1m} = \sqrt{2}I_1$, where I_1 is the r.m.s. value of the fundamental component, and $(I_{1m})^2 = (\sqrt{2}I_1)^2 = 2I_1^2$ Thus, from equation (18), r.m.s. current

$$I=\sqrt{\left(\frac{2I_1^2+2I_2^2+\cdots+2I_n^2}{2}\right)}$$

i.e. $$I=\sqrt{(I_1^2+I_2^2+\cdots+I_n^2)} \qquad (19)$$

where $I_1, I_2, \ldots, I_n$ are the r.m.s. values of the respective harmonics.

By similar reasoning, for a complex voltage waveform represented by

$$v = V_{1m}\sin(\omega t+\Psi_1)+V_{2m}\sin(2\omega t+\Psi_2) +\cdots+V_{nm}\sin(n\omega t+\Psi_n) \text{ volts}$$

the r.m.s. value of voltage, V, is given by

$$V=\sqrt{\left(\frac{V_{1m}^2+V_{2m}^2+\cdots+V_{nm}^2}{2}\right)} \qquad (20)$$

or $$V=\sqrt{\left(V_1^2+V_2^2+\cdots+V_n^2\right)} \qquad (21)$$

where V_1, V_2, ..., V_n are the r.m.s. values of the respective harmonics.

From equations (18) to (21) it is seen that the r.m.s. value of a complex wave is unaffected by the relative phase angles of the harmonic components. For a d.c. current or voltage, the instantaneous value, the mean value and the maximum value are equal. Thus, if a complex waveform should contain a d.c. component I_0, then the r.m.s. current I is given by

$$I=\sqrt{\left(I_0^2+\frac{I_{1m}^2+I_{2m}^2+\cdots+I_{nm}^2}{2}\right)}$$

or $$I=\sqrt{\left(I_0^2+I_1^2+I_2^2+\cdots+I_n^2\right)} \qquad (22)$$

Mean value

The mean or average value of a complex quantity whose negative half-cycle is similar to its positive half-cycle is given, for current, by

$$I_{av}=\frac{1}{\pi}\int_0^{\pi} i\,\mathrm{d}(\omega t) \qquad (23)$$

and for voltage by $$v_{av}=\frac{1}{\pi}\int_0^{\pi} v\,\mathrm{d}(\omega t) \qquad (24)$$

each waveform being taken over half a cycle.

Unlike r.m.s. values, mean values **are** affected by the relative phase angles of the harmonic components.

Form factor

The form factor of a complex waveform whose negative half-cycle is similar in shape to its positive half-cycle is defined as:

$$\textbf{form factor}=\frac{\textbf{r.m.s. value of the waveform}}{\textbf{mean value}} \qquad (25)$$

where the mean value is taken over half a cycle.

Changes in the phase displacement of the harmonics may appreciably alter the form factor of a complex waveform.

Problem 13. Determine the r.m.s. value of the current waveform represented by

$$i = 100\sin\omega t + 20\sin(3\omega t+\pi/6) + 10\sin(5\omega t+2\pi/3)\text{mA}$$

From equation (18), the r.m.s. value of current is given by

$$I=\sqrt{\left(\frac{100^2+20^2+10^2}{2}\right)}=\sqrt{\left(\frac{10\,000+400+100}{2}\right)}$$
$$=\mathbf{72.46\,mA}$$

Problem 14. A complex voltage is represented by

$$v = (10\sin\omega t+3\sin 3\omega t+2\sin 5\omega t) \text{ volts}$$

Determine for the voltage (a) the r.m.s. value, (b) the mean value and (c) the form factor.

(a) From equation (20), the r.m.s. value of voltage is given by

$$V=\sqrt{\left(\frac{10^2+3^2+2^2}{2}\right)}=\sqrt{\left(\frac{113}{2}\right)}=\mathbf{7.52\,V}$$

(b) From equation (24), the mean value of voltage is given by

$$V_{av} = \frac{1}{\pi}\int_0^{\pi}(10\sin\omega t + 3\sin 3\omega t + 2\sin 5\omega t)\,d(\omega t)$$

$$= \frac{1}{\pi}\left[-10\cos\omega t - \frac{3\cos 3\omega t}{3} - \frac{2\cos 5\omega t}{5}\right]_0^{\pi}$$

$$= \frac{1}{\pi}\left[\left(-10\cos\pi - \cos 3\pi - \frac{2}{5}\cos 5\pi\right) - \left(-10\cos 0 - \cos 0 - \frac{2}{5}\cos 0\right)\right]$$

$$= \frac{1}{\pi}\left[\left(10 + 1 + \frac{2}{5}\right) - \left(-10 - 1 - \frac{2}{5}\right)\right]$$

$$= \frac{22.8}{\pi} = \mathbf{7.26\,V}$$

(c) From equation (25), form factor is given by

$$\text{form factor} = \frac{\text{r.m.s. value of the waveform}}{\text{mean value}} = \frac{7.52}{7.26} = \mathbf{1.036}$$

Problem 15. A complex voltage waveform which has an r.m.s. value of 240 V contains 30% third harmonic and 10% fifth harmonic, both of the harmonics being initially in phase with each other. (a) Determine the r.m.s. value of the fundamental and each harmonic. (b) Write down an expression to represent the complex voltage waveform if the frequency of the fundamental is 31.83 Hz.

(a) From equation (21), r.m.s. voltage

$$V = \sqrt{(v_1^2 + V_3^2 + V_5^2)}$$

Since $V_3 = 0.30\,V_1$, $V_5 = 0.10\,V_1$ and $V = 240$ V, then

$$240 = \sqrt{[V_1^2 + (0.30\,V_1)^2 + (0.10\,V_1)^2]}$$

i.e. $240 = \sqrt{(1.10\,V_1^2)} = 1.049\,V_1$

from which the r.m.s. value of the fundamental,

$$V_1 = 240/1.049 = \mathbf{228.8\,V}$$

R.m.s. value of the third harmonic,

$$V_3 = 0.30\,V_1 = (0.30)(228.8) = \mathbf{68.64\,V}$$

and the r.m.s. value of the fifth harmonic,

$$V_5 = 0.10\,V_1 = (0.10)(228.8) = \mathbf{22.88\,V}$$

(b) Maximum value of the fundamental,

$$V_{1m} = \sqrt{2}V_1 = \sqrt{2}(228.8) = 323.6\,\text{V}$$

Maximum value of the third harmonic,

$$V_{3m} = \sqrt{2}V_3 = \sqrt{2}(68.64) = 97.07\,\text{V}$$

Maximum value of the fifth harmonic,

$$V_{5m} = \sqrt{2}V_5 = \sqrt{2}(22.88) = 32.36\,\text{V}$$

Since the fundamental frequency is 31.83 Hz, the fundamental voltage may be written as $323.6\sin 2\pi(31.83)t$, i.e. $323.6\sin 200t$ volts. The third harmonic component is $97.07\sin 600t$ volts and the fifth harmonic component is $32.36\sin 1000t$ volts. Hence an expression representing the complex voltage waveform is given by

$$\boldsymbol{v = (323.6\sin 200t + 97.07\sin 600t + 32.36\sin 1000t)\,\textbf{volts}}$$

Now try the following Practice Exercise

Practice Exercise 152 R.m.s. values, mean values and form factor of complex waves (Answers on page 830)

1. Determine the r.m.s. value of a complex current wave represented by
$$i = 3.5\sin\omega t + 0.8\sin\left(3\omega t - \frac{\pi}{3}\right) + 0.2\sin\left(5\omega t + \frac{\pi}{2}\right)\,\text{A}$$

2. Derive an expression for the r.m.s. value of a complex voltage waveform represented by
$$v = V_0 + V_{1m}\sin(\omega t + \phi_1) + V_{3m}\sin(3\omega t + \phi_3)\ \text{volts}$$
Calculate the r.m.s. value of a voltage waveform given by
$$v = 80 + 240\sin\omega t + 50\sin\left(2\omega t + \frac{\pi}{4}\right) + 20\sin\left(4\omega t - \frac{\pi}{3}\right)\ \text{volts}$$

3. A complex voltage waveform is given by
$$v = 150\sin 314t + 40\sin\left(942t - \frac{\pi}{2}\right) + 30\sin(1570t + \pi)\ \text{volts}$$
Determine for the voltage (a) the third harmonic frequency, (b) its r.m.s. value, (c) its mean value and (d) the form factor.

Part 4

4. A complex voltage waveform has an r.m.s. value of 220 V and it contains 25% third harmonic and 15% fifth harmonic. (a) Determine the r.m.s. value of the fundamental and each harmonic. (b) Write down an expression to represent the complex voltage waveform if the frequency of the fundamental is 60 Hz.
5. Define the term 'form factor' when applied to a symmetrical complex waveform. Calculate the form factor of an alternating voltage which is represented by

$$v = (50\sin 314t + 15\sin 942t + 6\sin 1570t)\text{ V}$$

39.7 Power associated with complex waves

Let a complex voltage wave be represented by

$$v = V_{1m}\sin\omega t + V_{2m}\sin 2\omega t + V_{3m}\sin 3\omega t + \cdots,$$

and when this is applied to a circuit let the resulting current be represented by

$$i = I_{1m}\sin(\omega t - \phi_1) + I_{2m}\sin(2\omega t - \phi_2) + I_{3m}\sin(3\omega t - \phi_3) + \cdots$$

(Since the phase angles are lagging, the circuit in this case is inductive.) At any instant in time the power p supplied to the circuit is given by $p = vi$, i.e.

$$p = (V_{1m}\sin\omega t + V_{2m}\sin 2\omega t + \cdots)(I_{1m}\sin(\omega t - \phi_1) + I_{2m}\sin(2\omega t - \phi_2) + \cdots)$$

$$= V_{1m}I_{1m}\sin\omega t\sin(\omega t - \phi_1) + V_{1m}I_{2m}\sin\omega t\sin(2\omega t - \phi_2) + \cdots \quad (26)$$

The average or active power supplied over one cycle is given by the sum of the average values of each individual product term taken over one cycle. It is seen from equation (17) that the average value of product terms involving harmonics of different frequencies is always zero. This means therefore that only products of voltage and current harmonics of the same frequency need be considered in equation (26).

Taking the first term, for example, the average power P_1 over one cycle of the fundamental is given by

$$P_1 = \frac{1}{2\pi}\int_0^{2\pi} V_{1m}I_{1m}\sin\omega t\sin(\omega t - \phi_1)\mathrm{d}(\omega t)$$

$$= \frac{V_{1m}I_{1m}}{2\pi}\int_0^{2\pi}\frac{1}{2}\{\cos\phi_1 - \cos(2\omega t - \phi_1)\}\mathrm{d}(\omega t)$$

since $\sin A\sin B = \frac{1}{2}\{\cos(A-B) - \cos(A+B)\}$,

$$= \frac{V_{1m}I_{1m}}{4\pi}\left[(\omega t)\cos\phi_1 - \frac{\sin(2\omega t - \phi_1)}{2}\right]_0^{2\pi}$$

$$= \frac{V_{1m}I_{1m}}{4\pi}\left[\left(2\pi\cos\phi_1 - \frac{\sin(4\pi - \phi_1)}{2}\right) - \left(0 - \frac{\sin(-\phi_1)}{2}\right)\right]$$

$$= \frac{V_{1m}I_{1m}}{4\pi}[2\pi\cos\phi_1] = \frac{V_{1m}I_{1m}}{2}\cos\phi_1$$

$V_{1m} = \sqrt{2}V_1$ and $I_{1m} = \sqrt{2}I_1$, where V_1 and I_1 are r.m.s. values, hence

$$P_1 = \frac{(\sqrt{2}V_1)(\sqrt{2}I_1)}{2}\cos\phi_1$$

i.e. $P_1 = V_1I_1\cos\phi_1$ watts

Similarly, the average power supplied over one cycle of the fundamental for the second harmonic is $V_2I_2\cos\Phi_2$, and so on. Hence the total power supplied by complex voltages and currents is the sum of the powers supplied by each harmonic component acting on its own. The average power P supplied for one cycle of the fundamental is given by

$$\boldsymbol{P = V_1I_1\cos\phi_1 + V_2I_2\cos\phi_2 + \cdots + V_nI_n\cos\phi_n} \quad (27)$$

If the voltage waveform contains a d.c. component V_0 which causes a direct current component I_0, then the average power supplied by the d.c. component is V_0I_0 and the total average power P supplied is given by

$$\boldsymbol{P = V_0I_0 + V_1I_1\cos\phi_1 + V_2I_2\cos\phi_2 + \cdots + V_nI_n\cos\phi_n} \quad (28)$$

Alternatively, if R is the equivalent series resistance of a circuit then the total power is given by

$$P = I_0^2R + I_1^2R + I_2^2R + I_3^2R + \cdots$$

i.e. $\boldsymbol{P = I^2R}$ (29)

where I is the r.m.s. value of current i.

Power factor

When harmonics are present in a waveform the overall circuit power factor is defined as overall power factor

$$= \frac{\text{total power supplied}}{\text{total r.m.s. voltage} \times \text{total r.m.s. current}}$$

$$= \frac{\text{total power}}{\text{volt amperes}}$$

i.e. $$\boldsymbol{p.f. = \frac{V_1 I_1 \cos\phi_1 + V_2 I_2 \cos\phi_2 + \cdots}{VI}} \quad (30)$$

Problem 16. Determine the average power in a $20\,\Omega$ resistance if the current i flowing through it is of the form

$$i = (12\sin\omega t + 5\sin 3\omega t + 2\sin 5\omega t)\text{ amperes}$$

From equation (18), r.m.s. current,

$$I = \sqrt{\left(\frac{12^2 + 5^2 + 2^2}{2}\right)} = 9.30\,\text{A}$$

From equation (29), average power,

$P = I^2 R = (9.30)^2(20) = \mathbf{1730\,W}$ or $\mathbf{1.73\,kW}$

Problem 17. A complex voltage v given by

$$v = 60\sin\omega t + 15\sin\left(3\omega t + \frac{\pi}{4}\right) + 10\sin\left(5\omega t - \frac{\pi}{2}\right)\text{ volts}$$

is applied to a circuit and the resulting current i is given by

$$i = 2\sin\left(\omega t - \frac{\pi}{6}\right) + 0.3\sin\left(3\omega t - \frac{\pi}{12}\right) + 0.1\sin\left(5\omega t - \frac{8\pi}{9}\right)\text{ amperes}$$

Determine (a) the total active power supplied to the circuit and (b) the overall power factor.

(a) From equation (27), total power supplied,

$$P = V_1 I_1 \cos\phi_1 + V_3 I_3 \cos\phi_3 + V_5 I_5 \cos\phi_5$$

$$= \left(\frac{60}{\sqrt{2}}\right)\left(\frac{2}{\sqrt{2}}\right)\cos\left(0 - \left(-\frac{\pi}{6}\right)\right) + \left(\frac{15}{\sqrt{2}}\right)\left(\frac{0.3}{\sqrt{2}}\right)\cos\left(\frac{\pi}{4} - \left(-\frac{\pi}{12}\right)\right) + \left(\frac{10}{\sqrt{2}}\right)\left(\frac{0.1}{\sqrt{2}}\right)\cos\left(-\frac{\pi}{2} - \left(-\frac{8\pi}{9}\right)\right)$$

$$= 51.96 + 1.125 + 0.171 = \mathbf{53.26\,W}$$

(b) From equation (18), r.m.s. current,

$$I = \sqrt{\left(\frac{2^2 + 0.3^2 + 0.1^2}{2}\right)} = 1.43\,\text{A}$$

and from equation (20), r.m.s. voltage,

$$V = \sqrt{\left(\frac{60^2 + 15^2 + 10^2}{2}\right)} = 44.30\,\text{V}$$

From equation (30),

$$\text{overall power factor} = \frac{53.26}{(44.30)(1.43)} = \mathbf{0.841}$$

(With a sinusoidal waveform,

$$\text{power factor} = \frac{\text{power}}{\text{volt-amperes}} = \frac{VI\cos\phi}{VI} = \cos\phi$$

Thus power factor depends upon the value of phase angle ϕ, and is lagging for an inductive circuit and leading for a capacitive circuit. However, with a complex waveform, power factor is not given by cos ϕ. In the expression for power in equation (27), there are n phase-angle terms, $\phi_1, \phi_2, \ldots, \phi_n$, all of which may be different. It is for this reason that it is not possible to state whether the overall power factor is lagging or leading when harmonics are present.)

Now try the following Practice Exercise

Practice Exercise 153 Power associated with complex waves (Answers on page 830)

1. Determine the average power in a $50\,\Omega$ resistor if the current i flowing through it is represented by

   $i = (140\sin\omega t + 40\sin 3\omega t + 20\sin 5\omega t)\,\text{mA}$

2. A voltage waveform represented by

   $$v = 100\sin\omega t + 22\sin\left(3\omega t - \frac{\pi}{6}\right) + 8\sin\left(5\omega t - \frac{\pi}{4}\right)\text{ volts}$$

   is applied to a circuit and the resulting current i is given by

   $$i = 5\sin\left(\omega t + \frac{\pi}{3}\right) + 1.91\sin 3\omega t + 0.76\sin(5\omega t - 0.452)\text{ amperes}$$

   Calculate (a) the total active power supplied to the circuit and (b) the overall power factor.

3. Determine the r.m.s. voltage, r.m.s. current and average power supplied to a network if the applied voltage is given by

$$v = 100 + 50\sin\left(400t - \frac{\pi}{3}\right) + 40\sin\left(1200t - \frac{\pi}{6}\right)\text{ volts}$$

and the resulting current is given by

$$i = 0.928\sin(400t + 0.424) + 2.14\sin(1200t + 0.756)\text{ amperes}$$

4. A voltage $v = 40 + 20\sin 300t + 8\sin 900t + 3\sin 1500t$ volts is applied to the terminals of a circuit and the resulting current is given by

$$i = 4 + 1.715\sin(300t - 0.540) + 0.389\sin(900t - 1.064) + 0.095\sin(1500t - 1.249)\text{A}$$

Determine (a) the r.m.s. voltage, (b) the r.m.s. current and (c) the average power.

39.8 Harmonics in single-phase circuits

When a complex alternating voltage wave, i.e. one containing harmonics, is applied to a single-phase circuit containing resistance, inductance and/or capacitance (i.e. linear circuit elements), then the resulting current will also be complex and contain harmonics.

Let a complex voltage v be represented by

$$\boldsymbol{v = V_{1m}\sin\omega t + V_{2m}\sin 2\omega t + V_{3m}\sin 3\omega t + \cdots}$$

(a) Pure resistance

The impedance of a pure resistance R is independent of frequency and the current and voltage are in phase for each harmonic. Thus the general expression for current i is given by

$$\boldsymbol{i = \frac{v}{R} = \frac{V_{1m}}{R}\sin\omega t + \frac{V_{2m}}{R}\sin 2\omega t + \frac{V_{3m}}{R}\sin 3\omega t + \cdots} \qquad (31)$$

The percentage harmonic content in the current wave is the same as that in the voltage wave. For example, the percentage second harmonic content from equation (31) is

$$\frac{V_{2m}/R}{V_{1m}/R}\times 100\%, \text{ i.e. } \frac{V_{2m}}{V_{1m}}\times 100\%$$

the same as for the voltage wave. The current and voltage waveforms will therefore be identical in shape.

(b) Pure inductance

The impedance of a pure inductance L, i.e. inductive reactance $X_L(=2\pi fL)$, varies with the harmonic frequency when voltage v is applied to it. Also, for every harmonic term, the current will lag the voltage by 90° or $\pi/2$ rad. The current i is given by

$$\boldsymbol{i = \frac{v}{X_L} = \frac{V_{1m}}{\omega L}\sin\left(\omega t - \frac{\pi}{2}\right) + \frac{V_{2m}}{2\omega L}\sin\left(2\omega t - \frac{\pi}{2}\right) + \frac{V_{3m}}{3\omega L}\sin\left(3\omega t - \frac{\pi}{2}\right) + \cdots} \qquad (32)$$

since for the nth harmonic the reactance is $n\omega L$.
Equation (32) shows that for, say, the nth harmonic, the percentage harmonic content in the current waveform is only $1/n$ of the corresponding harmonic content in the voltage waveform.
If a complex current contains a d.c. component then the direct voltage drop across a pure inductance is zero.

(c) Pure capacitance

The impedance of a pure capacitance C, i.e. capacitive reactance $X_C(=1/(2\pi fC))$, varies with the harmonic frequency when voltage v is applied to it. Also, for each harmonic term the current will lead the voltage by 90° or $\pi/2$ rad. The current i is given by

$$i = \frac{v}{X_C} = \frac{V_{1m}}{1/\omega C}\sin\left(\omega t + \frac{\pi}{2}\right) + \frac{V_{2m}}{1/2\omega C}\sin\left(2\omega t + \frac{\pi}{2}\right) + \frac{V_{3m}}{1/3\omega C}\sin\left(3\omega t + \frac{\pi}{2}\right) + \cdots,$$

since for the nth harmonic the reactance is $1/(n\omega C)$. Hence current,

$$\boldsymbol{i = V_{1m}(\omega C)\sin\left(\omega t + \frac{\pi}{2}\right) + V_{2m}(2\omega C)\sin\left(2\omega t + \frac{\pi}{2}\right) + V_{3m}(3\omega C)\sin\left(3\omega t + \frac{\pi}{2}\right) + \cdots} \qquad (33)$$

Equation (33) shows that the percentage harmonic content of the current waveform is n times larger for the nth harmonic than that of the corresponding harmonic voltage.

If a complex current contains a d.c. component then none of this direct current will flow through a pure capacitor, although the alternating components of the supply still operate.

Problem 18. A complex voltage waveform represented by

$$v = 100\sin\omega t + 30\sin\left(3\omega t + \frac{\pi}{3}\right) + 10\sin\left(5\omega t - \frac{\pi}{6}\right) \text{ volts}$$

is applied across (a) a pure 40 Ω resistance, (b) a pure 7.96 mH inductance and (c) a pure 25 μF capacitor. Determine for each case an expression for the current flowing if the fundamental frequency is 1 kHz.

(a) From equation (31),

$$\text{current } i = \frac{v}{R} = \frac{100}{40}\sin\omega t + \frac{30}{40}\sin\left(3\omega t + \frac{\pi}{3}\right) + \frac{10}{40}\sin\left(5\omega t - \frac{\pi}{6}\right)$$

i.e. $\boldsymbol{i = 2.5\sin\omega t + 0.75\sin\left(3\omega t + \frac{\pi}{3}\right) + 0.25\sin\left(5\omega t - \frac{\pi}{6}\right)}$ **amperes**

(b) At the fundamental frequency,

$$\omega L = 2\pi(1000)\,(7.96\times10^{-3}) = 50\,\Omega$$

From equation (19),

$$\text{current } i = \frac{100}{50}\sin\left(\omega t - \frac{\pi}{2}\right) + \frac{30}{3\times50}\sin\left(3\omega t + \frac{\pi}{3} - \frac{\pi}{2}\right) + \frac{10}{5\times50}\sin\left(5\omega t - \frac{\pi}{6} - \frac{\pi}{2}\right)$$

i.e.

current $\boldsymbol{i = 2\sin\left(\omega t - \frac{\pi}{2}\right) + 0.20\sin\left(3\omega t - \frac{\pi}{6}\right) + 0.04\sin\left(5\omega t - \frac{2\pi}{3}\right)}$ **amperes**

(c) At the fundamental frequency,

$$\omega C = 2\pi(1000)\,(25\times10^{-6}) = 0.157$$

From equation (33),

$$\text{current } i = 100(0.157)\sin\left(\omega t + \frac{\pi}{2}\right) + 30(3\times0.157)\sin\left(3\omega t + \frac{\pi}{3} + \frac{\pi}{2}\right) + 10(5\times0.157)\sin\left(5\omega t - \frac{\pi}{6} + \frac{\pi}{2}\right)$$

i.e. $\boldsymbol{i = 15.70\sin\left(\omega t + \frac{\pi}{2}\right) + 14.13\sin\left(3\omega t + \frac{5\pi}{6}\right) + 7.85\sin\left(5\omega t + \frac{\pi}{3}\right)}$ **amperes**

Problem 19. A supply voltage v given by

$$v = (240\sin 314t + 40\sin 942t + 30\sin 1570t) \text{ volts}$$

is applied to a circuit comprising a resistance of 12 Ω connected in series with a coil of inductance 9.55 mH. Determine (a) an expression to represent the instantaneous value of the current, (b) the r.m.s. voltage, (c) the r.m.s. current, (d) the power dissipated and (e) the overall power factor.

(a) The supply voltage comprises a fundamental, $240\sin 314t$, a third harmonic, $40\sin 942t$ (third harmonic since 942 is 3×314) and a fifth harmonic, $30\sin 1570t$

Fundamental

Since the fundamental frequency, $\omega_1 = 314$ rad/s, inductive reactance,

$$X_{L1} = \omega_1 L = (314)(9.55\times10^{-3}) = 3.0\,\Omega$$

Hence impedance at the fundamental frequency,

$$Z_1 = (12 + j3.0)\,\Omega = 12.37\angle 14.04°\,\Omega$$

Maximum current at fundamental frequency

$$I_{1m} = \frac{V_{1m}}{Z_1} = \frac{240\angle 0°}{12.37\angle 14.04°} = 19.40\angle -14.04°\text{ A}$$

$$14.04° = 14.04\times(\pi/180)\text{ rad} = 0.245\text{ rad},$$

thus

$$I_{1m} = 19.40\angle -0.245\text{ A}$$

Hence the fundamental current
$$i_1 = 19.40\sin(314t - 0.245)\text{ A}$$

(Note that with an expression of the form $R\sin(\omega t \pm \alpha)$, ωt is an angle measured in radians, thus the phase displacement, α, should also be expressed in radians.)

Third harmonic

Since the third harmonic frequency, $\omega_3 = 942$ rad/s, inductive reactance,

$$X_{L3} = 3X_{L1} = 9.0\,\Omega$$

Part 4

Hence impedance at the third harmonic frequency,

$$Z_3 = (12 + j9.0)\Omega = 15\angle 36.87°\Omega$$

Maximum current at the third harmonic frequency,

$$I_{3m} = \frac{V_{3m}}{Z_3} = \frac{40\angle 0°}{15\angle 36.87°}$$
$$= 2.67\angle -36.87°\,\text{A}$$
$$= 2.67\angle -0.644\,\text{A}$$

Hence the third harmonic current,
$$i_3 = 2.67\sin(942t - 0.644)\text{A}$$

Fifth harmonic

Inductive reactance, $X_{L5} = 5X_{L1} = 15\,\Omega$
Impedance
$$Z_5 = (12 + j15)\Omega = 19.21\angle 51.34°\,\Omega$$

$$\text{Current, } I_{5m} = \frac{V_{5m}}{Z_5} = \frac{30\angle 0°}{19.21\angle 51.34°}$$
$$= 1.56\angle -51.34°\,\text{A} = 1.56\angle -0.896\,\text{A}$$

Hence the fifth harmonic current,

$i_5 = 1.56\sin(1570t - 0.896)\text{A}$
Thus an expression to represent the instantaneous current, i, is given by $i = i_1 + i_3 + i_5$ i.e.

$$\mathbf{i = 19.40\sin(314t - 0.245) + 2.67\sin(942t - 0.644) + 1.56\sin(1570t - 0.896)\ A}$$

(b) From equation (20), r.m.s. voltage,

$$V = \sqrt{\left(\frac{240^2 + 40^2 + 30^2}{2}\right)} = \mathbf{173.35\ V}$$

(c) From equation (18), r.m.s. current,

$$I = \sqrt{\left(\frac{19.40^2 + 2.67^2 + 1.56^2}{2}\right)} = \mathbf{13.89\ A}$$

(d) From equation (29), power dissipated,

$$P = I^2R = (13.89)^2(12) = \mathbf{2315\ W} \text{ or } \mathbf{2.315\ kW}$$

(Alternatively, equation (27) may be used to determine power.)

(e) From equation (30),

$$\text{overall power factor} = \frac{2315}{(173.35)(13.89)} = \mathbf{0.961}$$

Problem 20. An e.m.f. is represented by

$$e = 50 + 200\sin\omega t + 40\sin\left(2\omega t - \frac{\pi}{2}\right) + 5\sin\left(4\omega t + \frac{\pi}{4}\right) \text{ volts,}$$

the fundamental frequency being 50 Hz. The e.m.f. is applied across a circuit comprising a 100 μF capacitor connected in series with a 50 Ω resistor. Obtain an expression for the current flowing and hence determine the r.m.s. value of current.

D.c. component

In a d.c. circuit no current will flow through a capacitor. The current waveform will not possess a d.c. component even though the e.m.f. waveform has a 50 V d.c. component. Hence $i_0 = 0$

Fundamental

Capacitive reactance,

$$X_{C1} = \frac{1}{2\pi fC} = \frac{1}{2\pi(50)(100\times 10^{-6})} = 31.83\,\Omega$$

Impedance $Z_1 = (50 - j31.83)\,\Omega$
$= 59.27\angle -32.48°\,\Omega$

$$I_{1m} = \frac{V_{1m}}{Z_1} = \frac{200\angle 0°}{59.27\angle -32.48°} = 3.374\angle 32.48°\,\text{A}$$
$$= 3.374\angle 0.567\,\text{A}$$

Hence the fundamental current,

$$i_1 = 3.374\sin(\omega t + 0.567)\text{A}$$

Second harmonic

Capacitive reactance,

$$X_{C2} = \frac{1}{2(2\pi fC)} = \frac{31.83}{2} = 15.92\,\Omega$$

Impedance $Z_2 = (50 - j15.92)\Omega = 52.47\angle -17.66°\,\Omega$

$$I_{2m} = \frac{V_{2m}}{Z_2} = \frac{40\angle -\pi/2}{52.47\angle -17.66°}$$
$$= 0.762\angle\left(-\frac{\pi}{2} - (-17.66°)\right) = 0.762\angle -72.34°\,\text{A}$$

Hence the second harmonic current,

$i_2 = 0.762\sin(2\omega t - 72.34°)\,\text{A}$
$= 0.762\sin(2\omega t - 1.263)\,\text{A}$

Fourth harmonic

Capacitive reactance, $X_{C4} = \frac{1}{4}X_{C1} = \frac{31.83}{4}$

$= 7.958\,\Omega$

Impedance, $Z_4 = (50 - j7.958)\Omega = 50.63\angle -9.04°\,\Omega$

$$I_{4m} = \frac{V_{4m}}{Z_4} = \frac{5\angle\pi/4}{50.63\angle -9.04°}$$
$$= 0.099\angle(\pi/4 - (-9.04°))$$
$$= 0.099\angle 54.04°\,\text{A}$$

Hence the fourth harmonic current,

$i_4 = 0.099\sin(4\omega t + 54.04°)\text{A}$

$= 0.099\sin(4\omega t + 0.943)\text{A}$

An expression for current flowing is therefore given by

$i = i_0 + i_1 + i_2 + i_4$

i.e.

$$\boldsymbol{i = 3.374\sin(\omega t + 0.567) + 0.762\sin(2\omega t - 1.263) + 0.099\sin(4\omega t + 0.943)\ \text{amperes}}$$

From equation (18), r.m.s. current,

$$I = \sqrt{\left(\frac{3.374^2 + 0.762^2 + 0.099^2}{2}\right)} = \mathbf{2.45\,A}$$

Now try the following Practice Exercise

Practice Exercise 154 Harmonics in single-phase circuits (Answers on page 830)

1. A complex voltage waveform represented by

$$v = 240\sin\omega t + 60\sin\left(3\omega t - \frac{\pi}{4}\right) + 30\sin\left(5\omega t + \frac{\pi}{3}\right)\ \text{volts}$$

is applied across (a) a pure $50\,\Omega$ resistance, (b) a pure $4.974\,\mu$F capacitor and (c) a pure 15.92 mH inductance. Determine for each case an expression for the current flowing if the fundamental frequency is 400 Hz.

2. A complex current given by

$$i = 5\sin\left(\omega t + \frac{\pi}{3}\right) + 8\sin\left(3\omega t + \frac{2\pi}{3}\right)\ \text{mA}$$

flows through a pure 2000 pF capacitor. If the frequency of the fundamental component is 4 kHz, determine (a) the r.m.s. value of current, (b) an expression for the p.d. across the capacitor and (c) the r.m.s. value of voltage.

3. A complex voltage, v, given by

$$v = 200\sin\omega t + 42\sin 3\omega t + 25\sin 5\omega t\ \text{volts}$$

is applied to a circuit comprising a $6\,\Omega$ resistance in series with a coil of inductance 5 mH. Determine, for a fundamental frequency of 50 Hz, (a) an expression to represent the instantaneous value of the current flowing, (b) the r.m.s. voltage, (c) the r.m.s. current, (d) the power dissipated and (e) the overall power factor.

4. An e.m.f. e is given by

$$e = 40 + 150\sin\omega t + 30\sin\left(2\omega t - \frac{\pi}{4}\right) + 10\sin\left(4\omega t - \frac{\pi}{3}\right)\ \text{volts}$$

the fundamental frequency being 50 Hz. The e.m.f. is applied across a circuit comprising a $100\,\Omega$ resistance in series with a $15\,\mu$F capacitor. Determine (i) the r.m.s. value of voltage, (ii) an expression for the current flowing and (iii) the r.m.s. value of current.

5. A circuit comprises a $100\,\Omega$ resistance in series with a 1 mH inductance. The supply voltage is given by

$$v = 40 + 200\sin\omega t + 50\sin\left(3\omega t + \frac{\pi}{4}\right) + 15\sin\left(5\omega t + \frac{\pi}{6}\right)\ \text{volts}$$

where $\omega = 10^5$ rad/s. Determine for the circuit (a) an expression to represent the current flowing, (b) the r.m.s. value of current and (c) the power dissipated.

39.9 Further worked problems on harmonics in single-phase circuits

Problem 21. A complex voltage v is represented by:

$$v = 25 + 100\sin\omega t + 40\sin\left(3\omega t + \frac{\pi}{6}\right) + 20\sin\left(5\omega t + \frac{\pi}{12}\right)\ \text{volts}$$

where $\omega = 10^4$ rad/s. The voltage is applied to a series circuit comprising a 5.0 Ω resistance and a 500 μH inductance.

Determine (a) an expression to represent the current flowing in the circuit, (b) the r.m.s. value of current, correct to two decimal places and (c) the power dissipated in the circuit, correct to three significant figures.

(a) **d.c. component**

Inductance has no effect on a steady current. Hence the d.c. component of the current, i_0, is given by

$$i_0 = \frac{v_0}{R} = \frac{25}{5.0} = 5.0\,\text{A}$$

Fundamental

Inductive reactance,

$$X_{L1} = \omega L = (10^4)(500 \times 10^{-6}) = 5\,\Omega$$

Impedance, $Z_1 = (5 + j5)\Omega = 7.071\angle 45°\,\Omega$

$$I_{1m} = \frac{V_{1m}}{Z_1} = \frac{100\angle 0°}{7.07\angle 45°} = 14.14\angle -45°\,\text{A}$$

$$= 14.14\angle -\pi/4\,\text{A or } 14.14\angle -0.785\,\text{A}$$

Hence fundamental current,

$$i_1 = 14.14\sin(\omega t - 0.785)\text{A}$$

Third harmonic

Inductive reactance at third harmonic frequency,

$$X_{L3} = 3X_{L1} = 15\,\Omega$$

Impedance, $Z_3 = (5 + j15)\,\Omega = 15.81\angle 71.57°\Omega$

$$I_{3m} = \frac{V_{3m}}{Z_3} = \frac{40\angle \pi/6}{15.81\angle 71.57°} = 2.53\angle -41.57°\text{A}$$

$$= 2.53\angle -0.726\,\text{A}$$

Hence the third harmonic current,

$$i_3 = 2.53\,\sin(3\omega t - 41.57°)\text{A}$$
$$= 2.53\sin(3\omega t - 0.726)\text{A}$$

Fifth harmonic

Inductive reactance at fifth harmonic frequency, $X_{L5} = 5X_{L1} = 25\,\Omega$

Impedance,
$Z_5 = (5 + j25)\Omega = 25.495\angle 78.69°\Omega$

$$I_5 = \frac{V_{5m}}{Z_5} = \frac{20\angle \pi/12}{25.495\angle 78.69°} = 0.784\angle -63.69°\text{A}$$

$$= 0.784\angle -1.112\,\text{A}$$

Hence the fifth harmonic current,

$$i_5 = 0.784\sin(5\omega t - 63.69°)\text{ A}$$
$$= 0.784\sin(5\omega t - 1.112)\text{ A}$$

Thus current, $i = i_0 + i_1 + i_3 + i_5$

i.e.

$$\boldsymbol{i = 5 + 14.14\sin(\omega t - 0.785) + 2.43\sin(3\omega t - 0.726) + 0.784\sin(5\omega t - 1.112)\,\text{A}}$$

(b) From equation (22), r.m.s. current,

$$I = \sqrt{\left(5.0^2 + \frac{14.14^2 + 2.53^2 + 0.784^2}{2}\right)}$$

$= 11.3348\,\text{A} = \mathbf{11.33\,A}$, correct to two decimal places.

(c) From equation (29), power dissipated,

$$P = I^2R = (11.3348)^2(5.0) = 642.4\,\text{W}$$

$= \mathbf{642\,W}$, correct to three significant figures

(Alternatively, from equation (28),

$$\text{power } P = (25)(5.0) + \left(\frac{100}{\sqrt{2}}\right)\left(\frac{14.14}{\sqrt{2}}\right)\cos 45° + \left(\frac{40}{\sqrt{2}}\right)\left(\frac{2.53}{\sqrt{2}}\right)\cos 71.57° + \left(\frac{20}{\sqrt{2}}\right)\left(\frac{0.784}{\sqrt{2}}\right)\cos 78.69°$$

$$= 125 + 499.92 + 16.00 + 1.54$$

$= 642.46\,\text{W}$ or $\mathbf{642\,W}$, correct to three significant figures, as above.)

Problem 22. The voltage applied to a particular circuit comprising two components connected in series is given by

$$v = (30 + 40\sin 10^3 t + 25\sin 2 \times 10^3 t + 15\sin 4 \times 10^3 t)\text{ volts}$$

and the resulting current is given by

$$i = 0.743\sin(10^3 t + 1.190) + 0.781\sin(2 \times 10^3 t + 0.896) + 0.636\sin(4 \times 10^3 t + 0.559)\text{ A}$$

Determine (a) the average power supplied, (b) the type of components present, and (c) the values of the components.

(a) From equation (28), the average power P is given by

$$P = (30)(0) + \left(\frac{40}{\sqrt{2}}\right)\left(\frac{0.743}{\sqrt{2}}\right)\cos 1.190$$
$$+ \left(\frac{25}{\sqrt{2}}\right)\left(\frac{0.781}{\sqrt{2}}\right)\cos 0.896$$
$$+ \left(\frac{15}{\sqrt{2}}\right)\left(\frac{0.636}{\sqrt{2}}\right)\cos 0.559$$

i.e. $P = 0 + 5.523 + 6.099 + 4.044 = \mathbf{15.67\,W}$

(b) The expression for the voltage contains a d.c. component of 30 V. However, there is no corresponding term in the expression for current. This indicates that one of the components is a **capacitor** (since in a d.c. circuit a capacitor offers an infinite impedance to a direct current). Since power is delivered to the circuit the other component is a **resistor**.

(c) From equation (8), r.m.s. current,

$$I = \sqrt{\left(\frac{0.743^2 + 0.781^2 + 0.636^2}{2}\right)} = 0.885\,\text{A}$$

Average power $P = I^2R$, from which,

$$\text{resistance } R = \frac{P}{I^2} = \frac{15.67}{(0.885)^2} = \mathbf{20\,\Omega}$$

At the fundamental frequency, $\omega = 10^3$ rad/s

$$\text{impedance } |Z_1| = \frac{V_{1m}}{I_{1m}} = \frac{40}{0.743} = 53.84\,\Omega$$

Impedance $|Z_1| = \sqrt{(R^2 + X_{C1}^2)}$, from which

$$X_{C1} = \sqrt{(Z_1^2 - R^2)} = \sqrt{(53.84^2 - 20^2)} = 50\,\Omega$$

Hence $1/\omega C = 50$, from which

$$\textbf{capacitance } C = \frac{1}{\omega(50)} = \frac{1}{10^3(50)} = \mathbf{20\,\mu F}$$

Problem 23. In the circuit shown in Figure 39.28 the supply voltage is given by $v = 300\sin 314t + 120\sin(942t + 0.698)$ volts. Determine (a) an expression for the supply current, i, (b) the percentage harmonic content of the supply current, (c) the total power dissipated, (d) an expression for the p.d. shown as v_1 and (e) an expression for current i_c.

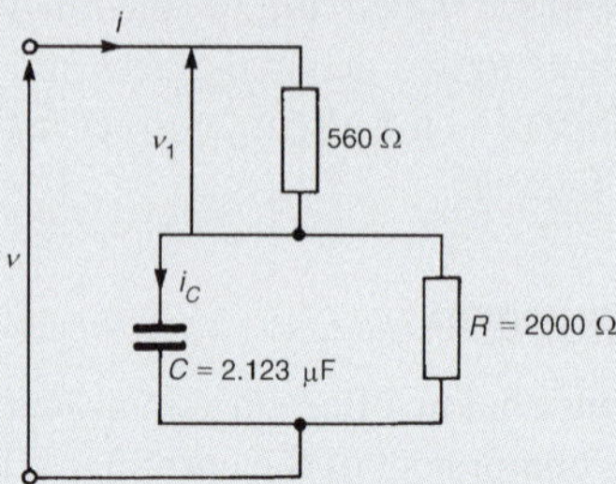

Figure 39.28

(a) Capacitive reactance of the 2.123 μF capacitor at the fundamental frequency is given by

$$X_{C1} = \frac{1}{(314)(2.123\times10^{-6})} = 1500\,\Omega$$

At the fundamental frequency the total circuit impedance, Z_1, is given by

$$Z_1 = 560 + \frac{(2000)(-j1500)}{(2000 - j1500)}$$
$$= 560 + \frac{3\times10^6\angle{-90^\circ}}{2500\angle{-36.87^\circ}}$$
$$= 560 + 1200\angle{-53.13^\circ} = 560 + 720 - j960$$
$$= (1280 - j960)\,\Omega = 1600\angle{-36.87^\circ}\,\Omega$$
$$= 1600\angle{-0.644}\,\Omega$$

Since for the nth harmonic the capacitive reactance is $1/(n\omega C)$, the capacitive reactance of the third harmonic is $\frac{1}{3}X_{C1} = \frac{1}{3}(1500) = 500\,\Omega$. Hence at the third harmonic frequency the total circuit impedance, Z_3, is given by

$$Z_1 = 560 + \frac{(2000)(-j500)}{(2000 - j500)}$$
$$= 560 + \frac{10^6\angle{-90^\circ}}{2061.55\angle{-14.04^\circ}}$$
$$= 560 + 485.07\angle{-75.96^\circ}$$
$$= 560 + 117.68 - j470.58$$
$$= (677.68 - j470.58)\,\Omega = 825\angle{-34.78^\circ}\,\Omega$$
$$= 825\angle{-0.607}\,\Omega$$

The fundamental current

$$i_1 = \frac{v_1}{Z_1} = \frac{300\angle 0}{1600\angle -0.644} = 0.188\angle 0.644\,\text{A}$$

The third harmonic current

$$i_3 = \frac{v_3}{Z_3} = \frac{120\angle 0.698}{825\angle -0.607} = 0.145\angle 1.305\,\text{A}$$

Thus, **supply current,**

$$\mathbf{i = 0.188\sin(314t + 0.644) + 0.145\sin(942t + 1.305)A}$$

(b) Percentage harmonic content of the supply current is given by

$$\frac{0.145}{0.188} \times 100\% = \mathbf{77\%}$$

(c) From equation (27), total active power

$$P = \left(\frac{300}{\sqrt{2}}\right)\left(\frac{0.188}{\sqrt{2}}\right)\cos 0.644 + \left(\frac{120}{\sqrt{2}}\right)\left(\frac{0.145}{\sqrt{2}}\right)\cos 0.607$$

i.e. $P = 22.55 + 7.15 = \mathbf{29.70\,W}$

(d) Voltage $v_1 = iR = 560[0.188\sin(314t + 0.644) + 0.145\sin(942t + 1.305)]$

i.e. $\mathbf{v_1 = 105.3\sin(314t + 0.644) + 81.2\sin(942t + 1.305)\,volts}$

(e) Current, $i_c = i_1\left(\frac{R}{R - jX_{C1}}\right) + i_3\left(\frac{R}{R - jX_{C3}}\right)$ by current division

$$= (0.188\angle 0.644)\left(\frac{2000}{2000 - j1500}\right) + (0.145\angle 1.305)\left(\frac{2000}{2000 - j500}\right)$$

$$= (0.188\angle 0.644)\left(\frac{2000}{2500\angle -0.644}\right) + (0.145\angle 1.305)\left(\frac{2000}{2061.55\angle -0.245}\right)$$

$$= 0.150\angle 1.288 + 0.141\angle 1.550$$

Hence $\mathbf{i_c = 0.150\sin(314t + 1.288) + 0.141\sin(942t + 1.550)\,A}$

Now try the following Practice Exercise

Practice Exercise 155 Harmonics in single-phase circuits (Answers on page 831)

1. The e.m.f. applied to a circuit comprising two components connected in series is given by

$$v = 50 + 150\sin(2 \times 10^3 t) + 40\sin(4 \times 10^3 t) + 20\sin(8 \times 10^3 t)\text{ volts}$$

and the resulting current is given by

$$i = 1.011\sin(2 \times 10^3 t + 1.001) + 0.394\sin(4 \times 10^3 t + 0.663) + 0.233\sin(8 \times 10^3 t + 0.372)\text{A}$$

Determine for the circuit (a) the average power supplied and (b) the value of the two circuit components.

2. A coil having inductance L and resistance R is supplied with a complex voltage given by

$$v = 240\sin\omega t + V_3\sin\left(3\omega t + \frac{\pi}{3}\right) + V_5\sin\left(5\omega t - \frac{\pi}{12}\right)\text{ volts}$$

The resulting current is given by

$$i = 4.064\sin(\omega t - 0.561) + 0.750\sin(3\omega t - 0.036) + 0.182\sin(5\omega t - 1.525)\text{ A}$$

The fundamental frequency is 500 Hz. Determine (a) the impedance of the circuit at the fundamental frequency, and hence the values of R and L, (b) the values of V_3 and V_5, (c) the r.m.s. voltage, (d) the r.m.s. current, (e) the circuit power and (f) the power factor.

3. An alternating supply voltage represented by

$$v = (240\sin 300t - 40\sin 1500t + 60\sin 2100t)\text{ volts}$$

is applied to the terminals of a circuit containing a 40 Ω resistor, a 200 mH inductor and a 25 μF capacitor in series. (a) Derive the expression for the current waveform and (b) calculate the power dissipated by the circuit.

4. A voltage v represented by

$$v = 120\sin 314t + 25\sin\left(942t + \frac{\pi}{6}\right)\text{ volts}$$

is applied to the circuit shown in Figure 39.29. Determine (a) an expression for current i, (b) the percentage harmonic content of the supply current, (c) the total power dissipated, (d) an expression for the p.d. shown as v_1 and (e) expressions for the currents shown as i_R and i_C

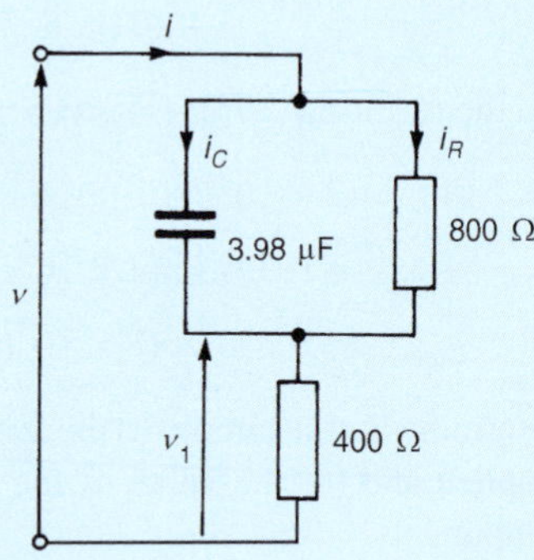

Figure 39.29

39.10 Resonance due to harmonics

In industrial circuits at power frequencies the typical values of L and C involved make resonance at the fundamental frequency very unlikely. (An exception to this is with the capacitor-start induction motor where the start-winding can achieve unity power factor during run-up.)

However, if the voltage waveform is not a pure sine wave it is quite possible for the resonant frequency to be near the frequency of one of the harmonics. In this case the magnitude of the particular harmonic in the current waveform is greatly increased and may even exceed that of the fundamental. The effect of this is a great distortion of the resultant current waveform so that dangerous volt drops may occur across the inductance and capacitance in the circuit.

When a circuit resonates at one of the harmonic frequencies of the supply voltage, the effect is called **selective or harmonic resonance**.

For resonance with the fundamental, the condition is $\omega L = 1/(\omega C)$; for resonance at, say, the third harmonic, the condition is $3\omega L = 1/(3\omega C)$; for resonance at the nth harmonic, the condition is

$$\mathbf{n\omega L = 1/(n\omega C)}$$

Problem 24. A voltage waveform having a fundamental of maximum value 400 V and a third harmonic of maximum value 10 V is applied to the circuit shown in Figure 39.30. Determine (a) the fundamental frequency for resonance with the third harmonic, and (b) the maximum value of the fundamental and third harmonic components of current.

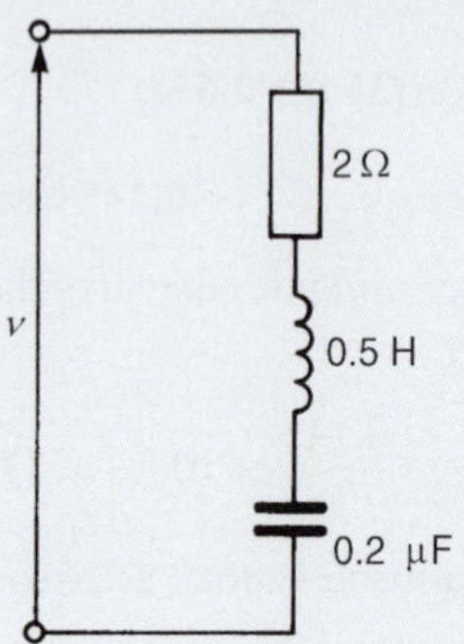

Figure 39.30

(a) Resonance with the third harmonic means that $3\omega L = 1/(3\omega C)$, i.e.

$$\omega = \sqrt{\left(\frac{1}{9LC}\right)} = \frac{1}{3\sqrt{(0.5)(0.2\times 10^{-6})}}$$

$$= 1054\,\text{rad/s}$$

from which, **fundamental frequency**, $f = \dfrac{\omega}{2\pi}$

$$= \frac{1054}{2\pi} = \mathbf{167.7\,Hz}$$

(b) At the fundamental frequency,

impedance

$$Z_1 = R + j\left(\omega L - \frac{1}{\omega C}\right)$$

$$= 2 + j\left[(1054)(0.5) - \frac{1}{(1054)(0.2\times 10^{-6})}\right]$$

$$= (2 - j4217)\,\Omega$$

i.e. $Z_1 = 4217\angle -89.97^\circ\,\Omega$

Maximum value of current at the fundamental frequency,

$$I_{1m} = \frac{V_{1m}}{Z_1} = \frac{400}{4217} = \mathbf{0.095\,A}$$

At the third harmonic frequency,

$$Z_3 = R + j\left(3\omega L - \frac{1}{3\omega C}\right) = R$$

since resonance occurs at the third harmonic, i.e. $Z_3 = 2\,\Omega$

Maximum value of current at the third harmonic frequency,

$$I_{3m} = \frac{V_{3m}}{Z_3} = \frac{10}{2} = \mathbf{5\,A}$$

(Note that the magnitude of I_{3m} compared with I_{1m} is 5/0.095, i.e. × **52.6 greater**)

Problem 25. A voltage wave has an amplitude of 800 V at the fundamental frequency of 50 Hz and its nth harmonic has an amplitude 1.5% of the fundamental. The voltage is applied to a series circuit containing resistance 5 Ω, inductance 0.369 H and capacitance 0.122 μF. Resonance occurs at the nth harmonic. Determine (a) the value of n, (b) the maximum value of current at the nth harmonic, (c) the p.d. across the capacitor at the nth harmonic and (d) the maximum value of the fundamental current.

(a) For resonance at the nth harmonic, $n\omega L = 1/(n\omega C)$, from which

$$n^2 = \frac{1}{\omega^2 LC} \text{ and } n = \frac{1}{\omega\sqrt{(LC)}}$$

Hence $$n = \frac{1}{2\pi 50\sqrt{(0.369)(0.122\times 10^{-6})}} = \mathbf{15}$$

Thus resonance occurs at the 15th harmonic.

(b) At resonance, impedance $Z_{15} = R = 5\,\Omega$. Hence the maximum value of current at the 15th harmonic,

$$I_{15m} = \frac{V_{15m}}{R} = \frac{(1.5/100)\times 800}{5} = \mathbf{2.4\,A}$$

(c) At the 15th harmonic, capacitive reactance,

$$X_{C15} = \frac{1}{15\omega C} = \frac{1}{15(2\pi 50)(0.122\times 10^{-6})} = 1739\,\Omega$$

Hence the p.d. across the capacitor at the 15th harmonic

$$= (I_{15m})(X_{C15}) = (2.4)(1739) = \mathbf{4.174\,kV}$$

(d) At the fundamental frequency, inductive reactance,

$$X_{L1} = \omega L = (2\pi 50)(0.369) = 115.9\,\Omega$$

and capacitive reactance,

$$X_{Cl} = \frac{1}{\omega C} = \frac{1}{(2\pi 50)(0.122\times 10^{-6})} = 26091\,\Omega$$

Impedance at the fundamental frequency,

$$|Z| = \sqrt{[R^2 + (X_C - X_L)^2]} = 25975\,\Omega$$

Maximum value of current at the fundamental frequency,

$$I_{1m} = \frac{V_{1m}}{Z_1} = \frac{800}{25975} = \mathbf{0.031\,A \text{ or } 31\,mA}$$

Now try the following Practice Exercise

Practice Exercise 156 Harmonic resonance (Answers on page 831)

1. A voltage waveform having a fundamental of maximum value 250 V and a third harmonic of maximum value 20 V is applied to a series circuit comprising a 5 Ω resistor, a 400 mH inductance and a 0.5 μF capacitor. Determine (a) the fundamental frequency for resonance with the third harmonic and (b) the maximum values of the fundamental and third harmonic components of the current.

2. A complex voltage waveform has a maximum value of 500 V at the fundamental frequency of 60 Hz and contains a 17th harmonic having an amplitude of 2% of the fundamental. The voltage is applied to a series circuit containing resistance 2 Ω, inductance 732 mH and capacitance 33.26 nF. Determine (a) the maximum value of the 17th harmonic current, (b) the maximum value of the 17th harmonic p.d. across the capacitor and (c) the amplitude of the fundamental current.

3. A complex voltage waveform v is given by the expression

$$v = 150\sin\omega t + 25\sin\left(3\omega t - \frac{\pi}{6}\right) + 10\sin\left(5\omega t + \frac{\pi}{3}\right) \text{ volts}$$

where $\omega = 314$ rad/s. The voltage is applied to a circuit consisting of a coil of resistance $10\,\Omega$ and inductance 50 mH in series with a variable capacitor.

(a) Calculate the value of the capacitance which will give resonance with the triple frequency component of the voltage. (b) Write down the corresponding equation for the current waveform. (c) Determine the r.m.s. value of current. (d) Find the power dissipated in the circuit.

4. A complex voltage of fundamental frequency 50 Hz is applied to a series circuit comprising resistance $20\,\Omega$, inductance $800\,\mu$H and capacitance $74.94\,\mu$F. Resonance occurs at the nth harmonic. Determine the value of n.

5. A complex voltage given by $v = 1200\sin\omega t + 300\sin 3\omega t + 100\sin 5\omega t$ volts is applied to a circuit containing a $25\,\Omega$ resistor, a $12\,\mu$F capacitor and a 37 mH inductance connected in series. The fundamental frequency is 79.62 Hz. Determine (a) the r.m.s. value of the voltage, (b) an expression for the current waveform, (c) the r.m.s. value of current, (d) the amplitude of the third harmonic voltage across the capacitor, (e) the circuit power and (f) the overall power factor.

39.11 Sources of harmonics

(i) Harmonics may be produced in the **output waveform of an a.c. generator**. This may be due either to 'tooth-ripple', caused by the effect of the slots that accommodate the windings, or to the non-sinusoidal airgap flux distribution.

Great care is taken to ensure a sinusoidal output from generators in large supply systems; however, non-linear loads will cause harmonics to appear in the load current waveform. Thus harmonics are produced in devices that have a **non-linear response** to their inputs. Non-linear circuit elements (i.e. those in which the current flowing through them is not proportional to the applied voltage) include rectifiers and any large-signal electronic amplifier in which diodes, transistors, valves or iron-cored inductors are used.

(ii) A **rectifier** is a device for converting an alternating or an oscillating current into a unidirectional or approximate direct current. A rectifier has a low impedance to current flow in one direction and a nearly infinite impedance to current flow in the opposite direction. Thus, when an alternating current is applied to a rectifier, current will flow through it during the positive half-cycles only; the current is zero during the negative half-cycles. A typical current waveform is shown in Figure 39.31. This 'half-wave rectification' is produced by using a single diode. The waveform is similar in shape to that shown in Figure 39.14, page 582, where the d.c. component brought the negative half-cycle up to the zero current point. The waveform shown in Figure 39.31 is typical of one containing a fairly large second harmonic.

(iii) **Transistors** and **valves** are non-linear devices in that sinusoidal input results in different positive and negative half-cycle amplifications. This means that the output half cycles have different amplitudes. Since they have a different shape, even harmonic distortion is suggested (see Section 39.3).

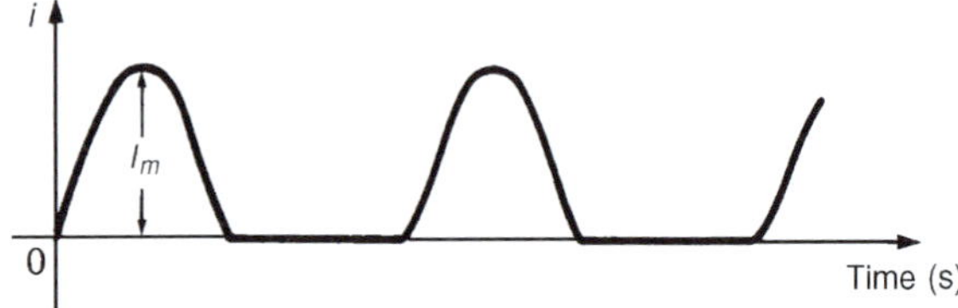

Figure 39.31 Typical current waveform containing a fairly large second harmonic

(iv) **Ferromagnetic-cored coils** are a source of harmonic generation in a.c. circuits because of the non-linearity of the B/H curve and the hysteresis loop, especially if saturation occurs. Let a sinusoidal voltage $v = V_m \sin\omega t$ be applied to a ferromagnetic-cored coil (having low resistance relative to inductive reactance) of cross-section area A square metres and possessing N turns.

If ϕ is the flux produced in the core then the instantaneous voltage is given by $\boldsymbol{v = N(d\phi/dt)}$.

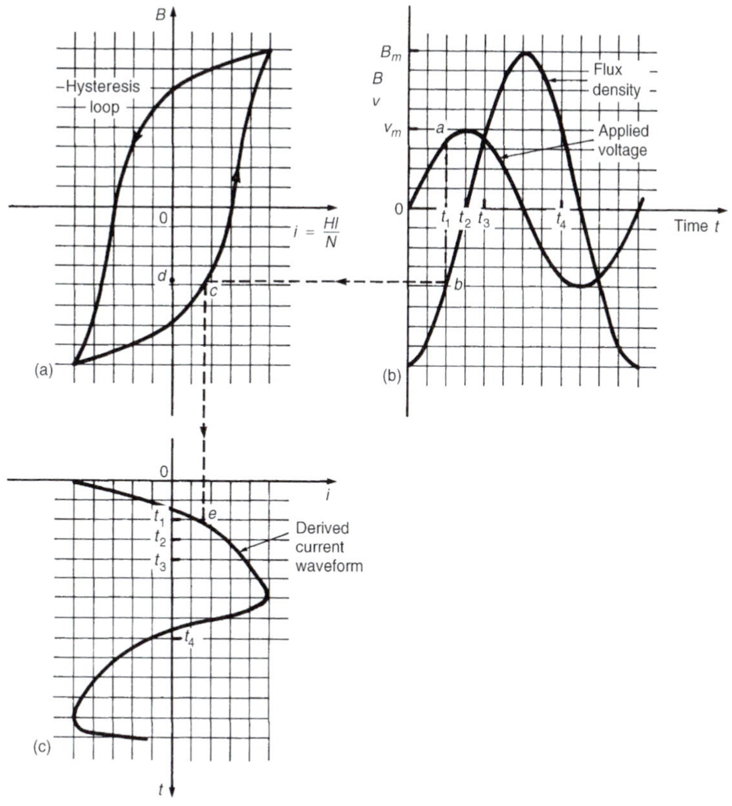

Figure 39.32

If B is the flux density of the core, then, since $\Phi = BA$,

$$v = N\frac{\mathrm{d}}{\mathrm{d}t}(BA) = \boldsymbol{NA\frac{\mathrm{d}B}{\mathrm{d}t}}$$

since area A is a constant for a particular core. Separating the variables gives

$$\int \mathrm{d}B = \frac{1}{NA}\int v\,\mathrm{d}t$$

i.e. $$B = \frac{1}{NA}\int V_m \sin\omega t\,\mathrm{d}t = \frac{-V_m}{\omega NA}\cos\omega t$$

Since $-\cos\omega t = \sin(\omega t - 90°)$

$$\boldsymbol{B = \frac{V_m}{\omega NA}\sin(\omega t - 90°)} \qquad (34)$$

Equation (34) shows that if the applied voltage is sinusoidal, the flux density B in the iron core must also be sinusoidal but lagging by 90°.

The condition of low resistance relative to inductive reactance, giving a sinusoidal flux from a sinusoidal supply voltage, is called **free magnetization**.

Consider the application of a sinusoidal voltage to a coil wound on a core with a hysteresis loop, as shown in Figure 39.32(a). The horizontal axis of a hysteresis loop is magnetic field strength H, but since $H = Ni/l$ and N and l (the length of the flux path) are constant, the axis may be directly scaled as current i (i.e. $i = Hl/N$). Figure 39.32(b) shows sinusoidal voltage v and flux density B waveforms, B lagging v by 90°.

The current waveform is shown in Figure 39.32(c) and is derived as follows. At time t_1, point a on the voltage curve corresponds to point b on the flux density curve and point c on the hysteresis loop. The current at time t_1 is given by the distance dc. Plotting this current on a vertical time-scale gives the derived point e on the current curve. A similar procedure is adopted for times t_2, t_3 and so on over one cycle of the voltage.

(Note that it is important to move around the hysteresis loop in the correct direction.) It is seen from the current curve that it is non-sinusoidal and that the positive and negative

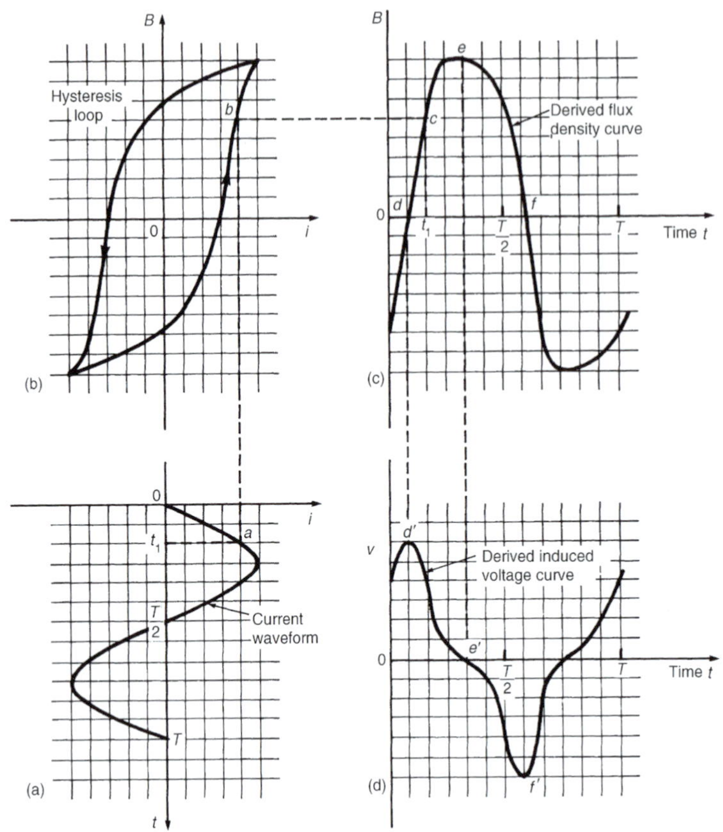

Figure 39.33

half-cycles are identical. This indicates that the waveform contains only odd harmonics (see Section 39.3).

(v) If, in a circuit containing a ferromagnetic-cored coil, the resistance is high compared with the inductive reactance, then the current flowing from a sinusoidal supply will tend to be sinusoidal. This means that the flux density B of the core cannot be sinusoidal since it is related to the current by the hysteresis loop. This means, in turn, that the induced voltage due to the alternating flux (i.e. $v = NA(dB/dt)$) will not be sinusoidal. This condition is called **forced magnetization**.

The shape of the induced voltage waveform under forced magnetization is obtained as follows. The current waveform is shown on a vertical axis in Figure 39.33(a). The hysteresis loop corresponding to the maximum value of circuit current is drawn as shown in Figure 39.33(b). The flux density curve which is derived from the sinusoidal current waveform is shown in Figure 39.33(c). Point a on the current wave at time t_1 corresponds to point b on the hysteresis loop and to point c on the flux density curve. By taking other points throughout the current cycle the flux density curve is derived as shown.

The relationship between the induced voltage v and the flux density B is given by $v = NA(dB/dt)$. Here dB/dt represents the rate of change of flux density with respect to time, i.e. the gradient of the B/t curve. At point d the gradient of the B/t curve is a maximum in the positive direction. Thus v will be maximum positive as shown by point d' in Figure 39.33(d). At point e the gradient (i.e. dB/dt) is zero, thus v is zero, as shown by point e'. At point f the gradient is maximum in a negative direction, thus v is maximum negative, as shown by point f'. If all such points are taken around the B/t curve, the curve representing induced voltage, shown in Figure 39.33(d), is produced. The resulting voltage waveform is non-sinusoidal. The positive and negative half cycles are identical in shape, indicating that the waveform

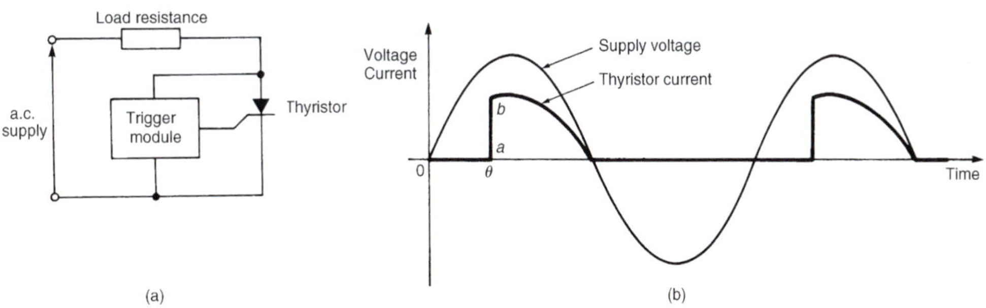

Figure 39.34

contains a fundamental and a prominent third harmonic.

(vi) The amount of power delivered to a load can be controlled using a **thyristor**, which is a semiconductor device. Examples of applications of controlled rectification include lamp and heater controls and the control of motor speeds. A basic circuit used for single-phase power control is shown in Figure 39.34(a). The trigger module contains circuitry to produce the necessary gate current to turn the thyristor on. If the pulse is applied at time θ/ω, where θ is the firing or triggering angle, then the current flowing in the load resistor has a waveform as shown in Figure 39.34(b). The sharp rise-time (shown as *ab* in Figure 39.34(b)), however, gives rise to harmonics.

(vii) In **microelectronic systems** rectangular waveforms are common. Again, fast rise-times give rise to harmonics, especially at high frequency. These harmonics can be fed back to the mains if not filtered.

There are thus a large number of sources of harmonics.

For fully worked solutions to each of the problems in Practice Exercises 148 to 156 in this chapter, go to the website:
www.routledge.com/cw/bird

Chapter 40

A numerical method of harmonic analysis

Why it is important to understand: **Harmonic analysis**

Harmonic analysis is a branch of mathematics concerned with the representation of functions or signals as the superposition of basic waves, and the study of and generalization of the notions of Fourier series and Fourier transforms. In the past two centuries, it has become a vast subject with applications in areas as diverse as signal processing, quantum mechanics and neuroscience. This chapter explains a tabular method of harmonic analysis and explains how to predict the probable harmonic content of a waveform on inspection.

At the end of this chapter you should be able to:

- use a tabular method to determine the Fourier series for a complex waveform
- predict the probable harmonic content of a waveform on inspection

40.1 Introduction

Many practical waveforms can be represented by simple mathematical expressions, and, by using Fourier series, the magnitude of their harmonic components determined. For waveforms not in this category, analysis may be achieved by numerical methods. **Harmonic analysis** is the process of resolving a periodic, non-sinusoidal quantity into a series of sinusoidal components of ascending order of frequency.

40.2 Harmonic analysis on data given in tabular or graphical form

A Fourier series is merely a trigonometric series of the form:

$$f(x) = a_0 + a_1 \cos x + a_2 \cos 2x + \cdots + b_1 \sin x + b_2 \sin 2x + \cdots$$

$$\text{i.e.} \quad f(x) = a_0 + \sum_{n=1}^{\infty}(a_n \cos nx + b_n \sin nx)$$

The Fourier coefficients a_0, a_n and b_n all require functions to be integrated, i.e.

$$a_0 = \frac{1}{2\pi}\int_{-\pi}^{\pi} f(x)\,dx = \frac{1}{2\pi}\int_{0}^{2\pi} f(x)\,dx$$

$= $ mean value of $f(x)$ in the range $-\pi$ to π or 0 to 2π

$$a_n = \frac{1}{\pi}\int_{-\pi}^{\pi} f(x)\cos nx\,dx = \frac{1}{\pi}\int_{0}^{2\pi} f(x)\cos nx\,dx$$

$= $ twice the mean value of $f(x)\cos nx$ in the range 0 to 2π

Electrical Circuit Theory and Technology. 978-1-138-67349-6, © 2017 John Bird. Published by Taylor & Francis. All rights reserved.

$$b_n = \frac{1}{\pi}\int_{-\pi}^{\pi} f(x)\sin nx\,\mathrm{d}x = \frac{1}{\pi}\int_{0}^{2\pi} f(x)\sin nx\,\mathrm{d}x$$
$$= \text{twice the mean value of } f(x)\sin nx \text{ in the range } 0 \text{ to } 2\pi$$

However, irregular waveforms are not usually defined by mathematical expressions and thus the Fourier coefficients cannot be determined by using calculus. In these cases, approximate methods, such as the **trapezoidal rule**, can be used to evaluate the Fourier coefficients.

Most practical waveforms to be analysed are periodic. Let the period of a waveform be 2π and be divided into p equal parts as shown in Figure 40.1. The width of each interval is thus $2\pi/p$. Let the ordinates be labelled $y_0, y_1, y_2, \ldots, y_p$ (note that $y_0 = y_p$). The trapezoidal rule states:

$$\text{Area} \approx \begin{pmatrix}\text{width of}\\ \text{interval}\end{pmatrix}\left[\frac{1}{2}\begin{pmatrix}\text{first + last}\\ \text{ordinate}\end{pmatrix} + \begin{matrix}\text{sum of}\\ \text{remaining}\\ \text{ordinates}\end{matrix}\right]$$
$$\approx \frac{2\pi}{p}\left[\frac{1}{2}(y_0 + y_p) + y_1 + y_2 + y_3 + \cdots\right]$$

Since $y_0 = y_p$, then $\frac{1}{2}(y_0 + y_p) = y_0 = y_p$

Hence area $\approx \frac{2\pi}{p}\sum_{k=1}^{p} y_k$

$$\text{Mean value} = \frac{\text{area}}{\text{length of base}}$$
$$\approx \frac{1}{2\pi}\left(\frac{2\pi}{p}\right)\sum_{k=1}^{p} y_k \approx \frac{1}{p}\sum_{k=1}^{p} y_k$$

However, a_0 = mean value of $f(x)$ in the range 0 to 2π.

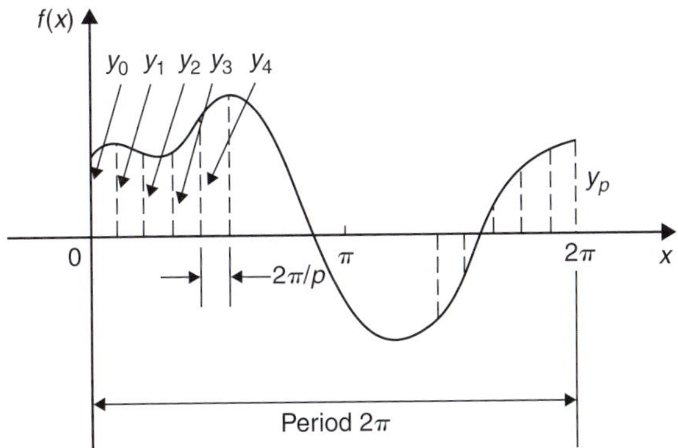

Figure 40.1

Thus

$$a_0 \approx \frac{1}{p}\sum_{k=1}^{p} y_k \qquad (1)$$

Similarly, a_n = twice the mean value of $f(x)\cos nx$ in the range 0 to 2π, thus,

$$a_n \approx \frac{2}{p}\sum_{k=1}^{p} y_k \cos nx_k \qquad (2)$$

and b_n = twice the mean value of $f(x)\sin nx$ in the range 0 to 2π, thus

$$b_n \approx \frac{2}{p}\sum_{k=1}^{p} y_k \sin nx_k \qquad (3)$$

Problem 1. The values of the voltage v volts at different moments in a cycle are given by:

θ degrees	30	60	90	120	150	180
v (volts)	62	35	−38	−64	−63	−52

θ degrees	210	240	270	300	330	360
v (volts)	−28	24	80	96	90	70

Draw the graph of voltage v against angle θ and analyse the voltage into its first three constituent harmonics, each coefficient correct to 2 decimal places.

The graph of voltage v against angle θ is shown in Figure 40.2. The range 0 to 2π is divided into 12 equal intervals giving an interval width of $2\pi/12$, i.e. $\pi/6$ or 30°. The values of the ordinates $y_1, y_2, y_3, \ldots$ are 62, 35, −38, … from the given table of values. If a larger number of intervals are used, results having a greater accuracy are achieved. The data is tabulated in the proforma shown in Table 40.1.

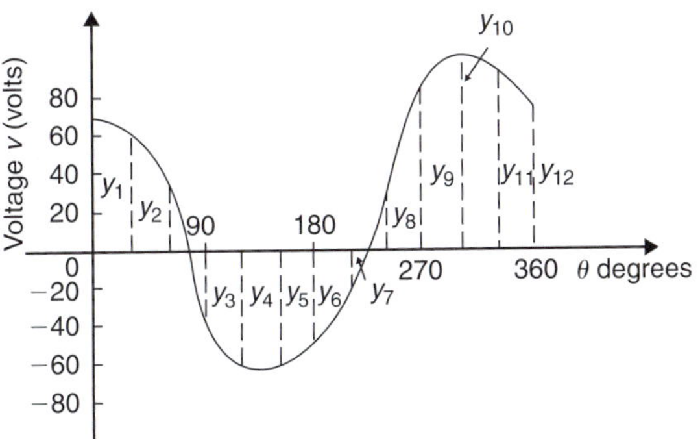

Figure 40.2

Part 4

Table 40.1

Ordinates	$\theta°$	v	$\cos\theta$	$v\cos\theta$	$\sin\theta$	$v\sin\theta$	$\cos 2\theta$	$v\cos 2\theta$	$\sin 2\theta$	$v\sin 2\theta$	$\cos 3\theta$	$v\cos 3\theta$	$\sin 3\theta$	$v\sin 3\theta$
y_1	30	62	0.866	53.69	0.5	31	0.5	31	0.866	53.69	0	0	1	62
y_2	60	35	0.5	17.5	0.866	30.31	−0.5	−17.5	0.866	30.31	−1	−35	0	0
y_3	90	−38	0	0	1	−38	−1	38	0	0	0	0	−1	38
y_4	120	−64	−0.5	32	0.866	−55.42	−0.5	32	−0.866	55.42	1	−64	0	0
y_5	150	−63	−0.866	54.56	0.5	−31.5	0.5	−31.5	−0.866	54.56	0	0	1	−63
y_6	180	−52	−1	52	0	0	1	−52	0	0	−1	52	0	0
y_7	210	−28	−0.866	24.25	−0.5	14	0.5	−14	0.866	−24.25	0	0	−1	28
y_8	240	24	−0.5	−12	−0.866	−20.78	−0.5	−12	0.866	−20.78	1	24	0	0
y_9	270	80	0	0	−1	−80	−1	−80	0	0	0	0	1	80
y_{10}	300	96	0.5	48	−0.866	−83.14	−0.5	−48	−0.866	−83.14	−1	−96	0	0
y_{11}	330	90	0.866	77.94	−0.5	−45	0.5	45	−0.866	−77.94	0	0	−1	−90
y_{12}	360	70	1	70	0	0	1	70	0	0	1	70	0	0
$\sum_{k=1}^{12} y_k = 212$				$\sum_{k=1}^{12} y_k \cos\theta_k = 417.94$		$\sum_{k=1}^{12} y_k \sin\theta_k = -278.53$		$\sum_{k=1}^{12} y_k \cos 2\theta_k = -39$		$\sum_{k=1}^{12} y_k \sin 2\theta_k = 29.43$		$\sum_{k=1}^{12} y_k \cos 3\theta_k = -49$		$\sum_{k=1}^{12} y_k \sin 3\theta_k = 55$

From equation (1), $a_0 \approx \frac{1}{p}\sum_{k=1}^{p} y_k = \frac{1}{12}(212)$

$= 17.67$ (since $p = 12$)

From equation (2), $a_n \approx \frac{2}{p}\sum_{k=1}^{p} \cos nx_k$

Hence $a_1 \approx \frac{2}{12}(417.94) = 69.66$

$a_2 \approx \frac{2}{12}(-39) = -6.50$

and $a_3 \approx \frac{2}{12}(-49) = -8.17$

From equation (3), $b_n \approx \frac{2}{p}\sum_{k=1}^{p} y_k \sin nx_k$

Hence $b_1 \approx \frac{2}{12}(-278.53) = -46.42$

$b_2 \approx \frac{2}{12}(29.43) = 4.91$

and $b_3 \approx \frac{2}{12}(55) = 9.17$

Substituting these values into the Fourier series:

$$f(x) = a_0 + \sum_{n=1}^{\infty}(a_n \cos nx + b_n \sin nx)$$

gives: $\boldsymbol{v = 17.67 + 69.66\cos\theta - 6.50\cos 2\theta}$

$\boldsymbol{- 8.17\cos 3\theta + \cdots - 46.42\sin\theta}$

$\boldsymbol{+ 4.91\sin 2\theta + 9.17\sin 3\theta + \cdots}$ (4)

Note that in equation (4), $(-46.42\sin\theta + 69.66\cos\theta)$ comprises the fundamental, $(4.91\sin 2\theta - 6.50\cos 2\theta)$ comprises the second harmonic, $(9.17\sin 3\theta - 8.17\cos 3\theta)$ comprises the third harmonic.

It is shown in *Higher Engineering Mathematics* that

$$a\sin\omega t + b\cos\omega t \equiv R\sin(\omega t + \alpha)$$

where $a = R\cos\alpha$, $b = R\sin\alpha$, $R = \sqrt{(a^2 + b^2)}$ and

$$\alpha = \tan^{-1}\frac{b}{a}$$

For the fundamental, $R = \sqrt{[(-46.42)^2 + (69.66)^2]}$

$= 83.71$

If $a = R\cos\alpha$, then $\cos\alpha = \frac{a}{R} = \frac{-46.42}{83.71}$, which is negative,

and if $b = R\sin\alpha$, then $\sin\alpha = \frac{b}{R} = \frac{69.66}{83.71}$, which is positive.

The only quadrant where $\cos\alpha$ is negative *and* $\sin\alpha$ is positive is the second quadrant.

Hence

$$\alpha = \tan^{-1}\frac{b}{a} = \tan^{-1}\frac{69.66}{-46.42} = 123.68° \text{ or } 2.16 \text{ rad}$$

Thus $(-46.42\sin\theta + 69.66\cos\theta) = 83.71\sin(\theta + 2.16)$

By a similar method it may be shown that the second harmonic

$(4.91\sin 2\theta - 6.50\cos 2\theta) \equiv 8.15\sin(2\theta - 0.92)$

and the third harmonic

$(9.17\sin 3\theta - 8.17\cos 3\theta) \equiv 12.28\sin(3\theta - 0.73)$

Hence equation (4) may be re-written as:

$$\boldsymbol{v = 17.67 + 83.71\sin(\theta + 2.16) + 8.15\sin(2\theta - 0.92) + 12.28\sin(3\theta - 0.73)\ \text{volts}}$$

which is the form used in Chapter 39 with complex waveforms.

Now try the following Practice Exercise

Practice Exercise 157 Harmonic analysis on data given in tabular form (Answers on page 831)

Determine the Fourier series to represent the periodic functions given by the tables of values in Problems 1 to 3, up to and including the third harmonics and each coefficient correct to 2 decimal places. Use 12 ordinates in each case.

1.

Angle $\theta°$	30	60	90	120	150	180
Displacement y	40	43	38	30	23	17

Angle $\theta°$	210	240	270	300	330	360
Displacement y	11	9	10	13	21	32

2.

Angle $\theta°$	0	30	60	90	120	150
Voltage v	−5.0	−1.5	6.0	12.5	16.0	16.5

Angle $\theta°$	180	210	240	270	300	330
Voltage v	15.0	12.5	6.5	−4.0	−7.0	−7.5

3.

Angle $\theta°$	30	60	90	120	150	180
Current i	0	−1.4	−1.8	−1.9	−1.8	−1.3

Angle $\theta°$	210	240	270	300	330	360
Current i	0	2.2	3.8	3.9	3.5	2.5

40.3 Complex waveform considerations

It is sometimes possible to predict the harmonic content of a waveform on inspection of particular waveform characteristics.

(i) If a periodic waveform is such that the area above the horizontal axis is equal to the area below then the mean value is zero. Hence $a_0 = 0$ (see Figure 40.3(a)).

(ii) An **even function** is symmetrical about the vertical axis and contains **no sine terms** (see Figure 40.3(b)).

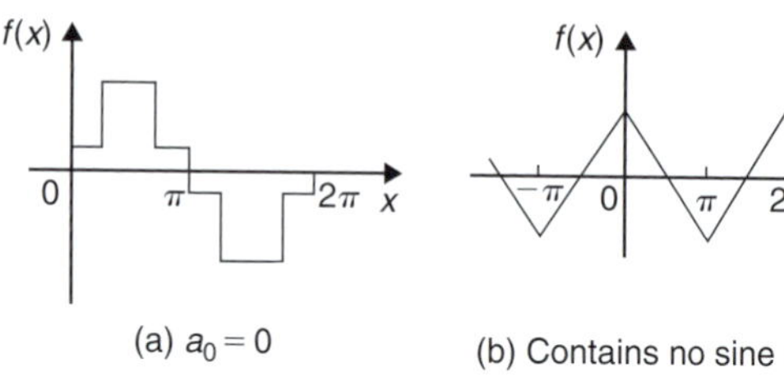

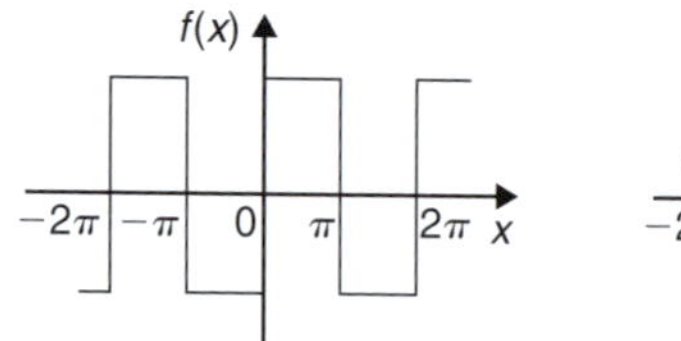

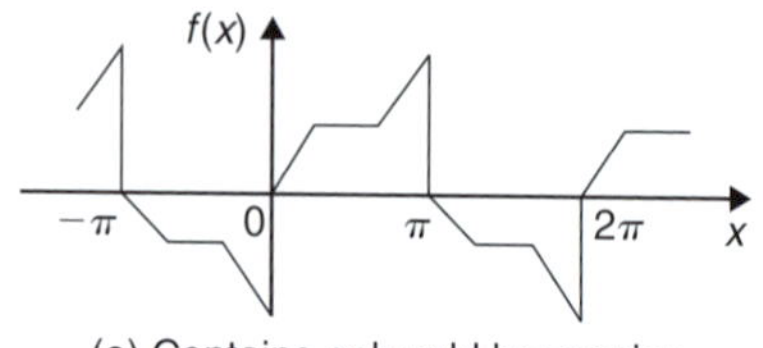

Figure 40.3

(iii) An **odd function** is symmetrical about the origin and contains no **cosine terms** (see Figure 40.3(c)).

(iv) $f(x) = f(x + \pi)$ represents a waveform which repeats after half a cycle and **only even harmonics** are present (see Figure 40.3(d)).

(v) $f(x) = -f(x + \pi)$ represents a waveform for which the positive and negative cycles are identical in shape and **only odd harmonics** are present (see Figure 40.3(e)).

Problem 2. Without calculating Fourier coefficients state which harmonics will be present in the waveforms shown in Figure 40.4.

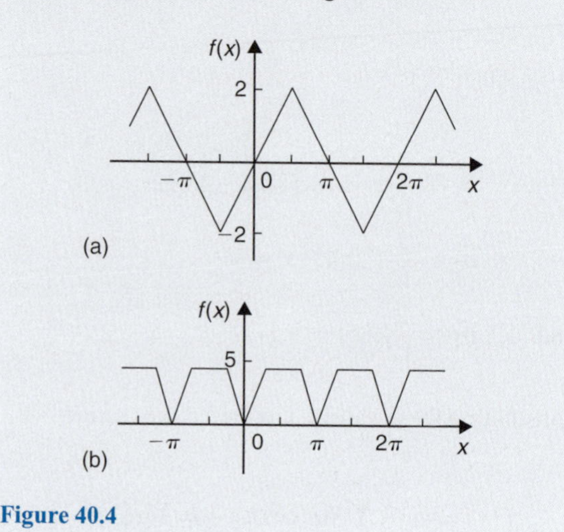

Figure 40.4

(a) The waveform shown in Figure 40.4(a) is symmetrical about the origin and is thus an odd function. An odd function contains no cosine terms. Also, the waveform has the characteristic $f(x) = -f(x + \pi)$, i.e. the positive and negative half-cycles are identical in shape. Only odd harmonics can be present in such a waveform. Thus the waveform shown in Figure 40.4(a) contains **only odd sine terms**. Since the area above the x-axis is equal to the area below, $a_0 = 0$

(b) The waveform shown in Figure 40.4(b) is symmetrical about the $f(x)$ axis and is thus an even function. An even function contains no sine terms. Also, the waveform has the characteristic $f(x) = f(x + \pi)$, i.e. the waveform repeats itself after half a cycle. Only even harmonics can be present in such a waveform. Thus the waveform shown in Figure 40.4(b) contains **only even cosine terms** (together with a constant term, a_0)

Problem 3. An alternating current i amperes is shown in Figure 40.5. Analyse the waveform into its constituent harmonics as far as and including the fifth harmonic, correct to 2 decimal places, by taking 30° intervals.

With reference to Figure 40.5, the following characteristics are noted:

(i) The mean value is zero since the area above the θ axis is equal to the area below it. Thus the constant term, or d.c. component, $a_0 = 0$

(ii) Since the waveform is symmetrical about the origin the function i is odd, which means that there are no cosine terms present in the Fourier series.

(iii) The waveform is of the form $f(\theta) = -f(\theta + \pi)$ which means that only odd harmonics are present.

Investigating waveform characteristics has thus saved unnecessary calculations and in this case the Fourier series has only odd sine terms present, i.e.

$$i = b_1 \sin\theta + b_3 \sin 3\theta + b_5 \sin 5\theta + \cdots$$

A proforma, similar to Table 40.1, but without the 'cosine terms' columns and without the 'even sine terms' columns is shown in Table 40.2 up to, and including, the fifth harmonic, from which the Fourier

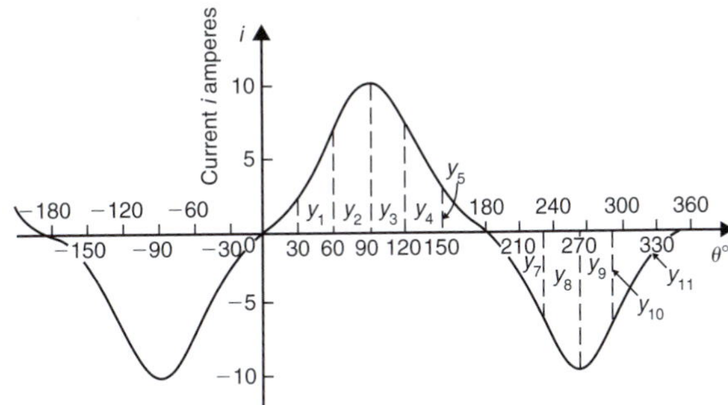

Figure 40.5

Table 40.2

Ordinate	$\theta°$	i	$\sin\theta$	$i\sin\theta$	$\sin 3\theta$	$i\sin 3\theta$	$\sin 5\theta$	$i\sin 5\theta$
y_1	30	2	0.5	1	1	2	0.5	1
y_2	60	7	0.866	6.06	0	0	−0.866	−6.06
y_3	90	10	1	10	−1	−10	1	10
y_4	120	7	0.866	6.06	0	0	−0.866	−6.06
y_5	150	2	0.5	1	1	2	0.5	1
y_6	180	0	0	0	0	0	0	0
y_7	210	−2	−0.5	1	−1	2	−0.5	1
y_8	240	−7	−0.866	6.06	0	0	0.866	−6.06
y_9	270	−10	−1	10	1	−10	−1	10
y_{10}	300	−7	−0.866	6.06	0	0	0.866	−6.06
y_{11}	330	−2	−0.5	1	−1	2	−0.5	1
y_{12}	360	0	0	0	0	0	0	0
				$\sum_{k=1}^{12} i_k \sin\theta_k = 48.24$		$\sum_{k=1}^{12} i_k \sin 3\theta_k = -12$		$\sum_{k=1}^{12} i_k \sin 5\theta_k = -0.24$

coefficients b_1, b_3 and b_5 can be determined. Twelve coordinates are chosen and labelled $y_1, y_2, y_3, \ldots y_{12}$ as shown in Figure 40.5.

From equation (3), Section 40.2, $b_n \approx \frac{2}{p}\sum_{k=1}^{p} i_k \sin n\theta_k$, where $p=12$.

Hence $b_1 \approx \frac{2}{12}(48.24) = 8.04$

$b_3 \approx \frac{2}{12}(-12) = -2.00$

and $b_5 \approx \frac{2}{12}(-0.24) = -0.04$

Thus the Fourier series for current i is given by:

$\boldsymbol{i = 8.04\sin\theta - 2.00\sin 3\theta - 0.04\sin 5\theta}$

Now try the following Practice Exercise

Practice Exercise 158 Harmonic wave considerations (Answers on page 831)

1. Without performing calculations, state which harmonics will be present in the waveforms shown in Figure 40.6.
2. Analyse the periodic waveform of displacement y against angle θ in Figure 40.4(a) into its constituent harmonics as far as and including the third harmonic, by taking 30° intervals.
3. For the waveform of current shown in Figure 40.7(b) state why only a d.c. component and even cosine terms will appear in the Fourier series and determine the series, using $\pi/6$ rad intervals, up to and including the sixth harmonic.

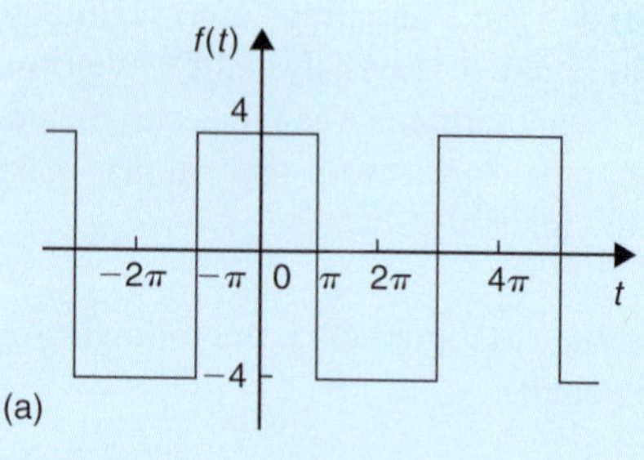

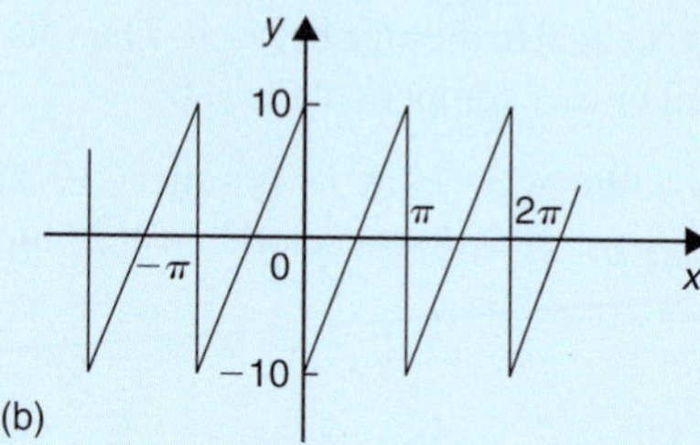

Figure 40.6

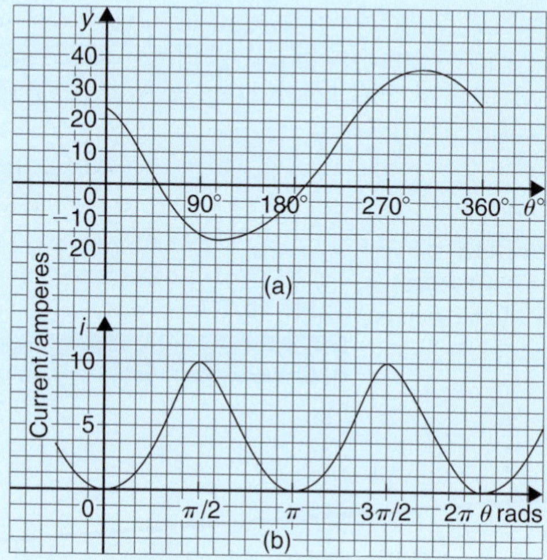

Figure 40.7

Part 4

For fully worked solutions to each of the problems in Practice Exercises 157 and 158 in this chapter, go to the website:
www.routledge.com/cw/bird

Chapter 41

Magnetic materials

***Why it is important to understand:* Magnetic materials**

Magnetism, the phenomenon by which materials assert an attractive or repulsive force or influence on other materials, has been known for thousands of years. However, the underlying principles and mechanisms that explain the magnetic phenomenon are complex and subtle, and their understanding has eluded scientists until relatively recent times. Many of our modern technological devices rely on magnetism and magnetic materials; these include electrical power generators and transformers, electric motors, radio, television, telephones, computers and components of sound and video reproduction systems. Iron, some steels and the naturally occurring mineral lodestone are well-known examples of materials that exhibit magnetic properties. Not so familiar, however, is the fact that all substances are influenced to one degree or another by the presence of a magnetic field. Permanent magnet materials are incorporated in a device for only one reason, to increase magnetic flux. The flux may be necessary for detection, as it is in sensors, to assist in creating a force or torque, as it does in motors, actuators and speakers, to generate a voltage as it does in generators and alternators, to create a magnetic field as it does in Magnetic Resonance Imaging (MRI) systems and electron beam devices. This chapter provides a brief description of terms associated with magnetic fields and briefly looks at the phenomena of diamagnetism, paramagnetism and ferromagnetism, explains hysteresis and eddy current losses and distinguishes between non-permanent and permanent magnetic materials.

At the end of this chapter you should be able to:

- recognize terms associated with magnetic circuits
- appreciate magnetic properties of materials
- categorize materials as ferromagnetic, diamagnetic and paramagnetic
- explain hysteresis and calculate hysteresis loss
- explain and calculate eddy current loss
- explain a method of separation of hysteresis and eddy current loss and determine the separate losses from given data
- distinguish between non-permanent and permanent magnetic materials.

Electrical Circuit Theory and Technology. 978-1-138-67349-6, © 2017 John Bird. Published by Taylor & Francis. All rights reserved.

41.1 Revision of terms and units used with magnetic circuits

In Chapter 9, page 125, a number of terms used with magnetic circuits are defined. These are summarized below.

(a) A **magnetic field** is the state of the space in the vicinity of a permanent magnet or an electric current throughout which the magnetic forces produced by the magnet or current are discernible.

(b) **Magnetic flux Φ** is the amount of magnetic field produced by a magnetic source. The unit of magnetic flux is the **weber***. If the flux linking one turn in a circuit changes by one weber in one second, a voltage of one volt will be induced in that turn.

(c) **Magnetic flux density B** is the amount of flux passing through a defined area that is perpendicular to the direction of the flux.

$$\text{Magnetic flux density} = \frac{\text{magnetic flux}}{\text{area}}$$

i.e. $\boldsymbol{B = \Phi/A}$ where A is the area in square metres. The unit of magnetic flux density is the **tesla***, $\boldsymbol{T}$, where $1\ T = 1\ \text{Wb/m}^2$.

(d) **Magnetomotive force (mmf)** is the cause of the existence of a magnetic flux in a magnetic circuit.

$$\textbf{mmf, } \boldsymbol{F_m = NI} \textbf{ amperes}$$

where N is the number of conductors (or turns) and I is the current in amperes. The unit of mmf is sometimes expressed as 'ampere-turns'. However, since 'turns' have no dimension, the S.I. unit of mmf is the ampere.

(e) **Magnetic field strength (or magnetizing force),**

$$\boldsymbol{H = NI/l} \textbf{ ampere per metre}$$

where l is the mean length of the flux path in metres.

Thus $\textbf{mmf} = \boldsymbol{NI = Hl}$ **amperes**

*Who was **Weber**? For image and resume of Weber, see page 124. To find out more go to **www.routledge.com/cw/bird**

*Who was **Tesla**? For image and resume of Tesla, see page 125. To find out more go to **www.routledge.com/cw/bird**

(f) μ_0 is a constant called the **permeability of free space** (or the magnetic space constant). The value of μ_0 is $4\pi \times 10^{-7}$ H/m.

For air, or any non-magnetic medium, the ratio

$$\boldsymbol{B/H = \mu_0}$$

(Although all non-magnetic materials, including air, exhibit slight magnetic properties, these can effectively be neglected.)

(g) μ_r is the **relative permeability** and is defined as

$$\frac{\text{flux density in material}}{\text{flux density in a vacuum}}$$

μ_r varies with the type of magnetic material and, since it is a ratio of flux densities, it has no unit. From its definition, μ_r for a vacuum is 1.

For all media other than free space, $\boldsymbol{B/H = \mu_0\mu_r}$

(h) Absolute permeability $\boldsymbol{\mu = \mu_0\mu_r}$

(i) By plotting measured values of flux density B against magnetic field strength H a **magnetization curve** (or B/H curve) is produced. For non-magnetic materials this is a straight line having the approximate gradient of μ_0. B/H curves for four materials are shown on page 126.

(j) From (g), $\mu_r = B/(\mu_0 H)$. Thus the relative permeability μ_r of a ferromagnetic material is proportional to the gradient of the B/H curve and varies with the magnetic field strength H.

(k) **Reluctance S** (or R_M) is the 'magnetic resistance' of a magnetic circuit to the presence of magnetic flux.

$$\text{Reluctance } S = \frac{F_m}{\Phi} = \frac{NI}{\Phi} = \frac{Hl}{BA} = \frac{l}{(B/H)A} = \frac{l}{\mu_0\mu_r A}$$

The unit of reluctance is $1/H$ (or H^{-1}) or A/Wb

(l) **Permeance** is the magnetic flux per ampere of total magnetomotive force in the path of a magnetic field. It is the reciprocal of reluctance.

Part 4

41.2 Magnetic properties of materials

The full theory of magnetism is one of the most complex of subjects. However, the phenomenon may be satisfactorily explained by the use of a simple model. Bohr and Rutherford, who discovered atomic structure, suggested that electrons move around the nucleus confined to a plane, like planets around the sun. An even better model is to consider each electron as having a surface, which may be spherical or elliptical or something more complicated.

Magnetic effects in materials are due to the electrons contained in them, the electrons giving rise to magnetism in the following two ways:

(i) by revolving around the nucleus

(ii) by their angular momentum about their own axis, called spin.

In each of these cases the charge of the electron can be thought of as moving round in a closed loop and therefore acting as a current loop.

The main measurable quantity of an atomic model is the **magnetic moment**. When applied to a loop of wire carrying a current,

$$\text{magnetic moment} = \text{current} \times \text{area of the loop}$$

Electrons associated with atoms possess magnetic moment which gives rise to their magnetic properties.

Diamagnetism is a phenomenon exhibited by materials having a relative permeability less than unity. When electrons move more or less in a spherical orbit around the nucleus, the magnetic moment due to this orbital is zero, all the current due to moving electrons being considered as averaging to zero. If the net magnetic moment of the electron spins were also zero then there would be no tendency for the electron motion to line up in the presence of a magnetic field. However, as a field is being turned on, the flux through the electron orbitals increases. Thus, considering the orbital as a circuit, there will be, by Faraday's laws, an e.m.f. induced in it which will change the current in the circuit. The flux change will accelerate the electrons in its orbit, causing an induced magnetic moment. By Lenz's law the flux due to the induced magnetic moment will be such as to oppose the applied flux. As a result, the net flux through the material becomes less than in a vacuum. Since relative permeability is defined as

$$\frac{\text{flux density in material}}{\text{flux density in vacuum}}$$

with diamagnetic materials the relative permeability is less than one.

Paramagnetism is a phenomenon exhibited by materials where the relative permeability is greater than unity. Paramagnetism occurs in substances where atoms have a permanent magnetic moment. This may be caused by the orbitals not being spherical or by the spin of the electrons. Electron spins tend to pair up and cancel each other. However, there are many atoms with odd numbers of electrons, or in which pairing is incomplete. Such atoms have what is called a permanent dipole moment. When a field is applied to them they tend to line up with the field, like compass needles, and so strengthen the flux in that region. (Diamagnetic materials do not tend to line up with the field in this way.) When this effect is stronger than the diamagnetic effect, the overall effect is to make the relative permeability greater than one. Such materials are called paramagnetic.

Ferromagnetic materials

Ferromagnetism is the phenomenon exhibited by materials having a relative permeability which is considerably greater than 1 and which varies with flux density. Iron, cobalt and nickel are the only elements that are ferromagnetic at ordinary working temperatures, but there are several alloys containing one or more of these metals as constituents, with widely varying ferromagnetic properties.

Consider the simple model of a single iron atom represented in Figure 41.1. It consists of a small, heavy central nucleus surrounded by a total of 26 electrons. Each electron has an orbital motion about the nucleus in a limited region, or shell, such shells being represented by circles K, L, M and N. The numbers in Figure 41.1 represent the number of electrons in each shell.

The outer shell N contains two loosely held electrons, these electrons becoming the carriers of electric current, making iron electrically conductive. There are 14 electrons in the M shell and it is this group that is responsible for magnetism. An electron carries a negative charge

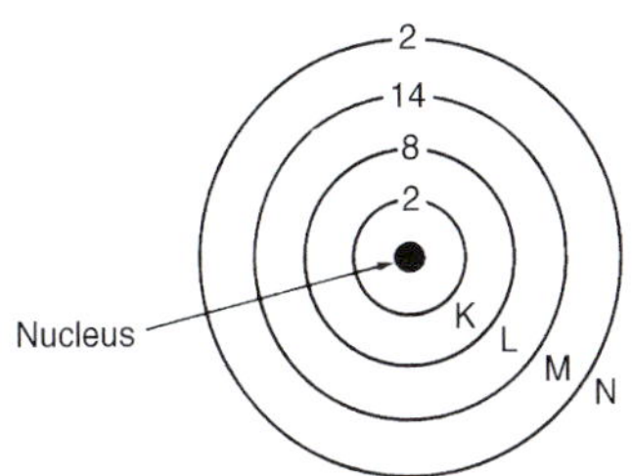

Figure 41.1 Single iron atom

and a charge in motion constitutes an electric current with which is associated a magnetic field. Magnetism would therefore result from the orbital motion of each electron in the atom. However, experimental evidence indicates that the resultant magnetic effect due to all the orbital motions in the metal solid is zero; thus the orbital currents may be disregarded.

In addition to the orbital motion, each electron spins on its own axis. A rotating charge is equivalent to a circular current and gives rise to a magnetic field. In any atom, all the axes about which the electrons spin are parallel, but rotation may be in either direction. In the single atom shown in Figure 41.1, in each of the K, L and N shells equal numbers of electrons spin in the clockwise and anticlockwise directions respectively, and therefore these shells are magnetically neutral. However, in shell M, nine of the electrons spin in one direction while five spin in the opposite direction. There is therefore a resultant effect due to four electrons.

The atom of cobalt has 15 electrons in the M shell, nine spinning in one direction and six in the other. Thus with cobalt there is a resultant effect due to 3 electrons. A nickel atom has a resultant effect due to 2 electrons. The atoms of the paramagnetic elements, such as manganese, chromium or aluminium, also have a resultant effect for the same reasons as that of iron, cobalt and nickel. However, in the diamagnetic materials there is an exact equality between the clockwise and anticlockwise spins.

The total magnetic field of the resultant effect due to the four electrons in the iron atom is large enough to influence other atoms. Thus the orientation of one atom tends to spread through the material, with atoms acting together in groups instead of behaving independently. These groups of atoms, called **domains** (which tend to remain permanently magnetized), act as units. Thus, when a field is applied to a piece of iron, these domains as a whole tend to line up and large flux densities can be produced. This means that the relative permeability of such materials is much greater than one. As the applied field is increased, more and more domains align and the induced flux increases.

The overall magnetic properties of iron alloys and materials containing iron, such as ferrite (ferrite is a mixture of iron oxide together with other oxides – lodestone is a ferrite), depend upon the structure and composition of the material. However, the presence of iron ensures marked magnetic properties of some kind in them. Ferromagnetic effects decrease with temperature, as do those due to paramagnetism. The loss of ferromagnetism with temperature is more sudden, however; the temperature at which it has all disappeared is called the **Curie*** temperature. The ferromagnetic properties reappear on cooling, but any magnetism will have disappeared. Thus a permanent magnet will be demagnetized by heating above the Curie temperature (1040 K for iron) but can be remagnetized after cooling. Above the Curie temperature, ferromagnetics behave as paramagnetics.

41.3 Hysteresis and hysteresis loss

Hysteresis loop

Let a ferromagnetic material which is completely demagnetized, i.e. one in which $B=H=0$ (either by heating the sample above its Curie temperature or by reversing the magnetizing current a large number of times while at the same time, gradually reducing the current to zero) be subjected to increasing values of magnetic field strength H and the corresponding flux density B measured. The domains begin to align and the resulting relationship between B and H is shown by the curve *0ab* in Figure 41.2. At a particular value of H, shown as *0y*, most of the domains will be aligned and it becomes difficult to increase the flux density any further. The material is said to be saturated. Thus *by* is the **saturation flux density**.

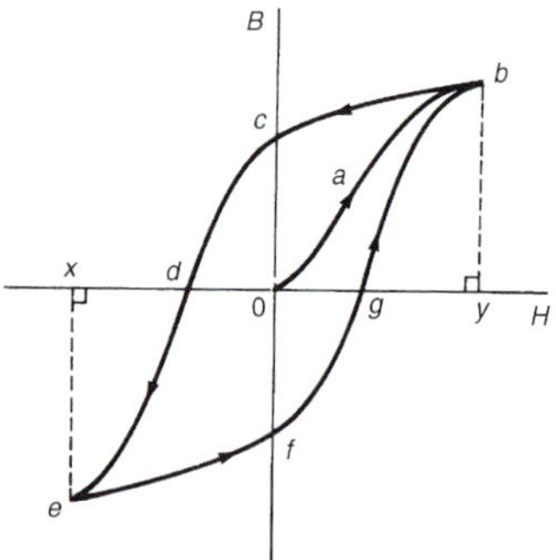

Figure 41.2

If the value of H is now reduced it is found that the flux density follows curve *bc*, i.e. the domains will tend to stay aligned even when the field is removed. When H is reduced to zero, flux remains in the iron. This **remanent flux density** or **remanence** is shown as *0c*

*Who was Curie? **Pierre Curie** (15 May 1859–19 April 1906) was a French polymath who received the Nobel Prize in Physics in 1903. To find out more go to www.routledge.com/cw/bird

in Figure 41.2. When H is increased in the opposite direction, the domains begin to realign in the opposite direction and the flux density decreases until, at a value shown as $0d$, the flux density has been reduced to zero. The magnetic field strength $0d$ required to remove the residual magnetism, i.e. reduce B to zero, is called the **coercive force**.

Further increase of H in the reverse direction causes the flux density to increase in the reverse direction until saturation is reached, as shown by curve de. If the reversed magnetic field strength $0x$ is adjusted to the same value of $0y$ in the initial direction, then the final flux density xe is the same as yb. If H is varied backwards from $0x$ to $0y$, the flux density follows the curve $efgb$, similar to curve $bcde$.

It is seen from Figure 41.2 that the flux density changes lag behind the changes in the magnetic field strength. This effect is called **hysteresis**. The closed figure $bcdefgb$ is called the **hysteresis loop** (or the B/H loop).

Hysteresis loss

A disturbance in the alignment of the domains of a ferromagnetic material causes energy to be expended in taking it through a cycle of magnetization. This energy appears as heat in the specimen and is called the **hysteresis loss**. Let the hysteresis loop shown in Figure 41.3 be that obtained for an iron ring of mean circumference l and cross-sectional area a m^2 and let the number of turns on the magnetizing coil be N.

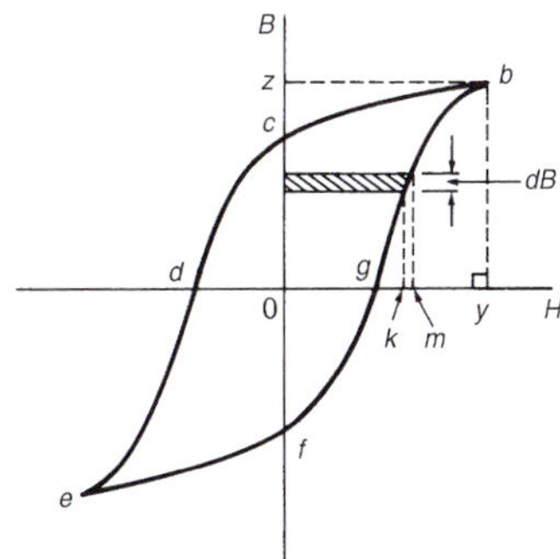

Figure 41.3

Let the increase of flux density be dB when the magnetic field strength H is increased by a very small amount km (see Figure 41.3) in time dt seconds, and let the current corresponding to $0k$ be i amperes. Thus since $H = NI/l$ then $0k = Ni/l$, from which,

$$i = \frac{l(0k)}{N} \qquad (1)$$

The instantaneous e.m.f. e induced in the winding is given by

$$e = -N\frac{d\Phi}{dt} = -N\frac{d(Ba)}{dt} = -aN\frac{dB}{dt}$$

The applied voltage to neutralize this e.m.f., $v = aN\dfrac{dB}{dt}$

The instantaneous power supplied to a magnetic field,

$$p = vi = i\left(aN\frac{dB}{dt}\right) \text{ watts}$$

Energy supplied to the magnetic field in time dt seconds

$$= \text{power} \times \text{time} = iaN\frac{dB}{dt}dt$$

$$= iaN\,dB \text{ joules} = \left(\frac{l(0k)}{N}\right)aN\,dB \text{ from equation (1)}$$

$$= (0k)\,dB(la) \text{ joules} = (\text{area of shaded strip}) \times (\text{volume of ring})$$

i.e. energy supplied in time dt seconds = (area of shaded strip) J/m^3.

Hence the energy supplied to the magnetic field when H is increased from zero to $0y$ = (area $fgbzf$) J/m^3.

Similarly, the energy returned from the magnetic field when H is reduced from $0y$ to zero = (area $bzcb$) J/m^3.

Hence net energy absorbed by the magnetic field = (area $fgbcf$) J/m^3.

Thus the hysteresis loss for a complete cycle

= area of loop $efgbcde$ J/m^3

If the hysteresis loop is plotted to a scale of 1 cm = α ampere/metre along the horizontal axis and 1 cm = β tesla along the vertical axis, and if A represents the area of the loop in square centimetres, then

hysteresis loss/cycle = $A\alpha\beta$ joules per metre^3 (2)

If hysteresis loops for a given ferromagnetic material are determined for different maximum values of H, they are found to lie within one another as shown in Figure 41.4.

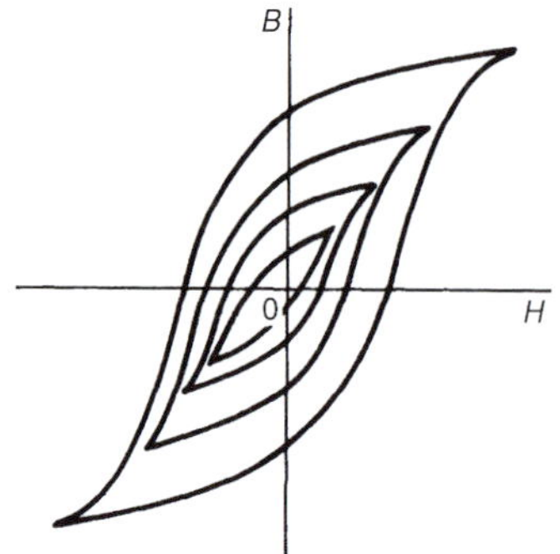

Figure 41.4

The maximum-sized hysteresis loop for a particular material is obtained at saturation. If, for example, the maximum flux density is reduced to half its value at saturation, the area of the resulting loop is considerably less than half the area of the loop at saturation. From the areas of a number of such hysteresis loops, as shown in Figure 41.4, the hysteresis loss per cycle was found by Steinmetz (an American electrical engineer) to be proportional to $(B_m)^n$, where n is called the **Steinmetz*** index and can have a value between about 1.6 and 3.0, depending on the quality of the ferromagnetic material and the range of flux density over which the measurements are made.

From the above it is found that the hysteresis loss is proportional to the volume of the specimen and the number of cycles through which the magnetization is taken. Thus

$$\textbf{hysteresis loss, } P_h = k_h v f (B_m)^n \textbf{ watts} \qquad (3)$$

where v = volume in cubic metres, f = frequency in hertz and k_h is a constant for a given specimen and given range of B.

The magnitude of the hysteresis loss depends on the composition of the specimen and on the heat treatment and mechanical handling to which the specimen has been subjected.

Figure 41.5 shows typical hysteresis loops for (a) hard steel, which has a high remanence $0c$ and a large coercivity $0d$, (b) soft steel, which has a large remanence and small coercivity and (c) ferrite, this being a ceramic-like magnetic substance made from oxides of iron, nickel, cobalt, magnesium, aluminium and manganese. The hysteresis of ferrite is very small.

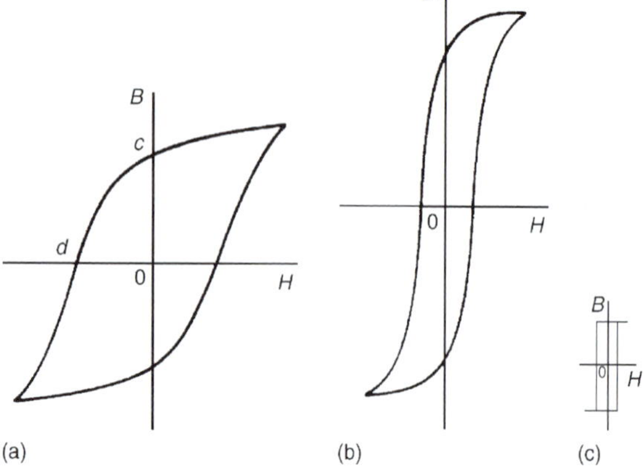

Figure 41.5

Problem 1. The area of a hysteresis loop obtained from a ferromagnetic specimen is 12.5 cm². The scales used were: horizontal axis 1 cm = 500 A/m; vertical axis 1 cm = 0.2 T. Determine (a) the hysteresis loss per m³ per cycle and (b) the hysteresis loss per m³ at a frequency of 50 Hz

(a) From equation (2), hysteresis loss per cycle

$$= A\alpha\beta = (12.5)(500)(0.2) = \mathbf{1250\ J/m^3}$$

(Note that, since $\alpha = 500$ A/m per centimetre and $\beta = 0.2$ T per centimetre, then 1 cm² of the loop represents

$$500\frac{\text{A}}{\text{m}} \times 0.2\,\text{T} = 100\,\frac{\text{A}}{\text{m}}\frac{\text{Wb}}{\text{m}^2} = 100\,\frac{\text{AVs}}{\text{m}^3}$$

$$= 100\,\frac{\text{Ws}}{\text{m}^3} = 100\,\text{J/m}^3$$

Hence 12.5 cm² represents

$$12.5 \times 100 = \mathbf{1250\ J/m^3})$$

(b) At 50 Hz frequency, hysteresis loss

$$= (1250\,\text{J/m}^3)(50\,1/\text{s}) = \mathbf{62\,500\ W/m^3}$$

Problem 2. If in Problem 1 the maximum flux density is 1.5 T at a frequency of 50 Hz, determine the hysteresis loss per m³ for a maximum flux density of 1.1 T and frequency of 25 Hz. Assume the Steinmetz index to be 1.6

*Who was Steinmetz? **Charles Proteus Steinmetz** (9 April 1865–26 October 1923) pushed forward the development of alternating current that made the expansion of the electrical power industry in the United States possible. To find out more go to **www.routledge.com/cw/bird**

Part 4

From equation (3), hysteresis loss $P_h = k_h v f (B_m)^n$

The loss at $f = 50$ Hz and $B_m = 1.5$ T is 62 500 W/m^3, from Problem 1.

Thus $62\,500 = k_h(1)(50)(1.5)^{1.6}$

from which, constant $k_h = \dfrac{62\,500}{(50)(1.5)^{1.6}} = 653.4$

When $f = 25$ Hz and $B_m = 1.1$ T,

$$\text{hysteresis loss, } P_h = k_h v f (B_m)^n = (653.4)(1)(25)(1.1)^{1.6} = \mathbf{19\,026\ W/m^3}$$

Problem 3. A ferromagnetic ring has a uniform cross-sectional area of 2000 mm^2 and a mean circumference of 1000 mm. A hysteresis loop obtained for the specimen is plotted to scales of 10 mm = 0.1 T and 10 mm = 400 A/m and is found to have an area of 10^4 mm^2. Determine the hysteresis loss at a frequency of 80 Hz.

From equation (2), hysteresis loss per cycle

$$= A\alpha\beta$$

$$= (10^4 \times 10^{-6}\,\text{m}^2)\left(\frac{400\ \text{A/m}}{10 \times 10^{-3}\,\text{m}}\right)\left(\frac{0.1\ \text{T}}{10 \times 10^{-3}\,\text{m}}\right)$$

$$= 4000\ \text{J/m}^3$$

At a frequency of 80 Hz,

$$\text{hysteresis loss} = (4000\ \text{J/m})(80\ 1/\text{s}) = 320\,000\ \text{W/m}^3$$

$$\text{Volume of ring} = (\text{cross-sectional area}) \times (\text{mean circumference}) = (2000 \times 10^{-6}\,\text{m}^2)(1000 \times 10^{-3}\,\text{m}) = 2 \times 10^{-3}\,\text{m}^3$$

Thus hysteresis loss $P_h = (320\,000\ \text{W/m}^3)(2 \times 10^{-3}\,\text{m}^3) = \mathbf{640\ W}$

Problem 4. The cross-sectional area of a transformer limb is 80 cm^2 and the volume of the transformer core is 5000 cm^3. The maximum value of the core flux is 10 mWb at a frequency of 50 Hz. Taking the Steinmetz constant as 1.7, the hysteresis loss is found to be 100 W. Determine the value of the hysteresis loss when the maximum core flux is 8 mWb and the frequency is 50 Hz.

When the maximum core flux is 10 mWb and the cross-sectional area is 80 cm^2,

$$\text{maximum flux density, } B_{m1} = \frac{\Phi_1}{A} = \frac{10 \times 10^{-3}}{80 \times 10^{-4}} = 1.25\ \text{T}$$

From equation (3), hysteresis loss, $P_{h1} = k_h v f (B_{m1})^n$

Hence $100 = k_h(5000 \times 10^{-6})(50)(1.25)^{1.7}$

$$\text{from which, constant } k_h = \frac{100}{(5000 \times 10^{-6})(50)(1.25)^{1.7}} = 273.7$$

When the maximum core flux is 8 mWb,

$$B_{m2} = \frac{8 \times 10^{-3}}{80 \times 10^{-4}} = 1\ \text{T}$$

Hence hysteresis loss,

$$P_{h2} = k_h v f (B_{m2})^n = (273.7)(5000 \times 10^{-6})(50)(1)^{1.7} = \mathbf{68.4\ W}$$

Now try the following Practice Exercise

Practice Exercise 159 Hysteresis loss (Answers on page 831)

1. The area of a hysteresis loop obtained from a specimen of steel is 2000 mm^2. The scales used are: horizontal axis 1 cm = 400 A/m; vertical axis 1 cm = 0.5 T. Determine (a) the hysteresis loss per m^3 per cycle, (b) the hysteresis loss per m^3 at a frequency of 60 Hz. (c) If the maximum flux density is 1.2 T at a frequency of 60 Hz, determine the hysteresis loss per m^3 for a maximum flux density of 1 T and a frequency of 20 Hz, assuming the Steinmetz index to be 1.7
2. A steel ring has a uniform cross-sectional area of 1500 mm^2 and a mean circumference of 800 mm. A hysteresis loop obtained for the

specimen is plotted to scales of 1 cm = 0.05 T and 1 cm = 100 A/m and it is found to have an area of 720 cm^2. Determine the hysteresis loss at a frequency of 50 Hz.

3. What is hysteresis? Explain how a hysteresis loop is produced for a ferromagnetic specimen and how its area is representative of the hysteresis loss.

   The area of a hysteresis loop plotted for a ferromagnetic material is 80 cm^2, the maximum flux density being 1.2 T. The scales of B and H are such that 1 cm = 0.15 T and 1 cm = 10 A/m. Determine the loss due to hysteresis if 1.25 kg of the material is subjected to an alternating magnetic field of maximum flux density 1.2 T at a frequency of 50 Hz. The density of the material is 7700 kg/m^3.

4. The cross-sectional area of a transformer limb is 8000 mm^2 and the volume of the transformer core is 4×10^6 mm^3. The maximum value of the core flux is 12 mWb and the frequency is 50 Hz. Assuming the Steinmetz constant is 1.6, the hysteresis loss is found to be 250 W. Determine the hysteresis loss when the maximum core flux is 9 mWb, the frequency remaining unchanged.

5. The hysteresis loss in a transformer is 200 W when the maximum flux density is 1 T and the frequency is 50 Hz. Determine the hysteresis loss if the maximum flux density is increased to 1.2 T and the frequency reduced to 32 Hz. Assume the hysteresis loss over this range to be proportional to $(B_m)^{1.6}$

6. A hysteresis loop is plotted to scales of 1 cm = 0.004 T and 1 cm = 10 A/m and has an area of 200 cm^2. If the ferromagnetic circuit for the loop has a volume of 0.02 m^3 and operates at 60 Hz frequency, determine the hysteresis loss for the ferromagnetic specimen.

41.4 Eddy current loss

If a coil is wound on a ferromagnetic core (such as in a transformer) and alternating current is passed through the coil, an alternating flux is set up in the core. The alternating flux induces an e.m.f. e in the coil given by $e = N(d\phi/dt)$. However, in addition to the desirable effect of inducing an e.m.f. in the coil, the alternating flux induces undesirable voltages in the iron core. These induced e.m.f.s set up circulating currents in the core, known as **eddy currents**. Since the core possesses resistance, the eddy currents heat the core, and this represents wasted energy.

Eddy currents can be reduced by laminating the core, i.e. splitting it into thin sheets with very thin layers of insulating material inserted between each pair of the laminations (this may be achieved by simply varnishing one side of the lamination or by placing paper between each lamination). The insulation presents a high resistance and this reduces any induced circulating currents.

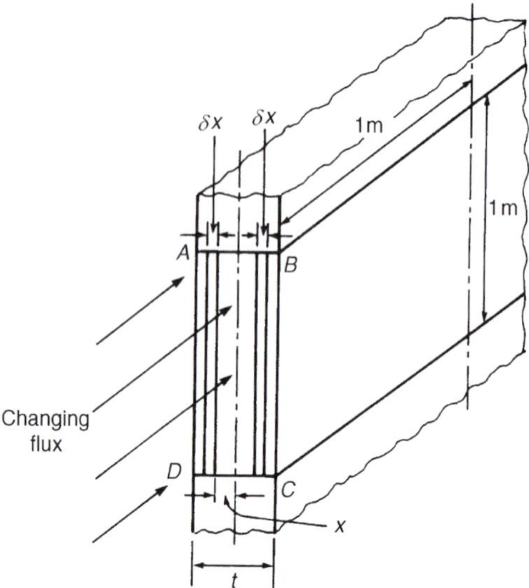

Figure 41.6

The eddy current loss may be determined as follows. Let Figure 41.6 represent one strip of the core, having a thickness of t metres, and consider just a rectangular prism of the strip having dimensions t m × 1 m × 1 m as shown. The area of the front face $ABCD$ is $(t \times 1)$ m^2 and, since the flux enters this face at right-angles, the eddy currents will flow along paths parallel to the long sides.

Consider two such current paths each of width δx and distance x m from the centre line of the front face. The area of the rectangle enclosed by the two paths, $A = (2x)(1) = 2x$ m^2. Hence the maximum flux entering the rectangle,

$$\Phi_m = (B_m)(A) = (B_m)(2x) \text{ weber} \quad (4)$$

Induced e.m.f. e is given by $e = N(\mathrm{d}\phi/\mathrm{d}t)$. Since the flux varies sinusoidally, $\phi = \Phi_m \sin \omega t$. Thus

e.m.f. $e = N\dfrac{\mathrm{d}}{\mathrm{d}t}(\Phi_m \sin \omega t) = N\omega\Phi_m \cos \omega t$

The maximum value of e.m.f. occurs when $\cos \omega t = 1$, i.e. $E_m = N\omega\Phi_m$

Rms value of e.m.f., $E = \dfrac{E_m}{\sqrt{2}} = \dfrac{N\omega\Phi_m}{\sqrt{2}}$

Now $\omega = 2\pi f$ hence

$$E = \left(\frac{2\pi}{\sqrt{2}}\right) fN\Phi_m = 4.44\, fN\Phi_m$$

i.e. $$E = 4.44\, fN(B_m)(A) \quad (5)$$

From equation (4), $\Phi_m = (B_m)(2x)$. Hence induced e.m.f. $E = 4.44 fN(B_m)(2x)$ and, since the number of turns $N = 1$,

$$E = 8.88 B_m f x \text{ volts} \quad (6)$$

Resistance R is given by $R = \rho l/a$, where ρ is the resistivity of the lamination material. Since the current set up is confined to the two loop sides (thus $l = 2$ m and $a = (\delta x \times 1)\,\mathrm{m}^2$), the total resistance of the path is given by

$$R = \frac{\rho(2)}{\delta x} = \frac{2\rho}{\delta x} \quad (7)$$

The eddy current loss in the two strips is given by

$$\frac{E^2}{R} = \frac{8.88^2 B_m^2 f^2 x^2}{2\rho/\delta x} \text{ from equations (6) and (7)}$$

$$= \frac{8.88^2 B_m^2 f^2 x^2 \delta x}{2\rho}$$

The total eddy current loss P_e in the rectangular prism considered is given by

$$P_e = \int_0^{t/2} \left(\frac{8.88^2 B_m^2 f^2}{2\rho}\right) x^2 \,\mathrm{d}x$$

$$= \left(\frac{8.88^2 B_m^2 f^2}{2\rho}\right)\left[\frac{x^3}{3}\right]_0^{t/2}$$

$$= \left(\frac{8.88^2 B_m^2 f^2}{2\rho}\right)\left(\frac{t^3}{24}\right) \text{ watts}$$

i.e. $$\boldsymbol{P_e = k_e(B_m)^2 f^2 t^3} \textbf{ watts} \quad (8)$$

where k_e is a constant.

The volume of the prism is $(t \times 1 \times 1)\,\mathrm{m}^3$. Hence the eddy current loss per m^3 is given by

$$\boldsymbol{P_e = k_e(B_m)^2 f^2 t^2} \textbf{ watts per m}^3 \quad (9)$$

From equation (9) it is seen that eddy current loss is proportional to the square of the thickness of the core strip. It is therefore desirable to make lamination strips as thin as possible. However, at high frequencies where it is not practicable to make very thin laminations, core losses may be reduced by using ferrite cores or dust cores. Ferrite is a ceramic material having magnetic properties similar to silicon steel, and dust cores consist of fine particles of carbonyl iron or permalloy (i.e. nickel and iron), each particle of which is insulated from its neighbour by a binding material. Such materials have a very high value of resistivity.

Problem 5. The eddy current loss in a particular magnetic circuit is $10\,\mathrm{W/m^3}$. If the frequency of operation is reduced from 50 Hz to 30 Hz with the flux density remaining unchanged, determine the new value of eddy current loss per cubic metre.

From equation (9), eddy current loss per cubic metre, $P_e = k_e(B_m)^2 f^2 t^2$ or $P_e = kf^2$, where $k = k_e(B_m)^2 t^2$, since B_m and t are constant.

When the eddy current loss is $10\,\mathrm{W/m^3}$, frequency f is 50 Hz. Hence $10 = k(50)^2$, from which

constant $k = \dfrac{10}{(50)^2}$

When the frequency is 30 Hz, eddy current loss,

$P_e = k(30)^2 = \dfrac{10}{(50)^2}(30)^2 = \mathbf{3.6\,W/m^3}$

Problem 6. The core of a transformer operating at 50 Hz has an eddy current loss of $100\,\mathrm{W/m^3}$ and the core laminations have a thickness of 0.50 mm. The core is redesigned so as to operate with the same eddy current loss but at a different voltage and at a frequency of 250 Hz. Assuming that at the new voltage the maximum flux density is one-third of its original value and the resistivity of the core remains unaltered, determine the necessary new thickness of the laminations.

From equation (9), $P_e = k_e(B_m)^2 f^2 t^2$ watts per m^3.

Hence, at 50 Hz frequency,

$100 = k_e(B_m)^2(50)^2\,(0.50\times 10^{-3})^2$, from which

$$k_e = \frac{100}{(B_m)^2(50)^2(0.50\times 10^{-3})^2}$$

At 250 Hz frequency, $100 = k_e\left(\frac{B_m}{3}\right)^2(250)^2(t)^2$ i.e.

$$100 = \left(\frac{100}{(B_m)^2(50)^2(0.50\times 10^{-3})^2}\right)\left(\frac{B_m}{3}\right)^2(250)^2(t)^2$$

$$= \frac{100(250)^2(t)^2}{(3)^2(50)^2(0.50\times 10^{-3})^2}$$

from which $t^2 = \dfrac{(100)(3)^2(50)^2(0.50\times 10^{-3})^2}{(100)(250)^2}$

i.e. **lamination thickness,** $\boldsymbol{t} = \dfrac{(3)(50)(0.50\times 10^{-3})}{250}$

$= 0.3\times 10^{-3}$ m or **0.30 mm**

Problem 7. The core of an inductor has a hysteresis loss of 40 W and an eddy current loss of 20 W when operating at 50 Hz frequency. (a) Determine the values of the losses if the frequency is increased to 60 Hz. (b) What will be the total core loss if the frequency is 50 Hz and the laminations are made one-half of their original thickness? Assume that the flux density remains unchanged in each case.

(a) From equation (3), hysteresis loss,

$P_h = k_h v f\,(B_m)^n = k_1 f$ (where $k_1 = k_h v(Bm)^n$), since the flux density and volume are constant. Thus when the hysteresis is 40 W and the frequency 50 Hz,

$$40 = k_1(50)$$

from which, $k_1 = \dfrac{40}{50} = 0.8$

If the frequency is increased to 60 Hz,

hysteresis loss, $\boldsymbol{P_h} = k_1(60) = (0.8)(60) = \mathbf{48\ W}$

From equation (8),

eddy current loss,

$$P_e = k_e(B_m)^2 f^2 t^3$$
$$= k_2 f^2 \text{ (where } (k_2 = k_e B_m)^2 t^3),$$

since the flux density and lamination thickness are constant.

When the eddy current loss is 20 W the frequency is 50 Hz. Thus $20 = k_2(50)^2$

from which $k_2 = \dfrac{20}{(50)^2} = 0.008$

If the frequency is increased to 60 Hz,

eddy current loss, $\boldsymbol{P_e} = k_2(60)^2 = (0.008)(60)^2$

$= \mathbf{28.8\ W}$

(b) The hysteresis loss, $P_h = k_h v f (B_m)^n$, is independent of the thickness of the laminations. Thus, if the thickness of the laminations is halved, the hysteresis loss remains at **40 W**

Eddy current loss $P_e = k_e(B_m)^2 f^2 t^3$,
i.e. $P_e = k_3 f^2 t^3$, where $k_3 = k_e(B_m)^2$

Thus $\quad 20 = k_3(50)^2 t^3$

from which $k_3 = \dfrac{20}{(50)^2 t^3}$

When the thickness is $t/2$,

$$P_e = k_3(50)^2(t/2)^3 = \left(\frac{20}{(50)^2 t^3}\right)(50)^2(t/2)^3$$
$$= \mathbf{2.5\ W}$$

Hence the **total core loss** when the thickness of the laminations is halved is given by hysteresis loss + eddy current loss = 40 + 2.5 = **42.5 W**

Problem 8. When a transformer is connected to a 500 V, 50 Hz supply, the hysteresis and eddy current losses are 400 W and 150 W, respectively. The applied voltage is increased to 1 kV and the frequency to 100 Hz. Assuming the Steinmetz index to be 1.6, determine the new total core loss.

From equation (3), the hysteresis loss, $P_h = k_h v f (B_m)^n$. From equation (5), e.m.f., $E = 4.44 fN(B_m)(A)$, from which, $B_m \alpha (E/f)$ since turns N and cross-sectional area, A are constants. Hence
$P_h = k_1 f (E/f)^{1.6} = k_1 f^{-0.6} E^{1.6}$

At 500 V and 50 Hz, $400 = k_1(50)^{-0.6}(500)^{1.6}$,

from which $k_1 = \dfrac{400}{(50)^{-0.6}(500)^{1.6}} = 0.20095$

At 1000 V and 100 Hz, hysteresis loss,

$$P_h = k_1(100)^{-0.6}(1000)^{1.6}$$
$$= (0.20095)(100)^{-0.6}(1000)^{-1.6} = 800\text{ W}$$

From equation (8)
eddy current loss, $P_e = k_e(B_m)^2 f^2 t^3 = k_2(E/f)^2 f^2$
$= k_2 E^2$

At 500 V, $150 = k_2(500)^2$, from which
$$k_2 = \frac{150}{(500)^2} = 6 \times 10^{-4}$$

At 1000 V,

eddy current loss, $P_e = k_2(1000)^2 = (6 \times 10^{-4})(1000)^2$
$= 600\text{ W}$

Hence the new **total core loss** $= 800 + 600 =$ **1400 W**

Now try the following Practice Exercise

Practice Exercise 160 Eddy current loss (Answers on page 831)

1. In a magnetic circuit operating at 60 Hz, the eddy current loss is 25 W/m^3. If the frequency is reduced to 30 Hz with the flux density remaining unchanged, determine the new value of eddy current loss per cubic metre.

2. A transformer core operating at 50 Hz has an eddy current loss of 150 W/m^3 and the core laminations are 0.4 mm thick. The core is redesigned so as to operate with the same eddy current loss but at a different voltage and at 200 Hz frequency. Assuming that at the new voltage the flux density is half of its original value and the resistivity of the core remains unchanged, determine the necessary new thickness of the laminations.

3. An inductor core has an eddy current loss of 25 W and a hysteresis loss of 35 W when operating at 50 Hz frequency. Assuming that the flux density remains unchanged, determine (a) the value of the losses if the frequency is increased to 75 Hz and (b) the total core loss if the frequency is 50 Hz and the laminations are 2/5 of their original thickness.

4. A transformer is connected to a 400 V, 50 Hz supply. The hysteresis loss is 250 W and the eddy current loss is 120 W. The supply voltage is increased to 1.2 kV and the frequency to 80 Hz. Determine the new total core loss if the Steinmetz index is assumed to be 1.6

5. The hysteresis and eddy current losses in a magnetic circuit are 5 W and 8 W, respectively. If the frequency is reduced from 50 Hz to 30 Hz, the flux density remaining the same, determine the new values of hysteresis and eddy current loss.

6. The core loss in a transformer connected to a 600 V, 50 Hz supply is 1.5 kW, of which 60% is hysteresis loss and 40% eddy current loss. Determine the total core loss if the same winding is connected to a 750 V, 60 Hz supply. Assume the Steinmetz constant to be 1.6

41.5 Separation of hysteresis and eddy current losses

From equation (3), hysteresis loss, $P_h = k_h v f (B_m)^n$.

From equation (8), eddy current loss, $P_e = k_e(B_m)^2 f^2 t^3$.

The total core loss P_c is given by $P_c = P_h + P_e$
If for a particular inductor or transformer, the core flux density is maintained constant, then $P_h = k_1 f$, where constant $k_1 = k_h v(B_m)^n$, and $P_e = k_2 f^2$, where constant $k_2 = k_e(B_m)^2 t^3$. Thus the total core loss $P_c = k_1 f + k_2 f^2$ and

$$\frac{P_c}{f} = k_1 + k_2 f$$

which is of the straight line form $y = mx + c$. Thus if P_c/f is plotted vertically against f horizontally, a straight line graph results having a gradient k_2 and a vertical-axis intercept k_1

If the total core loss P_c is measured over a range of frequencies, then k_1 and k_2 may be determined from the graph of P_c/f against f. Hence the hysteresis loss $P_h(=k_1 f)$ and the eddy current loss $P_e(=k_2 f^2)$ at a given frequency may be determined.

The above method of separation of losses is an approximate one since the Steinmetz index n is not a constant value but tends to increase with increase of frequency. However, a reasonable indication of the relative magnitudes of the hysteresis and eddy current losses in an iron core may be determined.

Problem 9. The total core loss of a ferromagnetic cored transformer winding is measured at different frequencies and the results obtained are:

Total core loss, P_c (watts)	45	105	190	305
Frequency, f (hertz)	30	50	70	90

Determine the separate values of the hysteresis and eddy current losses at frequencies of (a) 50 Hz and (b) 60 Hz.

To obtain a straight line graph, values of P_c/f are plotted against f.

f(Hz)	30	50	70	90
P_c/f	1.5	2.1	2.7	3.4

A graph of P_c/f against f is shown in Figure 41.7. The graph is a straight line of the form $P_c/f = k_1 + k_2 f$. The vertical axis intercept at $f = 0$, $\boldsymbol{k_1 = 0.5}$

The gradient of the graph, $\boldsymbol{k_2} = \dfrac{a}{b} = \dfrac{3.7 - 0.5}{100} = \mathbf{0.032}$

Since $P_c/f = k_1 + k_2 f$, then $P = k_1 f + k_2 f^2$, i.e. total core losses = hysteresis loss + eddy current loss.

(a) At a frequency of 50 Hz,

hysteresis loss $= k_1 f = (0.5)(50) = \mathbf{25\ W}$

eddy current loss $= k_2 f^2 = (0.032)(50)^2 = \mathbf{80\ W}$

(b) At a frequency of 60 Hz,

hysteresis loss $= k_1 f = (0.5)(60) = \mathbf{30\ W}$

eddy current loss $= k_2 f^2 = (0.032)(60)^2$

$= \mathbf{115.2\ W}$

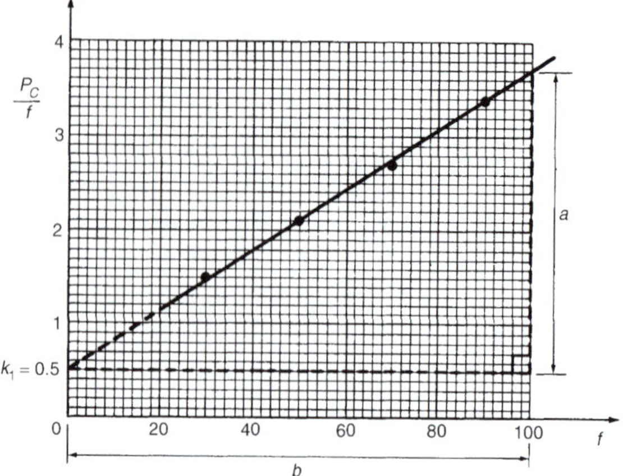

Figure 41.7

Problem 10. The core of a synchrogenerator has total losses of 400 W at 50 Hz and 498 W at 60 Hz, the flux density being constant for the two tests. (a) Determine the hysteresis and eddy current losses at 50 Hz. (b) If the flux density is increased by 25% and the lamination thickness is increased by 40%, determine the hysteresis and eddy current losses at 50 Hz. Assume the Steinmetz index to be 1.7

(a) From equation (3),

hysteresis loss, $P_h = k_h v f (B_m)^n = k_1 f$

(if volume v and the maximum flux density are constant).

From equation (8),

eddy current loss, $P_e = k_e (B_m)^2 f^2 t^3 = k_2 f^2$

(if the maximum flux density and the lamination thickness are constant).

Hence the total core loss $P_c = P_h + P_e$ i.e. $P_c = k_1 f + k_2 f^2$

At 50 Hz frequency, $400 = k_1(50) + k_2(50)^2$ (1)

At 60 Hz frequency, $498 = k_1(60) + k_2(60)^2$ (2)

Solving equations (1) and (2) gives the values of k_1 and k_2

$6 \times$ equation (1) gives: $2400 = 300k_1 + 15\,000k_2$ (3)

$5 \times$ equation (2) gives: $2490 = 300k_1 + 18\,000k_2$ (4)

Equation (4) – equation (3) gives: $90 = 3000k_2$ from which,

$k_2 = 90/3000 = \mathbf{0.03}$

Substituting $k_2 = 0.03$ in equation (1) gives
$400 = 50k_1 + 75$, from which $\boldsymbol{k_1 = 6.5}$
Thus, at 50 Hz frequency,

$$\text{hysteresis loss } P_h = k_1 f = (6.5)(50) = \mathbf{325\ W}$$

$$\text{eddy current loss } P_e = k_2 f^2 = (0.03)(50)^2 = \mathbf{75\ W}$$

(b) Hysteresis loss, $P_h = k_h v f (B_m)^n$. Since at 50 Hz the flux density is increased by 25%, the new hysteresis loss is $(1.25)^{1.7}$ times greater than 325 W,

i.e. $P_h = (1.25)^{1.7}(325) = \mathbf{474.9\ W}$

Eddy current loss, $P_e = k_e (B_m)^2 f^2 t^3$. Since at 50 Hz the flux density is increased by 25%, and the lamination thickness is increased by 40%, the new eddy current loss is $(1.25)^2(1.4)^3$ times greater than 75 W,

i.e. $P_e = (1.25)^2(1.4)^3(75) = \mathbf{321.6\ W}$

Now try the following Practice Exercise

Practice Exercise 161 Separation of hysteresis and eddy current losses (Answers on page 831)

1. Tests to determine the total loss of the steel core of a coil at different frequencies gave the following results:

Frequency (Hz)	40	50	70	100
Total core loss (W)	40	57.5	101.5	190

Determine the hysteresis and eddy current losses at (a) 50 Hz and (b) 80 Hz.

2. Explain why, when steel is subjected to alternating magnetization energy, losses occur due to both hysteresis and eddy currents.
The core loss in a transformer core at normal flux density was measured at frequencies of 40 Hz and 50 Hz, the results being 40 W and 52.5 W, respectively. Calculate, at a frequency of 50 Hz, (a) the hysteresis loss and (b) the eddy current loss.

3. Results of a test used to separate the hysteresis and eddy current losses in the core of a transformer winding gave the following results:

Total core loss (W)	48	96	160	240
Frequency (Hz)	40	60	80	100

If the flux density is held constant throughout the test, determine the values of the hysteresis and eddy current losses at 50 Hz.

4. A transformer core has a total core loss of 275 W at 50 Hz and 600 W at 100 Hz, the flux density being constant for the two tests. (a) Determine the hysteresis and eddy current losses at 75 Hz. (b) If the flux density is increased by 40% and the lamination thickness is increased by 20%, determine the hysteresis and eddy current losses at 75 Hz. Assume the Steinmetz index to be 1.6

41.6 Non-permanent magnetic materials

General

Nonpermanent magnetic materials are those in which magnetism may be induced. With the magnetic circuits of electrical machines, transformers and heavy current apparatus a high value of flux density B is desirable so as to limit the cross-sectional area A ($\Phi = BA$) and therefore the weight and cost involved. At the same time the magnetic field strength H ($= NI/l$) should be as small as possible so as to limit the I^2R losses in the exciting coils. The relative permeability ($\mu_r = B/(\mu_0 H)$) and the saturation flux density should therefore be high. Also, when flux is continually varying, as in transformers, inductors and armature cores, low hysteresis and eddy current losses are essential.

Silicon–iron alloys

In the earliest electrical machines the magnetic circuit material used was iron with low content of carbon and other impurities. However, it was later discovered that the deliberate addition of silicon to the iron brought about a great improvement in magnetic properties. The laminations now used in electrical machines and in transformers at supply frequencies are made of silicon-steel in which the silicon in different grades of the material varies in amounts from about 0.5% to 4.5% by weight. The silicon added to iron increases the resistivity. This in turn increases the resistance ($R = \rho l/A$) and thus helps to reduce eddy current loss. The hysteresis loss is also reduced; however, the silicon reduces the saturation flux density.

A limit to the amount of silicon which may be added in practice is set by the mechanical properties of the material, since the addition of silicon causes a material to become brittle. Also, the brittleness of a silicon–iron alloy depends on temperature. About 4.5% silicon is found to be the upper practical limit for silicon–iron sheets. Lohys is a typical example of a silicon–iron alloy and is used for the armatures of d.c. machines and for the rotors and stators of a.c. machines. Stalloy, which has a higher proportion of silicon and lower losses, is used for transformer cores.

Silicon steel sheets are often produced by a hot-rolling process. In these finished materials the constituent crystals are not arranged in any particular manner with respect, for example, to the direction of rolling or the plane of the sheet. If silicon steel is reduced in thickness by rolling in the cold state and the material is then annealed it is possible to obtain a finished sheet in which the crystals are nearly all approximately parallel to one another. The material has strongly directional magnetic properties, the rolling direction being the direction of highest permeability. This direction is also the direction of lowest hysteresis loss. This type of material is particularly suitable for use in transformers, since the axis of the core can be made to correspond with the rolling direction of the sheet and thus full use is made of the high permeability, low loss direction of the sheet.

With silicon–iron alloys a maximum magnetic flux density of about 2 T is possible. With cold-rolled silicon–steel, used for large machine construction, a maximum flux density of 2.5 T is possible, whereas the maximum obtainable with the hot-rolling process is about 1.8 T. (In fact, with any material, only under the most abnormal of conditions will the value of flux density exceed 3 T)

It should be noted that the term 'iron-core' implies that the core is made of iron; it is, in fact, almost certainly made from steel, pure iron being extremely hard to come by. Equally, an iron alloy is generally a steel and so it is preferred to describe a core as being a steel rather than an iron core.

Nickel–iron alloys

Nickel and iron are both ferromagnetic elements and when they are alloyed together in different proportions a series of useful magnetic alloys is obtained. With about 25%–30% nickel content added to iron, the alloy tends to be very hard and almost non-magnetic at room temperature. However, when the nickel content is increased to, say, 75%–80% (together with small amounts of molybdenum and copper), very high values of initial and maximum permeabilities and very low values of hysteresis loss are obtainable if the alloys are given suitable heat treatment. For example, Permalloy, having a content of 78% nickel, 3% molybdenum and the remainder iron, has an initial permeability of 20 000 and a maximum permeability of 100 000 compared with values of 250 and 5 000, respectively, for iron. The maximum flux density for Permalloy is about 0.8 T. Mumetal (76% nickel, 5% copper and 2% chromium) has similar characteristics. Such materials are used for the cores of current and a.f. transformers, for magnetic amplifiers and also for magnetic screening. However, nickel–iron alloys are limited in that they have a low saturation value when compared with iron. Thus, in applications where it is necessary to work at a high flux density, nickel–iron alloys are inferior to both iron and silicon–iron. Also nickel–iron alloys tend to be more expensive than silicon–iron alloys.

Eddy current loss is proportional to the thickness of lamination squared, thus such losses can be reduced by using laminations as thin as possible. Nickel–iron alloy strip as thin as 0.004 mm, wound in a spiral, may be used.

Dust cores

In many circuits high permeability may be unnecessary or it may be more important to have a very high resistivity. Where this is so, metal powder or dust cores are widely used up to frequencies of 150 MHz. These consist of particles of nickel–iron–molybdenum for lower frequencies and iron for the higher frequencies. The particles, which are individually covered with an insulating film, are mixed with an insulating, resinous binder and pressed into shape.

Ferrites

Magnetite, or ferrous ferrite, is a compound of ferric oxide and ferrous oxide and possesses magnetic properties similar to those of iron. However, being a semiconductor, it has a very high resistivity. Manufactured ferrites are compounds of ferric oxide and an oxide of some other metal such as manganese, nickel or zinc. Ferrites are free from eddy current losses at all but the highest frequencies (i.e. >100 MHz), but have a much lower initial permeability compared with nickel–iron alloys or silicon–iron alloys. Ferrites have typically a maximum flux density of about 0.4 T. Ferrite cores are used in audio-frequency transformers and inductors.

41.7 Permanent magnetic materials

A permanent magnet is one in which the material used exhibits magnetism without the need for excitation by a current-carrying coil. The silicon–iron and nickel–iron alloys discussed in Section 41.6 are ‘soft’ magnetic materials having high permeability and hence low hysteresis loss. The opposite characteristics are required in the ‘hard’ materials used to make permanent magnets. In permanent magnets, high remanent flux density and high coercive force, after magnetization to saturation, are desirable in order to resist demagnetization. The hysteresis loop should embrace the maximum possible area. Possibly the best criterion of the merit of a permanent magnet is its maximum energy product $(BH)_m$, i.e. the maximum value of the product of the flux density B and the magnetic field strength H along the demagnetization curve (shown as cd in Figure 41.2). A rough criterion is the product of coercive force and remanent flux density, i.e. $(0d)(0c)$ in Figure 41.2. The earliest materials used for permanent magnets were tungsten and chromium steel, followed by a series of cobalt steels, to give both a high remanent flux density and a high value of $(BH)_m$.

Alni was the first of the aluminium–nickel–iron alloys to be discovered, and with the addition of cobalt, titanium and niobium, the Alnico series of magnets was developed, the properties of which vary according to composition. These materials are very hard and brittle. Many alloys with other compositions and trade names are commercially available.

A considerable advance was later made when it was found that directional magnetic properties could be induced in alloys of suitable composition if they were heated in a strong magnetic field. This discovery led to the powerful Alcomex and Hycomex series of magnets. By using special casting techniques to give a grain-oriented structure, even better properties are obtained if the field applied during heat treatment is parallel to the columnar crystals in the magnet. The values of coercivity, the remanent flux density and hence $(BH)_m$ are high for these alloys.

The most recent and most powerful permanent magnets discovered are made by powder metallurgy techniques and are based on an intermetallic compound of cobalt and samarium. These are very expensive and are only available in a limited range of small sizes.

For fully worked solutions to each of the problems in Practice Exercises 159 to 161 in this chapter, go to the website:
www.routledge.com/cw/bird

Revision Test 12

This revision test covers the material in Chapters 39 to 41. *The marks for each question are shown in brackets at the end of each question.*

1. A voltage waveform represented by

$$v = 50\sin\omega t + 20\sin\left(3\omega t + \frac{\pi}{3}\right) + 5\sin\left(5\omega t + \frac{\pi}{6}\right) \text{ volts}$$

is applied to a circuit and the resulting current i is given by

$$i = 2.0\sin\left(\omega t - \frac{\pi}{6}\right) + 0.462\sin 3\omega t + 0.0756\sin(5\omega t - 0.71) \text{ amperes.}$$

Calculate (a) the r.m.s. voltage, (b) the mean value of voltage, (c) the form factor for the voltage, (d) the r.m.s. value of current, (e) the mean value of current, (f) the form factor for the current, (g) the total active power supplied to the circuit and (h) the overall power factor. (24)

2. Obtain a Fourier series to represent $f(t) = t$ in the range $-\pi$ to $+\pi$. (15)

3. (a) Sketch a waveform defined by:

$$f(x) = \begin{cases} 0 \text{ when } -4 \le x \le -2 \\ 3 \text{ when } -2 \le x \le 2 \\ 0 \text{ when } 2 \le x \le 4 \end{cases}$$

and is periodic outside of this range of period 8.

(b) State whether the waveform in (a) is odd, even or neither odd nor even.

(c) Deduce the Fourier series for the function defined in (a). (15)

4. The value of the current i (in mA) at different moments in a cycle are given by:

θ degrees	0	30	60	90	120	150	180
i mA	50	75	165	190	170	100	−150
θ degrees	210	240	270	300	330	360	
i mA	−210	−185	−90	−10	35	50	

Draw the graph of current i against θ and analyse the current into its first three constituent components, each coefficient correct to 2 decimal places. (30)

5. The cross-sectional area of a transformer limb is 8000 mm^2 and the volume of the transformer core is 4×10^6 mm^3. The maximum value of the core flux is 12 mWb at a frequency of 50 Hz. Taking the Steinmetz index as 1.6, the hysteresis loss is found to be 80 W. Determine the value of the hysteresis loss when the maximum core flux is 9 mWb and the frequency is 50 Hz. (6)

6. The core of an inductor has a hysteresis loss of 25 W and an eddy current loss of 15 W when operating at 50 Hz frequency. Determine (a) the values of the losses if the frequency is increased to 70 Hz, and (b) the total core loss if the frequency is 50 Hz and the laminations are made three-quarters of their original thickness. Assume that the flux density remains unchanged in each case. (10)

For lecturers/instructors/teachers, fully worked solutions to each of the problems in Revision Test 12, together with a full marking scheme, are available at the website:
www.routledge.com/cw/bird

Chapter 42

Dielectrics and dielectric loss

Why it is important to understand: **Dielectric and dielectric loss**

Dielectric loss is the loss of energy that goes into heating a dielectric material in a varying electric field. For example, a capacitor incorporated in an a.c. circuit is alternately charged and discharged each half-cycle. During the alternation of polarity of the plates, the charges must be displaced through the dielectric first in one direction and then in the other, and overcoming the opposition that they encounter leads to a production of heat through dielectric loss, a characteristic that must be considered when applying capacitors to electric circuits, such as those in radio and television receivers. Dielectric losses depend on frequency and the dielectric material. Heating through dielectric loss is widely employed industrially for heating thermosetting glues, for drying lumber and other fibrous materials, for preheating plastics before moulding and for fast jelling and drying of foam rubber. An important property of a dielectric is its ability to support an electrostatic field while dissipating minimal energy in the form of heat. The lower the dielectric loss (the proportion of energy lost as heat), the more effective is a dielectric material. The study of dielectric properties is concerned with the storage and dissipation of electric and magnetic energy in materials. It is important to explain various phenomena in electronics, optics and solid-state physics. This chapter assesses the dielectric properties of materials, and determines, via calculations, dielectric loss, loss angle, Q-factor and dissipation factor of capacitors.

At the end of this chapter you should be able to:

- understand electric fields, capacitance and permittivity
- assess the dielectric properties of materials
- determine dielectric loss, loss angle, Q-factor and dissipation factor of capacitors

42.1 Electric fields, capacitance and permittivity

Any region in which an electric charge experiences a force is called an electrostatic field. Electric fields, Coulomb's law, capacitance and permittivity are discussed in Chapter 8 – refer back to page 106. Summarizing the main formulae:

Electric field strength, $E = \frac{V}{d}$ **volts/metre**

Capacitance $C = \frac{Q}{V}$ **farads**

Electric flux density, $D = \frac{Q}{A}$ **coulombs/metre²**

$$\frac{D}{E} = \varepsilon_0 \varepsilon_r = \varepsilon$$

Relative permittivity $\varepsilon_r = \frac{\text{flux density in material}}{\text{flux density in vacuum}}$

The insulating medium separating charged surfaces is called a **dielectric**. Compared with conductors, dielectric materials have very high resistivities (and

Electrical Circuit Theory and Technology. 978-1-138-67349-6, © 2017 John Bird. Published by Taylor & Francis. All rights reserved.

hence low conductance, since $\rho = 1/\sigma$). They are therefore used to separate conductors at different potentials, such as capacitor plates or electric power lines.

For a parallel-plate capacitor,

$$\textbf{capacitance } C = \frac{\varepsilon_0 \varepsilon_r A(n-1)}{d}$$

42.2 Polarization

When a dielectric is placed between charged plates, the capacitance of the system increases. The mechanism by which a dielectric increases capacitance is called **polarization**. In an electric field the electrons and atomic nuclei of the dielectric material experience forces in opposite directions. Since the electrons in an insulator cannot flow, each atom becomes a tiny dipole (i.e. an arrangement of two electric charges of opposite polarity) with positive and negative charges slightly separated, i.e. the material becomes polarized.

Within the material this produces no discernible effects. However, on the surfaces of the dielectric, layers of charge appear. Electrons are drawn towards the positive potential, producing a negative charge layer, and away from the negative potential, leaving positive surface charge behind. Therefore the dielectric becomes a volume of neutral insulator with surface charges of opposite polarity on opposite surfaces. The result of this is that the electric field inside the dielectric is less than the electric field causing the polarization, because these two charge layers give rise to a field which opposes the electric field causing it. Since electric field strength, $E = V/d$, the p.d. between the plates, $V = Ed$. Thus, if E decreases when the dielectric is inserted, then V falls too and this drop in p.d. occurs without change of charge on the plates. Thus, since capacitance $C = Q/V$, capacitance increases, this increase being by a factor equal to ε_r above that obtained with a vacuum dielectric.

There are two main ways in which polarization takes place:

(i) The electric field, as explained above, pulls the electrons and nuclei in opposite directions because they have opposite charges, which makes each atom into an electric dipole. The movement is only small and takes place very fast since the electrons are very light. Thus, if the applied electric field is varied periodically, the polarization, and hence the permittivity due to these induced dipoles, is independent of the frequency of the applied field.

(ii) Some atoms have a permanent electric dipole as a result of their structure and, when an electric field is applied, they turn and tend to align along the field. The response of the permanent dipoles is slower than the response of the induced dipoles and that part of the relative permittivity which arises from this type of polarization decreases with increase of frequency.

Most materials contain both induced and permanent dipoles, so the relative permittivity usually tends to decrease with increase of frequency.

42.3 Dielectric strength

The maximum amount of field strength that a dielectric can withstand is called the dielectric strength of the material. When an electric field is established across the faces of a material, molecular alignment and distortion of the electron orbits around the atoms of the dielectric occur. This produces a mechanical stress which in turn generates heat. The production of heat represents a dissipation of power, such a loss being present in all practical dielectrics, especially when used in high-frequency systems where the field polarity is continually and rapidly changing.

A dielectric whose conductivity is not zero between the plates of a capacitor provides a conducting path along which charges can flow and thus discharge the capacitor. The resistance R of the dielectric is given by $R = \rho l/a$, l being the thickness of the dielectric film (which may be as small as 0.001 mm) and a being the area of the capacitor plates. The resistance R of the dielectric may be represented as a leakage resistance across an ideal capacitor (see Section 42.8 on dielectric loss). The required lower limit for acceptable resistance between the plates varies with the use to which the capacitor is put. High-quality capacitors have high shunt-resistance values. A measure of dielectric quality is the time taken for a capacitor to discharge a given amount through the resistance of the dielectric. This is related to the product CR.

$$\text{Capacitance, } C \propto \frac{\text{area}}{\text{thickness}} \quad \text{and} \quad \frac{1}{R} \propto \frac{\text{area}}{\text{thickness}}$$

thus CR is a characteristic of a given dielectric. In practice, circuit design is considerably simplified if the shunt conductance of a capacitor can be ignored (i.e. $R \to \infty$)

and the capacitor therefore regarded as an open circuit for direct current.

Since capacitance C of a parallel plate capacitor is given by $C=\varepsilon_0\varepsilon_r A/d$, reducing the thickness d of a dielectric film increases the capacitance, but decreases the resistance. It also reduces the voltage the capacitor can withstand without breakdown (since $V=Q/C$). Any material will eventually break down, usually destructively, when subjected to a sufficiently large electric field. A spark may occur at breakdown which produces a hole through the film. The metal film forming the metal plates may be welded together at the point of breakdown.

Breakdown depends on electric field strength E (where $E=V/d$), so thinner films will break down with smaller voltages across them. This is the main reason for limiting the voltage that may be applied to a capacitor. All practical capacitors have a safe working voltage stated on them, generally at a particular maximum temperature. Figure 42.1 shows the typical shapes of graphs expected for electric field strength E plotted against thickness and for breakdown voltage plotted against thickness. The shape of the curves depend on a number of factors, and these include:

(i) the type of dielectric material,

(ii) the shape and size of the conductors associated with it,

(iii) the atmospheric pressure,

(iv) the humidity/moisture content of the material,

(v) the operating temperature.

Dielectric strength is an important factor in the design of capacitors as well as transformers and high-voltage insulators, and in motors and generators. Dielectrics vary in their ability to withstand large fields. Some typical values of dielectric strength, together with resistivity and relative permittivity, are shown in Table 42.1. The ceramics have very high relative permittivities and they tend to be 'ferroelectric', i.e. they do not lose

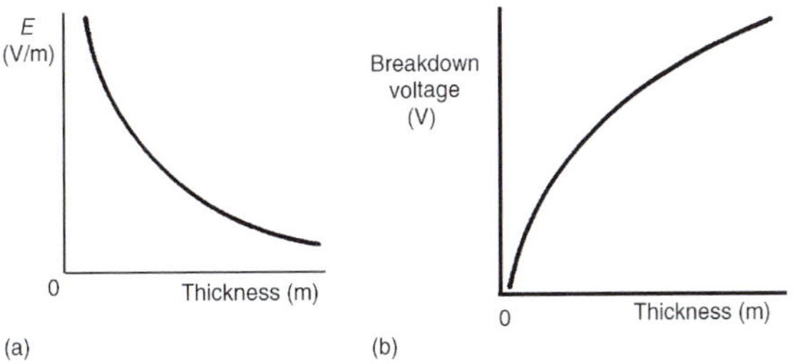

Figure 42.1

Table 42.1 Dielectric properties of some common materials

Material	Resistivity, $\rho(\Omega m)$	Relative permittivity, ε_r	Dielectric strength (V/m)
Air		1.0	3×10^6
Paper	10^{10}	3.7	1.6×10^7
Mica	5×10^{11}	5.4	10^8–10^9
Titanium dioxide	10^{12}	100	6×10^6
Polythene	$>10^{11}$	2.3	4×10^7
Polystyrene	$>10^{13}$	2.5	2.5×10^7
Ceramic (type 1)	4×10^{11}	6–500	4.5×10^7
Ceramic (type 2)	10^6–10^{13}	500–1000	2×10^6–10^7

their polarities when the electric field is removed. When ferroelectric effects are present, the charge on a capacitor is given by $Q=(CV)+$(remanent polarization). These dielectrics often possess an appreciable negative temperature coefficient of resistance. Despite this, a high permittivity is often very desirable and ceramic dielectrics are widely used.

42.4 Thermal effects

As the temperature of most dielectrics is increased, the insulation resistance falls rapidly. This causes the leakage current to increase, which generates further heat. Eventually a condition known as thermal avalanche or thermal runaway may develop, when the heat is generated faster than it can be dissipated to the surrounding environment. The dielectric will burn and thus fail.

Thermal effects may often seriously influence the choice and application of insulating materials. Some important factors to be considered include:

(i) the melting-point (for example, for waxes used in paper capacitors),

(ii) ageing due to heat,

(iii) the maximum temperature that a material will withstand without serious deterioration of essential properties,

(iv) flash-point or ignitability,

(v) resistance to electric arcs,

(vi) the specific heat capacity of the material,

(vii) thermal resistivity,

(viii) the coefficient of expansion,

(ix) the freezing-point of the material.

42.5 Mechanical properties

Mechanical properties determine, to varying degrees, the suitability of a solid material for use as an insulator: tensile strength, transverse strength, shearing strength and compressive strength are often specified. Most solid insulations have a degree of inelasticity and many are quite brittle, thus it is often necessary to consider features such as compressibility, deformation under bending stresses, impact strength and extensibility, tearing strength, machinability and the ability to fold without damage.

42.6 Types of practical capacitor

Practical types of capacitor are characterized by the material used for their dielectric. The main types include: variable air, mica, paper, ceramic, plastic, titanium oxide and electrolytic, together with supercapacitors. Refer back to Chapter 8, Section 8.12, page 117, for a description of each type.

42.7 Liquid dielectrics and gas insulation

Liquid dielectrics used for insulation purposes are refined mineral oils, silicone fluids and synthetic oils such as chlorinated diphenyl. The principal uses of liquid dielectrics are as a filling and cooling medium for transformers, capacitors and rheostats, as an insulating and arc-quenching medium in switchgear such as circuit breakers, and as an impregnant of absorbent insulations – for example, wood, slate, paper and pressboard, used mainly in transformers, switchgear, capacitors and cables.

Two **gases** used as insulation are nitrogen and sulphur hexafluoride. Nitrogen is used as an insulation medium in some sealed transformers and in power cables, and sulphur hexafluoride is finding increasing use in switchgear both as an insulant and as an arc-extinguishing medium.

42.8 Dielectric loss and loss angle

In capacitors with solid dielectrics, losses can be attributed to two causes:

(i) **dielectric hysteresis**, a phenomenon by which energy is expended and heat produced as the result of the reversal of electrostatic stress in a dielectric subjected to alternating electric stress – this loss is analogous to hysteresis loss in magnetic materials;

(ii) **leakage currents** that may flow through the dielectric and along surface paths between the terminals.

The total dielectric loss may be represented as the loss in an additional resistance connected between the plates. This may be represented as either a small resistance in series with an ideal capacitor or as a large resistance in parallel with an ideal capacitor.

Series representation

The circuit and phasor diagrams for the series representation are shown in Figure 42.2. The circuit phase angle is shown as angle ϕ. If resistance R_S is zero then current I would lead voltage V by 90°, this being the case of a perfect capacitor. The difference between 90° and the circuit phase angle ϕ is the angle shown as δ. This is known as the **loss angle** of the capacitor, i.e.

$$\textbf{loss angle, } \boldsymbol{\delta = (90° - \phi)}$$

For the equivalent series circuit,

$$\tan\delta = \frac{V_{R_S}}{V_{C_S}} = \frac{IR_S}{IX_{C_S}}$$

i.e. $$\tan\delta = \frac{R_S}{1/(\omega C_S)} = R_S\omega C_S$$

Since from Chapter 31, $Q = \frac{1}{\omega CR}$ then

$$\boldsymbol{\tan\delta = R_S\omega C_S = \frac{1}{Q}} \qquad (1)$$

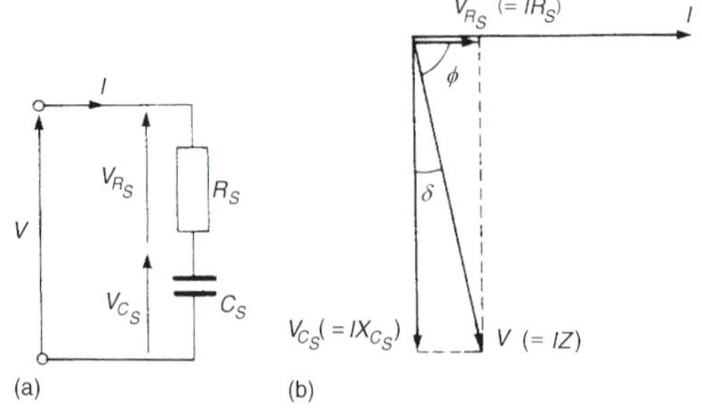

Figure 42.2 (a) Circuit diagram (b) phasor diagram

Power factor of capacitor,

$$\cos\phi = \frac{V_{R_S}}{V} = \frac{IR_S}{IZ_S} = \frac{R_S}{Z_S} \approx \frac{R_S}{X_{C_S}}$$

since $X_{C_S} \approx Z_S$ when δ is small. Hence **power factor = $\cos\phi \approx R_S\omega C_S$**, i.e.

$$\cos\phi \approx \tan\delta \qquad (2)$$

Dissipation factor, *D*, is defined as the reciprocal of Q-factor and is an indication of the quality of the dielectric, i.e.

$$D = \frac{1}{Q} = \tan\delta \qquad (3)$$

Parallel representation

The circuit and phasor diagrams for the parallel representation are shown in Figure 42.3. From the phasor diagram,

$$\tan\delta = \frac{I_{R_P}}{I_{C_P}} = \frac{V/R_P}{V/X_{C_P}} = \frac{X_{C_P}}{R_P}$$

i.e. $$\tan\delta = \frac{1}{R_P\omega C_P} \qquad (4)$$

Power factor of capacitor,

$$\cos\phi = \frac{I_{R_P}}{I} = \frac{V/R_P}{V/Z_P} = \frac{Z_P}{R_P} \approx \frac{X_{C_P}}{R_P}$$

since $X_{C_P} \approx Z_P$, when δ is small. Hence

$$\text{power factor} = \cos\phi \approx \frac{1}{R_P\omega C_P}$$

i.e. $$\cos\phi \approx \tan\delta$$

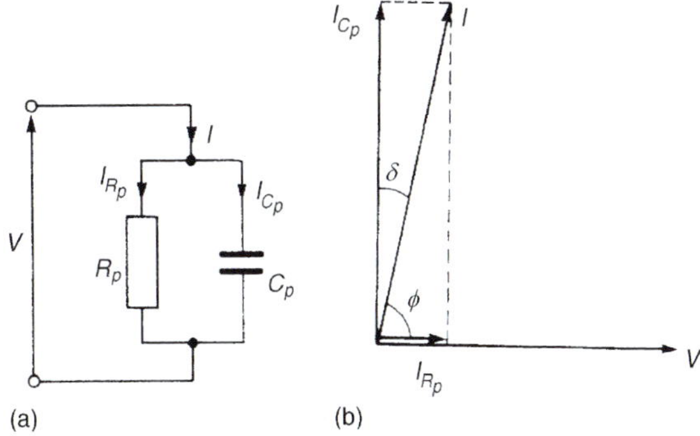

Figure 42.3 (a) Circuit diagram (b) phasor diagram

(For equivalence between the series and the parallel circuit representations,

$$C_S \approx C_P = C \quad \text{and} \quad R_S\omega C_S \approx \frac{1}{R_P\omega C_P}$$

from which $R_S \approx 1/R_P\omega^2C^2$)

Power loss in the dielectric $= VI\cos\phi$. From the phasor diagram of Figure 42.3

$$\cos\delta = \frac{I_{C_P}}{I} = \frac{V/X_{C_P}}{I} = \frac{V\omega C}{I} \text{ or } I = \frac{V\omega C}{\cos\delta}$$

$$\text{Hence power loss} = VI\cos\phi = V\left(\frac{V\omega C}{\cos\delta}\right)\cos\phi$$

However, $\cos\phi = \sin\delta$ (complementary angles), thus

$$\text{power loss} = V\left(\frac{V\omega C}{\cos\delta}\right)\sin\delta = V^2\omega C\tan\delta$$

(since $\sin\delta/\cos\delta = \tan\delta$)

Hence **dielectric power loss = $V^2\omega C\tan\delta$** (5)

Problem 1. The equivalent series circuit for a particular capacitor consists of a 1.5 Ω resistance in series with a 400 pF capacitor. Determine for the capacitor, at a frequency of 8 MHz, (a) the loss angle, (b) the power factor, (c) the Q-factor and (d) the dissipation factor.

(a) From equation (1), for a series equivalent circuit,

$$\tan\delta = R_S\omega C_S$$
$$= (1.5)(2\pi \times 8 \times 10^6)(400 \times 10^{-12})$$
$$= 0.030159$$

Hence **loss angle, δ** $= \tan^{-1}(0.030159)$ $=$ **1.727°** or **0.030 rad**

(b) From equation (2), **power factor** $= \cos\phi \approx \tan\delta =$ **0.030**

(c) From equation (1), $\tan\delta = \dfrac{1}{Q}$

hence $\boldsymbol{Q} = \dfrac{1}{\tan\delta} = \dfrac{1}{0.030159} =$ **33.16**

(d) From equation (3), **dissipation factor,**

$\boldsymbol{D} = \dfrac{1}{Q} = 0.030159$ or **0.030**, correct to 3 decimal places.

Problem 2. A capacitor has a loss angle of 0.025 rad, and when it is connected across a 5 kV, 50 Hz supply, the power loss is 20 W. Determine the component values of the equivalent parallel circuit.

From equation (5),

power loss $= V^2 \omega C \tan\delta$

i.e. $\quad 20 = (5000)^2(2\pi 50)(C)\tan(0.025)$

from which **capacitance** $\boldsymbol{C} = \dfrac{20}{(5000)^2(2\pi 50)\tan(0.025)}$

$= \mathbf{0.102\,\mu F}$

(Note tan (0.025) means 'the tangent of 0.025 rad')
From equation (4), for a parallel equivalent circuit,

$$\tan\delta = \frac{1}{R_P \omega C_P}$$

from which, parallel resistance,

$$R_P = \frac{1}{\omega C_P \tan\delta}$$

$$= \frac{1}{(2\pi 50)(0.102 \times 10^{-6})\tan 0.025}$$

i.e. $\boldsymbol{R_P = 1.248\,M\Omega}$

Problem 3. A 2000 pF capacitor has an alternating voltage of 20 V connected across it at a frequency of 10 kHz. If the power dissipated in the dielectric is 500 μW, determine (a) the loss angle, (b) the equivalent series loss resistance and (c) the equivalent parallel loss resistance.

(a) From equation (5), power loss $= V^2 \omega C \tan\delta$, i.e.

$$500 \times 10^{-6} = (20)^2(2\pi 10 \times 10^3) \times (2000 \times 10^{-12})\tan\delta$$

Hence $\tan\delta = \dfrac{500 \times 10^{-6}}{(20)^2(2\pi 10 \times 10^3)(2000 \times 10^{-12})}$

$= 9.947 \times 10^{-3}$

from which, **loss angle,** $\boldsymbol{\delta = 0.57°}$ or $\mathbf{9.95 \times 10^{-3}}$ **rad**.

(b) From equation (1), for an equivalent series circuit, $\tan\delta = R_S \omega C_S$, from which equivalent series resistance,

$$R_S = \frac{\tan\delta}{\omega C_S} = \frac{9.947 \times 10^{-3}}{(2\pi 10 \times 10^3)(2000 \times 10^{-12})}$$

i.e. $\boldsymbol{R_S = 79.16\,\Omega}$

(c) From equation (4), for an equivalent parallel circuit,

$$\tan\delta = \frac{1}{R_P \omega C_P}$$

from which equivalent parallel resistance,

$$R_P = \frac{1}{(\tan\delta)\omega C_P}$$

$$= \frac{1}{(9.947 \times 10^{-3})(2\pi 10 \times 10^3)(2000 \times 10^{-12})}$$

i.e. $\boldsymbol{R_P = 800\,k\Omega}$

Now try the following Practice Exercise

Practice Exercise 162 Dielectric loss and loss angle (Answers on page 831)

1. The equivalent series circuit for a capacitor consists of a 3 Ω resistance in series with a 250 pF capacitor. Determine the loss angle of the capacitor at a frequency of 5 MHz, giving the answer in degrees and in radians. Find also for the capacitor (a) the power factor, (b) the Q-factor and (c) the dissipation factor.
2. A capacitor has a loss angle of 0.008 rad and when it is connected across a 4 kV, 60 Hz supply the power loss is 15 W. Determine the component values of (a) the equivalent parallel circuit and (b) the equivalent series circuit.
3. A coaxial cable has a capacitance of 4 μF and a dielectric power loss of 12 kW when operated at 50 kV and frequency 50 Hz. Calculate (a) the value of the loss angle and (b) the equivalent parallel resistance of the cable.
4. What are the main reasons for power loss in capacitors with solid dielectrics? Explain the term 'loss angle'.

A voltage of 10 V and frequency 20 kHz is connected across a 1 nF capacitor. If the power dissipated in the dielectric is 0.2 mW, determine (a) the loss angle, (b) the equivalent series loss resistance and (c) the equivalent parallel loss resistance.

5. The equivalent series circuit for a capacitor consists of a 0.5 Ω resistor in series with a capacitor of reactance 2 kΩ. Determine for the capacitor (a) the loss angle, (b) the power factor and (c) the equivalent parallel resistance.

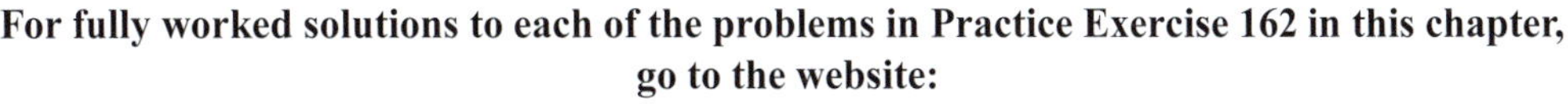
For fully worked solutions to each of the problems in Practice Exercise 162 in this chapter, go to the website:

www.routledge.com/cw/bird

Chapter 43

Field theory

***Why it is important to understand:* Field theory**

The conductor and sheath of coaxial cables, such as a telegraph cable and single-conductor lead-sheath power cable, are concentric cylinders. The capacitance between the inner cylinder, the conductor, and outer cylinder, the sheath or shield, has an important effect on the characteristics of such cables. Whenever two nearby conductors of any size or shape carry equal and opposite charges, the combination of these conducting bodies is called a capacitor. Because the isolated conducting bodies have equal but opposite charges on them, an electric field exists in the space between them. The importance of the capacitor lies in the fact that energy can be stored in the electric field between the two conducting bodies. The main properties of a co-axial cable are its inductance, capacitance, effective shunt conductance and series resistance per unit length. Taken together these influence the signal transmission and loss properties when a length of cable is employed as part of a system. This chapter looks at some aspects of field theory – field plotting, capacitance and inductance of concentric cylinders, capacitance and inductance of isolated twin line and energy stored in electric and electromagnetic fields.

At the end of this chapter you should be able to:

- understand field plotting by curvilinear squares
- show that the capacitance between concentric cylinders, $C=\dfrac{2\pi\varepsilon_0\varepsilon_r}{\ln(b/a)}$ and calculate C given values of radii a and b
- calculate dielectric stress $E=\dfrac{V}{r\ln(b/a)}$
- appreciate dimensions of the most economical cable
- show that the capacitance of an isolated twin line, $C=\dfrac{\pi\varepsilon_0\varepsilon_r}{\ln(D/a)}$ and calculate C given values of a and D
- calculate energy stored in an electric field
- show that the inductance of a concentric cylinder, $L=\dfrac{\mu_0\mu_r}{2\pi}\left(\dfrac{1}{4}+\ln\dfrac{b}{a}\right)$ and calculate L given values of a and b
- show that the inductance of an isolated twin line, $L=\dfrac{\mu_0\mu_r}{\pi}\left(\dfrac{1}{4}+\ln\dfrac{D}{a}\right)$ and calculate L given values of a and D
- calculate energy stored in an electromagnetic field

Electrical Circuit Theory and Technology. 978-1-138-67349-6, © 2017 John Bird. Published by Taylor & Francis. All rights reserved.

43.1 Field plotting by curvilinear squares

Electric fields, magnetic fields and conduction fields (i.e. a region in which an electric current flows) are analogous, i.e. they all exhibit similar characteristics. Thus they may all be analysed by similar processes. In the following the electric field is analysed.

Figure 43.1 shows two parallel plates, A and B. Let the potential on plate A be $+V$ volts and that on plate B be $-V$ volts. The force acting on a point charge of 1 coulomb placed between the plates is the electric field strength E. It is measured in the direction of the field and its magnitude depends on the p.d. between the plates and the distance between the plates. In Figure 43.1, moving along a line of force from plate B to plate A means moving from $-V$ to $+V$ volts. The p.d. between the plates is therefore 2 V volts and this potential changes linearly when moving from one plate to the other. Hence a potential gradient is followed which changes by equal amounts for each unit of distance moved.

Lines may be drawn connecting together all points within the field having equal potentials. These lines are called **equipotential lines** and these have been drawn in Figure 43.1 for potentials of $\frac{2}{3}$ V, $\frac{1}{3}$ V, 0, $-\frac{1}{3}$ V and $-\frac{2}{3}$ V. The zero equipotential line represents earth potential and the potentials on plates A and B are respectively above and below earth potential. Equipotential lines form part of an equipotential surface. Such surfaces are parallel to the plates shown in Figure 43.1 and the plates themselves are equipotential surfaces. There can be no current flow between any given points on such a surface since all points on an equipotential surface have the same potential. Thus a line of force (or flux) must intersect an equipotential surface at right-angles. A line of force in an electrostatic field is often termed a **streamline**.

An electric field distribution for a concentric cylinder capacitor is shown in Figure 43.2. An electric field is set up in the insulating medium between two good conductors. Any volt drop within the conductors can usually be neglected compared with the p.d.s across

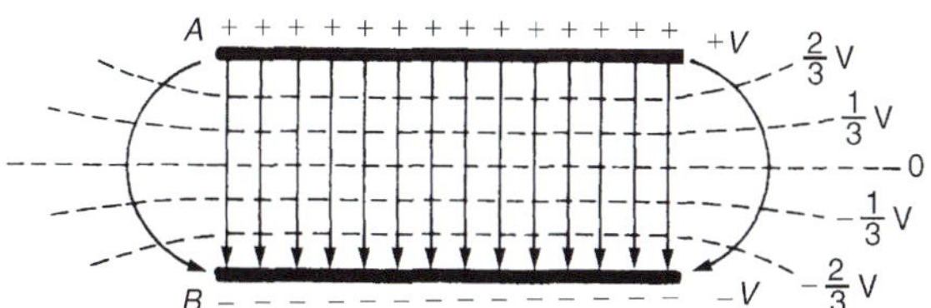

Figure 43.1 Lines of force intersecting equipotential lines in an electric field

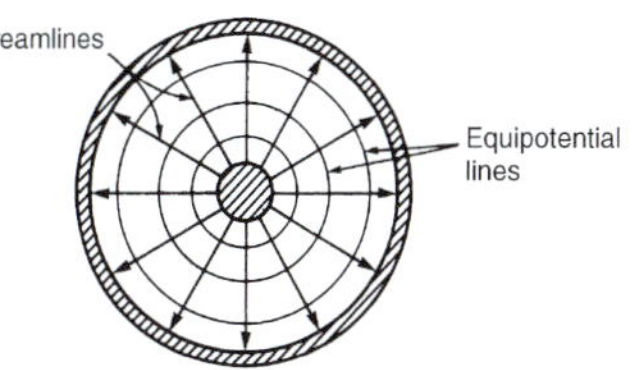

Figure 43.2 Electric field distribution for a concentric cylinder capacitor

the insulation since the conductors have a high conductivity. All points on the conductors are thus at the same potential so that the conductors form the boundary equipotentials for the electrostatic field. Streamlines (or lines of force) which must cut all equipotentials at right-angles leave one boundary at right-angles, pass across the field and enter the other boundary at right-angles.

In a magnetic field, a streamline is a line so drawn that its direction is everywhere parallel to the direction of the magnetic flux. An equipotential surface in a magnetic field is the surface over which a magnetic pole may be moved without the expenditure of work or energy.

In a conduction field, a streamline is a line drawn with a direction which is everywhere parallel to the direction of the current flow.

A method of solving certain field problems by a form of graphical estimation is available which may only be applied, however, to plane linear fields; examples include the field existing between parallel plates or between two long parallel conductors. In general, the plane of a field may be divided into a number of squares formed between the line of force (i.e. streamline) and the equipotential. Figure 43.3 shows a typical pattern. In most cases true squares will not exist, since the streamlines and equipotentials are curved. However, since the streamlines and the equipotentials intersect at right-angles, square-like figures are formed, and these are usually called **'curvilinear squares'**. The square-like figure shown in Figure 43.3 is a curvilinear square since, on successive sub-division by equal numbers of

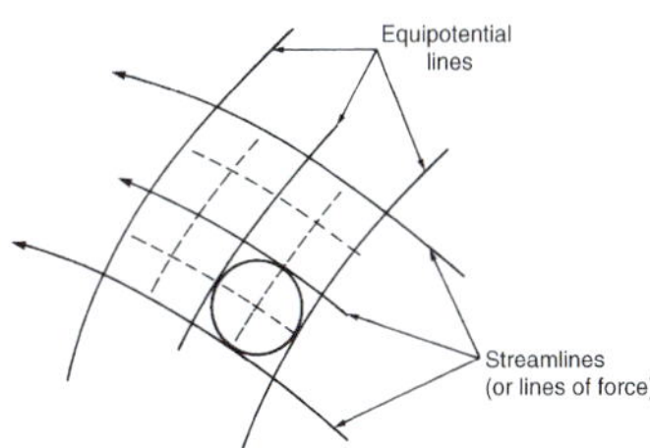

Figure 43.3 Curvilinear square

Part 4

intermediate streamlines and equipotentials, the smaller figures are seen to approach a true square form.

When sub-dividing to give a field in detail, and in some cases for the initial equipotentials, **'Moore's circle' technique** can be useful in that it tends to eliminate the trial and error process. If, say, two flux lines and an equipotential are given and it is required to draw a neighbouring equipotential, a circle tangential to the three given lines is constructed. The new equipotential is then approximately tangential to the circle, as shown in Figure 43.3.

Consider the electric field established between two parallel metal plates, as shown in Figure 43.4. The streamlines and the equipotential lines are shown sketched and are seen to form curvilinear squares. Consider a true square *abcd* lying between equipotentials *AB* and *CD*. Let this square be the end of x metres depth of the field forming a flux tube between adjacent equipotential surfaces *abfe* and *cdhg* as shown in Figure 43.5. Let l be the length of side of the squares. Then the capacitance C_1 of the flux tube is given by

$$C_1 = \frac{\varepsilon_0\varepsilon_r \text{ (area of plate)}}{\text{plate separation}}$$

$$\text{i.e. } \boldsymbol{C_1} = \frac{\varepsilon_0\varepsilon_r(lx)}{l} = \boldsymbol{\varepsilon_0\varepsilon_r x} \qquad (1)$$

Thus the capacitance of the flux tube whose end is a true square is independent of the size of the square.

Let the distance between the plates of a capacitor be divided into an exact number of parts, say n (in Figure 43.4, $n=4$). Using the same scale, the breadth of the plate is divided into a number of parts (which is not always an integer value), say m (in Figure 43.4, $m=10$, neglecting fringing). Thus between equipotentials *AB* and *CD* in Figure 43.4 there are m squares in parallel and so there are m capacitors in parallel. For m capacitors connected in parallel, the equivalent capacitance C_T is given by $C_T = C_1 + C_2 + C_3 + \cdots + C_m$. If the capacitors have the same value, i.e. $C_1 = C_2 = C_3 = \cdots = C_m = C_t$, then

$$C_T = mC_t \qquad (2)$$

Similarly, there are n squares in series in Figure 43.4 and thus n capacitors in series.

For n capacitors connected in series, the equivalent capacitance C_T is given by

$$\frac{1}{C_T} = \frac{1}{C_1} + \frac{1}{C_2} + \cdots + \frac{1}{C_n}$$

If $C_1 = C_2 = \cdots = C_n = C_t$ then $1/C_T = n/C_t$, from which

$$C_T = \frac{C_t}{n} \qquad (3)$$

Thus if m is the number of parallel squares measured along each equipotential and n is the number of series squares measured along each streamline (or line of force), then the total capacitance C of the field is given, from equations (1)–(3), by

$$\boldsymbol{C = \varepsilon_0\varepsilon_r x\frac{m}{n}} \textbf{ farads} \qquad (4)$$

For example, let a parallel-plate capacitor have plates 8 mm × 5 mm and spaced 4 mm apart (see Figure 43.6). Let the dielectric have a relative permittivity 3.5. If the distance between the plates is divided into, say, four equipotential lines, then each is 1 mm apart. Hence $n=4$.

Using the same scale, the number of lines of force from plate P to plate Q must be 8, i.e. $m=8$. This is, of course, neglecting any fringing. From equation (4), capacitance $C = \varepsilon_0\varepsilon_r x(m/n)$, where $x = 5$ mm or 0.005 m in this case. Hence

$$C = (8.85 \times 10^{-12})(3.5)(0.005)\left(\tfrac{8}{4}\right) = \mathbf{0.31\,pF}$$

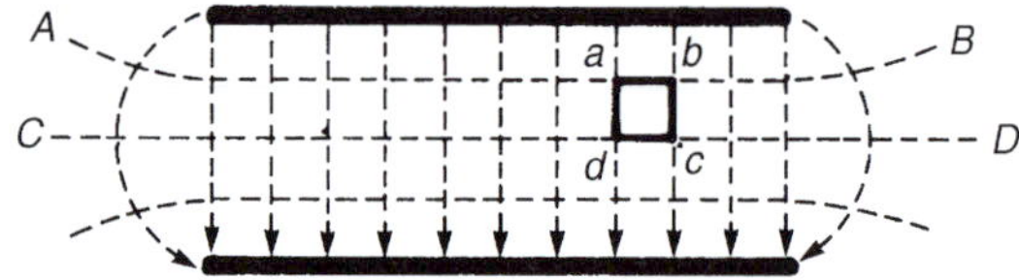

Figure 43.4

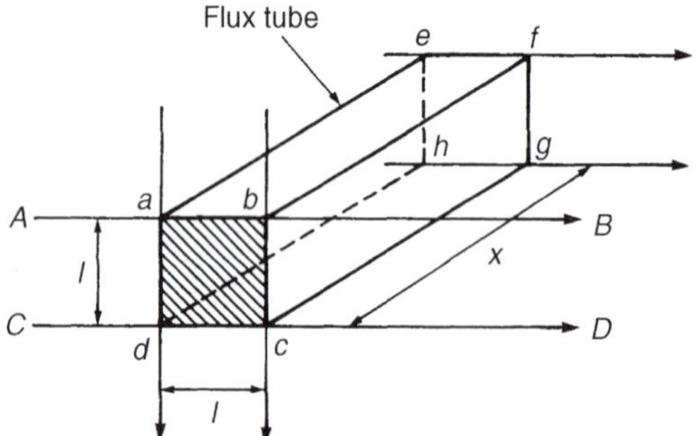

Figure 43.5

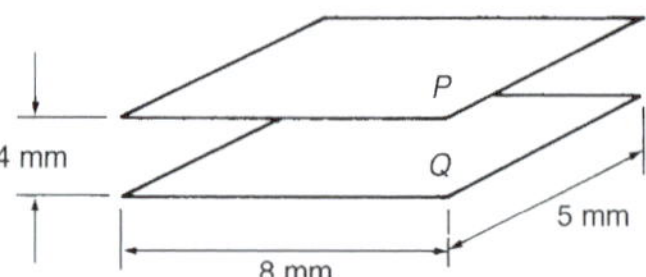

Figure 43.6

(Using the normal equation for capacitance of a parallel-plate capacitor,

$$C = \frac{\varepsilon_0\varepsilon_r A}{d} = \frac{(885 \times 10^{-12})(3.5)(0.008 \times 0.005)}{0.004}$$

$$= \mathbf{0.31\,pF}$$

The capacitance found by each method gives the same value; this is expected since the field is uniform between the plates, giving a field plot of true squares.)

The effect of fringing may be considered by estimating the capacitance by field plotting. This is described below.

In the side view of the plates shown in Figure 43.7, *RS* is the medial line of force or medial streamline, by symmetry. Also *XY* is the medial equipotential. The field may thus be divided into four separate symmetrical parts.

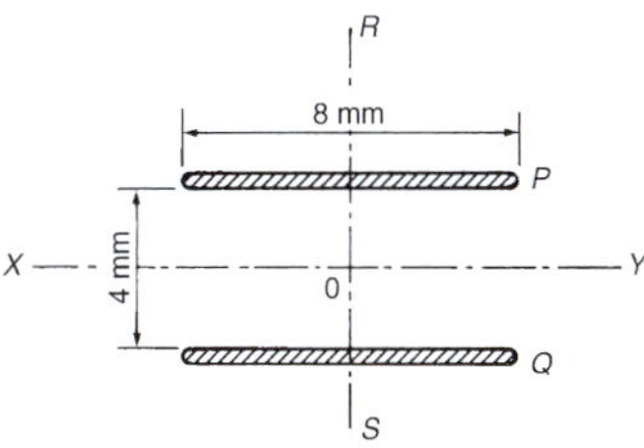

Figure 43.7

Considering just the top-left part of the field, the field plot is estimated as follows, with reference to Figure 43.8:

(i) Estimate the position of the equipotential *EF* which has the mean potential between that of the plate and that of the medial equipotential *X0*. F is not taken too far since it is difficult to estimate. Point E will lie slightly closer to point Z than point O.

(ii) Estimate the positions of intermediate equipotentials *GH* and *IJ*.

(iii) All the equipotential lines plotted are $\frac{2}{4}$, i.e. 0.5 mm apart. Thus a series of streamlines, cutting the equipotential at right-angles, are drawn, the streamlines being spaced 0.5 mm apart, with the object of forming, as far as possible, curvilinear squares.

It may be necessary to erase the equipotentials and redraw them to fit the lines of force. The field between the plates is almost uniform, giving a field plot of true squares in this region. At the corner of the plates the squares are smaller, this indicating a great stress in this region.

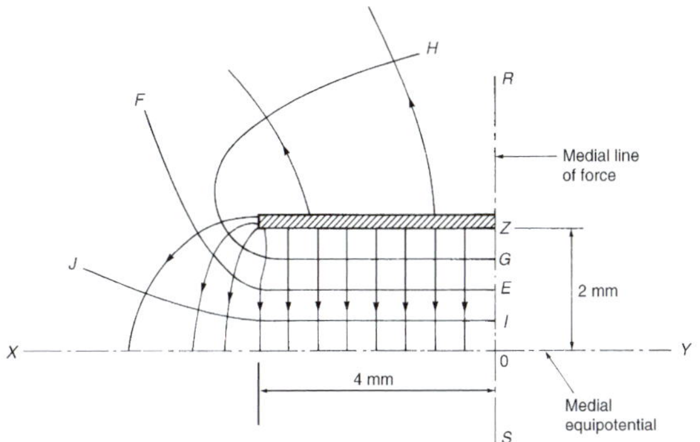

Figure 43.8

On the top of the plate the squares become very large, indicating that the main field exists between the plates. From equation (4),

$$\text{total capacitance, } C = \varepsilon_0\varepsilon_r x \frac{m}{n} \text{ farads}$$

The number of parallel squares measured along each equipotential is about 13 in this case and the number of series squares measured along each line of force is 4. Thus, for the plates shown in Figure 43.7, $m = 2 \times 13 = 26$ and $n = 2 \times 4 = 8$. Since x is 5 mm,

$$\text{total capacitance} = \varepsilon_0\varepsilon_r x \frac{m}{n}$$

$$= (8.85 \times 10^{-12})(3.5)(0.005)\frac{26}{8}$$

$$= \mathbf{0.50\,pF}$$

Problem 1. A field plot between two metal plates is shown in Figure 43.9. The relative permeability of the dielectric is 2.8. Determine the capacitance per metre length of the system.

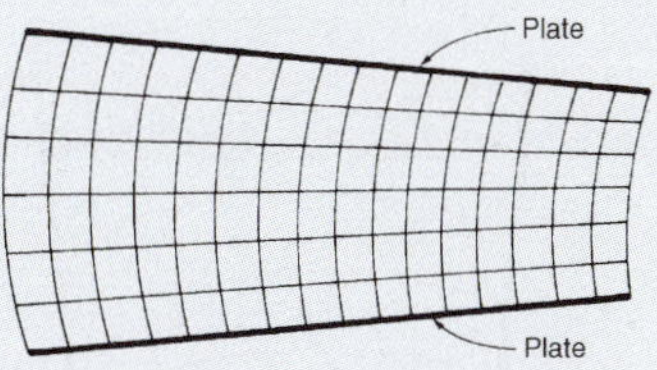

Figure 43.9

From equation (4), capacitance $C = \varepsilon_0\varepsilon_r x(m/n)$. From Figure 43.9, $m = 16$, i.e. the number of parallel squares measured along each equipotential, and $n = 6$, i.e. the number of series squares measured along each line of force. Hence capacitance for a 1 m length,

$$C = (8.85 \times 10^{-12})(2.8)(1)\frac{16}{6} = \mathbf{66.08\,pF}$$

Problem 2. A field plot for a cross-section of a concentric cable is shown in Figure 43.10. If the relative permeability of the dielectric is 3.4, determine the capacitance of a 100 m length of the cable.

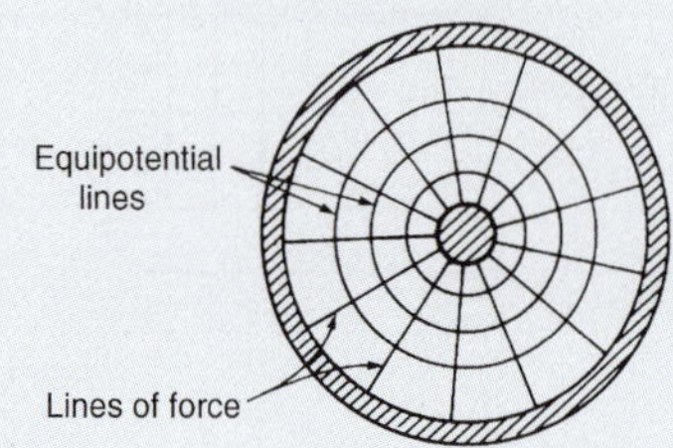

Figure 43.10

From equation (4), capacitance $C = \varepsilon_0 \varepsilon_r x(m/n)$. In this case, $m = 13$ and $n = 4$. Also $x = 100$ m. Thus

$$\text{capacitance } C = (8.85 \times 10^{-12})(3.4)(100)\frac{13}{4}$$

$$= \mathbf{9780\,pF} \text{ or } \mathbf{9.78\,nF}$$

Now try the following Practice Exercise

Practice Exercise 163 Field plotting by curvilinear squares (Answers on page 831)

1. (a) Explain the meaning of the terms (i) streamline and (ii) equipotential, with reference to an electric field.
   (b) A field plot between two metal plates is shown in Figure 43.11. If the relative permittivity of the dielectric is 2.4, determine the capacitance of a 50 cm length of the system.

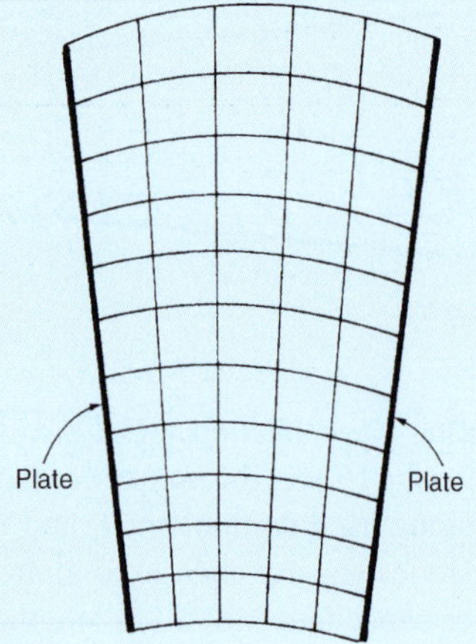

Figure 43.11

2. A field plot for a concentric cable is shown in Figure 43.12. The relative permittivity of the dielectric is 5. Determine the capacitance of a 10 m length of the cable.

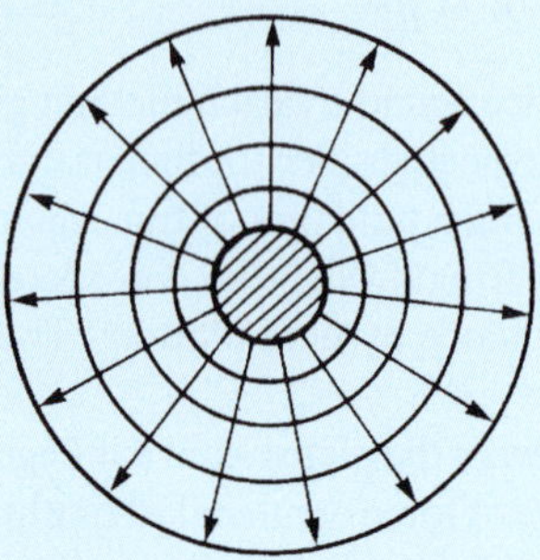

Figure 43.12

3. The plates of a capacitor are 10 mm long and 6 mm wide and are separated by a dielectric 3 mm thick and of relative permittivity 2.5. Determine the capacitance of the capacitor (a) when neglecting any fringing at the edges, (b) by producing a field plot taking fringing into consideration.

43.2 Capacitance between concentric cylinders

A **concentric cable** is one which contains two or more separate conductors, arranged concentrically (i.e. having a common centre), with insulation between them. In a **coaxial cable**, the central conductor, which may be either solid or hollow, is surrounded by an outer tubular conductor, the space in between being occupied by a dielectric. If air is the dielectric then concentric insulating discs are used to prevent the conductors touching each other. The two kinds of cable serve different purposes. The main feature they have in common is a complete absence of external flux and therefore a complete absence of interference with and from other circuits.

The electric field between two concentric cylinders (i.e. a coaxial cable) is shown in the cross-section of Figure 43.13. The conductors form the boundary equipotentials for the field, the boundary equipotentials in Figure 43.13 being concentric cylinders of radii a and b. The streamlines, or lines of force, are radial lines cutting the equipotentials at right-angles.

Let Q be the charge per unit length of the inner conductor. Then the total flux across the dielectric per unit

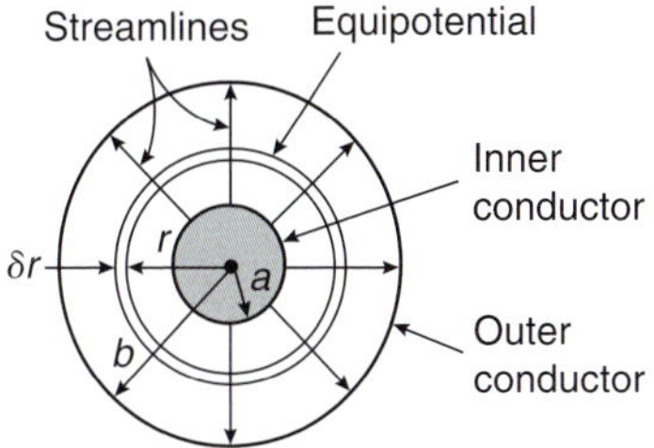

Figure 43.13 Electric field between two concentric cylinders

length is Q coulombs/metre. This total flux will pass through the elemental cylinder of width δr at radius r (shown in Figure 43.13) and a distance of 1 m into the plane of the paper.

The surface area of a cylinder of length 1 m within the dielectric with radius r is $(2\pi r \times 1)$ m^2. Hence the electric flux density at radius r,

$$D = \frac{Q}{A} = \frac{Q}{2\pi r}$$

The electric field strength or electric stress E, at radius r is given by

$$E = \frac{D}{\varepsilon_0 \varepsilon_r} = \frac{Q}{2\pi r \varepsilon_0 \varepsilon_r} \text{ volts/metre} \qquad (5)$$

Let the p.d. across the element be δV volts. Since

$$E = \frac{\text{voltage}}{\text{thickness}}$$

voltage $= E \times$ thickness. Therefore

$$\delta V = E\delta r = \frac{Q}{2\pi r \varepsilon_0 \varepsilon_r}\delta r$$

The total p.d. between the boundaries,

$$V = \int_a^b \frac{Q}{2\pi r \varepsilon_0 \varepsilon_r} \mathrm{d}r = \frac{Q}{2\pi \varepsilon_0 \varepsilon_r}\int_a^b \frac{1}{r}\mathrm{d}r$$

$$= \frac{Q}{2\pi \varepsilon_0 \varepsilon_r}[\ln r]_a^b = \frac{Q}{2\pi \varepsilon_0 \varepsilon_r}[\ln b - \ln a]$$

i.e. $$V = \frac{Q}{2\pi \varepsilon_0 \varepsilon_r} \ln \frac{b}{a} \text{ volts} \qquad (6)$$

The capacitance per unit length,

$$C = \frac{\text{charge per unit length}}{\text{p.d.}}$$

Hence capacitance,

$$C = \frac{Q}{V} = \frac{Q}{(Q/(2\pi \varepsilon_0 \varepsilon_r))\ln(b/a)}$$

i.e. $$\mathbf{C = \frac{2\pi \varepsilon_0 \varepsilon_r}{\ln(b/a)} \text{ farads/metre}} \qquad (7)$$

Problem 3. A coaxial cable has an inner core radius of 0.5 mm and an outer conductor of internal radius 6.0 mm. Determine the capacitance per metre length of the cable if the dielectric has a relative permittivity of 2.7.

From equation (7),

$$\text{capacitance } C = \frac{2\pi \varepsilon_0 \varepsilon_r}{\ln(b/a)} = \frac{2\pi(8.85 \times 10^{-12})(2.7)}{\ln(6.0/0.5)}$$

$$= \mathbf{60.4\,pF}$$

Problem 4. A single-core concentric cable has a capacitance of 80 pF per metre length. The relative permittivity of the dielectric is 3.5 and the core diameter is 8.0 mm. Determine the internal diameter of the sheath.

From equation (7), capacitance

$$C = \frac{2\pi \varepsilon_0 \varepsilon_r}{\ln(b/a)} \text{F/m}$$

$$\text{from which } \ln \frac{b}{a} = \frac{2\pi \varepsilon_0 \varepsilon_r}{C} = \frac{2\pi(8.85 \times 10^{-12})(3.5)}{(80 \times 10^{-12})}$$

$$= 2.433$$

Since the core radius, $a = 8.0/2 = 4.0$ mm,

$\ln(b/4.0) = 2.433$ and $b/4.0 = e^{2.433}$

Thus the internal radius of the sheath,

$b = 4.0e^{2.433} = 45.57$ mm.

Hence the internal diameter of the sheath

$= 2 \times 45.57 =$ **91.14 mm**.

Dielectric stress

Rearranging equation (6) gives:

$$\frac{Q}{2\pi \varepsilon_0 \varepsilon_r} = \frac{V}{\ln(b/a)}$$

However, from equation (5),

$$E = \frac{Q}{2\pi r \varepsilon_0 \varepsilon_r}$$

Thus dielectric stress,

$$\mathbf{E = \frac{V}{r \ln(b/a)} \text{ volts/metre}} \qquad (8)$$

From equation (8), the dielectric stress at any point is seen to be inversely proportional to r, i.e. $E \propto 1/r$
The dielectric stress E will have a maximum value when r is at its minimum, i.e. when $r = a$. Thus

$$\boldsymbol{E_{max} = \frac{V}{a \ln(b/a)}} \quad (9)$$

It follows that

$$\boldsymbol{E_{min} = \frac{V}{b \ln(b/a)}} \quad (9')$$

Problem 5. A concentric cable has a core diameter of 32 mm and an inner sheath diameter of 80 mm. The core potential is 40 kV and the relative permittivity of the dielectric is 3.5. Determine (a) the capacitance per kilometre length of the cable, (b) the dielectric stress at a radius of 30 mm and (c) the maximum and minimum values of dielectric stress.

(a) From equation (7), capacitance per metre length,

$$C = \frac{2\pi\varepsilon_0\varepsilon_r}{\ln(b/a)} = \frac{2\pi(8.85 \times 10^{-12})(3.5)}{\ln(40/16)}$$

$$= 212.4 \times 10^{-12}\text{ F/km}$$

$$= 212.4 \times 10^{-12} \times 10^3\text{ F/km}$$

$$= \mathbf{212\ nF/km}\text{ or }\mathbf{0.212\ \mu F/km}$$

(b) From equation (8), dielectric stress at radius r,

$$E = \frac{V}{r \ln(b/a)} = \frac{40 \times 10^3}{(30 \times 10^{-3})\ln(40/16)}$$

$$= \mathbf{1.46 \times 10^6\ V/m}\text{ or }\mathbf{1.46\ MV/m}$$

(c) From equation (9), maximum dielectric stress,

$$E_{max} = \frac{V}{a \ln(b/a)} = \frac{40 \times 10^3}{16 \times 10^{-3}\ln(40/16)}$$

$$= \mathbf{2.73\ MV/m}$$

From equation (9′), minimum dielectric stress,

$$E_{min} = \frac{V}{b \ln(b/a)} = \frac{40 \times 10^3}{40 \times 10^{-3}\ln(40/16)}$$

$$= \mathbf{1.09\ MV/m}$$

Dimensions of most economical cable

It is important to obtain the most economical dimensions when designing a cable. A relationship between a and b may be obtained as follows. If E_{max} and V are both fixed values, then, from equation (9),

$$\frac{V}{E_{max}} = a \ln\frac{b}{a}$$

Letting $V/E_{max} = k$, a constant, gives

$$a \ln\frac{b}{a} = k$$

from which $\ln(b/a) = k/a$, $b/a = e^{k/a}$ and $b = ae^{k/a}$ (10)

For the most economical cable, b will be a minimum value. Using the product rule of calculus,

$$\frac{db}{da} = (e^{k/a})(1) + (a)\left(-\frac{k}{a^2}e^{k/a}\right)$$

$$= 0 \text{ for a minimum value.}$$

(Note, to differentiate $e^{k/a}$ with respect to a, an algebraic substitution may be used, letting $u = 1/a$)

$$e^{k/a} - \frac{k}{a}e^{k/a} = 0$$

Therefore $e^{k/a}\left(1 - \frac{k}{a}\right) = 0$

from which $a = k$. Thus

$$\boldsymbol{a = \frac{V}{E_{max}}} \quad (11)$$

From equation (10), internal sheath radius, $b = ae^{k/a} = ae^1 = ae$, i.e.

$$\mathbf{b = 2.718a} \quad (12)$$

Problem 6. A single-core concentric cable is to be manufactured for a 60 kV, 50 Hz transmission system. The dielectric used is paper, which has a maximum permissible safe dielectric stress of 10 MV/m r.m.s. and a relative permittivity of 3.5. Calculate (a) the core and inner sheath radii for the most economical cable, (b) the capacitance per metre length and (c) the charging current per kilometre run.

(a) From equation (11),

$$\text{core radius, } a = \frac{V}{E_m} = \frac{60 \times 10^3\,\text{V}}{10 \times 10^6\,\text{V/m}}$$
$$= 6 \times 10^{-3}\,\text{m} = \mathbf{6.0\,mm}$$

From equation (12), internal sheath radius, $b = ae = 6.0\text{e} =$ **16.3 mm**

(b) From equation (7),

$$\text{capacitance } C = \frac{2\pi\varepsilon_0\varepsilon_r}{\ln(b/a)}\,\text{F/m}$$

Since $b = ae$,

$$C = \frac{2\pi\varepsilon_0\varepsilon_r}{\ln e} = 2\pi\varepsilon_0\varepsilon_r = 2\pi(8.85 \times 10^{-12})(3.5)$$
$$= \mathbf{195 \times 10^{-12}\,F/m} \text{ or } \mathbf{195\,pF/m}$$

(c) Charging current

$$= \frac{V}{X_C} = \frac{V}{1/(\omega C)} = \omega CV$$
$$= (2\pi 50)(195 \times 10^{-12}) \times (60 \times 10^3)$$
$$= 3.68 \times 10^{-3}\,\text{A/m}$$

Hence the charging current per kilometre **= 3.68 A**

Problem 7. A concentric cable has a core diameter of 25 mm and an inside sheath diameter of 80 mm. The relative permittivity of the dielectric is 2.5, the loss angle is 3.5×10^{-3} rad and the working voltage is 132 kV at 50 Hz frequency. Determine for a 1 km length of the cable (a) the capacitance, (b) the charging current and (c) the power loss.

(a) From equation (7), capacitance C

$$= \frac{2\pi\varepsilon_0\varepsilon_r}{\ln(b/a)}\,\text{F/m}$$
$$= \frac{2\pi(8.85 \times 10^{-12})(2.5)}{\ln(40/12.5)} \times 10^3\,\text{F/km}$$
$$= 0.120\,\mu\text{F/km}$$

Thus the capacitance for a 1 km length of the cable is **0.120 μF**

(b) Charging current

$$I = \frac{V}{X_C} = \frac{V}{1/(\omega C)} = \omega CV$$
$$= (2\pi 50)(0.120 \times 10^{-6})(132 \times 10^3)$$
$$= \mathbf{4.98\,A/km}$$

(c) From equation (5), Chapter 42, power loss

$$= V^2\omega C \tan\delta$$
$$= (132 \times 10^3)^2(2\pi 50)(0.120 \times 10^{-6}) \times \tan(3.5 \times 10^{-3})$$
$$= \mathbf{2300\,W}$$

Concentric cable field plotting

Figure 43.14 shows a cross-section of a concentric cable having a core radius r_1 and a sheath radius r_4. It was shown in Section 43.1 that the capacitance of a true square is given by $C = \varepsilon_0\varepsilon r$ farads/metre.

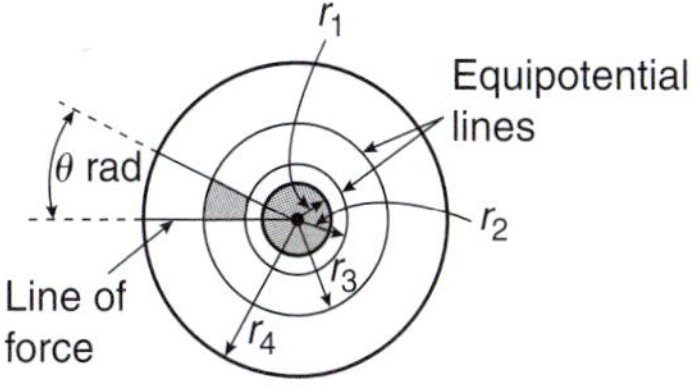

Figure 43.14

A curvilinear square is shown shaded in Figure 43.14. Such squares can be made to have the same capacitance as a true square by the correct choice of spacing between the lines of force and the equipotential surfaces in the field plot.

From equation (7), the capacitance between cylindrical equipotential lines at radii r_a and r_b is given by

$$C = \frac{2\pi\varepsilon_0\varepsilon_r}{\ln(r_b/r_a)}\ \text{farads/metre}$$

Thus for a sector of θ radians (see Figure 43.14) the capacitance is given by

$$C = \frac{\theta}{2\pi}\left(\frac{2\pi\varepsilon_0\varepsilon_r}{\ln(r_b/r_a)}\right) = \frac{\theta\varepsilon_0\varepsilon_r}{\ln(r_b/r_a)}\ \text{farads/metre}$$

Now if $\theta = \ln(r_b/r_a)$ then $C = \varepsilon_0\varepsilon_r$ F/m, the same as for a true square. If $\theta = \ln(r_b/r_a)$, then $e^\theta = (r_b/r_a)$. Thus if, say, two equipotential surfaces are chosen within the dielectric as shown in Figure 43.14, then $e^\theta = r_2/r_1$, $e^\theta = r_3/r_2$ and $e^\theta = r_4/r_3$. Hence

$$(e^\theta)^3 = \frac{r_2}{r_1} \times \frac{r_3}{r_2} \times \frac{r_4}{r_3}, \quad \text{i.e.} \quad \mathbf{e^{3\theta} = \frac{r_4}{r_1}} \qquad (13)$$

It follows that $e^{2\theta} = r_3/r_1$

Equation (13) is used to determine the value of θ and hence the number of sectors. Thus, for a concentric cable having a core radius 8 mm and inner sheath radius 32 mm, if two equipotential surfaces within the dielectric are chosen (and therefore form three capacitors in series in each sector).

$$e^{3\theta} = \frac{r_4}{r_1} = \frac{32}{8} = 4$$

Hence $3\theta = \ln 4$ and $\theta = \frac{1}{3}\ln 4 = 0.462$ rad (or 26.47°). Thus there will be $2\pi/0.462 = 13.6$ sectors in the field plot. (Alternatively, 360°/26.47° = 13.6 sectors.) From above,

$$e^{2\theta} = r_3/r_1, \text{ i.e. } r_3 = r_1 e^{2\theta} = 8e^{2(0.462)} = 20.15\,\text{mm}$$

$$e^{\theta} = \frac{r_2}{r_1}$$

from which

$$r_2 = r_1 e^{\theta} = 8e^{(0.462)} = 12.70\,\text{mm}$$

The field plot is shown in Figure 43.15. The number of parallel squares measured along each equipotential is 13.6 and the number of series squares measured along each line of force is 3. Hence in equation (4), where $C = \varepsilon_0 \varepsilon_r x(m/n)$, $m = 13.6$ and $n = 3$

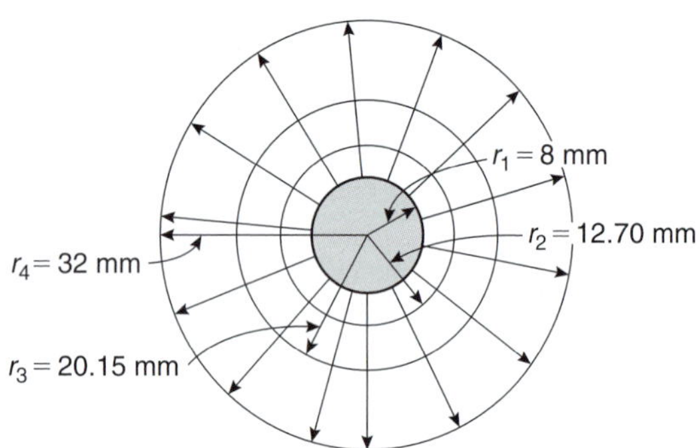

Figure 43.15

If the dielectric has a relative permittivity of, say, 2.5, then the capacitance per metre length,

$$C = (8.85 \times 10^{-12})(2.5)(1)\frac{13.6}{3} = \mathbf{100\,pF}$$

(From equation (7),

$$C = \frac{2\pi\varepsilon_0\varepsilon_r}{\ln(r_4/r_1)}\text{F/m} = \frac{2\pi(8.85 \times 10^{-12})(2.5)}{\ln(32/8)}$$
$$= \mathbf{100\,F/m)}$$

Thus field plotting using curvilinear squares provides an alternative method of determining the capacitance between concentric cylinders.

Problem 8. A concentric cable has a core diameter of 20 mm and a sheath inside diameter of 60 mm. The permittivity of the dielectric is 3.2. Using three equipotential surfaces within the dielectric, determine the capacitance of the cable per metre length by the method of curvilinear squares. Draw the field plot for the cable.

The field plot consists of radial lines of force dividing the cable cross-section into a number of sectors, the lines of force cutting the equipotential surfaces at right-angles. Since three equipotential surfaces are required in the dielectric, four capacitors in series are found in each sector of θ radians.

In Figure 43.16, $r_1 = 20/2 = 10$ mm and $r_5 = 60/2 = 30$ mm. It follows from equation (13) that $e^{4\theta} = r_5/r_1 = 30/10 = 3$, from which $4\theta = \ln 3$ and $\theta = \frac{1}{4}\ln 3 = 0.2747$ rad.

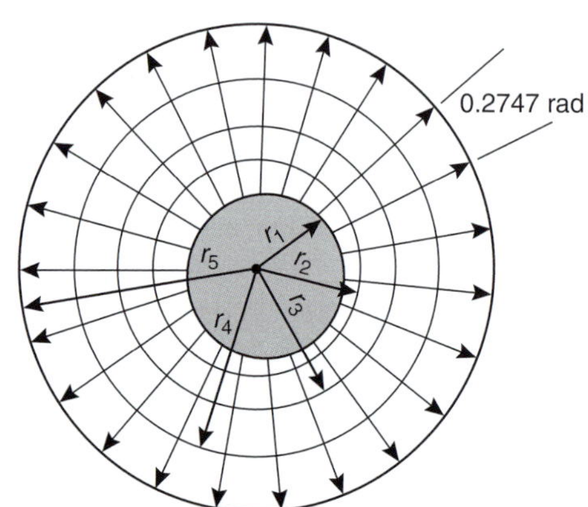

Figure 43.16

Thus the number of sectors in the plot shown in Figure 43.16 is $2\pi/0.2747 = \mathbf{22.9}$

The three equipotential lines are shown in Figure 43.16 at radii of r_2, r_3 and r_4

From equation (13),

$$e^{3\theta} = \frac{r_4}{r_1}, \text{ from which } r_4 = r_1 e^{3\theta} = 10e^{3(0.2747)}$$
$$= 22.80\,\text{mm}$$

$$e^{2\theta} = \frac{r_3}{r_1}, \text{ from which } r_3 = r_1 e^{2\theta} = 10e^{2(0.2747)}$$
$$= 17.32\,\text{mm}$$

$$e^{\theta} = \frac{r_2}{r_1}, \text{ from which } r_2 = r_1 e^{\theta} = 10e^{0.2747}$$
$$= 13.16\,\text{mm}$$

Thus the field plot for the cable is as shown in Figure 43.16.

From equation (4), capacitance $C=\varepsilon_0\varepsilon_r x(m/n)$. The number of parallel squares along each equipotential, $m=22.9$ and the number of series squares measured along each line of force, $n=4$. Thus

$$\text{capacitance } C = (8.85\times 10^{-12})(3.2)(1)\frac{22.9}{4} = \mathbf{162\,pF}$$

(Checking, from equation (7),

$$\text{capacitance } C = \frac{2\pi\varepsilon_0\varepsilon_r}{\ln(r_5/r_1)} = \frac{2\pi(8.85\times 10^{-12})(3.2)}{\ln(30/10)} = \mathbf{162\,pF})$$

Now try the following Practice Exercise

Practice Exercise 164 Capacitance between concentric cylinders (Answers on page 832)

1. A coaxial cable has an inner conductor of radius 0.4 mm and an outer conductor of internal radius 4 mm. Determine the capacitance per metre length of the cable if the dielectric has a relative permittivity of 2
2. A concentric cable has a core diameter of 40 mm and an inner sheath diameter of 100 mm. The relative permittivity of the dielectric is 2.5 and the core potential is 50 kV. Determine (a) the capacitance per kilometre length of the cable and (b) the dielectric stress at radii of 30 mm and 40 mm.
3. A coaxial cable has a capacitance of 100 pF per metre length. The relative permittivity of the dielectric is 3.2 and the core diameter is 1.0 mm. Determine the required inside diameter of the sheath.
4. A single-core concentric cable is to be manufactured for a 100 kV, 50 Hz transmission system. The dielectric used is paper, which has a maximum safe dielectric stress of 10 MV/m and a relative permittivity of 3.2. Calculate (a) the core and inner sheath radii for the most economical cable, (b) the capacitance per metre length and (c) the charging current per kilometre run.
5. A concentric cable has a core diameter of 30 mm and an inside sheath diameter of 75 mm. The relative permittivity is 2.6, the loss angle is 2.5×10^{-3} rad and the working voltage is 100 kV at 50 Hz frequency. Determine for a 1 km length of cable (a) the capacitance, (b) the charging current and (c) the power loss.
6. A concentric cable operates at 200 kV and 50 Hz. The maximum electric field strength within the cable is not to exceed 5 MV/m. Determine (a) the radius of the core and the inner radius of the sheath for ideal operation, and (b) the stress on the dielectric at the surface of the core and at the inner surface of the sheath.
7. A concentric cable has a core radius of 20 mm and a sheath inner radius of 40 mm. The permittivity of the dielectric is 2.5. Using two equipotential surfaces within the dielectric, determine the capacitance of the cable per metre length by the method of curvilinear squares. Draw the field plot for the cable.

43.3 Capacitance of an isolated twin line

The field distribution with two oppositely charged, long conductors, A and B, each of radius a is shown in Figure 43.17. The distance D between the centres of the two conductors is such that D is much greater than a. Figure 43.18 shows the field of each conductor separately.

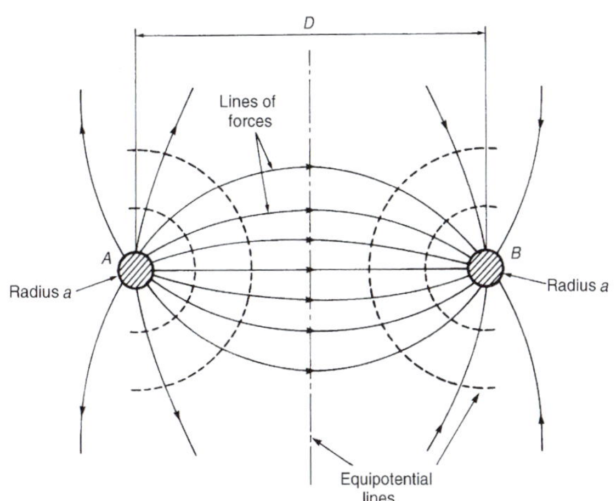

Figure 43.17

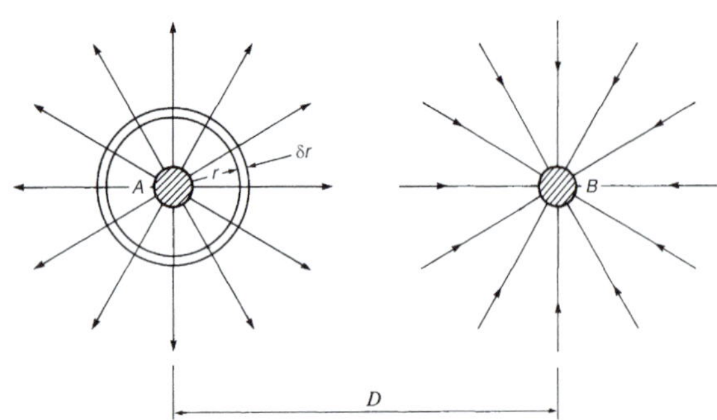

Figure 43.18

Initially, let conductor A carry a charge of $+Q$ coulombs per metre while conductor B is uncharged. Consider a cylindrical element of radius r about conductor A having a depth of 1 m and a thickness δr as shown in Figure 43.18.

The electric flux density D at the element (i.e. at radius r) is given by

$$D = \frac{\text{charge}}{\text{area}} = \frac{Q}{(2\pi r \times 1)} \text{ coulomb/metre}^2$$

The electric field strength at the element,

$$E = \frac{D}{\varepsilon_0 \varepsilon_r} = \frac{Q/2\pi r}{\varepsilon_0 \varepsilon_r} = \frac{Q}{2\pi r \varepsilon_0 \varepsilon_r} \text{ volts/metre}$$

Since $E = V/d$, potential difference, $V = Ed$. Thus

$$\text{p.d. at the element} = E\delta r = \frac{Q\delta r}{2\pi r \varepsilon_0 \varepsilon_r} \text{ volts}$$

The potential may be considered as zero at a large distance from the conductor. Let this be at radius R. Then the potential of conductor A above zero, V_{A_1}, is given by

$$V_{A_1} = \int_a^R \frac{Q\mathrm{d}r}{2\pi r \varepsilon_0 \varepsilon_r} = \frac{Q}{2\pi \varepsilon_0 \varepsilon_r} \int_a^R \frac{1}{r} \mathrm{d}r$$

$$= \frac{Q}{2\pi \varepsilon_0 \varepsilon_r} [\ln r]_a^R$$

$$= \frac{Q}{2\pi \varepsilon_0 \varepsilon_r} [\ln R - \ln a]$$

i.e. $$V_{A_1} = \frac{Q}{2\pi \varepsilon_0 \varepsilon_r} \ln \frac{R}{a}$$

Since conductor B lies in the field of conductor A, by reasoning similar to that above, the potential at conductor B above zero, V_{B_1}, is given by

$$V_{B_1} = \int_D^R \frac{Q\mathrm{d}r}{2\pi r \varepsilon_0 \varepsilon_r} = \frac{Q}{2\pi \varepsilon_0 \varepsilon_r} [\ln r]_D^R = \frac{Q}{2\pi \varepsilon_0 \varepsilon_r} \ln \frac{R}{D}$$

Repeating the above procedure, this time assuming that conductor B carries a charge of $-Q$ coulombs per metre, while conductor A is uncharged, gives

potential of conductor B below zero,

$$V_{B_2} = \frac{-Q}{2\pi \varepsilon_0 \varepsilon_r} \ln \frac{R}{a}$$

and the potential of conductor A below zero, due to the charge on conductor B, $V_{A_2} = \frac{-Q}{2\pi \varepsilon_0 \varepsilon_r} \ln \frac{R}{D}$

When both conductors carry equal and opposite charges, the total potential of A above zero is given by

$$V_{A_1} + V_{A_2} = \left(\frac{Q}{2\pi \varepsilon_0 \varepsilon_r} \ln \frac{R}{a}\right) + \left(\frac{-Q}{2\pi \varepsilon_0 \varepsilon_r} \ln \frac{R}{D}\right)$$

$$= \frac{Q}{2\pi \varepsilon_0 \varepsilon_r} \left(\ln \frac{R}{a} - \ln \frac{R}{D}\right)$$

$$= \frac{Q}{2\pi \varepsilon_0 \varepsilon_r} \left(\ln \frac{R/a}{R/D}\right) = \frac{Q}{2\pi \varepsilon_0 \varepsilon_r} \ln \frac{D}{a}$$

and the total potential of B below zero is given by

$$V_{B_1} + V_{B_2} = \frac{Q}{2\pi \varepsilon_0 \varepsilon_r} \left(\ln \frac{R}{D} - \ln \frac{R}{a}\right)$$

$$= \frac{Q}{2\pi \varepsilon_0 \varepsilon_r} \ln \frac{a}{D} = \frac{-Q}{2\pi \varepsilon_0 \varepsilon_r} \ln \frac{D}{a}$$

Hence the p.d. between A and B is

$$2\left(\frac{Q}{2\pi \varepsilon_0 \varepsilon_r} \ln \frac{D}{a}\right) \text{ volts/metre}$$

The capacitance between A and B per metre length,

$$C = \frac{\text{charge per metre}}{\text{p.d.}} = \frac{Q}{2(Q/(2\pi \varepsilon_0 \varepsilon_r)) \ln(D/a)}$$

i.e. $$\boldsymbol{C = \frac{1}{2} \frac{2\pi \varepsilon_0 \varepsilon_r}{\ln(D/a)}} \textbf{ farads/metre}$$

or $$\boldsymbol{C = \frac{\pi \varepsilon_0 \varepsilon_r}{\ln(D/a)}} \textbf{ farads/metre} \qquad (14)$$

Problem 9. Two parallel wires, each of diameter 5 mm, are uniformly spaced in air at a distance of 50 mm between centres. Determine the capacitance of the line if the total length is 200 m.

From equation (14), capacitance per metre length,

$$C = \frac{\pi\varepsilon_0\varepsilon_r}{\ln(D/a)} = \frac{\pi(8.85\times10^{-12})(1)}{\ln(50/(5/2))}$$

since $\varepsilon_r = 1$ for air,

$$= \frac{\pi(8.85\times10^{-12})}{\ln 20} = 9.28\times10^{-12}\,\text{F}$$

Hence the capacitance of a 200 m length is $(9.28\times10^{-12}\times 200)$ F = **1860 pF** or **1.86 nF**

Problem 10. A single-phase circuit is composed of two parallel conductors, each of radius 4 mm, spaced 1.2 m apart in air. The p.d. between the conductors at a frequency of 50 Hz is 15 kV. Determine, for a 1 km length of line, (a) the capacitance of the conductors, (b) the value of charge carried by each conductor and (c) the charging current.

(a) From equation (14),

$$\text{capacitance } C = \frac{\pi\varepsilon_0\varepsilon_r}{\ln(D/a)} = \frac{\pi(8.85\times10^{-12})(1)}{\ln(1.2/4\times10^{-3})}$$

$$= \frac{\pi(8.85\times10^{-12})}{\ln 300}$$

$$= 4.875\,\text{pF/m}$$

Hence the capacitance per kilometre length is $(4.875\times10^{-12})(10^3)$ F = **4.875 nF**

(b) Charge $Q = CV = (4.875\times10^{-9})(15\times10^3)$

$$= \mathbf{73.1\,\mu C}$$

(c) $$\text{Charging current} = \frac{V}{X_C} = \frac{V}{(1/\omega C)} = \omega CV$$

$$= (2\pi 50)(4.875\times10^{-9})(15\times10^3)$$

$$= \mathbf{0.023\,A} \text{ or } \mathbf{23\,mA}$$

Problem 11. The charging current for an 800 m run of isolated twin line is not to exceed 15 mA. The voltage between the lines is 10 kV at 50 Hz. If the line is air-insulated, determine (a) the maximum value required for the capacitance per metre length and (b) the maximum diameter of each conductor if their distance between centres is 1.25 m.

(a) $$\text{Charging current } I = \frac{V}{X_C} = \frac{V}{(1/\omega C)} = \omega CV$$

from which

$$\text{capacitance } C = \frac{I}{\omega V} = \frac{15\times10^{-3}}{(2\pi 50)(10\times10^3)}$$

farads per 800 metre run

$$= 4.775\,\text{nF}$$

Hence the required maximum value of capacitance

$$= \frac{4.775\times10^{-9}}{800}\,\text{F/m} = \mathbf{5.97\,pF/m}$$

(b) From equation (14)

$$C = \frac{\pi\varepsilon_0\varepsilon_r}{\ln(D/a)}$$

thus $$5.97\times10^{-12} = \frac{\pi(8.85\times10^{-12})(1)}{\ln(1.25/a)}$$

from which $$\ln\left(\frac{1.25}{a}\right) = \frac{\pi 8.85}{5.97} = 4.657$$

Hence $$\frac{1.25}{a} = e^{4.657} = 105.3$$

and radius $$a = \frac{1.25}{105.3}\,\text{m} = 0.01187\,\text{m}$$

or 11.87 mm

Thus **the maximum diameter of each conductor is** 2×11.87, i.e. **23.7 mm**.

Now try the following Practice Exercise

Practice Exercise 165 Capacitance of an isolated twin line (Answers on page 832)

1. Two parallel wires, each of diameter 5.0 mm, are uniformly spaced in air at a distance of 40 mm between centres. Determine the capacitance of a 500 m run of the line.
2. A single-phase circuit is comprised of two parallel conductors each of radius 5.0 mm and spaced 1.5 m apart in air. The p.d. between the conductors is 20 kV at 50 Hz. Determine (a) the capacitance per metre length of the conductors and (b) the charging current per kilometre run.

3. The capacitance of a 300 m length of an isolated twin line is 1522 pF. The line comprises two air conductors which are spaced 1200 mm between centres. Determine the diameter of each conductor.

4. An isolated twin line is comprised of two air-insulated conductors, each of radius 8.0 mm, which are spaced 1.60 m apart. The voltage between the lines is 7 kV at a frequency of 50 Hz. Determine for a 1 km length (a) the line capacitance, (b) the value of charge carried by each wire and (c) the charging current.

5. The charging current for a 1 km run of isolated twin line is not to exceed 30 mA. The p.d. between the lines is 20 kV at 50 Hz. If the line is air insulated and the conductors are spaced 1 m apart, determine (a) the maximum value required for the capacitance per metre length and (b) the maximum diameter of each conductor.

43.4 Energy stored in an electric field

Consider the p.d. across a parallel-plate capacitor of capacitance C farads being increased by $\mathrm{d}v$ volts in $\mathrm{d}t$ seconds. If the corresponding increase in charge is $\mathrm{d}q$ coulombs, then $\mathrm{d}q = C\mathrm{d}v$. If the charging current at that instant is i amperes, then $\mathrm{d}q = i\mathrm{d}t$. Thus $i\mathrm{d}t = C\mathrm{d}v$, i.e.

$$i = C\frac{\mathrm{d}v}{\mathrm{d}t}$$

(i.e. instantaneous current = capacitance × rate of change of p.d.)

The instantaneous value of power to the capacitor,

$$p = vi \text{ watts} = v\left(C\frac{\mathrm{d}v}{\mathrm{d}t}\right) \text{ watts}$$

The energy supplied to the capacitor during time $\mathrm{d}t$

$$= \text{power} \times \text{time} = \left(vC\frac{\mathrm{d}v}{\mathrm{d}t}\right)(\mathrm{d}t)$$

$$= Cv\,\mathrm{d}v \text{ joules}$$

Thus the total energy supplied to the capacitor when the p.d. is increased from 0 to V volts is given by

$$W_f = \int_0^V Cv\,\mathrm{d}v = C\left[\frac{v^2}{2}\right]_0^V$$

i.e. **energy stored in the electric field,**

$$\mathbf{W_f = \tfrac{1}{2}CV^2 \text{ joules}} \qquad (15)$$

Consider a capacitor with dielectric of relative permittivity ε_r, thickness d metres and area A square metres. Capacitance $C = Q/V$, hence energy stored

$$= \tfrac{1}{2}(Q/V)V^2 = \tfrac{1}{2}QV \text{ joules.}$$

The electric flux density, $D = Q/A$, from which $Q = DA$.

Hence the energy stored $= \frac{1}{2}(DA)V$ joules.

The electric field strength, $E = V/d$, from which $V = Ed$.

Hence the energy stored $= \frac{1}{2}(DA)(Ed)$ joules. However Ad is the volume of the field.

Hence **energy stored per unit volume,**

$$\mathbf{\omega_f = \frac{1}{2}DE \text{ joules/cubic metre}} \qquad (16)$$

Since $D/E = \varepsilon_0\varepsilon_r$, then $D = \varepsilon_0\varepsilon_r E$. Hence, from equation (16), the energy stored per unit volume,

$$\omega_f = \frac{1}{2}(\varepsilon_0\varepsilon_r E)E$$

i.e. $$\mathbf{\omega_f = \tfrac{1}{2}\varepsilon_0\varepsilon_r E^2 \text{ joules/cubic metre}} \qquad (17)$$

Also, since $D/E = \varepsilon_0\varepsilon_r$, then $E = D/(\varepsilon_0\varepsilon_r)$. Hence from equation (16), the energy stored per unit volume,

$$\omega_f = \frac{1}{2}D\left(\frac{D}{\varepsilon_0\varepsilon_r}\right)$$

i.e. $$\mathbf{\omega_f = \frac{D^2}{2\varepsilon_0\varepsilon_r} \text{ joules/cubic metre}} \qquad (18)$$

Summarizing,

energy stored in a capacitor $= \frac{1}{2}$ **CV^2 joules**

and energy stored per unit volume of dielectric

$$= \mathbf{\tfrac{1}{2}DE = \tfrac{1}{2}\varepsilon_0\varepsilon_r E^2}$$

$$= \mathbf{\frac{D^2}{2\varepsilon_0\varepsilon_r} \text{ joules/cubic metre}}$$

Problem 12. Determine the energy stored in a 10 nF capacitor when charged to 1 kV, and the average power developed if this energy is dissipated in 10 μs.

Part 4

From equation (15),

$$\text{energy stored, } W_f = \tfrac{1}{2}CV^2 = \tfrac{1}{2}(10\times10^{-9})(10^3)^2 = \mathbf{5\,mJ}$$

$$\text{average power developed} = \frac{\text{energy dissipated, } W}{\text{time, } t} = \frac{5\times10^{-3}\,\text{J}}{10\times10^{-6}\,\text{s}} = \mathbf{500\,W}$$

Problem 13. A capacitor is charged with 5 mC. If the energy stored is 625 mJ, determine (a) the voltage across the plates and (b) the capacitance of the capacitor.

(a) From equation (15),

$$\text{energy stored, } W_f = \frac{1}{2}CV^2 = \frac{1}{2}\left(\frac{Q}{V}\right)V^2 = \frac{1}{2}QV$$

from which voltage across the plates,

$$V = \frac{2\times\text{energy stored}}{Q} = \frac{2\times0.625}{5\times10^{-3}} = \mathbf{250\,V}$$

(b) Capacitance $C = \dfrac{Q}{V} = \dfrac{5\times10^{-3}}{250}\,\text{F} = 20\times10^{-6}\,\text{F} = \mathbf{20\,\mu F}$

Problem 14. A ceramic capacitor is to be constructed to have a capacitance of 0.01 μF and to have a steady working potential of 2.5 kV maximum. Allowing a safe value of field stress of 10 MV/m, determine (a) the required thickness of the ceramic dielectric, (b) the area of plate required if the relative permittivity of the ceramic is 10 and (c) the maximum energy stored by the capacitor.

(a) Field stress $E = V/d$, from which thickness of ceramic dielectric,

$$d = \frac{V}{E} = \frac{2.5\times10^3}{10\times10^6} = 2.5\times10^{-4}\,\text{m} = \mathbf{0.25\,mm}$$

(b) Capacitance $C = \varepsilon_0\varepsilon_r A/d$ for a two-plate parallel capacitor. Hence cross-sectional area of plate,

$$A = \frac{Cd}{\varepsilon_0\varepsilon_r} = \frac{(0.01\times10^{-6})(0.25\times10^{-3})}{(8.85\times10^{-12})(10)} = \mathbf{0.0282\,m^2} \text{ or } \mathbf{282\,cm^2}$$

(c) Maximum energy stored,

$$W_f = \frac{1}{2}CV^2 = \frac{1}{2}(0.01\times10^{-6})(2.5\times10^3)^2 = \mathbf{0.0313\,J} \text{ or } \mathbf{31.3\,mJ}$$

Problem 15. A 400 pF capacitor is charged to a p.d. of 100 V. The dielectric has a cross-sectional area of 200 cm^2 and a relative permittivity of 2.3. Calculate the energy stored per cubic metre of the dielectric.

From equation (18), energy stored per unit volume of dielectric,

$$\omega_f = \frac{D^2}{2\varepsilon_0\varepsilon_r}$$

Electric flux density

$$D = \frac{Q}{A} = \frac{CV}{A} = \frac{(400\times10^{-12})(100)}{200\times10^{-4}} = 2\times10^{-6}\,\text{C/m}^2$$

Hence energy stored,

$$\omega_f = \frac{D^2}{2\varepsilon_0\varepsilon_r} = \frac{(2\times10^{-6})^2}{2(8.85\times10^{-12})(2.3)} = \mathbf{0.0983\,J/m^3} \text{ or } \mathbf{98.3\,mJ/m^3}$$

Now try the following Practice Exercise

Practice Exercise 166 Energy stored in electric fields (Answers on page 832)

1. Determine the energy stored in a 5000 pF capacitor when charged to 800 V and the average power developed if this energy is dissipated in 20 μs.
2. A 0.25 μF capacitor is required to store 2 J of energy. Determine the p.d. to which the capacitor must be charged.
3. A capacitor is charged with 6 mC. If the energy stored is 1.5 J, determine (a) the voltage across the plates and (b) the capacitance of the capacitor.
4. After a capacitor is connected across a 250 V d.c. supply the charge is 5 μC. Determine (a) the capacitance and (b) the energy stored.

5. A capacitor consisting of two metal plates each of area 100 cm^2 and spaced 0.1 mm apart in air is connected across a 200 V supply. Determine (a) the electric flux density, (b) the potential gradient and (c) the energy stored in the capacitor.

6. A mica capacitor is to be constructed to have a capacitance of 0.05 μF and to have a steady working potential of 2 kV maximum. Allowing a safe value of field stress of 20 MV/m, determine (a) the required thickness of the mica dielectric, (b) the area of plate required if the relative permittivity of the mica is 5, (c) the maximum energy stored by the capacitor and (d) the average power developed if this energy is dissipated in 25 μs.

7. A 500 pF capacitor is charged to a p.d. of 100 V. The dielectric has a cross-sectional area of 200 cm^2 and a relative permittivity of 2.4. Determine the energy stored per cubic metre in the dielectric.

8. Two parallel plates each having dimensions 30 mm by 50 mm are spaced 8 mm apart in air. If a voltage of 40 kV is applied across the plates determine the energy stored in the electric field.

43.5 Induced e.m.f. and inductance

A current flowing in a coil of wire is accompanied by a magnetic flux linking with the coil. If the current changes, the flux linkage (i.e. the product of flux and the number of turns) changes and an e.m.f. is induced in the coil. The magnitude of the induced e.m.f. e in a coil of N turns is given by

$$e = N\frac{d\phi}{dt} \text{ volts}$$

where $d\phi/dt$ is the rate of change of flux.

Inductance is the name given to the property of a circuit whereby there is an e.m.f. induced into the circuit by the change of flux linkages produced by a current change. The unit of inductance is the **henry***, H. A circuit has an inductance of 1 H when an e.m.f. of 1 V is induced in it by a current changing uniformly at the rate of 1 A/s.

The magnitude of the e.m.f. induced in a coil of inductance L henry is given by

$$e = L\frac{di}{dt} \text{ volts}$$

where di/dt is the rate of change of current.

If a current changing uniformly from zero to I amperes produces a uniform flux change from zero to ϕ webers in t seconds then (from above) average induced e.m.f., $E_{av} = N\phi/t = LI/t$, from which

$$\text{inductance of coil, } L = \frac{N\phi}{I} \text{ henry}$$

Flux linkage means the product of flux, in webers, and the number of turns with which the flux is linked. Hence flux linkage $= N\phi$. Thus since $L = N\phi/I$, **inductance = flux linkages per ampere**.

43.6 Inductance of a concentric cylinder (or coaxial cable)

Skin effect

When a direct current flows in a uniform conductor the current will tend to distribute itself uniformly over the cross-section of the conductor. However, with alternating current, particularly if the frequency is high, the current carried by the conductor is not uniformly distributed over the available cross-section, but tends to be concentrated at the conductor surface. This is called **skin effect**. When current is flowing through a conductor, the magnetic flux that results is in the form of concentric circles. Some of this flux exists within the conductor and links with the current more strongly near the centre. The result is that the inductance of the central part of the conductor is greater than the inductance of the conductor near the surface. This is because of the greater number of flux linkages existing in the central region. At high frequencies the reactance ($X_L = 2\pi fL$)

*Who was **Henry**? For image and resume of Henry, see page 151. To find out more go to **www.routledge.com/cw/bird**

of the extra inductance is sufficiently large to seriously affect the flow of current, most of which flows along the surface of the conductor where the impedance is low, rather than near the centre where the impedance is high.

Inductance due to internal linkages at low frequency

When a conductor is used at high frequency the depth of penetration of the current is small compared with the conductor cross-section. Thus the internal linkages may be considered as negligible and the circuit inductance is that due to the fields in the surrounding space. However, at very low frequency the current distribution is considered uniform over the conductor cross-section and the inductance due to flux linkages has its maximum value.

Consider a conductor of radius R, as shown in Figure 43.19, carrying a current I amperes uniformly distributed over the cross-section. At all points on the conductor cross-section

$$\text{current density, } J = \frac{\text{current}}{\text{area}} = \left(\frac{I}{\pi R^2}\right) \text{amperes/metre}^2$$

Consider a thin elemental ring at radius r and width δr contained within the conductor, as shown in Figure 43.19. The current enclosed by the ring,

$$i = \text{current density} \times \text{area enclosed by the ring}$$

$$= \left(\frac{I}{\pi R^2}\right)(\pi r^2)$$

$$\text{i.e.} \quad i = \frac{Ir^2}{R^2} \text{ amperes}$$

Magnetic field strength, $H = Ni/l$ amperes/metre.

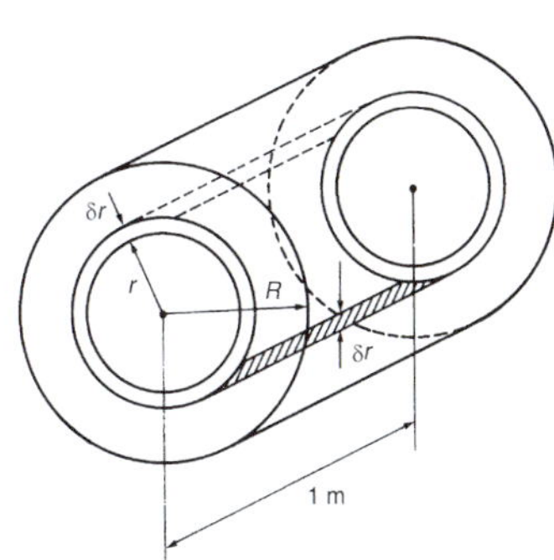

Figure 43.19

At radius r, the mean length of the flux path, $l = 2\pi r$ (i.e. the circumference of the elemental ring) and $N = 1$ turn.

Hence at radius r,

$$H_r = \frac{Ni}{l} = \frac{(1)(Ir^2/R^2)}{2\pi r} = \frac{Ir}{2\pi R^2} \text{ ampere/metre}$$

and the flux density, $B_r = \mu_0\mu_r H_r = \mu_0\mu_r\left(\frac{Ir}{2\pi R^2}\right)$ tesla. Flux $\phi = BA$ webers. For a 1 m length of the conductor, the cross-sectional area A of the element is $(\delta r \times 1)$ m^2 (see Figure 43.19). Thus the flux within the element of thickness δr,

$$\phi = \left(\frac{\mu_0\mu_r Ir}{2\pi R^2}\right)(\delta r) \text{ webers}$$

The flux in the element links the portion $\pi r^2/\pi R^2$, i.e. r^2/R^2 of the total conductor. Hence linkages due to the flux within radius r

$$= \left(\frac{\mu_0\mu_r Ir}{2\pi R^2}\delta r\right)\frac{r^2}{R^2} = \frac{\mu_0\mu_r Ir^3}{2\pi R^4}\delta r \text{ weber turns}$$

Total linkages per metre due to the flux in the conductor

$$= \int_0^R \frac{\mu_0\mu_r Ir^3}{2\pi R^4}\mathrm{d}r = \frac{\mu_0\mu_r I}{2\pi R^4}\int_0^R r^3\mathrm{d}r$$

$$= \frac{\mu_0\mu_r I}{2\pi R^4}\left[\frac{r^4}{4}\right]_0^R = \frac{\mu_0\mu_r I}{2\pi R^4}\left[\frac{R^4}{4}\right]$$

$$= \frac{1}{4}\left(\frac{\mu_0\mu_r I}{2\pi}\right) \text{ weber turns}$$

Inductance per metre due to the internal flux = internal flux linkages per ampere

$$= \mathbf{\frac{1}{4}\left(\frac{\mu_0\mu_r}{2\pi}\right)} \text{ or } \mathbf{\frac{\mu}{8\pi}} \textbf{ henry/metre}$$

It is seen that the inductance is independent of the conductor radius R.

Inductance of a pair of concentric cylinders

The cross-section of a concentric (or coaxial) cable is shown in Figure 43.20. Let a current of I amperes flow in one direction in the core and a current of I

amperes flow in the opposite direction in the outer sheath conductor.

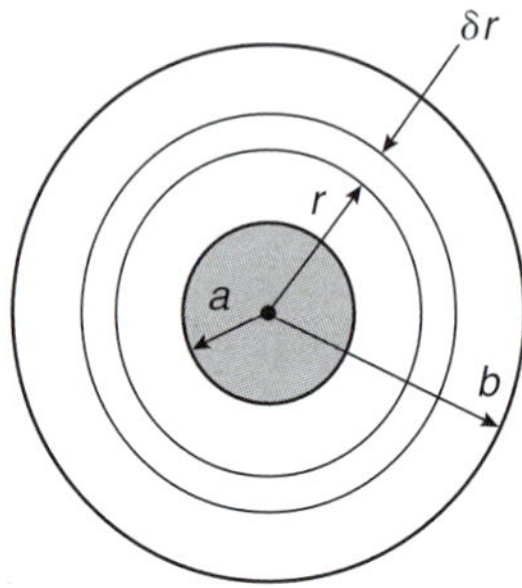

Figure 43.20 Cross-section of a concentric cable

Consider an element of width δr at radius r, and let the radii of the inner and outer conductor be a and b, respectively, as shown. The magnetic field strength at radius r,

$$H_r = \frac{Ni}{l} = \frac{(1)(I)}{2\pi r} = \frac{I}{2\pi r}$$

The flux density at radius r, $B_r = \mu_0\mu_r H_r = \dfrac{\mu_0\mu_r I}{2\pi r}$

For a 1 m length of the cable, the flux ϕ within the element of width δr is given by

$$\Phi = B_r A = \left(\frac{\mu_0\mu_r I}{2\pi r}\right)(\delta r \times 1) = \frac{\mu_0\mu_r I}{2\pi r}\delta r \text{ webers}$$

This flux links the loop of the cable formed by the core and the outer sheath. Thus the flux linkage per metre length of the cable is $(\mu_0\mu_r I/2\pi r)\delta r$ weber turns, and total flux linkages per metre

$$= \int_a^b \frac{\mu_0\mu_r I}{2\pi r}\mathrm{d}r = \frac{\mu_0\mu_r I}{2\pi}\int_a^b \frac{1}{r}\mathrm{d}r$$

$$= \frac{\mu_0\mu_r I}{2\pi}[\ln r]_a^b = \frac{\mu_0\mu_r I}{2\pi}[\ln b - \ln a]$$

$$= \frac{\mu_0\mu_r I}{2\pi}\ln\frac{b}{a} \text{ weber turns}$$

Thus inductance per metre length

$$= \text{flux linkages per ampere}$$

$$= \boldsymbol{\frac{\mu_0\mu_r}{2\pi}\ln\frac{b}{a}} \textbf{ henry/metre} \qquad (19)$$

At low frequencies the inductance due to the internal linkages is added to this result.
Hence the total inductance per metre at low frequency is given by

$$\boldsymbol{L = \frac{1}{4}\left(\frac{\mu_0\mu_r}{2\pi}\right) + \frac{\mu_0\mu_r}{2\pi}\ln\frac{b}{a}} \textbf{ henry/metre} \qquad (20)$$

or $$\boldsymbol{L = \frac{\mu}{2\pi}\left(\frac{1}{4} + \ln\frac{b}{a}\right)} \textbf{ henry/metre} \qquad (21)$$

Problem 16. A coaxial cable has an inner core of radius 1.0 mm and an outer sheath of internal radius 4.0 mm. Determine the inductance of the cable per metre length. Assume that the relative permeability is unity.

From equation (21),

$$\text{inductance } L = \frac{\mu}{2\pi}\left(\frac{1}{4} + \ln\frac{b}{a}\right) \text{ H/m}$$

$$= \frac{\mu_0\mu_r}{2\pi}\left(\frac{1}{4} + \ln\frac{4.0}{1.0}\right)$$

$$= \frac{(4\pi \times 10^{-7})(1)}{2\pi}(0.25 + \ln 4)$$

$$= \mathbf{3.27 \times 10^{-7}\ H/m} \text{ or } \mathbf{0.327\ \mu H/m}$$

Problem 17. A concentric cable has a core diameter of 10 mm. The inductance of the cable is 4×10^{-7} H/m. Ignoring inductance due to internal linkages, determine the diameter of the sheath. Assume that the relative permeability is 1

From equation (19),

$$\text{inductance per metre length} = \frac{\mu_0\mu_r}{2\pi}\ln\frac{b}{a}$$

where b = sheath radius and a = core radius. Hence

$$4 \times 10^{-7} = \frac{(4\pi \times 10^{-7})(1)}{2\pi}\ln\left(\frac{b}{5}\right)$$

from which $2 = \ln\left(\dfrac{b}{5}\right)$ and $e^2 = \dfrac{b}{5}$

Hence radius $b = 5e^2 = 36.95$ mm

Thus the diameter of the sheath is $2 \times 36.95 = \mathbf{73.9\ mm}$

Problem 18. A coaxial cable 7.5 km long has a core 10 mm diameter and a sheath 25 mm diameter, the sheath having negligible thickness. Determine for the cable (a) the inductance, assuming nonmagnetic materials, and (b) the capacitance, assuming a dielectric of relative permittivity 3.

(a) From equation (21),

inductance per metre length

$$= \frac{\mu}{2\pi}\left(\frac{1}{4} + \ln\frac{b}{a}\right)$$

$$= \frac{\mu_0\mu_r}{2\pi}\left[\frac{1}{4} + \ln\left(\frac{12.5}{5}\right)\right]$$

$$= \frac{(4\pi \times 10^{-7})(1)}{2\pi}(0.25 + \ln 2.5)$$

$$= 2.33 \times 10^{-7}\ \text{H/m}$$

Since the cable is 7500 m long,

the inductance $= 7500 \times 2.33 \times 10^{-7} = \mathbf{1.75\ mH}$

(b) From equation (7),

$$\text{capacitance, } C = \frac{2\pi\varepsilon_0\varepsilon_r}{\ln(b/a)} = \frac{2\pi(8.85\times10^{-12})(3)}{\ln(12.5/5)}$$

$$= 182.06\ \text{pF/m}$$

Since the cable is 7500 m long,

$$\text{the capacitance} = 7500 \times 182.06 \times 10^{-12}$$

$$= \mathbf{1.365\ \mu F}$$

Now try the following Practice Exercise

Practice Exercise 167 Inductance of concentric cylinders (Answers on page 832)

1. A coaxial cable has an inner core of radius 0.8 mm and an outer sheath of internal radius 4.8 mm. Determine the inductance of 25 m of the cable. Assume that the relative permeability of the material used is 1

2. A concentric cable has a core 12 mm diameter and a sheath 40 mm diameter, the sheath having negligible thickness. Determine the inductance and the capacitance of the cable per metre assuming nonmagnetic materials and a dielectric of relative permittivity 3.2

3. A concentric cable has an inner sheath radius of 4.0 cm. The inductance of the cable is 0.5 μH/m. Ignoring inductance due to internal linkages, determine the radius of the core. Assume that the relative permeability of the material is unity.

4. The inductance of a concentric cable of core radius 8 mm and inner sheath radius of 35 mm is measured as 2.0 mH. Determine (a) the length of the cable and (b) the capacitance of the cable. Assume that nonmagnetic materials are used and the relative permittivity of the dielectric is 2.5

43.7 Inductance of an isolated twin line

Consider two isolated, long, parallel, straight conductors A and B, each of radius a metres, spaced D metres apart. Let the current in each be I amperes but flowing in opposite directions. Distance D is assumed to be much greater than radius a. The magnetic field associated with the conductors is as shown in Figure 43.21. There is a force of repulsion between conductors A and B.

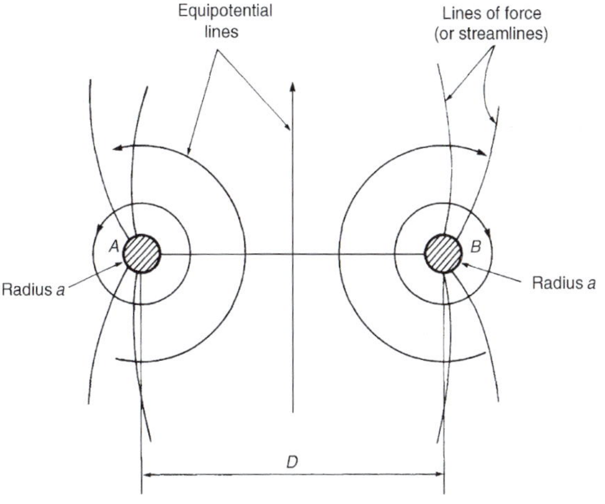

Figure 43.21

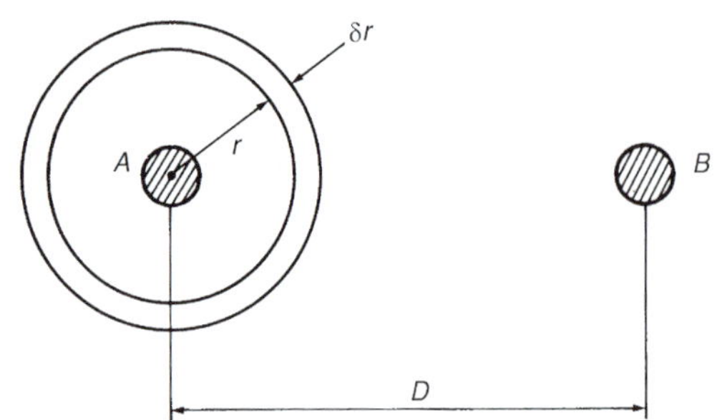

Figure 43.22

It is easier to analyse the field by initially considering each conductor alone (as in Section 43.3). At any radius r from conductor A (see Figure 43.22),

$$\text{magnetic field strength, } H_r = \frac{Ni}{l} = \frac{I}{2\pi r} \text{ ampere/metre}$$

$$\text{and flux density, } B_r = \mu_0\mu_r H_r = \frac{\mu_0\mu_r I}{2\pi r} \text{ tesla}$$

The total flux in 1 m of the conductor,

$$\Phi = B_r A = \left(\frac{\mu_0\mu_r I}{2\pi r}\right)(\delta r \times 1) = \frac{\mu_0\mu_r I}{2\pi r}\delta r \text{ webers}$$

Since this flux links conductor A once, the linkages with conductor A due to this flux $= \frac{\mu_0\mu_r I}{2\pi r}\delta r$ weber turns.

There is, in fact, no limit to the distance from conductor A at which a magnetic field may be experienced. However, let R be a very large radius at which the magnetic field strength may be regarded as zero. Then the total linkages with conductor A due to current in conductor A is given by

$$\int_a^R \frac{\mu_0\mu_r I}{2\pi r}\mathrm{d}r = \frac{\mu_0\mu_r I}{2\pi}\int_a^R \frac{\mathrm{d}r}{r} = \frac{\mu_0\mu_r I}{2\pi}[\ln r]_a^R$$

$$= \frac{\mu_0\mu_r I}{2\pi}[\ln R - \ln a] = \frac{\mu_0\mu_r I}{2\pi}\ln\left(\frac{R}{a}\right)$$

Similarly, the total linkages with conductor B due to the current in A

$$= \int_D^R \frac{\mu_0\mu_r I}{2\pi r}\mathrm{d}r = \frac{\mu_0\mu_r I}{2\pi}\ln\frac{R}{D}$$

Now consider conductor B alone, carrying a current of $-I$ amperes. By similar reasoning to above, total linkages with conductor B due to the current in B

$$= \frac{-\mu_0\mu_r I}{2\pi}\ln\left(\frac{R}{a}\right)$$

and total linkages with conductor A due to the current in B

$$= \frac{-\mu_0\mu_r I}{2\pi}\ln\frac{R}{D}$$

Hence total linkages with conductor A

$$= \left(\frac{\mu_0\mu_r I}{2\pi}\ln\frac{R}{a}\right) + \left(\frac{-\mu_0\mu_r I}{2\pi}\ln\frac{R}{D}\right)$$

$$= \frac{\mu_0\mu_r I}{2\pi}\left[\ln\frac{R}{a} - \ln\frac{R}{D}\right]$$

$$= \frac{\mu_0\mu_r I}{2\pi}\left[\ln\frac{R/a}{R/D}\right]$$

$$= \frac{\mu_0\mu_r I}{2\pi}\ln\frac{D}{a} \text{ weber-turns/metre}$$

Similarly, total linkages with conductor B

$$= -\frac{\mu_0\mu_r I}{2\pi}\ln\frac{D}{a} \text{ weber-turns/metre}$$

For a 1 m length of the two conductors,

$$\text{total inductance} = \text{flux linkages per ampere}$$

$$= 2\left(\frac{\mu_0\mu_r}{2\pi}\ln\frac{D}{a}\right) \text{ henry/metre}$$

i.e. **total inductance** $= \boldsymbol{\frac{\mu_0\mu_r}{\pi}\ln\frac{D}{a}}$ **henry/metre** (22)

Equation (22) does not take into consideration the internal linkages of each line.

From Section 43.6, inductance per metre due to internal linkages

$$= \frac{1}{4}\left(\frac{\mu_0\mu_r}{2\pi}\right) \text{ henry/metre}$$

Thus inductance per metre due to internal linkages of two conductors

$$= 2\left(\frac{1}{4}\left(\frac{\mu_0\mu_r}{2\pi}\right)\right) = \frac{\mu_0\mu_r}{4\pi} \text{ henry/metre}$$

Therefore, at low frequency, total inductance per metre of the two conductors

$$= \frac{\mu_0\mu_r}{4\pi} + \frac{\mu_0\mu_r}{\pi}\ln\frac{D}{a}$$

i.e. $\boldsymbol{L = \frac{\mu_0\mu_r}{\pi}\left(\frac{1}{4} + \ln\frac{D}{a}\right)}$ **henry/metre** (23)

(This is often referred to as the **'loop inductance'**).

In most practical lines the relative permeability, $\mu_r = 1$

Part 4

Problem 19. A single-phase power line comprises two conductors each with a radius 8.0 mm and spaced 1.2 m apart in air. Determine the inductance of the line per metre length ignoring internal linkages. Assume the relative permeability, $\mu_r = 1$

From equation (22), inductance

$$= \frac{\mu_0\mu_r}{\pi}\ln\frac{D}{a}$$

$$= \frac{(4\pi\times10^{-7})(1)}{\pi}\ln\left(\frac{1.2}{8.0\times10^{-3}}\right) = 4\times10^{-7}\ln 150$$

$$= \mathbf{20.0\times10^{-7}\ H/m} \quad \text{or} \quad \mathbf{2.0\,\mu H/m}$$

Problem 20. Determine (a) the loop inductance and (b) the capacitance of a 1 km length of single-phase twin line having conductors of diameter 10 mm and spaced 800 mm apart in air.

(a) From equation (23), total inductance per loop metre

$$= \frac{\mu_0\mu_r}{\pi}\left(\frac{1}{4}+\ln\frac{D}{a}\right)$$

$$= \frac{(4\pi\times10^{-7})(1)}{\pi}\left(\frac{1}{4}+\ln\frac{800}{10/2}\right)$$

$$= (4\times10^{-7})(0.25+\ln 160)$$

$$= 21.3\times10^{-7}\ \text{H/m}$$

Hence loop inductance of a 1 km length of line

$$= 21.3\times10^{-7}\ \text{H/m}\times10^3\ \text{m}$$

$$= \mathbf{21.3\times10^{-4}\ H} \quad \text{or} \quad \mathbf{2.13\ mH}$$

(b) From equation (14), capacitance per metre length

$$= \frac{\pi\varepsilon_0\varepsilon_r}{\ln(D/a)}$$

$$= \frac{\pi(8.85\times10^{-12})(1)}{\ln(800/5)}$$

$$= 5.478\times10^{-12}\ \text{F/m}$$

Hence capacitance of a 1 km length of line

$$= 5.478\times10^{-12}\ \text{F/m}\times10^3\ \text{m}$$

$$= \mathbf{5.478\ nF}$$

Problem 21. The total loop inductance of an isolated twin power line is 2.185 μH/m. The diameter of each conductor is 12 mm. Determine the distance between their centres.

From equation (23),

$$\text{total loop inductance} = \frac{\mu_0\mu_r}{\pi}\left(\frac{1}{4}+\ln\frac{D}{a}\right)$$

Hence

$$2.185\times10^{-6} = \frac{(4\pi\times10^{-7})(1)}{\pi}\left(\frac{1}{4}+\ln\frac{D}{6}\right)$$

where D is the distance between centres in millimetres.

$$\frac{2.185\times10^{-6}}{4\times10^{-7}} = \left(0.25+\ln\frac{D}{6}\right)$$

$$\ln\frac{D}{6} = 5.4625 - 0.25 = 5.2125$$

$$\frac{D}{6} = e^{5.2125}$$

from which, distance $D = 6e^{5.2125} = \mathbf{1100\ mm}$ or **1.10 m**

Now try the following Practice Exercise

Practice Exercise 168 Inductance of an isolated twin line (Answers on page 832)

1. A single-phase power line comprises two conductors each with a radius of 15 mm and spaced 1.8 m apart in air. Determine the inductance per metre length, ignoring internal linkages and assuming the relative permeability, $\mu_r = 1$
2. Determine (a) the loop inductance and (b) the capacitance of a 500 m length of single-phase twin line having conductors of diameter 8 mm and spaced 60 mm apart in air.
3. An isolated twin power line has conductors 7.5 mm radius. Determine the distance between centres if the total loop inductance of 1 km of the line is 1.95 mH.
4. An isolated twin line has conductors of diameter $d\times10^{-3}$ metres and spaced D millimetres apart in air. Derive an expression for the total loop inductance L of the line per metre length.

5. A single-phase power line comprises two conductors spaced 2 m apart in air. The loop inductance of 2 km of the line is measured as 3.65 mH. Determine the diameter of the conductors.

43.8 Energy stored in an electromagnetic field

Magnetic energy in a nonmagnetic medium

For a nonmagnetic medium the relative permeability, $\mu_r = 1$ and $B = \mu_0 H$

Thus the magnetic field strength H is proportional to the flux density B and a graph of B against H is a straight line, as shown in Figure 43.23.

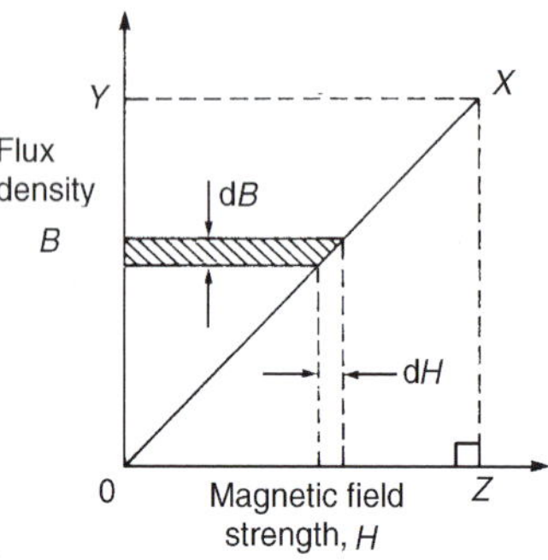

Figure 43.23

It was shown in Section 41.3 that, when the flux density is increased by an amount dB due to an increase dH in the magnetic field strength, then

energy supplied to the magnetic circuit

= area of shaded strip (in joules per cubic metre)

Thus, for a maximum flux density $0Y$ in Figure 43.23,

total energy stored in the magnetic field

$$= \text{area of triangle } 0YX$$

$$= \frac{1}{2} \times \text{base} \times \text{height}$$

$$= \frac{1}{2}(0Z)(0Y)$$

If $0Y = B$ teslas and $0Z = H$ ampere/metre, then the total energy stored in a non-magnetic medium,

$$\boldsymbol{\omega_f = \frac{1}{2} HB \text{ joules/metre}^3} \qquad (24)$$

Since $B = \mu_0$H for a non-magnetic medium, the energy stored, $\omega_f = \frac{1}{2} H(\mu_0 H)$

i.e. $$\boldsymbol{\omega_f = \frac{1}{2}\mu_0 H^2 \text{ joules/metre}^3} \qquad (25)$$

Alternatively, $H = B/\mu_0$, thus the energy stored,

$$\omega_f = \frac{1}{2}HB = \frac{1}{2}\left(\frac{B}{\mu_0}\right)B$$

i.e. $$\boldsymbol{\omega_f = \frac{B^2}{2\mu_0} \text{ joules/metre}^3} \qquad (26)$$

Magnetic energy stored in an inductor

Establishing a magnetic field requires energy to be expended. However, once the field is established, the only energy expended is that supplied to maintain the flow of current in opposition to the circuit resistance, i.e. the I^2R loss, which is dissipated as heat.

For an inductive circuit containing resistance R and inductance L (see Figure 43.24) the applied voltage V at any instant is given by $V = v_R + v_L$

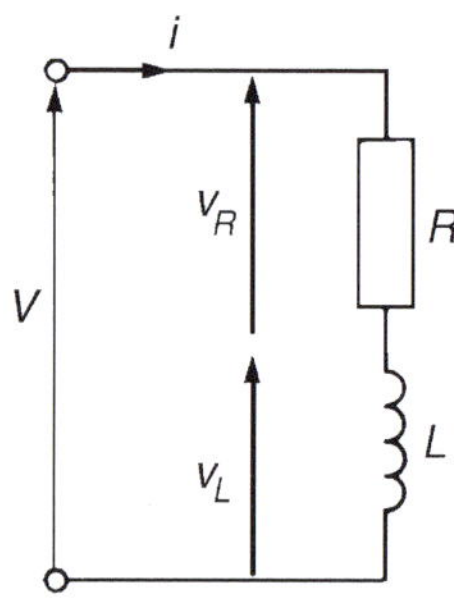

Figure 43.24

i.e. $$V = iR + L\frac{di}{dt}$$

Multiplying throughout by current i gives the power equation:

$$Vi = i^2R + Li\frac{di}{dt}$$

Multiplying throughout by time dt seconds gives the energy equation:

$$Vidt = i^2Rdt + Lidi$$

Vidt is the energy supplied by the source in time dt, i^2Rdt is the energy dissipated in the resistance and Lidi is the energy supplied in establishing the magnetic field

Part 4

or the energy absorbed by the magnetic field in time dt seconds.
Hence the total energy stored in the field when the current increases from 0 to I amperes is given by

$$\text{energy stored, } W_f = \int_0^I Li\,\mathrm{d}i = L\left[\frac{i^2}{2}\right]_0^I$$

i.e. **total energy stored,** $\mathbf{W_f = \frac{1}{2}LI^2}$ **joules** (27)

From Section 43.5, inductance $L = N\phi/I$, hence

$$\text{total energy stored} = \frac{1}{2}\left(\frac{N\phi}{I}\right)I^2 = \frac{1}{2}N\phi I \text{ joules}$$

Also $H = NI/l$, from which, $N = Hl/I$, and $\phi = BA$. Thus the total energy stored,

$$W_f = \frac{1}{2}N\phi l = \frac{1}{2}\left(\frac{Hl}{I}\right)(BA)I$$

$$= \frac{1}{2}HBlA \text{ joules}$$

$$\text{or } \omega_f = \frac{1}{2}HB \text{ joules/metre}^3$$

since lA is the volume of the magnetic field. This latter expression has already been derived in equation (24).
Summarizing, the energy stored in a nonmagnetic medium,

$$\boldsymbol{\omega_f = \frac{1}{2}BH = \frac{1}{2}\mu_0 H^2 = \frac{B^2}{2\mu_0}} \textbf{ joules/metre}^3$$

and the energy stored in an inductor,

$$\mathbf{W_f = \frac{1}{2}LI^2} \textbf{ joules}$$

Problem 22. Calculate the value of the energy stored when a current of 50 mA is flowing in a coil of inductance 200 mH. What value of current would double the energy stored?

From equation (27), energy stored in inductor,

$$W_f = \frac{1}{2}LI^2 = \frac{1}{2}(200\times10^{-3})(50\times10^{-3})^2$$

$$= \mathbf{2.5\times10^{-4}\,J} \text{ or } \mathbf{0.25\,mJ} \text{ or } \mathbf{250\,\mu J}$$

If the energy stored is doubled, then
$(2)(2.5\times10^{-4}) = \frac{1}{2}(200\times10^{-3})I^2$ from which

$$\text{current } I = \sqrt{\left(\frac{(4)(2.5\times10^{-4})}{(200\times10^{-3})}\right)} = \mathbf{70.71\,mA}$$

Problem 23. The airgap of a moving coil instrument is 2.0 mm long and has a cross-sectional area of 500 mm^2. If the flux density is 50 mT, determine the total energy stored in the magnetic field of the airgap.

From equation (26), energy stored,

$$\omega_f = \frac{B^2}{2\mu_0} = \frac{(50\times10^{-3})^2}{2(4\pi\times10^{-7})} = 9.95\times10^2\,\text{J/m}^3$$

Volume of airgap $= Al = (500\times2.0)$ mm^3
$= 500\times2.0\times10^{-9}$m^3

Hence the energy stored in the airgap,

$$W_f = 9.95\times10^2\,\text{J/m}^3\times500\times2.0\times10^{-9}\,\text{m}^3$$

$$= \mathbf{9.95\times10^{-4}\,J \equiv 0.995\,mJ \equiv 995\,\mu J}$$

Problem 24. Determine the strength of a uniform electric field if it is to have the same energy as that established by a magnetic field of flux density 0.8 T. Assume that the relative permeability of the magnetic field and the relative permittivity of the electric field are both unity.

From equation (26), energy stored in magnetic field,

$$\omega_f = \frac{B^2}{2\mu_0} = \frac{(0.8)^2}{2(4\pi\times10^{-7})} = 2.546\times10^5\,\text{J/m}^3$$

From equation (17), energy stored in electric field,

$$\omega_f = \frac{1}{2}\varepsilon_0\varepsilon_r E^2$$

Hence, if the current stored in the magnetic and electric fields is to be the same, then

$\frac{1}{2}\varepsilon_0\varepsilon_r E^2 = 2.546\times10^5$, i.e.

$$\frac{1}{2}(8.85\times10^{-12})(1)E^2 = 2.546\times10^5$$

from which electric field strength,

$$E = \sqrt{\left(\frac{(2)(2.546\times10^5)}{(8.85\times10^{-12})}\right)} = \sqrt{(5.75\times10^{16})}$$

$$= \mathbf{2.40\times10^8\,V/m}$$

or **240 MV/m**

Now try the following Practice Exercise

Practice Exercise 169 Energy stored in an electromagnetic field (Answers on page 832)

1. Determine the value of the energy stored when a current of 120 mA flows in a coil of 500 mH. What value of current is required to double the energy stored?
2. A moving-coil instrument has two airgaps, each 2.5 mm long and having a cross-sectional area of 8.0 cm^2. Determine the total energy stored in the magnetic field of the airgaps if the flux density is 100 mT.
3. Determine the flux density of a uniform magnetic field if it is to have the same energy as that established by a uniform electric field of strength 45 MV/m. Assume the relative permeability of the magnetic field and the relative permittivity of the electric field are both unity.
4. A long single-core concentric cable has inner and outer conductors of diameters D_1 and D_2, respectively. The conductors each carry a current of I amperes but in opposite directions. If the relative permeability of the material is unity and the inductance due to internal linkages is negligible, show that the magnetic energy stored in a 4 m length of the cable is given by
$$\frac{\mu_0 I^2}{\pi} \ln\left(\frac{D_2}{D_1}\right) \text{joules}$$
5. 1 mJ of energy is stored in a uniform magnetic field having dimensions 20 mm by 10 mm by 1.0 mm. Determine for the field (a) the flux density and (b) the magnetic field strength.

For fully worked solutions to each of the problems in Practice Exercises 163 to 169 in this chapter, go to the website:
www.routledge.com/cw/bird

Chapter 44

Attenuators

Why it is important to understand: **Attenuators**

Attenuation is the gradual loss in intensity of any kind of flux through a medium. For instance, sunlight is attenuated by dark glasses, X-rays are attenuated by lead and light and sound are attenuated by water. In electrical engineering and telecommunications, attenuation affects the propagation of waves and signals in electrical circuits, in optical fibres, as well as in air (radio waves). RF attenuators reduce the level of the signal; this may be required to protect a stage from receiving a signal level that is too high. An attenuator may be used to provide an accurate impedance match as most fixed attenuators offer a well-defined impedance, or attenuators may be used in a variety of areas where signal levels need to be controlled. Attenuation is a general term that refers to any reduction in the strength of a signal. Attenuation occurs with any type of signal, whether digital or analogue. Sometimes called loss, attenuation is a natural consequence of signal transmission over long distances. This chapter explains the function of attenuation, logarithmic units and the design of T, π and L attenuator sections. In addition, attenuation for two-port networks in cascade, and *ABCD* parameters for networks is explained.

At the end of this chapter you should be able to:

- understand the function of an attenuator
- understand characteristic impedance and calculate for given values
- appreciate and calculate logarithmic ratios
- design symmetrical T and symmetrical π attenuators given required attenuation and characteristic impedance
- appreciate and calculate insertion loss
- determine iterative and image impedances for asymmetrical T and π networks
- appreciate and design the L-section attenuator
- calculate attenuation for two-port networks in cascade
- understand and apply *ABCD* parameters for networks

Electrical Circuit Theory and Technology. 978-1-138-67349-6, © 2017 John Bird. Published by Taylor & Francis. All rights reserved.

44.1 Introduction

An **attenuator** is a device for introducing a specified loss between a signal source and a matched load without upsetting the impedance relationship necessary for matching. The loss introduced is constant irrespective of frequency; since reactive elements (L or C) vary with frequency, it follows that ideal attenuators are networks containing pure resistances. A fixed attenuator section is usually known as a 'pad'.

Attenuation is a reduction in the magnitude of a voltage or current due to its transmission over a line or through an attenuator. Any degree of attenuation may be achieved with an attenuator by suitable choice of resistance values, but the input and output impedances of the pad must be such that the impedance conditions existing in the circuit into which it is connected are not disturbed. Thus an attenuator must provide the correct input and output impedances as well as providing the required attenuation.

Attenuation sections are made up of resistances connected as T or π arrangements (as introduced in Chapter 37).

Two-port networks

Networks in which electrical energy is fed in at one pair of terminals and taken out at a second pair of terminals are called two-port networks. Thus an attenuator is a two-port network, as are transmission lines, transformers and electronic amplifiers. The network between the input port and the output port is a transmission network for which a known relationship exists between the input and output currents and voltages. If a network contains only passive circuit elements, such as in an attenuator, the network is said to be **passive**; if a network contains a source of e.m.f., such as in an electronic amplifier, the network is said to be **active**.

Figure 44.1(a) shows a T-network, which is termed **symmetrical** if $Z_A = Z_B$; Figure 44.1(b) shows a π-network which is symmetrical if $Z_E = Z_F$. If $Z_A \neq Z_B$ in Figure 44.1(a) and $Z_E \neq Z_F$ in Figure 44.1(b), the sections are termed **asymmetrical**. Both networks shown have one common terminal, which may be earthed, and are therefore said to be **unbalanced**. The **balanced** form of the T-network is shown in Figure 44.2(a) and the balanced form of the π-network is shown in Figure 44.2(b).

Symmetrical T- and π-attenuators are discussed in Section 44.4 and asymmetrical attenuators are discussed in Sections 44.6 and 44.7. Before this it is important to understand the concept of characteristic impedance, which is explained generally in Section 44.2 (characteristic impedances will be used again in Chapter 47), and logarithmic units, discussed in Section 44.3. Another important aspect of attenuators, that of insertion loss, is discussed in Section 44.5. To obtain greater attenuation, sections may be connected in cascade, and this is discussed in Section 44.8. Finally, in Section 44.9, $ABCD$ parameters are explained.

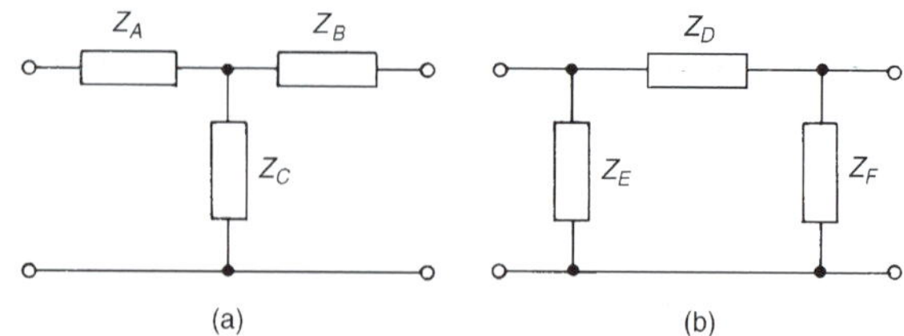

Figure 44.1 (a) T-network, (b) π-network

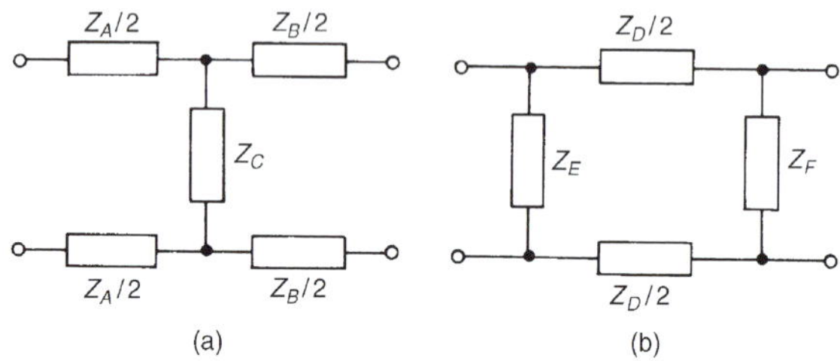

Figure 44.2 (a) Balanced T-network, (b) Balanced π-network

44.2 Characteristic impedance

The input impedance of a network is the ratio of voltage to current (in complex form) at the input terminals. With a two-port network the input impedance often varies according to the load impedance across the output terminals. For any passive two-port network it is found that a particular value of load impedance can always be found which will produce an input impedance having the same value as the load impedance. This is called the **iterative impedance** for an asymmetrical network and its value depends on which pair of terminals is taken to be the input and which the output (there are thus two values of iterative impedance, one for each direction). For a symmetrical network there is only one value for the iterative impedance and this is called the **characteristic impedance** of the symmetrical two-port network. Let the characteristic impedance be denoted by Z_0. Figure 44.3 shows a **symmetrical *T*-network** terminated in an impedance Z_0

Part 4

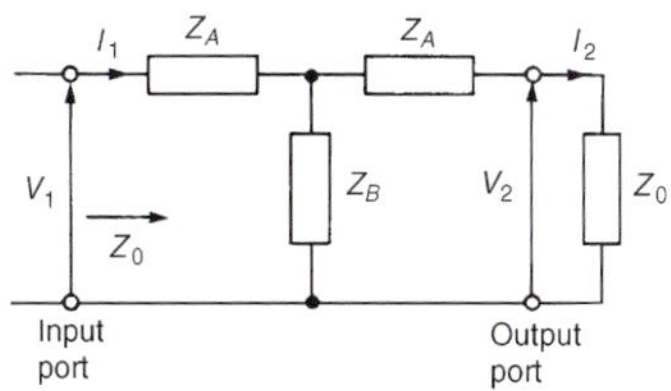

Figure 44.3

Let the impedance 'looking-in' at the input port also be Z_0. Then $V_1/I_1 = Z_0 = V_2/I_2$ in Figure 44.3. From circuit theory,

$$Z_0 = \frac{V_1}{I_1} = Z_A + \frac{Z_B(Z_A + Z_0)}{Z_B + Z_A + Z_0}, \quad \text{since } (Z_A + Z_0) \text{ is in parallel with } Z_B,$$

$$= \frac{Z_A^2 + Z_A Z_B + Z_A Z_0 + Z_A Z_B + Z_B Z_0}{Z_A + Z_B + Z_0}$$

i.e. $$Z_0 = \frac{Z_A^2 + 2Z_A Z_B + Z_A Z_0 + Z_B Z_0}{Z_A + Z_B + Z_0}$$

Thus

$$Z_0(Z_A + Z_B + Z_0) = Z_A^2 + 2Z_A Z_B + Z_A Z_0 + Z_B Z_0$$

$$Z_0 Z_A + Z_0 Z_B + Z_0^2 = Z_A^2 + 2Z_A Z_B + Z_A Z_0 + Z_B Z_0$$

i.e. $$Z_0^2 = Z_A^2 + 2Z_A Z_B, \text{ from which}$$

characteristic impedance, $\boldsymbol{Z_0 = \sqrt{(Z_A^2 + 2Z_A Z_B)}}$ (1)

If the output terminals of Figure 44.3 are open-circuited, then the open-circuit impedance, $Z_{OC} = Z_A + Z_B$. If the output terminals of Figure 44.3 are short-circuited, then the short-circuit impedance,

$$Z_{SC} = Z_A + \frac{Z_A Z_B}{Z_A + Z_B} = \frac{Z_A^2 + 2Z_A Z_B}{Z_A + Z_B}$$

Thus

$$Z_{OC} Z_{SC} = (Z_A + Z_B)\left(\frac{Z_A^2 + 2Z_A Z_B}{Z_A + Z_B}\right) = Z_A^2 + 2Z_A Z_B$$

Comparing this with equation (1) gives

$$\boldsymbol{Z_0 = \sqrt{(Z_{OC} Z_{SC})}} \quad (2)$$

Figure 44.4 shows a **symmetrical π-network** terminated in an impedance Z_0

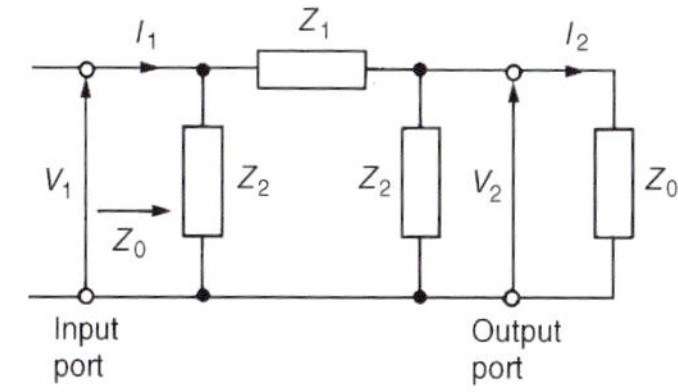

Figure 44.4

If the impedance 'looking in' at the input port is also Z_0, then

$$\frac{V_1}{I_1} = Z_0 = (Z_2) \text{ in parallel with } [Z_1 \text{ in series with } (Z_0 \text{ and } Z_2) \text{ in parallel}]$$

$$= (Z_2) \text{ in parallel with } \left[Z_1 + \frac{Z_0 Z_2}{Z_0 + Z_2}\right]$$

$$= (Z_2) \text{ in parallel with } \left[\frac{Z_1 Z_0 + Z_1 Z_2 + Z_0 Z_2}{Z_0 + Z_2}\right]$$

i.e. $$Z_0 = \frac{(Z_2)((Z_1 Z_0 + Z_1 Z_2 + Z_0 Z_2)/(Z_0 + Z_2))}{Z_2 + ((Z_1 Z_0 + Z_1 Z_2 + Z_0 Z_2)/(Z_0 + Z_2))}$$

$$= \frac{(Z_1 Z_2 Z_0 + Z_1 Z_2^2 + Z_0 Z_2^2)/(Z_0 + Z_2)}{(Z_2 Z_0 + Z_2^2 + Z_1 Z_0 + Z_1 Z_2 + Z_0 Z_2)/(Z_0 + Z_2)}$$

i.e. $$Z_0 = \frac{Z_1 Z_2 Z_0 + Z_1 Z_2^2 + Z_0 Z_2^2}{Z_2^2 + 2Z_2 Z_0 + Z_1 Z_0 + Z_1 Z_2}$$

Thus $$Z_0(Z_2^2 + 2Z_2 Z_0 + Z_1 Z_0 + Z_1 Z_2) = Z_1 Z_2 Z_0 + Z_1 Z_2^2 + Z_0 Z_2^2$$

and $$2Z_2 Z_0^2 + Z_1 Z_0^2 = Z_1 Z_2^2$$

from which

characteristic impedance, $\boldsymbol{Z_0 = \sqrt{\left(\frac{Z_1 Z_2^2}{Z_1 + 2Z_2}\right)}}$ (3)

If the output terminals of Figure 44.4 are open-circuited, then the open-circuit impedance,

$$Z_{OC} = \frac{Z_2(Z_1 + Z_2)}{Z_2 + Z_1 + Z_2} = \frac{Z_2(Z_1 + Z_2)}{Z_1 + 2Z_2}$$

If the output terminals of Figure 44.4 are short-circuited, then the short-circuit impedance,

$$Z_{SC} = \frac{Z_2 Z_1}{Z_1 + Z_2}$$

Thus

$$Z_{OC}Z_{SC} = \frac{Z_2(Z_1+Z_2)}{(Z_1+2Z_2)}\left(\frac{Z_2Z_1}{Z_1+Z_2}\right) = \frac{Z_1Z_2^2}{Z_1+2Z_2}$$

Comparing this expression with equation (3) gives

$$Z_0 = \sqrt{(Z_{OC}Z_{SC})} \qquad (2')$$

which is the same as equation (2).
Thus the characteristic impedance Z_0 is given by $Z_0 = \sqrt{(Z_{OC}Z_{SC})}$ whether the network is a symmetrical T or a symmetrical π
Equations (1) to (3) are used later in this chapter.

44.3 Logarithmic ratios

The ratio of two powers P_1 and P_2 may be expressed in logarithmic form as shown in Chapter 12.
Let P_1 be the input power to a system and P_2 the output power.
If logarithms to base 10 are used, then the ratio is said to be in **bels**, i.e. power ratio in bels $= \lg(P_2/P_1)$. The bel is a large unit and the **decibel (dB)** is more often used, where 10 decibels $= 1$ bel, i.e.

$$\textbf{power ratio in decibels} = 10\lg\frac{P_2}{P_1} \qquad (4)$$

The bel is named after **Alexander Graham Bell.***
For example:

P_2/P_1	power ratio (dB)
1	$10\lg 1 = 0$
100	$10\lg 100 = +20$ (power gain)
$\frac{1}{10}$	$10\lg\frac{1}{10} = -10$ (power loss or attenuation)

If **logarithms to base e** (i.e. natural or Napierian logarithms) are used, then the ratio of two powers is said to be in **nepers (Np)**, i.e.

$$\textbf{power ratio in nepers} = \frac{1}{2}\ln\frac{P_2}{P_1} \qquad (5)$$

*Who was **Bell**? For resume of Bell, see page 174. To find out more go to **www.routledge.com/cw/bird**

The neper is named after **John Napier**. *
Thus when the power ratio $P_2/P_1 = 5$, the power ratio in nepers $= \frac{1}{2}\ln 5 = 0.805$ Np, and when the power ratio $P_2/P_1 = 0.1$, the power ratio in nepers $= \frac{1}{2}\ln 0.1 = -1.15$ Np
The attenuation of filter sections and along a transmission line are of an exponential form and it is in such applications that the unit of the neper is used (see Chapters 45 and 47).
If the powers P_1 and P_2 refer to power developed in two equal resistors, R, then $P_1 = V_1^2/R$ and $P_2 = V_2^2/R$. Thus the ratio (from equation (4)) can be expressed by the laws of logarithms as

$$\textbf{ratio in decibels} = 10\lg\frac{P_2}{P_1} = 10\lg\left(\frac{V_2^2/R}{V_1^2/R}\right)$$

$$= 10\lg\frac{V_2^2}{V_1^2} = 10\lg\left(\frac{V_2}{V_1}\right)^2$$

i.e.
$$\textbf{ratio in decibels} = 20\lg\frac{V_2}{V_1} \qquad (6)$$

Although this is really a power ratio, it is called the **logarithmic voltage ratio**.

*Who was **Napier**? **John Napier** (1550–4 April 1617) is best known as the discoverer of logarithms and the inventor of 'Napier's bones'. To find out more go to **www.routledge.com/cw/bird**

Alternatively, (from equation (5)),

$$\text{ratio in nepers} = \frac{1}{2}\ln\frac{P_2}{P_1} = \frac{1}{2}\ln\left(\frac{V_2^2/R}{V_1^2/R}\right) = \frac{1}{2}\ln\left(\frac{V_2}{V_1}\right)^2$$

i.e. **ratio in nepers = ln $\frac{V_2}{V_1}$** (7)

Similarly, if currents I_1 and I_2 in two equal resistors R give powers P_1 and P_2, then (from equation (4))

ratio in decibels

$$= 10\lg\frac{P_2}{P_1} = 10\lg\left(\frac{I_2^2 R}{I_1^2 R}\right) = 10\lg\left(\frac{I_2}{I_1}\right)^2$$

i.e. **ratio in decibels = 20 lg $\frac{I_2}{I_1}$** (8)

Alternatively (from equation (5)),

$$\text{ratio in nepers} = \frac{1}{2}\ln\frac{P_2}{P_1} = \frac{1}{2}\ln\left(\frac{I_2^2 R}{I_1^2 R}\right)^2 = \frac{1}{2}\ln\left(\frac{I_2}{I_1}\right)^2$$

i.e. **ratio in nepers = ln $\frac{I_2}{I_1}$** (9)

In equations (4) to (9) the output-to-input ratio has been used. However, the input-to-output ratio may also be used. For example, in equation (6), the output-to-input voltage ratio is expressed as $20\lg(V_2/V_1)$ dB. Alternatively, the input-to-output voltage ratio may be expressed as $20\lg(V_1/V_2)$ dB, the only difference in the values obtained being a difference in sign.
If $20\lg(V_2/V_1) = 10$ dB, say, then $20\lg(V_1/V_2) = -10$ dB. Thus if an attenuator has a voltage input V_1 of 50 mV and a voltage output V_2 of 5 mV, the voltage ratio V_2/V_1 is 5/50 or 1/10. Alternatively, this may be expressed as **'an attenuation of 10'**, i.e. $V_1/V_2 = 10$.

Problem 1. The ratio of output power to input power in a system is

(a) 2, (b) 25, (c) 1000 and (d) $\frac{1}{100}$

Determine the power ratio in each case (i) in decibels and (ii) in nepers.

(i) From equation (4), power ratio in decibels $= 10\lg(P_2/P_1)$

(a) When $P_2/P_1 = 2$, power ratio $= 10\lg 2$ $= \mathbf{3\,dB}$

(b) When $P_2/P_1 = 25$, power ratio $= 10\lg 25$ $= \mathbf{14\,dB}$

(c) When $P_2/P_1 = 1000$, power ratio $= 10\lg 1000 = \mathbf{30\,dB}$

(d) When $P_2/P_1 = \frac{1}{100}$, power ratio $= 10\lg\frac{1}{100} = \mathbf{-20\,dB}$

(ii) From equation (5), power ratio in nepers $= \frac{1}{2}\ln(P_2/P_1)$

(a) When $P_2/P_1 = 2$, power ratio $= \frac{1}{2}\ln 2 = \mathbf{0.347\,Np}$

(b) When $P_2/P_1 = 25$, power ratio $= \frac{1}{2}\ln 25$ $= \mathbf{1.609\,Np}$

(c) When $P_2/P_1 = 1000$, power ratio $= \frac{1}{2}\ln 1000 = \mathbf{3.454\,Np}$

(d) When $P_2/P_1 = \frac{1}{100}$, power ratio $= \frac{1}{2}\ln\frac{1}{100}$ $= \mathbf{-2.303\,Np}$

The power ratios in (a), (b) and (c) represent power gains, since the ratios are positive values; the power ratio in (d) represents a power loss or attenuation, since the ratio is a negative value.

Problem 2. 5% of the power supplied to a cable appears at the output terminals. Determine the attenuation in decibels.

If P_1 = input power and P_2 = output power, then

$$\frac{P_2}{P_1} = \frac{5}{100} = 0.05$$

From equation (4), power ratio in decibels

$$= 10\lg(P_2/P_1) = 10\lg 0.05 = -13\,\text{dB}$$

Hence the attenuation (i.e. power loss) is 13 dB

Problem 3. An amplifier has a gain of 15 dB. If the input power is 12 mW, determine the output power.

From equation (4), decibel power ratio $= 10\lg(P_2/P_1)$
Hence $15 = 10\lg(P_2/12)$, where P_2 is the output power in milliwatts.

$$1.5 = \lg\left(\frac{P_2}{12}\right)$$

$$\frac{P_2}{12} = 10^{1.5}$$

from the definition of a logarithm. Thus the output power, $P_2 = 12(10)^{1.5} = \mathbf{379.5\,mW}$

Problem 4. The current output of an attenuator is 50 mA. If the current ratio of the attenuator is −1.32 Np, determine (a) the current input and (b) the current ratio expressed in decibels. Assume that the input and load resistances of the attenuator are equal.

(a) From equation (9),

current ratio in nepers $= \ln(I_2/I_1)$

Hence $-1.32 = \ln(50/I_1)$, where I_1 is the input current in mA.

$$e^{-1.32} = \frac{50}{I_1}$$

from which, **current input,** $I_1 = \frac{50}{e^{-1.32}} = 50e^{1.32}$

$$= \mathbf{187.2\,mA}$$

(b) From equation (8), current ratio in decibels

$$= 20\lg\frac{I_2}{I_1} = 20\lg\left(\frac{50}{187.2}\right)$$

$$= \mathbf{-11.47\,dB}$$

Now try the following Practice Exercise

Practice Exercise 170 Logarithmic ratios (Answers on page 832)

1. The ratio of two powers is (a) 3, (b) 10, (c) 30, (d) 10 000. Determine the decibel power ratio for each.
2. The ratio of two powers is (a) $\frac{1}{10}$, (b) $\frac{1}{3}$, (c) $\frac{1}{40}$, (d) $\frac{1}{1000}$. Determine the decibel power ratio for each.
3. An amplifier has (a) a gain of 25 dB, (b) an attenuation of 25 dB. If the input power is 12 mW, determine the output power in each case.
4. 7.5% of the power supplied to a cable appears at the output terminals. Determine the attenuation in decibels.
5. The current input of a system is 250 mA. If the current ratio of the system is (i) 15 dB, (ii) −8 dB, determine (a) the current output and (b) the current ratio expressed in nepers.

Part 4

44.4 Symmetrical T- and π-attenuators

(a) Symmetrical T-attenuator

As mentioned in Section 44.1, the ideal attenuator is made up of pure resistances. A symmetrical T-pad attenuator is shown in Figure 44.5 with a termination R_0 connected as shown. From equation (1),

$$\boldsymbol{R_0 = \sqrt{(R_1^2 + 2R_1R_2)}} \qquad (10)$$

and from equation (2) $\boldsymbol{R_0 = \sqrt{(R_{OC}R_{SC})}}$ (11)

With resistance R_0 as the termination, the input resistance of the pad will also be equal to R_0. If the terminating resistance R_0 is transferred to port A then the input resistance looking into port B will again be R_0.

The pad is therefore symmetrical in impedance in both directions of connection and may thus be inserted into a network whose impedance is also R_0. The value of R_0 is the characteristic impedance of the section.

As stated in Section 44.3, attenuation may be expressed as a voltage ratio V_1/V_2 (see Figure 44.5) or quoted in decibels as $20\lg(V_1/V_2)$ or, alternatively, as a power ratio as $10\lg(P_1/P_2)$. If a T-section is symmetrical, i.e. the terminals of the section are matched to equal impedances, then

$$10\lg\frac{P_1}{P_2} = 20\lg\frac{V_1}{V_2} = 20\frac{I_1}{I_2}$$

since $R_{IN} = R_{LOAD} = R_0$, i.e.

$$10\lg\frac{P_1}{P_2} = 10\lg\left(\frac{V_1}{V_2}\right)^2 = 10\lg\left(\frac{I_1}{I_2}\right)^2$$

from which $$\frac{P_1}{P_2} = \left(\frac{V_1}{V_2}\right)^2 = \left(\frac{I_1}{I_2}\right)^2$$

or $$\sqrt{\left(\frac{P_1}{P_2}\right)} = \left(\frac{V_1}{V_2}\right) = \left(\frac{I_1}{I_2}\right)$$

Let $N = V_1/V_2$ or I_1/I_2 or $\sqrt{(P_1/P_2)}$, where N is the attenuation. In Section 44.5, page 675, it is shown that,

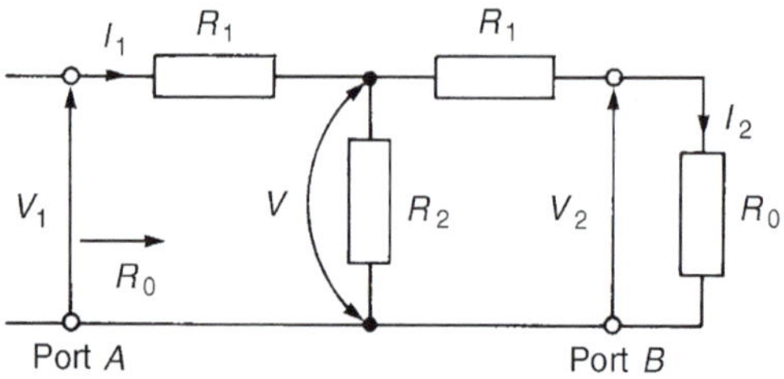

Figure 44.5 Symmetrical T-pad attenuator

for a matched network, i.e. one terminated in its characteristic impedance, N is in fact the insertion loss ratio. (Note that in an asymmetrical network, only the expression $N=\sqrt{(P_1/P_2)}$ may be used – see Section 44.7 on the L-section attenuator.)

From Figure 44.5,

current $I_1 = \dfrac{V_1}{R_0}$

Voltage $V = V_1 - I_1 R_1 = V_1 - \left(\dfrac{V_1}{R_0}\right) R_1$

i.e. $V = V_1 \left(1 - \dfrac{R_1}{R_0}\right)$

Voltage $V_2 = \left(\dfrac{R_0}{R_1 + R_0}\right) V$ by voltage division

i.e.
$$V_2 = \left(\frac{R_0}{R_1 + R_0}\right) V_1 \left(1 - \frac{R_1}{R_0}\right)$$
$$= V_1 \left(\frac{R_0}{R_1 + R_0}\right)\left(\frac{R_0 - R_1}{R_0}\right)$$

Hence $\dfrac{V_2}{V_1} = \dfrac{R_0 - R_1}{R_0 + R_1}$ or $\dfrac{V_1}{V_2} = N = \dfrac{R_0 + R_1}{R_0 - R_1}$ (12)

From equation (12) and also equation (10), it is possible to derive expressions for R_1 and R_2 in terms of N and R_0, thus enabling an attenuator to be designed to give a specified attenuation and to be matched symmetrically into the network. From equation (12),

$$N(R_0 - R_1) = R_0 + R_1$$
$$NR_0 - NR_1 = R_0 + R_1$$
$$NR_0 - R_0 = R_1 + NR_1$$
$$R_0(N-1) = R_1(1+N)$$

from which $\boldsymbol{R_1 = R_0 \dfrac{(N-1)}{(N+1)}}$ (13)

From equation (10), $R_0 = \sqrt{(R_1^2 + 2R_1R_2)}$

i.e. $R_0^2 = R_1^2 + 2R_1R_2$

from which, $R_2 = \dfrac{R_0^2 - R_1^2}{2R_1}$

Substituting for R_1 from equation (13) gives

$$R_2 = \frac{R_0^2 - [R_0(N-1)/(N+1)]^2}{2[R_0(N-1)/(N+1)]}$$
$$= \frac{[R_0^2(N+1)^2 - R_0^2(N-1)^2]/(N+1)^2}{2R_0(N-1)/(N+1)}$$

i.e.
$$= \frac{R_0^2[(N+1)^2 - (N-1)^2]}{2R_0(N-1)(N+1)}$$
$$= \frac{R_0[(N^2+2N+1) - (N^2-2N+1)]}{2(N^2-1)}$$
$$= \frac{R_0(4N)}{2(N^2-1)}$$

Hence $\boldsymbol{R_2 = R_0 \left(\dfrac{2N}{N^2-1}\right)}$ (14)

Thus if the characteristic impedance R_0 and the attenuation N $(=V_1/V_2)$ are known for a symmetrical T-network then values of R_1 and R_2 may be calculated. Figure 44.6 shows a T-pad attenuator having input and output impedances of R_0 with resistances R_1 and R_2 expressed in terms of R_0 and N.

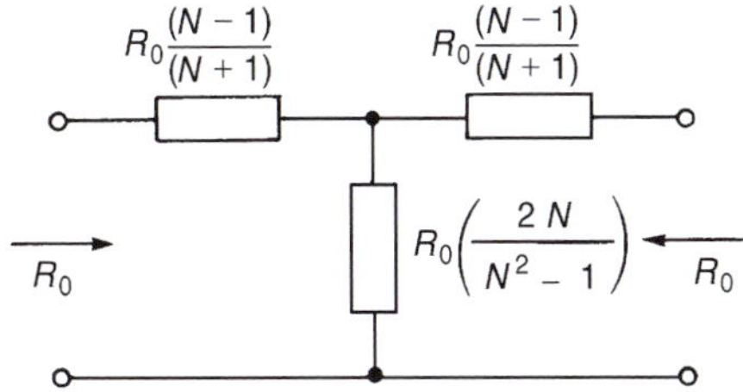

Figure 44.6

(b) Symmetrical π-attenuator

A symmetrical π-attenuator is shown in Figure 44.7, terminated in R_0.

From equation (3),

characteristic impedance $\boldsymbol{R_0 = \sqrt{\left(\dfrac{R_1R_2^2}{R_1 + 2R_2}\right)}}$ (15)

and from equation (2′), $\boldsymbol{R_0 = \sqrt{(R_{OC}R_{SC})}}$ (16)

Given the attenuation factor $N = \dfrac{V_1}{V_2}\left(=\dfrac{I_1}{I_2}\right)$

and the characteristic impedance R_0, it is possible to derive expressions for R_1 and R_2, in a similar way to the T-pad attenuator, to enable a π-attenuator to be effectively designed.

Since $N = V_1/V_2$ then $V_2 = V_1/N$. From Figure 44.7,

current $I_1 = I_A + I_B$ and current $I_B = I_C + I_D$. Thus

current
$$I_1 = \frac{V_1}{R_0} = I_A + I_C + I_D$$
$$= \frac{V_1}{R_2} + \frac{V_2}{R_2} + \frac{V_2}{R_0} = \frac{V_1}{R_2} + \frac{V_1}{NR_2} + \frac{V_1}{NR_0}$$

since $V_2 = V_1/N$,

i.e. $\dfrac{V_1}{R_0} = V_1\left(\dfrac{1}{R_2} + \dfrac{1}{NR_2} + \dfrac{1}{NR_0}\right)$

Part 4

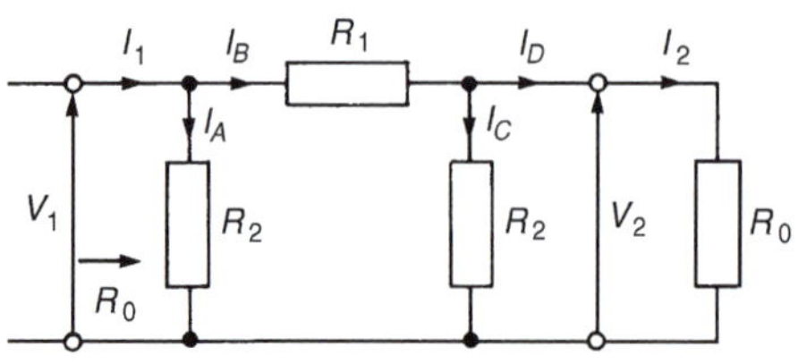

Figure 44.7 Symmetrical π-attenuator

$$\text{Hence}\ \frac{1}{R_0}=\frac{1}{R_2}+\frac{1}{NR_2}+\frac{1}{NR_0}$$

$$\frac{1}{R_0}-\frac{1}{NR_0}=\frac{1}{R_2}+\frac{1}{NR_2}$$

$$\frac{1}{R_0}\left(1-\frac{1}{N}\right)=\frac{1}{R_2}\left(1+\frac{1}{N}\right)$$

$$\frac{1}{R_0}\left(\frac{N-1}{N}\right)=\frac{1}{R_2}\left(\frac{N+1}{N}\right)$$

Thus $$\mathbf{R_2=R_0\frac{(N+1)}{(N-1)}} \qquad (17)$$

From Figure 44.7, current $I_1=I_A+I_B$, and since the p.d. across R_1 is (V_1-V_2),

$$\frac{V_1}{R_0}=\frac{V_1}{R_2}+\frac{V_1-V_2}{R_1}$$

$$\frac{V_1}{R_0}=\frac{V_1}{R_2}+\frac{V_1}{R_1}-\frac{V_2}{R_1}$$

$$\frac{V_1}{R_0}=\frac{V_1}{R_2}+\frac{V_1}{R_1}-\frac{V_1}{NR_1}\ \text{since}\ V_2=V_1/N$$

$$\frac{1}{R_0}=\frac{1}{R_2}+\frac{1}{R_1}-\frac{1}{NR_1}$$

$$\frac{1}{R_0}-\frac{1}{R_2}=\frac{1}{R_1}\left(1-\frac{1}{N}\right)$$

$$\frac{1}{R_0}-\frac{(N-1)}{R_0(N+1)}=\frac{1}{R_1}\left(\frac{N-1}{N}\right)$$

from equation (17),

$$\frac{1}{R_0}\left(1-\frac{N-1}{N+1}\right)=\frac{1}{R_1}\left(\frac{N-1}{N}\right)$$

$$\frac{1}{R_0}\left(\frac{(N+1)-(N-1)}{(N+1)}\right)=\frac{1}{R_1}\left(\frac{N-1}{N}\right)$$

$$\frac{1}{R_0}\left(\frac{2}{N+1}\right)=\frac{1}{R_1}\left(\frac{N-1}{N}\right)$$

$$R_1=R_0\left(\frac{N-1}{N}\right)\left(\frac{N+1}{2}\right)$$

Hence $$\mathbf{R_1=R_0\left(\frac{N^2-1}{2N}\right)} \qquad (18)$$

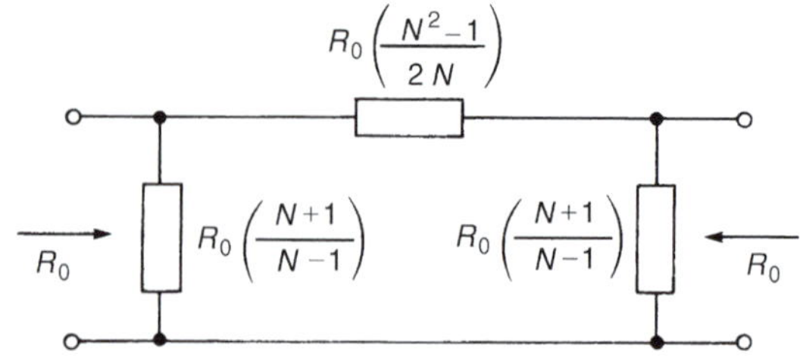

Figure 44.8

Figure 44.8 shows a π-attenuator having input and output impedances of R_0 with resistances R_1 and R_2 expressed in terms of R_0 and N.

There is no difference in the functions of the T- and π-attenuator pads and either may be used in a particular situation.

Problem 5. Determine the characteristic impedance of each of the attenuator sections shown in Figure 44.9.

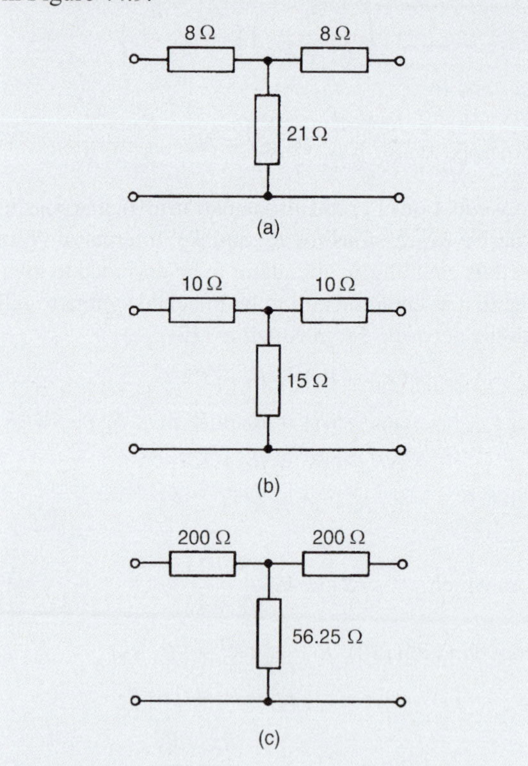

Figure 44.9

From equation (10), for a T-section attenuator the characteristic impedance,

$$R_0=\sqrt{(R_1^2+2R_1R_2)}$$

(a) $R_0=\sqrt{(8^2+(2)(8)(21))}=\sqrt{400}=\mathbf{20\ \Omega}$

(b) $R_0=\sqrt{(10^2+(2)(10)(15))}=\sqrt{400}=\mathbf{20\ \Omega}$

(c) $R_0=\sqrt{(200^2+(2)(200)(56.25))}=\sqrt{62\,500}=\mathbf{250\ \Omega}$

Part 4

It is seen that the characteristic impedance of parts (a) and (b) is the same. In fact, there are numerous combinations of resistances R_1 and R_2 which would give the same value for the characteristic impedance.

Problem 6. A symmetrical π-attenuator pad has a series arm of 500 Ω resistance and each shunt arm of 1 kΩ resistance. Determine (a) the characteristic impedance and (b) the attenuation (in dB) produced by the pad.

The π-attenuator section is shown in Figure 44.10, terminated in its characteristic impedance, R_0

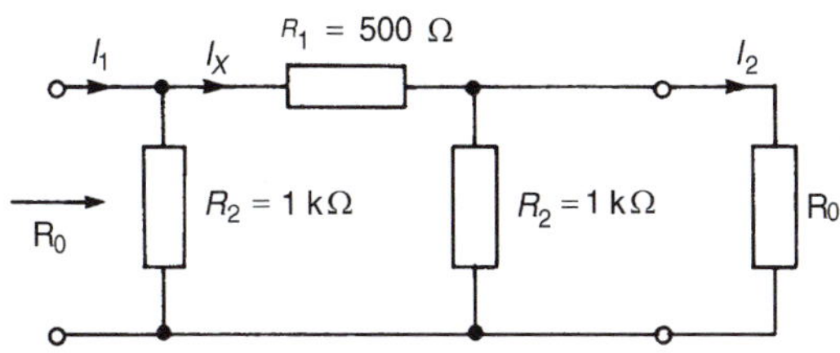

Figure 44.10

(a) From equation (15), for a symmetrical π-attenuator section,

$$\text{characteristic impedance, } R_0 = \sqrt{\left(\frac{R_1 R_2^2}{R_1 + 2R_2}\right)}$$

$$\text{Hence } \mathbf{R_0} = \sqrt{\left[\frac{(500)(1000)^2}{500 + 2(1000)}\right]} = \mathbf{447\ \Omega}$$

(b) Attenuation $= 20\lg(I_1/I_2)$ dB. From Figure 44.10,

$$\text{current } I_X = \left(\frac{R_2}{R_2 + R_1 + (R_2 R_0/(R_2 + R_0))}\right)(I_1),$$

by current division

i.e.

$$I_X = \left(\frac{1000}{1000 + 500 + ((1000)(447)/(1000 + 447))}\right) I_1$$

$$= 0.553 I_1$$

$$\text{and current } I_2 = \left(\frac{R_2}{R_2 + R_0}\right) I_X$$

$$= \left(\frac{1000}{1000 + 447}\right) I_X = 0.691 I_X$$

Hence $\quad I_2 = 0.691(0.553 I_1) = 0.382 I_1$

and $\quad I_1/I_2 = 1/0.382 = 2.617$

Thus **attenuation** $= 20\lg 2.617 = \mathbf{8.36\ dB}$

(Alternatively, since $I_1/I_2 = N$, then the formula

$$R_2 = R_0\left(\frac{N+1}{N-1}\right)$$

may be transposed for N, from which **attenuation = 20 lg** $\mathbf{N}$)

Problem 7. For each of the attenuator networks shown in Figure 44.11, determine (a) the input resistance when the output port is open-circuited, (b) the input resistance when the output port is short-circuited and (c) the characteristic impedance.

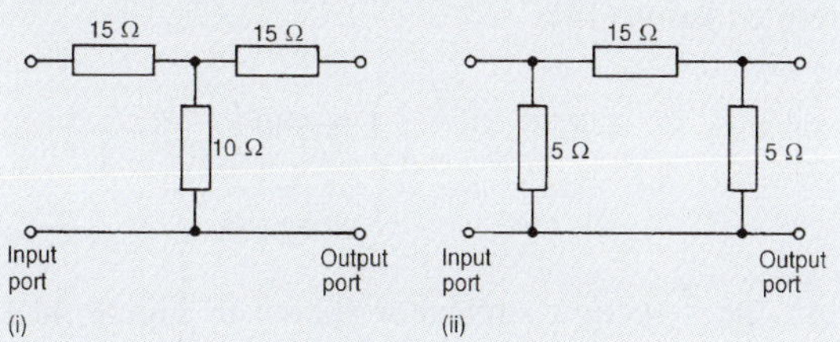

Figure 44.11

(i) For the T-network shown in Figure 44.11(i):

(a) $R_{OC} = 15 + 10 = \mathbf{25\ \Omega}$

(b) $R_{SC} = 15 + \dfrac{10 \times 15}{10 + 15} = 15 + 6 = \mathbf{21\ \Omega}$

(c) From equation (11),
$R_0 = \sqrt{R_{OC} R_{SC}} = \sqrt{[(25)(21)]} = \mathbf{22.9\ \Omega}$

(Alternatively, from equation (10),

$$R_0 = \sqrt{(R_1^2 + 2R_1 R_2)}$$
$$= \sqrt{(15^2 + (2)(15)(10))} = \mathbf{22.9\ \Omega})$$

(ii) For the π-network shown in Figure 44.11(ii):

(a) $R_{OC} = \dfrac{5 \times (15+5)}{5 + (15+5)} = \dfrac{100}{25} = \mathbf{4\ \Omega}$

(b) $R_{SC} = \dfrac{5 \times 15}{5 + 15} = \dfrac{75}{20} = \mathbf{3.75\ \Omega}$

(c) From equation (16),

$$R_0 = \sqrt{(R_{OC} R_{SC})} \text{ as for a T-network}$$
$$= \sqrt{[(4)(3.75)]} = \sqrt{15} = \mathbf{3.87\ \Omega}$$

(Alternatively, from equation (15),

$$R_0 = \sqrt{\left(\frac{R_1 R_2^2}{R_1 + 2R_2}\right)} = \sqrt{\left(\frac{15(5)^2}{15 + 2(5)}\right)}$$
$$= \mathbf{3.87\ \Omega})$$

Problem 8. Design a T-section symmetrical attenuator pad to provide a voltage attenuation of 20 dB and having a characteristic impedance of 600 Ω.

Voltage attenuation in decibels $= 20\lg(V_1/V_2)$. Attenuation, $N = V_1/V_2$, hence $20 = 20\lg N$, from which $N = 10$

Characteristic impedance, $R_0 = 600\,\Omega$

From equation (13),

$$\text{resistance } R_1 = \frac{R_0(N-1)}{(N+1)} = \frac{600(10-1)}{(10+1)} = \mathbf{491\,\Omega}$$

From equation (14),

$$\text{resistance } R_2 = R_0\left(\frac{2N}{N^2-1}\right) = 600\left(\frac{(2)(10)}{10^2-1}\right)$$
$$= \mathbf{121\,\Omega}$$

Thus the T-section attenuator shown in Figure 44.12 has a voltage attenuation of 20 dB and a characteristic impedance of 600 Ω. (Check: from equation (10))

$$R_0 = \sqrt{(R_1^2 + 2R_1R_2)} = \sqrt{[491^2 + 2(491)(121)]}$$
$$= 600\,\Omega)$$

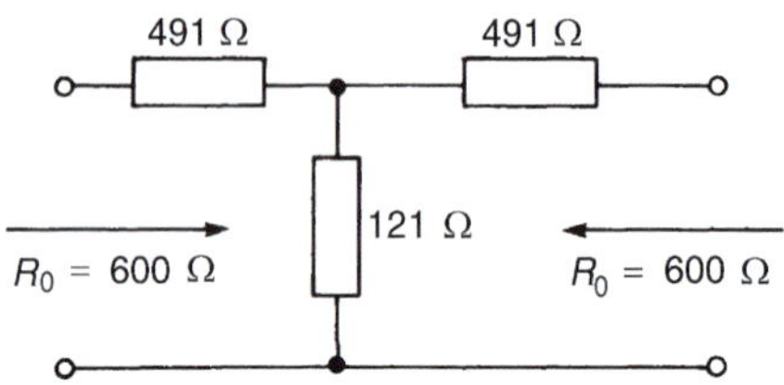

Figure 44.12

Problem 9. Design a π-section symmetrical attenuator pad to provide a voltage attenuation of 20 dB and having a characteristic impedance of 600 Ω.

From Problem 8, $N = 10$ and $R_0 = 600\,\Omega$

From equation (18),

$$\text{resistance } R_1 = R_0\left(\frac{N^2-1}{2N}\right) = 600\left(\frac{10^2-1}{(2)(10)}\right)$$
$$= \mathbf{2970\,\Omega} \text{ or } \mathbf{2.97\,k\Omega}$$

From equation (17),

$$R_2 = R_0\left(\frac{N+1}{N-1}\right) = 600\left(\frac{10+1}{10-1}\right) = \mathbf{733\,\Omega}$$

Thus the π-section attenuator shown in Figure 44.13 has a voltage attenuation of 20 dB and a characteristic impedance of 600 Ω.
(Check: from equation (15),

$$R_0 = \sqrt{\left(\frac{R_1R_2^2}{R_1+2R_2}\right)} = \sqrt{\left(\frac{(2970)(733)^2}{2970+(2)(733)}\right)}$$
$$= 600\,\Omega)$$

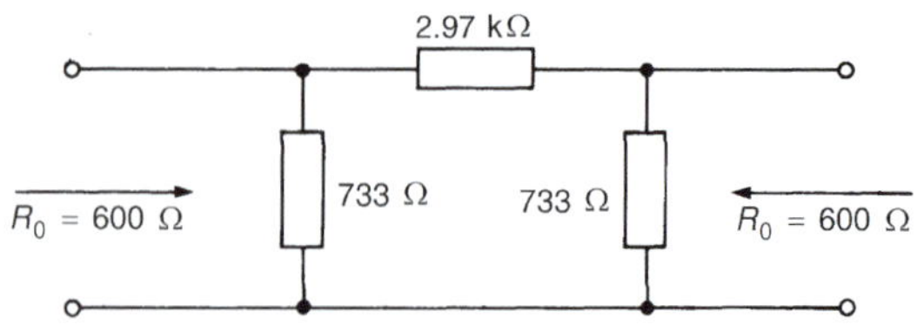

Figure 44.13

Now try the following Practice Exercise

Practice Exercise 171 Symmetrical T- and π-attenuators (Answers on page 832)

1. Determine the characteristic impedances of the T-network attenuator sections shown in Figure 44.14.

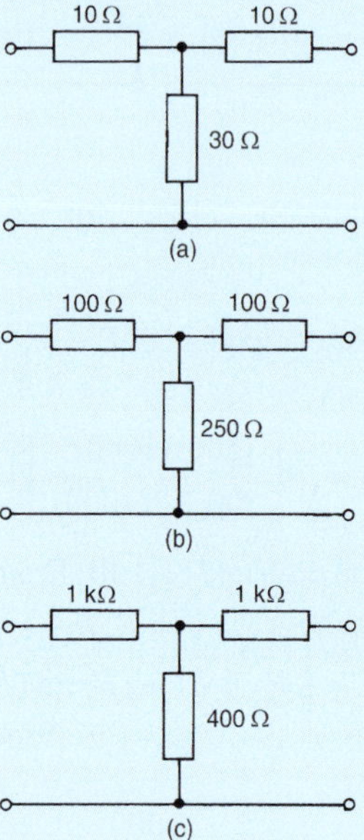

Figure 44.14

2. Determine the characteristic impedances of the π-network attenuator pads shown in Figure 44.15.
3. A T-section attenuator is to provide 18 dB voltage attenuation per section and is to match a 1.5 kΩ line. Determine the resistance values necessary per section.
4. A π-section attenuator has a series resistance of 500 Ω and shunt resistances of 2 kΩ. Determine (a) the characteristic impedance, and (b) the attenuation produced by the network.

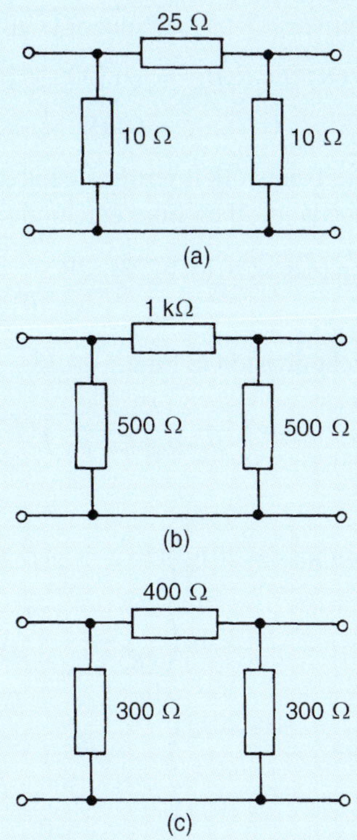

Figure 44.15

5. For each of the attenuator pads shown in Figure 44.16, determine (a) the input resistance when the output port is open-circuited, (b) the input resistance when the output port is short-circuited and (c) the characteristic impedance.

6. A television signal received from an aerial through a length of coaxial cable of characteristic impedance 100 Ω has to be attenuated by 15 dB before entering the receiver. If the input impedance of the receiver is also 100 Ω, design a suitable T-attenuator network to give the necessary reduction.

7. Design (a) a T-section symmetrical attenuator pad and (b) a π-section symmetrical attenuator pad, to provide a voltage attenuation of 15 dB and having a characteristic impedance of 500 Ω.

8. Determine the values of the shunt and series resistances for T-pad attenuators of characteristic impedance 400 Ω to provide the following voltage attenuations: (a) 12 dB, (b) 25 dB, (c) 36 dB.

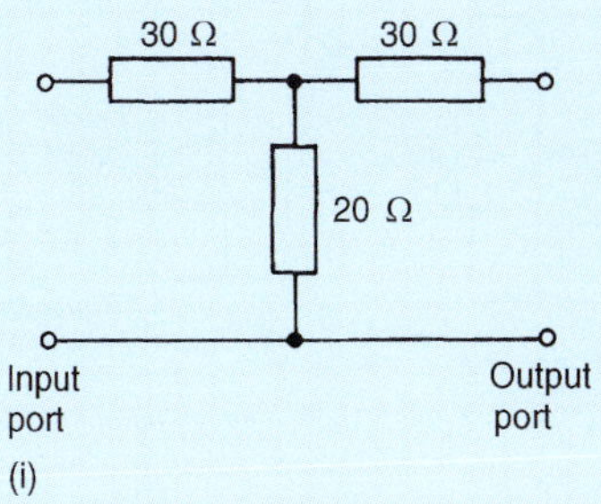

Figure 44.16 i

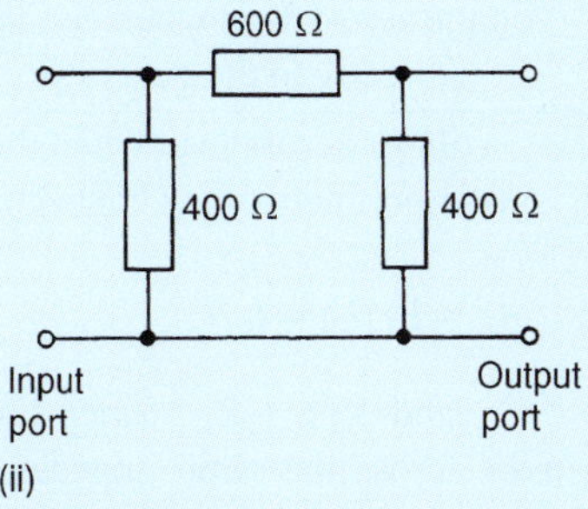

Figure 44.16 ii

9. Design a π-section symmetrical attenuator network to provide a voltage attenuation of 24 dB and having a characteristic impedance of 600 Ω.

10. A d.c. generator has an internal resistance of 600 Ω and supplies a 600 Ω load. Design a symmetrical (a) T-network and (b) π-network attenuator pad, having a characteristic impedance of 600 Ω which when connected between the generator and load will reduce the load current to $\frac{1}{4}$ its initial value.

44.5 Insertion loss

Figure 44.17(a) shows a generator E connected directly to a load Z_L. Let the current flowing be I_L and the p.d. across the load V_L. z is the internal impedance of the source

Figure 44.17(b) shows a two-port network connected between the generator E and load Z_L

The current through the load, shown as I_2, and the p.d. across the load, shown as V_2, will generally be less than current I_L and voltage V_L of Figure 44.17(a), as a result of the insertion of the two-port network between generator and load.

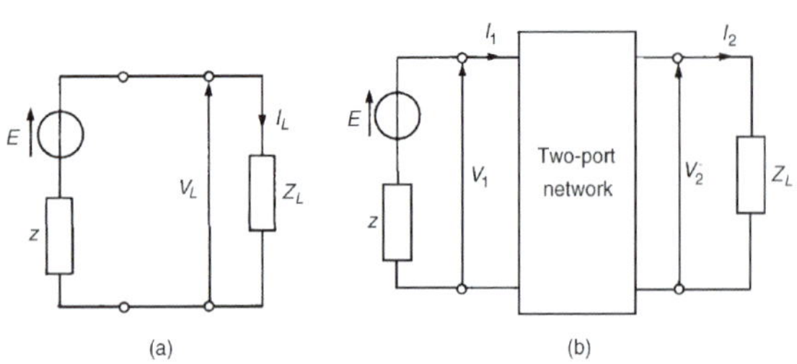

Figure 44.17

The **insertion loss ratio, A_L**, is defined as

$$A_L = \frac{\text{voltage across load when connected directly to the generator}}{\text{voltage across load when the two-port network is connected}}$$

i.e. $$\boldsymbol{A_L = V_L/V_2 = I_L/I_2} \qquad (19)$$

since $V_L = I_L Z_L$ and $V_2 = I_2 Z_L$. Since both V_L and V_2 refer to p.d.s across the same impedance Z_L, the insertion loss ratio may also be expressed (from Section 44.3) as

insertion loss ratio

$$= 20\,\text{lg}\left(\frac{V_L}{V_2}\right)\text{dB} \text{ or } 20\,\text{lg}\left(\frac{I_L}{I_2}\right)\text{dB} \qquad (20)$$

When the two-port network is terminated in its characteristic impedance Z_0 the network is said to be **matched**. In such circumstances the input impedance is also Z_0, thus the insertion loss is simply the ratio of input to output voltage (i.e. V_1/V_2). Thus, **for a network terminated in its characteristic impedance**,

$$\textbf{insertion loss} = 20\,\text{lg}\left(\frac{V_1}{V_2}\right)\text{dB or } 20\,\text{lg}\left(\frac{I_1}{I_2}\right)\text{dB} \qquad (21)$$

Problem 10. The attenuator shown in Figure 44.18 feeds a matched load. Determine (a) the characteristic impedance R_0 and (b) the insertion loss in decibels.

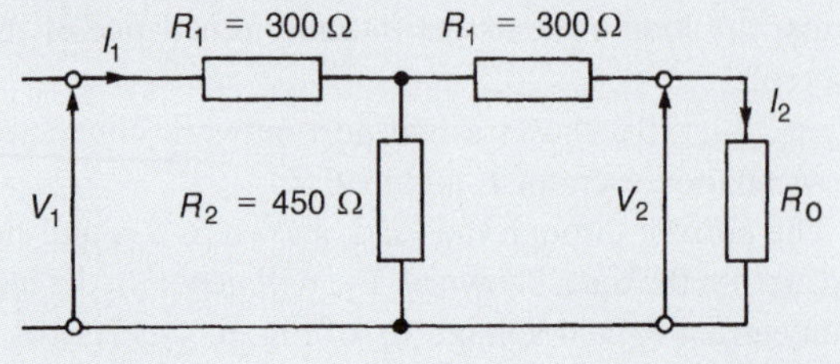

Figure 44.18

(a) From equation (10), the characteristic impedance of a symmetric T-pad attenuator is given by

$$R_0 = \sqrt{(R_1^2 + 2R_1R_2)} = \sqrt{[300^2 + 2(300)(450)]} = \mathbf{600\,\Omega}$$

(b) Since the T-network is terminated in its characteristic impedance, then from equation (21),

$$\text{insertion loss} = 20\,\text{lg}(V_1/V_2)\text{ dB} \text{ or } 20\,\text{lg}(I_1/I_2)\text{ dB}.$$

By current division in Figure 44.18,

$$I_2 = \left(\frac{R_2}{R_2 + R_1 + R_0}\right)(I_1)$$

Hence

$$\begin{aligned}\textbf{insertion loss} &= 20\,\text{lg}\frac{I_1}{I_2} \\ &= 20\,\text{lg}\left(\frac{I_1}{(R_2/(R_2+R_1+R_0))I_1}\right) \\ &= 20\,\text{lg}\left(\frac{R_2+R_1+R_0}{R_2}\right) \\ &= 20\,\text{lg}\left(\frac{450+300+600}{450}\right) \\ &= 20\,\text{lg}3 = \mathbf{9.54\,dB}\end{aligned}$$

Problem 11. A 0–3 kΩ rheostat is connected across the output of a signal generator of internal resistance 500 Ω. If a load of 2 kΩ is connected across the rheostat, determine the insertion loss at a tapping of (a) 2 kΩ, (b) 1 kΩ.

The circuit diagram is shown in Figure 44.19. Without the rheostat in the circuit the voltage across the 2 kΩ load, V_L (see Figure 44.20), is given by

$$V_L = \left(\frac{2000}{2000+500}\right)E = 0.8\text{E}$$

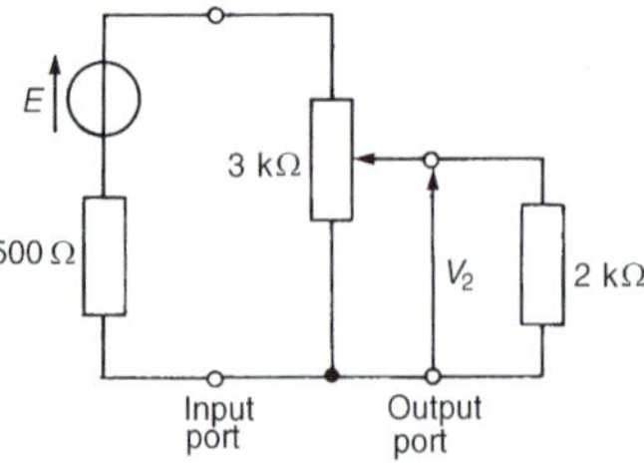

Figure 44.19

(a) With the 2 kΩ tapping, the network of Figure 44.19 may be redrawn as shown in Figure 44.21, which in turn is simplified as shown in Figure 44.22.

Part 4

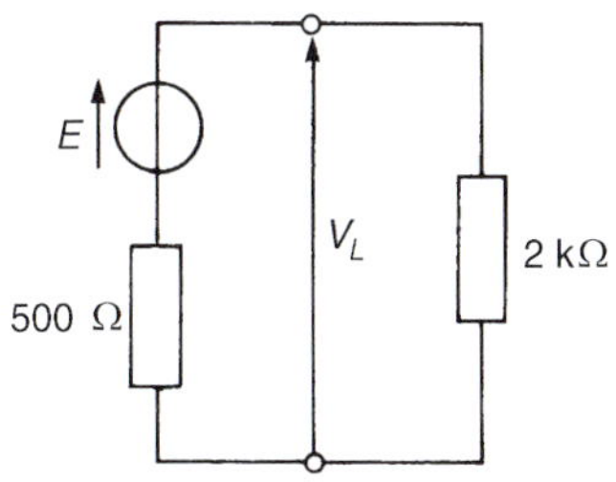

Figure 44.20

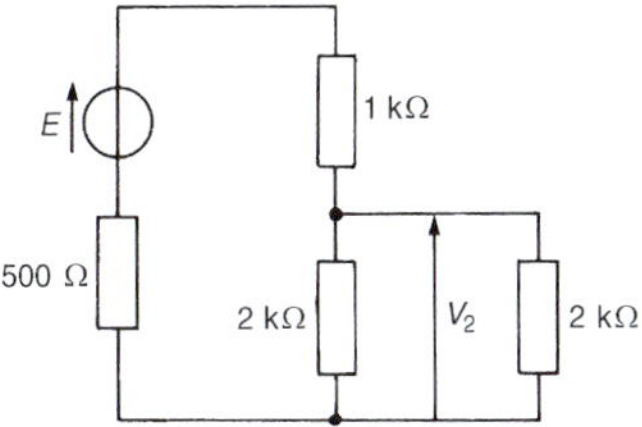

Figure 44.21

From Figure 44.22,

$$\text{voltage } V_2 = \left(\frac{1000}{1000+1000+500}\right)E = 0.4\,E$$

Hence, from equation (19), insertion loss ratio,

$$\boldsymbol{A_L} = \frac{V_L}{V_2} = \frac{0.8E}{0.4E} = \mathbf{2}$$

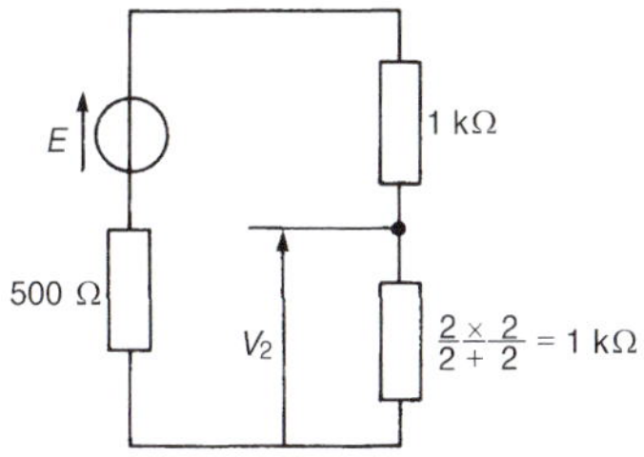

Figure 44.22

or, from equation (20),

$$\text{insertion loss} = 20\lg(V_L/V_2) = 20\lg 2$$

$$= \mathbf{6.02\,dB}$$

(b) With the 1 kΩ tapping, voltage V_2 is given by

$$V_2 = \left(\frac{\dfrac{(1000\times 2000)}{(1000+2000)}}{\dfrac{((1000\times 2000)}{(1000+2000))}+2000+500}\right)E$$

$$= \left(\frac{666.7}{666.7+2000+500}\right)E = 0.211\,E$$

Hence, from equation (19),

$$\text{insertion loss ratio } A_L = \frac{V_L}{V_2} = \frac{0.8E}{0.211E} = \mathbf{3.79}$$

or, from equation (20),

$$\text{insertion loss in decibels} = 20\lg\left(\frac{V_L}{V_2}\right)$$

$$= 20\lg 3.79 = \mathbf{11.57\,dB}$$

(Note that the insertion loss is not doubled by halving the tapping.)

Problem 12. A symmetrical π-attenuator pad has a series arm of resistance 1000 Ω and shunt arms each of 500 Ω. Determine (a) its characteristic impedance and (b) the insertion loss (in decibels) when feeding a matched load.

The π-attenuator pad is shown in Figure 44.23, terminated in its characteristic impedance, R_0

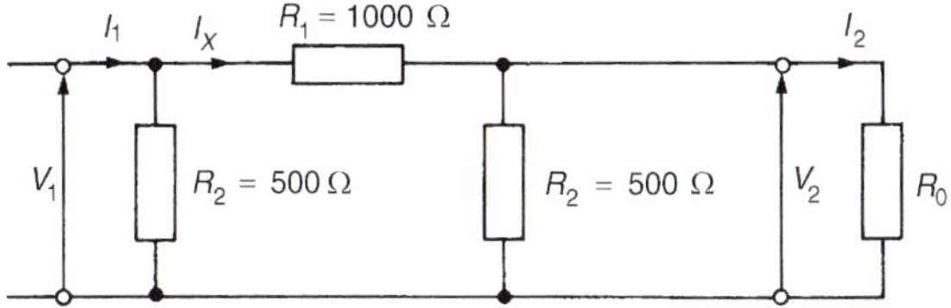

Figure 44.23

(a) From equation (15), the characteristic impedance of a symmetrical π-attenuator is given by

$$R_0 = \sqrt{\left(\frac{R_1R_2^2}{R_1+2R_2}\right)} = \sqrt{\left(\frac{(1000)(500)^2}{1000+2(500)}\right)}$$

$$= \mathbf{354\,\Omega}$$

(b) Since the attenuator network is feeding a matched load, from equation (21),

$$\text{insertion loss} = 20\lg\left(\frac{V_1}{V_2}\right)\text{ dB} = 20\lg\left(\frac{I_1}{I_2}\right)\text{ dB}$$

From Figure 44.23, by current division,

$$\text{current } I_X = \left\{\frac{R_2}{R_2+R_1+\dfrac{(R_2R_0}{(R_2+R_0))}}\right\}(I_1)$$

$$\text{and current } I_2 = \left(\frac{R_2}{R_2+R_0}\right)I_x$$

$$= \left(\frac{R_2}{R_2+R_0}\right)\left(\frac{R_2}{R_2+R_1+\dfrac{(R_2R_0}{(R_2+R_0))}}\right)I_1$$

i.e.

$$I_2=\left(\frac{500}{500+354}\right)\left(\frac{500}{500+1000+\dfrac{((500)(354)}{(500+354))}}\right)I_1$$

$$=(0.5855)(0.2929)I_1=0.1715I_1$$

Hence $I_1/I_2=1/0.1715=5.83$

Thus the insertion loss in decibels $=20\lg(I_1/I_2)$

$=20\lg 5.83$

$=\mathbf{15.3\,dB}$

Now try the following Practice Exercise

Practice Exercise 172 Insertion loss (Answers on page 832)

1. The attenuator section shown in Figure 44.24 feeds a matched load. Determine (a) the characteristic impedance R_0 and (b) the insertion loss.

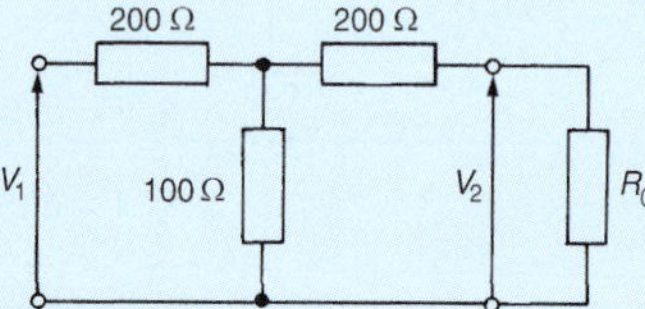

Figure 44.24

2. A 0–10 kΩ variable resistor is connected across the output of a generator of internal resistance 500 Ω. If a load of 1500 Ω is connected across the variable resistor, determine the insertion loss in decibels at a tapping of (a) 7.5 kΩ, (b) 2.5 kΩ.

3. A symmetrical π-attenuator pad has a series arm resistance of 800 Ω and shunt arms each of 250 Ω. Determine (a) the characteristic impedance of the section, and (b) the insertion loss when feeding a matched load.

44.6 Asymmetrical T- and π-sections

Figure 44.25(a) shows an asymmetrical T-pad section where resistance $R_1 \neq R_3$. Figure 44.25(b) shows an asymmetrical π-section where $R_2 \neq R_3$

When viewed from port A, in each of the sections, the output impedance is R_{OB}; when viewed from port B, the input impedance is R_{OA}. Since the sections are asymmetrical R_{OA} does not have the same value as R_{OB}

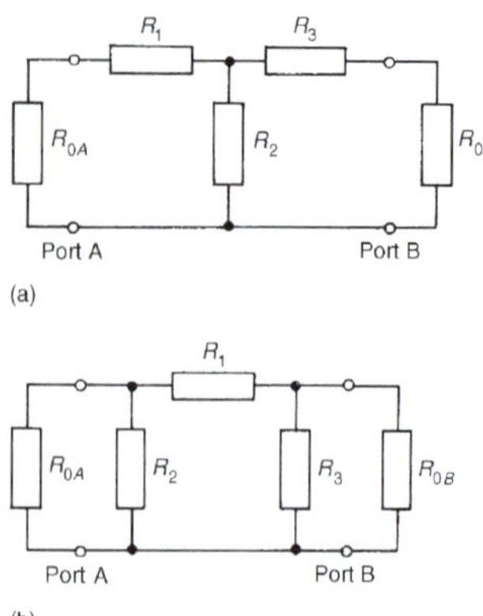

Figure 44.25 (a) Asymmetrical T-pad section, (b) asymmetrical π-section

Iterative impedance is the term used for the impedance measured at one port of a two-port network when the other port is terminated with an impedance of the same value. For example, the impedance looking into port 1 of Figure 44.26(a) is, say, 500 Ω when port 2 is terminated in 500 Ω and the impedance looking into port 2 of Figure 44.26(b) is, say, 600 Ω when port 1 is terminated in 600 Ω. (In symmetric T- and π-sections the two iterative impedances are equal, this value being the characteristic impedance of the section.)

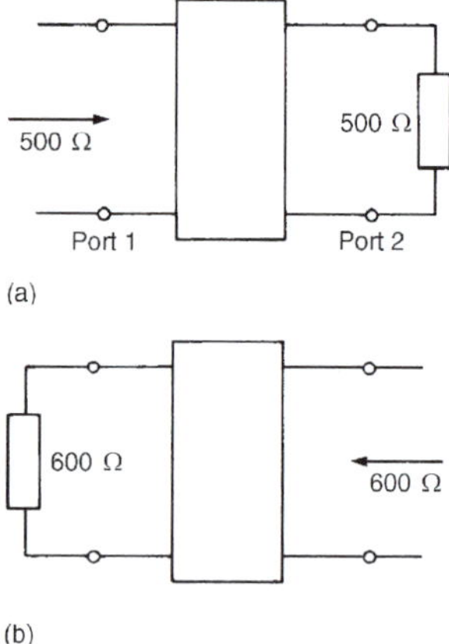

Figure 44.26

An **image impedance** is defined as the impedance which, when connected to the terminals of a network, equals the impedance presented to it at the opposite terminals. For example, the impedance looking into port 1 of Figure 44.27(a) is, say, 400 Ω when port 2 is terminated in, say 750 Ω, and the impedance seen looking into port 2 (Figure 44.27(b)) is 750 Ω when port 1 is terminated in 400 Ω. An asymmetrical network is correctly terminated when it is terminated in its image impedance. (If the image impedances are equal, the value is the characteristic impedance.)

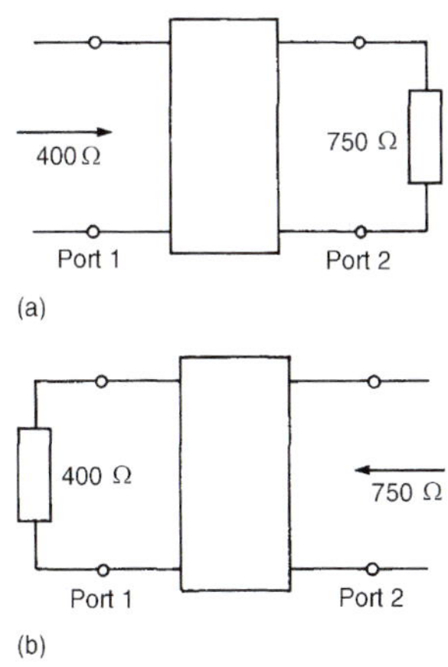

Figure 44.27

The following worked problems show how the iterative and image impedances are determined for asymmetrical T- and π-sections.

Problem 13. An asymmetrical T-section attenuator is shown in Figure 44.28. Determine for the section (a) the image impedances and (b) the iterative impedances.

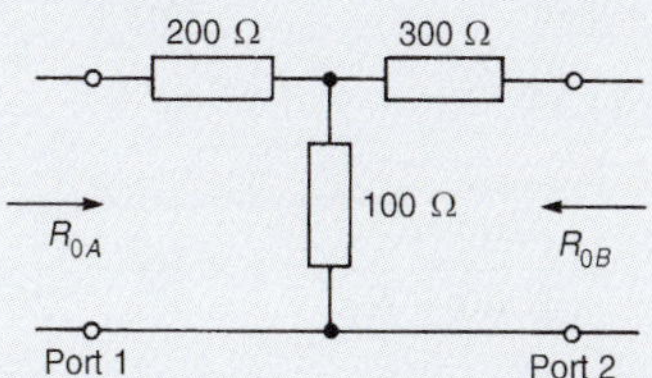

Figure 44.28

(a) The image impedance R_{OA} seen at port 1 in Figure 44.28 is given by equation (11):

$$R_{OA} = \sqrt{(R_{OC})(R_{SC})},$$

where R_{OC} and R_{SC} refer to port 2 being respectively open-circuited and short-circuited.

$$R_{OC} = 200 + 100 = 300\,\Omega$$

and $$R_{SC} = 200 + \frac{(100)(300)}{100 + 300} = 275\,\Omega$$

Hence $R_{OA} = \sqrt{[(300)(275)]} = \mathbf{287.2\,\Omega}$

Similarly, $R_{OB} = \sqrt{(R_{OC})(R_{SC})}$, where R_{OC} and R_{SC} refer to port 1 being respectively open-circuited and short-circuited.

$$R_{OC} = 300 + 100 = 400\,\Omega$$

and $$R_{SC} = 300 + \frac{(200)(100)}{200 + 100} = 366.7\,\Omega$$

Hence $R_{OB} = \sqrt{[(400)(366.7)]} = \mathbf{383\,\Omega}$

Thus the image impedances are 287.2 Ω and 383 Ω and are shown in the circuit of Figure 44.29

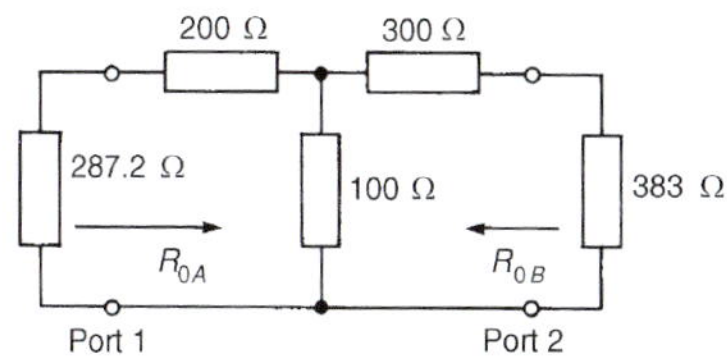

Figure 44.29

(Checking:

$$R_{OA} = 200 + \frac{(100)(300 + 383)}{100 + 300 + 383} = 287.2\,\Omega$$

and $$R_{OB} = 300 + \frac{(100)(200 + 287.2)}{100 + 200 + 287.2} = 383\,\Omega)$$

(b) The iterative impedance at port 1 in Figure 44.30 is shown as R_1. Hence

$$R_1 = 200 + \frac{(100)(300 + R_1)}{100 + 300 + R_1}$$

$$= 200 + \frac{30\,000 + 100R_1}{400 + R_1}$$

from which $400R_1 + R_1^2 = 80\,000 + 200R_1 + 30\,000 + 100R_1$

and $R_1^2 + 100R_1 - 110\,000 = 0$

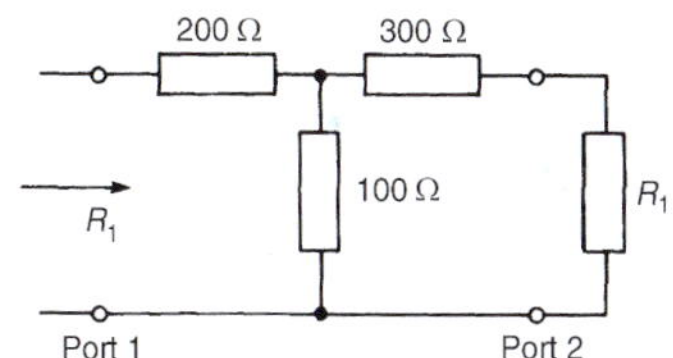

Figure 44.30

Solving by the quadratic formula gives

$$R_1 = \frac{-100 \pm \sqrt{[100^2 - (4)(1)(-110\,000)]}}{2}$$

$$= \frac{-100 \pm 670.8}{2} = \mathbf{285.4\,\Omega}$$

(neglecting the negative value).

The iterative impedance at port 2 in Figure 44.31 is shown as R_2. Hence

$$R_2 = 300 + \frac{(100)(200 + R_2)}{100 + 200 + R_2}$$

$$= 300 + \frac{20\,000 + 100R_2}{300 + R_2}$$

from which $300R_2 + R_2^2 = 90\,000 + 300R_2 + 20\,000 + 100R_2$

and $R_2^2 - 100R_2 - 110\,000 = 0$

Thus

$$R_2 = \frac{100 \pm \sqrt{[(-100)^2 - (4)(1)(-110\,000)]}}{2}$$

$$= \frac{100 \pm 670.8}{2} = \mathbf{385.4\,\Omega}$$

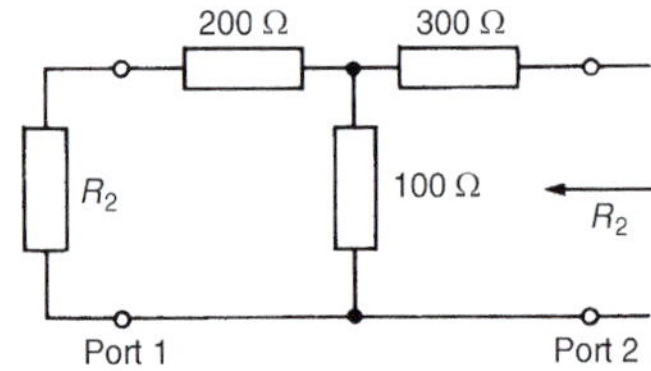

Figure 44.31

Thus the iterative impedances of the section shown in Figure 44.28 are 285.4 Ω and 385.4 Ω.

Problem 14. An asymmetrical π-section attenuator is shown in Figure 44.32. Determine for the section (a) the image impedances and (b) the iterative impedances.

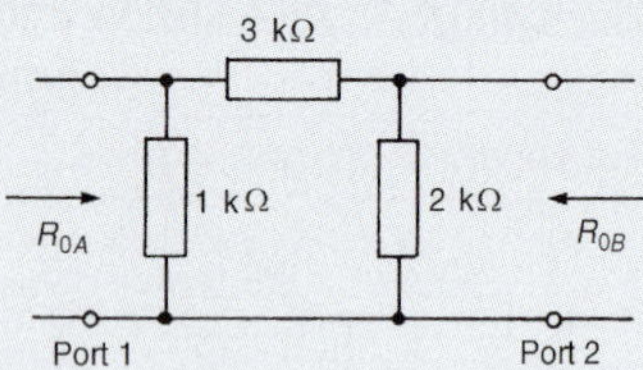

Figure 44.32

(a) The image resistance R_{OA} seen at port 1 is given by

$$R_{OA} = \sqrt{(R_{OC})(R_{SC})}$$

where the impedance at port 1 with port 2 open-circuited,

$$R_{OC} = \frac{(1000)(5000)}{1000 + 5000} = 833\,\Omega$$

and the impedance at port 1, with port 2 short-circuited,

$$R_{SC} = \frac{(1000)(3000)}{1000 + 3000} = 750\,\Omega$$

Hence $R_{OA} = \sqrt{[(833)(750)]} = \mathbf{790\,\Omega}$.
Similarly, $R_{OB} = \sqrt{(R_{OC})(R_{SC})}$, where the impedance at port 2 with port 1 open-circuited,

$$R_{OC} = \frac{(2000)(4000)}{2000 + 4000} = 1333\,\Omega$$

and the impedance at port 2 with port 1 short-circuited,

$$R_{SC} = \frac{(2000)(3000)}{2000 + 3000} = 1200\,\Omega$$

Hence $R_{OB} = \sqrt{[(1333)(1200)]} = \mathbf{1265\,\Omega}$
Thus the image impedances are 790 Ω and 1265 Ω.

(b) The iterative impedance at port 1 in Figure 44.33 is shown as R_1. From circuit theory,

$$R_1 = \frac{1000[3000 + (2000R_1/(2000 + R_1))]}{1000 + 3000 + (2000R_1/(2000 + R_1))}$$

i.e. $$R_1 = \frac{3 \times 10^6 + (2 \times 10^6 R_1/(2000 + R_1))}{4000 + (2000R_1/(2000 + R_1))}$$

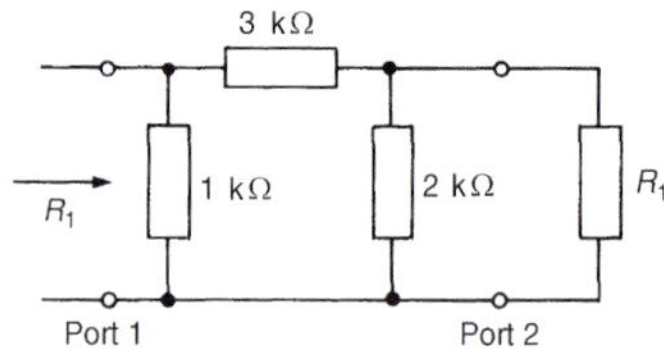

Figure 44.33

$$4000R_1 + \frac{2000R_1^2}{2000 + R_1} = 3 \times 10^6 + \frac{2 \times 10^6 R_1}{2000 + R_1}$$

$$8 \times 10^6 R_1 + 4000R_1^2 + 2000R_1^2 = 6 \times 10^9 + 3 \times 10^6 R_1 + 2 \times 10^6 R_1$$

$$6000R_1^2 + 3 \times 10^6 R_1 - 6 \times 10^9 = 0$$

$$2R_1^2 + 1000R_1 - 2 \times 10^6 = 0$$

Using the quadratic formula gives

$$R_1 = \frac{-1000 \pm \sqrt{[(1000)^2 - (4)(2)(-2 \times 10^6)]}}{4}$$

$$= \frac{-1000 \pm 4123}{4} = \mathbf{781\,\Omega}$$

(neglecting the negative value).
The iterative impedance at port 2 in Figure 44.34 is shown as R_2

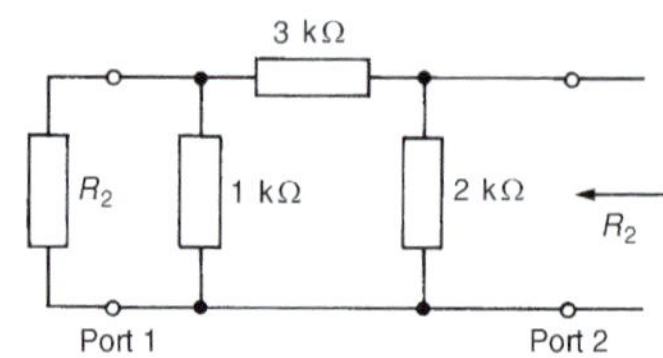

Figure 44.34

Part 4

$$R_2 = \frac{2000[3000 + (1000R_2/(1000 + R_2))]}{2000 + 3000 + (1000R_2/(1000 + R_2))}$$

$$= \frac{6 \times 10^6 + (2 \times 10^6 R_2/(1000 + R_2))}{5000 + (1000R_2/(1000 + R_2))}$$

Hence

$$5000R_2 + \frac{1000R_2^2}{1000 + R_2} = 6 \times 10^6 + \frac{2 \times 10^6 R_2}{1000 + R_2}$$

$$5 \times 10^6 R_2 + 5000R_2^2 + 1000R_2^2$$
$$= 6 \times 10^9 + 6 \times 10^6 R_2 + 2 \times 10^6 R_2$$

$$6000R_2^2 - 3 \times 10^6 R_2 - 6 \times 10^9 = 0$$
$$2R_2^2 - 1000R_2 - 2 \times 10^6 = 0$$

from which

$$R_2 = \frac{1000 \pm \sqrt{[(-1000)^2 - (4)(2)(-2 \times 10^6)]}}{4}$$

$$= \frac{1000 \pm 4123}{4} = \mathbf{1281\,\Omega}$$

Thus the iterative impedances of the section shown in Figure 44.32 are 781 Ω and 1281 Ω.

Now try the following Practice Exercise

Practice Exercise 173 Asymmetrical T- and π-attenuators (Answers on page 832)

1. An asymmetric section is shown in Figure 44.35. Determine for the section (a) the image impedances and (b) the iterative impedances.

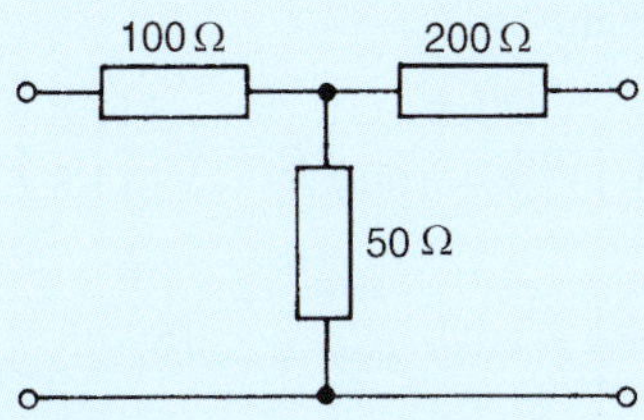

Figure 44.35

2. An asymmetric π-section is shown in Figure 44.36. Determine for the section (a) the image impedances and (b) the iterative impedances.

3. Distinguish between image and iterative impedances of a network. An asymmetric T-attenuator section has series arms of resistance 200 Ω and 400 Ω respectively, and a shunt arm of resistance 300 Ω. Determine the image and iterative impedances of the section.

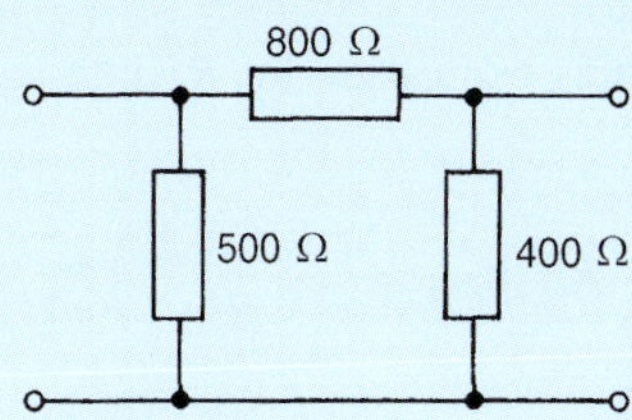

Figure 44.36

44.7 The L-section attenuator

A typical L-section attenuator pad is shown in Figure 44.37. Such a pad is used for matching purposes only, the design being such that the attenuation introduced is a minimum. In order to derive values for R_1 and R_2, consider the resistances seen from either end of the section.

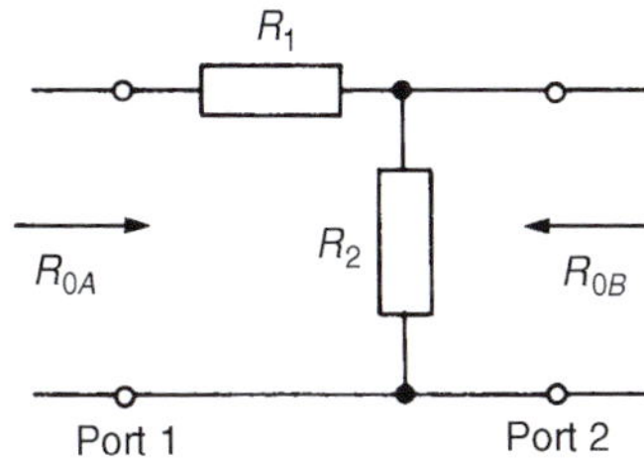

Figure 44.37 L-section attenuator pad

Looking in at port 1,

$$R_{OA} = R_1 + \frac{R_2 R_{OB}}{R_2 + R_{OB}}$$

from which

$$R_{OA}R_2 + R_{OA}R_{OB} = R_1R_2 + R_1R_{OB} + R_2R_{OB} \quad (22)$$

Looking in at port 2,

$$R_{OB} = \frac{R_2(R_1 + R_{OA})}{R_1 + R_{OA} + R_2}$$

from which

$$R_{OB}R_1 + R_{OA}R_{OB} + R_{OB}R_2 = R_1R_2 + R_2R_{OA} \quad (23)$$

Adding equations (22) and (23) gives

$$R_{OA}R_2 + 2R_{OA}R_{OB} + R_{OB}R_1 + R_{OB}R_2$$
$$= 2R_1R_2 + R_1R_{OB} + R_2R_{OB} + R_2R_{OA}$$

i.e. $2R_{OA}R_{OB}=2R_1R_2$

and $$R_1=\frac{R_{OA}R_{OB}}{R_2} \quad (24)$$

Substituting this expression for R_1 into equation (22) gives

$$R_{OA}R_2+R_{OA}R_{OB}=\left(\frac{R_{OA}R_{OB}}{R_2}\right)R_2 + \left(\frac{R_{OA}R_{OB}}{R_2}\right)R_{OB}+R_2R_{OB}$$

i.e. $$R_{OA}R_2+R_{OA}R_{OB}=R_{OA}R_{OB} + \frac{R_{OA}R_{OB}^2}{R_2}+R_2R_{OB}$$

from which $$R_2(R_{OA}-R_{OB})=\frac{R_{OA}R_{OB}^2}{R_2}$$

$$R_2^2(R_{OA}-R_{OB})=R_{OA}R_{OB}^2$$

and resistance, $$\boldsymbol{R_2=\sqrt{\left(\frac{R_{OA}R_{OB}^2}{R_{OA}-R_{OB}}\right)}} \quad (25)$$

Thus, from equation (24),

$$R_1=\frac{R_{OA}R_{OB}}{\sqrt{(R_{OA}R_{OB}^2/(R_{OA}-R_{OB}))}} = \frac{R_{OA}R_{OB}}{R_{OB}\sqrt{(R_{OA}/(R_{OA}-R_{OB}))}} = \frac{R_{OA}}{\sqrt{R_{OA}}}\sqrt{(R_{OA}-R_{OB})}$$

Hence resistance, $$\boldsymbol{R_1=\sqrt{[R_{OA}(R_{OA}-R_{OB})]}} \quad (26)$$

Figure 44.38 shows an L-section attenuator pad with its resistances expressed in terms of the input and output resistances, R_{OA} and R_{OB}

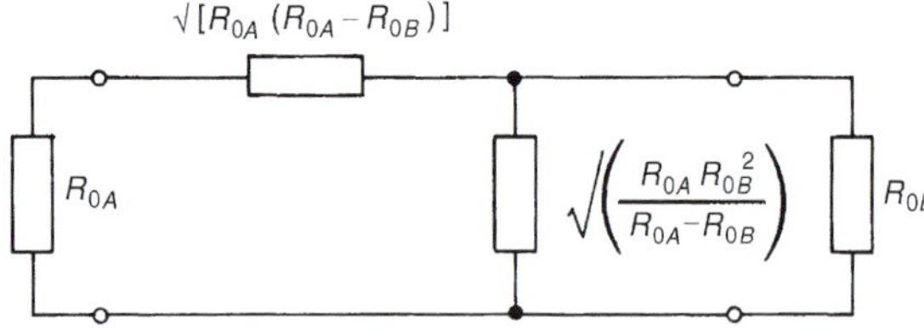

Figure 44.38

Problem 15. A generator having an internal resistance of 500 Ω is connected to a 100 Ω load via an impedance-matching resistance pad as shown in Figure 44.39. Determine (a) the values of resistance R_1 and R_2, (b) the attenuation of the pad in decibels and (c) its insertion loss.

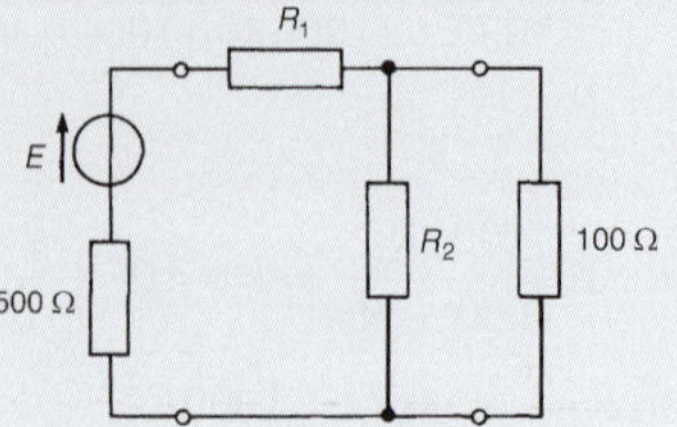

Figure 44.39

(a) From equation (26), $R_1=\sqrt{[500(500-100)]}$

$$=\mathbf{447.2\ \Omega}$$

From equation (25), $R_2=\sqrt{\left(\frac{(500)(100)^2}{500-100}\right)}$

$$=\mathbf{111.8\ \Omega}$$

(b) From Section 44.3, the attenuation is given by $10\lg(P_1/P_2)$ dB. Note that, for an asymmetrical section such as that shown in Figure 44.39, the expression $20\lg(V_1/V_2)$ or $20\lg(I_1/I_2)$ may **not** be used for attenuation since the terminals of the pad are not matched to equal impedances. In Figure 44.40,

$$\text{current } I_1=\frac{E}{500+447.2+\frac{(111.8\times 100}{(111.8+100))}} = \frac{E}{1000}$$

and current

$$I_2=\left(\frac{111.8}{111.8+100}\right)I_1=\left(\frac{111.8}{211.8}\right)\left(\frac{E}{1000}\right) = \frac{E}{1894.5}$$

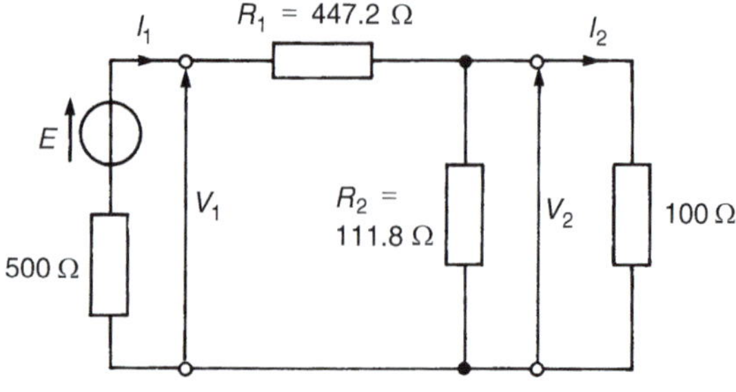

Figure 44.40

Part 4

Thus input power,

$$P_1 = I_1^2(500) = \left(\frac{E}{1000}\right)^2(500)$$

and output power,

$$P_2 = I_2^2(100) = \left(\frac{E}{1894.5}\right)^2(100)$$

Hence

$$\text{attenuation} = 10\lg\frac{P_1}{P_2}$$

$$= 10\lg\left\{\frac{[E/(1000)]^2(500)}{[E/(1894.5)]^2(100)}\right\}$$

$$= 10\lg\left\{\left(\frac{1894.5}{1000}\right)^2(5)\right\}\text{dB}$$

i.e. **attenuation = 12.54 dB**

(c) Insertion loss A_L is defined as

$$\frac{\text{voltage across load when connected directly to the generator}}{\text{voltage across load when the two-port network is connected}}$$

Figure 44.41 shows the generator connected directly to the load.

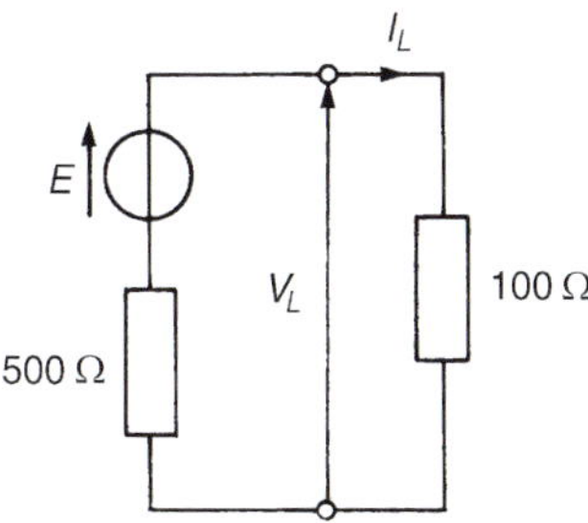

Figure 44.41

$$\text{Load current, } I_L = \frac{E}{500+100} = \frac{E}{600}$$

$$\text{and voltage, } V_L = I_L(100) = \frac{E}{600}(100) = \frac{E}{6}$$

From Figure 44.40 voltage,

$V_1 = E - I_1(500) = E - (E/1000)500$ from part (b)

i.e. $V_1 = 0.5\,E$

voltage, $V_2 = V_1 - I_1R_1$

$$= 0.5\,E - \left(\frac{E}{1000}\right)(447.2) = 0.0528\,E$$

$$\textbf{insertion loss, } A_L = \frac{V_L}{V_2} = \frac{E/6}{0.0528E} = \mathbf{3.157}$$

$$\textbf{In decibels, the insertion loss} = 20\lg\frac{V_L}{V_2}$$

$$= 20\lg 3.157 = \mathbf{9.99\,dB}$$

Now try the following Practice Exercise

Practice Exercise 174 L-section attenuators (Answers on page 832)

1. Figure 44.42 shows an L-section attenuator. The resistance across the input terminals is 250 Ω and the resistance across the output terminals is 100 Ω. Determine the values R_1 and R_2

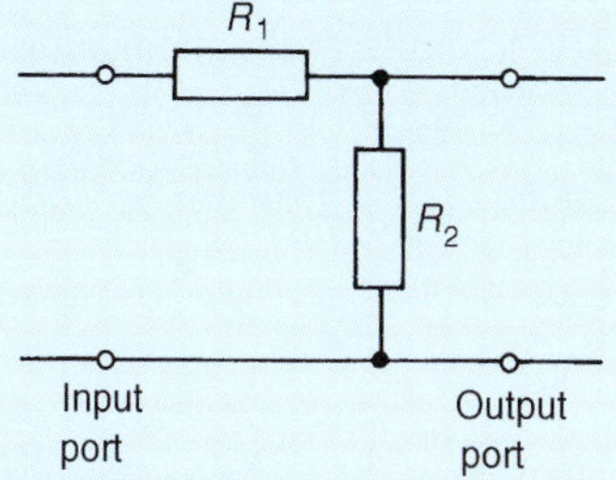

Figure 44.42

2. A generator having an internal resistance of 600 Ω is connected to a 200 Ω load via an impedance-matching resistive pad as shown in Figure 44.43. Determine (a) the values of resistances R_1 and R_2, (b) the attenuation of the matching pad and (c) its insertion loss.

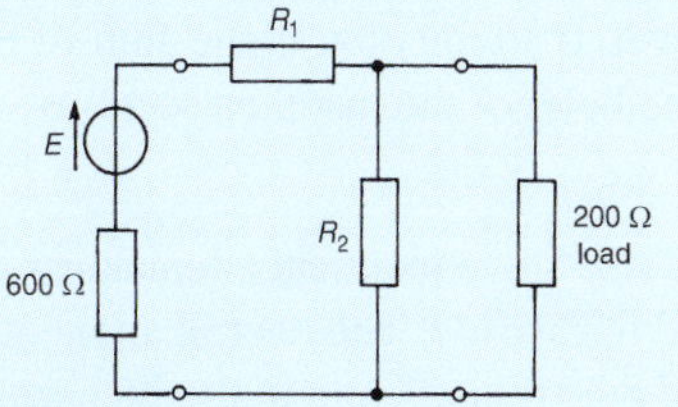

Figure 44.43

44.8 Two-port networks in cascade

Often two-port networks are connected in cascade, i.e. the output from the first network becomes the input to the second network, and so on, as shown in Figure 44.44. Thus an attenuator may consist of several cascaded

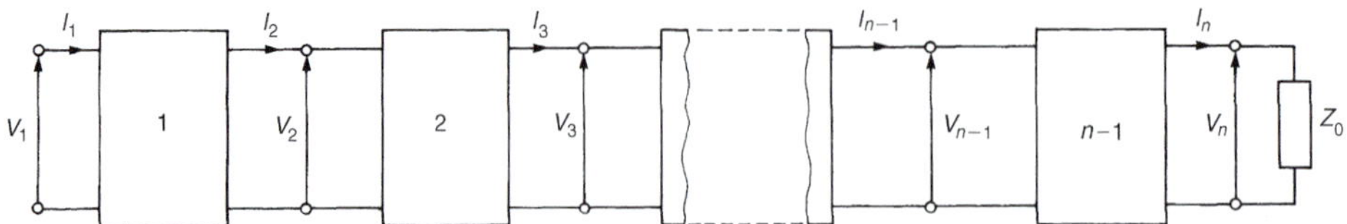

Figure 44.44 Two-port networks connected in cascade

sections so as to achieve a particular desired overall performance.

If the cascade is arranged so that the impedance measured at one port and the impedance with which the other port is terminated have the same value, then each section (assuming they are symmetrical) will have the same characteristic impedance Z_0 and the last network will be terminated in Z_0. Thus each network will have a matched termination and hence the attenuation in decibels of section 1 in Figure 44.44 is given by $a_1 = 20\lg(V_1/V_2)$. Similarly, the attenuation of section 2 is given by $a_2 = 20\lg(V_2/V_3)$, and so on.

The overall attenuation is given by

$$a = 20\frac{V_1}{V_n}$$

$$= 20\lg\left(\frac{V_1}{V_2} \times \frac{V_2}{V_3} \times \frac{V_3}{V_4} \times \cdots \times \frac{V_{n-1}}{V_n}\right)$$

$$= 20\lg\frac{V_1}{V_2} + 20\lg\frac{V_2}{V_3} + \cdots + 20\lg\frac{V_{n-1}}{V_n}$$

by the laws of logarithms, i.e.

overall attenuation, $a = a_1 + a_2 + \cdots + a_{n-1}$ (27)

Thus the overall attenuation is the sum of the attenuations (in decibels) of the matched sections.

Problem 16. Five identical attenuator sections are connected in cascade. The overall attenuation is 70 dB and the voltage input to the first section is 20 mV. Determine (a) the attenuation of each individual attenuation section, (b) the voltage output of the final stage and (c) the voltage output of the third stage.

(a) From equation (27), the overall attenuation is equal to the sum of the attenuations of the individual sections and, since in this case each section is identical, **the attenuation of each section** = 70/5 = **14 dB**.

(b) If V_1 = the input voltage to the first stage and V_0 = the output of the final stage, then the overall attenuation = $20\lg(V_1/V_0)$, i.e.

$$70 = 20\lg\left(\frac{20}{V_0}\right) \text{ where } V_0 \text{ is in millivolts}$$

$$3.5 = \lg\left(\frac{20}{V_0}\right)$$

$$10^{3.5} = \frac{20}{V_0}$$

from which **output voltage of final stage,**

$$\boldsymbol{V_0} = \frac{20}{10^{3.5}} = 6.32 \times 10^{-3}\,\text{mV}$$

$$= \mathbf{6.32\,\mu V}$$

(c) The overall attenuation of three identical stages is $3 \times 14 = 42$ dB. Hence $42 = 20\lg(V_1/V_3)$, where V_3 is the voltage output of the third stage. Thus

$$\frac{42}{20} = \lg\left(\frac{20}{V_3}\right) \text{ and } 10^{42/20} = \frac{20}{V_3}$$

from which **the voltage output of the third stage,**

$$\boldsymbol{V_3} = 20/10^{2.1} = \mathbf{0.159\,mV}$$

Problem 17. A d.c. generator has an internal resistance of 450 Ω and supplies a 450 Ω load.

(a) Design a T-network attenuator pad having a characteristic impedance of 450 Ω which, when connected between the generator and the load, will reduce the load current to $\frac{1}{8}$ of its initial value.

(b) If two such networks as designed in (a) were connected in series between the generator and the load, determine the fraction of the initial current that would now flow in the load.

(c) Determine the attenuation in decibels given by four such sections as designed in (a).

The T-network attenuator is shown in Figure 44.45 connected between the generator and the load. Since it is matching equal impedances, the network is symmetrical.

(a) Since the load current is to be reduced to $\frac{1}{8}$ of its initial value, the attenuation $N = 8$

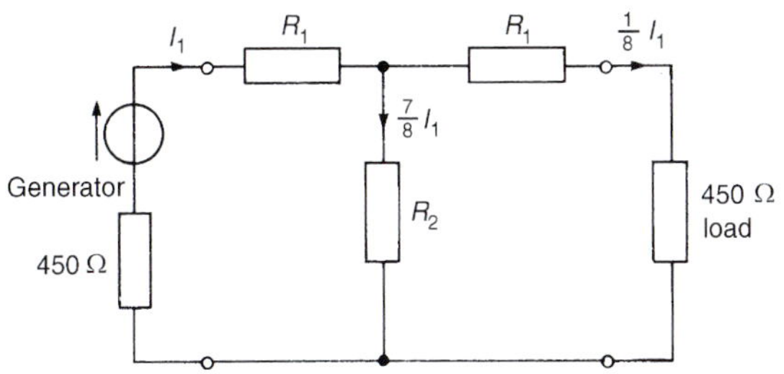

Figure 44.45

From equation (13),

$$\text{resistance, } \boldsymbol{R_1} = \frac{R_0(N-1)}{(N+1)} = 450\frac{(8-1)}{(8+1)}$$

$$= \mathbf{350\,\Omega}$$

and from equation (14),

$$\text{resistance, } \boldsymbol{R_2} = R_0\left(\frac{2N}{N^2-1}\right) = 450\left(\frac{2\times 8}{8^2-1}\right)$$

$$= \mathbf{114\,\Omega}$$

(b) When two such networks are connected in series, as shown in Figure 44.46, current I_1 flows into the first stage and $\frac{1}{8}I_1$ flows out of the first stage into the second.

Again, $\frac{1}{8}$ of this current flows out of the second stage,

i.e. $\frac{1}{8}\times\frac{1}{8}I_1$, i.e. $\frac{1}{64}$ of I_1 flows into the load.

Thus $\frac{1}{64}$ of the original current flows in the load.

(c) The attenuation of a single stage is 8. Expressed in decibels, the attenuation is
$20\lg(I_1/I_2) = 20\lg 8 = 18.06$ dB
From equation (27), the overall attenuation of four identical stages is given by $18.06 + 18.06 + 18.06 + 18.06$, i.e. **72.24 dB**

Now try the following Practice Exercise

Practice Exercise 175 Cascading two-port networks (Answers on page 832)

1. The input to an attenuator is 24 V and the output is 4 V. Determine the attenuation in decibels. If five such identical attenuators are cascaded, determine the overall attenuation.
2. Four identical attenuator sections are connected in cascade. The overall attenuation is 60 dB. The input to the first section is 50 mV. Determine (a) the attenuation of each stage, (b) the output of the final stage and (c) the output of the second stage.
3. A d.c. generator has an internal resistance of 300 Ω and supplies a 300 Ω load.
   (a) Design a symmetrical T network attenuator pad having a characteristic impedance of 300 Ω which, when connected between the generator and the load, will reduce the load current to $\frac{1}{3}$ its initial value.
   (b) If two such networks as in (a) were connected in series between the generator and the load, what fraction of the initial current would the load take?
   (c) Determine the fraction of the initial current that the load would take if six such networks were cascaded between the generator and the load.
   (d) Determine the attenuation in decibels provided by five such identical stages as in (a).

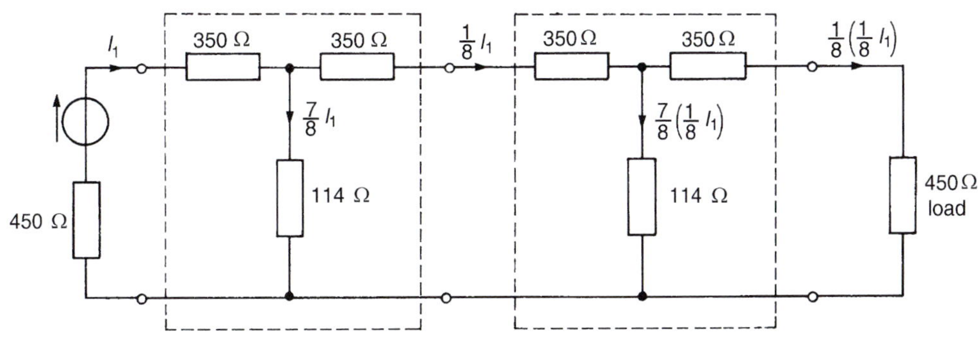

Figure 44.46

44.9 *ABCD* parameters

As mentioned earlier, a **two-port network** has a pair of input terminals, shown as *PQ* in Figure 44.47, and a pair of output terminals, shown as *RS*. When a voltage V_1 is applied to terminals *PQ*, the input and output currents, I_1 and I_2, flow and an output voltage V_2 is produced. There are therefore **four variable quantities** V_1, I_1, V_2 and I_2 for a two-port network. If the elements are assumed to be linear, then there are a number of ways in which the relationships can be written. One such method is termed ***ABCD*** **parameters** which we consider in this chapter; there are others, however, such as *z*-, *y*-, *h*- and *g*-parameters that are not considered here.

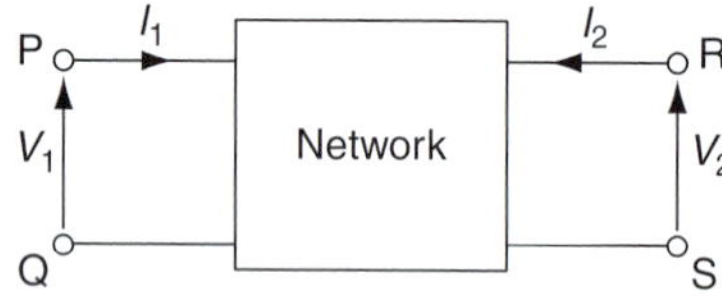

Figure 44.47

For *ABCD* parameters, the input voltage V_1 and the input current I_1 are specified in terms of the output voltage V_2 and current I_2 as follows:

$$V_1 = AV_2 - BI_2 \qquad (28)$$
$$I_1 = CV_2 - DI_2 \qquad (29)$$

A, B, C and D are constants for a particular network and called **parameters**.

In equation (28), when $I_2 = 0$, then: $A = \dfrac{V_1}{V_2}$

and when $V_2 = 0$, then: $B = -\dfrac{V_1}{I_2}$

In equation (29), when $I_2 = 0$, then: $C = \dfrac{I_1}{V_2}$

and when $V_2 = 0$, then: $D = -\dfrac{I_1}{I_2}$

Since A and D are ratios, they have no units, while B is an impedance in ohms, and C an admittance in siemens.

These parameters are often called the **general circuit** or **transmission parameters** and are generally used for the analysis of two-port networks which are heavy current circuits or power frequency transmission lines.

Problem 18. A two-port network has parameters $A=2+j$, $B=5\,\Omega$, $C=(1+j2)S$ and $D=4$. If the output is a current of 80 mA through a 20 Ω resistive load, determine (a) the input voltage and (b) the input current.

Since $I_2 = 80$ mA and $R_2 = 20\,\Omega$ then
$V_2 = I_2R_2 = 80\times10^{-3}\times20 = 1.6\,\text{V}$

(a) From equation (28), voltage,

$$\begin{aligned}V_1 &= AV_2 - BI_2\\ &= (2+j)(1.6) - (5)(80\times10^{-3})\\ &= 3.2 + j1.6 - 0.4 = 2.8 + j1.6\end{aligned}$$

i.e. $\boldsymbol{V_1 = 3.22\angle 29.74°\,\text{V}}$

(b) From equation (29), current,

$$\begin{aligned}I_1 &= CV_2 - DI_2\\ &= (1+j2)(1.6) - (4)(80\times10^{-3})\\ &= 1.6 + j3.2 - 0.32 = 1.28 + j3.2\end{aligned}$$

i.e. $\boldsymbol{I_1 = 3.45\angle 68.20°\,\text{A}}$

Problem 19. A two-port network has the following transmission parameters: $A=5$, $B=25\,\Omega$, $C=0.04S$ and $D=2$. Determine the values of the input impedance when the output is (a) open-circuited, (b) short-circuited.

(a) From equation (28), voltage, $V_1 = AV_2 - BI_2$

When the output is open-circuited, $I_2 = 0$ and $V_1 = AV_2$

From equation (29), current, $I_1 = CV_2 - DI_2$

When $I_2 = 0$, $I_1 = CV_2$

Thus, **input impedance**,

$$Z_{1oc} = \frac{V_1}{I_1} = \frac{AV_2}{CV_2} = \frac{A}{C} = \frac{5}{0.04} = \mathbf{125\,\Omega}$$

(b) From equation (28), voltage, $V_1 = AV_2 - BI_2$

When the output is short-circuited, $V_2 = 0$ and $V_1 = -BI_2$

From equation (29), current, $I_1 = CV_2 - DI_2$

When $V_2 = 0$, $I_1 = -DI_2$

Thus, **input impedance**,

$$Z_{1sc} = \frac{V_1}{I_1} = \frac{-BI_2}{-DI_2} = \frac{B}{D} = \frac{25}{2} = \mathbf{12.5\,\Omega}$$

Transmission matrix

Since $V_1 = AV_2 - BI_2$
and $I_1 = CV_2 - DI_2$

then in matrix form:

$$\begin{pmatrix} V_1 \\ I_1 \end{pmatrix} = \begin{pmatrix} A & B \\ C & D \end{pmatrix} \begin{pmatrix} V_2 \\ -I_2 \end{pmatrix}$$

$\begin{pmatrix} A & B \\ C & D \end{pmatrix}$ is called the ***ABCD* or transmission matrix**.

(For more on matrices and solving simultaneous equations using matrices see *Higher Engineering Mathematics 8th Edition*).

Problem 20. For a two-port network:

$$V_1 = 5V_2 - 3I_2$$
$$\text{and} \quad I_1 = 4V_2 - 2I_2$$

When $V_1 = 11\,\text{V}$ and $I_1 = 12\,\text{A}$, determine (a) the transmission matrix and (b) the values of V_2 and I_2

(a) The transmission matrix is: $\begin{pmatrix} \mathbf{5} & \mathbf{3} \\ \mathbf{4} & \mathbf{2} \end{pmatrix}$

(b) Since $V_1 = 11\,\text{V}$ and $I_1 = 12\,\text{A}$, then

$11 = 5V_2 - 3I_2$
$12 = 4V_2 - 2I_2$

In matrix form this becomes:

$$\begin{pmatrix} 11 \\ 12 \end{pmatrix} = \begin{pmatrix} 5 & 3 \\ 4 & 2 \end{pmatrix} \begin{pmatrix} V_2 \\ -I_2 \end{pmatrix} \quad (30)$$

The inverse of a matrix $\begin{pmatrix} P & Q \\ R & S \end{pmatrix}$ is:

$$\frac{1}{PS - RQ} \begin{pmatrix} S & -Q \\ -R & P \end{pmatrix}$$

Hence the inverse of $\begin{pmatrix} 5 & 3 \\ 4 & 2 \end{pmatrix}$ is:

$$\frac{1}{10-12} \begin{pmatrix} 2 & -3 \\ -4 & 5 \end{pmatrix} = \begin{pmatrix} -1 & 1.5 \\ 2 & -2.5 \end{pmatrix}$$

Multiplying both sides of equation (30) by this inverse matrix gives:

$$\begin{pmatrix} -1 & 1.5 \\ 2 & -2.5 \end{pmatrix} \begin{pmatrix} 11 \\ 12 \end{pmatrix} = \begin{pmatrix} -1 & 1.5 \\ 2 & -2.5 \end{pmatrix} \begin{pmatrix} 5 & 3 \\ 4 & 2 \end{pmatrix} \begin{pmatrix} V_2 \\ -I_2 \end{pmatrix}$$

i.e. $$\begin{pmatrix} (-11+18) \\ (22-30) \end{pmatrix} = \begin{pmatrix} (-5+6) & (-3+3) \\ (10-10) & (6-5) \end{pmatrix} \begin{pmatrix} V_2 \\ -I_2 \end{pmatrix}$$

i.e. $$\begin{pmatrix} 7 \\ -8 \end{pmatrix} = \begin{pmatrix} 1 & 0 \\ 0 & 1 \end{pmatrix} \begin{pmatrix} V_2 \\ -I_2 \end{pmatrix}$$

i.e. $$\begin{pmatrix} 7 \\ -8 \end{pmatrix} = \begin{pmatrix} V_2 \\ -I_2 \end{pmatrix}$$

from which $\boldsymbol{V_2 = 7\,\text{V}}$ and $\boldsymbol{I_2 = 8\,\text{A}}$

ABCD networks in cascade

Figure 44.48 shows two networks in cascade,

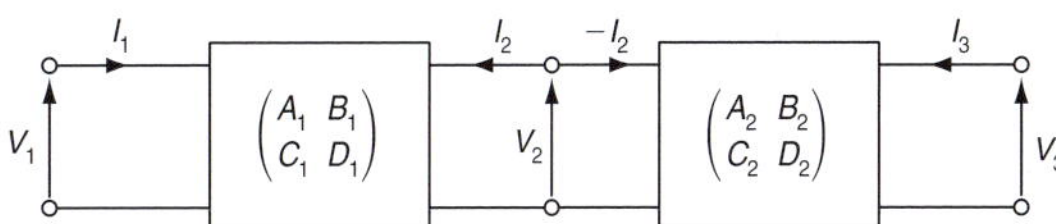

Figure 44.48

where for network 1,

$$\begin{pmatrix} V_1 \\ I_1 \end{pmatrix} = \begin{pmatrix} A_1 & B_1 \\ C_1 & D_1 \end{pmatrix} \begin{pmatrix} V_2 \\ -I_2 \end{pmatrix}$$

and for network 2,

$$\begin{pmatrix} V_2 \\ -I_2 \end{pmatrix} = \begin{pmatrix} A_2 & B_2 \\ C_2 & D_2 \end{pmatrix} \begin{pmatrix} V_3 \\ -I_3 \end{pmatrix}$$

Hence,

$$\begin{pmatrix} V_1 \\ I_1 \end{pmatrix} = \begin{pmatrix} A_1 & B_1 \\ C_1 & D_1 \end{pmatrix} \begin{pmatrix} A_2 & B_2 \\ C_2 & D_2 \end{pmatrix} \begin{pmatrix} V_3 \\ -I_3 \end{pmatrix}$$
$$= \begin{pmatrix} (A_1A_2 + B_1C_2) & (A_1B_2 + B_1D_2) \\ (C_1A_2 + D_1C_2) & (C_1B_2 + D_1D_2) \end{pmatrix} \begin{pmatrix} V_3 \\ -I_3 \end{pmatrix}$$

Thus the cascaded network behaves as a network with the parameters:

$$A = A_1A_2 + B_1C_2 \qquad B = A_1B_2 + B_1D_2$$
$$C = C_1A_2 + D_1C_2 \quad \text{and} \quad D = C_1B_2 + D_1D_2$$

Problem 21. A network has the transmission parameters of:

$$\begin{pmatrix} (2+j) & 200 \\ 0.002j & 1 \end{pmatrix}$$

Determine the parameters for two such networks in cascade.

Part 4

The parameters for two such networks in cascade will be

$$\begin{pmatrix}(2+j) & 200\\ 0.002j & 1\end{pmatrix}\begin{pmatrix}(2+j) & 200\\ 0.002j & 1\end{pmatrix}$$

$$=\begin{pmatrix}[(2+j)^2+200(0.002j)] & [(2+j)(200)+200]\\ [0.002j(2+j)+(1)(0.002j)] & [(0.002j)(200)+(1)(1)]\end{pmatrix}$$

$$=\begin{pmatrix}(4+j2+j^2+j0.4) & (400+j200+200)\\ (0.004j+j^2 0.002+0.002j) & (0.4j+1)\end{pmatrix}$$

$$=\begin{pmatrix}\mathbf{(3+j2.4)} & \mathbf{(600+j200)}\\ \mathbf{(-0.002+j0.006)} & \mathbf{(1+j0.4)}\end{pmatrix}$$

Passive networks and reciprocity theorem

As stated earlier, the $ABCD$ parameters for a two-port network are:

$$V_1 = AV_2 - BI_2 \quad (28)$$

$$I_1 = CV_2 - DI_2 \quad (29)$$

Figure 44.49(a) represents a two-port network whose terminals RS are short-circuited with an input voltage V across terminals PQ. Then V_2 in the above equations is zero and the equations become:

$$V = -BI_2 \quad (31)$$

and

$$I_1 = -DI_2 \quad (32)$$

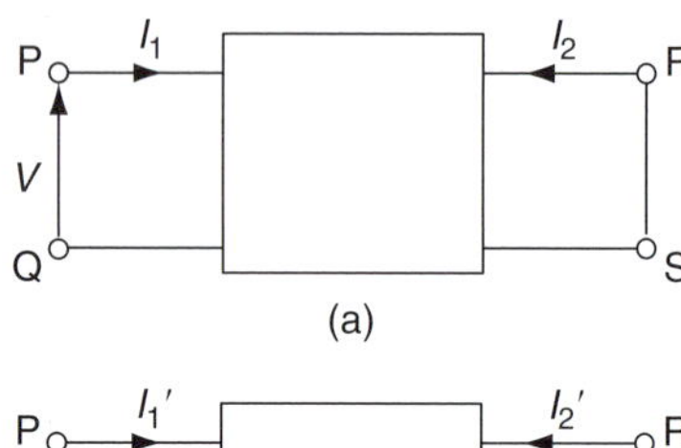

(a)

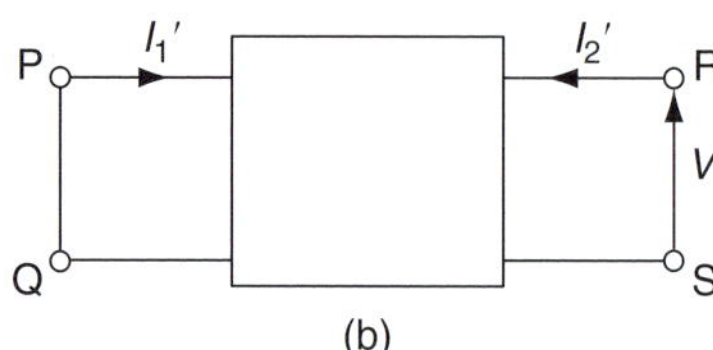

(b)

Figure 44.49

Figure 44.49(b) represents a two-port network whose terminals PQ are short-circuited with an input voltage V now applied across terminals RS, then V_1 in equation (28) is zero and the above equations become:

$$0 = AV - BI_2' \quad (33)$$

$$I_1' = CV - DI_2' \quad (34)$$

From equation (33),

$$I_2' = \left(\frac{A}{B}\right)V$$

Substituting in equation (34) gives:

$$I_1' = CV - D\left(\frac{A}{B}\right)V = \frac{(BC-AD)V}{B}$$

Substituting $V = -BI_2$ from equation (31) gives:

$$I_1' = \frac{(BC-AD)(-BI_2)}{B} = (BC-AD)(-I_2)$$

i.e. $\mathbf{I_1' = (AD-BC)I_2}$

This is a theorem called the **reciprocity theorem**, which states that if a voltage is applied to a linear passive network at one terminal and produces a current at another, then the same voltage applied to a second terminal will produce the same current at the first terminal.
Since the same voltage has been applied at each of the two terminals, then $I_1' = I_2$

Thus, $\mathbf{AD - BC = 1}$ (35)

This condition must be satisfied by any linear passive network.

Problem 22. Tests on a passive two-port network gave the following results:
Output open-circuited:
$V_1 = 20\,V$, $V_2 = 10\,V$, $I_1 = 0.5\,A$
Output short-circuited:
$V_1 = 20\,V$, $I_1 = 1\,A$, $I_2 = 1\,A$
(a) Determine the values of the $ABCD$ parameters and (b) confirm that the network is passive.

(a) From equation (28),

$$V_1 = AV_2 - BI_2$$

When the output is open-circuit,

$$V_1 = AV_2 \quad \text{since } I_2 = 0$$

Since $V_1 = 20\,V$ and $V_2 = 10\,V$, then $20 = A(10)$

from which, $\mathbf{A = 2}$

From equation (29),

$$I_1 = CV_2 - DI_2$$

When the output is open-circuit,

$$I_1 = CV_2 \quad \text{since } I_2 = 0$$

Since $V_2 = 10\,\text{V}$ and $I_1 = 0.5\,\text{A}$, then $0.5 = C(10)$

from which, $\boldsymbol{C} = \dfrac{0.5}{10} = \mathbf{0.05\,S}$

From equation (28),

$$V_1 = AV_2 - BI_2$$

When the output is short-circuit,

$$V_1 = -BI_2 \quad \text{since } V_2 = 0$$

Since $V_1 = 20\,\text{V}$ and $I_2 = -1\,\text{A}$, then $20 = -B(-1)$

from which, $\boldsymbol{B} = \mathbf{20\,\Omega}$

From equation (29),

$$I_1 = CV_2 - DI_2$$

When the output is short-circuit,

$$I_1 = -DI_2 \quad \text{since } V_2 = 0$$

Since $I_1 = 1\,\text{A}$ and $I_2 = -1\,\text{A}$, then $1 = -D(-1)$

from which, $\boldsymbol{D} = \mathbf{1}$

(b) For passive network, $AD - BC = 1$ from equation (35)

Thus, when $A = 2$, $B = 20$, $C = 0.05$ and $D = 1$:

$$\boldsymbol{AD - BC} = (2)(1) - (20)(0.05) = 2 - 1 = \mathbf{1}$$

Hence **the network is passive**.

Now try the following Practice Exercise

Practice Exercise 176 *ABCD* parameters (Answers on page 833)

1. A two-port network has parameters $A = (1 - j)$, $B = 10\,\Omega$, $C = (2 + j)S$ and $D = 5$. The output current is 100 mA through a 50 Ω load. Determine (a) the input voltage and (b) the input current.
2. A two-port network has the following parameters: $A = 10$, $B = 60\,\Omega$, $C = 5\,\text{mS}$ and $D = 4$. Determine the input impedance when the output is (a) open-circuited and (b) short-circuited.
3. A two-port network has the parameters: $A = 2\angle 30°$, $B = 50\angle 25°\,\Omega$, $C = 0.05\angle 45°\,\text{S}$ and $D = 1\angle 60°$. Determine the input impedance when the output is (a) open-circuited and (b) short-circuited.
4. For a two-port network:
$$V_1 = 6V_2 - 5I_2$$
and $$I_1 = 5V_2 - 3I_2$$
When $V_1 = 32\,\text{V}$ and $I_1 = 36\,\text{A}$, determine (a) the transmission matrix and (b) the values of V_2 and I_2
5. A network has the transmission matrix:
$$\begin{pmatrix} (1+j) & 2000 \\ j0.001 & 1 \end{pmatrix}$$
Determine the parameters for two such networks in cascade.
6. Tests on a passive two-port network gave the following results:
Output open-circuited:
$$V_1 = 40\,\text{V},\ V_2 = 20\,\text{V},\ I_1 = 20\,\text{mA}$$
Output short-circuited:
$$V_1 = 40\,\text{V},\ I_1 = 40\,\text{mA},\ I_2 = 40\,\text{mA}$$
(a) Determine the values of the *ABCD* parameters and (b) confirm that the network is passive.

44.10 *ABCD* parameters for networks

(a) Series impedance

A series impedance, Z, is shown in Figure 44.50.

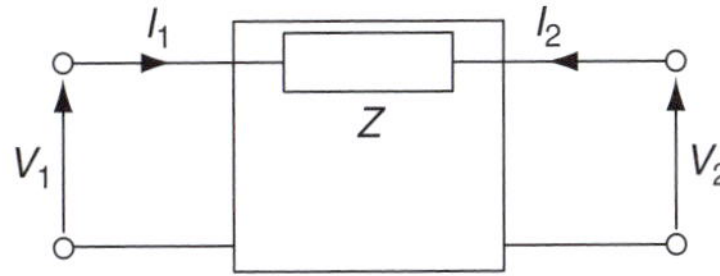

Figure 44.50

By Kirchhoff's voltage law:

$$V_1 = V_2 - ZI_2 \tag{36}$$

The input current I_1 must equal $-I_2$ for a passive network,

i.e. $$I_1 = 0 - I_2 \qquad (37)$$

Comparing equations (36) and (37) with equations (28) and (29),

i.e. $$V_1 = AV_2 - BI_2 \qquad (28)$$

$$I_1 = CV_2 - DI_2 \qquad (29)$$

shows that $A=1$, $B=Z$, $C=0$ and $D=1$

and the transmission matrix $$\begin{pmatrix} \boldsymbol{A} & \boldsymbol{B} \\ \boldsymbol{C} & \boldsymbol{D} \end{pmatrix} = \begin{pmatrix} \mathbf{1} & \boldsymbol{Z} \\ \mathbf{0} & \mathbf{1} \end{pmatrix} \qquad (38)$$

(b) Shunt admittance

A shunt admittance, Y, is shown in Figure 44.51.

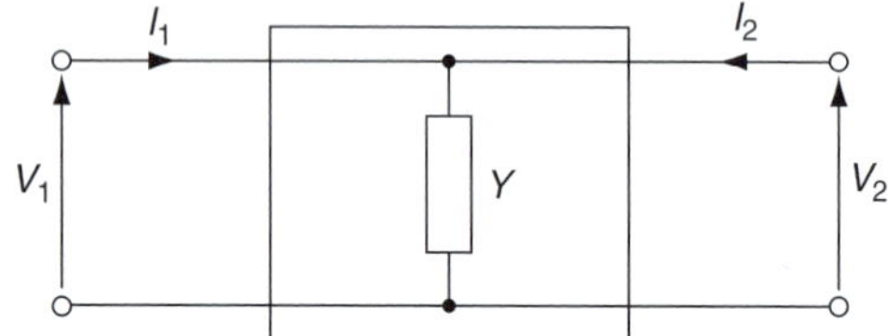

Figure 44.51

Hence, $$V_1 = V_2 + 0 \qquad (39)$$

and since the current through Y is $I_1 + I_2$

then $I_1 + I_2 = YV_2$

from which, $$I_1 = YV_2 - I_2 \qquad (40)$$

Comparing equations (39) and (40) with equations (28) and (29),

i.e. $$V_1 = AV_2 - BI_2 \qquad (28)$$

$$I_1 = CV_2 - DI_2 \qquad (29)$$

shows that $A=1$, $B=0$, $C=Y$ and $D=1$

and the transmission matrix $$\begin{pmatrix} \boldsymbol{A} & \boldsymbol{B} \\ \boldsymbol{C} & \boldsymbol{D} \end{pmatrix} = \begin{pmatrix} \mathbf{1} & \mathbf{0} \\ \boldsymbol{Y} & \mathbf{1} \end{pmatrix} \qquad (41)$$

(c) L-network

The L-network shown in Figure 44.52 can be considered to be a cascade connection of the series impedance and shunt admittance of (a) and (b) above.

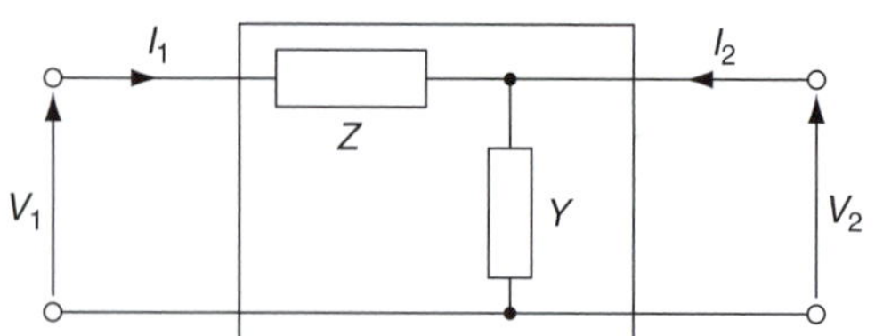

Figure 44.52

Thus, the transmission parameters,

$$\begin{pmatrix} A & B \\ C & D \end{pmatrix} = \begin{pmatrix} 1 & Z \\ 0 & 1 \end{pmatrix}\begin{pmatrix} 1 & 0 \\ Y & 1 \end{pmatrix}$$

from equations (38) and (41)

$$= \begin{pmatrix} (\mathbf{1} + \boldsymbol{YZ}) & \boldsymbol{Z} \\ \boldsymbol{Y} & \mathbf{1} \end{pmatrix} \qquad (42)$$

Problem 23. Determine the transmission parameters for the L-network shown in Figure 44.53.

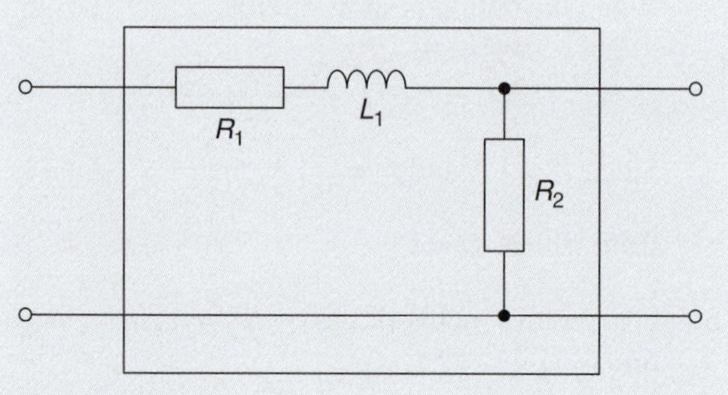

Figure 44.53

From equation (42), $\begin{pmatrix} A & B \\ C & D \end{pmatrix} = \begin{pmatrix} (1+YZ) & Z \\ Y & 1 \end{pmatrix}$

where shunt admittance, $Y = \dfrac{1}{R_2}$ and series impedance, $Z = R_1 + j\omega L_1$

Hence the transmission parameters,

$$\begin{pmatrix} A & B \\ C & D \end{pmatrix} = \begin{pmatrix} \left(1 + \left(\dfrac{1}{R_2}\right)(R_1 + j\omega L_1)\right) & (R_1 + j\omega L_1) \\ \dfrac{1}{R_2} & 1 \end{pmatrix}$$

$$= \begin{pmatrix} \left(\mathbf{1} + \dfrac{\boldsymbol{R_1 + j\omega L_1}}{\boldsymbol{R_2}}\right) & (\boldsymbol{R_1 + j\omega L_1}) \\ \dfrac{\mathbf{1}}{\boldsymbol{R_2}} & \mathbf{1} \end{pmatrix}$$

Part 4

(d) T-network

The T-network shown in Figure 44.54 can be considered to be a cascade connection of a series impedance, a shunt admittance and then another series impedance.

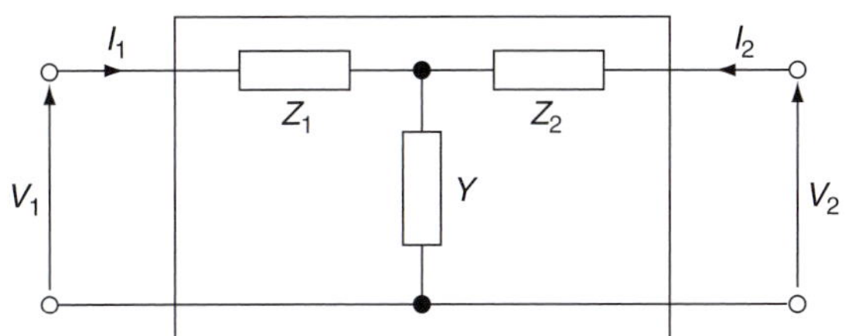

Figure 44.54

The transmission matrix for the T-network comprises the products of the matrices given by equations (38), (41) and (38),

i.e. $$\begin{pmatrix} A & B \\ C & D \end{pmatrix} = \begin{pmatrix} 1 & Z_1 \\ 0 & 1 \end{pmatrix}\begin{pmatrix} 1 & 0 \\ Y & 1 \end{pmatrix}\begin{pmatrix} 1 & Z_2 \\ 0 & 1 \end{pmatrix}$$

$$= \begin{pmatrix} (1+YZ_1) & Z_1 \\ Y & 1 \end{pmatrix}\begin{pmatrix} 1 & Z_2 \\ 0 & 1 \end{pmatrix}$$

$$= \begin{pmatrix} (1+YZ_1) & [(1+YZ_1)(Z_2)+Z_1] \\ Y & (YZ_2+1) \end{pmatrix}$$

$$= \begin{pmatrix} \mathbf{(1+YZ_1)} & \mathbf{(Z_1+Z_2+YZ_1Z_2)} \\ \mathbf{Y} & \mathbf{(1+YZ_2)} \end{pmatrix} \quad (43)$$

Problem 24. Determine the transmission parameters for a symmetrical T-network which has series arm impedances of $(50+j25)\,\Omega$ and a shunt arm of $-j50\,\Omega$.

In equation (43), $Z_1 = Z_2 = (50+j25)$ and $Y = \frac{1}{-j50} = \frac{j}{50}$ (by multiplying numerator and denominator by j)
Hence, transmission parameters,

$$\begin{pmatrix} A & B \\ C & D \end{pmatrix} = \begin{pmatrix} \left(1+\frac{j}{50}(50+j25)\right) & (50+j25+50+j25 + \frac{j}{50}(50+j25)(50+j25)) \\ \frac{j}{50}\left(1+\frac{j}{50}(50+j25)\right) & \end{pmatrix}$$

$$= \begin{pmatrix} (1+j+j^2 0.5) & (100+j50+j(1+j0.5)(50+j25)) \\ j0.02 & (1+j+j^2 0.5) \end{pmatrix}$$

$$= \begin{pmatrix} (0.5+j) & (100+j50+j50-25-25-j12.5) \\ j0.02 & (0.5+j) \end{pmatrix}$$

$$= \begin{pmatrix} \mathbf{(0.5+j)} & \mathbf{(50+j87.5)} \\ \mathbf{j0.02} & \mathbf{(0.5+j)} \end{pmatrix}$$

(e) π-network

The π-network shown in Figure 44.55 can be considered to be a cascade connection of a shunt admittance, a series impedance and another shunt admittance.

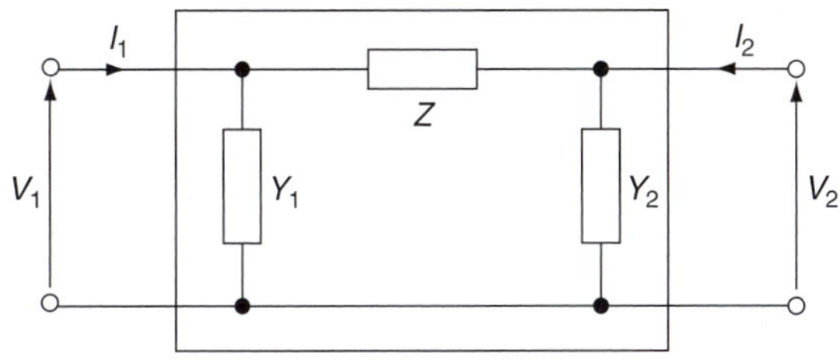

Figure 44.55

The transmission matrix for the π-network comprises the products of the matrices given by equations (41), (38) and (41),

i.e. $$\begin{pmatrix} A & B \\ C & D \end{pmatrix} = \begin{pmatrix} 1 & 0 \\ Y_1 & 1 \end{pmatrix}\begin{pmatrix} 1 & Z \\ 0 & 1 \end{pmatrix}\begin{pmatrix} 1 & 0 \\ Y_2 & 1 \end{pmatrix}$$

$$= \begin{pmatrix} 1 & Z \\ Y_1 & (Y_1Z+1) \end{pmatrix}\begin{pmatrix} 1 & 0 \\ Y_2 & 1 \end{pmatrix}$$

$$= \begin{pmatrix} (1+Y_2Z) & Z \\ (Y_1+Y_2(Y_1Z+1)) & (1+Y_1Z) \end{pmatrix}$$

$$= \begin{pmatrix} \mathbf{(1+Y_2Z)} & \mathbf{Z} \\ \mathbf{(Y_1+Y_2+Y_1Y_2Z)} & \mathbf{(1+Y_1Z)} \end{pmatrix} \quad (44)$$

Problem 25. A symmetrical π-network has a series impedance of $10\angle 60°\,\Omega$ and shunt admittances of $0.01\angle 75°$ S. Determine (a) the transmission matrix and (b) the input voltage and current when there is a load resistance of $20\,\Omega$ across the output and the input voltage produces a current of 1 mA through it.

(a) The transmission matrix is given by equation (44) where $Z = 10\angle 60°$ and $Y_1 = Y_2 = 0.01\angle 75°$. Thus,

$$\begin{pmatrix} A & B \\ C & D \end{pmatrix} = \begin{pmatrix} (1+Y_2 Z) & Z \\ (Y_1+Y_2+Y_1Y_2Z) & (1+Y_1Z) \end{pmatrix}$$

$$= \begin{pmatrix} (1+(0.01\angle 75°)(10\angle 60°)) & 10\angle 60° \\ \begin{pmatrix} (0.01\angle 75° + 0.01\angle 75° \\ +(0.01\angle 75°)(0.01\angle 75°) \\ \times(10\angle 60°)) \end{pmatrix} & \begin{pmatrix} (1+(0.01\angle 75°) \\ \times(10\angle 60°)) \end{pmatrix} \end{pmatrix}$$

$$= \begin{pmatrix} (1+0.1\angle 135°) & 10\angle 60° \\ (0.02\angle 75° + 0.001\angle 210°) & (1+0.1\angle 135°) \end{pmatrix}$$

$$= \begin{pmatrix} \mathbf{(0.929+j0.0707)} & \mathbf{(5+j8.660)} \\ \mathbf{(0.0043+j0.0188)} & \mathbf{(0.929+j0.0707)} \end{pmatrix} \text{ or}$$

$$\begin{pmatrix} \mathbf{0.932\angle 4.35°} & \mathbf{10\angle 60°} \\ \mathbf{0.0193\angle 77.12°} & \mathbf{0.932\angle 4.35°} \end{pmatrix}$$

(b) Output voltage,
$V_2 = I_2 R = 20 \times 1 \times 10^{-3} = 20\,\text{mV} = 0.02\,\text{V}$

Current, $I_2 = -1\,\text{mA} = -0.001\,\text{A}$

From equation (29),

$$\begin{aligned} V_1 &= AV_2 - BI_2 \\ &= (0.929 + j0.0707)(0.02) \\ &\quad - (5 + j8.660)(-0.001) \\ &= 0.02358 + j0.0880 \end{aligned}$$

i.e. **input voltage, $\mathbf{V_1 = 0.091\angle 75.00°\,V}$**

From equation (29),

$$\begin{aligned} I_1 &= CV_2 - DI_2 \\ &= (0.0043 + j0.0188)(0.02) \\ &\quad - (0.929 + j0.707)(-0.001) \\ &= 0.00523 + j0.000447 \end{aligned}$$

i.e. **input current, I_1** $= 0.00525\angle 4.89°\,\text{A}$
$= \mathbf{5.25\angle 4.89°\,mA}$

(f) Pure mutual inductance

A two-port network which is a pure mutual inductance is shown in Figure 44.56.

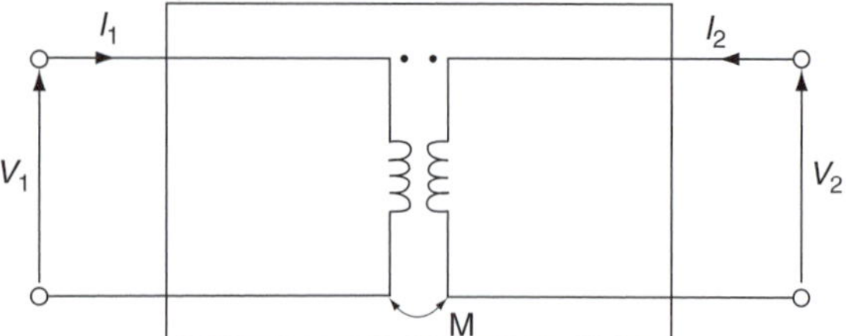

Figure 44.56

Applying Kirchhoff's voltage law to the primary circuit gives:

$V_1 + j\omega M I_2 = 0$ (this is explained in Chapter 46, page 734)

i.e. $$V_1 = -j\omega M I_2 \qquad (45)$$

For the secondary circuit:

$$V_2 = j\omega M I_1$$

from which,

$$I_1 = \left(\frac{1}{j\omega M}\right) V_2 \qquad (46)$$

Comparing equations (45) and (46) with equations (28) and (29),

i.e. $$V_1 = AV_2 - BI_2 \qquad (28)$$
$$I_1 = CV_2 - DI_2 \qquad (29)$$

shows that $A = 0$, $B = j\omega M$, $C = \dfrac{1}{j\omega M}$ and $D = 0$

and the transmission matrix $\begin{pmatrix} A & B \\ C & D \end{pmatrix} = \begin{pmatrix} \mathbf{0} & \mathbf{j\omega M} \\ \mathbf{\dfrac{1}{j\omega M}} & \mathbf{0} \end{pmatrix}$ (47)

(g) Symmetrical lattice

A symmetrical lattice two-port network is shown in Figure 44.57.

The circuit is shown redrawn as a bridge circuit in Figure 44.58.

With the output in Figure 44.58 **open-circuited**, $I_2 = 0$ and $I_1 = \dfrac{V_1}{Z_T}$ where Z_T is $(Z_1 + Z_2)$ in parallel with $(Z_1 + Z_2)$, i.e. $Z_T = \dfrac{1}{2}(Z_1 + Z_2)$ either from the

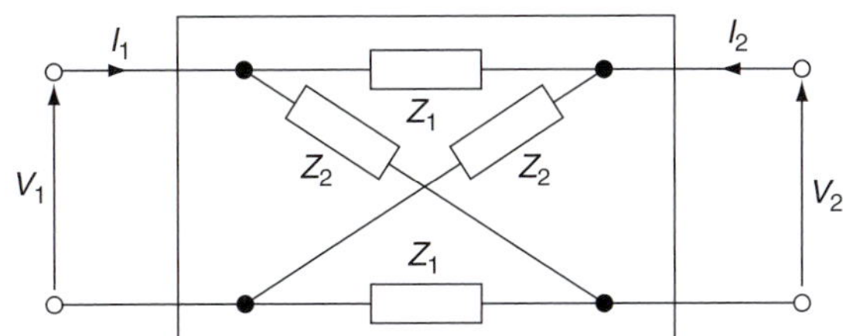

Figure 44.57

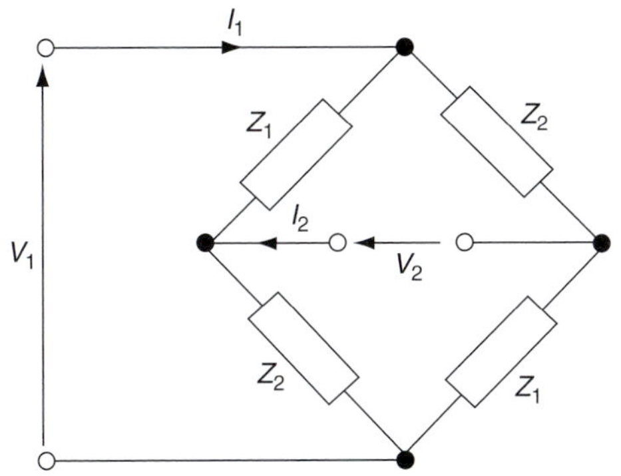

Figure 44.58

reciprocal formula or from product/sum.

Hence, $$I_1 = \frac{V_1}{Z_T} = \frac{V_1}{\frac{1}{2}(Z_1 + Z_2)}$$

In Figure 44.58, since the two arms have the same impedance, i.e. $(Z_1 + Z_2)$, the current through each will be $\frac{1}{2} I_1$

Thus, $$V_2 = \frac{1}{2} I_1 Z_2 - \frac{1}{2} I_1 Z_1$$
$$= \frac{1}{2} I_1 (Z_2 - Z_1) \quad (48)$$

Substituting for I_1 gives:

$$V_2 = \frac{1}{2} I_1 (Z_2 - Z_1)$$
$$= \frac{1}{2}\left[\frac{V_1}{\frac{1}{2}(Z_1 + Z_2)}\right](Z_2 - Z_1)$$

i.e. $$V_1 = \left(\frac{Z_1 + Z_2}{Z_2 - Z_1}\right) V_2 \quad (49)$$

From equation (28), $V_1 = AV_2 - BI_2$

and when $I_2 = 0$, $V_1 = AV_2$

Comparing this equation with equation (49) gives:

$$A = \left(\frac{Z_1 + Z_2}{Z_2 - Z_1}\right)$$

Rearranging equation (48) gives:

$$I_1 = \left(\frac{2}{Z_2 - Z_1}\right) V_2 \quad (50)$$

From equation (29), $I_1 = CV_2 - DI_2$

and when $I_2 = 0$, $I_1 = CV_2$

Comparing this equation with equation (50) gives:

$$C = \left(\frac{2}{Z_2 - Z_1}\right)$$

If the output in Figure 44.58 is now **short-circuited**, then $V_2 = 0$

The current I_2 will be given by:

$$I_2 = \frac{V}{Z_2} - \frac{V}{Z_1}$$

where V is the volt drop across both Z_1 and Z_2. If the voltages across each were different, then V_2 would not be equal to zero. Since the network is symmetrical, $V = \frac{1}{2} V_1$. Therefore,

$$I_2 = \frac{\frac{1}{2}V_1}{Z_2} - \frac{\frac{1}{2}V_1}{Z_1} = \frac{V_1}{2}\left(\frac{1}{Z_2} - \frac{1}{Z_1}\right) = \frac{V_1}{2}\left(\frac{Z_1 - Z_2}{Z_1 Z_2}\right)$$

Rearranging gives:

$$V_1 = \left(\frac{2Z_1 Z_2}{Z_1 - Z_2}\right) I_2 \quad (51)$$

From equation (28),

$$V_1 = AV_2 - BI_2$$

and when $V_2 = 0$,

$$V_1 = -BI_2$$

Comparing this equation with equation (51) gives:

$$B = -\left(\frac{2Z_1 Z_2}{Z_1 - Z_2}\right) = \left(\frac{2Z_1 Z_2}{Z_2 - Z_1}\right)$$

Input current, $$I_1 = \frac{V}{Z_1} + \frac{V}{Z_2} = \frac{V_1}{2}\left(\frac{1}{Z_1} + \frac{1}{Z_2}\right)$$
$$= \frac{V_1}{2}\left(\frac{Z_1 + Z_2}{Z_1 Z_2}\right)$$

and substituting for V_1 from equation (51) gives:

$$I_1 = \frac{V_1}{2}\left(\frac{Z_1+Z_2}{Z_1 Z_2}\right)$$
$$= \frac{I_2}{2}\left(\frac{2Z_1 Z_2}{Z_1-Z_2}\right)\left(\frac{Z_1+Z_2}{Z_1 Z_2}\right)$$

i.e. $$I_1 = \left(\frac{Z_1+Z_2}{Z_1-Z_2}\right) I_2 \qquad (52)$$

From equation (29), $I_1 = CV_2 - DI_2$

and when $V_2 = 0$, $I_1 = -DI_2$

Comparing this equation with equation (52) gives:

$$D = -\left(\frac{Z_1+Z_2}{Z_1-Z_2}\right) = \left(\frac{Z_1+Z_2}{Z_2-Z_1}\right)$$

Thus the transmission matrix for the symmetrical lattice is:

$$\begin{pmatrix} \left(\dfrac{\mathbf{Z_1+Z_2}}{\mathbf{Z_2-Z_1}}\right) & \left(\dfrac{\mathbf{2Z_1Z_2}}{\mathbf{Z_2-Z_1}}\right) \\ \left(\dfrac{\mathbf{2}}{\mathbf{Z_2-Z_1}}\right) & \left(\dfrac{\mathbf{Z_1+Z_2}}{\mathbf{Z_2-Z_1}}\right) \end{pmatrix} \qquad (53)$$

Problem 26. In the symmetrical lattice shown in Figure 44.57, $Z_1 = 25\,\Omega$ and $Z_2 = 35\,\Omega$. Determine the equivalent T-network.

From equation (43) the transmission matrix for a T-network is:

$$\begin{pmatrix} (1+YZ_1) & (Z_1+Z_2+YZ_1Z_2) \\ Y & (1+YZ_2) \end{pmatrix}$$

From equation (53) the transmission matrix for a symmetrical lattice is:

$$\begin{pmatrix} \left(\dfrac{Z_1+Z_2}{Z_2-Z_1}\right) & \left(\dfrac{2Z_1Z_2}{Z_2-Z_1}\right) \\ \left(\dfrac{2}{Z_2-Z_1}\right) & \left(\dfrac{Z_1+Z_2}{Z_2-Z_1}\right) \end{pmatrix}$$

$$= \begin{pmatrix} \left(\dfrac{25+35}{35-25}\right) & \left(\dfrac{2(25)(35)}{35-25}\right) \\ \left(\dfrac{2}{35-25}\right) & \left(\dfrac{25+35}{35-25}\right) \end{pmatrix}$$

when $Z_1 = 25\,\Omega$ and $Z_2 = 35\,\Omega$.

Since the lattice is symmetrical, the equivalent T-network must also be symmetrical. Thus

$$\begin{pmatrix} (1+YZ) & (2Z+YZ^2) \\ Y & (1+YZ) \end{pmatrix}$$

$$= \begin{pmatrix} \left(\dfrac{25+35}{35-25}\right) & \left(\dfrac{2(25)(35)}{35-25}\right) \\ \left(\dfrac{2}{35-25}\right) & \left(\dfrac{25+35}{35-25}\right) \end{pmatrix} = \begin{pmatrix} 6 & 175 \\ 0.2 & 6 \end{pmatrix}$$

Equating the terms gives:

$$\mathbf{Y = 0.2\,S},$$
$$1 + YZ = 6$$

i.e. $$1 + 0.2Z = 6$$

from which, $$\mathbf{Z} = \frac{6-1}{0.2} = \mathbf{25\,\Omega}$$

Hence, the T-network shown in Figure 44.59(b) is equivalent to the symmetrical lattice network shown in Figure 44.59(a).

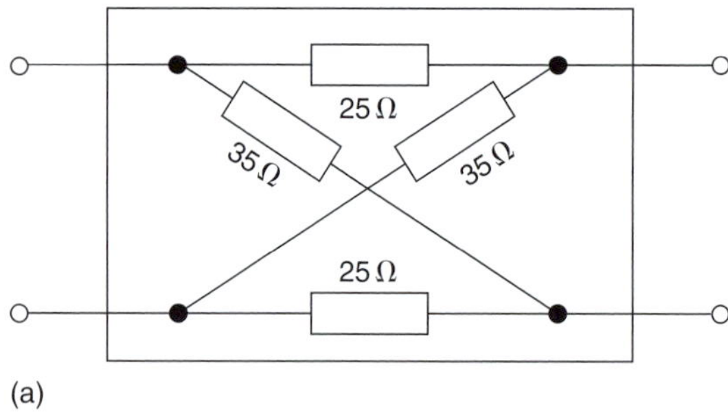

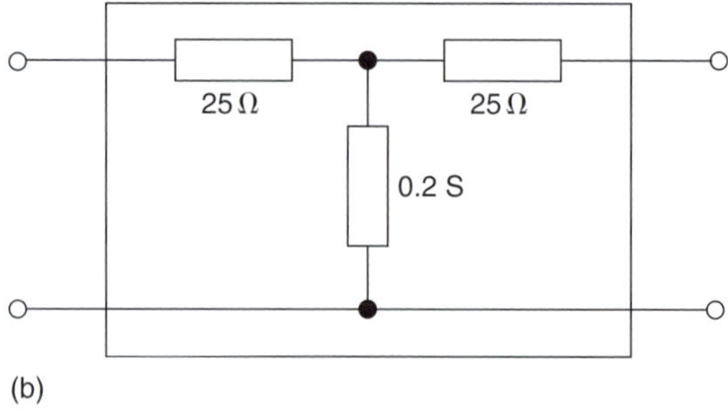

Figure 44.59

Now try the following Practice Exercise

Practice Exercise 177 *ABCD* parameters for networks (Answers on page 833)

1. Write down the transmission matrix for (a) a 5 Ω series impedance and (b) a 5 Ω shunt impedance.
2. Determine the transmission parameters for an L-network having a 5 Ω series impedance and a 5 Ω shunt impedance.
3. An L-network has a shunt arm admittance of 0.5 S and a series arm impedance of 10 Ω resistance and 318.3 μH inductor. If the frequency of the network is 5 kHz, determine its transmission matrix.
4. Determine the transmission parameters for a symmetrical T-section which has a series impedance of $(1+j2)$ kΩ and a shunt arm impedance of $j1$ kΩ.
5. Determine the transmission parameters for a symmetrical T-section which has a series impedance of $100\angle 30°$ Ω and a shunt arm impedance of $100\angle -90°$ Ω.
6. A symmetrical π-section has a series impedance of $(5+j10)$ Ω and shunt arm admittances of $(0.01+j0.02)$S. Determine the transmission parameters.
7. A symmetrical π-section has a series impedance of $100\angle 30°$ Ω and shunt arm admittances of $0.01\angle 60°$S. Determine (a) the transmission matrix and (b) the input voltage and current when there is a load resistance of 100 Ω across the output and the input voltage produces a current of 10 mA.
8. A two-port network has a pure mutual inductance of 3.979 mH at a frequency of 2 kHz. Determine the transmission matrix.
9. A symmetrical lattice (as shown in Figure 44.57) has $Z_1=10$ Ω and $Z_2=20$ Ω. Determine its transmission parameters.
10. For the symmetrical lattice of Problem 9, determine the equivalent symmetrical T-network.
11. A symmetrical π-network has a series impedance of 10 Ω and shunt arm impedances of 5 Ω. Determine (a) its transmission parameters and (b) the equivalent symmetrical T-network.

44.11 Characteristic impedance in terms of *ABCD* parameters

As stated previously, for a two-port network, the *ABCD* parameters are specified as follows:

$$V_1 = AV_2 - BI_2 \quad (28)$$

$$I_1 = CV_2 - DI_2 \quad (29)$$

From page 668, equation (2), characteristic impedance,

$$Z_0 = \sqrt{(Z_{OC} Z_{SC})}$$

When the output of a two-port network is open-circuited, $I_2 = 0$

and from equation (28), $V_1 = AV_2$

and from equation (29), $I_1 = CV_2$

Thus, input impedance, $Z_{1_{OC}} = \dfrac{V_1}{I_1} = \dfrac{AV_2}{CV_2} = \dfrac{A}{C}$

When the output of a two-port network is short-circuited, $V_2 = 0$

and from equation (28), $V_1 = -BI_2$

and from equation (29), $I_1 = -DI_2$

Thus, input impedance, $Z_{1_{SC}} = \dfrac{V_1}{I_1} = \dfrac{-BI_2}{-DI_2} = \dfrac{B}{D}$

It is noticed from the T- and π-networks that when they are symmetrical, $A = D$

Thus, $Z_{1_{SC}} = \dfrac{B}{D} = \dfrac{B}{A}$

Therefore, **the characteristic impedance,**

$$\mathbf{Z_0} = \sqrt{(Z_{OC} Z_{SC})} = \sqrt{\left[\frac{A}{C} \times \frac{B}{A}\right]} = \sqrt{\frac{\mathbf{B}}{\mathbf{C}}}$$

Problem 27. Find the characteristic impedance for the symmetrical T-network of Problem 24 on page 691.

From Problem 24, $B=(50+j87.5)=100.78\angle 60.26^\circ$, and $C=j0.02=0.02\angle 90^\circ$, from which characteristic impedance,

$$\boldsymbol{Z_0}=\sqrt{\frac{B}{C}}=\sqrt{\frac{100.78\angle 60.26^\circ}{0.02\angle 90^\circ}}$$

$$=\sqrt{5039\angle -29.74^\circ}$$

$$=\mathbf{71\angle -14.87^\circ\,\Omega}$$

using De Moivre's theorem, which for square roots states:

$$\sqrt{r\angle\theta}=[r\angle\theta]^{\frac{1}{2}}=r^{\frac{1}{2}}\angle\frac{1}{2}\theta=\sqrt{r}\angle\frac{\theta}{2}$$

Problem 28. Find the characteristic impedance for the symmetrical π-network of Problem 25 on page 691.

From Problem 25, $B=10\angle 60^\circ$, and $C=0.0193\angle 77.12^\circ$, from which characteristic impedance,

$$\boldsymbol{Z_0}=\sqrt{\frac{B}{C}}=\sqrt{\frac{10\angle 60^\circ}{0.0193\angle 77.12^\circ}}$$

$$=\sqrt{518.135\angle -17.12^\circ}$$

$$=\mathbf{22.76\angle -8.56^\circ\,\Omega}$$

Now try the following Practice Exercise

Practice Exercise 178 The characteristic impedance in terms of the *ABCD* parameters (Answers on page 833)

1. For the symmetrical T-network of Problem 5, Exercise 177, page 695, find its characteristic impedance.
2. For the symmetrical π-network of Problem 7, Exercise 177, page 695, find its characteristic impedance.
3. For the symmetrical lattice of Problem 9, Exercise 177, page 695, find its characteristic impedance.
4. For the symmetrical π-network of Problem 11, Exercise 177, page 695, find its characteristic impedance.

Revision Test 13

This revision test covers the material contained in Chapters 42 to 44. *The marks for each question are shown in brackets at the end of each question.*

1. The equivalent series circuit for a particular capacitor consists of a 2 Ω resistor in series with a 250 pF capacitor. Determine, at a frequency of 10 MHz, (a) the loss angle of the capacitor and (b) the power factor of the capacitor. (3)

2. A 50 V, 20 kHz supply is connected across a 500 pF capacitor and the power dissipated in the dielectric is 200 μW. Determine (a) the loss angle, (b) the equivalent series loss resistance and (c) the equivalent parallel loss resistance. (9)

3. A coaxial cable, which has a core of diameter 12 mm and a sheath diameter of 30 mm, is 10 km long. Calculate for the cable (a) the inductance, assuming non-magnetic materials and (b) the capacitance, assuming a dielectric of relative permittivity 5. (8)

4. A 50 km length single-phase twin line has conductors of diameter 20 mm and spaced 1.25 m apart in air. Determine for the line (a) the loop inductance and (b) the capacitance. (8)

5. Find the strength of a uniform electric field if it is to have the same energy as that established by a magnetic field of flux density 1.15 T. (Assume that the relative permeability of the magnetic field and the relative permittivity of the electric field are both unity.) (5)

6. 8% of the power supplied to a cable appears at the output terminals. Determine the attenuation in decibels. (3)

7. Design (a) a T-section attenuator and (b) a π-attenuator to provide a voltage attenuation of 25 dB and having a characteristic impedance of 620 Ω. (14)

8. Determine the transmission parameters for the following: (a) a symmetrical T-section which has a series impedance of $(2+j5)\,\Omega$ and a shunt arm impedance of $j2\,\Omega$, (b) a symmetrical π-section which has a series impedance of $100\angle 60^\circ\,\Omega$ and shunt arm admittances of $0.01\angle 30^\circ$ S. (c) For each of the above networks find their characteristic impedance. (20)

Part 4

For lecturers/instructors/teachers, fully worked solutions to each of the problems in Revision Test 13, together with a full marking scheme, are available at the website:
www.routledge.com/cw/bird

Chapter 45

Filter networks

***Why it is important to understand:* Filter networks**

In circuit theory, a filter is an electrical network that alters the amplitude and/or phase characteristics of a signal with respect to frequency. Ideally, a filter will not add new frequencies to the input signal, nor will it change the component frequencies of that signal, but it will change the relative amplitudes of the various frequency components and/or their phase relationships. Filters are often used in electronic systems to emphasize signals in certain frequency ranges and reject signals in other frequency ranges. Electronic filters are electronic circuits which perform signal processing functions, specifically to remove unwanted frequency components from the signal, to enhance wanted ones, or both. Filters are used in electronic music to alter the harmonic content of a signal, which changes its timbre. Many of the filters used in synthesizers are voltage-controlled filters, which allows the filter to be controlled by a signal generated elsewhere in the synthesizer. The purpose of filters and an explanation of the various types of filter is given in this chapter, including the design of '*m*-derived' filter sections.

At the end of this chapter you should be able to:

- appreciate the purpose of a filter network
- understand basic types of filter sections, i.e. low-pass, high-pass, band-pass and band-stop filters
- understand characteristic impedance and attenuation of filter sections
- understand low and high pass ladder networks
- design a low- and high-pass filter section
- calculate propagation coefficient and time delay in filter sections
- understand and design 'm-derived' filter sections
- understand and design practical composite filters

45.1 Introduction

A **filter** is a network designed to pass signals having frequencies within certain bands (called **pass-bands**) with little attenuation, but greatly attenuates signals within other bands (called **attenuation bands** or **stop-bands**).

As explained in the previous chapter, an attenuator network pad is composed of resistances only, the attenuation resulting being constant and independent of frequency. However, a filter is frequency sensitive and is thus composed of reactive elements. Since certain frequencies are to be passed with minimal loss, ideally the inductors and capacitors need to be pure

Electrical Circuit Theory and Technology. 978-1-138-67349-6, © 2017 John Bird. Published by Taylor & Francis. All rights reserved.

components since the presence of resistance results in some attenuation at all frequencies.

Between the passband of a filter, where ideally the attenuation is zero, and the attenuation band, where ideally the attenuation is infinite, is the **cut-off frequency**, this being the frequency at which the attenuation changes from zero to some finite value.

A filter network containing no source of power is termed **passive**, and one containing one or more power sources is known as an **active** filter network.

The filters considered in this chapter are symmetrical unbalanced T and π sections, the reactances used being considered as ideal.

Filters are used for a variety of purposes in nearly every type of electronic communications and control equipment. The bandwidths of filters used in communications systems vary from a fraction of a hertz to many megahertz, depending on the application.

45.2 Basic types of filter sections

(a) Low-pass filters

Figure 45.1 shows simple unbalanced T and π section filters using series inductors and shunt capacitors. If either section is connected into a network and a continuously increasing frequency is applied, each would have a frequency-attenuation characteristic as shown in Figure 45.2(a). This is an ideal characteristic and assumes pure reactive elements. All frequencies are seen to be passed from zero up to a certain value without attenuation, this value being shown as f_c, the cut-off frequency; all values of frequency above f_c are attenuated. It is for this reason that the networks shown in Figures 45.1(a) and (b) are known as **low-pass filters**. The electrical circuit diagram symbol for a low-pass filter is shown in Figure 45.2(b).

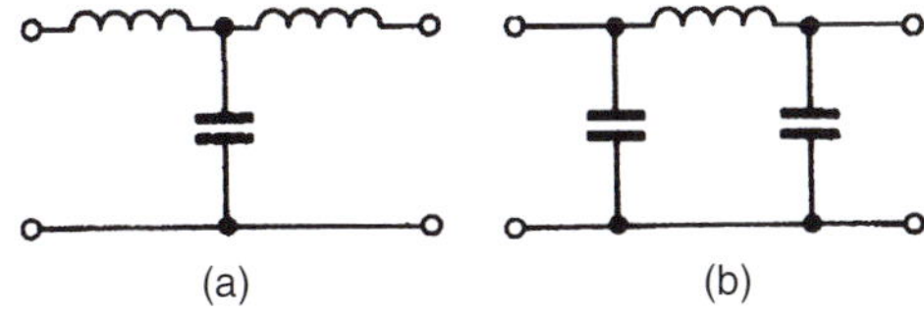

Figure 45.1

Summarizing, a low-pass filter is one designed to pass signals at frequencies below a specified cut-off frequency.

When rectifiers are used to produce the d.c. supplies of electronic systems, a large ripple introduces undesirable noise and may even mask the effect of the signal

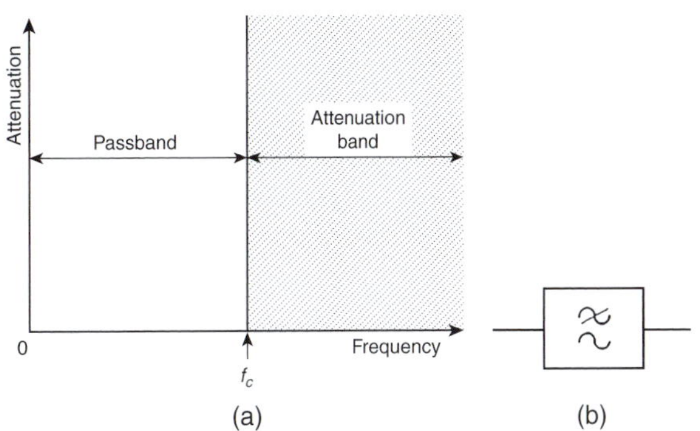

Figure 45.2

voltage. Low-pass filters are added to smooth the output voltage waveform, this being one of the most common applications of filters in electrical circuits.

Filters are employed to isolate various sections of a complete system and thus to prevent undesired interactions. For example, the insertion of low-pass decoupling filters between each of several amplifier stages and a common power supply reduces interaction due to the common power supply impedance.

(b) High-pass filters

Figure 45.3 shows simple unbalanced T and π section filters using series capacitors and shunt inductors. If either section is connected into a network and a continuously increasing frequency is applied, each would have a frequency-attenuation characteristic as shown in Figure 45.4(a).

Once again this is an ideal characteristic assuming pure reactive elements. All frequencies below the cut-off frequency f_c are seen to be attenuated and all frequencies above f_c are passed without loss. It is for this reason that the networks shown in Figures 45.3(a) and (b) are known as **high-pass filters**. The electrical circuit-diagram symbol for a high-pass filter is shown in Figure 45.4(b).

Summarizing, a high-pass filter is one designed to pass signals at frequencies above a specified cut-off frequency.

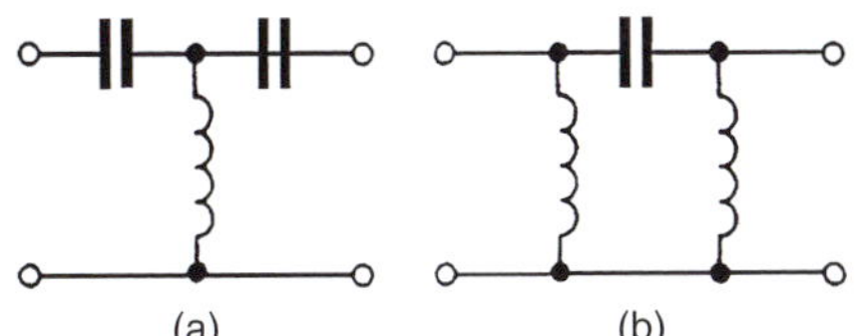

Figure 45.3

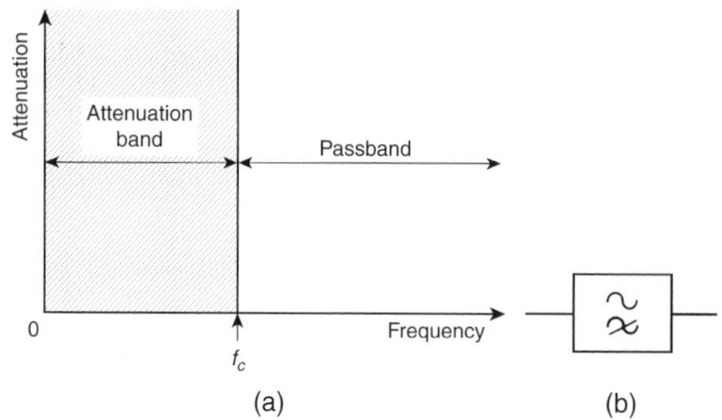

Figure 45.4

The characteristics shown in Figures 45.2(a) and 45.4(a) are ideal in that they have assumed that there is no attenuation at all in the passbands and infinite attenuation in the attenuation bands. Both of these conditions are impossible to achieve in practice. Due to resistance, mainly in the inductive elements, the attenuation in the passband will not be zero, and in a practical filter section the attenuation in the attenuation band will have a finite value. Practical characteristics for low-pass and high-pass filters are discussed in Sections 45.5 and 45.6. In addition to the resistive loss there is often an added loss due to mismatching. Ideally, when a filter is inserted into a network it is matched to the impedance of that network. However, the characteristic impedance of a filter section will vary with frequency and the termination of the section may be an impedance that does not vary with frequency in the same way. To minimize losses due to resistance and mismatching, filters are used under image impedance conditions as far as possible (see Chapter 44).

(c) Band-pass filters

A band-pass filter is one designed to pass signals with frequencies between two specified cut-off frequencies. The characteristic of an ideal band-pass filter is shown in Figure 45.5.

Such a filter may be formed by cascading a high-pass and a low-pass filter. f_{C_H} is the cut-off frequency of the high-pass filter and f_{C_L} is the cut-off frequency of the low-pass filter. As can be seen, $f_{C_L} > f_{C_H}$ for a band-pass filter, the passband being given by the difference between these values. The electrical circuit diagram symbol for a band-pass filter is shown in Figure 45.6.

Crystal and ceramic devices are used extensively as band-pass filters. They are common in the intermediate-frequency amplifiers of vhf radios where a precisely-defined bandwidth must be maintained for good performance.

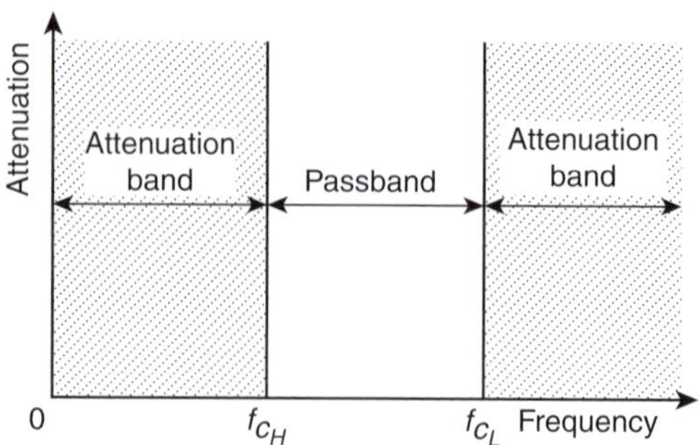

Figure 45.5

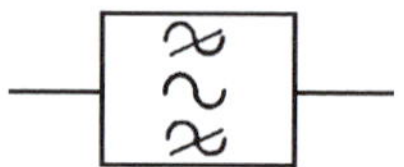

Figure 45.6

(d) Band-stop filters

A band-stop filter is one designed to pass signals with all frequencies except those between two specified cut-off frequencies. The characteristic of an ideal band-stop filter is shown in Figure 45.7. Such a filter may be formed by connecting a high-pass and a low-pass filter in parallel. As can be seen, for a band-stop filter $f_{C_H} > f_{C_L}$, the stop-band being given by the difference between these values. The electrical circuit diagram symbol for a band-stop filter is shown in Figure 45.8.

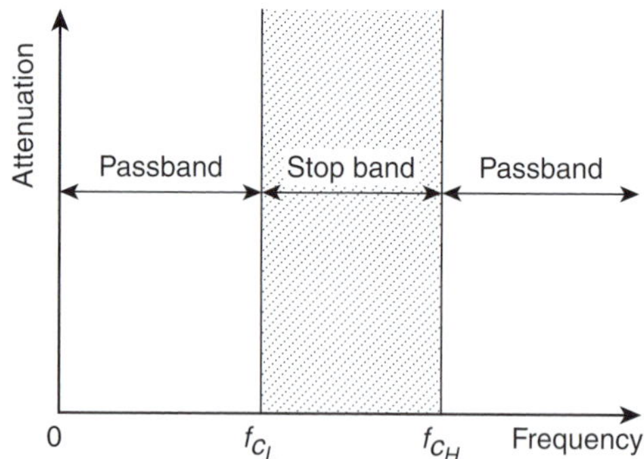

Figure 45.7

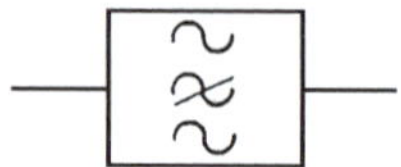

Figure 45.8

Sometimes, as in the case of interference from 50 Hz power lines in an audio system, the exact frequency of a spurious noise signal is known. Usually such interference is from an odd harmonic of 50 Hz, for example, 250 Hz. A sharply tuned band-stop filter, designed to attenuate the 250 Hz noise signal, is used to minimize the effect of the output. A high-pass filter with cut-off frequency greater than 250 Hz would also remove the interference, but some of the lower frequency components of the audio signal would be lost as well.

45.3 The characteristic impedance and the attenuation of filter sections

Nature of the input impedance

Let a symmetrical filter section be terminated in an impedance Z_O. If the input impedance also has a value of Z_O, then Z_O is the characteristic impedance of the section.

Figure 45.9 shows a T section composed of reactive elements X_A and X_B. If the reactances are of opposite kind, then the input impedance of the section, shown as Z_O, when the output port is open or short-circuited, can be either inductive or capacitive depending on the frequency of the input signal.

For example, if X_A is inductive, say jX_L, and X_B is capacitive, say, $-jX_C$, then from Figure 45.9,

$$Z_{OC} = jX_L - jX_C = j(X_L - X_C)$$

and
$$Z_{SC} = jX_L + \frac{(jX_L)(-jX_C)}{(jX_L) + (-jX_C)}$$

$$= jX_L + \frac{(X_L X_C)}{j(X_L - X_C)}$$

$$= jX_L - j\left(\frac{X_L X_C}{X_L - X_C}\right)$$

$$= \boldsymbol{j\left(X_L - \frac{X_L X_C}{X_L - X_C}\right)}$$

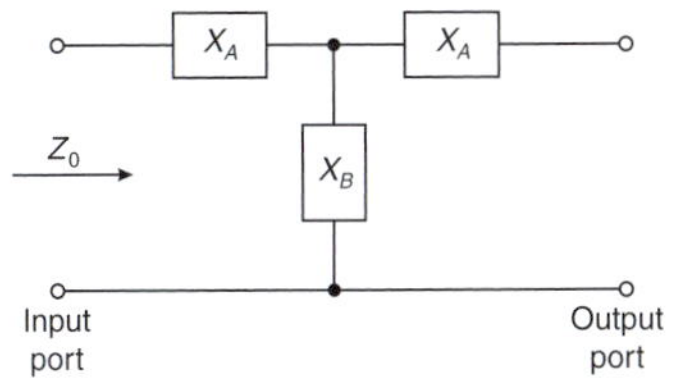

Figure 45.9

Since $X_L = 2\pi fL$ and $X_C = (1/2\pi fC)$ then Z_{OC} and Z_{SC} can be inductive (i.e. positive reactance) or capacitive (i.e. negative reactance) depending on the value of frequency, f

Let the magnitude of the reactance on open-circuit be X_{OC} and the magnitude of the reactance on short-circuit be X_{SC}. Since the filter elements are all purely reactive they may be expressed as jX_{OC} or jX_{SC}, where X_{OC} and X_{SC} are real, being positive or negative in sign. Four combinations of Z_{OC} and Z_{SC} are possible, these being:

(i) $Z_{OC} = +jX_{OC}$ and $Z_{SC} = -jX_{SC}$

(ii) $Z_{OC} = -jX_{OC}$ and $Z_{SC} = +jX_{SC}$

(iii) $Z_{OC} = +jX_{OC}$ and $Z_{SC} = +jX_{SC}$

and (iv) $Z_{OC} = -jX_{OC}$ and $Z_{SC} = -jX_{SC}$

From general circuit theory, input impedance Z_O is given by:

$$Z_O = \sqrt{(Z_{OC} Z_{SC})}$$

Taking either of combinations (i) and (ii) above gives:

$$Z_O = \sqrt{(-j^2 X_{OC} X_{SC})} = \sqrt{(X_{OC} X_{SC})},$$

which is real, thus the input impedance will be **purely resistive**.

Taking either of combinations (iii) and (iv) above gives:

$$Z_O = \sqrt{(j^2 X_{OC} X_{SC})} = +j\sqrt{(X_{OC} X_{SC})},$$

which is imaginary, thus the input impedance will be **purely reactive**.

Thus since the magnitude and nature of Z_{OC} and Z_{SC} depend upon frequency then so also will the magnitude and nature of the input impedance Z_O depend upon frequency.

Characteristic impedance

Figure 45.10 shows a low-pass T section terminated in its characteristic impedance, Z_O

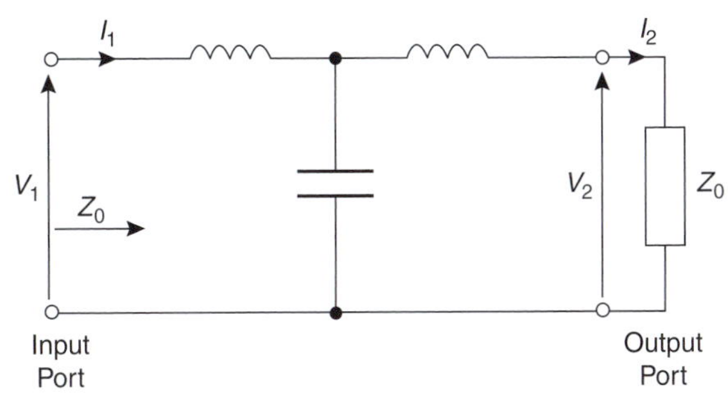

Figure 45.10

From equation (2), page 668, the characteristic impedance is given by $Z_O = \sqrt{(Z_{OC} Z_{SC})}$.
The following statements may be demonstrated to be true for any filter:

(a) *The attenuation is zero throughout the frequency range for which the characteristic impedance is purely resistive.*

(b) *The attenuation is finite throughout the frequency range for which the characteristic impedance is purely reactive.*

To demonstrate statement (a) above:

Let the filter shown in Figure 45.10 be operating over a range of frequencies such that Z_O is purely resistive.

From Figure 45.10, $Z_O = \dfrac{V_1}{I_1} = \dfrac{V_2}{I_2}$

Power dissipated in the output termination,
$P_2 = V_2 I_2 \cos\phi_2 = V_2 I_2$ (since $\phi_2 = 0$ with a purely resistive load).
Power delivered at the input terminals,

$$P_1 = V_1 I_1 \cos\phi_1 = V_1 I_1 \text{ (since } \phi_1 = 0)$$

No power is absorbed by the filter elements since they are purely reactive.

Hence $P_2 = P_1$, $V_2 = V_1$ and $I_2 = I_1$

Thus if the filter is terminated in Z_O and operating in a frequency range such that Z_O is purely resistive, then all the power delivered to the input is passed to the output and there is therefore no attenuation.

To demonstrate statement (b) above:

Let the filter be operating over a range of frequencies such that Z_O is purely reactive.

Then, from Figure 45.10, $\dfrac{V_1}{I_1} = jZ_O = \dfrac{V_2}{I_2}$

Thus voltage and current are at 90° to each other which means that the circuit can neither accept nor deliver any active power from the source to the load ($P = VI\cos\phi = VI\cos 90° = VI(0) = 0$). There is therefore infinite attenuation, theoretically. (In practice, the attenuation is finite, for the condition $(V_1/I_1) = (V_2/I_2)$ can hold for $V_2 < V_1$ and $I_2 < I_1$, since the voltage and current are 90° out of phase.)

Statements (a) and (b) above are important because they can be applied to determine the cut-off frequency point of any filter section simply from a knowledge of the nature of Z_O. **In the pass band, Z_O is real, and in the attenuation band, Z_O is imaginary.** The cut-off frequency is therefore at the point on the frequency scale at which Z_O changes from a real quantity to an imaginary one, or vice versa (see Sections 45.5 and 45.6).

45.4 Ladder networks

Low-pass networks

Figure 45.11 shows a low-pass network arranged as a ladder or repetitive network. Such a network may be considered as a number of T or π sections in cascade. In Figure 45.12(a), a T section may be taken from the ladder by removing $ABED$, producing the low-pass filter section shown in Figure 45.13(a). The ladder has been cut in the centre of each of its inductive elements hence giving $L/2$ as the series arm elements in Figure 45.13(a).

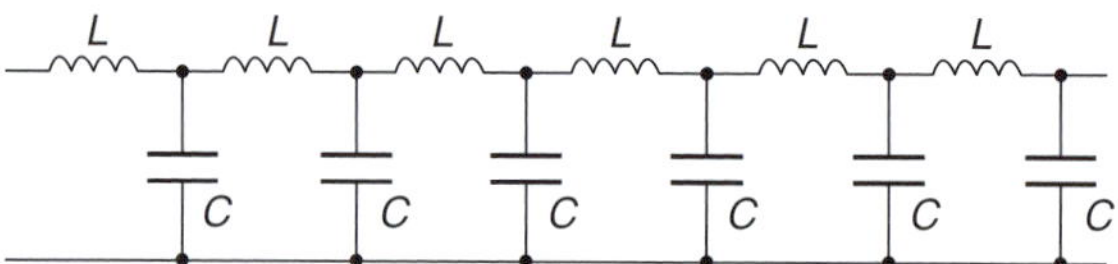

Figure 45.11

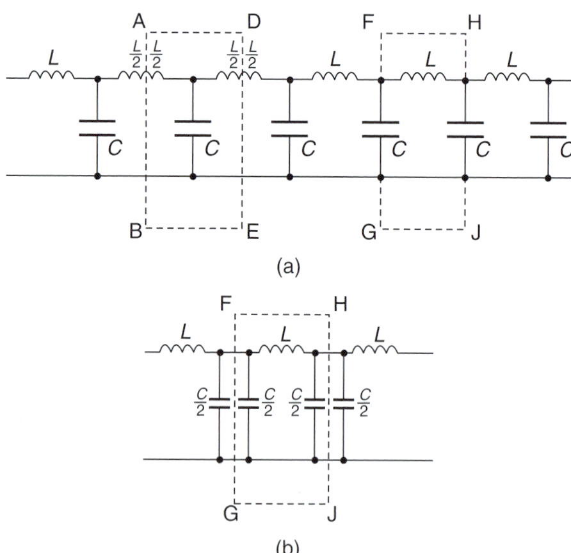

Figure 45.12

Similarly, a π section may be taken from the ladder shown in Figure 45.12(a) by removing $FGJH$, producing the low-pass filter section shown in Figure 45.13(b). The shunt element C in Figure 45.12(a) may be regarded as two capacitors in parallel, each of value $C/2$ as shown in the part of the ladder redrawn in Figure 45.12(b). (Note that for parallel capacitors, the total capacitance C_T is given by

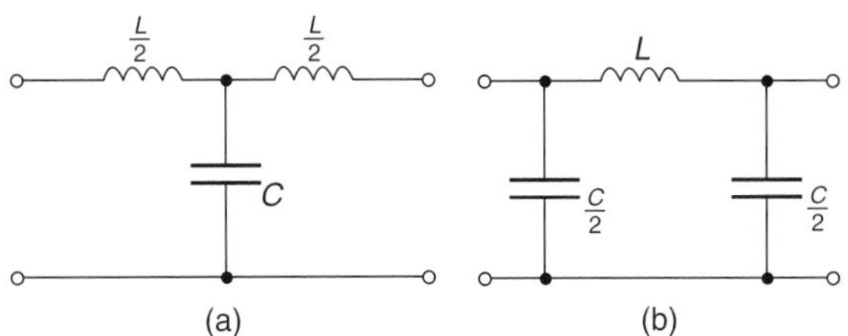

Figure 45.13

$C_T = C_1 + C_2 + \cdots$ In this case $\frac{C}{2} + \frac{C}{2} = C$)

The ladder network of Figure 45.11 can thus either be considered to be a number of the T networks shown in Figure 45.13(a) connected in cascade, or a number of the π networks shown in Figure 45.13(b) connected in cascade.

It is shown in Section 47.3, page 748, that an infinite transmission line may be reduced to a repetitive low-pass filter network.

High-pass networks

Figure 45.14 shows a high-pass network arranged as a ladder. As above, the repetitive network may be considered as a number of T or π sections in cascade.

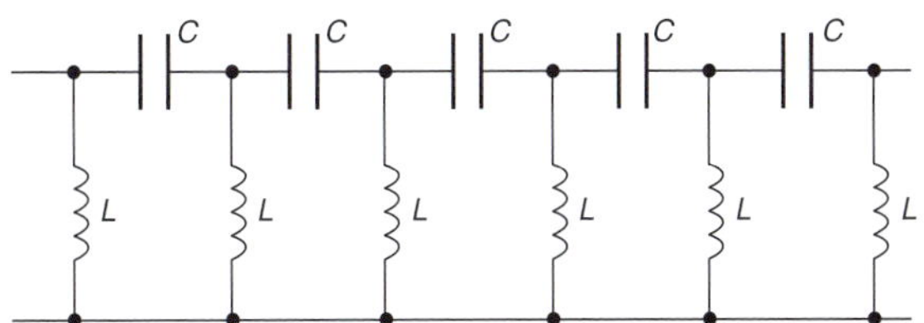

Figure 45.14

In Figure 45.15, a T section may be taken from the ladder by removing $ABED$, producing the high-pass filter section shown in Figure 45.16(a).

Note that the series arm elements are each $2C$. This is because two capacitors each of value $2C$ connected in series gives a total equivalent value of C (i.e. for series capacitors, the total capacitance C_T is given by

$$\frac{1}{C_T} = \frac{1}{C_1} + \frac{1}{C_2} + \cdots)$$

Similarly, a π section may be taken from the ladder shown in Figure 45.15 by removing $FGJH$, producing the high-pass filter section shown in Figure 45.16(b). The shunt element L in Figure 45.15(a) may be regarded as two inductors in parallel, each of value $2L$ as shown in the part of the ladder redrawn in Figure 45.15(b).

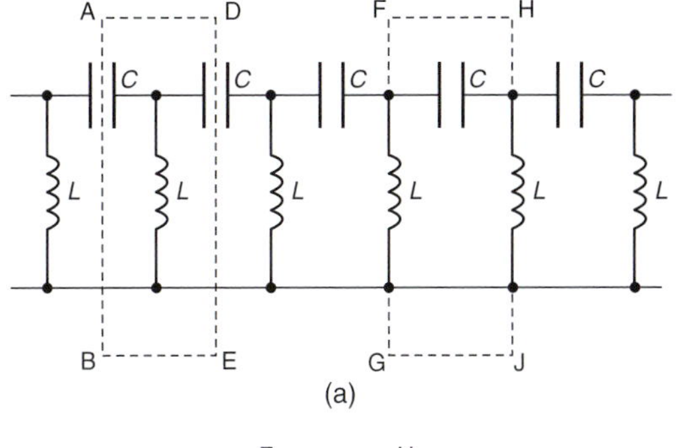

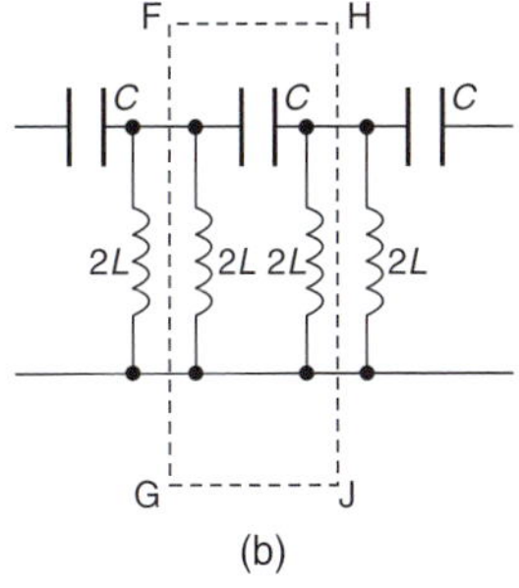

Figure 45.15

(Note that for parallel inductance, the total inductance L_T is given by

$$\frac{1}{L_T} = \frac{1}{L_1} + \frac{1}{L_2} + \cdots \text{ In this case, } \frac{1}{2L} + \frac{1}{2L} = \frac{1}{L})$$

The ladder network of Figure 45.14 can thus be considered to be either a number of T networks shown in Figure 45.16(a) connected in cascade, or a number of the π networks shown in Figure 45.16(b) connected in cascade.

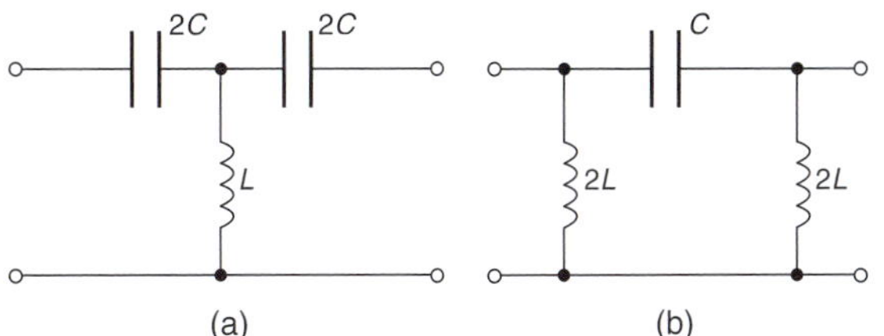

Figure 45.16

45.5 Low-pass filter sections

(a) The cut-off frequency

From equation (1), the characteristic impedance Z_0 for a symmetrical T network is given by: $Z_0 = \sqrt{(Z_A^2 + 2Z_A Z_B)}$. Applying this to the low-pass T section shown in Figure 45.17,

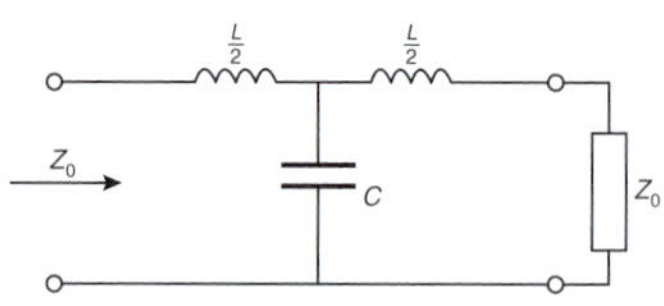

Figure 45.17

$$Z_A = \frac{j\omega L}{2} \text{ and } Z_B = \frac{1}{j\omega C}$$

Thus $$Z_0 = \sqrt{\left[\frac{j^2\omega^2L^2}{4} + 2\left(\frac{j\omega L}{2}\right)\left(\frac{1}{j\omega C}\right)\right]}$$

$$= \sqrt{\left(\frac{-\omega^2L^2}{4} + \frac{L}{C}\right)}$$

i.e. $$Z_0 = \sqrt{\left(\frac{L}{C} - \frac{\omega^2L^2}{4}\right)} \qquad (1)$$

Z_0 will be real if $\frac{L}{C} > \frac{\omega^2L^2}{4}$

Thus attenuation will commence when $\frac{L}{C} = \frac{\omega^2L^2}{4}$

i.e. when $\omega_c^2 = \frac{4}{LC}$ (2)

where $\omega_c = 2\pi f_c$ and f_c is the cut-off frequency.

Thus $$(2\pi f_c)^2 = \frac{4}{LC}$$

$$2\pi f_c = \sqrt{\left(\frac{4}{LC}\right)} = \frac{2}{\sqrt{(LC)}}$$

and $$f_c = \frac{2}{2\pi\sqrt{(LC)}} = \frac{1}{\pi\sqrt{(LC)}}$$

i.e. **the cut-off frequency,** $\mathbf{f_c = \frac{1}{\pi\sqrt{(LC)}}}$ (3)

The same equation for the cut-off frequency is obtained for the low-pass π network shown in Figure 45.18 as follows:

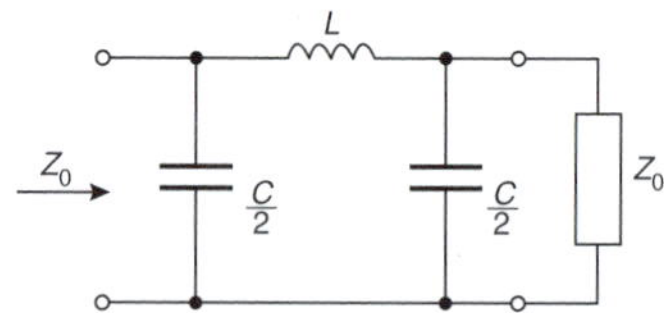

Figure 45.18

From equation (41.3), for a symmetrical π network,

$$Z_0 = \sqrt{\left(\frac{Z_1 Z_2^2}{Z_1 + 2Z_2}\right)}$$

Applying this to Figure 45.18,

$$Z_1 = j\omega L \text{ and } Z_2 = \frac{1}{j\omega\frac{C}{2}} = \frac{2}{j\omega C}$$

Thus $$Z_0 = \sqrt{\left\{\frac{(j\omega L)\left(\frac{2}{j\omega C}\right)^2}{j\omega L + 2\left(\frac{2}{j\omega C}\right)}\right\}}$$

$$= \sqrt{\left\{\frac{(j\omega L)\left(\frac{4}{-\omega^2C^2}\right)}{j\omega L - j\left(\frac{4}{\omega C}\right)}\right\}}$$

$$= \sqrt{\left\{\frac{-j\frac{4L}{\omega C^2}}{j\left(\omega L - \frac{4}{\omega C}\right)}\right\}} = \sqrt{\left\{\frac{\frac{4L}{\omega C^2}}{\frac{4}{\omega C} - \omega L}\right\}}$$

$$= \sqrt{\left\{\frac{4L}{\omega C^2\left(\frac{4}{\omega C} - \omega L\right)}\right\}}$$

$$= \sqrt{\left(\frac{4L}{4C - \omega^2 LC^2}\right)}$$

i.e. $$Z_0 = \sqrt{\left(\frac{1}{\frac{C}{L} - \frac{\omega^2C^2}{4}}\right)} \qquad (4)$$

Z_0 will be real if $\frac{C}{L} > \frac{\omega^2C^2}{4}$

Thus attenuation will commence when $\frac{C}{L} = \frac{\omega^2C^2}{4}$

i.e. when $\omega_c^2 = \frac{4}{LC}$

from which, **cut-off frequency,** $\mathbf{f_c = \frac{1}{\pi\sqrt{(LC)}}}$

as in equation (3).

(b) Nominal impedance

When the frequency is very low, ω is small and the term $(\omega^2L^2/4)$ in equation (1) (or the term $(\omega^2C^2/4)$ in equation (4)) may be neglected. The characteristic impedance then becomes equal to $\sqrt{(L/C)}$, which is purely resistive. This value of the characteristic impedance is known as the **design impedance** or the

nominal impedance of the section and is often given the symbol R_0,

i.e. $$R_0 = \sqrt{\frac{L}{C}} \qquad (5)$$

Problem 1. Determine the cut-off frequency and the nominal impedance of each of the low-pass filter sections shown in Figure 45.19.

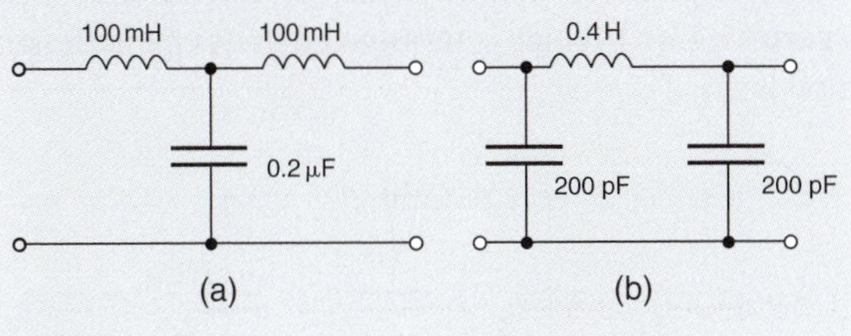

Figure 45.19

(a) Comparing Figure 45.19(a) with the low-pass T section in Figure 45.17 shows that $(L/2)=100\,\text{mH}$, i.e. inductance, $L=200\,\text{mH}=0.2\,\text{H}$ and capacitance, $C=0.2\,\mu\text{F}=0.2\times10^{-6}\,\text{F}$. From equation (3), cut-off frequency,

$$f_c=\frac{1}{\pi\sqrt{(LC)}}=\frac{1}{\pi\sqrt{(0.2\times0.2\times10^{-6})}}=\frac{10^3}{\pi(0.2)}$$

i.e. $\boldsymbol{f_c=1592\,\text{Hz}}$ or $\mathbf{1.592\,kHz}$

From equation (5), **nominal impedance,**

$$R_0=\sqrt{\left(\frac{L}{C}\right)}=\sqrt{\left(\frac{0.2}{0.2\times10^{-6}}\right)}=\mathbf{1000\,\Omega}\text{ or }\mathbf{1\,k\Omega}$$

(b) Comparing Figure 45.19(b) with the low-pass π section shown in Figure 45.18 shows that $(C/2)=200\,\text{pF}$, i.e. capacitance, $C=400\text{pF}=400\times10^{-12}\,\text{F}$ and inductance, $L=0.4\,\text{H}$.

From equation (3), **cut-off frequency**,

$$f_c=\frac{1}{\pi\sqrt{(LC)}}=\frac{1}{\pi\sqrt{(0.4\times400\times10^{-12})}}$$
$$=\mathbf{25.16\,kHz}$$

From equation (5), **nominal impedance,**

$$R_0=\sqrt{\left(\frac{L}{C}\right)}=\sqrt{\left(\frac{0.4}{400\times10^{-12}}\right)}=\mathbf{31.62\,k\Omega}$$

From equations (1) and (4) it is seen that the characteristic impedance Z_0 varies with ω, i.e. Z_0 varies with frequency. Thus if the nominal impedance is made to equal the load impedance into which the filter feeds then the matching deteriorates as the frequency increases from zero towards f_c. It is, however, convention to make the terminating impedance equal to the value of Z_0 well within the passband, i.e. to take the limiting value of Z_0 as the frequency approaches zero. This limit is obviously $\sqrt{(L/C)}$. This means that the filter is properly terminated at very low frequency but as the cut-off frequency is approached becomes increasingly mismatched. This is shown for a low-pass section in Figure 45.20 by curve (a). It is seen that an increasing loss is introduced into the passband. Curve (b) shows the attenuation due to the same low-pass section being correctly terminated at all frequencies. A curve lying somewhere between curves (a) and (b) will usually result for each section if several sections are cascaded and terminated in R_0, or if a matching section is inserted between the low-pass section and the load.

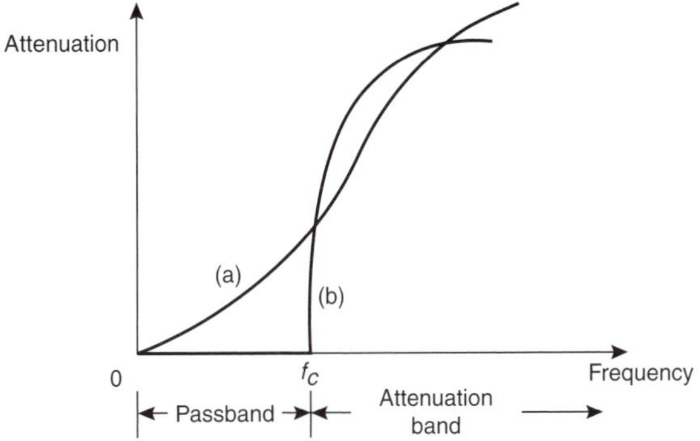

Figure 45.20

(c) To determine values of L and C given R_0 and f_c

If the values of the nominal impedance R_0 and the cut-off frequency f_c are known for a low-pass T or π section it is possible to determine the values of inductance and capacitance required to form the section.

From equation (5), $R_0=\sqrt{\dfrac{L}{C}}=\dfrac{\sqrt{L}}{\sqrt{C}}$ from which

$$\sqrt{L}=R_0\sqrt{C}$$

Substituting in equation (3) gives:

$$f_c=\frac{1}{\pi\sqrt{L}\sqrt{C}}=\frac{1}{\pi(R_0\sqrt{C})\sqrt{C}}=\frac{1}{\pi R_0 C}$$

from which, **capacitance** $\boldsymbol{C=\dfrac{1}{\pi R_0 f_c}}$ (6)

Similarly from equation (5), $\sqrt{C} = \frac{\sqrt{L}}{R_0}$

Substituting in equation (3) gives:

$$f_c = \frac{1}{\pi\sqrt{L}\left(\frac{\sqrt{L}}{R_0}\right)}$$

$$= \frac{R_0}{\pi L}$$

from which, **inductance,** $\boldsymbol{L = \frac{R_0}{\pi f_c}}$ (7)

Problem 2. A filter section is to have a characteristic impedance at zero frequency of 600 Ω and a cut-off frequency at 5 MHz. Design (a) a low-pass T section filter and (b) a low-pass π section filter to meet these requirements.

The characteristic impedance at zero frequency is the nominal impedance R_0, i.e. $R_0 = 600\,\Omega$; cut-off frequency, $f_c = 5\,\text{MHz} = 5 \times 10^6\,\text{Hz}$.

From equation (6),

$$\text{capacitance, } C = \frac{1}{\pi R_0 f_c} = \frac{1}{\pi(600)(5\times 10^6)}\,\text{F} = 106\,\text{pF}$$

and from equation (7),

$$\text{inductance, } L = \frac{R_0}{\pi f_c} = \frac{600}{\pi(5\times 10^6)}\,\text{H} = 38.2\,\mu\text{H}$$

(a) A low-pass T section filter is shown in Figure 45.21(a), where the series arm inductances are each $L/2$ (see Figure 45.17), i.e. $(38.2/2) = 19.1\,\mu\text{H}$

(b) A low-pass π section filter is shown in Figure 45.21(b), where the shunt arm capacitances are each $(C/2)$ (see Figure 45.18), i.e. $(106/2) = 53\,\text{pF}$

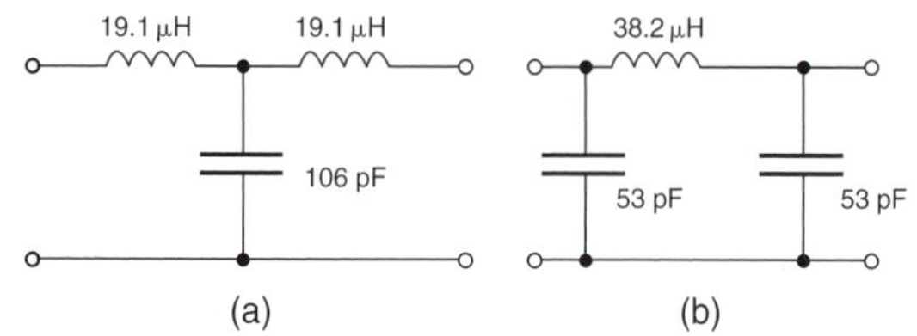

Figure 45.21

(d) 'Constant-k' prototype low-pass filter

A ladder network is shown in Figure 45.22, the elements being expressed in terms of impedances Z_1 and Z_2. The network shown in Figure 45.22(b) is equivalent to the network shown in Figure 45.22(a), where $(Z_1/2)$ in series with $(Z_1/2)$ equals Z_1 and $2Z_2$ in parallel with $2Z_2$ equals Z_2. Removing sections $ABED$ and $FGJH$ from Figure 45.22(b) gives the T section shown in Figure 45.23(a), which is terminated in its characteristic impedance Z_{OT}, and the π section shown in Figure 45.23(b), which is terminated in its characteristic impedance $Z_{0\pi}$

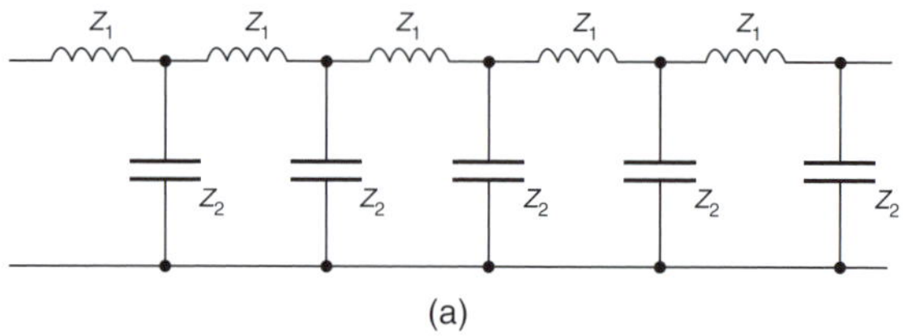

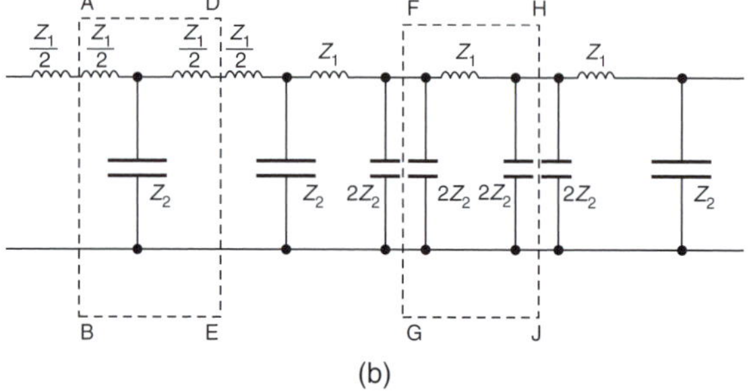

Figure 45.22

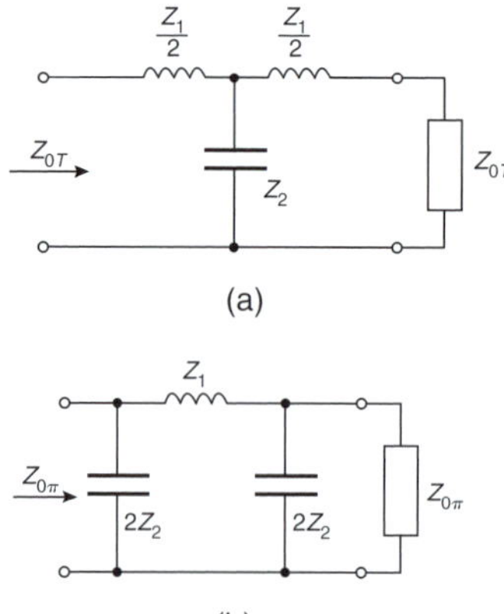

Figure 45.23

From equation (1), page 667,

$$Z_{OT} = \sqrt{\left[\left(\frac{Z_1}{2}\right)^2 + 2\left(\frac{Z_1}{2}\right)Z_2\right]}$$

Part 4

i.e. $$Z_{OT}=\sqrt{\left(\frac{Z_1^2}{4}+Z_1Z_2\right)} \qquad (8)$$

From equation (3), page 667

$$Z_{0\pi}=\sqrt{\left[\frac{(Z_1)(2Z_2)^2}{Z_1+2(2Z_2)}\right]}=\sqrt{\left[\frac{Z_1(Z_1)(4Z_2^2)}{Z_1(Z_1+4Z_2)}\right]}$$

$$=\frac{2Z_1Z_2}{\sqrt{(Z_1^2+4Z_1Z_2)}}=\frac{Z_1Z_2}{\sqrt{\left(\frac{Z_1^2}{4}+Z_1Z_2\right)}}$$

i.e. $Z_{0\pi}=\dfrac{Z_1Z_2}{Z_{OT}}$ from equation (8)

Thus $$\boldsymbol{Z_{0T}Z_{0\pi}=Z_1Z_2} \qquad (9)$$

This is a general expression relating the characteristic impedances of T and π sections made up of equivalent series and shunt impedances.
From the low-pass sections shown in Figures 45.17 and 45.18,

$$Z_1=j\omega L \text{ and } Z_2=\frac{1}{j\omega C}$$

Hence $$Z_{0T}Z_{0\pi}=(j\omega L)\left(\frac{1}{j\omega C}\right)=\frac{L}{C}$$

Thus, from equation (5), $$\boldsymbol{Z_{0T}Z_{0\pi}=R_0^2} \qquad (10)$$

From equations (9) and (10),

$$Z_{0T}Z_{0\pi}=Z_1Z_2=R_0^2=\text{constant (k)}.$$

A ladder network composed of reactances, the series reactances being of opposite sign to the shunt reactances (as in Figure 45.23) are called **'constant-k' filter sections**. Positive (i.e. inductive) reactance is directly proportional to frequency, and negative (i.e. capacitive) reactance is inversely proportional to frequency. Thus the product of the series and shunt reactances is independent of frequency (see equations (9) and (10)). The constancy of this product has given this type of filter its name.

From equation (10), it is seen that Z_{0T} and $Z_{0\pi}$ will either be both real or both imaginary together (since $j^2=-1$). Also, when Z_{0T} changes from real to imaginary at the cut-off frequency, so will $Z_{0\pi}$. The two sections shown in Figures 45.17 and 45.18 will thus have identical cut-off frequencies and thus identical passbands. Constant-k sections of any kind of filter are known as **prototypes**.

(e) Practical low-pass filter characteristics

From equation (1), the characteristic impedance Z_{0T} of a low-pass T section is given by:

$$Z_{0T}=\sqrt{\left(\frac{L}{C}-\frac{\omega^2L^2}{4}\right)}$$

Rearranging gives:

$$Z_{0T}=\sqrt{\left[\frac{L}{C}\left(1-\frac{\omega^2LC}{4}\right)\right]}=\sqrt{\left(\frac{L}{C}\right)}\sqrt{\left(1-\frac{\omega^2LC}{4}\right)}$$

$$=R_0\sqrt{\left(1-\frac{\omega^2LC}{4}\right)} \text{ from equation (5)}$$

From equation (2), $\omega_c^2=\dfrac{4}{LC}$, hence

$$Z_{0T}=R_0\sqrt{\left(1-\frac{\omega^2}{\omega_c^2}\right)}$$

i.e. $$\boldsymbol{Z_{0T}=R_0\sqrt{\left[1-\left(\frac{\omega}{\omega_c}\right)^2\right]}} \qquad (11)$$

Also, from equation (10),

$$Z_{0\pi}=\frac{R_0^2}{Z_{0T}}$$

$$=\frac{R_0^2}{R_0\sqrt{\left[1-\left(\frac{\omega}{\omega_c}\right)^2\right]}}$$

i.e. $$\boldsymbol{Z_{0\pi}=\frac{R_0}{\sqrt{\left[1-\left(\frac{\omega}{\omega_c}\right)^2\right]}}} \qquad (12)$$

(Alternatively, the expression for $Z_{0\pi}$ could have been obtained from equation (4), where

$$Z_{0\pi}=\sqrt{\left(\frac{\frac{1}{C}}{\frac{C}{L}-\frac{\omega^2C^2}{4}}\right)}=\sqrt{\left[\frac{\frac{L}{C}}{\frac{L}{C}\left(\frac{C}{L}-\frac{\omega^2C^2}{4}\right)}\right]}$$

$$=\frac{\sqrt{\frac{L}{C}}}{\sqrt{\left(1-\frac{\omega^2LC}{4}\right)}}=\frac{R_0}{\sqrt{\left[1-\left(\frac{\omega}{\omega_c}\right)^2\right]}} \text{ as above).}$$

From equations (11) and (12), when $\omega=0$ (i.e. when the frequency is zero),

$$Z_{0T}=Z_{0\pi}=R_0$$

At the cut-off frequency, f_c, $\omega = \omega_c$
and from equation (11), Z_{0T} falls to zero,
and from equation (12), $Z_{0\pi}$ rises to infinity.
These results are shown graphically in Figure 45.24, where it is seen that Z_{0T} decreases from R_0 at zero frequency to zero at the cut-off frequency; $Z_{0\pi}$ rises from its initial value of R_0 to infinity at f_c

(At a frequency, $f = 0.95 f_c$, for example,

$$Z_{0\pi} = \frac{R_0}{\sqrt{(1 - 0.95^2)}} = 3.2 R_0$$

from equation (12))

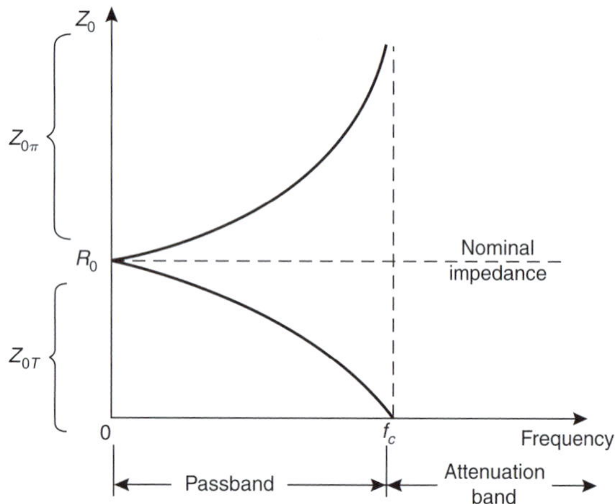

Figure 45.24

Note that since Z_0 becomes purely reactive in the attenuation band, it is not shown in this range in Figure 45.24.

Figure 45.2(a), on page 699, showed an ideal low-pass filter section characteristic. In practice, the characteristic curve of a low-pass prototype filter section looks more like that shown in Figure 45.25. The characteristic may be improved somewhat closer to the ideal by connecting two or more identical sections in cascade. This produces a much sharper cut-off characteristic, although the attenuation in the passband is increased a little.

Problem 3. The nominal impedance of a low-pass π section filter is 500 Ω and its cut-off frequency is at 100 kHz. Determine (a) the value of the characteristic impedance of the section at a frequency of 90 kHz and (b) the value of the characteristic impedance of the equivalent low-pass T section filter.

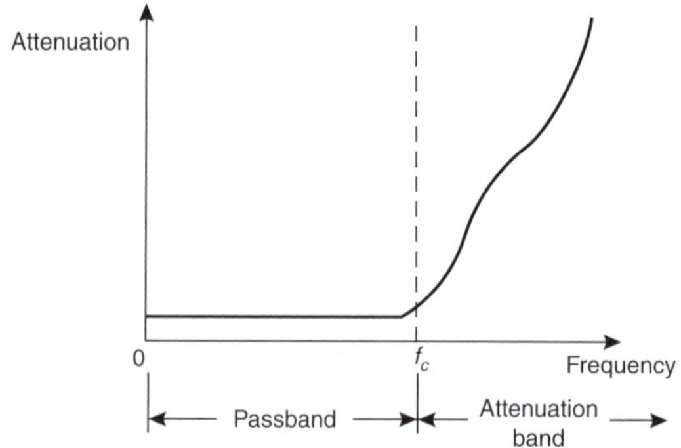

Figure 45.25

At zero frequency the characteristic impedance of the π and T section filters will be equal to the nominal impedance of 500 Ω.

(a) From equation (12), the characteristic impedance of the π section at 90 kHz is given by:

$$Z_{0\pi} = \frac{R_0}{\sqrt{\left[1 - \left(\frac{\omega}{\omega_c}\right)^2\right]}}$$

$$= \frac{500}{\sqrt{\left[1 - \left(\frac{2\pi 90 \times 10^3}{2\pi 100 \times 10^3}\right)^2\right]}}$$

$$= \frac{500}{\sqrt{[1 - (0.9)^2]}} = \mathbf{1147\,\Omega}$$

(b) From equation (11), the characteristic impedance of the T section at 90 kHz is given by:

$$Z_{0T} = R_0 \sqrt{\left[1 - \left(\frac{\omega}{\omega_c}\right)^2\right]} = 500\sqrt{[1 - (0.9)^2]}$$

$$= \mathbf{218\,\Omega}$$

(Check: from equation (10),

$$Z_{0T} Z_{0\pi} = (218)(1147) = 250\,000 = 500^2 = R_0^2)$$

Typical low-pass characteristics of characteristic impedance against frequency are shown in Figure 45.24.

Problem 4. A low-pass π section filter has a nominal impedance of 600 Ω and a cut-off frequency of 2 MHz. Determine the frequency at which the characteristic impedance of the section is (a) 600 Ω (b) 1 kΩ (c) 10 kΩ.

Part 4

From equation (12), $Z_{0\pi} = \dfrac{R_0}{\sqrt{\left[1-\left(\frac{\omega}{\omega_c}\right)^2\right]}}$

(a) When $Z_{0\pi} = 600\,\Omega$ and $R_0 = 600\,\Omega$, then $\omega = 0$, i.e. **the frequency is zero**.

(b) When $Z_{0\pi} = 1000\,\Omega$, $R_0 = 600\,\Omega$ and $f_c = 2\times10^6$ Hz

then $1000 = \dfrac{600}{\sqrt{\left[1-\left(\frac{2\pi f}{2\pi 2\times10^6}\right)^2\right]}}$

from which, $1-\left(\dfrac{f}{2\times10^6}\right)^2 = \left(\dfrac{600}{1000}\right)^2 = 0.36$

and $\left(\dfrac{f}{2\times10^6}\right) = \sqrt{(1-0.36)} = 0.8$

Thus when $Z_{0\pi} = 1000\,\Omega$,

frequency, f $= (0.8)(2\times10^6) =$ **1.6 MHz**

(c) When $Z_{0\pi} = 10\,\mathrm{k}\Omega$, then

$10000 = \dfrac{600}{\sqrt{\left[1-\left(\frac{f}{2}\right)^2\right]}}$

where frequency, f, is in megahertz.

Thus $1-\left(\dfrac{f}{2}\right)^2 = \left(\dfrac{600}{10000}\right)^2 = (0.06)^2$

and $\dfrac{f}{2} = \sqrt{[1-(0.06)^2]} = 0.9982$

Hence when $Z_{0\pi} = 10\,\mathrm{k}\Omega$,

frequency f $= (2)(0.9982) =$ **1.996 MHz**

The above three results are seen to be borne out in the characteristic of $Z_{0\pi}$ against frequency shown in Figure 45.24.

Now try the following Practice Exercise

Practice Exercise 179 Low-pass filter sections (Answers on page 833)

1. Determine the cut-off frequency and the nominal impedance of each of the low-pass filter sections shown in Figure 45.26.

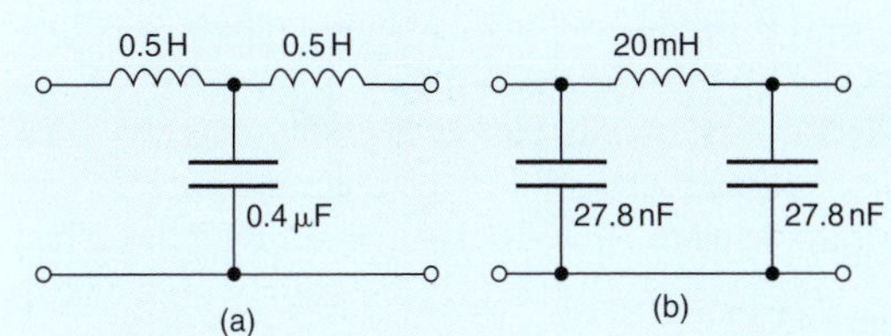

Figure 45.26

2. A filter section is to have a characteristic impedance at zero frequency of 500 Ω and a cut-off frequency of 1 kHz. Design (a) a low-pass T section filter and (b) a low-pass π section filter to meet these requirements.
3. Determine the value of capacitance required in the shunt arm of a low-pass T section if the inductance in each of the series arms is 40 mH and the cut-off frequency of the filter is 2.5 kHz.
4. The nominal impedance of a low-pass π section filter is 600 Ω and its cut-off frequency is at 25 kHz. Determine (a) the value of the characteristic impedance of the section at a frequency of 20 kHz and (b) the value of the characteristic impedance of the equivalent low-pass T section filter.
5. The nominal impedance of a low-pass π section filter is 600 Ω. If the capacitance in each of the shunt arms is 0.1 μF, determine the inductance in the series arm. Make a sketch of the ideal and the practical attenuation/frequency characteristic expected for such a filter section.
6. A low-pass T section filter has a nominal impedance of 600 Ω and a cut-off frequency of 10 kHz. Determine the frequency at which the characteristic impedance of the section is (a) zero, (b) 300 Ω, (c) 600 Ω.

45.6 High-pass filter sections

(a) The cut-off frequency

High-pass T and π sections are shown in Figure 45.27 (as derived in Section 45.4), each being terminated in their characteristic impedance.

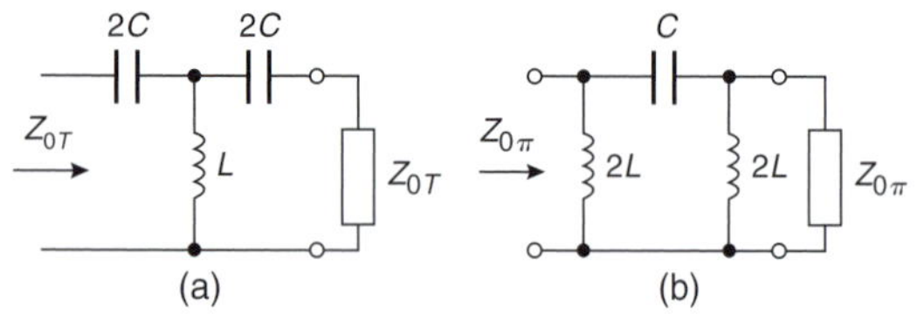

Figure 45.27

From equation (1), page 667, the characteristic impedance of a T section is given by:

$$Z_{0T} = \sqrt{(Z_A^2 + 2Z_A Z_B)}$$

From Figure 45.27(a), $Z_A = \dfrac{1}{j\omega 2C}$ and $Z_B = j\omega L$

Thus $$Z_{0T} = \sqrt{\left[\left(\frac{1}{j\omega 2C}\right)^2 + 2\left(\frac{1}{j\omega 2C}\right)(j\omega L)\right]}$$

$$= \sqrt{\left[\frac{1}{-4\omega^2C^2} + \frac{L}{C}\right]}$$

i.e. $$Z_{0T} = \sqrt{\left(\frac{L}{C} - \frac{1}{4\omega^2C^2}\right)} \qquad (13)$$

Z_{0T} will be real when $\dfrac{L}{C} > \dfrac{1}{4\omega^2C^2}$

Thus the filter will pass all frequencies above the point

where $\dfrac{L}{C} = \dfrac{1}{4\omega^2C^2}$

i.e. where $\omega_c^2 = \dfrac{1}{4LC}$ (14)

where $\omega_c = 2\pi f_c$, and f_c is the cut-off frequency.

Hence $(2\pi f_c)^2 = \dfrac{1}{4LC}$

and **the cut-off frequency,** $$\boldsymbol{f_c = \frac{1}{4\pi\sqrt{(LC)}}} \qquad (15)$$

The same equation for the cut-off frequency is obtained for the high-pass π network shown in Figure 45.27(b) as follows:

From equation (3), page 667, the characteristic impedance of a symmetrical π section is given by:

$$Z_{0\pi} = \sqrt{\left(\frac{Z_1 Z_2^2}{Z_1 + 2Z_2}\right)}$$

From Figure 45.27(b), $Z_1 = \dfrac{1}{j\omega C}$ and $Z_2 = j2\omega L$

Hence $$Z_{0\pi} = \sqrt{\left\{\frac{\left(\frac{1}{j\omega C}\right)(j2\omega L)^2}{\frac{1}{j\omega C} + 2j2\omega L}\right\}}$$

$$= \sqrt{\left\{\frac{j4\frac{\omega L^2}{C}}{j\left(4\omega L - \frac{1}{\omega C}\right)}\right\}}$$

$$= \sqrt{\left(\frac{\frac{4L^2}{C}}{4L - \frac{1}{\omega^2 C}}\right)}$$

i.e. $$Z_{0\pi} = \sqrt{\left(\frac{1}{\frac{C}{L} - \frac{1}{4\omega^2L^2}}\right)} \qquad (16)$$

$Z_{0\pi}$ will be real when $\dfrac{C}{L} > \dfrac{1}{4\omega^2L^2}$ and the filter will pass all frequencies above the point where $\dfrac{C}{L} = \dfrac{1}{4\omega^2L^2}$, i.e. where $\omega_c^2 = \dfrac{1}{4LC}$ as above.

Thus the cut-off frequency for a high-pass π network is also given by

$$\boldsymbol{f_c = \frac{1}{4\pi\sqrt{(LC)}}} \quad \text{(as in equation (15))} \qquad (15')$$

(b) Nominal impedance

When the frequency is very high, ω is a very large value and the term $(1/4\omega^2C^2)$ in equations (13) and (16) are extremely small and may be neglected.

The characteristic impedance then becomes equal to $\sqrt{(L/C)}$, this being the nominal impedance. Thus for a high-pass filter section the nominal impedance R_0 is given by:

$$\boldsymbol{R_0 = \sqrt{\left(\frac{L}{C}\right)}} \qquad (17)$$

the same as for the low-pass filter sections.

Problem 5. Determine for each of the high-pass filter sections shown in Figure 45.28 (i) the cut-off frequency and (ii) the nominal impedance.

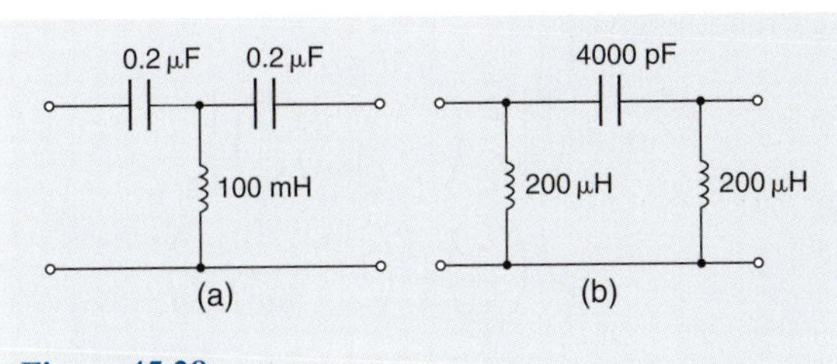

Figure 45.28

(a) Comparing Figure 45.28(a) with Figure 45.27(a) shows that:

$$2C = 0.2\,\mu\text{F, i.e. capacitance, } C = 0.1\,\mu\text{F} = 0.1\times 10^{-6}\,\text{F}$$

$$\text{and inductance, } L = 100\,\text{mH} = 0.1\,\text{H}$$

(i) From equation (15),

cut-off frequency,

$$f_c = \frac{1}{4\pi\sqrt{(LC)}} = \frac{1}{4\pi\sqrt{[(0.1)(0.1\times 10^{-6})]}}$$

i.e. $$f_c = \frac{10^3}{4\pi(0.1)} = \mathbf{796\,Hz}$$

(ii) From equation (17),

$$\textbf{nominal impedance, } R_0 = \sqrt{\left(\frac{L}{C}\right)} = \sqrt{\left(\frac{0.1}{0.1\times 10^{-6}}\right)} = \mathbf{1000\,\Omega} \text{ or } \mathbf{1\,k\Omega}$$

(b) Comparing Figure 45.28(b) with Figure 45.27(b) shows that:

$$2L = 200\,\mu\text{H, i.e. inductance, } L = 100\,\mu\text{H} = 10^{-4}\,\text{H}$$

$$\text{and capacitance } C = 4000\,\text{pF} = 4\times 10^{-9}\,\text{F}$$

(i) From equation (15′),

cut-off frequency,

$$f_c = \frac{1}{4\pi\sqrt{(LC)}} = \frac{1}{4\pi\sqrt{[(10^{-4})(4\times 10^{-9})]}} = \mathbf{126\,kHz}$$

(ii) From equation (17),

nominal impedance,

$$R_0 = \sqrt{\left(\frac{L}{C}\right)} = \sqrt{\left(\frac{10^{-4}}{4\times 10^{-9}}\right)} = \sqrt{\left(\frac{10^5}{4}\right)} = \mathbf{158\,\Omega}$$

(c) To determine values of L and C given R_0 and f_c

If the values of the nominal impedance R_0 and the cut-off frequency f_c are known for a high-pass T or π section, it is possible to determine the values of inductance L and capacitance C required to form the section.

From equation (17), $R_0 = \sqrt{\dfrac{L}{C}} = \dfrac{\sqrt{L}}{\sqrt{C}}$ from which,

$$\sqrt{L} = R_0\sqrt{C}$$

Substituting in equation (15) gives:

$$f_c = \frac{1}{4\pi\sqrt{L}\sqrt{C}} = \frac{1}{4\pi(R_0\sqrt{C})\sqrt{C}} = \frac{1}{4\pi R_0 C}$$

from which, **capacitance** $\mathbf{C = \dfrac{1}{4\pi R_0 f_c}}$ (18)

Similarly, from equation (17), $\sqrt{C} = \dfrac{\sqrt{L}}{R_0}$

Substituting in equation (15) gives:

$$f_c = \frac{1}{4\pi\sqrt{L}\left(\frac{\sqrt{L}}{R_0}\right)} = \frac{R_0}{4\pi L}$$

from which, **inductance,** $\mathbf{L = \dfrac{R_0}{4\pi f_c}}$ (19)

Problem 6. A filter is required to pass all frequencies above 25 kHz and to have a nominal impedance of 600 Ω. Design (a) a high-pass T section filter and (b) a high-pass π section filter to meet these requirements.

Cut-off frequency, $f_c = 25\times 10^3$ Hz and nominal impedance, $R_0 = 600\,\Omega$

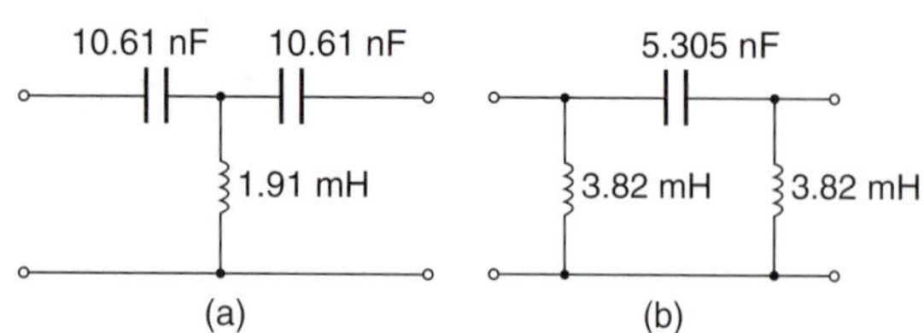

Figure 45.29

From equation (18),

$$C = \frac{1}{4\pi R_0 f_c} = \frac{1}{4\pi(600)(25 \times 10^3)} \text{F}$$

$$= \frac{10^{12}}{4\pi(600)(25 \times 10^3)} \text{pF}$$

i.e. $C = 5305\,\text{pF}$ or $5.305\,\text{nF}$

From equation (19), inductance,

$$L = \frac{R_0}{4\pi f_c} = \frac{600}{4\pi(25 \times 10^3)} \text{H} = 1.91\,\text{mH}$$

(a) A high-pass T section filter is shown in Figure 45.29(a), where the series arm capacitances are each $2C$ (see Figure 45.27(a)), i.e. $2 \times 5.305 = 10.61\,\text{nF}$.

(b) A high-pass π section filter is shown in Figure 45.29(b), where the shunt arm inductances are each $2L$ (see Figure 45.27(b)), i.e. $2 \times 1.91 = 3.82\,\text{mH}$.

(d) 'Constant-k' prototype high-pass filter

It may be shown, in a similar way to that shown in Section 45.5(d), that for a high-pass filter section:

$$\boldsymbol{Z_{0T} Z_{0\pi} = Z_1 Z_2 = R_0^2}$$

where Z_1 and Z_2 are the total equivalent series and shunt arm impedances. The high-pass filter sections shown in Figure 45.27 are thus 'constant-k' prototype filter sections.

(e) Practical high-pass filter characteristics

From equation (13), the characteristic impedance Z_{0T} of a high-pass T section is given by:

$$Z_{0T} = \sqrt{\left(\frac{L}{C} - \frac{1}{4\omega^2 C^2}\right)}$$

Rearranging gives:

$$Z_{0T} = \sqrt{\left[\frac{L}{C}\left(1 - \frac{1}{4\omega^2 LC}\right)\right]}$$

$$= \sqrt{\left(\frac{L}{C}\right)}\sqrt{\left(1 - \frac{1}{4\omega^2 LC}\right)}$$

From equation (14), $\omega_c^2 = \dfrac{1}{4LC}$

Thus $$\boldsymbol{Z_{0T} = R_0\sqrt{\left[1 - \left(\frac{\omega_c}{\omega}\right)^2\right]}} \qquad (20)$$

Also, since $Z_{0T} Z_{0\pi} = R_0^2$

then $$Z_{0\pi} = \frac{R_0^2}{Z_{0T}} = \frac{R_0^2}{R_0\sqrt{\left[1 - \left(\frac{\omega_c}{\omega}\right)^2\right]}}$$

i.e. $$\boldsymbol{Z_{0\pi} = \frac{R_0}{\sqrt{\left[1 - \left(\frac{\omega_c}{\omega}\right)^2\right]}}} \qquad (21)$$

From equation (20),

when $\omega < \omega_c$, Z_{0T} is reactive,

when $\omega = \omega_c$, Z_{0T} is zero,

and when $\omega > \omega_c$, Z_{0T} is real, eventually increasing to R_0 when ω is very large.

Similarly, from equation (21),

when $\omega < \omega_c$, $Z_{0\pi}$ is reactive,

when $\omega = \omega_c$, $Z_{0\pi} = \infty \left(\text{i.e. } \dfrac{R_0}{0} = \infty\right)$

and when $\omega > \omega_c$, $Z_{0\pi}$ is real, eventually decreasing to R_0 when ω is very large.

Curves of Z_{0T} and $Z_{0\pi}$ against frequency are shown in Figure 45.30.

Figure 45.4(a), on page 700, showed an ideal high-pass filter section characteristic of attenuation against frequency. In practice, the characteristic curve of a high-pass prototype filter section would look more like that shown in Figure 45.31.

Problem 7. A low-pass T section filter having a cut-off frequency of 15 kHz is connected in series with a high-pass T section filter having a cut-off frequency of 10 kHz. The terminating impedance of the filter is 600 Ω.

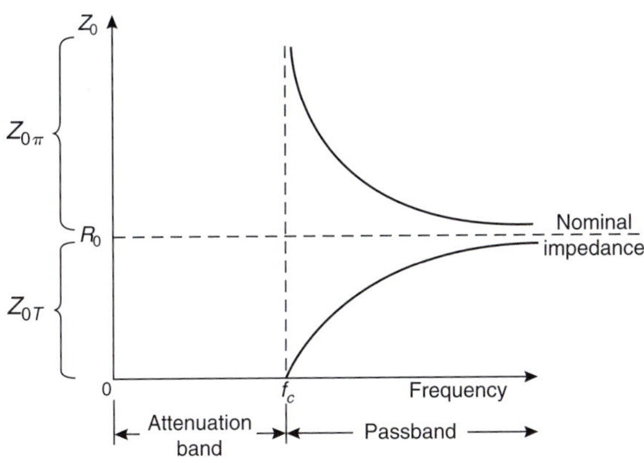

Figure 45.30

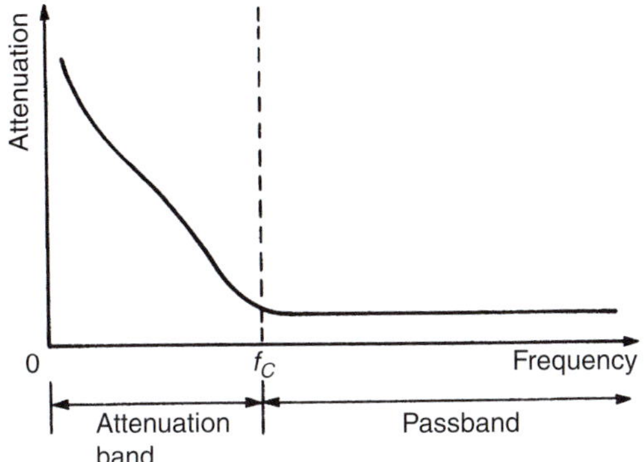

Figure 45.31

(a) Determine the values of the components comprising the composite filter.
(b) Sketch the expected attenuation against frequency characteristic.
(c) State the name given to the type of filter described.

(a) **For the low-pass *T* section filter:** $f_{cL} = 15\,000\,\text{Hz}$
From equation (6),

$$\text{capacitance, } C = \frac{1}{\pi R_0 f_c} = \frac{1}{\pi(600)(15\,000)} \equiv 35.4\,\text{nF}$$

From equation (7),

$$\text{inductance, } L = \frac{R_0}{\pi f_c} = \frac{600}{\pi(15\,000)} \equiv 12.73\,\text{mH}$$

Thus from Figure 45.17, the series arm inductances are each *L*/2, i.e. (12.73/2) = **6.37 mH** and the shunt arm capacitance is **35.4 nF**.

For a high-pass *T* section filter: $f_{CH} = 10\,000\,\text{Hz}$

From equation (18),

$$\text{capacitance, } C = \frac{1}{4\pi R_0 f_c} = \frac{1}{4\pi(600)(10\,000)} \equiv 13.3\,\text{nF}$$

From equation (19),

$$\text{inductance, } L = \frac{R_0}{4\pi f_c} = \frac{600}{4\pi 10\,000} \equiv 4.77\,\text{mH}$$

Thus from Figure 45.27(a), the series arm capacitances are each $2C$, i.e. $2 \times 13.3 = $ **26.6 nF**, and the shunt arm inductance is **4.77 mH**

The composite filter is shown in Figure 45.32.

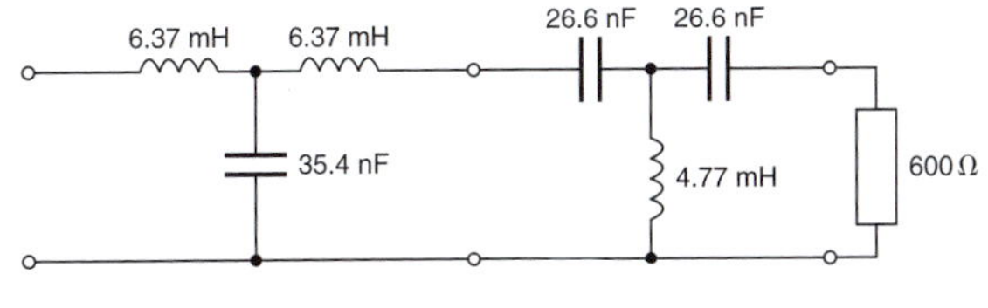

Figure 45.32

(b) A typical characteristic expected of attenuation against frequency is shown in Figure 45.33.

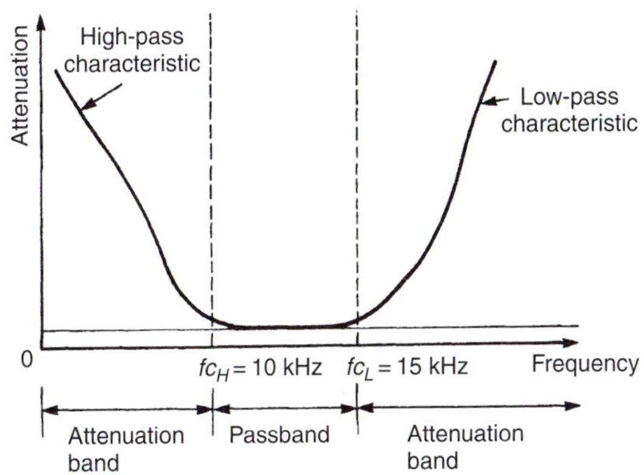

Figure 45.33

(c) The name given to the type of filter described is a **band-pass filter**. The ideal characteristic of such a filter is shown in Figure 45.5 on page 700.

Problem 8. A high-pass *T* section filter has a cut-off frequency of 500 Hz and a nominal impedance of 600 Ω. Determine the frequency at which the characteristic impedance of the section is (a) zero, (b) 300 Ω, (c) 590 Ω.

From equation (20), $Z_{0T} = R_0\sqrt{\left[1-\left(\frac{\omega_c}{\omega}\right)^2\right]}$

(a) When $Z_{0T}=0$, then $(\omega_c/\omega)=1$, i.e. **the frequency is 500 Hz**, the cut-off frequency.

(b) When $Z_{0T}=300\,\Omega$, $R_0=600\,\Omega$ and $f_c=500\,\text{Hz}$

$$300 = 600\sqrt{\left[1-\left(\frac{2\pi 500}{2\pi f}\right)^2\right]}$$

from which $\left(\frac{300}{600}\right)^2 = 1-\left(\frac{500}{f}\right)^2$

and $\frac{500}{f} = \sqrt{\left[1-\left(\frac{300}{600}\right)^2\right]} = \sqrt{0.75}$

Thus when $Z_{0T}=300\,\Omega$,

frequency, $\boldsymbol{f = \frac{500}{\sqrt{0.75}} = 577.4\,\text{Hz}}$

(c) When $Z_{0T}=590\,\Omega$, $590 = 600\sqrt{\left[1-\left(\frac{500}{f}\right)^2\right]}$

$$\frac{500}{f} = \sqrt{\left[1-\left(\frac{590}{600}\right)^2\right]} = 0.1818$$

Thus when $Z_{0T}=590\,\Omega$,

frequency, $\boldsymbol{f = \frac{500}{0.1818} = 2750\,\text{Hz}}$

The above three results are seen to be borne out in the characteristic of Z_{0T} against frequency shown in Figure 45.30.

Now try the following Practice Exercise

Practice Exercise 180 High-pass filter sections (Answers on page 833)

1. Determine for each of the high-pass filter sections shown in Figure 45.34 (i) the cut-off frequency and (ii) the nominal impedance.

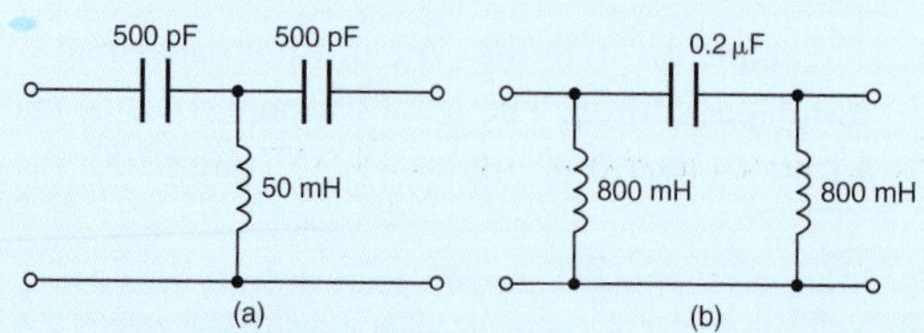

Figure 45.34

2. A filter is required to pass all frequencies above 4 kHz and to have a nominal impedance of 750 Ω. Design (a) an appropriate T section filter and (b) an appropriate π section filter to meet these requirements.

3. The inductance in each of the shunt arms of a high-pass π section filter is 50 mH. If the nominal impedance of the section is 600 Ω, determine the value of the capacitance in the series arm.

4. Determine the value of inductance required in the shunt arm of a high-pass T section filter if in each series arm it contains a 0.5 μF capacitor. The cut-off frequency of the filter section is 1500 Hz. Sketch the characteristic curve of characteristic impedance against frequency expected for such a filter section.

5. A high-pass π section filter has a nominal impedance of 500 Ω and a cut-off frequency of 50 kHz. Determine the frequency at which the characteristic impedance of the section is (a) 1 kΩ, (b) 800 Ω, (c) 520 Ω.

6. A low-pass T section filter having a cut-off frequency of 9 kHz is connected in series with a high-pass T section filter having a cut-off frequency of 6 kHz. The terminating impedance of the filter is 600 Ω.
   (a) Determine the values of the components comprising the composite filter.
   (b) Sketch the expected attenuation/frequency characteristic and state the name given to the type of filter described.

45.7 Propagation coefficient and time delay in filter sections

Propagation coefficient

In Figure 45.35, let A, B and C represent identical filter sections, the current ratios (I_1/I_2), (I_2/I_3) and (I_3/I_4) being equal.

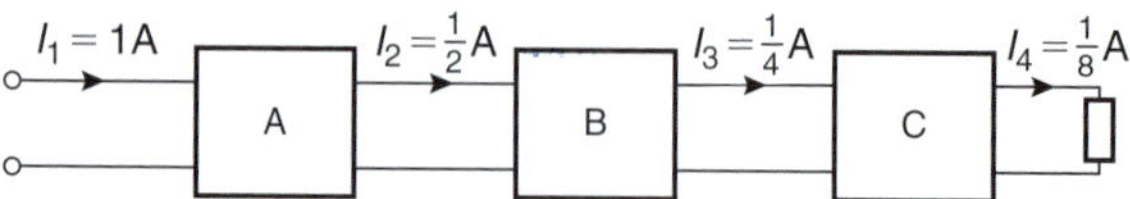

Figure 45.35

Although the rate of attenuation is the same in each section (i.e. the current output of each section is one half of the current input) the amount of attenuation in each is different (Section A attenuates by $\frac{1}{2}$ A, B attenuates by $\frac{1}{4}$ A and C attenuates by $\frac{1}{8}$ A). The attenuation is in fact in the form of a logarithmic decay and

$$\frac{I_1}{I_2} = \frac{I_2}{I_3} = \frac{I_3}{I_4} = e^{\gamma} \qquad (22)$$

where γ is called the **propagation coefficient** or the **propagation constant**.

From equation (22), propagation coefficient,

$$\gamma = \ln \frac{I_1}{I_2} \text{ nepers} \qquad (23)$$

(See Section 44.3, page 668, on logarithmic units.)

Unless Sections A, B and C in Figure 45.35 are purely resistive there will be a phase change in each section. Thus the ratio of the current entering a section to that leaving it will be a phasor quantity having both modulus and argument. The propagation constant which has no units is a complex quantity given by:

$$\boldsymbol{\gamma = \alpha + j\beta} \qquad (24)$$

where α is called the **attenuation coefficient**, measured in nepers, and β the **phase shift coefficient**, measured in radians. β is the angle by which a current leaving a section lags behind the current entering it.
From equations (22) and (24),

$$\frac{I_1}{I_2} = e^{\gamma} = e^{\alpha + j\beta} = (e^{\alpha})(e^{j\beta})$$

Since $\quad e^x = 1 + x + \frac{x^2}{2!} + \frac{x^3}{3!} + \frac{x^4}{4!} + \frac{x^5}{5!} + \cdots\cdots$

then $\quad e^{j\beta} = 1 + (j\beta) + \frac{(j\beta)^2}{2!} + \frac{(j\beta)^3}{3!} + \frac{(j\beta)^4}{4!} + \frac{(j\beta)^5}{5!} + \cdots\cdots$

$$= 1 + j\beta - \frac{\beta^2}{2!} - j\frac{\beta^3}{3!} + \frac{\beta^4}{4!} + j\frac{\beta^5}{5!} + \cdots\cdots$$

since $j^2 = -1$, $j^3 = -j$, $j^4 = +1$, and so on.

Hence $\quad e^{j\beta} = \left(1 - \frac{\beta^2}{2!} + \frac{\beta^4}{4!} - \cdots\right) + j\left(\beta - \frac{\beta^3}{3!} + \frac{\beta^5}{5!} - \cdots\right)$

$= \cos\beta + j\sin\beta$ from the power series for $\cos\beta$ and $\sin\beta$

Thus $\frac{I_1}{I_2} = e^{\alpha}e^{j\beta} = e^{\alpha}(\cos\beta + j\sin\beta) = e^{\alpha}\angle\beta$ in abbreviated polar form,

i.e. $$\boldsymbol{\frac{I_1}{I_2} = e^{\alpha}\angle\beta} \qquad (25)$$

Now $\quad e^{\alpha} = \left|\frac{I_1}{I_2}\right|$

from which

attenuation coefficient,

$$\boldsymbol{\alpha = \ln\left|\frac{I_1}{I_2}\right| \text{ nepers or } 20\lg\left|\frac{I_1}{I_2}\right| \text{ dB}}$$

If in Figure 45.35 current I_2 lags current I_1 by, say, 30°, i.e. $(\pi/6)$ rad, then the propagation coefficient γ of Section A is given by:

$$\gamma = \alpha + j\beta = \ln\left|\frac{1}{\frac{1}{2}}\right| + j\frac{\pi}{6}$$

i.e. $\quad \boldsymbol{\gamma = (0.693 + j0.524)}$

If there are n identical sections connected in cascade and terminated in their characteristic impedance, then

$$\frac{\boldsymbol{I_1}}{\boldsymbol{I_{n+1}}} = (e^{\gamma})^n = e^{n\gamma} = e^{n(\alpha + j\beta)} = \boldsymbol{e^{n\alpha}\angle n\beta} \ldots\ldots \qquad (26)$$

where I_{n+1} is the output current of the nth section.

Problem 9. The propagation coefficients of two filter networks are given by

(a) $\gamma = (1.25 + j0.52)$, (b) $\gamma = 1.794\angle -39.4°$

Determine for each (i) the attenuation coefficient and (ii) the phase shift coefficient.

(a) If $\gamma = (1.25 + j0.52)$
then (i) the attenuation coefficient, α, is given by the real part,

i.e. $\quad \boldsymbol{\alpha = 1.25\,\text{Np}}$

and (ii) the phase shift coefficient, β, is given by the imaginary part,

i.e. $\quad \boldsymbol{\beta = 0.52\,\text{rad}}$

(b) $\gamma = 1.794\angle -39.4° = 1.794[\cos(-39.4°) + j\sin(-39.4°)]$
$= (1.386 - j1.139)$

Part 4

Hence (i) the attenuation coefficient, $\boldsymbol{\alpha}=\mathbf{1.386\ Np}$ and (ii) the phase shift coefficient, $\boldsymbol{\beta}=\mathbf{-1.139\ rad}$

Problem 10. The current input to a filter section is $24\angle 10^\circ$ mA and the current output is $8\angle -45^\circ$ mA. Determine for the section (a) the attenuation coefficient, (b) the phase shift coefficient and (c) the propagation coefficient. (d) If five such sections are cascaded, determine the output current of the fifth stage and the overall propagation constant of the network.

Let $I_1=24\angle 10^\circ$ mA and $I_2=8\angle -45^\circ$ mA, then

$$\frac{I_1}{I_2}=\frac{24\angle 10^\circ}{8\angle -45^\circ}=3\angle 55^\circ=e^{\alpha}\angle\beta \text{ from equation (25).}$$

(a) Hence the attenuation constant, α, is obtained from $3=e^{\alpha}$, i.e. $\boldsymbol{\alpha}=\mathbf{\ln 3=1.099\ Np}$

(b) The phase shift coefficient $\beta=55^\circ\times\dfrac{\pi}{180}=\mathbf{0.960\ rad}$

(c) The propagation coefficient

$\gamma=\alpha+j\beta=\mathbf{(1.099+j0.960)}$ or $\mathbf{1.459\angle 41.14^\circ}$

(d) If I_6 is the current output of the fifth stage, then from equation (26),

$$\frac{I_1}{I_6}=(e^{\gamma})^n=[3\angle 55^\circ]^5=243\angle 275^\circ$$

(by De Moivre's theorem)

Thus the output current of the fifth stage,

$$I_6=\frac{I_1}{243\angle 275^\circ}=\frac{24\angle 10^\circ}{243\angle 275^\circ}$$

$$=\mathbf{0.0988\angle -265^\circ\ mA} \text{ or } \mathbf{98.8\angle 95^\circ\ \mu A}$$

Let the overall propagation coefficient be γ'

then $\dfrac{I_1}{I_6}=243\angle 275^\circ=e^{\gamma'}=e^{\alpha'}\angle\beta'$

The overall attenuation coefficient $\alpha'=\ln 243=5.49$

and the overall phase shift coefficient $\beta'=275^\circ\times\dfrac{\pi}{180^\circ}=4.80$ rad

Hence the overall propagation coefficient $\boldsymbol{\gamma'}=\mathbf{(5.49+j4.80)}$ or $\mathbf{7.29\angle 41.16^\circ}$

Problem 11. For the low-pass T section filter shown in Figure 45.36 determine (a) the attenuation coefficient, (b) the phase shift coefficient and (c) the propagation coefficient γ

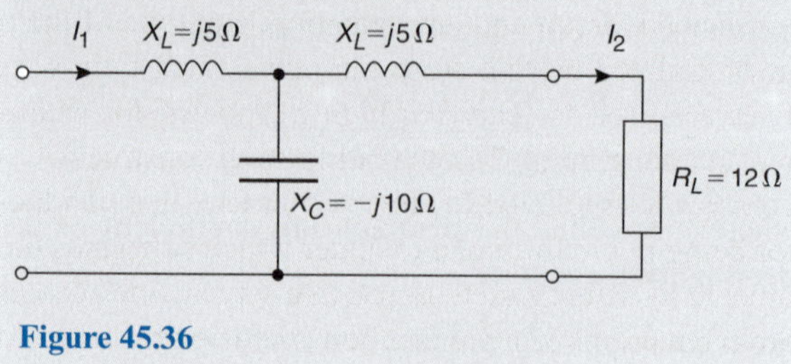

Figure 45.36

By current division in Figure 45.36,

$$I_2=\left(\frac{X_C}{X_C+X_L+R_L}\right)I_1$$

from which
$$\frac{I_1}{I_2}=\frac{X_C+X_L+R_L}{X_C}=\frac{-j10+j5+12}{-j10}$$
$$=\frac{-j5+12}{-j10}=\frac{-j5}{-j10}+\frac{12}{-j10}$$
$$=0.5+\frac{j12}{-j^2 10}=0.5+j1.2$$
$$=1.3\angle 67.38^\circ \text{ or } 1.3\angle 1.176$$

From equation (25), $\dfrac{I_1}{I_2}=e^{\alpha}\angle\beta=1.3\angle 1.176$

(a) The attenuation coefficient, $\alpha=\ln 1.3=\mathbf{0.262\ Np}$

(b) The phase shift coefficient, $\boldsymbol{\beta}=\mathbf{1.176\ rad}$

(c) The propagation coefficient, $\gamma=\alpha+j\beta=\mathbf{(0.262+j1.176)}$ or $\mathbf{1.205\angle 77.44^\circ}$

Variation in phase angle in the passband of a filter

In practice, the low- and high-pass filter sections discussed in Sections 45.5 and 45.6 would possess a phase shift between the input and output voltages which varies considerably over the range of frequency comprising the passband.

Let the **low-pass prototype *T* section** shown in Figure 45.37 be terminated as shown in its nominal impedance R_0. The input impedance for frequencies much less than the cut-off frequency is thus also equal to R_0 and is resistive. The phasor diagram representing Figure 45.37 is shown in Figure 45.38 and is produced as follows:

(i) V_1 and I_1 are in phase (since the input impedance is resistive)

Part 4

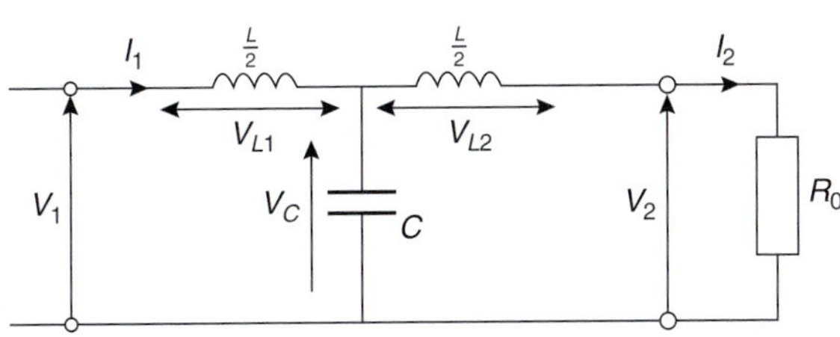

Figure 45.37

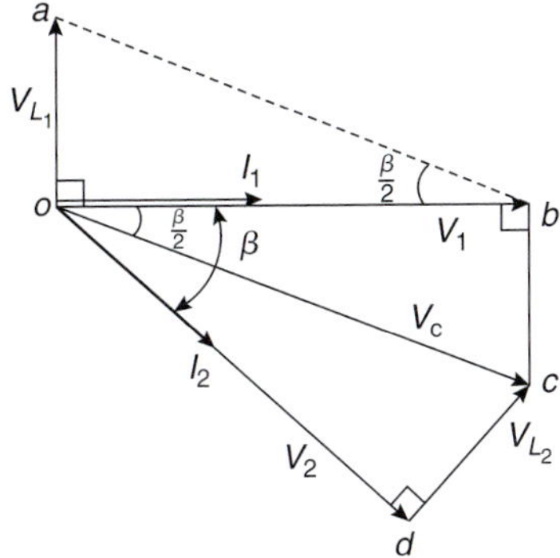

Figure 45.38

(ii) Voltage $V_{L1} = I_1 X_L = I_1\left(\frac{\omega L}{2}\right)$, which leads I by 90°

(iii) Voltage V_1 is the phasor sum of V_{L1} and V_C. Thus V_C is drawn as shown, completing the parallelogram *oabc*

(iv) Since no power is dissipated in reactive elements $V_1 = V_2$ in magnitude

(v) Voltage $V_{L2} = I_2\left(\frac{\omega L}{2}\right) = I_1\left(\frac{\omega L}{2}\right) = V_{L1}$

(vi) Voltage V_C is the phasor sum of V_{L2} and V_2 as shown by triangle *ocd*, where V_{L2} is at right-angles to V_2

(vii) Current I_2 is in phase with V_2 since the output impedance is resistive. The phase lag over the section is the angle between V_1 and V_2 shown as angle β in Figure 45.38,

where $\quad \tan\frac{\beta}{2} = \frac{oa}{ob} = \frac{V_{L1}}{V_1} = \frac{I_1\left(\frac{\omega L}{2}\right)}{I_1 R_0} = \frac{\frac{\omega L}{2}}{R_0}$

From equation (5), $R_0 = \sqrt{\frac{L}{C}}$

thus $\tan\frac{\beta}{2} = \frac{\frac{\omega L}{2}}{\sqrt{\frac{L}{C}}} = \frac{\omega\sqrt{(LC)}}{2}$

For angles of β up to about 20°,

$\tan\frac{\beta}{2} \approx \frac{\beta}{2}$ radians

Thus when $\beta < 20°$, $\frac{\beta}{2} = \frac{\omega\sqrt{(LC)}}{2}$

from which, **phase angle, $\beta = \omega\sqrt{(LC)}$ radian** (27)

Since $\beta = 2\pi f\sqrt{(LC)} = (2\pi\sqrt{(LC)})f$ then β is proportional to f and a graph of β (vertical) against frequency (horizontal) should be a straight line of gradient $2\pi\sqrt{(LC)}$ and passing through the origin. However, in practice this is only usually valid up to a frequency of about $0.7\,f_c$ for a low-pass filter and a typical characteristic is shown in Figure 45.39. At the cut-off frequency, $\beta = \pi$ rad. For frequencies within the attenuation band, the phase shift is unimportant, since all voltages having such frequencies are suppressed.

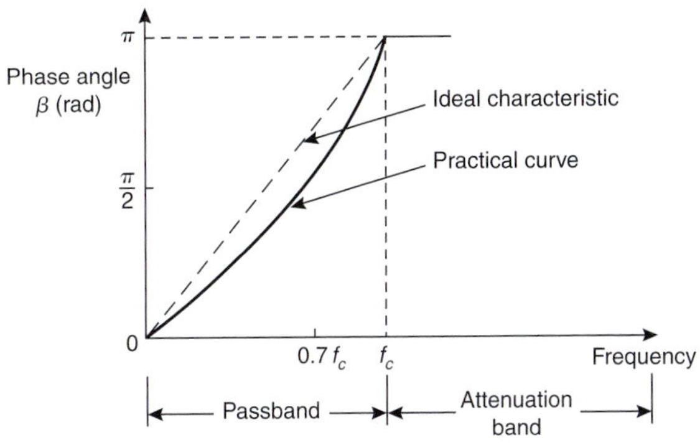

Figure 45.39

A **high-pass prototype *T* section** is shown in Figure 45.40(a) and its phasor diagram in Figure 45.40(b), the latter being produced by similar reasoning to above.

From Figure 45.40(b), $\tan\frac{\beta}{2} = \frac{V_{C1}}{V_1} = \frac{I_1\left(\frac{1}{\omega 2C}\right)}{I_1 R_0}$

$$= \frac{1}{2\omega C R_0} = \frac{1}{2\omega C\sqrt{\frac{L}{C}}}$$

$$= \frac{1}{2\omega\sqrt{(LC)}}$$

i.e. $\beta = \frac{1}{\omega\sqrt{(LC)}} = \frac{1}{(2\pi\sqrt{(LC)})f}$ for small angles.

Thus the phase angle is inversely proportional to frequency. The β/f characteristics of an ideal and a practical high-pass filter are shown in Figure 45.41.

Part 4

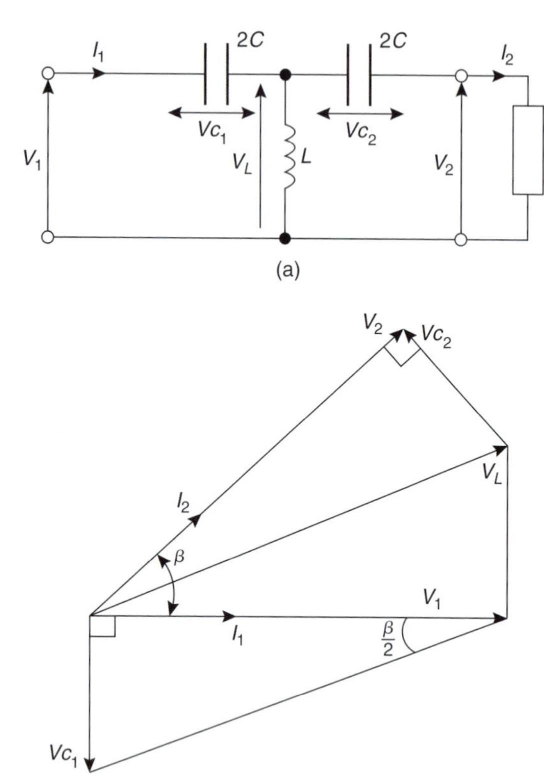

Figure 45.40

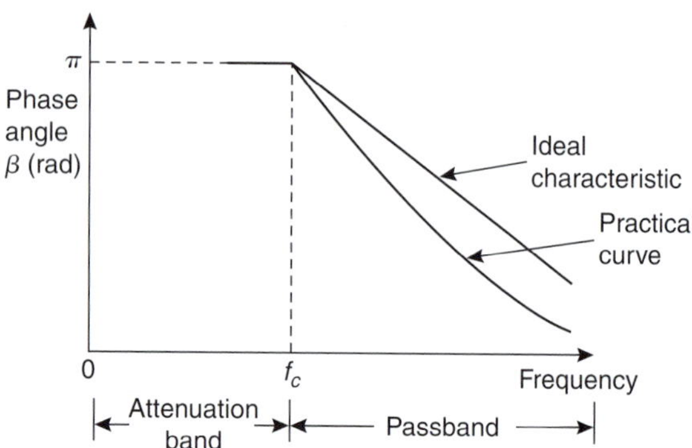

Figure 45.41

Time delay

The change of phase that occurs in a filter section depends on the time the signal takes to pass through the section. The phase shift β may be expressed as a time delay. If the frequency of the signal is f then the periodic time is $(1/f)$ seconds.

Hence the time delay $= \dfrac{\beta}{2\pi} \times \dfrac{1}{f} = \dfrac{\beta}{\omega}$

From equation (27), $\beta = \omega\sqrt{(LC)}$. Thus

$$\textbf{time delay} = \frac{\omega\sqrt{(LC)}}{\omega} = \sqrt{(LC)} \qquad (28)$$

when angle β is small.

Equation (28) shows that the time delay, or **transit time**, is independent of frequency. Thus a phase shift which is proportional to frequency (equation (27)) results in a time delay which is independent of frequency. Hence if the input to the filter section consists of a complex wave composed of several harmonic components of differing frequency, the output will consist of a complex wave made up of the sum of corresponding components all delayed by the same amount. There will therefore be no phase distortion due to varying time delays for the separate frequency components.

In practice, however, phase shift β tends not to be constant and the increase in time delay with rising frequency causes distortion of non-sinusoidal inputs, this distortion being superimposed on that due to the attenuation of components whose frequency is higher than the cut-off frequency.

At the cut-off frequency of a prototype low-pass filter, the phase angle $\beta = \pi$ rad. Hence the time delay of a signal through such a section at the cut-off frequency is given by

$$\frac{\beta}{\omega} = \frac{\pi}{2\pi f_c} = \frac{1}{2f_c} = \frac{1}{2\dfrac{1}{\pi\sqrt{(LC)}}} \text{ from equation (3),}$$

i.e. at f_c, **the transit time** $= \dfrac{\pi\sqrt{(LC)}}{2}$ **seconds** (29)

Problem 12. Determine for the filter section shown in Figure 45.42 (a) the time delay for the signal to pass through the filter, assuming the phase shift is small and (b) the time delay for a signal to pass through the section at the cut-off frequency.

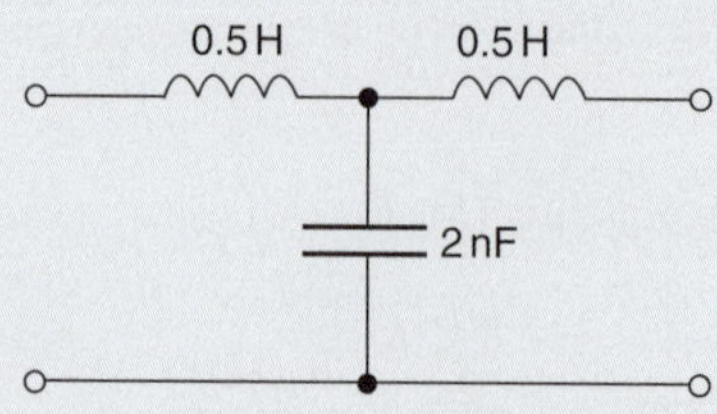

Figure 45.42

Comparing Figure 45.42 with the low-pass T section of Figure 45.13(a) shows that

$\frac{L}{2} = 0.5$ H, thus inductance $L = 1$ H, and capacitance $C = 2$ nF

(a) From equation (28),

$$\textbf{time delay} = \sqrt{(LC)} = \sqrt{[(1)(2\times 10^{-9})]} = \mathbf{44.7\,\mu s}$$

(b) From equation (29), at the cut-off frequency,

$$\textbf{time delay} = \frac{\pi}{2}\sqrt{(LC)} = \frac{\pi}{2}(44.7) = \mathbf{70.2\,\mu s}$$

Problem 13. A filter network comprising n identical sections passes signals of all frequencies up to 500 kHz and provides a total delay of 9.55 µs. If the nominal impedance of the circuit into which the filter is inserted is 1 kΩ, determine (a) the values of the elements in each section and (b) the value of n.

Cut-off frequency, $f_c = 500\times 10^3$ Hz and nominal impedance $R_0 = 1000\,\Omega$.

Since the filter passes frequencies up to 500 kHz then it is a **low-pass filter**.

(a) From equations (6) and (7), for a low-pass filter section,

$$\text{capacitance, } C = \frac{1}{\pi R_0 f_c} = \frac{1}{\pi(1000)(500\times 10^3)} \equiv \mathbf{636.6\,pF}$$

$$\text{and inductance, } L = \frac{R_0}{\pi f_c} = \frac{1000}{\pi(500\times 10^3)} \equiv \mathbf{636.6\,\mu H}$$

Thus if the section is a **low-pass T section** then the inductance in each series arm will be $(L/2) =$ **318.3 µH** and the capacitance in the shunt arm will be **636.6 pF**

If the section is a **low-pass π section** then the inductance in the series arm will be **636.6 µH** and the capacitance in each shunt arm will be $(C/2) =$ **318.3 pF**

(b) From equation (28), the time delay for a single section

$$= \sqrt{(LC)} = \sqrt{[(636.6\times 10^{-6})(636.6\times 10^{-12})]} = 0.6366\,\mu s$$

For a time delay of 9.55 µs therefore, the number of cascaded sections required is given by

$$\frac{9.55}{0.6366} = 15, \text{ i.e. } \boldsymbol{n = 15}$$

Problem 14. A filter network consists of 8 sections in cascade having a nominal impedance of 1 kΩ. If the total delay time is 4 µs, determine the component values for each section if the filter is (a) a low-pass T network and (b) a high-pass π network.

Since the total delay time is 4 µs then the delay time of each of the 8 sections is $\frac{4}{8}$, i.e. 0.5 µs.
From equation (28), time delay $= \sqrt{(LC)}$

Hence $0.5\times 10^{-6} = \sqrt{(LC)}$ (i)

Also, from equation (5), $\sqrt{\frac{L}{C}} = 1000$ (ii)

From equation (ii), $\sqrt{L} = 1000\sqrt{C}$

Substituting in equation (i) gives:

$0.5\times 10^{-6} = (1000\sqrt{C})\sqrt{C} = 1000\,C$

from which, capacitance $C = \frac{0.5\times 10^{-6}}{1000} = 0.5$ nF

From equation (ii), $\sqrt{C} = \frac{\sqrt{L}}{1000}$

Substituting in equation (i) gives:

$$0.5\times 10^{-6} = (\sqrt{L})\left(\frac{\sqrt{L}}{1000}\right) = \frac{L}{1000}$$

from which inductance, $L = 500\,\mu$H

(a) If the filter is a **low-pass T section** then, from Figure 45.13(a), each series arm has an inductance of $L/2$, i.e. **250 µH** and the shunt arm has a capacitance of **0.5 nF**

(b) If the filter is a **high-pass π network** then, from Figure 45.16(b), the series arm has a capacitance of **0.5 nF** and each shunt arm has an inductance of 2L, i.e. **1000 µH** or **1 mH**

Now try the following Practice Exercise

Practice Exercise 181 Propagation coefficient and time delay (Answers on page 833)

1. A filter section has a propagation coefficient given by (a) $(1.79 - j0.63)$, (b) $1.378\angle 51.6^\circ$.

Part 4

Determine for each (i) the attenuation coefficient and (ii) the phase angle coefficient.

2. A filter section has a current input of $200\angle 20°$ mA and a current output of $16\angle -30°$ mA. Determine (a) the attenuation coefficient, (b) the phase shift coefficient and (c) the propagation coefficient. (d) If four such sections are cascaded, determine the current output of the fourth stage and the overall propagation coefficient.

3. Determine for the high-pass T section filter shown in Figure 45.43, (a) the attenuation coefficient, (b) the phase shift coefficient and (c) the propagation coefficient.

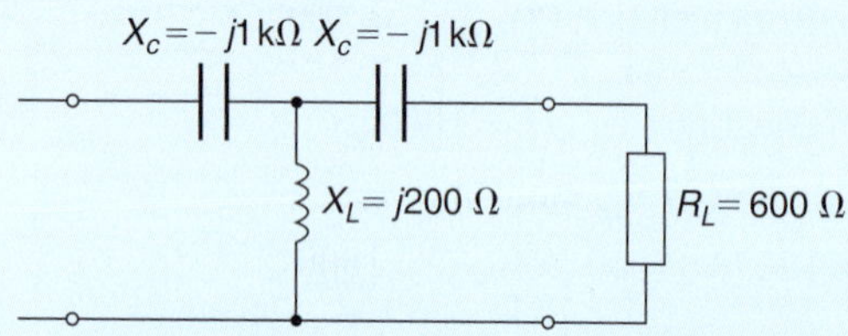

Figure 45.43

4. A low-pass T section filter has an inductance of 25 mH in each series arm and a shunt arm capacitance of 400 nF. Determine for the section (a) the time delay for the signal to pass through the filter, assuming the phase shift is small, and (b) the time delay for a signal to pass through the section at the cut-off frequency.

5. A T section filter network comprising n identical sections passes signals of all frequencies over 8 kHz and provides a total delay of 69.63 μs. If the characteristic impedance of the circuit into which the filter is inserted is 600 Ω, determine (a) the values of the components comprising each section and (b) the value of n

6. A filter network consists of 15 sections in cascade having a nominal impedance of 800 Ω. If the total delay time is 30 μs determine the component value for each section if the filter is (a) a low-pass π network, (b) a high-pass T network.

45.8 '*m*-derived' filter sections

(a) General

In a low-pass filter a clearly defined cut-off frequency followed by a high attenuation is needed; in a high-pass filter, high attenuation followed by a clearly defined cut-off frequency is needed. It is not practicable to obtain either of these conditions by wiring appropriate prototype constant-k sections in cascade. An equivalent section is therefore required having:

(i) the same cut-off frequency as the prototype but with a rapid rise in attenuation beyond cut-off for a low-pass type or a rapid decrease at cut-off from a high attenuation for the high-pass type,

(ii) the same value of nominal impedance R_0 as the prototype at all frequencies (otherwise the two forms could not be connected together without mismatch).

If the two sections, i.e. the prototype and the equivalent section, have the same value of R_0 they will have identical passbands.

The equivalent section is called an **'*m*-derived' filter section** (for reasons as explained below) and is one which gives a sharper cut-off at the edges of the passband and a better impedance characteristic.

(b) *T* sections

A prototype T section is shown in Figure 45.44(a). Let a new section be constructed from this section having a series arm of the same type but of different value, say mZ_1, where m is some constant. (It is for this reason that the new equivalent section is called an 'm-derived' section.) If the characteristic impedance Z_{0T} of the two sections is to be the same then the value of the shunt arm impedance will have to be different to Z_2. Let this be Z_2' as shown in Figure 45.44(b).

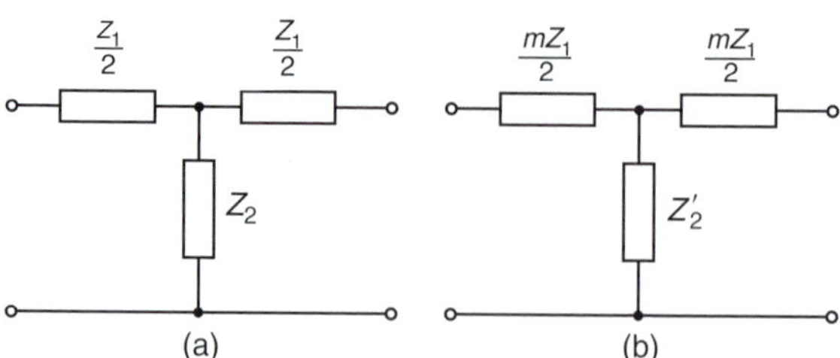

Figure 45.44

The value of Z_2' is determined as follows:

From equation (1), page 667, for the prototype shown in Figure 45.44(a):

$$Z_{0T} = \sqrt{\left[\left(\frac{Z_1}{2}\right)^2 + 2\left(\frac{Z_1}{2}\right)Z_2\right]}$$

i.e. $$Z_{0T} = \sqrt{\left(\frac{Z_1^2}{4} + Z_1 Z_2\right)} \qquad \text{(a)}$$

Similarly, for the new section shown in Figure 45.44(b),

$$Z_{0T} = \sqrt{\left[\left(\frac{mZ_1}{2}\right)^2 + 2\left(\frac{mZ_1}{2}\right)Z_2'\right]}$$

i.e. $$Z_{0T} = \sqrt{\left(\frac{m^2 Z_1^2}{4} + mZ_1 Z_2'\right)} \qquad \text{(b)}$$

Equations (a) and (b) will be identical if:

$$\frac{Z_1^2}{4} + Z_1 Z_2 = \frac{m^2 Z_1^2}{4} + mZ_1 Z_2'$$

Rearranging gives: $$mZ_1 Z_2' = Z_1 Z_2 + \frac{Z_1^2}{4}(1 - m^2)$$

i.e. $$\boldsymbol{Z_2' = \frac{Z_2}{m} + Z_1\left(\frac{1-m^2}{4m}\right)} \qquad (30)$$

Thus impedance Z_2' consists of an impedance Z_2/m in series with an impedance $Z_1((1-m^2)/4m)$. An additional component has therefore been introduced into the shunt arm of the m-derived section. The value of m can range from 0 to 1, and when $m=1$, the prototype and the m-derived sections are identical.

(c) π sections

A prototype π section is shown in Figure 45.45(a). Let a new section be constructed having shunt arms of the same type but of different values, say Z_2/m, where m is some constant. If the characteristic impedance $Z_{0\pi}$ of the two sections is to be the same then the value of the series arm impedance will have to be different to Z_1. Let this be Z_1' as shown in Figure 45.45(b).
The value of Z_1' is determined as follows:
From equation (9), $Z_{0T} Z_{0\pi} = Z_1 Z_2$
Thus the characteristic impedance of the section shown in Figure 45.45(a) is given by:

$$Z_{0\pi} = \frac{Z_1 Z_2}{Z_{0T}} = \frac{Z_1 Z_2}{\sqrt{\left(\frac{Z_1^2}{4} + Z_1 Z_2\right)}} \qquad \text{(c)}$$

from equation (a) above.

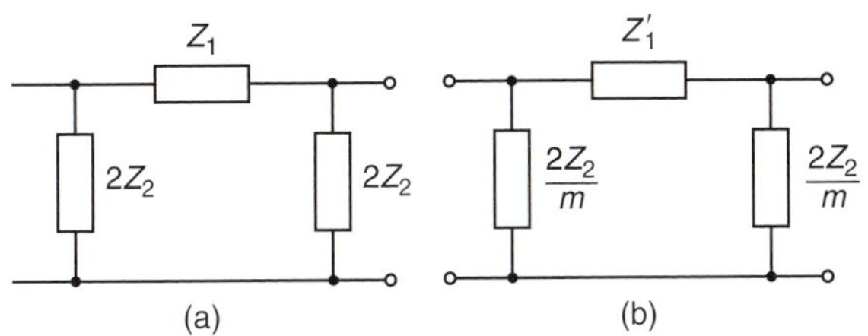

Figure 45.45

For the section shown in Figure 45.45(b),

$$Z_{0\pi} = \frac{Z_1' \frac{Z_2}{m}}{\sqrt{\left(\frac{(Z_1')^2}{4} + Z_1' \frac{Z_2}{m}\right)}} \qquad \text{(d)}$$

Equations (c) and (d) will be identical if

$$\frac{Z_1 Z_2}{\sqrt{\left(\frac{Z_1^2}{4} + Z_1 Z_2\right)}} = \frac{Z_1' \frac{Z_2}{m}}{\sqrt{\left(\frac{(Z_1')^2}{4} + Z_1' \frac{Z_2}{m}\right)}}$$

Dividing both sides by Z_2 and then squaring both sides gives:

$$\frac{Z_1^2}{\frac{Z_1^2}{4} + Z_1 Z_2} = \frac{\frac{(Z_1')^2}{m^2}}{\frac{(Z_1')^2}{4} + \frac{Z_1' Z_2}{m}}$$

Thus $$Z_1^2\left(\frac{(Z_1')^2}{4} + \frac{Z_1' Z_2}{m}\right) = \frac{(Z_1')^2}{m^2}\left(\frac{Z_1^2}{4} + Z_1 Z_2\right)$$

i.e. $$\frac{Z_1^2 (Z_1')^2}{4} + \frac{Z_1^2 Z_1' Z_2}{m} = \frac{(Z_1')^2 Z_1^2}{4m^2} + \frac{(Z_1')^2 Z_1 Z_2}{m^2}$$

Multiplying throughout by $4m^2$ gives:

$$m^2 Z_1^2 (Z_1')^2 + 4mZ_1^2 Z_1' Z_2 = (Z_1')^2 Z_1^2 + 4(Z_1')^2 Z_1 Z_2$$

Dividing throughout by Z_1' and rearranging gives:

$$4mZ_1^2 Z_2 = Z_1'(Z_1^2 + 4Z_1 Z_2 - m^2 Z_1^2)$$

Thus $$Z_1' = \frac{4mZ_1^2 Z_2}{4Z_1 Z_2 + Z_1^2(1-m^2)}$$

i.e. $$\boldsymbol{Z_1' = \frac{4mZ_1 Z_2}{4Z_2 + Z_1(1-m^2)}} \qquad (31)$$

An impedance mZ_1 in parallel with an impedance $(4mZ_2/1-m^2)$ gives (using (product/sum)):

$$\frac{(mZ_1)\dfrac{4mZ_2}{1-m^2}}{mZ_1+\dfrac{4mZ_2}{1-m^2}}=\frac{(mZ_1)4mZ_2}{mZ_1(1-m^2)+4mZ_2}$$

$$=\frac{4mZ_1Z_2}{4Z_2+Z_1(1-m^2)}$$

Hence the expression for Z_1' (equation (31)) represents an impedance mZ_1 in parallel with an impedance $(4m/1-m^2)Z_2$

(d) Low-pass '*m*-derived' sections

The '*m*-derived' low-pass T section is shown in Figure 45.46(a) and is derived from Figure 45.13(a), Figure 45.44 and equation (30). If Z_2 represents a pure capacitor in Figure 45.44(a), then $Z_2=(1/\omega C)$.

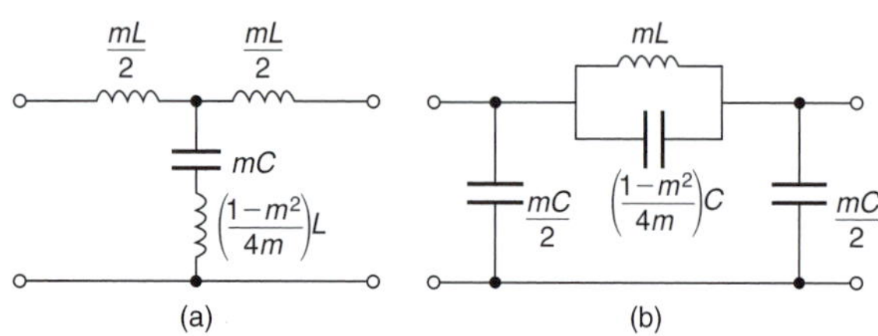

Figure 45.46

A capacitance of value mC shown in Figure 45.46(a) has an impedance

$$\frac{1}{\omega mC}=\frac{1}{m}\left(\frac{1}{\omega C}\right)=\frac{Z_2}{m} \text{ as in equation (30).}$$

The '*m*-derived' low-pass π section is shown in Figure 45.46(b) and is derived from Figure 45.13(b), Figure 45.45 and from equation (31).

Note that a capacitance of value $\left(\dfrac{1-m^2}{4m}\right)C$ has an impedance of

$$\frac{1}{\omega\left(\dfrac{1-m^2}{4m}\right)C}=\left(\frac{4m}{1-m^2}\right)\left(\frac{1}{\omega C}\right)=\left(\frac{4m}{1-m^2}\right)Z_2$$

where Z_2 is a pure capacitor.

In Figure 45.46(a), series resonance will occur in the shunt arm at a particular frequency – thus short-circuiting the transmission path. In the prototype, infinite attenuation is obtained only at infinite frequency (see Figure 45.25).

In the m-derived section of Figure 45.46(a), let the frequency of infinite attenuation be f_∞, then at resonance: $X_L=X_C$

i.e. $$\omega_\infty\left(\frac{1-m^2}{4m}\right)L=\frac{1}{\omega_\infty mC}$$

from which $$\omega_\infty^2=\frac{1}{(mC)\left(\dfrac{1-m^2}{4m}\right)L}=\frac{4}{LC(1-m^2)}$$

From equation (2),

$$\frac{4}{LC}=\omega_c^2,\ \text{thus } \omega_\infty^2=\frac{\omega_c^2}{(1-m^2)}$$

where $\omega_c=2\pi f_c$, f_c being the cut-off frequency of the prototype.

Hence $$\omega_\infty=\frac{\omega_c}{\sqrt{(1-m^2)}} \qquad (32)$$

Rearranging gives: $$\omega_\infty^2(1-m^2)=\omega_c^2$$

$$\omega_\infty^2-m^2\omega_\infty^2=\omega_c^2$$

$$m^2=\frac{\omega_\infty^2-\omega_c^2}{\omega_\infty^2}=1-\frac{\omega_c^2}{\omega_\infty^2}$$

i.e. $$\boldsymbol{m=\sqrt{\left[1-\left(\frac{f_c}{f_\infty}\right)^2\right]}} \qquad (33)$$

In the m-derived π section of Figure 45.46(b), resonance occurs in the parallel arrangement comprising the series arm of the section when

$$\omega^2=\frac{1}{mL\left(\dfrac{1-m^2}{4m}\right)C},\ \text{when } \omega^2=\frac{4}{LC(1-m^2)}$$

as in the series resonance case (see Chapter 31).
Thus equations (32) and (33) are also applicable to the low-pass m-derived π section.
In equation (33), $\boldsymbol{0<m<1}$, thus $\boldsymbol{f_\infty>f_c}$
The frequency of infinite attenuation f_∞ can be placed anywhere within the attenuation band by suitable choice of the value of m; the smaller m is made the nearer is f_∞ to the cut-off frequency, f_c

Problem 15. A filter section is required to have a nominal impedance of 600 Ω, a cut-off frequency of 5 kHz and a frequency of infinite attenuation at 5.50 kHz. Design (a) an appropriate '*m*-derived' T section and (b) an appropriate '*m*-derived' π section.

Part 4

Nominal impedance $R_0 = 600\,\Omega$, cut-off frequency, $f_c = 5000\,\text{Hz}$ and frequency of infinite attenuation, $f_\infty = 5500\,\text{Hz}$. Since $f_\infty > f_c$ the filter section is low-pass.
From equation (33),

$$m = \sqrt{\left[1 - \left(\frac{f_c}{f_\infty}\right)^2\right]} = \sqrt{\left[1 - \left(\frac{5000}{5500}\right)^2\right]} = 0.4166$$

For a low-pass prototype section:
from equation (6), capacitance,

$$C = \frac{1}{\pi R_0 f_c} = \frac{1}{\pi(600)(5000)}$$

$$\equiv 0.106\,\mu\text{F}$$

and from equation (7), inductance,

$$L = \frac{R_0}{\pi f_c} = \frac{600}{\pi(5000)}$$

$$\equiv 38.2\,\text{mH}$$

(a) For an 'm-derived' low-pass T section:
From Figure 45.46(a), the series arm inductances are each

$$\frac{mL}{2} = \frac{(0.4166)(38.2)}{2} = \mathbf{7.957\ mH},$$

and the shunt arm contains a capacitor of value mC, i.e. (0.4166)(0.106)=**0.0442 μF** or **44.2 nF**, in series with an inductance of

$$\text{value}\ \left(\frac{1-m^2}{4m}\right)L = \left(\frac{1-0.4166^2}{4(0.4166)}\right)(38.2)$$

$$= \mathbf{18.95\,mH}$$

The appropriate 'm-derived' T section is shown in Figure 45.47.

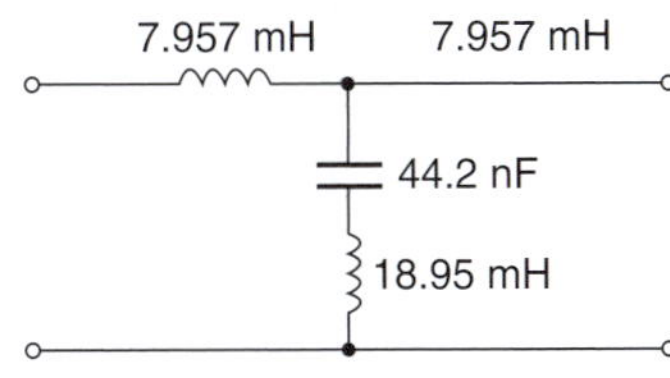

Figure 45.47

(b) For an 'm-derived' low-pass π section:
From Figure 45.46(b) the shunt arms each contain capacitances equal to $mC/2$

$$\text{i.e.}\quad \frac{(0.4166)(0.106)}{2} = \mathbf{0.0221\,\mu F}\ \text{or}\ \mathbf{22.1\,nF}$$

and the series arm contains an inductance of value mL, i.e. (0.4166)(38.2)=**15.91 mH** in parallel with a capacitance of

$$\text{value}\ \left(\frac{1-m^2}{4m}\right)C = \left(\frac{1-0.4166^2}{4(0.4166)}\right)(0.106)$$

$$= \mathbf{0.0526\,\mu F}\ \text{or}\ \mathbf{52.6\,nF}$$

The appropriate 'm-derived' π section is shown in Figure 45.48.

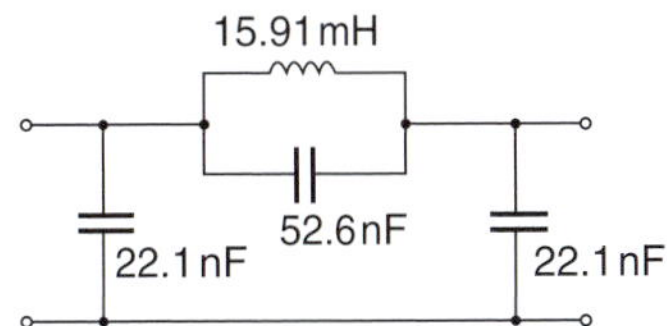

Figure 45.48

(e) High-pass '*m*-derived' sections

Figure 45.49(a) shows a high-pass prototype T section and Figure 45.49(b) shows the 'm-derived' high-pass T section which is derived from Figure 45.16(a), Figure 45.44 and equation (30).

Figure 45.50(a) shows a high-pass prototype π section and Figure 45.50(b) shows the 'm-derived' high-pass π section which is derived from Figure 45.16(b), Figure 45.45 and equation (31). In Figure 45.49(b), resonance occurs in the shunt arm when:

$$\omega_\infty \frac{L}{m} = \frac{1}{\omega_\infty\left(\dfrac{4m}{1-m^2}\right)C}$$

$$\text{i.e.}\quad \text{when } \omega_\infty^2 = \frac{1-m^2}{4LC} = \omega_c^2(1-m^2)$$

from equation (14)

$$\text{i.e.}\quad \omega_\infty = \omega_c\sqrt{(1-m^2)} \qquad (34)$$

$$\text{Hence}\quad \frac{\omega_\infty^2}{\omega_c^2} = 1 - m^2$$

$$\text{from which,}\quad \boldsymbol{m = \sqrt{\left[1 - \left(\frac{f_\infty}{f_c}\right)^2\right]}} \qquad (35)$$

For a high-pass section, $f_\infty < f_c$.
It may be shown that equations (34) and (35) also apply to the 'm-derived' π section shown in Figure 45.50(b).

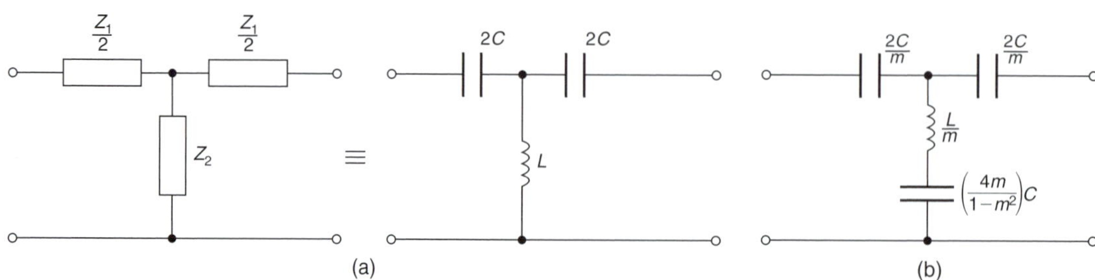

Figure 45.49

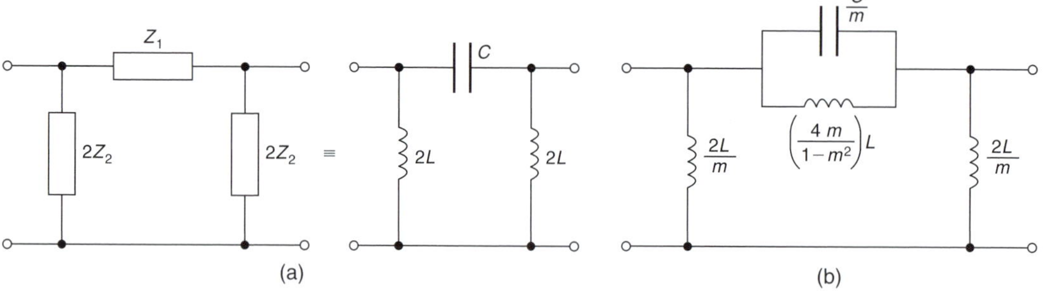

Figure 45.50

Problem 16. Design (a) a suitable 'm-derived' T section and (b) a suitable 'm-derived' π section having a cut-off frequency of 20 kHz, a nominal impedance of 500 Ω and a frequency of infinite attenuation 16 kHz.

Nominal impedance $R_0 = 500\,\Omega$, cut-off frequency, $f_c = 20$ kHz and the frequency of infinite attenuation, $f_\infty = 16$ kHz. Since $f_\infty < f_c$ the filter is high-pass.

From equation (35), $m = \sqrt{\left[1 - \left(\frac{f_\infty}{f_c}\right)^2\right]}$

$$= \sqrt{\left[1 - \left(\frac{16}{20}\right)^2\right]} = 0.60$$

For a high-pass prototype section:
From equation (18), capacitance,

$$C = \frac{1}{4\pi R_0 f_c} = \frac{1}{4\pi(500)(20\,000)} \equiv 7.958\,\text{nF}$$

and from equation (19), inductance,

$$L = \frac{R_0}{4\pi f_c} = \frac{500}{4\pi(20\,000)} \equiv 1.989\,\text{mH}$$

(a) For an 'm-derived' high-pass T section:

From Figure 45.49(b), each series arm contains a capacitance of value $2C/m$, i.e. 2(7.958)/0.60, i.e. **26.53 nF**, and the shunt arm contains an inductance of value L/m, i.e. (1.989/0.60) = **3.315 mH** in series with a capacitance of value

$$\left(\frac{4m}{1-m^2}\right)C \text{ i.e. } \left(\frac{4(0.60)}{1-0.60^2}\right)(7.958) = \mathbf{29.84\,nF}$$

A suitable 'm-derived' T section is shown in Figure 45.51.

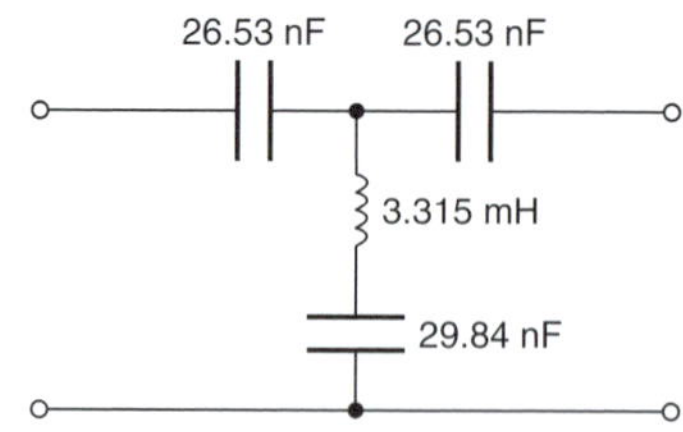

Figure 45.51

(b) For an 'm-derived' high-pass π section:
From Figure 45.50(b), the shunt arms each contain inductances equal to $2L/m$, i.e. (2(1.989)/0.60), i.e. **6.63 mH**, and the series arm contains a capacitance of value C/m, i.e. (7.958/0.60) = **13.26 nF** in parallel with an inductance of value $(4m/1-m^2)L$

i.e. $\left(\frac{4(0.60)}{1-0.60^2}\right)(1.989) \equiv \mathbf{7.459\,mH}$

A suitable 'm-derived' π section is shown in Figure 45.52.

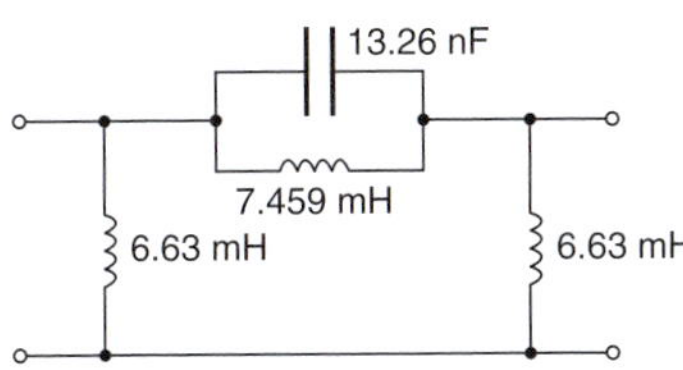

Figure 45.52

Now try the following Practice Exercise

Practice Exercise 182 'm-derived' filter sections (Answers on page 833)

1. A low-pass filter section is required to have a nominal impedance of 450 Ω, a cut-off frequency of 150 kHz and a frequency of infinite attenuation at 160 kHz. Design an appropriate 'm-derived' T section filter.

2. In a filter section it is required to have a cut-off frequency of 1.2 MHz and a frequency of infinite attenuation of 1.3 MHz. If the nominal impedance of the line into which the filter is to be inserted is 600 Ω, determine suitable component values if the section is an 'm-derived' π type.

3. Determine the component values of an 'm-derived' T section filter having a nominal impedance of 600 Ω, a cut-off frequency of 1220 Hz and a frequency of infinite attenuation of 1100 Hz.

4. State the advantages of an 'm-derived' filter section over its equivalent prototype. A filter section is to have a nominal impedance of 500 Ω, a cut-off frequency of 5 kHz and a frequency of infinite attenuation of 4.5 kHz. Determine the values of components if the section is to be an 'm-derived' π filter.

45.9 Practical composite filters

In practice, filters to meet a given specification often have to comprise a number of basic networks. For example, a practical arrangement might consist of (i) a basic prototype, in series with (ii) an 'm-derived' section, with (iii) terminating half-sections at each end. The 'm-derived' section improves the attenuation immediately after cut-off, the prototype improves the attenuation well after cut-off, whilst the terminating half-sections are used to obtain a constant match over the pass-band.

Figure 45.53(a) shows an 'm-derived' low-pass T section, and Figure 45.53(b) shows the same section cut into two halves through AB, each of the two halves being termed a 'half-section'. The 'm-derived' half section also improves the steepness of attenuation outside the pass-band.

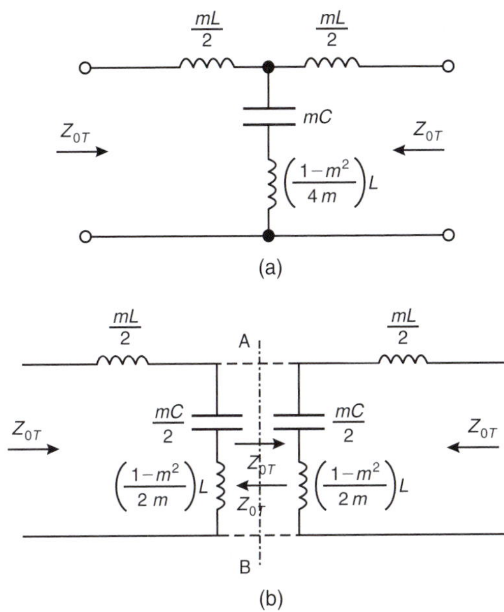

Figure 45.53

As shown in Section 45.8, the 'm-derived' filter section is based on a prototype which presents its own characteristic impedance at its terminals. Hence, for example, the prototype of a T section leads to an 'm-derived' T section and $Z_{0T} = Z_{0T(m)}$ where Z_{0T} is the characteristic impedance of the prototype and $Z_{0T(m)}$ is the characteristic impedance of the 'm-derived' section. It is shown in Figure 45.24 that Z_{0T} has a non-linear characteristic against frequency; thus $Z_{0T(m)}$ will also be non-linear.

Since from equation (9) $Z_{0\pi} = (Z_1 Z_2 / Z_{0T})$, then the characteristic impedance of the 'm-derived' π section,

$$Z_{0\pi(m)} = \frac{Z_1' Z_2'}{Z_{0T(m)}} = \frac{Z_1' Z_2'}{Z_{0T}}$$

where Z_1' and Z_2' are the equivalent values of impedance in the 'm-derived' section.

From Figure 45.44, $Z_1' = mZ_1$ and from equation (30),

$$Z_2' = \frac{Z_2}{m} + \left(\frac{1-m^2}{4m}\right) Z_1$$

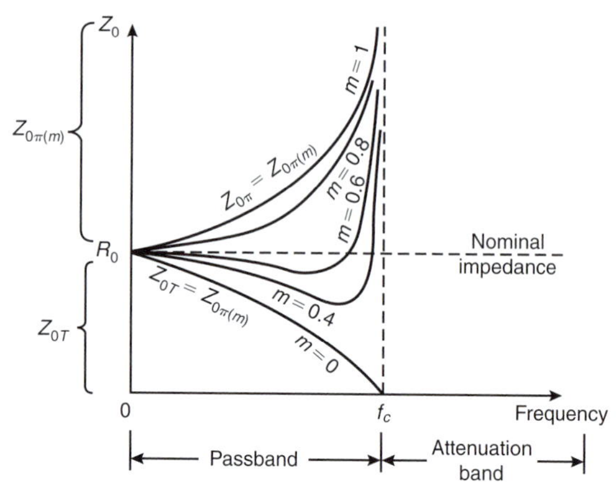

Figure 45.54

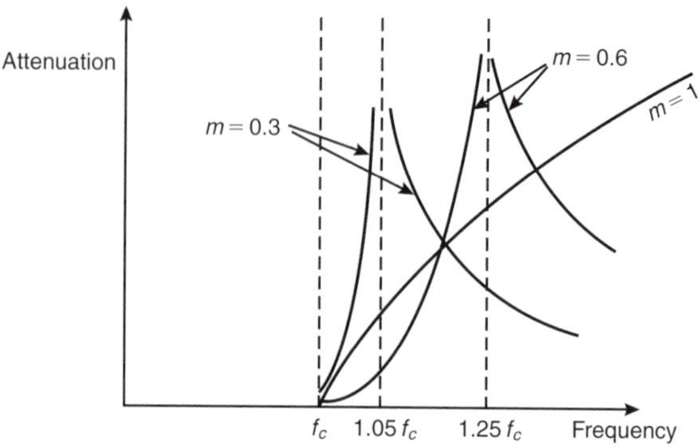

Figure 45.55

Thus $$Z_{0\pi(m)}=\frac{mZ_1\left[\frac{Z_2}{m}+\left(\frac{1-m^2}{4m}\right)Z_1\right]}{Z_{0T}}$$

$$=\frac{Z_1Z_2}{Z_{0T}}\left[1+\left(\frac{1-m^2}{4Z_2}\right)Z_1\right] \quad (36)$$

or $$\mathbf{Z_{0\pi(m)}=Z_{0\pi}\left[1+\left(\frac{1-m^2}{4Z_2}\right)Z_1\right]} \quad (37)$$

Thus the impedance of the 'm-derived' section is related to the impedance of the prototype by a factor of $[1+(1-m^2/4Z_2)Z_1]$ and will vary as m varies.

When $m=1,\ Z_{0\pi(m)}=Z_{0\pi}$

When $$m=0,\ Z_{0\pi(m)}=\frac{Z_1Z_2}{Z_{0T}}\left[1+\frac{Z_1}{4Z_2}\right]$$

from equation (36)

$$=\frac{1}{Z_{0T}}\left[Z_1Z_2+\frac{Z_1^2}{4}\right]$$

However, from equation (8), $Z_1Z_2+\frac{Z_1^2}{4}=Z_{0T}^2$

Hence, when $m=0$, $Z_{0\pi(m)}=\frac{Z_{0T}^2}{Z_{0T}}=Z_{0T}$

Thus the characteristic of impedance against frequency for $m=1$ and $m=0$ shown in Figure 45.54 are the same as shown in Figure 45.24. Further characteristics may be drawn for values of m between 0 and 1 as shown.

It is seen from Figure 45.54 that when $m=0.6$ the impedance is practically constant at R_0 for most of the passband. In a composite filter, 'm-derived' half-sections having a value of $m=0.6$ are usually used at each end to provide a good match to a resistive source and load over the passband.

Figure 45.54 shows characteristics of 'm-derived' low-pass filter sections; similar curves may be constructed for m-derived high-pass filters with the two curves shown in Figure 45.30 representing the limiting values of $m=0$ and $m=1$

The value of m needs to be small for the frequency of input attenuation, f_∞, to be close to the cut-off frequency, f_c. However, it is not practical to make m very small, below 0.3 being very unusual. When $m=0.3$, $f_\infty\approx 1.05f_c$ (from equation (32)) and when $m=0.6$, $f_\infty=1.25f_c$. The effect of the value of m on the frequency of infinite attenuation is shown in Figure 45.55, although the ideal curves shown would be modified a little in practice by resistance losses.

Problem 17. It is required to design a composite filter with a cut-off frequency of 10 kHz, a frequency of infinite attenuation of 11.8 kHz and nominal impedance of 600 Ω. Determine the component values needed if the filter is to comprise a prototype T section, an 'm-derived' T section and two terminating 'm-derived' half-sections.

$R_0=600\,\Omega$, $f_c=10\,\text{kHz}$ and $f_\infty=11.8\,\text{kHz}$. Since $f_c< f_\infty$ the filter is a low-pass T section.

For the prototype:

From equation (6), capacitance,

$$C=\frac{1}{\pi f_cR_0}=\frac{1}{\pi(10000)(600)}\equiv 0.0531\,\mu\text{F},$$

and from equation (7), inductance,

$$L=\frac{R_0}{\pi f_c}=\frac{600}{\pi(10000)}\equiv 19\,\text{mH}$$

Part 4

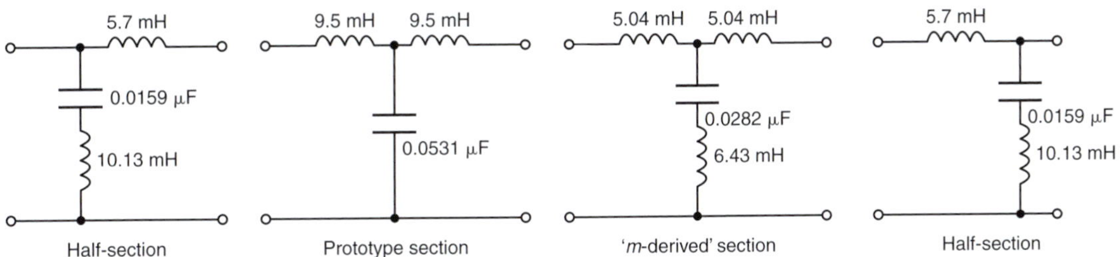

Figure 45.56

Thus, from Figure 45.13(a), the series arm components are $(L/2)=(19/2)=$ **9.5 mH** and the shunt arm component is **0.0531 μF**.

For the '*m*-derived' section:

From equation (33),

$$m=\sqrt{\left[1-\left(\frac{f_c}{f_\infty}\right)^2\right]}=\sqrt{\left[1-\left(\frac{10\,000}{11\,800}\right)^2\right]}$$
$$=0.5309$$

Thus from Figure 45.43(a), the series arm components are

$$\frac{mL}{2}=\frac{(0.5309)(19)}{2}=\mathbf{5.04\,mH}$$

and the shunt arm comprises $mC=(0.5309)(0.0531)=$ **0.0282 μF** in series with

$$\left(\frac{1-m^2}{4m}\right)L=\left(\frac{1-0.5309^2}{4(0.5309)}\right)(19)=\mathbf{6.43\,mH}$$

For the half-sections a value of $m=0.6$ is taken to obtain matching. Thus from Figure 45.53.

$$\frac{mL}{2}=\frac{(0.6)(19)}{2}=\mathbf{5.7\,mH},\ \frac{mC}{2}=\frac{(0.6)(0.0531)}{2}$$
$$\equiv\mathbf{0.0159\,\mu F}$$

and $\left(\frac{1-m^2}{2m}\right)L=\left(\frac{1-0.6^2}{2(0.6)}\right)(19)\equiv\mathbf{10.13\,mH}$

The complete filter is shown in Figure 45.56.

Now try the following Practice Exercise

Practice Exercise 183 Practical composite filter sections (Answers on page 834)

1. A composite filter is to have a nominal impedance of 500 Ω, a cut-off frequency of 1500 Hz and a frequency of infinite attenuation of 1800 Hz. Determine the values of components required if the filter is to comprise a prototype T section, an 'm-derived' T section and two terminating half-sections (use $m=0.6$ for the half-sections).

2. A filter made up of a prototype π section, an 'm-derived' π section and two terminating half-sections in cascade has a nominal impedance of 1 kΩ, a cut-off frequency of 100 kHz and a frequency of infinite attenuation of 90 kHz. Determine the values of the components comprising the composite filter and explain why such a filter is more suitable than just the prototype (use $m=0.6$ for the half-sections).

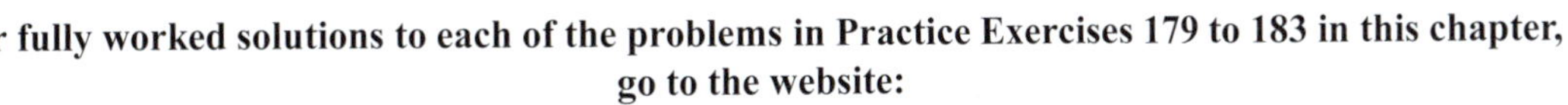
For fully worked solutions to each of the problems in Practice Exercises 179 to 183 in this chapter, go to the website:
www.routledge.com/cw/bird

Chapter 46

Magnetically coupled circuits

***Why it is important to understand:* Magnetically coupled circuits**

Magnetically coupled electric circuits are central to the operation of transformers and electric machines. In the case of transformers, stationary circuits are magnetically coupled for the purpose of changing the voltage and current levels. In the case of electric machines, circuits in relative motion are magnetically coupled for the purpose of transferring energy between mechanical and electrical systems. Because magnetically coupled circuits play such an important role in power transmission and conversion, it is important to establish the equations that describe their behaviour and to express these equations in a form convenient for analysis. This chapter reviews self and mutual inductance and performs calculations on mutually coupled circuits and describes and uses the dot rule in coupled circuit problems.

At the end of this chapter you should be able to:

- define mutual inductance
- deduce that $E_2 = -M\dfrac{dI_1}{dt}$, $M = N_2\dfrac{d\Phi_2}{dI_1}$, $M = N_1\dfrac{d\Phi_1}{dI_2}$ and perform calculations
- show that $M = k\sqrt{(L_1 L_2)}$ and perform calculations
- perform calculations on mutually coupled coils in series
- perform calculations on coupled circuits
- describe and use the dot rule in coupled circuit problems

46.1 Introduction

When the interaction between two loops of a circuit takes place through a magnetic field instead of through common elements, the loops are said to be inductively or **magnetically coupled.** The windings of a transformer, for example, are magnetically coupled (see Chapter 23).

46.2 Self-inductance

It was shown in Chapter 11, that the e.m.f. E induced in a coil of inductance L henrys is given by:

$\mathbf{E = -L\dfrac{di}{dt}}$ **volts** where $\dfrac{di}{dt}$ is the rate of change of current,

Electrical Circuit Theory and Technology. 978-1-138-67349-6, © 2017 John Bird. Published by Taylor & Francis. All rights reserved.

the magnitude of the e.m.f. induced in a coil of N turns is given by:

$$\mathbf{E = -N\frac{d\Phi}{dt}\ volts} \text{ where } \frac{d\Phi}{dt} \text{ is the rate of change of flux,}$$

and the inductance of a coil L is given by:

$$\mathbf{L = \frac{N\Phi}{I}\ henrys}$$

46.3 Mutual inductance

Mutual inductance is said to exist between two circuits when a changing current in one induces, by electro-magnetic induction, an e.m.f. in the other. An ideal equivalent circuit of a mutual inductor is shown in Figure 46.1.

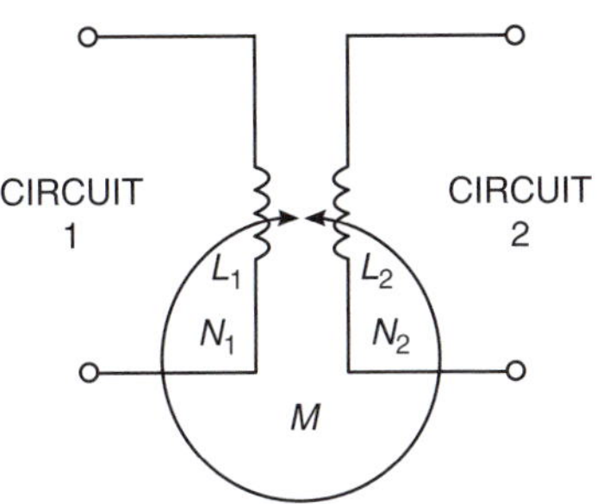

Figure 46.1

L_1 and L_2 are the self inductances of the two circuits and M the mutual inductance between them. The mutual inductance M is defined by the relationship:

$$\mathbf{E_2 = -M\frac{dI_1}{dt}} \text{ or } \mathbf{E_1 = -M\frac{dI_2}{dt}} \qquad (1)$$

where E_2 is the e.m.f. in circuit 2 due to current I_1 in circuit 1 and E_1 is the e.m.f. in circuit 1 due to the current I_2 in circuit 2.

The unit of M is the henry.

From Section 46.2, $E_2 = -N_2\frac{d\Phi_2}{dt}$ or

$$E_1 = -N_1\frac{d\Phi_1}{dt} \qquad (2)$$

Equating the E_2 terms in equations (1) and (2) gives:

$$-M\frac{dI_1}{dt} = -N_2\frac{d\Phi_2}{dt}$$

from which $\mathbf{M = N_2\frac{d\Phi_2}{dI_1}}$ (3)

Equating the E_1 terms in equations (1) and (2) gives:

$$-M\frac{dI_2}{dt} = -N_1\frac{d\Phi_1}{dt}$$

from which $\mathbf{M = N_1\frac{d\Phi_1}{dI_2}}$ (4)

If the coils are linked with air as the medium, the flux and current are linearly related and equations (3) and (4) become:

$$\mathbf{M = \frac{N_2\Phi_2}{I_1}} \text{ and } \mathbf{M = \frac{N_1\Phi_1}{I_2}} \qquad (5)$$

Problem 1. A and B are two coils in close proximity. A has 1200 turns and B has 1000 turns. When a current of 0.8 A flows in coil A a flux of 100 μWb links with coil A and 75% of this flux links coil B. Determine (a) the self inductance of coil A and (b) the mutual inductance.

(a) From Section 46.2,

self inductance of coil A,

$$\mathbf{L_A} = \frac{N_A\Phi_A}{I_A} = \frac{(1200)(100\times10^{-6})}{0.80} = \mathbf{0.15\,H}$$

(b) From equation (5),

mutual inductance, M

$$= \frac{N_B\Phi_B}{I_A} = \frac{(1000)(0.75\times100\times10^{-6})}{0.80} = \mathbf{0.09375\,H \text{ or } 93.75\,mH}$$

Problem 2. Two circuits have a mutual inductance of 600 mH. A current of 5 A in the primary is reversed in 200 ms. Determine the e.m.f. induced in the secondary, assuming the current changes at a uniform rate.

Secondary e.m.f., $E_2 = -M\frac{dI_1}{dt}$, from equation (1).

Since the current changes from +5 A to −5 A, the change of current is 10 A.

Hence $\frac{dI_1}{dt} = \frac{10}{200\times10^{-3}} = 50\,A/s$

Part 4

Hence **secondary induced e.m.f.,**

$$E_2 = -M\frac{dI_1}{dt} = -(600 \times 10^{-3})(50)$$
$$= \mathbf{-30\ volts}$$

Now try the following Practice Exercise

Practice Exercise 184 Mutual inductance (Answers on page 834)

1. If two coils have a mutual inductance of 500 μH, determine the magnitude of the e.m.f. induced in one coil when the current in the other coil varies at a rate of 20×10^3 A/s.
2. An e.m.f. of 15 V is induced in a coil when the current in an adjacent coil varies at a rate of 300 A/s. Calculate the value of the mutual inductance of the two coils.
3. Two circuits have a mutual inductance of 0.2 H. A current of 3 A in the primary is reversed in 200 ms. Determine the e.m.f. induced in the secondary, assuming the current changes at a uniform rate.
4. A coil, x, has 1500 turns and a coil, y, situated close to x has 900 turns. When a current of 1 A flows in coil x a flux of 0.2 mWb links with x and 0.65 of this flux links coil y. Determine (a) the self inductance of coil x and (b) the mutual inductance between the coils.

46.4 Coupling coefficient

The coupling coefficient k is the degree or fraction of magnetic coupling that occurs between circuits.

$$k = \frac{\text{flux linking two circuits}}{\text{total flux produced}}$$

When there is no magnetic coupling, $k = 0$. If the magnetic coupling is perfect, i.e. all the flux produced in the primary links with the secondary then $k = 1$. Coupling coefficient is used in communications engineering to denote the degree of coupling between two coils. If the coils are close together, most of the flux produced by current in one coil passes through the other, and the coils are termed **tightly coupled**. If the coils are spaced apart, only a part of the flux links with the second, and the coils are termed **loosely coupled**.

From Section 46.2, the inductance of a coil is given by $L = \frac{N\Phi}{I}$

Thus for the circuit of Figure 46.1, $L_1 = \frac{N_1\Phi_1}{I_1}$

$$\text{from which, } \Phi_1 = \frac{L_1 I_1}{N_1} \quad (6)$$

From equation (5), $M = (N_2\Phi_2/I_1)$, but the flux that links the second circuit, $\Phi_2 = k\Phi_1$

$$\text{Thus} \quad M = \frac{N_2\Phi_2}{I_1} = \frac{N_2(k\Phi_1)}{I_1} = \frac{N_2 k}{I_1}\left(\frac{L_1 I_1}{N_1}\right)$$

from equation (6)

$$\text{i.e. } M = \frac{kN_2L_1}{N_1} \text{ from which, } \frac{N_2}{N_1} = \frac{M}{kL_1} \quad (7)$$

Also, since the two circuits can be reversed,

$$M = \frac{kN_1L_2}{N_2} \text{ from which, } \frac{N_2}{N_1} = \frac{kL_2}{M} \quad (8)$$

Thus from equations (7) and (8),

$$\frac{N_2}{N_1} = \frac{M}{kL_1} = \frac{kL_2}{M}$$

from which,

$$M^2 = k^2L_1L_2 \text{ and } \mathbf{M = k\sqrt{(L_1L_2)}} \quad (9)$$

$$\text{or, } \quad \textbf{coefficient of coupling, } \mathbf{k = \frac{M}{\sqrt{(L_1L_2)}}} \quad (10)$$

Problem 3. Two coils have self inductances of 250 mH and 400 mH, respectively. Determine the magnetic coupling coefficient of the pair of coils if their mutual inductance is 80 mH.

From equation (10), coupling coefficient,

$$k = \frac{M}{\sqrt{(L_1L_2)}} = \frac{80 \times 10^{-3}}{\sqrt{[(250 \times 10^{-3})(400 \times 10^{-3})]}}$$
$$= \frac{80 \times 10^{-3}}{\sqrt{(0.1)}} = \mathbf{0.253}$$

Problem 4. Two coils, X and Y, having self inductances of 80 mH and 60 mH, respectively, are magnetically coupled. Coil X has 200 turns and coil Y has 100 turns. When a current of 4 A is reversed in coil X the change of flux in coil Y is 5 mWb.

Part 4

Determine (a) the mutual inductance between the coils and (b) the coefficient of coupling.

(a) From equation (3),

$$\textbf{mutual inductance, } M = N_Y \frac{d\Phi_Y}{dI_X} = \frac{(100)(5\times10^{-3})}{(4--4)} = \mathbf{0.0625\,H} \text{ or } \mathbf{62.5\,mH}$$

(b) From equation (10),

$$\textbf{coefficient of coupling, } k = \frac{M}{\sqrt{(L_X L_Y)}} = \frac{0.0625}{\sqrt{[(80\times10^{-3})(60\times10^{-3})]}} = \mathbf{0.902}$$

Now try the following Practice Exercise

Practice Exercise 185 Coupling coefficient (Answers on page 834)

1. Two coils have a mutual inductance of 0.24 H. If the coils have self inductances of 0.4 H and 0.9 H, respectively, determine the magnetic coefficient of coupling.
2. Coils A and B are magnetically coupled. Coil A has a self inductance of 0.30 H and 300 turns, and coil B has a self inductance of 0.20 H and 120 turns. A change of flux of 8 mWb occurs in coil B when a current of 3 A is reversed in coil A. Determine (a) the mutual inductance between the coils, and (b) the coefficient of coupling.

46.5 Coils connected in series

Figure 46.2 shows two coils, 1 and 2, wound on an insulating core with terminals B and C joined. The fluxes in each coil produced by current i are in the same direction and the coils are termed **cumulatively coupled.**

Let the self inductance of coil 1 be L_1 and that of coil 2 be L_2 and let their mutual inductance be M

If in dt seconds, the current increases by di amperes then the e.m.f. induced in coil 1 due to its self inductance is $L_1(di/dt)$ volts, and the e.m.f. induced in coil 2 due to its self inductance is $L_2(di/dt)$ volts. Also, the e.m.f. induced in coil 1 due to the increase of current in coil 2 is $M(di/dt)$ volts and the e.m.f. induced in coil 2 due to the increase of current in coil 1 is $M(di/dt)$

Hence the total e.m.f. induced in coils 1 and 2 is:

$$L_1\frac{di}{dt} + L_2\frac{di}{dt} + 2\left(M\frac{di}{dt}\right) \text{ volts}$$

$$= (L_1 + L_2 + 2M)\frac{di}{dt} \text{ volts}$$

If the winding between terminals A and D in Figure 46.2 are considered as a single circuit having a self inductance L_A henrys, then if the same increase in dt seconds is di amperes then the e.m.f. induced in the complete circuit is $L_A(di/dt)$ volts.

Hence $\quad L_A\frac{di}{dt} = (L_1 + L_2 + 2M)\frac{di}{dt}$

i.e. $\quad \mathbf{L_A = L_1 + L_2 + 2M} \quad (11)$

If terminals B and D are joined as shown in Figure 46.3 the direction of the current in coil 2 is reversed and the coils are termed **differentially coupled.** In this case, the total e.m.f. induced in coils 1 and 2 is:

$$L_1\frac{di}{dt} + L_2\frac{di}{dt} - 2M\frac{di}{dt}$$

The e.m.f. $M(di/dt)$ induced in coil 1 due to an increase di amperes in dt seconds in coil 2 is in the same direction as the current and is hence in opposition to the e.m.f. induced in coil 1 due to its self inductance. Similarly, the e.m.f. induced in coil 2 by mutual inductance is in opposition to that induced by the self inductance of coil 2.

If L_B is the self inductance of the whole circuit between terminals A and C in Figure 46.3 then:

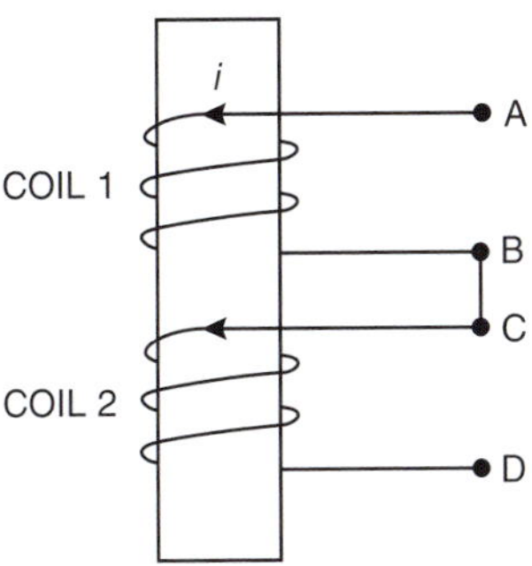

Figure 46.2

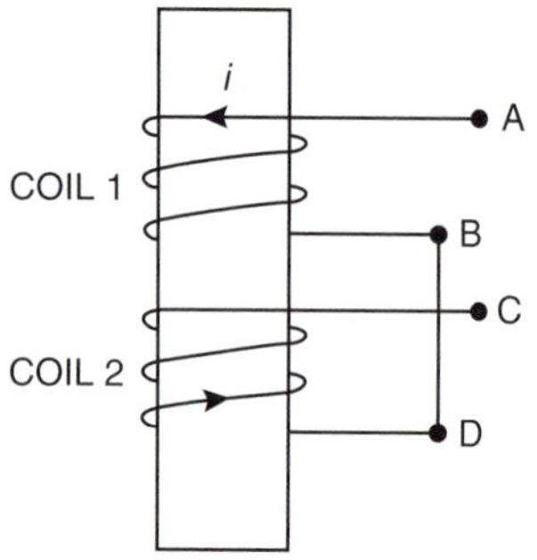

Figure 46.3

$$L_B\frac{di}{dt} = L_1\frac{di}{dt} + L_2\frac{di}{dt} - 2M\frac{di}{dt}$$

i.e. $\quad \boldsymbol{L_B = L_1 + L_2 - 2M} \qquad (12)$

Thus the total inductance L of inductively coupled circuits is given by:

$$\boldsymbol{L = L_1 + L_2 \pm 2M} \qquad (13)$$

Equation (11) – equation (12) gives:

$$L_A - L_B = (L_1 + L_2 + 2M) - (L_1 + L_2 - 2M)$$

i.e. $\quad L_A - L_B = 2M - (-2M) = 4M$

from which, **mutual inductance,** $\boldsymbol{M = \frac{L_A - L_B}{4}} \qquad (14)$

An experimental method of determining the mutual inductance is indicated by equation (14), i.e. connect the coils both ways and determine the equivalent inductances L_A and L_B using an a.c. bridge. The mutual inductance is then given by a quarter of the difference between the two values of inductance.

Problem 5. Two coils connected in series have self inductance of 40 mH and 10 mH, respectively. The total inductance of the circuit is found to be 60 mH. Determine (a) the mutual inductance between the two coils and (b) the coefficient of coupling.

(a) From equation (13), total inductance,

$$L = L_1 + L_2 \pm 2M$$

Hence $\quad 60 = 40 + 10 \pm 2M$

Since $(L_1 + L_2) < L$ then $60 = 40 + 10 + 2M$

from which $\quad 2M = 60 - 40 - 10 = 10$

and **mutual inductance,** $\boldsymbol{M = \frac{10}{2} = 5\,\text{mH}}$

(b) From equation (10), coefficient of coupling,

$$k = \frac{M}{\sqrt{(L_1L_2)}} = \frac{5 \times 10^{-3}}{\sqrt{[(40 \times 10^{-3})(10 \times 10^{-3})]}} = \frac{5 \times 10^{-3}}{0.02}$$

i.e. **coefficient of coupling,** $\boldsymbol{k = 0.25}$

Problem 6. Two mutually coupled coils, X and Y, are connected in series to a 240 V d.c. supply. Coil X has a resistance of 5 Ω and an inductance of 1 H. Coil Y has a resistance of 10 Ω and an inductance of 5 H. At a certain instant after the circuit is connected, the current is 8 A and increasing at a rate of 15 A/s. Determine (a) the mutual inductance between the coils and (b) the coefficient of coupling.

The circuit is shown in Figure 46.4.

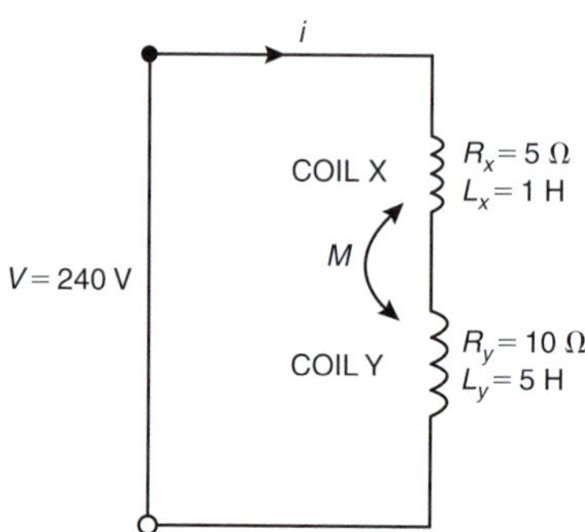

Figure 46.4

(a) From Kirchhoff's voltage law:

$$V = iR + L\frac{di}{dt}$$

i.e. $\quad 240 = 8(5 + 10) + L(15)$

i.e. $\quad 240 = 120 + 15L$

from which, $L = \frac{240 - 120}{15} = 8\,\text{H}$

From equation (11),

$$L = L_X + L_Y + 2M$$

Hence $\quad 8 = 1 + 5 + 2M$

from which, **mutual inductance,** $\boldsymbol{M = 1\,\text{H}}$

(b) From equation (10),

$$\textbf{coefficient of coupling, } k = \frac{M}{\sqrt{(L_X L_Y)}} = \frac{1}{\sqrt{[(1)(5)]}} = \mathbf{0.447}$$

Problem 7. Two coils are connected in series and their effective inductance is found to be 15 mH. When the connection to one coil is reversed, the effective inductance is found to be 10 mH. If the coefficient of coupling is 0.7, determine (a) the self inductance of each coil and (b) the mutual inductance.

(a) From equation (13),

total inductance, $L = L_1 + L_2 \pm 2M$

and from equation (9), $M = k\sqrt{(L_1 L_2)}$

hence $L = L_1 + L_2 \pm 2k\sqrt{(L_1 L_2)}$

Since in equation (11),

$$L_A = 15\,\text{mH}, \quad 15 = L_1 + L_2 + 2k\sqrt{(L_1 L_2)} \qquad (15)$$

and since in equation (12),

$$L_B = 10\,\text{mH}, \quad 10 = L_1 + L_2 - 2k\sqrt{(L_1 L_2)} \qquad (16)$$

Equation (15) + equation (16) gives:

$$25 = 2L_1 + 2L_2 \text{ and } 12.5 = L_1 + L_2 \qquad (17)$$

From equation (17), $L_2 = 12.5 - L_1$

Substituting in equation (15), gives:

$$15 = L_1 + (12.5 - L_1) + 2(0.7)\sqrt{[L_1(12.5 - L_1)]}$$

i.e. $15 = 12.5 + 1.4\sqrt{(12.5L_1 - L_1^2)}$

$$\frac{15 - 12.5}{1.4} = \sqrt{(12.5L_1 - L_1^2)}$$

and $$\left(\frac{15 - 12.5}{1.4}\right)^2 = 12.5L_1 - L_1^2$$

i.e. $3.189 = 12.5L_1 - L_1^2$

from which, $L_1^2 - 12.5L_1 + 3.189 = 0$

Using the quadratic formula:

$$L_1 = \frac{-(-12.5) \pm \sqrt{[(-12.5)^2 - 4(1)(3.189)]}}{2(1)}$$

i.e. $$L_1 = \frac{12.5 \pm (11.98)}{2} = \mathbf{12.24\,mH} \text{ or } \mathbf{0.26\,H}$$

From equation (17):

$$L_2 = 12.5 - L_1 = (12.5 - 12.24) = \mathbf{0.26\,mH}$$

$$\text{or } (12.5 - 0.26) = \mathbf{12.24\,mH}$$

(b) From equation (14),

$$\textbf{mutual inductance,} \quad M = \frac{L_A - L_B}{4} = \frac{15 - 10}{4} = \mathbf{1.25\,mH}$$

Now try the following Practice Exercise

Practice Exercise 186 Coils connected in series (Answers on page 834)

1. Two coils have inductances of 50 mH and 100 mH, respectively. They are placed so that their mutual inductance is 10 mH. Determine their effective inductance when the coils are (a) in series aiding (i.e. cumulatively coupled) and (b) in series opposing (i.e. differentially coupled).
2. The total inductance of two coils connected in series is 0.1 H. The coils have self inductance of 25 mH and 55 mH, respectively. Determine (a) the mutual inductance between the two coils and (b) the coefficient of coupling.
3. A d.c. supply of 200 V is applied across two mutually coupled coils in series, A and B. Coil A has a resistance of 2 Ω and a self inductance of 0.5 H; coil B has a resistance of 8 Ω and a self inductance of 2 H. At a certain instant after the circuit is switched on, the current is 10 A and increasing at a rate of 25 A/s. Determine (a) the mutual inductance between the coils and (b) the coefficient of coupling.
4. A ferromagnetic-cored coil is in two sections. One section has an inductance of 750 mH

and the other an inductance of 148 mH. The coefficient of coupling is 0.6. Determine (a) the mutual inductance, (b) the total inductance when the sections are connected in series aiding and (c) the total inductance when the sections are in series opposing.

5. Two coils are connected in series and their total inductance is measured as 0.12 H, and when the connection to one coil is reversed, the total inductance is measured as 0.04 H. If the coefficient of coupling is 0.8, determine (a) the self inductance of each coil and (b) the mutual inductance between the coils.

46.6 Coupled circuits

The magnitude of the secondary e.m.f. E_2 in Figure 46.5 is given by:

$$E_2 = M\frac{dI_1}{dt}, \text{ from equation (1)}$$

If the current I_1 is sinusoidal, i.e. $I_1 = I_{1m}\sin\omega t$

then $E_2 = M\frac{d}{dt}(I_{1m}\sin\omega t) = M\omega I_{1m}\cos\omega t$

Since $\cos\omega t = \sin(\omega t + 90°)$ then $\cos\omega t = j\sin\omega t$ in complex form.

Hence $E_2 = M\omega I_{1m}(j\sin\omega t) = j\omega M(I_{1m}\sin\omega t)$

i.e. $$\boldsymbol{E_2 = j\omega M I_1} \qquad (18)$$

If L_1 is the self inductance of the primary winding in Figure 46.5, there will be an e.m.f. generated equal to $j\omega L_1 I_1$ induced into the primary winding since the flux set up by the primary current also links with the primary winding.

(a) Secondary open-circuited

Figure 46.6 shows two coils, having self inductances of L_1 and L_2 which are inductively coupled together by a mutual inductance M. The primary winding has a voltage generator of e.m.f. E_1 connected across its terminals. The internal resistance of the source added to the primary resistance is shown as R_1 and the secondary winding which is open-circuited has a resistance of R_2

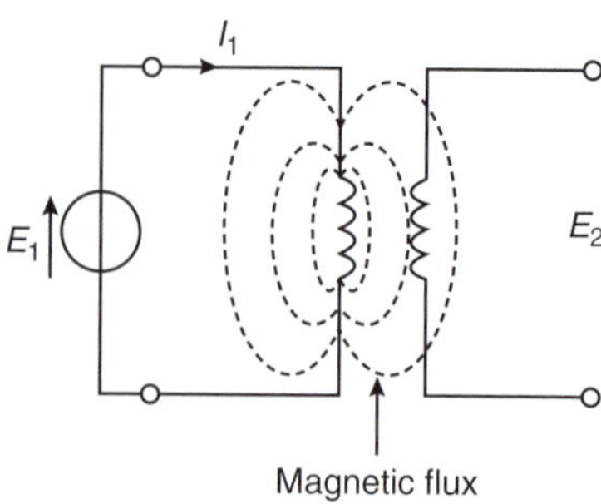

Figure 46.5

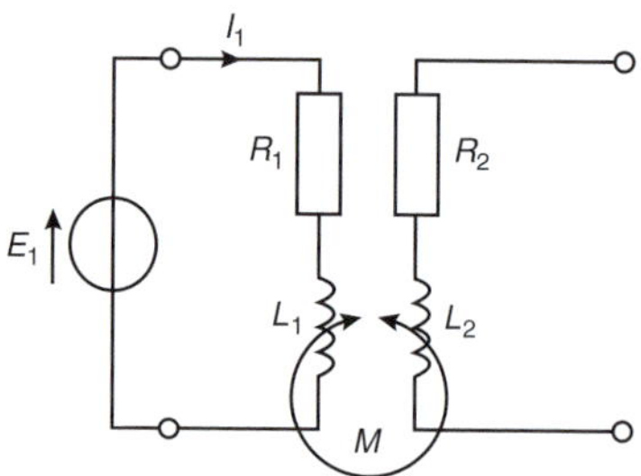

Figure 46.6

Applying Kirchhoff's voltage law to the primary circuit gives:

$$E_1 = I_1R_1 + L_1\frac{dI_1}{dt} \qquad (19)$$

If E_1 and I_1 are both sinusoidal then equation (19) becomes:

$$E_1 = I_1R_1 + L_1\frac{d}{dt}(I_{1m}\sin\omega t)$$
$$= I_1R_1 + L_1\omega I_{1m}\cos\omega t$$
$$= I_1R_1 + L_1\omega(I_{1m}\sin\omega t)$$

i.e. $E_1 = I_1R_1 + j\omega I_1L_1 = I_1(R_1 + j\omega L_1)$

i.e. $$I_1 = \frac{E_1}{R_1 + j\omega L_1} \qquad (20)$$

From equation (18), $E_2 = j\omega M I_1$

from which, $$I_1 = \frac{E_2}{j\omega M} \qquad (21)$$

Equating equations (20) and (21) gives:

$$\frac{E_2}{j\omega M} = \frac{E_1}{R_1 + j\omega L_1}$$

and $$\boldsymbol{E_2 = \frac{j\omega M E_1}{R_1 + j\omega L_1}} \qquad (22)$$

Problem 8. For the circuit shown in Figure 46.7, determine the p.d. E_2 which appears across the open-circuited secondary winding, given that $E_1 = 8\sin 2500t$ volts.

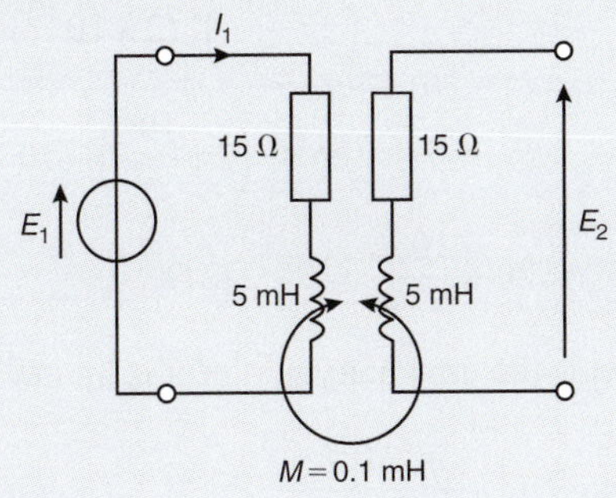

Figure 46.7

Impedance of primary,

$$Z_1 = R_1 + j\omega L_1 = 15 + j(2500)(5\times 10^{-3})$$
$$= (15 + j12.5)\Omega \text{ or } 19.53\angle 39.81°\Omega$$

Primary current $I_1 = \dfrac{E_1}{Z_1} = \dfrac{8\angle 0°}{19.53\angle 39.81°}$

From equation (18),

$$E_2 = j\omega M I_1 = \frac{j\omega M E_1}{(R_1 + j\omega L_1)}$$
$$= \frac{j(2500)(0.1\times 10^{-3})(8\angle 0°)}{19.53\angle 39.81°}$$
$$= \frac{2\angle 90°}{19.53\angle 39.81°} = \mathbf{0.102\angle 50.19°\,V}$$

Problem 9. Two coils, x and y, with negligible resistance, have self inductances of 20 mH and 80 mH, respectively, and the coefficient of coupling between them is 0.75. If a sinusoidal alternating p.d. of 5 V is applied to x, determine the magnitude of the open circuit e.m.f. induced in y.

From equation (9), mutual inductance,

$$M = k\sqrt{(L_x L_y)} = 0.75\sqrt{[(20\times 10^{-3})(80\times 10^{-3})]}$$
$$= 0.03\,\text{H}$$

From equation (22), the magnitude of the open circuit e.m.f. induced in coil y,

$$|E_y| = \frac{j\omega M E_x}{R_x + j\omega L_x}$$

When $R_1 = 0$, $|\boldsymbol{E_y}| = \dfrac{j\omega M E_x}{j\omega L_x} = \dfrac{M E_x}{L_x} = \dfrac{(0.03)(5)}{20\times 10^{-3}}$

$$= \mathbf{7.5\,V}$$

(b) Secondary terminals having load impedance

In the circuit shown in Figure 46.8 a load resistor R_L is connected across the secondary terminals. Let $R'_2 + R_L = R_2$

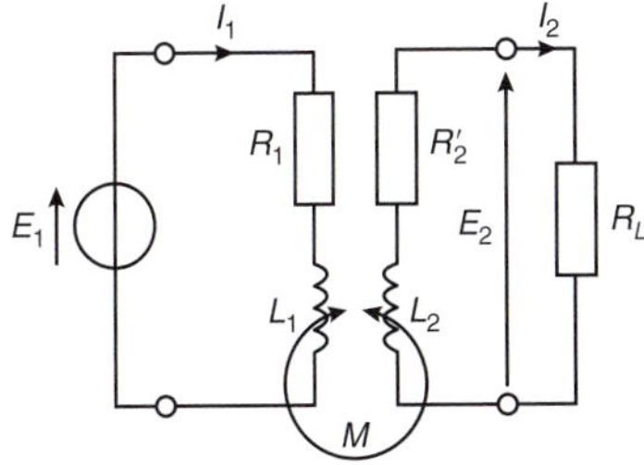

Figure 46.8

When an e.m.f. is induced into the secondary winding a current I_2 flows and this will induce an e.m.f. into the primary winding.

Applying Kirchhoff's voltage law to the primary winding gives:

$$E_1 = I_1(R_1 + j\omega L_1) \pm j\omega M I_2 \qquad (23)$$

Applying Kirchhoff's voltage law to the secondary winding gives:

$$0 = I_2(R_2 + j\omega L_2) \pm j\omega M I_1 \qquad (24)$$

From equation (24), $I_2 = \dfrac{\mp j\omega M I_1}{(R_2 + j\omega L_2)}$

Substituting this in equation (23) gives:

$$E_1 = I_1(R_1 + j\omega L_1) \pm j\omega M\left(\frac{\mp j\omega M I_1}{(R_2 + j\omega L_2)}\right)$$

i.e. $E_1 = I_1\left[(R_1 + j\omega L_1) + \dfrac{\omega^2 M^2}{(R_2 + j\omega L_2)}\right]$

since $j^2 = -1$

$$= I_1\left[(R_1 + j\omega L_1) + \frac{\omega^2 M^2 (R_2 - j\omega L_2)}{R_2^2 + \omega^2 L_2^2}\right]$$

$$= I_1\left[R_1 + j\omega L_1 + \frac{\omega^2 M^2 R_2}{R_2^2 + \omega^2 L_2^2} - \frac{j\omega^3 M^2 L_2}{R_2^2 + \omega^2 L_2^2}\right]$$

Part 4

The effective primary impedance $Z_{1(\text{eff})}$ of the circuit is given by:

$$\boldsymbol{Z_{1(\text{eff})} = \frac{E_1}{I_1}}$$

$$\boldsymbol{= R_1 + \frac{\omega^2 M^2 R_2}{R_2^2 + \omega^2 L_2^2} + j\left(\omega L_1 - \frac{\omega^3 M^2 L_2}{R_2^2 + \omega^2 L_2^2}\right)} \qquad (25)$$

In equation (25), the primary impedance is

$(R_1 + j\omega L_1)$. The remainder,

i.e. $\left(\frac{\omega^2 M^2 R_2}{R_2^2 + \omega^2 L_2^2} - j\frac{\omega^3 M^2 L_2}{R_2^2 + \omega^2 L_2^2}\right)$

is known as the **reflected impedance** since it represents the impedance reflected back into the primary side by the presence of the secondary current.
Hence reflected impedance

$$= \frac{\omega^2 M^2 R_2}{R_2^2 + \omega^2 L_2^2} - j\frac{\omega^3 M^2 L_2}{R_2^2 + \omega^2 L_2^2} = \omega^2 M^2\left(\frac{R_2 - j\omega L_2}{R_2^2 + \omega^2 L_2^2}\right)$$

$$= \omega^2 M^2 \frac{(R_2 - j\omega L_2)}{(R_2 + j\omega L_2)(R_2 - j\omega L_2)} = \frac{\omega^2 M^2}{R_2 + j\omega L_2}$$

i.e. **reflected impedance,** $\boldsymbol{Z_r = \frac{\omega^2 M^2}{Z_2}}$ (26)

Problem 10. For the circuit shown in Figure 46.9, determine the value of the secondary current I_2 if $E_1 = 2\angle 0°$ volts and the frequency is $\frac{10^3}{\pi}$ Hz.

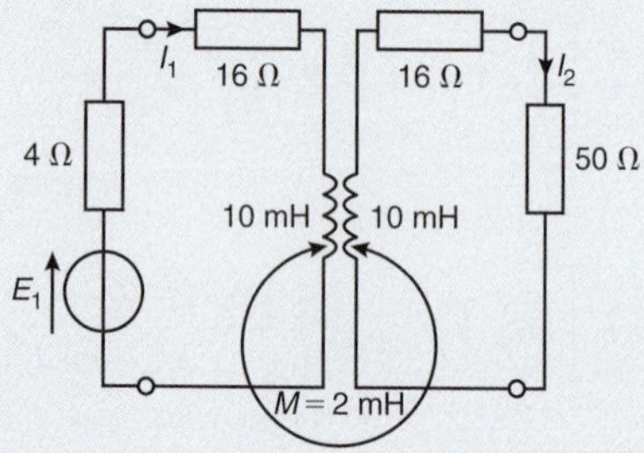

Figure 46.9

From equation (25), $R_{1(\text{eff})}$ is the real part of $Z_{1(\text{eff})}$,

i.e. $R_1(\text{eff}) = R_1 + \frac{\omega^2 M^2 R_2}{R_2^2 + \omega^2 L_2^2}$

$$= (4+16) + \frac{\left(2\pi\frac{10^3}{\pi}\right)^2 (2\times 10^{-3})^2(16+50)}{66^2 + \left(2\pi\frac{10^3}{\pi}\right)^2(10\times 10^{-3})^2}$$

$$= 20 + \frac{1056}{4756} = 20.222\,\Omega$$

and $X_{1(\text{eff})}$ is the imaginary part of $Z_{1(\text{eff})}$, i.e.

$$X_1(\text{eff}) = \omega L_1 - \frac{\omega^3 M^2 L_2}{R_2^2 + \omega^2 L_2^2}$$

$$= \left(2\pi\frac{10^3}{\pi}\right)(10\times 10^{-3}) - \frac{\left(2\pi\frac{10^3}{\pi}\right)^3 (2\times 10^{-3})^2(10\times 10^{-3})}{66^2 + \left(2\pi\frac{10^3}{\pi}\right)^2(10\times 10^{-3})^2}$$

$$= 20 - \frac{320}{4756} = 19.933\,\Omega$$

Hence primary current,

$$I_1 = \frac{E_1}{Z_{1(\text{eff})}} = \frac{2\angle 0°}{(20.222 + j19.933)}$$

$$= \frac{2\angle 0°}{28.395\angle 44.59°} = 0.0704\angle -44.59°\text{ A}$$

From equation (18),

$$E_2 = j\omega M I_1$$

$$= j\left(2\pi\frac{10^3}{\pi}\right)(2\times 10^{-3})(0.0704\angle -44.59°)$$

$$= (4\angle 90°)(0.0704\angle -44.59°)$$

$$= 0.282\angle 45.41°\text{ V}$$

Hence secondary current,

$$I_2 = \frac{E_2}{Z_2} = \frac{0.282\angle 45.41°}{66 + j\left(2\pi\frac{10^3}{\pi}\right)(10\times 10^{-3})}$$

$$= \frac{0.282\angle 45.41°}{(66 + j20)}$$

$$= \frac{0.282\angle 45.41°}{68.964\angle 16.86°} = 4.089\times 10^{-3}\angle 28.55°\text{ A}$$

i.e. $\boldsymbol{I_2 = 4.09\angle 28.55°\text{ mA}}$

Problem 11. For the coupled circuit shown in Figure 46.10, calculate (a) the self impedance of the primary circuit, (b) the self impedance of the secondary circuit, (c) the impedance reflected into the primary circuit, (d) the effective primary impedance, (e) the primary current and (f) the secondary current.

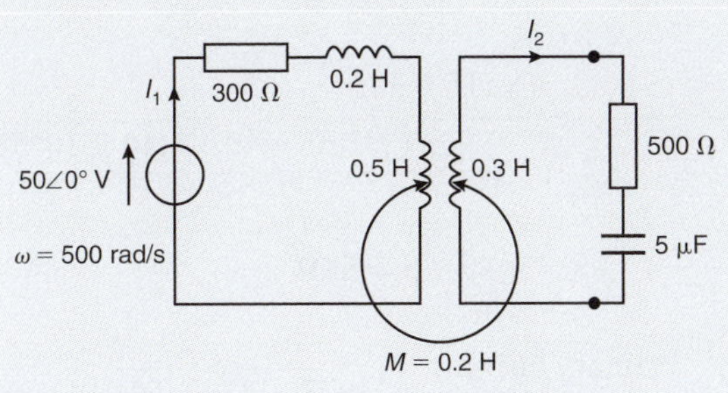

Figure 46.10

(a) Self impedance of primary circuit,

$$Z_1 = 300 + j(500)(0.2 + 0.5)$$

i.e. $\boldsymbol{Z_1 = (300 + j350)\,\Omega}$

(b) Self impedance of secondary circuit,

$$Z_2 = 500 + j\left[(500)(0.3) - \frac{1}{(500)(5 \times 10^{-6})}\right]$$

$$= 500 + j(150 - 400)$$

i.e. $\boldsymbol{Z_2 = (500 - j250)\,\Omega}$

(c) From equation (26),

$$\textbf{reflected impedance, } \boldsymbol{Z_r} = \frac{\omega^2 M^2}{Z_2}$$

$$= \frac{(500)^2(0.2)^2}{(500 - j250)}$$

$$= \frac{10^4(500 + j250)}{500^2 + 250^2}$$

$$= \boldsymbol{(16 + j8)\,\Omega}$$

(d) Effective primary impedance,

$$Z_{1(\text{eff})} = Z_1 + Z_r \text{ (note this is equivalent to equation (25))}$$

$$= (300 + j350) + (16 + j8)$$

i.e. $\boldsymbol{Z_{1(\text{eff})} = (316 + j358)\,\Omega}$

(e) Primary current, $\boldsymbol{I_1} = \dfrac{E_1}{Z_{1(\text{eff})}} = \dfrac{50\angle 0^\circ}{(316 + j358)}$

$$= \frac{50\angle 0^\circ}{477.51\angle 48.57^\circ}$$

$$= \mathbf{0.105\angle -48.57^\circ\ A}$$

(f) Secondary current, $I_2 = \dfrac{E_2}{Z_2}$, where

$E_2 = j\omega M I_1$ from equation (18)

Hence $\boldsymbol{I_2} = \dfrac{j\omega M I_1}{Z_2}$

$$= \frac{j(500)(0.2)(0.105\angle -48.57^\circ)}{(500 - j250)}$$

$$= \frac{(100\angle 90^\circ)(0.105\angle -48.57^\circ)}{559.02\angle -26.57^\circ}$$

$$= \mathbf{0.0188\angle 68^\circ\ A} \text{ or } \mathbf{18.8\angle 68^\circ\ mA}$$

(c) Resonance by tuning capacitors

Tuning capacitors may be added to the primary and/or secondary circuits to cause it to resonate at particular frequencies. These may be connected either in series or in parallel with the windings. Figure 46.11 shows each winding tuned by series-connected capacitors C_1 and C_2. The expression for the effective primary impedance $Z_{1(\text{eff})}$, i.e. equation (25), applies except that ωL_1 becomes $(\omega L_1 - (1/\omega C_1))$ and ωL_2 becomes $(\omega L_2 - (1/\omega C_2))$

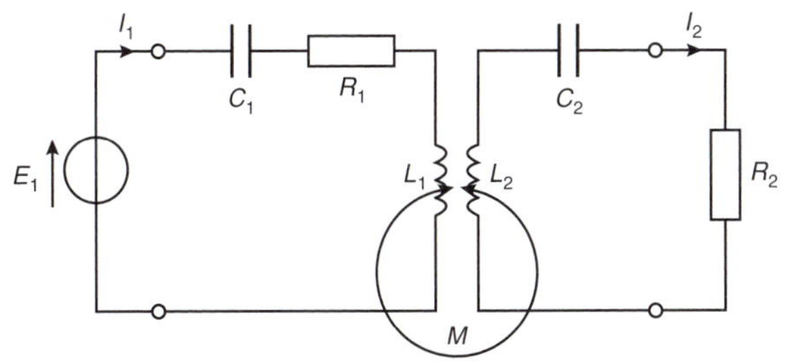

Figure 46.11

Problem 12. For the circuit shown in Figure 46.12 each winding is tuned to resonate at the same frequency. Determine (a) the resonant frequency, (b) the value of capacitor C_2, (c) the effective primary impedance, (d) the primary current, (e) the voltage across capacitor C_2 and (f) the coefficient of coupling.

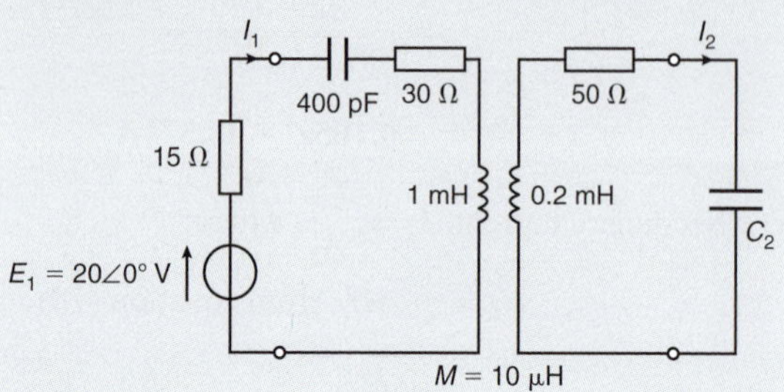

Figure 46.12

(a) For resonance in a series circuit, the resonant frequency, f_r, is given by:

$$f_r = \frac{1}{2\pi\sqrt{(LC)}}\text{ Hz}$$

Hence $\boldsymbol{f_r} = \dfrac{1}{2\pi\sqrt{(L_1C_1)}}$

$$= \frac{1}{2\pi\sqrt{(1\times10^{-3})(400\times10^{-12})}}$$

$= \mathbf{251.65\,kHz}$

(b) The secondary is also tuned to a resonant frequency of 251.65 kHz.

Hence $f_r = \dfrac{1}{2\pi\sqrt{(L_2C_2)}}$ i.e. $(2\pi f_r)^2 = \dfrac{1}{L_2C_2}$

and **capacitance,**

$$\boldsymbol{C_2} = \frac{1}{L_2(2\pi f_r)^2}$$

$$= \frac{1}{(0.2\times10^{-3})[2\pi(251.65\times10^3)]^2}$$

$= 2.0\times10^{-9}$ F or **2.0 nF**

(Note that since $f_r = \dfrac{1}{2\pi\sqrt{(L_1C_1)}} = \dfrac{1}{2\pi\sqrt{(L_2C_2)}}$

then $L_1C_1 = L_2C_2$

and $C_2 = \dfrac{L_1C_1}{L_2} = \dfrac{(1\times10^{-3})(400\times10^{-12})}{0.2\times10^{-3}}$

$= 2.0$ nF)

(c) Since both the primary and secondary circuits are resonant, the effective primary impedance $Z_{1(\text{eff})}$, from equation (25) is resistive,

i.e. $\boldsymbol{Z_{1(\text{eff})}} = R_1 + \dfrac{\omega^2M^2R_2}{R_2^2 + \left(\omega L_1 - \dfrac{1}{\omega C_1}\right)^2}$

$$= R_1 + \frac{\omega^2M^2R_2}{R_2^2}$$

$$= R_1 + \frac{\omega^2M^2}{R_2}$$

$$= (15+30) + \frac{[2\pi(251.65\times10^3)]^2(10\times10^{-6})^2}{50}$$

$= 45 + 5 = \mathbf{50\,\Omega}$

(d) **Primary current,** $\boldsymbol{I_1} = \dfrac{E_1}{Z_{1(\text{eff})}} = \dfrac{20\angle0°}{50}$

$= \mathbf{0.40\angle0°\,A}$

(e) From equation (18), secondary voltage

$E_2 = j\omega MI_1$

$= j(2\pi)(251.65\times10^3)(10\times10^{-6})(0.40\angle0°)$

$= 6.325\angle90°$ V

Secondary current, $I_2 = \dfrac{E_2}{Z_2} = \dfrac{6.325\angle90°}{50\angle0°}$

$= 0.1265\angle90°$ A

Hence **voltage across capacitor $\boldsymbol{C_2}$**

$= (I_2)(X_{C2}) = (I_2)\left(\dfrac{1}{\omega C_2}\right)$

$= (0.1265\angle90°)$

$\times\left(\dfrac{1}{[2\pi(251.65\times10^3)](2.0\times10^{-9})}\angle-90°\right)$

$= \mathbf{40\angle0°\,V}$

(f) From equation (10), the

coefficient of coupling, $\boldsymbol{k} = \dfrac{M}{\sqrt{(L_1L_2)}}$

$= \dfrac{10\times10^{-6}}{\sqrt{(1\times10^{-3})(0.2\times10^{-3})}} = \mathbf{0.0224}$

Now try the following Practice Exercise

Practice Exercise 187 Coupled circuits (Answers on page 834)

1. Determine the value of voltage E_2 which appears across the open-circuited secondary winding of Figure 46.13.

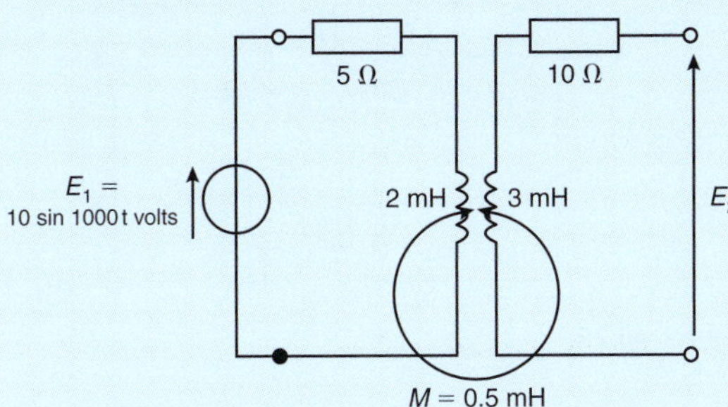

Figure 46.13

2. The coefficient of coupling between two coils having self inductances of 0.5 H and 0.9 H, respectively, is 0.85. If a sinusoidal alternating voltage of 50 mV is applied to the 0.5 H coil, determine the magnitude of the open-circuit e.m.f. induced in the 0.9 H coil.
3. Determine the value of (a) the primary current, I_1 and (b) the secondary current I_2 for the circuit shown in Figure 46.14.

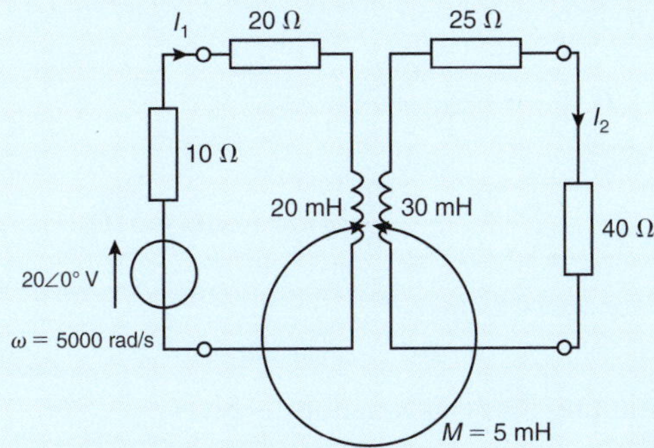

Figure 46.14

4. For the magnetically coupled circuit shown in Figure 46.15, determine (a) the self impedance of the primary circuit, (b) the self impedance of the secondary circuit, (c) the impedance reflected into the primary circuit, (d) the effective primary impedance, (e) the primary current and (f) the secondary current.

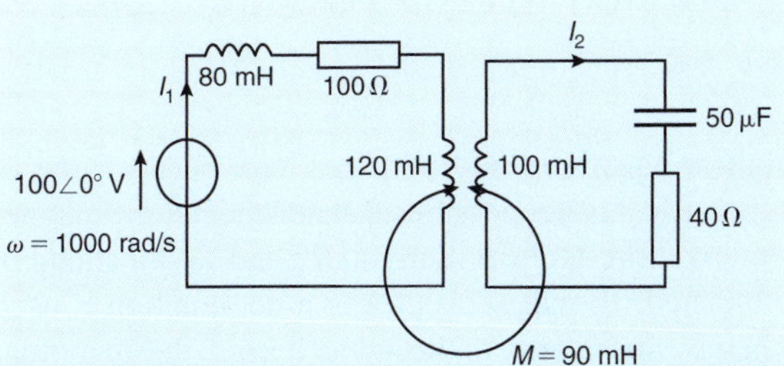

Figure 46.15

5. In the coupled circuit shown in Figure 46.16, each winding is tuned to resonance at the same frequency. Calculate (a) the resonant frequency, (b) the value of C_S, (c) the effective primary impedance, (d) the primary current, (e) the secondary current, (f) the p.d. across capacitor C_S and (g) the coefficient of coupling.

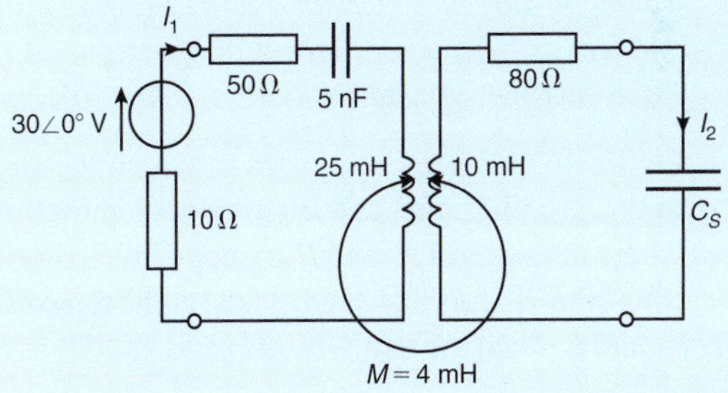

Figure 46.16

46.7 Dot rule for coupled circuits

Applying Kirchhoff's voltage law to each mesh of the circuit shown in Figure 46.17 gives:

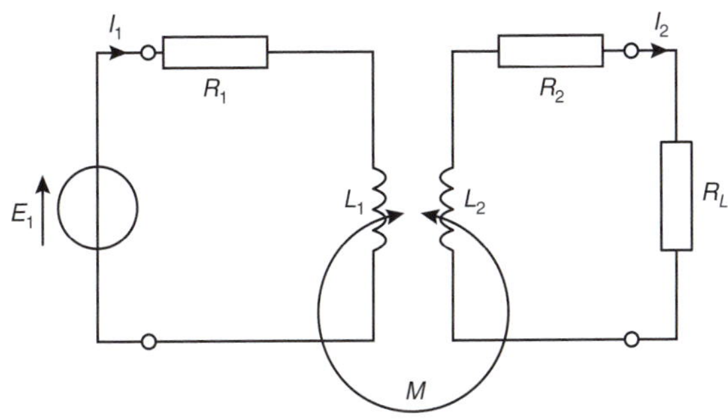

Figure 46.17

$$E_1 = I_1(R_1 + j\omega L_1) \pm j\omega M I_2$$

$$\text{and} \quad 0 = I_2(R_2 + R_L + j\omega L_2) \pm j\omega M I_1$$

In these equations the 'M' terms have been written as $\pm$ because it is not possible to state whether the magnetomotive forces due to currents I_1 and I_2 are added or subtracted. To make this clearer a dot notation is used whereby the polarity of the induced e.m.f. due to mutual inductance is identified by placing a dot on the diagram adjacent to that end of each equivalent winding which bears the same relationship to the magnetic flux.

The **dot rule** determines the sign of the voltage of mutual inductance in the Kirchhoff's law equations shown above, and states:

(i) *when both currents enter, or both currents leave, a pair of coupled coils at the dotted terminals, the signs of the 'M' terms will be the same as the signs of the 'L' terms, or*

(ii) *when one current enters at a dotted terminal and one leaves by a dotted terminal, the signs of the 'M' terms are opposite to the signs of the 'L' terms.*

Thus Figure 46.18 shows two cases in which the signs of M and L are the same, and Figure 46.19 shows two cases where the signs of M and L are opposite. In Figure 46.17, therefore, if dots had been placed at the top end of coils L_1 and L_2 then the terms $j\omega M I_2$ and $j\omega M I_1$ in the Kirchhoff's equation would be negative (since current directions are similar to Figure 46.19(a)).

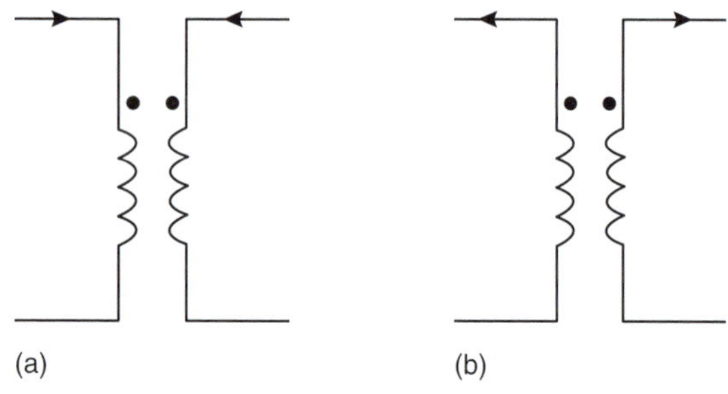

Figure 46.18

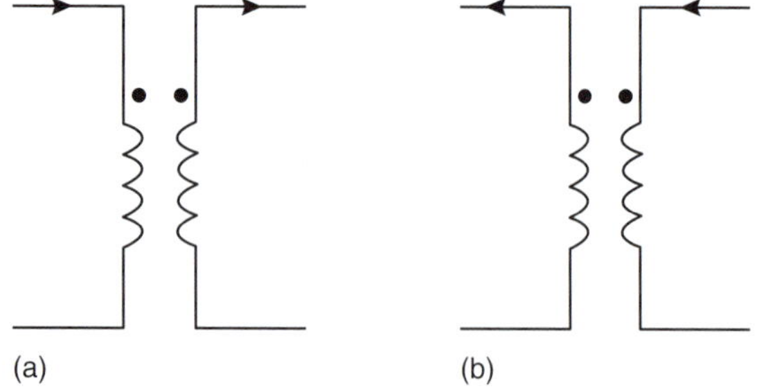

Figure 46.19

Problem 13. For the coupled circuit shown in Figure 46.20, determine the values of currents I_1 and I_2

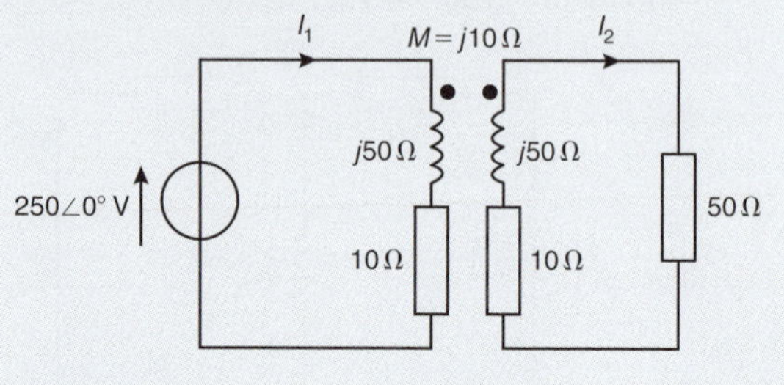

Figure 46.20

The position of the dots and the current directions correspond to Figure 46.19(a), and hence the signs of the M and L terms are opposite. Applying Kirchhoff's voltage law to the primary circuit gives:

$$250\angle 0° = (10 + j50)I_1 - j10I_2 \qquad (1)$$

and applying Kirchhoff's voltage law to the secondary circuit gives:

$$0 = (10 + 50 + j50)I_2 - j10I_1 \qquad (2)$$

From equation (2), $j10I_1 = (60 + j50)I_2$

$$\text{and} \quad I_1 = \frac{(60 + j50)I_2}{j10}$$

$$= \left(\frac{60}{j10} + \frac{j50}{j10}\right) I_2 = (-j6 + 5)I_2$$

$$\text{i.e.} \quad I_1 = (5 - j6)I_2 \qquad (3)$$

Substituting for I_1 in equation (1) gives:

$$250\angle 0° = (10 + j50)(5 - j6)I_2 - j10I_2$$

$$= (50 - j60 + j250 + 300 - j10)I_2$$

$$= (350 + j180)I_2$$

$$\text{from which,} \quad \boldsymbol{I_2} = \frac{250\angle 0°}{(350 + j180)} = \frac{250\angle 0°}{393.57\angle 27.22°}$$

$$= \mathbf{0.635\angle{-27.22}° \ A}$$

From equation (3),

$$I_1 = (5 - j6)I_2$$
$$= (5 - j6)(0.635\angle{-27.22°})$$
$$= (7.810\angle{-50.19°})(0.635\angle{-27.22°})$$

i.e. $\mathbf{I_1 = 4.959\angle{-77.41°}\ A}$

Problem 14. The circuit diagram of an air-cored transformer winding is shown in Figure 46.21. The coefficient of coupling between primary and secondary windings is 0.70. Determine for the circuit (a) the mutual inductance M, (b) the primary current I_1 and (c) the secondary terminal p.d.

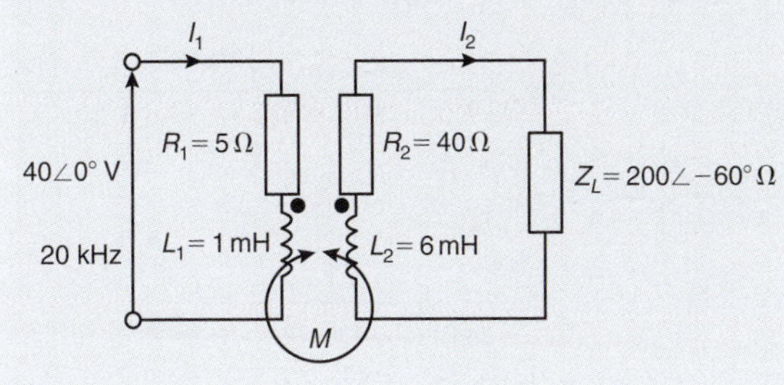

Figure 46.21

(a) From equation (9),

$$\textbf{mutual inductance, } M = k\sqrt{(L_1L_2)}$$
$$= 0.70\sqrt{[(1\times10^{-3})(6\times10^{-3})]}$$
$$= \mathbf{1.715\,mH}$$

(b) The two mesh equations are:

$$40\angle 0° = (R_1 + j\omega L_1)I_1 - j\omega MI_2 \quad (1)$$
$$\text{and} \quad 0 = (R_2 + j\omega L_2 + Z_L)I_2 - j\omega MI_1 \quad (2)$$

(Note that with the dots and current directions shown, the $j\omega MI$ terms are negative)

$$R_1 + j\omega L_1 = 5 + j2\pi(20\times10^3)(1\times10^{-3})$$
$$= (5 + j125.66)\,\Omega \text{ or } 125.76\angle 87.72°\,\Omega$$
$$j\omega M = j2\pi(20\times10^3)(1.715\times10^{-3})$$
$$= j215.51\,\Omega \quad\text{or}\quad 215.51\angle 90°\,\Omega$$

$$R_2 + j\omega L_2 + Z_L = 40 + j2\pi(20\times10^3) \times(6\times10^{-3}) + 200\angle{-60°}$$
$$= 40 + j753.98 + 100 - j173.21$$
$$= (140 + j580.77)\,\Omega \text{ or } 597.41\angle 76.45°\,\Omega$$

$$\text{Hence } 40\angle 0° = 125.76\angle 87.72° I_1 - 215.51\angle 90° I_2 \quad (3)$$
$$0 = -215.51\angle 90° I_1 + 597.41\angle 76.45° I_2 \quad (4)$$

From equation (4), $I_2 = \dfrac{215.51\angle 90°}{597.41\angle 76.45°}I_1$

$$= 0.361\angle 13.55° I_1 \quad (5)$$

Substituting for I_2 in equation (3) gives:

$$40\angle 0° = 125.76\angle 87.72° I_1 - (215.51\angle 90°)(0.361\angle 13.55° I_1)$$
$$= I_1(125.76\angle 87.72° - 77.80\angle 103.55°)$$
$$= I_1[(5 + j125.66) - (-18.23 + j75.63)]$$
$$= I_1(23.23 - j50.03)$$

i.e. $40\angle 0° = I_1(55.16\angle{-65.09°})$

Hence **primary current,** $\boldsymbol{I_1} = \dfrac{40\angle 0°}{55.16\angle{-65.09°}}$

$$= \mathbf{0.725\angle 65.09°\ A}$$

(c) From equation (5), $I_2 = 0.361\angle 13.55° I_1$

$$= (0.361\angle 13.55°)(0.725\angle 65.09°)$$
$$= 0.262\angle 78.64°\ \text{A}$$

Hence **secondary terminal p.d.**

$$= I_2Z_L$$
$$= (0.262\angle 78.64°)(200\angle{-60°})$$
$$= \mathbf{52.4\angle 18.64°\ V}$$

Problem 15. A mutual inductor is used to couple a 20 Ω resistive load to a $50\angle 0°$ V generator as shown in Figure 46.22. The generator has an internal resistance of 5 Ω and the mutual inductor parameters are $R_1 = 20\,\Omega$, $L_1 = 0.2\,\text{H}$, $R_2 = 25\,\Omega$, $L_2 = 0.4\,\text{H}$ and $M = 0.1\,\text{H}$. The supply frequency is $(75/\pi)$ Hz. Determine (a) the generator current I_1 and (b) the load current I_2

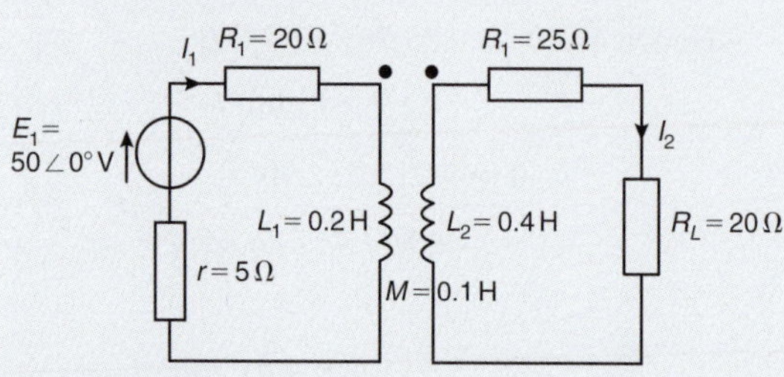

Figure 46.22

(a) Applying Kirchhoff's voltage law to the primary winding gives:

$$I_1(r + R_1 + j\omega L_1) - j\omega M I_2 = 50\angle 0°$$

i.e. $$I_1\left[5 + 20 + j2\pi\left(\frac{75}{\pi}\right)(0.2)\right] - j2\pi\left(\frac{75}{\pi}\right)(0.1)I_2 = 50\angle 0°$$

i.e. $$I_1(25 + j30) - j15I_2 = 50\angle 0° \qquad (1)$$

(b) Applying Kirchhoff's voltage law to the secondary winding gives:

$$-j\omega M I_1 + I_2(R_2 + R_L + j\omega L_2) = 0$$

i.e. $$-j2\pi\left(\frac{75}{\pi}\right)(0.1)I_1 + I_2\left[25 + 20 + j2\pi\left(\frac{75}{\pi}\right)(0.4)\right] = 0$$

i.e. $$-j15I_1 + I_2(45 + j60) = 0 \qquad (2)$$

Hence the equations to solve are:

$$(25 + j30)I_1 - j15I_2 - 50\angle 0° = 0 \qquad (1)'$$

and $$-j15I_1 + (45 + j60)I_2 = 0 \qquad (2)'$$

Using determinants:

$$\frac{I_1}{\begin{vmatrix} -j15 & -50\angle 0° \\ (45 + j60) & 0 \end{vmatrix}} = \frac{-I_2}{\begin{vmatrix} (25 + j30) & -50\angle 0° \\ -j15 & 0 \end{vmatrix}} = \frac{1}{\begin{vmatrix} (25 + j30) & -j15 \\ -j15 & (45 + j60) \end{vmatrix}}$$

i.e

$$\frac{I_1}{50(45 + j60)} = \frac{-I_2}{-50(j15)} = \frac{1}{(25 + j30)(45 + j60) - (j15)^2}$$

$$\frac{I_1}{50(75\angle 53.13°)} = \frac{I_2}{750\angle 90°} = \frac{1}{(39.05\angle 50.19°)(75\angle 53.13°) + 225}$$

$$\frac{I_1}{3750\angle 53.13°} = \frac{I_2}{750\angle 90°} = \frac{1}{2928.75\angle 103.32° + 225}$$

$$\frac{I_1}{3750\angle 53.13°} = \frac{I_2}{750\angle 90°} = \frac{1}{(-449.753 + j2849.962)}$$

$$\frac{I_1}{3750\angle 53.13°} = \frac{I_2}{750\angle 90°} = \frac{1}{2885.23\angle 98.97°}$$

(a) **Generator current,** $$I_1 = \frac{3750\angle 53.13°}{2885.23\angle 98.97°} = \mathbf{1.30\angle -45.84°\ A}$$

(b) **Load current,** $$I_2 = \frac{750\angle 90°}{2885.23\angle 98.97°} = \mathbf{0.26\angle -8.97°\ A}$$

Problem 16. The mutual inductor of Problem 15 is connected to the circuit of Figure 46.23. Determine the source and load currents for (a) the windings as shown (i.e. with the dots adjacent), and (b) with one winding reversed (i.e. with the dots at opposite ends).

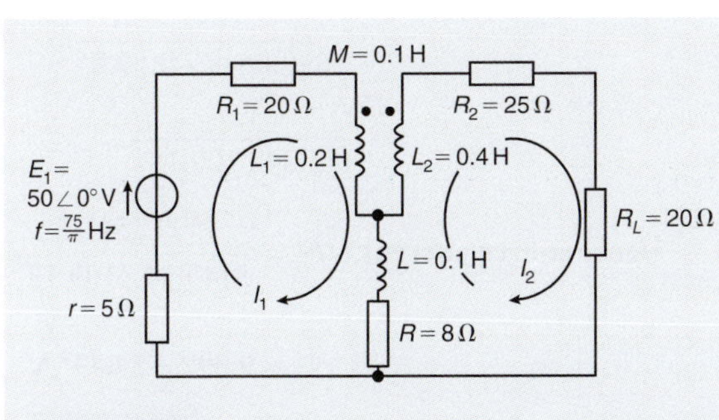

Figure 46.23

(a) The left-hand mesh equation in Figure 46.23 is:

$$E_1 = I_1(r + R_1 + R + j\omega L_1 + j\omega L) - j\omega M I_2 - I_2(R + j\omega L)$$

(Note that with the dots as shown in Figure 46.23, and the chosen current directions as shown, the $j\omega M I_2$ is negative – see Figure 46.19(a)). Hence

$$50\angle 0^\circ = I_1\left[5+20+8+j2\pi\left(\frac{75}{\pi}\right)(0.2) + j2\pi\left(\frac{75}{\pi}\right)(0.1)\right] - j2\pi\left(\frac{75}{\pi}\right)(0.1)I_2 - I_2\left[8+j2\pi\left(\frac{75}{\pi}\right)(0.1)\right]$$

i.e. $50\angle 0^\circ = I_1(33+j30+j15) - j15I_2 - I_2(8+j15)$ (i)

i.e. $50\angle 0^\circ = (33+j45)I_1 - (8+j30)I_2$ (1)

The right-hand mesh equation in Figure 46.23 is:

$$0 = I_2(R + R_2 + R_L + j\omega L_2 + j\omega L) - j\omega M I_1 - I_1(R + j\omega L)$$

i.e. $0 = I_2\left[8+25+20+j2\pi\left(\frac{75}{\pi}\right)(0.4) + j2\pi\left(\frac{75}{\pi}\right)(0.1)\right] - j2\pi\left(\frac{75}{\pi}\right)(0.1)I_1 - I_1\left[8+j2\pi\left(\frac{75}{\pi}\right)(0.1)\right]$

i.e. $0 = I_2(53+j60+j15) - j15I_1 - I_1(8+j15)$ (ii)

i.e. $0 = (53+j75)I_2 - (8+j30)I_1$ (iii)

Hence the simultaneous equations to solve are:

$$(33+j45)I_1 - (8+j30)I_2 - 50\angle 0^\circ = 0 \quad (1)$$

$$-(8+j30)I_1 + (53+j75)I_2 = 0 \quad (2)$$

Using determinants gives:

$$\frac{I_1}{\begin{vmatrix} -(8+j30) & -50\angle 0^\circ \\ (53+j75) & 0 \end{vmatrix}} = \frac{-I_2}{\begin{vmatrix} (33+j45) & -50\angle 0^\circ \\ -(8+j30) & 0 \end{vmatrix}} = \frac{1}{\begin{vmatrix} (33+j45) & -(8+j30) \\ -(8+j30) & (53+j75) \end{vmatrix}}$$

i.e.

$$\frac{I_1}{50(53+j75)} = \frac{-I_2}{-50(8+j30)} = \frac{1}{(33+j45)(53+j75)-(8+j30)^2}$$

i.e.

$$\frac{I_1}{50(91.84\angle 54.75^\circ)} = \frac{I_2}{50(31.05\angle 75.07^\circ)} = \frac{1}{\left[\begin{array}{c}(55.80\angle 53.75^\circ)(91.84\angle 54.75^\circ) \\ -(31.05\angle 75.07^\circ)^2\end{array}\right]}$$

$$\frac{I_1}{4592\angle 54.75^\circ} = \frac{I_2}{1552.5\angle 75.07^\circ} = \frac{1}{5124.672\angle 108.50^\circ - 964.103\angle 150.14^\circ}$$

$$\frac{I_1}{4592\angle 54.75^\circ} = \frac{I_2}{1552.5\angle 75.07^\circ} = \frac{1}{-789.97 + j4379.84} = \frac{1}{4450.51\angle 100.22^\circ}$$

Hence **source current,** $I_1 = \frac{4592\angle 54.75^\circ}{4450.51\angle 100.22^\circ}$ $= \mathbf{1.03\angle{-45.47}^\circ\,A}$

and **load current,** $I_2 = \frac{1552.5\angle 75.07^\circ}{4450.51\angle 100.22^\circ}$ $= \mathbf{0.35\angle{-25.15}^\circ\,A}$

(b) When one of the windings of the mutual inductor is reversed, with, say, the dots as shown in Figure 46.24, the $j\omega MI$ terms change sign, i.e. are positive. With both currents entering the dot ends of the windings as shown, it compares with Figure 46.18(a), which indicates that the 'L' and 'M' terms are of similar sign.

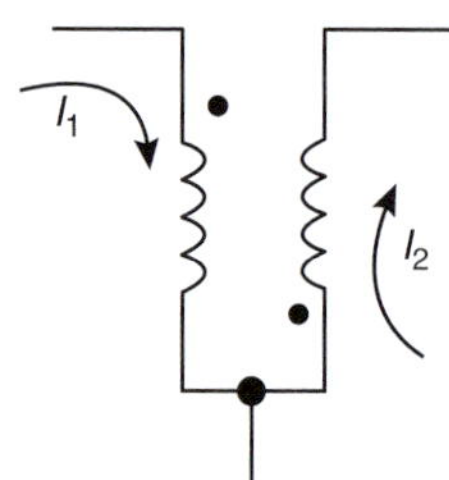

Figure 46.24

Thus equations (i) and (ii) of part (a) become:

$$50\angle 0^\circ = I_1(33 + j30 + j15) + j15I_2 - I_2(8 + j15)$$

and $$0 = I_2(53 + j60 + j15) + j15I_1 - I_1(8 + j15)$$

i.e. $I_1(33 + j45) - I_2(8) - 50\angle 0^\circ = 0$

and $-I_1(8) + I_2(53 + j75) = 0$

Using determinants:

$$\frac{I_1}{\begin{vmatrix} -8 & -50\angle 0^\circ \\ (53+j75) & 0 \end{vmatrix}} = \frac{-I_2}{\begin{vmatrix} (33+j45) & -50\angle 0^\circ \\ -8 & 0 \end{vmatrix}} = \frac{1}{\begin{vmatrix} (33+j45) & -8 \\ -8 & (53+j75) \end{vmatrix}}$$

i.e. $$\frac{I_1}{50(53+j75)} = \frac{-I_2}{-400\angle 0^\circ} = \frac{1}{(33+j45)(53+j75) - 64}$$

$$\frac{I_1}{4592\angle 54.75^\circ} = \frac{I_2}{400\angle 0^\circ} = \frac{1}{5124.672\angle 108.50^\circ - 64}$$

$$= \frac{1}{-1690.08 + j4859.85}$$

$$= \frac{1}{5145.34\angle 109.18^\circ}$$

Hence **source current,** $$I_1 = \frac{4592\angle 54.75^\circ}{5145.34\angle 109.18^\circ}$$

$$= \mathbf{0.89\angle{-54.43}^\circ\ A}$$

and **load current,** $$I_2 = \frac{400\angle 0^\circ}{5145.34\angle 109.18^\circ}$$

$$= \mathbf{0.078\angle{-109.18}^\circ\ A}$$

Now try the following Practice Exercise

Practice Exercise 188 The dot rule for coupled circuits (Answers on page 834)

1. Determine the values of currents I_p and I_s in the coupled circuit shown in Figure 46.25.

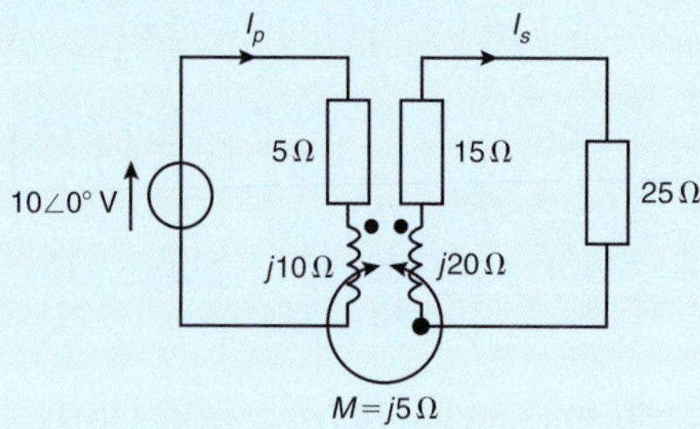

Figure 46.25

2. The coefficient of coupling between the primary and secondary windings for the air-cored transformer shown in Figure 46.26 is 0.84. Calculate for the circuit (a) the mutual inductance M, (b) the primary current I_p, (c) the secondary current I_s and (d) the secondary terminal p.d.

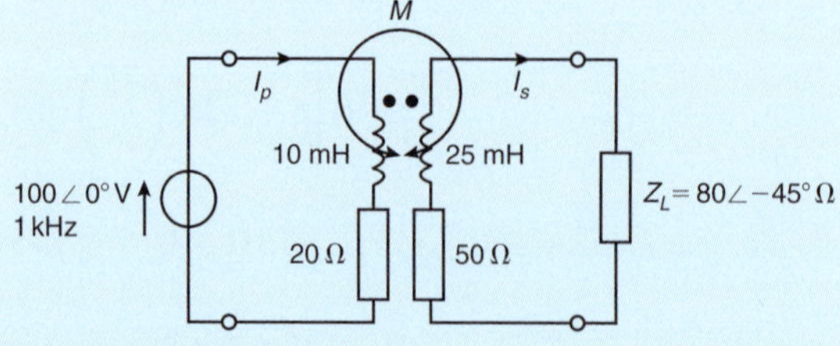

Figure 46.26

3. A mutual inductor is used to couple a $50\,\Omega$ resistive load to a $250\angle 0°$ V generator as shown in Figure 46.27. Calculate (a) the generator current I_g and (b) the load current I_L

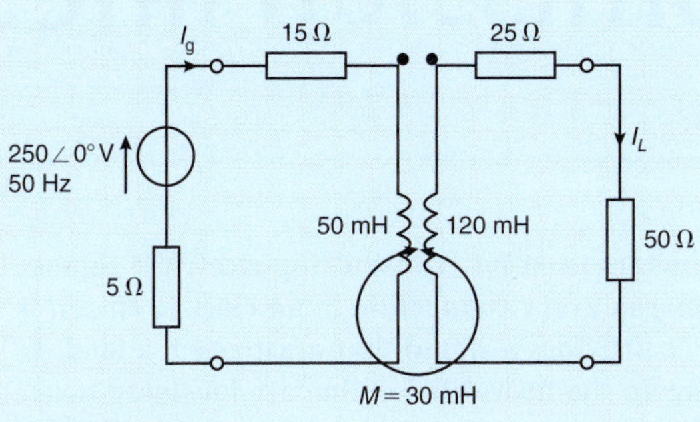

Figure 46.27

4. The mutual inductor of Problem 3 is connected to the circuit as shown in Figure 46.28. Determine (a) the source current and (b) the load current. (c) If one of the windings is reversed, determine the new value of source and load currents.

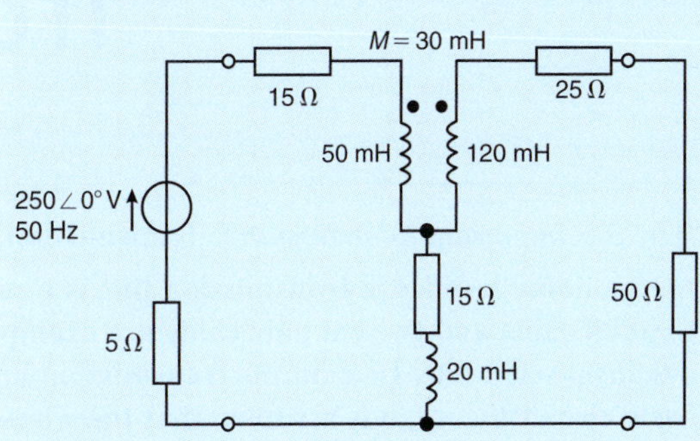

Figure 46.28

For fully worked solutions to each of the problems in Practice Exercises 184 to 188 in this chapter, go to the website:
www.routledge.com/cw/bird

Chapter 47

Transmission lines

***Why it is important to understand:* Transmission lines**

As the name implies, a transmission line is a set of conductors used for transmitting electrical signals. Coaxial cable and twisted pair cable are examples. In general, every connection in an electric circuit is a transmission line. In a simple transmission line, a source provides a signal that must reach a load. In basic circuit theory, it is assumed that the wires that make up the transmission line are ideal and hence that the voltage at all points on the wires is exactly the same. In reality, this situation is never quite true. Any real wire has series resistance, *R*, and inductance, *L*. Additionally, a capacitance, *C*, exists between any pair of real wires. Moreover, because all dielectrics exhibit some leakage, a small conductance, *G*, (i.e. a high shunt resistance) exists between the two wires. Understanding transmission lines is an important part of the art of radio – the proper selection and use of a feed line is often more important than the choice of radio. In this chapter the purpose of transmission lines is explained, as are the current and voltage relationships; calculations involving the characteristic impedance and propagation coefficient, distortion, wave reflection and standing waves are also explored.

At the end of this chapter you should be able to:

- appreciate the purpose of a transmission line
- define the transmission line primary constants R, L, C and G
- calculate phase delay, wavelength and velocity of propagation on a transmission line
- appreciate current and voltage relationships on a transmission line
- define the transmission line secondary line constants Z_0, γ, α and β
- calculate characteristic impedance and propagation coefficient in terms of the primary line constants
- understand and calculate distortion on transmission lines
- understand wave reflection and calculate reflection coefficient
- understand standing waves and calculate standing wave ratio

47.1 Introduction

A transmission line is a system of conductors connecting one point to another and along which electromagnetic energy can be sent. Thus telephone lines and power distribution lines are typical examples of transmission lines; in electronics, however, the term usually implies a line used for the transmission of radio-frequency (r.f.) energy such as that from a radio transmitter to the antenna.

An important feature of a transmission line is that it should guide energy from a source at the sending end to a load at the receiving end without loss by radiation.

Electrical Circuit Theory and Technology. 978-1-138-67349-6, © 2017 John Bird. Published by Taylor & Francis. All rights reserved.

One form of construction often used consists of two similar conductors mounted close together at a constant separation. The two conductors form the two sides of a balanced circuit and any radiation from one of them is neutralized by that from the other. Such twin-wire lines are used for carrying high r.f. power, for example, at transmitters. The coaxial form of construction is commonly employed for low power use, one conductor being in the form of a cylinder which surrounds the other at its centre, and thus acts as a screen. Such cables are often used to couple f.m. and television receivers to their antennas.

At frequencies greater than 1000 MHz, transmission lines are usually in the form of a waveguide which may be regarded as coaxial lines without the centre conductor, the energy being launched into the guide or abstracted from it by probes or loops projecting into the guide.

47.2 Transmission line primary constants

Let an a.c. generator be connected to the input terminals of a pair of parallel conductors of infinite length. A sinusoidal wave will move along the line and a finite current will flow into the line. The variation of voltage with distance along the line will resemble the variation of applied voltage with time. The moving wave, sinusoidal in this case, is called a voltage **travelling wave**. As the wave moves along the line the capacitance of the line is charged up and the moving charges cause magnetic energy to be stored. Thus the propagation of such an **electromagnetic wave** constitutes a flow of energy.

After sufficient time the magnitude of the wave may be measured at any point along the line. The line does not therefore appear to the generator as an open circuit but presents a definite load Z_0. If the sending-end voltage is V_S and the sending-end current is I_S then $Z_0 = V_S/I_S$. Thus all of the energy is absorbed by the line and the line behaves in a similar manner to the generator as would a single 'lumped' impedance of value Z_0 connected directly across the generator terminals.

There are **four parameters** associated with transmission lines, these being resistance, inductance, capacitance and conductance.

(i) **Resistance *R*** is given by $R = \rho l/A$, where ρ is the resistivity of the conductor material, A is the cross-sectional area of each conductor and l is the length of the conductor (for a two-wire system, l represents twice the length of the line). Resistance is stated in ohms per metre length of a line and represents the imperfection of the conductor. A resistance stated in ohms per loop metre is a little more specific since it takes into consideration the fact that there are two conductors in a particular length of line.

(ii) **Inductance *L*** is due to the magnetic field surrounding the conductors of a transmission line when a current flows through them. The inductance of an isolated twin line is considered in Section 43.7. From equation (23), page 660, the inductance L is given by

$$L = \frac{\mu_0 \mu_r}{\pi}\left\{\frac{1}{4} + \ln\frac{D}{a}\right\} \text{ henry/metre}$$

where D is the distance between centres of the conductor and a is the radius of each conductor. In most practical lines $\mu_r = 1$. An inductance stated in henrys per loop metre takes into consideration the fact that there are two conductors in a particular length of line.

(iii) **Capacitance *C*** exists as a result of the electric field between conductors of a transmission line. The capacitance of an isolated twin line is considered in Section 43.3. From equation (14), page 652, the capacitance between the two conductors is given by

$$C = \frac{\pi \varepsilon_0 \varepsilon_r}{\ln(D/a)} \text{ farads/metre}$$

In most practical lines $\varepsilon_r = 1$

(iv) **Conductance *G*** is due to the insulation of the line allowing some current to leak from one conductor to the other. Conductance is measured in siemens per metre length of line and represents the imperfection of the insulation. Another name for conductance is leakance.

Each of the four transmission line constants, R, L, C and G, known as the **primary constants**, are uniformly distributed along the line.

From Chapter 44, when a symmetrical T-network is terminated in its characteristic impedance Z_0, the input impedance of the network is also equal to Z_0. Similarly, if a number of identical T-sections are connected in cascade, the input impedance of the network will also be equal to Z_0

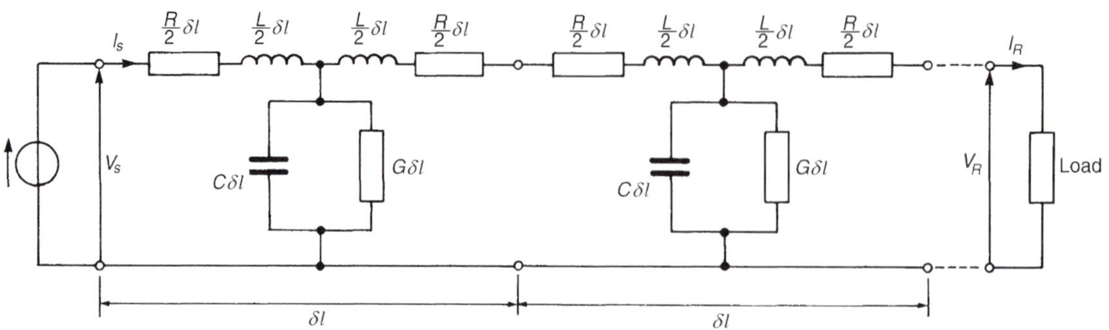

Figure 47.1

A transmission line can be considered to consist of a network of a very large number of cascaded T-sections each a very short length (δl) of transmission line, as shown in Figure 47.1. This is an approximation of the uniformly distributed line; the larger the number of lumped parameter sections, the nearer it approaches the true distributed nature of the line. When the generator V_S is connected, a current I_S flows which divides between that flowing through the leakage conductance G, which is lost, and that which progressively charges each capacitor C and which sets up the voltage travelling wave moving along the transmission line. The loss or attenuation in the line is caused by both the conductance G and the series resistance R

47.3 Phase delay, wavelength and velocity of propagation

Each section of that shown in Figure 47.1 is simply a low-pass filter possessing losses R and G. If losses are neglected, and R and G are removed, the circuit simplifies and the infinite line reduces to a repetitive T-section low-pass filter network, as shown in Figure 47.2. Let a generator be connected to the line as shown and let the voltage be rising to a maximum positive value just at the instant when the line is connected to it. A current I_S flows through inductance L_1 into capacitor C_1. The capacitor charges and a voltage develops across it. The voltage sends a current through inductance L_1' and L_2 into capacitor C_2. The capacitor charges and the voltage developed across it sends a current through L_2' and L_3 into C_3, and so on. Thus all capacitors will in turn charge up to the maximum input voltage. When the generator voltage falls, each capacitor is charged in turn in opposite polarity, and as before the input charge is progressively passed along to the next capacitor. In this manner voltage and current waves travel along the line together and depend on each other.

The process outlined above takes time; for example, by the time capacitor C_3 has reached its maximum voltage, the generator input may be at zero or moving towards its minimum value. There will therefore be a time, and thus a phase difference between the generator input voltage and the voltage at any point on the line.

Phase delay

Since the line shown in Figure 47.2 is a ladder network of low-pass T-section filters, it is shown in equation (27), page 717, that the phase delay, β, is given by:

$$\beta = \omega\sqrt{(LC)} \text{ radians/metre} \quad (1)$$

where L and C are the inductance and capacitance per metre of the line.

Wavelength

The wavelength λ on a line is the distance between a given point and the next point along the line at which the voltage is the same phase, the initial point leading

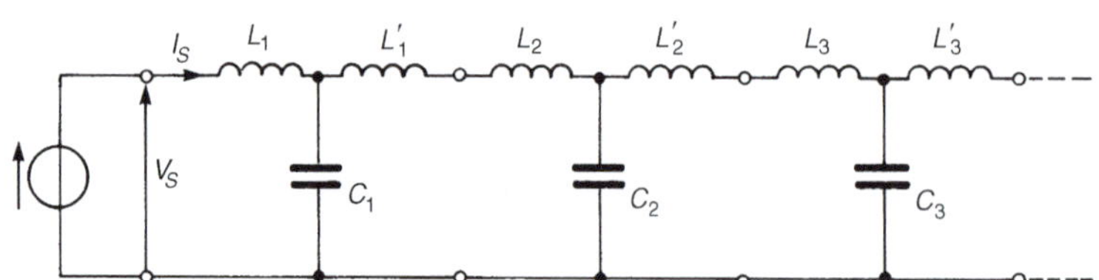

Figure 47.2

the latter point by 2π radian. Since in one wavelength a phase change of 2π radians occurs, the phase change per metre is $2\pi/\lambda$. Hence, phase change per metre, $\beta = 2\pi/\lambda$

or **wavelength,** $\lambda = \dfrac{2\pi}{\beta}$ **metres** (2)

Velocity of propagation

The velocity of propagation, u, is given by $u = f\lambda$, where f is the frequency and λ the wavelength. Hence

$$u = f\lambda = f(2\pi/\beta) = \frac{2\pi f}{\beta} = \frac{\omega}{\beta} \quad (3)$$

The velocity of propagation of free space is the same as that of light, i.e. approximately 300×10^6 m/s. The velocity of electrical energy along a line is always less than the velocity in free space. The wavelength λ of radiation in free space is given by $\lambda = c/f$ where c is the velocity of light. Since the velocity along a line is always less than c, the wavelength corresponding to any particular frequency is always shorter on the line than it would be in free space.

Problem 1. A parallel-wire air-spaced transmission line operating at 1910 Hz has a phase shift of 0.05 rad/km. Determine (a) the wavelength on the line and (b) the speed of transmission of a signal.

(a) From equation (2), wavelength $\lambda = 2\pi/\beta$

$= 2\pi/0.05$

$=$ **125.7 km**

(b) From equation (3), speed of transmission,

$u = f\lambda = (1910)(125.7)$

$=$ **240 x 10³ km/s** or **240 x 10⁶ m/s**

Problem 2. A transmission line has an inductance of 4 mH/loop km and a capacitance of 0.004 μF/km. Determine, for a frequency of operation of 1 kHz, (a) the phase delay, (b) the wavelength on the line and (c) the velocity of propagation (in metres per second) of the signal.

(a) From equation (1), phase delay,

$\beta = \omega\sqrt{(LC)}$

$= (2\pi 1000)\sqrt{[(4 \times 10^{-3})(0.004 \times 10^{-6})]}$

$=$ **0.025 rad/km**

(b) From equation (2), wavelength $\lambda = 2\pi/\beta$

$= 2\pi/0.025$

$=$ **251 km**

(c) From equation (3), velocity of propagation,

$u = f\lambda = (1000)(251)$ km/s $=$ **251 x 10⁶ m/s**

Now try the following Practice Exercise

Practice Exercise 189 Phase delay, wavelength and velocity of propagation (Answers on page 834)

1. A parallel-wire air-spaced line has a phase-shift of 0.03 rad/km. Determine (a) the wavelength on the line and (b) the speed of transmission of a signal of frequency 1.2 kHz.
2. A transmission line has an inductance of 5 μH/m and a capacitance of 3.49 pF/m. Determine, for an operating frequency of 5 kHz, (a) the phase delay (b) the wavelength on the line and (c) the velocity of propagation of the signal in metres per second.
3. An air-spaced transmission line has a capacitance of 6.0 pF/m and the velocity of propagation of a signal is 225×10^6 m/s. If the operating frequency is 20 kHz, determine (a) the inductance per metre, (b) the phase delay and (c) the wavelength on the line.

47.4 Current and voltage relationships

Figure 47.3 shows a voltage source V_S applied to the input terminals of an infinite line, or a line terminated in its characteristic impedance, such that a current I_S flows into the line. At a point, say, 1 km down the line, let the current be I_1. The current I_1 will not have the same magnitude as I_S because of line attenuation; also I_1 will lag I_S by some angle β. The ratio I_S/I_1 is therefore a phasor quantity. Let the current a further 1 km down the line be I_2, and so on, as shown in Figure 47.3. Each unit length of line can be treated as a section of a repetitive network, as explained in Section 47.2. The attenuation is in the form of a logarithmic decay and

$$\frac{I_S}{I_1} = \frac{I_1}{I_2} = \frac{I_2}{I_3} = e^{\gamma}$$

where γ is the **propagation constant**, first introduced in Section 45.7, page 714. γ has no unit.

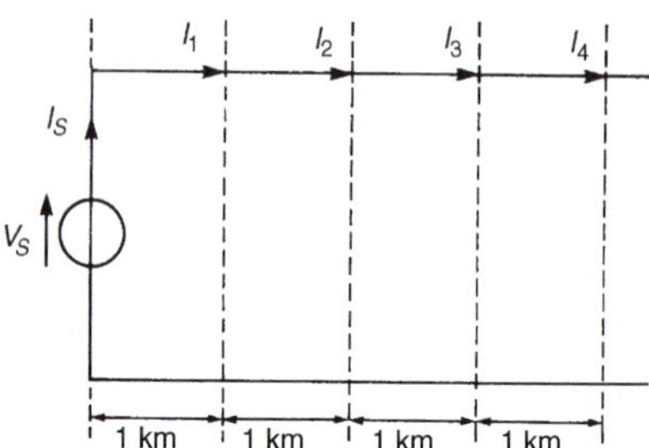

Figure 47.3

The propagation constant is a complex quantity given by $\gamma = \alpha + j\beta$, where α is the **attenuation constant**, whose unit is the neper, and β is the **phase shift coefficient**, whose unit is the radian. For n such 1 km sections, $I_S/I_R = e^{n\gamma}$ where I_R is the current at the receiving end.

Hence $\dfrac{I_S}{I_R} = e^{n(\alpha + j\beta)} = e^{(n\alpha + jn\beta)} = e^{n\alpha}\angle n\beta$

from which, $\boldsymbol{I_R = I_S e^{-n\gamma} = I_S e^{-n\alpha}\angle -n\beta}$ (4)

In equation (4), the attenuation on the line is given by $n\alpha$ nepers and the phase shift is $n\beta$ radians.

At all points along an infinite line, the ratio of voltage to current is Z_0, the characteristic impedance. Thus from equation (4) it follows that:

receiving end voltage,

$$\boldsymbol{V_R = V_S e^{-n\gamma} = V_S e^{-n\alpha}\angle -n\beta} \quad (5)$$

Z_0, γ, α, and β are referred to as the **secondary line constants** or **coefficients**.

Problem 3. When operating at a frequency of 2 kHz, a cable has an attenuation of 0.25 Np/km and a phase shift of 0.20 rad/km. If a 5 V r.m.s. signal is applied at the sending end, determine the voltage at a point 10 km down the line, assuming that the termination is equal to the characteristic impedance of the line.

Let V_R be the voltage at a point n km from the sending end, then from equation (5),
$V_R = V_S e^{-n\gamma} = V_S e^{-n\alpha}\angle -n\beta$

Since $\alpha = 0.25$ Np/km, $\beta = 0.20$ rad/km, $V_S = 5$ V and $n = 10$ km, then

$$V_R = (5)e^{-(10)(0.25)}\angle -(10)(0.20) = 5e^{-2.5}\angle -2.0\,\text{V}$$
$$= \mathbf{0.41\angle -2.0\,V} \quad \text{or} \quad \mathbf{0.41\angle -114.6^\circ\,V}$$

Thus the voltage 10 km down the line is 0.41 V r.m.s. lagging the sending end voltage of 5 V by 2.0 rad or 114.6°

Problem 4. A transmission line 5 km long has a characteristic impedance of $800\angle -25^\circ\,\Omega$. At a particular frequency, the attenuation coefficient of the line is 0.5 Np/km and the phase shift coefficient is 0.25 rad/km. Determine the magnitude and phase of the current at the receiving end, if the sending end voltage is $2.0\angle 0^\circ$ V r.m.s.

The receiving end voltage (from equation (5)) is given by:

$$V_R = V_S e^{-n\gamma} = V_S e^{-n\alpha}\angle -n\beta$$
$$= (2.0\angle 0^\circ)e^{-(5)(0.5)}\angle -(5)(0.25)$$
$$= 2.0e^{-2.5}\angle -1.25 = 0.1642\angle -71.62^\circ\,\text{V}$$

Receiving end current,

$$I_R = \frac{V_R}{Z_0} = \frac{0.1642\angle -71.62^\circ}{800\angle -25^\circ}$$
$$= 2.05 \times 10^4\angle(-71.62^\circ - (-25^\circ))\,\text{A}$$
$$= \mathbf{0.205\angle -46.62^\circ\,mA} \quad \text{or} \quad \mathbf{205\angle -46.62^\circ\,\mu A}$$

Problem 5. The voltages at the input and at the output of a transmission line properly terminated in its characteristic impedance are 8.0 V and 2.0 V r.m.s., respectively. Determine the output voltage if the length of the line is doubled.

The receiving-end voltage V_R is given by $V_R = V_S e^{-n\gamma}$.

Hence $2.0 = 8.0e^{-n\gamma}$, from which,
$e^{-n\gamma} = 2.0/8.0 = 0.25$

If the line is doubled in length, then

$$V_R = 8.0e^{-2n\gamma} = 8.0(e^{-n\gamma})^2$$
$$= 8.0(0.25)^2 = \mathbf{0.50\,V}$$

Now try the following Practice Exercise

Practice Exercise 190 Current and voltage relationships (Answers on page 834)

1. When the working frequency of a cable is 1.35 kHz, its attenuation is 0.40 Np/km and

Part 4

its phase-shift is 0.25 rad/km. The sending-end voltage and current are 8.0 V r.m.s., and 10.0 mA r.m.s. Determine the voltage and current at a point 25 km down the line, assuming that the termination is equal to the characteristic impedance of the line.

2. A transmission line 8 km long has a characteristic impedance $600\angle -30^\circ\ \Omega$. At a particular frequency the attenuation coefficient of the line is 0.4 Np/km and the phase-shift coefficient is 0.20 rad/km. Determine the magnitude and phase of the current at the receiving end if the sending-end voltage is $5\angle 0^\circ$ V r.m.s.

3. The voltages at the input and at the output of a transmission line properly terminated in its characteristic impedance are 10 V and 4 V r.m.s., respectively. Determine the output voltage if the length of the line is trebled.

47.5 Characteristic impedance and propagation coefficient in terms of the primary constants

Characteristic impedance

At all points along an infinite line, the ratio of voltage to current is called the characteristic impedance Z_0. The value of Z_0 is independent of the length of the line; it merely describes a property of a line that is a function of the physical construction of the line. Since a short length of line may be considered as a ladder of identical low-pass filter sections, the characteristic impedance may be determined from equation (2), page 667, i.e.

$$\mathbf{Z_0 = \sqrt{(Z_{OC} Z_{SC})}} \qquad (6)$$

since the open-circuit impedance Z_{OC} and the short-circuit impedance Z_{SC} may be easily measured.

Problem 6. At a frequency of 1.5 kHz the open-circuit impedance of a length of transmission line is $800\angle -50^\circ\ \Omega$ and the short-circuit impedance is $413\angle -20^\circ \Omega$. Determine the characteristic impedance of the line at this frequency.

From equation (6), characteristic impedance

$$Z_0 = \sqrt{(Z_{OC} Z_{SC})}$$
$$= \sqrt{[(800\angle -50^\circ)(413\angle -20^\circ)]}$$
$$= \sqrt{(330\,400\angle -70^\circ)} = \mathbf{575\angle -35^\circ\ \Omega}$$

by de Moivre's theorem.

The characteristic impedance of a transmission line may also be expressed in terms of the primary constants, R, L, G and C. Measurements of the primary constants may be obtained for a particular line and manufacturers usually state them for a standard length.

Let a very short length of line δl metres be as shown in Figure 47.4, comprising a single T-section. Each series arm impedance is $Z_1 = \frac{1}{2}(R + jwL)\delta l$ ohms, and the shunt arm impedance is

$$Z_2 = \frac{1}{Y_2} = \frac{1}{(G + j\omega C)\delta l}$$

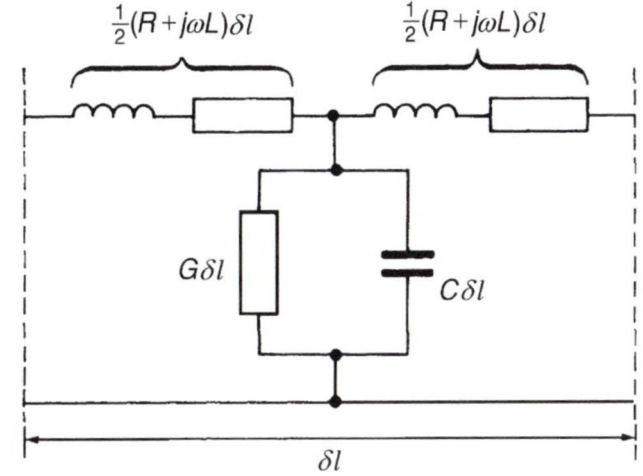

Figure 47.4

[i.e. from Chapter 28, the total admittance Y_2 is the sum of the admittance of the two parallel arms, i.e. in this case, the sum of

$$G\delta l \text{ and } \left(\frac{1}{1/(j\omega C)}\right)\delta l]$$

From equation (1), page 667, the characteristic impedance Z_0 of a T-section having in each series arm an impedance Z_1 and a shunt arm impedance Z_2 is given by: $Z_0 = \sqrt{(Z_1^2 + 2Z_1Z_2)}$

Hence the characteristic impedance of the section shown in Figure 47.4 is

$$Z_0 = \sqrt{\left\{\left[\frac{1}{2}(R + j\omega L)\delta l\right]^2 + 2\left[\frac{1}{2}(R + j\omega L)\delta l\right]\left[\frac{1}{(G + j\omega C)\delta l}\right]\right\}}$$

The term Z_1^2 involves δl^2 and, since δl is a very short length of line, δl^2 is negligible. Hence

Part 4

$$Z_0 = \sqrt{\frac{R+j\omega L}{G+j\omega C}} \text{ ohms} \qquad (7)$$

If losses R and G are neglected, then

$$Z_0 = \sqrt{(L/C)} \text{ ohms} \qquad (8)$$

Problem 7. A transmission line has the following primary constants: resistance $R = 15\,\Omega$/loop km, inductance $L = 3.4$ mH/loop km, conductance $G = 3\,\mu$S/km and capacitance $C = 10$ nF/km. Determine the characteristic impedance of the line when the frequency is 2 kHz.

From equation (7),

characteristic impedance $Z_0 = \sqrt{\dfrac{R+j\omega L}{G+j\omega C}}$ ohms

$$R + j\omega L = 15 + j(2\pi 2000)(3.4\times 10^{-3})$$
$$= (15 + j42.73)\,\Omega = 45.29\angle 70.66^\circ\,\Omega$$
$$G + j\omega C = 3\times 10^{-6} + j(2\pi 2000)(10\times 10^{-9})$$
$$= (3 + j125.66)10^{-6}\,\text{S}$$
$$= 125.7\times 10^{-6}\angle 88.63^\circ\,\text{S}$$

Hence $Z_0 = \sqrt{\dfrac{45.29\angle 70.66^\circ}{125.7\times 10^{-6}\angle 88.63^\circ}}$

$$= \sqrt{[0.360\times 10^6\angle -17.97^\circ]}\,\Omega$$

i.e. characteristic impedance, $\boldsymbol{Z_0 = 600\angle -8.99^\circ\,\Omega}$

Propagation coefficient

Figure 47.5 shows a T-section with the series arm impedances each expressed as $Z_A/2$ ohms per unit length and the shunt impedance as Z_B ohms per unit length. The p.d. between points P and Q is given by:

$$V_{PQ} = (I_1 - I_2)Z_B = I_2\left(\frac{Z_A}{2} + Z_0\right)$$

i.e. $$I_1 Z_B - I_2 Z_B = \frac{I_2 Z_A}{2} + I_2 Z_0$$

Hence $$I_1 Z_B = I_2\left(Z_B + \frac{Z_A}{2} + Z_0\right)$$

from which $$\frac{I_1}{I_2} = \frac{Z_B + (Z_A/2) + Z_0}{Z_B}$$

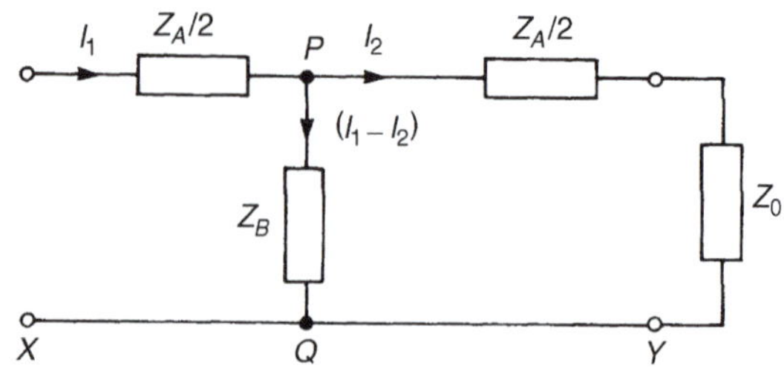

Figure 47.5

From equation (1), page 667, $Z_0 = \sqrt{(Z_1^2 + 2Z_1 Z_2)}$ In Figure 47.5, $Z_1 \equiv Z_A/2$ and $Z_2 \equiv Z_B$

Thus $$Z_0 = \sqrt{\left[\left(\frac{Z_A}{2}\right)^2 + 2\left(\frac{Z_A}{2}\right)Z_B\right]}$$
$$= \sqrt{\left(\frac{Z_A^2}{4} + Z_A Z_B\right)}$$

Hence $$\frac{I_1}{I_2} = \frac{Z_B + (Z_A/2) + \sqrt{(Z_A Z_B + (Z_A^2/4))}}{Z_B}$$
$$= \frac{Z_B}{Z_B} + \frac{(Z_A/2)}{Z_B} + \frac{\sqrt{(Z_A Z_B + (Z_A^2/4))}}{Z_B}$$
$$= 1 + \frac{1}{2}\left(\frac{Z_A}{Z_B}\right) + \sqrt{\left(\frac{Z_A Z_B}{Z_B^2} + \frac{(Z_A^2/4)}{Z_B^2}\right)}$$

i.e. $$\frac{I_1}{I_2} = 1 + \frac{1}{2}\left(\frac{Z_A}{Z_B}\right) + \left[\frac{Z_A}{Z_B} + \frac{1}{4}\left(\frac{Z_A}{Z_B}\right)^2\right]^{1/2} \qquad (9)$$

From Section 47.4, $I_1/I_2 = e^\gamma$, where γ is the propagation coefficient. Also, from the binomial theorem:

$$(a+b)^n = a^n + na^{n-1}b + \frac{n(n-1)}{2!}a^{n-2}b^2 + \cdots$$

Thus $$\left[\frac{Z_A}{Z_B} + \frac{1}{4}\left(\frac{Z_A}{Z_B}\right)^2\right]^{1/2}$$
$$= \left(\frac{Z_A}{Z_B}\right)^{1/2} + \frac{1}{2}\left(\frac{Z_A}{Z_B}\right)^{-1/2}\frac{1}{4}\left(\frac{Z_A}{Z_B}\right)^2 + \cdots$$

Hence, from equation (9),

$$\frac{I_1}{I_2} = e^\gamma = 1 + \frac{1}{2}\left(\frac{Z_A}{Z_B}\right) + \left[\left(\frac{Z_A}{Z_B}\right)^{1/2} + \frac{1}{8}\left(\frac{Z_A}{Z_B}\right)^{3/2} + \cdots\right.$$

Rearranging gives: $e^{\gamma} = 1 + \left(\frac{Z_A}{Z_B}\right)^{1/2} + \frac{1}{2}\left(\frac{Z_A}{Z_B}\right) + \frac{1}{8}\left(\frac{Z_A}{Z_B}\right)^{3/2} + \cdots$

Let length XY in Figure 47.5 be a very short length of line δl and let impedance $Z_A = Z\delta l$, where $Z = R + j\omega L$ and $Z_B = 1/(Y\delta l)$, where $Y = G + j\omega C$
Then

$$e^{\gamma\delta l} = 1 + \left(\frac{Z\delta l}{1/Y\delta l}\right)^{1/2} + \frac{1}{2}\left(\frac{Z\delta l}{1/Y\delta l}\right) + \frac{1}{8}\left(\frac{Z\delta l}{1/Y\delta l}\right)^{3/2} + \cdots$$

$$= 1 + (ZY\delta l^2)^{1/2} + \frac{1}{2}(ZY\delta l^2) + \frac{1}{8}(ZY\delta l^2)^{3/2} + \cdots$$

$$= 1 + (ZY)^{1/2}\delta l + \frac{1}{2}(ZY)(\delta l)^2 + \frac{1}{8}(ZY)^{3/2}(\delta l)^3 + \cdots$$

$$= 1 + (ZY)^{1/2}\delta l$$

if $(\delta l)^2$, $(\delta l)^3$ and higher powers are considered as negligible.

e^x may be expressed as a series:

$$e^x = 1 + x + \frac{x^2}{2!} + \frac{x^3}{3!} + \cdots$$

Comparison with $e^{\gamma\delta l} = 1 + (ZY)^{1/2}\delta l$ shows that

$\gamma\delta l = (ZY)^{1/2}\delta l$ i.e. $\gamma = \sqrt{(ZY)}$. Thus

propagation coefficient,

$$\boldsymbol{\gamma = \sqrt{[(R + j\omega L)(G + j\omega C)]}} \qquad (10)$$

The unit of γ is $\sqrt{(\Omega)(S)}$, i.e. $\sqrt{[(\Omega)(1/\Omega)]}$ thus γ is dimensionless, as expected, since $I_1/I_2 = e^{\gamma}$, from which $\gamma = \ln(I_1/I_2)$, i.e. a ratio of two currents. For a lossless line, $R = G = 0$ and

$$\boldsymbol{\gamma = \sqrt{(j\omega L)(j\omega C)} = j\omega\sqrt{(LC)}} \qquad (11)$$

Equations (7) and (10) are used to determine the characteristic impedance Z_0 and propagation coefficient γ of a transmission line in terms of the primary constants R, L, G and C. When $R = G = 0$, i.e. losses are neglected, equations (8) and (11) are used to determine Z_0 and γ.

Problem 8. A transmission line having negligible losses has primary line constants of inductance $L = 0.5$ mH/loop km and capacitance $C = 0.12\,\mu$F/km. Determine, at an operating frequency of 400 kHz, (a) the characteristic impedance, (b) the propagation coefficient, (c) the wavelength on the line and (d) the velocity of propagation, in metres per second, of a signal.

(a) Since the line is loss-free, from equation (8), the characteristic impedance Z_0 is given by

$$Z_0 = \sqrt{\frac{L}{C}} = \sqrt{\frac{0.5 \times 10^{-3}}{0.12 \times 10^{-6}}} = \mathbf{64.55\,\Omega}$$

(b) From equation (11), for a loss-free line, the propagation coefficient γ is given by

$$\gamma = j\omega\sqrt{(LC)}$$
$$= j(2\pi 400 \times 10^3) \times \sqrt{[(0.5 \times 10^{-3})(0.12 \times 10^{-6})]}$$
$$= j19.47 \text{ or } \mathbf{0 + j19.47}$$

Since $\gamma = \alpha + j\beta$, the attenuation coefficient, $\alpha = 0$ and the phase-shift coefficient, $\beta = 19.47$ rad/km.

(c) From equation (2), wavelength

$$\lambda = \frac{2\pi}{\beta} = \frac{2\pi}{19.47} = \mathbf{0.323\ km} \text{ or } \mathbf{323\ m}$$

(d) From equation (3), velocity of propagation

$$u = f\lambda = (400 \times 10^3)(323) = \mathbf{129 \times 10^6\ m/s}.$$

Problem 9. At a frequency of 1 kHz the primary constants of a transmission line are resistance $R = 25\,\Omega$/loop km, inductance $L = 5$ mH/loop km, capacitance $C = 0.04\,\mu$F/km and conductance $G = 80\,\mu$S/km. Determine for the line (a) the characteristic impedance, (b) the propagation coefficient, (c) the attenuation coefficient and (d) the phase-shift coefficient.

(a) From equation (7),

characteristic impedance $Z_0 = \sqrt{\frac{R + j\omega L}{G + j\omega C}}$ ohms

$$R + j\omega L = 25 + j(2\pi 1000)(5 \times 10^{-3})$$
$$= (25 + j31.42)$$
$$= 40.15\angle 51.49°\,\Omega$$

$$G + j\omega C = 80 \times 10^{-6} + j(2\pi 1000)(0.04 \times 10^{-6})$$
$$= (80 + j251.33)10^{-6}$$
$$= 263.76 \times 10^{-6}\angle 72.34°\,\text{S}$$

Thus characteristic impedance

$$Z_0 = \sqrt{\frac{40.15\angle 51.49^\circ}{263.76 \times 10^{-6}\angle 72.34^\circ}}$$

$$= \mathbf{390.2\angle{-10.43}^\circ\ \Omega}$$

(b) From equation (10), propagation coefficient

$$\gamma = \sqrt{[(R + j\omega L)(G + j\omega C)]}$$

$$= \sqrt{[(40.15\angle 51.49^\circ)(263.76 \times 10^{-6}\angle 72.34^\circ)]}$$

$$= \sqrt{(0.01059\angle 123.83^\circ)} = \mathbf{0.1029\angle 61.92^\circ}$$

(c) $\gamma = \alpha + j\beta = 0.1029(\cos 61.92^\circ + j \sin 61.92^\circ)$

i.e. $\gamma = 0.0484 + j0.0908$

Thus the attenuation coefficient,

$\boldsymbol{\alpha = 0.0484}$ **nepers/km**

(d) The phase-shift coefficient, $\boldsymbol{\beta = 0.0908}$ **rad/km**

Problem 10. An open wire line is 300 km long and is terminated in its characteristic impedance. At the sending end is a generator having an open-circuit e.m.f. of 10.0 V, an internal impedance of $(400 + j0)\ \Omega$ and a frequency of 1 kHz. If the line primary constants are $R = 8\ \Omega$/loop km, $L = 3$ mH/loop km, $C = 7500$ pF/km and $G = 0.25\ \mu$S/km, determine (a) the characteristic impedance, (b) the propagation coefficient, (c) the attenuation and phase-shift coefficients, (d) the sending-end current, (e) the receiving-end current, (f) the wavelength on the line and (g) the speed of transmission of signal.

(a) From equation (7),

characteristic impedance, $Z_0 = \sqrt{\dfrac{R + j\omega L}{G + j\omega C}}$ ohms

$$R + j\omega L = 8 + j(2\pi 1000)(3 \times 10^{-3})$$

$$= 8 + j6\pi = 20.48\angle 67.0^\circ\ \Omega$$

$$G + j\omega C$$

$$= 0.25 \times 10^{-6} + j(2\pi 1000)(7500 \times 10^{-12})$$

$$= (0.25 + j47.12)10^{-6}$$

$$= 47.12 \times 10^{-6}\angle 89.70^\circ\ \text{S}$$

Hence characteristic impedance

$$Z_0 = \sqrt{\frac{20.48\angle 67.0^\circ}{47.12 \times 10^{-6}\angle 89.70^\circ}} = \mathbf{659.3\angle{-11.35}^\circ\ \Omega}$$

(b) From equation (10), propagation coefficient

$$\gamma = \sqrt{[(R + j\omega L)(G + j\omega C)]}$$

$$= \sqrt{[(20.48\angle 67.0^\circ)(47.12 \times 10^{-6}\angle 89.70^\circ)]}$$

$$= \mathbf{0.03106\angle 78.35^\circ}$$

(c) $\gamma = \alpha + j\beta = 0.03106(\cos 78.35^\circ + j \sin 78.35^\circ)$

$= 0.00627 + j0.03042$

Hence the attenuation coefficient,

$\boldsymbol{\alpha = 0.00627}$ **Np/km**

and the phase shift coefficient,

$\boldsymbol{\beta = 0.03042}$ **rad/km**

(d) With reference to Figure 47.6, since the line is matched, i.e. terminated in its characteristic impedance, $V_S/I_S = Z_0$. Also

$$V_S = V_G - I_S Z_G = 10.0 - I_S(400 + j0)$$

Thus $I_S = \dfrac{V_S}{Z_0} = \dfrac{10.0 - 400 I_S}{Z_0}$

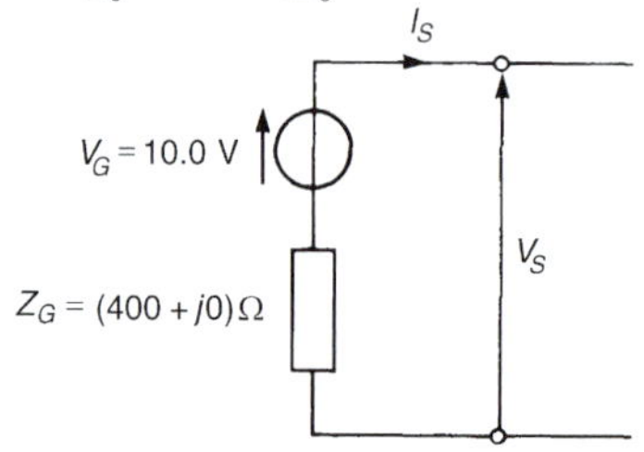

Figure 47.6

Rearranging gives: $I_S Z_0 = 10.0 - 400\ I_S$, from which,

$$I_S(Z_0 + 400) = 10.0$$

Thus the sending-end current,

$$I_S = \frac{10.0}{Z_0 + 400} = \frac{10.0}{659.3\angle{-11.35}^\circ + 400}$$

$$= \frac{10.0}{646.41 - j129.75 + 400} = \frac{10.0}{1054.4\angle{-7.07}^\circ}$$

$$= \mathbf{9.484\angle 7.07^\circ\ mA}$$

(e) From equation (4), the receiving-end current,

$$I_R = I_S e^{-n\gamma} = I_S e^{-n\alpha}\angle{-n\beta}$$

$$= (9.484\angle 7.07^\circ)e^{-(300)(0.00627)}$$

$$\angle{-(300)(0.03042)}$$

$= 9.484\angle 7.07°e^{-1.881}\angle{-9.13}\,\text{rad}$

$= 1.446\angle{-516}°\,\text{mA} = \mathbf{1.446\angle -156°\ mA}$

(f) From equation (2),

$$\text{wavelength, } \lambda = \frac{2\pi}{\beta} = \frac{2\pi}{0.03042} = \mathbf{206.5\ km}$$

(g) From equation (3),

speed of transmission, $u = f\lambda = (1000)(206.5)$

$= 206.5 \times 10^3$ km/s

$= \mathbf{206.5 \times 10^6\ m/s}$

Now try the following Practice Exercise

Practice Exercise 191 Characteristic impedance and propagation coefficients in terms of the primary constants (Answers on page 834)

1. At a frequency of 800 Hz, the open-circuit impedance of a length of transmission line is measured as $500\angle{-35}°\,\Omega$ and the short-circuit impedance as $300\angle{-15}°\Omega$. Determine the characteristic impedance of the line at this frequency.

2. A transmission line has the following primary constants per loop kilometre run: $R = 12\,\Omega$, $L = 3$ mH, $G = 4\,\mu$S and $C = 0.02\,\mu$F. Determine the characteristic impedance of the line when the frequency is 750 Hz.

3. A transmission line having negligible losses has primary constants: inductance $L = 1.0$ mH/loop km and capacitance $C = 0.20\,\mu$F/km. Determine, at an operating frequency of 50 kHz, (a) the characteristic impedance, (b) the propagation coefficient, (c) the attenuation and phase-shift coefficients, (d) the wavelength on the line and (e) the velocity of propagation of signal in metres per second.

4. At a frequency of 5 kHz the primary constants of a transmission line are: resistance $R = 12\,\Omega$/loop km, inductance $L = 0.50$ mH/loop km, capacitance $C = 0.01\,\mu$F/km and $G = 60\,\mu$S/km. Determine for the line (a) the characteristic impedance, (b) the propagation coefficient, (c) the attenuation coefficient and (d) the phase-shift coefficient.

5. A transmission line is 50 km in length and is terminated in its characteristic impedance. At the sending end a signal emanates from a generator which has an open-circuit e.m.f. of 20.0 V, an internal impedance of $(250 + j0)\,\Omega$ at a frequency of 1592 Hz. If the line primary constants are $R = 30\,\Omega$/loop km, $L = 4.0$ mH/loop km, $G = 5.0\,\mu$S/km, and $C = 0.01\,\mu$F/km, determine (a) the value of the characteristic impedance, (b) the propagation coefficient, (c) the attenuation and phase-shift coefficients, (d) the sending-end current, (e) the receiving-end current, (f) the wavelength on the line and (g) the speed of transmission of a signal, in metres per second.

47.6 Distortion on transmission lines

If the waveform at the receiving end of a transmission line is not the same shape as the waveform at the sending end, **distortion** is said to have occurred. The three main causes of distortion on transmission lines are as follows.

(i) The characteristic impedance Z_0 of a line varies with the operating frequency, i.e. from equation (7),

$$Z_0 = \sqrt{\frac{R + j\omega L}{G + j\omega C}}\ \text{ohms}$$

The terminating impedance of the line may not vary with frequency in the same manner.
In the above equation for Z_0, if the frequency is very low, ω is low and $Z_0 \approx \sqrt{(R/G)}$. If the frequency is very high, then $\omega L \gg R$, $\omega C \gg G$ and $Z_0 \approx \sqrt{(L/C)}$. A graph showing the variation of Z_0 with frequency f is shown in Figure 47.7.

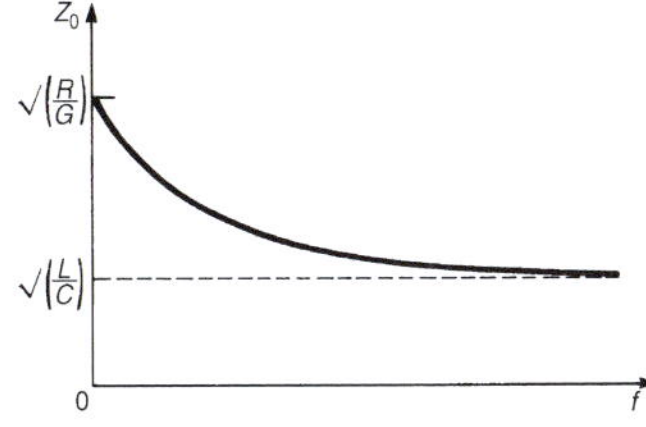

Figure 47.7

If the characteristic impedance is to be constant throughout the entire operating frequency range then the following condition is required:

$\sqrt{(L/C)} = \sqrt{(R/G)}$, i.e $L/C = R/G$, from which

$$\boldsymbol{LG = CR} \qquad (12)$$

Thus, in a transmission line, if $LG = CR$ it is possible to provide a termination equal to the characteristic impedance Z_0 at all frequencies.

(ii) The attenuation of a line varies with the operating frequency (since $\gamma = \sqrt{[(R + j\omega L)(G + j\omega C)]}$, from equation (10)), thus waves of differing frequencies and component frequencies of complex waves are attenuated by different amounts.

From the above equation for the propagation coefficient:

$$\gamma^2 = (R + j\omega L)(G + j\omega C)$$
$$= RG + j\omega(LG + CR) - \omega^2 LC$$

If $LG = CR = x$, then $LG + CR = 2x$ and $LG + CR$ may be written as $2\sqrt{x^2}$, i.e. $LG + CR$ may be written as $2\sqrt{[(LG)(CR)]}$

Thus $\gamma^2 = RG + j\omega(2\sqrt{[(LG)(CR)]}) - \omega^2 LC$

$$= [\sqrt{(RG)} + j\omega\sqrt{(LC)}]^2$$

and $\gamma = \sqrt{(RG)} + j\omega\sqrt{(LC)}$

Since $\gamma = \alpha + j\beta$,

attenuation coefficient, $\boldsymbol{\alpha = \sqrt{(RG)}}$ (13)

and **phase-shift coefficient,** $\boldsymbol{\beta = \omega\sqrt{(LC)}}$ (14)

Thus, in a transmission line, if $LG = CR$, $\alpha = \sqrt{(RG)}$, i.e. the attenuation coefficient is independent of frequency and all frequencies are equally attenuated.

(iii) The delay time, or the time of propagation, and thus the velocity of propagation, varies with frequency and therefore waves of different frequencies arrive at the termination with differing delays. From equation (14), the phase-shift coefficient, $\beta = \omega\sqrt{(LC)}$ when $LG = CR$

Velocity of propagation, $\boldsymbol{u} = \dfrac{\omega}{\beta} = \dfrac{\omega}{\omega\sqrt{(LC)}}$

i.e. $$\boldsymbol{u = \frac{1}{\sqrt{(LC)}}} \qquad (15)$$

Thus, in a transmission line, if $LG = CR$, the velocity of propagation, and hence the time delay, is independent of frequency.

From the above it appears that the condition $LG = CR$ is appropriate for the design of a transmission line, since under this condition no distortion is introduced. This means that the signal at the receiving end is the same as the sending-end signal except that it is reduced in amplitude and delayed by a fixed time. Also, with no distortion, the attenuation on the line is a minimum. In practice, however, $R/L \gg G/C$. The inductance is usually low and the capacitance is large and not easily reduced. Thus if the condition $LG = CR$ is to be achieved in practice, either L or G must be increased since neither CN or R can really be altered. It is undesirable to increase G since the attenuation and power losses increase. Thus the inductance L is the quantity that needs to be increased and such an artificial increase in the line inductance is called **loading**. This is achieved either by inserting inductance coils at intervals along the transmission line – this being called **'lumped loading'** – or by wrapping the conductors with a high-permeability metal tape – this being called **'continuous loading'**.

Problem 11. An underground cable has the following primary constants: resistance $R = 10\,\Omega$/loop km, inductance $L = 1.5$ mH/loop km, conductance $G = 1.2\,\mu$S/km and capacitance $C = 0.06\,\mu$F/km. Determine by how much the inductance should be increased to satisfy the condition for minimum distortion.

From equation (12), the condition for minimum distortion is given by $LG = CR$, from which,

$$\text{inductance } L = \frac{CR}{G} = \frac{(0.06 \times 10^{-6})(10)}{1.2 \times 10^{-6}}$$
$$= 0.5 \text{ H or } 500 \text{ mH.}$$

Thus the inductance should be increased by $(500 - 1.5)$ mH, i.e. **498.5 mH** per loop km, for minimum distortion.

Problem 12. A cable has the following primary constants: resistance $R = 80\,\Omega$/loop km, conductance, $G = 2\,\mu$S/km and capacitance $C = 5$ nF/km. Determine, for minimum distortion at a frequency of 1.5 kHz (a) the value of inductance per loop kilometre required, (b) the propagation coefficient, (c) the velocity of propagation of signal and (d) the wavelength on the line.

(a) From equation (12), for minimum distortion, $LG = CR$, from which, inductance per loop km,

$$L = \frac{CR}{G} = \frac{(5 \times 10^{-9})(80)}{(2 \times 10^{-6})} = \mathbf{0.20\,H} \text{ or } \mathbf{200\,mH}$$

(b) From equation (13), attenuation coefficient,

$$\alpha = \sqrt{(RG)} = \sqrt{[(80)(2 \times 10^{-6})]} = 0.0126\,\text{Np/km}$$

and from equation (14), phase shift coefficient,

$$\beta = \omega\sqrt{(LC)} = (2\pi 1500)\sqrt{[(0.20)(5 \times 10^{-9})]} = 0.2980\,\text{rad/km}$$

Hence the propagation coefficient,

$$\boldsymbol{\gamma = \alpha + j\beta = (0.0126 + j0.2980)} \text{ or } \mathbf{0.2983\angle 87.58^\circ}$$

(c) From equation (15), velocity of propagation,

$$u = \frac{1}{\sqrt{(LC)}} = \frac{1}{\sqrt{[(0.2)(5 \times 10^{-9})]}} = \mathbf{31\,620\,km/s} \text{ or } \mathbf{31.62 \times 10^6\,m/s}$$

(d) Wavelength, $\lambda = \frac{u}{f} = \frac{31.62 \times 10^6}{1500}\,\text{m} = \mathbf{21.08\,km}$

Now try the following Practice Exercise

Practice Exercise 192 Distortion on transmission lines (Answers on page 834)

1. A cable has the following primary constants: resistance $R = 90\,\Omega$/loop km, inductance $L = 2.0\,$mH/loop km, capacitance $C = 0.05\,\mu$F/km and conductance $G = 3.0\,\mu$S/km. Determine the value to which the inductance should be increased to satisfy the condition for minimum distortion.

2. A condition of minimum distortion is required for a cable. Its primary constants are: $R = 40\Omega$/loop km, $L = 2.0\,$mH/loop km, $G = 2.0\,\mu$S/km and $C = 0.08\,\mu$F/km. At a frequency of 100 Hz determine (a) the increase in inductance required, (b) the propagation coefficient, (c) the speed of signal transmission and (d) the wavelength on the line.

47.7 Wave reflection and the reflection coefficient

In earlier sections of this chapter it was assumed that the transmission line had been properly terminated in its characteristic impedance or regarded as an infinite line. In practice, of course, all lines have a definite length and often the terminating impedance does not have the same value as the characteristic impedance of the line. When this is the case, the transmission line is said to have a **'mismatched load'**.

The forward-travelling wave moving from the source to the load is called the **incident wave** or the sending-end wave. With a mismatched load the termination will absorb only a part of the energy of the incident wave, the remainder being forced to return back along the line toward the source. This latter wave is called the **reflected wave**.

Electrical energy is transmitted by a transmission line; when such energy arrives at a termination that has a value different from the characteristic impedance, it experiences a sudden change in the impedance of the medium. When this occurs, some reflection of incident energy occurs and the reflected energy is lost to the receiving load. (Reflections commonly occur in nature when a change of transmission medium occurs; for example, sound waves are reflected at a wall, which can produce echoes, and light rays are reflected by mirrors.)

If a transmission line is terminated in its characteristic impedance, no reflection occurs; if terminated in an open circuit or a short circuit, total reflection occurs, i.e. the whole of the incident wave reflects along the line. Between these extreme possibilities, all degrees of reflection are possible.

Open-circuited termination

If a length of transmission line is open-circuited at the termination, no current can flow in it and thus no power can be absorbed by the termination. This condition is achieved if a current is imagined to be reflected from the termination, the reflected current having the same magnitude as the incident wave but with a phase difference of 180°. Also, since no power is absorbed at the termination (it is all returned back along the line), the reflected voltage wave at the termination must be equal to the incident wave. Thus the voltage at the termination must be doubled by the open circuit. The resultant current (and voltage) at any point on the transmission line and at any instant of time is given by the sum of the currents (and voltages) due to the incident and reflected waves (see Section 47.8).

Short-circuit termination

If the termination of a transmission line is short-circuited, the impedance is zero, and hence the voltage developed across it must be zero. As with the open-circuit condition, no power is absorbed by the termination. To obtain zero voltage at the termination, the reflected voltage wave must be equal in amplitude but opposite in phase (i.e. 180° phase difference) to the incident wave. Since no power is absorbed, the reflected current wave at the termination must be equal to the incident current wave and thus the current at the end of the line must be doubled at the short circuit. As with the open-circuited case, the resultant voltage (and current) at any point on the line and at any instant of time is given by the sum of the voltages (and currents) due to the incident and reflected waves.

Energy associated with a travelling wave

A travelling wave on a transmission line may be thought of as being made up of electric and magnetic components. Energy is stored in the magnetic field due to the current (energy $=\frac{1}{2}LI^2$ – see page 663) and energy is stored in the electric field due to the voltage (energy $=\frac{1}{2}CV^2$ – see page 654). It is the continual interchange of energy between the magnetic and electric fields, and vice versa, that causes the transmission of the total electromagnetic energy along the transmission line.

When a wave reaches an open-circuited termination the magnetic field collapses since the current I is zero. Energy cannot be lost, but it can change form. In this case it is converted into electrical energy, adding to that already caused by the existing electric field. The voltage at the termination consequently doubles and this increased voltage starts the movement of a reflected wave back along the line. A magnetic field will be set up by this movement and the total energy of the reflected wave will again be shared between the magnetic and electric field components.

When a wave meets a short-circuited termination, the electric field collapses and its energy changes form to the magnetic energy. This results in a doubling of the current.

Reflection coefficient

Let a generator having impedance Z_0 (this being equal to the characteristic impedance of the line) be connected to the input terminals of a transmission line which is terminated in an impedance Z_R, where $Z_0 \neq Z_R$, as shown in Figure 47.8. The sending-end or incident current I_i flowing from the source generator flows along the line and, until it arrives at the termination Z_R, behaves as though the line were infinitely long or properly terminated in its characteristic impedance, Z_0

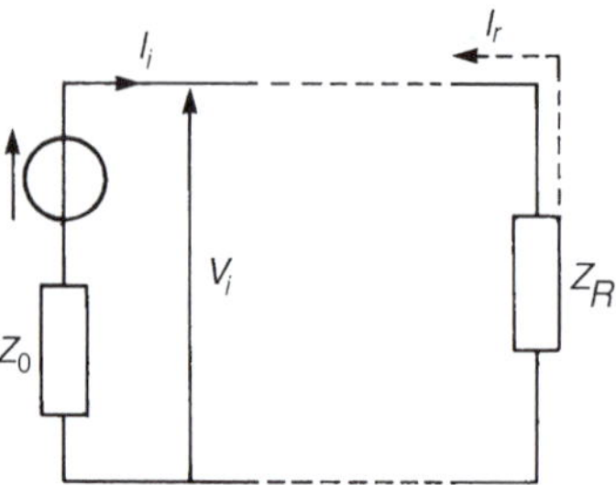

Figure 47.8

The incident voltage V_i shown in Figure 47.8 is given by:

$$V_i = I_i Z_0 \qquad (12)$$

from which $$I_i = \frac{V_i}{Z_0} \qquad (13)$$

At the termination, the conditions must be such that:

$$Z_R = \frac{\text{total voltage}}{\text{total current}}$$

Since $Z_R \neq Z_0$, part of the incident wave will be reflected back along the line from the load to the source. Let the reflected voltage be V_r and the reflected current be I_r Then

$$V_r = -I_r Z_0 \qquad (14)$$

from which $$I_r = -\frac{V_r}{Z_0} \qquad (15)$$

(Note the minus sign, since the reflected voltage and current waveforms travel in the opposite direction to the incident waveforms.)

Thus, at the termination,

$$Z_R = \frac{\text{total voltage}}{\text{total current}} = \frac{V_i + V_r}{I_i + I_r}$$

$$= \frac{I_i Z_0 - I_r Z_0}{I_i + I_r} \quad \text{from equations (12) and (14)}$$

i.e. $$Z_R = \frac{Z_0(I_i - I_r)}{(I_i + I_r)}$$

Hence $$Z_R(I_i + I_r) = Z_0(I_i - I_r)$$

$$Z_R I_i + Z_R I_r = Z_0 I_i - Z_0 I_r$$

$$Z_0 I_r + Z_R I_r = Z_0 I_i - Z_R I_i$$

$$I_r(Z_0 + Z_R) = I_i(Z_0 - Z_R)$$

from which $$\frac{I_r}{I_i} = \frac{Z_0 - Z_R}{Z_0 + Z_R}$$

The ratio of the reflected current to the incident current is called the **reflection coefficient** and is often given the symbol ρ, i.e.

$$\frac{I_r}{I_i} = \rho = \frac{Z_0 - Z_R}{Z_0 + Z_R} \quad (16)$$

By similar reasoning to above an expression for the ratio of the reflected to the incident voltage may be obtained. From above,

$$Z_R = \frac{V_i + V_r}{I_i + I_r} = \frac{V_i + V_r}{(V_i/Z_0) - (V_r/Z_0)}$$

from equations (13) and (15),

i.e. $$Z_R = \frac{V_i + V_r}{(V_i - V_r)/Z_0}$$

Hence $$\frac{Z_R}{Z_0}(V_i - V_r) = V_i + V_r$$

from which, $$\frac{Z_R}{Z_0}V_i - \frac{Z_R}{Z_0}V_r = V_i + V_r$$

Then $$\frac{Z_R}{Z_0}V_i - V_i = V_r + \frac{Z_R}{Z_0}V_r$$

and $$V_i\left(\frac{Z_R}{Z_0} - 1\right) = V_r\left(1 + \frac{Z_R}{Z_0}\right)$$

Hence $$V_i\left(\frac{Z_R - Z_0}{Z_0}\right) = V_r\left(\frac{Z_0 + Z_R}{Z_0}\right)$$

from which $$\frac{V_r}{V_i} = \frac{Z_R - Z_0}{Z_0 + Z_R} = -\left(\frac{Z_0 - Z_R}{Z_0 + Z_R}\right) \quad (17)$$

Hence $$\frac{V_r}{V_i} = -\frac{I_r}{I_i} = -\rho \quad (18)$$

Thus the ratio of the reflected to the incident voltage has the same magnitude as the ratio of reflected to incident current, but is of opposite sign. From equations (16) and (17) it is seen that when $Z_R = Z_0$, $\rho = 0$ and there is no reflection.

Problem 13. A cable which has a characteristic impedance of 75 Ω is terminated in a 250 Ω resistive load. Assuming that the cable has negligible losses and the voltage measured across the terminating load is 10 V, calculate the value of (a) the reflection coefficient for the line, (b) the incident current, (c) the incident voltage, (d) the reflected current and (e) the reflected voltage.

(a) From equation (16),

reflection coefficient, $$\rho = \frac{Z_0 - Z_R}{Z_0 + Z_R} = \frac{75 - 250}{75 + 250} = \frac{-175}{325} = \mathbf{-0.538}$$

(b) The circuit diagram is shown in Figure 47.9. Current flowing in the terminating load,

$$I_R = \frac{V_R}{Z_R} = \frac{10}{250} = 0.04\,\text{A}$$

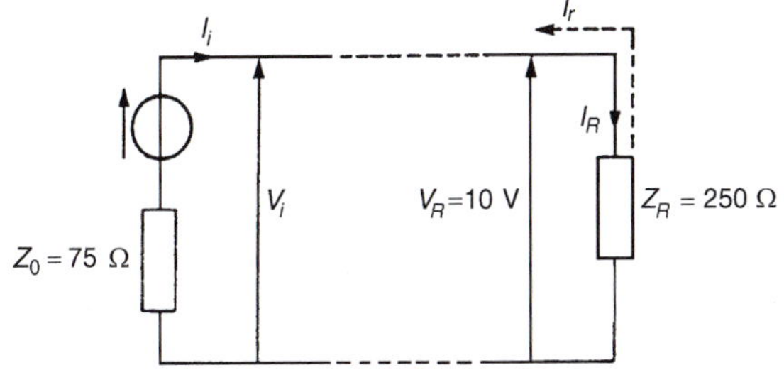

Figure 47.9

However, current $I_R = I_i + I_r$. From equation (16), $I_r = \rho I_i$

Thus $I_R = I_i + \rho I_i = I_i(1 + \rho)$

from which **incident current,** $$I_i = \frac{I_R}{(1+\rho)} = \frac{0.04}{1 + (-0.538)} = \mathbf{0.0866\,A} \text{ or } \mathbf{86.6\,mA}$$

(c) From equation (12),

incident voltage, $$V_i = I_i Z_0 = (0.0866)(75) = \mathbf{6.50\,V}$$

(d) Since $I_R = I_i + I_r$

reflected current, $$I_r = I_R - I_i = 0.04 - 0.0866 = \mathbf{-0.0466\,A} \text{ or } \mathbf{-46.6\,mA}$$

(e) From equation (14),

reflected voltage, $$V_r = -I_r Z_0 = -(-0.0466)(75) = \mathbf{3.50\,V}$$

Problem 14. A long transmission line has a characteristic impedance of $(500 - j40)$ Ω and is terminated in an impedance of (a) $(500 + j40)$ Ω and (b) $(600 + j20)$ Ω. Determine the magnitude of the reflection coefficient in each case.

(a) From equation (16), reflection coefficient,

$$\rho = \frac{Z_0 - Z_R}{Z_0 + Z_R}$$

When $Z_0 = (500 - j40)\,\Omega$ and $Z_R = (500 + j40)\,\Omega$

$$\rho = \frac{(500 - j40) - (500 + j40)}{(500 - j40) + (500 + j40)} = \frac{-j80}{1000}$$

$$= -j0.08$$

Hence the magnitude of the reflection coefficient, $|\rho| = \mathbf{0.08}$

(b) When $Z_0 = (500 - j40)\,\Omega$ and $Z_R = (600 + j20)\,\Omega$

$$\rho = \frac{(500 - j40) - (600 + j20)}{(500 - j40) + (600 + j20)}$$

$$= \frac{-100 - j60}{1100 - j20}$$

$$= \frac{116.62\angle -149.04^\circ}{1100.18\angle -1.04^\circ}$$

$$= 0.106\angle -148^\circ$$

Hence the magnitude of the reflection coefficient, $|\rho| = \mathbf{0.106}$

Problem 15. A loss-free transmission line has a characteristic impedance of $500\angle 0^\circ\,\Omega$ and is connected to an aerial of impedance $(320 + j240)\,\Omega$. Determine (a) the magnitude of the ratio of the reflected to the incident voltage wave and (b) the incident voltage if the reflected voltage is $20\angle 35^\circ$ V

(a) From equation (17), the ratio of the reflected to the incident voltage is given by:

$$\frac{V_r}{V_i} = \frac{Z_R - Z_0}{Z_R + Z_0}$$

where Z_0 is the characteristic impedance $500\angle 0^\circ\,\Omega$ and Z_R is the terminating impedance $(320 + j240)\,\Omega$.

$$\text{Thus} \quad \frac{V_r}{V_i} = \frac{(320 + j240) - 500\angle 0^\circ}{500\angle 0^\circ + (320 + j240)}$$

$$= \frac{-180 + j240}{820 + j240} = \frac{300\angle 126.87^\circ}{854.4\angle 16.31^\circ}$$

$$= 0.351\angle 110.56^\circ$$

Hence the magnitude of the ratio $V_r{:}V_i$ is **0.351**

(b) Since $V_r/V_i = 0.351\angle 110.56^\circ$,

$$\text{incident voltage, } V_i = \frac{V_r}{0.351\angle 110.56^\circ}$$

Thus, when $V_r = 20\angle 35^\circ$ V,

$$V_i = \frac{20\angle 35^\circ}{0.351\angle 110.56^\circ} = \mathbf{57.0\angle -75.56^\circ\ V}$$

Now try the following Practice Exercise

Practice Exercise 193 Wave reflection and the reflection coefficient (Answers on page 834)

1. A coaxial line has a characteristic impedance of $100\,\Omega$ and is terminated in a $400\,\Omega$ resistive load. The voltage measured across the termination is 15 V. The cable is assumed to have negligible losses. Calculate for the line the values of (a) the reflection coefficient, (b) the incident current, (c) the incident voltage, (d) the reflected current and (e) the reflected voltage.

2. A long transmission line has a characteristic impedance of $(400 - j50)\,\Omega$ and is terminated in an impedance of (i) $400 + j50)\,\Omega$, (ii) $(500 + j60)\,\Omega$ and (iii) $400\angle 0^\circ\,\Omega$. Determine the magnitude of the reflection coefficient in each case.

3. A transmission line which is loss-free has a characteristic impedance of $600\angle 0^\circ\,\Omega$ and is connected to a load of impedance $(400 + j300)\,\Omega$. Determine (a) the magnitude of the reflection coefficient and (b) the magnitude of the sending-end voltage if the reflected voltage is 14.60 V

47.8 Standing waves and the standing-wave ratio

Consider a loss-free transmission line **open-circuited** at its termination. An incident current waveform is completely reflected at the termination, and, as stated in Section 47.7, the reflected current is of the same magnitude as the incident current but is 180° out of phase. Figure 47.10(a) shows the incident and reflected current waveforms drawn separately (shown as I_i moving to the right and I_r moving to the left, respectively) at a time $t = 0$, with $I_i = 0$ and decreasing at the termination.

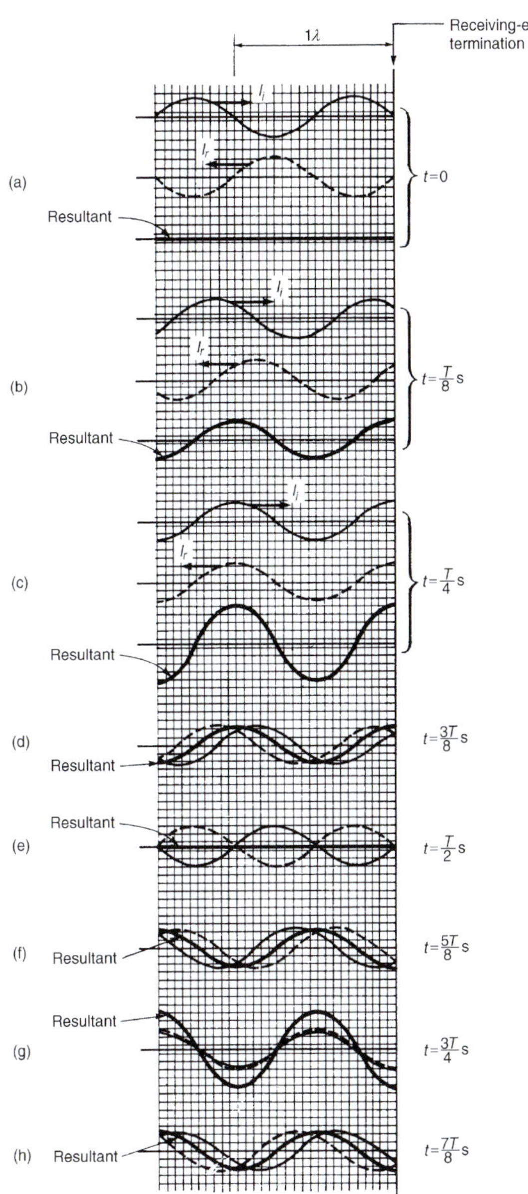

Figure 47.10 Current waveforms on an open-circuited transmission line

The resultant of the two waves is obtained by adding them at intervals. In this case the resultant is seen to be zero. Figures 47.10(b) and (c) show the incident and reflected waves drawn separately as times $t=T/8$ seconds and $t=T/4$, where T is the periodic time of the signal. Again, the resultant is obtained by adding the incident and reflected waveforms at intervals. Figures 47.10(d) to (h) show the incident and reflected current waveforms plotted on the same axis, together with their resultant waveform, at times $t=3T/8$ to $t=7T/8$ at intervals of $T/8$

If the resultant waveforms shown in Figures 47.10(a) to (g) are superimposed one upon the other, Figure 47.11 results. (Note that the scale has been increased for clarity.) The waveforms show clearly that waveform (a) moves to (b) after $T/8$, then to (c) after a further period of $T/8$, then to (d), (e), (f), (g) and (h) at intervals of $T/8$. It is noted that at any particular point the current varies sinusoidally with time, but the amplitude of oscillation is different at different points on the line.

Whenever two waves of the same frequency and amplitude travelling in opposite directions are superimposed on each other as above, interference takes place between the two waves and a **standing** or **stationary wave** is produced. The points at which the current is always zero are called **nodes** (labelled N in Figure 47.11). The standing wave does not progress to the left or right and the nodes do not oscillate. Those points on the wave that undergo maximum disturbance are called **antinodes** (labelled A in Figure 47.11). The distance between adjacent nodes or adjacent antinodes is $\lambda/2$, where λ is the wavelength. A standing wave is therefore seen to be a periodic variation in the vertical plane taking place on the transmission line without travel in either direction.

The resultant of the incident and reflected voltage for the open-circuit termination may be deduced in a similar manner to that for current. However, as stated in Section 47.7, when the incident voltage wave reaches the termination it is reflected without phase change. Figure 47.12 shows the resultant waveforms of incident and reflected voltages at intervals of $t=T/8$. Figure 47.13 shows all the resultant waveforms of Figure 47.12(a) to (h) superimposed; again, standing waves are seen to result. Nodes (labelled N) and antinodes (labelled A) are shown in Figure 47.13 and, in comparison with the current waves, are seen to occur 90° out of phase.

If the transmission line is short-circuited at the termination, it is the incident current that is reflected without phase change and the incident voltage that is

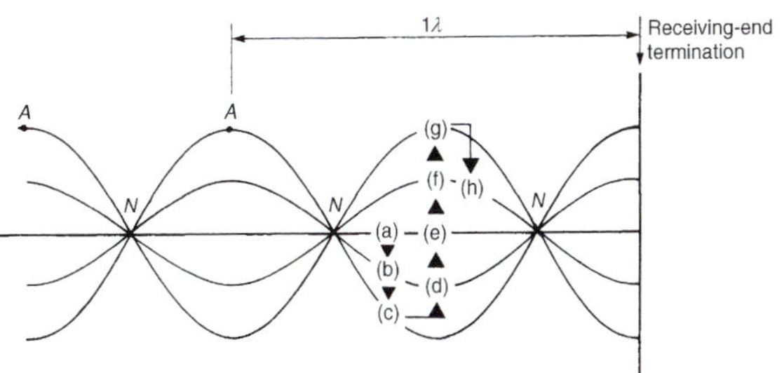

Figure 47.11

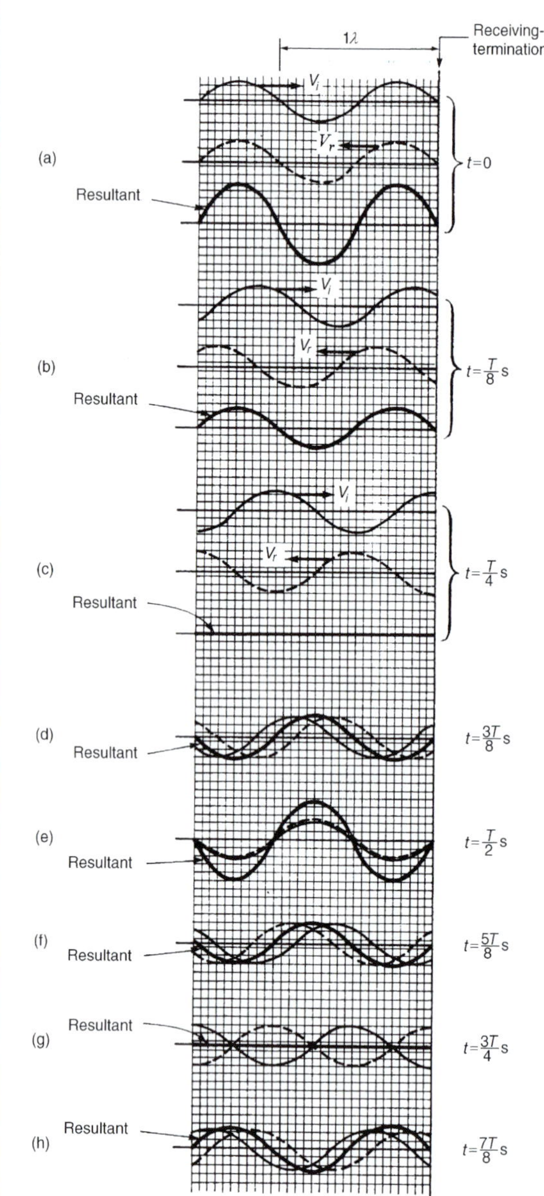

Figure 47.12 Voltage waveforms on an open-circuited transmission line

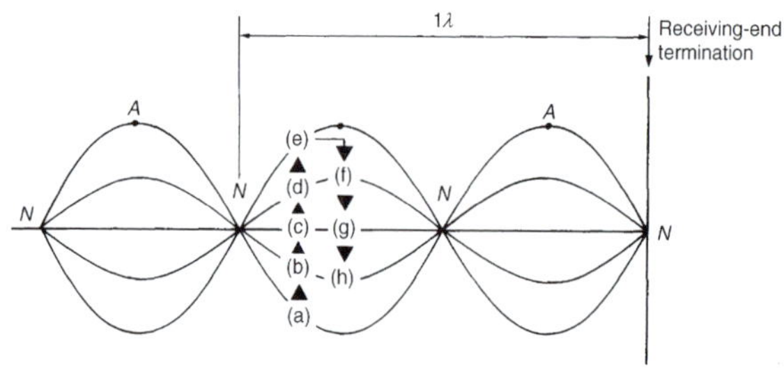

Figure 47.13

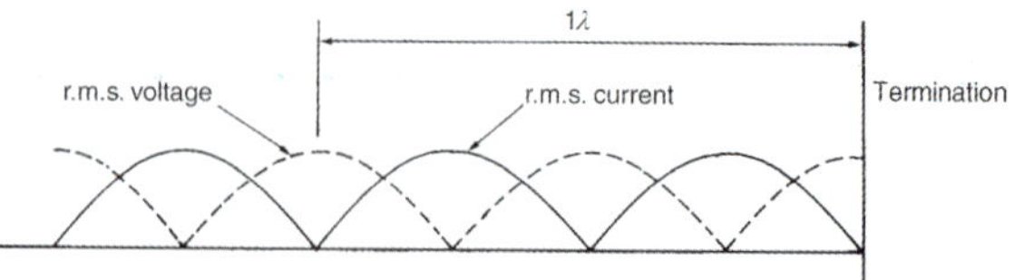

Figure 47.14

reflected with a phase change of 180°. Thus the diagrams shown in Figures 47.10 and 47.11 representing current at an open-circuited termination may be used to represent voltage conditions at a short-circuited termination and the diagrams shown in Figures 47.12 and 47.13 representing voltage at an open-circuited termination may be used to represent current conditions at a short-circuited termination.

Figure 47.14 shows the r.m.s. current and voltage waveforms plotted on the same axis against distance for the case of total reflection, deduced from Figures 47.11 and 47.13. The r.m.s. values are equal to the amplitudes of the waveforms shown in Figures 47.11 and 47.13, except that they are each divided by $\sqrt{2}$ (since, for a sine wave, r.m.s. value $=(1/\sqrt{2}) \times$ maximum value). With total reflection, the standing-wave patterns of r.m.s. voltage and current consist of a succession of positive sine waves with the voltage node located at the current antinode and the current node located at the voltage antinode. The termination is a current nodal point. The r.m.s. values of current and voltage may be recorded on a suitable r.m.s. instrument moving along the line. Such measurements of the maximum and minimum voltage and current can provide a reasonably accurate indication of the wavelength, and also provide information regarding the amount of reflected energy relative to the incident energy that is absorbed at the termination, as shown below.

Standing-wave ratio

Let the incident current flowing from the source of a mismatched low-loss transmission line be I_i and the current reflected at the termination be I_r. If I_{MAX} is the sum of the incident and reflected current, and I_{MIN} is their difference, then the **standing-wave ratio** (symbol s) on the line is defined as:

$$s = \frac{I_{MAX}}{I_{MIN}} = \frac{I_i + I_r}{I_i - I_r} \qquad (19)$$

Hence $s(I_i - I_r) = I_i + I_r$

$$sI_i - sI_r = I_i + I_r$$
$$sI_i - I_i = sI_r + I_r$$
$$I_i(s-1) = I_r(s+1)$$

i.e. $$\frac{I_r}{I_i} = \left(\frac{s-1}{s+1}\right) \qquad (20)$$

The power absorbed in the termination $P_t = I_i^2 Z_0$ and the reflected power, $P_r = I_r^2 Z_0$. Thus

$$\frac{P_r}{P_t} = \frac{I_r^2 Z_0}{I_i^2 Z_0} = \left(\frac{I_r}{I_i}\right)^2$$

Hence, from equation (20),

$$\frac{P_r}{P_t} = \left(\frac{s-1}{s+1}\right)^2 \qquad (21)$$

Thus the ratio of the reflected to the transmitted power may be calculated directly from the standing-wave ratio, which may be calculated from measurements of I_{MAX} and I_{MIN}. When a transmission line is properly terminated there is no reflection, i.e. $I_r = 0$, and from equation (19) the standing-wave ratio is 1. From equation (21), when $s = 1$, $P_r = 0$, i.e. there is no reflected power. In practice, the standing-wave ratio is kept as close to unity as possible. From equation (16), the reflection coefficient, $\rho = I_r/I_i$. Thus, from equation (20), $|\rho| = \dfrac{s-1}{s+1}$

Rearranging gives: $|\rho|(s+1) = (s-1)$

$$|\rho|s + |\rho| = s - 1$$

$$1 + |\rho| = s(1 - |\rho|)$$

from which $$s = \frac{1+|\rho|}{1-|\rho|} \qquad (22)$$

Equation (22) gives an expression for the standing-wave ratio in terms of the magnitude of the reflection coefficient.

Problem 16. A transmission line has a characteristic impedance of $600\angle 0°\,\Omega$ and negligible loss. If the terminating impedance of the line is $(400 + j250)\Omega$, determine (a) the reflection coefficient and (b) the standing-wave ratio.

(a) From equation (16),

$$\text{reflection coefficient, } \rho = \frac{Z_0 - Z_R}{Z_0 + Z_R} = \frac{600\angle 0° - (400 + j250)}{600\angle 0° + (400 + j250)} = \frac{200 - j250}{1000 + j250} = \frac{320.16\angle -51.34°}{1030.78\angle 14.04°}$$

Hence $\boldsymbol{\rho = 0.3106\angle -65.38°}$

(b) From above, $|\rho| = 0.3106$. Thus from equation (22),

$$\text{standing-wave ratio, } s = \frac{1+|\rho|}{1-|\rho|} = \frac{1+0.3106}{1-0.3106} = \mathbf{1.901}$$

Problem 17. A low-loss transmission line has a mismatched load such that the reflection coefficient at the termination is $0.2\angle -120°$. The characteristic impedance of the line is $80\,\Omega$. Calculate (a) the standing-wave ratio, (b) the load impedance and (c) the incident current flowing if the reflected current is 10 mA.

(a) From equation (22),

$$\textbf{standing-wave ratio, } s = \frac{1+|\rho|}{1-|\rho|} = \frac{1+0.2}{1-0.2} = \frac{1.2}{0.8} = \mathbf{1.5}$$

(b) From equation (16) reflection coefficient, $\rho = \dfrac{Z_0 - Z_R}{Z_0 + Z_R}$

Rearranging gives: $\rho(Z_0 + Z_R) = Z_0 - Z_R$

from which $Z_R(\rho + 1) = Z_0(1 - \rho)$

and $$\frac{Z_R}{Z_0} = \frac{1-\rho}{1+\rho} = \frac{1 - 0.2\angle -120°}{1 + 0.2\angle -120°} = \frac{1 - (-0.10 - j0.173)}{1 + (-0.10 - j0.173)} = \frac{1.10 + j0.173}{0.90 - j0.173} = \frac{1.1135\angle 8.94°}{0.9165\angle -10.88°} = 1.215\angle 19.82°$$

Hence load impedance $Z_R = Z_0(1.215\angle 19.82°)$
$= (80)(1.215\angle 19.82°)$
$= \mathbf{97.2\angle 19.82°\,\Omega}$
or $\mathbf{(91.4 + j33.0)\,\Omega}$

(c) From equation (20),

$$\frac{I_r}{I_i} = \frac{s-1}{s+1}$$

Hence $$\frac{10}{I_i} = \frac{1.5-1}{1.5+1} = \frac{0.5}{2.5} = 0.2$$

Thus the **incident current,** $\boldsymbol{I_i} = 10/0.2 = \mathbf{50\,mA}$

Part 4

Problem 18. The standing-wave ratio on a mismatched line is calculated as 1.60. If the incident power arriving at the termination is 200 mW, determine the value of the reflected power.

From equation (21),

$$\frac{P_r}{P_i} = \left(\frac{s-1}{s+1}\right)^2 = \left(\frac{1.60-1}{1.60+1}\right)^2 = \left(\frac{0.60}{2.60}\right)^2 = 0.0533$$

Hence the **reflected power,** $\boldsymbol{P_r} = 0.0533\,P_i$

$= (0.0533)(200)$

$= \mathbf{10.66\,mW}$

Now try the following Practice Exercise

Practice Exercise 194 Standing waves and the standing-wave ratio (Answers on page 835)

1. A transmission line has a characteristic impedance of $500\angle 0°\,\Omega$ and negligible loss. If the terminating impedance of the line is $(320 + j200)\,\Omega$, determine (a) the reflection coefficient and (b) the standing-wave ratio.
2. A low-loss transmission line has a mismatched load such that the reflection coefficient at the termination is $0.5\angle -135°$. The characteristic impedance of the line is $60\,\Omega$. Calculate (a) the standing-wave ratio, (b) the load impedance and (c) the incident current flowing if the reflected current is 25 mA.
3. The standing-wave ratio on a mismatched line is calculated as 2.20. If the incident power arriving at the termination is 100 mW, determine the value of the reflected power.
4. The termination of a coaxial cable may be represented as a $150\,\Omega$ resistance in series with a $0.20\,\mu$H inductance. If the characteristic impedance of the line is $100\angle 0°\,\Omega$ and the operating frequency is 80 MHz, determine (a) the reflection coefficient and (b) the standing-wave ratio.
5. A cable has a characteristic impedance of $70\angle 0°\,\Omega$. The cable is terminated by an impedance of $60\angle 30°\,\Omega$. Determine the ratio of the maximum to minimum current along the line.

For fully worked solutions to each of the problems in Practice Exercises 189 to 194 in this chapter, go to the website:
www.routledge.com/cw/bird

Chapter 48

Transients and Laplace transforms

Why it is important to understand: **Transients and Laplace transforms**

A transient state will exist in a circuit containing one or more energy storage elements (i.e. capacitors and inductors) whenever the energy conditions in the circuit change, until the new steady state condition is reached. Transients are caused by changing the applied voltage or current, or by changing any of the circuit elements; such changes occur due to opening and closing switches. In this chapter, such equations are developed analytically by using both differential equations and Laplace transforms for different waveform supply voltages. The solution of most electrical problems can be reduced ultimately to the solution of differential equations and the use of Laplace transforms provides an alternative method to those used previously. Laplace transforms provide a convenient method for the calculation of the complete response of a circuit. In this chapter, the technique of Laplace transforms is developed and then used to solve differential equations; in addition, Laplace transforms are used to analyse transient responses directly from circuit diagrams

At the end of this chapter you should be able to:

- determine the transient response of currents and voltages in R–L, R–C and L–R–C series circuits using differential equations
- define the Laplace transform of a function
- use a table of Laplace transforms of functions commonly met in electrical engineering for transient analysis of simple networks
- use partial fractions to deduce inverse Laplace transforms
- deduce expressions for component and circuit impedances in the s-plane given initial conditions
- use Laplace transform analysis directly from circuit diagrams in the s-plane
- deduce Kirchhoff's law equations in the s-plane for determining the response of R–L, R–C and L–R–C networks, given initial conditions
- explain the conditions for which an L–R–C circuit response is over, critical, under or zero-damped and calculate circuit responses
- predict the circuit response of an L–R–C network, given non-zero initial conditions

Electrical Circuit Theory and Technology. 978-1-138-67349-6, © 2017 John Bird. Published by Taylor & Francis. All rights reserved.

48.1 Introduction

A **transient state** will exist in a circuit containing one or more energy storage elements (i.e. capacitors and inductors) whenever the energy conditions in the circuit change, until the new **steady state** condition is reached. Transients are caused by changing the applied voltage or current, or by changing any of the circuit elements; such changes occur due to opening and closing switches. Transients were introduced in Chapter 19 where growth and decay curves were constructed and their equations stated for step inputs only. In this chapter, such equations are developed analytically by using both **differential equations** and **Laplace transforms** for different waveform supply voltages.

48.2 Response of *R–C* series circuit to a step input

Charging a capacitor

A series R–C circuit is shown in Figure 48.1(a).

A step voltage of magnitude V is shown in Figure 48.1(b). The capacitor in Figure 48.1(a) is assumed to be initially uncharged.

From Kirchhoff's voltage law, supply voltage,

$$V = v_C + v_R \qquad (1)$$

Voltage $v_R = iR$ and current $i = C\frac{dv_c}{dt}$, hence $v_R = CR\frac{dv_C}{dt}$

Therefore, from equation (1)

$$V = v_C + CR\frac{dv_C}{dt} \qquad (2)$$

This is a linear, constant coefficient, first-order differential equation. Such a differential equation may be solved, i.e. find an expression for voltage v_C, by separating the variables. (See *Higher Engineering Mathematics*).

Rearranging equation (2) gives:

$$V - v_C = CR\frac{dv_C}{dt}$$

and $$\frac{dv_C}{dt} = \frac{V - v_C}{CR}$$

from which, $$\frac{dv_C}{V - v_c} = \frac{dt}{CR}$$

and integrating both sides gives: $$\int \frac{dv_C}{V - v_C} = \int \frac{dt}{CR}$$

Hence $$-\ln(V - v_C) = \frac{t}{CR} + k \qquad (3)$$

where k is the arbitrary constant of integration

(To integrate $\int \frac{dv_C}{V - v_C}$ make an algebraic substitution, $u = V - v_C$ – see *Engineering Mathematics* or *Higher Engineering Mathematics*, J.O. Bird, 2017, 8th edition Taylor & Francis)

When time $t = 0$, $v_C = 0$, hence $-\ln V = k$

Thus, from equation (3), $-\ln(V - v_C) = \frac{t}{CR} - \ln V$

Rearranging gives:

$$\ln V - \ln(V - v_C) = \frac{t}{CR}$$

$$\ln\frac{V}{V - v_C} = \frac{t}{CR} \text{ by the laws of logarithms}$$

i.e. $$\frac{V}{V - v_C} = e^{t/CR}$$

and $$\frac{V - v_C}{V} = \frac{1}{e^{t/CR}} = e^{-t/CR}$$

$$V - v_C = Ve^{-t/CR}$$

$$V - Ve^{-t/CR} = v_C$$

i.e. capacitor p.d., $$\mathbf{v_c = V(1 - e^{-t/CR})} \qquad (4)$$

This is an exponential growth curve, as shown in Figure 48.2.

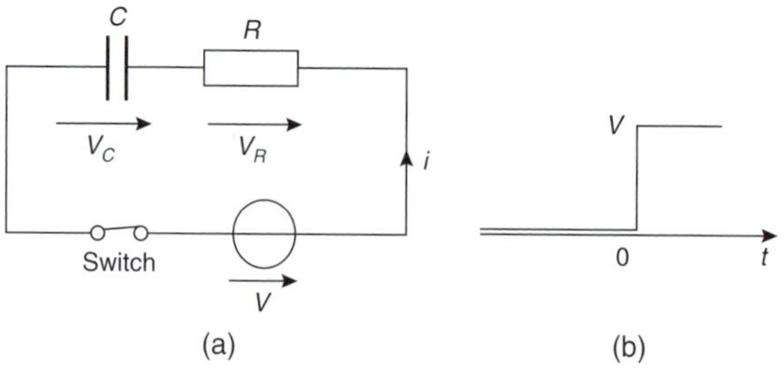

Figure 48.1

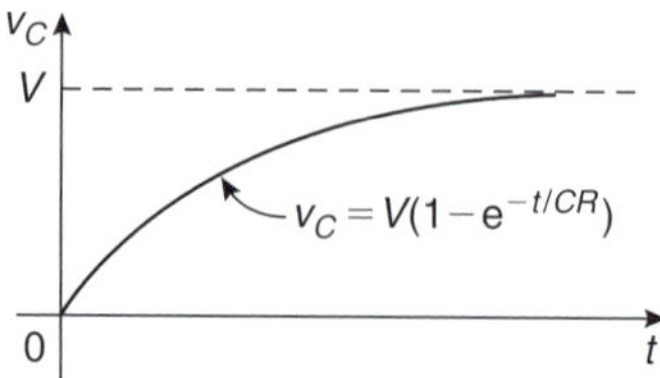

Figure 48.2

From equation (1),

$$v_R = V - v_C$$
$$= V - [V(1 - e^{-t/CR})] \text{ from equation (4)}$$
$$= V - V + Ve^{-t/CR}$$

i.e. resistor p.d., $\boldsymbol{v_R = Ve^{-t/CR}}$ (5)

This is an exponential decay curve, as shown in Figure 48.3.

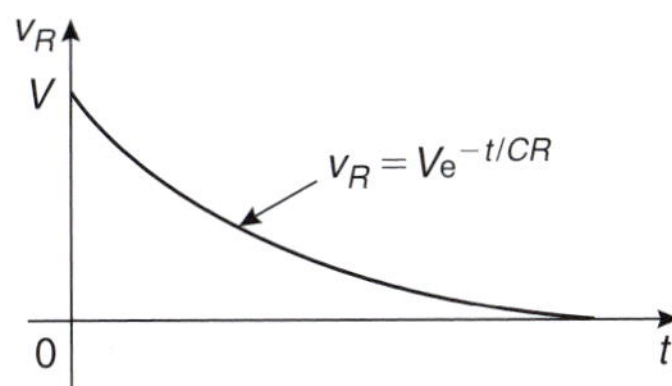

Figure 48.3

In the circuit of Figure 48.1(a), current $i = C\dfrac{dv_C}{dt}$

Hence $i = C\dfrac{d}{dt}[V(1 - e^{-t/CR})]$ from equation (4)

i.e. $i = C\dfrac{d}{dt}[V - Ve^{-t/CR}]$

$$= C\left[0 - (V)\left(\frac{-1}{CR}\right)e^{-t/CR}\right]$$

$$= C\left[\frac{V}{CR}e^{-t/CR}\right]$$

i.e. current, $\boldsymbol{i = \dfrac{V}{R}e^{-t/CR}}$ (6)

where $\dfrac{V}{R}$ is the steady state current, I.

This is an exponential decay curve as shown in Figure 48.4.

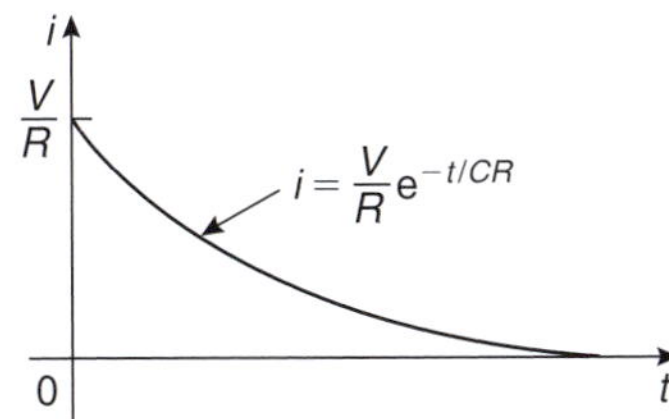

Figure 48.4

After a period of time it can be determined from equations (4) to (6) that the voltage across the capacitor, v_C, attains the value V, the supply voltage, whilst the resistor voltage, v_R, and current i both decay to zero.

Problem 1. A 500 nF capacitor is connected in series with a 100 kΩ resistor and the circuit is connected to a 50 V d.c. supply. Calculate (a) the initial value of current flowing, (b) the value of current 150 ms after connection, (c) the value of capacitor voltage 80 ms after connection and (d) the time after connection when the resistor voltage is 35 V.

(a) From equation (6), current, $i = \dfrac{V}{R}e^{-t/CR}$

Initial current, i.e. when $t = 0$

$$i_0 = \frac{V}{R}e^0 = \frac{V}{R} = \frac{50}{100 \times 10^3} = \mathbf{0.5\,mA}$$

(b) Current, $i = \dfrac{V}{R}e^{-t/CR}$ hence, when time $t = 150$ ms or 0.15 s,

$$i = \frac{50}{100 \times 10^3}e^{-0.5/(500\times10^{-9})(100\times10^3)}$$
$$= (0.5 \times 10^{-3})e^{-3} = (0.5 \times 10^{-3})(0.049787)$$
$$= \mathbf{0.0249\,mA} \text{ or } \mathbf{24.9\,\mu A}$$

(c) From equation (4), capacitor voltage, $v_C = V(1 - e^{-t/CR})$

When time $t = 80$ ms,

$$v_C = 50\left(1 - e^{-80\times10^{-3}/(500\times10^{-3}\times100\times10^3)}\right)$$
$$= 50(1 - e^{-1.6}) = 50(0.7981)$$
$$= \mathbf{39.91\,V}$$

(d) From equation (5), resistor voltage, $v_R = Ve^{-t/CR}$

When $v_R = 35$ V

then $35 = 50e^{-t/(500\times10^{-9}\times100\times10^3)}$

i.e. $\dfrac{35}{50} = e^{-t/0.05}$

and $\ln\dfrac{35}{50} = \dfrac{-t}{0.05}$

from which, **time,** $\boldsymbol{t} = -0.05\ln 0.7$
$= 0.0178$ s or **17.8 ms**

Discharging a capacitor

If after a period of time the step input voltage V applied to the circuit of Figure 48.1 is suddenly removed, by opening the switch, then

from equation (1), $v_R + v_C = 0$

or, from equation (2), $CR\dfrac{dv_C}{dt} + v_C = 0$

Rearranging gives: $\dfrac{dv_C}{dt} = \dfrac{-1}{CR}v_C$

and separating the variables gives: $\dfrac{dv_C}{v_C} = -\dfrac{dt}{CR}$

and integrating both sides gives: $\displaystyle\int \frac{dv_C}{v_C} = \int -\frac{dt}{CR}$

from which, $\ln v_C = -\dfrac{t}{CR} + k$ (7)

where k is a constant.

At time $t = 0$ (i.e. at the instant of opening the switch), $v_C = V$

Substituting $t = 0$ and $v_C = V$ in equation (7) gives:

$$\ln V = 0 + k$$

Substituting $k = \ln V$ into equation (7) gives:

$$\ln v_C = -\frac{t}{CR} + \ln V$$

and $$\ln v_C - \ln V = -\frac{t}{CR}$$

$$\ln\frac{v_C}{V} = -\frac{t}{CR}$$

and $$\frac{v_C}{V} = e^{-t/CR}$$

from which $$\boldsymbol{v_C = Ve^{-t/CR}} \quad (8)$$

i.e. the capacitor voltage, v_C, decays to zero after a period of time, the rate of decay depending on CR, which is the **time constant, τ** (see Section 19.3, page 296). Since $v_R + v_C = 0$ then the magnitude of the resistor voltage, v_R, is given by:

$$\boldsymbol{v_R = Ve^{-t/CR}} \quad (9)$$

and since $i = C\dfrac{dv_C}{dt} = C\dfrac{d}{dt}(Ve^{-t/CR})$

$$= (CV)\left(-\frac{1}{CR}\right)e^{-t/CR}$$

i.e. the magnitude of the current,

$$\boldsymbol{i = \frac{V}{R}e^{-t/CR}} \quad (10)$$

Problem 2. A d.c. voltage supply of 200 V is connected across a 5 μF capacitor as shown in Figure 48.5. When the supply is suddenly cut by opening switch S, the capacitor is left isolated except for a parallel resistor of 2 MΩ. Calculate the p.d. across the capacitor after 20 s.

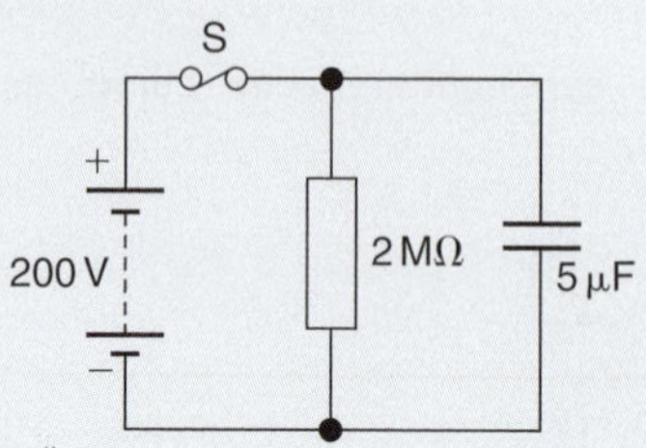

Figure 48.5

From equation (8), $v_C = Ve^{-t/CR}$

After 20 s, $v_C = 200e^{-20/(5\times10^{-6}\times2\times10^{6})}$

$= 200\,e^{-2}$

$= 200(0.13534)$

$= \mathbf{27.07\,V}$

48.3 Response of *R–L* series circuit to a step input

Current growth

A series R–L circuit is shown in Figure 48.6. When the switch is closed and a step voltage V is applied, it is assumed that L carries no current.

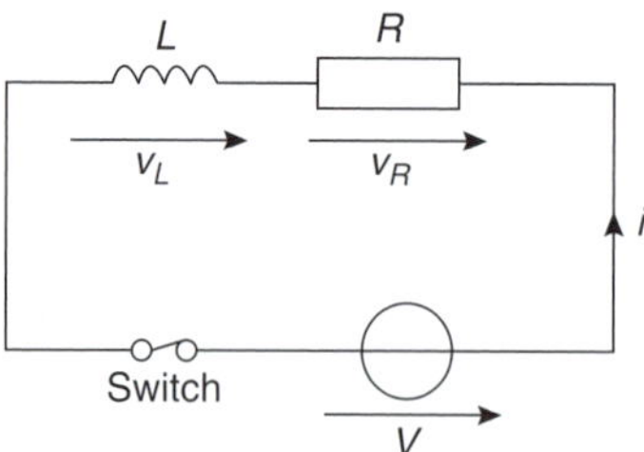

Figure 48.6

From Kirchhoff's voltage law, $V = v_L + v_R$

Voltage $v_L = L\dfrac{di}{dt}$ and voltage $v_R = iR$

Hence $V = L\dfrac{di}{dt} + iR$ (11)

This is a linear, constant coefficient, first-order differential equation. Again, such a differential equation may be solved by separating the variables.

Rearranging equation (11) gives: $\frac{\mathrm{d}i}{\mathrm{d}t}=\frac{V-iR}{L}$

from which, $\frac{\mathrm{d}i}{V-iR}=\frac{\mathrm{d}t}{L}$

and $\int\frac{\mathrm{d}i}{V-iR}=\int\frac{\mathrm{d}t}{L}$

Hence $$-\frac{1}{R}\ln(V-iR)=\frac{t}{L}+k \qquad (12)$$

where k is a constant

(Use the algebraic substitution $u=V-iR$ to integrate $\int\frac{\mathrm{d}i}{V-iR}$)

At time $t=0$, $i=0$, thus $-\frac{1}{R}\ln V=0+k$

Substituting $k=-\frac{1}{R}\ln V$ in equation (12) gives:

$$-\frac{1}{R}\ln(V-iR)=\frac{t}{L}-\frac{1}{R}\ln V$$

Rearranging gives: $\frac{1}{R}[\ln V-\ln(V-iR)]=\frac{t}{L}$

and $\ln\left(\frac{V}{V-iR}\right)=\frac{Rt}{L}$

Hence $\frac{V}{V-iR}=\mathrm{e}^{Rt/L}$

and $\frac{V-iR}{V}=\frac{1}{\mathrm{e}^{Rt/L}}=\mathrm{e}^{-Rt/L}$

$V-iR=V\mathrm{e}^{-Rt/L}$

$V-V\mathrm{e}^{-Rt/L}=iR$

and current, $$\boldsymbol{i=\frac{V}{R}(1-\mathrm{e}^{-Rt/L})} \qquad (13)$$

This is an exponential growth curve as shown in Figure 48.7.

The p.d. across the resistor in Figure 48.6, $v_R=iR$

Hence $v_R=R\left[\frac{V}{R}(1-\mathrm{e}^{-Rt/L})\right]$ from equation (13)

i.e. $$\boldsymbol{V_R=V(1-\mathrm{e}^{-Rt/L})} \qquad (14)$$

which again represents an exponential growth curve.

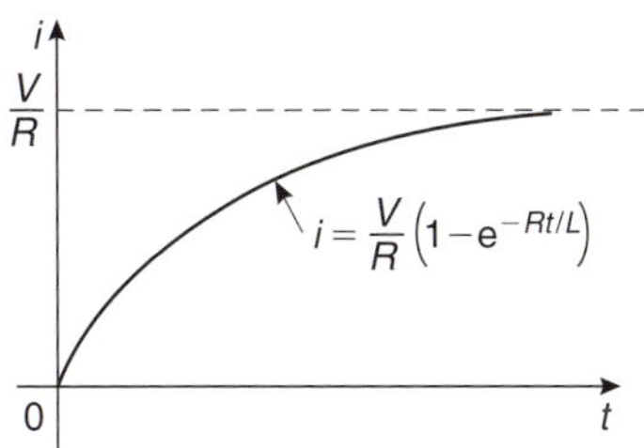

Figure 48.7

The voltage across the inductor in Figure 48.6, $v_L=L\frac{\mathrm{d}i}{\mathrm{d}t}$

i.e. $v_L=L\frac{\mathrm{d}}{\mathrm{d}t}\left[\frac{V}{R}(1-\mathrm{e}^{-Rt/L})\right]=\frac{LV}{R}\frac{\mathrm{d}}{\mathrm{d}t}[1-\mathrm{e}^{-Rt/L}]$

$=\frac{LV}{R}\left[0-\left(-\frac{R}{L}\right)\mathrm{e}^{-Rt/L}\right]=\frac{LV}{R}\left(\frac{R}{L}\mathrm{e}^{-Rt/L}\right)$

i.e. $$\boldsymbol{v_L=V\mathrm{e}^{-Rt/L}} \qquad (15)$$

Problem 3. A coil of inductance 50 mH and resistance 5 Ω is connected to a 110 V d.c. supply. Determine (a) the final value of current, (b) the value of current after 4 ms, (c) the value of the voltage across the resistor after 6 ms, (d) the value of the voltage across the inductance after 6 ms and (e) the time when the current reaches 15 A.

(a) From equation (13), when t is large, the final, or steady state current i is given by:

$$i=\frac{V}{R}=\frac{110}{5}=\mathbf{22\,A}$$

(b) From equation (13), current, $i=\frac{V}{R}(1-\mathrm{e}^{-Rt/L})$

When $t=4$ ms,

$$i=\frac{110}{5}(1-\mathrm{e}^{(-(5)(4\times10^{-3})/50\times10^{-3})})$$
$$=22(1-\mathrm{e}^{-0.40})=22(0.32968)$$
$$=\mathbf{7.25\,V}$$

(c) From equation (14), the voltage across the resistor,

$$v_R=V(1-\mathrm{e}^{-Rt/L})$$

When $t=6$ ms,

$$v_R=110(1-\mathrm{e}^{(-(5)(6\times10^{-3})/50\times10^{-3})})$$
$$=110(1-\mathrm{e}^{-0.60})=110(0.45119)$$
$$=\mathbf{49.63\,V}$$

Part 4

(d) From equation (15), the voltage across the inductance,

$$v_L = Ve^{-Rt/L}$$

When $t = 6$ ms,

$$\boldsymbol{v_L} = 110e^{(-(5)(6\times10^{-3})/50\times10^{-3})} = 110e^{-0.60}$$
$$= \mathbf{60.37\,V}$$

(Note that at $t = 6$ ms,

$$v_L + v_R = 60.37 + 49.63 = 110\,\text{V}$$
$$= \text{supply p.d., } V)$$

(e) When current i reaches 15 A,

$$15 = \frac{V}{R}(1 - e^{-Rt/L})$$

from equation (13)

i.e. $$15 = \frac{110}{5}(1 - e^{-5t/(50\times10^{-3})})$$

$$15\left(\frac{5}{110}\right) = 1 - e^{-100t}$$

and $$e^{-100t} = 1 - \frac{75}{110}$$

Hence $$-100t = \ln\left(1 - \frac{75}{110}\right)$$

and **time,** $$\boldsymbol{t} = \frac{1}{-100}\ln\left(1 - \frac{75}{100}\right)$$
$$= \mathbf{0.01145\,s \text{ or } 11.45\,ms}$$

Current decay

If after a period of time the step voltage V applied to the circuit of Figure 48.6 is suddenly removed by opening the switch, then from equation (11),

$$0 = L\frac{di}{dt} + iR$$

Rearranging gives: $L\frac{di}{dt} = -iR$ or $\frac{di}{dt} = -\frac{iR}{L}$

Separating the variables gives: $\frac{di}{i} = -\frac{R}{L}dt$

and integrating both sides gives:

$$\int \frac{di}{i} = \int -\frac{R}{L}dt$$

$$\ln i = -\frac{R}{L}t + k \qquad (16)$$

At $t = 0$ (i.e. when the switch is opened),

$$i = I\left(= \frac{V}{R}, \text{ the steady state current}\right)$$

then $\ln I = 0 + k$

Substituting $k = \ln I$ into equation (16) gives:

$$\ln i = -\frac{R}{L}t + \ln I$$

Rearranging gives: $$\ln i - \ln I = -\frac{R}{L}t$$

$$\ln\frac{i}{I} = -\frac{R}{L}t$$

$$\frac{i}{I} = e^{-Rt/L}$$

and **current,** $$\boldsymbol{i = Ie^{-Rt/L}} \text{ or } \boldsymbol{\frac{V}{R}e^{-Rt/L}} \qquad (17)$$

i.e. the current i decays exponentially to zero.

From Figure 48.6, $v_R = iR = R\left(\frac{V}{R}e^{-Rt/L}\right)$ from equation (17)

i.e. $$\boldsymbol{v_R = Ve^{-Rt/L}} \qquad (18)$$

The voltage across the coil,

$$v_L = L\frac{di}{dt} = L\frac{d}{dt}\left(\frac{V}{R}e^{-Rt/L}\right) \text{ from equation (17)}$$

$$= L\left(\frac{V}{R}\right)\left(-\frac{R}{L}\right)e^{-Rt/L}$$

Hence the magnitude of v_L is given by:

$$\boldsymbol{v_L = Ve^{-Rt/L}} \qquad (19)$$

Hence both v_R and v_L decay exponentially to zero.

Problem 4. In the circuit shown in Figure 48.8, a current of 5 A flows from the supply source. Switch S is then opened. Determine (a) the time for the current in the 2 H inductor to fall to 200 mA and (b) the maximum voltage appearing across the resistor.

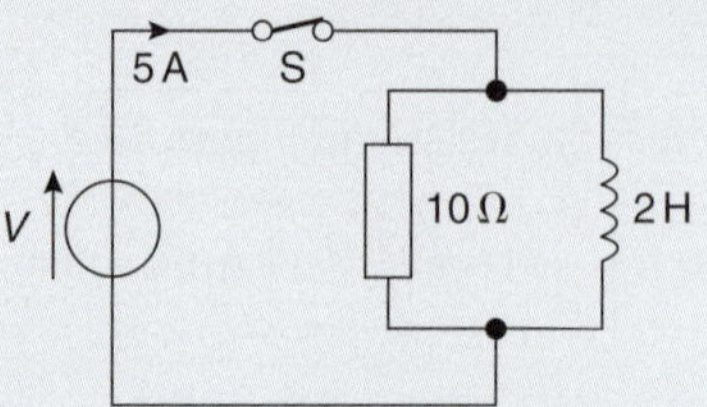

Figure 48.8

Part 4

(a) When the supply is cut off, the circuit consists of just the 10 Ω resistor and the 2 H coil in parallel. This is effectively the same circuit as Figure 48.6 with the supply voltage zero.

From equation (17), current $i = \frac{V}{R}e^{-Rt/L}$

In this case $\frac{V}{R} = 5$ A, the initial value of current.

When $i = 200$ mA or 0.2 A

$$0.2 = 5e^{-10t/2}$$

i.e. $$\frac{0.2}{5} = e^{-5t}$$

thus $$\ln\frac{0.2}{5} = -5t$$

and **time,** $t = -\frac{1}{5}\ln\frac{0.2}{5} = \mathbf{0.644\ s}$ or **644 ms**

(b) Since the current through the coil can only return through the 10 Ω resistance, the voltage across the resistor is a maximum at the moment of disconnection, i.e.

$$v_{R_m} = IR = (5)(10) = \mathbf{50\ V}$$

Now try the following Practice Exercise

Practice Exercise 195 *R–L* and *R–C* series circuits (Answers on page 835)

1. A 5 μF capacitor is connected in series with a 20 kΩ resistor and the circuit is connected to a 10 V d.c. supply. Determine (a) the initial value of current flowing, (b) the value of the current 0.4 s after connection, (c) the value of the capacitor voltage 30 ms after connection and (d) the time after connection when the resistor voltage is 4 V.

2. A 100 V d.c. supply is connected across a 400 nF capacitor as shown in Figure 48.9. When the switch S is opened the capacitor is left isolated except for a parallel resistor of 5 MΩ. Determine the p.d. across the capacitor 5 s after opening the switch.

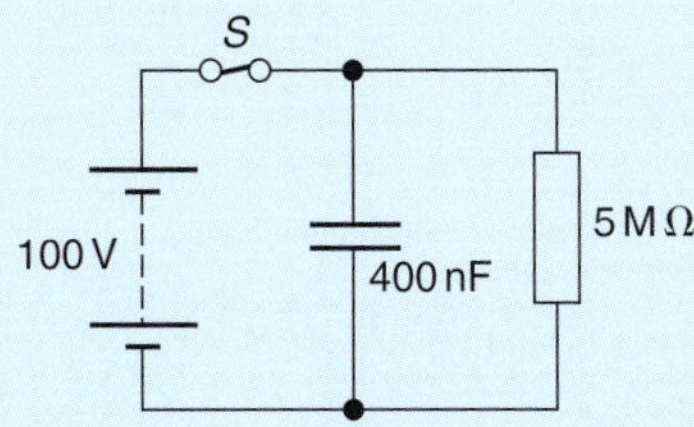

Figure 48.9

3. A 40 V d.c. supply is connected across a coil of inductance 25 mH and resistance 5 Ω. Calculate (a) the final value of current, (b) the value of current after 10 ms, (c) the p.d. across the resistor after 5 ms, (d) the value of the voltage across the inductance after 2 ms and (e) the time when the current reaches 3 A.

4. In the circuit shown in Figure 48.10 a current of 2 A flows from the source. If the switch S is suddenly opened, calculate (a) the time for the current in the 0.5 H inductor to fall to 0.8 A and (b) the maximum voltage across the resistor.

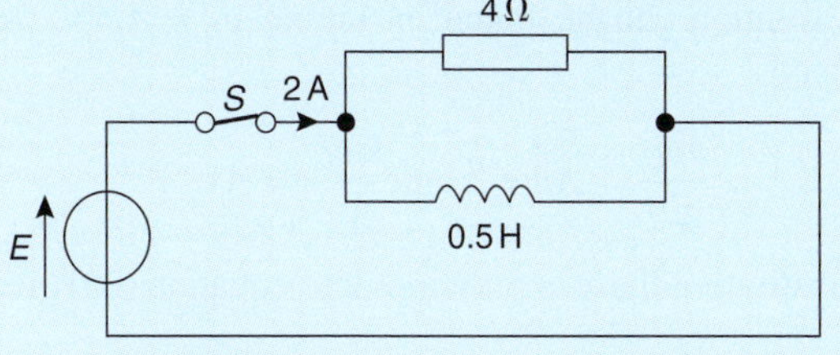

Figure 48.10

48.4 *L–R–C* series circuit response

L–R–C circuits are widely used in a variety of applications, such as in filters in communication systems, ignition systems in automobiles and defibrillator circuits in biomedical applications (where an electric shock is used to stop the heart, in the hope that the heart will restart with rhythmic contractions).

For the circuit shown in Figure 48.11, from Kirchhoff's voltage law,

$$V = v_L + v_R + v_C \qquad (20)$$

$$v_L = L\frac{di}{dt} \text{ and } i = C\frac{dv_C}{dt}, \text{ hence}$$

$$v_L = L\frac{d}{dt}\left(C\frac{dv_C}{dt}\right) = LC\frac{d^2v_C}{dt^2}$$

$$v_R = iR = \left(C\frac{dv_C}{dt}\right)R = RC\frac{dv_C}{dt}$$

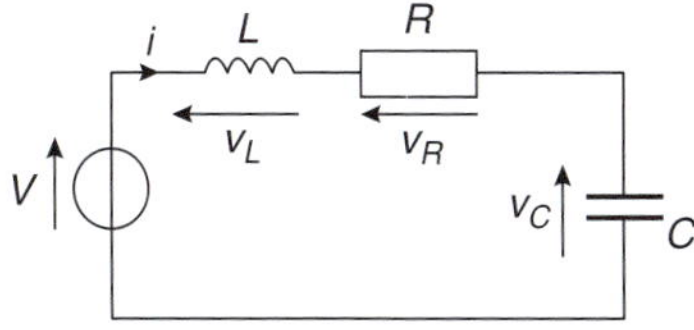

Figure 48.11

Hence from equation (20):

$$V = LC\frac{d^2 v_C}{dt^2} + RC\frac{dv_C}{dt} + v_C \quad (21)$$

This is a linear, constant coefficient, second-order differential equation. (For the solution of second-order differential equations, see *Higher Engineering Mathematics*.)

To determine the transient response, the supply p.d., V, is made equal to zero,

i.e. $$LC\frac{d^2 v_C}{dt^2} + RC\frac{dv_C}{dt} + v_C = 0 \quad (22)$$

A solution can be found by letting $v_C = Ae^{mt}$, from which,

$$\frac{dv_C}{dt} = Ame^{mt} \text{ and } \frac{dv_C}{dt^2} = Am^2e^{mt}$$

Substituting these expressions into equation (22) gives:

$$LC(Am^2e^{mt}) + RC(Ame^{mt}) + Ae^{mt} = 0$$

i.e. $$Ae^{mt}(m^2LC + mRC + 1) = 0$$

Thus $v_C = Ae^{mt}$ is a solution of the given equation provided that

$$m^2LC + mRC + 1 = 0 \quad (23)$$

This is called the **auxiliary equation**.

Using the quadratic formula on equation (23) gives:

$$m = \frac{-RC \pm \sqrt{[(RC)^2 - 4(LC)(1)]}}{2LC}$$

$$= \frac{-RC \pm \sqrt{(R^2C^2 - 4LC)}}{2LC}$$

i.e. $$m = \frac{-RC}{2LC} \pm \sqrt{\frac{R^2C^2 - 4LC}{(2LC)^2}}$$

$$= -\frac{R}{2L} \pm \sqrt{\left(\frac{R^2C^2}{4L^2C^2} - \frac{4LC}{4L^2C^2}\right)}$$

$$= -\frac{R}{2L} \pm \sqrt{\left[\left(\frac{R}{2L}\right)^2 - \frac{1}{LC}\right]} \quad (24)$$

This equation may have either:

(i) **two different real roots**, when $(R/2L)^2 > (1/LC)$, when the circuit is said to be **overdamped** since the transient voltage decays very slowly with time, or

(ii) **two real equal roots**, when $(R/2L)^2 = (1/LC)$, when the circuit is said to be **critically damped** since the transient voltage decays in the minimum amount of time without oscillations occurring, or

(iii) **two complex roots**, when $(R/2L)^2 < (1/LC)$, when the circuit is said to be **underdamped** since the transient voltage oscillates about the final steady state value, the oscillations eventually dying away to give the steady state value, or

(iv) if $\boldsymbol{R = 0}$ in equation (24), the oscillations would continue indefinitely without any reduction in amplitude – this is the **undamped** condition.

Damping is discussed again in Section 48.8 with typical current responses sketched in Figure 48.30 on page 795.

Problem 5. A series L–R–C circuit has inductance, $L = 2$ mH, resistance, $R = 1$ kΩ and capacitance, $C = 5$ μF. (a) Determine whether the circuit is over, critical or underdamped. (b) If $C = 5$ nF, determine the state of damping.

(a) $$\left(\frac{R}{2L}\right)^2 = \left[\frac{10^3}{2(2\times 10^{-3})}\right]^2 = \frac{10^{12}}{16} = 6.25\times 10^{10}$$

$$\frac{1}{LC} = \frac{1}{(2\times 10^{-3})(5\times 10^{6})} = \frac{10^9}{10} = 10^8$$

Since $\left(\frac{R}{2L}\right)^2 > \frac{1}{LC}$ the circuit is **overdamped**.

(b) When $C = 5$ nF, $\frac{1}{LC} = \frac{1}{(2\times 10^{-3})(5\times 10^{-9})} = 10^{11}$

Since $\left(\frac{R}{2L}\right)^2 < \frac{1}{LC}$ the circuit is **underdamped**.

Problem 6. In the circuit of Problem 5, what value of capacitance will give critical damping?

For critical damping: $\left(\frac{R}{2L}\right)^2 = \frac{1}{LC}$

from which, **capacitance,**

$$\boldsymbol{C} = \frac{1}{L\left(\frac{R}{2L}\right)^2} = \frac{1}{L\frac{R^2}{4L^2}} = \frac{4L^2}{LR^2} = \frac{4L}{R^2}$$

$$= \frac{4(2\times 10^{-3})}{(10^3)^2} = 8\times 10^{-9} \text{ F or } \mathbf{8\,nF}$$

Part 4

Roots of the auxiliary equation

With reference to equation (24):

(i) when the roots are **real and different**, say $m=\alpha$ and $m=\beta$, the general solution is

$$\boldsymbol{v_C = Ae^{\alpha t} + Be^{\beta t}} \qquad (25)$$

where $\alpha = -\frac{R}{2L} + \sqrt{\left[\left(\frac{R}{2L}\right)^2 - \frac{1}{LC}\right]}$ and

$$\beta = -\frac{R}{2L} - \sqrt{\left[\left(\frac{R}{2L}\right)^2 - \frac{1}{LC}\right]}$$

(ii) when the roots are **real and equal**, say $m=\alpha$ twice, the general solution is

$$\boldsymbol{v_C = (At+B)\,e^{\alpha t}} \qquad (26)$$

where $\alpha = -\frac{R}{2L}$

(iii) when the roots are **complex**, say $m = \alpha \pm j\beta$, the general solution is

$$\boldsymbol{v_C = e^{\alpha t}\{A\cos \beta t + B \sin \beta t\}} \qquad (27)$$

where $\alpha = -\frac{R}{2L}$ and $\beta = \sqrt{\left[\frac{1}{LC} - \left(\frac{R}{2L}\right)^2\right]}$ (28)

To determine the actual expression for the voltage under any given initial condition, it is necessary to evaluate constants A and B in terms of v_C and current i. The procedure is the same for each of the above three cases. Assuming in, say, case (iii) that at time $t=0$, $v_C=v_0$ and $i(=C(\mathrm{d}v_C/\mathrm{d}t))=i_0$ then substituting in equation (27):

$$v_0 = e^0\{A\cos 0 + B\sin 0\}$$

i.e. $\boldsymbol{v_0 = A}$ (29)

Also, from equation (27),

$$\frac{\mathrm{d}v_C}{\mathrm{d}t} = e^{\alpha t}[-A\beta\sin\beta t + B\beta\cos\beta t] + [A\cos\beta t + B\sin\beta t](\alpha e^{\alpha t}) \qquad (30)$$

by the product rule of differentiation

When $t=0$, $\frac{\mathrm{d}v_C}{\mathrm{d}t} = e^0[0+B\beta] + [A](\alpha e^0) = B\beta + \alpha A$

Hence at $t=0$, $i_0 = C\frac{\mathrm{d}v_C}{\mathrm{d}t} = C(B\beta + \alpha A)$

From equation (29), $A = v_0$ hence $i_0 = C(B\beta + \alpha v_0) = CB\beta + C\alpha v_0$

from which, $\boldsymbol{B = \frac{i_0 - C\alpha v_0}{C\beta}}$ (31)

Problem 7. A coil has an equivalent circuit of inductance 1.5 H in series with resistance 90 Ω. It is connected across a charged 5 μF capacitor at the moment when the capacitor voltage is 10 V. Determine the nature of the response and obtain an expression for the current in the coil.

$$\left(\frac{R}{2L}\right)^2 = \left[\frac{90}{2(1.5)}\right]^2 = 900 \text{ and } \frac{1}{LC} = \frac{1}{(1.5)(5\times 10^{-6})} = 1.333\times 10^5$$

Since $\left(\frac{R}{2L}\right)^2 < \frac{1}{LC}$ the circuit is **underdamped**.

From equation (28),

$$\alpha = -\frac{R}{2L} = -\frac{90}{2(1.5)} = -30$$

$$\text{and } \beta = \sqrt{\left[\frac{1}{LC} - \left(\frac{R}{2L}\right)^2\right]} = \sqrt{[1.333\times 10^5 - 900]} = 363.9$$

With $v_0=10$ V and $i_0=0$, from equation (29), $v_0=A=10$

and from equation (31),

$$B = \frac{i_0 - C\alpha v_0}{C\beta} = \frac{0-(5\times 10^{-6})(-30)(10)}{(5\times 10^{-6})(363.9)} = \frac{300}{363.9} = 0.8244$$

Current, $i = C\frac{\mathrm{d}v_C}{\mathrm{d}t}$, and from equation (30),

$$\begin{aligned} i &= C\{e^{-30t}[-10(363.9)\sin\beta t + (0.8244)(363.9)\cos\beta t] + (10\cos\beta t + 0.8244\sin\beta t)(-30e^{-30t})\} \\ &= C\{e^{-30t}[-3639\sin\beta t + 300\cos\beta t - 300\cos\beta t - 24.732\sin\beta t]\} \\ &= Ce^{-30t}[-3663.732\sin\beta t] \\ &= -(5\times 10^{-6})(3663.732)e^{-30t}\sin\beta t \end{aligned}$$

i.e. **current, $\boldsymbol{i = -0.0183e^{-30t}\sin 363.9t}$ amperes**

Part 4

Now try the following Practice Exercise

Practice Exercise 196 *L–R–C* series circuit response (Answers on page 835)

1. In a series L–R–C circuit the inductance, $L = 5\,\text{mH}$ and the resistance $R = 5\,\text{k}\Omega$. Determine whether the circuit is over, critical or underdamped when (a) capacitance $C = 500\,\text{pF}$ and (b) $C = 10\,\mu\text{F}$.
2. For the circuit in Problem 7 (page 773) calculate the value of capacitance C for critical damping.
3. A coil having an equivalent circuit of inductance 1 H in series with resistance 50 Ω is connected across a fully charged 0.4 μF capacitor at the instant when the capacitor voltage is 20 V. Determine the nature of the response and obtain an expression for the current in the coil.
4. If the coil in Problem 7 had a resistance of 500 Ω and the capacitance was 16 μF, determine the nature of the response and obtain an expression for the current flowing.

48.5 Introduction to Laplace* transforms

The solution of most electrical problems can be reduced ultimately to the solution of differential equations, and the use of **Laplace transforms** provides an alternative method to those used previously. Laplace transforms provide a convenient method for the calculation of the complete response of a circuit. In this section and in Section 48.5 the technique of Laplace transforms is developed and then used to solve differential equations. In Section 48.7 Laplace transforms are used to analyse transient responses directly from circuit diagrams.

Definition of a Laplace transform

The Laplace transform of the function of time $f(t)$ is defined by the integral

$$\int_0^\infty e^{-st} f(t)\,dt \quad \text{where } s \text{ is a parameter}$$

There are various commonly used notations for the Laplace transform of $f(t)$ and these include $\mathscr{L}\{f(t)\}$ or $L\{f(t)\}$ or $\mathscr{L}(f)$ or Lf or $\overline{f}(s)$.

Also the letter p is sometimes used instead of s as the parameter. The notation used in this chapter will be $f(t)$ for the original function and $\mathscr{L}\{f(t)\}$ for its Laplace transform,

$$\text{i.e. } \mathscr{L}\{f(t)\} = \int_0^\infty e^{-st} f(t)\,dt \qquad (32)$$

Laplace transforms of elementary functions

Using equation (32):

(i) when $\boldsymbol{f(t) = 1}$, $\mathscr{L}\{1\} = \int_0^\infty e^{-st}(1)\,dt = \left[\frac{e^{-st}}{-s}\right]_0^\infty$

$$= -\frac{1}{s}[e^{-s(\infty)} - e^0]$$

$$= -\frac{1}{s}[0-1]$$

$$= \boldsymbol{\frac{1}{s}} \quad (\text{provided } s > 0)$$

*Who was **Laplace**? Pierre-Simon, marquis de Laplace (23 March 1749–5 March 1827) was a French mathematician and astronomer who formulated Laplace's equation, and pioneered the Laplace transform which appears in many branches of mathematical physics. To find out more go to **www.routledge.com/cw/bird**

(ii) when $f(t)=k$, $\mathscr{L}\{k\}=k\mathscr{L}\{1\}=k\left(\frac{1}{s}\right)=\frac{k}{s}$ from (i) above

(iii) when $f(t)=e^{at}$, $\mathscr{L}\{e^{at}\}=\int_0^\infty e^{-st}(e^{at})\,dt$

$$=\int e^{-(s-a)t}\,dt$$

from the laws of indices

$$=\left[\frac{e^{-(s-a)t}}{-(s-a)}\right]_0^\infty$$

$$=\frac{1}{-(s-a)}(0-1)$$

$$=\frac{\mathbf{1}}{\boldsymbol{s-a}} \text{ (provided } s>a)$$

(iv) when $f(t)=t$, $\mathscr{L}\{t\}=\int_0^\infty e^{-st}t\,dt$

$$=\left[\frac{te^{-st}}{-s}-\int\frac{e^{-st}}{-s}\,dt\right]_0^\infty$$

$$=\left[\frac{te^{-st}}{-s}-\frac{e^{-st}}{s^2}\right]_0^\infty$$

by integration by parts

$$=\left[\frac{\infty e^{-s(\infty)}}{-s}-\frac{e^{-s(\infty)}}{s^2}\right]-\left[0-\frac{e^0}{s^2}\right]$$

$$=(0-0)-\left(0-\frac{1}{s^2}\right)$$

since $(\infty\times 0)=0$

$$=\frac{\mathbf{1}}{\boldsymbol{s^2}} \quad \text{(provided } s>0)$$

(v) when $f(t)=\cos\omega t$

$$\mathscr{L}\{\cos\omega t\}=\int_0^\infty e^{-st}\cos\omega t\,dt$$

$$=\left[\frac{e^{-st}}{s^2+\omega^2}(\omega\sin\omega t-s\cos\omega t)\right]_0^\infty$$

by integration by parts twice

$$=\frac{\boldsymbol{s}}{\boldsymbol{s^2+\omega^2}} \quad \text{(provided } s>0)$$

A list of standard Laplace transforms is summarized in Table 48.1. It will not usually be necessary to derive the transforms as above – but merely to use them.

The following worked problems only require using the standard list of Table 48.1.

Table 48.1 Standard Laplace transforms

Time function $f(t)$	**Laplace transform $\mathscr{L}\{f(t)\}=\int_0^\infty e^{-st}f(t)\,dt$**
1. δ (unit impulse)	1
2. 1 (unit step function)	$\frac{1}{s}$
3. e^{at} (exponential function)	$\frac{1}{s-a}$
4. unit step delayed by T	$\frac{e^{-sT}}{s}$
5. $\sin\omega t$ (sine wave)	$\frac{\omega}{s^2+\omega^2}$
6. $\cos\omega t$ (cosine wave)	$\frac{s}{s^2+\omega^2}$
7. t (unit ramp function)	$\frac{1}{s^2}$
8. t^2	$\frac{2!}{s^3}$
9. t^n $(n=1,2,3...)$	$\frac{n!}{s^{n+1}}$
10. $\cosh\omega t$	$\frac{s}{s^2-\omega^2}$
11. $\sinh\omega t$	$\frac{\omega}{s^2-\omega^2}$
12. $e^{at}t^n$	$\frac{n!}{(s-a)^{n+1}}$
13. $e^{-at}\sin\omega t$ (damped sine wave)	$\frac{\omega}{(s+a)^2+\omega^2}$
14. $e^{-at}\cos\omega t$ (damped cosine wave)	$\frac{s+a}{(s+a)^2+\omega^2}$
15. $e^{-at}\sinh\omega t$	$\frac{\omega}{(s+a)^2-\omega^2}$
16. $e^{-at}\cosh\omega t$	$\frac{s+a}{(s+a)^2-\omega^2}$

Problem 8. Find the Laplace transforms of (a) $1+2t-\frac{1}{3}t^4$ (b) $5e^{2t}-3e^{-t}$

(a) $$\mathscr{L}\left\{1+2t-\frac{1}{3}t^4\right\} = \mathscr{L}\{1\}+2\mathscr{L}\{t\}-\frac{1}{3}\mathscr{L}\{t^4\}$$
$$= \frac{1}{s}+2\left(\frac{1}{s^2}\right)-\frac{1}{3}\left(\frac{4!}{s^{4+1}}\right)$$
from 2, 7 and 9 of Table 48.1
$$= \frac{1}{s}+\frac{2}{s^2}-\frac{1}{3}\left(\frac{4\times 3\times 2\times 1}{s^5}\right)$$
$$= \mathbf{\frac{1}{s}+\frac{2}{s^2}-\frac{8}{s^5}}$$

(b) $$\mathscr{L}\{5e^{2t}-3e^{-t}\} = 5\mathscr{L}\{e^{2t}\}-3\mathscr{L}\{e^{-t}\}$$
$$= 5\left(\frac{1}{s-2}\right)-3\left(\frac{1}{s--1}\right)$$
from 3 of Table 48.1
$$= \frac{5}{s-2}-\frac{3}{s+1}$$
$$= \frac{5(s+1)-3(s-2)}{(s-2)(s+1)}$$
$$= \mathbf{\frac{2s+11}{s^2-s-2}}$$

Problem 9. Find the Laplace transform of $6\sin 3t-4\cos 5t$.

$$\mathscr{L}\{6\sin 3t-4\cos 5t\} = 6\mathscr{L}\{\sin 3t\}-4\mathscr{L}\{\cos 5t\}$$
$$= 6\left(\frac{3}{s^2+3^2}\right)-4\left(\frac{s}{s^2+5^2}\right)$$
from 5 and 6 of Table 48.1
$$= \mathbf{\frac{18}{s^2+9}-\frac{4s}{s^2+25}}$$

Problem 10. Use Table 48.1 to determine the Laplace transforms of the following waveforms:

(a) a step voltage of 10 V which starts at time $t=0$
(b) a step voltage of 10 V which starts at time $t=5$ s
(c) a ramp voltage which starts at zero and increases at 4 V/s
(d) a ramp voltage which starts at time $t=1$ s and increases at 4 V/s

(a) From 2 of Table 48.1,

$$\mathscr{L}\{\mathbf{10}\} = 10\mathscr{L}\{1\} = 10\left(\frac{1}{s}\right) = \mathbf{\frac{10}{s}}$$

The waveform is shown in Figure 48.12(a).

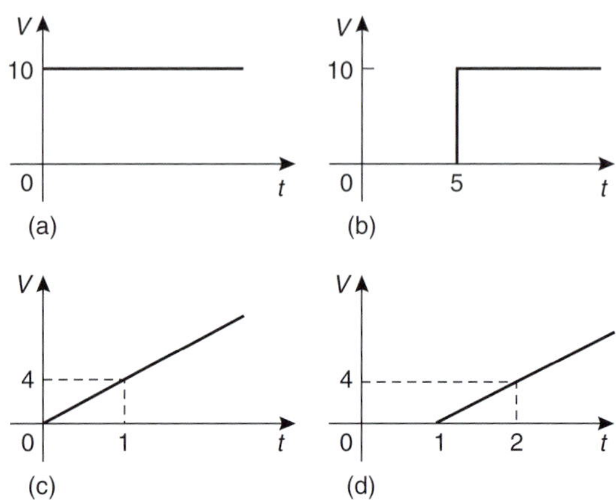

Figure 48.12

(b) From 4 of Table 48.1, a step function of 10 V which is delayed by $t=5$ s is given by:

$$10\left(\frac{e^{-sT}}{s}\right) = 10\left(\frac{e^{-5s}}{s}\right) = \mathbf{\frac{10}{s}e^{-5s}}$$

This is, in fact, the function starting at $t=0$ given in part (a), i.e. $(10/s)$ multiplied by e^{-sT}, where T is the delay in seconds.

The waveform is shown in Figure 48.12(b).

(c) From 7 of Table 48.1, the Laplace transform of the unit ramp, $\mathscr{L}\{t\}=(1/s^2)$.

Hence the Laplace transform of a ramp voltage increasing at 4 V/s is given by:

$$4\mathscr{L}\{t\} = \mathbf{\frac{4}{s^2}}$$

The waveform is shown in Figure 48.12(c).

(d) As with part (b), for a delayed function, the Laplace transform is the undelayed function, in this case $(4/s^2)$ from part (c), multiplied by e^{-sT} where T in this case is 1 s. Hence the Laplace transform is given by:

$$\mathbf{\left(\frac{4}{s^2}\right)e^{-s}}$$

The waveform is shown in Figure 48.12(d).

Problem 11. Determine the Laplace transforms of the following waveforms:

(a) an impulse voltage of 8 V which starts at time $t=0$
(b) an impulse voltage of 8 V which starts at time $t=2$ s
(c) a sinusoidal current of 4 A and angular frequency 5 rad/s which starts at time $t=0$

(a) An **impulse** is an intense signal of very short duration. This function is often known as the **Dirac*** **function**.

From 1 of Table 48.1, the Laplace transform of an impulse starting at time $t=0$ is given by $\mathscr{L}\{\delta\}=1$, hence an impulse of 8 V is given by: $8\mathscr{L}\{\delta\}=\mathbf{8}$

This is shown in Figure 48.13(a).

(b) From part (a) the Laplace transform of an impulse of 8 V is 8. Delaying the impulse by 2 s involves multiplying the undelayed function by $\mathbf{e^{-sT}}$ where $T=2$ s.

Hence the Laplace transform of the function is given by: $\mathbf{8\,e^{-2s}}$

This is shown in Figure 48.13(b).

(c) From 5 of Table 48.1, $\mathscr{L}\{\sin\omega t\}=\dfrac{\omega}{s^2+\omega^2}$

When the amplitude is 4 A and $\omega=5$, then

$$\mathscr{L}\{4\sin\omega t\}=4\left(\frac{5}{s^2+5^2}\right)=\mathbf{\frac{20}{s^2+25}}$$

The waveform is shown in Figure 48.13(c).

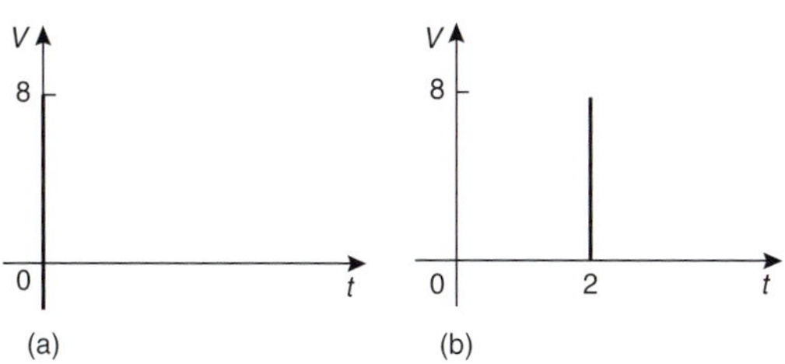

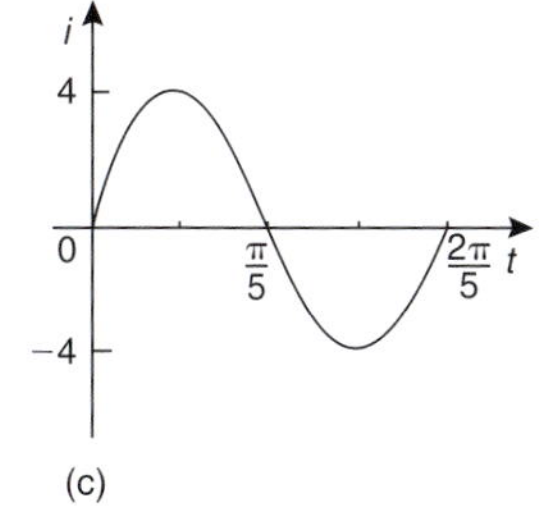

Figure 48.13

Problem 12. Find the Laplace transforms of (a) $2t^4e^{3t}$, (b) $4e^{3t}\cos 5t$

(a) From 12 of Table 48.1,

$$\mathscr{L}\{2t^4e^{3t}\}=2\mathscr{L}\{t^4e^{3t}\}=2\left[\frac{4!}{(s-3)^{4+1}}\right]=\frac{2(4\times3\times2\times1)}{(s-3)^5}=\mathbf{\frac{48}{(s-3)^5}}$$

(b) From 14 of Table 48.1,

$$\mathscr{L}\{4e^{3t}\cos 5t\}=4\mathscr{L}\{e^{3t}\cos 5t\}=4\left[\frac{s-3}{(s-3)^2+5^2}\right]=\frac{4(s-3)}{s^2-6s+9+25}=\mathbf{\frac{4(s-3)}{s^2-6s+34}}$$

Problem 13. Determine the Laplace transforms of (a) $2\cosh 3t$, (b) $e^{-2t}\sin 3t$

(a) From 10 of Table 48.1,

$$\mathscr{L}\{2\cosh 3t\}=2\mathscr{L}\cosh 3t=2\left[\frac{s}{s^2-3^2}\right]=\mathbf{\frac{2s}{s^2-9}}$$

(b) From 13 of Table 48.1,

$$\mathscr{L}\{e^{-2t}\sin 3t\}=\frac{3}{(s+2)^2+3^2}=\frac{3}{s^2+4s+4+9}=\mathbf{\frac{3}{s^2+4s+13}}$$

*Who was **Dirac**? Paul Adrien Maurice Dirac (8 August 1902–20 October 1984) was an English theoretical physicist who made fundamental contributions to the early development of both quantum mechanics and quantum electrodynamics. To find out more go to **www.routledge.com/cw/bird**

Laplace transforms of derivatives

Using integration by parts, it may be shown that:

(a) for the **first derivative:**

$$\mathscr{L}\{f'(t)\} = s\mathscr{L}\{f(t)\} - f(0)$$

or $$\mathscr{L}\left\{\frac{dy}{dx}\right\} = s\mathscr{L}\{y\} - y(0) \qquad (33)$$

where $y(0)$ is the value of y at $x=0$

(b) for the **second derivative:**

$$\mathscr{L}\{f''(t)\} = s^2\mathscr{L}\{f(t)\} - sf(0) - f'(0)$$

or $$\mathscr{L}\left\{\frac{d^2y}{dx^2}\right\} = s^2\mathscr{L}\{y\} - sy(0) - y'(0) \qquad (34)$$

where $y'(0)$ is the value of (dy/dx) at $x=0$

Equations (33) and (34) are used in the solution of differential equations in Section 48.5.

The initial and final value theorems

The initial and final value theorems can often considerably reduce the work of solving electrical circuits.

(a) The **initial value theorem** states:

$$\operatorname*{limit}_{t\to 0}[f(t)] = \operatorname*{limit}_{s\to\infty}[s\mathscr{L}\{f(t)\}]$$

Thus, for example, if

$$f(t) = v = Ve^{-t/CR} \text{ and if, say,}$$

$$V = 10 \text{ and } CR = 0.5, \text{ then}$$

$$f(t) = v = 10e^{-2t}$$

$$\mathscr{L}\{f(t)\} = 10\left[\frac{1}{s+2}\right]$$

from 3 of Table 48.1

$$s\mathscr{L}\{f(t)\} = 10\left[\frac{s}{s+2}\right]$$

From the initial value theorem, the initial value of $f(t)$ is given by:

$$10\left[\frac{\infty}{\infty+2}\right] = 10(1) = \mathbf{10}$$

(b) The **final value theorem** states:

$$\operatorname*{limit}_{t\to\infty}[f(t)] = \operatorname*{limit}_{s\to 0}[s\mathscr{L}\{f(t)\}]$$

In the above example of $f(t) = 10e^{-2t}$ the final value is given by:

$$10\left[\frac{0}{0+2}\right] = \mathbf{0}$$

The initial and final value theorems are used in pulse circuit applications where the response of the circuit for small periods of time, or the behaviour immediately after the switch is closed, are of interest. The final value theorem is particularly useful in investigating the stability of systems (such as in automatic aircraft-landing systems) and is concerned with the steady state response for large values of time t, i.e. after all transient effects have died away.

Now try the following Practice Exercise

Practice Exercise 197 Laplace transforms (Answers on page 835)

Determine the Laplace transforms in Problems 1 to 7.

1. (a) $2t - 3$ (b) $5t^2 + 4t - 3$
2. (a) $\frac{t^3}{24} - 3t + 2$ (b) $\frac{t^5}{15} - 2t^4 + \frac{t^2}{2}$
3. (a) $5e^{3t}$ (b) e^{-2t}
4. (a) $4\sin 3t$ (b) $3\cos 2t$
5. (a) $2te^{2t}$ (b) t^2e^t
6. (a) $4t^3e^{-2t}$ (b) $\frac{1}{2}t^4e^{-3t}$
7. (a) $5e^{-2t}\cos 3t$ (b) $4e^{-5t}\sin t$
8. Determine the Laplace transforms of the following waveforms:
   (a) a step voltage of 4 V which starts at time $t=0$
   (b) a step voltage of 5 V which starts at time $t=2$ s
   (c) a ramp voltage which starts at zero and increases at 7 V/s
   (d) a ramp voltage which starts at time $t=2$ s and increases at 3 V/s
9. Determine the Laplace transforms of the following waveforms:

Part 4

(a) an impulse voltage of 15 V which starts at time $t=0$
(b) an impulse voltage of 6 V which starts at time $t=5$
(c) a sinusoidal current of 10 A and angular frequency 8 rad/s

10. State the initial value theorem. Verify the theorem for the functions:
(a) $3-4\sin t$
(b) $(t-4)^2$
and state their initial values.

11. State the final value theorem and state a practical application where it is of use. Verify the theorem for the function $4+e^{-2t}(\sin t+\cos t)$ representing a displacement and state its final value.

48.6 Inverse Laplace transforms and the solution of differential equations

Since from 2 of Table 48.1, $\mathscr{L}\{1\}=\frac{1}{s}$ then

$$\mathscr{L}^{-1}=\left\{\frac{1}{s}\right\}=\mathbf{1}$$

where $\mathscr{L}^{-1}$ means the **inverse Laplace transform**.
Similarly, since from 5 of Table 48.1,

$$\mathscr{L}\{\sin\omega t\}=\frac{\omega}{s^2+\omega^2} \text{ then } \mathscr{L}^{-1}\left\{\frac{\omega}{s^2+\omega^2}\right\}=\mathbf{\sin\omega t}$$

Thus finding an inverse transform involves locating the Laplace transform from the right-hand column of Table 48.1 and then reading the function from the left-hand column. The following worked problems demonstrate the method.

Problem 14. Find the following inverse Laplace transforms:

(a) $\mathscr{L}^{-1}\left\{\frac{1}{s^2+9}\right\}$ (b) $\mathscr{L}^{-1}\left\{\frac{5}{3s-1}\right\}$

(a) $$\mathscr{L}^{-1}\left\{\frac{1}{s^2+9}\right\}=\mathscr{L}^{-1}\left\{\frac{1}{s^2+3^2}\right\}=\frac{1}{3}\mathscr{L}^{-1}\left\{\frac{3}{s^2+3^2}\right\}$$

and from 5 of Table 48.1,

$$\frac{1}{3}\mathscr{L}^{-1}\left\{\frac{3}{s^2+3^2}\right\}=\mathbf{\frac{1}{3}\sin 3t}$$

(b) $$\mathscr{L}^{-1}\left\{\frac{5}{3s-1}\right\}=\mathscr{L}^{-1}\left\{\frac{5}{3\left(s-\frac{1}{3}\right)}\right\}=\frac{5}{3}\mathscr{L}^{-1}\left\{\frac{1}{s-\frac{1}{3}}\right\}=\mathbf{\frac{5}{3}e^{\frac{1}{3}t}}$$

from 3 of Table 48.1

Problem 15. Determine the following inverse Laplace transforms:

(a) $\mathscr{L}^{-1}\left\{\frac{6}{s^3}\right\}$ (b) $\mathscr{L}^{-1}\left\{\frac{3}{s^4}\right\}$

(a) From 8 of Table 48.1, $\mathscr{L}^{-1}\left\{\frac{2}{s^3}\right\}=t^2$

Hence $\mathscr{L}^{-1}\left\{\frac{6}{s^3}\right\}=3\mathscr{L}^{-1}\left\{\frac{2}{s^3}\right\}=\mathbf{3t^2}$

(b) From 9 of Table 48.1, if s is to have a power of 4 then $n=3$.

Thus $\mathscr{L}^{-1}\left\{\frac{3!}{s^4}\right\}=t^3$, i.e. $\mathscr{L}^{-1}\left\{\frac{6}{s^4}\right\}=\mathbf{t^3}$

Hence $\mathscr{L}^{-1}\left\{\frac{3}{s^4}\right\}=\frac{1}{2}\mathscr{L}^{-1}\left\{\frac{6}{s^4}\right\}=\mathbf{\frac{1}{2}t^3}$

Problem 16. Determine

(a) $\mathscr{L}^{-1}\left\{\frac{7s}{s^2+4}\right\}$ (b) $\mathscr{L}^{-1}\left\{\frac{4s}{s^2-16}\right\}$

(a) $$\mathscr{L}^{-1}\left\{\frac{7s}{s^2+4}\right\}=7\mathscr{L}^{-1}\left\{\frac{s}{s^2+2^2}\right\}$$

$=\mathbf{7\cos 2t}$ from 6 of Table 48.1

(b) $$\mathscr{L}^{-1}\left\{\frac{4s}{s^2-16}\right\}=4\mathscr{L}^{-1}\left\{\frac{s}{s^2-4^2}\right\}$$

$=\mathbf{4\cosh 4t}$ from 10 of Table 48.1

Part 4

Problem 17. Find $\mathscr{L}^{-1}\left\{\frac{2}{(s-3)^5}\right\}$

From 12 of Table 48.1, $\mathscr{L}^{-1}\left\{\frac{n!}{(s-a)^{n+1}}\right\}=e^{at}t^n$

Thus $\mathscr{L}^{-1}\left\{\frac{1}{(s-a)^{n+1}}\right\}=\frac{1}{n!}e^{at}t^n$

and comparing with $\mathscr{L}^{-1}\left\{\frac{2}{(s-3)^5}\right\}$ shows that $n=4$ and $a=3$

Hence $\mathscr{L}^{-1}\left\{\frac{2}{(s-3)^5}\right\}=2\mathscr{L}^{-1}\left\{\frac{1}{(s-3)^5}\right\}$

$$=2\left[\frac{1}{4!}e^{3t}t^4\right]$$

$$=\mathbf{\frac{1}{12}e^{3t}t^4}$$

Problem 18. Determine

(a) $\mathscr{L}^{-1}\left\{\frac{3}{s^2-4s+13}\right\}$ (b) $\mathscr{L}^{-1}\left\{\frac{2(s+1)}{s^2+2s+10}\right\}$

(a) $\mathscr{L}^{-1}\left\{\frac{3}{s^2-4s+13}\right\}=\mathscr{L}^{-1}\left\{\frac{3}{(s-2)^2+3^2}\right\}$

$=\mathbf{e^{2t}\sin 3t}$ from 13 of Table 48.1

(b) $\mathscr{L}^{-1}\left\{\frac{2(s+1)}{s^2+2s+10}\right\}=\mathscr{L}^{-1}\left\{\frac{2(s+1)}{(s+1)^2+3^2}\right\}$

$=\mathbf{2e^{-t}\cos 3t}$ from 14 of Table 48.1

Note that in solving these examples the denominator in each case has been made into a perfect square.

Now try the following Practice Exercise

Practice Exercise 198 Inverse Laplace transforms (Answers on page 835)

Determine the inverse Laplace transforms in Problems 1 to 8.

1. (a) $\frac{7}{s}$ (b) $\frac{2}{s-5}$
2. (a) $\frac{3}{2s+1}$ (b) $\frac{2s}{s^2+4}$
3. (a) $\frac{1}{s^2+25}$ (b) $\frac{4}{s^2+9}$
4. (a) $\frac{5s}{2s^2+18}$ (b) $\frac{6}{s^2}$
5. (a) $\frac{5}{s^3}$ (b) $\frac{8}{s^4}$
6. (a) $\frac{15}{3s^2-27}$ (b) $\frac{4}{(s-1)^3}$
7. (a) $\frac{3}{s^2-4s+13}$ (b) $\frac{4}{2s^2-8s+10}$
8. (a) $\frac{s+1}{s^2+2s+10}$ (b) $\frac{3}{s^2+6s+13}$

Use of partial fractions for inverse Laplace transforms

Sometimes the function whose inverse is required is not recognizable as a standard type, such as those listed in Table 48.1. In such cases it may be possible, by using **partial fractions**, to resolve the function into simpler fractions which may be inverted on sight.

For example, the function $F(s)=\frac{2s-3}{s(s-3)}$ cannot be inverted on sight from Table 48.1. However, using partial fractions:

$$\frac{2s-3}{s(s-3)}\equiv\frac{A}{s}+\frac{B}{s-3}=\frac{A(s-3)+Bs}{s(s-3)}$$

from which, $2s-3=A(s-3)+Bs$

Letting $s=0$ gives: $-3=-3A$ from which $A=1$

Letting $s=3$ gives: $3=3B$ from which $B=1$

Hence $\frac{2s-3}{s(s-3)}\equiv\frac{1}{s}+\frac{1}{s-3}$

Thus $\mathscr{L}^{-1}\left\{\frac{2s-3}{s(s-3)}\right\}=\mathscr{L}^{-1}\left\{\frac{1}{s}+\frac{1}{(s-3)}\right\}$

$=\mathbf{1+e^{3t}}$ from 2 and 3 of Table 48.1

Partial fractions are explained in *Engineering Mathematics* and *Higher Engineering Mathematics*. The following worked problems demonstrate the method.

Part 4

Problem 19. Determine $\mathcal{L}^{-1}\left\{\dfrac{4s-5}{s^2-s-2}\right\}$

$$\frac{4s-5}{s^2-s-2} \equiv \frac{4s-5}{(s-2)(s+1)} \equiv \frac{A}{(s-2)} + \frac{B}{(s+1)}$$
$$= \frac{A(s+1)+B(s-2)}{(s-2)(s+1)}$$

Hence $4s-5=A(s+1)+B(s-2)$

When $s=2$, $3=3A$ from which, $A=1$

When $s=-1$, $-9=-3B$ from which, $B=3$

$$\text{Hence } \mathcal{L}^{-1}\left\{\frac{4s-5}{s^2-s-2}\right\} \equiv \mathcal{L}^{-1}\left\{\frac{1}{s-2}+\frac{3}{s+1}\right\}$$
$$= \mathcal{L}^{-1}\left\{\frac{1}{s-2}\right\} + \mathcal{L}^{-1}\left\{\frac{3}{s+1}\right\}$$
$$= \mathbf{e^{2t}+3e^{-t}} \text{ from 3 of Table 48.1}$$

Problem 20. Find $\mathcal{L}^{-1}\left\{\dfrac{3s^3+s^2+12s+2}{(s-3)(s+1)^3}\right\}$

$$\frac{3s^3+s^2+12s+12}{(s-3)(s+1)^3}$$
$$\equiv \frac{A}{s-3}+\frac{B}{s+1}+\frac{C}{(s+1)^2}+\frac{D}{(s+1)^3}$$
$$= \frac{A(s+1)^3+B(s-3)(s+1)^2 + C(s-3)(s+1)+D(s-3)}{(s-3)(s+1)^3}$$

Hence $3s^3+s^2+12s+2=A(s+1)^3+B(s-3)(s+1)^2 + C(s-3)(s+1)+D(s-3)$

When $s=3$, $128=64A$ from which, $A=2$

When $s=-1$, $-12=-4D$ from which, $D=3$

Equating s^3 terms gives: $3=A+B$ from which, $B=1$

Equating s^2 terms gives: $1=3A-B+C$ from which, $C=-4$

$$\text{Hence } \mathcal{L}^{-1}\left\{\frac{3s^3+s^2+12s+2}{(s-3)(s+1)^3}\right\}$$
$$\equiv \mathcal{L}^{-1}\left\{\frac{2}{s-3}+\frac{1}{s+1}-\frac{4}{(s+1)^2}+\frac{3}{(s+1)^3}\right\}$$
$$= \mathbf{2e^{3t}+e^{-t}-4e^{-t}t+\frac{3}{2}e^{-t}t^2}$$

from 3 and 12 of Table 48.1

Problem 21. Determine $\mathcal{L}^{-1}\left\{\dfrac{5s^2+8s-1}{(s+3)(s^2+1)}\right\}$

$$\frac{5s^2+8s-1}{(s+3)(s^2+1)} \equiv \frac{A}{s+3}+\frac{Bs+C}{s^2+1}$$
$$= \frac{A(s^2+1)+(Bs+C)(s+3)}{(s+3)(s^2+1)}$$

Hence $5s^2+8s-1=A(s^2+1)+(Bs+C)(s+3)$

When $s=-3$ $\quad 20=10A$ from which, $A=2$

Equating s^2 terms gives: $5=A+B$ from which, $B=3$

Equating s terms gives: $8=3B+C$ from which, $C=-1$

Hence

$$\mathcal{L}^{-1}\left\{\frac{5s^2+8s-1}{(s+3)(s^2+1)}\right\} \equiv \mathcal{L}^{-1}\left\{\frac{2}{s+3}+\frac{3s-1}{s^2+1}\right\}$$
$$= \mathcal{L}^{-1}\left\{\frac{2}{s+3}\right\} + \mathcal{L}^{-1}\left\{\frac{3s}{s^2+1}\right\} - \mathcal{L}^{-1}\left\{\frac{1}{s^2+1}\right\}$$
$$= \mathbf{2e^{-3t}+3\cos t-\sin t}$$

from 3, 6 and 5 of Table 48.1

Now try the following Practice Exercise

Practice Exercise 199 Inverse Laplace transforms using partial fractions (Answers on page 835)

Use partial fractions to find the inverse Laplace transforms of the functions in Problems 1 to 6.

1. $\dfrac{11-3s}{s^2+2s-3}$
2. $\dfrac{2s^2-9s-35}{(s+1)(s-2)(s+3)}$
3. $\dfrac{2s+3}{(s-2)^2}$
4. $\dfrac{5s^2-2s-19}{(s+3)(s-1)^2}$

5. $\dfrac{3s^2+16s+15}{(s+3)^3}$

6. $\dfrac{26-s^2}{s(s^2+4s+13)}$

Procedure to solve differential equations by using Laplace transforms

(i) Take the Laplace transform of both sides of the differential equation by applying the formulae for the Laplace transforms of derivatives (i.e. equations (33) and (34) on page 778) and, where necessary, using a list of standard Laplace transforms, such as Table 48.1 on page 775.

(ii) Put in the given initial conditions, i.e. $y(0)$ and $y'(0)$

(iii) Rearrange the equation to make $\mathscr{L}\{y\}$ the subject.

(iv) Determine y by using, where necessary, partial fractions, and taking the inverse of each term by using Table 48.1.

This procedure is demonstrated in the following problems.

Problem 22. Use Laplace transforms to solve the differential equation

$$2\frac{d^2y}{dx^2}+5\frac{dy}{dx}-3y=0$$

given that when $x=0$, $y=4$ and $\dfrac{dy}{dx}=9$

(i) $2\mathscr{L}\left\{\dfrac{d^2y}{dx^2}\right\}+5\mathscr{L}\left\{\dfrac{dy}{dx}\right\}-3\mathscr{L}\{y\}=\mathscr{L}\{0\}$

$$2[s^2\mathscr{L}\{y\}-sy(0)-y'(0)]+5[s\mathscr{L}\{y\}-y(0)]$$
$$-3\mathscr{L}\{y\}=0$$

from equations (33) and (34)

(ii) $y(0)=4$ and $y'(0)=9$

Thus

$$2[s^2\mathscr{L}\{y\}-4s-9]+5[s\mathscr{L}\{y\}-4]-3\mathscr{L}\{y\}=0$$

i.e.

$$2s^2\mathscr{L}\{y\}-8s-18+5s\mathscr{L}\{y\}-20-3\mathscr{L}\{y\}=0$$

(iii) Rearranging gives: $(2s^2+5s-3)\mathscr{L}\{y\}=8s+38$

i.e. $\mathscr{L}\{y\}=\dfrac{8s+38}{2s^2+5s-3}$

(iv) $y=\mathscr{L}^{-1}\left\{\dfrac{8s+38}{2s^2+5s-3}\right\}$

Let $\dfrac{8s+38}{2s^2+5s-3}=\dfrac{8s+38}{(2s-1)(s+3)}$

$$=\frac{A}{2s-1}+\frac{B}{s+3}$$

$$=\frac{A(s+3)+B(2s-1)}{(2s-1)(s+3)}$$

Hence $8s+38=A(s+3)+B(2s-1)$

When $s=\frac{1}{2}$, $42=3\frac{1}{2}A$ from which, $A=12$

When $s=-3$, $14=-7B$ from which, $B=-2$

Hence $y=\mathscr{L}^{-1}\left\{\dfrac{8s+38}{2s^2+5s-3}\right\}$

$$\equiv\mathscr{L}^{-1}\left\{\frac{12}{2s-1}-\frac{2}{s+3}\right\}$$

$$=\mathscr{L}^{-1}\left\{\frac{12}{2(s-\frac{1}{2})}\right\}-\mathscr{L}^{-1}\left\{\frac{2}{s+3}\right\}$$

Hence $\boldsymbol{y=6e^{(1/2)x}-2e^{-3x}}$ from 3 of Table 48.1.

Problem 23. Use Laplace transforms to solve the differential equation:

$$\frac{d^2y}{dx^2}+6\frac{dy}{dx}+13y=0$$

given that when $x=0$, $y=3$ and $\dfrac{dy}{dx}=7$

Using the above procedure:

(i) $\mathscr{L}\left\{\dfrac{d^2y}{dx^2}\right\}+6\mathscr{L}\left\{\dfrac{dy}{dx}\right\}+13\mathscr{L}\{y\}=\mathscr{L}\{0\}$

Hence $[s^2\mathscr{L}\{y\}-sy(0)-y'(0)]$
$$+6[s\mathscr{L}\{y\}-y(0)]+13\mathscr{L}\{y\}=0$$

from equations (33) and (34)

(ii) $y(0)=3$ and $y'(0)=7$

Thus $s^2\mathscr{L}\{y\}-3s-7+6s\mathscr{L}\{y\}-18$
$$+13\mathscr{L}\{y\}=0$$

(iii) Rearranging gives: $(s^2+6s+13)\mathscr{L}\{y\}=3s+25$

i.e. $\mathscr{L}\{y\}=\dfrac{3s+25}{s^2+6s+13}$

Part 4

(iv) $y = \mathscr{L}^{-1}\left\{\dfrac{3s+25}{s^2+6s+13}\right\}$

$$= \mathscr{L}^{-1}\left\{\frac{3s+25}{(s+3)^2+2^2}\right\} = \mathscr{L}^{-1}\left\{\frac{3(s+3)+16}{(s+3)^2+2^2}\right\}$$

$$= \mathscr{L}^{-1}\left\{\frac{3(s+3)}{(s+3)^2+2^2}\right\} + \mathscr{L}^{-1}\left\{\frac{8(2)}{(s+3)^2+2^2}\right\}$$

$$= 3e^{-3t}\cos 2t + 8e^{-3t}\sin 2t$$

from 14 and 13 of Table 48.1, page 775.

Hence $\boldsymbol{y = e^{-3t}(3\cos 2t + 8\sin 2t)}$

Problem 24. A step voltage is applied to a series C–R circuit. When the capacitor is fully charged the circuit is suddenly broken. Deduce, using Laplace transforms, an expression for the capacitor voltage during the transient period if the voltage when the supply is cut is V volts.

From Figure 48.1, page 766, $v_R + v_C = 0$ when the supply is cut

i.e. $\quad iR + v_c = 0$

i.e. $\left(C\dfrac{dv_c}{dt}\right)R + v_c = 0$

i.e. $\quad CR\dfrac{dv_c}{dt} + v_c = 0$

Using the procedure:

(i) $\mathscr{L}\left\{CR\dfrac{dv_c}{dt}\right\} + \mathscr{L}\{v_c\} = \mathscr{L}\{0\}$

i.e. $CR[s\mathscr{L}\{v_c\} - v_0] + \mathscr{L}\{v_c\} = 0$

(ii) $v_0 = V$, hence $CR[s\mathscr{L}\{v_c\} - V] + \mathscr{L}\{v_c\} = 0$

(iii) Rearranging gives: $CRs\mathscr{L}\{v_c\} - CRV + \mathscr{L}\{v_c\} = 0$

i.e. $(CRs+1)\mathscr{L}\{v_c\} = CRV$

hence $\quad \mathscr{L}\{v_c\} = \dfrac{CRV}{(CRs+1)}$

(iv) Capacitor voltage, $v_c = \mathscr{L}^{-1}\left\{\dfrac{CRV}{CRs+1}\right\}$

$$= CRV\mathscr{L}^{-1}\left\{\frac{1}{CR\left(s+\frac{1}{CR}\right)}\right\}$$

$$= \frac{CRV}{CR}\mathscr{L}^{-1}\left\{\frac{1}{s+\frac{1}{CR}}\right\}$$

i.e. $\boldsymbol{v_c = Ve^{(-t/CR)}}$ as previously obtained in equation (8) on page 768.

Problem 25. A series R–L circuit has a step input V applied to it. Use Laplace transforms to determine an expression for the current i flowing in the circuit given that when time $t = 0$, $i = 0$

From Figure 48.6 and equation (11) on page 768,

$v_R + v_L = V$ becomes $iR + L\dfrac{di}{dt} = V$

Using the procedure:

(i) $\mathscr{L}\{iR\} + \mathscr{L}\left\{L\dfrac{di}{dt}\right\} = \mathscr{L}\{V\}$

i.e. $R\mathscr{L}\{i\} + L[s\mathscr{L}\{i\} - i(0)] = \dfrac{V}{s}$

(ii) $i(0) = 0$, hence $R\mathscr{L}\{i\} + Ls\mathscr{L}\{i\} = \dfrac{V}{s}$

(iii) Rearranging gives: $(R+Ls)\mathscr{L}\{i\} = \dfrac{V}{s}$

i.e. $\mathscr{L}\{i\} = \dfrac{V}{s(R+Ls)}$

(iv) $i = \mathscr{L}^{-1}\left\{\dfrac{V}{s(R+Ls)}\right\}$

Let $\dfrac{V}{s(R+Ls)} \equiv \dfrac{A}{s} + \dfrac{B}{R+Ls} = \dfrac{A(R+Ls)+Bs}{s(R+Ls)}$

Hence $V = A(R+Ls) + Bs$

When $s = 0$, $V = AR$ from which, $A = \dfrac{V}{R}$

When $s = -\dfrac{R}{L}$, $V = B\left(-\dfrac{R}{L}\right)$ from which,

$$B = -\frac{VL}{R}$$

Hence $\mathscr{L}^{-1}\left\{\dfrac{V}{s(R+Ls)}\right\}$

$$= \mathscr{L}^{-1}\left\{\frac{V/R}{s} + \frac{-VL/R}{R+Ls}\right\}$$

$$= \mathscr{L}^{-1}\left\{\frac{V}{Rs} - \frac{VL}{R(R+Ls)}\right\}$$

$$= \mathscr{L}^{-1}\left\{\frac{V}{R}\left(\frac{1}{s}\right) - \frac{V}{R}\left(\frac{1}{\frac{R}{L}+s}\right)\right\}$$

Part 4

$$=\frac{V}{R}\mathscr{L}^{-1}\left\{\frac{1}{s}-\frac{1}{\left(s+\frac{R}{L}\right)}\right\}$$

Hence **current,** $\boldsymbol{i=\frac{V}{R}(1-e^{-Rt/L})}$ as previously obtained in equation (13), page 769.

Problem 26. If after a period of time, the switch in the R–L circuit of Problem 25 is opened, use Laplace transforms to determine an expression to represent the current transient response. Assume that at the instant of opening the switch, the steady state current flowing is I

From Figure 48.6, page 768, $v_L+v_R=0$ when the switch is opened,

i.e. $L\frac{di}{dt}+iR=0$

Using the procedure:

(i) $\mathscr{L}\left\{L\frac{di}{dt}\right\}+\mathscr{L}\{iR\}=\mathscr{L}\{0\}$

i.e. $L[s\mathscr{L}\{i\}-i_0]+R\mathscr{L}\{i\}=0$

(ii) $i_0=I$, hence $L[s\mathscr{L}\{i\}-I]+R\mathscr{L}\{i\}=0$

(iii) Rearranging gives: $Ls\mathscr{L}\{i\}-LI+R\mathscr{L}\{i\}=0$

i.e. $(R+Ls)\mathscr{L}\{i\}=LI$

and $\mathscr{L}\{i\}=\frac{LI}{R+Ls}$

(iv) Current, $i=\mathscr{L}^{-1}\left\{\frac{LI}{R+Ls}\right\}$

$$=LI\mathscr{L}^{-1}\left\{\frac{1}{L\left(\frac{R}{L}+s\right)}\right\}$$

$$=\frac{LI}{L}\mathscr{L}^{-1}\left\{\frac{1}{s+\frac{R}{L}}\right\}$$

i.e. $\boldsymbol{i=Ie^{(-Rt/L)}}$ from 3 of Table 48.1

Since $I=\frac{V}{R}$ then $\boldsymbol{i=\frac{V}{R}e^{-Rt/L}}$ as previously derived in equation (17), page 770.

Now try the following Practice Exercise

Practice Exercise 200 Solving differential equations using Laplace transforms (Answers on page 835)

In Problems 1 to 5, use Laplace transforms to solve the given differential equations.

1. $9\frac{d^2y}{dt^2}-24\frac{dy}{dt}+16y=0$, given $y(0)=3$ and $y'(0)=3$
2. $\frac{d^2x}{dt^2}+100x=0$, given $x(0)=2$ and $x'(0)=0$
3. $\frac{d^2i}{dt^2}+1000\frac{di}{dt}+250\,000i=0$, given $i(0)=0$ and $i'(0)=100$
4. $\frac{d^2x}{dt^2}+6\frac{dx}{dt}+8x=0$, given $x(0)=4$ and $x'(0)=8$
5. $\frac{d^2y}{dt^2}-2\frac{dy}{dt}+y=3e^{4t}$, given $y(0)=-\frac{2}{3}$ and $y'(0)=4\frac{1}{3}$
6. Use Laplace transforms to solve the differential equation: $\frac{d^2y}{dt^2}+\frac{dy}{dt}-2y=3\cos 3t-11\sin 3t$ given $y(0)=0$ and $y'(0)=6$

48.7 Laplace transform analysis directly from the circuit diagram

Resistor

At any instant in time $v=Ri$

Since v and i are both functions of time, a more correct equation would be $v(t)=Ri(t)$

However, this is normally assumed. The Laplace transform of this equation is:

$$V(s)=RI(s)$$

Hence, in the s-domain $R(s)=\frac{V(s)}{I(s)}=\boldsymbol{R}$

(Note that $V(s)$ merely means that it is the Laplace transform of v and $I(s)$ is the Laplace transform of i.

Whenever the Laplace transform of functions is taken it is referred to as the 's-domain' – as opposed to the 'time domain'.)

The resistor is shown in Figure 48.14 in both the time domain and the s-domain.

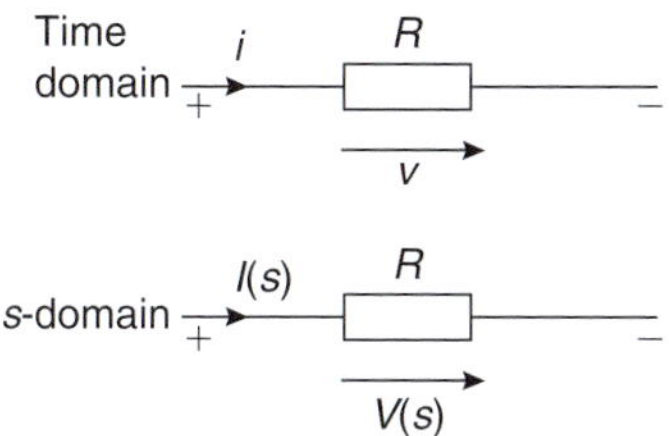

Figure 48.14

Inductor

If an inductor has no initial current, i.e. $i=0$ at time $t=0$, the normal equation is $v=L(\mathrm{d}i/\mathrm{d}t)$ where L is the inductance.

The Laplace transform of the equation is:

$V(s)=L[sI(s)-i(0)]$ from equation (33)

and as $i(0)=0$ then $V(s)=sLI(s)$

Thus the impedance of the inductor in the s-domain is given by:

$$Z(s)=\frac{V(s)}{I(s)}=\boldsymbol{sL}$$

The inductor is shown in Figure 48.15 in both the time domain and the s-domain.

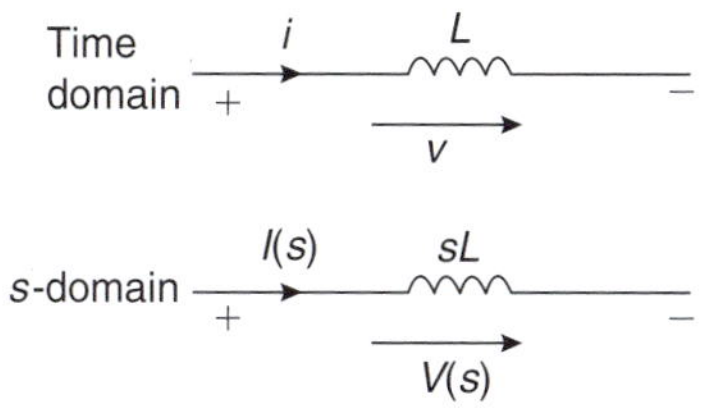

Figure 48.15

Capacitor

If a capacitor has no initial voltage, i.e. $v=0$ at time $t=0$, the normal equation is $i=C(\mathrm{d}v/\mathrm{d}t)$.

The Laplace transform of the equation is:

$$I(s)=C[sV(s)-v(0)]$$
$$=sCV(s) \text{ since } v(0)=0$$

Thus the impedance of the capacitor in the s-domain is given by:

$$Z(s)=\frac{V(s)}{I(s)}=\frac{V(s)}{sCV(s)}=\boldsymbol{\frac{1}{sC}}$$

The capacitor is shown in Figure 48.16 in both the time domain and the s-domain.

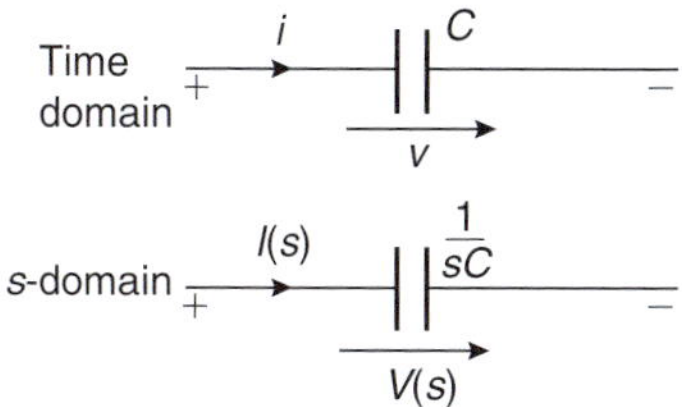

Figure 48.16

Summarizing, in the **time domain**, the circuit elements are ***R*, *L* and *C*** and in the **s-domain**, the circuit elements are ***R*, *sL* and (1/*sC*)**.

Note that the impedance of L is $X_L=j\omega L$ and the impedance of C is $X_c=(-j/\omega C)=(1/j\omega C)$.

Thus, just replacing $j\omega$ with s gives the s-domain expressions for L and C. (Because of this apparent association with j, s is sometimes called the **complex frequency** and the s-domain called the **complex frequency domain**.)

Problem 27. Determine the impedance of a 5 μF capacitor in the s-domain.

In the s-domain the impedance of a capacitor is $\frac{1}{sC}$, hence

$$Z(s)=\boldsymbol{\frac{1}{s(5\times10^{-6})}\ \Omega} \text{ or } \boldsymbol{\frac{1}{5\times10^{-6}s}\ \Omega} \text{ or } \boldsymbol{\frac{2\times10^{5}}{s}\ \Omega}$$

Problem 28. Determine the impedance of a 200 Ω resistor in series with an 8 mH inductor in the s-domain.

The impedance of the resistor in the s-domain is 200 Ω

The impedance of the inductor in the s-domain is

$sL=8\times10^{-3}s$

Since the components are in series,

$$Z(s)=\boldsymbol{(200+8\times10^{-3}s)\ \Omega}$$

Problem 29. A circuit comprises a 50 Ω resistor, a 5 mH inductor and a 0.04 μF capacitor. Determine, in the s-domain (a) the impedance when the components are connected in series and (b) the admittance when the components are connected in parallel.

(a) R, L and C connected in series in the s-domain give an impedance,

$$Z(s) = R + sL + \frac{1}{sC}$$

$$= \left(\mathbf{50 + 5 \times 10^{-3}s + \frac{1}{0.04 \times 10^{-6}s}}\right) \boldsymbol{\Omega}$$

(b) R, L and C connected in parallel gives:

$$\text{admittance } Y = Y_1 + Y_2 + Y_3 = \frac{1}{Z_1} + \frac{1}{Z_2} + \frac{1}{Z_3}$$

In the s-domain, admittance,

$$Y(s) = \frac{1}{R} + \frac{1}{sL} + \frac{1}{\frac{1}{sC}}$$

$$= \frac{1}{R} + \frac{1}{sL} + sC$$

i.e. $$\mathbf{Y(s) = \left(\frac{1}{50} + \frac{1}{5 \times 10^{-3}s} + 0.04 \times 10^{-6}s\right) S}$$

or $$Y(s) = \frac{1}{s}\left(\frac{s}{50} + \frac{1}{5 \times 10^3} + 0.04 \times 10^{-6}s^2\right) S$$

$$= \frac{0.04 \times 10^{-6}}{s}\left(\frac{s}{50(0.04 \times 10^{-6})} + \frac{1}{(5 \times 10^{-3})(0.04 \times 10^{-6})} + s^2\right) S$$

i.e. $$\mathbf{Y(s) = \frac{4 \times 10^{-8}}{s}(s^2 + 5 \times 10^5 s + 5 \times 10^9)\ S}$$

Kirchhoff's laws in the s-domain

Kirchhoff's* current and voltage laws may be applied to currents and voltages in the s-domain just as they can with normal time domain currents and voltages. To solve circuits in the s-domain using Kirchhoff's laws the procedure is:

(i) change the time domain circuit to an s-domain circuit,

(ii) apply Kirchhoff's laws in terms of s,

(iii) solve the equation to obtain the Laplace transform of the unknown quantity and

(iv) determine the inverse Laplace transform after rearranging into a form that can be recognized in a table of standard transforms.

This procedure is demonstrated in the following problems.

Problem 30. Determine an expression for (a) the current i through and (b) the voltage v_c across the capacitor for the circuit shown in Figure 48.17, after the switch is closed with a supply step voltage of V volts. Assume that the capacitor is initially uncharged.

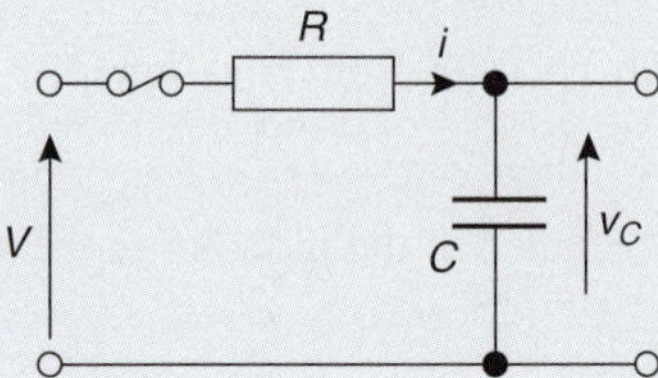

Figure 48.17

(a) Using the above procedure:

(i) In the s-domain the circuit impedance, $Z(s) = R + (1/sC)$ and the step input voltage is (V/s) and the circuit is as shown in Figure 48.18.

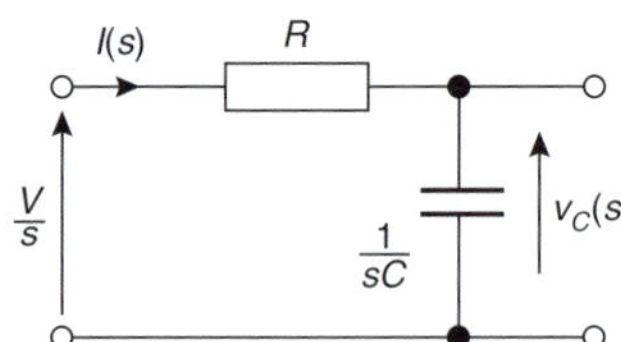

Figure 48.18

(ii) Applying Kirchhoff's voltage law:

$$\frac{V}{s} = RI(s) + v_c(s) \quad (35)$$

$$= RI(s) + \left(\frac{1}{sC}\right) I(s)$$

i.e. $$\frac{V}{s} = I(s)\left(R + \frac{1}{sC}\right)$$

*Who was **Kirchhoff**? For image and resume of Kirchhoff, see page 220. To find out more go to **www.routledge.com/cw/bird**

(iii) Rearranging gives:

$$I(s)=\frac{V/s}{\left(R+\frac{1}{sC}\right)}=\frac{V/s}{R\left(1+\frac{1}{RsC}\right)}$$

$$=\frac{V}{sR\left(1+\frac{1}{RsC}\right)}$$

i.e. $I(s)=\dfrac{V}{R\left(s+\frac{1}{RC}\right)}$ (36)

(iv) Hence current, $i=\mathscr{L}^{-1}\{I(s)\}$

$$=\mathscr{L}^{-1}\left\{\frac{V}{R\left(s+\frac{1}{RC}\right)}\right\}$$

$$=\frac{V}{R}\mathscr{L}^{-1}\left\{\frac{1}{s+\frac{1}{RC}}\right\}$$

since $\mathscr{L}^{-1}\left\{\frac{1}{s-a}\right\}=e^{at}$

then $\mathscr{L}^{-1}\left\{\frac{1}{s+a}\right\}=e^{-at}$ from 3 of Table 48.1. Hence

$$\textbf{current, } \boldsymbol{i=\frac{V}{R}e^{-(1/RC)t}=\frac{V}{R}e^{-t/RC}}$$

as previously obtained in equation (6), page 767.

(b) From equation (35), $v_c(s)=\dfrac{V}{s}-RI(s)$ and from equation (36),

$$v_c(s)=\frac{V}{s}-R\left(\frac{V}{R\left(s+\frac{1}{RC}\right)}\right)$$

$$=\frac{V}{s}-\frac{V}{\left(s+\frac{1}{RC}\right)}=V\left(\frac{1}{s}-\frac{1}{s+\frac{1}{RC}}\right)$$

Hence

$$v_c=\mathscr{L}^{-1}\{v_c(s)\}=\mathscr{L}^{-1}\left\{V\left(\frac{1}{s}-\frac{1}{s+\frac{1}{RC}}\right)\right\}$$

i.e. $\boldsymbol{v_c=V(1-e^{-t/RC})}$ from 2 and 3 of Table 48.1, as previously obtained in equation (4), page 766.

Alternatively, current, $i=C\dfrac{dv_c}{dt}$, hence

$$v_c=\int_0^t\frac{i}{C}dt=\int_0^t\frac{\frac{V}{R}e^{-t/RC}}{C}dt=\frac{V}{RC}\left[\frac{e^{-t/RC}}{\frac{-1}{RC}}\right]_0^t$$

$$=-V[e^{-t/RC}]_0^t=-V[e^{-t/RC}-e^0]$$

$$=-V[e^{-t/RC}-1]$$

i.e. $\boldsymbol{V_c=V(1-e^{-t/RC})}$

Problem 31. In the R–C series circuit shown in Figure 48.19, a ramp voltage V is applied to the input. Determine expressions for (a) current, i, and (b) capacitor voltage, v_c

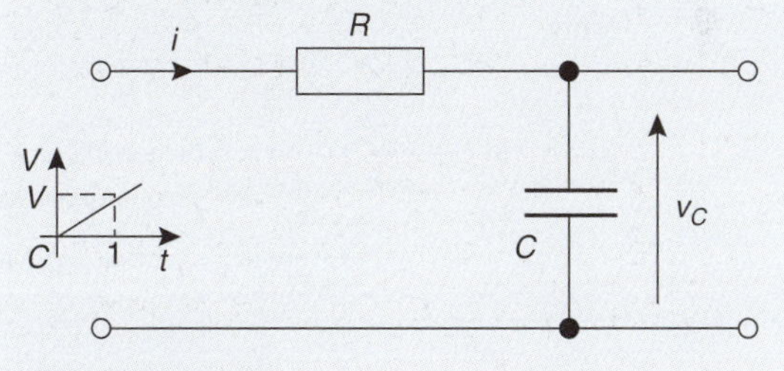

Figure 48.19

(a) Using the procedure:

(i) The time domain circuit of Figure 48.19 is changed to the s-domain as shown in Figure 48.20, where the ramp function is (V/s^2) from 7 of Table 48.1.

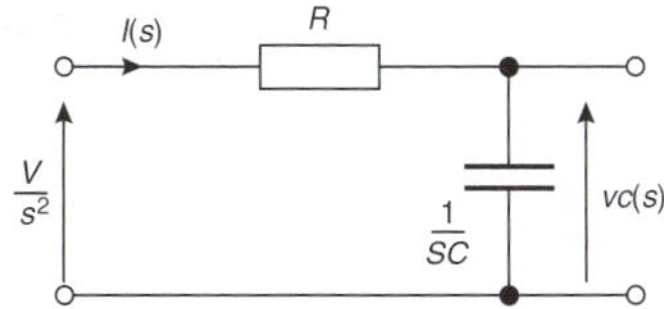

Figure 48.20

(ii) Applying Kirchhoff's voltage law gives:

$$\frac{V}{s^2}=RI(s)+\left(\frac{1}{sC}\right)I(s)=I(s)\left(R+\frac{1}{sC}\right)$$

Part 4

(iii) Hence

$$I(s)=\frac{V}{s^2\left(R+\frac{1}{sC}\right)}=\frac{V}{\frac{s^2}{sC}(RsC+1)}$$

$$=\frac{VC}{s(1+sRC)}$$

Using partial fractions:

$$\frac{VC}{s(1+sRC)}=\frac{A}{s}+\frac{B}{(1+sRC)}$$

$$=\frac{A(1+sRC)+Bs}{s(1+sRC)}$$

Thus $VC=A(1+sRC)+Bs$

When $s=0$ $\quad VC=A+0$ i.e. $A=VC$

When $s=\frac{-1}{RC}$ $\quad VC=0+B\left(\frac{-1}{RC}\right)$

i.e. $\quad B=-VC^2R$

Hence

$$I(s)=\frac{VC}{s(1+sRC)}=\frac{A}{s}+\frac{B}{(1+sRC)}$$

$$=\frac{VC}{s}+\frac{-VC^2R}{(1+sRC)}$$

$$=\frac{VC}{s}-\frac{VC^2R}{RC\left(\frac{1}{RC}+s\right)}$$

$$=\frac{VC}{s}-\frac{VC}{\left(s+\frac{1}{RC}\right)} \qquad (37)$$

(iv) Current,

$$i=\mathcal{L}^{-1}\left\{\frac{VC}{s}-\frac{VC}{\left(s+\frac{1}{RC}\right)}\right\}$$

$$=VC\mathcal{L}^{-1}\left\{\frac{1}{s}\right\}-VC\mathcal{L}^{-1}\left\{\frac{1}{s+\frac{1}{RC}}\right\}$$

$=VC-VC\mathrm{e}^{-t/RC}$ from 2 and 3 of Table 48.1

i.e. **current, $\boldsymbol{i=VC(1-\mathrm{e}^{-t/RC})}$**

(b) Capacitor voltage, $v_c(s)=I(s)\left(\frac{1}{sC}\right)$

$$=\frac{\frac{VC}{s}-\frac{VC}{s+\frac{1}{RC}}}{sC}$$

from equation (37)

$$=\frac{V}{s^2}-\frac{V}{s\left(s+\frac{1}{RC}\right)}$$

Using partial fractions:

$$\frac{V}{s\left(s+\frac{1}{RC}\right)}=\frac{A}{s}+\frac{B}{s+\frac{1}{RC}}$$

$$=\frac{A\left(s+\frac{1}{RC}\right)+Bs}{s\left(s+\frac{1}{RC}\right)}$$

hence $\quad V=A\left(s+\frac{1}{RC}\right)+Bs$

When $s=0$

$V=A\left(\frac{1}{RC}\right)+0$ from which, $A=VCR$

When $s=-\frac{1}{RC}$

$V=0+B\left(-\frac{1}{RC}\right)$ from which, $B=-VCR$

Thus

$$v_c(s)=\frac{V}{s^2}-\frac{V}{s\left(s+\frac{1}{RC}\right)}$$

$$=\frac{V}{s^2}-\left[\frac{VCR}{s}-\frac{VCR}{\left(s+\frac{1}{RC}\right)}\right]$$

$$=\frac{V}{s^2}-\frac{VCR}{s}+\frac{VCR}{\left(s+\frac{1}{RC}\right)}$$

Thus, capacitor voltage,

$$v_c=\mathcal{L}^{-1}\left\{\frac{V}{s^2}-\frac{VCR}{s}+\frac{VCR}{\left(s+\frac{1}{RC}\right)}\right\}$$

Part 4

$$= Vt - VCR + VCR\,e^{-t/RC}$$

from 7, 2 and 3 of Table 48.1

i.e. $\boldsymbol{v_c = Vt - VCR(1 - e^{-t/RC})}$

Problem 32. Determine for the R–L series circuit shown in Figure 48.21 expressions for current i, inductor voltage v_L and resistor voltage v_R when a step voltage V is applied to the input terminals.

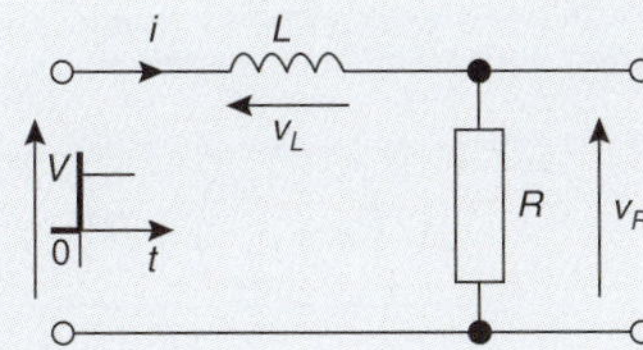

Figure 48.21

Using the procedure:

(i) The s-domain circuit is shown in Figure 48.22.

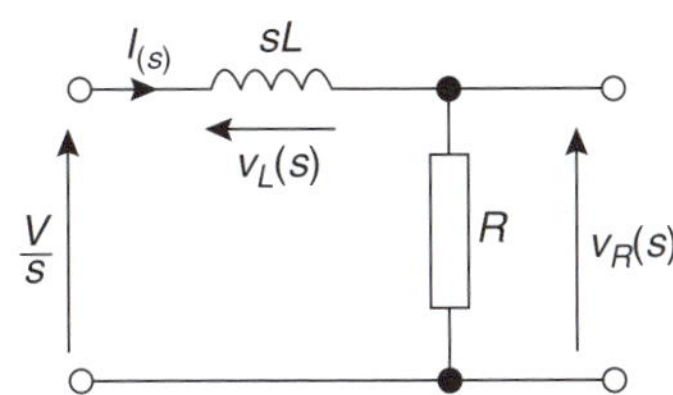

Figure 48.22

(ii) Using Kirchhoff's voltage law:

$$\frac{V}{s} = I(s)(sL) + I(s)R$$

(iii) Current $I(s) = \dfrac{V/s}{R + sL} = \dfrac{V}{s(R + sL)}$

$$= \frac{V}{sL\left(s + \frac{R}{L}\right)} = \frac{V/L}{s\left(s + \frac{R}{L}\right)}$$

Using partial fractions:

$$\frac{V/L}{s\left(s+\frac{R}{L}\right)} = \frac{A}{s} + \frac{B}{\left(s+\frac{R}{L}\right)} = \frac{A\left(s+\frac{R}{L}\right) + Bs}{s\left(s+\frac{R}{L}\right)}$$

Hence $\quad \dfrac{V}{L} = A\left(s + \dfrac{R}{L}\right) + Bs$

When $s = 0$: $\quad \dfrac{V}{L} = A\left(\dfrac{R}{L}\right) + 0,$

from which $\quad A = \dfrac{V}{R}$

When $s = -\dfrac{R}{L}$: $\quad \dfrac{V}{L} = 0 + B\left(-\dfrac{R}{L}\right)$, from which $B = -\dfrac{V}{R}$

$$\text{Hence } I(s) = \frac{V/L}{s\left(s+\frac{R}{L}\right)} = \frac{V/R}{s} - \frac{V/R}{\left(s+\frac{R}{L}\right)} \qquad (38)$$

(iv) Current $i = \mathcal{L}^{-1}\left\{\dfrac{V/R}{s} - \dfrac{V/R}{\left(s+\frac{R}{L}\right)}\right\}$

$$= \frac{V}{R} - \frac{V}{R}e^{-\frac{Rt}{L}}$$

from 2 and 3 of Table 48.1

i.e. $\boldsymbol{i = \dfrac{V}{R}(1 - e^{(-Rt/L)})}$

as previously obtained in equation (13), page 769, and in Problem 25, page 783.

In the s-domain, inductor voltage

$$v_L(s) = I(s)(sL)$$

$$= (sL)\left[\frac{V/R}{s} - \frac{V/R}{\left(s+\frac{R}{L}\right)}\right]$$

from equation (38)

$$= \frac{VL}{R} - \frac{VL}{R}\left(\frac{s}{\left(s+\frac{R}{L}\right)}\right)$$

$\dfrac{s}{\left(s+\frac{R}{L}\right)}$ needs to be divided out:

$$\begin{array}{r} 1 \\ \left(s+\frac{R}{L}\right)\overline{)\,s\qquad} \\ s+\frac{R}{L} \\ \hline -\frac{R}{L} \end{array}$$

Thus $\dfrac{s}{\left(s+\frac{R}{L}\right)} \equiv 1 - \dfrac{R/L}{\left(s+\frac{R}{L}\right)}$

$$\text{Hence } v_L(s) = \frac{VL}{R} - \frac{VL}{R}\left[1 - \frac{R/L}{\left(s+\frac{R}{L}\right)}\right]$$

$$= \frac{VL}{R} - \frac{VL}{R} + \frac{VL}{R}\left[\frac{R/L}{\left(s+\frac{R}{L}\right)}\right]$$

$$= \frac{VL}{R}\frac{R}{L}\left[\frac{1}{\left(s+\frac{R}{L}\right)}\right] = V\left[\frac{1}{\left(s+\frac{R}{L}\right)}\right]$$

$$\text{Thus inductor voltage } v_L = \mathscr{L}^{-1}\left\{V\left[\frac{1}{s+\frac{R}{L}}\right]\right\}$$

$$= V\mathscr{L}^{-1}\left\{\frac{1}{s+\frac{R}{L}}\right\}$$

i.e. $\boldsymbol{v_L = V\mathrm{e}^{-Rt/L}}$

from 3 of Table 48.1, as previously obtained in equation (15), page 769.

Since $V = v_L + v_R$ in Figure 48.21,

resistor voltage, $v_R = V - v_L = V - V\mathrm{e}^{-Rt/L}$

$$= \boldsymbol{V(1-\mathrm{e}^{-Rt/L})}$$

as previously obtained in equation (14), page 769.

Problem 33. For the circuit of Figure 48.21 of Problem 32, a ramp of V volts/s is applied to the input terminals, instead of a step voltage. Determine expressions for current i, inductor v_L and resistor voltage v_R

(i) The circuit for the s-domain is shown in Figure 48.23.

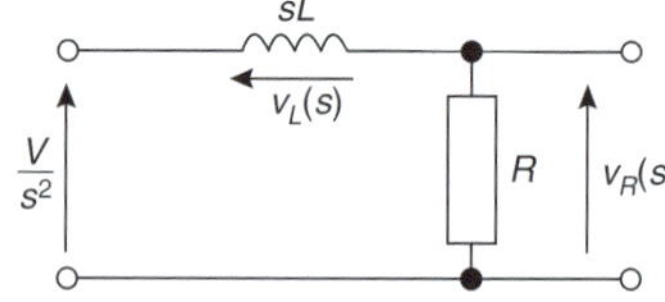

Figure 48.23

(ii) From Kirchhoff's voltage law:

$$\frac{V}{s^2} = I(s)(R+sL)$$

(iii) Current $I(s) = \dfrac{V}{s^2(R+sL)} = \dfrac{V}{s^2L\left(s+\frac{R}{L}\right)}$

$$= \frac{V/L}{s^2\left(s+\frac{R}{L}\right)}$$

Using partial fractions:

$$\frac{V/L}{s^2\left(s+\frac{R}{L}\right)} = \frac{A}{s} + \frac{B}{s^2} + \frac{C}{s+\frac{R}{L}}$$

$$= \frac{As\left(s+\frac{R}{L}\right) + B\left(s+\frac{R}{L}\right) + Cs^2}{s^2\left(s+\frac{R}{L}\right)}$$

from which,

$$\frac{V}{L} = As\left(s+\frac{R}{L}\right) + B\left(s+\frac{R}{L}\right) + Cs^2$$

when $s = 0$,

$$\frac{V}{L} = 0 + B\left(\frac{R}{L}\right) + 0, \text{ from which } B = \frac{V}{R}$$

when $s = -\dfrac{R}{L}$,

$$\frac{V}{L} = 0 + 0 + C\left(-\frac{R}{L}\right)^2, \text{ from which}$$

$$C = \frac{V}{L}\left(\frac{L^2}{R^2}\right) = \frac{VL}{R^2}$$

Equating s^2 coefficients: $0 = A + C$, from which

$$A = -C = -\frac{VL}{R^2}$$

$$\text{Thus } I(s) = \frac{V/L}{s^2\left(s+\frac{R}{L}\right)}$$

$$= \frac{A}{s} + \frac{B}{s^2} + \frac{C}{s+\frac{R}{L}}$$

$$= \frac{-VL/R^2}{s} + \frac{V/R}{s^2} + \frac{VL/R^2}{\left(s+\frac{R}{L}\right)} \qquad (39)$$

(iv) Current,

$$i = \mathscr{L}^{-1}\{I(s)\}$$

$$= -\frac{VL}{R^2}\mathscr{L}^{-1}\left\{\frac{1}{s}\right\} + \frac{V}{R}\mathscr{L}^{-1}\left\{\frac{1}{s^2}\right\} + \frac{VL}{R^2}\mathscr{L}^{-1}\left\{\frac{1}{s+\frac{R}{L}}\right\}$$

$$= -\frac{VL}{R^2}(1) + \frac{V}{R}(t) + \frac{VL}{R^2}(e^{(-Rt/L)})$$

from 2, 7 and 3 of Table 48.1

i.e. $\boldsymbol{i = \frac{V}{R}t - \frac{VL}{R^2}(1 - e^{(-Rt/L)})}$

Inductor voltage,

$$v_L(s) = I(s)(sL) = sL\left[\frac{-VL/R^2}{s} + \frac{V/R}{s^2} + \frac{VL/R^2}{s+\frac{R}{L}}\right]$$

from equation (39) above

$$= -\frac{VL^2}{R^2} + \frac{VL/R}{s} + \frac{(VL^2/R^2)s}{s+\frac{R}{L}}$$

$$= -\frac{VL^2}{R^2} + \frac{VL}{sR} + -\frac{VL^2}{R^2}\left(\frac{s}{s+\frac{R}{L}}\right)$$

The division $\frac{s}{s+\frac{R}{L}}$ was shown on page 789,

and is equivalent to $1 - \frac{R/L}{s+\frac{R}{L}}$

Hence $v_L(s) = -\frac{VL^2}{R^2} + \frac{VL}{sR} + \frac{VL^2}{R^2}\left[1 - \frac{R/L}{s+\frac{R}{L}}\right]$

$$= \frac{VL}{sR} - \frac{VL^2}{R^2}\frac{R}{L}\left(\frac{1}{s+\frac{R}{L}}\right)$$

$$= \frac{VL}{sR} - \frac{VL}{R}\left(\frac{1}{s+\frac{R}{L}}\right)$$

Thus $v_L = \mathscr{L}^{-1}\{v_L(s)\}$

$$= \frac{VL}{R}\mathscr{L}^{-1}\left\{\frac{1}{s}\right\} - \frac{VL}{R}\mathscr{L}^{-1}\left\{\frac{1}{s+\frac{R}{L}}\right\}$$

i.e. inductor voltage, $\boldsymbol{v_L} = \frac{VL}{R} - \frac{VL}{R}e^{-(Rt/L)}$

$$\boldsymbol{= \frac{VL}{R}(1 - e^{(-Rt/L)})}$$

Resistor voltage,

$$v_R(s) = I(s)R = R\left[\frac{-VL/R^2}{s} + \frac{V/R}{s^2} + \frac{VL/R^2}{\left(s+\frac{R}{L}\right)}\right]$$

from equation (39)

$$= -\frac{VL}{sR} + \frac{V}{s^2} + \frac{VL}{R\left(s+\frac{R}{L}\right)}$$

hence $v_R = \mathscr{L}^{-1}\left\{-\frac{VL}{sR} + \frac{V}{s^2} + \frac{VL}{R\left(s+\frac{R}{L}\right)}\right\}$

$$= -\frac{VL}{R} + Vt + \frac{VL}{R}e^{(-Rt/L)}$$

from 2, 7 and 3 Table 48.1

i.e. $\boldsymbol{v_R = Vt - \frac{VL}{R}(1 - e^{(-Rt/L)})}$

Problem 34. At time $t = 0$, a sinusoidal voltage $10\sin\omega t$ is applied to an L–R series circuit. Determine an expression for the current flowing.

(i) The circuit is shown in Figure 48.24 and the s-domain circuit is shown in Figure 48.25, the $10\sin\omega t$ input voltage becoming $10\left(\frac{\omega}{s^2+\omega^2}\right)$ in the s-domain from 5 of Table 48.1

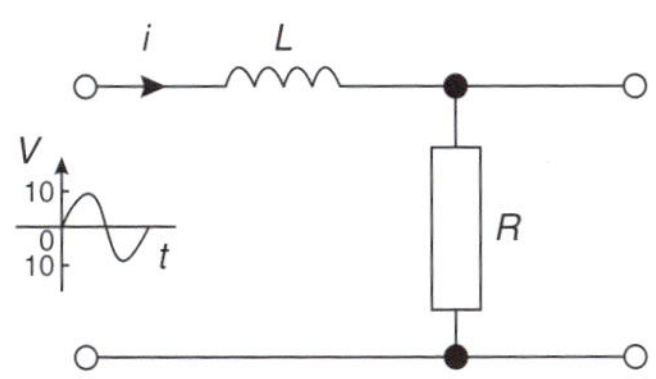

Figure 48.24

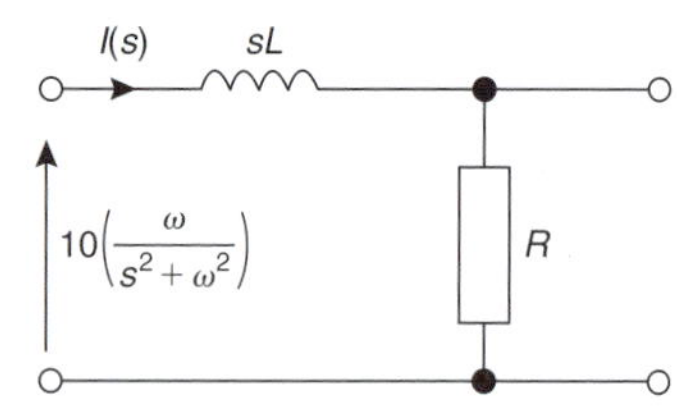

Figure 48.25

(ii) From Kirchhoff's voltage law:

$$\frac{10\omega}{s^2+\omega^2} = I(s)(sL) + I(s)R$$

(iii) Hence current, $I(s) = \dfrac{10\omega}{(s^2+\omega^2)(R+sL)}$

$$= \frac{10\omega}{(s^2+\omega^2)L\left(s+\frac{R}{L}\right)}$$

$$= \frac{10\omega/L}{(s^2+\omega^2)\left(s+\frac{R}{L}\right)}$$

Using partial fractions:

$$\frac{10\omega/L}{(s^2+\omega^2)\left(s+\frac{R}{L}\right)} = \frac{As+B}{(s^2+\omega^2)} + \frac{C}{\left(s+\frac{R}{L}\right)}$$

$$= \frac{(As+B)\left(s+\frac{R}{L}\right) + C(s^2+\omega^2)}{(s^2+\omega^2)\left(s+\frac{R}{L}\right)}$$

hence $\dfrac{10\omega}{L} = (As+B)\left(s+\dfrac{R}{L}\right) + C(s^2+\omega^2)$

When $s = -\dfrac{R}{L}$: $\dfrac{10\omega}{L} = 0 + C\left[\left(-\dfrac{R}{L}\right)^2 + \omega^2\right]$

from which, $C = \dfrac{10\omega}{L\left(\frac{R^2}{L^2}+\omega^2\right)} = \dfrac{10\omega}{\frac{L}{L^2}(R^2+L^2\omega)}$

$$= \frac{10\omega L}{(R^2+\omega^2 L^2)}$$

Equating s^2 coefficients,

$0 = A + C$, from which, $A = -C = -\dfrac{10\omega L}{(R^2+\omega^2 L^2)}$

Equating constant terms, $\dfrac{10\omega}{L} = B\left(\dfrac{R}{L}\right) + C\omega^2$

$$\frac{10\omega}{L} - C\omega^2 = B\left(\frac{R}{L}\right)$$

from which, $B = \dfrac{L}{R}\left(\dfrac{10\omega}{L} - C\omega^2\right)$

$$= \frac{10\omega}{R} - \frac{L\omega^2}{R}\left(\frac{10\omega L}{(R^2+\omega^2 L^2)}\right)$$

$$= \frac{10\omega(R^2+\omega^2L^2) - L\omega^2(10\omega L)}{R(R^2+\omega^2L^2)}$$

$$= \frac{10\omega R^2 + 10\omega^3L^2 - 10\omega^3L^2}{R(R^2+\omega^2L^2)}$$

$$= \frac{10\omega R}{(R^2+\omega^2L^2)}$$

Hence

$$I(s) = \frac{10\omega/L}{(s^2+\omega^2)\left(s+\frac{R}{L}\right)}$$

$$= \frac{As+B}{(s^2+\omega^2)} + \frac{C}{\left(s+\frac{R}{L}\right)}$$

$$= \frac{\left(\frac{-10\omega L}{R^2+\omega^2L^2}\right)s + \left(\frac{10\omega R}{R^2+\omega^2L^2}\right)}{(s^2+\omega^2)} + \frac{\left(\frac{10\omega L}{R^2+\omega^2L^2}\right)}{\left(s+\frac{R}{L}\right)}$$

$$= \frac{10\omega}{R^2+\omega^2L^2}\left[\frac{L}{\left(s+\frac{R}{L}\right)} - \frac{sL}{(s^2+\omega^2)} + \frac{R}{(s^2+\omega^2)}\right]$$

(iv) Current, $i = \mathscr{L}^{-1}\{I(s)\}$

$$= \boldsymbol{\frac{10\omega}{R^2+\omega^2L^2}\left\{Le^{(-Rt/L)} - L\cos\omega t + \frac{R}{\omega}\sin\omega t\right\}}$$

from 3, 6 and 5 of Table 48.1

Problem 35. In the series–parallel network shown in Figure 48.26, a 5 V step voltage is applied at the input terminals. Determine an expression to show how current i varies with time

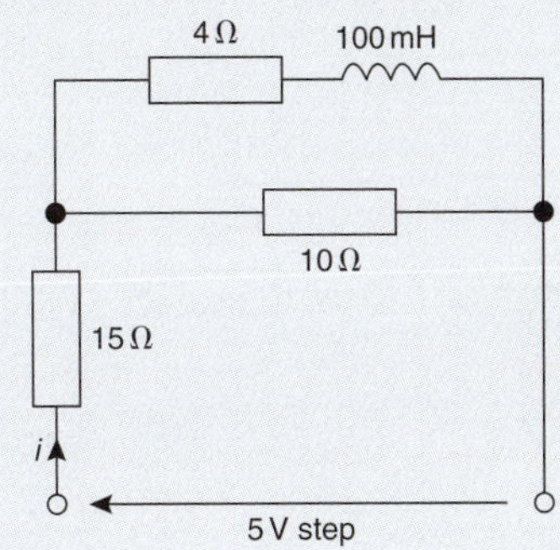

Figure 48.26

In the s-domain, $Z(s) = 15 + \dfrac{10(4+0.1s)}{10+4+0.1s}$

$$= 15 + \frac{40+s}{14+0.1s}$$

$$= \frac{15(14+0.1s)+(40+s)}{14+0.1s}$$

$$= \frac{210+1.5s+40+s}{14+0.1s}$$

$$= \frac{250+2.5s}{14+0.1s}$$

Since in the s-domain the input voltage is (V/s) then

$$I(s) = \frac{V(s)}{Z(s)} = \frac{5/s}{\left(\dfrac{250+2.5s}{14+0.1s}\right)} = \frac{5(14+0.1s)}{s(250+2.5s)}$$

$$= \frac{70+0.5s}{s(250+2.5s)}$$

$$= \frac{70}{s(250+2.5s)} + \frac{0.5s}{s(250+2.5s)}$$

$$= \frac{70}{2.5s(s+100)} + \frac{0.5}{2.5(s+100)}$$

i.e. $I(s) = \dfrac{28}{s(s+100)} + \dfrac{0.2}{(s+100)}$

Using partial fractions: $\dfrac{28}{s(s+100)} = \dfrac{A}{s} + \dfrac{B}{(s+100)}$

$$= \frac{A(s+100)+Bs}{s(s+100)}$$

from which $\quad 28 = A(s+100)+Bs$

When $s=0 \quad 28 = 100A$ and $A = 0.28$

When $s=-100 \quad 28 = 0 - 100B$ and $B = -0.28$

Hence $I(s) = \dfrac{0.28}{s} - \dfrac{0.28}{(s+100)} + \dfrac{0.2}{(s+100)}$

$$= \frac{0.28}{s} - \frac{0.08}{(s+100)}$$

and **current,** $\boldsymbol{i = \mathscr{L}^{-1}\{I(s)\} = 0.28 - 0.08e^{-100t}}$

from 2 and 3 of Table 48.1

Now try the following Practice Exercise

Practice Exercise 201 Circuit analysis using Laplace transforms (Answers on page 835)

1. Determine the impedance of a 2000 pF capacitor in the s-domain.
2. Determine the impedance of a 0.4 H inductor in series with a 50 Ω resistor in the s-domain.
3. Determine the circuit impedance in the s-domain for the following:
   (a) a resistor of 100 Ω in series with a 1 μF capacitor
   (b) an inductance of 10 mH, a resistance of 500 Ω and a capacitance of 400 nF in series
   (c) a 10 Ω resistance in parallel with a 10 mH inductor
   (d) a 10 mH inductor in parallel with a 1 μF capacitor.
4. An L–R–C network comprises a 20 Ω resistor, a 20 mH inductor and a 20 μF capacitor. Determine in the s-domain (a) the impedance when the components are connected in series and (b) the admittance when the components are connected in parallel.
5. A circuit consists of a 0.5 MΩ resistor in series with a 0.5 μF capacitor. Determine how the voltage across the capacitor varies with time when there is a step voltage input of 5 V. Assume the initial conditions are zero.

6. An exponential voltage, $V = 20e^{-50t}$ volts is applied to a series R–L circuit, where $R = 10\,\Omega$ and $L = 0.1$ H. If the initial conditions are zero, find the resulting current.

7. If in Problem 6 a supply of $20.2\sin(10t + \phi)$ volts is applied to the circuit, find the resulting current. Assume the circuit is switched on when $\phi = 0$

8. An R–C series network has (a) a step input voltage E volts and (b) a ramp voltage E volts/s, applied to the input. Use Laplace transforms to determine expressions for the current flowing in each case. Assume the capacitor is initially uncharged.

9. An R–L series network has (a) a step input of E volts, (b) a ramp input of 1 V/s, applied across it. Use Laplace transforms to develop expressions for the voltage across the inductance L in each case.
Assume that at time $t = 0$, current $i = 0$

10. A sinusoidal voltage $5\sin t$ volts is applied to a series R–L circuit. Assuming that at time $t = 0$, current $i = 0$, derive an expression for the current flowing.

11. Derive an expression for the growth of current through an inductive coil of resistance $20\,\Omega$ and inductance 2 H using Laplace transforms when a d.c. voltage of 30 V is suddenly applied to the coil.

Part 4

48.8 L–R–C series circuit using Laplace transforms

An L–R–C series circuit is shown in Figure 48.27 with a step input voltage V. In the s-domain, the circuit components are as shown in Figure 48.28 and if the step is applied at time $t = 0$, the s-domain supply voltage is V/s

Hence $\dfrac{V}{s} = I(s)\left(R + sL + \dfrac{1}{sC}\right)$

from which, current,

$$I(s) = \frac{V/s}{R + sL + (1/sC)} = \frac{V/s}{(1/s)(sR + s^2L + (1/C))}$$

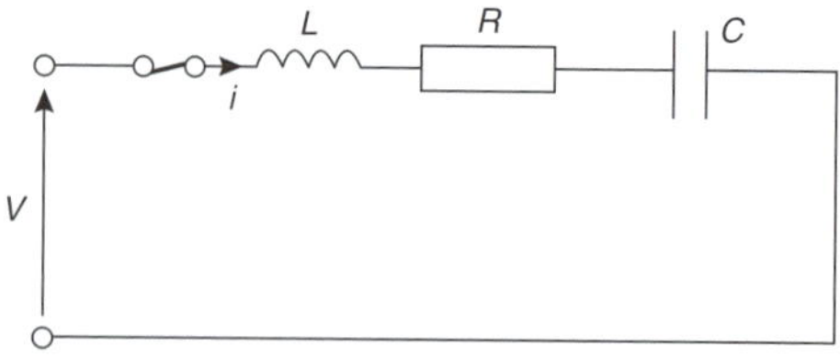

Figure 48.27

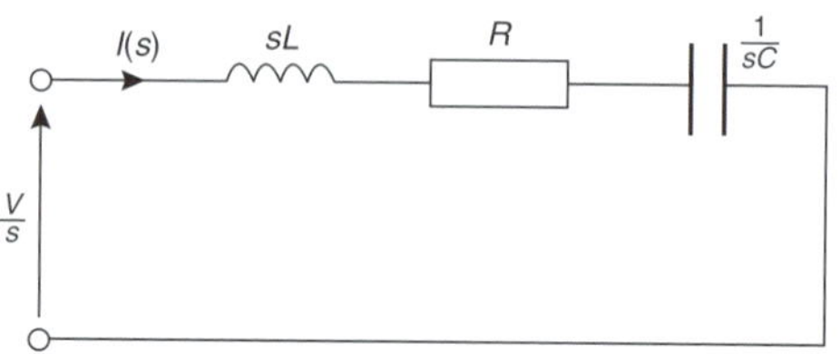

Figure 48.28

$$= \frac{V}{sR + s^2L + (1/C)} = \frac{V}{L(s^2 + s(R/L) + (1/LC))}$$

$$= \frac{V/L}{(s^2 + (R/L)s + (1/LC))}$$

The denominator is made into a perfect square (as in Problem 18):

Hence $I(s)$

$$= \frac{V/L}{\left\{s^2 + \frac{R}{L}s + \left(\frac{R}{2L}\right)^2\right\} + \left\{\frac{1}{LC} - \left(\frac{R}{2L}\right)^2\right\}}$$

$$= \frac{V/L}{\left(s + \frac{R}{2L}\right)^2 + \sqrt{\left(\frac{1}{LC} - \left(\frac{R}{2L}\right)^2\right)}^{\,2}} \tag{40}$$

or $I(s)$

$$= \frac{V/L}{\sqrt{\left(\frac{1}{LC} - \left(\frac{R}{2L}\right)^2\right)}} \; \frac{\sqrt{\left(\frac{1}{LC} - \left(\frac{R}{2L}\right)^2\right)}}{\left(s + \frac{R}{2L}\right)^2 + \sqrt{\left(\frac{1}{LC} - \left(\frac{R}{2L}\right)^2\right)}^{\,2}}$$

and current, $i = \mathscr{L}^{-1}\{I(s)\}$

From 13 of Table 48.1,

$$\mathscr{L}^{-1}\left\{\frac{\omega}{(s+a)^2 + \omega^2}\right\} = e^{-at}\sin\omega t, \text{ hence}$$

current,

$$i = \frac{V/L}{\sqrt{\left(\frac{1}{LC} - \left(\frac{R}{2L}\right)^2\right)}} e^{(-R/2L)t} \sin\sqrt{\left(\frac{1}{LC} - \left(\frac{R}{2L}\right)^2\right)} t \qquad (41)$$

Problem 36. For the circuit shown in Figure 48.29 produce an equation which shows how the current varies with time. Assume zero initial conditions when the switch is closed.

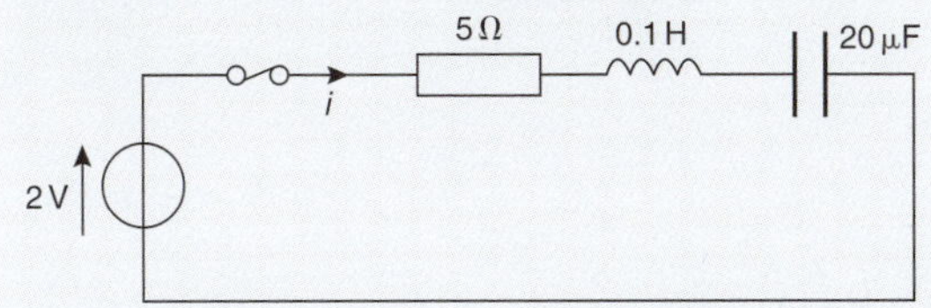

Figure 48.29

In the s-domain, applying Kirchhoff's voltage law gives:

$$\frac{2}{s} = I(s)\left[5 + 0.1s + \frac{1}{20 \times 10^{-6}s}\right]$$

and current

$$I(s) = \frac{2}{s\left(5 + 0.1s + \frac{5 \times 10^4}{s}\right)}$$

$$= \frac{2}{5s + 0.1s^2 + 5 \times 10^4}$$

$$= \frac{2}{0.1\left(s^2 + \frac{5}{0.1}s + \frac{5 \times 10^4}{0.1}\right)}$$

$$= \frac{20}{(s^2 + 50s + 5 \times 10^5)}$$

$$= \frac{20}{\{s^2 + 50s + (25)^2\} + \{5 \times 10^5 - (25)^2\}}$$

$$= \frac{20}{(s+25)^2 + \sqrt{(499375)}^2}$$

$$= \frac{20}{\sqrt{(499375)}} \frac{\sqrt{(499375)}}{(s+25)^2 + \sqrt{(499375)}^2}$$

$$= \frac{20}{706.7} \frac{706.7}{(s+25)^2 + (706.7)^2}$$

Hence current,

$$i = \mathscr{L}^{-1}\{I(s)\} = 0.0283e^{-25t} \sin 706.7t,$$

from 13 of Table 48.1,

i.e. $\boldsymbol{i = 28.3e^{-25t} \sin 706.7t}$ **mA**

Damping

The expression for current $I(s)$ in equation (40) has four possible solutions, each dependent on the values of R, L and C.

Solution 1. When $R = 0$, the circuit is **undamped** and, from equation (40),

$$I(s) = \frac{V/L}{\left(s^2 + \frac{1}{LC}\right)}$$

From Chapter 31, at resonance, $\omega_r = \frac{1}{LC}$ hence

$$I(s) = \frac{V/L}{(s^2 + \omega_r^2)} = \frac{V}{\omega_r L} \frac{\omega_r}{(s^2 + \omega_r^2)}$$

Hence **current**, $\boldsymbol{i = \mathscr{L}^{-1}\{I(s)\} = \frac{V}{\omega_r L} \sin \omega_r t}$ from 5 of Table 48.1 which is a sine wave of amplitude $\frac{V}{\omega_r L}$ and angular velocity ω_r rad/s.

This is shown by curve A in Figure 48.30.

Solution 2. When $\left(\frac{R}{2L}\right)^2 < \frac{1}{LC}$, the circuit is **under-damped** and the current i is as in equation (41). The current is oscillatory which is decaying exponentially. This is shown by curve B in Figure 48.30.

Solution 3. When $\left(\frac{R}{2L}\right)^2 = \frac{1}{LC}$, the circuit is **critically damped** and from equation (40),

$$I(s) = \frac{V/L}{\left(s + \frac{R}{2L}\right)^2}$$

and **current**,

$$i = \mathscr{L}^{-1}\left\{\frac{V/L}{\left(s + \frac{R}{2L}\right)^2}\right\}$$

$$= \frac{V}{L} te^{-(Rt/2L)} \qquad (42)$$

from 12 of Table 48.1

Part 4

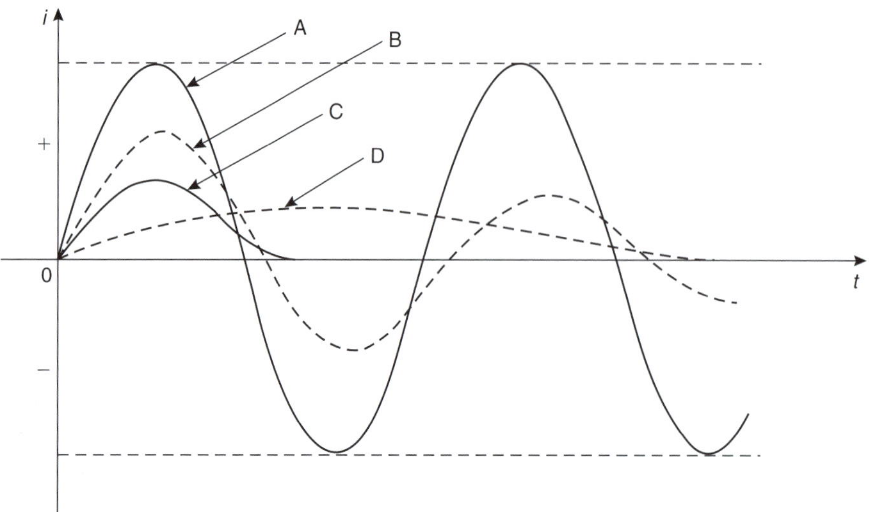

Figure 48.30

The current is non-oscillatory and is as shown in curve C in Figure 48.30.

Solution 4. When $\left(\frac{R}{2L}\right)^2 > \frac{1}{LC}$, the circuit is **over-damped** and from equation (40),

$$I(s) = \frac{V/L}{\left(s+\frac{R}{2L}\right)^2 - \sqrt{\left[\left(\frac{R}{2L}\right)^2 - \frac{1}{LC}\right]}^2}$$

$$= \frac{V/L}{\sqrt{\left[\left(\frac{R}{2L}\right)^2 - \frac{1}{LC}\right]}} \frac{\sqrt{\left[\left(\frac{R}{2L}\right)^2 - \frac{1}{LC}\right]}}{\left(s+\frac{R}{2L}\right)^2 - \sqrt{\left[\left(\frac{R}{2L}\right)^2 - \frac{1}{LC}\right]}^2}$$

and **current** $\boldsymbol{i = \mathscr{L}^{-1}\{I(s)\}}$

$$= \frac{V}{L\sqrt{\left[\left(\frac{R}{2L}\right)^2 - \frac{1}{LC}\right]}} e^{-(Rt/2L)} \sinh \sqrt{\left[\left(\frac{R}{2L}\right)^2 - \frac{1}{LC}\right]}t$$

from 15 of Table 48.1

This curve is shown as curve D in Figure 48.30.

Problem 37. An L–R–C series circuit contains a coil of inductance 1 H and resistance 8 Ω and a capacitor of capacitance 50 μF. Assuming current $i = 0$ at time $t = 0$, determine (a) the state of damping in the circuit and (b) an expression for the current when a step voltage of 10 V is applied to the circuit.

(a) $\left(\frac{R}{2L}\right)^2 = \left(\frac{8}{2(1)}\right)^2 = 16$ and

$$\frac{1}{LC} = \frac{1}{(1)(50\times10^{-6})} = 20\,000$$

Since $\left(\frac{R}{2L}\right)^2 < \frac{1}{LC}$ the circuit is **underdamped**

(b) When $\left(\frac{R}{2L}\right)^2 < \frac{1}{LC}$, equation (41) applies,

$$\text{i.e. } i = \frac{V/L}{\sqrt{\frac{1}{LC} - \left(\frac{R}{2L}\right)^2}} e^{-(Rt/2L)} \sin \sqrt{\left[\frac{1}{LC} - \left(\frac{R}{2L}\right)^2\right]}t$$

$$= \frac{10/1}{\sqrt{(20\,000-16)}} e^{-4t} \sin \sqrt{(20\,000-16)}t$$

i.e. $\boldsymbol{i = 0.0707e^{-4t} \sin 141.4t}$ **A**

Problem 38. Values of R, L and C in a series R–L–C circuit are $R = 100$ Ω, $L = 423$ mH and $C = 169.2$ μF. A step voltage of 2 V is applied to the circuit. Assuming current $i = 0$ at the instant of applying the step, determine (a) the state of damping, and (b) an expression for current i.

(a) $\left(\frac{R}{2L}\right)^2 = \left[\frac{100}{2(0.423)}\right]^2 = 13\,972$

and $\frac{1}{LC} = \frac{1}{(0.423)(169.2\times10^{-6})} = 13\,972$

Since $\left(\frac{R}{2L}\right)^2 = \frac{1}{LC}$ the circuit is **critically damped**.

(b) From equation (42), current

$$i = \frac{V}{L} t e^{-(Rt/2L)} = \frac{2}{0.423} t e^{(-100/2(0.423))t}$$

i.e. $\boldsymbol{i = 4.73\, te^{-118.2t}}$ **A**

48.9 Initial conditions

In an L–R–C circuit it is possible, at time $t = 0$, for an inductor to carry a current or a capacitor to possess a charge.

(a) For an inductor: $v_L = L\frac{di}{dt}$

The Laplace transform of this equation is:

$$v_L(s) = L[sI(s) - i(0)]$$

If, say, $i(0) = I_0$ then $v_L(s) = sLI(s) - LI_0$ (43)

The p.d. across the inductor in the s-domain is given by: $(sL)I(s)$

Equation (43) would appear to comprise two series elements, i.e.

$v_L(s) =$ (p.d. across L) + (voltage generator of $-LI_0$)

An inductor can thus be considered as an impedance sL in series with an independent voltage source of $-LI_0$ as shown in Figure 48.31.

Series equivalent circuit for an inductor in the s-domain

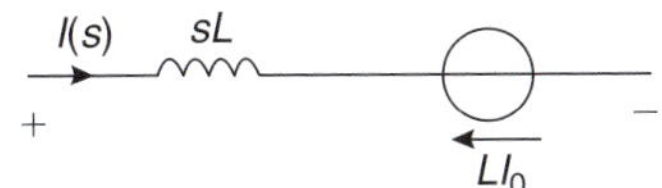

Figure 48.31

Transposing equation (43) for $I(s)$ gives:

$$I(s) - \frac{v_L(s) + LI_0}{sL} = \frac{v_L(s)}{sL} + \frac{I_0}{s}$$

Thus, alternatively, an inductor can be considered to be an impedance sL in parallel with an independent current source I_0/s

i.e. $I(s) =$ (current through L) + $\left(\text{current source } \frac{I_0}{s}\right)$

as shown in Figure 48.32.

Parallel equivalent circuit for an inductor in the s-domain

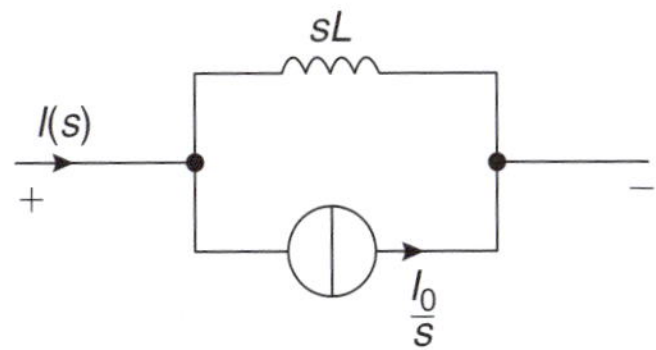

Figure 48.32

Problem 39. An L–R series circuit has a step voltage V applied to its input terminals. If after a period of time the step voltage is removed and replaced by a short-circuit, determine the expression for the current transient.

The L–R circuit with a step voltage applied to the input is shown in Figure 48.33.

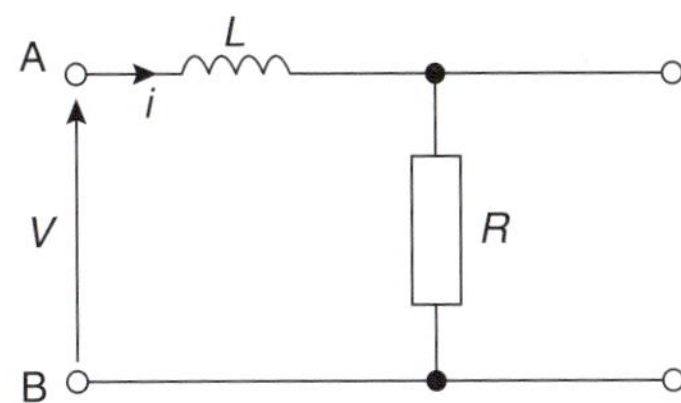

Figure 48.33

Using Kirchhoff's voltage law:

$$V = iR + L\frac{di}{dt} \qquad (44)$$

If the step voltage is removed the circuit of Figure 48.34 results.

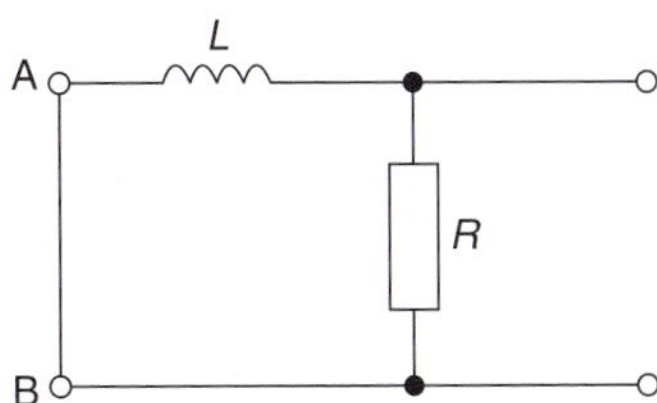

Figure 48.34

The s-domain circuit is shown in Figure 48.35, where the inductor is considered as an impedance sL in series with a voltage source LI_0 with its direction as shown.

If $V = 0$ in equation (44) then $0 = iR + L\dfrac{di}{dt}$

i.e. $0 = I(s)R + L[sI(s) - I_0]$

$= I(s)R + sLI(s) - LI_0$ which verifies Figure 48.35

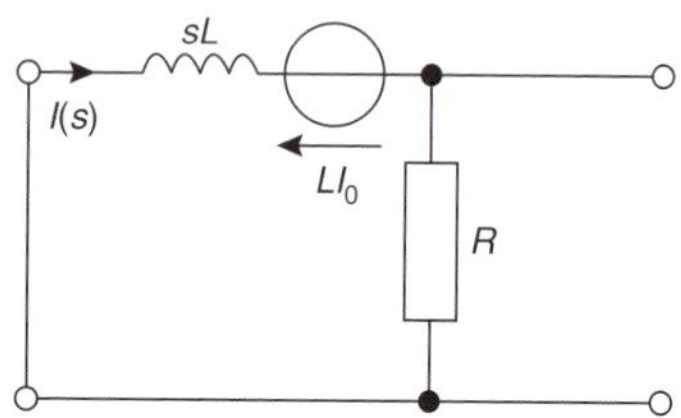

Figure 48.35

In this case $I_0 = (V/R)$, the steady state current before the step voltage was removed.

Hence $\quad 0 = I(s)R + sLI(s) - L\dfrac{V}{R}$

i.e. $\quad \dfrac{LV}{R} = I(s)(R + sL)$

and $\quad I(s) = \dfrac{VL/R}{R + sL}$

$$= \frac{VL/R}{L(s + (R/L))} = \frac{V/R}{(s + (R/L))}$$

Hence **current** $\boldsymbol{i} = \mathscr{L}^{-1}\left\{\dfrac{V/R}{s + (R/L)}\right\} = \boldsymbol{\dfrac{V}{R}e^{(-Rt/L)}}$

(b) For a capacitor: $i = C\dfrac{dv}{dt}$

The Laplace transform of this equation is:

$$I(s) = C[sV(s) - v(0)]$$

If, say, $v(0) = V_0$ then $\quad I(s) = CsV(s) - CV_0 \quad (45)$

Rearranging gives: $\quad CsV(s) = I(s) + CV_0$

from which $\quad V(s) = \dfrac{I(s)}{Cs} + \dfrac{CV_0}{Cs}$

$$= \left(\frac{1}{sC}\right)I(s) + \frac{V_0}{s}$$

i.e. $\quad V(s) = $ (p.d. across capacitor)

$$+\left(\text{voltage source } \frac{V_0}{s}\right) \quad (46)$$

as shown in Figure 48.36.

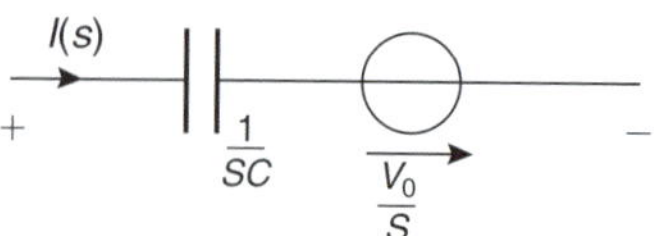

Figure 48.36

Thus the equivalent circuit in the s-domain for a capacitor with an initial voltage V_0 is a capacitor with impedance $(1/sC)$ in series with an impedance source (V_0/s). Alternatively, from equation (45),

$I(s) = sCV(s) - CV_0$

$=$ (current through C) + (a current source of $-CV_0$)

as shown in Figure 48.37.

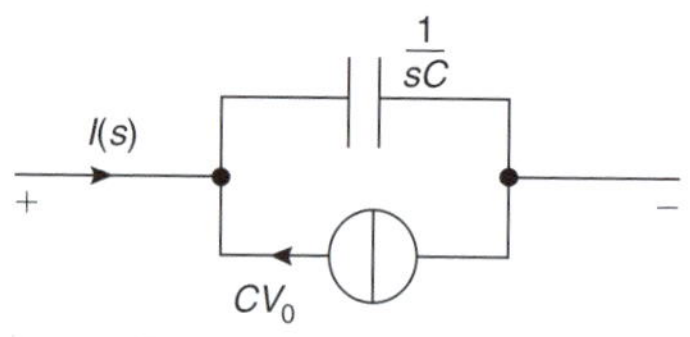

Figure 48.37

Problem 40. A C–R series circuit is shown in Figure 48.38. The capacitor C is charged to a p.d. of V_0 when it is suddenly discharged through the resistor R. Deduce how the current i and the voltage v vary with time.

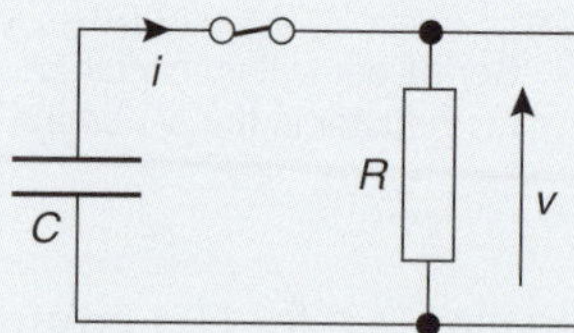

Figure 48.38

The s-domain equivalent circuit, from equation (46), is shown in Figure 48.39.

Applying Kirchhoff's voltage law:

$$\frac{V_0}{s} = I(s)\left(\frac{1}{sC}\right) + RI(s)$$

$$= I(s)\left(R + \frac{1}{sC}\right)$$

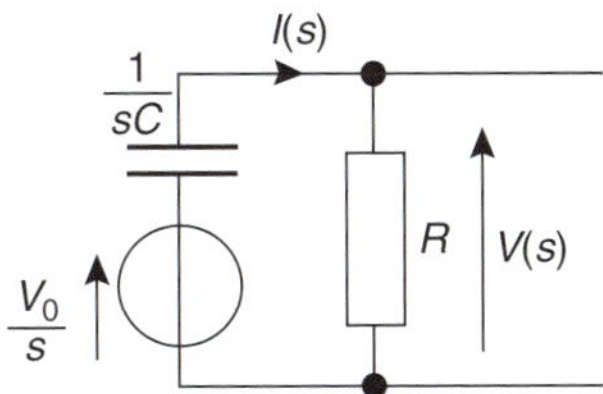

Figure 48.39

from which,

$$I(s) = \frac{V_0/s}{\left(R + \frac{1}{sC}\right)} = \frac{V_0}{s\left(R + \frac{1}{sC}\right)} = \frac{V_0}{sR + \frac{1}{C}}$$

$$= \frac{V_0}{R\left(s + \frac{1}{RC}\right)}$$

and current $i = \mathscr{L}^{-1}\{I(s)\} = \frac{V_0}{R}\mathscr{L}^{-1}\left\{\frac{1}{s + \frac{1}{RC}}\right\}$

$$= \frac{V_0}{R}e^{(-t/CR)}$$

i.e. $\boldsymbol{i = \frac{V_0}{R}e^{(-t/CR)}}$

Since $v = iR$, then $v = \left(\frac{V_0}{R}e^{(-t/CR)}\right)R$

i.e. $\boldsymbol{v = V_0 e^{(-t/CR)}}$

Problem 41. Derive an equation for current i flowing through the 1 kΩ resistor in Figure 48.40 when the switch is moved from x to y. Assume that the switch has been in position x for some time.

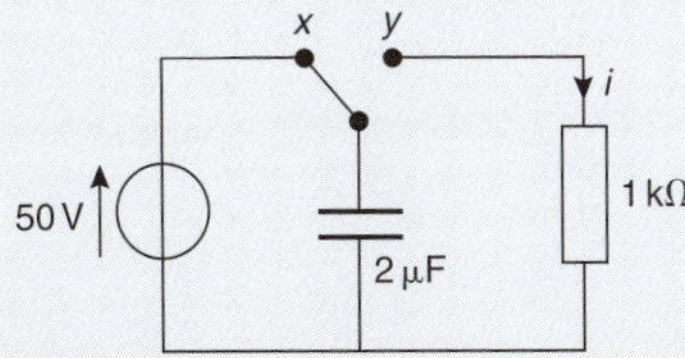

Figure 48.40

The 2 μF capacitor will have become fully charged to 50 V after a period of time. When the switch is changed from x to y the charged capacitor can be considered to be a voltage generator of voltage $(50/s)$. The s-domain circuit is shown in Figure 48.41.

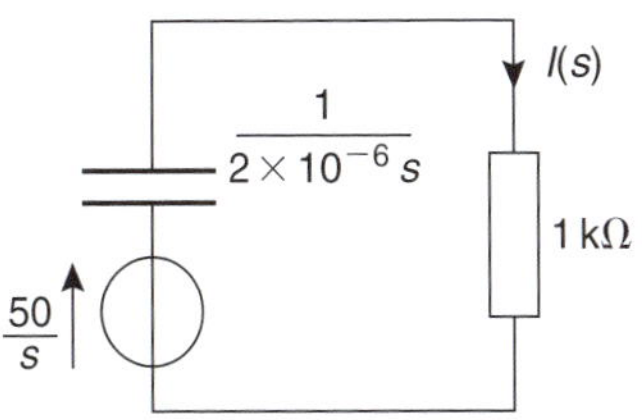

Figure 48.41

Applying Kirchhoff's voltage law in the s-domain gives:

$$\frac{50}{s} = I(s)\left(10^3 + \frac{1}{2\times 10^{-6}s}\right)$$

from which, $I(s) = \dfrac{50/s}{\left(10^3 + \frac{5\times 10^5}{s}\right)}$

$$= \frac{50}{(10^3 s + 5\times 10^5)}$$

$$= \frac{50/10^3}{\left(\frac{10^3 s}{10^3} + \frac{5\times 10^5}{10^3}\right)} = \frac{0.05}{(s+500)}$$

$$= 0.05\left(\frac{1}{s+500}\right)$$

Hence **current, $\boldsymbol{i = 0.05e^{-500t}}$ A**

Now try the following Practice Exercise

Practice Exercise 202 Circuit analysis using Laplace transforms (Answers on page 835)

1. For the circuit shown in Figure 48.42, derive an equation to represent the current i flowing. Assume zero conditions when the switch is closed. Is the circuit over, critical or under damped?

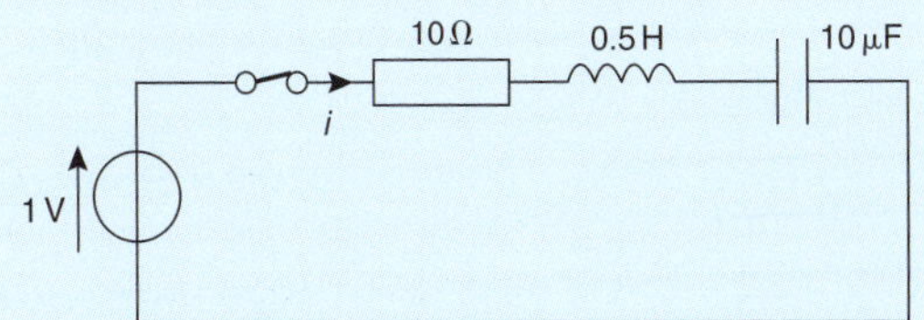

Figure 48.42

2. If for the circuit of Problem 1, $R=100\,\Omega$, $L=0.5\,H$ and $C=200\,\mu F$, derive an equation to represent current.

3. If for the circuit of Problem 1, $R=1\,k\Omega$, $L=0.5\,H$ and $C=200\,\mu F$, derive an equation to represent current.

4. In a C–R series circuit the $5\,\mu F$ capacitor is charged to a p.d. of 100 V. It is then suddenly discharged through a $1\,k\Omega$ resistor. Determine, after 10 ms, (a) the value of the current and (b) the voltage across the resistor.

5. In the circuit shown in Figure 48.43 the switch has been connected to point a for some time. It is then suddenly switched to point b. Derive an expression for current i flowing through the $20\,\Omega$ resistor.

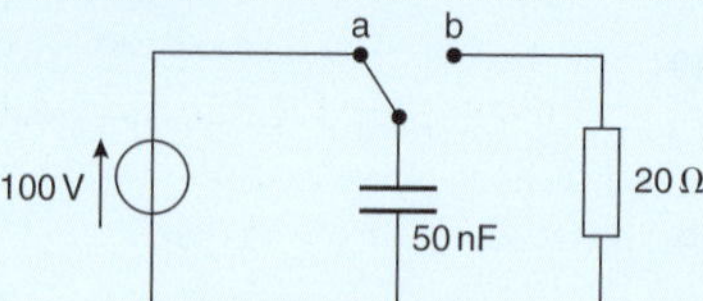

Figure 48.43

Part 4

For fully worked solutions to each of the problems in Practice Exercises 195 to 202 in this chapter, go to the website:
www.routledge.com/cw/bird

Revision Test 14

This revision test covers the material contained in Chapters 45 to 48. *The marks for each question are shown in brackets at the end of each question.*

1. A filter section is to have a characteristic impedance at zero frequency of 720 Ω and a cut-off frequency of 2 MHz. To meet these requirements, design (a) a low-pass T section filter and (b) a low-pass π section filter. (8)

2. A filter is required to pass all frequencies above 50 kHz and to have a nominal impedance of 620 Ω. Design (a) a high-pass T section filter and (b) a high-pass π section filter to meet these requirements. (8)

3. Design (a) a suitable 'm-derived' T section and (b) a suitable 'm-derived' π section having a cut-off frequency of 50 kHz, a nominal impedance of 600 Ω and a frequency of infinite attenuation 30 kHz. (14)

4. Two coils, A and B, are magnetically coupled; coil A has 400 turns and a self inductance of 20 mH and coil B has 250 turns and a self inductance of 50 mH. When a current of 10 A is reversed in coil A, the change of flux in coil B is 2 mWb. Determine (a) the mutual inductance between the coils and (b) the coefficient of coupling. (4)

5. Two mutually coupled coils P and Q are connected in series to a 200 V d.c. supply. Coil P has an inductance of 0.8 H and resistance 2 Ω; coil Q has an inductance of 1.2 H and a resistance of 5 Ω. Determine the mutual inductance between the coils if, at a certain instant after the circuit is connected, the current is 5 A and increasing at a rate of 7.5 A/s. (5)

6. For the coupled circuit shown in Figure RT14.1, calculate the values of currents I_P and I_S (9)

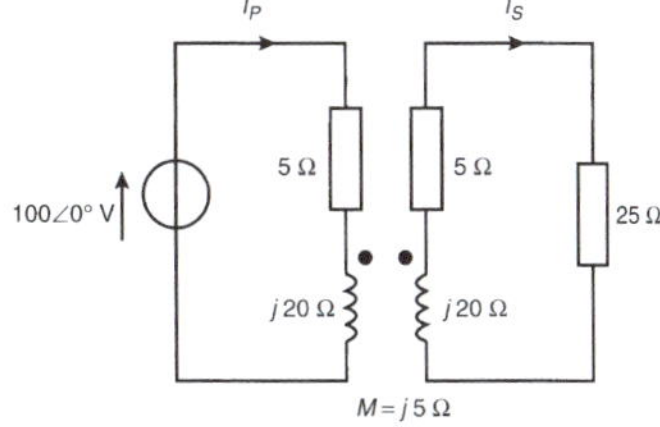

Figure RT14.1

7. A 4 km transmission line has a characteristic impedance of $600\angle -30°$ Ω. At a particular frequency, the attenuation coefficient of the line is 0.4 Np/km and the phase-shift coefficient is 0.20 rad/km. Calculate (a) the magnitude and phase of the voltage at the receiving end if the sending end voltage is $5.0\angle 0°$ V and (b) the magnitude and phase of the receiving end current. (5)

8. The primary constants of a transmission line at a frequency of 5 kHz are: resistance, $R = 20$ Ω/loop km, inductance, $L = 3$ mH/loop km, capacitance, $C = 50$ nF/km and conductance, $G = 0.4$ mS/km. Determine for the line (a) the characteristic impedance, (b) the propagation coefficient, (c) the attenuation coefficient, (d) the phase-shift coefficient, (e) the wavelength on the line and (f) the speed of transmission of signal. (13)

9. A loss-free transmission line has a characteristic impedance of $600\angle 0°$ Ω and is connected to an aerial of impedance $(250 + j200)$ Ω. Determine (a) the magnitude of the ratio of the reflected to the incident voltage wave and (b) the incident voltage if the reflected voltage is $10\angle 60°$ V. (5)

10. A low-loss transmission line has a mismatched load such that the reflection coefficient at the termination is $0.5\angle -150°$. The characteristic impedance of the line is 200 Ω. Determine (a) the standing wave ratio, (b) the load impedance and (c) the incident current flowing if the reflected current is 15 mA. (11)

11. Determine, in the s-domain, the impedance of (a) an inductance of 1 mH, a resistance of 100 Ω and a capacitance of 2 μF connected in series, and (b) a 50 Ω resistance in parallel with an inductance of 5 mH. (6)

12. A sinusoidal voltage $9 \sin 2t$ volts is applied to a series R–L circuit. Assuming that at time $t = 0$, current $i = 0$ and that resistance, $R = 6$ Ω and inductance $L = 1.5$ H, determine using Laplace transforms an expression for the current flowing in the circuit. (12)

For lecturers/instructors/teachers, fully worked solutions to each of the problems in Revision Test 14, together with a full marking scheme, are available at the website:
www.routledge.com/cw/bird

Main formulae for Part 4 Advanced circuit theory and technology

Complex numbers

$z = a + jb = r(\cos\theta + j\sin\theta) = r\angle\theta$,
where $j^2 = -1$ Modulus, $r = |z| = \sqrt{(a^2 + b^2)}$

Argument, $\theta = \arg z = \tan^{-1}\frac{b}{a}$

Addition: $(a + jb) + (c + jb) = (a + c) + j(b + d)$

Subtraction: $(a + jb) - (c + jd) = (a - c) + j(b - d)$

Complex equations: If $a + jb = c + jd$, then $a = c$ and $b = d$

If $z_1 = r_1\angle\theta_1$ and $z_2 = r_2\angle\theta_2$ then

Multiplication: $z_1 z_2 = r_1 r_2 \angle(\theta_1 + \theta_2)$

and Division: $\frac{z_1}{z_2} = \frac{r_1}{r_2}\angle(\theta_1 - \theta_2)$

De Moivre's theorem:

$[r\angle\theta]^n = r^n\angle n\theta = r^n(\cos n\theta + j\sin n\theta)$

General

$Z = \frac{V}{I} = R + j(X_L - X_C) = |Z|\angle\phi$

where $|Z| = \sqrt{[R^2 + (X_L - X_C)^2]}$ and

$\phi = \tan^{-1}\frac{X_L - X_C}{R}$

$X_L = 2\pi fL \quad X_C = \frac{1}{2\pi fC} \quad Y = \frac{I}{V} = \frac{1}{Z} = G + jB$

Series: $Z_T = Z_1 + Z_2 + Z_3 + \cdots$

Parallel: $\frac{1}{Z_T} = \frac{1}{Z_1} + \frac{1}{Z_2} + \frac{1}{Z_3} + \cdots$

$P = VI\cos\phi$ or $P = I_R^2 R \quad S = VI \quad Q = VI\sin\phi$

Power factor $= \cos\phi = \frac{R}{Z}$

If $V = a + jb$ and $I = c + jd$ then $P = ac + bd$

$Q = bc - ad \quad S = VI^* = P + jQ$

R–L–C series circuit

$f_r = \frac{1}{2\pi\sqrt{(LC)}} \qquad Q = \frac{\omega_r L}{R} = \frac{1}{\omega_r CR} = \frac{1}{R}\sqrt{\frac{L}{C}}$

$= \frac{V_L}{V} = \frac{V_C}{V} = \frac{f_r}{f_2 - f_1}$

$f_r = \sqrt{(f_1 \ f_2)}$

LR–C network

$f_r = \frac{1}{2\pi}\sqrt{\left(\frac{1}{LC} - \frac{R^2}{L^2}\right)} \quad R_D = \frac{L}{CR} \quad Q = \frac{I_C}{I_r} = \frac{\omega_r L}{R}$

LR–CR network

$f_r = \frac{1}{2\pi\sqrt{(LC)}}\sqrt{\left(\frac{R_L^2 - L/C}{R_C^2 - L/C}\right)}$

Determinants

$\begin{vmatrix} a & b \\ c & d \end{vmatrix} = ad - bc$

$\begin{vmatrix} a & b & c \\ d & e & f \\ g & h & j \end{vmatrix} = a\begin{vmatrix} e & f \\ h & j \end{vmatrix} - b\begin{vmatrix} d & f \\ g & j \end{vmatrix} + c\begin{vmatrix} d & e \\ g & h \end{vmatrix}$

Delta–star

$Z_1 = \frac{Z_A Z_B}{Z_A + Z_B + Z_C}$, etc.

Star–delta

$Z_A = \frac{Z_1 Z_2 + Z_2 Z_3 + Z_3 Z_1}{Z_2}$, etc.

Impedance matching

$|z| = \left(\frac{N_1}{N_2}\right)^2 |Z_L|$

Complex waveforms

$I = \sqrt{\left(I_0^2 + \frac{I_{1m}^2 + I_{2m}^2 + \cdots}{2}\right)}$

$i_{AV} = \frac{1}{\pi}\int_0^{\pi} i\,d(\omega t) \qquad$ form factor $= \frac{\text{r.m.s.}}{\text{mean}}$

$P = V_0 I_0 + V_1 I_1\cos\phi_1 + V_2 I_2\cos\phi_2 + \cdots$ or $P = I^2 R$

power factor $= \frac{P}{VI}$

Harmonic resonance: $n\omega L = \frac{1}{n\omega C}$

Fourier series

If $f(x)$ is a periodic function of period 2π then its Fourier series is given by:

$$\boldsymbol{f(x) = a_0 + \sum_{n=1}^{\infty}(a_n \cos nx + b_n \sin nx)}$$

where, for the range $-\pi$ to $+\pi$:

$$a_0 = \frac{1}{2\pi}\int_{-\pi}^{\pi} f(x)\,dx$$

$$a_n = \frac{1}{\pi}\int_{-\pi}^{\pi} f(x)\cos nx\,dx \quad (n = 1,2,3,\ldots)$$

$$b_n = \frac{1}{\pi}\int_{-\pi}^{\pi} f(x)\sin nx\,dx \quad (n = 1,2,3,\ldots)$$

If $f(x)$ is a periodic function of period L then its Fourier series is given by:

$$\boldsymbol{f(x) = a_0 + \sum_{n=1}^{\infty}\left\{a_n \cos\left(\frac{2\pi nx}{L}\right) + b_n \sin\left(\frac{2\pi nx}{L}\right)\right\}}$$

where for the range $-\frac{L}{2}$ to $+\frac{L}{2}$:

$$a_0 = \frac{1}{L}\int_{-L/2}^{L/2} f(x)\,dx$$

$$a_n = \frac{2}{L}\int_{-L/2}^{L/2} f(x)\cos\left(\frac{2\pi nx}{L}\right)dx \quad (n = 1,2,3,\ldots)$$

$$b_n = \frac{2}{L}\int_{-L/2}^{L/2} f(x)\sin\left(\frac{2\pi nx}{L}\right)dx \quad (n = 1,2,3,\ldots)$$

Harmonic analysis

$$a_0 \approx \frac{1}{p}\sum_{k=1}^{p} y_k \quad a_n \approx \frac{2}{p}\sum_{k=1}^{p} y_k \cos nx_k$$

$$b_n \approx \frac{2}{p}\sum_{k=1}^{p} y_k \sin nx_k$$

Hysteresis and eddy current

Hysteresis loss/cycle $= A\alpha\beta$ J/m^3
or hysteresis loss $= k_h v f (B_m)^n$ W

Eddy current loss/cycle $= k_e (B_m)^2 f^2 t^3$ W

Dielectric loss

Series representation: $\tan\delta = R_S \omega C_S = 1/Q$

Parallel representation: $\tan\delta = \dfrac{1}{R_p \omega C_p}$

Loss angle $\delta = (90° - \phi)$

Power factor $= \cos\phi \approx \tan\delta$

Dielectric power loss $= V^2 \omega C \tan\delta$

Field theory

Coaxial cable: $C = \dfrac{2\pi\varepsilon_0\varepsilon_r}{\ln\frac{b}{a}}$ F/m $\quad E = \dfrac{V}{r\ln\frac{b}{a}}$ V/m

$$L = \frac{\mu_0\mu_r}{2\pi}\left(\frac{1}{4} + \ln\frac{b}{a}\right) \textbf{ H/m}$$

Twin line: $C = \dfrac{\pi\varepsilon_0\varepsilon_r}{\ln\frac{D}{a}}$ F/m

$$L = \frac{\mu_0\mu_r}{\pi}\left(\frac{1}{4} + \ln\frac{D}{a}\right) \text{ H/m}$$

Energy stored: in a capacitor, $W = \frac{1}{2}CV^2$ J;

in an inductor $W = \frac{1}{2}LI^2$ J

in electric field per unit volume,

$$\omega_f = \tfrac{1}{2}DE = \tfrac{1}{2}\varepsilon_0\varepsilon_r E^2 = \frac{D^2}{2\varepsilon_0\varepsilon_r} \text{ J/m}^3$$

in a non-magnetic medium,

$$\omega_f = \tfrac{1}{2}BH = \tfrac{1}{2}\mu_0 H^2 = \frac{B^2}{2\mu_0} \text{ J/m}^3$$

Attenuators

Logarithmic ratios:

in decibels $= 10\lg\dfrac{P_2}{P_1} = 20\lg\dfrac{V_2}{V_1} = 20\lg\dfrac{I_2}{I_1}$

in nepers $= \frac{1}{2}\ln\dfrac{P_2}{P_1} = \ln\dfrac{V_2}{V_1} = \ln\dfrac{I_2}{I_1}$

Symmetrical T-attenuator:

$$R_0 = \sqrt{(R_1^2 + 2R_1R_2)} = \sqrt{(R_{OC}R_{SC})}$$

$$R_1 = R_0\left(\frac{N-1}{N+1}\right) \quad R_2 = R_0\left(\frac{2N}{N^2-1}\right)$$

Symmetrical π-attenuator:

$$R_0 = \sqrt{\left(\frac{R_1R_2^2}{R_1 + 2R_2}\right)} = \sqrt{(R_{OC}R_{SC})}$$

$$R_1 = R_0\left(\frac{N^2-1}{2N}\right) \qquad R_2 = R_0\left(\frac{N+1}{N-1}\right)$$

L-section attenuator: $\quad R_1 = \sqrt{[R_{OA}(R_{OA} - R_{OB})]}$

$$R_2 = \sqrt{\left(\frac{R_{OA}R_{OB}^2}{R_{OA} - R_{OB}}\right)}$$

ABCD parameters

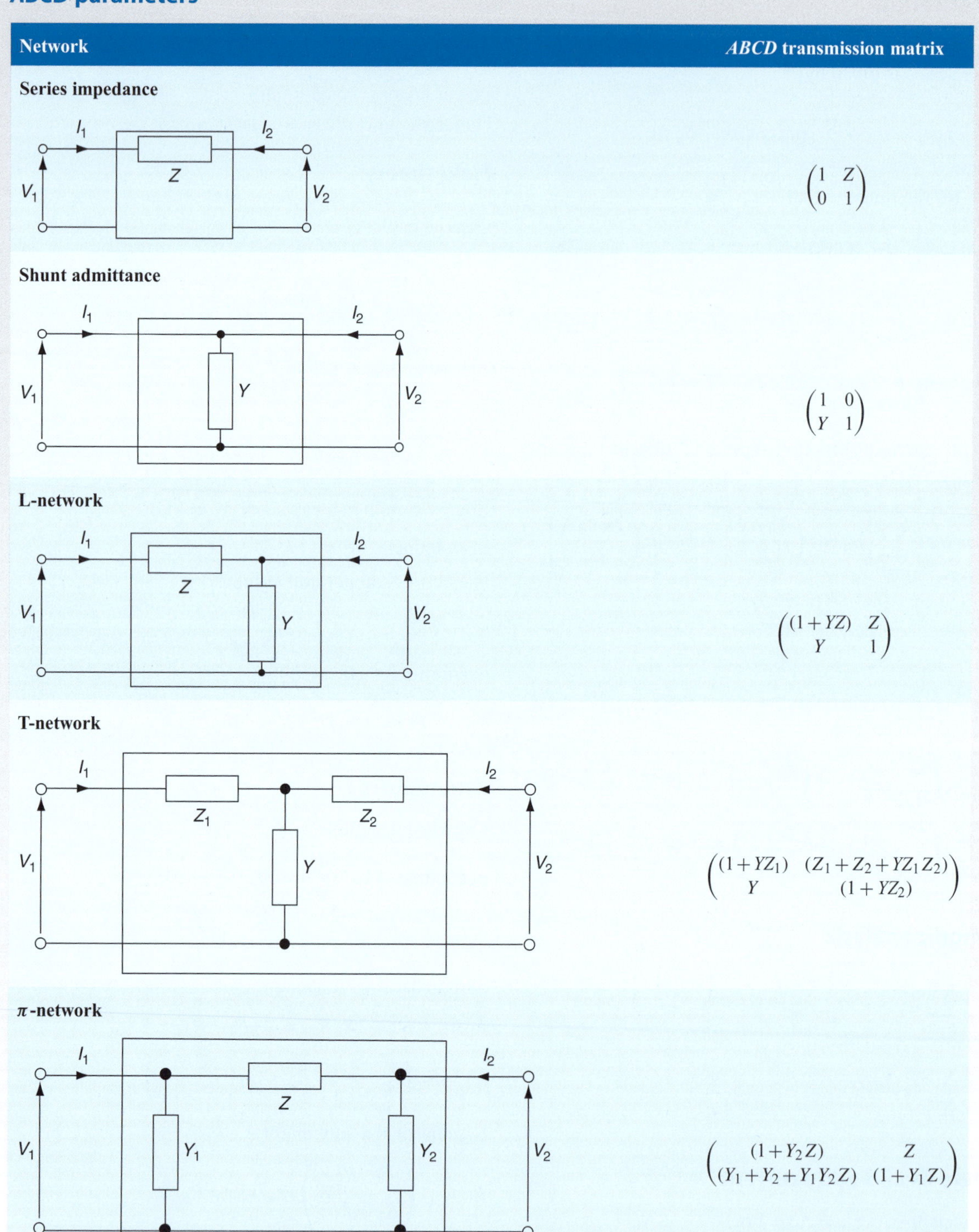

Network	*ABCD* transmission matrix
Series impedance	$\begin{pmatrix} 1 & Z \\ 0 & 1 \end{pmatrix}$
Shunt admittance	$\begin{pmatrix} 1 & 0 \\ Y & 1 \end{pmatrix}$
L-network	$\begin{pmatrix} (1+YZ) & Z \\ Y & 1 \end{pmatrix}$
T-network	$\begin{pmatrix} (1+YZ_1) & (Z_1+Z_2+YZ_1Z_2) \\ Y & (1+YZ_2) \end{pmatrix}$
π-network	$\begin{pmatrix} (1+Y_2Z) & Z \\ (Y_1+Y_2+Y_1Y_2Z) & (1+Y_1Z) \end{pmatrix}$

(*Continued*)

ABCD parameters (*Continued*)

Network	*ABCD* transmission matrix
Pure mutual inductance 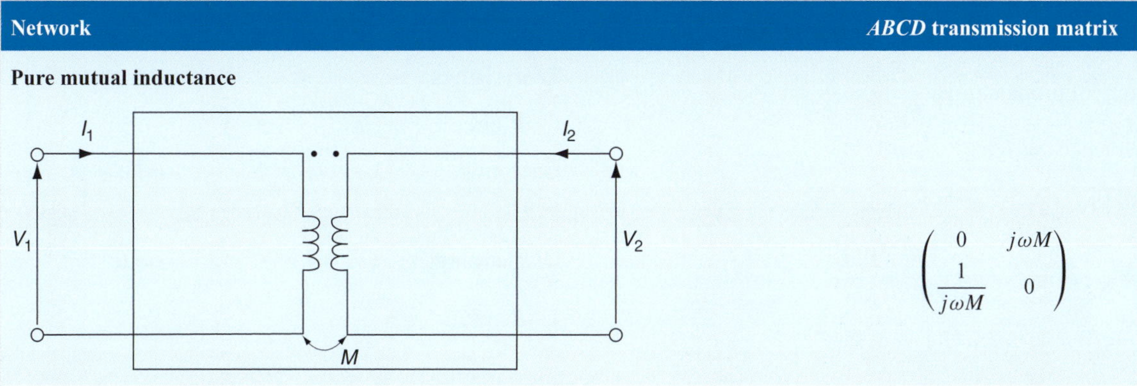	$\begin{pmatrix} 0 & j\omega M \\ \frac{1}{j\omega M} & 0 \end{pmatrix}$
Symmetrical lattice 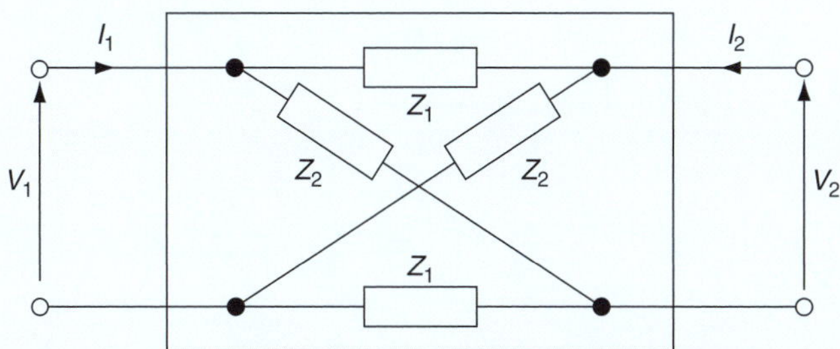	$\begin{pmatrix} \left(\frac{Z_1+Z_2}{Z_2-Z_1}\right) & \left(\frac{2Z_1Z_2}{Z_2-Z_1}\right) \\ \left(\frac{2}{Z_2-Z_1}\right) & \left(\frac{Z_1+Z_2}{Z_2-Z_1}\right) \end{pmatrix}$

Characteristic impedance, $Z_0 = \sqrt{\frac{B}{C}}$

Filter networks

Low-pass T or π: $f_C = \dfrac{1}{\pi\sqrt{(LC)}}$ $R_0 = \sqrt{\dfrac{L}{C}}$

$$C = \frac{1}{\pi R_0 f_C} \quad L = \frac{R_0}{\pi f_C}$$

$$Z_{0T} = R_0\sqrt{\left[1 - \left(\frac{\omega}{\omega_C}\right)^2\right]}$$

$$Z_{0\pi} = \frac{R_0}{\sqrt{\left[1 - \left(\frac{\omega}{\omega_C}\right)^2\right]}}$$

High-pass T or π: $f_C = \dfrac{1}{4\pi\sqrt{(LC)}}$ $R_0 = \sqrt{\dfrac{L}{C}}$

$$C = \frac{1}{4\pi R_0 f_C} \quad L = \frac{R_0}{4\pi f_C}$$

$$Z_{0T} = R_0\sqrt{\left[1 - \left(\frac{\omega_C}{\omega}\right)^2\right]}$$

$$Z_{0\pi} = \frac{R_0}{\left[1 - \left(\frac{\omega_C}{\omega}\right)^2\right]}$$

Low- and high-pass:

$$Z_{0T} Z_{0\pi} = Z_1 Z_2 = R_0^2$$

$$\frac{I_1}{I_2} = \frac{I_2}{I_3} = \frac{I_3}{I_4} = e^{\gamma} = e^{\alpha + j\beta} = e^{\alpha}\angle\beta$$

Phase angle $\beta = \omega\sqrt{(LC)}$

time delay $= \sqrt{(LC)}$

m-derived filter sections:

$$\text{Low-pass} \quad m = \sqrt{\left[1 - \left(\frac{f_C}{f_\infty}\right)^2\right]}$$

$$\text{High-pass} \quad m = \sqrt{\left[1 - \left(\frac{f_\infty}{f_C}\right)^2\right]}$$

Magnetically coupled circuits

$$E_2 = -M\frac{dI_1}{dt} = \pm j\omega M I_1$$

$$M = N_2\frac{d\phi_2}{dI_1} = N_1\frac{d\phi_1}{dI_2} = k\sqrt{(L_1 L_2)} = \frac{L_A - L_B}{4}$$

Transmission lines

Phase delay $\beta = \omega\sqrt{(LC)}$ wavelength $\lambda = \dfrac{2\pi}{\beta}$

velocity of propagation $u = f\lambda = \dfrac{\omega}{\beta}$

$$I_R = I_S e^{-n\gamma} = I_S e^{-n\alpha} \angle -n\beta$$

$$V_R = V_S e^{-n\gamma} = V_S e^{-n\alpha} \angle -n\beta$$

$$Z_0 = \sqrt{(Z_{OC} Z_{SC})} = \sqrt{\frac{R + j\omega L}{G + j\omega C}}$$

$$\gamma = \sqrt{[(R + j\omega L)(G + j\omega C)]}$$

Reflection coefficient, $\rho = \dfrac{I_r}{I_i} = \dfrac{Z_O - Z_R}{Z_O + Z_R} = -\dfrac{V_r}{V_i}$

Standing-wave ratio, $s = \dfrac{I_{max}}{I_{min}} = \dfrac{I_i + I_r}{I_i - I_r} = \dfrac{1 + |\rho|}{1 - |\rho|}$

$$\frac{P_r}{P_t} = \left(\frac{s-1}{s+1}\right)^2$$

Transients

C–R circuit $\tau = CR$

Charging: $v_C = V(1 - e^{-(t/CR)})$ $v_r = Ve^{-(t/CR)}$

$i = Ie^{-(t/CR)}$

Discharging: $v_c = v_R = Ve^{-(t/CR)}$ $i = Ie^{-(t/CR)}$

L–R circuit $\tau = \dfrac{L}{R}$

Current growth: $v_L = Ve^{-(Rt/L)}$

$v_R = V(1 - e^{-(Rt/L)})$

$i = I(1 - e^{-(Rt/L)})$

Current decay: $v_L = v_R = Ve^{-(Rt/L)}$ $i = Ie^{-(Rt/L)}$

These formulae are available for download at the website:
www.routledge.com/cw/bird

Part 5

General reference

Standard electrical quantities – their symbols and units

QUANTITY	QUANTITY SYMBOL	UNIT	UNIT SYMBOL
Admittance	Y	siemen	S
Angular frequency	ω	radians per second	rad/s
Area	A	square metres	m^2
Attenuation coefficient (or constant)	α	neper	Np
Capacitance	C	farad	F
Charge	Q	coulomb	C
Charge density	σ	coulomb per square metre	C/m^2
Conductance	G	siemen	S
Current	I	ampere	A
Current density	J	ampere per square metre	A/m^2
Efficiency	η	per unit or per cent	p.u. or %
Electric field strength	E	volt per metre	V/m
Electric flux	Ψ	coulomb	C
Electric flux density	D	coulomb per square metre	C/m^2
Electromotive force	E	volt	V
Energy	W	joule	J
Field strength, electric	E	volt per metre	V/m
Field strength, magnetic	H	ampere per metre	A/m
Flux, electric	Ψ	coulomb	C
Flux, magnetic	Φ	weber	Wb
Flux density, electric	D	coulomb per square metre	C/m^2
Flux density, magnetic	B	tesla	T
Force	F	newton	N
Frequency	f	hertz	Hz
Frequency, angular	ω	radians per second	rad/s

(Continued)

QUANTITY	QUANTITY SYMBOL	UNIT	UNIT SYMBOL
Frequency, rotational	n	revolutions per second	rev/s
Impedance	Z	ohm	Ω
Inductance, mutual	M	henry	H
Inductance, self	L	henry	H
Length	l	metre	m
Loss angle	δ	radian or degrees	rad or °
Magnetic field strength	H	ampere per metre	A/m
Magnetic flux	Φ	weber	Wb
Magnetic flux density	B	tesla	T
Magnetic flux linkage	Ψ	weber	Wb
Magnetizing force	H	ampere per metre	A/m
Magnetomotive force	F_m	ampere	A
Mutual inductance	M	henry	H
Number of phases	m	–	–
Number of pole-pairs	p	–	–
Number of turns (of a winding)	N	–	–
Period, Periodic time	T	second	s
Permeability, absolute	μ	henry per metre	H/m
Permeability of free space	μ_0	henry per metre	H/m
Permeability, relative	μ_r	–	–
Permeance	Λ	weber per ampere or per henry	Wb/A or /H
Permittivity, absolute	ε	farad per metre	F/m
Permittivity of free space	ε_0	farad per metre	F/m
Permittivity, relative	ε_r	–	–
Phase-change coefficient	β	radian	rad
Potential, Potential difference	V	volt	V
Power, active	P	watt	W
Power, apparent	S	volt ampere	VA
Power, reactive	Q	volt ampere reactive	var
Propagation coefficient (or constant)	γ	–	–

(Continued)

QUANTITY	QUANTITY SYMBOL	UNIT	UNIT SYMBOL
Quality factor, magnification	Q	–	–
Quantity of electricity	Q	coulomb	C
Reactance	X	ohm	Ω
Reflection coefficient	ρ	–	–
Relative permeability	μ_r	–	–
Relative permittivity	ε_r	–	–
Reluctance	S or R_m	ampere per weber or per henry	A/Wb or /H
Resistance	R	ohm	Ω
Resistance, temperature coefficient of	α	per degree Celsius or per kelvin	/°C or /K
Resistivity	ρ	ohm metre	Ωm
Slip	s	per unit or per cent	p.u. or %
Standing-wave ratio	s	–	–
Susceptance	B	siemen	S
Temperature coefficient of resistance	α	per degree Celsius or per kelvin	/°C or /K
Temperature, thermodynamic	T	kelvin	K
Time	t	second	s
Torque	T	newton metre	Nm
Velocity	v	metre per second	m/s
Velocity, angular	ω	radian per second	rad/s
Volume	V	cubic metres	m^3
Wavelength	λ	metre	m

(Note that m/s may also be written as ms^{-1}, C/m^2 as Cm^{-2}, /K as K^{-1}, and so on.)

Greek alphabet

LETTER	UPPER CASE	LOWER CASE
Alpha	A	α
Beta	B	β
Gamma	Γ	γ
Delta	Δ	δ
Epsilon	E	ε
Zeta	Z	ζ
Eta	H	η
Theta	Θ	θ
Iota	I	ι
Kappa	K	κ
Lambda	Λ	λ
Mu	M	μ

LETTER	UPPER CASE	LOWER CASE
Nu	N	ν
Xi	Ξ	ξ
Omicron	O	o
Pi	Π	π
Rho	P	ρ
Sigma	Σ	σ
Tau	T	τ
Upsilon	Υ	υ
Phi	Φ	ϕ
Chi	X	χ
Psi	Ψ	ψ
Omega	Ω	ω

Common prefixes

PREFIX	NAME	MEANING: multiply by
E	exa	10^{18}
P	peta	10^{15}
T	tera	10^{12}
G	giga	10^{9}
M	mega	10^{6}
k	kilo	10^{3}
h	hecto	10^{2}
da	deca	10^{1}

PREFIX	NAME	MEANING: multiply by
d	deci	10^{-1}
c	centi	10^{-2}
m	milli	10^{-3}
μ	micro	10^{-6}
n	nano	10^{-9}
p	pico	10^{-12}
f	femto	10^{-15}
a	atto	10^{-18}

Resistor colour coding and ohmic values

Colour code for fixed resistors

COLOUR	SIGNIFICANT FIGURES	MULTIPLIER	TOLERANCE
Silver	–	10^{-2}	±10%
Gold	–	10^{-1}	±5%
Black	0	1	–
Brown	1	10	±1%
Red	2	10^2	±2%
Orange	3	10^3	–
Yellow	4	10^4	–
Green	5	10^5	±0.5%
Blue	6	10^6	±0.25%
Violet	7	10^7	±0.1%
Grey	8	10^8	–
White	9	10^9	–
None	–	–	±20%

Thus, for a **five-band fixed resistor** (i.e. resistance values with two significant figures):

yellow-violet-orange-red indicates 47 kΩ with a tolerance of ±2%

orange-orange-silver-brown indicates 0.33 Ω with a tolerance of ±1%

and brown-black-brown indicates 100 Ω with a tolerance of ±20%

(Note that the first band is the one nearest the end of the resistor).

For a **five-band fixed resistor** (i.e. resistance values with three significant figures):

red-yellow-white-orange-brown indicates 249 kΩ with a tolerance of ±1%

(Note that the fifth band is 1.5 to 2 times wider than the other bands).

Letter and digit code for resistors

RESISTANCE VALUE	MARKED AS
0.47 Ω	R47
1 Ω	1R0
4.7 Ω	4R7
47 Ω	47R
100 Ω	100R
1 kΩ	1K0
10 kΩ	10K
10 MΩ	10M

Tolerance is indicated as follows:

$F = \pm 1\%$, $G = \pm 2\%$, $J = \pm 5\%$, $K = \pm 10\%$ and $M = \pm 20\%$

Thus, for example,
R33M = 0.33 Ω ± 20%
4R7K = 4.7 Ω ± 10%
390RJ = 390 Ω ± 5%
6K8F = 6.8 kΩ ± 1%
68KK = 68 kΩ ± 10%
4M7M = 4.7 MΩ ± 20%

Answers to Practice Exercises

Chapter 1

Exercise 1, Page 4

1. 30.797 2. 53.832 3. 1.0944
4. 50.330 5. 36.45 6. 46.923
7. 1.296×10^{-3} 8. 2.197 9. 0.0549
10. 5.832×10^{-6} 11. 137.9 12. 14.96
13. 19.4481 14. 515.36×10^{-6} 15. 1.0871
16. 52.70 17. 185.82 18. 2.571
19. 1.068 20. 3.5×10^6 21. 4.2×10^{-6}
22. 202.767×10^{-3} 23. 18.32×10^6 24. 0.4667
25. $\frac{13}{14}$ 26. 4.458 27. 2.732
28. $-\frac{9}{10}$ 29. $3\frac{1}{3}$ 30. 2.567
31. 0.0776 32. 0.9205 33. 2.9042
34. 0.4424 35. 0.0321 36. 0.4232
37. −0.6992 38. 5.8452 39. 4.995
40. 5.782 41. 25.72 42. 0.6977
43. 591.0 44. 3.520 45. 0.3770

Exercise 2, Page 6

1. $A = 66.59\,cm^2$ 2. C = 52.78 mm 3. R = 37.5
4. 159 m/s 5. 0.407 A 6. 5.02 mm
7. 0.144 J 8. $628.8\,m^2$ 9. 224.5
10. $14230\,kg/m^3$ 11. $2.526\,\Omega$ 12. V = 2.61 V
13. I = 3.81 A 14. E = 3.96 J 15. I = 12.77 A

Exercise 3, Page 8

1. $\frac{7}{12}$ 2. $\frac{9}{20}$ 3. $\frac{7}{15}$
4. $\frac{1}{21}$ 5. $-\frac{9}{10}$ 6. $\frac{17}{24} = 0.7083$
7. $3\frac{1}{3}$ 8. 2.567 9. $\frac{3x + 2y}{xy}$

Exercise 4, Page 10

1. 5.7% 2. 37.4% 3. 0.20
4. 68.75% 5. 38.462%
6. (b) 52.9% (d) 54.5% (c) 55.6% (a) 57.1%
7. $\frac{13}{20}$ 8. 21.8 kg 9. 9.72 m
10. (a) 496.4 t (b) 8.657 g (c) 20.73 s
11. (a) 14% (b) 15.67% (c) 5.36%
12. £624 13. 37.49% 14. 38.7%
15. 2.7%

Exercise 5, Page 11

1. 36:1 2. 3.5:1 or 7:2 3. 96 cm, 240 cm
4. £3680, £1840, £920 5. 12 cm
6. 1:15 7. 76 ml 8. 25%
9. 12.6 kg 10. 14.3 kg

Exercise 6, Page 12

1. £556 2. £66
3. 264 kg 4. (a) 0.00025 (b) 48 MPa

Electrical Circuit Theory and Technology. 978-1-138-67349-6, © 2017 John Bird. Published by Taylor & Francis. All rights reserved.

5. (a) 440 K (b) 5.76 litre **6.** (a) 2 mA (b) 25 V
7. 83 lb 10 oz **8.** (a) 159.1 litres (b) 16.5 gallons

Exercise 7, Page 13

1. 3.5 weeks **2.** 20 days
3. (a) 9.18 (b) 6.12 (c) 0.3375
4. 50 minutes **5.** (a) 300×10^3 (b) $0.375\ m^3$ (c) 24×10^3 Pa

Exercise 8, Page 15

1. $2^7 = 128$ **2.** 3^9 **3.** $2^4 = 16$
4. $3^{-2} = \frac{1}{3^2} = \frac{1}{9}$ **5.** 1 **6.** $2^3 = 8$
7. $10^2 = 100$ **8.** $10^3 = 1000$ **9.** $10^{-2} = \frac{1}{10^2} = \frac{1}{100} = 0.01$
10. 5 **11.** 7^6 **12.** $3^6 = 729$
13. 3^6 **14.** 81 **15.** 1
16. $5^2 = 25$ **17.** $3^{-5} = \frac{1}{3^5} = \frac{1}{243}$ **18.** $7^2 = 49$
19. z^8 **20.** a^8 **21.** n^3
22. b^{11} **23.** $b^{-3} = \frac{1}{b^3}$ **24.** c^4
25. m^4 **26.** $x^{-3} = \frac{1}{x^3}$ **27.** x^{12}
28. $y^{-6} = \frac{1}{y^6}$ **29.** t^8 **30.** c^{14}
31. $a^{-9} = \frac{1}{a^9}$ **32.** $\frac{1}{b^{12}} = b^{-12}$ **33.** b^{10}
34. $\frac{1}{s^9} = s^{-9}$ **35.** $p^6q^7r^5$ **36.** $x^{-2}yz^{-2}$ or $\frac{y}{x^2z^2}$

Exercise 9, Page 16

1. 21 **2.** 72 **3.** $2x - 4y + 6$
4. $7x - y - 4z$ **5.** 0 **6.** $11b - 2a$
7. $abc + abd - abef + abeg - abehi - abehj$

Exercise 10, Page 18

1. 1 **2.** 2 **3.** 6 **4.** -4
5. 2 **6.** $\frac{1}{2}$ **7.** 0 **8.** 3
9. 2 **10.** -10 **11.** 6 **12.** -2
13. 2.5 **14.** 2 **15.** 6 **16.** -3
17. 5 **18.** -2 **19.** $-4\frac{1}{2}$ **20.** 2
21. 12 **22.** 15 **23.** $5\frac{1}{3}$ **24.** 2
25. 13 **26.** -10 **27.** 2 **28.** 3
29. -11 **30.** 9 **31.** $6\frac{1}{4}$ **32.** 10
33. ± 12 **34.** $-3\frac{1}{3}$ **35.** ± 3

Exercise 11, Page 21

1. $d = c - e - a - b$ **2.** $x = \frac{y}{7}$
3. $v = \frac{c}{p}$ **4.** $a = \frac{v - u}{t}$
5. $R = \frac{V}{I}$ **6.** $y = \frac{1}{3}(t - x)$
7. $r = \frac{c}{2\pi}$ **8.** $x = \frac{y - c}{m}$
9. $T = \frac{I}{PR}$ **10.** $L = \frac{X_L}{2\pi f}$
11. $R = \frac{E}{I}$ **12.** $x = a(y - 3)$
13. $C = \frac{5}{9}(F - 32)$ **14.** $f = \frac{1}{2\pi CX_C}$
15. $r = \frac{S - a}{S}$ or $1 - \frac{a}{S}$ **16.** $x = \frac{d}{\lambda}(y + \lambda)$ or $d + \frac{yd}{\lambda}$
17. $f = \frac{3F - AL}{3}$ or $f = F - \frac{AL}{3}$ **18.** $D = \frac{AB^2}{5Cy}$
19. $t = \frac{R - R_0}{R_0\alpha}$ **20.** $R = \frac{E - e - Ir}{I}$ or $R = \frac{E - e}{I} - r$
21. $b = \sqrt{\left(\frac{y}{4ac^2}\right)}$ **22.** $L = \frac{gt^2}{4\pi^2}$
23. $u = \sqrt{v^2 - 2as}$ **24.** $a = N^2y - x$
25. $L = \frac{\sqrt{Z^2 - R^2}}{2\pi f}$, 0.080

Exercise 12, Page 22

1. apple = 8p, banana = 35 p
2. apple = 28p, orange = 17 p
3. car = £15000, van = £12000
4. $F_1 = 1.5$, $F_2 = -4.5$
5. $a = 5$, $b = 2$
6. $a = 6$, $b = -1$

Chapter 2

Exercise 13, Page 25

1. (a) 34.377° (b) 45.837° (c) 114.592° (d) 180°

2. (a) $\frac{\pi}{4}$ rad or 0.7854 rad (b) $\frac{\pi}{2}$ rad or 1.5708 rad (c) $\frac{2\pi}{3}$ rad or 2.0944 rad (d) π rad or 3.1416 rad

Exercise 14, Page 26

1. (a) 0.5000, 0.8660, 1.7321 (b) 0, 1, ∞ (c) −0.8660, 0.5000, −0.5774 (d) −1, 0, 0 (e) −0.8660, −0.5000, 0.5774 (f) 0, −1, $-\infty$ (g) 0.8660, −0.5000, −0.5774 (h) 0.8660, −0.5000, −0.5774 (i) 0.5000, 0.8660, 1.7321 (j) 0, 1, ∞ (k) −0.8660, 0.5000, −0.5774

Exercise 15, Page 27

1. 35.54° **2.** 62.61° **3.** 28.97°
4. 7.08 cm **5.** 4.65 m

Exercise 16, Page 28

1. A = 38.96°, C = 41.04°, a = 3.83 m

2. b = 22.01 m, A = 38.86°, C = 74.14°

3. 28.96°

4. A = 83.33°, B = 52.62°, C = 44.05°

5. P = 39.73°, QR = 7.38 m

Exercise 17, Page 29

1. log 6 **2.** log 15 **3.** log 2
4. log 3 **5.** log 12 **6.** log 500
7. log 100 **8.** log 6 **9.** log 10
10. log 1 = 0 **11.** x = 2.5 **12.** t = 8
13. b = 2 **14.** x = 2

Exercise 18, Page 30

1. 1.690 **2.** 3.170 **3.** 6.058
4. 2.251 **5.** 2.542 **6.** −0.3272
7. 316.2

Exercise 19, Page 31

1. −0.4904 **2.** −0.5822 **3.** 2.197
4. 816.2 **5.** 0.8274 **6.** 1.962
7. 3 **8.** 4 **9.** 500
10. $W = PV \ln\left(\frac{U_2}{U_1}\right)$

Exercise 20, Page 32

1. (a) 150°C (b) 100.5°C

2. (a) 29.32 volts (b) 71.31×10^{-6} s

3. (a) 1.993 m (b) 2.293 m

4. (a) 3.04 A (b) 1.46 s

5. (a) 7.07 A (b) 0.966 s

Exercise 21, Page 34

1. (a) −1 (b) −8 (c) −1.5 (d) 4 **2.** 14.5
3. (a) −1.1 (b) −1.4
4. 1010 rev/min should be 1070 rev/min; (a) 1000 rev/min (b) 167 V

Exercise 22, Page 36

1. Missing values: −0.75, 0.25, 0.75, 1.75, 2.25, 2.75, Gradient = $\frac{1}{2}$ or 0.5

2. (a) 4, −2 (b) −1, 0 (c) −3, −4 (d) 0, 4

3. (2, 1)

4. (a) 89 cm (b) 11 N (c) 2.4 (d) $\ell = 2.4W + 48$

Exercise 23, Page 38

1. (a) 40°C (b) 128 Ω

2. (a) 0.25 (b) 12 (c) F = 0.25L + 12 (d) 89.5 N (e) 592 N (f) 212 N

3. (a) 850 rev/min (b) 77.5 V

4. (a) $\frac{1}{5}$ or 0.2 (b) 6 (c) E = 0.2 L + 6 (d) 12 N (e) 65 N

Exercise 24, Page 41

1. (a) 80 m (b) 170 m **2.** 27.2 cm^2
3. 18 cm **4.** 1200 mm
5. (a) 29 cm^2 (b) 650 mm^2 **6.** 560 m^2
7. 6750 mm^2 **8.** 43.30 cm^2

9. 32 **10.** 482 m^2
11. (a) 50.27 cm^2 (b) 706.9 mm^2 (c) 3183 mm^2
12. (a) 53.01 cm^2 (b) 129.9 mm^2
13. 5773 mm^2 **14.** 1.89 m^2
15. (a) 0.698 rad (b) 804.2 m^2
16. 10.47m^2

Exercise 25, Page 43

1. 4.563 square units (using 6 intervals)

2. 54.5 square units (using 8 intervals)

Chapter 3

Exercise 26, Page 52

1. 600 N **2.** 5.1 kN **3.** 8 kg
4. 14.72 N **5.** 8 J **6.** 12.5 kJ
7. 4.5 W **8.** (a) 29.43 kN m (b) 981 W
9. 1000 C **10.** 30 s **11.** 900 C
12. 5 minutes
13. (a) 1 nF (b) 20 000 pF (c) 5 MHz (d) 0.047 MΩ (e) 320 μA

Exercise 27, Page 55

1. (a) 0.1 S (b) 0.5 mS (c) 500 S
2. 20 kΩ **3.** 1 kW **4.** 7.5 W
5. 1 V **6.** 7.2 kJ **7.** 10 kW, 40 A

Chapter 4

Exercise 28, Page 58

1. 5 s **2.** 3600 C **3.** 13 min 20 s

Exercise 29, Page 61

1. 7 Ω **2.** (a) 0.25 A (b) 960 Ω
3. 2 mΩ, 5 mΩ **4.** 30 V
5. 50 mA

Exercise 30, Page 63

1. 0.4 A, 100 W **2.** (a) 2 kΩ (b) 0.5 MΩ
3. 20 Ω, 2.88 kW, 57.6 kWh
4. 0.8 W **5.** 9.5 W
6. 2.5 V **7.** (a) 0.5 W (b) 1.33 W (c) 40 W
8. 9 kJ **9.** £26.70
10. 3 kW, 90 kWh, £12.15

Exercise 31, Page 64

1. 3 A, 5 A

Chapter 5

Exercise 32, Page 68

1. (a) 8.75 Ω (b) 5 m **2.** (a) 5 Ω (b) 0.625 mm^2
3. 0.32 Ω **4.** 0.8 Ω **5.** 1.5 mm^2
6. 0.026 μΩm **7.** 0.216 Ω

Exercise 33, Page 70

1. 69 Ω **2.** 24.69 Ω **3.** 488 Ω
4. 26.4 Ω **5.** 70°C **6.** 64.8Ω
7. 5.95 Ω

Exercise 34, Page 72

1. 68 kΩ±2% **2.** 4.7 Ω±20%
3. 690 Ω±5% **4.** Green-brown-orange-red
5. brown-black-green-silver **6.** 1.8 MΩ to 2.2 MΩ
7. 39.6 kΩ to 40.4 kΩ **8.** (a) 0.22 Ω±2% (b) 4.7 kΩ±1%
9. 100KJ **10.** 6M8M

Chapter 6

Exercise 35, Page 78

1. (a) 18 V, 2.88 Ω (b) 1.5 V, 0.02 Ω

2. (a) 2.17 V (b) 1.6 V (c) 0.7 V

3. 0.25 Ω

4. 18 V, 1.8 Ω

5. (a) 1 A (b) 21

6. (i)(a) 6 V (b) 2 V (ii)(a) 4 Ω (b) 0.25 Ω
7. 0.04 Ω, 51.2 V

Chapter 7

Exercise 36, Page 93

1. (a) 22 V (b) 11 Ω (c) 2.5 Ω, 3.5 Ω, 5 Ω
2. 10 V, 0.5 A, 20 Ω, 10 Ω, 6 Ω
3. 4 A, 2.5 Ω
4. 45 V
5. (a) 1.2 Ω (b) 12 V
6. 6.77 Ω
7. (a) 4 Ω (b) 48 V

Exercise 37, Page 98

1. (a) 3 Ω (b) 3 A (c) 2.25 A, 0.75 A
2. 2.5 A, 2.5 Ω
3. (a)(i) 5Ω (ii) 60 kΩ (iii) 28 Ω (iv) 6.3 kΩ (b)(i) 1.2Ω (ii) 13.33 kΩ (iii) 2.29 Ω (iv) 461.54 Ω
4. 8 Ω **5.** 27.5 Ω **6.** 2.5 Ω, 6 A
7. (a) 1.6 A (b) 6 Ω
8. $I_1 = 5\text{ A}, I_2 = 2.5\text{ A}, I_3 = 1\frac{2}{3}\text{ A}, I_4 = \frac{5}{6}\text{ A}, I_5 = 3\text{ A}, I_6 = 2\text{ A}\ V_1 = 20\text{ V}, V_2 = 5\text{ V}, V_3 = 6\text{ V}$
9. 1.8 A **10.** 7.2 Ω **11.** 30 V

Exercise 38, Page 102

1. 44.44 V, potentiometer
2. (a) 0.545 A, 13.64 V (b) 0.286 A, 7.14 V, rheostat
3. 136.4 V **4.** 9.68 V **5.** 63.40 V

Exercise 39, Page 103

1. (a) +40 V, +29.6 V, +24 V (b) +10.4 V, +16 V (c) −5.6 V, −16 V
2. (a) 1.68 V (b) 0.16 A (c) 460.8 mW (d) +2.88 V (e) +2.88 V
3. (a) 10 V, 10 V (b) 0 V

Exercise 40, Page 105

1. 400 Ω **2.** (a) 70 V (b) 210 V

Chapter 8

Exercise 41, Page 109

1. 2.5 mC **2.** 2 kV **3.** 2.5 μF
4. 1.25 ms **5.** 7.5μF

Exercise 42, Page 111

1. 750 kV/m **2.** 50 kC/m^2 **3.** 312.5 μC/m^2, 50 kV/m
4. 226 kV/m
5. 250 kV/m (a) 2.213 μC/m^2 (b) 11.063 μC/m^2

Exercise 43, Page 112

1. 885 pF **2.** 0.885 mm **3.** 65.14 pF
4. 7 **5.** 2.97 mm **6.** 1.67
7. (a) 0.005 mm (b) 10.44 cm^2

Exercise 44, Page 115

1. (a) 8 μF (b) 1.5 μF
2. 15 μF
3. 2.4 μF, 2.4 μF
4. (a)(i) 14 μF (ii) 0.17 μF (iii) 500 pF (iv) 0.0102 μF (b)(i) 1.143 μF (ii) 0.0125 μF (iii) 45 pF (iv) 196.1 pF
5. (a) 1.2 μF (b) 100 V
6. (a) 4 μF (b) 3 mC (c) 250 V
7. (a) 150 V, 90 V (b) 0.45 mC on each
8. 4.2 μF each
9. (a) 0.857 μF (b) 1.071 mJ (c) 42.85 μC on each

Exercise 45, Page 117

1. (a) 0.02 μF (b) 0.4 mJ **2.** 20 J
3. 550 V **4.** (a) 1.593 μJ (b) 5.31 μC/m^2 (c) 600 kV/m
5. (a) 0.04 mm (b) 361.6 cm^2 (c) 0.02 J (d) 1 kW

Chapter 9

Exercise 46, Page 126

1. 1.5 T **2.** 2.7 mWb **3.** 32 cm
4. (a) 5000 A (b) 6631 A/m

Exercise 47, Page 128

1. (a) 262600 A/m (b) 3939 A **2.** 23870 A
3. 960 A/m **4.** 1 A **5.** 7.85 A
6. 1478 **7.** (a) 79580 /H (b) 1 mH/m
8. 1000 **9.** (a) 466000 /H (b) 233 **10.** 0.60 A
11. (a) 110 A (b) 0.25 A

Exercise 48, Page 131

1. 0.195 mWb **2.** (a) 270 A (b) 1860 A **3.** 0.83 A
4. 550 A, 1.83×10^6/H **5.** 2970 A

Chapter 10

Exercise 49, Page 142

1. 21.0 N, 14.8 N **2.** 4.0 A **3.** 0.80 T
4. 0.582 N **5.** (a) 14.2 mm (b) towards the viewer
6. (a) 2.25×10^{-3}N (b) 0.9 N

Exercise 50, Page 144

1. 8×10^{-19} N **2.** 10^6 m/s

Chapter 11

Exercise 51, Page 149

1. 0.135 V **2.** 25 m/s **3.** (a) 0 (b) 0.16 A
4. 1.56 mV **5.** (a) 48 V (b) 33.9 V (c) 24 V
6. (a) 10.21 V (b) 0.408 N

Exercise 52, Page 151

1. 72.38 V **2.** 47.50 V
3. (a) 1243 rev/min, 568 rev/min (b) 1.33 T, 1.77 T

Exercise 53, Page 152

1. −150 V **2.** 144 ms **3.** 0.8 Wb/s
4. 3.5 V

Exercise 54, Page 153

1. 0.18 mJ **2.** 40 H

Exercise 55, Page 155

1. (a) 7.2 H (b) 90 J (c) 180 V **2.** 4 H
3. 40 ms **4.** 12.5 H, 1.25 kV **5.** 4.8 A
6. 12500 **7.** 0.1 H, 80 V
8. (a) 1.492 A (b) 33.51 mH (c) −50 V

Exercise 56, Page 156

1. 4.5 V **2.** 1.6 mH **3.** 250 V
4. (a) −180 V (b) 5.4 mWb
5. (a) 0.30 H (b) 320 kA/Wb (c) 0.18 H

Chapter 12

Exercise 57, Page 161

1. (a) 2 kΩ (b) 10 kΩ (c) 25 kΩ
2. (a) 18.18 Ω (b) 10.00 mΩ (c) 2.00 mΩ
3. 39.98 kΩ
4. (a) 50.10 mΩ in parallel (b) 4.975 kΩ in series

Exercise 58, Page 164

1. (a) 0.250 A (b) 0.238 A (c) 2.832 W (d) 56.64 W
2. (a) 900 W (b) 904.5 W **3.** 160 V, 156.7 V
4. (a) 24 mW, 576 W (b) 24 mW, 57.6 mW

Exercise 59, Page 169

1. (a) 41.7 Hz (b) 176 V **2.** (a) 0.56 Hz (b) 8.4 V
3. (a) 7.14 Hz (b) 220 V (c) 77.78 V

Exercise 60, Page 176

1. (a) 4.77 dB (b) 10 dB (c) 13 dB (d) 40 dB
2. (a) −10 dB (b) −4.77 dB (c) −16.02 dB (d) −20 dB
3. 13.98 dB **4.** 13 dB **5.** 2.51 W
6. 0.39 mV
7. (a) 0.775 V (b) 0.921 V (c) 0.138 V (d) −3.807 dB

Exercise 61, Page 178

1. 3 kΩ **2.** 1.525 V

Exercise 62, Page 180

1. 6.25 mA ± 1.3% or 6.25 ± 0.08 mA
2. 4.16 Ω± 6.08% or 4.16 ± 0.25 Ω
3. 27.36 Ω± 2.6% or 27.36 ± 0.71 Ω

Chapter 13

Exercise 63, Page 189

1. to **10.** Descriptive answers may be found from within the text on pages 182 to 188
11. (a) Germanium (b) 17 mA (c) 0.625 V (d) 50 Ω

Exercise 64, Page 194

1. to **5.** Descriptive answers may be found from within the text on pages 189 to 194
6. (a) 5.6 V (b) −5.8 V (c) −5 mA (d) 195 mW

Chapter 14

Exercise 65, Page 203

1. to **7.** Descriptive answers may be found from within the text on pages 195 to 203
8. 1.25 A **9.** 24 **10.** (a) 32.5 μA (b) 20 kΩ (c) 3 kΩ
11. 98

Exercise 66, Page 211

1. (a) false (b) true (c) false (d) true (e) true (f) true (g) true
2. to **5.** Descriptive answers may be found from within the text on pages 204 to 211
6. (a) 5V, 7 mA (b) 8.5 V
7. (a) 12.2 V, 6.1 mA (b) 5.5 V (c) 2.75 **8.** 1200
9. (a) 5.2 V, 3.7 mA (b) 5.1 V (c) 106 (d) 87 (e) 9222

Chapter 15

Exercise 67, Page 223

1. $I_3 = 2$ A, $I_4 = -1$ A, $I_6 = 3$ A

2. (a) $I_1 = 4$ A, $I_2 = -1$ A, $I_3 = 13$ A (b) $I_1 = 40$ A, $I_2 = 60$ A, $I_3 = 120$ A, $I_4 = 100$ A, $I_5 = -80$ A

3. $I_1 = 0.8$ A, $I_2 = 0.5$ A

4. 2.162 A, 42.07 W

5. 2.715 A, 7.410 V, 3.948 V

6. (a) 60.38 mA (b) 15.09 mA (c) 45.29 mA (d) 34.20 mW

7. $I_1 = 1.259$ A, $I_2 = 0.752$ A, $I_3 = 0.153$ A, $I_4 = 1.412$ A, $I_5 = 0.599$ A

Exercise 68, Page 226

1. $I_1 = 2$ A, $I_2 = 3$ A, $I_3 = 5$ A **2.** 0.385 A
3. 10 V battery discharges at 1.429 A, 4 V battery charges at 0.857 A, current through 10 Ω resistor is 0.571 A
4. 24 V battery charges at 1.664 A, 52 V battery discharges at 3.280 A, current in 20 Ω resistor is 1.616 A

Exercise 69, Page 232

1. 0.434 A, 2.64 W **2.** 2.162 A, 42.07 W
3. See answers for Exercise 68 above
4. 0.918 A **5.** 0.153 A from B to A

Exercise 70, Page 235

1. See answers for Exercise 68 above
2. See answers for questions 1, 2, 4 and 5 of Exercise 69 above
3. 2.5 mA

Exercise 71, Page 238

1. (a) $I_{SC} = 25$ A, r = 2Ω (b) $I_{SC} = 2$ mA, r = 5Ω
2. (a) E = 20 V, r = 4Ω (b) E = 12 mV, r = 3Ω
3. (a) E = 18 V, r = 1.2Ω (b) 6 A
4. $E = 9\frac{1}{3}$ V, r = 1Ω, $1\frac{1}{3}$ A

Exercise 72, Page 240

1. 2 Ω, 50 W **2.** 147 W **3.** 30 W
4. (a) 5 A (b) r = 4Ω (c) 40 V
5. $R_L = 1.6\Omega$, P = 57.6 W

Chapter 16

Exercise 73, Page 245

1. (a) 0.4 s (b) 10 ms (c) 25 μs
2. (a) 200 Hz (b) 20 kHz (c) 5 Hz **3.** 800 Hz

Exercise 74, Page 248

1. (a) 50 Hz (b) 5.5 A, 3.1 A (c) 2.8 A (d) 4.0 A
2. (a)(i) 100 Hz (ii) 2.50 A (iii) 2.87 A (iv) 1.15 (v) 1.74
(b)(i) 250 Hz (ii) 20 V (iii) 20 V (iv) 1.0 (v) 1.0
(c)(i) 125 Hz (ii) 18 A (iii) 19.56 A (iv) 1.09 (v) 1.23
(d)(i) 250 Hz (ii) 25 V (iii) 50 V (iv) 2.0 (v) 2.0
3. (a) 150 V (b) 170 V **4.** 212.1 V **5.** 282.9 V, 180.2 V **6.** 84.8 V, 76.4 V **7.** 23.55 A, 16.65 A

Exercise 75, Page 251

1. (a) 20 V (b) 25 Hz (c) 0.04 s (d) 157.1 rad/s
2. (a) 90 V, 63.63 V, 200 Hz, 5 ms, 0°
(b) 50 A, 35.35 A, 50 Hz, 0.02 s, 17.19° lead
(c) 200 V, 141.4 V, 100 Hz, 0.01 s, 23.49° lag
3. $i = 30 \sin 120\pi t$ A
4. $v = 200 \sin(10\pi t - 0.384)$ V
5. (a) 200 V, 25 Hz, 0.04 s, 29.97° lagging (b) −49.95 V (c) 66.96 V (d) 7.426 ms, 19.23 ms (e) 25.95 ms, 40.71 ms (f) 13.33 ms

Exercise 76, Page 254

1. (a) $v_1 + v_2 = 12.6 \sin(\omega t - 0.32)$V
(b) $v_1 - v_2 = 4.4 \sin(\omega t + 2)$V
2. (a) $v_1 + v_2 = 12.58 \sin(\omega t - 0.324)$V
(b) $v_1 - v_2 = 4.44 \sin(\omega t + 2.02)$V
3. $i = 23.43 \sin(\omega t + 0.588)$ A
4. (a) $13.14 \sin(\omega t + 0.217)$V
(b) $94.34 \sin(\omega t + 0.489)$V
(c) $88.88 \sin(\omega t + 0.751)$V
5. (a) $229 \sin(314.2t - 0.233)$V (b) 161.9 V (c) 50 Hz
6. (a) $12.96 \sin(628.3t + 0.762)$V (b) 100 Hz (c) 10 ms
7. (a) $97.39 \sin(300\pi t + 0.620)$ V (b) 150 Hz (c) 6.67 ms (d) 68.85 V

Chapter 17

Exercise 77, Page 261

1. (a) 62.83 Ω (b) 754 Ω (c) 50.27 kΩ **2.** 4.77 mH
3. 0.637 H **4.** (a) 628 Ω (b) 0.318 A
5. (a) 397.9 Ω (b) 15.92 Ω (c) 1.989 Ω **6.** 39.79 μF
7. 15.92 μF, 0.25 A **8.** 0.25 μF

Exercise 78, Page 264

1. 20 Ω **2.** 78.27 Ω, 2.555 A, 39.95° lagging
3. (a) 40 Ω (b) 1.77 A (c) 56.64 V (d) 42.48 V
4. (a) 4 Ω (b) 8 Ω (c) 22.05 mH
5. (a) 200Ω (b) 223.6 Ω (c) 1.118 A (d) 111.8 V, 223.6 V (e) 63.43° lagging

Exercise 79, Page 265

1. 28 V **2.** (a) 93.98 Ω (b) 2.128 A (c) 57.86° leading
3. (a) 39.05 Ω (b) 4.526 A (c) 135.8 V (d) 113.2 V (e) 39.81° leading
4. 225 kHz

Exercise 80, Page 269

1. (a) 13.18 Ω (b) 15.17 A (c) 52.63° lagging (d) 772.1 V (e) 603.6 V
2. R = 131Ω, L = 0.545 H
3. (a) 11.12 Ω (b) 8.99 A (c) 25.92° lagging (d) 53.92 V, 78.53 V, 76.46 V
4. $V_1 = 26.0$ V at 67.38° lagging, $V_2 = 67.05$ V at 72.65° leading, V = 50 V, 53.14° leading

Exercise 81, Page 271

1. 3.183 kHz, 10 A **2.** 1.25 kΩ, 63.3 μH
3. (a) 1.453 kHz (b) 8 A (c) 36.51
4. 0.158 mH **5.** 100, 150 V

Exercise 82, Page 275

1. 13.33 W **2.** 0, 628.3 VA **3.** 1875 W, £38.43
4. 2.016 kW **5.** 132 kW, 0.66
6. 62.5 kVA, 37.5 kvar **7.** 5.452 W
8. (a) 154.9 Ω (b) 0.968 A (c) 75 W **9.** 60 Ω, 255 mH
10. (a) 7 A (b) 53.13° lagging (c) 4.286 Ω (d) 7.143 Ω (e) 9.095 mH
11. 37.5 Ω, 28.61 μF

Chapter 18

Exercise 83, Page 279

1. (a) $I_R = 3.67$A, $I_L = 2.92$A (b) 4.69 A (c) 38.51° lagging (d) 23.45 Ω (e) 404 W (f) 0.782 lagging
2. 102 mH

Exercise 84, Page 280

1. (a) $I_R = 0.625A, I_C = 0.943A$ (b) 1.131 A (c) 56.46° leading (d) 8.84 Ω (e) 6.25 W (f) 11.31 VA (g) 0.553 leading
2. $R = 125\Omega, C = 9.55\mu F$

Exercise 85, Page 281

1. (a) $I_L = 1.194A, I_C = 0.377A$ (b) 0.817 A (c) 90° lagging (d) 73.44 Ω (e) 0 W
2. (a) $I_L = 0.597A, I_C = 0.754A$ (b) 0.157 A (c) 90° leading (d) 382.2 Ω (e) 0 W

Exercise 86, Page 284

1. (a) 1.715 A (b) 0.943 A (c) 1.028 A at 30.88° lagging (d) 194.6 Ω (e) 176.5 W (f) 205.6 VA (g) 105.5 var
2. (a) 18.48 mA (b) 62.83 mA (c) 46.17 mA at 81.49° leading (d) 2.166 kΩ (e) 0.683 W

Exercise 87, Page 288

1. (a) 4.11 kHz (b) 38.73 mA **2.** (a) 37.68 Ω (b) 2.94 A (c) 2.714 A
3. (a) 127.2 Hz (b) 600 Ω (c) 0.10 A (d) 4.80
4. (a) 1561 pF (b) 106.8 kΩ (c) 93.66 μA
5. (a) 2.533 pF (b) 5.264 MΩ (c) 418.9 (d) 11.94 kHz (e) $I_C = 15.915\angle 90°$ mA, $I_{LR} = 15.915\angle -89.863°$ mA (f) 38 μA (g) 7.60 mW

Exercise 88, Page 293

1. (a) 84.6 kVA (b) 203.9 A(c) 84.6 kVA
2. (a) 22.80 A (b) 19.50 A
3. (a) 25.13 Ω (b) $32.12\angle 51.49°\Omega$ (c) $6.227\angle -51.49°A$ (d) 0.623 (e) 775.5 W (f) 77.56 μF (g) 47.67 μF
4. (a) 50 A (b) 34.59 A (c) 25.28 A (d) 268.2 μF (e) 6.32 kvar
5. 21.74 A, 0.966 lagging, 21.74 μF

Chapter 19

Exercise 89, Page 301

1. 39.35 V **2.** (a) 0.309 μF (b) 14.47 V **3.** 105.0 V, 23.53 s
4. 55.90 μF **5.** 1.08 MΩ
6. (a) 0.10 A (b) 50 ms (c) 36.78 mA (d) 30.12 V
7. 150 ms, 3.67 mA, 1.65 mA
8. 80 ms, 0.11 A (a) 66.7 mA (b) 40.5 mA
9. (a) 0.60 s (b) 200 V/s (c) 12 mA (d) 0.323 s
10. (a) 35.95 μA (b) 89.87 V (c) 12.13 mJ
11. (a)(i) 80 μA (ii) 18.05 V (iii) 0.892 s (b)(i) 40 μA (ii) 48.30 μJ

Exercise 90, Page 306

1. 4.32 A **2.** 1.95 A (1.97 A by calculation)
3. (a) 0.984 A (b) 0.183 A
4. (a) 25 ms (b) 6.32 A (c) 8.65 A
5. (a) 0.15 s (b) 2.528 A (c) 0.75 s (d) 0.147 s (e) 26.67 A/s
6. (a) 64.38 ms (b) 0.20 s (c) 0.20 J (d) 7.67 ms

Chapter 20

Exercise 91, Page 314

1. 22.5 V **2.** 6×10^4 **3.** 3.75×10^{-3}, 92.04 dB
4. (a) −1.0 V (b) +1.5 V
5. (a) −80 (b) 1.33 mV **6.** (a) 3.56 MΩ (b) 1.78 MHz

Exercise 92, Page 319

1. (a) 3.21 (b) −1.60 V **2.** (a) −10 V (b) +5 V
3. −3.9 V **4.** 0.3 V
5. (a) −60 mV (b) +90 mV (c) −150 mV (d) +225 mV

Chapter 22

Exercise 93, Page 341

1. (a) 231 V (b) 4.62 A (c) 4.62 V
2. (a) 400 V (b) 8 A (c) 13.86 A
3. (a) 212 V (b) 367 V
4. (a) 165.4 μF (b) 55.13 μF
5. (a) 16.78 mH (b) 73.84 mH
6. $I_R = 64.95A, I_Y = 86.60A, I_B = 108.25A, I_N = 37.50A$
7. (a) 219.4 V (b) 65 A (c) 37.53 A
8. 8 μF

Exercise 94, Page 347

1. (a) 9.68 kW (b) 29.04 kW
2. 1.35 kW **3.** 5.21 kW

4. (a) 0.406 (b) 10 A (c) 17.32 A (d) 98.53 V
5. 0.509 **6.** (a) 13.39 kW (b) 21.97 A (c) 12.68 A
7. (a) 14.7 kW (b) 0.909
8. 5.431 kW, 2.569 kW
9. (a) 17.15 kW, 5.73 kW (b) 51.46 kW, 17.18 kW
10. (a) 27.71 A (b) 11.52 kW (c) 19.20 kVA
11. (a) 4.66 A (b) 8.07 A (c) 2.605 kW (d) 5.80 kVA

Chapter 23

Exercise 95, Page 352

1. 96 **2.** 990 V **3.** 400 V
4. 16 V **5.** 12 V, 60 A **6.** 50 A
7. 16 V, 45 A **8.** (a) 50 A (b) 4 Ω (c) 4.17 A
9. 225 V, 3:2

Exercise 96, Page 354

1. (a) 20 A (b) 2 kVA **2.** 0.786 A, 0.152 A
3. (a) 0.40 A (b) 0.40 (c) 0.917 A

Exercise 97, Page 356

1. (a) 37.5 A, 600 A (b) 800 (c) 9.0 mWb
2. (a) 1.25 T (b) 3.90 kV **3.** 464, 58
4. (a) 150, 5 (b) 792.8 cm^2

Exercise 98, Page 357

1. 23.26 A, 0.73

Exercise 99, Page 359

1. (a) 0.92 Ω, 3.0 Ω, 3.14 Ω (b) 72.95°

Exercise 100, Page 360

1. 2.5% **2.** 106.7 volts

Exercise 101, Page 362

1. (a) 2.7 kVA (b) 2.16 kW (c) 5 A
2. 96.10% **3.** 95.81% **4.** 97.56%
5. (i) 96.77% (ii) 96.84% (iii) 95.62%
6. (a) 204.1 kVA (b) 97.61%

Exercise 102, Page 365

1. 3.2 kΩ **2.** 3:1 **3.** (a) 30 A (b) 4.5 kW
4. 1:8 **5.** (a) 78.13 Ω (b) 5 mW

Exercise 103, Page 366

1. $I_1 = 62.5$ A, $I_2 = 100$ A, $(I_2 - I_1) = 37.5$ A
2. (a) 80% (b) 25%

Exercise 104, Page 368

1. (a) 649.5 V (b) 216.5 V

Exercise 105, Page 369

1. (a) 5 A (b) 1 V (c) 7.5 VA

Chapter 24

Exercise 106, Page 375

1. 270 V **2.** 15 rev/s or 900 rev/min
3. (a) 400 V (b) 200 V **4.** 50%

Exercise 107, Page 379

1. 238 V **2.** (a) 500 volts (b) 505 V
3. (a) 240 V (b) 112.5 V (c) 270 V
4. (a) 425 volts (b) 431.68 V **5.** 304.5 volts
6. (a) 315 V (b) 175 V (c) 381.2 V **7.** 270 V

Exercise 108, Page 381

1. 82.14%

Exercise 109, Page 381

1. 326 V **2.** (a) 175 volts (b) 235 V
3. (a) 224 V (b) 238 V

Exercise 110, Page 383

1. 123.1 V **2.** 65.2 N m **3.** 203.7 N m
4. 167.1 N m
5. (a) 5.5 rev/s or 330 rev/min (b) 152.8 N m
6. (a) 83.4% (b) 748.8 W

Exercise 111, Page 389

1. (a) 78 A (b) 208.8 V **2.** 559 rev/min
3. 212 V, 15.85 rev/s **4.** 81.95%
5. 30.94 A **6.** 78.5% **7.** 92%
8. 80% **9.** 21.2 A, 1415 rev/min

Exercise 112, Page 392

1. (a) 11.83 rev/s (b) 16.67 rev/s **2.** 2 Ω
3. 1239 rev/min

Chapter 25

Exercise 113, Page 397

1. 120 Hz **2.** 2 **3.** 100 rev/s

Exercise 114, Page 399

1. 3% **2.** (a) 750 rev/min (b) 731 rev/min (c) 1.25 Hz
3. 1800 rev/min **4.** (a) 1500 rev/min (b) 4% (c) 2 Hz

Exercise 115, Page 400

1. (a) 25 rev/min (b) 5% (c) 2.5 Hz
2. (a) 0.04 or 4% (b) 960 rev/min

Exercise 116, Page 402

1. (a) 1.92 kW (b) 46.08 kW (c) 45.08 kW (d) 90.16%
2. (a) 28.80 kW (b) 36.40%

Exercise 117, Page 404

1. (a) 50 rev/s or 3000 rev/min (b) 0.03 or 3%
(c) 22.43 N m (d) 6.34 kW (e) 40.74 N m
(f) 45 rev/s or 2700 rev/min (g) 8.07 N m
2. (a) 13.27 A (b) 211.3 W (c) 45.96 A
3. (a) 7.57 kW (b) 83.75% (c) 13.0 A
4. 4.0 Ω

Chapter 26

Exercise 118, Page 417

1. (a) $-1+j7$ (b) $4+j3$
2. (a) $18+j$ (b) $12+j$
3. (a) $21+j38$ (b) $3+j5$
4. (a) $-\frac{6}{25}+j\frac{17}{25}$ (b) $-\frac{2}{3}+j$
5. (a) $\frac{80}{26}+j\frac{23}{26}$ (b) $-\frac{11}{5}-j\frac{12}{5}$
6. 4 **7.** $x=0.188, y=0.216$
8. (a) -4 (b) $\frac{3}{5}-j\frac{5}{5}$ (c) $\frac{2}{13}-j\frac{3}{13}$
9. $0-j2$ **10.** $Z=21.62+j8.39$
11. (a) $0.0472+j0.0849$ (b) $0.1321-j0.0377$
(c) $0.1793+j0.0472$ (d) $5.2158-j1.3731$

Exercise 119, Page 418

1. $a=\frac{7}{4}, b=-\frac{3}{4}$ **2.** $x=18, y=-1$
3. $a=18, b=-14$ **4.** $e=21, f=20$
5. $R_x=\frac{R_3C_4}{C_1}, C_x=\frac{C_1R_4}{R_3}$

Exercise 120, Page 419

1. (a) 5, 53°8′ (b) 5.385, −68°12′
2. (a) 4.123, 165°58′ (b) 5.831, −149°2′
3. (a) $\sqrt{61}\angle 39°48'$ (b) $\sqrt{13}\angle -33°41'$
(c) $3\angle 180°$ or $3\angle\pi$
4. (a) $\sqrt{26}\angle 168°41'$ (b) $5\angle -143°8'$
(c) $2\angle -90°$ or $2\angle -\pi/2$
5. (a) $5.195+j3.000$ (b) $2.000+j3.464$
(c) $2.121+j2.121$
6. (a) $0+j2.000$ (b) $-3.000+j0$ (c) $-4.330+j2.500$
7. (a) $-6.928+j4.000$ (b) $-2.100-j3.637$
(c) $3.263-j1.521$
8. (a) $10\angle 60°$ (b) $11.18\angle 117°$
9. (a) $2.9\angle 45°$ (b) $6\angle 115°$
10. (a) $10.93\angle 168°$ (b) $7.289\angle -24°35'$
11. (a) $x=4.659, y=-17.387$ (b) $r=30.52$,
$\theta=81°31'$
12. $6.61\angle 37.24°$
13. $(4-j12)\Omega$ or $12.65\angle -71.57°\Omega$
14. $6.36\angle 11.46°$ A
15. $(10+j20)\Omega$, $22.36\angle 63.43°\Omega$

Exercise 121, Page 422

1. (a) $-5+j12$; $13\angle 112.62°$
(b) $-9-j40$; $41\angle -102.68°$
2. (a) $597+j122$; $609.3\angle 11.55°$
(b) $-2-j11$; $11.18\angle -100.30°$

3. (a) $-157.6+j201.7; 256\angle 128^\circ$
(b) $-2.789-j31.88; 32\angle -95^\circ$
4. (a) $-27+j0; 27\angle -\pi$
(b) $0.8792+j4.986; 5.063\angle 80^\circ$
5. (a) $\pm(1.455+j0.344)$ (b) $\pm(1.818-j0.550)$
6. (a) $\pm(1+j2)$ (b) $\pm(1.040-j1.442)$
7. (a) $\pm(2.127+j0.691)$ (b) $\pm(-2.646+j2.646)$
8. $\sqrt{5}\angle -26.57^\circ; 279.5\angle 174.04^\circ$

Chapter 27

Exercise 122, Page 429

1. (a) $R=4\Omega, L=22.3$ mH (b) $R=3\Omega, C=159.2\mu F$
(c) $R=0, L=31.8$ mH (d) $R=0, C=1.061\mu F$
(e) $R=7.5\Omega, L=41.3$ mH
(f) $R=4.243\Omega, C=0.750$ nF
2. $1.257\angle 90^\circ$ A or $j1.257$ A
3. 27.66 V
4. $246.8\angle 0^\circ$ V
5. (a) $Z=(8.61-j2.66)\Omega$ or $9.01\angle -17.19^\circ\Omega$
(b) $R=8.61\Omega, C=59.83\mu F$
6. 45.95 Ω
7. (a) $(68+j127.5)$ V or $144.5\angle 61.93^\circ$ V (b) $68\angle 0^\circ$ V
(c) $127.5\angle 90^\circ$ V (d) 61.93° lagging
8. (a) 20 Ω (b) 106.1 μF (c) 2.774 A (d) 56.31° leading
9. (a) 61.28 Ω (b) 16.59 μF (c) $59.95\angle -130^\circ$ V
(d) $242.9\angle 10.90^\circ$ V
10. $1.69\angle -28.36^\circ$ A; $(1.49-j0.80)$ A

Exercise 123, Page 433

1. (a) $44.53\angle -63.31^\circ\Omega$ (b) $19.77\angle 52.62^\circ\Omega$
(c) $113.5\angle -58.08^\circ\Omega$
2. $(1.85+j6.20)\Omega$ or $6.47\angle 73.39^\circ\Omega$
3. 8.17 A lagging V by 35.20°
4. $4.44\angle 42.31^\circ$ mA
5. (a) 50 V (b) 40.01° lagging (c) $16.8\angle 30^\circ$ V
(d) $33.6\angle 45^\circ$ V
6. (a) $150\angle 45^\circ$ V (b) $86.63\angle -54.75^\circ\Omega$
(c) $1.73\angle 99.75^\circ$ A (d) 54.75° leading
(e) $86.50\angle 99.75^\circ$ V (f) $122.38\angle 9.75^\circ$ V
7. (a) $39.95\angle -28.82^\circ\Omega$ (b) $3.00\angle 28.82^\circ$ A (c) 28.82° leading (d) $V_1=44.70\angle 86.35^\circ$ V,
$V_2=36.00\angle 28.82^\circ$ V, $V_3=105.56\angle -35.95^\circ$ V
8. $V_1=164\angle -12.68^\circ$ V or $(160-j\,36)$ V,
$V_2=104\angle 67.38^\circ$ V or $(40+j\,96)$ V,
$V=208.8\angle 16.70^\circ$ V or $(200+j60)$ V
Phase angle = 16.70° lagging
9. 46.04 μF
10. (i) $3.71\angle -17.35^\circ$ A (ii) $V_1=55.65\angle 12.65^\circ$ V,
$V_2=37.10\angle -77.35^\circ$ V, $V_3=44.52\angle 32.65^\circ$ V

Chapter 28

Exercise 124, Page 438

1. (a) $0.1\angle -90^\circ$ S, 0, 0.1 S (b) $0.025\angle 90^\circ$ S, 0, 0.025 S
(c) $0.03125\angle 30^\circ$ S, 0.0271 S, 0.0156 S
(d) $0.0971\angle -60.93^\circ$ S, 0.0472 S, 0.0849 S
(e) $0.0530\angle 32.04^\circ$ S, 0.0449 S, 0.0281 S
2. (a) $20\angle -40^\circ\Omega$ (b) $625\angle 25^\circ\Omega$
(c) $2.425\angle -75.96^\circ\Omega$ (d) $21.20\angle 57.99^\circ\Omega$
3. $R=86.21\Omega, L=109.8$ mH
4. (a) $R=5.62\Omega, X_L=8.99\Omega$
(b) $R=20\Omega, X_L=12.5\Omega$
5. $R=50\Omega, C=7.958\mu F$
6. (a) $(0.0154-j0.0231)$ S or $0.0277\angle -56.31^\circ$ S
(b) $(0.132-j0.024)$ S or $0.134\angle -10.30^\circ$ S
(c) $(0.08+j0.01)$ S or $0.0806\angle 7.13^\circ$ S
(d) $(0.0596-j0.0310)$ S or $0.0672\angle -27.49^\circ$ S

Exercise 125, Page 442

1. (a) $(4-j8)\Omega$ or $8.94\angle -63.43^\circ\Omega$
(b) $(7.56+j1.95)\Omega$ or $7.81\angle 14.46^\circ\Omega$
(c) $(14.04-j0.74)\Omega$ or $14.06\angle -3.02^\circ\Omega$
2. $I_1=8.94\angle -10.30^\circ$ A, $I_2=17.89\angle 79.70^\circ$ A
3. (a) $10\angle 36.87^\circ\Omega$ (b) $3\angle -16.87^\circ$ A
4. (a) $10.33\angle -6.31^\circ\Omega$ (b) $4.84\angle 6.31^\circ$ A
(c) 6.31° leading (d) $I_1=0.953\angle -73.38^\circ$ A,
$I_2=4.765\angle 17.66^\circ$ A
5. (a) $15.08\angle 90^\circ$ A (b) $3.39\angle -45.15^\circ$ A
(c) $12.90\angle 79.33^\circ$ A (d) $9.30\angle -79.33^\circ\Omega$
(e) 79.33° leading
6. (a) $0.0733\angle 43.39^\circ S$ (b) $13.64\angle -43.39^\circ\Omega$
(c) $1.833\angle 43.39^\circ$ A (d) 43.39° leading
(e) $I_1=0.455\angle -43.30^\circ$ A, $I_2=1.863\angle 57.50^\circ$ A,
$I_3=1\angle 0^\circ$ A, $I_4=1.570\angle 90^\circ$ A
7. $32.63\angle 43.55^\circ$ A
8. 7.53 A

Exercise 126, Page 444

1. (a) $(9.73+j1.81)\Omega$ or $9.90\angle 10.56^\circ\Omega$
2. (a) $1.632\angle -17.10^\circ$ A (b) $5.412\angle -8.46^\circ$ A
3. $7.65\angle -33.63^\circ$ A

4. (a) $11.99\angle -31.81°$ A (b) $8.54\angle 20.56°\Omega$ (c) $102.4\angle -11.25°$ V (d) 20.56° lagging (e) $86.0\angle 17.91°$ V (f) $I_1 = 7.37\angle -13.05°$ A, $I_2 = 5.54\angle -57.16°$ A
5. (a) $6.25\angle 52.34°\Omega$ (b) $16.0\angle 7.66°$ A (c) $R = 3.819\Omega$, $L = 0.394$ mH
6. (a) $39.31\angle -61.84°\Omega$ (b) $3.816\angle 61.84°$ A (c) 61.84° leading (d) $2.595\angle 60.28°$ A (e) $1.224\angle 65.14°$ A
7. (a) $10.0\angle 36.87°\Omega$ (b) $150\angle 66.87°$ V (c) $90\angle 51.92°$ V (d) $2.50\angle 18.23°$ A

Chapter 29

Exercise 127, Page 453

1. 36 W
2. (a) 2.90 kW (b) 1.89 kvar lagging (c) 3.46 kVA
3. 480 W
4. $(4.0 + j3.0)\Omega$ or $5\angle 36.87°\Omega$
5. (a) $25\angle -60°$ A (b) 4.64 Ω (c) 12.4 mH
6. 1550 W; 1120 var lagging
7. (a) 2469 W (b) 2970 VA (c) 0.83 lagging
8. (a) $24\angle 60°\Omega$ (b) 337.5 W (c) 584.6 var lagging (d) 675 VA
9. (a) 600 VA (b) 360 var leading (c) $(3 - j3.6)\Omega$ or $4.69\angle -50.19°\Omega$
10. (a) $I_1 = 6.20\angle 29.74°$ A, $I_2 = 19.86\angle -8.92°$ A (b) 981 W (c) 153.9 var leading (d) 992.8 VA
11. (a) 49.34 W (b) 28.90 var leading (c) 0.863 leading
12. (a) 254.1 W (b) 0 (c) 65.92 W

Exercise 128, Page 457

1. 334.8 kvar leading
2. (a) $10\angle -27.38°$ A (b) 500 W, 1300 VA, 1200 var lagging (c) 1018.5 var leading (d) 4.797 μF
3. (a) $3.51\angle 58.40°\Omega$ (b) 35.0 A (c) 4300 VA (d) 3662 var lagging (e) 0.524 lagging (f) 542.3 μF
4. (a) 184 kW (b) 108.4 kvar lagging (c) 0.862 lagging
5. (a) 23.56 Ω (b) $27.93\angle 57.52°\Omega$ (c) $7.16\angle -57.52°$ A (d) 0.537 (e) 769 W (f) 70 μF

Chapter 30

Exercise 129, Page 469

1. $L = 0.20$ H, $R = 10\Omega$
2. 6 μF
3. $R = 4.00\Omega$, $L = 3.96$ mH
4. $C_x = 38.77$ nF, $R_x = 3.27$ kΩ
5. $R_x = 10$ kΩ, $C_x = 100$ pF
6. $R_x = 2$ kΩ, $L_x = 0.2$ H
7. $R_x = 500\Omega$, $C_x = 4\mu$ F
8. (a) 250 Ω (b) 9.95 kHz
9. (a) 3 Ω (b) 0.4 μF (c) 0.0075 (d) 0.432°
10. $R_x = 2$ kΩ, $L_x = 1.5$ H
11. $R = 59.41\Omega$, $L = 37.6$ mH
12. (a) $R_x = 6.25$ kΩ, $C_x = 240$ pF (b) 0.81°
13. (a) 649.7 Hz (b) 375 Ω

Chapter 31

Exercise 130, Page 474

1. (a) 142.4 Hz (b) 12.5 A
2. (a) 200 Ω (b) 316.6 pF
3. (a) 6.50 kHz (b) 102.1 V (c) 102.2 V
4. (a) 26 Ω , 154.8 nF (b) 639.6 Hz
5. (a) 100 pF (b) 1.60 mH

Exercise 131, Page 479

1. (a) 200 mH (b) $V_R = 10\angle 0°$ V, $V_L = 500\angle 90°$ V, $V_C = 500\angle -90°$ V (c) 50
2. (a) 5033 Hz (b) 10.54 V (c) 4741 Hz (d) 10.85 V
3. 60.61
4. (a) 9 V (b) 50 Ω, 0.179 H, 8.84 nF
5. 158.3 μH, 16.58

Exercise 132, Page 484

1. (a) 400 Hz (b) 100 (c) 4 Hz (d) 398 Hz and 402 Hz
2. (a) 95.5 mH (b) 66.3 nF (c) 50 Hz (d) 1975 Hz and 2025 Hz (e) 42.43 Ω
3. 238.7 Hz
4. (a) 79.06 (b) 6366 Hz
5. (a) 31.66 pF (b) 99.46 μA (c) 250 mV (d) 40 kHz (e) 2.02 MHz, 1.98 MHz
6. (a) $0.25\angle 0°$ A (b) $0.223\angle 27.04°$ A (c) $12.25\angle -11.54°\Omega$

Chapter 32

Exercise 133, Page 492

1. (a) 63.66 Hz (b) 100 Ω (c) 0.30 A (d) 2
2. (a) 48.73 nF (b) 6.57 kΩ (c) 3.81 mA (d) 5.03 (e) 497.3 Hz (f) 2761 Hz, 2264 Hz (g) 4.64 kΩ

3. (a) 3.56 kHz (b) 26.83 mA
4. (i)(a) 7958 Hz (b) 16.67 (c) 300 Ω (ii)(a) 7943 Hz (b) 16.64 (c) 83.33 kΩ
5. (a) 32.5 Hz (b) 25.7 Hz
6. 2.30 μF
7. (a) 355.9 Hz (b) 318.3 Hz

Exercise 134, Page 494

1. 10 Ω
2. 2.50 mH or 0.45 mH
3. 667 Hz
4. 11.87 Ω
5. 928 Hz, 5.27 Ω
6. 73.8
7. (a) 250 (b) 4 kΩ (c) 1.486 kΩ

Chapter 33

Exercise 135, Page 504

1. 50 V source discharges at 2.08 A, 20 V source charges at 0.62 A, current through 20 Ω resistor is 1.46 A
2. $I_A = 5.38$ A, $I_B = 4.81$ A, $I_C = 0.58$ A
3. (a) 4 A (b) 1 A (c) 7 A
4. (a) 1.59 A (b) 3.72 V (c) 3.79 W
5. $40\angle 90°$ V source discharges at $4.40\angle 74.48°$ A, $20\angle 0°$ V source discharges at $2.94\angle 53.13°$ A, current in 10 Ω resistance is $1.97\angle 107.35°$ A (downwards)
6. 1.58 A
7. (a) 0.14 A (b) 10.1 A (c) 2.27 V (d) 1.81 W
8. $I_A = 2.80\angle -59.59°$ A, $I_B = 2.71\angle -58.78°$ A, $I_C = 0.097\angle 97.13°$ A
9. (a) 0.275 A (b) 0.700 A (c) 0.292 A discharging (d) 1.607 W
10. 11.37 V

Chapter 34

Exercise 136, Page 510

1. The answers to problems 1 to 10 of Exercise 135 are given above.
2. $6.96\angle -49.94°$ A, 644 W
3. $I_1 = 8.73\angle -1.37°$ A, $I_2 = 7.02\angle 17.25°$ A, $I_3 = 3.05\angle -48.67°$ A
4. 0
5. (a) 14.5 A (b) 11.5 A (c) 71.8 V (d) 2499 W
6. $I_R = 7.84\angle 71.19°$ A, $I_Y = 9.04\angle -37.49°$ A, $I_B = 9.89\angle -168.81°$ A
7. (a) 1.03 A (b) 1.48 A (c) 16.28 W (d) 3.47 V
8. $I_1 = 83\angle 173.13°$ A, $I_2 = 83\angle 53.13°$ A, $I_3 = 83\angle -66.87°$ A, $I_R = 143.8\angle 143.13°$ A, $I_Y = 143.8\angle 23.13°$ A, $I_B = 143.8\angle -96.87°$ A
9. $I_A = 2.40\angle 52.52°$ A, $I_B = 1.02\angle 46.19°$ A, $I_C = 1.39\angle 57.17°$ A, $I_D = 0.67\angle 15.57°$ A, $I_E = 0.996\angle 83.74°$ A

Exercise 137, Page 517

1. The answers to problems 1, 2, 5, 8 and 10 of Exercise 135 are given above.
2. The answers to problems 2, 3, 5 and 9 of Exercise 136 are given above.
3. $V_1 = 59.0\angle -28.92°$ V, $V_{AB} = 45.3\angle 10.89°$ V
4. $V_{PQ} = 55.87\angle 50.60°$ V
5. $I_A = 1.21\angle 150.95°$ A, $I_B = 1.06\angle -56.31°$ A, $I_C = 0.56\angle 32.01°$ A
6. (a) $V_1 = 88.12\angle 33.86°$ V, $V_2 = 58.72\angle 72.28°$ V (b) $2.20\angle 33.86°$ A away from node 1 (c) $2.80\angle 118.65°$ A away from node 1 (d) 226 W
7. $V_{AB} = 54.23\angle -102.52°$ V

Chapter 35

Exercise 138, Page 523

1. The answers to problems 1, 5, 8 and 9 of Exercise 135 are given above.
2. The answers to problems 3 and 5 of Exercise 136 are given above.
3. The answers to problem 5 of Exercise 137 are given above.
4. 6.86 A, 2.57 A, 4.29 A
5. 2.584 A
6. (a) $1.30\angle 38.65°$ A downwards (b) $20\angle 90°$ V source discharges at $1.58\angle 120.97°$ A, $30\angle 0°$ V source discharges at $1.90\angle -16.50°$ A
7. $0.529\angle 5.72°$ A

Exercise 139, Page 526

1. (a) 1.28 A (b) 0.74 A (c) 3.02 V (d) 2.91 W
2. (a) 0.375 A, 8.0 V, 57.8 W (b) 0.625 A
3. (a) 3.97 A (b) 28.7 V (c) 36.4 W (d) 371.6 W

Chapter 36

Exercise 140, Page 535

1. 0.85 A **2.** $I_1 = 2.8$ A, $I_2 = 4.8$ A, $I_3 = 7.6$ A
3. $E = 15.37\angle -38.66°$ V, $z = (3.20 + j4.00)\Omega$, 1.97 A, 15.5 W
4. 1.17 A **5.** $E = 2.5$ V, $r = 5\Omega$, 0.10 A

Exercise 141, Page 539

1. $E = 14.32\angle 6.38°$ V, $z = (3.99 + j0.55)\Omega$, 1.29 A
2. 1.157 A
3. 0.24 W
4. 0.12 A from Q to P
5. The answers to problems 1 to 10 of Exercise 135 are given above.
6. The answers to problems 2 and 3 of Exercise 136 are given above.
7. The answers to problems 3 to 7 of Exercise 137 are given above.
8. The answers to problems 4 to 7 of Exercise 138 are given above.
9. The answers to problems 1 to 3 of Exercise 139 are given above.

Exercise 142, Page 545

1. The answers to problems 1 to 4 of Exercise 140 are given above.
2. The answers to problems 1 to 3 of Exercise 141 are given above.
3. 3.13 A
4. 1.08 A
5. $I_{SC} = 2.185\angle -43.96°$ A, $z = (2.40 + j1.47)\Omega$, 0.88 A
6. The answers to problems 1 to 10 of Exercise 135 are given above.
7. The answers to problems 2 and 3 of Exercise 136 are given above.
8. The answers to problems 3 to 6 of Exercise 137 are given above.

Exercise 143, Page 549

1. (a) $I_{SC} = 2.5$ A, $z = 2\Omega$ (b) $I_{SC} = 2\angle 30°$ A, $z = 5\Omega$
2. (a) $E = 20$ V, $z = 4\Omega$ (b) $E = 12\angle 50°$ V, $z = 3\Omega$
3. (a) $E = 18$ V, $z = 1.2\Omega$ (b) 3.6 A
4. $E = 9\frac{1}{3}$ V, $z = 1\Omega$, $1.87\angle -53.13°$ A
5. $E = 4.82\angle -41.63°$ V, $z = (0.8 + j0.4)k\Omega$, 4.15 V
6. (a) $E = 6.71\angle -26.57°$ V, $z = (4.50 + j3.75)\Omega$
(b) $I_{SC} = 1.15\angle -66.38°$ A, $z = (4.50 + j3.75)\Omega$
(c) 0.60 A

Chapter 37

Exercise 144, Page 560

1. (a) $Z_1 = 0.4\Omega$, $Z_2 = 2\Omega$, $Z_3 = 0.5\Omega$
(b) $Z_1 = -j100\Omega$, $Z_2 = j100\Omega$, $Z_3 = 100\Omega$
2. $Z_1 = 5.12\angle 78.35°\Omega$, $Z_2 = 6.82\angle -26.65°\Omega$, $Z_3 = 10.23\angle -11.65°\Omega$
3. (a) $7.32\angle 24.06°$ A (b) 668 W
4. Each impedance = $8\angle 60°\Omega$
5. 131 mA
6. 31.25 V
7. 4.47 A
8. (a) 9.73 A (b) 98.6 W

Exercise 145, Page 563

1. (a) $Z_{12} = 18\Omega$, $Z_{23} = 9\Omega$, $Z_{31} = 13.5\Omega$
(b) $Z_{12} = (10 + j0)\Omega$, $Z_{23} = (5 + j5)\Omega$, $Z_{31} = (0 - j10)\Omega$
2. $Z_{12} = 35.93\angle 40.50°\Omega$, $Z_{23} = 53.89\angle -19.50°\Omega$, $Z_{31} = 26.95\angle -49.50°\Omega$
3. Each impedance = $(6 + j9)\Omega$
4. (a) 5 Ω, 6 Ω, 3 Ω (b) 21 Ω, 12.6 Ω, 10.5 Ω

Chapter 38

Exercise 146, Page 570

1. $r = 9\Omega$, $P = 208.4$ W
2. (a) 11.18 Ω (b) 151.1 W
3. (a) $(10 + j5)\Omega$ (b) 160 W
4. (a) 26.83 Ω (b) 35.4 W
5. $R_L = 2.1\Omega$, $P = 23.3$ W
6. (a) 16 Ω (b) 48 W
7. (a) $R = 1.706\Omega$, $X = 0.177\Omega$ (b) 269 W
8. (a) 7.07 Ω (b) 596.5 Ω
9. $R = 3.47\Omega$, $X = -0.93\Omega$, 13.6 W
10. (a) $R = 0.80\Omega$, $X = -1.40\Omega$, $P = 225$ W
(b) $R = 1.61\Omega$, $P = 149.2\Omega$

Exercise 147, Page 573

1. 4:1 **2.** 7 Ω **3.** 2.70:1
4. (a) 15 A (b) 2.81 kW
5. (a) 40 Ω (b) 9.77 mW

Chapter 39

Exercise 148, Page 577

1. (a) 2nd and 4th (b) 106.1 V (c) 100 Hz (d) 10 ms (e) 200 Hz, 400 Hz
2. (a) 80 Hz (b) 40% (c) 10% (d) 5.656 A (e) 3rd harmonic leading by $\pi/2$ rad (i.e. leading by 90°), 5th harmonic lagging by $\pi/5$ rad (i.e. lagging by 36°) (f) 1.274 A
3. $30\sin 800\pi t + 12\sin(2400\pi t + \pi/3)$ volts

Exercise 149, Page 584

1. (a) $i = (70.71\sin 628.3t + 16.97\sin 1885t$ A
2. (a) $v = 300\sin 314.2t + 90\sin(628.3t - \pi/2) + 30\sin(1256.6t + \pi/3)$ volts
3. Sketch
4. $i = (16\sin 2\pi 10^3 t + 3.2\sin 6\pi 10^3 t + 1.6\sin \pi 10^4 t)$ A
5. (a) Fundamental and even harmonics, or all harmonics present, initially in phase with each other (b) Fundamental and odd harmonics only (c) Fundamental and even harmonics, initially out of phase with each other, or all harmonics present, some being initially out of phase with each other
6. (a) 60 Hz, 180 Hz, 300 Hz (b) 40% (c) 10%

Exercise 150, Page 589

1. $f(x) = \frac{8}{\pi}\left(\sin x + \frac{1}{3}\sin 3x + \frac{1}{5}\sin 5x + ...\right)$
2. $\frac{2}{3\pi}\sin 3x$
3. $f(t) = \frac{2}{\pi}\left(\cos t - \frac{1}{3}\cos 3t + \frac{1}{5}\cos 5t - .... + \sin 2t + \frac{1}{3}\sin 6t + \frac{1}{5}\sin 10t + ...\right)$
4. $f(x) = \pi - 2\left(\sin x + \frac{1}{2}\sin 2x + \frac{1}{3}\sin 3x + \frac{1}{4}\sin 4x + ...\right)$
5. $f(x) = \frac{\pi}{2} - \frac{4}{\pi}\left(\cos x + \frac{\cos 3x}{3^2} + \frac{\cos 5x}{5^2} + ...\right)$
6. $f(x) = \pi + 2\left(\sin x - \frac{1}{2}\sin 2x + \frac{1}{3}\sin 3x - \frac{1}{4}\sin 4x + \frac{1}{5}\sin 5x - ....\right)$

Exercise 151, Page 593

1. $f(x) = \frac{4}{\pi}\left(\cos x - \frac{1}{3}\cos 3x + \frac{1}{5}\cos 5x - \frac{1}{7}\cos 7x + ...\right)$
2. $f(t) = -2\left(\sin t + \frac{1}{2}\sin 2t + \frac{1}{3}\sin 3t + \frac{1}{4}\sin 4t + ...\right)$
3. $f(x) = \frac{\pi}{2} + 1 - \frac{4}{\pi}\left(\cos x + \frac{1}{3^2}\cos 3x + \frac{1}{5^2}\cos 5x + ...\right)$
4. $f(t) = \frac{5}{2} + \frac{10}{\pi}\left(\sin\left(\frac{\pi t}{10}\right) + \frac{1}{3}\sin\left(\frac{3\pi t}{10}\right) + \frac{1}{5}\sin\left(\frac{5\pi t}{10}\right) + ...\right)$
5. $f(x) = \frac{5}{2} - \frac{5}{\pi}\left(\sin\left(\frac{2\pi x}{5}\right) + \frac{1}{2}\sin\left(\frac{4\pi x}{5}\right) + \frac{1}{3}\sin\left(\frac{6\pi x}{5}\right) + ...\right)$
6. $f(x) = \frac{12}{\pi}\left(\sin\left(\frac{\pi x}{2}\right) + \frac{1}{3}\sin\left(\frac{3\pi x}{2}\right) + \frac{1}{5}\sin\left(\frac{5\pi x}{2}\right) + ...\right)$

Exercise 152, Page 596

1. 2.54 A
2. 191.4 V
3. (a) 150 Hz (b) 111.8 V (c) 91.7 V (d) 1.22
4. (a) 211.2 V, 52.8 V, 31.7 V (b) $v = (298.7\sin 377t + 74.7\sin 1131t + 44.8\sin 1885t)$ volts
5. 1.038

Exercise 153, Page 598

1. 0.54 W
2. (a) 146.1 W (b) 0.526
3. 109.8 V, 1.65 A, 14.60 W
4. (a) 42.85 V (b) 4.189 A (c) 175.5 W

Exercise 154, Page 602

1. (a) $i = 4.8\sin\omega t + 1.2\sin(3\omega t - \pi/4) + 0.6\sin(5\omega t + \pi/3)$ A
(b) $i = 3\sin(\omega t + \pi/2) + 2.25\sin(3\omega t + \pi/4) + 1.875\sin(5\omega t + 5\pi/6)$ A
(c) $i = 6\sin(\omega t - \pi/2) + 0.5\sin(3\omega t - 3\pi/4) + 0.15\sin(5\omega t - \pi/6)$ A
2. (a) 6.671 mA (b) $v = 99.47\sin(\omega t - \pi/6) + 53.05\sin(3\omega t + \pi/6)$ V (c) 79.71 V

3. (a) $i = 32.25\sin(314t - 0.256) + 5.50\sin(942t - 0.666) + 2.53\sin(1570t - 0.918)$ A (b) 145.6 V (c) 23.20 A (d) 3.23 kW (e) 0.956

4. (i) 115.5 V (ii) $i = 0.639\sin(\omega t + 1.130) + 0.206\sin(2\omega t + 0.030) + 0.088\sin(4\omega t - 0.559)$ A (iii) 0.479 A

5. (a) $i = 0.40 + 1.414\sin(\omega t - \pi/4) + 0.158\sin(3\omega t - 0.464) + 0.029\sin(5\omega t - 0.850)$ A (b) 1.08 A (c) 117 W

Exercise 155, Page 605

1. (a) 49.3 W (b) $R = 80\Omega$, $C = 4.0\mu F$

2. (a) 59.06 Ω, $R = 50\Omega$, $L = 10$ mH (b) 80 V, 30 V (c) 180.1 V (d) 2.93 A (e) 427.8 W (f) 0.811

3. (a) $i = 2.873\sin(300t + 1.071) - 0.145\sin(1500t - 1.425) + 0.149\sin(2100t - 1.471)$ A (b) 166 W

4. (a) $i = 0.134\sin(314t + 0.464) + 0.047\sin(942t + 0.988)$ A (b) 35.07% (c) 7.72 W (d) $v_l = 53.6\sin(314t + 0.464) + 18.8\sin(942t + 0.988)$ V (e) $i_R = 0.095\sin(314t - 0.321) + 0.015\sin(942t - 0.261)$ A, $i_C = 0.095\sin(314t + 1.249) + 0.045\sin(942t + 1.310)$ A

Exercise 156, Page 607

1. (a) 118.6 Hz (b) 0.105 A, 4 A

2. (a) 5 A (b) 23.46 kV (c) 6.29 mA

3. (a) 22.54 μF (b) $i = 1.191\sin(314t + 1.491) + 2.500\sin(942t - 0.524) + 0.195\sin(1570t - 0.327)$ A (c) 1.963 A (d) 38.56 W

4. 13

5. (a) 877.5 V (b) $i = 7.991\sin(\omega t + 1.404) + 12\sin 3\omega t + 1.555\sin(5\omega t - 1.171)$ A (c) 10.25 A (d) 666.4 V (e) 2626 W (f) 0.292

Chapter 40

Exercise 157, Page 615

1. $y = 23.92 + 7.81\cos\theta + 14.61\sin\theta + 0.17\cos 2\theta + 2.31\sin 2\theta - 0.33\cos 3\theta + 0.50\sin 3\theta$

2. $v = 5.00 - 10.78\cos\theta + 6.83\sin\theta + 0.13\cos 2\theta + 0.79\sin 2\theta + 0.58\cos 3\theta - 1.08\sin 3\theta$

3. $i = 0.64 + 1.58\cos\theta - 2.73\sin\theta - 0.23\cos 2\theta - 0.42\sin 2\theta + 0.27\cos 3\theta + 0.05\sin 3\theta$

Exercise 158, Page 618

1. (a) only odd cosine terms present (b) only even sine terms present

2. $y = 9.4 + 13.2\cos\theta - 24.1\sin\theta + 0.92\cos 2\theta - 0.14\sin 2\theta + 0.83\cos 3\theta + 0.67\sin 3\theta$

3. $i = 4.00 - 4.67\cos 2\theta + 1.00\sin 4\theta - 0.66\cos 6\theta$

Chapter 41

Exercise 159, Page 625

1. (a) 4 kJ/m^3 (b) 240 kW/m^3 (c) 58.68 kW/m^3
2. 216 W **3.** 0.974 W **4.** 157.8 W
5. 171.4 W **6.** 9.6 W

Exercise 160, Page 629

1. 6.25 W/m^3 **2.** 0.20 mm
3. (a) $P_h = 52.5$ W, $P_e = 56.25$ W (b) 36.6 W
4. 2173.6 W **5.** 3 W, 2.88 W **6.** 2090 W

Exercise 161, Page 631

1. (a) 20 W, 37.5 W (b) 32 W, 96 W
2. (a) 40 W (b) 12.5 W **3.** 20 W, 50 W
4. (a) 375 W, 56.25 W (b) 624.4 W, 190.5 W

Chapter 42

Exercise 162, Page 640

1. 1.35° or 0.024 rad (a) 0.0236 (b) 42.4 (c) 0.0236

2. (a) 0.311 μF, 1.066 MΩ (b) 0.311 μF, 68.24 Ω

3. (a) 0.219° or 3.82×10^{-3} rad (b) 208.3 kΩ

4. (a) 0.912° or 0.0159 rad (b) 126.7Ω (c) 0.5 MΩ

5. (a) 0.014° or 2.5×10^{-4} (b) 2.5×10^{-4} (c) 8 MΩ

Chapter 43

Exercise 163, Page 646

1. (a) See textbook pages 643 to 646 (b) 23.4 pF

2. 1.66 nF

3. (a) 0.44 pF (b) 0.60–0.70 pF, depending on the accuracy of the plot

Exercise 164, Page 651

1. 48.30 pF
2. (a) 0.1517 μF (b) 1.819 MV/m, 1.364 MV/m
3. 5.93 mm
4. (a) 10 mm, 27.2 mm (b) 177.9 pF (c) 5.59 A
5. (a) 0.1578 μF (b) 4.957 A (c) 1239 W
6. (a) 40 mm, 108.7 mm (b) 5 MV/m, 1.84 MV/m
7. 200.6 pF

Exercise 165, Page 653

1. 5.014 nF
2. (a) 4.875 pF (b) 30.63 mA
3. 10 mm
4. (a) 5.248 nF (b) 36.74 μC (c) 11.54 mA
5. (a) 4.775 pF (b) 5.92 mm

Exercise 166, Page 655

1. 1.6 mJ, 80 W
2. 4 kV
3. (a) 500 V (b) 12 μF
4. (a) 20 nF (b) 0.625 mJ
5. (a) 17.7 μC/m^2 (b) 2 MV/m (c) 17.7 μJ
6. (a) 0.1 mm (b) 0.113m^2 (c) 0.1 J (d) 4 kW
7. 0.147 J/m^3
8. 1.328 mJ

Exercise 167, Page 659

1. 10.2 μH **2.** 0.291 μH/m, 147.8 pF/m
3. 3.28 mm **4.** (a) 5.794 km (b) 0546 μF

Exercise 168, Page 661

1. 1.915 μH/m
2. (a) 0.592 mH (b) 5.133 nF
3. 765 mm
4. $L = \frac{\mu_0}{\pi}\left(\frac{1}{4} + \ln\frac{2D}{d}\right)$
5. 53.6 mm

Exercise 169, Page 664

1. 3.6 mJ, 169.7 mA **2.** 15.92 mJ **3.** 0.15 T
4. $\frac{\mu_0 I^2}{\pi}\ln\left(\frac{D_2}{D_1}\right)$ joules **5.** (a) 0.112 T (b) 89200 A/m

Chapter 44

Exercise 170, Page 670

1. (a) 4.77 dB (b) 10 dB (c) 14.77 dB (d) 40 dB
2. (a) −10 dB (b) −3 dB (c) −16 dB (d) −30 dB
3. (a) 3.795 mW (b) 37.9 μW
4. 11.25 dB
5. (i)(a) 1.406 A (b) 1.727 Np (ii)(a) 99.53 mA (b) −0.921 Np

Exercise 171, Page 674

1. (a) 26.46 Ω (b) 244.9 Ω (c) 1.342 kΩ
2. (a) 7.45 Ω (b) 353.6 Ω (c) 189.7 Ω
3. $R_1 = 1165\Omega, R_2 = 384\Omega$
4. (a) 667 Ω (b) 6 dB
5. (i)(a) 50 Ω (b) 42 Ω (c) 45.83 Ω (ii)(a) 285.7 Ω (b) 240 Ω (c) 261.9 Ω
6. $R_1 = 69.8\Omega, R_2 = 36.7\Omega$
7. (a) $R_1 = 349\Omega, R_2 = 184\Omega$ (b) $R_1 = 1.36k\Omega$, $R_2 = 716\Omega$
8. (a) $R_1 = 239.4\Omega, R_2 = 214.5\Omega$ (b) $R_1 = 357.4\Omega$, $R_2 = 45.13\Omega$ (c) $R_1 = 387.5\Omega, R_2 = 12.68\Omega$
9. $R_1 = 4.736\ k\Omega, R_2 = 680.8\Omega$
10. (a) $R_1 = 360\Omega, R_2 = 320\Omega$ (b) $R_1 = 1125\Omega$, $R_2 = 1000\Omega$

Exercise 172, Page 678

1. (a) 282.8 Ω (b) 15.31 dB
2. (a) 8.13 dB (b) 17.09 dB
3. (a) 196.1 Ω (b) 18.36 dB

Exercise 173, Page 681

1. (a) 144.9 Ω, 241.5 Ω (b) 143.6 Ω, 243.6 Ω
2. (a) 329.5 Ω, 285.6 Ω (b) 331.2 Ω, 284.2 Ω
3. 430.9 Ω, 603.3 Ω; 419.6 Ω, 619.6 Ω

Exercise 174, Page 683

1. $R_1 = 193.6\Omega, R_2 = 129.1\Omega$
2. (a) $R_1 = 489.9\Omega, R_2 = 244.9\Omega$ (b) 9.96 dB (c) 8.71 dB

Exercise 175, Page 685

1. (a) 15.56 dB (b) 77.80 dB
2. (a) 15 dB (b) 50 μV (c) 1.58 mV
3. (a) $R_1 = 150\Omega, R_2 = 225\Omega$ (b) $\frac{1}{9}$ (c) $\frac{1}{729}$ (d) 47.71 dB

Exercise 176, Page 689

1. (a) (4 − j5) V or $6.403\angle -51.34°$ V (b) (9.5 + j5) A or $10.74\angle 27.76°$ A

2. (a) 2 kΩ (b) 15 Ω

3. (a) $40\angle -15°\Omega$ (b) $50\angle -35°\Omega$

4. (a) $\begin{pmatrix} 6 & 5 \\ 5 & 3 \end{pmatrix}$ (b) $V_2 = 12$ V, $I_2 = 8$ A

5. $\begin{pmatrix} j4 & (4000+j2000) \\ (-0.001+j0.002) & (1+j2) \end{pmatrix}$

6. (a) A = 2, B = 1 kΩ, C = 1 mS and D = 1
(b) AD − BC = 1, hence the network is passive

Exercise 177, Page 695

1. (a) $\begin{pmatrix} 1 & 5 \\ 0 & 1 \end{pmatrix}$ (b) $\begin{pmatrix} 1 & 0 \\ 0.2 & 1 \end{pmatrix}$

2. $\begin{pmatrix} 2 & 5 \\ 0.2 & 1 \end{pmatrix}$

3. $\begin{pmatrix} (6+j5) & (10+j10) \\ 0.5 & 1 \end{pmatrix}$

4. $\begin{pmatrix} (3-j) & (6+j7)10^3 \\ -j0.001 & (3-j) \end{pmatrix}$

5. $\begin{pmatrix} 1\angle 60° & 173.2\angle 60° \\ 0.01\angle 90° & 1\angle 60° \end{pmatrix}$

6. $\begin{pmatrix} (0.85+j0.20) & (5+j10) \\ (0.0145+j0.039 & (0.85+j0.20 \end{pmatrix}$

7. (a) $\begin{pmatrix} 1.414\angle 45° & 100\angle 30° \\ 0.0224\angle 86.57° & 1.414\angle 45° \end{pmatrix}$
(b) $V_1 = 2.394\angle 38.79°$ V, $I_1 = 34.25\angle 70.67°$ mA

8. $\begin{pmatrix} 0 & j50 \\ -j0.02 & 0 \end{pmatrix}$

9. $\begin{pmatrix} 3 & 40 \\ 0.2 & 3 \end{pmatrix}$

10. Y = 0.2 S, Z = 10Ω

11. (a) $\begin{pmatrix} 3 & 10 \\ 0.8 & 3 \end{pmatrix}$ (b) Y = 0.8 S, Z = 2.5Ω

Exercise 178, Page 696

1. $131.6\angle -15°\Omega$ **2.** $66.82\angle -28.29°\Omega$
3. 14.14 Ω **4.** 3.536 Ω

Chapter 45

Exercise 179, Page 709

1. (a) 1592 Hz, 5 kΩ (b) 9545 Hz, 600 Ω
2. (a) Each series arm 79.60 mH, shunt arm 0.6366 μF
(b) Series arm 159.2 mH, each shunt arm 0.3183 μF
3. 0.203 μF **4.** (a) 1 kΩ (b) 360 Ω
5. 72 mH **6.** (a) 10 kHz (b) 8.66 kHz (c) 0

Exercise 180, Page 714

1. (a)(i) 22.51 kHz (ii) 14.14 kΩ (b)(i) 281.3 Hz (ii) 1414 Ω
2. (a) Each series arm = 53.06 nF, shunt arm = 14.92 mH
(b) Series arm = 26.53 nF, each shunt arm = 29.84 mH
3. 69.44 nF **4.** 11.26 mH
5. (a) 57.74 kHz (b) 64.05 kHz (c) 182 kHz
6. (a) Low-pass T section: each series arm 10.61 mH, shunt arm 58.95 nF; high-pass T section: each series arm 44.20 nF, shunt arm 7.96 mH (b) Band-pass filter

Exercise 181, Page 719

1. (a)(i) 1.79 Np (ii) −0.63 rad
(b)(i) 0.856 Np (ii) 1.08 rad
2. (a) 2.526 Np (b) 0.873 rad (c) (2.526 + j0.873) or $2.673\angle 19.07°$ (d) $8.19\angle -180°$μ A, (10.103 + j3.491) or $10.69\angle 19.06°$
3. (a) 1.61 Np (b) −2.50 rad (c) (1.61 − j2.50) or $2.97\angle -57.22°$
4. (a) 141.4 μs (b) 222.1 μs
5. (a) Each series arm 33.16 nF; shunt arm 5.97 mH
(b) 7
6. (a) Series arm 1.60 mH, each shunt arm 1.25 nF
(b) Each series arm 5 nF; shunt arm 1.60 mH

Exercise 182, Page 725

1. Each series arm 0.166 mH; shunt arm comprises 1.641 nF capacitor in series with 0.603 mH inductance
2. Each shunt arm 85.0 pF; series arm contains 61.21 μH inductance in parallel with 244.9 pF capacitor
3. Each series arm 0.503 μF; shunt arm comprises 90.49 mH inductance in series with 0.231 μF capacitor
4. Each shunt arm 36.51 mH inductance; series arm comprises 73.02 nF capacitor in parallel with 17.13 mH inductance

Exercise 183, Page 727

1. Prototype: Each series arm 53.1 mH; shunt arm comprises 0.424 μF capacitor

'm-derived': Each series arm 29.3 mH; shunt arm comprises 0.235 μF capacitor in series with 33.32 mH inductance

Half-sections: Series arm 31.8 mH; shunt arm comprises 0.127 μF capacitor in series with 56.59 mH inductance

2. Prototype: Series arm 795.8 pF; each shunt arm 1.592 mH

'm-derived': Each shunt arm 3.651 mH; series arm comprises 1.826 nF capacitor in parallel with 1.713 mH inductance

Half-sections: Shunt arm 2.653 mH; series arm comprises 2.653 nF capacitor in parallel with 1.492 mH inductance

Chapter 46

Exercise 184, Page 730

1. 10 V **2.** 50 mH **3.** −6 V
4. (a) 0.30 H (b) 0.117 H

Exercise 185, Page 731

1. 0.40 **2.** (a) 0.16 H (b) 0.653

Exercise 186, Page 733

1. (a) 170 mH (b) 130 mH
2. (a) 10 mH (b) 0.270
3. (a) 0.75 H (b) 0.75
4. (a) 200 mH (b) 1.298 H (c) 0.498 H
5. (a) $L_1 = 71.22$ mH or 8.78 mH, $L_2 = 8.78$ mH or 71.22 mH (b) 20 mH

Exercise 187, Page 739

1. $0.93\angle 68.20°$ V
2. 57 mV
3. (a) $0.197\angle -71.91°$ A (b) $0.030\angle -48.48°$ A
4. (a) $(100 + j200)\Omega$ (b) $(40 + j80)\Omega$ (c) $(40.5 - j81.0)\Omega$ (d) $(140.5 + j119)\Omega$ (e) $0.543\angle -40.26°$ A(f) $0.546\angle -13.69°$ A
5. (a) 14.235 kHz (b) 12.5 nF (c) 1659.9 Ω (d) $18.07\angle 0°$ mA (e) $80.94\angle 90°$ mA (f) $72.40\angle 0°$ V (g) 0.253

Exercise 188, Page 744

1. $I_P = 893.3\angle -60.57°$ mA, $I_S = 99.88\angle 2.86°$ mA
2. (a) 13.28 mH (b) $1.603\angle -28.97°$ A (c) $0.913\angle 17.71°$ A (d) $73.04\angle -27.29°$ V
3. (a) $I_g = 9.653\angle -36.03°$ A (b) $I_L = 1.084\angle 27.28°$ A
4. (a) $6.658\angle -28.07°$ A (b) $1.444\angle -7.79°$ A (c) $6.087\angle -35.38°$ A, $0.931\angle -73.25°$ A

Chapter 47

Exercise 189, Page 749

1. (a) 209.4 km (b) 251.3×10^6m/s
2. (a) 0.131 rad/km (b) 48 km (c) 240×10^6m/s
3. (a) 3.29 μH/m (b) 0.558×10^{-3} rad/m (c) 11.25 km

Exercise 190, Page 750

1. $V_R = 0.363\angle -6.25$ mV or $0.363\angle 1.90°$ mV, $I_R = 0.454\angle -6.25\mu$ A or $0.454\angle 1.90° \mu$ A
2. $0.340\angle -61.67°$ mA
3. 0.64 V

Exercise 191, Page 755

1. $387.3\angle -25°\Omega$
2. $443.4\angle -18.95°\Omega$
3. (a) 70.71 Ω (b) j4.443 (c) 0; 4.443 rad/km (d) 1.414 km (e) 70.70×10^6m/s
4. (a) $248.6\angle -13.29°\Omega$ (b) $0.0795\angle 65.91°$ (c) 0.0324 Np/km (d) 0.0726 rad/km
5. (a) $706.6\angle -17°\Omega$ (b) $0.0708\angle 70.14°$ (c) 0.024 Np/km; 0.067 rad/km (d) $21.1\angle 12.58°$ mA (e) $6.35\angle -178.21°$ mA (f) 94.34 km (g) 150.2×10^6m/s

Exercise 192, Page 757

1. 1.5 H
2. (a) 1.598 H (b) $(8.944 + j225)10^{-3}$ (c) 2.795×10^6m/s (d) 27.93 km

Exercise 193, Page 760

1. (a) −0.60 (b) 93.75 mA (c) 9.375 V (d) −56.25 mA (e) 5.625 V
2. (i) 0.125 (ii) 0.165 (iii) 0.062
3. (a) 0.345 (b) 42.32 V

Exercise 194, Page 764

1. (a) $0.319\angle -61.72°$ (b) 1.937
2. (a) 3 (b) $113.93\angle 43.32°\Omega$ (c) 50 mA
3. 14.06 mW
4. (a) $0.417\angle -138.35°$ (b) 2.43
5. 1.77

Chapter 48

Exercise 195, Page 771

1. (a) 0.5 mA (b) 9.16 μA (c) 2.59 V (d) 91.63 ms
2. 8.21 V
3. (a) 8 A (b) 6.92 A (c) 25.28 V (d) 26.81 V (e) 2.35 ms
4. (a) 114.5 ms (b) 8 V

Exercise 196, Page 774

1. (a) Underdamped (b) Overdamped
2. 800 pF
3. Underdamped; $i = -0.0127e^{-25t}\sin 1580.9\,t$ A
4. Critically damped; $i = -20te^{-250t}$ A

Exercise 197, Page 778

1. (a) $\frac{2}{s^2} - \frac{3}{s}$ (b) $\frac{10}{s^3} + \frac{4}{s^2} - \frac{3}{s}$
2. (a) $\frac{1}{4s^4} - \frac{3}{s^2} + \frac{2}{s}$ (b) $\frac{8}{s^6} - \frac{48}{s^5} + \frac{1}{s^3}$
3. (a) $\frac{5}{s-3}$ (b) $\frac{2}{s+2}$
4. (a) $\frac{12}{s^2+9}$ (b) $\frac{3s}{s^2+4}$
5. (a) $\frac{2}{(s-2)^2}$ (b) $\frac{2}{(s-1)^3}$
6. (a) $\frac{24}{(s+2)^4}$ (b) $\frac{12}{(s+3)^5}$
7. (a) $\frac{5(s+2)}{s^2+4s+13}$ (b) $\frac{4}{s^2+10s+26}$
8. (a) $\frac{4}{s}$ (b) $\frac{5}{s}e^{-2s}$ (c) $\frac{7}{s^2}$ (d) $\frac{3}{s^2}e^{-2s}$
9. (a) 15 (b) $6e^{-5s}$ (c) $\frac{80}{s^2+64}$
10. (a) 3 (b) 16 **11.** 4

Exercise 198, Page 780

1. (a) 7 (b) $2e^{5t}$ **2.** (a) $\frac{3}{2}e^{-\frac{1}{2}t}$ (b) $2\cos 2t$
3. (a) $\frac{1}{5}\sin 5t$ (b) $\frac{4}{3}\sin 3t$ **4.** (a) $\frac{5}{2}\cos 3t$ (b) 6t
5. (a) $\frac{5}{2}t^2$ (b) $\frac{4}{3}t^3$ **6.** (a) $\frac{5}{3}\sinh 3t$ (b) $2e^t t^2$
7. (a) $e^{2t}\sin 3t$ (b) $2e^{2t}\sin t$ **8.** (a) $e^{-t}\cos 3t$ (b) $\frac{3}{2}e^{-3t}\sin 2t$

Exercise 199, Page 781

1. $2e^t - 5e^{-3t}$ **2.** $4e^{-t} - 3e^{2t} + e^{-3t}$
3. $2e^{2t} + 7te^{2t}$ **4.** $2e^{-3t} + 3e^t - 4te^t$
5. $e^{-3t}\left(3 - 2t - 3t^2\right)$ **6.** $2 - 3e^{-2t}\cos 3t - \frac{2}{3}e^{-2t}\sin 3t$

Exercise 200, Page 784

1. $y = (3-t)\,e^{\frac{4}{3}t}$ **2.** $x = 2\cos 10t$
3. $i = 100te^{-500t}$ **4.** $x = 4\left(3e^{-2t} - 2e^{-4t}\right)$
5. $y = (4t-1)\,e^t + \frac{1}{3}e^{4t}$ **6.** $y = e^t - e^{-2t} + \sin 3t$

Exercise 201, Page 793

1. $\frac{1}{2\times 10^{-9}s}\Omega$ or $\frac{5\times 10^8}{s}\Omega$
2. $(50 + 0.4s)\Omega$
3. (a) $\left(100 + \frac{10^6}{s}\right)\Omega$ (b) $\left(500 + 0.01s + \frac{10^7}{4s}\right)\Omega$
(c) $\frac{0.1s}{10+0.01s}\Omega$ or $\frac{10s}{s+1000}\Omega$ (d) $\frac{10^4 s}{0.01s^2 + 10^6}\Omega$ or $\frac{10^6 s}{s^2 + 10^8}\Omega$
4. (a) $\left(20 + 0.02s + \frac{1}{2\times 10^{-5}s}\right)\Omega$
(b) $\left(0.05 + \frac{50}{s} + 2\times 10^{-5}s\right)\Omega$
5. $v_C = 5\left(1 - e^{-4t}\right)$ volts
6. $i = 4e^{-50t} - 4e^{-100t}$ A
7. $i = 2\sin 10t - 0.2\cos 10t + 0.2e^{-100t}$ A

8. (a) $\frac{E}{R}e^{-\frac{t}{CR}}$ (b) $EC\left(1-e^{-\frac{t}{CR}}\right)$

9. (a) $Ee^{-\frac{Rt}{L}}$ (b) $\frac{L}{R}\left(1-e^{-\frac{Rt}{L}}\right)$

10. $i=\frac{5}{R^2+L^2}\left\{Le^{-\frac{Rt}{L}}-L\cos t+R\sin t\right\}$

11. $i=1.5-1.5e^{-10t}$ A

Exercise 202, Page 799

1. $i=4.47e^{-10t}\sin 447t$ mA; underdamped

2. $i=2te^{-100t}$ A

3. $2.01e^{-1000t}\sinh 995t$ mA

4. (a) 13.53 mA (b) 13.53 V

5. $i=5e^{-10^6 t}$ A

Index

ABCD networks in cascade, 687
  parameters, 686
    for networks, 689
Absolute permeability, 126, 620
  permittivity, 110
  voltage, 103
A.c. bridges, 178, 460
  generator, 243, 608
  values, 245
Acceptor circuit, 269
Active networks, 666, 699
  power, 274, 449
Admittance, 437
Air capacitor, 117
Alkaline cell, 74
Alternating current, 242
  waveforms, 244
Alternative energy sources, 85
Aluminium, 184
Ammeter, 58, 159
Ampere, 50, 57
Amplifier gain, 312, 314
Amplitude, 166, 245, 249
Analogue instrument, 159, 165
Analogue-to-digital conversion, 320
Anderson bridge, 557
Angles, 25
Angular velocity, 249
Annulus, 41
Anode, 75, 190
Antimony, 184
Antinode, 761
Apparent power, 274, 449
Applications of resonance, 286
Areas of common shapes, 38
  irregular figures, 43
Argand diagram, 414
Argument, 418
Armature, 373
  reaction, 374
Arsenic, 184
Asymmetrical π-section, 678
  T-section, 666, 678
Atom, 57, 622
Attenuation, 175, 666
  bands, 698
  coefficient, 715, 756
  constant, 715, 750
Attenuators, 665, 666
  asymmetrical π, 678
  asymmetrical T, 678
  cascade, 683
  L-section, 681
  symmetrical π, 671
  symmetrical T, 670
Audio-frequency transformer, 357
Auto transformer, 365
  starting, 406
Auxiliary equation, 772
Avalanche effect, 188
  breakdown, 191
Average value, 245
Avometer, 68

Back e.m.f., 381
Band-pass filter, 700, 713
  -stop filter, 700
Bandwidth, 272, 311, 479
Barrier potential, 186
Base, 14
Batteries, 73, 74, 77, 82
  disposal of, 84
Bell, Alexander, 174, 668
  electric, 137
B-H curves, 127, 621
Bias, 199
Biomass, 85, 333
Bipolar junction transistor, 196, 199
  characteristics, 200, 203
BM 80, 59
Boron, 184
Brackets, 16
Breakdown voltage, 188
Bridge, a.c., 178, 460
Bridge megger, 59
  rectifier, 190, 255
Brush contact loss, 380
Brushes, 372
Buffer amplifier, 315

Calculator use, 4
Calibration accuracy, 179
Camera flash, 302
Candela, 50
Capacitance, 106, 108, 635
  between concentric cylinders, 646
  of isolated twin line, 651
Capacitive reactance, 260
Capacitors, 106, 109, 638
  charging, 295
  discharging, 121, 300
  energy stored in, 117
  in parallel, 112
  in series, 113
  parallel plate, 111
  practical types, 117, 638
  reservoir, 256
Carbon resistors, 66
Car ignition, 64
Cartesian axes, 33
Cartesian complex numbers, 414
Cathode, 75
Cell, 75
  capacity, 84
Ceramic capacitor, 118
Characteristic impedance, 666
  in terms of ABCD parameters, 695
  in terms of primary line constants, 751
  of filters, 701
  of transmission lines, 751
Characteristics, d.c. generator, 376–379
  d.c. motor, 314–318
  of semiconductors, 190
  of transistor, 200
Charge, 50, 109
  density, 110
Charged particles, 74
Charging of cell, 77
  capacitor, 295, 766
Chemical effect of electric current, 64, 74
  electricity, 74
Choke, 152
Circuit magnification factor, 475
  theory, 226
CIVIL, 261
Class A amplifier, 208
Closed-loop gain, 313
Coal, for electricity generation, 324
Coaxial cable, 646
Coefficient of coupling, 730
  proportionality, 13

Coercive force, 132, 623
Cofactor, 499
Coils in series, cumulatively coupled, 731
differentially coupled, 731
Collector, 199
Combination of waveforms, 251
Commercial bridge, 179
Common logarithms, 28
Common-mode rejection ratio, 311
Commutation, 372
Commutator, 372
Complex conjugate, 415
equations, 417
Complex numbers, 413
applications to parallel networks, 435
applications to series circuits, 421
Cartesian form, 415
De Moivres theorem, 420
determination of power, 450
equations, 417
operations involving, 415
polar form, 418
Complex wave, 173, 575
form factor, 595
frequency, 785
general equation, 576
harmonics in single-phase circuits, 599
mean value, 595
power associated, 597
resonance due to harmonics, 606
r.m.s. value, 594
sources of harmonics, 608
waveform considerations, 616
waveforms, 574
Composite filters, 725
series magnetic circuits, 129
Compound motor, 373, 387
winding, 373
wound generator, 378
Concentric cable, 646, 656
field plotting, 649
Conductance, 53, 436
Conductor, 57, 61
Constant-current source, 233
Constant-k high-pass filter, 712
low-pass filter, 706
Contact potential, 185, 186
Continuity tester, 59, 162
Continuous function, 585
loading, 756
Cooker, 64
Co-ordinates, 33
Copper loss, 360, 380
Core loss, 380
component, 353
Core type transformer construction, 357
Corrosion, 76
Cosine, 26
rule, 27
Coulomb, 50, 57
Coulomb's law, 107
Coupled circuits, 734
dot rule, 739
Coupling coefficient, 730
Covalent bonds, 184
Critically damped circuit, 772, 795
Cumulative compound motor, 387
Curie temperature, 622
Current, 50, 55
decay in L-R circuit, 305
division, 96, 439
gain, 202
growth in L-R circuit, 302, 768
magnification, 286, 489
main effects of, 64
transformer, 368
Current gain in transistors, 202
Curvilinear squares, 643
Cut-off frequency, 699
Cycle, 244

Damping, 143, 795
device, 159
D.c. circuit theory, 219
generators, 375
machine construction, 373
machine losses, 380
machines, 371
motors, 142, 371, 381
motor starter, 389
potentiometer, 177
transients, 294
Descartes, 414
Decibel, 174, 668
meter, 175
Degrees, 24
Delta connection, 340, 552
Delta-star comparison, 348
transformation, 553–561
De Moivre's theorem, 420
Denominator, 7
Depletion layer, 185, 186
Derived units, 49
De Sauty bridge, 178, 465
Design impedence, 704
Detector types, a.c. bridges, 462
Determinants, 488
Deviation from resonant frequency, 483
Diamagnetism, 621
Dielectric, 109, 110, 635
hysteresis, 638
liquid, 638
loss, 638
strength, 116, 636
stress, 647
Differential amplifier, 317, 318
compound motor, 387
equation solution, 782
voltage amplifier, 310
Differentiator circuit, 308
Diffusion, 186
Digital multimeter, 162
oscilloscopes, 159, 166
voltmeter, 161
Digital-to-analogue conversion, 320
Dimensions of most economical cable, 648
Diode characteristics, 190
Diodes, 182
Dirac function, 777
Direct proportion, 11
Discharging of capacitors, 121, 300
cells, 77
Disposal of batteries, 84
Dissipation factor, 639
Distortion on transmission line, 755
Diverter, 391
Domains, 622
Doping, 184
Dot rule for coupled circuits, 739
Double beam oscilloscope, 166
cage induction motor, 407
layer capacitance, 119
Drift, 57
Dust core, 632
Dynamic current gain, 202, 211
forward transfer conductance, 205
resistance, 285, 488

Earth, 103
point, 103
potential, 104
Eddy current loss, 360, 626
Edison cell, 80
Effective value, 245
Efficiency of d.c. generator, 380
d.c. motor, 387
induction motor, 401
transformer, 360
Electrical energy, 62
measuring instruments and measurements, 58, 158
potential, 53
power, 54, 61
quantities and units, 809

safety, 248
symbols, 57, 809
Electric bell, 137
cell, 75
circuit, 56
current, 50
field strength, 109, 635
fire, 64
flux density, 110, 635
force, 107
potential, 53
Electricity generation, 323
using biomass, 333
coal, 324
oil, 326
hydro power, 329
natural gas, 327
nuclear energy, 328
pumped storage, 330
solar energy, 333
tidal power, 331
wind, 331
Electrochemical series, 75
Electrodes, 74
Electrolysis, 74
Electrolyte, 74, 80
Electrolytic capacitors, 119
Electromagnetic induction, 145, 146
laws of, 147
Electromagnetic wave, 747
Electromagnetism, 135
Electromagnets, 137
Electromotive force, 53, 76
Electron, 57, 74, 183
Electronic instruments, 161
Electroplating, 75
Electrostatic field, 107
E.m.f., 53, 76
equation of transformer, 354
in armature winding, 374
Emitter, 199
follower, 199
Energy, 62
associated with travelling wave, 758
electrical, 54, 62
stored in capacitor, 117
stored in electric field, 153, 654
stored in electromagnetic field, 662
stored in magnetic field of inductor, 153, 662
Equations, complex, 417
indicial, 29
involving exponential functions, 30
of a straight line graph, 35
simultaneous, 21
sinusoidal waveforms, 248
solving, 16
Equipotential lines, 643
Equivalent circuit, transformer, 358
Errors, measurement, 179
Evaluating formulae, 5
Even function, 590, 616
Exponential growth and decay curves, 295
Exponentials, 30, 31
Extrapolation, 34

Farad, 108
Faraday, Michael, 108
Faraday's laws of electromagnetic induction, 147
Ferrites, 132, 632
Ferromagnetic-cored coils, 608
materials, 127, 621
Field effect transistor, 196, 204, 208
amplifiers, 206
characteristics, 205
Field plotting, 643
theory, 642
Filter, 256, 688
band-pass, 700
band-stop, 700
composite, 725
high-pass, 699, 703, 709
low-pass, 699, 702, 703
'm derived', 720
networks, 698
time delay, 718
Final value theorem, 778
Finite discontinuities, 585
Fleming's left-hand rule, 140
right-hand rule, 147
Fluke, 58, 162
Flux density, 125, 620
electric, 110
linkage, 656
Flux, magnetic, 125, 620
Force, 50
on a charge, 143
current-carrying conductor, 139
Forced magnetization, 610
resonant frequency, 489
Form factor, 245, 595
Formulae, 5, 215, 409, 802
transposing, 19
Forward bias, 186
characteristics, 187
transconductance, 205
Fossil fuels, 324
Fourier coefficients, 585
Fourier cosine series, 590
Fourier series, 578, 585
for non-periodic functions, 587
for periodic functions, 585
over any range, 591, 592
Fourier sine series, 591
Fractions, 7
Free magnetization, 609
Frequency, 166, 244, 249
resonant, 269, 285
Friction and windage losses, 380
Fuel cell, 84
Full wave rectification, 190, 255
Fundamental, 173, 576
Furnace, 64
Fuse, 57, 64, 248

Gallium arsenide, 184
Galvanometer, 177
Gas insulation, 638
Generating electricity, 323
using biomass, 333
coal, 324
oil, 326
hydro power, 329
natural gas, 327
nuclear energy, 328
pumped storage, 330
solar energy, 333
tidal power, 331
wind, 331
Generator, 64, 372, 375
a.c., 243
efficiency of, 380
Geothermal energy, 85
Germanium, 184
Gradient, of a graph, 35
Graphs, 33
Gravitational force, 51
Greek alphabet, 812
Grip rule, 137
Growth and decay, laws of, 31

Half-power points, 272, 479
–wave rectification, 254
–wave rectifier, 190
Harmonic analysis, 577
numerical method, 612
Harmonic resonance, 606
synthesis, 577
Harmonics, 173, 576
in single phase circuits, 599
sources of, 608
Hay bridge, 178, 463
Heating effect of electric current, 64
Heaviside bridge, 178
Henry, 151, 656

Hertz, 244
High-pass filter, 699, 703, 709
ladder, 703
'm derived', 723
Hole, 184, 185
Hydroelectricity, 85
Hydrogen cell, 84
Hydro power, for electricity generation, 329
Hyperbolic logarithms, 28
Hysteresis, 132, 623
loop, 132, 622
loss, 132, 360, 623

Image impedance, 678
Imaginary numbers, 413
Impedance, 261, 265
matching, 571
triangle, 261, 265
Improper fraction, 7
Impulse, 777
Impurity, 184
Incident wave, 757
Indices, 13
Indicial equations, 29
Indium, 184
Indium arsenide, 184
Induced e.m.f., 147, 656
Inductance, 151, 656, 747
mutual, 151, 155
of a coil, 153
of a concentric cylinder, 656
of an isolated twin line, 659
Induction motor, three-phase, 393
construction, 397
copper loss, 400
double cage, 407
impedance and current, 400
losses and efficiency, 401
principle of operation, 397
production of rotating magnetic field, 394
rotor e.m.f. and frequency, 399
slip, 398
starting, 405
synchronous speed, 396
torque equation, 402
–speed characteristic, 404
uses, 407
Inductive circuit, switching, 307
reactance, 259
Inductors, 152
Initial conditions, 797
value theorem, 778
Initial slope and three-point method, 296

Input bias current, 311
impedance, 313
offset current, 311
voltage, 311
Insertion loss, 675
ratio, 676
Instantaneous values, 245
Instrument 'loading' effect, 162
Insulated gate field effect transistor (IGFET), 204
Insulation, 248
Insulation and dangers of high current, 64
Insulation resistance tester, 59
Insulator, 57, 61, 183
Integrated circuit, 193
Integrator circuit, 307
op amp, 317
Internal resistance of a cell, 76
Interpolation, 34
Interpoles, 373
Intrinsic semiconductors, 186
Inverse Laplace transforms, 779
Inverse proportion, 13
Inverting op amp, 312
Ion, 57, 74
Iron, 64
loss, 360, 380
Irregular figures, 43
Isolating transformer, 367
Iterative impedance, 666, 678

Joule, 51, 54, 62
Junction gate field effect transistor (JFET), 204

Kelvin, 50
Kettle, 64
kilo, 51
Kilowatt-hour, 54, 62
Kirchhoff's laws, a.c., 498
d.c., 220
in the s-domain, 786
network analysis, a.c., 498, 499

Ladder networks, 702
Lag, angle of, 249
Lamps in series and parallel, 104
Laplace transforms, 774
by partial fractions, 780
capacitor, 785
definition of, 774
elementary functions, 774
final value theorem, 778
inductor, 785
initial conditions, 797
initial value theorem, 778
inverse, 779
L-R-C circuit, 794
of derivatives, 778
resistor, 784
to solve differential equations, 779
Lap winding, 373
Laws of electromagnetic induction, 147
growth and decay, 31
indices, 13
logarithms, 28
L-C parallel network, 280
Lead-acid cell, 79
Lead, angle of, 249
Leakage currents, 628
Leclanche cell, 79
Lenz's law, 147
Letter and digit code for resistors, 814
Level compounded machine, 379
Lifting magnet, 138
Light emitting diodes, 190, 193
Li-ion battery, 81
Linear device, 59
scale, 159
Lines of electric force, 107
magnetic force, 124
Liquid dielectrics, 678
Lithium-ion battery, 74, 81–84
Loading effect, 99, 162
Load line, 208
Local action, 75
Logarithmic ratios, 174, 668
Logarithms, 28
laws of, 28
Long shunt compound generator, 378
motor, 387
Loop currents, 507
inductance, 660
Loss angle, 466, 638
Losses in d.c. machines, 380
induction motor, 401
transformers, 360
Loudspeaker, 139
Low-pass filter, 699, 702, 703
ladder, 702
'm derived', 722
L-R-C circuit using Laplace transforms, 794
LR-C parallel network, 282
resonance, 487
LR-CR parallel network resonance, 488
L-R-C series circuit, 794
L-section attenuator, 681
Luminous intensity, 50
Lumped loading, 756

Magnetically coupled circuits, 728
Magnetic effect of electric current, 64
circuit, 122
field, 124, 136, 620
strength, 126, 620
flux, 124, 125, 620
density, 125, 620
force, 124
materials, 619
moment, 621
properties of materials, 621
screens, 127
space constant, 126
Magnetic field due to electric current, 136
Magnetization curves, 126, 620
Magnetizing component, 353
force, 126
Magnetomotive force, 125, 620
Magnification factor, 476
Majority carriers, 185, 186
Manganese battery, 74
Matched network, 676
Matching transformer, 363, 571
Mathematics revision, 3–45
Maximum efficiency, transformer, 362
Maximum power transfer theorems, d.c., 239
a.c., 565, 566
Maximum repetitive reverse voltage, 190
Maximum value, 245, 249
Maxwell bridge, 178, 463
Maxwell's theorem, 508
Maxwell-Wien bridge, 464
'm derived' filter, 720
Mean value, 245
of complex wave, 595
Measurement errors, 179
of angles, 25
of power in three phase system, 343
Mechanical analogy of parallel resonance, 286
Mega, 51
Megger, 59
Mercury cell, 79
Mesh-connection, 340, 552
-current analysis, 507
Metal film resistors, 66
Mica capacitor, 118
Micro, 51
Microelectronic systems, 611
Mid-ordinate rule, 43
Milli, 51
Minor, 499
Minority carriers, 186
Mismatched load, 757
Mixed number, 8
Modulus, 418
Mole, 50
Moore's circle technique, 644
Motor, 64, 381
compound, 387
cooling, 392
d.c., principle of operation, 142
efficiency, 387
series wound, 385, 391
shunt wound, 383, 390
speed control, 390
starter, 389
Moving coil instrument, principle of, 143
Multimeter, 58, 162
Multiples, 51, 59, 60
Multiplier, 159, 160
Mutual inductance, 151, 155, 729

Nano, 51
Napier, John, 668
Napierian logarithms, 28, 668
National grid, 243
Natural frequency, 489
Natural gas, for electricity generation, 327
logarithms, 28, 668
Negative feedback, 311
Nepers, 668
Network analysis, 497
Neutral point, 337
Neutron, 57
Newton, 50, 51
Nickel cadmium cells, 80
Nickel-iron alloys, 632
Nickel metal cells, 80
Nife cell, 80
Nodal analysis, 511
Node, 511, 761
No load phasor diagram, transformer, 352
Nominal impedance, 704
Non inverting amplifier, 314
Non-linear device, 59
scale, 159
Nonpermanent magnetic materials, 631
Norton and Thevenin equivalent networks, a.c., 546–550
d.c., 236
Norton's theorem, a.c., 539–545
d.c., 233
n-p-n transistor, 198
n-type material, 184, 186
Nuclear energy, for electricity generation, 328
Nucleus, 57
Null method of measurement, 176
Numerator, 7
Numerical methods, 578
of harmonic analysis, 612
Nyquist, 170

Odd functions, 591, 616
Ohm, 53, 58
Ohmic values, 70, 71, 814
Ohmmeter, 58, 161
Ohm's law, 59
Oil, for electricity generation, 326
On load phasor diagram, transformer, 356
Operating point, 208
Operational amplifiers, 309
differential, 318
integrator, 317
inverting amplifier, 312
non-inverting amplifier, 314
parameters, 311
summing amplifier, 315
voltage comparator, 316
voltage follower, 315
Oscilloscope, 58, 164
analogue, 165
digital, 165, 166
double beam, 166
Over-compounded machine, 379
Over-damped circuit, 772, 796
Owen bridge, 178, 464

Paper capacitor, 118
Parallel networks, 94, 435, 439
plate capacitor, 111
resonance, 285, 486
Paramagnetism, 621
Partial fractions, 780
Passbands, 698
Passive network, 666, 688, 691
Peak factor, 245
Peak inverse voltage, 190
value, 166, 245
Peak-to-peak value, 245, 249
Pentavalent impurity, 184
Percentages, 8
Period, 244, 576
Periodic function, 576
time, 244, 249
Permanent magnet, 124
magnetic materials, 633
Permeability, absolute, 126

of free space, 125, 620
relative, 126, 127, 620
Permeance, 620
Permittivity, 110
absolute, 110
Permittivity of free space, 110
relative, 110
Phase delay, 748
shift coefficient, 715, 750, 756
Phasor, 248
Phosphorus, 184
Photovoltaic cells, 85, 87
π-attenuator, 667, 671, 678
Pico, 51
π-connection, 667
π-section m-derived filter, 721
Plastic capacitors, 119
p-n junction, 185
p-n-p transistor, 197, 198
Polar form of complex number, 418
Polarization, 75, 636
Poles, 373
Potential difference, 53, 58
divider, 92
gradient, 108
Potentiometer, 92, 100, 177
Power, 52, 54, 61
associated with complex waves, 597
factor, 274, 450, 597
improvement, 289, 454
gain, 211
in a.c. circuits, 272, 446
in three phase systems, 342
loss, 639
transformer, 357
triangle, 274, 449
Practical straight line graphs, 37
types of capacitor, 117
Prefixes, 51, 813
Primary cell, 74, 78
constants, 747
Principal node, 511
Principle of operation, d.c. motor, 142
m.c. instrument, 143
three-phase induction motor, 397
transformer, 350
Product-arm bridge, 462
Propagation coefficient, 714, 752
constant, 715, 750
Proper fraction, 7
Proportion, 10
direct, 11
inverse, 13
Protons, 57, 183
Prototype filter, 707
Pseudocapacitance, 120
p-type material, 184
Pumped storage, for electricity generation, 330
Pythagoras' theorem, 26

Q-factor, 270, 286, 475, 489
in series, 478
Quantity of electric charge, 57
Quiescent point, 209

Radians, 24
Radio frequency transformer, 358
Rating of a machine, 449
transformer, 351
Ratio, 10
Ratio-arm bridge, 462
R-C parallel network, 279
series circuit, 264, 426, 766
Reactive power, 274, 449
Real number, 413
Reciprocity theorem, 688
Rectangular axes, 33
complex number, 414
Rectification, 190, 254
Rectifier, 190, 608
Rectifier diode, 190
Reference level, 175
Reflected impedance, 736
wave, 757
Reflection coefficient, 758
Regulation of a transformer, 359
Rejector circuit, 286
Relative permeability, 127, 620
permittivity, 110, 635
voltage, 103
Relay, 138
Reluctance, 127, 620
Remanence, 132, 622
Renewable energy sources, 85
Reservoir capacitor, 256
Residual flux density, 132, 623
Resistance, 53, 58, 66
dynamic, 285, 488
internal, 76
matching, 363
variation, 65
Resistivity, 66
Resistor colour coding, 70, 814
construction, 66
Resolution, 131
Resonance, applications of, 286
Resonance, by tuning capacitors, 737
due to harmonics, 606
parallel, 285
series, 266, 269, 427, 471
Reverse bias, 186
characteristics, 187
Rheostat, 100, 101
Ripple, 256
R-L-C circuit using Laplace transforms, 794
R-L-C series circuit, 266
R-L parallel network, 278
series circuit, 261, 425
R.m.s. value, 166, 245
complex wave, 594
Roots of auxiliary equation, 773
Rotation of loop in magnetic field, 150
Rotor copper loss, 400

Saturation flux density, 132, 622
Schering bridge, 178, 465
Schottky diodes, 193
Screw rule, 136, 137
s-domains, 785
Kirchhoff's laws, 786
Secondary cell, 74, 79
line constants, 750
Selective resonance, 606
Selectivity, 272, 481
Self-excited generator, 375
Self inductance, 151, 728
Semiconductor diodes, 182, 189
materials, 183, 184
Semiconductors, 183
Separately-excited generator, 375, 376
Separation of hysteresis and eddy current losses, 629
Series circuit, 91
a.c., 259, 423
Series magnetic circuit, 129
resonance, 266, 269, 427, 471
winding, 373
wound motor, 385
generator, 378
Shells, 57, 183
Shell type transformer construction, 357
Short circuits, 104
Short-shunt compound generator, 378
motor, 387
Shunt, 159, 160
field regulator, 390
winding, 373
wound generator, 377
motor, 387
Siemens, 53
Silicon, 184
Silicon controlled rectifiers, 192
Silicon-iron alloys, 631
Silver oxide batteries, 74

Simple cell, 75
equations, 16
Simultaneous equations, 21
using determinants, 498
Sine, 26
rule, 27
Sine wave, 244
general equation, 248
Single-phase parallel a.c. network, 277
series a.c. circuit, 258
supply, 337
S.I. units, 49
Skin effect, 656
Slew rate, 312
Slip, 398
Smoothing of rectified output, 255
Solar cars, 87
power satellites, 88
updraft tower, 88
Solar energy, 85–88, 333
advantages of, 86, 374
applications of, 87
disadvantages of, 86, 334
heating systems, 87
panels, 85
Soldering iron, 64
Solenoid, 136
Sources of harmonics, 608
Spectrum analysis, 173
Speed control of d.c. motors, 390
Squirrel-cage induction motor, 397, 405
applications of, 406
rotor, 397, 405
Standing wave, 761
ratio, 762
Star connection, 337, 552
Star-delta comparison, 348
starting, 406
transformation, 561–564
Star point, 337
Stator, 373
Steady state, 295, 766
Steinmetz index, 624
Step input, L-R-C circuit, 771
R-C circuit, 766
R-L circuit, 768
Stopbands, 698
Straight line graphs, 33
practical, 37
Streamline, 643
Stroboscope, 59
Sub-multiples, 59, 60
Summing amplifier, 315, 320
point, 316
Supercapacitors, 119
applications of, 120
Superposition theorem, a.c., 518–527
d.c., 224
Surface mount technology, 66
Susceptance, 436
Switched-mode power supplies, 193
Switching inductive circuits, 307
Symmetrical lattice, 693
π-attenuator, 671
T-attenuator, 666, 670
Synchronous speed, 396

Tachometer, 59
Tangent, 26
method, 296
T-attenuator, 670
T-connection, 666
Telephone receiver, 138
Temperature coefficient of resistance, 68
Tesla, 125
Thermal effects of dielectrics, 637
generation of electron-hole pairs, 186
Thermodynamic temperature, 50
Thevenin and Norton equivalent networks, a.c., 546–550
d.c., 236
Thevenin's theorem, a.c., 528–539
d.c., 228
Three-phase induction motors, 393
construction, 397
copper loss, 400
double cage, 407
impedance and current, 400
losses and efficiency, 401
principle of operation, 397
production of rotating magnetic field, 394
rotor e.m.f. and frequency, 399
slip, 398
starting, 405
synchronous speed, 396
torque equation, 402
–speed characteristics, 404
uses, 407
Three-phase systems, 336
advantages of, 348
power, 342
transformers, 367
Thyristor, 611
Tidal power, 85, 331
Time constant, C-R circuit, 296
L-R circuit, 303
Time delay, 718
Titanium oxide capacitor, 119
Torque of a d.c. machine, 382
Torque-speed characteristic of induction motor, 404
Transfer characteristics, 200, 310
Transformation ratio, 351
Transformer, 156, 349
a.f., 357
auto, 365
cooling, 358
construction, 357
current, 368
e.m.f. equation, 354
equivalent circuit, 358
isolating, 367
losses and efficiency, 360
maximum efficiency, 362
no-load phasor diagram, 352
on-load phasor diagram, 356
power, 357
principle of operation, 350
rating, 351
regulation, 359
r.f., 358
three-phase, 367
voltage, 369
windings, 358
Transient curves, C-R, 296
L-R, 303
Transients, 295, 765
Transistor, 195, 608
action, 197, 198
amplifier, 206
characteristics, 200, 205
connections, 196, 197
maximum ratings, 206
symbols, 197
Transistor classification, 196
operating configurations, 199
parameters, 201
Transit time, 718
Transmission lines, 746
current and voltage relationships, 749
distortion, 755
primary constants, 747
secondary constants, 750
standing waves, 760
wave reflection, 757
Transmission matrix, 687
parameters, 686
Transposing formulae, 19
Trapezoidal rule, 613
Travelling wave, 747
Trigonometry, 26
Trivalent impurity, 184
True power, 274, 449
T-section m-derived filter, 720

Two port networks, 666
in cascade, 683
Types of a.c. bridge circuits, 462
capacitor, 117
material, 183

UK supply voltage, 338
Ultracapacitors, 119
Under compounded machine, 379
Underdamped circuit, 772, 795
Unidirectional waveforms, 244
Unit of electricity, 62
Units, 49, 55, 809
S.I., 49
Universal bridge, 179
instrument, 162

Vacuum, 110
Valence electrons, 184
shell, 183
Valves, 608
Varactor diodes, 190, 193
Variable air capacitor, 117
Velocity of propagation, 749, 756
Vertical-axis intercept, 34
Virtual digital storage oscilloscope, 170
earth, 312
test and measuring instruments, 169
Volt, 53
Voltage, 57, 58
absolute, 103
comparator, 316
follower op amp, 315
gain, 211
magnification at resonance, 270, 476
regulator, 191
relative, 103
ratios, 351
transformer, 369
triangle, 261, 264
Voltmeter, 58, 159

Water heater, 64
Watt, 52, 61, 273
Wattmeter, 59, 162
Waveform analyser, 578
considerations, 616
harmonics, 173
Waveforms, 244
combination of, 251
Wavelength, 748
Wave reflection, 757
winding, 373
Weber, 125
Weight, 51
Wheatstone bridge, 177, 460
Wien bridge, 178, 466
Wind, for electricity generation, 331
Wind power, 85, 331
Wire wound resistors, 66
Work, 51
Wound rotor, 397
induction motor, 405
advantages of, 407

y-axis intercept, 35
Yoke, 373

Zener diode, 190, 191
effect, 188